Geological Survey of Canada

Geology of Canada, no. 1

QUATERNARY GEOLOGY OF CANADA AND GREENLAND

edited by

R.J. Fulton

1989

This is volume K-1 of the Geological Society of America's Geology of North America series produced as part of the Decade of North American Geology project.

© Minister of Supply and Services Canada 1989

Available in Canada through
authorized bookstore agents and other bookstores
or by mail from

Canadian Government Publishing Centre
Supply and Services Canada
Ottawa, Canada K1A 0S9

and from

Geological Survey of Canada offices:

601 Booth Street
Ottawa, Canada K1A 0E8

3303-33rd Street N.W.
Calgary, Alberta T2L 2A7

100 West Pender Street,
Vancouver, B.C. V6B 1R8

A deposit copy of this publication is also available for reference
in public libraries across Canada

Cat. No. M40-49/1E
ISBN 0-660-13114-5

Price subject to change without notice

Cette publication est aussi disponible en français

Technical editor
H. Dumych

Design and layout
M.J. Kiel
P.A. Melbourne
D.A. Busby

Cartography
Cartographic Section, GSC, Ottawa

Cover

*Kaskawulsh Glacier, Kluane National Park, Yukon Territory.
Photo by M. Poirel, SSC Photo Centre.*

Printed in Canada

PREFACE

The Geology of North America series has been prepared to mark the Centennial of The Geological Society of America. It represents the cooperative efforts of more than 1000 individuals from academia, state and federal agencies of many countries, and industry to prepare syntheses that are as current and authoritative as possible about the geology of the North American continent and adjacent oceanic regions.

This series is part of the Decade of North American Geology (DNAG) Project which also includes eight wall maps at a scale of 1:5 000 000 that summarize the geology, tectonics, magnetic and gravity anomaly patterns, regional stress fields, thermal aspects, seismicity, and neotectonics of North America and its surroundings. Together, the synthesis volumes and maps are the first coordinated effort to integrate all available knowledge about the geology and geophysics of a crustal plate on a regional scale.

The products of the DNAG Project present the state of knowledge of the geology and geophysics of North America in the 1980s, and they point the way toward work to be done in the decades ahead.

From time to time since its foundation in 1842 the Geological Survey of Canada has prepared and published overviews of the geology of Canada. This volume represents a part of the seventh such synthesis and besides forming part of the DNAG Project series is one of the nine volumes that make up the latest *Geology of Canada*.

J.O. Wheeler
General Editor for the volumes
published by the
Geological Survey of Canada

A.R. Palmer
General Editor for the volumes
published by the
Geological Society of America

ACKNOWLEDGMENTS

Although the *Geology of Canada* is produced and published by the Geological Survey of Canada, additional support from the following contributors through the Canadian Geological Foundation assisted in defraying special costs related to the volume on the Appalachian Orogen in Canada and Greenland.

Alberta Energy Co. Ltd.
Bow Valley Industries Ltd.
B.P. Canada Ltd.
Canterra Energy Ltd.
Norcen Energy Resources Ltd.
Petro-Canada
Shell Canada Ltd.
Westmin Resources Ltd.

J.J. Brummer
D.R. Derry (deceased)
R.E. Folinsbee

CONTENTS

Foreword 1

PART 1: REGIONAL QUATERNARY GEOLOGY OF CANADA

Introduction 13
1. Canadian Cordillera 15
2. Canadian Interior Plains 97
3. Canadian Shield 175
4. St. Lawrence Lowlands of Canada 319
5. Atlantic Appalachian region of Canada 391
6. Queen Elizabeth Islands 441

PART 2: APPLIED QUATERNARY GEOLOGY IN CANADA

Introduction 479
7. Quaternary environments in Canada as documented by paleobotanical case histories 481
8. Quaternary geodynamics in Canada 541
9. A survey of geomorphic processes in Canada 573
10. Terrain geochemistry in Canada 645
11. Quaternary resources in Canada 665
12. Influence of the Quaternary geology of Canada on man's environment 699

PART 3: QUATERNARY GEOLOGY OF GREENLAND

Introduction 739
13. Quaternary geology of the ice-free areas and adjacent shelves of Greenland 741
14. Dynamic and climatic history of the Greenland Ice Sheet 793

Index 823

FOREWORD

R.J. Fulton

The Quaternary is the geological time period in which we live. The internationally recommended time for the beginning of the Quaternary is 1.64 Ma (Pliocene/Pleistocene boundary). The predominant characteristics of the Quaternary are marked climatic change, glaciation, and the activity of other processes fuelled by climatic oscillations. Climatic change was the cause of periods of buildup and decay of ice sheets of continental scale and also resulted in transgression and regression of oceans, advance and retreat of deserts, and migration and extinction of plants and animals. From our own narrow perspective this dynamic period is important because it is the age of man.

Because the Quaternary is the latest geological time period, it is the one we know best and is the time during which geological processes fashioned the surface of the Earth as we know it today. Quaternary events are largely responsible for the soil in which we grow our food and the forests that we harvest; they are responsible for deposition of the surface materials with which we must contend in constructing buildings, roads, and pipelines; and they are responsible for the complex succession of units from which much of our groundwater is drawn, in which we bury solid wastes and into which we pump liquid wastes. Because patterns of climatic change similar to those which occurred in the past will, in all likelihood, occur in the future, we need an understanding of the mechanisms and patterns of climatic change during the Quaternary in order to provide a means of predicting future climatic changes and the probable regional distribution of climatically dependent phenomena such as permafrost.

The objective of this volume is to provide a synthesis and overview of the Quaternary geology of Canada and Greenland. It is hoped that this volume will serve as a source of information on those regions not particularly familiar to the reader and that the syntheses of areas with which the reader is familiar will stimulate new ideas. For those who are not Quaternary specialists, for readers from other countries, and for students, this volume is intended as a reference, a starting point for developing an understanding of the Quaternary of Canada and Greenland and an introduction to the Quaternary scientific literature.

This volume is organized in three parts: Part 1 covers regional aspects of the Quaternary geology of Canada (Fig. 1) and is aimed at supplying descriptive information on the nature of Quaternary materials, stratigraphy and events, and a review of Quaternary history. Part 2 considers some topical aspects of the Quaternary geology which in general apply to all regions of Canada. The topical chapters are aimed at providing an understanding of some of the processes and changes which have characterized the Quaternary and an understanding of the importance of Quaternary geology and events in our day to day lives.

Certain of the regional chapters also consider aspects of these same topics but at varying levels of detail. Part 3 concerns the Quaternary of Greenland; regional aspects are covered in sections dealing with eastern, western, and northern ice-free areas, and with offshore areas. The Greenland Ice Sheet is covered in a separate chapter.

THE QUATERNARY

The Quaternary is often referred to as the glacial age. The Quaternary is not the only geological time period during which glaciation occurred; records of glaciation as old as 2.5 Ga and glacial deposits of other glacial ages have been reported from many parts of the world. Nor was the Quaternary record exclusively glacial because there were many times when conditions were as warm as or warmer than present. Studies of marine and terrestrial sediments indicate that beginning about 2 million years ago the earth's climate began a series of cold-warm cycles that apparently caused growth and decay of continent scale ice sheets as many as eight times. These oscillations were irregular and many additional ones occurred that were not great enough to lead to full glacial conditions and the climate during each glaciation was not always the same because the limits of ice sheets varied from glaciation to glaciation. The net result, however, is that climatic swings and associated advances and retreats of the glaciers and seas that occurred during the Quaternary left unique signatures on the distribution of plants and animals and in geological deposits.

The dating and correlation of deposits that can be clearly related to early Quaternary glaciations are difficult. The defined base of the Quaternary used in this report is 1.65 Ma. This approximates the time when the main climatic oscillations began in the Mediterranean area and is marked by a major influx of cold climate marine and terrestrial fauna. The boundary stratotype adopted by the International Commission on Stratigraphy is the Vrica section, Calabria, Italy where the Quaternary/Pliocene boundary is placed 3-6 m above the top of the Olduvai Normal Polarity Subchronozone (Basset, 1985; Aguirre and Pasini, 1985). Paleomagnetic and vertebrate fossil information has been used to identify deposits of this approximate age in one section in southwestern Saskatchewan, but in general, this boundary has not been recognized as marking a significant stratigraphic or faunal break in Canada.

The Quaternary Period has been subdivided into Pleistocene and Holocene epochs with Holocene defined as the last 10 ka (Fig. 2). The Pleistocene Epoch is divided into Early, Middle, and Late subdivisions. The proposed boundary between Early and Middle Pleistocene is at the Brunhes/Matuyama paleomagnetic reversal (about 790 ka, Johnson, 1982; see also Richmond and Fullerton, 1986) and the boundary between Middle and Late is commonly placed at the beginning of the last interglaciation (in this report 130 ka, the approximate beginning of oxygen isotope stage 5 of deep sea cores). In Canada the latest part of the Middle Pleistocene is referred to as the Illinoian Stage and the Late Pleistocene is subdivided into Sangamonian and

---

Fulton, R.J.
1989: Foreword to the Quaternary Geology of Canada and Greenland; *in* Geology of Canada and Greenland, R.J. Fulton, (ed.); Geological Survey of Canada, Geology of Canada, no. 1 (*also* Geological Society of America, The Geology of North America, v. K-1).

Wisconsinan stages with 80 ka as the boundary (the approximate age of the boundary between oxygen isotope stages 4 and 5). The Wisconsinan is further subdivided into Early, Middle, and Late substages with boundaries at 65 ka (approximate age of the end of oxygen isotope stage 4) and 23 ka (the approximate time at which most of Canada was under or coming under the influence of the last major glacial advance).

The subdivision outlined above is a hybrid system and while it does not strictly follow the *North American Stratigraphic Code* (North American Commission on Stratigraphic Nomenclature, 1983), it is an attempt to adapt rational, consistent terminology while making the smallest possible changes from generally accepted usages. The Illinoian, Sangamonian, and Wisconsinan (chronostratigraphic units) are defined by type sections in the midwestern United States (Willman and Frye, 1970). In this report, the Sangamonian and Wisconsinan are equated with oxygen isotope stage 5 and stages 4-2 of the deep-sea record as defined in equatorial Pacific cores V28-238 and V28-239 by Shackleton and Opdyke (1973, 1976). Despite the fact that the subdivisions of the Wisconsinan are called substages, a chronostratigraphic unit, this system follows Dreimanis and Karrow (1972) in referring to them as early, middle, and late (designations used for geochronological units).

Ideally all boundaries of a stratigraphic system should be defined in type sections, based on similar criteria, and dated by similar methods; however the availability of suitable deposits and sections and the techniques that can be used for dating varies with age of the deposits. The Holocene, Late Wisconsinan, and part of the Middle Wisconsinan stratigraphic framework are based on terrestrial deposits that are correlated mainly by radiocarbon dating. This is because terrestrial deposits for this part of the

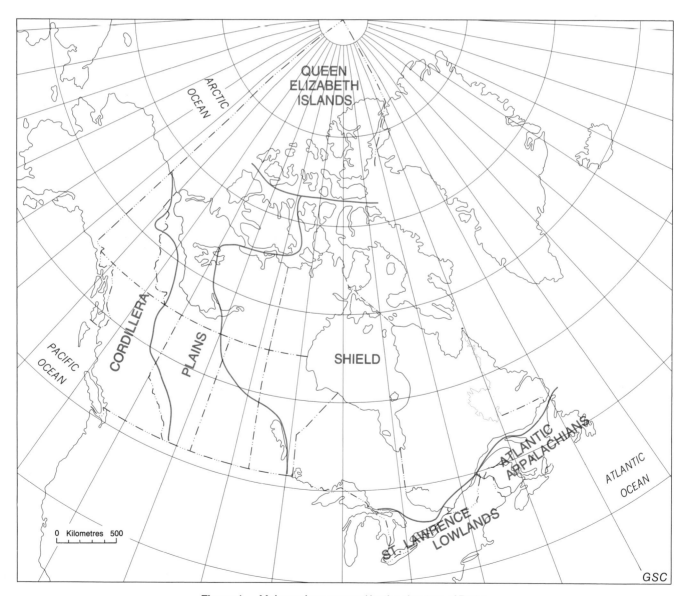

Figure 1. Major regions covered by the chapters of Part 1.

stratigraphic record are abundant and many contain materials suitable for radiocarbon dating. Older terrestrial records are fragmentary and because there is no reliable dating technique similar to radiocarbon dating which can be universally applied, it is difficult to erect a regional framework based on terrestrial deposits. Marine deposits, however, provide a relatively continuous record, materials can be dated by a variety of techniques other than radiocarbon, and climatic proxy units can be derived by detailed study of oxygen isotope content of fossils. The terrestrial Middle Wisconsinan, Early Wisconsinan, Sangamonian, and Illinoian units, which correspond to geoclimatic units, can be correlated with the marine isotope record (proxy climate units), and where material for dating is lacking in the terrestrial units, the ages of the correlated climatic proxy units in the marine record can be used. The record of terrestrial deposits older than Illinoian is even more fragmentary and more difficult to correlate with the marine proxy climate record than is the younger record. Dating is possible by fission track and potassium argon techniques where deposits include tephra or are associated with volcanic deposits, but these materials are present only locally. The common parameter most readily available and present in both terrestrial and marine materials is paleomagnetism. If a good paleomagnetic record can be obtained from a deposit, then it may be correlated with the worldwide paleomagnetic signature and in this way the isolated deposit can be "dated". Paleomagnetism is the main criterion used to define Middle Pleistocene, Early Pleistocene, and the Pleistocene-Pliocene boundaries.

DEVELOPMENT OF QUATERNARY GEOLOGY IN CANADA

Development of an understanding of regional aspects of the Quaternary of Canada began shortly after founding of the Geological Survey of Canada in 1842. Before that time explorers and visiting scientists made a variety of observations on Quaternary geology, along with notes on other aspects of geology, fauna, flora, and the landscape; however, there was no attempt at making systematic observations to solve specific problems or to gain a complete understanding of an area. Following founding of the Geological Survey of Canada, geologists, who in reality were natural scientists, made systematic observations of Quaternary phenomena and all other aspects of geology. By 1863 an adequate body of data was available so that a map of the "superficial deposits" of Canada was prepared and a chapter on "superficial geology" was included in *Geology of Canada*.

During the early stages of Quaternary studies in Canada the theory of a glacial origin for drift was gaining dominance over the marine origin theory. Canadian workers for the most part favoured a glacial origin although in early work authors, such as G.M. Dawson (1849-1901), made some concessions to the older hypothesis. It appears that evidence of deep abrasion, the carving of rounded landforms in hard rock, and the consistent unidirectional transport of materials provided overwhelming evidence of glacier action in Canada. Only in areas submerged during deglaciation and where floating ice was a strong contemporary agent was a case made for a marine origin of drift. The last and strongest advocate of a marine origin for drift was J.W. Dawson (1820-1898), the father of G.M. Dawson. He continued to his death to point to modern floating ice or shore ice processes in the estuary of St. Lawrence River as evidence of the marine origin of drift.

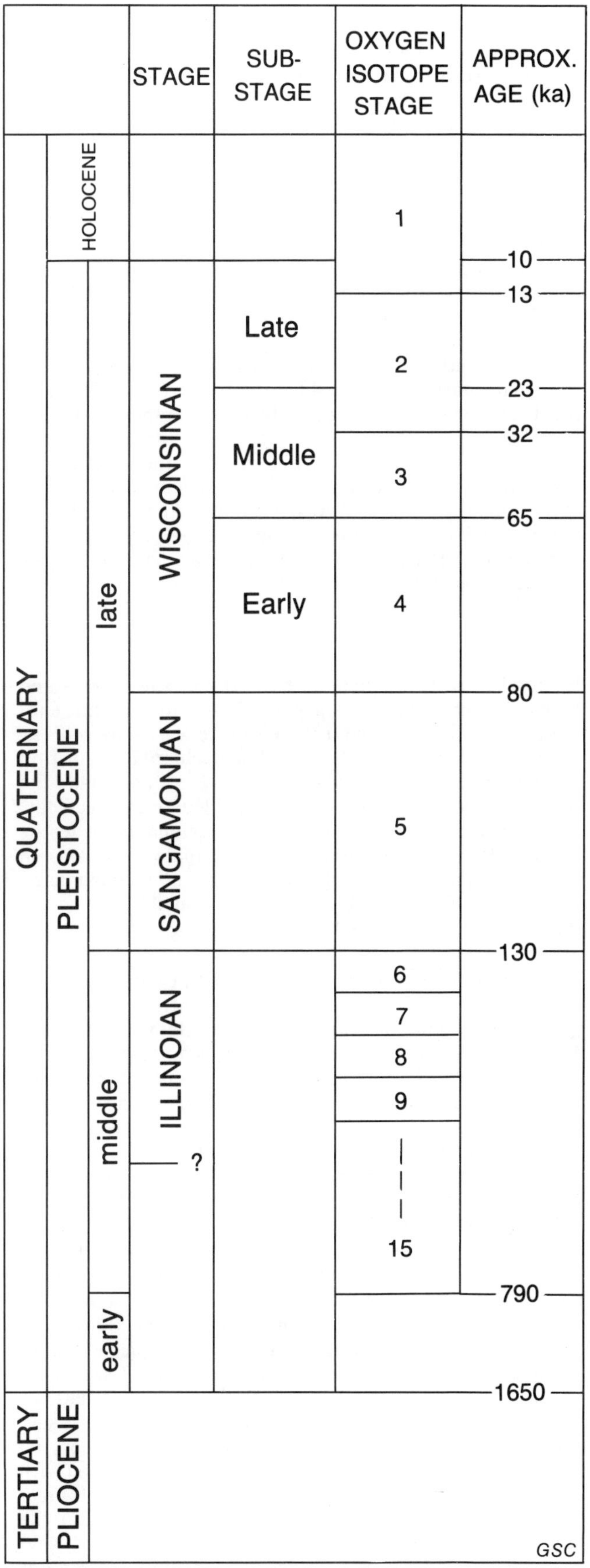

			STAGE	SUB-STAGE	OXYGEN ISOTOPE STAGE	APPROX. AGE (ka)
QUATERNARY	PLEISTOCENE		HOLOCENE		1	
		late	WISCONSINAN	Late	2	10 — 13 — 23
				Middle	3	32 — 65
				Early	4	80
			SANGAMONIAN		5	130
		middle	ILLINOIAN		6	
					7	
					8	
					9	
			?			
					15	
		early				790 — 1650
TERTIARY	PLIOCENE					

GSC

Figure 2. Subdivision of the Quaternary used in Canada.

FOREWORD

Many prominent geologists gathered Quaternary data as part of regional survey work during early geological surveys. R. Bell (1841-1917) studied the stratigraphy of Quaternary sediments in the Hudson Bay Lowlands, made early observations on the extent and origin of Lake Agassiz, contributed to the paleoecology of Champlain Sea deposits, and prepared material for the chapter on "Superficial Geology of Canada" for the 1863 *Geology of Canada* volume. G.M. Dawson recognized the existence of several major glacial flow patterns in the area between Lake of the Woods and the Rocky Mountains, named the Laurentide Ice Sheet, developed the concept of a "Great Cordilleran Glacier", studied the stratigraphy of Quaternary sediments in southwestern Alberta, and provided many astute observations on the Quaternary of the Cordillera. J.B. Tyrrell (1858-1957) has been generally credited as being the father of the dynamic ice sheet theory. Drawing on his extensive observations from northern Manitoba and the District of Keewatin, he put together a picture of an ice sheet with several regional centres that waxed and waned at slightly different times and migrated as the ice sheet expanded and contracted. A.P. Low (1861-1942) found evidence indicating the existence and migration of similar regional centres east of Hudson Bay.

R. Chalmers (1833-1908) concentrated his work on surficial geology. Most of his surveys were conducted in New Brunswick, eastern Quebec, and parts of Nova Scotia and Prince Edward Island. He mapped the extent of marine submergence and recognized the importance of local ice centres, originating the concept of an Appalachian ice sheet in southern Quebec. A.P. Coleman (1852-1939) also made the bulk of his contributions in the field of Quaternary geology. He used information gained through regional work in the Atlantic Provinces and eastern Quebec to develop ideas on the extent and thickness of Labrador ice and through many years of careful research made major contributions to the stratigraphy and paleoecology of deposits in the Toronto area.

During the first half of the Twentieth Century workers from outside Canada made important contributions to the Quaternary knowledge of Canada. J.W. Goldthwait (1880-1947), an American, did extensive work in the Maritimes and St. Lawrence Lowlands and made major contributions to an understanding of the Champlain Sea and correlations of Champlain Sea events with those of glacial lakes to the south. E. Antevs (1888-1974), a Scandinavian, conducted mapping and varve correlation studies in several parts of Canada and produced reports on the retreat of the last ice sheet in eastern Canada and in Manitoba.

Because the Quaternary ice sheets extended south from Canada into the United States, a flow of ideas and work across the border has influenced Quaternary research developments in Canada. The concepts of W. Upham (1850-1934) on the development of Lake Agassiz were picked up in Canada and Upham extended his work into Canada by working for a time with the Geological Survey of Canada in southern Manitoba. R.F. Flint (1902-1976) influenced several generations of Quaternary workers with his theory on development of the Laurentide Ice Sheet. He disagreed with the emphasis placed by Tyrrell on local flow centres. He recognized the existence of distinct flow centres within the Laurentide Ice Sheet but felt that, because there were several, they must have been only of local significance. He put forward a theory on the growth of Laurentide ice which suggested a westward expansion of the ice mass from highlands in eastern Canada with movement from local centres in marginal areas but with the dominant feature being a thick ice mass centred on Hudson Bay. The ideas of R.P. Goldthwait, N.E. Odell, and M.L. Fernald, among others, influenced the development of ideas concerning the presence of nunataks and refugia and the position of the last glacial limits in eastern Canada. The stratigraphic work of T.C. Chamberlin, F. Leverett, L. Horberg, J.C. Frye, W.C. Alden, P. MacClintock, G.M. Richmond, and B. Willis, influenced the Quaternary stratigraphic framework developed in Canada. Ideas on landform development, timing and pattern of ice advance and retreat, and development of glacial lakes have been influenced by the work of J.L. Hough, R.F. Flint, S.R. Moran, L. Clayton, and D.S. Fullerton.

A Canadian who made notable contributions to Quaternary geology during the first half of this century was W.A. Johnston (1874-1949); he joined the Geological Survey of Canada as the first geologist since Chalmers to work exclusively on surficial geology problems. Johnston mapped surficial deposits and worked on Champlain Sea problems in the Ottawa area, contributed to the growing body of data on tilting of shorelines in the Great Lakes area, mapped much of southern Manitoba and Saskatchewan, and provided insight into the nature of the Fraser River delta.

During the middle to late 1930s a new phase of surficial geology work began as regional surficial geology mapping and Quaternary stratigraphy were initiated for use in groundwater studies. Following the Second World War this work was expanded within the Geological Survey of Canada and several provinces added surficial geology components to their 'Surveys'. This work resulted in sound systematic mapping and stratigraphic studies which built on the Quaternary framework provided by earlier workers

Geologists mapping bedrock have continued to make casual observations of Quaternary features (primarily measurements of striations) and generally include one or two pages on "glaciation" in their reports. Several bedrock geologists have made significant contributions to Quaternary geology. H.S. Bostock, while mapping in the Cordillera, gained many important insights into the Quaternary of the area. He used striations and transport of erratics to define the direction of ice flow; noted the upper limit of erratics in many areas; described the pattern of deglaciation and development of lakes in response to neoglacial advances; traced the limits of glaciation of different ages in the Yukon; and presented a physiographic classification of the Cordillera. J.E. Armstrong, who also conducted bedrock mapping in the Cordillera, made many astute observations on Quaternary features in northern and central British Columbia before switching to Quaternary research in the lower Fraser Valley where he mapped most of the Fraser Lowland and conducted and inspired significant stratigraphic studies. Other bedrock geologists who made significant contributions to the Quaternary of the Cordillera by mapping glacial features are H.W. Tipper, R.B. Campbell, H. Gabrielse, W.H. Mathews, J.E. Muller, A. Sutherland-Brown, and J.O. Wheeler. R.L. Christie, in the course of bedrock studies in the Arctic Islands published a report on the surficial geology of northeastern Ellesmere Island. J. Tuzo Wilson reported on drumlins and ice flow in Nova Scotia and conducted some of the earliest airphoto interpretation of glacial features in the District of Keewatin.

Systematic regional mapping and Quaternary stratigraphic studies continued and were expanded during the

FOREWORD

1950s and 1960s. With much of the settled part of the country being covered by the combined effort of the Geological Survey of Canada and provincial 'Surveys'. The rationale for much of this work was collection of data which would aid: soil studies of agricultural and forestry land, groundwater studies, and engineering development and land use planning.

The first *Glacial Map of Canada* (Wilson et al., 1958) showed orientation of ice flow features, limits of marine and lacustrine submergence, eskers and morainal belts. This map, put together by a committee of the Geological Association of Canada under the direction of J. Tuzo Wilson, was the first published, relatively complete picture of glaciation in Canada.

During the 1950s and 1960s Quaternary specialists were able to take advantage of the logistics provided by Geological Survey of Canada reconnaissance mapping projects in the North. In this way it was possible for B.G. Craig, J.G. Fyles, H.A. Lee, and O.L. Hughes to gather reconnaissance data for a large part of Canada's North. The scattered observations resulting from this work, combined with systematically collected data from southern Canada, made it possible to produce a second "glacial map" (Prest et al., 1968) which accompanied a summary of the *Quaternary Geology of Canada* published in the Geological Survey of Canada's *Geology and Economic Minerals of Canada* (Douglas, 1970). This map, inspired by V.K. Prest, showed location and orientation of glacial features, limits of glaciation, and marine and lacustrine submergence areas in greater detail than the map produced ten years earlier. In addition Prest produced a second map that showed the changing glacial configuration during ice retreat. This map, the first to show the retreat picture of the entire Laurentide Ice Sheet, has profoundly influenced our ideas of ice sheet dynamics and since its publication has been used in virtually every reconstruction of deglaciation

During the early 1960s, the Fisheries Research Board began a program of sampling and mapping the surface sediments off the east coast of Canada. This program expanded under the auspices of the Bedford Institute of Oceanography and later, under the Atlantic Geoscience Centre, became a full scale study of Quaternary deposits. What started as a task of plotting the texture of grab samples and interpreting bottom conditions from echo sounding tracks, became a full three-dimensional mapping program using geophysical surveys to map the extent and thickness of deposits, and the stratigraphy of cores from critical localities used to calibrate the signature of seismic records. By 1980 the Scotian Shelf and much of the Grand Banks had been covered and work had been extended northward as far as Baffin Bay. The Fisheries Research Board also began research in the fiords of the Pacific coast of Canada with surface sampling and echo sounding surveys. The Department of Oceanography at the University of British Columbia added geophysics to the bottom studies and extended this work to the sound, strait, and shelf areas off the west coast. With the establishment of the Pacific Geoscience Centre in 1977, the Geological Survey of Canada began regional studies of Quaternary geology and surface materials, similar to those being carried on the east coast. In response to exploration for offshore petroleum resources and the need to locate possible pipeline routes in the 1970s, surficial geology studies were conducted in the straits and sounds of the Arctic Islands and the Beaufort Sea. In general this work was oriented towards gathering information on parameters that affect engineering use of the sediments and on presently active processes.

Beginning in 1970, the Geological Survey of Canada fielded its own large systematic Quaternary mapping operations in poorly accessible parts of Canada. The first work in Labrador was prompted by a request for information that would aid in assessing forestry potential; later projects in the Mackenzie Valley and the Arctic Islands were aimed at gathering data that could be used in assessing the viability of constructing oil and gas pipelines and assessing the environmental impact of hydrocarbon exploration in the North; programs mounted in the District of Keewatin, and northern Manitoba, and on Banks and Victoria islands were aimed at developing techniques for drift prospecting, at systematically determining the geochemical composition of surface materials, and at systematically obtaining Quaternary geological information that could be used in assessing the impact of future developments.

Three main developments, made since 1950 have profoundly influenced Quaternary work in Canada. The development of radiocarbon dating made it possible to obtain absolute ages of organic materials that are younger than 50 ka. This made it possible to set up a late Quaternary stratigraphy based on absolute ages. Research into amino acid racemization showed that the breakdown of amino acids is, among other things, age dependent. This has made it possible to use the ratios of different amino acids present in organic substances as a correlation tool and as an indication of relative age. A third development which has strongly affected the direction of Quaternary research is the study of oxygen isotope content of marine shells. This work has led to the setting up in marine Quaternary sediments of a series of crudely dated oxygen isotope stages which in theory are related to cycles of glaciation. Because this work was done in deep marine basins, the sediment record is assumed to be continuous so that the signature of climatic change developed is considered to be complete. Interpretation of this signature has generated a myriad of papers on the cyclicity of glaciations, relationships between glaciation and the Earth's orbital parameters, and causes of glaciation. Possibly the aspect of this work most significant to terrestrial studies is that it provides a continuous record to which the discontinuous Quaternary land record can be compared.

Several Quaternary scientists have made outstanding contributions to the present foundation of Quaternary knowledge in Canada over the past three decades. J.G. Fyles developed the framework for Quaternary stratigraphy that is used in southern British Columbia, made many valuable observations and put forward many of the concepts embodied in the Quaternary framework of Arctic Canada; moreover, he was instrumental in establishing the Quaternary Division of the Geological Survey of Canada. J.R. Mackay has single-handedly pushed back the frontiers of our understanding of permafrost processes and has inspired many students, who continue to advance our understanding of this significant northern process. V.K. Prest, the first head of the Pleistocene Section of the Geological Survey of Canada, is responsible for many of our concepts of glaciation in the Maritime Provinces and has profoundly influenced our view of Laurentide Ice Sheet dynamics with his glacial and ice retreat maps of Canada. Probably the greatest contributor to Quaternary geology in Canada over the past century is A. Dreimanis. He has made substantial contributions in the fields of Quaternary stratigraphy, glacial history, drift prospecting, genesis of till, and classification of glacial deposits. Through his own work and that of his students and collaborators, he has developed a Quaternary stratigraphic framework for the Great Lakes - St. Lawrence Valley; he has fostered an interest in the study of the composition of tills and use of this information to determine till provenance and to develop an understanding of the transport of debris by ice; he has fostered an interest in structure and texture of till and in the use of this information to interpret and understand the genesis of glacial deposits.

The most recent trend influencing Quaternary geology work is a demand for surficial geology information that would aid mineral development. This requires greater emphasis on systematic collection of samples and extensive chemical analyses in order to learn more about the geochemistry of surficial materials. Hence Quaternary studies, which began as casual observations by geologists who were mapping bedrock in an attempt to find mineral deposits, developed into a specialty which is itself used to locate mineral deposits. This has resulted in a progression from the 1863 surface deposit map of Canada showing highly generalized units developed from widely scattered observations, through several generations of feature maps showing largely ice flow features and limits, to a surface materials map indicating the general nature of deposits over the entire country. Similarly, from a chapter prepared in 1863 by Robert Bell, limited to specific observations made largely in eastern Canada, we have progressed to a volume synthesizing the results of hundreds of papers reporting on data from all parts of the country.

PERSPECTIVES ON CANADIAN ICE SHEETS

The dominant factor in the Quaternary history of Canada is glaciation and most of the country has been covered by ice at least once (Fig. 3). The number of times ice sheets have expanded and contracted across the country is unknown because erosion during subsequent ice advances and during nonglacial periods removes evidence of earlier events. In several parts of the country, however, evidence of at least four major glacial periods is present. In general terms the pattern of ice buildup and retreat was similar from glaciation to glaciation with ice accumulating and moving outward from the highlands during each glaciation and late ice remaining longest in these same areas. In detail, however, each glaciation probably differed from the others in extent of ice sheets, dominance of specific ice centres, and timing of advance and retreat from individual centres. During periods of glaciation the ice cover consisted of a major core ice sheet surrounded by areas dominated by smaller ice sheets, ice sheet complexes, and local ice caps. What follows is a thumbnail introduction to the major regions of Canada and to the ice masses that so profoundly affected their Quaternary geology; details are discussed in the regional chapters of this volume. Most of this is based on our understanding of the ice cover during the last glaciation (Fig. 4).

Canadian Cordillera

The Canadian Cordillera consists of the mountainous region that lies to the west of the plains of Canada (Fig. 1, 3). It consists of several belts of general northwest-trending mountain ranges and highlands with local plains and plateaus which extend from the 49th Parallel to the Arctic Ocean. The Pacific Ocean and Alaska lie to the west, the Beaufort Sea to the north, and the Interior Plains to the east. It is assumed that during glacial periods moisture was delivered from the Pacific by the prevailing westerly winds. Accumulation, therefore, was greatest in the belt of coastal mountains with subsidiary buildup of ice in the interior mountains. Because of the nature of the topography, no single area would have acted as a dominant accumulation centre but rather there were chains of accumulation centres along the west coast and in various mountain ranges of the interior. During initiation of glacial periods local glaciers and ice caps expanded and coalesced to form major ice sheets and ice caps. Maximum accumulation probably was on the ocean side of the western mountains but much of the ice from this area was also funnelled to fiords and ablated mainly by calving into the Pacific Ocean. Ice moving towards the east, however, accumulated in the relatively low interior of the Cordillera and coalesced with glacier complexes flowing from local ice centres, and mountainous areas on the eastern side of the Cordillera to form a major Cordilleran ice sheet. The Rocky Mountains, on the eastern side of the Cordillera, are farthest from the Pacific moisture source, contained the least active ice centres, and probably even at glacial maxima were marked by nunataks and flanked by local unglaciated areas. Ice from the Rocky Mountains advanced down major valleys onto the plains and coalesced with the main core of the inland ice sheet to the east. Ice from the more active accumulation areas in the interior of the Cordillera moved eastward into the Rocky Mountains and, in several areas, pushed through the mountains and flowed onto the plains. A few local glaciers developed in the northern Yukon, but moisture was available in such small quantities that this area remained largely unglaciated.

At glacial maxima the Cordilleran region was dominated by an ice sheet or complex of coalescing domes surrounded by less coalesced domes. The central core occupied the interior of the Cordillera and flowed to the north and south, and also east and west in places where it overtopped lower parts of the fringing mountain ranges. Even though the general picture was relatively simple, in detail it was complex, with flow in upper levels of the ice controlled by the location of many local domes and ice centres and with ice flow at the base influenced by the complex buried mountainous topography. The north and south termini of the ice sheet formed a series of lobes and tongues; in the west, the ice extended into the Pacific as a series of ice shelves; in the east, lobes and valley glaciers of Cordilleran ice coalesced with Laurentide ice, and its flow was diverted southerly parallel to the margin of the continental ice sheet.

Because ice in the central part of the Cordillera was surrounded by mountains, the sheet was relatively stable. With the ice sheet buttressed by the surrounding mountains reaching 2500 m, snowline had to rise to near this elevation before recession of the margin of the core ice sheet could begin. Consequences of this were: local ice caps did not persist on highlands during deglaciation; topographic highs were deglaciated before adjacent low areas; and the longest lasting ice remnants occupied main valleys and the lowest parts of the interior (areas where ice was thickest).

The ice sheet in the Cordillera developed on the western side of the continent in a belt of prevailing westerly winds and consequently it was little affected by development of other ice sheets; however, it probably influenced the supply of moisture to Laurentide ice sheets which lay to the east.

Stratigraphic evidence from the zone of confluence between Cordilleran and Laurentide ice suggests that either the Cordilleran ice reached its limit and was retreating when Laurentide ice reached its maximum or that the two coalesced. It is not surprising that Laurentide ice would have reached the western edge of the Interior Plains later than Cordilleran ice because the western margin of the inland ice sheet was about 1500 km from the area of its inception whereas accumulation areas for Cordilleran ice were no more than a few hundred kilometres from the eastern edge of the Cordilleran ice sheet. Even though this was apparently true for the eastern margin of the Cordilleran ice sheet, during the Late Wisconsinan the southern margin of the Cordilleran ice sheet did not reach its maximum position until about 15 ka, several thousand years after the Laurentide Ice Sheet had begun to retreat.

Queen Elizabeth Islands

The Queen Elizabeth Islands area is shaped roughly like a right triangle with the northwest-facing hypotenuse adjoining the Arctic Ocean, the east-facing side adjacent to northwest Greenland and Baffin Bay, and the base separated from Banks, Victoria, and Baffin islands to the south by Parry Channel (Fig. 3). The eastern third of the Queen Elizabeth Islands consists of mountains and highlands with intervening lowlands and interisland channels culminating along its eastern margin in a rim of mountains. The remainder of the area grades from hills, ridges, and plateaus in the east to gently sloping coastal plains in the west.

Distribution of ice in the Queen Elizabeth Islands during periods of glaciation is poorly known. One interpretation for the Late Wisconsinan is that observed crustal downwarping can be accounted for only if the area was covered by an ice sheet (Innuitian Ice Sheet) which joined Laurentide and Greenland ice sheets. At the other extreme is the interpretation that there was limited ice cover during the last gla-

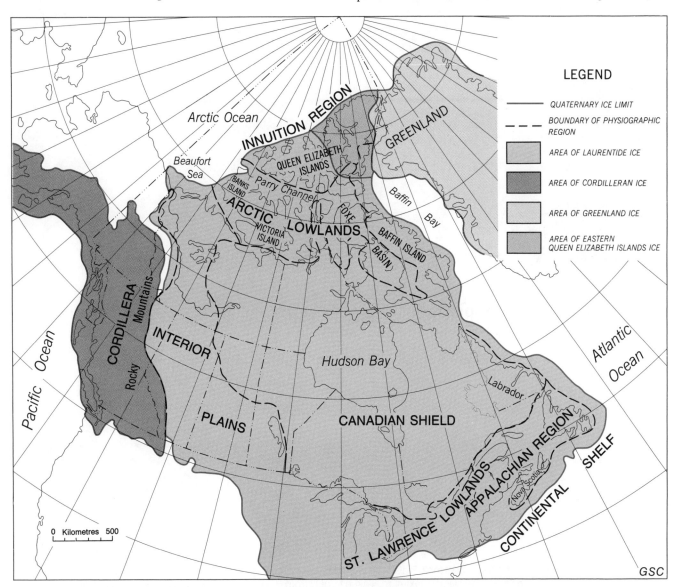

Figure 3. Major physiographic subdivisions of Canada and areal coverage of ice from different sources.

ciation in the Queen Elizabeth Islands. This is based on the observations that extensive glacial deposits that clearly can be assigned to the Late Wisconsinan are not present. Proponents of this interpretation attribute the observed crustal depression to the combined loading effect of local ice caps and the adjacent Laurentide and Greenland ice. They agree that the presence of erratics and glacial landforms suggests that parts of the Queen Elizabeth Islands were covered by Laurentide and Greenland ice but say this did not occur during the last ice advance. The proponents of an Innuitian Ice Sheet explain the lack of glacial deposits by proposing that this ice was cold based (frozen to its bed) and consequently extensive glacial deposits were not produced and glacial erosion did not greatly modify the preglacial surface. Those arguing against an Innuitian Ice Sheet counter by saying that even a cold based ice sheet could not disintegrate without leaving evidence, at least in the form of meltwater channels.

That there should be two different glaciation hypotheses is understandable because of the contradictory nature of evidence of Quaternary glacial coverage in the region. The distribution of erratics suggests that much of the Queen Elizabeth Islands was at one time covered by a major extension of Laurentide ice and a limited extension of Greenland ice. The pattern of fiords and interisland channels leading away from topographic basins suggests former existence of lowland ice centres. There are, however, problems with finding a moisture source capable of nourishing a regional ice sheet at these high latitudes. Extensive modern ice fields occupy eastern highland areas where moisture carried

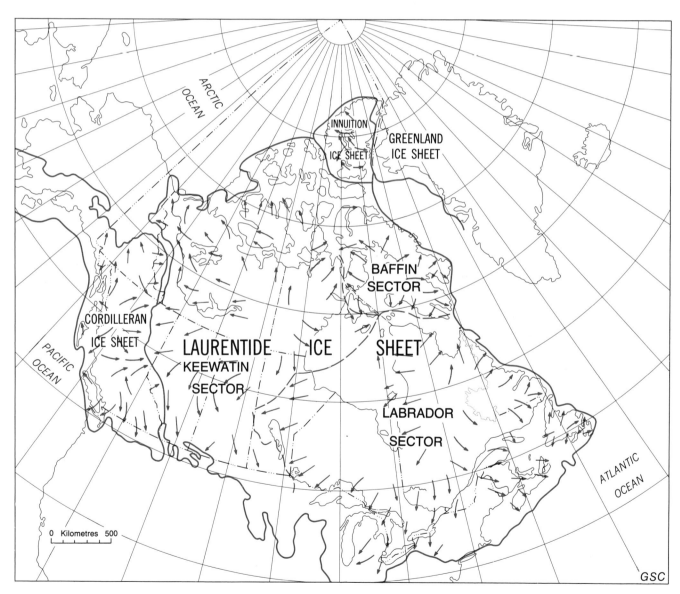

Figure 4. Approximate extent of main ice sheets during the last (Wisconsin) glaciation; arrows indicate probable directions of ice flow at the glacial maximum. Taken largely from Prest et al. (1968) and Prest (1984) but with some modifications.

northwestward from Baffin Bay supplies about 300 mm of annual precipitation. On the other hand the low western islands are true arctic deserts and in general receive less than 30 mm annually. Hence construction of a regional ice sheet from current levels of precipitation would be difficult and slow. If one assumes that both Baffin Bay and the Arctic Ocean were ice covered during glacial periods, moisture would have been even less readily available and ice sheet nourishment even more difficult at the culminations of glaciations. Consequently, it is difficult to imagine how there could have been a Late Wisconsinan Innuitian Ice Sheet in the Queen Elizabeth Islands unless through some quirk, glacial conditions resulted in greater quantities of moisture in the area than through present climatic conditions. To get around this problem one theory proposes that development of major ice sheets in the Queen Elizabeth Islands could only occur during warm interglaciations when the Arctic Ocean was not ice covered and relatively warm water in Baffin Bay could supply abundant precipitation. If this is true, a major ice sheet was not built over the Queen Elizabeth Islands during each glacial period but only following interglaciations warm enough for destruction of the Arctic Ocean ice cover. Evidence from cores taken in the Arctic Ocean suggests that this might not have occurred within the last 700 ka.

The current assumption is that when the northern hemisphere entered a glacial period, nourishment of glaciers in the Queen Elizabeth Islands decreased as ice cover in the Arctic Ocean became permanent, as the northward expanding Laurentide ice filled the channels and bays to the south, and as construction of ice shelves and development of a permanent, pack ice cover closed off Baffin Bay. These considerations have led to the interpretation that Queen Elizabeth Islands glaciers achieved maximum size at the start of glaciation and retreated slowly during the remainder of the glacial cycle with possibly a readvance during deglaciation when moisture sources again became available. An early (late interglacial) buildup of ice in the Queen Elizabeth Islands might help explain the rapid fall of sea level which oxygen isotope ratios of marine cores suggest occurred during the inception of glaciation before extensive glaciers appeared in the south. It also fits the story of the last glaciation during which extensive ice apparently occupied the region early in the cycle but late advance did not occur until early Holocene, after much of the Laurentide Ice Sheet had disintegrated.

Atlantic Appalachian area

The Atlantic Appalachian area consists of Nova Scotia, New Brunswick, Gaspésie, and Newfoundland, a series of islands in the Gulf of St. Lawrence (Prince Edward Island, Magdalen Islands, and Anticosti Island), and adjacent shelf areas (Fig. 3). The Atlantic Ocean lies to the east and south, the Canadian Shield to the north, and the Appalachian Mountains of the United States to the west. The region consists of low hills, hilly uplands, lowlands, and shallowly submerged shelves. The Atlantic Ocean provides abundant precipitation to the region and consequently many areas, including parts of the emerged shelves, supported local ice caps during intervals of cooler climate. In addition to local ice centres, Laurentide ice invaded the area from the north during glaciation. Because of the presence of many local competing ice centres, the glacier flow picture was complicated and it varied during glacial cycles.

Local terrestrial ice centres apparently developed before the end of the Sangamonian Stage but were overwhelmed by southward flowing Laurentide ice which overrode the area during the early part of the last glacial cycle. General recession of Laurentide ice during the middle of the last glaciation permitted local glaciers to again become dominant and ice centres developed on parts of the emerged shelves. Flow of ice from land to sea again was dominant during final stages of the glaciation but apparently Laurentide ice did not override the area at this time.

Central Canada

In the context of this Foreword, central Canada includes the parts of Canada covered by Laurentide ice (the Canadian Shield, the Interior Plains, the Arctic Lowlands, and the St. Lawrence Lowlands; Fig. 3). The core part of Canada is dominated by Hudson Bay and Foxe Basin, two low relief features which lie below present sea level and are underlain largely by Paleozoic sedimentary rocks. The region around these basins consists of hilly to low relief areas underlain largely by Precambrian crystalline rocks and metasediments. The northern segment of the central core contains marine channels and bays separating peninsulas and islands. The northwestern, western, and southern part of the area consists of low relief to hilly plains underlain by Paleozoic and poorly consolidated Mesozoic sedimentary rocks. The northern and eastern edge of the area is fringed by mountains and highlands. The St. Lawrence Lowlands, a major valley and chain of large lake basins, lie just inside the southeastern margin of the region.

During glacial periods the central part of Canada was covered by inland or Laurentide ice which has been subdivided into Labrador, Keewatin, and Baffin sectors (Fig. 4). This consisted of a large complex of ice domes, divides, saddles, and lobes. The different elements of this vast complex reacted semi-independently in response to a variety of regional and local, physiographic, climatic, and glacier bed conditions. The pattern of ice flow features and distribution of dispersed debris that must be used in reconstructing the configuration of Laurentide ice sheets are complex, disjunct, contradictory, and, most significantly, of different ages in different areas. In some regions glacier flow may have been maintained in one direction throughout an entire glaciation; in other regions flow patterns during inception and disintegration may have been similar but different at glacial maxima; other areas apparently had continually shifting flow patterns. In addition, flow features developed during an early glaciation that have different orientations from those developed during later glaciations may be preserved; materials transported in one direction by an early ice flow could be moved in different directions by later ones or could be carried farther by a later flow. Because of the complicated array of evidence of largely undated, nonsynchronous flow, it is not surprising that there is no consensus on the paleogeography and flow dynamics of the Laurentide Ice Sheet. (Throughout this volume the name Laurentide Ice Sheet is reserved for the Wisconsinan inland ice sheet. The ice cover of central Canada of other ages is referred to as Laurentide ice or inland ice). One further factor which complicates discussion of inception, growth, and decay of Laurentide ice is that one school bases its ideas on the aggregate pattern of drift dispersal developed during all glaciations, another works on the basis of ice flow features which might or might not be contemporaneous, and a third school bases its story on dynamic glaciological models which may be valid for only short periods of the existence of a Laurentide ice cover.

Moisture to nourish Laurentide ice was available on all sides. The Atlantic Ocean to the east was undoubtedly a prime moisture source and all models depicting inception of glaciation recognize this. Abundant moisture was also available from the Gulf of Mexico but this probably had maximum influence only as the ice approached its southern limit. Some moisture was probably available from the Pacific Ocean but this was limited when Cordilleran ice was present. The Arctic Ocean was a potential source of moisture early during a glacial cycle, particularly if the preceding interglacial was warm enough to destroy the sea ice cover; but if most interglaciations were similar to the present one, a sea ice cover would have been present so only limited moisture for ice sheet nourishment would have been available from the north and northwest.

Two main theories for inception of Laurentide ice have been considered. The oldest is based on mountain ice sheet analogies and assumes that ice buildup began in the mountainous regions at the eastern edge of the area and that coalescing alpine glaciers became an ice sheet which grew towards the west and eventually covered all central Canada. The second theory, referred to as instantaneous glacierization, hypothesizes that lowering of snow line permitted preservation of snow throughout the summer on sizable upland areas of Baffin Island and Labrador, this increased the albedo (ability to reflect solar radiation), cooled the atmosphere, and led to additional lowering of snow line. By this method large areas could be covered with ice in a short time.

Under the theory of expansion from mountain glaciers, the centre of ice dispersal migrated from highlands in the east towards Hudson Bay and remained there until final stages of deglaciation when the invading sea carved out the heart of the sheet, leaving remnant ice divides on the east and west shores of Hudson Bay. One problem with this theory is that buildup is a slow process, requiring tens of thousands of years, and removal of water from the oceans would have been slow at first followed by a relatively steady acceleration. By contrast, the oxygen isotope records from marine cores have been interpreted as suggesting that large amounts of moisture were removed from the sea early during glaciation and variable amounts at later times. A second problem is that ice is supposed to have flowed into Hudson Bay from the east and the west during only a relatively short period of the glaciation. Patterns of drift dispersal, however, indicate long term flow of ice into Hudson Bay from the east and west rather than flow outward from Hudson Bay. Finally, this theory predicts a relatively stable ice sheet but stratigraphic and ice flow data suggest that it was dynamic.

Computer modelling of glacier inception, based on the instantaneous glacierization theory, has shown that sizable ice sheets could be grown on Baffin Island and northern Labrador in 10 ka. Hence this theory seems to fit requirements of the rate of ice sheet growth. It also would produce a complex, multidomed feature similar to that postulated from field studies.

At its maximum extent during each glaciation, Laurentide ice probably had the same general size and shape as at its maximum during Wisconsinan (Fig. 4). In the north and east the ice sheet terminated in the sea; in the south and west most of the ice ended on land. As stated earlier, locations of domes, saddles, and lobes were probably somewhat different during each glacial maximum but may have borne some similarity to the Wisconsinan picture.

It is possible to locate certain glacial limits and to correlate these from region to region with some degree of confidence. These are generally assumed to relate to the maximum of particular glaciations but in several cases may reflect readvance or stillstand positions developed during glacial retreats. Our concrete knowledge of the age of these limits is meagre and in only a few places can dated ice marginal deposits be assigned to specific glacial stages. Even with the last glacial advance, ages are normally based on a few radiocarbon dates on overridden interstadial sediments which yield a maximum but not the true age of the glacial event.

Our knowledge of the style of inland ice retreat is based largely on the retreat pattern of Late Wisconsinan ice. During this period ice retreat appears to have begun earliest in the west and south with remnants being preserved longest in the east and north (Baffin Island and Central Quebec). Parts of the ice sheet in the east and northwest retreated in contact with the sea and at times large segments of the southern and western ice margins retreated in contact with glacial lakes. The ice apparently was extremely dynamic during retreat with evidence of readvances, development of lobes, shifts in centres of flow, and surges. These late movements produced a complicated record of flow which obscured earlier regional flow patterns and probably is responsible for causing much of the confusion and controversy over the structure of the Laurentide Ice Sheet.

ACKNOWLEDGMENTS

As associate editors, J.A. Heginbottom (Geological Survey of Canada) had general responsibility for Part 2 of this volume and S. Funder (Geological Museum, University of Copenhagen) for Part 3. They were active in general planning of their respective parts, in selecting and contacting authors, in maintaining contact with chapter co-ordinators and authors during first draft and review stages, and in choosing and acting as liaisons with critical reviewers.

The successful completion of this volume has resulted from the efforts of many people. All 53 direct contributors are credited with authorship of particular sections of this volume. Critical readers are acknowledged in each chapter but special thanks must go to W.H. Mathews (University of British Columbia), and to J.B. Bird (McGill University), who supplied helpful comments on the entire volume, N.R. Gadd, (Geological Survey of Canada) who reviewed Part 1, and B.R. Pelletier (Geological Survey of Canada) who reviewed Part 2. L. Maurice (Geological Survey of Canada) checked many of the references for this volume and provided invaluable support by conducting extensive proofing of several versions of each chapter. Particular mention also must be made of the efforts of H. Dumych (Geological Survey of Canada), who, in addition to correcting the English, insisting on consistency in use of terminology, and improving the readability, pointed out errors in fact and picked up faults in scientific reasoning. This volume was indexed by Patricia Sheahan.

REFERENCES

Aguirre, E. and Pasini, G.
1985: The Pliocene-Pleistocene boundary; Episodes, v. 8, p. 116-120.

Bassett, M.G.
1985: Towards a "Common Language" in stratigraphy; Episodes, v. 8, p. 87-92.

Douglas, R.J.W. (editor)
1970: Geology and economic minerals of Canada; Geological Survey of Canada, Economic Geology Report, 5th edition.

Dreimanis, A. and Karrow, P.F.
1972: Glacial history of the Great Lakes-St. Lawrence Region, the classification of the Wisconsin(an) Stage, and its correlatives; 24th International Geological Congress (Montréal), Section 2, Quaternary Geology, p. 5-15.

Johnson, R.G.
1982: Brunhes-Matuyama magnetic reversal dated at 790,000 yr B.P. by marine-astronomical correlations; Quaternary Research, v. 17, p. 135-92.

North American Commission on Stratigraphic Nomenclature
1983: North American stratigraphic code; The American Association of Petroleum Geologists Bulletin, v. 67, p. 841-875.

Prest, V.K.
1984: The Late Wisconsinan glacier complex; in Quaternary Stratigraphy of Canada — A Canadian Contribution to IGCP Project 24, R.J. Fulton (ed.); Geological Survey of Canada, Paper 84-10, p. 21-36, Map 1584A, scale 1:7 500 000.

Prest, V.K., Grant, D.R. and Rampton, V.N.
1968: Glacial Map of Canada; Geological Survey of Canada, Map 1253A, scale 1:5 000 000.

Wilson, J.T., Falconer, G., Mathews, W.H., and Prest, V.K. (compilers)
1958: Glacial Map of Canada; Geological Association of Canada, Toronto. (out of print)

Richmond, G.M. and Fullerton, D.S.
1986: Introduction to Quaternary glaciation in the United States of America; in Quaternary Glaciations in the Northern Hemisphere, V. Sibrava, D.Q. Bowen, and G.M. Richmond (ed.); Quaternary Science Reviews, Special Volume 5, Pergamon Press, Oxford, p. 3-10.

Shackleton, N.J. and Opdyke, N.D.
1973: Oxygen isotope and paleomagnetic stratigraphy of equatorial Pacific core V28-238: Oxygen isotope temperatures and ice volumes on a 10^5 year and 10^6 year scale; Quaternary Research, v. 3, p. 39-55.

1976: Oxygen isotope and paleomagnetic stratigraphy of Pacific core V28-239, late Pliocene to latest Pleistocene; in Investigations of Late Quaternary Paleoceanography and Paleoclimatology, R.M. Cline and J.D. Hays (ed.); Geological Society of America, Memoir 145, p. 449-464.

Willman, H.B. and Frye, J.C.
1970: Pleistocene stratigraphy of Illinois; Illinois State Geological Survey, Bulletin 94, 204 p.

Printed in Canada

Part 1
REGIONAL QUATERNARY GEOLOGY OF CANADA

Part 1 of the *Quaternary Geology of Canada and Greenland* consists of descriptions of the nature, distribution, stratigraphy, and history of the Quaternary deposits of Canada. It is presented in six chapters, each covering a major natural region. Facets which give unity to the Quaternary geology of each region are: bedrock geology, gross physiography, style of glaciation, and position in relation to major ice sheets. These factors have resulted in roughly similar Quaternary deposits, stratigraphic successions, and Quaternary histories throughout each natural region. For example: In the Canadian Shield region, tills are largely derived from crystalline and metamorphic rocks and hence tend to be nonplastic and have a sandy to silty texture. In the Cordilleran region, lakes in overdeepened valleys are the main sediment traps so that nonglacial deposits predominantly consist of successions of silts, sands, and gravels representing valley fill sequences. In the Interior Plains region the substrate consists largely of poorly lithified shale and siltstone and consequently glacial deposits are thick, highly plastic, and over large areas are characterized by hummocky ice disintegration and ice thrust features. The Atlantic region has many local highlands and lies at the periphery of Laurentide ice coverage and, as a consequence, its Quaternary glacial history is a record of competition between local ice centres and regional Laurentide ice.

The location of natural regions is shown in Figure 1 (in Foreword). The Cordilleran region corresponds largely to the Cordilleran physiographic province, is underlain by rocks of mixed lithologies of the Cordilleran Orogen, and was glaciated by Cordilleran ice, local glaciers, or not at all. The Interior Plains region is underlain by sedimentary rock, mainly poorly lithified Mesozoic age shales and siltstones, consists largely of plains and low relief uplands, and lies in the peripheral zone of Laurentide ice. The Canadian Shield region occupies the stable cratonic core of the continent, is underlain mainly by Precambrian igneous, metasedimentary, and sedimentary rocks, has variable topography but in general consists of plains and low relief hills; it occupies the core zone of the inland ice sheet. The St. Lawrence Lowlands region is underlain by moderately competent Paleozoic sedimentary rocks, consists in general of plains and low relief hills, and lies in the peripheral zone of Laurentide ice. The Atlantic Appalachian region is underlain by rocks of mixed lithologies of the Appalachian Orogen, has variable physiography, lies in the peripheral zone of Laurentide ice, but was also strongly affected by local glaciers. The Queen Elizabeth Islands region is underlain mainly by sedimentary rocks of Paleozoic and Mesozoic age, which are highly deformed in the east but are largely flat lying and unlithified in the west, and varies in physiographic expression from mountainous in the east to a flat coastal plain in the west; in addition to local glaciation, the region was affected by ice sheet glaciation of Laurentide ice in the south, and of Greenland ice in the east.

The nature and distribution of Quaternary deposits, style of glaciation, and the level of knowledge of Quaternary geology vary from region to region and from area to area within regions. Consequently, the emphasis and level of detail varies from chapter to chapter and between subsections within chapters. Also, each chapter was approached slightly differently by chapter co-ordinators and authors. The Cordilleran chapter consists of contributions of several authors which have been combined as a relatively homogeneous entity. The Interior Plains, Canadian Shield, and St. Lawrence Lowlands chapters consist of subregion sections that essentially stand alone. The Queen Elizabeth Islands chapter consists of a main regional description and an essay on glaciers of the area. The Atlantic Appalachians chapter was prepared by a single contributor.

The general order or treatment of regions and of subregions within chapters is from west to east and from south to north (counter-clockwise).

Most chapters were prepared during the period 1984-1986 with some information added in 1987.

R.J. Fulton

Chapter 1

QUATERNARY GEOLOGY OF THE CANADIAN CORDILLERA

Summary

Introduction — *J.J. Clague*

Bedrock geology — *J.J. Clague*

Quaternary tectonic setting — *J.J. Clague*

Physiography and drainage — *J.J. Clague*

Climate — *J.M. Ryder*

Development of Cordilleran landscapes during the Quaternary — *W.H. Mathews*

Character and distribution of Quaternary deposits — *J.J. Clague*

Controls on Quaternary deposition and erosion — *J.J. Clague*

Cordilleran Ice Sheet — *J.J. Clague*

Relationship of Cordilleran and Laurentide glaciers — *J.J. Clague*

Quaternary sea levels — *J.J. Clague*

Regional Quaternary stratigraphy and history

    Area of Cordilleran Ice Sheet — *J.J. Clague, J.M. Ryder, O.L. Hughes, and N.W. Rutter*

    Glaciated fringe — *L.E. Jackson, Jr., N.W. Rutter, O.L. Hughes, and J.J. Clague*

    Unglaciated areas — *O.L. Hughes, N.W. Rutter, J.V. Matthews, Jr., and J.J. Clague*

Paleoecology and paleoclimatology — *J.J. Clague and G.M. MacDonald*

Holocene glacier fluctuations — *J.M. Ryder*

Economic implications of Quaternary geology — *J.J. Clague*

Acknowledgments

References

Tiedemann Glacier, British Columbia. Tiedemann Glacier heads in the highest part of the Coast Mountains and is flanked by large lateral moraines of neoglacial age. The lakes at the bottom of the photo lie between two sets of these moraines. Province of British Columbia BC1413-55.

Chapter 1

QUATERNARY GEOLOGY OF THE CANADIAN CORDILLERA

Compiled by

John J. Clague

SUMMARY

The Canadian Cordillera, the westernmost of the major physiographic and geological regions of Canada, is an area of rugged mountains, plateaus, lowlands, valleys, and seaways. This region extends from the International Boundary on the south to Beaufort Sea on the north and from Pacific Ocean and Alaska on the west to the Interior Plains on the east, a land area in excess of 1 500 000 km^2.

The Canadian Cordillera is located at the edge of the America lithospheric plate and consists of the deformed western margin of the North American craton and a collage of crustal fragments, or terranes, that were accreted to the craton and subsequently fragmented and displaced northward along major strike-slip faults. These processes have given the Cordillera a strong northwest-southeast structural grain and are largely responsible for the present complex distribution of rocks, faults, and other structures in the region.

The distribution of earthquakes, active and recently active faults, and young volcanoes in the Canadian Cordillera is controlled by the motions of the Pacific, America, Juan de Fuca, and Explorer plates and Winona Block which are in contact in the northeast Pacific Ocean west of British Columbia. Most large earthquakes and active faults are associated with offshore plate boundaries beyond the British Columbia continental margin; some, however, occur on the continental shelf and on land. Most Quaternary volcanoes are located in four narrow belts that constitute part of the circum-Pacific "Ring of Fire". Several of these volcanoes have erupted during the Holocene, one less than 150 years ago.

Lithospheric plate interactions are also responsible for recent vertical crustal movements in the Canadian Cordillera. Some outer coastal areas of British Columbia apparently are being uplifted at rates of up to 2 mm/a. In contrast, the inner coast is either stable or subsiding. This pattern continues southward into Washington but is interrupted to the north by considerable uplift in southeastern Alaska and bordering areas of British Columbia and Yukon Territory.

The character of the Cordilleran landscape at the beginning of the Quaternary has been inferred from the distribution of lavas, sediments, and relict surfaces of late Tertiary age, and by estimating late Cenozoic denudation rates from fission-track ages of unroofed plutons and uranium-series dates on speleothems in relict phreatic caves. The evidence indicates that the distribution of high and low ground at the beginning of the Quaternary was probably much as it is today, but relief was much less. The landscape had not yet been extensively modified by glaciation and may have resembled that of northern Yukon Territory today.

Quaternary deposits in the Canadian Cordillera are extremely varied and have complex distributions controlled largely by physiography and glacial history. Most of the surface sediments were deposited during the last glaciation and during postglacial time. Older materials occur at the surface in central and northern Yukon Territory, along the eastern margin of the Cordillera, and on parts of the continental shelf; they also underlie Late Wisconsinan glacial deposits in some parts of British Columbia and southern Yukon Territory.

During Pleistocene nonglacial periods, as at present, sedimentation was concentrated in valleys, lakes, and the sea. Mountains were areas of erosion during nonglacial periods, whereas most plateaus and some lowlands were little affected by either erosion or deposition at these times.

During the early phase of each Pleistocene glaciation, as ice spread from mountains into low-lying areas, streams aggraded their valleys with outwash. Advancing glaciers also impounded large lakes in which fine sediments were deposited. At the climax of each major glaciation, most of British Columbia and southern Yukon Territory was covered by ice. At these times, till was deposited in places at the base of the Cordilleran Ice Sheet and its satellite glaciers. However, mountain areas, fiords, and valleys parallel to the direction of ice flow were subject to intense scour by glaciers, and the older unconsolidated fills of many valleys were thus partly or completely removed. Deglacial intervals, like periods of glacier growth, were times of rapid valley aggradation and extensive drainage change. Glacial lacustrine sediments accumulated in lakes that formed and evolved as the ice sheet decayed. Glacial marine sediments were laid down on the continental shelf and on isostatically depressed lowlands vacated by retreating glaciers and covered by the sea. Many valleys became choked with glacial fluvial and fluvial sediments that were eroded from poorly vegetated, unstable drift deposits mantling upland slopes and valley walls.

Clague, J.J. (compiler)
1989: Quaternary geology of the Canadian Cordillera; Chapter 1 in Quaternary Geology of Canada and Greenland, R.J. Fulton (ed.); Geological Survey of Canada, Geology of Canada, no. 1 (also Geological Society of America, The Geology of North America, v. K-1).

Glaciations were initiated by the expansion of alpine glaciers during periods of global climatic cooling. With continued cooling, these glaciers coalesced to form piedmont complexes and mountain ice sheets. Eventually, piedmont complexes from separate mountain source areas joined to cover most of British Columbia and adjacent areas. Throughout this period, the major mountain systems remained the principal source areas of glaciers, and ice flow continued to be controlled by topography. Occasionally, ice thickened to such an extent (approximately 2500 m) that one or more domes with surface flow radially away from their centres became established over the interior of British Columbia.

Each glacial cycle terminated with rapid climatic amelioration. Deglaciation occurred by complex frontal retreat in peripheral glaciated areas and by downwasting accompanied by widespread stagnation throughout much of the interior. Along the western periphery of the ice sheet, glaciers calved back in contact with eustatically rising seas. In areas of moderate relief, the pattern of deglaciation was more complex, with uplands becoming ice-free first and dividing the ice sheet into a series of valley tongues that decayed in response to local conditions.

Growth and decay of the Cordilleran Ice Sheet triggered isostatic adjustments in the crust and mantle of western Canada. In combination with related eustatic and diastrophic effects, these adjustments produced complex sea level changes along the coast of British Columbia. Gradual growth of glaciers at the beginning of each glacial cycle led to progressive isostatic depression of the land surface. Initially, shorelines may have fallen as a proglacial forebulge migrated through coastal areas and as water was transferred from oceans to expanding ice sheets. However, as glaciers continued to grow, the coastal region began to subside, and the sea eventually rose far above its present level relative to the land in most areas. At the climax of each major glaciation, the entire glaciated Cordillera was isostatically depressed, with areas near the centre of the ice sheet displaced downward more than areas near the periphery. During deglaciation, isostatic uplift in most coastal areas was greater than the coeval eustatic rise, thus the sea fell rapidly relative to the land. Uplift occurred at different times during deglaciation due to diachronous retreat of the Cordilleran Ice Sheet.

Although there probably were glaciers in the high mountains of the Canadian Cordillera during late Tertiary time, the oldest glaciations for which there is reasonable stratigraphic and landform evidence are Pleistocene in age. Westlynn Drift in southwestern British Columbia and drift of the Nansen, Klaza, Shakwak, and "old" glaciations in Yukon Territory predate the last (Sangamonian) interglaciation, some by a considerable margin (i.e., Nansen and "old" glaciations are older than 1 Ma). The Great Glaciation (Glacial Episode 1) in the southern Rocky Mountains is probably Middle Pleistocene in age, and there is some evidence for several older glaciations in this region.

The Muir Point Formation and correlative Highbury Sediments in southwestern British Columbia underlie two drift sheets and are either Sangamonian or pre-Illinoian in age. Westwold Sediments of south-central British Columbia are thought to correlate in part with these units.

Units deposited by the Cordilleran Ice Sheet during the penultimate glaciation (Early Wisconsinan or Illinoian) include Semiahmoo, Dashwood, Muchalat River, and Okanagan Centre drifts in British Columbia, and Reid, Mirror Creek, and (?)Icefield drifts in Yukon Territory. Correlative units deposited by mountain glaciers at the periphery of the ice sheet include: Waterton II, Albertan, and Maycroft tills and associated glacial lacustrine and glacial fluvial sediments in the Rocky Mountains (perhaps also Hummingbird, Baseline, and Early Cordilleran tills in this same region); drift of the "intermediate" glaciation in the Southern Ogilvie Mountains; and drift of the penultimate glaciation on the Queen Charlotte Islands.

A lengthy nonglacial interval, known as the Olympia in British Columbia and the Boutellier in Yukon Territory, began before 60 ka and persisted through the Middle Wisconsinan. In British Columbia, glaciers probably were confined to the major mountain ranges throughout the Olympia Nonglacial Interval, and in general the physical environment was similar to that of postglacial time. Sediments accumulated mainly in large intermontane valleys, on coastal lowlands, and in the sea. Two well defined stratigraphic units of Olympia age are the Cowichan Head Formation of southwestern British Columbia and Bessette Sediments of south-central British Columbia.

The Olympia-Boutellier nonglacial interval ended about 25-29 ka with climatic deterioration and glacier growth at the onset of the last glaciation (locally termed Fraser, McConnell, Macauley, and Kluane glaciations). Glacier growth was slow at first, with ice confined to mountain ranges until 20-25 ka, depending on the locality. The Cordilleran Ice Sheet attained its maximum extent in the south about 14-14.5 ka. Ice sheet growth was accompanied by widespread aggradation of outwash in coastal lowlands and interior valleys and by the formation of proglacial lakes.

Late Wisconsinan drift in many parts of the Canadian Cordillera comprises one till and bounding stratified units (e.g., Kamloops Lake, Gold River, Port McNeill, McConnell, and Macauley drifts). In some areas, however, Late Wisconsinan drift is more complex and includes two or more till units. For example, in Fraser Lowland, three main suites of Fraser Glaciation deposits are recognized: (1) Quadra Sand and Coquitlam Drift deposited during the early part of the Fraser Glaciation; (2) Vashon Drift deposited at the climax of this glaciation; and (3) Capilano Sediments, Fort Langley Formation, and Sumas Drift deposited during deglaciation. Each of these suites contains, among other things, one or more tills.

Late Wisconsinan ice cover in many mountain ranges at the periphery of the Cordilleran Ice Sheet was limited. Most glaciers in the Mackenzie, Wernecke, and Southern Ogilvie mountains terminated inside mountain fronts. Parts of the Queen Charlotte Islands also apparently escaped glaciation during Late Wisconsinan time. In the Rocky Mountains, however, many glaciers flowed onto the Interior Plains and coalesced with the Laurentide Ice Sheet. Drift units of Late Wisconsinan age in peripheral glaciated areas include: Waterton III and IV, Hidden Creek, Bow Valley, Canmore, Eisenhower Junction, Ernst, Lamoral, Jackfish Creek, Marlboro, Obed, and Drystone tills and associated outwash in the Rocky Mountains; the surface drift in the Fort St. John area; and drift of the "last" glaciation in the Southern Ogilvie Mountains.

The southwestern sector of the Late Wisconsinan Cordilleran Ice Sheet began to decay about 14 ka. Parts of the coastal lowlands of southwestern British Columbia were ice-free by 13 ka, and the ice sheet and most satellite glaciers had completely disappeared by 10 ka or shortly thereafter. During deglaciation, the sea transgressed isostatically depressed lowlands along the British Columbia coast, and large lakes formed behind masses of decaying ice and drift in the interior.

Late Wisconsinan deglaciation initiated a period of redistribution of glacial materials by fluvial and mass wasting processes that continued into the Holocene. A period of rapid aggradation in many valleys was followed by downcutting as the supply of sediment to streams decreased. By middle or late Holocene time, streams were flowing near their present levels. The main sedimentation sites in the Cordillera during the Holocene have been lake and seafloor basins, fans, and deltas.

Northern Yukon Territory was not glaciated during the Quaternary. The only surficial deposits in most of this region are weathered rock and colluvium on slopes, organic sediments in depressions, and fluvial deposits along streams. However, thick fluvial, lacustrine, and glacial lacustrine sediments underlie Old Crow, Bluefish, and Bell flats. During much of the Pleistocene, as at present, alluvium and lake sediments accumulated in these areas. On at least one occasion, however, the Laurentide Ice Sheet advanced against the Richardson Mountains and blocked easterly drainage at McDougall Pass, thus forming an extensive proglacial lake in the lowlands to the west. Large amounts of glacial lacustrine silt and clay were deposited in this lake.

Quaternary paleoenvironmental information for the Canadian Cordillera comes mainly from studies of fossil pollen, plant macrofossils, paleosols, speleothems, and vertebrate and invertebrate faunal remains. Information for early and middle Quaternary time is sparse, although recent paleoecological studies in northern Yukon Territory have shed new light on the character and evolution of circumpolar floras and faunas during this period. In contrast, there is abundant paleoenvironmental information for Wisconsinan and Holocene time. Environmental conditions during the long Olympia-Boutellier interval were variable in both space and time; the climate at times was colder and at times similar to that of the present. During the Fraser-McConnell Glaciation, very cold and probably arid conditions prevailed at the eastern and northern margins of the Cordilleran Ice Sheet; the vegetation in these areas was tundra. Along the Pacific coast, however, the climate was more moderate, and diverse floras may have persisted in refugial areas.

Major floral changes in the Canadian Cordillera during and immediately following deglaciation are attributable in part to climatic change and in part to different rates of plant migration and plant succession on freshly deglaciated terrain. Nonarboreal plant communities adapted to cold and probably dry conditions were the first to appear. In the southern Cordillera, these plant communities were rapidly replaced by forests as climate ameliorated. In the Yukon, herb tundra was replaced by shrub tundra at this time; later, various tree species expanded into the region in a complex fashion. The general warming trend which accompanied deglaciation continued well into the Holocene, although it may have been interrupted by brief intervals of climatic deterioration. The warm Hypsithermal interval was followed during the latter half of the Holocene by generally cooler and wetter conditions. Vegetation in most areas has changed little in the last few thousand years.

Alpine glaciers in the Canadian Cordillera have fluctuated in response to Holocene climatic change. Cirque moraines of latest Pleistocene or early Holocene age occur in several mountain ranges and were constructed during a minor expansion of glaciers at the end of the Fraser Glaciation or soon thereafter. Glacier advances also occurred between 3.3 ka and 1.9 ka, during the last millennium, and probably between 6 ka and 5 ka. In most areas, glaciers achieved their maximum Holocene extent between about 1500 AD and 1850 AD during the Little Ice Age.

INTRODUCTION
J.J. Clague

The geology of Quaternary deposits in the Canadian Cordillera and various aspects of the landscape of the region attributable to Quaternary events are discussed in this chapter. In Canada in general, and in the Cordilleran region in detail, Quaternary glaciation has profoundly altered the landscape and left a mantle of unconsolidated sediments on bedrock that has significantly affected economic development and other human activities. Repeatedly during the Quaternary Period and perhaps also in late Tertiary time, glaciers enveloped large areas of the Cordillera; only parts of central and northern Yukon Territory, the eastern Mackenzie Mountains, and scattered peaks were never covered by ice. Glaciations were separated by nonglacial or interglacial periods during which the Canadian Cordillera was largely free of ice.

This chapter highlights the history of the last nonglacial-glacial cycle which is so well documented in the Canadian Cordillera. The relationship between topography, climate, and glacier buildup and decay are discussed, and depositional and erosional responses to major climatic changes are assessed. Special consideration is given to the Quaternary history of Yukon Territory north of the limit of glaciation using evidence obtained from sedimentary deposits in Porcupine River basin. Other significant topics include: (1) the contemporary tectonic setting of the Canadian Cordillera; (2) the evolution of Cordilleran landscapes during the Quaternary; (3) interrelationships of past Cordilleran and Laurentide ice masses; (4) land-sea relationships during the advance and retreat of the last Pleistocene glaciers; (5) the Quaternary paleoecology and paleoclimatology of the Cordillera; and (6) the activity of glaciers during the Holocene.

Description of region

The Canadian Cordillera is part of the great belt of mountains that forms the western margin of North America and South America. Within Canada, this belt is up to 900 km

---

Clague, J.J.
1989: Introduction (Quaternary geology of the Canadian Cordillera); in Chapter 1 of Quaternary Geology of Canada and Greenland, R.J. Fulton (ed.); Geological Survey of Canada, Geology of Canada, no. 1 (also Geological Society of America, The Geology of North America, v. K-1).

CHAPTER 1

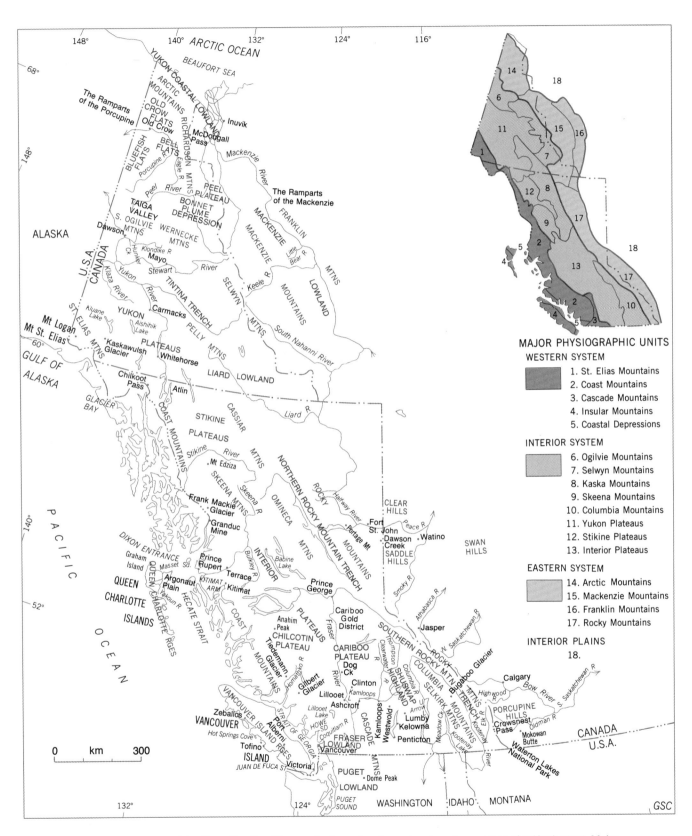

Figure 1.1. Map of the Canadian Cordillera and adjacent regions showing place names cited in the text. Major physiographic units (after Mathews, 1986) are shown on the small-scale inset map.

wide and more than 2500 km long. It extends north from the International Boundary along the forty-ninth parallel and Juan de Fuca Strait to Beaufort Sea, and east from the Pacific Ocean and Alaska to the Interior Plains (Fig. 1.1). The Canadian Cordillera includes most of British Columbia and Yukon Territory, the continental shelf and slope west of British Columbia, and parts of Alberta and District of Mackenzie, an area of about 2 000 000 km^2.

The region is one of extremely diverse topography, geology, climate, and vegetation. Rugged glacier-clad mountains that receive more than 2500 mm of precipitation per year are within sight of broad semiarid valleys. The general form of the Cordillera resembles that of a great wall flanking the Interior Plains and consisting of an elevated platform rimmed by mountain battlements (Bostock, 1948). The eastern battlement of this wall comprises the Rocky, Mackenzie, and Arctic mountains, and the western battlement, the Coast and St. Elias mountains. The platform between the two consists of plateaus, lowlands, valleys, and mountains which form much of the interior of British Columbia and the Yukon. West of the Coast Mountains are the lower ranges of Vancouver Island and the Queen Charlotte Islands, and lowlands bordering the Pacific Ocean.

The high mountains of the Cordillera today support valley glaciers and small ice caps. During major Pleistocene cold periods, glaciers advanced from these mountains to bury most of British Columbia and parts of the Yukon, District of Mackenzie, and Alberta beneath as much as 2.5 km of ice. In the process, the drainage network of western Canada was repeatedly disrupted and rearranged, mountains were sculpted by glaciers, and prodigious quantities of sediment were deposited in valleys and on plateaus, coastal lowlands, and the seafloor. Those parts of the Yukon that completely escaped glaciation have a somewhat muted topography that contrasts sharply with the rugged mountains and fresh drift-covered lowlands of formerly glaciated areas. The landscape of both glaciated and unglaciated areas has been modified since the end of the Pleistocene by fluvial, mass wasting, and other processes, but the resulting changes have been minor in comparison to those induced by repeated growth and decay of Cordilleran glaciers over a period of millions of years.

History of development of Quaternary ideas

Our present understanding of the Quaternary geology of the Canadian Cordillera is based on observations made by earth scientists over about the last 100 years. During this time, ideas concerning the Quaternary have gradually evolved and become more sophisticated.

The first significant studies of the Quaternary of the Canadian Cordillera were made in the late Nineteenth Century by G.M. Dawson. In a series of papers that were remarkable for their time, Dawson (1877, 1878, 1879, 1881, 1886, 1888, 1889, 1890, 1891) established that British Columbia and adjacent areas at one time had been covered by glaciers. Strongly influenced by ideas prevailing elsewhere in North America and Europe at the time, he initially invoked a marine submergence of the Cordilleran interior to account for gravel deposits, or "beaches", high on valley walls, "boulder clay" (till), and bedded silt deposits (subsequently shown by Daly, 1912, to be glacial lacustrine in origin). Dawson abandoned the idea of a marine transgression after it received widespread criticism (Chamberlin, 1894) and devoted more of his energies to understanding the geometry and flow patterns of former ice masses that buried the Cordillera.

In his most comprehensive report on glacial geology, Dawson (1891) suggested that an ice sheet developed in the Canadian Cordillera when mountain glaciers (especially those in the Coast Mountains) expanded and coalesced to form a continuous cover over plateaus and coastal lowlands. Once fully established, this ice sheet had a dome-like central area in northern British Columbia from which ice flowed northwesterly into the Yukon and southeasterly into the United States.

In the late Nineteenth Century, Dawson's "Cordilleran glacier" was the subject of considerable debate among geologists working in British Columbia and Yukon Territory. The existence of such a glacier continued to be questioned into the Twentieth Century (e.g., Tyrrell, 1919). However, careful observations from many parts of the Canadian Cordillera and adjacent areas provided increasing support for an all-encompassing Pleistocene ice sheet (e.g., Willis, 1898; Tyrrell, 1901; Gwillim, 1902; McConnell, 1903a, b; Coleman, 1910; Daly, 1912, 1915; Bretz, 1913; Stewart, 1913; Read, 1921; Johnston, 1926b).

During the early Twentieth Century, attention shifted from questions relating to the existence of a former ice sheet to questions concerning the pattern of ice sheet growth (Kerr, 1934, 1936; Davis and Mathews, 1944) and the sequence of glacial and nonglacial intervals during the Quaternary (e.g., Clapp, 1913; Burwash, 1918; Berry and Johnston, 1922; Johnston, 1923, 1926a; Johnston and Uglow, 1926; Cockfield and Walker, 1933; Bostock, 1934; Kerr, 1934). In addition, a wide variety of other Quaternary phenomena began to attract the attention of scientists. For example, during the late Nineteenth and early Twentieth centuries, the first substantive contributions were made in the following fields in the Canadian Cordillera: Quaternary paleontology and paleoecology (Crickmay, 1925, 1929; Hansen, 1940; Cowan, 1941); recent glacier fluctuations (Munday, 1931, 1936; Wheeler, 1931); Quaternary sea level change (Lamplugh, 1886; Newcombe, 1914; Johnston, 1921a); slope stability (Stanton, 1898; McConnell and Brock, 1904; Daly et al., 1912); the sedimentology and genesis of Quaternary sediments (Johnston, 1921b, 1922; Hanson, 1932; Flint, 1935); and landform development (Peacock, 1935; Beach and Spivak, 1943). Notwithstanding these contributions, there were relatively few scientists working on Quaternary problems in the Canadian Cordillera during this period, and interest was correspondingly low.

The development of Quaternary ideas entered a new era in the middle and late 1940s when an increasing number of geologists developed interests in the surficial geology of the region. Some of these individuals (notably J.E. Armstrong, H.S. Bostock, and W.H. Mathews) recognized that a better understanding of the Quaternary Period could only be achieved if surficial sediments and landforms were investigated in their own right, rather than as adjuncts to bedrock mapping programs, as generally had been the case before. As a result of their efforts, new insights were gained into Quaternary geomorphic change in the Canadian Cordillera and into the pattern and history of glaciation in the region. These new insights came at a time of accelerated economic development and hydrocarbon and mineral exploration, which created a large demand for information on the distribution and character of unconsolidated deposits. This, in turn, led to a rapid growth in Quaternary research by government agencies and universities, a process that has continued unabated to the present. The explosion in Quaternary scientific endeavours in the last four decades has been accompanied by an increase in diversity and sophistication of research. Before World War II, many Quaternary studies were limited to observations of former ice flow directions, upper limits of glaciation, and changes in drainage patterns. In contrast, in recent years there has been increasing emphasis on, among other things, stratigraphy, former sedimentary environments, paleoecology, neotectonics, and hydrogeology. This chapter is a summary and synthesis of information collected in these and other fields.

Organization and authorship

These introductory remarks are followed by an overview of the Canadian Cordillera, which includes brief summaries of the following: bedrock geology; Quaternary tectonic setting; physiography and drainage; climate; Quaternary landscape development; the regional character and distribution of Quaternary deposits; controls on Quaternary deposition and erosion; the character and extent of the Cordilleran Ice Sheet; patterns of ice sheet growth and decay; relationships of Cordilleran and Laurentide ice masses; and Quaternary sea levels. Following this regional overview are sections concerned with the Quaternary stratigraphy, history, and paleoecology of various parts of the Canadian Cordillera. These are followed, in turn, by a section on Holocene glacier fluctuations and by one on the economic implications of Quaternary deposits and contemporary geologic processes.

The authorship of each major section is indicated after the corresponding section heading. Clague edited all contributions to ensure uniformity of style and completeness of coverage.

BEDROCK GEOLOGY

J.J. Clague

The Canadian Cordillera comprises part of the deformed western margin of the North American craton and a collage of far-travelled crustal fragments or "terranes"; the latter have become accreted to the craton in response to oblique convergence of North American and Pacific Ocean lithosphere during Mesozoic and Cenozoic time (Coney et al., 1980; Jones et al., 1982a, b; Monger et al., 1982; Chamberlain and Lambert, 1985). Each terrane is separated from its neighbours by major faults, intrusions, or a cover of younger rocks, and comprises a unique assemblage of rocks that formed in a particular tectonic setting (Tipper et al., 1981). By analogy with modern examples, most terranes in the Cordillera can be interpreted to be relics of island arcs, oceanic plateaus and islands, continental margin fragments, and complex accretionary terranes, the latter including melange belts, ophiolite fragments, and thrust-faulted forearc provinces (Silver and Smith, 1983).

Clague, J.J.
1989: Bedrock geology (Canadian Cordillera); in Chapter 1 of Quaternary Geology of Canada and Greenland, R.J. Fulton (ed.); Geological Survey of Canada, Geology of Canada, no. 1 (also Geological Society of America, The Geology of North America, v. K-1).

Accretion and related faulting have given the Canadian Cordillera a strong northwest-southeast structural grain. The region comprises five major northwest-trending geological belts, each with its own distinctive stratigraphy, metamorphism, plutonism, volcanism, and structure (Fig. 1.2, 1.3; Tipper et al., 1981; Monger et al., 1982; Monger and Berg, 1984; Gabrielse and Yorath, in press):

1. The Foreland Belt borders undeformed rocks of the Interior Platform on the east and consists of a northeasterly thinning wedge of middle Proterozoic to Upper Jurassic miogeoclinal and platform carbonates and craton-derived clastics, and Upper Jurassic to Paleogene exogeoclinal clastics derived from the Cordillera. This wedge is foreshortened about 200 km along a series of west-dipping thrust faults and northwest-trending folds that developed during the late Mesozoic and early Cenozoic.

2. The Omineca Belt straddles the boundary between ancestral North America and allochthonous terranes to the west, and is a major tectonic welt characterized by deformation, regional metamorphism, and granitic plutonism. The eastern part of the Omineca Belt comprises middle Proterozoic to middle Paleozoic miogeoclinal rocks, whereas the western part consists of accreted Paleozoic and lower Mesozoic volcanic and sedimentary rocks. Granitic plutons are common throughout the belt and are mainly of Jurassic and Cretaceous age. The intense metamorphism and complex structure of the Omineca Belt were produced when a large composite terrane collided with and became accreted to continental crust to the east, probably in Early and Middle Jurassic time.

3. The Intermontane Belt is a composite of three major exotic terranes formed of upper Paleozoic to middle Mesozoic, allochthonous, marine volcanic and sedimentary rocks. These rocks are superposed by autochthonous Jurassic and Cretaceous clastic wedges and generally flat-lying Tertiary volcanic and sedimentary rocks. The belt also contains granitic intrusions which are comagmatic with the various volcanics. The three major terranes of the Intermontane Belt amalgamated by the Middle Jurassic prior to colliding with continental crust to the east.

4. The Coast Belt, west of the Intermontane Belt, is the second large metamorphic and plutonic welt in the Canadian Cordillera. It consists of Jurassic to Tertiary granitic rocks and variably metamorphosed sedimentary and volcanic strata ranging in age from Paleozoic to early Tertiary. The Coast Belt formed mainly during Cretaceous and early Tertiary time, probably as a result of the collision of amalgamated terranes of the Intermontane Belt with a large composite terrane to the west (the Insular Belt). Compressional thickening and tectonic overlap resulting from this collision have been invoked to explain 5 to 25 km of uplift and erosion that occurred along the axis of the belt in Cenozoic time.

5. The Insular Belt, like the Intermontane Belt, is a composite allochthonous terrane. It comprises Upper Cambrian to Tertiary volcanic and sedimentary rocks, and granitics in part comagmatic with the volcanics. Its major elements probably amalgamated in Middle Jurassic time and collided with terranes of the Intermontane Belt during the Cretaceous. Accretionary

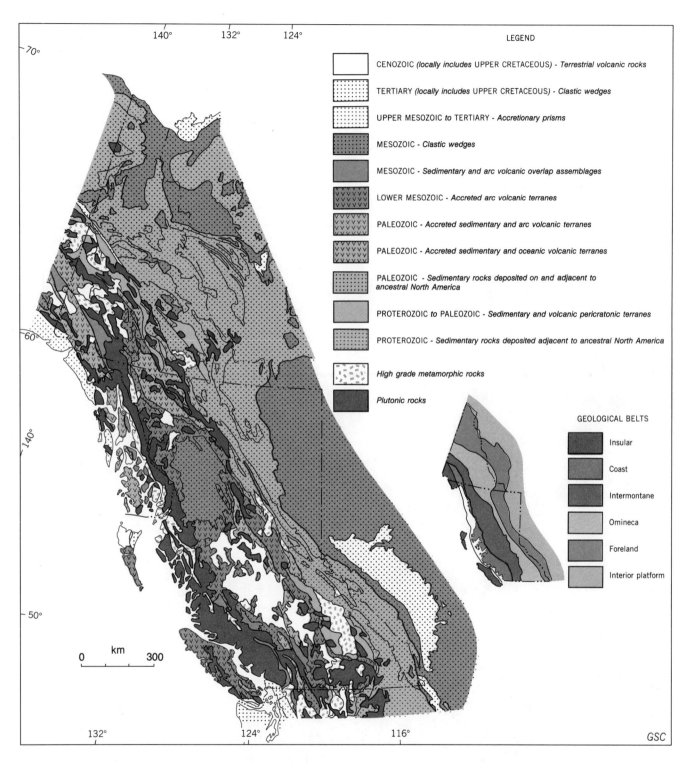

Figure 1.2. Simplified bedrock geology map of the Canadian Cordillera (adapted from Tipper et al., 1981).

prisms at the western margin of the Insular Belt, however, were added to the continent later, and, in fact, the Yakutat terrane is still accreting as the Pacific plate underthrusts Alaska.

Accretion of allochthonous terranes to North America was accompanied by compression and by northwestward displacements of accreted rocks. Displacements are manifested, in part, in the great right-lateral strike-slip faults of the region, such as the Tintina in Yukon Territory, active in Late Cretaceous-early Tertiary time, the Denali in Alaska and the Yukon, active since middle Tertiary time, and the offshore Queen Charlotte-Fairweather system which is

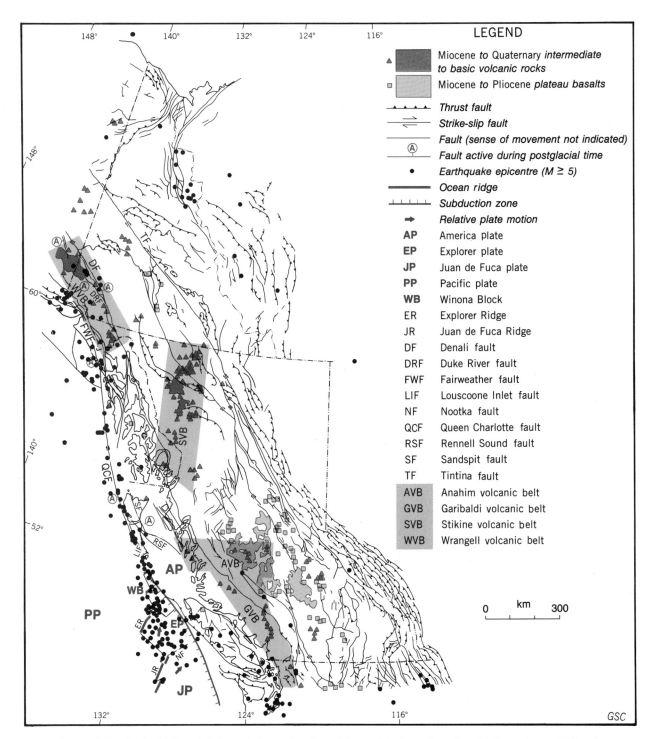

Figure 1.3. Major faults, plate boundaries, epicentres of large historic earthquakes (to December 1980) and young volcanic rocks in the Canadian Cordillera. Sources of information: Riddihough and Hyndman (1976), Souther (1976), Tipper et al. (1981), Geological Survey of Canada (unpublished).

active today. In the southeastern Canadian Cordillera, early Tertiary displacements were accompanied by crustal extension and uplift along listric normal faults, resulting in tectonic denudation of cover rocks above metamorphic core complexes.

QUATERNARY TECTONIC SETTING
J.J. Clague

The present tectonic regime of western North America is controlled mainly by the motions of the Pacific, America, and Juan de Fuca plates (Fig. 1.3; Atwater, 1970). In addition, there are two small lithospheric blocks (Explorer plate, Winona Block) at the north end of the Juan de Fuca plate; these may be moving as independent units (Riddihough, 1977). The Pacific and America plates and Winona Block intersect off the north end of Vancouver Island (Riddihough et al., 1980). North of this triple junction, the Queen Charlotte-Fairweather transform fault separates the Pacific and America plates. Displacements along this fault are right-lateral and average about 5.5 cm/a (Chase and Tiffin, 1972; Keen and Hyndman, 1979). South of the triple junction, a system of spreading ridges and short transform faults forms the boundary between the Pacific plate on the west and the Juan de Fuca and Explorer plates and Winona Block on the east (Barr and Chase, 1974; Riddihough et al., 1983). Spreading rates on Juan de Fuca and Explorer ridges range from about 4 to 6 cm/a (Keen and Hyndman, 1979). The Juan de Fuca and Explorer plates are separated by the Nootka transform fault which extends in a northeasterly direction from Juan de Fuca Ridge to the continental shelf of north-central Vancouver Island (Hyndman et al., 1979). Motion along this fault probably is left-lateral at a rate of about 2 cm/a.

The boundary between the America and Juan de Fuca plates is thought to be a zone of convergence or subduction. It is well established that subduction has occurred along the coasts of British Columbia, Washington, and Oregon during the past few million years (Riddihough and Hyndman, 1976), but there has been some debate as to whether or not it is continuing at present (Crosson, 1972). The doubt arises primarily from the absence of a deep marginal trench characteristic of most active subduction zones, the lack of deep or thrust-type earthquakes in an eastward-dipping Benioff zone, and the relatively low level of historic volcanism in the mountains bordering the Pacific Ocean. Riddihough and Hyndman (1976) reviewed relevant geological and geophysical data bearing on this problem and concluded that subduction probably is continuing south of 50°N, although at a very low rate in southwestern British Columbia.

The distribution of earthquakes, active and recently active faults, and Quaternary volcanoes in the Canadian Cordillera (Fig. 1.3) is related to the plate tectonic regime outlined above. The main earthquake areas are (1) the Queen Charlotte-Fairweather fault, (2) the offshore spreading ridge and associated fracture zones, (3) the southern Strait of Georgia-Puget Sound region, (4) the St. Elias Mountains, and (5) part of eastern Yukon Territory and westernmost District of Mackenzie (Basham et al., 1977; Milne et al., 1978). The first two areas are seismically active because they are plate boundaries. Focal mechanisms and fault-plane solutions of earthquakes in these two areas are in good agreement with postulated relative plate motions (Hyndman and Weichert, 1983). The southern Strait of Georgia-Puget Sound seismic area may correspond to the zone beneath which the Juan de Fuca plate remains essentially coherent as it is subducted beneath North America (Keen and Hyndman, 1979). Seismicity in this area may be related to lateral compression or overlapping in the sinking plate due to the major change in trend of the North American continental margin at this latitude. Earthquake activity in the Strait of Georgia decreases northward and is very low north of the presumed Explorer-America plate boundary at about 51°N. Seismicity in the St. Elias Mountains probably is linked in some way to convergence of the Pacific and America plates in the northern Gulf of Alaska. The relationship of seismicity in eastern Yukon Territory and western District of Mackenzie to the present-day tectonic regime is unknown, and the same is true for scattered earthquakes that occur elsewhere in the Cordillera, for example in south-central and southeastern British Columbia.

Many faults in the Canadian Cordillera are presently active or have been active sometime during the Holocene. On the basis of high seismicity, the Queen Charlotte fault and some of the transform faults connecting segments of Juan de Fuca and Explorer ridges are known to be active at present (Milne et al., 1978). Other major, predominantly strike-slip faults for which Holocene and recent movements are known or suspected include the Denali, Duke River, and associated faults in southwestern Yukon Territory (Clague, 1979; Horner, 1983), and the Sandspit, Rennell Sound, and Louscoone Inlet faults on the Queen Charlotte Islands (Sutherland-Brown, 1968; Young, 1981). Other major faults also may be active, but studies required to document recent displacements have not been undertaken. Finally, there probably are many small active faults in the Cordillera that have not been recognized by geologists. Hamilton and Luternauer (1983), however, identified numerous small Holocene faults of possible tectonic origin on the seafloor of the southern and central Strait of Georgia, and Eisbacher (1977) described fresh fault scarps in the Mackenzie Mountains.

Most Quaternary volcanoes in the Canadian Cordillera are located in four linear belts (Fig. 1.3; Souther, 1970, 1977). The southernmost, or Garibaldi, belt is the northern extension of a well defined chain of late Cenozoic volcanoes in northern California, Oregon, and Washington. It contains more than 30 Quaternary cones and domes of andesite, dacite, and minor basalt, which probably formed in response to subduction of the Juan de Fuca and Explorer plates. The Anahim volcanic belt, which extends in an east-west direction across western British Columbia at about 52°N, includes about 30 Quaternary volcanoes of mainly basaltic composition. This belt may be the product of progressive movement of the America plate over one or more hot spots in the mantle (Bevier et al., 1979; Rogers, 1981; Souther, 1986; Souther et al., 1987). The Stikine volcanic belt of northwestern British Columbia and southeastern Alaska contains more than 50 Quaternary eruptive centres, at least one of which is younger than 150 years old. The volcanoes

---

Clague, J.J.
1989: Quaternary tectonic setting (Canadian Cordillera); in Chapter 1 of Quaternary Geology of Canada and Greenland, R.J. Fulton (ed.); Geological Survey of Canada, Geology of Canada, no. 1 (also Geological Society of America, The Geology of North America, v. K-1).

are mainly basaltic in composition like those of the Anahim belt. Souther (1977) suggested that this belt may have formed in response to tensional stresses related to right-lateral shear between the Pacific and America plates. However, the position of the Stikine belt far to the east of the Pacific-America plate boundary and its termination near 60°N are difficult to explain in the context of the present plate tectonic regime. The northernmost, or Wrangell, volcanic belt defines a broad arc of calc-alkaline eruptives that extends through the St. Elias Mountains from southwestern Yukon Territory to central Alaska. Although the Canadian part of the belt is dominated by Tertiary lavas, the Alaskan part includes numerous Quaternary volcanoes which probably have formed in response to subduction of the Pacific plate at Aleutian Trench (Van Wormer et al., 1974).

Finally, lithospheric plate interactions are also responsible for recent vertical crustal movements in the Canadian Cordillera. Evidence from tidal records and geodetic relevelling indicates that there is a consistent pattern of uplift in outer coastal areas of British Columbia (up to 2 mm/a) and subsidence or no change on the inner coast (0-2 mm/a); the zero-uplift contour or "hinge line" runs through Hecate Strait and either the Strait of Georgia or Vancouver Island (Clague et al., 1982b; Riddihough, 1982). This pattern continues southward into Washington (Ando and Balazs, 1979), but is interrupted to the north by considerable uplift in southeastern Alaska and bordering areas of British Columbia and Yukon Territory (Hicks and Shofnos, 1965; Vaníček and Nagy, 1980, 1981). Parts of the rugged St. Elias Mountains and northern Coast Mountains are presently rising at rates of several millimetres to a few centimetres per year. Contemporary uplift also has been postulated for other areas. For example, an area in southeastern British Columbia, purportedly above a mantle hot spot, may be rising at a rate of several millimetres per year (Vaníček and Nagy, 1980; Riddihough, 1982). The pattern of contemporary uplift and subsidence in the Cordillera indicates that the movements are mainly tectonic in origin (Clague et al., 1982b; Riddihough, 1982). However, some uplift in heavily glacierized areas such as the St. Elias Mountains (e.g., at Glacier Bay) may be an isostatic response to recent deglaciation (Hicks and Shofnos, 1965).

PHYSIOGRAPHY AND DRAINAGE
J.J. Clague

The Canadian Cordillera comprises four main landscape elements: mountains (including "highlands"), plateaus, lowlands, and valleys (including fiords) (Fig. 1.4 to 1.7). Local physiographic units consisting of one or more of these landscape elements may be grouped into three major "systems" that extend the length of British Columbia and Yukon Territory: the Western System, Interior System, and Eastern System (Fig. 1.1; Bostock, 1948, 1970; Holland, 1964).

The Western System includes: (1) the great mountain ranges of southwestern Yukon Territory (St. Elias Mountains) and western British Columbia (Coast, Cascade, and Insular mountains); (2) narrow strips of coastal lowland bordering the British Columbia coast; (3) intramontane valleys and fiords; and (4) the seafloor of the British Columbia continental margin. The Western System also contains the largest icefields in the Cordillera and the highest peaks in Canada, including Mount Logan (ca. 5951 m) and Mount St. Elias (5489 m) (Fig. 1.8).

The Interior System is dominated by the large plateaus of interior British Columbia and Yukon Territory and by bordering mountain ranges. The main plateaus are the Yukon, Stikine, and Interior plateaus. Major mountain systems include the Ogilvie, Selwyn, Pelly, Cassiar, Omineca, Skeena, and Columbia mountains (some of these comprise two or more named ranges). The boundary between the Interior and Eastern systems in British Columbia is Rocky Mountain Trench, a remarkable topographic and structural feature that extends, with one major gap, over 1400 km in a northwesterly direction from the forty-ninth parallel to near the British Columbia-Yukon border. Tintina Trench, a similar structurally controlled valley, crosses the Interior System of Yukon Territory along the same trend.

The Eastern System is a region of rugged mountains, plateaus, lowlands, and valleys bordering the Interior Plains on the east. Its main elements are the Arctic, Mackenzie, Franklin, and Rocky mountains, and Peel Plateau.

Parts of the watersheds of the Pacific Ocean, Bering Sea, Arctic Ocean, and Hudson Bay occur within the Canadian Cordillera, and most of the large rivers of western Canada flow through or head in this region (Fig. 1.1). Fraser River and Columbia River drain most of the southern half of British Columbia and empty into the Pacific Ocean. Skeena River and Stikine River drain much of northwestern British Columbia and also flow to the Pacific. Saskatchewan River, which heads in the southern Rocky Mountains of Alberta, is within the Hudson Bay watershed. Athabasca River to the north also heads in the Rocky Mountains, but is part of the Mackenzie River system emptying into the Arctic Ocean. Most of north-central and northeastern British Columbia is drained by Peace and Liard rivers which are major tributaries of Mackenzie River. Western District of Mackenzie and parts of northern Yukon Territory are the sources of several streams which also empty into the Mackenzie. Mackenzie River itself is outside the Cordillera, except for a 500 km reach northwest of the mouth of Liard River (Fig. 1.1). Much of Yukon Territory is drained by Yukon River and its tributaries; this large river flows across Alaska into Bering Sea.

CLIMATE
J.M. Ryder

The climate of the Canadian Cordillera is controlled primarily by: (1) the location of the region in middle to high latitudes along a continental margin adjacent to the Pacific Ocean and (2) topography, with several mountain belts trending parallel to the coastline. The dominant atmospheric movement in this region is eastward, with moist air masses and cyclonic storms generated over the Pacific

---

Clague, J.J.
1989: Physiography and drainage (Canadian Cordillera); in Chapter 1 of Quaternary Geology of Canada and Greenland, R.J. Fulton (ed.); Geological Survey of Canada, Geology of Canada, no. 1 (also Geological Society of America, The Geology of North America, v. K-1).

Ryder, J.M.
1989: Climate (Canadian Cordillera); in Chapter 1 of Quaternary Geology of Canada and Greenland, R.J. Fulton (ed.); Geological Survey of Canada, Geology of Canada, no. 1 (also Geological Society of America, The Geology of North America, v. K-1).

Ocean moving inland across successive mountain ranges, plateaus, and valleys (Hare and Hay, 1974; Hare and Thomas, 1974).

The coastal region, west of the crest of the Coast, Cascade, and St. Elias mountains, is dominated by warm moist Pacific air. Much of this region receives more than 2500 mm of precipitation per year and thus is by far the wettest part of Canada (Hare and Hay, 1974). Winter precipitation along the coast results from a succession of frontal systems associated with cyclonic storms in the Gulf of Alaska; frontal precipitation is enhanced orographically so that amounts increase markedly with elevation. Snowfall accounts for only a small fraction of precipitation near sea level, but thick snowpacks accumulate during winter in mountains adjacent to the coast. In summer, the coastal region comes under the influence of a large anticyclone and there are spells of fair weather with infrequent convective storms. Air temperatures vary much less on a daily and seasonal basis than farther inland due to the moderating influence of maritime air masses. Winters are mild; mean January temperatures range from about 1°C to 5°C, decreasing rapidly inland from the open coast (Atmospheric Environment Service, n.d.). Along the coast, the mean annual temperature range (i.e., the difference between the mean temperatures of the warmest and coldest months) is only 10-15°C, the lowest in Canada.

The southern interior region, which extends eastward from the crest of the Coast and Cascade mountains to the

Figure 1.4. Coast Mountains west of Kitimat, British Columbia. Province of British Columbia BC528-21.

Interior Plains and northward to about 58°N, is characterized by strong climatic contrasts. Temperatures and humidity are controlled partly by modified maritime air masses which have lost much of their moisture during their passage across the coastal mountains. However, incursions of continental arctic air in winter and continental tropical air in summer produce extreme and variable conditions. Precipitation is more evenly distributed throughout the year in this region than on the coast, although the proportion of snow to rain is greater. Local rain shadows and orographic effects determine the regional distribution of precipitation. Some valleys in the south receive less than 250 mm of precipitation per year, whereas valleys in the northern part of this region receive 400-600 mm (Atmospheric Environment Service, n.d.). Greater amounts of precipitation fall on plateaus (700 mm/a or more) and in the Columbia and Rocky mountains (up to 1500-2000 mm/a on the western flanks). In general, air temperatures are warmest in the south and decrease both northward and with increasing elevation. Mean valley floor temperatures in

Figure 1.5. Fraser River valley west of Clinton, British Columbia. Province of British Columbia BC1087-46.

January range from −3°C to −6°C in the south to −10°C to −17°C in the north; corresponding mean July temperatures are 18-21°C and 14-17°C (Atmospheric Environment Service, n.d.). The annual temperature range in the southern interior climatic region is about 26°C, twice that of the coast.

The northern Cordilleran region, which includes British Columbia north of 58°N, Yukon Territory, and western District of Mackenzie, experiences a climate that is similar to other high latitude boreal forest regions of Canada. Winters are long (October-April) and are dominated by extremely cold dry continental arctic air. Precipitation amounts are low (300-400 mm/a on plateaus and in valleys and lowlands) due to the position of this region on the lee side of the St. Elias and Coast mountains and due to the low water-vapour holding capacity of the cold air masses (Kendrew and Kerr, 1955; Atmospheric Environment Service, 1982). There is no pronounced wet or dry season, although spring is relatively dry and many localities have modest summer precipitation maxima.

Figure 1.6. Drumlinized plateau northeast of Prince George, British Columbia; view northeast towards the Rocky Mountains. Province of British Columbia BC761-71.

Permafrost is continuous in northern Yukon Territory and is common although discontinuous elsewhere in the Yukon and in the mountainous part of District of Mackenzie. In British Columbia, permafrost has a patchy distribution above 1200 m elevation in the north and above 2100 m in the south (Brown, 1967; Hare and Thomas, 1974).

Figure 1.7. Strandflat and mountain fringe, western Vancouver Island. Province of British Columbia BC666-93.

Figure 1.8. Mount St. Elias, the second highest mountain in Canada. Courtesy of Austin Post.

DEVELOPMENT OF CORDILLERAN LANDSCAPES DURING THE QUATERNARY

W.H. Mathews

An understanding of landscape evolution in the Canadian Cordillera involves two more or less interrelated questions: (1) what was the landscape like at the beginning of the Quaternary (or how did it differ from that of the present) and (2) what geomorphic processes were effective during the Quaternary in bringing about physiographic change.

The landscape at the beginning of the Quaternary

Data to answer the first question are limited because of the scarcity of well dated deposits from this part of geological time. Nevertheless, some generalizations can be made regarding the physiography of the Canadian Cordillera at the start of the Quaternary Period. These generalizations are based largely on observations of the distribution and character of late Tertiary sedimentary and volcanic rocks and landforms.

The Cordilleran landscape at the beginning of the Quaternary was dominated by mountainous uplands, dissected plateaus, and alluvial plains; it thus probably resembled the landscape of present-day northern Yukon Territory. The distribution of high and low ground was probably much as it is today, but the relief was almost certainly significantly less. Valleys, particularly those of the interior, were not cut to anything like their present depths nor were they yet extensively modified by glaciation. The streams that occupied them may have differed in size and even in direction of flow from those now existing. The mountains were almost certainly lower than their modern counterparts, and their glacial modification, though probably present, may have been much more restricted than now in both distribution and degree. The coastline was probably less indented than at present, looking perhaps somewhat like the northeast shores of the Queen Charlotte Islands or those bordering the northern Gulf of Alaska today. The possibilities of sea level changes associated with Pliocene glaciations in Antarctica could have led to incision of river mouths followed by drowning to form estuaries such as Masset Sound on the Queen Charlotte Islands. It is questionable whether any local Pliocene glacier reached the sea to create the fiords which are such a distinctive feature of the present west coast. The existence at that time of fiord lakes in the interior valleys is also in doubt.

Geomorphic processes

Major geomorphic processes that have shaped the surface of the Cordillera during Quaternary time include diastrophism, volcanism, subaerial erosion and deposition, and glaciation.

Mathews, W.H.
1989: Development of Cordilleran landscapes during the Quaternary; in Chapter 1 of Quaternary Geology of Canada and Greenland, R.J. Fulton (ed.); Geological Survey of Canada, Geology of Canada, no. 1 (also Geological Society of America, The Geology of North America, v. K-1).

Diastrophism

Diastrophism has resulted in uplift and subsidence of various parts of the Cordillera during the Quaternary, as illustrated by the following examples:

1. Miocene lavas in the northeastern St. Elias Mountains and southern Coast Mountains have been tilted, faulted, and uplifted (Tipper, 1963; Souther and Stanciu, 1975). On the basis of apatite fission-track dates, Parrish (1981) concluded that there has been sustained uplift and denudation of Mount Logan in the St. Elias Mountains averaging about 0.3 m/ka for the past 15 Ma. Somewhat slower uplift, generally >0.2 m/ka, is recorded in the southern Coast Mountains over the last 10 Ma (Fig. 1.9; Parrish, 1983). Tilting and deep incision of Pliocene and early Quaternary lavas along Fraser River between 51°N and 52°N suggest that this uplift has continued in the Quaternary.

2. Dissected gravel-capped pediplains are locally present over the folded and faulted rocks of the eastern foothills of the Cordillera from Montana north to the Richardson Mountains (Alden, 1932; Ross, 1959; Roed, 1975; Rampton, 1982); some also occur on the Interior Plains to the east. These relict surfaces range in age from Early Oligocene to late Tertiary (Russell, 1957; Russell and Churcher, 1972; Mathews, 1978; Rampton, 1982). They were uplifted and incised during middle to late Tertiary and Quaternary time as the mountains to the west rose.

3. Additional evidence for relatively recent uplift of the Rocky Mountains comes from a study of speleothems in relict phreatic caves preserved in valley walls. Uranium-series dates from such speleothems provide information on rates of valley deepening; data from the southern Rockies (Ford et al., 1981) indicate downcutting at rates of >0.04-2.07 m/ka over the past several hundred thousand years, and by extrapolation suggest that the mean relief of these mountains, some 1340 m, could have developed over a span of 0.65-33.4 Ma, most probably 1.2-12 Ma.

4. In the Klondike area of central Yukon Territory, the "White Channel gravels", assigned a Pliocene and early Pleistocene age (Hughes et al., 1972; Naeser et al., 1982), mark a series of river courses since incised to depths of about 100 m. Hughes et al. (1972) concluded that tectonic tilting contributed to the aggradation of these and associated Yukon River gravels which are now 150 m higher downstream from Dawson than at the mouth of Stewart River, 150 km upstream.

5. Tectonic subsidence or faulting has been offered as the explanation for Old Crow, Bluefish, and Bell flats in northern Yukon Territory (Hughes, 1972), and for the Strait of Georgia (Mathews, 1972) and Hecate Strait (Yorath and Hyndman, 1983). Downwarping of late Miocene strata beneath the floor of Hecate Strait (Fig. 1.9) indicates that much of the subsidence there is Plio-Pleistocene in age.

Volcanism

Volcanism, which generated widespread basaltic lavas in the British Columbia interior in the late Tertiary, continued on a reduced scale during the Quaternary. Most Quaternary volcanoes in the Canadian Cordillera are localized in four relatively narrow belts (Fig. 1.3; see also Quaternary tectonic setting). The eruptive products are mainly basalts, except in the Garibaldi belt in the south where andesites and dacites of calc-alkaline affiliation are dominant. The most common volcanic landforms are steep-sided pyroclastic cones, large eroded stratovolcanoes, and intra-valley flows. Anomalous landforms at several centres have been ascribed to eruption under, through, or onto glacier ice or into glacier-dammed lakes. A flat-topped version, the "tuya" (Mathews, 1947), consists of a cap of gently sloping, subaerially cooled lava flows crowning a pile of outward-dipping brecciated basalt and/or pillow lava laid down within a glacier-dammed lake. Another form, most closely resembling a giant sawn-off tree stump, is believed to be a slightly modified plug that was intruded through the Cordilleran Ice Sheet (Mathews, 1951b). Yet other forms, which are eskerlike in shape, probably were produced when lava solidified while flowing through trenches or tunnels in the ice (Mathews, 1958).

Subaerial erosion and deposition

Subaerial erosion accompanying and following late Cenozoic uplift did much to develop the major valleys of the Cordillera. Fluvial incision, particularly in its early stages, could be expected to form V-shaped valleys. Subsequent occupation by valley glaciers could then lead to their widening and deepening. Although the relative importance of these two processes in glaciated parts of the Cordillera cannot be completely evaluated, the common localization of valleys along narrow fracture systems or belts of nonresistant rocks is more consistent with the selective action of streams than with the less precise scouring of glaciers. Accordingly, most Cordilleran valleys probably were initiated by steams prior to glaciation (some exceptions are noted in the next section). In unglaciated northern Yukon Territory, streams alone controlled erosion. However, even here periodic glacial activity to the east and south could have influenced stream behaviour by adding to the sediment load and encouraging aggradation of such streams as Klondike and Yukon rivers (Hughes et al., 1972).

Though streams may have been instrumental in cutting the major valleys of the Cordillera, they did not necessarily flow along them in the same direction as their modern counterparts. Indeed, stream diversions and reversals took place frequently during the Quaternary. Many diversions resulted from the blockage of stream courses by glaciers, but damming and diversion of waters by glacial and fluvial deposits or more locally by lava flows or landslides were also important in rearranging the drainage pattern.

Mass movement processes have contributed to the denudation of mountains and the widening of valleys during the Quaternary by transferring material from oversteepened mountainsides to sites where the debris could be carried away by streams. Weathering and soil formation accompanied by removal of the alteration products have been less important in this respect. Periglacial processes, such as solifluction, cryoplanation, and thermokarst development, limited today to high latitudes and high altitudes, were more widespread during cooler nonglacial stages of the Pleistocene and during early parts of glacial cycles. Such processes were probably more important at inland sites, with a relatively dry continental climate, than near the coast where a blanket of winter snow or permanent glacier ice would inhibit periglacial activity.

A strandflat (i.e., low coastal rock platform abutting against a sharply rising mountain slope) is recognizable along the Pacific coast from Vancouver Island northward

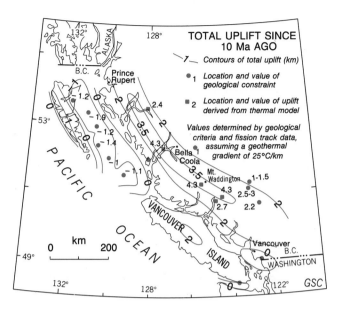

Figure 1.9. Total uplift of western British Columbia during the last 10 Ma (adapted from Parrish, 1983, Fig. 11).

(Fig. 1.7; Holland, 1964). It extends as much as 10 km inland and to an elevation of about 150 m; locally it comprises swarms of small islets and rocky reefs more or less modified by wave action. The origin of the strandflat is debated, but accelerated Pleistocene frost weathering in the presence of abundant moisture and poor drainage likely contributed to its formation (Holtedahl, 1960); marine planation and glaciation seem to have played a lesser role.

Subaerial depositional processes have had a significant, although localized, impact on the Cordilleran landscape. Deltas, floodplains, and alluvial fans are the most important features produced by fluvial deposition; landslides and colluvial fans are the main products of mass wasting.

Glaciation

Glaciers first formed in the northern Cordillera more than 9 Ma ago (Denton and Armstrong, 1969) and since then have profoundly altered the landscape of the region. In the high mountains, the gathering ground for the Cordilleran Ice Sheet and satellite glaciers, where summits and ridges stood above névé surfaces, classic alpine landforms were created, including cirques and overdeepened valley heads, horns, and comb ridges. In the major valleys, especially along the Pacific shore, the effects of glaciation are even more conspicuous. The coastline is highly indented by fiords extending as much as 150 km inland and to water depths of as much as 755 m (Peacock, 1935; Pickard, 1961). Most fiords have reaches with relatively flat floors stemming from partial infilling by sediments up to several hundred metres thick. Nevertheless, longitudinal profiles at the sea-sediment interface are generally highly irregular, and even more so on the sediment-bedrock interface. The major fiords trend more or less perpendicular to the coastline, but oblique and cross channels linking these fiords to one another are common.

The continental shelf between Vancouver Island and the Queen Charlotte Islands is crossed by three major streamlined U-shaped troughs 20-30 km wide and 100-300 m deep (Luternauer and Murray, 1983). These share many of the characteristics of the Laurentian Channel west and south of Newfoundland and others off the Labrador coast. The similarity in pattern and scale of these troughs to those carrying fast-moving ice streams within the glacier apron draining westerly to the Ross Ice Shelf, Antarctica (Rose, 1979) supports the idea that they owe their origin to glacial scour. Juan de Fuca Strait, the Strait of Georgia, and Dixon Entrance also have been extensively modified by streams of ice at the western margin of the Cordilleran Ice Sheet.

East of the axis of the Coast and St. Elias mountains, the erosional effects of glaciation, though not as spectacular as on the coast, are still evident. Deep elongate lakes take the place of the coastal inlets. These fiord lakes are most common at the margins of major mountain systems, are up to 125 km long, and are known to have depths of as much as 418 m. Most are dammed at their outlets by either glacial deposits or alluvial fans.

Some valleys in the Cordillera, such as those of Homathko River crossing the southern Coast Mountains, Kootenay River between Kootenay Lake and Columbia River, and Mackenzie River near The Ramparts (Mackay and Mathews, 1973), show no signs of channel fills predating the last glaciation. It is likely that these valleys mark the sites of former bedrock saddles which have been deepened by glacial scour, aided perhaps by meltwater streams, so that in late glacial time they could be occupied by throughgoing rivers which had formerly discharged by other routes.

Smaller-scale glacial landforms, almost all of which were produced during the last (Late Wisconsinan) glaciation, are conspicuous in many parts of the Canadian Cordillera. Drumlins, flutings, and other streamlined forms are products of subglacial erosion and/or deposition and are found on perhaps 50% of the plateau surface of the Cordilleran interior (Armstrong and Tipper, 1948; Tipper 1971a, b). In contrast, morainal ridges are rare except in front of modern glaciers. Glacial fluvial landforms, including abandoned meltwater channels, compound eskers, kames, kame terraces, outwash plains, and valley trains, are striking features of the landscape and record the decay of the Cordilleran Ice Sheet at the close of the Pleistocene. Incised depositional plains of glacial lacustrine and glacial marine origin are found in some interior valleys and along the coast, respectively, but are subdued and inconspicuous except in relatively arid grassland areas.

CHARACTER AND DISTRIBUTION OF QUATERNARY DEPOSITS

J.J. Clague

Quaternary deposits in the Canadian Cordillera are extremely varied in character and have complex distributions controlled largely by physiography and glacial history. Most of the surface sediments were deposited during the last glaciation and during postglacial time. Older materials occur at the surface in central and northern Yukon Territory, District of Mackenzie, the eastern Rocky Mountains, and on parts of the continental shelf; they also underlie Late Wisconsinan glacial deposits in some valleys and lowlands in British Columbia and the southern Yukon. Surficial geology maps prepared by the Geological Survey of Canada, British Columbia Ministry of Environment, and Alberta Research Council show the distribution of surface Quaternary deposits and landforms for many parts of the Canadian Cordillera. Map 1704A shows the areas covered by these maps and gives a list of references.

Morainal deposits (till)

In the Canadian Cordillera, sediments deposited directly from glacier ice in subglacial and supraglacial settings typically are poorly sorted, massive to weakly stratified, and very stony; they have matrixes of mixed sand, silt, and clay (Fig. 1.10). Morainal sediments vary in character both locally and regionally because of differences in source materials, complexities in the pattern of glacier flow, the presence of diverse depositional environments, and the effects of water and gravity during deposition (Keser, 1970; Scott, 1976). In general, there is a direct relationship between local bedrock type and till composition and texture. Till derived from volcanic rocks, carbonates, mudstone, shale, or slate typically has a matrix rich in silt and clay; in contrast, till derived from granitic rocks, schist, gneiss, sandstone, or conglomerate has a sandy matrix. Some till is composed of particles eroded from older Quaternary sediments and commonly exhibits the textural and compositional properties of these materials. For example, in the Strait of Georgia region of southwestern British Columbia, the surface till is very sandy and contains abundant, well rounded granitic pebbles and cobbles. It was deposited by glaciers that flowed across sandy outwash rich in granitic detritus.

Till in high-relief areas generally is more gravelly and has less silt and clay than till in lowlands and on plateaus. Some coarse till on steep mountain slopes is similar to poorly sorted glacial fluvial and colluvial sediments and, in fact, there is a complete spectrum of materials ranging from till to glacial fluvial gravel on one hand, and from till to colluvium on the other.

Till probably is the most extensive of all surficial materials in the Canadian Cordillera. It covers most plateaus in British Columbia and southern Yukon Territory and also is common in some lowlands. In mountain areas, it occurs on the floors of some valleys and on adjacent slopes. At high elevations and on steep slopes, however, till generally is either thin and patchy, or completely absent.

Glacial fluvial deposits

Glacial fluvial sediments include both ice contact and outwash materials deposited in subaerial environments and in lakes and the sea. Ice contact fluvial materials consist of gravel and/or sand which exhibit variable sorting and stratification. Where present, bedding commonly is distorted and broken due to melt of associated ice during and shortly after deposition. Ice contact sediments in the Canadian

---

Clague, J.J.
1989: Character and distribution of Quaternary deposits (Canadian Cordillera); in Chapter 1 of Quaternary Geology of Canada and Greenland, R.J. Fulton (ed.); Geological Survey of Canada, Geology of Canada, no. 1 (also Geological Society of America, The Geology of North America, v. K-1).

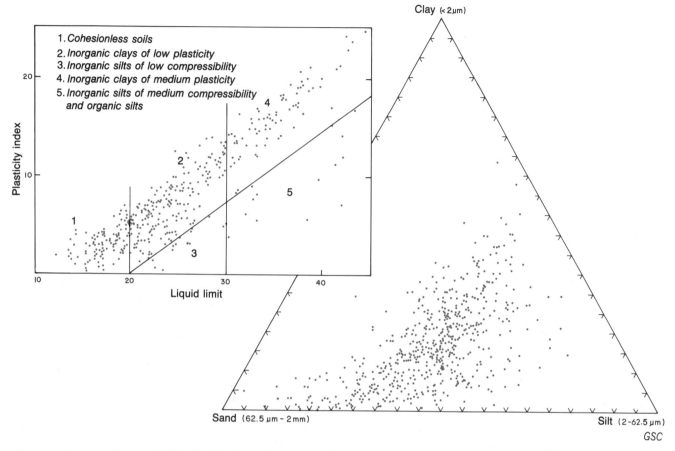

Figure 1.10. Particle-size distribution (<2 mm) and engineering properties of representative till samples from the Canadian Cordillera. Plasticity chart and nomenclature after Casagrande (1932). Data, in part, from Fyles (1963), Clague (1973), Jackson (1977, 1987), Howes (1981b), Ryder (1981a, 1985), Clague and Luternauer (1983), and Klassen (1987); unpublished data provided by J.J. Clague, D.E. Howes, R.W. Klassen, L. Lacelle, J.M. Ryder, and B. Thomson.

Cordillera are preserved in kames, kame terraces, knob-and-kettle topography, and eskers, and also occur as thin discontinuous sheets of gravel and sand overlying till. These sediments are widely distributed, except in rugged alpine terrain and, of course, in areas that escaped glaciation. Major esker and kame complexes and associated large meltwater channels are present on the Yukon and Interior plateaus and in Liard Lowland (Armstrong and Tipper, 1948; Prest et al., 1968; Tipper, 1971a, b).

Outwash consists of stratified, well sorted gravel and sand and occurs as valley trains, plains, terraces, and deltas. These materials generally lack deformation structures characteristic of most ice contact deposits. In the Canadian Cordillera, large amounts of outwash were deposited at the close of the last glaciation on proglacial floodplains and deltas and in subaqueous fans that prograded outward from glaciers terminating in the sea or in lakes. Outwash is common in most large valleys and lowlands and also is present locally on the British Columbia continental shelf. Late Pleistocene valley trains, outwash plains, and deltas were incised by streams during the Holocene, and thus now occur as terraces above present-day floodplains. In contrast, active outwash trains extending downvalley from large glaciers in the high mountains of the Cordillera are graded to existing streams.

Glacial fluvial sediments also occur beneath other types of Quaternary sediments and have little or no surface expression. Some of these subsurface glacial fluvial materials predate the Late Wisconsinan glacial maximum.

Glacial lacustrine deposits

Glacial lacustrine deposits consist mainly of fine sand, silt, and clay carried into glacier-dammed lakes by meltwater streams and deposited from overflows, interflows, and underflows; coarser sediments generally are restricted to subaqueous outwash fans, deltas, and beaches. Most glacial lacustrine sediments are well stratified, and many are laminated or varved, suggesting deposition under relatively stable conditions; others have irregular or disturbed stratification and apparently were deposited as ice contact sediments.

These materials are sufficiently thick in many areas to completely mask the morphology of underlying sediments or bedrock; in these areas the old lake bed forms a level or undulating surface dissected by streams. Thick glacial lacustrine sediments deposited in contact with ice may slump or settle as the ice melts, giving rise to irregular or kettled topography.

Glacial lacustrine sediments deposited at the close of the last glaciation are common in several large valleys in British Columbia and southern Yukon Territory (Mathews, 1944; Kindle, 1952; Prest et al., 1968; Fulton, 1969; Tipper, 1971a; Clague, 1987). Similar sediments accumulated in lakes in unglaciated northern Yukon Territory when the Laurentide Ice Sheet advanced to the front of the Richardson Mountains and dammed east-flowing streams (Hughes, 1972; Morlan, 1979; Hughes et al., 1981). Glacial lacustrine silt and clay also are found in many places beneath till; these sediments probably accumulated in lakes impounded by advancing Late Wisconsinan glaciers.

Glacial marine deposits

Glacial marine sediments in the Canadian Cordillera were deposited on the sea floor in front of temperate tidewater glaciers. They consist of material that was introduced into the sea by meltwater streams, by flows of supraglacial debris directly from ice, and by the melting of icebergs. Fine detritus settled out of suspension from overflows, interflows, and underflows; some was redeposited by mass movement processes. Some coarse material was dropped from icebergs and the calving fronts of debris-laden glaciers; some was deposited at the mouths of meltwater streams.

The most common type of glacial marine sediment in the Cordillera is massive to stratified mud. Most of this material contains less than 5% gravel-size material, and some is stone-free. In contrast, some mud is very gravelly and is not easily distinguished from till. Stratified gravel and sand of glacial marine origin are restricted to deltas, subaqueous outwash fans, and beaches. These sediments are fairly well sorted, except where laid down directly against ice.

Glacial marine sediments occur locally on the coastal lowlands of western British Columbia (Fyles, 1963; Armstrong, 1981; Clague, 1981, 1985) and on the adjacent continental shelf (Luternauer and Murray, 1983). Most of the sediments were deposited at the close of the last glaciation as the Cordilleran Ice Sheet decayed and retreated east and north in contact with the sea. At that time, coastal lowlands in most areas were submerged because the crust was isostatically depressed by the decaying ice sheet. Glacial marine sedimentation ceased when glaciers disappeared from coastal areas. This was accompanied by the emergence of coastal lowlands due to isostatic rebound. Older glacial marine sediments, deposited during one or more glaciations preceding the Late Wisconsinan, have been identified on eastern Vancouver Island and in Fraser Lowland (Fyles, 1963; Armstrong, 1981; Howes, 1981a; Hicock and Armstrong, 1983).

Colluvial deposits

Sediments produced by mass movement processes are common at the surface throughout the Canadian Cordillera and also occur locally in the subsurface beneath Late Wisconsinan drift. There are four main types of colluvium in this region: (1) Landslide deposits. Deposits produced by landslides range from fine textured masses of displaced glacial lacustrine and glacial marine sediments to coarse blocky accumulations of broken rock. Their physical characteristics are determined by the geology of the source materials, the topography of the failure site, and the type of landslide motion (e.g., fall, topple, slide, spread, flow; Varnes, 1978). Most landslides have irregular hummocky surfaces, although highly fluid, channelized flows of debris and mud characteristically build cones or fans similar to alluvial fans. Although widely distributed in the Canadian Cordillera, landslide deposits cover a very small part of the total surface area ($<1\%$). (2) Talus. Cones, aprons, and thin mantles of broken rock occur below steep bedrock slopes in all mountain areas. Some thick talus accumulations have glacier-like forms resulting from the deformation of interstitial ice or the flow of subjacent glacier ice (i.e., rock glaciers). (3) Colluviated drift. Sheet erosion and creep have gradually transformed some Late Wisconsinan glacial deposits into colluvium. Thin mantles of diamicton and poorly sorted gravel formed in this way are common on moderate and steep slopes in most parts of the Cordillera. The role, if any, of freeze-thaw activity in the genesis of these sediments is unknown. (4) Solifluction deposits. Masses of weathered rock and unconsolidated sediments which have flowed slowly downhill in response to cyclic freezing and thawing are common both in alpine areas and at lower elevations on slopes with impervious (commonly frozen) substrates.

Fluvial (alluvial) deposits

Stratified sediments deposited by streams are associated with Holocene and contemporary channels, floodplains, terraces, fans, and deltas, and also occur beneath Late Wisconsinan drift.

Streams with steep gradients and high sediment loads tend to have shallow braided channels and gravelly floodplains. They are most common in mountain areas. Streams with low gradients and low to intermediate sediment loads typically have sinuous or meandering channels. Active channels of low gradient streams commonly are floored by sand and/or fine gravel, whereas their floodplains are underlain mainly by sand and silt deposited during floods. Streams of this sort occur in broad, low gradient valleys, mostly outside mountain ranges.

Fluvial terraces are underlain by gravel similar to that found on floodplains of braided streams (note: a thin veneer of sand and silt commonly overlies the gravel). They formed during the Holocene when streams incised Pleistocene valley fills.

Alluvial fans consist of gravel, sand, and minor diamicton. Sediments generally are coarser at the apex of a fan than at the toe. Many alluvial fans are inactive relict features that formed during and shortly after deglaciation at the close of the Pleistocene, whereas others have accumulated sediment throughout the Holocene and remain active today.

Deltas are gently sloping alluvial surfaces built out into lakes or the sea by streams. Deposition occurs mainly on subaqueous foreslopes that drop off from the delta top to the lake or sea floor beyond. The texture of deltaic sediments depends to a large extent on the gradient of the source stream and the calibre of the material carried by that stream. Deltas of steep mountain torrents consist of coarse gravel, whereas those of large, low gradient streams such as the Fraser, Columbia, and Skeena rivers consist mainly of sand and silt. Fine deltaic sediments are overlain by coarser alluvium at the mouths of many rivers and for some distance upstream. Some active deltas in the Canadian Cordillera began to grow during deglaciation at the close of the Pleistocene; others were initiated during the Holocene. In addition to active deltas, there are terraced relict deltas that were built into former lakes and the sea and later incised by streams.

Lacustrine and marine deposits

Lacustrine and marine sediments resemble their glacial counterparts (glacial lacustrine and glacial marine sediments), but unlike them, accumulated in environments free of glacier ice. These nonglacial materials include offshore silt and clay deposited from suspension and locally redistributed by sediment gravity flows, and nearshore gravel and sand formed by wave and current action. In addition, coarse lag deposits occur in areas of strong bottom currents on the British Columbia continental shelf. Holocene lacustrine sediments are thickest and most extensive in large lakes with high sediment influx, for example, Upper Arrow, Lillooet, and Kamloops lakes in British Columbia (Fulton and Pullen, 1969; Gilbert, 1975; Pharo and Carmacks, 1979) and Kluane Lake in Yukon Territory (Terrain Analysis and Mapping Services Ltd., 1978). Marine sediments are widespread on the seafloor off British Columbia (e.g., Pickard, 1961; Carter, 1973; Clague, 1977b; Bornhold, 1983; Luternauer and Murray, 1983), and also occur in a narrow fringe inland from the present shore in a few areas, for example on parts of the Queen Charlotte Islands (Clague et al., 1982a, b). Lacustrine and marine sediments predating the last glaciation are also present in the Cordillera, but they typically occur beneath younger sediments, consequently their extent is not well known.

Other Quaternary deposits

Organic materials in the Canadian Cordillera include peat, gyttja, and marl. Deposits of these materials are generally less than 5 m thick and are found in shallow closed depressions, at the margins of shallow lakes, and on gently to moderately sloping, poorly drained surfaces. They occur in favourable topographic situations throughout the Cordillera, but attain their greatest extent in areas of low relief and high rainfall and in low-lying areas underlain by permafrost. Organic deposits record past vegetation changes and thus are an important source of paleoenvironmental information (see Paleoecology and paleoclimatology).

Eolian sediments, mainly sand and silt, form localized dunes and thin mantles on older deposits. In addition, fine wind blown particles are present in the surface soil throughout the Cordillera. Thick eolian sediments are restricted to valleys containing sandy floodplain sediments. Most surface eolian sediments were deposited at the close of the Pleistocene when large outwash plains and valley trains were active. Today, significant eolian sedimentation is restricted to small widely scattered sites.

Volcanic materials, mainly lavas and coarse pyroclastics of basaltic, andesitic, and dacitic composition, are found in more than 100 Quaternary eruptive centres in British Columbia and southwestern Yukon Territory (Fig. 1.3; Souther, 1970, 1977). In addition, layers of tephra, consisting of sand-, silt-, and clay-size particles, occur within nonvolcanic sedimentary deposits in many areas.

Although volumetrically small, these tephras are invaluable for correlation purposes because they are widespread and of known age (Table 1.1, Fig. 1.11). Those in the southern Cordillera are the products of volcanic eruptions in western Oregon, western Washington, and southwestern British Columbia, whereas those in the Yukon were erupted from Alaskan volcanoes.

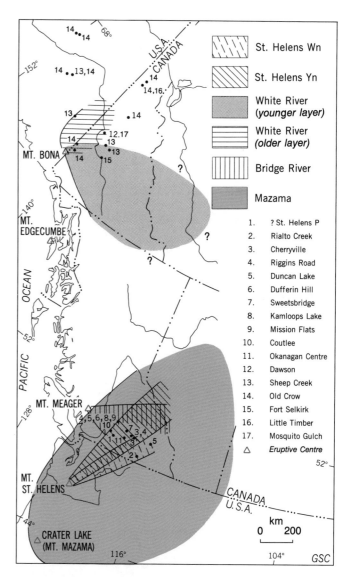

Figure 1.11. Distribution of Quaternary tephra deposits found in the Canadian Cordillera. Sources of information: Bostock (1952), Nasmith et al. (1967), Rampton (1969), Westgate et al. (1970, 1985), Fulton (1971), Hughes et al. (1972), Mullineaux et al. (1975), Westgate and Fulton (1975), Smith et al. (1977), Westgate (1977, 1982), Mathewes and Westgate (1980), Naeser et al. (1982). Tephras derived from Mount Edgecumbe are present locally in the northern Coast Mountains near the Alaska-British Columbia boundary.

Table 1.1. Documented Quaternary tephras in the Canadian Cordillera

Tephra	Age (ka BP)[1]	Source	Known occurrence in Canadian Cordillera	References
HOLOCENE				
St. Helens Wn	0.508[2]	Mt. St. Helens, Washington	south-central B.C.	Fulton, 1971; Smith et al., 1977; Yamaguchi, 1983
White River (younger layer)	1.2	Mt. Bona, Alaska	southern Yukon, western District of Mackenzie	Bostock, 1952; Lerbekmo and Campbell, 1969; Hughes et al., 1972; Lerbekmo et al., 1975
White River (older layer)	>1.5 <1.9	Mt. Bona, Alaska	west-central Yukon	Bostock, 1952; Lerbekmo and Campbell, 1969; Hughes et al., 1972; Lerbekmo et al., 1975
Bridge River[3]	2.4	Mt. Meager, B.C.	southern B.C., southwestern Alberta	Nasmith et al., 1967; Westgate and Dreimanis 1967; Westgate et al., 1970; Fulton, 1971; Westgate, 1977; Mathewes and Westgate, 1980
St. Helens Yn	3.4	Mt. St. Helens, Washington	southern B.C., southwestern Alberta	Westgate and Dreimanis, 1967; Westgate et al., 1970; Fulton, 1971; Mullineaux et al., 1975; Westgate, 1977
Mazama	6.8	Crater Lake, Oregon	southern B.C., southwestern Alberta	Powers and Wilcox, 1964; Fryxell, 1965; Westgate and Dreimanis, 1967; Westgate et al., 1970; Fulton, 1971; Bacon, 1983
PLEISTOCENE				
Rialto Creek	20	Mt. St. Helens, Washington?	south-central B.C.	Westgate and Fulton, 1975
Cherryville	25	Mt. St. Helens, Washington?	south-central B.C.	Westgate and Fulton, 1975
Riggins Road	30 ±	Mt. St. Helens, Washington?	south-central B.C.	Westgate and Fulton, 1975
Duncan Lake	34	Mt. St. Helens, Washington?	south-central B.C.	Westgate and Fulton, 1975
Dufferin Hill[4]	?	?	south-central B.C.	Westgate and Fulton, 1975
Sweetsbridge[4]	>22	?	south-central B.C.	Westgate and Fulton, 1975
Kamloops Lake	34 +[5]	?	south-central B.C.	Westgate and Fulton, 1975
Mission Flats	>35	Mt. St. Helens, Washington	south-central B.C.	Westgate and Fulton, 1975
Okanagan Centre	?[6]	?	south-central B.C.	Westgate and Fulton, 1975
Dawson	<52[7,8]	?	west-central Yukon	Naeser et al., 1982
Sheep Creek	80 ±[8,9]	Wrangell Mountains?	central Yukon	Schweger and Matthews, 1985
Old Crow	>80 ≤130[8,10]	Aleutians?	western Yukon	Naeser et al., 1982; Westgate, 1982; Westgate et al., 1983, 1985; Schweger and Matthews 1985; Wintle and Westgate 1986; Berger, 1987
Surprise Creek	?	?	northern Yukon	Matthews, 1987
Coutlee	>670[8,11]	?	south-central B.C.	Westgate and Fulton, 1975; Berger, 1985
Fort Selkirk	>840 <940[7,8]	?	south-central Yukon	Naeser et al., 1982
Little Timber	>1200[7,8]	?	northern Yukon	Matthews, 1987
Mosquito Gulch	1220 ±[7,8]	?	west-central Yukon	Naeser et al., 1982

[1] All ages are based on radiocarbon dates unless otherwise indicated.
[2] Age based on dendrochronology (1988 datum).
[3] Bridge River tephra at Otter Creek bog was thought by Westgate (1977) to date about 1.9-2.0 ka and thus to be distinct from the slightly older Bridge River tephra occurring more widely in south-central British Columbia. However, more recent work (Mathewes and Westgate, 1980) has raised the possibility that the former is, in fact, the 2.4 ka tephra.
[4] This tephra is perhaps close in age to the Duncan Lake tephra and, in fact, the two may be equivalent.
[5] Kamloops Lake tephra is only slightly older than Duncan Lake tephra.
[6] Okanagan Centre tephra probably was deposited sometime between the last glaciation (Late Wisconsinan) and the penultimate glaciation (Early Wisconsinan or Illinoian).
[7] Fission-track age.
[8] Age is provisional and subject to revision or refinement as a result of ongoing work.
[9] Age is based on radiocarbon, thorium-uranium, and protactinium-uranium dates and paleomagnetic data.
[10] Age is based on fission-track, thermoluminescence, and radiocarbon dates and paleomagnetic data.
[11] Thermoluminescence age.

CONTROLS ON QUATERNARY DEPOSITION AND EROSION

J.J. Clague

Physiography probably has been the most important factor affecting deposition and erosion in the Canadian Cordillera during Quaternary time. During nonglacial periods, sedimentation occurred mainly in valleys, lakes, and the sea. The principal subaerial depositional sites were floodplains and fans. Deltas were constructed where streams emptied into lakes and the sea. Large amounts of fine sediment were deposited from suspension and turbidity flows in lakes and fiords, on parts of the British Columbia continental shelf and slope, and on North Pacific abyssal plains and basins.

Mountains were areas of erosion during nonglacial periods. Glacial, periglacial, and fluvial processes and mass wasting contributed to the denudation of mountain slopes. Sediment produced by these denudational processes initially may have been transported only a short distance, for example to a position lower on the slope or to the adjacent valley floor. Ultimately, however, much of the sediment was reentrained and carried to lowland floodplains, lakes, and the sea. Quaternary deposits and bedrock outside mountain areas also were eroded during nonglacial periods. At these times, many streams flowed in valleys incised into Quaternary deposits; escarpments in these materials supplied large amounts of sediment to the streams. Similarly, bluffs of unconsolidated deposits along the British Columbia coast were eroded by currents and waves and thus contributed material directly to the sea.

Large areas of the Canadian Cordillera were relatively stable during nonglacial periods. For example, much of the plateau surfaces in British Columbia and southern Yukon Territory experienced no significant erosion or deposition for tens of thousands of years between glaciations. Support for this supposition is provided by the fact that unmodified glacial landforms dating back to the close of the Pleistocene are widespread on low-relief surfaces in the interior of the Cordillera.

Denudation during Quaternary nonglacial periods also has been affected by climate. Rates of mass wasting and fluvial erosion at these times varied in relation to precipitation, air temperature, and vegetation. The mechanical breakdown of rock due to repeated freezing and thawing was most effective in humid areas where air temperatures regularly fluctuated around 0°C. In regions where mean annual temperature was at or below freezing (i.e., most of Yukon Territory, District of Mackenzie, and the high mountains of British Columbia and Alberta), there was significant downslope movement of sediment and shattered rock due to solifluction and shallow slope failures involving water-saturated material overlying frozen ground. Such denudation was much reduced or did not take place at all in areas lacking permafrost or seasonally frozen ground.

The pattern of erosion and deposition during glaciations is quite different from that outlined above. Climatic deterioration at the onset of each glaciation led to increased sediment production in the mountains of the Cordillera. As glaciers advanced out of mountain areas, many interior and coastal valleys became aggraded with outwash, and thick deltaic and marine sediments accumulated locally in offshore areas. At the same time, the nonglacial drainage system was disrupted and rearranged. Lakes were impounded by advancing glaciers, and thick deposits of sand, silt, and clay accumulated in them. Those within the limits of glaciation eventually were overridden and obliterated, and their fills either eroded or buried by drift. In contrast, in northern Yukon Territory large lakes dammed by the Laurentide Ice Sheet were not subsequently overridden by glaciers, and thus their sedimentary fills are better preserved than those farther south. Advancing glaciers also forced streams out of their nonglacial valleys. Swollen with meltwater and laden with sediment, these streams followed complex paths along constantly shifting glacier margins. This situation contrasts with that in unglaciated Yukon Territory where streams generally maintained their courses during each glaciation, although they carried large amounts of meltwater and additional sediment.

At climaxes of major glaciations, a variety of sediments continued to accumulate in unglaciated parts of the Yukon, District of Mackenzie, and offshore British Columbia, and till was deposited locally beneath the Cordilleran Ice Sheet and satellite glaciers. However, mountains, fiords, and valleys parallel to the direction of ice flow were subject to intense scour by glaciers, and the unconsolidated fills of many valleys were partly or completely removed. The eroded materials, in part, were redeposited as till and, in part, carried englacially to the periphery of the ice sheet and laid down in ice contact, proglacial, and extraglacial environments.

Deglacial intervals, like periods of glacier growth, were times of rapid valley aggradation and extensive drainage change. Large glacial lakes formed and evolved as the Cordilleran Ice Sheet downwasted and separated into valley tongues that retreated in response to local conditions. In coastal areas, thick glacial marine sediments were laid down on isostatically depressed lowlands vacated by retreating glaciers and covered by the sea. Similar sediments also accumulated on the adjacent continental shelf. Throughout the glaciated Cordillera, unconsolidated sediments (mainly till) were eroded from uplands and valley walls and transported to lower elevations where they were redeposited on floodplains and fans and in lakes and the sea. Streams at first rapidly aggraded their valleys because they were unable to cope with the large amounts of sediment made available during deglaciation (Church and Ryder, 1972). However, the supply of sediment decreased as slopes stabilized and vegetation became established. Together with a fall in base level due to isostatic uplift, this reduction in sediment supply forced streams to deeply incise their late glacial floodplains, and the drainage network thus became locked into a relatively stable pattern that prevailed until the next glaciation.

Clague, J.J.
1989: Controls on Quaternary deposition and erosion (Canadian Cordillera); in Chapter 1 of Quaternary Geology of Canada and Greenland, R.J. Fulton (ed.); Geological Survey of Canada, Geology of Canada, no. 1 (also Geological Society of America, The Geology of North America, v. K-1).

CORDILLERAN ICE SHEET

J.J. Clague

Character and extent

Several times during the Pleistocene, large parts of the Canadian Cordillera were covered by an interconnected mass of valley and piedmont glaciers and mountain ice sheets, collectively known as the Cordilleran Ice Sheet (Fig. 1.12; Flint, 1971). At its maximum, this ice sheet and its satellite glaciers buried almost all of British Columbia, southern Yukon Territory, and southern Alaska, and extended south into the northwestern United States. To a considerable degree, the ice sheet was confined between the high bordering ranges of the Cordillera, although large areas on the east flank of the Rockies and west of the Coast Mountains were also covered by ice. Glaciers in several mountain ranges were more or less independent of the ice sheet, even at the climax of glaciation (Fig. 1.12).

The Cordilleran Ice Sheet attained its maximum size in British Columbia where it was up to 900 km wide and at the climax of glaciation extended to 2000-3000 m elevation over much of the Stikine and Interior plateaus (Wilson et al., 1958). When fully formed, the ice sheet probably had the shape of an elongate inverted dish, with gentle slopes in the broad interior region and steeper slopes at the periphery. At such times, it closely resembled the present-day Greenland Ice Sheet.

In western British Columbia, ice streamed down fiords and valleys in the coastal mountains and covered large areas of the continental shelf; in places ice lobes extended to the shelf edge where they calved into deep water (Prest, 1969, 1970; Luternauer and Murray, 1983). Ice from the southern Coast Mountains and Vancouver Island coalesced to produce a great piedmont lobe that flowed far into Puget Lowland in Washington (Armstrong et al., 1965; Waitt and Thorson, 1983). Glaciers streaming down valleys in south-central and southeastern British Columbia likewise terminated as large lobes on plateaus and in broad valleys in eastern Washington, Idaho, and Montana.

The Cordilleran Ice Sheet in southern Alaska consisted of ice fields and large valley and piedmont glaciers that flowed from montane source areas across the continental shelf to the south and into the broad low country drained by Yukon River to the north (Hamilton and Thorson, 1983). Ice was thicker and more extensive on the seaward side of the mountains than on the interior side, reflecting the predominant Pacific source of snowfall (Flint, 1971). Most of interior Alaska and central and northern Yukon Territory were dry throughout the Quaternary and consequently remained unglaciated. The Cordilleran Ice Sheet in southern Yukon Territory was fed principally from the Selwyn and Cassiar mountains; some ice also was supplied from the St. Elias Mountains (Hughes et al., 1969).

On the east, ice flowed out of the Rocky Mountains and locally coalesced with the Keewatin portion of the Laurentide Ice Sheet. Some valley glaciers in the eastern Mackenzie Mountains also came into contact with Laurentide ice. Coalescence probably occurred only at the climaxes of major glaciations; at other times, an ice-free zone existed between Cordilleran and Laurentide glaciers on the westernmost Interior Plains (see Relationship of Cordilleran and Laurentide glaciers).

Growth

At the end of each major Pleistocene nonglacial period, glaciers were restricted to the high mountains of the Canadian Cordillera. As climate deteriorated during the early part of each glaciation, small mountain ice fields grew and alpine glaciers advanced (alpine phase of glaciation; Kerr, 1934; Davis and Mathews, 1944; Flint, 1971; Fig. 1.13). With continued cooling and perhaps increased precipitation, glaciers expanded and coalesced to form a more extensive cover of ice in mountain areas (intense alpine phase). During long sustained cold periods, these glaciers advanced across plateaus and lowlands and eventually grew into large confluent masses of ice that covered most of British Columbia and adjacent regions (mountain ice sheet phase). Throughout this period, the major mountain systems remained the principal source areas of glaciers, and ice flow was controlled by topography. During the final phase of glaciation, which probably was infrequently achieved, ice thickened to such an extent that one or more domes with surface flow radially away from their centres became established over the interior of British Columbia (continental ice sheet phase; Dawson, 1881, 1891; Kerr, 1934; Mathews, 1955; Wilson et al., 1958; Fulton, 1967; Flint, 1971).

In this general sequence of glacier growth, the transition from the third to the fourth phase was accompanied by a local reversal of glacier flow in the Coast Mountains as the ice divide (i.e., the axis of outflow) shifted from the mountain crest eastward to a position over the British Columbia interior (Kerr, 1934; Flint, 1971). A comparable westward shift and reversal of flow also may have occurred locally in the Rocky Mountains. These flow reversals resulted from the buildup of ice in the interior to levels higher than the main accumulation areas in the flanking mountains.

Drumlins, flutings, and other streamlined forms indicate that there were at least three ice divides over the Interior System at the climax of the last glaciation or shortly thereafter (Fig. 1.12). This suggests that the Cordilleran Ice Sheet did not constitute a single monolithic ice dome as has often been suggested. These ice divides, however, probably were subordinate to the main divide along the axis of the Coast Mountains which may have persisted throughout the last glaciation. If this is true, the continental ice sheet phase of glaciation was not completely achieved in the Canadian Cordillera during Late Wisconsinan time.

The successional four-phase model outlined above provides a useful framework for conceptualizing the growth of the Cordilleran Ice Sheet; however, the actual history of glacier growth in the Cordillera during the Pleistocene is more complicated, because ice did not build up in a uniform fashion. Rather, periods of growth were interrupted, at least in some areas, by intervals during which glaciers stabilized or receded. These complex glacier fluctuations were controlled mainly by global climatic changes and, to a lesser extent, by secondary local and regional factors induced by glaciation but only indirectly related to climate (e.g., eustatic sea level lowering, ocean cooling, and changes in local atmospheric circulation due to glacier growth).

Clague, J.J.
1989: Cordilleran Ice Sheet; in Chapter 1 of Quaternary Geology of Canada and Greenland, R.J. Fulton (ed.); Geological Survey of Canada, Geology of Canada, no. 1 (also Geological Society of America, The Geology of North America, v. K-1).

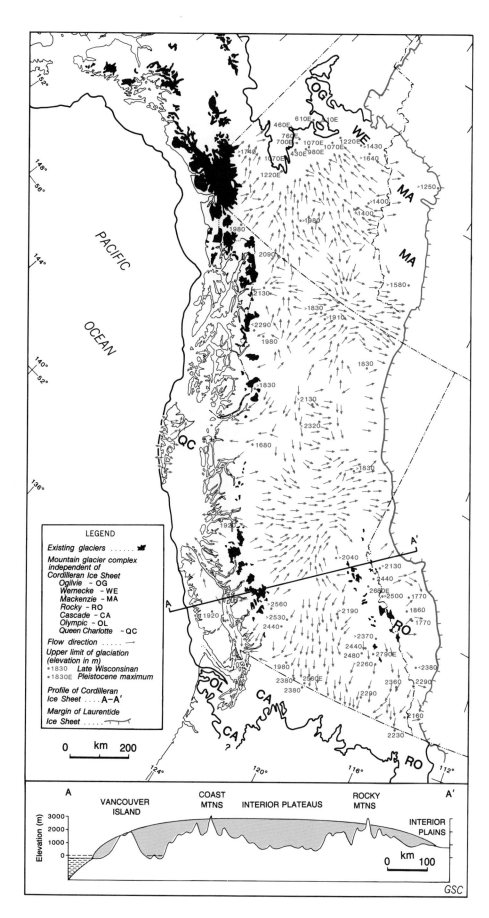

Figure 1.12. Maximum extent of Pleistocene glaciation in the Canadian Cordillera and adjacent areas, and ice flow pattern during the Late Wisconsinan. The glacier complex shown in this figure includes the Cordilleran Ice Sheet and independent and nearly independent glacier systems in some peripheral mountain ranges. Nunataks were small in ice sheet areas at the climax of glaciation. In contrast, there were large ice-free areas in some peripheral glaciated mountain ranges. Extent of glaciation, in part, from Crandell (1965), Lemke et al. (1965), Richmond et al. (1965), Prest et al. (1968), Hamilton and Thorson (1983), and Porter et al. (1983); ice flow pattern from Prest et al. (1968); data on upper limit of glaciation from Wilson et al. (1958).

Decay

Each glacial cycle terminated with rapid climatic amelioration. Deglaciation was characterized by complex frontal retreat in peripheral glaciated areas and by downwasting accompanied by widespread stagnation throughout much of the interior (Fig. 1.13; Rice, 1936; Fulton, 1967; Tipper, 1971a, b; Clague, 1981).

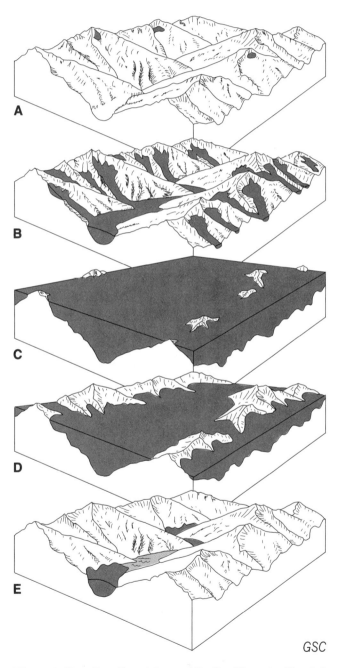

Figure 1.13. Growth and decay of the Cordilleran Ice Sheet. A. Mountain area at the beginning of a glaciation. B. Development of a network of valley glaciers. C. Coalescence of valley and piedmont lobes to form an ice sheet. D. Decay of ice sheet by downwasting; upland areas are deglaciated before adjacent valleys. E. Residual dead ice masses confined to valleys.

At the close of each glaciation, the western periphery of the Cordilleran Ice Sheet became unstable, probably due to a eustatic rise in sea level. The British Columbia continental shelf was rapidly deglaciated as glaciers calved back to fiord heads and shallow protected coastal embayments. Frontal retreat also occurred elsewhere along the periphery of the ice sheet, for example in northeastern Washington and southern Yukon Territory.

A different pattern of deglaciation has been documented for areas of low and moderate relief nearer the centre of the ice sheet. On the basis of observations of late Pleistocene landforms and sediments in south-central British Columbia, Fulton (1967) suggested that deglaciation in such areas occurred mainly by downwasting and stagnation and proceeded through four stages: (1) active ice phase — regional flow continued but diminished as ice thinned; (2) transitional upland phase — highest uplands appeared through the ice sheet, but regional flow continued in major valleys; (3) stagnant ice phase — ice was confined to valleys but was still thick enough to flow; and (4) dead ice phase — valley tongues thinned to the point where plasticity was lost. This model has been used with minor modifications to describe deglaciation of other parts of the British Columbia interior (Tipper, 1971a, b; Heginbottom, 1972; Ryder, 1976, 1981b; Howes, 1977).

The first areas to become ice-free at the end of each glaciation were those near the periphery of the ice sheet, for example the British Columbia continental shelf. Active glaciers probably persisted longest in some mountain valleys; however, these glaciers may have coexisted with large remnant dead ice masses on the plateaus of the Cordilleran interior. In general, retreat in the interior proceeded from both southern and northern peripheral areas towards the centre of the ice sheet. In detail, however, the pattern of retreat was complex, with uplands in each region becoming ice-free before adjacent valleys and other low-lying areas.

Decay of the Cordilleran Ice Sheet at the close of each glacial cycle was interrupted repeatedly by glacier stillstands and readvances. Most of these fluctuations affected relatively small areas and were not synchronous from one region to another; thus, they may have resulted from local factors rather than global climatic change.

RELATIONSHIP OF CORDILLERAN AND LAURENTIDE GLACIERS

J.J. Clague

Rocky Mountain glaciers on occasion advanced onto the Interior Plains and coalesced with the Laurentide Ice Sheet. In the Athabasca, Peace, and Liard river areas and perhaps elsewhere, these glaciers were augmented by streams of ice flowing from the interior of British Columbia and southern Yukon Territory. The Laurentide Ice Sheet spread into the region from the north and east, at times reaching into the foothills and mountains at the eastern edge of the Cordillera. During lesser glaciations, Rocky Mountain glaciers failed to reach the Interior Plains and the Laurentide

---

Clague, J.J.
1989: Relationship of Cordilleran and Laurentide glaciers; in Chapter 1 of Quaternary Geology of Canada and Greenland, R.J. Fulton (ed.); Geological Survey of Canada, Geology of Canada, no. 1 (also Geological Society of America, The Geology of North America, v. K-1).

Ice Sheet terminated short of the mountain front. At such times, montane outwash trains ended as deltas in lakes impounded by Laurentide ice.

Some glaciers in the Mackenzie Mountains also reached the eastern edge of the Cordillera on one or more occasions during the Pleistocene. Within this region, there are small areas that were affected by both montane and Laurentide ice masses; locally these ice masses coalesced.

Evidence on the relationship of Cordilleran and Laurentide glaciers comes from studies of till and associated stratified drift along the eastern margin of the Canadian Cordillera. Deposits of the Laurentide Ice Sheet and of Cordilleran glaciers are very different in lithology and thus are easily distinguished. The former are rich in plutonic and metamorphic detritus derived from the Canadian Shield, whereas the latter consist dominantly of detritus of sedimentary rocks eroded from the Rocky and Mackenzie mountains. Using lithological criteria as well as stratigraphic and geomorphic evidence, scientists have recognized several advances of the Laurentide Ice Sheet and broadly correlative episodes of mountain glaciation in the eastern Cordillera (see Regional Quaternary stratigraphy and history). However, there probably were other glaciations that have not been recognized because of their antiquity and because their deposits have been eroded or covered with younger sediments.

In general, older known glaciations in the eastern Cordillera were more extensive than younger ones. Differences in the extent of glaciation in this region are reflected in the degree of contact between Cordilleran and Laurentide glaciers and the width of the ice-free zone between them. The pattern of glaciation might be likened to "the periodical closing and opening of a gigantic zipper along the Albertan mountain front, with the zipper closing a lesser distance during each successive glaciation. During each glaciation the zipper would close from north to south as the Cordilleran and Laurentide glaciers spread farther east and southwest respectively. Then, when they began to wane, the zipper would part from south to north, opening up an ever-widening corridor between them" (Stalker and Harrison, 1977, p. 887). Although various scientists differ in opinion as to the amount of contact between Cordilleran and Laurentide glaciers during each glaciation, they all visualize the ice-free corridor as narrowing northward in western Alberta. During the most extensive glaciation, there may have been fairly complete contact from Waterton Lakes to beyond the British Columbia-Yukon border. In contrast, at the climax of the last glaciation, closure did not extend as far south, and many intervening high areas in the eastern Rocky Mountains were left unglaciated.

In some areas, Cordilleran and Laurentide advances were out-of-phase. For example, during the pre-Wisconsinan "Great Glaciation" of Waterton Lakes National Park, Cordilleran glaciers advanced and retreated before the Laurentide Ice Sheet invaded the area (Stalker and Harrison, 1977). The main evidence for this is the occurrence of Laurentide Labuma Till over Cordilleran Albertan Till deposited during the same glaciation. The two tills do not interfinger, nor do they exhibit a transitional zone of mixed lithology as might be expected if the two ice masses had coalesced. Locally, the two tills are separated by Cordilleran outwash or by glacial lacustrine sediments deposited in lakes dammed by advancing Laurentide ice. Altogether, the evidence points to the fact that Cordilleran glaciers had largely receded from the region before the arrival of the Laurentide Ice Sheet. Similar arguments have been made for nonsynchronous advances in Oldman River basin directly north of Waterton Lakes (Alley, 1973). In addition, mountain and Laurentide glaciers may have been out-of-phase in the Mackenzie Mountains and on the adjacent Interior Plains (Ford, 1976).

In some cases, however, Cordilleran and Laurentide advances were broadly synchronous, or at least the Laurentide Ice Sheet began to retreat along its western margin at about the same time as mountain glaciers and lobes of the Cordilleran Ice Sheet. General synchroneity is suggested by the fact that Cordilleran and Laurentide ice coalesced in several areas during the last glaciation. Till of mixed Cordilleran and Keewatin provenance and associated quartzose conglomerate erratics derived from the Rocky Mountains near Jasper (Foothills Erratics Train; Stalker, 1956; Mountjoy, 1958) define the zone of coalescence of Cordilleran ice flowing from the west and Laurentide ice flowing from the north and east at the climax of this glaciation. The till of mixed provenance extends southeastward at the edge of the Rocky Mountains from Athabasca River on the north to near Oldman River on the south. There is no evidence within this region for overriding of the mixed provenance till or the Foothills Erratics Train by either Cordilleran or Laurentide ice; rather, the two ice masses apparently retreated at about the same time. It thus is clear that the Cordilleran and Laurentide advances during the last glaciation were broadly synchronous, although there undoubtedly were some differences in timing.

There is no satisfactory explanation for the apparent nonsynchroneity of Cordilleran and Laurentide advances in some areas and for synchroneity in others. However, it seems reasonable that mountain glaciers, with their shorter travel distance, would reach positions at or beyond the mountain front before the more massive Laurentide Ice Sheet entered the region. The early retreat of mountain glaciers in some areas may have resulted from a reduction in precipitation in the eastern Cordillera due to growth of the Cordilleran Ice Sheet to the west. Ice covering the British Columbia interior may have depleted or diverted moist air masses that previously had flowed across the Rocky Mountains, making the air reaching that area rather dry. This, in turn, may have caused some local glaciers in the Rocky Mountains to retreat at a time when both the Cordilleran and Laurentide ice sheets were growing.

QUATERNARY SEA LEVELS
J.J. Clague

Changes in sea level along the Pacific coast of Canada during Quaternary time are attributable mainly to diastrophism, eustasy, and isostasy (Mathews et al., 1970; Clague, 1975a, 1981, 1983; Fladmark, 1975; Andrews and Retherford, 1978; Clague et al., 1982b; Riddihough, 1982). The net effect on past sea levels of these three factors can be determined by studying former shorelines and associated

---

Clague, J.J.
1989: Quaternary sea levels (Canadian Cordillera); in Chapter 1 of Quaternary Geology of Canada and Greenland, R.J. Fulton (ed.); Geological Survey of Canada, Geology of Canada, no. 1 (also Geological Society of America, The Geology of North America, v. K-1).

deposits, but it is difficult to assess the individual effect of each. This arises because eustasy and isostasy are interdependent and because there is disagreement about the nature of past eustatic sea level changes (e.g., Mörner, 1976; Clark et al., 1978). Notwithstanding these problems, it is possible to summarize late Quaternary sea level fluctuations on the British Columbia coast and to comment on their likely causes.

Sea levels during the last glaciation

The position of the sea relative to the land at the beginning of the last glaciation (i.e., Fraser Glaciation in British Columbia) is uncertain because there is no consensus concerning eustatic sea levels at this time and because subsequent vertical tectonic movements are poorly known. It seems most likely, however, that British Columbia shorelines at the beginning of the Fraser Glaciation were somewhat lower than, although perhaps close to, the present.

Loading of British Columbia by glacier ice during the Fraser Glaciation caused major isostatic adjustments in the crust and mantle of the region (Clague, 1983). Gradual growth of glaciers led to progressive isostatic depression of the land surface. At first, this depression was localized beneath major mountain ranges that served as loci of glacier growth. Lateral movement of material in the asthenosphere away from these areas probably produced outward migrating forebulges. Initially, relative sea levels may have fallen as these forebulges passed through coastal areas, as oceans cooled and contracted, and as water was transferred from oceans to expanding Late Wisconsinan ice sheets. Eustatic sea level lowering, in turn, may have caused hydro-isostatic uplift of the continental shelf, resulting in a further fall of the sea relative to the land. As glaciers advanced from mountains onto plateaus and into lowlands, however, isostatically depressed areas grew in size, and the coastal region began to subside. Eventually, glacial isostatic depression became dominant in most areas, and the sea rose above its present level relative to the land.

At the climax of the Fraser Glaciation, the entire glaciated Cordillera was isostatically depressed, with areas near the centre of the ice sheet displaced downward more than areas near the periphery. Although the magnitude of isostatic depression at this time is unknown, limits are provided by levels of shorelines that formed at the end of the Pleistocene as glaciers withdrew from the British Columbia coast. Relict shorelines occur on the mainland coast up to about 200 m above present sea level. Taking into account a eustatic sea level lowering of perhaps 50-100 m at the time the highest shorelines formed, local glacial isostatic depression of more than 250 m is indicated. In fact, this is a minimum value for isostatic depression in this area because the Cordilleran Ice Sheet had decreased in size before the highest shorelines formed, and consequently isostatic rebound probably had commenced earlier.

The highest shorelines on the coasts of Vancouver Island and mainland British Columbia date to the time of glacier retreat at the close of the Pleistocene. The elevation of the marine limit in these areas varies in relation to the distance from main centres of ice accumulation and, to a lesser extent, to the timing of retreat. In general, the marine limit is highest (ca. 200 m) on the mainland coast and declines towards the west and southwest (Fig. 1.14; Mathews et al., 1970; Clague, 1975a, 1981). It is less than 50 m in elevation on the west coast of Vancouver Island near the margin of the former ice sheet (Clague, 1981; Howes, 1981b). In contrast, relative sea level on the Queen Charlotte Islands and in parts of Dixon Entrance, Hecate Strait, and Queen Charlotte Sound was lower at the end of the last glaciation than at present (Fladmark, 1975; Clague, 1981; Clague et al., 1982a, b; Warner et al., 1982). This indicates that glacial isostatic depression of these areas was minor.

Rapid deglaciation at the close of the Pleistocene triggered isostatic movements in the Canadian Cordillera that were opposite in direction to those that occurred during ice sheet growth. Isostatic uplift of Vancouver Island and the mainland was greater than the coeval eustatic rise, thus the sea fell from the marine limit in these areas as deglaciation progressed (Fig. 1.15). The fall in sea level was extremely rapid. For example, at Courtenay on eastern Vancouver Island, the highest marine delta, 150 m above the present shore, formed about 12 500 ± 450 BP (I(GSC)-9, Table 1.2). In contrast, deltas 60-90 m below marine limit about 65 km to the southeast have yielded radiocarbon dates ranging from 12 400 ± 200 BP to 12 000 ± 450 BP (GSC-1, I(GSC)-1). Similarly, at Kitimat on the northern mainland coast, the sea fell 85 m with respect to the land between 10 100 ± 160 BP and 9300 ± 90 BP (GSC-2492, GSC-2425). The error terms associated with these and other radiocarbon dates are sufficiently large and the dates themselves so closely spaced in time that it is not possible to assign precise figures to rates of uplift. Nevertheless, most recorded uplift at each site occurred within a time interval of less than 2 ka.

Figure 1.14. Maximum late Quaternary marine overlap in British Columbia; extent of overlap inland from the heads of most mainland fiords is unknown (adapted from Clague, 1981, Fig. 5).

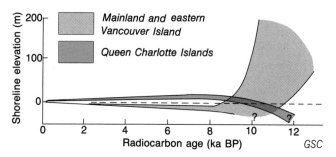

Figure 1.15. Generalized patterns of sea level change on the British Columbia coast since the end of the last glaciation. Each envelope depicts a range of sea level positions, reflecting regional and local differences in ice loads and time of deglaciation as well as uncertainties in locating former shorelines.

Isostatic uplift occurred at different times along the British Columbia coast due to diachronous retreat of the Cordilleran Ice Sheet. In general, regions that were deglaciated first rebounded earlier than those that were deglaciated at a late date. For example, the sea had fallen to its present level at Victoria on southern Vancouver Island by about 11.5 ka, about 1.5 ka after initial deglaciation of the area. In contrast, the sea was at its upper limit at Kitimat about 11 ka when the area was first deglaciated, and most isostatic uplift there occurred between 10.5 ka and 9 ka. Such data indicate that there were regional differences in the timing of the isostatic response to deglaciation and show that the crust deformed in a complex nonuniform manner.

Even within a region, the sea probably did not fall uniformly relative to the land at the end of the Fraser Glaciation. There are at least three reasons for this: (1) the rate and direction of eustatic change during this period varied; (2) the rate of isostatic uplift may have varied due to local stillstands and readvances of glaciers; and (3) some vertical displacements may have occurred instantaneously and sporadically along faults during earthquakes. Evidence for nonuniform emergence has been found at several sites on the British Columbia coast. At Kitimat, for example, shorelines apparently were relatively stable for hundreds of years after initial deglaciation of the area; about 10-10.5 ka, however, the sea began to fall rapidly relative to the land (Clague, 1984, 1985). In some areas, emergence probably was interrupted by short periods of submergence. Mathews et al. (1970) and Armstrong (1981), for example, proposed that emergence of Fraser Lowland at the close of the Fraser Glaciation was interrupted by a strong short-lived submergence that culminated about 11.5 ka.

The record of sea level changes during deglaciation on the Queen Charlotte Islands is different from that outlined above. Shorelines there were lower than at present from at least 15 ka until 9.5-10 ka; this contrasts sharply with the situation on the mainland coast directly to the east (Fig. 1.15; Clague, 1981; Clague et al., 1982a, b; Mathewes and Clague, 1982). This period of low sea levels was followed by a marine transgression that culminated about 7.5-8.5 ka with shorelines in some areas about 15 m higher than at present.

The opposing character of late Quaternary sea level changes on the Queen Charlotte Islands and the adjacent mainland is best explained in terms of differing ice loads and forebulge migration during deglaciation. Clague (1983) proposed that shorelines on the Queen Charlotte Islands were low at the close of the last glaciation because ice loads there were insufficient to depress the crust below lowered eustatic water levels prevailing at that time. He further suggested that mantle material flowed away from peripheral glaciated areas as the Cordilleran Ice Sheet decayed. As a result, the asthenosphere was depleted beneath the Queen Charlotte Islands, the crust subsided, and lowland areas were transgressed by the sea. In contrast, eastward forebulge migration beneath the mainland coast caused uplift and a marine regression there.

Sea levels during postglacial time

The coastal lowlands of eastern Vancouver Island and mainland British Columbia continued to emerge during the early Holocene, although generally more slowly than during deglaciation (Fig. 1.15; Mathews et al., 1970). This indicates that glacial isostatic uplift must have persisted into the early Holocene, because seas were still rising eustatically at that time due to final disintegration of the Laurentide and Fennoscandian ice sheets (Curray, 1965; Hopkins, 1967; Denton and Hughes, 1981). Emergence culminated during early or middle Holocene time when shorelines in some areas were considerably lower than they are today. The best evidence for low shorelines at this time comes from southern Vancouver Island and Fraser Lowland where Holocene terrestrial peats occur below present sea level (Mathews et al., 1970; Clague and Luternauer, 1982, 1983; Clague et al., 1982b); supporting evidence has been found on eastern Vancouver Island (Mathews et al., 1970) and on the central mainland coast (Andrews and Retherford, 1978). In these areas, the sea apparently rose during middle Holocene time, inundating some coastal archeological sites and terrestrial vegetation. Shorelines have deviated no more than a few metres during the last 5 ka, thus late Holocene sea level fluctuations have been relatively minor in comparison to those accompanying and immediately following deglaciation (Clague et al., 1982b).

In contrast to eastern Vancouver Island and the mainland, the Queen Charlotte Islands experienced sea levels higher than at present throughout the Holocene (Fig. 1.15; Fladmark, 1975; Clague, 1981; Clague et al., 1982a, b; Hebda and Mathewes, 1986). At the northeast end of this island chain, the sea has fallen about 15 m relative to the land since 7.5-8.5 ka. Features that formed during this regression and are now elevated above sea level include wave-cut scarps, wave-cut benches overlain by littoral gravel and sand, bars, spits, and beach and dune ridges parallel to the modern shoreline (Fig. 1.16). Sea level changes on westernmost Vancouver Island during the last 4 ka perhaps were similar to those on the Queen Charlotte Islands during the same period (Clague et al., 1982b).

In conclusion, significant isostatic uplift continued at gradually decreasing rates in most coastal areas until the early Holocene. The subsequent marine transgression on eastern Vancouver Island and the British Columbia mainland probably resulted from the combined effects of the global eustatic rise in sea level and forebulge collapse. In contrast, the recent regression on the Queen Charlotte Islands and western Vancouver Island is due largely to tectonic uplift, although some late Holocene sea level change in these areas may be eustatic in nature or even a residual isostatic response to deglaciation at the close of the Pleistocene (Riddihough, 1982; Clague et al., 1982b).

Table 1.2. Radiocarbon dates

Age (years BP)	Laboratory dating no.[1]	Locality	Location lat.	Location long.	Material	Reference	Comment
5 260 ± 200	Y-140bis	Mount Garibaldi	49°52'	122°59'	wood	Stuiver et al., 1960	treeline higher than at present
7 390 ± 250	GX-4039	Dunn Peak	51°27'	119°55'	charcoal	Duford and Osborn, 1978	minimum date for Dunn Peak advance
7 985 ± 125	I-9162	Dunn Peak	51°25'	119°57'	wood	Alley, 1980	minimum date for Harper Creek advance
8 460 ± 120[2]	GSC-2605	Old Crow Flats	68°13'	140°00'	wood	Lowdon and Blake, 1979	postdates proglacial lake
9 070 ± 130	GSC-3173	Bridge Glacier	40°48.3'	123°25.4'	wood	Blake, 1983	treeline higher than at present
9 300 ± 90	GSC-2425	Kitimat River	54°06.0'	128°36.8'	marine shells	Lowdon and Blake, 1979	sea level higher than at present
9 350 ± 80[3]	GSC-3120	Graham Island	53°41.7'	131°52.8'	marine shells	Blake, 1982	sea level higher than at present
9 670 ± 140[4]	I-5677	Rocky Mountain House	52°28.3'	114°32.0'	bone	Boydell, 1972	minimum date for deglaciation
9 780 ± 80	Y-1483	Kaskawulsh Glacier	60°49'	138°35'	grass	Denton and Stuiver, 1966	Kaskawulsh Glacier less extensive
9 960 ± 170	GSC-1548	Dawson Creek	55°58.8'	120°14.6'	freshwater shells	Lowdon and Blake, 1973	minimum date for deglaciation
10 100 ± 160	GSC-2492	Hirsch Creek	54°03.8'	128°35.6'	marine shells	Lowdon and Blake, 1979	sea level higher than at present
10 200 ± 280	GSC-3065	Calgary	51°02.9'	114°04.4'	bone	Blake, 1986	minimum date for deglaciation
10 250 ± 165[4]	I-5675	Rocky Mountain House	52°28.3'	114°32.0'	freshwater shells	Boydell, 1972	minimum date for deglaciation
10 400 ± 110[4]	GSC-2965	Kananaskis valley	50°52'	115°10'	wood	Lowdon and Blake, 1980	minimum date for deglaciation
10 400 ± 170[4]	GSC-1654	Dawson Creek	55°59.0'	120°15.7'	freshwater shells	Lowdon and Blake, 1973	minimum date for deglaciation
10 850 ± 320[2]	I-4224	Old Crow Flats	67°51'	139°48'	bone	Harington, 1977	Old Crow River near present level
11 100 ± 90[3]	GSC-3337	Graham Island	53°41.7'	131°52.8'	peat	Warner et al., 1982	sea level less than 8 m
11 370 ± 170	GSC-613	Cochrane	51°10.7'	114°27.5'	bone	Lowdon et al., 1967	minimum date for deglaciation
11 900 ± 120[4]	GSC-3885	Pocahontas	53°13'	117°55'	freshwater shells	Blake, 1986	minimum date for deglaciation
12 000 ± 450	I(GSC)-1	Parksville	ca. 49°17'	124°16'	wood	Walton et al., 1961	sea level higher than at present
12 400 ± 200	GSC-1	Parksville	ca. 49°17'	124°16'	wood	Dyck and Fyles, 1962	sea level higher than at present
12 460 ± 440[2,5]	I-3574	Old Crow Flats	67°50'	139°50'	bone	Harington, 1977	minimum date for deglaciation
12 500 ± 200[5]	Y-1386	Kluane Lake	61°03'	138°21'	plant detritus	Denton and Stuiver, 1966	sea level higher than at present
12 500 ± 450	I(GSC)-9	Courtenay	49°38.7'	125°00.3'	marine shells	Walton et al., 1961	minimum date for deglaciation
13 660 ± 180[5]	GSC-495	Macauley Ridge	62°17'	140°42.5'	organic silt	Lowdon et al., 1967	minimum date for deglaciation
13 740 ± 190[5]	GSC-515	East Blackstone R	64°38'	138°24'	organic silt	Lowdon and Blake, 1968	minimum date for intermediate glaciation
13 870 ± 180	GSC-296	Chapman Lake	64°51.5'	138°19'	organic silt	Dyck et al., 1966	minimum date for Hungry Creek Glaciation
16 000 ± 420	GSC-2690	Doll Creek	66°02'	135°42'	plant detritus	Lowdon and Blake, 1981	minimum date for deglaciation
16 000 ± 570[3]	GSC-3340	Graham Island	53°41.7'	131°52.8'	moss	Clague et al., 1982a	ice-free conditions in Rocky Mountains
18 300 ± 380[6]	GSC-2668	Chalmer's bog	50°39.5'	114°33.5'	moss	Lowdon and Blake, 1979	ice-free conditions in Rocky Mountains
18 400 ± 1090[6]	GSC-2670	Chalmer's bog	50°39.5'	114°33.5'	plant detritus	Lowdon and Blake, 1979	maximum date for last ice advance
19 100 ± 240	GSC-913	Bessette Creek	50°17.9'	118°51.8'	plant detritus	Lowdon and Blake, 1970	maximum date for proglacial lake
20 800 ± 200	GSC-3946	Bluefish Flats	67°23.1'	140°21.7'	wood	Blake, 1987	Coquitlam Drift
21 500 ± 240[7]	GSC-2536	Coquitlam valley	49°18.7'	122°46.8'	wood	Lowdon and Blake, 1978	Quadra Sand (above Coquitlam Drift)
21 700 ± 130[7]	GSC-2416	Coquitlam valley	49°18.8'	122°46.6'	wood	Lowdon and Blake, 1978	maximum date for last glaciation
22 700 ± 1000[4]	GaK-2336	Eagle Cave	49°37'	114°38'	bone	Kigoshi et al., 1973	Cowichan Head Formation
23 840 ± 300	GSC-518	Mill Bay	48°37'	123°31'	wood	Lowdon et al., 1967	Tom Creek Silt
23 900 ± 1140[5,8]	GSC-2811	Tom Creek	60°13.7'	129°00.4'	twigs	Lowdon and Blake, 1981	Quadra Sand
23 920 ± 400	GSC-59	Sidney Island	48°38.7'	123°19.7'	wood	Dyck and Fyles, 1963	Quadra Sand
24 400 ± 900[9]	L-502	Spanish Banks	49°17'	123°13'	wood	Olson and Broecker, 1961	maximum date for proglacial lake
25 170 ± 630[5]	NMC-1232	Cadzow Bluff	67°33.5'	138°53.5'	mammoth tusk	Morlan, 1986	Bessette Sediments
25 840 ± 320[10]	GSC-715	Meadow Creek	50°15.1'	116°59.0'	wood	Lowdon and Blake, 1968	Olympia Nonglacial Interval sediments
25 940 ± 3804[11]	GSC-573	Finlay River	56°18'	124°21'	plant detritus	Lowdon et al., 1971	Quadra Sand
26 100 ± 320[9]	GSC-1635	Point Grey	49°15.9'	123°15.8'	wood	Lowdon and Blake, 1973	Olympia Nonglacial Interval sediments
26 800 +1200/-1000[11]	GX-2032	Sand Creek	49°21.4'	115°17.1'	wood	Clague, 1973	

Table 1.2 (cont.)

Age (years BP)	Laboratory dating no.[1]	Locality	Location lat.	Location long.	Material	Reference	Comment
27 400 ± 580[4]	GSC-2034	Taylor	56°09'	120°42'	tooth	Lowdon and Blake, 1979	maximum date for last ice advance
27 400 ± 850	I-4878	Watino	55°43'	117°38'	peat	Westgate et al., 1972	maximum date for last ice advance
27 500 ± 400	GSC-3530	Yakoun valley	53°31.2'	132°11.9'	wood	Blake, 1984	maximum date for last ice advance
28 800 ± 740	GSC-95	Willemar Bluff	49°40.2'	124°53.8'	wood	Dyck and Fyles, 1963	Quadra Sand
29 100 ± 560	GSC-3792	Jasper	52°53'	118°06'	wood	Blake, 1986	maximum date for Marlboro (?) advance
29 600 ± 460[5,12]	GSC-769	Silver Creek	61°00'	138°19'	plant detritus	Lowdon and Blake, 1970	maximum date for Kluane Glaciation
30 100 ± 600[12]	Y-1385	Silver Creek	61°00'	138°19'	plant detritus	Denton and Stuiver, 1967	maximum date for Kluane Glaciation
31 300 ± 640[2]	GSC-1191	Old Crow Flats	68°03'	139°49'	plant detritus	Lowdon and Blake, 1979	maximum date for proglacial lake
31 400 ± 660[2]	GSC-2739	Old Crow Flats	67°50.0'	139°51.8'	peat	Lowdon and Blake, 1979	maximum date for proglacial lake
32 710 ± 800[10]	GSC-493	Meadow Creek	50°15.1'	116°59.0'	wood	Lowdon and Blake, 1968	Bessette Sediments
33 400 ± 800[12]	Y-1488	Silver Creek	61°00'	138°19'	plant detritus	Denton and Stuiver, 1967	maximum date for Kluane Glaciation
33 700 ± 300[10]	GSC-542	Meadow Creek	50°15.1'	116°59.0'	wood	Lowdon and Blake, 1968	Bessette Sediments
36 900 ± 300	GSC-2422	Hungry Creek	64°34.5'	135°30'	wood	Hughes et al., 1981	maximum date for Hungry Creek Glaciation
37 700 +1500/-1300[5,12]	Y-1356	Silver Creek	61°00'	138°19'	plant detritus	Denton and Stuiver, 1967	minimum date for Icefield Glaciation
40 900 ± 200[11]	GSC-2591	Gold River	49°50.7'	126°07'	wood	Lowdon and Blake, 1981	minimum age of Muchalat River Drift
41 100 ± 1650[2]	GSC-2574	Old Crow Flats	67°51.5'	139°49.6'	wood	Blake, 1984	predates proglacial lake
41 500 ± 520[10]	GSC-1017	Meadow Creek	50°15.1'	116°59.0'	peat	Lowdon and Blake, 1970	Bessette Sediments
41 800 ± 600[10]	GSC-716	Meadow Creek	50°15.1'	116°59.0'	wood	Lowdon and Blake, 1968	Bessette Sediments
41 900 ± 600[10]	GSC-733	Meadow Creek	50°15.1'	116°59.0'	wood	Lowdon and Blake, 1968	Bessette Sediments
42 300 ± 65[10]	GSC-1015	Meadow Creek	50°15.1'	116°59.0'	wood, moss	Lowdon and Blake, 1970	Bessette Sediments
42 300 ± 700[10]	GSC-720	Meadow Creek	50°15.1'	116°59.0'	peat	Lowdon and Blake, 1968	Bessette Sediments
43 000 ± 600[10]	GSC-740-2	Meadow Creek	50°15.1'	116°59.0'	wood	Lowdon et al., 1971	Bessette Sediments
43 600 ± 700[10]	GSC-1017-2	Meadow Creek	50°15.1'	116°59.0'	peat	Lowdon et al., 1971	Bessette Sediments
43 800 ± 800[10,11]	GSC-740	Meadow Creek	50°15.1'	116°59.0'	wood	Lowdon and Blake, 1968	Bessette Sediments
48 000 ± 1300[5]	GSC-732	White River	62°00'	140°34'	wood	Lowdon and Blake, 1970	minimum date for Mirror Ck. Glaciation
58 800 +2900/-2100	QL-195	East Delta	49°09.3'	122°55.6'	wood	Clague, 1977a	Cowichan Head Formation
>30 000[8]	GSC-2949	Tom Creek	60°13.7'	129°00.4'	twigs	Lowdon and Blake, 1981	Tom Creek Silt
>33 000	GSC-3208	Graham Island	53°42.5'	131°52.4'	peat	Blake, 1982	minimum date for penultimate glaciation
>38 000[11]	I-9772	Port McNeill	50°34.5'	127°02.5'	wood	Howes, 1983	older drift
>42 900	GSC-524	Stewart River	63°30.2'	137°16'	wood	Lowdon and Blake, 1968	minimum date for Reid Glaciation
>46 400[12]	Y-1355	Silver Creek	61°00'	138°19'	plant detritus	Denton and Stuiver, 1967	Shakwak Drift
>46 580[5]	GSC-331	Mayo	63°36'	135°56'	wood	Dyck et al., 1966	minimum date for Reid Glaciation
>49 000[12]	Y-1486	Silver Creek	61°00'	138°19'	peat	Denton and Stuiver, 1967	Icefield Drift
>49 000[7]	GSC-2094-2	Coquitlam valley	49°18.8'	122°46.8'	wood	Lowdon et al., 1977	Cowichan Head Formation?
>51 000	GSC-94-2	Cowichan Head	48°34'	123°22'	wood	Fulton and Halstead, 1972	Dashwood Drift
>52 000	GSC-3151-2	Graham Island	53°39.7'	131°54.3'	peat	Blake, 1982	minimum date for penultimate glaciation
>53 900[5]	GSC-527	Hunker Creek	63°58'	138°57'	wood	Lowdon and Blake, 1968	minimum date for intermediate glaciation
>62 000[11]	QL-194	Mary Hill	49°13.8'	122°46.5'	wood	Armstrong and Hicock, 1976	Semiahmoo Drift

[1] Laboratories: GaK — Gakushuin University; GSC — Geological Survey of Canada; GX — Geochron Laboratories; I — Teledyne Isotopes; L — Lamont; NMC — National Museums of Canada (AMS age determination for National Museums of Canada by Atomic Energy of Canada); QL — Quaternary Isotope Laboratory; Y — Yale

[2] See Figure 1.32.
[3] See Figure 1.31.
[4] See Figure 1.30.
[5] See Figure 1.26.
[6] Contamination by dead carbon suspected.
[7] See Figure 1.18.
[8] See Figure 1.27.
[9] See Figure 1.20.
[10] See Figure 1.19.
[11] See Figure 1.17.
[12] See Figure 1.28.

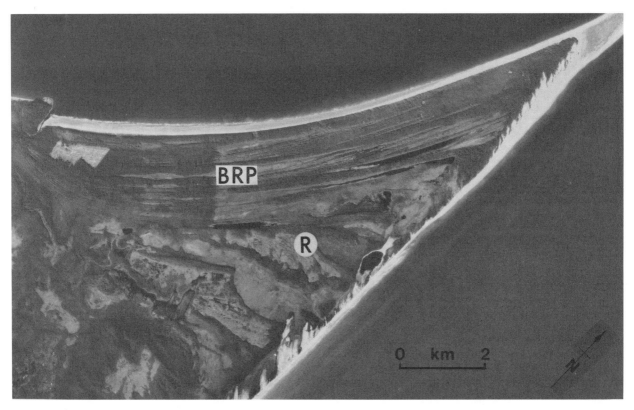

Figure 1.16. Elevated beach ridge plain (BRP) and older recurves (R) at the northeast corner of Graham Island, British Columbia. These features formed during a middle and late Holocene marine regression. Province of British Columbia BC5630-198.

QUATERNARY STRATIGRAPHY AND HISTORY
Area of Cordilleran Ice Sheet

Introduction
J.J. Clague

Most of British Columbia and southern and central Yukon Territory were repeatedly enveloped by the Cordilleran Ice Sheet during the Pleistocene. Surface sediments and landforms in these areas are mainly products of Late Wisconsinan glaciation (locally termed Fraser, McConnell, Macauley, and Kluane glaciations) and of postglacial time; Middle Wisconsinan and older sediments, in general, are covered by these younger materials. Our understanding of the Quaternary history of the area affected by the Cordilleran Ice Sheet thus is based on studies of relatively young surface sediments and associated landforms and of older subsurface deposits. At the northern limit of ice sheet glaciation in Yukon Territory, however, pre-Late Wisconsinan glacial and proglacial sediments occur at the surface beyond the McConnell and Macauley limits. Here, surface studies similar to those conducted on younger materials farther south have provided information on events predating the last glaciation.

British Columbia
J.M. Ryder and J.J. Clague

Thick Quaternary sediments representing several glacial and nonglacial intervals are present in some valleys and lowlands in British Columbia and also occur on the continental shelf. These sediments are most common at sites where ice flow was transverse to valley axes, where bedrock projections sheltered downstream areas from scour, and where ice flow was sluggish, as for example in some lowlands near the margin of the Cordilleran Ice Sheet and in some valleys near its centre.

The stratigraphic relationships and internal characteristics of Quaternary sedimentary units in British Columbia are complex. Valleys and lowlands experienced

---

Clague, J.J.
1989: Introduction (Quaternary stratigraphy and history, Cordilleran Ice Sheet); in Chapter 1 of Quaternary Geology of Canada and Greenland, R.J. Fulton (ed.); Geological Survey of Canada, Geology of Canada, no. 1 (also Geological Society of America, The Geology of North America, v. K-1).

Ryder, J.M. and Clague, J.J.
1989: British Columbia (Quaternary stratigraphy and history, Cordilleran Ice Sheet); in Chapter 1 of Quaternary Geology of Canada and Greenland, R.J. Fulton (ed.); Geological Survey of Canada, Geology of Canada, no. 1 (also Geological Society of America, The Geology of North America, v. K-1).

several cycles of infilling and partial or complete reexcavation, thus younger sediments commonly are inset into older deposits or are draped over an irregular older landscape. Younger sediments typically have been reworked from older Quaternary materials, thus units of different age within a given area may be physically similar. On the other hand, individual units may display abrupt facies changes due to the local variability of depositional environments within rugged terrain. Consequently, physical characteristics generally cannot be used to correlate discontinuous stratigraphic units. The firm correlation of major glacial and nonglacial units from valley to valley and from region to region has depended largely on radiocarbon dating and the presence of dated tephra layers.

Most Quaternary stratigraphic studies in British Columbia have been carried out in the southern part of the province. Chronologies that have been established and local names for the principal stratigraphic units are summarized in Figure 1.17 (see also Fulton et al., 1984). The longest record comes from Fraser Lowland near Vancouver, where three major glaciations and three nonglacial intervals have been recognized (Armstrong, 1981). Two major glaciations and three nonglacial intervals have been described from several areas, for example eastern Vancouver Island (Fyles, 1963) and south-central British Columbia (Fulton and Smith, 1978). Late Wisconsinan drift and postglacial sediments are common everywhere in British Columbia.

Old glaciations

There was glaciation in British Columbia in the Middle and Early Pleistocene and probably during the late Tertiary as well. Anahim Peak volcano on Chilcotin Plateau may have erupted into or against glacier ice in late Miocene or Pliocene time. On Mount Edziza in northwestern British Columbia, till and glacial fluvial sediments containing clasts foreign to the mountain are interbedded with lava flows and pyroclastic deposits dating back to perhaps 3-4 Ma (Souther et al., 1984). Till, proglacial sediments, and a glacial pavement at Dog Creek on Cariboo Plateau date to about 1.2 Ma (Mathews and Rouse, 1986). Tuyas and other ice contact volcanic landforms in the Clearwater River area of east-central British Columbia are at least 0.3 Ma old (Hickson and Souther, 1984). Finally, basal lavas about 0.4 Ma old in the Grand Canyon of Stikine River (P.B. Read, Geotex Consultants Ltd., personal communication, 1983) overlie both glacially scoured rock surfaces and glacial fluvial gravel containing clasts derived from the Coast Mountains to the west (Kerr, 1948).

Exposures of Early and Middle Pleistocene glacial deposits are rare outside of volcanic areas, in part because most of these materials occur below present base level. On southern Vancouver Island and in Fraser Lowland, however, there are exposures of drift beneath Sangamonian or older nonglacial deposits (Armstrong, 1975, 1981; Hicock and Armstrong, 1983). In Fraser Lowland, these old glacial deposits are termed Westlynn Drift and consist of a complex of till and glacial marine, glacial fluvial, and (?) glacial lacustrine sediments similar to materials in younger drift sheets. Deposits which correlate with, or are older than, Westlynn Drift may be present in other parts of British Columbia. Armstrong and Leaming (1968), for example, noted the presence of three or four till sheets in central British Columbia, and Ryder (1976) recognized two tills beneath Fraser Glaciation drift along Fraser River north of Lillooet.

Old nonglacial intervals

Outside of volcanic areas, the oldest known Pleistocene nonglacial deposits in British Columbia occur in Fraser Lowland (Armstrong, 1975, 1981; Hicock and Armstrong, 1983). These deposits, which underlie Westlynn Drift and

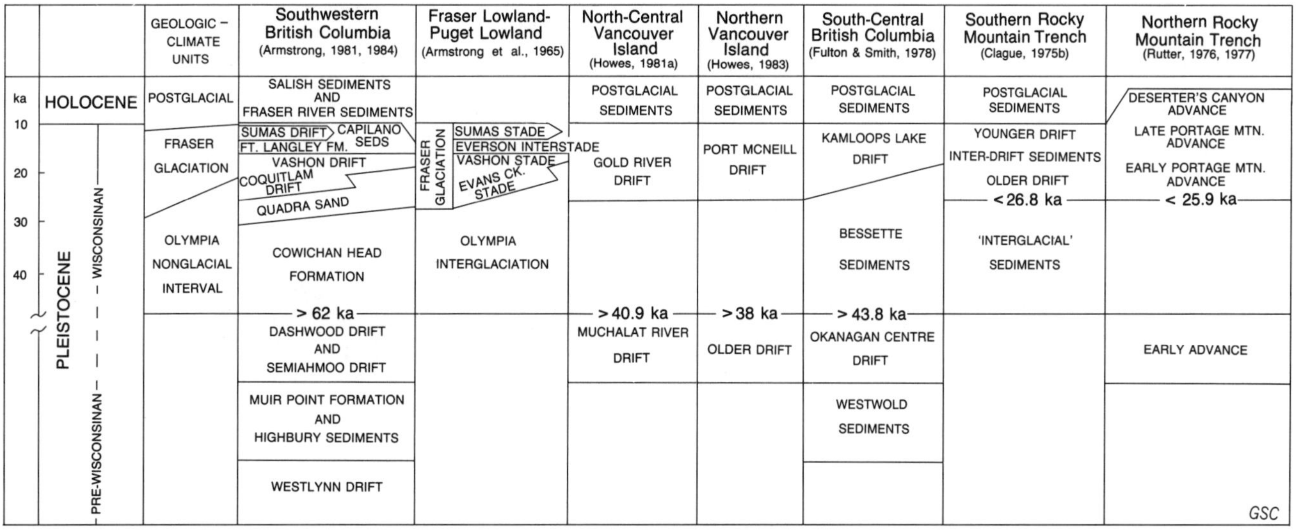

Figure 1.17. Subdivisions of Quaternary events and deposits in British Columbia. Different ages have been assigned to Northern Rocky Mountain Trench events by Bobrowsky et al. (1987).

are pre-Sangamonian in age, consist of more than 60 m of sand and silt encountered in drill holes. No outcrops are known.

Possible Sangamonian deposits have been described for only a few localities. In Fraser Lowland, Highbury Sediments consist of fluvial, deltaic, marine, and organic sediments, mainly sand and silt. Highbury deltaic and marine sediments grade upward into fluvial sediments, recording seaward progradation of floodplains, a process repeated during subsequent nonglacial intervals. The Muir Point Formation on southern Vancouver Island, which is Sangamonian or older in age and may correlate with Highbury Sediments, comprises gravel, sand, silt, peat, and diamicton, presumably deposited as alluvium and colluvium on a coastal floodplain (Hicock and Armstrong, 1983; Alley and Hicock, 1986). At the type section, the Muir Point Formation is more than 30 m thick and is unconformably overlain by Middle Wisconsinan nonglacial sediments and Late Wisconsinan drift.

Strongly oxidized, fluvial gravel and sand containing fragments of wood and blocks of peat and marl occur beneath the oldest till in Northern Rocky Mountain Trench (Rutter, 1976, 1977). These sediments, which commonly are more than 15 m thick, are beyond the range of radiocarbon dating and may correlate with Highbury Sediments and/or the Muir Point Formation.

Two well defined stratigraphic units which underlie till of the penultimate glaciation in southern British Columbia may record the transition from the Sangamonian Stage to the Wisconsinan Stage; alternatively, they may be pre-Sangamonian in age. Mapleguard Sediments, exposed at the base of some sea cliffs on eastern Vancouver Island, consist mainly of bedded sand, silt, and minor gravel of fluvial or perhaps deltaic or marine origin (Fyles, 1963; Hicock, 1980). The relationship of Mapleguard Sediments to the Muir Point Formation is unknown, although Hicock and Armstrong (1983) proposed that the former are younger than, and thus stratigraphically separate from, the latter. Mapleguard Sediments may be strictly nonglacial in origin, although they more likely are outwash deposits laid down during the early part of the penultimate glaciation. Westwold Sediments, which probably correlate with Mapleguard Sediments, are exposed at two sites in south-central British Columbia (Fulton and Smith, 1978). At the type section, Westwold Sediments consist of 16.5 m of cross-bedded gravelly sand capped by 1.8 m of sand, silt, clay, and marl. Features resembling ice-wedge pseudomorphs are present near the top of the gravelly sand unit, suggesting that permafrost may have been present at one time during the final stages of deposition of Westwold Sediments. However, the thin, fine grained upper unit lacks such features and contains molluscan shells, plant impressions, and fragments of bison bones, fish, beetles, and rodents, suggesting subsequent climatic warming.

Penultimate glaciation

Drift of the penultimate glaciation has been identified in most parts of British Columbia where detailed stratigraphic work has been carried out, although it generally is confined to major valleys and coastal lowlands (Fig. 1.18).

In the interior of British Columbia and on Vancouver Island, this drift consists of a single till bounded by stratified sediments. Okanagan Centre Drift in south-central British Columbia, for example, comprises a till, an underlying unit of glacial lacustrine silt and glacial fluvial gravel, and an overlying unit of glacial lacustrine silt and minor glacial fluvial and beach gravel (Fulton and Smith, 1978). A similar succession has been found at other sites in the southern interior (Ryder, 1976, 1981b) and in Northern Rocky Mountain Trench (Rutter, 1976, 1977). On north-central Vancouver Island, poorly exposed Muchalat River Drift consists of till and overlying glacial lacustrine silt (Howes, 1981a). Along the coast, the same till is overlain by fossiliferous glacial marine mud (Howes, 1983). On eastern Vancouver Island south of Howes' study area, Dashwood Drift comprises till and an overlying unit of glacial marine silt and silty sand (Fyles, 1963). Hicock and Armstrong (1983) included the previously mentioned Mapleguard Sediments, which they consider to be glacial in origin, in Dashwood Drift.

In Fraser Lowland, drift of the penultimate glaciation (Semiahmoo Drift) apparently is more complex than in other areas. Semiahmoo Drift consists of two or more tills interlayered with glacial marine, glacial fluvial, and possibly glacial lacustrine sediments (Armstrong, 1975; Hicock and Armstrong, 1983). Some of the "glacial fluvial" materials were deposited subaqueously as outwash fans and deltas. The complexity of this unit probably results from the fact that tidewater glaciers fluctuated in this area during the decay phase of the penultimate glaciation. Semiahmoo and Dashwood drifts are similar in character and complexity to Late Wisconsinan drift in the same area, thus the pattern of glaciation in the Strait of Georgia region during the last two glacial periods probably was similar.

Olympia Nonglacial Interval

Sediments assigned to the Olympia Nonglacial Interval (or Olympia Interglaciation) are common in southern British Columbia and also occur locally in the central and northern parts of the province. These sediments were deposited during a lengthy nonglacial period that preceded the Late Wisconsinan Fraser Glaciation.

The Cowichan Head Formation underlies lowlands adjacent to the Strait of Georgia (Armstrong and Clague, 1977). The marine lower member, which consists mainly of sand and mud, conformably overlies Dashwood and Semiahmoo glacial marine sediments deposited during the transition from glacial to nonglacial conditions at the end of the penultimate glaciation. The upper member comprises interbedded organic-rich gravel, sand, and silt of fluvial and estuarine origin. The marine member is less widely distributed on land than the terrestrial upper member, but may occur extensively beneath the seafloor of the Strait of Georgia. Where the marine member is absent, upper Cowichan Head sediments overlie an irregular erosion surface developed on older drift. The Cowichan Head Formation is less than 10 m thick in most exposures and is much thinner than bounding drift units.

Deposition of the Cowichan Head Formation probably commenced while the sea was falling relative to the land at the end of the Semiahmoo Glaciation. At many coastal sites, shallow marine environments were gradually replaced by estuarine and fluvial environments as streams prograded seaward or extended across the emergent sea floor. Upper Cowichan Head sediments were deposited in channel, overbank, and swamp settings on coastal floodplains; they possibly also accumulated along prograding delta shorelines in bars, dunes, lagoons, and marshes (Fyles, 1963; Armstrong and Clague, 1977). Nonglacial sedimentation ceased with the deposition of outwash (Quadra Sand) from advancing glaciers at the onset of the Fraser Glaciation.

Bessette Sediments of south-central and southeastern British Columbia are time-equivalents of the Cowichan Head Formation (Fulton, 1968; Fulton and Smith, 1978). They consist chiefly of fluvial, deltaic, lacustrine, and colluvial materials deposited in a physiographic setting similar to that prevailing today in the southern interior. At the type section on Bessette Creek near Lumby, Bessette Sediments comprise 22 m of interbedded fluvial gravel, sand, and silt containing plant remains and two tephra layers. These sediments are sharply overlain by laminated silt deposited when the regional drainage was impounded by glaciers during the Fraser Glaciation. Another section of Bessette Sediments at Meadow Creek in southeastern British Columbia illustrates the wide range of Olympia-age materials in the southern interior (Fig. 1.19). Here, Bessette Sediments overlie a sloping surface developed on Okanagan Centre Drift; they consist of (1) thin colluvial and eolian sediments containing organic matter and well developed soil horizons, and (2) fluvial gravel, sand, and silt containing at least one tephra layer, plant fragments, and peat beds. The fluvial succession, which extends through a stratigraphic interval of about 15 m, intertongues with and overlies the colluvial and eolian sediments which are 1-2 m thick. At some sites in south-central British Columbia, Bessette Sediments include thick (up to about 100 m) sand and silt of probable deltaic origin.

Depositional analogues for Bessette Sediments exist under present-day conditions in the valleys of southern British Columbia (Fulton, 1975a; Fulton and Smith, 1978):

Figure 1.18. Exposure in Coquitlam River valley, southwestern British Columbia. Glacial fluvial gravel of the penultimate glaciation (1) is overlain successively by thin sand and silt of the Cowichan Head Formation (2), upward-coarsening glacial fluvial or fluvial sediments (3), Fraser Glaciation till (4), and Fraser Glaciation gravelly and sandy subaqueous (?) outwash (5). Wood from units 2, 4, and 5 yielded radiocarbon dates of >49 000 BP, 21 500 ± 240 BP, and 21 700 ± 130 BP, respectively (GSC-2094-2, GSC-2536, GSC-2416; Table 1.2).

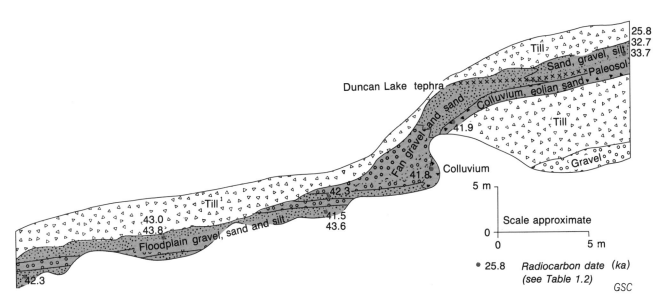

Figure 1.19. Exposure near Meadow Creek, southeastern British Columbia, showing varied Middle Wisconsinan nonglacial sediments (Bessette Sediments, shaded) overlain and underlain by till (adapted from Fulton, 1975b, Fig. I-1).

gravel and sand with cut-and-fill structures are present in the channels of most streams; finer sediments occur on many floodplains, on delta foreslopes, and elsewhere on the floors of lakes; and soil and colluvium cover valley walls. However, the present-day situation is not completely analogous to that of the Olympia Nonglacial Interval because Holocene fluvial deposits comparable in thickness to those of Olympia age are rare in interior valleys. During the Holocene, most streams in this region degraded their valleys, and they are more or less at equilibrium today. In contrast, Bessette Sediments in part record significant aggradation by major streams in southern British Columbia during the Olympia Nonglacial Interval. Fulton (1975a) proposed that aggradation occurred at the end of this period, probably in response to climatic change heralding the Fraser Glaciation. At that time, base level rose, and fluvial and lacustrine sediments accumulated up to 180 m above present valley floors.

This late aggradational phase was preceded by a period during which streams either were in equilibrium or were incising their valleys. This period may include most of the Olympia Nonglacial Interval, for in many areas where drift of both the penultimate glaciation and the Fraser Glaciation is present, the Olympia interval is represented only by an unconformity. This is the case, for example, on northern Vancouver Island (Howes, 1981a, 1983) and in Thompson River valley near Ashcroft (Ryder, 1976). Again, comparison with present-day conditions suggests that erosion was probably the dominant process in many valleys during the Olympia Nonglacial Interval.

Olympia-age sediments also have been reported at a few localities in central and northern British Columbia, but have not been studied as thoroughly as those in the southern part of the province. Lacustrine sediments at Babine Lake (Harington et al., 1974), intertill sediments in Northern Rocky Mountain Trench (Rutter, 1976, 1977) and near Atlin (Miller, 1976), and fluvial gravel and sand along Peace River and its tributaries near Fort St. John (Mathews, 1978) probably were deposited during the Olympia Nonglacial Interval. The fluvial sediments near Fort St. John may record the transition from the Olympia interval to the Fraser Glaciation because they are conformably overlain by glacial lacustrine silt and clay laid down in lakes dammed by the advancing Laurentide Ice Sheet. Aggradation of these sediments followed a period of erosion during which Peace River and its tributaries cut trenches comparable in size to present-day valleys in the region. This pattern of erosion followed by fluvial aggradation is identical to that reconstructed for many valleys in southern British Columbia during the Olympia Nonglacial Interval (Clague, 1986, 1987).

Fraser Glaciation

The Fraser Glaciation is the last period of ice sheet glaciation in British Columbia. In general, Fraser Glaciation drift consists of till and underlying and overlying glacial fluvial and glacial lacustrine sediments; in addition, glacial marine sediments are common in some coastal areas. At many localities, there is only a single Fraser Glaciation till, but at others two or more tills are present.

Early Fraser Glaciation stratigraphy and chronology are best documented in southwestern British Columbia where a prominent unit of advance outwash, Quadra Sand, has been described in detail (Fig. 1.20; Clague, 1976, 1977a; Armstrong and Clague, 1977). Quadra Sand is widely distributed in the Strait of Georgia region below an elevation of about 100 m and also occurs at higher elevations in some valleys extending into the bordering mountains. It overlies the Cowichan Head Formation and Semiahmoo and Dashwood drifts and underlies younger Fraser Glaciation deposits, including till. The unit, which locally is more than 50 m thick, consists mainly of well sorted, horizontally and cross-stratified sand. It includes small amounts of silt and clay in discrete beds and locally coarsens to gravel (e.g., Saanichton Gravel of Halstead, 1968). Paleocurrent indicators and provenance studies show that Quadra Sand was derived largely from the southern Coast Mountains and was deposited as outwash aprons in front of, and perhaps along the margins of, glaciers moving southward into the Strait of Georgia region and Puget Lowland during Late Wisconsinan time. Deposition occurred in channels of braided streams, on adjacent floodplains, and at the fronts of deltas that were prograding into the sea. Quadra Sand was progressively overridden by glaciers and partly eroded, giving rise to the present patchy distribution of the unit. Radiocarbon dates show that Quadra Sand is markedly diachronous and overlaps the Cowichan Head Formation in age (Clague, 1976, 1977a).

Thick subtill gravel and sand in interior valleys, mentioned in the preceding section, may correlate with Quadra Sand. These sediments are horizontally bedded, well sorted, and typically occur far above present base level. Although some scientists have grouped these materials with Olympia nonglacial units, they probably are outwash deposited in response to the expansion of glaciers during the early part of the Fraser Glaciation. The gravel and sand commonly grade up into, or are sharply overlain by, silt and clay deposited in lakes impounded by advancing glaciers or drift (Clague, 1987).

Thick and extensive glacial lacustrine sediments underlie Fraser Glaciation till in Stikine River basin. These sediments accumulated in a lake that formed when glaciers in the Coast Mountains advanced across and blocked Stikine River near its mouth. At its maximum, this lake extended at least 200-300 km upriver from the ice dam. Glaciers today reach almost to river level in this area, so the lake must have developed very early in the Fraser Glaciation and probably persisted for thousands of years. Similar lakes developed in the basins of other rivers that flow through the Coast Mountains, for example Skeena River basin (Clague, 1984).

In areas near the centre of the former Cordilleran Ice Sheet, Fraser Glaciation drift generally includes one till unit deposited during a single glacial event. This till typically is overlain and underlain by stratified sediments of glacial fluvial or glacial lacustrine origin. For example, Kamloops Lake Drift in south-central British Columbia is divisible into three units: (1) a lower stratified unit deposited during glacier expansion, mainly in proglacial lakes; (2) a middle till unit; and (3) an upper stratified unit deposited during deglaciation on floodplains and other subaerial surfaces and in glacier-dammed lakes (Fulton, 1975a; Fulton and Smith, 1978). Fraser Glaciation drift exhibiting this same tripartite stratigraphy is present elsewhere in the British Columbia interior (Ryder, 1976, 1981b; Clague, 1984) and on north-central Vancouver Island (Howes, 1981a).

In many areas, especially near the margins of the former ice sheet, Fraser Glaciation drift is more complex

Figure 1.20. Quadra Sand, Vancouver, British Columbia. The lower darker part of the unit contains interbeds of silt and has yielded radiocarbon dates ranging from 26 100 ± 320 BP at the bottom to 24 400 ± 900 BP at the top (GSC-1635, L-502; Table 1.2).

and includes two or more till units. Such is the case, for example, in Fraser Lowland. There, Coquitlam Drift (Hicock and Armstrong, 1981) was deposited early during the Fraser Glaciation by valley and piedmont glaciers that expanded into lowland areas from the Coast Mountains. These glaciers receded from at least the western part of Fraser Lowland before 18.7 ka when forests became reestablished in the area. Glaciers subsequently readvanced and deposited Vashon Drift at the climax of the Fraser Glaciation (Armstrong et al., 1965; Hicock and Armstrong, 1985). During the Vashon Stade, the piedmont glacier in the Strait of Georgia advanced south into Puget Lowland and west into Juan de Fuca Strait; at its maximum, the Puget lobe reached 250 km south of the International Boundary (Waitt and Thorson, 1983). As this lobe retreated, a variety of sediments were deposited in Fraser Lowland in glacial marine environments. These sediments include till, subaqueous outwash, glacial marine mud and gravelly mud, and deltaic and beach gravel and sand (Fort Langley Formation and Capilano Sediments of Armstrong, 1981). Sumas Drift (Armstrong, 1981) is the youngest unit in the Fraser Glaciation drift sequence in this area and consists of till and outwash deposited during a local readvance at the end of the Pleistocene.

Another area with evidence for more than one ice advance during the Fraser Glaciation is Southern Rocky Mountain Trench (Clague, 1975b). During an early advance, some time after 27 ka, till and glacial fluvial gravel were deposited on the floor of the trench. This "older drift" is overlain by "inter-drift sediments" which consist mainly of glacial lacustrine and (?) lacustrine sand and silt deposited in one or more lakes that formed on the trench floor during an interval of glacier recession. Inter-drift sediments, in turn, are overlain by "younger drift" comprising two tills and associated glacial fluvial and glacial lacustrine sediments. Younger drift is the product of two glacier advances separated by a weak, short-lived interstade during which the floor of Southern Rocky Mountain Trench was only partially deglaciated.

Multiple Fraser Glaciation advances also have been reported for Northern Rocky Mountain Trench adjacent to the eastern margin of the Cordilleran Ice Sheet. Rutter (1976, 1977) recognized three advances in this area during Late Wisconsinan time: Early Portage Mountain (oldest and most extensive), Late Portage Mountain, and Deserter's Canyon. These advances were separated by periods of retreat during which glacial fluvial and glacial lacustrine sediments were deposited on the floor of the trench. This interpretation, however, has recently been questioned by Bobrowsky et al. (1987) who suggested that the Early Portage Mountain advance is Early Wisconsinan in age and that there was no Deserter's Canyon advance. According to them, the only Late Wisconsinan event in this area is the Late Portage Mountain advance.

Minor readvances of glaciers at the end of the Fraser Glaciation, such as the Sumas, have been recognized in many parts of British Columbia. For example, Alley and Chatwin (1979) presented evidence for resurgence of ice in Juan de Fuca Strait and on adjacent southern Vancouver Island during deglaciation of that area. Armstrong (1966) found stratigraphic evidence for a local readvance in the Coast Mountains near Terrace at the end of the Fraser Glaciation (see also Clague, 1985). Most of these advances occurred at different times and probably were controlled more by local factors than by global or regional climatic change (Clague, 1981).

The distribution of different types of surface sediments and the relationships of these materials to subglacial, ice marginal, and proglacial landforms enable a reconstruction to be made of the sequence and pattern of late Fraser Glaciation events in British Columbia. Such an approach was used by Fulton (1967), for example, to document deglaciation in south-central British Columbia. He showed that deglaciation took place largely by downwasting accompanied by stagnation and frontal retreat. Uplands appeared through the ice cover first, dividing the ice sheet into a series of valley tongues that retreated in response to local conditions.

As deglaciation progressed, glacier- and drift-dammed lakes formed in valleys and lowlands, trapping large quantities of fine sediment (Fig. 1.21). The most extensive lakes were in central and south-central British Columbia (Mathews, 1944; Armstrong and Tipper, 1948; Fulton, 1965, 1969; Tipper, 1971a). Some of these have been documented in considerable detail by studying shorelines, outlet channels, glacial lacustrine deposits, and ice marginal features (Fig. 1.22). Smaller lakes were present in many other areas during deglaciation, for example in parts of Rocky Mountain Trench and adjacent valleys (Clague, 1975b; Rutter, 1976, 1977; Ryder, 1981a), in Bulkley River valley (Clague, 1984), and on northern Vancouver Island (Howes, 1981a, 1983).

Postglacial

During and immediately following deglaciation, streams established courses across drift-covered terrain, including the floors of former glacial lakes and isostatically emergent coastal lowlands. Lakes remained in some basins dammed by drift and in basins previously occupied by stagnant ice.

Deglaciation initiated a period of redistribution of glacial materials by fluvial and mass wasting processes that continued into the Holocene (i.e., "paraglacial sedimentation"; Church and Ryder, 1972; Jackson et al., 1982). Freshly deglaciated drift was removed from slopes by landslides and running water and transported to lower elevations. As a result, large fluvial and debris-flow fans were constructed, and streams aggraded their valleys. Much of the fine detritus eroded from drift was carried through the fluvial system and contributed to the rapid growth of deltas in lakes and the sea (Clague et al., 1983).

Paraglacial sedimentation commenced as soon as local areas became ice-free. Processes were initially rapid due to plentiful meltwater and the lack of vegetation; they continued at decreasing rates as slopes stabilized and the supply of available drift declined. Mazama tephra commonly occurs within the uppermost 2-3 m of debris-flow fans in the southern Canadian Cordillera, indicating that most deposition took place before 6.8 ka (Ryder, 1971).

Eolian sedimentation likewise occurred mainly during latest Pleistocene and early Holocene time (Fulton, 1975a). During deglaciation, silt and clay were blown from unvegetated surfaces, especially floodplains, and deposited in protected areas as loess. At the same time, wherever there was an abundant supply of sand, dunes were constructed. There was a marked reduction in loess deposition and dune formation when bare surfaces became stabilized by vegetation.

Aggradation due to paraglacial effects was followed by degradation as the supply of sediment to streams decreased. Streams entrenched paraglacial fills and older deposits, thus producing terraces and dissected fans. As a consequence of this degradation, many streams in British Columbia today flow in trenches cut into broad sediment-filled valleys. For example, Fraser and Thompson rivers locally occupy trenches incised up to 300 m below late glacial valley floors (Fig. 1.5; Ryder, 1971).

This period of degradation ended in the middle to late Holocene with streams near their present levels. Since then, most fluvial systems have been, more or less, in dynamic equilibrium; consequently, Holocene alluvium beneath floodplains generally is thin. Streams issuing from present-day glaciers, however, have deposited thick bodies of outwash in some mountain valleys during late Holocene time.

The main sedimentation sites in British Columbia during the Holocene have been fans, deltas, and offshore basins (both lacustrine and marine). A variety of sediments have accumulated at other sites during this period, but most of the deposits are small.

Chronology and correlation

In southwestern British Columbia, all units below the Cowichan Head Formation (e.g., Highbury Sediments, Muir Point Formation, and Semiahmoo and Dashwood drifts) have consistently yielded "infinite" radiocarbon dates (i.e., dates beyond the limit of the radiocarbon dating method) (Fig. 1.17). On the basis of relative stratigraphic position and palynological data, Hicock and Armstrong (1983) suggested that the Muir Point Formation was deposited during the Sangamonian Stage and that it might correlate with the Whidbey Formation in nearby Puget Lowland, which has yielded amino acid dates of about 100 ka (Easterbrook and Rutter, 1982). However, there still seems to be some uncertainty about the age(s) of the Muir Point Formation and Whidbey Formation; Alley and Hicock (1986) recently stated that the former could be either Sangamonian or pre-Sangamonian. Correlation of the Muir Point Formation and Highbury Sediments is based largely on lithological similarity and relative stratigraphic position. If this correlation is valid, Westlynn Drift must be Illinoian (stage 6 of the marine isotope record) or older.

Figure 1.21. Glacial lacustrine sediments in Thompson River valley near Kamloops, British Columbia. These sediments were deposited over a period of 100-200 years at the close of the Fraser Glaciation (Fulton, 1965). The layers in the lower photograph are varves (person indicated by arrow provides scale). 204640-B, 204640

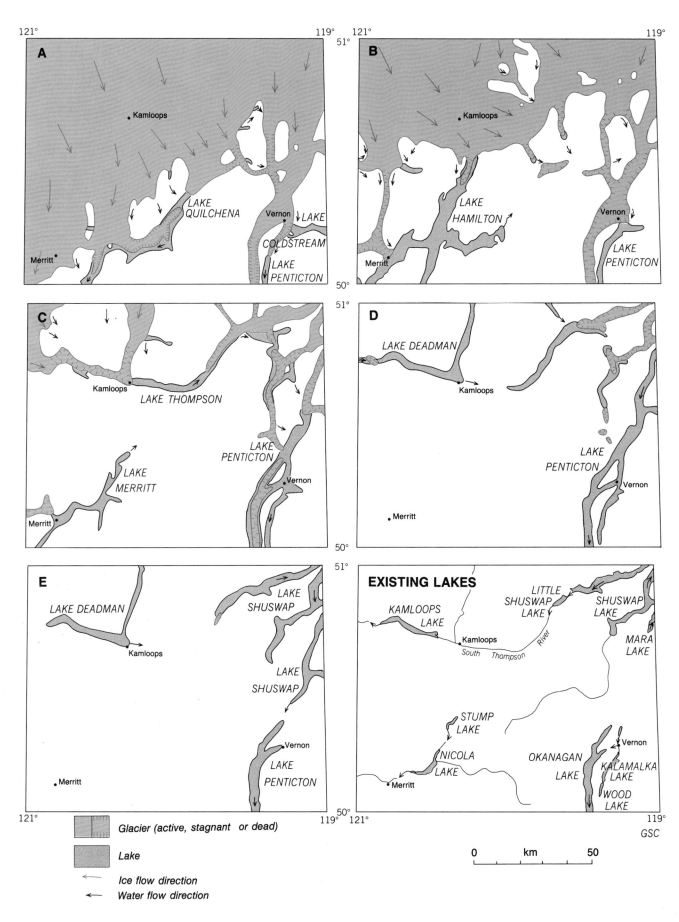

The absolute age of Westwold Sediments is unknown, although it too exceeds the limit of radiocarbon dating. On the basis of relative stratigraphic position, Fulton and Smith (1978) correlated this unit with Highbury and Mapleguard sediments and suggested that it most likely is of Sangamonian and perhaps Early Wisconsinan age.

Most workers have assigned the penultimate glaciation in British Columbia to the Early Wisconsinan Substage. However, dating control is poor, and it is possible that this glaciation is Illinoian. Semiahmoo and Dashwood drifts, correlated by Hicock and Armstrong (1983), have yielded radiocarbon dates of >62 000 BP and >51 000 BP, respectively (QL-194, GSC-94-2; Table 1.2). Dashwood Drift may correlate with Possession Drift in Puget Lowland, which has yielded wood and shell amino acid dates of 50-80 ka (Easterbrook and Rutter, 1981). Okanagan Centre Drift, which almost certainly correlates with Semiahmoo and Dashwood drifts, is known to be older than 43 800 ± 800 BP (GSC-740).

A large number of radiocarbon dates have been obtained from the Cowichan Head Formation, Bessette Sediments, and correlative deposits. These dates indicate that the Olympia Nonglacial Interval commenced more than 59 ka and persisted until the beginning of the Fraser Glaciation, 25-29 ka. Finite dates from the Cowichan Head Formation range from 58 800 +2900/−2100 BP to 23 840 ± 300 BP (QL-195, GSC-518; see also Clague, 1980). Bessette Sediments have yielded dates from 43 800 ± 800 BP to 19 100 ± 240 BP (GSC-740, GSC-913; Clague, 1980). Independent evidence of the duration of the Olympia Nonglacial Interval is provided by $^{230}Th/^{234}U$ dates on speleothems from a cave on central Vancouver Island (Gascoyne et al., 1981). The main period of speleothem growth in this cave, from 67 ka to 28 ka, corresponds to at least part of the Olympia Nonglacial Interval. It is possible, however, that the Olympia interval began well before 67 ka and may, in fact, include part or all of the Early Wisconsinan. Stratigraphic evidence and radiocarbon dates indicate that glaciers probably were confined to mountain areas throughout the Olympia interval (Fulton et al., 1976; Clague, 1978).

Climatic deterioration marking the end of the Olympia Nonglacial Interval may have begun as early as 29 ka on the Pacific coast, based on radiocarbon dates from Quadra Sand in the Strait of Georgia area (Clague, 1976, 1977a, 1980; Alley, 1979). Glacier growth was slow during the early stages of the Fraser Glaciation, and many thousands of years elapsed before ice reached lowland areas outside of the major mountain systems (Fig. 1.23). This is indicated, in part, by the long period of time during which Quadra Sand was deposited and by the presence of relatively young nonglacial sediments beneath till in the vicinity of some mountain

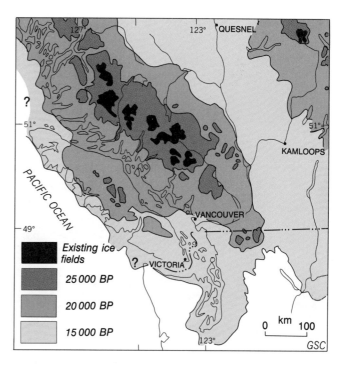

Figure 1.23. Growth of the Cordilleran Ice Sheet in southern British Columbia and northern Washington during the Fraser Glaciation (from Clague, 1981, Fig. 3). Approximate glacier margins at 25 ka, 20 ka, and 15 ka are depicted. Unglaciated areas within the confines of the ice sheet are not shown.

ranges in southern British Columbia. Radiocarbon dates on that part of Quadra Sand predating Coquitlam Drift range from 28 800 ± 740 BP to 23 920 ± 400 BP (GSC-95, GSC-59; Clague, 1980). Dates on other subtill stratified sediments in southern British Columbia indicate that most plateaus and coastal lowlands remained ice-free until after 21 ka and that some areas were not overridden until after 17 ka (Clague et al., 1980). Chronological control on the growth of the Cordilleran Ice Sheet in northern British Columbia is sparse. However, glaciers in the Omineca Mountains and northern Rocky Mountains did not extend into bordering lowlands until some time after 25 940 ± 380 BP (GSC-573).

The Coquitlam and Vashon stades are fairly well dated. Coquitlam Drift has yielded a radiocarbon date of 21 500 ± 240 BP (GSC-2536). The interval of glacier recession separating the Coquitlam and Vashon stades lasted from 19-20 ka to 18 ka, based on radiocarbon dates from organic beds in western Fraser Lowland. The Puget lobe of the Cordilleran Ice Sheet achieved its maximum extent about 14-14.5 ka, and other lobes along the southern periphery of the ice sheet also may have reached their Late Wisconsinan limits at about the same time (Waitt and Thorson, 1983). The Puget lobe advanced 200-250 km between 17 ka and 14.5 ka, thus ice sheet growth during this period was relatively rapid.

Decay of the Cordilleran Ice Sheet at the end of the Fraser Glaciation was equally rapid. Radiocarbon dates from glacial marine sediments in the Strait of Georgia region and on northern Vancouver Island indicate that deglaciation was in progress in these areas about 13-13.5 ka

Figure 1.22. Glacial lake evolution and ice retreat in part of south-central British Columbia (adapted from Fulton, 1969, Fig. 3). Five stages in the development of glacial lakes in this region are shown, A being the oldest and E the youngest. Existing lakes are shown in the lower right panel. Although the age of each stage is unknown, deglaciation of this area probably began about 11-12 ka, and the modern drainage pattern was established before 9 ka.

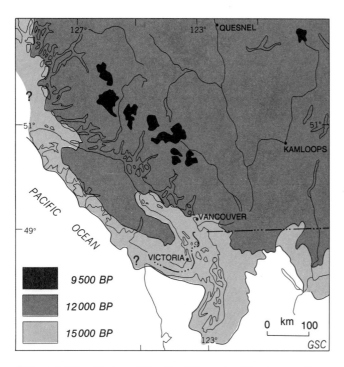

Figure 1.24. Decay of the Cordilleran Ice Sheet in southern British Columbia and northern Washington at the close of the Fraser Glaciation (from Clague, 1983, Fig. 5). Approximate glacier margins at 15 ka, 12.5 ka, and 9.5 ka are depicted. Unglaciated areas within the confines of the ice sheet are not shown.

(Fig. 1.24; Clague, 1980). Fraser Lowland and fiords bordering the Strait of Georgia were completely ice-free about 11 ka at the end of the Sumas Stade (Armstrong, 1981; Saunders et al., 1987). Some upland areas in the southern interior may have become deglaciated as early as 13-14 ka; low plateaus and intermontane valleys remained covered until about 10.5-11 ka (Fulton, 1971; Clague, 1981). The Terrace-Kitimat area in the northern Coast Mountains still supported glaciers as late as 10 ka (Clague, 1984, 1985), but by 9.5 ka glaciers throughout British Columbia were no more extensive than they are today (Clague, 1981).

Yukon Territory
O.L. Hughes, N.W. Rutter, and J.J. Clague

Our understanding of the Quaternary history of that part of Yukon Territory covered by the Cordilleran Ice Sheet has come mainly from studies of glacial sediments and landforms, thus the main emphasis in this section is on the glacial record.

Hughes, O.L., Rutter, N.W., and Clague, J.J.
1989: Yukon Territory (Quaternary stratigraphy and history, Cordilleran Ice Sheet); in Chapter 1 of Quaternary Geology of Canada and Greenland, R.J. Fulton (ed.); Geological Survey of Canada, Geology of Canada, no. 1 (also Geological Society of America, The Geology of North America, v. K-1).

Old glaciations

Bostock (1966) inferred four advances of the Cordilleran Ice Sheet in southern and central Yukon Territory, with each successive advance being less extensive than its predecessor: Nansen (oldest), Klaza, Reid, and McConnell (Fig. 1.25, 1.26). Few readily recognizable landforms remain from the Nansen and Klaza glaciations, hence their limits are poorly known. In contrast, ice marginal features of the Reid Glaciation are moderately well preserved and those of the McConnell Glaciation very well preserved, permitting airphoto interpretation of their limits across much of central Yukon Territory (Hughes et al., 1969). Reid and McConnell deposits also may be distinguished from each other and from older glacial deposits by conspicuous differences in soil development (Hughes et al., 1972; Foscolos et al., 1977; Rutter et al., 1978; Tarnocai et al., 1985). Soils on pre-Reid drift at well drained sites are Luvisols with thick Bt horizons; soils on Reid Drift are moderately developed Luvisols, or Brunisols with thick Bm horizons; and soils on McConnell Drift are Brunisols with thin Bm horizons (see Canada Soil Survey Committee, 1978, for definition of soil terms).

The Nansen Glaciation was named for occurrences of weathered till and weathered gravel of presumed glacial origin in the upper reaches of Nansen Creek and other streams west of Carmacks (Bostock, 1966). A later glaciation, the Klaza, was inferred by Bostock on the basis of evidence for northward diversion of the headwaters of south-flowing Lonely Creek into the Klaza River system. Although it cannot be proven that this diversion was not accomplished during the more extensive Nansen Glaciation, there is evidence elsewhere for two pre-Reid glaciations. For example, at Fort Selkirk outside the Reid limit, a sequence of till, gravel, sand, and silt deposited during an earlier (Nansen?) glaciation is overlain by about 100 m of basaltic lava flows. Glacial striae on the surface of the flows are the product of a later glaciation, presumably the Klaza.

Deposits of two pre-Reid glaciations also are exposed in sections in Liard Lowland, southeastern Yukon Territory (Fig. 1.27; Klassen, 1978, 1987). The older glaciation is represented by a till (Till A) which overlies Tertiary sediments and volcanics (Liard Formation) and is overlain by thick sand and silt. Later ice sheet glaciation of Liard Lowland is indicated by a second till (Till B) which underlies dense lacustrine clay containing plant detritus and gastropod shells. Basalt flows occur directly above Till A and between Till B and the dense lacustrine clay (Fig. 1.27).

In southwestern Yukon Territory, which was affected by successive advances of glaciers from the St. Elias Mountains, there are no surface deposits comparable either in soil or landform development to Nansen or Klaza drifts. However, the lowest of the three drift units exposed at Silver Creek near Kluane Lake probably predates the Reid Glaciation (Fig. 1.28; Denton and Stuiver, 1967). This unit, termed Shakwak Drift, consists of till and outwash gravel that have been oxidized throughout their exposed thickness

Figure 1.25. Maximum extent of glaciers in Yukon Territory and western District of Mackenzie during various Pleistocene glaciations (adapted from Hughes et al., 1983, Fig. 2).

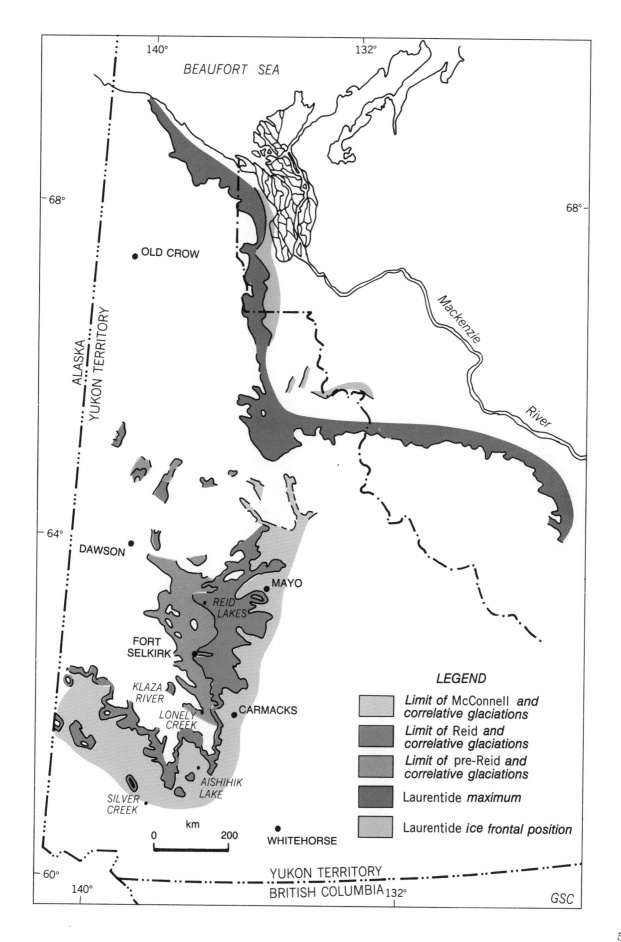

	Yukon Plateaus (Bostock, 1966; Hughes et al., 1969)	Snag-Klutlan area (Rampton, 1971a)	Silver Creek (Denton and Stuiver, 1967)	Liard Lowland (Klassen, 1978, 1987)	Southern Ogilvie Mtns (Vernon and Hughes, 1966)	Old Crow Flats (Morlan, 1980)
HOLOCENE	POSTGLACIAL	NEOGLACIATION	NEOGLACIATION	POSTGLACIAL		
			SLIMS NONGLACIAL INTERVAL			
		—13.7 ka—	—12.5 ka—		—13.7 ka—	—12.5 ka—
PLEISTOCENE — WISCONSINAN	McCONNELL GLACIATION	MACAULEY GLACIATION	KLUANE GLACIATION	TILL D	LAST GLACIATION	GLACIAL-LACUSTRINE CLAY
			—< 29.6 ka—	—< 23.9 ka—		—< 25.2 ka—
	REID SOIL SHEEP CK TEPHRA (ca. 80 ka)	OLD CROW TEPHRA (80-130 ka)	BOUTELLIER NONGLACIAL INTERVAL	INTERTILL UNIT C-D		ALLUVIUM "DISCONFORMITY A" OLD CROW TEPHRA (80-130 ka)
	—> 46.6 ka—	—> 48 ka—	—> 37.7 ka—		—> 53.9 ka—	
	REID GLACIATION	MIRROR CREEK GLACIATION	ICEFIELD GLACIATION	TILL C	INTERMEDIATE GLACIATION	
PLEISTOCENE — PRE-WISCONSINAN	PRE-REID SOIL		SILVER NONGLACIAL INTERVAL	INTERTILL UNIT B-C?		ALLUVIUM
				—> 0.23 Ma—		
	KLAZA GLACIATION	?	SHAKWAK GLACIATION?	TILL B?	OLD GLACIATION?	
	—< 1.08 Ma—					
	FORT SELKIRK TEPHRA (0.84-0.94 Ma)			INTERTILL UNIT A-B?	MOSQUITO GULCH TEPHRA (1.22 Ma)	LITTLE TIMBER TEPHRA (>1.2 Ma)
				—> 0.76 Ma—		
	NANSEN GLACIATION		SHAKWAK GLACIATION?	TILL A?	OLD GLACIATION	LACUSTRINE CLAY

Figure 1.26. Subdivisions of Quaternary events and deposits in Yukon Territory.

(24 m). At Silver Creek, Shakwak Drift is overlain by two younger drift units (Icefield and Kluane drifts), each of which comprises lower and upper outwash gravels and one or more intervening tills.

Reid Glaciation

The Reid Glaciation was named by Bostock (1966) for prominent moraines south and west of Reid Lakes in central Yukon Territory. These mark the limit of an advance of the Cordilleran Ice Sheet in Stewart River valley that was less extensive than the older Nansen and Klaza advances. Moraines and other ice marginal features of Reid age have been modified by mass wasting processes, but in general they are readily recognizable on airphotos.

Reid ice marginal features can be traced with some breaks across the Yukon Plateaus into the St. Elias area of southwestern Yukon Territory. Here, they appear to be continuous with features marking the limit of the Mirror Creek Glaciation (Hughes et al., 1969, 1983; Rampton, 1971a).

Units of possible Reid age exposed in sections in southern Yukon Territory include Till C in Liard Lowland and Icefield Drift at Silver Creek (Fig. 1.27, 1.28). Till C is underlain by lacustrine clay and is overlain successively by Intertill unit C-D and Till D, the youngest till in Liard Lowland. Intertill unit C-D comprises a lower clay layer, a middle gravel, and an upper fossiliferous silt (Tom Creek Silt) which contains an interstadial pollen assemblage. Icefield Drift is overlain by Kluane Drift, the youngest Pleistocene glacial deposit in southwestern Yukon Territory.

McConnell Glaciation

Bostock (1966) applied the name McConnell to a prominent moraine loop that crosses Stewart River about 18 km southwest of Mayo. The moraine and associated features can be traced almost continuously across the Yukon Plateaus to delineate the highly digitate northwestern margin of the Cordilleran Ice Sheet at the climax of the last glaciation (Fig. 1.29). The McConnell limit is continuous with the Macauley glacial limit in the St. Elias area of southwestern Yukon Territory (Hughes et al., 1969, 1983; Rampton, 1971a). Other deposits of McConnell age include Till D and associated glacial fluvial sediments in Liard Lowland and Kluane Drift in the Kluane Lake area (Fig. 1.27, 1.28).

Chronology and correlation

Evidence bearing on the age of pre-Reid glaciations has been obtained at sections near Fort Selkirk and in Liard Lowland. Till at the base of the Fort Selkirk section, thought by Bostock (1966) to be Nansen in age or possibly older, is overlain by stratified sediments containing Fort Selkirk tephra and by lava flows with a striated upper surface. The tephra has yielded glass fission-track ages of 0.84 ± 0.13 Ma and 0.86 ± 0.18 Ma and a zircon fission-track age of 0.94 ± 0.40 Ma (Naeser et al., 1982). A sample of basalt near the base of the flows gave a whole-rock K-Ar age of 1.08 ± 0.05 Ma (Naeser et al., 1982), which does not differ significantly from the fission-track ages at the 2σ level. The K-Ar age is compatible with the reversed magnetic polarity of the dated lava flow. It follows that the till at the base of the section is older than about 1 Ma and that the advance

(Klaza?) that left striae on the surface of the lava flows is younger than 1 Ma.

In Liard Lowland, basalt flows below Till C but above the Liard Formation range in age from 0.765 ± 0.049 Ma to 0.232 ± 0.021 Ma (Fig. 1.27). The flow that yielded the 0.765 Ma date is thought to overlie Till A and underlie Till B. If so, Till B is Middle Pleistocene and Till A Middle or Early Pleistocene in age. Till C, which presumably was deposited during the Reid Glaciation, is younger than about 0.232 Ma.

Radiometric age determinations and continuity of surface deposits indicate that the Reid and Mirror Creek glaciations are equivalent. The Cordilleran Ice Sheet advanced to the Reid limit near the type locality and began to retreat >42 900 BP (GSC-524, Table 1.2) and probably before 80 ka. The former date was obtained on wood collected from a layer of Sheep Creek tephra overlying Reid Drift; the tephra is thought to be about 80 ka old (Hamilton and Bischoff, 1984). A radiocarbon date of >46 580 BP (GSC-331) on wood collected from beneath McConnell till is also a minimum for the Reid Glaciation. Radiocarbon dates from organic material above Mirror Creek Drift and below Macauley Drift in southwestern Yukon Territory are all infinite, except for one of 48 000 ± 1300 BP (GSC-732) for which contamination by modern rootlets is considered possible (Rampton, 1971a). At one site near Kluane Lake, Mirror Creek Drift is overlain by Old Crow tephra, which may be older than 100 ka (Schweger and Matthews, 1985; Westgate et al., 1985; Wintle and Westgate, 1986; Berger, 1987). It thus is clear that the Mirror Creek and Reid glaciations are no younger than Early Wisconsinan; many workers favour an Illinoian age for these events.

The Silver Creek section of Denton and Stuiver (1967) and a section in Liard Lowland studied by Klassen (1978, 1987) have yielded finite radiocarbon dates from sediments beneath the surface drift. At Silver Creek, Icefield till is overlain by gravel containing organic-rich silt beds dating from 37 700 +1500/−1300 BP to 29 600 ± 460 BP (Y-1356, GSC-769; Fig. 1.28). Twig fragments from the lower and upper parts of Tom Creek Silt in Liard Lowland have yielded radiocarbon dates of >30 000 BP and 23 900 ± 1140 BP, respectively (GSC-2949, GSC-2811). These dates suggest that Intertill unit C-D (which includes Tom Creek Silt)

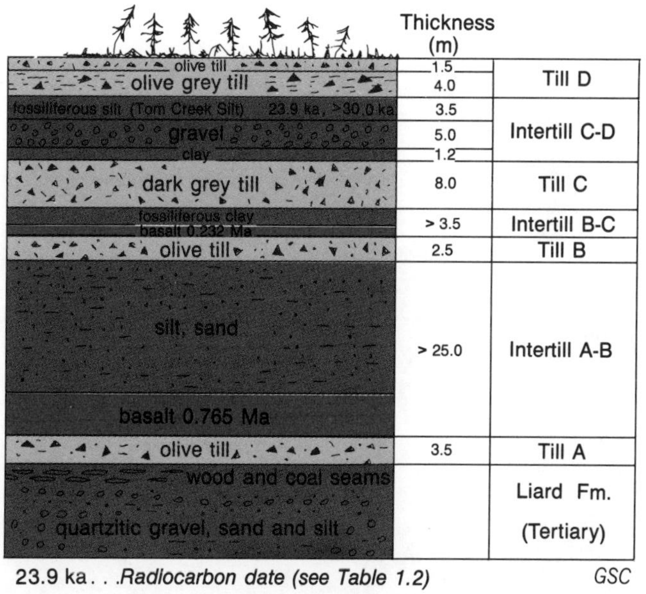

Figure 1.27. Composite Tertiary-Quaternary stratigraphic section, Liard Lowland, Yukon Territory (adapted from Klassen, 1978, Fig. 2, and 1987, Fig. 14).

is Middle Wisconsinan in age and that southern Yukon Territory was deglaciated during the Boutellier Nonglacial Interval.

The date of 29 600 ± 460 BP (GSC-769) at Silver Creek is a maximum for the initial advance of ice out of the St. Elias Mountains during the Kluane Glaciation. The date of 23 900 ± 1140 BP (GSC-2811) from Liard Lowland likewise is a maximum for the last glacier incursion into that area. Deglaciation began on the east side of the St. Elias Mountains before 13 660 ± 180 BP (GSC-495), and Kaskawulsh Glacier, one of the large valley glaciers in the St. Elias Mountains, was less extensive than at present by 9780 ± 80 BP (Y-1483).

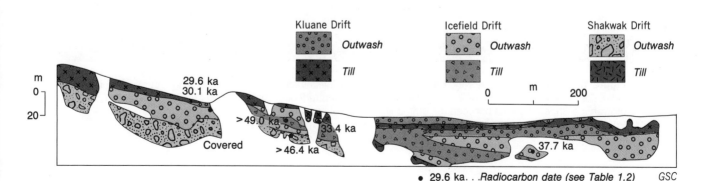

Figure 1.28. Exposure along Silver Creek, Yukon Territory, showing deposits of three glaciations (adapted from Denton and Stuiver, 1967, Plate 4A).

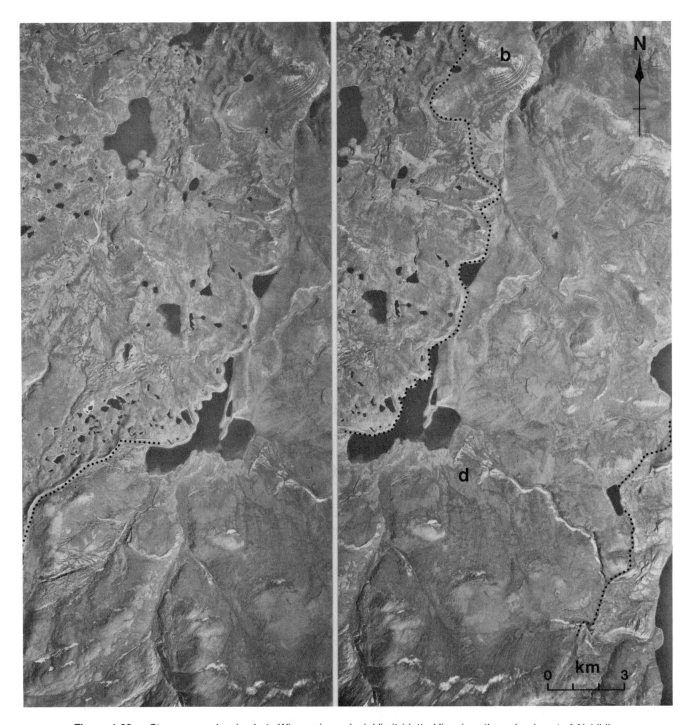

Figure 1.29. Stereogram showing Late Wisconsinan glacial limit (dotted lines) on the upland east of Aishihik Lake, southwestern Yukon Territory. The upland formed a reentrant between two north-flowing lobes of the Cordilleran Ice Sheet; the area between the dotted lines was not glaciated during the Late Wisconsinan. The delta (d) and beaches (b) formed in a glacial lake impounded at the ice margin. NAPL A15739-76, -77

Glaciated fringe

L.E. Jackson, Jr., N.W. Rutter, O.L. Hughes, and J.J. Clague

Introduction

Parts of the Canadian Cordillera were incompletely glaciated during the Pleistocene or escaped glaciation altogether. This section summarizes the Quaternary history of those areas affected mainly by glaciers that were independent or nearly independent of the Cordilleran Ice Sheet. These areas are located at the fringe of the former ice sheet and include parts of the Rocky Mountains and Mackenzie Mountains on the east, the Richardson, Southern Ogilvie, and Wernecke mountains on the north, and the Queen Charlotte Ranges on the west.[1] During Pleistocene glaciations, major valleys in these ranges were occupied by tongues of ice, and summit areas supported ice caps. In addition, the Laurentide Ice Sheet on a few occasions overrode the eastern foothills of the Rocky, Mackenzie, and Richardson mountains.

The Quaternary history of the glaciated fringe has been inferred mainly from stratigraphic and geomorphic evidence. Most deposits that have been studied are glacial in origin, thus Quaternary historical reconstructions for this region emphasize glacial events. The paucity of relevant absolute age determinations, however, has made it extremely difficult to date these events, and the lack of continuity of individual drift sheets from one drainage basin to the next and from one mountain range to another has made regional correlations inferential and tentative at best. Even in a single basin, stratigraphic relationships may not be straightforward, and consequently different conclusions as to the sequence of glacial events have been reached by different workers. As a result of these problems, there is considerable disagreement and confusion regarding the number and age of glaciations in the eastern Cordillera, even in the most intensively studied areas. The Quaternary stratigraphy and glacial history of the Cordilleran glaciated fringe are summarized on the following pages and in the accompanying correlation chart (Fig. 1.30). Undoubtedly, some of the interpretations and correlations that are made here will have to be modified as additional data become available.

Old glaciations

Drift deposited by mountain glaciers during various pre-Wisconsinan glaciations is found in several areas on the eastern flank of the Rocky Mountains and in the Southern Ogilvie Mountains.

Stalker and Harrison (1977), working in Waterton Lakes National Park, recognized four major glaciations, at least one of which is pre-Wisconsinan in age. During the oldest, or Great, glaciation, Rocky Mountain ice extended to high elevations in the foothills and far out onto the Interior Plains, where it deposited Albertan Till. Laurentide ice dropped erratics derived from the Canadian Shield high on the mountain front and spread Labuma Till over Albertan Till in lower areas. Mountain glaciers and the Laurentide Ice Sheet were out-of-phase during the Great Glaciation, the former having withdrawn before the latter reached its climax position. Whether the two ever coalesced in the southern Rocky Mountains remains uncertain, but there probably was at least local contact.

A study of paleosols by Karlstrom (1981, 1987) complements the work of Stalker and Harrison (1977) in the Waterton Lakes area. Karlstrom found five diamicton layers, each with a capping paleosol, on Mokowan Butte, a plateau-like hill rising over 400 m above surrounding terrain at the eastern boundary of Waterton Lakes National Park. According to Karlstrom, the diamictons probably are tills deposited during five separate glaciations, the youngest of which presumably correlates with the Great Glaciation.

Farther north, in Oldman and Crowsnest river valleys, Alley (1973) and Alley and Harris (1974) found evidence for a four-fold sequence of glaciation similar to that proposed by Stalker and Harrison (1977). The oldest advance, termed Glacial Episode 1, is probably equivalent to the Great Glaciation in Waterton Lakes National Park. The Laurentide Ice Sheet impounded lakes during the advance and recessional stages of Glacial Episode 1 and subsequent glaciations. Sediments deposited in these lakes locally separate tills of Rocky Mountain and Keewatin provenance and thus provide stratigraphic evidence that mountain glaciers and the Laurentide Ice Sheet were out-of-phase in the Oldman River area.

Jackson (1980), working in Highwood River basin north of Oldman River, identified an old glaciation equivalent to Alley's (1973) Glacial Episode 1. Evidence consists of erratics and scattered patches of till above 1400 m elevation in Porcupine Hills at the mountain front. This glaciation involved nonsynchronous advances of Rocky Mountain and Laurentide glaciers.

Farther north, Ford (1973, 1976) and Mathews (1978) documented pre-Wisconsinan advances of the Laurentide Ice Sheet to the western edge of the Interior Plains. Mathews (1978), working in the Peace River area, found Canadian Shield pebbles in an aggradational gravel unit deposited before the penultimate incursion of ice into the area. This gravel was laid down by streams flowing from the Rocky Mountains, thus Mathews inferred that some time prior to the penultimate glaciation, Laurentide ice extended west of the location of the shield pebbles (i.e., west of the British Columbia-Alberta boundary). Ford (1973, 1976) identified shield erratics in the foothills of the Mackenzie Mountains near South Nahanni River and assigned them to the oldest and most extensive of three major glaciations in the region (First Canyon Glaciation). This glaciation and the oldest glaciation in the Peace River area thus record major expansions of the Laurentide Ice Sheet. Presumably, mountain glaciers and the Cordilleran Ice Sheet grew at the same time, although sedimentary and other evidence for this has not been found.

Valley glaciers elsewhere in the Mackenzie Mountains reached Mackenzie Lowland and flowed onto Peel Plateau

---

[1] Glaciers in the St. Elias Mountains and Vancouver Island Ranges are not discussed here because they were confluent with, and thus part of, the Cordilleran Ice Sheet at the climaxes of most glaciations.

Jackson, L.E., Jr., Rutter, N.W., Hughes, O.L., and Clague, J.J.
1989: Glaciated fringe (Quaternary stratigraphy and history, Canadian Cordillera); in Chapter 1 of Quaternary Geology of Canada and Greenland, R.J. Fulton (ed.); Geological Survey of Canada, Geology of Canada, no. 1 (also Geological Society of America, The Geology of North America, v. K-1).

on at least one occasion. Extensive interfluve areas in the eastern ranges of the Mackenzie Mountains, however, were never covered by these glaciers, nor by the Laurentide Ice Sheet which abutted against the mountain front.

During one or more "old" glaciations, glaciers flowed from the Southern Ogilvie Mountains into Taiga Valley on the north and Tintina Trench on the south (Vernon and Hughes, 1966). These glaciers deposited more than 200 m of glacial lacustrine sediments, outwash gravel, and till in Tintina Trench east of Dawson (Flat Creek beds of McConnell, 1905). West of the trench, outwash gravel derived from the Southern Ogilvie Mountains (Klondike gravels of McConnell, 1907) lies on a high bedrock terrace along the lower reaches of Klondike River. Near Dawson, the Klondike gravels overlie and intertongue with the White Channel gravels (McConnell, 1907) of probable late Pliocene-early Pleistocene age.

Penultimate glaciation

Deposits of the penultimate glaciation are much more common and better preserved than those of older glaciations. In this section, the record of the penultimate glaciation is summarized, starting first in the Rocky Mountains, proceeding to the Mackenzie and Southern Ogilvie mountains, and ending on the Queen Charlotte Islands.

In Waterton Lakes National Park, the Great Glaciation was followed by the less extensive Waterton II advance, during which Rocky Mountain ice extended east onto the Interior Plains and the Laurentide Ice Sheet deposited till as far west as the mountain front (Stalker and Harrison, 1977). As was the case during the Great Glaciation, mountain glaciers receded before the Laurentide Ice Sheet achieved its maximum extent. With present information, it is not clear if Waterton II and Waterton III, the next major Rocky Mountain advance, were separated by a major nonglacial interval or were part of one large glaciation. The former interpretation is favoured here and is shown in Figure 1.30.

Maycroft Till, the oldest widespread montane till in Oldman, Crowsnest, and Highwood river valleys, was deposited during Glacial Episode 2 when Rocky Mountain ice extended east onto the Interior Plains (Alley, 1973; Jackson, 1980). This ice perhaps coalesced locally with Laurentide ice which deposited Maunsell Till. Glacial lacustrine sediments (Chain Lake Clays and Silts) accumulated in lakes at the margins of these ice masses.

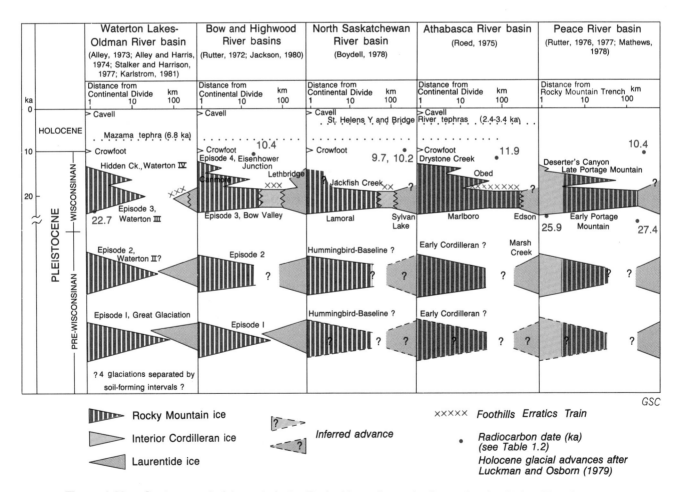

Figure 1.30. Quaternary glacial events in the Rocky Mountains and adjacent Interior Plains. The diagram shows the approximate extent of glaciers, the names of glaciations, and proposed inter-basin correlations. Some age assignments differ from those of original authors.

Fragmentary deposits attributable either to the penultimate glaciation or to an earlier glaciation include outwash underlying the oldest till in Bow River valley (Rutter, 1972), the montane Baseline and Hummingbird tills in North Saskatchewan River basin (Boydell, 1978), and the montane Early Cordilleran and Laurentide Marsh Creek tills in Athabasca River basin (Roed, 1975).

The penultimate glaciation in the Peace River area is recorded by a stratified valley fill capped by till of Keewatin provenance (Mathews, 1978). The valley fill consists of fluvial or glacial fluvial gravel and overlying glacial lacustrine sand, silt, and clay deposited when Peace River was ponded by advancing Laurentide ice. This ice eventually entered the area from the east or north and flowed an unknown distance to the west. The extent of Cordilleran ice at this time is unknown.

Undated moraines comparable in preservation to those of the penultimate (Reid) glaciation in southern and central Yukon Territory are present in many valleys in the eastern Mackenzie Mountains. The glaciers that deposited these moraines were smaller than those of at least one earlier glaciation, but larger than those of the last glaciation. At several localities along the northern front of the Mackenzie Mountains, Laurentide ice marginal features truncate moraines deposited by mountain glaciers during the penultimate glaciation. Elsewhere, however, these moraines are situated upvalley of the later limit of the Laurentide Ice Sheet. The trunk glacier in South Nahanni River valley, for example, terminated about 100 km short of the mountain front at the climax of the penultimate (Flat River-Clausen) glaciation; during the same glaciation, Laurentide ice extended into the eastern foothills of the Mackenzie Mountains, damming a lake (glacial Lake Nahanni) in which thick silt and clay accumulated (Ford, 1976). These sediments overlie the montane Flat River Till, suggesting that the South Nahanni valley glacier may have retreated prior to the incursion of Laurentide ice into the area.

Glaciers in the Southern Ogilvie Mountains during the penultimate ("intermediate") glaciation were similar to, although less extensive than, those of the "old" glaciation(s) (Vernon and Hughes, 1966). During the "intermediate" glaciation, ice in North Klondike River valley reached the east margin of Tintina Trench where it deposited drift on surfaces incised into deposits of an "old" glaciation. Apparently, these two glaciations were separated by a lengthy period of time during which Klondike River and its tributaries eroded their valleys to near present-day levels, about 200 m below "old" glaciation valley floors.

During the penultimate glaciation, the Queen Charlotte Islands supported valley and piedmont glaciers that flowed west and east from local mountain ice caps (Clague et al., 1982a). On eastern Graham Island, these glaciers probably came into contact with a westward-flowing lobe of the Cordilleran Ice Sheet. Deposits of this glaciation underlie a younger drift succession and are well exposed in coastal bluffs bordering northern Hecate Strait. These deposits comprise: (1) massive to weakly stratified, glacial marine mud containing foraminifera, molluscan shells, and scattered stones derived from the British Columbia mainland; (2) till of Queen Charlotte Islands provenance; and (3) stratified sand and gravel.

Last glaciation

Deposits and landforms of the last Pleistocene glaciation in the glaciated fringe are conspicuous and relatively easy to correlate, although in some areas their age is in dispute. In this section, we briefly review the record of the last glaciation, again proceeding in a counterclockwise direction around the Canadian Cordillera.

Sediments deposited at the maximum of the last glaciation in the southern and central Rocky Mountains have been recognized and correlated on the basis of their association with the Foothills Erratics Train, a band of distinctive erratics derived from the mountains near Jasper (Stalker, 1956; Mountjoy, 1958; Morgan, 1966, 1969). The Foothills Erratics Train and associated "mixed provenance" till were deposited by a south-flowing ice stream bordered on the west by montane ice and on the east by the Laurentide Ice Sheet. In the Highwood and Bow river areas, the montane Ernst and Bow Valley tills are in contact with the mixed provenance Erratics Train Till near the front of the Rocky Mountains. These tills were deposited when Rocky Mountain and Laurentide ice last coalesced in this region (Glacial Episode 3 of Jackson, 1980; Bow Valley advance of Rutter, 1972). Farther north in North Saskatchewan River basin, mountain glaciers coalesced with the Laurentide Ice Sheet during the Jackfish Creek-Sylvan Lake glacial episode (Boydell, 1978). There, Jackfish Creek Till of Rocky Mountain provenance grades eastward into mixed provenance Athabasca Till capped by the Foothills Erratics Train; Athabasca Till, in turn, is bordered on the east by Sylvan Lake Till of Keewatin provenance. Still farther north in Athabasca River basin, Marlboro Till, deposited in part by mountain glaciers and in part by ice streaming out of the British Columbia interior, grades laterally into mixed provenance till overlain by Foothills erratics. In this area, the mixed provenance till is bordered on the east by Edson Till, deposited by the Laurentide Ice Sheet. Thus, during the last glaciation, Rocky Mountain (and interior Cordilleran) ice coalesced with the Laurentide Ice Sheet along the eastern edge of the Canadian Cordillera from Highwood River on the south to at least Athabasca River on the north.

South of Highwood River, in Oldman River basin and in Waterton Lakes National Park, Rocky Mountain and Laurentide ice apparently did not coalesce at the climax of the last glaciation (Glacial Episode 3 of Alley, 1973; Waterton III of Stalker and Harrison, 1977). Mountain glaciers flowed east to the front of the Rockies and deposited Ernst Till, while at about the same time mixed provenance till, the Foothills Erratics Train, and Laurentide Buffalo Lake Till were being deposited on the adjacent Interior Plains. Lakes dammed by the Laurentide Ice Sheet covered parts of this region during Glacial Episode 3.

There were one or more readvances in the southern Rocky Mountains during the closing stages of the last glaciation. During the Waterton IV-Hidden Creek advance, glaciers in Waterton Lakes National Park and Oldman and Crowsnest valleys terminated well inside the mountain front and thus were much less extensive than at the last glacial maximum.[1] In Highwood, Bow, and Athabasca valleys,

---

[1] Some workers have questioned this interpretation, claiming that the Waterton IV and Hidden Creek advances are the climactic advances of the last glaciation and that the Foothills Erratics Train and associated mixed provenance till were deposited during the penultimate glaciation (Alley, 1973; Stalker and Harrison, 1977; Jackson, 1980; Rutter, 1980, 1984; Reeves, 1983).

glaciers reached the mountain front during the Canmore and Obed advances (probable correlatives of the Waterton IV — Hidden Creek event; Rutter, 1972; Roed, 1975); a subsequent advance (Eisenhower Junction, Drystone Creek) was much less extensive.

The Cordilleran and Laurentide ice sheets and Rocky Mountain glaciers coalesced in northeastern British Columbia and adjacent Alberta during the last glaciation. Drumlins and other streamlined forms in the Peace River area indicate that southwest-flowing Laurentide ice was deflected in a broad sweeping arc south and southeast by Cordilleran ice flowing east and northeast (Mathews, 1978). The surface till in this area is Cordilleran in provenance in the west and Keewatin in the east, and locally overlies an aggradational valley fill similar to that deposited during the early part of the penultimate glaciation (see preceding section).

The pattern of deglaciation in the northern Rocky Mountains and adjacent Interior Plains at the end of the last glaciation has been reconstructed by Mathews (1980) from geomorphic and stratigraphic evidence. Mathews was able to relate the retreat of the Cordilleran Ice Sheet to that of the Laurentide Ice Sheet by tracing the evolution of glacial Lake Peace which was ponded by Laurentide ice. He showed that the end moraine near Portage Mountain marking the terminus of the Late Portage Mountain advance (Rutter, 1976, 1977) was built out into a high-level lake, probably the Bessborough stage of Lake Peace. Cordilleran ice remained near the mountain front through the Bessborough stage and an unnamed stage that followed. During the next (Clayhurst) stage, Cordilleran ice retreated more than 100 km from the Portage Mountain area, while the Laurentide Ice Sheet continued to impound Lake Peace from positions east of the British Columbia-Alberta boundary. There was a minor readvance of glaciers in Halfway River valley on the east side of the Rocky Mountains during or shortly after the Clayhurst stage.

The last glaciation in the eastern Mackenzie and Southern Ogilvie mountains was restricted, with glaciers in many areas confined to valleys near the crests of the ranges. In South Nahanni National Park, for example, glaciers apparently reached no more than 30 km from cirques during the last (Hole-in-the-Wall) glaciation (Ford, 1976). Some valley glaciers in the eastern Mackenzie Mountains, however, extended into the foothills and coalesced with the Laurentide Ice Sheet at the climax of the last glaciation. These glaciers may have achieved their maximum extent shortly after Laurentide ice began to recede (A. Duk-Rodkin, Geological Survey of Canada, personal communication, 1987).

During the last glaciation, there were significant ice-free areas on the Queen Charlotte Islands, including intervalley ridges, some mountains and coastal lowlands, and possibly shallow offshore platforms (Warner et al., 1982). Glaciers flowed from the Queen Charlotte Ranges and terminated near the present shoreline; those on eastern Graham Island briefly coalesced with the Hecate lobe of the Cordilleran Ice Sheet. Drift of the last glaciation is particularly well exposed in sea cliffs bordering Argonaut Plain on northeastern Graham Island. In this area, the drift consists of: (1) thick, well sorted, crossbedded sand; and (2) overlying till and associated ice contact gravel of local provenance (Clague et al., 1982a). A unit of massive to weakly stratified, gravelly mud, which in places directly underlies the crossbedded sand, may be part of this drift sequence; alternatively, it may predate the last glaciation. Over much of Argonaut Plain, unit 1 sand extends to the surface, and unit 2 is absent. The sand, which typically is a few tens of metres thick, probably was deposited at the western edge of the Cordilleran Ice Sheet just before glaciers achieved their maximum extent in this area during the last glaciation. West of Argonaut Plain, thin local provenance till veneers the sand in an area of low hills and swales oriented in a northerly to northwesterly direction. The linear topography was produced when glaciers flowing east and northeast from the Queen Charlotte Ranges were deflected northward by the margin of the Cordilleran Ice Sheet.

On eastern Graham Island, drift of the last glaciation is conformably overlain by fossiliferous sediments that have yielded a detailed record of Late Pleistocene and Holocene paleoenvironments and sea level change (Clague et al., 1982a, b; Mathewes and Clague, 1982; Warner et al., 1982; Clague, 1983; Warner, 1984). These sediments are thickest in coastal areas below about 15 m elevation and consist of fluvial, shallow marine, littoral, and slopewash deposits, and surface and buried peats (Fig. 1.31).

Chronology and correlation

Old glaciations

There is very little chronological control on events predating the penultimate glaciation. However, because the penultimate glaciation is no younger than Early Wisconsinan, older glaciations must be pre-Sangamonian in age. Estimates of soil age based on degree of soil formation and paleomagnetic data suggest that the youngest of five diamictons on Mokowan Butte in southwestern Alberta is Middle Pleistocene and that the oldest is late Pliocene in age (Karlstrom, 1987). If these diamictons are tills, as argued by Karlstrom (1981, 1987), there were several episodes of late Tertiary and early Quaternary glaciation in the southern Canadian Cordillera. ^{230}Th/^{234}U age determinations on speleothems in caves in South Nahanni River valley suggest that the earliest incursion of the Laurentide Ice Sheet into that area (First Canyon Glaciation) occurred before 320 ka (Ford, 1976; Harmon et al., 1977). A fission track date of 1.22 ± 0.49 Ma on Mosquito Gulch tephra near Dawson provides a minimum age for deposition of the Klondike gravels and hence the "old" glaciation(s) of the Southern Ogilvie Mountains (Naeser et al., 1982; Hughes et al., 1983). This tephra occurs on a terrace cut into the White Channel gravels which, at least in part, are correlative with the Klondike gravels. This date and similar fission-track dates on Fort Selkirk tephra indicate at least comparable antiquity for the Nansen Glaciation of central Yukon Territory and the "old" glaciation(s) of the Southern Ogilvie Mountains. This is supported by the fact that pre-Reid drift, the Flat Creek beds, and Klondike gravels all support Luvisolic soils with thick Bt horizons (Tarnocai et al., 1985).

Because of the lack of adequate chronological control, the number of old glaciations is uncertain. There is some evidence for several in the Waterton Lakes area (Fig. 1.30), but elsewhere only one has been recognized. The correlation of old glaciations from region to region also is uncertain; those shown in Figure 1.30 are tentative.

Penultimate glaciation

Most workers consider the penultimate glaciation in the fringe areas of the Canadian Cordillera to be Early Wisconsinan in age. This event, however, is not well dated, and it could just as well be Illinoian. Radiocarbon dates indicate only that it is older than Middle Wisconsinan.

^{230}Th/^{234}U dates on cave speleothems indicate that the penultimate (Clausen-Flat River) glaciation in South Nahanni River valley occurred sometime after 190 ka (Ford, 1976; Harmon et al., 1977). This glaciation may correspond to the Reid advance of the Cordilleran Ice Sheet in southern Yukon Territory. In terms of degree of preservation of glacial morphology and soil development, drift of the "intermediate" glaciation of the Southern Ogilvie Mountains is similar to Reid deposits (Tarnocai et al., 1985). Organic silt on the floor of Hunker Creek valley near Dawson has yielded a radiocarbon date of >53 900 BP (GSC-527). This is a minimum age for incision of Klondike River and its tributaries preceding the "intermediate" glaciation. The silt may be reworked loess deposited during or after this glaciation; if so, the advance is older than 53.9 ka. The oldest radiocarbon date from sediments overlying drift of the "intermediate" glaciation in the Southern Ogilvie Mountains is 13 870 ± 180 BP (GSC-296).

Radiocarbon dates ranging from >52 000 BP to >33 000 BP (GSC-3151-2, GSC-3208; Clague et al., 1982a) have been obtained on sediments underlying drift of the last glaciation on eastern Graham Island. The penultimate glaciation on the Queen Charlotte Islands thus is older than 52 ka.

Figure 1.31. Late Quaternary sediments exposed in a sea cliff on eastern Graham Island, British Columbia: (1) till; (2) outwash gravel; (3) fluvial and ponded water sand and silt; (4) peat; (5) marine and littoral sand and mud; and (6) peat. Plant detritus at the base of unit 3 yielded a radiocarbon date of 16 000 ± 570 BP (GSC-3340); peat at the base of unit 4 dated 11 100 ± 90 BP (GSC-3337); and shells in unit 5 dated 9350 ± 80 BP (GSC-3120, Table 1.2). Courtesy of R.L. Long.

Last glaciation

The Foothills Erratics Train and associated mixed provenance till provide a basis for correlating deposits of the last glaciation in the eastern Rocky Mountains from Waterton Lakes to Athabasca River. Drift units associated with the Foothills Erratics Train and thus assignable to the last glaciation include Ernst, Bow Valley, Lamoral, Jackfish Creek, and Marlboro tills of Rocky Mountain provenance, and Buffalo Lake, Sylvan Lake, and Edson tills of Keewatin provenance.

Boydell (1978) proposed that Sylvan Lake Till and, by inference, the mixed provenance Athabasca Till and Foothills Erratics Train in the North Saskatchewan River area are Late Wisconsinan in age. He cited as evidence radiocarbon dates of $10\ 250 \pm 165$ BP and 9670 ± 140 BP (I-5675, I-5677) obtained on gastropods and bison bone from diamicton-capped lake sediments in an area of Sylvan Lake dead-ice moraine. Boydell attributed the burial of the lake sediments by diamicton to melting and collapse of underlying ice and to flowage of supraglacial drift into ponds.

In contrast, some workers favour an Early Wisconsinan or older age for the Foothills Erratics Train and associated deposits. Jackson (1980), for example, assigned an Early Wisconsinan age to the Erratics Train on the basis of radiocarbon dates of $18\ 400 \pm 1090$ BP and $18\ 300 \pm 380$ BP (GSC-2670, GSC-2668) from a peat bog in a meltwater channel incised into Bow Valley Till (time equivalent of Erratics Train Till). Recent work, however, has shown that these dates probably are several thousand years too old (MacDonald et al., 1987). This suggests that Bow Valley Till may be Late Wisconsinan in age.

The last glaciation in Athabasca River valley probably occurred sometime after $29\ 100 \pm 560$ BP (GSC-3792; Levson and Rutter, 1986), which is a date on wood in gravel below a single till near Jasper. A date of $22\ 700 \pm 1000$ BP (GaK-2336; Kigoshi et al., 1973) on bone from a cave in Crowsnest Pass is a maximum for the last glaciation in that area.

Radiocarbon dates on ungulate remains from a terraced postglacial gravel fill along Bow River range from $11\ 370 \pm 170$ BP to $10\ 200 \pm 280$ BP (GSC-613, GSC-3065; Wilson, 1981; Jackson et al., 1982). On the basis of these and older dates from Elk River valley to the south (Harrison, 1976; Ferguson and Osborn, 1981), it appears that the Canmore advance ended before 12 ka and perhaps before 13 ka.

Several radiocarbon dates help fix the time of the last glaciation in the Peace River region. A tooth from gravel thought by Mathews (1978) to have been deposited before the last incursion of glaciers into the Fort St. John area gave a date of $27\ 400 \pm 580$ BP (GSC-2034). This gravel seems to correlate with similar sediments occupying a former valley of Smoky River at Watino, Alberta, 200 km east of Fort St. John (Westgate et al., 1972; Mathews, 1978). At Watino, fluvial gravel is overlain by lacustrine sediments that have yielded radiocarbon dates as young as $27\ 400 \pm 850$ BP (I-4878). These dates suggest that the surface tills in the Fort St. John area, shown by Rutter (1976, 1977) and Mathews (1978, 1980) to have been deposited by coalescing Cordilleran and Laurentide ice, are Late Wisconsinan in age. Deglaciation of this area was in progress long before 10 ka. Freshwater shells from silts near Dawson Creek that were deposited either in a late stage of glacial Lake Peace or in local ponds that postdate the lake have yielded radiocarbon dates of $10\ 400 \pm 170$ BP and 9960 ± 170 BP (GSC-1654, GSC-1548). Other dates from bog and lake bottoms, proglacial lacustrine sediments, and younger fossiliferous silt in the Swan Hills area about 240 km southeast of Fort St. John indicate that much of north-central Alberta was deglaciated before 11.5 ka, with some areas ice-free perhaps as early as 13.5 ka (St-Onge, 1972).

The beginning of the "last" glaciation in the Southern Ogilvie Mountains is unknown. Glaciers in these mountains, however, probably were in retreat before $13\ 740 \pm 190$ BP (GSC-515).

Peat beneath outwash gravel and till in Yakoun River valley on the Queen Charlotte Islands has yielded a radiocarbon date of $27\ 500 \pm 400$ BP (GSC-3530). The last advance of glaciers in the Queen Charlotte Ranges thus occurred after 28 ka. Recent work on sediments exposed in coastal bluffs on northern and eastern Graham Island indicate that Late Wisconsinan glaciers achieved their maximum extent in this area after 21 ka. Glaciers retreated from their climax positions on eastern Graham Island before 15 ka (Clague et al., 1982a; Warner et al., 1982).

Unglaciated areas

O.L. Hughes, N.W. Rutter, J.V. Matthews, Jr., and J.J. Clague

Unglaciated central and northern Yukon Territory includes diverse terrains that range from flat lowlands through rolling plateaus to locally rugged mountains. This area is unified only by its lack of glacial deposits.

Cryoplanation (altiplanation) terraces and tors are common features in this region (Hughes et al., 1972). Cryoplanation terraces exhibit all the characteristics described by Demek (1969) for European, Asian, and other North American examples and likely are products of parallel scarp retreat in a periglacial environment. They are absent in areas of readily recognizable glacial features and therefore are pre-Reid in age. Some tors may have formed by cryoplanation of larger rock masses. However, the common occurrence of these features independently of cryoplanation terraces, their presence on western Klondike Plateau at a lower elevation than the terraces, and their occurrence in some glaciated areas where there are no terraces suggest a different origin for many tors. One possible formative mechanism is downwasting by solifluction; the tors would be left where the rock is more resistant to weathering and erosion.

Throughout much of the unglaciated Yukon, the only surficial deposits are weathered rock and colluvium on slopes, organic silt and peat in depressions, and fluvial sediments along major streams (Hughes, 1972). Thick and extensive unconsolidated sediments have been identified in only three lowland areas within Porcupine River basin: Old Crow, Bluefish, and Bell flats. The sediments in Old Crow and Bluefish flats, and probably those in Bell Flats, extend

---

Hughes, O.L., Rutter, N.W., Matthews, J.V., Jr., and Clague, J.J.
1989: Unglaciated areas (Quaternary stratigraphy and history, Canadian Cordillera); in Chapter 1 of Quaternary Geology of Canada and Greenland, R.J. Fulton (ed.); Geological Survey of Canada, Geology of Canada, no. 1 (also Geological Society of America, The Geology of North America, v. K-1).

below bedrock thresholds, therefore these lowlands cannot be entirely erosional in origin; Tertiary or Pleistocene warping or faulting must have contributed to their formation (Hughes, 1972).

Quaternary deposits in Old Crow, Bluefish, and Bell flats are broadly similar in character and are mainly of fluvial, lacustrine, and glacial lacustrine origin. The flats once were drained by streams that flowed east through the Richardson Mountains at McDougall Pass. These streams deposited a variety of fluvial, lacustrine, and deltaic sediments in a nonglacial setting. At least once during the Pleistocene, however, the Laurentide Ice Sheet advanced against the Richardson Mountains and blocked east-flowing streams to form a vast lake that inundated low-lying areas to the west (Hughes, 1972; Thorson and Dixon, 1983). This lake discharged through a canyon at the Alaska-Yukon boundary (The Ramparts of the Porcupine). Waters escaping from the lake gradually eroded the Ramparts; by the time McDougall Pass last became ice-free, the outlet was lower than the pass and westward drainage was permanently established. Drainage of the glacial lake was followed by incision, local deposition of fluvial and shallow water lacustrine sediments, and widespread peat deposition.

Quaternary deposits in Old Crow, Bluefish, and Bell flats have been extensively studied, in part because bones thought to be human-modified have been found at some exposures (Morlan, 1980, 1986; Jopling et al., 1981). A composite section of the deposits in Old Crow Flats, shown in Figure 1.32, illustrates the complex history of sedimentation and erosion in this area during the Quaternary.[1] The lowest unit (1a) commonly extends to a few metres above river level and consists of massive clay, formerly thought to be glacial lacustrine in origin (Hughes, 1972; Matthews, 1975b; Jopling et al., 1981), but now suspected of having accumulated in a lake impounded by tectonic warping. Unit 1b consists of clay reworked from unit 1a. At one locality, it contains a layer of volcanic ash, termed the Little Timber tephra. Small channels at the contact between units 1a and 1b contain concentrations of wood, plant macrofossils, insects, mammal and fish bones, and molluscs (Clarke and Harington, 1978; Cumbaa et al., 1981; Bobrowsky, 1982; Morlan and Matthews, 1983). Unit 1b is overlain by 20-30 m of well bedded sand, silt, and minor gravel (unit 2) deposited intermittently by streams over an extended period of time. Large channel cut-and-fill structures occur in unit 2, as do ice-wedge pseudomorphs, cryoturbation structures, paleosols, peat layers, wood, molluscs, and bone. Unconformities demarcated by concentrations of bone, truncated cryoturbation structures, and peat are also present, especially in the upper part of the succession. The most conspicuous of these unconformities (= boundary between units 2a and 2b in Fig. 1.32, "Disconformity A." of Morlan, 1980) merges with a paleosol and peat that formed on a coniferous forest floor (Morlan and Matthews, 1983). Some of the bones found on this surface may be artifacts (see Morlan, 1986, for a critique). Two tephras occur in unit 2: Surprise Creek tephra near the base and Old Crow tephra (Westgate et al., 1983, 1985; Schweger and Matthews, 1985) in the upper part, 1-2 m below "Disconformity A". Unit 2 is sharply overlain

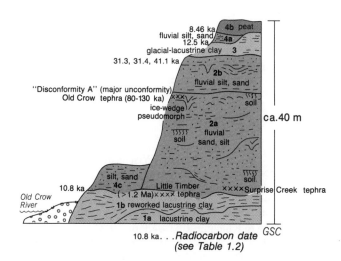

Figure 1.32. Composite stratigraphic section of Old Crow River bluffs, northern Yukon Territory (adapted from Morlan, 1980, Fig. 2.2).

by unfossiliferous glacial lacustrine clay (unit 3) up to 7 m thick. This clay, commonly with a capping of peat (unit 4b), forms the upper surface of much of Old Crow Flats. In places, however, the peat is underlain by fluvial sand and silt (unit 4a) which fill channels cut into the glacial lacustrine sediments.

Chronological control on the deposits in Old Crow Flats is provided by radiocarbon dates on wood, peat, and bone, fission-track and thermoluminescence dates on tephras, uranium-series dates on bone, and paleomagnetic stratigraphy. Part of unit 1b is reversely magnetized (J.A. Westgate, University of Toronto, personal communication, 1986) and may have been deposited during the Matuyama Reversal Epoch, in which case it is older than 790 ka but younger than 2.5 Ma. This is supported by fission-track dates of >1.2 Ma on Little Timber tephra and uranium-series dates of >460 ka on bone from the lower part of unit 2a. Estimates of the age of Old Crow tephra range from about 130 ka to 80 ka (Schweger and Matthews, 1985; Wintle and Westgate, 1986; Berger, 1987; J.A. Westgate, University of Toronto, personal communication, 1987). Finite radiocarbon dates ranging from 41 100 ± 1650 BP to 31 300 ± 640 BP (GSC-2754, GSC-1191; Table 1.2) have been obtained on plant material near the top of unit 2b, 4-5 m above Old Crow tephra. Organic detritus at about the same stratigraphic level in an exposure in Bluefish Flats yielded a radiocarbon date of 20 800 ± 200 BP (GSC-3946); mammoth tusk from this level in another exposure yielded an AMS (accelerator) radiocarbon date of 25 170 ± 630 BP (NMC-1232). Finally, bone from a channel fill incised into unit 3 has been dated at 12 460 ± 440 BP (I-3574).

Unit 2 thus spans a long period of time, perhaps much of the Middle and Late Pleistocene. During this period, intervals of fluvial deposition alternated with intervals of soil formation and permafrost degradation (Schweger and Matthews, 1985).

Glacial lacustrine sediments of unit 3 were deposited during the Late Wisconsinan Hungry Creek Glaciation when the Laurentide Ice Sheet blocked a major east-flowing

---

[1] The lowest sediments in this section, however, may be pre-Pleistocene in age. A section of similar sediments exposed along Porcupine River in adjacent Bluefish Flats is described by Schweger (1989).

river at McDougall Pass. At the climax of this glaciation, some time after 36 900 ± 300 BP (GSC-2422) but before 16 000 ± 420 BP (GSC-2690), Laurentide ice filled Bonnet Plume Depression and diverted Peel River northward via Eagle River into the Porcupine drainage (Hughes et al., 1981; Vincent, 1989). The ice sheet also backed up against the eastern Mackenzie and Richardson mountains, reaching an elevation of about 1500 m near Keele River and 885 m just south of Yukon Coastal Lowland (Hughes, 1987). McDougall Pass (314 m) thus was blocked and a lake impounded to the west. A Late Wisconsinan age for the Hungry Creek Glaciation, however, conflicts with evidence from the Interior Plains to the north and east which suggests that part of this area was ice-free during this period (Hughes et al., 1981; Rampton, 1982; see also Vincent, 1989).

The lake in which unit 3 was deposited had drained by 12 460 ± 440 BP (I-3574). By 10 850 ± 320 BP (I-4224), Old Crow River had cut 40 m through the Pleistocene fill to near its present level and was integrated with the Alaskan Porcupine River system.

Thorson and Dixon (1983) studied the alluvial history of Porcupine River in Alaska and suggested that there were at least three glacial lacustrine inundations of Old Crow Flats and Bluefish Flats during Wisconsinan time, two of which occurred after about 31 ka. In contrast, Morlan (1980) and Hughes et al. (1981) suggested, on the basis of stratigraphic and geochronological studies in Old Crow Flats and Bonnet Plume Depression, that there was only one lake episode during the Late Wisconsinan, represented by unit 3. However, because of the high ice content of unit 3, there are no sections where good exposures of the entire glacial lacustrine sequence may be seen; it thus may represent two or more ponding events.

PALEOECOLOGY AND PALEOCLIMATOLOGY

J.J. Clague and G.M. MacDonald

Paleoecological and paleoclimatic information for the Canadian Cordillera has come from studies of fossil pollen, plant macrofossils, paleosols, speleothems, and vertebrate and invertebrate faunal remains (Fig. 1.33). Most of this information is from the southern part of the region, but even here an understanding of Quaternary paleoecology remains fragmentary. This section provides a brief synopsis of Quaternary climates and vegetation in the Canadian Cordillera. Additional information is given in Chapter 7 of this volume.

Coastal British Columbia

Studies of plant microfossils from Middle Wisconsinan sediments in south-coastal British Columbia and adjacent Washington indicate that the climate of the Olympia Nonglacial Interval was variable, but generally cooler than the present-day climate (Hansen and Easterbrook, 1974; C.J. Heusser, 1977, 1983; Clague, 1978; Alley, 1979; Heusser

Clague, J.J. and MacDonald, G.M.
1989: Paleoecology and paleoclimatology (Canadian Cordillera); in Chapter 1 of Quaternary Geology of Canada and Greenland, R.J. Fulton (ed.); Geological Survey of Canada, Geology of Canada, no. 1 (also Geological Society of America, The Geology of North America, v. K-1).

et al., 1980; Hicock, 1980; Hebda et al., 1983). Some beds examined by these authors are dominated by trees that presently grow at low elevations; others contain assemblages indicative of subalpine forests or grass-herb meadows, implying a cooler climate. Slightly cooler and wetter conditions also prevailed on the Queen Charlotte Islands during part of the Olympia Nonglacial Interval (Warner et al., 1984).

Palynological investigations of early Fraser Glaciation sediments in the Strait of Georgia region indicate a gradual deterioration of climate after about 29 ka (Alley, 1979; Mathewes, 1979). By 21 ka, subalpine parkland and possibly alpine plant communities were dominant on the lowlands of eastern Vancouver Island, and large vertebrates, including *Mammuthus imperator* (mammoth) and *Symbos cavifrons* (muskox), were relatively common in the region (Harington, 1975; Hicock et al., 1982b).

Diverse floras existed in some British Columbia coastal areas near the climax of the Fraser Glaciation. About 18 ka, *Abies lasiocarpa-Picea* cf. *engelmannii* (subalpine fir-Engelmann spruce) forest and parkland grew under cold humid continental conditions at sea level near Vancouver (Hicock et al., 1982a). Mean annual temperature was depressed about 8°C, and treeline was 1200-1500 m lower than today. About 15 ka, the lowlands of the Queen Charlotte Islands supported a varied nonarboreal terrestrial and aquatic flora characteristic of alpine herbaceous and shrub meadows (Warner et al., 1982; Warner, 1984).

Vegetation became reestablished in coastal British Columbia soon after the area was deglaciated at the end of the Fraser Glaciation (Mathewes, 1973, 1985; Hebda, 1983; Warner, 1984). The earliest fossil pollen assemblages, dating from about 13-13.5 ka, are dominated by herbs and shrubs adapted to a cold, relatively dry climate. About 13 ka, forest became established and began to change in composition as plants migrated into the region from extraglacial areas and as climate ameliorated. There was rapid warming and drying between 10.5 ka and 10 ka in the Strait of Georgia region, marked by the appearance of *Pseudotsuga menziesii* (Douglas-fir) and *Tsuga heterophylla* (western hemlock). There was also climatic amelioration on the Queen Charlotte Islands at this time.

Pseudotsuga and *Alnus* (alder) dominated southcoastal forests during the early Holocene (Mathewes, 1985), and it is likely that the climate at that time was similar to or warmer than the present. Maximum dry and warm conditions during this "early Holocene xerothermic interval" (Mathewes and Heusser, 1981) occurred between 10 ka and 7 ka. The xerothemic (Hypsithermal) interval was followed by generally wetter and cooler conditions. By 4-5 ka, modern forests dominated by *Tsuga heterophylla* and *Thuja plicata* (western red cedar) had become established on the south coast (Mathewes, 1985). Clear evidence for the present cool wet climate on the Queen Charlotte Islands is not apparent in the fossil record until about 3 ka when *Thuja plicata* became dominant in lowland forests. Since then, there has been little change in vegetation along the British Columbia coast (Warner, 1984).

Interior British Columbia

Fossil animal and plant remains from Bessette Sediments suggest that the climate of south-central British Columbia during at least part of the Olympia Nonglacial Interval was similar to the present-day climate of the region (Fulton,

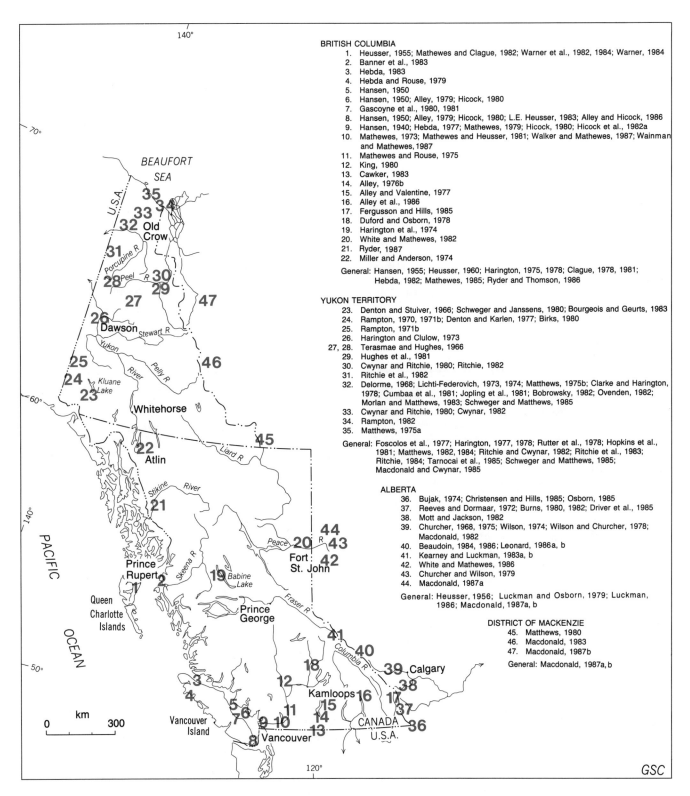

Figure 1.33. Selected Quaternary paleoecological and paleoclimatic studies in the Canadian Cordillera.

1975a; Alley et al., 1986). Major climatic deterioration accompanying the onset of the Fraser Glaciation began about 25 ka, but a full glacial climate was not attained until 19 ka (Alley and Valentine, 1977).

Early deglacial plant assemblages in the southern interior, as on the coast, were dominated by nonarboreal taxa, indicating an open landscape and generally cold and possibly dry climatic conditions (Hebda, 1982). A cold, relatively dry climate also prevailed in the Atlin area in northwestern British Columbia during deglaciation (Miller and Anderson, 1974); there, shrub tundra apparently grew at sites now covered by spruce forest.

Geological and botanical evidence pertaining to early and middle Holocene climatic conditions in the British Columbia interior is complex and in some cases contradictory. Pollen evidence generally supports the existence of a xerothermic interval during the Holocene, although the proposed dating varies considerably from locality to locality (Clague, 1981; Mathewes, 1985). For example, Hansen (1955) identified a warm dry period in south-central British Columbia between 7.5 ka and 3.5 ka, whereas Alley (1976b), working in the same region, concluded that the Hypsithermal interval extended from about 8.4 ka to 6.6 ka. King (1980) presented evidence for low lake levels and, by inference, xerothermic conditions near Lillooet from 8 ka until well after 7 ka. Miller and Anderson (1974) proposed that the period of maximum Holocene warmth at Atlin lasted from about 8 ka until 2.5 ka. Precipitation at Atlin during this period apparently equaled or exceeded present levels, in contrast to the situation in the southern part of the province.

Fossil tree remains found above present treeline in the mountains of southern British Columbia provide additional evidence for a Holocene xerothermic interval. Some of these remains are of early and middle Holocene age and have yielded radiocarbon dates ranging from 9070 ± 130 BP to 5260 ± 200 BP (GSC-3173, Y-140bis; Clague, 1980; Ryder and Thomson, 1986).

The xerothermic interval in southern British Columbia was followed by a generally cooler wetter period marked by sharp, but relatively minor fluctuations in climate. Hebda (1982) concluded that essentially modern climatic conditions were attained in the southern interior about 4.5 ka after a period of cooling and increased precipitation that began some time between 8 ka and 6 ka. Results of paleoecological studies by Alley (1976b), Hazell (1979), and King (1980) at various sites in southern British Columbia generally support this conclusion.

Yukon Territory

The fact that a large part of Yukon Territory was ice-free or glaciated only infrequently during the Quaternary has made this an important region for paleoecological research. Evidence for early and middle Quaternary environmental conditions has been obtained from sediment sections in river valleys, on plateaus, and along the coast (see Schweger, 1989, for pollen diagram of Twelvemile Bluff, Porcupine River). In contrast, evidence for late Quaternary environmental conditions comes mainly from sediment cores taken from extant lakes and bogs.

Fossil pollen and plant macrofossils from the Porcupine River area indicate that a forest dominated by *Picea* (spruce), *Pinus* (pine), and *Betula* (birch), but also containing *Larix* (tamarack) and *Corylus* (hazel), was present in northern Yukon Territory during the Pliocene and perhaps the Early Pleistocene (Lichti-Federovich, 1974; Pearce et al., 1982; Ritchie, 1984). The climate probably was warmer and moister than the present. The later part of the Quaternary in northern Yukon Territory appears to have been characterized by an alternation of tundra, represented in the fossil record by Cyperaceae (sedges), Gramineae (grasses), and various herbs, and forest, dominated by *Picea* (Lichti-Federovich, 1973, 1974; Matthews, 1975b; Schweger and Janssens, 1980; Hughes et al., 1981).

Additional climatic information comes from studies of paleosols developed on drift of the Cordilleran Ice Sheet in central Yukon Territory. Well developed Luvisolic soils on pre-Reid drift indicate a temperate and humid climate prior to the Klaza and/or Nansen glaciations (Foscolos et al., 1977; Rutter et al., 1978; Tarnocai et al., 1985). Moderately developed Luvisols and Brunisols on Reid Drift are evidence of a cool subhumid climate during the Boutellier Nonglacial Interval. Boreal forest conditions prevailed during part of the Boutellier interval, but by about 37.7 ka conditions were colder than at present (Denton and Stuiver, 1967; Schweger and Janssens, 1980). The paleosols contain periglacial and cryogenic features such as sand wedges, sand involutions, and oriented stones, which developed under cold dry conditions during the Reid and McConnell glaciations.

Abundant Pleistocene vertebrate remains have been found at several sites in unglaciated northern Yukon Territory, for example near Dawson and Old Crow (Harington, 1970, 1978). More than 60 species of vertebrates have been recovered, including *Equus* (horse), *Bison* (bison), and *Mammuthus* (mammoth). Unfortunately, most of the collections that have been described are reworked and lack stratigraphic context (Morlan, 1980), thus their paleoecological usefulness is limited.

Radiocarbon-dated cores from lakes and bogs have provided a wealth of paleoenvironmental information for the Late Pleistocene and Holocene of Yukon Territory (e.g., Terasmae and Hughes, 1966; Rampton, 1971b; Birks, 1980; Cwynar, 1982; Ovenden, 1982; Bourgeois and Geurts, 1983; MacDonald and Cwynar, 1985). Late Quaternary pollen, plant macrofossils, insect fossils, and vertebrate remains also have been obtained from cave deposits (Ritchie et al., 1982) and from sediment sections along streams and the Yukon coast (Matthews, 1975a; Cumbaa et al., 1981; Rampton, 1982; Morlan and Matthews, 1983). These studies collectively provide a detailed record of environmental change that spans the last 30 ka.

During Late Wisconsinan time, Yukon Territory was occupied by tundra, characterized in the pollen and plant macrofossil record by Cyperaceae, Gramineae, *Artemisia* (sage), and various herbs. This vegetation was widely established in the Yukon by 30 ka (Rampton, 1971b; Cwynar, 1982). It has been suggested that the vegetation from 30 ka to 14 ka resembled a sparse, fell-field tundra typical of the modern vegetation of the Arctic islands (Cwynar and Ritchie, 1980; Cwynar, 1982; Ritchie and Cwynar, 1982; Ritchie, 1984). This interpretation runs counter to the view that a highly productive "steppe tundra" existed in the Yukon during this period (Guthrie, 1968, 1982; Matthews, 1976, 1982). Although there is a continuing dispute over this issue, it generally is agreed that a reasonably large mammal population was supported by the available plant cover and that the climate was very cold and arid (Guthrie, 1982; Ritchie, 1984).

The climate in northern Yukon Territory ameliorated slightly between about 23 ka and 18.5 ka. The evidence for this is an increase in *Picea, Alnus,* and *Betula* pollen at some sites and the presence of fossil beetle and ostracode species typical of modern shrub tundra and the forest-tundra boundary (Cwynar, 1982; Rampton, 1982).

This interval was followed by a herb and willow tundra phase which ended about 12 ka with the establishment of birch tundra over most of northern Yukon Territory (Matthews, 1975a; Ovenden, 1982; Ritchie, 1982; Ritchie et al., 1982). The change from herb-willow to birch tundra generally has been attributed to climatic amelioration, although the diachronous nature of birch expansion in northwestern Canada remains an enigma (Ritchie, 1984). *Populus* (poplar), *Myrica* (bog myrtle), and *Typha* (cat-tail) grew north of their present limits in this region at 10-11 ka (Cwynar, 1982; Rampton, 1982; Ritchie et al., 1983), and the climate was warmer than it is today.

Picea pollen increased significantly in the northern Yukon as early as 10 ka, possibly reflecting the invasion of spruce from the east (Hopkins et al., 1981) and south (Ritchie and MacDonald, 1986; MacDonald, 1987b). The dispersal of *Picea* and, later, *Alnus* apparently was diachronous: *Picea* did not reach some sites in south and central Yukon Territory until 9 ka or later, and *Alnus* did not achieve its present distribution and density in this region until after 6 ka (Rampton, 1971b; Hopkins et al., 1981). Since the establishment of *Picea* and *Alnus,* the vegetation appears to have changed little, except at climatically stressed sites near treeline and in parts of southern Yukon Territory where *Pinus* only recently (<0.5 ka) has reached its present northern limit (MacDonald and Cwynar, 1985).

There exists a rich record of middle and late Holocene climatic change in the St. Elias Mountains. Stumps above present treeline in this region have been dated at 5.35 ka, 5.25 ka, 3.0-3.6 ka, and 1.23-2.1 ka (Rampton, 1971b; Denton and Karlén, 1977); climate probably was warmer at these times than it is today. Times of cooler climate (2.9-2.1 ka, 1.23-1.05 ka, and 0.5-0.1 ka) are recorded by glacier advances beyond present limits. These and other data (e.g., Bourgeois and Geurts, 1983) indicate that there have been minor adjustments of vegetation at sensitive sites in Yukon Territory during the last half of the Holocene.

Eastern Cordillera

Pollen records from the Rocky and Richardson mountains indicate that the Late Wisconsinan climate of the eastern Canadian Cordillera was cold and probably arid. A sparse herbaceous tundra, dominated by Cyperaceae, Gramineae, *Salix, Artemisia,* and various herbs, existed in the foothills of the Rocky Mountains early during deglaciation (Mott and Jackson, 1982; MacDonald, 1982). The vegetation in this area changed to shrub tundra near the end of the Pleistocene. Herb tundra existed in the Richardson Mountains between 15 ka and 12 ka (Ritchie, 1982). Shortly after 12 ka, it was replaced by shrub tundra dominated by *Betula glandulosa* (resin birch).

A number of Holocene paleoecological records are available for the southern Rocky Mountains and adjacent Interior Plains, Peace River region, Mackenzie Mountains, and Richardson Mountains:

Southern Rocky Mountains and adjacent Interior Plains. Coniferous forests dominated by *Picea* and *Pinus* invaded lower mountain valleys in southwestern Alberta and southeastern British Columbia between about 12 ka and 10 ka, replacing nonarboreal vegetation and, in places, pioneering stands of *Populus* (poplar-aspen) (Harrison, 1976; Schweger et al., 1981; MacDonald, 1982; Mott and Jackson, 1982; Fergusson and Hills, 1985). *Pinus contorta* (lodgepole pine) was growing near present treeline by 9.7 ka (Kearney and Luckman, 1983b; Beaudoin, 1984). Since then, significant changes in vegetation in the southern Rocky Mountains probably have been restricted to sites at the upper and lower limits of forests.

Studies of pollen, plant macrofossils, and paleosols have shown that there were shifts in treeline in the southern Rocky Mountains during the Holocene, presumably in response to changes in temperature and precipitation (Reeves and Dormaar, 1972; Reeves, 1975; Kearney and Luckman, 1983a, b; Beaudoin, 1984, 1986). During most of early and middle Holocene time, treeline was higher than at present, and the climate thus warmer. A cool climate, similar to the present, was achieved after 5 ka. There were additional complex, but minor, timberline fluctuations during the late Holocene associated with neoglacial climatic oscillations.

Holocene malacological data from the southern Rocky Mountains and Interior Plains support these conclusions (Harris and Pip, 1973; MacDonald, 1982). They show that the regional molluscan fauna contained a high complement of south-ranging species from approximately 9 ka to 6 ka; after 6 ka, north-ranging taxa increased in abundance.

Fossil vertebrates in 10-11 ka terrace gravels flanking Bow River near the front of the Rocky Mountains represent a diverse ungulate fauna, including *Camelops* cf. *hesternus* (camel), *Bison bison antiquus* (bison), *Equus conversidens* (Mexican half-ass), *Mammuthus, Cervus canadensis* (wapiti), *Rangifer tarandus* (caribou), and *Ovis canadensis* (bighorn sheep) (Churcher, 1968, 1975; Wilson, 1974; Wilson and Churcher, 1978). This fauna may have lived in a parkland vegetational setting similar to that of present-day southern and central Alberta (Wilson and Churcher, 1978), although paleobotanical information to confirm or negate this hypothesis is lacking.

In summary, climate had ameliorated sufficiently by 10 ka for the establishment of regional vegetation in the southern Rocky Mountains similar to that of the present. The general warming trend seems to have continued to the middle Holocene, although there may have been brief periods of climatic deterioration. Warmer conditions terminated about 4-5 ka.

Peace River region. Holocene paleobotanical information has been obtained from sediment cores taken from lakes in the Clear Hills (MacDonald, 1987a), Saddle Hills (White and Mathewes, 1986), and near Fort St. John (White and Mathewes, 1982). The Clear Hills pollen record shows that the vegetation was dominated by *Artemisia,* Cyperaceae, Gramineae, *Salix,* and *Populus* from 10.7 ka to 9.9 ka. Spruce forest expanded into the Clear Hills between 9.9 ka and 9 ka, and *Pinus* appeared about 7.5 ka. *Pinus* and *Picea* may have arrived in the Saddle Hills as early as 11 ka. The Fort St. John pollen record indicates that there have been no major changes in the boreal forest of this region during the last 7.25 ka.

Early postglacial vertebrate fossils from the eastern Peace River region include *Mammuthus primigenius, Equus,* three species of *Bison,* a camelid, *Cervus canadensis,* and *Ovis,* a fauna indicative of an environmental setting similar to the modern parkland of southern and central Alberta (Churcher and Wilson, 1979).

Neither the pollen nor faunal evidence from this part of the eastern Cordillera provides a clear climatic signal. The transition from parkland to coniferous forest during the early Holocene may reflect differing rates of migration of the dominant plant species rather than climatic change. However, widespread development of small lakes and peatlands between 7 ka and 5 ka may reflect moister conditions during the middle and late Holocene (White and Mathewes, 1986; MacDonald, 1987a).

Mackenzie and Richardson mountains. Two Holocene pollen and plant macrofossil studies have been reported for the Mackenzie Mountains, one from the Boreal Forest zone in the southeast (Matthews, 1980), and another from a small bog above treeline near the crest of the range (MacDonald, 1983). In the former area, a spruce-dominated forest was established by 9.6 ka, and *Alnus* and *Pinus* appeared at 8.7 ka and 6.7 ka, respectively. In the latter area, shrub birch tundra prevailed from at least 8.6 ka until 7.7 ka, after which *Picea* became established. Treeline at this site was higher than at present, and the climate thus warmer, between about 7.7 ka and 5 ka.

Farther north in the Richardson Mountains, birch shrub tundra existed from 11 ka to 9 ka (Ritchie, 1982). At about 9 ka, *Picea* expanded into the region and within 1 ka was well established. *Alnus* appeared at about 7.5 ka and achieved its present abundance by 6.5 ka. *Betula neoalaskanis* (Alaskan paper birch) apparently attained its modern extent and abundance by 6 ka. While all of these changes may be the result of migrational lags, it is worth noting that pollen, plant macrofossil, and invertebrate fossil data from sites on the Mackenzie River delta and Yukon Coastal Lowland indicate that the climate in areas bordering the Richardson Mountains was significantly warmer than at present from about 11 ka until 5-6 ka (Ritchie et al., 1983).

HOLOCENE GLACIER FLUCTUATIONS
J.M. Ryder

Fluctuations of glaciers in the Canadian Cordillera during the Holocene (Table 1.3) have been recognized from detailed mapping of moraines and from stratigraphic investigations. These fluctuations have been dated by dendrochronology, lichenometry, radiocarbon ages, and stratigraphic relationships of drift to various tephras.

Moraines of latest Pleistocene or early Holocene age have been described from Shuswap Highland in south-central British Columbia (Alley, 1976a; Duford and Osborn, 1978), the southern Rocky Mountains (Luckman and Osborn, 1979; Osborn, 1985), and the southern Coast Mountains (Ricker, 1983). In Shuswap Highland, the Harper Creek and Dunn Peak moraines occur in cirques beyond late Holocene moraines and are older than 7985 ± 125 BP and 7390 ± 250 BP, respectively (I-9162, GX-4039; Table 1.2). In the Rocky Mountains, moraines and rock glaciers of the Crowfoot advance generally lie within 1 km of present ice margins and are older than 6.8 ka. Both the Dunn Peak and Crowfoot advances occurred after substantial or complete retreat of Late Wisconsinan glaciers and thus are younger than 12 ka. Davis and Osborn (1987) argued that these and comparable advances elsewhere in the North American Cordillera are probably latest Pleistocene rather than early Holocene in age.

Glaciers expanded during middle and late Holocene time after the Hypsithermal warm interval. This period of glacier expansion has been referred to as "Neoglaciation" by Porter and Denton (1967). Early neoglacial advances are recognized in the southern Coast Mountains ("Garibaldi phase" of Ryder and Thomson, 1986) and in nearby Washington (Miller, 1969; Beget, 1984). The Garibaldi phase is poorly dated, but probably occurred between 6 ka and 5 ka (Ryder and Thomson, 1986). An early neoglacial advance on Dome Peak in northern Washington took place shortly after 5 ka (Miller, 1969).

Glaciers grew in many parts of the Canadian Cordillera and elsewhere in western North America between about 3.3 ka and 1.9 ka (Porter and Denton, 1967; Denton and Karlén, 1973, 1977; Burke and Birkeland, 1983; Ryder and Thomson, 1986). Advances during this period have been documented for three glaciers in the Coast Mountains (Ryder and Thomson, 1986): (1) Tiedemann Glacier achieved its maximum Holocene extent at 2.3 ka; (2) an advance of Gilbert Glacier began before 2.2 ka and culminated shortly after 2.0 ka; (3) Frank Mackie Glacier had about the same extent at 2.7 ka as it does today. A major neoglacial advance of Bugaboo Glacier in the Columbia Mountains began before 2.4 ka (Osborn, 1986). The Battle Mountain advance in Shuswap Highland culminated between 3.4 ka and 2.4 ka (Alley, 1976a). During this same general period, glaciers also advanced in the St. Elias Mountains, with a culmination about 2.6-2.8 ka (Denton and Karlén, 1977). Glaciers in the southern Rocky Mountains began to advance from retracted middle Holocene positions before 3 ka; by 2.7 ka, they may have been as extensive as at present (Leonard, 1986b).

A short-lived advance occurred locally in the St. Elias Mountains about 1.23-1.05 ka (Denton and Karlén, 1977). This advance has not been documented elsewhere in the Canadian Cordillera and thus likely was a minor event. However, Alley (1976b) proposed on the basis of palynological evidence that the present relatively moist phase at Kelowna began about this time.

In many parts of the Cordillera, glaciers attained their maximum Holocene extent during the last several centuries, destroying or burying evidence of earlier advances. This interval of glacier growth has been referred to as the "Little Ice Age" and is recognized in mountain ranges throughout the world (for a summary and references, see Porter and Denton, 1967; Denton and Karlén, 1973). Little Ice Age advances produced the fresh, sparsely vegetated morainal ridges adjacent to most existing glaciers (Fig. 1.34). Dates from in situ stumps and roots exposed by recent glacier recession in the southern Coast Mountains show that the Little Ice Age began there before 0.9 ka and continued without notable interruption until 0.1-0.2 ka (Mathews, 1951a;

Ryder, J.M.
1989: Holocene glacier fluctuations (Canadian Cordillera); in Chapter 1 of Quaternary Geology of Canada and Greenland, R.J. Fulton (ed.); Geological Survey of Canada, Geology of Canada, no. 1 (also Geological Society of America, The Geology of North America, v. K-1).

Table 1.3. Holocene glacier fluctuations in the Canadian Cordillera

	Coast Mountains	Shuswap Highland		Rocky Mountains	St. Elias Mountains
	Ryder and Thomson, 1986; Ryder, 1987	Alley, 1976a	Duford and Osborn, 1978	Luckman and Osborn, 1979; Luckman, 1986	Denton and Karlén, 1977
Late Neoglacial phase	Late neoglacial advance maximum - 0.1-0.3 ka began - >0.9 ka	Mammoth Creek advance late phase - <0.4 ka early phase - 0.8-1.0 ka	Raft Mountain advance "last few centuries"	Cavell advance maximum - <0.1-0.4 ka began - >0.6 ka	Late Neoglacial advances late phase - max. - 0.1-0.5 ka late phase - began - >0.5 ka early phase - 1.05-1.23 ka
Middle Neoglacial phase	Tiedemann advance ended - <1.9 ka maximum - 1.9-2.3 ka[1] began - ≥3.3 ka	Battle Mountain advance >2.4 ka <3.4 ka			Middle Neoglacial advance ended - 2.1 ka maximum - 2.6-2.8 ka began - 2.9 ka
Early Neoglacial phase	Garibaldi phase ca. 5-6 ka				
Early Holocene-Late Pleistocene phase		Dunn Peak advance[2]	Dunn Peak advance >7.4 ka <11 ka	Crowfoot advance >6.8 ka <12 ka	

[1] 2.3 ka for Tiedemann Glacier, 1.9 ka for Gilbert Glacier.
[2] Alley (1976a) assigned an age of 4-5 ka to the Dunn Peak Advance, but Duford and Osborn (1978, 1980) later showed that this advance is older than 7.4 ka.

Figure 1.34. "Little Ice Age" moraines bordering Tiedemann Glacier, southern Coast Mountains, British Columbia.

Ryder and Thomson, 1986). Dates from similar materials in the northern Coast Mountains indicate that glaciers began to advance there before 0.6 ka (Ryder, 1987). The earliest Little Ice Age advances in the Rocky Mountains may have occurred before 0.6 ka (Luckman and Osborn, 1979; Leonard, 1986a; Luckman, 1986), and in the St. Elias Mountains before 0.5 ka (Denton and Stuiver, 1966; Rampton, 1970; Denton and Karlén, 1977).

Dates obtained by dendrochronology and lichenometry indicate that most glaciers in the Canadian Cordillera began to recede from their maximum Little Ice Age positions at various times during the Eighteenth, Nineteenth, and early Twentieth centuries (Mathews, 1951a; Denton and Karlén, 1977; Duford and Osborn, 1978; Luckman and Osborn, 1979; Osborn, 1985; Luckman, 1986; Ryder, 1987). Since then, glaciers generally have undergone sporadic retreat interrupted by stillstands and readvances. Today, almost all glaciers are much less extensive than they were at the climax of the Little Ice Age.

In the Coast and Rocky mountains, most glaciers reached their greatest Holocene extent during the Little Ice Age. In contrast, early Holocene advance(s) in Shuswap Highland were more extensive than subsequent advances. In the St. Elias Mountains, some glaciers climaxed about 2.6-2.8 ka, whereas others attained maximum Holocene positions during the Little Ice Age.

ECONOMIC IMPLICATIONS OF QUATERNARY GEOLOGY

J.J. Clague

Quaternary deposits and geological processes are fundamental to the very character of the Canadian Cordillera and have played a major role in economic development of the region. The significance of these deposits and processes to the economy of the region is briefly summarized in this section, with emphasis on urban and industrial development, resource exploitation, and natural hazards.

Foundations and excavations

Most towns and cities in the Canadian Cordillera are located in areas of low relief, low elevation, and thick Quaternary sediments. Excavation in such areas is relatively easy and inexpensive, facilitating construction of roads, sewers, and water lines. On the other hand, special care must be taken to provide protection against landslides and to ensure the stability of structures built on permafrost and loose, water-saturated or highly compressible materials. For example, compressible fine grained sediments of the Fraser River delta were preloaded at considerable expense prior to the expansion of the Vancouver International Airport in the 1960s (Meyerhof and Sebastyan, 1970). As another example, governments have restricted construction on unstable glacial lacustrine silt benches in Kamloops, Penticton, and some other interior communities to prevent or minimize property damage and loss of life due to landsliding.

Clague, J.J.
1989: Economic implications of Quaternary geology (Canadian Cordillera); in Chapter 1 of Quaternary Geology of Canada and Greenland, R.J. Fulton (ed.); Geological Survey of Canada, Geology of Canada, no. 1 (also Geological Society of America, The Geology of North America, v. K-1).

Agriculture

Land suitable for farming in the Canadian Cordillera is restricted to valley bottoms, coastal and intermontane lowlands, and some plateaus, and represents less than 2% of the total land area of region (Barker, 1977). The most fertile and productive lands are those underlain by silt and clay of glacial lacustrine and glacial marine origin (old lake basins and coastal lowlands formerly submerged by the sea), and by sand and silt of fluvial origin (river terraces and present-day floodplains). Till is less suited for cultivation because it is gravelly and commonly is associated with irregular terrain; nevertheless, large areas of rolling ground moraine in the British Columbia interior are used as range land for livestock and also yield forage crops.

Forestry

About 60% of the Canadian Cordillera is forested, but commercial timber covers only one-third of British Columbia and one-tenth of Yukon Territory (Barker, 1977). Forest productivity is highest in the wet coastal belt of British Columbia, especially in lowland areas. In relatively dry areas of the interior, trees tend to be smaller and productivity lower. Productivity also is low where the water table is near the surface (e.g., bogs), in areas subject to sporadic flooding (e.g., floodplains), and throughout most of the northern Cordillera. Rugged mountains support little commercial timber because climatic conditions there are generally severe and because steep bedrock slopes mantled by thin bouldery drift and colluvium provide a poor substrate for trees and are subject to avalanching. Geological materials exert less of an influence on forest productivity outside major mountain ranges, although fine sediments with high moisture-holding capacities (e.g., clay-rich till, glacial lacustrine silt and clay) are more productive than coarse, well drained sediments (e.g., outwash gravel) in areas of summer drought (the opposite is true in areas of discontinuous permafrost). Substrate materials also may affect forest composition. Within a specific region, for example, the forest on rolling till surfaces may differ significantly from that on gravelly outwash terraces.

Geology and topography also have important economic implications for timber harvesting. In general, the costs of road construction and logging are much higher on steep rock slopes than on low-relief surfaces underlain by Quaternary sediments. Economic and environmental costs may be so high in some areas, for example on slopes prone to landsliding, that logging is not justified.

Mining

Glaciation has played a vital, although largely negative, role in mineral development in the Canadian Cordillera. Pleistocene glaciers have eroded the weathered supergene mantles of many orebodies, dispersing valuable secondary ores. Many primary orebodies thus exposed by glacier action have been buried by younger drift and are not amenable to discovery by traditional prospecting methods. Instead, they remain to be found by geophysical, geochemical, and aeromagnetic surveys, and by drilling. Some surface and subsurface orebodies may be found by tracing mineralized boulders along former glacier flow lines back to their sources. Others may be located by identifying and mapping areas of till containing elevated concentrations of base metals ("dispersal fans").

More than 50% of the gold and 4% of the silver produced in the Canadian Cordillera have come from placers in Quaternary and late Tertiary sediments (Fig. 1.35, 1.36; Debicki, 1983; see also Morison, 1989). Some of these placers predate one or more glaciations and either were not eroded by glaciers or were never overridden by them; others postdate the last glaciation. Most placers in the Cordillera, irrespective of age, are gravels of fluvial or glacial fluvial origin. The richest deposits occur in stream valleys and gulches in hilly terrain. Some are associated with and clearly derived from lodes (e.g., the placers of the Cariboo gold fields), but others have no obvious bedrock sources.

Three examples illustrate the range of gold placers in the Cordillera. The Klondike placers were discovered in 1896 and have yielded about 285 000 kg of gold, more than half of the placer production of the Canadian Cordillera

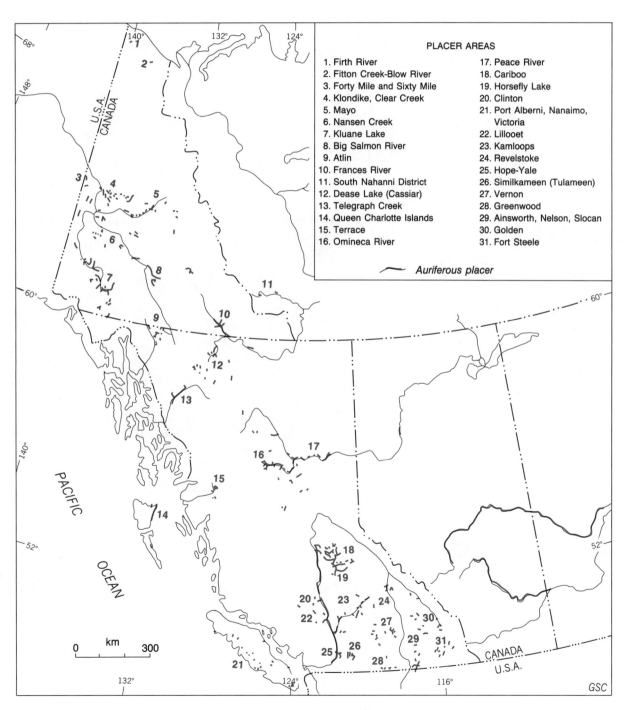

Figure 1.35.   Placer gold deposits in the Canadian Cordillera (adapted from Boyle, 1979, Fig. 79).

(Fig. 1.36; McConnell, 1905; Boyle, 1979; Morison, 1987). The gold occurs in several valleys dissecting an unglaciated upland near Dawson, Yukon Territory. These valleys are flat, wide, and locally terraced in their lower reaches, but they narrow into steep-sided gulches towards their heads and end abruptly in broad amphitheatres. The Klondike placers owe their existence to uplift and erosion of deeply weathered, gold-bearing schists in late Tertiary time. Auriferous White Channel gravels accumulated in broad valleys cut into the uplifted upland. Later, during the Pleistocene, streams cut down through the White Channel gravels into underlying bedrock, and a series of successively lower terraces were formed. Streams completed the incision of their valleys and deposited reworked White Channel gravels on the eroded valley floors some time before 50 ka (Naeser et al., 1982). Most Klondike gold has been recovered

Figure 1.36. The Klondike gold district near Dawson, Yukon Territory. Large amounts of placer gold have been recovered from Plio-Pleistocene gravel which overlies high bedrock benches in this area; arrows indicate old hydraulic workings in these materials. Placer gold also is present in younger gravels underlying terraces and present-day valley floors. Valley floor deposits have been extensively mined by dredges (lower right), giving rise to the conspicuous, worm-like pattern of tailings along Klondike River and Bonanza Creek. NAPL A17155-99, 204655.

from basal White Channel deposits on benches along valley sides and from gravel underlying adjacent valley floors.

The Cariboo placers, discovered in 1860, have supplied over 70 000 kg of gold, more than any other placer district in British Columbia (Johnston and Uglow, 1926; Boyle, 1979). Several types of auriferous sediments are found in the district, and the gold has had a complex history following its release during the Tertiary from quartz veins and stringers in metamorphosed sedimentary rocks. First, there was deposition during the Tertiary of gold-bearing gravel and colluvium in valleys cut into the upland plateau. This was followed during the Pleistocene by repeated glaciation of the region. Some of the Tertiary placers were scoured away by Pleistocene glaciers and reworked by meltwater streams, but others were preserved beneath a thick mantle of drift. Gold placers also formed during Pleistocene nonglacial intervals when streams reworked older gold-bearing gravels. Finally, auriferous stream gravel was deposited in some valleys at the close of the Pleistocene and during the Holocene. These deposits underlie present-day floodplains and stream terraces, and contain variable although generally low amounts of gold.

The Atlin placers, discovered in 1898, occur in valleys cut into an upland that was repeatedly glaciated during the Pleistocene. Placer gold is found in stream gravels within a region containing base metal sulphide deposits and auriferous quartz veins (Boyle, 1979). There are two main types of auriferous gravels in the Atlin district: (1) highly decomposed yellow gravels of Tertiary and Pleistocene age; and (2) undecomposed gravel eroded from drift and old gold-bearing deposits during and following Late Wisconsinan deglaciation of the region.

Construction materials

Quaternary deposits are economically important as sources of construction aggregate. Gravel and sand are in great demand in all urban areas of the Canadian Cordillera and are used extensively in the construction of highways. The annual value of these materials in recent years has been of the order of 50-100 million dollars (British Columbia Ministry of Energy, Mines and Petroleum Resources, 1980). Production is concentrated near urban centres where most of the material is used in concrete, asphalt, and as fill. Supplies near Vancouver, Victoria, and a few other cities, however, are rapidly being depleted or covered with houses, necessitating lengthy transport at high cost. In some areas, transport costs are prohibitive, and crushed rock is substituted for gravel for some purposes.

A wide variety of Quaternary sediments are quarried for construction aggregate. Most of the gravel and sand produced in the Canadian Cordillera comes from outwash and ice contact deposits that were laid down during the advance and recessional phases of the last glaciation. These materials commonly occur above present base level and are terraced. They are thus relatively easy to mine, except where covered by thick till, silt, or clay. Large amounts of construction aggregate and fill also are obtained from Holocene fluvial deposits. In a few areas, Pleistocene beach deposits and even till and rubbly colluvium are mined for aggregate. Till, colluvium, and clay also are used where construction fill of low permeability is required, for example in earth-fill dams. Finally, gravel is dredged from the seafloor to meet the needs of some communities lacking suitable deposits on land (e.g., Prince Rupert).

Clay deposits of glacial marine origin were once exploited in the manufacture of brick, tile, and related products in British Columbia. This practice has been discontinued, despite large reserves of suitable material, due to competition from imported and substitute products.

Water resources

Abundant lakes and ponds which owe their origin mainly to Pleistocene glacial erosion and deposition play a major role in regulating runoff. Quaternary sediments also modulate the flow of streams by holding large amounts of water that would otherwise be lost through rapid runoff. In this manner, Quaternary sediments help maintain lake levels and enhance stream discharge during the dry summer season, both of which are important in hydroelectric power generation, irrigation, and the supply of water to communities. Glaciers and seasonal snowpack play a similar role in the seasonal storage of water in the high mountains of the Canadian Cordillera. In northern Yukon Territory where glaciers, lakes, and thick Quaternary deposits are generally lacking and where the subsoil is perennially frozen, drainage basins have only limited water storage capacity.

Thick Quaternary sediments in valleys and other lowland areas in the Canadian Cordillera contain large amounts of groundwater. At present, about 10% of the water used in the region is groundwater; the remainder is surface water from streams, natural lakes, and reservoirs (Barker, 1977). People in rural areas and in some communities rely almost exclusively on groundwater for domestic and farm needs. In some areas, however, groundwater use is limited by high concentrations of dissolved solids, by contamination from septic tanks, sanitary landfills, feed lots, fertilizers, and pesticides, and by the lack of adequate aquifers.

Although a variety of Quaternary materials contain groundwater, the best aquifers in the Cordillera are thick permeable glacial fluvial and fluvial deposits that occur in areas of high surface recharge such as valleys and lowlands adjacent to mountain slopes. Many of these deposits underlie one or more till sheets and thus are confined aquifers. In contrast, Holocene fluvial deposits directly underlie present-day floodplains and adjacent stream terraces and generally are unconfined.

Natural hazards

Landslides, floods, earthquakes, and volcanic eruptions have played a major role in shaping the landscape of the Canadian Cordillera. These and other natural processes also pose hazards to people and property in this region. Some of these hazards can be mitigated, but only after sufficient geological, geophysical, and historical data have been collected to appraise the probable location, magnitude, and frequency of future destructive events. Floods, earthquakes, and snow avalanches occur repeatedly, thus risk can be assessed by making observations over a long period of time. Areas susceptible to landslides and volcanic eruptions also can be identified, although predicting exactly when and where such events will occur is not yet possible. While in some cases the only option is to avoid a hazardous area, development may be possible after appropriate steps have been taken to reduce risk.

Landslides

A wide variety of potentially destructive mass movements occur in the Canadian Cordillera (Eisbacher, 1979; Evans, 1982; Cruden, 1985; Carson and Bovis, 1989; Evans and Gardner, 1989). Most are complex and are related to instabilities in steep bedrock slopes and to high-gradient mountain streams. They are controlled by unique combinations of topographic, geological, meteorological, and seismic factors and by human activity.

Landslides involving bedrock range from small falls and topples to very large ($>10^6$ m^3) slumps and slides.[1] Rockfalls are common on steep bedrock slopes in areas of intensely fractured or jointed rocks and are triggered by freeze-thaw activity, intense precipitation, and earthquakes (Peckover and Kerr, 1977; Piteau, 1977; Mathews, 1979). They occur frequently throughout the Cordillera, disrupt road and rail traffic, and occasionally take human lives. The small size of these landslides (typically 10-1000 m^3) belies the fact that they are among the most costly in the region. Aside from economic losses due to traffic delays, there is an enormous cost involved in scaling, blasting, and grouting threatening faces and in removing debris from roads and railways.

Rapid bedrock failures involving a sliding or flowing type of movement (rockslides and rock avalanches) are most common in areas of high relief where geological discontinuities (e.g., fractures, faults, bedding planes, cleavage, intrusive contacts) dip in the direction of the slope. They are triggered by excessive porewater pressure, seismic shaking, and human activity, among other things. Fundamental natural causes include erosion of slopes by streams and glaciers and the gradual destruction of cohesion along discontinuities by physical weathering and solution. At least 17 large rockslides and rock avalanches have occurred in historical time in the Canadian Cordillera, and 4 of these have claimed lives (Kerr, 1948; Mokievsky-Zubok, 1977; Cruden and Krahn, 1978; Mathews and McTaggart, 1978; Moore and Mathews, 1978; Clague and Souther, 1982; Cruden, 1982; Eisbacher, 1983; Evans, 1987; Evans et al., 1987). Although uncommon in comparison to small rockfalls and debris flows, rockslides and rock avalanches can be extremely destructive; they are responsible for about 40% (140) of the recorded landslide deaths in the Canadian Cordillera.

Rotational slides (slumps) of a range of sizes occur both in bedrock and Quaternary sediments, and are common in most parts of the Cordillera. Rotational sliding often takes place in association with other types of mass movements. For example, many landslides involving Quaternary sediments are retrogressive slumps at their heads but translational slides or flows at their distal ends. Also, some debris flows are initiated by the blockage of rain-swollen torrents by slumps.

Large deep-seated slope failures characterized by the slow downslope movement of internally broken rock masses along poorly defined rupture surfaces ("sagging slopes") are found in some areas of foliated metamorphic rocks, for example parts of the Columbia Mountains. An especially well documented example of such a failure is the Downie slide in Columbia River valley (Piteau et al., 1978). Although initiated thousands of years ago, this and many other large slope sags in the Cordillera are still active, and any permanent structures located on them ultimately will be damaged or destroyed. In addition, parts of some sags may detach from the main mass of creeping debris and move rapidly downslope; such a possibility should be considered when siting communities and major developments such as hydroelectric reservoirs.

Flows of unconsolidated sediment or weathered bedrock involve the displacement of a mass of material as a viscous fluid. Most such failures in the Canadian Cordillera are debris flows (Fig. 1.37) resulting from the failure of water-saturated, heterogeneous Quaternary sediments (e.g., till, glacial fluvial deposits, pyroclastics, colluvium). Debris flows are most common in mountain areas which receive abundant precipitation. They are triggered principally by intense fall, winter, and spring rainstorms, often accompanied by rapid snowmelt (Eisbacher and Clague, 1984; VanDine, 1985; Church and Miles, 1987); some, however, are initiated by the sudden draining of moraine- or glacier-dammed lakes (Jackson, 1979; Blown and Church, 1985; Clague et al., 1985). Most debris flows are funnelled along steep valleys and ravines and debouch onto fans or cones which often are sites of death and destruction (e.g., Nasmith and Mercer, 1979; Eisbacher, 1983; Jackson et al., 1985; VanDine, 1985). Debris flows are much smaller than most rockslides and rock avalanches, but occur more frequently and are responsible for over one-third (ca. 115) of all recorded landslide deaths in the Canadian Cordillera.

Flows and complex slides consisting mainly of silt and clay occur in areas of Pleistocene glacial lacustrine and glacial marine sediments (Evans, 1982). Failure commonly results from high pore water pressures attributable to both natural causes such as heavy rains and rapid snowmelt, and to human activity, for example irrigation. In drier parts of the Cordillera, special care must be taken with surface water in areas where glacial lacustrine silt and clay form bluffs. Several cities and towns (e.g., Vancouver, Victoria, Kamloops, Prince George, Kitimat, Whitehorse) are located partly on glacial lacustrine or glacial marine deposits, thus flows and related slides and lateral spreads pose a significant hazard to people and property. At least 25 people have died in landslides in these materials in the Twentieth Century, and property damage has been extensive.

Sediment flows also are common on slopes in permafrost areas. Failure occurs in zones of high porewater pressure at the base of the active layer during summer thaw (Hughes et al., 1973; Mackay and Mathews, 1973; Rutter et al., 1973; McRoberts and Morgenstern, 1974). These flows are found on slopes as gentle as about 7° and are especially common along river and coastal bluffs underlain by ice-rich Quaternary sediments.

Large slow-moving flows in weathered volcanic and sedimentary rocks of Cretaceous and Tertiary age are present in Fraser River basin in south-central British Columbia (Eisbacher, 1979; VanDine, 1980; Bovis, 1985). Bentonites and other clay-rich rocks and high porewater pressures are implicated in their movement. Many of these flows are presently active, but they move very slowly and consequently pose a hazard only to certain structures located on them (e.g., roads and rail lines).

Subaqueous mass movements similar to some terrestrial sediment flows are relatively common in British Columbia fiords, at several other sites along the coast, and in some lakes. Most occur at the fronts of active deltas and on steep slopes underlain by unconsolidated sediments (Prior

---

[1] Landslide terminology after Varnes (1978).

and Bornhold, 1984, 1986; Prior et al., 1982, 1984, 1986). Several historical submarine slope failures have occurred in Howe Sound (Terzaghi, 1956; Prior et al., 1981) and Kitimat Arm (Luternauer and Swan, 1978; Swan and Luternauer, 1978; Prior et al., 1982, 1984); one in Kitimat Arm in 1975 generated a train of sea waves that damaged port facilities at Kitimat (see Tsunamis).

Snow avalanches

Avalanches of dry and wet snow, in some cases with dispersed sediment or freshly broken rock, are common in the mountains of western Canada during winter and spring. They occur both on open slopes and in ravines and gullies. Areas of frequent avalanche activity on forested slopes are easily identified because they lack trees; in contrast, avalanche areas above treeline may be more difficult to recognize.

Although they can occur almost any time on slopes that are moderately steep and snow-covered, avalanches are favoured by certain weather conditions (Fitzharris and Schaerer, 1980). They are especially common during thaws or periods of rain after heavy snowfalls, and when thick dry snow accumulates on old icy snow surfaces. In the latter case, the buried icy surface is a plane of weakness along which overlying snow may easily slide.

Figure 1.37. Locomotives derailed by debris-flow deposits in the southern Rocky Mountains, British Columbia, September 7, 1978. Photo by Calgary Herald, courtesy of Glenbow Archives, Calgary, Alberta.

In recent decades, avalanches have taken the lives of many people engaged in recreation activities in remote areas (Stethem and Schaerer, 1979, 1980). Avalanches also were responsible for two disastrous mining-related accidents (Chilkoot Pass, 1898; Granduc Mine, 1965), and regularly disrupt road and rail traffic in some mountain passes and valleys (Schaerer, 1962; Avalanche Task Force, 1974; Barker, 1977).

Floods

Although there has been little loss of life due to floods in the Canadian Cordillera, property damage has been extensive and the cost of flood-control measures high. The most destructive historical flood in this region occurred in southwestern British Columbia in May and June of 1948. At that time, Fraser River broke its dykes in Fraser Lowland and flooded large tracts of agricultural land. As a result, 16 000 people were evacuated and 2000 homes damaged; total losses amounted to more than $20 000 000 (Barker, 1977).

Flooding in the southern Cordillera occurs during extended periods of warm weather in the spring and during periods of heavy rainfall. Large rivers generally crest in late spring as the winter snowpack melts. In contrast, many medium-size rivers have maximum flows during rainy periods in the fall. Streams with small catchments may flood at any time of the year.

Although rainstorm and snowmelt floods also occur in the northern Cordillera, another type of flood perhaps is more common there. In winter, rivers in the north freeze over, often to a depth of more than 1 m. During spring thaw, these rivers swell and their ice crusts are broken into sheets that float downstream. Channel constrictions and river bars may obstruct the free passage of the ice blocks and thus create large jams that force the river over its banks. When the flow becomes powerful enough to break the ice jams, the backed-up waters flood downstream. Considerable damage may be caused in this manner both by the water and by the debris and ice carried along with it. Similar floods occasionally occur when aufeis impedes flow sufficiently to force a stream over its banks during spring thaw.

Some low-lying areas along the coast of British Columbia are flooded by the sea during severe storms, unusually high tides, or as a result of tsunamis. For example, exceptionally high spring tides raised the level of Fraser River during the 1948 flood, causing additional damage to parts of Fraser Lowland.

Finally, lakes impounded by present-day glaciers and by bulky neoglacial end moraines may drain catastrophically to produce severe downstream floods. There have been numerous historical jökulhlaups (glacier-outburst floods) in the Coast, Rocky, and St. Elias mountains (Marcus, 1960; Mathews, 1965, 1973; Gilbert, 1972; Jackson, 1979; Clague, 1982b; Clague and Rampton, 1982; Clarke, 1982; Clarke and Waldron, 1984; Blown and Church, 1985; Jones et al., 1985; Ryder, 1985; Gilbert and Desloges, 1987). Floods from moraine-dammed lakes have been recognized in the Selkirk Mountains and southern Coast Mountains (S.G. Evans, Geological Survey of Canada, personal communication, 1984; Blown and Church, 1985).

Erosion

During periods of high discharge, streams may erode their banks and, in some cases, occupy new channels through avulsion. Erosion may be particularly severe at meander bends and other inflection points and opposite the mouths of aggrading tributaries. Rapid lateral erosion by streams may create steep banks prone to landsliding. Braided streams, which are common in and around glaciated mountain ranges, are especially susceptible to channel migration and avulsion; these processes, in conjunction with normal flooding, render the floodplains of these streams hazardous sites for development.

Parts of the shoreline of British Columbia have receded rapidly (up to a few metres per year) in historical time, necessitating the abandonment or relocation of homes and roads. Erosion has been most severe along those parts of the coast bordered by thick Quaternary sediments, for example eastern Graham Island, eastern and southern Vancouver Island, and near Vancouver (Clague and Bornhold, 1980; Clague, 1982a).

Tsunamis

The most destructive historical tsunami in British Columbia resulted from the Alaska earthquake of March 27, 1964 (Wigen and White, 1964; Thomson, 1981). Waves up to 7 m high swept into the lower parts of Port Alberni, damaging 260 homes, 60 extensively. Hot Springs Cove and Zeballos also suffered wave damage. Seismic sea waves comparable to those generated by the 1964 earthquake are rare. Of the 176 tsunamis recorded in the Pacific Ocean between 1900 and 1970, 35 caused damage near their sources, but only 9 resulted in widespread destruction (Thomson, 1981). Even fewer produced significant runups on British Columbia shores.

Tsunamis generated by distant earthquakes do not affect all parts of the British Columbia coast equally. The 1964 tsunami, although highly destructive at Port Alberni, caused little damage at Tofino only 65 km away and had little effect on protected waterways such as the Strait of Georgia.

Not all tsunamis in British Columbia are the result of distant earthquakes; some are produced by local quakes and submarine landslides. For example, the Vancouver Island earthquake of June 23, 1946 (M = 7.2) produced a local tsunami that affected shores along the Strait of Georgia and nearby inlets (Murty, 1977). In April 1975, sea waves generated by a large submarine landslide near the head of Kitimat Arm slammed into shore installations at Kitimat, causing about $600 000 damage (Campbell and Skermer, 1975; Murty, 1979).

Earthquakes

Although the Canadian Cordillera is seismically active (Milne et al., 1978), damage from earthquakes in the region has been extremely low. The main reason for this is that most large quakes have occurred far offshore at lithospheric plate boundaries. Also, with the exception of the 1946 Vancouver Island earthquake, the few large earthquakes that have occurred on land have been sufficiently far from urban centres that damage has been minor. Nevertheless, some time in the future, a large earthquake probably will occur near a major population centre in western British Columbia. In general, earthquake risk decreases towards the east and is moderate to low in most of the interior of the Canadian Cordillera. High levels of seismicity, however, have been recorded in parts of Yukon Territory and District of Mackenzie, and earthquake risk in these remote, sparsely populated areas is considered to be relatively high (Basham et al., 1977; Horner, 1983).

Earthquake damage is produced both by direct ground motion and by secondary causes such as landslides and fires. The extent of damage is closely related to magnitude, epicentral proximity, and focal depth. In addition, earth materials respond differently to seismic shaking, thus geological factors are important determinants of earthquake damage. Structures located on Quaternary sediments generally are more prone to damage from shaking than structures on bedrock, although some dense compact sediments perform as well as rock in this respect. Saturated, fine grained sediments may liquefy during a major earthquake, giving rise to destructive flows and spreads. Delta-front sediments off the mouth of Fraser River presumably could fail in this way (Luternauer and Finn, 1983). Other materials prone to failure during earthquakes are artificial fills and fine glacial lacustrine and glacial marine deposits.

Volcanic eruptions

There have been many volcanic eruptions in the Canadian Cordillera during the last several thousand years, although none during this century. Quaternary volcanoes occur in clusters and narrow belts in the western Cordillera (Fig. 1.3; Souther, 1970, 1977), and it is likely that any future eruption will also take place in these areas.

Most Quaternary eruptive centres have been the locus of a single pulse of activity during which one or more small pyroclastic cones were built and a small volume of basalt erupted to form thin blocky flows. The past record indicates that an eruption of this type is likely to occur somewhere in the Canadian Cordillera in the next several centuries. Such an eruption probably would affect only a small area and would not be hazardous, unless it occurred in an inhabited region.

In contrast, a few volcanic centres in Alaska, British Columbia, Washington, and Oregon have erupted explosively on several occasions during the Holocene, with far-reaching effects (Table 1.1, Fig. 1.11). The most recent significant eruption of this sort occurred about 1.2 ka when a layer of tephra was deposited over a large area of southern Yukon Territory and western District of Mackenzie from a vent near the Alaska-Yukon boundary. Future eruptions of this sort would likely damage property and crops and disrupt transportation in the Canadian Cordillera. This would result mainly from fallout of ash and dust over large areas and from flooding and aggradation in stream valleys surrounding the volcano. Lahars, pyroclastic flows, and pyroclastic surges would produce additional damage near the volcano.

ACKNOWLEDGMENTS

Unpublished till textural data presented in Figure 1.10 were provided by D.E. Howes, L. Lacelle, B. Thomson (British Columbia Ministry of Environment), R.W. Klassen (Geological Survey of Canada), and J.M. Ryder (University of British Columbia). Locations of earthquake epicentres plotted in Figure 1.3 were supplied by R.B. Horner (Geological Survey of Canada). The photograph of Mount St. Elias (Fig. 1.8) was taken by A. Post and supplied by D.R. Hirst (United States Geological Survey). The photo of the Graham Island section (Fig. 1.31) was taken by R.L. Long and provided by R.W. Mathewes (Simon Fraser University). The photo showing debris flow deposits in the Rocky Mountains (Fig. 1.37) was supplied by Glenbow Archives (Glenbow-Alberta Institute) from a Calgary Herald negative. B.D. Bornhold, J.L. Luternauer, R.P. Riddihough, and J.O. Wheeler (Geological Survey of Canada) provided information and useful suggestions. Drafts of the chapter were reviewed by S.R. Hicock (University of Western Ontario), N.R. Gadd, O.L. Hughes (Geological Survey of Canada), and W.H. Mathews (University of British Columbia).

REFERENCES

Alden, W.C.
1932: Physiography and glacial geology of eastern Montana and adjacent areas; United States Geological Survey, Professional Paper 174, 133 p.

Alley, N.F.
1973: Glacial stratigraphy and the limits of the Rocky Mountain and Laurentide ice sheets in southwestern Alberta, Canada; Canadian Petroleum Geology, Bulletin, v. 21, p. 153-177.

1976a: Post Pleistocene glaciations in the interior of British Columbia; in Geomorphology of the Canadian Cordillera and Its Bearing on Mineral Deposits; Geological Association of Canada, Cordilleran Section, Vancouver, British Columbia, p. 6-7.

1976b: The palynology and palaeoclimatic significance of a dated core of Holocene peat, Okanagan Valley, southern British Columbia; Canadian Journal of Earth Sciences, v. 13, p. 1131-1144.

1979: Middle Wisconsin stratigraphy and climatic reconstruction, southern Vancouver Island, British Columbia; Quaternary Research, v. 11, p. 213-237.

1980: Holocene and latest Pleistocene cirque glaciations in the Shuswap Highland, British Columbia: Discussion; Canadian Journal of Earth Sciences, v. 17, p. 797-798.

Alley, N.F. and Chatwin, S.C.
1979: Late Pleistocene history and geomorphology, southwestern Vancouver Island, British Columbia; Canadian Journal of Earth Sciences, v. 16, p. 1645-1657.

Alley, N.F. and Harris, S.A.
1974: Pleistocene glacial lake sequences in the Foothills, southwestern Alberta, Canada; Canadian Journal of Earth Sciences, v. 11, p. 1220-1235.

Alley, N.F. and Hicock, S.R.
1986: The stratigraphy, palynology, and climatic significance of pre-middle Wisconsin Pleistocene sediments, southern Vancouver Island, British Columbia; Canadian Journal of Earth Sciences, v. 23, p. 369-382.

Alley, N.F. and Valentine, K.W.G.
1977: Palaeoenvironments of the Olympia Interglacial (mid-Wisconsin) in southeastern British Columbia, Canada (abstract); International Union for Quaternary Research, 10th International Congress (Birmingham, England), Abstracts, p. 12.

Alley, N.F., Valentine, K.W.G., and Fulton, R.J.
1986: Paleoclimatic implications of middle Wisconsinan pollen and a paleosol from the Purcell Trench, south central British Columbia; Canadian Journal of Earth Sciences, v. 23, p. 1156-1168.

Ando, M. and Balazs, E.I.
1979: Geodetic evidence for aseismic subduction of the Juan de Fuca plate; Journal of Geophysical Research, v. 84, p. 3023-3028.

Andrews, J.T. and Retherford, R.M.
1978: A reconnaissance survey of late Quaternary sea levels, Bella Bella/Bella Coola region, central British Columbia coast; Canadian Journal of Earth Sciences, v. 15, p. 341-350.

Armstrong, J.E.
1966: Glaciation along a major fiord valley in the Coast Mountains of British Columbia (abstract); Geological Society of America, Program, 1966 Annual Meetings (San Francisco, California), p. 7.

1975: Fraser Lowland-Canada; in The Last Glaciation, D.J. Easterbrook (ed.); IUGS-Unesco International Geological Correlation Program, Project 73-1-24, Guidebook for Field Conference (Western Washington University, Bellingham, Washington), p. 74-97.

1981: Post-Vashon Wisconsin glaciation, Fraser Lowland, British Columbia, Canada; Geological Survey of Canada, Bulletin 322, 34 p.

1984: Environmental and engineering applications of the surficial geology of the Fraser Lowland, British Columbia; Geological Survey of Canada, Paper 83-23, 54 p.

Armstrong, J.E. and Clague, J.J.
1977: Two major Wisconsin lithostratigraphic units in southwest British Columbia; Canadian Journal of Earth Sciences, v. 14, p. 1471-1480.

Armstrong, J.E. and Hicock, S.R.
1976: Quaternary multiple valley development of the lower Coquitlam Valley, Coquitlam, British Columbia (92G/7c); in Report of Activities, Part B, Geological Survey of Canada, Paper 76-1B, p. 197-200.

Armstrong, J.E. and Leaming, S.F.
1968: Surficial geology, Prince George map-area, British Columbia (93G); in Report of Activities, Part A, Geological Survey of Canada, Paper 68-1A, p. 151-152.

Armstrong, J.E. and Tipper, H.W.
1948: Glaciation in north central British Columbia; American Journal of Science, v. 246, p. 283-310.

Armstrong, J.E., Crandell, D.R., Easterbrook, D.J., and Noble, J.B.
1965: Late Pleistocene stratigraphy and chronology in southwestern British Columbia and northwestern Washington; Geological Society of America, Bulletin, v. 76, p. 321-330.

Atmospheric Environment Service
n.d.: Canadian climatic normals, 1951-1980, temperature and precipitation, British Columbia; Canada Department of Environment, Atmospheric Environment Service, Downsview, Ontario, 268 p.

1982: Canadian climatic normals, 1951-1980, temperature and precipitation, the North — Y.T. and N.W.T.; Canada Department of Environment, Atmospheric Environment Service, Downsview, Ontario, 55 p.

Atwater, T.
1970: Implications of plate tectonics for the Cenozoic tectonic evolution of western North America; Geological Society of America, Bulletin, v. 81, p. 3513-3536.

Avalanche Task Force
1974: Report on findings and recommendations to the Honorable Graham R. Lea, Ministry of Highways, September 30, 1974; British Columbia Department of Highways, Victoria, British Columbia, 33 p.

Bacon, C.R.
1983: Eruptive history of Mount Mazama and Crater Lake caldera, Cascade Range, U.S.A.; Journal of Volcanology and Geothermal Research, v. 18, p. 57-115.

Banner, A., Pojar, J., and Rouse, G.E.
1983: Postglacial paleoecology and successional relationships of a bog woodland near Prince Rupert, British Columbia; Canadian Journal of Forest Research, v. 13, p. 938-947.

Barker, M.L.
1977: Natural resources of British Columbia and the Yukon; Douglas, David & Charles, North Vancouver, British Columbia, 155 p.

Barr, S.M. and Chase, R.L.
1974: Geology of the northern end of Juan de Fuca Ridge and sea-floor spreading; Canadian Journal of Earth Sciences, v. 11, p. 1384-1406.

Basham, P.W., Forsyth, D.A., and Wetmiller, R.J.
1977: The seismicity of northern Canada; Canadian Journal of Earth Sciences, v. 14, p. 1646-1667.

Beach, H.H. and Spivak, J.
1943: The origin of Peace River canyon, British Columbia; American Journal of Science, v. 241, p. 366-376.

Beaudoin, A.B.
1984: Holocene environmental change in the Sunwapta Pass area, Jasper National Park; unpublished PhD thesis, University of Western Ontario, London, Ontario, 487 p.

1986: Using *Picea/Pinus* ratios from the Wilcox Pass Core, Jasper National Park, Alberta, to investigate Holocene timberline fluctuations; Géographie physique et Quaternaire, v. 40, p. 145-152.

Beget, J.E.
1984: Tephrochronology of late Wisconsin deglaciation and Holocene glacier fluctuations near Glacier Peak, North Cascade Range, Washington; Quaternary Research, v. 21, p. 304-316.

Berger, G.W.
1985: Thermoluminescence dating of volcanic ash; Journal of Volcanology and Geothermal Research, v. 25, p. 333-347.

1987: Thermoluminescence dating of the Pleistocene Old Crow tephra and adjacent loess, near Fairbanks, Alaska; Canadian Journal of Earth Sciences, v. 24, p. 1975-1984.

Berry, E.W. and Johnston, W.A.
1922: Pleistocene interglacial deposits in the Vancouver region, British Columbia; Royal Society of Canada, Transactions, ser. 3, v. 16, sec. 4, p. 133-139.

Bevier, M.L., Armstrong, R.L., and Souther, J.G.
1979: Miocene peralkaline volcanism in west-central British Columbia — Its temporal and plate tectonics setting; Geology, v. 7, p. 389-392.

Birks, H.J.B.
1980: Modern pollen assemblages and vegetational history of the moraines of the Klutlan Glacier and its surroundings, Yukon Territory, Canada; Quaternary Research, v. 14, p. 101-129.

Blake, W., Jr.
1982: Geological Survey of Canada radiocarbon dates XXII; Geological Survey of Canada, Paper 82-7, 22 p.

1983: Geological Survey of Canada radiocarbon dates XXIII; Geological Survey of Canada, Paper 83-7, 34 p.

1984: Geological Survey of Canada radiocarbon dates XXIV; Geological Survey of Canada, Paper 84-7, 35 p.

1986: Geological Survey of Canada radiocarbon dates XXV; Geological Survey of Canada, Paper 85-7, 32 p.

1987: Geological Survey of Canada radiocarbon dates XXVI; Geological Survey of Canada, Paper 86-7, 60 p.

Blown, I. and Church, M.
1985: Catastrophic lake drainage within the Homathko River basin, British Columbia; Canadian Geotechnical Journal, v. 22, p. 551-563.

Bobrowsky, P.T.
1982: The quantitative and qualitative significance of fossil and subfossil gastropod remains in archaeology; unpublished MA thesis, Simon Fraser University, Burnaby, British Columbia, 294 p.

Bobrowsky, P.T., Levson, V., Liverman, D.G.E., and Rutter, N.W.
1987: Quaternary geology of northwestern Alberta and northeastern British Columbia; International Union for Quaternary Research, 12th International Congress (Ottawa, Ontario), Guidebook, Field Excursion A-24, 62 p.

Bornhold, B.D.
1983: Sedimentation in Douglas Channel and Kitimat Arm; in Proceedings of a Workshop on the Kitimat Marine Environment, R.W. Macdonald (ed.); Canadian Technical Report of Hydrography and Ocean Sciences 18, p. 88-114.

Bostock, H.S.
1934: The mining industry of Yukon, 1933, and notes on the geology of Carmacks map-area; Geological Survey of Canada, Summary Report 1933, Pt. A, p. 1A-8A.
1948: Physiography of the Canadian Cordillera, with special reference to the area north of the fifty-fifth parallel; Geological Survey of Canada, Memoir 247, 106 p.
1952: Geology of northwest Shakwak Valley, Yukon Territory; Geological Survey of Canada, Memoir 267, 54 p.
1966: Notes on glaciation in central Yukon Territory; Geological Survey of Canada, Paper 65-36, 18 p.
1970: Physiographic regions of Canada; Geological Survey of Canada, Map 1254A.

Bourgeois, J.C. et Geurts, M-A.
1983: Palynologie et morphogenèse récente dans le bassin du Grizzly Creek (Territoire du Yukon); Journal Canadien des sciences de la terre, vol. 20, p. 1543-1553.

Bovis, M.J.
1985: Earthflows in the Interior Plateau, southwest British Columbia; Canadian Geotechnical Journal, v. 22, p. 313-334.

Boydell, A.N.
1972: Multiple glaciation in the Foothills, Rocky Mountain House area, Alberta; unpublished PhD thesis, University of Calgary, Calgary, Alberta, 128 p.
1978: Multiple glaciations in the Foothills, Rocky Mountain House area, Alberta; Alberta Research Council, Bulletin 36, 35 p.

Boyle, R.W.
1979: The geochemistry of gold and its deposits (together with a chapter on geochemical prospecting for the element); Geological Survey of Canada, Bulletin 280, 584 p.

Bretz, J.H.
1913: Glaciation of the Puget Sound region; Washington Geological Survey, Bulletin 8, 244 p.

British Columbia Ministry of Energy, Mines and Petroleum Resources
1980: Minister of Energy, Mines and Petroleum Resources annual report 1979; British Columbia Ministry of Energy, Mines and Petroleum Resources, Victoria, British Columbia, 189 p.

Brown, R.J.E.
1967: Permafrost in Canada; Geological Survey of Canada, Map 1246A.

Bujak, C.A.
1974: Recent palynology of Goat Lake and Lost Lake, Waterton Lakes National Park; unpublished MSc thesis, University of Calgary, Calgary, Alberta, 60 p.

Burke, R.M. and Birkeland, P.W.
1983: Holocene glaciation in the mountain ranges of the western United States; in Late-Quaternary Environments of the United States, Volume 2, the Holocene, H.E. Wright, Jr. (ed.); University of Minnesota Press, Minneapolis, Minnesota, p. 3-11.

Burns, J.A.
1980: The brown lemming, *Lemmus sibiricus* (Rodentia, Aruicolidae) in the Late Pleistocene of Alberta and its postglacial dispersal; Canadian Journal of Zoology, v. 58, p. 1507-1511.
1982: Water vole *Microtus richardsoni* (Mammalia, Rodentia) from the Late Pleistocene of Alberta; Canadian Journal of Earth Sciences, v. 19, p. 628-631.

Burwash, E.M.J.
1918: The geology of Vancouver and vicinity; University of Chicago Press, Chicago, Illinois, 106 p.

Campbell, D.B. and Skermer, N.A.
1975: Report to British Columbia Water Resources Service on investigation of seawave at Kitimat, B.C.; Golder Associates, Vancouver, British Columbia, 9 p.

Canada Soil Survey Committee
1978: The Canadian system of soil classification; Canada Department of Agriculture, Publication 1646, 164 p.

Carson, M.A. and Bovis, M.J.
1989: Slope processes; in Chapter 9 of Quaternary Geology of Canada and Greenland, R.J. Fulton (ed.); Geological Survey of Canada, Geology of Canada, no. 1 (also Geological Society of America, The Geology of North America, v. K-1).

Carter, L.
1973: Surficial sediments of Barkley Sound and the adjacent continental shelf, west coast Vancouver Island; Canadian Journal of Earth Sciences, v. 10, p. 441-459.

Casagrande, A.
1932: Research on the Atterberg limits of soils; Public Roads, v. 13, p. 121-136.

Cawker, K.B.
1983: Fire history and grassland vegetation change: three pollen diagrams from southern British Columbia; Canadian Journal of Botany, v. 61, p. 1126-1139.

Chamberlain, T.C.
1894: Glacial phenomena of North America; in The Great Ice Age and Its Relation to the Antiquity of Man (3rd edition), J. Geikie (ed.); Edward Standford, London, England, p. 724-775.

Chamberlain, V.E. and Lambert, R. St.J.
1985: Cordilleria, a newly defined Canadian microcontinent; Nature, v. 314, p. 707-713.

Chase, R.L. and Tiffin, D.L.
1972: Queen Charlotte fault-zone, British Columbia; in Section 8, Marine Geology and Geophysics; 24th International Geological Congress (Montreal, Quebec), p. 17-28.

Christensen, O.A. and Hills, L.V.
1985: Postglacial pollen and paleoclimate in southwestern Alberta and southeastern British Columbia. Part 1: palynologic and paleoclimatic interpretation of Holocene sediments, Waterton Lakes National Park, Alberta; in Climatic Change in Canada 5, Critical Periods in the Quaternary Climatic History of Northern North America, C.R. Harington (ed.); National Museums of Canada, National Museum of Natural Sciences, Syllogeus Series, no. 55, p. 345-354.

Church, M. and Miles, M.J.
1987: Meteorological antecedents to debris flow in southwestern British Columbia; Some case studies; in Debris Flows/Avalanches: Process, Recognition, and Mitigation, J.E. Costa and G.F.Wieczorck (ed.); Geological Society of America, Reviews in Engineering Geology, v. 7, p. 63-79.

Church, M. and Ryder, J.M.
1972: Paraglacial sedimentation: a consideration of fluvial processes conditioned by glaciation; Geological Society of America, Bulletin, v. 83, p. 3059-3072.

Churcher, C.S.
1968: Pleistocene ungulates from the Bow River gravels at Cochrane, Alberta; Canadian Journal of Earth Sciences, v. 5, p. 1467-1488.
1975: Additional evidence of Pleistocene ungulates from the Bow River gravels at Cochrane, Alberta; Canadian Journal of Earth Sciences, v. 12, p. 68-76.

Churcher, C.S. and Wilson, M.
1979: Quaternary mammals from the eastern Peace River District, Alberta; Journal of Paleontology, v. 53, p. 71-76.

Clague, J.J.
1973: Late Cenozoic geology of the southern Rocky Mountain Trench, British Columbia; unpublished PhD thesis, University of British Columbia, Vancouver, British Columbia, 274 p.
1975a: Late Quaternary sea-level fluctuations, Pacific coast of Canada and adjacent areas; in Report of Activities, Part C, Geological Survey of Canada, Paper 75-1C, p. 17-21.
1975b: Late Quaternary sediments and geomorphic history of the southern Rocky Mountain Trench, British Columbia; Canadian Journal of Earth Sciences, v. 12, p. 595-605.
1976: Quadra Sand and its relation to the late Wisconsin glaciation of southwest British Columbia; Canadian Journal of Earth Sciences, v. 13, p. 803-815.
1977a: Quadra Sand: a study of the late Pleistocene geology and geomorphic history of coastal southwest British Columbia; Geological Survey of Canada, Paper 77-17, 24 p.
1977b: Holocene sediments in northern Strait of Georgia, British Columbia; in Report of Activities, Part A, Geological Survey of Canada, Paper 77-1A, p. 51-58.
1978: Mid-Wisconsinan climates of the Pacific Northwest; in Current Research, Part B, Geological Survey of Canada, Paper 78-1B, p. 95-100.
1979: The Denali fault system in southwest Yukon Territory — a geologic hazard?; in Current Research, Part A, Geological Survey of Canada, Paper 79-1A, p. 169-178.

1980: Late Quaternary geology and geochronology of British Columbia. Part 1: radiocarbon dates; Geological Survey of Canada, Paper 80-13, 28 p.
1981: Late Quaternary geology and geochronology of British Columbia. Part 2: summary and discussion of radiocarbon-dated Quaternary history; Geological Survey of Canada, Paper 80-35, 41 p.
1982a: Erosion at Point Grey, British Columbia; Geoscience Canada, v. 9, p. 129-131.
1982b: The role of geomorphology in the identification and evaluation of natural hazards; in Applied Geomorphology, R.G. Craig and J.L. Craft (ed.); George Allen & Unwin, London, England, p. 17-43.
1983: Glacio-isostatic effects of the Cordilleran Ice Sheet, British Columbia, Canada; in Shorelines and Isostasy, D.E. Smith and A.G. Dawson (ed.); Academic Press, London, England, p. 321-343.
1984: Quaternary geology and geomorphology, Smithers-Terrace-Prince Rupert area, British Columbia; Geological Survey of Canada, Memoir 413, 71 p.
1985: Deglaciation of the Prince Rupert-Kitimat area, British Columbia; Canadian Journal of Earth Sciences, v. 22, p. 256-265.
1986: The Quaternary stratigraphic record of British Columbia — evidence for episodic sedimentation and erosion controlled by glaciation; Canadian Journal of Earth Sciences, v. 23, p. 885-894.
1987: Quaternary stratigraphy and history, Williams Lake, British Columbia; Canadian Journal of Earth Sciences, v. 24, p. 147-158.

Clague, J.J. and Bornhold, B.D.
1980: Morphology and littoral processes of the Pacific coast of Canada; in The Coastline of Canada, S.B. McCann (ed.); Geological Survey of Canada, Paper 80-10, p. 339-380.

Clague, J.J. and Luternauer, J.L.
1982: Excursion 30A: late Quaternary sedimentary environments, southwestern British Columbia; International Association of Sedimentologists, 11th International Congress on Sedimentology (Hamilton, Ontario), Field Excursion Guide Book, 167 p.
1983: Late Quaternary geology of southwestern British Columbia; Geological Association of Canada, Mineralogical Association of Canada, Canadian Geophysical Union, Joint Annual Meeting (Victoria, British Columbia), Field Trip Guidebook, no. 6, 112 p.

Clague, J.J. and Rampton, V.N.
1982: Neoglacial Lake Alsek; Canadian Journal of Earth Sciences, v. 19, p. 94-117.

Clague, J.J. and Souther, J.G.
1982: The Dusty Creek landslide on Mount Cayley, British Columbia; Canadian Journal of Earth Sciences, v. 19, p. 524-539.

Clague, J.J., Armstrong, J.E., and Mathews, W.H.
1980: Advance of the late Wisconsin Cordilleran Ice Sheet in southern British Columbia since 22,000 yr B.P.; Quaternary Research, v. 13, p. 322-326.

Clague, J.J., Evans, S.G., and Blown, I.
1985: A debris flow triggered by the breaching of a moraine-dammed lake, Klattasine Creek, British Columbia; Canadian Journal of Earth Sciences, v. 22, p. 1492-1502.

Clague, J.J., Harper, J.R., Hebda, R.J., and Howes, D.E.
1982b: Late Quaternary sea levels and crustal movements, coastal British Columbia; Canadian Journal of Earth Sciences, v. 19, p. 597-618.

Clague, J.J., Luternauer, J.L., and Hebda, R.J.
1983: Sedimentary environments and postglacial history of the Fraser Delta and lower Fraser Valley, British Columbia; Canadian Journal of Earth Sciences, v. 20, p. 1314-1326.

Clague, J.J., Mathewes, R.W., and Warner, B.G.
1982a: Late Quaternary geology of eastern Graham Island, Queen Charlotte Islands, British Columbia; Canadian Journal of Earth Sciences, v. 19, p. 1786-1795.

Clapp, C.H.
1913: Geology of the Victoria and Saanich map-areas, Vancouver Island, B.C.; Geological Survey of Canada, Memoir 36, 143 p.

Clark, J.A., Farrell, W.E., and Peltier, W.R.
1978: Global changes in postglacial sea level: a numerical calculation; Quaternary Research, v. 9, p. 265-287.

Clarke, A.H. and Harington, C.R.
1978: Asian freshwater mollusks from Pleistocene deposits in the Old Crow Basin, Yukon Territory; Canadian Journal of Earth Sciences, v. 15, p. 45-51.

Clarke, G.K.C.
1982: Glacier outburst floods from "Hazard Lake", Yukon Territory, and the problem of flood magnitude prediction; Journal of Glaciology, v. 28, p. 3-21.

Clarke, G.K.C. and Waldron, D.A.
1984: Simulation of the August 1979 sudden discharge of glacier-dammed Flood Lake, British Columbia; Canadian Journal of Earth Sciences, v. 21, p. 502-504.

Cockfield, W.E. and Walker, J.F.
1933: Geology and placer deposits of Quesnel Forks area, Cariboo District, British Columbia; Geological Survey of Canada, Summary Report 1932, pt. AI, p. 76AI-143AI.

Coleman, A.P.
1910: The drift of Alberta and the relations of the Cordilleran and Keewatin ice sheets; Royal Society of Canada, Transactions (1909), ser. 3, v. 3, sec. 4, p. 3-12.

Coney, P.J., Jones, D.L., and Monger, J.W.H.
1980: Cordilleran suspect terranes; Nature, v. 288, p. 329-333.

Cowan, I. McT.
1941: Fossil and subfossil mammals from the Quarternary (sic) of British Columbia; Royal Society of Canada, Transactions, ser. 3, v. 35, sec. 4, p. 39-50.

Crandell, D.R.
1965: The glacial history of western Washington and Oregon; in The Quaternary of the United States, H.E. Wright, Jr. and D.G. Frey (ed.); Princeton University Press, Princeton, New Jersey, p. 341-353.

Crickmay, C.H.
1925: A Pleistocene fauna from the southwestern mainland of British Columbia; Canadian Field-Naturalist, v. 39, p. 140-141.
1929: A Pleistocene fauna from British Columbia; Canadian Field-Naturalist, v. 43, p. 205-206.

Crosson, R.S.
1972: Small earthquakes, structure, and tectonics of the Puget Sound region; Seismological Society of America, Bulletin, v. 62, p. 1133-1171.

Cruden, D.M.
1982: The Brazeau Lake slide, Jasper National Park, Alberta; Canadian Journal of Earth Sciences, v. 19, p. 975-981.
1985: Rock slope movements in the Canadian Cordillera; Canadian Geotechnical Journal, v. 22, p. 528-540.

Cruden, D.M. and Krahn, J.
1978: Frank rockslide, Alberta, Canada; in Rockslides and Avalanches, 1, Natural Phenomena, B. Voight (ed.); Elsevier Scientific Publishing Company, New York, p. 97-112.

Cumbaa, S.L., McAllister, D.E., and Morlan, R.E.
1981: Late Pleistocene fish fossils of *Coregonus, Stenodus, Thymallus, Catostomus, Lota,* and *Cottus* from the Old Crow basin, northern Yukon, Canada; Canadian Journal of Earth Sciences, v. 18, p. 1740-1754.

Curray, J.R.
1965: Late Quaternary history, continental shelves of the United States; in The Quaternary of the United States, H.E. Wright, Jr. and D.G. Frey (ed.); Princeton University Press, Princeton, New Jersey, p. 723-735.

Cwynar, L.C.
1982: A late-Quaternary vegetation history from Hanging Lake, northern Yukon; Ecological Monographs, v. 52, p. 1-24.

Cwynar, L.C. and Ritchie, J.C.
1980: Arctic steppe-tundra: a Yukon perspective; Science, v. 208, p. 1375-1377.

Daly, R.A.
1912: Geology of the North American Cordillera at the forty-ninth parallel; Geological Survey of Canada, Memoir 38, 3 pt., 857 p.
1915: A geological reconnaissance between Golden and Kamloops, B.C., along the Canadian Pacific Railway; Geological Survey of Canada, Memoir 68, 260 p.

Daly, R.A., Miller, W.G., and Rice, G.S.
1912: Report of the commission appointed to investigate Turtle Mountain, Frank, Alberta, 1911; Geological Survey of Canada, Memoir 27, 34 p.

Davis, N.F.G. and Mathews, W.H.
1944: Four phases of glaciation with illustrations from southwestern British Columbia; Journal of Geology, v. 52, p. 403-413.

Davis, P.T. and Osborn, G.
1987: Age of pre-Neoglacial cirque moraines in the central North American Cordillera; Géographie physique et Quaternaire, v. 41, p. 365-375.

Dawson, G.M.
1877: Note on some of the more recent changes in level of the coast of British Columbia and adjacent regions; Canadian Naturalist, new ser., v. 8, p. 241-248.
1878: On the superficial geology of British Columbia; Geological Society of London, Quarterly Journal, v. 34, p. 89-123.
1879: Notes on the glaciation of British Columbia; Canadian Naturalist, new ser., v. 9, p. 32-39.
1881: Additional observations on the superficial geology of British Columbia and adjacent regions; Geological Society of London, Quarterly Journal, v. 37, p. 272-285.
1886: Preliminary report on the physical and geological features of that portion of the Rocky Mountains between latitudes 49° and 50°30′; Geological Survey of Canada, Annual Report, new ser., v. 1 (1885), report B, 169 p.

1888: Recent observations on the glaciation of British Columbia and adjacent regions; Geological Magazine, new ser., v. 5, p. 347-350.

1889: Glaciation of high points in the southern interior of British Columbia; Geological Magazine, new ser., v. 6, p. 350-352.

1890: On the glaciation of the northern part of the Cordillera, with an attempt to correlate the events of the glacial period in the Cordillera and Great Plains; American Geologist, v. 6, p. 153-162.

1891: On the later physiographical geology of the Rocky Mountain region in Canada, with special reference to changes in elevation and to the history of the glacial period; Royal Society of Canada, Transactions (1890), v. 8, sec. 4, p. 3-74.

Debicki, R.L.
1983: Placer gold deposits in the Canadian Cordillera (abstract); Geological Association of Canada, Program with Abstracts, v. 8, p. A16.

Delorme, L.D.
1968: Pleistocene freshwater Ostracoda from Yukon, Canada; Canadian Journal of Zoology, v. 46, p. 859-876.

Demek, J.
1969: Cryoplanation terraces, their geographical distribution, genesis and development; Československá Akademia Věd, Rozpravy, Rada Matematických a Přírodních Věd, v. 79, no. 4, 80 p.

Denton, G.H. and Armstrong, R.L.
1969: Miocene-Pliocene glaciations in southern Alaska; American Journal of Science, v. 267, p. 1121-1142.

Denton, G.H. and Hughes, T.E. (editors)
1981: The Last Great Ice Sheets; John Wiley & Sons, New York, 484 p.

Denton, G.H. and Karlén, W.
1973: Holocene climatic variations — their pattern and possible cause; Quaternary Research, v. 3, p. 155-205.

1977: Holocene glacial and tree-line variations in the White River valley and Skolai Pass, Alaska and Yukon Territory; Quaternary Research, v. 7, p. 63-111.

Denton, G.H. and Stuiver, M.
1966: Neoglacial chronology, northeastern St. Elias Mountains, Canada; American Journal of Science, v. 264, p. 577-599.

1967: Late Pleistocene glacial stratigraphy and chronology, northeastern St. Elias Mountains, Yukon Territory, Canada; Geological Society of America, Bulletin, v. 78, p. 485-510.

Driver, J.C., Hills, L.V., and Reeves, B.O.K.
1985: Postglacial pollen and paleoclimate in southwestern Alberta and southeastern British Columbia. Part 3: Holocene palynology of Crowsnest Lake, Alberta, with comments on Holocene paleoenvironments of the southern Alberta Rockies and surrounding areas; in Climatic Change in Canada 5, Critical Periods in the Quaternary Climatic History of Northern North America, C.R. Harington (ed.); National Museums of Canada, National Museum of Natural Sciences, Syllogeus Series, no. 55, p. 370-396.

Duford, J.M. and Osborn, G.D.
1978: Holocene and latest Pleistocene cirque glaciations in the Shuswap Highland, British Columbia; Canadian Journal of Earth Sciences, v. 15, p. 865-873.

1980: Holocene and latest Pleistocene cirque glaciations in the Shuswap Highland, British Columbia: Reply; Canadian Journal of Earth Sciences, v. 17, p. 799-800.

Dyck, W. and Fyles, J.G.
1962: Geological Survey of Canada radiocarbon dates I; Radiocarbon, v. 4, p. 13-26. Reprinted (1963) in Geological Survey of Canada Radiocarbon Dates I and II; Geological Survey of Canada, Paper 63-21, p. 1-14.

1963: Geological Survey of Canada radiocarbon dates II; Radiocarbon, v. 5, p. 39-55. Reprinted (1963) in Geological Survey of Canada Radiocarbon Dates I and II; Geological Survey of Canada, Paper 63-21, p. 15-31.

Dyck, W., Lowdon, J.A., Fyles, J.G., and Blake, W., Jr.
1966: Geological Survey of Canada radiocarbon dates V; Radiocarbon, v. 8, p. 96-127. Reprinted (1966) as Geological Survey of Canada, Paper 66-48.

Easterbrook, D.J. and Rutter, N.W.
1981: Amino acid ages of Pleistocene glacial and interglacial sediments in western Washington (abstract); Geological Society of America, Abstracts with Programs, v. 13, p. 444.

1982: Amino acid analyses of wood and shells in development of chronology and correlation of Pleistocene sediments in the Puget Lowland, Washington (abstract); Geological Society of America, Abstracts with Programs, v. 14, p. 480.

Eisbacher, G.H.
1977: Rockslides in the Mackenzie Mountains, District of Mackenzie; in Report of Activities, Part A, Geological Survey of Canada, Paper 77-1A, p. 235-241.

1979: First-order regionalization of landslide characteristics in the Canadian Cordillera; Geoscience Canada, v. 6, p. 69-79.

1983: Slope stability and mountain torrents, Fraser Lowlands and southern Coast Mountains, British Columbia; Geological Association of Canada, Mineralogical Association of Canada, Canadian Geophysical Union, Joint Annual Meeting (Victoria, British Columbia), Field Trip Guidebook, no. 15, 46 p.

Eisbacher, G.H. and Clague, J.J.
1984: Destructive mass movements in high mountains: hazard and management; Geological Survey of Canada, Paper 84-16, 230 p.

Evans, S.G.
1982: Landslides and surficial deposits in urban areas of British Columbia: a review; Canadian Geotechnical Journal, v. 19, p. 269-288.

1987: A rock avalanche from the peak of Mount Meager, British Columbia; in Current Research, Part A, Geological Survey of Canada, Paper 87-1A, p. 929-933.

Evans, S.G. and Gardner, J.S.
1989: Geological hazards in the Canadian Cordillera; in Chapter 12 of Quaternary Geology of Canada and Greenland, R.J. Fulton (ed.); Geological Survey of Canada, Geology of Canada, no. 1 (also Geological Society of America, The Geology of North America, v. K-1).

Evans, S.G., Aitken, J.D., Wetmiller, R.J., and Horner, R.B.
1987: A rock avalanche triggered by the October 1985 North Nahanni earthquake, District of Mackenzie, N.W.T.; Canadian Journal of Earth Sciences, v. 24, p. 176-184.

Ferguson, A. and Osborn, G.
1981: Minimum age of deglaciation of upper Elk Valley, British Columbia; Canadian Journal of Earth Sciences, v. 18, p. 1635-1636.

Fergusson, A. (sic) and Hills, L.V.
1985: Postglacial pollen and paleoclimate in southwestern Alberta and southeastern British Columbia. Part 2: a palynological record, upper Elk Valley, British Columbia; in Climatic Change in Canada 5, Critical Periods in the Quaternary Climatic History of Northern North America, C.R. Harington (ed.); National Museums of Canada, National Museum of Natural Sciences, Syllogeus Series, no. 55, p. 355-369.

Fitzharris, B.B. and Schaerer, P.A.
1980: Frequency of major avalanche winters; Journal of Glaciology, v. 26, p. 43-52.

Fladmark, K.R.
1975: A paleoecological model for Northwest Coast prehistory; National Museums of Canada, National Museum of Man, Mercury Series, Archaeological Survey of Canada, Paper 43, 328 p.

Flint, R.F.
1935: "White-silt" deposits in the Okanagan Valley, British Columbia; Royal Society of Canada, Transactions, ser. 3, v. 29, sec. 4, p. 107-114.

1971: Glacial and Quaternary Geology; John Wiley & Sons, New York, 892 p.

Ford, D.C.
1973: Development of the canyons of the South Nahanni River, N.W.T.; Canadian Journal of Earth Sciences, v. 10, p. 366-378.

1976: Evidences of multiple glaciation in South Nahanni National Park, Mackenzie Mountains, Northwest Territories; Canadian Journal of Earth Sciences, v. 13, p. 1433-1445.

Ford, D.C., Schwarcz, H.P., Drake, J.J., Gascoyne, M., Harmon, R.S., and Latham, A.G.
1981: Estimates of the age of the existing relief within the southern Rocky Mountains of Canada; Arctic and Alpine Research, v. 13, p. 1-10.

Foscolos, A.E., Rutter, N.W., and Hughes, O.L.
1977: The use of pedological studies in interpreting the Quaternary history of central Yukon Territory; Geological Survey of Canada, Bulletin 271, 48 p.

Fryxell, R.
1965: Mazama and Glacier Peak volcanic ash layers: relative ages; Science, v. 147, p. 1288-1290.

Fulton, R.J.
1965: Silt deposition in late-glacial lakes of southern British Columbia; American Journal of Science, v. 263, p. 553-570.

1967: Deglaciation studies in Kamloops region, an area of moderate relief, British Columbia; Geological Survey of Canada, Bulletin 154, 36 p.

1968: Olympia Interglaciation, Purcell Trench, British Columbia; Geological Society of America, Bulletin, v. 79, p. 1075-1080.

1969: Glacial lake history, southern Interior Plateau, British Columbia; Geological Survey of Canada, Paper 69-37, 14 p.

1971: Radiocarbon geochronology of southern British Columbia; Geological Survey of Canada, Paper 71-37, 28 p.

1975a: Quaternary geology and geomorphology, Nicola-Vernon area, British Columbia (82L W1/2 and 92I E1/2); Geological Survey of Canada, Memoir 380, 50 p.

1975b: Quaternary stratigraphy south central British Columbia; in The Last Glaciation, D.J. Easterbrook (ed.); IUGS-Unesco International Geological Correlation Program, Project 73-1-24, Guidebook for Field Conference (Western Washington University, Bellingham, Washington), p. 98-124.

Fulton, R.J. and Halstead, E.C.
1972: Quaternary geology of the southern Canadian Cordillera; 24th International Geological Congress (Montreal, Quebec), Guidebook, Field Excursion A02, 49 p.

Fulton, R.J. and Pullen, M.J.L.T.
1969: Sedimentation in Upper Arrow Lake, British Columbia; Canadian Journal of Earth Sciences, v. 6, p. 785-790.

Fulton, R.J. and Smith, G.W.
1978: Late Pleistocene stratigraphy of south-central British Columbia; Canadian Journal of Earth Sciences, v. 15, p. 971-980.

Fulton, R.J., Armstrong, J.E., and Fyles, J.G.
1976: Stratigraphy and palynology of late Quaternary sediments in the Puget Lowland, Washington: Discussion; Geological Society of America, Bulletin, v. 87, p. 153-155.

Fulton, R.J., Fenton, M.M., and Rutter, N.W.
1984: Summary of Quaternary stratigraphy and history, western Canada; in Quaternary Stratigraphy of Canada — A Canadian Contribution to IGCP Project 24, R.J. Fulton (ed.); Geological Survey of Canada, Paper 84-10, p. 69-83.

Fyles, J.G.
1963: Surficial geology of Horne Lake and Parksville map-areas, Vancouver Island, British Columbia; Geological Survey of Canada, Memoir 318, 142 p.

Gabrielse, H. and Yorath, C.J. (editors)
in press: The Cordilleran Orogen: Canada; Geological Survey of Canada, Geology of Canada, no. 4 (also Geological Society of America, The Geology of North America, v. G-2).

Gascoyne, M., Ford, D.C., and Schwarcz, H.P.
1981: Late Pleistocene chronology and paleoclimate of Vancouver Island determined from cave deposits; Canadian Journal of Earth Sciences, v. 18, p. 1643-1652.

Gascoyne, M., Schwarcz, H.P., and Ford, D.C.
1980: A palaeotemperature record for the mid-Wisconsin in Vancouver Island; Nature, v. 285, p. 474-476.

Gilbert, R.
1972: Drainings of ice-dammed Summit Lake, British Columbia; Canada Department of Environment, Inland Waters Directorate, Water Resources Branch, Scientific Series, no. 20, 17 p.
1975: Sedimentation in Lillooet Lake, British Columbia; Canadian Journal of Earth Sciences, v. 12, p. 1697-1711.

Gilbert, R. and Desloges, J.R.
1987: Sediments of ice-dammed, self-draining Ape Lake, British Columbia; Canadian Journal of Earth Sciences, v. 24, p. 1735-1747.

Guthrie, R.D.
1968: Paleoecology of the large-mammal community in interior Alaska during the late Pleistocene; American Midland Naturalist, v. 79, p. 346-363.
1982: Mammals of the mammoth steppe as paleoenvironmental indicators; in Paleoecology of Beringia, D.M. Hopkins, J.V. Matthews, Jr., C.E. Schweger, and S.B. Young (ed.); Academic Press, New York, p. 307-326.

Gwillim, J.C.
1902: Glaciation in the Atlin district, Journal of Geology, v. 10, p. 182-185.

Halstead, E.C.
1968: The Cowichan Ice tongue, Vancouver Island; Canadian Journal of Earth Sciences, v. 5, p. 1409-1415.

Hamilton, T.D. and Bischoff, J.L.
1984: Uranium-series dating of fossil bones from the Canyon Creek vertebrate locality in central Alaska; in The United States Geological Survey in Alaska: Accomplishments during 1982, K.M. Reed and S. Bartsch-Winkler (ed.); United States Geological Survey, Circular 939, p. 26-29.

Hamilton, T.D. and Thorson, R.M.
1983: The Cordilleran ice sheet in Alaska; in Late-Quaternary Environments of the United States, Volume 1, the Late Pleistocene, S.C. Porter (ed.); University of Minnesota Press, Minneapolis, Minnesota, p. 38-52.

Hamilton, T.S. and Luternauer, J.L.
1983: Evidence for seafloor instability in the south-central Strait of Georgia, British Columbia: a preliminary compilation; in Current Research, Part A, Geological Survey of Canada, Paper 83-1A, p. 417-421.

Hansen, B.S. and Easterbrook, D.J.
1974: Stratigraphy and palynology of late Quaternary sediments in the Puget Lowland, Washington; Geological Society of America, Bulletin, v. 85, p. 587-602.

Hansen, H.P.
1940: Paleoecology of two peat bogs in southwestern British Columbia; American Journal of Botany, v. 27, p. 144-149.
1950: Pollen analysis of three bogs on Vancouver Island, Canada; Journal of Ecology, v. 38, p. 270-276.
1955: Postglacial forests in south central and central British Columbia; American Journal of Science, v. 253, p. 640-658.

Hanson, G.
1932: Varved clays of Tide Lake, British Columbia; Royal Society of Canada, Transactions, ser. 3, v. 26, sec. 4, p. 335-339.

Hare, F.K. and Hay, J.E.
1974: The climate of Canada and Alaska; in Climates of North America, R.A. Bryson and F.K. Hare (ed.); Elsevier Scientific Publishing Company, Amsterdam, Netherlands, p. 49-192.

Hare, F.K. and Thomas, M.K.
1974: Climate Canada; Wiley Publishers of Canada, Toronto, Ontario, 256 p.

Harington, C.R.
1970: Ice age mammal research in the Yukon Territory and Alaska; in Early Man and Environments in Northwest North America, R.A. Smith and J.W. Smith (ed.); University of Calgary, Archaeological Association, Calgary, Alberta, p. 35-51.
1975: Pleistocene muskoxen (*Symbos*) from Alberta and British Columbia; Canadian Journal of Earth Sciences, v. 12, p. 903-919.
1977: Pleistocene mammals of the Yukon Territory; unpublished PhD thesis, University of Alberta, Edmonton, Alberta, 1060 p.
1978: Quaternary vertebrate faunas of Canada and Alaska and their suggested chronological sequence; National Museums of Canada, National Museum of Natural Sciences, Syllogeus Series, no. 15, 105 p.

Harington, C.R. and Clulow, F.V.
1973: Pleistocene mammals from Gold Run Creek, Yukon Territory; Canadian Journal of Earth Sciences, v. 10, p. 697-759.

Harington, C.R., Tipper, H.W., and Mott, R.J.
1974: Mammoth from Babine Lake, British Columbia; Canadian Journal of Earth Sciences, v. 11, p. 285-303.

Harmon, R.S., Ford, D.C., and Schwarcz, H.P.
1977: Interglacial chronology of the Rocky and Mackenzie Mountains based upon ^{230}Th-^{234}U dating of calcite speleothems; Canadian Journal of Earth Sciences, v. 14, p. 2543-2552.

Harris, S.A. and Pip, E.
1973: Molluscs as indicators of late- and post-glacial climatic history in Alberta; Canadian Journal of Zoology, v. 51, p. 209-215.

Harrison, J.E.
1976: Dated organic material below Mazama (?) tephra: Elk Valley, British Columbia; in Report of Activities, Part C, Geological Survey of Canada, Paper 76-1C, p. 169-170.

Hazell, S.D.
1979: Late Quaternary vegetation and climate of Dunbar Valley, British Columbia; unpublished MSc thesis, University of Toronto, Toronto, Ontario, 101 p.

Hebda, R.J.
1977: The paleoecology of a raised bog and associated deltaic sediments of the Fraser River delta; unpublished PhD thesis, University of British Columbia, Vancouver, British Columbia, 202 p.
1982: Postglacial history of grasslands of southern British Columbia and adjacent regions; in Grassland Ecology and Classification Symposium Proceedings, A.C. Nicholson, A. McLean, and T.E. Baker (ed.); British Columbia Ministry of Forests, Victoria, British Columbia, p. 157-191.
1983: Late-glacial and postglacial vegetation history at Bear Cove bog, northeast Vancouver Island, British Columbia; Canadian Journal of Botany, v. 61, p. 3172-3192.

Hebda, R.J. and Mathewes, R.W.
1986: Radiocarbon dates from Anthony Island, Queen Charlotte Islands, and their geological and archaeological significance; Canadian Journal of Earth Sciences, v. 23, p. 2071-2076.

Hebda, R.J. and Rouse, G.E.
1979: Palynology of two Holocene cores from the Hesquiat Peninsula, Vancouver Island; Syesis, v. 12, p. 121-130.

Hebda, R.J., Hicock, S.R., Miller, R.F., and Armstrong, J.E.
1983: Paleoecology of mid-Wisconsin sediments from Lynn Canyon, Fraser Lowland, British Columbia; Geological Association of Canada, Program with Abstracts, v. 8, p. A31.

Heginbottom, J.A.
1972: Surficial geology of Taseko Lakes map-area British Columbia; Geological Survey of Canada, Paper 72-14, 9 p.

Heusser, C.J.
1955: Pollen profiles from the Queen Charlotte Islands, British Columbia; Canadian Journal of Botany, v. 33, p. 429-449.
1956: Postglacial environments in the Canadian Rocky Mountains; Ecological Monographs, v. 26, p. 263-302.

1960: Late-Pleistocene environments of North Pacific North America; American Geographical Society, Special Publication 35, 308 p.
1977: Quaternary palynology of the Pacific slope of Washington; Quaternary Research, v. 8, p. 282-306.
1983: Vegetational history of the northwestern United States including Alaska; in Late-Quaternary Environments of the United States, Volume 1, the Late Pleistocene, S.C. Porter (ed.); University of Minnesota Press, Minneapolis, Minnesota, p. 239-258.

Heusser, C.J., Heusser, L.E., and Streeter, S.S.
1980: Quaternary temperatures and precipitation for the north-west coast of North America; Nature, v. 286, p. 702-704.

Heusser, L.E.
1983: Palynology and paleoecology of postglacial sediments in an anoxic basin, Saanich Inlet, British Columbia; Canadian Journal of Earth Sciences, v. 20, p. 873-885.

Hicks, S.D. and Shofnos, W.
1965: The determination of land emergence from sea level observations in southeast Alaska; Journal of Geophysical Research, v. 70, p. 3315-3320.

Hickson, C.J. and Souther, J.G.
1984: Late Cenozoic volcanic rocks of the Clearwater-Wells Gray area, British Columbia; Canadian Journal of Earth Sciences, v. 21, p. 267-277.

Hicock, S.R.
1980: Pre-Fraser Pleistocene stratigraphy, geochronology, and paleoecology of the Georgia Depression, British Columbia; unpublished PhD thesis, University of Western Ontario, London, Ontario, 230 p.

Hicock, S.R. and Armstrong, J.E.
1981: Coquitlam Drift: a pre-Vashon glacial formation in the Fraser Lowland, British Columbia; Canadian Journal of Earth Sciences, v. 18, p. 1443-1451.
1983: Four Pleistocene formations in southwest British Columbia: their implications for patterns of sedimentation of possible Sangamonian to early Wisconsinan age; Canadian Journal of Earth Sciences, v. 20, p. 1232-1247.
1985: Vashon Drift: definition of the formation in the Georgia Depression, southwest British Columbia; Canadian Journal of Earth Sciences, v. 22, p. 748-757.

Hicock, S.R., Hebda, R.J., and Armstrong, J.E.
1982a: Lag of Fraser glacial maximum in the Pacific Northwest: pollen and macrofossil evidence from western Fraser Lowland, British Columbia; Canadian Journal of Earth Sciences, v. 19, p. 2288-2296.

Hicock, S.R., Hobson, K., and Armstrong, J.E.
1982b: Late Pleistocene proboscideans and early Fraser glacial sedimentation in eastern Fraser Lowland, British Columbia; Canadian Journal of Earth Sciences, v. 19, p. 899-906.

Holland, S.S.
1964: Landforms of British Columbia, a physiographic outline; British Columbia Department of Mines and Petroleum Resources, Bulletin 48, 138 p. Reprinted 1976.

Holtedahl, H.
1960: Mountain, fiord, strandflat; geomorphology and general geology of parts of western Norway; 21st International Geological Congress (Norden), Guide to Excursions A6 and C3, 29 p.

Hopkins, D.M.
1967: The Cenozoic history of Beringia — a synthesis; in The Bering Land Bridge, D.M. Hopkins (ed.); Stanford University Press, Stanford, California, p. 451-484.

Hopkins, D.M., Smith, P.A., and Matthews, J.V., Jr.
1981: Dated wood from Alaska and the Yukon: implications for forest refugia in Beringia; Quaternary Research, v. 15, p. 217-249.

Horner, R.B.
1983: Seismicity in the St. Elias region of northwestern Canada and southeastern Alaska; Seismological Society of America, Bulletin, v. 73, p. 1117-1137.

Howes, D.E.
1977: Terrain inventory and late Pleistocene history of the southern part of the Nechako Plateau; British Columbia Ministry of Environment, Resource Analysis Branch, Bulletin 1, 27 p.
1981a: Late Quaternary sediments and geomorphic history of north-central Vancouver Island; Canadian Journal of Earth Sciences, v. 18, p. 1-12.
1981b: Terrain inventory and geological hazards: northern Vancouver Island; British Columbia Ministry of Environment, Assessment and Planning Division, APD Bulletin 5, 105 p.
1983: Late Quaternary sediments and geomorphic history of northern Vancouver Island, British Columbia; Canadian Journal of Earth Sciences, v. 20, p. 57-65.

Hughes, O.L.
1972: Surficial geology of northern Yukon Territory and northwestern District of Mackenzie, Northwest Territories; Geological Survey of Canada, Paper 69-36, 11 p.
1987: Late Wisconsinan Laurentide glacial limits of northwestern Canada: The Tutsieta Lake and Kelly Lake phases; Geological Survey of Canada, Paper 85-25, 19 p.

Hughes, O.L., Campbell, R.B., Muller, J.E., and Wheeler, J.O.
1969: Glacial limits and flow patterns, Yukon Territory, south of 65 degrees north latitude; Geological Survey of Canada, Paper 68-34, 9 p.

Hughes, O.L., Harington, C.R., Janssens, J.A., Matthews, J.V., Jr., Morlan, R.E., Rutter, N.W., and Schweger, C.E.
1981: Upper Pleistocene stratigraphy, paleoecology, and archaeology of the northern Yukon interior, eastern Beringia. 1. Bonnet Plume Basin; Arctic, v. 34, p. 329-365.

Hughes, O.L., Rampton, V.N., and Rutter, N.W.
1972: Quaternary geology and geomorphology, southern and central Yukon (northern Canada); 24th International Geological Congress (Montreal, Quebec), Guidebook, Field Excursion A11, 59 p.

Hughes, O.L., van Everdingen, R.O., and Tarnocai, C.
1983: Regional setting — physiography and geology; in Guidebook to Permafrost and Related Features of the Northern Yukon Territory and Mackenzie Delta, Canada, H.M. French and J.A. Heginbottom (ed.); Alaska Division of Geological and Geophysical Surveys, Fairbanks, Alaska, p. 5-34.

Hughes, O.L., Veillette, J.J., Pilon, J., Hanley, P.T., and van Everdingen, R.O.
1973: Terrain evaluation with respect to pipeline construction, Mackenzie Transportation Corridor, central part, lat. 64° to 68°N.; Environmental-Social Program, Northern Pipelines, Task Force on Northern Oil Development, Report 73-37, 74 p.

Hyndman, R.D. and Weichert, D.H.
1983: Seismicity and rates of relative motion on the plate boundaries of western North America; Geophysical Journal of the Royal Astronomical Society, v. 72, p. 59-82.

Hyndman, R.D., Riddihough, R.P., and Herzer, R.
1979: The Nootka fault zone: a new plate boundary off western Canada; Geophysical Journal, v. 58, p. 667-683.

Jackson, L.E., Jr.
1977: Quaternary stratigraphy and terrain inventory of the Alberta portion of the Kananaskis Lakes 1:250 000 sheet (82-J); unpublished PhD thesis, University of Calgary, Calgary, Alberta, 480 p.
1979: A catastrophic glacial outburst flood (jökulhlaup) mechanism for debris flow generation at the Spiral Tunnels, Kicking Horse River basin, British Columbia; Canadian Geotechnical Journal, v. 16, p. 806-813.
1980: Glacial history and stratigraphy of the Alberta portion of the Kananaskis Lakes map area; Canadian Journal of Earth Sciences, v. 17, p. 459-477.
1987: Terrain inventory and Quaternary history of the Nahanni map area, Yukon Territory and Northwest Territories; Geological Survey of Canada, Paper 86-18, 23 p.

Jackson, L.E., Church, M., Clague, J.J., and Eisbacher, G.H.
1985: Slope hazards in the southern Coast Mountains of British Columbia, Field Trip 4; in Field Guides to Geology and Mineral Deposits in the Southern Canadian Cordillera, D.J. Tempelman-Kluit (ed.); Geological Society of America, Cordilleran Section, 1985 Annual Meeting (Vancouver, British Columbia), Field Trip Guidebook, p. 4-1 - 4-34.

Jackson, L.E., Jr., MacDonald, G.M., and Wilson, M.C.
1982: Paraglacial origin for terraced river sediments in Bow Valley, Alberta; Canadian Journal of Earth Sciences, v. 19, p. 2219-2231.

Johnston, W.A.
1921a: Pleistocene oscillations of sea-level in the Vancouver region, British Columbia; Royal Society of Canada, Transactions, ser. 3, v. 15, sec. 4, p. 9-19.
1921b: Sedimentation of the Fraser River delta; Geological Survey of Canada, Memoir 125, 46 p.
1922: The character of the stratification of the sediments in the Recent delta of Fraser River, British Columbia, Canada; Journal of Geology, v. 30, p. 115-129.
1923: Geology of Fraser River delta map-area; Geological Survey of Canada, Memoir 135, 87 p.
1926a: Gold placers of Dease Lake area, Cassiar district, B.C.; Geological Survey of Canada, Summary Report 1925, pt. A, p. 33A-74A.
1926b: The Pleistocene of Cariboo and Cassiar districts, British Columbia, Canada; Royal Society of Canada, Transactions, ser. 3, v. 20, sec. 4, p. 137-147.

Johnston, W.A. and Uglow, W.L.
1926: Placer and vein gold deposits of Barkerville, Cariboo district, British Columbia; Geological Survey of Canada, Memoir 149, 246 p.

Jones, D.L., Cox, A., Coney, P., and Beck, M.
1982a: The growth of western North America; Scientific American, v. 247, no. 5, p. 70-84.

Jones, D.L., Howell, D.G., Coney, P.J., and Monger, J.W.H.
1982b: Recognition, character, and analysis of tectonostratigraphic terranes in western North America; in Accretion Tectonics in the Circum-Pacific Regions, M. Hashimoto and S. Uyeda (ed.); Advances in Earth and Planetary Sciences, Terra Scientific Publishing Company, Tokyo, Japan, p. 21-35.

Jones, D.P., Ricker, K.E., Desloges, J.R., and Maxwell, M.
1985: Glacier outburst flood on the Noeick River: the draining of Ape Lake, British Columbia, October 20, 1984; Geological Survey of Canada, Open File 1139, 81 p.

Jopling, A.V., Irving, W.N., and Beebe, B.F.
1981: Stratigraphic, sedimentological and faunal evidence for the occurrence of pre-Sangamonian artefacts in northern Yukon; Arctic, v. 34, p. 3-33.

Karlstrom, E.T.
1981: Late Cenozoic soils of the Glacier and Waterton Parks area, northwestern Montana and southwestern Alberta, and paleoclimatic implications; unpublished PhD thesis, University of Calgary, Calgary, Alberta, 358 p.
1987: Stratigraphy and genesis of five superposed paleosols in pre-Wisconsinan drift on Mokowan Butte, southwestern Alberta; Canadian Journal of Earth Sciences, v. 24, p. 2235-2253.

Kearney, M.S. and Luckman, B.H.
1983a: Holocene timberline fluctuations in Jasper National Park, Alberta; Science, v. 221, p. 261-263.
1983b: Postglacial vegetational history of Tonquin Pass, British Columbia; Canadian Journal of Earth Sciences, v. 20, p. 776-786.

Keen, C.E. and Hyndman, R.D.
1979: Geophysical review of the continental margins of eastern and western Canada; Canadian Journal of Earth Sciences, v. 16, p. 712-747.

Kendrew, W.G. and Kerr, D.
1955: The climate of British Columbia and the Yukon Territory; Queen's Printer, Ottawa, Ontario, 222 p.

Kerr, F.A.
1934: Glaciation in northern British Columbia; Royal Society of Canada, Transactions, ser. 3, v. 28, sec. 4, p. 17-31.
1936: Quaternary glaciation in the Coast Range, northern British Columbia and Alaska; Journal of Geology, v. 44, p. 681-700.
1948: Lower Stikine and western Iskut River areas, British Columbia; Geological Survey of Canada, Memoir 246, 94 p.

Keser, N.
1970: Soil and forest growth in the Sayward Forest, British Columbia; unpublished PhD thesis, University of British Columbia, Vancouver, British Columbia, 302 p.

Kigoshi, K., Suzuki, N., and Fukatsu, H.
1973: Gakushuin natural radiocarbon measurements VIII; Radiocarbon, v. 15, p. 42-67.

Kindle, E.D.
1952: Dezadeash map-area, Yukon Territory; Geological Survey of Canada, Memoir 268, 68 p.

King, M.
1980: Palynological and macrofossil analyses of lake sediments from the Lillooet area, British Columbia; unpublished MSc thesis, Simon Fraser University, Burnaby, British Columbia, 125 p.

Klassen, R.W.
1978: A unique stratigraphic record of late Tertiary-Quaternary events in southeastern Yukon; Canadian Journal of Earth Sciences, v. 15, p. 1884-1886.
1987: The Tertiary-Pleistocene stratigraphy of the Liard Plain, southeastern Yukon Territory; Geological Survey of Canada, Paper 86-17, 16 p.

Lamplugh, G.W.
1886: On glacial shell-beds in British Columbia; Geological Society of London, Quarterly Journal, v. 42, p. 276-286.

Lemke, R.W., Laird, W.M., Tipton, M.J., and Lindvall, R.M.
1965: Quaternary geology of northern Great Plains; in The Quaternary of the United States, H.E. Wright, Jr. and D.G. Frey (ed.); Princeton University Press, Princeton, New Jersey, p. 15-27.

Leonard, E.M.
1986a: Varve studies at Hector Lake, Alberta, Canada, and the relationship between glacial activity and sedimentation; Quaternary Research, v. 25, p. 199-214.
1986b: Use of lacustrine sedimentary sequences as indicators of Holocene glacial history, Banff National Park, Alberta, Canada; Quaternary Research, v. 26, p. 218-231.

Lerbekmo, J.F. and Campbell, F.A.
1969: Distribution, composition, and source of the White River ash, Yukon Territory; Canadian Journal of Earth Sciences, v. 6, p. 109-116.

Lerbekmo, J.F., Westgate, J.A., Smith, D.G.W., and Denton, G.H.
1975: New data on the character and history of the White River volcanic eruption, Alaska; in Quaternary Studies, R.P. Suggate and M.M. Cresswell (ed.); Royal Society of New Zealand, Bulletin 13, p. 203-209.

Levson, V. and Rutter, N.W.
1986: A facies approach to the stratigraphic analysis of Late Wisconsinan sediments in the Portal Creek area, Jasper National Park, Alberta; Géographie physique et Quaternaire, v. 40, p. 129-144.

Lichti-Federovich, S.
1973: Palynology of six sections of late Quaternary sediments from the Old Crow River, Yukon Territory; Canadian Journal of Botany, v. 51, p. 553-564.
1974: Palynology of two sections of late Quaternary sediments from the Porcupine River, Yukon Territory; Geological Survey of Canada, Paper 74-23, 6 p.

Lowdon, J.A. and Blake, W., Jr.
1968: Geological Survey of Canada radiocarbon dates VII; Radiocarbon, v. 10, p. 207-245. Reprinted (1968) as Geological Survey of Canada, Paper 68-2B.
1970: Geological Survey of Canada radiocarbon dates IX; Radiocarbon, v. 12, p. 46-86. Reprinted (1970) as Geological Survey of Canada, Paper 70-2B.
1973: Geological Survey of Canada radiocarbon dates XIII; Geological Survey of Canada, Paper 73-7, 61 p.
1978: Geological Survey of Canada radiocarbon dates XVIII; Geological Survey of Canada, Paper 78-7, 20 p.
1979: Geological Survey of Canada radiocarbon dates XIX; Geological Survey of Canada, Paper 79-7, 58 p.
1980: Geological Survey of Canada radiocarbon dates XX; Geological Survey of Canada, Paper 80-7, 28 p.
1981: Geological Survey of Canada radiocarbon dates XXI; Geological Survey of Canada, Paper 81-7, 22 p.

Lowdon, J.A., Fyles, J.G., and Blake, W., Jr.
1967: Geological Survey of Canada radiocarbon dates VI; Radiocarbon, v. 9, p. 156-197. Reprinted (1967) as Geological Survey of Canada, Paper 67-2B.

Lowdon, J.A., Robertson, I.M., and Blake, W., Jr.
1971: Geological Survey of Canada radiocarbon dates XI; Radiocarbon, v. 13, p. 255-324. Reprinted (1971) as Geological Survey of Canada, Paper 71-7.
1977: Geological Survey of Canada radiocarbon dates XVII; Geological Survey of Canada, Paper 77-7, 25 p.

Luckman, B.H.
1986: Reconstruction of Little Ice Age events in the Canadian Rocky Mountains; Géographie physique et Quaternaire, v. 40, p. 17-28.

Luckman, B.H. and Osborn, G.D.
1979: Holocene glacier fluctuations in the middle Canadian Rocky Mountains; Quaternary Research, v. 11, p. 52-77.

Luternauer, J.L. and Finn, W.D.L.
1983: Stability of the Fraser River delta front; Canadian Geotechnical Journal, v. 20, p. 603-616.

Luternauer, J.L. and Murray, J.W.
1983: Late Quaternary morphologic development and sedimentation, central British Columbia continental shelf; Geological Survey of Canada, Paper 83-21, 38 p.

Luternauer, J.L. and Swan, D.
1978: Kitimat submarine slump deposit(s): a preliminary report; in Current Research, Part A, Geological Survey of Canada, Paper 78-1A, p. 327-332.

MacDonald, G.M.
1982: Late Quaternary paleoenvironments of the Morley Flats and Kananaskis Valley of southwestern Alberta; Canadian Journal of Earth Sciences, v. 19, p. 23-35.
1983: Holocene vegetation history of the Upper Natla River area, Northwest Territories, Canada; Arctic and Alpine Research, v. 15, p. 169-180.
1987a: Postglacial development of the subalpine-boreal transition forest of western Canada; Journal of Ecology, v. 75, p. 303-320.
1987b: Postglacial vegetation history of the Mackenzie River basin; Quaternary Research, v. 28, p. 245-262.

MacDonald, G.M. and Cwynar, L.C.
1985: A fossil pollen based reconstruction of lodgepole pine (*Pinus contorta* ssp. *latifolia*) in the western interior of Canada; Canadian Journal of Forestry Research, v. 15, p. 1039-1044.

MacDonald, G.M., Beukens, R.P., Keiser, W.E., and Vitt, D.H.
1987: Comparative radiocarbon dating of terrestrial plant macrofossils and aquatic moss from the "ice-free corridor" of western Canada; Geology, v. 15, p. 837-840.

Mackay, J.R. and Mathews, W.H.
1973: Geomorphology and Quaternary history of the Mackenzie River valley near Fort Good Hope, N.W.T., Canada; Canadian Journal of Earth Sciences, v. 10, p. 26-41.

Marcus, M.G.
1960: Periodic drainage of glacier-dammed Tulsequah Lake, British Columbia; Geographical Review, v. 50, p. 89-106.

Mathewes, R.W.
1973: A palynological study of postglacial vegetation changes in the University Research Forest, southwestern British Columbia; Canadian Journal of Botany, v. 51, p. 2085-2103.
1979: A paleoecological analysis of Quadra Sand at Point Grey, British Columbia, based on indicator pollen; Canadian Journal of Earth Sciences, v. 16, p. 847-858.
1985: Paleobotanical evidence for climatic change in southern British Columbia during late-glacial and Holocene time; in Climatic Change in Canada 5, Critical Periods in the Quaternary Climatic History of Northern North America, C.R. Harington (ed.); National Museums of Canada, National Museum of Natural Sciences, Syllogeus Series, no. 55, p. 397-422.

Mathewes, R.W. and Clague, J.J.
1982: Stratigraphic relationships and paleoecology of a late-glacial peat bed from the Queen Charlotte Islands, British Columbia; Canadian Journal of Earth Sciences, v. 19, p. 1185-1195.

Mathewes, R.W. and Heusser, L.E.
1981: A 12 000 year palynological record of temperature and precipitation trends in southwestern British Columbia; Canadian Journal of Botany, v. 59, p. 707-710.

Mathewes, R.W. and Rouse, G.E.
1975: Palynology and paleoecology of postglacial sediments from the lower Fraser River canyon of British Columbia; Canadian Journal of Earth Sciences, v. 12, p. 745-756.

Mathewes, R.W. and Westgate, J.A.
1980: Bridge River tephra: revised distribution and significance for detecting old carbon errors in radiocarbon dates of limnic sediments in southern British Columbia; Canadian Journal of Earth Sciences, v. 17, p. 1454-1461.

Mathews, W.H.
1944: Glacial lakes and ice retreat in south-central British Columbia; Royal Society of Canada, Transactions, ser. 3, v. 38, sec. 4, p. 39-57.
1947: "Tuyas," flat-topped volcanoes in northern British Columbia; American Journal of Science, v. 245, p. 560-570.
1951a: Historic and prehistoric fluctuations of alpine glaciers in the Mount Garibaldi map-area, southwestern British Columbia; Journal of Geology, v. 59, p. 357-380.
1951b: The Table, a flat-topped volcano in southern British Columbia; American Journal of Science, v. 249, p. 830-841.
1955: Late Pleistocene divide of the Cordilleran Ice Sheet (abstract); Geological Society of America, Bulletin, v. 66, p. 1657.
1958: Geology of the Mount Garibaldi map-area, southwestern British Columbia, Canada. Part II: geomorphology and Quaternary volcanic rocks; Geological Society of America, Bulletin, v. 69, p. 179-198.
1965: Two self-dumping ice-dammed lakes in British Columbia; Geographical Review, v. 55, p. 46-52.
1972: Geology of Vancouver area of British Columbia; 24th International Geological Congress (Montreal, Quebec), Guidebook, Field Excursion A05-C05, 47 p.
1973: Record of two jökullhlaups (sic); International Association of Scientific Hydrology, Publication 95, p. 99-110.
1978: Quaternary stratigraphy and geomorphology of Charlie Lake (94A) map-area, British Columbia; Geological Survey of Canada, Paper 76-20, 25 p.
1979: Landslides of central Vancouver Island and the 1946 earthquake; Seismological Society of America, Bulletin, v. 69, p. 445-450.
1980: Retreat of the last ice sheets in northeastern British Columbia and adjacent Alberta; Geological Survey of Canada, Bulletin 331, 22 p.
1986: Physiographic map of the Canadian Cordillera; Geological Survey of Canada, Map 1701A.

Mathews, W.H. and McTaggart, K.C.
1978: Hope rockslides, British Columbia, Canada; in Rockslides and Avalanches, 1, Natural Phenomena, B. Voight (ed.); Elsevier Scientific Publishing Company, New York, p. 259-275.

Mathews, W.H. and Rouse, G.E.
1986: An Early Pleistocene proglacial succession in south-central British Columbia; Canadian Journal of Earth Sciences, v. 23, p. 1796-1803.

Mathews, W.H., Fyles, J.G., and Nasmith, H.W.
1970: Postglacial crustal movements in southwestern British Columbia and adjacent Washington state; Canadian Journal of Earth Sciences, v. 7, p. 690-702.

Matthews, J.V., Jr.
1975a: Incongruence of macrofossil and pollen evidence: a case from the late Pleistocene of the northern Yukon coast; in Report of Activities, Part B, Geological Survey of Canada, Paper 75-1B, p. 139-146.
1975b: Insects and plant macrofossils from two Quaternary exosures in the Old Crow-Porcupine region, Yukon Territory, Canada; Arctic and Alpine Research, v. 7, p. 249-259.
1976: Arctic-steppe: an extinct biome (abstract); American Quaternary Association, 4th Biennial Conference (Tempe, Arizona), Abstracts, p. 73-79.
1980: Paleoecology of John Klondike bog, Fisherman Lake region, southwest District of Mackenzie; Geological Survey of Canada, Paper 80-22, 12 p.
1982: East Beringia during late Wisconsin time: a review of the biotic evidence; in Paleoecology of Beringia, D.M. Hopkins, J.V. Matthews, Jr., C.E. Schweger, and S.B. Young (ed.); Academic Press, New York, p. 127-150.
1984: Synthesis of environmental history and stratigraphy of the northern Yukon: iteration 1; Joint Canadian-American Workshop on Correlation of Quaternary Deposits and Events in the Area around the Beaufort Sea (Calgary, Alberta), unpublished report.
1987: Late Cenozoic history of northern Yukon basins: an update (abstract); International Union for Quaternary Research, 12th International Congress (Ottawa, Ontario), Programme with Abstracts, p. 221.

McConnell, R.G.
1903a: Yukon district; Geological Survey of Canada, Annual Report, new ser., v. 13 (1900), p. 37A-52A.
1903b: The MacMillan River, Yukon district; Geological Survey of Canada, Annual Report, new ser., v. 15 (1902), p. 22A-38A.
1905: Report on the Klondike gold fields; Geological Survey of Canada, Annual Report, new ser., v. 14 (1901), p. 1B-71B.
1907: Report on gold values in the Klondike high level gravels; Geological Survey of Canada, Publication 979, 34 p.

McConnell, R.G. and Brock, R.W.
1904: Report on the great landslide at Frank, Alta.; Canada Department of Interior, Annual Report 1903, pt. 8, 17 p.

McRoberts, E.C. and Morgenstern, N.R.
1974: The stability of thawing slopes; Canadian Geotechnical Journal, v. 11, p. 447-469.

Meyerhof, G.G. and Sebastyan, G.Y.
1970: Settlement studies on air terminal building and apron, Vancouver International Airport, British Columbia; Canadian Geotechnical Journal, v. 7, p. 433-456.

Miller, C.D.
1969: Chronology of Neoglacial moraines in the Dome Peak area, North Cascade Range, Washington; Arctic and Alpine Research, v. 1, p. 49-65.

Miller, M.M.
1976: Quaternary erosional and stratigraphic sequences in the Alaska-Canada Boundary Range; in Quaternary Stratigraphy of North America, W.C. Mahaney (ed.); Dowden, Hutchinson & Ross, Stroudsburg, Pennsylvania, p. 463-492.

Miller, M.M. and Anderson, J.H.
1974: Out-of-phase Holocene climatic trends in the maritime and continental sectors of the Alaska-Canada Boundary Range; in Quaternary Environments: Proceedings of a Symposium, W.C. Mahaney (ed.); York University, Geographical Monographs, no. 5, p. 33-58.

Milne, W.G., Rogers, G.C., Riddihough, R.P., McMechan, G.A., and Hyndman, R.D.
1978: Seismicity of western Canada; Canadian Journal of Earth Sciences, v. 15, p. 1170-1193.

Mokievsky-Zubok, O.
1977: Glacier-caused slide near Pylon Peak, British Columbia; Canadian Journal of Earth Sciences, v. 14, p. 2657-2662.

Monger, J.W.H. and Berg, H.C.
1984: Part B — Lithotectonic terrane map of western Canada and southeastern Alaska; in Lithotectonic Terrane Maps of the North American Cordillera, N.J. Siberling and D.L. Jones (ed.); United States Geological Survey, Open-File Report 84-523, p. B-1-B-31 and Map Sheet 2.

Monger, J.W.H., Price, R.A., and Tempelman-Kluit, D.J.
1982: Tectonic accretion and the origin of the two major metamorphic and plutonic welts in the Canadian Cordillera; Geology, v. 10, p. 70-75.

Moore, D.P. and Mathews, W.H.
1978: The Rubble Creek landslide, southwestern British Columbia; Canadian Journal of Earth Sciences, v. 15, p. 1039-1052.

Morgan, A.V.
1966: Lithological and glacial geomorphological studies near the Erratics Train, Calgary area, Alberta; unpublished MSc thesis, University of Calgary, Calgary, Alberta, 178 p.
1969: Lithology of the Erratics Train in the Calgary area; in Geomorphology: Selected Readings, J.G. Nelson and M.J. Chambers (ed.); Methuen, Toronto, Ontario, p. 165-182.

Morison, S.R.
1987: Stop 26: White Channel placer deposits in the Klondike area; in Quaternary Research in Yukon, S.R. Morison and C.A.S. Smith (ed.); International Union for Quaternary Research, 12th International Congress (Ottawa, Ontario), Guidebook, Field Excursions A-20a and A-20b, p. 68-71.
1989: Placer deposits in Canada; in Chapter 11 of Quaternary Geology of Canada and Greenland, R.J. Fulton (ed.); Geological Survey of Canada, Geology of Canada, no. 1 (also Geological Society of America, The Geology of North America, v. K-1).

Morlan, R.E.
1979: A stratigraphic framework for Pleistocene artifacts from Old Crow River, northern Yukon Territory; in Pre-Llano Cultures of the Americas: Paradoxes and Possibilities, R.L. Humphrey and D. Stanford (ed.); Anthropological Society of Washington, Washington, D.C., p. 125-145.
1980: Taphonomy and archaeology in the Upper Pleistocene of the northern Yukon Territory: a glimpse of the peopling of the New World; National Museums of Canada, National Museum of Man, Mercury Series, Archaeological Survey of Canada, Paper 94, 398 p.
1986: Pleistocene archaeology in Old Crow Basin: a critical reappraisal; in New Evidence for the Pleistocene Peopling of the Americas, A.L. Bryan (ed.); University of Maine, Center for the Study of Early Man, Orono, Maine, p. 27-48.

Morlan, R.E. and Matthews, J.V., Jr.
1983: Taphonomy and paleoecology of fossil insect assemblages from Old Crow River (CRH-15) northern Yukon Territory, Canada; Géographie physique et Quaternaire, v. 37, p. 147-157.

Mörner, N.-A.
1976: Eustasy and geoid changes; Journal of Geology, v. 84, p. 123-151.

Mott, R.J. and Jackson, L.E., Jr.
1982: An 18 000 year palynological record from the southern Alberta segment of the classical Wisconsinan "Ice-free Corridor"; Canadian Journal of Earth Sciences, v. 19, p. 504-513.

Mountjoy, E.W.
1958: Jasper area Alberta, a source of the Foothills Erratics Train; Alberta Society of Petroleum Geology, Journal, v. 6, p. 218-226.

Mullineaux, D.R., Hyde, J.H., and Rubin, M.
1975: Widespread late glacial and postglacial tephra deposits from Mount St. Helens volcano, Washington; United States Geological Survey, Journal of Research, v. 3, p. 329-335.

Munday, D.
1931: Retreat of Coast Range glaciers; Canadian Alpine Journal, v. 20, p. 140-142.
1936: Glaciers of Mt. Waddington region; Canadian Alpine Journal, v. 23, p. 68-75.

Murty, T.S.
1977: Seismic sea waves — tsunamis; Fisheries Research Board of Canada, Bulletin 198, 337 p.
1979: Submarine slide-generated water waves in Kitimat Inlet, British Columbia; Journal of Geophysical Research, v. 84, p. 7777-7779.

Naeser, N.D., Westgate, J.A., Hughes, O.L., and Péwé, T.L.
1982: Fission-track ages of late Cenozoic distal tephra beds in the Yukon Territory and Alaska; Canadian Journal of Earth Sciences, v. 19, p. 2167-2178.

Nasmith, H.W. and Mercer, A.G.
1979: Design of dykes to protect against debris flows at Port Alice, British Columbia; Canadian Geotechnical Journal, v. 16, p. 748-757.

Nasmith, H., Mathews, W.H., and Rouse, G.E.
1967: Bridge River ash and some other Recent ash beds in British Columbia; Canadian Journal of Earth Sciences, v. 4, p. 163-170.

Newcombe, C.F.
1914: Pleistocene raised beaches at Victoria, B.C.; Ottawa Naturalist, v. 28, p. 107-110.

Olson, E.A. and Broecker, W.S.
1961: Lamont natural radiocarbon measurements VII; Radiocarbon, v. 3, p. 141-175.

Osborn, G.
1985: Holocene tephrostratigraphy and glacier fluctuations in Waterton Lakes and Glacier national parks, Alberta and Montana; Canadian Journal of Earth Sciences, v. 22, p. 1093-1101.
1986: Lateral-moraine stratigraphy and Neoglacial history of Bugaboo Glacier, British Columbia; Quaternary Research, v. 26, p. 171-178.

Ovenden, L.
1982: Vegetation history of a polygonal peatland, northern Yukon; Boreas, v. 11, p. 209-224.

Parrish, R.R.
1981: Uplift rates of Mt. Logan, Y.T., and British Columbia's Coast Mountains using fission track dating methods (abstract); Eos, v. 62, p. 59-60.
1983: Cenozoic thermal evolution and tectonics of the Coast Mountains of British Columbia, 1. Fission track dating, apparent uplift rates, and patterns of uplift; Tectonics, v. 2, p. 601-631.

Peacock, M.A.
1935: Fiord-land of British Columbia; Geological Society of America, Bulletin, v. 46, p. 633-696.

Pearce, G.W., Westgate, J.A., and Robertson, S.
1982: Magnetic reversal history of Pleistocene sediments at Old Crow, northwestern Yukon Territory; Canadian Journal of Earth Sciences, v. 19, p. 919-929.

Peckover, F.L. and Kerr, J.W.G.
1977: Treatment and maintenance of rock slopes on transportation routes; Canadian Geotechnical Journal, v. 14, p. 487-507.

Pharo, C.H. and Carmacks, E.C.
1979: Sedimentation processes in a short residence-time intermontane lake, Kamloops Lake, British Columbia; Sedimentology, v. 26, p. 523-541.

Pickard, G.L.
1961: Oceanographic features of inlets in the British Columbia mainland coast; Fisheries Research Board of Canada, Journal, v. 18, p. 907-999.

Piteau, D.R.
1977: Regional slope-stability controls and engineering geology of the Fraser Canyon, British Columbia; in Landslides, D.R. Coates (ed.); Geological Society of America, Reviews in Engineering Geology, v. 3, p. 85-111.

Piteau, D.R., Mylrea, F.H., and Blown, I.G.
1978: Downie slide, Columbia River, British Columbia; in Rockslides and Avalanches, 1, Natural Phenomena, B. Voight (ed.); Elsevier Scientific Publishing Company, New York, p. 365-392.

Porter, S.C. and Denton, G.H.
1967: Chronology of Neoglaciation in the North American Cordillera; American Journal of Science, v. 265, p. 177-210.

Porter, S.C., Pierce, K.L., and Hamilton, T.D.
1983: Late Wisconsin mountain glaciation in the western United States; in Late-Quaternary Environments of the United States, Volume 1, the Late Pleistocene, S.C. Porter (ed.); University of Minnesota Press, Minneapolis, Minnesota, p. 71-111.

Powers, H.A. and Wilcox, R.E.
1964: Volcanic ash from Mount Mazama (Crater Lake) and from Glacier Peak; Science, v. 144, p. 1334-1336.

Prest, V.K.
1969: Retreat of Wisconsin and Recent ice in North America; Geological Survey of Canada, Map 1257A.
1970: Quaternary geology of Canada; in Geology and Economic Minerals of Canada, R.J.W. Douglas (ed.); Geological Survey of Canada, Economic Geology Report 1, p. 675-764.

Prest, V.K., Grant, D.R., and Rampton, V.N.
1968: Glacial map of Canada; Geological Survey of Canada, Map 1253A.

Prior, D.B. and Bornhold, B.D.
1984: Geomorphology of slope instability features of Squamish Harbour, Howe Sound, British Columbia; Geological Survey of Canada, Open File 1095.
1986: Sediment transport on subaqueous fan delta slopes, Britannia Beach, British Columbia; Geo-Marine Letters, v. 5, p. 217-224.

Prior, D.B., Bornhold, B.D., Coleman, J.M., and Bryant, W.R.
1982: Morphology of a submarine slide, Kitimat Arm, British Columbia; Geology, v. 10, p. 588-592.

Prior, D.B., Bornhold, B.D., and Johns, M.W.
1984: Depositional characteristics of a submarine debris flow; Journal of Geology, v. 92, p. 707-727.
1986: Active sand transport along a fjord-bottom channel, Bute Inlet, British Columbia; Geology, v. 14, p. 581-584.

Prior, D.B., Wiseman, W.J., and Gilbert, R.
1981: Submarine slope processes on a fan delta, Howe Sound, British Columbia; Geo-Marine Letters, v. 1, p. 85-90.

Rampton, V.N.
1969: Pleistocene geology of the Snag-Klutlan area, southwestern Yukon, Canada; unpublished PhD thesis, University of Minnesota, Minneapolis, 237 p.
1970: Neoglacial fluctuations of the Natazhat and Klutlan Glaciers, Yukon Territory, Canada; Canadian Journal of Earth Sciences, v. 7, p. 1236-1263.
1971a: Late Pleistocene glaciations of the Snag-Klutlan area, Yukon Territory; Arctic, v. 24, p. 277-300.
1971b: Late Quaternary vegetational and climatic history of the Snag-Klutlan area, southwestern Yukon Territory, Canada; Geological Society of America, Bulletin, v. 82, p. 959-978.
1982: Quaternary geology of the Yukon Coastal Plain; Geological Survey of Canada, Bulletin 317, 49 p.

Read, L.C.
1921: The Cordilleran ice sheet; Natural History, v. 21, p. 251-254.

Reeves, B.O.K.
1975: Early Holocene (ca. 8000 to 5500 B.C.) prehistoric land/resource utilization patterns in Waterton Lakes National Park, Alberta; Arctic and Alpine Research, v. 7, p. 237-248.
1983: Bergs, barriers and Beringia: reflections on the peopling of the New World; in Quaternary Coastlines and Marine Archaeology: Towards the Prehistory of Land Bridges and the Continental Shelves, P.M. Masters and N.C. Flemming (ed.); Academic Press, London, England, p. 389-411.

Reeves, B.O.K. and Dormaar, J.F.
1972: A partial Holocene pedological and archaeological record from the southern Alberta Rocky Mountains; Arctic and Alpine Research, v. 4, p. 325-336.

Rice, H.M.A.
1936: Glacial phenomena near Cranbrook, British Columbia; Journal of Geology, v. 44, p. 68-73.

Richmond, G.M., Fryxell, R., Neff, G.E., and Weis, P.L.
1965: The Cordilleran Ice Sheet of the northern Rocky Mountains, and related Quaternary history of the Columbia Plateau; in The Quaternary of the United States, H.E. Wright, Jr. and D.G. Frey (ed.); Princeton University Press, Princeton, New Jersey, p. 231-242.

Ricker, K.
1983: Preliminary observations on a multiple morainal sequence and associated periglacial features on the Mt. Tatlow area, Chicotin Ranges, Coast Mountains; Canadian Alpine Journal, v. 66, p. 61-67.

Riddihough, R.P.
1977: A model for recent plate interactions off Canada's west coast; Canadian Journal of Earth Sciences, v. 14, p. 384-396.
1982: Contemporary movements and tectonics on Canada's west coast: a discussion; Tectonophysics, v. 86, p. 319-341.

Riddihough, R.P. and Hyndman, R.D.
1976: Canada's active western margin — the case for subduction; Geoscience Canada, v. 3, p. 269-278.

Riddihough, R.P., Beck, M.E., Chase, R.L., Davis, E.E., Hyndman, R.D., Johnson, S.H., and Rogers, G.C.
1983: Geodynamics of the Juan de Fuca Plate; in Geodynamics of the Eastern Pacific Region, Caribbean and Scotia Arcs, R. Cabre (ed.); American Geophysical Union, Geodynamics Series, v. 9, p. 5-21.

Riddihough, R.P., Currie, R.G., and Hyndman, R.D.
1980: The Dellwood Knolls and their role in triple junction tectonics off northern Vancouver Island; Canadian Journal of Earth Sciences, v. 17, p. 577-593.

Ritchie, J.C.
1982: The modern and late Quaternary vegetation of the Doll Creek area, north Yukon, Canada; New Phytologist, v. 90, p. 563-603.
1984: Past and present vegetation of the far northwest of Canada; University of Toronto Press, Toronto, Ontario, 251 p.

Ritchie, J.C. and Cwynar, L.C.
1982: The late Quaternary vegetation of the north Yukon; in Paleoecology of Beringia, D.M. Hopkins, J.V. Matthews, Jr., C.E. Schweger, and S.B. Young (ed.); Academic Press, New York, p. 113-126.

Ritchie, J.C. and MacDonald, G.M.
1986: The patterns of postglacial spread of white spruce; Journal of Biogeography, v. 13, p. 527-540.

Ritchie, J.C., Cinq-Mars, J., et Cwynar, L.C.
1982: L'environnement tardiglaciaire du Yukon septentrional, Canada; Géographie physique et Quaternaire, vol. 36, p. 241-250.

Ritchie, J.C., Cwynar, L.C., and Spear, R.W.
1983: Evidence from north-west Canada for an early Holocene Milankovitch thermal maximum; Nature, v. 305, p. 126-128.

Roed, M.A.
1975: Cordilleran and Laurentide multiple glaciation, west central Alberta; Canadian Journal of Earth Sciences, v. 12, p. 1493-1515.

Rogers, G.C.
1981: McNaughton Lake seismicity — more evidence for an Anahim hotspot?; Canadian Journal of Earth Sciences, v. 18, p. 826-828.

Rose, K.E.
1979: Characteristics of ice flow in Marie Byrd Land, Antarctica (includes Discussion); Journal of Glaciology, v. 24, p. 63-75.

Ross, C.P.
1959: Geology of Glacier National Park and the Flathead region, northwestern Montana; United States Geological Survey, Professional Paper 296, 125 p.

Russell, L.S.
1957: Tertiary plains of Alberta and Saskatchewan; Geological Association of Canada, Proceedings, v. 9, p. 17-19.

Russell, L.S. and Churcher, C.S.
1972: Vertebrate paleontology, Cretaceous to Recent, Interior Plains, Canada; 24th International Geological Congress (Montreal, Quebec), Guidebook, Field Excursion A21, 46 p.

Rutter, N.W.
1972: Geomorphology and multiple glaciation in the area of Banff, Alberta; Geological Survey of Canada, Bulletin 206, 54 p.
1976: Multiple glaciation in the Canadian Rocky Mountains with special emphasis on northeastern British Columbia; in Quaternary Stratigraphy of North America, W.C. Mahaney (ed.); Dowden, Hutchinson & Ross, Stroudsburg, Pennyslvania, p. 409-440.
1977: Multiple glaciation in the area of Williston Lake, British Columbia; Geological Survey of Canada, Bulletin 273, 31 p.
1980: Late Pleistocene history of the western Canadian ice-free corridor; Canadian Journal of Anthropology, v. 1, no. 1, p. 1-8.
1984: Pleistocene history of the western Canadian ice-free corridor; in Quaternary Stratigraphy of Canada — a Canadian Contribution to IGCP Project 24, R.J. Fulton (ed.); Geological Survey of Canada, Paper 84-10, p. 49-56.

Rutter, N.W., Boydell, A.N., Savigny, K.W., and van Everdingen, R.O.
1973: Terrain evaluation with respect to pipeline construction, Mackenzie Transportation Corridor, southern part, lat. 60° to 64°N; Environmental-Social Program, Northern Pipelines, Task Force on Northern Oil Development, Report 73-36, 135 p.

Rutter, N.W., Foscolos, A.E., and Hughes, O.L.
1978: Climatic trends during the Quaternary in central Yukon based upon pedological and geomorphological evidence; in Quaternary Soils, W.C. Mahaney (ed.); Geo Abstracts, Norwich, England, p. 309-359.

Ryder, J.M.
1971: Some aspects of the morphometry of paraglacial alluvial fans in south-central British Columbia; Canadian Journal of Earth Sciences, v. 8, p. 1252-1264.
1976: Terrain inventory and Quaternary geology Ashcroft, British Columbia; Geological Survey of Canada, Paper 74-49, 17 p.
1981a: Biophysical resources of the East Kootenay area: terrain; British Columbia Ministry of Environment, Assessment and Planning Division, APD Bulletin 7, 152 p.
1981b: Terrain inventory and Quaternary geology, Lytton, British Columbia; Geological Survey of Canada, Paper 79-25, 20 p.
1985: Terrain inventory for the Stikine-Iskut area; British Columbia Ministry of Environment, MOE Technical Report 11, 85 p.
1987: Neoglacial history of the Stikine-Iskut area, northern Coast Mountains, British Columbia; Canadian Journal of Earth Sciences, v. 24, p. 1294-1301.

Ryder, J.M. and Thomson, B.
1986: Neoglaciation in the southern Coast Mountains of British Columbia: chronology prior to the late Neoglacial maximum; Canadian Journal of Earth Sciences, v. 23, p. 273-287.

Saunders, I.R., Clague, J.J., and Roberts, M.C.
1987: Deglaciation of Chilliwack River valley, British Columbia; Canadian Journal of Earth Sciences, v. 24, p. 915-923.

Schaerer, P.
1962: The avalanche hazard evaluation and prediction at Rogers Pass; National Research Council of Canada, Division of Building Research, DBR Technical Paper 142 (NRC 7051), 49 p.

Schweger, C.E.
1989: Paleocology of the western Canadian ice-free corridor; in Chapter 7 of Quaternary Geology of Canada and Greenland, R.J. Fulton (ed.); Geological Survey of Canada, Geology of Canada, no. 1 (also Geological Society of America, The Geology of North America, v. K-1).

Schweger, C.E. and Janssens, J.A.P.
1980: Paleoecology of the Boutellier nonglacial interval, St. Elias Mountains, Yukon Territory, Canada; Arctic and Alpine Research, v. 12, p. 309-317.

Schweger, C.E. and Matthews, J.V., Jr.
1985: Early and Middle Wisconsinan environments of eastern Beringia: stratigraphic and paleoecological implications of the Old Crow tephra; Géographie physique et Quaternaire, v. 39, p. 275-290.

Schweger, C.E., Habgood, T., and Hickman, M.
1981: Late glacial-Holocene climatic changes of Alberta — the record from lake sediment studies; in The Impacts of Climatic Fluctuations on Alberta's Resources and Environment; Canada Department of Environment, Atmospheric Environment Service, western region (Edmonton, Alberta), Report WAES-1-81, p. 47-60.

Scott, J.S.
1976: Geology of Canadian tills; in Glacial Till, an Inter-Disciplinary Study, R.F. Legget (ed.); Royal Society of Canada, Special Publication 12, p. 50-66.

Silver, E.A. and Smith, R.B.
1983: Comparison of terrane accretion in modern Southeast Asia and the Mesozoic North American Cordillera; Geology, v. 11, p. 198-202.

Smith, H.W., Okazaki, R., and Knowles, C.R.
1977: Electron microprobe analysis of glass shards from tephra assigned to set W, Mount St. Helens, Washington; Quaternary Research, v. 7, p. 207-217.

Souther, J.G.
1970: Volcanism and its relationship to recent crustal movements in the Canadian Cordillera; Canadian Journal of Earth Sciences, v. 7, p. 553-568.
1976: Geothermal potential of western Canada; 2nd United Nations Symposium on the Development and Use of Geothermal Resources (San Francisco, California), Proceedings, v. 1, p. 259-267.
1977: Volcanism and tectonic environments in the Canadian Cordillera — a second look; in Volcanic Regimes in Canada, W.R.A. Baragar, L.C. Coleman, and J.M. Hall (ed.); Geological Association of Canada, Special Paper 16, p. 3-24.
1986: The western Anahim Belt: root zone of a peralkaline magma system; Canadian Journal of Earth Sciences, v. 23, p. 895-908.

Souther, J.G. and Stanciu, C.
1975: Operation Saint Elias, Yukon Territory: Tertiary volcanic rocks; in Report of Activities, Part A, Geological Survey of Canada, Paper 75-1A, p. 63-70.

Souther, J.G., Armstrong, R.L., and Harakal, J.
1984: Chronology of the peralkaline, late Cenozoic Mount Edziza Volcanic Complex, northern British Columbia, Canada; Geological Society of America, Bulletin, v. 95, p. 337-349.

Souther, J.G., Clague, J.J., and Mathewes, R.W.
1987: Nazko cone, a Quaternary volcano in the eastern Anahim Belt; Canadian Journal of Earth Sciences, v. 24, p. 2477-2486.

Stalker, A.MacS.
1956: The Erratics Train, foothills of Alberta; Geological Survey of Canada, Bulletin 37, 28 p.

Stalker, A.MacS. and Harrison, J.E.
1977: Quaternary glaciation of the Waterton-Castle River region of Alberta; Canadian Petroleum Geology, Bulletin, v. 25, p. 882-906.

Stanton, R.B.
1898: The great land-slides on the Canadian Pacific Railway in British Columbia (includes Discussion); Institution of Civil Engineers, Proceedings, v. 132, pt. 2, p. 1-48.

Stethem, C.J. and Schaerer, P.A.
1979: Avalanche accidents in Canada. I. A selection of case histories of accidents, 1955 to 1976; National Research Council of Canada, Division of Building Research, DBR Paper 834 (NRCC 17292), 114 p.
1980: Avalanche accidents in Canada. II. A selection of case histories of accidents, 1943 to 1978; National Research Council of Canada, Division of Building Research, DBR Paper 926 (NRCC 18525), 75 p.

Stewart, C.A.
1913: The extent of the Cordilleran ice-sheet; Journal of Geology, v. 21, p. 427-430.

St-Onge, D.A.
1972: Sequence of glacial lakes in north-central Alberta; Geological Survey of Canada, Bulletin 213, 16 p.

Stuiver, M., Deevey, E.S., and Gralenski, L.J.
1960: Yale natural radiocarbon measurements V; Radiocarbon, 2, p. 49-61.

Sutherland Brown, A.
1968: Geology of the Queen Charlotte Islands, British Columbia; British Columbia Department of Mines and Petroleum Resources, Bulletin 54, 226 p.

Swan, D. and Luternauer, J.L.
1978: Mosaic of side scan sonar records, northern Kitimat Arm, B.C.; Geological Survey of Canada, Open File 579.

Tarnocai, C., Smith, S., and Hughes, O.L.
1985: Soil development on Quaternary deposits of various ages in the central Yukon Territory; in Current Research, Part A, Geological Survey of Canada, Paper 85-1A, p. 229-238.

Terasmae, J. and Hughes, O.L.
1966: Late-Wisconsinan chronology and history of vegetation in the Ogilvie Mountains, Yukon Territory, Canada; Palaeobotanist, v. 15, p. 235-242.

Terrain Analysis and Mapping Services Ltd.
1978: Geology and limnology of Kluane Lake, I. Preliminary assessment; Geological Survey of Canada, Open File 527, 51 p., 5 appendixes.

Terzaghi, K.
1956: Varieties of submarine slope failures; 8th Texas Conference on Soil Mechanics and Foundation Engineering, Proceedings, 41 p.

Thomson, R.E.
1981: Oceanography of the British Columbia coast; Canada Department of Fisheries and Oceans, Canadian Special Publication of Fisheries and Aquatic Sciences 56, 291 p.

Thorson, R.M. and Dixon, E.J., Jr.
1983: Alluvial history of the Porcupine River, Alaska: role of glacial-lake overflow from northwest Canada; Geological Society of America, Bulletin, v. 94, p. 576-589.

Tipper, H.W.
1963: Geology, Taseko Lakes, British Columbia; Geological Survey of Canada, Map 29-1963.
1971a: Glacial geomorphology and Pleistocene history of central British Columbia; Geological Survey of Canada, Bulletin 196, 89 p.
1971b: Multiple glaciation in central British Columbia; Canadian Journal of Earth Sciences, v. 8, p. 743-752.

Tipper, H.W., Woodsworth, G.J., and Gabrielse, H. (co-ordinators)
1981: Tectonic assemblage map of the Canadian Cordillera and adjacent parts of the United States of America; Geological Survey of Canada, Map 1505A.

Tyrrell, J.B.
1901: Yukon district; Geological Survey of Canada, Annual Report, new ser., v. 11 (1898), p. 36A-46A.
1919: Was there a "Cordilleran glacier" in British Columbia?; Journal of Geology, v. 27, p. 55-60.

Van Wormer, J.D., Davies, J., and Gedney, L.
1974: Seismicity and plate tectonics in south central Alaska; Seismological Society of America, Bulletin, v. 64, p. 1467-1475.

VanDine, D.F.
1980: Engineering geology and geotechnical study of Drynoch landslide, British Columbia; Geological Survey of Canada, Paper 79-31, 34 p.
1985: Debris flows and debris torrents in the southern Canadian Cordillera; Canadian Geotechnical Journal, v. 22, p. 44-68.

Vaníček, P. and Nagy, D.
1980: Report on the compilation of the map of vertical crustal movements in Canada; Canada Department of Energy, Mines and Resources, Earth Physics Branch, Open File 80-2, 59 p.
1981: On the compilation of the map of contemporary vertical crustal movements in Canada; Tectonophysics, v. 71, p. 75-86.

Varnes, D.J.
1978: Slope movement types and processes; in Landslides, Analysis and Control, R.L. Schuster and R.J. Krizek (ed.); National Research Council, Transportation Research Board, Special Report 176, p. 11-33.

Vernon, P. and Hughes, O.L.
1966: Surficial geology, Dawson, Larsen Creek, and Nash Creek map-areas, Yukon Territory (116B and 116C E1/2, 116A and 106D); Geological Survey of Canada, Bulletin 136, 25 p.

Vincent, J-S.
1989: Quaternary geology of the northern Interior Plains of Canada; in Chapter 2 of Quaternary Geology of Canada and Greenland, R.J. Fulton (ed.); Geological Survey of Canada, Geology of Canada, no. 1 (also Geological Society of America, The Geology of North America, v. K-1).

Wainman, N. and Mathewes, R.W.
1987: Forest history of the last 12 000 years based on plant macrofossil analysis of sediment from Marion Lake, southwestern British Columbia; Canadian Journal of Botany, v. 65, p. 2179-2187.

Waitt, R.B., Jr. and Thorson, R.M.
1983: The Cordilleran ice sheet in Washington, Idaho, and Montana; in Late-Quaternary Environments of the United States, Volume 1, the Late Pleistocene, S.C. Porter (ed.); University of Minnesota Press, Minneapolis, Minnesota, p. 53-70.

Walker, I.R. and Mathewes, R.W.
1987: Chironomidae (Diptera) and postglacial climate at Marion Lake, British Columbia, Canada; Quaternary Research, v. 27, p. 89-102.

Walton, A., Trautman, M.A., and Friend, J.P.
1961: Isotopes, Inc. radiocarbon measurements I; Radiocarbon, v. 3, p. 47-59.

Warner, B.G.
1984: Late Quaternary paleoecology of eastern Graham Island, Queen Charlotte Islands, British Columbia, Canada; unpublished PhD thesis, Simon Fraser University, Burnaby, British Columbia, 190 p.

Warner, B.G., Clague, J.J., and Mathewes, R.W.
1984: Geology and paleoecology of a mid-Wisconsin peat from the Queen Charlotte Islands, British Columbia, Canada; Quaternary Research, v. 21, p. 337-350.

Warner, B.G., Mathewes, R.W., and Clague, J.J.
1982: Ice-free conditions on the Queen Charlotte Islands, British Columbia, at the height of late Wisconsin glaciation; Science, v. 218, p. 675-677.

Westgate, J.A.
1977: Identification and significance of late Holocene tephra from Otter Creek, southern British Columbia, and localities in west-central Alberta; Canadian Journal of Earth Sciences, v. 14, p. 2593-2600.
1982: Discovery of a large-magnitude, late Pleistocene volcanic eruption in Alaska; Science, v. 218, p. 789-790.

Westgate, J.A. and Dreimanis, A.
1967: Volcanic ash layers of Recent age at Banff National Park, Alberta, Canada; Canadian Journal of Earth Sciences, v. 4, p. 155-162.

Westgate, J.A. and Fulton, R.J.
1975: Tephrostratigraphy of Olympia Interglacial sediments in south-central British Columbia, Canada; Canadian Journal of Earth Sciences, v. 12, p. 489-502.

Westgate, J.A., Fritz, P., Matthews, J.V., Jr., Kalas, L., Delorme, L.D., Green, R., and Aario, R.
1972: Geochronology and palaeoecology of mid-Wisconsin sediments in west-central Alberta, Canada (abstract); 24th International Geological Congress (Montreal, Quebec), Abstracts, p. 380.

Westgate, J.A., Hamilton, T.D., and Gorton, M.P.
1983: Old Crow tephra: a new late Pleistocene stratigraphic marker across north-central Alaska and western Yukon Territory; Quaternary Research, v. 19, p. 38-54.

Westgate, J.A., Smith, D.G.W., and Tomlinson, M.
1970: Late Quaternary tephra layers in southwestern Canada; in Early Man and Environments in Northwest North America, R.A. Smith and J.W. Smith (ed.); University of Calgary, Archaeological Association, Calgary, Alberta, p. 13-33.

Westgate, J.A., Walter, R.C., Pearce, G.W., and Gorton, M.P.
1985: Distribution, stratigraphy, petrochemistry, and palaeomagnetism of the late Pleistocene Old Crow tephra in Alaska and the Yukon; Canadian Journal of Earth Sciences, v. 22, p. 893-906.

Wheeler, A.O.
1931: Glacial change in the Canadian Cordillera, the 1931 expedition; Canadian Alpine Journal, v. 20, p. 120-137.

White, J.M. and Mathewes, R.W.
1982: Holocene vegetation and climatic change in the Peace River district, Canada; Canadian Journal of Earth Sciences, v. 19, p. 555-570.

1986: Postglacial vegetation and climatic change in the upper Peace River district, Alberta; Canadian Journal of Botany, v. 64, p. 2305-2318.

Wigen, S.O. and White, W.R.
1964: Tsunami of March 27-29, 1964, west coast of Canada; Canada Department of Mines and Technical Surveys, Ottawa, Ontario, 6 p.

Willis, B.
1898: Drift phenomena of Puget Sound; Geological Society of America, Bulletin, v. 9, p. 111-162.

Wilson, J.T., Falconer, G., Mathews, W.H., and Prest, V.K. (compilers)
1958: Glacial map of Canada; Geological Association of Canada, Toronto, Ontario. (Out of print)

Wilson, M.
1974: Fossil bison and artifacts from the Mona Lisa Site, Calgary, Alberta. Part I: stratigraphy and artifacts; Plains Anthropologist, v. 19, p. 34-45.

1981: Once upon a river: archaeology and geology of the Bow River valley at Calgary, Alberta, Canada; unpublished PhD thesis, University of Calgary, Calgary, Alberta, 464 p.

Wilson, M. and Churcher, C.S.
1978: Late Pleistocene *Camelops* from the Gallelli Pit, Calgary, Alberta: morphology and geologic setting; Canadian Journal of Earth Sciences, v. 15, p. 729-740.

Wintle, A.G. and Westgate, J.A.
1986: Thermoluminescence age of Old Crow tephra in Alaska; Geology, v.14, p. 594-597.

Yamaguchi, D.K.
1983: New tree-ring dates for recent eruptions of Mount St. Helens; Quaternary Research, v. 20, p. 246-250.

Yorath, C.J. and Hyndman, R.D.
1983: Subsidence and thermal history of Queen Charlotte Basin; Canadian Journal of Earth Sciences, v. 20, p. 135-159.

Young, I.
1981: Structure of the western margin of the Queen Charlotte Basin, British Columbia; unpublished MSc thesis, University of British Columbia, Vancouver, British Columbia 380 p.

Author's Addresses

J.J. Clague
Geological Survey of Canada
100 West Pender Street
Vancouver, British Coumbia
V6B 1R8

W.H. Mathews
Department of Geological Sciences
University of British Columbia
Vancouver, British Columbia
V6T 1W5

J.M. Ryder
Department of Geography
University of British Columbia
Vancouver, BritishColumbia
V6T 1W5

O.L. Hughes
Geological Survey of Canada
3303-33rd Street N.W.
Calgary, Alberta
T2L 2A7

N.W. Rutter
Department of Geology
University of Alberta
Edmonton, Alberta
T6G 2E3

L.E. Jackson, Jr.
Geological Survey of Canada
100 West Pender Street
Vancouver, British Columbia
V6B 1R8

J.V. Matthews, Jr.
Geological Survey of Canada
601 Booth Street
Ottawa, Ontario
K1A 0E8

G.M. MacDonald
Department of Geography
McMaster University
Hamilton, Ontario
L8S 4K1

Printed in Canada

Chapter 2

QUATERNARY GEOLOGY OF THE CANADIAN INTERIOR PLAINS

Summary

Introduction — *J-S. Vincent and R.W. Klassen*

Quaternary geology of the northern Canadian Interior Plains — *J-S. Vincent*

Quaternary geology of the southern Canadian Interior Plains — *R.W. Klassen*

Acknowledgments

References

Limit of Hungry Creek Glaciation (Late Wisconsinan) against the north flank of Knorr Range, Wernecke Mountains. Although generally not as clearly marked as in this area, the Canadian Plains contain the limits of several glaciations. At this site hummocky moraine deposited at the southern margin of a small lobe of Laurentide ice is separated from colluviated slopes to the south by ice marginal channels and glaciofluvial deposits. NAPL T4-100R

Chapter 2

QUATERNARY GEOLOGY OF THE CANADIAN INTERIOR PLAINS

compiled by
R.J. Fulton

SUMMARY

The Canadian Interior Plains are a low relief area lying between the Canadian Shield and the western Cordillera. They are underlain by flat-lying rocks which over much of the region consist of poorly consolidated shales, siltstones, and sandstones. The thickness and nature of the glacial deposits and the types of glacial processes that occurred are in part a function of the nature of the bedrock. Bedrock relief plays a role in large-scale physiographic features but small-scale elements are largely related to Quaternary glaciation.

This region contains the oldest record of Quaternary deposits in Canada with deposits at one site in the south and several in the north apparently reaching back to the beginning of the Epoch. Stratigraphy in the north records at least three full glaciations whereas that in the south records as many as four.

In the north the Banks Glaciation, which occurred before the start of the Brunhes Normal Polarity Chron (older than 790 ka), was the most extensive glaciation. This was followed by a Middle Pleistocene glaciation (the Thomsen) and by a Late Pleistocene glaciation (the Amundsen). Each of these recorded glaciations was less extensive than the preceding one. Nonglacial deposits, possibly of Plio-Pleistocene age have been found underlying the oldest till and interglacial deposits separate deposits of the three glaciations; all three nonglacial successions contain fossils indicating a climate warmer than present.

In the south there is evidence of glacial activity prior to deposition of sediments containing a paleomagnetic record believed to represent the Olduvai Normal Subchron of the Matuyama Reversed Chron which occurred between 1.7 and 1.67 Ma. Other glacial activity apparently occurred before deposition of Pearlette Ash, type 0 between 700 and 600 ka, but age and extent of the glacial units developed during these events is uncertain. The most extensive glaciation in the south resulted in deposition of the Labuma Till and occurred during the Middle Pleistocene. This was followed by two other Middle Pleistocene ice advances and several advances of Late Pleistocene age. Nonglacial sediments in the southern Canadian Interior Plains have yielded a large variety of vertebrate remains. These provide important data on absolute and relative ages of deposits and supply information pertinent to paleoenvironments.

The timing of Late Pleistocene glaciation is controversial in both the northern and southern Canadian Interior Plains. Similar successions of events have been proposed in both regions; one group of workers propose extensive glaciation early during the Wisconsinan and limited glaciation later and a second group propose the reverse. Both groups cite radiocarbon ages and stratigraphic and geomorphologic evidence to support their positions but in general draw their supporting evidence from different parts of the regions. In general this report supports extensive Early and limited Late Wisconsinan glaciation although reference is made to the growing support for limited Early Wisconsinan and extensive Late Wisconsinan glaciation.

INTRODUCTION
J-S. Vincent and R.W. Klassen

The Interior Plains area of Canada encompasses the region between the Canadian Shield and the western Cordillera, and includes the western Arctic islands (Fig. 2.1). It is distinguished by wide expanses of flat-lying sedimentary bedrock, much of which is poorly consolidated or even unconsolidated. As it is a region into which glaciers advanced, rather than originated, deposition of drift was more important in general than glacial erosion; this has resulted in thick Quaternary successions.

The Interior Plains region occupies the western margin of the Keewatin Sector of the Laurentide Ice Sheet (see Dyke et al., 1989). In fact it contains the limits of advance of many of the Quaternary glaciers, both Laurentide and Cordilleran. Apparently the ice sheet dynamics were different on the Plains than on the Canadian Shield, probably, because of the contrast in the composition of the glacial bed. It is recognized (Fisher et al., 1985) that in the Interior

---

Fulton, R.J. (compiler)
1989: Quaternary geology of the Canadian Interior Plains; Chapter 2 in Quaternary Geology of Canada and Greenland, R.J. Fulton (ed.); Geological Survey of Canada, Geology of Canada, no. 1 (also Geological Society of America, The Geology of North America, v. K-1).

Vincent, J-S. and Klassen, R.W.
1989: Introduction (Quaternary geology of the Canadian Interior Plains); in Chapter 2 of Quaternary Geology of Canada and Greenland, R.J. Fulton (ed.); Geological Survey of Canada, Geology of Canada, no. 1 (also Geological Society of America, The Geology of North America, v. K-1).

Plains, the beds are deformable and the glaciers were extremely mobile and had low gradient surface profiles. Morphological features such as hummocky moraines and ice thrusted sheets of bedrock and Quaternary materials are better developed there than in any other region of Canada.

The Interior Plains region is discussed in two parts: the northern part, which consists of the western Arctic islands and the Interior Plains north of 60°N and the part south of 60°N, which consists of the Interior Plains of Manitoba, Saskatchewan, Alberta, and northeastern British Columbia. The Quaternary deposits of the southern part of the area have been studied at a variety of levels of detail; the surficial geology of Mackenzie Valley and Banks and Victoria islands has been covered by several inventory studies; the Quaternary of the boreal forest area of northern Alberta however is virtually unknown. Consequently the two areas emphasized in this chapter are separated by "terra incognita". Due to constraints of space, the writers could not always provide all of the abundant background details and supporting evidence for statements made, nor could conflicting theories be discussed adequately. In addition, it was necessary to be selective of the facets of the Quaternary treated, and the text may be superficial in the level of treatment of many topics. In general, emphasis is on features that are characteristic of, important to, or best displayed in this region, or are important to an overall knowledge of the Quaternary. Many features that are well displayed in other parts of the country or that are comprehensively described in other chapters are ignored or given only cursory treatment here. In this chapter, emphasis is given to stratigraphy, chronology, older glacial deposits whose known occurrences are largely confined to this area, the vertebrate record, drainage development, topographic forms developed on weak plastic bedrock by ice moving upslope, and the relationships between mountain (Cordilleran) and continental (Laurentide) glaciers.

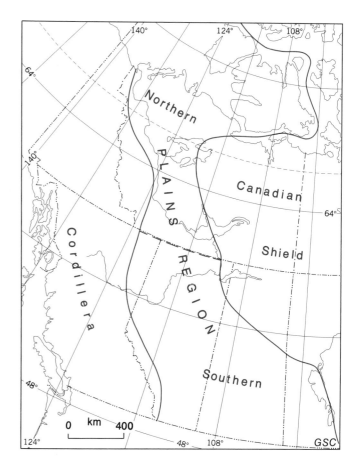

Figure 2.1 Regional location map.

QUATERNARY GEOLOGY OF THE NORTHERN CANADIAN INTERIOR PLAINS

J-S. Vincent

The northern Canadian Interior Plains region includes a large "mainland" area as well as Banks and Victoria islands (Fig. 2.2a). This area lies north of 60°N, west of the Canadian Shield, and east of the western limit reached by continental ice (Fig. 2.2b).

Continental glaciers, which likely originated west of Hudson Bay repeatedly covered the northern Interior Plains during the Quaternary. An exceptional record of the various ice advances and nonglacial intervals is documented by sediments and landforms of various ages lying on the surface and by glacial, marine, and terrestrial deposits in numerous sections.

The Quaternary deposits and history of the area were initially synthesized by Craig and Fyles (1960, 1965), Bird (1967), Prest (1970), and Prest et al. (1968). They mainly summarized the field work of Craig (1960, 1965) and Mackay (1958, 1963) in the District of Mackenzie, and of Fyles (1962, 1963) on Banks and Victoria islands. Map 1704A, which accompanies this volume, shows the areas covered by Quaternary or surficial geology maps. Most of this work has been done since the early 1970s, in response to needs related to oil and gas exploration. Recent major regional syntheses include those of Catto (1986b), Hughes (1972, 1987), Hughes et al. (1981) and Rampton (1988) for the District of Mackenzie; Rampton (1982) for the Yukon Coastal Plain; and Vincent (1983) for Banks Island. A proposed correlation for the deposits and events in northwestern North America, is shown in Table 2.1.

Vincent, J-S.
1989: Quaternary geology of the northern Canadian Interior Plains; in Chapter 2 of Quaternary Geology of Canada and Greenland, R.J. Fulton (ed.); Geological Survey of Canada, Geology of Canada, no. 1 (also Geological Society of America, The Geology of North America, v. K-1).

PHYSICAL SETTING

Bedrock geology

The main geological provinces of the northern Interior Plains are shown on an inset in Figure 2.3. The platform areas are generally underlain by flat-lying to gently inclined sedimentary rocks (carbonates, shales, and sandstones) of Paleozoic and Mesozoic age. Rocks in the Minto Uplift Province are Proterozoic carbonates, shales, siltstones, and sandstones on the mainland, and carbonates, shales, sandstones, and diabase on Victoria and Banks islands. In the Mackenzie Delta area and on central and western Banks Island, unconsolidated to poorly consolidated terrestrial and marine silts, sands, and gravel of Cenozoic age are present. The widespread fossil-bearing alluvial sands and gravels of the Beaufort Formation, in the Arctic Coastal Plain Geological Province, are Miocene in age (Hills et al., 1974).

Relief and drainage

Topography of the region is shown in Figure 2.2b. Typically the area is low and flat. On the mainland, the ground rises east of Mackenzie River towards the Canadian Shield and high areas such as Cameron Hills, Horn Plateau, Franklin

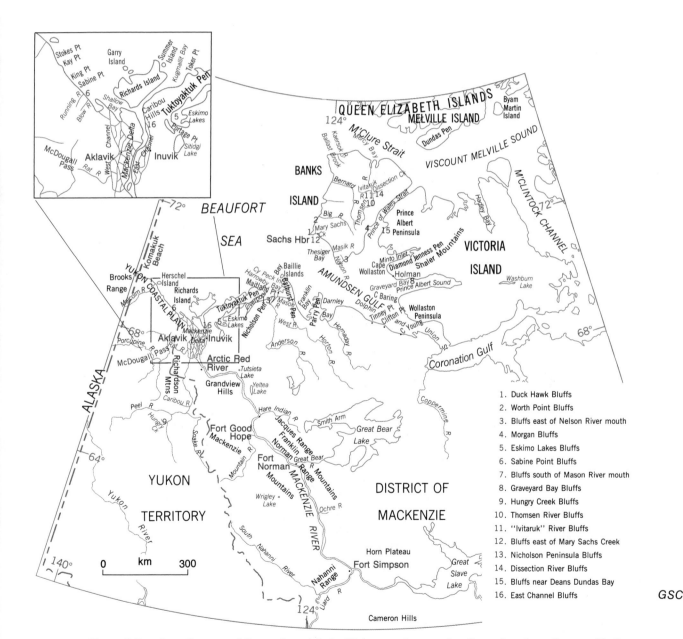

Figure 2.2a Location map of the northern Interior Plains; numbers are locations of sections discussed in the text.

Mountains, and Brock Upland south of Melville Hills dominate the landscape. West of Mackenzie River the surface rises rapidly to the Mackenzie and Richardson mountains. Mackenzie River valley between the Franklin and Mackenzie mountains is a striking feature of the relief. Some time before the last glacial advance, Mackenzie River flowed through the Franklin Mountains, in a deep trench cut in bedrock along the present Hare Indian River and between the Jacques and Norman ranges (Mackay and Mathews, 1973). The physiography clearly shows how ice moving westward and downhill from the District of Keewatin would be deflected north as it reached the Mackenzie River Basin and the mountain barrier to the west. Most of the mainland part of the northern Interior Plains region is drained by Mackenzie River and its tributaries, but in the northeast Anderson, Horton, and Hornaday rivers drain areas north of Great Bear Lake. The Mackenzie Delta is one of the largest in the world with a total length of about 200 km and a width of 65 km. Its surface is a complex network of thermokarst lakes and shifting anastomosing channels.

On Victoria Island the flat landscape is broken by the deeply dissected Shaler Mountains in the northwest, and on Banks Island by the southern uplands and the northeastern Devonian plateau. Spectacular cliffs are present where the coast intersects the high areas. The major elements of relief, which controlled glacial flow, are the interisland channels (Amundsen Gulf — Dolphin and Union Strait, Prince of

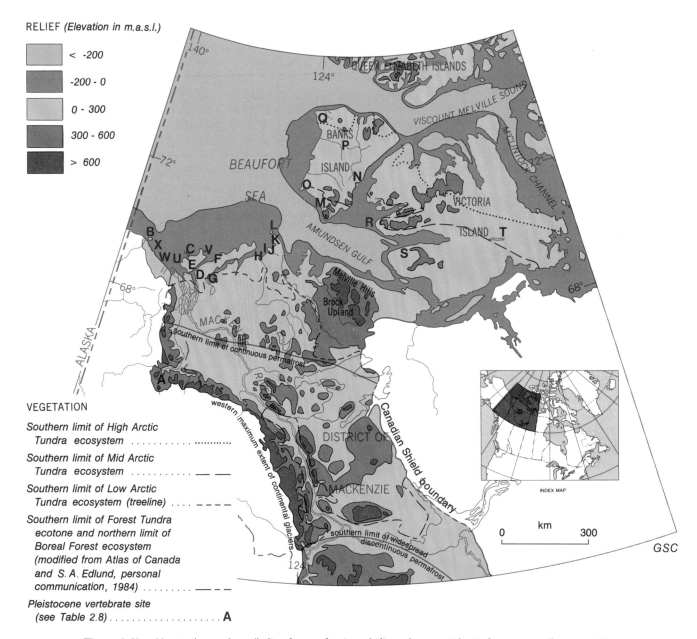

Figure 2.2b Vegetation regions, limits of permafrost, and sites where vertebrate fauna were discovered (see Table 2.8).

Table 2.1. Correlation of Quaternary deposits and events in the area adjacent to the Beaufort Sea[1] (compiled by J-S. Vincent from data provided by the researchers indicated)

General Chronostratigraphy (age in ka)		Oxygen Isotope Stage	Yukon Cordilleran Ice Sheets (O.L. Hughes)	Yukon Basins (O.L. Hughes, J.V. Matthews, Jr., N.W. Rutter and C. Schweger)	Brooks Range and Basins to South (T.D. Hamilton)	Alaskan Arctic Coastal Plain (J.K. Brigham-Grette, L.D. Carter, D.A. Dinter, D.M. Hopkins, S.E. Rawlinson, and P.A. Smith)	Yukon Coastal Plain and Mackenzie Delta Valley (N. Catto, O.L. Hughes, V.N. Rampton and J-S. Vincent)	Mackenzie Delta Offshore (P.R. Hill and S.M. Blasco)	Western Arctic Islands (J-S. Vincent)
LATE PLEISTOCENE	LATE WISCONSINAN STAGE	2	McCONNELL GLACIATION = MACAULEY GLACIATION	Upper glacio-lacustrine (12-30 ka - 14C) Interstadial fluctuation? Hanging Lake interval (at about 18-20 ka - 14C)	WALKER LAKE GLACIATION (13-24 ka - 14C)	Put River outwash and alluvium (9-15 ka - 14C) UNIT A marine wedge on middle and outer shelf (9-15 ka - 14C) Ikpikpuk sand sea Canning gravel	STIDJI STADE = Tutsieta Lake phase (13 ka - 14C) (incl. AKLAVIK FM, PARSONS LAKE MEMBER, ESKIMO LAKES MEMBER, SITIDJI POINT SANDS, SITIDJI MEMBER) HUNGRY CREEK GLACIATION (16-25 ka - 14C)	Sea level drop or standstill-Mackenzie Delta progradation in west and outwash plain in east (8.8 to 21.6 ka - 14C) Sea level rise from mid or early(?) Wisconsinan minimum Delta progradation in west (21.6 to 27.4 ka - 14C)	AMUNDSEN GLACIATION (RUSSELL STADE) = PRINCE OF WALES FM (incl. SCHUYTER POINT SEA SEDS - 12.6 to 9 ka - 14C, and PASSAGE POINT SEDIMENTS)
	――23―― MIDDLE WISCONSINAN	3	THOM CREEK INTERSTADIAL (28 ka - 14C) BOUTELLIER NONGLACIAL INTERVAL (29, >37 ka - 14C)	Alternating warm and cold intervals within stage 5d to 3	Unnamed paleosol (24-34 ka - 14C)	Paleosols in Put River outwash and alluvium (24 and 43 ka - 14C) Ugnuravik sand; marine UNIT B on middle and outer shelf	Nonglacial beds (33.8 and 36.9 ka - 14C)	Outwash plain in east, off Tuktoyaktuk Peninsula	Unnamed interstadial (>41 and >49 ka - 14C and 71.8 - Th/U)
	――64―― EARLY WISCONSINAN	5d-4	Sheep Creek Tephra (>42 ka - 14C; 73, 78 ka - Th/U)		ITKILLIK GLACIATION (Chebanika advance) Forest beds (>55 ka - 14C) Old Crow Tephra? ITKILLIK GLACIATION (maximum advance)	SIMPSONIAN TRANSGRESSION = FLAXMAN MEMBER of GUBIK FM (75 ka - Tl) = Cross Island Unit on inner shelf? = mid shelf deltas = UNIT C marine wedge (?)	TOKER POINT STADE = Buckland Glaciation[2] = Franklin Bay Stade = Deposition deglaciation (>35 and >39 ka - 14C) (incl. Cape Dalhousie Sands, Garry Island, Turnabout and Toker Point members, North Star Outwash, Malloch Till) Deformed ground ice Sabine grey member?		AMUNDSEN GLACIATION (M'CLURE STADE) = WALKER BAY STADE = MEEK POINT SEA SEDIMENTS and EAST COAST SEA SEDIMENTS - 46 ka - Th/U and 24.7 ka - 14C; incl. CARPENTER, BAR HARBOUR, MERCY, SACHS and JESSE tills; and PRE-AMUNDSEN SEA SEDIMENTS - 92.4 ka 14C and >37 ka 14C)
	――115―― SANGAMONIAN STAGE	5e		Koy-Yukon Interglaciation	Bettes gravel	PELUKIAN TRANSGRESSION (125 ka - Tl) = Walakpa Member of GUBIK FM = McGUIRE ISLAND UNIT on inner shelf = UNIT E marine wedge Ugnuravik gravel	LIVERPOOL BAY INTER-GLACIATION (incl. Mailland Sea Sediments) (>36; >38 & >39 ka - 14C) = IKPISUGYUK FORMATION Sabine oxidized member? Peel fluvial deposits?		CAPE COLLINSON INTER-GLACIATION = CAPE COLLINSON FM (>61 ka - 14C and 69.9 ka - Th/U)
	――128―― MIDDLE PLEISTOCENE		Old Crow Tephra MIRROR CREEK GLACIATION = REID GLACIATION	Old Crow Tephra Interlacustrine alluvium with multiple paleosols	Old Crow Tephra? SAGAVANIRKTOK RIVER GLACIATION Long interglacial	WAINWRIGHTIAN TRANSGRESSION (210 ka - Tl) = Karmuk Member of GUBIK FM = Beaded Lower Complex = KOTZEBUAN TRANSGRESSION = LEFFINGWELL LAGOON UNIT on inner shelf (?) = UNIT I marine wedge (?)	MASON RIVER GLACIATION (incl. Harrowby Sea Sediments, Stanton Sea Sediments, Thinly Bedded Lower Complex, Nicholson Peninsula Sea Sediments, Baillie Clay?) = KITTIGAZUIT FM. = KIDLUIT FORMATION Hooper clay Peel gravels Kendall sediments		THOMSEN GLACIATION = NELSON RIVER FM (incl. BIG SEA SEDIMENTS >30 ka - 14C, KELLETT, BAKER and KANGE tills, and PRE-THOMSEN SEA SEDIMENTS 104.6 ka - Th/U) MORGAN BLUFFS INTER-GLACIATION = MORGAN BLUFFS FM (>200 ka - Th/U) Brunhes/Matuyama boundary within unit
	――730―― EARLY PLEISTOCENE		KLAZA GLACIATION Fort Selkirk Tephra (0.94 Ma-KAr) NANSEN GLACIATION Klondike gravels Flat Creek beds White Channel gravels	Little Timber Tephra (>1.2 Ma-Ft) Lower lacustrine (in Old Crow Basin) Sands containing permafrost structures (in Bluefish Basin)	ANAKTUVUK RIVER GLACIATION High terraces	FISHCREEKIAN TRANSGRESSION = Tuapaktushak Member of GUBIK FM (-?) BIGBENDIAN TRANSGRESSION = Big Bend Member of GUBIK FM (+1.7-2.2 Ma - Aa, mammals) ANVILIAN TRANSGRESSION = Nulavik Member of GUBIK FM OLDUVAI GEOMAGNETIC EVENT?			BANKS GLACIATION = DUCK HAWK BLUFFS FM (incl. POST BANKS SEA SEDIMENTS, BERNARD PLATEAU SEDIMENTS, DURHAM HEIGHTS tills and PRE BANKS SEA SEDIMENTS - magnetically reversed in Duck Hawk Bluffs Old erratics
LATE TERTIARY				Paleosol with extinct Larix minuta type, Picea and Pinus?	Gunsight Mountain erratics	COLVILLIAN TRANSGRESSION (<3.5 Ma - Pacific molluscs) = BERINGIAN TRANSGRESSION? = Nulavik Member of GUBIK FM Erratics in Kuparuk gravel NUWOK FM. = Papigak Clay			WORTH POINT FORMATION (preglacial) - Late Pliocene? BEAUFORT FORMATION - Miocene to Pliocene

[1] This table updates the one initially published in Heginbottom and Vincent (1986) by the participants of the Canadian-American workshop on the Correlation of Quaternary deposits and events around the margin of the Beaufort Sea.
  a) Names in upper case letters are published and in the Alaskan column, are formal names that are published and/or have the approval of the USGS Geologic Names Committee.
  b) Names in lower case are informal and in the Alaskan column, if formal, have not yet been published and do not have the approval of the USGS Geologic Names Committee.
  c) It should be stressed that the correlation chart is a working document. Readers will note the lack of consistency in the nature of the units discussed. Few formally defined names of lithological units are used. Geologic-climate units (glaciations, interglaciations, stades, and interstades) are used even though these have been abandoned by the North American Commission on Stratigraphic Nomenclature, and are now recognized only as informal units.

[2] According to Hughes, Buckland Glaciation is correlative with Hungry Creek Glaciation.

Aa age estimate from amino acid analyses
14C age estimate from radiocarbon analyses
Ft age estimate from fission track analyses
Tl age estimate from thermoluminescence analyses
Th/U age estimate from uranium-thorium analyses
KAr age estimate from potassium-argon analyses
(+) magnetically normal
(−) magnetically reversed

Wales Strait, and Viscount Melville Sound — M'Clure Strait). The origin and age of the channels have been the subject of much discussion. Carsola (1954) attributed the development of at least Amundsen Gulf to glacial overdeepening. Fortier and Morley (1956) and Pelletier (1966) suggested that the network of interisland channels is a submerged Tertiary fluvial system. Kerr (1980), on the other hand, showed that at least M'Clure Strait and Prince of Wales Strait opened through rifting in Miocene time. The rifting postdates deposition of the Beaufort Formation as paleocurrent measurements indicate that streams on the Arctic Coastal Plain flowed westward across areas that are now the site of the channels (L.V. Hills, University of Calgary, personal communication, 1984). It predates the Early Pleistocene, as deposits of a marine transgression of that age are found along central Prince of Wales Strait on Banks Island. Although rifting is undoubtedly the main agent, possibly rivers and certainly glaciers enlarged the channels. The general physiography during the late Tertiary, probably resembled that of today.

Climate, permafrost, and vegetation

The climate of the northern Interior Plains is polar continental. Winters are long and cold and summers short. Precipitation is low. Table 2.2 shows how temperatures and precipitation decrease northward. Winter temperatures do not vary significantly but summers are distinctly cooler in the north.

Permafrost underlies most of the area. The southern limits of continuous and of widespread discontinuous permafrost are shown in Figure 2.2b. Permafrost is generally less than 150 m thick south of the Mackenzie River Delta, but it attains thicknesses of more than 600 m on northern Richards Island and Tuktoyaktuk Peninsula (Judge, 1986). Offshore on the Beaufort Sea Shelf, permafrost thicknesses up to 780 m have been measured (Weaver and Stewart, 1982). Some of the thickest permafrost (up to 1000 m — A.S. Judge, Geological Survey of Canada, personal communication, 1984) is present on western Banks Island, probably because deep permafrost growth has not been inhibited by a cover of glacial ice or sea water since the Early Pleistocene. Large bodies of massive ground ice are widespread, particularly south of the Beaufort Sea, on eastern Banks Island, and on Prince Albert Peninsula of Victoria Island. On the mainland (Mackay, 1986), as well as on Banks and Victoria islands (Lorrain and Demeur, 1985), some of these ice bodies may be buried Wisconsinan glacier ice but most probably developed as segregated ice after deglaciation (Rampton, 1974a).

As shown in Figure 2.2b, vegetation within the region varies from dense boreal forest in the south to sparsely vegetated High Arctic tundra in the north. The northern limit of trees was as much as 70 km north of its present position during the early Holocene (Spear, 1983), and reached at least southern Banks Island during the Early Pleistocene (Kuc, 1974; Vincent, 1984, Matthews et al., 1986).

NATURE AND DISTRIBUTION OF QUATERNARY DEPOSITS

The nature and distribution of Quaternary deposits will be portrayed on the map of Surficial Materials of Canada (R.J. Fulton, Geological Survey of Canada, in preparation), which forms the basis for the following discussion. Figures 2.4a,b, and c, which in many places update the Glacial Map of Canada (Prest et al., 1968), and which were compiled from recent surficial geology maps (Map 1704A), show the distribution of the main glacial landforms of the northern Interior Plains.

On the mainland of the region, Quaternary deposits vary somewhat with topographic position. In very low areas along Mackenzie and Liard rivers, floodplain, terrace and delta deposits (including the Mackenzie Delta), and widespread flat-lying expanses of glacial lake silts and sands are present. Glacial lake deposits are also found near Great Slave and Great Bear lakes. Somewhat higher, in the low-lying interfluve areas, peat-mantled till plains, commonly fluted and drumlinized, are extensive. On higher ground, hummocky moraine with dispersed glaciofluvial deposits covers much of the plateaus and hills. Such regions are mostly on the plateaus south of Mackenzie River and east of Liard River, on Horn Plateau, in the area south and southeast of the Mackenzie Delta, and over large areas between Great Bear Lake and Amundsen Gulf. In the highest areas, including the glaciated, mountainous eastern edge of the Cordillera, Franklin Mountains, and Brock Upland, Quaternary materials are typically thin and surfaces are mantled with felsenmeer or bedrock-derived colluvium. Alluvium, deltaic, and eolian deposits cover much of the Arctic Coastal Plain. The hummocky nature of the terrain in the Coastal Plain results mainly from thermokarst activity rather than from stagnation of glaciers (Rampton, 1988).

Much of the low-lying part of Banks Island is a nondescript till plain broken by widespread trains of outwash, particularly in Big River and Bernard River valleys. The eastern part of Banks Island contains numerous lakes of glacial (kettle) and thermokarst origin. The higher extreme southern and northern parts display large areas of dissected bedrock, and Quaternary materials are sparse. Well developed glacial landforms are rare because they never developed, or were modified by periglacial processes over a long period of time, or were obliterated by waves in glacial lakes and seas. Apart from the Shaler Mountains, Victoria Island is, like Banks Island, low-lying. Much of the terrain is drumlinized till plain or ice scoured and felsenmeer-covered bedrock with only a thin mantle of glacial deposits. Smooth and

Table 2.2. Temperature and precipitation for five selected stations in northwestern Canada (1951-1980)

Station	Annual	January	July	Average
	Mean Daily Temperature (°C)			Annual Precipitation (mm)
Sachs Harbour (71°59'N, 125°17'W)	-14.1	-30.4	5.9	114.1
Holman (70°44'N, 117°47'W)	-12.2	-29.2	7.4	177.7
Inuvik (68°18'N, 133°29'W)	-9.8	-29.6	13.6	266.1
Fort Norman (64°53'N, 125°34'W)	-6.3	-28.6	16.0	324.9
Fort Simpson (61°52'N, 121°21'W)	-4.2	-28.2	16.6	355.1

Source: Atmospheric Environment Service, 1982

hummocky till plains, with well developed ice contact landforms and outwash trains, cover much of Wollaston, southern Diamond Jenness, and Prince Albert peninsulas. The postglacial sea inundated much of the southern and eastern parts of the island, leaving little debris but reworking the glacial deposits into spectacular flights of raised beaches.

Over this large area the matrix texture of the tills varies markedly but generally it is finer than in the sandier tills of the Canadian Shield (Fig. 2.5; cf. Fig. 3.47 of Vincent, 1989). The tills are not very stony, contain much carbonate, and reflect closely the substratum over which the glaciers flowed. On Banks Island, for example, Bernard Till is dark with a fine matrix and contains only few clasts, much montmorillonite, and little carbonate, because the glacier flowed over broad areas of Cretaceous and early Tertiary fine grained marine sediments. A significant property of the tills is their content of shield erratics. In areas more than 800 km away from the Canadian Shield, such as Banks Island or the Yukon Coastal Plain, tills still have at least 1% shield clasts. This property is used to determine the extent of Laurentide glaciation — particularly along the eastern Cordilleran margin.

QUATERNARY HISTORY
Introduction

Sediments in the northern Interior Plains and western Arctic Islands contain a long Quaternary record. The sequence of events in the area has been derived by mapping the areal distribution of deposits associated with distinct glacial and marine events, and by stratigraphic studies.

Glacier and marine submergence limits of distinct ages have been mapped throughout the area (Fig. 2.4). For example, on Banks Island, ice limits of three separate glaciations have been mapped, along with limits of associated glacial lakes and marine transgressions (Vincent, 1982, 1983). The

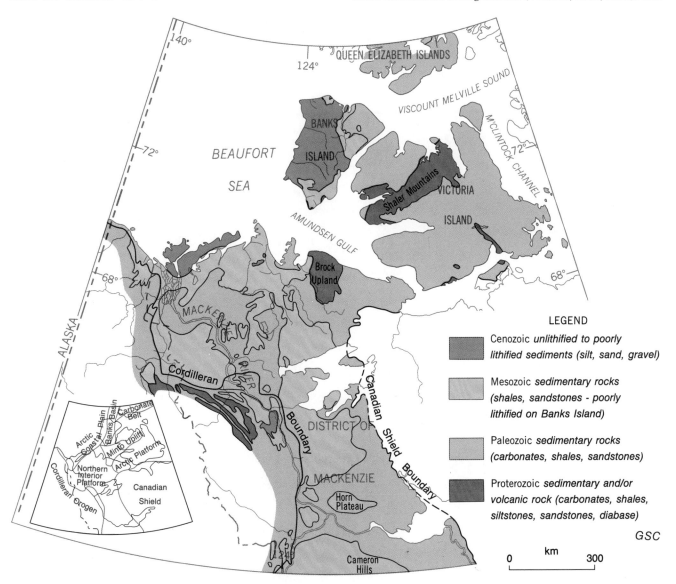

Figure 2.3 Geological provinces and geology of the northern Interior Plains (after Douglas, 1969; Fraser et al., 1978; Miall, 1979; Tipper et al., 1981).

Figure 2.4 Glacial features map of the northern Interior Plains.

a. Lower Mackenzie River area;

b. Banks and Victoria islands and Brock Upland area;

c. Upper Mackenzie River area.

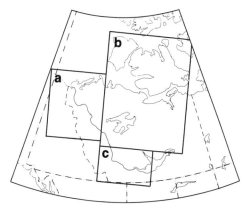

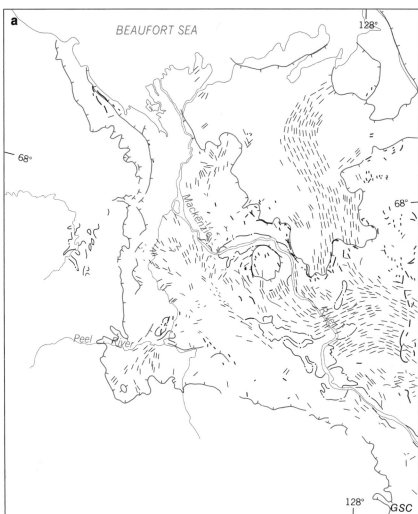

Glacial limit
Striae
Drumlins or fluting
End moraine
Esker
Glacial lake limit

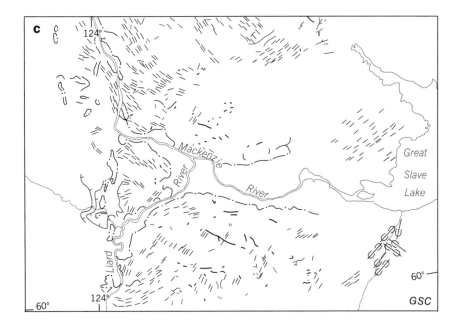

QUATERNARY GEOLOGY — CANADIAN INTERIOR PLAINS

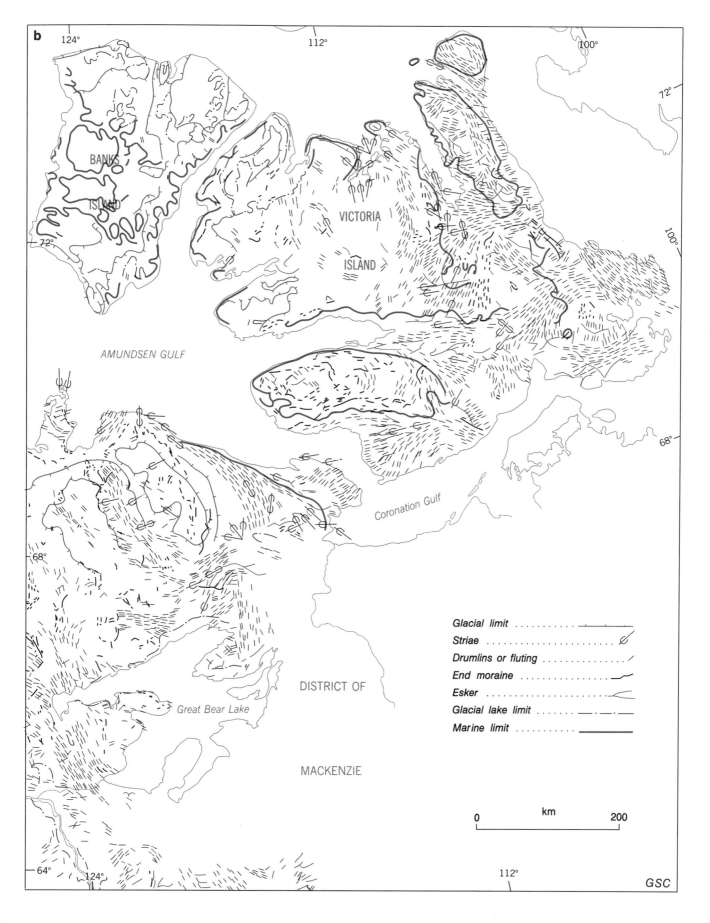

establishment of the lateral stratigraphic relationship between the till sheets and the marine and lacustrine sediments and features allows, in turn, the determination of the relative sequence of events.

Figures 2.6a-i describe some of the main Quaternary sections. These are found chiefly along the larger rivers, the Beaufort Sea coast from Herschel Island to Bathurst Peninsula, and along the coast of Banks Island. The last area provides by far the longest and fullest record, indicating terrestrial preglacial, interglacial, and glacial events. In many places a suite of sediments bears witness to transgressive marine events resulting from the buildup of ice, then glacial overlap followed by regressive marine events related to isostatic recovery during and following ice retreat. The extensive record, also present on the Arctic Coastal Plain of the mainland, is difficult to unravel for three reasons. Firstly, mainly nonglacial sediment suites are stacked on each other without the intervening glacial or marine sediments that elsewhere serve as marker units. Secondly, widespread ice thrusting along the mainland Beaufort Sea coast (Mackay, 1956 and 1959) has made the correct relative position of sediments difficult to establish. Finally, widespread thermokarst activity and the growth of extensive bodies of ground ice have strongly disturbed the sediments.

Chronological control. Even though distinct events can be identified and their relative timing established, their absolute age is rarely known. However, hundreds of radiocarbon ages on faunal and floral remains are now available and help to establish some of the Late Wisconsinan and Holocene chronology. Pertinent dates mentioned in the

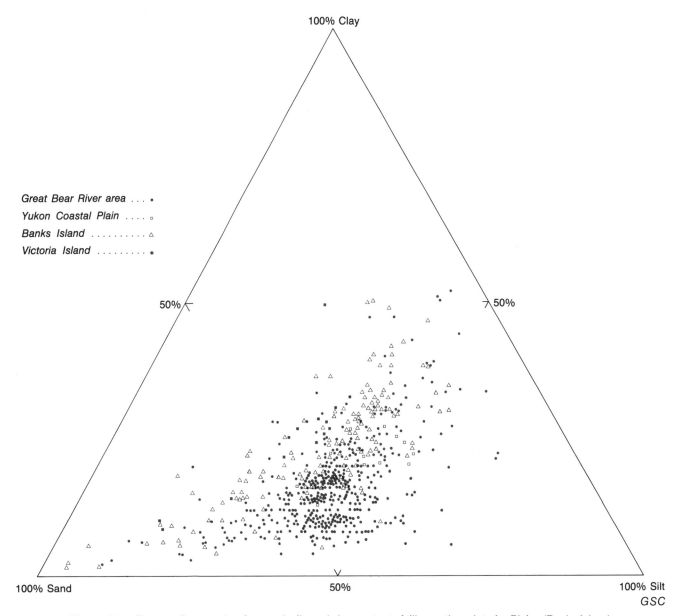

Figure 2.5 Ternary diagram showing sand, silt, and clay content of tills, northern Interior Plains (Banks Island — Vincent, 1983; Victoria Island — D.A. Hodgson, D.R. Sharpe, and J-S. Vincent, unpublished; Yukon Coastal Plain — J-S. Vincent, unpublished; Great Bear River — Savigny, 1989).

figures and in the text are listed in Table 2.3 and the sites are located on Figure 2.7. For older events, thorium/uranium ages on shells and wood have provided some chronological control. Th/U ages have proved useful in the area since permafrost limits leaching and contamination by circulating water (Causse and Vincent, 1985). In addition, numerous amino acid racemization analyses on shell and wood remains have enabled correlation of Quaternary units, mainly within local areas. Finally, paleomagnetism studies on Banks Island have indicated that part of the sediment suite there is in the Matuyama Reversed-Polarity Chron (>790 ka), because some units are magnetically reversed (Vincent et al., 1984; Vincent and Barendregt, 1987).

Nomenclature. Outside of Banks Island and Tuktoyaktuk coastlands, only a few stratigraphic units have been named. On the other hand, names for the main geological, particularly glacial, events have been proposed. Even though there is much inconsistency, the established terminology is used here unmodified. Most names are listed in Table 2.4, which summarizes Quaternary events on Banks Island, and in Table 2.5, which presents a tentative correlation between various parts of the northern Interior Plains.

Glacial limits. In many areas, various authors have mapped what they considered to be the ice sheet limits during distinct glacial stages or stages; these limits are shown in Figures 2.4a-c without any age connotation. Prest (1984)

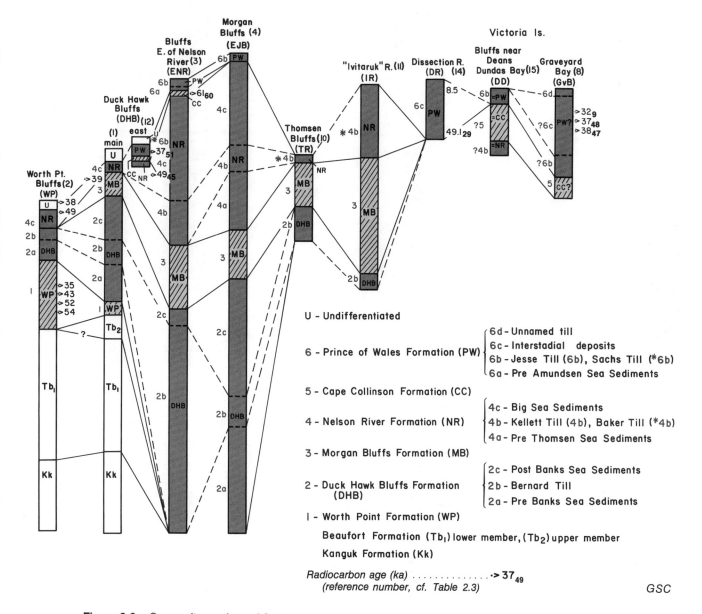

Figure 2.6 Composite sections of Quaternary sediments. Numbers in brackets refer to section location numbers which are shown in Figure 2.2a. Light red shading denotes nonglacial units, darker red shading denotes glacial units.

CHAPTER 2

Figure 2.6. Continued.

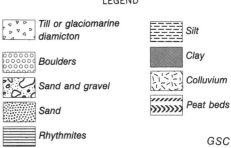

LEGEND

- Till or glaciomarine diamicton
- Boulders
- Sand and gravel
- Sand
- Rhythmites
- Silt
- Clay
- Colluvium
- Peat beds

GSC

Figure 2.6. Continued.

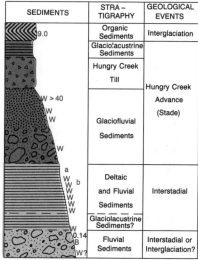

a. Duck Hawk Bluffs, Banks Island (1) (modified from Vincent et al., 1983).

b. Bluffs east of mouth of Nelson River, Banks Island (3) (modified from Vincent, 1983).

c. Morgan Bluffs, Banks Island (4) (modified from Vincent, 1983).

d. Outer Fingers of Eskimo Lakes, District of Mackenzie (5) (V.N. Rampton, Terrain Analysis and Mapping Services Ltd., Carp, Ontario, personal communication, 1984).

e. Sabine Point Bluffs, Yukon Coastal Plain (6) (V.N. Rampton and J-S. Vincent, unpublished).

f. Bluffs south of Mason River mouth, Bathurst Peninsula, District of Mackenzie (7) (V.N. Rampton and J-S. Vincent, unpublished).

g. River bluffs northwest of Graveyard Bay, Victoria Island (8) (J-S. Vincent, unpublished).

h. Hungry Creek composite section, District of Mackenzie (9) (Hughes, et al., 1981).

i. Composite sections and probable correlation of the main exposures on Banks Island and Victoria Island (modified from Matthews et al., 1986). The number in brackets, near the section name, refers to the section locations shown in Figure 2.2a.

Table 2.3. Significant radiocarbon dates from the northern Interior Plains of Canada

Reference number (cf. Fig. 2.7)	Laboratory number	Date (years BP)	Locality	Reference or Collector	Material	Significance
1	B-6059	27 380 ± 470	Tarsuit N44, Beaufort Sea	Hill et al. (1985)	peat	Continental shelf exposed to -141 m about 27 ka
2	B-2676	21 620 ± 630	Uviluk P-66, Beaufort Sea	Hill et al. (1985)	wood	Continental shelf exposed to -100 m about 22 ka
3	B-6280	28 360 ± 630	Uviluk P-66, Beaufort Sea	Hill et al. (1985)	wood	Continental shelf exposed to -70 m about 28 ka
4	GSC-42	9 710 ± 150	East of Prince Albert Sound, Victoria Island	Dyck and Fyles, 1963	marine shells	Minimum age for deglaciation and marine inundation in area
5	GSC-54	6 900 ± 110	Mackenzie River Delta	Dyck and Fyles, 1963; Mackay, 1963	wood	Sea level between -33 m and -18 m
6	GSC-151-2	>51 100	King Point, Yukon	Dyck et al., 1966; Rampton, 1982	wood and peat	Minimum age for pre-Buckland Glaciation sediments
7	GSC-278	10 340 ± 150	Cape Phipps, Melville Island	Lowdon et al., 1967; Hodgson and Vincent, 1984b	marine shells	"Youngest" maximum age on Melville Island for Viscount Melville Sound Ice Shelf and deposition of Winter Harbour Till
8	GSC-329	>50 900	Portage Point, Eskimo Lakes District of Mackenzie	Dyck et al., 1966; Rampton, 1988	peat	Minimum age for interglacial-like sediments beneath Toker Point Stade sediments (Sangamonian?)
9	GSC-388	>32 400	Graveyard Bay, Victoria Island	Blake, 1974; Vincent, 1984	undra plants	Interstadial deposits (between Early and Late Wisconsinan tills?)
10	GSC-481	17 860 ± 260	Ibyuk Pingo, Tuktoyaktuk District of Mackenzie	Lowdon and Blake, 1973; Rampton, 1988	peat	Minimum age for Toker Point Stade (Early Wisconsinan?)
11	GSC-562	>35 000	Garry Island, District of Mackenzie	Lowdon et al., 1971; Rampton, 1988	marine shells	Minimum age for marine transgression after the Toker Point Stade (Early Wisconsinan)
12	GSC-690	>37 000	South of Kendall Island, District of Mackenzie	Lowdon and Blake, 1968; Rampton, 1988	marine shells	Minimum age for marine transgression after the Toker Point Stade (Early Wisconsinan?)
13	GSC-727	>33 000	Parry Point, Melville Island	Lowdon and Blake, 1968; Hodgson et al., 1984	marine shells	Minimum age for marine transgression after deposition of Bolduc Till (Early Wisconsinan?)
14	GSC-787	42 400 ± 1900	Cape Phipps, Melville Island	Lowdon and Blake, 1968; Hodgson et al., 1984	marine shells	Minimum age for marine transgression after deposition of Bolduc Till (Early Wisconsinan?)
15	GSC-1088	>41 000	Kaersok River, Banks Island	Vincent, 1983, 1984	moss	Minimum age for Prince Alfred Lobe (M'Clure Stade of Amundsen Glaciation), Bar Harbour Till and lakes Ballast and Ivitaruk (Early Wisconsinan?)
16	GSC-1100	>41 000	Horton River, District of Mackenzie	Lowdon et al., 1971	peat	Minimum age for interglacial-like sediments predating the Amundsen Gulf ice lobe
17	GSC-1139	10 800 ± 150	Erly Lake, District of Mackenzie	Lowdon et al., 1971	moss	Minimum age for Tutsieta Lake Stade deglaciation outh of Parry Peninsula
18	GSC-1262	22 400 ± 240	Stokes Point, Yukon	Lowdon and Blake, 1976; Rampton, 1982	peat	Minimum age for Buckland Glaciation (Early Wisconsinan?)
19	GSC-1281	>36 000	Turnabout Point, Liverpool Bay District of Mackenzie	Lowdon and Blake, 1978; Rampton, 1988	wood	Minimum age for the upper grey sand sequences in area
20	GSC-1478	>19 000	Worth Point, Banks Island	Lowdon and Blake, 1973; Vincent, 1983, 1984	marine shells	"Oldest" minimum age for marine transgression before advance of Viscount Melville Sound Ice Shelf
21	GSC-1707	12 600 ± 140	Peel Point, Victoria Island	Lowdon and Blake, 1976; Hodgson et al., 1984	marine shells	Age of Tutsieta Lake Stade at its maximum stand in Sitidgi Lake area
22	GSC-1784-2	12 900 ± 150	Eskimo Lakes, District of Mackenzie	Blake, 1987; Rampton, 1988	grass	Age of Tutsieta Lake Stade at its maximum stand in Sitidgi Lake area
23	GSC-1792	14 400 ± 180	Sabine Point, Yukon	Lowdon and Blake, 1976; Rampton, 1982	peat	Minimum age for Sabine Phase of Buckland Glaciation (Early Wisconsinan?)
24	GSC-1798-2	>51 000	King Point, Yukon	Lowdon and Blake, 1976; Rampton, 1982	wood	Minimum age for pre-Buckland Glaciation sediments
25	GSC-1900	34 600 ± 1480	Malcolm River, Yukon	Lowdon and Blake, 1976; Rampton, 1982	plant and moss	Interstadial deposits postdating (?) Buckland Glaciation
26	GSC-1974	33 800 ± 880	Cy Peck Inlet, Bathurst Peninsula, District of Mackenzie	Lowdon and Blake, 1978; Rampton, 1988	wood and bark	Minimum age for the Amundsen Gulf ice lobe and the Toker Point Stade (Early Wisconsinan?)
27	GSC-1995	13 000 ± 130	Eskimo Lakes, District of Mackenzie	Blake, 1987; Rampton, 1988	plants	Age of Tutsieta Lake Stade at maximum stand in Sitidgi Lake area
28	GSC-2328	10 600 ± 260	Great Bear River, District of Mackenzie	Lowdon and Blake, 1979; Hughes, 1987	organic detritus	Minimum age for deglaciation and for drainage of glacial Lake McConnell in area
29	GSC-2375-2	49 100 ± 980	Dissection River, Banks Island	Vincent, 1983, 1984; Blake, 1987	peat	Minimum age for M'Clure Stade of Amundsen Glaciation (Early Wisconsinan?)
30	GSC-2422	36 900 ± 300	Hungry Creek, District of Mackenzie	Hughes et al., 1981	wood	Maximum age for the Hungry Creek Glaciation
31	GSC-2545	11 200 ± 100	South of Jesse Harbour, Banks Island	Lowdon and Blake, 1980; Vincent, 1983	marine shells	Schuyter Point Sea 21 m shoreline (marine limit ca. 25 m)
32	GSC-2690	16 000 ± 420	Doll Creek, District of Mackenzie	Lowdon and Blake, 1981; Ritchie, 1982	organic lake mud	Minimum age for Hungry Creek Glaciation
33	GSC-2758	15 200 ± 230	Lateral Pond, District of Mackenzie	Lowdon and Blake, 1981; Ritchie, 1982	organic lake mud	Minimum age for Hungry Creek Glaciation

QUATERNARY GEOLOGY — CANADIAN INTERIOR PLAINS

Table 2.3. (cont'd.)

Reference number (cf. Fig. 2.7)	Laboratory number	Date (years BP)	Locality	Reference or Collector	Material	Significance
34	GSC-2819	>39 000	Dissection River, Banks Island	Vincent 1983, 1984; Blake, 1987	peat	Minimum age for M'Clure Stade of Amundsen Glaciation (Early Wisconsinan?)
35	GSC-3249	11 700 ± 100	Cape Clarendon, Melville Island	Hodgson and Vincent, 1984b; Blake, 1984	marine shells	Maximum age for Viscount Melville Sound Ice Shelf and Deposition of Winter Harbour Till
36	GSC-3359	>43 000	Rat River, District of Mackenzie	Catto, 1986b	fecal remains	Minimum age for interglacial sediments beneath till of probable Early Wisconsinan age
37	GSC-3371	21 300 ± 270	Rat River, District of Mackenzie	Catto, 1986b	organic detritus	Minimum age for deglaciation of Upper Rat River
38	GSC-3376	11 300 ± 210	Deans Dundas Bay, Victoria Island	J-S. Vincent	marine shells	Minimum age for marine inundation of area
39	GSC-3387	13 100 ± 150	Twin Tamarack Lake, District of Mackenzie	Blake, 1983; Ritchie et al., 1983	organic lake mud	Minimum age for retreat of Tutsieta Lake Stade ice in lower Mackenzie Valley
40	GSC-3511	11 800 ± 100	Natkusiak Peninsula, Victoria Island	Blake, 1984	marine shells	Minimum age for deglaciation of area, predates Viscount Melville Sound Ice Shelf advance
41	GSC-3524	10 500 ± 200	Horn Plateau, District of Mackenzie	Blake, 1986; MacDonald, 1987	organic lake mud	Minimum age for deglaciation of eastern end of Horn Plateau
42	GSC-3527	9 880 ± 150	Worksop Point, Victoria Island	Hodgson and Vincent, 1984b; Blake, 1984	marine shells	"Oldest" minimum age on Victoria Island for Viscount Melville Sound Ice Shelf and deposition of Winter Harbour Till
43	GSC-3536	11 000 ± 340	Franklin Mountains, District of Mackenzie	Blake, 1986	organic lake mud	Minimum age for deglaciation of Franklin Mountains
44	GSC-3558	11 000 ± 100	Holman, Victoria Island	J-S. Vincent	marine shells	Approximate age of marine limit in area
45	GSC-3560-2	>49 000	Sachs Harbour, Banks Island	Vincent et al., 1983; Blake, 1987	peat	Minimum age of Cape Collinson Formation (Sangamonian) deposits lying below M'Clure Stade (Amundsen Glaciation) glacial deposits
46	GSC-3566	10 700 ± 100	Cape Baring, Victoria Island	D.R. Sharpe	marine shells	Minimum age for deglaciation of western Wollaston Peninsula
47	GSC-3592	>38 000	Graveyard Bay, Victoria Island	Vincent, 1984	wood	Interstadial deposits between Early and Late Wisconsinan tills?
48	GSC-3613	>37 000	Graveyard Bay, Victoria Island	Vincent, 1984	willow leaves	Interstadial deposits between Early and Late Wisconsinan tills?
49	GSC-3691	12 400 ± 120	Caribou River, District of Mackenzie	Catto, 1986b	peat	Minimum age for deglaciation of Upper Caribou River basin
50	GSC-3697	>39 000	Snake River, District of Mackenzie	Catto, 1986b	wood	Minimum age for interglacial-like sediments beneath till of Wisconsinan age
51	GSC-3698	>37 000	Sachs Harbour, Banks Island	Vincent, 1984; Blake, 1987	marine shells	Minimum age of Thesiger Lobe (M'Clure Stade of Amundsen Glaciation is Early Wisconsinan)
52	GSC-3722	>39 000	Maitland Point, Bathurst Peninsula, District of Mackenzie	J-S. Vincent and V.N. Rampton	wood	Driftwood postdating Mason River Stade (Sangamonian?)
53	GSC-3759	>38 000	Mason River, District of Mackenzie	J-S. Vincent and V.N. Rampton	wood	Driftwood postdating Mason River Stade (Sangamonian?)
54	GSC-3801	10 300 ± 100	East of Diamond Jenness Peninsula, Victoria Island	D.A. Hodgson	marine shells	Minimum age for deglaciation of eastern Prince Albert Sound area
55	GSC-3813	21 200 ± 240	Rat River, District of Mackenzie	Catto, 1986b	organic detritus	Minimum age for deglaciation of Upper Rat River
56	GSC-4075	>36 000	Mason River, District of Mackenzie	J-S. Vincent and V.N. Rampton	wood	Driftwood postdating Mason River Stade (Sangamonian?)
57	I(GSC)-25	10 530 ± 260	Harding River, District of Mackenzie	Craig, 1960	marine shells	Minimum age for deglaciation of area northeast of Melville Hills
58	I-3734	11 530 ± 170	Mountain River, District of Mackenzie	Mackay and Mathews, 1973	wood	Minimum age for deglaciation of area west of Franklin Mountains and for drainage of glacial lake in Mackenzie Valley
59	L-522 A	>42 000	East Channel, District of Mackenzie	Mackay, 1963	peat	Minimum age of interglacial-like sediments (Sangamonian?)
60	QL-1230	>61 000	Nelson River, Banks Island	Vincent, 1983, 1984	wood	Minimum age of interglacial Cape Collinson Formation sediments
61	TO-217	11 280 ± 100	Between Brock River and Hornaday River deltas	J-S. Vincent	marine shells	Minimum age for deglaciation of area at the head of Darnley Bay and for low marine inundation of area
62	TO-650	24 730 ± 260	South of Jesse Harbour, Banks Island	J-S. Vincent	marine shells	Minimum age for East Coast Sea and M'Clure Stade on eastern Banks Island
63	TO-796	43 550 ± 470	Garry Island, District of Mackenzie	J-S. Vincent	marine shells	Minimum age for marine transgression after the Toker Point Stade (Early Wisconsinan?), should be considered nonfinite
64	RIDDL-801	48 200 ± 1100	South of Kendall Island, District of Mackenzie	J-S. Vincent	marine shells	Minimum age for marine transgression after the Toker Point Stade (Early Wisconsinan?), should be considered nonfinite
65	GSC-3646	15 500 ± 440	"Kate's Pond", District of Mackenzie	Blake, 1987	organic lake mud	Possible minimum age for retreat of Tutsieta Lake Stade ice in lower Mackenzie Valley

CHAPTER 2

Table 2.4. Correlation of Quaternary events on Banks Island

Geological Events		Stratigraphy[1]			Aminostratigraphy			Age (ka)			Paleomagnetism	
		North Zone	East Zone	West Zone	Shells[2] Free	Shells[2] Total	Wood[3]	^{14}C	Th/U			
Postglacial		Organic, eolian, alluvial, marine, and colluvial sediments					0.14	7.8				
AMUNDSEN GLACIATION	RUSSELL STADE	Schuyter Point Sea Sediments	Schuyter Point Sea Sediments									
			Passage Point Sediments (Viscount Melville Sound Ice Shelf)					10.6				
			Schuyter Point Sea Sediments			0.02		11.2				
	M'CLURE STADE	Investigator Sea and Meek Point sediments	East Coast Sea Sediments	Carpenter Till (Sand Hills Advance)						Brunhes Normal-Polarity Chron		
				Meek Point Sea Sediments	0.42-0.51	0.04-0.09		>49.1, >19 24.7 >41	46		N	
		Lake Ivitaruk and Lake Ballast sediments	Lake Cardwell, Lake De Salis, and Lake Sarfarssuk sediments	Lake Rufus, Lake Masik, and Lake Raddi sediments								
		Bar Harbour Till and Mercy Till (Prince Alfred Lobe)	Jesse Till (Prince of Wales Lobe)	Sachs Till (Thesiger Lobe)								
			Pre-Amundsen Sea Sediments			0.03		>37	92.4		N	
CAPE COLLINSON INTERGLACIATION			CAPE COLLINSON FORMATION					>49 >61	69.9		N	
THOMSEN GLACIATION		Big Sea Sediments	Big Sea Sediments	Big Sea Sediments	0.54-0.73	0.16-0.20 (0.19)	0.22	>30	104.6		N / R	
		Baker Till	Lake Parker and Lake Dissection sediments							Matuyama Reversed-Polarity Chron		
			Kellett Till, Baker Till, and Kange Till	Kellett Till								
			Pre-Thomsen Sea Sediments									
MORGAN BLUFFS INTERGLACIATION			MORGAN BLUFFS FORMATION					>39	>200 >200		R	
			Post-Banks Sea Sediments	Post-Banks Sea Sediments			0.32-0.35					
BANKS GLACIATION		Bernard Till and Plateau Till		Lake Egina and Lake Storkerson sediments								
			Bernard Till and Plateau Till	Bernard Till and Durham Heights Till								
			Pre-Banks Sea Sediments	Pre-Banks Sea Sediments							R	
INTERGLACIATION OR PREGLACIAL				WORTH POINT FORMATION				>54			N	

Stratigraphic formations (left column sub-labels): Prince of Wales Formation; Nelson River Formation; Duck Hawk Bluffs Formation

1 As well as the formally named stratigraphic units, the position of glacial lakes (in italics) is indicated in the stratigraphic sequence.
2 Ratios of the amino acids D-alloisoleucine to L-isoleucine in the free and total (free and peptide-bound) fractions in *Hiatella arctica*.
3 Ratios of the amino acids D-aspartic to L-aspartic in *Salix* and *Betula*.

plotted the maximum extent of Pleistocene glaciation and the two limits for Late Wisconsinan time suggested by proponents of minimum and maximum ice extent models (Fig. 2.7, inset map).

Because of the lack of reliable chronological data and detailed mapping in certain critical areas, together with conflicting interpretations of the existing data, there is no agreement on the age of recognized glacial limits in the northern Interior Plains. Indeed one worker's glaciation limit can be a stillstand or readvance position to a colleague (Dyke and Prest, 1987; Hughes, 1987; Sharpe, 1984). Obviously these limits are still tentative.

An attempt is made here to assign ages to the limits suggested by various researchers (Fig. 2.7). Where the ages are controversial, the rationale for the interpretation chosen is stated.

Preglacial events

In the northern Interior Plains sediments that predate those of the earliest known glaciation have been described from Duck Hawk Bluffs(1)[1] and Worth Point Bluffs(2) of southwestern Banks Island. They have been assigned to the Worth Point Formation (Vincent, 1983; Vincent et al., 1983), which includes sediments that overlie the Miocene age Beaufort Formation and underlie deposits of the oldest recognized glaciation (Duck Hawk Bluffs Formation; Table 2.4). At the Worth Point type section, which is in a channel cut in sediments of the Beaufort Formation and the Cretaceous Kanguk Formation, more than 5 m of woody

---

[1] Number in brackets indicates site location shown in Figure 2.2a.

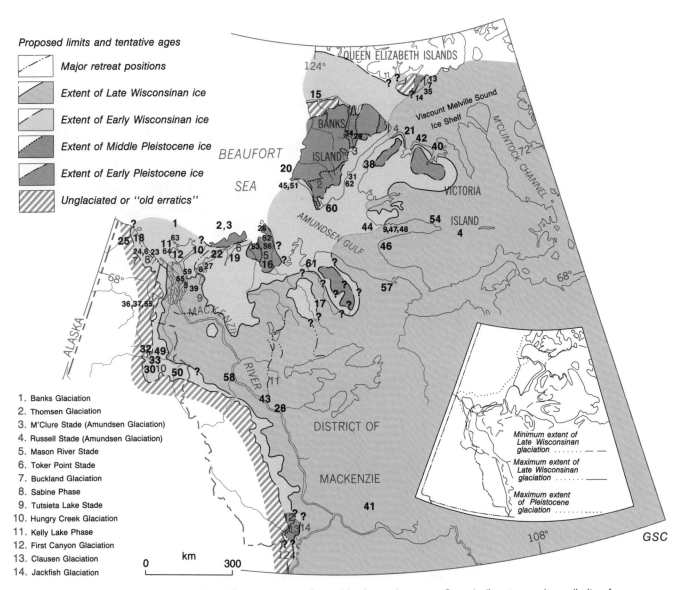

Figure 2.7 Proposed limits of Quaternary continental ice in northwestern Canada (inset map shows limits of Prest, 1984). Numbers refer to the site locations for radiocarbon dated material in Table 2.3.

peat underlies Bernard Till of Banks Glaciation. On the basis of macrofossils of larch *(Larix laricina)* and of various shrubs and herbaceous plants (Table 2.6), Kuc (1974) concluded that the peats accumulated in an open, subarctic forest-tundra environment resembling that of the northern part of the boreal forest today. Duck Hawk Bluffs (Fig. 2.6a) show terrestrial, organic-bearing sediments of fluvial, lacustrine, and eolian origin in the same stratigraphic position (Vincent et al., 1983). The organic deposits contain plant macrofossils and arthropod remains (Table 2.6) of numerous species not living on Banks Island today.

Paleomagnetic data indicate that the Worth Point Formation is older than 790 ka, because the sediments are magnetically reversed (Vincent et al., 1984; Vincent and Barendregt, 1987; Table 2.4). In general the sediments record the last time that trees grew on Banks Island and they show a much warmer climate than during subsequent interglacials (Matthews et al., 1986). However, the climate was still much cooler than in Miocene time, when deciduous hardwood forests with trees such as walnut *(Juglans eocinera)* and boreal forests dominated by spruce *(Picea)* and larch *(Larix)* were present on the island (Hills, 1975). The

Table 2.5. Correlation of Quaternary deposits and events in the northern Interior Plains of Canada

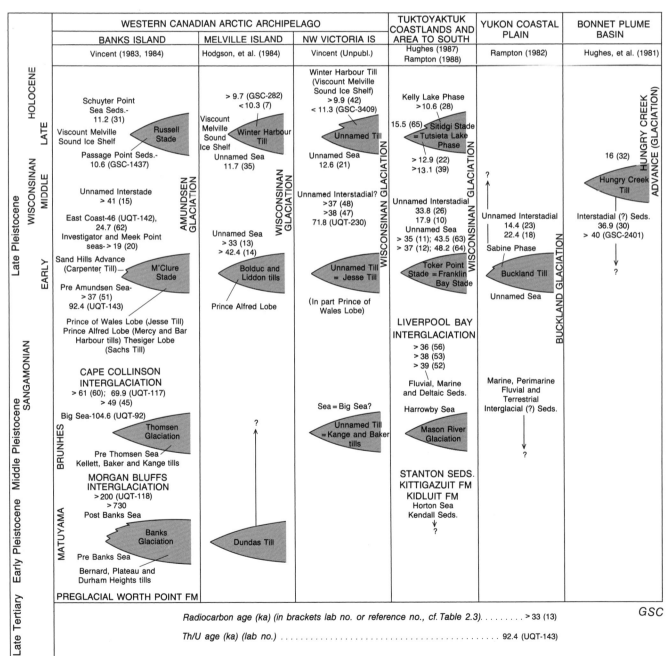

Table 2.6. Plant and arthropod macrofossils from the Beaufort and Worth Point Formations, Banks Island (modified from Matthews et al., 1986)[1]

Taxa	Beaufort lower (Tb$_1$)	Beaufort upper (Tb$_2$)	Worth Point WP[2]	Worth Point DHB[2]
PLANTS				
Bryophyta	r[3]		+ +[3]	+ +
Equisetaceae				
Equisetum sp.				
Spermatophyta				
Pinaceae				
Abies grandis type	+ +			
Larix sp.	+[3]			
Larix laricina (DuRoi) K.			+	
Pinus sp. (5-needle type)	+			
Picea sp.	+	+		
Tsuga sp.	+			
Taxodiaceae				
Metasequoia type	+			
Sparganiceae				
Sparganium sp.			+	+
Potamogetonaceae				
Potamogeton sp.	+	+	+	+
Alismaceae				
Sagittaria sp.	?			
Cyperaceae				
Carex aquatilis Wahlenb.			+	
Carex sp.	+	+	+	+
Scirpus sp.	+			
Juncaceae				
Luzula sp.			+	
Salicaceae				
Salix sp.			+	
S. niphoclada Rydb.			+	
S. alaxensis (And.) Cor.			+	
S. ovalifolia Trautv.			?	
Betulaceae				
Alnus crispa type	+			
A. crispa (Ait.)			+	
Alnus incanalrugosa type	+			
Betula sp.			+	
Betula (arboreal)				
Caryophyllaceae				
Stellaria sp.			+	
Arenaria humifusa Wahlenb.			?	
Ranunculaceae				
Ranunculus sp.			+	
R. trichophyllus type		+	+	+
R. lapponicus L.			+	
R. Macounii-pensylvanicus typ.			+	
Capparidaceae				
Polanisia sp.	+	+		
Crassulaceae				
Sedum sp.	+			
Rosaceae				
Dryas integrifolia Vahl			+	
Potentilla sp.		+	+	+
Rubus sp.	+			
Hypericaceae				
Hypericum sp.	+	+		
Lythraceae				
Decodon sp.	+	+		
Araliaceae				
Aralia sp.	+	?		
Haloragaceae				
Hippuris sp.		+		
Hippuris vulgaris L.			+	
Pyrolaceae				
Pyrola grandiflora Rad.			+	
Empetraceae				
Empetrum nigrum L.			+	
Ericaceae				
Ledum decumbens (Ait.) Lodd			+	
Andromeda sp.		+		
Arctostaphylos alpina/ruba type			+	
Vaccinium uliginosum L.				
var. *uliginosum*			+	
V.u. spp. *microphyllum* Lange.			+	
V.u. var. *alpinum* Big.			+	
V. vitis-idea var. *minus* Lodd			+	
Gentianaceae				
Menyanthes trifoliata L.			+	
Labiatae				
Teucrium sp.	+			
Solanaceae				
Genus?			?	
Caprifoliaceae				
Sambucus		+		
ARTHROPODS				
COLEOPTERA				
Carabidae				
Genus?				
Carabus truncaticollis Eschz.			+	+
Notiophilus sp.			+	
Elaphrus sp.			+	
Elaphrus lapponicus Gyll.			+	
Diacheila polita Fald.			+	
Dyschirius sp.		+		
Bembidion umiatense Lth.			+	
Bembidion sp.			+	
Pterostichus sp.			+	
P. nearcticus Lth.			+	
P. (Cryobius) ventricosus grp.			+	
Amara alpina Payk.			+	
Amara sp.			+	
Trichocellus mannerheimi Sahlb.		+		
Dytiscidae				
Hydroporus sp.			+	
Agabus sp.			+	
Staphylinidae				
Genus?			+	
Olophrum sp.			+	
Micralymma type			+	
Tachinus brevipennis type			+	
Byrrhidae				
Genus?			+	
Chrysomelidae				
Chrysolina sp.			+	
Curculionidae				
Genus?	+		+	
Apion sp.			+	
Lepidophorus lineaticollis Kirby		+		
Vitavitus thulius Kiss.			+	
Notaris sp.			+	
Cleonus sp.			+	
TRICHOPTERA				
Family?				
DITERA				
Family?				
Chironomidae			+	+
Genus?				+
Xylophagidae				
Xylophagus sp.		+		
HYMENOPTERA				
Ichneumonidea			+	
CRUSTACEA				
Cladocera				
Daphina sp.			+	
Notostraca				
Lepiduris sp.			+	+
ARACHNIDA				
Acari-Mesostigmata				
Trachytes type			+	
Acari-Orbatei				
Ceratozetidae			+	
Damaeidae				
Epidamaeus sp.			+	
Family?		+		

[1] Compiled from unpublished GSC Macrofossil and Arthropod Identification Reports by J.V. Matthews, Jr.; from unpublished GSC Wood Identification Reports by R.J., Mott and L.D. Farley-Gill; and from Kuc (1974).

[2] WP = Worth Point Bluffs; DHB = Duck Hawk Bluffs; location of bluffs shown in Figure 2.2a.

[3] + = present; + + = abundant; r = rare.

[4] * = taxa which probably does not occur on Banks Island at present. Absence of vascular plants confirmed by S.A. Edlund, Geological Survey of Canada.

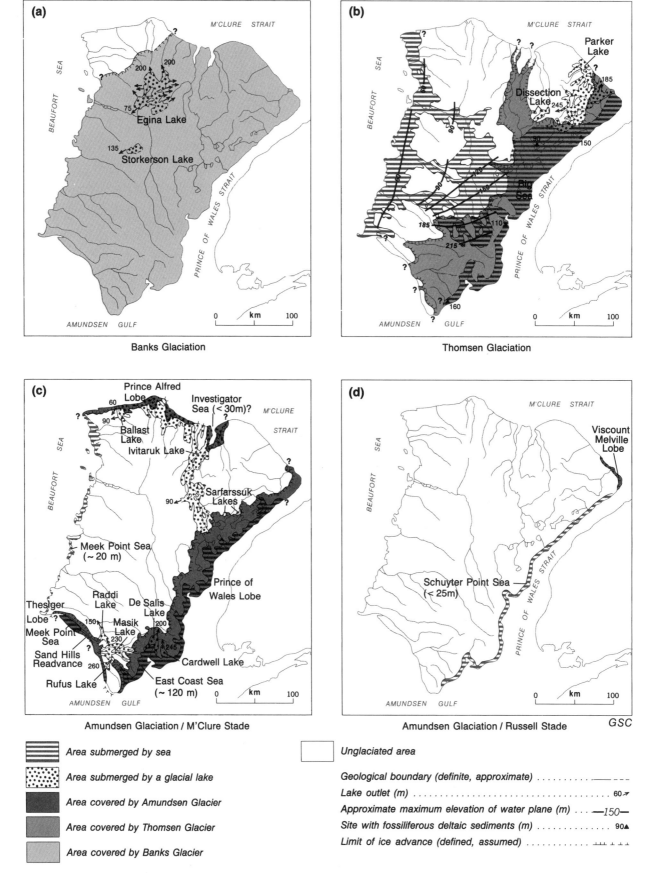

Figure 2.8 Quaternary paleogeography of Banks Island (modified from Vincent, 1983).

Worth Point Formation may span Late Miocene to early Quaternary time and therefore be associated with the old marine, continental and tephra units of Greenland (Kap København Formation; Funder et al.,1985; Funder, 1989), Yukon (Old Crow basin; Pearce et al., 1982; Naeser et al., 1982) and Alaska (Gubik Formation; Nelson and Carter, 1985; Carter et al., 1986). Participants at a 1987 workshop on circum-Arctic late Tertiary/Early Pleistocene stratigraphy and environments compared the Worth Point Formation flora and fauna with those of the above mentioned Arctic localities and in fact concluded that the beds are of likely Late Pliocene age.

In the northern Interior Plains, only the landscapes of the northwestern corner of Banks Island (on the distal side of the Banks Glaciation limit, see next section) show nonglacial characteristics (Vincent, 1982, 1983). There, V-shaped valleys and terraces cut into sands and gravels of the Beaufort Formation record post-Miocene fluvial events. These valleys and terraces are old because meltwater channels of the Banks Glaciation cut across them. Nevertheless, large diabase erratics have been observed on the surface (J.G. Fyles, Geological Survey of Canada, personal communication, 1984). These erratics may be from the Beaufort Formation or, like the old erratics of the western Queen Elizabeth Islands (Tozer and Thorsteinsson, 1964; Hodgson, 1989) and the "Gunsight Mountain" erratics of the Brooks Range in Alaska (Hamilton, 1979), they may represent a pre-Pleistocene (?) glaciation of which little evidence remains.

On the mainland, preglacial conditions are recorded in the unglaciated areas of interior Yukon (see Hughes et al., 1989), in the major pediments of the Yukon Coastal Plain (Rampton, 1982), and in river valleys east of the Richardson Mountains where preglacial gravels are present (Catto, 1986b).

Pre-Wisconsinan events

Record in southwestern Arctic Archipelago

Following the work of Fyles (1962, 1963) on Banks and Victoria islands, Vincent (1978, 1982, 1983, 1984) provided evidence for two full pre-Wisconsinan glaciations, (Banks and Thomsen) and for two interglaciations (Morgan Bluffs and Cape Collinson) in the area. The limits reached by the Banks and Thomsen glaciers are shown in Figures 2.7, 2.8a, and 2.8b, together with the limits of associated glacial lakes and marine transgressions. Figures 2.6a,b, and c are composite sections for the three areas that have provided the best information and Table 2.4 is a correlation chart for the events on Banks Island. These figures and table are used in the following chronology of events based, unless otherwise stated, on the work of Vincent (1978, 1982, 1983, 1984) and Vincent et al. (1983).

Banks Glaciation

The oldest known glacial advance onto the southwestern Arctic Archipelago by a continental glacier is recorded by sediments of the Banks Glaciation. Banks glacier flowed northwest from an ice centre on the mainland, mostly overriding fine grained Cretaceous and Tertiary sediments. It covered all of Banks Island except for the northwest sector. This likely was the only glaciation, apart from the one that left the possible "old erratics", strong enough to have covered the high southern tip of the island and the northeastern Devonian Plateau as well as large parts of the western Queen Elizabeth Islands (Fig. 2.7, 2.8a). Figure 2.9 shows part of the limit on northwestern Banks Island.

Till plains, left by Banks glacier, cover much of western and northern Banks Island. In numerous sections (Fig. 2.6a, b, c; Table 2.4) marine and glaciomarine sediments immediately predate and postdate the ice advance that is recorded by the till. Marine and glaciomarine sediments of Pre-Banks Sea overlie the nonglacial deposits, discussed earlier, at Duck Hawk Bluffs (1) (Fig. 2.6a), and underlie Banks Glaciation till at both Duck Hawk and Morgan (4) bluffs (Fig. 2.6c). They likely were laid down when the advancing ice depressed the land surface. As the glacier retreated southeastward, glacial lakes formed between the ice and high ground (Fig. 2.8a). At the same time a sea covered the west and east coasts of the island. Marine and glaciomarine deposits of this Post- Banks Sea are found in both Duck Hawk (1) and Morgan (4) bluffs, and in bluffs east of the mouth of Nelson River (3) (Fig. 2.6a-c).

Banks Glaciation is more than 790 ka old and therefore likely of Early Pleistocene age, since both Bernard Till and Post-Banks Sea sediments at Duck Hawk Bluffs are magnetically reversed (Vincent et al., 1984; Vincent and Barendregt, 1987). Clark et al. (1984) have suggested that Banks Glaciation correlates with unit J of the Central Arctic Ocean basin (Clark et al., 1980), which is the coarsest glaciomarine unit included in the Matuyama Reversed-Polarity Chron. In Table 2.5 and Figure 2.7, the unnamed glaciation responsible for deposition of the Dundas Till on Dundas Peninsula of Central Melville Island (Hodgson et al., 1984) is correlated with Banks Glaciation. First Canyon Glaciation discussed by Ford (1976) in the South Nahanni River basin may also be equivalent. All three glaciations represent the strongest Early Pleistocene continental glacier advance recorded by a till sheet in the northern Interior Plains. Northern Bathurst Peninsula should also have been glaciated at this time, considering the extent of ice elsewhere (Fig. 2.7), but Rampton (1988) found no evidence of glacial overriding there.

Morgan Bluffs Interglaciation

At five locations on Banks Island, plant- and shell-bearing, nonglacial deposits overlie marine and glacial sediments associated with Banks Glaciation, and underlie marine and glacial sediments associated with Thomsen Glaciation (Fig. 2.6a-c, i; Tables 2.4, 2.5). These sediments are considered to have been laid down during the long Morgan Bluffs Interglaciation. Correlation of the various beds is mainly by stratigraphic position but amino acid ratios from fossil wood corroborate the correlation of the sediments in the bluffs east of Nelson River (3) with those of the Morgan Bluffs (4) (Fig. 2.6b, c). In those two groups of bluffs the sediments are mainly interstratified, shallow water marine, fluvial, and peat deposits laid down in a perimarine environment. Some of the sediments in the Duck Hawk Bluffs (1) may also be perimarine, but beds of colluvial, fluvial, lacustrine (tundra ponds), and eolian origin are also recognized.

The organic deposits contain plant macrofossils and arthropod remains (Table 2.7) of species not living on Banks Island today. The paleoecology of beds other than Duck Hawk Bluffs (1) (Table 2.7) indicates a climate slightly warmer than that of today during Morgan Bluffs Interglaciation. At Duck Hawk Bluffs, some of the plant material is definitely in situ but some may be reworked from preglacial Worth Point Formation sediments. Assuming that the macrofossils found there, and particularly the alder and larch, represent the flora of the time, then the climate was significantly warmer than that of today (Matthews et al., 1986). Sea level during Morgan Bluffs Interglaciation, as recorded by the elevation of the perimarine sediments in the Duck Hawk (1) and Morgan (4) bluffs and in the bluffs east of the mouth of Nelson River (3), would have been about 30 m higher than it is today. The thick sequence of perimarine sediments indicates that the area was not affected by glacial isostatic movement. Sea level must therefore have been higher during the interglaciation, a situation akin to that of the record on the Alaskan Arctic Coastal Plain (Carter et al., 1986), or perhaps either faulting or uplift of tectonic origin affected much of Banks Island. Two nonfinite (>200 ka) uranium disequilibrium series age determinations[1] (UQT-118 and UQT-229) on autochthonous wood from Morgan (4) and Duck Hawk (1) bluffs have been obtained. A better age definition is provided by the fact that the Brunhes-Matuyama boundary is now known to lie within the Morgan Bluffs interglacial sequence at Duck Hawk Bluffs (1) (Vincent et al., 1984; Vincent and Barendregt, 1987). These interglacial deposits may correlate (Clark et al., 1984) with unit K of the Central Arctic Ocean

---

[1] The several uranium disequilibrium series age determinations, mentioned here and later, have been completed on material provided by the author to C. Causse of the Université du Québec à Montréal.

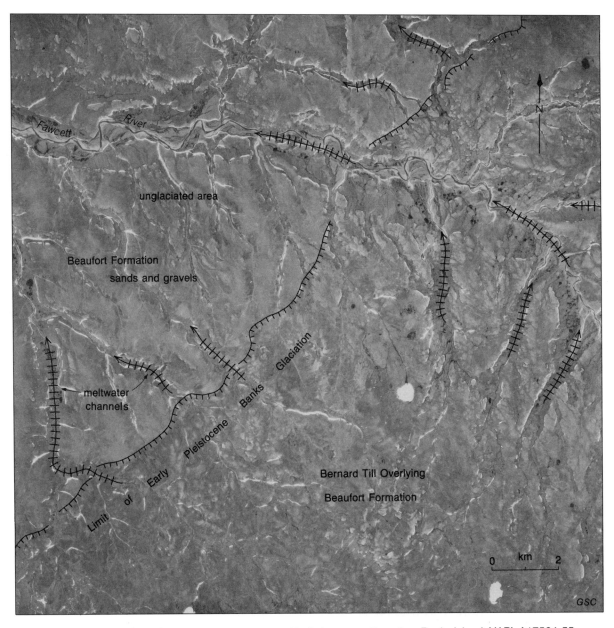

Figure 2.9 Limit of the Early Quaternary Banks Glaciation on northwestern Banks Island. NAPL A17564-55

basin (Clark et al., 1980), a unit considered to have been deposited during reduced glacial activity. The Morgan Bluffs interglacial beds record a long nonglacial period of perhaps more than 0.5 Ma, a situation akin to that of Beringia (Table 2.1). Some of the plant-bearing silt and sand in sections on the west coast of Bathurst Peninsula (Stanton Sediments), or the Tuktoyaktuk coastlands (Kidluit Formation; Kendall sediments; Rampton, 1988) may be equivalent in age (Table 2.1) to the Morgan Bluffs interglacial sediments. The Grotte Valérie I and II interglaciations identified by Ford (1976) in the South Nahanni River basin may also be of the same age.

Thomsen Glaciation

Thomsen Glacier, like Banks Glacier, flowed northwestward from the mainland to southeast Banks Island. It overrode much of southern and eastern parts of that island and the Thomsen River basin (Fig. 2.7, 2.8b; Table 2.4). The limits were controlled by high ground at the southern extremity of Banks Island, the drainage divide, and the northeastern Devonian Plateau. However, the ice was able to advance north in the Thomsen River basin. Extensive areas depressed by the ice were submerged by the Big Sea before, during, and after the ice advance (Fig. 2.8b). On the unglaciated west coast of the island the Big Sea limit is at about 60 m. Arms of the sea extended into valleys to the east and reached the exceptionally high levels of over 160 m at the margin of the ice in central Banks Island and over 250 m in a reentrant at the head of the Big and Bernard drainage basins. When the Thomsen Glacier reached its limit in the northeast, glacial lakes were dammed between the ice and the Devonian Plateau (Fig. 2.8b). The lakes finally drained when lower outlets opened to the sea. Till of the Thomsen Glacier and thick marine deposits, which both predate and postdate the ice advance, are particularly well displayed along the east coast of Banks Island (Fig. 2.6b-c). At Morgan Bluffs (4), Pre-Thomsen Sea sediments, including a glacial marine facies, underlie Kellett Till. Big Sea sediments, which overlie Kellett and Baker tills and underlie younger Cape Collinson interglacial and Wisconsinan deposits, are chiefly glacial marine bedded silts, sands, and clays. Where rivers flowed into the sea, the fine sediments are capped by shell-bearing deltaic deposits. Amino acid analyses of those shells has helped to correlate deposits from different localities on Banks Island. Total amino acid ratios of D-alloisoleucine to L-isoleucine (aIle:Ile) were approximately 0.19 in three samples, including in situ shells (Table 2.4) which also gave a thorium/uranium age of 104.6 ka(UQT-80).

The absolute age of Thomsen Glaciation is not known. However, it predates the last interglaciation and, as Big Sea sediments at Duck Hawk Bluffs (1) are normally polarized, it took place in the Brunhes Normal-Polarity Chron (<790 ka). On this basis, a Middle Pleistocene age is suggested for Thomsen Glaciation.

Figure 2.7 shows a large area of central Prince Albert Peninsula and the Shaler Mountains on Victoria Island as having escaped the Wisconsin Glaciation. The till at the surface of those areas probably was deposited during the Thomsen Glaciation and hence is assigned a Middle Pleistocene age (Table 2.5). Stratified marine deposits that underlie Wisconsinan till along Prince of Wales Strait, on Prince Albert Peninsula, resemble those on Banks Island and may also have been deposited in the Big Sea. On the mainland, as portrayed in Figure 2.7, areas affected by Clausen Glaciation in the South Nahanni River basin (Ford, 1976), Mason River Glaciation on Bathurst Peninsula (Rampton, 1988), and the glaciation that overrode Brock Upland are assigned the same age as Thomsen Glaciation. If this is the case, Harrowby Sea (the marine transgression postdating Mason River Glaciation; Rampton, 1988) could be the mainland equivalent of the Big Sea. The Kittigazuit Formation sediments (glaciofluvial outwash related to Mason River Glaciation or eolian deposits from the exposed Beaufort Sea Shelf?) in the Tuktoyaktuk coastlands (Rampton, 1988) could also be of Thomsen Glaciation age.

Cape Collinson Interglaciation

At two sites on Banks Island, plant-bearing, nonglacial deposits of the Cape Collinson Formation overlie and underlie marine and glacial sediments associated with Thomsen and Amundsen glaciations, respectively (Table 2.4). Correlation of the beds is based on stratigraphic position. In the type section of this unit, east of Nelson River (3) (Fig. 2.6b), in situ organic remains collected from a fossilized tundra surface have given radiocarbon ages of >61 000 BP (QL-1230) and thorium/uranium ages of 69.9 ka(UQT-117) that, like other such determinations, should be considered minimal. Those deposits overlie Big Sea sediments and underlie both Pre-Amundsen Sea sediments and Jesse Till of the M'Clure Stade of Amundsen Glaciation (Table 2.4). In the bluffs east of Mary Sachs Creek (12) (Fig. 2.6a), peat beds dated at >49 000 BP (GSC-3560-2) possibly overlie Big Sea sediments and underlie pre-Amundsen Sea glaciomarine sediments and Sachs Till.

The climate when these beds were deposited was distinctly warmer than that of today, as indicated by the dominance of tundra species such as dwarf birches, (Betula, including B. glandulosa) and a few species of coleoptera (Table 2.7) that presently are restricted to the mainland near treeline (Matthews et al., 1986). As a result, the Cape Collinson Formation is believed to date from the last (Sangamon) interglaciation (Table 2.5). The widespread plant-bearing, bedded, lacustrine silts and the fluvial sands and gravels with ice-wedge casts of the Yukon Coastal Plain (Rampton, 1982), along with the organic- and driftwood-bearing sediments on low benches of Bathurst Peninsula (Rampton, 1988), are also probably interglacial and are tentatively correlated with the Cape Collinson Formation.

Record on mainland

Pre-Wisconsinan glacial and interglacial deposits are present in the District of Mackenzie and the Yukon. These deposits lie beneath Wisconsinan sediments in sections mainly in the Mackenzie River basin and on the Arctic Coastal Plain from Yukon Territory to Bathurst Peninsula. Presently many of the deposits cannot be fitted into a chronological or even a geological event framework. Many of the widespread vertebrate remains (Table 2.8) collected in the study area must have come from these beds, even though few have been collected in place.

East and north slope of Mackenzie Mountains

Although continental ice likely extended into the Mackenzie Mountains several times during the Quaternary, limits associated with specific glaciations generally have not been traced. A maximum limit of ice advance of presumably

Table 2.7. Plant and arthropod macrofossils from Pleistocene interglacial and interstadial sediments, Banks and Victoria islands (modified from Matthews et al., 1986)[1]

Taxa	Morgan Bluffs Fm.					Cape Collinson Fm.			Prince of Wales Fm.	
	MB[2]	ENR	DHB	IR	TR	ENR	DHB	DD	DR	GB
PLANTS										
Characeae										
Chara/Nitella sp.			+[3]							+
Bryophyta			+					+		+
Equisetaceae										
Equisetum sp.	+	+			+	+				+
Selaginellaceae										
Selaginella sp.										+
Pinaceae										
Larix sp.			+							
Sparganiaceae										
Sparganium sp.	+	+				+				
Potamogetonaceae										
Potamogeton filliformis Pers.				+		+		+		+ +
Potamogeton sp.	+			+	+	+				
Graminaeae										
Genus?								+		
Cyperaceae										
Carex aquatilis Wahlenb.	+	+	+		+	+				+
Carex sp.	+	+	+		+	+	+	+	+	+
Carex maritima Gunn.								+		
Kobresia sp.										?
Scirpus sp.	+	+				+				
Juncaceae										
Luzula sp.	+	+				+				
Salicaceae										
Salix sp.	+	+	+		+	+		+		+ +
Betulaceae										
Alnus sp.			?							
Betula glandulosa type			+		+	+				
Betula sp.	+	+	+			+				
Polygonaceae										
Oxyria digyna (L.) Hill										+
Caryophyllaceae										
Cerastium sp.								+		
Melandrium apetalum (L.) Fenzl.										+
Melandrium sp.				?						
Stellaria sp.	?	?				?				
Ranunculaceae										
Ranunculus sp.	+	+	+			+		+		
R. trichophyllus type			+							
R. lapponicus L.	+	+				+	+			
Cruciferae										
Draba type								+		
Genus?					+					
Rosaceae										
Dryas sp.								+		
Dryas integrifolia Vahl	+	+		+	+	+		+		+
Sibbaldia procumbens L.			+							
Potentilla palustris (L.) Scop.			+							
Potentilla sp.			+		+					
Rubus chamaemorus L.								+		
Leguminosae										
Genus?								+		
Haloragaceae										
Myriophyllum sp.			+							
Hippuris sp.		+			+	+	+			
Empetraceae										
Empetrum nigrum L.			+							
Ericaceae										
Cassiope sp.										+
Arctostaphylos alpina/ruba type			+							
Gentianaceae										
Menyanthes trifoliata L.	+	+	+		+	+				
Compositae										
Taraxacum sp.								+		
ARTHROPODS										
BRYOZOA										
Cristatella mucedo L.			+							
Genus?								+		
ARTHROPODA										
INSECTA										
HEMIPTERA										
Miridae								+		
HOMOPTERA										
Cicadellidae										
Genus?								+		
COLEOPTERA										
Carabidae										
Genus?			+							
Carabus truncaticollis Eschz.[3]	+									
Notiophilus sp.	+	+		+		+			+	
Elaphrus sp.	+	+		+		+			+	
Elaphrus lapponicus Gyll.	+	+		+		+				
Diacheila polita Fald.							+		+	
Dyschirius sp.	+	+	+	+		+			+	+
Dyschirius frigidus type										
Bembidion (Platophodes) sp.	+		+							
Bembidion umiatense Lth.	+	+			+	+				
Bembidion sp.	+	+		+	+	+	+			
Pterostichus sp.	+	+	+	+		+				
P. nearcticus Lth.	+			+		+				
P. (Cryobius) cf. *kotzebuei* Ball				+						
P. (Cryobius) ventricosus Eschz.	+		+	+						
P. (Cryobius) brevicornis Kby.	+		+	+					+	
P. (Cryobius) sp.	+	+	+	+		+				+
Pterostichus haematopus Dej.										+
Amara alpina Payk.	+	+		+	+	+				

Table 2.7 (cont'd.)

Taxa	Morgan Bluffs Fm.					Cape Collinson Fm.			Prince of Wales Fm.	
	MB²	ENR	DHB	IR	TR	ENR	DHB	DD	DR	GB
Amara sp.	+	+		+		+		+		
Harpalus sp.			?							
*Trichocellus mannerheimi Sahlb	+	+		+		+				
Dytiscidae										
Genus?	+				+			+		+
Hudroporus sp.	+	+		+		+				
Agabus sp.	?	?		?		?				
Ilybius sp.	?	?		?		?				
Colymbetes sp.			+							
Gyrinidae										
*Gyrinus sp.										+
Hydrophilidae										
Hydrobius sp.										+
Hydraenidae										
Genus?	+									
Staphylinidae										
Genus?	+	+	+	+		+				
Bledius sp.			+							
Omalinae			+							
Olophrum sp.	+	+		+		+				
Olophrum latum Makl.	+	+		+		+				
Macralymma brevilinque type	+	+		+	+	+				+
Stenus sp.	+	+		+		+			+	
*Euaesthetus sp.		+							+	+
Tachinus sp.	+		+							
Tachinus apterus Maklin					+					
Tachinus brevipennis Sahlb.	+	+		+		+				
*Tachinus instabilis Maklin	+									
Aleocharinae	+									
Genus?		+	+						+	
Silphidae										
*Silpha sp.										+
Leiodidae										
Agathidium sp.										+
Genus?	+									
Byrrhidae										
Genus?	+			+		+				
Simplocaria sp.				+						
Simplocaria remota Brown									+	
*Morychus sp.								+		
*Byrrhus sp.									+	
Coccinellidae										
Scymnini								+		
Ceratomegilla sp.								+		
Curculionidae										
Genus?			+							
Apion sp.	+	+		+		+				+
*Lepidophorus lineaticollis Kirby	+	+	+	+		+				
*Vitavitus sp.			+							
*Vitavitus thulius Kiss.	+	+	+	+		+				+
Hypera sp.			+					+		
Hypera diversipunctata Schrank			+							
*Sitona sp.										?
*Lepyrus sp.								+		
*Notaris sp.	+	+		+		+				
Rhynchaenus sp.			+							
*Cleonus sp.	+	+		+		+				
Ceutorhunchus sp.					+					
TRICHOPTERA										
Family?			+					+		+
LEPIDOPTERA										
Genus?			+		+					
DIPTERA										
Family?			+					+		+
Tipulidae										
Tipila sp.								+		
Chironomidae										
Genus?	+	+	+	+		+		+		
Calliphoridae								+		
HYMENOPTERA										
Family?			+							
Symphyta										
Tenthredinidae					+			+		
Ichneumonoidea										
Genus?		+			+					
Diapriidae					+					
Cynipoidea										
Genus?								+		
CRUSTACEA										
Cladocera										
Daphina sp.		+		+		+				+
Notostraca										
Lepiduris sp.		+		+						
Ostracoda										
Genus								+		
ARACHNIDA										
Acari										
Oribatei										
Genus?			+						+	
Araneae										
Lycosidae?			+							

1. Compiled from unpublished GSC Macrofossil and Arthropod Identification Reports by J.V. Matthews, Jr.; and from unpublished GSC Wood Identification Reports by R.J. Mott and L.D. Farley-Gill.
2. MB = Morgan Bluffs; ENR = Bluffs East of Nelson River mouth; DHB = Duck Hawk Bluffs; IR = "Ivitaruk" River Bluffs; TR = Thomsen River Bluffs; DD = Bluffs near Deans Dundas Bay; DR = Dissection River Bluffs; GB = Graveyard Bay Bluffs; location of bluffs shown in Figure 2.2a.
3. + = present.
4. * = taxa which probably does not occur on Banks or Victoria islands at present. Absence of vascular plants confirmed by S.A. Edlund, Geological Survey of Canada.

Table 2.8. Quaternary vertebrate remains of the northern Interior Plains

	Locality (cf. Fig. 2.2b)	Fauna recorded*	Age	^{14}C age (years BP)	Reference or Collector
A.	Hungry Creek	Spermophilus cf. S. parryi Dicrostonyx sp. Lemmus sp.? Microtus sp.? Equus (Asinus) cf. E. lambei	Pleistocene " " " "		Hughes et al., 1981 " " " "
B.	Herschel Island	Canis familiaris Mammuthus sp. Equus (Asinus) lambei Bison cf. crassicornis ?Bootherium sp. Ovibos moschatus Cetacea (genus and sp. indeterminate)	Pleistocene " " " " " Pre-Early Wisconsinan	16 200 ± 150 (RIDDL-765)	Harington, 1978 " " " " " Harington, 1977
C.	Beaufort Sea (Immerk)	Equus (Asinus) lambei	Pleistocene		Harington, 1978
D.	East Channel (Mackenzie River) – Tununuk	Equus (Asinus) lambei Mammuthus cf. primigenius Mammuthus sp.	Pleistocene " "	19 440 ± 290 (I-8578)	Harington, 1978 Mackay, 1958 "
E.	Richards Island	Equus (Asinus) lambei Bison sp.	Pleistocene "		Harington, 1978 J-S. Vincent
F.	Tuktoyaktuk	Bison crassicornis?	Pleistocene		Mackay, 1958
G.	Southern Eskimo Lakes	Mammuthus sp. Alces cf. latifrons	Middle Pleistocene "		J-S. Vincent and S.R. Dallimore "
H.	Nicholson Peninsula	Mammuthus sp.? Bison sp.	Pleistocene "		Mackay, 1958 C.R. Harington
I.	Maitland Point	Bison sp. Proboscidea cf. Mammuthus sp.	Pleistocene "		J-V. Matthews "
J.	Inlet east of Maitland Point	Proboscidea cf. Mammuthus sp.	Pleistocene		J-S. Vincent
K.	Bathurst Peninsula	Mammuthus sp.?	Pleistocene		Mackay, 1958
L.	Baillie Islands	Mammuthus primigenius Equus sp. Rangifer tarandus Bison sp. Saiga tatarica Ovibos moschatus Balaena mysticetus Phoca (Pusa) hispida Ursus maritimus	Pleistocene " " " " " Holocene Pleistocene "	1810 ± 90 (I-5407)	Harington, 1978 " " " " " Harington, 1980 " "
M.	Masik River (Banks Island)	Rangifer tarandus	Pleistocene		Harington, 1978
N.	Morgan Bluffs (Banks Island)	Dicrostonyx torquatus Lagopus sp.?	Middle Pleistocene "		Vincent, 1983 "
O.	Duck Hawk Bluffs (Banks Island)	Indeterminate mammal bone	Middle Pleistocene		J-S. Vincent
P.	Bernard River (Banks Island)	Ovibos moschatus	Pleistocene	>34 000 (S-288)	Maher, 1968
Q.	Ballast Brook (Banks Island)	Proboscidea cf. Mammuthus sp.	Pleistocene		L.V. Hills
R.	Cape Wollaston (Victoria Island)	Balaena mysticetus	Early Holocene	9285 ± 140 (S-2729)	J-S. Vincent
S.	East of Cape Baring (Victoria Island)	Cetacea (genus and sp. indeterminate)	Early Holocene	9780 ± 250 (S-2686) 8565 ± 220 (S-2687)	D.R. Sharpe
T.	East of Washburn Lake (Victoria Island)	Cetacea (genus and sp. indeterminate)	Early Holocene		D.R. Sharpe
U.	Garry Island	Ovibos moschatus Mammuthus sp.	Pleistocene "		J-S. Vincent
V.	Summer Island	Equus (Asinus) lambei	Pleistocene		J-S. Vincent
W.	Kay Point	Mammuthus sp.	Pleistocene		S.R. Dallimore
X.	Phillips Bay	Mammuthus sp.	Pleistocene		S.R. Dallimore

*Most fauna has been identified by C.R. Harington, National Museum of Canada

Wisconsinan age at about 1500 m on the south flank and 1400 m on the north flank of the mountains generally has been the only one portrayed (Hughes, 1972, 1987; Prest, 1983; Figure 2.7). Nevertheless, in some areas older deposits probably lie beyond the limit of Wisconsinan ice advances. Rutter and Minning (1972) and Rutter and Boydell (1973) recognized two advances on the southern flank of the Mackenzie Mountains, the oldest of which reached elevations of about 1500 m. On the basis of work by Ford (1976) and Harmon et al. (1977) in the South Nahanni River basin, this limit is Middle Pleistocene or older. Ford (1976) mapped three continental glacial limits in the South Nahanni, two of which, on the basis of $^{230}Th/^{234}U$ age determinations of speleothems, he attributed to pre-Sangamonian glaciations (Fig. 2.7) — the pre-Illinoian First Canyon Glaciation (>350 ka), which was most extensive, and the Illinoian Clausen Glaciation (>145 ka). They are separated by the Grotte Valérie I and II interglaciations and followed by the Sangamon Interglaciation.

Upper Mackenzie River basin

Along Mackenzie River and its tributaries, Craig (1965) and Rutter et al. (1973) identified numerous sections with thick intertill units of sand and gravel. Where Ochre River crosses the Franklin Mountains, three tills separated by sands and silt were observed by Rutter et al. (1973). It is not known if all the deposits that underlie a Wisconsinan age surface are also Wisconsinan, or if, as could be the case for those at the Ochre River site, they predate the Wisconsinan.

East flank of Richardson Mountains

Catto (1986a,b) described sediments in the Rat, Peel, Caribou, and Snake river basins that likely predate the Wisconsinan. Weathered fluvial gravels, which contain no Canadian Shield stones, occur along Snake, Peel, and Rat rivers and are assigned a preglacial age. Fresher fluvial gravels stratigraphically above these and with shield stones probably record an early glaciation. Along Snake River, a diamicton, informally named the "Snake River till", underlies fluvial sediments that contains wood dated at >39 000 BP (GSC-3697). In the Rat River basin, sediments deposited in a lake dammed by Laurentide ice standing to the east also record an early ice advance. These sediments are overlain by organic-bearing fluvial sediments considered to be of interglacial origin. They record a boreal forest to open tundra environment, with a climate warmer than today's. The age of these deposits is not known, but autochthonous fecal remains in the upper part of the fluvial sediments gave a radiocarbon age of >43 000 BP (GSC-3359).

Lower Mackenzie River

Along Hare Indian and Mackenzie rivers downstream from Fort Good Hope, Mackay and Mathews (1973) observed sections where valley fill gravel, sand, silt, and till underlie Late Wisconsinan deposits; some of these may predate the Wisconsinan. Along Hare Indian River, for example, more than 20 m of sand and gravel underlies till and sand overridden by ice. Wood in the intertill sediments has provided four nonfinite radiocarbon ages. Savigny (1989) also noted the presence of alluvial clay and sand, and deltaic sands and gravel, all below till, along Great Bear River.

On the north flank of the Caribou Hills, along East Channel (16) of Mackenzie River, up to 50 m of fluvial sands, silts and clays, interbedded with peat containing wood, is exposed (Mackay, 1963; Rampton, 1970, 1988). As expected in a fluvial sequence, the buried organics are typically allochthonous. A radiocarbon age of >42 000 BP (L-522A) on peat and a thorium/uranium age of >200 ka(UQT-149) on wood from the peat were obtained.

Mackay (1963), Fyles (1966), and Rampton (1972, 1988) described sections, at the south end of Eskimo Lakes (5) in the Portage Point area, in which about 25 m of interstratified silts, sands, and minor clay with peat beds underlies outwash gravels, and in places till. An age of >50 900 BP (GSC-329) for the peat and a thorium/uranium age of >200 ka(UQT-188) on wood from the peat were obtained. These beds resemble the East Channel beds north of Caribou Hills. The sediments at East Channel and Portage Point may have been deposited by an ancestral Mackenzie River, possibly during the Middle Pleistocene.

Yukon Coastal Plain

Deposits underlying drift of the presumably Early Wisconsinan Buckland Glaciation (Rampton, 1982) have been investigated along the Beaufort Sea coast by Fyles (1966), Naylor et al. (1972), Bouchard (1974), Johnson et al. (1976), and Rampton (1982). Between Herschel Island and Sabine Point, thick and complex sequences of sands, bedded silts, clays, and gravels lie below till (Fig. 2.6e and Table 2.5). These sediments, commonly ice-thrusted and deformed (Mackay, 1959), contain marine shells, allochthonous and autochthonous peats, and wood which dated >51 000 BP (GSC-151-2 and GSC-1798-2). These deposits were laid down in rivers, estuaries, and shallow and moderately deep seas; they typically represent deposition of sediments derived from the mountains to the south.

Beneath Buckland Glaciation till, east of Sabine Point, and along the lower stretches of Blow and Running rivers, unoxidized organic-bearing sands and gravels are widespread over oxidized sands and gravels containing ice-wedge casts (Rampton, 1982). Both the oxidized and unoxidized materials probably are the fluvial equivalents of the thick marine and estuarine deposits predating the till west of Sabine Point. These sediments of the Yukon Coastal Plain are important, as they may record interglacial and interstadial(?) conditions and events in pre-Wisconsinan time (Table 2.5). Five wood samples found beneath Buckland Till have given thorium/uranium ages in excess of 120 ka.

Liverpool Bay-Eskimo Lakes

Thick sands with minor gravel and clay underlie till, presumably of the Early Wisconsinan Toker Point Stade (Rampton, 1988; Table 2.5), over an extensive region that includes islands fronting Mackenzie Delta, Tuktoyaktuk Peninsula, and areas surrounding Eskimo Lakes and Liverpool Bay. These deposits have been described in various localities by Mackay (1963), Fyles (1966, 1967), Fyles et al. (1972), Kerfoot (1969), Rampton (1970, 1971a, 1972, 1974b), Rampton and Bouchard (1975), and Mackay and Matthews (1983).

Recently Rampton (1988) proposed a lithostratigraphic framework for these sediments. Kendall sediments and Hooper clay, found west of East Channel and probably on

Nicholson Peninsula (13), consist of fossiliferous marine clays, silts, and sands. Overlying these are beds of Kidluit Formation, which consist of fluvial, crossbedded, grey sands, with much organic detritus. Paleoecological studies indicate that they likely were deposited during an interglacial period with a climate similar or slightly warmer than today. Finally, beds of Kittigazuit Formation, consisting of brown sands, are well exposed on the south side of Liverpool Bay and on Richards Island. Rampton (1988) believed that the sands were likely glaciofluvial sediments deposited by glacial meltwater entering a 50 m-high sea (equivalent to Harrowby Sea postdating Mason River Glaciation(?) discussed below). On the other hand it could be argued that Kittigazuit Formation sands are eolian and that they are part of a fossilized dune complex built during glacial times when the Beaufort Sea Shelf was subaerially exposed. A composite section (Fig. 2.6d), made up from observations in the Eskimo Lakes area, shows units traceable from Nicholson Peninsula (13) to western Richards Island. All these sediments likely were deposited on the coastal plain by the ancestral Mackenzie River and rivers farther east, either as outwash or as nonglacial deposits when sea level was higher or lower. They thus resemble the stream-derived sediments of the Yukon Coastal Plain. Abundant allochthonous wood has been found in these beds, but all dates have been nonfinite (e.g., >36 000 BP, GSC-1281).

Following deposition, many of these sediments were ice-thrust, particularly on Nicholson Peninsula (13) (Mackay, 1956) and on Richards Island (Rampton, 1971a; Mackay and Matthews, 1983); terraces were incised into them by Early and Late Wisconsinan meltwaters; and they have been affected by the growth and disintegration of massive ice bodies within them (Mackay, 1971; Rampton 1971b, 1974a).

Bathurst Peninsula and lower Horton River basin

As shown in Figure 2.7 a Middle Pleistocene glacial advance, named the Mason River Glaciation by Rampton (1988), is recorded on southwestern Bathurst Peninsula, and much of the remainder of the Peninsula is believed to have escaped later glacial advances. The Mason River glacial limit clearly truncates the up to 75 m-high marine wave-cut plain of northern Bathurst Peninsula (Horton Sea). A thick suite of sediments bordering the coasts and inlets of Bathurst Peninsula, which were named the Stanton Sediments by Rampton (1988) (Fig. 2.6f), for the most part appears to underlie the glaciated surface. A younger marine event (Harrowby Sea) is in turn responsible for reworking the Mason River glacial surface up to 50 m a.s.l., and deposits, interglacial in character, are inset into the surface.

Unless tectonic forces have been active, Horton Sea is likely a glacial isostatic sea caused by glacial events for which other evidence is not presently available. The Kendall sediments and Hooper clay, in the Tuktoyaktuk Coastlands, may have been deposited in the same glacial sea. The basal marine unit of Stanton Sediments was probably laid down in the Horton Sea. This is overlain by alluvial sands, silts, and clays, containing organic material, soil horizons and ice-wedge casts, which were deposited on an alluvial plain in a perimarine environment and in tundra ponds during an interglacial. These are capped by channel sands and beach gravels. This complex may record a long interglacial period, with fluctuating sea levels and intense thermokarst activity, predating Mason River Glaciation. Alternatively, Rampton (1988) recognized a thinly bedded organic-rich fine sand (his thinly bedded member) which he stated could have been deposited during Mason River Glaciation. This interpretation makes the channel sands and beach gravels at the top post-Mason River Glaciation.

During Mason River Glaciation (Rampton, 1988), northeast flowing ice wrapped around southwestern Bathurst Peninsula (Fig. 2.7) and Harrowby Sea submerged the part of the area now lying below 50 m a.s.l. Following this event, low benches, up to 7 m a.s.l., were cut into the older surfaces likely during the Sangamon Interglaciation. Thick driftwood mats found on these were dated at >36 000 BP (GSC-4075), >38 000 BP (GSC-3759), and >39 000 BP (GSC-3722). They record a time when sea level was higher than today and when Mackenzie River or Horton River basin must have been ice-free in order to produce the driftwood.

In the lower Horton River area, pre-Wisconsinan beds are also present, but have received little attention. Fulton and Klassen (1969) reported at least three tills lying above Tertiary gravels along a tributary of the Horton River. Paleomagnetic reversal of the lowermost tills suggests these are Early Pleistocene (J-S. Vincent and R.W. Barendregt, unpublished). Peat, possibly interglacial, lying beneath the upper till was dated at >41 000 BP (GSC-1100).

Wisconsin Glaciation

Most Quaternary geologists agree as to what are Wisconsinan glacial deposits in the northern Interior Plains, but disagree on their ages within the Wisconsinan. Some contend that the strongest advance was during the Early Wisconsinan and that Late Wisconsinan ice was less extensive; this is the point of view portrayed in Figure 2.7. On the other hand, others consider that the maximum Wisconsinan advance occurred in the Late Wisconsinan, and was followed locally by a readvance. Accordingly, they consider what is shown as the Early Wisconsinan limit in Figure 2.7 to be Late Wisconsinan and the younger limit to be a readvance. For example, Hughes (1987) assigned what in this text is described as the Late Wisconsinan limit to the Tutsieta Lake Phase — a Late Wisconsinan readvance (Fig. 2.7).

In this report the Early and Late Wisconsinan histories are discussed separately and alternative interpretations of ages of the glacial limits are given where controversy exists. It should be noted that because of the absence of nonequivocal age control, age assignments are commonly tentative and much depend on the individual author's biases. This situation in the northern Interior Plains is analogous to that of all other areas at the periphery of the Laurentide Ice Sheet (Vincent and Prest, 1987).

Early Wisconsinan glaciation

Figure 2.7 shows the tentative limit proposed for Early Wisconsinan ice advance and Table 2.5 show the correlation of glacial events assigned this age. Bathurst Peninsula, Tuktoyaktuk Peninsula, most of Banks (Fig. 2.8c) and Melville islands, central Prince Albert Peninsula, and possibly part of the Shaler Mountains and the Brock Upland escaped glaciation. Offshore extensions of ice margins and the horizontal nature of terminal moraines on land indicate that ice shelves and/or ice streams existed in Amundsen Gulf and M'Clure Strait. On Banks Island (Vincent, 1983) and in the lower Mackenzie River basin (Beget, 1987) reconstructed ice profiles indicate that both Early and Late Wisconsinan ice was thin and characterized by low profiles.

Banks Island

Deposits on Banks Island assigned to the M'Clure Stade of Amundsen Glaciation (Tables 2.4 and 2.5) are thought to be of Early Wisconsinan age (Vincent, 1978, 1982, 1983, 1984). Lobes of ice from Amundsen Gulf, Victoria Island, and Viscount Melville Sound advanced onto the coastal areas of the island. In all cases ice flow was controlled by the inter-island channels and the ice surface gradients were low.

The Prince of Wales Lobe from Amundsen Gulf and Victoria Island impinged on the southern part of Banks Island and flowed northwestward in Prince of Wales Strait to lap onto the east coast of the island (Fig. 2.8c). The gradient of the ice surface was <1 m/km. Jesse Till, characterized by its pinkish grey colour, by a high carbonate content, and by a dense network of high-centred polygons, was deposited. Terminal moraines, and spectacular outwash plains that extend onto nonglaciated areas, mark the glacial limit. Small glacial lakes were dammed and, in the course of deglaciation, the eastern coastal areas below about 120 m were submerged by the East Coast Sea. The trimline of the sea cuts into Jesse Till and can be followed continuously for 350 km.

The Thesiger Lobe, coming from Amundsen Gulf, impinged on the southwest coast after rounding the high southern tip of Banks Island (Fig. 2.8c). The lobe, with a gradient of <3 m/km, left Sachs Till as well as terminal and end moraines and dammed glacial lakes in large valleys at the margin of the ice. During deglaciation, the area along Thesiger Bay and all the unglaciated western coast of Banks Island were submerged to about 20 m by the Meek Point Sea. The limit of the sea, typically marked by a wave-cut cliff, can be traced for 250 km. Southeast of Sachs Harbour, the well developed Sand Hills Moraine extends along the coast for 25 km and overlaps Sachs Till at its southeastern extremity. The moraine was likely constructed by a local ice readvance in Thesiger Bay after the first retreat of the Thesiger Lobe.

The Prince Alfred Lobe impinged on the north coast as it progressed into M'Clure Strait from Viscount Melville Sound (Fig. 2.8c). East of Mercy Bay the lobe was confined to the marine channel by high cliffs, whereas to the west ice overlapped the land and built terminal and end moraines. The limit of ice extent stands consistently at about 100 m between Mercy Bay and Ballast Brook, indicating that the Prince Alfred Lobe was likely an ice stream in western M'Clure Strait. Much of the north slope was inundated by proglacial lakes, and the postglacial sea flooded coastal areas to about 30 m following ice retreat.

Prince of Wales and Thesiger lobes were contemporaneous, for meltwaters of the first built deltas into a glacial lake (Lake Masik) which was dammed by the latter. Similarly, Prince of Wales and Prince Alfred lobes were contemporaneous, for the extent of the glacial lake in Thomsen River valley (Lake Ivitaruk) was controlled by the simultaneous presence of the two different lobes in M'Clure Strait and in the upper Thomsen River basin (Table 2.4).

Some researchers (Mayewski et al., 1981; Denton and Hughes, 1983; Dyke, 1987) have argued that the M'Clure Stade, as portrayed here, is Late Wisconsinan. Although Vincent has opted for an Early Wisconsinan age assignment, he has stated (Vincent, 1982, p. 228; 1983, p. 87) that the possibility still remains that both the M'Clure Stade and later Russell Stade occurred in the Late Wisconsinan. Evidence for an Early Wisconsinan age is perhaps best provided by a >37 000 BP (GSC-3698) radiocarbon age and a 92.4 ka(UQT-143) thorium/uranium age on in situ shells of *Portlandia arctica* collected from coastal bluffs east of Mary Sachs Creek (12). These came from ice contact, glaciomarine sediments of the pre-Amundsen Sea, and lay immediately below Sachs Till of the Thesiger Lobe (M'Clure Stade of Amundsen Glaciation; Tables 2.4 and 2.5). As these fossiliferous sediments contain inclusions of the overlying till, the ice reached its farthest extent, at 1 km north of site, well before the Late Wisconsinan. In Kaersok River valley of northern Banks Island, autochthonous mosses overlie glacial lake silts and sands deposited when M'Clure Strait was filled by the Prince Alfred Lobe. These gave a radiocarbon age of >41 000 BP (GSC-1088). Shells of *Astarte* sp., from a raised spit built in Meek Point Sea, gave an age of >19 000 BP (GSC-1478). Amino acid ratios of aIle:Ile (total), for six fragments of *Hiatella arctica* shells from a delta in the East Coast Sea, vary between 0.04 and 0.09. These ratios differ markedly from the average of 0.19 for shells of the Middle Pleistocene Big Sea and of 0.02 from Late Wisconsinan-Holocene Schuyter Point Sea. The ratios indicate that the East Coast Sea, and Prince of Wales Lobe, which it immediately postdates, are post-Sangamonian but distinctly older than Late Wisconsinan. An accelerator radiocarbon analysis of these shells provided an age of 24 730 ± 260 BP (TO-650) whereas a thorium/uranium determination gave an age of 46 ka(UQT-142). Both these determinations should probably be considered minimum ages. These lines of evidence, among others, have been used to infer that, on Banks Island, the M'Clure Stade was pre-Late Wisconsinan, but it is not known if the glaciation occurred during oxygen isotope stage 4 or 5 (Vincent and Prest, 1987).

Melville Island

Bolduc Till of continental origin on southern Dundas Peninsula and Liddon Till on the northern part of the peninsula were assigned to the Early Wisconsinan M'Clure Stade by Hodgson et al. (1984) and Vincent (1984). These tills likely were deposited on the north side of Prince Alfred Lobe, which on its south side deposited Mercy and Bar Harbour tills of northern Banks Island. Shells on the surface of Bolduc Till gave ages of >33 000 BP (GSC-727) and 42 400 ± 1900 BP (GSC-787); the latter is best regarded as a minimum age. The oldest marine sediments that definitely overlap the till are dated 11.7 ka(Hodgson, 1989). As indicated by Hodgson et al. (1984), both samples may be autochthonous and hence date a marine event subsequent to till deposition, or they may be older shells that were brought in by the ice as erratics. Whatever the case, if Bolduc Till correlates with the tills of Prince Alfred Lobe, these dates are a further indication that the M'Clure Stade predates the Late Wisconsinan.

Victoria Island

Figure 2.7 shows that ice, mostly from the southeast, covered Victoria Island except for central Prince Albert

Peninsula and possibly part of the Shaler Mountains (Vincent 1982, 1983, 1984). A well defined glacial limit can be traced around central Prince Albert Peninsula, but not in the more rugged terrain of the Shaler Mountains. At one time there was a strong northwesterly flow across the Shaler Mountains; this may be Early Wisconsinan or older. No evidence for local glaciers has yet been found in the Shaler Mountains (D.A. Hodgson, Geological Survey of Canada, personal communication, 1984). As much of northwestern Victoria Island was possibly not covered by Late Wisconsinan ice, Early Wisconsinan glacial deposits should lie at the surface in these areas (Fig. 2.7).

The composition and surface morphology of the till near Prince of Wales Strait resemble those of Jesse Till (M'Clure Stade) of eastern Banks Island, and that till obviously was laid down by the Prince of Wales Lobe, discussed above.

Plant-bearing sediments, which gave radiocarbon ages of >32 400 BP (GSC-388), >37 000 BP (GSC-3613), and >38 000 BP (GSC-3592) and a thorium/uranium age of 71.8 ka (UQT-230), separate the Late Wisconsinan till from an older one in a river section near Graveyard Bay (8) on southern Diamond Jenness Peninsula (Fig. 2.6g). Both tills are reddish brown, a colour that is characteristic of Wisconsinan tills in the southwestern Arctic Archipelago. The lower till could therefore be Early Wisconsinan and the organic-rich unit Middle Wisconsinan. Further work is needed to confirm the chronology of this crucial site.

Area south of Amundsen Gulf

The lobe of ice that advanced in Amundsen Gulf from the mainland, and impinged on the eastern and southwestern coast of Banks Island during the M'Clure Stade, also impinged on the mainland coast and merged with northwestwardly flowing ice from the Great Bear Lake area. These glaciers probably were too thin to cover the Brock Upland and Bathurst Peninsula (Fig. 2.7; Vincent, 1984).

The tentative Early Wisconsinan limit around the Brock Upland is the upper limit of a moraine belt at an elevation of about 600 m a.s.l. (Fulton and Klassen, 1969). The area above this belt consists of rubble-strewn hills which, as mentioned before, may have last been covered by ice in Middle Pleistocene time. The limit north of 69°N was initially traced by Klassen (1971a) as the "approximate limit of Late Wisconsinan Glaciation"; however, Hughes (1987) assigned the limit on the west side of the plateau to the Late Wisconsinan Tutsieta Lake phase. Based on interpretation of air photographs, Vincent (1984) considered that two distinct glacial limits existed around the high plateau area. The highest, at about 600 m, was tentatively assigned an Early Wisconsinan age, whereas the lower (<300 m in the north) was thought to mark the limit of a much fresher looking till sheet and was considered to be Late Wisconsinan in age. St-Onge (1987), however, following a survey of the area east of the Brock Upland, believed that the two morphosedimentary zones described above are likely of Late Wisconsinan age, and that the observed differences are related to varying flow regimes rather than age.

Cliffs on the east coast of Bathurst Peninsula effectively blocked further westerly progress of the lobe of ice (ice shelf?) in Amundsen Gulf, whereas the ice flowing northwesterly from Great Bear Lake did not progress much north of West River and was restricted to the west of the Anderson River basin (Toker Point Stade of Rampton, 1988; Fig. 2.7).

The possible Early Wisconsinan glacial limit is considered to be the one that Klassen (1971a) mapped, south of 70°N, as the "approximate limit of Wisconsinan Glaciation". This is the limit that Fyles et al. (1972) considered to represent the "maximum extent of the Laurentide Glaciation", and that Rampton (1981a) initially mapped as the "Early Wisconsin (?) glacial limit" and later (Rampton, 1988) as the Franklin Bay Stade limit on the east side of Bathurst Peninsula north of 70°N. West of the peninsula the limit is placed on the west side of the Anderson River basin. On the west side of Franklin Bay the limit is classic. There, a narrow belt of till and glaciofluvial deposits mantles the surface along the sheer coastal cliffs; west of this belt wide proglacial outwash channels, incised in the unglaciated area, are graded to terraces along the river valleys. Later, during its retreat in Amundsen Gulf, the ice built extensive end moraine systems that trend north on the eastern part of Parry Peninsula (Mackay, 1958).

Sands and gravels on Bathurst Peninsula and at the mouth of Mason River, which likely are distal outwash sediments from the ice in Amundsen Gulf, overlie sediments containing driftwood which gave the nonfinite radiocarbon ages mentioned above in the discussion of pre-Wisconsinan events. Fragile willow twigs in crossbedded terrace sands postdating the glacial advance have been dated at 33 800 ± 880 BP (GSC-1974). The dates appear to confirm the Early Wisconsinan age assigned to the last ice along these coastal areas.

Area south of Beaufort Sea and east of Mackenzie River

Ice from north of Great Bear Lake reached the Beaufort Sea at the mouth of Mackenzie River and impinged on the southern part of Tuktoyaktuk Peninsula (Fig. 2.7); there it caused ice thrusting of older sediments in the Nicholson Peninsula area (Mackay, 1956). Rampton (1981b, c) mapped the ice limit on Tuktoyaktuk Peninsula and assigned it (Rampton, 1988) to the Toker Point Stade. Generally, till veneers and glaciofluvial deposits are set into the older fluvial (Kidluit Formation) and deltaic or eolian (Kittigazuit Formation) sediments of the Tuktoyaktuk coastlands. Due to intense thermokarst activity, the glacial drift is typically disturbed or completely reworked, but glacial erratics from it are conspicuous in an area where all underlying sediments are fine. From Toker Point, just northeast of Kugmallit Bay, the ice crossed into the Beaufort Sea, thus covering the area of the Mackenzie River mouth. During deglaciation, the area was isostatically depressed so that shell-bearing marine sediments (Garry Island Member of Rampton, 1988) are now found locally along the Tuktoyaktuk coastlands. Dates on shell fragments have been used to indicate that the advance was pre-Late Wisconsinan (>35 000 BP (GSC-562) on Garry Island, and >37 000 BP (GSC-690) south of Kendall Island). Recent age determinations on likely in situ shells, collected by the author, which overlie Toker Point Till on Garry Island (43 550 ± 470 BP, TO-796; aIle/Ile = 0.017) and from south of Kendall Island (48 000 ± 1100 BP, RIDDL-801; aIle/Ile = 0.032) provide added support to the pre-Late Wisconsinan age assignment of the Toker Point Stade ice advance. The elevation of marine limit is difficult to estimate, but Kerfoot (1969) noted possible strandlines up to 46 m a.s.l. on Garry Island. Proglacial outwash from this stade, particularly on Tuktoyaktuk Peninsula, was assigned to the Cape Dalhousie sands by Rampton (1988). Much allochthonous wood, in many cases contained in fluvial and other sediments postdating Toker Point Stade, has provided nonfinite radio-carbon ages. In addition, many Late Wisconsinan dates from sediments postdating Toker Point Stade deposits are available, the oldest being 17 860 ± 260 BP (GSC-481) for autochthonous peat overlying till that was tilted during growth of the Ibyuk Pingo near Tuktoyaktuk.

Yukon Coastal Plain

West of Mackenzie River, ice of the Toker Point Stade advanced from Mackenzie Valley into the Beaufort Sea and impinged on the Yukon Coastal Plain (Fig. 2.7). The limit and deposits of this event, termed the Buckland Glaciation, have been mapped by Rampton (1982). The glacial limit falls from about 1000 m southwest of Aklavik, in the Richardson Mountains, to near sea level west of Herschel Island. The glacial limit, in many places marked by kame deltas and terraces, is well marked (Fig. 2.10). Fieldwork in 1985 has shown that the 75 ka glaciomarine Flaxman Member of the Gubik Formation, found on the Alaskan Coastal Plain (Carter et al., 1986), extends east along the Yukon Coastal Plain at least to Komakuk Beach (35 km east of the Alaska border). The Flaxman Member could represent a distal glacial marine facies of the Buckland ice advance. If this is the case, and on the basis of the Alaskan age assignment, a late stage 5 age for Buckland Glaciation is obtained. The oldest minimum ages for the glaciation are 22 400 ± 240 BP (GSC-1262) based on fragmented autochthonous plant remains in a pond developed on glacial sediments near Stokes Point and 14 400 ± 180 BP (GSC-1792) from peat collected from a pond deposit in a coastal bluff west of Sabine Point.

According to Rampton (1982), a major stillstand or readvance, termed Sabine Phase of the Buckland Glaciation, occurred during deglaciation. Evidence for this is the major, commonly ice-thrusted moraine-outwash complexes along the coast between Kay Point and King Point.

Area west of Mackenzie River

Glacial limits on the lower slopes of the Mackenzie and Richardson mountains are the subject of much controversy (Hughes et al., 1981; Prest, 1984). They are not readily traced on the steep slopes where significant changes in ice thicknesses cause only limited change in extent. As there are indications that continental ice reached its all time limit along the mountain front, at least locally, during the latter part of the Wisconsin Glaciation (Hughes et al., 1981; Catto, 1986b), this area is discussed below under Late Wisconsinan ice. Because of the evidence for limited Late Wisconsinan ice on the Arctic Coastal Plain, however, Figure 2.7 arbitrarily portrays the most extensive glacial limit in this area as Early Wisconsinan.

Wisconsinan interstade

Radiocarbon-dated organic deposits from a few localities in the northern Interior Plains (Table 2.3), overlying materials thought to be Early Wisconsinan, indicate that an interstadial period preceded the Late Wisconsinan (>23 ka).

On Banks Island, in Kaersok River valley, autochthonous mosses younger than Bar Harbour Till have been dated at >41 000 BP (GSC-1088). At Dissection Creek (14), autochthonous peats, also thought to postdate the M'Clure Stade advance of the Prince Alfred Lobe, gave radiocarbon ages of >39 000 BP (GSC-2819) and 49 100 ± 980 BP (GSC-2375-2). Arthropod remains in these peats indicate a warmer climate than now (Matthews et al., 1986). On Victoria Island, the organic deposits underlying a Late Wisconsinan till, in a section (8) previously discussed (Fig. 2.6g), may also be interstadial. The macrofloral and arthropod remains in the various organic deposits indicate conditions similar to the present.

East of Mackenzie River, on the Arctic Coastal Plain, much allochthonous wood has been radiocarbon dated. The only finite date (33 800 ± 880 BP, GSC-1974) on non-redeposited material, however, is from a terrace younger than the Toker Point Stade along Cy Peck Inlet on Bathurst Peninsula.

On the unglaciated part of the Yukon Coastal Plain, west of Herschel Island, plant and moss fragments from a terrace on Malcolm River dated as 34 600 ± 1480 BP (GSC-1900); macrofloral remains were of extant plants. Finally, in a section along Hungry Creek (9) (Fig. 2.6h), fluvial and deltaic sediments found under till are likely interstadial. Plant macroremains and insects from those beds, which contain wood dated 36 900 ± 300 BP (GSC-2422), indicate climatic conditions similar to or warmer than those of today (Hughes et al., 1981).

Late Wisconsinan glaciation

The limit proposed for Late Wisconsinan ice is shown in Figure 2.7, glacial events assigned this age are correlated in Table 2.5, and features developed during this glacial interval are portrayed in Figures 2.4a-c. The limit shown is a minimum position (see Prest, 1984) and, consequently, most of Banks Island and northwestern Victoria Island, along with significant areas of the mainland south of the Beaufort Sea and Amundsen Gulf, is shown as having escaped glaciation at this time. However, an ice shelf existed in Viscount Melville Sound (Hodgson and Vincent, 1984b) and perhaps in eastern Amundsen Gulf. Both on the mainland and Victoria Island, ice flowed westward and northwestward from the Keewatin Sector of the Laurentide Ice Sheet.

Upper Mackenzie River basin

In the Upper Mackenzie River basin ice advanced westwardly to the Mackenzie Mountains, which deflected it north along Mackenzie River valley and south along Liard River valley (Craig, 1965; Rutter and Boydell, 1973). A silty to clayey, stony till covers most of the area, generally as a till plain or fluted till plain. According to Rutter (1974), the Late Wisconsinan glacial limit, in many places marked by meltwater channels, is about 1525 m a.s.l. west of the Nahanni Range and 1295 m in the Wrigley Lake area. On the other hand, in the South Nahanni River basin Ford (1976) equated the limit of Rutter to that of his Illinoian Clausen Glaciation, and places the limit of his Late Wisconsinan Jackfish Glaciation west of Nahanni Range at 800 m. The minimum limit of Ford (1976) is shown in Figure 2.7.

Recent detailed airphoto interpretation in Mackenzie Mountains is contributing to a better understanding of Laurentide ice extent and relationship to montane glaciers.

Hughes and Duk-Rodkin (1987) have shown that as continental ice pressed on the Canyon Ranges of Mackenzie Mountains, ice tongues extended more than 40 km up the Keele, Mountain, and Arctic Red River valleys. Montane glaciers in the Keele and Mountain river valleys did not merge with the Laurentide Ice Sheet, but evidence now indicates that valley glaciers in the Arctic Red, North Redstone, and Redstone river basins merged with continental ice after it had begun to retreat from its maximum stand.

The glacier generally retreated northeastward and eastward (Craig, 1965) allowing glacial lakes dammed in mountain valleys at the glacial maximum to spread eastward. Ice stagnated on the higher Horn Plateau and on plateaus south of Mackenzie River, forming hummocky topography and leaving end moraines. Glacial Lake McConnell (Craig, 1965) enveloped much of the Liard and southern Mackenzie River basins, extended south of 60°N in the valleys of Peace and Slave rivers and into the extensive lowlands around and

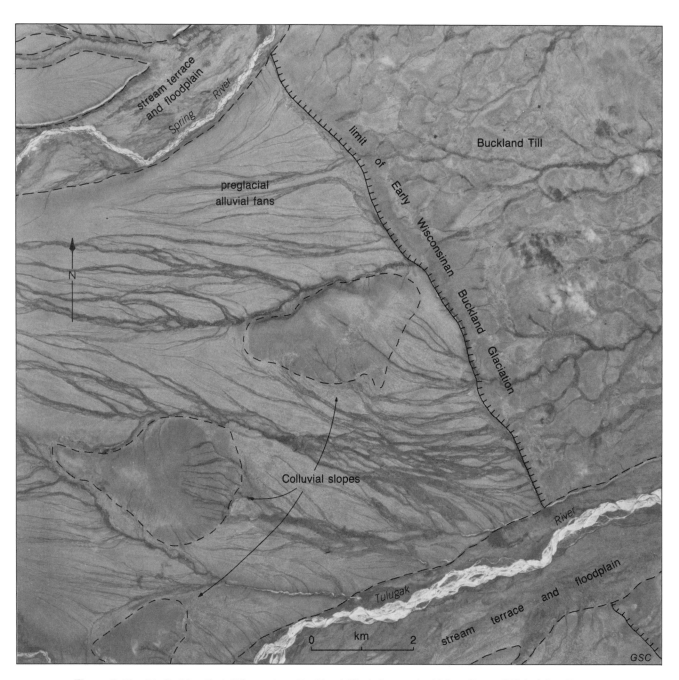

Figure 2.10 Limit of the Early Wisconsinan Buckland Glaciation on the Yukon Coastal Plain (after Rampton, 1982) NAPL A24123-129

between Great Slave, Great Bear, and Athabasca lakes (Fig. 2.4a), locally leaving thick glacial lake deposits (see also Dyke and Dredge, 1989).

Craig did not show the lake as extending much north of the mouth of Liard River in Mackenzie River valley, but Rutter et al. (1973), Hughes et al., (1973), Mackay and Mathews (1973), and Savigny (1989) reported extensive lacustrine deposits throughout the valley. The relationship of these "ponding events" to glacial Lake McConnell, however, is unknown. The mechanism for ponding glacial Lake McConnell in the Mackenzie River basin is not completely understood, but Craig (1965) suggested that isostatic tilting might have played a role. His shoreline elevation data appear to verify this hypothesis, because they indicate upward tilting to the east. Relations between glacial Lake McConnell and the other lakes that formed along the western and southern margin of the retreating Late Wisconsinan glacier, in the Peace River area to the south (Prest, 1970) and Coppermine River area to the east (St-Onge, 1980), are unknown. Elson (1967) suggested that water from his Campbell Phase of Lake Agassiz flowed westward into Lake McConnell, but Christiansen (1979) stated Lake Agassiz did not reach the Clearwater Spillway because lake sediment do not extend that far west.

Organic detritus overlying glacial lake sediments and till from near the mouth of Great Bear River dated 10 600 ± 260 BP (GSC-2328). This is the oldest minimal age available for deglaciation and drainage of glacial Lake McConnell in this area.

Area between Richardson Mountains and Anderson Plain

In this area, ice flowed north to northwest down Mackenzie Valley (Fig. 2.7). West of Mackenzie River and north of Mackenzie Mountains, its limit is placed at what Hughes (1972) initially mapped as the "limit of Wisconsinan Laurentide Ice". North of 68°N in the West Channel area, it follows the margin of Late Wisconsinan Glaciation (Sitidgi Lake Stade) shown by Rampton (1982, 1988). The limit drops from about 500 m in the Snake River area to 120 m west of Aklavik and thence to sea level near Shallow Bay. According to Hughes (1972), these limits "are marked by well-preserved moraines that contrast sharply with subdued glacial landforms beyond". East of Mackenzie River, an ice margin with many well developed terminal moraines mirrors the margin west of Mackenzie River. Again north of 68°N (Fig. 2.7), the limit follows the "Late Wisconsin glacial limit" of Rampton (1980). From 68°N to Yeltea Lake it coincides with the Tutsieta Lake Moraine (Fig. 2.11), which was named and assigned to the Tutsieta Lake Phase by Hughes (1987). Hughes et al. (1981) do not agree that this is the Late Wisconsinan limit, but consider it a stillstand during retreat of Laurentide ice in Late Wisconsinan time. They place the Late Wisconsinan limit farther west at the all time glacial margin, and at what Rampton (1988) considered the Early Wisconsinan Toker Point Stade limit on the Arctic Coastal Plain. Hughes et al. (1981) and Catto (1986b) placed the Late Wisconsinan margin far to the west on the east flank of the Cordillera in Upper Rat River. This is done to allow blockage of drainage from the Porcupine River basin through McDougall Pass, and so to divert it west to create Late Wisconsinan lakes in the basins of the unglaciated Yukon (for a discussion of these lakes see Hughes et al., 1989). In Upper Rat River valley, proglacial lake sediments overlie till assigned to the Late Wisconsinan (Catto, 1986a). Organic detritus within the lake sediments have given radiocarbon ages of 21 300 ± 270 BP (GSC-3371) and 21 200 ± 240 BP (GSC-3813). According to Catto, these indicate that the lake, which was dammed by major advance of continental ice against the Richardson Mountain front, existed at least until 21.2 ka. On Hungry Creek (9) (Fig. 2.6h), well west of the Late Wisconsinan limit shown in Figure 2.7, allochthonous wood from below till gave a radiocarbon age of 36 900 ± 300 BP (GSC-2422), indicating that the overlying till and its glacier (Hungry Creek Glaciation of Hughes et al., 1981; Table 2.5) are Late Wisconsinan. The age of the Hungry Creek Glaciation is likely earliest Late Wisconsinan, as it appears that the ice reached its maximum between 25 ka (maximum age obtained by dating a mammoth tusk found in alluvium underlying glacial lake clay in the Yukon Bluefish Basin; see Morlan, 1986) and 16 000 ± 420 BP (GSC-2690) or 15 200 ± 230 BP (GSC-2758) (basal dates on lake sediment cores inside the maximum limit of Hungry Creek ice; see Ritchie, 1982). This extreme position of Late Wisconsinan ice is seemingly not compatible with data from the Arctic Coastal Plain, which corroborate the limit shown in Figure 2.7. Reasons for assigning the Toker Point Stade — Franklin Bay Stade to a pre-Late Wisconsinan stade have already been mentioned (nonfinite radiocarbon dates for shells and wood postdating the advance; see Table 2.3). Also, the coast has not experienced glacial isostatic depression since at least 48 200 ± 1100 BP (RIDDL-801). If Late Wisconsinan ice had reached the limit of Hughes et al. (1981), it would also have covered much of the Beaufort Sea coast and so caused isostatic submergence of the coastline. However, the offshore record shows that the shelf was exposed to at least 140 m below present sea level, about 27 ka ago (radiocarbon dates on core material; see Table 2.3), and that it has progressively been submerging since that time (Hill et al., 1985). The above discussions show that having abundant ice on the east slope of the Cordillera creates problems with Arctic Coastal Plain reconstructions, whereas less ice on the Arctic Coastal Plain would create problems in the reconstruction of events on the east side of the Cordillera. A possible solution is to have three Wisconsinan advances, with the oldest (Early Wisconsinan) corresponding to the Toker Point Stade and Buckland Glaciation, the youngest to the Tutsieta Lake Phase being "late" Late Wisconsinan, and with an intermediate advance late in the Middle Wisconsinan or early in the Late Wisconsinan. This middle advance would correspond to the limits of Hungry Creek Glaciation, and perhaps the Sabine Phase of Rampton (1982) on the Yukon Coastal Plain, and the limit just north of the southern Eskimo Lakes that was mapped by Fyles et al. (1972) as an "alternate (sic) possible limit of classical Wisconsin glaciation east of Caribou Hills". Unfortunately, dating control is not adequate to decide the matter.

The Late Wisconsinan lobe in lower Mackenzie Valley (Fig. 2.7) receded to the southeast after reaching its limit, leaving ice marginal features such as major meltwater channels or moraines like the Tutsieta Lake Moraine or those around the Grandview Hills. Lakes were dammed in valleys at the margin of this lobe and also in Mackenzie River valley (Hughes et al., 1973; Mackay and Mathews, 1973). Belts of hummocky moraine, such as the one north of Arctic Red River, were also formed. Radiocarbon dates at the north end of the lobe indicate that ice retreat was underway at least by 13 ka (GSC-1784-2, -1995, and -3387, Table 2.3) and possibly as early as 15 500 ± 440 BP (GSC-3646, Table 2.3) if a recent age determination is deemed reliable. By 12 400 ± 120 BP (GSC-3691) ice had disappeared from the upper Caribou River basin (Catto, 1986b). At the southern end of the lobe, the area near the mouth of Mountain River was free of ice and the glacial lake in Mackenzie Valley drained before 11 530 ± 170 BP (I-3734). Speculative ice margins during retreat are portrayed in Figure 2.12.

Area east of Mackenzie River and south of Amundsen Gulf

The Late Wisconsinan Sitidgi Stade limit east of Mackenzie River and south of Amundsen Gulf is portrayed in Figure 2.7. Ice advanced northwesterly from Great Bear Lake; topographic depressions, such as upper Anderson River and Hornaday River valleys and Dolphin and Union Strait, along with higher plateaus in the upper Horton River basin and south of Brock Upland, directed the ice flow. From east

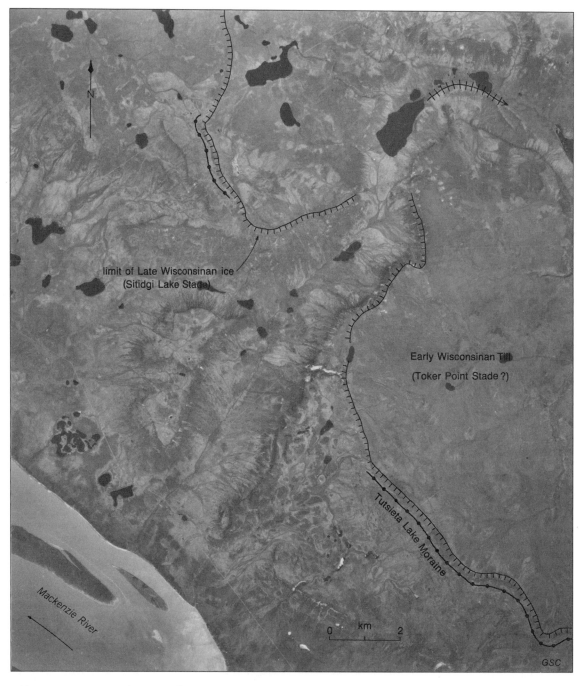

Figure 2.11 Limit of ice advance during the Late Wisconsinan Sitidgi Lake Stade in the Tutsieta Lake area, District of Mackenzie (modified from Hughes, 1987). NAPL A21585-191

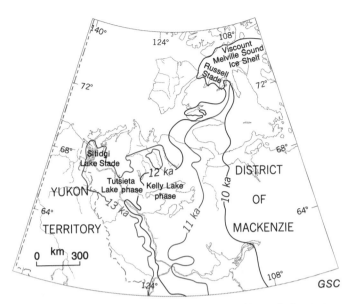

Figure 2.12 Speculative ice marginal positions during Late Wisconsinan retreat in the northern Interior Plains (at 10, 11, 12, and 13 ka).

Victoria Island

Ice advanced northwesterly and westerly on Victoria Island from the M'Clintock Ice Divide (Keewatin Sector) of the Laurentide Ice Sheet (Dyke et al., 1982; Dyke, 1984; Dyke and Prest, 1987). Ice was channelled along major depressions in Coronation Gulf-Dolphin and Union Strait, Prince Albert Sound, Minto Inlet, and Hadley Bay. The ice limits in Figure 2.7 are consistent with a minimum Late Wisconsinan ice advance point of view (Vincent, 1984), but the extent of ice remains uncertain. The westernmost part of Diamond Jenness Peninsula and most of Prince Albert Peninsula and the Shaler Mountains are shown as remaining ice free. Sharpe (1984 and 1988) convincingly argued for complete Late Wisconsinan coverage of Wollaston Peninsula, but he believed that the ice that overrode the top of the peninsula was thin. On westernmost Diamond Jenness Peninsula, preliminary mapping of well marked glacial limits north of Prince Albert Sound and south of Minto Inlet indicates that Late Wisconsinan ice wrapped around the westernmost part of the peninsula. The Prince of Wales Lobe that impinged on eastern Banks Island during M'Clure Stade also impinged on Prince Albert Peninsula, leaving, as on Banks Island, a characteristic continuous cover of reddish brown till now covered with high-centre polygons. Because evidence for a younger ice cover is lacking on Prince Albert Peninsula, the equivalent Russell Stade limit must lie farther south and east. No precise limit has been traced in the dissected bedrock there, but a tentative limit, which wraps around the higher Shaler Mountains, is portrayed in Figure 2.7. If ice was thin and cold-based over the upland areas, as suggested for Wollaston Peninsula (Sharpe, 1984), then distinct landforms representing an ice limit may not have resulted. More information is obviously needed to confirm the position of ice limits for northwest Victoria Island.

The pattern of deglaciation was complex on Victoria Island (Fyles, 1963). On Wollaston Peninsula, Sharpe (1988) outlined a pattern of late glacial landforms that suggests normal ice marginal retreat (a series of moraines and drainage channels), followed by possible surges (shear moraines) and regional stagnation (broad areas with no marginal forms). Streamlined landforms occur in broad depressions whereas hummocky moraine occurs against topographic highs. Glacial lakes were dammed between thick ice in major channels and uplands prior to marine incursion. Low profiles (1-2 m/km) for the Colville Moraines on Wollaston Peninsula may relate to surging induced by abundant subglacial water (Sharpe, 1988). Low profiles on drift limits on northern Victoria Island have been related to the development of an ice shelf in Viscount Melville Sound (Hodgson and Vincent, 1984). On eastern Victoria Island crosscutting streamlined landforms are common but they are not generally bounded by moraines. The crosscutting features appear to represent rapid surges triggered by ice profiles responding to calving in the postglacial sea (Hodgson, 1987; see also Dyke and Dredge, 1989). These events may be analogous to multiple ice marginal surges that have been inferred for James Bay Lowlands (Dredge and Cowan, 1989; Vincent, 1989). Alternatively, the streamlined forms have been suggested to represent bedforms produced by subglacial meltwater floods as their internal structure reflects passive rather than deformed sedimentation (Sharpe, 1985). Others have suggested that the stratified sediments may have been proglacial deposits overridden during the surges. Furthermore, the crosscutting landforms could have been

of Mackenzie River northeastward to south of Parry Peninsula, and around upper Horton River basin, the limit follows the Tutsieta Lake Phase limit of Hughes (1987). Southeast of Parry Peninsula the Late Wisconsinan limit is linked with the lower limit around the Brock Upland and the ice is shown as not having reached Amundsen Gulf. This picture is similar to that of Mackay (1958) who plotted the limit of the "last ice advance", which he termed "Great Bear Lake Ice Lobe". As described by Craig (1960), the ice front generally retreated southeasterly towards the Canadian Shield while building spectacular systems of end moraines and leaving widespread hummocky moraine and ice contact deposits (Fig. 2.4a,b). Dead ice topography is particularly prominent north and northwest of Smith Arm of Great Bear Lake (Mackay, 1960; Hughes, 1987). The best developed moraine system is the Kelly Lake Moraine, which extends northeastward from the Norman Range (Fig. 2.7; Hughes, 1987). The area west of this moraine in the Franklin Mountains was certainly deglaciated by 11 000 ± 340 BP (GSC-3536) as indicated by a basal lake sediment age determination. Ice retreated from the Kelly Lake Moraine by 10 600 ± 260 BP (GSC-2328). In the Hornaday River basin, a ^{14}C date of 10 800 ± 150 BP (GSC-1139) on peaty moss from a pingo provides a minimum age for deglaciation in the north, whereas by 10 500 ± 200 BP (GSC-3524) the eastern end of the Horn Plateau in the south was deglaciated. As the ice withdrew, the sea flooded the coastal area (Fig. 2.13). The marine limit, according to Craig (1960), decreases from about 145 m above present sea level in the lower Coppermine River area to less than 45 m near Tinney Point at the west end of Dolphin and Union Strait. The most westerly record of emergence is found at the head of Darnley Bay (J-S. Vincent, unpublished). Shells from Darnley Bay dated at 11 280 ± 100 BP (TO-217) and from south of Cape Young dated at 10 530 ± 260 BP (I(GSC)-25) provide the oldest minimum ages for deglaciation of the coast south of Amundsen Gulf. Postulated ice margins during retreat are portrayed in Figure 2.12.

CHAPTER 2

formed by short lived ice streams surrounded by slower moving ice rather than by outlet flows such as those of surges (D.R. Sharpe, Geological Survey of Canada, personal communication, 1988). Ice streaming may explain the lack of ice marginal landforms. Esker patterns that transect streamlined fields appear to define former multiple ice streams. These eskers, and the lack of ice marginal landforms, suggest regional ice stagnation followed the ice thinning events in much of eastern as well as western Victoria Island (Sharpe, 1988). The interpretation of complex flow patterns and their inferred ice dynamics models may allow researchers to apply this new understanding on Victoria Island to other areas covered by the Laurentide Ice Sheet.

The sea inundated isostatically depressed areas outside the Late Wisconsinan glacial limit. Marine limit reached 25 m on eastern Banks Island (Schuyter Point Sea), between 64 and 90 m on Prince Albert Peninsula, and between 70 and 81 m on western Diamond Jenness Peninsula. The sea also submerged large tracts of land as the ice receded (Fig. 2.13). On southern Victoria Island marine limit rises eastward from 115 m at the western tip of Wollaston Peninsula (Sharpe, 1984) to 175 m on the southeast coast. Radiocarbon dates indicate that submergence of the unglaciated areas took place as early as 12 600 ± 140 BP (GSC-1707). By 10 300 ± 100 BP (GSC-3801) eastern Prince Albert Sound, and by about 9 ka most of Victoria Island were free of ice.

Viscount Melville Sound region

In the Viscount Melville Sound region, Hodgson and Vincent (1984b) have shown that an ice shelf, with a surface area of at least 60 000 km^2, spread northwestward at the end of the last glaciation. The edge of the shelf marked the limit of Late Wisconsinan Laurentide Ice (Fig. 2.7). The ice shelf impinged on coasts and deposited Winter Harbour Till up to 125 m above present sea level on the southern coast of Melville Island, 135 m on Byam Martin Island, possibly 90 m on the northeast tip of Banks Island (Fig. 2.8d), and to 150 m on the north coast of Victoria Island (Fig. 2.14). Shells from marine sediments older than the till left by the shelf provide a maximum age of 10 340 ± 150 BP (GSC-278) for the ice advance whereas shells from marine sediments above the till give a minimum age of 9880 ± 150 BP (GSC-3527; Table 2.3). The major advance of shelf ice into Viscount Melville Sound may have been connected with the disintegration of the M'Clintock Dome, which in turn was probably associated with climatic amelioration in the Western Arctic (Hodgson and Vincent, 1984b; Hodgson, 1987).

The Viscount Melville Sound region is the only part of the northern Interior Plains where it has been possible to determine precisely when Laurentide ice reached its Late Wisconsinan limit. The maximum in this region (about 10 ka) is two to three thousand years younger than that of lower Mackenzie River valley (about 13 ka) or of the

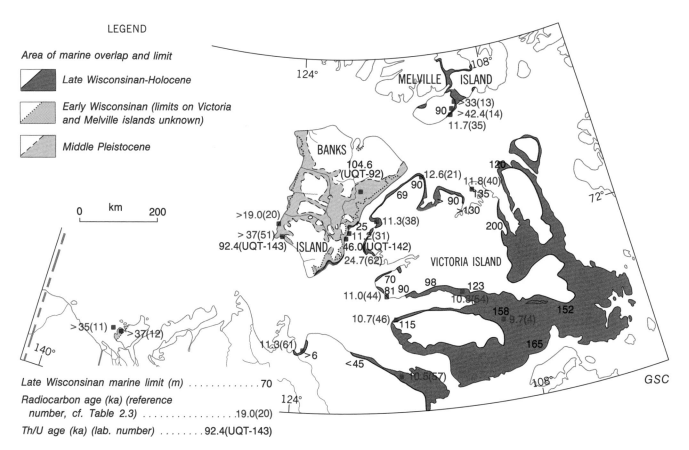

Figure 2.13 Middle and Late Pleistocene marine transgressions in the northern Interior Plains (modified from Craig, 1960; Fyles, 1963; Hodgson et al., 1984; Sharpe, 1984; J-S. Vincent, 1983, unpublished data).

QUATERNARY GEOLOGY — CANADIAN INTERIOR PLAINS

Amundsen Gulf area (about 12 ka). This younger age of the ice limit to the east apparently continues to Somerset Island where the ice did not retreat from its Late Wisconsinan maximum until about 9.3 ka (Dyke, 1983; Dyke and Dredge, 1989).

Discussion of Late Wisconsinan limit problem

In this synthesis of the Wisconsin Glaciation, an attempt has been made to correlate and assign ages to various deposits and glacial limits documented by several authors. As is

Figure 2.14 Limit of the Late Wisconsinan Viscount Melville Sound Ice Shelf, northern Victoria Island (D.A. Hodgson, unpublished data). NAPL A17372-170

also the case with the southern Interior Plains part of this chapter, age assignments have to be considered tentative because of the lack of firm chronological control (see also Vincent and Prest, 1987). As more data become available, particularly for northwestern Victoria Island, the Brock Upland, Bathurst Peninsula, and the flanks of the Cordillera, a more accurate picture should evolve.

Two distinct Wisconsinan glacial stades are definitely recorded: the oldest by the Hungry Creek Glaciation, Buckland Glaciation, Toker Point Stade, Franklin Bay Stade, M'Clure Stade and the youngest by the Tutsieta Lake Phase, Sitidgi Lake Stade, Russell Stade. The evidence of various authors for stating that the former is separated from the latter by a Middle Wisconsinan interstadial has been presented as has the evidence for suggesting that the two stades followed each other closely during the Late Wisconsinan and that the youngest ice advance was merely a readvance of previously more extensive ice. A definitive solution of this problem has an important bearing on our understanding of the Laurentide Ice Sheet. Was there or was there not a major Laurentide Glaciation in the northwest at some time during isotopic stage 5 or 4? If one occurred, when did it occur? If there was not a glaciation at this time, when did ice first cover the northern Interior Plains during the last glaciation?

SEA LEVEL, CLIMATE, AND PROCESSES
Sea level history in the Beaufort Sea area
In the Beaufort Sea area north of the Mackenzie Delta and the Tuktoyaktuk Peninsula, a continuous record of Late Wisconsinan and Holocene submergence exists. Following the work of Richards (1950), Mackay (1963), and Forbes (1980) who discussed the evidence for submergence, Hill et al. (1985) have prepared a sea level curve for the Beaufort Shelf showing a relative rise of sea level of 140 m since 27 ka BP. On land the evidence for Late Wisconsinan-Holocene submergence includes drowned valleys, lagoons, large barrier and spit complexes, terrestrial organic beds uplifted from below present sea level during growth of pingos, and radiocarbon dated wood from a core in the Mackenzie Delta that indicates a sea level between -33 m and -18 m at 6900 ± 110 BP (GSC-54) (Johnson and Brown, 1965, and Mackay, 1963). On the continental shelf drowned river valleys, channelled levees and delta mouth bars, a well defined shoreline at about -70 m, submarine permafrost and ground ice, relict ice scour tracts in deep water, numerous radiocarbon-dated organic sediments from drill cores and, a previously exposed land surface identified by acoustic data give clear indication of submergence. Despite this, the causes of the submergence and the precise times of the low sea levels are not clear. Hill et al. (1985) estimated that about 35 m of the 140 m rise of relative sea level was from subsidence due to basin deformation, subsidence along shallow growth faults, and sediment loading or consolidation. They attributed the rest of the submergence to migration of a forebulge associated with "late" Middle Wisconsinan and Late Wisconsinan ice. If the minimum reconstruction and timing of the Late Wisconsinan ice emphasized in this section are correct, little glacial isostatic effect would be expected on the continental shelf. As on the Alaskan coastal shelf (Dinter, 1984), the relative rise of sea level in the Late Wisconsinan and Holocene may be due largely to worldwide eustatic rise of sea level associated with the melting of the continental ice sheets and with hydro-isostatic loading.

Late Wisconsinan and Holocene climate
Climatic reconstructions based on pollen studies of lake cores are available for the northern Interior Plains (Ritchie and Hare, 1971; Ritchie, 1972, 1977, 1980, 1981, 1982, 1984; Hyvarinen and Ritchie, 1975; Cwynar and Ritchie, 1980; Cwynar, 1982; Ritchie and Cwynar, 1982; Ritchie et al., 1983; Spear, 1983; Slater, 1985; MacDonald, 1987). Ritchie (1989) and Schweger (1989) have discussed the development and changes of vegetation during the Late Wisconsinan and Holocene in northern Yukon and the boreal forest part of this region. From about 30 to 14 ka the unglaciated northern Yukon was cold and the vegetation resembled the modern herb fell-field communities. About 14 ka ago northern Yukon Territory, and about 12 ka ago lower Mackenzie Valley, became warmer and wetter with an increase in dwarf birch indicating a shift from herb to shrub tundra. The time between about 11 ka and 9 ka ago was much warmer (perhaps reflecting the predicted Milankovich thermal maximum, Ritchie et al., 1983) and saw northward migration of treeline with increases, first in spruce and then in alder. About 5.5 ka ago treeline started to retreat southward, and by 4 ka the climate had cooled substantially. Shrub tundra, similar to that of today, became established on the mainland coastal plain.

Little information exists on vegetation and climate change in the Western Arctic Archipelago. Numerous radiocarbon dates on Banks Island and on the mainland coastal plain indicate that peat growth reached its maximum around 10 ka ago.

The Late Wisconsinan-early Holocene general warming on the islands and the mainland may explain the rapid disintegration of the M'Clintock Dome, the rapid advance of the Viscount Melville Sound Ice Shelf (Hodgson and Vincent, 1984b) and the well dated maximum of thermokarst activity on the Arctic Coastal Plain (Rampton, 1973). The late Holocene deterioration of climate may have initiated geological processes, such as the growth of pingos (Mackay, 1962) or eolian activity (Pissart et al., 1977).

Permafrost history
The presence of ice-wedge pseudomorphs and other frost forms in pre-Wisconsinan deposits along the Yukon coast (Rampton, 1982), in bluffs on the west side of Bathurst Peninsula, and in many sediments including late Tertiary-early Quaternary Worth Point Formation on Banks Island (Vincent, 1983; Matthews et al., 1986), bears witness to a long history of permafrost in the northern Interior Plains. Particularly good evidence for Wisconsinan permafrost is found on the Arctic Coastal Plain of the mainland. Ground ice is preserved at several localities in pre-Wisconsinan sediments that were deformed by the Early Wisconsinan glaciers (Mackay et al., 1972; Mackay and Matthews, 1983). Widespread and thick, undeformed bodies of ground ice postdating the Early Wisconsinan glaciers are also found along the Beaufort Sea coast (Mackay, 1971; Rampton, 1971b). The source of the water may have been glacial (Mackay, 1983), which would support the hypothesis that

the large bodies of segregated ice grew during the various deglaciations (Rampton, 1974a). During the early Holocene thermal maximum, bodies of ground ice partly thawed, giving a thicker active layer that truncated ice wedges both on the mainland (Rampton, 1973) and in the Western Archipelago (French et al., 1982). Much of the permafrost found on the floor of the Beaufort Sea probably grew when the shelf was exposed before the Late Wisconsinan-Holocene submergence.

Geomorphic processes

Numerous reports, particularly those of Mackay and French, deal with specific recent geomorphic processes in the northern Interior Plains. These reports are listed in summary regional works, such as those of Mackay (1958, 1963), Rampton (1982, 1988), Vincent (1983), and Washburn (1947), and in syntheses such as those of French (1976) and Carter et al. (1987). As in most areas, rivers are the chief landscape modifier. The recent Horton River breakthrough (Mackay, 1981) is an example of the more spectacular impact of rivers; the breakthrough of the Horton, which formerly flowed into Harrowby Bay, to Franklin Bay shortened the river by some 100 km and caused rejuvenation of the basin. Mass movements on slopes are also important. Gelifluction and creep are common on all slopes with fine grained material, as shown by the widespread presence of stripes and solifluction lobes and sheets. Rapid mass movements are less common, but active layer detachment failures in the Cretaceous marine formations and rockfalls and debris slides in the mountains are important. On the mainland Arctic Coastal Plain and in the Arctic Archipelago frost processes have caused sorting, mixing, heaving, tension cracking, and shattering of both consolidated and unconsolidated surface materials. Patterned ground with hummocks, high and low centre polygons, and mudboils reflect this frost activity. As mentioned before, the growth or decay of ground ice bodies, such as massive segregated ice, ice wedges, and pingo ice, has substantially altered large tracts of land. Coastal processes that modify seashores and shores of the larger lakes, as well as limited eolian, nival, and human activity, have contributed to a lesser extent in modifying landscapes.

ECONOMIC GEOLOGY

Sources of aggregate materials can be identified from the surficial geology maps that cover much of the northern Interior Plains (Map 1704A). General comments on the aggregate sources on the mainland can be found in Rutter et al. (1973), Hughes et al. (1973), and Rampton (1974b), and on Banks Island in Vincent (1983). Granular materials can generally be recovered from outwash and ice contact deposits, Quaternary and Tertiary alluvium (floodplains and terraces), and nearshore marine and glacial lake deposits. The large amount of ground ice on both the mainland Arctic Coastal Plain and on Banks and Victoria islands can cause problems in utilizing some of these deposits. In the lower Mackenzie River area many of those sources generally lack material coarser than sand. Consequently, finding readily available aggregate in that area can be a problem.

Melting of ground ice leads to subsidence and loss of strength in sediments; aggradation of permafrost can lead to buildup of ground ice and heaving. Therefore, any construction activity must be preceded by careful assessment of subsurface conditions and be conducted with the least possible disruption of the natural thermal regime. References to studies on permafrost and processes related to it are given in the sections on *Permafrost history* and *Geomorphic processes*. Geotechnical investigations related to exploitation of petroleum resources have been carried out throughout Mackenzie Valley, Mackenzie River Delta, and Beaufort Sea Shelf. The majority of the information gathered during these studies remains in the hands of private companies or is on file with government agencies; some of these references are given in Heginbottom et al. (1977).

SUMMARY

The northern Interior Plains contain one of the longest records of Quaternary glacial and interglacial events in Canada. In the western Arctic Archipelago, Banks Island reveals three full glaciations followed or preceded by warm, nonglacial intervals. The oldest, Banks Glaciation, is Early Pleistocene or older, Thomsen Glaciation is Middle Pleistocene, and Amundsen Glaciation, with two distinct stades, is Late Pleistocene. The coastal plain of the mainland records one or two full glaciations preceding the last interglaciation. During the Late Pleistocene, ice apparently advanced twice towards the Beaufort Sea. Finally, in the upper or middle Mackenzie River basin two Late Pleistocene stadials are recorded in some areas, whereas elsewhere up to three full glaciations are possibly documented.

Large areas of the northern Interior Plains remain relatively unexplored. Further work and better dating methods should make it possible to correlate glacial limits with more certainty and determine with more precision the ages of the glacial and nonglacial events. At present, correlation of the terrestrial events with the marine isotopic stages or with terrestrial events in neighbouring areas, such as the Alaskan Brooks Range and Arctic Coastal Plain, the Cordillera, or the unglaciated part of northern Yukon Territory, can only be tentative (Table 2.1).

CHAPTER 2

QUATERNARY GEOLOGY OF THE SOUTHERN CANADIAN INTERIOR PLAINS

R.W. Klassen

Prest (1970) provided a good summary of our knowledge of the Quaternary of the southern Interior Plains of Canada. Studies since then have little changed our knowledge of the broad Quaternary framework or our understanding of the surface features. The focus of work has changed, however, particularly in southern regions, to a more three-dimensional approach based on stratigraphic drilling and mapping. This has extended our knowledge of glaciation back in time and provided a better understanding of the Quaternary history. Studies that exemplify this approach include those of Stalker (1963, 1969), Klassen and Wyder (1970), Christiansen (1971a, 1972, 1979), Stalker and Churcher (1972), Klassen (1979), Stalker and Wyder (1983), and Shetsen (1984). In addition to this work in well settled parts of the area, several reconnaissance mapping projects touching on the northern part of the region have added to our knowledge of regional Quaternary history (Dredge, 1983; Klassen, 1983a, 1986; Schreiner, 1983, 1984). Other important investigations include studies in southern Manitoba by Fenton and Anderson (1971), McPherson et al., (1971), Teller (1976), and Teller and Fenton (1980); studies in Saskatchewan by Christiansen (1968a, 1968b); and those farther west in the plains and along the mountain front by Reeves (1971), Alley (1973), Stalker (1973, 1976b, 1983), Stalker and Harrison (1977), and Jackson (1979, 1980a, b). Further, the increasing number of radiocarbon dates available, along with vertebrate paleontology and other studies, has helped establish a chronological framework and provide a record of climate changes. In particular, the radiocarbon dates and the studies of postglacial vegetation history (Mott, 1973; Ritchie, 1976; Ritchie and Yarranton, 1978) have laid the basis for deciphering the late glacial and Holocene history of the Plains.

PHYSICAL SETTING
Bedrock geology

Bedrock geology of the area is shown in Figure 2.15. Except for a narrow belt of Paleozoic, mainly carbonate, rock found along the southwest margin of the Precambrian Shield and minor exposures of shield rock in this same area, the surface rock consists of clastic sediments. These largely fine grained rocks are poorly consolidated, and only in the foothills and near the mountains are they moderately to well consolidated.

The most extensive rocks are of Cretaceous age. These consist typically of shales and siltstones, laid down in shallow seas, that interfinger westward with deltaic and fluvial siltstone and sandstones which include some coal. An important component of these units is bentonite which consists mainly of highly expansive montmorillonite. The presence of this clay imparts distinctive properties to these rocks.

Lower Tertiary rocks are widespread in southern and western parts of the Canadian Interior Plains (Fig. 2.15). These consist largely of terrestrial sandstones and siltstones. Although poorly consolidated, they are generally coarser and so more resistant than the Cretaceous beds. As a consequence they commonly persist as caps on hills and plateaus that rise above lowlands cut in Cretaceous shales. They represent the outpouring of debris that was shed as the Rocky Mountains rose during Paleocene through into Oligocene time. Subsequent erosion has reduced their extent to remnants found largely near the foothills, where they were thickest (e.g. Porcupine Hills), and close to the United States border (Cypress Hills, Wood Mountain, and Turtle Mountain). These latter remnants were preserved chiefly because they occupied a broad, interfluve area between drainage to the ancestral Missouri River and that to rivers farther north.

Younger Tertiary deposits are found locally over much of the southern Canadian Plains (Vonhoff, 1965, 1969) either exposed in unglaciated areas such as Cypress Hills and the Del Bonita area or covered by drift as in areas north of Brooks, Alberta and in the Swan Hills. These deposits typically are less than 10 m thick and consist chiefly of coarse to fine sand and quartzite pebble gravel, derived from the western mountains, along with some local material. They lack materials from the Precambrian Shield. The coarseness of these younger Tertiary deposits distinguishes them readily from the older ones, which contain little coarse sand and lack gravel. The reason for the change from fine to coarse deposits is unknown; changes in relative rates of uplift of the Rocky Mountains and adjoining plains or in climate may have been factors. Deposition of these distinctive gravels and sands evidently continued from the Oligocene right through the Pliocene, and the deposits are well known for their vertebrate fossils.

Gravels and sands, which are of similar lithology to the coarse Tertiary materials that occur largely on the uplands, are found in the bottoms of preglacial valleys. These deposits were first named "South Saskatchewan Gravels" (McConnell, 1886, p. 70C) but have generally been referred to as the Saskatchewan Gravels and Sands (Stalker, 1968a). This unit makes up the lower part of the Empress Group in Saskatchewan (Whitaker and Christiansen, 1972) and is largely equivalent to the Souris Formation of Manitoba (Klassen, 1969). Saskatchewan Gravels and Sands span the Tertiary-Quaternary boundary and are defined as those deposits laid down within the broad confines of preglacial valleys before the first glaciation of any region on the Canadian Interior Plains. They generally form a belt 1-2 km wide along the deepest part of those valleys and are typically 3-10 m thick but may reach 65 m. Deposition of Saskatchewan Gravels and Sands ended when ice advancing

---

Klassen, R.W.
1989: Quaternary geology of the southern Canadian Interior Plains; in Chapter 2 of Quaternary Geology of Canada and Greenland, R.J. Fulton (ed.); Geological Survey of Canada, Geology of Canada, no. 1 (also Geological Society of America, The Geology of North America, v. K-1).

from the direction of the Canadian Shield disrupted the local drainage. Because proglacial lakes were formed in many preglacial valleys during ice advance, the coarse unit in many places is overlain by clay and silt. Because early glaciers from the Canadian Shield did not reach the western edge of the Plains, the top of the unit is time transgressive. Therefore the upper part of this unit in the western part of the area is younger than in the east. Gravels of a similar nature were deposited by eastward flowing rivers during subsequent nonglacial intervals but these may be distinguished from the preglacial gravels by their higher stratigraphic position and content of debris from eastern and northern sources.

The nature of the bedrock influenced topographic development, affected the way in which the rock reacted at the base of the glaciers, and played a role in determining the composition of glacial deposits. For example, the coarse grained and porous rock in general resisted fluvial erosion better than the finer ones. As a consequence, paleovalleys in which Tertiary rivers deposited thick gravel units later became uplands. In addition, the montmorillonite-rich Cretaceous beds were weak and, because of this, these rocks in many places were thrust and deformed by the overriding glaciers. Also, the low permeability of the Cretaceous beds, and to a lesser extent of the lower Tertiary ones, slowed subglacial drainage and probably caused elevated subglacial water pressure. This affected processes acting at the base of the ice and is probably in part responsible for the low surface slopes reported for glaciers on the plains (Mathews, 1974).

Physiography related to bedrock

The Southern Canadian Interior Plains rise generally to the west-southwest, from about 250 m near the edge of the

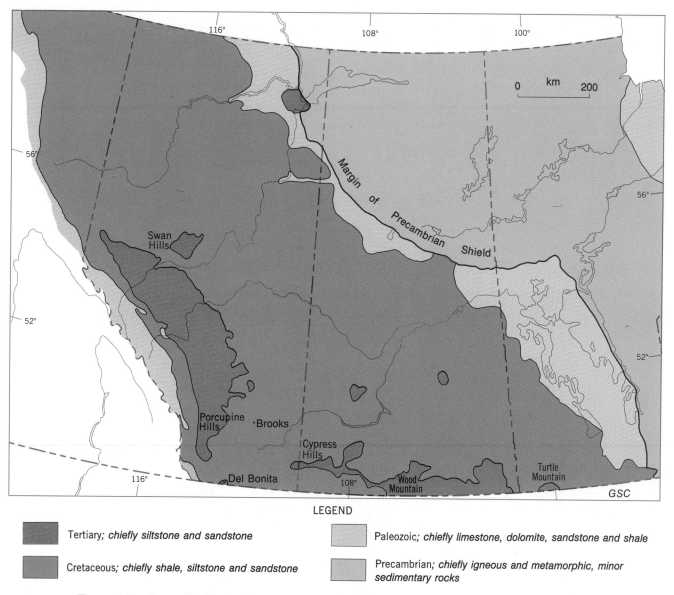

Figure 2.15 Generalized bedrock geology of the southern Interior Plains of Canada (after Douglas, 1969).

Precambrian Shield to about 1200 m near the Rocky Mountain Front. The major physiographic features are the result of differential fluvial erosion of the weak, largely flat-lying sedimentary bedrock that characterizes most of the plains, whereas most of the smaller, superimposed features, described later, were produced during Quaternary glaciations.

At onset of glaciation, the Interior Plains had undergone a prolonged spell of erosion. This had produced a smooth surface which was broken by remnant uplands where more resistant rock outcropped, and by valleys of the preglacial rivers. These valleys were broad, typically 5 to 40 km from lip to lip with gently sloping valley walls, and deep. Though the uplands generally remain salient features of the skyline, the valleys now are less prominent, because they have been largely filled with Quaternary sediments.

The gradual westward rise of the Interior Plains is punctuated by a number of steep ascents, called the Prairie "steps", which separate broad areas of more gradually rising surfaces. These sharp rises are a legacy from the preglacial surface and developed where weak bedrock is overlain by more resistant beds. In general, these steps trend northwest, are most prominent in the south, and become lower and disappear towards the northwest. The two most prominent are the Manitoba Escarpment and the Missouri Coteau (Fig. 2.16), but a third marked step is found west of the Coteau, close to where the Tertiary beds overlap the less resistant Cretaceous formations (Fig. 2.15). The Manitoba Escarpment rises about 300 m to separate the Manitoba Plain on the east from the Saskatchewan Plain (Bostock, 1970; Fig. 2.16). It lies at the eastern margin of a series of uplands, including Riding and Duck mountains and the Porcupine Hills. These heights are separated by broad embayments that mark the positions of former preglacial valleys. Farther west, the Missouri Coteau rises about 100 m from the Saskatchewan Plain on the east to the Alberta Plain. The central and northern parts of the Interior Plains lack the distinctive natural boundaries found in the south, but Bostock (1970) designated certain plateau-like interfluve areas as the Alberta Plateau and the separating broad lowlands as Peace River Lowland and Fort Nelson Lowland.

Manitoba Plain has low relief, with neither prominent uplands or preglacial valleys; however, in the north it includes low bedrock hills and cuestas. Saskatchewan Plain

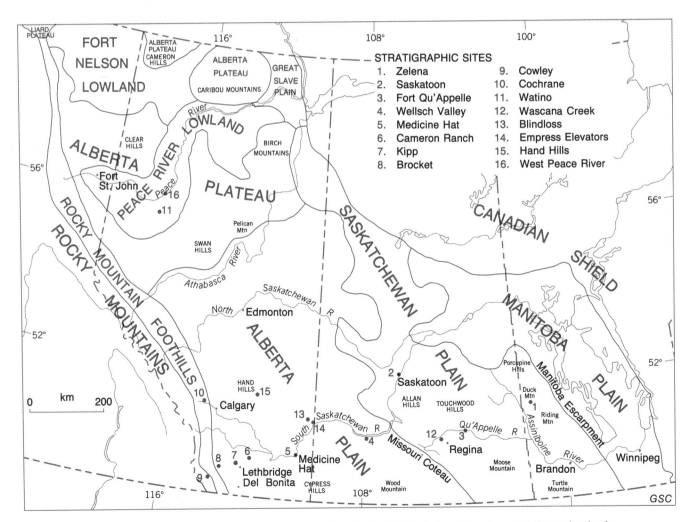

Figure 2.16 Location map showing places mentioned in the text, stratigraphic sites, and the main physiographic elements of the southern Interior Plains of Canada.

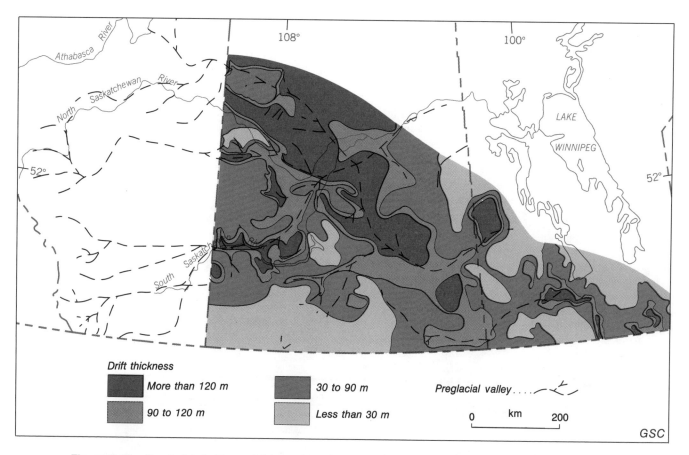

Figure 2.17 Preglacial drainage of the southern Interior Plains of Canada and drift thickness in southern Saskatchewan and Manitoba. Preglacial drainage information for southern Alberta modified from Stalker (1961); for northern Alberta mainly based on information from Alberta Research Council; for Saskatchewan from Christiansen (1967). Drift thickness information compiled from various sources.

lies mainly between 400 and 800 m a.s.l., and contains several higher areas such as Moose Mountain, Touchwood and Allan hills, and Wapawekka and the Waskesiu uplands. As the drift blanket on the Saskatchewan Plain is substantially thicker (30 to 120 m) than that on the adjoining Manitoba and Alberta plains, features of the bedrock topography, such as preglacial valleys, are less well reflected at the surface (Fig. 2.17). Alberta Plain is generally 800 to 1000 m a.s.l. It typically is broadly rolling with interspersed uplands such as Wood Mountain, Cypress Hills, and Hand Hills; local badlands occur mainly along valleys in its southern part. Alberta Plateau is 600 to 900 m a.s.l., and includes Pelican, Birch, and Caribou mountains, and the Swan, Clear, and Cameron hills. These uplands are separated by the Peace River and Fort Nelson lowlands lying at 300 to 600 m a.s.l. The southernmost part of the Great Slave Plain, mostly below 300 m, extends into northeastern Alberta.

Glaciation was responsible for most of the lesser features superimposed on the broader bedrock physiography. This aspect of the geomorphology is discussed in a later section but mention should be made here of the deep, narrow valleys cut as meltwater channels and glacial spillways. These are steeper walled and narrower (typically <2 km wide) than preglacial and interglacial valleys and typically trend east or southeast in contrast with the large nonglacial valleys that typically trend northeastward (Fig. 2.18). In many places the glacial valleys cut across bedrock uplands or along their margins, but they also may be incised into the Quaternary fill of buried valleys. In some places they form series of subparallel channels that probably were cut at the margin of ice that was retreating downslope. In other places these valleys drained glacial lakes and formed as spillways (Kehew and Lord, 1986). These valleys are interesting aspects of the southern Interior Plains because they impart variety to a generally flat landscape.

Drainage evolution

The land surface slopes to the northeast — from the mountains on the west towards the margin of the Canadian Shield. The preglacial drainage system developed on this slope, with rivers carrying debris northeastward from the mountains. The latter part of the Tertiary was a general period of erosion and rivers moved laterally away from their coarse channel deposits and deepened their valleys in adjacent more easily eroded bedrock. At the time of the first glaciation the drainage system of the southern Interior Plains of Canada probably consisted of a series of major northeast to east trending, regularly spaced valleys which headed in the mountains and, at least in the west, were separated by uplands capped with Tertiary sands and gravels (Fig. 2.17).

The main effects of glaciation relative to drainage were: (1) an infilling of many of the pre-existing valleys with till and other sediments; (2) the carving of new basins by glacial erosion; (3) the cutting of new valleys across former uplands; and (4) development of vast tracks of rolling to hummocky terrain which lacked integrated drainage. During interglacial times, glacially carved basins were occupied by lakes and some segments of streams were re-established in old valleys whereas others followed glacial valleys.

This general sequence of filling old valleys and of cutting new ones occurred during each glacial-interglacial cycle. Consequently the present drainage system consists of interconnected segments that originated as preglacial valleys, glacial spillways cut during one or more glaciations, interglacial valleys developed during one or more nonglacial periods, and new valleys cut since the last glaciation.

Climate, soils, and vegetation

The climate of the southern Canadian Interior Plains is continental and for most of the year the area is under the influ-

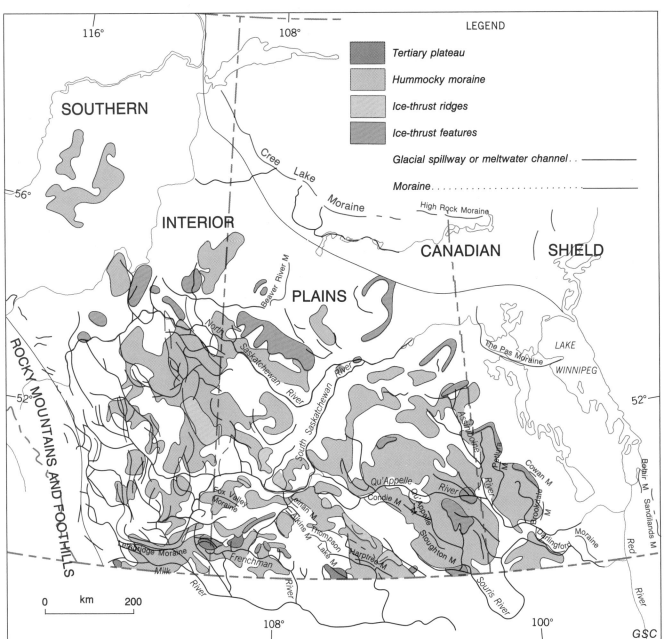

Figure 2.18 Various types of moraines and meltwater drainage features of the southern Interior Plains. Most meltwater channels originated during early glaciations but were reoccupied during subsequent advances or retreats. Compiled from Prest et al. (1968); Christiansen and Whitaker (1976); Moran et al. (1980).

ence of Polar continental air. Modifications to this continental climate occur in southern Alberta and southwest Saskatchewan as a result of the warming effects of downslope-flowing Pacific air, in northern Alberta as a result of precipitation from Pacific air, and in southeastern Manitoba from somewhat random incursions of warm and moist tropical air from the Gulf of Mexico (Laycock, 1972). Average mean temperatures, over a 30 year period, are about $-10°C$ in January and $+20°C$ in July in southeastern Alberta and southwestern Saskatchewan, and about $-25°C$ in January and $+15°C$ in July in northern Alberta (Laycock, 1972; Longley, 1972). Mean annual precipitation varies from more than 500 mm in southeast Manitoba to less than 300 mm in southwest Saskatchewan and southeast Alberta (Putnam and Putnam, 1979). About 70% of the precipitation falls between April and September (Laycock, 1972). According to the climatic divisions of Putnam and Putnam (1979), which are based on both precipitation and temperature, southwestern Saskatchewan and southeastern Alberta are semiarid; southwest Manitoba, along with much of southern Saskatchewan and Alberta, is in the sub-humid prairie subdivision; most of southern Manitoba, central Saskatchewan, and central Alberta are sub-boreal; northernmost Alberta is boreal; and a small part of southeastern Manitoba lies in the humid subdivision.

The regional soil and vegetation pattern within the southern Interior Plains of Canada is controlled mainly by climate. The major soil zones are from south to north: Brown, Dark Brown, Black, and Grey Wooded (Putnam and Putnam, 1979). The first three constitute the so-called grassland (Chernozemic) soils of the drier areas (Laycock, 1972) whereas the last is related to cooler, largely forested areas. In addition to these zonal soils, large areas of northern Alberta are underlain by organic soils (Clayton et al., 1977).

The southern Canadian Interior Plains can also be divided into prairie, parkland, and boreal forest zones. The prairie zone comprises the grasslands of the south, which are characteristically treeless save for poplars, willows, and smaller brush growing along the rivers. Almost all this area is devoted to agriculture with ranching and dry farming the main pursuits. Parkland forms an intermediate zone north of the prairie, where the grasslands become punctuated by bluffs or clumps of trees, mostly poplar with some spruce; like the prairie zone, it is extensively farmed. In the cooler and damper regions farther north lies the boreal forest zone, consisting largely of spruce and jack pine. Parts of this zone have been cleared for farming but forestry is the prime use of much of this land.

NATURE AND DISTRIBUTION OF QUATERNARY DEPOSITS

Quaternary deposits are the surface materials and form the local landforms over virtually all of this region. Bedrock, which controls the broad elements of the physiography, rarely outcrops. Most of the surficial materials that blanket or veneer the bedrock were deposited during Pleistocene glaciations. Only a small proportion of the Quaternary materials are of nonglacial origin, partly because little sediment was deposited in the Interior Plains area during nonglacial periods and partly because the thin nonglacial sediments which were deposited were readily destroyed during subsequent glaciations.

The nature of the Quaternary deposits reflects broad aspects of bedrock lithologies and patterns of glacial and fluvial transport. Clay and silt, derived from the fine grained Cretaceous beds which underlie most of this region, and sand from Cretaceous and Tertiary sandstones are the main components of the Quaternary sediments. Minor and locally important components of the sediments come from: igneous and metamorphic lithologies of the Precambrian Shield, carbonates at the margin of the shield and the Cordillera, and quartzites and other clastic rocks of the Cordillera. Competent lithologies, which form pebble and larger sized clasts, are mainly carbonates and igneous and metamorphic lithologies carried into the area by glaciers moving from the shield; carbonates and quartzite derived from the Rocky Mountains and carried onto the Interior Plains by glacial and fluvial transport; and ironstone concretions, petrified wood, and coal eroded from the local bedrock.

The distribution of Quaternary sediments in the southern Interior Plains bears a general relationship to bedrock topography, pattern of ice flow, and the glacial processes which were active when the ice margin lay in certain areas. Areas where these factors combined to produce thick drift are the Manitoba Escarpment and the Missouri Coteau. At these sites the Cretaceous shale in the east-facing escarpments provided southwest flowing glaciers with materials to construct thick (100 to 300 m) hummocky moraines.

Subsurface studies in southern Saskatchewan and Manitoba over the past 20 years indicate that tills, which make up about 70% of the Quaternary deposits, have considerable continuity whereas the stratified deposits, with several notable exceptions, are discontinuous and thinner. In southern Alberta and southwestern Saskatchewan, the drift is generally less than 20 m thick and thick deposits of till and stratified sediments are restricted to buried valleys and belts of hummocky moraines.

Glacial deposits

In this report glacial deposits are subdivided into three general groups — till, glaciofluvial deposits, and glacial lake deposits. The discussion of these three categories is based mainly on deposits of the last glaciation but those of older glaciations undoubtedly were similar.

Till

Till is here used in the sense of Dreimanis and Lundqvist (1984) "a genetic term applied to a sediment that has been transported by glacier ice and is subsequently deposited by or from it with little or no sorting by water". The tills of this region are remarkably uniform because they were largely derived from the shale, siltstone, and sandstone which underlie most of this region. They contain roughly equal amounts of clay, silt, and sand with minor coarser material. Most stones in the till, and a large part of the heavy minerals, were derived from the Canadian Shield and the fringe of Paleozoic carbonate rock along its margin. Some of the large clasts came from Tertiary gravels of the preglacial valleys and the upland surfaces, and tills of the western part of the region include carbonates and quartzites from the Rocky Mountains.

Important characteristics of the tills of the southern Canadian Plains, including stickiness when wet and low permeability, are due to the large clay component, which includes much swelling montmorillonite derived from bentonitic shale (Scott, 1976). Differences between till units are in many cases subtle and reflect glacial loading, changes in regional ice flow, and weathering. Field and laboratory studies of thousands of drill core and outcrop samples in southeastern Manitoba (Fenton and Anderson, 1971; McPherson et al., 1971; Fenton, 1974; Teller and Fenton, 1980); southwestern Manitoba (Klassen, 1966, 1969, 1971b, 1979; Klassen and Wyder, 1970); southern Saskatchewan (Christainsen, 1968a,b, 1971a,b, 1972; Mahaney and Stalker, 1988) and southern Alberta (Stalker, 1963, 1969, 1976b, 1983; Stalker and Wyder, 1983; Jackson, 1980b; Bayrock, 1962; Bayrock and Pauluk, 1967; Westgate, 1968) have verified these differences on a regional basis. The criteria used to characterize different tills are: carbonate content, grain size distribution, colour; degree of weathering (oxidation, carbonate leaching, soil development), stone lithology, structure (jointing, fissility), and mineralogy.

The general conclusions of the till studies are that tills of southeastern Manitoba are sandier and contain more carbonate in their silt component (50%) than the tills west of the Manitoba Escarpment. Carbonate contents decreases westward from about 15 to 20% in southwestern Manitoba to between 5 and 15% in southwestern Saskatchewan and southern Alberta. In contrast with these tills deposited by ice moving from the Precambrian Shield, tills of Cordilleran provenance, are sandy, consist of up to 20% stones, and contain about 20% carbonate.

Composition of coarse clasts has proven to be useful in distinguishing tills from different source areas. Igneous and crystalline metamorphic stones from the Precambrian Shield are a reliable indicator of Laurentide till (from the north and east), except near the western edge of the area where Cordilleran tills (from the west) include some shield lithologies which were derived from older Laurentide till. In addition to identifying till of western and northern source, stone lithology has been used to differentiate tills related to different lobes of ice (Shetsen, 1984).

The distribution of erratics on the southern Canadian Plains has been used to provide information on former ice flow patterns. The Foothills Erratics Train and several other trains of erratics in southern Alberta have been used to trace ice flow in the zone of coalescence between Cordilleran and Laurentide ice (Stalker, 1956; Jackson, 1980b; Fig. 2.19). In addition, the distribution of two distinctive types of erratics suggests they were emplaced by ice which was following different paths from that of the last advance (Fig. 2.20). One of these is fine grained greywacke, containing small carbonate concretions (Omaraluk Formation), suggested to come from southeastern Hudson Bay (Prest and Nielsen, 1987). This is found in the tills of Manitoba, Saskatchewan, and locally in southeastern Alberta, almost 2200 km south-southwest of their presumed source area. The other is oolitic iron formation which occurs as erratics at the western end of the Cypress Hills in southeastern Alberta about 1100 and 1500 km due south of possible source areas north of Lake Athabasca and near the east end of Great Slave Lake (Fig. 2.19).

Successions of till units have been identified in boreholes and outcrops and correlated on the basis of stratigraphic position and material attributes. Fenton (1974) identified 8 till units in southeastern Manitoba; Klassen (1979) and Christiansen (1968a,b) identified at least four distinct tills in southwestern Manitoba and southern Saskatchewan respectively; and Stalker (1983) identified 8 till units in the Cameron Ranch section on Old Man River in south-central Alberta. Caution is necessary in assigning an ice advance to each till unit of the southern Interior Plains however, because glacial thrusting has in places caused repetition of units. In many cases this repetition is only recognized where slices of bedrock occur in the Quaternary succession.

Glacial lake deposits

Glacial lake deposits are the second most common Quaternary glacial deposit on the southern Canadian Plains. They occur in thicknesses of up to 10 m or more in the former basins of large glacial lakes, such as Lake Peace (northwestern Alberta and adjacent British Columbia), Lake Regina (south-central Saskatchewan), and Lake Agassiz (eastern and central Manitoba and east-central Saskatchewan). They also occur in numerous smaller basins and locally can be a significant component of the drift in hummocky moraine complexes. They consist dominantly of clay and silt, are well stratified to massive, and in some areas

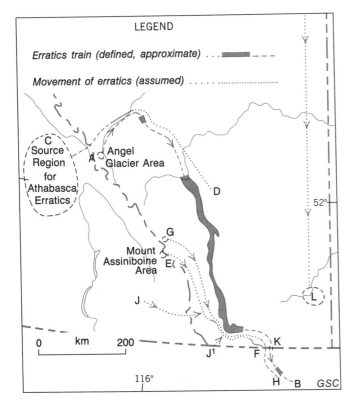

Figure 2.19 Movement of some glacial erratics in southern Alberta. A-B, Foothills Erratics Train (Stalker, 1956); C-D, Athabasca Valley Erratics Train (Roed et al., 1967); E-F, McNeill (Twin River) Erratics Train (dolomites); G-H, vuggy dolomite; J-K, shale from Purcell Supergroup (J and J¹ are two possible source areas); and L, oolitic iron formation. Where no specific reference is given, erratic distributions are from A. MacS. Stalker, unpublished data.

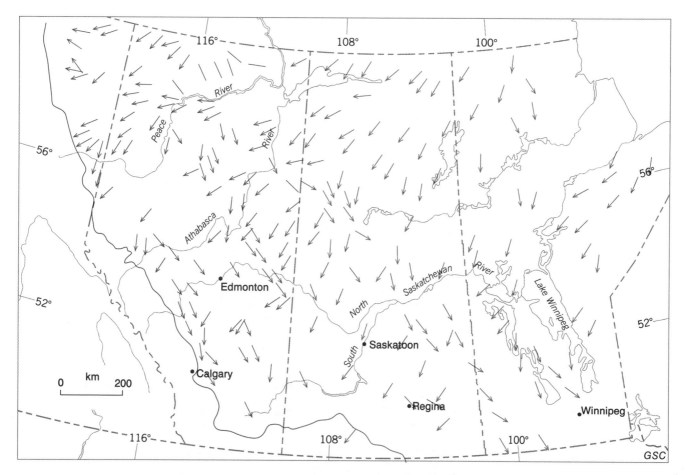

Figure 2.20 General flow directions of latest Late Wisconsinan ice.

dropstones are so abundant that the lake sediments can be mistaken for till. Sand is an important component of glacial lake deposits in some areas, particularly in the west and in hummocky moraine complexes. Beach ridges, which contain mainly gravel derived from the underlying till, formed mainly along shorelines where large lakes were stable for considerable periods. Wave eroded (washed or bevelled) till marks the former margins of small lakes or those that did not achieve extended periods of stability.

Laurentide ice in general advanced up the regional slope and receded down it. Consequently, glacial lakes formed as each ice sheet advanced, and again, as each ice sheet retreated. Glacial lake sediments deposited during the last ice retreat are widespread at the surface and older glacial lake sediments are an important component of Quaternary valley fills and are commonly found as intertill sediments in boreholes. In most cases, the buried glacial lake sediments are interpreted as deposits formed as the ice advanced. This might indicate that the glacial lake sediments that formed during deglaciation were deposited in the shallow basins which were left as the ice retreated and hence were readily destroyed by the next ice advance. On the other hand, glacial lake sediments that were deposited in river valleys by advancing ice probably stood a better chance of surviving subsequent glaciations.

Glaciofluvial deposits

Glaciofluvial deposits of the last glaciation are widespread. They consist of sands and gravels which are typically well sorted, moderately well washed, and well stratified. Locally, however, they are massive or are slumped and have deformed bedding. Compositions of large clasts are similar to those of adjacent till sheets, and in general their content of shale is greater than that of preglacial and interglacial gravels. Glaciofluvial deposits at the surface occur in most places as distinct landforms, such as kames, eskers, deltas, fans, outwash trains, and kame and kettle complexes. Commonly their occurrence can be associated with well defined glacial drainage systems but, in some places, in particular where they are part of a hummocky moraine complex, it is difficult to relate them to specific meltwater flow systems.

In section, glaciofluvial deposits contain few features that permit easy differentiation from preglacial and interglacial sands and gravels. Two criteria that have been used to suggest a glaciofluvial origin are high content of shale and other weak stones and deformed structure. In many circumstances, gravels on the southern Canadian Interior Plains have been assumed to be of glaciofluvial origin unless they consist largely of quartzite and other resistant stones or contain Quaternary fossils. As a consequence gravels reported from boreholes that do not have a stratigraphic context to suggest they are preglacial or interglacial, generally have been referred to as glaciofluvial.

Nonglacial deposits

The main nonglacial Quaternary deposits that can be studied on the southern Canadian Plains are those that have been deposited during the Holocene. The sites where Holocene deposition has largely occurred are closed depressions, valleys, and areas where vegetation cover and source sediment supply have led to formation of eolian sediments. Pre-Holocene nonglacial deposits are rarely exposed, and consequently few general comments can be made on their mode and environment of deposition. In the following discussion it is assumed that past nonglacial periods were roughly similar to the present and that most Pleistocene nonglacial deposits have an analogue in the Holocene.

Depressions of various sizes, resulting mainly from glaciation, are distributed across the plains. Most of these contain permanent or ephemeral lakes and ponds. Some basins, particularly larger ones such as Lake Winnipeg, are fed by major rivers and hence are sinks for large amounts of clastic sediment. Most basins however receive only small amounts of clastic material derived from local streams, slope wash, and shoreline erosion, but in the driest part of the area evaporites are a significant sediment in many basins (Last, 1984). Small depressions in the dry southern region generally do not hold water all year and receive little sediment other than fines washed from the nearby slopes during snowmelt or infrequent heavy rains. Basins that do retain water or at least remain moist all year can be sites of organic sediment accumulation. This sediment can vary from an organic-rich mud in southern areas to gyttja and peat in more humid northern areas.

Although basins are widespread, Holocene lake sediments are generally thin (up to 2 m), occupy shallow basins, and hence would be unlikely to be preserved if there were another glaciation. Buried sediment of this type, however, has been described from southwestern Manitoba, where an organic-rich clay (Roaring River Clay) contains pollen and invertebrate fossils that indicates deposition in a small pond during temperate conditions (Klassen et al., 1967).

Valleys on the Plains may presently receive sediments from four sources: (1) Cordilleran mountains in the west — quartzite, carbonate and other debris; (2) local bedrock — soft shale and siltstone with minor ironstone concretions, petrified wood, and coal; (3) preglacial gravels and sands — rounded quartzite and chert pebbles and a variety of other sedimentary rock types; and (4) glacial deposits — materials from the other three sources plus metamorphic and igneous debris derived from the Precambrian Shield. Major river valleys that traverse the southern Canadian Plains contain Holocene fills up to 60 m thick that consist of channel gravels and sands, quiet water silts and clays, and colluvium (Klassen, 1975); the valleys of small streams contain fills up to 10 m thick that mainly consist of colluvium. Coarse deposits of Pleistocene age, which accumulated as fluvial sediments, include stratigraphic units such as: the previously described Saskatchewan Gravels and Sands; Riddell member of the Floral Formation at Saskatoon (SkaraWoolf, 1981); Echo Lake gravel at Fort Qu'Appelle (site 3 of Fig. 2.16; Khan, 1970; Christiansen, 1972); Mitchell Bluff Formation at Medicine Hat (Stalker, 1976a); and the extensive valley fill gravel and sand that underlies the surface till in the Fort St. John area of east-central British Columbia (Mathews, 1978). Fine grained fluvial sediments of Pleistocene age are the "Evilsmelling Band" of Middle Wisconsinan age at Medicine Hat (Stalker, 1976b) and the fine grained, fossiliferous sediment of Middle Wisconsinan age at Watino in central Alberta (site 11 of Fig. 2.16; Westgate et al., 1972). Colluvium apparently is not a significant component of the fill in most valleys but it makes up part of the basal unit at the Wellsch Valley site of southwestern Saskatchewan (site 4 of Fig. 2.16; Stalker and Churcher, 1972; Stalker, 1976b).

Holocene eolian sediments consist of sand and silt. The sand occurs mainly as dunes associated with deltas or other coarse glacial lake deposits; the loess is in many places downwind from dune fields. The largest areas of Holocene dune sands in the southern Canadian Plains are the Great Sand Hills between the Cypress Hills and South Saskatchewan River in southwestern Saskatchewan and the Carberry Sand Hills on the Assiniboine River east of Brandon in south-central Manitoba. Holocene loess is widespread in the southwestern part of the region but is generally too thin to map. Pleistocene loess occurs on those uplands of southern Alberta and southern Saskatchewan that were not overridden during the last ice advance.

Volcanic ash or tephra locally occurs in nonglacial deposits of the southern Interior Plains. It is volumetrically insignificant but its identification is an important means of determining the age of enclosing sediment. The tephras found in this region are grey to white, clay to sand in texture, and consist dominantly of glass shards. The tephra units are typically discrete beds 1-2 cm thick but locally are much thicker. Mazama ash from an eruption that occurred about 6.6 ka at Crater Lake in the Cascade Mountains of southern Oregon (Powers and Wilcox, 1964; Westgate et al., 1970) is widespread in the southwestern corner of the region; Glacier Peak ash from an eruption that occurred in the northern Cascade Mountains of Washington about 12 ka (Powers and Wilcox, 1964; Westgate and Evans, 1978) occurs locally in the vicinity of Cypress Hills in southeastern Alberta; and Pearlette ash, type O, from an eruption about 600 ka in the Yellowstone area of northern Wyoming (Izett and Wilcox, 1970) is exposed on Wascana Creek northwest of Regina (site 12 of Fig. 2.16; Westgate et al., 1977). Single tephra units have been reported from pre-Holocene deposits at a number of other sites but their sources and ages have not been precisely determined.

Geomorphology of Quaternary deposits

Regional geomorphology of the southern Interior Plains is largely a reflection of the underlying bedrock surface; however, the smaller features over most of the region are of glacial origin.

Abundant published information on the nature and distribution of glacial features is available in many regional reports and accompanying maps (see Map 1704A). The regional distribution of major glacial features is presented in Prest et al. (1968) and summary descriptions of their main characteristics and nomenclature are available in Prest (1968, 1984). In this report discussion is limited to hummocky moraine, ice-thrust features, and corrugated moraine, which are uniquely well developed in the southern Interior Plains.

Hummocky moraine and ice-thrust features are dominant aspects of many upland areas with intervening low areas characterized by gently irregular ground moraine and flat lake plains. End moraines are associated in many areas with hummocky moraine, but the most prominent end moraines — the Cree Lake and the Pas moraines — occur in the north, near the margin of the Precambrian Shield, well separated from hummocky moraine areas (Fig. 2.18). Kames and eskers are scattered here and there but are far less common here than they are in the Canadian Shield region. Sand dunes are also prominent features, locally.

Hummocky moraine

Hummocky moraine is strongly developed over the higher parts of most uplands and in somewhat more subdued form occurs in many other places throughout the region (Fig. 2.18). Morainal hills, knolls, randomly oriented ridges and intervening depressions characterize both high relief (10-30 m) and low relief (5-10 m) hummocky moraine. Surface patterns range from regularly spaced knob and kettle topography to irregular, rounded or flat-topped hills and knolls with intervening depressions, and a variety of ridge forms ranging from sinuous to nearly circular (Fig. 2.21). Hummocky moraine is commonly subdivided into two classes: low relief (5-10 m) and high relief (10 - 30 m). The nature and origin of hummocky moraine and associated features of the southern Canadian Plains have been studied by numerous workers, including Gravenor (1955), Gravenor and Kupsch (1959), Stalker (1960a), Parizek (1969), and Bik (1969). The general consensus is that they reflect deglaciation by stagnation of segments of the glacier margin, although disagreement exists concerning the degree to which specific glacial processes were involved. The extent of this disagreement is clearly evident from studies of the broad, elevated flat-topped hills known as moraine plateaus (Stalker, 1960a) or moraine-lake-plateaus (Parizek, 1969) that are so common in hummocky moraine.

The hummocky moraine of this region is formed mainly of till which is similar to the till in adjacent ground moraine. Sediments deposited or reworked by water, ranging from clay to gravel, may form part or all of some knolls, and in places entire tracts of hummocky terrain are formed of glaciolacustrine silt and clay.

Ice-thrust features

Ice-thrust features occur in upland areas, in association with hummocky moraine and, in several limited areas, on their own, (Fig. 2.22). These features formed as a result of glacial entrainment and deformation of discrete masses of bedrock and/or pre-existing drift. Evidence of ice-thrusting was first recognized in the hummocky moraine belt of east-central Alberta (Hopkins, 1923; Slater, 1927) and later, similar distinctive features were identified in hummocky moraine along the Missouri Coteau in southern Saskatchewan (Byers, 1959; Kupsch, 1962; David, 1964; St-Onge, 1967). Outcrop and subsurface studies have indicated that large masses of bedrock were displaced and that disruption by thrusting was not limited only to areas with distinct thrust surface features (Wickenden, 1945; Christiansen and Whitaker, 1976; Stalker, 1976a; Klassen, 1979; Fenton, 1983, 1984a). Several other studies have contributed to our understanding of the origin and significance of ice-thrust features

Figure 2.21 Hummocky moraine along the southern slopes of the Cypress Hills upland viewed north towards the unglaciated Cypress Hills plateau. The gap between the west and centre blocks of the plateau forms the notch on the horizon to the left. ISPG 2519-8

(Mathews and Mackay, 1960; Moran, 1971; Moran et al., 1980; Bluemle and Clayton, 1984).

The most prominent type of ice-thrust feature occurs along the northwest-southeast trending Missouri Coteau in southwestern Saskatchewan (Fig. 2.18). Hummocky moraine there is bordered by a succession of broadly arcuate to nearly straight ridges with local relief from 15 to 60 m and spaced at 150 to 300 m intervals. Individual ridges, typically several kilometres long, have sharp crests or somewhat rounded profiles. Source depressions located up-ice from the ridges have been noted in places (Christiansen and Whitaker, 1976). The terms ice-thrust moraine (Prest, 1968), transverse compressional ridges (Clayton and Moran, 1974), glaciotectonic terrain (Fenton, 1984a), and transverse ridge forms (Bluemle and Clayton, 1984) have also been applied to this type of feature.

Other forms of ice-thrust features consist of an individual hill with a source depression or a train of hills, decreasing in size down-glacier from the source depression. Fenton (1983, 1984a) referred to isolated hills and source depressions as glaciotectonic terrain and to a series of hills as "rubble terrain". Similar features in North Dakota are referred to as hill-depression forms and irregular forms, respectively (Bluemle and Clayton, 1984).

Structural disruptions associated with ice-thrust features include folding, faulting, and stacking of discrete slabs or blocks of drift and bedrock (Fenton, 1984a). Some ice-thrust features, however, were emplaced in a virtually undeformed state (Stalker, 1976a). In this situation, recognition of ice-thrust material requires noting a unit out of its correct stratigraphic position. Recognition of ice-thrusting of till sheets is difficult unless bedrock masses are included. The maximum known thickness of materials affected by ice-thrusting is in the range of 60 to 180 m (Wickenden, 1945; Kupsch, 1962; Christiansen and Whitaker, 1976).

Ice-thrust features along the Yukon coast described by Mackay (1959) are virtual models for similar features in the southern Canadian Interior Plains region. Mackay (1959)

Table 2.9. Quaternary stratigraphic units recognized in south Alberta and southwest Saskatchewan

Unit Name	Description	References	Remarks
Hidden Creek Till	Light grey, sandy, calcareous, stony Cordilleran till	Alley, 1973	This unit is restricted to large valleys in the Rocky Mountains and rarely reaches the mountain front; it is equivalent to the Canmore advance of Rutter (1980).
Buffalo Lake Till	Light grey to brown, silty to sandy Laurentide till	Stalker, 1960b; Stalker, 1983	This is the youngest till of the southwest Canadian Plains; it extends as far south as the Lethbridge Moraine; near its southern limit (Cameron Ranch site) it consists of three till units with lake silt and sand separating the upper two tills.
Evilsmelling Band	Silts and clays containing disseminated organic material, including wood fragments and vertebrate fossils	Stalker and Churcher, 1982; Stalker, 1983	This unit lies on the floor of a filled valley in the Medicine Hat area; is thought to have been deposited in part under permafrost conditions.
Ernst Till	Bouldery Cordilleran till	Alley, 1973	Composition varies with underlying bedrock; correlated with Bow Valley Till of Rutter (1972).
Maycroft Till	Light brown, silty Cordilleran till	Stalker, 1963; Alley, 1973	This is considered equivalent to the middle subunit of the Cameron Ranch Formation and was deposited during Waterton II advance of Stalker and Harrison (1977).
Cameron Ranch Formation	Light grey to dark brown, stony Laurentide till	Stalker, 1983	Generally poorly consolidated and locally includes three till subunits; gravel and sand containing fossils overlie the lowest subunit and lake and stream deposits underlie the upper subunit; upper till correlative with Erratics Train Till of Jackson (1980b).
Mitchell Bluff Formation	Discontinuous sand, silt, and clay with abundant vertebrate fossils	Stalker, 1976b	Lies on unconformity which marked the floor of an interglacial valley; upper part of unit strongly contorted in many places.
Brocket Till	Dark brown, silty and clayey Laurentide till	Stalker, 1963	Moderately indurated, massive, and forms cliffs locally with columnar structures; readily confused with Maunsell Till but in preglacial valleys overlies Labuma or Maunsell till, and in interglacial valleys overlies bedrock or Maunsell till.
Maunsell Till	Light to dark grey, silty clayey Laurentide till	Stalker, 1960b	Plastic and sticky when wet, forms cliffs with rectangular or hexagonal columns.
Labuma Till	Dark brown to nearly black, clayey and silty Laurentide till	Stalker, 1960b; Stalker and Harrison, 1977	The dark colour and distinctive fracturing into small cubes when dry makes this till distinctive; deposited by the most extensive glaciation to cover this part of the Plains and is the oldest till at Medicine Hat and areas to the west; extends to 1700 m in the Foothills.
Albertan Formation	Silty, sandy, stony Cordilleran till	Dawson and McConnell, 1896; Stalker, 1963; Alley, 1973	This is the earliest and most widespread of Cordilleran tills and is found in preglacial valleys and perhaps in interfluve areas.
Wellsch Farm Formation	Silt, clayey silt, sand, and gravel containing vertebrate fossils and a tephra	Stalker, 1976b	Fossils indicate an age of about 1.8 Ma and paleomagnetic data tend to collaborate this (Barendregt, 1985); the age of the tephra is inconclusive (Westgate et al., 1978).
Saskatchewan Gravels and Sands	Silt, sand, gravel, and clay	McConnell, 1886; Rutherford, 1937; Stalker, 1968a; Westgate, 1968	Fluvial and lacustrine sediments deposited in bottoms of preglacial valleys before and during early Laurentide glacial advances.

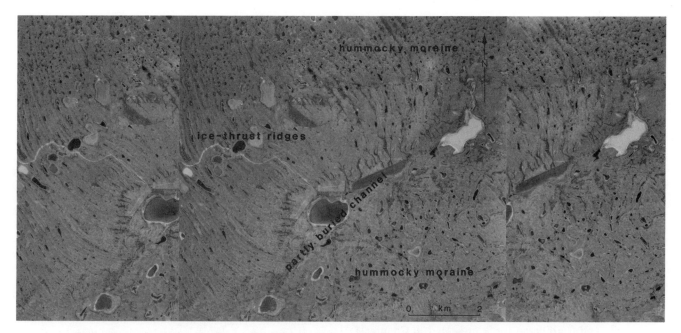

Figure 2.22 Ice-thrust ridges in the Matador area, southwestern Saskatchewan. The northwest-trending ridges are truncated by a meltwater channel which is partly buried by hummocky moraine; the hummocky moraine and channel are thus contemporaneous or younger than the ridges (Christiansen and Whitaker, 1976). NAPL A25126-71, 72, 73

and Mathews and Mackay (1960) suggested that the thrusting mechanisms were dependent upon specific hydrological and ice flow conditions in the marginal zone of continental glaciers. These studies have provided the basis for a better understanding of the origin and significance of ice-thrust features in the Interior Plains region (Moran, 1971; Christiansen and Whitaker, 1976; Moran et al., 1980; Bluemle and Clayton, 1984).

Corrugated moraine

Corrugated moraine consists of aligned ridges, knolls, and depressions that retain the trends of lobate ice margins. A variety of terms have been applied to these features, such as minor end moraine, minor recessional moraine, washboard moraine, ridged moraine, corrugated ground moraine, and corrugated moraine (Christiansen, 1956, 1961; Elson, 1957; Gravenor and Kupsch, 1959; Prest, 1968; Klassen, 1979), but the term washboard moraine is most common. However, Prest (1968) pointed out that it was earlier applied to morainal ridges of a different type and origin and proposed the name corrugation ground moraine for these distinctly patterned areas. Klassen (1979) proposed adoption of the name corrugated moraine.

Corrugated moraine typically consists of successive aligned ridges, 3 to 9 m high and 300 to 1200 m long, spaced at 120 to 200 m intervals. Shorter ridges and knolls linking the aligned ridges form myriads of elongate and irregular closed depressions. The characteristic aspect of this type of terrain is a corrugated to waffle-like pattern that has a broadly lobate outline. The ridges and knolls are typically composed of the last till deposited in the area although some features have cores of older till (Klassen, 1979). Ice contact sand and gravel, in the form of minor eskers and outwash, are locally associated with corrugated moraine, especially in the re-entrants between two lobes. Corrugated moraine appears to reflect patterns of fracture systems or belts of weakness in the marginal zones of stagnating ice lobes (Elson, 1957; Gravenor and Kupsch, 1959; Prest, 1968).

QUATERNARY HISTORY

The record of Quaternary events on the southern Interior Plains is long, with evidence of deposits reaching back to the beginning of the Quaternary (Stalker, 1976b). Stratigraphic units, which are described in many regional reports, are summarized in Stalker (1976b), Fenton (1984b), and Fulton et al. (1984). The nature of the main categories of materials that make up the Quaternary units of the area was described in the earlier section *Nature and distribution of Quaternary deposits*. The emphasis in this report, however, is on Quaternary history and the reader is referred to regional reports and the summaries mentioned above for the stratigraphy on which this history is based. Table 2.9 briefly describes stratigraphic units recognized in southern Alberta and southwestern Saskatchewan and Table 2.10 lists units used in southern Saskatchewan and Manitoba. Figure 2.23 shows the relative southwestern extent of till units in southern Alberta and the relationship of these to tills of Cordilleran provenance.

One important aspect of the glacial history of this region is that throughout the Pleistocene it was overridden by ice that came from two distinct areas. The Canadian Shield was the source area for what in general terms is referred to as Laurentide ice, which flowed onto the Interior Plains from the north and east; the Rocky Mountains were primarily the source area for Cordilleran ice which entered the area from the west. Laurentide source glacial deposits contain Precambrian Shield debris that is absent in Cordilleran drift, except locally where it is recycled from older Laurentide drift. In most places where stratigraphic relationships are clear, Cordilleran ice was either coeval with Laurentide ice or had reached its eastern limit and was retreating when correlative Laurentide ice achieved its maximum. This is not too surprising because ice from the west was moving downslope whereas ice from the east was moving upslope; and the eastern limit of Cordilleran ice lay within 200 km of source areas in the mountains, whereas the western limit of Laurentide ice was at least 1500 km from its source area on the west side of Hudson Bay. In addition to direct contact of the two ice masses during several glaciations, Cordilleran ice probably indirectly affected the Laurentide ice sheets by intercepting moisture from air masses moving eastward from the Pacific Ocean. The emphasis in this report is on the history of that part of the region affected primarily by ice from the east. Additional information on the zone of overlap and interaction of Laurentide and Cordilleran ice is presented in Chapter 1 (Jackson et al., 1989).

Laurentide ice reached the Rocky Mountain Foothills at the western edge of the Plains during several advances with the strongest of these reaching elevations of 1500-1650 m. The only parts of the area that were not covered by ice at least once were the highest parts of the Cypress Hills and Wood Mountain and the uplands around Del Bonita (Fig. 2.16). Laurentide ice moved up the regional slope with the consequences that (1) ice did not advance far across the Canadian Interior Plains during weak glaciations and (2) glacial lakes formed during both ice advance and retreat. The glacial events recorded in any area are dependant upon the number of glaciers that affected the area and the erosive power of each succeeding glacier. Near the margin of the Canadian Shield, the area would have been ice covered during most Laurentide ice advances, the glaciers were strongly erosive, and most older drift was removed. Farther southwest, where the ice sheets neared their limits, deposition dominated over erosion and older drift was preserved in valleys or was beyond the reach of later glaciers. Thus the most complete record of Laurentide glacial events is preserved in the southwestern part of the Plains even though a greater number of glacial events occurred in lower areas to the northeast.

Ice from the Cordilleran region probably never advanced far onto the Plains. This is because during weaker glaciations, the glaciers did not advance east of the foothills, and during stronger glaciations, Cordilleran ice was confined to the western edge of the Interior Plains by westward advancing Laurentide ice. South of 55°N, mountain ice formed valley glaciers and piedmont lobes in the mountains or at the edge of the plains and coalesced with Laurentide ice only during the strongest glaciations. North of 55°N, however, mountain ice apparently coalesced with plains ice during most major glaciations. Under some circumstances streams of ice from the mountains joined Laurentide ice and flowed southward depositing distinctive Cordilleran debris (Stalker, 1956; Jackson, 1980b; Fig. 2.19). The record of Cordilleran glacial events is less complete than of Laurentide events. Two reasons for this are: (1) valleys,

Table 2.10. Correlation of lithostratigraphic units in southern Saskatchewan and southern Manitoba

Lithostratigraphy			Biostratigraphy	Chronostratigraphy	
Southern Saskatchewan (Christiansen, 1968b, 1971a, and personal communication, 1984)	Southern Manitoba (Klassen, 1969, 1979)	Southwestern Manitoba (Fenton, 1974; Teller and Fenton, 1980)			
Saskatoon Group — Battleford Formation	Arran Formation / Zelena Formation / Lennard Formation	Marchand Formation / Whitemouth Formation / Roseau Formation / Whiteshell Formation / Senkiw Formation		Late	Wisconsinan
			Unnamed deposits (Klassen, 1969) / Vita Formation (Fenton, 1974) / Units XX and XXI (Stalker, 1976b)	Middle	
	Minnedosa Formation	Tolstoi Formation / Stuartburn Formation		Early	
Saskatoon Group — Floral Formation			Riddell Member (SkaraWoolf, 1981) / Echo Lake Gravel (Khan, 1970) / Roaring River clay (Klassen et al., 1967) / St. Malo Formation (Fenton, 1974)	Sangamonian	
	Shell Formation	Woodmore Formation		Pre-Sangamonian	
Sutherland Group — Upper till / Middle till / Lower till	Tee Lakes Formation / Largs Formation	Rosa Formation			
Empress Group					

where thick drift could be deposited, are also the areas where glacial erosion would be most concentrated and (2) glacial deposits in areas outside valleys generally consisted of thin, easily erodible drift. As a general rule the record of mountain glaciation has a better chance of being preserved in the plains than in the mountains. This is because even if an ice advance was not extensive enough to deposit till beyond the mountain front, the associated outwash train could be preserved in valley fills. Unfortunately, the provenance of both glacial and nonglacial sediments in the mountains is the same so that it is seldom possible to differentiate Cordilleran outwash from other Cordilleran source channel sediments.

The dating and correlation of Quaternary events on the southern Interior Plains have been accomplished by various techniques, but vertebrate paleontology has been the main technique used. The span of time encompassed by Quaternary deposits of the region is long enough so that some evolution of organisms occurred and therefore faunal composition can be used to indicate relative age.

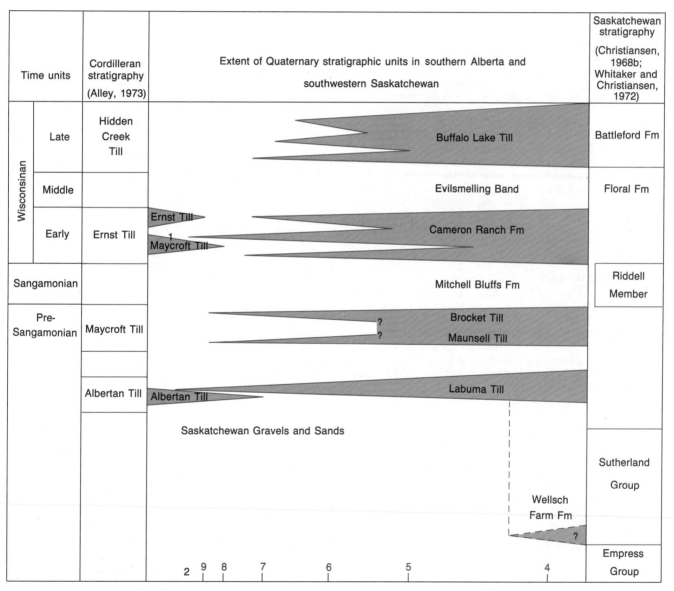

1 There is disagreement in the position of the Maycroft Till; Alley believes it is equivalent to a single till which encompasses Maunsell and Brocket tills; Stalker believes it overlies Brocket Till and was deposited during the next glaciation.

2 Numbers indicate approximate locations of sites that are listed and located in Figure 2.16.

GSC

Figure 2.23 Extent of named Quaternary stratigraphic units in southern Alberta and southwestern Saskatchewan and correlation of these units with those of the Cordillera and the standard Quaternary stratigraphic units of Saskatchewan.

Furthermore, the regional history of animal migration and extinction is well enough known so that the presence of certain species can provide limiting ages. Also, the fauna of a unit can indicate climate during deposition and this can be useful in making regional correlations. Despite all these positive points, vertebrate paleontology is a relatively coarse dating and correlating tool. One of the more promising techniques is paleomagnetic studies. To date, insufficient data are available for construction of secular variation curves that could be used in correlation, but studies have shown that most Quaternary sediments on the south Canadian Interior Plains are normally polarized. In one case, however, paleomagnetic results indicate that sediments in one section were deposited during the Olduvai Normal-Polarity Subchron of the Matuyama Reversed-Polarity Chron which occurred about 1.75 Ma (Foster and Stalker, 1976; Barendregt, 1985). Radiocarbon dating is a good technique for dating events younger than 50 ka. Unfortunately, few of the Pleistocene deposits on the Canadian Interior Plains that fall in this age range contain sufficient quantities of organic materials for dating. Tephra beds are present locally but only the Wascana Creek ash has been dated and correlated with a known tephra unit (Westgate et al., 1977). Paleosols and weathered horizons locally record weathering events which help decipher the history of the area. These have been used in estimating the relative length of nonglacial intervals but the Prelate paleosol (David, 1966), is the only one that has been fully described and named. Attempts have been made to date sediments by obtaining uranium-series and electron spin resonance ages on included vertebrate remains, but these so far have not provided consistent results (Szabo et al., 1973; Zymela et al., 1988).

Early Pleistocene

Early Pleistocene is the oldest subdivision of the Quaternary and began at the end of the Olduvai Normal-Polarity Subchron, about 1.65 Ma, and lasted until the start of the Brunhes Normal-Polarity Chron, about 0.79 Ma (Johnson, 1982; Aguire and Pasini, 1985; Bowen et al., 1986). Few of the Quaternary events recorded on the southern Plains of Canada have been ascribed with a high degree of certainty to this period.

The oldest deposits which have been assigned to this period are the sediments that underlie four tills at the Wellsch Valley section (site 4 of Fig. 2.16; Table 2.9; Stalker and Churcher, 1972). The middle unit of these sediments consists of thin bedded grey silt with local lenses of fine gravel and contains a vertebrate fossil assemblage that suggests an early Irvingtonian Land Mammal age (Fig. 2.24; Stalker, 1976b; Churcher, 1984). In addition, the lowest sediments of this section contain a pattern of paleomagnetic polarity which has been correlated with the Olduvai Normal-Polarity Subchron (Foster and Stalker, 1976; Barendregt, 1984). Consequently, the age of the base of this unit has been estimated at about 1.7 to 1.8 Ma. The sediments at Wellsch Valley include badlands deposits at the base and pond deposits overlain by glacial deposits at the top. This succession appears to indicate a marked cooling trend, probably accompanied by an increase in moisture. However, the climate probably fluctuated widely because the fauna from the central part of the unit contains both cold and warm elements, which suggests a climate like that of today. Deposition of this formation is considered to have taken between 10 and 20 ka.

Substantial numbers of shield stones are present in several places within the sediments at Wellsch Valley and the valley in which the sediments were deposited is, on the basis of morphology, interpreted as an interglacial valley (Stalker and Churcher, 1972). These factors lead to the conclusion that the Wellsch Valley site was overridden by ice, which originated on the Canadian Shield, at least once prior to deposition of the sediments that lie at the base of the succession. Consequently a possible till unit is shown underlying Wellsch Farm Formation in Fig. 2.23. Till from this early ice advance has not been described from the Wellsch Valley site but possibly one of the lower tills of the Sutherland Group (Christiansen, 1968b; E.A. Christiansen, E.A. Christiansen Consulting Ltd., Saskatoon, Saskatchewan, personal communication, 1984) was deposited during this early glaciation.

Middle Pleistocene

The Middle Pleistocene is the time between the beginning of the Brunhes Normal-Polarity Chron and the start of the last (Sangamon) interglaciation (about 790-130 ka; Bowen et al., 1986). Nonglacial conditions are known to have prevailed in part of the southern Interior Plains during an early part of this interval because tephra (correlated with Pearlette ash, type O) was deposited in lake clay at a site on Wascana Creek, Saskatchewan, at some time between 700 and 600 ka (Westgate et al., 1977). The next younger unit from this period for which an absolute age is available is the Saskatchewan Gravels and Sands at Medicine Hat (Table 2.9). These nonglacial fluvial sediments at this site contain a tephra which on the basis of fission-track dating is 435 ka (Westgate et al., 1978) and on the basis of vertebrate fauna is mid-Kansan to Yarmouthian in age (Stalker, 1976b). Apparently ice sheets from the east had not reached this area before this time because the Saskatchewan Gravels and Sands do not contain shield stones. However, incorporation of mountain outwash into the upper part of the Saskatchewan Gravels and Sands, and deposition of proglacial lake sediment on top of this unit in front of advancing Laurentide ice in the valleys on the plains, signalled the start of an episode of glaciation.

Little information is available on this Laurentide advance because its drift is rarely exposed (Stalker, 1983, p. 10, unit CRD). Stalker regarded it as probably an early phase of the Labuma advance that succeeded it. The earliest advance of the first glaciers to reach Medicine Hat appears to have

Figure 2.24 North American time spans of some of the mammalian taxa recovered from Quaternary deposits of the southern Plains of Canada (modified from Stalker and Churcher, 1982). Site locations are shown in Figure 2.16. Site names: 2. Saskatoon area; 3. Fort Qu'Appelle (Bliss); 4. Wellsch Valley; 5. Medicine Hat area; 10. Cochrane; 11. Watino; 13. Bindloss; 14. Empress Elevators; 15. Hand Hills; 16. West Peace River; and * = general Edmonton area but definite location unknown.

Note: Question marks on the chart for taxa numbers 5 and 69 are for sites at Hand Hills and in the Edmonton area where the ages of the deposits from which the bones were recovered are unknown but are assumed to be Sangamonian. The fauna represented in these two areas has a long possible time range.

QUATERNARY GEOLOGY — CANADIAN INTERIOR PLAINS

Taxa Number	Taxa Scientific Names	Common Names	Suggested Ages (ka BP) according to central North America time scale	Sites from which taxa were recovered (Numbers refer to sites shown in Fig 2.16)
1.	Megalonyx sp.	Ground sloth		4, 5,
2.	Nothrotherium cf. shastense	Shasta ground sloth		5,
3.	Paramylodon harlani	Harlan's ground sloth		5,
4.	?Hypolagus limnetis	Gazin's marsh rabbit		4,
5.	Leporidae indet	Hare or rabbit		15,
6.	Lepus cf. townsendii	Townsend hare		5,
7.	Sylvilagus floridianus	Eastern cottontail rabbit		5,
8.	Cynomys cf. leucurus	White-tailed prairie dog		2, 5, 15,
9.	Cynomys cf. meadensis	Meade prairie dog		4,
10.	Citellus cf. meadensis	Meade ground squirrel		5,
11.	Citellus richardsonii	Ground squirrel		2, 5, 15,
12.	Thomomys sp.	Undescribed pocket gopher		4,
13.	Thomomys talpoides	Northern pocket gopher		5,
14.	Castor canadensis	Canadian beaver		5,
15.	Microtus sp.	Field vole		2, 5,
16.	Microtus paroperarius	Extinct vole		4,
17.	Microtus cf. pennsylvanicus	Meadow vole		2, 5, 15,
18.	Allophaiomys sp.	Undescribed extinct vole		4,
19.	Pliophenacomys osborni	Osborn's extinct tree vole		4,
20.	Synaptomys kansasensis	Kansas southern bog lemming		4,
21.	Ondatra zibethicus	Muskrat		5,
22.	Erethizon dorsatum	Porcupine		5,
23.	Mustela nigripes	Black-footed ferret		5,
24.	Spilogale cf. putorius	Eastern spotted skunk		5,
25.	Vulpes vulpes	Red fox		2, 5,
26.	Borophagus diversidens	Cope's bone-eating dog		4,
27.	Taxidea taxus	Badger		2, 3,
28.	Canis cf. latrans	Coyote		2, 5,
29.	Canis lupus	Gray wolf		5,
30.	Canis dirus	Dire wolf		5,
31.	Canis sp.	Dog or wolf		3, 5, 11,
32.	Procyon lotor	Raccoon		5,
33.	Arctodus simus	Short-faced bear		3,
34.	Smilodon fatalis	Sabre-toothed cat		5,
35.	Felis leo atrox	Pleistocene lion		5, 13,
36.	Lynx canadensis	Canada lynx		5,
37.	Lynx cf. rufus	Bobcat		4,
38.	Mammuthus sp.	Mammoth		2, 5, 11, 15, *
39.	M. imperator haroldcooki	Cook's mammoth		4,
40.	Mammuthus imperator	Imperial mammoth		5, 13, 14,
41.	Mammuthus columbi	Columbian mammoth		2, 3, 5,
42.	Mammuthus primigenius	Siberian or northern mammoth		14, 16,
43.	Equus complicatus	Eastern horse		4,
44.	Equus calobatus	Stilt-legged ass		5,
45.	Equus conversidens	Mexican ass		2, 5, 10, 11, 13, 14, 15, *
46.	Equus pacificus	Pacific horse		4,
47.	Equus niobrarensis	Niobrara horse		2, 5, 11,
48.	Equus scotti	Scott's horse		2, 3, 5, 14, *
49.	Equus giganteus	Giant horse		5,
50.	Amerhippus sp.	Neogene horse		5,
51.	Platygonus ?bicalcaratus	Cope's peccary		4,
52.	Camelidae indet.	Unidentified camel		5,
53.	Hemiauchenia sp.	Plains llama		5,
54.	Hemiauchenia stevensi	Steven's plains llama		5,
55.	Hemiauchenia hollomani	Holloman's plains llama		5,
56.	Camelops sp.	Camel		2, 4,
57.	Camelops minidokae	Irvingtonian camel		5,
58.	Camelops hesternus	Western camel		2, 3, 5, 14, *
59.	Odocoileus sp.	Nearctic deer		2, 5,
60.	Odocoileus cf. virginianus	White-tailed deer		5,
61.	Cervus canadensis	Wapiti or elk		2, 5, 16,
62.	Rangifer tarandus	Caribou		5, 10,
63.	Rangifer sp.	Small caribou		14,
64.	?Cervalces sp.	Moose deer		3, 5,
65.	?Tetrameryx sp.	Four-horned pronghorn		4,
66.	Antilocapridae indet.	Unidentified pronghorn		5,
67.	Antilocapra cf. americana	Pronghorn		2, 5,
68.	Ovis canadensis	Mountain sheep		5, 10,
69.	Ovibos moschatus	Tundra muskox		11, *
70.	Symbos cavifrons	Woodland muskox		3, 5, *
71.	?Euceratherium sp.	Woodland shrubox		4,
72.	Biscon cf. bison	Bison		5, 10, 11, 14, 16,
73.	Bison cf. latifrons	Giant long-horned bison		2, 3, 5, 11,
74.	Bison sp.	Large extinct bison		5, 11,

Recovered from deposits of indicated age ... ──────

Existance known during indicated age ...... ───────

Existance assumed during indicated age .... ─ ─ ─ ─

Extant before Nebraskan in North America ..... <

Extant at present in North America .... >

Extant at present on southern Interior Plains ... ×

GSC

reached an elevation of about 900 m in the southwestern Plains but failed to extend as far west as Lethbridge.

The term "Great Glaciation" was coined by Stalker and Harrison (1977) for the glaciation during which both Laurentide and Cordilleran ice reached their all time limits in this area. Labuma Till was deposited by Laurentide ice and Albertan Till was deposited by Cordilleran ice (Table 2.9; Stalker and Harrison, 1977). The oldest glacial deposits at Medicine Hat overlie the Saskatchewan Gravels and Sands which contain a fauna as young as late Yarmouthian (Stalker, 1976b) and therefore the glaciation during which the Labuma Till was deposited is post-Yarmouthian in age. This differs from the age of the most extensive glaciation in the northern Canadian Interior Plains (Banks Glaciation) which appears to have occurred prior to the end of the Matuyama Reversed-Polarity Chron or during Early Pleistocene (Vincent, 1989).

The ice that deposited Labuma Till covered all of the southern Canadian Plains with the exception of small areas in the Wood Mountain Upland, in Cypress Hills, and at Del Bonita (Prest et al., 1968). In the foothills and at the front of the Rocky Mountains it extended to elevations of about 1555 m near the border with the United States and to about 1680 m in the Porcupine Hills, 80 km to the north (Stalker, 1959, 1962).

When this ice commenced its eastward retreat, lakes undoubtedly formed along its margin, but spillways cut by outflow from these are the main evidence for existence of these lakes. Most of the large spillway valleys found in the southwestern part of this region likely were cut at this time (Fig. 2.18), even though they were reused and modified during subsequent glaciations.

The ice that deposited Albertan Till flowed out of the mountains to form a large piedmont lobe. This was the Waterton I Advance and the Great Cordilleran Glacier of Stalker and Harrison (1977, p. 887-892). In contrast with later glaciations, Laurentide ice did not block the flow of ice out from the valleys except during late stages. The ability of the ice on the plains to advance to the mountain front, and the local occurrence of sands and silts between the Albertan and Labuma tills, indicate that the Cordilleran ice had undergone substantial retreat before the Laurentide ice arrived on the scene. Some contact between the two ice sheets, however is suggested by evidence that Laurentide ice reached its greatest elevations in areas between the major valleys. Evidently Cordilleran ice in the mountain valleys fended off the advancing Laurentide glacier (Stalker, 1959; Stalker and Harrison, 1977, p. 892-897).

Maunsell and Brocket tills, which overlie Labuma Till, represent two ice advances of the latter part of the Middle Pleistocene (Table 2.9). They were large enough to approach the mountain front and to rank as third and fourth most extensive of all the Laurentide ice advances (Fig. 2.23). However, little is known about them, because their area of coverage was completely overrun by ice of Late Pleistocene age. Alley (1973, p. 166) considered the Maunsell and Brocket to be the same till but here they are referred to two separate ice advances. This interpretation is based chiefly on the fact that Maunsell Till typically overlies bedrock in the oldest set of interglacial valleys, whereas Brocket Till overlies bedrock in the second oldest set and overlies bedrock and Maunsell Till in the oldest. Undoubtedly, there were Cordilleran ice advances during those times, but of them we have no record.

Our knowledge of Quaternary deposits in Saskatchewan and Manitoba that might relate to the Middle Pleistocene is based on borehole stratigraphy. As a consequence, accurate correlation with the outcrop record of southern Alberta is difficult. In Saskatchewan, if the strongly weathered zone at the top of the Sutherland Group is assumed to predate the Middle Pleistocene, and the Riddell Member of the Floral Formation is Sangamonian in age (SkaraWoolf, 1981), then the part of the Floral Formation below the Riddell Member represents deposition during the Middle Pleistocene (Table 2.10). In Manitoba, the units that underlie deposits referred to as Sangamonian and which therefore might have been deposited during the Middle Pleistocene (but could also be older) are the Shell, Tee Lakes, and Largs formations (Klassen, 1969, 1979) and the Rosa and Woodmore formations (Fenton, 1974).

Late Pleistocene

Late Pleistocene includes the Sangamonian and Wisconsinan stages and in this volume is considered to have begun at 130 ka and to have ended at 10 ka. It includes the last interglaciation and the last glaciation. Events of the early part of this time are poorly known because exposures of deposits of this age are scarce and this period is beyond the limit of radiocarbon dating. Events of the latter part of this interval are relatively well known because their deposits form the surface of much of the southern Canadian Interior Plains and fall within the limit of radiocarbon dating.

Sangamonian

Sangamonian time in western Canada included a nonglacial period during which climate is inferred to have been similar or slightly warmer than at present. Deposits at several sites have been assigned to this stage because of their content of vertebrate fossils (for lists of fossils, see Sangamonian in Fig. 2.24; Harington, 1978, p. 27-30, 37-43). The main units referred to the Sangamonian based on vertebrate content are: Echo Lake Gravels at Fort Qu'Appelle (site 3 of Fig. 2.16; Khan, 1970; Christiansen, 1972), Riddell Member of the Floral Formation at Saskatoon (Skarawoolf, 1980, 1981), and Mitchell Bluffs Formation at Medicine Hat (Stalker, 1976b). Other local nonglacial deposits which have been correlated with this interglaciation are: Roaring River Clay of southwestern Manitoba (Klassen et al., 1967) and St. Malo Formation of southeastern Manitoba (Fenton, 1974). These deposits are referred to the Sangamonian because they apparently are older than a nonglacial of Middle Wisconsinan age and they contain plant fossil evidence of conditions as warm or warmer than present. Because there are no fossils that are uniquely Sangamonian and methods of absolute dating are lacking, difficulty arises in verifying the Sangamonian age of all of these deposits. The age assignment of these deposits has been challenged by Fullerton and Colton (1986); they place these units in a late Middle Pleistocene position saying that none of the fossil vertebrates found in these deposits is demonstrably diagnostic of a post-Illinoian pre-Wisconsinan age.

The length of the Sangamon Interglaciation on the Plains is unknown, but it is estimated to have lasted roughly from 125 to 85 ka BP. During this time, which is much longer than postglacial time, rivers re-established themselves and formed a drainage network that, apparently was better developed than the present one. In general, trunk valleys were slightly deeper and broader than their present day equivalents, but were not nearly as wide as the preglacial ones. Most deposits are from the latter part of Sangamonian time, and were laid down as valley fill after the valleys had been incised and enlarged; hence, they do not provide an accurate assessment of conditions during the Sangamonian as a whole. It is likely they stem from a cool and humid part of this nonglacial period and possibly some of them were deposited as glaciers were starting to form in the east, north, and west.

Wisconsinan

During the Wisconsinan, the southern Canadian Interior Plains were glaciated by the Laurentide Ice Sheet (Dyke et al., 1989, contains an explanation of the nomenclature used for this major Wisconsinan feature). The main components of the Laurentide Ice Sheet which affected the Plains were: Hudson Ice of the Labrador Sector, which was active at the southeastern margin of the area in Late Wisconsinan time and which probably transported "dark erratics" across much of the area at some earlier time(s); Keewatin Ice of the Keewatin Sector, which was responsible for southerly and southwesterly flow across north and south-central parts of the region; and Plains Ice of the Keewatin Sector which left southerly and southeasterly flow patterns across western parts of the area. Figure 2.20, however, suggests that this is a gross simplification of a flow pattern that is extremely complex in detail. This flow pattern is considerably more complex than that of the same ice sheet where it was overriding the Canadian Shield. Dyke and Prest (1987) speculated that this might be due to a change in glacier flow conditions which occurred when the ice moved from competent rocks of the Canadian Shield to the more easily deformed rocks of the Interior Plains. The significance of degree of deformability of bed on glacier movement is discussed by Fisher et al. (1985).

Early Wisconsinan glaciation

If the Middle Wisconsinan of the southern Canadian Interior Plains was a time of general ice retreat (Fenton, 1984b) then where was the glacial limit during Early Wisconsinan In setting up regional stratigraphies, workers as a general rule have assigned tills directly underlying Middle Wisconsinan deposits to the Early Wisconsinan. This is generally based on the assumption that deposits assigned to Middle Wisconsinan were insufficient to encompass Middle Wisconsinan through Sangamonian time and because the till underlying the sediments lacked a paleosol that could be referred to as the Sangamon Soil. Using these assumptions, a series of Early Wisconsinan tills have been named as outlined in Table 2.10 and Figure 2.23 (see also Fenton, 1984b).

There is a lack of agreement on the probable position of the Early Wisconsinan limit. Stalker and Harrison (1977) placed the limit in the Rocky Mountain Foothills well beyond that of Late Wisconsinan. This is based in part on Stalker's (1977, 1983) observations that the tills that overlie his Middle Wisconsinan unit at Medicine Hat end at the Lethbridge Moraine, whereas the tills that lie between his Middle Wisconsinan and Sangamonian units at Medicine Hat (Cameron Ranch Formation) continue west to the foothills (Fig. 2.23). This makes the Early Wisconsinan the second most extensive Quaternary ice advance on the Interior Plains, rising to elevations of 1380 m in the Rocky Mountain Foothills (limit 2 of Stalker and Harrison, 1977). During this advance, large quartzite boulders were added to the margin of the Laurentide Ice Sheet by glaciers advancing from the Cordillera. The train of these boulders became the Foothills Erratics Train of Stalker (1956), and the till associated with them has been referred to as the Erratics Train Till (Stalker and Harrison, 1977; Jackson, 1980b).

Fullerton and Colton (1986) rejected the stratigraphic evidence for Early Wisconsinan glaciation in the western Plains and interpreted the till containing the Erratics Train (Cameron Ranch Formation) as Late Wisconsinan. Clayton and Moran (1982) disagreed with Stalker (1977) and placed their Late Wisconsinan limit in Canada approximately on Stalker's Early Wisconsinan limit but also disagreed with Fullerton and Colton by showing the Early Wisconsinan limit to lie slightly beyond that of the Late Wisconsinan in the Plains of United States.

At the moment all that can be said is that there are no deposits inside or outside the limit of the last ice advance that can be proven to be Early Wisconsinan on the basis of unquestioned stratigraphic evidence or absolute dates.

Nonglacial periods of the Wisconsinan

One of the contentious points of the Quaternary history of the southern Canadian Interior Plains is the number of nonglacial periods (interstades) that occurred during the Wisconsinan. Only one fossiliferous nonglacial unit has been assigned to the Early Wisconsinan. This is a 2-3 m thick deposit of gravel, sand, and silt that contains wood and bones of horse, mammoth, and pronebuck (Fig. 2.24, Nos. 45, 42) and is exposed at Medicine Hat (Stalker, 1976b). The content of large mammal bones suggests these sediments represent a significant nonglacial period. The regional significance of this unit however is difficult to evaluate because the deposits have been recognized at only one place.

Sediments representing an interstade have provided radiocarbon ages ranging from 22 ka to 43 ka (14-20 of Table 2.11). At Watino, Alberta (site 11 of Fig. 2.16; Westgate et al., 1972) and at Zelena, Manitoba (site 1 of Fig. 2.16; Klassen, 1969) evidence points to a climate during this time that apparently was little different from that of present. At Medicine Hat, however, there is evidence of permafrost near the end of this period (Stalker, 1976b). Most of the southern Canadian Interior Plains apparently were ice free at this time because sediments from several sites scattered throughout the area date to this interval and Stalker (1976b) referred to a nearly continuous sediment record 40 to 24 ka at Medicine Hat. Paraphrasing Stalker in describing deposits related to this interval at Medicine Hat: Lag gravel and alluvial silt and clay at the base contain an extensive fauna which indicates a not overly cold climate. Radiocarbon ages of $37\,900 \pm 1100$ BP and $38\,700 \pm 1100$ BP (different size fractions of the same sample; 19 of Table 2.11) were obtained for plant fragments 1 m above the main bone bed. The next younger unit is a mixture of silt, clay, sand, stones, and organic material that is severely churned and was formed under permafrost conditions. A radiocarbon date from near the base of this unit (Evil-smelling Band) gave an age of $28\,630 \pm 800$ BP, one 5 m higher an age of $25\,000 \pm 800$ BP, and a third close to the top of the unit an age of $24\,490 \pm 200$ BP (see 18, 17, 16 of Table 2.11 for more information on dates). The Prelate

Table 2.11. Significant radiocarbon dates from the southern Interior Plains of Canada

Reference number (cf. Fig. 2.26)	Laboratory number	Date (years BP)	Material	Location	Reference	Comments
1	GSC-2294	7030 ± 170	marine shells	Limestone River, Man. 56°31'N 94°05'W	Teller, 1980	From 1.2 m depth in surficial marine clay; records post-Lake Agassiz marine inundation
2	GSC-1825	7970 ± 150	gyttja	Flin Flon, Man. 54°45'N 101°41'W	Lowdon and Blake, 1975	From 3.7 m depth in clayey lake sediments
3	GSC-3402	9910 ± 90	wood	Fort McMurray, Alta. 58°15'N 111°25'W	Blake, 1982	From about 10 m depth in deltaic sand of glacial Lake McConnell or early Lake Athabasca
4	GSC-391	9990 ± 160	wood	Buffalo Point, Man. 49°00'N 95°14'W	Lowdon, et al., 1967	From about 2 m depth in lower Campbell Beach
5	GSC-677	10 690 ± 190	wood	Shellmouth, Man. 50°58'N 101°24'W	Lowdon and Blake, 1968	From alluvium below slump debris about 3 m above modern Assiniboine River floodplain
6	S-2077	10 840 ± 355	wood	Petrafka, Sask. 52°39'N 106°51'W	E.A. Christiansen, personal communication, 1984	From about 18 m depth in North Saskatchewan River alluvium
7	S-1374	10 875 ± 660	wood	Denholm, Sask. 52°36'N 108°05'W	E.A. Christiansen, personal communication, 1984	From about 24 m depth in North Saskatchewan River alluvium
8	S-553	12 025 ± 205	wood	Marieval, Sask. 50°35'N 102°39'W	Christiansen, 1979	From about 50 m depth in Qu'Appelle alluvium
9	GSC-1319	12 100 ± 160	peat	Rossendale, Man. 49°47'N 98°36'W	Lowdon et al., 1971	From about 4 m depth in fill within gulley in Assiniboine delta
10	TO-216	12 630 ± 80	freshwater shells	Ham site, Sask. 49°04'05" 108°45'35"	Klassen and Vreeken, 1987	Shells from silt and sand deposited in an ice-walled lake
11	I-1682	12 800 ± 350	gyttja	Glenora, Man. 49°26'N 99°17'W	Ritchie and Lichti-Federovich, 1968	From about 10 m depth in pollen-rich Kettle Lake sediments
12	I-3476	13 900 ± 240	organic detritus	Brandon, Man. 49°50'N 99°35'W	Ritchie, 1976	From 4.2 m depth in pollen-rich sediment in postglacial lake on Assiniboine delta
13	S-176	20 000 ± 850	humus	Leader, Sask. 50°59'N 109°22'W	McCallum and Wittenberg, 1965	From about 36 m depth in paleosol below two tills (David, 1966)
14	GX-3530	22 260 ± 1000	organic detritus and wood fragments	Roseau River, Man. 49°11'N 96°39'W	Teller, 1980	From about 20 m depth in silt below two tills (Whitemouth and Marchand formations); same unit at about 30 m depth dated >39 ka
15	GSC-1279	23 700 ± 290	charcoal	Zelena, Man. 51°24'N 101°14'W	Lowdon and Blake, 1968	From about 3 m depth from within a silt less than 1 m thick between Minnedosa and Zelena tills (Klassen, 1979)
16	GSC-205	24 490 ± 200	plant fragments	Medicine Hat, Alta. 50°06'N 110°38'W	Dyck et al., 1965	From 18 m depth; provides maximum age for overlying till and age of top part of "Evilsmelling Band"
17	GSC-1370	25 000 ± 800	plant fragments	Medicine Hat, Alta. 50°06'N 110°38'W	Lowdon et al., 1971	Provides maximum age for overlying till and age of material about 5 m below top of "Evilsmelling Band"
18	GSC-578	28 630 ± 800	plant fragments	Medicine Hat, Alta. 50°06'N 110°38'W	Lowdon et al., 1967	From 24 m depth; provides age of bottom of "Evilsmelling Band" about 9 m below its top
19	GSC-1442 GSC-1442-2	37 900 ± 1100 38 700 ± 1100	plant fragments	Redcliff, Alta. 54°05'N 110°49'W	Lowdon and Blake, 1975	From 30 m depth; provides minimum age for vertebrate fossil site and maximum age for onset of Late Wisconsin glaciation
20	GSC-1020	43 500 ± 620	wood	Watino, Alta. 55°43' 117°38'	Lowdon and Blake, 1970	Wood from peat- and mollusc-bearing unit which overlies quartzite gravel and underlies Late Wisconsinan glacial deposits
21	GSC-2052	>29 000	humus	Prelate, Sask. 51°1.5' 109°15.3'	David, 1987	Humus collected from Prelate Ferry Paleosol 14.5 km east of the reference section

Ferry Paleosol, originally dated as 20 000 ± 850 BP (David, 1966; 13 of Table 2.11) but redated as >29 000 BP (21 of Table 2.11) may also have formed during this period (David, 1987). This Middle Wisconsinan nonglacial interval was referred to as the Watino Interval by Fenton (1984b).

Late Wisconsinan glaciation

Late Wisconsinan deposits make up much of the surface of the southern Interior Plains and outcrops are abundant. Consequently, materials of this last ice advance are easily studied. One would therefore expect that the stratigraphy and chronology of Late Wisconsinan deposits would be well known and that the limit would be precisely located and dated. This is not the case and much controversy exists over the nature and chronology of Late Wisconsinan events, and both location and age of the Late Wisconsinan limit remain speculative.

In the western part of the southern Interior Plains, Stalker (1977) located the Late Wisconsinan limit at the Lethbridge Moraine and many have accepted this position (Jackson, 1980b; Reeves, 1973; Rutter, 1980). Fullerton and Colton (1986) disputed Stalker's evidence and placed the limit in Montana, as much as 250 km south of the Lethbridge Moraine. Clayton and Moran (1982) and Christiansen (1979) likewise showed Late Wisconsinan ice to have extended well beyond the Lethbridge Moraine. The case for the limit lying at the Lethbridge Moraine is based on Late Wisconsinan tills apparently ending there and on surface nonglacial materials from areas some distance beyond the Lethbridge Moraine being older than the age of the Late Wisconsinan maximum (Karlstrom, 1981, 1986; Jackson, 1983; Jackson and Pawson, 1984). The case for a more southerly Late Wisconsinan limit is based mainly on the apparent continuity of fresh surface forms from north of the Lethbridge Moraine to the limit drawn in Montana (Fullerton and Colton, 1986). The most recent evidence available in Canada appears to also favour a limit beyond the Lethbridge Moraine. Klassen and Vreeken (1987) have dated freshwater shells from deposits of a former ice-walled lake about 75 km south of the Lethbridge Moraine position as 12 630 ± 80 BP (10 of Table 2.11); MacDonald et al. (1987) have suggested that the old radiocarbon ages that have been used to indicate areas that were not covered by Late Wisconsinan ice are of questionable value because one of the mosses commonly present in the dated samples of postglacial pond sediments can ingest carbon from lake water to such an extent that 9 ka old material will provide a radiocarbon date of 16 ka.

Finite radiocarbon dates have been obtained for a variety of organic materials enclosed in deposits underlying Late Wisconsinan till. These include humus dated 20 000 ± 850 BP, organic detritus dated 22 260 ± 1000 BP, charcoal dated 23 700 ± 290 BP, and wood dated 24 490 ± 200 BP (see 13-16, Table 2.11). Stalker (1980) stated that ice stood at the Lethbridge Moraine 22 ka; Clayton and Moran (1982) suggested that the ice reached the Late Wisconsinan limit about 20 ka; and Christiansen (1979) gave an age of 17 ka to the ice margin shown on his earliest Late Wisconsinan retreat map.

As indicated in the discussion above, a precise position and age cannot be given for the Late Wisconsinan limit in the southern Interior Plains area. This report follows Prest's (1984) proposal that the Lethbridge Moraine is the minimum position of the limit and that the maximum possible position of the limit lies in Montana. Late Wisconsinan ice in this area reached its limit some time after 23 ka but may not have begun its final retreat from that limit until as late as 17 ka.

Late Wisconsinan deglaciation

Even though the age and position of the Late Wisconsinan limit are disputed, there is general agreement on the pattern of Late Wisconsinan deglaciation and the picture presented in Figure 2.25 differs little from that presented by Prest (1969). There is however considerable controversy about the age of the various ice margin positions. This controversy mostly involves regions in Manitoba, eastern Saskatchewan, and the adjacent states that lie in and near the basin of glacial Lake Agassiz. Elson (1967), Prest (1969), Klassen (1972, 1975) and Christiansen (1979) proposed a deglaciation chronology for the Lake Agassiz basin that begins 13-14 ka. This is based on radiocarbon ages for a variety of organic materials contained in alluvial sediments that largely postdate early glacial Lake Agassiz sediments. Recently, Teller and Fenton (1980), Clayton and Moran (1982), Fenton et al. (1983), and Mickelson et al. (1983, p. 9) used dates on wood overlain by till in North and South Dakota and Iowa to propose that final deglaciation did not begin until after 12 ka. In order to explain the discrepancy, they suggested that most radiocarbon dates from southern Canada, on materials other than wood, are erroneous due to contamination (Teller et al., 1980). Other studies have raised similar doubts about validity of nonwood dates (Karrow and Anderson, 1975; Mathewes and Westgate, 1980; Nambudiri et al., 1980). This controversy is not discussed in detail here, and the discrepancies can largely be explained by retaining ice in the glacial Lake Agassiz basin after it had retreated from the adjacent plains.

Ice retreat pattern. The pattern of retreat of Late Wisconsinan ice for the southern Interior Plains as shown in Figure 2.25 is based on recognition of ice marginal landforms. These include corrugated ground moraine, ice thrust moraine, end and interlobate moraines, and ice marginal channels. Caution must be exercised in using morainic features and meltwater channels in working out the pattern of ice retreat, because not all such features necessarily formed at the ice margin or during the last glaciation. Although linear belts of knob-and kettle topography are considered to be end or recessional moraines, the extensive areas of hummocky moraine associated with most of the upland areas do not necessarily denote important Late Wisconsinan ice marginal positions, contrary to suggestions by early workers (Johnston, 1934; Johnston et al., 1948). Rather, they were the loci of glacial deposition during successive glaciations. Thus, much of the thick drift associated with the large belts of hummocky moraine found on the southeast-trending Missouri Coteau and Manitoba Escarpment predates the last glaciation (Christiansen, 1971a; Klassen, 1979). Similarly, the Pas Moraine of northern Manitoba marks an ice marginal position that predates the last advance (Klassen, 1965, 1983a). In addition, many meltwater channels, such as segments of Qu'Appelle and Assiniboine valleys (Klassen, 1972, 1975), and probably Frenchman Valley of southwest Saskatchewan (Christiansen, 1979, p. 934), were also established before Late Wisconsinan glaciation.

Retreat of Late Wisconsinan ice from the southern Interior Plains appears to have proceeded largely by melting

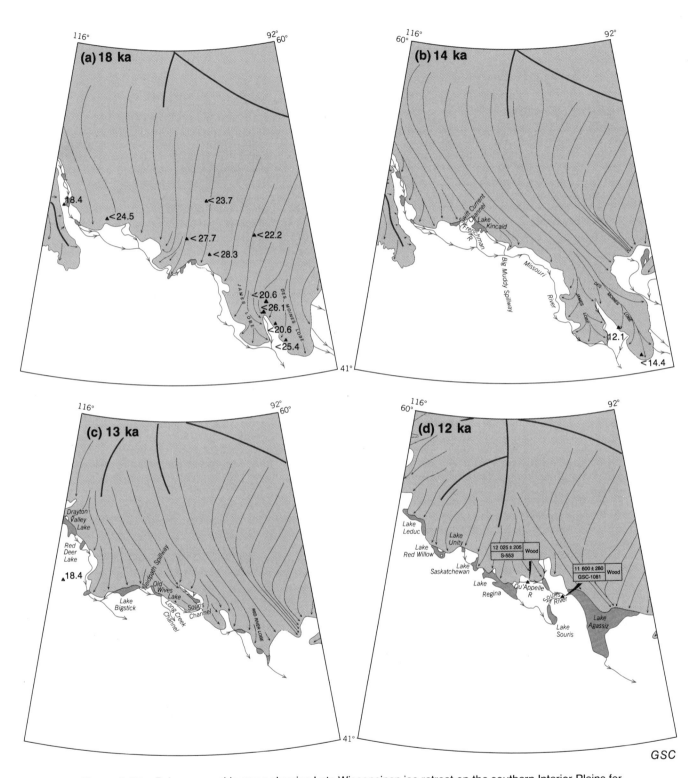

Figure 2.25 Paleogeographic maps showing Late Wisconsinan ice retreat on the southern Interior Plains for (a) 18 ka, (b) 14 ka, (c) 13 ka, (d) 12 ka, (e) 11 ka, (f) 10 ka, and (g) 9 ka.

QUATERNARY GEOLOGY — CANADIAN INTERIOR PLAINS

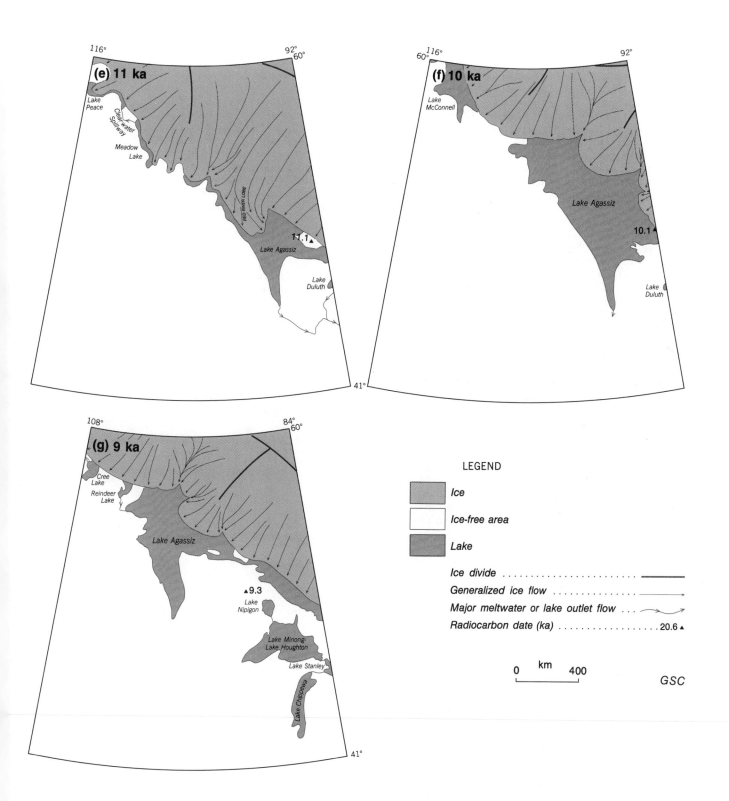

of stagnant ice along broad marginal belts. This is in contrast with retreat along a sharply demarcated ice front, by the melting of still active and advancing ice, as envisioned in the traditional model of retreat deduced from study of valley glaciers. However, this does not substantially alter the traditional approach for deciphering the general "phases" of deglaciation on the Interior Plains of Canada (Klassen, 1972, 1983a, b; Christiansen, 1979), which is based largely on interpreting ice marginal features, flow indicators, drainage patterns, and stratigraphy. Mickelson et al. (1983, p. 13) used the "glacial-process-form" model of Clayton and Moran (1974) in interpreting Late Wisconsinan events of the north-central United States for glaciated terrain similar to that of the southern Canadian Plains. This model relates suites of landforms (fringe, marginal, transitional, inner) to basal ice regimes (melting or freezing) and to glacial processes (erosion, transportation, deposition). This concept provides a better approach to deciphering the Late Wisconsinan deglaciation than the more traditional approach followed in this report. Both, however are subjective and strongly dependent on accurate mapping and correct interpretation of the origin of landforms and deposits.

A number of ice retreat rates have been proposed for the Late Wisconsinan ice on the southern Canadian Plains. Christiansen (1979, p. 934), proposed an average of 60 m/a for the early stages and 275 m/a for final stages of retreat; Ritchie (1976, p. 1809) suggested 200 to 300 m/a for northward encroachment of late glacial forests, and Klassen (1983a) derived retreat rates of 250 m/a and 300 m/a in northern Manitoba from varve counts and radiocarbon dates respectively.

Phases of deglaciation. Phase 1. The 14 ka ice margin location assumed for the southern Canadian Plains as shown in Figure 2.25b is approximately the minimum limit position of the Late Wisconsinan. It is largely a modification of phase 2 of Christiansen (1979). It correlates with Stalker's (1977) margin associated with the Lethbridge Moraine in southern Alberta (Fig. 2.18) and approximately coincides with a major end moraine in northeastern Montana (Colton et al., 1961). End moraines associated with this margin in southwestern Saskatchewan include the Aikins Moraine (Christiansen, 1959), Thomson Moraine (Whitaker, 1965), and Harptree Moraine (Parizek, 1964). Glacial lake basins related to phase 1 include a number of small lakes in the Rocky Mountain Foothills at the western margin of the ice, several unnamed basins at about 900 m elevation along the north-facing slopes of the Cypress Hills, and glacial Lake Kincaid (Whitaker, 1965) at about 800 m. Spillways in Saskatchewan include Swift Current channel, Braddock channel, and the Big Muddy spillway that joined the Missouri River in Montana. Meltwater was also diverted southward into the Frenchman channel via Swift Current channel early in this phase.

The oldest postglacial ages behind this ice marginal position are in the 12 to 13 ka range (8,9,10,11 of Table 2.11; Fig. 2.26) and hence assigning a 14 ka age to this position is reasonable. One age that does not agree with this is the 13 900 ± 240 BP date from the surface of the Assiniboine delta near Brandon (12 of Table 2.11; Fig. 2.26). This sample of organic detritus may be contaminated or contain moss which ingested old carbon as mentioned above.

The moraines associated with this position are considered to be readvance positions of Late Wisconsinan ice by Christiansen (1959), Whitaker (1965), and Parizek (1964). Parizek (1964) placed the Late Wisconsinan ice limit at the northern edge of the unglaciated part of the Wood Mountain Upland, some 35 km southwest of the phase 1 margin shown here, whereas Whitaker (1965) placed the Late Wisconsinan margin some 60 km south of the phase 1 margin. The ice margin position of Colton et al. (1961) in northeastern Montana, which ties in with the phase 1 position, is considered to be between 14 and 13.5 ka by Clayton and Moran (1982, Fig. 6). They suggested that it marks the limit of a major readvance of Keewatin Ice that deposited shale-rich tills above boulder pavements in North and South Dakota. Similar tills above boulder pavements mapped in southern Manitoba (Lennard Formation) and east-central Saskatchewan (Battleford Formation) are considered to be associated with the main Late Wisconsinan ice advance (Christiansen, 1968a; Klassen, 1969).

Phase 2. The ice margin position in the southern Canadian Interior Plains shown for 13 ka in Figure 2.25c is based largely on Christiansen's (1979) phase 3 in southern Saskatchewan. It differs from Christiansen's phase 3 in Manitoba in that the ice is shown as completely occupying all of the basin later occupied by glacial Lake Souris. In Alberta this position joins St-Onge's (1972) phase 3, rather than Westgate's (1968) Etzikom advance as proposed by Christiansen (1979). If the connection with phase 3 of St-Onge (1972) is correct, then it also correlates with phase 2 of Mathews (1980) in northwestern British Columbia. The nearest correlative in Clayton and Moran (1982) appears to be phase L.

End moraines associated with phase 2 in southern Saskatchewan include Fox Valley, Leinan, Dirt Hills, and

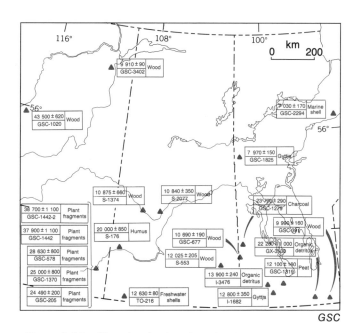

Figure 2.26 Map showing location of radiocarbon dates listed in Table 2.11. Note: After this manuscript was prepared, an age of 9510 ± 90 BP (GSC-4490) was obtained for wood from the peat that yielded GSC-1319, indicating it was associated with the Campbell beach phase.

Stoughton (Christiansen, 1979). Several glacial lakes, including glacial Lake Bigstick, formed at about 760 m elevation between the glacier and the Cypress Hills. These lakes drained eastward, via Neidpath spillway, into glacial Old Wives Lake at about 685 m elevation, and then into Missouri River via Big Muddy spillway (Christiansen, 1979). Farther east, meltwater flowed along the margin of the Weyburn and Souris lobes, by Long Creek and Souris channels.

The age of about 13 ka here proposed for this position is slightly younger than the age of about 13.5 ka that St-Onge (1972) assigned to his phase 3 ice margin in adjacent east-central Alberta. The 11.7 ka age assigned to this margin in the adjoining part of North Dakota (Clayton and Moran, 1982) reflects the problems of Late Wisconsinan chronology discussed earlier.

A number of events occurred between when the ice was at the 13 ka position and when it reached the 12 ka position. In the far west, glacial Lake Iosegun formed and drained (St-Onge, 1972); glacial Lake Souris, draining via the Pembina spillway, formed at the margin of the Souris lobe (Clayton et al., 1980); large tracks of hummocky moraine were formed in south-central Alberta and southwestern Manitoba; and ice pushed into the re-entrant in the Manitoba escarpment occupied by Assiniboine Valley and deposited the Darlingford Moraine.

Phase 3. The 12 ka ice margin for the southern Canadian Plains (Fig. 2.25d) is after phase 4 of Christiansen (1979) in Saskatchewan and phase 1 of Klassen (1972) in Manitoba. It is approximately equivalent to Elson's (1967) Herman phase, the middle Lochart Phase of Fenton et al. (1983), and the phase Teller and Fenton (1980) showed in their Figure 10-f. It correlates with St-Onge's (1972) phase 4 margin in east-central Alberta, and corresponds approximately to phase N of Clayton and Moran (1982).

End moraines along this ice margin include the Condie and Qu'Appelle (Christiansen, 1979) in Saskatchewan, and the Petlura (Klassen, 1979) and Cowan (Elson, 1961) in southern Manitoba (Fig. 2.18).

Herman beach, at 400 m a.s.l., marks the highest level of glacial Lake Agassiz in southern Manitoba; it formed when Assiniboine River started to build a delta where it entered Lake Agassiz. Lake Regina and other glacial lakes to the west, including early Lake Saskatchewan (600 m a.s.l.), Lake Unity (655 m a.s.l.), and Lake Leduc, drained into Lake Agassiz via Souris Valley. At least one catastrophic flood occurred in this system (Kehew, 1982). Subsequent drainage of glacial Lake Regina took place along Qu'Appelle Valley when the glacier withdrew from the Condie and Qu'Appelle moraines. The bulk of the Assiniboine delta was built near the end of this phase, when the South Saskatchewan-Qu'Appelle-Assiniboine system drained most of the deglaciated southern Canadian Interior Plains into Lake Agassiz. Three radiocarbon dates (8,9,10 of Table 2.11; Fig. 2.26) from fluvial deposits overlying the delta and within Assiniboine and Qu'Appelle valleys suggest that the ice had withdrawn from this position before the 12 ka age that is indicated (Klassen, 1983b). St-Onge's (1972) chronology in east-central Alberta is compatible with this age. However, Clayton and Moran's (1982) proposed age of about 11.3 ka, for ice margins in South Dakota that are continuous with this position, again reflects the problems of regional chronology.

Between 12 ka and 11 ka, Laurentide ice retreated from much of Alberta and central Saskatchewan. All ice marginal lakes in Alberta, with the exception of glacial Lake Peace, were drained. In central Saskatchewan, glacial lakes Regina and Saskatchewan drained, and glacial lakes Lost Mountain, Saltcoats, and Melfort formed and were drained. The ice margin retreated rapidly northward in the western part of the glacial Lake Agassiz basin, and the Bélair and Sandilands interlobate moraine complex (Fig. 2.18) formed between this lobe and Hudson Ice which lay on the Canadian Shield to the northeast.

Phase 4. The ice front position shown for 11 ka (Fig. 2.25e) is an arbitrary one slightly modified from Christiansen's (1979) phase 8, Schreiner's (1983) phase 1, and Elson's (1967) Campbell phase in Manitoba. It coincides with the late Lochart phase of Clayton (1983) in the southern parts of the Lake Agassiz basin.

This position is derived mainly from stratigraphic and chronological evidence from the Assiniboine and Saskatchewan deltas in Lake Agassiz, rather than from geomorphology. Beaver River Moraine in western Saskatchewan is the only major end moraine associated with it. Glacial Meadow Lake, at 480 m a.s.l., at this time drained north through Clearwater spillway to glacial Lake Peace which still existed in Peace River valley in front of an ice lobe pushing southwest from the general vicinity of Lake Athabasca.

The opening of outlets east of the Canadian Interior Plains into Lake Superior basin shortly after 11 ka (Kashabowie-Seine valleys of Teller and Thorleifson, 1983) lowered Lake Agassiz substantially. This caused valley incision by Assiniboine and Saskatchewan rivers across and upstream from their Lake Agassiz deltas (Elson, 1967; Klassen, 1972, 1983b; E.A. Christiansen, E.A. Christiansen Consulting Ltd., Saskatoon, personal communication, 1984). The oldest radiocarbon dates from fill deposited in the channels cut by the Assiniboine and Saskatchewan systems are between 11 and 10.5 ka (5,6,7 of Table 2.11; Klassen, 1983b; E.A. Christiansen, personal communication, 1984).

Phase 5. The ice margin shown for 10 ka (Fig. 2.25f) lies on the Canadian Shield east of the Canadian Interior Plains at the Cree Lake Moraine (Christiansen, 1979, phase 9; Schreiner, 1983, phase 2). In Manitoba, the limit lies at one of the end and interlobate moraine complexes that marks the contact between Keewatin Ice and Hudson Ice (Klassen, 1983a). This phase approximates ice front positions proposed by Elson (1967, phase 1), Prest (1969, 10 ka margin), and the Emerson phase of Clayton (1983) and Teller and Thorleifson (1983) in the southern part of the Lake Agassiz basin.

At this time the Marquette ice advance closed the eastern outlets of glacial Lake Agassiz (Teller and Thorleifson, 1983) and the lake reached its greatest extent and stability, as indicated by the Campbell beaches (Elson, 1971). The lower part of Assiniboine delta (separated from the main delta by a scarp at 365 to 320 m a.s.l.; Klassen, 1983b) and the highest shoreline features above the Saskatchewan delta near Nipawin (371 m a.s.l.) formed at this time. Glacial Lake Agassiz stood at this high level from 10 to 9.5 ka. This age is based on a radiocarbon age of 9990 ± 160 BP from Campbell beach in southeastern Manitoba (4 of Table 2.11) and dates from the upper valley fill underlying the

Campbell terrace both in Assiniboine delta (Klassen, 1983b) and in Saskatchewan delta near Nipawin (E.A. Christiansen, personal communication, 1984). An age of 9910 ± 90 BP (3 of Table 2.11) from an early Lake Athabasca or late glacial Lake McConnell delta at about 245 m elevation (Smith in Blake, 1982) aids in correlation between this stage of Lake Agassiz and deltas built into lakes farther north.

In addition to draining south, Lake Agassiz may have spilled into the Clearwater-Mackenzie drainage system (Elson, 1967) and thence into glacial Lake McConnell, which occupied the northeastern corner of Alberta at this time (Craig, 1965). According to Christiansen (1979) and Schreiner (1983), however, the fact that Lake Agassiz deposits do not extend west to the Churchill-Clearwater spillway indicates that Lake Agassiz did not drain north during this phase. Anastomosing channels, however, join Clearwater Valley at a point where it deepens and widens markedly to the west. These features may be due to ice marginal drainage or to drainage from a northwest arm of Lake Agassiz.

Retreat of ice from the Marquette position in the Lake Superior Basin caused the level of Lake Agassiz to fall rapidly. Discharge to the east was catastrophic although periods of lake level stability permitted a series of weakly developed beaches to be constructed as outflow was shunted into a series of progressively lower channels (Teller and Thorleifson, 1983).

Phase 6. The ice margin position shown for 9 ka (Fig. 2.25 g) corresponds approximately to phase 5 of Schreiner (1983) in Saskatchewan and is midway between phases 3 and 4 of Klassen (1983a) in Manitoba. During this phase, which is approximately equivalent to Elson's (1967) Pipun stage, glacial Lake Agassiz built the Ponton beach. Drainage was southeast to glacial Lake Ojibway which covered much of deglaciated Hudson Bay Lowland in Ontario and Quebec (Dredge and Cowan, 1989).

Radiocarbon ages on surface organic deposits from near this ice margin in Saskatchewan and from Ponton beach in Manitoba indicate that the ice had retreated from this position by 8 ka (1, 2 of Table 2.11; Fig. 2.26). Ritchie and Koivo (1975) identified a transition from deep to shallow water diatoms that occurred at 7.3 ka. This is considered to be due to Lake Agassiz dropping from its Ponton level prior to final drainage. Klassen (1983a) estimated disintegration of Hudson Ice occurred about 7.5 ka; this permitted drainage of Lake Agassiz and inundation of the Hudson Bay Lowland by the Tyrrell Sea. Dredge and Cowan (1989) and Dyke and Dredge (1989) however, estimated that Tyrrell Sea entered the area about 8 ka; this is based on radiocarbon ages on marine shells.

Holocene

Holocene, the last epoch of the Quaternary Period, began 10 ka and continues at present. The Laurentide Ice Sheet had retreated from the Canadian Interior Plains by the beginning of the Holocene, however, glacial Lake Agassiz occupied the southeast corner of the area, and glacial Lake McConnell extended into the northwestern corner (Fig. 2.25f). By 9 ka drainage was almost that of present although glacial Lake Agassiz continued to occupy an area slightly larger than modern Lake Winnipeg until the opening of Hudson Bay at about 8 ka.

Conditions similar to those of present were established soon after deglaciation. Immediately following deglaciation, vast areas of unvegetated drift supplied abundant sediment which promoted aggradation of streams and development of terraces such as those on the Bow River (Stalker, 1968b; Jackson et al., 1982). In addition, valleys that carried glacial meltwater and those with gradients reduced by glacial isostatic tilting went through several brief phases of downcutting and alluviation before present drainage regimes were established (see for example, Kugler and St-Onge, 1973; and evolution of Assiniboine drainage system in Klassen, 1975). Once the area was vegetated, slopes cut to stable angles, and glacial isostatic adjustments completed, processes stabilized at approximately their present rate of activity.

Postglacial evolution of vegetation in this area is briefly described in Chapter 7 (Ritchie, 1989; Schweger, 1989). In general, spruce forest colonized the region as ice retreated and was shortly replaced by grassland in the drier southern part of the area. Climatic fluctuations have caused some migration of the boundary between forest and grasslands but in general the vegetation structure developed early during Holocene is the same as that present today.

Holocene climate has been deduced largely from studies of pollen and of paleosols. Anderson et al. (1989) summarized Holocene climatic trends as indicated by pollen studies; data on Holocene paleosols are found in a number of publications, such as Westgate et al. (1972), Waters (1979), Vreeken (1986), and Klassen and Vreeken (1987). In addition, studies of salinity variation, sedimentation, and paleoecology of lakes have provided a picture of fluctuating Holocene climate (for example, Delorme, 1971; Teller and Last, 1981). The general story is that climate was relatively warm early during the Holocene and a soil, as well developed as a present one, had formed before 6.6 ka (time of deposition of Mazama tephra; Powers and Wilcox, 1964). The period of maximum warmth and aridity (Hypsithermal, Altithermal, or climatic optimum) occurred between about 8 and 6 ka. This was the time of maximum northward extension of grasslands and of lowest lake levels. Cooler and more moist conditions began sometime after 6 ka with forest reinvading the northern margins of the grasslands and water levels rising in lakes. Climate since then has fluctuated but generally has not been too different from present.

Process activity has varied with climate and vegetation cover. Eolian activity accelerated when drought reduced the vegetation cover. This has resulted in eolian sands being deposited in areas where there was an appropriate sand source and loess being deposited downwind from sand dune areas and as a thin layer in many other regions. In addition, reduction of the vegetation cover undoubtedly led to accelerated sheet wash, increased production of colluvium, and a cycle of alluviation. The conditions of accelerated eolian processes were best developed during the Hypsithermal but the reactivation of dunes and the dust storms induced by the drought of the 1930s indicate that even short dry periods can quickly lead to accelerated eolian activity.

PALEOECOLOGY

Despite recent advances in knowledge about the Quaternary, the paleoecology of the southern Interior Plains during most of that period remains largely unknown. Most is known about the interval between retreat of Wisconsinan ice and the present, with detail of knowledge increasing with nearness to the present. In addition, vertebrate remains from the Saskatchewan Gravels and Sands supply information for preglacial conditions, and records from the northern plains of the United States can be extrapolated to the Canadian Plains for parts of the Quaternary. However, the record is sketchy to practically nil for much of the Quaternary. Further, the scattered fossil records we do have are probably not representative of general conditions during the nonglacials, because deposits preserved probably formed during cool, final stages of those intervals.

Our knowledge of the flora is restricted mainly to Late Wisconsinan and Holocene times, whereas knowledge of the fauna, and particularly the vertebrate fauna, is spread more evenly throughout the Quaternary. One thing is certain: both flora and fauna were vastly richer and much more varied at the beginning of the Quaternary than in historic times, despite the intermittent introduction of fresh species of animals, and perhaps of plants, from Asia via Bering Land bridges. Each large ice advance extinguished or drove most life from the Interior Plains of Canada. When each glacier withdrew, apparently fewer genera and species returned from the south to reinhabit the region. Though depletion of vertebrate animals was general throughout the Quaternary, the greatest impoverishment took place at the end of the last glaciation. On the other hand, loss of plant taxa probably was spread more evenly throughout the Quaternary, partly because plants could not spread as quickly and repopulate the region as readily as animals.

Flora

Pollen studies have been the chief source of information about the vegetational history of the southern Canadian Plains. Palynology has also provided valuable information about Late Wisconsinan deglaciation, chronology, and postglacial climate changes (Ritchie, 1969, 1976; Mott, 1973; Ritchie and Yarranton, 1978). Further, much of the radiocarbon chronology available for the Plains was obtained from lake cores collected for palynology studies. The pollen sequence of the western part of this area is described in Chapter 7 (Schweger, 1989).

Only sparse information is available concerning the vegetation of earlier times. Data collected from Medicine Hat sites indicate that, in Yarmouthian time, the trees growing along the river resembled those found there now, with the addition of ash (*Fraxinus* sp.) and Manitoba maple (*Acer negundo* sp.) (R.J. Mott, Geological Survey of Canada Wood Identification Report No. 71-68, 1971). The limited data available for Middle Wisconsinan also indicate vegetation similar to that of present (Westgate et al., 1972; Delorme in Fenton, 1974).

Considerable work has been done on the vegetation succession following deglaciation. In general, the studies show a predominantly spruce forest along the ice margin in Manitoba (Mott, 1973; Ritchie, 1976). That spruce forest spread north and east as deglaciation proceeded and covered the southern part of the region by 14 to 12 ka. With warmer and drier conditions, by about 10.5 ka, grassland replaced the spruce forest in the southwest part of that region (Ritchie, 1976, Fig. 3, 4), and by the height of the Altithermal, about 6.5 ka, the spruce forest was largely restricted to the Canadian Shield, and northern Saskatchewan and Alberta, as it is at present. This early spread of forest, however, appears incompatible with the concept of active glaciation in southern Manitoba as late as 11 ka (Teller and Fenton, 1980; Clayton and Moran, 1982; Mickelson et al., 1983).

Since disappearance of the forest, the grasslands have undergone a series of vicissitudes, including prairie fires, overgrazing by bison and other herbivores, and prolonged warm and cold spells; however, none equalled the drastic events of the Altithermal or climatic optimum from about 9 to 4.5 ka. That episode brought widespread drought, soil erosion, and wind drifting with burial of extensive areas under loess and windblown sand and must have caused great stress to the vegetation of the Interior Plains, to the animals that depended on it, and to early man who depended on both.

Fauna

Only recently it was generally accepted that, outside the Yukon, Quaternary vertebrate fossils were rare in Canada, and too scarce to be used for stratigraphic purposes. Finds on the southwestern Canadian Plains during the last twenty years have dispelled that concept and now mammal fossils provide the leading means of correlation between sections and of determining the relative ages of the Quaternary deposits in that region. They have also established the great time span covered by the surficial deposits of the southern Interior Plains and put to rest the once common notion that the glacial deposits of the regions were all of Wisconsinan Age. Further, in several cases, they have indicated that certain successive till sheets represented distinct ice advances, rather than being different facies of the same advance or else representing merely changes in flow direction.

Most of the Quaternary vertebrate taxa from within the area are listed in Figure 2.24. Most of the fossils have come from southeastern Alberta and southwest and central Saskatchewan because exposures are good in this region and because most collection work has been done in that area. Harington (1978) listed, by geographic areas, the Quaternary vertebrate faunas described in Canada until 1978 and gave brief descriptions of the sites and paleoenvironments. Stalker and Churcher (1982) depicted the fossils recovered from Wellsch Valley and Medicine Hat sites, indicated the ages of the fossiliferous beds, and showed the North American time ranges of various taxa. Figure 2.24 is based on this earlier work with a few additions and deletions.

In general, the Canadian Interior Plains supported a rich diversity of taxa through the Quaternary until the great extinction towards the end of the last glaciation. There were constant changes, but the extinctions that took place were largely offset by evolutionary change and by new taxa that migrated across the Bering Land Bridge from Asia and, in the early Quaternary by the Isthmus of Panama from South America. Indeed, additions may themselves have led to some of the extinctions. The picture presented by the fossils, however, is probably a false representation of the overall fauna, for there were many biases involved in determining which animal fossils would be preserved and which discovered. As prompt burial after death was required to preserve the bones, animals that congregated along rivers or browsed on the brush growing in the large valleys would be overrepresented whereas those smaller animals that lived on the high plains and uplands would be underrepresented.

Further, if most of the deposits represent only cooler parts of the nonglacials, as suggested previously, further bias is introduced.

Faunal development

At inception of the Quaternary Period, the fauna of the Interior Plains had had a prolonged, uninterrupted term of development, and so was well adapted to the environment. Fossils from the Saskatchewan Gravels and Sands give intermittent glimpses of that fauna, with those from Medicine Hat telling a great deal about the fauna that had survived there during early glacier advances farther east. The sediments from near the base of the Quaternary section at the Wellsch Valley Site, however, supply the oldest abundant record of specific age in the Quaternary (1, 4, 9, 12, 16, 18, 19, 20, 26, 37, 39, 43, 46, 51, 56, 65, 71 of Fig. 2.24). That fauna is considered to have been a typical preglacial one for the region, and it was drastically different from the modern fauna. Ground sloths, horses, and camels were prominent amongst the bigger animals, while mammoths had recently arrived on the scene and were becoming firmly established. Among the mid-size animals early rabbits, peccaries, pronghorns, and shruboxen were present, though nothing is known about relative numbers. Carnivores are poorly represented, for only bobcat and an early, heavily-built dog (*Borophagus*) have been recovered. Rodents, especially voles, are common; many of them are primitive types.

Nearly a million years elapsed before the next good view of the fauna. This glimpse is afforded by the basal beds at Medicine Hat which represent the final deposits of the Saskatchewan Gravels and Sands laid down prior to the Labuma Glaciation. They are of Kansan and Yarmouthian ages, and their fossils (2, 3, 8, 10, 14, 31, 41, 44, 48, 55, 57, 58, 59, 66 of Fig. 2.24) depict a scene with camels, horses, and elephants apparently dominant, though mostly of different species from the earlier Wellsch Valley ones. Ground sloths were still prominent, and beaver, deer, pronghorn, and prairie dog — this last perhaps indicating a climate warmer than now — were present. Carnivores are poorly represented, perhaps due to biased preservation, and *Borophagus* had become extinct. The poor representation of small rodents shown (Fig. 2.24) is a result of the coarseness of the fossiliferous deposits, in turn an indication of river currents strong enough to destroy or carry away small bones.

Some 300 ka elapsed before the next extensive fauna is revealed in the Mitchell Bluff Formation (Stalker, 1976b) and in other deposits considered to be of Sangamonian age from Medicine Hat, Saskatoon, and Fort Qu'Appelle sites (Fig. 2.16). These deposits are considered to be between 125 and 85 ka. Once again, drastic changes had taken place, and at this time the fauna attains some modern aspects (Fig. 2.24, Sangamonian column). The most striking difference from the earlier faunas was the importance of bison, which had migrated from Asia some time previously, had begun to speciate, and was soon to become the dominant grazer of the Interior Plains. According to numbers of bones recovered, elephants and horses — the latter mainly new species — were other prominent large mammals. The camels, though still common, apparently were declining in numbers. Deer, wapiti, elk, caribou, and mountain sheep are recorded for the first time, as are the smaller porcupine, muskrat, raccoon and hare, while ground sloth, rabbit, prairie dog, and ground squirrel carried on successfully. Due to conditions of preservation, carnivores are well represented for the first time, with lion, lynx, wolf, fox, and ferret resident in the region. Short-faced bear and muskox are found at the Fort Qu'Appelle site, and badger there and at Saskatoon.

The next younger deposits to afford a comprehensive glimpse at the vertebrate record are thought to be of Middle Wisconsinan age (Fig. 2.24). The restricted fauna recovered appears more Sangamonian than modern because it still includes many of the earlier forms that subsequently became extinct at the end of the last glaciation. It displays the first coyote and skunk found in the region, although both taxa had appeared much earlier elsewhere in North America and had probably been present here too, though their fossils have not been recovered from older deposits.

Towards the end of Wisconsinan time (roughly between 15 and 9 ka) came what has been referred to as the great extinction. The rodents, rabbits, hares, beaver, smaller carnivores, pronghorn, mountain sheep, bison, and some of the deer, caribou, and elk survived without much change, but many previously prominent taxa, including elephants, horses, camels, ground sloths, disappeared, as did the lion, saber-toothed cat, and dire wolf that preyed on them. Only the smaller carnivores, such as fox, coyote, ferret, and skunk, remained widespread on the southern Interior Plains, with bobcat, lynx, and bear putting in occasional appearances. Altogether, the great extinction left behind a most impoverished fauna, which was dominated by bison until these were replaced with modern cattle and other farm animals.

Climate

Few details are known about the climate of the southern Canadian Interior Plains before Late Wisconsinan time. Certainly it was a period of drastic climate fluctuations, with cold spells during which glaciers developed, alternating with nonglacial intervals with moderate climate, at times warmer than at present. Information about precipitation and other climatic detail is almost entirely lacking. Even the relative lengths of the glacial and nonglacial episodes remain as enigmas, though it is probable that the nonglacial periods were by far the longer. This was particularly true in the southwest, where the high land enjoyed much longer ice-free periods than those lower areas farther to the east and north.

In general, the vertebrate record gives little information about climate. Fossil beds only point to conditions prevailing during their deposition, and their very preservation may denote that those times were abnormal. The much longer periods without fossils could have been drastically different. Furthermore, most of the animals identified, and particularly the large ones, could tolerate a broad range of conditions. In addition, the admixture of bones from warm and cold climate species of large animals, found in many of the deposits, may merely denote seasonal migrations, with bones from animals that travelled north in summer and south in winter intermingled, during deposition and burial with those of the permanent residents. Only a few of the smaller animals preserved provide useful clues as to past climates; thus the prairie dog, which now does not live as far north as Medicine Hat, is found in some of the deposits and may signify past warm spells. Altogether, the climate during much of nonglacial time probably resembled the present climate, but there were also cool times and, probably, shorter spells of warmer temperatures.

ECONOMIC GEOLOGY

The great importance of the Quaternary to the economics of the southern Interior Plains is rarely appreciated. It has affected practically all facets of Plains life, from impoverishment of fauna and flora to enrichment of soils. This section gives brief summaries of its contribution to water supply, agriculture and forestry, and to construction materials; additional information on some of these aspects and its significance to engineering problems and slope stabilities is described in Chapter 12 (Scott, 1989).

Groundwater

Over much of the southern Interior Plains the Quaternary deposits are a major source of groundwater, although they commonly are not a fully satisfactory source. They are drawn upon mainly because they are the best available as bedrock over the greater part of the area yields little water, due to its clayey and silty, commonly bentonitic, nature. Further, the quality of the water that is available from the bedrock typically is poor, because most of that water contains much dissolved matter. In addition, surface water commonly is not available or else provides an unreliable source, for large rivers are few and widely separated and many of the ponds and lakes dry up in years of low precipitation. Further, the water in many of those ponds, especially in southern Saskatchewan and southeast Alberta, is too salty for most uses. As a result, resort must be made to aquifers in the surface deposits.

In general, the Quaternary deposits yield higher quality water than either the bedrock or that found in ponds on the surface, but quantity is a major problem. Tills typically yield small quantities, sufficient for household and some farm purposes, and the interglacial silt and sand beds found locally, along with the gravel and sand deposits found at the surface, provide more satisfactory supplies. The best potential sources, however, are the Saskatchewan Gravels and Sands and the other gravels and sands found towards the base of the buried preglacial and interglacial valleys (Fig. 2.17). These materials are tapped in a few places, as at Medicine Hat where they yield water of fairly constant temperature in sufficient quantities for industrial purposes. In general, however, the potential of these aquifers remains relatively unknown.

Agriculture and forestry soils

Fertility. Practically all the soil of the three Prairie Provinces is developed on Quaternary deposits. This is a good thing, for over much of the Plains, soils developed on bedrock are deficient in certain elements, such as calcium, sulphur, and phosphorous, and over enriched in sodium and certain salts. It was the glaciers spreading over the Plains from northeast, north, and northwest that brought in fresh, unweathered debris and laid down nutrient-rich deposits that constitute the parent material of much of the soil of the southern Interior Plains of Canada. Another material, whose economic significance is rarely recognized, is the Holocene loess that forms a thin blanket over extensive areas of the Plains. It was deposited chiefly in the Altithermal episode, but is added to by each dust storm that envelopes the Prairies. The loess supplies valuable nutrients, and much of the richest soil on the Plains has developed on that windblown silt. Even where other materials form the main parent material, the addition of limited amounts of loess has greatly enriched the soil. The glacial drift and loess are between them largely responsible for the fertility of the soils found on the southern Interior Plains, and so are responsible for the importance of agriculture and forestry there. Further information on soils of the southern Interior Plains is given in Chapter 11 (Acton, 1989).

Drainage. In both agriculture and forestry the ability of precipitation to permeate into the soil, along with subsurface drainage, is important. Without adequate permeability and drainage, water would either remain on or near the surface until it evaporated, water-logging the soil and causing leaching and salinity problems, or else run off quickly and so be lost to the soil, while at the same time causing erosion problems. These features and difficulties are found in various parts of the southern Canadian Interior Plains, and they are especially acute where rather impermeable bedrock is near surface. The most extreme examples are the badlands and areas of salinity of southeast Alberta. Though permeability and drainage of areas underlain by lake clay and till may often be poor, they are typically much better than in most areas of shallow bedrock. There are exceptions; for example, the good drainage of coarse sandstone of the Porcupine Hills Formation, in the Porcupine Hills, permits the growth of valuable stands of Douglas fir, which support a lumbering industry. Over most of the area, however, without the parent material laid down during the Quaternary, poor surface drainage and impermeability of the bedrock would have strongly deterred development of economic woodlands and a viable farming industry. Good subsurface drainage is especially important to forestry, for certain species of trees are largely excluded from areas that lack it.

Salinity. Salinity problems in the soils of the Interior Plains of southern Canada would be considerably worse without the blanket of drift and loess. Such problems arise chiefly through the high content of salts found in much of the bedrock; the troubles come about where groundwater carries the dissolved salts to near the surface, there to be deposited when the water evaporates. The lesser salt content in the blanket of surficial deposits helps cushion this effect and makes agriculture possible in many areas where it would otherwise be marginal. Even with that buffering blanket, however, salinity is a major problem, especially in irrigated areas. In those areas, the extra water added by irrigation sinks a short distance into the ground, collects various salts, rises and then evaporates and deposits those salts in the soil zone. Over time, the accumulation of salts can render land useless for farming and cause abandonment of formerly irrigated areas. This is a difficult problem to solve, and commonly the only treatment is addition of copious quantities of water to leach the salts from the soil. Such vast amounts of water, however, are rarely available in irrigated areas. In general, irrigation should largely be restricted to areas where original salt content in the drift is low and the subsoil material sufficiently porous to allow downward percolation of water.

Construction materials

Gravels and sands laid down by rivers during the last half of the Tertiary Period and still found on some remnant high areas, such as at Del Bonita and in the Cypress, Swan, and Hand hills (Fig. 2.16), are valuable sources of gravel and sand for parts of the Interior Plains. Over most of the area, however, such deposits are lacking and resort must be made to Quaternary deposits for those materials. The best quality and largest Quaternary deposits of gravel and sand are found in the preglacial and, to a lesser extent, interglacial valleys. However, those deposits commonly are too deeply buried to be economic at present, and so ice contact deposits such as kames and eskers, outwash, and the bars and terraces of present day rivers and glacial spillways supply the construction materials for much of the region. The sand and gravel deposits, and especially those in the ice contact deposits, are commonly poorly sorted and may be of poor quality, with an excess of weak or rotted clasts. However, ease of access and development help compensate for those deficiencies.

Glacial lake clay and clayey tills have, in the past, been used locally to make bricks and tiles. In general, such clay is of too poor quality to be entirely satisfactory for those purposes.

ACKNOWLEDGMENTS

Discussions with J.G. Fyles, O.L. Hughes (Geological Survey of Canada), V.N. Rampton (Terrain Analysis and Mapping Services Ltd., Carp, Ontario), and N.W. Rutter (University of Alberta) and peer review by A.S. Dyke (Geological Survey of Canada) were important elements in compiling the northern Interior Plains section. The critical review of the entire report by M.M. Fenton (Alberta Research Council), W.O. Kupsch (University of Saskatchewan), E. Nielsen (Manitoba Mineral Resources Division), B.T. Schreiner (Saskatchewan Research Council), A. MacS. Stalker, N.R. Gadd (Geological Survey of Canada), W.H. Mathews (University of British Columbia), and J.B. Bird (McGill University), is gratefully acknowledged.

Many of the ideas and some of the material in the section dealing with the southern Interior Plains come directly from an unpublished manuscript by A. MacS. Stalker. Parts which have been taken almost without modification from the original Stalker manuscript are: *Fauna, Faunal development, Climate,* and *Economic geology.* A shortened version of the Stalker manuscript is being published as part of Volume 6 (*Sedimentary Cover of the Craton: Canada*) of this series (Stalker and Vincent, in press).

REFERENCES

Acton, D.F.
1989: Soils of the Interior Plains; in Chapter 11 of Quaternary Geology of Canada and Greenland, R.J. Fulton (ed.); Geological Survey of Canada, Geology of Canada, no. 1 (also Geological Society of America, The Geology of North America, v. K-1).

Aguirre, E. and Pasini, G.
1985: The Pliocene-Pleistocene boundary; Episodes, v. 8, p. 116-120.

Alley, N.F.
1973: Glacial stratigraphy and the limits of the Rocky Mountain and Laurentide ice sheets in Southwestern Alberta, Canada; Canadian Petroleum Geology, Bulletin, v. 21, p. 153-177.

Anderson, T., Mathewes, R., and Schweger, C.
1989: Holocene climatic trends in Canada; in Chapter 7 of Quaternary Geology of Canada and Greenland, R.J. Fulton (ed.); Geological Survey of Canada, Geology of Canada, no. 1 (also Geological Society of America, The Geology of North America, v. K-1).

Atmospheric Environment Service
1982: Canadian Climate Normals, Temperature and Precipitation 1951-1980, The North, Y.T. and N.W.T.; Canada Department of Environment, Downsview, 55 p.

Barendregt, R.W.
1984: Correlation of Quaternary chronologies using paleomagnetism — examples from southern Alberta and Saskatchewan; in Correlation of Quaternary Chronologies, W.C. Mahaney (ed.); Geo Books, Norwich, England, p. 59-71.
1985: Correlation of Quaternary chronologies using paleomagnetism — examples from southern Alberta and Saskatchewan; in Correlation of Quaternary Chronologies, W.C. Mahaney (ed.); Geo Books, Norwich, England, p. 120-136.

Bayrock, L.A.
1962: Heavy minerals in till of central Alberta; Alberta Society of Petroleum Geologist Journal, v. 10, p. 171-184.

Bayrock, L.A. and Pauluk, S.
1967: Trace elements in tills of Alberta; Canadian Journal of Earth Sciences, v. 4, p. 597-607.

Beget, J.
1987: Low profile of the northwest Laurentide Ice Sheet; Arctic and Alpine Research, v. 19, p. 81-88.

Bik, M.J.J.
1969: The origin and age of the prairie mounds of southern Alberta, Canada; Biuletyn Peryglacjainy, no. 19, Lodz, p. 85-130.

Bird, J.B.
1967: The Physiography of Arctic Canada; The John Hopkins Press, Baltimore, 336 p.

Blake, W. Jr.
1974: Studies of glacial history in Arctic Canada. II. Interglacial peat deposits on Bathurst Island; Canadian: Journal of Earth Sciences, v. 11, p. 1025-1042.
1982: Geological Survey of Canada radiocarbon dates XXII; Geological Survey of Canada, Paper 82-7, 22 p.
1983: Geological Survey of Canada radiocarbon dates XXIII; Geological Survey of Canada, Paper 83-7, 34 p.
1984: Geological Survey of Canada radiocarbon dates XXIV; Geological Survey of Canada, Paper 84-7, 35 p.
1986: Geological Survey of Canada radiocarbon dates XXV; Geological Survey of Canada, Paper 85-7, 32 p.
1987: Geological Survey of Canada radiocarbon dates XXVI; Geological Survey of Canada, Paper 86-7, 60 p.

Bluemle, J.P. and Clayton, L.
1984: Large-scale glacial thrusting and related processes in North Dakota; Boreas, v. 13, p. 279-299.

Bostock, H.S.
1970: Physiographic regions of Canada; Geological Survey of Canada, Map 1254A, scale 1:5,000,000.

Bouchard, M.
1974: Géologie des dépôts meubles de l'île Herschel, Territoire du Yukon; Thèse de MSc non publiée, Université de Montréal, Montréal, 125 p.

Bowen, D.Q., Richmond, G.M., Fullerton, D.S., Šibrava, V., Fulton, R.J., and Velichko, A.A.
1986: Correlation of Quaternary glaciations in the Northern Hemisphere; in Glaciations in the Northern Hemisphere, v. Šibrava, D.Q. Bowen, and G.M. Richmond (ed.); Quaternary Science Reviews Special Volume 5, Pergamon Press, Oxford, England, Chart 1.

Byers, A.R.
1959: Deformation of the Whitemud and Eastend Formations near Claybank, Saskatchewan; Royal Society of Canada, Transactions, v. 53, ser. 3, sec. 4, p. 1-11.

Carsola, A.J.
1954: Extent of glaciation on the continental shelf in the Beaufort Sea; American Journal of Science, v. 252, p. 366-371.

Carter, L.D., Brigham-Grette, J., and Hopkins, D.M.
1986: Late Cenozoic marine transgressions of the Alaskan Arctic Coastal Plain; in Correlation of Quaternary Deposits and Events Around the Margin of the Beaufort Sea: Contributions from a Joint Canadian-American Workshop, April 1984; J.A. Heginbottom and J-S. Vincent (ed.); Geological Survey of Canada, Open File 1237, p. 21-26.

Carter, L.D., Heginbottom, J.A., and Woo, M.
1987: Arctic Lowlands; in Geomorphic Systems of North America, W.L. Graf (ed.); Geological Society of America, Centennial Special Volume 2, p. 583-628.

Catto, N.R.
1986a: Quaternary stratigraphy and chronology in the Richardson Mountains — Peel Plateau region, Yukon, N.W.T.; in Correlation of Quaternary Deposits and Events Around the Margin of the Beaufort Sea: Contributions from a Joint Canadian-American Workshop, April 1984, J.A. Heginbottom and J-S. Vincent (ed.); Geological Survey of Canada, Open File 1237, p. 18-19.
1986b: Quaternary sedimentology and stratigraphy, Peel Plateau and Richardson Mountains, Yukon and N.W.T.; unpublished PhD thesis, University of Alberta, Edmonton, 728 p.

Causse, C. and Vincent, J-S.
1985: U/Th disequilibrium dating of Middle and Late Pleistocene wood in northwestern Canada; Geological Association of Canada and Mineralogical Association of Canada, Joint Annual Meeting, Fredericton, Program with Abstracts, v. 10, p. A8.

Christiansen, E.A.
1956: Geology of the Moose Mountain area, Saskatchewan; Saskatchewan Department of Mineral Resources, Report 21.
1959: Glacial geology of the Swift Current area, Saskatchewan; Saskatchewan Department of Mineral Resources, Report no. 32, 62 p.
1961: Geology and groundwater resources of the Regina area, Saskatchewan; Saskatchewan Research Council, Geology Division Report no. 2, 72 p.
1967: Preglacial valleys in southern Saskatchewan; Saskatchewan Research Council, Geology Division, Map no. 3, scale 1:520 640.
1968a: A thin till in west-central Saskatchewan; Canadian Journal of Earth Sciences, v. 5, p. 329-336.
1968b: Pleistocene stratigraphy of the Saskatoon area, Saskatchewan, Canada; Canadian Journal of Earth Sciences, v. 5, p. 1167-1173.
1971a: Tills in southern Saskatchewan, Canada; in Till: A Symposium, R.P. Goldthwait (ed.); The Ohio State University Press, Columbus, p. 167-183.
1971b: Geology of the Crater Lake collapse structure in southeastern Saskatchewan; Canadian Journal of Earth Sciences, v. 8, p. 1505-1513.
1972: Stratigraphy of the Fort Qu'Appelle vertebrate fossil locality, Saskatchewan; Canadian Journal of Earth Sciences, v. 9, p. 212-218.
1979: The Wisconsinan glaciation of southern Saskatchewan and adjacent areas; Canadian Journal of Earth Sciences, v. 16, p. 913-938.

Christiansen, E.A. and Whitaker, S.H.
1976: Glacial thrusting of drift and bedrock; in Glacial Till, An Interdisciplinary Study, R.F. Leggett (ed.); Royal Society of Canada, Special Publication, no. 12, p. 121-130.

Churcher, C.S.
1984: Faunal correlations of Pleistocene deposits in Western Canada; in Correlation of Quaternary Chronologies, W.C. Mahaney (ed.); Geo Books, Norwich, England, p. 145-158.

Clark, D.L., Vincent, J-S., Jones, G.A., and Morris, W.A.
1984: Correlation of marine and continental glacial and interglacial events, Arctic Ocean and Banks Island; Nature, v. 311, no. 5982, p. 147-149.

Clark, D.L., Whitman, R.R., Morgan, K.A., and Mackey, S.D.
1980: Stratigraphy and glacial-marine sediments of the Amerasian Basin, Central Arctic Ocean; Geological Society of America, Special Paper 181, 57 p.

Clayton, J.S., Ehrlich, W.A., Cann, D.B., Day, J.H., and Marshall, I.B.
1977: Soils of Canada; Research Branch, Canada Department of Agriculture, v. I, 243 p., v. II, 239 p., map scale 1:5 000 000.

Clayton, L.
1983: Chronology of Lake Agassiz drainage to Lake Superior; in Glacial Lake Agassiz, J.T. Teller and L. Clayton (ed.); Geological Association of Canada, Special Paper 26, p. 291-307.

Clayton, L. and Moran, S.R.
1974: A glacial process-form model; in Glacial Geomorphology, D.R. Coates, (ed.); Publications in Geomorphology, State University of New York, Binghamton, New York.
1982: Chronology of Late Wisconsinan glaciation in middle North America; Quaternary Science Reviews, v. 1, p. 55-82.

Clayton, L., Moran, S.R., and Bluemle, J.P.
1980: "Explanatory text to accompany the Geologic Map of North Dakota"; North Dakota Geological Survey, Report of Investigation no. 69, 93 p.

Colton, R.B., Lemke, R.W., and Lindvall, R.M.
1961: Glacial map of Montana east of the Rocky Mountains; United States Geological Survey, Miscellaneous Geological Investigations, Map I-327.

Craig, B.G.
1960: Surficial geology of north-central District of Mackenzie, Northwest Territories; Geological Survey of Canada, Paper 60-18, 8 p.
1965: Glacial Lake McConnell, and the surficial geology of parts of Slave River and Redstone River map-areas, District of Mackenzie; Geological Survey of Canada, Bulletin 122, 33 p.

Craig, B.G. and Fyles, J.G.
1960: Pleistocene geology of Arctic Canada; Geological Survey of Canada, Paper 60-10, 21 p.
1965: Quaternary of Arctic Canada; in Anthropogen Period in Arctic and Subarctic; Scientific Research Institute of the Geology of the Arctic, Transactions, State Geological Committee; U.S.S.R. Moscow, v. 143, p. 5-33 (In Russian with English summary).

Cwynar, L.C.
1982: A Late-Quaternary vegetation history from Hanging Lake, Northern Yukon; Ecological Monographs, v. 52, p. 1-24.

Cwynar, L.C. and Ritchie, J.C.
1980: Arctic steppe-tundra: a Yukon perspective; Science, v. 208, no. 4450, p. 1375-1377.

David, P.P.
1964: Surficial geology and ground water resources of the Prelate area (72K), Saskatchewan; unpublished PhD thesis, Department of Geological Sciences, McGill University, Montreal.
1966: The Late Wisconsin Prelate Ferry paleosol of Saskatchewan; Canadian Journal of Earth Sciences, v. 3, p. 685-696.
1987: The Prelate Ferry Paleosol: a new date; Geological Association of Canada, Program with Abstracts, v. 12, p. 36.

Dawson, G.M. and McConnell, R.G.
1896: Glacial deposits of southwestern Alberta in the vicinity of the Rocky Mountains; Geological Survey of America, Bulletin, v. 7, p. 31-66.

Delorme, L.D.
1971: Paleoecology of Holocene sediments from Manitoba using freshwater ostracodes; in Geoscience Studies in Manitoba, A.C. Turnock (ed.); Geological Association of Canada, Special Paper 9, p. 301-304.

Denton, G.H. and Hughes, T.J.
1983: Milankovitch theory of ice ages: hypothesis of ice-sheet linkage between regional insolation and global climate; Quaternary Research, v. 20, p. 125-144.

Dinter, D.A.
1984: Quaternary sedimentation on the Alaskan Beaufort Shelf: sediment sources, glacio-eustatism and regional tectonics; in United States Geological Survey Polar Research Symposium — Abstracts with Program, United States Geological Survey Circular 911, p. 37-38.

Douglas, R.J.W.
1969: Geological map of Canada; Geological Survey of Canada, Map 1250A, scale 1:5 000 000.

Dredge, L.A.
1983: Character and development of northern Lake Agassiz and its relation to Keewatin and Hudsonian ice regimes; in Glacial Lake Agassiz, J.T. Teller and L. Clayton (ed.); Geological Association of Canada, Special Paper 26, p. 117-131.

Dredge, L.A. and Cowan, W.R.
1989: Quaternary geology of the southwestern Canadian Shield; in Chapter 3 of Quaternary Geology of Canada and Greenland, R.J. Fulton (ed.); Geological Survey of Canada, Geology of Canada, no. 1 (also Geological Society of America, The Geology of North America, v. K-1).

Dreimanis, A. and Lundquist, J.
1984: What should be called till?; in 10 years at Nordic Research, L.K. Konigsson (ed.); Striae, v. 20, p. 5-10.

Dyck, W. and Fyles, J.G.
1963: Geological Survey of Canada radiocarbon dates II; Radiocarbon, v. 5, p. 39-55.

Dyck, W., Lowdon, J.A., Fyles, J.G., and Blake, W., Jr.
1965: Geological Survey of Canada radiocarbon dates IV; Radiocarbon, v. 7, p. 24-46.
1966: Geological Survey of Canada radiocarbon dates V; Radiocarbon, v. 8, p. 96-127.

Dyke, A.S.
1983: Quaternary geology of Somerset Island, District of Franklin; Geological Survey of Canada, Memoir 404, 32 p.
1984: Quaternary geology of Boothia Peninsula and northern District of Keewatin, central Canadian Arctic; Geological Survey of Canada, Memoir 407, 26 p.
1987: A reinterpretation of glacial and marine limits around the northwestern Laurentide Ice Sheet; Canadian Journal of Earth Sciences, v. 24, p. 591-601.

Dyke, A.S. and Dredge, L.A.
1989: Quaternary geology of the northwestern Canadian Shield; in Chapter 3 of Quaternary Geology of Canada and Greenland, R.J. Fulton (ed.); Geological Survey of Canada, Geology of Canada, no. 1 (also Geological Society of America, The Geology of North America, v. K-1).

Dyke, A.S. and Prest, V.K.
1987: The Late Wisconsinan and Holocene history of the Laurentide Ice Sheet; Géographie physique et Quaternaire, v. 41, p. 237-263.

Dyke, A.S., Dredge, L.A., and Vincent, J-S.
1982: Configuration and dynamics of the Laurentide Ice Sheet during the Late Wisconsin maximum; Géographie physique et Quaternaire, v. 36, p. 5-14.

Dyke, A.S., Vincent, J-S., Andrews, J.T., Dredge, L.A., and Cowan, W.R.
1989: The Laurentide Ice Sheet and an introduction to the Quaternary geology of the Canadian Shield; in Chapter 3 of Quaternary Geology of Canada and Greenland, R.J. Fulton (ed.); Geological Survey of Canada, Geology of Canada, no. 1 (also Geological Society of America, The Geology of North America, v. K-1).

Elson, J.A.
1957: Origin of washboard moraine; Geological Society of America, Bulletin, v. 68, p. 1721.
1961: Soils of the Lake Agassiz region; in Soils in Canada, Geological, Pedological and Engineering Studies, R.F. Legget (ed.); The Royal Society of Canada, Special Publications, no. 3; University of Toronto Press, p. 51-79.
1967: Geology of glacial Lake Agassiz; in Life, Land and Water, W.J. Mayer-Oakes (ed.); Occasional Papers, Department of Anthropology, University of Manitoba, no. 1, University of Manitoba Press, Winnipeg, p. 37-95.
1971: Roundness of Lake Agassiz beach pebbles; in Geoscience Studies in Manitoba, A.C. Turnock (ed.); Geological Association of Canada, Special Paper 9, p. 285-291.

Fenton, M.M.
1974: The Quaternary stratigraphy of a portion of southeastern Manitoba, Canada; unpublished PhD thesis, University of Western Ontario, Department of Geology, London, Ontario, 285 p.
1983: Deformation terrain mid-continent region: properties subdivision, recognition; Geological Society of America, Annual Meeting, North-central Section, Madison, Wisconsin, Abstracts with Programs, v. 15, p. 250.
1984a: Deformation terrain in Canadian Prairies: morphology and sediment facies; INQUA Commission on Genesis and Lithology of Quaternary Deposits, Symposium on the relationship between glacial terrain and glacial sediment facies, Medicine Hat-Lethbridge, Alberta, Abstracts and Program, p. 7-8.
1984b: Quaternary stratigraphy of the Canadian Prairies; in Quaternary Stratigraphy of Canada — A Canadian Contribution to IGCP Project 24, R.J. Fulton (ed.); Geological Survey of Canada, Paper 84-10, p. 57-68.

Fenton, M.M. and Anderson, D.T.
1971: Pleistocene stratigraphy of the Portage La Prairie area, Manitoba; in Geoscience Studies in Manitoba, A.C. Turnock (ed.); Geological Association of Canada, Special Paper 9, p. 271-276.

Fenton, M.M., Moran, S.R., Teller, J.T., and Clayton, L.
1983: Quaternary stratigraphy and history in the southern part of the Lake Agassiz Basin; in Glacial Lake Agassiz, J.T. Teller and L. Clayton (ed.); Geological Association of Canada, Special Paper 26, p. 49-74.

Fisher, D., Reeh, N., and Langley, K.
1985: Objective reconstructions of the Late Wisconsinan Laurentide Ice Sheet and the significance of deformable beds; Géographie physique et Quaternaire, v. 39, p. 229-238.

Forbes, D.L.
1980: Late Quaternary sea levels in the southern Beaufort Sea; in Current Research, Part B, Geological Survey of Canada, Paper 80-1B, p. 75-87.

Ford, D.C.
1976: Evidences of multiple glaciations in South Nahanni National Park, Mackenzie Mountains, Northwest Territories; Canadian Journal of Earth Sciences, v. 13, p. 1433-1445.

Fortier, Y.O. and Morley, L.W.
1956: Geological unity of the Arctic Islands; Royal Society of Canada, Transactions, v. 50, Series 3, p. 3-12.

Foster, J.H. and Stalker, A. MacS.
1976: Paleomagnetic stratigraphy of the Wellsch Valley site, Saskatchewan; in Report of Activities, Part C, Geological Survey of Canada, Paper 76-1C, p. 191-193.

Fraser, J.A., Heywood, W.W., and Mazurski, M.A.
1978: Metamorphic map of the Canadian Shield; Geological Survey of Canada, Map 1475A.

French, H.M.
1976: The Periglacial Environment; Longman, London, 309 p.

French, H.M., Harry, D.G., and Clark, M.J.
1982: Ground ice stratigraphy and Late-Quaternary events, south-west Banks Island, Canadian Arctic; in Proceedings Fourth Canadian Permafrost Conference, National Research Council of Canada, p. 81-90.

Fullerton, D.S. and Colton, R.B.
1986: Stratigraphy and correlation of the glacial deposits of the Montana plains; in Quaternary Glaciations in the Northern Hemisphere, v. Sibrava, D.Q. Bowen, and G.M. Richmond (ed.); Pergamon Press, Oxford, Quaternary Science Reviews, Special Volume 5, p. 69-82.

Fulton, R.J. and Klassen, R.W.
1969: Quaternary geology, Northwest District of Mackenzie; in Report of Activities, Part A, Geological Survey of Canada, Paper 69-1A, p. 193-194.

Fulton, R.J., Fenton, M.M., and Rutter, N.W.
1984: Summary of Quaternary stratigraphy and history, Western Canada; in Quaternary Stratigraphy of Canada — A Canadian contribution to IGCP Project 24, R.J. Fulton (ed.); Geological Survey of Canada, Paper 84-10, p. 69-83.

Funder, S.
1989: Quaternary geology of North Greenland; in Chapter 13 of Quaternary Geology of Canada and Greenland, R.J. Fulton (ed.); Geological Survey of Canada, Geology of Canada, no. 1 (also Geological Society of America, The Geology of North America, v. K-1).

Funder, S., Abrahamsen, N., Bennike, O., and Feyling-Hanssen, R.W.
1985: Forested arctic, evidence from North Greenland; Geology, v. 13, p. 542-546.

Fyles, J.G.
1962: Physiography; in Banks, Victoria and Stefansson islands, Arctic Archipelago; Geological Survey of Canada, Memoir 330, p. 8-17.
1963: Surficial geology of Victoria and Stefansson islands, District of Franklin; Geological Survey of Canada, Bulletin 101, 38 p.
1966: Quaternary stratigraphy, Mackenzie Delta and Arctic Coastal Plain; in Report of Activities, Part 1, Geological Survey of Canada, Paper 66-1, p. 30-31.
1967: Mackenzie Delta and Arctic Coastal Plain; in Report of Activities Part A, Geological Survey of Canada, Paper 67-1A, p. 34-35.

Fyles, J.G., Heginbottom, J.A., and Rampton, V.N.
1972: Quaternary geology and geomorphology, Mackenzie Delta to Hudson Bay; XXIV International Geological Congress, Montreal, Canada, Guidebook A-30, 23 p.

Gravenor, C.P.
1955: The origin and significance of prairie mounds; American Journal of Science, v. 253, p. 475-481.

Gravenor, C.P. and Kupsch, W.O.
1959: Ice-disintegration features in western Canada; Journal of Geology, v. 67, p. 48-64.

Hamilton, T.D.
1979: Late Cenozoic glaciations and erosion intervals, north-central Brooks Range; United States Geological Survey, Circular 804-B, p. B-27-B-29.

Harington, C.R.
1977: Pleistocene mammals of the Yukon Territory; unpublished PhD thesis, University of Alberta, Edmonton, 1060 p.
1978: Quaternary vertebrate faunas of Canada and Alaska and their suggested chronological sequence; National Museum of Natural Sciences, Syllogeus, no. 15, 105 p.
1980: Radiocarbon dates on some Quaternary mammals and artifacts from northern North America; Arctic, v. 33, p. 815-832.

Harmon, R.S., Ford, D.C., and Schwarcz, H.P.
1977: Interglacial-chronology of the Rocky and Mackenzie Mountains based upon ^{230}Th-^{234}U dating of calcite speleothems; Canadian Journal of Earth Sciences, v. 14, p. 2543-2552.

Heginbottom, J.A. and Vincent, J.A. (editors)
1986: Correlation of Quaternary deposits and events around the margin of the Beaufort Sea: Contributions from a Joint Canadian-American Workshop, April 1984; Geological Survey of Canada, Open File 1237, 60 p.

Heginbottom, J.A., Kurfurst, P.J., and Lau, J.S.O.
1977: Evaluation of regional occurrence of ground ice and frozen ground, Mackenzie Valley, District of Mackenzie; in Report of Activities, Part C, Geological Survey of Canada, Paper 77-1C, p. 35-38.

Hill, P.R., Mudie, P.J., Moran, K., and Blasco, S.M.
1985: A sea-level curve for the Canadian Beaufort Shelf; Canadian Journal of Earth Sciences, v. 22, p. 1383-1393.

Hills, L.V.
1975: Late Tertiary floras Arctic Canada: an interpretation; in Proceedings of the Circumpolar Conference on Northern Ecology, Ottawa, Canada, p. I-61 to I-71.

Hills, L.V., Klovan, J.E., and Sweet, A.R.
1974: *Juglans eocinerea* n. sp., Beaufort Formation (Tertiary), south-western Banks Island, Arctic Canada; Canadian Journal of Botany, v. 52, p. 65-90.

Hodgson, D.A.
1987: Episodic retreat of the Late Wisconsinan Laurentide Ice Sheet over northeast Victoria Island, and the source of the Viscount Melville Sound Ice Shelf (abstract); 12th INQUA Congress, National Research Council of Canada, Ottawa, Programme with Abstracts, p. 188.
1989: Quaternary stratigraphy and chronology (Queen Elizabeth Islands); in Chapter 6 of Quaternary Geology of Canada and Greenland, R.J. Fulton (ed.); Geological Survey of Canada, Geology of Canada, no. 1 (also Geological Society of America, The Geology of North America, v. K-1).

Hodgson, D.A. and Vincent, J-S.
1984a: Surficial geology, central Melville Island; Geological Survey of Canada, Map 1583A, scale 1:250 000.
1984b: A 10 000 yr. B.P. extensive ice shelf over Viscount Melville Sound, Arctic Canada; Quaternary Research, v. 22, p. 18-30.

Hodgson, D.A., Vincent, J-S., and Fyles, J.G.
1984: Quaternary geology of central Melville Island, Northwest Territories; Geological Survey of Canada, Paper 83-16, 25 p.

Hopkins, D.B.
1923: Some structural features of the plains and of Alberta caused by Pleistocene glaciation; Geological Society of America, Bulletin, v. 34, p. 419-420.

Hughes, O.L.
1972: Surficial geology of northern Yukon Territory and northwestern District of Mackenzie, Northwest Territories; Geological Survey of Canada, Paper 69-36, 11 p.

1987: The Late Wisconsinan Laurentide glacial limits of northwestern Canada: the Tutsieta Lake and Kelly Lake phases; Geological Survey of Canada, Paper 85-25, 19 p.

Hughes, O.L. and Duk-Rodkin, A.
1987: Relationships of Late Pleistocene landforms and deposits along Mackenzie and Richardson Mountains (abstract); 12th INQUA Congress, National Research Council of Canada, Ottawa, Programme with Abstracts, p. 190.

Hughes, O.L., Harington, C.R., Janssens, J.A., Matthews, J.V., Jr., Morlan, R.E., Rutter, N.W., and Schweger, C.E.
1981: Upper Pleistocene stratigraphy, paleoecology, and archaeology of the northern Yukon interior, Eastern Beringia 1. Bonnet Plume Basin; Arctic, v. 34, p. 329-365.

Hughes, O.L., Rutter, N.W., Matthews, J.V., Jr., and Clague, J.J.
1989: Unglaciated areas (Quaternary stratigraphy and history, Canadian Cordillera); in Chapter 1 of Quaternary Geology of Canada and Greenland, R.J. Fulton (ed.); Geological Survey of Canada, Geology of Canada, no. 1 (also Geological Society of America, Geology of North America, v. K-1).

Hughes, O.L., Veillette, J.J., Pilon, J., Hanley, P.T., and van Everdingen, R.O.
1973: Terrain evaluation with respect to pipeline construction, Mackenzie transportation corridor, central part, 64° to 68°N; Environmental-Social Program, Task Force on Northern Oil Development, Report 73-77, 74 p.

Hyvarinen, H. and Ritchie, J.C.
1975: Pollen stratigraphy of Mackenzie pingo sediments, N.W.T., Canada; Arctic and Alpine Research, v. 7, p. 261-272.

Izett, G.A. and Wilcox, R.E.
1970: The Bishop ash bed, a Pleistocene marker bed in the Western United States; Quaternary Research, v. 1, p. 121-132.

Jackson, L.E., Jr.
1979: New evidence for the existence of an ice free corridor in the Rocky Mountain Foothills near Calgary, Alberta, during late Wisconsinan Time; in Current Research, Part A, Geological Survey of Canada, Paper 79-1A, p. 107-111.

1980a: Quaternary stratigraphy and history of the Alberta portion of the Kananaskis map area (82J) and its implications for the existence of an ice free corridor during Wisconsinan time; Canadian Journal of Anthropology, v. 1, p. 9-10.

1980b: Glacial history and stratigraphy of the Alberta portion of the Kananaskis Lakes map area; Canadian Journal of Earth Sciences, v. 17, p. 459-477.

1983: Comments on chronology of Late Wisconsinan glaciation in middle North America; Quaternary Science Reviews, v. 1, p. vii-xiv.

Jackson, L.E., Jr. and Pawson, M.
1984: Alberta radiocarbon dates; Geological Survey of Canada, Paper 83-25, 27 p.

Jackson, L.E., Jr., MacDonald, G.M., and Wilson, M.C.
1982: Paraglacial origin for terraced river sediments in Bow Valley, Alberta; Canadian Journal of Earth Sciences, v. 19, no. 12, p. 2219-2231.

Jackson, L.E., Rutter, N.W., Hughes, O.L., and Clague, J.J.
1989: Glaciated fringe (Quaternary stratigraphy and history, Canadian Cordillera): in Chapter 1 of Quaternary Geology of Canada and Greenland, R.J. Fulton (ed.); Geological Survey of Canada, Geology of Canada, no. 1 (also Geological Society of America, The Geology of North America, v. K-1).

Johnson, B., Calverley, A.E. and Pendlebury, D.C.
1976: Pleistocene foraminifera from northern Yukon, Canada; in 1st International Symposium on Benthonic Foraminifera of Continental Margins, Part B, Paleoecology and Biostratigraphy, Maritime Sediments, Special Publication 1, p. 393-400.

Johnson, G.H. and Brown, R.J.E.
1965: Stratigraphy of the Mackenzie River Delta, Northwest Territories, Canada; Geological Society of America, Bulletin, v. 76, p. 103-112.

Johnson, R.G.
1982: Brunhes-Matuyama magnetic reversal dated at 790 000 yr. B.P. by marine-astronomical correlations; Quaternary Research, v. 17, p. 135-147.

Johnston, W.A.
1934: Surface deposits and groundwater supply of Winnipeg map-area, Manitoba; Geological Survey of Canada, Memoir 174, 110 p.

Johnston, W.A., Wickenden, R.T.D., and Weir, J.D.
1948: Preliminary map, surface deposits, southern Saskatchewan; Geological Survey of Canada, Paper 48-18, scale 1 in to 6 mi.

Judge, A.S.
1986: Permafrost distribution and the Quaternary history of the Mackenzie Beaufort Region: a geothermal perspective; in Correlation of Quaternary Deposits and Events Around the Margin of the Beaufort Sea: Contributions from a Joint Canadian-American Workshop, April 1984, J.A. Heginbottom and J-S. Vincent (ed.); Geological Survey of Canada, Open File 1237, p. 41-45.

Karlstrom, E.
1981: Late Cenozoic soils of the Glacier and Waterton Parks area, northwestern Montana and southwestern Alberta, and paleoclimatic implications; unpublished PhD thesis, University of Calgary, Calgary, 358 p.

1986: Late Cenozoic glacial and pedological history of Glacier-Waterton Park area, southwestern Alberta and northwestern Montana; in Developments in Palaeontology and Stratigraphy, R.W. Barendregt (ed.), v. 8, Elsevier, New York.

Karrow, P.F. and Anderson, T.W.
1975: Palynological study of lake sediment profiles from southwestern New Brunswick: Discussion; Canadian Journal of Earth Sciences, v. 12, p. 1808-1812.

Kehew, A.E.
1982: Catastrophic flood hypothesis for origin of Souris spillway, Saskatchewan and North Dakota; Geological Society of America, Bulletin, v. 93, p. 1051-1058.

Kehew, A.E. and Lord, M.L.
1986: Origin and large-scale erosional features of glacial-lake spillways in the northern Great Plains; Geological Society of America, Bulletin, v. 97, p. 162-177.

Kerfoot, D.E.
1969: The geomorphology and permafrost conditions of Garry Island, N.W.T.; unpublished PhD thesis, University of British Columbia, Vancouver, 308 p.

Kerr, J.W.
1980: Evolution of the Canadian Arctic Islands: a transition between the Atlantic and Arctic oceans; in The Ocean Basins and Margins, A.E.M. Nairn, M. Churkin, Jr., and F.G. Stehli (ed.); v. 5, The Arctic Ocean; Plenum Press, New York, p. 105-199.

Khan, E.
1970: Biostratigraphy and paleontology of a Sangamon deposit at Fort Qu'Appelle, Saskatchewan; National Museums of Canada, Publications in Palaeontology, no. 5, 82 p.

Klassen, R.W.
1965: Surficial geology of the Waterhen-Grand Rapids area, Manitoba; Geological Survey of Canada, Paper 66-36, 6 p.

1966: The surficial geology of the Riding Mountain area, Manitoba-Saskatchewan; unpublished PhD thesis, University of Saskatchewan, Saskatoon, Saskatchewan.

1969: Quaternary stratigraphy and radiocarbon chronology of southwestern Manitoba; Geological Survey of Canada, Paper 69-27, 19 p.

1971a: Surficial geology, Franklin Bay (97C) and Brock River (97D); Geological Survey of Canada, Open File 48, scale 1:250 000.

1971b: Nature, thickness, and subsurface stratigraphy of the drift in southwestern Manitoba; in Geoscience Studies in Manitoba, A.C. Turnock (ed.); Geological Association of Canada, Special Paper 9, p. 253-261.

1972: Wisconsin events and the Assiniboine and Qu'Appelle valleys of Manitoba and Saskatchewan; Canadian Journal of Earth Sciences, v. 9, p. 544-560.

1975: Quaternary geology and geomorphology of Assiniboine and Qu'Appelle valleys of Manitoba and Saskatchewan; Geological Survey of Canada, Bulletin 228, 61 p.

1979: Pleistocene geology and geomorphology of the Riding Mountain and Duck Mountain areas, Manitoba-Saskatchewan; Geological Survey of Canada, Memoir 396, 52 p.

1983a: Lake Agassiz and the late glacial history of northern Manitoba; in Glacial Lake Agassiz, J.T. Teller and L. Clayton (ed.); Geological Association of Canada, Special Paper 26, p. 97-115.

1983b: Assiniboine Delta and the Assiniboine-Qu'Appelle valley system — implications concerning the history of Lake Agassiz in southwestern Manitoba; in Glacial Lake Agassiz, J.T. Teller and L. Clayton (ed.); Geological Association of Canada, Special Paper 26, p. 212-229.

1986: Surficial geology of north-central Manitoba; Geological Survey of Canada, Memoir 419, 57 p.

Klassen, R.W. and Vreeken, W.J.
1987: The nature and chronological implications of surface tills and post-till sediments in the Cypress Lake area, Saskatchewan; in Current Research, Part A, Geological Survey of Canada, Paper 87-1A, p. 111-125.

Klassen, R.W. and Wyder, J.E.
1970: Bedrock topography, buried valleys and nature of the drift, Virden map- area, Manitoba; Geological Survey of Canada, Paper 70-56, 11 p.

Klassen, R.W., Delorme, L.O., and Mott, R.J.
1967: Geology and paleontology of Pleistocene deposits in southwestern Manitoba; Canadian Journal of Earth Sciences, v. 4, p. 433-447.

Kuc, M.
1974: The interglacial flora of Worth Point, western Banks Island; in Report of Activities, Part B, Geological Survey of Canada, Paper 74-1B, p. 227-231.

Kugler, M. et St-Onge, D.A.
1973: Composantes du mouvement de rebondissement isostatique d'après des données de remblaiements alluviaux (exemple de la rivière Saskatchewan Sud); Journal canadien des sciences de la terre, vol. 4, p. 551-556.

Kupsch, W.O.
1962: Ice-thrust ridges in western Canada; Journal of Geology, v. 70, p. 582-594.

Last, W.M.
1984: Sedimentology of playa lakes of the northern Great Plains; Canadian Journal of Earth Sciences, v. 21, p. 107-125.

Laycock, A.H.
1972: The diversity of the physical landscape; in Studies in Canadian Geography — The Prairie Provinces, P.J. Smith (ed.); University of Toronto Press, Toronto.

Longley, R.W.
1972: The climate of the Prairie Provinces; Canada Department of Environment, Climatological Studies no. 13, Toronto.

Lorrain, R.D. and Demeur, P.
1985: Isotopic evidence for relic Pleistocene glacier ice on Victoria Island, Canadian Arctic Archipelago; Arctic and Alpine Research, v. 17, no. 1, p. 89-98.

Lowdon, J.A. and Blake, W., Jr.
1968: Geological Survey of Canada radiocarbon dates VII; Radiocarbon, v. 10, p. 207-245.
1970: Geological Survey of Canada radiocarbon dates IX; Radiocarbon v. 12, p. 46-86
1973: Geological Survey of Canada radiocarbon dates XIII; Geological Survey of Canada, Paper 73-7, 61 p.
1975: Geological Survey of Canada radiocarbon dates XV; Geological Survey of Canada Paper 75-7, 32 p.
1976: Geological Survey of Canada radiocarbon dates XVI; Geological Survey of Canada, Paper 76-7, 21 p.
1978: Geological Survey of Canada radiocarbon dates XVIII; Geological Survey of Canada, Paper 78-7, 20 p.
1979: Geological Survey of Canada radiocarbon dates XIX; Geological Survey of Canada, Paper 79-7, 58 p.
1980: Geological Survey of Canada radiocarbon dates XX; Geological Survey of Canada, Paper 80-7, 28 p.
1981: Geological Survey of Canada radiocarbon dates XXI; Geological Survey of Canada, Paper 81-7, 22 p.

Lowdon, J.A., Fyles, J.G., and Blake, W., Jr.
1967: Geological Survey of Canada radiocarbon dates VI; Radiocarbon, v. 9, p. 156-197.

Lowdon, J.A., Robertson, I.M., and Blake, W., Jr.
1971: Geological Survey of Canada radiocarbon dates XI; Radiocarbon, v. 13, p. 255-324.

MacDonald, G.M.
1987: Postglacial vegetation history of the Mackenzie River basin, Quaternary Research, v. 28, p. 245-262.

MacDonald, G.M., Beukens, B.P., Keiser, W.E., and Vitt, D.H.
1987: Comparative radiocarbon dating of terrestrial plant microfossils and aquatic moss from the "Ice-free Corridor" of western Canada; Geology, v. 15, p. 837-840.

Mackay, J.R.
1956: Deformation by glacier-ice at Nicholson Peninsula, N.W.T., Canada; Arctic v. 9, p. 219-228.
1958: The Anderson River map-area, N.W.T.; Geographical Branch, Memoir 5, 137 p.
1959: Glacier ice-thrust features of the Yukon Coast; Geographical Bulletin, no. 13, p. 5-21.
1960: Crevasse fillings and ablation slide moraines, Stopover Lake area, N.W.T.; Geographical Bulletin, no. 14, p. 89-99.
1962: Pingos of the Pleistocene Mackenzie Delta area; Geographical Bulletin no. 18, p. 21-63.
1963: The Mackenzie delta area, N.W.T.; Geographical Branch, Memoir no. 8, 202 p.
1971: The origin of massive icy beds in permafrost, Western Arctic Coast, Canada; Canadian Journal of Earth Sciences, v. 8, p. 397-422.
1981: Dating the Horton River breakthrough, District of Mackenzie; in Current Research, Part B, Geological Survey of Canada, Paper 81-1B, p. 129-132.
1983: Oxygen isotope variations in permafrost, Tuktoyaktuk Peninsula area, Northwest Territories; in Current Research, Part B, Geological Survey of Canada, Paper 83-1B, p. 67-74.
1986: The permafrost record and Quaternary history of northwestern Canada; in Correlation of Quaternary Deposits and Events Around the Margin of the Beaufort Sea: Contribution from a Joint Canadian-American Workshop, April 1984, J.A. Heginbottom and J-S. Vincent (ed.); Geological Survey of Canada, Open File 1237, p. 38-40.

Mackay, J.R., and Mathews, W.H.
1973: Geomorphology and Quaternary history of the Mackenzie River valley near Fort Good Hope, N.W.T., Canada; Canadian Journal of Earth Sciences, v. 10, p. 26-41.

Mackay, J.R., and Matthews, J.V., Jr.
1983: Pleistocene ice and sand wedges, Hooper Island, Northwest Territories; Canadian Journal of Earth Sciences, v. 20, p. 1087-1097.

Mackay, J.R., Rampton, V.N., and Fyles, J.G.
1972: Relic Pleistocene permafrost, Western Arctic, Canada; Science, v. 176, p. 1321-1323.

Mahaney, W.C. and Stalker, A. MacS.
1988: Stratigraphy of the North Cliff section, Wellsch Valley site, Saskatchewan; Canadian Journal of Earth Sciences, v. 25, p. 206-214.

Maher, W.J.
1968: Musk-ox bone of possible Wisconsin age from Banks Island, Northwest Territories; Arctic, v. 21, p. 260-266.

Mathewes, R.W. and Westgate, J.A.
1980: Bridge River tephra; revised distribution and significance for detecting old carbon errors in radiocarbon dates of limnic sediments in southern British Columbia; Canadian Journal of Earth Sciences, v. 17, p. 1454-1461.

Mathews, W.H.
1974: Surface profiles of the Laurentide Ice Sheet in its marginal areas; Journal of Glaciology, v. 13, p. 37-43.
1978: Quaternary stratigraphy and geomorphology of Charlie Lake (94 A) map-area, British Columbia; Geological Survey of Canada, Paper 76-20, 25 p.
1980: Retreat of last ice sheet in northeastern British Columbia and adjacent Alberta; Geological Survey of Canada, Bulletin 331, 22 p.

Mathews, W.H. and Mackay, J.R.
1960: Deformation of soils by glacier ice and the influence of pore pressures and permafrost; Royal Society of Canada, Transactions, third series, v. 54, p. 27-36.

Matthews, J.V., Jr., Mott, R.J., and Vincent, J-S.
1986: Preglacial and interglacial environments of Banks Island: pollen and macrofossils from Duck Hawk Bluffs and related sites; Géographie physique et Quaternaire, v. 40, p. 279-298.

Mayewski, P.A., Denton, G.H., and Hughes, T.J.
1981: Late Wisconsin ice sheets of North America; in The Last Great Ice Sheets, G.H. Denton and T.J. Hughes (ed.); John Wiley and Sons, New York, p. 67-178.

McCallum, K.J. and Wittenberg, J.
1965: University of Saskatchewan radiocarbon dates IV; Radiocarbon, v. 7, p. 229-235.

McConnell, R.G.
1886: Report on the Cypress Hills, Wood Mountain, and adjacent country; Geological Survey of Canada, Annual Report 1, Part C, p. 1C-169C.

McPherson, R.A., Leith, E.I., and Anderson, D.T.
1971: Pleistocene stratigraphy of a portion of southeastern Manitoba; in Geoscience Studies in Manitoba, A.C. Turnock (ed.); Geological Association of Canada, Special Paper 9, p. 277-283.

Miall, A.D.
1979: Mesozoic and Tertiary geology of Banks Island, Arctic Canada: The history of an unstable craton margin; Geological Survey of Canada, Memoir 387, 235 p.

Mickelson, D.M., Clayton, L., Fullerton, D.S., and Barns, H.W.
1983: The Late Wisconsin glacial record of the Laurentide ice sheet in the United States; in The Late Pleistocene, S.C. Porter, (ed.), volume 1 of Late-Quaternary Environments of the United States, H.E. Wright, Jr. (ed.); University of Minnesota Press, Minneapolis, p. 3-37.

Moran, S.R.
1971: Glaciotectonic structures in drift; in Till: A Symposium, R.P. Goldthwait (ed.); Ohio State University Press, Columbus, Ohio, p. 127-148.

Moran, S.R., Clayton, L., Hooke, R.L., Fenton, M.M., and Andriashek, L.D.
1980: Glacier-bed landforms of the prairie region of North America; Journal of Glaciology, v. 25, p. 457-476.

Morlan, R.E.
1986: Pleistocene archaeology in Old Crow Basin: a critical reappraisal; in New Evidence for the Pleistocene Peopling of the Americas, A.L. Bryan (ed.); Center for Study of Early Man, University of Maine, p. 27-48.

Mott, R.J.
1973: Palynological studies in central Saskatchewan — pollen stratigraphy from lake sediment sequences; Geological Survey of Canada, Paper 72-49, 18 p.

Naeser, N.D., Westgate, J.A., Hughes, O.L., and Péwé, T.L.
1982: Fission-track ages of late Cenozoic distal tephra beds in the Yukon Territory and Alaska; Canadian Journal of Earth Sciences, v. 19, p. 2167-2178.

Nambudiri, E.M.V., Teller, J.T., and Last, W.M.
1980: Pre-Quaternary microfossils — A guide to errors in radiocarbon dating; Geology, v. 8, p. 123-126.

Naylor, D., McIntyre, D.J., and McMillan, N.F.
1972: Pleistocene deposits exposed along the Yukon Coast; Arctic, v. 25, p. 49-55.

Nelson, R.E. and Carter, L.D.
1985: Pollen analysis of a Late Pliocene and Early Pleistocene section from the Gubik Formation of Arctic Alaska; Quaternary Research, v. 24, p. 295-306.

Parizek, R.R.
1964: Geology of the Willow Bunch Lake area (72H), Saskatchewan; Saskatchewan Research Council, Geology Division, Report no. 4, 47 p.
1969: Glacial ice-contact rings and ridges; in United States Contributions to Quaternary Research, S.A. Schumm and W.C. Bradley (ed.); Geological Society of America Special Paper 123, p. 49-102.

Pearce, G.W., Westgate, J.A., and Robertson, S.
1982: Magnetic reversal history of Pleistocene sediments at Old Crow, northwestern Yukon Territory; Canadian Journal of Earth Sciences, v. 19, p. 919-929.

Pelletier, B.R.
1966: Development of submarine physiography in the Canadian Arctic and its relation to crustal movements; in Continental Drift, G.D. Garland (ed.); The Royal Society of Canada, Special Publication no. 9, University of Toronto Press, p. 77-101.

Pissart, A., Vincent, J-S., et Edlund, S.A.
1977: Dépôts et phénomènes éoliens sur l'île Banks, Territoires du Nord-Ouest, Canada; Journal canadien des sciences de la terre, vol. 14, p. 2462-2480.

Powers, H.A. and Wilcox, R.E.
1964: Volcanic ash from Mount Mazama (Crater Lake) and from Glacier Park; Science, v. 144, p. 1334-1336.

Prest, V.K.
1968: Nomenclature of moraines and ice-flow features as applied to the glacial map of Canada; Geological Survey of Canada, Paper 67-57, 32 p.
1969: Retreat of Wisconsin and recent ice in North America; Geological Survey of Canada, Map 1257A, 1:5 000 000.
1970: Quaternary geology of Canada; in Geology and Economic Minerals of Canada, R.J.W. Douglas (ed.); Geological Survey of Canada, Economic Geology Report 1, 5th edition, p. 676-764.
1983: Canada's heritage of glacial features; Geological Survey of Canada, Miscellaneous Report 28, 119 p.
1984: The Late Wisconsinan glacier complex; in Quaternary Stratigraphy of Canada — A Canadian Contribution to IGCP Project 24, R.J. Fulton (ed.); Geological Survey of Canada, Paper 84-10, p. 21-36.

Prest, V.K. and Nielsen, E.
1987: The Laurentide Ice Sheet and long distance transport; Geological Survey of Finland, Special Paper 3, p. 91-101.

Prest, V.K., Grant, D.R., and Rampton, V.N.
1968: Glacial map of Canada; Geological Survey of Canada, Map 1253A, scale 1:5 000 000.

Putnam, D.F. and Putnam, R.G.
1979: Canada: A Regional Analysis; J.M. Dent and Sons, Canada.

Rampton, V.N.
1970: Quaternary geology, Mackenzie Delta and Arctic Coastal Plain, District of Mackenzie (Parts of 107B, 117A, C, D); in Report of Activities, Part A, Geological Survey of Canada, Paper 70-1A, p. 181-182.
1971a: Quaternary geology, Mackenzie Delta and Arctic Coastal Plain, District of Mackenzie; in Report of Activities, Part A, Geological Survey of Canada, Paper 71-1A, p. 173-177.
1971b: Massive ice and icy sediments throughout the Tuktoyaktuk Peninsula, Richards Island, and nearby areas, District of Mackenzie; Geological Survey of Canada, Paper 71-21, 16 p.
1972: Quaternary geology, Arctic Coastal Plain, District of Mackenzie and Herschel Island, Yukon Territory; in Report of Activities, Part A, Geological Survey of Canada, Paper 72-1A, p. 171-175.
1973: The history of thermokarst in the Mackenzie-Beaufort region, Northwest Territories, Canada; in Abstracts, 9th Congress, International Union of Quaternary Research, p. 299.
1974a: The influence of ground ice and thermokarst upon the geomorphology of the Mackenzie-Beaufort area; in Research in Polar and Alpine Geomorphology, B.D. Fahey and R.D. Thompson (ed.), Proceedings, 3rd Guelph Symposium on Geomorphology, Guelph, p. 43-59.
1974b: Terrain evaluation with respect to pipeline construction, Mackenzie transportation corridor, northern part, Lat. 68°N to coast; Environmental-Social Program, Task Force on Northern Oil Development, Report 73-47, 44 p.
1980: Surficial geology, Aklavik, District of Mackenzie; Geological Survey of Canada, Map 31-1979, scale 1:250 000.
1981a: Surficial geology, Malloch Hill, District of Mackenzie; Geological Survey of Canada, Map 30-1979, scale 1:250 000.
1981b: Surficial geology, Mackenzie Delta, District of Mackenzie; Geological Survey of Canada, Map 32-1979, scale 1:250 000.
1981c: Surficial geology, Stanton, District of Mackenzie; Geological Survey of Canada, Map 33-1979, scale 1:250 000.
1982: Quaternary geology of the Yukon Coastal Plain; Geological Survey of Canada, Bulletin 317, 49 p.
1988: Quaternary geology of the Tuktoyaktuk Coastlands, Northwest Territories; Geological Survey of Canada, Memoir 423, 98 p.

Rampton, V.N. and Bouchard, M.
1975: Surficial geology of Tuktoyaktuk, District of Mackenzie; Geological Survey of Canada, Paper 74-53, 17 p.

Reeves, B.O.K.
1971: On the coalescence of the Laurentide and Cordilleran ice sheets in the western interior of North America with particular reference to the southern Alberta area; in Aboriginal Man and Environments on the Plateau of North America, A.H. Styrd and R.A. Smith (ed.); The Student's Press, Calgary, p. 205-227.
1973: The nature and age of the contact between the Laurentide and Cordilleran ice sheets in the western interior of North America; Arctic and Alpine Research, v. 5, no. 1, p. 1-16.

Richards, H.G.
1950: Postglacial marine submergence of Arctic North America with special reference to the Mackenzie Delta; Proceedings, American Philosophical Society, v. 94, p. 31-37.

Ritchie, J.C.
1969: Absolute pollen frequencies and carbon-14 age of a section of Holocene lake sediment from the Riding Mountain area of Manitoba; Canadian Journal of Botany, v. 47, no. 9, p. 1345-1349.
1972: Pollen analysis of late-Quaternary sediments from the Arctic treeline of the Mackenzie River Delta region, Northwest Territories, Canada; in Climatic Changes in Arctic Areas During the Last Ten Thousand Years, Y. Vasari, H. Hyvarinen and S. Hicks (ed.); Acta Universitatis Ouluensis, Serie A, Geologica 1, p. 239-252.
1976: The Late-Quaternary vegetational history of the western interior of Canada; Canadian Journal of Botany, v. 54, no. 15, p. 1793-1818.
1977: The modern and late Quaternary vegetation of the Campbell-Dolomite Uplands, near Inuvik, N.W.T., Canada; Ecological Monographs, v. 47, p. 401-423.
1980: Towards a late-Quaternary paleoecology of the ice-free corridor; in The Ice-Free Corridor and Peopling of the New World, N.W. Rutter and C.E. Schweger (ed.); Canadian Journal of Anthropology, v. 1, p. 15-28.
1981: Problems of interpretation of the pollen stratigraphy of northwest North America; in Quaternary Paleoclimates, W.C. Mahaney (ed.); Geo Abstracts, Norwich, p. 377-391.
1982: The modern and Late-Quaternary vegetation of the Doll Creek area, North Yukon, Canada; The New Phytologist, v. 90, p. 563-603.
1984: Past and present vegetation of the far northwest of Canada; University of Toronto Press, Toronto, 251 p.
1989: History of the boreal forest in Canada; in Chapter 7 of Quaternary Geology of Canada and Greenland, R.J. Fulton (ed.); Geological Survey of Canada, Geology of Canada, no. 1 (also Geological Society of America, The Geology of North America, v. K-1).

Ritchie, J.C., and Cwynar, L.C.
1982: The Late Quaternary vegetation of the north Yukon; in Paleoecology of Beringia, D.M. Hopkins, J.V. Matthews Jr., C.E. Schweger, and S.B. Young (ed.); Academic Press, New York, p. 113-126.

Ritchie, J.C., and Hare, F.K.
1971: Late-Quaternary vegetation and climate near the Arctic tree line of northwestern North America; Quaternary Research, v. 1, p. 331-342.

Ritchie, J.C., and Koivo, L.K.
1975: Postglacial diatom stratigraphy in relation to the recession of Glacial Lake Agassiz; Quaternary Research, v. 5, p. 529-540.

Ritchie, J.C. and Lichti-Federovich, S.
1968: Holocene pollen assemblages from the Tiger Hills, Manitoba; Canadian Journal of Earth Sciences, v. 5, p. 873-880.

Ritchie, J.C. and Yarranton, G.A.
1978: Patterns of change in the late-Quaternary vegetation of the western interior of Canada; Canadian Journal of Botany, v. 56, no. 17, p. 2177-2183.

Ritchie, J.C., Cwynar, L.C., and Spear, R.W.
1983: Evidence from north-west Canada for an early Holocene Milankovitch thermal maximum; Nature, v. 305, no. 5930, p. 126-128.

Roed, M.A., Mountjoy, E.W., and Rutter, N.W.
1967: The Athabasca Valley erratics train, Alberta and Pleistocene ice movements across the continental divide; Canadian Journal of Earth Sciences, v. 4, p. 625-632.

Rutherford, R.L.
1937: Saskatchewan gravels and sands in central Alberta; Royal Society of Canada, Transactions, section IV, series 3, v. 31, p. 81-95.

Rutter, N.W.
1972: Geomorphology and multiple glaciation in the area of Banff, Alberta; Geological Survey of Canada, Bulletin 206, 54 p.
1974: Surficial geology and land classification, Mackenzie Valley transportation corridor (85D, E, 95A, B, G, H, I, J, K, N, O); in Report of Activities, Part A, Geological Survey of Canada, Paper 74-1A, p. 285.
1980: Late Pleistocene history of the western Canadian ice-free corridor; in The Ice-Free Corridor and Peopling of the New World, N.W. Rutter and C.E. Schweger (ed.); Canadian Journal of Anthropology, v. 1, p. 1-8.

Rutter, N.W. and Boydell, A.N.
1973: Surficial geology and land classification, Mackenzie Valley transport corridor (85D, 95B (north half), 95G, I, K (east half), N, O); in Report of Activities, Part A, Geological Survey of Canada, Paper 73-1A, p. 239-240.

Rutter, N.W. and Minning, G.V.
1972: Surficial geology and land classification, Mackenzie Valley transportation corridor (85E, 95A, B (south half), H, J); in Report of Activities, Part A, Geological Survey of Canada, Paper 72-1A, p. 178.

Rutter, N.W., Boydell, A.N., Savigny, K.W., and van Everdingen, R.O.
1973: Terrain evaluation with respect to pipeline construction, Mackenzie transportation corridor, southern part, Lat. 60° to 64°; Environmental-Social Program, Task Force on Northern Oil Development, Report no. 73-36, 145 p.

Savigny, K.W.
1989: Engineering geology of the Great Bear River area, Northwest Territories; Geological Survey of Canada, Paper 88-23.

Schreiner, B.T.
1983: Lake Agassiz in Saskatchewan; in Glacial Lake Agassiz, J.T. Teller and L. Clayton (ed.); Geological Association of Canada, Special Paper 26, p. 75-96.
1984: Quaternary geology of the Precambrian Shield, Saskatchewan; Saskatchewan Geological Survey, Report 221, 106 p.

Schweger, C.E.
1989: Paleoecology of the western Canadian ice-free corridor; in Chapter 7 of Quaternary Geology of Canada and Greenland, R.J. Fulton (ed.); Geological Survey of Canada, Geology of Canada, no. 1 (also Geological Society of America, The Geology of North America, v. K-1).

Scott, J.S.
1976: Geology of Canadian tills; in Glacial Till, R.F. Leggett (ed.); Royal Society of Canada, Special Publication 12, p. 50-68.
1989: Engineering geology and land use planning in the Prairie region of Canada; in Chapter 12 of Quaternary Geology of Canada and Greenland, R.J. Fulton (ed.); Geological Survey of Canada, Geology of Canada, no. 1 (also Geological Society of America, The Geology of North America, v. K-1).

Sharpe, D.R.
1984: Late Wisconsinan glaciation and deglaciation of Wollaston Peninsula, Victoria Island, Northwest Territories; in Current Research, Part A, Geological Survey of Canada, Paper 84-1A, p. 259-269.
1985: The stratified nature of deposits in streamlined glacial landforms on southern Victoria Island, District of Franklin; in Current Research, Part A, Geological Survey of Canada, Paper 85-1A, p. 365-371.
1988: Late glacial landforms of Wollaston Peninsula, Victoria Island, Northwest Territories: product of ice-marginal retreat, surge, and mass stagnation; Canadian Journal of Earth Sciences, v. 25, p. 262-279.

Shetsen, I.
1984: Application of till pebble lithology to the differentiation of glacial lobes in southern Alberta; Canadian Journal of Earth Sciences, v. 21, p. 920-933.

SkaraWoolf, T.
1980: Mammals of the Riddell Local Fauna (Floral Formation, Pleistocene, Late Rancholabrean), Saskatoon, Canada; Saskatchewan Museum of Natural History, Regina, Natural History Contributions, Number 2, 129 p.

1981: Biostratigraphy and paleoecology of Pleistocene deposits (Riddell Member, Floral Formation, Late Rancholabrean), Saskatoon, Canada; Canadian Journal of Earth Sciences, v. 18, p. 311-322.

Slater, D.S.
1985: Pollen analysis of postglacial sediments from Eildun Lake, District of Mackenzie, N.W.T., Canada; Canadian Journal of Earth Sciences, v. 22, p. 663-674.

Slater, G.
1927: Structure of the Mud Buttes and Tit Hills in Alberta; Geological Society of America, Bulletin, v. 38, p. 721-730.

Spear, R.W.
1983: Paleoecological approaches to a study of tree-line fluctuation in the Mackenzie Delta region, Northwest Territories: preliminary results; Collection Nordicana, no. 47, p. 61-72.

Stalker, A. MacS.
1956: The Erratics Train; Foothills of Alberta; Geological Survey of Canada, Bulletin 37, 28 p.
1959: Surficial geology, Fort Macleod, Alberta; Geological Survey of Canada, Map 21-1958 (with marginal notes), scale 1:250 000.
1960a: Ice-pressed drift forms and associated deposits in Alberta; Geological Survey of Canada, Bulletin 57, 38 p.
1960b: Surficial geology of the Red Deer — Stettler map-area, Alberta; Geological Survey of Canada, Memoir 306, 140 p.
1961: Buried valleys in central and southern Alberta; Geological Survey of Canada, Paper 60-32, 13 p.
1962: Surficial geology, Fernie, 82G, Alberta portion; Geological Survey of Canada, Map 31-1961, scale 1:250 000.
1963: Quaternary stratigraphy in southern Alberta; Geological Survey of Canada, Paper 62-34, 52 p.
1968a: Identification of Saskatchewan Gravels and Sands; Canadian Journal of Earth Sciences, v. 5, p. 155-163.
1968b: Geology of the terraces at Cochrane, Alberta; Canadian Journal of Earth Sciences, v. 5, no. 6, p. 1455-1466.
1969: Quaternary stratigraphy in southern Alberta; Report II: Sections near Medicine Hat, Alberta; Geological Survey of Canada, Paper 69-26, 28 p.
1973: Surficial geology of the Drumheller area, Alberta; Geological Survey of Canada, Memoir 370, 122 p.
1976a: Mega blocks, or the enormous erratics of the Albertan Prairies; in Report of Activities, Part C, Geological Survey of Canada, Paper 76-1C, p. 185-188.
1976b: Quaternary stratigraphy of the southwestern Canadian Prairies; in Quaternary Stratigraphy of North America, W.C. Mahaney (ed.); Dowden, Hutchinson & Ross, Stroudsburg, Pa., p. 381-407.
1977: The probable extent of Classical Wisconsin ice in southern and central Alberta; Canadian Journal of Earth Sciences, v. 14, p. 2614-2619.
1980: The geology of the ice-free corridor: the southern half; Canadian Journal of Anthropology, v. 1, p. 11-13.
1983: Quaternary stratigraphy in southern Alberta; Report III: The Cameron Ranch Section; Geological Survey of Canada, Paper 83-10, 20 p.

Stalker, A. MacS. and Churcher, C.S.
1972: Glacial stratigraphy of the southwestern Canadian Prairies; the Laurentide record; 24th International Geological Congress (Montreal), Section 12, p. 110-119.
1982: Ice Age deposits and animals from the southwestern part of the Great Plains of Canada; Geological Survey of Canada, Miscellaneous Report no. 31.

Stalker, A. MacS. and Harrison, J.E.
1977: Quaternary glaciation of the Waterton-Castle River region of Alberta; Canadian Petroleum Geology, Bulletin, v. 25, p. 882-906.

Stalker, A.MacS. and Vincent, J-S.
in press: The Quaternary geology of the Interior Plains; Chapter 10 in Sedimentary Cover of the Craton: Canada, D.F. Stott and J.D. Aitken (ed.); Geological Survey of Canada, Geology of Canada no. 6 (also Geological Society of America, The Geology of North America, v. D-1).

Stalker, A. MacS. and Wyder, J.E.
1983: Borehole and outcrop stratigraphy compared, with illustrations from the Medicine Hat area of Alberta; Geological Survey of Canada, Bulletin 296, 28 p.

St-Onge, D.A.
1967: Geomorphology, Lancer, Saskatchewan; Department of Energy, Mines and Resources, Geographical Branch, scale 1:50 000.
1972: Sequence of glacial lakes in north-central Alberta; Geological Survey of Canada, Bulletin 213, 16 p.
1980: Glacial Lake Coppermine, north-central District of Mackenzie, Northwest Territories; Canadian Journal of Earth Sciences, v. 17, p. 1310-1315.
1987: Morphosedimentary zones and Wisconsinan glacial limit, Bluenose Lake Region, Northwest Territories; Geological Association of Canada, Program with Abstracts, v. 12, p. 92.

Szabo, B.J., Stalker, A. MacS., and Churcher, C.S.
1973: Uranium series ages of some Quaternary deposits near Medicine Hat, Alberta; Canadian Journal of Earth Sciences, v. 10, p. 1464-1469.

Teller, J.T.
1976: Lake Agassiz deposits in the main offshore basin of southern Manitoba; Canadian Journal of Earth Sciences, v. 13, p. 27-43.
1980: Radiocarbon dates in Manitoba; Manitoba Department of Energy and Mines, Mineral Resources Division, Geological Report GR 80-4.

Teller, J.T. and Fenton, M.M.
1980: Late Wisconsinan glacial stratigraphy and history of southeastern Manitoba; Canadian Journal of Earth Sciences, v. 17, p. 19-35.

Teller, J.T. and Last, W.M.
1981: Late Quaternary history of Lake Manitoba, Canada; Quaternary Research, v. 16, p. 97-116.

Teller, J.T. and Thorleifson, L.H.
1983: The Lake Agassiz-Lake Superior connections; in Glacial Lake Agassiz, J.T. Teller and L. Clayton (ed.); Geological Association of Canada, Special Paper 26, p. 261-290.

Teller, J.T., Moran, S.R., and Clayton, L.
1980: The Wisconsinan deglaciation of southern Saskatchewan and adjacent areas: Discussion; Canadian Journal of Earth Sciences, v. 17, p. 539-541.

Tipper, H.W., Woodsworth, G.J., and Gabrielse, H.
1981: Tectonic assemblage map of the Canadian Cordillera and adjacent parts of the United States of America; Geological Survey of Canada, Map 1505A.

Tozer, E.T. and Thorsteinsson, R.
1964: Western Queen Elizabeth Islands, Arctic Archipelago; Geological Survey of Canada, Memoir 332, 242 p.

Vincent, J-S.
1978: Limits of ice advance, glacial lakes, and marine transgressions on Banks Island, District of Franklin: a preliminary interpretation; in Current Research, Part C, Geological Survey of Canada, Paper 78-1C, p. 53-62
1982: The Quaternary history of Banks Island, N.W.T., Canada; Géographie physique et Quaternaire, v. 36, p. 209-232.
1983: La géologie du Quaternaire et la géomorphologie de l'île Banks, Arctique canadien; Commission géologique du Canada, Mémoire 405, 118 p.
1984: Quaternary stratigraphy of the western Canadian Arctic Archipelago; in Quaternary Stratigraphy of Canada — A Canadian Contribution to IGCP Project 24, R.J. Fulton (ed.); Geological Survey of Canada, Paper 84-10, p. 87-100.
1989: Quaternary geology of the southeastern Canadian Shield; in Chapter 3 of Quaternary Geology of Canada and Greenland, R.J. Fulton (ed.); Geological Survey of Canada, Geology of Canada, no. 1 (also Geological Society of America, The Geology of North America, v. K-1).

Vincent, J-S. and Barendregt, R.W.
1987: Late Cenozoic paleomagnetic record of the Duck Hawk Bluffs, Banks Island, Canadian Arctic Archipelago (abstract); 12th INQUA Congress, National Research Council of Canada, Ottawa, Programme with Abstracts, p. 282.

Vincent, J-S. and Prest, V.K.
1987: The Early Wisconsinan history of the Laurentide Ice Sheet; Géographie physique et Quaternaire, v. 41, p. 199-213.

Vincent, J-S., Morris, W.A., and Occhietti, S.
1984: Glacial and nonglacial sediments of Matuyama paleomagnetic age on Banks Island, Canadian Arctic Archipelago; Geology, v. 12, p. 139-142.

Vincent, J-S., Occhietti, S., Rutter, N.W., Lortie, G., Guilbault, J.P., and Boutray, B. de
1983: The late-Tertiary-Quaternary stratigraphic record of the Duck Hawk Bluffs, Banks Island, Canadian Arctic Archipelago; Canadian Journal of Earth Sciences, v. 20, p. 1694-1712.

Vonhoff, J.A.
1965: The Cypress Hills Formation and its reworked deposits in southwestern Saskatchewan; in Alberta Society of Petroleum Geologists Guidebook, 15th Annual Field Conference Guidebook, Part 1, Cypress Hills Plateau, p. 142-161.
1969: Tertiary gravels and sands in the Canadian Great Plains; unpublished PhD thesis, Department of Geological Sciences, University of Saskatchewan, Saskatoon.

Vreeken, W.J.
1986: Quaternary events in the Elkwater Lake area of southeastern Alberta; Canadian Journal of Earth Sciences, v. 23, p. 2024-2038.

Washburn, A.L.
1947: Reconnaissance geology of portions of Victoria Island and adjacent regions, Arctic Canada; Geological Society of America; Memoir 22, 142 p.

Waters, P.L.
1979: Postglacial environment of southern Alberta; unpublished MSc thesis, University of Alberta, Department of Geology, Edmonton, 107 p.

Weaver, J.S. and Stewart, J.M.
1982: In situ hydrates under the Beaufort Sea Shelf; in Proceedings of the Fourth Canadian Permafrost Conference, H.M. French (ed.); National Research Council, Ottawa, p. 312-319.

Westgate, J.A.
1968: Surficial geology of the Foremost-Cypress Hills area, Alberta; Research Council of Alberta, Bulletin 22, 122 p.

Westgate, J.A. and Evans, M.E.
1978: Compositional variability of Glacier Peak tephra and its stratigraphic significance; Canadian Journal of Earth Sciences, v. 15, p. 1554-1567.

Westgate, J.A., Bonnichsen, R., Schweger, C., and Dormaar, J.F.
1972: The Cypress Hills; in Quaternary Geology and Geomorphology between Winnipeg and the Rocky Mountains, N.W. Rutter and E.A. Christiansen (ed.); Field Excursion C-22, 24th International Geological Congress (Montreal), p. 50-62.

Westgate, J.A., Briggs, N.D., Stalker, A. MacS., and Churcher, C.S.
1978: Fission-track age of glass from tephra beds associated with Quaternary vertebrate assemblages in the southern Canadian plains; Geological Society of America, Abstracts with Program, v. 10, no. 7, p. 514-515.

Westgate, J.A., Christiansen E.A., and Boellstorff, J.D.
1977: Wascana Creek Ash (Middle Pleistocene) in southern Saskatchewan: characterization, source, fission track age, palaeomagnetism and stratigraphic significance; Canadian Journal of Earth Sciences, v. 14, no. 3, p. 357-374.

Westgate, J.A., Fritz, P., Matthews, J.V., Jr., Klalas, L., Delorme, L.D., Green, R., and Aario, R.
1972: Geochronology and paleoecology of mid-Wisconsinan sedimentation in west-central Alberta, Canada; 24th International Geological Congress (Montreal), Abstracts, p. 380.

Westgate, J.A., Smith, D.G.W., and Tomlinson, M.
1970: Late Quaternary tephra layers in southwestern Canada; in Early Man and Environment in Northwest North America, R.A. Smith and J.W. Smith (ed.); Proceedings of the Second Annual Paleoenvironmental Workshop, University of Calgary Archaeological Association, The Student Press, Calgary, Alberta, p. 13-34.

Whitaker, S.H.
1965: Geology of the Wood Mountain area (72 G), Saskatchewan; unpublished PhD thesis, University of Illinois, University Microfilms, Inc., Ann Arbor, Michigan.

Whitaker, S.H. and Christiansen, E.A.
1972: The Empress Group in southern Saskatchewan; Canadian Journal of Earth Sciences, v. 9, p. 353-360.

Wickenden, R.T.D.
1945: Mesozoic stratigraphy of the eastern plains, Manitoba and Saskatchewan; Geological Survey of Canada, Memoir 239, 87 p.

Zymela, S., Schwarcz, H.P., Grün, R., Stalker, A. MacS., and Churcher, C.S.
1988: ERS dating of Pleistocene fossil teeth from Alberta and Saskatchewan; Canadian Journal of Earth Sciences, v. 25, p. 235-245.

Authors' addresses

R.J. Fulton
Geological Survey of Canada
601 Booth Street
Ottawa, Ontario K1A 0E8

J-S. Vincent
Geological Survey of Canada
601 Booth Street
Ottawa, Ontario K1A 0E8

R.W. Klassen
Geological Survey of Canada
3303-33rd Street N.W.
Calgary, Alberta T2L 2A7

Printed in Canada

Chapter 3

QUATERNARY GEOLOGY OF THE CANADIAN SHIELD

The Laurentide Ice Sheet and an introduction to the Quaternary geology of the Canadian Shield — *A.S. Dyke, J-S. Vincent, J.T. Andrews, L.A. Dredge, and W.R. Cowan*

Quaternary geology of the northwestern Canadian Shield — *A.S. Dyke and L.A. Dredge*

Quaternary geology of the southwestern Canadian Shield — *L.A. Dredge and W.R. Cowan*

Quaternary geology of the southeastern Canadian Shield — *J-S. Vincent*

Quaternary geology of the northeastern Canadian Shield — *J.T. Andrews*

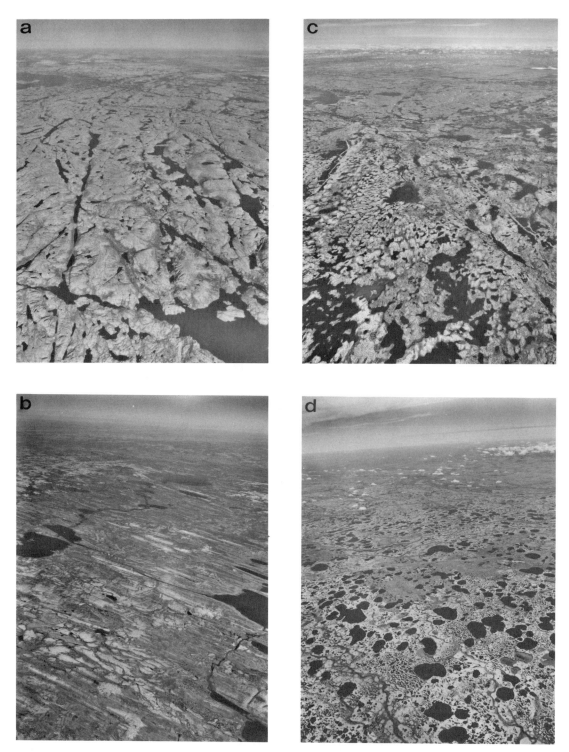

Aerial photographs of typical terrain and features in the Canadian Shield region. a) Area of abundant outcrops with a small esker and patch of streamlined drift (lower right). Differential erosion has accentuated bedrock structures which control the shape and distribution of most lakes. Labrador Plateau, Mistastin Lake area, Quebec / Labrador. NAPL T383L-200. b) Fluted drift area with individual ridges several kilometres long; areas of outcrop break up the fluted pattern. Several short meltwater channels occur in the lower right of the photo. Caniapiskau River area, northern Quebec. NAPL T201L-164. c) Moderately thick drift with fluted ribbed moraine in the foreground; several eskers cut diagonally across the area. Rock and unpatterned drift occur in the background. Central Quebec-Labrador. NAPL T189R-55. d) Muskeg terrain in the Hudson Bay Lowlands showing bog and fenland, and shallow thermokarst ponds. The Hudson Bay railway crosses the central part of the picture. South of Churchill, Manitoba. NAPL T169R-27.

Chapter 3

QUATERNARY GEOLOGY OF THE CANADIAN SHIELD

Co-ordinated by
R.J. Fulton

SUMMARY

This chapter summarizes the Quaternary geology of more than half of Canada — the Precambrian (Canadian) Shield, as well as the central cratonic basins of Hudson Bay and Foxe Basin. It thus encompasses the supposed areas of inception of the Laurentide Ice Sheet and its precursors, the centres of final retreat of the Late Wisconsinan ice sheet, and large areas of marine inundation that record the isostatic response of the crust to glacial unloading in areas at and locally beyond the limit of ice advance (eastern Baffin Island) and all the way back to the geographic centre of the ice sheet (Hudson Bay).

Quaternary glaciations, especially the last, have created a concentric landscape zonation on the Canadian Shield. The entire shield has been extensively scoured by ice sheets and this process has created hundreds of thousands of lake basins, perhaps cumulatively throughout the Quaternary. Overprinted on this is a zonation of landform-sediment assemblages centred on the final areas of ice retreat in District of Keewatin and Quebec-Labrador. The retreat centres are characterized by large fields of ribbed moraine formed in association with drumlins and flutings. Beyond those are broad zones of thick and nearly continuous drift composed mostly of streamlined till and large, long eskers. The outer shield is mostly bare rock but what drift is there is largely streamlined or oriented along ice flow (e.g., eskers). Hence glacial bedforms and eskers form two giant systems radiating from central Keewatin and central Quebec-Labrador. They do not represent the pattern of either ice or meltwater flow at any single instant in time but rather reflect the ice retreat pattern. No simple concentric zonation of landscapes is associated with the Baffin Sector of the ice sheet.

Over most of the region Quaternary deposits record only events of the last glaciation and of postglacial time because of the self-erasing nature of glaciation. The important exceptions to this are the widespread stratigraphic exposures along the rivers of the Hudson Bay Lowlands, areas beyond the Late Wisconsinan glacial limit along the mountainous eastern rim of the shield on Baffin Island, and exposures at other widely scattered localities.

Fulton, R.J. (co-ordinator)
1989: Quaternary geology of the Canadian Shield; Chapter 3 in Quaternary Geology of Canada and Greenland, R.J. Fulton (ed); Geological Survey of Canada, Geology of Canada, no. 1 (also Geological Society of America, The Geology of North America, v. K-1).

Pre-Sangamonian glaciations, involving both regional and alpine glaciers, are recorded by weathered and morphologically degraded surface tills and erratics beyond the limit of little weathered and morphologically fresh Wisconsinan drift on Bylot Island, along eastern Baffin Island, and in northern Labrador. Probable Illinoian till and glaciomarine sediments are exposed along Pasley River on Boothia Peninsula, and at least three tills occur beneath interglacial sediments in the Hudson Bay Lowlands.

Deposits assigned to the Sangamon Interglaciation are exposed at several localities in the Hudson Bay Lowlands and are known as the Missinaibi Formation. The formation contains marine, fluvial, peat, and lacustrine deposits that collectively record a complete interglacial cycle. Organic detritus in marine sediments on Baffin Island and in glacially deformed beds near the Barnes Ice Cap are thought to date from Sangamonian time and contain <u>Betula</u> (dwarf birch) at sites well north of its present range. On Boothia Peninsula a Sangamonian fluvial gravel also contains macrofossils of <u>Betula</u> beyond its present range.

The initial advances of the Laurentide Ice Sheet after the Sangamon Interglaciation are not dated closely. The largest advances along the eastern Baffin fiords, recorded by fresh lateral moraines extending to the fiord mouths, occurred at a time beyond the range of radiocarbon dating and may date to middle Sangamonian or Early Wisconsinan time. Amino acid ratios and species composition of marine molluscs in sediments related to these advances indicate that glaciation in the eastern Arctic may have been triggered by warmer than present conditions associated with a warming influence of the West Greenland Current on the Canadian side of Baffin Bay. Amino acid ratios also indicate that, if the advances occurred after marine oxygen isotopic stage 5e, the integrated mean annual temperatures in the eastern Arctic since stage 5e have been 2° to 5°C warmer than present.

Except for alpine nunataks in the northeast, the Laurentide Ice Sheet probably covered all of the Canadian Shield during the Early Wisconsinan as it is thought to have extended across St. Lawrence Valley by 75 ka or shortly after. As yet, unlike areas south and west of the Canadian Shield, no subtill nonglacial sediments have yielded finite Middle Wisconsinan ages. As many as four tills, however, overlie the Missinaibi Formation. Erratic shells in the younger of these tills have amino acid ratios indicative of Middle Wisconsinan ages. Hence, it is possible that ice cover was not continuous throughout the

Wisconsinan at the geographic centre of the ice sheet. Indeed, as many as three openings of Hudson Bay may have occurred since oxygen isotopic stage 5e but before the postglacial. Analogy with West Antarctica may suggest an inherent instability of the Laurentide Ice Sheet, which was marine based at its centre, due to absence of buttressing ice shelves in the Labrador Sea and Baffin Bay.

At the Late Wisconsinan glacial maximum the ice sheet consisted of three sectors, each with a complex of ice divides radiating from central domes. By 11 ka the margin of the ice sheet was nearly coincident with the edge of the Canadian Shield. At approximately 8 ka the sea penetrated Hudson Bay and the Laurentide Ice Sheet ceased to exist as a contiguous mass. Keewatin Ice remnants disappeared shortly after 8 ka; Foxe Ice held a near-maximum configuration until shortly before 7 ka when the sea extended into Foxe Basin, while final remnants of Labrador Ice persisted until about 6.5 ka.

Families of relative sea level curves arranged along transects from centres of glacial loading to the glacial limits indicate the general form of postglacial crustal delevelling and isostatic relaxation. The relative sea level record is still poorly dated, especially in Hudson Bay and Foxe Basin, but apparently each of these major regional centres of the ice sheet has a corresponding uplift centre. Centres of uplift seem to have migrated in delayed response to migration of ice divides but precise cause-effect relationships are obscured by an ambiguous data base.

THE LAURENTIDE ICE SHEET AND AN INTRODUCTION TO THE QUATERNARY GEOLOGY OF THE CANADIAN SHIELD

A.S. Dyke, J-S. Vincent, J.T. Andrews, L.A. Dredge, and W.R. Cowan

DEFINITION OF REGION

The Canadian Shield forms the stable cratonic core of North America and occupies about 4 828 000 km², more than half of Canada. It consists of Precambrian rocks, with an extensive cover of flat-lying Phanerozoic sedimentary rocks in the Hudson Bay, Foxe Basin, and Hudson Strait areas, which are overlain by a variable thickness of Quaternary deposits, mainly of glacial and marine origin. In this chapter, the Canadian Shield is subdivided into four areas, each thought to have been affected primarily by dynamically different parts of the Laurentide Ice Sheet during and following the last glacial maximum (Fig. 3.1). The southeastern Canadian Shield area comprises Quebec and Labrador (region situated east of the Harricana Interlobate Moraine). The southwestern shield area comprises Ontario, eastern Manitoba, and adjacent Hudson Bay (region situated between the Harricana Interlobate Moraine and the Burntwood-Knife Interlobate Moraine). The northwestern shield area comprises the northern Prairie Provinces, District of Keewatin, and parts of the districts of Mackenzie and Franklin (area situated north and northwest of the Burntwood-Knife Interlobate Moraine and west of the Gulf of Boothia). Finally, the northeastern shield area comprises Baffin Island, Melville Peninsula, Southampton Island, and Foxe Basin.

CLIMATE, PERMAFROST, AND VEGETATION

The climate of this region is highly variable due to the large extent of the region, to topography, to effects of surrounding water bodies, and to air mass tracking. The climate falls into three major divisions: continental temperate, boreal, and tundra. In all seasons, average temperatures in general decrease from south to north. In July, the average daily maximum temperatures are about 25°C in the Ottawa area in

Figure 3.1. Location map of the Canadian Shield and subdivisions as used in this report.

Dyke, A.S., Vincent, J-S., Andrews, J.T., Dredge, L.A., and Cowan, W.R.
1989: The Laurentide Ice Sheet and an introduction to the Quaternary geology of the Canadian Shield; in Chapter 3 of Quaternary Geology of Canada and Greenland, R.J. Fulton (ed.); Geological Survey of Canada, Geology of Canada, no. 1 (also Geological Society of America, The Geology of North America, v. K-1).

the southeast, 10°C on Victoria Island in the northwest, and 5°C on the coast facing Baffin Bay. In January average daily minimum temperatures in the same areas are, respectively, −5°C, −30°C, and −25°C. Average precipitation generally decreases from southeast (1000 mm/a) to northwest and northeast (100 mm/a).

Permafrost conditions prevail over much of the region (Brown, 1967; Fig. 3.2). Permafrost extends to depths of more than 500 m in northern Ungava Peninsula, Somerset Island, and Bathurst Inlet (Taylor and Judge, 1979). Southwards, the thickness of permafrost decreases to about 60 m in the Churchill area (mid-north), which lies near the limit of continuous permafrost, and gradually thins out in the peatlands and forested bogs farther south. Towards the southernmost limit of discontinuous permafrost, small islands of thin permafrost are found in peatbogs, on north-facing slopes, and on high summits. Palsas, peat plateaus, thermokarst lakes, patterned ground (mudboils, polygons, stripes), solifluction lobes, rock heave features, and blockfields are the principal manifestations of permafrost. The role of frost action is particularly evident on frost shattered bedrock and in areas of thick, fine grained sediments. In bedrock areas, presence of permafrost is indicated by occurrence of ice in rock pores, and of lenses and veinlets of ice in fissures.

Four main vegetation zones occur on the Canadian Shield (Hare, 1959; Rowe, 1972; Payette, 1983; Fig. 3.2): a southern mixed coniferous and deciduous forest zone, a closed boreal forest zone, a forest/tundra zone, and an arctic tundra zone. In addition, much of the Hudson Bay Lowlands and adjacent Canadian Shield is mantled extensively with peat and constitutes one of the world's largest wetlands. (For additional information on peat and wetlands see Tarnocai, 1989 and National Wetlands Working Group, 1981).

BEDROCK GEOLOGY

Precambrian rocks of the Canadian Shield have been subdivided into structural provinces based on internal structural trends and style of folding. Each structural province consists of numerous belts of stratified rocks, metamorphosed and deformed to varying degrees, and areas of crystalline intrusive and metamorphic rocks (see no. 7 of the *Geology of Canada*). Figure 3.3 shows the Precambrian rocks subdivided into four major groups: belts of greenstones and metamorphosed sediments which include gneisses, schists, and volcanic rocks; high grade metamorphic and basic intrusive rocks which include granulites, charnockites, gabbros, diorites, and anorthosites; granitic rocks which include granites per se, allied plutonic rocks, granitic gneisses, and syenites; and Precambrian sediments, which include variably metamorphosed Proterozoic sandstones, greywackes, carbonates, and shales. Phanerozoic limestones and related sedimentary rocks (undifferentiated sediments in Fig. 3.3), which surround the Canadian Shield and occupy several basins within it, possibly at one time completely covered the shield.

Only a few rock types are sufficiently distinctive to make them useful in determining directions of ice flow and loci of ice dispersal. Red erratics of the Proterozoic Dubawnt Group in western Keewatin have been traced over a distance of 800 km to Coates Island in northern Hudson Bay (Shilts, 1980). Another train of Proterozoic erratics from the southeastern part of Hudson Bay has been traced at least 1500 km to southern Manitoba and North Dakota and possibly occurs as far west as southeastern Alberta. Paleozoic and younger rocks bordering the shield are also valuable ice flow indicators where carried onto Precambrian rocks. Major dispersal trains of these have been traced from Hudson Bay to Lake Superior (this chapter, Dredge and Cowan, 1989), from Foxe Basin across Baffin Island (Sim, 1964; Andrews and Miller, 1979) and across Melville Peninsula (Sim, 1960a; Andrews and Sim, 1964), or from M'Clintock Channel across Prince of Wales Island and onto Somerset Island and Boothia Peninsula (Dyke, 1983, 1984). Smaller Paleozoic outliers near Lac Saint-Jean (Dionne, 1973) and Lac Témiscamingue (Veillette, 1986a) provide distinctive dispersal trains which have been traced onto Precambrian rocks. Although erratic tracing has been useful for understanding configuration of ice sheets, problems in interpretation exist. The main difficulties arise where erratics may be recycled from older glacial deposits, where transport may have been accomplished by ice flow in more than one direction during a single glaciation, and where erratics might come from still unmapped rock formations.

PHYSIOGRAPHY AS A CONTROL ON GLACIATION

The broad physiographic elements of the Canadian Shield have been mapped by Bostock (1970a) and described and analyzed in various reports, among them, Bird (1967) and Ambrose (1964). As well, physiographic evolution is treated in the regional reports that form the body of this chapter. The features of the Canadian Shield that are of primary glaciological importance are those that asserted a major influence on the pattern of growth and decay of the Laurentide Ice Sheet (and of its precursors) and on the functioning of the ice sheets. Those physiographic elements of import are: (1) the uptilted eastern rim, (2) the high plateaus of Labrador-Ungava and Baffin Island, (3) the lower plateau of Keewatin-Mackenzie, and Ontario, (4) the large cratonic basins of Hudson Bay and Foxe Basin, and (5) the deep channel of Hudson Strait (Fig. 3.2).

The scenic uplifted eastern rim resulted from topographic development during the Tertiary and Cretaceous. Consequent upon the rifting-open of the North Atlantic Ocean, the rim was tectonically raised as much as 2000 m in the region northward of the middle Labrador coast. Deep fluvial dissection of the rim resulted from rejuvenation of an integrated dendritic stream system, the ancestors of our east coast fiords. The fluvial valleys were then overdeepened by outlet glaciers which drained a succession of ice sheets that were unable to overtop the high interfluves. The high interfluves are heavily glacierized today, primarily by thousands of small ice caps, cirque glaciers, and transection glaciers averaging a few kilometres long, although many cirques, particularly south-facing ones, today stand empty. Slow erosion by these cirque and valley glaciers throughout the Quaternary, and likely the late Tertiary, gave rise to the spectacular alpine scenery of northeastern Canada.

The plateaus of Labrador-Ungava, Baffin Island, District of Keewatin, and Ontario (Fig. 3.2) played a vital role in initiation of the Laurentide Ice Sheet and its precursors. It is now generally accepted that ice sheet inception was triggered by a lowering of the regional snowline to the level of the plateaus (Ives, 1957; Ives et al., 1975). In fact, such a lowering during the Little Ice Age (about 300 years

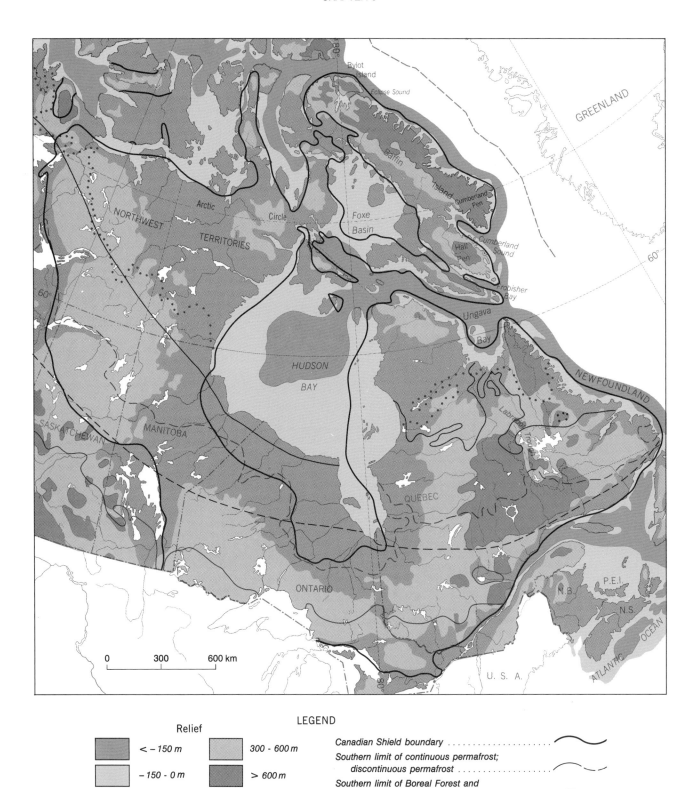

Figure 3.2. Map of the Canadian Shield showing relief, limits of permafrost, and vegetation zones.

QUATERNARY GEOLOGY — CANADIAN SHIELD

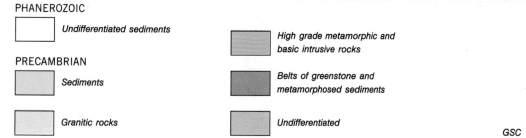

Figure 3.3. Bedrock geology of the Canadian Shield.

ago) saw permanent snow and thin ice fields extend over much of the plateau of Baffin Island (Ives, 1962); this provided an intriguing preview of the initial stage of ice sheet inception. Lamb and Woodroffe (1970) referred to this event as an "abortive glaciation". The Keewatin-Mackenzie plateau is lower than the plateaus adjacent to the North Atlantic and is farther removed from sources of moisture. For these reasons it is probably less close to the glaciation threshold. Nevertheless, there is strong evidence that the area acted as the initial gathering ground for the western half of the Laurentide Ice Sheet (Tyrrell, 1898b). Although Flint (1943) challenged Tyrrell's conclusions and suggested instead that the ice sheet had spread westward across Hudson Bay, Keewatin and western Canada from original kernel areas (cirques) in the eastern mountain rim, Tyrrell's evidence stands today with little essential change whereas Flint's concept stands without any geological substantiation.

The two large shallow epicontinental seas of Hudson Bay and Foxe Basin are striking features of the Canadian Shield. These shallow seas were considerably deeper and larger during early postglacial time and have been constantly shrinking since deglaciation as the land rebounds, following removal of ice load. A considerable amount of rebound has yet to occur, an estimated 150 to 300 m (Andrews, 1970; Walcott, 1970), and upon completion these inland basins will be largely dry assuming sea level remains constant (Fig. 3.2). The point of importance here is that at the beginning of a glacial cycle (inception) these basins may have been largely dry land, making it relatively easy for ice to spread into and fill them from surrounding plateaus. These structures apparently have maintained the form of broad shallow basins since early Paleozoic time because they are partly filled with Ordovician to Devonian sediments and Cretaceous deposits which dip radially inward towards the centres of the basins. The epicontinental seas, however, appear to be ephemeral Quaternary features; the bottom of Hudson Bay contains a drowned, integrated dendritic drainage system of probable Tertiary age, showing that most of the basin was emergent for a long interval prior to the Quaternary. These drainage systems would have been reactivated during any interglaciation of sufficient duration to allow emergence to be completed. The location of these glacial isostatic seas near the central region of the Laurentide Ice Sheet profoundly influenced the process and timing of the last deglaciation and in all likelihood all previous deglaciations. Hudson Strait, which connects the twin basins to the Labrador Sea also played a pivotal role. Because the central parts of the ice sheet were grounded below sea level, a plentiful supply of ice to Hudson Strait was essential in preventing destabilization of the ice mass (see below). With the possible exception of the Gulf of Boothia, Hudson Strait is the only corridor of access along which a marine calving bay could allow deep water to penetrate to the central zone of the ice sheet.

Another striking feature of the Canadian Shield, a series of troughs and large elongate depressions, occurs at its perimeter. Proceeding clockwise from the southeast, these features include St. Lawrence Valley, Great Lakes, Lake Winnipeg, Lake Athabasca, Great Slave Lake, Great Bear Lake, Coronation Gulf, and Queen Maud Gulf. An equivalent submerged feature — the Labrador Marginal Trough — occurs at the Canadian Shield edge off the coast of Labrador (Piper et al., 1989). These major topographic features exerted substantial control on the style of glacial advance and retreat. To some extent there is a danger of confusing cause and effect here because the basins are in some measure the result of glacial overdeepening and consequently their configuration and the role they played as controls on ice flow would have evolved with each glaciation.

The extent to which the Canadian Shield and its cratonic basins have been modified by glacial erosion has been a subject of disagreement and debate, particularly during the past decade. On the one hand, White (1972) has proposed an eloquently simple hypothesis that accounts for most of the geography of Canada sketched above. The starting point of White's hypothesis is that it is not a matter of coincidence that the major northern hemisphere ice sheets were centred on Precambrian shields. He proposed that the central basins (Hudson Bay, Foxe Basin) are glacial features produced by excavation of about 1000 m of bedrock from beneath the ice sheet centre. He further proposed that the shield had been covered by sedimentary rocks until the Quaternary, when it was exhumed by deep glacial erosion, in this light making the shield a Quaternary feature. On the other hand, Sugden (1976a) countered by summarizing most earlier views on the subject. He pointed out the great antiquity of Hudson and Foxe basins, the fact that they are still floored by Paleozoic strata, the existence of Tertiary drainage systems on the floor of Hudson Bay, widespread pre-Quaternary erosion surfaces on both the Canadian Shield and adjacent cover rocks, and the fact that the oldest Quaternary tills contain shield clasts, all of which are incompatible with White's or any other hypothesis of deep glacial erosion. In Sugden's view, and that of most others including the authors of this chapter, the Canadian Shield has experienced net Quaternary glacial erosion of only some tens of metres on average, a depth which perhaps did not greatly exceed that of the Tertiary weathering zone (regolith mantle). This, in turn, suggests the possibility that the first glaciation may have accomplished a greatly disproportionate amount of total Quaternary glacial erosion.

The effects of erosion and deposition, at least that accomplished by the last major ice sheet, were not spatially uniform. If we take the distribution of bare or only thinly veneered bedrock (Fig. 3.4) to represent areas of glacial erosion, the distribution of such terrain suggests that more erosion has occurred near the outer margin of the Canadian Shield than near the central regions. This pattern is particularly clear in the belt extending from Coronation Gulf to Ottawa Valley. The Quebec-Labrador Peninsula, however, displays its own eroded perimeter, inset by a central zone of drift cover. The main areas of drift cover in Keewatin and in Quebec-Labrador coincide with areas that lay beneath central zones of major component parts of the Wisconsinan Laurentide and earlier ice sheets, although we cannot assume that there is a causal relationship here. Outside of the Interior Plains, the thickest and most continuous drift cover in Canada underlies the Hudson and James Bay lowlands. That region has experienced net and incremental deposition for perhaps much of Quaternary time, resulting in a stacking of sediments of glacial and nonglacial character. Some of the debris composing these sediments is known to have come from east of Hudson Bay, but the sediments are also charged with debris, mainly calcareous, from sedimentary rocks of the basin. Hence, although the now emergent part of the lowlands has experienced net Quaternary deposition, deposition of successively younger layers of drift, each calcareous, during successive glacial events indicates either substantial but incomplete recycling of older drift layers or existence of a major region of net erosion of Paleozoic strata offshore.

The Canadian Shield is speckled with innumerable small lakes (Fig. 3.4) which can be used as a measure of glacial erosion, provided that basins in bedrock are distinguished from basins in drift. Most of those in the zone of thin or no drift cover occupy bedrock basins and are products of glacial erosion. Those within the zone of drift cover include both erosional rock basin lakes and lakes occupying basins within drift, which resulted primarily from uneven deposition from ice. Although it is difficult to estimate the

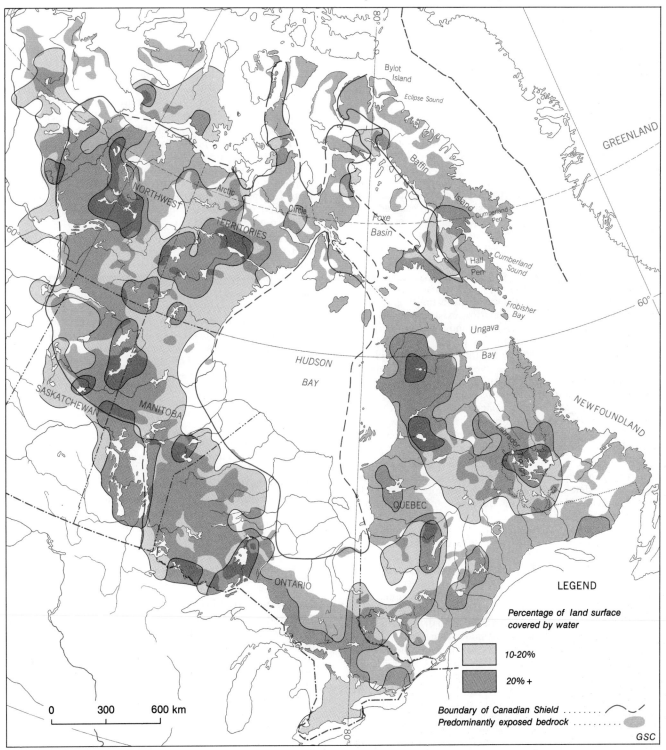

Figure 3.4. Percentage of the Canadian Shield covered by water (lakes) and areas underlain predominantly by bare rock or thin drift.

proportions of these two types of lakes, we know that lakes are drift-floored over certain broad areas. For example, the lakes of southern Victoria Island occur within drumlin fields and belts of hummocky disintegration moraine. Many of the lakes of Keewatin, northern Manitoba, and central Quebec-Labrador occupy inter-rib depressions in vast tracts of ribbed moraine, whose distribution is highlighted on the Glacial Map of Canada (Prest et al., 1968).

The small lakes of Canada have received less attention than most other mesoscale glacial landforms although they are no less important, either academically or culturally. Sugden (1978) was the first to attempt to use the spatial density of these small basins as an index of intensity of glacial erosion. He derived a map, using satellite imagery, which differs little from that portrayed here (Fig. 3.4). Sugden recognized that practically all of the shield comprised a vast landscape of glacial scour and proposed that zones of greatest lake basin density indicated those areas that had experienced most glacial erosion. He employed a computer model to determine the arrangement of thermal regimes at the base of a hypothetical Pleistocene maximum Laurentide ice mass which had a single dome centred over Hudson Bay. He found that, provided the ice sheet held its maximum or near maximum configuration for a period of 50 to 100 ka, basal thermal regimes were arranged in a concentric pattern such that a large circular central zone of warm-melting ice, was followed outward in succession by a narrow ring of regelation (warm-freezing), by a broad ring of cold-based ice, and by a marginal zone of warm-melting. Sugden considered that a spatial correlation existed between the zone of maximum lake basin density and the zone of regelation, hence offering support for his theoretical model and lending some credence to the assumption upon which his experiment was predicated — namely, that most erosion was accomplished during maximum ice conditions. Whether Sugden's conclusions are correct or not, the concept that the subcontinental scale zones of glacial erosion imprinted on the Canadian Shield in the form of lake basins may have been shaped long before inception of the Wisconsinan Laurentide Ice Sheet has intriguing implications.

It is questionable whether the zones of erosion suggested by Sugden do, in fact, support his experimental results, as Sugden freely admits. For example, most of the zone of high lake density on the Quebec-Labrador Peninsula coincides with the wide zone of cold-based ice in the hypothetical ice sheet. Sugden suggests that this is possibly the product of erosion by cold-based ice which was carrying a considerable thickness of debris inherited from the warm-freezing (regelation) zone up ice. If such is the case, however, there clearly are few constraints on interpreting zones of intensity of erosion in terms of basal thermal regime. Still there remains great potential in Sugden's approach, particularly if rock basin lakes are separated from drift basin lakes in the process of mapping the actual erosion zones, and the method has been pursued in more detail in the Foxe Basin region (this chapter, Andrews, 1989). At the scale of the ice sheet, however, we contend that it is presently not possible to use patterns such as those shown on Figure 3.4 to support or refute any particular ice sheet configuration or any particular arrangement of basal thermal regimes. Indeed, it is possible that the patterns are simply the net result of total Quaternary erosion and deposition, that the surface was substantially affected by the ice sheet during multiple buildups and recessions, as well as maxima, and that it will not be possible to separate these various effects.

THE LAURENTIDE ICE SHEET (WISCONSINAN)

Nomenclature of the Laurentide Ice Sheet

The ice sheet that was centred on the Canadian Shield during the last glaciation and its component parts have been referred to in a variety of ways and the spatial hierarchy of terms is not consistent from author to author. Ice sheet nomenclature used in this chapter is specified below.

Laurentide Ice Sheet: an ice sheet of complex morphology, centred on the Canadian Shield during the last glaciation but at times also covering large areas peripheral to the shield. Precursors during earlier glaciations could be referred to as Laurentide ice but not as the Laurentide Ice Sheet.

Labrador Sector, Keewatin Sector, Baffin Sector: major components of the ice sheet that reflect its likely pattern of inception, buildup, and in some cases, maximum configuration (for locations see Fig. 3 of Introduction to this Volume). These subdivisions follow Prest (1984) but here the name Labrador has been substituted for the adjectival term Labradorean and the name of the northern sector is simplified to Baffin (the area of northern ice sheet inception) from Foxe-Baffin.

Labrador Ice, Hudson Ice: components of the Labrador Sector that resulted from bipartition of the sector late in the buildup phase, as discussed below.

Keewatin Ice, Plains Ice: components of the Keewatin Sector that resulted from bipartition of the sector during buildup or maximum; Plains Ice produced regional southeastward flow over Alberta, Saskatchewan, and western Manitoba and northwestward flow over western District of Mackenzie. The western ice mass probably spread from an ice divide that branched off the main divide of Keewatin Ice and extended westward across northern Alberta and southern Mackenzie. Dyke et al. (1982b) referred to this as Caribou Dome because they inferred that the ice divide lay over the Caribou Hills, northern Alberta. The concept is discussed further below.

Foxe Ice, Amadjuak Ice, Penny Ice, Barnes Ice: components of the Baffin Sector that represent dynamically distinct units flowing from secondary centres or ice divides or that became distinct units during retreat (i.e. Barnes Ice).

Ice divide: topographic ridge on an ice mass which induces a divergence of flow lines from a linear axis; may include saddles (i.e.: Keewatin, M'Clintock, and Labrador ice divides).

Ice dome: topographic dome on an ice mass which induces a radial divergence of flow lines.

Inception and Buildup

As stated above, the process of glacial inception is controlled by topography and availability of snow. We accept the major tenets of Ives et al. (1975) regarding the process of inception (instantaneous glacierization, Ives, 1957) and hence the hypothesis that the Wisconsinan Laurentide Ice Sheet grew from plateau-based centres in Baffin Island, Keewatin, Quebec-Labrador, and possibly northern Ontario. This hypothesis is fully in accordance with some of the earliest views on the subject (Tyrrell, 1898b).

Details of the pattern of buildup are practically unknown, but logic and a limited amount of field data suggest that Hudson Bay became infilled by ice from both Keewatin and Quebec-Labrador. If either the Keewatin or the

Labrador ice mass accomplished more infilling than the other, it was probably the one growing from Quebec-Labrador as this was nearer the major moisture source and had a less continental climate, that is, cooler summers. Computer modelling of ice sheet inception and growth by Andrews and Mahaffy (1976), utilizing a three-dimensional ice flow model and plausible snowline lowerings and mass balances, supports the possible predominant role of the Quebec-Labrador centre in infilling the southern and eastern part of Hudson Bay. Hence, it is likely that during buildup phases the Labrador Sector of the ice sheet filled at least southeastern Hudson Bay and covered northeastern Ontario.

Skinner's (1973) work in the James Bay area apparently confirms that Labrador Ice crossed into northeastern Ontario during the Early Wisconsinan. Similar studies on till composition and fabric from northeastern Manitoba and observations from northern Ontario (this chapter, Dredge and Cowan, 1989) suggest that Keewatin Ice also occupied substantial parts of the Hudson Bay Lowlands during the inception phase of the last glaciation. Of considerable controversy is whether Labrador Ice was sufficiently overpowering to cross extensive areas of northern Ontario and central Manitoba. The main geological data favouring extensive Labrador Ice are "dark erratics" which are found in tills of all ages throughout the Hudson Bay Lowlands and are generally considered to have been derived from rocks of the Circum-Ungava Geosyncline underlying Richmond Gulf and outcropping on the Belcher Islands in the middle of the Gulf. These erratics occur in greatest concentrations in a belt between the middle James Bay coast of Ontario and the mouth of Nelson River, Manitoba. There are several plausible interpretations of this set of data, the extremes being: (1) that a single sustained flow of Labrador Ice crossing southern Hudson Bay delivered abundant dark erratics to northern Ontario and Manitoba at a time represented by the oldest till known to contain these erratics, and that these erratics were later incorporated into younger drifts due to successive recycling of sediment, regardless of subsequent ice flow directions; (2) that each and every ice flow event of any considerable duration in that region resulted in a nearly identical pattern of dispersal of dark erratics and recycling of erratics from older drifts was minimal or served only to augment dispersal along the same paths; and (3) that there are multiple sources for the dark erratics. This problem is explored further below.

Stable or unstable Wisconsinan ice cover

A substantial part of the Laurentide Ice Sheet lay over the interior marine basins of Hudson Bay and Foxe Basin and was connected to the Atlantic Ocean by the deep trough of Hudson Strait. Research in Antarctica (cf. Thomas, 1977; Denton and Hughes, 1981) has suggested that ice sheets which are grounded below sea level are inherently unstable if the interior marine-based portion is connected to the ocean via major ice streams. Such ice sheets, it is argued, can be stable only if the ice stream flows into a buttressing ice shelf. The ice shelf provides resistance to forward flow and allows the ice sheet to thicken. If the ice shelf starts to thin or break up, however, the ice stream can accelerate and cause rapid "drawdown" in the marine-based section of the ice sheet.

Until recently it was assumed that ice occupied Hudson Bay and Foxe Basin continuously throughout the Wisconsin Glaciation (Prest, 1970; Andrews and Barry, 1978). The Missinaibi Formation nonglacial beds were considered by most workers to be of Sangamonian age, and no marine sediments or fossils of Wisconsinan age were recognized. This implied a stable Wisconsinan ice cover in the central region of the ice sheet over Hudson Bay. If the Laurentide Ice Sheet extended across Hudson Bay and Foxe Basin during the entire Wisconsin Glaciation, analogy with the current West Antarctic Ice Sheet requires that the ice sheet was buttressed by a large ice shelf which extended across Baffin Bay and the northern Labrador Sea (Denton and Hughes, 1981, 1983). However, evidence from marine cores (Mudie and Aksu, 1984) and from inshore pollen and mollusc data (Mode, 1980, 1985) indicates that an ice shelf was not present throughout the Wisconsin Glaciation. If there was no ice shelf, then the Laurentide Ice Sheet should have been, in theory, inherently unstable and its central marine-based parts may have disintegrated.

Andrews et al. (1980a, 1983b), Shilts et al. (1981), and Shilts (1982a,b) have reported results of amino acid analyses of marine shells collected throughout the Hudson Bay Lowlands and have interpreted these results as indicating, to quote the title of their culminating report, "Multiple deglaciation of the Hudson Bay Lowlands...since deposition of the Missinaibi (Last-interglacial?) Formation". This interpretation suggests an unstable central core to the ice sheet on the time scale of 40 ka and emphasizes the critical importance of the stratigraphic exposures in the Hudson Bay Lowlands to understanding the history and dynamics of the world's largest former ice sheet. The lithostratigraphy and aminostratigraphy of the region are discussed below in the section on the southwestern shield (this chapter, Dredge and Cowan, 1989).

Late Wisconsinan configuration and flow

The surface morphology, and hence flow pattern, of the Laurentide Ice Sheet during the last glacial maximum is a topic of debate. Two major schools of thought exist proposing that (1) the ice sheet had a steady-state configuration with a single central dome located over Hudson Bay, and that (2) the ice sheet was multidomed.

The idea of a steady-state, single-domed ice sheet was first clearly proposed by Flint (1943) and was raised in contradistinction to the ice sheet complex espoused by Tyrrell (1898b; 1913), involving centres in Keewatin, Patricia (northern Ontario), Labrador, and Baffin Island. Tyrrell's views were generally accepted prior to Flints' influential arguments against them. The arguments are inextricably linked to each author's views on ice sheet inception. As outlined above, Tyrrell believed that the ice sheet grew from centres located west, east, and north of Hudson Bay and that these centres remained the primary centres of outflow during each glaciation, including the last. Flint, on the other hand, felt that Tyrrell's "gathering grounds" were unlikely sites of glacier inception because of their low elevation, distance from moisture sources, and warm summers. He proposed instead, in his "hypothesis of highland origin and windward growth", that the ice sheet started with expansion of cirque glaciers in the mountains of eastern Baffin Island and northern Labrador to form piedmont lobes on their western flanks. These were then nourished by "snowfall from relatively warm moist air masses moving across the continent from the south and west" leading to expansion in these directions, until ultimately the margin reached the

Cordillera and its centre lay over Hudson Bay. Flint's view was in wide acceptance for about 35 years, partly because he reiterated his conclusions in a series of highly influential publications (Flint, 1947, 1957, 1971).

Flint's model was almost entirely conceptually based and also based in part on a misunderstanding of the topography of the eastern Canadian Arctic. Ives (1957) pointed out that a mountain slope exists only to the east and that in almost every case, expansion of cirque glaciers would lead simply to calving at tide water and would not produce a vast coalescent system of west-flowing piedmont lobes, as required by Flint's model. He also showed that the receding hemicycle of Flint's model was invalid. Further work in Labrador and Baffin Island led Ives and his colleagues (Ives et al., 1975) to return consistently to a model of inception similar to that of Tyrrell, though more elaborately developed. Also growing out of that work, Ives and Andrews (1963) proposed a major independent centre of dispersal over Foxe Basin and in so doing departed substantially from any view of a single domed ice sheet. Later, however, Andrews (1973) clearly abandoned the concept of an independent centre of dispersal over Foxe Basin during the last glacial maximum. Ives et al. (1975) portrayed a single domed form centred on Hudson Bay along a profile from Labrador to Keewatin and ascribed all land-based centres of outflow to deglacial phases. This was in line with what was then the most recent interpretation of the glacial history of Keewatin (Lee, 1959a). Thus, by this point, an intriguing double irony had developed: Flint had rejected Tyrrell's ice sheet configuration because he did not find his pattern of inception to be tenable; Ives et al. (1975) had shown Flint's concept of inception and growth to be untenable and had proposed concepts very similar to those of Tyrrell. However, they did not find new favour in Tyrrell's concept of ice sheet configuration. They continued to favour a single domed configuration despite even their own proposition of a major independent dome over Foxe Basin and Andrews' (1970) delineation of a double celled maxima in postglacial uplift.

A model of the Late Wisconsinan Laurentide Ice Sheet similar to Flint's was proposed by Denton and Hughes (1981). Theirs is a steady-state, integrated, two dimensional model using as input the position of the ice margin, positions of major ice streams, and positions of ice divides, all of which are assumed to be known. Their model shows a radial flow outward from Hudson Bay. The primary difficulty with it lies in the fact that there is no geological evidence of either westward or eastward dispersal of drift from Hudson Bay. Instead, strong dispersal patterns towards the bay from both sides are well known (see regional reports below).

Because the single-dome model predicts a pattern of drift dispersal contrary to known patterns, a multi-dome model is more realistic. Such models have been proposed by Shilts et al. (1979), Andrews and Miller (1979), Shilts (1980), and Dyke et al. (1982b) based on geological data. There is general agreement that domes or ice divides existed over Foxe Basin, Keewatin and adjacent arctic archipelago, and Quebec-Labrador. There is, however, controversy over ice sheet configuration over Hudson Bay. Dyke et al. proposed an additional dome over southwestern Hudson Bay and northwestern Ontario during the Late Wisconsinan maximum and during retreat, in contrast to the model of Shilts (1980) which showed ice flowing from an ice divide over Ungava, across southern Hudson Bay to Lake Winnipeg, and presumably to the glacial limit on the American Plains. The reasons for proposing a Hudson centre were outlined by Dyke et al. (1982b) and are addressed again in the section on the southwestern shield. The idea has not been widely accepted, however, primarily because of the claim that a model involving a Hudson centre of outflow cannot explain dispersal of "dark erratics" into northern Ontario and Manitoba from the Belcher Islands and vicinity (Shilts, 1985).

As mentioned above, a possible explanation of the dark erratics train is that it was produced by ice flow during glacier buildup whereby ice growing from a centre on the Quebec-Labrador plateau crossed southern Hudson Bay and invaded Ontario and Manitoba. Hence, in that sense, the Labrador Sector of the Laurentide Ice Sheet could have covered an area identical to that proposed by Prest (1984). We argue, however, that after Hudson Bay (and presumably Foxe Basin) was sufficiently infilled to generate an eastward flowing ice stream in Hudson Strait, that ice stream produced widespread drawdown of the ice surface over the northern part of Hudson Bay, and consequently a northeastward ice flow. This convergent northeastward flow eventually carried ice from Keewatin on its western side, from southern Hudson Bay through its central part, and from Ungava on its eastern side. The predominant, and apparently longest sustained ice flow across the Ottawa Islands (Andrews and Falconer, 1969) was northeastward, recording flow from southern Hudson Bay. Sugden (1978) commenting on the data of Andrews and Falconer (1969) makes this point quite clearly: " ...the streamlined outline of the islands reflects maximum ice flow conditions out of Hudson Bay, and yet overlapping striations of completely different orientation tell of the phases of deglaciation". If the southerly to southwesterly flow across Ontario and Manitoba operated at the same time as the convergent but generally northeasterly flow beneath the ice-surface trough over northern Hudson Bay, then there must have been an ice divide over the southern part of Hudson Bay or over northern Ontario (Dyke and Prest, 1987; Map 1702A).

In summary, we propose invasion of northeastern Ontario and parts of northwestern Ontario by the Labrador Sector of the ice sheet during buildup, followed by a bipartition of the sector to produce Hudson and Labrador centres, essentially as a product of development of the Hudson Strait ice stream. Through this sequence of events dark erratics could have been carried from the Belcher Islands to a final site of deposition anywhere within the "Labrador Sector", but still could have been deposited during retreat of Hudson Ice. It remains possible, however, that some of these erratics were recycled from older drift sheets, and there is a possibility that they may have come from more than one bedrock source (this chapter, Dredge and Cowan, 1989).

Shilts (1985 and earlier) has been the most vociferous opponent of an ice divide over southern Hudson Bay. He maintains that there is no geological evidence of such a divide. This is an inappropriate dismissal of a hypothesis which can as easily account for the major patterns of drift dispersal, especially when extended to include westward dispersal across southern Hudson Basin during glacial buildup episodes, as can his own model. However, in his latest conceptualization of the "possible ice flow configuration early or late in development of Laurentide Ice Sheet" Shilts (1985, his Fig. 3.3) shows an ice divide (divergence of flow lines) extending across southern Hudson Bay as an extension of the Keewatin Ice Divide.

Experiments by Fisher et al. (1985) are useful in examining the influence of bed conditions and topography on ice sheet configuration, and in particular in examining the range of possible ice surface configurations in the critical Hudson Bay area (cf. Boulton et al., 1985). Because there is empirical evidence for low gradient ice surface profiles and for substrate deformation in the Interior Plains region (Mathews, 1974), Fisher et al. simulated deforming (soft) bed conditions there by specifying very low basal yield stresses in the ice. Over the Canadian Shield they specified "normal" yield stresses (nondeforming beds); over Hudson Bay they tested the possible influence of bed conditions by specifying normal yield stresses in one experiment and low yield stresses in another. Several results are of critical interest to the present discussion:

(1) Both experiments yielded multidomed ice sheets with primary ice divides or domes over Foxe Basin, Quebec-Labrador, southwestern Hudson Bay/northern Ontario, Keewatin, and the northern Prairies.

(2) The ice divide over the northern Prairies separated regional southeastward flow across the Prairie Provinces from northwestward flow across District of Mackenzie. This regional flow pattern on the Interior Plains resulted from enhanced sensitivity of low yield stress ice to regional topography, not to deflection of westward flowing Keewatin Ice by the Cordillera (cf. Klassen, 1989; Vincent, 1989a). It is proposed that this dynamically distinct ice mass, part of the Keewatin Sector of the Laurentide Ice Sheet, be referred to as Plains Ice (Dyke and Prest, 1987) as this is more in keeping with the nomenclature system outlined above than the term Caribou Dome suggested by Dyke et al. (1982b).

(3) The experiment incorporating a nondeforming bed in Hudson Bay generated an ice dome located over southwestern Hudson Bay, linked by primary ice divides to domes over Keewatin and Quebec-Labrador. Introduction of a deforming bed condition in Hudson Bay created an ice surface trough over the bay and moved all ice centres landward. It did not eliminate the controversial Hudson centre but merely shifted it onto northern Ontario. This resilience of the Hudson centre to changes in bed condition is highly significant because one of the purposes of the experiment was to see if very low yield stresses in the Hudson Bay area would produce the east-west asymmetry of Labrador Ice suggested by the model of Shilts (1980). It failed to do so because of the overwhelming importance of drainage through Hudson Strait on configuration of the central parts of the Laurentide Ice Sheet. It must however be noted that even though large volumes of ice are assumed to have flowed through Hudson Strait, sedimentation rates and influx of sand-sized particles during oxygen isotope stage 2 are rather low (Fillon, 1985).

Deglaciation

The details of the record of deglaciation are treated separately in four regional sections of this chapter. However, the main points of these discussions are brought together here in the form of a series of maps. Map 1702A is a summary map depicting approximate extent of all Canadian ice sheets at the Late Wisconsinan maximum and the total retreat picture of the Laurentide Ice Sheet; Map 1703A is a series of maps, each showing the paleogeography at a specific time.

The nomenclature applied to components and features of the Laurentide Ice Sheet on these maps follows the hierarchy and definitions set out above.

The retreat record of concern to this chapter is that part which occurred after 10 ka in the northeast and after 12 ka in the northwest, southwest, and southeast. Earlier retreat, mostly in the interval 14 to 10 ka involved deglaciation of the Interior Plains and St. Lawrence Lowlands. Interpretation of deep sea plankton abundance and oxygen isotope records (Ruddiman and McIntyre, 1981; Duplessy et al., 1981) have been viewed as forecasting that deglaciation occurred in two discrete steps with the first step dated either between 15 and 13 ka (Duplessy et al., 1981) or 13 and 11.5 ka (see discussion in Ruddiman and Duplessy, 1985) and the second step between 10 and 8 ka. The oxygen isotope evidence suggests that 50% of the volume of northern hemisphere ice sheets disappeared during the 2-3 ka of the first deglaciation step. Ruddiman and McIntyre (1981) suggested that because relatively little ice marginal recession occurred during that interval, the first deglaciation step required a massive downdraw and collapse of the Laurentide Ice Sheet, primarily focused on drainage through Hudson Strait. Although others (e.g., Andrews and Peltier, 1976) have also suggested a "collapse" of the hypothetical central dome of the ice sheet at about the same time, we can only repeat that there is no known direct geological evidence of such a single central dome.

In attempting to understand deglaciation dynamics and to trace the flow paths of transported debris, one important consideration is the role and frequency of surges. A surge is an unusually rapid though short-lived flow event leading to an advance. Large, and in some cases repeated, surges have been invoked to account for the formation and advance of the Viscount Melville Sound Ice Shelf (Hodgson and Vincent, 1984; Dyke, 1987), for the Cochrane readvances south of James Bay (Hardy, 1977), for many other advances into glacial lakes Agassiz and Ojibway (this chapter, Dredge and Cowan, 1989), and for low-profile extremely lobate ice margins on the southern Interior Plains (Clayton et al., 1985).

Surges have been postulated both for ice margins retreating in deep water and for those retreating on land. Mechanisms responsible for triggering surges along dry margins would likely operate as effectively along aquatic margins. However, calving and buoyancy are additional mechanisms that operate along aquatic margins and may trigger surges. Calving can produce margins steeper than can be supported by the shear strength of ice and hence lead to accelerated flow. If a calving bay forms in the margin, flow from a wide area of the ice sheet is drawn into it, leading to ice streaming and readvance. Readvance produces general thinning and if the ice should thin to the point where the margin becomes buoyant, resistance to flow is eliminated and the rate of advance is restrained only by rate of supply in the feeding ice stream. The result is a surging ice shelf which floats in deeper water and grounds on rises where it remoulds (flutes) the substrate surface, as did the Cochrane surges.

Surges along dry margins are thought to be triggered by episodic deformation of the glacier bed or accumulation of a subglacial water layer. Many saturated sediments have shear strengths less than that of ice and will deform under an ice sheet before the ice surface attains a normal profile.

However, before a surge can develop, some means of preventing substrate deformation is required in order to allow the profile to "oversteepen". This temporary stiffening of the substrate could be caused by permafrost. The main evidence for surges of this sort that has been cited is low gradient, extremely lobate ice marginal features of the Interior Plains (Mathews, 1974; Clayton et al., 1985). Occurrence of similar surges have yet to be postulated for the Canadian Shield. Shaw and Kvill (1984), however, suggested that the Athabasca drumlin field was formed by giant subglacial sheet floods that triggered surging.

QUATERNARY GEOLOGY AND LANDUSE ON THE CANADIAN SHIELD

Extractive industries and transportation are the main landuse activities conducted in the Canadian Shield region. Consequently the main properties of Quaternary material of economic importance are those that control their suitability as foundations and their suitability and availability as fill and aggregate. The presence of certain Quaternary deposits can be a hindrance to construction; but absence of unconsolidated materials creates problems where construction activity requires installation of costly footings in bedrock, extensive cuts in rock, or long distance transport of aggregate and fill.

Tills, sands, and gravels — the most abundant Quaternary deposits on the Canadian Shield — generally provide stable, well drained foundation conditions. However, some sandy ablation tills, which occur in northern Ontario and Manitoba and central Quebec, have low density and high void ratio and consequently low bearing strength and are subject to liquefaction when disturbed while saturated. Fine grained materials found in former basins of glacial lakes and areas submerged by the sea are generally poorly drained, have high water content and, particularly in the case of marine silts and clays, may be sensitive to disturbance. Thick organic sediments, which are abundant south of Hudson Bay and locally in other parts of the forested area of the shield, provide particularly difficult foundation conditions. Ice content of all sediments may be high in areas of permafrost, but fine grained sediments and organic deposits create particularly difficult conditions in this situation.

Granular aggregate is abundant in most parts of the Canadian Shield but may be covered by fine grained sediments in areas of marine and lacustrine sedimentation and be very coarse grained and widely distributed in areas characterized by abundant outcrop. Glaciofluvial, morainic (till), fluvial, and nearshore glacial lacustrine and marine Quaternary deposits are commonly used aggregate sources. Despite a general abundance of granular deposits in the shield area, materials of suitable grading and quality are not always bountiful. The dominant by-products of glacial processes on Precambrian rocks are fine sand and boulders. Consequently many landforms that would appear suitable for granular aggregates (eskers or moraines) have a short fall of medium and coarse sand and fine pebbles, and an excess of fine sand. Beach deposits locally cover extensive areas and in many cases consist of coarse sand and fine pebbles but they generally are thin so extensive areas must be scraped to obtain appreciable volumes of aggregate. In most areas coarse textured till is a source of low grade aggregate but must have fines removed and the coarse fraction crushed if it is to be used as a higher grade product.

Precambrian rocks normally provide sound, nonreactive materials but in some areas aggregates may contain biotite-rich lithologies which are subject to granular disintegration, soft limonitic rocks, or siltstones and slaty lithologies.

Quaternary deposits which mask ore bodies are often considered an inconvenience to bedrock mapping and mineral exploration and exploitation. In recent years, glacial deposits, particularly tills, have been receiving considerable attention as a means of locating ore bodies (Nichol and Bjorklund, 1973; Shilts, 1976, 1984b). Detailed accounts of the use of drift prospecting for locating ore bodies on the Canadian Shield can be found in Bradshaw (1975) and DiLabio (1989).

Due to the short length of the growing season in most parts of the Canadian Shield, agriculture is generally limited to the growing of forage and feed grain for local dairy and meat production. In addition, the texture of most soil parent material in the Canadian Shield is too coarse to permit development of soils with good moisture retention characteristics, and stoniness and topography are considerable hindrances to cultivation. In limited southern parts of the shield soils developed on fine grained marine or glacial lake deposits are suitable for cultivation. The main areas where these occur are in the Lac Saint-Jean area and in narrow valleys of the southern Laurentian Highlands, on the glacial lake deposits in the Clay Belt of Quebec and Ontario, and in limited areas north of Lake Huron and northwest of Lake Superior. Additional general information on the nature of soils in the shield area may be obtained from Clayton et al. (1977), Canada Soil Survey Committee (1978), and Acton (1989).

The Canadian Shield occupies a large part of the Boreal Forest Zone of Canada and forestry is an important industry in all southern parts of the area. In more northerly parts of this forest belt productivity is adversely affected by a short growing season and the presence of permafrost which limits the depth of rooting and impedes soil drainage. The presence and nature of Quaternary deposits are critical to tree growth and forest productivity. Within any climatic zone, the best stands of trees are found on thick, moderately well drained Quaternary deposits. In the areas where Quaternary materials are absent or thin, trees are not present or their growth is slowed. Excessively stony drift reduces forest productivity, the presence of large boulders impedes harvesting activities, and bouldery drift and presence of near surface rock can make construction of forestry roads difficult and expensive.

A subject of considerable concern in recent years is the capacity of soils to buffer the effects of acid solutions created in the natural environment by industrial pollutants (Shilts, 1981). Quaternary deposits derived from quartz-rich granitic and gneissic shield rocks have a low buffering capacity, while those derived from metasedimentary (marbles) or volcanic rocks have a somewhat higher capacity. The presence of Paleozoic carbonate material eroded from the Hudson Bay region in tills on the shield in Ontario and Manitoba, and the greater availability of carbonates and other substances with buffering capabilities in fine grained glacial lake and marine deposits have imparted to some regions a high buffering potential relative to other areas of the shield.

Landslides and flowslides are common along the steeply and deeply incised valley walls developed in marine clays of the Tyrrell, Champlain, Laflamme, and Goldthwait seas or the glacial lake clays of lakes Barlow and Ojibway (see regional chapters for extent of these water bodies). The loss of life and property associated with the Saint-Jean-Vianney flow slide in the Lac Saint-Jean area, is an example of the catastrophic nature of the process (Tavenas et al., 1971). In addition, gulleying of fine grained waterlain sediments can reach intense levels. Perhaps the most susceptible deposits to this process are the widespread glacial marine silts that extend from Bathurst Inlet to Boothia Peninsula. River channel migration and bank erosion can cause problems in areas underlain by sand and finer sediments. For example, the severe erosion of the banks of Île du Gouverneur, at the mouth of La Grande Rivière, has forced the displacement of the Fort George community to the safer site of Chisasibi, and many other communities constructed on floodplains or terraces adjacent to large rivers have similar problems. The frost processes, particularly in the part of the shield underlain by permafrost, are very disruptive; they are responsible for the sorting, mixing, heaving, cracking, and frost shattering of both consolidated and unconsolidated surface materials. In addition, thermokarst has locally been active. The growth or decay of ground ice bodies, such as massive segregated ice and ice wedges, have substantially altered tracks of land particularly in areas of fine grained sediments on Prince of Wales and Somerset islands (Dyke, 1983) and areas of organic sediments southwest of Hudson Bay (Dredge and Nixon, 1979).

In summary, terrain conditions on the Canadian Shield are quite variable. Areas of abundant glaciofluvial deposits, and hence good supplies of aggregate, lie adjacent to areas where the surface material consists of a cover of peat or fine grained glacial lacustrine sediment. Areas where rock or lodgment till provide solid foundation conditions can contain local seemingly bottomless bogs which can supply no foundation support. Areas where ice flow off tracts of carbonate rock has provided tills with good buffering capacity against acid precipitation can occur adjacent to areas where till was derived solely from felsic rocks and has virtually no buffering capacity. These examples are given to point out that in the Canadian Shield, as elsewhere, it is difficult to predict correctly how a particular area will react to a given land use and that thorough investigations of terrain conditions must be undertaken before initiating landuse activities.

QUATERNARY GEOLOGY OF THE NORTHWESTERN CANADIAN SHIELD

A.S. Dyke and L.A. Dredge

INTRODUCTION

This section deals with the area of the Canadian Shield west of Hudson Bay and the Gulf of Boothia (Fig. 3.1, 3.5). During the last glacial maximum, this region was covered by a large ice mass which has been called the Keewatin Ice Sheet (Tyrrell, 1898b), the Keewatin Sector of the Laurentide Ice Sheet (Prest, 1970, 1984; Dyke and Prest, 1987), and the M'Clintock Dome of the Laurentide Ice Sheet (Dyke et al., 1982b). The ice sheet nomenclature used here follows that set down in the introduction to Chapter 3.

Information on the glacial landforms and glacial history of this region that predates 1968 is summarized on the Glacial Map of Canada (Prest et al., 1968), the ice retreat map of Prest (1969), and in Prest (1970). In the northern and central part of the report area, fieldwork conducted since 1968 was concentrated along a proposed gas pipeline corridor across Somerset Island (Netterville et al., 1976; Dyke, 1983), Boothia Peninsula (Boydell et al., 1975; Dyke, 1984), the central tract of Keewatin (Arsenault et al., 1981, 1982; Aylsworth et al., 1981a-d; Thomas, 1982a,b; Thomas and Dyke, 1982a-f; Dyke, 1984; Aylsworth, 1986a-c; Aylsworth et al., 1986a-c), with additional data collected on Prince of Wales and King William islands (Hélie, 1984; A.S. Dyke, unpublished). Intensive work conducted in southern Keewatin focused on drift geochemistry (see DiLabio, 1989) and periglacial geomorphology, but the general conclusions contribute substantially to this report (Shilts et al., 1979; Shilts, 1980). In the southern part of the report area, mapping projects have been completed in northern Manitoba (Dredge et al., 1981a,b; Nixon et al., 1981; Dredge and Nixon, 1982a-g; Richardson et al., 1982; Dredge et al., 1986), north-central Manitoba (Klassen, 1986), and northern Saskatchewan (Schreiner, 1984). Recent work on Victoria Island is summarized by Sharpe (1984) and additional information supplied by J-S. Vincent and D.A. Hodgson (Geological Survey of Canada, personal communication 1986). Victoria Island is discussed in Chapter 2 of this volume (Vincent, 1989a), and only information pertaining to deglaciation of the island is treated here. St-Onge et al. (1981) and St-Onge and Bruneau (1982) have reported on the middle and lower Coppermine River valley. Few ground observations are available for much of the region dealt with here, particularly southeastern District of Mackenzie and northeastern District of Keewatin.

Although substantial areas remain largely unmapped, recent work along with re-evaluation of earlier work have generated new or revised hypotheses regarding the nature, extent, and dynamics of the ice cover during the last glacial maximum. In addition, lateral facies changes in the regional till sheet have been delineated, thus defining major dispersal trains; a more detailed reconstruction of the pattern and chronology of the last ice retreat has been made; new sea level data have been added, and some stratigraphic sections have been studied. At present, however, pre-Late Wisconsinan events are known only in their scantiest outline.

Dyke, A.S. and Dredge, L.A.
1989: Quaternary geology of the northwestern Canadian Shield; in Chapter 3 of Quaternary Geology of Canada and Greenland, R.J. Fulton (ed.); Geological Survey of Canada, Geology of Canada, no. 1 (also Geological Society of America, The Geology of North America, v. K-1).

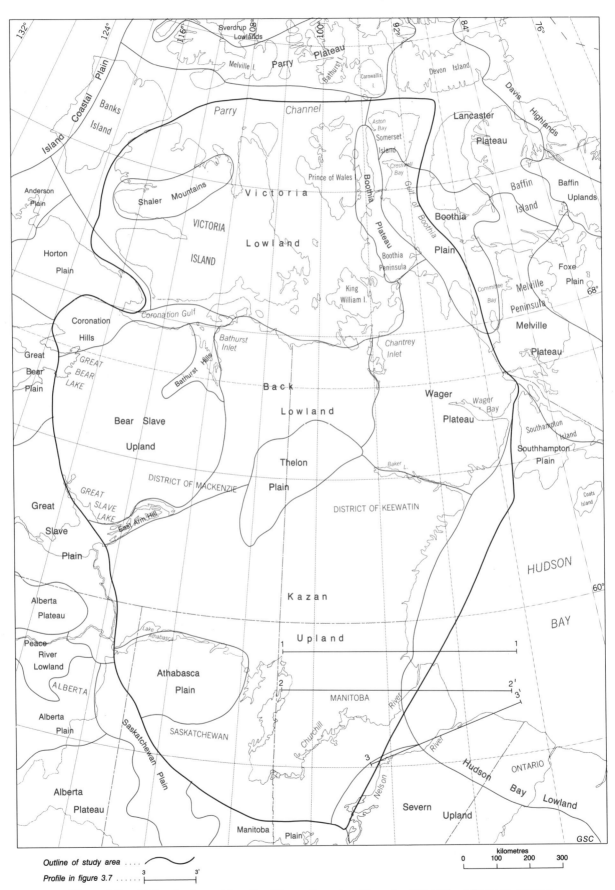

Figure 3.5. Major physiographic units of the northwestern Canadian Shield (from Bostock, 1970b).

BEDROCK AND PHYSIOGRAPHY

For the purpose of this report the region can be divided into four simple bedrock units: belts of greenstone and metamorphosed sediments; crystalline Precambrian terrane, largely underlain by granitic gneiss and intrusions; Proterozoic sedimentary and volcanogenic rocks; and the Paleozoic sedimentary cover, predominantly carbonates (Fig. 3.3). Glacial erosion, transport, and deposition of these materials have produced sharply contrasting glacial sediments and landscapes.

The physiography is more complex than the simple treatment given here would imply. Many physiographic aspects of the Canadian Shield may have been inherited from fluvial planation events of Precambrian age, although a major post-Devonian erosion surface bevels the Precambrian surfaces over broad regions. Contacts between these geomorphic features are indistinct and the various erosion surfaces throughout most of the northwestern shield have not been greatly disturbed by faulting. The major post-Devonian surface in the northern part of the region, however, was severely fragmented during the Eurekian Rifting Episode, which led to formation of the interisland channels during the middle and late Tertiary (Bird, 1967; Kerr, 1980).

The northern part of the region consists of relatively high plateaus, medium elevation plains, and lowlands (Fig. 3.6). The contacts between these topographic elements in places, such as on Boothia Peninsula, coincide with fault-line scarps suggesting that the different levels may have resulted from tectonic displacement of a once continuous erosion surface rather than development of multiple erosion surfaces (Bird, 1967). Other areas, however, display impressive continuity of erosion surfaces across bedrock contacts and across structural zones; the most striking example is the continuity of the Barrow Surface (Bird, 1967) across Precambrian gneiss and across steeply dipping and flat-lying carbonate rocks on Somerset Island. Bird (1967) recognized correlative surfaces on adjacent islands and argued that the surface developed on a once continuous plain which was later fractured to produce the interisland channels and straits. Bird's suggested origin of the interisland channels has since been corroborated by an analysis of the tectonic history of the area (e.g., Kerr, 1980).

On Somerset Island the Barrow Surface is crossed by canyons with meandering channels incised as much as 100 m into bedrock. Some of these canyons are truncated by the fault scarp, which forms the east coast of the island. On the graben floor, which forms the bottom of Barrow Strait north of Somerset Island, similar canyons are incised into Paleozoic rocks (Bornhold et al., 1976). The similarity of the submerged canyons to those on land supports the suggestion that the canyons on Somerset Island formed prior to fracturing of the Barrow Surface and formation of the interisland channels (Dyke, 1983).

In all likelihood, the Barrow Surface probably predates separation of Canada from Greenland, which culminated during the Eocene (37 Ma; Grant, 1975), and hence predates opening of Baffin Bay. The stream incision recorded on Somerset Island may signify rejuvenation engendered by uplift which accompanied the rifting that resulted in opening of Baffin Bay. Flow in the fossil river system on the floor of Barrow Strait apparently was to the east.

Whereas Somerset Island is interesting for its uniformity of elevation and beautifully preserved erosion surface, which crosses diverse rock types, Prince of Wales Island is remarkable for its stepped topography developed on a single bedrock formation, the Peel Sound Formation. The Prince of Wales Lowland, extends between blocks of the Beams Plateau in a series of broad, flat-floored, curving to sinuous "channels", best exemplified as the inland extensions of bays along the east coast. These channels have not been recognized as grabens. It is possible, therefore, that they are fluvial forms predating rifting. If so, they differ from the pre-rifting, deeply entrenched canyons of Somerset Island, and this difference in form is still not explained.

The physiographic elements of northern Keewatin, eastern Mackenzie, and King William Island are neither spectacular nor easy to interpret. Vast areas are occupied by lowlands, here given the local names Rasmussen, Back, and Bathurst lowlands (Fig. 3.6), conforming largely to the Victoria and Back lowlands of Bostock (1970a). These seem to define collectively, along with Prince of Wales Lowland and Pasley Plain, a vast lowland terrain, the central third of which is presently submerged by M'Clintock Channel, Queen Maud Gulf, and Coronation Gulf. Possibly the two submerged central zones of the lowland represent drowned fluvial systems, although it is difficult to substantiate that hypothesis. Nevertheless, the marine channels crossing that lowland do not have the same appearance as the tectonically formed channels farther north.

Back Lowland extends southward between the Arctic mainland coast and the Hudson Bay coast. Its margins rise gently and without significant break to the adjacent plateaus. There is no sharp boundary and it is certainly not a fault-bounded feature. The lowland is possibly an ancient pre-Quaternary erosion form, and because it passes gently beneath the Paleozoic sedimentary cover, it may well be a Precambrian geomorphic element.

The southern part of the northwestern Canadian Shield forms a series of slightly and variously inclined peneplains lowering from 450 m at the height of land east of Lake Athabasca to below sea level under Hudson Bay. Within these plains local topography and bedrock type are closely harmonized. Granitic and arkosic rocks form bold rounded hills; pelitic metasediments and volcanic rocks form valleys. Extensive plains have developed over areas of uniform lithology, as in the case of Athabasca Plain (Fig. 3.5), developed on sandstone, and the granitoid batholith in northernmost Manitoba (Nejanilini Plain, Fig. 3.7), where relief rarely exceeds 5 m. The plains are distinguished from each other by their different relief and slope (Fig. 3.7). Topographic and bathymetric information, as well as data on basement depths, from examination of outcrops and seismic refraction profiles (Hobson, 1968), suggests three to five periods of planation in development of the present landscape (Fig. 3.7; Bird, 1967, p. 77). The oldest surface (facet 1, of Fig. 3.7), which has broad undulations, underlies the Paleozoic rocks of Hudson Bay, and dips towards the centre of the bay with a slope of 5 m/km. A second surface, resolved into two surfaces in some profiles, truncates and bevels the older surface. This surface is best preserved in the Caribou River area (Caribou slope, transect 2, Fig. 3.7). It is a mature, regular profile which dips underneath Paleozoic rocks with a uniform dip of 2.5 m/km. Alcock (1920), Wright (1932), and Ambrose (1964) examined the character of this surface and noted that the gentle rolling relief extends beneath the Paleozoic rock. Nelson River valley may have formed during this phase of planation.

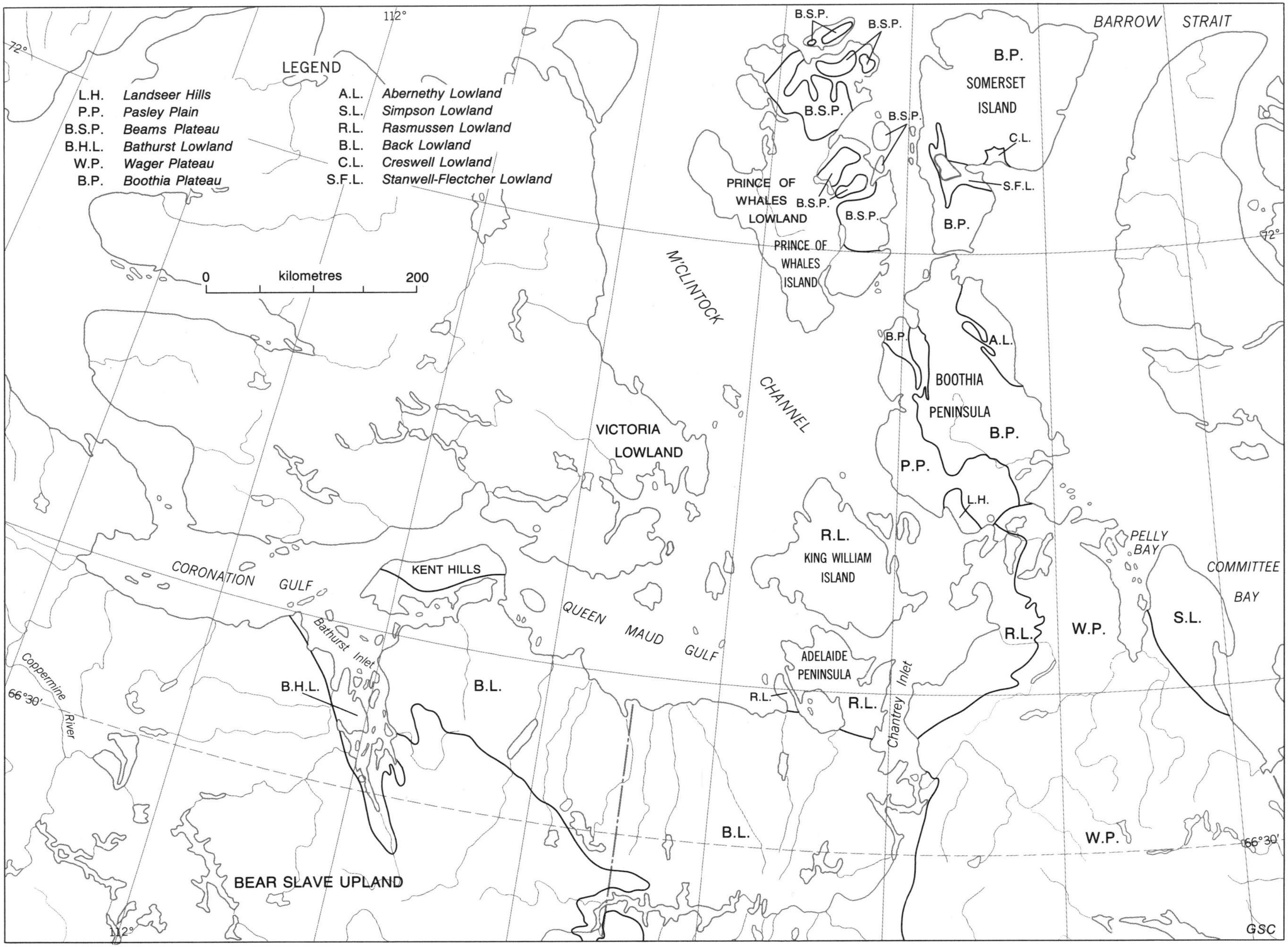

CHAPTER 3

Figure 3.6. Physiographic subdivisions used for the northern part of the study area.

The third and fourth facets (surfaces) consist of the hills and plain of the western uplands and the truncated Paleozoic beds in Hudson Bay. Both surfaces have irregularities, including broad river valleys. The regional slope of both surfaces is 0.6 m/km. They are post-Devonian in age, and may have developed during the Tertiary. The proto-Seal River developed during this phase, and its drainage basin included much of the present Churchill River watershed (Dredge et al., 1986). Although the two surfaces have fairly equal slopes they are separated in elevation by about 300 m and thus may have developed during different episodes. Locally along North Knife River deeply weathered rock occurs as remnants underlying drift (Dredge et al., 1986).

Although major elements of the landscape were shaped in pre-Quaternary times, many details of the terrain are products of Quaternary glaciation. Glaciations have scoured and freshened rock surfaces, producing an intricate second order relief consisting of ridges, valleys, and shallow basins (e.g., Schreiner, 1984) and leaving deeply altered regolith at only a few sites. Ice is estimated to have removed 5.5-8.0 m of rock in the part of central Keewatin underlain by Proterozoic sedimentary and felsic volcanic rocks of the Dubawnt Group (Kaszycki and Shilts, 1980). Glacial drift in places reaches 100 m in thickness; its pattern of flutes, transverse ridges, and eskers radiates outwards from late centres of outflow. Proglacial lake deposits and fine grained marine deposits have infilled the underlying topography creating extensive wetlands — a flat muskeg area with disrupted drainage, high ground ice contents, and an abundance of thermokarst lakes.

NATURE AND DISTRIBUTION OF QUATERNARY MATERIALS

The character and distribution of Quaternary deposits have been determined by reconnaissance ground mapping and airphoto interpretation. The main types of surface materials here are bare rock, thick and thin till, glacial marine, glaciofluvial, and glacial lake deposits, and an extensive cover of organic materials. Drift cover is bulkier and much more continuous in the eastern half of the region than in the western half (Fig. 3.4).

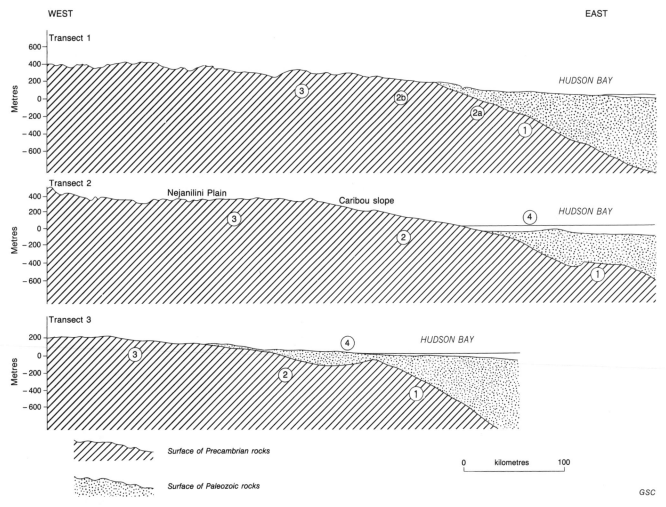

Figure 3.7. Physiographic transects across northern Manitoba showing landscape facets (1-4), each of which may have been produced during an ancient period of peneplanation. Location of profiles is shown in Figure 3.5.

Nonglacial materials

Nonglacial materials of two ages occur at the surface: (1) extensive areas of residuum (weathered bedrock) and associated colluvium, which developed prior to the last glaciation, occur on Somerset and locally on Prince of Wales islands, and in parts of Keewatin and northern Manitoba and (2) postglacial marine, fluvial, and organic sediments and blockfields occur throughout the region. In addition, pre-Holocene marine and fluvial sediments are exposed beneath the surface till along Pasley River on Boothia Peninsula and along North Knife River in Manitoba; several river sections on the Canadian Shield of northern Manitoba expose marine, fluvial, and organic materials, presumably of Missinaibi interglacial age; and subtill organic material, beyond the range of radiocarbon dating, and paleosols are exposed at several sites in northern Saskatchewan (Schreiner, 1984, p. 38-43).

The extensive covers of residuum and associated colluvium on Somerset Island are among the most interesting Quaternary materials in the region (Dyke, 1983; Hélie and Elson, 1984). The residuum consists of blocks with substantial quantities of interstitial grus and finer weathering residues on Precambrian rocks and forms a diamicton (stony, sandy, clay-silt) on Paleozoic rocks. On the Precambrian rocks the only distinctive landforms within the areas of residuum are cryoplanation terraces and associated tors. On Paleozoic rocks, with the exception of an area underlain by sandstone and conglomerate on northwest Somerset Island, even these minor landforms are absent and the terrain consists of smooth, graded hillslopes, interrupted in a few places by low cuestas.

The residuum mantles middle to upper hillslopes and summits but is buried beneath several metres of fine grained, slightly stony to stone-free colluvium on lower slopes and in broad valleys. These sheets of colluvium are continuous across many tens of square kilometres and must have accumulated over a long interval. Although accumulation is still occurring, only a minor proportion is of post-glacial age; glacial meltwater channels, which are incised into the colluvium, have been neither infilled nor extensively modified by mass movement. Hence, the bulk of it seems to have accumulated during previous nonglacial intervals. Ice cover in these areas of Somerset Island probably was always cold-based and protective, although it did deposit scattered erratics and small patches of till.

Blockfields interpreted as being of both interglacial and postglacial age extensively cover rocky areas in northern Manitoba (Dredge et al., 1986) and Saskatchewan (Schreiner, 1984, p. 22) and are most visible beyond treeline, where some cover areas of about 1000 km^2. Blockfields are also abundant farther south and inland, but are masked by forest cover. Their genesis involves rock heaving which has occurred preferentially in exposed rocks that have well developed vertical and horizontal joint systems. Heaving results from expansion due to excessive pore water pressures created during annual freezeback (L.D. Dyke, 1979) and to ice-riving. Although most of the blockfields in Manitoba are Holocene in age and frost-heaving is presently a widespread process, some may have developed subglacially or during a nonglacial interval since some areas of ribbed moraine and some larger morainic forms are composed entirely or mainly of frost-riven blocks.

In order of decreasing areal importance, postglacial deposits on the northwestern shield consist of raised beach gravel and sand, terraced and active alluvium, terraced and active delta sediments, and eolian sands. Throughout most of the region postglacial organic deposits cover small areas and are thin, but the southernmost part of the Canadian Shield, within the zone of discontinuous permafrost, is mantled by a continuous organic blanket, which is up to 4 m thick, and which contains substantial quantities of ground ice. The organic blanket consists of forested bog, pocked with thermokarst depressions and small lakes, and of fenlands (Dredge and Nixon, 1979).

Raised gravel beaches form continuous flights of ridges along the north and southeast coasts of Somerset Island, along much of the west coast of Boothia Peninsula, and on steeper sloping segments of the Prince of Wales, Simpson, and Rasmussen lowlands. These beaches developed from wave erosion of highly calcareous stony till and are generally less than 1 m thick because the till became armoured as the fines were washed out. Continuous flights of raised beaches are uncommon on the Canadian Shield except along the flanks of glaciofluvial deposits. The best examples of such beaches occur along the Hudson Bay coast. Isolated beaches, mostly composed of sand and fine gravel, but in places of boulders, were formed by proglacial lakes. Although the volume of sediment in these features is small, they are important features in reconstructing the extent and evolution of lakes.

Postglacial alluvium and delta deposits cover small areas and are locally extensive only where abundant sediment was supplied from erosion of glacial marine sediments or fine grained till, and where streams were superimposed on glaciofluvial sediments.

Small areas of active wind erosion and deposition occur throughout the region, particularly on and adjacent to eskers in Keewatin. Large areas of eolian features occur on outwash deposits and on sandy raised marine sediments. Several large conspicuous areas of wind erosion, 100 km^2 and more in area, occur in eastern and northwestern Prince of Wales Island, on eastern Adelaide Peninsula, and near the head of Wager Bay. The only large eolian deposits occur in northern Saskatchewan (David, 1977; Smith, 1978; Schreiner, 1984, p. 28). Large active sand dune fields around Lake Athabasca were derived from deltas deposited into glacial Lake McConnell. Inactive dunes are widespread north and south of the Cree Lake Moraine (see Map 1702A for location of named moraines).

Glacial materials

On the regional scale of this report, only three kinds of glacigenic sediments — till, glacial lacustrine sediment, and glacial marine sediment — are areally significant. However, glaciofluvial deposits, primarily in the form of eskers and flanking outwash materials are both common and voluminous throughout the shield areas of Keewatin, Mackenzie, Manitoba, and Saskatchewan.

Till

Till is by far the most widespread glacial deposit and is also the most varied in form and composition. In general till is thickest where it overlies Paleozoic or Proterozoic sedimentary rocks and where large amounts of debris from these rocks have been carried onto the shield. Hence, in the north, thick till occurs on the Prince of Wales Lowland, throughout most of the Rasmussen and Simpson lowlands, and on parts of the western flank of the Boothia Plateau. In the south, thick till overlies the Hudson Bay Lowlands and adjacent shield terrain to the west (this chapter, Dredge and Cowan, 1989). In the west, till thickness over the Athabasca Sandstone averages about 20 m and reaches up to 37 m, and thick till occurs down ice from the sandstone (Schreiner, 1984, p. 16). These areas of thick till commonly form extensive drumlin fields with local relief of 10 to 40 m, but large end moraines and end moraine complexes composed of till with relief of 30 to 80 m also occur, particularly on northwestern and east-central Prince of Wales Island, on King William Island, and on Simpson Lowland.

Till is either thin or absent on the sedimentary rock of northeastern Somerset Island, presumably because that area was covered by a cold-based local ice cap during the last glaciation. Till is also thin on the sedimentary rocks of the Abernethy Lowland of northeastern Boothia Peninsula, probably because the lowland lies in the lee of the resistant rock of the Boothia Plateau.

Much Precambrian rock terrain is veneered by thin patchy till or rock is exposed at the surface, but there are also large areas where till is thick enough to form drumlinized and fluted terrain and less commonly large fields of ribbed moraine (Aylsworth and Shilts, 1985). The till in these areas is probably of the order of 5-20 m thick because that is the usual relief of these landforms.

Landforms

The till of the region exhibits a variety of landforms including streamlined drift forms, major and minor end moraine, and ribbed moraine. Although these landforms have been mapped in some detail and have been used extensively in defining deglacial history, their internal compositions and structures remain largely unstudied.

Among the most interesting landforms composed of till in this region is ribbed moraine (Fig. 3.8). Ribs are asymmetric and trend transverse to ice flow, with steeper slopes on the down-ice side. In Keewatin and Manitoba ribbed moraine occurs in trains alternating with trains of drumlins. In places, particularly near the sides of the rib trains, drumlins and flutings are superimposed on the ribs, and in many places eskers, which postdate the ribs, wander through the rib trains. Both Lee (1959a) and Shilts (1977) have suggested that the ribs resulted from stacking of shear plates charged with englacial debris. Ribs are composed of coarse, exceedingly bouldery till, and both Lee (1959a) and Dredge et al. (1986) have observed interrib depressions floored with felsenmeer and adjacent ribs composed of thrust up ridges of felsenmeer. The felsenmeer, in these cases, obviously predates rib formation and may have formed either subglacially during the Wisconsinan or subaerially during the preceding interglaciation.

The alternating belts of ribs and drumlins could reflect alternating conditions in the basal ice just prior to deposition. Reasonably, the drumlin trains formed under unimpeded (extending) flow whereas the rib trains formed under impeded (compressive) flow which resulted in shearing and stacking of debris zones. The impedance to flow could have resulted from either a frozen bed or an excessive and coarse debris load. In fact, the debris load could have triggered the frozen bed condition by a reduction of flow rate and hence a reduction in the amount of frictional heat produced. It seems more likely that the coarseness of the debris load, rather than the volume of debris, triggered flow reduction and compression over the rib trains because both drumlins and ribs appear to have been deposited by ice that was heavily charged with debris. Gilchrist (1982, p. 184) has suggested that there is a textural control on the formation of ribs versus drumlins: "I suggest that ribs are representative of potential drumlins that were without a significant clay component to allow them to be moulded into a spoon shape". We are of the opinion that the parallel alternating trains of ribbed and drumlinized till require a glaciological (process) rather than a petrological or textural explanation.

On a regional scale, the ribbed moraine fields are concentrated in a U-shaped zone outlining the southern half of the Keewatin Ice Divide (Lee, 1959a; Aylsworth and Shilts, 1985). This association of the ribbed moraine and drumlin fields with the final ice divide position limits the possible time and duration of their formation to late glacial time. Ribbed moraine is also associated with late glacial deposition in the vicinity of ice divides in Quebec-Labrador and Newfoundland (Henderson, 1959; Hughes, 1964; Cowan, 1968; Shilts et al., 1987) and is rare in areas well removed from ice divides (Prest et al., 1968).

Little work has been done on the internal composition and structure of drumlins and other streamlined glacial landforms in the region. Field sampling at thousands of sites, however, demonstrates that they are composed of till at least in their upper parts. Generally, that till is not different from nonstreamlined till other than in thickness; most tills more than 5-10 m thick are drumlinized or fluted, or else are ribbed.

Shaw and Kvill (1984) examined three sections through drumlinoid forms over the Athabasca Sandstone and found that the features are composed primarily of water-sorted deposits. They concluded that the drumlins were deposited in subglacial cavities and that the regional drumlin field may be a gigantic glaciofluvial feature deposited by enormous meltwater floods involving water depths of several metres to tens of metres. They pointed out that such thick meltwater sheets under vast areas of the Laurentide Ice

Figure 3.8. Areal oblique of ribbed moraine field. 204035-C

Sheet would trigger giant surges. Sharpe (1985) has observed similar sediments in drumlins and flutings on Victoria Island and concluded that crosscutting drumlin fields there resulted from redirection of subglacial meltwater sheets under stagnant ice. Hence, Sharpe's mechanism differs substantially from that of Shaw and Kvill in that theirs invokes enhanced activity (surging) while his invokes stagnation. Either model, however, would predict the release of enormous plumes of sediment-laiden meltwater at the ice front. This should have produced striking ice marginal features and have lead to deposition over wide areas of either glacial marine sediment (e.g., on Victoria Island), glacial lacustrine sediment (e.g., in glacial lakes Agassiz, McConnell, Cree, or Wollaston in northern Saskatchewan), or glaciofluvial sediment where the water debouched subaerially.

In places, drumlin fields are intimately associated with large end moraines composed of till, for example, the Chantrey Moraine System of northern Keewatin (Dyke, 1984). It seems unlikely that end moraines of this sort would be deposited along ice margins debouching giant sheet floods because the morainal debris required to build the moraine would be swept away. It should be noted as well that Schreiner's (1984) investigation of the drumlins overlying the Athabasca Sandstone has revealed till at most, if not all, sites. In other places, drumlins and flutings grade laterally into ribbed moraine or are superimposed on the ribs. Hence, the meltwater hypothesis either should be expanded to include rib formation accompanied by drumlin formation, with the two forming alternating belts, or it has to accommodate preservation of ribs during subsequent drumlin/fluting formation.

Schreiner (1984, p. 21) drew attention to a particularly long drumlinoid feature, which Tyrrell and Dowling (1896) referred to as Ispatinow, Cree for 'conspicuous hill'. These features are "related to very sandy sediments and associated with the glacial lakes. They are concentrated primarily within the basins of the glacial lakes" in northern Saskatchewan.

Another remarkable feature of the till sheet of northern Keewatin and northeastern Mackenzie is the nested set of end moraines which arc across the entire region and continue eastward onto Melville Peninsula and Baffin Island. These moraines, known locally as the MacAlpine Moraine (Falconer et al., 1965; Blake, 1963) and the Chantrey Moraine System (Dyke, 1984), were first recognized as a coherent set of features by Falconer et al. (1965) and probably are the world's longest system of end moraines (Map 1702A). Although younger ice marginal features, mainly ice dammed lakes, can be recognized up-ice of this moraine belt, the moraines apparently represent the last climatically controlled stillstand or readvance of Keewatin Ice (Dyke, 1984). Other than the large Cree Lake Moraine (Map 1702A), no substantial end moraines were formed by the retreating western, southern, or southeastern margins of Keewatin Ice within the boundary of the Canadian Shield. Most of the time these margins lay in the deep waters of either Lake Agassiz or Tyrrell Sea and marginal deposition resulted primarily in esker nodes and esker deltas. Cree Lake Moraine (Schreiner, 1984, p. 16) can be traced for about 800 km across the Canadian Shield of Saskatchewan, east into Manitoba, and northwest into Alberta. Over much of its length the moraine consists of till but large kames and kame deltas occupy gaps in the moraine and occur at ends of sections composed of till. A number of flutings on the north side of the moraine suggest that moraine building was accompanied by a readvance. The moraine marks the last well delineated ice marginal position of this segment of the Keewatin Ice margin.

De Geer moraines were deposited in several parts of the study area. They are well formed along the southern part of the Hudson Bay coast where they indicate retreat rates of about 240 m/a (Lee, 1959a) and in northern Manitoba where they indicate retreat of about 1.5 km/a for an interval of 132 years. De Geer moraines also formed in the northern part of the region. The moraines are best developed south and west of Chantrey Inlet where they form a series of closely spaced parallel ridges that protrude 1-2 m above a glacial marine silt plain or ornament the flanks and tops of drumlins. Most individual ridges are a few metres wide and can be traced laterally for only a few hundred metres. Several ridges, however, are developed to a height of 5 to 10 m and can be traced across tens of kilometres. These larger ridges have scalloped proximal faces and distal bulges similar, except in scale, to ice-pushed ramparts. Likely, these larger ridges represent marginal positions that were maintained for more than a single season and they probably correlate with ridges within the Chantrey Moraine System. Schreiner (1984, his Plates 13 and 14) has mapped De Geer moraines in the southern glacial Lake McConnell basin as well as closely spaced minor moraine ridges deposited subaerially. Both types of moraines have similar spacings.

Composition

The composition of till has been examined in some detail along a corridor extending from Somerset Island to north-central Manitoba and in northern Saskatchewan. Most work has been done on the surface till but in some sections colour, texture, lithological differences, and paleosols suggest that multiple tills are present. In most cases, however, there is no firm evidence that tills of different appearance are of distinctly different ages and some variations can be attributed to facies changes within the till or to changing ice flow directions.

The till derives its textural and lithic character to a large degree from its source materials, and lateral variations in the till reveal regional ice flow patterns. In the shield region, most till is derived primarily from medium textured igneous and metamorphic rock. Till derived from the Precambrian basement has a small amount ($<2\%$) but a large variety of sand-sized heavy mineral species, and heavy mineral components vary radically over short distances because of complex structure, contrasting lithologies, and metamorphic grades of the bedrock; the clay-sized fraction has a strong quartz-feldspar component mixed with significant amounts of chlorite and illite (Shilts, 1977, p. 203). Till derived from the large sandstone bodies, such as the Proterozoic Athabasca Sandstone of Saskatchewan, the Thelon Formation of the eastern District of Mackenzie and western District of Keewatin area, or the Paleozoic Peel Sound Sandstone of Prince of Wales and Somerset islands, are more uniform both in terms of clast petrology and matrix composition.

The carbonate content of the till matrix and the limestone/dolomite content of the granule fraction define three large dispersal trains of calcareous till extending across the crystalline rocks of the Canadian Shield (Fig. 3.9a, 3.10).

QUATERNARY GEOLOGY — CANADIAN SHIELD

Two of these dispersal trains occur on the Boothia Plateau (Dyke, 1984). They are 150 km and 210 km wide and debris within them has been transported at least 100 km eastward. They have sharply defined edges and central plumes of farthest travelled material. What is most likely an upstream extension of the northern dispersal train can be recognized on the satellite imagery and airphotos of Prince of Wales Island, where limestone- and dolomite-charged grey till has been spread eastward as a distinct plume across red sandstones and conglomerates (Fig. 3.11). The third large carbonate dispersal train extends southwestward across the shield from Hudson Basin. That train has a sharply defined northern side marking the juncture of ice flowing from Hudson Bay and ice flowing from Keewatin. The zone of

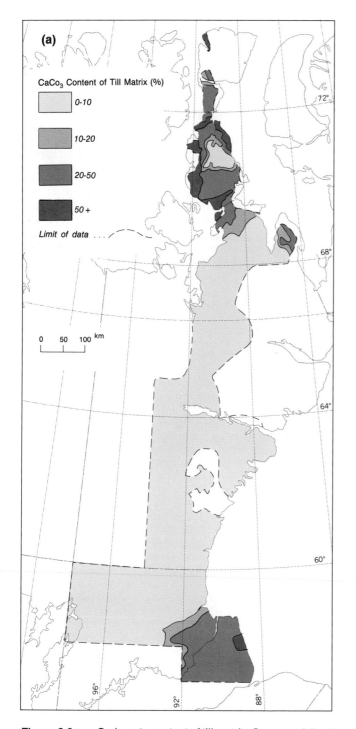

Figure 3.9a. Carbonate content of till matrix, Somerset Island to Manitoba.

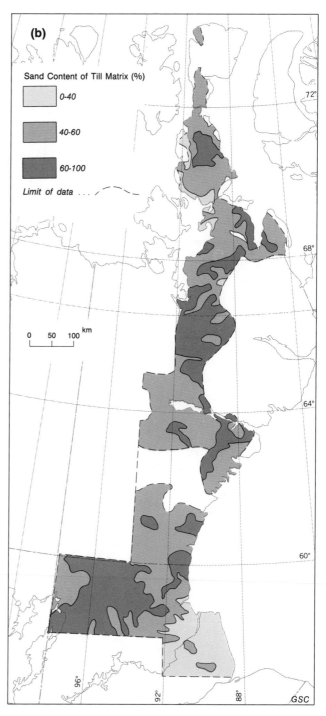

Figure 3.9b. Sand content of till matrix, Somerset Island to Manitoba.

197

confluence of Hudson and Keewatin ice was not stationary, and in places Keewatin Ice overrode older calcareous silty till of Hudson Bay provenance and produced a till with much higher silt and carbonate content than is typical of Keewatin till.

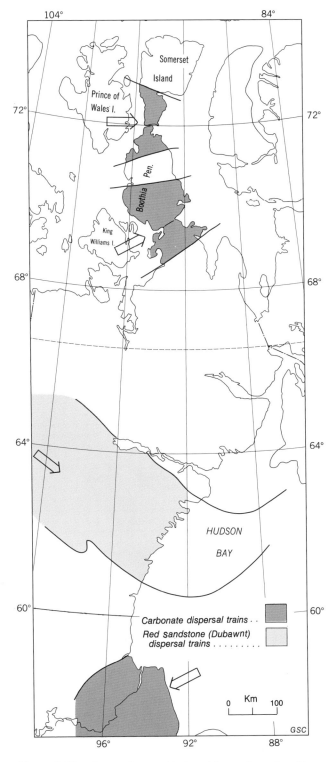

Figure 3.10. Large dispersal trains of the northwestern Canadian Shield.

Perhaps the longest dispersal train yet recognized consists of till with a red matrix and red erratics derived from the Dubawnt Group of central District of Keewatin (Fig. 3.10). The red pigment is finely ground hematite which resides in the clay fraction so that when the clay is removed the residue is pink or grey (Shilts, 1978, p. 1058). The red till is generally low in trace elements so areas within this train have depressed heavy metal contents (DiLabio, 1989). This till spreads southeastward and extends into Hudson Bay; it defines a dispersal train 150 km wide and at least 300 km long (Shilts et al., 1979). In fact, distinctive cobble and boulder-sized fragments of red Pitz and Christopher Island formations of the Dubawnt Group are numerous on Coats Island in north-central Hudson Bay (Shilts, 1980, p. 3), and hence, the dispersal train can be recognized over a length of at least 800 km.

Another large dispersal train extends southwestward from the Athabasca Sandstone (underlies Athabasca Plain, Fig. 3.8), in the same direction as the trend of drumlins on the surface till; its exact dimensions have not been mapped, but pebbles of Athabasca Sandstone have been reported from southern Alberta (I. Shetsen, Alberta Research Council, personal communication, 1984). The till is light grey, grey, or pinkish depending on the colour of the parent sandstone, which varies, depending on hematite content, from white, to pink, to purple (Schreiner, 1984, p. 14, 30). Down-ice tracing of the Athabasca dispersal train across the Phanerozoic sediment of the Interior Plains would be very helpful in interpreting the distance and relative duration of various ice flows.

Although the large dispersal trains discussed above are the most conspicuous and most useful in regional ice sheet reconstruction, numerous other dispersal trains exist, of which only a small fraction has been mapped. Smaller trains that have been defined usually consist of material that has been transported only a few kilometres or even a few tens of metres, but in the case of dispersal trains of economically valuable minerals, they do provide an enhanced target area for prospectors. Among the small trains recognized are a 16-km-long, 1.5-km wide, ribbon-shaped garnet train near Kaminak Lake, central Keewatin (Shilts, 1975); similar garnet trains, less than 10 km long, in northern Manitoba (Dredge, 1981); and a zinc, copper, nickel train about 10 km long on Boothia Peninsula (Dyke, 1980b). Obviously, within the same region till of local provenance can exist side by side with, and above or below, till that has travelled much farther. These contrasts can be caused by lateral and vertical differences in ice flow rates, by differences in duration or direction of flow, or by early or late entrainment of debris.

The till varies widely in chemical composition and clast petrology and, therefore, varies widely in texture (Fig. 3.9b). Fine grained tills overlie carbonate bedrock and comprise the carbonate dispersal trains. Till of the Dubawnt dispersal train is also rich in silt and clay (Shilts, 1973, p. 3). Small deposits of fine grained till were laid down following readvances into glacial lakes, as in northern Saskatchewan (Schreiner, 1984, p. 30). Extremely sandy tills, with more than 60% of the matrix in the sand fraction, cover large areas of central Boothia Peninsula, north-central Keewatin, northern Manitoba and areas underlain by Athabasca and Thelon sandstones. These were derived from local gneisses and loosely cemented Proterozoic sandstones. The till of central and southern District of Keewatin is somewhat less sandy. The extensive boulder cover of the central Boothia till led Boydell et al. (1975) to describe the material as felsenmeer, and it is possible that this till was derived from an interglacial felsenmeer that underwent little glacial modification. Likewise, exceedingly bouldery tills that comprise some ribbed moraine fields may have been produced from felsenmeer.

There is reason to believe, however, that the sandy till of northern Keewatin does not simply represent an immature till but that it is a result of depletion of fines from debris-rich basal ice. The sandy till occurs up-ice from thick and extensive glacial marine silt and clay. Had these materials been deposited with the till, it would have contained up to twice as much silt and much more clay than it does. The main conduits that delivered the fines to the sea were undoubtedly the numerous esker corridors that cross the region. The sandy till, however, occurs regionally across hundreds of square kilometres. This would require some mechanism comparable to sheetwash operating pervasively at the base of the ice sheet and delivering the fine material to the esker corridors.

Unlike the sandy till of central Boothia Peninsula, which has an extensive boulder cover almost everywhere, the sandy till of northern Keewatin has a variable boulder cover. Some areas tens of square kilometres in extent are

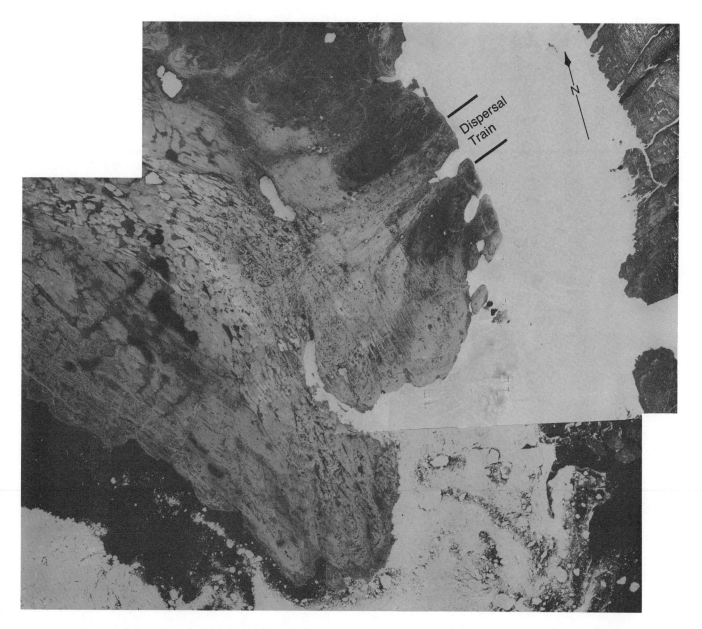

Figure 3.11. Landsat image of a carbonate dispersal train on Prince of Wales Island. Note the crosscutting relationship between this feature and the regional northward flow pattern.

nearly completely boulder covered, but more commonly boulder cover is in the 20-30% range. The exceedingly bouldery tills, which resemble felsenmeer, possibly represent the end product of subglacial sheetwash carried to its ultimate development. Even in these areas, however, ghosts of drumlinoids are recognizable and demonstrate that these are not simply areas of frost shattered bedrock that have developed in postglacial time or that have survived glaciation.

Glaciofluvial sediment

Glaciofluvial sediments occur throughout the region, although their distribution is by no means uniform. Eskers are relatively rare and short on Somerset, Prince of Wales, and King William islands, and on Boothia Peninsula, and at a regional scale the features do not form a coherent pattern. Over much of Keewatin and eastern Mackenzie however, numerous and much longer eskers describe a giant radial flow pattern emanating from the Keewatin Ice Divide (Lee, 1959a). At a slightly smaller scale eskers on the northwestern Canadian Shield are parallel and seldom, if ever, are there more than two orders of tributaries, in the sense of Horton (1945) (Banerjee and McDonald, 1975). The eskers are regularly spaced, some 12-15 km apart (Banerjee and McDonald, 1975; St-Onge, 1984), probably a function of the minimum discharge required to maintain the tunnel from closing and the roughly equal size of each esker catchment, a function of the uniform erodibility and slope of the ice surface. Many eskers are more than 100 km long; the Thelon esker, more than 800 km long, is probably the longest in the world. We do not mean to imply, however, that the entire length of an esker or that the whole network of eskers formed at the same time, for at more local scales different sets of eskers display directional changes corresponding to changes in drumlin and fluting directions, and hence, to shifts in ice divide position. Although the abundance and continuity of eskers indicate either rather sluggish ice flow or, more likely, large amounts of meltwater within the Keewatin Ice , the fact that successively younger esker segments are accompanied by successively younger drumlins and flutings suggests that the ice was not stagnant (cf. Shilts, 1985).

The most conspicuous component of esker systems is the esker ridge. This may be straight, meandering, or braided, depending upon the hydraulic gradient or the sediment load. In many places eskers are flanked by water-scoured bedrock (e.g., St-Onge, 1984); this was formed at the base of the glacier. Although composed predominantly of sand and gravel, numerous eskers contain short segments composed almost entirely of boulders. St-Onge suggested that these mark points at which englacial tunnels plunged to the glacier bed. Eskers that formed above sea level or lake level are intermittently flanked by terraced outwash which is commonly severely kettled. Shilts (1984c) suggested that much of this outwash was deposited over thin, debris-covered ice. Where esker streams discharged into standing water, the sudden decrease in competence led to deposition of esker nodes or esker deltas. This imparted a beaded appearance to many eskers in the glacial lake basins and below marine limit. The lack of ice marginal (delta, fan) deposition between nodes is not easy to account for. It may imply discrete jumps in position of the margin due to calving or an intermittent sediment supply, perhaps regulated by intermittent sediment plugging of the esker tunnel.

The largest glaciofluvial feature in this region is the Burntwood-Knife Interlobate Moraine which marks the convergence of southward flowing Keewatin Ice and westward flowing Hudson Ice during deglaciation (Map 1702A). The moraine is composed of broad, ridged segments of stratified sand, gravelly sand, and till. Individual segments vary in length (3-12 km), height (10-60 m), and width (1-4 km). The feature extends northward from Burntwood River, then swings eastward north of North Knife Lake, where it forms a massive arcuate mound 60 m high.

Radial (kame) moraines also occur in northern Manitoba. They are bulky, esker-like features similar to interlobate moraines which developed in zones of relatively sluggish flow between ice streams. Ice flow patterns are slightly convergent towards these features. They are commonly 20-30 m high, 50-200 m wide and 10-100 km long. Frontal kame moraines spread as wings from these ridges. The two largest radial moraines, originally identified by Alcock (1921) as terminal moraines, are located east and south of South Seal River.

A third variety of interlobate glaciofluvial deposit occurs in northern Saskatchewan. Large, severely kettled, outwash plains with minor esker ridges and straight to gently curving, parallel margins follow the regional ice flow trend. Schreiner (1984, p. 19) concluded that these sediments were deposited in ice-walled channels open to the sky in a stagnant marginal zone of the ice sheet. A trench in one of these deposits revealed more than 60 m of stratified sand with some gravel layers.

Several elongate belts of conjugate or reticulate ridges, each belt 15-20 km long and 1-2 km wide, composed of gravelly sand with slugs of till, lie between ribbed moraine belts near Hudson Bay in northern Manitoba and southern Keewatin. These features are presumed to be crevasse fillings and identify areas where Keewatin Ice became stagnant, perhaps detached from the main mass.

In many areas glaciofluvial waters have cut meltwater channels, which are a particularly common feature on both till and bedrock throughout the region and have been used extensively in reconstructing the pattern of ice marginal recession described below. Lateral meltwater channels are common on Somerset and Prince of Wales islands and on central Boothia Peninsula. They are rare in Keewatin and Manitoba where subglacial and proglacial channels are common.

Glacial lake sediment

Dozens of small glacial lakes formed on Boothia Peninsula and northeastern Keewatin during deglaciation (Thomas and Dyke, 1982a-f; Dyke, 1984) but these lakes left trivial quantities of sediment and are recognized mainly by their weakly developed strandlines. A lake with a more complex history formed in the valley of Coppermine River (St-Onge, 1980) and thick deltaic and deeper water rhythmites accumulated in it. The eastern part of glacial Lake McConnell (Map 1703A) overlapped the western margin of the Canadian Shield (Craig, 1965a; St-Onge and Dredge, 1985) and other large lakes formed in the Contwoyto, Dubawnt, and Thelon and Kazan basins (Map 1703A; Prest et al., 1968) but little sediment accumulated in any of them.

The largest glacial lake, Lake Agassiz (Map 1703A), rimmed much of the southern margin of receding Keewatin

Ice and influenced both the dynamics and the pattern of deglaciation (Dredge, 1983a; Schreiner, 1984). Over lower and more southern parts of the Agassiz basin the principal deposits are slightly calcareous, reddish or grey silts, locally varved but typically massive, which represent a mixture of material from Keewatin and Hudson Bay sources. The lacustrine sequence is generally sandy at the base and siltier and more clay-rich towards the top. At higher elevations and along the northern periphery of the lake basin, material was derived largely from debris washed directly from Keewatin Ice and by reworking of sandy till. These deposits form beaches or a noncalcareous sandy blanket, typically less than 2 m thick. The thickest sediments accumulated where meltwater channels debouched from the ice and successive ice marginal positions are marked by delta pads and by kame moraines. The surface of glacial lake deposits is scarred by crisscrossing networks of furrows caused by grounding of icebergs against sublacustrine rises (Dredge, 1982).

Glacial marine sediment

Little glacial marine sediment accumulated in the northwestern Tyrrell Sea (Hudson Bay basin), but thick and extensive glacial marine sediments extend nearly unbroken for almost 350 km from the south end of Committee Bay westward to Chantrey Inlet. They are at least 100 m thick in the larger valleys. In these areas badlands topography has developed through extensive postglacial fluvial erosion. Glacial marine sediments are present over much of the lowland between Chantrey Inlet and Bathurst Inlet but are only a few metres thick and are interrupted in many places by till and bedrock knobs. In Bathurst Lowland these sediments again become much thicker and badland topography has developed on them. The distribution of glacial marine sediments west of Bathurst Inlet is not well known but rhythmically bedded glacial marine sediments attain thicknesses of 70 m in the Coppermine River region (St-Onge and Bruneau, 1982).

Glacial marine sediments, in most places no more than a few metres thick, occupy many local basins, from a few square kilometres to a few tens of square kilometres in extent on King William and Prince of Wales islands. On northwestern Prince of Wales Island, glacial marine silts and clays discontinuously cover several hundred square kilometres and the largest deposit is thick enough to have developed badland topography and large thermokarst features. The deposits on Prince of Wales Island are commonly fossiliferous and range from very stony muds to stone-free silts and clays.

Little research has been done on these sediments. They consist predominantly of horizontally stratified silts, clays, and fine sands and are barren of marine macrofossils over much of their extent, but are locally fossiliferous. Most of these sediments are grey though Craig (1964a) reported that they are brick red in northwestern Keewatin.

The deposits described above are predominantly distal glacial marine sediments. Coarser ice contact sediment (glaciomarine) occurs mostly in the form of deltas which are common at marine limit in northern Keewatin, on northeastern Boothia Peninsula, and in the western Coronation Gulf region (St-Onge and Bruneau, 1982). They are less common on Somerset and Prince of Wales islands. Esker nodes are a related form of ice contact glaciomarine sediment that are common in areas of marine submergence of Keewatin, Mackenzie, and Manitoba. They differ from deltas mainly in that sedimentation was interrupted before the features could aggrade to sea level and in that they are interconnected by the eskers. Along many eskers in northern Keewatin, nodes are spaced at roughly regular intervals a few kilometres apart. In the space between nodes several De Geer moraines commonly are present on the adjacent till sheet; if the moraines are annual features, the nodes must have formed at longer than annual intervals.

Iceberg scour marks occur on deepwater glacial marine sediment as well as on till below marine limit on King William Island (Hélie, 1983) and on Prince of Wales Island. Most grooves are about 1 km long but reach lengths of 10 km. They occur in profusion in some areas but are absent or rare in many others. In places their orientations are random or multimodal; in others they have strongly preferred unimodal orientations.

QUATERNARY HISTORY

In this region, events prior to the Late Wisconsinan are represented in stratigraphic sections in few places. Where these older sediments are present, their ranks (i.e., glaciation, interglaciation) are usually apparent but only minimum ages can be assigned. Similarly, the time interval spanned by the last glaciation is not known. We know that deglaciation started during the latest Pleistocene or earliest Holocene and was completed shortly after 8 ka but estimates of the timing of the beginning of the last glacial buildup remain speculative because that event lies beyond the range of radio-carbon dating. Because the time interval represented by the uppermost tills remains so uncertain, other than that they likely span the interval from early Holocene to more than 40 ka (upper limit of radiocarbon dating), formally defined time terms such as "Late Wisconsinan" and "Wisconsinan" can be used only in a restricted sense. Hence, the surface till sheet is considered to have been deposited during the "last glaciation", and events predating deposition of the till are assigned to either the "last glacial buildup" or are considered to predate the last glaciation.

Events prior to deposition of the surface till sheet

Several events are recognized as predating deposition of the surface till but region-wide correlation is not possible. Therefore, the discussion below is largely an inventory and proceeds from Somerset Island in the north to Manitoba and Saskatchewan in the south.

Somerset Island

Early glaciation of Somerset Island is recorded by granite and gneiss erratics scattered sparsely on the Paleozoic carbonate plateau and by carbonate erratics scattered on the gneissic terrane in areas not overridden by Late Wisconsinan Laurentide ice. Precambrian Shield erratics are scattered similarly over Prince of Wales Island, Cornwallis Island, Bathurst Island, Melville Island, and other islands lying north and west of Somerset Island (Prest et al., 1968), and collectively they record the most extensive

Laurentide advance in Arctic Canada. On Somerset Island, the nearest source of Precambrian erratics lies on the western side of the island. The erratics, however, are no more abundant near the Precambrian contact than on the farthest parts of the island, suggesting that they came from elsewhere.

The possible age of this regional glaciation could lie anywhere between early Quaternary and Early Wisconsinan. Dyke (1983, p. 23) suggested that the glaciation occurred during the early part of Foxe Glaciation (see Andrews and Miller, 1984; and this chapter, Andrews, 1989, for definition) because amino acid ratios on erratic shells transported onto Somerset Island during the regional glaciation were comparable to ratios on shells then thought to be of last interglacial age on Baffin Island (e.g., Dyke et al., 1982a). Reinterpretations of the Baffin Island stratigraphy (e.g., Szabo et al., 1981; Nelson, 1982; Brigham, 1983) have assigned ages of 300 to 600 ka to these deposits which previously were considered to represent the last interglaciation. As a consequence, the "early glaciation" on Somerset Island also could be this old. Nevertheless, the recognition of moraines of Early Foxe age on northern Bylot Island, which were formed by a large outlet glacier flowing east in Lancaster Sound (Klassen, 1982), makes it likely that Somerset Island and all nearby islands were covered by thick ice during Early Foxe time.

Residuum, colluvium, and associated cryoplanation terraces and tors that cover large parts of Somerset Island were described above. The old erratics occur throughout the weathered rock terrains, but whether the processes that produced these terrains occurred before or after the "early glaciation" is not clear. The weathering must have occurred prior to the last glaciation, however, because meltwater channels formed during the last ice retreat are incised into both the residuum and the colluvium. The weathering interval, thus could have occurred at any time between early Quaternary (or even Tertiary) and the last glaciation. The fact that weathering and movement of colluvium have resumed during the postglacial and that the last glaciation left a record of only local meltwater erosion suggests that the thick colluvial deposits probably accumulated during more than one nonglacial interval.

Evidence of a period of ice retreat prior to the last advance occurs in two areas of Somerset Island. Driftwood collected from an ice contact delta is more than 38 ka, and high level beaches near the base of the delta contain shells that appear to predate the last glaciation on the basis of their amino acid ratios. On southern Somerset Island, the regional till was emplaced by ice flowing onshore and contains abundant marine shell fragments. Dyke (1983) concluded that these erratic shells indicated Middle Wisconsinan deglaciation based on comparison with shells from Baffin Island. However, reinterpretation of the age of Baffin Island stratigraphy (this chapter, Andrews, 1989) renders that conclusion invalid, and no Middle Wisconsinan nonglacial interval is presently recognized.

Boothia Peninsula

Sediments are exposed beneath the surface till at two places on Boothia Peninsula. At a site on northeastern Boothia Peninsula, foreset-bedded marine sand outcrops both at the surface and beneath till. Amino acid analysis suggests that the deposit is among the oldest recognized in this region. The high elevation of the deposit and its position on the side of an isolated plateau require that the sediment must have been deposited from a nearby ice front.

Along Pasley River on west-central Boothia Peninsula, 30 to 40 m-high bluffs expose a sequence of sediments below till (Fig. 3.12, Dyke and Matthews, 1987). These sections contain, from the base up, a lower marine deltaic sand, a lower glacigenic assemblage (till and various glaciomarine facies), a mid-section fluvial gravel, an upper marine deltaic sand, and an upper glacigenic assemblage (till and glaciomarine sediment). The mid-section fluvial gravel is >55 ka and is thought to represent the Sangamon Interglaciation separating Illinoian from Wisconsinan glacial sediments. The upper marine deltaic sand extends to nearly 100 m above present sea level and records crustal depression prior to arrival of ice at the site (slow glacial buildup), and the upper till probably represents most of Wisconsinan time.

Prince of Wales Island

Bluffs eroded by Fisher River expose gravels, containing transported marine shell fragments, beneath the surface till. Limited amino acid analyses compared with ratios from Baffin Island prompted G.H. Miller (Institute of Arctic and Alpine Research, personal communication, 1978) to suggest that they are >40 ka old, and Dyke and Matthews (1987) suggested that they are probably Sangamonian. A section on the northeastern part of the island exposes fluvial gravel between tills and glacial marine sediment. The gravel occurs at low elevation and has a well developed paleosol, possibly indicative of interglacial conditions.

Keewatin

The limited information available on events predating deposition of the surface till has been summarized by Shilts (1984a, p. 124). "Multiple till sections are known from exposures along Kazan and Thelon rivers near Baker Lake and in the vicinity of Kaminak Lake (Shilts, 1971; Ridler and Shilts, 1974). In these areas two tills of sharply contrasting geochemical composition, lithology, and texture are superposed but are not separated by intervening waterlaid sediments or weathering zones. Although these sections may represent deposition during separate glaciations, it is just as likely that they represent either deposition from a single ice sheet that shifted ice flow trajectory with time or that they represent a single basal meltout till, the layers of varying composition representing debris bands of varying composition that were stacked vertically in the ice." The Kaminak Lake till exposure was described by Shilts (1971, p. 48) as follows: "The till consists of a lower massive, sandy, grey unit which is covered by a zone of shear or thrust plates of alternating grey and brick-red till. The intersheared zone is overlain by massive red till."

The lower grey till at Kaminak Lake has been interpreted as the product of ice flowing southward from a centre "100 or more miles north of the region" (Ridler and Shilts, 1974, p. 17) and has been correlated with the inscription of southward oriented striations that record a flow predating establishment of the Keewatin Ice Divide (Lee, 1959a; Ridler and Shilts, 1974). This earlier southward flow has been recognized in the area at the west end of Baker Lake, by the distribution of quartzite erratics "of presumed Aphebian age...A preliminary interpretation of these observations could be that the Wisconsinan (or Late

Wisconsinan) ice sheet grew with its centre of outflow in northern Keewatin and that the centre migrated to the south and east to the vicinity of Baker Lake in the final phase of glaciation" (Cunningham and Shilts, 1977, p. 313).

An early southward flow predating establishment of the Keewatin Ice Divide is recognized throughout southern Keewatin, where "widespread Dubawnt erratics southward at least to the Manitoba border... support the concept that prior to the southeastward flow, a general southerly ice flow affected most of the area, an idea first suggested by Tyrrell (1898a) and later confirmed by Lee (1959a)" (Shilts et al., 1979, p. 539). Tyrrell suggested that the flow emanated from a major centre northwest or north of Dubawnt Lake, and Gilchrist (1982, p. 224) interpreted these older striae as products of early Wisconsinan ice advancing from a north-northwestern centre in the Districts of Keewatin and Mackenzie. In fact, all that can be said about the age of the early southward flow is that it predates flow from the Keewatin Ice Divide and it may, as suggested by Cunningham and Shilts (1977), be as young as Late Wisconsinan.

Multiple-till units of contrasting geochemical and lithological composition have also been recognized in boreholes in southern Keewatin and northern Manitoba (Shilts, 1984a, p. 124). Shilts pointed out that intertill gravel units, whether of fluvial or glaciofluvial origin, in Keewatin must represent deposition during a period when the glacier that deposited the underlying till was either confined to the Keewatin mainland west of the Hudson Bay depression or had melted away altogether. This can no longer be taken to mean that these gravels are of Sangamonian age or older because the Laurentide Ice Sheet may have been reduced to that state one or more times during the Wisconsinan (Andrews, et al., 1983b).

Manitoba

Although interpretation of the stratigraphic record in Manitoba also suffers from a lack of absolute age control, the sequence of sediments suggests that there is a relatively long Quaternary record preserved, and that there was an interplay of both Keewatin and Hudson or Labrador ice across northern Manitoba during the Wisconsin Glaciation.

The pre-Wisconsinan record is best preserved along Nelson River, which lies marginally within the Hudson Bay Lowlands (see this chapter, Dredge and Cowan, 1989). The principal elements of the stratigraphic record (base to top) are: (1) a sandy granitic till of northern provenance, with a paleosol developed on its upper metre, (2) a silty calcareous till of eastern provenance, (3) clastic sediments and organic

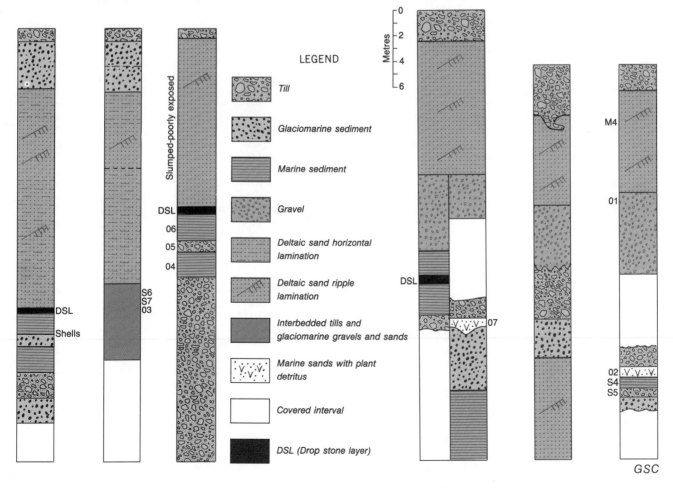

Figure 3.12. Pasley River stratigraphic sections.

beds, and (4) several silty calcareous till sheets of eastern provenance. The organic beds have been assigned an interglacial rank and a Sangamonian age on the basis of the stratigraphic position of the sediments, the pollen assemblages, and the similarity of amino acid ratios from wood within this unit to those from wood from the Missinaibi type section in Ontario (Nielsen and Dredge, 1982; Dredge and Nielsen, 1985). The stratigraphic column records an early extensive glaciation of northern provenance, a weathering interval during which a paleosol developed, a succeeding glaciation when the area was covered by ice of eastern provenance, the Sangamon Interglaciation, and a succeeding glaciation of eastern provenance.

Over much of the shield in Manitoba few stratigraphic units are exposed, and nothing is dated. On Churchill River till of northern provenance underlies interglacial (?) sediments containing twigs dated as >32 ka (Dredge and Nielsen, 1985). Other than this there is no record of pre-Missinaibi glaciation north of Nelson River, although some interglacial deposits may be exposed. An exposure along Seal River containing imprints of leaves and twigs (Taylor, 1961) has been assigned a Missinaibi age. Bedded sands and silts between till sheets, but without plant remains, farther west along Seal River may be of the same age. Sections along North Knife River reveal fluvial silt and indurated gravel below till. These fluvial deposits also have been tentatively assigned a Missinaibi age (Dredge et al., 1986) because of their similarity to interglacial deposits along Nelson River.

Ice flow direction in northern Manitoba changed several times prior to establishment of the main Late Wisconsinan flow pattern, but it is difficult to assign ages to these events. Near Chippewyan Lakes, in northwestern Manitoba, for example, the last major ice movement was southward, but below the surface, sandy, granitic till of northern provenance lies a silty, calcareous till containing Paleozoic fossils and limestone pebbles of eastern provenance. This silty till was emplaced when ice from the east extended westward well beyond Burntwood-Knife Interlobate Moraine (Map 1702A). This flow is of post-Missinaibi age and could be as young as Late Wisconsinan maximum. In contrast, near the mouth of Churchill River, sections and buried moraines indicate that Wisconsinan age Keewatin Ice once extended substantially south of the Burntwood-Knife Interlobate Moraine and that its drift sheet was subsequently overridden by ice of eastern provenance (Dredge and Nielsen, 1985).

A final problematic set of data is the early, eastward trending striations on high ground near the Keewatin-Saskatchewan-Manitoba border (Dredge, et al., 1986). These indicate a glacial dispersal centre in northern Saskatchewan or southern District of Mackenzie, possibly part of the same ice mass that formed older southerly striations in Keewatin (see above).

Saskatchewan

Schreiner (1984, p. 39-40) reported evidence of interglacial or interstadial conditions at three localities in northern Saskatchewan. Intertill organics, including wood, were recovered from two drillholes and proved to be beyond the range of radiocarbon dating (>34 ka). In both cases the organic layers are overlain by a single till. Three exposures of weathered till beneath the surface till occur along Churchill River; two of these have upper oxidized zones interpreted as paleosols and the other displays oxidation along fractures and joints. A till with an upper oxidized zone, interpreted as a paleosol, was exposed beneath the surface till in a trench. In all cases there is evidence of only one glacial event following the nonglacial interval.

At two other localities Schreiner (1984, p. 38-39) reported the presence of more than one till sheet. Although these could represent different glaciations, they could also represent intermittent deposition or shifts in ice flow direction during the last period of ice cover. Two tills with an erosional contact are exposed in cuts through drumlins. The lower is greyish pink and sandy, the upper is pinkish grey and sandy, perhaps suggesting a slight difference in provenance. Three tills were exposed in an artificial trench. The lower till, pink to pinkish white, was derived from Athabasca Sandstone and is interpreted as lodgment till. The middle till is either a different lodgment till of the same provenance or an ablation facies of the lower till. The upper till is much more mafic-rich and was derived from the basement rocks.

Johnston (1978) and Schreiner (1984, p. 11) reported west to northwest trending striae at several localities. These seem to record an ice movement that predates the regional southwesterly flow recorded by the drumlin fields and Athabasca Sandstone dispersal. If they correlate with the early easterly striae in northern Manitoba, they indicate an ice divide near the Saskatchewan/Manitoba border.

Last glacial buildup and maximum

The limit of Late Wisconsinan Keewatin Ice in the north and northwest cannot be fixed with certainty and this led Prest (1984) to present two possibilities. The maximum extent model essentially retains the limit shown on the Glacial Map of Canada (Prest et al., 1968). This limit crosses southern, eastern, and northern Banks Island and southern Melville Island and then extends eastward in Parry Channel to Baffin Bay. Dyke et al. (1982b) and Dyke (1983) mapped a limit of Laurentide ice on Somerset Island, coinciding roughly with the limit proposed by Jenness (1952) and by Craig and Fyles (1960). The limit on Banks Island was interpreted as being of Early Wisconsinan age and a new less extensive Late Wisconsinan limit was proposed (Vincent, 1982, 1983, 1989). That limit was adopted by Dyke et al. (1982b) and Dyke (1984) and modified by Vincent (1984) to conform to the boundary of Winter Harbour Till deposited by the Viscount Melville Sound Ice Shelf (Hodgson and Vincent, 1984). More recently, however, Dyke (1987) has argued that the earlier proposed limit on Banks Island is of Late Wisconsinan age. That limit, therefore, is portrayed here (Map 1702A), while the minimum model, which places the limit at the edge of the Viscount Melville Sound Ice Shelf, is portrayed in Chapter 2 (Vincent, 1989a).

Keewatin Ice was flanked on the northeast by Foxe Ice (Ives and Andrews, 1963; Andrews and Miller, 1979; Dyke et al., 1982b) and on the southeast by Hudson Ice of the Labrador Sector (Dyke et al., 1982b; this chapter, Dredge and Cowan, 1989). The zone of confluence of Keewatin and Hudson ice shifted during the course of the last glaciation, as discussed above. Initially Hudson Ice was more extensive (see discussion of Chippewyan Lakes area, above), but its northern and western limits were pushed back by ice flowing out of Keewatin. The final confluence of these two ice masses in northern Manitoba is marked by the Burntwood-

Knife Interlobate Moraine. In southern Manitoba the final convergence zone is marked by the Belaire-Bedford Hills interlobate moraines (Map 1702A; Manitoba Mineral Resources Division, 1981; Klassen, 1989).

During the last glacial maximum ice flowed eastward and northeastward across the entire region extending from northwestern Somerset Island to north-central Keewatin, southeastward into Hudson Bay across central and eastern Keewatin, southwestward and westward across the Canadian Shield of District of Mackenzie, and westward and northwestward across Victoria Island (Dyke, 1984). At its last maximum this ice extended southward beyond the limits of the shield into the central Interior Plains (Klassen, 1989). The main ice divide, the M'Clintock Ice Divide, trended north-south (Map 1703A, Sheet 1). The position of the ice divide in Keewatin or eastern Mackenzie is only vaguely known but it lay north and west of the well known Keewatin Ice Divide (Lee, 1959a). The term "Keewatin Ice Divide" as used here refers to the axis of the late glacial Keewatin Ice, essentially as defined by Lee (1959a).

The eastward flow from the M'Clintock Ice Divide across Prince of Wales Island, Somerset Island, and Boothia Peninsula generated two large carbonate dispersal trains (Fig. 3.10). At the head of the northernmost train (on Prince of Wales Island), drumlins and flutings clearly crosscut larger and older drumlins that indicate northward to northwestward flow (Fig. 3.11). To the east, they have been obliterated by the younger flow, but to the west older drumlins are perfectly preserved, and the subsequent eastward flow is either not recorded or is indicated by a weak fluting of the till surface. Possibly the older drumlins record a flow direction from a centre to the south during the buildup phase of the last glaciation. Their preservation can be accounted for by their proximity to the M'Clintock Ice Divide (horizontal component of ice flow is low and basal sliding is minimal or zero). Hence, flow patterns can be radically different during buildup phases than during later phases, and the location of ice domes and ice divides during the last glacial maximum and during retreat is not necessarily a reliable indicator of the location of the initial accumulation centres.

Dyke (1984) suggested that the two large dispersal trains on Somerset Island and Boothia Peninsula could have only been generated by vigorously flowing ice streams bounded laterally by zones of more sluggish ice. The most intriguing aspects of these ice streams are that they crossed the regional topographic grain at right angles and that they had sharply defined edges that do not coincide with topographic breaks. Since the ice streams were not topographically generated, the question arises of what did generate them. The head of the northern ice stream on Prince of Wales Island is marked by a drumlin field showing strong convergence of flow lines (Map 1703A, Sheet 1). The drumlin fields on King William Island, at the head of the southern dispersal train, show a similar strong convergence (Map 1703A, Sheet 1). Hence, the ice streams could have been generated by convergent flow in the vicinity of the divide.

Hodgson and Vincent (1984) have concluded that the northwestern margin of Keewatin Ice floated and formed the Viscount Melville Sound Ice Shelf, but no definite evidence of an ice shelf has been found along the northeastern margin of Keewatin Ice. However, Klassen (1982) has concluded that a low elevation, horizontal moraine on northern Bylot Island was formed by an ice shelf extending along Lancaster Sound during the late part of Foxe Glaciation (Late Wisconsinan?). If Klassen is correct, that ice shelf likely extended westward to the grounding line (or margin) of Keewatin Ice (Map 1702A).

An ice cap with a radius of about 50 km covered most of Somerset Island that escaped inundation by Keewatin Ice (Dyke, 1983). The ice cap was cold-based beneath its central region but had enough erosive power to severely scour its peripheral zone. The contact between scoured and non-scoured terrain is abrupt and controlled by both small and large scale topographic changes. The ice cap coalesced with Keewatin Ice in Aston Bay and Creswell Bay; elsewhere it was separated from the main ice sheet by nunataks (Dyke, 1983, Fig. 34).

Deglaciation

Retreat of Keewatin Ice and accompanying shifts in ice divide positions and ice flow paths is shown in Figure 3.13. The ice marginal positions and ages shown on these maps are derived or modified from Christiansen (1979), Manitoba Mineral Resources Division (1981), Klassen (1983b), Dredge (1983a), Dredge et al. (1986), and Schreiner (1983) south of the territorial boundary; from Craig (1965a, 1964a,b), St.Onge et al. (1981), and Vincent (1989) for District of Mackenzie and northwestern Keewatin; and from Dyke (1983, 1984, unpublished; Vincent, 1989) for northern Keewatin and District of Franklin. In areas where local retreat patterns have not been published, the patterns shown by Prest (1969) have been largely retained.

There is close agreement among workers on the spatial patterns of retreat for the areas south of the boundary between the provinces and territories but considerable differences in the ages assigned to the ice marginal positions. The ages we suggest for ice marginal positions in that area are slightly older than those suggested by Klassen (1983b) and Schreiner (1983). We arrived at older ages for two reasons: (1) older ages (by 400-1000 years) are more compatible with the radiocarbon-dated chronology of the Arctic coast and (2) the sea entered the southwestern Hudson Bay earlier than suggested by Klassen (1983b) and by Manitoba Mineral Resources Division (1981).

Over most of the area our ice retreat pattern differs only in detail from that portrayed by Prest (1969) and our departures from Prest's interpretation cannot be shown to be correct in every case. The main differences between our ice retreat history and that of Prest are:

(1) Prest showed a major zone of uncoupling of Laurentide Ice and an ice sheet over the Queen Elizabeth Islands (the Innuitian Ice Sheet of Blake, 1970) propagating eastward from the Arctic Ocean and westward from Baffin Bay along Parry Channel. We do not recognize any evidence of confluence of Laurentide and Innuitian ice.

(2) Deglaciation occurred earlier in the Gulf of Boothia and in Coronation Gulf than suggested by Prest and initial ice retreat across Boothia Peninsula and southern Somerset Island was westward rather than southward.

(3) Prest showed Keewatin Ice separating from Labrador Ice along a line passing through central Hudson Bay southwestward from Hudson Strait. Hence, in his model, final deglaciation of coastal Manitoba and adjacent Hudson Bay involved a northwestward retreating Keewatin Ice margin. We propose that Hudson Ice retreated northeastward to or beyond the Manitoba coast

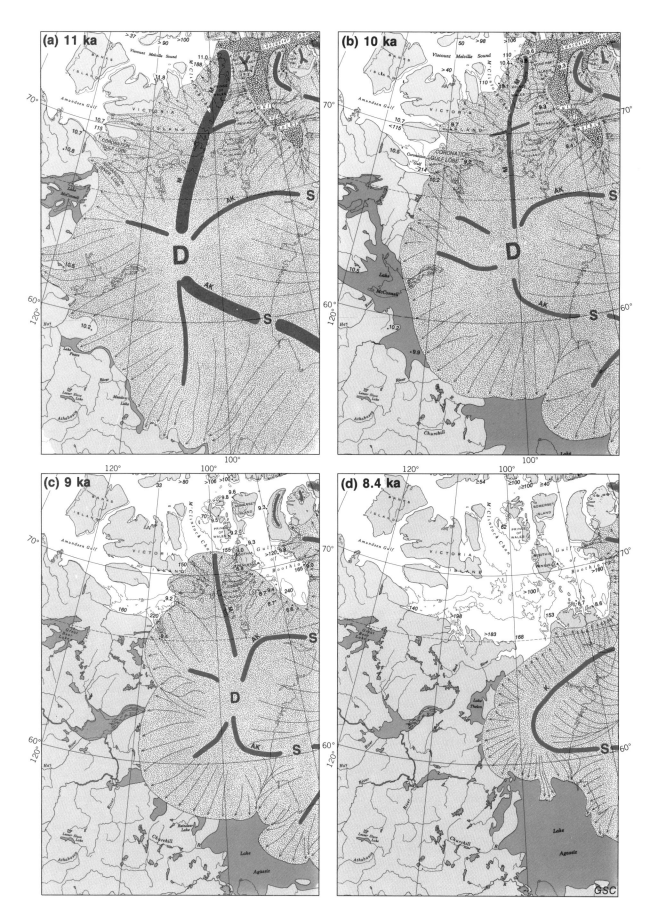

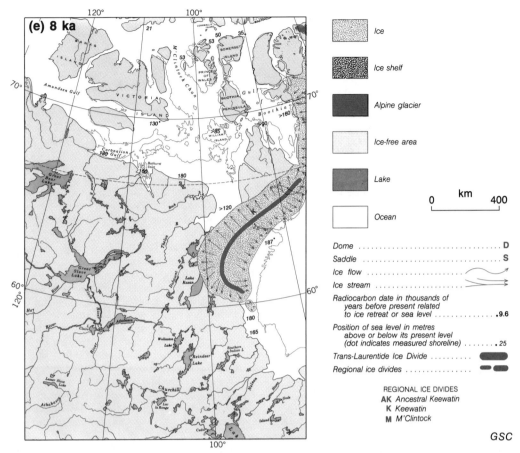

Figure 3.13. Paleogeographic maps of the northwestern Canadian Shield showing the pattern of ice retreat for (a) 11 ka, (b) 10 ka, (c) 9 ka, (d) 8.4 ka, and (e) 8 ka.

and that Hudson Ice finally separated from Keewatin Ice along a line (calving bay) in western Hudson Bay.

(4) We propose that the final ice remnants in the vicinity of the Keewatin Ice Divide disappeared shortly after 8.0 ka, whereas Prest showed them persisting until possibly as late as 7.0 ka.

The discussion below outlines the local details of the ice retreat record and comments on the likely controls on the pattern of deglaciation.

11 ka and earlier

By 11 ka retreat of Keewatin Ice was well underway everywhere except apparently along its northeastern margin (Fig. 3.13a). Glacial Lake Agassiz had come into existence along the southern margin, and drained southward via Mississippi River. Glacial Meadow Lake abutted the ice front in western Saskatchewan and spilled northward into one of the later phase of glacial Lake Peace (Mathews, 1980). Lake Peace also spilled northward with meltwater eventually finding its way into Mackenzie River. Parts of the headwaters of the Mackenzie, particularly near Great Bear Lake, were captured by isostatic tilting to form glacial Lake McConnell (Craig, 1965a).

By 11 ka parts of southwestern Victoria Island were probably ice-free (Sharpe, 1984, p. 264) and the sea had penetrated to northwestern Prince of Wales Island, immediately following breakup of the Viscount Melville Sound Ice Shelf (Hodgson and Vincent, 1984; Dyke, 1987). The ice shelf had resulted from a rapid and extensive readvance, probably a surge which started after 11.3 ka.

At 11 ka Keewatin Ice had a north-south oriented primary divide that extended from northeastern Saskatchewan to northern Prince of Wales Island (Fig. 3.13a). This divide had earlier been located farther west because the initial retreat of Keewatin Ice was entirely from the north, west, and south. The northwestern margin of Keewatin Ice began to retreat before 12.6 ka (Dyke, 1987) or at about the same time as the southern margin (Map 1703A).

Hence retreat of the north, west, and south margins of Keewatin Ice were more or less synchronous. This differed from the Baffin and Labrador sectors of the Laurentide Ice Sheet where retreat of the Labrador and Baffin Island margin began in the 8 to 10 ka range.

11 ka to 10 ka

The southern margin of Keewatin Ice continued to retreat in contact with glacial Lake Agassiz and by 10 ka the southern end of the Burntwood-Knife Interlobate Moraine had formed along the uncoupling line of Keewatin Ice and Hudson Ice (Fig. 3.13b). Some time between 11 and 10 ka The Pas Moraine was constructed near the north end of Lake Winnipeg and later was overridden during an extensive but brief readvance which carried the margin back to a location near its 11 ka position. Glacial Lake McConnell expanded southward into the Great Slave Lake basin and into the lower valleys draining into Lake Athabasca and by 10 ka nearly matched the size of glacial Lake Agassiz. Between about 11 and 10.5 ka a glacial lake was dammed in Coppermine River valley and another small lake was dammed in Richardson River valley by an ice lobe that extended westward from Coronation Gulf (St-Onge and Bruneau, 1982; Mercier, 1984). After the ice retreated east of Forcier Moraine (St-Onge, 1984), about 11 ka, the western end of the remarkably long Thelon esker formed, apparently along a substantial indentation in the ice front, as indicated by a strong flow convergence (Craig, 1964a; Prest et al., 1968; Prest, 1970).

At 10 ka Keewatin Ice still had a north-south oriented primary ice divide but it had shifted 100 km or so east of its 11 ka position (Fig. 3.13b). Very slow retreat on Prince of Wales Island had nearly stabilized the northern end of the divide. The ice flow pattern at 10 ka, derived primarily from the Glacial Map of Canada (Prest et al., 1968), suggests that several secondary divides branched off the main divide. Two of these were connected by saddles to divides in Foxe Ice and Hudson Ice.

10ka to 8.4 ka

Keewatin Ice retreated rapidly between 10 ka and 8.4 ka (Fig. 3.13c, 3.13d). Most of the Burntwood-Knife Interlobate Moraine was constructed during this interval as Keewatin and Hudson ice continued to retreat in deep water of glacial Lake Agassiz. The western margin retreated well east of the former glacial Lake McConnell basin. The Thelon esker continued to form along a major confluence, and late in this interval, Lake Thelon (Craig, 1964a) formed. Between about 9.6 ka and 8.6 ka, large glacial lakes formed southwest of Bathurst Inlet (Blake, 1963; Prest et al., 1968). During this interval most of the remarkably continuous eskers of the northwestern Canadian Shield formed.

The most rapid retreat occurred in the northern part of the region, most of which subsequently was flooded by a high postglacial sea. The ice disintegrated largely by calving, producing abundant iceberg scours on Prince of Wales and King William islands. Despite the rapidity of retreat, large end moraines were deposited on eastern Victoria Island, on eastern Prince of Wales Island, on Boothia Peninsula, and in northern Keewatin and Mackenzie. On the other hand, few end moraines were constructed south of the Thelon esker, despite somewhat slower retreat there. Some of the end moraines represent re-equilibration of the ice sheet profiles following massive calving as the margin temporarily became securely grounded at anchor points, either on shallow banks or on islands in the high postglacial sea or on dry ground south of marine limit.

The most continuous system of end moraines formed during this interval is the Chantrey Moraine System, a set of nested end moraine ridges stretching from Committee Bay to Chantrey Inlet (Falconer et al., 1965; Dyke, 1984). The youngest moraine in this system is estimated to date about 8.4 ka; MacAlpine Moraine, south of Bathurst Inlet (Blake, 1963; Falconer et al., 1965), is estimated to have formed about 8.5 ka. These periods of end moraine construction correspond to intervals of end moraine construction around the perimeter of Foxe Ice, from Melville Peninsula in the west to Frobisher Bay in the east (Cockburn Moraines, Ives and Andrews, 1963; this chapter, Andrews, 1989). Because this part of the Holocene may have been as warm as or warmer than the present, Andrews (1989) has suggested that the Cockburn Moraines signify a period of enhanced snowfall from a moisture source in Baffin Bay. Similarly, Dyke (1984) suggested that the Chantrey Moraine System recorded a period of enhanced snowfall on Keewatin Ice. The likely moisture source was the vast sea that occupied newly deglaciated areas to the north.

An interesting aspect of the ice retreat pattern during this interval is the contrast in retreat rates on opposite sides of Boothia Peninsula. Despite massive disintegration of the northern third of Keewatin Ice, including the thick central parts in the vicinity of the M'Clintock Ice Divide, the margin at the south end of the Gulf of Boothia retreated slowly, or perhaps remained quasi-stationary, even though it too stood in deep water. The near stability of the margin in the Gulf of Boothia requires a sustained and vigorous supply of ice from Foxe Basin, thus drawing a sharp contrast between the mass balance of Foxe and Keewatin ice. The vigorous flow from Foxe Basin probably indicates that the sea had not yet penetrated west of Hudson Strait and sapped the southern margin of Foxe Ice.

The southern and western margin of Keewatin Ice retreated in an orderly and parallel fashion without major readvances, except for the Quinn Lake readvance, possibly a surge into glacial Lake Agassiz in northern Manitoba (Dredge et al., 1986; Fig. 3.13d), and a similar readvance into glacial Lake Agassiz in Saskatchewan (Schreiner, 1984, p. 30). Systematic changes in the orientations of eskers and streamlined drift features indicate that ice flow lines swung gradually in response to an orderly eastward migration of the ancestral Keewatin Ice Divide. Farther north, however, where the margin was more lobate and where retreat involved rapid shortening of the M'Clintock Ice Divide, fields of crosscutting drumlins and flutes formed in response to development of calving bays, to rapid shifts in ice divide positions, and to readvances triggered by oversteepening of the ice front. Thus, crosscutting drumlin fields are a conspicuous element of the landscape of eastern Victoria Island (Fyles, 1963) and of King William Island; on Prince of Wales Island as well, deglacial flows in places followed paths very different from those followed at the last glacial maximum.

Shifts in ice flow direction are well recorded on King William Island and in northern Keewatin where five generations of drumlin and fluting fields occur. The oldest ice flow features on King William Island probably represent the flow pattern that operated at 10 ka and earlier (Fig. 3.13b). A younger drumlin field on northern King William Island includes features that in places cross the older drumlins; the younger drumlins describe a strong convergence of flow lines indicating an embayment in the ice front just off the present coast. A third drumlin field crosses southwestern King William Island and covers nearly all of Adelaide Peninsula.

These drumlins are strongly asymmetric and indicate northwestward flow. The margin of the drumlin field on King William Island is marked by a 20 km-long, broad, low, hummocky ridge of till, with the same orientation as the drumlins (Hélie, 1984). Because of its length and its position at the edge of the field, it is taken as a lateral moraine. Beyond the moraine, and apparently truncated by it, is a nested series of minor moraines indicating that the ice margin had retreated before readvancing to form the drumlin field. This readvance is interpreted as a surge involving a substantial area of ice, and a marginal fluctuation of at least 40 km. The southern part of that drumlin field was remoulded by ice flowing north-northeastward along Chantrey Inlet. Drumlins with that orientation can be traced up-ice about 200 km where they are clearly crossed by a still younger set of features indicating northwesterly flow.

By 8.4 ka the Keewatin Ice Divide had migrated to near its final position, and its south end hooked eastward in the manner portrayed by Shilts et al. (1979) and by Shilts (1980). We would like to stress here that the M'Clintock Ice Divide (Fig. 3.13e) was the primary divide of Keewatin Ice as late as 10 ka and as such was ancestral to the Keewatin Ice Divide. Although the point was made in the past by referring to the entire Keewatin ice mass as the M'Clintock Dome (Dyke et al., 1982b), unfortunately the M'Clintock Ice Divide (Dome) and Keewatin Ice Divide have been portrayed as entirely separate coeval features (Andrews et al., 1984; Shilts, 1984a).

At 8.4 ka we show Keewatin Ice entirely confluent with a dynamically separate ice mass centred over Hudson Bay (Fig. 3.13d). There is, however, no consensus on this point. For example, Shilts (1980, p. 5) felt that at this time Keewatin, not Hudson, Ice occupied most of the Hudson Bay, and that the Cochrane readvances south of James Bay emanated from Keewatin Ice: "It is tempting to speculate that within the central part of the Laurentide ice sheet the Labrador-Quebec centre of outflow predominated in the early and maximum stages of glaciation and that the Keewatin centre became relatively stronger during the later stages. Certainly during the last stages of glaciation there was a significant break in the vicinity of the Quebec-Ontario border between ice retreating towards a centre in Quebec and ice retreating generally northward or northwestward towards a Keewatin centre. The Cochrane readvances seems to have been independent of the Quebec ice and may, therefore, have emanated from an ice mass shrinking towards a Keewatin centre". More recently, Shilts (1984a, p. 123) has presented other evidence in support of a late extension of Keewatin Ice: "Recent geochemical and mineralogical evidence (Shilts, 1980; Paré, 1982) suggests a northerly or northwesterly provenance for the upper part of the uppermost till north of the Ontario-Manitoba border. The latter mineralogical data are consistent with observations of northwest-southeast oriented striae reported by McDonald (1969) on boulder pavements and bedrock at several places in the Hudson Bay Lowlands". This dichotomy of interpretation is fundamental for if Shilts and Paré are correct, the entire framework of our paleogeographic reconstruction is invalid and major questions are posed. Among these questions are: (1) If Keewatin Ice extended southeast of James Bay at approximately the same time as the youngest ridges of the Chantrey Moraine System were being formed, where was the Keewatin Ice Divide at that time? Could it still have been located over Keewatin? (2) How can we reconcile the interlobate interpretation of the Burntwood-Knife Moraine with the fact that Keewatin Ice flowed across all of northern Manitoba and into Ontario after it was formed? We feel that the Cochrane readvances were a set of surge events, which along with many others, defines a system of radial surges from Hudson Ice (this chapter, Dredge and Cowan, 1989).

The 8.4 ka paleogeography shown here also presents a modified interpretation of the extent of marine submergence south of the Arctic mainland coast. Figure 3.13d shows a broad arm of the sea extending southward along Back Lowland at 8.4 ka. Previously marine limit was thought to lie well to the north (Craig, 1961; Prest et al., 1968). Although a body of water was known to have formed strandlines in upper Back River valley, it was concluded to have been a glacial lake. There are several reasons for this change of interpretation: 1) the proposed lake is not higher than the regional marine limit and the terrain to the east is even lower; 2) there is nothing in the ice retreat pattern that necessitates damming of a lake in upper Back River valley; 3) recently determined marine limits are substantially higher than previously supposed, and the area shown as submerged lies below these; and 4) a veneer of water laid sand overlies till throughout much of Back Lowland nearly to the height of land. Thus, although marine fossils have not been found yet to prove that the sea extended as far south as suggested, this interpretation opens up the important possibility that the postglacial sea may have entered the Tyrrell Sea Basin through a corridor across Back Lowland at approximately the same time as it entered through Hudson Strait.

After 8.4 ka

The precise time when the sea entered Hudson Basin and isolated Keewatin Ice from Hudson Ice remains unknown. The oldest date on Tyrrell Sea shells is on a sample from northern Manitoba — 8530 ± 220 BP (GSC-896, age uncorrected for ^{13}C fractionation; Craig in Lowdon and Blake, 1970). These shells were considered to be contaminated, that is, mixed with shells of infinite age, because (a) they dated some centuries older than other Tyrrell Sea shells that had been dated from the region, (b) the shell fragments were well worn, and (c) old shells recycled from Bell Sea sediments are fairly abundant in the Hudson Bay Lowlands. Several corroborating dates are now available to show that the sea probably occupied the area by 7.8-8.2 ka (this chapter, Dredge and Cowan, 1989).

The pattern of retreat of Keewatin Ice after 8.4 ka is in part obvious and in part problematical. Obviously, the margin continued to retreat southward, eastward, and northward from the 8.4 ka position. In so doing, glacial Lake Kazan (Lee, 1959a) and Lake Hyper Dubawnt (Tyrrell, 1898a), and other small lakes in northern Manitoba (Dredge et al., 1986) and southern Keewatin (Gilchrist, 1982), were held up along the west and south sides of the ice. Obviously as well, the ice retreated in a simple orderly fashion inland from Hudson Bay to remnants in the vicinity of the final ice divide position (Fig. 3.13e), as indicated by De Geer moraines as well as by the esker patterns in Keewatin (Lee, 1959a; Ayslworth et al., 1981a-d; Arsenault et al., 1981, 1982). De Geer moraines and small, short eskers in northern Manitoba indicate a counterclockwise shift in ice flow direction as the ice front retreated in the vicinity of the present coast. There may have been a minor northwestward shift in the ice divide position during the final phase of deglaciation brought on by the more rapid retreat on the Tyrrell Sea side of the ice remnant.

The problematic part involves the manner in which Keewatin Ice finally separated from other ice masses in Hudson Bay. Two widely contrasting alternative interpretations exist: (1) Keewatin Ice separated from Labrador Ice along a major uncoupling zone in eastern Hudson Bay following the Cochrane readvances, as discussed above or (2) Keewatin Ice separated from Hudson Ice along an uncoupling zone in western Hudson Bay, while at the same time Hudson and Labrador ice separated along an eastern zone (Andrews and Falconer, 1969; Dyke et al., 1982b; Vincent, this chapter). In both cases these uncouplings are thought to have occurred through the continued northward extension of glacial lakes Agassiz and Ojibway along the zones of confluence between the major ice masses (as is well documented on land) and the simultaneous propagation of marine calving bays along either side of Hudson Ice. Both marine calving bays could have started in Hudson Strait, although it is possible that the western calving bay extended southward from the Gulf of Boothia. It is even possible, though not necessary, that a calving bay extending westward from Hudson Strait would stimulate development of another calving bay southward from the Gulf of Boothia. As pointed out in the preceding section, sustained vigorous flow of Foxe Ice into the Gulf of Boothia had prevented development of a calving bay there during the interval 10-8.4 ka. However, once the Hudson Strait calving bay had extended westward across the mouth of Foxe Channel, between Southampton Island and Baffin Island, drawdown of central Foxe Ice through Foxe Channel would greatly diminish supply of ice to the Gulf of Boothia.

The extent of Keewatin Ice at 7.9-8 ka is not well known. Keewatin Ice may have covered much of Keewatin and still have been connected to Foxe Ice or Keewatin Ice may have been reduced to three or more much smaller remnants (Map 1702A). One of these remnants probably was located southwest of Baker Lake. We have drawn its margin at the upstream ends of eskers depicted on the surficial geology maps referred to above. That remnant apparently degenerated into even smaller remnants, perhaps localized by individual lake basins, as suggested by Cunningham and Shilts (1977, p. 313). Another remnant of stagnant ice has been recognized in the vicinity of the Keewatin-Manitoba border and was responsible for deposition of reticulate moraines (Gilchrist, 1982; Dredge et al., 1986; Dredge and Nixon, 1986).

The case for more extensive ice at 7.9-8 ka rests on the assertion that the Kaminak Lake area did not become ice free until 6600 ± 230 BP (GSC-1434; Shilts in Lowdon and Blake, 1979, p. 37). This date is on marine shells from the base of a deltaic deposit overlying till at 61 m a.s.l. Shilts commented: "The date also closely approximates the age of the marine limit which is well marked at 170 ± 1 m on a hill 15 km from the exposure. This date should probably be used to estimate times of deglaciation and maximum marine submergence in this region as it is the only one that can be tied confidently to a stratigraphic break — the till-marine contact." If that date is correctly interpreted, then Keewatin Ice must have retreated much more slowly during the interval 8.4-6.6 ka than it did during the interval 10-8.4 ka. Considerations that suggest that retreat following 8.4 ka was as rapid as before are: (1) no substantial ice marginal deposition occurred after deposition of the Chantrey Moraine System; (2) the entire southeastern margin of Keewatin Ice terminated in the Tyrrell Sea following initial marine incursion 7.8-8.2 ka; (3) retreat rates following incursion of Tyrrell Sea were 120 to 330 m/a according to the spacing of De Geer moraines (Lee, 1959a, p. 13); (4) the Keewatin esker swarms indicate that enormous quantities of meltwater were being produced throughout the period of deglaciation, including that part which occurred after 8.4 ka; and (5) during the period in question, the climate was as warm as or warmer than at present.

On the other hand, if we assume that retreat after 8.4 ka continued at about the same rate as it did during the 10-8.4 ka interval, then very little ice would have been left in Keewatin by 7.9-8 ka. Therefore, it is quite possible that the arm of the sea extending southward from the Arctic Ocean along Back Lowland at 8.4 ka (Fig. 3.13d, 3.13e) eventually submerged the low continental divide (ca. 120-135 m a.s.l.) and extended southward into Hudson Bay. If the early deglaciation that we suggest here is correct, the sea may have entered Hudson Bay through Back Lowland and from the Gulf of Boothia as early as it did through Hudson Strait. By this means, it is possible for the sea to have reached Manitoba before it reached James Bay or even northwestern Ungava.

Activity of retreating ice

The discussion above has outlined what we know or do not know regarding the pattern and chronology of ice retreat. We have made only passing mention of the state of activity of the ice during deglaciation. Here again, however, there is no consensus. One school of thought favours massive stagnation of the retreating Keewatin Ice, either to account for preservation of ribbed moraine (Gilchrist, 1982, p. 186) or because of the presence of integrated esker systems (Shilts et al., 1979, p. 540; Shilts, 1985). We feel that parts of the retreating marginal zone of the glacier became crevassed through to its base, and hence stagnant, resulting in deposition of crevasse fillings and reticulate moraines (Gilchrist, 1982; Dredge, et al., 1986; Dredge and Nixon, 1986), such as those in northernmost coastal Manitoba and adjacent southeastern Keewatin, and that areas of ice-cored moraine may have become detached from the main glacier front during retreat, but that the main residual ice mass remained active until very near the end. The crux of our interpretation is the multiple generations of drumlin and fluting fields, each displaying a different direction of glacier flow in a successively reduced ice mass, and each associated with a companion generation of eskers and in some cases ribbed moraine. Perhaps the youngest generation of streamlined drift forms (younger than the five generations described above) is the one extending westward from near the southwest corner of Baker Lake (Prest et al., 1968; Cunningham and Shilts, 1977, p. 313). These likely were formed when the ice was reduced to the final remnants indicated on Map 1702A.

Postglacial emergence
Emergence curves

A continuing weakness of the Quaternary data base in this region is the paucity of radiocarbon-dated paleosea level positions. In addition, many of the existing dates are on marine shells that cannot be tied confidently to a given sea

(Churchill area, northern Manitoba). These curves are nested in the theoretically expected manner. It has not always been clear, however, that Keewatin Ice left its own distinct isostatic signature, as is indicated by the fact that the central Keewatin curve lies well to the left of the Churchill and Bathurst Inlet curves. In other words, more rebound has occurred since a given time beneath the central Keewatin Ice than has occurred either to the northwest or southeast.

The Churchill and central Keewatin curves can also be used to attempt a preliminary answer to the question of the date of deglaciation of these areas. The marine limit near Churchill lies in the 164-180 m range (Prest et al., 1968; Dredge and Nixon, 1986). Extrapolation of the Churchill curve suggests that the age of that feature is about 8.25 ka, and therefore the date of 8530 ± 220 BP (GSC-896) should not be disregarded without better reasons than have been advanced. The marine limit near Chesterfield Inlet, central Keewatin lies at 187 m (Lee, 1959a). Extrapolation of the Keewatin curve suggests that the limit may be as old as 8.3 ka. The marine limit near Kaminak Lake is at least 170 m a.s.l. (Shilts in Lowdon and Blake, 1979, p. 37). By extrapolation of the Keewatin curve, it could be as old as 7.9 ka. The suggested age of 6.6 ka for that feature cannot be taken as unequivocal because the dated sample came from a delta at only 61 m a.s.l. Hence, although these curves do not clinch the argument for relatively early disappearance of the last remnants of Keewatin Ice, they do add substance to it.

Early emergence rates, when averaged over the first 1 to 2 ka of record, are similar throughout most of the region, being approximately 10 m/century in Bathurst Inlet, 10 m/century in northernmost Keewatin, and 9-11 m/century along northern Somerset Island. When averaged over a shorter interval, however, specifically the brief interval of local deglaciation (200 years), early (deglacial) emergence rates on Boothia Peninsula were more than 30 m/century (Dyke, 1979a, 1984). In all likelihood, the early steepest parts of the central Keewatin and Churchill curves have been eclipsed by the late deglaciation of these sites relative to the other sites shown here, and by the fact that these sites were deglaciated at the very end of the process of regional unloading. Late Holocene and ongoing emergence rates are expectedly highest closest to the centre of the former ice load; they are about 1.3 m/century in central Keewatin, 0.66 m/century in Bathurst Inlet, at least 0.53 m/century in northernmost Keewatin, and 0.46 to 0.28 m/century along northern Somerset Island, declining eastward. Where the littoral zone slopes at 1:600, this translates to a contemporary shoreline regression rate of 3.18 m/a in northeastern Keewatin and 7.8 m/a in central Keewatin.

Marine limits

It follows from the discussion of both deglaciation and postglacial emergence curves that the marine limit in the region is a strongly diachronous feature ranging in age from 12.6 ka or older in the northwest (Victoria Island) to ca. 8 ka in the western Tyrrell Sea. Interpretation of the pattern of marine limit elevation (Fig. 3.15), therefore, must include this characteristic. The pattern of highs and lows is a result of both the pattern of original crustal depression and the timing of deglaciation.

The marine limit surface is dominated by two pronounced highs — one centred south of Bathurst Inlet and one forming a ridge over northeastern Keewatin and

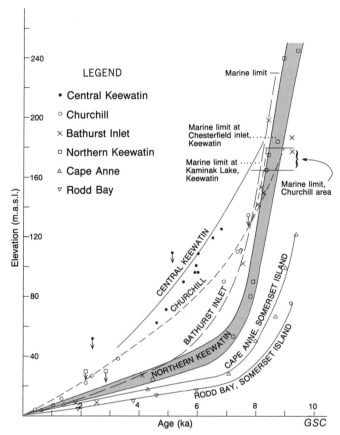

Figure 3.14. Family of emergence curves for Bathurst Inlet, northern Keewatin, Somerset Island, and Churchill area, Manitoba.

level. The largest groups of radiocarbon dates are concentrated in the Bathurst Inlet area (Blake, 1963), in northernmost Keewatin (Dyke, 1984), on Somerset Island (Dyke, 1979b, 1980a, 1983), and in the Churchill area of northern Manitoba. Of these four areas, only the Somerset Island data are of sufficient quality to construct a set of emergence curves; an emergence envelope, which describes the sea level history of a sizeable region, can also be constructed for northernmost Keewatin; and minimum sea level curves can be constructed for Bathurst Inlet and northern Manitoba (Fig. 3.14). In addition, part of a minimum emergence curve can be constructed for central Keewatin (in the vicinity of the Keewatin Ice Divide).

This family of curves (Fig. 3.14) can be used to illustrate several important points or to quantify theoretically predictable relationships. The nesting of the curves illustrates the relative isostatic response of: 1) an area beyond the limit of Keewatin Ice, but near the limit of a local ice cap on Somerset Island, (Rodd Bay); 2) an area near the limits of both Keewatin and Somerset ice (Cape Anne); 3) two areas well behind the ice limit, with similar dates of deglaciation, but on opposite sides of M'Clintock Ice Divide (northeastern Keewatin and Bathurst Inlet); 4) a site near the centre of the Keewatin Ice load (central Keewatin); and 5) a site near the zone of confluence of Keewatin and Hudson ice,

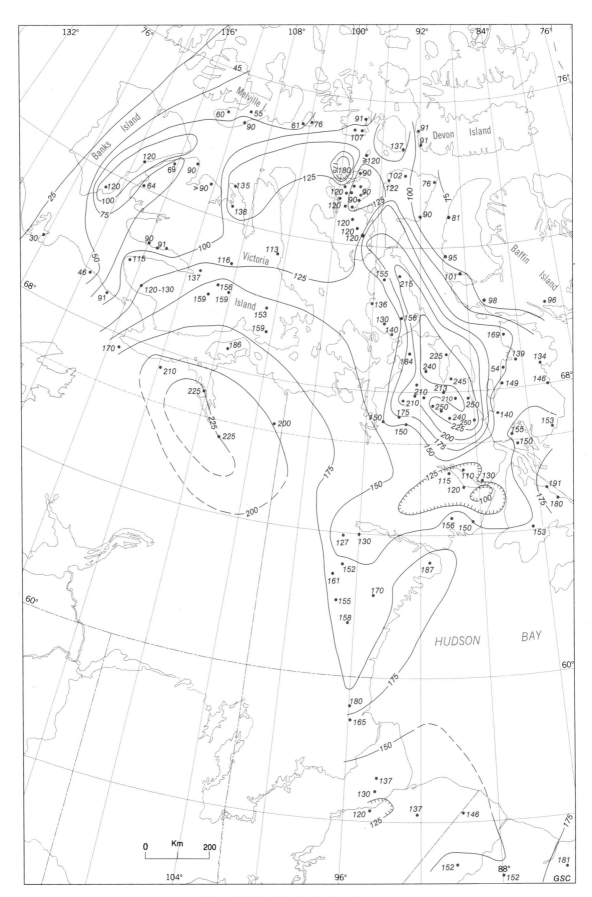

Figure 3.15. Marine limit isoline map.

Boothia Peninsula. In the area between these two highs, marine limit declines by nearly 100 m. Thus, although the lowest contours (100 m and less) describe a rise in elevation of marine limit inland from the Late Wisconsinan glacial limit, there is a trough in the surface situated approximately beneath the former M'Clintock Ice Divide. This reflects the ice retreat pattern described above with ice persisting relatively late in the vicinity of the ice divide. The marine limit elevations in northeastern Keewatin and Bathurst Inlet, up to 250 m, are among the highest in Canada, exceeded only by those in the southeastern Hudson Bay. However, these high marine limits require a different interpretation than that which has been offered traditionally for the high marine limit in the southeastern Hudson Bay basin. They represent not a centre of loading but regions into which the postglacial sea entered relatively early in the deglacial sequence.

Marine limit in the western part of Tyrrell Sea Basin is more uniform in elevation than in areas to the north, perhaps reflecting near synchronous establishment following the rapid or instantaneous disappearance of Hudson Ice, perhaps because it became buoyant and floated. The most pronounced feature of the marine limit surface in that region is the closed depression in the vicinity of Wager Bay. This could indicate the presence of a late ice remnant, similar to the one already proposed for the area southwest of Baker Lake, where marine limit appears to be low as well. We show another low on the marine limit surface centred on the Nelson River area of northern Manitoba. This feature does not appear to relate to a late ice remnant and the low, therefore, may reflect the location of the boundary between the isostatic domains of Keewatin and Hudson ice. Alternatively, the actual marine limit might not have been found.

Marine limit elevations increase eastward across Melville Peninsula and across Southampton Island, a result of the isostatic response to Foxe Ice.

Isobases

At present the best dated and surveyed shoreline in the central Arctic is the feature formed about 9300 BP (Fig. 3.16). Within our region the best control on its elevation lies

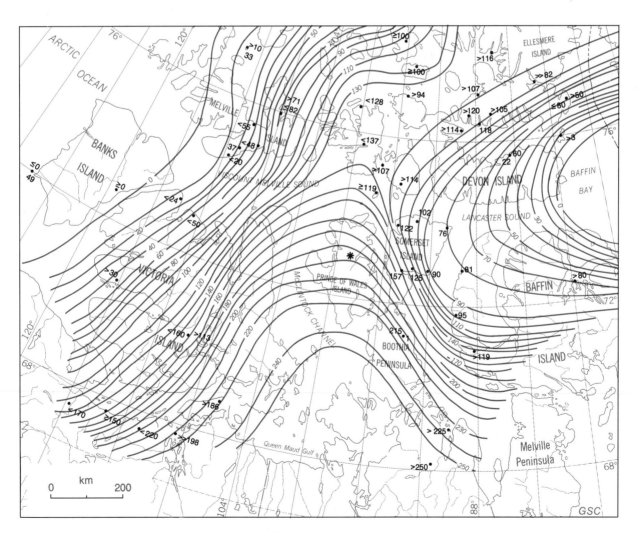

Figure 3.16. Isobase map for 9.3 ka (from Dyke, 1984). The * shows location of ice contact marine delta dated 9470 ± 100 BP (GSC-3679) which is discussed in the text.

within the Somerset Island/Boothia Peninsula area (e.g., Dyke, 1979a,b), where it is distinctly tilted up towards the west. Dyke (1984) contoured the elevation of that paleoshoreline based on 49 control points, and concluded that its gross configuration reflected the location of the M'Clintock Ice Divide.

Additional data were gathered on Prince of Wales Island after Figure 3.16 was published. Marine limit lies at about 95 to 100 m along the entire northeast coast of Prince of Wales Island and on adjacent eastern Russell Island (Fig. 3.15). Shells from a glaciomarine delta on Prince of Wales Island (Fig. 3.16) were dated 9470 ± 100 BP (GSC-3679). Figure 3.16 predicts that the 9300 BP shoreline at that site should lie between 180-200 m a.s.l.; instead it lies at or slightly below 95 m. The only reasonable explanation of the anomalously low elevation of the marine limit and the 9300 BP shoreline on Prince of Wales Island that can be suggested at present is differential tectonic movement between Prince of Wales Island and Somerset Island and Boothia Peninsula.

Postglacial climate and vegetation changes

The northwestern Canadian Shield is roughly bisected by northern treeline. Early studies (Terasmae and Craig, 1958; Nichols, 1967; Sorenson et al., 1971) demonstrated that treeline, a major bioclimatic boundary controlled by the mean July position of the Arctic Front (Bryson, 1966), fluctuated substantially during the Holocene. Unfortunately these few studies have not been augmented, and our present interpretation still rests on only three primary sites.

Nichols (1972) has summarized palynological data from the region. The most instructive record is that from Ennadai Lake in southern Keewatin. There the climate was generally warmer and wetter than present between 2.5 ka and the start of record about 6 ka. Although fluctuations of climate occurred during that interval, involving a relatively cool period between 5 and 4.5 ka and a relatively cool and variable period between 3.5 and 2.5 ka, treeline was displaced northward of its present position by as much as 250 km and may have extended to Dubawnt Lake. Coker and DiLabio (1979) have suggested that there was no permafrost in northwestern Manitoba during that period. An abrupt climatic deterioration at 2.5 ka involved a drop in mean July temperature of about 2°C and a retreat of treeline south of its present position. This cool period, which corresponds with the onset of Neoglacial conditions globally, was interrupted by warmer than present conditions during the interval 1.4 ka to about 0.7 ka when treeline advanced about 100 km north of its present position. Cooler than present conditions, with a treeline south of its present position, characterized the last 700 years, with temperature returning to present levels during the historical warming of the present century.

There are only limited proxy climatic data for the rest of the region. Interestingly, vegetation, including *Salix arctica*, was established on northern Somerset Island immediately upon local deglaciation, when most of the island was still ice covered (Dyke, 1983); buried soil with *Salix* macrofossils have been dated there at 9.5 ka. Somerset Island also has a minor Neoglacial record. In the very recent past, probably during the Little Ice Age, 12 small ice caps grew on the plateau area of the island in response to a depression in the glaciation level to about 300-350 m a.s.l. (Dyke, 1983, p. 29).

QUATERNARY GEOLOGY OF THE SOUTHWESTERN CANADIAN SHIELD

L.A. Dredge and W.R. Cowan

INTRODUCTION

This section provides a brief overview of the Quaternary geology and history of the northern Canadian Shield in Ontario, that part of the shield in northeastern Manitoba affected primarily by the Hudson Ice domain of the Laurentide Ice Sheet, and the Hudson Bay Lowlands (Fig. 3.1). Essentially the area is bounded by the Harricana interlobate moraine on the east, the Burntwood-Knife Interlobate Moraine system on the west, northern Hudson Bay on the north, and Lake Superior on the south. Of necessity some discussion will exceed the limits of these boundaries, particularly in the Lake Superior region. Much of the region has been mapped only at reconnaissance scales (Map 1704A). Although the area is a key one for Quaternary glacial stratigraphy in North America, a great deal of basic work remains to be done; many questions are yet to be answered and asked.

REGIONAL SETTING
Bedrock geology

The area is underlain by Precambrian granites, gneisses, metasedimentary, and metavolcanic rocks of the Canadian Shield, which are covered by sedimentary rocks of Phanerozoic age in the Hudson Bay area (Fig. 3.3). The glacially dispersed by-products of these rock types have been used to determine patterns of ice flow.

The Precambrian rocks consist mainly of east-west trending belts of metasediments and greenstones interspersed with granitic intrusives and gneisses. Glaciers crossing these belts commonly carried mixed suites of erratics, but in a few cases, erratics can be traced to unique source areas. Such rocks have been used to reconstruct local ice flow patterns.

Dredge, L.A. and Cowan, W.R.
1989: Quaternary geology of the southwestern Canadian Shield; in Chapter 3 of Quaternary Geology of Canada and Greenland, R.J. Fulton (ed.); Geological Survey of Canada, Geology of Canada, no. 1 (also Geological Society of America, The Geology of North America, v. K-1).

Southward ice flow is indicated by dispersal trains of jasper-bearing quartzites and tillites from source rocks in the Sault Ste. Marie-Haileybury region, and of the Grenville marbles farther east. Southward flow from the James Bay region is also suggested by the dispersal of erratic diamonds from lamprophyric or kimberlitic intrusions (e.g., Gunn, 1968; Satterly, 1971; Wolfe et al., 1975; Brummer, 1978, 1984). The Proterozoic dolostone, iron formation, greywacke, argillite, quartzite, chert, conglomerate, and diabase (Sanford et al., 1968; Bostock, 1969) indicate more widespread dispersal to the west, southwest, and southeast.

Belcher Island rocks contain oolitic jasper (Kipalu Formation), black to dark grey siltstone and greywacke, and concretionary siltstone and greywacke associated with the Omarolluk Formation (Ricketts and Donaldson, 1981). These rock types have been identified in southwestern Ontario (E.V. Sado, Ontario Geological Survey, personal communication, 1984), throughout much of the area north of Lake Superior (Prest, 1963), and even as far west as Alberta.

The Paleozoic carbonate rocks of the Hudson Bay Lowlands provide the best source rocks for detecting regional dispersal patterns. Carbonate drift from the lowland fans westwards across the Canadian Shield into north-central Manitoba, and southwards to Lake Superior (Geddes et al., 1985) and central Ontario.

Physiography

Principal physiographic subdivisions of the region under discussion are the Hudson Bay Lowlands on the north, the Great Lakes basins on the south, and the intervening Precambrian uplands. Bostock (1970a, b) referred to the lowlands as the Hudson Region, and divided the Canadian Shield into the James and Laurentian regions, which he further subdivided according to elevation and bedrock type (Fig. 3.17).

The Hudson Bay Lowlands comprise a low-lying bedrock plain, formed by subaerial denudation in post-Devonian times (this chapter, Dyke and Dredge, 1989) and mantled to a variable degree by glacial and marine sediments. The area is low and almost flat; elevations rise gently from sea level to 120 m a.s.l. except for the Sutton Hills inlier of Proterozoic rocks, which stands about 150 m above the surrounding surface.

The Severn, Abitibi, and Kazan uplands are low and rolling with local relief exceeding 60-90 m only in deeply incised canyons. Elevations rise from near sea level to 500-600 m. Most of these areas are underlain by Archean rocks of varied lithologies but are highlighted by numerous greenstone belts traversing granitic or gneissic terrain. These areas have been subjected to several long periods of subaerial erosion during the late Archean and early Proterozoic, late Proterozoic, much of the Mesozoic, and probably throughout much of the Tertiary. Through interpretation of the structural level of Proterozoic metamorphic rocks in the Southern Province, Young (in Card et al., 1972) has estimated that at least 6000 m of erosion has occurred since the last mountain building episode. Ambrose (1964) catalogued exhumed paleoplains of the shield areas and concluded that much of the present day shield topography is pre-Paleozoic, perhaps as old as mid-Proterozoic, and that in some cases paleoplains have been buried and exhumed several times without much topographic change. In addition, scientists have concluded that overall erosion of the shield area by Pleistocene glaciers was relatively modest, perhaps only a few metres (e.g., Lawson, 1890; Jahns, 1943; Smith, 1948; Flint, 1971; Gravenor, 1975; Mathews, 1975; Sugden, 1976a, b; Kaszycki and Shilts, 1980).

Nipigon Plain, Port Arthur Hills, Penokean Hills, and Cobalt Plain (Bostock, 1970b) are diverse subregions within the James Region. Nipigon Plain, surrounding Lake Nipigon, is underlain by nearly flat lying Keweenawan gabbro sills and sediments which have been deeply dissected by glacial meltwaters. It occurs at an elevation of about 270 m and has local relief of the order of 50-200 m. Port Arthur Hills consist of Proterozoic diabase sills and comprise a series of south-dipping cuestas and ridges which reach a maximum elevation of 482 m a.s.l., 299 m above Lake Superior. Penokean Hills have a local relief of up to 100 m and are composed of folded Huronian sediments with quartzite ridges generally reaching elevations of 240-300 m a.s.l. Cobalt Plain is composed of nearly flat-lying Huronian clastic sediments with some ridges and hills formed by inliers of Archean rocks or gabbro sills.

That part of the area lying south of the Grenville Front includes the Frontenac axis and Haliburton Highlands, parts of Bostock's (1970b) Laurentian Highlands. This subregion generally presents a rounded landscape of low knobs and ridges, frequently referred to as mammillated topography, with maximum elevation of 600 m (similar to that of the Severn and Abitibi uplands).

Flanking the Canadian Shield area on the south are the two largest of the Great Lakes — Superior and Huron. The Lake Superior basin occupies a syncline which may have originated as a result of crustal rifting and plate separation between 1.3-0.6 Ga (Card et al., 1972; Wallace, 1981). Although the Lake Superior basin is primarily tectonic in origin, it has been further modified by fluvial and glacial erosion. The Lake Huron basin (and Lake Michigan basin to the south) are structurally controlled and have their long axes associated with more easily eroded units of the Michigan Structural Basin (Hough, 1958). These lake basins are believed to have been the locus of preglacial river systems which were subsequently modified by Pleistocene glacial erosion (Hough, 1958).

SURFICIAL MATERIALS AND LANDFORMS

Figure 3.18 shows the general distribution of surface materials and some of the main glacial landforms. The text draws attention to the nature and distribution of surface materials within the southwestern Canadian Shield and material-landform associations. Most discussion centres around distributions of materials of Holocene and Late Wisconsinan age, but reference is also made to older Quaternary stratigraphic units.

Till and till forms

Distribution and thickness. Till mantles much of the bedrock within the region (Fig. 3.18). On the Canadian Shield, till seldom exceeds 4 m in thickness, although thicknesses of more than 30 m have been noted in some places (e.g., the Geraldton and Hemlo areas of Ontario). Commonly only one till sheet is recognized, and there are substantial areas without significant drift cover. In contrast,

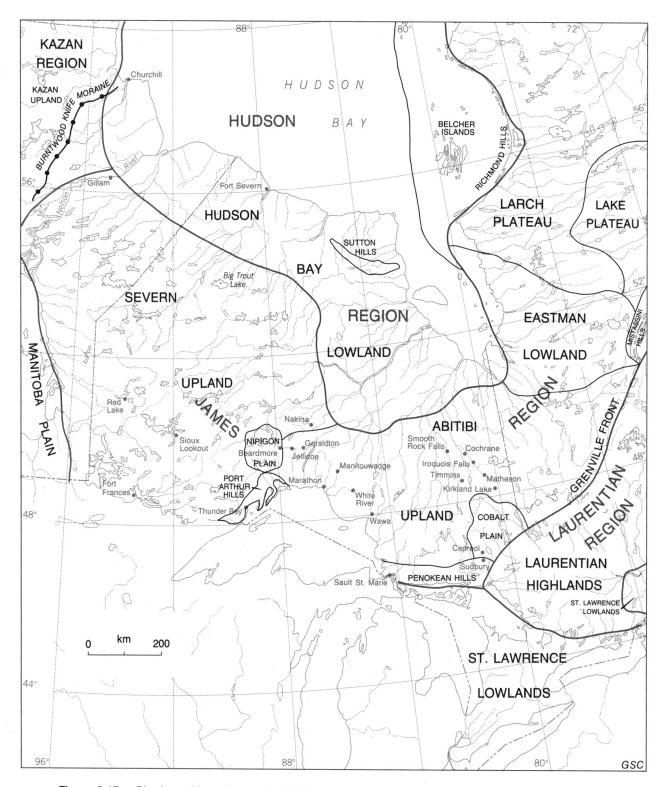

Figure 3.17. Physiographic regions and subdivisions of the southwestern Canadian Shield (from Bostock, 1970a).

QUATERNARY GEOLOGY — CANADIAN SHIELD

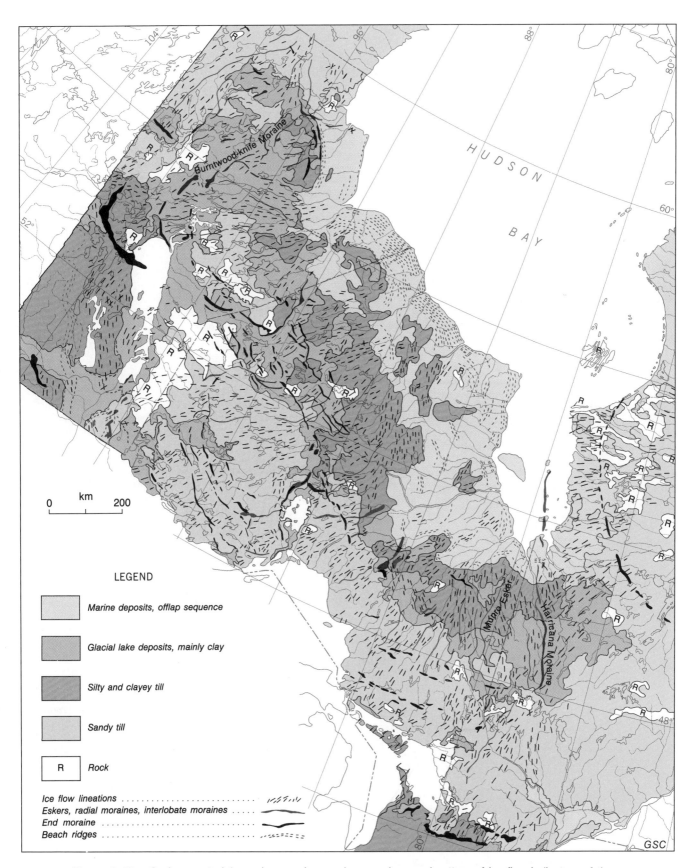

Figure 3.18. Surface materials, major moraines, eskers, and general pattern of ice flow indicators of the southwestern Canadian Shield.

in the Hudson Bay Lowlands, the till is almost continuous and is considerably thicker than on the shield. Broad preglacial valleys (e.g., Nelson River, Manitoba) have been infilled with more than 60 m of till, and a thickness of 145 m has been logged during drilling programs sponsored by the Ontario Geological Survey. In the lowlands as many as seven separate till sheets, either stacked one upon another or separated by nonglacial beds, have been recognized.

Compositional types. Two basic types of till occur in the region. The first is derived from Precambrian Shield terrain, where rocks are predominantly of igneous and metamorphic lithologies; the tills derived from these source rocks contain crystalline clasts, are sandy textured (average grain size of the matrix is 75% sand, 20% silt, 5% clay), and have matrix carbonate contents of less than 4% by weight. The second type of till is largely derived from Paleozoic carbonate rocks of the Hudson platform. These tills have an abundance of limestone clasts (up to 80% of the pebbles), a clayey silt matrix texture (typical grain size of the matrix is about 33% sand, 45% silt, 22% clay), and high matrix carbonate contents (35% by weight: range 9-52%). Where ice has crossed more than one rock type or has overrun older glacial deposits of different provenance, tills with intermediate characteristics are common.

Figures 3.19 and 3.20 show the regional distribution of carbonate in tills, and the progressive dilution of Paleozoic carbonate (by incorporation of noncalcareous Precambrian lithologies) southwards and westwards from Hudson Bay. The Nipigon and Chapleau moraines mark the southern limit of calcareous tills in Ontario (Karrow and Geddes, 1987). No carbonates were identified in tills from Sault Ste. Marie, or from the Lake Superior basin (Dell, 1975); however, Sado (1975; Ontario Geological Survey, personal communication, 1984), Geddes (1984), and Geddes et al. (1985) identified carbonate-enriched drift in the Geraldton-Hemlo-White River area. They found that carbonate clast contents were highest in the supraglacial (ablation) till facies (up to 38% carbonate), and lowest in lodgment and basal melt-out tills. The carbonate contents in the Ontario tills decrease abruptly west of Geraldton; and Hudson Bay carbonates have not been found west of the Nipigon Moraine (Zoltai, 1965). The pattern of carbonate dispersal for northern Manitoba (Fig. 3.20) shows regional dilution of Hudson Bay carbonates westwards as far as a line extending from Tadoule Lake to the northwest corner of Lake Winnipegosis. Local dispersal trains, indicative of ice streams, are superposed onto this general pattern. In addition to the carbonates delivered to the shield area from the Hudson Bay Lowlands, there is an area near the southern Manitoba/Ontario border that has had carbonate till emplaced by ice from the St. Louis Sublobe (Nielsen et al., 1982; Zoltai, 1962), which flowed eastwards across carbonate bedrock in southern Manitoba. Similarly, carbonates, derived from lower Ottawa River valley have been mixed with tills deposited on the Frontenac Arch (Kettles and Shilts, 1987).

A third type of till, which is clay-rich and stone-poor, overlies the regional sandy and silty till sheets fringing Hudson Bay. This till — known as till of the Cochrane Formation (Hughes, 1965) after the locality where it was first recognized — was generated by the glacial overriding of glacial lake deposits and is considered to have been produced by surging.

Till landforms. The predominant landforms related to till deposition in this area are featureless plains which commonly mimic the underlying bedrock topography, fluted terrain, and end moraines. Drumlinized or fluted terrain occurs in extensive fields whose edges either converge or crosscut one another. Convergent fields are separated by moraines oriented parallel to ice flow. The fluted fields are assumed to reflect regional flow zones within the Hudson Ice mass ice. Crosscutting patterns indicate the changes in ice flow which accompanied deglaciation.

Although normally fluted forms are associated with regional flow patterns, a second type of streamlined terrain covers parts of northernmost Ontario and northeastern Manitoba. These features are parallel, delicate, highly elongated streamlined forms, that occur in sharp-edged swaths roughly 10-25 km wide and 50-75 km long. The southernmost extremities of some flutings within a single swath curve and intersect one another. Flutings are developed preferentially on topographic rises and are absent in intervening low areas. The flutes are formed of clayey till derived from glacial lake deposits, in particular, overridden varves. The flutes therefore are assumed to record ice marginal fluctuations, occurring as local readvances into the proglacial lakes. Near Cochrane, Ontario where the Cochrane till was first examined, the varve record indicates that the fluctuations there lasted 25 years (Hughes, 1965). The streamlined swaths are therefore thought to have been produced by surging. Other features support this idea: their limited extent indicates that some parts of the ice margin abruptly gave way, while others remained intact; the sharp lateral edges and parallel nature of the streamlining suggest that these areas had been functioning as ice streams — zones of faster moving ice possibly generated by local calving of the ice front. Other aspects suggest that these readvances occurred under conditions of low basal shear stress. The location of flutes on topographic highs and their absence in intervening lows suggest that the ice sheet was partially floating near the margin. What is not clear is what caused these (surge) events, which, from the distribution of the features mentioned above, appear to have occurred at the periphery of the Hudson Ice mass during the last phases of deglaciation. The presence of the impermeable substrate (glacial lake clay and silty till) may have created surge-prone conditions by inhibiting drainage, thereby creating a pressurized basal water layer which acted as a slippage plane at the base of the ice. Also, local concavities along the ice margin (small calving bays) may have caused local profile oversteepening, which in turn led to forward flow, drawdown, and the generation of ice streams (a situation analogous to ice streaming observed within the Antarctic ice sheet today; Hughes et al., 1981). The threshold between normal advance and surging probably occurred where (or when) the ice was thin enough, or water depth was deep enough, for the ice front to float. The yield stress along the ice margin would then decrease to zero and allow the margin to surge forward. The extent of the surge would be limited by the driving force supplying the ice shelf, that is, the generation of feeder ice streams, and by buttressing by the ice shelf. The observed streamlined forms mark locations where the advancing ice shelf grounded on rises. The arcing and crosscutting termini of some of these streamlines suggest that extending, unconfined flow of the shelving margin caused the ice to break up into ice bergs. There may have been numerous surging shelves which did not scrape topographic highs and are thus recorded only as rain-out diamictons within the varve sequences.

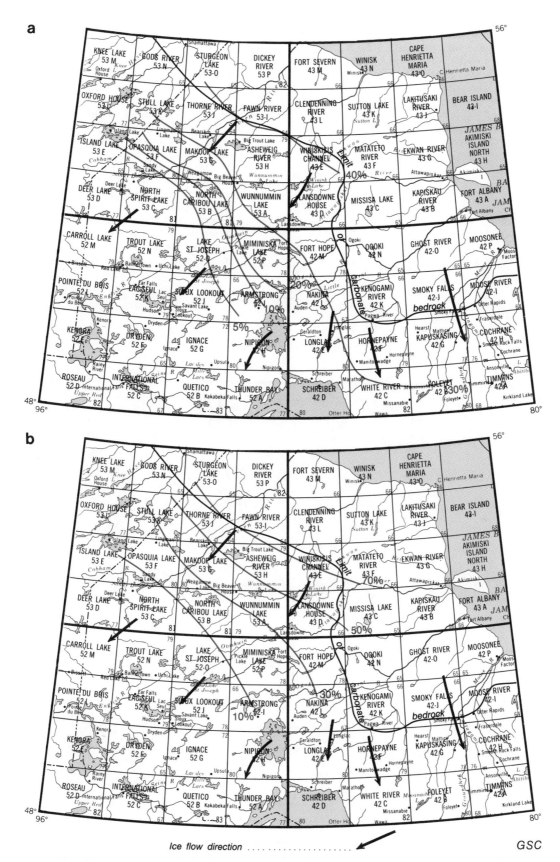

Figure 3.19. Dispersal of Paleozoic carbonates derived from Hudson Bay Lowlands over northwestern Ontario: a) carbonate in till matrix; b) carbonate pebble content.

Major end moraines in the northwestern Ontario shield are composed of till and sorted ice contact debris. Some workers have contended that these features mark regional readvances (e.g., Zoltai, 1965; Prest, 1970), a view not held by the present authors who regard them as primarily recessional features. The clayey moraines in northern Manitoba, which form gentle, broad swells in the landscape, are considered to be subaquatic grounding line moraines. De Geer moraines in north-central Ontario are associated with backwasting of the Windigo Lobe. These and ribbed moraine are relatively minor features in the region.

Glaciofluvial deposits

Most glaciofluvial deposits in the southwestern region of the Canadian Shield are of ice contact origin and display fluviodeltaic facies characteristics. Because the ice retreated downslope in most of the area and the ice margins lay against glacial lakes, proglacial outwash generally did not develop. The principal glaciofluvial landforms include interlobate moraines, radial moraines, eskers, and the end moraines mentioned above. Interlobate and radial moraines are broad, massive ridges consisting of till and coarse gravel, separated by zones of predominantly fine sand. They are thought to have formed along sutures between confluent ice masses. The Harricana and Burntwood-Knife interlobate moraines are regionally significant features which developed between major ice domains (Fig. 3.21). The term interlobate moraine has also been used to denote kame moraines, radial to ice flow, which have developed between ice lobes within a major ice domain; e.g., the Hayes Lobe in Manitoba (Fig. 3.22) is bounded by the Limestone and Sachigo interlobate moraines. Radial moraines mark the junction between smaller scale flow discontinuities. Eskers vary greatly in size from major features, such as the Munro Esker

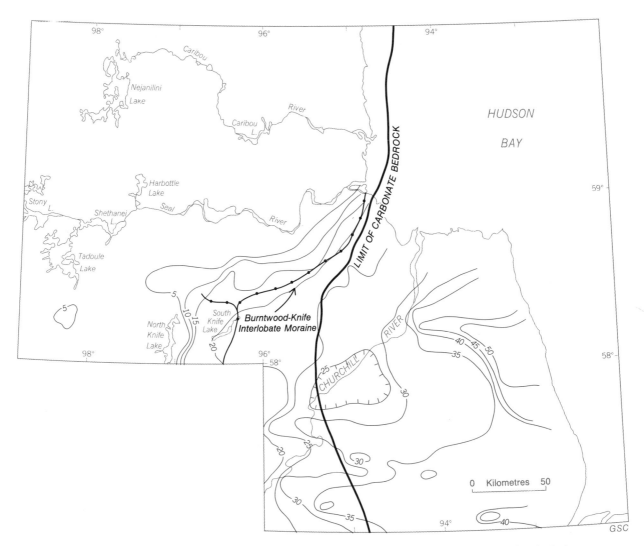

Figure 3.20. Carbonate dispersal in northern Manitoba: carbonate content (% of the < 2 mm fraction) of surface till. General pattern shows the transport westward from the area of Paleozoic carbonates; anomalously high values within the area of carbonate-rich till mark the sites of ice streams within Hudson Ice (after Dredge, 1983c).

(Fig. 3.18), to narrow ridges. They are fairly uniformly scattered over the area but are generally abundant on the proximal side of major end moraines. The eskers display sequences of deltaic sedimentation; fine sand is the dominant material in many of these features.

Two types of minor buried glaciofluvial deposits may be seen in sections. The first consists of crossbedded gravels which separate till sheets and are commonly associated with other nonglacial beds; these are interpreted as fluvial or glaciofluvial deposits and are associated with receding ice sheets. The second comprise seams of massive sorted sand, or sand and gravel, commonly separating till sheets of a single glaciation. These occur most commonly in areas that lay at the edges of ice streams or ice lobes. They are believed to represent either subglacial sheet flow or incipient short lived interlobate or radial moraines. Shifting ice flow led to glacial overriding and destruction of internal sedimentary structures in these relatively thin deposits.

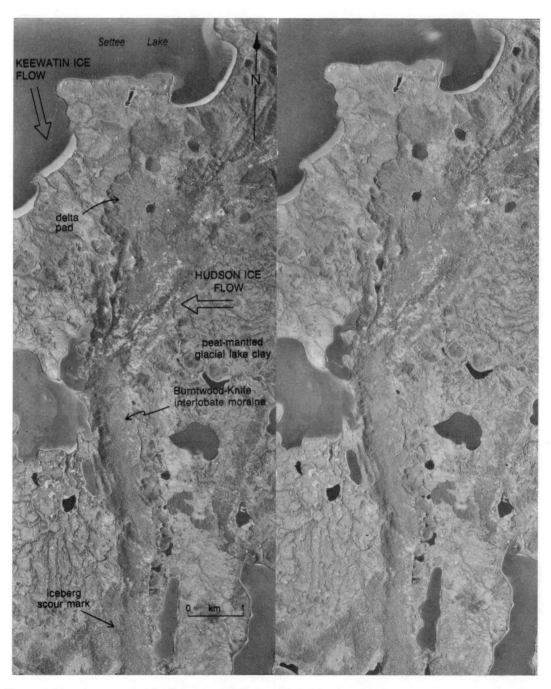

Figure 3.21. A segment of the Burntwood-Knife Interlobate Moraine near Settee Lake in northern Manitoba. The stereopair shows the main ridge with attached deltas. The moraine consists predominantly of sand whereas adjacent areas are glacial lake silts mantled by bogs and fens. NAPL A14988-17, 18

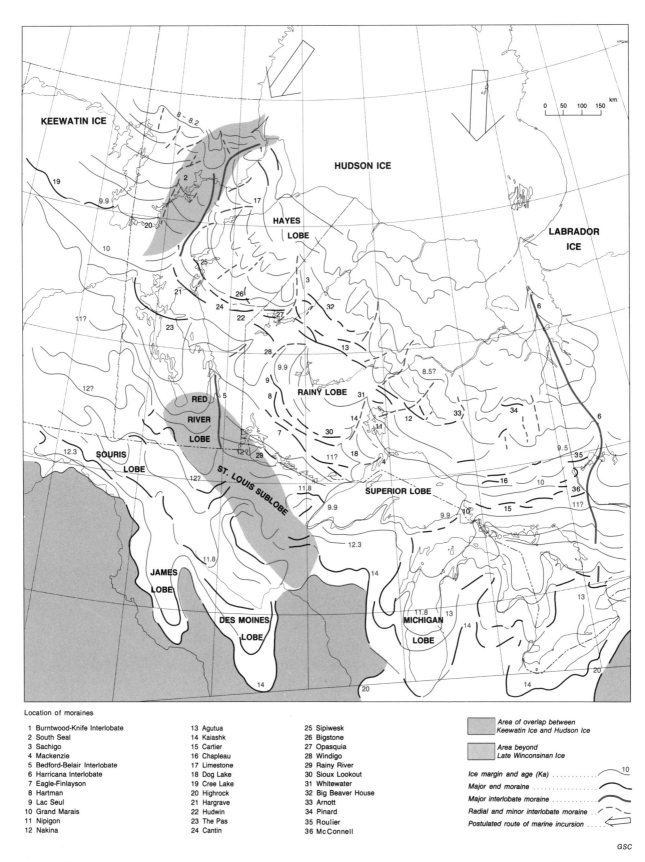

Figure 3.22. Ice retreat pattern for the area south of Hudson Bay. The digitate pattern of the former ice margins is characteristic of a highly mobile ice mass. Ice margins modified from Prest (1969).

Glacial lake deposits

Distribution. Glacial lake sediments were deposited in the proglacial lakes that developed during deglaciation while the land either was isostatically depressed or drainage into Hudson Bay was blocked by the continental ice sheet (Fig. 3.23). The largest lakes were glacial lakes Agassiz (western part of area) and Barlow-Ojibway (eastern part of area), which covered thousands of square kilometres, though not concurrently. Glacial lake sediments cover much of the till plains in the areas submerged by lakes Agassiz and Barlow-Ojibway and extend beneath marine sediment at least as far north as the present coast of Hudson Bay (Fig. 3.18). In northeastern Ontario, glacial Lake Barlow developed in the Lake Timiskaming basin and expanded northward against the receding ice front. The lake was called Lake Ojibway where it occupied areas north of the Hudson Bay-Great Lakes drainage divide (this chapter, Vincent, 1989b). Deposits range in thickness from a thin sand or clay veneer over till or rock, to clay deposits more than 60 m thick (locally in Back River valley near Marathon, on the north shore of Lake Superior, glacial lake sediments exceed 100 m in thickness).

Sediment facies. Five sediment facies have been recognized: stony clay, sandy substratified turbidite units, varves, massive clay, and littoral sand. Glaciolacustrine sediment at the base of the sequence consists of a stony clayey diamicton, similar to the regional till sheet, but less dense and finer grained. This deposit is considered to have been emplaced by rainout, possibly from beneath an ice ramp extending beyond the grounded glacier, into the glacial lake. The next facies is a sandy diamicton (Nielsen and Dredge, 1982), consisting of turbidity flow, massive, and laminar flow subunits. This material was deposited from turbidity currents where subglacial conduits debouched into the glacial lake.

Varved silts and clays are commonly made up of alternating beds of red-brown and grey sediments (Fig. 3.24).

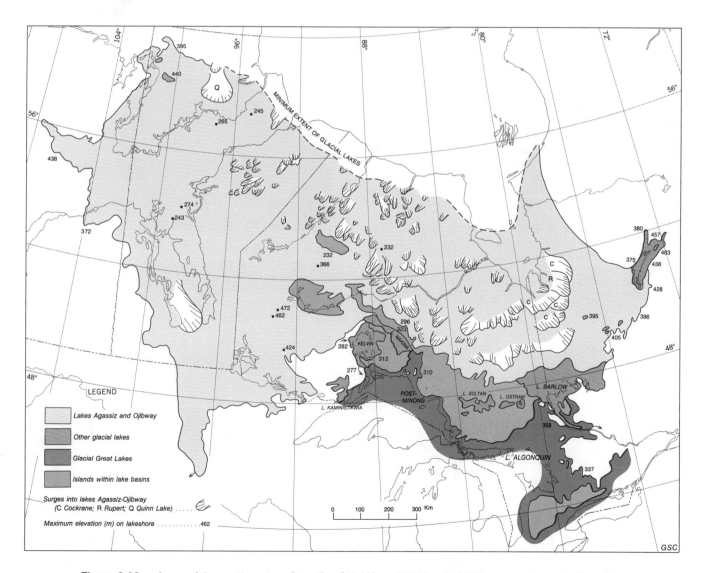

Figure 3.23. Areas of the southwestern Canadian Shield inundated by glacial lakes; also shown is where the retreating ice margin surged.

Some 2075 varves were counted from glacial Lake Barlow-Ojibway (Hughes, 1965). The varves thin upwards, without substantial changes in grain size. The thinning upward sequences relate to ice recession and hence increased distance from sediment source areas. In the Cochrane area, Hughes (1965) identified three thinning upward varve cycles, which he called the Lower, Frederick House, and Connaught sequences. He related the abrupt change in thickness between the upper part of the Frederick House sequence and the Connaught sequence to a glacial readvance (one of the Cochrane readvances) some distance farther north. Massive silts and clays are found overlying the varved sediments and represent deposition beyond the influence of turbidity currents. Layers of clayey till and rainout diamicton occur locally within the varved and massive units. These probably represent the proximal and distal deposits of readvances (or surges) into proglacial lakes such as Ojibway and Agassiz (Dredge, 1982).

Littoral sand deposits are rare, probably because lake levels fell from one level to another too rapidly for extensive beach development; exceptions occur where large eskers or moraines such as the Agutua Moraine or Munro Esker provided ample sorted sediment for reworking. In such environments well developed beaches formed over short periods of time (Fig. 3.25). In areas of shallow drift, washed bedrock is common within the major lake basins.

Landforms. Over extensive areas in Manitoba and northern Ontario lacustrine deposits form a flat clay plain inscribed with iceberg scour marks (Fig. 3.26). Northern areas are poorly drained and covered with muskeg. Other features which are associated locally with glacial lake deposits are small areas of De Geer moraines (near Windigo Moraine) and deltaic pads along the margins of eskers. Multiple strandlines and large deltas occur along the former southwest shore of glacial Lake Agassiz in southern Manitoba (and northern Ontario), where land, rather than ice, confined the lake and where large rivers carried sediment into the lake basin. Strandlines and large deltas are also well developed locally along the north shore of Lake Superior where south flowing rivers debouched into high level lakes.

Marine and glaciomarine deposits

Distribution. At the time of deglaciation of the Hudson Bay Lowlands, a high sea (Tyrrell Sea) invaded the depressed region around Hudson Bay, and extended inland as much as 300 km beyond the present coast (Fig. 3.27). In the

Figure 3.24. Varved lake sediments (A) overlying laminar and diamictic sands (B and C) which probably were deposited by turbidity flows into a glacial lake; Long Spruce Dam on Nelson River. 204315-V

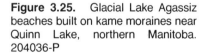

Figure 3.25. Glacial Lake Agassiz beaches built on kame moraines near Quinn Lake, northern Manitoba. 204036-P

study region its general outline was similar to the present coastline, although major offshore islands existed along the Manitoba-Ontario border and the coastline was much more indented than the present shoreline, which has very few irregularities. Deposits are generally 3-4 m thick along the present coast, thinning inland to a veneer near the upper limits of marine incursion. Deposits up to 30 m thick are found, however, where major rivers flowed into the sea (e.g., along Nelson River which drained much of the deglaciated prairie region).

Sediment facies. The lowermost facies associated with the marine episode is a stony mud with abundant broken shells. This deposit is best exposed near the mouth of Churchill River, in the general vicinity of the zone of confluence between Keewatin and Hudson ice; it is considered to be a glaciomarine deposit (deposited in the sea by direct sedimentation from glacier ice). In the James Bay area, lacustrine deposits are separated from overlying marine units by a coarse horizon composed of boulders, pebble gravel and rip-up blocks of clay. Skinner (1973) interpreted these as indicators of subaqueous erosion by density underflows generated during the marine incursion. In Quebec similar horizons have been interpreted as resulting from the rapid drainage of glacial Lake Ojibway (e.g., Hardy, 1982b). Across the remainder of the lowlands, however, the sediment sequence is a coarsening upward clay to sand sequence, which suggests that the Tyrrell Sea began as a high level sea and then continually regressed to present sea level.

Main surface features. Across the Hudson Bay Lowlands, the Tyrrell Sea sediment mantles till and lake plains, filling irregularities. At several places this featureless plain is interrupted by beach ridges. The most prominent single feature is the Great Beach, which is 200-800 m wide and consists of multiple beach ridges and spits. This beach complex can be traced from the north end of the Burntwood-Knife Interlobate Moraine, across northeastern Manitoba and into Northern Ontario as far as Winisk River, a distance of about 800 km. Another more extensive flight of beach ridges lies across northern Ontario between Nelson River and the Sutton Hills area, between 60 m elevation and present sea level (Fig. 3.28). The occurrence of these voluminous sandy and gravelly deposits is somewhat anomalous considering that most of the source material in the area is glacial lake clay and silty till. The massive nature and location of these ridges suggest that they are the reworked remains of sandy, ice contact sediment. The Great Beach possibly formed along a former grounding line moraine. The widespread northern Ontario beaches are more problematic: they might mark either successive ice grounding margins or an area where ice contact debris was dumped from the base of the ice sheet.

Organic deposits

The Hudson Bay Lowlands and adjacent shield support the largest area of organic terrain in the world. Swamp, fen, forested bog, and tundra peat have accumulated on the flat, poorly drained glacial lake and marine plains. During the last 8000 years up to 4 m of organic sediment has accumulated in peatland areas formerly covered by glacial lakes Agassiz and Ojibway. The peatlands that overlie younger marine deposits become increasingly thinner towards the modern coast.

Figure 3.26. Criss-crossing pattern of iceberg scour marks on the floor of glacial Lake Agassiz east of Lake Winnipeg. NAPL A21815-40

Peat forms a semi-continuous blanket throughout the poorly drained areas both within and beyond the boreal forest (Fig. 3.29). Thermokarst depressions, palsa mounds, and ice-wedge polygons form mesorelief features within the peatlands.

Many organic deposits in the northern Hudson Bay Lowlands contain abundant ice which occurs as interstitial ice, thin icy layers, and as segregated masses (palsas and ice wedges). The thermal insulation provided by the peat has led to the development of ground ice conditions in substrates which climatically would not otherwise be within the permafrost zone.

LITHOSTRATIGRAPHIC RECORD OF THE HUDSON BAY LOWLANDS

The lowlands adjacent to Hudson and James bays harbour an extensive record of multiple glacial units punctuated by nonglacial deposits. These deposits have long been recognized as being critical to understanding the Quaternary glacial history of North America since they record events that affected the heart of the main continental ice sheets. The region is the focus of debate on centres of glacier buildup and ice dispersal. A relatively complete description of the known lithostratigraphic record is given below because of the controversy surrounding its interpretation.

The stratigraphic record of the Hudson Bay Lowlands consists of three or more old tills (believed to be pre-Sangamonian), an interglacial formation, several younger

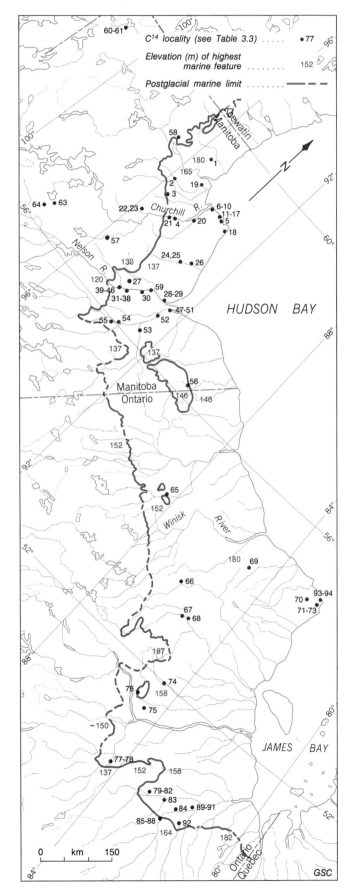

tills (Wisconsinan) separated by sand and silt beds of uncertain and hence controversial origin, and a postglacial lacustrine and marine sequence.

No section contains all units and there are commonly substantial differences between adjacent exposures but Figure 3.30 shows three generalized, regional, composite, stratigraphic columns. Generally the sequences are more complex and contain more till units at the eastern and western extremities of the lowlands than in the central parts. This is probably related to the fact that these two areas were interlobate zones experiencing flow from two different ice centres, at least during the Wisconsin Glaciation. Figure 3.31 shows individual stratigraphic columns from throughout the lowlands, which form the basis of our interpretation of events. Difficulties have arisen over the interpretation of the record because of its incompleteness in any given section; the lack of distinctiveness of some tills, which makes correlation difficult; and problems concerning the rank and genesis of units, particularly the intertill gravelly, sandy, and silty beds which do not contain organic materials. Throughout this discussion the main subtill or intertill organic beds (discussed below as the Missinaibi Formation and Gods River Sediments) are assumed to represent the Sangamon Interglaciation and are used as a marker bed. Problems with this assumption are discussed below.

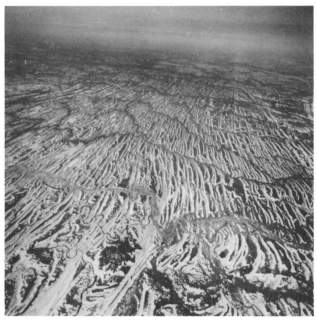

Figure 3.28. Flights of raised beaches in northwestern Ontario with Hudson Bay on the right side of the photograph. NAPL T131L-22

Figure 3.27. Limit of postglacial marine submergence on the southwestern Canadian Shield (elevation in m a.s.l.) and location of radiocarbon date sites listed in Table 3.3.

QUATERNARY GEOLOGY — CANADIAN SHIELD

Figure 3.29. Organic terrain in the Hudson Bay Lowlands: (a) tundra peat with ice-wedge polygons and thermokarst lakes; (b) forested (spruce) bog with peat about 4 m thick. 1-16-78, 204035-Q

Manitoba Dredge and Nielsen (1985); Klassen (1986)		James Bay Skinner (1973)[2]		pan-Lowlands Shilts (1984a)		
Peat and stream deposits Tyrrell Sea deposits	present interglaciation	Terrestrial unit Marine unit	present interglaciation	Peat and fluvial deposits Tyrrell Sea deposits	present interglaciation	0.036[5]
varves clay till varves and sandy diamicton	Ice recession against glacial Lake Agassiz, with surges into the lake	Glaciolacustrine unit Kipling Till Friday Creek sediments	(a) ice recession and surges into Lake Ojibway / (b) Lake Ojibway glacial stade interstade	varves	Lake Agassiz-Ojibway	
				Cochrane till Kipling Till	partial readvance glaciation	
				Friday Creek varves	Wisconsin deglaciation *	0.065[5]
				Till	glaciation	
Wigwan Creek Fm. till sand and gravel till sand and gravel till	Wisconsin Glaciation with shifting ice flow	Adam Till	Wisconsin Glaciation	Fawn River gravel	Wisconsin deglaciation *	0.14[5]
				Adam Till	glaciation	
Gods River Sediments	Sangamon Interglaciation[1]	Missinaibi Formation[3]	Sangamon Interglaciation	Missinaibi beds[4]	Sangamon Interglaciation *	0.25[5]
Amery Till	glaciation	Till III sand and silt Till II sand and silt Till I	glaciation with an oscillating ice margin against a proglacial lake	till	glaciation	
(soil)	cool interglaciation			sand and gravel, silt	interstade	
Sundance Till	glaciation			till	glaciation	
Mountain Rapids diamicts and gravels	glacial and nonglacial intervals			sand and varves	interstade	
				till	glaciation	
					interglaciation *	
				till	glaciation	

[1] assuming the silts at Port Nelson are Sangamonian age deposits
[2] Skinner's alternative (b) is essentially the same as McDonald (1969)
[3] including marine beds
[4] the stratigraphic position of marine beds is uncertain
[5] amino acid ratios from J.T. Andrews, University of Colorado, personal communication, 1986

* ice-free Hudson Bay

Figure 3.30. Composite stratigraphic sections and interpretations of the lithostratigraphic record in the Hudson Bay Lowlands.

Pre-Missinaibi record

Tills and intertill sediments underlying the Missinaibi Formation are well exposed on the Missinaibi and other tributaries of Moose River in Ontario, in numerous sections along Nelson River and on Hayes River and tributaries in Manitoba, and at isolated places across northwestern Ontario. Terasmae and Hughes (1960a, b) and Skinner (1973) identified three sandy tills, labelled I, II, III from bottom to top, below the Missinaibi marker beds on Missinaibi River (31 of Fig. 3.32). A fourth till lying directly below till I has been reported by Shilts (1984a). Laminated silts and clays, oxidized sands whose structures indicate current directions opposite to present river flow, and intercalated diamictons separate tills I, II, and III. Skinner (1973) concluded that these tills represented oscillations into a proglacial lake dammed by southward flowing ice advancing from Labrador. Shilts (1984a), however, noted that the gravel beds are substantially weathered and that they may represent major nonglacial intervals rather than oxidation zones related to groundwater fluctuations. In the James Bay area ice depositing the pre-Missinaibi tills flowed towards the southwest (Skinner, 1973), while to the northwest of James Bay the ice that deposited these tills flowed due south (McDonald, 1971).

In Manitoba, at the Henday site along Nelson River (6 of Fig. 3.32; Fig. 3.33; Nielsen and Dredge, 1982; Dredge and Nielsen, 1985; Nielsen et al., 1986), a dark grey silty calcareous till containing shell fragments (Amery Till) lies directly below an interglacial unit. Nearby, at Sundance (11 of Fig. 3.32), the Amery Till is underlain by a second, sandier till derived from granitic source rocks. An oxidized and bleached zone containing pollen separates the two tills; this zone is thought to be the remains of a paleosol developed during ice-free conditions when a tundra environment prevailed. The Nelson River sections therefore appear to record two pre-Missinaibi glaciations separated by a cool interglaciation. Striae, fabric, and lithological data indicate that the first ice advance was from the north (Keewatin); the second came from the east (Hudson Bay or Quebec-Labrador). The ice centre related to the second, later till might have been located in Hudson Bay because in the Gods River area ice flow was from the northeast (Netterville, 1974).

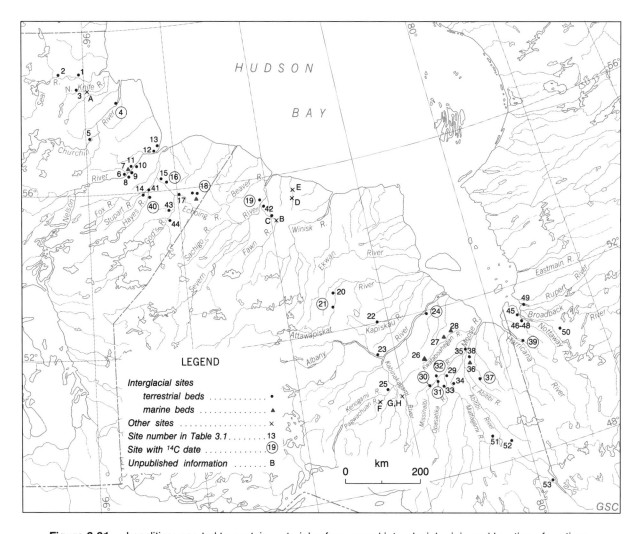

Figure 3.31. Localities reported to contain materials of presumed interglacial origin and location of sections used in Figure 3.32.

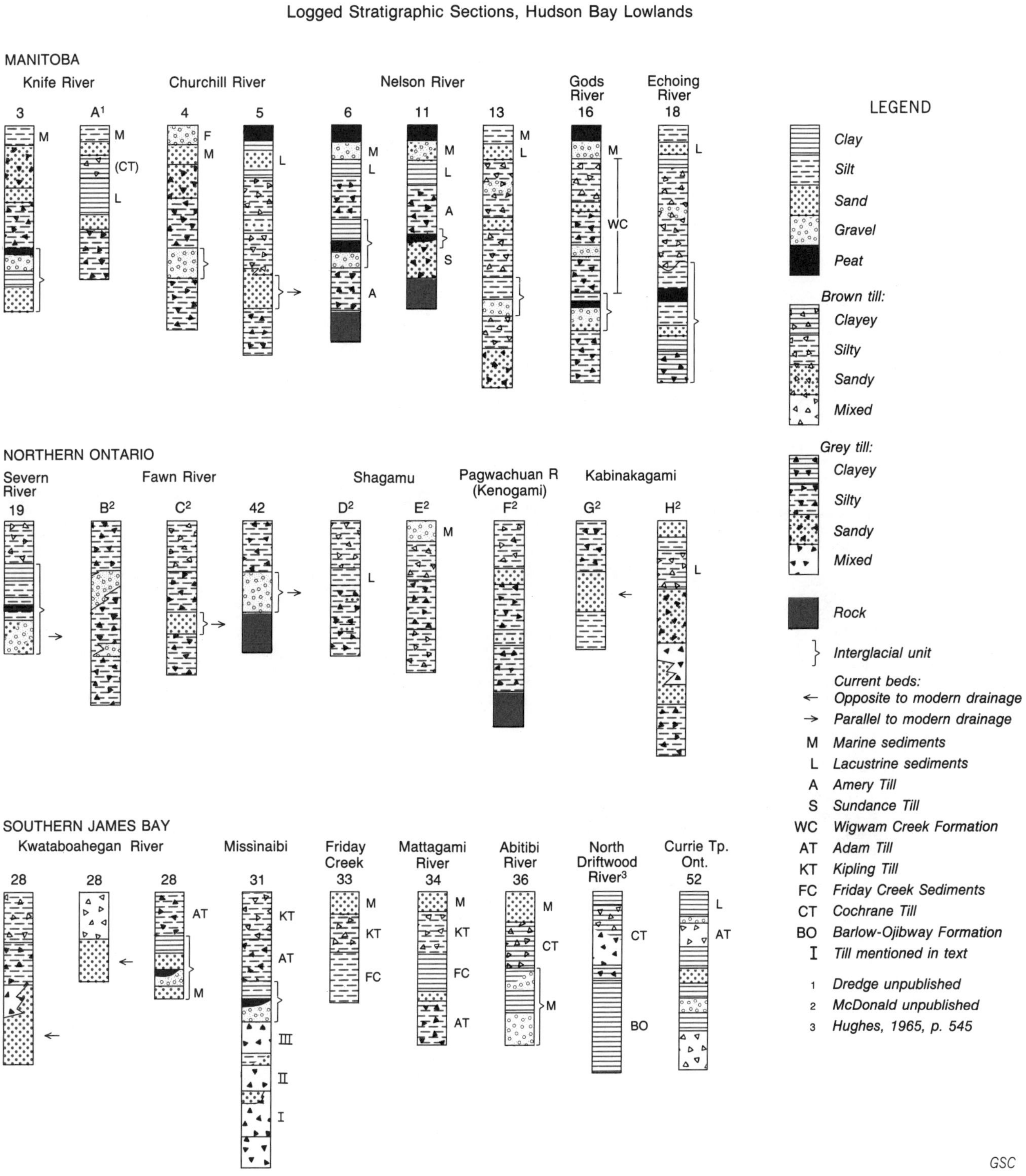

QUATERNARY GEOLOGY — CANADIAN SHIELD

Logged Stratigraphic Sections, Hudson Bay Lowlands

Figure 3.32. Logged stratigraphic sections, Hudson Bay Lowlands. Numbers or letters over stratigraphic columns refer to sites located in Figure 3.31 and referenced in Table 3.1.

Missinaibi and Gods River records (Sangamon Interglaciation and isotope stage 5a-5e)

Missinaibi Formation and Gods River Sediments

The Missinaibi beds, named by Terasmae and Hughes (1960a, b), and renamed Missinaibi Formation by Skinner (1973), are a series of subtill peats and inorganic sediments now considered to represent the Sangamon Interglaciation. The beds along Missinaibi and Kwataboahegan rivers were first described and identified by R. Bell (1887) and J. Bell (1904). Terasmae and Hughes (1960a,b) completed a palynological analysis of sections along Missinaibi and Opasatika rivers; McDonald (1969) logged similar sections across the lowlands and initially concluded that they related to an interglacial environment; Skinner (1973) summarized previous work in the Moose River basin and developed a facies model to explain the Missinaibi sediment sequence. In Manitoba Netterville (1974) studied similar sediment packages along Gods River (Gods River Sediments); Klassen (1985) and Dredge and Nielsen (1985; Nielsen and Dredge, 1982; Nielson et al., 1986) discussed these sediments and others across northeastern Manitoba, particularly along

Table 3.1. Selected interglacial sites, Hudson Bay Lowlands

Site Number (cf. Fig. 3.31)	Name/Locality	Reference	^{14}C Age (years BP)
1	Great Island/Seal River	Taylor, 1961	
2	Seal/Seal River	Dredge et al., 1986	
3	Knife/N. Knife River	Dredge et al., 1986	
4	Limestone Rapids/Churchill River	Dredge and Nielsen, 1985	
5	Mountain Rapids/Churchill River	Dredge and Nielsen, 1985	>32 000 (GSC-3074)
6	Henday/Nelson River	Nielsen and Dredge, 1982	
7	Limestone/Nelson River	Nielsen and Dredge, 1982	
8	Cofferdam/Nelson River	L.A. Dredge, unpublished	
9	Nelson River	Nielsen and Dredge, 1982	
10	Eagle Bluff/Nelson River	Nielsen and Dredge, 1982	
11	Sundance/Nelson River	Nielsen and Dredge, 1982	
12	Flamborough/Nelson Estuary	Dredge and Nielsen, 1985	
13	Port Nelson/Nelson Estuary	Dredge and Nielsen, 1985	
14	Twisty Creek/Stupart River	Dredge and Nielsen, 1985	
15	Gods River	Netterville, 1974	
16	Gods River	Netterville, 1974	>41 000 (GSC-1736)
17	Shamattawa	McDonald, 1969	
18	Echoing Creek	Dredge and Nielsen, 1985	>37 000 (GSC-892)
19	Severn River	Tyrrell, 1913; McDonald, 1969	>41 000 (GSC-1011)
20	Attawapiskat River	McDonald, 1969	
21	Attawapiskat River	Prest, 1963	>35 800 (GSC-83)
22	Kapiskau River	McDonald, 1969	
23	Albany River	Williams, 1921	
24	Albany River	Martison, 1953; Skinner, 1973	>54 000 (GSC-1185)
25	Kenogami River	Skinner, 1973	
26	Kwataboahegan River	McDonald, 1969	
27	Kwataboahegan River	Bell, 1904; Wilson, 1906	
28	Kwataboahegan River	McDonald, 1969	
29	Opasatika River	McLearn, 1927; Terasmae and Hughes, 1960a	
30	Missinaibi River	Bell, 1879b; Terasmae and Hughes, 1960a	>42 000 (GrN-1921)
31	Missinaibi River	Wilson, 1906; Terasmae and Hughes, 1960a	>50 000 (GrN-1435)
32	Soweska River	Bell, 1904; McLearn, 1927	>42 600 (L-369b)
33	Opasatika River	Terasmae and Hughes, 1960a; Skinner 1973	
34	Smoky Falls/Mattagami River	Keele, 1921	
35	Moose River	Bell, 1904; Keele, 1921	
36	Otter Rapids/Abitibi River	Prest, 1966	
37	Little Abitibi River	Prest, 1970	>43 600 (GSC-435)
38	Onakawan	Dyer and Crozier, 1933	
39	Harricana	Stuiver et al., 1963	>42 000 (Y-1165)
40	Hayes River	Tyrrell, 1913; McDonald, 1969(?)	
41	Stupart River Junction	Dredge and Nielsen, 1985	
42	Fawn River	McDonald, 1969	
43	Gods River	Tyrrell, 1913	
44	Gods River	Tyrrell (?)	
45	Nottaway River, Quebec	Hardy, 1982b	
46	Nottaway River, Quebec	Hardy, 1982b	
47	Nottaway River, Quebec	Hardy, 1982b	
48	Nottawa River, Quebec	Hardy, 1982b	
49	Rupert River, Quebec	Hardy, 1982b	
50	SW of Lac Dionne	Hillaire-Marcel and Vincent, 1980	
51	Kidd Creek	Lowdon and Blake, 1979	>28 000 (GSC-1633)
52	Matheson	Brereton and Elson, 1979	>37 000 (GSC-2148)
53	Montreal River	Lowdon et al., 1971	>42 000 (GSC-1299)

Figure 3.33. Tills and interglacial deposits along Nelson River, Manitoba: (A) Amery Till (Illinoian), (B) interglacial sediments, (C) Wisconsinan grey and brown tills, and (D) Lake Agassiz and Tyrrell Sea sediments. Courtesy of E. Nielsen.

Nelson and Echoing (Fig. 3.34) rivers. Similar but less complete sediment packages are found at many localities across the Hudson Bay Lowlands (Fig. 3.31, Table 3.1).

Skinner (1973) summarized and interpreted the sediment sequence in the Moose River basin and divided the beds of the Missinaibi Formation into marine, fluvial, forest peat, and glacial lake members. In northern Manitoba additional glaciofluvial and lacustrine members lie stratigraphically below the marine member (Dredge and Nielsen, 1985), and additional pond and stream deposits are associated with the forest peat member. Also in Manitoba, massive and poorly stratified gravels form the base of the interglacial sequence and have been interpreted both as a recessional facies associated with the Amery Till and as a terrestrial stream deposit.

The marine member is present at only a few localities in the Hudson Bay Lowlands. On Kwataboahegan River (Bell, 1904; McDonald, 1969) it consists of fossil-rich sand and silt, and along Abitibi River (Prest, 1966, 1970) it is a fossiliferous stony blue clay. Shells from the Abitibi River site were dated at >19 000 BP (GSC-1535; Blake, 1988). Along Nelson River, Johnston (1918) (below the mouth of the Limestone River) and Dredge and Nielsen (1985) (at Eagle Bluff) found massive blue clay below several tills; the clay may be a marine unit although it is not necessarily correlative with the Gods River Sediments. The clay, which forms the base of the Gods River Sediments where they were examined by Dredge and Nielsen (1985), may also be a marine deposit, since it contains salt water microfauna and abnormally high amounts of boron. The marine member is thought to represent a glacial isostatic marine incursion which reached at least 90 m a.s.l. in the James Bay region and about 120 m at Echoing River in Manitoba. Skinner (1973) called this water body the Bell Sea. Fossils indicate cold water conditions similar to those found in the postglacial Tyrrell Sea and present Hudson Bay.

The next unit upward in the interglacial sequence consists of crossbedded stream gravels, and beds of sand and silt with organic detritus. These deposits are present in major valleys, including the Fawn and Severn, but have not been seen in other areas. Current directions indicate flow towards Hudson Bay. The unit represents an interval of subaerial stream deposition. Because these deposits are found at relatively low elevations, base level (sea level) must have been similar to, or lower than present. In Manitoba the gravel contains abundant organic detritus, including reworked peat and beaver-chewed sticks (Netterville, 1974). At Echoing River, Manitoba the fluvial unit is replaced by laminated silt and clay and a calcareous marl containing abundant freshwater gastropods and small molluscs indicating the former existence of a pond or small lake.

Peaty members are found in many sections and range in thickness from 5 cm to 2 m. They include moss, sedge, and woody peat, including in situ stumps. The tree species are mainly spruce. Peat types, tree species (based on wood identifications), pollen assemblages, seeds, and insects suggest a boreal environment similar to present, although in Manitoba the uppermost part of the peat contains fossil material indicative of a succeeding period of tundra conditions (Netterville, 1974; Nielsen et al., 1986).

The peat is overlain by massive or laminated silts and clays. In some places, these sediments have been interpreted as proglacial lake deposits, which developed when northward drainage was blocked by ice advancing out of the Hudson Bay basin, although it is also possible that in other places these deposits accumulated in nonglacial lakes. Indeed, J.H. McAndrews (Royal Ontario Museum, unpublished report to W.R. Cowan, 1985) analyzed the pollen content of 30.4 m of rhythmically laminated silt and clay (Missinaibi Formation) from a borehole in the Moose River basin and concluded that the rhythmite sequence represented about 813 years of sedimentation into an interglacial freshwater lake. It should be noted however, that Skinner

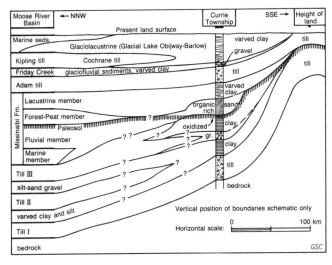

Figure 3.34. Correlation of stratigraphy of Moose River basin with that of the area to the south (from Brereton and Elson, 1979).

(1973) attributed most organic material in this unit to reworking of a land surface by a transgressing ice marginal lake. The uppermost part of this unit is commonly intercalated with till, which includes clasts of massive and laminated clay near the contact.

In summary, the sediment record of the Missinaibi Formation and Gods River Sediments represents an early glacial lake (receding ice); a high level sea which later regressed to levels similar to or below present; a subaerial unit of peat, stream, and pond deposits; and a succeeding lake which may represent a proglacial lake and ice advance over the area.

Rank

Terasmae and Hughes (1960a, b) initially concluded from palynological evidence that the Missinaibi beds represented an "interstadial" because the pollen species indicated boreal conditions no warmer than present, although they pointed out that slight climatic differences may be difficult to detect in boreal regions. However, because pollen from the Missinaibi type section (Terasmae and Hughes, 1960a, b), the Gods River Sediments (Netterville, 1974), the sediments along Nelson River (Nielsen et al., 1986) and Echoing River (R.J. Mott, unpublished Geological Survey of Canada Palynological Report 85-10) all indicate climate at least as warm as or warmer than present, most palynologists now assign an interglacial rank to the pollen-bearing deposits.

The Missinaibi Formation and Gods River Sediments are assigned an interglacial rank on the basis of the sediment sequence as well as bioclimatic criteria. The existence of marine beds and the direction of streamflow indicated by crossbedding in the fluvial units demand an ice-free Hudson Bay. Similarly, the existence of terrestrial peat precludes the presence of a proglacial lake, which would have occupied the region if Hudson Bay were filled with ice. Furthermore, the existence of subaerial fluvial and organic deposits at low elevations necessitates a relatively low sea level and a substantial amount of isostatic recovery.

Age and correlation

The Missinaibi Formation and Gods River Sediments have been correlated with one another on the basis of the similarity of their sedimentary and bioclimatic records, and their stratigraphic position. At each exposure they are the first (and only) organic-bearing beds below till. In addition, aspartic acid ratios on wood samples from most sites in Manitoba are similar and correlate with ratios from wood at the Missinaibi type section in Ontario.

Though there is some debate as to the age of the Missinaibi Formation and Gods River Sediments, it is our opinion and that of many others (e.g., Prest, 1970; Skinner, 1973; Andrews et al., 1983b; Klassen, 1985) that these deposits represent the last interglaciation — the Sangamon. Organic materials from these units, however, are beyond the ^{14}C dating limit (Table 3.1); the oldest exceeding 72.5 ka (Stuiver et al., 1978). Because of the lack of absolute age control, it is possible that not all these beds are correlative and/or that not all beds are from the same part of the Sangamon Interglaciation. This situation would not be unexpected since oxygen isotope stage 5, which encompasses the Sangamonian stage (130 to 75 ka), contains two relatively warm climatic intervals. In addition, none of the confirmed marine units reported above occur in section with organics from the Missinaibi Formation and Gods River Sediments; therefore they could belong in entirely different stratigraphic intervals and represent different interglaciations.

Post-Missinaibi record

The succession of deposits overlying the Missinaibi Formation consists of tills and intertill sediments, capped by glacial lake, marine, and terrestrial deposits. The discussion here is concerned only with the Wisconsin Glaciation part of this record.

As many as four till sheets, with intertill deposits, overlie the Missinaibi Formation and Gods River Sediments (Fig. 3.32). The tills all contain shell fragments and are of three different types: 1) Grey compact calcareous silty till sheets occur as the lower tills in sections. 2) Compact buff brown silty till forms the uppermost till in many areas; the brown colour may relate to reddish Devonian source rocks in Hudson Bay or (less likely) to oxidation of iron colloids in grey calcareous tills during deposition into glacial lakes. This till commonly grades upwards into sandy glaciolacustrine diamictons. 3) Relatively loose brown clayey till which consists in part of contorted varves and clasts of varved clay and commonly is enclosed by glaciolacustrine deposits. It forms the uppermost till sheet where parts of the receding Hudson Ice margin readvanced (by surging) into glacial lakes Agassiz and Ojibway.

The intertill sediments are also of several varieties: 1) Lenses of sand and gravel, which are irregularly intercalated with till. 2) Sorted sand, or sand and gravel beds, which locally form horizontal seams or wet bands traceable for several hundred metres along sections, and which commonly separate tills of different colour, fabric, or clast content. In Manitoba, these beds are generally massive. In other places, small, poorly defined crossbeds indicate current flow opposite to the flow direction of present streams, which suggests that they are meltwater deposits related to a Hudson Ice mass. We believe that the gravel bodies represent subglacial deposits during a single glacial phase, but others interpret them as subaerial deposits marking periods of substantial ice retreat. 3) Sandy diamictons, reddish varved silts, and clays form proximal and distal deposits of glacial lakes; some of the sands have crossbedding indicating flow away from Hudson Bay. Where these deposits outcrop beneath a brown clay till (till 3 above) but have been traced to surface deposits nearby, they are interpreted as glacial Lake Agassiz and Lake Ojibway deposits. However, in a number of places, particularly in northern Ontario where they underlie brown till (i.e., either till 3 or 2), they could alternatively relate to an earlier Wisconsinan ice recession. The interpretation of the sand beds and the silt units underlying till is crucial to arguments concerning possible Middle Wisconsinan openings of Hudson Bay that are outlined in the following sections.

Northern Manitoba

In Manitoba, the post-Gods River Sediments (post-Missinaibi) record consists of the Wigwam Creek Formation, the deposits of glacial Lake Agassiz, a Cochrane-type till which was deposited by surges into Lake Agassiz, and the deposits of the Tyrrell Sea. The Wigwam Creek Formation (Nielsen and Dredge, 1982; Klassen, 1985; Dredge and Nielsen, 1985) consists of up to four till sheets which either lie directly over each other or are separated from one another by sand and gravel beds. The lower tills are compact grey tills; the upper is a compact buff brown till that grades into glaciolacustrine diamictons at the top. Major lithic and fabric differences in the sandy and silty tills exposed in section in the Churchill River area appear to reflect the interplay of Hudson and Keewatin ice. In other parts of Manitoba minor variations in striae, fabric, and lithic composition of the silty tills record shifts in flow within the Hudson Ice mass. The grey till was emplaced by ice moving either across (from Labrador), or out of, Hudson Bay. The brown colour of the upper till (red Devonian source rocks) and systematic fabric changes across the region reflect a radial ice flow westward and southward out from Hudson Bay. Minor intertill gravelly lenses are considered to be melt-out deposits. More extensive massive intertill sand, and sand and gravel, form thin horizontal seams which commonly rest unconformably on the till underneath but are intercalated with or grade into the till resting above them. They are not related to other nonglacial sediments. Their position between slightly different till sheets, and their massive structure and contact relations with the till above and below suggest that they may be subglacial (basal) meltwater deposits, laid down either as sheets or shallow channel fills (Dredge and Nielsen, 1985). They may mark former confluences of ice streams or ice lobes in the same manner as radial moraines seen at the surface delimit confluent ice streams which developed during final glacial recession. The sand beds are prevalent where drumlin fields converge at acute angles (Gods River) and where interlobate moraines (Mountain Rapids) provide independent evidence of the presence of convergent ice lobes.

If these particular structureless intertill sand and gravel beds are interpreted as subglacial or englacial in origin, then there are no stratified units that might be assigned to partial or total deglaciation in northern Manitoba between deposition of the Gods River Sediments and the present interglaciation.

A sandy diamicton or crossbedded sands, grading into red-brown varves and massive clay, overlie the main till sequence. These constitute the surficial units over substantial parts of northern Manitoba and are considered to be proximal and distal facies of glacial Lake Agassiz sediment. In places these deposits are interrupted by a till which, in its classic form (North Knife River), consists largely of contorted varves with clasts of varved clay at its base and grades upward into brown or grey till. The streamlined surface expression of this till suggests that it was deposited where the retreating ice margin surged into Lake Agassiz, and that it is genetically equivalent to the Cochrane till in Ontario. The sequence is capped by glacial lake clay or by the postglacial marine and peat deposits.

In summary, Dredge and Nielsen (1985) have suggested that there may have been continuous ice cover over the region during the entire Wisconsin Glaciation and that the existence of thin intertill sand seams need not represent nonglacial intervals. North of Churchill River, major differences between till sheets record the interplay of Keewatin Ice and Hudson Ice and indicate that the last ice movement was from the north. The area south of Churchill River was continuously covered by ice which flowed across, or out of, Hudson Bay and deposited a silty till. Colour and fabric differences within the silty till sheets, and intertill sediments, suggest that accumulation zones and ice flow patterns (hence ice divide positions) shifted during the course of glaciation. The uppermost sheet suggests flow from a dispersal centre in Hudson Bay. The receding ice margin periodically surged into the glacial lake which expanded as the ice sheet retreated.

Netterville (1974), Shilts (1982a,b), and Andrews et al. (1983b), on the basis of the presence of the sand and gravel layers, and included detrital shell chips found in all tills and intertill sediments, have interpreted the stratigraphic record differently and have suggested that all intertill sands in this region (and others across the lowlands) represent nonglacial events during the Wisconsinan. This concept is discussed further below.

Moose River Basin

In the James Bay region Skinner (1973) has described a sequence of two tills separated by beds of cross-stratified sand, silty rhythmites, and clay. The lower grey Adam Till (Fig. 3.30) was deposited by southwestward flowing ice. Skinner was unable to determine how much of the Wisconsinan Stage was spanned by Adam Till because the age of the overlying (Friday Creek) sediments was unknown. The Friday Creek Sediments consist of glacial lake deposits containing some pollen but no organic matter suitable for radiocarbon dating. Analysis of a sample of Friday Creek silt and clay from the junction of Adam Creek and Mattagami River shows it to contain pollen similar to that of the Erie and Mackinaw interstadial silts of southwestern Ontario. Typically this pollen was deposited in deep sterile proglacial lakes, with the pollen largely derived from glacial meltwater (J.H. McAndrews, Royal Ontario Museum, unpublished report to R. Cowan, 1985). Skinner (1973) raised the possibility that Friday Creek Sediments are Middle Wisconsinan; however, there is no concrete evidence to suggest that the Friday Creek Sediments are anything other than glacial Lake Barlow-Ojibway sediments overlain by a Cochrane till; we consider these sediments to be Holocene.

The brownish, less compact Kipling Till, which overlies the Friday Creek Sediments, represents a glacial event of unknown magnitude; some researchers (e.g., Shilts, 1982a,b) believe it may represent a Late Wisconsinan stade following a Friday Creek "interstade", whereas we believe that it only represents a fluctuation during deglaciation. Hardy (1976, 1977) suggested that Kipling Till and Cochrane till are correlative. Cochrane till is a clay-rich till formed by the glacial incorporation of glacial Lake Ojibway sediment as the receding Late Wisconsinan ice margin surged (Hughes, 1965). Stratigraphically it usually overlies (or is included within) glacial lake sediment of the Barlow-Ojibway Formation and correlates with the Connaught varve sequence farther south.

Northern Ontario

B.C. McDonald (1969, and unpublished field notes), B.G. Craig (1969, and unpublished field notes) and Q.H.J. Gwyn (unpublished field notes) have logged many sections across northern Ontario (Fig. 3.32). McDonald's excellently recorded observations and included discussions show the range and complexity of the sediment record, and the difficulties of correlation in an area where one section differs substantially from the next (e.g., Fawn River sections, Fig. 3.32). In some sections an upper unit of brown silty clay till overlies contorted silty rhythmites, or beds of fine sand whose crossbeds indicate flow opposite to the direction of modern streams (e.g., D of Fig. 3.32). These closely resemble Cochrane till and sediments of glacial lake Agassiz and Ojibway, and are probably of late glacial age. However, in other places, such as along Kwataboahegan and Kabinakagami rivers the uppermost till, which also overlies silt, is more compact and was considered by McDonald (1969, 1971) to be different from Cochrane till. This interpretation would make the silts below not postglacial, but rather interstadial glacial lake deposits created during partial retreat of the ice sheet during the Wisconsinan. Andrews et al. (1983b), on the basis of amino acid ratios, interpreted subtill sands and gravels on Kabinakagami River (G,H of Fig. 3.32) as representing a major Wisconsinan deglaciation during which Hudson Bay became ice free. Along other rivers in the James Bay region and along the Hudson Bay coast, McDonald found interbedded grey and brown tills (E of Fig. 3.32), or grey tills separated by gravels (F, B of Fig. 3.32). McDonald thought that in some places (section B) the gravels were lenses within a single till sheet, and at other sites (section F) that they might lie between tills relating either to shifts in ice flow during a single glacial event or possibly to interstadial intervals.

The degree to which differences in correlation affect our interpretation is exemplified by the Fawn River and Severn River sections. At a site on Severn River (19 of Fig. 3.32) brown till overlies a sequence of clays, peaty sand, and crossbedded gravels. B.C. McDonald (1969 and unpublished field notes) interpreted these as interglacial gravels and correlated them with the Missinaibi Formation farther south. Nearby, on Fawn River (C of Fig. 3.32) extensive gravel beds without organics, but with current bedding towards Hudson Bay, were shown to underlie, not overlie, a grey till, which in turn underlies the brown surface till. McDonald correlated the Fawn River gravels with the Severn River deposits, and with the Missinaibi Formation — that is, the Sangamon Interglaciation. Andrews et al. (1983b), however, found that shell fragments within the Fawn River gravels had amino acid ratios that were lower than those assigned to the Sangamonian. They therefore assigned the gravel units to a separate Wisconsin interstade and moved them to a stratigraphic position above the grey till. Wyatt and Thorleifson (1986) similarly concluded that the Severn River gravels were younger than the interglacial deposits in the James Bay region. Radiocarbon dates of 37 040 ± 1660 BP (WAT-1378), >38 000 BP (GSC-4146), and >43 000 BP (GSC-4154) were used to assign a Middle Wisconsinan age to peat between the brown and a grey till on nearby Beaver River. We concur with McDonald's (1969) interpretation.

McDonald's (1969) interpretation of the regional stratigraphy of northern Ontario is shown in his composite stratigraphic sections. Resting on recognized interglacial beds there is a two or three till sequence. Till fabrics and striated boulder pavements indicate that the lowermost (grey) till was emplaced by ice from the northeast, McDonald interpreted the glacial lake deposits above this till as indicating a partial deglaciation in the James Bay region. A succeeding glacial advance from the north-northeast and northwest deposited the upper buff till or layered buff and grey tills. McDonald's interstadial deposits do not occur in sections having glacial Lake Ojibway deposits and Cochrane till; therefore it is possible that McDonald's interstadial deposits and the Holocene glacial Lake Ojibway sediments are stratigraphically equivalent.

Ice flow summary

Although for many areas of the Canadian Shield the only ice flow patterns preserved are those associated with final deglaciation, a much longer record of ice flow is preserved in the Hudson Bay Lowlands. Former ice movements can be partially reconstructed from the fabric and lithic studies on the various till sheets, and from striated boulder pavements.

The tills directly below the major organic (Sangamonian) marker beds, and overlying older paleosols, indicate ice flow towards the southwest in the James Bay area (Skinner, 1973), due south in the Fawn-Severn River area (McDonald, 1971), southwest in the Hayes-Gods River area of northeastern Manitoba (Netterville, 1974), and roughly westwards along Nelson River (Nielsen and Dredge, 1982) and at Churchill (Dredge and Nielsen, 1985). It appears, therefore that one ice dispersal centre for these tills lay in Labrador, with another probably in Hudson Bay. Scattered evidence in Manitoba indicates that prior to this glaciation there was a major dispersal centre in Keewatin from which ice flowed across Churchill at least as far south as Nelson River (Dredge and Nielsen, 1985) and probably extended well into Hudson Bay.

The grey tills (Early and Middle Wisconsinan?) directly above the interglacial organic marker beds were also emplaced by ice flowing in several directions. Skinner (1973) and McDonald (1971) reported southwesterly ice flow in the James Bay area and northeastern Ontario, presumably from a Labrador centre. Farther west Tyrrell (1913; based on striae) and Netterville (1974; based on fabric) reported flow due south in northwestern Ontario and adjacent Manitoba, while flow was westwards in northern Manitoba, both along Nelson River and at Churchill (Dredge and Nielsen, 1985). The south flow reported by Netterville and Tyrrell suggests a secondary dispersal centre in Hudson Bay lying somewhat west of the earlier Hudson centre, but the other ice flow patterns could be accommodated by a single major dispersal centre in Labrador. Early Wisconsinan Keewatin Ice flow is not recorded in the Hudson Bay Lowlands.

Fabrics from the uppermost regional till sheet (Late Wisconsinan?) indicate that the ice crossing the southern James Bay area came from a Labrador centre. The brown colour, fabric, and clast types from the Severn River area, Gods River, and Nelson River indicate flow radiating from Hudson Bay. Flow in the Severn River area varies from south-southwest to southeast, flow in the Gods River area is southwest, and in the Nelson River area is west. These patterns indicate a major dispersal centre in Hudson Bay. The Churchill area was covered by southward flowing ice from a centre in Keewatin. The Keewatin/Hudson convergence thus initially lay between Churchill and Nelson rivers and moved northwards towards the end of the last glacial cycle.

Conclusion

The range of interpretations which arise from lithostratigraphic examination of sections in the Hudson Bay Lowlands is shown in Figure 3.30. There were at least two, and possibly as many as four, pre-Wisconsinan glaciations and possibly two pre-Sangamonian interglaciations. Opinions on the events related to the Wisconsinan Stage vary from having shifts in ice flow during a period of continuous ice cover (Dredge and Nielsen, 1985), to one or more interstadial events during which Hudson Bay remained ice covered (MacDonald, 1971), to multiple regional

Wisconsinan deglaciations with openings of Hudson Bay (Andrews et al., 1983b). These different interpretations hinge on the interpretation and correlation of the lithostratigraphic units. It should be reiterated however that no single section contains the complete stratigraphic record. At no place has more than one organic-bearing bed been logged in a single section; nor have (interglacial) organic beds and pre-Holocene subtill silt deposits (interstadial) been seen in one exposure; nor have "interstadial" lake deposits and postglacial pre-surge lake deposits been observed in the same section (are there two events or only one?). No Middle Wisconsinan marine beds have been identified, even where there are good cliff sections at low elevations near present Hudson Bay (e.g., Port Nelson); and the "Sangamon" marine units cited from the eastern lowlands do not occur in section with organic deposits: they could belong to an earlier interglaciation. There are clearly many problems to be resolved. Specifically, we need to identify a diagnostic signature for the various till sheets so that they can be correlated; determine the provenance and glaciological significance of the various till sheets; determine the genesis and rank of the various intertill units; and determine absolute ages for various events. Amino acid stratigraphy has allowed us to resolve some of the problems involved with these various lithostratigraphic interpretations; but the method itself has generated additional problems (Dyke, 1984), and the results from the Hudson Bay Lowlands indicate that the data have to be extensively statistically manipulated before apparent patterns emerge.

AMINO ACID STRATIGRAPHY
J.T. Andrews

Amino acid stratigraphy involves the recognition of regional sedimentary units that contain fossils, usually marine molluscs, with similar amino acid ratios (Miller, 1985a; this chapter, Andrews, 1989). In the Hudson Bay Lowlands, amino acid stratigraphy is based on the ratio of D-alloisoleucine/L-isoleucine (henceforth aIle/Ile) in the total fraction. The diagenetic reaction involved is mainly a function of temperature, time since death of the animal, and the genus (Miller, 1985a). Our results are primarily for the genus *Hiatella*. We assume that within the Hudson Bay Lowlands past temperature gradients (horizontal and vertical) have been low and that the area constitutes a homogeneous climatic region (with accrued data, such an assumption is testable). Thus changes in ratios largely reflect the age of the samples, and deposits with shells of similar aIle/Ile ratios are coeval. This statement only applies to in situ marine sediments and thus could only apply to certain nonglacial units within the stratigraphic sections. However, transported shells can also supply important information. This situation is common within the lowlands where many tills contain shell fragments (McDonald, 1969; Shilts, 1982b, 1984a; Andrews et al., 1983b). In such cases the lowest (i.e.,youngest) ratios provide minimum "ages" for the nonglacial event preceding deposition of the till. Thus the youngest till in a sequence might contain shells with a variety of aIle/Ile ratios but stratigraphically older tills cannot contain the lowest ratios.

As of 1985 more than 300 samples have been run; however, much data are unpublished and only the published information is used here. The upper Quaternary amino stratigraphy is delimited on one end by ratios from Bell Sea deposits and on the other by ratios from shells from the Holocene Tyrrell Sea sediments, although many shells give intermediate values. Since initial publication (Andrews et al., 1983b), many samples have been rerun to ensure that the ratios of these end member units are reproducible (cf. Miller, 1985a). The results indicate that the ratios published by Andrews et al. (1983b) are broadly correct. Ratios from Bell Sea deposits cluster around 0.2 ± 0.03 and are radically different from Tyrrell Sea molluscs which have a mean ratio of 0.036 ± 0.015. Two intermediate amino stratigraphy units were recognized and have been identified on the basis of transported shells. Shilts (1982b) called these nonglacial units the Fawn River gravel and Kabinakagami sediments. Typical ratios for these two units were $0.14 \pm .02$ and 0.065 ± 0.01 (Andrews et al., 1983b, Table 1). An independent age estimate for the Fawn River gravel unit comes from thermoluminescence dates on correlative silty clay which contained shells in growth position. The average age was 74 ± 10 ka and the shells associated with the dated material had aIle/Ile ratios of 0.13 to 0.14 (Forman et al., 1987). These data suggest an intermediate age between Bell Sea and Tyrrell Sea episodes. The collection site is about 77 m above sea level, indicating considerable isostatic depression at the time of deposition. This information points to an interstadial marine incursion and apparently confirms at least one of the deglacial events previously suggested by Andrews et al. (1983b) and Shilts (1982b).

Andrews et al. (1983b) suggested a chronology for the units discussed earlier on the assumption that the Bell Sea is early marine isotope stage 5e and dates from 130 ka. On that assumption, and using a kinetic model developed by Miller (see Miller, 1985a), they predicted the ages of the Fawn River gravel units as 76 ka and the Kabinakagami sediments as 35 ka. The 76 ka age estimate is close to the more recent independent age estimates noted above. It is possible, however, that the thermoluminescence dates are minimum ages and that Bell Sea sediments date from an interglaciation previous to the Sangamon as stated in the lithostratigraphic section, above.

LITHOSTRATIGRAPHIC RECORD ON THE ONTARIO SHIELD
L.A. Dredge and W.R. Cowan

The sedimentary record of Quaternary events in parts of this area underlain by Precambrian rocks is generally similar to that found elsewhere on the Precambrian Shield. Till is the dominant deposit, and commonly only one till is recognized in section. Where additional tills or other deposits are present, it is not possible to determine the age of the older units.

Subtill organic sediments

In addition to the Hudson Bay-James Bay nonglacial sediments described above, numerous subtill organic deposits have been reported within the Canadian Shield area to the

---

Andrews, J.T.
1989: Amino acid stratigraphy (southwestern Canadian Shield); in Chapter 3 of Quaternary Geology of Canada and Greenland, R.J. Fulton (ed.); Geological Survey of Canada, no. 1 (also Geological Society of America, The Geology of North America, v. K-1).

south (e.g., Lowdon et al., 1971; Brereton and Elson, 1979; DiLabio, 1982; Steele and Baker, 1985), many of them containing allochthonous wood. All samples dated have proven to be beyond ^{14}C dating limit. Suggested correlations for these materials include the last interglaciation (Sangamon) as well as possible Wisconsinan interstades. Perhaps the most complete study to date of such material is provided by Brereton and Elson (1979) who correlated allochthonous wood dated at greater than 37 ka with the Missinaibi Formation of the James Bay area. Their correlation of a borehole sequence from Currie Township near Matheson, Ontario is given in Figure 3.34. The interpretation indicates that a pre-Missinaibi till is present, as are sediments correlated with the Missinaibi and the ubiquitous Matheson Till (here correlated with the Adam Till of the lowlands). Though this and other examples leave many questions, it is possible that both interglacial and interstadial organic deposits exist within the southwestern Canadian Shield area.

Late Wisconsinan and Holocene stratigraphy

Much of the southwestern Canadian Shield region contains only one late glacial till sheet. Sections that do consist of multiple Late Wisconsinan or Holocene till sheets occur mainly around the margin of Lake Superior, particularly near Thunder Bay, along the northern edges of the glacial Great Lakes, and near the margins of the Hudson Bay Lowlands. These multiple successions are probably associated with ice marginal lakes and the tendency of the ice sheets to fluctuate (sometimes rapidly) within the lacustrine environment, and with the interplay of various ice lobes.

Burwasser (1977) reported two Late Wisconsinan tills near Thunder Bay, the second having as many as four textural facies. The oldest till is a stony sand till deposited during the main Late Wisconsinan advance by the Laurentide Ice Sheet. A subsequent readvance to the Dog Lake and Hartman moraines resulted in the deposition of a second till of similar texture and composition. At the same time a readvance of the Superior Lobe (Marquette readvance) to the Marks-Mackenzie moraine system deposited fine grained tills derived from glacial lake sediment; these vary in texture with the underlying sediments and are about 10 ka in age.

Along the northern margin of the southwestern Canadian Shield region, and adjoining the Hudson Bay Lowlands, thin calcareous clay till overlying older till was deposited by the Holocene glacial surges which deposited the Cochrane till and its stratigraphic equivalents. Near its attenuated western border the upper till is commonly less than 1 m thick and can be penetrated with a hand auger or shovel. The underlying till may be calcareous and silty, depending on the Phanerozoic component derived from the Hudson Bay Lowlands.

For the Cochrane-Timmins area of northern Ontario, Hughes (1959, 1965) provided detailed descriptions of the glacial Lake Barlow-Ojibway sediments and Cochrane Till and subsequent glacial lake sediments. For the Barlow-Ojibway sediments (Barlow-Ojibway Formation, Hughes, 1965) north of the Hudson Bay-St. Lawrence drainage divide he recognized three sequences of varved sediments: "these are informally termed Lower sequence, consisting of varves No. 1527 and below of the Timiskaming Series of Antevs (1925, 1928), Frederick House sequence, consisting of varves 1528 to 2014 of the Timiskaming Series, and Connaught sequence, consisting of 60 or more varves that overlie Frederick House sequence, which are numbered separately. Varves of the Connaught sequence are known only from north of Hudson Bay-St. Lawrence divide (i.e. glacial Lake Ojibway, where varves 1 to 800 (glacial Lake Barlow) of the Lower sequence are missing; hence the entire varved sediments of the Barlow-Ojibway Formation are exposed nowhere at a single locality" (Hughes, 1965, p. 542-543). Each sequence consists of a thinning upward set of varves. The basal units of each sequence correspond to glacial readvances of the Hudson Ice margin which lay farther north. The Barlow-Ojibway sediments form a nearly continuous sheet from the mouth of Montreal River on Lake Timiskaming northward beneath the Cochrane till to at least Island Falls on Abitibi River (49°35′N) Hughes (1965).

Hughes (1965) defined the Cochrane Formation to include "the stony clay till that overlies varved sediments of the Barlow-Ojibway Formation; it also includes thin glaciolacustrine sediments that locally overlie the till" (Hughes, 1965, p. 544). As such, the Cochrane till and associated sediments represent the Cochrane readvance, one of the late glacial, ice margin surges into glacial Lake Ojibway.

GROWTH OF THE LAST ICE SHEET

As discussed in the introduction to this chapter, there is considerable controversy over the nature of inception, and dynamics of growth and retreat of the last ice sheet which covered this area. The discussion below touches on the two most important points of this controversy: the location of centres of inception (was there a Patrician growth centre in north-central Ontario?) and the possibility of an ice dome (Hudson Ice) independent of Keewatin Ice and Labrador Ice in southern Hudson Bay.

Patrician ice centre

In addition to identifying Labradorean and Keewatin centres of ice accumulation, Tyrrell (1913) proposed a third centre of glacial outgrowth in the District of Patricia (north-central Ontario). The orientation of striae and roches moutonnées at many sites throughout northeastern Manitoba and across Ontario north of Albany River, showed an ice flow radiating northwards from a centre that lay between Big Trout Lake (Fig. 3.17) and Windigo Lake 150 km to the southwest. Crosscutting relationships indicated that the radiating striae predated a later set produced by ice flow from the northeast. Tyrrell's observations in the Wunnummin, Red Lake, and Landsdowne House areas were substantiated by Prest (1963), although it was not clear that the northwesterly sense of the flow was correct in all cases.

South of Albany River the existence of a Patrician ice centre (vis-à-vis other source areas) cannot be proven by the pattern of striae, although various workers have expressed support for the Patrician concept. Burwash (1934) and Zoltai (1962) reported multiple generations of striae in northern Ontario between Lake Superior and the Manitoba border. Burwash identified an early ice flow from the north ("Keewatin"), based on striae preserved below oxidized till on lee slopes of rock outcrops. He considered the next ice flow, which produced roches mountonnées and striae trending southwest and west-southwest near Lake of the Woods, to be from a Patrician centre, and a final flow towards the southwest to be "Labradorean". Zoltai (1962), however, thought that this last flow might rather be from a "Patrician" centre 1) because the striae appeared to be radiating from an area between Hudson Bay and Lake Superior and 2) because of the recessional pattern of moraines.

Although glacial erosion forms apparently suggest there may have been an ice dispersal centre in north-central Ontario at some time, the composition and fabric of the various till sheets (see stratigraphy section) in northern Ontario and Manitoba indicate that those tills were emplaced by ice flowing from centres in Keewatin, central Quebec, and southern Hudson Bay, rather than from northern Ontario. In addition, the presence of glacial lake sediments directly below Early Wisconsinan tills in the Manitoba and James Bay areas demands a blockage of drainage by ice flowing out of or across Hudson Bay, rather than ice flow from a Patrician centre south of the bay. Many sections in northern Ontario, however, have yet to be investigated. In conclusion, the District of Patricia may have been a centre of ice flow during the initial growth of the last ice sheet, but the area was probably not a centre of outflow during subsequent glacial phases.

Hudson Ice Divide

Today there are three contrasting views on the configuration of the Laurentide Ice Sheet during the last glacial maximum. (1) Flint (1943), Walcott (1970), and Denton and Hughes (1981) proposed a model which today would be described as a single-domed equilibrium ice sheet centred over Hudson Bay. The single-domed model portrays outward movement of erratics from Hudson Bay in all directions. However, because we know that ice flowed into the bay from the northwest and the east, this model is discounted and will not be discussed further. (2) Tyrrell (1898b) and later Shilts (1980) envisaged a two-domed model with centres in Keewatin and central Quebec (in 1913, Tyrrell proposed a third centre, the Patrician Centre). This model has Labrador Ice flowing southwestward across Ontario and the southern half of Hudson Bay, and coalescing with Keewatin Ice somewhere west of Lake Winnipeg. Shilts' (1985) latest model shows Keewatin Ice flowing southeastward across Manitoba and Ontario instead, but he does not present evidence to support his conclusions. (3) Dyke et al. (1982b) proposed a multidomed configuration for the Laurentide Ice Sheet, in which there was an ice divide over southern Hudson Bay; this divide produced a centre of outflow and an ice mass (Hudson Ice) that was distinct from Keewatin and Labrador Ice. The Hudson Ice Divide was part of a complex of branching ice divides within the Laurentide Ice Sheet (e.g., Map 1703A). Dyke et al. suggested that the Hudson Ice Divide became operative as soon as there was considerable drawdown into Hudson Strait, although during early ice sheet expansion, ice from Keewatin, Labrador, or Patrician centres may have flowed into Hudson Basin. The validity of the last two models has been tested by examining the glacial deposits (items 1 to 3 below) and features (items 4 to 6) in the area surrounding Hudson Bay, since the nature of glacial sediments beneath Hudson Bay is largely unknown.

(1) The major zones of carbonate dispersal in Manitoba (Fig. 3.20) and northern Ontario (Fig. 3.19) can be accounted for by all models except that of Shilts (1985), as can flow features that are oriented generally south and southwest across the region (Fig. 3.18; Prest et al., 1968).

(2) The upper till sheet in Manitoba and much of northern Ontario is a brown till containing reddish mudstones and dolomites which originate in Devonian formations in the middle of Hudson Bay. The occurrence of the brown till in northwestern Ontario requires that ice flow be southwards from Hudson Bay, rather than westwards or southwestwards from a Labrador source. The underlying grey tills, however, may have been emplaced by ice from other flow centres.

(3) The interpretation of "dark erratics" in tills has been the subject of considerable controversy. Their distribution was initially used to argue against a Hudson ice centre on the basis that their presence in tills demanded ice flow out of Quebec across southeastern Hudson Bay; this line of reasoning has recently been questioned. The problematic erratics appear as pebbles and cobbles of dark greywacke with diagnostic yellowish 'eyes' (concretions), and in the granule-sized fraction as nondistinctive dark grey clasts.

Granule-sized dark grey rock fragments have been retrieved from tills in northern Ontario and Manitoba by Shilts et al. (1979). Greywacke pebbles and cobbles with yellowish eyes have been discovered in northern Manitoba east of a line joining Churchill, Southern Indian Lake, and The Pas (Nielsen, 1982); they are present in all tills of eastern provenance and some (Mountain Rapids, northern Manitoba) that also contain "Keewatin" erratics. They are exposed in the surface tills of central Saskatchewan (E. Nielsen, Manitoba Department of Energy and Mines, personal communication, 1985), in both Wisconsinan and pre-Wisconsinan tills in southwestern Manitoba (Nielsen, 1982), and as far west as southern Alberta (H. Groom, Manitoba Department of Energy and Mines, personal communication, 1984). They have been reported in northwestern Ontario (Prest, 1963), the James Bay Lowlands (Bell, 1877; V.K. Prest, Geological Survey of Canada, personal communication, 1985), and as far south as Leamington, Ontario (E.V. Sado, Ontario Geological Survey, personal communication, 1985). One of the first observers of these erratics in Manitoba was Bell (1879a), who noted that they resembled the Proterozoic rocks of the circum-Ungava geosyncline which underlie southeastern Hudson Bay and outcrop as the Omaralluk Formation on the Belcher Islands and east side of Hudson Bay. Since then, all similar rocks have been assumed to originate from that area or from Sutton Hills. The principal questions as far as Hudson Ice Divide is concerned are: (1) are there other sources for the greywackes (and jaspers)? and (2) when were the erratics transported?

Prest (1963) concluded that the greywackes in northwestern Ontario had been transported by ice crossing Hudson Bay from a centre in Labrador; Shilts et al. (1979) have further argued that the distribution of granule sized "dark erratics" demands ice flow across Hudson Bay from a Labrador centre throughout most of the last glacial stage, and that a Hudson Ice Divide was therefore impossible. On the basis of heavy mineral studies of tills, marine sediments, and stream deposits, Paré (1982), Henderson (1983), and Adshead (1983a,b) have also concluded that the ice mass affecting the Hudson Bay Lowlands was centred in Quebec, although they mentioned that their samples may have been derived from recycled sediment. However, Dyke et al. (1982b) and the authors of this chapter feel that the distribution of greywackes cannot be used to argue against a Hudson Ice Divide, and that these erratics were emplaced by means other than by sustained flow of Labrador Ice across Ontario and onto the Plains. The reasons are as follows:

(a) There may be alternative source areas for these rocks, particularly for those found in northern Manitoba and Saskatchewan. Greywacke is exposed in the Fox River basin of northeastern Manitoba and also constitutes part of the metavolcanic and metasedimentary suites along Seal River farther north. Greywacke cobbles are particularly abundant in tills exposed along Churchill River due south of the Seal River metavolcanic suites. These, and possibly some of the erratics in the Plains region, might therefore have been derived from source areas to the west of Hudson Bay and carried by Hudson, Labrador, or Keewatin ice. Also, unrecognized source areas may exist in Hudson Bay (e.g., offshore from Churchill) and protrude through shallow Paleozoic cover rocks.

(b) In some cases the erratics are thought to be polycyclic, since they occur in pre-Wisconsinan tills and all Wisconsinan tills of eastern provenance. Either the erratics were recycled into younger tills, or else source rocks were re-eroded during each glacial event. The polycyclic alternative is favoured because, in some cases, the erratics are associated with deposits of ice lobes which are not of eastern provenance (e.g., Red River Lobe in southern Manitoba; Nielsen, 1982), and in northern Manitoba they are found in till sheets which also contain erratics from Keewatin sources.

(c) In the introduction to Chapter 3 we have also suggested that even if all the distinctive concretionary greywackes were derived from the Omaralluk Formation in eastern Hudson Bay, they may have been initially transported westwards into Hudson Bay by Labrador Ice and then redispersed by Hudson Ice as the Hudson Ice Divide developed.

(4) Crossed striae on the Ottawa Islands (eastern Hudson Bay) show a major ice flow towards the northeast, followed by a flow west-southwest (Andrews and Falconer, 1969). In the model of Dyke et al. (1982b), the general northeastern flow over the north half of Hudson Bay is said to be due to converging flow from centres in Labrador, southern Hudson Bay, and Keewatin, which developed when an ice stream began to operate in Hudson Strait. This was followed by a late, deglacial flow into Hudson Bay from central Quebec. Shilts' (1980) model, however, shows maximum flow towards the west across the Ottawa Islands, which conflicts with the observed data.

(5) Major landforms and ice retreat patterns are best explained by ice flow radial to a centre in Hudson Bay, rather than from a centre in Labrador. (a) The development of large interlobate moraine systems both southwest and southeast of Hudson Bay (cf. Fig. 3.22) implies that an ice mass dynamically distinct from Keewatin and Labrador ice occupied southern Hudson Bay. The Burntwood-Knife and Harricana interlobate moraines formed during and since the last glacial maximum. Although the Burntwood-Knife Moraine, which separates southward and westward flow, could have marked the convergence of Keewatin Ice with either Hudson Ice or Labrador Ice, the location of the Harricana Moraine demands that ice flow in Hudson Bay be independent from Labrador Ice flow. (b) The pattern of late glacial surges in northern Ontario and Manitoba, as determined by mapping of glacial-fluting patterns, suggests ice in the Hudson Bay Lowlands receded towards a centre in Hudson Bay, not towards outflow centres in Labrador (Hardy, 1976; Prest, 1970) or central Keewatin (Shilts, 1981, 1985).(c)The development and geographic distribution of glacial lakes along the southern periphery of the ice sheet suggests ice recession and lake expansion controlled by an ice divide in Hudson Bay; otherwise the progressive expansion of waterbodies northwards (especially glacial Lake Ojibway) would have interrupted the regional Labrador Ice flow into Manitoba and northeastern Ontario. (d) Radiocarbon dates on shells indicate that the postglacial sea occupied the entire lowlands more or less instantaneously. A synchronous marine incursion cannot be explained by recession towards a Labrador ice centre.

(6) Dyke et al. (1982b) discussed the problem of ice sheet asymmetry and pointed out that if flow from the Labrador Ice Divide extended westwards, at least as far as Winnipeg, but only extended eastwards to about the Labrador coast, then the profile of the ice sheet was grossly asymmetrical. They pointed out that unless the basal shear stress on the western side was radically lower than that on the eastern side, then Labrador Ice could not have flowed much farther west than the east coast of Hudson Bay. The recent glaciological models of the Laurentide Ice Sheet developed by Denton and Hughes (1983) and Fisher et al. (1985) support this contention. Their models show a multi-domed ice sheet with one ice divide lying in southern Hudson Bay. Fisher et al. (1985) have also shown that this ice divide persists even when lower shear stresses are assumed to have existed at the base of the ice in the Hudson Bay and Prairie regions.

In conclusion, we contend that if there was an ice stream in Hudson Strait, which generated a northeast flow over the northern part of Hudson Bay while ice continued to flow southwards and westwards over Ontario and Manitoba, then there should have been an ice divide over the southern part of Hudson Bay. The existence of this divide is supported by glacial deposits and landforms and by deglaciation patterns. The position of the divide would have been extremely sensitive to the rate of flow in the Hudson Strait ice stream (Denton and Hughes, 1983). Rapid flow rates there would have displaced the divide to the south; slow flow caused by grounding at the mouth of the Strait would have displaced the divide to the north. Similarly, widespread surging into the fringing glacial lakes on the south side of the ice mass would have displaced the position of the ice divide and change its shape (i.e. straight to curved to serpentine).

Variations in ice flow patterns

Although regional ice flow patterns and till fabrics suggest outflow from an ice divide in Hudson Bay, there is considerable variation in the direction of ice flow across eastern

Manitoba and northern Ontario, and multiple generations of striae and streamlined forms have been reported (Zoltai, 1962; Prest, 1963; see *Ice flow summary* in *Lithostratigraphic record of the Hudson Bay Lowlands*). Although the existence of weathered striae on "lee sides" (Burwash, 1934) implies some striations may predate the last glaciation, only relative ages can be assigned to most ice surface flow features. Most intersecting patterns have been assigned to ice flow changes accompanying deglaciation.

Ice sheet limits

From the Wisconsinan maximum, once an ice stream was active in Hudson Strait, ice flowed northward from a divide over southern Hudson Bay. Hudson Ice coalesced with Foxe Ice on its northern margin and streamed into Hudson Strait. The exact location of the confluence is unknown, but it is presumed to lie north of Mansel Island. In Hudson Bay, Keewatin Ice and Hudson Ice coalesced well offshore, as Keewatin Ice flowed strongly into the bay before being deflected northward to cross Coats Island (Shilts, 1980; Map 1703A). The western margin and zone of final confluence between Hudson Ice and Keewatin Ice on land is marked by the Burntwood-Knife Interlobate Moraine in northern Manitoba, and the Bedford Hills-Belair Moraine complex south of Lake Winnipeg (Fig. 3.22). The position of the confluence varied somewhat throughout the last glaciation as is indicated by regional dispersal of carbonates west of the Burntwood-Knife Interlobate Moraine and east of the Belair Moraine, and the superposition of striae. The southern limit of Hudson Ice lay as an unconstrained margin in Minnesota, Wisconsin, and south of the Great Lakes. Hudson Ice reached as far east as the Harricana Interlobate Moraine, but did not extend substantially beyond it (cf. this chapter, Vincent, 1989b). To the northeast, Hudson Ice probably did not extend far east of the Ottawa Islands. The northward striae found there may mark the area of convergence of eastward flowing Hudson Ice and westward flowing Labrador Ice.

DEGLACIATION

Perspective and general summary

Many essential aspects of the deglacial history have been synthesized by Prest (1970). Since then, stratigraphic and regional mapping studies have added numerous details to our understanding of the late glacial history of this region, and in some cases, have led to new concepts about the style and pattern of ice retreat. Some of the main ideas resulting from recent syntheses are:

(1) An ice divide was present in Hudson Bay during deglaciation. Hudson Ice was a distinctive functioning entity (i.e. an ice domain), within the Laurentide Ice Sheet, and had its own style and pattern of ice flow.

(2) Along the southern extremities of Hudson Ice, deglaciation began in the zones of convergence between Hudson and Keewatin ice, and Hudson and Labrador ice. The Burntwood-Knife and Harricana interlobate moraines are major landforms marking the zones of convergence. These features developed because meltwater was funnelled down the hydraulic gradient towards areas of thinner, relatively inactive ice at the margins of the main ice masses, and because of the persistence of flow towards these areas (i.e. they were zones of debris accumulation).

(3) Proglacial lakes developed where meltwater was impounded. The first to form were in the southern part of the Great Lakes basins, followed by the largest — glacial Lake Agassiz — and later by glacial Lake Barlow-Ojibway which developed along the east side of the ice sheet (Map 1703A).

(4) The Hudson Ice margin was highly mobile, due to frequent shifting of the ice divide and centres of outflow, the existence of major glacial lakes along its southern periphery, its location in a marine basin, and at some stages, probably due to the presence of a water layer between the ice sheet and impermeable substrate. Consequently, flow zones and ice streams developed within the ice mass. The locations of these are marked by fields of streamlined forms, and by the radial moraines which developed along their margins (e.g., Sachigo and Limestone moraines). Where flow zones approached the unconfined southern boundary of the ice sheet, they created lobes (e.g., Rainy Lobe, Hayes Lobe). The distribution of radial features suggests that many small ice streams developed within Hudson Ice as deglaciation proceeded.

(5) Short readvances or stabilizations of the ice, related either to changing lake levels or to topographic obstructions, are marked by ice front moraines. Many of these were deposited subaqueously.

(6) Proglacial lakes continued to expand against the receding ice front until, eventually, the entire southern periphery of the ice sheet was rimmed by glacial lakes (Agassiz-Ojibway). Major embayments developed along the interlobate convergence zones.

(7) During the final stages of deglaciation, parts of the ice margin surged into the glacial lakes. The Cochrane surges are a part of these events, and airphoto mapping of northern Ontario suggests that surging occurred along much of the ice margin (Fig. 3.23). The mapping also suggests that the Cochrane "readvances" (Prest, 1970), and similar events, may have consisted of many short surges along a continually receding ice front, rather than one or two major readvances.

(8) Surging, downwasting, sapping of the northern margins of this marine-based ice sheet, and possibly eventual uncoupling of ice from its base created an unstable ice mass whose existence ended catastrophically with the incursion of the Tyrrell Sea. The gravelly debris (reworked into beaches) near the Hudson Bay coast in Ontario may mark the southern edge of the ice sheet at the time of breakup.

(9) Radiocarbon dates from marine shells indicate that the Tyrrell Sea invasion was essentially a synchronous event throughout the Hudson Bay Lowlands, resulting from marine incursions along both the east and west sides of Hudson Bay.

Early stages of deglaciation (20 ka - 10 ka)

Recession of the Rainy (northwestern Ontario) Lobe and Superior Lobe

The southwest margin of Hudson Ice reached its maximum about 20 ka ago (Prest, 1984), when its terminus reached into Minnesota and Wisconsin (Fig. 3.22 and Map 1702A). During early phases of deglaciation there was slow, steady retreat of the Superior Lobe from that position despite major readvances of the adjacent Lake Michigan, and Des

Moines lobes. By the time of the Two Creeks Interstade (12 ka) ice had retreated into the confines of the Lake Superior basin, but at about 11.8 ka it readvanced to the southern margin of the basin and created the Nickerson and Two Rivers moraines in northern United States (Farrand and Drexler, 1985). This readvance, called the Great Lakean (formerly Valders), is not recorded by glacial sediments in Ontario or Manitoba, but forms a datum for subsequent events in the Lake Superior basin. At about the same time, the St. Louis Sublobe of the Red River Lobe advanced into northern Minnesota.

After 11.8 ka both the Rainy Lobe and Superior Lobe retreated. Glacial Lake Johnston (Antevs, 1951) developed in northwest Ontario in the area between the Rainy and St. Louis lobes. With additional retreat, the small ice marginal lakes coalesced with glacial Lake Agassiz, which formed in Red River valley as the Des Moines Lobe retreated north of the Mississippi drainage divide (Elson, 1967). This is the early Cass Phase of Fenton et al. (1983), shown on Figure 3.35a. As the ice front receded, the Bedford Hills and Belair interlobate moraines developed between an ice lobe flowing south in the Lake Winnipeg basin and one flowing westward from the Canadian Shield. At about this same time the Assiniboine delta formed where meltwaters from western Canada debouched into the western margin of the glacial lake (Klassen, 1989); this Lockhart phase of Lake Agassiz lasted until ice retreated east of the Eagle-Findlayson moraines (Fig. 3.35b).

By 10.8 ka the Rainy Lobe may have backwasted at least as far east as Thunder Bay. Exposure of low areas permitted glacial Lake Agassiz to drain eastwards into the Lake Superior basin (Zoltai, 1965; Teller and Thorliefson, 1983) and water level fell below the level of the southern outlet. The Herman, Ojata, and The Pas beaches developed during this low phase, called the Moorhead Phase. There is some debate as to how far east the ice had receded at this time and whether the spillways near Lake Nipigon were operative (e.g., as shown by Clayton, 1983); exposing these Nipigon outlets necessitates a major readvance during the next glacial event. Alternatively, or additionally, it is possible that the lake drained northwestwards through Clearwater Valley in Saskatchewan during this phase (Elson, 1967; Christiansen, 1979). In any case the Moorhead phase ended about 9.9 ka when a readvance of the Rainy and Superior lobes closed the eastern outlets.

North-central Ontario

East of Sault Ste. Marie, events between 12 and 11 ka are not well documented; however, the ice margin was probably near the north shore of the Lake Huron basin throughout much of this period (Fig. 3.35a, b). Throughout the Laurentian Highlands area, grounded ice was retreating over relatively uniform Precambrian terrain. Northward ice retreat was accompanied by rising water levels in the Huron and Michigan basins (probably the result of isostatic uplift of the Kirkfield outlet), and by 11.2 ka glacial Lake Algonquin inundated the north shore of Lake Huron (Karrow et al., 1975) and entered the easternmost part of Lake Superior (Prest, 1970; Cowan, 1985). South of North Bay several east-west trending outlets were uncovered (Harrison, 1972; Chapman and Putnam, 1984) and a series of lake levels below the Main Algonquin level resulted (Hough, 1958; Prest, 1970; Karrow, 1989, Table 3.2). Isostatic rebound also played a significant role in development of this sequence.

Ottawa-Mattawa valleys

Ice retreat was sufficient for the Champlain Sea (entering Ottawa Valley from the east) to reach Ottawa by 12 ka (Richard, 1980b), although age assignments are the subject of much controversy; it possibly reached its most westerly extension at Chalk River by 11.3 ka (Catto et al., 1981). In the Upper Ottawa River area, the Post-Algonquin lakes spread northeasterly into the Lake Timiskaming area while a lobe of ice to the east of Mattawa blocked drainage towards lower Ottawa Valley (Vincent and Hardy, 1979; Veillette, 1988; this chapter, Vincent, 1989b). The lake followed the suture between Hudson and Labrador ice, which is marked by the Harricana Moraine. The relationships between Hudson and Labrador ice, glacial lakes Barlow and Ojibway, and the Harricana Interlobate Moraine are discussed by Vincent (this chapter, 1989b).

Between 10.5 ka and 10 ka the ice lobe blocking Mattawa River valley retreated, allowing drainage of the last phases of Lake Algonquin through the Mattawa and Ottawa rivers (Lewis, 1969; Karrow et al., 1975; Chapman and Putnam, 1984; Karrow, 1989). This opening of the "North Bay outlet" created the low level Lake Stanley Phase in the Lake Huron basin, and Lake Hough in Georgian Bay and permitted Lake Barlow to develop in the Lake Timiskaming area (Fig. 3.35c).

Hudson ice margin about 10 ka

By 10 ka there had been extensive retreat of Hudson Ice, accompanied by expansion of proglacial lakes, although the Lake Superior basin was still ice-filled (Fig. 3.35c). Glacial Lake Agassiz occupied an extensive area between the Manitoba Escarpment on the west, the regional drainage divide on the south, and the combined Keewatin and Hudson ice in the north and east. Large rivers from deglaciated areas deposited deltas in the southwest. Glacial debris was deposited into the lake as blanket sands and clays, esker deltas, and radial moraines. Small ice marginal lakes, Duluth and Minong, also occupied both western and eastern parts of the Lake Superior basin. Lake Barlow developed in the upper Ottawa River drainage basin along the eastern margin of Hudson Ice in the vicinity of the contact between Hudson and Labrador ice.

Table 3.2 Algonquin and post-Algonquin Lakes in the Lake Huron and Eastern Lake Superior basins

Name	Elevation (m) at Sault Ste. Marie	Original level (m) (from Farrand and Drexler, 1985)
Main Algonquin	309	184
Upper Group		
Ardtrea	302 ?	170
Upper Orillia	295	
Lower Orillia	?	
Wyebridge	275 ?	165
Penetang	265	155
Cedar Point	257	150
Payette	247	142
Sheguiandah	~233	134
Korah (Minong)	210	122
Nipissing	198	184

Major segments of both the Burntwood-Knife Interlobate Moraine, along the convergence zone between Hudson-Keewatin ice and the Harricana Interlobate Moraine, separating Hudson Ice and Labrador Ice, formed between 10 ka and 8.5 ka. Large quantities of sand and gravel comprising these moraines were deposited where meltwater debouched into ice marginal lakes as the ice sheet retreated. The Hudson Ice margin was distinctly scalloped

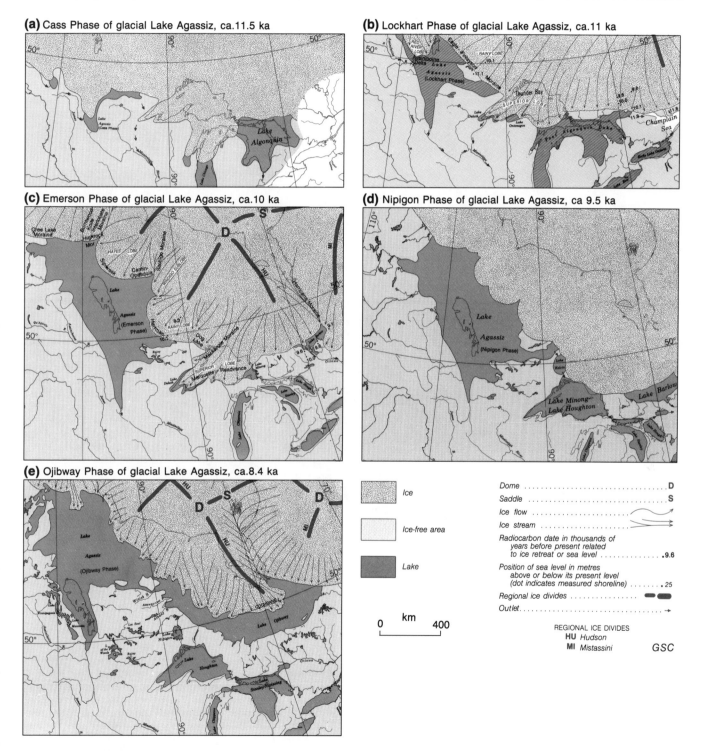

Figure 3.35. Deglaciation of the southwestern Canadian Shield region. Ice margin position: (a) Cass phase of glacial Lake Agassiz, about 11.5 ka; (b) Lockhart phase of glacial lake Agassiz, about 11 ka; (c) Emerson phase of glacial lake Agassiz, about 10 ka; (d) Nipigon phase of glacial lake Agassiz, about 9.5 ka; (e) Ojibway phase of glacial lake Agassiz, about 8.4 ka.

241

in character, particularly on its western edge where the ice abutted glacial Lake Agassiz. Large lobes occupied northeast Manitoba (Hayes Lobe) where ice may have been streaming through the Hayes River-Nelson River topographic lows; a small lobe existed in the vicinity of Trout Lake in northern Ontario; a large lobe with several sublobes occupied northwestern Ontario (Rainy Lobe); and another major lobe occupied the Lake Superior basin. The fronts of these lobes are delimited by arcing morainic systems. Some of these features may mark readvance positions, but others probably represent minor halts or pulses related to changes in lake level, especially where the moraines consist of stratified debris. The structural character of Hudson Ice at this time is shown by systems of ice flow features, particularly streamlined forms, and by striae. Lobes and sublobes were set off from each other by areas of thinner or slower flowing ice whose existence is marked by esker-like interlobate or radial moraines and minor areas of converging striae.

Hayes Lobe

By about 10 ka the south margin of Keewatin Ice had retreated to a position marked by the Cree Lake Moraine and Highrock Moraine (Fig. 3.35c). This age for the Cree Lake Moraine follows the chronology of Christiansen (1979). The Hayes ice lobe filled Nelson-Hayes valley. This lobe was separated from Keewatin Ice by a re-entrant of Lake Agassiz and by the Burntwood-Knife Interlobate Moraine. Its eastern limit is demarked by the Sachigo Moraine, which continually separated this lobe from other areas of Hudson Ice. Flow within the lobe was splayed outwards from the central part of an ice stream which occupied the Hayes River-Nelson River lowlands. The ice position shown in Figure 3.35c follows the Sipiwesk-Cantin moraines, but because of the absence of absolute age control in this area, it is equally possible that the more southerly system composed of the Hargrave, Hudwin, and Pasquia moraines marks the margin at this time. These moraines are composed largely of substratified ice contact debris and are considered to represent halts or short forward pulses of the ice margin, rather than significant readvances.

Northwest Ontario (Rainy) Lobe and Superior Lobe

The Rainy Lobe occupied northwestern Ontario and generated a recessional southwestward ice flow which crossed an earlier more southerly regional flow. The boundaries of this lobe are the Sachigo Moraine on the north and Mackenzie radial moraine to the south (Fig. 3.22). Several sublobes are further recognized by flow patterns and are demarcated by both end moraines and radial moraines. At 10 ka the margin of the Rainy Lobe lay at the Hartman and Dog Lake moraines which mark a readvance (Zoltai, 1965).

In the Lake Superior basin a major readvance, called the Marquette, formed the Marks-Mackenzie moraine system near Thunder Bay (Zoltai, 1965; Burwasser, 1977) and constructed the Grand Marais Moraines on the south side of the lake about 9.9 ka (Black, 1979; Nielsen et al., 1982; Drexler et al., 1983; Clayton, 1983; Farrand and Drexler, 1985). The Hartman, Dog Lake, and Marquette readvances blocked the eastern outlets of glacial Lake Agassiz, marked the end of the Moorhead (low) phase, and caused glacial Lake Agassiz to drain southward through Red River valley.

Levels in glacial Lake Agassiz rose during the subsequent period called the Emerson phase (approximately 9.9-9.5 ka; Fig3.35c); it was during this high level phase that the upper Campbell beach (350 m) formed in western Manitoba (Nielsen et al., 1982) and the beaches at 470 m (Prest, 1963) developed on the Lac Seul Moraine. A distinctive marker deposit of red clay, derived from the west end of Lake Superior (Zoltai, 1962; Nielsen, et al., 1982), was deposited in eastern glacial Lake Agassiz during this time. Glacial Lake Minong occupied the eastern part of the Lake Superior basin and glacial Lake Duluth lay in the southwestern part. The extensive readvances involved with the Marquette and related events apparently caused a considerable drawdown of ice north of Lake Superior and thus considerably influenced deglaciation events thereafter.

Early Holocene ice recession (9.9 ka-8.2 ka)

As the ice front receded north of the Hudson Bay-St. Lawrence River drainage divide, glacial lakes continued to expand. Hudson Ice retreated about 220 m/a between Lake Superior and James Bay (Prest, 1968), and slightly faster farther west.

The partitioning of the ice sheet, which had developed by about 10 ka, persisted and was accentuated, with progressive development of secondary ice "lobes" separated by small radial moraines (Fig. 3.22). Although the reason for development of lobes is unclear, it possibly relates to the development of ice streams or flow zones within Hudson Ice, which in turn may have been driven by shifts in the late glacial ice divide, and to ice dynamics along glacial lake margins. Processes related to the proglacial lakes that might have promoted ice streaming are lowering of basal shear stresses by buoyancy along particular parts of the ice margin and accelerated marginal retreat by calving (i.e. unstable frontal profiles). The abundance of crisscrossing iceberg scour marks across Manitoba and northern Ontario (Dredge and Grant, 1982) indicates that icebergs were common in the glacial lakes, as was calving. The many small areas characterized by flutings diverging from regional flow patterns suggest that local surges were common during this phase of deglaciation (Fig. 3.23).

The major role of meltwater during deglaciation is abundantly clear. Most of the radial moraines are glaciofluvial in nature. Many of the major end moraines are also largely substratified debris dumped into glacial lakes, rather than till or push moraines. Since many are not composed of till, these moraines may mark halts or minor pulses of the ice sheet rather than major readvances; the position and creation of many of the largest moraines may relate to glacial adjustments to changing lake levels.

Retreat of the northwest Ontario (Rainy) Lobe

After the 10 ka readvances, the northwest Ontario Lobe and associated sublobes receded, forming several end moraines. The later readvance to the Lac Seul Moraine overran lake sediments and truncated part of the Hartman Moraines. The Lac Seul Moraine is considered by Zoltai (1965) to represent a readvance of a least 40 km. De Geer moraines mark the pattern of ice recession from the Lac Seul Moraine to the Nipigon Moraine. During this time there were minor shunts of the various sublobes of the Rainy Lobe and development of the Kaishak and other radial moraines.

Recession of ice to the position of the Nipigon Moraine exposed a series of successively lower outlets which permitted drainage of glacial Lake Agassiz into Lake Superior, ended the Emerson phase of glacial Lake Agassiz and began the Nipigon phase (Fig. 3.35d). This change probably occurred about 9.5 ka (Teller and Thorleifson, 1983). The drop in lake level may have been responsible for the minor stillstands or readvances which created the Whitewater and Nipigon moraines.

The general pattern of ice recession was later interrupted by readvances that formed the Nakina (Zoltai, 1965) and Agutua (Prest, 1963) moraines. Prest considered the Agutua Moraine to be the product of a readvance of 40 km based on his observations of overridden lacustrine sediment. It is also possible, however, that the moraine was formed by a less extensive readvance and that the buried clays resulted from a series of individual small pulses during general retreat. Glacial Lake Nakina formed in front of the Nakina Moraines in the area between the ice sheet and the Great Lakes-Hudson Bay divide. Retreat beyond the Nakina and Agutua moraines probably permitted the coalescence of glacial Lake Agassiz with glacial Lake Ojibway (Fig. 3.35e) which was expanding from the east towards the northwest (see below). This ended the Nipigon Phase and began an Ojibway Phase (possibly correlative with Elson's (1967) Gimli Phase) of Lake Agassiz.

Superior Lobe

During the recession which followed the Marquette readvance, glacial Lake Minong expanded to occupy the entire Lake Superior basin. Lake level was maintained by the sill at Sault Ste. Marie until about 9.5 ka (Saarnisto, 1974). Subsequent lowering of lake level was accomplished by erosion of the St. Marys River sill, possibly caused by inflow of glacial Lake Agassiz waters after the Nipigon outlets were reopened (Teller and Thorliefson, 1983). A red clay may have been flushed into the Great Lakes system during this event. The lowest lake level recognized in the basin, Lake Houghton, developed later, when lakes Chippewa and Stanley occupied the Michigan and Huron basins (Fig. 3.35e).

Numerous intermediate lake levels developed between the glacial Lake Duluth phase and the extension of glacial Lake Minong into the western part of the basin (Farrand, 1960; Farrand and Drexler, 1985).

North-central Ontario

From about 10 ka to 8.2 ka the grounded receding ice margin was generally straight and trended in an east-west direction in north-central Ontario. Successive marginal positions are recorded by a series of ill defined moraines which all trend east-west (Fig. 3.22). A series of recessional moraines at Capreol (a short distance north of Sudbury) have been designated Cartier I-III by Boissonneau (1968); he linked the southernmost of these, Cartier I, with the McConnell Lake Moraine (which is part of the Harricana Interlobate Moraine, Veillette, 1988) in the upper Ottawa valley. If these relate to the Korah phase of the Post Algonquin lakes then the age of this morainic belt is probably in excess of 10 ka. Farther north, Boissonneau (1968) described another series of moraines called Chapleau I, II, and III; of these he suggested that only Chapleau I (the northernmost) represented a readvance. Small proglacial lakes formed south of Chapleau (glacial Lake Sultan), at Gogama (glacial Lake Ostram), and at Gowganda (glacial Lake Ogilvie).

North of the Great Lakes-Hudson Bay drainage divide there are few moraines and deglacial history is contained primarily within the sediments of glacial Lake Ojibway (see below).

Upper Ottawa Valley

Radiocarbon dates presented by Veillette (1983a) indicate that the Lake Timiskaming basin was deglaciated by 10 ka. Baker (1980) had earlier placed the 10 ka margin about 100 km north of Lake Timiskaming, near Kirkland Lake, on the basis of a date from basal organics in a bog above the highest levels of Lake Barlow-Objiway (9990 ± 260 BP, BGS-552).

Glacial Lake Barlow occupied the Lake Timiskaming basin, expanded northward to become glacial Lake Barlow-Ojibway, and finally glacial Lake Ojibway. Glacial Lake Barlow was once thought to have been contained by a drift barrier about 15 km north of Témiscaming, Quebec (Prest, 1970). However, Vincent and Hardy (1979) have suggested that early phases of the lake were held up by drift barriers near Deux Rivières and Mattawa, Ontario, with differential glacial isostatic depression controlling the size of the lake basin. The separation of glacial Lake Ojibway from Barlow occurred when isostatic rebound restricted the glacial lake to the north side of the present Hudson Bay- Great Lakes drainage divide.

Retreat northward from the Cartier-McConnell Lake moraines was accompanied by several stillstands or minor readvances which resulted in morainic features (Veillette, 1983a), including the Roulier Moraine (Vincent and Hardy, 1979) which is aligned with Boissoneau's (1968) Chapleau III features. Glacial Lake Ojibway spread westward along the retreating ice front and eventually included a large area of northern Ontario and western Quebec and possibly joined with glacial Lake Agassiz (see above). The lake drained southward through Ottawa Valley during all stages.

Final stages of deglaciation and late glacial surges

The northern margin of Hudson Ice was receding along calving bays extending in from Hudson Strait (this chapter, Andrews, 1989), and possibly from the Gulf of Boothia (this chapter, Dyke and Dredge, 1989). Reinterpretation of the northern extremities of glacial Lake Agassiz (Dredge and Grant, 1982), based on distribution of iceberg scour marks and glacial lake features, suggests that the southern periphery of Hudson Ice was rimmed by glacial lakes prior to deglaciation. Glacial Lake Agassiz drained through northeastern Manitoba into glacial Lake Ojibway after ice had retreated from the Agutua Moraine. The main drain for this enlarged lake system was southward through Ottawa valley.

Ice marginal positions during the final phase of retreat are not marked by end moraines. There was, however, continued development of radial moraines, including the Limestone Moraine in northern Manitoba (Fig. 3.22), which split the Hayes lobe. The sublobe between the Limestone Moraine and Burntwood-Knife Interlobate Moraine changed configuration as deglaciation proceeded (Dredge, 1985) and switched from a convex ice lobe to a large concave calving bay, whose recessional positions are marked by broad arcuate rises. These are considered to be grounding line moraines, marking the inner edge of a floating ice ramp. The Beaverhouse Moraine (Prest, 1963) developed during this phase. Several other radial moraines developed, including the Arnott Moraine west of Hearst (Boissoneau, 1966). Both this moraine and the Pinard Moraine to the east were formed as ice contact deltas. Prest (1970) suggested that they correlate with the Nakina Moraines to the west. In the ice marginal configuration shown in this chapter, however, the Nakina Moraines are linked to radial features south of the Pinard, in which case the Pinard might correlate better with the Beaverhouse or later events.

Cochrane and related surges

During final stages of deglaciation the southern margin of Hudson Ice appears to have been relatively unstable, with irregular marginal fluctuations occurring at several places between Manitoba and Quebec. Evidence for these is seen in the lake sediment record, by the character of the till they generated, and by truncating swaths of streamlined landforms. Varve counts suggest that individual fluctuations lasted about 25 years (Hughes, 1965). These events are therefore considered to be surges, which consisted of an advancing ice stream fronted by a floating shelf (see section on till landforms). The location and extent of surges have been mapped from airphotos (Fig. 3.23). Individual surges extended 50 to 75 km into the lakes and in many places successive surges crosscut both the regional ice retreat pattern and earlier surges. Commonly the ice skimmed across the top of topographic rises leaving a swath of parallel, extremely delicate flutes (Fig. 3.36). Some of the delicately fluted forms terminate in crossing arcs, which suggest that the leading edge of the advancing ice shelf broke up into icebergs (D.R. Grant, Geological Survey of Canada, personal communication, 1984). Where the ice was grounded, it scoured the bedrock and fluted the surface of till or varved glacial lake deposits, and incorporated or contorted both types of material to form Cochrane-type till (Hughes, 1965; Hardy, 1976, 1977). In intervening low areas there was little alteration of the landforms, although a rainout sediment facies is present in some varve records. In areas beyond the ice margin, surges resulted in sudden increases in varve thicknesses (Hughes, 1965; Hardy, 1976), interrupting generally thinning upwards varve sequences.

Although recent airphoto mapping and ground investigations have shown that surging was prevalent all around the southern margin of Hudson Ice, the best documented surges were those in the James Bay area, first recognized in the district of Cochrane, Ontario. Boissonneau (1966) identified two phases of the Cochrane readvance which Prest (1970) later referred to as Cochrane I and II: Cochrane I moved southerly and was restricted to the James Bay area; Cochrane II occurred at a somewhat later date and was a southeasterly thrust. In Quebec, Hardy (1976, 1977) recognized three "Cochrane type" readvances, Cochrane I, Rupert, and Cochrane II. He correlated the latest of these, Cochrane II, with the main Cochrane events of Prest (1970). Hardy (1982b) however dropped the Rupert readvance.

Prest (1969) depicted these events as extensive readvances of 800 to 1200 km of an ice mass retreating towards Labrador via the Hudson Bay Lowlands; Hardy (1976, 1977) showed them as emanating from an ice mass centred in Hudson Bay; and Shilts (1980, 1985) considered the Cochrane to be a major readvance from ice retreating back to a Keewatin centre. We view the Cochrane readvances as a series of small (50-75 km) surges along a continually retreating Hudson Ice front; this is more in keeping with the evidence of distribution of Cochrane flutings determined from airphoto mapping, available ground data across the ice margin, and with the limits reported by Hughes (1965) and Hardy (1976, 1977). The Cochrane surges are only one set of many similar surges now mapped along the entire south margin of the ice sheet (Fig. 3.23). They collectively define radial surging from the Hudson centre.

Hughes (1965) estimated the minimum age of the Cochrane readvance at 8275 BP using a date of 7875 ± 200 BP (GSC-14) for initiation of the Tyrrell Sea and adding 400 years based on the varve chronology for the area. Hardy (1982b) has suggested that his earliest Cochrane phases culminated at 8.2 and 7.9 ka, but these ages may be too young in light of other radiocarbon dates for the Tyrrell Sea (see *Tyrrell Sea Episode*; and this chapter, Vincent, 1989b). Dredge et al. (1986) reported an age of 7770 ± 140 BP (GSC-3070) near North Knife River in Manitoba on

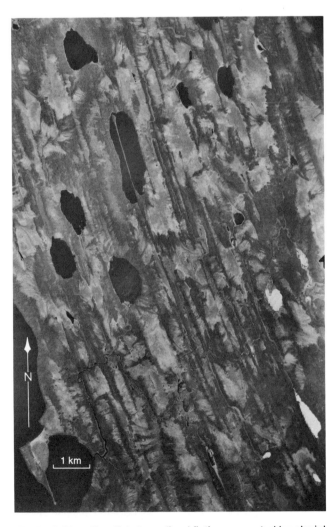

Figure 3.36. Parallel streamlined fluting generated by glacial surging into ice marginal lakes. View from Kattawagami Lake, Ontario, near Cochrane (50°N, 80°W). NAPL A13390-75

marine shells in deposits directly overlying a Cochrane-type surge till, and suggested that surging there is of roughly comparable age to events in the Cochrane area.

Demise of Hudson Ice and Tyrrell Sea invasion

Glacial lake sediments have been observed as far north as the Hudson Bay coast along Nelson River, where they are covered by marine deposits, but it is not known how far the glacial lake extended into Hudson Bay.

During the last phases of glaciation, Hudson Ice was sapped along its southern margin by glacial lakes, particularly in the interlobate zones, as has been shown by the extension of the Harricana Interlobate Moraine well into Hudson Bay (Hardy, 1976, 1977). It was also calving rapidly on its northern sides both through Hudson Strait, Boothia Strait west of Southampton Island, and possibly along basins in Hudson Bay. Lowering of the ice sheet profile by melting and marginal sapping, together with the buoyant forces of the invading sea and the lake to the south may have eventually caused the entire ice remnant to float. Such an event would have led to a rapid invasion of the Tyrrell Sea across the Hudson Bay Lowlands. Evidence for sudden lowering of the level of the glacial lakes to the level of the Tyrrell Sea is identified in "drainage horizons" consisting of rounded varved clay pebbles overlain by pebbly sands in Quebec (Hardy, 1976, 1982b). Similar deposits were observed in the Moose River basin by Skinner (1973). Marine waters entered the James Bay area along a suture between Hudson Ice and Labrador Ice in the Great Whale River-Lac Guillaume-Delisle area (Hardy, 1976, 1977; this chapter, Vincent, 1989b). On the western side of Hudson Bay the sea first reached the basin of glacial Lake Agassiz in the vicinity of Churchill via a calving bay along the suture between Keewatin Ice and Hudson Ice (see this chapter, Dyke and Dredge, 1989; Fig. 3.13e).

TYRRELL SEA EPISODE

The Tyrrell Sea (Lee, 1960) is the glacial isostatic marine water body that occupied Hudson Bay Lowlands at the close of and following deglaciation. Prest (1969, 1970) thought that the sea may have invaded the lowlands via a calving bay which could have developed along the deeps in central Hudson Bay. Andrews and Falconer (1969) proposed a calving bay along eastern Hudson Bay to account for the ice flow sequence on the Ottawa Islands. This idea was later corroborated by Hardy (1976, 1977), who studied the field relationships between glacial, lacustrine, and marine deposits in eastern Hudson Bay. Dyke et al. (1982b) suggested that the calving bay exploited the suture between Hudson Ice and Labrador Ice. Field studies and ^{14}C dates from northern Manitoba now suggest that there was also a calving bay in western Hudson Bay along the convergence of Hudson Ice and Keewatin Ice. This calving bay could have penetrated either from Hudson Strait or from the Gulf of Boothia (this chapter, Dyke and Dredge, 1989). The sea was at its maximum level at the time of initial invasion and subsequently regressed to the present level of Hudson Bay.

Age of marine incursion

The oldest date on marine shells in the James Bay area is 7800 ± 160 BP (QU-122; Hardy, 1982b). Until recently it was thought that the Tyrrell Sea entered the area at about that time, or slightly earlier since the dated materials lie below the marine limit. Craig (1969) proposed that the sea gradually encroached westwards across Hudson Bay Lowlands. Dates from northwest Manitoba (Wagner, 1967; 8010 ± 95 BP, GX-1063; 8000 ± 200 BP, BGS-812; 8200 ± 300 BP, BGS-813, Table 3.3), however, suggest that the Tyrrell Sea had invaded that area by 8 ka or possibly earlier. If these dates and the pattern of Hudson Ice disintegration are correct, then the sea entered the entire Hudson Bay Lowlands at about the same time. The oldest and highest date on marine shells (8530 ± 220 BP, GSC-896) was obtained from beach deposits in northeast Manitoba. This sample was small and consisted of worn shell fragments; because it was about 700 years older than the oldest date in the James Bay area, and 1200 years older than what had been the oldest date in Manitoba, Craig (1969) thought that it contained old reworked shells. Accelerator dates of 6980 ± 60 BP (TO-220) and 39 140 ± 520 BP (TO-221) from fresh and worn shells in a beach ridge on Nelson River support Craig's position.

The apparent synchroneity of the marine incursion in areas within the Hudson ice domain, which had a marine-based ice divide, is a marked contrast to the time-transgressive incursion in the adjacent Keewatin Sector, which reflects a sea expanding against a glacier which downwasted and backwasted towards a terrestrial ice divide.

Configuration and maximum extent

Figure 3.27 shows the area covered by the Tyrrell Sea, elevations for marine limit (most extrapolated from contours on topographic maps), and locations of radiocarbon dated material. Marine limit was approximated by the elevation of the highest beaches or wave-cut notches visible on aerial photographs. Marine limits can be closely estimated by this method beyond treeline. Farther south, where no beach ridges are visible, marine limit was approximated by the lowermost limit of iceberg scour marks. These marks, created by icebergs in glacial Lake Agassiz, were probably ubiquitous on the lake floor but apparently were subsequently eradicated by waves and currents in areas covered by the Tyrrell Sea (Dredge and Grant, 1982). More precise limits are difficult to determine because of the extensive peat cover which masks deposits and small relief features. The elevation of marine limit is lowest in the Nelson River area (122 m) and rises both to the north and to the southeast. It reaches a maximum elevation of about 180 m in the Sutton Hills area and at the south end of James Bay. North of Nelson River, marine limit also rises to a maximum of 180 m near Seal River (which lies within the area covered by Keewatin Ice). North of that point it declines in an irregular manner. The differences in elevation and gradient of the marine limit in northern Manitoba reflect differences in glacial and deglacial history between the Keewatin and Hudson ice masses. Because the marine incursion is thought to have been almost synchronous throughout the Hudson Bay Lowlands, elevations of marine limit can be used to reconstruct crustal tilting in that area. The high marine limits in the southeast Hudson Bay area suggest that an ice loading centre lay in that vicinity, a conclusion reached by Andrews and Peltier (1976). Similarly the rising marine limits north of Nelson River reflect loading by Keewatin Ice. The relatively low marine limits in the Nelson River area support the concept of lower ice load and partial recovery prior to establishment of the Tyrrell Sea.

Landform development and sediment deposition

Tyrrell Sea deposits are generally thin (see materials section) within the area covered by Hudson Ice, because during Tyrrell Sea time there was no ice mass feeding sediment-laden meltwater into this part of the Tyrrell Sea.

There was little beach ridge development during the early stages of emergence. An exception is the Great Beach, a remarkable feature which is 200-800 m wide. It can be traced from North Knife River across northeastern Manitoba and into northern Ontario as far as Winisk River, a distance of about 800 km. The varied elevations at which this feature is found suggest that it is not the product of a single shoreline, but rather that beaches developed on a pre-existing sandy morainic feature. Similarly, the massive set of beach ridges, between Nelson River and the Sutton Hills, between an elevation of about 60 m and present sea level, may be the reworked remains of coarse debris from a stranded remnant of the Hudson Ice mass (mentioned earlier, Fig. 3.28). The beach sequences appear to be regularly spaced and could relate to periodic Holocene climatic cycles as has been interpreted for similar sequences from eastern Hudson Bay (Hillaire-Marcel and Fairbridge, 1978).

Sea level change

Figure 3.37 shows emergence curves for the Hudson Bay Lowlands and adjacent areas. The graphs show minimum emergence, since they are based to a large degree on shell dates which do not necessarily date water planes at the elevation at which they were found. Emergence has been represented by simple curves; actual emergence may in fact have been much more complex. Also, little data exist for the middle parts of most curves.

Despite these drawbacks, the curves illustrate several interesting facets of glacial isostasy: The curves from southern Hudson Bay (James Bay, Churchill, Nelson River, and

Table 3.3. Selected radiocarbon ages pertaining to the Holocene of the Hudson Bay Lowlands, southwestern Canadian Shield

Site (cf. Fig. 3.27)	Age (years BP)	Lab No.	Elevation (m)	Material	Latitude/Longitude		Reference	Comments
Churchill and west area								
1. Caribou River	6790 ± 100	GSC-2579	90	shells	59°19'	95°23'	Blake, 1982	deltaic
2. N. Knife River	7770 ± 140	GSC-3070	110	shells	58°33'	95°50'	Blake, 1982	from sands, early emergence
3. Gt. Beach	7270 ± 120	GSC-92	142	shells	58°11'	95°03'	Craig, 1969	elev. may be 110 m
4. Churchill River	8010 ± 95	GX-1063	67	shells	58°06'	94°42'	Wagner, 1967	location approximate
5. Watson Pt.	385 ± 80	GX-1073	4	shells	58°45'	93°21'	Wagner, 1967	location approximate
6. Churchill	3190 ± 80	GX-1065	38	shells	59°45'	94°16.5'	Wagner, 1967	location approximate
7. Seahorse	3560 ± 105	S-738	35	shells	58°45'	94°15'	Rutherford et al., 1975	
8. Churchill	3080 ± 130	GSC-245	5	peat	58°45.2'	94°08'	Lowdon et al., 1971	
9. Churchill	3040 ± 130	GSC-261	23	shells	58°44.5'	94°04.8'	Lowdon et al., 1971	from littoral sand
10. Goose Ck.	3430 ± 140	GSC-735	5	shells	58°40.3'	94°10.2'	Craig, 1969	in stony clay
11. Churchill	5020 ± 140	GSC-1549	22	shells	58°44.4'	94°04.6'	Craig, 1969	
12. Rocket range	2120 ± 130	GSC-723	22	shells	58°45.4'	93°58.8'	Craig, 1969	littoral sands
13. Rocket range	1020 ± 140	GSC-684	7	shells	58°45.6'	93°57.0'	Craig, 1969	littoral gravel
14. Rocket range	1240 ± 130	GSC-682	11	shells	58°44.8'	93°50.4'	Craig, 1969	littoral gravel
15. Twin Lakes	2320 ± 130	GSC-683	27	shells	58°42.2'	93°50.6'	Craig, 1969	littoral gravel
16. Twin Lakes	3180 ± 140	GSC-685	39	shells	58°37.1'	93°48.7'	Craig, 1969	gravel
17. Twin Lakes	3190 ± 80	GX-1072	30	shells	58°36.4'	93°49.4'	Wagner, 1967	location approximate
18. Cape Churchill	0 ± 130	GSC-1226	0	shells	58°45'	93°12'	Lowdon et al., 1971	intertidal
19. N. Knife R.	4000 ± 90	GSC-3851	30	shells	58°53.5'	94°58.0'	Blake, 1988	in silty clay
20. Deer River	5150 ± 110	BGS-796	.30	shells	58°23.4'	94°14.2'	Nielsen, unpub.	
21. Churchill R.	3530 ± 100	BGS-793	.10	wood	57°58'	95°00'	Nielsen, unpub.	
22. Churchill R.	7670 ± 370	GSC-3348	134	shells	57°40'	95°25'	Blake, 1982	freshwater shells
23. Churchill R.	5960 ± 100	BGS-980	134	wood	57°40'	95°25'	Nielsen and Dredge, unpub.	basal peat, log
24. Owl River	6880 ± 130	GSC-3855	59	shells	57°36'	93°32'	Blake, 1988	
25. Owl River	5310 ± 80	GSC-3856	49	shells	57°41.5'	93°21.5'	Blake, 1988	
26. Owl River	5290 ± 70	GSC-3896	37	shells	57°46.3'	93°11'	Blake, 1988	sand below marine diamicton
57. Recluse Lake	6490 ± 170	GSC-1738	185	peat	56°52'	95°47'	Lowdon et al., 1977	basal peat
58. Spruce Lake	4850 ± 60	GSC-2567	210	wood	59°05'	96°45'	Lowdon and Blake, 1978	
60. Moorby Lake	6040 ± 80	GSC-2759	190	peat	59°28.3'	101°12.5'	Lowdon and Blake, 1979	basal peat
61. Moorby Lake	4450 ± 60	GSC-2803	190	peat	59°28.3'	101°12.5'	Lowdon and Blake, 1979	
63. Baldock Lake	5430 ± 210	GSC-1782	305	peat	56°21'	97°57.5'	Lowdon and Blake, 1975	moss
64. Baldock Lake	6920 ± 150	GSC-1818	305	peat	56°21'	97°57.5'	Lowdon and Blake, 1975	basal organic debris
Nelson River area								
27. Charlebois	6280 ± 80	GSC-2760	100	peat	56°40'	95°05'	Roggensack, unpub.	basal peat
28. Port Nelson	7020 ± 100	GSC-3921	16	shells	57°1.6'	92°38.0'	Blake, 1988	
29. Flamborough Hd.	7370 ± 90	GSC-3930	30	shells	56°58.1'	92°47.0'	Blake, 1988	
30. Angling R.	7250 ± 80	GSC-3904	15	shells	56°45.0'	93°36.5'	Blake, 1988	surface at 35 m
31. Sundance	7180 ± 70	GSC-3326	76	shells	56°32.0'	94°05'	Nielson and Dredge, 1982	re-collection of BGS 714
32. Sundance	6900 ± 150	BGS-714	76	shells	56°32.0'	94°05'	Nielsen and Dredge, 1982	
33. Sundance	6900 ± 100	BGS-798	76	shells	56°32.0'	94°05'	Nielsen and Dredge, 1982	
34. Sundance	7030 ± 170	GSC-2294	90	shells	56°32.0'	94°05'	Klassen, 1985	
35. Nelson Rt. bk	6760 ± 80	GSC-3367	67	shells	56°32.5'	94°08'	Nielsen and Dredge, 1982	re-collection of BGS 791
36. Nelson Rt. bk	6760 ± 100	BGS-791	67	shells	56°32.5'	94°00'	Nielsen and Dredge, 1982	
37. Nelson Rt. bk	6280 ± 180	BGS-711	67	shells	56°32.5'	94°00'	Nielsen and Dredge, 1982	re-collection of BGS 791
38. Nelson Rt. bk	6500 ± 140	BGS-797	67	shells	56°32.5'	94°00'	Nielsen and Dredge, 1982	upper sand unit
39. 5-mile hole	7120 ± 120	BGS-906	84	shells	56°31.7'	94°15'	Nielsen, unpublished	
40. Nelson L. bk (418)	6990 ± 130	BGS-712	82	shells	56°25.8'	94°12'	Nielsen and Dredge, 1982	
41. Nelson L. bk	7050 ± 150	BGS-815	90	shells	56°26.1'	94°10.5'	Nielsen, unpub.	
42. Nelson L. bk	6750 ± 150	BGS-713	90	shells	56°25.0'	94°13'	Nielsen, unpub.	
43. Nelson L. bk	7300 ± 200	BGS-814	90	shells	56°25.0'	94°14'	Nielsen, unpub.	
44. Long Spruce	8000 ± 200	BGS-812	105	shells	56°25.0'	94°20.5'	Nielsen, unpub.	
45. Long Spruce	7760 ± 80	GSC-3916	100	shells	56°24.9'	94°20.2'	Nielsen and Dredge, 1982	re-collection of BGS-812
46. Nelson R.	8200 ± 300	BGS-813	90	shells			Nielsen, unpub.	
59. Island	6900 ± 130	GSC-3928	15	shells	56°50'	93°30'	Blake, 1988	

Hayes River) lack the initial rapid emergence which is typical of most curves (e.g., La Grande Rivière), and have lower marine limits than for other parts of the central Laurentide Ice Sheet. These two aspects suggest that substantial rebound may have occurred beneath Lake Agassiz prior to the marine incursion. The lower marine limit elevations alone might only have indicated that this area was farther from early loading centres. Emergence for southern Hudson Bay is expressed as "flattened", rather than simple exponential, curves. This form possibly reflects unloading influenced by multiple ice centres, whose deglaciation histories were not in phase.

Since marine limit was apparently achieved at about the same time at Churchill, Nelson River, Hayes River, James Bay, and Cape Henrietta Maria, the differently shaped curves within the Hudson Sector of the Laurentide Ice Sheet reflect loading differences. The highest load or combined loads in the Hudson Bay Lowlands were apparently near Cape Henrietta Maria in southeast Hudson Bay, near one of the proposed divides of Hudson Ice. Moderately high loads are suggested from the rest of the James Bay area and for Churchill (which was affected by both Hudson Ice and Keewatin Ice), while the lowest loading occurred in the Nelson River and Hayes River areas which were either farthest away from the main loading centres, or were substantially unloaded prior to establishment of marine limit (about 8.0 ka).

The crossing of some sets of curves with others indicates either that the curves are poorly constrained or that centres of rebound shifted during the mid Holocene. The 7 ka shoreline is highest in the La Grande region (it could be even higher at Henrietta Maria, but the elevation is unknown). It remains high over the Ottawa Islands (see southeastern Canadian Shield section of this chapter), is considerably lower in western James Bay, and lower still over the Nelson River area. The shoreline shows a slight rise towards Churchill. By 5 ka, shorelines remain highest at La Grande Rivière but the western James Bay and south Hudson Bay shoreline is higher than that in northern Keewatin and the Ottawa Islands. As was the case with earlier shorelines, the shoreline still drops towards the Nelson River area and rises towards Churchill. The changes in these shoreline elevations indicate that the eastern centres of rebound shifted from Ottawa Islands-La Grande Rivière area to La Grande Rivière-Cape Henrietta Maria area. It implies that substantial rebound may still occur along the south shore of Hudson

Table 3.3. (cont'd.)

Site (cf. Fig. 3.27)	Age (years BP)	Lab No.	Elevation (m)	Material	Latitude/Longitude		Reference	Comments
Hayes River area								
47. York Factory	660 ± 190	GSC-1468	2	wood	57°03'	92°14'	Simpson, 1972	
48. York Factory	1930 ± 130	GSC-1305	5	wood	56°59'	93°39'	Simpson, 1972	
49. York Factory	2065 ± 125	GX-2061	9	wood	57°00'	92°15'	Simpson, 1972	location approximate
50. York Factory	1055 ± 125	GX-2062	4	wood	57°00'	92°15'	Simpson, 1972	location approximate
51. York Factory	1250 ± 105	GX-2063	2.3	wood	57°00'	92°15'	Simpson, 1972	
52. Hayes R.	4890 ± 140	GSC-1745	21	wood	56°48'	92°42'	Klassen and Netterville, 1973	in alluvial deltaic sand
53. Gods R.	6610 ± 100	GSC-1955	75	shells	56°15'	92°45'	Lowdon et al., 1977	in sand
54. Hayes R.	7570 ± 140	GSC-878	115	shells	56°2.3'	93°17'	Lowdon and Blake, 1970	in silty clay, marine limit at 130 m
55. Stupart R.	7110 ± 90	GSC-3926	90	shells	55°59'	93°23'	Blake, 1988	upper sand
56. Kaskattama	8530 ± 220	GSC-896	125	shells	56°18'	90°24'	Lowdon and Blake, 1970	marine limit at 140 m
Northern Ontario								
65. Fawn R.	7400 ± 140	GSC-877	137	shells	54°29'	88°16'	Lowdon and Blake, 1970	paired valves 45 m
66. Ekwan R.	7220 ± 140	GSC-872	120	shells	53°32'	86°03'	Lowdon and Blake, 1970	
67. Attawapiskat	5670 ± 110	GSC-31	140	peat	53°07'	85°25'	Dyck and Fyles, 1963	
68. Attawapiskat	4940 ± 80	GrN-1925	-	peat	53°08'	85°18'	Terasmae and Hughes, 1960a	
69. Hawley Lake	5580 ± 150	GSC-247	129	peat	54°34'	84°40'	Dyck et al., 1965	
70. C. Henrietta Maria	2310 ± 200	I-3909	41	shells	54°36'	82°36'	Webber et al., 1970	location approximate
71. C. Henrietta Maria	1430 ± 190	I-3983	23	shells	54°52'	82°25'	Webber et al., 1970	location approximate
72. C. Henrietta Maria	2410 ± 200	I-3907	7	shells	54°52'	82°18'	Webber et al., 1970	location approximate
73. C. Henrietta Maria	1210 ± 130	GSC-231	14	peat	54°58'	82°20'	Dyck et al., 1965	
74. Kapiskau	7720 ± 140	GSC-880	120	shells	51°56'	84°32'	Lowdon and Blake, 1970	
75. Albany Forks	7140 ± 170	GSC-831	158	peat	51°23'	84°31'	Lowdon and Blake, 1970	
76. Albany Forks	5820 ± 150	GSC-885	168	peat	51°28'	84°48'	Lowdon and Blake, 1970	
77. Kabinakagami	7540 ± 140	GSC-915	99	shells	50°13'	84°14'	Lowdon and Blake, 1970	
78. Nagagami	7760 ± 160	GSC-897	105	shells	50°13'	84°18'	Craig, 1969	
79. Pivabiska R.	7630 ± 170	GSC-1309	116	shells	50°14'	83°04'	Skinner, 1973	in coarse sand, location approximate
80. Missinaibi	7500 ± 160	GSC-1418	109	shells	50°11'	83°00'	Skinner, 1973	in sand and silt, location approximate
81. Missinaibi	7875 ± 200	GSC-14	121	shells	50°13'	82°54'	Hughes, 1965, Lee 1960	elev. may be 106 m
82. Opasatika R.	7520 ± 80	GrN-1698	102	shells	50°05'	82°30'	Vogel and Waterbolk, 1972	gravels at top of section,
83. Opasatika R.	6890 ± 220	GSC-1489	86	shells	50°22'	82°22'	Skinner, 1973	in basal marine gravel, location approximate
84. Opasatika R.	6620 ± 240	GSC-1499	100	shells	50°12'	82°20'	Skinner, 1973	in gravels, location approximate
85. Opasatika R.	7280 ± 150	GSC-1436	102	shells	50°05'	82°30'	Skinner, 1973	re-collection of GrN 1698
86. Mattagami	7523 ± 400	I-1256	112	shells	50°00'	82°12'	Skinner, 1973	location approximate
87. Little Long Rapids	6910 ± 120	GSC-7	112	wood	50°01'	82°12'	Dyck and Fyles, 1963	base of peat
88. Mattagami	7380 ± 160	GSC-1430	124	shells	50°08'	82°11'	Skinner, 1973	in gravels, Kipling Dam
89. Mattagami	4900 ± 130	GSC-1396	46	shells	50°34'	81°33'	Skinner, 1973	location approximate
90. Mattagami	4160 ± 140	GSC-1323	46	wood	50°34'	81°33'	Skinner, 1973	in alluvium
91. Mattagami	4120 ± 130	GSC-1531	46	wood	50°34'	81°33'	Skinner, 1973	in gravel
92. Otter Rapids	7100 ± 160	GSC-1241	133	shells	50°10'	81°40'	Skinner, 1973	in beach, location approximate
93. C. Henrietta Maria	235 ± 190	I-3908	7	wood	55°	82°	Webber et al., 1970	driftwood
94. C. Henrietta Maria	2450 ± 190	I-3982	22	wood	55°	82°	Webber et al., 1970	driftwood

Bay and that parts of Hudson Bay may eventually become dry land.

The curve for Southampton Island shows an initial rapid emergence and persistently higher rebound than for the Ottawa Islands; this may reflect a cell of uplift related to late Foxe Ice. Shorelines on Southampton Island that are younger than 5 ka are as low or lower than contemporaneous ones along the south coast of Hudson Bay.

Present day rates of emergence (Table 3.4), as determined by tide gauge and historical records from Churchill (39 cm/century; Barnett 1970) and historical records from Fort Albany in the James Bay area (90-120 cm/century; Hunter 1970), support the evidence of loading centres shown by the more generalized emergence curves. Lake records (Cowan, 1978; Table 3.4) suggest that the limit of postglacial rebound lies near Sault Ste. Marie; areas south of the Sault are subsiding.

NIPISSING GREAT LAKES

As described previously, deglaciation of the isostatically depressed Mattawa River valley between 10.5 ka and 10 ka allowed the Great Lakes to drain to very low levels, forming lakes Stanley and Hough in Lake Huron and Georgian Bay basins (Fig. 3.35c). The low level Houghton phase of the Lake Superior basin did not occur until ca. 8-7 ka (Farrand

Table 3.4. Estimates of present day emergence or submergence in the Hudson Bay and Lake Superior regions

Location	Estimated rate of emergence (+) or submergence (-)	Reference
Churchill, Manitoba	+0.39 m/century	Barnett, 1970 (tide gauge analysis)
Cape Henrietta Maria, Ontario	+1.2 m/century	Webber et al.,1970 (radiocarbon controlled emergence curves)
Fort Albany, James Bay, Ontario	+0.9 to +1.2 m/century	Hunter, 1970 (historical data)
Richmond Gulf, Quebec	+1.1 m/century	Hillaire-Marcel and Fairbridge, 1978 (emergence curves)
Michipicoten area, Ontario	+0.27 m/century	Farrand and Drexler, 1985 (U.S. Corps of Engineers Lake gauge data)
Pt. Iroquois (St. Marys River), Michigan	0	Farrand and Drexler, 1985 (U.S. Corps of Engineers Lake gauge data)
Duluth, Minnesota	-0.21 m/century	Farrand and Drexler, 1985 (U.S. Corps of Engineers Lake gauge data)

and Drexler, 1985) when the St. Marys River sill was lowered by the downcutting waters of Lake Minong. During these low water phases, large areas of the modern lake basins were poorly drained swampy areas.

Uplift of the North Bay outlet (Mattawa River valley) caused lake levels to rise, eventually water bodies in the Superior, Michigan, Huron, and Georgian Bay basins became confluent, and the Nipissing Great Lakes were initiated (see Map 1703A, 7 ka). Fluvial and deltaic deposition into the rising Nipissing waters along the north shores of lakes Superior and Huron occurred and numerous dates between 7.5 and 5 ka (Lewis, 1969, 1970; Prest, 1970; Cowan, 1978) have been obtained for these deposits. When the rising North Bay outlet attained the same level as the Chicago and Port Huron outlets (three outlet phase), lake level stabilized at 184 m a.s.l. (Nipissing Great Lakes I), and a prominent shoreline was constructed along the north shore of lakes Huron and Superior. Present day elevations of this isostatically tilted shoreline are: 213 m at North Bay, 198 m at Sault Ste.-Marie, and 200-204 m at Thunder Bay. The age of this main beach-forming phase is considered to be about 4.6 ka (Cowan, 1978). Subsequent uplift raised the North Bay outlet, creating the two-outlet (Chicago and Port Huron) Nipissing Great Lakes II, and causing withdrawal of the Nipissing waters from the north shores of lakes Huron and Superior.

With further downcutting of the Port Huron outlet, a new level — Algoma — formed a weak shore feature in the Superior, Michigan, and Huron basins about 3.2 ka (Saarnisto, 1974). Uplift of the St. Marys River sill at the east end of Lake Superior then isolated the Lake Superior basin about 2.2 ka (Farrand 1962; Saarnisto 1975). This brought into being the Sault level of Lake Superior (190 m at Thunder Bay) and later the sub-Sault level; on the north shore the latter is at a level similar to the storm wave level of present day Lake Superior (Farrand, 1960).

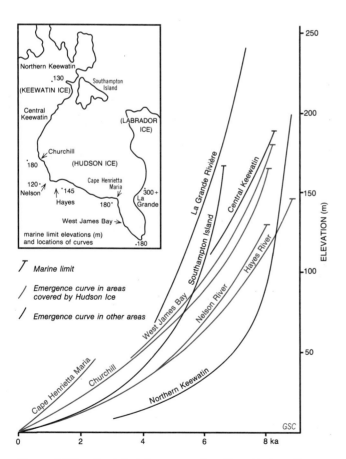

Figure 3.37. Postglacial recovery of southern Hudson Bay as shown by sea level emergence curves.

Larsen (1985) saw the Nipissing and Algoma Great Lakes in a different light. He viewed these as largely climatically controlled water levels during a period of relatively steady outlet incision rather than being due to isostatic adjustment.

Modern levels of both lakes Huron and Superior were substantially achieved by about 2 ka. Further discussion of Great Lakes history is provided by Karrow (1989).

VEGETATION HISTORY

Revegetation of the study region following disappearance of glacier ice, glacial lake waters, and marine waters spans the period from 12 ka to present. McAndrews (1981) summarized fossil pollen assemblages along a north-south transect across Ontario. Another report on the vegetation history of the Hudson Bay Lowlands is based on a pollen profile from Sutton Ridge (McAndrews et al., 1982). Chapter 7 of this volume reports on the general development of the Boreal Forest in Canada (Ritchie, 1989), the Holocene changes in vegetation (Anderson et al., 1989) and patterns of vegetation colonization in adjacent Quebec (Richard, 1989).

In general, as deglaciation proceeded in Ontario, a brief period of tundra vegetation existed. This was followed by spruce forest, forests in which pine pollen dominated, and then the modern zonal vegetation. The pine forest period was more subject to latitudinal effects than was the spruce forest period. In central Ontario the spruce-pine transition occurred between 10.5 and 10.0 ka. Northward of 46°30′N the spruce minimum and white pine peak are related to the Hypsithermal period. At this time, about 5 ka, temperatures were 1-2 °C higher than present. White pine macrofossils dated at 5.0 ka indicate that this species existed nearly 100 km north of its present range during this episode. Sutton Ridge vegetation history includes a succession from sparse coastal tundra through shrub tundra to modern spruce woodland between the time of emergence and 6.5 ka, as the Tyrrell Sea maritime effects decreased (McAndrews et al., 1982). Plant macrofossils indicate the presence of *Najas flexilis* between 6.5 and 3.0 ka, suggesting that the Hypsithermal range for this species was 300 km north of the known modern range. Post-Hypsithermal cooling contracted the ranges of species such as white pine to their present range about 2.5-3.0 ka.

Throughout much of the region the postglacial growth of muskeg was an important bioclimatic and economic development. Large scale muskeg expansion in northern Ontario did not occur until the climatic decline following the Hypsithermal, when more favourable colder, moister conditions prevailed (Terasmae, 1977). The oldest date on basal peat in the Manitoba part of the Hudson Bay Lowlands is 6490 ± 70 BP (GSC-1738) at Recluse Lake and the oldest date on spruce trunks is 5960 ± 100 BP (BGS-980).

QUATERNARY GEOLOGY OF THE SOUTHEASTERN CANADIAN SHIELD

J-S. Vincent

INTRODUCTION

The southeastern Canadian Shield lies west of the Labrador Sea; north of the Gulf of St. Lawrence, Saint-Narcisse Moraine, and Ottawa River; east of the Harricana Interlobate Moraine and Hudson Bay; and south of Hudson Strait (Fig. 3.1). The Quaternary deposits of this region bear witness, with few exceptions, to events that occurred during and since the Wisconsin Glaciation. The southeastern Canadian Shield area is particularly interesting since it is presumably there that ice from the Labrador Sector of the Laurentide Ice Sheet first accumulated. After coalescing with other ice masses, it subsequently extended to the Canadian Interior Plains and the northern United States. It is also in the centre of the southeastern Canadian Shield that one of the last remnants of continental ice finally disappeared about 6.5 ka ago.

Detailed or reconnaissance surficial geology maps have been completed for only about 10% of the area (Fig. 3.38).

Vincent, J-S.
1989: Quaternary geology of the southeastern Canadian Shield; in Chapter 3 of Quaternary Geology of Canada and Greenland, R.J. Fulton (ed.); Geological Survey of Canada, Geology of Canada, no. 1 (also Geological Society of America, The Geology of North America, v. K-1).

The principal references for the mapped areas and for regional or topical reports are listed in Table 3.5. Several comprehensive studies deal specifically with particular topics such as glacial limits and weathering zones in Labrador (Ives, 1978; Clark, 1984a), glacier movements of different ages (Veillette, 1986a), the Quebec North Shore morainic systems (Dubois and Dionne, 1985), the Sakami and Harricana moraines and the Cochrane surges (Hardy, 1977, 1982b), glacial lakes Barlow and Ojibway (Vincent and Hardy, 1977, 1979; Veillette, 1988), and postglacial seas (Hillaire-Marcel, 1979). Finally, Prest (1970), Prest et al. (1968), Hillaire-Marcel and Occhietti (1980), Mayewski et al. (1981), Peltier and Andrews (1983), and Parent et al. (1985) provide syntheses on the Quaternary of the area.

Relief and drainage

The present physiographic aspect of the southeastern Canadian Shield results from a long series of events. In Precambrian time at least one erosion surface was developed. This surface was later submerged by Paleozoic seas in which thick sequences of sediments were deposited. During the Mesozoic and Cenozoic, the sedimentary cover was stripped by running water, and during the Quaternary, glaciers modified the old erosion surface.

Bostock (1970b) described the area as a vast undulating lake-studded plateau. The elevation of the plateau gradually rises towards the central interior from the peripheral lowlands of the St. Lawrence River basin and Hudson Bay, Hudson Strait, and Labrador Sea coasts (Fig. 3.2). Local relief is generally low, rarely exceeding 100 m, with isolated hills standing above the monotonous plateau surface. Exceptionally high terrains, with elevations in excess of 1000 m, are restricted to summit areas of the Laurentian Highlands and of the Mealey and Torngat mountains. The presence of fault-line scarps, cut here and there by deeply incised valleys in the southern part of the southeastern sector of the shield gives a mountainous aspect to the Laurentian Highlands. The most spectacular scenery is found on the Atlantic side of the Torngat Mountains. There, the ancient erosion surface, dipping west, abruptly terminates in a series of cliffs, in places more than 1000 m high, incised by valleys and fiords.

Many features of the modern physiography of the Canadian Shield relate to the presence or absence of Quaternary deposits (Surficial Materials Map of Canada, Geological Survey of Canada, in preparation). Large areas of the Laurentian Highlands and Ungava Peninsula have a

Table 3.5. References dealing with the Late Wisconsinan deposits and history of the various areas of the Canadian Shield in Quebec-Labrador

Surficial Geology Maps[1]	Regional Reports	Topical Reports
Northern Labrador and northeastern New Quebec		
(58) Barré (1984)[*2]	Clark (1984a); Evans (1984)	Andrews (1963a), Barnett (1967), Barnett and Peterson (1964), Evans and Rogerson (1986), Ives (1957; 1958a, 1960a, b, 1976), Johnson (1969), Løken (1962a, b, 1964), Matthew (1961), McCoy (1983), Peterson (1965), Short (1981), Smith (1969), Tomlinson (1963)
Central and southern Labrador		
(15) [3]Fulton and Hodgson (1980), (8, 11-13, 16-20)[3] Fulton et al. (1979, 1980a,b,c,d, 1981a,b,c), (9-10, 14) [3]Fulton et al. (1980e,f, 1981d), (21) [3]Fulton et al. (1981e),(22) Grant (1986)	Fulton and Hodgson (1979)	Blake (1956), Fitzhugh (1973), Grant (1969, 1971), Gray (1969), Hodgson and Fulton (1972), Klassen (1983a, 1984),Morrison (1963), Rogerson (1977)
Quebec North Shore		
(24) Dredge (1983b), (25) Dubois and Desmarais (1983), (60) Dubois et al. (1984)	Dredge (1983b), Dubois (1980), Tremblay (1975)	Boutray and Hillaire-Marcel (1977), Dionne (1977), Dredge (1976a, b), Dubois (1976, 1977, 1979), Dubois and Dionne (1985), Dubois et al. (1984)
Lac Saint-Jean		
(29) Dionne (unpublished), (30-36), LaSalle and Tremblay (1978)	LaSalle and Tremblay (1978)	Dionne (1968, 1973), Dionne and Laverdière (1969), LaSalle et al. (1977a, b), Laverdière and Mailloux (1956), G. Tremblay (1971, 1973)
Laurentian Highlands		
(52) Chagnon (1969), (49) Denis (1976), (56) Gadd (unpublished), (54) Gadd and Veillette[*4], (42) Fulton (unpublished), (46) Lamothe (1977), (51) LaSalle (1978), (55) LaSalle and Gadd[*4], (50) Occhietti (1980), (48) Pagé (1977), (44-45) S.H. Richard (in press, 1984), (43) S.H. Richard et al. (1977), (47) Tremblay (1977)	Denis (1976), Occhietti (1980), Tremblay (1977)	Dadswell (1974), Denis (1974, 1976), Hardy (1970), Lamothe (1977), LaSalle et al. (1977a, b), Laverdière and Courtemanche (1960), 1980b), Romanelli Pagé (1977), Parry (1963), Richard (1978),(1975)
Upper Ottawa River basin		
(37-41) Veillette (1986b,c, 1987a,b,c), (53) Veillette and Daigneault (1987)		Antevs (1925), Veillette (1982, 1983a,b, 1986a, 1988), Vincent (1975), Vincent and Hardy (1977, 1979).
Quebec Clay Belt and area southeast of James Bay		
(24) Baker and Storrison (1979), (61) Bisson (1987), (23) Bouchard et al. (1974), (26) Chauvin (1977), (59) Hardy (1976)[*],(2) Lee et al. (1960), (28a,b), G. Tremblay (1972; 1974), (3-5) Vincent (1985a,b,c)	Bisson (1987), Bouchard (1980, 1986), Chauvin (1977), Hardy (1976), Martineau (1984a), G. Tremblay (1974), Vincent (1977)	Allard (1974), Antevs (1925), Bouchard and Martineau (1984, 1985), Bouchard et al. (1984), DiLabio (1981), Dionne (1974), Hardy (1977, 1982a,b), Hillaire-Marcel el al. (1981), Hillaire-Marcel and Vincent (1980), Ignatius (1958), Lee (1959b, 1960, 1962, 1968), Martineau (1984b), Mawdsley (1936), Norman (1938, 1939), Prichonnet et al. (1984), Shaw (1944), Vincent and Hardy (1977, 1979), Wilson (1938)
Area east of Hudson Bay		
	Allard and Séguin (1985), Hillaire-Marcel (1976)	Allard and G. Tremblay (1983), Andrews and Falconer (1969), Archer (1968), Fairbridge and Hillaire-Marcel (1977), Hillaire-Marcel (1979, 1980), Hillaire-Marcel and Boutray (1975), Hillaire-Marcel and Fairbridge (1978), Hillaire-Marcel and Vincent (1980), Plumet (1974), Portman (1972),Walcott and Craig (1975)
Ungava Peninsula		
(1) Lauriol (1982)[*]	Lauriol (1982), Gray and Lauriol (1985)	Blake (1966), Bouchard and Marcotte (1986), Drummond (1965), Gangloff et al. (1976), Gray et al. (1980), Lauriol and Gray (1983), Lauriol and Gray (1987),Lauriol et al. (1979), Løken (1978), Matthews (1966, 1967), Taylor (1982)
Central New-Quebec and western Labrador		
(57) Guimont and Laverdière (1982)[*], (7) Henderson (1959)[*], (6) Hughes (1964)[*]	Henderson (1959), Hughes (1964)	Barr (1969), Derbyshire (1962b), Ives (1959, 1960a,c), Kirby (1961a,b), Richard et al. (1982)

[1] The number in brackets, before the author's name, refers to the number shown in Figure 3.38.
[2] The reference accompanied by an asterix indicates that the map portrays major landforms not deposits.
[3] Synthesis maps at a scale of 1:500 000 are also available in Fulton (1986a, b).
[4] Manuscript maps prepared as part of the Quaternary Geologic Atlas of the United States of the United States Geological Survey (scale 1:1 000 000).

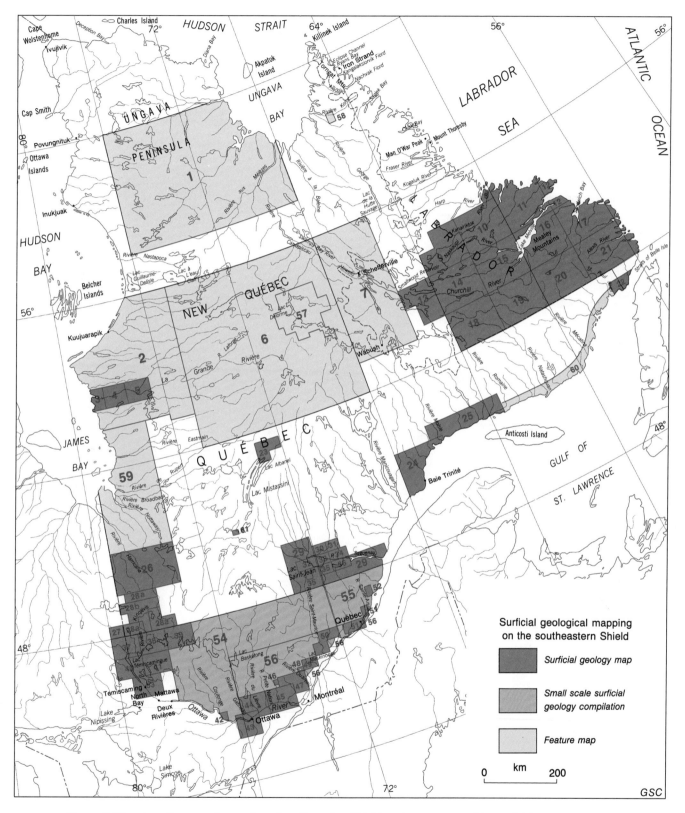

Figure 3.38. Location map for southeastern Canadian Shield and areas covered by surficial geology maps listed in Table 3.5 and by maps used as data source for Figure 3.39.

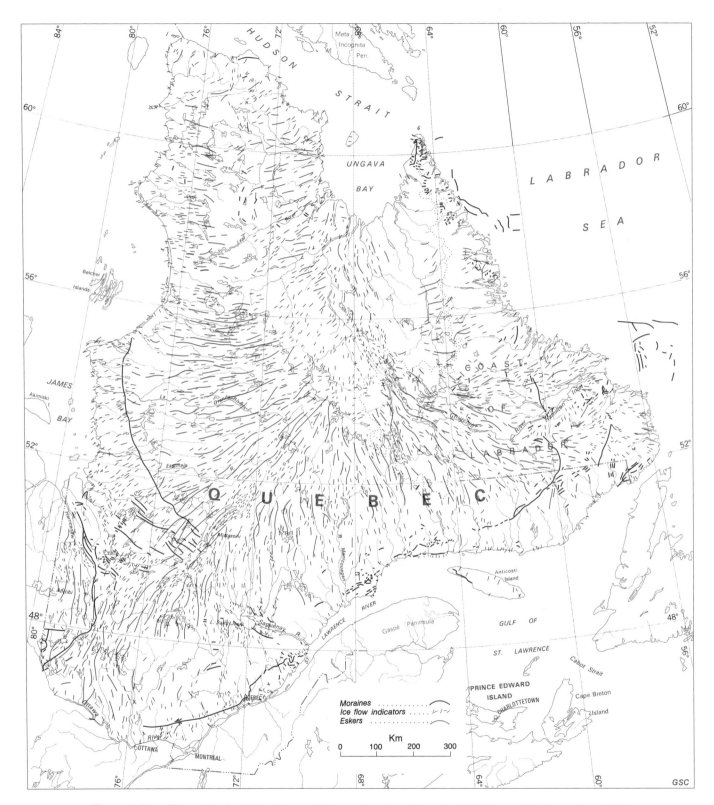

Figure 3.39. The main glacial landforms of the southeastern Canadian Shield compiled from all available sources of information.

sparse cover of Quaternary materials (Fig. 3.4), whereas the central part of the southeastern shield has a thick till sheet moulded into drumlins and ribbed moraine. Coastal areas are underlain by flat-lying marine or glacial marine sediments.

The drainage of the southeastern shield is obviously related to the slope of the land and thus is essentially radial from the higher south-central areas. Rivers with the highest average annual discharges are the Saguenay, Ottawa, Nottaway, La Grande, Caniapiscau, and Churchill. River courses are generally dictated by the structural elements of the Precambrian Shield although locally they are controlled by glacial deposits.

NATURE AND DISTRIBUTION OF QUATERNARY DEPOSITS

Figure 3.39 shows the distribution of the principal glacial landforms of the southeastern Canadian Shield. The general distribution of Quaternary deposits is shown on the Surficial Materials Map of Canada (Geological Survey of Canada, in preparation). The definition and descriptions of glacial landforms mentioned here can be found in Prest (1968, 1983a).

Glacial deposits

Till and related landforms

The texture, structure, composition, and morphology of the tills of the southeastern shield are variable. Generally derived from metamorphic and igneous rocks, tills in this area are sandy, stony (Fig. 3.40) and noncalcareous. Although thicknesses in excess of 10 m are reported locally, the till is generally thin, probably less than 2 m in average. The Cochrane till southeast of James Bay is finer grained, highly calcareous, and nearly stone free; it was formed by a glacier advancing across glacial lake sediments. Finer grained till, with significant carbonate content ($>5\%$) is locally associated with Proterozoic dolomitic sources in the Lac Mistassini area (DiLabio, 1981) and in northwestern Ungava Peninsula (Delisle et al., 1984). Figure 3.41 shows matrix textures of tills from various locations in the area. In the oxidized zone, up to 3 m thick, the tills are brown to yellowish brown whereas in the unoxidized zone they are greyish. In the Labrador Trough (Fig. 3.3) area, some tills are reddish due to incorporation of hematite. Except in the lowland southeast of James Bay and Torngat Mountains, only one till is recognized.

In central New Quebec there is a gradational series of belts of glacial depositional landform assemblages. Proceeding from Hudson Bay towards the ice divide, drumlin fields give way to ribbed moraine fields, which in turn grade into hummocky moraine. As recently portrayed by Bouchard et al. (1984) in the Lac Mistassini area (Fig. 3.42), tills in regions of ground moraine or fluted ground moraine are massive, found mainly in interfluve areas, and deposited by ice near the pressure melting temperature. Tills comprising ribbed moraine are commonly stratified, generally occur in low areas, and are thought to be formed by shearing of basal ice. Finally, tills comprising the hummocky moraine areas are massive, generally interstratified with sand and gravel, and thought to have formed at the margin of cold-based ice. Some hummocky moraines show conspicuous glacial lineations suggesting that their genesis is more related to shearing and ice-flow rather than to marginal ablation of cold-based ice. Clasts in fluted till or ribbed moraine are dominantly local whereas clasts in hummocky moraine have more distant sources. Whether these landform-sediment associations in the Lac Mistassini area are representative of all areas remains to be shown.

Figure 3.40. Bouldery till surface in southwestern Labrador. Courtesy of R.J. Fulton. 161608

Extensive tracts of land are made up of ground moraine moulded into drumlins (Fig. 3.43) or plastered on the lee side of bedrock hills (crag-and-tail). In the Lac Mistassini area drumlins are 100-3000 m long, 30-600 m wide, and 10-25 m high (Bouchard, 1980). In the La Grande Rivière area drumlins are 200-2000 m long, 100-400 m wide, and 3-25 m high (Vincent, 1977). Well developed glacial lineaments are common except in the rugged terrains of the southern Laurentian Highlands, the Torngat Mountains, and in the area of final ice retreat (Fig. 3.39). Lauriol (1982) and Gray and Lauriol (1985) have shown that the ice divide area in Ungava is not characterized by hummocky moraine but instead by an unmoulded till sheet flanked distally first by thin and discontinuous till and then by fluted till. The northern portion of the unmoulded till sheet is made up of rounded solifluected slopes that contrast with fresh morainic surfaces elsewhere. Perhaps this is a remnant till surface only slightly affected by the last ice advance.

Ribbed moraine (Fig. 3.44) and hummocky moraine (Fig. 3.45) occur mainly where Labrador Ice persisted latest (Hughes, 1964; Prest et al., 1968). The ribbed moraine has been described in some detail by Hughes (1964), Cowan (1968), and Bouchard (1980). Ribs are up to 1600 m long, 200 m wide, and 27 m high; generally the rib crests are spaced 90-300 m apart. In many areas radial belts of ribbed moraine alternate with belts of drumlins as they do in

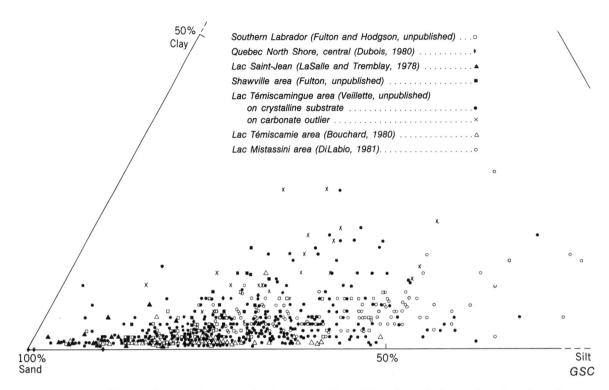

Figure 3.41. Ternary diagram showing grain size composition of till matrixes in the southeastern Canadian Shield

Keewatin and Manitoba (Shilts et al., 1987; this chapter, Dyke and Dredge, 1989). This attests to the complexity of ice flow regimes even over small areas. Large areas of hummocky moraine, which consist of irregularly shaped mounds of melt-out till, and "drift pressed features" such as moraine plateau, are common particularly in the final retreatal ground of Labrador Ice (Fig. 3.45). A belt of hummocky moraine, 100 km long, also occurs in southeastern Labrador (Fulton et al., 1981c, e). No detailed studies have been completed on these in the southeastern Canadian Shield.

De Geer moraines (Fig. 3.46) are restricted to low relief areas where the margin of the retreating ice sheet was in contact with lacustrine or marine water. They are remarkably well formed and abundant in the northeastern glacial Lake Ojibway basin and in the Tyrrell Sea basin. The moraines are composed largely of till and are largest and most consistently aligned where the ice sheet retreated in deep water. Where formed in shallow water, the moraines are smaller and their alignment is in places erratic. In La Grande Rivière area, De Geer moraines are 1-10 m high, 5-150 m wide, and 50-1500 m long. Spacing between moraine crests varies between 50 and 400 m but is most commonly between 100-200 m, presumably indicating annual retreat rates (Vincent, 1977). East of Hudson Bay, Lauriol and Gray (1987) measured a similar average spacing of 100-200 m between moraine crests.

Glaciofluvial deposits

Glaciofluvial deposits are widespread on the southeastern Canadian Shield. They occur as major end and interlobate moraines (Fig. 3.39, 3.47, 3.48), as eskers (Fig. 3.39), as kames, and as subaqueous and subaerial outwash deposits. Glaciofluvial deposits consist mostly of stratified sands and gravels.

The Saint-Narcisse (see Occhietti, 1989), Quebec North Shore, and Sakami end moraines mark major halts during recession (Fig. 3.39, 3.47). The Quebec North Shore

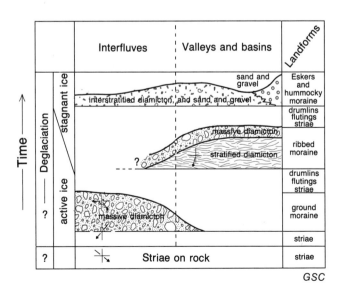

Figure 3.42. Relative stratigraphic and topographic occurrences of till lithofacies in the Lac Albanel area (after Bouchard et al., 1984).

Moraine system extends for 800 km (including gaps) between Rivière Manicouagan in Quebec and south of Lake Melville in Labrador (Dubois and Dionne, 1985). It consists of discontinuous till and ice contact ridges, up to 4 km wide and 50 m thick (Fig. 3.48), with outwash deltas and plains on its distal side. The Sakami Moraine extends over a distance of 630 km between Kuujjuarapik (Great Whale, Poste-de-la-Baleine) on the southeastern coast of Hudson Bay to Lac Mistassini (Hillaire-Marcel et al., 1981; Hardy, 1982a). Asymmetric ridges of ice contact and proglacial deposits, up to 6 km wide and 40 m thick, make up the moraine. The Harricana Interlobate Moraine (Hardy 1977, 1982b; Veillette, 1986a) marks the final zone of convergence of Labrador Ice and Hudson Ice. It can be traced for 1000 km from islands in eastern James Bay to Lake Simcoe in Ontario (Veillette, 1986a). The moraine is composed of a complex series of ridges, up to 10 km wide and 100 m high, of glaciofluvial deposits profusely pocked with elongated kettles.

Eskers are conspicuous features of the region everywhere except in the more rugged Torngat Mountains and the central and eastern parts of the Laurentian Highlands. Eskers form a radiating pattern from the central divide area. A characteristic feature of the eskers is that they commonly cross the grain of the landscape, ascending and descending hills and ridges. Where the ice retreated in contact with water bodies, eskers consist of subaqueous outwash deposits and may be buried by thick sequences of fine grained sediments.

Subaerial outwash deposits are less common because over much of the area the ice retreated in contact with either glacial lakes or the sea. Outwash terraces or delta complexes are present along some major valleys, such as those draining southward from the Laurentian Highlands. Meltwater channels are particularly abundant near the final retreat centre (Fig. 3.49); they were used by Ives (1959) to designate the sites of final ice sheet disintegration.

Figure 3.43. Drumlin field northeast of Lac Mistassini area. Courtesy of J-C. Dionne.

Figure 3.45. Hummocky moraine in the area northeast of Lac Mistassini. Courtesy of C. Laverdière.

Figure 3.44. Ribbed moraine field northeast of Lac Mistassini. Courtesy of C. Laverdière.

Figure 3.46. Suite of De Geer moraines in the area north of La Grande Rivière and east of Sakami Moraine. 204061-M

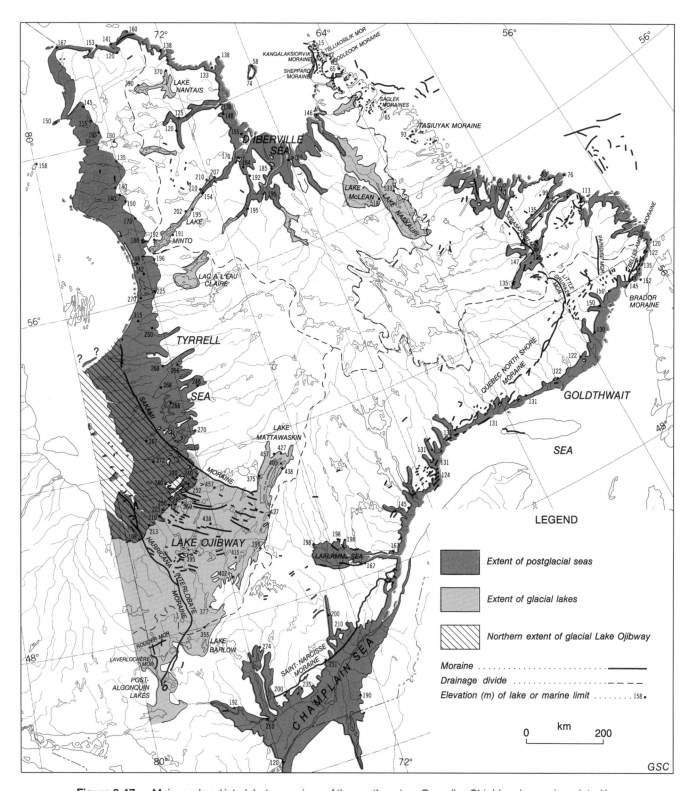

Figure 3.47. Major end and interlobate moraines of the southeastern Canadian Shield and areas inundated by glacial lakes and marine waters during the last deglaciation.

Glacial lake and marine deposits

Extensive areas were flooded by glacial lakes and seas during deglaciation (Fig. 3.47). Glacial Lake Ojibway and the Tyrrell Sea were by far the largest water bodies.

Fossiliferous silts and clays, generally massive, were deposited offshore in the seas, whereas varved silts and clays were laid down in glacial lakes. Glaciomarine facies (rainout sediment from glacial ice) are recognized locally but in most places glacial marine rhythmites occur at the base of the marine offshore sediment suite especially in proximal locations or in estuaries. At the shoreline of basins, littoral facies were also developed resulting in beaches, spits, and other related features. As the seas or lakes regressed or were drained, nearshore facies were produced in places where waves and currents could rework glacial, particularly ice contact, deposits. Hence, thick sequences of nearshore sands, overlying the offshore facies, are found near moraines and eskers. Flights of raised beaches (Fig. 3.50) can be spectacular, particularly in the Tyrrell Sea basin and in many locations along the Labrador coast. Estuarine and deltaic sediments were deposited at mouths of rivers. Since rivers entered the sea at progressively lower elevations as isostatic uplift progressed, there is downstream overlapping of younger coarser estuarine and deltaic deposits onto older fine grained deposits. Most older and higher delta deposits, which generally overlie fine grained offshore materials, were later incised by rivers; thus they occur as terrace remnants capping fine grained sediments.

Eolian, fluvial, and organic deposits

Eolian deposits are found in small areas where the wind has reworked sands and silts of glaciofluvial, marine, or deltaic origin. Large dune fields are found on outwash terraces and deltas in the Lac Saint-Jean area, the Quebec North Shore area, and in lower Churchill River valley (David, 1977); on reworked glacial lake materials near the Harricana Interlobate Moraine (Fig. 3.51); and locally on glaciofluvial and deltaic deposits elsewhere. The largest occurrence is south of Harp River in east-central Labrador.

Present day rivers are reworking older, mainly glacial and marine sediments and depositing materials on their

Figure 3.48. Quebec North Shore Moraine in the Rivière Natashquan area. Courtesy of J-C. Dionne.

Figure 3.49. Meltwater channels 70 km north of Schefferville, near the retreat centre of Labrador Ice. Courtesy of J.D. Ives.

Figure 3.50. Flights of raised beaches of the Tyrrell Sea in the Lac Guillaume-Delisle area. Courtesy of C. Hillaire-Marcel.

Figure 3.51. Parabolic dunes outlining bog areas on the east flank of the Harricana Interlobate Moraine in the area east of Lac Témiscamingue. Courtesy of J.J. Veillette. 203506-C

floodplains and as deltas in lakes and the sea. Most drainage basins contain an abundance of lakes which trap sediments entrained by streams. As a consequence, floodplains of more than limited extent are found only on the lower reaches of some of the larger rivers.

Organic deposits have accumulated in wetlands on glacial lake and marine plains, low relief till areas, and in depressions on bedrock surfaces. The peat cover is widespread in the area adjacent to James Bay and according to the National Wetlands Working Group (1981) wetlands in which organic sediments accumulate cover more than 50% of the surface in an area extending roughly 250 km inland. This same group also shows a belt of wetlands about 500 km wide stretching from James Bay to the Labrador Coast. Because climate varies from north to south, the type of wetland in which organic deposits accumulates also varies from north to south. At the southern fringe of the Canadian Shield, wetlands are dominantly domed raised bogs and ladder fens. Farther north, and throughout much of the central part of the region, organic deposits are being laid down in domed, flat, basin and string bogs and ribbed fens. In the northern part of the area, peat plateaus and patterned fens give way to wetlands characterized by low and high centred polygons containing ice wedges and lenses and underlain by permafrost. Palsas are common in coastal and northern bogs and thermokarst depressions are locally common in ice-rich organic deposits.

QUATERNARY HISTORY

Sediments on the southeastern Canadian Shield mostly record late Quaternary deglacial events. Older deposits are exposed along rivers to the southeast of James Bay and are found in the Schefferville area. Older tills have also been reported at the surface in the Torngat Mountains and coastal summits to the south of these, in the Mealy Mountains, possibly in southeastern Labrador, and in northern Abitibi. Age control on deglacial events is based mainly on radiocarbon dates on marine shells and basal lake sediments (Table 3.6). Major moraines, glacial lakes, postglacial seas and morphostratigraphic (weathering) zones have been named, but few lithostratigraphic units have been carefully defined and formally named.

In this section the limited information on pre-Late Wisconsinan events is discussed first. Then, deglacial events and chronology are discussed for a series of subregions and correlations are suggested (Fig. 3.52).

Pre-Late Wisconsinan events
Southeast James Bay area

Sediments recording pre-Late Wisconsinan events are known from eight locations (Hardy, 1982b; Bouchard et al, 1986; Fig. 3.31; Table 3.1). Thin laminae of peat underlying till, interbedded with compact clayey silt rhythmites, in a 10 m sequence, have been dated at >42 000 BP (Y-1165, Stuiver et al, 1963; Prest, 1970) along Rivière Harricana. In four sections along the banks of lower Rivière Nottaway, compact shell-bearing silts, wood-bearing clayey silts, and organic-bearing rhythmites and silty sands underlie the surface till. Near the mouth of Rivière de Rupert, compact organic-bearing silts and sandy silts underlying till have been observed in a core. All these sediments record marine or lacustrine facies similar to those of the interglacial Missinaibi

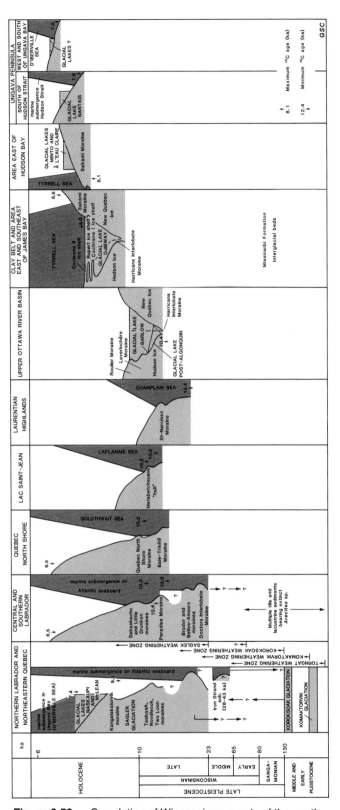

Figure 3.52. Correlation of Wisconsinan events of the southeastern Canadian Shield area. Each column is constructed from references given in the text.

Formation in the Hudson Bay Lowlands of Ontario (Skinner, 1973; this chapter, Dredge and Cowan, 1989). Compact waterlain sediments, similar to those described above, occur beneath glacial deposits in a borrow pit situated 115 km southeast of the mouth of Rivière Nottaway (50°39′N, 77°34.5′W; Hillaire-Marcel and Vincent, unpublished). The location and elevation (275 m) of this site suggest that one or the other of the water bodies recorded by deposits of the Missinaibi Formation may have extended far inland in Quebec. Finally, at the Selbaie Mine in northern Abitibi, Bouchard et al. (1986) have noted the presence of sand, gravel, and till under definitely Late Wisconsinan tills.

Schefferville-Wabush area

Recent investigations by Klassen and Thompson (1987) and Klassen et al. (1988), have indicated that several lithologically distinct till units are present in the Wabush area. In one section glaciofluvial and organic-bearing lacustrine sediments were identified. The significance of these finds is still not known but the deposits may well record pre-Sangamonian events since fruits of the extinct *Aracites* have been identified in lake sediments buried by three distinctively different tills.

During a drilling operation 25 km northwest of Schefferville a twig was obtained from a peat layer which was overlain by 18 m of glaciofluvial materials and till. This small piece of wood gave a radiocarbon age of 24 250 ± 600 BP (Granberg and Krishnan, 1984); the date, if valid, may record a nonglacial pre-Late Wisconsinan event. This is a surprising discovery because it may suggest that an area of the southeastern shield near the heart of the Laurentide Ice Sheet may have been ice free during the Middle Wisconsinan.

Torngat Mountains and adjacent coastal areas

Surface materials on summits of the Torngat Mountains, and on coastal summits as far south as Fraser River, apparently record glacial and nonglacial events of pre-Late Wisconsinan and pre-Wisconsinan age. A long lasting controversy exists between researchers who feel extensive nunatak areas existed during Late Wisconsinan and those who feel the areas were ice covered. The former link weathering zones to different glacial events; the latter believe that weathering zones result from progressive deglaciation of summit areas during the Late Wisconsinan, represent zones developed under different subglacial thermal regimes, or record subaerially weathered terrains that were preserved under cold-based ice. The argument between the two groups of researchers persists because of equivocal field evidence, poor chronological control of events, and a tendency to fit the facts to theory (Ives, 1974; Clark, 1984a). Comprehensive reviews of this equivocal evidence can be found in Ives (1974, 1978), Mayewski et al. (1981, p. 106-108), Brookes (1982), Clark (1984a), Evans (1984), and Evans and Rogerson (1986).

Ives (1978), mainly using his own work (Ives, 1957, 1958a,b, 1963, 1974, 1975, 1976; Ives et al., 1975) and that of Tomlinson (1958 and 1963), Wheeler (1958), Løken (1962a, b), Andrews (1963a), and Johnson (1969), proposed that three regional ice sheet glacial events are recorded in the Torngat Mountains, on the basis of clearly defined weathering zones. The weathering zone limits (or glacial boundaries, Figure 3.53) slope seaward and ice was progressively thinner and less extensive during successive glaciations.

During the oldest **Komaktorvik Glaciation,** all of northern Labrador, except for higher summits in the central and northern Torngat, was ice covered and glaciers extended far out onto the continental shelf. The nunataks of this glaciation were assigned to the Torngat Weathering Zone which is characterized by mature felsenmeer, tors, and deep weathering pits, and by the assumed absence of glacial erratics. Areas affected by the Komaktorvik Glaciation, and not later covered by ice, comprise the Komaktorvik Weathering Zone. This zone is similar in character to the Torngat Weathering Zone but contains erratics.

During the **Koroksoak Glaciation** ice again covered most of Labrador and extended out onto the shelf, but nunataks were larger than during the Komaktorvik Glaciation. Areas covered by the ice during Koroksoak Glaciation, and not later covered by ice, comprise the Koroksoak Weathering Zone. In this zone there is abundant evidence of glaciation and only limited (incipient) felsenmeer development. Weathering characteristics are intermediate between those of the Komaktorvik Weathering Zone above and those of the areas covered by ice during Saglek Glaciation. The boundary between the Koroksoak and Komaktorvik zones is a trimline which is obscured in places by periglacial slope processes.

Abundant and fresh looking glacial landforms characterize the area covered by the **Saglek Glaciation.** Presumably during this glaciation ice reached the outer coast adjacent to the Torngat Mountains, only at mouths of fiords. The Saglek glacial limit, is marked by fresh lateral and end moraines (Ives, 1976; Clark, 1984a). Felsenmeer and advanced weathering features are generally absent in the area covered by Saglek ice. Locally, due to lithology or preservation beneath the Saglek ice, small areas of extensive weathering have been reported (Gangloff, 1983).

The absence of detailed studies comparing the weathering characteristics of each zone and the lack of chronological control makes it difficult to assign ages to the different glaciations. Andrews (1974) argued on the basis of comparisons with the Baffin Island weathering zones, the development of clay minerals, and increase in ferric oxides, that the Koroksoak and older terrains on Baffin Island must be at least mid-Quaternary in age. Even though many workers, including J.D. Ives (University of Colorado, personal communication, 1985), believe the Saglek Moraines were built during the Late Wisconsinan maximum, the absolute age of the Saglek Glaciation is still not known (Andrews, 1977; Mayewski et al., 1981), although a Late Wisconsinan age is likely. Clark (1984a) corrrelated his Late Wisconsinan Two Loon drift with the deposits of Saglek Glaciation; Clark and Josenhans (1985), after studying the Saglek Moraines in their type area, argued that they are Late Wisconsinan.

Interpretations based on weathering zones are still being challenged. Gangloff (1983) saw no significant differences between the matrix texture, clay content, mineralogy, or surface morphology of quartz sand grains in Saglek Glaciation till and in felsenmeer of the Koroksoak Weathering Zone. Gangloff also documented the presence of tors in valley bottoms which were not destroyed by Saglek Glaciation ice, and of weathering pits and incipient tafoni which developed in Holocene time. Systematic mapping of

CHAPTER 3

Table 3.6. Pertinent radiocarbon ages of the Canadian Shield area of Quebec-Labrador

Laboratory Dating no.	Age (years BP)	Locality	Reference	Material	Significance
Beta-9516	6 600 ± 100	Rivière Laforge, Que.	P.J.H. Richard (personal communication, 1985)	gyttja	"Oldest" minimum age for deglaciation of Rivière Laforge area (LG 4 Reservoir).
Beta-11121	9 800 ± 220	Rivière Deception, Que.	Gray and Lauriol (1985)	marine shells	With I-488, oldest shell date for deglaciation of the south shore of Hudson Strait.
DIC-517	42 730 +6680/-9970	Iron Strand, Labrador	Ives (1977) Short (1981)	marine shells	Date implies that Iron Strand site records a Middle Wisconsinan Interstade.
Gif-424	10 250 ± 350	Metabetchouan, Que.	LaSalle and Rondot (1967)	marine shells	"Oldest" minimum age for deglaciation of Lac Saint-Jean area.
Gif-3770	10 230 ± 180	Rivière-à-la-Chaloupe, Que.	Dubois (1980)	marine shells	"Oldest" minimum age for deglaciation of area west of Rivière Romaine, Quebec North Shore.
GSC-672	7 970 ± 250	Sugluk Inlet, Que.	Matthews (1967) Lowdon and Blake (1968)	marine shells	"Oldest" reliable minimum age for deglaciation of the south shore of Hudson Strait.
GSC-706	7 430 ± 180	Ottawa Islands, N.W.T.	Andrews and Falconer (1969) Lowdon and Blake (1968)	marine shells	"Oldest" minimum age for deglaciation of the Ottawa Islands and best approximation for the age of marine limit.
GSC-1337	9 140 ± 200	Rivière Moisie, Que.	Dredge (1983b) Lowdon et al. (1971)	marine shells	"Oldest" minimum age for deglaciation and age of the Quebec North Shore Moraine in the Rivière Moisie area.
GSC-1533	12 400 ± 160	Charlesbourg, Que.	LaSalle et al. (1977a) Lowdon and Blake (1973)	marine shells	"Oldest" minimum age for deglaciation of St. Lawrence estuary east of Québec City and begining of Champlain Sea Episode.
GSC-1592	6 460 ± 200	Michikamau Lake	Fulton and Hodgon (1979) Lowdon and Blake (1973)	peat	"Oldest" minimum age for deglaciation of the Upper Churchill River area.
GSC-1646	12 200 ± 160	Cantley, Que.	Romanelli (1975) Lowdon and Blake (1973)	marine shells	"Oldest" minimum age for deglaciation of lower Rivière Gatineau valley and incursion of Champlain Sea there.
GSC-1772	11 900 ± 160	Martindale, Que.	Romanelli (1975) Lowdon and Blake (1973)	marine shells	"Oldest" minimum age for deglaciation of middle Rivière Gatineau valley and incursion of Champlain Sea there.
GSC-2101	10 300 ± 100	Shawinigan, Que.	Occhietti (1980)	marine shells	"Oldest" minimum age for deglaciation in area between Montrealand Québec City.
GSC-2825	10 900 ± 140	Pinware, Labrador	Grant (1986) Lowdon and Blake (1979)	marine shells	"Oldest" minimum age for deglaciation of area north of the Straits of Belle Isle and for Goldthwait Sea on the north shore of the St. Lawrence.
GSC-2946	9 120 ± 480	Hudson Strait	Fillon and Harmes (1982) Lowdon and Blake (1980)	marine shells	"Oldest" minimum age for deglaciation of eastern Hudson Strait.
GSC-2970	7 600 ± 100	Northwest River, Labrador	Lowdon and Blake (1980)	marine shells	"Oldest" minimum age for deglaciation of the western end of Lake Melville.
GSC-3022	10 400 ± 140	Lake Hope Simpson, Labrador	Blake (1982); Engstrom and Hansen (1985)	gyttja	"Oldest" likely reliable minimum age for deglaciation of Alexis River area.
GSC-3067	9 640 ± 170	Moraine Lake, Labrador	Blake (1982) Engstrom and Hansen (1985)	gyttja	"Oldest" minimum age for deglaciation of upper St. Paul River area.
GSC-3094	6 320 ± 180	Lac Delorme, Que.	Richard et al. (1982) Blake (1982)	gyttja	"Oldest" minimum age for deglaciation of the ice devide area in the Lac Caniapiscau region of central New Quebec.
GSC-3241	6 500 ± 100	Border Beacon, Labrador	Blake (1982)	gyttja	"Oldest" minimum age for deglaciation of area northeast of Smallwood Reservoir.
GSC-3460	10 400 ± 200	Montreal River, Ont.	Veillette (1988)	gyttja	"Oldest" minimum age for deglaciation of southern portion of Lac Témiscamingue area and minimum age for McConnell Lake Moraine (part of Harricana Interlobate Moraine) in that area.
GSC-3467	10 100 ± 180	Lac Kipawa, Que.	Veillette (1988)	gyttja	With GSC-3460 "oldest" minimum age for deglaciation of area just south of Lac Témiscamingue.
GSC-3615	6 510 ± 110	Lac Gras, Que.	King (1985)	gyttja	"Oldest" minimum age for deglaciation of upper Rivière Moisie area.

Table 3.6. (cont.)

Laboratory Dating no.	Age (years BP)	Locality	Reference	Material	Significance
GSC-3644	6 200 ± 100	Lac Starkel, Que.	King (1985)	gyttja	"Oldest" minimum age for deglaciation of the wide-mouthed U-shaped retreat center northeast of Lac Opiscotéo.
GSC-3947	7 130 ± 100	Deception Bay, Que.	B. Lauriol (personal communication, 1986)	marine shells	Shells dated are from immediate vicinity and about at same altitude as other collection of Matthews (I-488) which gave an age of 10 450 ± 250 BP.
GX-5522	11 160 ± 520	Moraine Lake, Labrador	Short (1981)	lake mud (low organic content)	Lakes situated on distal site of Saglek moraines. With GX-6362 date provides support for ice free areas in northern Labrador in the Late Wisconsinan?
GX-6345	10 275 ± 225	Makkovik Harbour, Labrador	Barrie and Piper (1982)	foraminifera in marine muds	"Oldest" minimum age for deglaciation of area north of Lake Melville.
GX-6362	18 210 ± 1900	Square Lake, Labrador	Short (1981); Clark et al. (1986)	lake mud (low organic content)	Lake situated on distal side of Saglek moraines. With GX-5522 date provides support for ice free areas in northern Labrador in the Late Wisconsinan?
GX-8240	34 200 +2100/-1600	Iron Strand, Labrador	Clark (1984a)	marine shells	Date implies that Iron Strand site records a Middle Wisconsinan Interstade.
GX-8241	28 200 +1200/-1000	Iron Strand, Labrador	Clark (1984a)	marine shells	Date implies that Iron Strand site records a Middle Wisconsinan Interstade.
GX-9293	9 110 ± 470	Shoal Cove, Labrador	Clark (1984a)	marine shells	With L-642 and TO-305 "oldest" minimum age for deglaciation of the Labrador coast east of the Torngat Mountains.
I-488	10 450 ± 250	Deception Bay, Que.	Matthews (1967)	marine shells	Oldest shell date for deglaciation of the south shore of Hudson Strait, but is likely erroneous (see GSC-3947).
I-5922	10 400 ± 150	Sacre-Coeur-de-Saguenay, Que.	Dionne (1977)	marine shells	Oldest age for deglaciation of the mouth of Rivière Saguenay area. Oldest age determination for Laflamme Sea.
I-8363	8 230 ± 135	Kuujjuarapik, Que.	Hillaire-Marcel (1976)	concretion	Maximum age for construction of Sakami Moraine and incursion of Tyrrell Sea east of Hudson Bay.
I-9632	6 990 ± 150	Payne Bay, Que.	Gray et al. (1980)	marine shells	"Oldest" minimum age for deglaciation of south-western Ungava Bay and incursion of D'Iberville Sea.
L-642	9 000 ± 200	Eclipse Channel, Labrador	Løken (1962b)	marine shells	With GX-9293 and TO-305 "oldest" minimum age for deglaciation of the Labrador coast east of the Torngat Mountains.
QU-122	7 880 ± 160	Rivière La Grande, Que.	Hardy (1976)	marine shells	Maximum age for drainage of Lake Ojibway and oldest available minimum age for construction of Sakami Moraine and incursion of Tyrrell Sea.
QU-574	9 970 ± 130	Rivière des Anglais, Que.	Dubois (1980)	marine shells	"Oldest" minimum age for deglaciation of the Rivière Manicouagan area.
SI-1737	10 240 ± 1240	Saint-John Island, Labrador	Jordan (1975)	gyttja	"Oldest" minimum age for deglaciation of eastern Lake Melville area.
SI-1959	6 815 ± 125	Pyramid Hills Lake, Que.	Short (1981)	lake mud	"Oldest" minimum age for deglaciation of area southeast of Ungava Bay.
SI-3139	10 550 ± 290	Eagle River, Labrador	Lamb (1980)	gyttja	"Oldest" minimum age for deglaciation of area south of the Mealy Mountains and for construction of the Paradise Moraine.
SI-4131	34 360 ± 850	Iron Strand, Labrador	Ives (1977); Short (1981); Vilks et al. (1987)	marine shells	Date implies that Iron Strand site records a Middle Wisconsinan Interstade.
TO-200	7 970 ± 90	Central Lake Melville, Labrador		marine shells	"Oldest" reliable minimum age for the deglaciation of central Lake Melville.
TO-305	9 830 ± 70	Nachvak Fiord, Labrador	R.J. Rogerson (personal communication, 1986)	marine shells	"Oldest" minimum age for deglaciation of the Labrador coast east of the Torngat Mountains.
UQ-547	6 700 ± 100	Rivière Nastapoca, Que.	Allard and Seguin (1985)	marine shells	"Oldest" minimum age for deglaciation of the Nastapoka Sound area.
Y-1165	>42 000	Rivière Harricana, Que.	Stuiver et al. (1963)	peat	Minimum age for Missinaibi Formation deposits in the Quebec James Bay Lowlands.

glacial deposits and weathering zones, augmented by weathering and stratigraphic studies, must be completed before the Quaternary geology of the Torngat Mountains can be more clearly interpreted.

Recent studies by Clark (1984a) in the Iron Strand area of northern Labrador, and by Evans (1984), Evans and Rogerson (1986), and by Bell et al. (1987) in the Nachvak Fiord area, provide further insight into the Quaternary history of the Torngat Mountains. On the basis of field mapping and ice sheet profile reconstructions, Clark (1984a) and Clark and Josenhans (1986) recorded two glacial advances. The older, which they correlate with the Koroksoak Glaciation of Ives (1978), deposited the Iron Strand drift, left nunataks in the Torngats, and extended onto Saglek Bank. The younger advance, which they correlate with the Saglek Glaciation of Ives (1978), deposited Two Loon drift, left large ice-free areas, and extended according to them out to the shelf edge through the deep areas on the shelf (Clark and Josenhans, 1986). Clark (1984b) estimated that the ice surface on the drainage divide was 818 m a.s.l. west of Kangalaksiorvik Fiord and 606 m a.s.l. west of Eclipse Channel.

At Iron Strand, in an area known to have been covered by the ice that deposited Iron Strand drift but not by the younger ice, massive marine clayey sands are overlain by a silty sand till or glaciomarine diamicton containing abraded marine shells, and by fossiliferous nearshore marine sands. South of this site, a soil thicker than that developed during the Holocene is present above similar beds. Shells from the diamicton dated 34 200 +2100/−1600 BP (GX-8240); likely in situ shells from the overlying marine sands dated 28 200 +1200/−1000 BP (GX-8241); and shells from colluvium derived from the same bluffs are 42 730 +6680/−9970 BP (DIC-517) and 34 360 ± 850 BP (SI- 4131; see Ives, 1977; Short, 1981). The shell dates indicate that the diamicton records a pre-Late Wisconsinan advance of continental ice which reached the sea. Based on radiocarbon dating and amino acid analyses, Clark (1984a) tentatively corelated the Iron Strand drift with the Loks Land Member of the Clyde Foreland Formation of Baffin Island and suggested that it represents an event which occurred between 100 ka and 34 ka. Andrews and Miller (1984) suggested that the Iron Strand fossiliferous deposits represent a Middle Wisconsinan nonglacial interval because the amino acid racemization of shells was significantly less than those which typify the Kogalu aminozone, possibly of Sangamonian age (this chapter, Andrews, 1989). If the radio-carbon dates are considered to lie beyond the limit of the method, as is commonly done with shell dates in this age range, the Iron Strand drift could be older than Middle Wisconsinan. Despite the problem of absolute age, the record in the Iron Strand area and the minimum ages support the Quaternary framework established on the basis of weathering zones and confirm that parts of the north coast of Labrador remained ice-free throughout the Late Wisconsinan (Ives, 1977).

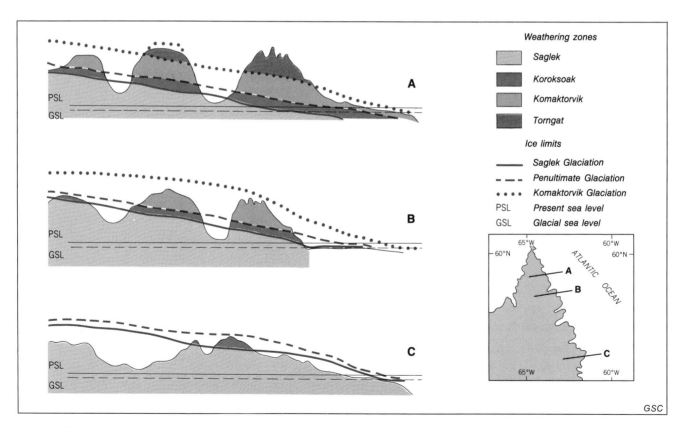

Figure 3.53. Cross-sections, drawn perpendicular to the general trend of the Laurentide Ice Sheet margin, showing ice profiles and weathering zones of the northern and central Labrador highlands. Inset shows locations of A, B, and C (after Ives, 1978).

Based on detailed mapping, Evans (1984), Evans and Rogerson (1986), Rogerson and Bell (1986), and Bell et al. (1987) recognized two advances of Laurentide ice in the Nachvak Fiord area. Their Ivitak phase is the oldest and most extensive; nunataks were present but the ice extended onto the shelf. As in the Iron Strand area, shells from a glacial marine diamicton assigned to the Ivikak phase have provided four finite radiocarbon ages in the Middle Wisconsinan time range (Bell et al, 1987). The subsequent Nachvak phase was less extensive: the ice only extended to the mouths of fiords and is marked west of the head of Nachvak Fiord by the Tinutyaruik Moraine. On the basis of soil development studies and the older radiocarbon dates, the workers believed that deposits laid down during Ivitak phase glaciation are >40 ka old and are equivalent in age to the Iron Strand drift of Clark (1984a). They further believed that their younger Nachvak phase is equivalent in age to the Saglek Glaciation. Even though correlations are tentative, the studies also confirm that parts of northern Labrador remained ice free at least during the Late Wisconsinan and that the ice may not have extended onto the Labrador Shelf at that time.

Additional support for ice-free areas during the Late Wisconsinan may be provided by a radiocarbon age of 18 210 ± 1900 BP (GX-6362) on organic sediments at the base of a core from a lake on the distal side of and dammed by the Saglek Moraines (Short, 1981; Clark et al., 1986). However, since the organic content in the dated core was very low (<0.1% total organic matter) the accuracy of the age determination can be questioned (see discussions in Short, 1981). Whatever the case the date probably provides a maximum age for the Saglek Moraines (Clark et al., 1986). Another age determination of 11 160 ± 520 BP (GX-5522) dates the recession of the ice front from the Saglek Moraines but may also be too old.

Mealy Mountains and southeastern Labrador
Various researchers (Gray, 1969; Rogerson, 1977; Fulton and Hodgson, 1979) have speculated that summits of the Mealy Mountains were not overtopped by Late Wisconsinan ice, and Fulton and Hodgson (1979) have suggested that the Paradise Moraine (Fig. 3.47) marks the Late Wisconsinan glacial limit. Differences are readily apparent in the character of the terrains separated by moraine systems. Radiocarbon ages, to 21 ka obtained from sediments in a lake core near Alexis River by H.E. Wright, Jr. (University of Minnesota, personal communication, 1979) and in a core taken offshore from Lake Melville (Vilks and Mudie, 1978) appear to lend support to the idea of a Late Wisconsinan ice-free area in southeastern Labrador. The reliability of the dates can be questioned, however, since the offshore dates are based on total organic content (Fillon et al., 1981) and contamination from older carbon could account for the old gyttja dates on land.

Late Wisconsinan buildup and limit
The entire southeastern Canadian Shield, except for some nunataks in northern Labrador, possibly summits of the Mealy Mountains, and possibly the region lying beyond the Paradise Moraine, was covered by Late Wisconsinan ice of the Labrador Sector of the Laurentide Ice Sheet. In this section the sparse evidence on glacial buildup and the problems of defining the glacial limit on the Atlantic seaboard are discussed.

It is generally agreed that initial development of the Labrador Sector of the Laurentide Ice Sheet in the Late Wisconsinan (or much earlier?) was on the interior uplands and that glacial flow was radial to limits on the Labrador Shelf and in the Gulf of St. Lawrence, and was radial to zones of confluence with adjacent ice masses in the west and in the north. In two areas of the southeastern Canadian Shield, ice flow indicators record movements different from those thought to have been produced during the last flow phase. Whether these flows are related to Late Wisconsinan dispersal centres or whether they are older is not known.

In southern Labrador, Klassen (1983a, 1984) and Klassen and Bolduc (1984), postulated a dispersal centre situated between Churchill River and the Gulf of St. Lawrence. This is based on striae indicating northward and north-northeastward flow in the area north of Lake Melville and in the upper Churchill River region. The striae predate the Late Wisconsinan regional northeastward, eastward, and southeastward flow. The age of this dispersal centre is not known but it could be as young as "early" Late Wisconsinan. Occhietti (1982) argued for an Early Wisconsinan accumulation centre on the central Laurentian Highlands to explain the Quaternary record in the St. Lawrence Lowlands, and Quinlan and Beaumont (1982) have suggested independently, on the basis of the sea level record in Atlantic Canada, that a major ice dome must have existed in the same area during the "early" Late Wisconsinan. The observations of Klassen and Bolduc may be field evidence for Occhietti's and Quinlan and Beaumont's suggestions. Farther north in the Schefferville area, Klassen and Thompson (1987) recently identified five distinctive phases of ice flow requiring different or major shifts of dispersal centres. Because, according to Klassen and Thompson (1987, p. 65) "the ice flow features recorded on outcrop surfaces do not show differential weathering, the phases ..... appear to have been produced during a period of continuous ice cover and may represent changes during one or more stadials of Wisconsin Glaciation".

In Abitibi, Lac Témiscamingue, and adjoining regions of western Quebec, striae and glacially transported materials indicate that ice flowing southwest from New Quebec extended west of the Harricana Interlobate Moraine (the feature that marks the final contact between Labrador Ice and Hudson Ice; Chauvin, 1977, p.16; Kish et al., 1979, p. 8, Veillette, 1982, 1986a). Time of this event is not known. There is also evidence that at one time ice flowing southeast out of James Bay extended at least 20 km east of the Harricana Moraine in the area between Rivière Harricana and Rivière Nottaway (L. Hardy, Poly-Géo. Inc., Longueil, Quebec, personal communication, 1985). In the area east of James Bay, the existence of eastward moving ice proposed by Lee (1959b) was later rejected by Dionne (1974) who attributed the presence of erratics from a western provenance, noted by Lee, to sea ice rafting. On Ungava Peninsula, Bouchard and Marcotte (1986) saw no evidence for ice flowing eastward from Hudson Bay. Much farther inland, in the Lac Mistassini area, Martineau (1984b), Prichonnet et al. (1984), and Bouchard and Martineau (1985) have documented southeastward ice flow which preceded the regional southwestern ice flow from New Quebec recorded on the eastern and western side of the Harricana Moraine. At this time it is impossible to say which flow directions relate to the Late Wisconsinan maximum and which ones might be older. Whatever the case, at some time before final deglaciation, ice from central New Quebec carried materials west of the Harricana Interlobate Moraine, whereas ice from a dispersal centre west of the east coast of James Bay flowed southeastwardly as far as Lac Mistassini. The identification along the middle St. Lawrence estuary of stromatolitic dolomite from the Lac Mistassini and Lac Albanel area (Dionne, 1986) indicates that the southeasterly flows could have been even more extensive.

The Late Wisconsinan extent of Laurentide ice on the Atlantic seaboard of Labrador is not definitely known and is the subject of much controversy (Fig. 3.54). Even though early workers such as Bell (1884), Daly (1902), and Coleman (1921) suggested that parts of northeastern and southeastern Labrador had not been glaciated, the tendency until recently, as expressed by Flint (1957) and Prest (1969), has been to portray the Late Wisconsinan ice sheet as extending well offshore (Fig. 3.54). Based on the weathering zone concept, areas of eastern Labrador have more recently been shown as having escaped Late Wisconsinan glaciation (Ives, 1960a, 1978; Hughes et al., 1981; Fig. 3.54). Based in part on the location of the Paradise Moraine, which Fulton and Hodgson (1979) speculated might mark the limit of Late Wisconsinan ice in southeastern Labrador, and on data from offshore investigations (Vilks and Mudie, 1978), Rogerson (1981, 1982), and Prest (1984) have shown relatively large ice-free areas on land (Fig. 3.54). Finally, using the mapped extent of a distinctive offshore till sheet assigned to the "early" Late Wisconsinan on the basis of total organic radiocarbon ages (deemed anomalously old by Fillon et al., 1981), Josenhans et al. (1986) and Clark and Josenhans (1986) have reasserted that Late Wisconsinan ice extended well offshore onto the continental shelf (Fig. 3.54). They proposed that outlet glaciers in the Torngat Mountains in fact advanced onto the shelf, coalesced and deposited a till sheet.

None of the models portrayed in Figure 3.54 can be validated at this time, but the best interpretation at the moment is that areas of northern Labrador, summit areas of the Mealy Mountains, and possibly areas distal to the Paradise Moraine were not covered by Late Wisconsinan ice, and that the ice that covered the rest of the landmass, flowed onto the shelf. Until (1) glacial limits and weathering zones are mapped everywhere in detail, (2) in depth studies are undertaken, as on Baffin Island, to document the relative differences between all weathering zones, and (3) stratigraphic studies accompanied by unequivocal radiometric dates are undertaken, the location of the Late Wisconsinan limit will remain in question.

Deglaciation history

Ice initially retreated in water (Fig. 3.47) in all peripheral areas of the southeastern Canadian Shield. This circumstance greatly influenced the style of deglaciation. A series of paleogeographic maps (Fig. 3.55) portrays the ice cover and location of major moraines, glacial lakes, and postglacial seas for the period 12 ka to 7 ka. Figure 3.52 provides a tentative correlation of the events in each of the regions discussed and is referred to in the following discussions.

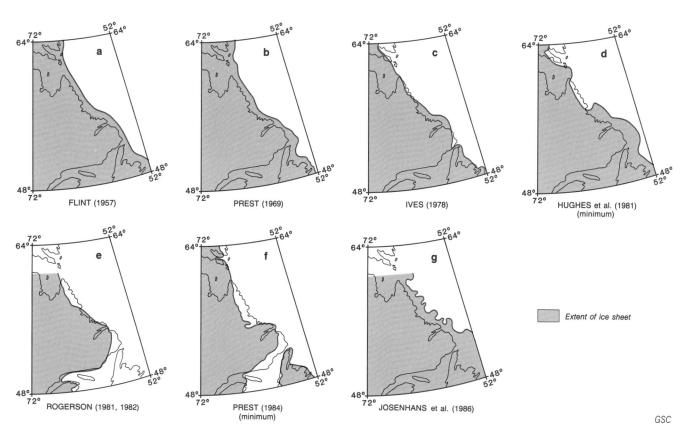

Figure 3.54. Speculative Late Wisconsinan limits of the Laurentide Ice Sheet on the Atlantic seaboard of Labrador as proposed by various authors.

Northern Labrador and northeastern New Quebec

During the Late Wisconsinan, ice of the Labrador Sector advanced east northeasterly from central New Quebec and easterly from Ungava Bay (Ives, 1957, 1958a; Løken, 1962a) towards the Labrador Shelf over northeastern New Quebec and northern Labrador (Saglek Glaciation in the Torngat Mountains; Andrews, 1963a). Higher coastal summit areas from Fraser River north and much of the more northerly Torngat Mountains and adjacent coastal forelands were not overtopped by ice. Clark (1984a; Fig. 3.56) recently illustrated that ice from west of the watershed crossed the Torngat Mountains in the form of outlet glaciers, extended to fiord mouths, and did not override large nunatak areas comprising the Koroksoak and higher weathering zones. The probable upper limit of extent of Late Wisconsinan ice is documented in several local studies but, except in Andrews (1963a), Clark (1984a), Evans (1984) and Evans and Rogerson (1986), no attempt has been made at mapping precisely the limit of ice cover. Generally the upper trimline of Saglek Glaciation not only slopes from the divide to the ocean but also slopes down towards the north. In the area between Fraser River and Okak Bay, ice reached a maximum elevation of about 700 m (Andrews, 1963a). Mount Thoresby and Man O'War Peak are the most southerly nunataks (Andrews, 1963a; Johnson, 1969). The limit in the Saglek Fiord area is 615 m (Smith, 1969); in the Ryans Bay region, Clark (1984a) stated that the ice passing through the Torngat Mountains did not reach elevations of more than 800 m on the drainage divide. In summit areas lying above the Saglek Glaciation level, cirque glaciers or small local ice caps existed which were independent of the Laurentide Ice Sheet (Fig. 3.56; Ives, 1960a; Løken, 1962a; Evans, 1984; Clark, 1984a; Evans and Rogerson, 1986; Bell et al., 1987). The question of the actual extent of ice on the shelf is the subject of much controversy. Initial studies by Clark (1984a) and the studies of Evans (1984), Evans and Rogerson (1986), and Rogerson and Bell (1986) indicated that the Late Wisconsinan ice only extended to the mouths of the fiords. In opposition other studies, as mentioned above (Josenhans et al., 1986; Clark and Josenhans, 1986), stated that the same ice extended well offshore to near the shelf edge.

Deglacial events in rugged coastal areas are complex and have been discussed for various locations particularly by Ives (1958a, 1960a), Løken (1962b, 1964), Tomlinson (1963), Andrews (1963a), Johnson (1969), Clark (1984a), Evans (1984), and Evans and Rogerson (1986) . At the limit reached by the Late Wisconsinan ice, extensive systems of lateral moraines and kame complexes were built (Saglek Moraines of Ives, 1976; Fig. 3.52). Several end and lateral moraines mark positions of halts or local readvances of the ice front during retreat from the Saglek Moraines towards an ice mass located west of the Atlantic/Ungava Bay watershed (Fig. 3.47). Notable examples are the Tasiuyak Moraines in the Fraser River-Okak Bay area (Andrews, 1963a), and the well correlated Noodleook, Two Loon, and Kangalaksiorvik (=Sheppard) moraines of Løken (1962b, 1964) on Torngat Peninsula. Andrews (1977) suggested that the Kangalaksiorvik Moraines, which extend from Killinek Island to Ryans Bay may be correlative with moraines of Cockburn age elsewhere in Arctic Canada (Andrews and Ives, 1978). J.D. Ives (University of Colorado, personal communication, 1985) has observed ice flow indicators between the southwestern Torngat Mountains and lower Rivière George, and along the east coast of Ungava Bay which clearly indicate a late reversal of ice flow into Ungava Bay. He interpreted this as resulting from the development of a calving bay in northern Ungava Bay which extended rapidly southward until the entire coastline was ice free.

Because coastal areas were glacial isostatically depressed, waters of the Atlantic Ocean submerged both ice-free and newly deglaciated low-lying areas. Little is known about sea level history in the area, but generally marine limit progressively decreases in elevation towards the north. Marine waters may have reached elevations of about 93 m south of Okak Bay (Andrews, 1963a), 65 m in the Saglek Bay and Ryans Bay areas (Løken, 1962b; Smith, 1969), 73 m at the head of Nachvak Fiord (Bell et al., 1987), 42 m north of Ryans Bay (Løken, 1964), and 16 m on Killinek Island (Løken, 1964). Shells collected by R.J. Rogerson (Memorial University personal communication, 1986) from the head of Nachvak Fiord gave an age of 9820 ± 70 BP (TO-305), the oldest Holocene age so far obtained on the north coast. Other shells collected by Løken (1962b) north of Ryans Bay gave an age of 9000 ± 200 BP (L-642) and by Clark (1984a) from south of Iron Strand gave an age of 9820 ± 470 (GX-9293). The presence in northernmost Labrador of tilted shorelines truncated by a 15 m high horizontal shoreline is considered by Løken (1962b) as recording an early Holocene transgression and also suggests that thick continental ice did not overlie the northern tip of Labrador.

During deglaciation of the Torngat Mountains, numerous small and possibly ephemeral glacial lakes were ponded on the east side of the drainage divide in tributary valleys blocked by ice tongues. As the ice margin receded downslope on the west side of the mountains, larger and longer lasting glacial lakes were created. Lakes in tributary valleys of Rivière Alluviaq drained into a fiord south of Iron Strand (Ives, 1957), whereas other lakes, in the Rivière Koroc basin drained towards Saglek Bay (Ives, 1958a; Fig. 3.47). The most important lakes were the Naskaupi and McLean glacial lakes (Ives, 1960a,b; Matthew, 1961; Barnett and Peterson, 1964; Barnett, 1964, 1967; Peterson, 1965; Fig. 3.47), which extended over large areas in the upper Rivière George and Rivière à la Baleine drainage basins. These lakes were dammed between the watershed crest on the east, the southwestward retreating main body of Labrador Ice on the west, and the inferred late ice ridge or dome over Ungava Bay. The configuration of the ice required to hold the lakes is discussed by Prest (1970, 1984). Glacial Lake Naskaupi cut a series of well defined strandlines, some of which are incised in bedrock. Probable outlets, which permitted drainage of water to the Labrador Coast, are the headwaters of Fraser, Kogaluk, Harp, Kanairiktok, and Naskaupi rivers. Glacial Lake McLean, in upper Rivière à la Baleine basin, was separate from Lake Naskaupi but drained into it by a channel just west of Lac de la Hutte Sauvage. Both lakes finally drained into the D'Iberville Sea (postglacial Ungava Bay) when ice had retreated sufficiently, perhaps through the development of a calving bay, to allow penetration of marine waters.

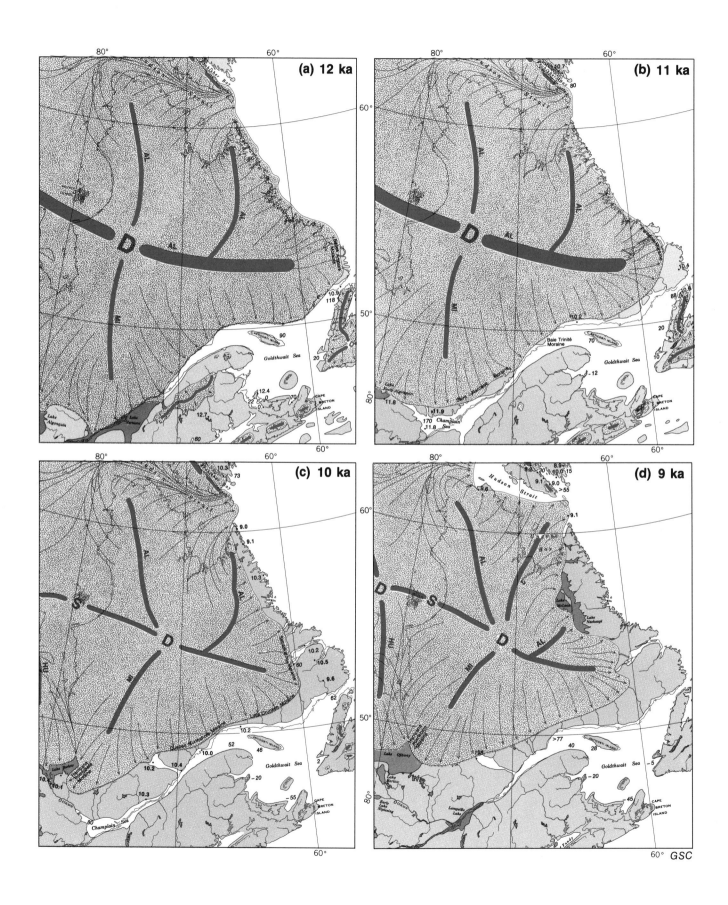

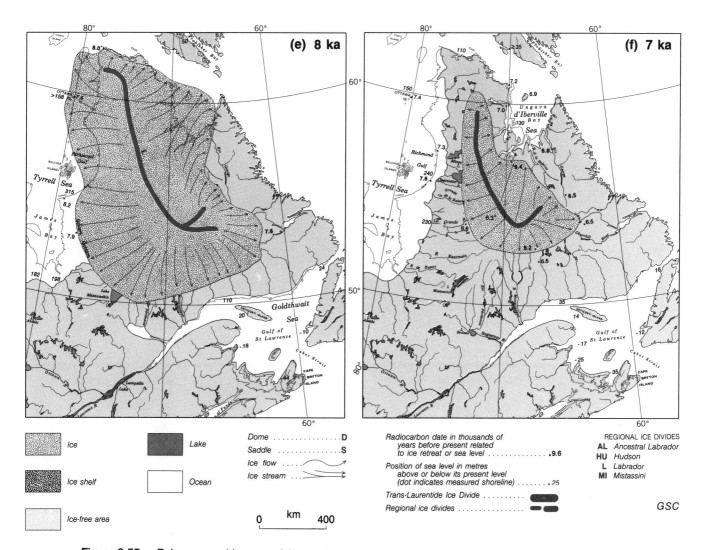

Figure 3.55. Paleogeographic maps of the southeastern Canadian Shield area showing ice cover, location of major end moraines, glacial lakes, and postglacial seas for (a) 12 ka, (b) 11 ka, (c) 10 ka, (d) 9 ka, (e) 8 ka, and (f) 7 ka.

Figures 3.55a to 3.55f present a speculative reconstruction for the deglaciation of northern Labrador and northeastern Quebec. Ice is shown as remaining for a long time east of Ungava Bay in such a way that the damming of glacial lakes Naskaupi and McLean and other lakes to the north can be accounted for. The time when the ice receded from the Saglek Moraines is unknown. Numerous radiocarbon age determinations on lake cores or bogs are reported by Short (1981). Many of the dates appear suspect because of anomalously old ages, the very low organic carbon content of many dated samples, age reversals in the cores, and the possibility of incorporation of old organic carbon. It seems preferable to rely solely on the few shell dates that give a minimum age for deglaciation of the coastal areas of about 9 ka. Inland, in the lower Rivière George basin, an age of 6815 ± 125 BP (SI-1959; Short, 1981) may provide a minimum age for deglaciation and drainage of glacial Lake Naskaupi.

Abandoned moraines and lichen-kill areas adjacent to present cirque glaciers, and glacier-free cirques provide a record of neoglacial expansion of glaciers in the Torngat Mountains (McCoy, 1983; Clark, 1984a; Evans, 1984, Evans and Rogerson, 1986).

Central and southern Labrador

Ice from the Labrador Sector advanced towards the coast of central and southern Labrador. As previously mentioned, summit areas in the Mealy Mountains and a large area on the distal side of the Paradise Moraine may not have been overridden by Laurentide Ice during the Late Wisconsinan. Based on lateral moraines sloping to the east at the limit of what appears to be different weathering zones, Late Wisconsinan ice may have reached elevations of only 710-555 m in the Mealy Mountains south of Lake Melville

(Gray, 1969), 500-300 m in the Mealy Mountains north of Sandwich Bay, and 275-122 m near the coast east southeast of Sandwich Bay (Rogerson, 1977).

Ice flow during deglaciation was generally northeastward in central Labrador, eastward in the Lake Melville area, and southeastward in southeastern Labrador. Flow was controlled to a great extent by the Mealy Mountains which parted the flow, and by Lake Melville where a calving bay may have existed (Fulton and Hodgson, 1979). Major end moraines were built during the retreat phase. If Laurentide Ice covered all of southeastern Labrador and joined with the Newfoundland Ice Cap, then the Bradore and Belles-Amours moraines (D.R. Grant, unpublished; Fig. 3.47) are likely Late Wisconsinan retreatal features. If, on the other hand, southeastern Labrador remained ice free, these moraines must be older and possibly date from a Middle or Early Wisconsinan stade, with the Paradise Moraine farther west representing the Late Wisconsinan stadial maximum. As shown in Figure 3.47 and 3.55a-f, many moraines were built during retreat of ice from the area. The longest of these is the Sebaskachu-Little Drunken Moraine System (Blake, 1956; Fulton and Hodgson, 1979), which apparently is part of the Quebec-North Shore System of Dubois and Dionne (1985). Later retreat towards central New Quebec was marked largely by construction of numerous eskers and fluted landforms (Fig. 3.39). Generally free drainage to the Atlantic prevented major glacial lake development. Rhythmites, observed in the Naskaupi River basin (Blake, 1956) and elsewhere (surficial geology maps of Fulton, 1986a, b and Fulton et al., 1979-1981), may be marine rhythmites developed in estuaries with low salinity.

Atlantic Ocean waters submerged glacial isostatically depressed coastal areas of central and southern Labrador during deglaciation. Marine limit has been traced by R.J. Fulton (Geological Survey of Canada, personal communication, 1984) and is shown on Figure 3.47. Along the coast, north of Straits of Belle Isle it may have reached 150 m (D.R. Grant, Geological Survey of Canada, personal communication, 1985), between Sandwich Bay and Lake Melville 113 m (Rogerson, 1977), and northeast of Lake Melville 80-85 m (Hodgson and Fulton, 1972). In the Lake Melville area marine limit increases in elevation inland from about 75 m on the outer coast to 150 m west of the lake (Fitzhugh, 1973).

Based on few radiocarbon dates, and on the location of major moraines, the paleogeographic maps (Figs. 3.55a-f) provide a simple reconstruction of deglaciation. On the basis of radiocarbon ages of shells, the area north of the Straits of Belle Isle was ice free by at least 10 900 ± 140 BP (GSC-2825), the coastal area east of Kanairiktok River by 10 275 ± 225 BP (GX-6345), central Lake Melville by 7970 ± 90 BP (TO-200), and the west end of Lake Melville by 7600 ± 100 BP (GSC-2970)(Table 3.6). On the basis of radiocarbon-dated lake cores, the area south of Alexis River was ice free by at least 10 400 ± 140 BP (GSC-3022). The age determinations of 9640 ± 170 BP (GSC-3067), 10 550 ± 290 BP (SI-3139), and 10 240 ± 1240 BP (SI-1737) on lake sediments provide minimum ages for the deglaciation of the upper St. Paul River area, the southern Mealy Mountains, and the eastern end of Lake Melville, respectively, and provide a minimum age for the construction of the Paradise Moraine. Farther inland a date of 6460 ± 200 BP (GSC-1592) on peat provides a minimum age for the deglaciation of upper Churchill River area and a 6500 ± 100 BP (GSC-3241) date on lake sediments, a minimum age for the deglaciation of the upper Harp River basin of central Labrador.

Quebec North Shore

The Quaternary history of the coastal fringe north of the Gulf of St. Lawrence between the mouth of Saguenay River and the Quebec/Labrador border is relatively well known but little data are available for the area farther inland.

During the Late Wisconsinan, south flowing ice of the Labrador Sector covered the Quebec North Shore and extended to its limit in the Gulf of St. Lawrence (see Grant, 1989), without apparently overrunning the eastern tip of Anticosti Island (Gratton et al, 1984). Ice retreated generally northwestward during deglaciation (Fig. 3.55a-f). The eastern extremity of the North Shore and the headland in the Baie-Trinité region were likely the first areas to become ice free. Moraines were built in the Baie-Trinité area between 13.5 and 9 ka (Dredge, 1976b, 1983b). The Baie-Trinité Moraines are correlated with the about 11 ka old Paradise and St-Narcisse moraines (Fig 3.52, 3.55b).

The marine submergence (Fig. 3.47) in the Gulf of St. Lawrence east of Québec City (Gadd, 1964, p. 1253) was named Goldthwait Sea by Elson (1969) and details on its development in Quebec can be found particularly in Dionne (1977), Hillaire-Marcel (1979), and Dubois (1980). Maps showing the extent of the sea can be found in Dubois et al. (1984). Marine limit varies from 150 m in the Quebec/

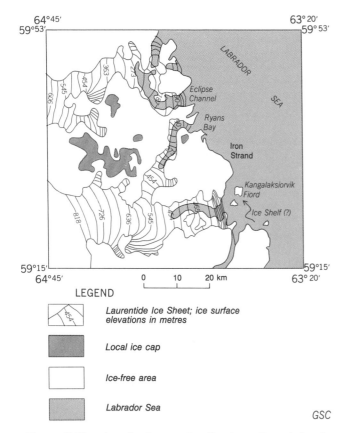

Figure 3.56. Ice sheet reconstruction in northern Labrador during the Late Wisconsinan (after Clark, 1984a).

Labrador border area (Boutray and Hillaire-Marcel, 1977) to 130 m and 122 m in the Rivière Mécatina and Rivière Natashquan areas, respectively (Dubois et al., 1984). Farther west in the Rivière Romaine to Rivière Moisie area it lies between 128-131 m (Dubois, 1977, 1980), in the Rivière Moisie to Baie Trinité between 100-130 m (Dredge, 1976a, 1983b), in the Rivière Manicouagan area between 138-145 m (J-M. Dubois, Université de Sherbrooke, personal communication, 1985), and finally near the mouth of Rivière Saguenay it is at about 167 m (J-M. Dubois, personal communication, 1985).

Radiocarbon dates on shells provide the best available minimum ages for deglaciation of the Quebec North Shore. In addition to the $10\,900 \pm 140$ BP (GSC-2825) date in Labrador near the Quebec border, the oldest reliable age determinations are from the area west of Rivière Romaine ($10\,230 \pm 180$ BP, Gif-3770), the Rivière Moisie area (9140 ± 200 BP, GSC-1337), and the Rivière Manicouagan area (9970 ± 130 BP, QU-574). Three radiocarbon dates 10.0 to 10.6 ka obtained by Tremblay in the Rivière Moisie area are erroneous since redated samples gave ages 1000 years younger (Dubois, 1980).

As ice retreated farther north onto the Canadian Shield, the Quebec North Shore Moraine (Dubois and Dionne, 1985), more than 800 km long, was built in areas generally lying between 200 and 400 m elevation (Fig. 3.47, 3.48). It extends from Rivière Manicouagan to south of Lake Melville in Labrador (Dubois, 1979, 1980; Fulton and Hodgson, 1979). The feature includes segments in the Rivière Manicouagan area (Sauvé and LaSalle, 1968), west of Rivière Moisie (Lac Daigle Moraine of Dredge, 1976b, 1983b), in the Rivière Moisie to Rivière Romaine area (the Manitou-Matamec Moraine of Dubois, 1976, 1977, 1979, 1980), and in the Rivière Natashquan to south of Lake Melville area (the Little Drunken Moraine of Fulton and Hodgson, 1979, and the Aguanus-Kenamiou Moraine of Dionne and Dubois, 1980). The moraine is considered by Dubois and Dionne (1985) to represent a halt of the ice sheet during a cooler climatic phase. Another possible explanation is that this is a re-equilibration moraine (Andrews, 1973; Hillaire-Marcel et al., 1981) which was built as the ice established a new equilibrium profile after retreating onto land from a major topographic depression (the Gulf of St. Lawrence) where it had retreated by calving. The age of the moraine is uncertain. The Goldthwait Sea was in contact with the moraine only locally in the Rivière Moisie and Rivière Manicouagan areas. On the basis of the paleogeographic setting of Goldthwait Sea radiocarbon-dated shells, Dubois (1980) and Dubois and Dionne (1985) suggested an age of 9.5 to 9.7 ka.

The ice margin retreated northward and northwestward from the Quebec North Shore Moraine towards central New Quebec leaving eskers and fluted landforms (Fig. 3.39). Rhythmites in river valleys were either deposited in lakes dammed by ice or ice contact deposits (Dubois, 1980) or more likely in low salinity estuaries of the Goldthwait Sea extending up the valleys. According to Dubois (1980), the Goldthwait Sea in the middle North Shore area attained its inland limit (128-131 m) at about 9.5 ka and had regressed to 106 m by 9.1 ka, to 75-76 m by 7.7 ka, to 45-46 m by 7.2 ka, and to 15 m by 5.2 ka. Inland the oldest minimum age for deglaciation, 6510 ± 110 BP (GSC-3615), was obtained from a lake core in the upper Rivière Moisie area (King, 1985).

Lac Saint-Jean area

Generally, south flowing ice of the Labrador Sector covered all the region and extended south of the Gulf of St. Lawrence. Tremblay (1971), Dionne (1973), and LaSalle and Tremblay (1978) have shown that the ice moved in a southerly direction except along Rivière Saguenay, where ice flow was generally southeasterly along the axis of the valley (Fig. 3.39). As the ice retreated from the uplands south of Lac Saint-Jean, small glacial lakes were dammed and De Geer moraines were built (LaSalle and Tremblay, 1978). A late ice lobe occupied the Lac Saint-Jean depression and ice contact materials were deposited at its receding margin (LaSalle et al., 1977a; LaSalle and Tremblay, 1978; Fig. 3.55c). Some of these deposits, such as those on the south side of Lac Saint-Jean, have been assigned to the Metabetchouane "halt" by LaSalle et al. (1977a) and should be considered as end moraines.

As the ice receded northwesterly up Saguenay Valley and into the Lac Saint-Jean basin, marine waters from the Gulf of St. Lawrence extended over lower lying, newly deglaciated land. This arm of the Goldthwait Sea was referred to as the Laflamme Sea by Laverdière and Mailloux (1956). Marine limit lies at 167 m (J-M. Dubois, Université de Sherbrooke, personal communication, 1985) at the mouth of Rivière Saguenay and at about 167 m and 198 m south and north of Lac Saint-Jean, respectively (LaSalle and Tremblay, 1978). The "oldest" minimum ages for marine invasion and deglaciation are $10\,400 \pm 150$ BP (I-5922) at the mouth of Rivière Saguenay and $10\,250 \pm 350$ BP (Gif-424) south of Lac Saint-Jean. Following deglaciation of the Lac Saint-Jean basin, the ice retreated northward leaving behind numerous eskers and fluted landforms (Fig. 3.39, 3.55d,e,f).

Western Laurentian Highlands

The western Laurentian Highlands are situated north of St. Lawrence River and Ottawa River between Rivière Saguenay and Lac Témiscamingue. Only events of the area situated north of the St. Narcisse Moraine (Fig. 3.47) are discussed here. The events that preceded and accompanied construction of the moraine are discussed by Parent and Occhietti (1988) and Occhietti (1989).

Ice flow during advance and retreat was controlled by topography with ice funnelled into the Rivière Saint-Maurice basin, and into the low general area centred on basins of Petite Nation, du Lièvre, Gatineau, and Coulonge rivers. The characteristic deglacial deposits are short segments of end moraines, isolated ice contact deposits, and outwash trains (Parry, 1963; Hardy, 1970; Denis, 1974; Lamothe, 1977; Pagé, 1977; Tremblay, 1977; Occhietti, 1980). In this moderate relief area higher summits first became ice free; at places this led to formation of small isolated glacial lakes perched on the flanks of higher hills, dammed by ice still occupying depressions. Such lakes are documented by Laverdière and Courtemanche (1960), Parry (1963), Lamothe (1977), and Pagé (1977). This style of deglaciation also led to development of typical kame and kettle topography in the Lac Maskinongé area where a remnant ice mass disintegrated (Denis, 1974). Fluted till terrains and extensive esker complexes are not as common as elsewhere on the southeastern Canadian Shield.

As ice retreated north of the St. Narcisse Moraine, the

Champlain Sea, which already flooded the glacial isostatically depressed St. Lawrence River basin, invaded the lower parts of the main valleys between about Rivière Sainte-Anne and Rivière Ouareau, in Rivière Petite Nation, and in the du Lièvre and Gatineau river basins (Fig. 3.47). The marine limit on the north shore of the Champlain Sea lies at about 192 m in the western end of the basin (R.J. Fulton, Geological Suvey of Canada, personal communication, 1984), 200 m in Rivière Petite Nation valley (Richard, 1980b), 235 m north of Montréal (Lamothe, 1977; Pagé, 1977), and 200 m in the Saint-Maurice valley (S. Occhietti, Université du Québec à Montréal, personal communication, 1984). Waterlain deposits of problematic (marine, estuarine, or freshwater) origin are found in the upper parts of Saint-Maurice valley (Occhietti, 1980), as well as in Gatineau and du Lièvre valleys. Wilson (1924) initially interpreted waterlain sediments in the latter valleys as Champlain Sea sediments. Gadd (1972), mainly because of the presence of rhythmites, proposed that the valleys were occupied by a glacial lake. Dadswell (1974) used the distribution of modern day crustaceans to define a former water body connected to the Champlain Sea which extended to north of Lac Baskatong. There is no sill or morainic dam that might have produced a lake of this extent but northward tilting of 0.4 m/km would be sufficient to extend the Champlain Sea from its limit of 210 m near Ottawa to the 274 m upper limit of the water body north of Lac Baskatong (Fig. 3.47). It is therefore likely that the rhythmites were in fact deposited in a marine estuary rather than a glacial lake.

On the basis of radiocarbon ages on Champlain Sea and Goldthwait Sea shells the southern periphery of the Laurentian Highlands was ice free by at least 12 400 ± 160 BP (GSC-1533) in the Québec City area, and possibly as early as 12 200 ± 160 BP (GSC-1646) or 11 900 ± 160 BP (GSC-1772) in lower Rivière Gatineau valley (Romanelli, 1975; Parent and Occhietti, 1988; Occhietti, 1989.) As discussed by Occhietti (1989), the timing of the initiation of Champlain Sea is controversial but it probably occurred about 12 ka. Figure 3.55a shows the paleogeography just before the ice dam near Québec City was breached. Several Goldthwait Sea shell samples have given radiocarbon ages that vary between 11 ka and 12 ka in the area east of Québec City. Likewise, several Champlain Sea shell samples have been dated in that age range in the area between Montréal and the western end of the basin. The oldest shell date between Montréal and Québec City however is 10 300 ± 100 BP (GSC-2101). This unusual distribution of dates has led Hillaire-Marcel (1981) to question the validity of the old dates near Ottawa and has, in part, prompted Gadd (1980) to propose a calving bay mechanism for deglaciation of the western basin of the Champlain Sea. The deglacial chronology of the adjacent Lac Saint-Jean and upper Ottawa River areas indicates that the Laurentian Highlands were completely ice free by 10 ka.

Upper Ottawa River basin

In the Late Wisconsinan, ice from the Labrador Sector flowed southward to southwestward across the upper basin of Ottawa River (Veillette, 1983a, 1986a). Ice flow during deglaciation was complex because it involved the uncoupling of two major ice masses along the Harricana Interlobate Moraine (Fig. 3.47). South of the Hudson Bay-St. Lawrence divide, the moraine has been mapped and identified mostly on the basis of ice flow indicators by Veillette (1983b, 1986a), who argued that the McConnell Lake Moraine (Boissonneau, 1968) southwest of Lac Témiscamingue, and the Boulter "esker" and other extensive glaciofluvial deposits (Chapman, 1975) from south of North Bay to the vicinity of Lake Simcoe constitute an extension of the Harricana Moraine. Vincent and Hardy (1979) inferred a similar extension of the interlobate zone to North Bay based on regional ice flow features.

Ice retreated initially in a north-northeasterly direction but in the northwestern part of the area retreat was to the northwest. The region northeast of Témiscaming and northwest towards Lac Témiscamingue became ice free first (Veillette, 1983b, 1986a, 1988). Later a lobe of ice occupied the northern Lac Témiscamingue trough. The Laverlochère Moraine (Veillette, 1983a,b, 1986a, 1988) was built on the borders of this lobe (Fig. 3.47). West of the Harricana Interlobate Moraine and north of Lac Témiscamingue, but south of the Hudson Bay drainage divide, the Roulier Moraine (Fig. 3.47; Vincent and Hardy, 1977, 1979) was formed as a result of a halt or as a re-equilibration moraine formed when ice halted on the edge of higher ground near the height of land after retreating in a glacial lake basin (Hillaire-Marcel et al., 1981).

Glacial lakes abutted the ice margin in low-lying western and northern areas (Vincent and Hardy, 1977, 1979; Veillette, 1983b, 1988). The earliest glacial lake phase is thought to be related to the northeast extension of the Post-Algonquin glacial lake from the Great Lakes basin (Harrison, 1972). Possibly during the Sheguiandah and certainly during the Korah Phase, the lake was still dammed by ice blocking drainage down Ottawa River in the Mattawa area, and extended northeast of Témiscaming along a re-entrant in the ice front along the Harricana Interlobate Moraine and on the west side of Ottawa River to Lac Témiscamingue (Vincent and Hardy, 1977, 1979; Veillette, 1988; Fig. 3.57a). When ice withdrew from Ottawa River valley, east of Mattawa, water levels dropped, and glacial Lake Barlow (Wilson, 1918) occupied Lac Témiscamingue basin. Glacial Lake Barlow was initially controlled first by a morainic dam at Deux-Rivières (the Aylen Phase) and later, as differential uplift occurred, by a rock and morainic sill at Témiscaming (the Témiscaming Phase, Fig. 3.57b; Vincent and Hardy, 1979). Higher elevations of maximum water planes west of the Harricana Interlobate Moraine indicate that the area west of the moraine became ice free first. Maximum glacial Lake Barlow levels in the area east of Lac Témiscamingue were at about 300 m and rise to the northeast to about 380 m in the vicinity of the present Hudson Bay watershed where the Harricana Moraine crosses it (Veillette, 1988). The northeastern tilt of glacial Lake Barlow water planes suggests thicker ice in New Quebec rather than in Hudson Bay (Hillaire-Marcel et al., 1980).

Numerous minimum ages for deglaciation are available from dates on lake cores. Age determinations of 10 400 ± 200 BP (GSC-3460) and 10 100 ± 180 BP (GSC-3467) in the area east of Témiscaming provide the best estimate for the time of deglaciation in the Harricana Interlobate Moraine re-entrant (Veillette, 1988).

Quebec Clay Belt/James Bay area

During deglaciation, southwest flowing Labrador Ice and southeast flowing Hudson Ice separated along the Harricana Interlobate Moraine. The moraine, named by Hardy (1976; Fig. 3.47) and originally recognized by Low (1888) as an end

QUATERNARY GEOLOGY — CANADIAN SHIELD

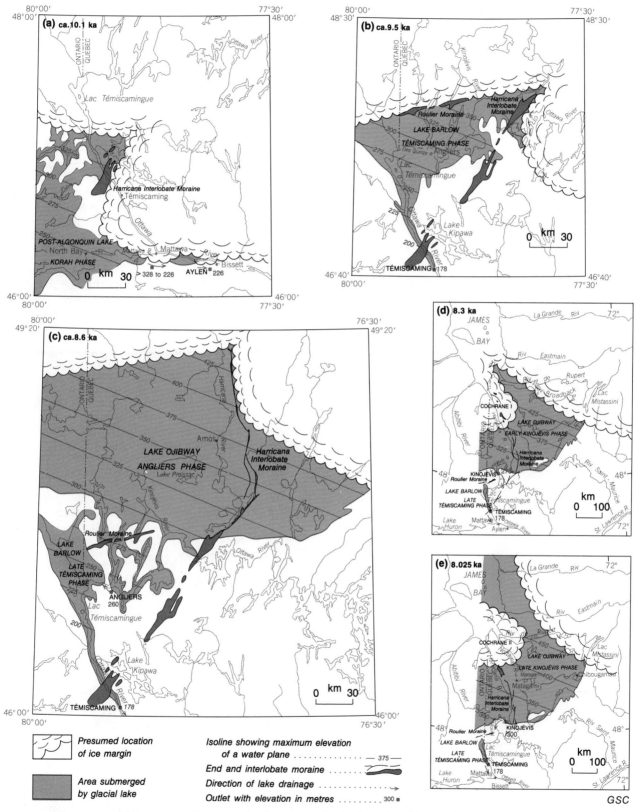

Figure 3.57. Paleogeographic maps showing successive ice frontal positions and extent of water bodies during different phases of the Post-Algonquin, Barlow, and Ojibway glacial lakes in the western part of the southeastern Canadian Shield (after Vincent and Hardy, 1979).

moraine and by Wilson (1938) as "a moraine between two ice-sheets" was studied by L.P. Tremblay (1950), Allard (1974), G. Tremblay (1974), Hardy (1976, 1977, 1982b), and Veillette (1982, 1983a,b, 1986a, 1988). The latter two authors carefully documented the two converging ice flows.

East of the Harricana Interlobate Moraine ice flow during deglaciation was towards the south-southwest near the continental divide and it swung progressively to the west until, in the area north of La Grande Rivière, it was distinctly westerly (Fig. 3.39; Hardy, 1976). In the vicinity of the interlobate moraine, the ice flow was everywhere deflected to the south, and in many areas ice flowed parallel to the moraine axis. In the La Grande Rivière area, superposed drumlins and striae (Lee et al., 1960) show that Labrador Ice was free to move in a more westerly direction following separation of the ice masses (Vincent, 1977). West of the Harricana Interlobate Moraine, in Abitibi and Lac Témiscamingue region, Veillette (1986a) has carefully documented late glacial southeasterly flow.

As Labrador Ice and Hudson Ice retreated, glacial Lake Ojibway (Coleman, 1909) was dammed between the ice front and the continental divide to the south (Fig. 3.47). The extent, as well as the history, of this lake in Quebec is discussed in detail by Vincent and Hardy (1977, 1979). The maximum lake limit rises northward from about 380 m at the divide to more than 450 m in the interfluve area between Rivière de Rupert and Rivière Broadback. The maximum depth of Lake Ojibway was more than 500 m on the east coast of James Bay. Lake waters flooded the lowlands east of James Bay as far as the Sakami Moraine (Hardy, 1976) and as far north as just east of Kuujjuarapik (Hillaire-Marcel, 1976; Fig. 3.57e). Glacial Lake Obijway became separated from glacial Lake Barlow with the emergence of the Angliers sill east of the head of Lac Témiscamingue (Vincent and Hardy, 1977, 1979; Angliers Phase, Fig. 3.57c). This sill was then the lowest point on the Hudson Bay-St. Lawrence drainage divide, which was displaced south by isostatic tilting of the crust. With differential uplift, the sill of the outlet migrated north along Rivière Kinojévis (Early Kinojévis Phase, Fig. 3.57d) until it reached the most northerly lowest point on the present divide between the upper Kinojévis and upper Harricana drainage basins (Late Kinojévis Phase, Fig. 3.57e). Large channels on the divide and wide and deeply incised Rivière Kinojévis valley were cut by overflowing glacial lake waters.

Extensive fields of De Geer moraines were built in glacial Lake Ojibway (Mawdsley, 1936; Norman, 1938; Shaw, 1944; Ignatius, 1958). Norman (1938) calculated that the rate of ice front retreat was 173-239 m/a.

A varve chronology has been established for lakes Barlow and Ojibway. Antevs (1925) counted 2027 varves beginning with varve 1 in a river cut southwest of Lac Témiscamingue. Hughes (1965) confirmed the upper part of Antevs varve diagrams, in the Clay Belt of Ontario, and added 58 varves to the sequence. Hardy (1976) correlated some 625 varves in the James Bay Lowlands of Quebec with Hughes' sequence and added 25 varves which register the most northerly and northeasterly extent of Lake Ojibway. Hardy's varve sequence extends to the time of final drainage of the lake into Tyrrell Sea. This varve chronology indicates that a total of 2110 years elapsed from the time varve 1 was laid down southwest of Lac Témiscamingue to the time Lake Ojibway drained. Using this, Hardy (1976) calculated that the rate of ice retreat increased from about 320 m/a southeast of James Bay to possibly 900 m/a in the La Grande Rivière area in the deep lake basin just before it drained.

As in Ontario (Prest, 1970; this chapter, Dredge and Cowan, 1989), late glacial ice surges occurred into glacial Lake Objiway. Hardy (1976) has documented three "Cochrane" surges in the lowlands southeast of James Bay. His reconstruction is based on: (1) mapping the extent of the clayey and carbonate-rich Cochrane till; (2) measuring ice flow features related to each movement; (3) logging sections in which Cochrane tills are interstratified with Lake Ojibway sediments, and (4) studying the effect of the Cochrane surges on the sedimentology of the varves. His work enabled him to reconstruct ice profiles by showing precisely where the ice was grounded or floating in the lake basin. The extent of ice during Cochrane I surge is shown in Figure 3.57d and during Cochrane II in Figure 3.57e. A third surge, intermediate between the Cochrane I and II was called the Rupert surge. These originated to the northwest and, hence, came from Hudson Ice. The Cochrane I of Hardy (1976) in Quebec is probably somewhat younger than the Cochrane I of Prest (1970) in Ontario. During Cochrane I and II surges, Lake Ojibway varves became coarser and thicker (effect of the ice advance) and richer in carbonate (effect of overriding Paleozoic beds south and southeast of James Bay) (Hardy, 1976). The maximum Cochrane I and II surges occurred 300 years (varve year 1810 and correlative with Hughes' (1965) Frederick House varve sequence) and 75 years (varve year 2035 and correlative with Hughes' (1965) Connaught varve sequence) before glacial Lake Ojibway drained. Changes in sedimentology of the varves can be attributed entirely to the effect of the surges. Hence there is no need to invoke fluctuating lake levels, as suggested by Hughes (1965) or different lake phases such as the Antevs or Opemiska lakes suggested by Prest (1970).

When Labrador Ice had retreated to the approximate position of the Sakami Moraine, waters from Hudson Strait penetrated Hudson and James bays and flooded the isostatically depressed lowlands. Opening to the sea allowed drainage of glacial Lake Ojibway and led to formation of the Sakami Moraine (Hardy, 1976). The moraine is a major feature extending over a distance of 630 km from the vicinity of Kuujjuarapik on the southeastern coast of Hudson Bay to southern Lac Mistassini (Fig. 3.47). Hardy (1976) has shown that the Sakami Moraine marks the position of Labrador Ice at the time glacial Lake Ojibway drained and the postglacial sea, called the Tyrrell Sea by Lee (1960), submerged coastal areas. Hillaire-Marcel et al. (1981) considered the Sakami Moraine as a re-equilibration moraine, that is, a moraine that results from the stabilization of the ice front when the glacier, which was previously calving and in part floating in a deep water basin, grounded. Ice did not resume its retreat until a new equilibrium profile had been established. Locally the ice may have readvanced. Drainage of glacial Lake Ojibway led to a slight readvance in the Lac Mistassini area (Bouchard, 1980), previously recognized as the Waconichi ice advance by DiLabio (1981). Drainage of glacial Lake Ojibway and submergence by the Tyrrell Sea was likely a catastrophic event; it is recorded in sections where post-Cochrane II Lake Ojibway varves are overlain by a thin diamicton with carbonate debris, marking the drainage horizon, and by fossiliferous marine clays (Hardy, 1976). The sequence is similar to that described by Skinner (1973) in the Hudson Bay Lowlands of Ontario. He interpreted the diamicton as resulting from a sudden widespread density flow. Higher hills in the Lake Ojibway basin also record the seemingly sudden drainage; on these hills both upper and lower wave washing limits exist (Norman, 1939; Hardy, 1976). Below the lower limit of washing, no strandlines are present because the lake level fell too rapidly for strandlines to form. The lower washing limit represents a long synchronous shoreline that is now tilted, and it offers an excellent opportunity for measuring the amount of delevelling accomplished since it formed.

The Tyrrell Sea followed the retreating ice front after construction of the Sakami Moraine. Marine limit is at about 198 m in the southern part of its basin (Hardy, 1976) and rises northward to the Kuujjuarapik area where it reaches elevations of 315 m (Hillaire-Marcel, 1976). In La Grande Rivière area marine limit decreases eastward from close to 270 m in the area just east of the Sakami Moraine to 246 m farther up river (Vincent, 1977). Extensive swarms of De Geer moraines, many of which overlie drumlins and eskers, were built east of the Sakami Moraine (Fig. 3.46). Rates of ice retreat averaged 217 m/a (Vincent, 1977) . In the Lac Mistassini area, a relatively shallow glacial lake, named glacial Lake Mattawaskin by Bouchard (1986) followed the retreating ice front (Fig. 3.55e). Discharge of lake waters was first via Rivière Broadback, and then via Rivière de Rupert and Rivière Eastmain as lower outlets became free of ice (Vincent and Hardy, 1977, 1979; Bouchard, 1980, 1986). Rates of ice recession in this lake basin were estimated at 220 to 260 m/a (Bouchard, 1980).

The late Quaternary chronological framework for the Quebec Clay Belt and the area southeast and east of James Bay rests on the age assigned the Sakami Moraine. Shells collected in clay lying on the proximal flank of the Sakami Moraine date 7880 ± 160 BP (QU-122). This date led Hardy (1976) to assign an age of 7.9 ka, or slightly older, to the moraine. Hillaire-Marcel (1976) assigned a 8.1-8.0 ka maximum age to the moraine on the basis of a 8230 ± 135 BP (I-8363) age (corrected to 8.1-8.0 ka to account for isotopic fractionation) on concretions in glacial Lake Ojibway clays closely associated with formation of the moraine near Kuujjuarapik. The approximate age of ca. 8.0 ka for the Sakami Moraine (therefore also for the drainage of glacial Lake Ojibway and incursion of the Tyrrell Sea) agrees well with the age proposed for the drainage of glacial lakes Agassiz and Ojibway and for the incursion of Tyrrell Sea on the west side of Hudson Bay (see this chapter, Dyke and Dredge, 1989; Dredge and Cowan, 1989). Assuming an 8.0 ka age for the Sakami Moraine, and using the varve chronology, the area west of Lac Témiscamingue was deglaciated some 10 ka ago and the height of land at the Quebec-Ontario border, 9.2 ka ago; the Cochrane I reached its maximum 8.3 ka ago and Cochrane II 8025BP. East of the Sakami Moraine, based on the De Geer moraine chronology (Vincent, 1977), the Tyrrell Sea reached its eastern limit in the La Grande Rivière area about 7.5 ka ago. Farther inland the oldest radiocarbon age of 6600 ± 100 BP (B-9516) was obtained by P.J.H. Richard (Université de Montréal, personal communication, 1984) on basal gyttja from a lake in the Rivière Laforge area.

Area east of Hudson Bay

Ice from central New Quebec and central Ungava Peninsula flowed generally westward and southwestward into Hudson Bay (Hillaire-Marcel, 1979; Gray and Lauriol, 1985). This ice flow direction is also recorded on the Belcher Islands (Jackson, 1960). On the Ottawa Islands the direction of last ice movement was west-southwest, but Andrews and Falconer (1969) have documented an earlier northeasterly flow, associated by Dyke et al. (1983) with Hudson Ice.

When Labrador Ice finally separated from Hudson Ice in Hudson Bay, only the area west of the Sakami Moraine in the Kuujjuarapik area, and perhaps the extreme northwestern part of Ungava Peninsula, was ice free (Fig. 3.55e). On the basis of the progressively shifting ice flow directions recorded by striae on the Ottawa Islands , the drainage corridor is inferred to have been located west of the islands (Andrews and Falconer, 1969).

As the ice margin retreated to the east, the Tyrrell Sea covered the newly deglaciated areas and swarms of De Geer moraines were built at the ice front in a belt extending northward from Inukjuak to south of Ivujivik. Marine limit decreases northward from about 315 m on the southeastern coast of Hudson Bay to possibly as low as 105 m east of Povungnituk. From there it rises northerly to about 170 m near Hudson Strait (Gray and Lauriol, 1985). As in the area east of James Bay, marine limit also declines inland; it falls from 248 m to 196 m along Rivière Nastapoca (Allard and Seguin, 1985), and from 158 m on the Ottawa Islands (Andrews and Falconer, 1969) to 105 m east of Povungnituk (Gray and Lauriol, 1985) (Fig. 3.47). Rates of uplift, as measured in the Lac Guillaume-Delisle (Richmond Gulf) area, were 9.6-10 m/century at the time of deglaciation (Allard and Seguin, 1985). According to Hillaire-Marcel (1976), by 7 ka the rate of uplift was 6.5 m/century and it decreased linearly to the present rates estimated to be of the order of 1.1 m/century. Fairbridge and Hillaire-Marcel (1977) and Hillaire-Marcel and Fairbridge (1978) recognized a 45 year cycle in beach building in the Lac Guillaume-Delisle area that they related to the "double Hale" solar cycle; thus they deduced a cyclic record of storminess.

East of marine limit, ice continued its retreat towards central Ungava Peninsula in the north or towards central New Quebec farther south. Beyond the Tyrrell Sea limit, a shallow glacial lake was formed in the Lac à l'Eau Claire area between an uplifted sill west of the lake and the ice front (Allard and Seguin, 1985; Fig. 3.47, 3.55f). At the head of Rivière aux Mélèzes, glacial Lake Minto formed between the Hudson Bay-Ungava Bay drainage divide and the receding ice front (Lauriol, 1982; Lauriol and Gray, 1983; Fig. 3.47, 3.55f).

The time of deglaciation for the area west of the Sakami Moraine, in the Kuujjuarapik area, is estimated at 8.1 ka by Hillaire-Marcel (1976). In the Rivière Nastapoca area, the oldest dated shells, 45 m below marine limit, were 6700 ± 100 years old (UQ-547). Shells 17 m below marine limit on the Ottawa Islands (Andrews and Falconer, 1969) were dated at 7430 ± 180 BP (GSC-706), indicating rather late deglaciation of western Ungava. Marine shells from the Cape Smith area originally provided radiocarbon ages of ca. 8 ka but were later redated at ca. 6.8 BP (Lauriol and Gray, 1987).

Ungava Peninsula

Both northerly flow into southern Ungava Bay, from central New Quebec, and flow towards Hudson Bay, western Hudson Strait, and western Ungava Bay, from a central north-south ice divide on Ungava Peninsula are recorded

(Gray and Lauriol, 1985; Bouchard and Marcotte, 1986). Ice flowing westward and northward from this ice flow centre, called the Payne center by Bouchard and Marcotte (1986), apparently coalesced with ice moving northeastward in Hudson Bay and eastward in Hudson Strait (flow in offshore area from Andrews and Falconer, 1969; Shilts, 1980; Laymon, 1984; Gray and Lauriol, 1985).

The northwestern and northern extremities of the area were deglaciated first (Fig. 3.55e). Marine waters submerged the Ungava coast along Hudson Strait where marine submergence generally decreases from west to east (167 m at Cape Wolstenholme, Matthews, 1967; 138 m near Diana Bay, Gray et al., 1980), and from north to south (170 m on Charles Island and 120 m at the head of Deception Bay, Gray and Lauriol, 1985). Rates of uplift at about 8 ka were estimated at 7.9 m/century (Matthews, 1967). The age of deglaciation of the south shore of Hudson Strait has been the subject of some controversy. Most researchers would agree that some areas were ice free at least 7970 ± 250 BP (GSC-672), but some, on the basis of three "older" radiocarbon age determinations, postulate a much earlier deglaciation. A $10 450 \pm 250$ BP (I-488; Matthews, 1966, 1967) age may in fact be erroneous since shells collected by B. Lauriol (University of Ottawa, personal communication, 1985) from the immediate vicinity of the Deception Bay site originally sampled by Matthews were dated at 7130 ± 100 BP (GSC-3947). Notwithstanding this, two other age determinations, the oldest of which is 9800 ± 220 BP (Beta-11121), have been obtained from the Deception Bay area by dating in situ *Portlandia arctica* and *Nuculana minuta* shells collected in glacial marine sediments overlying till (Gray and Lauriol, 1985; Lauriol and Gray, 1987). This conflicts with evidence on Meta Incognita Peninsula of Baffin Island which requires that ice extend to the mouth of Hudson Strait until 8.6 ka or later (this chapter, Andrews, 1989). Miller et al. (1988) have suggested that a late readvance of Labrador Ice across Hudson Strait, between 9 ka and 8.2 ka, could account for the late presence of ice on southern Baffin Island. Upon further retreat of the ice front towards the interior of Ungava Peninsula, glacial lakes were dammed between the ice front and higher ground on the Hudson Bay-Hudson Strait and Hudson Bay-Ungava Bay drainage divides (Fig. 3.47; Prest et al., 1968; Prest, 1970). Of these lakes, the best documented is glacial Lake Nantais (Lauriol and Gray, 1987; Fig. 3.47).

Ice filling Ungava Bay retreated both westward towards central Ungava Peninsula and south towards central New Quebec perhaps through the development of a calving bay. Eastern Hudson Strait was deglaciated by 9.1 ka on the basis of a radiocarbon date on shells collected from a seabed core (9120 ± 480 BP, GSC-2946). The D'Iberville Sea (Laverdière and Bernard, 1969) followed the retreating ice front. Blake (1976), Gangloff et al. (1976), Lauriol et al. (1979), Gray et al. (1980), Lauriol (1982), and Gray and Lauriol (1985) provide detailed accounts of sea level history in this sector. On the west coast of Ungava Bay marine limit rises from 138 m near Diana Bay to 195 m in upper Rivière aux Mélèzes drainage basin (Gray and Lauriol, 1985). On Akpatok Island, 75 km offshore, marine limit is much lower (58-74 m; Løken, 1978). Several radiocarbon dates on shells indicate that the west coast of Ungava Bay was ice free by 7 ka (e.g., 6990 ± 150 BP, I-9632; Fig. 3.55e). Standing water bodies extending far inland in the lower parts of the valleys of Rivière aux Mélèzes (Gray and Lauriol, 1985) and Rivière Caniapiscau (Drummond, 1965), thought to be glacial lakes, could well have been brackish estuaries of the D'Iberville Sea.

Central New Quebec and western Labrador

Since Low (1896) first recognized central Labrador-Ungava Peninsula as one of the final centres of ice disintegration, much controversy has surrounded the identification of the exact locations of the last glacial masses. As portrayed in Wilson et al. (1958), Ives (1960a), Prest et al. (1968), and Prest (1969), ice flow indicators provide evidence that Labrador Ice flowed radially from the horseshoe-shaped ice divide extending from northwest of Lac Delorme in the west, to northern Smallwood Reservoir in the east. Whether the ice divide had been stable for a considerable part of the Late Wisconsinan, or whether its location fluctuated considerably, is not definitely established. For example, Hughes (1964) and Richard et al. (1982) suggested that the ice divide must at one time have been situated well to the northeast of its final location because rocks from the Labrador Trough were transported to areas west of the ice divide. Equally, whether or not the final location of the divide coincides with the location of the last ice remnants is not clearly established. Recent fieldwork by Klassen and Thompson (1987), who have recognized up to five distinct phases of flow in the area, should help clarify these points.

Field investigations by Perrault (1955), Grayson (1956), Henderson (1959), Ives (1959, 1960a, c, 1968, 1979), Kirby (1961a, b), Derbyshire (1962a), in the Schefferville area of the Labrador Trough, brought conclusive evidence for the presence of small ice remnants in the low-lying basins of Howells River (Kivivic ice divide; just west of Schefferville), and Swampy Bay River valleys (north-northwest of Schefferville). Other authors, basing their arguments on glacial ice flow indicators and landforms (Low, 1896; Hughes, 1964; Laverdière, 1967; Richard et al., 1982) and on the intersection of projected strandline tilt directions of glacial lakes as an indicator of the location of the maximum ice thickness (Ives, 1960b; Harrison, 1963; Barnett, 1964; Barnett and Peterson, 1964), proposed that the final ice masses disintegrated in the general area of the "horseshoe-shaped" ice divide. Much controversy has ensued between the different proponents (Ives, 1968; Barnett and Peterson, 1968; Bryson et al., 1969; Laverdière, 1969a,b; Laverdière and Guimont, 1982), but it is likely that there were numerous retreat centres both along the final location of the divide and in adjacent low-lying basins where discrete ice masses finally melted.

Dates from numerous lake cores have been used to date the final disappearance of ice (Grayson, 1956; Morisson, 1970; McAndrews and Samson, 1977; Short, 1981; Stravers, 1981; Richard et al., 1982; King, 1985). On or near the final location of the ice divide, basal dates of 6320 ± 180 BP (GSC-3094; Richard et al., 1982) in the Lac Delorme area, and 6200 ± 100 BP (GSC-3644; King, 1985) in the Lac Stakel area provide the best minimal estimates. In the Schefferville area of the Labrador Trough numerous dates have been obtained (Short, 1981; Stravers, 1981), including some as old as 16 ka; however, these dates are suspect because of the possible presence of redeposited older organic matter in samples containing less than 1% organic carbon (Stravers, 1981). Apart perhaps from small remnant ice masses in depressions, it is probably safe to assume that Labrador Ice had completely melted by 6.5 ka.

Postglacial history
Postglacial emergence

Much data on sea level history have been presented in the sections dealing with the deglaciation of each major region. Minimum sea level curves have been constructed for six areas (Fig. 3.58) and are discussed here. Although a considerable number of shell samples has been dated, few can be tied to specific sea level positions. Also, in most cases, only part of an emergence curve can be plotted since no material has been found to date both the lower and the upper parts of curves. The curves are nevertheless useful since they provide a value for the minimal sea level which could have existed at any given time in different areas.

The family of curves in Figure 3.58 illustrates the relative isostatic response of different areas to the ice load. As would be expected, the middle Quebec North Shore and Deception Bay curves, farthest away from the centres of loading, exhibit the least glacial isostatic recovery, whereas those which are closest (La Grande Rivière and Lac Guillaume-Delisle) exhibit most. The intersection of the La Grande Rivière and Lac Guillaume-Delisle curves at about 6 ka may indicate the switch from the initial dominant influence of the Hudson Ice load to a later dominant influence of the Labrador Ice load as proposed by Hillaire-Marcel (1980).

The shape of the La Grande Rivière curve and its extrapolation to marine limit from two dates on sediments associated with the construction of the Sakami Moraine interestingly support the age estimate, of about 8 ka, for the drainage of glacial Lake Ojibway and incursion of Tyrrell Sea east of James Bay.

Late Wisconsinan and Holocene vegetational history
P.J.H. Richard

Thanks to data from pollen analyses, the vegetation landscapes are known to have been diversified at the margin of the retreating Laurentide Ice Sheet, but in the central part of the southeastern Canadian Shield, since about 5 ka, as well as in peripheral areas, since about 8 ka, changes have been generally minor. Detailed accounts of vegetational history and lists of the numerous available studies are given in Richard (1977a, 1981, 1985, 1989).

At about 11 ka in the south (Richard, 1977b; Savoie and Richard, 1979), and at about 7.5 ka in Ungava Peninsula (Richard, 1981), nearly unvegetated and then tundra environments existed. In Labrador, shrub-tundra occupied the newly deglaciated terrain until about 8 ka in the south, and 5 ka in the north-central part of the area (Short, 1978; Lamb, 1980, 1984). Elsewhere, in the central portion of the southeastern Canadian Shield, trees colonized the land soon after ice retreat, the drainage of glacial lakes (Richard, 1980a), or emergence from postglacial seas (Richard, 1979). *Larix, Populus, Picea,* and *Betula,* accompanied by *Alnus,* were the first colonizers. In southern Labrador and on the

Richard, P.J.H.
1989: Late Wisconsinan and Holocene vegetational history; in Chapter 3 of Quaternary Geology of Canada and Greenland, R.J. Fulton (ed.); Geological Survey of Canada, Geology of Canada, no. 1 (also Geological Society of America, The Geology of North America, v. K-1)

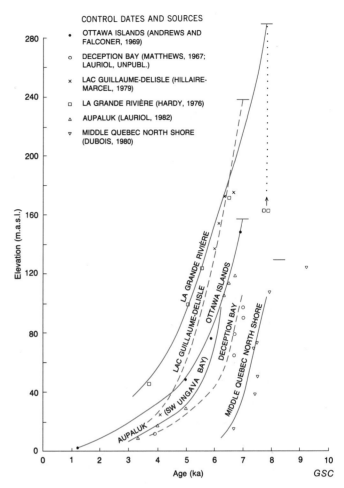

Figure 3.58. Minimum emergence curves for six areas of the southeastern Canadian Shield.

Quebec North Shore, *Picea glauca* and *Abies balsamea* were the first trees present (Lamb, 1980, 1984). In the Laurentian Highlands tree colonization of the higher areas was slow; *Populus* played a dominant role there for more than a millennium (Richard, 1977b).

During the early Holocene, a complex history occurred on the southern margin of the Canadian Shield. Large fluctuations in the abundance of *Pinus divaricata, Pinus strobus,* and *Tsuga canadensis* on one hand, and the northward migration of *Acer saccharum, Fagus grandiflora,* and other thermophilous species on the other hand, characterize this period. The presence of spruce, then fir, led the way to the present day mixed forests. *Pinus strobus* was 50 km farther north than today about 5 ka ago in the middle Holocene (Terasmae and Anderson, 1970). In New Quebec, *Pinus divaricata* migrated into the spruce dominated area east of James Bay about 3 ka (Richard, 1979) barely reaching Rivière Caniapiscau to the east (Richard et al., 1982). In the north, however, the position of treeline has fluctuated little since deglaciation (Gagnon and Payette, 1981; Richard, 1981). In the last 1000 years the boreal forest has progressively opened up, possibly in response to climatic cooling in the late Holocene (Short, 1978).

CHAPTER 3

QUATERNARY GEOLOGY OF THE NORTHEASTERN CANADIAN SHIELD

J.T. Andrews

INTRODUCTION

This section provides a description of the Quaternary geology of the part of the Canadian Shield lying between Baffin Bay and Melville Peninsula, and between Hudson Strait and Lancaster Sound (Fig. 3.1). The region lies in part within the Baffin Sector of the Laurentide Ice Sheet (Fig. 3) and was covered by Foxe Ice and a series of small independent ice caps during the last glaciation. Because the region lies at and extends beyond the northeastern limit of the last glaciation, abundant older Quaternary deposits are present. As a consequence, development of methods for recognizing, subdividing, and dating Quaternary sediments that predate the last ice advance has been a major facet of Quaternary studies in this region. Marine deposits have figured prominently in chronostratigraphic studies and there has been considerable research on Quaternary land and sea movements. Finally, because the area includes alpine glaciers and ice caps and parts of the area lie near the present limit of glacierization, neoglacial glacier activity has been an additional subject of research.

Bedrock geology and physiography

The bedrock geology of the region is critical to many facets of the Quaternary geology. Figure 3.3 illustrates the major bedrock units exposed on land and under adjacent marine basins and shelves. The physiography of the region clearly is associated in part with the broad structural framework provided by a variety of Precambrian shield rocks forming a structural high and ringing a cratonic basin (Foxe Basin) delimited by the outcrop of younger sedimentary rocks.

Much of Baffin Island is composed of a variety of granites, granite gneisses, migmatites, gneisses, and schists. Ages range from Archean (>2500 Ma) to Aphebian (~1600 Ma). Northern Baffin Island is geologically the most complex part of the region — Tertiary/Cretaceous rocks outcrop on southern Bylot Island and on the floor of Eclipse Sound. Below this in the sequence is a series of Paleozoic rocks, predominantly limestones and dolostones. Proterozoic volcanic and metasedimentary rocks also outcrop.

Central Baffin Island consists nearly entirely of undifferentiated Archean granite gneiss. Around the eastern shore of Foxe Basin the Paleozoic limestone, which underlies this feature, is locally preserved on land (Fig. 3.3). A thin bed of early Tertiary sediment has been located on Precambrian granite gneiss just north of the Barnes Ice Cap (Central Baffin Island, Andrews et al., 1972).

Southern Baffin Island is dominated by Precambrian shield rocks. A major feature of the bedrock geology is the tongue of Paleozoic limestone that extends from Foxe Basin eastward towards Frobisher Bay (Fig. 3.3). Recent geophysical and geological surveys in Frobisher Bay (MacLean, 1985) have shown that the Paleozoic rocks occur beneath the outer part of Frobisher Bay. North of Frobisher Bay, Paleozoic limestone underlies part of the continental shelf off Hall and Cumberland peninsulas and extends into Cumberland Sound where the limestone is at least partly overlain by sediments of Cretaceous/Tertiary age (MacLean and Williams, 1983).

Seaward from the east coast of Baffin Island, Precambrian and Paleozoic rocks are onlapped by Cretaceous and Tertiary marine sediments that form a thick accretionary wedge along the length of the Baffin Island continental shelf. Quaternary sediments on the shelf are relatively thin and range between 0 and 500 m thick (Grant, 1975; MacLean, 1985).

The geology of Hudson Strait is complicated by a series of major faults. Paleozoic limestone is present beneath the floor of the Strait and extends southward into Ungava Bay where it forms the surface rock in Atpatok Island.

Melville Peninsula stands as a structural high or arch and divides Paleozoic rocks in Foxe Basin from similar sediments around the Gulf of Boothia.

There is a broad relationship between the bedrock geology and the physiographic expression with younger sedimentary rocks underlying coastal plains and sounds and old resistant igneous and metamorphic rocks making up the higher elevations. The relationships break down on northern Baffin Island where relatively unresistant Paleozoic sediments underlie high upland surfaces. Figure 3.59 shows the major physiographic regions; they are largely based on elevation, degree of preservation of summit levels, and bedrock geology. Bird (1967) suggested that upland surfaces throughout Arctic Canada were formed during major erosional cycles. One of these, which was also recognized by Ambrose (1964), is an exhumed erosion surface produced by the stripping of Paleozoic rocks (cf. Higgs, 1978). These surfaces are defined as extensive areas of similar summit elevation, and as Bird (1967, p. 78) remarked "there can be no doubt that considerable areas of arctic Canada are composed of surfaces of denudation unrelated to contemporary erosive processes."

Fronting parts of Baffin Bay are a series of low coastal forelands. These extend intermittently from Bylot Island to near Cape Dyer and represent Quaternary and possibly late Tertiary accretions of glacial, marine, and terrestrial sediments. The coastal forelands are backed by steeped bedrock slopes that rise rapidly to a dissected plateau. In places the higher summits are eroded by individual cirque basins but near the outer coast the summits are broad and rounded.

The middle stretches of the fiords on eastern Baffin Island consist of nearly vertical cliffs descending to sea level. Fiord depths vary between 300 and 800 m (Løken and Hodgson, 1971; Syvitski and Blakeney, 1983). Mountain

---

Andrews, J.T.
1989: Quaternary geology of the northeastern Canadian Shield; in Chapter 3 of Quaternary Geology of Canada and Greenland, R.J. Fulton (ed.); Geological Survey of Canada, Geology of Canada, no. 1 (also Geological Society of America, The Geology of North America, v. K-1).

summits commonly reach elevations of 1300-1500 m and near Pangnirtung elevations of close to 2000 m occur.

A few major rivers cut into the Baffin Surface and flow eastward towards Baffin Bay, but in most cases the drainage into the fiords is short and steep, and there is a rapid ascent from the fiords to the high upland surface that constitutes an extensive physiographic region across much of central Baffin Island (Fig. 3.59; Ives, 1962; Ives and Andrews, 1963; Bird, 1967, Fig. 67). The plateau has a relative relief averaging 200-300 m.

The coast of eastern Foxe Basins, commonly a low, swampy lowland with elevations below 100 m, is largely associated with the outcrop of Paleozoic limestones, and forms a zone of low relief that encircles Foxe Basin (Fig. 3.59). High land close to the coast occurs in a few places. The Foxe Basin lowlands penetrate through the higher upland surfaces as the ice-scoured lowland that extends between Nettling Lake and the head of Cumberland Sound, and as a narrower lowland that extends to the head of Frobisher Bay from Amadjuak Lake.

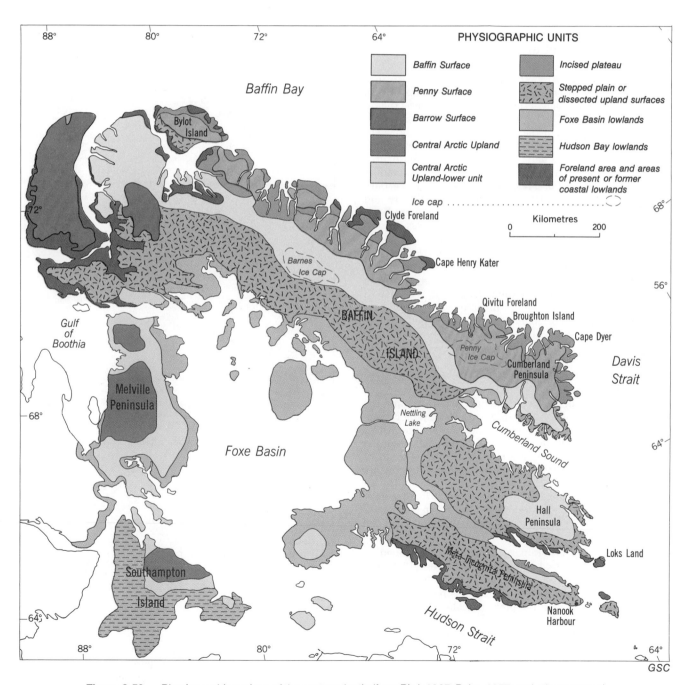

Figure 3.59. Physiographic regions of the eastern Arctic (from Bird, 1967; Dyke, 1977; and other sources).

The physiographic regions of Southampton Island/ Melville Peninsula/north Baffin Island consist of a Central Arctic Upland (lower unit); the Barrow Surface; and the Foxe Basin lowlands (Bird, 1967; Fig. 3.59). Bird (1967, p. 69) observed that in Baffin Island the Barrow Surface, which has a summit elevation of 350-380 m, ends abruptly against the higher Baffin Surface. On Melville Peninsula the main upland surface lies between 410 and 460 m a.s.l., although Bird (1967, p. 70) noted that hills rise to 520 m. Relief is low and the valleys are broad. Towards both coasts the drainage becomes more incised. On Southampton Island the upper surface extends to about 440 m elevation with a local relief of only 30-45 m (Bird, 1967), but in places isolated hills rise to 600 m.

Physiographic and drainage evolution

There has been considerable speculation on the physiographic evolution of Arctic Canada. Extreme views range from the largely discredited hypothesis of White (1972), who would explain much of the present landscape as the result of "deep erosion" by Laurentide ice sheets, to hypotheses that explain much of the present relief and drainage as being a response to regional and continental tectonism during the late Cretaceous and early Tertiary. It is reasonably certain that the large features of the eastern Canadian Arctic are the product of late Cretaceous and early Tertiary tectonism. Cretaceous rocks are known from offshore and inshore Baffin Island (e.g., MacLean and Williams, 1983) and indicate that a seaway of some dimensions was present between Greenland and Baffin Island at that time. Early Tertiary rifting that occurred between West Greenland and Baffin Island resulted in uplift, arching, and faulting.

Dowdeswell and Andrews (1985) have speculated on the relationship between the tectonic setting of the eastern Canadian Arctic and the origin of fiords. They have suggested that the fiords originated as canyons incised into the flank of the rift margin. Although not denying some glacial component in the development of the fiords, they also observed that the uplift/submergence model of rift development might also be invoked to explain some facets of fiord morphology.

The relationships between the upland surfaces and the tectonic evolution of the region are not clear. The critical questions are whether each surface intersects another with a significant topographic break; and if they do, what is the origin of the break. The occurrence of a small outcrop of Paleocene (lake?) sediments on a hilltop north of the Barnes Ice Cap (Andrews et al., 1972) suggests that much of the relief of the Baffin Surface is post-Paleocene; it further indicates that the Paleozoic limestone of Foxe Basin did not extend across north-central Baffin Island at the onset of the Tertiary.

The evolution of drainage in the eastern Canadian Arctic has been discussed by Fortier and Morley (1956) and Bird (1967) amongst others. The channels and sounds of the Canadian Arctic have been interpreted as part of an early drainage network. Given the present geological knowledge (e.g., Kerr, 1980; Grant and Manchester, 1970; MacLean and Williams, 1983), however, it appears that many of these features represent large-scale horst and graben structures floored by marine sediments of various ages.

Climate and glaciology

The topography and the proximity to marine environments cause significant climatic variability across the region. Maxwell (1981) divided the area into a series of climatic subregions based on these influences. July temperatures throughout the region vary between about 3° and 8°C with the lower temperatures being recorded along the eastern coast of Baffin Island. These cold summer conditions are caused by the continued presence of fast ice and pack ice offshore until late July or August. Temperatures rise inland over short distances but these in turn are modulated by the increase in elevation away from the outer coast. Precipitation is generally low but may exceed 50 cm in the high mountains of Cumberland Peninsula. About one quarter of the precipitation falls as rain during the months of June to September; snow is possible during all months.

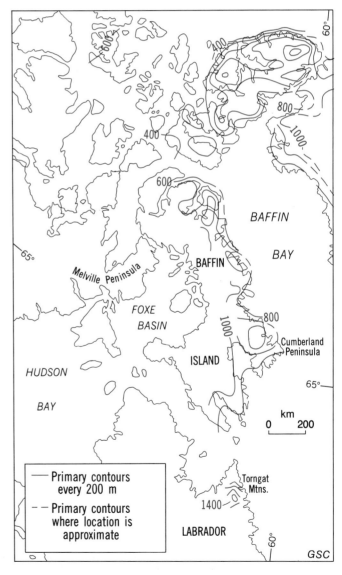

Figure 3.60. Glaciation limit or threshold map of the land around Baffin Bay and the northern Labrador Sea (from Jacobs et al., 1985).

Ice masses are a fundamental feature of the landscape of the eastern Canadian Arctic. These range in size from the 6000 km² Barnes and Penny ice caps, to extensive mountain ice complexes, valley glaciers, cirque glaciers, and permanent snow and ice patches (Falconer, 1962). The "Glaciation Level" (GL) (Andrews and Miller, 1972) lies between 600-700 m along the east coast of Baffin Island and rises inland to 900-1000 m in the vicinity of the Barnes and Penny ice caps (Fig. 3.60). Across the eastern Canadian Arctic the present GL lies very close to the upland plateaus and there is abundant evidence that during the late neoglaciation it was lower and thus permanent snowcover was significantly more extensive (Ives, 1962; Andrews et al., 1976; Locke and Locke, 1977; Williams, 1978, Dyke, 1978a,b).

Permafrost is continuous throughout the region but its extent offshore in Foxe Basin is not known. Offshore permafrost has not been reported from the eastern Arctic shelf. Permafrost thicknesses are largely unknown but probably exceed 100 m throughout most of the region.

Nature and distribution of Quaternary deposits and landforms

The surficial deposits in this region are generally thin. Only along the outer coast of Baffin Island and Bylot Island are there extensive, relatively thick sequences of Quaternary sediment (Miller et al., 1977; Klassen, 1981, 1985; Mode et al., 1983), but even here total exposed thickness rarely exceeds 60 m.

Landforms

The eastern Canadian Arctic is dominated by landscapes that reflect varying degrees of glacial erosion (Bird, 1967; Sugden, 1978). Areas of glacial scour (Sugden, 1977) have been delimited by measuring the number of small, bedrock-controlled lakes per unit area. This approach has been used to map aspects of the glacial landforms using 1:1 million Landsat imagery and 1:500 000 topographic maps (Fig. 3.61; Sugden, 1978; Andrews et al., 1985). Figure 3.62 is an example of the significant differences in the degree of bedrock erosion that occur around the head of Cumberland Sound.

Analysis of the Landsat imagery indicates that there is a distinct boundary that runs parallel to the western edge of the Baffin Surface. The boundary is close to 380-400 m elevation (Andrews et al., 1985). To the west of this boundary, the landscape is dominated by ice-smoothed bedrock, little till, and many small lakes. The scouring increases towards Foxe Basin and the zone of moderate glacial scouring penetrates seaward towards Baffin Bay and the Davis Strait through the low divides that link Foxe Basin to Cumberland Sound and Frobisher Bay (Fig. 3.61, 3.62). The Baffin Surface is largely coincident with an area that has few lakes (<5%) and this, plus the presence on the surface of tors and angular blockfields, suggests that this area has not been scoured by glacial ice. Farther east the relatively unglaciated terrain of the Baffin Surface gives way to the landscape of selective linear erosion. This is the region of fiords where fast moving ice streams were postulated responsible for the erosion of these features (Sugden, 1977). Above the fiords, and on the broad rolling hills of the interfiord regions, there are tors, extensive blockfields, and only limited evidence of active glaciation in the form of rare erratics (Boyer and Pheasant, 1974; Ives, 1974, 1978; Isherwood, 1975; Dyke, 1977; Sugden and Watts, 1977; Locke, 1980). Higher mountains are glacially modified by cirque erosion and form an "alpine landscape" (Sugden, 1978; Dyke et al., 1982a). Some of these landform divisions can also be recognized on Melville Peninsula and on Southampton Island, although neither area has "alpine landforms".

An explanation for the distribution of surface features as defined in Figure 3.61 must be sought in the basal thermal regimes of the Quaternary glaciers. Reasonable models of basal thermal regimes (Sugden, 1977; Andrews et al. 1985) indicate that the 400 m boundary lies close to the elevation at which a transition from warm-freezing to cold-based ice can be predicted. Within the area of moderate to high glacial scour (areas below 400 m), the dominant landforms are ice-moulded bedrock knolls and small scoured lake basins. Within the zone of little or no glacial erosion, the surface landforms are nonglacial except where broad, deep valleys are incised into the high plateau surface. Along the shoulders of these valleys and at the heads of fiords there are abundant signs of active glacial erosion in the form of ice-moulded bedrock bosses.

The fiords of Baffin Island are a major feature of the landscape. Recent investigations into the geomorphology, sedimentology, and oceanography of the fiords have been conducted as part of the Sedimentology of Arctic Fiords Experiment project (S.A.F.E.) (Syvitski and Blakney, 1983). Gilbert (1983), Dowdeswell and Andrews (1985), and Gilbert and MacLean (1983) have described many aspects of fiord morphometry. It is impossible to measure the relative contribution of tectonics, fluvial, and glacial erosion to the fiord morphology. The floors of the fiords have significant topographic irregularity which is smoothed over by rather thick accumulations of sediment (Fig. 3.63). As sediment rates in the fiords over the last 10 to 15 ka are not excessively high, and are one to two orders of magnitude less than in the fiords of southwest Alaska (Molnia, 1978; Gilbert, 1982; Andrews and Jull, 1984), these thick sediments represent many tens of thousands of years of accumulation.

Reconstruction of fiord glaciers, based on the mapping of lateral and terminal moraine segments (Smith, 1966; Ives and Buckley, 1969; Buckley, 1969; Boyer, 1972; Locke, 1980) indicates that glaciers in the fiords were relatively thin with basal shear stress much lower than 100 kPa. Seismic profiles in several fiords recorded few diamicts.

On the regional scale, another important landform is the zone of moraines that runs along the junction between the Baffin Surface and the incised plateau (Ives and Andrews, 1963; Falconer et al., 1965; Craig, 1965b; Blake, 1966; Hodgson and Hazelton, 1974; Miller and Dyke, 1974). The moraines form a belt of low, broad ridges where they occur on the high ground between fiords, whereas within fiords they are narrower and higher. The moraine belt between fiords is from 5-25 km wide and individual moraine loops are commonly strikingly well developed (Fig. 3.64).

One final set of landforms that is distinctive is the cross-valley or sublacustrine moraines (Fig. 3.65; Andrews, 1963b; Barnett and Holdsworth, 1974). These moraines are restricted to former glacial lake basins. Genetically similar forms, De Geer moraines, occur in large numbers below marine limit in certain areas of Canada (Prest et al., 1968; Prest, 1983b). The cross-valley moraines on Baffin Island are associated with deglaciation from the Late Wisconsinan maximum when large proglacial lakes were dammed between the ice margin and the drainage divides of many

CHAPTER 3

major rivers that flowed towards Foxe Basin (Ives and Andrews, 1963; Hodgson and Hazelton, 1974). Although Andrews and Smithson (1966) and Barnett and Holdsworth (1974) have proposed different mechanisms for the genesis of the cross-valley moraines, it is certain that their origin is intimately connected with retreat of this ice margin in a body of water.

On Southampton Island, Melville Peninsula, and throughout much of Baffin Island, retreat of the Late Wisconsinan (Foxe) ice sheet and smaller residual ice masses is well demarcated by flights of lateral and sublateral drainage channels (e.g., Sim, 1960b, 1964; Ives and Andrews, 1963; Bird, 1970; Hodgson and Hazelton, 1974). These are good indications of the shape of the retreating ice margin and extremely useful in correlating ice margin positions between different areas.

An important component in understanding the Quaternary history of the region is the mapping of ice directional indicators. The presence of striations and other flow indicators (Fig. 3.66) indicates that some erosion, albeit limited, has occurred on the Baffin Surface within the zone of

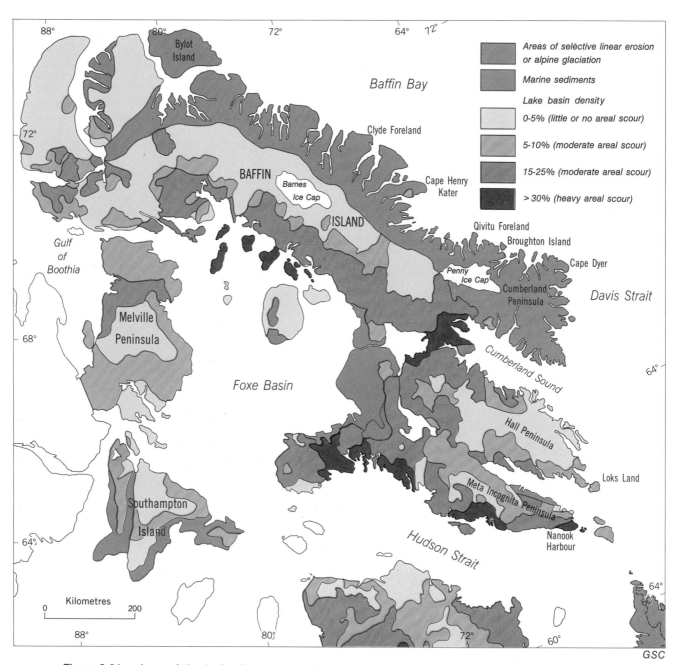

Figure 3.61. Areas of 1) selective linear erosion, 2) glacial scour as indicated by density of lake basins, and 3) marine sediment cover. Map based on analysis of Landsat imagery and of 1:500 000 topographic (NTS) maps (from Andrews et al., 1985).

QUATERNARY GEOLOGY — CANADIAN SHIELD

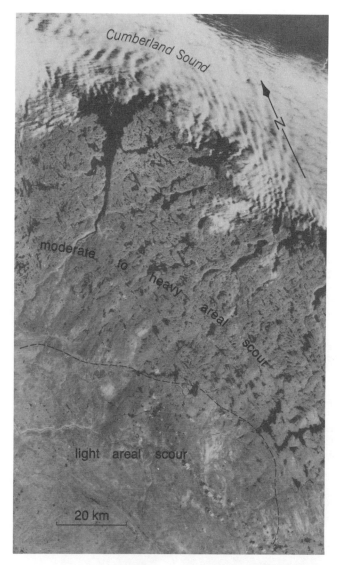

Figure 3.62. Landsat image near the head of Cumberland Sound showing the profound difference between areas of moderate glacial scour and areas of light glacial scour at higher elevations which are mantled with deposits and have few lakes.

little glacial scour. Crossing striations have not been reported in many localities, although they have been reported from north-central Baffin Island (Ives, 1962; Ives and Andrews, 1963), west Baffin Island (Andrews, 1970), southeast Baffin Island, and offshore islands.

Over the eastern Canadian Arctic, the areas covered by drumlins, ice contact/dead-ice deposits, and till plains are limited (e.g. Sim, 1964; Bird, 1970). The eastern Arctic is, however, the site for a series of large inactive and active sandurs (outwash trains). These have been studied by Church (1972) and are a major topic of study in the S.A.F.E. project (Syvitski and Blakney, 1983).

Materials

Glacigenic and other Quaternary sediments are normally thin throughout the area. Locally at the heads of fiords and within side entry valleys there are thick (20-100 m) sequences of deltaic and outwash sediments exposed because of glacial isostatic rebound. The forelands of the outer east coast of Baffin Island consist primarily of sand, gravel, mud, and diamicts associated with glacial and glacial marine deposition during intervals of marine transgression and regression (Miller et al., 1977; Nelson, 1981; Mode et al., 1983). This complex of sediments outcrops in wave-cut, 20-60 m-high exposures along tens of kilometres of coastline.

Andrews (1985a) presented information on more than 600 sediment samples collected from a variety of environments on Baffin Island; Church (1972, 1978) outlined the nature of some sandurs (outwash plains) on Baffin Island; and Gilbert (1982), Mode et al. (1983), and Osterman and Andrews (1983) described a variety of glacial marine facies from raised and offshore deposits on eastern Baffin Island.

The tills of the region consist of two principal lithologies. Those derived from Precambrian rocks are sandy gravels with a matrix dominated by sand containing only small amounts of silt and clay (Fig. 3.67). In contrast, tills that incorporated a significant fraction of Paleozoic limestone and dolostone are significantly enriched in silt and clay (Fig. 3.68). The large volume of suspended sediment in glacier meltwater streams (Church, 1972; Dowdeswell, 1982) indicates that silts and clays are being preferentially transported from the subglacial environment.

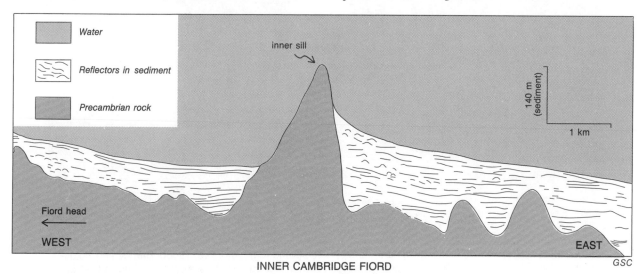

Figure 3.63. Interpretation of an air gun record of the sediment fill in Cambridge Fiord, eastern Baffin Island. The fiord consists of a number of basins which contain acoustically laminated sediments.

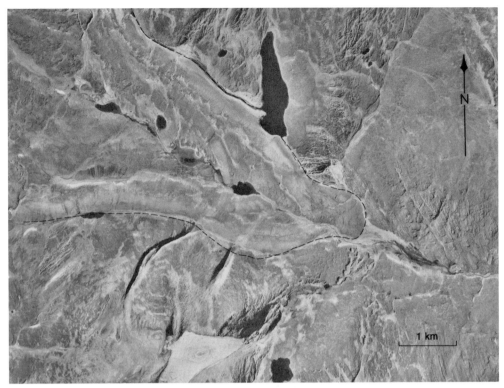

Figure 3.64. Large end moraines on the uplands west of Itirbilung Fiord, eastern Baffin Island. These moraines can be traced on the aerial photographs to the south where they link with the Ekalugad Moraine in Home Bay (Andrews et al., 1970) which is radiocarbon-dated in the fiord south of Itirbilung Fiord at about 8.4 ka. NAPL A17019-149

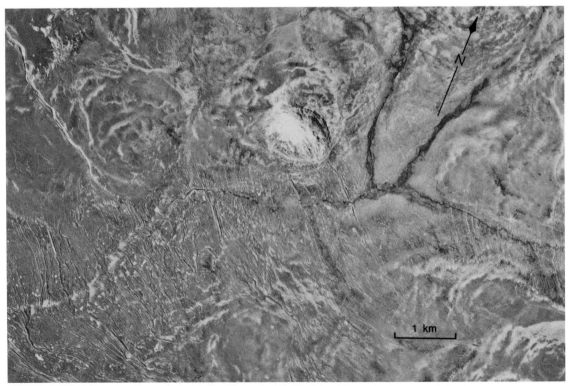

Figure 3.65. Cross-valley moraines of north-central Baffin Island (e.g., Andrews, 1963b; Barnett and Holdsworth, 1974). The moraines delimit the margin of a calving bay in Isortoq Valley just west of the Barnes Ice Cap. They can be traced up to the highest proglacial lake shorelines. NAPL A16293-100

Large volumes of gravels and sands are stored in the outwash trains and in raised deltaic sequences along the coasts of the region (e.g., Sim, 1960b, 1964; Andrews, 1966; Andrews et al., 1970; Bird, 1970; Church, 1972; Hodgson and Haselton, 1974; Dyke et al., 1982a). These deposits are well to moderately sorted (Fig. 3.67). Sandy diamictons are common on the iceberg ploughed shelf (MacLean, 1985). Figure 3.67 illustrates grain size histograms from raised glacial marine mud and offshore marine silt facies (Andrews, 1985a). The tail of very fine clay (finer than 11 phi) is noteworthy.

The availability of sand and finer sediments on active outwash plains, plus the occurrence of strong summer and fall winds, has resulted in a significant addition of eolian material to several sedimentary environments. Bird (1953), Dyke (1977), Miller et al. (1977), and Andrews et al. (1979), have drawn attention to the presence of eolian sediments in parts of this region. In glaciated valleys large dune fields of thick (1-6 m) windblown sand are not uncommon. Gilbert (1982, 1983) noted that at the head of Maktak Fiord, Cumberland Peninsula, sand is blown onto the sea ice in winter and later deposited in the fiord as sand laminae during melting of the sea ice. The sediment in the dunes can be well sorted, but in the crudely stratified "layered sands" (Thompson, 1954), the sediment is poorly sorted sand (Fig. 3.67). Silt and clay transported by the strong fiord and valley winds are not deposited extensively at lower elevations. However, silt content of Baffin Island soils increases with age (Locke, 1980, 1985), suggesting an eolian addition to soils on fiord slopes.

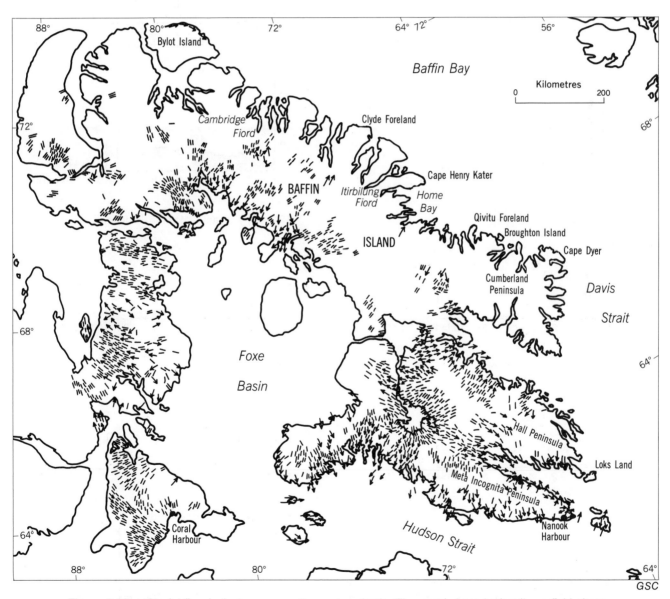

Figure 3.66. Glacial flow indicators across the eastern Arctic. The map is based primarily on field observations of the direction of striations and large bedrock forms plus the spread of carbonate drift onto areas of shield rock (see Fig. 3.3). Arrows indicate areas where the direction of movement has been determined.

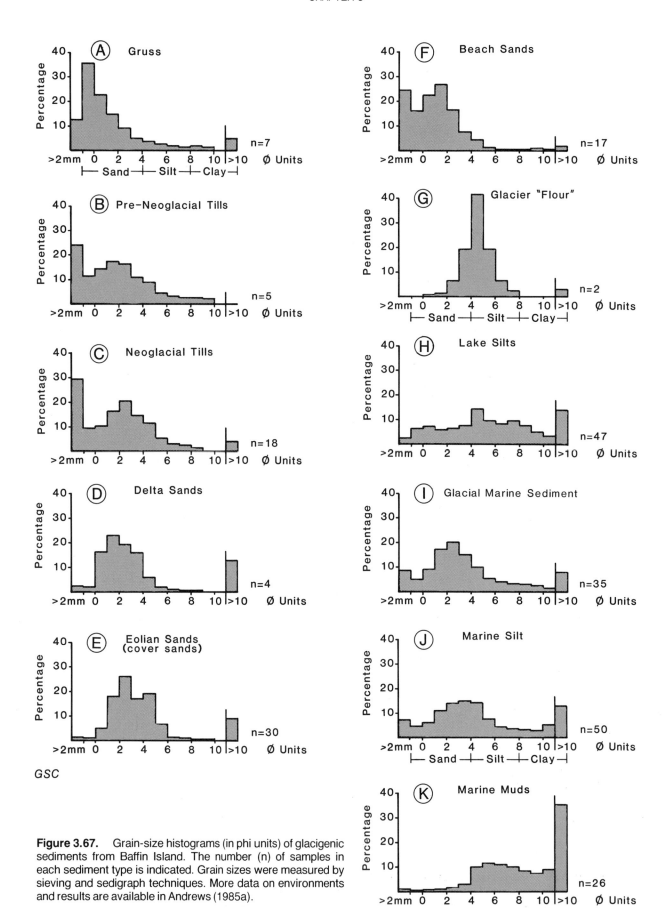

Figure 3.67. Grain-size histograms (in phi units) of glacigenic sediments from Baffin Island. The number (n) of samples in each sediment type is indicated. Grain sizes were measured by sieving and sedigraph techniques. More data on environments and results are available in Andrews (1985a).

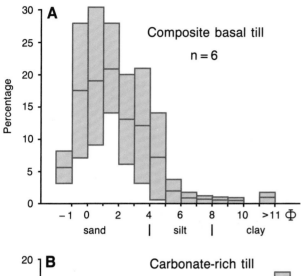

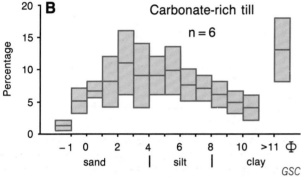

Figure 3.68. Grain size distribution of carbonate and noncarbonate tills from Meta Incognita Peninsula, Baffin Island. Scale in phi units (from Andrews, 1985a); the boxes represent ± one standard deviation about the mean of each grain size.

Cliffed coastlines and fault-guided escarpments commonly have extensive development of scree or talus slopes (Church et al., 1979). In places the sediment has been mobilized by interstitial ice into valley side rock glaciers (Dyke et al., 1982a) although a few valley head rock glaciers also occur. The talus can occur either as cones or sheets, depending on the nature of the rockfall source.

In terms of geomorphic processes it must be noted that the area is dry and cold with a limited melt season. Runoff peaks rapidly in June and by August many rivers are low. Glacier melt is similarly restricted to a period of between one and two months. Eolian activity and rockfall may be very important processes at present but, again, the total volume of sediment involved is low. Given the extremely long coastline, the presence of fast ice, high to moderate tidal ranges, and extensive areas of low shore gradient, the most effective geomorphic agent might be sea ice. Bird (1953, p. 38) noted that around Coral Harbour, Southampton Island, in June "the debris of the surface (of the sea ice) at that time was estimated to be 10 tons per acre of ice." Thus erosion and transportation by sea ice may be a vastly underestimated geological process in the eastern Arctic.

QUATERNARY HISTORY

Over the last two decades research on the Quaternary stratigraphy of the eastern Canadian Arctic has greatly increased our knowledge of the Quaternary history of the area although many perplexing problems have not been solved. The presence of diamictons on the Baffin Island shelf (MacLean, 1985) stands in apparent contrast to the absence of till in the exposed sections of the east coast (e.g., Mode et al., 1983) and may suggest that offshore areas were influenced by an ice shelf at various times during the Quaternary. Løken (1966) demonstrated that parts of the outer coast of Baffin Island were not glaciated during the last glaciation, whereas interglacial deposits have been preserved particularly in the central and northern areas once covered by Quaternary ice sheets (Terasmae et al., 1966).

Nature of the stratigraphic record

The stratigraphic record consists of 1) the stratified and complex sediment packages exposed along the west coast of Baffin Bay; 2) the surfaces of moraines and terraces; 3) the deposits of the zone of contact between the glacier ice and the sea during late glacial advance and final deglaciation; and 4) neoglacial moraines and associated sediments and landforms. A variety of stratigraphic procedures and dating methods have been applied to the coastal sediment complexes, including ^{14}C dating, U-series dating, amino acid racemization, paleomagnetism, macro and micropaleontology, and analysis of the lithology and mineralogy of the sediments (e.g., Miller et al., 1977; Mode, 1980; Andrews et al., 1981; Nelson, 1981, 1982; Osterman, 1982; Brigham, 1983; Mode et al., 1983; Miller, 1985a). In the case of surfaces, a variety of relative weathering indexes and soil development measurements have been used (Isherwood, 1975; Dyke, 1977; Birkeland, 1978; Locke, 1979, 1980, 1985). With ice contact deposits, the emphasis has been largely on ^{14}C dating and on tracing former ice margins by connecting associated glacial and glaciofluvial landforms (e.g., Ives and Andrews, 1963; Falconer et al., 1965; Bird, 1970). Some neoglacial events are ^{14}C dated (Miller, 1973, 1975; Stuckenrath et al., 1979) but in the majority of cases these events are dated by lichenometry (e.g., Miller, 1973; Andrews and Barnett, 1979; Davis, 1980, 1985).

The emphasis of Quaternary research in the eastern Arctic has changed over the past two decades. Much of the work in the 1960s concentrated on delimiting events at the end of the last glaciation (Sim, 1960a; Ives and Andrews, 1963; Craig, 1965b; Andrews, 1966; Blake, 1966; Bird, 1970) and determining the maximum extent of glaciation (e.g., Ives, 1974). However, starting in 1950 with Goldthwait's observations on the Quaternary sediments exposed near Clyde (R.P. Goldthwait, personal communication to J.D. Ives, 1960-1961), the value of these extensive outcrops of Quaternary sediments was steadily appreciated as research concentrated on these sections (Løken, 1966; Feyling-Hanssen, 1976a,b, 1980, 1985; Miller et al., 1977; Nelson, 1981; Brigham, 1983; Mode et al., 1983; Mode, 1985; Klassen, 1985).

Westward from the outer east coast, surficial mapping (Hodgson and Haselton, 1974; Dyke et al., 1982a) indicated that within the zone of "selective linear erosion" (Sugden, 1978) most fiord slopes were either covered by till, colluvium, or simply bare rock. In this situation, relative weathering studies (Mercer, 1956; Pheasant and Andrews, 1972;

Boyer and Pheasant, 1974; Ives, 1974, 1975; Isherwood, 1975; Dyke, 1977; Locke, 1980) indicated that the fiord slopes could be divided into a series of vertically arranged "weathering zones" which inclined inland. In several cases the boundaries between different zones coincided with lateral moraine systems. The significance of the weathering zones in the Quaternary stratigraphy of the eastern Arctic has been a contentious issue (Ives, 1975, 1978; Sugden and Watts, 1977; Dyke, 1977; Denton and Hughes, 1981; Brookes, 1982) and it has proven difficult to correlate the glacial events recorded in the outer coast sections with the valleyside surfaces delimited as weathering zones (e.g., Dyke, 1977; Nelson, 1980; Mode, 1980).

Prior to the mid 1960s no attempt was made to develop a rational Quaternary stratigraphy; since that period, several attempts have been made although these have usually included discussion about the problems attendant upon such an exercise (Pheasant and Andrews, 1973; Andrews and Ives, 1978; Andrews et al., 1981; Andrews and Miller, 1984; Miller, 1985a).

Aminostratigraphy

Aminozones are one type of unit that has been used in investigations of the stratigraphy of the east coast forelands (Miller et al., 1977; Nelson, 1982; Miller, 1985a). Aminozones represent sediment sequences that have similar amino acid ratios, in particular similar aIle:Ile ratios (D-alloisoleucine: L-isoleucine). Aminostratigraphy has been used to subdivide the Clyde Foreland Formation into several members which are not lithologically distinct (Mode et al., 1983), but which do have distinct amino acid ratios in the free and/or total fraction. The Clyde Foreland Formation encompasses the exposed sediments of the outer coast (Feyling-Hanssen, 1976a) and on the basis of amino acid kinetics and foraminiferal biostratigraphy the base of the section may be as old as Pliocene (Feyling-Hanssen, 1980, 1985; Miller, 1985a). Transitional probability analysis (Nelson, 1981; Mode et al., 1983) of measured sections from Clyde Foreland, Qivitu Foreland, and Broughton Island indicates that each aminozone represents a sequence of sediments associated with glacial isostatically driven changes in relative sea level (Fig. 3.69). Age estimates for the members (Table 3.7) are based on ^{14}C dates, U-series dates, and on amino acid kinetics. As Miller (1985a) notes there is an intriguing suggestion in these data for an ca 100 ka periodicity.

Weathering zones

For several decades researchers have noted that physical weathering increases with elevation in the eastern Arctic (Mercer, 1956). Likewise soil development on moraine sequences also increases with elevation (Locke, 1979, 1980). These observations are the inverse of what would be expected if the surfaces were all of the same age because summer temperatures at low elevations are noticeably warmer than at the higher elevations. Pheasant and Andrews (1973) suggested that surfaces with similar relative weathering states be designated as "weathering zones" and they defined three major weathering zones on northern Cumberland Peninsula. Weathering Zone III (Boyer and Pheasant, 1974) most probably records weathering between the maximum of the last glaciation (see below) and the present. Dyke (1977) subdivided Weathering Zone III into other units based on

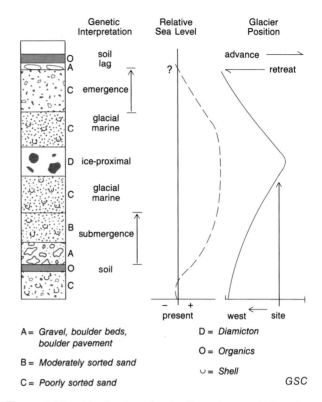

Figure 3.69. Idealized section for the outer coast of eastern Baffin Island showing the "typical" sediment package associated with a marine transgression and regression which are correlated with ice loading and unloading. For more information and a complete description of facies see Mode et al. (1983).

surface weathering parameters. Weathering Zone II may contain deposits that span more than 1 million years but additional techniques and methods of analysis will be required if it is to be further subdivided. The age and nature of glaciation (if any) in Zone I is difficult to assess (Ives and Borns, 1971; Ives, 1975; Sugden and Watts, 1977) and the evidence for little or any active glaciation has to be balanced against the finding, for example, of "pink" granites on many mountain summits in northern Cumberland Peninsula (Pheasant, 1971; Boyer, 1972), as well as the notion that the summits may have been covered by inactive cold-based ice.

Geologic-climate units

Geologic-climate units have been proposed for the Eastern Canadian Arctic (Fig. 3.70). The last glaciation of the region is called the Foxe Glaciation (Andrews, 1965; Andrews and Miller, 1984). The start of the event is heralded in the outer coast stratigraphy by a marine transgression that buried organics which had accumulated on the surface of the regressive phase of the Cape Broughton aminozone. During the early phase (Ayre Lake Stade) glaciers flowed seaward from the ice sheet which covered Baffin Island (and Foxe Basin?) and reached the outer coast. The Ayr Lake Stade is coeval with the Kogalu aminozone. In other parts of the region local "glaciations" have been defined (Locke, 1980; Klassen, 1981, 1985; Fig. 3.70). The Baffinland Stade of the

Foxe Glaciation (Andrews and Ives, 1978) represents the regional Late Wisconsinan glaciation equivalent in the eastern Arctic. In north-central Baffin Island the Foxe Glaciation is still occurring because the Barnes Ice Cap contains basal ice of probable Pleistocene age (Hooke, 1982). However, the Foxe Glaciation can be considered to have ended over most of the island between 6 and 5 ka when the residual ice on Baffin Island split into discrete units (Ives and Andrews, 1963; Blake, 1966; Bryson et al., 1969; Dyke, 1974) and Foxe Basin, Southampton Island, and Melville Peninsula were largely deglaciated (e.g., Bird, 1970; Andrews, 1970).

Late Holocene (i.e. neoglacial) glacial activity of mountain ice caps, valley and cirque glaciers is recorded throughout the eastern Arctic as one or more ice-cored moraines. Miller (1975) and Davis (1985) have given these various glacial advances an informal nomenclature (Table 3.8) based on lichenometry.

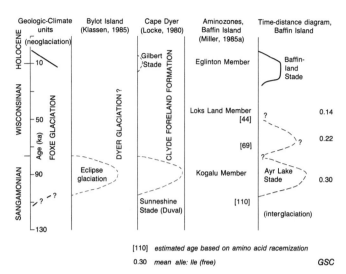

Figure 3.70. Stratigraphic nomenclature and postulated ages, Baffin and Bylot islands.

Table 3.7. Aminostratigraphy and chronological interpretation of the Clyde Foreland Formation, eastern Baffin Island

Clyde Foreland Formation aminozones	Mean alle:Ile		Age (ka)			Comments
	Total	Free	^{14}C	U-series[1]	Amino acid[2]	
Modern	0.011	N.D.				
Eglinton	0.013	N.D.	8-11	≥6	13	Limited glacial advance, panarctic fauna overlain by nonglacial sediment with nondiagnostic fauna
Loks Land	0.020	0.14	≥40		44	Deglacial sediments, panarctic fauna
Kogalu late	0.027	0.22	≥54		69	Nonglacial sediment, warm subarctic fauna
Kogalu ice-proximal	0.029	0.30	≥54	>70	110	Ice proximal to deglacial sediment with panarctic to subarctic fauna; overlies interglacial soil, overlain by marine subarctic sediment with warm fauna
Cape Broughton	0.045	0.43	≥40	≥140	210	Ice proximal sediment, warm subarctic fauna
Kuvinilk	0.060	0.54		≥140	305	Includes ice distal:ice proximal: ice distal sediment, warm subarctic fauna throughout
Cape Christian	0.092	0.62	≥50	≥190	500	Includes a till overlain by warm subarctic fauna in marine sands on which is developed an interglacial soil
youngest pre-Cape Christian	0.116	0.72			650	Undivided multiple glacial and marine events (e.g., Nelson, 1978)
oldest pre-Cape Christian	0.55	1.02		>300	3500	

[1] From Szabo et al., 1981
[2] From EDT-9°C column of Table 14-11 of Miller, 1985

Table 3.8. Neoglacial terminology, Baffin Island (from Davis, 1985)

Chronostratigraphy (ka)		Neoglacial advances (estimated age BP)	
0			
		Cumberland advance I	<100
		Cumberland advance II	200-400
		Cumberland advance III	500-650
1		Pangnirtung advance	900-1150
	Windy Lake		
2		Kingnait advance I	1900-2000
		Kingnait advance II	2200-2400
3		Snow Creek advance	2900-3100
	Iglutalik		
4		unnamed	3800
5			
6	Kangilo		
7			
8			

Chronostratigraphy

The application of chronostratigraphy within the eastern Arctic has been discussed by Andrews and Ives (1978) and Andrews (1982a,b). Because of the wide availability of ^{14}C dates and lichen dates, the Holocene series is divided into five intervals whose boundaries coincide with the Norden chronostratigraphy (Table 5 of Mangerud et al., 1974).

Pre-Foxe glaciations

There is abundant evidence for several glaciations which occurred previous to Foxe Glaciation. The evidence consists of highly weathered lateral moraines and tills that lie above the lateral moraines of the Ayr Lake Stade (early Foxe), such as the Pre-Duval moraines in southern Cumberland Peninsula (Dyke, 1977), the Mooneshine Drift near Cape Dyer (Locke, 1980), and the Pre-Eclipse glacial deposits on Bylot Island (Klassen, 1982). These deposits and landforms are not directly dated but because of their position and relative weathering they must predate Foxe Glaciation.

The Clyde Foreland Formation consists of gravels, sands, silty clays, and diamictons deposited during glacial isostatic cycles of sea level rise and fall (Fig. 3.69). Within the Clyde Foreland Formation there are seven or eight aminozones that predate the Kogalu aminozone (Mode, 1980). Feyling-Hanssen (1985) noted that assemblages of benthic foraminifera from the lower strata at Qivitu and Clyde forelands are similar to Pliocene assemblages from the North Sea. Miller's assessment of the amino acid kinetics of shells from these deposits is such that an age of >1 million years for these units is probable. It appears, therefore, that the sections exposed by wave action along the eastern coast of Baffin Island may contain a relatively complete record of the Quaternary. However, there are unconformities within the sections (Mode et al., 1983) which represent the planing of sediments by renewed marine transgression associated with glacial isostatic depression of the outer coast.

Mode (1985) reported on the palynology and malacology of pre-Foxe glacial and nonglacial units encompassing the aminozones of eastern Baffin Island. Figure 3.71 illustrates the variations in climatically sensitive taxa within each aminozone. Although most of the interglacial soils and other indicators of warmer climates have been recovered from sites on the east coast of Baffin Island (e.g., Miller et al, 1977; Mode, 1985), interglacial peats have been reported from sites close to the Barnes Ice Cap (Terasmae et al., 1966). Radiocarbon ages indicate that these sites are at least 40 ka, and based on their pollen and macrofossils they clearly represent a climate significantly warmer than present.

Detailed stratigraphic description and interpretation of amino acid racemization ratios, ^{14}C, and U-series dates on pre-Foxe deposits are contained in reports and theses (e.g., Feyling-Hanssen, 1976a,b; Miller et al., 1977; Mode, 1980; Nelson, 1981; Szabo et al., 1981; Brigham, 1983; and Miller, 1985a). A critical site that has been examined by a number of investigators occurs at Cape Broughton, at the north end of Broughton Island (England and Andrews, 1973; Feyling-Hanssen, 1976b; Brigham, 1983). The type stratum of the Cape Broughton Member of the Clyde Foreland Formation is exposed in a terrace at 32 m (site 13, Fig. 2 of Brigham, 1983) where 18 m of deltaic sediment overlies the Platform Till. The Cape Broughton Member is highly fossiliferous

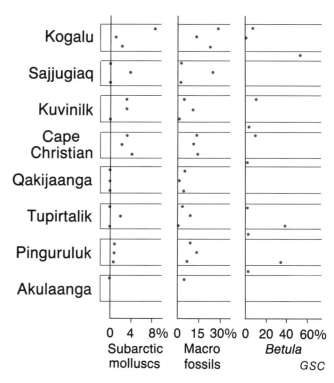

Figure 3.71. Diagram showing variation in the numbers or percentage of indicator taxa for the aminozones of eastern Baffin Island (from Mode, 1985). *Betula* is a low arctic shrub that currently grows only in southeastern Baffin Island. The position of the circles refers to the position of samples within each aminozone (i.e. lower, middle, upper). Qakijaanga is the local equivalent of the Cape Broughton aminozone.

with both foraminifera and molluscs. Finite ^{14}C analysis has provided several dates of between 28 and 47 ka but the amino acid ratios indicate that the sediments are significantly older than the Kogalu sediments of early part of Foxe Glaciation (Wisconsinan/Sangamonian) (Szabo et al., 1981; Brigham, 1983). U-series dates on shells of *Mya truncata* and *Hiatella arctica* gave concordant ages of 142 and 156 ka. The literature shows that U-series ages on marine shells are always <u>minimum</u> age estimates, thus the Cape Broughton site represents an important section within the eastern Arctic. Although the deltaic sediments overlie a till, there is no evidence that the site has been subsequently overridden by an ice lobe moving seaward down Broughton Channel.

Foxe Glaciation

Foxe Glaciation is divided into early, middle, and late stages. The relative and absolute chronology of these subdivisions is based on a multiparameter approach. Early Foxe Glaciation is represented by the Ayr Lake Stade which occurred more than 54 ka ago and most probably represents regional glaciation during parts of marine oxygen isotope stages 5/4. Middle Foxe events are poorly known but may date from between 50 and 30 ka. Late Foxe, or Baffinland Stade deglaciation, is well dated between 10.7 and 8.0 ka, but the age of the onset of this stade is not known.

Early Foxe Glaciation

Early Foxe Glaciation is here considered to include the Eclipse Glaciation of Bylot Island and northern Baffin Bay (Klassen, 1982), and the Sunneshine Stade of Dyer Glaciation (Locke, 1980). The limit of Foxe Glaciation is mapped at the outer "fresh" lateral moraines that decline seaward towards Baffin Bay and which have moderate weathering and soil development. In places, such as at Cape Aston (Løken, 1966), Remote Peninsula (Ives and Buckley, 1969), and near Cape Dyer (Locke, 1980), lateral moraines of

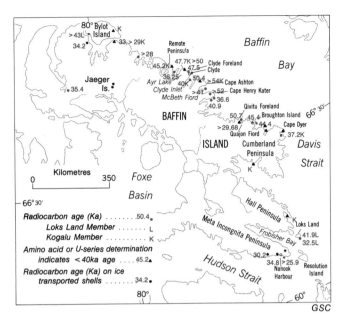

Figure 3.72. Distribution of "old" radiocarbon dates throughout the eastern Arctic. In addition the map shows sites where ice has transported old shells and incorporated them into tills, and sites where amino acid analysis indicates old (> 40 ka) materials exist. See also Table 3.10.

Foxe Glaciation can be associated directly with shell-bearing glacial marine deltas (e.g., Nelson, 1980) or ice proximal sediments. Table 3.9 gives aIle:Ile ratios from some of these key sites and compares them with ratios from the Kogalu aminozone. The comparison is such that the correlation between the Kogalu aminozone and the early Foxe Glaciation appear assured.

The precise age of early Foxe Glaciation is not well known (Szabo et al., 1981; Andrews and Miller, 1984). On the basis of infinite ^{14}C dates (Fig. 3.72 and Table 3.10), it is known that it dates from prior to 54 ka. Because the beginning of this event cannot be precisely determined, it is possible that a stade, or stades, of early Foxe Glaciation are pre-Wisconsinan in age and represent glacial events of Sangamonian age. An additional point of note is that the frequency of amino acid ratios in the Free fraction (Mode, 1980; Brigham, 1980) suggests that the Kogalu aminozone represents two discrete submergence/glacial events.

It is suggested that the Ayr Lake Stade and early Foxe Glaciation date from within marine oxygen isotope stages 4 and/or 5. This might be more rigorously defined through an examination of the deep sea record from Baffin Bay and correlation to the Eclipse glaciation of northern Baffin Island and Bylot Island.

In piston cores from Baffin Bay, Aksu (1981, 1985) delimited facies rich in detrital carbonate and associated these with deglaciation in the sounds and channels of the High Arctic. In core 040 off Clyde Foreland (Fig. 3.73), intervals with more than 50% carbonate occurred at depths of about 0.5 and ~2 m. When these lithologies are compared with the planktonic formainiferal oxygen isotopic record (Fig. 3.73) these two occurrences of carbonate detrital muds are associated with marine oxygen isotope stage 1/2 and 4/5 with boundary dates in core 040 of 12 and 77 ka, respectively.

Table 3.9. Free D-alloisoleucine:L-isoleucine ratios from deposits considered broadly correlative to the Kogalu Member (Ayr Lake Stade, Foxe Glaciation) (from Miller, 1985a)

	Free alle:Ile
Kogalu Member	0.30 ± 0.024 (n = 134)
Alikdjuak Moraine (N. Cumberland Peninsula; Nelson, 1980)	0.35-0.30
Duval Stade (S. Cumberland Peninsula)	0.28 (n = 3)
Sunneshine Stade (Cape Dyer)	0.29 (n = 7)
Harbour Till (Broughton Island)	0.29 (n = 3) 0.33 (n = 27)
McBeth Fiord, moraine	0.29 (n = 3)
Hall Peninsula	0.30 (n = 3)

Klassen (1982, 1985) has mapped an extensive system of lateral moraines that impinged upon Bylot Island from ice flowing seaward in northern Baffin Bay. These lateral moraines can be easily traced on air photographs because the till is heavily charged with light toned carbonates. This event has been termed the Eclipse glaciation. On the basis of limiting ^{14}C dates and amino acid ratios, this glaciation is correlated with the Kogalu aminozone and hence with the Ayr Lake Stade of the Foxe Glaciation. Thus the breakup of the ice at the end of the Eclipse Glaciation is equated here with the high detrital carbonate facies in Baffin Bay (Aksu, 1981).

A site in north-central Baffin Island provides some additional constraints on the timing of the onset of the Foxe Glaciation. Shells from a glacial marine diamicton on Jaeger Island (north of Foxe Basin) were collected by Falconer and dated at $18\,700 \pm 1200$ BP (I-1314). This site was revisited by Miller and Klassen in 1981 and shells collected (*Mya truncata*) had aIle:Ile ratios of about 0.054 (total). If the shells had been transported in (as suggested in Fig. 3.72) or covered by basal ice layers, a temperature of $-2°C$ might be assumed for their thermal history (Andrews et al., 1985); this temperature would provide an estimated age of 80 ± 40 ka (G.H. Miller, Institute of Arctic and Alpine Research, Boulder, Colorado, personal communication, 1983).

The pattern that has emerged from several studies of glacial and marine deposits at the mouths of several Baffin Island fiords (e.g., Løken, 1966; Ives and Buckley, 1969; King, 1969; Dyke, 1977; Miller et al., 1977; Nelson, 1978; Mode, 1980; Locke, 1980; Klassen, 1982) is that during the Ayr Lake Stade outlet glaciers flowed seaward through the fiords and terminated on the low coastal forelands as broad piedmont lobes with relatively shallow gradients. Soils and the surface weathering of boulders in lateral moraine ridges easily distinguish deposits of the Ayr Lake and Baffinland stades (Dyke, 1977; Birkeland, 1978; Bockheim, 1979). During the Ayr Lake Stade large ice contact deltas were deposited at the ice/marine interface and lie at elevations of between 70 and 90 m a.s.l. These deltaic surfaces extend some 40 to 50 m **above** the well dated (9 to 10 ka) early Holocene marine sediments.

As noted above, dating the Ayr Lake event directly has proven difficult because it is older than 54 ka. Szabo et al. (1981) argued that a close minimum age is provided by a U-series age on marine shells of 68 ka from a recessional deltaic deposit in Quajon Fiord, northern Cumberland Peninsula. The best glacial and marine record of the Ayr Lake Stade, however, is preserved along the outer reaches of Clyde Inlet, Inugsuin, and McBeth fiords (between Clyde Foreland and Cape Henry Kater) and adjacent forelands (e.g., Løken, 1966; Miller, 1976; Mode, 1980; J.E. Smith, unpublished). The major feature along this coast is the massive Cape Aston delta which was first described by Løken (1966). The delta front is about 10 km in length and between 80 and 88 m a.s.l. A shell date of $>54\,000$ BP (Y-1703) from in situ valves at 61.3 m a.s.l. in the delta provides a limiting date on deposition of the delta. Across McBeth Fiord, Miller (1979) has described and dated another important raised deltaic sequence, and both he and King (1969) have provided a series of ^{14}C dates on both Holocene and older marine units on the south side of outer McBeth Fiord and

Table 3.10. "Old" radiocarbon ages

Location (cf. Fig. 3.72)	Dated Material	Age (years BP)	Lab no.	Reference
Loks Land	Shell	41 900 + 7100/-3700	QC-446	Miller, 1979, p. 17
Loks Land	Shell	32 500 + 2400/-1800	GX-8591	Andrews and Short, 1983, p. 32
Cape Dyer	Shell	37 200 ± 800	QL-979	Miller, 1979, p. 38
Broughton Island	Shell	44 400 ± 1000	QL-974	Miller, 1979, p. 43
Qivitu Foreland	Peat	50 700 + 2000/-1600	QL-1179	Miller, 1979, p. 45
McBeth Fiord	Whalebone	>52 000	QL-976-2	Miller, 1979, p. 50
Clyde Cliffs	Shell	47 500 ± 1200	QL-973	Miller, 1979, p. 56
Cape Broughton	Shell	45 400 ± 600	QL-179	Andrews, 1976, p. 24
Cape Henry Kater	Shell	36 600 ± 350	QL-185	Andrews, 1976, p. 32
Clyde Foreland	Soil	50 400 + 1000/-900	QL-188	Andrews, 1976, p. 34
Clyde Foreland	Shell	47 700 ± 700	QL-183	Andrews, 1976, p. 34
Kogalu River	Shell	33 600 ± 300	QL-136	Miller, 1976
Scott Inlet	Shell	45 200 ± 800	QL-177	Andrews, 1976, p. 20
Clyde Foreland	Shell	>50 000	Y-1702	Andrews and Drapier, 1967, p. 124
Clyde Fiord	Shell	40 000 ± 3000	QL-184	Andrews, 1976, p. 35
Cape Aston	Shell	>54 000	Y-1703	Andrews and Drapier, 1967, p. 124
McBeth Fiord	Shell	>41 000	I-1829	Andrews and Drapier, 1967, p. 124
Sam Ford Fiord	Shell	36 250 + 3600/-2000	I-2581	Andrews and Drapier, 1967, p. 148
Henry Kater Pen.	Shell	40 900 ± 2000	Y-1985	King, 1969
Cape Jameson	Shell	>28 000	GSC-1090	Hodgson and Haselton, 1974
Pond Inlet	Shell	33 100 ± 900	GSC-1153	Hodgson and Haselton, 1974
Nannuk Harbour	Shell	>25 900	GSC-468	Blake, 1966, p. 6
Navy Board Inlet	Shell	34 200 + 3400/-2400	GSC-184	Dyck et al., 1966
Jungerson Bay	Shell	35 400 + 2100/-1600	GSC-188	Dyck et al., 1966
Clyde River	Shell	>39 000	GSC-2797	Mode, 1980

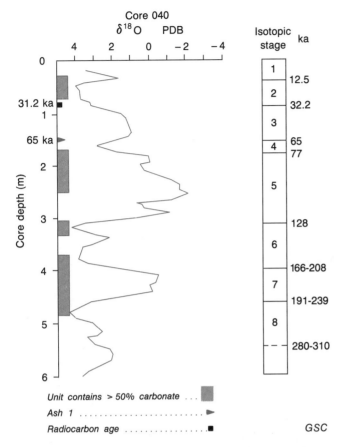

Figure 3.73. The δ^{18}O record from core 040 off Clyde Foreland (cf. Fig. 3.72) and segments of core where carbonate content is > 50% (Facies B of Aksu, 1981, 1985). Facies B represents the input of sediment high in detrital carbonate.

across Henry Kater Peninsula. In an ice contact delta on the south side of McBeth Fiord, Miller (1979) dated a whalebone (>52 000 BP, QL-976-2) from 65 m a.s.l. which is related to the delta surface at 73 m a.s.l.. The sandy foreset beds of the delta are overlain by a diamicton 1.5 m thick. The sedimentary sequence is interpreted by Miller (1979, p. 50-51) as indicating the approach and overriding of the site by a glacier in McBeth Fiord. A lower, distinct shoreline in outer McBeth Fiord is of early Holocene age. For example, on Aulitivik Island (mouth of McBeth Fiord) shells were dated at 8650±80 BP (DIC-332) and are associated with a shoreline at 17 m a.s.l., whereas from 10 km west of the "old" whalebone site, King (1969) reported a series of dates between 8160±135 BP and 8760±140 BP associated with a sea level close to 25 m a.s.l. Moraines extend across the same valley at elevations between 143 and 64 m a.s.l. and King (1969) interpreted them as being of late glacial age; however, no relative dating has been undertaken on deposits in McBeth Fiord and hence it is uncertain whether these moraines date from early, middle, or late Foxe time.

To the north, between Clyde Inlet and Ayr Lake, Miller (1976) described the surficial geology and presented a radiocarbon date from marine shells in a delta 11 km inland along Kogalu River (which drains Ayr Lake to Baffin Bay) and at an elevation of 42 m. The date of 33 600±300 BP (QL-136) is considered a minimum date for deglaciation. The delta surface shows no evidence for glacier overriding. In outer Clyde Inlet shells that date from the early Holocene are associated with shorelines that lie below 25 m a.s.l. (Andrews, 1980; Mode, 1980). Mode (1980) collected a sample of shells from near the settlement of Clyde, including the subarctic mollusc *Mya pseudoarenaria*, from distal foreset beds of a shoreline at 25 m a.s.l. The shells gave an age of >39 000 BP (GSC-2797) and Mode interpreted the local landforms and stratigraphy to indicate that ice has not reached the outer coast since that date.

These short descriptions illustrate the essence of the outer coastal glacial record of the eastern Canadian Arctic. There are abundant fossiliferous marine sediments dated >39 to >54 ka which lie well above fossiliferous marine sediments that date between 8 and 10 ka. The former are associated with lateral moraines and ice proximal deposits whereas the Holocene shorelines are distal (i.e., not ice contact) normal littoral sediments.

Middle Foxe events

The stratigraphic record from the outer central east coast of Baffin Island has failed to document any depositional events between the termination of the late Kogalu regression and the onset (at about 10 ka) of the earliest Holocene Eglinton aminozone (cf. Miller et al., 1977; Nelson, 1982; Brigham, 1983). However, marine and glacial marine sediments of a "middle" Foxe event have been described at sites along the west coast of Baffin Bay and the Labrador Sea (Ives, 1977; Blake, 1980; Clark, 1982, 1984a; Miller, 1985a; Klassen, 1985). On Baffin Island, Miller (1985a) called these marine sediments the Loks Land Member of the Clyde Foreland Formation. In northern Labrador possible correlative units have been described from Iron Strand (Ives, 1977; Clark, 1984a). Radiocarbon dates on this event vary between 42 and 28 ka and amino acid ratios are significantly lower than in the Kogalu aminozone (Table 3.7). Miller's assessment of the ratios suggests that the Loks Land Member is probably between 40 and 50 ka old. Little is known about middle Foxe glacial events although Dugdale (1972) and Dyke (1977) have suggested that there are local glacial deposits that may be ascribed to such an event.

Late Foxe glacial events

The Baffinland Stade (Andrews and Ives, 1978) is the geologic-climate unit that embraces glacial events correlative with Late Wisconsinan (Fig. 3.70). Our knowledge of glacial events, particularly the deglaciation from the last major ice advance, is vastly superior to that of the earlier episodes. We know very little, however, about the onset of late Foxe Glaciation. Shells incorporated into tills during readvances of fiord glaciers generally date from the last 8 to 9 ka (Falconer et al., 1965; Smith, 1966). Marine shells in ice contact deltaic deposits from along the mapped margin of Baffinland Stade deposits have been dated between 10.7 ka and ca. 8 ka (Andrews and Ives, 1978; Miller, 1980; Klassen, 1985), but nowhere have in situ shells been found which date from the "classical" Late Wisconsinan glaciation of 20 to 15 ka. The absence of ^{14}C ages between 30 and 11 ka has been the cause of much discussion and concern (Blake, 1966; Miller and Dyke, 1974; Hodgson and Haselton, 1974; Andrews and Ives, 1978; Denton and Hughes, 1981).

Recent attempts to provide a chronology for late Foxe Glaciation, especially the onset of this event, have focused on two main strategies. First, attempts have been made to model the glacial isostatic response of the region to Baffinland Stade loads and compare various scenarios with relative sea level history (Andrews, 1975, 1980; Dyke, 1979c; Clark, 1980; Quinlan, 1981, 1985). These studies all concluded that the eastern margin of late Foxe ice did not fluctuate significantly between about 18 and 8 ka. Thus the Baffinland Stade moraines represent either a period of long, slow till accumulation or a series of relatively short-lived oscillations of a margin that remained near the same position for 10 ka.

The second approach has been to examine marine cores on the continental shelf and in the fiords for lithostratigraphic units that could be tied to the glacial record (Osterman, 1982). This work is still in progress. A major problem in this approach is how to judge the reliability of ^{14}C dates on organic materials from such cores (Fillon et al., 1981; Andrews and Jull, 1984). Although radiocarbon dates on the acid insoluble fraction give basal age estimates from various cores between 27 and 12 ka (e.g., Andrews et al., 1983a) these are maximum age estimates because of contamination from older units. The oldest reliable date so far obtained from Baffin Island fiord cores (Cape Aston) is 11 770 ± 550 BP (GX-6280) on shells from 0.62 m depth. At present, only core, HU77-156, from outer Frobisher Bay (north of Resolution Island) (Osterman, 1982) may extend back to the beginning of the Baffinland Stade; however, a basal date of 27 255 ± 1250 BP (GX-7883) on a 2.45 m piston core appears anomalously old. Amino acid racemization data on foraminifera from the core indicate that a date of 18 ka is more probable (Osterman et al., 1985). We should also note that not all laboratories report radiocarbon dates corrected to the same standard. This could mean a difference of 400 to 700 years for the same shell date reported by different laboratories. Consequently care is necessary in comparing the ^{14}C data from different laboratories.

In evaluating the late Foxe Glaciation of the eastern Canadian Arctic we must remember that not only the Laurentide Ice Sheet affected the region, but also local ice caps, valley and cirque glaciers existed in the highlands adjacent to the eastern margin of the ice sheet. Studies on the glacial response of these local ice masses indicates that they did not fluctuate in phase with the Laurentide Ice Sheet (Miller, 1975, 1976; Dyke, 1977; Birkeland, 1978; Locke, 1980; Hawkins, 1980; Klassen, 1985). Figure 3.74 illustrates the time/distance history for some of these areas (Andrews, 1982b). Two contrasting responses are evident: 1) neoglacial extent of local ice as extensive as it has been for the last 30 to 40 ka (e.g., Miller, 1976) and 2) extensive Baffinland Stade expansion of local glaciers (Miller, 1973) and a subsequent extensive early to middle Holocene retreat.

The position of the margin of the ice sheet at the maximum of the Baffinland Stade has been discussed by several workers (Falconer et al., 1965; Craig, 1965b; Blake, 1966; Andrews and Ives, 1972; Miller and Dyke, 1974; Hodgson and Haselton, 1974; Miller, 1980; Denton and Hughes, 1981), although the position of the margin of this stadial event is not known with certainty in many localities.

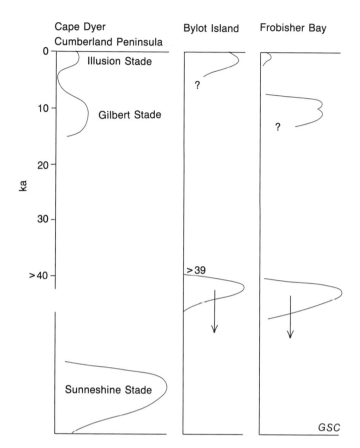

Figure 3.74. Schematic time-distance diagram of variations in local cirque and valley glaciers on Baffin and Bylot islands over the course of the last glaciation.

Mapping of lateral moraines that line the inner slopes of most fiords has only been carried out in a few areas (Smith, 1966; Pheasant, 1971; Dyke, 1977) — in particular, the tracing of the fiord moraines around and onto the high plateau has not been systematically carried out. Thus we do not know if breaks in surface weathering and soil development occur within this transect. Figure 3.75 shows the area in which moraines associated with the maximum of the Baffinland Stade occur and the locations of critical ^{14}C dates (Table 3.11). Most dates are from marine shells collected from foreset or bottomset sequences associated with moraines (cf. Andrews et al., 1970). Figure 3.75 also shows the location of a suite of radiocarbon dates on organics and shells that are associated with the early Holocene Cape Adair transgression; this was more or less contemporaneous with the Baffinland maximum (Table 3.12; Andrews, 1980). This transgression is represented by a well preserved terrace along much of the outer east coast (Fig. 3.76).

The work of Miller (1980, 1985b) from Frobisher Bay indicates that the ice sheet in southeast Baffin Island attained its maximum extent 10.7 ka. Klassen (1985) has reported a ^{14}C date from Cape Hatt, northern Baffin Island on in situ shells from an ice proximal delta which indicates that ice filled Milne Inlet at 9.9 ka. Falconer (in Andrews and Drapier, 1967) discovered a site near the settlement of

Table 3.11. Radiocarbon ages for ice proximal Baffinland Stade deposits

Location (cf. Fig. 3.75)	Dated Material	^{14}C age (years BP)	Lab no.	Reference
Frobisher Bay	Shell	10 720 ± 140	QC-480A	Miller, 1979, p. 17
Frobisher Bay	Shell	10 100 ± 110 (10 510 ± 110)*	GSC-2725	Miller, 1979, p. 21
Cornelius Grinnel Bay	Shell	8890 ± 100 (9300 ± 100)*	GSC-2568	Miller, 1979, p. 27
Chidlick Bay	Shell	8660 ± 160 (9070 ± 160)*	GSC-2466	Miller, 1979, p. 28-29
Quajon Fiord	Shell	8980 ± 180	Gak-5479	Andrews, 1976, p. 27
S. Cumberland Peninsula	Shell	8660 ± 110 (9070 ± 110)*	GSC-2183	Andrews, 1976, p. 6
Scott Inlet	Shell	10 095 ± 95	SI-2612	Andrews, 1976, p. 40
Lewis Bay	Shell	8450 ± 190	GX-8159	Andrews and Short, 1983, p. 25
Eggleston Bay	Shell	8690 ± 120 (9100 ± 120)*	GSC-3157	Andrews and Short, 1983, p. 24
Pugh Island	Shell	9875 ± 130	QC-903	Andrews and Short, 1983, p. 26
Pugh Island	Shell	8590 ± 100 (9000 ± 100)*	GSC-3660	Unpublished
York Sound	Shell	8820 ± 110	SI-4368	Andrews and Short, 1983, p. 35
Jackman Sound	Shell	9845 ± 175	SI-5172	Andrews and Short, 1983, p. 37
SE Meta Incognita	Shell	9190 ± 195	SI-8194	Andrews and Short, 1983, p. 43
Tay Sound	Shell	8350 ± 300	I-724	Andrews and Drapier, 1967, p. 127
Sam Ford Fiord	Shell	8210 ± 130	I-1933	Andrews and Drapier, 1967, p. 132
Clyde Fiord	Shell	7940 ± 130	I-1932	Andrews and Drapier, 1967, p. 136
Inugsuin Fiord	Shell	7970 ± 340	I-1673	Andrews and Drapier, 1967, p. 136
Inugsuin Fiord	Shell	7900 ± 210	I-1602	Andrews and Drapier, 1967, p. 136
Sam Ford Fiord	Shell	8000 ± 150 (8410 ± 150)*	GSC-630	Andrews and Drapier, 1967, p. 136
Tingin Fiord	Shell	8430 ± 140	Y-1830	Andrews and Drapier, 1967, p. 149
Tingin Fiord	Shell	8300 ± 135	I-2611	Andrews and Drapier, 1967, p. 149
"Foxe C Bay"	Shell	9180 ± 1140 (9590 ± 1140)*	GSC-707	Andrews and Drapier, 1967, p. 150
Kangok Fiord	Shell	8435 ± 105	GX-930	Andrews and Drapier, 1967, p. 152
Kangok Fiord	Shell	7820 ± 140	Y-1834	Andrews and Drapier, 1967, p. 152
Okoa Bay	Shell	8760 ± 350	St-3816	Andrews and Drapier, 1967, p. 265
McBeth Fiord	Shell	8760 ± 140	I-3211	King, 1969, p. 207
Bernier Bay	Shell	8830 ± 170 (9240 ± 170)*	GSC-182	Craig, 1965b, p. 6
York Sound	Shell	8840 ± 160 (9250 ± 160)*	GSC-463	Blake, 1966, p. 6
Cape Hatt	Shell	9510 ± 180 (9920 ± 180)*	GSC-3318	Klassen, 1985
Rannock Arm	Shell	7890 ± 160 (8300 ± 160)*	GSC-1064	Hodgson and Haselton, 1974
Kentra Bay	Shell	8090 ± 160 (8500 ± 160)*	GSC-1060	Hodgson and Haselton, 1974

*Age in parentheses is GSC date + 410 years to make the GSC date comparable to the other dates in the table

Arctic Bay, northern Baffin Island (cf. Fig. 3.75) that is important in deciphering the extent of Baffinland Stade ice. He obtained a date of ca 9.2 ka for the middle of a 5 m thick peat bed. The site was revisited by Short who collected a series of samples which dated between modern and 16 ka. This peat lies on the distal side of moraines that have been associated with the maximum advance position of the late Foxe ice sheet (e.g., Falconer et al, 1965; Miller and Dyke, 1974). The situation within other Baffin Island fiords is currently unresolved. Sites that lay beyond the margin of the Baffinland Stade ice can be delimited in several areas where in situ marine shells (e.g. Løken, 1966; Mode et al., 1983) give minimum ^{14}C and U-series dates of >40.0 ka (Miller and Dyke, 1974; Miller et al., 1977; Locke, 1980; Nelson, 1982; Brigham, 1983; Klassen, 1985).

Late Foxe deglaciation

Maps showing the deglaciation of the region have been prepared by Ives and Andrews (1963), Falconer et al. (1965), Bryson et al. (1969), Prest (1969), and Dyke (1974). Since these publications there has been a significant increase in the number of radiocarbon dates, particularly for parts of eastern and southern Baffin Island (e.g., Andrews, 1976; Miller, 1979; Andrews and Short, 1983). Figure 3.77a-f is an up to date series of paleogeographic maps showing deglaciation of the area.

Of particular concern for the deglacial chronology is the timing of ice retreat within Hudson Strait because events

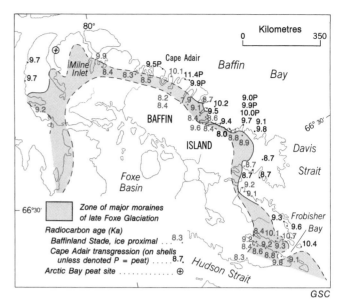

Figure 3.75. Map showing (1) the distribution of radiocarbon ages associated with ice contact deposits, many of which are of Cockburn age (Andrews and Ives, 1978); (2) the location of radiocarbon ages on peats and shells associated with an early Holocene marine transgression (the Cape Adair transgression); and (3) the generalized distribution of end and lateral moraines of late Foxe Glaciation. For further information on the ^{14}C dates see Tables 3.11 and 3.12.

Table 3.12. Radiocarbon ages associated with Cape Adair transgression

Location (cf. Fig. 3.75)	Dated Material	^{14}C age (years BP)	Lab no.	Reference
? Loks Land	Shell	9960 ± 230 (10 370 ± 230)*	GSC-2752	Miller, 1979, p. 22
Allen Island	Shell	9230 ± 110 (9640 ± 110)*	GSC-2618	Miller, 1979, p. 26
Kingnait Fiord	Shell	8680 ± 160 (9090 ± 160)*	GSC-2478	Miller, 1979, p. 31
Qivitu Foreland	Peat	9950 ± 185	QC-453	Miller, 1979, p. 31
Qivitu Foreland	Shell	9280 ± 120 (9690 ± 120)*	GSC-2479	Miller, 1979, p. 31-32
Qivitu Foreland	Peat	9935 ± 165	QC-451	Miller, 1979, p. 47
Qivitu Foreland	Peat	9092 ± 150	QC-454	Miller, 1979, p. 47-48
Itiribilung Fiord	Shell	9110 ± 160 (9520 ± 160)*	GSC-2215	Andrews, 1976, p. 33
Itiribilung Fiord	Shell	8670 ± 140	I-3136	King, 1969, p. 207
Clyde Foreland	Plant	9880 ± 200	GSC-2201	Andrews, 1976, p. 36
Clyde Foreland	Plant	11 360 ± 320	SI-2614	Andrews, 1976, p. 42
Cape Adair	Peat	9480 ± 165	DIC-374	Andrews, 1976, p. 42
Broughton Island	Seaweed	9100 ± 140	GSC-1969	Andrews, 1975, p. 80
Pangnirtung Fiord	Shell	8690 ± 90 (9100 ± 90)*	GSC-2001	Andrews, 1975, p. 84
Irkalulik	Shell	8410 ± 340 (8820 ± 340)*	GSC-1638	Andrews and Miller, 1972, p. 263
Broughton Island	Shell	9850 ± 250	GaK-2573	Andrews and Miller, 1972, p. 274
Henry Kater Peninsula	Shell	10 210 ± 180	Y-1986	King, 1969, p. 207
Cape Kater	Shell	9260 ± 150 (9670 ± 150)*	GSC-392	Dyck et al., 1966
McBean Bay	Shell	9280 ± 150 (9690 ± 150)*	GSC-241	Dyck et al., 1966

*Age in parentheses is GSC date +410 to make GSC date comparable to other dates in the table

there exert a control on the timing of deglaciation in Hudson Bay, Ungava Bay, Foxe Basin, and the eastern margin of Keewatin Ice. Within Frobisher Bay, immediately north of Hudson Strait, the Hall Moraines were still being constructed at 11 ka (Miller, 1980; Colvill, 1982; Lind, 1983), the sea entered the area behind the moraines at 10.7 ka, and readvance to the moraines occurred at 10 ka (Osterman et al., 1985). Between 10 and 9 ka the ice retreated rapidly to the Frobisher Bay Moraines at the head of Frobisher Bay (cf. Blake, 1966). In contrast, radiocarbon dates on ice contact deltas along the extreme southeast coast of Baffin Island (Andrews and Miller, 1985) indicate that ice was still offshore on the shelf as late as 8.5 ka. However dates of 9100 ± 480 BP (GSC-2946) and 8730 ± 250 BP (GSC-2698) from a core at the mouth of Hudson Strait and dates of 9.6 to 9.8 ka from northern Ungava Peninsula (Gray and Lauriol, 1985) indicate that most of Hudson Strait was ice free well before 8.5 ka. Perhaps the ice contact deltas on southeastern Baffin Island record a readvance of Labrador Ice across Hudson Strait after initial deglaciation of the Strait (Fig. 3.77c). To the south, the west coast of Ungava Bay was not ice free until ca 7.0 ka (Fig. 3.77e; Hillaire-Marcel, 1979; Lauriol et al., 1979) and Foxe Basin was deglaciated about this same time (Ives and Andrews, 1963; Blake, 1966). The deglaciation of Southampton Island did not occur until between 7.5 and 6.6 ka (Bird, 1970, Fig. 3). The pattern of deglaciation on Melville Peninsula (Sim, 1960a; Craig, 1965b) suggests that after deglaciation of the west coast some time prior to 8 ka and the east coast around 7 ka, residual ice masses decayed on the uplands.

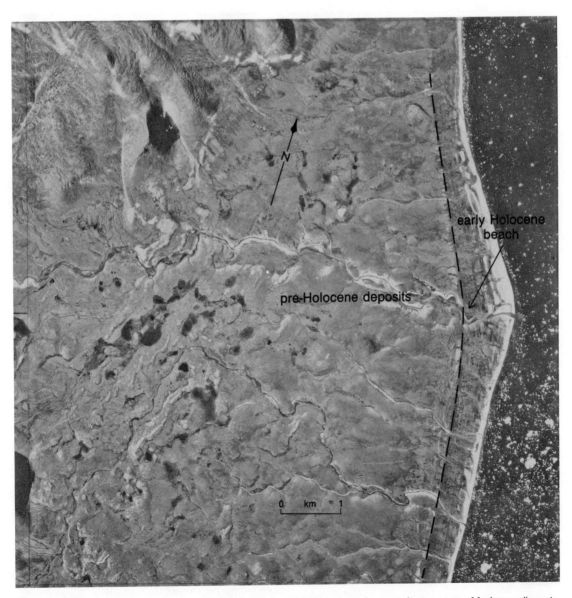

Figure 3.76. Upper limit of the early Holocene Cape Adair transgression near its type area. Marine sediments extend above the upper limit of the transgression, and amino acid analysis indicates that these sediments are coeval with the Kogalu Member of the Clyde Foreland Formation. NAPL A16301-29

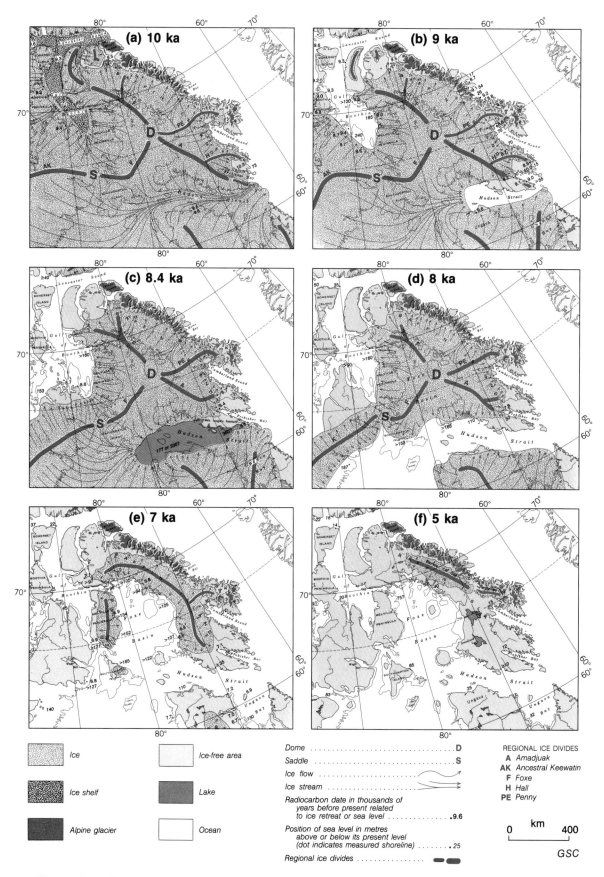

Figure 3.77. Paleogeographic maps of deglaciation of the eastern Arctic for (a) 10 ka, (b) 9 ka, (c) 8.4 ka, (d) 8 ka, (e) 7 ka, and (f) 5 ka.

The slow deglaciation of many fiords in eastern Baffin Island between 8 and 5 ka (cf. Andrews, 1982b) needs to be commented on. These slow rates of deglaciation occur at a time when pollen and macrofossil evidence suggests that the marine and atmospheric climate was at least as warm as present (Andrews, 1985b). It is possible that increased snowfall, resulting from an open (seasonal ice) Baffin Bay, compensated for the increased summer temperatures and associated increase in glacier ablation.

A significant aspect of the deglaciation, particularly on Baffin Island, was the creation of extensive systems of glacial lakes (Ives, 1962; Dyke et al., 1982a; Ives and Andrews, 1963). For example, Miller (1985b) recognized a series of major glacial lakes that existed between the margin of the ice sheet and the rising terrain of Hall Peninsula during construction of the Hall and Frobisher Bay moraines.

Although the history of the ice sheet and local glaciers between 9 and 5.5 ka was dominated by deglaciation, episodes of readvance or stillstand punctuated the overall retreat. Within the fiords of eastern Baffin Island several fiord glaciers readvanced during the Cockburn Substage (Falconer et al., 1965; Smith, 1966; Andrews et al., 1970). In addition the ice sheet readvanced to build the Frobisher Bay Moraines between 8.4 and 8.6 ka (Blake, 1966; Miller, 1980; Lind, 1983) and at about the same time local glaciers advanced northward from the Terra Nivea and Grinnell ice caps (near the tip of Meta Incognita Peninsula; Muller, 1980). Between 8 and 5 ka, during the Kangilo Substage (Table 3.8) there is evidence that moraines were deposited along parts of the ice sheet. On west Baffin Island the Isortoq Moraine (Andrews, 1966) was formed immediately after deglaciation and hence may not reflect a climatic pulse but a readjustment to a change in ice dynamics (e.g., Andrews, 1973; Hillaire-Marcel et al., 1981).

Neoglaciation

By 5 ka the only remaining remnants of the Laurentide Ice Sheet were on Baffin Island (Fig. 3.77f). Miller (1973) suggested that at about 5 ka the northern margin of the Penny Ice Cap was behind its current margin and it is probable that local valley and cirque glaciers were significantly less extensive than they are now. Shortly after 5 ka, however, the western margins of the Penny and Barnes ice caps advanced (Dyke, 1977; Andrews and Barnett, 1979; Dyke et al., 1982a), and by about 3.2 ka (lichen date) local glaciers had retreated from a series of outer Neoglacial moraines (Miller, 1973; Davis, 1985). Moraines that are dated by lichenometry consist of a layer of supraglacial debris covering an ice core; thus moraines that are dated by lichenometry date the onset of glacial thinning and retreat. Most neoglacial moraines are characterized by small diameter lichens (Fig. 3.78) and hence are young. Also, the most extensive neoglacial advance generally occurred within the last 100 to 300 years. In places the magnitude of neoglacial advances equals or exceeds any that have occurred in the last 30 to 40 ka (Miller, 1976; Dyke, 1977; Klassen, 1982; Locke, 1980).

The onset of neoglaciation can be associated with evidence for cooler conditions as seen in the palynology of peat and lake sediments across eastern Baffin Island (Mode, 1980; Davis, 1980; Andrews et al., 1980b; Short et al., 1985).

There is some evidence that during the Little Ice Age permanent snowfields occupied some of the cirques on the rim of Ungava Plateau and it is thus conceivable that glacierettes existed in cirques on Southampton Island and elsewhere throughout the area. The global warming of the early and mid Twentieth Century has resulted in most glaciers showing evidence of limited to extensive retreat. However, since about 1963 the summer climate of the eastern Arctic has become on average more severe and in 1983 it appeared that some tidewater glaciers might even be advancing. Ives (1962) postulated that during the Little Ice Age extensive areas of north-central Baffin Island were mantled by a thin permanent snow cover. Subsequently, these ideas were corroborated through the analysis of aerial photographs and Landsat imagery (Wright, 1975; Andrews et al., 1976; Locke and Locke, 1977). On imagery and on the ground, these lichen-free areas have sharp margins where they abut more densely lichen-covered terrain. Radiocarbon dates on dead mosses collected from within the lichen-free areas suggest that the moss had been killed some time during the last 300 years (e.g., Falconer, 1966). Although regional mapping has been restricted to Baffin Island, it is apparent that a Little Ice Age snow cover probably existed on the higher surfaces of Southampton Island and Melville Peninsula.

Ives (1962) suggested that the Little Ice Age snow cover on the Baffin plateau might serve as a suitable analog for "instantaneous glacierization" of the eastern Canadian Arctic. Williams' (1978, 1979) reconstructions of snow cover during the onset of the last glaciation support this analog, although Koerner (1980a,b) has been critical of both the evidence and the interpretation of the lichen-free areas.

Relative sea level changes

One of the recurring themes in Quaternary research in the eastern Arctic is the history of sea level and its association with glacial and deglacial events. Mapping the Holocene marine limit and dating both it and lower marine shorelines

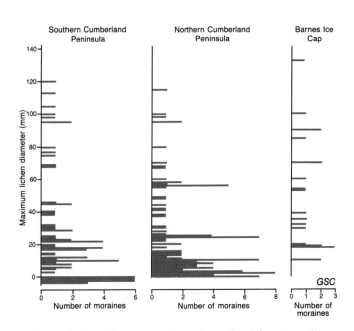

Figure 3.78. Frequency of moraines with different maximum lichen diameters for three localities on Baffin Island (from Davis, 1985).

have been major focuses (Bird, 1953, 1967; Sim, 1960b; Ives, 1964; Blake, 1966; Andrews, 1966, 1980; Matthews, 1967; Dyke, 1974, 1979c; Andrews and Miller, 1985; Quinlan, 1985). In addition, research on the outer coast of east Baffin Island has resulted in the mapping of marine limits and sediments that were more than 40 ka old (Løken, 1966; Ives and Buckley, 1969; King, 1969; Miller and Dyke, 1974; Miller et al., 1977).

Figure 3.79 illustrates the elevation and ages of shorelines of Cumberland Peninsula, Baffin Island, but can be considered "representative" for the outer coast of the eastern Arctic. In the area covered by the diagram, shorelines occur up to 105 m a.s.l. with shells from associated sediments providing amino acid ratios correlative with the Kogalu aminozone (Ayr Lake Stade). Below these high shorelines are landforms that date close to 10 ka and that represent a marine transgression to a Holocene marine limit. Figure 3.76 shows the marked terrace which forms the upper limit of the early Holocene marine transgression along the eastern Arctic outer coast; this terrace is eroded into sediments dated at >40 ka (e.g., Nelson, 1980, 1982). At, and close to the fiord heads, early Holocene shorelines grade towards ice margins of Cockburn and Remote Lake ages (Dyke, 1979c, Fig. 7). In contrast, towards the eastern limit of the transect, fiords west-northwest of Cape Dyer contain large glacial marine deltas, associated with terminal moraine loops, that are submerged to depths of 20-40 m **below** present sea level (Miller, 1975; Andrews, 1980). Much of the outer east coast is presently undergoing submergence and, in the vicinity of Cape Dyer and westward for 50-70 km, the present sea level is the Holocene "marine limit".

The elevation of marine limit increases from east to west across the region (Fig. 3.80). Along Hudson Strait the limit increases westward from 35 to 172 m a.s.l. and on Southampton Island reaches elevations of 190 m (Bird, 1970). Marine limit increases westward from 90-110 m on the outer coast of west Baffin Island to 104-152 m along the east coast of Melville Peninsula. Marine limit elevations vary a great deal on the local scale, and indeed these changes in elevation are very useful in discussing the relative rate of glacial retreat (cf. Andrews, 1970; Andrews et al., 1970). In Frobisher Bay, for example, marine limit on the distal side of the Frobisher Bay Moraine is between 100 and 122 m whereas on the proximal side of the moraine marine limit is between 30-42 m a.s.l. (Blake, 1966; Miller, 1980; Colvill, 1982; Lind, 1983). A similar decrease in marine limit occurs across the Ekalugad Moraine in Home Bay (Andrews et al., 1970).

Relative sea level across much of the region was dominated by a regression cycle throughout the middle and late Holocene (Fig. 3.81). On the outer coast of Baffin Island the evidence suggests that sea level was lower than present over the last 6 ka (Pheasant and Andrews, 1973; Dyke, 1979c; Andrews and Miller, 1985). In terms of glacial isostatic modelling, the eastern Arctic falls into two zones (Fig. 3.80; Clark et al., 1978). Zone I is dominated by glacial rebound and, with adequate materials and favourable slopes and exposures, sites in this zone have abundant raised beaches. Most of the study area lies in this zone. Zone II is characterized by an absence of raised beaches; such areas lie well beyond the limit of Foxe (Wisconsinan) Glaciation. Zone II (Fig. 3.80, site 12) is restricted to the outer tip of

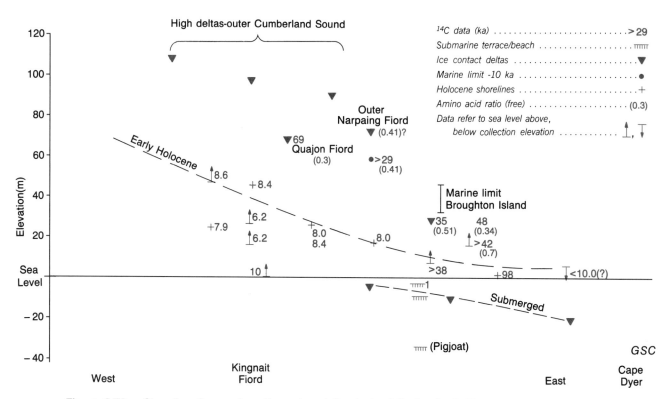

Figure 3.79. Shoreline diagram from the region of Cumberland Peninsula, Baffin Island, showing the elevation of shorelines along a transect (shown in Fig. 3.80) from Cape Dyer inland (from Andrews, 1980).

Cumberland Peninsula (Miller, 1975; Dyke, 1979c; Locke, 1980; Hawkins, 1980). There is also an important transition zone between zones I and II. In the transition zone initial emergence is followed by a subsequent submergence as the collapsing forebulge moves across a region. The transition zone (Fig. 3.80) has been recognized on northern and southern Cumberland Peninsula and in outer Frobisher Bay (Pheasant and Andrews, 1973; Dyke, 1979c; Andrews and Miller, 1985). The delimitation of the transition zone provides additional evidence on the extent of late Foxe Glaciation across the eastern Arctic.

Most emergence curves from Arctic Canada have shown a simple exponential decrease in sea level throughout the Holocene (Zone I; Fig. 3.81, sites 5 and 6) or an emergence/submergence cycle (Zone I/II). There is growing evidence from the outer coast of eastern Baffin Island that the history of sea level between 10 and 8 ka was more complex and consisted of at least one, possibly two, emergence/submergence events prior to the onset of regional emergence ca. 8 ka (Fig. 3.81, sites 10 and 11). At the outer tip of Meta Incognita Peninsula (site 10) there is stratigraphic evidence for a rapid fall in sea level after 8.6 ka, followed by a transgression of 10-15 m about 8.2 ka (Andrews and Miller, 1985). Evidence of similar stratigraphic sequences has been noted along several of the outer forelands of Baffin Island (e.g., Nelson, 1978; Andrews, 1980). In addition, comparable data come from the coast of northernmost Labrador where Løken (1962b) determined that what he referred to as shoreline S1-4 represented a major transgression as it crosscut older, tilted shorelines. Shells associated with the S1-4 shoreline gave an age of 8 ka. A limiting age on the next older (S1-3) strandline was given as 9 ka.

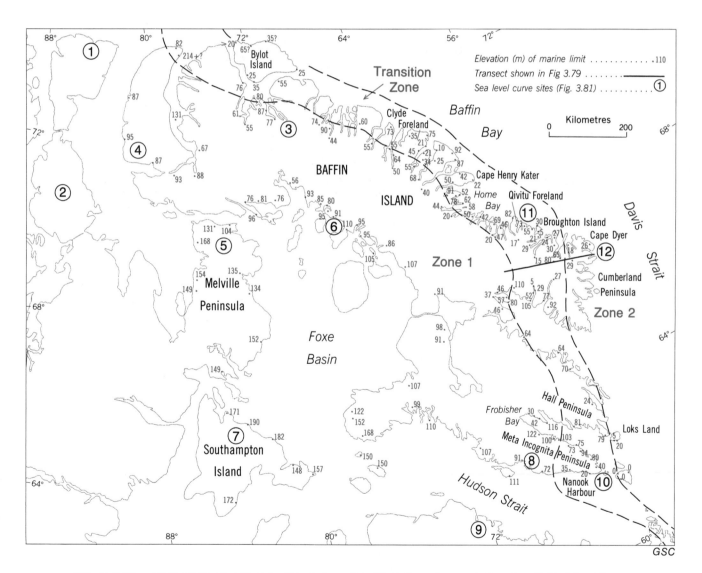

Figure 3.80. Marine limit elevations for the study region. Along the outer coast of eastern Baffin Island the marine limit is dated at > 40 to > 80 ka and is commonly associated with the Ayr Lake Stade. Over most of the region, however, marine limit is a Holocene feature which dates between 10 and 5 ka; in the vicinity of Cape Dyer marine limit is present sea level. The zones of sea level response after Clark et al. (1978) and Quinlan (1981).

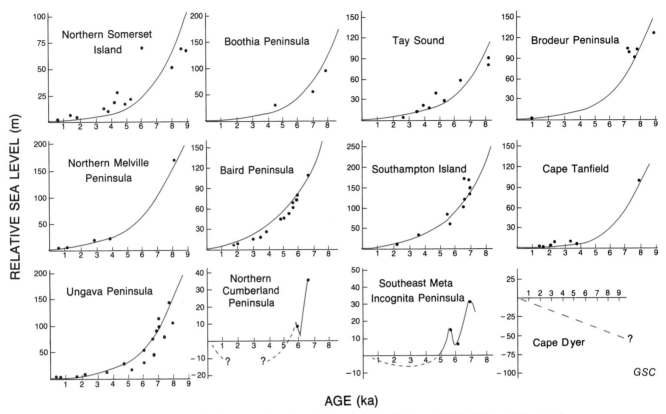

Figure 3.81. Relative sea level curves from the study area (after Quinlan, 1981, 1985; additional references for these curves include Andrews, 1970; Dyke, 1974; Quinlan, 1981, 1985). See Figure 3.80 for site locations. Curves are generally based on the ^{14}C age of marine shells. In several cases, the curves are based on a variety of materials and may synthesize two or three more specific relative sea level curves.

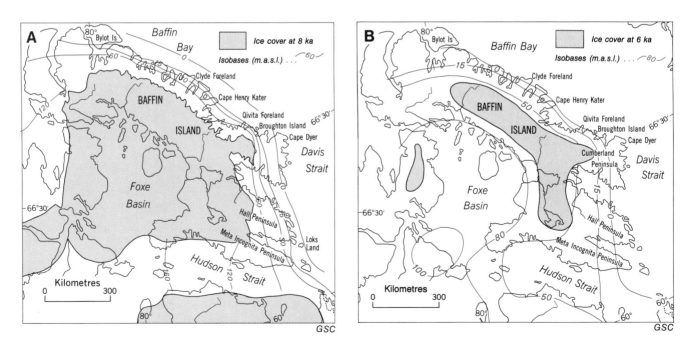

Figure 3.82. Ice extent and isobases (lines of equal sea level elevation relative to present) for (a) 8 ka and (b) 6 ka.

Relative sea level curves (Fig. 3.81) provide a series of site specific pictures of the Holocene sea level history of the eastern Arctic. A regional perspective can be provided by constructing isobase maps based on radiocarbon-dated samples associated with specific or limiting sea levels (Dyke, 1974). Such maps, if combined with isochrone maps of the ice margins, provide critical information on the interaction between glacial unloading and glacial isostatic recovery (Andrews and Peltier, 1989). Figure 3.82 presents two scenarios for 8 and 6 ka. They are based on the maps presented by Dyke (1974) but have been modified to take into account new sea level and age determinations.

Debate on reconstructions of Laurentide Ice Sheet

Over the last several years there has been an increased debate over the geometry of the Laurentide Ice Sheet during the Late Wisconsinan glaciation. Tyrrell (1898b) produced a map which suggested that the ice sheet was composed of several domes, one of which was located over Baffin Island. Ives and Andrews (1963) also suggested that a major dome of the ice sheet existed over Foxe Basin. Since then the debate has polarized into a discussion on whether the Laurentide Ice Sheet during the Late Wisconsinan consisted of a restricted and thin ice sheet with several major ice divides (e.g., Andrews and Miller, 1979; Shilts et al., 1979; Shilts, 1980, 1982b; Dyke et al., 1982b; Prest, 1983b) or whether the ice sheet conformed to the 1943 model of Flint with a central zone or single dome of dispersal located astride Hudson Bay. Denton and Hughes (1981) have produced models for both "minimum" and "maximum" ice sheets. It is worth noting that the "minimum" model can be represented as a logical phase in deglaciation from a single-domed ice sheet. Peltier and Andrews (1983) attempted to evaluate the glacial geological, the glacial isostatic, and the free air gravity anomaly data from around Hudson Bay to see if a common ground can be established between the different lines of evidence — at the moment the answer is that no single ice sheet model explains <u>all</u> aspects of the field evidence.

Figure 3.83 presents a picture of the northeastern corner of the Laurentide Ice Sheet at the late glacial maximum. This reconstruction attempts to take into account the information that has been developed from field evidence and from glacial isostatic models (e.g., Quinlan, 1985). Although weight is attached to the dispersal patterns of erratics (e.g., Andrews and Sim, 1964; Shilts et al., 1979; Andrews and Miller, 1979, Tippett, 1985), it is difficult to fairly interpret this data because the rate of transportation of distinctive erratics from their source to ultimate point of deposition involves many complex variables, including transportation over several glacial cycles. However, the presence of large bedforms which indicate ice movement direction; the distribution of landforms of glacial erosion (e.g., Sugden, 1978; Andrews et al, 1985); and the occurrence of extensive areas of carbonate drift plastered onto the Precambrian terrain indicate that there have been prolonged periods of what might be termed "characteristic" glacial flow. Thus Figure 3.83 represents a notion of the ice sheet geometry and flow regime integrated over some considerable period of time. The main glaciological constituents of Figure 3.83 can be listed as: 1) a large dome situated over Foxe Basin which was probably linked via a ridge to the Keewatin Sector of the

Figure 3.83. Reconstruction of the Laurentide Ice Sheet in the region of Foxe Basin, Baffin Island, and Hudson Strait. This figure is based largely on paleoflow indicators and on the transport vectors of glacial erratics.

Laurentide Ice Sheet; 2) subsidiary areas of glacial outflow located over the uplands of Cumberland Peninsula (Dyke, 1977), Meta Incognita Peninsula (Muller, 1980), and possibly over northern Baffin Island; 3) an ice stream within Hudson Strait that was separated from ice flowing into Hudson Strait from Baffin Island and Ungava Peninsula by zones of intense shearing such as has been proposed by Hughes (1975) for ice streams in the West Antarctic Ice Sheet; and 4) local valley and cirque glaciers which existed in the mountains beyond the margin of the main ice sheet, although in several cases these ice bodies were less extensive during the late Foxe Glaciation than they are at present.

The major problem areas that exist with this static reconstruction are:

(1) what were the glaciological conditions in Ungava Bay and outer Hudson Strait? Did a separate ice divide exist over Ungava Bay and do the Labrador-Ungava erratics on southern Baffin Island (Osterman et al, 1985; Stravers, 1986) reflect a dominant flow regime across Hudson Strait or the reworking of ice rafted sediments?

(2) Can specific erratic trains, related to each component of the Laurentide Ice Sheet, be recognized at the entrance into Hudson Strait where ice from Foxe Basin, Keewatin, Hudson Bay, and the Ungava Peninsula converged?

(3) Is there any evidence for the existence of a fringing ice shelf against the eastern Arctic coast (e.g., Osterman, 1982; Fillon and Harmes, 1982; Andrews, 1985b) during late glaciation and is such an ice shelf, or a much larger one, truly a necessary condition for the existence of the Laurentide Ice Sheet (e.g., Denton and Hughes, 1981, 1983).

ACKNOWLEDGMENTS

We thank W.H. Mathews (University of British Columbia), J.B. Bird (McGill University), J.D. Ives (University of Colorado), and N.R. Gadd (Geological Survey of Canada) for reviewing the entire chapter; V.K. Prest, D.A. St-Onge, D.R. Grant, and D.A. Hodgson for reviews of the Introduction and the section on the northwestern Canadian Shield; V.K. Prest, D.R. Grant, W.W. Shilts (Geological Survey of Canada), R. Geddes, E.V. Sado (Ontario Geological Survey), and E. Nielson (Manitoba Mineral Resources Division), for reviews of the section on the southwestern Canadian Shield; M. Bouchard (Université de Montréal), J-M. Dubois (Université de Sherbrooke), S. Occhietti (Université du Québec), and V.K. Prest (Geological Survey of Canada) for reviews of the section on the southeastern Shield; and B. McLean (Atlantic Geoscience Centre) for review of the section on the northeastern Canadian Shield. P.J.H. Richard (Université de Montréal) provided the section on the vegetational history of the southeastern Canadian Shield. J.A. Stravers and C.A. Layman (University of Colorado) assisted in preparing materials for the section on the northeastern Canadian Shield. The work of J.T. Andrews on Baffin Island has been supported by the United States National Science Foundation (Geology and Polar Program Office).

REFERENCES

Acton, D.F.
1989: Shield region (Soils of Canada); in Chapter 11 of Quaternary Geology of Canada and Greenland, R.J. Fulton (ed.); Geological Survey of Canada, Geology of Canada, no. 1 (also Geological Society of America, The Geology of North America, v. K-1).

Adshead, J.D.
1983a: Hudson Bay river sediments and regional glaciation: I. Iron and carbonate dispersal trains southwest of Hudson and James Bay; Canadian Journal of Earth Sciences, v. 20, p. 290-304.
1983b: Hudson Bay river sediments and regional glaciation: III. Implications of mineralogical studies for Wisconsinan and earlier ice-flow patterns; Canadian Journal of Earth Sciences, v. 20, p. 313-321.

Aksu, A.
1981: Late Quaternary stratigraphy, paleoenvironmentology and sedimentation history of Baffin Bay and Davis Strait; unpublished PhD thesis, Dalhousie University, Halifax, 771 p.
1985: Climatic and oceanographic changes over the past 400,000 years: evidence from deep-sea cores on Baffin Bay and Davis Strait; in Quaternary Environments: Eastern Canadian Arctic, Baffin Bay, and West Greenland, J.T. Andrews (ed.); Allen and Unwin, London, p. 181-209.

Alcock, F.J.
1920: The Reed-Wakusko map-area, northern Manitoba; Geological Survey of Canada, Memoir 119, 47 p.
1921: The terminal moraine of the Seal-Churchill divide; Geological Survey of Canada, Summary Report, 1920, Part C, p. 13-18

Allard, M.
1974: Géomorphologie des eskers abitibiens; Cahiers de géographie de Québec, vol. 18, p. 271-276.

Allard, M. et Seguin, M.K.
1985: La déglaciation d'une partie du versant hudsonien québécois; bassin des rivières Nastapoca, Sheldrake et à l'eau Claire; Géographie physique et Quaternaire, vol. 39, p. 13-24.

Allard, M. et Tremblay, G.
1983: La dynamique littorale des îles Manitounuk durant l'Holocène; Zeitschrift fur geomorphologie, Supp. 47, p. 61-95.

Ambrose, J.W.
1964: Exhumed paleoplains of the Precambrian Shield of North America; American Journal of Sciences, v. 262, p. 817-857.

Anderson, T.W., Mathewes, R.W., and Schweger, C.E.
1989: Holocene climatic trends with specific reference to the Hypsithermal interval; in Chapter 7 of Quaternary Geology of Canada and Greenland, R.J. Fulton (ed.); Geological Survey of Canada, Geology of Canada, no. 1 (also Geological Society of America, The Geology of North America, v. K-1).

Andrews, J.T.
1963a: End moraines and late-glacial chronology in the northern Nain Okak section of the Labrador Coast; Geografiska Annaler, v. 45A, p. 158-171.
1963b: The cross-valley moraines of the Rimrock and Isortoq river valleys, Baffin Island, N.W.T.: A descriptive analysis; Geographical Bulletin, v. 19, p. 49-77.
1965: Glacial geomorphological studies on north-central Baffin Island, Northwest Territories, Canada; unpublished PhD thesis, University of Nottingham, England, 476 p.
1966: Pattern of coastal uplift and deglacierization, west Baffin Island, N.W.T.; Geographical Bulletin, v. 8, p. 174-193.
1970: Differential crustal recovery and glacial chronology (6,700-0 BP), west Baffin Island, N.W.T., Canada; Arctic and Alpine Research, v. 2, p. 115-134.
1973: The Wisconsin Laurentide Ice Sheet: dispersal centers, problems of rates of retreat and climatic implications; Arctic and Alpine Research, v. 5, p. 185-200.
1974: Cainozoic glaciations and crustal movements in the Arctic; in Arctic and Alpine Environments, J.D. Ives and R.G. Barry (ed.); Methuen and Co. Ltd., London, p. 277-317.
1975: Support for a stable late Wisconsin ice margin (14,000 to ca. 9000 BP): a test based on glacial rebound; Geology, v. 3, no. 11, p. 617-620.
1976: Radiocarbon date list III, Baffin Island, N.W.T., Canada; institute of Arctic and Alpine Research, University of Colorado, Occasional Paper 21, 50 p.
1977: Status of Late Quaternary correlation <125,000 BP along the eastern Canada seaboard-latitude 45°N to 82°N; in IGCP Project 73/1/24, Quaternary Glaciations in the Northern Hemisphere; - Report no. 4 on sessions in Stuttgart, p. 180-195.
1980: Progress in relative sea level and ice sheet reconstructions, Baffin Island, N.W.T. for the last 125,000 years; in Earth Rheology, Isostasy and Eustasy, N-A Mörner (ed.); John Wiley and Sons, London, p. 175-200.
1982a: Chronostratigraphic division of the Holocene, Arctic Canada; Striae, v. 16, p. 56-64.
1982b: Holocene glacier variations in the eastern Canadian Arctic: A review; Striae, v. 18, p. 9-14.
1985a: Grain size characteristics of Quaternary sediments, Baffin Island region; in Quaternary Environments: Eastern Canadian Arctic, Baffin Bay, and West Greenland, J.T. Andrews (ed.); Allen and Unwin, London, p. 124-153.
1985b: Reconstruction of environmental conditions in the eastern Canadian Arctic during the last 11,000 years; in Climatic Change in Canada, 5th edition, C.R. Harington (ed.); Syllogeus, National Museums of Canada, Ottawa, p. 423-451.
1989: Quaternary geology of the northeastern Canadian Shield; in Chapter 3 of Quaternary Geology of Canada and Greenland, R.J. Fulton (ed.); Geological Survey of Canada, Geology of Canada, no. 1 (also Geological Society of America, The Geology of North America, v. K-1).

Andrews, J.T. and Barnett, D.M.
1979: Holocene (neoglacial) moraine and proglacial lake chronology, Barnes Ice Cap, Canada; Boreas, v. 8, p. 339-356.

Andrews, J.T. and Barry, R.G.
1978: Glacial inception and disintegration during the last glaciation; Annual Review, Earth and Planetary Sciences, v. 6, p. 205-228.

Andrews, J.T. and Drapier, L.
1967: Radiocarbon dates obtained through Geographical Branch field observations; Geographical Bulletin, v. 9, p. 115-162.

Andrews, J.T. and Falconer, G.
1969: Late glacial and post-glacial history and emergence of the Ottawa Islands, Hudson Bay, N.W.T.: Evidence on the deglaciation of Hudson Bay; Canadian Journal of Earth Sciences, v. 6, p. 1263-1276.

Andrews, J.T. and Ives, J.D.
1972: Late- and postglacial events (less than 10,000 BP) in the eastern Canadian Arctic with particular reference to the Cockburn moraines and break-up of the Laurentide Ice Sheet; in Climatic Change in Arctic Areas During the Last 10,000 Years, Y. Vasari, H. Hyvarinen, and S. Hicks (ed.); Acta Universitatis Oulensis, Series A, Scientiae rerum naturalium no. 3, Geologicam no. 1, p. 149-174.
1978: "Cockburn" nomenclature and the Late Quaternary history of the Eastern Canadian Arctic; Arctic and Alpine Research, v. 10, p. 617-633.

Andrews, J.T. and Jull, T.
1984: Rates of fiord and shelf sediment accumulation, Baffin Island, Canada, based on ^{14}C accelerator dates on in situ molluscs (abstract); in Abstracts with Program, Geological Society of America, v. 16, p. 431.

Andrews, J.T. and Mahaffy, M.A.W.
1976: Growth rate of the Laurentide Ice sheet and sea level lowering (with emphasis on the 115,000 BP sea level low); Quaternary Research v. 6, p. 167-183.

Andrews, J.T. and Miller, G.H.
1972: Maps of the present glaciation limits and lowest equilibrium line altitude for north and south Baffin Island; Arctic and Alpine Research, v. 4, p. 45-60.
1979: Glacial erosion and ice sheet divides, northeastern Laurentide ice sheet, on the basis of the distribution of limestone erratics; Geology, v. 7, p. 592-596.
1984: Quaternary glacial and nonglacial correlations for the eastern Canadian Arctic; in Quaternary Stratigraphy of Canada — A Canadian Contribution to IGCP Project 24, R.J. Fulton (ed.); Geological Survey of Canada, Paper 84-10, p. 101-116.
1985: Holocene sea level variations within Frobisher Bay; in Quaternary Environments: Eastern Canadian Arctic, Baffin Bay, and West Greenland, J.T. Andrews (ed.); Allen and Unwin, London, p. 585-606.

Andrews, J.T. and Peltier, W.R.
1976: Collapse of the Hudson Bay ice center and glacio-isostatic rebound; Geology, v.4, p. 73-75.
1989: Quaternary geodynamics; Chapter 8 in Quaternary Geology of Canada and Greenland, R.J. Fulton (ed.); Geological Survey of Canada, Geology of Canada, no. 1 (also Geological Society of America, The Geology of North America, v. K-1).

Andrews, J.T. and Short, S.K.
1983: Radiocarbon date list V from Baffin Island; Radiocarbon date list II from Labrador; institute of Arctic and Alpine Research, University of Colorado, Occasional Paper 40, 71 p.

Andrews, J.T. and Sim, V.W.
1964: Examination of the carbonate content of drift in the area of Foxe Basin, N.W.T.; Geographical Bulletin, v. 8, p. 174-193.

Andrews, J.T. and Smithson, B.B.
1966: Till fabrics of the cross-valley moraines of north-central Baffin Islands, N.W.T., Canada; Geological Society of America, Bulletin, v. 77, p. 271-290.

Andrews, J.T., Buckley, J.T., and England, J.H.
1970: Late-glacial chronology and glacio-isostatic recovery, Home Bay, east Baffin Island; Geological Society of America, Bulletin, v. 81, p. 1123-1148.

Andrews, J.T., Clark, P.U., and Stravers, J.
1985: The pattern of glacial erosion across the eastern Canadian Arctic; in Quaternary Environments: Eastern Canadian Arctic, Baffin Bay, and West Greenland, J.T. Andrews, (ed.); Allen and Unwin, London, p. 69-92.

Andrews, J.T., Davis, P.T., and Wright, C.
1976: Little Ice Age permanent snow cover in the eastern Canadian Arctic: Extent mapped from LANDSAT-1 satellite imagery; Geografiska Annaler, v. 58, p. 71-81.

Andrews, J.T., Guennel, G.K., Wray, J.L., and Ives, J.D.
1972: An early Tertiary outcrop in north central Baffin Island, Northwest Territories, Canada: environment and significance; Canadian Journal of Earth Sciences, v. 9, p. 233-238.

Andrews, J.T., Miller, G.H., Nelson, A.R., Mode, W.N., and Locke, W.W. III
1981: Quaternary near-shore environments on eastern Baffin Island, N.W.T.; in Quaternary Paleoclimate, W.C. Mahaney (ed.); Geoabstracts, Norwich, England, p. 13-44.

Andrews, J.T., Miller, G.H., and Shilts, W.W.
1980a: History of Hudson Bay during the Wisconsin Glaciation, based on amino acid geochronology; Fourth Symposium on the Quaternary of Quebec, Quebec City.

Andrews, J.T., Miller, G.H., Vincent, J-S., and Shilts, W.W.
1984: Quaternary correlations in Arctic Canada; in Quaternary Stratigraphy of Canada — A Canadian Contribution to IGCP Project 24, R.J. Fulton (ed.); Geological Survey of Canada, Paper 84-10, p. 127-134.

Andrews, J.T., Mode, W.N., and Davis, P.T.
1980b: Holocene climate based on pollen transfer function equations, eastern Canadian arctic; Arctic and Alpine Research, v. 12, p. 41-64.

Andrews, J.T., Osterman, L., Kravitz, J., Jennings, A., Williams, K., and Mothersill, J.
1983a: Quaternary piston cores; in Sedimentology of Arctic Fjords Experiment: HU82-031 Data Report, J.P.M. Syvitski and C.P. Blakeney (ed.); Canadian Data Report of Hydrography and Ocean Sciences, Supply and Services Canada, no. 12, 70 p.

Andrews, J.T., Shilts, W.W., and Miller, G.H.
1983b: Multiple deglaciations of the Hudson Bay Lowlands, Canada, since deposition of the Missinaibi (last-interglacial?) Formation; Quaternary Research, v. 19, p. 18-37.

Andrews, J.T., Webber, P.J., and Nichols, H.
1979: A late Holocene pollen diagram from Pangnirtung Pass, Baffin Island, N.W.T., Canada; Review of Palaeobotany and Palynology, v. 27, p. 1-28.

Antevs, E.
1925: Retreat of the last ice-sheet in eastern Canada; Geological Survey of Canada, Memoir 146, 142 p.
1928: The last glaciation; American Geographical Society, Research Series no. 17, 292 p.
1951: Glacial clays in Steep Rock Lake, Ontario; Geological Society of America, Bulletin, v. 62, p. 1223-1262.

Archer, D.R.
1968: The upper marine limit in the Little Whale River area, New Quebec; Arctic, v. 21, p. 153-160.

Arsenault, L., Aylsworth, J.M., Cunningham, C.M., Kettles, I.M., and Shilts, W.W.
1982: Surficial geology, Eskimo Point, District of Keewatin; Geological Survey of Canada, Map 8-1980, scale 1:125 000.

Arsenault, L., Aylsworth, J.M., Kettles, I.M., and Shilts, W.W.
1981: Surficial geology, Kaminak Lake, District of Keewatin; Geological Survey of Canada, Map 7-1979, scale 1:125 000.

Aylsworth, J.M.
1986a: Surficial geology, Ennadai Lake, District of Keewatin, Northwest Territories; Geological Survey of Canada, Map 5-1985, scale 1:125 000.
1986b: Surficial geology, Nueltin Lake, District of Keewatin, Northwest Territories; Geological Survey of Canada, Map 6-1985, scale 1:125 000.
1986c: Surficial geology, Kamilukuak Lake, District of Keewatin, Northwest Territories; Geological Survey of Canada, Map 4-1985, scale 1:125 000.

Aylsworth, J.M. and Shilts, W.W.
1985: Glacial features of the west-central Canadian Shield; in Current Research, Part B, Geological Survey of Canada, Paper 85-1B, p. 375-381.

Aylsworth, J.M., Boydell, A.N., Cunningham, C.M., and Shilts, W.W.
1981a: Surficial geology, Macquoid Lake, District of Keewatin; Geological Survey of Canada, Map 11-1980, scale 1:125 000.

Aylsworth, J.M., Boydell, A.N., and Shilts, W.W.
1986a: Surficial geology, Chesterfield Inlet, District of Keewatin; Geological Survey of Canada, Map 1-1985, scale 1:125 000.

Aylsworth, J.M., Cunningham, C.M., Kettles, I.M., and Shilts, W.W.
1986b: Surficial geology, Henik Lakes, District of Keewatin; Geological Survey of Canada, Map 2-1985, scale 1:125 000.

Aylsworth, J.M., Cunningham, C.M., and Shilts, W.W.
1981b: Surficial geology, Ferguson Lake, District of Keewatin; Geological Survey of Canada, Map 2-1979, scale 1:125 000.
1981c: Surficial geology, Hyde Lake, District of Keewatin; Geological Survey of Canada, Map 8-1979, scale 1:125 000.
1986c: Surficial geology, Baker Lake, District of Keewatin; Geological Survey of Canada, Map 3-1985, scale 1:125 000.

Aylsworth, J.M., Kettles, I.M., and Shilts, W.W.
1981d: Surficial geology, Dawson Inlet, District of Keewatin; Geological Survey of Canada, Map 9-1979, scale 1:125 000.

Baker, C.L.
1980: Quaternary geology of the Kirkland Lake area, District of Timiskaming; Ontario Geological Survey, Preliminary Map 2382, scale 1:50,000.

Baker, C.L. and Storrison, D.J.
1979: Quaternary geology, Larder Lake Area, District of Timiskaming, Ontario; Ontario Geological Survey, Preliminary Map 2290, scale 1:50 000.

Banerjee, I. and McDonald, B.C.
1975: Nature of esker sedimentation; in Glaciofluvial and Glaciolacustrine Sedimentation, A.V. Jopling and B.C. McDonald (ed.); Society of Economic Paleontologists and Mineralogists, Special Publication 23, p. 132-154.

Barnett, D.M.
1964: Some aspects of the deglaciation of the Indian House Lake area, with particular reference to the former proglacial lakes; unpublished MSc thesis, McGill University, Montréal, 175 p.
1967: Glacial Lake McLean and its relationships with Glacial Lake Naskaupi; Geographical Bulletin, v. 9, p. 96-101.
1970: An amendment and extension of tide gauge data analysis for Churchill, Manitoba; Canadian Journal of Earth Sciences, v. 7, p. 626-627.

Barnett, D.M. and Holdsworth, G.
1974: Origin, morphology, and chronology of sublacustrine moraines, Generator Lake, Baffin Island, Northwest Territories, Canada; Canadian Journal of Earth Sciences, v. 11, p. 380-408.

Barnett, D.M. and Peterson, J.A.
1964: The significance of glacial Lake Naskaupi 2 in the deglaciation of Labrador-Ungava; Canadian Geographer, v. 8, p. 173-181.
1968: Comments on "Sur le lieu de fonte sur place de la calotte glaciaire de Scheffer"; Canadian Geographer, v. 12, p. 53-54.

Barr, W.
1969: Structurally-controlled fluvioglacial erosion features near Schefferville, Quebec; Cahiers de géographie de Québec, no. 30, p. 295-320.

Barré, D.
1984: Cartographie géomorphologique détaillée appliquée à la région du mont Nuvulialuk et du Koroc-aval, Nouveau-Québec; thèse de MSc non publiée, Université de Montréal, Montréal, 173 p.

Barrie, C.Q. and Piper, D.J.W.
1982: Late Quaternary marine geology of Makkovik Bay, Labrador; Geological Survey of Canda, Paper 81-17, 37 p.

Bell, J.M.
1904: Economic resources of Moose River basin; Ontario Bureau of Mines, v. 13, Part 1, p. 135-179.

Bell, R.
1877: Report on exploration in 1875 between James Bay and lakes Superior and Huron; Report of Progress for 1875-76, Geological Survey of Canada, p. 294-342.
1879a: Report on the country between Lake Winnipeg and Hudson's Bay, 1878; Geological Survey of Canada, Report of Progress, 1877-78, Part. CC, p. 1CC-31CC.
1879b: Report on an exploration of the east coast of Hudson Bay in 1877; Geological Survey of Canada, Report of Progress 1877-78, Part. C, p. 1C-37C.
1884: Observations on geology, mineralogy, zoology and botany of the Labrador coast, Hudson's Strait and Bay; Geological Survey of Canada, Report of Progress 1882-83-84, v.1, Part D, p. 5-62.
1887: Report on an exploration of portions of the Attawapiskat and Albany Rivers; Geological Survey of Canada, Annual Report 1886, Part G, p. 5-38.

Bell, T., Rogerson, R.J., Klassen, R.A., and Dyer, A.
1987: Acoustic survey and glacial history of Adam Lake, outer Nachvak Fiord, northern Labrador; in Current Research, Part A, Geological Survey of Canada, Paper 87-1A, p. 101-110.

Bird, J.B.
1953: Southampton Island; Geographical Branch, Department of Mines and Technical Surveys, Ottawa, Memoir 1, 84 p.
1967: The Physiography of Arctic Canada (with special reference to the area south of Parry Channel); The Johns Hopkins Press, Baltimore, 336 p.
1970: The final phase of the Pleistocene ice sheet north of Hudson Bay; Acta Geographica Lodziensia, no. 24, p. 75-89.

Birkeland, P.W.
1978: Soil development as an indication of relative age of Quaternary deposits, Baffin Island, N.W.T., Canada; Arctic and Alpine Reserach, v. 10, p. 733-747.

Bisson, L.
1987: Géologie des dépôts quaternaires du canton de Scott, Chibougamau, Québec; avec applications à la prospection minérale; Thèse de maîtrise, Université du Québec à Montréal, Montréal, 182 p.

Black, R.F.
1979: Quaternary geology of Wisconsinin and contiguous upper Michigan; in Quaternary Stratigraphy of North America, W.C. Mahaney (ed.); Dowden, Hutchison & Ross Inc., Stroudsburg, Pa., 512 p.

Blake, W., Jr.
1956: Landforms and topography of the Lake Melville area, Labrador, Newfoundland; Geographical Bulletin, no. 9, p. 75-100.
1963: Notes on glacial geology, northeastern District of Mackenzie; Geological Survey of Canada, Paper 63-28, 12 p.
1966: End moraines and deglaciation chronology in northern Canada with special reference to southern Baffin Island; Geological Survey of Canada, Paper 66-26, 31 p.
1970: Studies of glacial histories in Arctic Canada I: Pumice, radiocarbon dates, and differential postglacial uplift in the eastern Queen Elizabeth Islands; Canadian Journal of Earth Sciences, v. 7, p. 634-664.
1976: Postglacial marine submergence at Lac Ford, northern Ungava, Quebec; in Report of Activities, Part C, Geological Survey of Canada, Paper 76- 1C, p. 171-174.
1980: Mid-Wisconsinan interstadial deposits beneath Holocene beaches, Cape Storm, Ellesmere Island, arctic Canada (abstract); AMQUA 6th Biennial Meeting, Orono, Maine, Abstracts, p. 26-27.
1982: Geological Survey of Canada radiocarbon dates XXII; Geological Survey of Canada, Paper 82-7, 22 p.
1988: Geological Survey of Canada radiocarbon dates XXVII; Geological Survey of Canada, Paper 87-7, 100 p.

Bockheim, J.G.
1979: Properties and relative age of soils of southwestern Cumberland Peninsula, Baffin Island, N.W.T., Canada; Arctic and Alpine Research, v. 11, p. 289-306.

Boissonneau, A.N.
1966: Glacial history of northeastern Ontario, I. The Cochrane-Hearst area; Canadian Journal of Earth Sciences, v. 3, p. 559-578.
1968: Glacial history of northeastern Ontario, II. The Timiscaming-Algoma area; Canadian Journal of Earth Sciences, v. 5, p. 97-109.

Bornhold, B.D., Finlayson, N.M., and Monahan, D.
1976: Submerged drainage patterns in Barrow Strait, Canadian Arctic; Canadian Journal of Earth Sciences, v. 13, p. 305-311.

Bostock, H.H.
1969: Precambrian sedimentary rocks of the Hudson Bay Lowlands; in Earth Science Symposium on Hudson Bay, P.J. Hood (ed.); Geological Survey of Canada, Paper 68-53, 386 p.

Bostock, H.S.
1970a: Physiography of Canada, Geological Survey of Canada, Map 1254A, scale 1:5 000 000.
1970b: Physiographic subdivisions of Canada; in Geology and Economic Minerals of Canada, R.J.W. Douglas (ed.); Geological Survey of Canada, Economic Geology Report no. 1, p. 10-30.

Bouchard, M.A.
1980: Late Quaternary geology of the Témiscamie area, Central Québec, Canada; unpublished PhD thesis, McGill University, Montréal, 284 p.
1986: Géologie des dépôts meubles de la région de Témiscamie (territoire du Nouveau-Québec); Ministère de l'Energie et des Ressources, Rapport MM-83- 03, 88 p.

Bouchard, M.A. and Marcotte, C.
1986: Regional glacial dispersal patterns in Ungava, Noveau-Québec; in Current Research, Part B, Geological Survey of Canada, Paper 86-1B, p. 295-304.

Bouchard, M.A. et Martineau, G.
1984: Les aspects régionaux de la dispersion glaciaire, Chibougamau, Québec; Canadian Institute of Mining and Metallurgy, Special volume 34, p. 431-441.
1985: Southeastern ice flow in central Québec and its paleogeographic significance; Canadian Journal of Earth Sciences, v. 22, p. 1536-1541.

Bouchard, M.A., Cadieux, B., et Goutier, F.
1984: L'origine et les caractéristiques des lithofaciès du till dans le secteur nord du lac Albanel, Québec: une étude de la dispersion glaciaire clastique; Canadian Institute of Mining and Metallurgy, Special volume 34, p. 244-261.

Bouchard, M.A., LaSalle, P., Lamothe, M., David, P.P.,and Bouillon, J.J.
1986: Pleistocene stratigraphy of northwestern Abitibi from boreholes and excavations at Selbaie Mine, Quebec (abstract); in Program with Abstracts, Geological Association of Canada, v. 11, p. 47.

Bouchard, M.A., St-Jacques, G., et Hamel, M.
1974: Géologie du Quaternaire, Lac Clary (32 /7), Rivière Pepeshquasati (32 P/10) et Rivière Témiscamie (32 P/9), Territoire de Mistassini; Ministère des Richesses naturelles, Dossier public 322, échelle 1:50 000.

Boulton, G.S., Smith, G.D., Jones, A.S., and Newsome, J.
1985: Glacial geology and glaciology of the last mid-latitude ice sheets; Journal of the Geological Society of London, v. 142, p. 447-474.

Boutray, B. de et Hillaire-Marcel, C.
1977: Aperçu géologique du substratum et des dépôts quaternaires dans la région de Blanc-Sablon, Québec; Géographie physique et Quaternaire, vol. 32, p. 207-215.

Boydell, A.N., Drabinsky, K.A., and Netterville, J.A.
1975: Terrain inventory and land classification, Boothia Peninsula and northern Keewatin; in Report of Activities, Part A, Geological Survey of Canada, Paper 75-1A, p. 393-396.

Boyer, S.J.
1972: Pre-Wisconsin, Wisconsin and Neoglacial ice limits in Maktak Fiord, Baffin Island: a statistical analysis; unpublished MSc thesis, University of Colorado, Boulder, 117 p.

Boyer, S.J. and Pheasant, D.R.
1974: Delimination of weathering zones in the fiord area of eastern Baffin Island; Geological Society of America, Bulletin, v. 85, p. 805-810.

Bradshaw, P.M.D. (ed.)
1975: Conceptual models in exploration geochemistry — The Canadian Cordillera and Canadian Shield; Association of Exploration Geochemists, Special Publication no. 3, 213 p.

Brereton, W.E. and Elson, J.A.
1979: A late Pleistocene plant-bearing deposit in Currie Township, near Matheson, Ontario; Canadian Journal of Earth Sciences, v. 16, p. 1130-1136.

Brigham, J.K.
1980: Stratigraphy, amino acid geochronology and genesis of Quaternary sediments, east Baffin Island, Canada; unpublished MSc thesis, University of Colorado, Boulder, 199 p.
1983: Stratigraphy, amino acid geochronology, and correlation of Quaternary sea-level and glacial events, Broughton Island, Arctic Canada; Canadian Journal of Earth Sciences, v. 20, p. 577-598.

Brookes, I.A.
1982: Dating methods of Pleistocene deposits and their problems; VIII. Weathering; Geoscience Canada, v. 9, p. 188-199.

Brown, R.J.E.
1967: Permafrost in Canada; Geological Survey of Canada, Map 1246A, scale 1:7 063 200.

Brummer, J.J.
1978: Diamonds in Canada; Canadian Mining and Metallurgical Bulletin, October 1978, p. 64-79.
1984: Diamonds in Canada; in The Geology of Industrial Minerals in Canada, G.R. Guillet and W. Martin (ed.); Canadian Institute of Mining and Metallurgy, Special Volume 29, 350 p.

Bryson, R.A.
1966: Air masses, streamlines and the boreal forest; Geographical Bulletin, v. 8, p. 228-269.

Bryson, R.A., Wendland, W.M., Ives, J.D., and Andrews, J.T.
1969: Radiocarbon isochrones and the disintegration of the Laurentide Ice Sheet; Arctic and Alpine Research, v. 1, p. 1-14.

Buckley, J.T.
1969: Gradients of past and present outlet glaciers; Geological Survey of Canada, Paper 69-29, 13 p.

Burwash, E.M.
1934: Geology of the Kakagi Lake area; Ontario Department of Mines, v. 42, part 4, 1933.

Burwasser, G.J.
1977: Quaternary geology of the City of Thunder Bay and vicinity; District of Thunder Bay; Ontario Geological Survey, Report GR164, 70 p.

Canada Soil Survey Committee
1978: The Canadian System of Soil Classification; Canada Department of Agriculture, Research Branch, Publication 1646, 164 p.

Card, K.D., Church, W.R., Franklin, J.M., Frarey, M.J., Robertson, J.A., West, G.F., and Young, G.M.
1972: The Southern Province; in Variations in Tectonic Style in Canada; Geological Association of Canada, Special Paper 11, 688 p.

Catto, M.R., Patterson, R.J., and Gorman, W.A.
1981: Late Quaternary marine sediments at Chalk River, Ontario; Canadian Journal of Earth Sciences, v. 18, p. 1261-1267.

Chagnon, J-Y.
1969: Étude des phénomènes d'érosion et des dépôts de surface dans la région de Baie-Saint-Paul — Saint-Urbain; Ministère des Richesses naturelles, Étude spéciale 3, 31 p.

Chapman, L.J.
1975: The physiography of the Georgian Bay-Ottawa Valley area of southern Ontario; Ontario Division of Mines, Geoscience Report 128, 33 p.

Chapman, L.J. and Putnam, D.F.
1984: The physiography of southern Ontario, 3rd edition; Ontario Geological Survey, Special Volume 2, 270 p.

Chauvin, L.
1977: Géologie des dépôts meubles de la région de Joutel-Matagami; Ministère des Richesses naturelles, Dossier public 539, 106 p.

Christiansen, E.A.
1979: The Wisconsinan deglaciation of southern Saskatchewan and adjacent areas; Canadian Journal of Earth Sciences, v. 16, p. 913-938.

Church, M.
1972: Baffin Island sandurs: a study of arctic fluvial processes; Geological Survey of Canada, Bulletin 216, 208 p.
1978: Palaeohydrological reconstructions from a Holocene valley fill; in Fluvial Sedimentology, A.D. Miall (ed.); Canadian Society of Petroleum Geologists, Memoir 5, p. 743-772.

Church, M., Stock, R.F., and Ryder, J.M.
1979: Contemporary sedimentary environments on Baffin Island, N.W.T., Canada: debris slope accumulation; Arctic and Alpine Research, v. 11, p. 371-402.

Clark, J.A.
1980: A numerical model of world wide sea level changes on a visco-elastic earth; in Earth Rheology, Isostasy, and Eustasy, N-A. Mörner (ed.); John Wiley, New York, p. 525-534.

Clark, J.A., Farrell, W.E., and Peltier, W.R.
1978: Global changes in postglacial sea level: a numerical calculation; Quaternary Research, v. 9, p. 265-287.

Clark, P.U.
1982: Late Quaternary history of the Iron Strand region, Torngat Mountains, northernmost Labrador, Canada (abstract); in Program and Abstracts, American Quaternary Association, Seventh Biennial Conference, Seattle, p. 82.
1984a: Glacial geology of the Kangalaksiorvik-Abloviak region, northern Labrador, Canada; unpublished PhD thesis, University of Colorado, Boulder, 240 p.
1984b: Wisconsinan glacial history of northern Labrador, Canada (abstract); in Program and Abstracts, Vth Congress, Association québécoise pour l'étude du Quaternaire, p. 24-25.

Clark, P.U. and Josenhans, H.
1985: Late Quaternary land-sea correlations, northern Labrador, Canada (abstrct); in Abstracts with Program, Geological Society of America, Annual Meeting,v. 17, p. 548.
1986: Late Quaternary land-sea correlations, northern Labrador and Labrador Shelf; in Current Research, Part B, Geological Survey of Canada, Paper 86-1B, p. 171-178.

Clark, P.U., Andrews, J.T., Short, S.K., Williams, K., and Melcer, A.
1986: Late-glacial and Holocene paleoenvironmental record from Square Lake, Torngat Mountains, Labrador; in Program and Abstracts, Ninth Biennial Meeting, American Quaternary Association, p. 73.

Clayton, J.S., Ehrlich, W.A., Cann, D.B., Day, J.H., and Marshall, I.B.
1977: Soils of Canada; Canada Department of Agriculture, Research Branch, v. 1, 243 p., v. 2, 239 p.

Clayton, L.
1983: Chronology of Lake Agassiz drainage to Lake Superior; in Glacial Lake Agassiz, J.T. Teller and L. Clayton, (ed.) Geological Association of Canada, Special Paper 26, p. 291-307.

Clayton, L., Teller, J.T., and Attig, J.W.
1985: Surging of the southwestern part of the Laurentide Ice Sheet; Boreas, v. 14, p. 235-241.

Coker, W.B. and DiLabio, R.N.W.
1979: Initial geochemical results and exploration significance of two uraniferous peat bogs, Kasmere Lake, Manitoba; in Current Research, Part B, Geological Survey of Canada, Paper 79-1B, p. 199-206.

Coleman, A.P.
1909: Lake Ojibway; last of the great glacial lakes; Ontario Bureau of Mines, 18th Annual Report, v. 18, Part 1, p. 284-293.
1921: Northeastern part of Labrador, and New Quebec; Geological Survey of Canada, Memoir 124, 68 p.

Colvill, A.
1982: Glacial landforms at the head of Frobisher Bay, Baffin Island, Canada; unpublished MA thesis, University of Colorado, Boulder, 202 p.

Cowan, W.R.
1968: Ribbed moraine: till-fabric analysis and origin; Canadian Journal of Earth Sciences, v. 5, p. 1145-1159.
1978: Radiocarbon dating of Nipissing Great Lakes events near Sault Ste. Marie, Ontario; Canadian Journal of Earth Sciences, v. 15, p. 2026-2030.
1985: Deglacial Great Lakes shorelines at Sault Ste. Marie, Ontario; in Quaternary Evolution of the Great Lakes, P.F. Karrow (ed.); Geological Association of Canada, Special Paper 30, p. 33-37.

Craig, B.G.
1961: Surficial geology of northeastern district of Keewatin, Northwest Territories; Geological Survey of Canada, Paper 61-5.
1964a: Surficial geology of east-central District of Mackenzie; Geological Survey of Canada, Bulletin 99, 41 p.
1964b: Surficial geology of Boothia Peninsula and Somerset, King William, and Prince of Wales Islands, District of Franklin; Geological Survey of Canada, Paper 63-44, 10 p.
1965a: Glacial Lake McConnell, and the surficial geology of part of Slave River and the Redstone River map-area, District of Mackenzie; Geological Survey of Canada, Bulletin 122, 33 p.
1965b: Note on moraines and radiocarbon dates in northwest Baffin Island, Melville Peninsula, and northeast District of Keewatin; Geological Survey of Canada, Paper 65-20, 7 p.
1969: Late glacial and postglacial history of the Hudson Bay region; Geological Survey of Canada, Paper 68-53, p. 63-77.

Craig, B.G. and Fyles, J.G.
1960: Pleistocene geology of Arctic Canada; Geological Survey of Canada, Paper 60-12, 21 p.

Cunningham, C.M. and Shilts, W.W.
1977: Surficial geology of the Baker Lake area, District of Keewatin; in Report of Activities, Part B, Geological Survey of Canada, Paper 77-1B, p. 311-314.

Dadswell, M.J.
1974: Distribution, ecology, and postglacial dispersal of certain crustaceans and fishes in eastern North America; National Museums of Canada Publications in Zoology, no. 11, 110 p.

Daly, R.A.
1902: The geology of the northeast coast of Labrador; Harvard University Museum of Comparative Zoology, Bulletin 38, p. 205-270.

305

David, P.P.
1977: Sand dune occurrences of Canada; Canada Department of Indian and Northern Affairs, National Parks Branch, Contract 74-230, 183 p.

Davis, P.T.
1980: Late Holocene glacial, vegetational, and climatic history of Pangnirtung and Kingnait fiord areas, Baffin Island, N.W.T., Canada; unpublished PhD thesis, University of Colorado, Boulder, 366 p.
1985: Neoglacial moraines on Baffin Island; in Quaternary Environments: Eastern Canadian Arctic, Baffin Bay and West Greenland, J.T. Andrews (ed.); Allen and Unwin, London, p. 682-718.

Delisle, C.E., Bouchard,M.A., et André, P.
1984: Les précipitations acides et leurs effets au nord du 55 ème parallèle du Québec; Rapport présenté à la fondation canadienne Donner, Section Environnement École Polytechnique de Montréal, Département de géologie Université de Montréal, 134 p.

Dell, C.I.
1975: Relationships of till to bedrock in the Lake superior region; Geology, v. 13, p. 563-564.

Denis, R.
1974: Late Quaternary geology and geomorphology in the Lake Maskinongé area, Quebec; Uppsala Universitet Naturgeografiska Institutionen, Report 28, 125 p.
1976: Région de Saint-Gabriel-de-Brandon; Ministère des Richesses naturelles, Rapport géologique 168, 56 p.

Denton, G.H. and Hughes, T.J.
1983: Milankovitch theory of ice ages: Hypothesis of ice-sheet linkage between regional insolation and global climate; Quaternary Research, v. 20, p. 125-144.

Denton, G.H. and Hughes, T.J. (ed.)
1981: The Last Great Ice Sheets; John Wiley and Sons, New York, 484 p.

Derbyshire, E.
1962a: The deglaciation of the Howells River valley and the adjacent parts of the watershed region, Central Labrador-Ungava; University of McGill, McGill Sub-Arctic Research Papers no. 14, 23 p.
1962b: Fluvioglacial erosion near Knob Lake, central Quebec-Labrador, Canada; Geological Society of America, Bulletin, v. 73, p. 1111-1126.

DiLabio, R.N.W.
1981: Glacial dispersal of rocks and minerals at the south end of Lac Mistassini, Québec, with special reference to the Icon dispersal train; Geological Survey of Canada, Bulletin 323, 46 p.
1982: Wood in Quaternary sediments near Timmins, Ontario; in Current Research, Part A, Geological Survey of Canada, Paper 82-1A, p. 433-434.
1989: Terrain geochemistry in Canada; Chapter 10 in Quaternary Geology of Canada and Greenland, R.J. Fulton (ed.); Geological Survey of Canada, Geology of Canada, no. 1 (also Geological Society of America, The Geology of North America, v. K-1).

Dionne, J-C.
1968: Fossiles marins pléistocènes dans la partie nord du lac Saint-Jean; Le Naturaliste canadien, vol. 95, p. 1401-1408.
1973: La dispersion des cailloux ordoviciens dans les formations quaternaires, au Saguenay/Lac Saint-Jean, Québec; Revue de géographie de Montréal, vol. 27, p. 339-364.
1974: The eastward transport of erratics in James Bay area, Québec; Revue de géographie de Montréal, v. 28, p. 453-457.
1977: La mer de Goldthwait au Québec; Géographie physique et Quaternaire, vol. 31, p. 61-80.
1986: Blocs de dolomie à stromatolites sur les rives de l'estuaire du Saint-Laurent, Québec; Géographie physique et Quaternaire, vol. 40, p. 93-98.

Dionne, J-C. et Dubois, J-M.
1980: Le complexe morainique frontal d'Aquanus-Kénamiou, Basse Côte Nord du Saint-Laurent (résumé); in Résumés et programmes, 4è Colloque sur le Quaternaire du Québec, Québec, p. 13.

Dionne, J-C. et Laverdière, C.
1969: Sites fossilifères du golfe de Laflamme; Revue de géographie de Montréal, vol. 23, p. 259-270.

Dowdeswell, E.K. and Andrews, J.T.
1985: The fiords of Baffin Island: description and classification; in Quaternary Environments: Eastern Canadian Arctic, Baffin Bay and West Greenland, J.T. Andrews (ed.); Allen and Unwin, London, p. 93-123.

Dowdeswell, J.A.
1982: Debris transport paths and sediment flux through the Grinnel Ice Cap, Frobisher Bay, Baffin Island, N.W.T., Canada; unpublished MA thesis, University of Colorado, Boulder, 176 p.

Dredge, L.A.
1976a: The Goldthwait Sea and its sediments: Godbout-Sept-Îles region, Quebec North Shore; in Report of Activities, Part C, Geological Survey of Canada, Paper 76-1C, p. 179-181.
1976b: Moraines in the Godbout-Sept-Îles area, Quebec North Shore; in Report of Activities, Part C, Geological Survey of Canada, Paper 76-1C, p. 183-184.
1981: Trace elements in till and esker sediments in northeastern Manitoba; in Current Research, Part A, Geological Survey of Canada, Paper 81-1A, p. 377-381.
1982: Relict ice-scour marks and late phases of Lake Agassiz in northernmost Manitoba; Canadian Journal of Earth Sciences, v. 19, p. 1079-1087.
1983a: Character and development of northern Lake Agassiz and its relation to Keewatin and Hudsonian ice regimes; in Glacial Lake Agassiz, J.T. Teller and L. Clayton (ed.); Geological Association of Canada, Special Paper 26, p. 117-131.
1983b: Surficial geology of the Sept-Îles area, Québec North Shore; Geological Survey of Canada, Memoir 408, 40 p.
1983c: Uranium and base metal concentrations in till, northern Manitoba; Geological Survey of Canada, Open File 931.

Dredge, L.A. and Cowan, W.R.
1989: Quaternary geology of the southwestern Canadian Shield; in Chapter 3 of Quaternary Geology of Canada and Greenland, R.J. Fulton (ed.); Geological Survey of Canada, Geology of Canada, no. 1 (also Geological Society of America, The Geology of North America, v. K-1).

Dredge, L.A. and Grant, D.R.
1982: Glacial Lake Agassiz and Laurentide ice domains (abstract); in Program with Abstracts; Geological Association of Canada, v. 7, p. 46.

Dredge, L.A. and Nielsen, E.
1985: Glacial and interglacial deposits in the Hudson Bay Lowlands: A summary of sites in Manitoba; in Current Research, Part A, Geological Survey of Canada, Paper 85-1A, p. 247-257.

Dredge, L.A. and Nixon, F.M.
1979: Thermal sensitivity and the development of tundra ponds and thermokarst lakes in the Manitoba portion of the Hudson Bay Lowland; in Current Research, Part C, Geological Survey of Canada, Paper 79-1C, p. 23-26.
1982a: Surficial geology, Cape Churchill, Manitoba; Geological Survey of Canada, Map 3- 1980, scale 1:250 000.
1982b: Surficial geology, Churchill, Manitoba; Geological Survey of Canada, Map 4-1980, scale 1:250 000.
1982c: Surficial geology, Nejanilini Lake, Manitoba; Geological Survey of Canada, Map 7-1980, scale 1:250 000.
1982d: Surficial geology, Shethanei Lake, Manitoba; Geological Survey of Canada, Map 6-1980, scale 1:250 000.
1982e: Surficial geology, Caribou River, Manitoba; Geological Survey of Canada, Map 5-1980, scale 1:250 000.
1982f: Surficial geology, York Factory, Manitoba; Geological Survey of Canada, Map 2-1980, scale 1:250 000.
1982g: Surficial geology, Herchmer, Manitoba; Geological Survey of Canada, Map 1-1980, scale 1:250 000.
1986: Surficial geology, northeastern Manitoba; Geological Survey of Canada, Map 1617A, scale 1:500 000.

Dredge, L.A., Nixon, F.M., and Richardson, R.J.
1981a: Surficial geology, Kasmere Lake, Manitoba; Geological Survey of Canada, Map 19-1981, scale 1:250 000.
1981b: Surficial geology, Tadoule Lake, Manitoba; Geological Survey of Canada, Map 17-1981, scale 1:250 000.
1986: Quaternary geology and geomorphology of northwestern Manitoba; Geological Survey of Canada, Memoir 418, 38 p.

Drexler, W., Farrand, W.R., and Hughes, J.D.
1983: Correlation of glacial lakes in the Superior basin with eastward discharge events from Lake Agassiz; in Glacial Lake Agassiz, J.T. Teller and L. Clayton (ed.); Geological Association of Canada, Special Paper 26, p. 309-329.

Drummond, R.N.
1965: Glacial geomorphology of the Cambrian Lake area, Labrador-Ungava; unpublished PhD thesis, McGill University, Montréal, 222 p.

Dubois, J-M.
1976: Levé préliminaire du complexe morainique du Manitou-Matamek sur la Côte Nord de l'estuaire maritime du Saint-Laurent; dans Report of Activities, Partie B, Commission géologique du Canada, Etude 76-1B, p. 89-93.
1977: La déglaciation de la Côte Nord du Saint-Laurent, analyse sommaire; Géographie physique et Quaternaire, vol. 31, p. 229-246.
1979: Télédétection, cartographie et interprétation des fronts glaciaires sur la Côte Nord du Saint-Laurent entre le lac Saint-Jean et le Labrador; Département de géographie, Université de Sherbrooke, Bulletin de recherche n° 42, 33 p.
1980: Environnements quaternaires et évolution littorale d'une zone côtière en émersion en bordure sud du bouclier canadien: la Moyenne Côte Nord du Saint-Laurent, Québec; thèse de PhD non publiée, Université d'Ottawa, 754 p.

Dubois, J-M. et Desmarais, G.
1983: Géomorphologie et formations meubles quaternaires de la Moyenne Côte Nord du Saint-Laurent; Commission géologique du Canada, Dossier public 958.

Dubois, J-M. and Dionne, J-C.
1985: The Québec North Shore frontal moraine system: a major feature of Late Wisconsinan deglaciation; Geological Society of America, Special Paper 197, p. 125-133.

Dubois, J-M., Desmarais, G., Brouillette, D., Perras, S., Tremblay, G.L., Larivière, L., Denis, F., et Lessard, G.
1984: La mer de Goldthwait sur la Côte Nord du Saint-Laurent: Harrington Harbour (12 J), Havre-St-Pierre (12 L), Musquaro (12 K), St-Augustin (12 O) et Blanc Sablon (12 P); Commission géologique du Canada, Dossier public 1048.

Dugdale, R.E.
1972: The Quaternary history of northern Cumberland Peninsula, Baffin Island, N.W.T., Part III. The late glacial deposits of Sulung Valley and adjacent parts of the Maktak trough; Canadian Journal of Earth Sciences, v. 9, p. 366-374.

Duplessis, J-C., Delibrias, G., Turon, J.L., Pujol, C., and Duprat, J.
1981: Deglacial warming of the northeastern North Atlantic: correlation with the paleoclimatic evolution of the European continent; Palaeogeography, Palaeoclimatology, Palaeoecology, v. 35, p. 121-144.

Dyck, W. and Fyles, J.G.
1963: Geological Survey of Canada radiocarbon dates I and II; Geological Survey of Canada, Paper 63-21, 31 p.

Dyck, W., Fyles, J.G., and Blake, W., Jr.
1965: Geological Survey of Canada radiocarbon dates IV; Geological Survey of Canada, Paper 65-4, 23 p.

Dyck, W., Lowdon, J.A., Fyles, J.G., and Blake, W., Jr.
1966: Geological Survey of Canada radiocarbon dates V; Geological Survey of Canada, Paper 66-48, 32 p.

Dyer, W.S. and Crozier, A.R.
1933: Lignite and refractory clay deposits of the Onakawana lignite field; Ontario Department of Mines, 42nd Annual Report, v. 42, part 3, p. 46-78.

Dyke, A.S.
1974: Deglacial chronology and uplift history: northeastern sector, Laurentide Ice Sheet; institute of Arctic and Alpine Research, University of Colorado, Occasional Paper, v. 12, 73 p.
1977: Quaternary geomorphology, glacial chronology, and climatic and sea-level history of southwestern Cumberland Peninsula, Baffin Island, Northwest Territories, Canada; unpublished PhD thesis, University of Colorado, Boulder, 185 p.
1978a: Indications of neoglacierization of Somerset Island, District of Franklin; Geological Survey of Canada, Paper 78-18, p. 215-217.
1978b: Glacial and marine limits on Somerset Island, Northwest Territories (abstract); in Abstracts with Programs; Geological Society of America, v. 10, p. 394.
1979a: Glacial geology of northern Boothia Peninsula, District of Franklin; in Current Research, Part B, Geological Survey of Canada, Paper 79-1B, p. 385-394.
1979b: Radiocarbon-dated Holocene emergence of Somerset Island, central Canadian Arctic; in Current Research, Part B, Geological Survey of Canada, Paper 79-1B, p. 307-318.
1979c: Glacial and sea level history of the southwestern Cumberland Peninsula, Baffin Island, N.W.T., Canada; Arctic and Alpine Research, v. 11, p. 179-202.
1980a: Redated Holocene whale bones from Somerset Island, District of Franklin; in Current Research, Part B, Geological Survey of Canada, Paper 80-1B, p. 269-270.
1980b: Base metal and uranium concentrations in till, northern Boothia Peninsula, District of Franklin; in Current Research, Part C, Geological Survey of Canada, Paper 80-1C, p. 155-159.
1983: Quaternary geology of Somerset Island, District of Franklin; Geological Survey of Canada, Memoir 404, 32 p.
1984: Quaternary geology of Boothia Peninsula and northern District of Keewatin, Central Canadian Arctic; Geological Survey of Canada, Memoir 407, 26 p.
1987: A reinterpretation of glacial and marine limits around the northwestern Laurentide Ice Sheet; Canadian Journal of Earth Sciences, v. 24, p. 591-601.

Dyke, A.S. and Dredge, L.A.
1989: Quaternary geology of the northwestern Canadian Shield; in Chapter 3 of Quaternary Geology of Canada and Greenland, R.J. Fulton (ed.); Geological Survey of Canada, Geology of Canada, no. 1 (also Geological Society of America, The Geology of North America, v. K-1).

Dyke, A.S. and Matthews, J.V., Jr.
1987: Stratigraphy and paleoecology of Quaternary sediments exposed along Pasley River, Boothia Peninsula, Central Canadian Arctic; Géographie physique et Quaternaire, v. 41, p. 323-344.

Dyke, A.S. and Prest, V.K.
1987: Late Wisconsinan and Holocene history of the Laurentide Ice Sheet; Géographie physique et Quaternaire, v. 41, p. 237-263.

Dyke, A.S., Andrews, J.T., and Miller, G.H.
1982a: Quaternary geology of Cumberland Peninsula, Baffin Island, District of Franklin; Geological Survey of Canada, Memoir 403, 32 p.

Dyke, A.S., Dredge, L.A., and Vincent, J-S.
1982b: Configuration of the Laurentide ice sheet during the Late Wisconsin maximum; Géographie physique et Quaternaire, v. 36, p. 5-14.
1983: Reply to comments by J.T. Andrews on "Configuration and dynamics of the Laurentide Ice Sheet during the Late Wisconsin maximum"; Géographie physique et Quaternaire, v. 37, p. 119-120.

Dyke, L.D.
1979: Bedrock heave in the central Canadian Arctic; in Current Research, Part A, Geological Survey of Canada, Paper 79-1A, p. 241-246.

Elson, J.A.
1967: The geology of Lake Agassiz; in Life, Land and Water, W.J. Mayer-Oakes (ed.); University of Manitoba Press, p. 37-95.
1969: Late Quaternary marine submergence of Quebec; Revue de géographie de Montréal, v. 23, p. 247-258.

England, J.H. and Andrews, J.T.
1973: Broughton Island — a reference area for Wisconsin and Holocene chronology and sea level changes on eastern Baffin Island; Boreas, v. 2, p. 17-32.

Engstrom, D.R., and Hansen, B.C.S.
1985: Postglacial vegetational change and soil development in southeastern Labrador as inferred from pollen and chemical stratigraphy; Canadian Journal of Botany, v. 63, p. 543-561.

Evans, D.J.A.
1984: Glacial geomorphology and chronology in the Selamiut Range/Nachvak Fiord area, Torngat Mountains, Labrador; unpublished MSc thesis, Memorial University of Newfoundland, St. John's, 138 p.

Evans, D.J.A. and Rogerson, R.J.
1986: Glacial geomorphology and chronology in the Selamiut Range/Nachvak Fiord area, Torngat Mountains, Labrador; Canadian Journal of Earth Sciences, v. 23, p. 66-76.

Fairbridge, R.W. and Hillaire-Marcel, C.
1977: An 8,000-yr paleoclimatic record of the "Double-Hale' 45-yr solar cycle; Nature, v. 268, no. 5619, p. 413-416.

Falconer, G.
1962: Glaciers of northern Baffin and Bylot Islands, N.W.T.; Canadian Geographical Branch, Geographical Paper 33, 31 p.
1966: Preservation of vegetation and patterned ground under thin ice body in northern Baffin Island, N.W.T.; Canadian Geographical Branch, Geographical Bulletin 8, p. 194-200.

Falconer, G., Ives, J.D., Løken, O.H., and Andrews, J.T.
1965: Major end moraines in eastern and central Arctic Canada; Geographical Bulletin, v. 7, p. 137-153.

Farrand, W.R.
1960: Former shorelines in western and northern Lake Superior basin; unpublished PhD thesis, University of Michigan, Ann Arbour, 226 p.
1962: Postglacial uplift in North America; American Journal of Science, v. 260, p. 181-199.

Farrand, W.R. and Drexler, C.W.
1985: Late Wisconsinan and Holocene history of the Lake Superior Basin; in Quaternary Evolution of the Great Lakes, P.F. Karrow (ed.); Geological Association of Canada, Special Paper 30, p. 17-32.

Fenton, M.M., Moran, S.R., Teller, J.T., and Clayton, L.
1983: Quaternary stratigraphy and history in the southern part of the Lake Agassiz basin; in Glacial Lake Agassiz, J.T. Teller and L. Clayton (ed.); Geological Association of Canada, Special Paper 26, p. 49-74.

Feyling-Hanssen, R.W.
1976a: The Clyde Foreland Formation, a micropaleontological study of Quaternary stratigraphy; Maritime Sediments, Special Publication no. 1, Part B, p. 315-377.
1976b: A mid-Wisconsin Interstadial on Broughton Island, Arctic Canada, and its foraminifera; Arctic and Alpine Reserach, v. 8, p. 161-182.
1980: Microbiostratigraphy of young Cenozoic marine deposits at the Qivituq Peninsula, Baffin Island; Marine Micropaleontology, p. 153-184.
1985: Late Cenozoic marine deosits of East Baffin Island and East Greenland: Microbiostratigraphy, Correlation, Age; in Quaternary Environments: Eastern Canadian Arctic, Baffin Bay and West Greenland, J.T. Andrews (ed.); Allen and Unwin, London, p. 354-393.

Fillon, R.H.
1985: Northwest Labrador Sea stratigraphy, sand input and paleooceanography during the last 160 000 years; in Quaternary Environments: Eastern Canadian Arctic, Baffin Bay, and West Greenland, J.T. Andrews (ed.); Allen and Unwin, London, p. 210-247.

Fillon, R.H. and Harmes, R.A.
1982: Northern Labrador shelf glacial chronology and depositional environments; Canadian Journal of Earth Sciences, v. 19, p. 162-192.

Fillon, R.H., Hardy, I.A., Wagner, F.J.E., Andrews, J.T., and Josenhans, H.W.
1981: Labrador shelf: Shell and total organic matter ^{14}C date discrepancies; in Current Research, Part B, Geological Survey of Canada, Paper 81-1B, p. 105-111.

Fisher, D.A., Reeh, N., and Langley, K.
1985: Objective reconstruction of the Late Wiconsinan Ice Sheet; Géographie physique et Quaternaire, v. 39, p. 229-238.

Fitzhugh, W.
1973: Environmental approaches to the prehistory of the north; Journal of the Washington Academy of Sciences, v. 63, p. 39-53.

Flint, R.F.
1943: Growth of the North American Ice Sheet during the Wisconsin Age; Geological Society of America, Bulletin, v. 54, p. 325-362,.
1947: Glacial Geology of the Pleistocene Epoch; John Wiley, New York.
1957: Glacial and Pleistocene Geology; Chapman and Hall Ltd., London, 553 p.
1971: Glacial and Quaternary Geology; John Wiley and Sons, New York, 892 p.

Forman, S.L., Wintle, A., Thorleifson, L.H., and Wyatt, P.H.
1987: Thermoluminescent properties and dates from Holocene and Pleistocene raised marine sediments, Hudson Bay Lowland, Canada; Canadian Journal of Earth Sciences, v. 24, p. 2405-2411.

Fortier, Y.O. and Morley, L.W.
1956: Geological unit of the Arctic Islands; Transactions of the Royal Society of London, v. 50, p. 3-12.

Fulton, R.J.
1986a: Surficial geology, Red Wine River, Labrador, Newfoundland; Geological Survey of Canada, Map 1621A, scale 1:500 000.
1986b: Surficial geology, Cartwright, Labrador, Newfoundland; Geological Survey of Canada, Map 1620A, scale 1:500 000.

Fulton, R.J. and Hodgson, D.A.
1979: Wisconsin glacial retreat, Southern Labrador; in Current Research, Part C, Geological Survey of Canada, Paper 79-1C, p. 17-21.
1980: Surficial materials, Goose Bay, Newfoundland; Geological Survey of Canada, Map 22-1979, scale 1:250 000.

Fulton, R.J., Hodgson, D.A., and Minning, G.V.
1979: Surficial materials, Lac Brûlé, Newfoundland-Québec; Geological Survey of Canada, Map 1-1978, scale 1:250 000.
1980a: Surficial materials, Rigolet, Newfoundland; Geological Survey of Canada, Map 26-1979, scale 1:250 000.
1980b: Surficial materials, Groswater Bay, Newfoundland; Geological Survey of Canada, Map 25-1979, scale 1:250 000.
1980c: Surficial materials, Lake Melville, Newfoundland; Geological Survey of Canada, Map 23-1979, scale 1:250 000.
1980d: Surficial materials, Cartwright, Newfoundland; Geological Survey of Canada, Map 24-1979, scale 1:250 000.
1981a: Surficial materials, Ossokmanuan, Newfoundland; Geological Survey of Canada, Map 29-1979, scale 1:250 000.
1981b: Surficial materials, Minipi Lake, Newfoundland; Geological Survey of Canada, Map 1531A, scale 1:250 000.
1981c: Surficial materials, upper Eagle River, Newfoundland; Geological Survey of Canada, Map 20-1979, scale 1:250 000.

Fulton, R.J., Hodgson, D.A., Minning, G.V., and Thomas, R.D.
1980e: Surficial materials, Kasheshibaw Lake, Newfoundland-Québec; Geological Survey of Canada, Map 28-1979, scale 1:250 000.
1980f: Surficial materials, Snegamook Lake, Newfoundland; Geological Survey of Canada, Map 27-1979, scale 1:250 000.

Fulton, R.J., Hodgson, D.A., Thomas, R.D., and Minning, G.V.
1981d: Surficial materials, Winokapau Lake, Newfoundland; Geological Survey of Canada, Map 21-1979, scale 1:250 000.

Fulton, R.J., Minning, G.V., and Hodgson, D.A.
1981e: Surficial materials, Battle Harbour, Newfoundland; Geological Survey of Canada, Map 19-1979, scale 1:250 000.

Fyles, J.G.
1963: Surficial geology of Victoria and Stefansson Islands, District of Franklin; Geological Survey of Canada, Bulletin 101, 38 p.

Gadd, N.R.
1964: Moraines in the Appalachian region of Quebec; Geological Society of America, Bulletin, v. 75, p. 1249-1254.
1972: Marine deposits, Gatineau Valley, Quebec; in Report of Activities, Part A, Geological Survey of Canada, Paper 72-1A, p. 156-157.
1980: Late-glacial regional ice-flow patterns in eastern Ontario; Canadian Journal of Earth Sciences, v. 17, p. 1439-1453.

Gagnon, R. et Payette, S.
1981: Fluctuations holocènes de la limite des forêts de mélèze, Rivière aux Feuilles, Nouveau-Québec: une analyse macrofossile en milieu tourbeux; Géographie physique et Quaternaire, vol. 35, p. 57-72.

Gangloff, P.
1983: Les fondements géomorphologiques de la théorie des paléonunataks: le cas des monts Torngats; Zeitschrift fur Geomorphologie, Supplement 47, p. 109-136.

Gangloff, P., Gray, J.T., et Hillaire-Marcel, C.
1976: Reconnaissance géomorphologique sur la côte ouest de la baie d'Ungava, Nouveau-Québec; Revue de géographie de Montréal, vol. 30, p. 339-348.

Geddes, R.S.
1984: An exotic till in the Hemlo area, northern Ontario (abstract); in Program with Abstracts, Geological Association of Canada, p. 66.

Geddes, R.S., Bajc, A.F., and Kristjansson, F.J.
1985: Quaternary geology of the Hemlo Region, District of Thunder Bay; in Summary of Field Work and Other Activities, J. Wood, O.L. White, R.B. Barlow, and A.C. Colvine (ed.); Ontario Geological Survey, Miscellaneous Paper 126, p. 151-154.

Gilbert, R.
1982: Contemporary sedimentary environments on Baffin Island, N.W.T., Canada: Glaciomarine processes in fiords of eastern Cumberland Peninsula; Arctic and Alpine Research, v. 14, p. 1-12.
1983: Sedimentary processes of Canadian arctic fiords; Sedimentary Geology, v. 36, p. 147-175.

Gilbert, R., and MacLean, B.
1983: Geophysical studies based on conventional shallow and Huntec high resolution seismic surveys of fiords on Baffin Island; in Sedimentology of Arctic Fjords Experiment:HU82-031 Data Report. Volume 1 J.P.M. Syvitski and C.P. Blakeney (ed.); Canadian Data Report of Hydrography and Ocean Sciences, Supply and Services Canada, no. 12, 90 p.

Gilchrist, C.M.
1982: The glacial geology of the southeastern area of the District of Keewatin, Northwest Territories, Canada; unpublished PhD thesis, University of Massachusetts, Boston, 307 p.

Granberg, H.B. and Krishnan, T.K.
1984: Wood remnants 24 250 years old in Central Labrador (abstract); in Program and Abstracts, 5th AQQUA Congress, Sherbrooke, p. 30.

Grant, A.C.
1975: Geophysical results from the continental margin off southern Baffin Island; in Canada's Continental Margins and Offshore Petroleum Exploration, C.J. Yorath, E.R. Parker, and D.J. Glass (ed.); Canadian Society of Petroleum Geologists, Memoir 6, p. 411-431.

Grant, A.C. and Manchester, K.S.
1970: Geophysical investigation in the Ungava Bay-Hudson Strait region of northern Canada; Canadian Journal of Earth Sciences, v. 7, p. 1062-1076.

Grant, D.R.
1969: Surficial deposits Lake Melville area, Labrador (13 F and G); in Report of Activities, Part A, Geological Survey of Canada, Paper 69-1A, p. 197- 198.
1971: Geomorphology, Lake Melville area, Labrador (13 F and G); in Report of Activities, Part B, Geological Survey of Canada, Paper 71-1A, p. 114- 117.
1986: Surficial geology, St. Anthony-Blanc-Sablon, Newfoundland-Québec; Geological Survey of Canada, Map 1610A, scale 1:125 000.
1989: Quaternary geology of the Atlantic Appalachian region of Canada; Chapter 5 in Quaternary Geology of Canada and Greenland, R.J. Fulton (ed.); Geological Survey of Canada, Geology of Canada, no. 1 (also Geological Society of America, The Geology of North America, v. K-1).

Gratton, P., Gwyn, Q.H.T., et Dubois, J.M.M.
1984: Les paléoenvironnements sédimentaires au Wisconsinien moyen et supérieur, île d'Anticosti, golfe du Saint-Laurent, Québec; Géographie physique et Quaternaire, vol. 38, p. 229-242.

Gravenor, C.P.
1975: Erosion by continental ice sheets; American Journal of Science, vol. 275, p. 594-604.

Gray, J.T.
1969: Glacial history of the eastern Mealy Mountains, southern Labrador; Arctic, v. 22, p. 106-111.

Gray, J.T. and Lauriol, B.
1985: Dynamics of the Late Wisconsin Ice Sheet in the Ungava Peninsula interpreted from geomorphological evidence; Arctic and Alpine Research, v. 17, p. 289-310.

Gray, J., Boutray, B. de, Hillaire-Marcel, C., and Lauriol, B.
1980: Postglacial emergence of the west coast of Ungava Bay, Québec; Arctic and Alpine Research, v. 12, p. 19-30.

Grayson, J.T.
1956: The post-glacial history of vegetation and climate in the Labrador-Quebec region as determined by palynology; unpublished PhD thesis, University of Michigan, 252 p.

Guimont, P. et Laverdière, C.
1982: Le réservoir du Caniapiscau esquisse géomorphologique; Rapport, Société de développement de la Baie James, 5 cartes au 1/60 000.

Gunn, C.B.
1968: Relevance of the Great Lakes discoveries to Canadian Diamond prospecting; Canadian Mining Journal, v. 39, p. 39-42.

Hardy, L.
1970: Géomorphologie glaciaire et post-glaciaire de St-Siméon à St-François d'Assises (Comtés de Charlevoix est et de Chicoutimi); thèse de maîtrise non publiée, Université Laval, 112 p.
1976: Contribution à l'étude géomorphologique de la portion québécoise des basses terres de la baie de James; thèse de doctorat non publiée, Université McGill, Montréal, 264 p.
1977: La déglaciation et les épisodes lacustre et marin sur le versant québécois des basses-terres de la baie de James; Géographie physique et Quaternaire, vol. 31, p. 261-273.
1982a: La moraine frontale de Sakami, Québec subarctique; Géographie physique et Quaternaire, vol. 36, p. 51-61.
1982b: Le Wisconsinien supérieur à l'est de la baie James (Québec); Naturaliste canadien, vol. 109, p. 333-351.

Hare, F.K.
1959: A photo-reconnaissance survey of Labrador-Ungava; Geographical Branch, Memoir 6, 63 p.

Harrison, D.A.
1963: The tilt of the abandoned lake shorelines in the Wabush-Shabogamo Lake area, Labrador; in Geographical Studies in Labrador, Annual Report 1961-1962; McGill Sub-Arctic Research Paper no. 15, p. 14-22.

Harrison, J.E.
1972: Quaternary geology of the North Bay-Mattawa region; Geological Survey of Canada, Paper 71-26, 37 p.

Hawkins, F.F.
1980: Glacial geology and late Quaternary paleoenvironment in the Merchants Bay area, Baffin Island, N.W.T., Canada; unpublished MSc thesis, University of Colorado, Boulder, 145 p.

Hélie, R.G.
1983: Relict iceberg scours, King William Island, Northwest Territories; in Current Research, Part B, Geological Survey of Canada, Paper 83-1B, p. 415-418.
1984: Surficial geology, King William Island and Adelaide Peninsula, Districts of Keewatin and Franklin, Northwest Territories; Geological Survey of Canada, Map 1618A, scale 1:250 000.

Hélie, R.G. and Elson, J.A.
1984: Discrimination between glacigenic and weathering residue diamictons, Somerset Island, Northwest Territories; in Current Research, Part A, Geological Survey of Canada, Paper 84-1A, p. 339-344.

Henderson, E.P.
1959: A glacial study of central Quebec-Labrador; Geological Survey of Canada, Bulletin 50, 94 p.

Henderson, P.J.
1983: A study of the heavy mineral distribution in the bottom sediments of Hudson Bay; in Current Research, Part A, Geological Survey of Canada, Paper 83-1A, p. 347-351.

Higgs, R.
1978: Provenance of Mesozoic and Cenozoic sediments from the Labrador and western Greenland continental margins; Canadian Journal of Earth Sciences, v. 15, p. 1850-1860.

Hillaire-Marcel, C.
1976: La déglaciation et le relèvement isostatique sur la côte est de la baie d'Hudson; Cahiers de géographie de Québec, vol. 20, p. 185-220.
1979: Les mers post-glaciaires du Québec: quelques aspects; thèse de doctorat d'état non publiée, Université Pierre et Marie Curie, Paris VI, vol. 1, 293 p., vol. 2, 249 p.
1980: Multiple component postglacial emergence, eastern Hudson Bay, Canada; in Earth Rheology, Isostasy and Eustasy, N-A. Mörner (ed.) John Wiley and Sons, Chichester, p. 215-230.
1981: Late-glacial regional ice-flow patterns in eastern Ontario: Discussion; Canadian Journal of Earth Sciences, v. 18, p. 1385-1386.

Hillaire-Marcel, C. et Boutray, B. de
1975: Les dépôts meubles de la région du Poste-de-la-Baleine (Nouveau-Québec); Nordicana, n° 38, 47 p.

Hillaire-Marcel, C. and Fairbridge, R.W.
1978: Isostasy and eustasy of Hudson Bay; Geology, v. 6, p. 117-122.

Hillaire-Marcel, C. and Occhietti, S.
1980: Chronology paleogeography and paleoclimatic significance of the late and post-glacial events in eastern Canada; Zeitschrift fur geomorphologie, v. 24, p. 373-392.

Hillaire-Marcel, C. and Vincent, J-S.
1980: Holocene stratigraphy and sea level changes in southeastern Hudson Bay, Canada; Paleo-Québec, no. 11, 165 p.

Hillaire-Marcel, C., Grant, D.R., and Vincent, J-S.
1980: Comment and reply on "Keewatin Ice Sheet-re-evaluation of the traditional concept of the Laurentide Ice Sheet" and "Glacial erosion and ice sheet divides, northeastern Laurentide Ice Sheet, on the basis of the distribution of limestone erratics"; Geology, v. 8, p. 466-468.

Hillaire-Marcel, C., Occhietti, S., and Vincent, J-S.
1981: Sakami moraine, Quebec: a 500-km-long-moraine without climatic control; Geology, v. 9, p. 210-214.

Hobson, G. D.
1968: Seismic refraction results from the Hudson region; in Earth Science Symposium on Hudson Bay, P.J. Hood (ed.); Geological Survey of Canada, Paper 68-53, p. 227-246.

Hodgson, D.A. and Fulton, R.J.
1972: Site description, age and significance of a shell sample from the mouth of the Michael River, 30 km south of Cape Harrison, Labrador; in Report of Activities, Part B, Geological Survey of Canada, Paper 72-1B, p. 102-105.

Hodgson, D.A. and Haselton, G.M.
1974: Reconnaissance glacial geology, north-eastern Baffin Island; Geological Survey of Canada, Paper 74-20, 10 p.

Hodgson, D.A. and Vincent, J-S.
1984: A 10 000 yr BP extensive ice shelf over Viscount Melville Sound, Arctic Canada; Quaternary Research, v. 22, p. 18-30.

Hooke, R. LeB.
1982: Wisconsin and Holocene delta 18O variations, Barnes Ice Cap, Canada; Geological Society of America, Bulletin, v. 93, p. 784-789.

Horton, R.E.
1945: Erosional development of streams and their drainage basins: Hydrophysical approach to quantitative morphology; Geological Society of America, Bulletin, v. 56, p. 275-370.

Hough, J.L.
1958: Geology of the Great Lakes; University of Illinois Press, Urbana, 313 p.

Hughes, O.L.
1959: Surficial geology of Smooth Rock and Iroquois Falls map-areas, Cochrane District, Ontario; unpublished PhD thesis, University of Kansas, Lawrence, 190 p.
1964: Surficial geology, Nichicun-Kaniapiskau map-area, Quebec; Geological Survey of Canada, Bulletin 106, 20 p.
1965: Surficial geology of part of the Cochrane District, Ontario, Canada; in International Studies on the Quaternary, H.E. Wright and D.G. Frey (ed.); Geological Society of America, Special Paper 84, p. 535-565.

Hughes, T.J.
1975: West Antarctic ice streams; Reviews of Geophysics and Space Physics, v. 15, p. 1-46.

Hughes, T.J., Denton, G.H., Andersen, B.G., Schilling, D.G., Fastook, J.L., and Lingle, C.S.
1981: The last great ice sheets: a global view; in The Last Great Ice Sheets, G.H. Denton and T.J. Hughes (ed.); John Wiley and Sons, New York, p. 263-317.

Hunter, G.T.
1970: Postglacial uplift at Fort Albany, James Bay; Canadian Journal of Earth Sciences, v. 7, p. 547-548.

Ignatius, H.
1958: On the Late-Wisconsin deglaciation in eastern Canada; Acta Geographica, v. 16, p. 1-34.

Isherwood, D.J.
1975: Soil geochemistry and rock weathering in an arctic environment; unpublished PhD thesis, University of Colorado, Boulder, 173 p.

Ives, J.D.
1957: Glaciation of the Torngat Mountains, Northern Labrador; Arctic, v. 10, p. 67-87.
1958a: Glacial geomorphology of the Torngat Mountains, Northern Labrador; Geographical Bulletin, no. 12, p. 47-75.
1958b: Mountain-top detritus and the extent of the last glaciation in northeastern Labrador-Ungava; The Canadian Geographer, no. 12, p. 25-31.
1959: Glacial drainage channels as indicators of late-glacial conditions in Labrador-Ungava: a discussion; Cahiers de géographie de Québec, v. 3, p. 57-72.
1960a: The deglaciation of Labrador-Ungava, an outline; Cahiers de géographie de Québec, v. 4, p. 323-343.
1960b: Former ice-dammed lakes and the deglaciation of the middle reaches of the George River, Labrador-Ungava; Geographical Bulletin, v. 14, p. 44-70.
1960c: Glaciation and deglaciation of the Helluva Lake area, Central Labrador-Ungava; Geographical Bulletin, no. 15, p. 46-64.
1962: Indications of recent extensive glacierization in north central Baffin Island, N.W.T.; Journal of Glaciology, v. 4, p. 197-205.
1963: Field problems in determining the maximum extent of Pleistocene Glaciation along the eastern Canadian seaboard — A geographer's point of view; in North Atlantic Biota and Their History, A. Love and D. Love (ed.), The Macmillan Company, New York, p. 337-354.
1964: Deglaciation and land emergence in northeastern Foxe Basin, N.W.T.; Geographical Bulletin, v. 21, p. 54-65.

1968: Late-Wisconsin events in Labrador-Ungava: an interim commentary; Canadian Geographer, v. 12, p. 192-203.
1974: Biological refugia and Nunatak hypothesis; in Arctic and Alpine Environments, J.D. Ives and R.G. Barry (ed.); Methuen, London, p. 605-636.
1975: Delimitation of surface weathering zones in eastern Baffin Island, Northern Labrador and Arctic Norway: a discussion; Geological Society of America, Bulletin, v. 86, p. 1096-1100.
1976: The Saglek moraines of northern Labrador: a commentary; Arctic and Alpine Research, v. 8, p. 403-408.
1977: Were parts of the north coast of Labrador ice-free at the Wisconsin glacial maximum?; Géographie physique et Quaternaire, v. 31, p. 401- 403.
1978: The maximum extent of the Laurentide Ice Sheet along the east coast of North America during the last glaciation; Arctic, v. 31, p. 24-53.
1979: A proposed history of permafrost development in Labrador-Ungava; Géographie physique et Quaternaire, v. 33, p. 233-244.

Ives, J.D. and Andrews, J.T.
1963: Studies in the physical geography of north central Baffin Island; Geographical Bulletin, v. 19, p. 5-48.

Ives, J.D. and Borns, J.W., Jr.
1971: Thickness of the Wisconsin ice sheet in southeast Baffin Island; Zeitschrift fur Gletscher und Glacialgeologie, v. 7, p. 17-21.

Ives, J.D. and Buckley, J.T.
1969: Glacial geomorphology of Remote Peninsula, Baffin Island, N.W.T.; Arctic and Alpine Research, v. 1, p. 83-96.

Ives, J.D., Andrews, J.T., and Barry, R.G.
1975: Growth and decay of the Laurentide Ice Sheet and comparisons with Fenno-Scandinavia; Naturwissenchaften, v. 62, p. 118-125.

Jackson, G.D.
1960: Belcher Islands, Northwest Territories; Geological Survey of Canada, Paper 60-20, 13 p.

Jacobs, J.D., Andrews, J.T., and Funder, S.
1985: Environmental background; in Quaternary Environments: Eastern Canadian Arctic, Baffin Bay, and West Greenland, J.T. Andrews (ed.); Allen and Unwin, London, p. 26-68.

Jahns, R.H..
1943: Sheet structure in granites: its origin and use as a measure of glacial erosion in New England; Journal of Geology, v. 51, p. 71-98.

Jenness, J.L.
1952: Problem of glaciation in the western islands of Arctic Canada; Geological Society of America, Bulletin, v. 63, p. 939-952.

Johnson, J.P., Jr.
1969: Deglaciation of the central Nain-Okak Bay section of Labrador; Arctic, v. 22, p. 373-394.

Johnston, W.A.
1918: Reconnaissance soil survey of the area along the Hudson Bay railway; Geological Survey of Canada, Summary Report 1917, Part D, p. 25-36.

Johnston, W.G.Q.
1978: Lake Agassiz's northernmost and other features of the Quaternary geology in the region around the southern part of Reindeer Lake, northern Saskatchewan; University of Saskatchewan, Musk-Ox, no. 21, p. 39-50.

Jordan, R.
1975: Pollen diagrams from Hamilton Inlet, Central Labrador, and their environmental implications for the northern Maritime Archaic; Arctic Anthropology, v. 12, p. 92-116.

Josenhans, H.W., Zevenhuizen, J., and Klassen, R.A.
1986: The Quaternary geology of the Labrador Shelf; Canadian Journal of Earth Sciences, v. 23, p. 1190-1213.

Karrow, P.F.
1989: Quaternary geology of the Great Lakes subregion; in Chapter 4 of Quaternary Geology of Canada and Greenland, R.J. Fulton (ed.); Geological Survey of Canada, Geology of Canada, no. 1 (also Geological Society of America, The Geology of North America, v. K-1).

Karrow, P.F. and Geddes, R.S.
1987: Drift carbonate on the Canadian Shield; Canadian Journal of Earth Sciences, v. 24, p. 365-369.

Karrow, P.F., Anderson, T.W., Clarke, A.H., Delorme, L.D., and Sreenivasa, M.R.
1975: Stratigraphy, paleontology, and age of Lake Algonquin sediments in southwestern Ontario, Canada; Quaternary Research, v. 5, p. 49-87.

Kaszycki, C.A. and Shilts, W.W.
1980: Glacial erosion of the Canadian Shield — calculation of average depths; Atomic Energy of Canada Limited Research Company, Technical Record, TR-106, 37 p.

Keele, J.
1921: Mesozoic clays and sands in northern Ontario; Geological Survey of Canada, Summary Report 1920, Part. D, p. 35-39.

Kerr, J.W.
1980: Structural framework of Lancaster Aulacogen, Arctic Canada; Geological Survey of Canada, Bulletin 319, 24 p.

Kettles, I.M. and Shilts, W.W.
1987: Tills of the Ottawa region; in Quaternary Geology of the Ottawa region, Ontario and Quebec, R.J. Fulton (ed.); Geological Survey of Canada, Paper 86-23, p. 10-13.

King, C.A.M.
1969: Glacial geomorphology and chronology of Henry Kater Peninsula, east Baffin Island, N.W.T.; Arctic and Alpine Research, v. 1, p. 195-212.

King, G.A.
1985: A standard method for evaluating radiocarbon dates of local deglaciation: Application to the deglaciation history of southern Labrador and adjacent Quebec; Géographie physique et Quaternaire, v. 39, p. 163-182.

Kirby, R.P.
1961a: Deglaciation in central Labrador-Ungava as interpreted from glacial deposits; Geographical Bulletin, no. 16, p. 4-23.
1961b: Movements of ice in Central Labrador-Ungava; Cahiers de géographie de Québec, no. 10, p. 205-218.

Kish, L., La Salle, P., et Szoghy, I.M.
1979: Rb, Sr, Y, Zr, Nb, Mo dans les tills de base de l'Abitibi; Ministère des Richesses naturelles, DPV-662, 8 p.

Klassen, R.A.
1981: Aspects of the glacial history of Bylot Island, District of Franklin; in Current Research, Part A, Geological Survey of Canada, Paper 81-1A, p. 317-326.
1982: Quaternary stratigraphy and glacial history of Bylot Island, N.W.T., Canada; unpublished PhD thesis, University of Illinois, Urbana, 161 p.
1983a: A preliminary report on drift prospecting studies in Labrador; in Current Research, Part A, Geological Survey of Canada, Paper 83-1A, p. 353- 355.
1984: A preliminary report on drift prospecting studies in Labrador Part II; in Current Research, Part A, Geological Survey of Canada, Paper 84-1A, p. 247-254.
1985: An outline of glacial history of Bylot Island, District of Franklin, N.W.T.; in Quaternary Environments: Eastern Canadian Arctic, Baffin Bay and West Greenland, J.T. Andrews (ed.); Allen and Unwin, London, p. 428-460.

Klassen, R.A. and Bolduc, A.
1984: Ice flow directions and drift composition, Churchill Falls, Labrador; in Current Research, Part A, Geological Survey of Canada, Paper 84-1A, p. 255-258.

Klassen, R.A. and Thompson, F.J.
1987: Ice flow history and glacial dispersal in the Labrador Trough; in Current Research, Part A, Geological Survey of Canada, Paper 87-1A, p. 61-71.

Klassen R.A., Matthews, J.V., Jr., Mott, R.J., Ovenden, L., and Thompson, F.J.
1988: The stratigraphic and paleobotanical record of interglaciation in Wabush region of western Labrador (abstract); in Program and Abstracts, Third Annual Meeting, Canadian Committee on Climatic Fluctuations and Man, Ottawa, p. 24-26.

Klassen, R.W.
1983b: Lake Agassiz and the glacial history of northern Manitoba; in Glacial Lake Agassiz, J.T. Teller and L. Clayton (ed.); Geological Association of Canada, Special Paper 26, p. 97-115.
1986: Surficial geology of north-central Manitoba; Geological Survey of Canada, Memoir 419, 57 p.
1989: Quaternary geology of the southern Canadian Interior Plains; in Chapter 2 of Quaternary Geology of Canada and Greenland, R.J. Fulton (ed.); Geological Survey of Canada, Geology of Canada, no. 1 (also Geological Society of America, The Geology of North America, v. K-1).

Klassen, R.W. and Netterville, J.A.
1973: Quaternary geology inventory, lower Nelson River basin (53M, 54C,D, 63O,P, 64A,B); in Report of Activities, Part A, Geological Survey of Canada, Paper 73-1A, p. 204-205.

Koerner, R.M.
1980a: Instantaneous glacierization, the rate of albedo change, and feedback effects at the beginning of an ice age; Quaternary Research, v. 13, p. 153-159.
1980b: The problem of lichen-free zones in arctic Canada; Arctic and Alpine Research, v. 12, p. 87-94.

Lamb, H.F.
1980: Late Quaternary vegetational history of northeastern Labrador; Arctic and Alpine Research, v. 12, p. 117-135.
1984: Modern pollen spectra from Labrador and their use in reconstructing Holocene vegetational history; Journal of Ecology, v. 72, p. 37-59.

Lamb, H.H. and Woodroffe, A.
1970: Atmospheric circulation during the last ice-age; Quaternary Research, v. 1, p. 29-58.

Lamothe, M.
1977: Les dépôts meubles de la région de Saint-Faustin, Saint-Jovite, Québec: cartographie, sédimentologie et stratigraphie; thèse de maîtrise non publiée, Université du Québec à Montréal, Montréal, 118 p.

Larsen, C.E.
1985: Lake level, uplift, and outlet incision, the Nipissing and Algoma Great Lakes; in Quaternary Evolution of the Great Lakes, P.F. Karrow and P.E. Calkin (ed.); Geological Association of Canada, p. 61-77.

LaSalle, P.
1978: Géologie des sédiments de surface de la région de Québec; Ministère des Richesses naturelles, DPV 565, 22 p.

LaSalle, P. and Rondot, J.
1967: New ^{14}C dates from the Lake St. John area, Québec; Canadian Journal of Earth Sciences, v. 4, p. 568-571.

LaSalle, P. et Tremblay, G.
1978: Dépôts meubles, Saguenay Lac Saint-Jean;Ministère des Richesses naturelles, Rapport géologique 191, 61 p.

LaSalle, P., Martineau, G., et Chauvin, L.
1977a: Morphologie, stratigraphie et déglaciation dans la région de Beauce — Monts Notre-Dame — Parc des Laurentides; Ministère des Richesses naturelles, DPV 516, 74 p.

1977b: Dépôts morainiques et stries glaciaires dans la région de Beauce — Monts Notre-Dame — Parc des Laurentides; Ministère des Richesses naturelles, DPV 515, 22 p.

Lauriol, B.
1982: Géomorphologie quaternaire du sud de l'Ungava; PaléoQuébec, n° 15, 174 p.

Lauriol, B. et Gray, J.T.
1983: Un lac glaciaire dans la région du lac Minto-Nouveau Québec; Journal canadien des sciences de la terre, vol. 10, p. 1488-1492.

Lauriol, B. and Gray, J.T.
1987: The decay and disappearence of the Late Wisconsinan Ice Sheet in the Ungava Peninsula, northern Quebec, Canada; Arctic and Alpine Research, v. 19, p. 109-126.

Lauriol,B.,Gray, J.T., Hétu, B., et Cyr, A.
1979: Le cadre chronologique et paléogéographique de l'évolution marine depuis la déglaciation dans la région d'Aupaluk, Nouveau-Québec, Géographie physique et Quaternaire, vol. 33, p. 189-203.

Laverdière, C.
1967: Sur le lieu de fonte sur place de la calotte glaciaire de Scheffer; Le géographe canadien, vol. 11, p. 87-95.

1969a: The Scheffer Ice-Sheet: a reply to Ives' comments; Canadian Geographer, v. 13, no. 3, p. 269-283.

1969b: Le retrait de la calotte glaciaire de Scheffer: du Témiscamingue au Nouveau-Québec; Revue de géographie de Montréal, vol. 23, p. 233-246.

Laverdière, C. et Bernard, C.
1969: Sur quelques néochronymes (Mer d'Iberville); dans Le vocabulaire de la géomorphologie glaciaire (V^e article); Revue de géographie de Montréal, vol. 23, p. 355-358.

Laverdière, C. et Courtemanche, A.
1960: La géomorphologie glaciaire de la région du mont Tremblant, 2e partie, La région de Saint-Faustin — Saint-Jovite; Cahiers de géographie de Québec, vol. 5, p. 5-32.

Laverdière, C. et Guimont, P.
1982: Le réservoir de Caniapiscau, étude du milieu physique; Rapport interne, Société de développement de la baie James, 125 p.

Laverdière, C. et Mailloux, A.
1956: État de nos connaissances d'une transgression marine post-glaciaire dans les régions du haut Saguenay et du lac Saint-Jean; Revue canadienne de géographie, vol. 10, p. 201-220.

Lawson, A.C.
1890: Notes on the pre-Paleozoic surface of the Archean terranes of Canada; Geological Society of America, Bulletin no. 1, p. 163-173.

Laymon, C.
1984: Glacial geology of western Hudson Strait with reference to Laurentide Ice Sheet dynamics (abstract); in Abstracts with Program, Geological Society of America, v. 16, Annual Meeting, Reno, Nevada, p. 571.

Lee, H.A.
1959a: Surficial geology of southern District of Keewatin and the Keewatin Ice Divide, Northwest Territories; Geological Survey of Canada, Bulletin 51, 42 p.

1959b: Eastward transport of erratics from Hudson Bay; Geological Society of America, Bulletin, v. 70, p. 219-222.

1960: Late glacial and postglacial Hudson Bay sea episode; Science, v. 131, no. 3413, p. 1609-1611.

1962: Method of deglaciation, age of submergence, and rate of uplift west and east of Hudson Bay, Canada; Biuletyn Peryglacjalny, no. 11, p. 239-245.

1968: Quaternary geology; in Science, History and Hudson Bay, Volume 2, C.S. Beals and D.A. Shenstone (ed.); Queen's Printer, Ottawa, p. 503-543.

Lee, H.A., Eade, K.E., and Heywood, W.W.
1960: Surficial geology, Sakami Lake (Fort George-Great Whale Area, New Quebec); Geological Survey of Canada, Map 52-1959, scale 1:506 880.

Lewis, C.F.M.
1969: Late Quaternary history of lake levels in the Huron and Erie basins; in Proceedings 12th Conference Great Lakes Research, International Association Great Lakes Research, Ann Arbour, Michigan, p. 250-270.

1970: Recent uplift of Manitoulin Island, Ontario; Canadian Journal of EarthSciences, v. 7, p. 665-675.

Lind, E.K.
1983: Sedimentology and paleoecology of the Cape Rammelsberg area, Baffin Island, Canada; unpublished MSc thesis, University of Colorado, Boulder.

Locke, C. and Locke, W.W. III
1977: Little Ice Age snow-cover extent and paleoglacierization thresholds: north-central Baffin Island, N.W.T., Canada; Arctic and Alpine Research, v. 9, p. 291-300.

Locke, W.W. III
1979: Etching of Hornblende grains in Arctic soils: an indicator of relative age and paleoclimate; Quaternary Research, v. 11, p. 197-212.

1980: The Quaternary geology of the Cape Dyer area, southeasternmost Baffin Island, Canada; unpublished PhD thesis, University of Colorado, Boulder, 331 p.

1985: Weathering and soil development on Baffin Island; in Quaternary Environments: Eastern Canadian Arctic, Baffin Bay, and West Greenland, J.T. Andrews (ed.); Allen and Unwin, London, p. 331-353.

Løken, O.H.
1962a: On the vertical extent of glaciation in northeastern Labrador-Ungava; Canadian Geographer, v. 6, p. 106-119.

1962b: The late-glacial and postglacial emergence and the deglaciation of northernmost Labrador; Geographical Bulletin, v. 17, p. 23-56.

1964: A study of the late and postglacial changes of sea level in northernmost Labrador; Report to Arctic Institute of North America, Montréal, 74 p.

1966: Baffin Island refugia older than 54,000 years; Science, v. 153, p. 1378-1380.

1978: Postglacial tilting of Akpatok Island, Northwest Territories; Canadian Journal of Earth Sciences, v. 15, p. 1547-1553.

Løken, O.H. and Hodgson, D.A.
1971: On the submarine geomorphology along the east coast of Baffin Island; Canadian Journal of Earth Sciences, v. 8, p.185-195.

Low, A.P.
1888: Report on explorations in James Bay and country east of Hudson Bay drained by the Big Great Whale and Clearwater rivers; Geological Survey of Canada, Annual Report 1887 and 1888, v. 3, Part J, 94 p.

1896: Report on exploration in the Labrador Peninsula along the East Main, Koksoak, Hamilton, Manicuagan and portions of other rivers in 1892-93-94-95; Geological Survey of Canada, Annual Report 1895, v. 8, Part L, 387 p.

Lowdon, J.A. and Blake, W., Jr.
1968: Geological Survey of Canada radiocarbon dates VII; Radiocarbon, v. 10, p. 207-245.

1970: Geological Survey of Canada radiocarbon dates IX; Geological Survey of Canada, Paper 70-2, Part B, 41 p.

1973: Geological Survey of Canada radiocarbon dates XIII; Geological Survey of Canada, Paper 73-7, 61 p.

1979: Geological Survey of Canada radiocarbon dates XIX; Geological Survey of Canada, Paper 79-7, 58 p.

1980: Geological Survey of Canada radiocarbon dates XX; Geological Survey of Canada, Paper 80-7, 28ecp.

Lowdon, J.A., Robertson, I.M., and Blake, W., Jr.
1971: Geological Survey of Canada radiocarbon dates XI; Radiocarbon, v. 13, p. 255-324.

1977: Geological Survey of Canada radiocarbon dates XVII; Geological Survey of Canada, Paper 77-7, 25 p.

MacLean, B.
1985: Geology of the Baffin Island Shelf; in Quaternary Environments: Eastern Canadian Arctic, Baffin Bay, and West Greenland, J.T. Andrews (ed.); Allen and Unwin, London, p. 154-177.

MacLean, B. and Williams, G.L.
1983: Geological investigations of Baffin Island shelf in 1982; in Current Research, Part B, Geological Survey of Canada, Paper 83-1B, p. 309-315.

Mangerud, J., Andersen, S.T., Berglund, B.E., and Donner, J.J.
1974: Quaternary stratigraphy of Norder, a proposal for terminology and classification; Boreas, v. 3, p. 109-128.

Manitoba Mineral Resources Division
1981: Surficial geological map of Manitoba; Manitoba Mineral Resources Division, Map 81-1, scale 1:1 000 000

Martineau, G.
1984a: Géologie du Quaternaire de la région de Chibougamau; Ministère de l'Énergie et des Ressources, Rapport ET 83-20, 15 p.
1984b: Aspects de la géologie du Quaternaire, région de Chibougamau, Ministère de l'Énergie et des Ressources, Manuscrits Bruts 84-13, 24 p.

Martison, N.W.
1953: Petroleum possibilities of the James Bay Lowland area; Ontario Department of Mines, Annual Report 1952, v. 61, Part 6, p. 1-58.

Mathews, W.H.
1974: Surface profiles of the Laurentide Ice Sheet in its marginal areas; Journal of Glaciology, v. 13, p. 37-43.
1975: Cenozoic erosion and erosion surfaces of eastern North America; American Journal of Science, v. 275, p. 818-824.
1980: Retreat of the last ice sheets in northeastern British Columbia and adjacent Alberta; Geological Survey of Canada, Bulletin 331, 22 p.

Matthew, E.M.
1961: Deglaciation of the George River basin Labrador-Ungara; in Field Research in Labrador-Ungava, Annual Report 1959-1960, McGill Sub-Arctic Research Paper no. 11, p. 29-45.

Matthews, B.
1966: Radiocarbon dated postglacial land uplift in Northern Ungava, Canada; Nature, v. 211, no. 5054, p. 1164-1166.
1967: Late Quaternary land emergence in northern Ungava, Quebec; Arctic, v. 20, p. 176-201.

Mawdsley, J.B.
1936: The wash-board moraines of the Opawica-Chibougamau area, Quebec; Royal Society of Canada, Transactions, Serie 3, Section IV, v. 30, p. 9-12.

Maxwell, J.B.
1981: Climatic regions of the Canadian Arctic Islands; Arctic, v. 34, p. 225-240.

Mayewski, P.A., Denton, G.H., and Hughes, T.J.
1981: Late Wisconsin ice sheets of North America; in The Last Great Ice Sheets, G.H. Denton and T.J. Hughes (ed.); John Wiley and Sons, New York, p. 67-178.

McAndrews, J.H.
1981: Late Quaternary climate of Ontario: temperature trends from the fossil pollen record; in Quaternary Paleoclimate, W.C. Mahaney (ed.); Geo Abstracts Ltd., Norwich, England, p. 319-333.

McAndrews, J.H. et Samson, G.
1977: Analyse pollinique et implications archéologiques et géomorphologiques, lac de la Hutte Sauvage (Mushuau Nipi), Nouveau-Québec; Géographie physique et Quaternaire, vol. 31, p. 177-183.

McAndrews, J.H., Riley, J.L., and Davis, A.M.
1982: Vegetation history of the Hudson Bay lowland: a postglacial pollen diagram from the Sutton Ridge; Le Naturaliste Canadien, v. 190, p. 597-608.

McCoy, W.D.
1983: Holocene glacier fluctuations in the Torngat Mountains, Northern Labrador; Géographie physique et Quaternaire, v. 37, p. 211-216.

McDonald, B.C.
1969: Glacial and interglacial stratigraphy, Hudson Bay lowland; in Earth Science Symposium on Hudson Bay, P.J. Hood (ed.); Geological Survey of Canada, Paper 68-53, p. 78-99.
1971: Late Quaternary stratigraphy and deglaciation in eastern Canada; in The Late Cenozoic Glacial Ages, K.K. Turekian (ed.); Yale University Press, p. 331-353.

McLearn, F.H.
1927: The Mesozoic and Pleistocene deposits of the Lower Missinaibi, Opazatika, and Mattagami rivers, Ontario; Geological Survey of Canada, Summary Report 1926, Part C, p. 16-47.

Mercer, J.H.
1956: Geomorphology and glacial history of southernmost Baffin Island; Geological Society of America, Bulletin, v. 67, p. 553-570.

Mercier, A.L.
1984: Glacial lake in the Richardson and Rae River basins, District of Mackenzie, N.W.T.; Géographie physique et Quaternaire, vol. 38, p. 75-80.

Miller, G.H.
1973: Late Quaternary glacial and climatic history of northern Cumberland Peninsula, Baffin Island, N.W.T., Canada; Quaternary Research, v. 3, p. 561-583.
1975: Quaternary glacial and climatic history of northern Cumberland Peninsula, Baffin Island, Canada, with particular reference to fluctuations during the last 20,000 years; unpublished PhD dissertation, University of Colorado, Boulder, 226 p.
1976: Anomalous local glacier activity, Baffin Island, Canada: Paleoclimatic implications; Geology, v. 4, p. 502-504.
1979: Radiocarbon date list IV: Baffin Island, N.W.T., Canada; institute of Arctic and Alpine Research, University of Colorado, Occasional Paper 29, 61 p.
1980: Late Foxe glaciation of southern Baffin Island, N.W.T., Canada; Geological Society of America, Bulletin, v. 91, p. 399-405.
1985a: Aminostratigraphy of Baffin Island shell-bearing deposits; in Quaternary Environments: Eastern Canadian Arctic, Baffin Bay, and West Greenland, J.T. Andrews (ed.); Allen and Unwin, London, p. 394-427.
1985b: Moraines and proglacial lake shorelines, Hall Peninsula, Baffin Island; in Quaternary Environments: Eastern Canadian Arctic, Baffin Bay and West Greenland, J.T. Andrews (ed.); Allen and Unwin, London, p. 546-557.

Miller, G.H. and Dyke, A.S.
1974: Proposed extent of late Wisconsin Laurentide ice on eastern Baffin Island; Geology, v. 2, p. 125-130.

Miller, G.H., Andrews, J.T., and Short, S.K.
1977: The last interglacial-glacial cycle, Clyde foreland, Baffin Island, N.W.T.: Stratigraphy, biostratigraphy, and chronology; Canadian Journal of Earth Sciences, v. 14, p. 2824-2857.

Miller, G.H., Andrews, J.T., Stravers, J.A., and Laymon, C.A.
1988: The Cockburn Readvance in northeastern Canada: a Younger Dryas style regional climatic oscillation during the last deglaciation (abstract); in Program and Abstracts, American Quaternary Association, 10th biennial meeting, Amherst, p. 139.

Mode, W.N.
1980: Quaternary stratigraphy and palynology of the Clyde Foreland, Baffin Island, N.W.T., Canada; unpublished PhD thesis, University of Colorado, Boulder, 219 p.
1985: Pre-Holocene pollen and molluscan records from eastern Baffin Island, Canada; in Quaternary Environments: Eastern Canadian Arctic, Baffin Bay and West Greenland, J.T. Andrews (ed.); Allen and Unwin, London, p. 502-519.

Mode, W.N., Nelson, A.R., and Brigham, J.K.
1983: Sedimentologic evidence for Quaternary glaciomarine cyclic sedimentation along eastern Baffin Island, Canada; in Glacial-marine Sedimentation, B.F. Molnia (ed.); Plenum Press, New York, p. 495-534.

Molnia, B.F.
1978: Surface sedimentary units of northern Gulf of Alaska continental shelf; American Association of Petroleum Geologists, Bulletin, v. 62, p. 633-643.

Morrison, A.
1963: Landform studies in the middle Hamilton River area, Labrador; Arctic, v. 16, p. 272-275.
1970: Pollen diagrams from interior Labrador; Canadian Journal of Botany, v. 48, p. 1957-1975.

Mudie, P.J. and Aksu, A.E.
1984: Paleoclimate of Baffin Bay from 300,000-year record of foraminifera, dinoflagellates and pollen; Nature, v. 312, p. 630-634.

Muller, D.S.
1980: Glacial geology and Quaternary history of southeast Meta Incognita Peninsula, Baffin Island, Canada; unpublished MSc thesis, University of Colorado, Boulder, 211 p.

National Wetlands Working Group
1981: Wetlands of Canada; Environment Canada, Ecological Land Classification Series no. 14, 2 maps, scale 1:7500.

Nelson, A.R.
1978: Quaternary glacial and marine stratigraphy of the Qivitu Peninsula, northern Cumberland Peninsula, Baffin Island, Canada; unpublished PhD dissertation, University of Colorado, Boulder, 215 p.
1980: Chronology of Quaternary landforms, Qivitu Peninsula, northern Cumberland Peninsula, N.W.T., Canada; Arctic and Alpine Research, v. 12, p. 256-286.
1981: Quaternary glacial and marine stratigraphy of the Qivitu Peninsula, northern Cumberland Peninsula, Baffin Island; Geological Society of America, Bulletin, v. 92 Part I, p. 512-518; Part II, p. 1143-1261.
1982: Aminostratigraphy of Quaternary marine and glaciomarine sediments, Qivitu Peninsula, Baffin Island; Canadian Journal of Earth Sciences, v. 19, p. 945-961.

Netterville, J.A.
1974: Quaternary stratigraphy of the lower Gods River region, Hudson Bay Lowlands; unpublished MSc thesis, University of Calgary, Calgary, 79 p.

Netterville, J.A., Dyke, A.S., Thomas, R.D., and Drabinsky, K.A.
1976: Terrain inventory and Quaternary geology, Somerset, Prince of Wales and adjacent islands; in Report of Activities, Part A, Geological Survey of Canada, Paper 76-1A, p. 145-154.

Nichol, I. and Bjorklund, A.
1973: Glacial geology as a key to geochemical exploration in areas of glacial overburden with particular reference to Canada; Journal of Geochemical Exploration, v. 2, p. 133-170.

Nichols, H.
1967: The post-glacial history of vegetation and climate at Ennadei Lake, Keewatin, and Lynn Lake, Manitoba (Canada); Eiszeitalter Gegenwart, Band 18, p. 1-76-97.
1972: Summary of the palynological evidence for Late-Quaternary vegetational and climatic change in the central and eastern Canadian Arctic; in Climatic Changes in Arctic Areas During the Last 10,000 Years, Y. Vasari, H. Hyvarinen, and S. Hicks (ed.); Acta Universitatis Ouluensis, Ser A., no. 3, Geologica no. 2, p.309-340.

Nielsen, E.
1982: Observation on the distribution of Proterozoic erratics in Manitoba (abstract); in Program with Abstracts, Geological Association of Canada, v. 7, p. 70.

Nielsen, E. and Dredge, L.A.
1982: Trip 5: Quaternary stratigraphy and geomorphology of a part of the Lower Nelson River; Geological Association of Canada, Field Trip Guidebook no. 5, Winnipeg, Manitoba, 56 p.

Nielsen, E., McKillop, W.B., and McCoy, J.P.
1982: The age of the Hartman moraine and the Campbell beach of Lake Agassiz in northwestern Ontario; Canadian Journal of Earth Sciences, v. 719, p. 1933-1937.

Nielsen, E., Morgan, A.V., Morgan, A., Mott, R,J., Rutter, N.W., and Causse, C.
1986: Stratigraphy, paleoecology, and glacial history of the Gillam area, Manitoba; Canadian Journal of Earth Sciences, v. 23, p. 1641-1661.

Nixon, F.M., Dredge, L.A., and Richardson, R.J.
1981: Surficial geology, Munroe Lake, Manitoba; Geological Survey of Canada, Map 10-1981, scale 1:250 000.

Norman, G.W.H.
1938: The last Pleistocene ice-front in Chibougamau District, Quebec; Royal Society of Canada, Transactions, Series 3, Section IV, v. 32, p. 69-86.
1939: The south-eastern limit of Glacial Lake Barlow-Ojibway in the Mistassini Lake region, Quebec; Royal Society of Canada, Transactions, Section IV, p. 59-65.

Occhietti, S.
1980: Le Quaternaire de la région de Trois-Rivières-Shawinigan, Québec, Contribution à la paléogéographie de la vallée moyenne du St-Laurent et corrélations stratigraphiques; Paléo-Québec, n° 10, 218 p.
1982: Synthése lithostratigraphique et paléoenvironnements du Quaternaire au Québec méridional; Géographie physique et Quaternaire, vol. 36, p. 15- 49.
1989: Quaternary geology of St. Lawrence Valley and adjacent Appalachian subregion; in Chapter 4 of Quaternary Geology of Canada and Greenland, R.J. Fulton (ed.); Geological Survey of Canada, Geology of Canada, no. 1 (also Geological Society of America, The Geology of North America, v. K-1).

Osterman, L.E.
1982: Late Quaternary history of southern Baffin Island, Canada: A study of foraminifera and sediments from Frobisher Bay; unpublished PhD dissertation, University of Colorado, Boulder, 380 p.

Osterman, L.E. and Andrews, J.T.
1983: Changes in glacial-marine sedimentation in core HU77-159, Frobisher Bay, Baffin Island, N.W.T.: a record of proximal, distal and ice-rafting glacial-marine environments; in Glacial-marine sedimentation, B.J. Molnia (ed.); Plenum Press, New York, p. 451-494.

Osterman, L.E., Miller, G.H., and Stravers, J.A.
1985: Late and mid-Foxe glaciation of southern Baffin Island; in Quaternary Environments: Eastern Canadian Arctic, Baffin Bay and West Greenland, J.T. Andrews (ed.); Allen and Unwin, London, p. 520-545.

Pagé, P.
1977: Les dépôts meubles de la région de Saint-Jean de Matha — Sainte Émilie de l'Énergie, Québec — cartographie, sédimentologie et stratigraphie; thèse de maîtrise non publiée, Université du Québec à Montréal, 118 p.

Paré, D.
1982: Application of heavy mineral analysis to problems of till provenance along a transect from Longlac, Ontario to Somerset Island; unpublished MA thesis, Carleton University, Ottawa, 76 p.

Parent, M., Dubois, J.M.M., Bail, P., Larocque, A., et Larocque, G.
1985: Paléogéographie du Québec méridional entre 12,500 et 8000 BP; Recherches amérindiennes au Québec, vol. 15, p. 17-37.

Parent, M. and Occhietti, S.
1988: Late Wisconsinan deglaciation and Champlain Sea invasion in the St. Lawrence Valley, Quebec; Géographie physique et Quaternaire, v. 42, p. 215-246.

Parry, J.T.
1963: The Laurentians: a study in geomorphological development; unpublished PhD thesis, McGill University, Montréal, 222 p.

Payette, S.
1983: The forest tundra and present tree-lines of the Northern Québec Labrador Peninsula; in Tree-Line Ecology, Proceedings of the Northern Québec Tree-Line Conference, P. Morisset and S. Payette (ed.); Collection Nordicana, no. 47, p. 3-23.

Peltier, W.R. and Andrews, J.T.
1983: Glacial geology and glacial isostasy of the Hudson Bay region; in Shorelines and Isostasy, D.E. Smith and A.G. Dawson (ed.); institute of British Geographers, Special Publication no. 16, Academic Press, London, p. 285-319.

Perrault, G.
1955: Geology of the western margin of the Labrador Trough, Part I: General geology of the western margin of the Labrador Trough, Part II: The Sokoman Iron Formation, Part III: Some data on iron silicate minerals occuring in iron rich sedimentary rocks; unpublished PhD thesis, University of Toronto, Toronto, 300 p.

Peterson, J.A.
1965: Deglaciation of the White Gull Lake area, Labrador-Ungava; Cahiers de géographie de Québec, v. 9, p. 183-196.

Pheasant, D.R.
1971: The glacial chronology and glacio-isostasy of the Narping-Quajon fiord area, Cumberland Peninsula, Baffin Island; unpublished PhD thesis, University of Colorado, Boulder, 232 p.

Pheasant, D.R. and Andrews, J.T.
1972: The Quaternary history of northern Cumberland Peninsula, Baffin Island, N.W.T. Part VIII: Chronology of Narping and Quajon fiords during the past 120,000 years; Proceedings, 24th International Geological Congress, Montréal, Section 12, p. 81-88.
1973: Wisconsin glacial chronology and relative sea level movements, Narpaing Fiord Broughton Island area, eastern Baffin Island, N.W.T.; Canadian Journal of Earth Sciences, v. 10, p. 1621-1641.

Piper, D.J.W., Mudie, P.J., Fader, G.B., Josenhans, H., MacLean, B., and Vilks, G.
1989: Quaternary geology; Chapter 10 in Geology of the Continental Margin of Eastern Canada, M.J. Keen and G.L. Williams (ed.); Geological Survey of Canada, Geology of Canada, no. 2 (also Geological Society of America, The Geology of North America, v. I-2).

Plumet, P.
1974: L'archéologie et le relèvement glacio-isostatique de la région de Poste-de-la-Baleine, Nouveau-Québec; Revue de géographie de Montréal, vol. 28, p. 443-447.

Portmann, J-P.
1972: Les dépôts quaternaires de l'estuaire de la Grande-Rivière de la Baleine, Nouveau-Québec; Revue de géographie de Montréal, vol. 26, p.208-214.

Prest, V.K.
1963: Surficial geology, Red Lake - Lansdowne House area, northwestern Ontario; Geological Survey of Canada, Paper 63-6, 23 p.
1966: Glacial studies, northeastern Ontario and northwestern Quebec; Geological Survey of Canada, Paper 66-1, p. 202-203.
1968: Nomenclature of moraines and ice-flow features as applied to the Glacial Map of Canada; Geological Survey of Canada, Paper 67-57, 32 p.
1969: Retreat of Wisconsin and recent ice in North America; Geological Survey of Canada, Map 1257A, scale 1:5 000 000.
1970: Quaternary geology of Canada; in Geology and Economic Minerals of Canada, R.J.W. Douglas (ed.); Geological Survey of Canada, Economic Geology Report 1, 5th ed., p. 676-764.
1983a: Canada's heritage of glacial features; Geological Survey of Canada, Miscellaneous Report 28, 119 p.
1983b: The Wisconsinan glacial complex; in Quaternary Glaciations in the Northern Hemisphere, A. Billard, D. Conchon, and F.W. Shutton (ed.); International Geological Correlation Project 73/1/24, Report no. 9, Prague, p. 90-102.
1984: The Late Wisconsinan glacier complex; in Quaternary Stratigraphy of Canada — A Canadian Contribution to IGCP Project 24, R.J. Fulton (ed.); Geological Survey of Canada, Paper 84-10, p. 21-36 Map 1584A, scale 1:7 500 000.

Prest, V.K., Grant, D.R., and Rampton, V.N.
1968: Glacial Map of Canada; Geological Survey of Canada, Map 1253A, scale 1:5 000 000.

Prichonnet, G., Martineau, G., et Bisson, L.
1984: Les dépôts quaternaires de la région de Chibaugamau, Québec; Géographie physique et Quaternaire, vol. 38, p. 287-304.

Quinlan, G.
1981: Numerical models of postglacial relative sea level change in Atlantic Canada and the eastern Canadian arctic; unpublished PhD thesis, Dalhousie University, Halifax, 499 p.
1985: A numerical model of postglacial relative sea level change near Baffin Island; in Quaternary Environments: Eastern Canadian Arctic, Baffin Bay and West Greenland, J.T. Andrews (ed.); Allen and Unwin, London, p. 560-584.

Quinlan, G. and Beaumont, C.
1982: The deglaciation of Atlantic Canada as reconstructed from the postglacial relative sea-level record; Canadian Journal of Earth Sciences, v. 19, p. 2232-2246.

Richard, P.J.H.
1977a: Histoire post-wisconsinienne de la végétation du Québec méridional par l'analyse pollinique; Service à la recherche, Direction générale des forêts, Ministère des terres et forêts du Québec, tome 1, 312 p. et tome 2, 142 p.

1977b: Végétation tardiglaciaire au Québec méridional et implications paléoclimatiques; Géographie physique et Quaternaire, vol. 31, p. 161-176.

1979: Contribution à l'histoire postglaciaire de la végétation au nord-est de la Jamésie; Géographie physique et Quaternaire, vol. 33, p. 93-112.

1980a: Histoire postglaciaire de la végétation au sud du lac Abitibi, Ontario et Québec; Géographie physique et Quaternaire; vol. 34, p. 77-94.

1981: Paléophytogéographie postglaciaire en Ungava par l'analyse pollinique; Paléo-Québec, n° 13, 153 p.

1985: Couvert végétal et paléoenvironnements du Québec entre 12 000 et 8000 ans BP — l'habitabilité dans un milieu changeant; Recherches amérindiennes au Québec, vol. 15, p. 39-56.

1989: Patterns of post-Wisconsinan plant colonization in Quebec-Labrador; in Chapter 7 of Quaternary Geology of Canada and Greenland, R.J. Fulton (ed.); Geological Survey of Canada, Geology of Canada, no. 1 (also Geological Society of America, The Geology of North America, v. K-1).

Richard, P.J.H., Larouche, A., et Bouchard, M.A.
1982: Âge de la déglaciation finale et histoire postglaciaire de la végétation dans la partie centrale du Nouveau-Québec; Géographie physique et Quaternaire, v. 36, p. 63-90.

Richard, S.H.
1978: Surficial geology, Lachute-Montebello area, Quebec; in Current Research, Part B, Geological Survey of Canada, Paper 78-1B, p. 115-119.

1980b: Surficial geology, Papineauville-Wakefield region, Quebec; in Current Research, Part C, Geological Survey of Canada, Paper 80-1C, p. 121-128.

1984: Surficial geology, Lachute-Arundel, Quebec-Ontario; Geological Survey of Canada, Map 1577A, scale 1:100 000.

in press: Surficial geology, Buckingham, Quebec-Ontario; Geological Survey of Canada, Map 1678A, scale 1:100 000.

Richard, S.H., Gadd, N.R., and Vincent, J-S.
1977: Surficial materials and terrain features, Ottawa-Hull, Ontario-Quebec; Geological Survey of Canada, Map 1425A, scale 1:125 000.

Richardson, R.J., Dredge, L.A., and Nixon, F.M.
1982: Surficial geology, Whiskey Jack Lake, Manitoba; Geological Survey of Canada, Map 18-1981, scale 1:250 000.

Ricketts, B.D. and Donaldson, J.A.
1981: Sedimentary history of the Belcher Group of Hudson Bay; in Proterozoic Basins of Canada, F.H.A. Campbell (ed.), Geological Survey of Canada, Paper 81-10, p. 235-254.

Ridler, R.H. and Shilts, W.W.
1974: Exploration for Archean polymetallic sulphide deposits in permafrost terrains: An integrated geological/geochemical technique; Kaminak Lake area, District of Keewatin; Geological Survey of Canada, Paper 73-34, 33 p.

Ritchie, J.C.
1989: History of the boreal forest in Canada; in Chapter 7 of Quaternary Geology of Canada and Greenland, R.J. Fulton (ed.); Geological Survey of Canada, Geology of Canada, no. 1 (also Geological Society of America, The Geology of North America, v. K-1).

Rogerson, R.J.
1977: Glacial geomorphology and sediments of the Porcupine Strands, Labrador, Canada; unpublished PhD thesis, Macquarie University, 276 p.

1981: The tectonic evolution and surface morphology of Newfoundland; in The Natural Environment of Newfoundland, Past and Present; Department of Geography, University of Newfoundland, p. 24-55.

1982: The glaciation of Newfoundland and Labrador; in Prospecting in Areas of Glaciated Terrain, P.H. Davenport (ed.), The Canadian Institute of Mining and Metallurgy, p. 37-56.

Rogerson, R.J. and Bell, T.
1986: The Late-Wisconsin maximum in the Nachavak Fiord area of northern Labrador (abstract); in Abstracts, 15th Arctic Workshop, Boulder, p. 57-60.

Romanelli, R.
1975: The Champlain Sea episode in the Gatineau River valley and Ottawa area; Canadian Field Naturalist, v. 89, p. 356-360.

Rowe, J.S.
1972: Forest Regions of Canada; Canada Department of the Environment, Canadian Forestry Service, Publication no. 1300, 172 p.

Ruddiman, W.R. and Duplessy, J-C.
1985: Conference on the last deglaciation: timing and mechanisms; Quaternary Research, v. 23, p. 1-17.

Ruddiman, W.R. and McIntyre, I.
1981: The mode and mechanism of the last deglaciation; Quaternary Research, v. 16, p. 125-134.

Rutherford, A.A., Wittenberg, J., and McCallum, K.J.
1975: University of Saskatchewan radiocarbon dates VI; Radiocarbon, v. 17, p. 328-353.

Saarnisto, M.
1974: The deglaciation history of the Lake Superior region and its climatic implications; Quaternary Research, v. 4, p. 316-339.

1975: Stratigraphical studies on the shoreline displacement of Lake Superior; Canadian Journal of Earth Sciences, v. 12, p. 300-319.

Sado, E.V.
1975: Quaternary geology of the Wildgoose Lake area, District of Thunder Bay; in Summary of Field Work 1975, by the Geological Branch, V.G. Milne, D.F. Hewitt, K.D. Card, and J.A. Robertson (ed.); Ontario Division of Mines, Miscellaneous Paper 63, p. 128-129.

St-Onge, D.A.
1980: Glacial Lake Coppermine, north-central District of Mackenzie, Northwest Territories; Canadian Journal of Earth Sciences, v. 17, p. 1310-1314.

1984: Surficial deposits of the Redrock Lake area, District of Mackenzie; in Current Research, Part A, Geological Survey of Canada, Paper 84-1A, p. 271-278.

St-Onge, D.A. and Bruneau, H.
1982: Dépôts meubles du secteur aval de la rivière Coppermine, Territoires du North-Ouest; in Recherches en cours, Partie B, Commission géologique du Canada, Étude 82-1B, p. 51-55.

St-Onge, D.A. and Geurts, L.A.
1985: Northeast extension of glacial Lake McConnell in the Dean River basin, District of Mackenzie; in Current Research, Part A, Geological Survey of Canada, Paper 85-1A, p. 181-186.

St-Onge, D.A., Geurts, M.A., Guay, F., Dewez, V., Landriault, F., and Léveillé, P.
1981: Aspects of the deglaciation of the Coppermine River region, District of Mackenzie; in Current Research, Part A, Geological Survey of Canada, Paper 81-1A, p. 327-331.

Sanford, B.V., Norris, A.W., and Bostock, H.H.
1968: Geology of the Hudson Bay Lowlands (Operation Winisk); Geological Survey of Canada, Paper 67-60, 118 p. (includes Bibliography on Hudson Bay Lowlands by A.W. Norris, B.V. Sanford, and R.T. Bell).

Satterly, J.
1971: Diamond, U.S.S.R. and North America, a target for exploration in Ontario; Ontario Department of Mines, Miscellaneous Paper 48, 43 p.

Sauvé, P. et LaSalle, P.
1968: Notes sur la géeologie glaciaire de la région de Manic 2; Naturaliste canadien, vol. 95, p. 1293-1300.

Savoie, L. et Richard, P.J.H.
1979: Paléophytogéographie de l'épisode de Saint-Narcisse dans la région de Sainte-Agathe, Québec; Géographie physique et Quaternaire, vol. 33, p. 175-188.

Schreiner, B.
1983: Lake Agassiz in Saskatchewan; in Glacial Lake Agassiz, J.T. Teller and L. Clayton (ed.), Geological Association of Canada, Special Paper 26, p. 75-96.

1984: Quaternary geology of the Precambrian Shield, Saskatchewan; Saskatchewan Geological Survey, Report 221, 106 p.

Sharpe, D.R.
1984: Late Wisconsinan glaciation and deglaciation of Wollaston Peninsula, Victoria Island, Northwest Territories; in Current Research, Part A, Geological Survey of Canada, Paper 84-1A, p. 259-269.

1985: The stratified nature of deposits in streamlined glacial landforms on southern Victoria Island, District of Franklin; in Current Research, Part A, Geological Survey of Canada, Paper 85-1A, p. 365-371.

Shaw, G.
1944: Moraines of Late Pleistocene ice fronts near James Bay, Quebec; Royal Society of Canada, Transactions, Section IV, p. 79-85.

Shaw, J. and Kvill, D.
1984: A glaciofluvial origin for drumlins of the Livingstone Lake area, Saskatchewan; Canadian Journal of Earth Sciences, v. 21, p. 1442-1459.

Shilts, W.W.
1971: Till studies and their application to regional drift prospecting; Canadian Mining Journal, v. 92, p. 45-50.

1973: Drift prospecting: Geochemistry of eskers and till in permanently frozen terrain, District of Keewatin, Northwest Territories; Geological Survey of Canada, Paper 72-45, 34 p.

1975: Principles of geochemical exploration for sulphide deposits using shallow samples of glacial drift; Canadian Mining and Metallurgical Bulletin, v. 68, p. 1-8.

1976: Glacial till and mineral exploration; in Glacial Till, R.F. Legget (ed.); Royal Society of Canada, Special Publication no. 12, p. 205-224.

1977: Geochemistry of till in perennially frozen terrain of the Canadian Shield — application to prospecting; Boreas v. 5, p. 203-212.

1978: Nature and genesis of mudboils, central Keewatin, Canada; Canadian Journal of Earth Sciences, v. 15, p. 1053-1068.

1980: Flow patterns in the central North American ice sheet; Nature, v. 286, p. 213-218.

1981: Sensitivity of bedrock to acid precipitation: modification by glacial processes; Geological Survey of Canada, Paper 81-14, 7 p.

1982a: Implications of mid Wisconsin marine episodes for stratigraphy of Hudson Bay Lowlands (abstract); in Program with Abstracts, Geological Association of Canada, v. 7, p. 81.

1982b: Quaternary evolution of the Hudson/James Bay region; Naturaliste Canadien, v. 109, p. 309-332.

1984a: Quaternary events — Hudson Bay Lowland and southern District of Keewatin; in Quaternary Stratigraphy of Canada — A Canadian Contribution to IGCP Project 24, R.J. Fulton (ed.); Geological Survey of Canada, Paper 84-10, p. 117-126.

1984b: Till geochemistry in Finland and Canada; Journal of Geochemical Exploration, v. 21, p. 95-117.

1984c: Esker sedimentation models, Deep Rose Lake map area, District of Keewatin; in Current Research, Part B, Geological Survey of Canada, Paper 84-1B, p. 217-222.

1985: Geological models for the configuration, history and style of disintegration of the Laurentide Ice Sheet; in Models in Geomorphology, M.J. Waldenberg (ed.); Binghampton Symposium in Geomorphology, International Series no. 12, George Allen and University, London, p. 73-91.

Shilts, W.W., Aylsworth, J.M., Kaszycki, C.A., and Klassen, R.A.
1987: Geomorphology of the Canadian Shield; in Geomorphic Systems of North America, W. Graf (ed.); Geological Society of America, The Geology of North America, v. CSV-2.

Shilts, W.W., Cunningham, C.M., and Kaszycki, C.A.
1979: Keewatin Ice Sheet — re-evaluation of the traditional concept of the Laurentide Ice Sheet; Geology, v. 7, p. 537-541.

Shilts, W.W., Miller, G.H., and Andrews, J.T.
1981: Glacial flow indicators and Wisconsin glacial chronology, Hudson Bay/James Bay Lowlands: evidence against a Hudson Bay ice divide (abstract); in Abstracts with Program, Geological Society of America, v. 13, p. 553.

Short, S.K.
1978: Palynology: a Holocene environmental perspective for archaeology in Labrador-Ungava; Arctic Anthropology, v. 15, p. 9-35.

1981: Radiocarbon date list 1, Labrador and northern Quebec, Canada; Institute of Arctic and Alpine Research, University of Colorado, Occasional Paper 36, 33 p.

Short, S.K., Mode, W.N., and Davis, P.T.
1985: The Holocene record from Baffin Island; modern and fossil pollen studies; in Quaternary Environments: Eastern Canadian Arctic, Baffin Bay and West Greenland, J.T. Andrews (ed.); Allen and Unwin, London, p. 608-642.

Sim, V.W.
1960a: A preliminary account of late Wisconsin glaciation in Melville Peninsula, Northwest Territories; Canadian Geographer, v. 17, p. 21-34.

1960b: Maximum post-glacial marine submergence in northern Melville Peninsula; Arctic, v. 13, p. 178-193.

1964: Terrain analysis of west-central Baffin Island; Geographical Bulletin, v. 21, p. 66-92.

Simpson, S.J.
1972: An account of the environment and the evolution of the tract of land between the mouths of the Nelson and Hayes rivers; unpublished PhD thesis, University of Manitoba, Winnipeg.

Skinner, R.G.
1973: Quaternary stratigraphy of the Moose River basin, Ontario; Geological Survey of Canada, Bulletin 225, 77 p.

Smith, D.G.
1978: The Athabasca sand dunes: a physical inventory; Canada Department of Indian and Northern Affairs, National Parks Branch, Contract 77-31, Report, 104 p.

Smith, H.T.U.
1948: Giant glacial grooves in northern Canada; American Journal of Science, v. 246, p. 503-514.

Smith, J.E.
1966: Sam Ford Fiord: a study in deglaciation; unpublished MSc thesis, McGill University, Montréal, 93 p.

Smith, P.A.W.
1969: Glacial geomorphology of the Saglek Fjord area of northeast Labrador; in Field Research in Labrador-Ungava, McGill Sub-Arctic Research, Paper no. 24, p. 115-123.

Sorenson, C.J., Knox, J.C., Larsen, J.A., and Bryson, R.A.
1971: Paleosols and the recent forest border in Keewatin, Northwest Territories; Quaternary Research, v. 1, p. 468-473.

Steele, K.G. and Baker, C.L.
1985: Sonic drilling: Framework for Quaternary stratigraphy and till geochemistry (abstract); in Program with Abstracts, Geological Association of Canada, v. 10, p. A59.

Straver, J.A.
1986: Glacial geology of outer Meta Incognita Peninsula and adjacent Frobisher Bay and Hudson Strait, southern Baffin Island, N.W.T., Canada; unpublished PhD thesis, University of Colorado, Boulder.

Stravers, L.K.S.
1981: Palynology and deglaciation history of the central Labrador-Ungava Peninsula; unpublished MSc thesis, University of Colorado, Boulder, 171 p.

Stuckenrath, R., Miller, G.H., and Andrews, J.T.
1979: Problems of radiocarbon dating Holocene organic-poor sediments, Cumberland Peninsula; Arctic and Alpine Research, v. 11, p. 109-120.

Stuiver, M., Deevey, E.S., Jr., and Rouse, I.
1963: Yale Natural Radiocarbon Measurements VIII; Radiocarbon, v. 5, p. 312-341.

Stuiver, M., Heusser, C.J., and Yang, I.C.
1978: North American glacial history extended to 75,000 years ago; Science, v. 200, p. 16-21.

Sugden, D.E.
1976a: A case against deep erosion of shields by ice sheets; Geology, v. 4, p. 580-582.

1976b: Glacial erosion by the Laurentide ice sheet and its relationship to ice, topographic and bedrock conditions; Department of Geography, University of Aberdeen, Aberdeen, 86 p.

1977: Reconstruction of the morphology, dynamics and thermal characteristics of the Laurentide Ice Sheet at its maximum; Arctic and Alpine Research, v. 9, p. 21-47.

1978: Glacial erosion by the Laurentide Ice Sheet; Journal of Glaciolgoy, v. 20, p. 367-391.

Sugden, D.E. and Watts, S.H.
1977: Tors, felsenmeer, and glaciation in northern Cumberland Peninsula, Baffin Island; Canadian Journal of Earth Sciences, v. 3, p. 243-263.

Syvitski, J.P.M. and Blakeney, C.P. (ed.)
1983: Sedimentology of Arctic Fjords Experiment: HU82-031 Data Report. Volume I; Canadian Data Report of Hydrography and Ocean Sciences, Supply and Services Canada, no. 12, 935 p.

Szabo, B.J., Miller, G.H., Andrews, J.T., and Stuiver, M.
1981: Comparison of uranium-series, radiocarbon, and amino acid data from marine molluscs, Baffin Island, Arctic Canada; Geology, v. 9, p. 451-457.

Tarnocai, C.
1989: Peat resources of Canada; in Chapter 11 of Quaternary Geology of Canada and Greenland, R.J. Fulton (ed.); Geological Survey of Canada, Geology of Canada, no. 1 (also Geological Society of America, The Geology of North America, v. K-1).

Tavenas, F., Chagnon, J-Y., and Larochelle, P.
1971: The Saint-Jean Vianney landslide: observations and eyewitness accounts; Canadian Geotechnical Journal, v. 8, p. 463-478.

Taylor, A. and Judge, A.
1979: Permafrost studies in northern Quebec; Géographie physique et Quaternaire, v. 33, p. 245-251.

Taylor, F.C.
1961: Interglacial conglomerate in northern Manitoba, Canada; Geological Society of America, Bulletin, v. 72, p. 167-168.

1982: Reconnaissance geology of a part of the Canadian Shield, northern Quebec and Northwest Territories; Geological Survey of Canada, Memoir 399, 32 p.

Teller, J.T. and Thorleifson, L.H.
1983: The Lake Agassiz — Lake Superior connection; in Glacial Lake Agassiz, J.T. Teller and L. Clayton (ed.); Geological Association of Canada, Special Paper 26, p. 261-290.

Terasmae, J.
1977: Postglacial history of Canadian muskeg; in Muskeg and The Northern Environment in Canada, N.W. Radforth and C.O. Brawner (ed.); University of Toronto Press, p. 9-30.

Terasmae, J. and Anderson, T.W.
1970: Hypsithermal range extension of white pine (Pinus strobus L.) in Quebec, Canada; Canadian Journal of Earth Sciences, v. 7, p. 406-413.

Terasmae, J. and Craig, B.G.
1958: Discovery of fossil Ceratophyllum demersum L. in Northwest Territories, Canada; Canadian Journal of Botany, v. 36, p. 567-569.

Terasmae, J. and Hughes, O.L.,
1960a: A palynological and geological study of Pleistocene deposits in the James Bay Lowlands, Ontario; Geological Survey of Canada, Bulletin 62, 15 p.
1960b: Glacial retreat in the North Bay area, Ontario; Science, v. 131, p. 1444-1446.

Terasmae, J., Webber, P.J., and Andrews, J.T.
1966: A study of late-Quaternary plant-bearing beds in north-central Baffin Island, Canada; Arctic, v. 19, p. 296-318.

Thomas, R.D.
1982a: Surficial geology, Amex Lake, District of Keewatin; Geological Survey of Canada, Map 9-1981, scale 1:250 000.
1982b: Surficial geology, Montresor River, District of Keewatin; Geological Survey of Canada, Map 10-1981, scale 1:250 000.

Thomas, R.D. and Dyke, A.S.
1982a: Surficial geology, Woodburn Lake, District of Keewatin; Geological Survey of Canada, Map 3-1981, scale 1:250 000.
1982b: Surficial geology, Penningotn Lake, District of Keewatin; Geological Survey of Canada, Map 4-1981, scale 1:250 000.
1982c: Surficial geology, Laughland Lake, District of Keewatin; Geological Survey of Canada, Map 5-1981, scale 1:250 000.
1982d: Surficial geology, Mistake River, District of Keewatin; Geological Survey of Canada, Map, 6-1981, scale 1:250 000.
1982e: Surficial geology, Lower Hayes River, District of Keewatin; Geological Survey of Canada, Map 7-1981, scale 1:250 000.
1982f: Surficial geology, Darby Lake, District of Keewatin; Geological Survey of Canada, Map 8-1981, scale 1:250 000.

Thomas, R.H.
1977: Calving bay dynamics and ice sheet retreat up the St. Lawrence Valley system; Géographie physique et Quaternaire, v. 31, p. 347-356.

Thompson, H.R.
1954: Pangnirtung Pass: an explanatory geomorphology; unpublished PhD thesis, McGill University, Montréal, 227 p.

Tippett, C.R.
1985: Glacial dispersal train of Paleozoic erratics, central Baffin Island, N.W.T., Canada; Canadian Journal of Earth Sciences, v. 22, p. 1818-1826.

Tomlinson, R.F.
1958: Geomorphological fieldwork in the Kaumajet Mountains and Okak Bay area of the Labrador Coast; Arctic, v. 11, p. 254-256.
1963: Pleistocene evidence related to glacial theory in northeastern Labrador; The Canadian Geographer, v. 7, p. 83-90.

Tremblay, G.
1971: Glaciation et déglaciation dans la région Saguenay — Lac Saint-Jean, Québec, Canada; Cahiers de géographie de Québec, vol. 15, p. 467-494.
1972: Géologie du Quaternaire, région d'Abitibi, centre ouest, Comtés d'Abitibi-ouest, d'Abitibi-est et de Rouyn-Noranda — Polmarolle (32 D/11), Amos (32 D/9), Konasuta River (32 D/10); Ministère des Richesses naturelles, GM 28573, échelle 1:50 000.
1973: Caractéristiques sédimentologiques des dépôts morainiques et fluvioglaciaires dans la région Saguenay — Lac Saint-Jean, Québec, Canada; Zeitschrift fur Geomorphologie, vol. 17, p. 405-427.
1974: Géologie du Quaternaire - régions de Rouyn-Noranda et d'Abitibi - Comtés d'Abitibi-est et d'Abitibi-ouest; Ministère des Richesses naturelles, Dossier public 236, 100 p.
1975: Géologie du Quaternaire, région de Sept-Îles — Port-Cartier; Ministère des Richesses naturelles, Dossier public 304, 43 p.
1977: Géologie du Quaternaire, région de Rawdon-Laurentides-Shawbridge-Ste-Agathe-des-Monts; Ministère des Richesses naturelles, Dossier public 555, 28 p.

Tremblay, L.P.
1950: Fiedmont map-area, Abitibi County, Quebec; Geological Survey of Canada, Memoir 253, 113 p.

Tyrrell, J.B.
1898a: Report on the Doobaunt, Kazan, and Ferguson Rivers, and the northwest coast of Hudson Bay; and on two overland routes from Hudson Bay to Lake Winnipeg; Geological Survey of Canada, Annual Report (new series), v. 9, p. 1-218.
1898b: The glaciation of north-central Canada; Journal of Geology, v. 6, p. 147-160.
1913: Hudson Bay exploring expedition, 1912; Ontario Bureau of Mines, Annual Report 22, Part 1, p. 161-209.

Tyrrell, J.B. and Dowling, D.G.
1896: Report on the country between Athabasca Lake and Churchill River, with notes on two routes between the Churchill and Sakatchewan rivers; Geological Survey of Canada, Annual Report (new series), v. 8, Report D, p. 1D-120D.

Veillette, J.J.
1982: Ice flow patterns, Lake Temiscaming area, Quebec; Geological Survey of Canada, Open File 841.
1983a: Les polis glaciaires au Témiscamingue: une chronologie relative; in Recherches en cours, Partie A, Commission géologique du Canada, Étude 83-1A, p. 187-196.
1983b: Déglaciation de la vallée supérieure de l'Outaouais, le lac Barlow et le sud du lac Ojibway, Québec; Géographie physique et Quaternaire, vol. 37, p. 67-84.
1986a: Former southwesterly ice flows in Abitibi-Timiskaming region: implications for the configuration of the Late Wisconsinan ice sheet; Canadian Journal of Earth Sciences, v. 23, p. 1724-1741.
1986b: Surficial geology, New Liskeard, Ontario-Quebec; Geological Survey of Canada, Map 1639A, Scale 1:100 000.
1986c: Surficial geology, Haileybury, Ontario-Quebec; Geological Survey of Canada, Map 1642A, scale 1:100 000.
1987a: Surficial geology, Lac Simard, Quebec; Geological Survey of Canada, Map 1640A, scale 1:100 000.
1987b: Surficial geology, Grand Lake Victoria North, Quebec; Geological Survey of Canada, Map 1641A, scale 1:100 000.
1987c: Surficial geology, Belleterre, Quebec; Geological Survey of Canada, Map 1643A, scale 1:100 000.
1988: Déglaciation et évolution des lacs proglaciaires Post-Algonquin et Barlow au Témiscamingue, Québec et Ontario; Géographie physique et Quaternaire, vol. 42, p. 7-31.

Veillette, J.J. and Daigneault, R.A.
1987: Surficial geology, Lac Kipawa, Quebec-Ontario; Geological Survey of Canada, Map 1644A, scale 1:100 000.

Vilks, G. and Mudie, P.J.
1978: Early deglaciation of the Labrador Shelf; Science, v. 202, no. 4373, p. 1181-1183.

Vilks, G., Deomarine, B., and Winters, G.
1987: Late Quaternary marine geology of Lake Melville, Labrador; Geological Survey of Canada, Paper 87-22, 50 p.

Vincent, J-S.
1975: Le glaciaire et le postglaciaire de la région à l'est du lac Témiscamingue, Québec; Revue de géographie de Montréal, vol. 29, p. 109-122.
1977: Le Quaternaire récent de la région du cours inférieur de La Grande Rivière, Québec; Commission géologique du Canada, Étude 76-19, 20 p.
1982: The Quaternary history of Banks Island, Northwest Territories, Canada; Géographie physique et Quaternaire, vol. 36, p. 209-232.
1983: La géologie du Quaternaire et la géomorphologie de l'île Banks, Arctique Canadien; Commission géologique du Canada, Mémoire 405, 118 p.
1984: Quaternary stratigraphy of the western Canadian Arctic Archipelago; in Quaternary Stratigraphy of Canada — A Canadian Contribution to IGCP Project 24, R.J. Fulton (ed.); Geological Survey of Canada, Paper 84-10, p. 87-100.
1985a: Géologie des formations en surface, Chisasibi, Québec; Commission géologique du Canada, Carte 1492A, échelle 1/100 000.
1985b: Géologie des formations en surface, Radisson, Québec; Commission géologique du Canada, Carte 1491A, échelle 1/100 000.
1985c: Géologie des formations en surface, Réservoir La Grande 2; Commission géologique du Canada, Carte 1490A, échelle 1/100 000.
1989a: Quaternary geology of the northern Canadian Interior Plains; in Chapter 2 of Quaternary Geology of Canada and Greenland, R.J. Fulton (ed.); Geological Survey of Canada, Geology of Canada, no. 1 (also Geological Society of America, The Geology of North America, v. K-1).
1989b: Quaternary geology of the southeastern Canadian Shield; in Chapter 3 of Quaternary Geology of Canada and Greenland, R.J. Fulton (ed.); Geological Survey of Canada, Geology of Canada, no. 1 (also Geological Society of America, The Geology of North America, v. K-1).

Vincent, J-S. et Hardy, L.
1977: L'évolution et l'extension des lacs glaciaires Barlow et Ojibway en territoire québécois; Géographie physique et Quaternaire, vol. 31, p. 357-372.
1979: The evolution of glacial lakes Barlow and Ojibway, Quebec and Ontario; Geological Survey of Canada, Bulletin 316, 18 p.

Vogel, J.C. and Waterbolk, H.T.
1972: Groningen radiocarbon dates X; Radiocarbon, v. 14, p. 6-110.
Wagner, F.J.E.
1967: Additional radiocarbon dates, Tyrrell Sea area; Maritime Sediments, v. 3, p. 100-104.
Walcott, R.I.
1970: Isostatic response to loading of the crust in Canada; Canadian Journal of Earth Sciences, v. 7, p. 716-726.
Walcott, R.I. and Craig, B.G.
1975: Uplift studies, southeastern Hudson Bay; in Report of Activities, Part A, Geological Survey of Canada, Paper 75-1A, p. 455-456.
Wallace, H.
1981: Keweenawan geology of the Lake Superior basin; in Proterozoic Basins of Canada, F.H.A. Campbell (ed.), Geological Survey of Canada, Paper 81-10, p. 399-417.
Webber, P.J., Richardson, J.W., and Andrews, J.T.
1970: Postglacial uplift and substrate age at Cape Henrietta, southeastern Hudson Bay, Canada; Canadian Journal of Earth Sciences, v. 7, p. 317-325.
Wheeler, E.P.
1958: Pleistocene glaciation in northern Labrador; Geological Society of America, Bulletin, v. 69, p. 343-344.
White, W.A.
1972: Deep erosion by continental ice sheets; Geological Society of America, Bulletin, v. 83, p. 1037-1056.
Williams, L.D.
1978: The Litle Ice Age glaciation level on Baffin Island, Arctic Canada; Palaeogeography, Palaeoclimatology, Palaeoecology, v. 25, p. 199-207.
1979: An energy balance model of potential glacierization of northern Canada; Arctic and Alpine Research, v. 11, p. 443-456.
Williams, M.Y.
1921: Palaeozoic stratigraphy of Pagwachuan, Lower Kenogami, and Lower Albany rivers, Ontario; Geological Survey of Canada, Summary Report 1920, Part D, p. 18-25.

Wilson, J.T.
1938: Glacial geology of part of northwestern Quebec; Royal Society of Canada, Transactions, Section 4, vol. 32, p. 49-59.
Wilson, J.T., Falconer, G., Mathews, W.H., and Prest, V.K.
1958: Glacial Map of Canada; Geological Association of Canada, scale 1:3 801 600.
Wilson, M.E.
1918: Timiskaming County, Quebec; Geological Survey of Canada, Memoir 103, 197 p.
1924: Arnprior-Quyon and Maniwaki areas, Ontario and Quebec; Geological Survey of Canada, Memoir 136, 152 p.
Wilson, W.J.
1906: Reconnaissance surveys of four rivers southwest of James Bay; Geological Survey of Canada, Annual Report 1902-03, p. 222A-243A.
Wolfe, W.J., Lee, H.A., and Hicks, W.D.
1975: Heavy mineral indicators in alluvial and esker gravels in the Moose River Basin, James Bay Lowlands, District of Cochrane; Ontario Geological Survey, Geoscience Report 16, 60 p.
Wright, C.
1975: Lichen-free areas as indicators of recent extensive glacierization in north-central Baffin Island, N.W.T., Canada; unpublished MA thesis, University of Colorado, Boulder.
Wright, J.F.
1932: Geology and mineral deposits of a part of southwestern Manitoba; Geological Survey of Canada, Memoir 169, p. 5-7.
Wyatt, P.H. and Thorleifson, L.H.
1986: Provenance and geochronology of Quaternary glacial deposits in the central Hudson Bay Lowland, northern Ontario (abstract); in Program with Abstracts, GAC-MAC-CGU Joint Annual Meeting, Ottawa 1986, p. 147.
Zoltai, S.C.
1962: Glacial history of part of northwestern Ontario; Geological Association of Canada, Proceedings, v. 13, p. 61-83.
1965: Glacial features of the Quetico - Nipigon area; Canadian Journal of Earth Sciences, v. 2, p. 247-269.

Authors' addresses

A.S. Dyke
Geological Survey of Canada
601 Booth Street
Ottawa, Ontario K1A 0E8

J-S. Vincent
Geological Survey of Canada
601 Booth Street
Ottawa, Ontario K1A 0E8

J.T. Andrews
Department of Geological Sciences
University of Colorado
Boulder, Colorado USA 80309

L.A. Dredge
Geological Survey of Canada
601 Booth Street
Ottawa, Ontario K1A 0E8

W.R. Cowan
Ministry of Northern Development and Mines
Mining Lands Section
99 Wellesley St. W.
Toronto, Ontario M7A 1W3

J.T. Andrews
Department of Geological Sciences and INSTAAR
University of Colorado
Boulder, Colorado USA 80309

P.J.H. Richard
Département de Géographie
Université de Montréal
Montréal, Québec H3C 3J7

Chapter 4

QUATERNARY GEOLOGY OF THE ST. LAWRENCE LOWLANDS OF CANADA

Summary

Introduction — *P.F. Karrow and S. Occhietti*

Quaternary geology of Great Lakes subregion — *P.F. Karrow*

Quaternary geology of St. Lawrence Valley and adjacent Appalachian subregion — *S. Occhietti*

Acknowledgments

References

Cliffs cut in Quaternary sediments, Port Talbot, north shore of Lake Erie. Most of the Canadian shore of Lake Erie is cut in unconsolidated sediments and consequently shoreline erosion is a major problem. Courtesy of P.F. Karrow.

Chapter 4

QUATERNARY GEOLOGY OF THE ST. LAWRENCE LOWLANDS OF CANADA

P.F. Karrow and S. Occhietti

SUMMARY

The St. Lawrence Lowlands region includes the Lake Superior basin, underlain by Precambrian igneous, sedimentary, and metamorphic rocks; the other Great Lakes basins and St. Lawrence Lowlands proper, underlain by nearly horizontal Paleozoic carbonate and fine clastic sedimentary rocks; and the Appalachian area of southeastern Quebec, underlain by folded and faulted Paleozoic sedimentary and metamorphic rocks. Relief is generally low, but in parts of the Lake Superior basin and in the Appalachian area it is moderate. Topography and bedrock lithology affected the style of glaciation, with coarse discontinuous till on the Precambrian Shield, extensive uniform till sheets deposited by markedly lobate ice bodies in the Great Lakes area, and variable tills in the Appalachian area.

The oldest deposits, of probable Illinoian age, include the York Till at Toronto and till underlying stratified sediments of southeastern Quebec. Sangamonian time is represented by fossiliferous fluvial sands and clays of the Don Formation at Toronto and by the lower stratified unit at Pointe-Fortune. Early Wisconsinan tills (Bradtville and Sunnybrook of Ontario and Bécancour (?) and Johnville of Quebec) are known from several sites. In Middle Wisconsinan time much of the Great Lakes area and southeasternmost Quebec were deglaciated. Fossiliferous interstadial deposits are known from a dozen sites in southwestern Ontario and from several other sites in southeastern Quebec.

A complex history is known from Late Wisconsinan glaciation, when the last major ice advance covered the whole region. Multiple till sequences have been described from much of southwestern Ontario, the first part of the region to be uncovered during the general ice recession after about 15 ka. These represent fluctuations in the ice, which after 14 ka created successions of glacial lakes in each of the Great Lakes basins. Numerous local lakes developed concurrently in the northward-sloping valleys of southeastern Quebec. In southwestern Ontario, end moraine trends roughly parallel the margins of the Great Lakes basins and mark the shrinking extent of the melting ice. Early formed moraines in southeastern Quebec were irregular and discontinuous but later became nearly straight, paralleling the margins of St. Lawrence Valley.

Deglaciation of St. Lawrence Valley allowed marine waters to enter, creating the Champlain Sea about 12 ka, which extended to the head of St. Lawrence River and up Ottawa Valley. Fossiliferous marine sediments include extrasensitive clays subject to characteristic earth flow landslides. Glacial isostatic uplift caused regression of the Champlain Sea and tilting of the glacial lake shorelines in southwestern Ontario.

Economic products of the region include abundant glaciofluvial sand and gravel, clay, and placer gold. Engineering problems include the marine clay landslides and shore erosion.

INTRODUCTION

P.F. Karrow and S. Occhietti

General description of the region

The St. Lawrence Lowlands (Fig. 4.1) is that part of the St. Lawrence River drainage basin underlain by gently dipping Paleozoic bedrock. For the purposes of this report, the boundary of the region has been chosen to embrace a part of southeastern Quebec extending into the Appalachian Mountains, which consist of folded and faulted Paleozoic sedimentary rocks, and the Lake Superior basin, underlain and bordered by Precambrian metasediments and crystalline intrusive rocks. The region is long and narrow trending southwest-northeast, with its central parts characterized by generally low relief and extensive drift cover which is commonly 30 to 60 m thick. The Appalachian area and the Lake Superior area are characterized by moderate relief and a drift cover of variable thickness.

The region is positioned near the southern and eastern edge of the main North American ice sheet. As a result, a relatively long period intervened between initiation of glaciation and coverage of the area. Also, the area was affected by fluctuations that occurred while the ice was near its limit, and deglaciation occurred while the ice sheet was still sufficiently active to initiate major readvances. Glaciation and

---

Karrow, P.F. and Occhietti, S.
1989: Quaternary geology of the St. Lawrence Lowlands; Chapter 4 in Quaternary Geology of Canada and Greenland, R.J. Fulton (ed.); Geological Survey of Canada, Geology of Canada, no. 1 (also Geological Society of America, The Geology of North America, v. K-1).

Karrow, P.F. and Occhietti, S.
1989: Introduction (Quaternary geology of the St. Lawrence Lowlands of Canada); in Chapter 4 of Quaternary Geology of Canada and Greenland, R.J. Fulton (ed.); Geological Survey of Canada, Geology of Canada, no. 1; (also Geological Society of America, The Geology of North America, v. K-1).

deglaciation were dominated by major lobes which developed in the large basins of the Great Lakes area. This differed from the area to the north where the ice margin was apparently more linear. The contrast in style may be attributed to: (1) differences in topography — broad, low relief lowlands, contrasted with moderate relief uplands, (2) differences in glacier bed material — softer sedimentary rock vs. harder crystalline rock, (3) the presence of large water bodies in the basins of the St. Lawrence Lowlands, and (4) climatic influence of the low latitudes and marginal water bodies.

During each glaciation the Quaternary ice sheets generally flowed south, southeast, or southwest from the Canadian Shield carrying a load of coarse debris derived from crystalline rocks. The glacial load was modified rapidly by the incorporation of Paleozoic shale and carbonate rocks. In the southeast the ice encountered higher ground which diverted and complicated its flow patterns, but in the southwest lower ground and relief allowed the ice to advance in a more regular pattern to its most southerly position at 38°N in the mid-West United States.

The region is divided into two major subregions (Fig. 4.1) comprising the *Great Lakes subregion*, which encompasses the upper part of the St. Lawrence River drainage basin, and the *St. Lawrence Valley and adjacent Appalachian subregion*, which in addition to the lowland along the valley includes part of the Eastern Townships extending south into the Appalachian Mountains. The boundary between the Great Lakes and St. Lawrence Valley subregions is placed at the Frontenac Arch, a low ridge of exposed Precambrian rock which joins the Canadian Shield of eastern Ontario to the Adirondack Mountains of New York State. This boundary not only separates two Paleozoic sedimentary basins but also approximates the western limit of submergence by the Champlain Sea in late glacial time.

The St. Lawrence Lowlands have the highest population concentration in Canada with the country's two largest cities, Toronto and Montréal, located in the Great Lakes and St. Lawrence Valley subregions, respectively. The St. Lawrence drainage system served as a major access route to the continental interior for early European explorers and settlers, who encountered native population densities in the Great Lakes area among the highest in the Americas.

History of Quaternary studies

General descriptions of the people, fauna, flora, and landscape were provided by early explorers and the first published geological account was that of Guettard (1752). This praised the quality of sand at Trois-Rivières as a source of lime. It was not until the early Nineteenth Century, along with the development of the science of geology itself, that other geological descriptions began to appear. Some of the important contributions, particularly in early years, were made by visitors from England and the United States. Because of its location next to the International Boundary, the St. Lawrence Lowlands region has particularly benefited from and been influenced by studies by Americans in nearby portions of the United States.

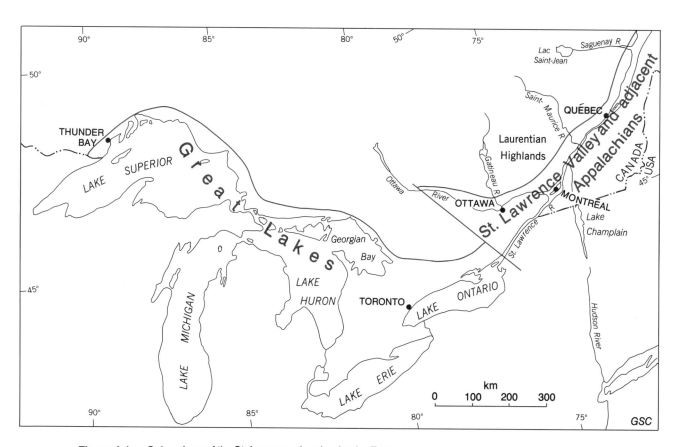

Figure 4.1. Subregions of the St. Lawrence Lowlands; the Frontenac Arch separates the two subregions.

The earliest geological description of Scarborough Bluffs was by Bigsby (1829), who described the erosional forms and materials and speculated that it would be a promising place to search for fossils. Later work by others has abundantly demonstrated his prescience.

Lyell, perhaps the most prominent English geologist of the Nineteenth Century, visited the area in 1841-42. Evidence of former higher water levels north of Toronto reported by Roy (1837) and of raised marine deposits in the St. Lawrence Valley reported by Bayfield (1837) were reviewed by Lyell and described in the two volume account of his travels published in 1845. Lyell also noted the significant role of the St. Davids buried valley in the evolution of the present Niagara River gorge, thus being one of the first to recognize that earlier drainage differed from present drainage.

The Geological Survey of Canada was founded in 1842 and undertook the mapping of the geology of Canada, including considerable attention to the Pleistocene deposits. Knowledge was sufficient by 1863 to enable Logan and his colleagues to present a map of the "superficial deposits" of eastern Canada on which were shown several named stratigraphic units. This type of map has not been attempted since, in spite of subsequent great advances in knowledge.

Study of the glacial and marine deposits of St. Lawrence Valley was a special interest of J.W. Dawson, a strong proponent of the floating ice theory of glaciation. During the second half of the Nineteenth Century, he greatly advanced the study of the Champlain Sea marine faunas and produced a major monograph entitled *The Canadian ice age* in 1893.

An important description of the stratigraphy of Scarborough Bluffs (Hinde, 1878) and some of the contained fossils, the presence of which had been anticipated by Bigsby (1829), included recognition of multiple glaciation, glacial transport of erratics from eastern Ontario, and reference to present time as interglacial. The significance of the report was later emphasized by Coleman, who gave the first description of true interglacial Don beds in 1894 and culminated the work of many decades with his 1941 monograph *The last million years*. Coleman is best known for his work on the Toronto sequence (Coleman, 1933), which he interpreted as representing three glaciations separated by two interglacial formations.

A major reconnaissance study of southern Ontario by Chapman and Putnam (1951) provided a standard reference on the physiography and history of ice retreat. In it the various moraines, shorelines, and drumlin fields were named and described. Revised versions of this were published in 1966 and 1984.

Evidence of former higher lake levels attracted the interest of many workers in the late nineteenth and early twentieth centuries. Several lake stages and various outlets were recognized in each of the Great Lakes basins. This work resulted in numerous reports (Leverett and Taylor, 1915; Coleman, 1936); interest in and publications on this subject have continued to be important up to the present (Hough, 1958; Fullerton, 1980; Karrow and Calkin, 1985).

Although a few areas had been mapped in detail earlier, systematic mapping at a scale of 1:63 360 (1 inch to 1 mile) or greater got under way in the 1940s and 1950s. Rapid expansion of these programs by the Geological Survey of Canada and provincial agencies has resulted in large areas now being covered by published maps and reports, which have provided a wealth of detail on the thickness, extent, composition, age, and provenance of till sheets and related deposits. Concurrently, radiocarbon dating was developed and led to major insights into the age and correlation of deposits.

Since about 1950, there has been considerable emphasis on regional stratigraphy. Gadd (1971) described the St. Pierre interstadial deposits in St. Lawrence Valley and McDonald and Shilts (1971) defined the Quaternary sequence in the southern Appalachians of Quebec. Dreimanis (1958) and Dreimanis et al. (1966) described the Plum Point and Port Talbot interstadial deposits near Lake Erie. This work culminated in a three-fold subdivision of the last glaciation and a time classification was proposed by Dreimanis and Karrow (1972). In conjunction with mapping and other stratigraphic studies, drilling was used to supplement and provide subsurface information, beginning in the 1960s and accelerating in the 1970s. The work of many on the mineralogy and petrography of tills — data used for correlation and provenance studies based on methods largely developed in the Great Lakes area — was summarized by Karrow (1976). The Trois-Rivières congress of l'Association québécoise pour l'étude du Quaternaire in 1976 was used as an opportunity to summarize the Quaternary of Quebec and to make correlations with the rest of eastern Canada (Occhietti, 1977a). The probable correlation of the main stratigraphic units used in this area is presented in Table 4.1.

With the rapid growth of university geology departments in the 1960s and 1970s, the scope of research broadened and activity intensified. New dating methods have been or are being developed and additional interstadial organic deposits have been found. Specific reference to some of the recent work will be made in later sections of this report.

Economic importance of Quaternary geology

The important role played by Quaternary deposits and their history in the economy of southern Ontario and the central lowlands of Quebec is immediately evident when traversing the Paleozoic-Precambrian contact, which extends from southeastern Georgian Bay to Kingston and along the south edge of the Laurentian Highlands of Quebec. North of the line there is little agriculture and a sparse population, whereas south of it agriculture is an important part of the economy and cities and towns are numerous. Thick and nearly ubiquitous Quaternary deposits located in the southernmost part of Canada have fostered unique agricultural development. Proximity to the transportation and water resources of the Great Lakes and St. Lawrence Valley has encouraged industrial and urban development. With this development came the need for large quantities of construction materials (Fig. 4.2) and other industrial minerals and rocks. Consideration of mineral resources usually emphasizes materials extracted for use elsewhere, yet the nature of the ground for bearing loads under highways and buildings, and for excavation of tunnels and canals is just as important. Compared to many other areas, construction in this area is relatively easy and cheap. Topography is important too as it affects slopes, drainage, and cut-and-fill requirements in construction.

Figure 4.2. Glacial erratics provided fieldstone for construction of many farm buildings, such as this house near Lucknow. 200300-T

Engineering geology

The Great Lakes area is a region of generally stable soils. There is an abundance of rather ordinary problems associated with compressible soils such as peat, saturated sands and silts, high water table, and difficult excavation in till, but difficulties have usually arisen through human failure to take precautions that should be normal practice. Only a few problems that are peculiar to or typify the features of the area can be mentioned here.

The Catfish Creek Till is commonly very hard and causes difficulty for well drillers. Fortunately, it is generally covered by younger softer tills and is not often encountered in the course of excavation. Quarrying of Devonian limestone near Woodstock, Ontario, however, involves the removal of 20 m or more of overburden, including Catfish Creek Till, which requires the use of heavy rippers on which teeth need frequent replacement.

Much of the surface of southern Ontario is exposed till, which is usually preconsolidated from former ice loads. Because of the presence of extensive proglacial lakes, in many areas sediments could not drain freely under load. As

Table 4.1. Summary correlation chart for the St. Lawrence Lowlands region

Classification following Dreimanis and Karrow (1972)			LAKE ERIE	TORONTO	TROIS-RIVIÈRES [1]	SOUTHEAST QUEBEC	Classification following Fulton (1984)
WISCONSINAN	LATE	PORT HURON STADE		Halton Till	Gentilly Till	Lennoxville Till	LATE / WISCONSINAN
		MACKINAW INTERSTADE		sand			
		PORT BRUCE STADE	Port Stanley Till	sandy till			
		ERIE INTERSTADE	Malahide Formation				
		NISSOURI STADE	Catfish Creek Till				
	MIDDLE	PLUM POINT INTERSTADE	Wallacetown Formation	Thorncliffe Formation	?	Gayhurst Formation	MIDDLE
		CHERRYTREE STADE					
		PORT TALBOT INTERSTADE	Tyrconnell Formation				
	EARLY	GUILDWOOD STADE	Bradtville Till	Sunnybrook Till		Chaudière Till	EARLY
		ST. PIERRE INTERSTADE		Pottery Road Formation	St. Pierre Formation	Massawippi Formation	SANGAMONIAN
		NICOLET STADE		Scarborough Formation	Bécancour Till	Johnville Till	
SANGAMONIAN				Don Formation		pre-Johnville sediments	
ILLINOIAN				York Till			

GSC

[1] Table 4.14 shows an interpretation of this stratigraphy that was added during editing

a result, some till is only partly preconsolidated. Till, however, has been a useful material for earth-fill dam construction (Legget, 1942). There are also extensive areas of fine lacustrine sediments, deposited in proglacial lakes, which are normally consolidated.

By contrast, lower Ottawa and upper and central St. Lawrence valleys are extensively mantled by marine clay and silt deposited in the Champlain Sea. These deposits, as thick as 100 m in Ottawa Valley (Fransham et al., 1976) and 30 m or more in St. Lawrence Valley (Gadd, 1971), cause many engineering problems. Most spectacular are the numerous earth flow landslides which have caused extensive rural and urban property damage and often loss of life. The properties of Leda Clay, as it is known in engineering usage, have been reviewed by Crawford (1968) and are discussed by Locat and Chagnon (1989).

Excavations for the St. Lawrence Seaway west of Montréal in the 1950s encountered extremely hard till. The unexpected hardness of the till necessitated blasting in some places (Cleaves, 1963). Costs were increased markedly and some contractors went bankrupt.

Shoreline erosion

The southern and western boundary of the region is lake or river shoreline. Water levels in the Great Lakes fluctuate about 2 m over periods of a decade or two, and it has been noticed that in general shore erosion rates increase during high water periods (Matyas et al., 1974). Such widespread problems were experienced during high water levels in the early 1970s that a comprehensive study of the shorelines of Lake Huron, Erie, and Ontario was undertaken (Haras and Tsui, 1976). Active erosion is present along much of the lakeshores, as indicated by the shorecliffs, which are particularly high and impressive at Newcastle and Scarborough (up to 100 m, Fig. 4.3) on the north shore of Lake Ontario, and which are 20 to 50 m high along Lake Erie and Huron shoreline. Erosion rates may reach several metres per year, and some areas have experienced severe property damage. The most severe problems occur where the bluffs include materials other than till; till is generally more resistant to erosion and often forms small headlands along the shore.

A subtle factor contributing to the shore erosion problem is crustal tilting, which is raising water level in the western Lake Ontario basin and lowering it in the northern Huron basin by about 0.3 m per century (Clark and Persoage, 1970). Present crustal tilting is variously attributed to tectonism and to residual glacial isostatic effects. The geological record clearly indicates much higher tilting rates in the early postglacial period. Such tilting is of great concern too because of its effect on water depths and transport on the Great Lakes.

Mineral resources

Mineral resources of Quaternary origin include clay for brick and tile manufacture, and sand and gravel for concrete aggregate. Clay was formerly more important than now, as shale from the bedrock has largely replaced it for brick making and plastic has largely replaced clay for tile. Nevertheless, the extensive use of brick in building construction in this area has depended to a large extent on lacustrine clays in many parts of the region. Although these have usually been of late glacial age, in former years several brickyards in Toronto used clay of the Early Wisconsinan Scarborough Formation (Coleman, 1933). Partial weathering gave it a low carbonate content and yielded red brick. Deschaillons varves and some marine clays are used in Quebec.

The chief Quaternary mineral product at present is gravel. It is distributed unevenly, being most abundant in association with coarse-textured tills, and is found as kames, eskers, and outwash. Extensive outwash deposits are associated with the Grand, Thames, and Saugeen valleys of Ontario. Large volumes are also found in kame moraines (Waterloo, Orangeville, Oak Ridges, and Highland Front). Substantial volumes of gravel also occur in some glacial lake beaches such as those of Lake Warren and Lake Algonquin. Large pits in Lake Iroquois deposits at Hamilton and Toronto formerly provided a convenient supply, but remaining material has been built over and hence has become too costly to remove. Deltaic and outwash deposits, as well as marine beach deposits, are widely used in eastern Ontario and Quebec. Present day rivers and beaches yield only minor amounts for local use. Large volumes of gravel remain in some areas, but their removal has become a matter of increasing controversy because of the disruption to agriculture and the scenic value of the rural landscape.

Peat moss for horticulture is harvested mainly in Quebec, particularly on Île aux Coudres, at Rivière-Ouelle on the lower St. Lawrence, and in central St. Lawrence Valley (Risi et al., 1953; Gauthier, 1971). Peat is little exploited in Ontario — Welland bog near Niagara Falls being a notable exception. Due to their size and thickness, peat bogs as yet unexploited have great economic potential.

The presence of bog iron ore in the marshes of the lower Saint-Maurice was responsible for the founding of the Saint-Maurice ironworks, active between 1733 and 1883, and the Radnor ironworks, which operated until 1910. This ore, a limonite containing manganese and alumina, was extracted from peat bogs and shallow lakes by hand and by

Figure 4.3. Scarborough Bluffs; upper blocky unit is Sunnybrook Drift, which overlies stratified sand and clay of the Scarborough Formation. 200300-N

dredging. Near Champlain, at Red Mill, the iron oxide found in a peat bog has been used in the production of red ochre for paint pigment (Terasmae, 1960a; Gadd, 1971).

Placer gold deposits have been mined in the glacial and Holocene sediments of the Beauce region (Chalmers, 1898; Morison, 1989). Tills and stream sediments are currently utilized in geochemical prospecting in the Appalachians (Shilts, 1973, 1976; DiLabio, 1989).

Planning

The nature of the landscape, which is largely glacial in origin, has influenced development in many ways. The large valleys in Toronto, whose trend is roughly parallel to the last ice movement and the resulting flutings and drumlins, created barriers to east-west transportation in early years and affected the pattern of development (Taylor, 1936). Other transportation routes in southern Ontario have followed ready-made easy gradients along meltwater channels, such as the Trent Canal along the outlet of Lake Algonquin. Lake terraces have been similarly used, for example, Queen Elizabeth Way from Toronto to the mouth of Niagara River largely follows the Iroquois plain, and Highway 401 follows it east of Toronto. River valleys incised into the flat marine plain in Quebec have similarly influenced the road pattern. Last, but not least, the combination of climate and fertile soil has provided a pleasant environment for human habitation, as shown by the large fraction of Canada's population that lives in this area.

QUATERNARY GEOLOGY OF THE GREAT LAKES SUBREGION

P.F. Karrow

The Great Lakes subregion is the western part of the St. Lawrence Lowlands region. It consists largely of the area known as southern Ontario, includes the basins of lakes Ontario, Erie, Huron, and Superior, and extends eastward to the Frontenac Arch (Fig. 4.1). The southern limit is the boundary with the United States and the northern limit is the northern shores of lakes Superior and Huron and the southern edge of the Precambrian Shield. This area is one of the most fully economically developed areas of Canada and as a consequence our knowledge of the Quaternary geology of this area surpasses in detail that of any area of similar size in Canada.

BEDROCK GEOLOGY AND PHYSIOGRAPHY

The Lake Superior basin is the only part of the Great Lakes subregion underlain by Precambrian rocks. These include a variety of igneous, metamorphic, volcanic, and sedimentary rocks that are mainly of relatively young Precambrian age. Much of this Canadian Shield area has its typical rounded hill form, whereas near Thunder Bay basic sills have been eroded into mesas resembling parts of the arid southwestern United States. The geology of the floor of Lake Superior is still poorly known but young Precambrian redbeds are apparently extensive, with a varied topography of moderate relief. The Precambrian geology of the basin has recently been compiled by Wold and Hinze (1982).

The Great Lakes subregion is underlain mostly by gently dipping rocks of Cambrian through Devonian age (Fig. 4.4). Paleozoic rocks are draped over a hilly Precambrian surface, with dips that diminish upward in the sequence. The Precambrian-Paleozoic contact is irregular, with numerous Paleozoic outliers and Precambrian inliers in a zone a few kilometres wide.

The Precambrian surface slopes steeply southward into the Allegheny Basin and westward into the Michigan Basin, where sedimentary accumulations are more than 4000 m thick. Separating these basins is the Algonquin Arch (Fig. 4.4), a broad rise in the Precambrian surface with an axis trending southwestward under the interlake peninsula of southwestern Ontario.

The Paleozoic succession is characterized by alternating sequences of fine clastic, chiefly shale, and carbonate sedimentary rocks. Massive carbonate formations form bedrock highs and escarpments (Fig. 4.5), whereas shale underlies the lowlands and basins. Through a combination of long-continued stream erosion and later modification by glacial erosion, the present cuesta-vale topography developed. The basins of the lower Great Lakes developed around the Michigan Basin and along the edge of the Allegheny Basin mainly in areas of shale. Even within the lakes, secondary basins are found over shale, with submerged carbonate rock escarpments commonly separating them.

As a result of their gentle dips, the Paleozoic formations are exposed, or subcrop, as broad belts trending northwest-southeast (Fig. 4.4). The composition of glacial deposits reflects that of the bedrock and similarly shows broad uniformity over large areas, facilitating the correlation of till sheets within individual lake basins but making more difficult the task of distinguishing between tills from different basins.

The surface of the Paleozoic rocks is commonly buried by 30-60 m (but may exceed 250 m) of Quaternary deposits; compilation of water, oil, and gas well records and engineering borings has provided a general picture of the buried topography (White and Karrow, 1971; Karrow, 1973; Eyles et al., 1985; Flint and Lolcama, 1986). Over large areas the bedrock surface is flat or gently sloping. Narrow valleys 30-60 m deep are widely distributed and probably were eroded in preglacial and interglacial time.

Karrow, P.F.
1989: Quaternary geology of the Great Lakes subregion; in Chapter 4 of Quaternary Geology of Canada and Greenland, R.J. Fulton (ed.); Geological Survey of Canada, Geology of Canada, no. 1 (also Geological Society of America, The Geology of North America, v. K-1).

The edge of Ordovician limestones is commonly marked by a low escarpment between Georgian Bay and eastern Lake Ontario, but except for Laurentian Valley (Fig. 4.6), which extends between Georgian Bay and Toronto (Spencer, 1890; White and Karrow, 1971), there has been only limited study of bedrock topography between it and the Niagara Escarpment. Laurentian Valley is the largest buried valley in southern Ontario and, where it passes under the Oak Ridges interlobate moraine (Fig. 4.6) north of Toronto, drift thickness exceeds 250 m. This valley is believed to be of preglacial origin and to have carried the drainage of the upper Great Lakes at least until the last glaciation (White and Karrow, 1971).

The Niagara Escarpment (Silurian) is the most prominent exposed bedrock topographic feature in southern Ontario (Fig. 4.5, 4.6). It commences in New York State south of Lake Ontario and enters Ontario north of Niagara Falls. Niagara River has cut a gorge 100 m deep which has extended about 11 km upstream from the edge of the Niagara Escarpment in the 13 ka since deglaciation. The escarpment forms the crest of Bruce Peninsula, separating Georgian Bay and Lake Huron, and is the principal topographic feature of Manitoulin and Cockburn islands. Along part of its course in Ontario, it is a multiple escarpment (Hewitt, 1971). Relief along the escarpment varies from 100 m in the Niagara Peninsula to over 200 m south of Georgian Bay.

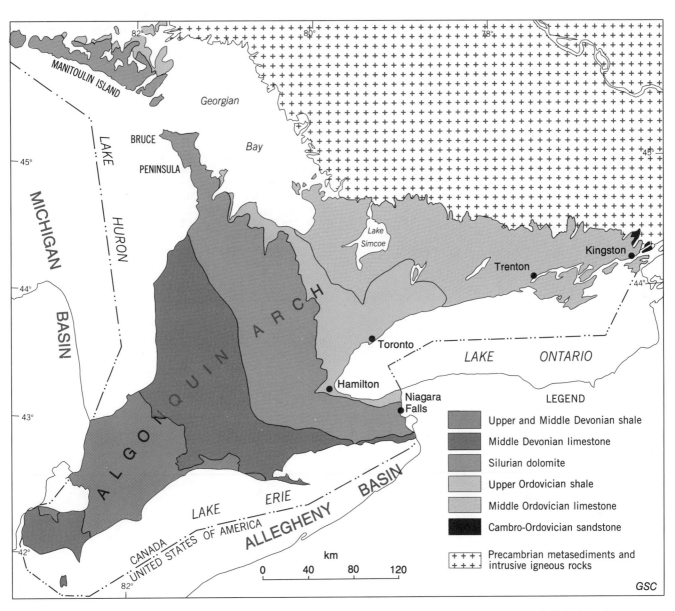

Figure 4.4. Bedrock geology of southern Ontario (from map DDM 4114A compiled 1976 by P.G. Teleford, Ontario Geological Survey).

The Devonian Onondaga Escarpment lies south of the Niagara Escarpment and extends from Fort Erie at least to Brantford (Fig. 4.6, 4.7) before disappearing under thick overburden. It reappears near Kincardine, jogs sharply south under Lake Huron, then continues northwest under the lake (Sly and Lewis, 1972) and into Michigan. The Ipperwash Escarpment (Devonian) is best known in the southern Lake Huron basin; it is believed to extend eastward under Lake Erie into New York State.

Major buried valleys (Laurentian and Wingham-Dundas valleys, Fig. 4.6) occur between the escarpments and follow the trend of lowlands underlain by shale. Parts of numerous lesser tributary valleys 10-60 m deep have been identified but their drainage connections and history of formation are largely unknown (Karrow, 1973). Many of these buried valleys are steep-walled and gorge-like. In a few cases drilling has revealed the presence of interstadial deposits of probable Middle Wisconsinan age. All these valleys probably represent numerous episodes of stream erosion in preglacial, interglacial, interstadial, and postglacial time. The bedrock surface between the valleys seems to be generally flat or gently sloping.

Figure 4.5. View looking southward to Niagara Escarpment, north of Hamilton. 200300-Q

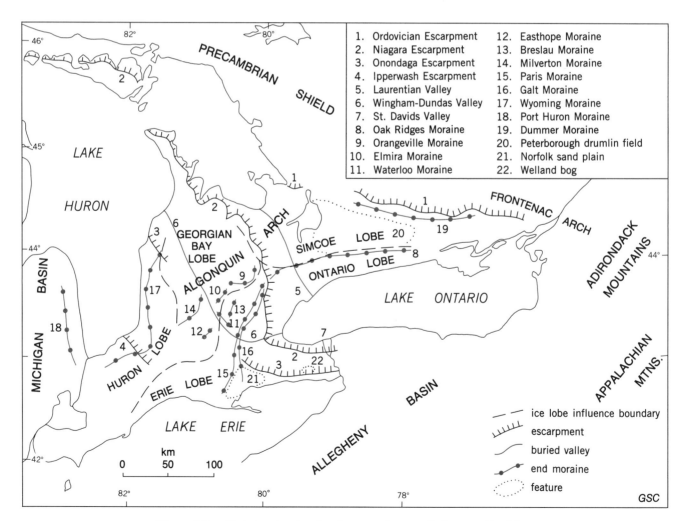

Figure 4.6. Buried and surface landforms and areas of influence of ice lobes.

There is little evidence of tectonic disruption of the Paleozoic rocks. Scattered occurrences of faults, with displacements of 1 to 2 m, have been noted (Liberty, 1969), but at least some local folds and faults have been attributed to recent stress-relief phenomena and are referred to as "popups" (White et al., 1974; Russell et al., 1982). Aside from these, the area has been subjected to repeated flexure as a result of glacial isostatic tilting, the postglacial record of which will be described later.

PHYSIOGRAPHY AND CLIMATE

Chapman and Putnam (1951, 1966, 1984) have described and portrayed on 1:250 000 scale maps the landform features of southern Ontario. Their work quickly became a standard and now classic reference and provided an important reconnaissance base for the more detailed mapping of Quaternary geology that followed it (see also Map 1704A). The following summary is largely based on the work of Chapman and Putnam (1984). Figure 4.7 shows most of the place names used in this section of the chapter.

Bedrock features

As already mentioned, bedrock topography influences surface landforms in part of southern Ontario; bedrock outcrops in several escarpments, the most prominent being the Niagara Escarpment. Adjacent to the Ordovician Escarpment extensive areas of limestone plain are concentrated northeast of Lake Simcoe, in the Prince Edward County peninsula of eastern Lake Ontario near Trenton, and northwest of Kingston (Fig. 4.6, 4.7). Limestone plains are also extensive along the Niagara Escarpment northwest of Hamilton, in the Bruce Peninsula, and on Manitoulin Island.

Erosional landforms developed in bedrock contribute to the scenic beauty of the area. Rock gorges are a notable feature along the Niagara Escarpment, the most spectacular being the Niagara River gorge. Variations in the width of the gorge have been related to drainage changes in the Great Lakes as discussed at length by Spencer (1907) and many others (Kindle and Taylor, 1913; Calkin and Brett, 1978).

Solution features are minor on the carbonate terrains; disappearing streams, caves, and sinkholes have been noted in only a few places on Silurian dolostones in the Niagara Peninsula, near Guelph, near Beaver Valley southeast of Owen Sound, and near Hepworth northwest of Owen Sound. Numerous sinkholes have been noted in areas of shallow overburden south of Seaforth (northeast of Grand Bend; Karrow, 1977) and near Lake Erie east of Tillsonburg (Barnett, 1978); this area is underlain by limestone of the Devonian Dundee Formation. Presumably more extensive karst development was present in preglacial times but has been covered, filled, and collapsed as a result of glaciation.

Small areas of shale plain lie northwest and south of Lake Ontario and southwest of Georgian Bay. These areas are underlain by grey and red shales of Late Ordovician age and vary from flat areas planed by glacial lake waters to steeply sloping dissected areas of badlands.

Glacial landforms

Most other landforms in southern Ontario owe their origin directly or indirectly to the effects of glaciation. They will be described by type, generally in their presumed order of formation.

Till plains are an extensive landform and consist of nearly flat to undulating areas underlain by till. They are the predominant landform over most of the southwestern peninsula of southern Ontario. Extensive areas near Windsor, Sarnia, and Toronto, are bevelled till plain and are believed to have been affected by glacial lake planation, leading to typically flat terrain. Because of the genesis of these features, discontinuous veneers of lake sediment are in places present. Other parts of the till plains are fluted to varying degrees by glacier flow. Fluting is seldom visible on the ground but is obvious on airphotos (Fig. 4.8). Fluting has been noted on till plains between Toronto and Hamilton and around Conestogo Lake (between Guelph and Lake Huron).

Flutings, mostly found on silty to clayey tills, grade into drumlins, which usually consist of sandy till (Karrow, 1981a). In many places drumlins occur in several "fields", the largest being the Peterborough drumlin field (20 of Fig. 4.6), with several thousand drumlins in central southern Ontario, which demonstrate ice movement from the northeast. Other prominent drumlin fields are located near Woodstock, Guelph, Wingham, and Owen Sound. A drumlin field on St. Joseph Island, northwestern Lake Huron, extends offshore under Lake Huron, and onto Drummond Island in the State of Michigan. In many places drumlins rest on limestone plains such as northwest of Hamilton and Owen Sound, and on Manitoulin Island, but elsewhere occur in areas of moderately deep overburden. They vary greatly in size and axial proportions, and a few contain cores of older tills or glaciofluvial deposits.

Ridge-like, hummocky, ice marginal accumulations are here grouped into three categories of moraines: end moraines, composed predominantly of till with variable amounts of associated kame sand and gravel; kame or interlobate moraines, composed predominantly of sorted sand and gravel; and minor moraines, which are just small end moraines that occur in series.

About 30 end moraines have been identified in southern Ontario; they have played a major role in hypotheses about the history of ice retreat (Taylor, 1913; Chapman and Putnam, 1951). During deglaciation the higher part of the interlake peninsula was uncovered first, with major ice lobes remaining in the lake basins (Fig. 4.6). The spasmodic retreat of ice lobe margins downslope toward the lake basins formed a series of subparallel moraines encircling the higher interlake peninsula. Some moraines are only a few kilometres long and mark local events, whereas others are nearly continuous from lobe to lobe (Fig. 4.9). Among the larger end moraines are the 135 km-long Paris and Galt moraines, formed by the Erie and Ontario lobes, and the 190 km-long Wyoming Moraine of the Huron Lobe (Fig. 4.6, 4.10). The large moraines may be several kilometres in width and stand up to 30 m above adjacent plains. Those formed of clayey till (Wyoming Moraine) commonly have more subdued local relief, whereas those formed of sandy till (Paris Moraine) exhibit much steeper slopes and rougher surface. Also, in many cases the moraine material is coarser on the flanks of the lobe, but as the moraine is traced toward the axis of the lobe, the influence of proglacial lakes becomes more pronounced and the material becomes finer and relief more subdued. In addition, the moraine may become concealed under a cover of lake sediments. Such variations occur along the Paris Moraine.

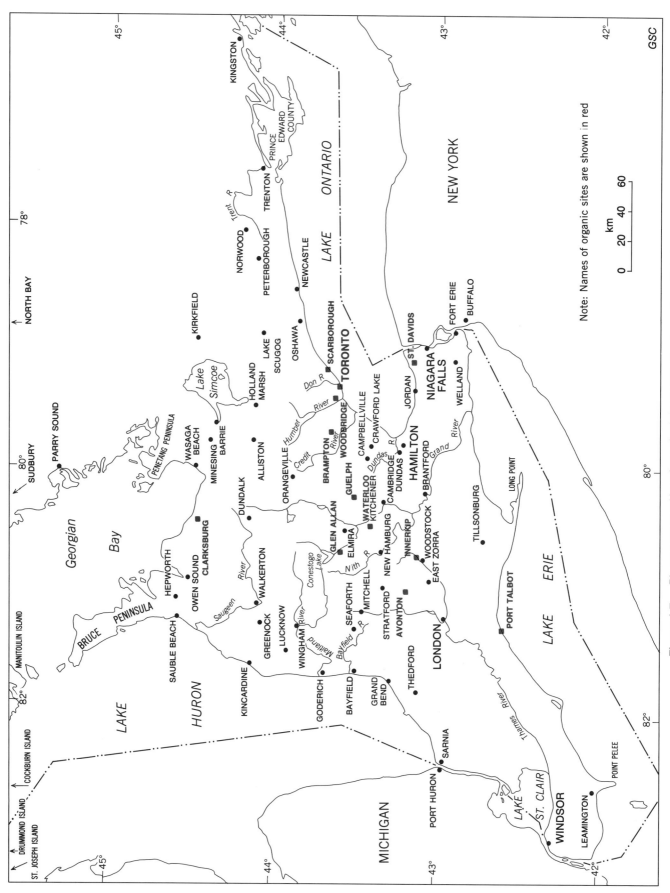

Figure 4.7. Place names and locations of subtill organic sites in southern Ontario.

Kame or interlobate moraines are not as common as end moraines, but are among the largest masses of glacial accumulations in the area. Their surface is marked by rough irregularity with numerous kame hills and ice block depressions. They formed as the result of concentration of meltwater along the junction of two ice lobes and consequently waterlaid sediments predominate. The principal kame moraines are the Waterloo, Elmira, and Orangeville moraines, formed between the Huron-Georgian Bay and Erie-Ontario lobes, and the Oak Ridges Moraine, formed between the Georgian Bay-Simcoe and the Ontario lobes (Fig. 4.6). The moraines are complex in structure and yield conflicting evidence as to their origin. They generally mark an interlobate zone of interaction between opposing lobes with strong evidence of final overriding by ice of one of the lobes. Duckworth (1979) has described braided stream, deltaic, and lacustrine deposits from the western part of the Oak Ridges Moraine and related them to alternating advances and retreats of the Ontario and Simcoe lobes.

Minor moraines may be only 1 or 2 m or up to 5 m high and extend for only a few kilometres. A series of about 20 such moraines has been recognized in the vicinity of Stratford, where they apparently were deposited at the ice margin while it stood in shallow glacial lakes, since lacustrine silts largely cover the till in the troughs between the gentle ridges. Other minor moraines occur west of Conestogo Lake and where they cross eskers, which trend perpendicularly, a node of kame sand and gravel occurs. These moraines evidently mark short pauses during which meltwater streams built kames at the mouths of ice tunnels in which eskers were forming (Karrow, 1977, in press).

Moraines of all types commonly show evidence of multi-stage history, and some were obviously formed prior to the last ice advance; a partial listing of these features has been given by Karrow (1974).

Study of lake bottom topography and sediments has revealed the presence of cross-lake moraines under lakes Erie and Ontario (Lewis et al., 1966; Wall, 1968; Sly and Lewis, 1972) which form continuations of some of the end moraines known from adjacent land areas. Similar features have been noted under Lake Superior (Landmesser et al., 1982).

Glaciofluvial landforms

Eskers are numerous in southern Ontario, although many are small and even the largest are only of moderate size (Fig. 4.9). The smallest eskers are 1 or 2 m high and 2 or 3 km long west of Guelph (Karrow, in press); the longest one extends with gaps from near Trenton northeast about 135 km. Most eskers are 10 to 25 km long and 5 to 15 m high; the highest eskers exceed 30 m (Norwood esker near Peterborough). The Brampton esker, the sedimentology of which has been studied by Saunderson (1975), is palimpsest, having probably been formed during melting of the penultimate ice and overridden by the last ice advance which covered it with till (Karrow et al., 1977).

Eskers are most common in areas characterized by coarse textured till, because meltwater washing of the till in these areas provided an abundance of coarse clastic material. Eskers trend in the general direction of ice flow but follow or trend toward low ground. Because drumlins are also

Figure 4.8. Flutings formed by Georgian Bay lobe ice flowing southeastward, north of Conestogo Lake. NAPL A18273-139

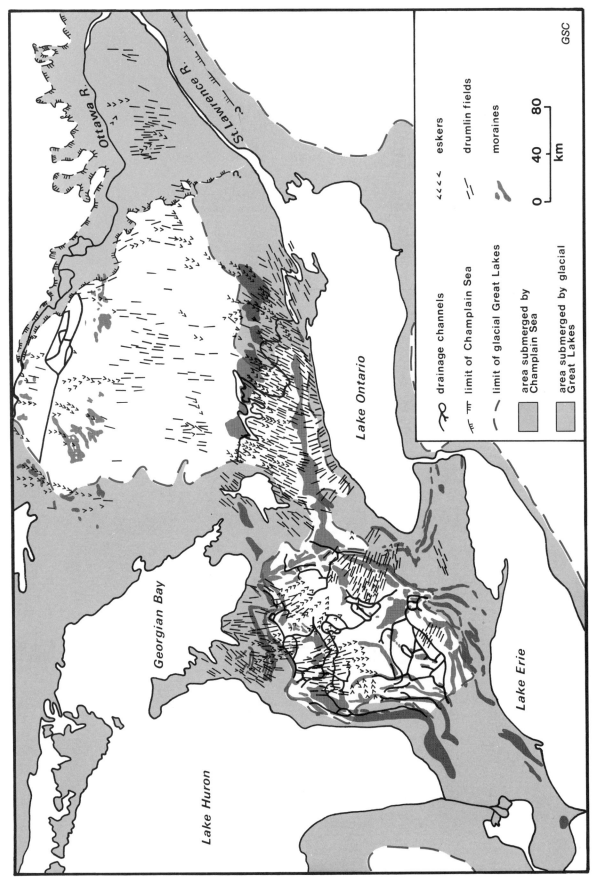

Figure 4.9. Glacial features of southern Ontario.

preferentially formed of coarse till, eskers are common in the drumlin fields. At least one case is known, in the Guelph drumlin field (Karrow, 1968), of an esker extending from both sides of a drumlin, with a channel eroded across its crest joining the esker ridges. This provides evidence of sequential development, with the drumlin formed by actively flowing ice and the esker formed later as the ice stagnated and melted.

Other meltwater deposits are outwash fans, plains, and terraces. Here too the glaciofluvial deposits are related to till texture, with few deposits associated with the finer tills of the Huron Lobe (Karrow, 1977; Cooper, 1979) but many more associated with the coarse tills of the Georgian Bay (Cowan, 1979) and Ontario (Cowan, 1972; Karrow, 1987) lobes. Outwash can commonly be traced to specific sources in outwash fans along end moraine fronts (Karrow, 1968) where the outwash plains appear to have formed through a coalescence of outwash fans. Such arrangements of landforms provide clear indication of the position and direction of movement of ice. Outwash was transported many kilometres along or away from the ice front, in places forming extensive deposits along the flanks of present stream valleys, and in places accumulating in deltaic deposits at the margins of glacial lakes, such as near Brantford (Cowan, 1972). Kettles are common and in places occur in large numbers to form pitted outwash. Kettle lakes are numerous southeast of Kitchener, south of Owen Sound, and north of Toronto in the Oak Ridges Moraine.

Many of the present streams originated as meltwater streams, but experienced changes in their course during ice retreat as lower paths were made available. Thus upper Grand River north of Guelph traces an ice marginal path, but formerly it flowed southwest across part of Nith River valley, and entered upper Thames River valley upstream from Woodstock (Karrow, 1987). Later, ice retreat opened the present middle Grand Valley, beheading what had been part of the Thames drainage. In addition to changes in the drainage pattern, disappearance of ice from the area has left many smaller streams underfit in former meltwater valleys. Hence, the present drainage pattern evolved through the complex interaction of many events associated with glaciation.

Chapman and Putnam (1951, 1966, 1984) classified all meltwater channels as spillways, and outwash as spillway deposits. True spillways, which are channels draining lakes and ponds, are comparatively rare in southern Ontario, but some cross the Milverton Moraine (Fig. 4.6) north of Stratford (Karrow, 1977) and a major one which was an outlet for glacial Lake Algonquin, follows approximately the Precambrian-Paleozoic boundary along Trent River valley (the present route of the Trent Canal). Meltwater channels are numerous and vary from small and short to those of moderately large dimensions. They form erosional counterparts of outwash deposits, with which they are closely associated. Although temporarily significant to drainage evolution, these features are small and form only a minor part of the landscape. Some particularly conspicuous examples developed as ice marginal channels along the Niagara Escarpment (White, 1975).

Glaciolacustrine landforms

Features of glacial lake origin were mapped by Chapman and Putnam (1951) as clay plain, sand plain, and beaches;

Figure 4.10. Hummocky topography of the Galt Moraine, south of Galt. 200300-S

they will be considered together here because of their common origin and association.

During ice retreat downslope into the Great Lakes basins, meltwaters were dammed between higher ground and the ice margin. As the ice retreated northward from the United States, it exposed the St. Lawrence-Mississippi drainage divide and formed ice marginal glacial lakes which grew in extent as the ice retreated. The former presence of a series of lakes is shown by widespread deep water clay plains, shallow water sand plains, and numerous raised shorecliffs and beaches. Most of these features are marginal to the present lakes and, in general, the highest features, which are farthest from the present shore, are the oldest, and the lowest features, which are nearest the present shore, are youngest. Shoreline features rise in elevation to the north and northeast as a result of isostatic uplift.

Because of the pattern of ice retreat, the Erie basin was deglaciated first and a belt some tens of kilometres wide, extending north to London and Brantford and covering most of the Niagara Peninsula, was affected by glacial lakes Maumee, Whittlesey (Fig. 4.11), and Warren. Thick clay deposits underlie this area. A large complex delta plain underlain by gravel and sand extends south from Brantford to Lake Erie, forming the chief tobacco-growing area of Ontario today. Similar deltaic sand plains are to be found along Thames River valley southwest of London.

The belt of lake plain along Lake Huron is much narrower (1-5 km) than that near Lake Erie and lacustrine deposits are generally thin, but the Warren shoreline is well developed. Younger shorelines have been removed by recent shore erosion but the former presence of Lake Algonquin is indicated by prominent hanging terraces along streams flowing into Lake Huron (Karrow, 1986). North of Kincardine the Algonquin plain is still present (Fig. 4.12), marked by large gravel bars at several places. Bruce

Figure 4.11. Lake Whittlesey shoreline eroded into a drumlin east of Galt. 200300-P

Figure 4.12. Lake Algonquin shorebluff, east of Douglas Point on Lake Huron. Terrace at mouth of Underwood Creek (in distant trees) and a notch in the bluff mark the Main Algonquin shoreline. 200300-M

Peninsula was mostly submerged by Lake Algonquin, but lacks extensive lacustrine deposits and consists largely of exposed rock.

Sand plains related to Lake Algonquin are extensive southeast of Georgian Bay. The terrain is irregular and so is the extent of the Algonquin sand plain; these deposits underlie the potato-growing district of the Alliston embayment, Canadian Forces Base Borden, the Penetang Peninsula, and a fringe around Lake Simcoe. This is a classic area for the Algonquin beaches, as described by Deane (1950) and Stanley (1936, 1937).

The principal glacial lake in the Ontario basin was Lake Iroquois. The associated sand and clay plain is only 1 to 10 km wide but shore cliffs and gravel bars are prominent features. In Niagara Peninsula the shoreline extended along the base of the Niagara Escarpment. Large bars formed across Dundas Valley at Hamilton and Humber and Don valleys at Toronto. Multiple shorelines have been described by Mirynech (1962) in the area northwest of Kingston and all shorelines disappear, presumably at the damming ice front, to the north on the Canadian Shield.

In addition to the lake plains of the major glacial lakes, numerous smaller glacial lakes existed for short periods during ice retreat. Their former presence is usually indicated by clay plains, sand plains, or deltas. Shore features are rare, probably because of the brevity of the lakes' existence and their limited extent, which hindered shore erosion. Lake Peel (White, 1975; Karrow, 1987) was one of these shallow temporary lakes which formed northeast of Hamilton; most others have not been named but examples have been noted in many areas.

Eolian landforms

Eolian landforms are widespread but are prominent in only a few areas. The largest active sand dunes are those developed in areas of abundant sand downwind (east) of present shorelines. Parabolic dunes are well developed south of Grand Bend and at Sauble Beach on Lake Huron, Wasaga Beach on Georgian Bay, and in western Prince Edward County in eastern Lake Ontario. Some dunes are more than 30 m high but more commonly they are 10-15 m high. Stabilized dunes occur along former glacial lake shorelines and on deltaic or shallow water lacustrine sand plains, such as south of Brantford and southwest of London. Parts of the interlobate kame moraines, such as the Waterloo and Oak Ridges moraines, show evidence of wind reworking with small irregular sand dunes in some places. Pockets of loess up to 1 or 2 m thick are known from these moraines but are not of mappable extent.

Wetlands

Wetlands include bogs, swamps, and marshes. The deepest organic accumulations are found in kettles. Most such deposits are not areally extensive but are numerous in the irregular topography of the kame moraines (Oak Ridges, Waterloo), end moraines (Paris, Galt), and in pitted outwash. Some of these bogs preserve relict vegetational assemblages from former colder climates. Some bogs and swamps are the filled remnants of former glacial lakes or meltwater channels and are more extensive, covering several tens of square kilometres but only 1 to 3 m deep of peat and muck (Welland bog, Fig. 4.6; Minesing swamp, Greenock swamp). Other bogs and marshes reflect isostatic crustal tilting and drainage interference; examples are the Holland Marsh, south of Lake Simcoe, and southern parts of nearby Lake Scugog. Lagoons behind baymouth bars of former lakes are also sites of bogs at Thedford, Kincardine, and Wasaga Beach. Because of the difference in local relief, bogs are more abundant in areas of coarse grained till, with their abundant associated ice contact and outwash deposits, than in areas of fine grained till.

Fluvial landforms

Alluvial features are found flanking former and present streams. Although the overall degree of drainage development is youthful, because of local resistant bedrock or rising lake levels for example, some stream reaches have developed meanders with floodplains; examples include Grand River south of Kitchener and near Brantford, Thames River southwest of London, and Saugeen River downstream from Walkerton. Many streams have a history of flooding, in part reflecting human interference by land clearance and urbanization, but also reflecting natural factors such as extensive clay soils in headwaters which promote rapid runoff. The geomorphology of Ontario's valleys has received little attention, but the studies of Gardner (1977) and Martini (1977) on flood effects in Grand Valley are of interest.

Terraces are well developed in many stream valleys and in some areas a close relationship with former lake levels is evident (Karrow, 1986). The continuing historical rise in the level of Lake Ontario on its southwest shore has led to the drowning of river mouths (Jordan Harbour, west of the mouth of Niagara River) and accumulation of thick sediment fills.

Climate

This subregion includes the most southerly part of Canada with the southwestern peninsula projecting well into the region of the Mid-West United States and hence sharing many aspects of its climate.

The Great Lakes themselves have a notable effect on climate, in that they tend to moderate winter and summer temperatures and steepen temperature gradients around individual basins. Moisture from the Great Lakes also causes additional cloudiness and in winter, prevailing westerly winds contribute to "snow belts" in lee areas of the lakes, such as the Parry Sound district east of Georgian Bay and Goderich-Owen Sound east of Lake Huron. Precipitation is between 650 and 900 mm/a and is rather evenly distributed throughout the year with higher areas generally having greater precipitation than lower areas. The Niagara Escarpment, with relief varying from 100 to 200 m, also affects local climate, with temperatures being lower on top of the escarpment.

Table 4.2 adapted from Brown et al. (1968), illustrates some typical parameters for climate in southern Ontario.

NATURE AND DISTRIBUTION OF QUATERNARY DEPOSITS

Information on the nature and distribution of deposits has been obtained mostly from government mapping programs, with important contributions from university research workers. Almost all of the land areas of the Great Lakes subregion have been mapped at 1:50 000 or larger scale while the distribution of deposits under the Great Lakes has been mapped only on a reconnaissance basis (Map 1704A, in pocket, shows surficial geology mapping coverage). For most land areas, at least preliminary maps and reports are available, with final reports available for some areas only (see Map 1704A).

Deposits and landforms are predominantly glacial, with extensive areas of modification by glacial lakes. Limited areas were subject to meltwater action along channels and over outwash plains; small areas of alluvial environment occur along present-day streams. In general, most sediments have been deposited in association with glaciation, while during interstadial or interglacial time the landscape was primarily subject to erosion, as it is now.

Table 4.2. Typical climatic parameters for southern Ontario (from Brown et al., 1968)

	Leamington	Lake Ontario Shore	Dundalk Upland
Elevation (m)	183	91	488
Mean annual temperature (°C)	9	7	6
Extreme low temperature (°C)	-29	-34	-33
Extreme high temperature (°C)	40	40	34
Mean annual frost free days	170	150	115
Mean annual growth days	220	205	190
Mean annual precipitation (mm)	762	864	686
Mean annual snow (cm)	89	165	254

Nonglacial deposits

Few deposits in southern Ontario were formed in the absence of glacial activity because, as mentioned above, erosional regimes prevailed in interglacial and recent times. Only in low-lying, poorly drained depressions, along stream floodplains, and in lake basins is sedimentation generally occurring now, and similar conditions are believed to have prevailed in comparable times in the past. Thus, the nonglacial record of the past (Fig. 4.13) is limited. The chief postglacial sediment record lies under the Great Lakes where, because of relative inaccessibility, it has received only modest attention. Numerous smaller sedimentary basins under lakes and bogs have received attention sufficient in only a few places to establish a general picture of sediment history. A detailed record documenting local variations within basins or between basins has not been attempted.

The various deposits will be considered from oldest to youngest. In this section emphasis is on the physical nature of the deposits, with their historical significance discussed in the next section.

The Don Formation

This unit includes fossiliferous gravel, sand, silt, and clay averaging about 8 m thick and underlies much of Metropolitan Toronto. Because of the indications of warm climate (wood, leaves, pollen, molluscs) and its position above and below glacial deposits, its deposition is attributed to an interglaciation, presumably the last one, about 125 ka. The Don Formation has only one "permanent" exposure at the Don Valley Brickyard (Fig. 4.14) in central Toronto, but

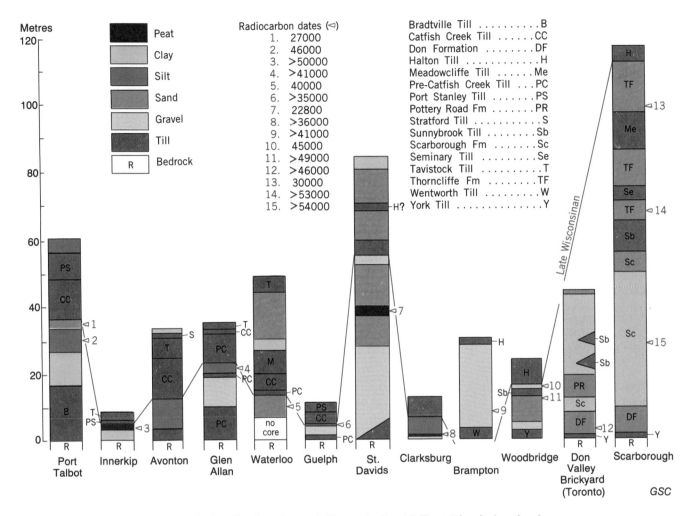

Figure 4.13. Stratigraphy at subtill organic sites (cf. Fig. 4.7 for site locations).

it has been exposed temporarily in numerous building, subway, and service trench or tunnel excavations in downtown Toronto. A possible extension northward along the buried Laurentian Valley has been indicated by deep borings (Coleman, 1933; Karrow, 1970; Sado et al., 1983). Borings at Scarborough Bluffs in eastern Toronto reveal its presence there below the level of Lake Ontario, and the recent discovery of molluscs in intertill sands at the Woodbridge railway cut suggests its presence in northwestern Toronto.

Although Gray (1950) attempted a study of the Don Formation sediments, it was based on the one limited exposure at the brickyard; new studies are currently underway by N. Eyles, University of Toronto (Eyles and Clark, 1986; Eyles, 1987), and R. Kelly, Ontario Geological Survey. Sediments and fossils suggest deposition in a fluvial environment grading into a shallow lacustrine environment.

The paleontology of the Don Formation has received considerably more attention than the sedimentology (Coleman, 1933; Terasmae, 1960a; Duthie and Mannada Rani, 1967; Terasmae et al., 1972; Williams and Morgan, 1977; Poplawski and Karrow, 1981; Hann and Karrow, 1984; Kerr-Lawson, 1985). These studies have been the basis of the interpretation that deposition took place under conditions of a climate up to 3°C warmer than that at present. The fossils in the lower part generally indicate the warmest conditions and a substantial missing record between the Don Formation and the immediately subjacent York Till has been inferred; the plant record indicates an upward cooling trend through the Don Formation.

Altogether some seven taxa of vertebrates (including extinct *Castoroides ohioensis*), 200 diatom taxa, 48 taxa of plant macrofossils, 29 pollen taxa, 37 taxa of molluscs, 12 taxa of ostracodes, 22 Trichoptera taxa, and 34 Cladocera taxa have been described from the Don Formation. Insect and microvertebrate faunas are currently under study.

The age of the Don Formation remains uncertain. Its age is beyond the range of radiocarbon dating, based on minimum ages of over 50 ka from overlying units. Molluscs are too decomposed for dating by the uranium series method. Preliminary results from amino acid analysis of wood from the Don Formation are consistent with assignment to the Sangamon Interglaciation (N.W. Rutter, University of Alberta, personal communication, 1980).

Figure 4.14. Don Valley Brickyard exposure, Toronto. Units exposed are: 1. Ordovician shale, 2. York Till, 3. Don Formation, 4. Scarborough Formation, 5. Pottery Road Formation, 6. Sunnybrook Drift, 7. Lake Iroquois sand. 200300-L

Pottery Road Formation.

At Toronto several prominent channel fill deposits of sand and gravel are inset into the top of the Early Wisconsinan Scarborough Formation. The deposits are up to a few tens of metres thick and are grouped as the Pottery Road Formation (Karrow, 1969, 1974). Possible glacial affinities at the Don Valley Brickyard (Fig. 4.14) are suggested by graben-type fault structures in this formation which could have resulted from melting ice blocks. Otherwise, molluscs and a vertebrate fauna (deer, bear, bison, elephant; Coleman, 1933) suggest nonglacial conditions, at a time when St. Lawrence Valley was ice free and lake level had dropped from that of the underlying Scarborough Formation. The tentative correlation of the Pottery Road Formation with the St. Pierre interstadial sediments of Quebec would suggest an age of about 75 ka (Stuiver et al., 1978). Sharpe and Barnett (1985), however, argued for a subaqueous or subglacial origin for these channel fills and question correlation with the St. Pierre interval and events.

Middle Wisconsinan interstadial deposits

Nonglacial deposits of Middle Wisconsinan age are not well known, but have been encountered at several sites. Probably such deposits are relatively widespread as alluvial valley fills, lacustrine deposits, and paludal deposits, but because they are generally deeply buried, they are seldom exposed, and because they may be unfossiliferous their nonglacial affinities go unrecognized. In most cases the sediments appear to represent only a short time span and they are generally thin. Nevertheless, their importance in establishing chronology and paleoenvironments cannot be overemphasized.

The best known interstadial deposit is that at Port Talbot (Fig. 4.15) in the middle of the north shore of Lake Erie (Dreimanis et al., 1966; Berti, 1975), where exposures just above lake level and borings below lake level reveal the presence of two interstadial units separated by glacial lake clay. The lower (Port Talbot I) consists of up to 5 m of weathered green lacustrine clay and the upper (Port Talbot II) consists of up to 3 m of shallow water lacustrine sand, silt, and a thin gyttja layer, which has provided numerous radiocarbon dates averaging about 46 ka. Pollen was the only fossil material found in the lower green clay, which is interpreted as an accretion-gley pond deposit containing predominantly pine and spruce with up to 40% oak pollen (Quigley and Dreimanis, 1972). The upper unit contains molluscs, ostracodes, seeds, wood (spruce and tamarack), and pollen (also dominantly pine and spruce with a maximum of only 3% oak). Waterworn balls of peat are often found on the present beach near the site. Radiocarbon dates and pollen analysis indicate that these are of Port Talbot II age and are presumed to be derived from an undiscovered underwater source.

Near Woodstock, Ontario, the Innerkip interstadial site (Cowan, 1975; Karrow et al., 1978) reveals more than 1 m of compressed peat between beds of lacustrine silt; the upper silts contain freshwater molluscs. The peat is dated at >50 000 BP (GSC-2010-2; Table 4.3). This site is the best exposure of interstadial peat in southern Ontario. A nearby conservation dam impounds Pittock Lake which partly submerges the section, but the base of the peat may be seen at times of low water. Pollen analysis reveals predominantly pine (75%) and spruce (23%), amounts typical of interstadial assemblages in southern Ontario. Recent study (Pilny and Morgan, 1987) has revealed a fossil insect and vertebrate fauna suggestive of interglacial conditions. This peat bed is probably not areally extensive.

Figure 4.15. Tilted gyttja layer at the Port Talbot Interstade site on Lake Erie. 200300-R

At Guelph, the Victoria Road underpass excavations (Karrow et al., 1982) revealed a 1 m-thick lens of peat dated at >35 000 BP (WAT-367; previously published as >45 000 BP but revised by the laboratory; Table 4.3) with pine-spruce pollen, plant macrofossils, insects, molluscs, and ostracodes indicating dry and cool conditions. An underlying well developed paleosol showed up to 1.3 m of leaching in dolostone-rich outwash gravel. Soil development was judged to be intermediate between typical interglacial Sangamonian soil and the Sidney interstadial soil of Ohio (Forsyth, 1965). Overlying lacustrine silts (1 m thick) contain freshwater and terrestrial molluscs.

A subsurface interstadial site was encountered in engineering borings at Waterloo (Karrow and Warner, 1984). About 6 m of stratified gravel, sand, silt, and diamicton, evidently an alluvial deposit, occurs under several tills at a depth of over 30 m. Pollen, plant macrofossils, molluscs, ostracodes, and insects have been identified and indicate cool dry conditions. An accelerator radiocarbon date of 40 085 ± 1280 BP (Chalk River accelerator, Table 4.3) was obtained on a small wood fragment, suggesting correlation with the Port Talbot interstadial deposits.

About 25 km northwest of Waterloo, near Glen Allan, a thin paleosol is overlain by 1 or 2 m of clay and silt, which contain pine and spruce pollen, and two lenses of charred plant material which yielded a date of >41 000 BP (GSC-2141; Table 4.3). Several till layers overlie the deposit. The enclosing sediments appear to be of glacial lake origin and the plant debris may be reworked, but all available facts are consistent with a Middle Wisconsinan (Port Talbot) age.

Along Grier Creek (Fig. 4.16), 3 km southwest of Clarksburg, and near the southwestern shore of Georgian Bay, is the northernmost interstadial site (Pinch, 1979). Basal stream gravels are overlain by about 0.3 m of peat, beneath about 5 m of lacustrine clay, then till. The peat has an age of >36 000 BP (GSC-2053; Table 4.3). Pollen and insect assemblages show a colder environment than the other sites (Warner et al., 1988).

Figure 4.16. Clarksburg interstadial site along Grier Creek, southwest of Georgian Bay. 200300-O

Table 4.3. Cited radiocarbon ages

^{14}C age (years BP)	Lab No.	Material	Site	Reference
5 109 ± 131	BGS-79	Charcoal	Sauble beach	Terasmae et al., 1972
10 500 ± 150	GSC-1126	Woody peat	Eighteen Mile River	Karrow et al., 1975
10 600 ± 160	GSC-1127	Spruce wood	Eighteen Mile River	Karrow et al., 1975
10 800 ± 100	GSC-1904	*Picea mariana* cone	Penetangore River	Miller et al., 1979
11 300 ± 140	GSC-1842	*Picea* or *Larix* wood	Penetangore River	Miller et al., 1979
12 650 ± 170	I-4040	Plant debris	Lake Erie	Lewis, 1969
22 800 ± 450	GSC-816	Wood	St. Davids gorge	Hobson and Terasmae, 1969
28 300 ± 600	GSC-1082	Plant debris	Thorncliffe Formation	Berti, 1975
32 000 ± 690	GSC-1221	Plant debris	Thorncliffe Formation	Berti, 1975
40 080 ± 1200	Chalk River accelerator	Wood	Waterloo	Karrow and Warner, 1984
>35 000	WAT-367	Peat	Guelph	Karrow et al., 1982
>36 000	GSC-2053	Peat	Clarksburg	Pinch, 1979
>41 000	GSC-2141	Charred wood	Glen Allan	Karrow, in press
>50 000	GSC-2010-2	Peat	Innerkip	Karrow et al., 1978
>53 000	GSC-1228	Plant debris	Thorncliffe Formation	Terasmae et al., 1972
>54 300	GrN-4817	Plant debris	Scarborough Formation	Karrow, 1969

At other scattered sites in southern Ontario, pollen-bearing sediments have been encountered in borings below several till layers. Borings in buried valleys 60 m deep at Fergus (20 km north of Guelph) and Rockwood (12 km northeast of Guelph) are among these. Stratigraphy indicates a minimum age of Middle Wisconsinan for these valley fills, which consist of silt and sand. The origin of the deposits is unknown.

Postglacial deposits

Postglacial deposits include alluvial, paludal, eolian, and lacustrine sediments. Although limited in extent, older alluvium occurs at scattered locations along some stream valleys. Sediments consist of abandoned floodplain deposits of gravel, sand, and silt on terraces intermediate between glacial outwash and modern floodplains. Molluscs, wood, and other plant matter may be present, but in general there has been little study of valley terrace history. Valley terraces graded to glacial Lake Algonquin in the Huron basin have yielded plant remains, pollen (Karrow et al., 1975), molluscs (Miller et al., 1979, 1985), and insects (Ashworth, 1977) dated between 10 and 11 ka (GSC-1126, 1127, 1842, 1904; Table 4.3); sediments are fluvial gravel and sand upstream, and transitional fluvial-lacustrine silts up to several metres thick downstream.

Because valley development is still youthful, floodplains are generally not extensive. Among the more prominent rivers with floodplains, which generally have some history of flooding, are the Humber, Grand, and Thames rivers, which flow into lakes Ontario, Erie, and St. Clair respectively. Alluvial deposits are commonly thin (1 to 3 m) veneers which underlie terraces or floodplains, although thicker accumulations are present in the lower reaches of streams which flow into lakes where isostatic uplift has raised the water level. Such conditions are encountered in the Ontario and Erie basins. Drowned river mouths with baymouth bars are prominently developed along the Lake Ontario shore in the Niagara Peninsula, and thick sediment fills in buried bedrock gorges are present between Hamilton and Toronto adjacent to and under Lake Ontario. Similar conditions existed in the Huron basin leading up to the Nipissing phase about 5 ka, as uplift raised the level of the North Bay outlet to that of the sill at Sarnia, so that entrenched stream valleys of the previous low water stage were flooded and infilled. Rivers subsequently became entrenched again as outlet downcutting dropped the water level in Lake Huron to the present. Remnants of those floodplains are present today as elevated terraces in the larger valleys such as the Saugeen, Lucknow, Maitland and Bayfield.

Paludal deposits include swamps, bogs, and marshes, which occur in poorly drained depressions or at the shallow margins of ponds and lakes. Swamps are widespread in parts of the more level till plains and glacial lake plains, but organic accumulations are usually no more than 1 or 2 m thick. Thicker accumulations are found along former meltwater channels. Deepest of all are those formed in kettles or ice block depressions, where accumulations of 6 to 10 m are not uncommon. Sediment sequences may begin with lacustrine clay or silt, overlain in some places by marl (commonly less than 1 m, but up to 3 m or more), then gyttja and peat to the surface. These organic deposits have been of major interest for palynological and paleoecological study. Terasmae (in Karrow, 1968; 1987) and Karrow et al. (1975) have described the pollen record of some of these sites, the vegetative succession, and the interpreted paleoenvironments. Results of such studies are presented in Chapter 7 of this volume dealing with Quaternary paleoenvironments.

Eolian deposits are widespread in southern Ontario, but substantial deposits are limited to areas near present or past lakeshores. Other deposits occur sporadically on former delta plains and in the interlobate kame moraine tracts, where local pockets of loess have also been identified.

Dominant among the eolian deposits are dune sands, typified by those near Grand Bend and Sauble Beach on Lake Huron, Wasaga Beach on Georgian Bay, west of Long Point on Lake Erie, and in Prince Edward County on Lake Ontario. In these areas parabolic coastal dune complexes up to 20 m high are typical, with large stabilized dunes inland and smaller active dunes near the present shore. There has been little study of eolian deposits in Ontario, but a short description by Martini (1975) of those at Wasaga Beach may be mentioned. Several paleosols are encountered in the eolian deposits; a dated paleosol in the Sauble Beach dunes established their age as Nipissing or younger (less than 5.5 ka; Terasmae et al., 1972), and the Grand Bend dunes are believed to be of similar age.

Deltaic sediments deposited in glacial lakes, such as by Thames River southwest of London and Grand River southwest of Brantford, have been modified by wind action, but eolian deposits are generally thin (less than 6 m). Small dunes near Alliston (southwest of Lake Simcoe) were developed on lacustrine sands of glacial Lake Algonquin.

Nonglacial lacustrine sediments occur in numerous small postglacial lakes as well as in the Great Lakes. While the basins in which sedimentation occurs are mostly of glacial origin (kettles, meltwater channels, glacially scoured basins), sedimentation has been largely continuous in postglacial time (the last 10 to 15 ka). Postglacial sediments are greatest in areal extent, thickness, and volume under the Great Lakes, and these will be discussed first.

Geophysical surveys combined with coring show that postglacial sediments cover about half of the lake floor area, with older glacial sediments and bedrock exposed over the other half (Sly and Lewis, 1972). The postglacial sediments are up to 20 m in thickness but are found mainly on the floors of the deeper basins, where they form a nearly level plain (Thomas et al., 1973). The sediments consist of soft, compressible, grey muds (silty clay or clay), commonly showing black banding which vanishes soon after exposure to air (Lewis and McNeely, 1967; Thomas et al., 1973). Unconformities and buried veneers of shallow water sediments at the base of the postglacial muds indicate former low water stages in each of the basins. In the western Lake Erie basin plant debris marks an unconformity about 30 m below present lake level that has been dated at 12 650 ± 170 BP (I-4040; Table 4.3).

Thin veneers (a few centimetres) of sand, gravel, and shells characterize shallow water areas but thicker accumulations are found near submerged moraine crests (Lewis et al., 1966; Wall, 1968; Coakley and Lewis, 1985) as the result of wave washing of coarser glacial materials. Coakley (1976) has suggested that the Point Pelee cuspate foreland developed from the Pelee-Lorraine cross-lake moraine as it was reworked by rising Lake Erie waters since about 4 ka. The surface sand ridges enclose a marsh and overlie a platform of till, superimposed on the moraine crest.

Fossil mollusc and ostracode shells in Lake Erie sediments have been the subject of carbon and oxygen isotope studies (Fritz et al., 1975) which detected changes in the hydrology of the basin. The lake sediments have also been of interest for their paleomagnetic record; Mothersill and Brown (1982) and Creer and Tucholka (1982) have observed that cyclic inclination and declination variations are present which can be used for correlation and dating when calibrated by palynostratigraphy and radiocarbon dates.

The sedimentary record of small inland lakes has been given less attention than that of the Great Lakes, even though human activity comes into more direct contact with them. They too preserve important records of postglacial time. They are generally much shallower and their history is commonly closely related to those of paludal environments — lakes often evolve into bogs or marshes by organic growth and filling. The present concern, however, is mainly with the inorganic record.

Few studies on the sediments of small lakes have been published, but two types of lake sediments have been of particular interest — laminated muds and marl. Tippett (1964) studied eastern Ontario lakes containing banded sediments and from detailed study of microfossils interpreted an annual sedimentation cycle. Light layers are carbonate rich and dark layers are organic. Similar layering was observed at meromictic Crawford Lake, near Hamilton, by Boyko-Diakonow (1979; see also McAndrews and Boyko-Diakonow, 1989) and was correlated with the pollen record.

Several marl depositing lakes near Kingston have been the subject of theses; some of this work has been summarized by Vreeken (1981). Marl deposition was found to have occurred in areas of carbonate-rich till or limestone bedrock following deposition of lacustrine clay. Marl was deposited generally in the early postglacial period and in most cases ceased on upland sites before 8 ka and on lowland sites 2 to 6 ka. Marl deposition was related to soil development and was commonly followed by peat or muck formation. Deposition rates averaged near 4 m per century with thicknesses ranging up to about 3 m.

Lewis (1969, 1970) studied several small lakes on Manitoulin Island to assess the effects of isostatic uplift and the Nipissing transgression on sedimentation. The lakes selected were near the Nipissing level and recorded small lake — large lake — small lake sequences from which the Nipissing phase was dated around 5 ka and uplift rates were estimated at about 3 mm per year over the last 5 ka.

Several lakes have been the subject of paleontological study, with palynology the leading topic of interest. Only a few studies have been published (e.g., Karrow et al., 1975; Mott and Farley-Gill, 1978) with more work in theses. The chief aim of this work is paleoecological and it will be described more fully in Chapter 7 of this volume.

Glacial deposits

Glacial deposits include sediments deposited in direct association with glacial ice and those deposited by or in its meltwaters, such as glaciofluvial and glaciolacustrine deposits. The boundary with nonglacial deposits is commonly arbitrary and associations with glaciers must usually be inferred. There are also gradational boundaries between glacial, glaciofluvial, and glaciolacustrine deposits, and controversy has arisen about the interpretation of the depositional environments associated with some materials. A simple definition is here adopted that till is material believed to have been deposited by ice; it will be used in a general sense to include all till facies, however deposited.

The development of major ice lobes was favoured in the various basins of the Great Lakes (Fig. 4.6). The Huron, Erie, and Ontario lobes had their axes close to the present International Boundary and as a result only about half of the deposits associated with each is in Canada. The International Boundary has reinforced the natural barrier formed by the water bodies that inhibits correlation and comparison between deposits on the two sides of each lobe. Study of glacial lake history, however, has had a particularly unifying effect on studies across the International Boundary.

Southern Ontario was invaded by ice moving through the lake basins. It expanded outward from each basin to surround the highest ground in the peninsula — the "Ontario Island" of Taylor (1913) — and a roughly concentric pattern of end moraines was formed during ice retreat. While ice flow was, broadly speaking, southwestward through the Great Lakes region, only north of the Oak Ridges Moraine, southwest of Georgian Bay, and in the Niagara Peninsula, was local ice movement in that direction, as indicated by landforms and till fabrics. Ice moved east or southeast out of the Huron basin, but nearly north, northwest, and west out of the Ontario basin, and northwest out of the Erie basin.

Ice flow direction is indicated by bedrock striae, trends of drumlins, flutings, eskers, end moraines, ice contact faces on kames and outwash fans, and by till fabrics. Provenance of glacial deposits, particularly till, has been the subject of many studies of the mineralogy, such as heavy minerals and carbonates, and this has provided more generalized information on former ice flow directions. Few indicator lithologies have been identified but Grenville (Precambrian) marble and Potsdam (Cambrian) sandstone are found in the deposits of the Ontario Lobe while Huronian (Precambrian) quartzite and conglomerate (both tillite and jasper) are widely distributed in the deposits of the Huron and Georgian Bay lobes. In the interlobate zone along the junction between lobes, however, mixing occurs, making it difficult to separate the deposits of the various lobes.

Till

Till study has been emphasized in the Quaternary stratigraphy of the region, in contrast to other kinds of deposits, because of its volume and extent, its climatic implications, and its relatively constant characteristics which allow till sheets to be used as stratigraphic marker beds over scores of kilometres. The consistency of characteristics is a reflection of the uniformity of nearly flat-lying bedrock formations and the homogenizing effect of glacial transport.

The complex mixture of materials in the tills has inspired many kinds of study of its composition. Texture is an attribute almost universally observed in the field and refined in the laboratory. Other field characteristics including colour, cracking while drying, stoniness, fabric, and erosion features usually allow identification of individual till sheets (Karrow, 1974), while pebble lithology, matrix carbonate, heavy mineral composition, trace elements, and geophysical properties provide additional information for correlation and provenance studies (Dreimanis, 1961). Stratigraphy in the region has been heavily dependent on such data; the development of these studies has been reviewed by Karrow (1976).

Although distribution patterns are often complex, some generalizations can be made and examples given of trends and variations. The ice arrived in southern Ontario carrying debris from the Canadian Shield. This created a "parent till" which is coarse textured (Scott, 1976), generally carbonate free, and characterized by a heavy mineral suite related to specific parts of the Canadian Shield (Gwyn and Dreimanis, 1979). Heavy minerals are consequently useful in distinguishing tills from the different lobes. Magnetic susceptibility, which is dependent on the amount of magnetic minerals present, is also useful in distinguishing tills from different lobes but becomes progressively less distinctive as one moves farther away from the Canadian Shield (Gravenor and Stupavsky, 1974).

After the ice moved onto Paleozoic rock, the till became more silty from the incorporation of carbonates and fine clastics (Dreimanis, 1961); carbonate contents rose for all size classes, with the matrix from 30 to 50% carbonate. Crushing of minerals proceeds to their terminal grades, which is coarse silt for dolomite and fine silt for calcite (Dreimanis and Vagners, 1971) and consequently these grain sizes are abundantly represented in St. Lawrence Lowlands area tills. Clayey tills, such as the Port Stanley Till of the Erie Lobe, apparently represent the incorporation of lacustrine clays during ice advance over proglacial lake plains (Dreimanis, 1961). On the northern flank of the Erie Lobe, however, the ice overrode bedrock and outwash leaving sandy Port Stanley Till near Kitchener, (Karrow, 1974). Incorporation of underlying coarse Catfish Creek Till created a coarse marginal zone in Tavistock Till (Cowan, 1978) and clayey Wartburg Till and Mornington Till (Cowan, 1979) caused a fining of the overlying Elma Till near its margin (Karrow, 1977).

Tills in southern Ontario range from stony and sandy to clayey in texture (Dreimanis, 1961; Karrow, 1974). As shown in Table 4.4, each of the lobes deposited tills of various textures. A major contrast is, however, present between coarse grained Nissouri Stade Catfish Creek Till and fine grained Port Bruce Stade Port Stanley tills, attributed to ice readvance over proglacial lacustrine sediments (Fig. 4.17). A correlative change is widely recognized across the Great Lakes region (Bleuer, 1974; Schneider, 1983).

The genesis of till in southern Ontario recently has been given more detailed study. Although the many genetic types of till have been classified and defined (Dreimanis, 1982a), consensus is still lacking on criteria to be used in recognizing them in the field. The growing literature on sedimentological study of active glaciers is providing additional ideas for the interpretation of field evidence, and new approaches and viewpoints are being applied. A major point of contention is the role of water in the deposition of several Great Lakes area tills. Dreimanis (1982b) has provided evidence that both subaquatic flow till and basal meltout till are present in stratified Catfish Creek Till along the Lake Erie shore. Eyles and Eyles (1983) and Eyles and Westgate (1987) have described sedimentary structures and faunal content at Scarborough Bluffs which lead them to conclude that the lowest three till units there were deposited by floating ice and that grounded ice did not cover the area. At present this opinion is contrary to other evidence from provenance studies, with basic disagreement on the significance of respective field observations. Sharpe and Barnett (1985) presented other evidence and reasoning which disagrees with the Eyles and Westgate interpretations.

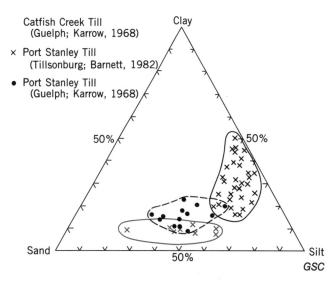

Figure 4.17. Some typical till textures; Catfish Creek Till is of Nissouri Stade age whereas Port Stanley Till is of Port Bruce Stade age.

Till is widely distributed at the surface in southern Ontario as till plain (undulating to level, fluted, drumlinized) and as end moraine. In the subsurface, stratigraphic sequences are commonly mostly till, with a single till sheet resting on bedrock (Halton Till on Queenston Shale; Karrow, 1987) or sequences of several till sheets (nine superimposed tills at a quarry near Woodstock; Westgate and Dreimanis, 1967). Till is the surface material over significant areas of the floors of the Great Lakes, chiefly in shallower zones which are swept free of sediment cover by currents. It is also sometimes encountered as the oldest sediment in lake bottom cores, underlying glacial lake and postglacial sediments, but little is known of the distribution of various till sheets under the lakes.

Till sheets are commonly 2 to 6 m thick, but may be 20 m or more thick, particularly in drumlins, buried valley fills, and in end moraines. Discontinuities in buried till sheets are known to exist, although are seldom documented (Catfish Creek Till missing at the Innerkip Interstadial site; Cowan, 1975). Windows in surface tills are also known, allowing buried tills to show at the surface (e.g., Maryhill Till through Port Stanley Till in the Breslau Moraine, Karrow, 1968; Catfish Creek Till through Tavistock Till north of Woodstock, Cowan, 1975; Karrow, in press).

Presumably because of general bedrock competence, there is little evidence of movement and deformation of bedrock blocks. Although 1 or 2 m of disturbed bedrock may be present under till, particularly where the bedrock is shale, till-bedrock contacts are generally fairly abrupt. Bedrock boulders and slabs are commonly seen down-ice from escarpments (in the Dummer Moraine at the Ordovician Escarpment, and west of the Niagara Escarpment).

Glaciofluvial deposits

Glaciofluvial deposits are usually distinguished and classified on the basis of their landforms, and to a lesser extent on internal structures, into ice contact deposits such as eskers, kames, and crevasse fillings, and outwash such as in fans, plains, and terraces.

Table 4.4. Texture and associated end moraines for named till units associated with ice lobes in southern Ontario

Lobe	Till (reference)	Texture	Associated end moraines
Erie	Lower Bradtville	sandy	-
	Upper Bradtville (Dreimanis et al., 1966)	clayey	-
	Dunwich (deVries and Dreimanis, 1960)	sandy	-
	Canning (Karrow, 1963)	clayey	-
	Catfish Creek (deVries and Dreimanis, 1960)	sandy	-
	Maryhill (Karrow, 1974)	clayey	Breslau
	Port Stanley (deVries and Dreimanis, 1960)	clayey, sandy	Blenheim, Ingersoll, Westminster, St. Thomas, Norwich, Tillsonburg Courtland, Maybee
	Wentworth (Karrow, 1959)	sandy	Paris, Galt, Moffat
Ontario	York (Terasmae, 1960a)	silty	-
	Sunnybrook (Terasmae, 1960a)	clayey	
	Seminary (Karrow, 1967)	silty	
	Meadowcliffe (Karrow, 1967)	clayey	-
	Halton (Karrow, 1959)	silty	Port Maitland, Crystal Beach, Fort Erie, Niagara Falls, Vinemount, Waterdown, Trafalgar
Huron-Georgian Bay	Stirton (Karrow, 1974)	clayey	-
	Tavistock (Karrow, 1974)	silty	-
	Mornington (Karrow, 1974)	clayey	Macton
	Stratford (Karrow, 1974)	silty	Gadshill
	Wartburg (Karrow, 1977)	clayey	Milverton
	Elma (Karrow, 1974)	sandy	-
	Rannock (Karrow, 1974)	silty	Michell, Dublin, Lucan, Seaforth, Centralia
	St. Joseph (Cooper and Clue, 1974)	clayey	Wyoming, Banks
Simcoe	Bogarttown (Gwyn, 1972)	silty	-
	Newmarket (Gwyn, 1972)	sandy	Singhampton, Gibraltar
	Kettleby (Gwyn, 1972)	clayey	-

Ice contact deposits are commonly bouldery; show rapid changes in sorting, size, and bedding; and may be bounded by steep slopes which represent the shape of the ice margin at the time of deposition. Till lenses may be present within or on top of the waterlaid sediments. Faults may be present, indicating collapse over melting ice blocks. There have been few sedimentological studies of ice contact deposits but those of Saunderson on the Brampton esker (1975, 1976, 1977a) and Guelph esker (1977b), of Martini (Sanford et al., 1972) on the Campbellville kame terrace outwash fan, and of Sharpe (1982) on the Allan Park kame delta are relevant. There is common evidence in most of these sediments of deposition at the ice margin or in a re-entrant in the ice front, with debris flows, braiding, channelling, and local ponding. In the Guelph esker the presence of poorly sorted massive deposits was interpreted by Saunderson (1977b) to suggest mass movement along a sliding bed under high hydrostatic pressure.

Outwash deposits are more sheet-like and uniform. Even though they are more sought out for aggregate than ice contact deposits, still fewer studies of them have been made. Outwash in the Oak Ridges Moraine was studied by Duckworth (1979) who described a westward-fining sequence, supported by westward paleocurrent indicators, with braided stream deposition and bar formation grading westward into lacustrine sediments of glacial Lake Schomberg. Costello and Walker (1972) described coarsening upward silt to gravel sequences deposited as a channel fill in outwash along the Niagara Escarpment, and Eynon and Walker (1974) studied part of a major outwash deposit at Paris, where they identified several facies associated with bar growth in a braided outwash stream.

Some studies have been made of sand and gravel composition, particularly related to their use as aggregate sources. In this respect, sand mineralogy has been analyzed in many deposits, as well as pebble lithology (Hewitt and Karrow, 1963). There is a close relationship between composition and subjacent bedrock, and in crossing bedrock boundaries a rapid change in composition occurs downstream from the contact. One of the most obvious tracers is chert. In the vicinity of the headwaters of Nith River, ice movement was to the southeast parallel to the buried Silurian-Devonian contact and the cherty Bois Blanc Formation. Esker gravels in the vicinity commonly contain more than 10% chert. The effect of the bedrock contact can be seen down ice to the margin of the Tavistock Till of the Georgian Bay Lobe at Waterloo, where buried outwash also contains abundant chert, in contrast to its absence in outwash along Grand River only a few kilometres to the east.

Sand mineralogy was studied by Dell (1959) in relation to soil formation. Chapman and Dell (1963) inferred that sands in the Norfolk sand plain (south of Brantford) were derived from east of the Niagara Escarpment because of their high calcite content.

Glaciolacustrine sediments

Glacial lake sediments are widespread, particularly in a fringing belt around the present Great Lakes, as well as in numerous smaller sedimentary basins on till plains, and interbedded with till sequences. Shallow water sediments include beach sand and gravel and nearshore sand sheets; deep water sediments include massive, laminated, and varved silt and clay. A postulated wave base at a depth of about 10 m in glacial lakes seems to coincide with a rough boundary between the two groups.

Beach deposits typically consist of beach ridges, bars, and spits of rounded and well sorted stratified sand and gravel. They have received little sedimentological study. As with glaciofluvial gravels, lithology has been of interest for aggregate resources, but only a few beach gravels are of large enough volume to be considered important resources. Beach gravels occur along the shorelines of most of the large former lakes, such as Warren, Algonquin, and Nipissing in the Huron-Georgian Bay basins; Maumee, Whittlesey, and Warren in the Erie basin; and Iroquois in the Ontario basin. Large baymouth bars and spits are prominent features at Toronto, Hamilton, Kincardine, Barrie, and near the mouth of Saugeen River, where gravels may be more than 5 m thick and extend for several kilometres. Buried beach gravels attributed to glacial lakes subsequently overridden by ice have been identified in a few places (Mörner and Dreimanis, 1973). Beach deposits contribute particularly important information to the history of glacial retreat and isostatic uplift.

Deep water lacustrine sediments are predominantly silt and clay. Those that are varved have attracted the most interest because of their earlier use as a geochronometer, and of all varved clays in southern Ontario, those at Toronto have been given the most study. Antevs (1928), by detailed varve counting, established correlation between the varves at the Don Valley Brickyard and at Scarborough Bluffs. More recent interest has emphasized the sedimentology and genesis of varve stratification. Lajtai (1967) and Banerjee (1973) described many details relating varves at Toronto to turbidity currents. Eyles and Eyles (1983) and Eyles and Clark (1988) have described features associated with storm action, ice scouring, ice rafting, slumping, and loading in glacial lake sediments at Scarborough Bluffs.

Glacial lake sediments are widely distributed as subsurface units interbedded with till. Glacial lake sand, silt, and clay, commonly varved, are extensively exposed in shorecliffs along Lake Erie (Dreimanis, 1958; Barnett, 1978) and Lake Ontario (Karrow, 1967; Brookfield et al., 1982; Eyles and Eyles, 1983) and occurrences are known from many sites along valleys. Although these deposits are generally unfossiliferous, the Scarborough Formation is a prominent exception. It is included here with glacial lake sediments because it is believed to have been deposited in a lake having a level near that of Lake Iroquois, which requires an ice dam blocking St. Lawrence Valley, and thus is a glacial lake. The Scarborough Formation consists of some 50 m of stratified sand overlying clay in a deltaic assemblage which extends under much of Metropolitan Toronto. Fossils present include diatoms (Duthie and Mannada Rani, 1967), pollen, plant debris (Terasmae, 1960a), molluscs, ostracodes (Poplawski and Karrow, 1981), and insects (Williams et al., 1981). Their study indicates deposition under cooler conditions than those at present. Sedimentological study has confirmed its deltaic origin (Kelly and Martini, 1986) at the mouth of a large river, which was likely Laurentian River flowing south from Georgian Bay. The delta was of similar size to modern deltas in the St. Lawrence system today in Lake St. Clair and Lac Saint-Pierre.

Late glacial lake sediments have also locally yielded fossils generally from areas distant from the ice front, and in protected lagoons (lacustrine) behind baymouth bars. Karrow et al. (1972) described molluscs from Lake Iroquois deposits in southwestern Lake Ontario basin and Karrow et al. (1975) described molluscs from Lake Algonquin deposits in the Alliston embayment. Older glacial lake deposits of Lake Schomberg, south of Georgian Bay, have been reported to contain molluscs (Gwyn, 1972) but these have not yet been studied. Still more remarkable is the presence of trace fossils on the bedding planes of varved clays; Gibbard and Dreimanis (1978) described evidence of several species from three sites in southwestern Ontario, and trails have recently been found on bedding planes of varved clay at the Don Valley Brickyard in Toronto.

Glacial lake clay and silt occurring as thin veneers or more locally as deposits up to a few tens of metres thick between till and postglacial muds are widely distributed under the Great Lakes. They occur at depths of 40 to 120 m in Lake Ontario (Lewis and McNeely, 1967) and typically show varving and greater stiffness than the overlying postglacial muds. For Lake Superior, Dell (1972) has attributed red and grey glacial lake clay to differences in bedrock and till sediment sources. As with the postglacial lake sediments, glacial lake sediments flooring the Great Lakes have been little studied.

QUATERNARY HISTORY

The stratigraphic record in southern Ontario is based on natural exposures in river valleys and in bluffs along the Great Lakes (at Port Talbot and Scarborough; Fig. 4.3), and on man-made exposures in pits and quarries (East Zorra), cuts for roads (Guelph), railroads (Woodbridge), and subways (Toronto), building excavations (Hamilton), and soil borings for engineering and stratigraphic purposes. Numerous cities and towns are in the area but, unfortunately, the vast majority of the temporary construction exposures are not seen by geologists. Toronto is the most tragic example because it is known to have a rich stratigraphic record, and the greatest concentration of construction activity in Canada, yet no organized effort to record the information exists.

Subsurface information from water and oil drilling (thousands of wells each year) has contributed greatly to defining bedrock topography, but because of the unreliability of many drill logs, seldom provide useful Quaternary stratigraphic information. Engineering boring records are more dependable, but seldom do they go deeper than 10 to 20 m. In the past decade, drilling for stratigraphic purposes has become increasingly important, although because of its cost, the number of holes and density of sampling are limited. Geophysical borehole logging shows promise of reducing sampling costs in some areas (e.g., Eyles et al., 1985; Farvolden et al., 1987).

Chronological control has depended almost entirely on radiocarbon dating, mainly on wood or peat, and the reproducibility of dates between 45 and 48 ka on the Port Talbot II Interstade deposits has been one of the striking successes of the method. There exists, however, a frustrating lack of time control for the early Middle Wisconsinan, a time interval, 50 to 70 ka, in which several deposits are believed to fall. Although isotopic enrichment makes it possible to reach back to that time range, samples seldom are of high enough quality to make enrichment dating worthwhile. Accelerator dating has enlarged the scope of ^{14}C dating because now it is possible to obtain dates from much smaller samples, such as the very small pieces of wood that are obtained from time to time from borehole samples.

Other dating methods, such as uranium series, amino acid racemization, and thermoluminescence (Berger, 1984), have had little application in the area, although the latter two methods have not been fully exploited.

At present, chronological control by radiocarbon dating is available for only a few Middle Wisconsinan sites containing materials dating between 23 and 50 ka, and for late glacial and postglacial sites younger than about 13 ka. No material that dates from the interval 23 to 13 ka has yet been found, presumably because of glacier cover and lack of plant growth. As more Middle Wisconsinan sites are dated, the extent of inferred Middle Wisconsinan ice advances (presumed age 30 to 45 ka) will become clearer. It must be emphasized that chronology is based on the dating of plant material formed under nonglacial conditions. Because glacial deposits commonly lack datable organic material, most glacial fluctuations are not directly dated but their ages are interpolations between dated nonglacial events.

Stratigraphic nomenclature has largely developed through the influence of workers in the Mid-West United States. Up until 1960, the evolving time classification of Leighton (1933, 1958) had held sway for many years. The publication of a new classification by Frye and Willman (1960) marked the beginning of a divergence of opinion between Ontario and many American workers and led to the classification of Dreimanis and Karrow (1972; Fig. 4.18). Although time boundaries and some details have been modified, it remains generally in use in this area today and is followed in this report. It should, however, be noted that this scheme restricts Sangamonian to the approximate time of oxygen isotope stage 5e and consequently includes the latter part of stage 5 in the Early Wisconsinan rather than in Sangamonian as is the case for other parts of this volume.

Lithostratigraphic names have generally been applied following North American stratigraphic practice. As rules have evolved through the several versions of stratigraphic codes, practice has become more refined. Because of the rapid growth of mapping and stratigraphic work in the 1960s and 1970s, it was the 1961 code that was most influential (American Commission on Stratigraphic Nomenclature, 1961).

The continuity and consistency of till sheets has led to emphasis on the naming of tills (Fig. 4.18) and at present more than a score of named tills exist in the area. Waterlaid deposits, on the other hand, are seldom traceable for substantial distances and only those that are notable for such things as thickness, extent, and fossil content or age have been named.

The various units will be discussed in stratigraphic order.

Pre-Wisconsinan

In southern Ontario, only at Toronto is there positive evidence of deposits of pre-Wisconsinan age.

Illinoian. The oldest named deposit at Toronto is the York Till (Terasmae, 1960a), which overlies Ordovician shale in downtown Toronto, at Scarborough, and at the Don Valley Brickyard (Fig. 4.14, type section). At Woodbridge it overlies interbedded gravel and till of unknown age. Excavations in downtown Toronto (Watt, 1954; Lajtai, 1969) have exposed up to 6 m of shale-rich, clayey silt till and up to 5 m of overlying associated varved clay (York Drift of Lajtai, 1969). About 5 m of till are also exposed at Woodbridge but at the Don Valley Brickyard and at Scarborough the till is less than 1 m thick, or is absent. It is evident at the brickyard that the till surface is an erosional one as it is in places overlain by a lag of cobbles. Rare striae on the underlying shale indicate ice movement out of the Ontario Lake basin (Terasmae, 1960a). The age of the York Till is undetermined, but it is assumed to be Illinoian.

Sangamonian. Directly overlying the York Till are the sands and clays of the Don Formation, previously described in the section on nonglacial deposits. Although 8 m or less in thickness, it is a nearly ubiquitous sediment sheet under the Toronto area. It has long been known from downtown excavations, the Don Valley Brickyard (its type section), and from below lake level at Scarborough Bluffs (Coleman, 1933; Karrow, 1969). A correlative unit may also be present at Woodbridge where several centimetres of sand overlying York Till contain a varied molluscan assemblage.

The fossils of the Don Formation indicate that an interglacial climate existed. The record of the early part of the interglaciation is apparently missing because the warmest conditions are inferred at the base of the sediments directly overlying the York Till. The age of the sediments is presumed to be about 125 ka, the age of the warmest part of the Sangamon Interglaciation. During deposition of the alluvial-lacustrine Don Formation, waters in the Ontario basin probably stood several metres above the present level of Lake Ontario. Prior to burial, water levels apparently fell; the surface of these sediments is somewhat weathered (Gray, 1950; Terasmae, 1960a).

Early Wisconsinan

About 50 m of sand over clay forms the lower half of the cliff exposures at Scarborough Bluffs, type section of the Scarborough Formation (Karrow, 1967; Kelly and Martini, 1986; Fig. 4.3). The unit, which has been described in the section on glaciolacustrine deposits, also extends under much of Metropolitan Toronto. This deltaic sediment mass is attributed to a large river probably carrying drainage from the Upper Great Lakes during the last use of the buried Laurentian Valley which extends from Georgian Bay to Lake Ontario. The abundant fossils indicate much colder conditions than those at present. The high level lake into which it was deposited has been attributed to advancing ice blocking drainage in St. Lawrence Valley so that an outlet into Mohawk Valley in New York was used instead. Only minimum ages have been determined by radiocarbon dating (Table 4.3; >54 300 BP, GrN-4817) and its age is estimated to be around 100 ka (Early Wisconsinan as used in this report).

Lenticular masses of stream sand and gravel inset into underlying Scarborough and Don formations are referred to as the Pottery Road Formation with a type section at the Don Valley Brickyard (Karrow, 1974). Scattered vertebrate fossils and molluscs are not diagnostic as to climate. The Christie Street sands (Coleman, 1933) are considered to be correlative, as are stream sediments at the Dutch Church section at Scarborough Bluffs (Churcher and Karrow, 1977). These high energy valley fills indicate a lowered water level

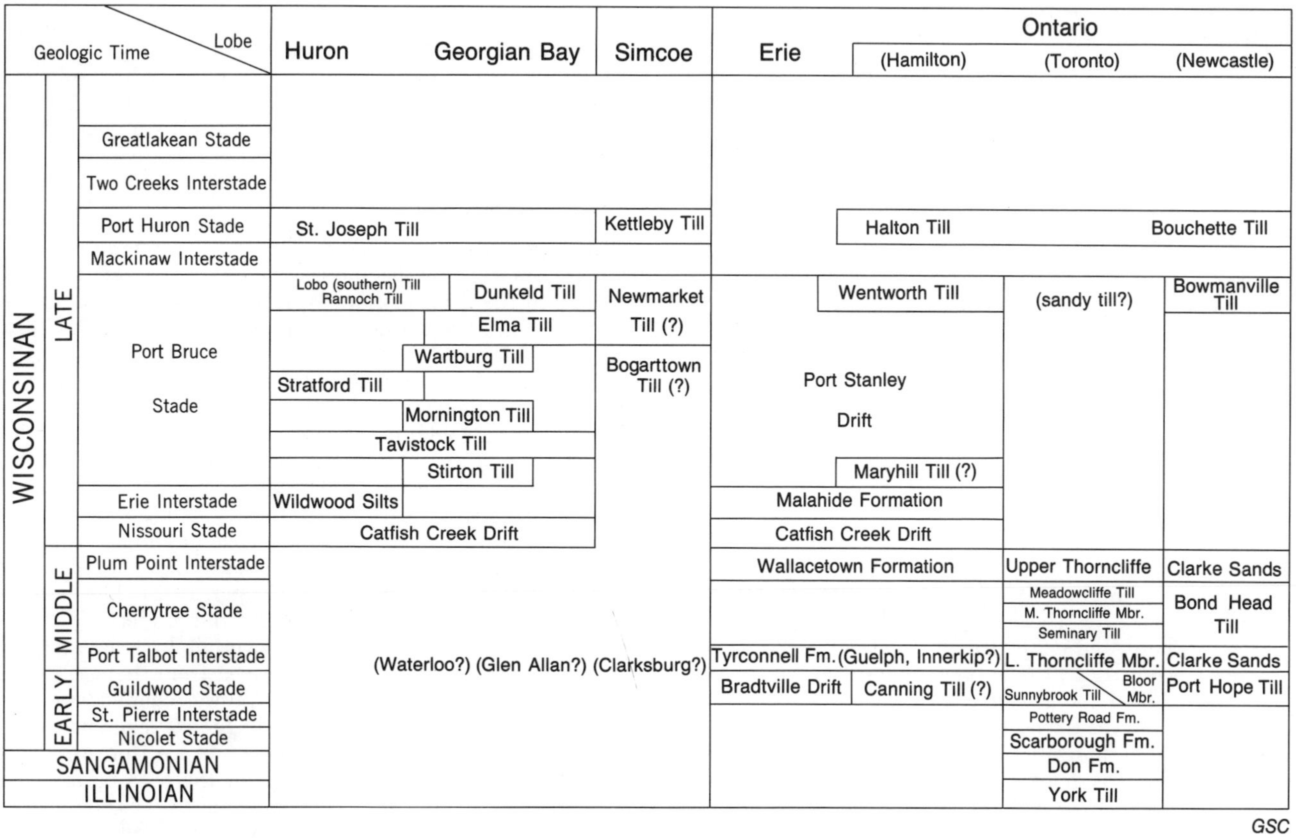

Figure 4.18. Correlation chart for southwestern Ontario; time stratigraphic units follow Dreimanis and Karrow (1972).

in the Lake Ontario basin which has been interpreted to signify drainage through St. Lawrence Valley as a result of ice retreat. The age of the deposits is unknown, but correlation with the St. Pierre fluvial deposits of Quebec, which have an age of about 75 ka has been suggested (Karrow, 1974; Stuiver et al., 1978). They are thus considered to be interstadial deposits of Early Wisconsinan age. As previously mentioned, Sharpe and Barnett (1985) suggested a subaqueous or subglacial origin and disputed correlation with the St. Pierre.

The Pottery Road Formation is overlain by Sunnybrook Till (Terasmae, 1960a), which along with associated varved clay of the Bloor Member, comprises Sunnybrook Drift (Karrow, 1969). Although a subglacial origin has been disputed by Eyles and Eyles (1983), Sharpe and Barnett (1985) supported the subglacial interpretation. Whatever its genesis, it represents a glacial advance in Early Wisconsinan time (Eyles and Westgate, 1987). This drift sheet is characterized by a low matrix-carbonate ratio and a notable proportion of Potsdam sandstone among the pebbles, both of which are derived from the St. Lawrence Lowlands in eastern Ontario (Karrow, 1967). Sunnybrook Till is typically clay rich, although it is in places sandy from the incorporation of underlying Scarborough Formation sand. Its composition is the most distinctive of the tills in the Toronto area. It is found along most of the Scarborough Bluffs, in numerous valley exposures, in downtown excavations, and in the Woodbridge cut. It is commonly about 6 to 9 m thick but where it occupies valleys in underlying units, it thickens considerably. Its high clay content, associated varved clay, and sedimentary features indicate that the ice advanced and retreated in a glacial lake in the Ontario basin.

Few correlative deposits have been found in other parts of Ontario (this is in part because valley incision seldom extends below Late Wisconsinan deposits). The Bradtville Drift (Dreimanis, 1964) is also considered to be of Early Wisconsinan age and may be correlative. This occurs at Port Talbot below the level of Lake Erie where two till units underlie the Port Talbot interstadial beds. Along Nith River, downstream from New Hamburg, the Canning Till (Karrow, 1987; in press) may be of similar age, but no associated organic deposits have been found to test this correlation. Brookfield et al. (1982) described the lowest till in the Lake Ontario Bluffs at Newcastle as a possible correlative of the Sunnybrook Till. In many boreholes, tills underlie deposits known or inferred to be of Middle Wisconsinan age; some of these occurrences may be of Early Wisconsinan age.

Middle Wisconsinan

At Toronto, the Middle Wisconsinan is represented by the Thorncliffe Formation, consisting of stratified clay (varved in part), silt, sand, and locally till, up to a thickness of 30 m or more (Karrow, 1967, 1974). The sediments, which are largely glaciofluvial and glaciolacustrine, were deposited in proglacial lakes in the Lake Ontario basin. The elevations of these sediments indicate that drainage through St. Lawrence Valley was dammed by ice for much of the time. Two till layers in Scarborough Bluffs, the silty Seminary Till and the clayey Meadowcliffe Till, suggest expansions of the ice lobe in the Lake Ontario basin which did not extend far inland from the present shore. These tills divide the Thorncliffe Formation at Scarborough Bluffs into three members, the upper and lower containing peat lenses dated around 30 ka and >53 ka, respectively, whereas inland the Thorncliffe Formation is undivided. At least part of this unit probably extends into or correlates with parts of the sediments at the core of the Oak Ridges Moraine, north of Toronto. These till units are considered to be lacustrine deposits by Eyles and Eyles (1983), Eyles (1987), and Eyles and Westgate (1987) but glacial deposition has been suggested by Sharpe and Barnett (1985). Possibly equivalent till has been described from the bluffs at Newcastle by Brookfield et al. (1982).

Several interstadial deposits (Fig. 4.13), generally less than 6 m thick, date from the Middle Wisconsinan in southwestern Ontario, as already described. They indicate that Middle Wisconsinan sediments occur widely as thin, discontinuous sheets, lenses, and thicker fills in buried valleys. Fluvial, lacustrine, and paludal sediments are included and suggest that most of southern Ontario was deglaciated for most of Middle Wisconsinan time. Fossil assemblages suggest cool climate and boreal vegetation. In the Lake Erie basin, glacial deposits referred to as Middle Wisconsinan have been the basis for suggesting brief glacial advances. At the Port Talbot section intervals of climate moderation and ice retreat are recorded — the Port Talbot I (undated), Port Talbot II (46 ka), and Plum Point (25 ka) (Dreimanis et al., 1966) — whereas elsewhere single organic zones, commonly with only minimum ages or undated, are believed to be of Middle Wisconsinan age.

Late Wisconsinan

The Late Wisconsinan is generally well known through much of southern Ontario. It was dominated by glacial activity with the earliest advance overriding sediments containing organics at St. Davids gorge near Niagara Falls shortly after 22.8 ka (Hobson and Terasmae, 1969; GSC-816, Table 4.3).

Nissouri Stade

The beginning of the Nissouri Stade is commonly recognized as marked by the major ice advance out from the lake basins to the limit of Late Wisconsinan glaciation, and is represented in southwestern Ontario by the Catfish Creek Till (deVries and Dreimanis, 1960). The Dunwich Drift, which was at one time considered Middle Wisconsinan in age, is now thought to be the lowermost deposit of the Nissouri Stade in the Erie basin (Dreimanis, 1987). The Southwold Drift, which was also thought to be Middle Wisconsinan, is now looked upon as part of the Catfish Creek Drift (Dreimanis and Barnett, 1985). Some tills underlying Catfish Creek Till in the region around Kitchener might represent slightly earlier Nissouri Stade fluctuations, but they could be much older (Cooper, 1975). The Catfish Creek Till contains wood picked up from underlying Plum Point interstadial deposits and is overlain by Erie interstadial clays or Port Stanley Till derived from them. The ice that deposited the Catfish Creek Till was characterized by several different patterns of flow, including regional southwestward flow at the time of maximum ice thickness and extent, and later lobate flows out of the lake basins (see 18 ka front of Map 1703A). In many areas it is the lowest stratigraphic unit in valley wall exposures and marks the beginning of the exposed stratigraphy. In small areas near Woodstock it is the surface till, and apparently was not overridden subsequently by younger ice advances. As noted by Cowan (1978) and widely observed by others, the till is markedly uniform in its characteristics through much of southwestern Ontario and commonly can be used as a marker bed in till sequences. It is characteristically stony, sandy to silty, very hard (well drillers' "hardpan"), with a low calcite-dolomite ratio.

Erie Interstade

Following deposition of Catfish Creek Till, the ice retreated at least as far as the east end of the Erie basin and well to the north in the Huron basin, before the readvances of the Port Bruce Stade. This time of ice retreat, the Erie Interstade, saw the development of extensive proglacial lake deposits with the result that tills of the following Port Bruce Stade, where derived from the basins, are clayey or silty and are commonly interbedded with lacustrine sediments. No datable material of this age has been found but its age is estimated to be around 16 ka (Mörner and Dreimanis, 1973; Sigleo and Karrow, 1977).

Port Bruce Stade

Various ice advances during the Port Bruce Stade almost completely covered southern Ontario again. The ice advance through the Erie basin deposited the clayey Port Stanley Till, with several till layers interbedded with lacustrine clay (Dreimanis and Goldthwait, 1973); the advance through the Huron basin deposited the Stirton, Tavistock, Mornington, Stratford, and Wartburg tills (Karrow, 1974). Some of these till units may be the result of local surges; they are of limited distribution and certainly the time available for the many fluctuations is short. In central Grand River Valley clayey Maryhill Till and sandy Port Stanley Till are equivalent in age to the above mentioned Huron lobe tills (Fig. 4.18). Later in the Port Bruce Stade the Georgian Bay Lobe became independent of the Huron Lobe and deposited the Elma Till, while the Huron Lobe deposited the silty Rannoch Till, and the Ontario-Erie lobe deposited the Wentworth Till (Fig. 4.18). These were the last tills deposited before another substantial retreat occurred.

Fluctuating ice retreat of Port Bruce time developed several prominent moraines concentrically arranged around the first deglaciated land in central southwestern Ontario, known as "Ontario Island" (Taylor, 1913). Moraine crests are recognized under Lake Erie and suggest some correlations with moraines of similar age in Ohio and New York (Sly and Lewis, 1972). Sedimentological study by Barnett (1984) and Sharpe and Barnett (1985) revealed complexities of the interaction of the fluctuating ice margin and the proglacial lakes of the Erie basin during the formation of some of these moraines, as seen in the bluff exposures on the north shore of Lake Erie.

During Port Bruce fluctuations, glacial lakes, which were initially of local extent along the ice margins, coalesced and became major lakes. The first of these to affect southern Ontario — Lake Maumee (Fig. 4.19a) — was formed in the Erie basin about 14 ka, and had outlets which can be correlated with the three stages of Maumee, to the west across Indiana and Michigan (Hough, 1963). Maumee shorelines in Ontario are restricted to the area between London and Tillsonburg (Barnett, 1982).

During the next few thousand years, a complex assemblage of glacial lakes formed in the Great Lakes basins (Fig. 4.19). This interval of time is comparatively short, and the history of these lakes has attracted the attention of some of the leading geologists of each generation of this century. The lake history is closely tied to the fluctuations of the retreating ice sheet and it is in this period of time that most of the present landscape was formed. Evidence of former higher water levels had been noted early in the 19th Century, and names were applied to some of them late in the century; at about the same time, outlets were identified. All of the details of this complex study cannot be covered here, but the reader is referred to classic work by Leverett and Taylor (1915), Hough (1958), and more recent summaries by Prest (1970), Fullerton (1980), and Karrow and Calkin (1985).

Mackinaw Interstade

As the ice withdrew into the Erie and Huron basins, lower outlets for proglacial lakes were opened to the east and water levels dropped. The Ontario basin was at least partly deglaciated at this time. This was a time of presumably milder conditions — the Mackinaw Interstade — dated at just over 13 ka from a few widely scattered sites. Also, Warner and Barnett (1986) reported detrital plant material of about this age from north-central Lake Erie bluffs. Large masses of glaciofluvial and glaciolacustrine deposits were formed in the Oak Ridges Moraine between the Ontario and Simcoe lobes, which were further augmented during the subsequent, and last, ice advance to and over the moraine (Duckworth, 1979). Land areas at this time were apparently subjected to rigorous climate; Morgan (1972, 1982) has recorded widespread evidence of patterned ground or soil polygons associated with permafrost conditions which apparently disappeared by about 13 ka (Fig. 4.20).

Port Huron Stade

The subsequent readvance during the Port Huron Stade was a significant event in the central Great Lakes; the advance limit of the Huron Lobe is marked by the massive Port Huron Moraine of Michigan and the Wyoming-Banks moraines of Ontario (Fig. 4.19b). The Erie Lobe entered only the east end of that basin, representing a spillover of the Ontario Lobe southward beyond the Niagara and Onondaga escarpments (Barnett, 1979). Several small moraines were formed in the Niagara Peninsula and north of Hamilton along the Niagara Escarpment. This advance is inferred to have occurred about 13 ka (Dreimanis and Goldthwait, 1973). The advance of the Huron and Ontario lobes once again blocked drainage to the east and created glacial lakes in the basins. Related to this readvance is Lake Whittlesey, a major lake of the Erie basin, lower than the Maumee levels, which extended into the southern Huron basin. It was followed closely in time by three phases of Lake Warren at still lower levels, whose shorelines are prominent to the north in the Huron basin along the west edge of the Wyoming Moraine, and by the formation of much less distinct shorelines of lakes Grassmere and Lundy (Calkin and Feenstra, 1985). Most of these lakes used various outlets across Michigan.

Two Creeks Interstade

By 12.5 ka, the ice lobes were again retreating and the opening of lower outlets to the east caused water levels in the Erie

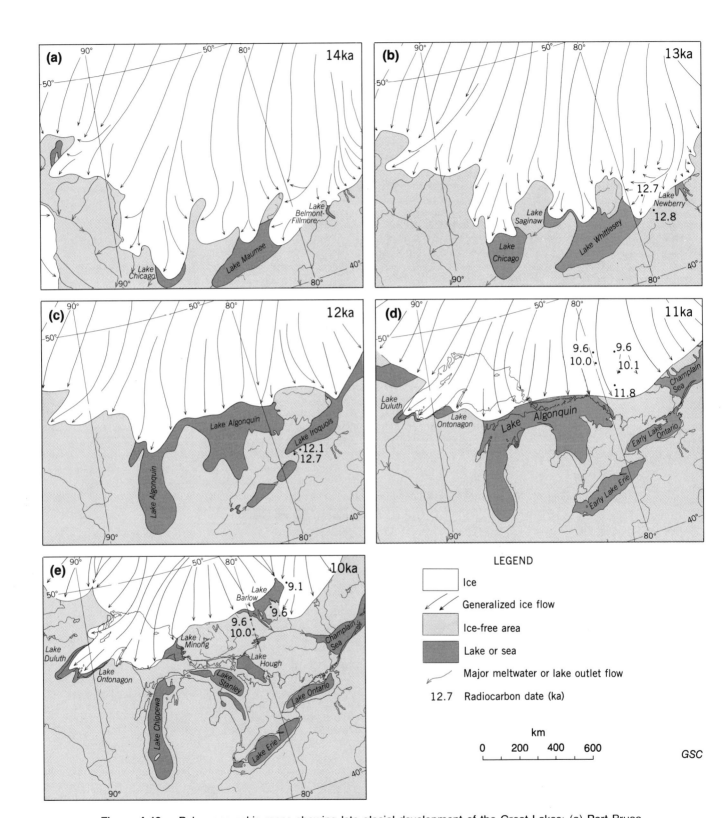

Figure 4.19. Paleogeographic maps showing late glacial development of the Great Lakes: (a) Port Bruce Stade (about 14 ka); (b) Port Huron readvance (about 13 ka); (c) Stage of deglaciation during Two Creeks Interstade (about 12 ka); (d) Lake and ice configuration (about 11 ka), shortly before opening of North Bay outlet; and (e) Great Lakes shortly after opening of North Bay outlet and shortly before culmination of Marquette advance in the Lake Superior basin (about 10 ka).

basin to fall to low levels. At this time, the shallow western part of the Erie basin was mostly land with swamps which were subsequently resubmerged because of isostatic uplift of the outlet near Buffalo, New York (Lewis et al., 1966). Meanwhile, retreat of the Huron Lobe had initiated an Early Lake Algonquin, which at first may have received inflow from the Erie basin, with outflow at Chicago. With ice retreat north from the Niagara Escarpment south of Georgian Bay and deglaciation of Trent Valley (the Fenelon Falls outlet of Lake Algonquin), a low water phase was established. It discharged eastward into the major proglacial Lake Iroquois in the Ontario basin about 12 ka (Fig. 4.19c; Karrow et al., 1961). Lake Iroquois had been created when the retreating Ontario Lobe opened an outlet into Mohawk Valley at Rome, New York (Muller and Prest, 1985). This outlet must have been relatively stable for a period of time because a prominent shoreline formed all around the basin, except in the northeast.

These events correspond approximately with formation of the well known Two Creeks forest bed in the Lake Michigan basin (Broecker and Farrand, 1963), but equivalent deposits have not been found in Ontario. The subsequent readvance of the Lake Michigan lobe — the Two Rivers advance — has been traced across Michigan (Burgis and Eschman, 1981) but has not been identified in Ontario.

When the ice retreated northward from the north slope of the Adirondack Mountains, Lake Iroquois drained by a series of steps to a level much lower than that of present Lake Ontario (Anderson and Lewis, 1985). Just prior to final drainage this lake extended north into Ottawa Valley and east in upper St. Lawrence Valley to join glacial Lake Vermont in the Lake Champlain Valley (Muller and Prest, 1985; Fig. 4.19c). Raising of the outlet by isostatic uplift during subsequent time is clearly indicated by the tilt of the Lake Iroquois shoreline from 110 m at Hamilton today, to near 300 m in northern New York State. Lake Iroquois is the earliest glacial lake from which a molluscan fauna has been documented in Ontario (Karrow et al., 1972).

With deglaciation, lower parts of St. Lawrence and Ottawa valleys were occupied by the Champlain Sea. Marine waters may even have entered the Lake Ontario basin (Clark and Karrow, 1984; Pair et al., in press) but would have regressed quickly due to rapid isostatic uplift. Gadd (1980b) has proposed that a calving bay advancing up Ottawa Valley permitted marine waters to reach Ottawa while a remnant ice mass remained at the east end of the Ontario basin. He felt that this was necessary to explain marine shell dates of about 12.7 ka in the Ottawa area which conflicted with Lake Iroquois wood dates from the Ontario basin that indicated ice in St. Lawrence Valley continued to support Lake Iroquois until later than 12 ka. Karrow (1981b) has argued that old carbon in sea water has resulted in unreliable shell dates and that the wood dates should take precedence. In addition, preservation of a residual ice mass east of Kingston, as suggested by Gadd to explain the younger Iroquois dates, appears to the author to be unreasonable. Anderson et al. (1985) have suggested, largely on the basis of palynological evidence, that the Champlain Sea did not occupy its western basin until about 11.7 ka.

---

* Editor's note: The 11 ka panel of Map 1703A shows a Post-Algonquin Lake present rather than Main Algonquin Lake. This scenario is based largely on a lake bottom date of 11.8 ka from southeast of North Bay (Harrison, 1972, p. 21). Deglaciation of the dated site would have occurred after the Main Algonquin Lake phase. Karrow does not accept this date as a real age.

Figure 4.20. Polygonal patterned ground developed on Port Stanley Till at Muir, near Woodstock. Polygons reflect moisture differences in sand wedges penetrating the till. Courtesy of A.V. Morgan.

Continued retreat of the main ice front from the Huron, Georgian Bay, and Muskoka areas resulted in uplift of the Kirkfield outlet, raising water levels in the Huron basin and causing a transgression southward. When the rising waters again reached the level of the Port Huron outlet (south end of Lake Huron) Main or Late Algonquin came into existence (Fig. 4.19d)*. Dated sequences of lacustrine estuarine sediments in valleys near Lake Huron (Karrow et al., 1975) document this rise and record the lake history between 11.3 and 10.5 ka. Combined uplift and uncovering of a succession of outlets south of North Bay initiated post-Algonquin lakes and lowered water levels through many named phases to the low water Lake Stanley of the Huron basin (Hough, 1955) and Lake Hough of the Georgian Bay basin (Lewis, 1969). The falling water levels left a profusion of shoreline features around the western and southern margins of present Lake Huron and Georgian Bay. The shorelines become difficult to trace, however, in the Precambrian Shield area east of Georgian Bay and north of Lake Huron. The lowest channel in the North Bay area, the Mattawa, eventually opened near 10 ka and emptied into the regressing Champlain Sea in Ottawa Valley (Prest, 1970; Fig. 4.19e).

Meanwhile, in the Superior basin, ice recession began in the southwest near Duluth, Minnesota, and progressed northeastward. Local ice fluctuations near Thunder Bay have been described by Burwasser (1977) and an unknown but substantial area of the Superior basin was apparently deglaciated before a final rapid readvance (Marquette advance). This advance covered most of the basin again and came to a halt about 10 ka on the south edge of the Superior basin (Farrand and Drexler, 1985). The equivalent margin in Ontario has not been found, but these events took place at about the time the ice margin had retreated to near North Bay. High level glacial lakes in the Superior basin at first were confined to the southwest end but expanded with ice retreat, and outflow changed from southwestward into Mississippi River, to southward into the Lake Michigan basin. Lake Algonquin spread into the southeast corner of the Superior basin for a brief interval, and with the opening of the North Bay outlets, water levels dropped through a series of levels. A high moraine ridge at Gros Cap, northwest of Sault Ste. Marie, temporarily retained Lake Minong in the Superior basin, which expanded throughout the basin with retreat of the Marquette ice. With the opening of outlets northwest of Lake Superior from the Agassiz basin, surges of water spilled into the Superior basin and caused spasmodic downcutting of the Gros Cap moraine sill to bedrock, and the lowest level — Lake Houghton — was reached (Farrand and Drexler, 1985).

Deglaciation of the North Bay area marks the end of the glacial history of southern Ontario. The effect of residual isostatic upwarp continued to play a role, however, and as the North Bay outlet rose, waters again transgressed southward until they once more spilled through the Port Huron and Chicago outlets, creating the Nipissing phase by about 5.5 ka (Lewis, 1969). Rich fossil assemblages of molluscs and plants are often found in or under Nipissing deposits. In the north these distinguish them from barren Algonquin deposits, and in the south they contrast with species assemblages found in Algonquin deposits (Miller et al., 1985). With downcutting of the Port Huron outlet, the Chicago outlet was abandoned and the water level gradually lowered in the Huron and Georgian Bay basins to the present 177 m. In the Superior basin, Lake Houghton was succeeded by the Nipissing phase; as the Nipissing phase gave way to lower levels and the rock sill at Sault Ste. Marie rose isostatically, a separate Lake Superior was created at 183 m (Farrand and Drexler, 1985). While the postglacial sedimentation record in the Great Lakes has been studied in several deep bottom cores, that of the land area has been studied in bogs and small lakes. This record has been largely the focus of palynological studies, which are described elsewhere in this volume.

QUATERNARY GEOLOGY OF ST. LAWRENCE VALLEY AND ADJACENT APPALACHIAN SUBREGION[1]

S. Occhietti

St. Lawrence Valley and adjacent Appalachian subregion is bounded on the north by the St. Narcisse Moraine, on the south by the Canada — United States Boundary, on the west by the limit of the Champlain Sea in the Ottawa Valley and the Frontenac Arch, and on the east by Saguenay Valley. It includes the Appalachians of southern Quebec, the lowlands of Ottawa and St. Lawrence valleys and of Lake Champlain in Ontario and Quebec, and the southern margin of the Laurentian Highlands (Fig. 4.1). These lowlands, referred to here collectively as St. Lawrence Valley, occupy the central part of the St. Lawrence sedimentary platform, and overlap the eroded margins of the Appalachians and the Canadian Shield.

BEDROCK GEOLOGY

The greatest part of this subregion is situated on the St. Lawrence sedimentary platform. The St. Lawrence sedimentary platform occupies an ancient rift system, bounded on the north by a system of en echelon faults (Wilson, 1964) or by an unconformity, and on the south terminates where the Appalachian Orogen has been thrust over the platform along Logan's Line (Fig. 4.21). The sedimentary cover which underlies this platform is composed of Cambrian and Ordovician sandstones, dolomites, limestones, and shales (Houde and Clark, 1961, Fig. 4.21), with a total thickness of 2300 m in the central Quebec basin. The northwest-southeast Oka-Beauharnois arch southwest of Montréal isolates a thinner sedimentary sequence in the Ottawa Embayment. The lower units, Upper Cambrian sandstones and Lower Ordovician (Beekmantown) dolomitic limestones are exposed on the flanks of the Frontenac Arch, the north edge of the Adirondack Mountains, which lie in the United States to the south and on the Beauharnois arch. In the central part of the Quebec basin, only Middle Ordovician formations are exposed. Due to the thickness of Quaternary cover, Paleozoic outcrops are confined to the beds and banks of streams and to a few structural rises. Downstream from Québec, the extent of Paleozoic rocks is limited largely to a narrow band of rock terraces which extend into the Charlevoix Astrobleme.

The Monteregian Hills are a distinct geological unit, consisting of Lower Cretaceous basic intrusive rocks. These ten plutonic plugs extend in a general east-west line between Mont Mégantic, in the Appalachians, and the Oka complex at the edge of the Precambrian Shield. These hills are composed mainly of gabbros and syenites that were resistant to the erosion that removed the sedimentary rock cover.

The low margins of the Canadian Shield are composed of intrusives, metasedimentary rocks, and gneisses of the Precambrian Grenville Province. In addition to the main Canadian Shield area to the north and west, they include the Frontenac Arch and Rigaud Mountain, and the Oka hills which lie west of Montréal. In the Ottawa Valley reentrant, slices of Precambrian rock are uplifted along a series of WNW-ESE faults which delimit anticlinal folds in Paleozoic sedimentary rocks. North of Ottawa River the structural limit of the Canadian Shield is marked by fault-line scarps. From north of Montréal to Québec, vertical southwest-northeast faults delimit four bedrock terraces inclined towards the southwest (Occhietti, 1980). These terraces are remnants of the ancient, pre-Ordovician, erosion surface. Downstream from Québec, the uplifted Precambrian rocks are separated from the Paleozoic cover

---

Occhietti, S.
1989: Quaternary geology of St. Lawrence Valley and adjacent Appalachian subregion; in Chapter 4 of Quaternary Geology of Canada and Greenland, R.J. Fulton (ed.); Geological Survey of Canada, Geology of Canada, no. 1 (also Geological Society of America, The Geology of North America, v. K-1).

[1] Translation of a French text.

of the middle estuary of the St. Lawrence as far east as the Saguenay, by a fault-line scarp. This structural configuration is disrupted at the Charlevoix Astrobleme, where a circular depression was produced by the impact of a meteorite during the Devonian (Rondot, 1968).

The Appalachians of southern Quebec are characterized by regional northeast-southwest oriented structures that separate a variety of terranes (Fig. 4.21; Lamarche in Dubois, 1973). The foreland at the southern margin of the St. Lawrence Valley Lowlands is composed of a sediment complex, consisting of flysch, carbonate, and shale that lies between the Quebec basin and Logan's Line. The extensions of the Green Mountain anticlinorium structures of Vermont in Quebec are the Sutton Mountains, composed of Cambrian phyllites and quartzites, and the Serpentine Belt. The latter includes a Lower Ordovician ophiolite complex which contains asbestos deposits. The rock to the southeast consists of a variety of slates, greywackes, limestones, and volcano-sedimentary lithologies.

Upper Saint-François Valley and middle and lower Chaudière Valley provide gaps transverse to the regional structure. In the Chaudière drainage basin, the Beauce, one of the natural subdivision of the southern Appalachians of Quebec, separates the Bois-Francs and the Eastern Townships, in the southwest, from the Notre-Dame Mountains and Lower St. Lawrence in the northeast.

In the eastern part of St. Lawrence Valley, Logan's Line runs close to the Canadian Shield. Because of this structural constriction, the lower part of the upper estuary and the middle estuary (Dionne, 1963a) cross segments of Appalachian basement. The middle estuary valley is narrow, of the order of 25 km wide.

MORPHOLOGY AND DRAINAGE EVOLUTION

To date, no general medium-scale synthesis has been prepared on the geomorphology of St. Lawrence Valley. MacPherson (1967) described the development of the St. Lawrence terraces in the Montréal region, Gadd (1971) described the morphology of a major sector of central St. Lawrence Valley, and Le Menestral (1969) published a geomorphic map of the Blackburn Hamlet area near Ottawa. Dumont prepared a series of maps of the area north of Ottawa River, only one of which has been published (Dumont et al., 1980).

Bedrock morphology

Looking at the overall geomorphology, the gross morphological features of St. Lawrence Valley are due to bedrock topography with low areas filled with Quaternary deposits. The underlying bedrock topography was reconstructed using data from borings made for oil, water, or engineering purposes (Prévôt, 1972; J. Schroeder, unpublished). This topography is characterized by the main St. Lawrence paleovalley, with tributary valleys either buried or reoccupied by present streams. The orientation of buried valleys indicates a preglacial flow towards the northeast, as at present. Upstream from Québec, the river passes through a rocky gorge 55 km long and 2 to 3 km wide. This constriction is possibly due to recent regional upwarping (Gale, 1970). In the Quebec basin, rock escarpments face the drainage axes, suggesting an origin by fluvial erosion. This contrasts with the morphology of the Great Lakes basin where escarpments face up the regional dip, suggesting that they originated as cuestas and more completely mirror bedrock structure.

At the margins of the Canadian Shield, the Laurentian Highlands are cut by fracture lines and faults. These structural discontinuities have been exploited by glacial erosion so that a knob and depression topography has been developed.

The substratum morphology of the southern Appalachians has been dealt with in general descriptions (Bird, 1972), summaries (Dubois, 1973), and detailed regional studies (Shilts, 1981). The area appears to have undergone several phases of peneplanation (Blanchard, 1947), which were responsible for surfaces ranging in elevation from 550 to 240 m (Stalker, 1948). Second-order relief features are controlled by the bedrock lithology and structure. Consequently the area is characterized by ridges and trenches which have been modified by glacial action. In the middle of the St. Lawrence estuary, this "Appalachian" relief emerges in the form of long, parallel, and asymmetrical islands.

Karst landforms have locally developed in carbonate rocks (Schroeder et al., 1980). North of Ottawa River, small caves are present in Grenville marble, notably La Flèche, Lusk, Bear, and Pointe Confort caverns. Postglacial karstic phenomena, including lapiés, sinkholes, caverns, seeps, resurgences, and canyons, have developed in the Chazy, Black River, and Trenton limestones. Many caverns are located near major rivers or escarpments, for example, the 900 m-long Saint-Casimir cavern, and the karstic cavities cut into the Trenton limestone during the deepening of the Sainte-Anne gorges, northeast of Trois-Rivières. At Boischatel, in the Québec City suburbs, karst features have developed in these same flat-lying limestones. At one point Laval River is captured by a system of caverns at least 2500 m long.

The substratum is locally affected by glacial tectonics. When excavations were being carried out in Montréal for the Metro system and the Olympic site, Durand and Ballivy (1974) and M. Durand (Université du Québec à Montréal, personal communication, 1984) discovered Trenton limestone thrust sheets more than 900 m long. At Saint-Léonard, in the northern part of Montréal, a system of cavities developed in this limestone by glaciotectonic thrusting (Schroeder and Beaupré, 1985; Schroeder et al., 1986).

Pockets of saprolites and rock weathered during pre-Wisconsinan time have survived the passage of the last ice sheet. On the margin of the Canadian Shield, Bouchard and Godard (1984) reported grus at a number of sites. At Château-Richer, near Québec, anorthosite is kaolinized to a depth of 20 m (Cimon, 1969; Dejou et al., 1982), indicating long-term weathering. Also in the Québec City area at Charlesbourg a saprolite containing kaolinite, smectite, vermiculite, and gibbsite has developed from biotite hornblende gneiss and mylonite (LaSalle et al., 1983). The saprolites of Mount Mégantic (Clément and De Kimpe, 1977; Dejou et al., 1982) and Mount Orford (LaSalle et al., 1985) represent the base of ancient deep regoliths considerably truncated by glacial erosion (see Fig. 4.22 for location map). It is highly probable that all these alteration products reflect pre-Quaternary weathering reactivated during interglaciations and perhaps the Holocene.

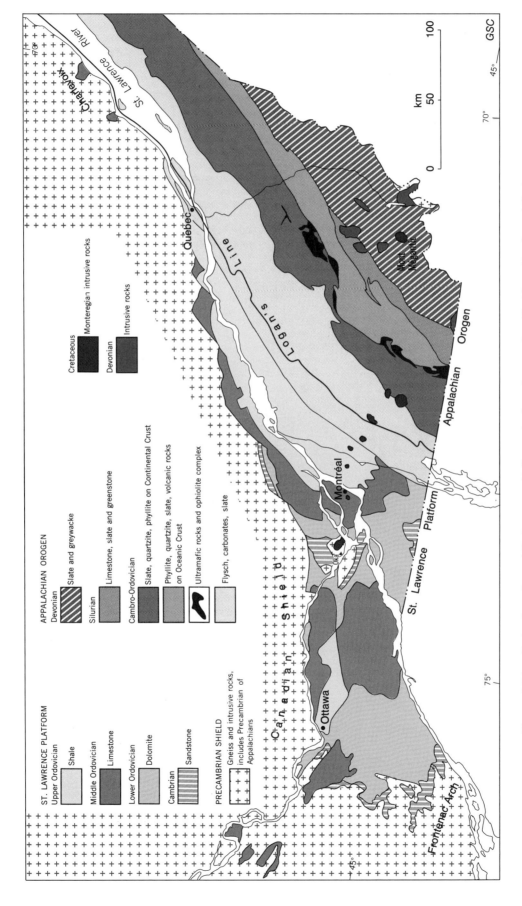

Figure 4.21. Bedrock geology of St. Lawrence Valley and adjacent areas (modified from Houde and Clark, 1961, and Lamarche in Dubois, 1973).

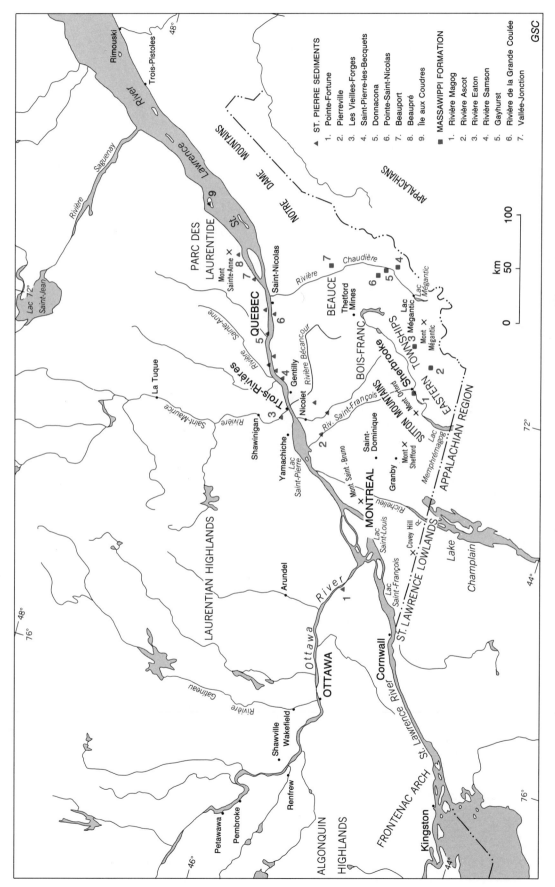

Figure 4.22. Place names used in text and locations of sites containing St. Pierre Sediments, Massawippi Formation, and equivalent deposits.

CHAPTER 4

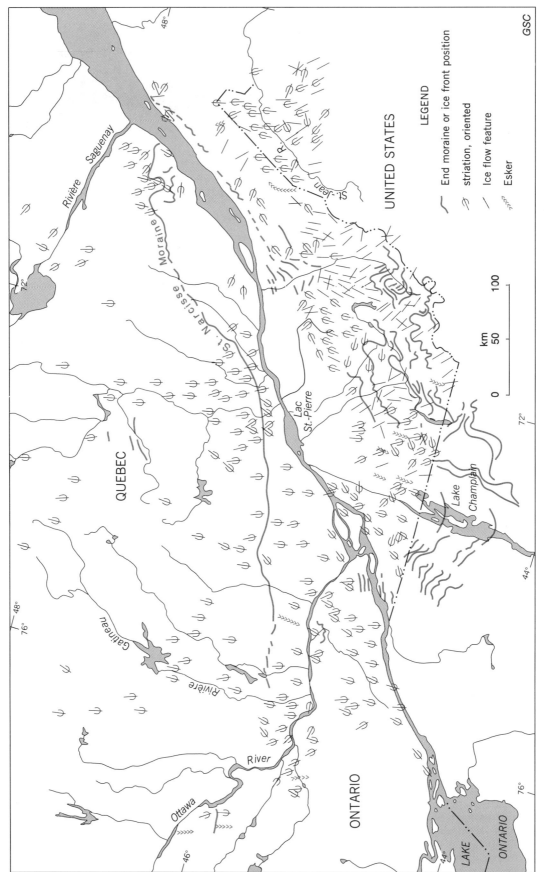

Figure 4.23. Ice flow features and main moraines.

Morphology of Quaternary deposits

Quaternary deposits cover the greater part of St. Lawrence Valley and the main low areas in the Appalachians but are discontinuous on the high ground of the Appalachians and of the Canadian Shield margin. The deposits are more than 100 m thick in the central part of St. Lawrence Valley, in Ottawa Valley, as well as in several buried valleys; however, in most areas these deposits are no more than a few metres thick and are somewhat thinner on valley slopes and hilltops.

The floor of St. Lawrence Valley consists of a series of terraces and stepped plains formed by littoral and fluvial processes during regression of the Champlain Sea at the end of the Wisconsinan. The channels of St. Lawrence and Ottawa rivers are incised in these plains. A few constructional glacial forms interrupt this generally level terrain. Constructed earlier than the terraces, moraines and other recessional features were partly or totally submerged and reworked by waves. On the Appalachians and the Canadian Shield margin, glacial landforms are relatively unmodified by postglacial processes.

Moraines in this area (Fig. 4.23, 4.24) commonly are composed of till ridges and discontinuous masses of ice contact sediments. They are generally tens of metres in height and width, and several hundreds of metres to several tens of kilometres in length. They parallel contours in the Appalachians, and are more or less straight in St. Lawrence Valley. Glaciofluvial constructional features such as kames, perched deltas, and outwash fans are associated with them. Because till sheets are generally thin, till plains and other till surfaces reflect the form of the substrate. Till-cored forms, measuring tens of metres long, have been noted in the Montréal region by Prichonnet (1977) and isolated drumlins occur between Ottawa and Cornwall, but the drumlin fields of the Great Lakes subregion to the west, end at the western limit of St. Lawrence Valley. Eskers of the order of 1 to 10 km long are relatively common (Fig. 4.23). Kame terraces, valley fills, deltas, and raised glacial lake beaches are abundant in the Appalachians.

Marine clay plains and terraces; sand terraces of marine, lacustrine, and fluvial origin; and incised deltas are the most extensive landforms in St. Lawrence Valley. They carry the imprint of ancient tidal channels and abandoned channels of the St. Lawrence and its tributaries, many of which are now occupied by swamps and bogs. Eolian activity remodelled the surfaces of sandy terraces and raised deltas. Locally parabolic dunes attain 20 m height and 300 m length and form "crêtes-de-coq" networks (Gadd, 1971).

Drainage evolution

The main valleys of the present drainage system (St. Lawrence and Ottawa rivers and Lake Champlain) are a reflection of major structural features. Conversely, tributary stream courses developed in valleys almost entirely filled by glacial, marine, lacustrine, and deltaic deposits under the influence of postglacial isostatic uplift. Despite this, many streams have managed to reoccupy their preglacial valleys. In the Appalachians, the orthogonal drainage system, governed by bedrock structure, converges towards three major drainage systems — the Saint-François in the Eastern Townships, the Chaudière in the Beauce, and the Saint John on the Atlantic slope of the Notre-Dame Mountains. Glacial filling and overdeepening are responsible for

Figure 4.24. Morainic ridge associated with the St. Narcisse Moraine, southeast of Lac Simon, Quebec; the more gently sloping left side of the ridge is the distal side of the moraines. 200300-Y

changes to the pre-Quaternary system. This is well displayed in the basins of lakes such as Memphrémagog and Saint-François (Bird, 1972; Dubois, 1973). The abundant gorges, rapids, and falls in the Appalachians and St. Lawrence Valley indicate, according to local circumstance, either superposition or renewal of erosion in the paleohydrological system, during the Holocene.

CLIMATE

St. Lawrence Valley and adjacent Appalachian subregion has a temperate continental climate. Winters are long and rigorous, with heavy precipitation. Trois-Rivières appears to be the western limit of the main oceanic influence with a markedly greater precipitation at Québec than at Montréal. During the summer, Montréal has the same monthly average temperatures as Toronto, where temperature extremes are somewhat dampened by the influence of Lake Ontario. At Québec, however, summers are cooler and winters colder. In the Appalachians, temperatures are lower and precipitation higher than in the lowland to the west.

One of the more noteworthy phenomena resulting from influence of climate on fluvial activity is the "glaciel" process (Dionne, 1977). Streams are frozen during the winter; flooding, stream and tidal flow, during spring breakup results in transport of debris frozen in ice and shore erosion by seasonal ice. Other important climate related phenomena are thawing and heavy precipitation, which may be responsible for triggering landslides in marine clays and mass movements in Appalachian tills.

NATURE AND DISTRIBUTION OF QUATERNARY DEPOSITS

Quaternary deposits of much of St. Lawrence Valley and the adjacent Appalachians have been mapped by geologists of the Geological Survey of Canada, Ministère de l'Énergie et des Ressources du Québec, Ontario Geological Survey, and universities in Quebec and Ontario. Early surficial deposits maps were published by Gadd and Karrow (1959) and McDonald (1966). Gadd (1971) prepared a map and report on the east-central part of St. Lawrence Valley. Reports and maps prepared by the various government agencies constitute the main body of descriptive information on

Quaternary deposits, but considerable additional information is available in journal articles, unpublished theses, and field trip guidebooks.

During Quaternary glaciations, ice entered St. Lawrence Valley from the major Laurentide distribution centres to the north and first blocked the valley at the constriction between the highlands of Parc des Laurentides and the Appalachians. This damming resulted in a large lake which extended into the Great Lakes basin to the southwest and Lake Champlain Valley to the south. Once the ice reached St. Lawrence Valley it flowed southwestward towards the Lake Ontario basin, southward into the Lake Champlain basin, and northeastward into the Gulf of St. Lawrence. Loading by the ice during glaciation, isostatically depressed the area; during the last glaciation this glacial isostatic subsidence was great enough to allow the area to be submerged by marine water at the time of deglaciation. Because of the relief, greater elevation, and marginal position in relation to Laurentide ice, the glacial pattern of the Appalachians is more complex than that of St. Lawrence Valley. In addition to the southward push of Laurentide ice, there was at times an input of Appalachian ice and consequently a vying for dominance between Laurentide ice and northward-moving Appalachian ice. Due to the moderate relief of the area, ice retreat was characterized by the development of local stagnant ice tongues in many valleys and by numerous glacial lakes.

Several episodes of fluvial deposition and erosion have affected St. Lawrence Valley during the Quaternary; glacial sediments have been interleaved with the fluvial sediments and glacial erosion has further complicated the stratigraphic picture. Early workers apparently oversimplified the stratigraphic succession and recent workers have suggested revisions (Lamothe, 1985, 1987; Occhietti et al., 1987).

Nonglacial terrestrial deposits

In St. Lawrence Valley and the adjacent Appalachians, terrestrial nonglacial deposits can be divided into three categories: deposits of the Sangamonian optimum, deposits predating the Late Wisconsinan glacial maximum, and postglacial deposits dating from the end of the Late Wisconsinan and the beginning of the Holocene. The older deposits are of fluviolacustrine, fluvial, and organic origin, and locally outcrop in lower parts of sections. The postglacial deposits are extensive and are of lacustrine, fluvial, eolian, and organic origin.

Deposits of the Sangamon optimum

Subtill deposits at Pointe-Fortune, on the Quebec-Ontario border, consist of unfossiliferous sand overlying organic-bearing sand and silty sand, massive clay, sand-clay, and till (Veillette and Nixon, 1984). Pollen and macrofossils from the organic-bearing sediments are characterized by high pine content and by deciduous taxa such as elm, oak, beech, and hickory. This suggests a climate slightly warmer than that of today (T.W. Anderson, J.V. Matthews Jr., and R.J. Mott, Geological Survey of Canada, personal communication, 1987). The samples which contained the warm climate, organic remains came from below the floor of a gravel pit, so stratigraphic relationships are not clear. However, because this is the first horizon below the surface till that contains evidence of warm (interglacial) climate, it is assumed to date from the optimum of the Sangamon Interglaciation.

Deposits predating last glacial maximum

All other nonglacial deposits recorded in St. Lawrence Valley that contain organic matter and predate the last glacial maximum have been included in a formation called the St. Pierre Sediments (Gadd, 1960, 1971). These sediments outcrop sporadically over an area extending from Ottawa Valley (Gadd et al., 1981; Veillette and Nixon, 1984) to the middle estuary of the St. Lawrence (Brodeur and Allard, 1985; Fig. 4.22). They include gravels, sands, silts, silty clays, and accumulations of organic matter (Table 4.5) and vary in thickness from less than 1 m to more than 8 m. The type section is located west of Saint-Pierre-les-Becquets, in the ravine of an intermittent stream (Site 4 of Fig. 4.22; Gadd, 1960, 1971); at this site, the St. Pierre Sediments consist of three beds of compacted peat containing pieces of flattened wood, interbedded with stratified sediments comprising sand, silty sand, and silt with disseminated organic matter. The overall thickness is about 4 m. At other sections nearby the units are underlain by a till, the lower of two tills in the area. Exposures of clean gravels and gravelly sands, underlying the upper till, have been mentioned by Gadd (1971) in the Bécancour and Saint-François valleys. These fluvial sediments, of local lithological composition, possibly are coarse stream facies that were deposited along the margin of the basin during the main St. Pierre Interval.

Later research has questioned this relatively simple succession. Lamothe (1987) suggested that a second organic-rich sand (St. Pierre I event), overlain by varves and till, underlies the unit described by Gadd (1971).

The St. Pierre Sediments is a heterogeneous group of fluvial, lacustrine, and paludal sediments (Fig. 4.25). The extent of the complex basins in which the sediments were deposited is as yet unknown but the sedimentation framework during the St. Pierre Interval apparently was closely analogous to the present situation. The St. Lawrence flows through several large shallow lakes (such as Saint-François and Saint-Pierre). They are generally less than 3 m deep; their shores are either sandy or muddy and are enclosed by impenetrable alder thickets; spring flooding produces ephemeral channels and associated deposits; marshes and bogs are common. During deposition of the St. Pierre Sediments the river channel meandered over a valley floor that was at least 50 km wide opposite Rivière Saint-François; the elevation of St. Pierre Sediments varies from 5 m at Les Vieilles-Forges, 25 m on Île aux Coudres, and 42 m at Pointe-Fortune.

In the southern Appalachians of Quebec, two nonglacial units predate the Wisconsinan glacial deposits — the pre-Johnville sediments and the Massawippi Formation. The pre-Johnville lacustrine and fluvial succession which outcrops on the banks of Rivière de la Grande Coulée underlies a till correlated with Johnville Till (McDonald and Shilts, 1971). At its base are clayey sands containing fragments of organic matter, overlain by 150 laminations of clayey silt. This is overlain by sands and gravels encrusted with iron oxide and containing clasts from the Canadian Shield (Shilts, 1981). McDonald and Shilts (1971) attributed this oxidation to an interglacial pedogenesis and hence propose a pre-Wisconsinan age for the sediments.

The Massawippi Formation predates Chaudière Till and appears to lie on Johnville Till. In the type section on Rivière Ascot (Site 2 of Fig. 4.22) noncalcareous laminated silts, 5.8 m thick, contain fragments of organic matter and vivianite (Fig. 4.25). Lacustrine and fluvial sediments in the same stratigraphic position outcrop on the banks of the Magog, Eaton, Samson, and Grande Coulée (McDonald and Shilts, 1971) and at Vallée-Jonction (LaSalle et al., 1979). Pollen in the lacustrine silts indicates a climate cooler than that of today. The organic has provided nonfinite ^{14}C ages (Table 4.5). Because of these considerations and its lithostratigraphic position, the unit has been correlated with the St. Pierre Sediments (McDonald and Shilts, 1971).

Postglacial deposits

Following the retreat of the last ice sheet and of marine waters, lacustrine, fluvial, paludal, eolian, estuarine, and littoral sediments were deposited.

In St. Lawrence Valley, marine clays (described in a later section) grade upwards into stratified silt and silty sand of a former lake characterized by *Lampsilis siliquoidea* (Elson and Elson, 1959). Such sediments presently fill the basins of lakes Saint-François, Saint-Louis, and Saint-Pierre and make up much of the shore deposits of the middle estuary. Sand and silty sand of the lacustrofluvial system of the proto-St. Lawrence and the modern St. Lawrence overlie the lake sediments. Marine, lacustrine, and fluvial sandy and silty facies in the area are virtually identical. This is in part because older marine silt and sand have been continually remobilized and redeposited in fresh water during isostatic uplift of the area. As a consequence, these facies can be differentiated with a degree of certainty only where they contain autochthonous fossils.

Table 4.5. Main occurrences of fossiliferous St. Pierre, Massawippi and possibly correlative sediments

	Locality	Nature of sediment	Plant macrofossils	References	Selected dates and references
ST. PIERRE SEDIMENTS	Pierreville	peat	*Larix laricina, Picea*	R.J. Mott, unpublished GSC Pollen Identification Report 83-27	enriched ^{14}C: 74 700 +2700/-2000 BP, QL-198, Stuiver et al., 1978; 67 000 ± 2000 BP GrN-1711, Gadd, 1971
	Les Vieilles-Forges	peat	*Picea* stumps; branches, trunks, roots of *Larix laricina* and probably *Picea*	R.J. Mott, unpublished GSC Pollen Identification Report 83-21; A. Larouche, Université de Montréal, unpublished report 23-2-1983	>30 840 BP, Y-255, Gadd, 1971; 32 200 ± 2800 BP, UQ-588, unpublished
	Pointe Saint-Nicolas	sands and silts, (Anse aux Hirondelles Sediments)	*Larix laricina* and possibly *Picea*	A. Larouche, Université de Montréal, unpublished report 23-2-1983	38 600 ± 2000 BP, UQ-388, Occhietti, 1982; >42 000 BP, GSC-3420, LaSalle, 1984
	Beaupré	varves with interbedded sands	mosses: *Sphagnum, Drepanocladus revolvens, Aulacomnium palustre, Ditrichum flexicaule, Polytrichum juniperinum*	M. Kuc in LaSalle et al., 1977b	>39 000 BP, GSC-1539, LaSalle et al., 1977b
	Saint-Pierre-les-Becquets	peat	*Picea*, possibly *Larix*, fruits of *Menyanthes*	R.J. Mott, unpublished GSC Pollen Identification Report 83-20	65 300 ± 1400 BP, GrN-1799, Gadd, 1971
	Beauport	sandy beds of upper varves	*Picea* sp. or *Larix* sp.	LaSalle et al., 1977b	>37 000 BP, GSC-1473, LaSalle et al., 1977b
	Donnacona	silt and silty sands			>44 470 BP, Y-463, Karrow, 1957; >35 000 BP, UQ-678, Clet et al., 1986
	Pointe-Fortune	sand and silty sand			>42 000 BP, GSC-2932, Gadd et al., 1981; Veillette and Nixon, 1984
MASSAWIPPI FORMATION	Rivière Ascot	laminated silts			>54 000 BP, Y-1683, McDonald and Shilts, 1971
	Rivière Magog	lacustrine sediments			>41 500 BP, GSC-507, McDonald and Shilts, 1971
	Rivière de la Grande Coulée	medium to coarse sand			>40 000 BP, GSC-1084, McDonald and Shilts, 1971
	Vallée-Jonction	laminated sandy sediments	Bryophytae: *Aulacomnium turgidum, A. palustre, Racomitrium canescens* var. *ericoides*	W.C. Steere in LaSalle et al., 1977b	>39 000 BP, QU-327, LaSalle et al., 1977b
	Île aux Coudres	peat and wood in sand			34 430 ± 1770 BP, UL-11; 28 170 ± 800 BP, I-13549; Brodeur and Allard, 1985; >39 000 BP, GSC-4252, >35 000 BP, UL-11-2, M. Allard, personal communication, 1986

Sandy deltaic and channel deposits between 5 and 30 m thick, with horizontal and crossbedded sedimentary structures, have been deposited at the base of the escarpment at the edge of the Laurentian Highlands (Gadd, 1971; Occhietti, 1980). Similar deposits, with associated silts, also underlie low terraces and islands in Ottawa Valley and occur at the head of Lac Saint-Pierre. In addition to fluvial deposits, abandoned channels of the Ottawa-St. Lawrence River system have been progressively filled by peat bogs or swamps. These organic deposits are up to several metres thick (Terasmae, 1965).

Eolian and niveo-eolian sands — sand-snow mixtures resulting from eolian and periglacial activities — are scattered over the high and middle terraces. These deposits are up to 30 m thick and form series of fixed dunes, associated with deflation areas that are commonly occupied by peat bogs (e.g., on the high terraces east of Shawinigan and south of Québec; Gadd, 1971; Dubé, 1971; Occhietti, 1980). Silty eolian sands, and sandy silts, 0.3 to 1 m thick, locally form a discontinuous blanket on till and glaciofluvial deposits (e.g., near Shawville, R.J. Fulton, Geological Survey of Canada, personal communication, 1984; and in the vicinity of Shawinigan, Occhietti, 1980).

The upper and middle estuary are subject to a tidal range of 0.3 to 5 m (Fig. 4.26). At low tide, in the middle estuary the strand in places is several hundred metres wide. The tidal zone deposits commonly consist of alluvial mud and coarser sediments, from 0.1 m to several metres in

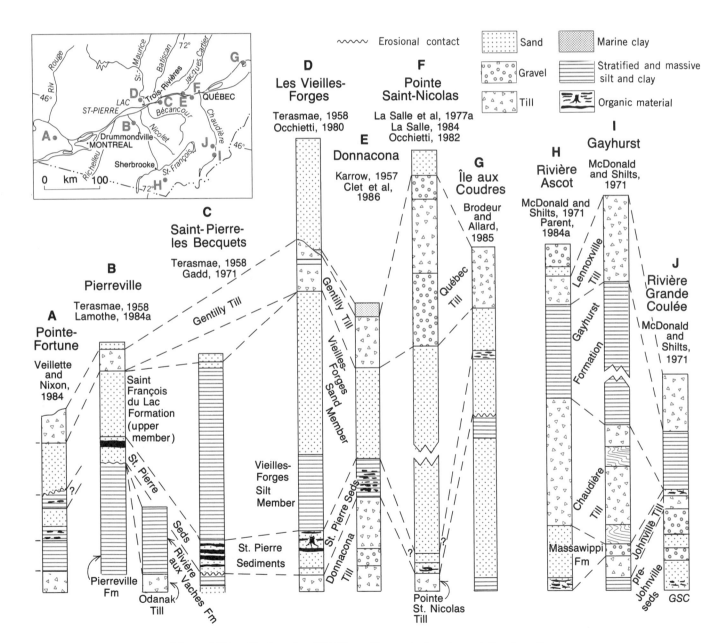

Figure 4.25. Stratigraphy of St. Pierre, Massawippi, and correlative deposits.

Figure 4.26. Tidal flats of the middle estuary of the St. Lawrence at mid tide; view from south shore of Île aux Coudres towards the Appalachians. 200300-U

thickness. Seasonal ice rafted or "glaciel" boulders are common. They indicate longshore movement and transport of clasts from the north shore to the south shore (Dionne, 1968, 1969, 1972b, 1977).

Glacial deposits

Glacial deposits include tills, ice contact deposits, proximal and distal glaciofluvial deposits, glaciolacustrine, and glacial lake deposits. Glaciomarine deposits are discussed in the next section along with marine deposits.

Tills

Three general types of regional tills and of till successions are present in this subregion. On the Canadian Shield, the till is usually loose with a grey, noncalcareous, sandy matrix. Thickness is variable, between a few decimetres in ground moraine, to several metres in morainic ridges and locally in the lee of rock knobs. The composition of these consists solely of Precambrian materials reflecting the regional geology. The till cover is discontinuous and preferentially occurs around the flanks and on the lee side of hills. Locally this till is deposited as small moraine ridges and also occurs within the St. Narcisse Moraine (Fig. 4.27). Occhietti (1982) includes the shield till in his Matawin Formation (Table 4.6A).

On the St. Lawrence sedimentary platform till occurs in places as several superimposed sheets. The thickness of individual sheets varies between 0.5 and 6.6 m; texture is generally clayey, and the matrix calcareous (Tables 4.6B, C); and the degree of compaction of older tills is more pronounced than that of the younger till. Clast composition is divided between Precambrian lithologies and local sedimentary rocks, but the matrix consists largely of materials abraded from the local sedimentary rocks and is the main determinant of colour. The sheets, of different ages, are relatively continuous and are more extensive and thicker south of Ottawa and St. Lawrence rivers than to the north. Downstream from the Ottawa-St. Lawrence confluence, and in lower Ottawa Valley, the tills are commonly buried by marine deposits; they do, however, outcrop on the northern edge of the Quebec Appalachians. West of the confluence, the upper till protrudes through younger overlying material in several areas. It has been reworked during marine regression but in many areas displays a rolling surface that is locally drumlinized.

Three till sheets have been described in the Appalachians but have not been described from a single section. The evidence for the lowest till is limited to information from bore holes and in two sections. The intermediate till sheet is common in depressions but seldom occurs at the surface. The upper till is relatively extensive and continuous in the Appalachian valleys and discontinuous on the ridges and monadnocks. All Appalachian tills are grey, have silty to silty sand matrixes, are relatively compact, distinctly Appalachian in lithology, but contain some shield debris. A facies, transitional between tills of the St. Lawrence Lowlands and those of the Appalachians, occurs at the northern margin of the Appalachians.

According to their lithostratigraphy, the tills of St. Lawrence Valley and southern Appalachians can be divided into two groups according to stratigraphic position: tills younger than St. Pierre Sediments (Table 4.6A, C) and older tills (Table 4.6B). Stratigraphic position is commonly the main criterion used in identifying the old tills, and is the only criterion in the Appalachians, where the Johnville Till has the same visual characteristics as the surface tills (Tables 4.6B, C). In places the lithological composition of the old tills reflects that of the local bedrock, suggesting active glacial erosion of the bedrock and only moderate transportation. The old tills are commonly oxidized but the clays do not show significant alteration (Gadd, 1971; Shilts, 1981). The red colour as well as degree of compaction have been used as identifying criteria but are not a means of recognizing old tills.

Differences between the young tills largely reflect the regional variation of glacial transport patterns after the St. Pierre Interstade. Three regional lithostratigraphic successions have been identified: a succession with two glacial units in upper St. Lawrence Valley and Montréal region (Table 4.6C); an Appalachian succession with two glacial units (Table 4.6C); and apparently a single glacial unit younger than the St. Pierre Interstade in the remainder of

Figure 4.27. Cut in the St. Narcisse Moraine north of Portneuf, Quebec. 200300-X

CHAPTER 4

Table 4.6. Description of named till units

Name	Locality and extent	Description	Direction of glacial flow	Stratigraphic position	References
A. Single tills younger than St. Pierre Sediments (central St. Lawrence and southern edge of Laurentian Highlands)					
Gentilly Till	Gentilly, central St. Lawrence Valley and Appalachian Lowlands	grey sandy till, calcareous, slightly compacted; mixed lithology; Precambrian and sedimentary; thickness 3 m	towards the SE	equivalent of Early, Middle, and Late Wisconsinan if St. Pierre Sediments are late Sangamonian or early Early Wisconsinan	Gadd, 1971; Occhietti, 1980
Québec Till	Québec area	calcareous till, very compact, sandy to clayey; mixed lithology: sedimentary, Precambrian and Appalachian		equivalent to Gentilly Till	LaSalle 1984
Matawin Formation (member: till)	Saint-Joseph-de-Mékinac, Laurentians, shield lowlands, and Saint-Narcisse Moraine	sandy diamicton, grey, variable compactness, fissile, Precambrian lithology, noncalcareous discontinuous; dominantly ablation facies, thickness generally less than 1 m	towards the SE but locally towards the S and the SSW (Prichonnet, 1977)	Late Wisconsinan	Occhietti 1980, 1982; Scott 1976
Rochette glacial deposits	southern area of Charlevoix between Baie Saint-Paul and la Malbaie	diamicton and ice marginal deposits; Precambrian lithology, charateristic anorthosite and ilmenite blocks	towards the ESE	glacial flow interpolated between Late Wisconsinan glacial maximum, with flow towards the SE, and the St. Narcisse readvance, with flow towards the S	Rondot 1974
B. Tills older than St. Pierre Sediments					
Bécancour Till	Rivière Bécancour; south shore, central St. Lawrence Valley	sandy clayey diamicton, usually calcareous and red, with medium compaction; consisting of sedimentary and Precambrian rocks fragments; thickness 3 m	S to W (?)	older than St. Pierre Sediments	Gadd, 1960, 1971
Pointe Saint-Nicolas Till	Pointe Saint-Nicolas; Québec area; on the south shore of St. Lawrence River	greenish diamicton, with silty matrix, moderate compaction; consisting of shield, sedimentary platform, and Appalachians debris; thickness 2.5 m	unknown	older than Anse aux Hirondelles Sediments (equivalent to St. Pierre Sediments)	LaSalle et al. 1977b; LaSalle 1984
Odanak Till	Pierreville; Rivière Saint-François	brick red, sandy clayey diamicton, calcareous, very compact, with two units divided by a gravel pavement or a sand bed; consisting of sedimentary and Precambrian lithologies; thickness 4 m	NNW-SSE	older than Pierreville Varves and St. Pierre Sediments	Lamothe 1984a
Pointe-Fortune till	Pointe-Fortune	sandy silty diamicton, dark greyish brown, calcareous; consisting of shield Paleozoic sedimentary lithologies; thickness 2.25 m	towards the SW likely	older than clays and silts containing organic matter that might be of Sangamonian age	Veillette and Nixon, 1984
Donnacona Till	Donnacona, north shore of the St. Lawrence between Rivière Yamachiche and Québec	grey sa...y silty diamicton, calcareous, medium to very compact, divided locally into three units by fine grained sediments; consisting of Precambrian and sedimentary debris; thickness 2-12 m	towards the SSW	older than St. Pierre Sediments	Coleman 1941; Karrow 1957; Occhietti 1980
Johnville Till	Grande Coulée River; in southern Appalachians	grey diamicton, very compact, noncalcareous, Appalachian lithology, oxidized; thickness up to 1.6 m	towards the SE	questionable evidence of a till older than Massawippi Formation (correlated with St. Pierre Sediments)	McDonald and Shilts 1971; Shilts, 1981; Parent, 1984a

Table 4.6. (cont.)

C. Two tills younger than St. Pierre Sediments (upper St. Lawrence, Montréal area, Appalachian region)

Name	Locality and extent	Description	Direction of glacial flow	Stratigraphic position	References
Till B (ex-Fort-Covington Till)	St. Lawrence Seaway; upper St. Lawrence Valley and margins of the valley	sandy diamicton, grey, moderately compacted; shield and Paleozoic debris; thickness to 9 m	towards the SSE	younger than stratified sterile sediments; either Late Wisconsinan (usual interpretation) or Middle and Late Wisconsinan	MacClintock, 1958; Fullerton, 1980; Clark and Karrow, 1984; Dreimanis, 1985
Till A (ex-Malone Till)	St. Lawrence Seaway; upper St. Lawrence Valley and margins of the valley	clayey diamicton, very compact, dark blue grey, dominantly Paleozoic debris, 10% shield debris; locally two units divided by stratified silts; thickness to 15 m	towards the SW (240°)	older than stratified sterile sediments; younger than the St. Pierre Interstade; equivalent to Chaudière Till (usual interpretation), so Early Wisconsinan	
Upper till	Montréal Island	grey diamicton with variable texture, fissile		same position as Till B (above)	Prest and Hode-Keyser, 1962, 1977
Lower till	Montréal Island	basal till, silty to sandy, with large Paleozoic and Precambrian blocks, compact		same position as Till A; with the lower complex apparently equivalent to the upper unit of Till A	
Saint-Jacques Till	Saint-Jacques-le-Mineur; area to the S and E of Montréal	clayey till, grey or reddish, Paleozoic and Precambrian debris		same position as Till B (above)	LaSalle, 1981
Upper till	Granby area, southern Piedmont of the Appalachians	sandy diamicton, grey, largely noncalcareous with Appalachian lithologies; thickness 1-3 m, and 10-15 m	towards the SE-SEE during the peniglacial, late flows towards the S and the SSW	same as Till B; could be equivalent to Lennoxville Till	Prichonnet et al., 1982a; Prichonnet, 1982a, 1984a
Ange-Gardien Till	Granby area, southern Piedmont of the Appalachians	grey diamicton, pebbly, slightly calcareous; thickness 1.5 m	towards the SW (220°)	same position as Till A; could be equivalent to Chaudière Till	
Lennoxville Till and Thetford Mines Till	Rivière Ascot, southern Appalachians	sandy and clayey silty diamicton, dark olive to dark grey, moderately compact; slightly calcareous, dominantly appalachian lithologies with some shield debris; units of lacustrine sediments locally intercalated, generally leached; thickness 3.5 m	towards the SE (110°, 130°, 140°)	surface till, Late Wisconsinan, Laurentide Ice Sheet deposit	McDonald and Shilts, 1971; Shilts, 1970, 1978, 1981; Chauvin, 1979a; Parent, 1984a
Chaudière Till and Norbestos Till		olive grey to olive black diamicton, calcareous, variable texture, highly compacted, exclusively Appalachian lithologies; thickness 1 m and more	fabric towards the W, WSW, and N	middle till; younger than Massawippi Formation, Appalachian ice cap deposit	

St. Lawrence Valley (Table 4.6A). Differentiating between these tills in the field, where no clear stratigraphic relationships are available, often requires detailed analysis. Distinguishing parameters that have been used are fabric, lithological composition, heavy minerals, alteration and percentage of clay-silt and of carbonates in the matrix (Shilts, 1981; Prichonnet, 1984a; Parent, 1984a; Lamothe, 1985). In association with the last till sheet, ablation tills and covers of coarse angular clasts occur throughout the Appalachians and the Laurentians.

Ice contact deposits

Ice contact and ice marginal deposits constitute considerable parts of frontal ridges, interlobate moraines, and other recessional deposits. These deposits are characterized by collapse structures, slumps, faults, and sedimentary discordances. The texture is extremely varied and units may contain diamictons and flowtills, boulder beds, massive and stratified gravels, sands and silts, and silts and sands interbedded with the coarser sediments (Denis and Prichonnet, 1973; Rondot, 1974; Lamothe, 1977; Pagé, 1977; LaSalle et al., 1977a; Occhietti, 1980; Prichonnet 1982a, b). They occur in a variety of geomorphic forms. An outwash plain was described north of Saint-Raymond-de-Portneuf on Rivière Sainte-Anne by Faessler (1948) and deltaic fans associated with the St. Narcisse Moraine system are well developed at Arundel (Laverdière and Courtemanche, 1961) and other locations (Gadd, 1971; Denis, 1976; Occhietti, 1980). Ice contact deltas are abundant in the Appalachians and are invaluable in helping to outline positions of the retreating ice front (McDonald, 1968; Prichonnet et al., 1982a; Boissonnault and Gwynn, 1983; Larocque et al., 1983b). At least one segment of the Saint-Antonin Moraine is formed of fluvioglacial deposits of exclusively Appalachian origin (Martineau and Corbeil, 1983). The great majority of deglaciation ridges in the Granby region (Prichonnet, 1984a) are formed of ice contact and outwash deposits. In areas below marine limit most of these types of deposits have been reworked (Hillaire-Marcel, 1974; Harrison, 1977; Richard, 1982). Subaqueous outwash deposits and features are common in areas inundated by the Champlain Sea (Rust, 1977; Chell, 1982) although there is some dispute over the exact nature and origin of some of these sediments (Gadd, 1978b; Rust, 1978).

Glacial lake and glaciolacustrine deposits

Glaciers periodically blocked St. Lawrence Valley and adjacent Appalachian valleys during ice advance and retreat, creating ice-dammed lakes. Exposure of deposits associated with these pondings are, however, of limited extent but may be 10 to 30 m thick. These glacial lake sediments include several distinct facies, true varves, rhythmites, interstratified or massive silts and sands, and turbidites, and are found in association with several distinct lithostratigraphic units of different ages (Table 4.7).

Sediments interpreted as glaciolacustrine (deposited in contact with or adjacent to the ice margin) locally contain materials that are interpreted as indicative of glacial activity. These materials include dropstones, diamictons, stratified silts, sands, and gravels, and turbidites. Sands with a suggested proglacial significance lie between the Les Vieilles-Forges silts above St. Pierre Sediments and Gentilly Till of the central St. Lawrence Valley (Fig. 4.25). In the Montréal and upper St. Lawrence Valley areas,

Table 4.7. Glaciolacustrine, glacial lake and lacustrine sediments (St. Lawrence Valley and southern Appalachians)

	Ottawa Valley	Upper St. Lawrence Valley	Montréal area	Central St. Lawrence Valley	St. Lawrence Estuary	Appalachians
Lacustrine units younger than Champlain Sea			*Lampsilis* Lake silts and sands (Elson and Elson, 1959)			
lacustrine units immediately predating marine invasion	varved silts of Ottawa-upper St. Lawrence area (Anderson et al., 1985; Fransham and Gadd, 1976)	laminated sediments (Rodrigues, 1987)	Côte-Saint-Luc rhythmites (Prest and Hode-Keyser, 1977) Lake Chambly sediments (LaSalle, 1981)	Nicolet and Saint-François varves (Gadd, 1971)	Saint-Féréol clays (LaSalle, 1978), Saint-Maxime varved silts (Gadd, 1978a) in lower Chaudière Valley	lacustrine sediments of glacial lakes Vermont and Memphrémagog (MacClintock, 1954; McDonald, 1968; Parent, 1984a)
lacustrine units immediately predating final glacial advance		lower and upper rhythmites of the intermediate complex (MacClintock and Stewart, 1965)	Châteauguay sediments (LaSalle, 1981), lower and upper rhythmites of the intermediate complex (Prest and Hode-Keyser, 1962)			Gayhurst Formation (McDonald and Shilts, 1971), Ruisseau Perry Formation (LaSalle, 1984; Chauvin, 1979a)
glaciolacustrine units associated with post-St. Pierre ice advance		? silts lower than till (Johnston, 1917)	rhythmites under lower till (MacClintock and Stewart, 1965)	laminated (turbidites) silt and clay underlying Gentilly Till (Occhietti, unpublished data)	Beauport and Beaupré varves (LaSalle et al., 1977a)	
lacustrine and glaciolacustrine units associated with St. Pierre Interval		Pointe-Fortune sediments associated with St. Pierre Sediments (Veillette and Nixon, 1984)		Saint-François-du-Lac Formation (Lamothe, 1985); silt and sand at top of St. Pierre Sediments at Les Vieilles-Forges (Occhietti et al., 1987)	Donnacona sediments (Karrow, 1957), Pointe Saint-Nicolas sediments (Occhietti, 1982; LaSalle et al., 1977a), Île aux Coudres sediments (Brodeur and Allard, 1985)	Massawippi Formation (McDonald and Shilts, 1971)
glaciolacustrine units intercalated between older tills and St. Pierre Sediments				Rivière aux Vaches Formation and Pierreville Formation (Lamothe, 1985); Deschaillons Varves (Lamothe, 1987)	Intermediate rhythmites at Île aux Coudres (Brodeur and Allard, 1985)	
glaciolacustrine and lacustrine units older than one of the older tills				varves mixed with Bécancour Till (Karrow, 1957)		pre-Johnville sediments (McDonald and Shilts, 1971)

rhythmically bedded glaciolacustrine silt and sand interstratified with gravel and till is inferred to indicate ice marginal fluctuations in a glacial lake (Middle Till Complex of Prest and Hode-Keyser, 1962, 1977).

In the Quebec Appalachians, a variety of deposits have been related to lakes formed during deglaciation. These include rhythmically bedded silts, nearshore sands and gravels, bouldery shore zone lags, and deltas perched above the limit of Champlain Sea. These have aided in the reconstruction of various phases of glacial Lake Vermont, which was blocked by the front of the Laurentide Ice Sheet and also of various lakes formed by the complicated pattern of ice retreat in the Appalachians (MacClintock and Terasmae, 1960; McDonald, 1968; Prichonnet, 1982b). In addition, the Gayhurst Formation, rhythmically bedded silt and sand which underlies the surface till, occurs in this area (McDonald and Shilts, 1971; Shilts, 1978; Parent, 1984a). Locally it contains small amounts of fine grained organic materials but this is thought to have been reworked from older deposits and the unit is interpreted as glacial lake sediment deposited when the Appalachians were partly deglaciated during the Middle Wisconsinan.

Marine sediments

At the end of the last glacial stage, a sea occupied St. Lawrence Valley. The valley, isostatically depressed by the ice sheet, gradually opened to the waters of the Atlantic as the ice sheet retreated. The marine basin has been subdivided into the Goldthwait Sea, which occupied the valley downstream from Québec City, and the Champlain Sea, which occupied St. Lawrence Valley upstream from Québec and lower Ottawa Valley (Elson, 1969a). The basin acted as an immense estuary with large quantities of meltwater depositing a cover of fine debris which in places exceeds 100 m thickness. At the same time as deposition was taking place, isostatic recovery was causing the sea to regress from the basin.

These Champlain Sea sediments, the most extensive Quaternary deposits in St. Lawrence Valley, extend from Pembroke in Ottawa Valley to the St. Lawrence estuary, and from the south end of Lake Champlain to La Tuque, in the valley of Rivière Saint-Maurice 130 km north of St. Lawrence River. The lithological and mineralogical composition of the marine deposits is related directly to the glacial deposits. The gravels and sands have the same composition as the regional till; and the clays and silts consist mainly of rock flour that includes quartz, feldspar, amphibole, illite, chlorite, and some interstratified clays. Gadd (1986) described these sediments as they occur in the Ottawa area and treated them as a series of lithofacies. This report groups the marine sediment into facies related to incursion, inundation, and regression. Locally the incursion facies is underlain by "true" glaciomarine or possibly glaciolacustrine deposits and in many areas the regression facies is overlain by deposits of the lacustrofluvial systems of the St. Lawrence basin.

Glaciomarine deposits

The glaciomarine deposits in this report include only materials deposited by processes directly linked to glacial ice. These materials are of limited extent. They include gravels and sands deposited in the sea as ice contact deposits and submarine fans (Rust, 1977; Hillaire-Marcel, 1979). These deposits in general resemble normal subaqueous outwash but locally contain marine fossils. Glaciomarine diamictons, or materials interpreted as deposited directly from debris melting out of floating ice, are rare. They have been described in the Trois-Rivières region at the position of the St. Narcisse Moraine and downstream at Saint-Alban on Rivière Sainte-Anne (Occhietti, 1977b, 1980); at the base of the Saint-Nicolas section south of Québec City (Gadd et al., 1972a; Occhietti and Hillaire-Marcel, 1982); and near Wakefield in Gatineau Valley (Fulton et al., 1986). They are as much as 15 m thick and massive, resemble till, but contain shells or shell fragments. Other deposits referred to as glaciomarine are marine clays similar to those described below but containing clasts and masses of diamicton dropped from icebergs.

Incursion facies

Deposits formed during marine incursion are found at the base of sections but are of limited extent. They are thin (a few decimetres) and mark the arrival of the sea water. They include pseudo-varved facies, rhythmites, and facies with parallel stratification comprising gravel, sand, and silt under the clay of definite marine origin (Gadd, 1971; Occhietti, 1980). Similar appearing deposits which contain the freshwater ostracode assemblage with *Candona subtriangulata* have been referred to as glaciolacustrine (Anderson et al., 1985; Parent, 1987). Rodrigues (1987) confirmed the presence of freshwater fauna in laminated sediments below the marine sequence.

Inundation facies

The marine inundation facies forms a cover from several to 100 m thick over much of the lowland. The sediments were deposited in quiet marine waters from a few tens to 250 m deep. These sediments are commonly called deep water marine deposits but they are not comparable to deposits of deep marine environments. They consist of clay, silt, and locally even fine sand, and vary from massive to finely stratified; fossils are rare. Sedimentation rates were high in ice proximal areas, reaching 200 mm per year in the para-marine basin of Rivière Saint-Maurice (Occhietti, 1980). Occhietti (1980) subdivided these sediments, as they occur at the Laurentian margin of the Champlain Sea, into finely laminated or massive 'decantation facies', nonfossiliferous stratified sediment 'para-marine facies', and 'prodeltaic dispersion facies', represented by sparsely fossiliferous silts with parallel stratification or with contorted structure.

Regression facies

Marine offlap deposits are locally extensive. They generally consist of stratified sands and silts, from some decimetres to some metres thick and are not everywhere fossiliferous. Their mineralogical composition is uniformly quartzitic, and hence they are easily distinguished from glaciofluvial sediments (de Boutray, 1975). Stratification is horizontal or crossbedded. The finest example of this facies is illustrated in the extremely fossiliferous gravel pit at Saint-Nicolas (Gadd et al., 1972a; Occhietti and Hillaire-Marcel, 1982; Fig. 4.28). At the same site, another facies, 4 to 5 m thick, with alternating sand and silt beds, marks the final emergent phase with deposition influenced by tidal fluctuations in a small bay (Hillaire-Marcel, 1974, 1979).

Figure 4.28. Fossiliferous sand of the regression facies of the Champlain Sea at Saint-Nicolas, southwest of Québec City. 200300-W

The littoral facies varies according to paleogeography, duration of shoreline stability, sources of detrital material, and mode of transport of material (Hillaire-Marcel, 1979). Raised shorelines are represented by horizontal bands of washed boulders and pebbles on the rocky slopes, particularly at the base of the Laurentian Highlands. Extensive sandy beach ridges were built on deltas and outwash terraces, and extend from glaciofluvial ridges. Spits were built in sheltered locations, for example, the fossil-bearing spits constructed in the shelter of the Monteregian Hills (Prichonnet, 1977). The most common littoral deposit is a bouldery or gravelly lag produced by marine reworking of the underlying materials. These either occur as coarse unconformable surface layers or as a fossiliferous coarse surface unit in gradational contact with the underlying deposits (Gadd et al., 1972b; Hillaire-Marcel, 1979, 1981a; Occhietti and Hillaire-Marcel, 1982).

Deltaic deposits are abundant at the margin of and within the Champlain Sea basin. These consist largely of sand with characteristic foreset and topset beds, and lie most commonly on well bedded silty prodelta facies. The locus of delta sedimentation migrated towards the centre of the basin as the sea regressed, producing strip deltas that parallel the modern water courses.

Marine fossils

The marine deposits contain a wide variety of marine and terrestrial fossils (Table 4.8). Marine macrofossils have been catalogued by Wagner (1970) and Elson (1969b). Cronin (1977a) used microfossil assemblages in his studies of paleoecology of the Champlain Sea and recognized three environmentally distinct marine phases. Hillaire-Marcel (1977, 1980, 1981a) described the various types of communities found in all the postglacial seas of Quebec and using available ecological data and ^{18}O and ^{13}C isotope analysis, distinguished seven communities and six subcommunities, characteristic of different marine environments (Fig. 4.29, Table 4.8). Rodrigues and Richard (1983, 1986) identified seven macrofaunal assemblages in their studies in the western basin of the Champlain Sea. Some of these correspond to the communities described by Hillaire-Marcel.

QUATERNARY HISTORY

St. Lawrence Valley and adjacent Appalachians occupy a position intermediate between the heart of the Laurentide ice in central Quebec and the terminal zone on Long Island, New York. In addition, the morphological variability of the area resulted in different styles of glaciation in different areas: the main St. Lawrence Valley was generally a conduit carrying ice flowing off the Canadian Shield, into the Gulf of St. Lawrence to the east, into the Great Lakes to the west, and into the Champlain-Hudson trough to the south; the Appalachians were probably a centre of accumulation during glacial inception, were overridden by ice overflowing southward from St. Lawrence Valley at glacial maxima, and may have again been the site of local accumulations during retreat of Laurentide ice. These various factors produced a variety of different and yet synchronous units which are difficult to fit into a consistent Quaternary history.

Several regional and interregional Quaternary history syntheses are available for St. Lawrence Valley and adjacent Appalachian area (Prest, 1970, 1977; McDonald and Shilts, 1971; McDonald, 1971; Gadd et al., 1972a, b; Dreimanis and Karrow, 1972; Dreimanis and Goldthwait, 1973; Dreimanis, 1977; Occhietti, 1982; Karrow, 1984; LaSalle, 1984). The main unit used in making these regional and interregional correlations is the St. Pierre Sediments; it was the only nonglacial unit that could be recognized with some degree of certainty in most parts of the subregion. Because of this, the age of the St. Pierre Sediments is critical to the Quaternary history of the subregion. Samples of plant material, calcareous concretions, and the sediments themselves have been subjected to conventional and enriched ^{14}C dating, thermoluminescence dating, uranium/thorium dating, and relative dating by the amino acid technique. The results have not provided a single chronological framework acceptable to all workers (Gadd et al., 1972b; Lamothe et al., 1983; LaSalle, 1984).

Pre-last glacial maximum

Nonglacial events predating the St. Pierre Interval

The Mic-Mac terrace, identified by Goldthwait (in Gadd, 1971), has been described on the southern shore of the middle estuary by Dionne (1963b) and Locat (1977), and its existence on the north shore has been noted by Brodeur and Allard (1985). This terrace has been correlated with a littoral abrasion platform observed in the Atlantic Provinces and attributed to the Sangamon Interglaciation (Grant, 1977). This feature implies a period of littoral erosion associated with a marine water plane about 6 m above modern sea level but there is no direct measure of its age. In the Québec City area, however, it is overlain by the rhythmites that underlie sediments of the St. Pierre Interval (LaSalle, 1984).

Table 4.8. Marine fossils from the Champlain Sea and western Goldthwait Sea

Marine shell communities Pelecypods, Gastropods Cirripeds, Brachiopods (Hillaire-Marcel, 1977, 1981a)	Other invertebrates (Wagner, 1970; Fulton, 1987)	Fish (Harington, 1978; Fulton, 1987)	Mammals (Harington, 1971, 1972, 1977, 1978; Harington and Occhietti, 1988) Birds (Harington and Occhietti, 1980)	Plants Marine algae (Mott, 1968; Illman et al., 1970) Plants (Fulton, 1987)	
Intertidal or shallow zones: 7 *Hemithyris psittacea* 6 *Mya arenaria* 5 *Macoma balthica* 4 *Mytilus edulis* Coarse features 3 *Hiatella arctica* a. upper sub-community: *Macoma balthica* b. middle sub-community: *Mytilus edulis, Balanus crenatus, B. balanus* c. deep sub-community: *Mya truncata, Balanus hameri* Ultrahaline cold waters 2 *Macoma calcarea* a. upper sub-community: *Astarte* sp. b. middle sub-community: *Chlamys islandicus, Serripes groenlandicus* c. deep sub-community: *Balanus hameri, Nuculana* sp. *Nucula tenuis* Near ice 1 *Portlandia arctica*	Echinoderms: *Crossaster papposus Ophiura* sp. *Ophiocoma* sp. or *Amphiura* sp. *Strongylocentrus drobachienois* Demosponge: *Tethya logani* Marine worms: *Nereis pelagica* Crustaceans: *Mesidotea sabini* Euphausiacea (near *Meganyctiphanes*) *Estheria dawsonii* xx Insects: March fly: *Bibio* sp. May fly: Ephemeridae Beetle: *Fornax ledensis* x Beetle: *Tenebrio calculensis* x Beetle: *Byrrhus ottawaensis* x Pill Beetle: Byrrhidae (*Cytilus* or *Byrrhus*) Caddisfly: *Phryganea ejecta* x	-Capelin: *Mallotus villosus* -Lake trout: *Salvelinus namaycush* -Rainbow smelt: *Osmerus mordax* -Lump fish: *Cyclopterus lumpus* -Threespine stickleback: *Gasterosteus aculeatus* (trachurus form) -Atlantic cod: *Gadus morhua* -Lake cisco: *Coregonus artedii* -Longnose sucker: *Catostomus catostomus* -Atlantic tomcod: *Microgadus tomcod* -Spoonhead sculpin: *Cottus ricei* -Deepwater sculpin: *Myoxocephalus thompsoni* -Blenny-like fish: Blennioidea	Marine mammals: Whales: -*Delphinapterus leucas* -*Megaptera novaengliae* -*Balaena mysticetus* -*Balaenoptera physalus* Seals: -*Phoca groenlandica* -*Phoca hispida* -*Phoca vitulina* -*Erignathus barbatus* Sea-elephant: -*Odobenus rosmarus* Terrestrial mammals: hare: -*Lepus americanus* Marten: -*Martes americana* Chipmunk: -*Tamias striatus* Birds: -*Somateria* cf. *mollissima*	*Laminaria Rodymenia* probable *Audouinella membranacea Acrochaetium* sp. Sugar maple: -*Acer saccharinum* Alder: -*Alnus* sp. Yellow birch: -*Betula alleghaniensis* (*B. lutea*) Water-shield: -*Brasenia schreberi* (*B. peltata*) Brome grass: -*Bromus ciliatus* Sedge: -*Carex magellanica* Round-leaved sundew: -*Drosera rotundifolia* Water-weed: -*Elodea canadensis* Algae: -*Encyonema prostratum* (*Cymbella prostratum*) Water horsetail: -*Equisetum fluviatile* (*E. limosum*) Dwarf horsetail: -*Equisetum scirpoides* Wood horsetail: -*Equisetum sylvaticum* Aquatic moss -*Fontinalis* sp. Rockweed -*Fucus digitalis*	Huckleberry -*Gaylussacia baccata* (*G. resinosa*) Bog moss -*Hypnum fluitans* Rice grass -*Orizopsis asperifolia* Balsam poplar -*Populus balsamifera* Large toothed aspen -*Populus grandidentata* Pondweed -*Potamogeton pectinatus* Pondweed -*Potamogeton perfoliatus* Pondweed -*Potamogeton pusillus* Pondweed -*Potamogeton rutilans* xx Silverweed -*Potentilla anserina* Willow -*Salix* sp. common cattail -*Typha latifolia* American eel-grass -*Valisneria americana* (*V. spiralis*)

x Record requires verification
xx Taxonomic position uncertain

Ostracode assemblages (Cronin, 1977a, b)	Foraminifer biozones from Lake Champlain cores (Fillon and Hunt, 1974; Corliss et al., 1982)	Foraminifer assemblages (Guilbault, 1980)	Diatom assemblages Goldthwait Sea (Lortie, 1983)	Shell assemblages and communities, Foraminifers and Ostracodes, in Ottawa Valley and western Champlain Sea (Rodrigues and Richard, 1983, 1986; Anderson et al., 1985)	Biostratigraphic and paleohydrological synthesis: diachronic composite vertical sequence
Upper stage, cold to temperate, brackish: *Cytheromorpha fuscata Cytherura gibba Leptocythere castanea* Intermediate phase, cold, euhaline to polyhaline: *Cytheropteron* sp. *Finmarchinella curvicosta Baffinicythere emarginata Cythere lutea Palmanella limicola Cytheromorpha macchesneyi* Lower transitional phase, cold, fresh to brackish water: *Candona subtriangulata Cytheromorpha macchesneyi*	5 *Elphidium clavatum* 4 *Elphidium clavatum* and *Protelphidium orbiculare* 3 *Protelphidium orbiculare* and *Elphidium bartletti* 2 *Islandiella islandica* and *Islandiella teretis* 1 *Islandiella teretis*	Shallow units: *Elphidium hallandense* or *Elphidium albiumbilicatum* Zone C mesohaline temperate: *Elphidium excavatum* Zone B mesohaline: *Elphidium clavatum* forma clavata Zone A arctic: *Cassidulina reniforme Islandiella helenae* or *Elphidium excavatum Islandiella norcrossi* Pre-A transitional zone, deep waters: *Cassidulina reniforme Elphidium* sp.	Upper phase, brackish and shallow waters: *Cyclotella caspia Cocconeis costata C. scutellum* var. *stauroneiformis* Regressive phase, shallower and more brackish waters: *Cocconeis scutellum* var. *stauroneiformis Cocconeis costata Nitzschia cylindrus Chaetoceros* sp. Full-sea phase, deep and saline waters: *Stephanopyxis furris* var. *intermedia Coscinodiscus* sp.	Upper lacustrine phase: *Lampsilis siliquoidea* Shallow water assemblages: *Elphidium albiumbilicatum Mya arenaria Elphidium clavatum Protelphidium orbiculare Macoma balthica Mytilus edulis* Rare assemblage: *Mya truncata* Deep, cold, high or variable salinity waters assemblages: *Elphidium clavatum Hiatella arctica Eoeponidella* sp. *Protelphidium orbiculare Astronomion gallowayi Balanus hameri Cassidulina reniforme Cibicides lobatulus Islandiella norcrossi Portlandia arctica* Glacial lake phase, fresh, deep water: *Candona subtriangulata Cytherissa lacustris Lymnocythere friabilis*	g) *Lampsilis* freshwater lake or fluvial estuary f) temperate, shallow, shore or coastal, brackish waters (summer temperature 20°C; salinity 3 to 18°/₀₀) e) arctic to subarctic, mesohaline, moderately shallow waters (18 to 24°/₀₀, summer temperature 10-12°C) d) arctic, cold, polyhaline, moderately deep waters (salinity 22 to 28°/₀₀) c) arctic, cold, euhaline, deep waters (0-4°C, salinity 30 to 33°/₀₀) b) transitional, deep and cold waters, fresh to brackish a) *Candona subtriangulata* fresh waters glacial lake

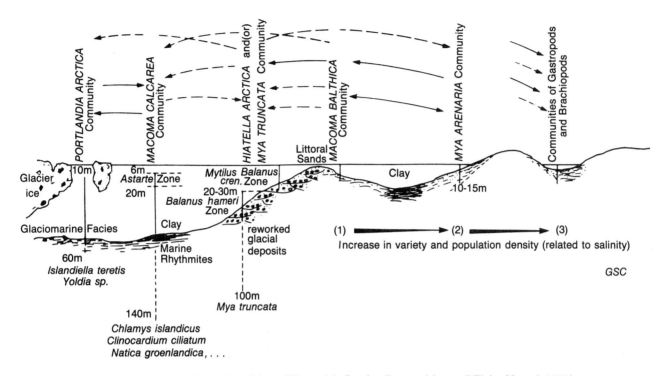

Figure 4.29. Ecological niches of Champlain Sea fossil assemblages, (Hillaire-Marcel, 1979)

The bedrock floor that underlies the pre-St. Pierre Sediments is at or below present base level. Even if residual glacial isostatic compensation is taken into account, it appears that base level of St. Lawrence River during the Sangamon Interglaciation (and possibly older ones) was as low as it is today.

The Sangamonian climatic optimum (equivalent to oxygen isotope stage 5e) apparently is represented by the lower organic-rich sediments at Pointe-Fortune (T.W. Anderson, J.V. Matthews, Jr., and R.J. Mott, Geological Survey of Canada, personal communication, 1987; Veillette and Nixon, 1984; Anderson et al., 1987). Few details are available on these or on the 30 to 40 m of sediments that underlies St. Pierre Sediments near the type section or the more than 60 m that underlies St. Pierre equivalent sediments at Île aux Coudres (J.P. Leroux, Transports Québec, unpublished report, 1986). The discovery of several units underlying the classical St. Pierre Sediments in the Saint-Pierre-les-Becquets area (Lamothe, 1987; Occhietti et al., 1987), however, suggests that a long sequence of events predating the St. Pierre Interval remains to be described.

In the Appalachians, the pre-Johnville sediments are of nonglacial origin and yet contain pebbles from the Canadian Shield; consequently, in addition to being a product of a pre-St. Pierre nonglacial period, they provide evidence of at least one regionally significant glaciation which must have occurred before they were deposited (McDonald, 1971).

Glacial events predating the St. Pierre Interval

There is scattered evidence for as many as three major glaciations prior to the St. Pierre (Ford et al., 1984); however, the nature, extent, and flow pattern of these older glaciers are unknown.

For the Appalachians, the evidence for a glaciation predating the pre-Johnville sediments has already been mentioned. Johnville Till is related to a glaciation immediately predating the St. Pierre but exposures of Johnville Till are extremely rare (McDonald and Shilts, 1971; Shilts, 1981; Parent, 1987).

In an area lying between Montréal and Québec City and between Les Vieilles-Forges and the Appalachian piedmont, there are several tills that may be placed reliably in a stratigraphic position below the St. Pierre Sediments. These are Bécancour, Odanak, Donnacona, Pointe Saint-Nicolas, the lower till at Pointe-Fortune (Table 4.6B), and the till at the base of borings near Saint-Pierre-les-Becquets (Lamothe, 1987). Correlation of these tills, however, is only speculative.

In St. Lawrence Valley varved clay units, in addition to tills, are used as evidence of past glaciations. This is based on the assumption that each glacial lake deposit represents a time when either Laurentide or Appalachian ice was sufficiently extensive to seal off the lower end of St. Lawrence Valley. Glacial events that were extensive enough to fill St. Lawrence Valley with ice were accompanied by lacustrine episodes that both predated and postdated the glacial maximum. However, it can be difficult to distinguish one unit of glacial lake sediments from another and consequently a degree of ambiguity has entered the Quaternary stratigraphy of the lowlands. For example, in the Pierreville section, varved sediments of Pierreville Formation (Lamothe, 1985) underlie St. Pierre Sediments. If a thermoluminescence date of 135 ka (Table 4.9) for these varved sediments is correct, these deposits are late Illinoian (Lamothe, 1985). Gadd (1971) assumed that finely laminated clays of the Rivière aux Vaches Formation exposed on the opposite bank of Rivière Saint-François were correlative. However, thermoluminescence dates of ca. 80 ka on these (Lamothe, 1984a; Table 4.9) suggest that these sediments, which are overlain by nonfossiliferous sand, represent a glacial lake episode that postdates the warmest part of the Sangamonian (oxygen isotope stage 5e), are younger than the Pierreville varves, and directly predate the St. Pierre Sediments. Gadd (1971) interpreted his single varved succession the result of a glacial lake that formed during retreat of the ice which deposited the Bécancour Till. He referred both the till and the varved sediments to the Bécancour Stade which immediately predates the St. Pierre Interval. Dreimanis and Karrow (1972) renamed this stade the Nicolet Stade. Lamothe (1984b, 1985), basing his interpretation on thermoluminescence dates and stratigraphic data, suggested that the Pierreville varves are Illinoian and are separated from what he referred to as the Rivière aux Vaches clays by a hiatus that represents the Sangamon Interglaciation. Hence there is a lack of agreement on the number and timing of glacial and lacustrine units in St. Lawrence Valley.

St. Pierre Interval events

In St. Lawrence Valley, a group of sediments deposited in glacial lacustrine, lacustrine, and fluvial environments is interbedded between the Bécancour Till (or old tills) and the Gentilly Till (or a group of equivalent units) (Table 4.10). The succession of these units was examined and defined in the central part of the valley by Gadd (1960, 1971), partially extended by Occhietti (1980), and re-examined by Lamothe (1984a, b, 1985, 1987), Occhietti (1982), and Occhietti et al. (1987). Similar lithostratigraphic suites are identified at a number of other sites (Fig. 4.22) which can be placed in five groups: Pointe-Fortune (site 1), Saint-François Valley (site 2), Les Vieilles-Forges-Saint-Pierre-les-Becquets (sites 3 and 4), Donnacona-Pointe Saint-Nicolas (sites 5 and 6), and Beauport-Île aux Coudres (sites 7, 8, and 9). Palynostratigraphic studies and absolute dating are currently insufficient to establish definite correlations between the units of these five areas; however, based on the sequence of deposits (Fig. 4.30), the interval appears to have been characterized by two main phases: 1) a nonglacial fluvial episode: St. Pierre Sediments (Gadd, 1960); 2) a lacustrine and deltaic episode: (Gadd, 1971; Occhietti, 1982; Lamothe, 1985).

Nonglacial fluvial episode. The St. Pierre Sediments, characterized by fluvial sand, peat, silt, and clay, suggest a relatively low energy fluvial environment. It appears that St. Lawrence Valley at this time was a broad, flat-floored basin occupied by peat bogs and shallow lakes connected by a relatively large, but low gradient fluvial system.

The study of pollen, plant, and insect macrofossils suggests interstadial climatic conditions, with temperatures generally cooler than at present (Terasmae, 1958). Terasmae's pollen diagram for Les Vieilles-Forges, shows *Picea* and *Pinus* as the dominant genera with temperate deciduous trees, such as *Quercus*, *Fagus*, *Tilia*, and *Carya*, representing the optimum climatic period. The occasional

Table 4.9. Radiocarbon ages for concretions and thermoluminescence ages for lake sediments

Lithostratigraphic unit	Locality	^{14}C age, lab no.	Characteristics of ^{14}C dated material	Thermoluminescence dates	Reference	Comments
Gentilly Till	Pierreville	34 000 +1800/-1400 BP, UQ-312	Striated calcareous concretions, carried by ice		Lamothe, 1985	Concretions older than ice invasion; age of concretion formation and possibly maximum age for glacier advance
		36 400 +3000/-2400 BP, UQ-494				
		37 200 +2400/-2500 BP, I-12 894				
		26 000 ± 2300 BP, UQ-484				
Pierreville Formation varves	Pierreville	34 000 ± 1050 BP, UQ-406 δ^{13}C = -13.13‰	calcareous ovoid concretions	135 ± 26 ka >82 ± 15 ka	Lamothe, 1984a, 1985	34 ka could be the age of concretion formation; the sediment is late Illinoian according to thermoluminescence dates
		36 400 +3000/-2400 BP, UQ-494				
Deschaillons Formation varves	Deschaillons	36 280 ± 2410 BP, QU-279	deformed, elliptic syngenetic concretions		Hillaire-Marcel and Pagé, 1981	age of concretion formation and probably minimum age for sediments
		37 500 +2300/-1800 BP, QC-357				
		34 900 +1625/-1350 BP, QU-559				
Gayhurst Formation varves	Old Gayhurst Dam on Rivière Chaudière	32 900 +1450/-1225 BP, QC-508	calcareous concretions		Hillaire-Marcel, 1979	age of concretion formation and possibly minimum age for sediment
		20 640 ± 640 BP, QC-558				
		20 600 ± 350 BP, UQ-556				
Rivière aux Vaches Formation varves	Pierreville	28 030 ± 760 BP, UQ-130	disc-shaped concretions	86.3 ± 17 ka >76.7 ± 15 ka	Lamothe, 1984a, 1985	28 ka could be the age of concretion formation; sediment age 86 ka
Stratified silts and sands of Saint-François-du-Lac Formation	Pierreville			61.2 ± 11 ka 61.1 ± 9.2 ka >53.7 ± 8.1 ka	Lamothe, 1984a, 1985	sediment age about 60 ka

abundance of birch (*Betula*) and alder (*Alnus*), and of herbaceous plants such at *Typha*, indicates local variations, particularly in drainage. Because of presence of eastern hemlock (*Tsuga canadensis*) in the flora at the Les Vieilles-Forges site, it is suggested that palynology indicates conditions similar to those of today's southern boreal forest (Clet and Occhietti, 1988).

In the Appalachians, the Massawippi Formation, which has generally been correlated with the St. Pierre Sediments (McDonald and Shilts, 1971) appears to be mainly lacustrine. It contains pollen grains of *Picea*, *Pinus*, and *Betula* — evidence of a boreal forest — and also pollen of some subarctic species.

Considerable controversy has surrounded the age and duration of the nonglacial St. Pierre Interval (see discussion in Occhietti, 1982). The earliest radiocarbon date of ca. 11 ka led Gadd (1953) to correlate this nonglacial with the Allerod and Two Creeks Interval. This date, obtained when the dating method was in its infancy, was soon discounted by a large number of dates beyond the range of radiocarbon dating. The oldest apparently acceptable date is 74 700 +2700/-2000 BP (QL-198; Stuiver et al., 1978) obtained by isotopic enrichment procedures. This date agrees relatively well with thermoluminescence dates of ca. 61 ka (Lamothe, 1984a; Table 4.9). These old dates have led to the designation of the St. Pierre as an Early Wisconsinan interstade which lasted for only a few thousand years (Terasmae, 1958; Dreimanis and Karrow, 1972). Because only one till was found above the St. Pierre near the type section, it was hypothesized that ice advanced over the site in the Early Wisconsinan and did not retreat until the end of the Wisconsinan. Finite radiocarbon dates in the 28 to 38 ka

Figure 4.30. Sediments exposed at the Pierreville site: A, Pierreville varves; B, St. Pierre Sediments; C, lacustrine silts and sands; D, Gentilly Till. 200300-V

Table 4.10. Lithostratigraphic correlations

		Upper St. Lawrence Valley (MacClintock, 1958; Clark and Karrow, 1983)	Montréal area and southern Appalachian piedmont (LaSalle, 1981; Prichonnet, 1982a; Veillette and Nixon, 1984)	Central St. Lawrence Valley (Gadd, 1971; Occhietti, 1982; Lamothe, 1985; Parent, 1987)	Québec City and middle St. Lawrence estuary (Gadd, 1971; Karrow, 1957; LaSalle, 1984)	Southern Appalachians Eastern Townships, upper Beauce (McDonald and Shilts, 1971)	Bois-Francs, Beauce (Chauvin, 1979a)
QUATERNARY	Holocene	Lacustrine, fluvial, eolian, and organic sediments					
	Late Wisconsinan		Marine sediments				
		varved sediments	Lake Chambly varves	varved sediments		glacial Lake Vermont and Memphrémagog sediments	
		Trois-Rivières Stade					
		Till B (ex-Fort-Covington Till)	Saint-Jacques Till	Gentilly Till (several units)	Gentilly Till, Québec Till	Lennoxville Till	Thetford Mines Till
	Middle Wisconsinan	-varved sediment -intermediate till complex -varved sediment	Lac Châteauguay varves			Gayhurst Formation	Ruisseau Perry Formation (varved sediment)
	Early Wisconsinan	Till A (ex-Malone Till)	Ange-Gardien Till			Chaudière Till	Norbestos Till
		varved sediment		turbidites	Beaupré varves		
		St. Pierre Interstade		St. Pierre Sediments	Saint-François-du-Lac Formation, St. Pierre Sediments	St. Pierre, Anse aux Hirondelles, and Donnacona sediments	Massawippi Formation
	Early Wisconsinan and/or Sangamonian and Illinoian			Rivière aux Vaches Formation, Pierreville Formation, Deschaillons varves			
		Nicolet Stade		Pointe-Fortune old till	Bécancour Till Odanak Till	Bécancour Till, Pointe Saint-Nicolas Till	Johnville Till
					pre-Bécancour varved sediment		pre-Johnville sediments

range on wood and peat from units correlated with the St. Pierre (Table 4.5) have been used to suggest that the St. Pierre Interval extended through much of the Middle Wisconsinan. LaSalle (1984) argued that finite dates from the Québec City area should be disregarded because they were obtained on samples contaminated by modern rootlets and also that the Québec Laboratory (QU) dates and other dates determined by means of the benzene method are unreliable in this dating range. As indicated in Table 4.5, finite radiocarbon dates have been obtained for wood and peat enclosed in nonglacial sediments exposed on Île aux Coudres. Brodeur and Allard (1985) correlated these sediments with the St. Pierre Sediments and, although they were concerned about the validity of the finite dates, tentatively suggested that the St. Pierre Interval might have extended into the Middle Wisconsinan. A nonfinite age has now been accepted for these deposits (M. Allard, Département de géographie, Université Laval, personal communication 1987). Consequently the deposits on Île aux Coudres can be correlated with the St. Pierre Sediments without requiring a change in the generally accepted time span of the St. Pierre Interval.

Lacustrine episode. Stratified silts and clayey silts deposited in a lake (lower part Saint-François-du-Lac formation; Lamothe, 1985) and sediments containing thermophilous pollen at Les Vieilles-Forges (Clet and Occhietti, 1988) overlie the fluvial St. Pierre Sediments. Pollen is abundant in these sediments with 7.6% thermophilous trees (*Tsuga, Tilia, Carya*) at Les Vieilles-Forges (Clet and Occhietti, 1988). The stratified silts grade upward into sands deposited as shallow water fans at Pierreville (upper part of Saint-François-du-Lac formation; Lamothe, 1985) and to prodelta and delta sands at Les Vieilles-Forges (Occhietti, 1980). Going by the highest occurrence of these sediments at Les Vieilles-Forges, this lake reached a level of at least 36 m above present sea level (Occhietti, 1982). The possible extent of this lake is shown (Fig. 4.31), but the reason for this rise in water level is at present not known.

Environmental context of the St. Pierre Interval. During the St. Pierre Interval, the drainage system in which the sediments were deposited was graded to a base level that was between 5 and 27 m above present sea level. The maximum elevation is based principally on the elevation of organic sediments on Île aux Coudres. The paleoenvironmental context implies cool "interstadial"? conditions which may have occurred during the latter cool part of the Sangamonian or early part of the Wisconsinan (Occhietti, 1980, 1982). During these times world sea level was 13 to 18 m below present. The easiest way to explain the discrepancy between the high base level of the sediments and the probable low sea level is isostatic downwarping of St. Lawrence Valley due to an ice load on the Canadian Shield and possibly accentuated by residual downwarping of the possible earlier stade.

In conclusion, cool climate, sedimentation on the floor of a broad valley, isostatic depression, and low relative sea level are all conditions postulated for the St. Pierre Interval. Conditions were possibly similar to those of the early Holocene (ca. 8 ka). At that time the Champlain Sea had regressed from St. Lawrence Valley and the river system was superposed on the flat valley floor which was covered by an open spruce forest (Richard, 1977). Ice still covered about half of Quebec and world mean sea level was around -13 to -15 m, according to the curve of Clark (1980).

Last glacial maximum

The succession of glacial events following the St. Pierre Interval was referred to the Gentilly Stade by Gadd (1971) and renamed Trois-Rivières Stade by Occhietti (1982). This latter is the term used throughout the rest of this report. Table 4.11 illustrates the events that occurred in different parts of the subregion during this episode.

The Trois-Rivières Stade ended at the time of construction of the St. Narcisse Moraine ca. 10.8 ka (LaSalle and Elson, 1975; Occhietti, 1977b, 1982), but there is no consensus on when this stade began. If sedimentation of the lacustrine sediments at Les Vieilles-Forges, which ended deposition of St. Pierre Sediments and heralded advance of the subsequent ice sheet, occurred ca. 75 ka, then this stade began either in the Early Wisconsinan (Dreimanis and Karrow, 1972; Dreimanis, 1977) at the beginning of oxygen isotope stage 4 or in the Middle Wisconsinan during oxygen isotope stage 3 (see discussion in Occhietti, 1982).

The Trois-Rivières Stade is represented by a single till in Ottawa Valley, middle and lower St. Lawrence Valley, and at the margin of the Laurentian Highlands. This unit has been referred to as the Gentilly Till, the Québec Till, and the Matawin Formation (Tables 4.6A, 4.10). Locally this unit may include a till of local lithology at the base and deposits of limited distribution that suggest local glacial fluctuations (Gadd, 1971; Occhietti, 1977b, 1980), but in general it is interpreted to represent a single glacial advance and retreat. In the Appalachians, the Trois-Rivières Stade has three parts: an early glaciation which deposited the Chaudière Till, a period of recession related to the Gayhurst Formation, and a final glaciation that deposited the Lennoxville Till (Tables 4.6C, 4.10). The nature of this stade in the upper St. Lawrence Valley is uncertain because more than one till has been recognized in several places (Table 4.6C) but, because St. Pierre Sediments generally are not present, it is not known whether the lower till predates the St. Pierre or whether it is equivalent to the Chaudière Till.

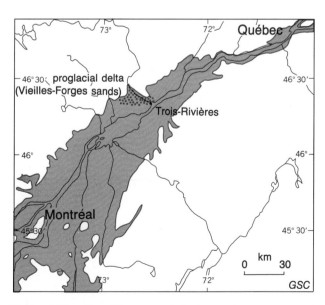

Figure 4.31. Possible extent of glacial lake in which Les Vieilles-Forges sands and silts were deposited at the end of the St. Pierre Interval.

CHAPTER 4

During the Trois-Rivières Stade, it is likely that the dispersal centre of Laurentide ice in Quebec shifted westward between the initiation of glaciation and the major final glacial advances (Occhietti, 1983). It is also possible that one or more Appalachian ice sheets, first autonomous and then coalescent, developed on the high plateaus of Quebec, Maine, and probably Vermont (Fig. 4.32). St. Lawrence Valley functioned as a channel way for a number of glacial lobes and ice streams which moved in different directions at different times. The general flow pattern is apparently as follows: upstream from Québec City early Laurentide ice flow was towards the southwest, up St. Lawrence Valley. Appalachian ice probably also entered the valley at this time flowing north in the east and west in the west (Fig. 4.32). At the time of glacial maximum, flow from the Canadian Shield extended across St. Lawrence Valley into and over the Appalachians, through Lake Champlain Valley, and into the Adirondacks of northern New York. During deglaciation, drawdown into a calving bay, in the Gulf of St. Lawrence, caused reversal of flow in the Appalachians and St. Lawrence Valley and a flow pattern influenced by local topography at the southern edge of the Canadian Shield.

Early Trois-Rivières and Chaudière glaciations

Prior to the maximum of the glacial advance that preceded the Gayhurst lacustrine episode, the Laurentide Ice Sheet covered the entire St. Lawrence Valley and probably abutted against an ice sheet advancing out of the Appalachians. Several till fabrics from the lower part of the Gentilly Till and striations, possibly of this age, suggest a southwesterly flow (up St. Lawrence Valley) for at least the early part of this episode (Gadd, 1971; Occhietti, 1977b; M. Lamothe, Université du Québec à Montréal, personal communication, 1984; Fig.4.32). There is, however, no defined till unit in central St. Lawrence Valley that clearly can be correlated with an early advance as distinct from the maximum Late Wisconsinan advance.

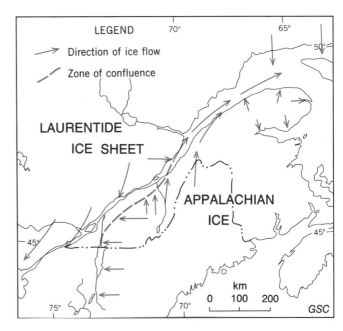

Figure 4.32. Ice flow during Early Wisconsinan buildup.

Table 4.11. Correlation of post-St. Pierre Interval events

Conventional Climatostratigraphy	Age	Southern Laurentians and central St. Lawrence Valley (Gadd, 1971; Occhietti, 1982)	Southern Quebec Appalachians (McDonald and Shilts, 1971; Lortie, 1975; Chauvin, 1979a, b; Shilts, 1981; Parent, 1987)	Upper St. Lawrence Valley, Montreal (MacClintock and Stewart, 1965; Prest and Hode-Keyser, 1977; Prichonnet, 1982a; Clark and Karrow 1983; LaSalle, 1984)	Vermont, Lake Champlain (Stewart and MacClintock, 1969)	Middle estuary, Lower St. Lawrence, Chaudière Valley (Gadd, 1978a; LaSalle et al., 1979; Martineau et Corbeil, 1983)
	6 ka					
Shawinigan Stade		?Shawinigan Stade				Goldthwait Sea
		Saint-Narcisse episode	Champlain Sea			
	11 ka		glacial Lake Vermont / glacial Lake Memphrémagog / Bois-Francs reversal	glacial Lake Chambly	glacial Lake Vermont	Notre-Dame Mountains reversal / Lake Chaudière Valley
Trois-Rivières Stade		Matawin Formation Till episode / Gentilly Till episode	Lennoxville Till episode, *Laurentide*	Till B (ex-Fort Covington Till) episode	Burlington Till episode	Lennoxville Till episode, *Laurentide*
			glacial Lake Gayhurst	glacial Lake Châteauguay	glacial lake	?
		?	Asbestos reversal / *Laurentide*	intermediary Till complex		Chaudière Till episode, *Appalachian*
		turbidite episode	Chaudière Till episode *Appalachian*	Till A (ex-Malone Till) / Ange-Gardien Till episode	Shelburne Till episode	
St. Pierre Interstade	40 or 70 ka ?	fluvial and lacustrine episode (St. Pierre)	(Massawippi)			(Vallée-Jonction)

In the Appalachians on the other hand, Chaudière Till overlies the Massawippi Formation (St. Pierre equivalent) and underlies glacial lake sediments (Gayhurst Formation), all of which are overlain by the Late Wisconsinan Lennoxville Till. This package of two tills and intervening stratified sediments is correlated with the single Gentilly Till of central St. Lawrence Valley because both apparently occupy the same stratigraphic position, that is, overlie deposits of the St. Pierre Interval and underlie postglacial deposits. Lower parts of the Chaudière Till apparently were deposited by ice flowing northwestward and westward from an ice cap in the upper Beauce and northern Maine (Fig. 4.32; McDonald, 1967; Shilts, 1976, 1981; Parent, 1984a). Later flow, however, was from the northwest suggesting that the Appalachian ice was overwhelmed by ice moving southward from the Canadian Shield into the Eastern Townships (McDonald and Shilts, 1971). In contrast with this, Parent (1984a, 1987), reported evidence for occupation only by Appalachian ice in part of this region.

Partial deglaciation episode

During an interval of undetermined age, which is normally placed before the last glacial maximum, a large part of the southern Appalachians in Quebec was apparently deglaciated (McDonald and Shilts, 1971; Fig. 4.33). At this time the Laurentide Ice Sheet apparently remained in central St. Lawrence Valley blocking drainage from the Appalachians and creating glacial Lake Gayhurst. This lake existed for at least 4 ka because this number of varves have been counted in the Gayhurst Formation (Shilts, 1981). This interval may be equivalent to the Plum Point Interstade, between 32 and 23 ka, or less probably to the Port Talbot Interstade, between 53 and 36 ka (Dreimanis and Karrow, 1972). A nonfinite date on disseminated organic matter of >20 ka (McDonald and Shilts, 1971) provides a minimum age for the Gayhurst Formation.

The southern Great Lakes basins are generally considered to have been ice free at this time (Karrow, 1984, 1989). Upper St. Lawrence Valley and the Lake Champlain depression, are also generally considered to have been ice free, but there is no direct stratigraphic proof of this. Deposits in this region that might be related to a period of Middle Wisconsinan deglaciation are the glaciolacustrine sediments intercalated between Shelbourne and upper Burlington tills in the Lake Champlain trough; Lac Châteauguay sediments south of Montréal (LaSalle, 1984); the sediments between the Ange-Gardien Till and the upper till in the Granby area (Prichonnet, 1984a); and varves lying in the upper part of the "middle till complex" of the Montréal area (Prest and Hode-Keyser, 1977). These correlations give a paleogeographic picture for this deglacial episode which may have been similar to that at the end of the Late Wisconsinan just prior to incursion of the Champlain Sea (Fig. 4.33).

Late Wisconsinan glacial maximum

The surface tills of the area (upper part of Gentilly, Lennoxville, and their equivalents, Tables 4.6A, C) are correlated with the Late Wisconsinan maximum and the deglaciation which followed. These units all contain evidence of a general flow from the Canadian Shield towards the southeast and south (Fig. 4.34a). During this advance Laurentide ice covered Gaspésie (David and Lebuis, 1985) and filled the Great Lakes basin (Karrow, this chapter, 1989); the ice margin lay off the coast of Maine and on Long Island. The ice may have reached its maximum extent on Long Island before 21.3 ka (Sirkin, 1981) and had begun to retreat from its margin in northern New Jersey by 18.5 ka (Cotter et al., 1985).

Two main factors, topography and development of a calving bay (or of a major ice stream) in the Gulf of St. Lawrence, influenced ice dynamics during retreat. The calving bay cut into the Laurentide Ice Sheet, producing a saddle over St. Lawrence Valley (Thomas, 1977; Lebuis and David, 1977; LaSalle et al., 1977a, b), and induced a northerly gradient on the part of the Laurentide Ice Sheet covering the Appalachians of Quebec and northwestern Maine (Lowell, 1985; Fig. 4.34b). This gradient caused a brief reversal of flow which transported little debris but produced an abundance of north and westward trending flow features in an area inside the Appalachian front (Lamarche, 1971, 1974; Gauthier, 1975; Lortie, 1976; Rencz and Shilts, 1980) reaching from Rivière Saint-François to Gaspésie. Ice thickness was appreciably greater over the valleys of the Appalachians than it was over the adjacent highlands. Consequently, late flow followed the valleys, and during downwasting, highlands appeared through the ice first, leaving thick stagnant ice tongues in the valleys (Gerath et al., 1985). This resulted in a complex pattern of ice marginal features, stagnation deposits, and local glacial lakes which are difficult to correlate from one valley to the next. This in turn has left considerable leeway for interpretation and has led to controversy about the paleogeography during deglaciation (Gadd, 1983, 1984; Parent, 1984b; Dubois et al., 1984).

As much as 5 ka may have passed between retreat from the terminal moraine and the start of the final deglaciation of the Appalachians of Quebec; this interval included the Erie Interstade, Port Bruce Stade, Mackinaw Interstade, and much of the Huron Stade (Karrow, 1989). Many recessional and advance moraines were built in Hudson-Champlain valley (Connally and Sirkin, 1973); the calving bay in the Gulf of St. Lawrence had separated ice in Gaspésie from the main ice sheet on the Quebec north shore; and ice flow in the Quebec Appalachians had reversed and

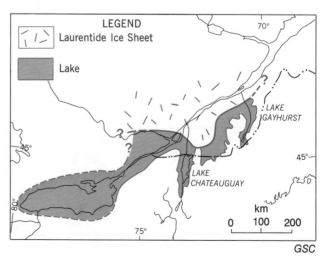

Figure 4.33. Paleogeography during maximum Middle Wisconsinan ice recession.

the Appalachian remnant had stagnated as this ice became too thin to sustain regional flow. After this time, ice retreat involved: extension of the Gulf of St. Lawrence calving bay, retreat of stagnant ice in the Appalachians, retreat of the ice from the Lake Ontario basin and the upper St. Lawrence Valley, and retreat across central St. Lawrence Valley.

Progress of the calving bay

Considerable controversy surrounds timing of development and the ultimate extension of the Gulf of St. Lawrence calving bay. Most workers agree that such a feature occupied St. Lawrence Valley as far upstream as Québec City but there is no unanimity on its speed and direction of propagation from that point. The chronology of this major feature of deglaciation is based on the earliest dates of marine incursion of various parts of the St. Lawrence system. These dates indicate that by 13.4 ka the calving bay had passed Sainte-Marthe-de-Gaspé and had reached Trois-Pistoles (Locat, 1977; David and Lebuis, 1985). By 12.4 ka it had passed Québec City and, according to Gadd (1980b) and Richard (1975b), reached Ottawa by 12.7 ka. This is, however, based on a controversial date from Clayton, Ontario (GSC-1859) which has been discounted by Karrow et al. (1975), Karrow (1981b), and Hillaire-Marcel (1981b); a subsequent accelerator date on this same material gave an age of 12 180 ± 90 BP (TO-245). In addition, Karrow (1981b) disputed the concept of a calving bay entering Ottawa Valley and, based on Great Lakes chronology, stated that marine waters could not have reached the western basin of the Champlain Sea before the end of Lake Iroquois (later than 12 ka). Additional discussion of this controversy is included in the *Chronology* subsection of the *Postglacial marine episode*.

Ice retreat in the Quebec Appalachians. As mentioned above, the ice retreat picture in the Appalachians was complex and lacks good chronological control. In the Upper Beauce region, ice contact features indicate a retreat of the glacial front towards the north-northwest, transverse to Chaudière Valley (Gadd, 1978a). West of this valley, on the low plateaus of the Beauce and Bois-Francs regions, the ice mass in the Appalachians appears to have acted independently and, according to glacial striation patterns, to have dissipated in a complex fashion (Fig. 4.34b; Lortie, 1976). This probably explains the ablation till in the Thetford Mines region (Chauvin, 1979b), which suggests in situ melting of a stagnant ice mass, and the absence of well differentiated recessional moraines as has been suggested by M. Parent (Geological Survey of Canada, personal communication, 1985). Farther south, in the Lac Mégantic area and in the Sherbrooke region of the Appalachians, the northward retreating ice front was oriented more or less east-west but was strongly controlled by local topography (Shilts, 1981; Fig. 4.23). Ice flowed towards the south-southeast and, between 13.5 and 12.5 ka, built the Cherry River Moraine (McDonald, 1967), Stoke Mountain interlobate moraine (McDonald, 1967; Clément and Parent, 1977), Ditchfield, Mégantic, La Guadeloupe (Shilts, 1981), and Tingwick-Ulverton (Parent, 1987) morainic systems. Following this time topography strongly controlled flow of the remaining ice masses, notably on the Appalachian piedmont to the east of the Lake Champlain-Rivière Richelieu axis (Prichonnet, 1977, 1982, b, 1984a). A series of small glacial lakes (Clément and Parent, 1977; Larocque et al., 1983a, b) and the larger glacial Lake Memphrémagog formed between the ice front and divides in west and north draining valleys during retreat of the glacier (McDonald, 1967; Boissonnault and Gwyn, 1983; Parent, 1984b).

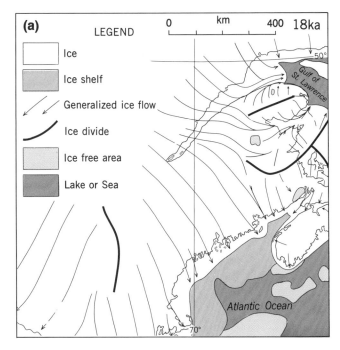

Figure 4.34a. Ice flow at Late Wisconsinan maximum.

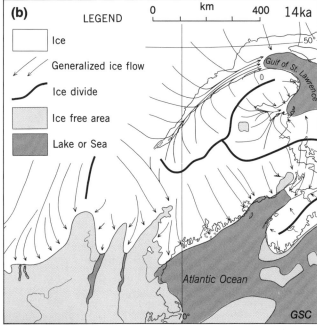

Figure 4.34b. Ice flow during inception of a calving bay in the Gulf of St. Lawrence.

Retreat from Lake Ontario basin and upper St. Lawrence Valley. Distinct patterns of flow features in the upper St. Lawrence Valley indicate that several ice lobes developed during deglaciation (Gadd, 1980a, b). One lobe flowed southward into the Lake Champlain-Hudson trough. Glacial lakes, draining southward into Hudson River, occupied the valley south of this lobe and expanded northward as the ice retreated (Connally and Sirkin, 1973; Fig. 4.35a). The last glacial lake was the Fort Ann phase of glacial Lake Vermont; this ended with invasion of the Champlain Sea which probably occurred ca. 12 ka (Cronin, 1979b).

Late glacial south and southwest flowing ice occupied upper St. Lawrence Valley between Montréal and the Frontenac Arch. This ice abutted against the Adirondack Mountains, to the south, precluding glacial lakes from this segment of the valley during most of deglaciation. In addition, it separated lakes in the Ontario basin from those in the Lake Champlain basin almost until the time of marine invasion.

A major flow of ice extended south from the vicinity of Ottawa and swung southwest into the Lake Ontario basin. As this lobe retreated, Lake Iroquois developed and expanded to the northeast (Muller and Prest, 1985). Lake Iroquois, which emptied east into Mohawk River at Rome, New York, was in existence ca. 12.5 ka. Ice retreat from the Covey Hill area permitted drawdown of Lake Iroquois and developed a single, southward draining, water body joining the Lake Ontario basin with the Lake Champlain basin. This confluent phase combined the Fort Ann phase of glacial Lake Vermont in the Champlain trough, Lake Chambly south of Montréal (LaSalle, 1981), and Belleville phase of the eastern Ontario basin (Fig. 4.35b). Anderson et al. (1985) used the presence of freshwater ostracodes, mostly *Candona subtriangulata*, in sediments overlying glacial deposits to suggest that this lake extended into the Ottawa area. Rodrigues and Richard (1986) are not certain that a single lake occupied the entire Ottawa-upper St. Lawrence region. The water level was lowered and the fresh water was replaced by marine when ice retreat in middle St. Lawrence Valley permitted the Champlain Sea to invade the area, probably about 12 ka.

To the west of Ottawa ice flowed southward across Ottawa Valley and over the Algonquin Highlands towards the Ontario basin (Fig. 4.35a). Early during retreat this ice separated from that in the Lake Ontario basin, and the Oak Ridges Moraine formed between the two lobes. When this ice thinned to the point where it could no longer overtop the Algonquin Highlands, flow was diverted southeastward to form a local Ottawa Valley ice lobe which left a series of small moraines as it retreated up valley from Renfrew (Gadd, 1980b; Barnett and Kennedy, 1987). The chronology of deglaciation of Ottawa Valley is not well dated but the Champlain Sea had reached Westmeath (20 km east of Pembroke) by 11 000 ± 160 BP (GSC-1664, Lowdon and Blake, 1979), and ice had receded from Ottawa Valley downstream from Lake Timiskaming so that drainage from the Great Lakes could use the North Bay outlet before 10.1 ka (Harrison, 1972).

Retreat from central St. Lawrence Valley. In central St. Lawrence Valley several of the Monteregian Hills became nunataks prior to 12.5 ka and the Champlain Sea apparently invaded the south shore of St. Lawrence Valley a few centuries later (Prichonnet, 1977). As the ice retreated across the lowlands there is no evidence of significant stillstands or readvances. Scattered glaciofluvial deposits and moraine ridges, however, suggest deposition was occurring at the margin of a north to northwest retreating arcuate ice

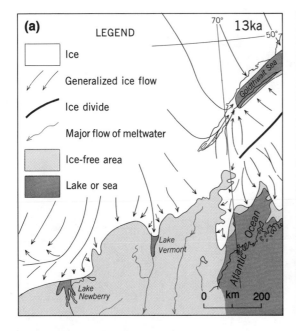

Figure 4.35a. Pattern of ice flow when front had receded to approximately Quebec-United States border.

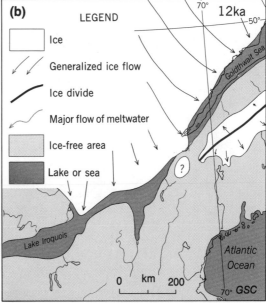

Figure 4.35b. Paleogeography immediately prior to invasion of Champlain Sea into upper St. Lawrence Valley.

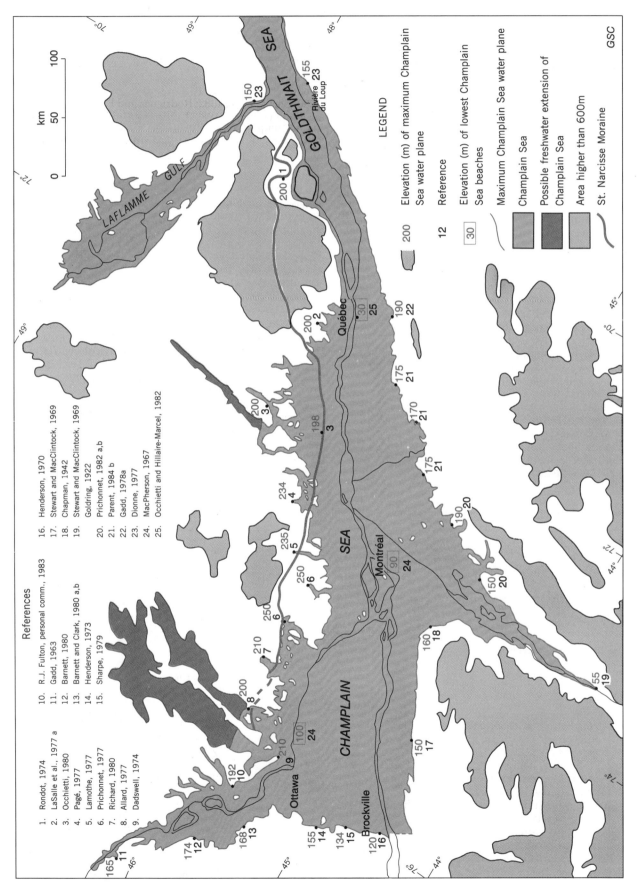

Figure 4.36. Extent and maximum water plane of the Champlain Sea.

front (Prichonnet et al., 1982b; Prichonnet, 1984b). Ice recession halted near the margin of the Canadian Shield north and northwest of the valley and at least locally readvanced, overriding marine sediments, to build the St. Narcisse Moraine (Parry and Macpherson, 1964; Fig. 4.36) between 11 and 10.6 ka (Occhietti, 1980). This period of moraine building and the following deglaciation has been referred to as the Shawinigan Stade (Occhietti, 1980; Table 4.11). According to LaSalle and Elson (1975) this moraine may represent a period of ice margin stability that was climatically controlled. Hillaire-Marcel et al. (1981) on the other hand referred to the St. Narcisse Moraine as a re-equilibrium feature formed where the ice became anchored following rapid retreat within the Champlain Sea. The ice left St. Lawrence Valley for the last time when it retreated from the St. Narcisse Moraine. The ice had apparently retreated from the St. Narcisse Moraine and from the Lac Saint-Jean area to the north, where marine waters had breached the St. Narcisse Moraine and occupied the Lac Saint-Jean basin by 10.3 ka (LaSalle and Tremblay, 1978; Vincent, 1989).

Postglacial marine episode

As described above, areas adjacent to the Gulf of St. Lawrence, St. Lawrence Valley, and lower Ottawa Valley were submerged by the sea at the time of the Late Wisconsinan deglaciation. The postglacial sea in the Gulf of St. Lawrence downstream from Québec City is referred to as the Goldthwait Sea. The ephemeral sea which occupied as much as 55 000 km² of valley system upstream from Québec City between 12 and 9.5 ka is called the Champlain Sea (Elson, 1969a; Fig. 4.36).

Lithostratigraphy

The sediments of the Champlain Sea have already been described in the section on marine sediments. The lithostratigraphic suite within the Champlain Sea basin includes sediments deposited during ice retreat, glaciolacustrine and glaciomarine sediments, followed by quiet or deep water sediments and by marine regression sediments. The glacial retreat and quiet water sediments are time transgressive from east to west and from south to north; the regression sequence is time transgressive from the margins towards the centre of the basin. Mixed facies of the glacial margin and of marine incursion were deposited in a narrow zone near the retreating ice front. Littoral and offlap sediments were deposited at the retreating margins of the sea as the basin shrank because of glacial isostatic uplift and sediment deposition.

Biostratigraphy

The Champlain Sea sediments contain a wide variety of marine vertebrates and invertebrates which attest to the variety of ecological niches present (Table 4.8). Cold water communities tolerant of salinity variations followed the retreat of the ice; cold euryhaline communities were associated with the main period of submergence; littoral, sublittoral, and brackish communities developed along the margins of the basin. This variety of communities is a complete reconsideration of the *Hiatella arctica* and *Mya arenaria* phases proposed by Elson and Elson (1959) which do not reflect the total biostratigraphy of the basin (Hillaire-Marcel, 1979; Occhietti, 1980; Rodrigues and Richard, 1983, 1985, 1986). The picture of evolving biostratigraphic zonation in response to environmental change in the basin as established by macrofossils (Hillaire-Marcel, 1980; Rodrigues and Richard, 1983), by ostracodes (Gunther and Hunt, 1977; Cronin, 1977a, b, 1979a, 1981; Rodrigues and Richard, 1986), by foraminifera (Cronin, 1977a, 1979 a, b; Guilbault, 1980; Corliss et al., 1982; Rodrigues and Richard, 1986), and by diatoms (Lortie, 1983; Lortie and Guilbault, 1984) are all similar. This evolution follows the sequence of type communities defined by Hillaire-Marcel (1980; Fig. 4.29, Table 4.8) and indicates the gradual change from ice proximal facies to deep water facies to shallow water facies.

Water conditions in the Champlain Sea

Analysis of the stable isotopes of ^{18}O and ^{13}C, supported by biostratigraphy, have permitted an analysis of the water conditions of the marine basin (Hillaire-Marcel, 1977, 1981a; Corliss et al., 1982). The closed basin of the Champlain Sea received substantial amounts of water from land and was characterized by steep vertical temperature and salinity gradients (Hillaire-Marcel, 1979). During the middle part of the marine episode, the water of the basin was stratified, with cold, saline water at depth and brackish water with seasonal temperature variations near the surface. Species living in the deep water, particularly *Balanus hameri* and *Portlandia arctica*, have high (positive) isotopic values (Fig. 4.37). Littoral shells, such as *Macoma balthica*, *Mya arenaria* and *Mytilus edulis*, on the other hand, have much lower (negative) isotopic values, which are related to temperature and to dilution of ocean water by glacial meltwater depleted in ^{18}O (Hillaire-Marcel, 1981a). As uplift occurred, the depth of water decreased and the proportion of fresh water increased. The general oceanographic conditions are interpreted as being similar to those of James Bay today, with salinity values of 10 to 30‰ and temperatures between -1 and $8°C$.

Chronology

Based on fossil-bearing diamicton at Petite-Matane (south shore of the St. Lawrence about 100 km northeast of Rimouski), Lebuis and David (1977) have postulated that a glaciomarine phase existed between Laurentide and Appalachian ice sheets in an area downstream from Trois-Pistoles. The age of the shells from this diamicton, and from shells in marine deposits at other sites between Petite-Matane and Trois-Pistoles-Saint-Fabien (13.4 and 13.4 ka,

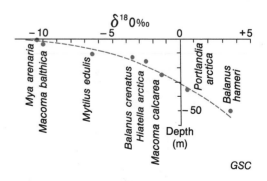

Figure 4.37. Oxygen isotope content of Champlain Sea shells related to water depth (Hillaire-Marcel, 1979).

Table 4.12; Locat, 1977) suggests inception of marine conditions in that area around 13.5 ka. The opening of the western arm of the Goldthwait Sea into the middle estuary, upstream from Trois-Pistoles, must have occurred by around 12.8 ka, according to a date from shells of *Portlandia arctica* from the Saint-Antonin Moraine at Trois-Pistoles (Table 4.12; Lee, 1962).

The timing of invasion of the Champlain Sea is controversial. Radiocarbon dates of 12.7 to 12.1 ka on marine shells in the Ottawa area (Table 4.12; Rodrigues and Richard, 1986), a shell date of 12.5 ka from near Saint-Dominique (Prichonnet, 1982a), and two dates on gyttja from Mont Saint-Bruno (LaSalle, 1966; Terasmae, 1968) are from 200 to 800 years older than dates on similar materials from surrounding areas. Gadd (1980b) proposed rapid extension of a calving bay, limited to the deepest part of the valley, as a means of getting marine waters into a relatively limited area near Ottawa at a time significantly earlier than they arrived in the Montréal and the upper St. Lawrence area. Hillaire-Marcel et al. (1979) and Karrow (1981b) proposed that the ^{14}C equilibrium of the Champlain Sea waters was upset by incorporation of old carbon derived from meltwaters, and consequently the shells provide dates older than their actual ages. It is also proposed that aging of the sea water during the 400 to 600 km flow from the Atlantic into the closed basin may in part be responsible for "old" dates (Mangerud and Gulliksen, 1975; Hillaire-Marcel, 1981b).

It is the author's view that marine incursion at ca. 12 ka best fits available evidence. Several authors have proposed extensive development of glacial lakes in the southern part of Ottawa and upper St. Lawrence valleys prior to Champlain Sea incursion (Prest, 1970; Anderson et al., 1985; Parent et al., 1985; Muller and Prest, 1985). They placed the ice margin at the time of invasion on a line running approximately from Ottawa to Bois-Francs (Fig. 4.35b). Anderson (1987) used pollen stratigraphy of Champlain Sea sediments to suggest that the Champlain Sea probably reached the Ottawa area about 11.7 ka. In addition, Karrow (1981b) and Karrow et al. (1975) argued that the Champlain Sea formed in the upper St. Lawrence after Lake Iroquois had drained, and because Lake Iroquois probably did not drain until shortly after 12 ka, the Champlain Sea could not have reached Ottawa earlier. Dates on shells from the upper St. Lawrence-Lake Champlain valleys suggest that the Champlain Sea reached its southern limit about 12 ka (Table 4.12).

Table 4.12. Radiocarbon ages for marine shells referred to in text. (A more complete list of dates for the western basin of the Champlain Sea may be found in Rodrigues and Richard, 1985.)

Area	Age years BP	Laboratory number	Species	Elevation m	Locality	Reference	Comments
Lower estuary	13 360 ± 320	QU-264	Hiatella arctica	98-126	Saint-Donat, Quebec	Locat, 1977	
	13 390 ± 690	QU-271	Hiatella arctica	138-155	Saint-Fabien, Quebec	Locat, 1977	
Middle estuary	12 720 ± 170	GSC-102	Portlandia arctica	167	Trois-Pistoles, Quebec	Gadd et al., 1972a	Minimum age for Goldthwait Sea in middle estuary
Québec City	12 400 ± 160	GSC-1533	Portlandia arctica	109	Charlesbourg, Quebec	Gadd et al., 1972a	Beginning of Champlain Sea in Québec City area
	12 230 ± 250	QU-93			Saint-Henri-de-Lévis, Quebec	LaSalle, in Richard, 1978b	
South shore of central St. Lawrence Valley	12 000 ± 230	GSC-936	several species	121	L'Avenir, Quebec	Lowdon and Blake, 1970	Beginning of Champlain Sea in main basin
	12 480 ± 240	QC-475	Mya sp.	90	Saint-Dominique, Quebec	Prichonnet, 1982a	Date not consistent with species and elevation: new date 11 250 ± 100 BP, UQ-1429 (Occhietti, unpub.) on Mya arenaria same site but slightly different position
Western part of Champlain Sea basin	12 700 ± 100	GSC-2151	Macoma balthica	168	Clayton, Lanark County, Ontario	Richard, 1978b	Date verified twice but on same material as T0-245 (below)
	12 180 ± 90	T0-245	Macoma balthica	168	Clayton	unpublished	
	12 200 ± 160	GSC-1646	Macoma balthica	192	Cantley, Gatineau County, Quebec	Lowdon and Blake, 1973	
	12 100 ± 100	GSC-3110	Macoma balthica	170	White Lake, Renfrew County, Ontario	Rodrigues and Richard, 1983	
	12 000 ± 200	—			Massena, New York	Kirkland and Coates, 1977	Lab number not available
	11 900 ± 200	GSC-1772	Macoma balthica	176	Martindale, Gatineau County, Quebec	Lowdon and Blake, 1973	
	11 900 ± 120	GSC-2338	Macoma balthica	101	Peru, New York	Lowdon and Blake, 1979; Cronin, 1979b	
	11 900 ± 100	GSC-3767	Portlandia arctica	76	Sparrowhawk Point, New York	Rodrigues and Richard, 1985	
	11 800 ± 210	GSC-1013	several species	104	Maitland, Ontario	Lowdon and Blake, 1970	
	11 800 ± 100	GSC-3523	Macoma balthica	120	Merrickville, Grenville County, Ontario	Rodrigues and Richard, 1985	
	11 800 ± 100	GSC-2366	Macoma balthica	96	Plattsburg, New York	Lowdon and Blake, 1979; Cronin, 1979b	
North shore of central St. Lawrence Valley	11 300 ± 160	GSC-1729	Portlandia arctica	81	Rivière la Fourche, Quebec	Occhietti, 1976	Glaciomarine sediments predating St. Narcisse event
Québec	9 355 ± 185	UQ-64	Hiatella arctica	64	Saint-Nicolas	Occhietti and Hillaire-Marcel, 1982	LaSalle (1984) reported dates as young as 9730 ± 190 BP (GSC-1726), from this same site

The oldest marine shell date at Yamachiche on the north side of central St. Lawrence Valley is 11.3 ka (Occhietti, 1980). The distance across valley from Bois-Francs to Yamachiche is about 65 km. If the Champlain Sea breached the ice dam at Bois-Francs at 12 ka, then 700 years were required for retreat across the valley (a rate of ca. 90 m/a).

Isostatic uplift caused a progressive shoaling which in turn pushed the lower highly saline waters from the Champlain Sea basin. This led to the formation of a shallow lake upstream from Québec ca. 10 ka (*Lampsilis* Lake Phase of Elson, 1962, 1969b) but the youngest marine fauna in the Québec region at Saint-Nicolas has provided dates as young as 9.4 ka (Occhietti and Hillaire-Marcel, 1982).

Marine limits and emergence

The age and level of marine limit vary throughout the basin but there is insufficient chronological information to establish consistent patterns (Fig. 4.36). About the only general comment that can be made is that the highest limits are on the northern margin of the basin, nearest the centre of ice loading.

Old shorelines, marine terraces, and terrace scarps provide evidence that emergence rates were variable and that uplift included periods of little relative sea level change. Unfortunately few of the small number of regional emergence curves available are complete (Elson, 1969a; Dionne, 1972a; Hillaire-Marcel, 1974, 1979; Locat, 1977; Fulton and Richard, 1987; Fig. 4.38). What data are available suggest that emergence was at times rapid, reaching rates of 115 m/ka.

Postglacial and postmarine episode

The beginning of the postglacial nonmarine history of St. Lawrence Valley and the adjacent Appalachians is time transgressive. It began as areas were deglaciated and as glacial lakes and postglacial seas drained. The ice disappeared from areas of the Appalachians as early as 13 ka but remained in some valleys north of the St. Narcisse Moraine until after 10.5 ka. Glacial lakes in the area drained when ice retreat opened St. Lawrence Valley to the Champlain Sea ca. 12 ka. Crustal rebound caused emergence of marine inundated areas, beginning at the time of marine submergence of the Gaspésie coast about 13.5 ka and ending with retreat of marine waters from the Québec City area before 9 ka.

Vegetation colonization and evolution

The postglacial history of the vegetation of St. Lawrence Valley and adjacent Appalachians has been reconstructed by analysis of pollen and macrofossils from the bottom of lakes and peat bogs (Potzger and Courtemanche, 1956; LaSalle, 1966; Terasmae and LaSalle, 1968; Richard, 1971, 1973, 1975a, 1978a; Richard and Poulin, 1976; Mott, 1977; Savoie and Richard, 1979; Mott and Farley-Gill, 1981; Anderson, 1987; Table 4.13). Plant colonization prior to the Holocene was influenced by the proximity of the ice front, by the dampening effect that the large, cold Champlain Sea had on climate, by the migration barrier formed by the Champlain Sea, and to a lesser degree by changes in climate. Subsequent vegetation development depended on climate and regional and local conditions.

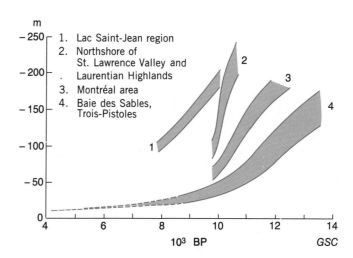

Figure 4.38. Shoreline relation diagrams for Champlain Sea.

The same general vegetation sequence may be observed in all regions. The Mont Shefford succession (Richard, 1978a) may be taken as an example of southern regions. It comprises a periglacial desert at the base, then a shrub tundra which seems to have lasted 400 years and ended 11 ka. Afforestation led to open spruce woods, dominated by *Picea mariana* and *Populus tremuloides*, lasting about a thousand years. This was succeeded by a spruce forest including white birch, which persisted from the early Holocene to around 7.5 ka, followed by a climax maple forest. Despite what is indicated by this example, locally the tundra period may have lasted for some time, due to the persistence of stagnant ice, the presence of glacial lakes, and the slow migration of vegetation (Richard, 1977). The modern climax vegetation zones in St. Lawrence Valley and the adjacent Appalachians are distributed according to a latitudinal gradient, ranging from maple forest with hickory in the south, to maple forest with basswood, to maple forest with yellow birch and, farther north, spruce forest with yellow birch and white birch (Grandtner, 1966).

Landforms and sediments related to evolution of drainage

Since ice retreat, drainage systems have become incised. In areas where streams have been superimposed on Quaternary sediments, series of terraces have been formed. Where streams have been superimposed on bedrock, channels cut in rock with abundant rapids and waterfalls have resulted.

In the southern Quebec Appalachians region, notable fluvial terraces have been formed in the Chaudière and Saint-François river valleys. Charny Falls on Chaudière River is one example of a locality where the river has been superimposed on rock. In other places old valleys, such as the buried valley of Rivière Eaton northeast of Sherbrooke (Dubois, 1973), have not been reoccupied by streams.

In the Montréal area of St. Lawrence Valley, Lampsilis Lake (Elson, 1962) developed as the Champlain Sea regressed. This lake existed ca. 9.8 ka (Richard, 1978b) and was part of an extensive, evolving lacustrine and fluvial system that was slowly but continually downcutting and prograding seaward. Present day lakes Saint-Louis, Saint-François, and Saint-Pierre are lake basin remnants of this system. Extensive terrace systems developed as the lakes were filled and paleochannel systems were left behind by incising streams (MacPherson, 1967). Peat bogs and marshes developed in many of the abandoned channels. Dating of the lowest organic bed of peat bogs on high terraces puts the minimum age for marine regression and high terrace abandonment in central St. Lawrence Valley at ca. 9.5 ka (Terasmae, 1960b).

Extensive and locally thick sandy and silty fluvial sediments overlie marine deposits in lower Ottawa Valley (Gadd, 1986). These sediments were deposited in a deltaic system that prograded from Petawawa through to St. Lawrence River. The deposited materials were of several origins: glacial meltwaters flowing directly from the receding ice front, drainage from the glacial Great Lakes, drainage from glacial lakes to the north (Vincent, 1989), and erosion of marine and deltaic sediments exposed as isostatic uplift raised recently deposited sediments above base level. *Lampsilis* shells locally occurring near the top of this succession indicate that freshwater conditions dominated in later periods of delta building (Rodrigues and Richard, 1983). Dates on *Lampsilis* from Ottawa Valley suggest freshwater conditions in that area as early as ca. 10.3 ka. Marine shells at higher elevations in this same area have, however, been dated as young as 10 ka and consequently there appears to be a problem with shell dates in this region (Fulton and Richard, 1987).

Several major deltas are still being formed in the region. Several rivers join the St. Lawrence, at the head of Lac Saint-Pierre and have built a delta consisting of several islands. Rivière Saint-Maurice is building a delta of three islands at its mouth, from which the city of Trois-Rivières takes its name. In addition at low tide, mud flats as much as 1000 m wide occupy the shore zone of the present middle estuary of the St. Lawrence and clay, silt, and sandy mud are being transported within the St. Lawrence estuary (D'Anglejean, 1971, 1981). All shores of the present system of lakes and rivers are affected by the "glaciel" phenomena (Dionne, 1969). Spring ice breakup sweeps the banks bare of vegetation and the ice transports blocks and other debris downstream.

Periglacial processes

Ice-wedge casts, cryoturbation, and niveo-eolian accumulations have been identified in the Appalachians and St. Lawrence Valley (Hamelin, 1961; Gangloff, 1970, 1973; Dionne, 1971; Cailleux, 1972). These structures and accumulations are characteristic of periglacial processes developed under conditions in which there was discontinuous permafrost. This activity is compatible with the rigorous climate, indicated by the presence of tundra vegetation as late as 11 ka on Mont Shefford and 10 ka in the Charlevoix area (Richard and Poulin, 1976; Richard, 1978).

Mass movements

During recession of the sea and during development of drainage channels in the Champlain Sea basin massive landslides occurred in saturated marine clays (Fransham and

Table 4.13. Late glacial and Holocene pollen assemblage zones (Richard, 1977)

Age ka	Southern Laurentians				Lowlands			Age ka
	PAZ	Sub-PAZ	Taxon guide		PAZ	sub-PAZ	Taxon guide	
0 0.3	Betula-Pinus	(-*Ambrosia*)			Betula-Pinus / Betula-Picea (in peat bogs)	(-*Ambrosia*)		0 0.4
4		-*Fagus*				-*Fagus*		4
		-*Tsuga*	*Acer saccharum*	4.8 minimum *Tsuga*		-*Tsuga*	*Acer saccharum*	
5						*Betula-Pinus*		5
	Pinus-Betula	-*Tsuga*				(-*Tsuga*)		
7		-*Quercus*				(-*Quercus*)		7
		-*Picea*				(-*Picea*)		
		Betula arborescent — *Picea*			Pinus-Betula or Pinus (west and central)	*Betula* arborescent — *Picea*		
		Betula arborescent — (*Alnus crispa*)				(*Picea*)		10
10		*Populus* — *Picea* — *Juniperus*				(Cyperaceae)		
		Betula (shrub)						
		Cyperaceae-Gramineae						
11.5		Trees — minimal pollen						

( ) may be absent
PAZ pollen assemblage zone

Gadd, 1976). Downcutting enhanced by glacial isostatic uplift, steadily deepened channels and gullies causing slope failures and many individual and multiphase landslides (Karrow, 1972). The triggering of landslides in these thixotropic clays can be caused by several different means. Earthquakes triggered the 1663 landslides in Saint-Maurice Valley described in the Jesuit Relations and confirmed by ^{14}C dating (Desjardins, 1980). Heavy spring rains were a major factor in the Saint-Alban landslide (on Rivière Sainte-Anne) in 1894 (Chalmers, 1900; Occhietti et al., 1975). The catastrophic Nicolet landslide in 1955 showed that landslides constitute a permanent and continuing threat in the densely populated St. Lawrence Valley (Hurtubise et al., 1957).

In the Quebec Appalachians, small slumps and solifluction flows occur in the soft rocky slopes and debris slopes following periods of heavy rain. These are insignificant on a regional scale but locally cause road maintenance problems.

Eolian phenomena

Immediately following glacial retreat or emergence, and before vegetation colonization, the surface of unconsolidated deposits was vulnerable to eolian processes. Osborne (1950) and Clark and Elson (1961) reported finding ventifacts, which probably date from this time, on the top of the St. Narcisse Moraine at Mont-Carmel. All the high and middle terraces and perched deltas of St. Lawrence Valley are covered with stabilized dunes. Parabolic dune systems have formed "crête de coq" structures on the south (Gadd, 1971; Dubé, 1971) and north shores of the St. Lawrence (Occhietti, 1980). The deflation areas are at present commonly filled in by peat bogs. Niveo-eolian sediments (Cailleux, 1972) and loess have been deposited by eolian processes in the lower Saint-Maurice region (Occhietti, 1980) and in the Montréal region (Gangloff, 1973).

Earthquakes and neotectonics

St. Lawrence Valley is frequently affected by earthquakes of low, medium or, rarely, high magnitude (Basham et al., 1979). These shocks are concentrated in the area of the Charlevoix Astrobleme, in the area of Montréal, and in Gatineau Valley. It has been suggested that the Charlevoix activity may be related to residual stress release associated with a Devonian impact feature (Rondot, 1968). In addition, recent vertical crustal movements have been proposed for the southern Laurentians (Gale, 1970). There is other evidence of neotectonic, possibly residual glacial isostatic uplift, in the area. For example, we know that Samuel de Champlain would not now be able to launch his ships in Rivière Saint-Charles (at Québec), as he was able to in 1603 and that postglacial faulting has been reported in southern Quebec (Oliver et al., 1970).

ACKNOWLEDGMENTS

This chapter has been critically reviewed by W.H. Mathews (University of British Columbia), J.B. Bird (McGill University), N.R. Gadd, V.K. Prest (Geological Survey of Canada), P. LaSalle (Ministère de l'Énergie et des Ressources), J.A. Elson (McGill University), G.P. Prichonnet (Université du Québec à Montréal), E.V. Sado and P.J. Barnett (Ontario Geological Survey).

Table 4.14. An alternative interpretation of central St. Lawrence Lowlands stratigraphy.

Approximate age (Ka)	Oxygen Isotope Stages	Chronostratigraphic Classification (as used in this volume)		Climatic Interpretation	Stratigraphic units
	1	Holocene			
13	2	W I S C O N S I N A N	Late	Cool	Gentilly Till (with stratified interbeds)
32	3		Middle	Cold	
65				Cool	
	4		Early		
80					
	5a	S A N G A M O N I A N		Cool	St. Pierre Sediments St. Francois-du-Lac Formation[2] silts and sands at Les Vieilles Forges[3]
	5b			Cold	unnamed till?[1] Lévrard Till[2]
	5c			Cool	Deschaillons Varves[1] pre-Deschaillons sand[1]
	5d			Cold	
130	5e			Warm	lower sand at Pointe-Fortune[4]
	6	Illinoian			unnamed varves[1] and clay[4,5] Bécancour Till?[1] lower till at Pointe-Fortune[5]

(1) Lamothe, 1987
(2) Lamothe, 1985
(3) Occhietti et al, 1987
(4) Anderson, T.W., Matthews, J.V.Jr. and Mott, R.J., personal communication 1987
(5) Veillette and Nixon, 1984

Editor's Note. In the time between completion of the first draft of this chapter (1984) and final editing (1987), a number of papers were published which questioned some of the earlier Quaternary stratigraphy. Some of this preliminary work has been incorporated into the chapter and Table 4.14 has been added to give a clear picture of new ideas on the Quaternary stratigraphy of the central St. Lawrence Lowlands.

REFERENCES

Allard, M.
1977: Le rôle de la géomorphologie dans les inventaires biophysiques: l'exemple de la région Gatineau-Lièvre; thèse de PhD, non publiée, McGill University, Montréal, 274 p.

American Commission on Stratigraphic Nomenclature
1961: Code of Stratigraphic Nomenclature; American Association of Petroleum Geologists, Bulletin, v. 45, p. 645-660.

Anderson, T.W.
1987: Terrestrial environments and age of the Champlain Sea based on pollen stratigraphy of the Ottawa-Lake Ontario region; in Quaternary Geology of the Ottawa Region, Ontario and Quebec, R.J. Fulton (ed.); Geological Survey of Canada, Paper 86-23.

Anderson, T.W. and Lewis, C.F.M.
1985: Postglacial water-level history of the Lake Ontario basin; in Quaternary Evolution of the Great Lakes, P.F. Karrow and P.E. Calkin (ed.); Geological Association of Canada, Special Paper 30, p. 231-253.

Anderson, T.W., Matthews, J.V., Jr., Mott, R.J., and Richard, S.H.
1987: Stratigraphy, pollen, plant, and insect macrofossils at Pointe-Fortune site, Ontario: A Possible St. Pierre Interstadial equivalent; XII INQUA Congress, Ottawa, 1987, Program with abstracts, National Research Council of Canada, Ottawa, p. 121.

Anderson, T.W., Mott, R.J., and Delorme, L.D.
1985: Evidence for a pre-Champlain Sea glacial lake phase in Ottawa valley, Ontario, and its implications; in Current Research, Part A, Geological Survey of Canada, Paper 85-1A, p. 239-245.

Antevs, E.
1928: The last glaciation; American Geographical Society, Research Series no. 17, 292 p.

Ashworth, A.C.
1977: A late Wisconsinan coleopterous assemblage from southern Ontario and its environmental significance; Canadian Journal of Earth Sciences, v. 14, p. 1625-1634.

Banerjee, I.
1973: Part A: Sedimentology of Pleistocene glacial varves in Ontario, Canada; Geological Survey of Canada, Bulletin 226, p. 1-44.

Barnett, P.J.
1978: Quaternary geology of the Simcoe area, southern Ontario; Ontario Division of Mines, Geoscience Report 162, 74 p.
1979: Glacial Lake Whittlesey: the probable ice frontal position in the eastern end of the Erie basin; Canadian Journal of Earth Sciences, v. 16, p. 568-574.
1980: Quaternary geology of Fort Coulonge (31F/15) area, Renfrew County, Ontario; Ontario Geological Survey, Preliminary Map 2367, scale 1:50 000.
1982: Quaternary geology of the Tillsonburg area, southern Ontario; Ontario Geological Survey, Report 220, 87 p.
1984: Glacial stratigraphy and sedimentology central north shore area, Lake Erie, Ontario; Geological Association of Canada, Annual Meeting, London, Guidebook for Field Trip no. 12, 42 p.

Barnett, P.J. and Clarke, W.S.
1980a: Quaternary geology of the Renfrew area, Renfrew County; Ontario Geological Survey, Preliminary Map 2365, scale 1:50 000.
1980b: Quaternary geology of the Cobden (31F/10) area, Renfrew County, Ontario; Ontario Geological Survey, Preliminary Map 2366, scale 1:50 000.

Barnett, P.J. and Kennedy, C.C.
1987: Deglaciation, marine inundation and archeology of the Renfrew-Pembroke area; in Quaternary Geology of the Ottawa Region and Guides for Day Excursions, R.J. Fulton (ed.); INQUA XIIth International Congress, Ottawa, National Research Council.

Basham, P.W., Weichert, D.H., and Berry, M.J.
1979: Regional assessment of seismic risk in eastern Canada; Seismological Society of Canada, Bulletin, v. 69, p. 1567-1602.

Bayfield, H.W.
1837: Notes on the geology of the north coast of the St. Lawrence; Geological Society of London, Transactions (2), v. 5, p. 89-102.

Berger, G.W.
1984: Thermoluminescence dating studies of glacial silts from Ontario; Canadian Journal of Earth Sciences, v. 21, p. 1393-1399.

Berti, A.A.
1975: Paleobotany of Wisconsinan interstadials, eastern Great Lakes region, North America; Quaternary Research, v. 5, p. 591-619.

Bigsby, J.J.
1829: A sketch of the topography and geology of Lake Ontario; Philosophical Magazine, new series, v. 5, p. 1-15, 81-87, 263-274, 339-347, 424-431.

Bird, B.J.
1972: The Natural Landscapes of Canada, a Study in Regional Earth Sciences; Wiley, Toronto, 191 p.

Blanchard, R.
1947: Le Centre du Canada français, Province de Québec; Beauchemin, Montréal, 577 p.

Bleuer, N.K.
1974: Buried till ridges in the Fort Wayne area, Indiana, and their regional significance; Geological Society of America, Bulletin, v. 85, p. 917-920.

Boissonnault, P. et Gwyn, Q.H.J.
1983: L'évolution du lac proglaciaire Memphrémagog, sud du Québec; Géographie physique et Quaternaire, vol. 37, p. 197-204.

Bouchard, M. et Godard, A.
1984: Les altérites du bouclier canadien: premier bilan d'une campagne de reconnaissance; Géographie physique et Quaternaire, vol. 38, p. 149-163.

Boutray, B. de
1975: Minéralogie et pétrographie des dépôts meubles pléistocènes de la région de Montréal; Revue de géographie de Montréal, vol. 29, p. 347-356.

Boyko-Diakonow, M.
1979: The laminated sediments of Crawford Lake, southern Ontario, Canada; in Moraines and Varves, C. Schluchter (ed.); Balkema, p. 303-308.

Brodeur, D. et Allard, M.
1985: Stratigraphie et Quaternaire à l'Île aux Coudres, estuaire moyen du Saint-Laurent, Québec; Géographie physique et Quaternaire, vol. 39, p. 183-198.

Broecker, W.S. and Farrand, W.R.
1963: Radiocarbon age of the Two Creeks Forest Bed, Wisconsin; Geological Society of America, Bulletin, v. 74, p. 795-802.

Brookfield, M.E., Gwyn, Q.H.J., and Martini, I.P.
1982: Quaternary sequences along the north shore of Lake Ontario; Oshawa-Port Hope; Canadian Journal of Earth Sciences, v. 19, p. 1836-1850.

Brown, B.M., McKay, G.A., and Chapman, L.J.
1968: The climate of southern Ontario; Canada Department of Transport, Climatological Studies no. 5, 50 p.

Burgis, W.A. and Eschman, D.F.
1981: Late-Wisconsinan history of northeastern Lower Michigan; Midwest Friends of the Pleistocene, 30th Annual Field Conference Guidebook, Ann Arbor, Michigan, 110 p.

Burwasser, G.J.
1977: Quaternary geology of the City of Thunder Bay and vicinity; Ontario Geological Survey, Report 164, 70 p.

Cailleux, A.
1972: Les formes et dépôts nivéo-éoliens actuels en Antarctique et au Nouveau-Québec; Cahiers de géographie de Québec, n° 39, p. 377-409.

Calkin, P.E. and Brett, C.E.
1978: Ancestral Niagara River drainage: stratigraphic and paleontologic setting; Geological Society of America, Bulletin, v. 89, p. 1140-1154.

Calkin, P.E. and Feenstra, B.H.
1985: Evolution of the Erie-basin Great Lakes; in Quaternary Evolution of the Great Lakes, P.F. Karrow and P.E. Calkin (ed.); Geological Association of Canada, Special Paper 30, p. 149-170.

Chalmers, R.
1898: Report on the surface geology and auriferous deposits of southeastern Quebec; Geological Survey of Canada, Annual Report, v. 10, Part J, 160 p.
1900: Notes on the Pleistocene marine shore-lines and landslips (sic) of the north side of the St. Lawrence Valley; Geological Survey of Canada, Annual Report, v. 11, Part J, Appendix 1, p. 63-70.

Chapman, D.H.
1942: Late glacial and post-glacial history of the Champlain Valley; American Journal of Science, 5th series, v. 34, p. 89-124.

Chapman, L.J. and Dell, C.I.
1963: Revisions in the early history of the retreat of the Wisconsinan glacier in Ontario based on the calcite content of sands; Geological Association of Canada, Proceedings, v. 15, p. 103-108.

Chapman, L.J. and Putnam, D.F.
1951: The Physiography of Southern Ontario; Ontario Research Foundation, Toronto, 284 p.
1966: The Physiography of Southern Ontario (second edition); Ontario Research Foundation, Toronto, 386 p.
1984: The Physiography of Southern Ontario (third edition); Ontario Geological Survey, Special Volume 2, 270 p.

Chauvin, L.
1979a: Géologie des dépôts meubles, région d'Asbestos-Disraeli, rapport préliminaire; Ministère de l'Énergie et des Ressources du Québec; DPV-716, 10 p.
1979b: Dépôts meubles de la région de Thetford Mines-Victoriaville; Ministère des Richesses naturelles du Québec, DPV-622, 20 p.

Cheel, R.J.
1982: The depositional history of an esker near Ottawa, Canada; Canadian Journal of Earth Sciences, v. 19, p. 1417-1427.

Churcher, C.S. and Karrow, P.F.
1977: Late Pleistocene muskox (Ovibos) from the Early Wisconsin at Scarborough Bluffs, Ontario; Canadian Journal of Earth Sciences, v. 14, p. 326-331.

Cimon, J.
1969: Étude de la kaolinisation d'une anorthosite à Château-Richer, comté de Montmorency, Québec; thèse de maîtrise en sciences non publiée, Département de géologie, Université Laval, Québec, 82 p.

Clark, J.A.
1980: A numerical model of worldwide sea level changes on a visco-elastic earth since 18,000 BP; in Earth Rheology and Isostasy, N.A. Mörner(ed.); Wiley and Sons, New York, p. 175-200.

Clark, P. and Karrow, P.F.
1983: Till stratigraphy in the St. Lawrence Valley near Malone, New York. Revised glacial history and stratigraphic nomenclature; Geological Society of America, Bulletin, v. 94, p. 1306-1318.
1984: Late Pleistocene water bodies in the St. Lawrence Lowland, New York, and regional correlations; Geological Society of America, Bulletin, v. 95, p. 805-813.

Clark, R.H. and Persoage, N.P.
1970: Some implications of crustal movement in engineering planning; Canadian Journal of Earth Sciences, v. 7, p. 628-633.

Clark, T.H. and Elson, J.A.
1961: Ventifacts and eolian sand at Charette, P.Q.; Royal Society of Canada, Transactions, v. 55, series III, p. 1-11.

Cleaves, A.B.
1963: Engineering geology characteristics of basal till, St. Lawrence Seaway project; P.D. Trask and G.A. Kiersch (ed.); Engineering Geology Case Histories, no. 4, Geological Society of America, p. 51-57.

Clément, P. and De Kimpe, C.R.
1977: Geomorphological conditions of gabbro weathering at Mount Mégantic, Quebec; Canadian Journal of Earth Sciences, v. 14, p. 2262-2273.

Clément, P. et Parent, M.
1977: Contribution à l'étude de la déglaciation wisconsinienne dans le centre des Cantons de l'Est, Québec; Géographie physique et Quaternaire, vol. 31, p. 217-228.

Clet, M. et Occhietti, S.
1988: Palynologie des sediments attribués à l'intervalle nonglaciaire de Saint-Pierre, Québec, Canada: étude préliminaire; Actes du symposium de l'Association des palynologues de langue française, Institut française de Pondichéry, Travaux de la section scientifique et technique, T. XXV, p. 185-196.

Clet, M., Occhietti, S., et Richard, P.
1986: Stratigraphie et palynologie du Wisconsinien de la coupe de Donnacona, Québec; ACFAS, 54e congrès, Recueil des resumés de communications, Annales de l'ACFAS, vol. 54, p. 216.

Coakley, J.P.
1976: The formation and evolution of Point Pelee, western Lake Erie; Canadian Journal of Earth Sciences, v. 13, p. 136-144.

Coakley, J.P. and Lewis, C.F.M.
1985: Postglacial lake levels in the Erie basin; in Quaternary Evolution of the Great Lakes, P.F. Karrow and P.E. Calkin (ed.); Geological Association of Canada, Special Paper 30, p. 195-212.

Coleman, A.P.
1894: Interglacial fossils from the Don Valley, Toronto; American Geologist, v. 13, p. 85-93.
1933: The Pleistocene of the Toronto region; Ontario Department of Mines, Annual Report, v. 41, part 7, 69 p.
1936: Lake Iroquois; Ontario Department of Mines Annual Report, v. 45, part 7, 36 p.
1941: The last million years; a history of the Pleistocene in North America; Toronto, University of Toronto Press, 216 p.

Connally, G. and Sirkin, L.A.
1973: The Wisconsinan history of the Hudson-Champlain lobe. The Wisconsinan stage; Geological Society of America, Memoir 136, p. 47-69.

Cooper, A.J.
1975: Pre-Catfish Creek tills of the Waterloo area; unpublished MSc thesis, University of Waterloo, Waterloo, Ontario 178 p.
1979: Quaternary geology of the Grand Bend-Parkhill area, southern Ontario; Ontario Geological Survey, Report 188, 70 p.

Cooper, A.J. and Clue, J.
1974: Quaternary geology of the Grand Bend area, southern Ontario; Ontario Division of Mines, Preliminary Map P-974.

Corliss, B.H., Hunt, A.S., and Keigwin, L.D., Jr.
1982: Benthonic foraminiferal faunal and isotopic data for the postglacial evolution of the Champlain Sea; Quaternary Research, v. 17, p. 325-338.

Costello, W.R. and Walker, R.G.
1972: Pleistocene sedimentology, Credit River, southern Ontario; a new component of the braided river model; Journal of Sedimentary Petrology, v. 42, p. 389-400.

Cotter, J.F., Ridge, J.C., Evenson, E.B., Sevon, W.D., Sirkin, L., and Stuckenrath, R.
1985: The Wisconsinan history of the Great Valley, Pennsylvania and New Jersey, and the age of the "Terminal Moraine"; in Woodfordian Deglaciation of the Great Valley, New Jersey, E.B. Evenson (ed.); 48th Friends of the Pleistocene, Field Conference Guidebook, Department of Geology, Lehigh University, Bethlehem, Pennsylvania.

Cowan, W.R.
1972: Pleistocene geology of the Brantford area, southern Ontario; Ontario Department of Mines and Northern Affairs, Industrial Minerals Report 37, 66 p.
1975: Quaternary geology of the Woodstock area; Ontario Division of Mines, Geological Report 119, 91 p.
1978: Trend surface analysis of major late Wisconsinan till sheets, Brantford-Woodstock area, southern Ontario; Canadian Journal of Earth Sciences, v. 15, p. 1025-1036.
1979: Quaternary geology of the Palmerston area, southern Ontario; Ontario Geological Survey, Report 187, 64 p.

Crawford, C.B.
1968: Quick clays of eastern Canada; Engineering Geology, v. 2, p. 239-265.

Creer, K.M. and Tucholka, P.
1982: Construction of type curves of geomagnetic secular variation for dating lake sediments from east central North America; Canadian Journal of Earth Sciences, v. 19, p. 1106-1115.

Cronin, T.M.
1977a: Champlain sea foraminifera and ostracoda: a systematic and paleoecological synthesis; Géographie physique et Quaternaire, v. 31, p. 107-122.
1977b: Late-Wisconsin marine environments of the Champlain Valley (New York, Quebec); Quaternary Research, v. 7, p. 238-253.
1979a: Foraminifer and ostracode species diversity in the Pleistocene Champlain Sea of the St. Lawrence Lowlands; Journal of Paleontology, v. 53, p. 233-244.
1979b: Late Pleistocene benthic foraminifers from the St. Lawrence Lowlands; Journal of Paleontology, v. 53, p. 781-814.
1981: Paleoclimatic implications of late Pleistocene marine ostracodes from the St. Lawrence Lowlands; Micropaleontology, v. 27, p. 384-418.

Dadswell, M.J.
1974: Distribution, ecology, and postglacial dispersal of certain crustaceans and fishes in eastern North America; National Museums of Canada, Publications Zoology, no. 11, 110 p.

D'Anglejan, B.F.
1971: Submarine sand dunes in the St. Lawrence estuary; Canadian Journal of Earth Sciences, v. 8, p. 1480-1486.
1981: Evolution post-glaciaire et sédiments récents de la plate-forme infralittorale, baie de Sainte-Anne, estuaire du Saint-Laurent, Québec; Géographie physique et Quaternaire, vol. 35, p. 253-260.

David, P.P. and Lebuis, J.
1985: Glacial maximum and deglaciation of western Gaspé, Québec, Canada; in Late Pleistocene History of Northeastern New England and Adjacent Quebec, H.J.W. Borns, P. LaSalle, and W.B. Thompson (ed.); Geological Society of America, Special Paper 197, p. 85-109.

Dawson, J.W.
1893: The Canadian Ice Age; William v. Dawson, Montreal, 301 p.

Deane, R.E.
1950: Pleistocene geology of the Lake Simcoe district, Ontario; Geological Survey of Canada, Memoir 256, 108 p.

Dejou, J., De Kimpe, C.R., et LaSalle, P.
1982: Evolution géochimique superficielle naissante des anorthosites. Cas du profil de Château-Richer, Québec, Canada; Journal canadien des sciences de la terre, vol. 19, p. 1697-1706.

Dell, C.I.
1959: A study of the mineralogical composition of sand in southern Ontario; Canadian Journal of Soil Sciences, v. 39, p. 185-196.
1972: The origin and characteristics of Lake Superior sediments; Proceedings, 15th Conference on Great Lakes Research, Ann Arbor, Michigan, p. 361-370.

Denis, R.
1976: Région de Saint-Gabriel; Ministère des Richesses naturelles du Québec, Rapport géologique 168, 56 p., carte 1772, échelle 1:63 360.

Denis, R. et Prichonnet, G.
1973: Aspects du Quaternaire dans la région au nord de Joliette; 2éme colloque sur le Quaternaire du Québec, Montréal, Livret-guide d'excursion, 53 p.

Desjardins, R.
1980: Tremblement de terre et glissement de terrain: corrélation entre des datations au ^{14}C et des données historiques à Shawinigan, Québec; Géographie physique et Quaternaire, vol. 34, p. 359-362.

deVries, H. and Dreimanis, A.
1960: Finite radiocarbon dates of the Port Talbot Interstadial deposits in southern Ontario; Science, v. 131, p. 1738-1739.

DiLabio, R.N.W.
1989: Terrain geochemistry in Canada; Chapter 10 in Quaternary Geology of Canada and Greenland, R.J. Fulton (ed.); Geological Survey of Canada, Geology of Canada, no. 1 (also Geological Society of America, The Geology of North America, v. K-1).

Dionne, J-C.
1963a: Vers une définition plus adéquate de l'Estuaire du Saint-Laurent; Annales de géomorphologie, vol. 7, p. 36-47.
1963b: Le problème de la terrasse et de la falaise Mic Mac (Côte sud de l'Estuaire maritime du Saint-Laurent); Revue canadienne de géographie, vol. 17, p. 9-25.
1968: Schorre morphology on the south shore of the St. Lawrence Estuary; American Journal of Science, v. 266, p. 380-388.
1969: Erosion glacielle littorale, estuaire du St-Laurent; Revue de géographie de Montréal, vol. 23, p. 5-20.
1971: Fentes de cryoturbation tradiglaciaires dans la région de Québec; Géographie physique et Quaternaire, vol. 25, p. 245-264.
1972a: La dénomination des mers du post-glaciaire au Québec; Cahiers de géographie de Québec, v. 16, p. 483-487.
1972b: Caractéristiques des blocs erratiques des rives de l'estuaire du Saint-Laurent; Revue de géographie de Montréal, vol. 26, p. 125-152.
1977: La mer de Goldthwait au Québec; Géographie physique et Quaternaire, vol. 31, p. 61-80.

Dreimanis, A.
1958: Wisconsin stratigraphy at Port Talbot on the north shore of Lake Erie, Ontario; Ohio Journal of Science, v. 58, p. 64-84.

1961: Tills of southern Ontario; in Soils in Canada, R.F. Legget (ed.); Royal Society of Canada, Special Publication no. 3, p. 80-96.

1964: Notes on the Pleistocene time-scale in Canada; Royal Society of Canada, Special Publication no. 8, p. 139-156.

1977: Correlation of Wisconsin glacial events between the eastern Great Lakes and the St. Lawrence lowlands; Géographie physique et Quaternaire, v. 31, p. 37-51.

1982a: Genetic classification of tills and criteria for their differentiation: progress report on activities 1977-1982, and definitions of glacigenic terms; in INQUA Commission on Genesis and Lithology of Quaternary Deposits, C. Schluchter (ed.); ETH-Honggerberg, Zurich p. 12-31.

1982b: Two origins of the stratified Catfish Creek Till at Plum Point, Ontario, Canada; Boreas, v. 11, p. 173-180.

1985: Till stratigraphy in the St. Lawrence Valley near Malone, New York: Revised glacial history and stratigraphic nomenclature: Discussion and reply; Geological Society of America, Bulletin, v. 96, p. 155-156.

1987: The Port Talbot interstadial site, southwestern Ontario; Geological Society of America, Centennial Field Guide — Northeastern Section, CFG-5, p. 345-348.

Dreimanis, A. and Barnett, P.J.
1985: Quaternary geology, Port Stanley area, southern Ontario; Ontario Geological Survey Geological Series — Preliminary Map, Map P-2827, scale 1:50 000.

Dreimanis, A. and Goldthwait, R.P.
1973: Wisconsin glaciation in the Huron, Erie and Ontario Lobes; in The Wisconsinan Stage, R.F. Black, R.P. Goldthwait, and H.B. Willman (ed.); Geological Society of America, Memoir 136, p. 71-106.

Dreimanis, A. and Karrow, P.F.
1972: Glacial history of the Great Lakes-St. Lawrence Region, the classification of the Wisconsin(an) Stage, and its correlatives; 24th International Geological Congress, Montreal, Section 12, p. 5-15.

Dreimanis, A. and Vagners, U.J.
1971: Bimodal distribution of rock and mineral fragments in basal tills; in Till: a Symposium, R.P. Goldthwait (ed.); Ohio State University Press, p. 237-250.

Dreimanis, A., Terasmae, J., and McKenzie, G.D.
1966: The Port Talbot Interstade of the Wisconsin glaciation; Canadian Journal of Earth Sciences, v. 3, p. 305-325.

Dubé, J.C.
1971: Géologie des dépôts meubles, région de Lyster, Comtés de Mégantic, Lotbinière, Nicolet, et Arthabaska; Ministère des Richesses naturelles du Québec, Rapport Préliminaire 596, 11 p., carte 1732, échelle 1:63 360.

Dubois, J.M.M.
1973: Les caractéristiques naturelles des Cantons de l'Est; Centre de Recherches en Aménagement Régional, Université de Sherbrooke, Sherbrooke, 130 p.

Dubois, J-M., Larocque, A., Boissonnault, P., Dubé, C., Poulin, A., Gwyn, Q.H.J., Larocque, G., et Morissette, A.
1984: Discussion de "Notes on the deglaciation of southeastern Quebec"; dans Recherches en cours, Partie B, Commission géologique du Canada, Étude 84-1B, p. 391-394.

Duckworth, P.B.
1979: The late depositional history of the western end of the Oak Ridges Moraine, Ontario; Canadian Journal of Earth Sciences, v. 16, p. 1094-1107.

Dumont, A., Allard, M., et Soucy, J.M.
1980: Wakefield — Géomorphologie; Office de planification et de développement du Québec, carte avec texte, échelle 1:50 000.

Durand, M. et Ballivy, G.
1974: Particularités rencontrées dans la région de Montréal résultant de l'arrachement d'écailles de roc par la glaciation; Revue canadienne de géotechnique, v. 11, p. 302-306.

Duthie, H.C. and Mannada Rani, R.G.
1967: Diatom assemblages from Pleistocene interglacial beds at Toronto, Ontario; Canadian Journal of Botany, v. 45, p. 2249-2261.

Elson, J.A.
1962: Pleistocene geology between Montreal and Covey Hill; Intercollegiate Geological Conference Guidebook, T.H. Clark (ed.); 54th Annual Meeting, Montreal, p. 61-66.

1969a: Late Quaternary marine submergence of Québec; Revue de géographie de Montréal, v. 23, p. 247-270.

1969b: Radiocarbon dates, *Mya arenaria* phase of the Champlain Sea; Canadian Journal of Earth Sciences, v. 6, p. 367-372.

Elson, J.A. and Elson, J.B.
1959: Phases of the Champlain Sea indicated by littoral mollusks; Geological Society of America, Bulletin, v. 70, p. 1596.

Eyles, C.H. and Eyles, N.
1983: Sedimentation in a large lake: a reinterpretation of the late Pleistocene stratigraphy at Scarborough Bluffs, Ontario, Canada; Geology, v. 11, p. 146-152.

Eyles, N.
1987: Late Pleistocene depositional systems of Metropolitan Toronto and their engineering and glacial geological significance; Canadian Journal of Earth Sciences, v. 24, p. 1009-1021.

Eyles, N. and Clark, B.M.
1986: Significance of hummocky and swaley cross-stratification in late Pleistocene lacustrine sediments of the Ontario basin, Canada; Geology, v. 14, p. 679-682.

1988: Storm-influenced deltas and ice scouring in a late Pleistocene glacial lake; Geological Society of America, Bulletin, v. 100, p. 793-809.

Eyles, N. and Westgate, J.A.
1987: Restricted regional extent of the Laurentide Ice Sheet in the Great Lakes basins during early Wisconsin glaciation; Geology, v. 15, p. 537-540.

Eyles, N., Clark, B.M., Kaye, B.G., Howard, K.W.F., and Eyles, C.H.
1985: The application of basin analysis techniques to glaciated terrains: An example from the Lake Ontario Basin, Canada; Geoscience Canada, v. 12, p. 22-32.

Eynon, G. and Walker, R.G.
1974: Facies relationships in Pleistocene outwash gravels, southern Ontario: a model for bar growth in braided rivers; Sedimentology, v. 21, p. 43-70.

Faessler, C.
1948: L'extension maximum de la Mer Champlain au nord du St-Laurent, de Trois-Rivières à Moisie; Université Laval, Géologie et Minéralogie, Contribution no. 88, p. 16-28.

Farrand, W.R. and Drexler, C.W.
1985: Late Wisconsinan and Holocene history of the Lake Superior basin; in Quaternary Evolution of the Great Lakes, P.F. Karrow and P.E. Calkin (ed.); Geological Association of Canada, Special Paper 30, p. 17-32.

Farvolden, R.N., Greenhouse, J.P., Karrow, P.F., Pehme, P.E., and Ross, L.C.
1987: Ontario Geoscience Research Grant Program Grant no. 128, Subsurface Quaternary stratigraphy of the Kitchener-Waterloo area using borehole geophysics; Ontario Geological Survey, Open File Report 5623, 76 p.

Fillon, R.F. and Hunt, A.
1974: Late Pleistocene benthonic Foraminifera of the southern Champlain Sea: paleotemperature and paleosalinity indications; Maritime Sediments, v. 10, p. 14-18.

Flint, J.J. and Lolcama, J.
1986: Buried ancestral drainage between lakes Erie and Ontario; Geological Society of America, Bulletin, v. 97, p. 75-84.

Ford, D.C., Andrews, J.T., Day, T.E., Harris, S.E., Macpherson, J.B., Occhietti, S., Rannie, W.F., and Slaymaker, H.O.
1984: Canada: how many glaciations?; Canadian Geographer, v. 28, p. 205-225.

Forsyth, J.L.
1965: Age of the buried soil in the Sidney, Ohio, area; American Journal of Science, v. 263, p. 571-597.

Fransham, P.B. and Gadd, N.R.
1976: Geological and geomorphological controls of landslides in Ottawa Valley, Ontario; Proceedings of the 29th Canadian Geotechnical Conference, Vancouver, session V, p. V-I-V-II.

Fransham, P.B., Gadd, N.R., and Carr, P.A.
1976: Geological variability of marine deposits, Ottawa-St. Lawrence Lowlands; in Report of Activities, Part A, Geological Survey of Canada, Paper 76-1A, p. 37-41.

Fritz, P., Anderson, T.W., and Lewis, C.F.M.
1975: Late-Quaternary climatic trends and history of Lake Erie from stable isotope studies; Science, v. 190, p. 267-269.

Frye, J.C. and Willman, H.B.
1960: Classification of the Wisconsinan Stage in the Lake Michigan glacial lobe; Illinois State Geological Survey, Circular 285, 16 p.

Fullerton, D.S.
1980: Preliminary correlation of post-Erie Interstadial events (16 000-10 000 radiocarbon years before present), central and eastern Great Lakes region, and Hudson, Champlain, and St. Lawrence Lowlands, United States and Canada; United States Geological Survey, Professional Paper 1089, 52 p.

Fulton, R.J.
1984: Summary: Quaternary stratigraphy of Canada; in Quaternary Stratigraphy of Canada — A Canadian Contribution to IGCP Project 24, R.J. Fulton (ed.); Geological Survey of Canada, Paper 84-10, p. 1-5.

Fulton, R.J. (editor)
1987: Quaternary geology of the Ottawa region and guides for day excursions; XII INQUA Congress, Ottawa 1987, National Research Council of Canada, Ottawa, NRC 27536, 125 p.

Fulton, R.J. and Richard, S.H.
1987: Chronology of late Quaternary events in the Ottawa region; in Quaternary geology of the Ottawa region, Ontario and Quebec, R.J. Fulton (ed.); Geological Survey of Canada, Paper 86-23.

Fulton, R.J., Gadd, N.R., Rodrigues, C.G., and Rust, B.R.
1986: Quaternary geology of the western Champlain Sea basin; Geological Association of Canada, Ottawa '86, Guidebook for Field Trip 7, 36 p.

Gadd, N.R.
1953: Interglacial deposits at St. Pierre, Quebec (abstract); Geological Society of America, Bulletin, v. 64, p. 1426.
1960: Géologie de la région de Bécancour, Québec (dépôts meubles); Commission géologique du Canada, Étude 59-8, 33 p.
1963: Géologie des dépôts meubles, Chalk-River, Ontario-Québec; Commission géologique du Canada, carte 1132A, échelle 1:63 360.
1971: Pleistocene geology of the central St. Lawrence Lowlands, with selected passages from an unpublished manuscript: The St. Lawrence Lowlands, by J.W. Goldthwait; Geological Survey of Canada, Memoir 359, 153 p.
1978a: Surficial geology of Saint-Sylvestre map-area, Quebec; Geological Survey of Canada, Paper 77-16, 9 p.
1978b: Mass flow deposits in a Quaternary succession near Ottawa, Canada: Diagnostic criteria for subaqueous outwash: Discussion; Canadian Journal of Earth Sciences, v. 15, p. 327-328.
1980a: Iceflow patterns, Montreal-Ottawa Lowland areas; in Current Research, Part A, Geological Survey of Canada, Paper 80-1A, p. 375-376.
1980b: Late-glacial regional ice-flow patterns in eastern Ontario; Canadian Journal of Earth Sciences, v. 17, p. 1439-1453.
1983: Notes on the deglaciation of southeastern Quebec; in Current Research, Part B, Geological Survey of Canada, Paper 83-1B, p. 403-412.
1984: Notes on the deglaciation of southeastern Quebec: Reply; in Current Research, Part B, Geological Survey of Canada, Paper 84-1B, p. 399-400.
1986: Lithofacies of Leda clay in the Ottawa basin of the Champlain Sea; Geological Survey of Canada, Paper 85-21, 44 p.

Gadd, N.R. and Karrow, P.F.
1959: Surficial geology, Trois-Rivières, St-Maurice, Champlain, Maskinongé, and Nicolet counties, Quebec; Geological Survey of Canada, Map 54-1959, scale 1:63 360.

Gadd, N.R., LaSalle, P., MacDonald, B.C., Shilts, W.W., et Dionne, J-C.
1972a: Géologie et géomorphologie du Quaternaire dans le Sud du Québec; 24e Congrés international de géologie, Montréal, Livret-guide d'excursion C-44, 74 p.

Gadd, N.R., McDonald, B.C., and Shilts, W.W.
1972b: Deglaciation of southern Quebec; Geological Survey of Canada, Paper 71-47, 19 p., Map 10-1971, scale 1:253 440.

Gadd, N.R., Richard, S.H., and Grant, D.R.
1981: Pre-last-glacial organic remains in Ottawa valley; in Current Research, Part C, Geological Survey of Canada, Paper 81-1C, p. 65-66.

Gale, L.A.
1970: Geodetic observations for the detection of vertical crustal movement; Canadian Journal of Earth Sciences, v. 7, p. 602-606.

Gangloff, P.
1970: Structures de gélisols reliques dans la région de Montréal; Revue de géographie de Montréal, vol. 24, p. 241-253.
1973: Le milieu morphoclimatique tardiglaciaire dans la région de Montréal; Cahiers de géographie de Québec, vol. 17, p. 415-448.

Gardner, J.S.
1977: Some geomorphic effects of a catastrophic flood on the Grand River, Ontario; Canadian Journal of Earth Sciences, v. 14, p. 2294-2300.

Gauthier, C.R.
1975: Déglaciation d'un secteur des rivières Chaudière et Etchemin; thèse de maîtrise non publiée McGill Université, Montréal, 180 p.

Gauthier, R.
1971: Étude de cinq tourbières du bas Saint-Laurent. I: Ecologie, II: Tourbe litière; Ministère des Richesses naturelles, Québec, étude spéciale 10, 25 p.

Gerath, R.F., Fowler, B.K., and Haselton, G.M.
1985: The deglaciation of the northern White Mountains of New Hampshire; in Late Pleistocene History of Northeastern New England and Adjacent Quebec, H.W. Borns Jr., P. LaSalle, and W.B. Thompson (ed.); Geological Society of America, Special Paper 197, p. 21-28

Gibbard, P.L. and Dreimanis, A.
1978: Trace fossils from Late Pleistocene glacial lake sediments in southwestern Ontario, Canada; Canadian Journal of Earth Sciences, v. 15, p. 1967-1976.

Goldring, W.
1922: The Champlain Sea, evidence of its decreasing salinity southward as shown by the character of its fauna; New York State Museum Bulletin, no. 232-240, p.153-194.

Grandtner, M.M.
1966: La végétation forestière du Québec méridional; Presses de l'Université Laval, Québec, 216 p.

Grant, D.R.
1977: Glacial style and ice limits, the Quaternary stratigraphic record, and changes of land and ocean level in the Atlantic Provinces, Canada; Géographie physique et Quaternaire, v. 31, p. 247-260.

Gravenor, C.P. and Stupavsky, M.
1974: Magnetic susceptibility of the surface tills of southern Ontario; Canadian Journal of Earth Sciences, v. 11, p. 658-663.

Gray, A.B.
1950: Sedimentary facies of the Don member (Toronto Formation); unpublished MA thesis, University of Toronto, Toronto, Ontario.

Guettard, M.
1752: Mémoire dans lequel on compare le Canada à la Suisse par rapport à ses minéraux; Mémoire de l'Académie Royale des Sciences de Paris, p. 189-220, Suite du mémoire, p. 323-360, Addition au mémoire, p. 524-538.

Guilbault, J.P.
1980: A stratigraphic approach to the study of the late-glacial Champlain Sea deposits with the use of Foraminifera; unpublished PhD thesis, Aarhus University, Denmark, 294 p.

Gunther, F.J. and Hunt, A.S.
1977: Paleoecology of marine and freshwater Holocene ostracodes from Lake Champlain, Nearctic North America; Proceedings of the Sixth International Ostracode Symposium, Saalfelden, p. 327-334.

Gwyn, Q.H.J.
1972: Quaternary geology of the Alliston-Newmarket area, southern Ontario; Ontario Division of Mines, Miscellaneous Paper 53, p. 144-146.

Gwyn, Q.H.J. and Dreimanis, A.
1979: Heavy mineral assemblages in tills and their use in distinguishing glacial lobes in the Great Lakes region; Canadian Journal of Earth Sciences, v. 16, p. 2219-2235.

Hamelin, L-E.
1961: Périglaciaire du Canada: idées nouvelles et perspectives globales; Cahiers de géographie de Québec, vol. 5, p. 141-203.

Hann, B.J. and Karrow, P.F.
1984: Pleistocene paleoecology of the Don and Scarborough Formations, Toronto, Canada, based on cladoceran microfossils at the Don Valley Brickyard; Boreas, v. 13, p. 377-391.

Haras, W. S. and Tsui, K.K.
1976: Canada/Ontario Great Lakes shore damage survey coastal zone atlas; Canada, Department of Environment — Ontario Ministry of Natural Resources, 97 p.

Harington, C.R.
1971: The Champlain Sea and its vertebrate fauna; Part I, The history and environment of the Champlain Sea; Trail & Landscape, v. 5, p.137-141.
1972: The Champlain Sea and its vertebrate fauna; Part II, Vertebrates of the Champlain Sea; Trail & Landscape, v. 6, p. 33-39.
1977: Marine mammals in the Champlain Sea and the Great Lakes; New York Academy of Sciences, Annals, v. 288, p. 508-537.
1978: Quaternary vertebrate faunas of Canada and Alaska and their suggested chronological sequence; National Museum of Natural Sciences, Ottawa, Syllogeus no. 15, 105 p.

Harington, C.R. and Occhietti, S.
1980: Pleistocene eider duck (*Somateria cf. mollissima*) from Champlain Sea deposits near Shawinigan, Québec; Géographie physique et Quaternaire, v. 34, p. 239-245.
1988: Inventaire systématique et paléoécologie des mammifères marins de la Mer de Champlain (fin du Wisconsinien) et de ses voies d'accès; Géographie physique et Quaternaire, vol. 42, no. 1, p. 45-64.

Harrison, J.E.
1972: Quaternary geology of the North Bay — Mattawa region; Geological Survey of Canada, Paper 71-26, 37 p.
1977: Coastal studies in the Ottawa area; in Report of Activities, Part A, Geological Survey of Canada, Paper 77-1A, p. 59-60.

Henderson, E.P.
1970: Surficial geology of Brockville and Mallorytown map-areas, Ontario; Geological Survey of Canada, Paper 70-18, Map 6-1970, scale 1:50 000.
1973: Surficial geology of Kingston map-area, Ontario; Geological Survey of Canada, Paper 72-48, Maps 7-1972, 8-1972, scale 1:125 000.

Hewitt, D.F.
1971: The Niagara Escarpment; Ontario Department of Mines and Northern Affairs, Industrial Minerals Report 35, 71 p.

Hewitt, D.F. and Karrow, P.F.
1963: Sand and gravel in southern Ontario; Ontario Department of Mines, Industrial Minerals Report 11, 151 p.

Hillaire-Marcel, C.
1974: État actuel des connaissances sur le relèvement glacio-isostatique dans la région de Montréal (Québec) entre moins 13 000 et moins 9 000 ans; Compte rendu de l'Académie des Sciences de Paris, T. 298, Série D, p. 1939-1941.
1977: Les isotopes du carbone et de l'oxygène dans les mers post-glaciaires du Québec; Géographie physique et Quaternaire, vol. 31, p. 81-106.
1979: Les mers post-glaciaires du Québec, quelques aspects; thèse de doctorat d'état non publiée, Université Pierre et Marie Curie, Paris VI, 293 p.
1980: Les faunes des mers post-glaciaires du Québec: quelques considérations paléoécologiques; Géographie physique et Quaternaire, vol. 34, p. 3-59.
1981a: Paléo-océanographie isotopique des mers post-glaciaires du Québec; Palaeogeography, Palaeoclimatology, Palaeoecology, v. 35, p. 35-119.
1981b: Late-glacial regional ice-flow patterns in eastern Ontario: Discussion; Canadian Journal of Earth Sciences, v. 18, p. 1385-1386.

Hillaire-Marcel, C. et Pagé, P.
1981: Paléotempératures isotopiques du Lac Glaciaire de Deschaillons; in Quaternary Paleoclimatc, W.C. Mahaney (ed.); Geo Books, University of East Anglia, Norwich, England, p. 273-298.

Hillaire-Marcel, C., Occhietti, S., and Vincent, J-S.
1981: The late glacial Sakami Moraine (New-Québec): an example of a reequilibration moraine without climatic control; Geology, v. 9, p. 210-214.

Hillaire-Marcel, C., Soucy, J.M., et Cailleux, A.
1979: Analyses isotopiques de concrétions sous-glaciaires de l'inlandsis laurentidien et teneur en oxygène 18 de la glace; Journal canadien des sciences de la terre, vol. 16, p. 1494-1498.

Hinde, G.J.
1878: The glacial and interglacial strata of Scarboro Heights and other locations near Toronto, Ontario; Canadian Journal, new series, v. 15, p. 388-413.

Hobson, G.D. and Terasmae, J.
1969: Pleistocene geology of the buried St. Davids Gorge, Niagara Falls, Ontario: Geophysical and palynological studies; Geological Survey of Canada, Paper 68-67, 16 p.

Houde, M. et Clark, T.H.
1961: Carte géologique des Basses-Terres du Saint-Laurent; Ministère des Richesses naturelles du Québec, carte 1407, échelle 1:253 440.

Hough, J.L.
1955: Lake Chippewa, a low stage of Lake Michigan indicated by bottom sediments; Geological Society of America, Bulletin, v. 66, p. 957-968.
1958: Geology of the Great Lakes; University of Illinois Press, 313 p.
1963: The prehistoric Great Lakes of North America; American Scientist, v. 51, p. 84-109.

Hurtubise, J.E., Gadd, N.R., et Meyerhoff, G.G.
1957: Les éboulements de terrain dans l'est du Canada; Proceedings 4th International Conference on Mechanics and Foundation Engineering, Tome II, London, Butterworths Scientific Publications, August, 1957, p. 325-329. Reprinted 1958 as: Associate Committee on Soil and Snow Mechanics, National Research Council of Canada, Technical Memoir no. 52, Ottawa.

Illman, W.I., McLachlan, J., and Edelstein, T.
1970: Marine algae of the Champlain Sea episode near Ottawa; Canadian Journal of Earth Sciences, v. 7, p. 1583-1585.

Johnston, W.A.
1917: Pleistocene and Recent Deposits in the vicinity of Ottawa, with a description of the soils, Ottawa; Geological Survey of Canada, Memoir 101, 69 p, Map 1662.

Karrow, P.F.
1957: Pleistocene geology of the Grondines map-area, Québec; unpublished PhD dissertation, University of Illinois, Urbana 97 p.
1959: Pleistocene geology of the Hamilton map-area; Ontario Department of Mines, Geological Circular no. 8, 6 p.
1961: The Champlain Sea and its sediments; in Soils in Canada, R.F. Legget (ed.); Royal Society of Canada, Special Publication no. 3, p. 97-108.
1967: Pleistocene geology of the Scarborough area; Ontario Department of Mines, Geological Report 46, 108 p.
1968: Pleistocene geology of the Guelph area; Ontario Department of Mines, Geological Report 61, 38 p.
1969: Stratigraphic studies in the Toronto Pleistocene; Geological Association of Canada, Proceedings, v. 20, p. 4-16.
1970: Pleistocene geology of the Thornhill area; Ontario Department of Mines, Industrial Mineral Report 32, 51 p.
1972: Earth flows in the Grondines and Trois-Rivières areas of Québec; Canadian Journal of Earth Sciences, v. 9, p. 561-573.
1973: Bedrock topography in southwestern Ontario: A progress report; Geological Association of Canada, Proceedings, v. 25, p. 67-77.
1974: Till stratigraphy in parts of southwestern Ontario; Geological Society of America, Bulletin, v. 85, p. 761-768.
1976: The texture, mineralogy, and petrography of North American tills; in Glacial Till, R.F. Legget (ed.); Royal Society of Canada, Special Publication no. 12, p. 83-98.
1977: Quaternary geology of the St. Mary's area, southern Ontario; Ontario Division of Mines, Geoscience Report 148, 59 p.
1981a: Till texture in drumlins; Journal of Glaciology, v. 27, p. 497-502.
1981b: Late-glacial regional ice-flow patterns in eastern Ontario: Discussion; Canadian Journal of Earth Sciences, v. 18, p. 1386-1390.
1984: Quaternary stratigraphy and history, Great Lakes-St. Lawrence region; in Quaternary Stratigraphy of Canada — A Canadian Contribution to IGCP Project 24, R.J. Fulton (ed.); Geological Survey of Canada, Paper 84-10, p. 137-153.
1986: Valley terraces and Huron basin water levels; Geological Society of America, Bulletin, v. 97, p. 1089-1097.
1987: Quaternary geology of the Hamilton-Cambridge area southern Ontario; Ontario Geological Survey; Report 255, 94 p.
1989: Quaternary geology of the Great Lakes subregion; in Chapter 4 of Quaternary Geology of Canada and Greenland, R.J. Fulton (ed.); Geological Survey of Canada, Geology of Canada, no. 1 (also Geological Society of America, The Geology of North America, v. K-1).
in press: Quaternary geology of the Stratford-Conestogo area, Ontario; Ontario Geological Survey, Report.

Karrow, P.F. and Calkin, P.E. (editors)
1985: Quaternary evolution of the Great Lakes; Geological Association of Canada, Special Paper 30, 258 p.

Karrow, P.F. and Warner, B.G.
1984: A subsurface Middle Wisconsinan interstadial site at Waterloo, Ontario, Canada; Boreas, v. 13, p. 67-85.

Karrow, P.F., Anderson, T.W., Clarke, A.H., Delorme, L.D., and Sreenivasa, M.R.
1975: Stratigraphy, paleontology, and age of Lake Algonquin sediments in southwestern Ontario, Canada; Quaternary Research, v. 5, p. 49-87.

Karrow, P.F., Clarke, A.H., and Herrington, H.B.
1972: Pleistocene molluscs from Lake Iroquois deposits in Ontario; Canadian Journal of Earth Sciences, v. 9, p. 589-595.

Karrow, P.F., Clark, J.R., and Terasmae, J.
1961: The age of Lake Iroquois and Lake Ontario; Journal of Geology, v. 69, p. 659-667.

Karrow, P.F., Cowan, W.R., Dreimanis, A., and Singer, S.N.
1978: Middle Wisconsinan stratigraphy in southern Ontario; Toronto '78 Field Trips Guidebook, A.L. Currie and W.O. Mackasey (ed.); Geological Society of America, Geological Association of Canada Annual Meeting, p. 17-27.

Karrow, P.F., Harrison, W., and Saunderson, H.C.
1977: Reworked Middle Wisconsinan(?) plant fossils from the Brampton esker, southern Ontario; Canadian Journal of Earth Sciences, v. 14, p. 426-430.

Karrow, P.F., Hebda, R.J., Presant, E.W., and Ross, G.J.
1982: Late Quaternary inter-till paleosol and biota at Guelph, Ontario; Canadian Journal of Earth Sciences, v. 19, p. 1857-1872.

Kelly, R.I. and Martini, I.P.
1986: Pleistocene glacio-lacustrine deltaic deposits of the Scarborough Formation, Ontario, Canada; Sedimentary Geology, v. 47, p. 27-52.

Kerr-Lawson, L.J.
1985: Gastropods and plant microfossils from the Quaternary Don Formation (Sangamonian Interglacial), Toronto, Ontario; unpublished MSc thesis, University of Waterloo, Waterloo, Ontario, 202 p.

Kindle, E.M. and Taylor, F.B.
1913: Description of the Niagara quadrangle; United States Geological Survey, Geological Atlas Folio 190, 26 p.

Kirkland, J.T. and Coates, D.R.
1977: The Champlain Sea and Quaternary deposits in the St. Lawrence Lowland, New York; New York Academy of Sciences, Annals, p. 498-507.

Lajtai, E.Z.
1967: The origin of some varves in Toronto, Canada; Canadian Journal of Earth Sciences, v. 4, p. 633-639.
1969: Stratigraphy of the University subway, Toronto, Canada; Geological Association of Canada, Proceedings, v. 20, p. 17-23.

Lamarche, R.Y.
1971: Northward moving ice in the Thetford Mines area of southern Quebec; American Journal of Science, v. 271, p. 383-388.

1974: Southeastward, northward and westward ice movement in the Asbestos area of southern Quebec; Geological Society of America, Bulletin, v. 85, p. 465-470.

Lamothe, M.
1977: Les dépôts meubles de la région de Saint-Faustin-Saint-Jovite, Québec; Cartographie, sédimentologie et stratigraphie; thèse de maîtrise non publiée, Université du Québec à Montréal, 118 p.

1984a: Le Quaternaire de la région de Pierreville, basses terres du Saint-Laurent; Association québécoise pour l'étude du Quaternaire, Ve Congrès, Sherbrooke, Livret guide d'excursion, p. 26-38.

1984b: Apparent thermoluminescence ages of St. Pierre sediments at Pierreville, Quebec, and the problem of anomalous fading; Canadian Journal of Earth Sciences, v. 21, p. 1406-1409.

1985: Lithostratigraphy and geochronology of the Quaternary deposits of the Pierreville and St-Pierre-les-Becquets area, Quebec; unpublished PhD thesis, University of Western Ontario, London, 240 p.

1987: Lithostratigraphic superposition and geochronology of the Pleistocene sediments in the St-Pierre-les-Becquets area, Quebec (abstract); XII INQUA Congress, Ottawa 1987, Program with abstracts, National Research Council of Canada, Ottawa, p. 206.

Lamothe, M., Hillaire-Marcel, C., et Pagé, P.
1983: Découverte de concrétions calcaires striées dans le till de Gentilly, basses terres du Saint-Laurent, Québec; Journal canadien des sciences de la terre, vol. 20, p. 500-505.

Landmesser, C.W., Johnson, T.C., and Wold, R.J.
1982: Seismic reflection study of recessional moraines beneath Lake Superior and their relationship to regional deglaciation; Quaternary Research v. 17, p. 173-190.

Larocque, A., Gwyn, Q.J.H., et Poulin, A.
1983a: Développement des lacs proglaciaires et déglaciation des hauts bassins des rivières au Saumon et Chaudière, sud du Québec; Géographie physique et Quaternaire, vol. 37, p. 93-105.

1983b: Évolution des lacs proglaciaires et déglaciation du haut Saint-François, sud du Québec; Géographie physique et Quaternaire, vol. 37, p. 85-92.

LaSalle, P.
1966: Late Quaternary vegetation and glacial history in the St. Lawrence Lowlands, Canada; Leidse Geologische Mededelingen, v. 38, p. 91-128.

1978: Géologie des sédiments de surface de la région de Québec; Ministère des Richesses naturelles du Québec, DPV-565, 2 p., 22 cartes, échelle 1:50 000.

1981: Géologie des dépôts meubles de la région Saint-Jean-Lachine; Ministère de l'Énergie et des Ressources du Québec, DPV-780, 13 p.

1984: Quaternary stratigraphy of Quebec: A review; in Quaternary Stratigraphy of Canada — A Canadian Contribution to IGCP Project 24, R.J. Fulton (ed.); Geological Survey of Canada, Paper 84-10, p. 155-171.

LaSalle, P. and Elson, J.A.
1975: Emplacement of the St. Narcisse Moraine as a climatic event in eastern Canada; Quaternary Research, v. 5, P. 621-625.

LaSalle, P. et Tremblay, G.
1978: Dépôts meubles, Saguenay-Lac St-Jean; Ministère des Richesses naturelles du Québec, Rapport géologique 191, 61 p.

LaSalle, P., De Kimpe, C., et Laverdière, M.
1983: Altération profonde préglaciaire d'un gneiss à Charlesbourg, Québec; Geoderma, vol. 31, p. 117-132.

LaSalle, P., De Kimpe, C.R., and Laverdière, M.R.
1985: Sub-till saprolites in southeastern Quebec and adjacent New England: Erosional, stratigraphic, and climatic significance; in Late Pleistocene History of Northeastern New England and Adjacent Quebec, H.J.W. Borns, P. LaSalle, and W.B. Thompson (ed.); Geological Society of America, Special Paper 194, p. 13-20.

LaSalle, P., Martineau, G., et Chauvin, L.
1977a: Dépôts morainiques et stries glaciaires dans la région de Beauce-Monts Notre-Dame-parc des Laurentides; Ministère des Richesses naturelles du Québec, DPV-515, 22 p.

1977b: Morphologie, stratigraphie, et déglaciation dans la région de Beauce-Monts Notre-Dame-Parc des Laurentides; Ministère des Richesses naturelles du Québec, DPV-516, 74 p.

1979: Lits de bryophites du Wisconsin moyen, Vallée-Jonction, Québec; Journal canadien des sciences de la terre, vol. 16, p. 593-598.

Laverdière, C. et Courtemanche, A.
1961: La géomorphologie glaciaire de la région de Mont-Tremblant: 2e partie: La région de Saint-Faustin — Saint-Jovite; Cahiers de géographie de Québec, no. 9, p. 5-32.

Lebuis, J. et David, P.P.
1977: La stratigraphie et les événements géologiques du Quaternaire de la partie occidentale de la Gaspésie; Géographie physique et Quaternaire, vol. 31, p. 275-296.

Lee, H.A.
1962: Géologie de la région de Rivière-du-Loup — Trois-Pistoles, Québec; Commission géologique du Canada, Étude 61-32, 2 p., carte 43-1961, échelle 1:63 630.

Legget, R.F.
1942: An engineering study of glacial drift for an earth dam, near Fergus, Ontario; Economic Geology, v. 37, p. 531-536.

Leighton, M.M.
1933: The naming of the subdivisions of the Wisconsin glacial age; Science, v. 77, p. 168.

1958: Important elements in the classification of the Wisconsin glacial stage; Journal of Geology, v. 66, p. 288-309.

LeMenestral, J.
1969: Blackburn (Geomorphology), Ontario-Quebec; Geological Survey of Canada, Map 1264A, scale 1:25 000.

Leverett, F. and Taylor, F.B.
1915: The Pleistocene of Indiana and Michigan and the history of the Great Lakes; United States Geological Survey, Monograph 53, 529 p.

Lewis, C.F.M.
1969: Late Quaternary history of lake levels in the Huron and Erie basins; Proceedings of the 12th Conference on Great Lakes Research, Ann Arbor, Michigan, p. 250-270.

1970: Recent uplift of Manitoulin Island, Ontario; Canadian Journal of Earth Sciences, v. 7, p. 665-675.

Lewis, C.F.M. and McNeely, R.N.
1967: Survey of Lake Ontario bottom deposits; Proceedings of the 10th Conference on Great Lakes Research, p. 133-142.

Lewis, C.F.M., Anderson, T.W., and Berti, A.A.
1966: Geological and palynological studies of early Lake Erie deposits; Great Lakes Research Division, University of Michigan, Publication 15, p. 176-191.

Liberty, B.A.
1969: Paleozoic geology of the Lake Simcoe area, Ontario; Geological Survey of Canada, Memoir 355, 201 p.

Locat, J.
1977: L'émersion des terres dans la région de Baie-des-Sables/Trois-Pistoles, Québec; Géographie physique et Quaternaire, vol. 31, p. 297-306.

Locat, J. and Chagnon, J-Y.
1989: Geological hazards in eastern Canada; in Chapter 12 of Quaternary Geology of Canada and Greenland, R.J. Fulton (ed.); Geological Survey of Canada, Geology of Canada, no. 1 (also Geological Society of America, The Geology of North America, v. K-1).

Logan, W.E.
1863: Geology of Canada; Geological Survey of Canada, Report of Progress to 1863, 983 p.

Lortie, G.
1975: Direction d'écoulement des glaciers du Pléistocène des Cantons de l'Est, Québec; in Report of Activities, Part A, Commission géologique du Canada, Étude 75-1A, p. 415-416.

1976: Les écoulements glaciaires wisconsiniens dans les Cantons de l'Est et la Beauce, Québec; thèse de MA, non publiée, Université de McGill, Montréal, 219 p.

1983: Les diatomées de la mer de Goldthwait dans la région de Rivière-du-Loup, Québec; Géographie physique et Quaternaire, vol. 37, p. 279-296.

Lortie, G. et Guilbault, J-P.
1984: Les diatomées et les foraminifères de sédiments marins post-glaciaires du Bas-Saint-Laurent (Québec): une analyse comparée des assemblages; Naturaliste canadien, vol. 3, p. 297-310.

Lowdon, J.A. and Blake, W., Jr.
1970: Geological Survey of Canada radiocarbon dates IX; Radiocarbon, v. 12, p. 46-86.

1973: Geological Survey of Canada radiocarbon dates XIII; Geological Survey of Canada, Paper 73-7, 61 p.

1979: Geological Survey of Canada radiocarbon dates XIX; Geological Survey of Canada, Paper 79-7, 58 p.

Lowell, T.V.
1985: Late Wisconsin ice-flow reversal and deglaciation, northwestern Maine; in Late Pleistocene History of Northeastern New England and Adjacent Quebec, H.J.W. Borns, P. LaSalle, and W.B. Thompson (ed.); Geological Society of America, Special Paper 197, p. 71-83.

Lyell, C.
1845: Travels in North America in the Years 1841-1842, with Geological Observations on the United States, Canada, and Nova Scotia; Wiley and Putnam, New York, v. 1, 231 p., v. 2, 251 p.; map scale approximately 1 inch to 100 miles, John Murray, London.

MacClintock, P.
1954: Pleistocene geology of the St. Lawrence Lowland; New York State Museum and Science Service, Report 10, 20 p.

1958: Glacial geology of the St. Lawrence Seaway and power projects; New York State Museum and Science Service, Report, 26 p.

MacClintock, P., and Stewart, D.P.
1965: Pleistocene geology of the St. Lawrence Lowland; New York State Museum and Science Service, Bulletin, no. 394, 152 p.

MacClintock, P.P. and Terasmae, J.
1960: Glacial history of Covey Hill; Journal of Geology, v. 68, p. 232-241.

MacPherson, J.B.
1967: Raised shorelines and drainage evolution in the Montréal Lowland; Cahiers de géographie de Québec, no. 23, p. 343-360.

McAndrews, J.H. and Boyko-Diakonow, M.
1989: Pollen analysis of varved sediment at Crawford Lake, Ontario: Evidence of Indian and European farming; in Quaternary Geology of Canada and Greenland, R.J. Fulton (ed.); Geological Survey of Canada, Geology of Canada, no. 1 (also Geological Society of America, The Geology of North America, v. K-1).

McDonald, B.C.
1966: Surficial geology, Richmond-Dudswell, Quebec; Geological Survey of Canada, Map 4-1966, scale 1:63 360.
1967: Surficial geology, Sherbrooke-Orford-Memphrémagog, Quebec; Geological Survey of Canada, Map 5-1966, scale 1:63 360.
1968: Deglaciation and differential post-glacial rebound in the Appalachian region of southeastern Quebec; Journal of Geology, v. 76, p. 664-677.
1971: Late Quaternary stratigraphy and deglaciation in eastern Canada, in The Late Cenozoic Glacial Ages, K.K. Turekian (ed.); Yale University Press, p. 331-353.

McDonald, B.C. and Shilts, W.W.
1971: Quaternary stratigraphy and events in southeastern Quebec; Geological Society of America, Bulletin, v. 82, p. 683-698.

Mangerud, J. and Gulliksen, S.
1975: Apparent radiocarbon ages of recent marine shells from Norway, Spitsbergen and Arctic Canada; Quaternary Research, v. 5, p. 263-274.

Martineau, G. et Corbeil, P.
1983: Réinterprétation d'un segment de la moraine de Saint-Antonin, Québec; Géographie physique et Quaternaire, vol. 37, p. 217-221.

Martini, I.P.
1975: Sedimentology of a lacustrine barrier system at Wasaga Beach, Ontario, Canada; Sedimentary Geology, v. 14, p. 169-190.
1977: Gravelly flood deposits of Irvine Creek, Ontario, Canada; Sedimentology, v. 24, p. 603-622.

Matyas, E.L., White, O.L., and LeLievre, B.
1974: A study of shoreline erosion in western Lake Ontario; in Report of Activities, Part B, Geological Survey of Canada, Paper 74-1B, p. 80-81.

Miller, B.B., Karrow, P.F., and Kalas, L.L.
1979: Late Quaternary mollusks from Glacial Lake Algonquin, Nipissing, and transitional sediments from southwestern Ontario, Canada; Quaternary Research, v. 11, p. 93-112.

Miller, B.B., Karrow, P.F., and Mackie, G.L.
1985: Late Quaternary molluscan faunal changes in the Huron basin; in Quaternary evolution of the Great Lakes, P.F. Karrow and P.E. Calkin (ed.); Geological Association of Canada, Special Paper 30, p. 95-107.

Mirynech, E.
1962: Pleistocene geology of the Trenton-Campbellford map area, Ontario; unpublished PhD thesis, University of Toronto, Toronto, Ontario, 197 p.

Morgan, A.V.
1972: Late Wisconsinan ice-wedge polygons near Kitchener, Ontario, Canada; Canadian Journal of Earth Sciences, v. 9, p. 607-617.
1982: Distribution and probable age of relict permafrost features in southwestern Ontario; in Proceedings, 4th Canadian Permafrost Conference (The Roger J.E. Brown Memorial Volume), H.M. French (ed.); National Research Council, p. 91-100.

Morison, S.R.
1989: Placer deposits in Canada; in Chapter 11 of Quaternary Geology of Canada and Greenland, R.J. Fulton (ed.); Geological Survey of Canada, Geology of Canada, no. 1 (also Geological Society of America, The Geology of North America, v. K-1).

Mörner, N.-A. and Dreimanis, A.
1973: The Erie Interstade; Geological Society of America, Memoir 136, p. 107-134.

Mothersill, J.S. and Brown, H.
1982: Late Quaternary stratigraphic sequence of the Goderich basin: texture, mineralogy, and paleomagnetic record; Journal of Great Lakes Research, v. 8, p. 578-586.

Mott, R.J.
1968: A radiocarbon-dated marine algal bed of the Champlain Sea episode near Ottawa, Ontario; Canadian Journal of Earth Sciences, v. 5, p. 319-323.
1977: Late Pleistocene and Holocene palynology in southeastern Québec; Géographie physique et Quaternaire, v. 31, p. 139-149.

Mott, R.J. and Farley-Gill, L.D.
1978: A Late-Quaternary pollen profile from Woodstock, Ontario; Canadian Journal of Earth Sciences, v. 15, p. 1101-1111.
1981: Two late Quaternary pollen profiles from Gatineau Park, Quebec; Geological Survey of Canada, Paper 80-31, 10 p.

Muller, E.H. and Prest, V.K.
1985: Glacial lakes in the Ontario Basin; in Quaternary Evolution of the Great Lakes, P.F. Karrow and P.E. Calkin (ed.); Geological Association of Canada, Special Paper 30, p. 213-229.

Occhietti, S.
1976: Dépôts et faits quaternaires du Bas St-Maurice, Québec (2e partie); in Report of Activities, Part C, Commission géologique du Canada, Étude 76-1C, p. 217-220.
1977a: Troisième colloque sur le Quaternaire du Québec; Géographie physique et Quaternaire, vol. 31, 408 p.
1977b: Stratigraphie du Wisconsinien de la région de Trois-Rivières — Shawinigan, Québec; Géographie physique et Quaternaire, vol. 31, p. 307-322.
1980: Le Quaternaire de la région de Trois-Rivières-Shawinigan, Québec. Contribution à la paléogéographie de la vallée moyenne du Saint-Laurent et corrélations stratigraphiques; Université du Québec à Trois-Rivières, Paléo-Québec, vol. 10, 227 p.
1982: Synthèse lithostratigraphique et paléoenvironnements du Quaternaire au Québec méridional. Hypothèse d'un centre d'englacement wisconsinien au Nouveau-Québec; Géographie physique et Quaternaire, vol. 36 p. 15-49.
1983: The Laurentide Ice-Sheet: climatic and oceanic implications; Palaeogeography, Palaeoclimatology, Palaeoecology, v. 44, p. 1-22.

Occhietti, S. et Hillaire-Marcel, C.
1982: Les paléoenvironnements de la Mer de Champlain dans la région de Québec, entre 11 500 et 9 000 BP (résumé); 50e Congrès ACFAS, Montréal, colloque AQQUA: Milieux glacio-marins actuels et passés.

Occhietti, S., Clet, M., et Richard, P.
1987: Lithostratigraphie et biostratigraphie du Pléistocène de la vallée du Saint-Laurent; XIIe Congrès de l'INQUA, Ottawa 1987, Programme et Résumés, Le Conseil national de recherche du Canada, Ottawa, p. 234.

Occhietti, S., Gadd, N.R., LaSalle, P., et Tremblay, G.
1975: Infiltration latérale par des dépôts morainiques: particularité du glissement argileux de St-Alban (1894) (résumé); 43e Congrès ACFAS, Moncton.

Oliver, J., Johnson, T., and Dorman, J.
1970: Postglacial faulting and seismicity in New York and Quebec; Canadian Journal of Earth Sciences, v. 7, p. 579-590.

Osborne, F.F.
1950: Ventifacts at Mont Carmel, Quebec; Royal Society of Canada, Transactions, v. 44, Series III, sect. IV, p. 41-49.

Pagé, P.
1977: Les dépôts meubles de la région de Saint-Jean-de-Matha — Sainte-Émélie-de-l'Énergie, Québec, cartographie, sédimentologie et stratigraphie; thèse de maîtrise non publiée, Université du Québec à Montréal, 118 p.

Pair, D., Karrow, P.F., and Clark, P.U.
in press: History of the Champlain Sea in the central St-Lawrence Lowland, New York and its relationship to water levels in the Lake Ontario basin; in Late Quaternary Development of the Champlain Sea Basin, N.R. Gadd (ed.); Geological Association of Canada, Special Paper 35.

Parent, M.
1984a: Le Quaternaire de la région d'Asbestos-Valcourt: Aspects stratigraphiques; Stratigraphie Quaternaire des Appalaches du Québec méridional: La coupe-type de la rivière Ascot; Association québécoise pour l'étude du Quaternaire, 5e Congrès, Sherbrooke, Livret-guide d'excursion, p. 2-25, p. 39-46.
1984b: Notes on the deglaciation of southeastern Québec: Discussion; in Current Research, Part B, Geological Survey tof Canada, Paper 84-1B, p. 395-397.
1987: Late Pleistocene stratigraphy and events in the Asbestos-Valcourt region, southeastern Quebec; unpublished PhD thesis, University of Western Ontario, London, Ontario, 320 p.

Parent, M., Dubois, J.M.M., Bail, P., Larocque, A., et Larocque, G.
1985: Paléogéographie du Québec méridional entre 12 500 et 8000 ans B.P.; Recherches amérindiennes au Québec, vol. 15, 26 p.

Parry, J.T. and Macpherson, J.C.
1964: The St. Faustin — St. Narcisse Moraine and the Champlain Sea; Revue de géographie de Montréal, v. 18, p. 235-248.

Pilny, J. and Morgan, A.V.
1987: Paleoentomology and paleoecology of a possible Sangamonian site near Innerkip, Ontario; Quaternary Research, v. 28, p. 157-174.

Pinch, J.J.
1979: Sedimentology and stratigraphy of Wisconsinan deposits in the McKittrick site and Beaver River gorge, Clarksburg, Ontario; unpublished MSc thesis, University of Western Ontario, London, 103 p.

Poplawski, S. and Karrow, P.F.
1981: Ostracodes and paleoenvironments of the late Quaternary Don and Scarborough Formations, Toronto, Ontario; Canadian Journal of Earth Sciences, v. 18, p. 1497-1505.

Potzger, J.E. and Courtemanche, A.
1956: A series of bogs across Quebec from the St. Lawrence Valley to James Bay; Canadian Journal of Botany, v. 34, p. 473-500.

Prest, V.K.
1970: Quaternary geology of Canada; Chapter XII in Geology and Economic Minerals of Canada, R.J.W. Douglas (ed.); Geological Survey of Canada, Economic Geology Report no. 1, 5th edition, p. 676-764.
1977: General stratigraphic framework of the Quaternary in Eastern Canada; Géographie physique et Quaternaire, v. 31, p. 7-14.

Prest, V.K. and Hode Keyser, J.
1962: Géologie des dépôts meubles et sols de la région de Montréal-Québec; Cité de Montréal, Service des travaux publics, 35 p.
1977: Geology and engineering characteristics of surficial deposits, Montréal Island and vicinity, Québec; Geological Survey of Canada, Paper 75-27, 29 p., Maps 1426A, 1427A, scale 1:50 000.

Prévôt, J.M.
1972: Carte hydrogéologique des Basses Terres du Saint-Laurent; Ministère des Richesses naturelles du Quéebec, carte 1748, échelle 1:250 000.

Prichonnet, G.
1977: La déglaciation de la vallée du Saint-Laurent et l'invasion marine contemporaine; Géographie physique et Quaternaire, vol. 31 p. 323-345.
1982a: Quelques données nouvelles sur les dépôts quaternaires du Wisconsinien et de l'Holocène dans le piedmont appalachien, Granby, Québec; in Recherches en cours, Partie B, Commission géologique du Canada, Étude 82-1B, p. 225-238.
1982b: Résultats préliminaires sur la géologie quaternaire de la région de Cowansville, Québec; in Recherches en cours, Partie B, Commission géologique du Canada, Étude 82-1B, p. 297-300.
1984a: Dépôts quaternaires de la région de Granby, Québec (31 H/7); Commission géologique du Canada, Étude 83-30, 8 p., Carte 4-1983, échelle 1:50 000.
1984b: Réévaluation des systèmes morainiques du sud du Québec (Wisconsinien supérieur); Commission géologique du Canada, Étude 83-29, 20 p.

Prichonnet, G., Cloutier, M., et Doiron, A.
1982a: Données récentes lithostratigraphiques. Nouveaux concepts sur la déglaciation wisconsinienne, en bordure des Appalaches au sud du Québec; 50e Congrès, ACFAS, Montréal, Livret guide d'excursion, 51 p.

Prichonnet, G., Doiron, A., et Cloutier, M.
1982b: Le mode de retrait glaciaire tardiwisconsinien sur la bordure appalachienne au sud du Québec; Géographie physique et Quaternaire, vol. 36, p. 125-137.

Quigley, R.M. and Dreimanis, A.
1972: Weathered interstadial green clay at Port Talbot, Ontario; Canadian Journal of Earth Sciences, v. 9, p. 991-1000.

Rencz, A.N. and Shilts, W.W.
1980: Nickel in soils and vegetation of glaciated terrains; in Nickel in the Environment, J.D. Nriagu (ed.); John Wiley and Sons, New York, p. 151-188.

Richard, P.
1971: Two pollen diagrams from the Québec city area, Canada; Pollen et Spores, v. 13, p. 523-559.
1973: Histoire postglaciaire comparée de la végétation dans deux localités au nord du parc des Laurentides, Québec; Naturaliste Canadien, vol. 100, p. 577-590.
1975a: Contribution à l'histoire postglaciaire de la végétation dans la plaine du Saint-Laurent: Lotbinière et Princeville; Revue de géographie de Montréal, vol. 29, p. 95-107.
1977: Végétation tardiglaciaire au Québec méridional et implications paléoclimatiques; Géographie physique et Quaternaire, vol. 31, p. 161-176.
1978a: Histoire tardiglaciaire et postglaciaire de la végétation au mont Shefford, Quebec; Géographie physique et Quaternaire, vol. 32, p. 80-93. Richard, p. et Poulin, P.
1976: Un diagramme pollinique au Mont des Eboulements, région de Charlevoix, Québec; Journal canadien des sciences de la terre, vol. 13, p. 145-156.

Richard, S.H.
1975b: Surficial geology mapping: Ottawa Valley Lowlands (Parts of 31 G,B and F); in Report of Activities, Part B, Geological Survey of Canada, Paper 75-1B, p. 113-117.
1978b: Age of Champlain Sea and "Lampsilis Lake" episode in the Ottawa-St. Lawrence Lowlands; in Current Research, Part C, Geological Survey of Canada, Paper 78-1C, p. 23-28.
1980: Surficial geology: Papineau-Wakefield region, Québec; in Current Research, Part C, Geological Survey of Canada, Paper 80-1C, p. 121-128.
1982: Géologie de surface, Vaudreuil, Québec-Ontario; Commission géologique du Canada, Carte 1488A, échelle 1:50 000.

Risi, J., Brunette, C.E., Spence, D., et Girard, H.
1953: Étude chimique des tourbes du Québec. II-Tourbière du lac-à-la-Tortue, comté de Laviolette; Ministère des mines du Québec, Rapport Préliminaire 281, 31 p.

Rodrigues, C.G.
1987: Late Pleistocene invertebrate macrofossils, microfossils and depositional environments of the western basin of the Champlain Sea; in Quaternary Geology of the Ottawa Region, Ontario and Quebec, R.J. Fulton (ed.); Geological Survey of Canada, Paper 86-23.

Rodrigues, C.G. and Richard, S.H.
1983: Late glacial and postglacial macrofossils from the Ottawa-St. Lawrence Lowlands, Ontario and Quebec; in Current Research, Part A, Geological Survey of Canada, Paper 83-1A, p. 371-379.
1985: Temporal distribution and significance of Late Pleistocene fossils in the western Champlain Sea basin, Ontario and Quebec; in Current Research, Part B, Geological Survey of Canada, Paper 85-1B, p. 401-411.
1986: An ecostratigraphic study of late Pleistocene sediments of the western Champlain Sea basin, Ontario and Quebec; Geological Survey of Canada, Paper 85-22, 33 p.

Rondot, J.
1968: Nouvel impact météoritique fossile: la structure semi-circulaire de Charlevoix; Journal canadien des sciences de la terre, vol. 5, p. 1305-1317.
1974: L'épisode glaciaire de Saint-Narcisse dans Charlevoix, Québec; Revue de géographie de Montréal, vol. 28 p. 375-388.

Roy, T.
1837: On the ancient state of the North American continent; Geological Society of London, Proceedings, v. 2, p. 537-538.

Russell, D.J., Graham, J., and White, O.L.
1982: Study of surface stress-release phenomena, southern Ontario; Ontario Geological Survey, Miscellaneous Paper 106, p. 115-116.

Rust, B.R.
1977: Mass flow deposits in a Quaternary succession near Ottawa, Canada: Diagnostic criteria for subaqueous outwash; Canadian Journal of Earth Sciences, v. 14, p. 175-184.
1978: Mass flow deposits in a Quaternary succession near Ottawa, Canada: diagnostic criteria for subaqueous outwash: Reply; Canadian Journal of Earth Sciences, v. 15, p. 329-330.

Sado, E.V., White, O.L., Barnett, P.J., and Sharpe, D.R.
1983: The glacial geology, stratigraphy, and geomorphology of the North Toronto area: a field excursion; Symposium, Correlation of Quaternary Chronologies, York University, Abstracts with Program and Field Guide, p. 505-517.

Sanford, J.T., Martini, I.P., and Mosher, R.E.
1972: Niagaran stratigraphy: Hamilton, Ontario; Michigan Basin Geological Society Annual Field Excursion, Guidebook, 89 p.

Saunderson, H.C.
1975: Sedimentology of the Brampton esker and its associated deposits: an empirical test of theory; Society of Economic Paleontologists and Mineralogists, Special Publication no. 23, p. 155-176.
1976: Paleocurrent analysis of large-scale cross-stratification in the Brampton esker, Ontario; Journal of Sedimentary Petrology, v. 46, p. 761-769.
1977a: Grain size characteristics of sands from the Brampton esker; Zeitschrift für Geomorphologie, v. 221, p. 44-56.
1977b: The sliding-bed facies in esker sands and gravels: a criterion for full-pipe (tunnel) flow?; Sedimentology, v. 24, p. 623-638.

Savoie, L. et Richard, P.
1979: Paléophytogéographie de l'épisode de Saint-Narcisse dans la région de Sainte-Agathe, Québec; Géographie physique et Quaternaire, vol. 33, p. 175-188.

Schneider, A.F.
1983: Lithologic and stratigraphic evidence for a late mid-Woodfordian proglacial lake in the Lake Michigan basin (abstract); Abstracts with Program, Geological Society of America, v. 15, p. 680.

Schroeder, J. et Beaupré, M.
1985: Impacts des cavités glaciotectoniques sur l'aménagement urbain de Montréal, Canada; Annales de la Société géologique de Belgique, T. 108, p. 69-75.

Schroeder, J., Beaupré, M., and Cloutier, M.
1986: Ice-push caves in platform limestones of the Montréal area; Canadian Journal of Earth Sciences, v. 23, p. 1842-1850.

Schroeder, J., Roberge, J., Beaupré, M.E., Harington, C.R., Bélanger, Y., Boily, P., et Caron, D.
1980: Le Karst de la plate-forme de Boischâtel et le karst barré de la Rédemption, état des connaissances; Société Québécoise de spéléologie, Livret guide de l'excursion de l'Association Québécoise pour l'étude du Quaternaire, 110 p.

Scott, J.S.
1976: Geology of Canadian Tills; in Glacial Till, R.F. Leggett (ed.); Royal Society of Canada, Special Publication no. 12, p. 50-66.

Sharpe, D.R.
1979: Quaternary geology of the Merrickville area, southern Ontario; Ontario Geological Survey, Report 180, 54 p.
1982: Allan Park kame-delta; in International Association of Sedimentologists Guidebook for Excursion 11A: p. 53-60.

Sharpe, D.R. and Barnett, P.J.
1985: Significance of sedimentological studies on the Wisconsinan stratigraphy of southern Ontario; Géographie physique et Quaternaire, v. 39, p. 255-273.

Shilts, W.W.
1970: Pleistocene geology of the Lac-Mégantic region, southeastern Quebec, Canada; unpublished PhD dissertation, Department of Geology, Syracuse University, Syracuse, 154 p.
1973: Glacial dispersal of rocks, minerals, and trace elements in Wisconsinan Till, southeastern Quebec, Canada; in The Wisconsinan Stage, R.F. Black, R.P. Goldthwait, and H.B. Willman (ed.); Geological Society of America, Memoir 136, p. 189-219.
1976: Mineral exploration and till; in Glacial Till, R.F. Legget (ed.); Royal Society of Canada, Special Publication no. 12, p. 205-224.
1978: Detailed sedimentological study of till sheets in a stratigraphic section, Samson River, Quebec; Geological Survey of Canada, Bulletin 285, 30 p.
1981: Surficial geology of the Lac-Mégantic area, Quebec; Geological Survey of Canada, Memoir 397, 102 p.

Sigleo, W.R. and Karrow, P.F.
1977: Pollen-bearing Erie Interstadial sediments from near St. Mary's, Ontario; Canadian Journal of Earth Sciences, v. 14, p. 1888-1896.

Sirkin, L.
1981: Wisconsinan glaciation of Long Island, New York, to Block Island, Rhode Island; in Late Wisconsinan Glaciation of New England, G.J. Larson and B.D. Stone (ed.); Kendall/Hunt, Dubuque, Iowa, p. 35-59.

Sly, P.G. and Lewis, C.F.M.
1972: The Great Lakes of Canada: Quaternary geology and limnology; 24th International Geological Congress, Montréal, Guidebook for Excursion A43, 92 p.

Spencer, J.W.
1890: Origin of the basins of the Great Lakes of America; American Geologist, v. 7, p. 86-97.
1907: The falls of Niagara, their evolution and varying relations to the Great Lakes, characteristics of the power and the effect of its diversion; Geological Survey of Canada, Report 970, 490 p.

Stalker, A. MacS.
1948: A study of erosion surfaces in the southern part of the Eastern Townships of Quebec; unpublished MSc thesis, Department of Geology, McGill University, Montréal, Quebec.

Stanley, G.M.
1936: Lower Algonquin beaches of Penetanguishene Peninsula; Geological Society of America, Bulletin, v. 47, p. 1933-1960.
1937: Lower Algonquin beaches of Cape Rich, Georgian Bay; Geological Society of America, Bulletin, v. 48, p. 1665-1686.

Stewart, D.P. and MacClintock, P.
1969: The surficial geology and Pleistocene history of Vermont; Vermont Geological Survey, Bulletin 31, 251 p.

Stuiver, M., Heusser, C.J., and Yang, I.C.
1978: North American glacial history extended to 75 000 years ago; Science, v. 200, no. 4337, p. 16-21.

Taylor, F.B.
1913: The moraine systems of southwestern Ontario; Canadian Institute Transactions, v. 10, p. 57-79.

Taylor, G.
1936: Topographic control in the Toronto region; Canadian Journal of Economic and Political Science, v. 2, p. 1-19.

Terasmae, J.
1958: Contributions to Canadian palynology, Part II: non-glacial deposits in the St. Lawrence Lowlands, Quebec; Geological Survey of Canada, Bulletin 46, p. 13-28.
1960a: Contributions to Canadian palynology no. 2, Part I — A palynological study of post-glacial deposits in the St. Lawrence Lowlands, Part II — A palynological study of Pleistocene interglacial beds at Toronto, Ontario; Geological Survey of Canada, Bulletin 56, 40 p.
1960b: Géologie des dépôts meubles de la région de Cornwall, Ontario et Québec; Commission géologique du Canada, Étude 60-28, 4 p., carte 4-1960, échelle 1:63 360.
1965: Geological Survey palynological studies; Geological Survey of Canada, Paper 61-1, p. 158-159.
1968: A discussion of deglaciation and the boreal forest history in the northern Great Lakes region; Proceedings of the Entomological Society of Ontario, v. 99, p. 31-43.

Terasmae, J. and LaSalle, P.
1968: Notes on late-glacial palynology and geochronology at St-Hilaire, Québec; Canadian Journal of Earth Sciences, v. 5, p. 249-257.

Terasmae, J., Karrow, P.F., and Dreimanis, A.
1972: Quaternary stratigraphy and geomorphology of the eastern Great Lakes region of southern Ontario; 24th International Geological Congress, Montréal, Guidebook, Excursion A42, 79 p.

Thomas, R.H.
1977: Calving bay dynamics and ice sheet retreat up the St. Lawrence valley system; Géographie physique et Quaternaire, v. 31, p. 347-356.

Thomas, R.L., Kemp, A.L., and Lewis, C.F.M.
1973: The surficial sediments of Lake Huron; Canadian Journal of Earth Sciences, v. 10, p. 226-271.

Tippet, R.
1964: An investigation into the nature of the layering of deep-water sediments in two eastern Ontario lakes; Canadian Journal of Botany, v. 42, p. 1693-1709.

Veillette, J.J. and Nixon, F.M.
1984: Sequence of Quaternary sediments in the Bélanger sand pit, Pointe-Fortune, Québec-Ontario; Géographie physique et Quaternaire, v. 38, p. 59-68.

Vincent, J-S.
1989: Quaternary geology of the southeastern Canadian Shield; in Chapter 3 of Quaternary Geology of Canada and Greenland, R.J. Fulton (ed.); Geological Survey of Canada, Geology of Canada, no. 1 (also Geological Society of America, The Geology of North America, v. $\overline{K-1}$).

Vreeken, W.J.
1981: Distribution and chronology of freshwater marls between Kingston and Belleville, Ontario; Canadian Journal of Earth Sciences, v. 18, p. 1228-1239.

Wagner, F.J.E.
1970: Faunas of the Pleistocene Champlain Sea; Geological Survey of Canada, Bulletin 181, 104 p.

Wall, R.E.
1968: A sub-bottom reflection survey in the central basin of Lake Erie; Geological Society of America, Bulletin, v. 79, p. 91-106.

Warner, B.G. and Barnett, P.J.
1986: Transport, sorting, and reworking of late Wisconsinan plant microfossils from Lake Erie, Canada; Boreas, v. 15, p. 323-329.

Warner, B.G., Morgan, A.V., and Karrow, P.F.
1988: A Wisconsinan interstadial Arctic flora and insect Fauna from Clarksburg, southwestern Ontario, Canada; Palaeogeography, Palaeoclimatology Palaeoecology, v.

Watt, A.K.
1954: Correlation of the Pleistocene geology as seen in the subway, with that of the Toronto region, Canada; Geological Association of Canada, Proceedings, v. 6, part II, p. 68-81.

Westgate, J.A. and Dreimanis, A.
1967: The Pleistocene sequence at Zorra, southwestern Ontario; Canadian Journal of Earth Sciences, v. 4, p. 1127-1143.

White, O.L.
1975: Quaternary geology of the Bolton area, southern Ontario; Ontario Division of Mines, Geological Report 117, 118 p.

White, O.L. and Karrow, P.F.
1971: New evidence for Spencer's Laurentian River; Proceedings of the 14th Conference on Great Lakes Research, Ann Arbor, Michigan, p. 394-400.

White, O.L., Karrow, P.F., and Macdonald, J.R.
1974: Residual stress relief phenomena in southern Ontario; Proceedings of the 9th Canadian Rock Mechanics Symposium, Mines Branch, Department of Energy, Mines and Resources, Ottawa, p. 323-348.

Williams, N.E. and Morgan, A.V.
1977: Fossil caddisflies (Insecta: Trichoptera) from the Don Formation, Toronto, Ontario, and their use in paleoecology; Canadian Journal of Zoology, v. 55, p. 519-527.

Williams, N.E., Westgate, J.A., Williams, D.D., Morgan, A., and Morgan, A.V.
1981: Invertebrate fossils (Insecta: Trichoptera, Diptera, Coleoptera) from the Pleistocene Scarborough Formation at Toronto, Ontario, and their paleoenvironmental significance; Quaternary Research, v. 16, p. 146-166.

Wilson, A.E.
1964: Geology of the Ottawa-St. Lawrence Lowland, Ontario and Quebec; Geological Survey of Canada, Memoir 241, 66 p., Map 852A scale 1:253 440, Maps 414A, 413A scale 1:63 360.

Wold, R.J. and Hinze, W.J. (ed.)
1982: Geology and tectonics of the Lake Superior basin; Geological Society of America, Memoir 156, 280 p.

Authors' Addresses

P.F. Karrow
Department of Earth Sciences
University of Waterloo
Waterloo, Ontario
H2L 3G1

S. Occhietti
Département de géographie
Université du Québec à Montréal
C.P. 8888, Montréal, Québec
H3C 3P8

Printed in Canada

Chapter 5
QUATERNARY GEOLOGY OF THE ATLANTIC APPALACHIAN REGION OF CANADA

Tree stumps partly buried by marine sand, Louis Head, Nova Scotia. A forest occupied this site when sea level was lower; recent crustal movement is causing sea level to rise so the site is being drowned by the sea. Courtesy of D.R. Grant. 200183-H.

Chapter 5

QUATERNARY GEOLOGY OF THE ATLANTIC APPALACHIAN REGION OF CANADA

Douglas R. Grant

SUMMARY

Knowledge of the Quaternary geology and sequence of events has developed spasmodically, more or less dependent on the pace of systematic mapping. A century ago in the early days of the glacial theory, and in the initial reconnaissance phase of geological exploration, attention was focused mainly on determining the direction and sequence of glacier movements. Although this approach was prompted by economic considerations, specifically the search for gold by means of displaced erratics, it did serve to successfully elucidate the general pattern which showed the activity of local glacial centres. Study of uplifted shorelines supported the concept of two independent ice domains on Newfoundland and the Maritimes, separated by a largely ice-free Gulf of St. Lawrence which Laurentide glaciers infrequently invaded.

Then, for several decades in this century, while regional fieldwork declined in popularity, the notion became entrenched that the last event was an invasion by a single northern ice sheet, as in the mid-continent region, and that all radial patterns were late, local, short-lived diversions consequent on marine calving of the super ice sheet. In the last 15 years resumption of systematic surveys and special stratigraphic studies have revalidated the original concept that while the Wisconsinan glacial stage may have included an intermediate regional ice sheet phase, it began and ended as a complex of coalescent local ice caps, some of which were at times situated on the emergent shelf. Most deposits and features in Gaspésie, Maritime Provinces, and Newfoundland are the product of the radial regime. Only in the northern Gulf of St. Lawrence where Laurentide ice was the last event, and on the shelf where the outer margins of the large adjacent ice complexes terminated, is the pattern one of simpler unidirectional movement.

Lately the pattern of delevelled paleoshores has played a decisive role in corroborating glaciological reconstructions of the last glacial maximum. Numerical simulations by computer show a two-centred ice system satellitic to the Laurentide Ice Sheet. Vestiges of earlier major advances are detected geomorphically on mountain tops in Newfoundland. Some summits there and possibly on Cape Breton Island were never overridden. The seaward slope of the limits reinforces the belief that Newfoundland has always had its own ice caps; Laurentide ice was largely barred from reaching it. Two major early glaciations are thus inferred and are speculated to date from oxygen isotope stages 6 and 12. Deep sea sediments point to the same periods as times of maximal erosion.

The low glacial limits may place constraints on estimates of the extent of advances on the shelf where deposits are less definitive of actual glacier presence. However, better appreciation of glaciological style in major marine-based ice masses, as learned from Antarctica, introduces new components to the dynamic model. The most important is the realization that cold-based conditions must have prevailed at times in certain parts of the area. Fixed, frozen, protective basal ice is the only condition that can explain the juxtaposition of scoured rock, fluted till, and ancient regolith on the highlands of Cape Breton, Cobequids, New Brunswick, and Gaspésie. Frozen-bed plastic shearing is the only process to deform and interfold bedrock and sediment in southwestern Nova Scotia and Magdalen Islands. During maximal cover, ice streams linked to marginal calving bays may have swept the region, creating the till plumes and erratics trains across Nova Scotia. Offshore, ice shelves were a probable fixture where the ice sheet went afloat in shelf basins. These were capable of bridging to regrounded portions on outer banks where possibly semi-independent small spreading centres or ice rises became localized. By this system, till was deposited over and around prominences as grounding line moraines which intertongue with coeval subshelf/marine drift. No more can a submarine moraine signify an ice margin. The problem of delimiting glacier extent means defining ice state. Thus the land-based Appalachian ice, which included transient ice streams and patches of frozen base, was fringed by a marine-based component characterized by ice shelves, rises, and calving bays. The general succession of changes from one regime to the other, and the broad lines of glacial evolution onshore and offshore as the Wisconsinan glacier complex waxed and waned, is fairly well understood, but the exact timing is poorly controlled and there are contradictions between the marine and terrestrial sequences.

Grant, D.R.
1989: Quaternary geology of the Atlantic Appalachian region of Canada; Chapter 5 *in* Quaternary Geology of Canada and Greenland, R.J. Fulton (ed.); Geological Survey of Canada, Geology of Canada, no.1 (<u>also</u> Geological Society of America, The Geology of North America, v. K-1).

The stratigraphic succession for the Middle and Late Pleistocene is based on fragmentary, scattered, partial sequences. A provisional correlation of lithostratigraphic units links sedimentary successions in widely separated areas. As a simplified summary it correlates the main elements in the rock record with the inferred climatic events and with sea-surface temperature variations based on oxygen isotope ratios. The apparent agreement in the number and timing of events lends support to the notion that the 23 ka astronomic insolation cycle, which largely determines oceanic temperature, may also have governed the regime of the maritime ice cap complex. Thus the last three largest glaciations of the 100 ka insolation cycles are oxygen isotope stages 12, 6, and 4/3/2. The Wisconsinan Stage is divisible into three pulses; the last, which culminated about 15 ka, was evidently the smallest. The final general ice retreat, climatic warming, and forest succession was interrupted by a severe cooling 11-10 ka which caused readvance of glacial remnants, reversion to tundra, and reactivation of slope processes so that organic deposits became buried. The reversal, which is linked to a southward shift of the oceanic Polar Front, further supports the belief that the Appalachian regional climate is more sensitive to maritime oceanographic conditions than to continental atmospheric dynamics.

The record of Quaternary geological evolution in the Canadian Appalachian region is thus primarily one of broad climatic change which caused glacier variations in five major cycles, and attendant sea level fluctuations which continue towards ultimate isostatic and geoidal equilibrium. A variety of other crustal movements and rock distortions have more deep seated causes. The history of the recent past thus serves to give a perspective for the present condition and to provide a basis for future prognosis.

In the coming years, a major preoccupation will be the mechanisms and processes of global change, both natural and anthropogenic, and their impact on environment and development. Understanding global change and its local expression in a maritime region will depend on improving our knowledge of recent past changes so as to provide a framework for evaluating contemporary man made perturbations. Multidisciplinary stratigraphic study, backed by adequate chronometric control, holds the key for defining process/response relationships.

A well documented record of recent Quaternary evolution, coupled with numerical modelling, is the soundest basis for predicting the future effects of current or expected changes.

INTRODUCTION

This report deals with the southeastern part of Canada, and comprises separate accounts of six distinct geographic divisions: the island of Newfoundland, the Maritime Provinces (Nova Scotia, Prince Edward Island, New Brunswick), Gaspésie of southern Quebec, and the Gulf of St. Lawrence including Magdalen Islands, Anticosti Island, and the Quebec North Shore (Fig. 5.1). The division is necessary because of the isolation of each area and because studies in each have been carried out largely independently. Although a certain degree of commonality in the results is emerging, correlations remain tenuous. Adjacent offshore areas of Gulf of Maine and the Continental Shelf are integral parts of the region, but they are the subject of a separate volume (Piper et al., 1989). Results from submarine areas, therefore, are mentioned briefly only where pertinent to onshore conditions. The interpretation of regional events such as glaciation and sea level change, and reconstruction of paleogeographic conditions is given for the entire area. A condensed version has been given by Grant (1987a).

As a thematic review, this explanatory description embodies an analysis of the deposits and features which are currently referred to the Quaternary Epoch, viewed in the context of regional geological and geomorphic evolution. It proceeds from an outline of pre-Quaternary geology insofar as bedrock distributions have determined the lithology and texture of Quaternary deposits, and through their influence on physiography, have affected the extent and vigour of Quaternary processes in general. Physiography is discussed in terms of geomorphic evolution to show how an interpretation of the origin and age of landscapes and landforms reveals a complex history of crustal warping and stratigraphic superposition, of which the Quaternary is the latest phase.

Bedrock geology and tectonics

Knowledge of lithological distributions and crustal movements is pertinent to Quaternary interpretations which seek to explain the composition and derivation of deposits, the juxtaposition of contrasting erosional and depositional glacial regimes, the structural control of landscapes, and the imprint of neotectonics on isostasy. Summaries by Poole (1976) and Williams (1979) form the basis of the following outline and the reader is referred to Keen and Williams (1989) for the most recent synthesis of the geology of submarine areas and the shelf and to Williams (in press) for the most recent synthesis of the land area.

The region comprises four main tectonostratigraphic divisions: a craton, a platform, a fold belt, and a shelf prism (Fig. 5.2). The North Shore of Gulf of St. Lawrence is the Grenville Province of the Precambrian Shield, which consists of Precambrian granite plutons and gneisses with infolded metasedimentary and volcanic belts. A fringe of relatively flat-lying lower Paleozoic cratonic sedimentary rock of the St. Lawrence Platform lies on crystalline basement along the Precambrian Shield margin. When the proto-Atlantic Ocean closed and American and African plates collided, former continental terraces were crumpled to produce a chain of fold mountains, the Appalachian Orogen. It consists mainly of highly deformed lower Paleozoic sedimentary and volcanic sequences intruded by Devonian and Carboniferous granite. Overthrust remnants of older crystalline crust stand as higher massifs within the sedimentary terrane. The present uplands and highlands are the worn remnants of the orogen. Block faulting and broad downwarping in Carboniferous time created narrow grabens in Nova Scotia and Newfoundland and a deep basin under Gulf of St. Lawrence which was filled with mainly clastic sediments and evaporites. By the end of the Permian and locally into Triassic time, denudation had reduced the region to a low level.

In Triassic time prior to inception of the present Atlantic basin, renewed tension created the Fundy Epigeosyncline which comprises narrow basins through southern Newfoundland, central Nova Scotia and Bay of Fundy. Broad subsidence produced a sag of the Paleozoic basement unconformity under the Gulf of Maine which persists as a deep reentrant separating the aligned coastal segments of Nova Scotia and Massachusetts. Redbeds and volcanic sheets filled the depressions. Plate separation prior to Early Jurassic time was followed by growth of the present continental terrace as a sedimentary prism on the trailing edge of the American plate. Rejuvenation of drainage systems maintained weak rock areas as lowlands, while hinterlands became elevated and dissected. The resulting detritus was delivered to the continental margin throughout Mesozoic and Cenozoic time.

Cenozoic episodes of uplift and fluctuations of base level can be inferred from several late-stage degradation surfaces on land and from facies changes and unconformities within the Cenozoic part of the shelf sequence. The latter records several broad swings of relative sea level from which periods of continental uplift and dissection may be inferred (Jansa and Wade, 1975). Shelf strata accumulated as the continental margin subsided and tilted; deposition was interrupted periodically by planation and dissection. Onshore there is no sedimentary record of conditions during the Tertiary and early Quaternary, so it is assumed that this area was undergoing erosion and isostatic uplift. The resulting planation developed several surfaces but correlation of these is speculative and is dealt with further in the section on geomorphic evolution.

After an early Tertiary fall of sea level and fluvial incision, there were two major marine incursions into fluvial lowlands. The present configuration of land and sea constitutes a third deep submergence of the continental margin. Evidence of Quaternary age crustal warping, and sea level change is given in a later section.

Physiography and geomorphic evolution

Consideration of landscape distributions and their character leads to an appreciation of what features of the present topographic makeup have been inherited from pre-Quaternary times and hence what the imprint of Quaternary processes has been. Conversely, the arrangement of relief elements has had a marked impact on the nature and location of the major processes of glaciation, marine transgression, and weathering. Hence, a series of glaciations of comparable magnitude will be progressively more confined to lower areas as each erosional episode enlarges and deepens the valleys that carry ice from the highlands. Thus even though glaciations may have been of similar size, each younger one would cover a smaller part of the highland. Further, the greater the volume of localized excavation thus resulting, the greater the possible differential isostatic uplift of pre-Quaternary reference surfaces such as paleoplains and peneplains.

Relief map and physiographic divisions

No systematic regional synthesis or interpretation of geomorphic evolution has yet been attempted, although physiographic units have been proposed by Bostock (1964) and Sanford and Grant (1976). Several interpretations of smaller areas with seemingly correlatable elements can be brought together to provide a general overview which gives new insight into Cenozoic geomorphic and tectonic evolution. To illustrate the topographic diversity, a simple relief diagram shows four elevational intervals above and below sea level (Fig. 5.1). Comparison with the generalized geological map (Fig. 5.2) shows that the region is characterized by flat-topped resistant rock uplands of varying size and elevation, rising more or less abruptly from weak rock lowlands. Many of the lowlands continue offshore as shallow banks cut by numerous basins and long submarine channels. The mosaic of different levels which developed as the result of a lengthy history of uplift and dissection, beginning in the Precambrian and continuing through to present, is discussed below. Some of the earlier diastrophic movements may have continued into the Quaternary, and some additional distortions may be of neotectonic origin.

Most of the physiographic divisions originate from major differences in tectonic-stratigraphic makeup. Differential erosion and uplift have accentuated the structural and lithological contrasts, and Quaternary glaciation has superimposed a small-scale superficial imprint. Six major recognized divisions are thus present (Sanford, in press). On the north the Precambrian Shield consists of the Laurentian Region which is subdivided into Laurentian Highlands and Mecatina Plateau — vast rolling expanses with ancient south-trending fluvial systems that are incised 200-500 m below the general level. Glacial scouring has further etched out the small-scale structural elements. Bordering the Precambrian Shield in the central part of the region is the East St. Lawrence Lowland which has developed on a lower Paleozoic platformal succession of carbonate, shale, and sandstone. The low and generally flat terrain reflects the relatively undisturbed character of the underlying strata. Gross lithology, however, determined the position of major topographic elements which developed by fluvial and glacial erosion. For example, resistant sandstone and carbonate beds underlie Anticosti, Mingan, and Dorset Ridges; Bank Beaugé is a structural drape of carbonates over a Precambrian basement ridge. Conversely, weaker shale belts underlie the various troughs and channels. Even in small scale, localized folding and fracture patterns are set in relief by solution and selective glacial scouring.

The lowland merges southward with the Maritime Plain, a gently undulating surface which slopes eastward from 200 to 100 m across New Brunswick and under Magdalen Shallows where it is truncated by Laurentian Channel. The area is underlain by gently deformed Paleozoic clastic rocks, and fold axes are weakly manifest in the major promontories and bays. Northumberland Strait marks a belt of weaker rocks, and Prince Edward Island is a cuesta of resistant Permian clastics. Magdalen Islands project from the plain partly because of the occurrence of resistant volcanic necks and partly because of salt diapirism.

The submerged lowlands continue farther offshore as the present continental shelf, which may be divided into Grand Banks of Newfoundland and Scotian Shelf including Bay of Fundy. The shelf is a mosaic of smooth plateaus and mesas separated by channels and basins partly relict from pre-Quaternary fluvial dissection and locally modified by glaciation. Piper et al. (1989) treat this region in detail and some additional interpretation is given below under planation surfaces and the degree of Quaternary erosion.

The bordering land areas of Newfoundland, Maritime Provinces, and Gaspésie are locally fringed by the submarine lowlands but are mainly a complex of rolling uplands, and highland plateaus (Fig. 5.1). The Nova Scotia Upland and Newfoundland Upland, the New Brunswick Highlands,

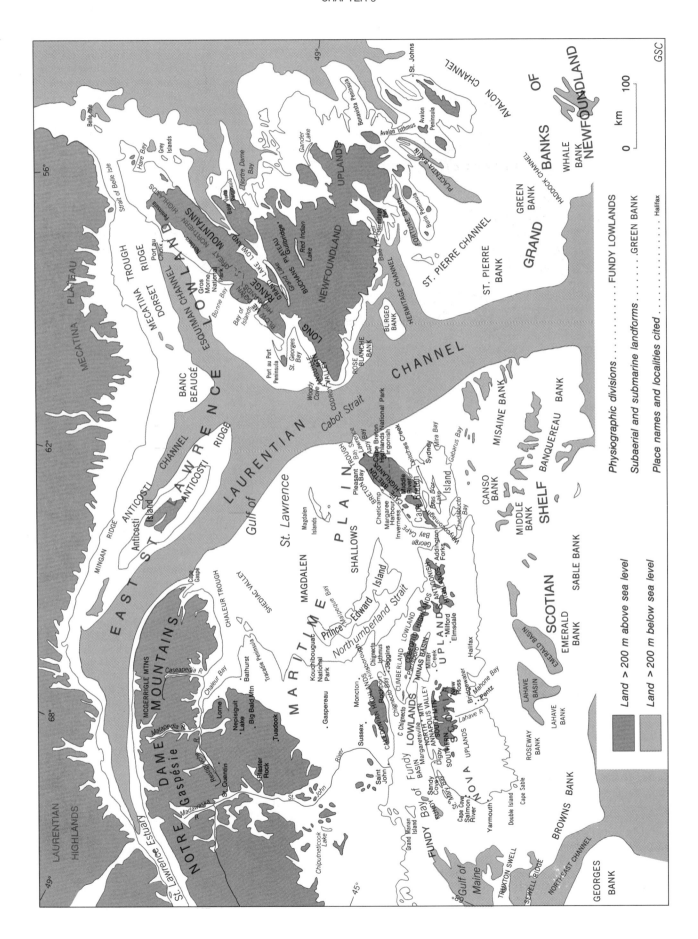

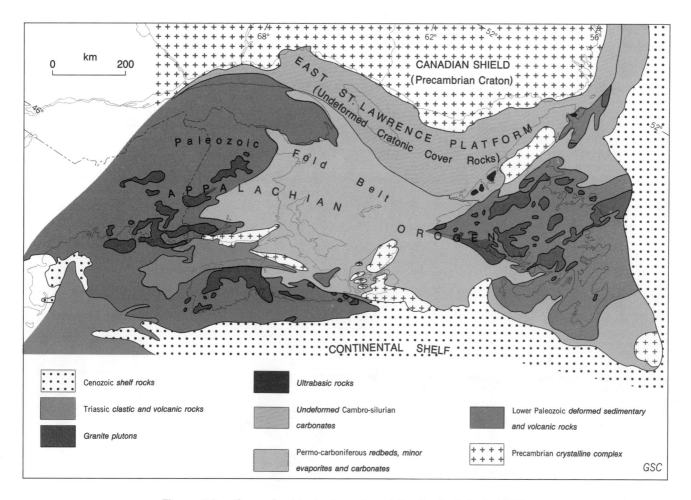

Figure 5.2. Generalized bedrock geology (after Sanford et al., 1979).

and the Notre Dame Mountains are the worn remnants of the older, more resistant rocks of the Appalachian Orogen. Locally, the relief fabric reflects the dominant northeast trend of folding, but for the most part the drainage network is consequent on intersecting tilted planation surfaces which form facets around the periphery of the emergent areas. In bare rock terrains, glacial scouring has cut lineaments along fault and joint patterns, and given a ribbed surface to foliated terranes. Finally, the steep-sided plateaus of Nova Scotia Upland and Newfoundland Upland are the truncated remnants of overthrust sheets composed of Grenville age crystalline complexes. Local relief ranges from a few tens to hundreds of metres depending on the degree of later dissection after uplift. Surface condition varies from completely denuded fracture-lineated landscapes of areal scouring to gently undulating mantles of till and residuum.

Figure 5.1. Location map of study area showing land areas above 200 m elevation (red shading) and submarine areas greater than 200 m deep (grey shading), physiographic divisions, subaerial and submarine landforms (based on Sanford and Grant, 1976) and place names and localities referred to.

There is thus a close adjustment of relief and physiography to rock type and structure, while topographic differentiation results from episodes of uplift, dissection, and planation. The degree of Quaternary modification is treated more fully below.

Denudation surfaces

The relief diagram highlights the major positive and negative physiographic elements (Fig. 5.1). Closer inspection of hypsometry on detailed topographic maps and in the field reveals two groups of distinct denudation surfaces of regional extent which form important facets of the topography. One is a set of three or four peneplains that have been recognized in all land areas of the region. They cut across or bevel older surfaces which are known to disappear beneath cover rocks of different ages. The latter are exhumed unconformities or resurrected peneplains, which have been termed "paleoplains" by Ambrose (1964). Figure 5.3 shows these elevated and partly submerged peneplains and paleoplains together with other smaller-scale relief features. Their distribution, age and origin are treated bleow.

Tilted paleoplains (exhumed peneplains)

The oldest paleoplain forms the North Shore of Gulf of St. Lawrence. With a slope of 5 m/km, it cuts a facet on the southern flank of the Hamilton and Mecatina plateaus on the Canadian Shield, and represents the essentially unmodified sub-Cambrian unconformity. More steeply inclined, tectonically tilted, portions of this surface are being exhumed along the west flank of the northern Long Range Mountains in Newfoundland. The similarity of level and the glacial-basin character of the buried and exhumed portions have led several authors to argue that much of the present glacial character is not Quaternary but has been inherited from a Late Precambrian glaciation (e.g. Lawson, 1890; Ambrose, 1964; White, 1972; Sugden, 1976; Swett, 1981).

More localized stripped unconformities form coastal facets along northern Chaleur Bay, Burin Peninsula, southwestern Newfoundland, and western Nova Scotia. These cut across or pass beneath Carboniferous and Triassic rocks. Modern coastal cliffs cut in the weak Carboniferous and Triassic rocks do not obscure the fact that the onshore paleoplains continue offshore beneath covers of basinal rock sequences (e.g., King and MacLean, 1976, Map 812H).

Perhaps the best known example of an exhumed unconformity is the Fall Zone Peneplain which truncates pre-Cretaceous basement rocks along the United States Atlantic Coast from Georgia to Massachusetts (Flint, 1963). It may be represented in this region by the gentle ramp along the outer Nova Scotian coast (sub-Cretaceous surface of

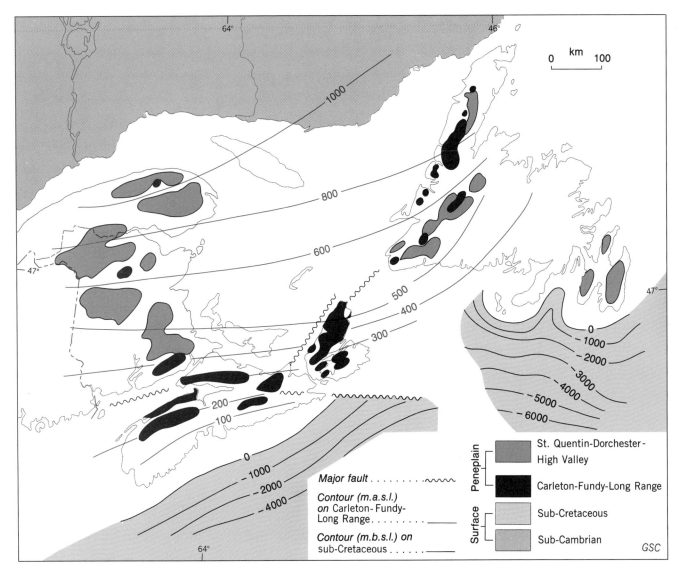

Figure 5.3. Denudation surfaces and tilted peneplains (after Gray et al., 1981, for Gaspésie; Bird, 1972, for New Brunswick; Goldthwait, 1924, and Roland, 1982, for Nova Scotia; Twenhofel and MacClintock, 1940, and Rogerson, 1981, for Newfoundland).

Fig. 5.3). However, it is not clear how this surface, with an onshore gradient of 5 m/km, can continue offshore as the sub-Jurassic unconformity with a slope of 40 m/km (McIver, 1972). More probably, it correlates with the Lower Cretaceous unconformity recognized by King et al. (1974) within the shelf prism. In eastern Nova Scotia the paleoplain is offset by the Chedabucto Fault and is evidently downdropped at least 200 m where it continues along the Cape Breton Island coast. This disturbance may date from the period of Late Cretaceous-early Tertiary deformation along the fault (King and Maclean, 1970) which may also explain why Cretaceous rocks occur downfaulted in valleys below the level of the paleoplain in Nova Scotia (Davies et al., 1984).

Another possible Cretaceous(?) paleoplain may explain the two gently shelving coastal segments along Bonavista Bay and Notre Dame Bay, Newfoundland. Many of the larger valleys in those areas are thought to be relict, which raises questions about the amount of Quaternary glacial remodelling there.

Uplifted peneplains

Bevelling the older exhumed peneplains at much gentler seaward slopes (2 m/km) are a series of three major relict peneplains (Fig. 5.4). The oldest and highest is restricted to remnants on high-standing resistant crystalline terranes; the lower two are developed mainly over intervening weaker sedimentary terranes. Submarine banks, basins, and channels may be submerged equivalents or additional younger degradational levels.

Recognition of these inset relict erosion surfaces by many authors in all parts of the Appalachian region serves as a basis for regional synthesis. Opinion has varied on the exact number of distinct surfaces but the general consensus is that there are three regional peneplains. The location of the two highest of these is shown in Figure 5.3. The ages are largely unknown because Quaternary sediments are the only deposits seen covering these surfaces. Tertiary and Cretaceous dates are generally inferred. Recently, fresh insight has come from studies of the Mesozoic-Cenozoic offshore sequence in which several unconformities representing regional degradation cycles are seen; however, links to the onshore peneplains remain tenuous.

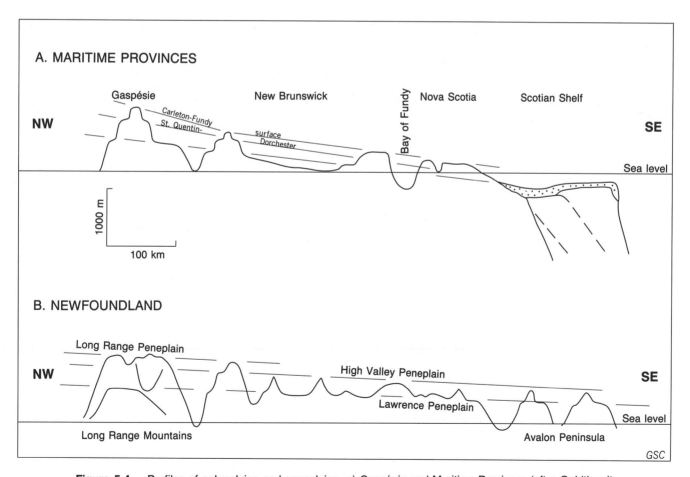

Figure 5.4. Profiles of paleoplains and peneplains. a) Gaspésie and Maritime Provinces (after Goldthwait, 1924; Bird, 1972; King et al., 1974; Gray et al., 1981; Roland, 1982). b) Island of Newfoundland (after Twenhofel and MacClintock, 1940; Rogerson, 1982).

In the Maritime Provinces, Daly (1901) Goldthwait (1913, 1924), Alcock (1935), Bird (1972), Brookes (1972), and Roland (1982) have recognized two and locally three major levels rising northwestward from the Atlantic coast to Gaspésie where an apparently correlative suite has been described by Coleman (1922), Alcock (1926), McGerrigle (1959), and Gray et al. (1981). These surfaces, some of which have been named, are shown schematically in Figure 5.4a. In Newfoundland three similar levels also rising northwestward (Fig. 5.4b) were identified by Twenhofel and MacClintock (1940). In addition, Brookes (1964) noted subordinate intermediate benches on the upland flanks. The spacing, elevation, and slope of the three main surfaces are in harmony with those in the Maritimes and the two sets are therefore tentatively assumed to be coeval. There is a slight suggestion that correlative levels are disjunctive, warped, or inflected across major crustal breaks such as the Chedabucto-Cobequid Fault, across certain tectonostratigraphic divisions like the Triassic graben, and across submarine deeps like the Laurentian Channel. If true, this suggests postplanation disturbance as alluded to by Bird (1972, p. 429). Only detailed definition and correlation of the surfaces will reveal whether there has been Cenozoic faulting, and whether there has been isostatic adjustment to deep erosion.

Dating of the younger planation surfaces is important in order to assess the amount of Quaternary erosion and because they reflect spasmodic upwarp and subsidence which may have continued into Quaternary time. Possible age can be inferred from offshore geology. In one approach, Mathews (1975) compared the volume of offshore sediments to the 2 km relief of the upland or oldest peneplain, and derived a mid-Tertiary (Miocene?) or younger age for the beginning of its uplift. This date is, to a certain extent, corroborated by King et al. (1974) who discerned three major unconformities, each reflecting periods of dissection, within the shelf prism, the youngest being Pliocene(?). Although no onshore peneplain has yet been directly traced to any one of the shelf unconformities, they suggest that the deeps of both Gulf of St. Lawrence and Gulf of Maine are fluvial lowlands which were buried and exhumed twice from beneath a cover of Cretaceous and Tertiary sediment. The outer shelf-edge banks thus appear to be cuestas and mesas which developed during the process of stripping the soft Mesozoic-Cenozoic cover rock from basement. The interbank channels and saddles are remnants of drainage systems which operated at various times when base level was lowered, first to as much as 700 m in the Late Eocene phase, then again to about 300 m below present sea level (King and Maclean, 1970).

Quaternary modification

The degree of Quaternary reshaping of Tertiary and older landscapes, by erosion, deposition, or by change of level and elevation, ranges from nil to great; this is portrayed in Figure 5.5. At the zero end of the scale, parts of the exhumed paleoplains such as adjacent to Strait of Belle Isle are essentially unaltered. Most of the uplifted Tertiary peneplain remnants appear, from their extreme flatness and almost total lack of glacial features, to be in essentially their preglacial condition (Fig. 5.6a). Indeed remnants of deeply altered regolith and saprolite may be found on these surfaces (e.g. Wang et al., 1981; McKeague et al., 1983), thus confirming that there has been little if any glacial erosion or deposition — a point to be further discussed in relation to glacial history and ice limits. At the other end of the scale, judging from the distribution and depth of basins which are the hallmark of glacial action, some areas have suffered considerable glacial degradation. The most obvious indications are the U-shaped glacial troughs which drain the northern Cape Breton Island and Gaspésie tablelands and the spectacular fiords of southern and western Newfoundland (Fig. 5.6b), for example, Bay of Islands, depth 300 m and Bay D'Espoir, depth 762 m — the deepest in eastern North America. Comparison with adjacent V-shaped gorges incised into the upland margins shows that the troughs are localized glacial overdeepenings of former fluvial valleys. On a broader areal basis, the weak rock tracts have generally been glacial depositional areas, whereas most of the resistant rock areas are denuded and pocked with lake basins up to a few kilometres across and a few tens of metres deep. Exceptionally large glacial basins are the Bras d'Or Lakes (depth 282 m) of Nova Scotia and the three serpentine lakes of Newfoundland: Grand, Red Indian, Gander (depth 300 m). Offshore, glacial basins are widespread, broader, and have closures ranging from 100-150 m. This difference in shape is probably due to the fact that the offshore basins are cut in semi-consolidated rocks. The closure (generally in excess of 100 m) of these basins, is clearly the minimum amount of glacial erosion below hypothetical Tertiary fluvial base levels. Glacial modification of preglacial landscapes thus ranges from zero to more than 100 m, and is highly localized. On balance, glacial erosion is generally slight over much of the area. The degree of Quaternary modification and the distribution of glacial basins, both onshore and offshore, provide a useful approximation of overall average extent and vigour of glaciers in the region and can be compared with inferred ice limits based on stratigraphy.

Previous work

Quaternary geology investigations in the region have spanned more than a century and generally reflect changing concepts of global Quaternary history. Although work has proceeded semi-independently in the various areas, and has been based on several approaches at different scales and for a variety of purposes, a common thread in the evolution of concepts is discernible. A large number of reports and maps, both topical and local and regional, provide a fairly consistent basis for an interpretation of Quaternary history and reconstruction of paleogeography (see Map 1704A for surficial geology map coverage and Fig. 5.7b for locations of sites where specific data have been reported). General frameworks of Quaternary events have emerged, although correlations are tenuous and ages speculative. Some aspects remain controversial, diametrically opposed views are still held, and onshore/offshore correlations are uncertain. Nonetheless, opinions are converging on the general framework of Quaternary history.

Newfoundland

Summaries of Quaternary events and the history of Quaternary studies are given by Tucker (1976), Grant (1977b), and Rogerson (1982). Brookes (1982) traced the development of concepts in relation to the traditional wisdom of their day. In these may be recognized three major stages ("Pre-cognitive", "Drift", and "Glacial" phases), which link the progress of understanding to improving accessibility, systematic scientific surveys, importation of applicable concepts, emerging chronometry, and glaciological modelling. In many respects, this evolution of ideas holds true for the Atlantic Appalachian region as a whole.

Inquiry about the glacial history of Newfoundland began a century ago with reports by Kerr (1870), Milne (1874, 1876), and Murray (1883), who first became aware of the evidence for ice action during the early days of the glacial theory in North America. Murray (1883) was the first to propound a hypothesis of glaciation for the whole island; it incorporated both external and local ice sources, using striations, drift sequences, and paleoshores. Thereafter debate centred on the relative roles of Labrador versus local ice, the areal and vertical extent of ice sheets, and the number and ages of separate advances. These controversies arose and many still persist largely because of incomplete knowledge of surface deposits, limited exposure and stratigraphic analysis, opposing interpretations of differently weathered glacial terrains, paucity of radiometric control, and ambiguity of crustal warping patterns as a proxy for ice disposition.

In brief, the major contributors and their findings may be reviewed as follows. De Geer (1892), Daly (1902, 1921), and Fairchild (1918) used isobase maps showing the upwarp of postglacial shores to judge the relative importance of a

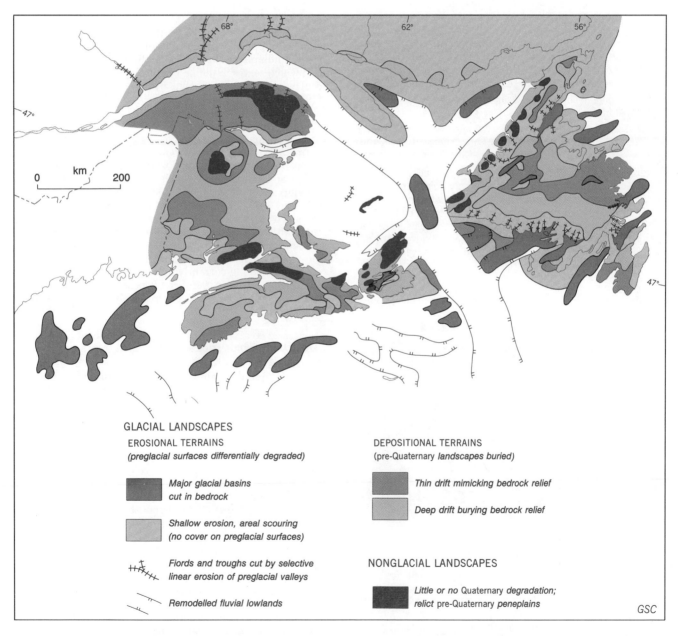

Figure 5.5. Degrees of glacial modification of pre-Quaternary landscapes.

supposed separate Newfoundland ice cap, versus the regional influence of the Laurentide Ice Sheet. Evidence that more than one glaciation was recorded, and that the last advance was from local sources which failed to cover the higher and outer parts of the island was discussed by Coleman (1926) and figured by Antevs (1929). The concept of incomplete glaciation was based on a few exposures of weathered till underlying fresh till (Van Alstine, 1948), and the existence of degraded glacial terrains, or lack of glacial features, on certain high western summits and outlying eastern peninsulas (Coleman, 1926). Fernald (1925) and Lindroth (1963) advanced biological arguments in favour of the inferred nunataks. Flint (1940) and MacClintock and Twenhofel (1940) dismissed the evidence and took the opposing view that only Laurentide ice had overwhelmed the island. Some authors concerned with regional glaciation still hold this view (e.g., Mayewski et al., 1981), notwithstanding the fact that the original observations have been further substantiated by systematic mapping (Grant, 1977a; Brookes, 1977a).

Systematic mapping has led to delimiting of four or more successive ice advances. The latter two are known to be Late Wisconsinan and the earlier ones probably predate the last interglaciation (Grant 1977a; Brookes, 1977b; Tucker and McCann, 1980; Rogerson,1982). Information which supports the concept of an independent Newfoundland ice cap, made up of several separate centres, and which serves to define the history of interaction of flow centres, has been presented by Summers (1949), Widmer (1950), Murray (1955), Jenness (1960), Lundqvist (1965), Grant (1969a, 1974b), Brookes (1970, 1977b), Henderson (1972), Grant and Tucker (1976), and Leckie (1979). Studies of stratigraphic successions of deposits from various competing and coalescent flow centres are hampered by lack of exposures, but detailed studies in two areas, the Red Indian Lake and Deer Lake lowlands (Sparkes 1981, 1983; Vanderveer and Sparkes 1982) and Burin Peninsula (Tucker, 1979; Tucker and McCann, 1980) corroborate the sequence and extent of ice movements that can be inferred from crossing striations and weathering contrasts. Evidence of invasion by Laurentide ice seems limited to the northernmost part of the island (Grant, in press). In summary, the consensus among currently active workers on the island is that glaciers from a number of local centres expanded at least four times to reach progressively lower and less extended limits.

Figure 5.6a. Landscape contrasts, Aspy Valley, northern Cape Breton Island. On the left the Tertiary summit peneplain, only lightly glaciated by external ice sheets, stands at 450 m on Precambrian crystalline rocks which are truncated by the Aspy Fault. On the right Carboniferous mudstone and gypsum underlie a late Tertiary fluvial lowland that is deeply blanketed by tills of a local ice cap. The lowland has been upthrown 14 m along the fault in recent time. 203673-X

Figure 5.6b. Landscape of selective linear erosion and glacial limits, Bakers Brook Pond, Gros Morne National Park, Newfoundland. Outlet glaciers debouched from a highlands plateau ice cap and spread onto the coastal lowland as tidewater piedmont lobes which built lateral moraines (foreground) in the sea at +90 m. Plateau summits adjacent to trough were nunataks during Late Wisconsinan time but were ice covered in pre-Wisconsinan time. 202188-B

Maritime Provinces

Overviews for the Maritime Provinces vary in depth and scope. A detailed summary and interpretation for New Brunswick by Rampton et al. (1984) contains an exhaustive source list on the lengthy history of Quaternary geological exploration and presents many new ideas on the sequence of Quaternary events. For Nova Scotia and surrounding areas, Greenwood and Davidson-Arnott (1972) outlined the evolution of ideas and their factual basis, while Roland (1982) presented a lengthy popular account of the geology and physiography. Systematic mapping in Prince Edward Island has been recast on one map, together with an explanatory description by Prest (1973). From these, the following generalized statements may be drawn.

Observations on the "superficial" geology began with reconnaissance surveys of the Province of Canada when the Geological Survey was founded in 1842. Logan (1863) and Dawson (1855) were the first to report on Quaternary glacial and associated marine deposits. Prompted then, as now, by economics, the first main concerns were to locate the centres of glacial outflow and to map their ice flow trends in order to trace transported mineralized boulders to their sources.

New Brunswick was surveyed at an early stage by Matthew (1872, 1879, 1894a, b), but it was Chalmers (1881, 1882, 1885, 1886, 1887a, b, 1888, 1890, 1893, 1895, 1901, 1902) who prepared surficial geology maps for most of the province, and for part of Nova Scotia. That knowledge formed the basis for his ideas on the interplay of local and external ice masses and the role of floating ice. His interpretations remain essentially valid today. Ganong also made abundant observations and drew attention to many glacial and nonglacial features which remain crucial to current paleogeographic reconstructions (cf. Ganong, 1896, 1913). More recent work has resulted in many maps and reports (cf. Lee, 1953, 1957; Gadd, 1973; Brinsmead, 1974, 1979; Bélanger, 1978; Thibault, 1978, 1982; Rampton and Paradis, 1981a, 1981b, 1981c; Gauthier, 1982; Seaman, 1982a, 1982b) that have been summarized by Rampton et al. (1984). In essence, the current understanding is that glaciation of New Brunswick was mainly by local ice caps located over lowlands as well as uplands; Laurentide ice invaded only upper Saint John River valley and Chaleur Bay. An offshore centre on Magdalen Shallows is postulated. Several phases of glaciation are recognized; the last extended more or less to the present shore.

In Nova Scotia early detailed bedrock mapping of the entire province by H. Fletcher and E.R. Faribault, between 1877 and 1921, provided much precise information on glaciation in terms of ice flow directions based on striations and erratics. Between 1874 and 1890 Honeyman produced a score of reports on drift dispersal and iceflow trends (e.g. Honeyman, 1882). Goldthwait's (1924) monograph on physiography and glaciation remains a standard source of information and interpretations of general aspects. Reconnaissance surveys of the mainland part of the province by R.H. MacNeill for the Nova Scotia Research Foundation between 1950 and 1970 produced maps of surface materials and landforms. Mapping in central Nova Scotia by Stevenson (1959) and Hickox (1962) provided evidence for local ice caps. Stratigraphic outlines by Prest (1970) and a hypothesis of glacier behaviour by the shifting and coalescence of local ice caps strongly influenced by calving (Prest and Grant, 1969; Prest et al., 1972) were the first tentative steps toward unifying concepts. Studies of till lithology over the Atlantic Uplands by Grant (1963) and of till mineralogy by Nielsen (1974) revealed stacked drift sheets and divergent ice flow sequences which began to suggest a lengthy and complex history. A sequence of separate glacial advances was revealed through geochemical mapping of tills (Stea and Fowler, 1978, 1979, 1981; Stea, 1982a; Stea and Grant, 1982; Stea et al., 1986) and by topical stratigraphic studies of complex till/organic sections in the central lowlands (Stea and Hemsworth, 1979; Stea, 1982b). Detailed mapping and stratigraphic studies along the Yarmouth-Digby coast and of Cape Breton Island by Grant (1971a, b, 1972b, 1974a, 1975a, 1980b, 1987b) revealed three regional glacial advances since the last interglaciation. From this information regional correlation charts were prepared by Grant (1976), Prest (1977), and Grant and King (1984); the latter also correlates terrestrial sequences with submarine sediments.

Much attention has also been given to the regional variation in postglacial sea level change in order to deduce the extent and timing of the glacier advances responsible for isostatic deformation. Regional maps and reports on the upwarped late glacial paleoshore (marine limit) were prepared by De Geer (1892), Fairchild (1918), Daly (1921), Goldthwait (1924), Flint (1940), Welsted (1976), Wightman and Cooke (1978), and Grant (1980a). In these, the general consensus is that the overall deformation pattern during the latest glacial stage can be explained by separate glacier complexes over Newfoundland and the Maritimes, abutted on the north by the Laurentide Ice Sheet, and with Gulf of St. Lawrence largely free of grounded glaciers. Inverse mathematical modelling of former ice loads using sea level histories at eleven widely separated sites in the region seems to corroborate the hypothesis of limited Late Wisconsinan glaciers (Quinlan and Beaumont, 1982).

Gulf of St. Lawrence

Relatively little is known about the Quaternary geology of this vast inland sea. Analysis of the mainly sandy and muddy surface sediment suggests that Laurentide ice moved southward across parts of the area (Loring and Nota 1969, 1973). The Magdalen Islands on the other hand have been the subject of lengthy debate about the evidence for and against any sort of glaciation (e.g., Prest et al., 1977). Any history imputed to this large central part of the region based on findings on the Magdalen Islands must harmonize with that deduced for surrounding land areas. Hence the problem of whether and when the Magdalen Islands were glaciated becomes a central one. Anticosti Island, which has been largely overlooked until recently and generally assumed to have little Quaternary data of importance, has been the subject of studies by Painchaud et al. (1984) among others, who have outlined a complex sequence and a lengthy record of Wisconsinan nonglacial conditions.

Gaspésie

The surface deposits and Quaternary history of this rugged mountainous area attracted attention at an early date for much the same reason as did Newfoundland. Evidence of glaciation on the summits is absent or obscure; there are intermediate levels of weathered and subdued drift forms above lower, fresher glacial terrains. Coleman (1922) believed that the weathered zones were degraded glacial terrains representing former glacial expansions, but Flint et al. (1942) discounted the weathering as a measure of time. Mapping of the peninsula reveals a sequence of four till sheets and shows the paths of several ice streams. Local ice caps as well as the Laurentide Ice Sheet have affected the area and an interpretation of their interplay and timing has been presented (David and Lebuis, 1985).

Quebec North Shore

This southern segment of the Canadian Shield margin is largely unexplored. It was covered by Laurentide ice and its major feature, a series of end moraines, has been reported on by Dredge (1983) and Dubois and Dionne (1985). On the basis of a few radiocarbon dates, these have been assigned to the period 9.5-12 ka. The chronology of glaciation in this area remains the crucial missing link between the separate histories being developed for Newfoundland and Gaspésie both of which lay at the extremity of Laurentide ice. The Quaternary geology of this area is not discussed further in this chapter but it is summarized in Chapter 3 (Vincent, 1989).

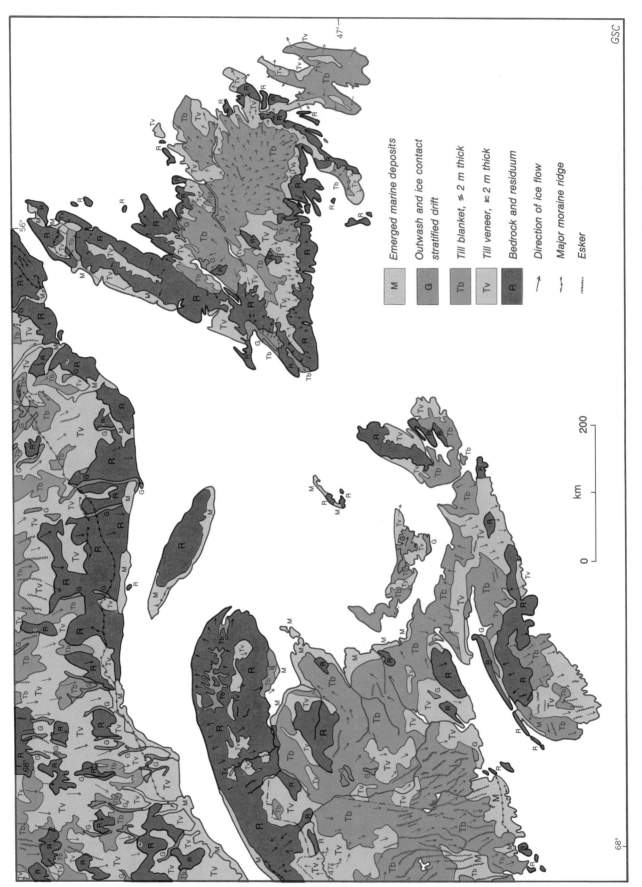

Figure 5.7a. Surficial materials and geomorphic features, compiled from various sources (cf. Fig. 5.7b).

Summary

Over the last hundred years or more, a large number of Quaternary studies, ranging from localized topical investigations to regional overviews, have been conducted. Systematic areal surficial geological mapping is incomplete and stratigraphic studies are just beginning. Radiometric control is poor and correlations are speculative. Nonetheless parallel schemes of paleogeographic evolution are emerging in four areas: western Newfoundland, Nova Scotia, southern New Brunswick, and Gaspésie. The recurrent theme is weak low-level maritime glaciation by local land-based ice cap complexes which spread radially seaward, while Laurentide ice intruded only infrequently and locally. Preliminary findings offshore generally support with this model (Piper et al., 1989).

SURFICIAL GEOLOGY AND QUATERNARY HISTORY

Knowledge of the distribution and composition of surface materials is based on systematic mapping supplemented, on land areas, by airphoto interpretation. From this, a generalized small-scale map can be derived (Fig. 5.7a) which shows the disposition of rocky areas, thick and thin till areas, marine and glaciofluvial deposits. Trends of ice flow indicators are added to convey an idea of ice trajectory and recessional pattern. Figure 5.7b also shows locations of extensive exposure where several significant units are in contact, location of sites containing pre-Late Wisconsinan organic remains and the locations of isotopically dated stratigraphic successions.

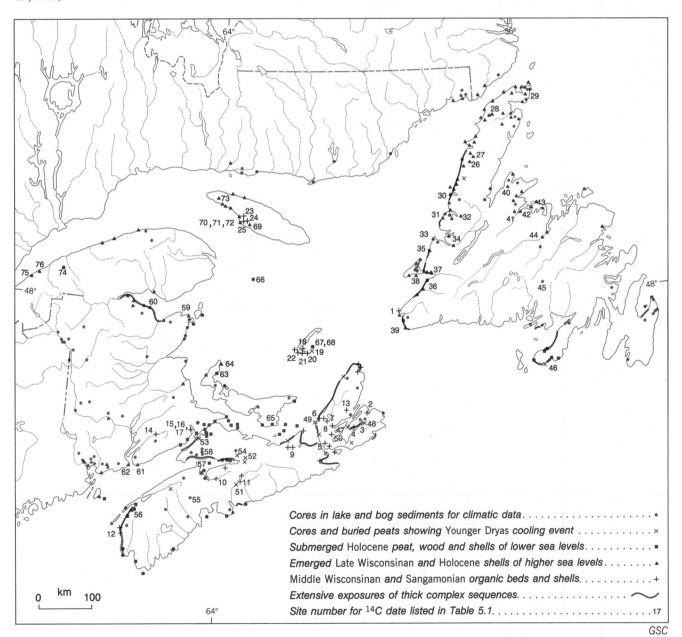

Figure 5.7b. Location map for main sources of data; symbols indicate the type of data available at each site.

In Atlantic Canada, much has been learned about the distribution of surficial deposits and their sequence as a result of systematic mapping, geochemical analysis, and stratigraphic study. However, the understanding of their origin, sequence, and age remains imperfect because of inadequate radiometric dating control and correlation problems. As well, reconstruction of a coherent geological history is, as ever, dependent on changing concepts of the mechanisms of glacierization and the underlying climatic controls. The history of this maritime area seems to indicate that glaciers spread from a number of accumulation areas on land and on emergent shelf areas and then moved into submarine areas and were transitional to floating ice shelves. These in turn impinged on submarine prominences and behaved again as grounded glaciers. Depending on the size and vigour of the ice system, basal conditions ranged from wet and sliding, to dry and frozen. Therefore the Appalachian glacier complex, in the course of a single glaciation, is thought to have evolved through several configurations which produced locally varying features and deposits. Hence, it is only through description and understanding of the resulting surficial geological make up that a history can be developed using whatever theoretical concepts seem appropriate. The Quaternary geology of the Atlantic region is presented from that point of view, beginning with Nova Scotia where the most work has been done and the succession of events is best documented.

Nova Scotia

Systematic surficial geological mapping (Map 1704A) is being completed for the entire province and has provided a regionally consistent first-stage reconnaissance of virtually all the exposed sedimentary units. This, together with stratigraphic studies, notably in the central region by Stea (1980, 1982b), in southwestern Nova Scotia by Grant (1980b), and in Cape Breton by Mott and Grant (1985), Vernal and Mott (1986), and Vernal et al. (1986), has led to development of a "standard" sedimentary succession and to various correlation schemes (Grant, 1963, p. 123, 1976; Prest, 1977; Grant and King, 1984). A recent bibliography by Stea and Brisco (1984) is an important document for further work.

Preglacial and/or early Quaternary weathered bedrock

The oldest material which might possibly be of Quaternary age comprises various forms of regolith or subaerially altered bedrock which may underlie any of the younger deposits or may outcrop at the surface. It is developed in granite in the New Ross area of Southern Uplands, in a variety of rocks on Cobequid Mountain (Stea, 1983), and in gneiss on Cape Breton Highlands (McKeague et al., 1983). At the latter site it consists of a gibbsite-bearing saprolite that underlies a pre-Quaternary peneplain. Pedological analysis is incomplete so it is unclear whether these weathering products represent the cumulative effect of a succession of warm interglacial periods, or whether, as LaSalle et al. (1985a) have proposed for these deposits in adjacent Quebec, they may be relicts of preglacial soils that formed in the warm humid climates that prevailed before the Quaternary. The main significance of these and of similar occurrences in New Brunswick, Gaspésie, and Newfoundland is that they could only have been preserved if the site areas were subject to only limited and/or weak glacial action.

Bridgewater Conglomerate

The oldest deposit of possible Quaternary age is a deeply oxidized, iron-cemented glacial diamicton and well sorted sediment, that rests on bedrock and occurs sporadically under, and as clasts within, younger Quaternary deposits. Discovered by Prest(1898), the Bridgewater Conglomerate is a goethite-cemented rock usually found as small patches on slate bedrock along the Atlantic coast (Stea, 1978). Grant (1963, p. 130) recognized glacial and glaciofluvial phases and, because they seemed to occur in a belt aligned along the coast, he thought they might be remnants of a terminal moraine. MacNeill (in Prest et al., 1972) described it as a till cemented to glacially striated bedrock at Pentz, near the type locality. In the Yarmouth area, the conglomerate is said to rest on a marine platform at Cape Cove (Grant, 1980b, p. 7) and is in a similar position on Double Island. Clearly the material predates other Quaternary sediments and its consolidation and degree of alteration suggest a considerable age. Until the conglomerate is better analyzed, however, its age and origin will remain conjectural. The Main Brook Till in Newfoundland (Tucker and McCann, 1980) might be a possible correlative.

Older buried organic beds

The most important stratigraphic units in the Canadian Atlantic region are a group of organic beds that underlie tills and postglacial sediments (Fig. 5.8). Paleoecological interpretation and stratigraphic assignment is given by Mott and Grant (1985), Vernal and Mott (1986), Vernal et al. (1986),

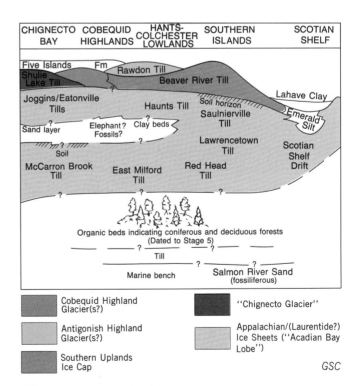

Figure 5.8. Nova Scotia till stratigraphy shown by diagrammatic north-south cross-section from Northumberland Strait to Scotian Shelf (modified after Stea et al., 1986).

and by Mott (1989). As has been reported in earlier studies (Mott and Prest, 1967; Livingstone, 1968; Mott, 1971; Stea and Hemsworth, 1979; Stea, 1982b; Mott et al., 1982; Vernal et al., 1983), these beds represent short intervals with climates ranging from tundra to thermophilous forest. A few have pollen spectra that represent unique forest associations with no Holocene analogues: anomalously high oak, beech, and balsam fir; no hemlock maximum. In general however the pollen sequences resemble those during the last deglaciation and the present interglacial period. That similarity and the fact that the inferred temperature at times exceeded the postglacial maximum, together with ^{14}C ages beyond the limit of the method (2, 4, 5, 7, 9, 10 of Table 5.1), led Mott and Grant (1985) to assign them all to the Sangamon Interglaciation, which is broadly equated with oxygen isotope stage 5 (125-75 ka). Moreover, from distinctive pollen assemblages and from successions of assemblages in different beds at separate sites, they concluded that three separate warm intervals were represented, namely substages 5e, 5c, and 5a. Vernal et al. (1986) studied three key sequences and, on the basis of Th/U ages (3, 6, 8, 11 of Table 5.1) generally supported the correlation, except for Bay St. Lawrence which they assign to Middle Wisconsinan time, as did Vernal et al. (1983).

A special problem is the Salmon River Sand — a grey marine mud that is intercalated in a till sequence near Yarmouth. The bed is 6 m above tide level and contains a temperate-water molluscan assemblage which typically lives at depths of 10-40 m. Paleoecologically the bed represents conditions warmer than those in the area today, and, despite a finite ^{14}C date of 38.6 ka (12 of Table 5.1) was originally assigned to the Sangamonian (Clarke et al., 1972). Miller (1973) and Nielsen (1974), however, argued on sedimentological grounds that this unit was deposited ice marginally, notwithstanding the temperate fauna, and Grant (1980b) accordingly reassigned it to a Middle Wisconsinan time. Reexamination has shown however that the bed is glacial tectonically emplaced within the lower till (Grant, 1987b). Its stratigraphic position and paleodepth are thus meaningless. Considering that similar oyster-bearing shelly deposits in the same area are deeply oxidized (Grant, 1980a) and lie undisturbed beneath the till sequence, Grant and King (1984) reaffirmed that the Salmon River Sand is last interglacial age or older. Gustavson (1976) proposed a last interglacial age for similar molluscan beds in New England.

A link between the temperate marine shell beds and the supposed interglacial peat beds is a stratum of littoral sand and gravel and an intertidal abrasion platform that is

Figure 5.9. Tills over an emerged littoral rock platform, Morrison Brook, northern Cape Breton Island. Cut across Carboniferous arkose, the platform ranges 4-6 m above tide level and is assigned to the Sangamon Interglaciation (stage 5e). A south-flowing regional ice sheet deposited the lower granitic till (pink), a local northeast-flowing ice lobe deposited the upper gabbroic till (grey). 203752-P

Table 5.1. Isotopic ages cited in text

Site No.[1]	Locality	Material[2]	Age* (years BP)	Laboratory[3] Dating No.	Location latitude	Location longitude	Elev.[4] (m)	Reference	Collector[5]	Stratigraphic Context; Comments
Newfoundland										
			*Dates pertaining to pre-Late Wisconsinan events*							
1	Woody Cove	wood	> 40 000	I-10203	47°51.67'	59°22.20'	15	Brookes et al., 1982	IAB	subtill; littoral; temperate "interglacial"[6]
Nova Scotia										
2	Leitches Creek	peat	52 000	GSC-2678	46°09.20'	60°22.87'	35	Blake, 1984	NJZ	subtill; boreal; late interglacial?
3	East Bay	wood	126 400 ± 15 000	UQT-175	45°59.08'	60°27.83'	1-2	Vernal et al., 1986	RJM	subtill; littoral?; boreal; "interglacial"
4	Castle Bay	mud	> 42 000	GSC-1577	45°55.22'	60°38.75'	5	Grant, 1972b	DRG	subtill; forest-tundra; late "interglacial"
5	River Inhabitants	wood	> 49 000	GSC-1406	45°40.57'	61°19.58'	53	Mott, 1971	DRG	subtill; littoral; temperate; full "interglacial"
6	Green Point	wood	117 400 ± 10 000	UQT-181	46°05.37'	61°28.58'		Vernal et al., 1986	DRG	subtill; karst filling; temperate; "interglacial"
7	Hillsboro	wood	> 51 000	GSC-570	46°04.37'	61°22.06'		Mott and Prest, 1967	VKP	subtill; temperate; "interglacial"
8	Bay St. Lawrence	wood	47 000 ± 4700	UQT-178	46°58.45'	61°07.80'	10	Vernal et al., 1986	ADV	intertill; temperate; "interglacial"
9	Addington Forks	wood	> 42 000	GSC-1598	46°34.06'	62°06.05'	15	MacNeill, 1969	VKP	subtill; temperate/boreal; "interglacial"
10	Miller Creek	wood	52 000	GSC-2694	45°00.85'	64°02.65'	20	Stea, 1982b	RRS	under two tills; temperate, "interglacial"
11	Milford	wood	84 900 ± 6500	UQT-185	45°00.33'	63°25.16'	25	Mott and Grant, 1985	VKP	intertill; ice-thrust bed; temperate fauna, "interglacial"?
12	Salmon River	shell	38 600 ± 1300	GSC-1440	44°03.50'	66°10.50'	4	Clarke et al., 1972	DRG	mastodon; boreal; "interglacial"?
13	Middle River	bone	31 900 ± 630	GSC-1220	46°08.1'	60°55.2'	15	Grant and King, 1984	DRG	
New Brunswick										
14	Sussex	peat	> 35 000	BGS-806	45°44.6'	65°32.2'	40	Rampton et al., 1984	VNR	subtill; temperate; "interglacial"
15	Hillsborough	bone	13 600 ± 220	GSC-1222	45°54.6'	64°39.8'	15	Schroeder and Arsenault, 1978	WMI	mastodon in karst; boreal; interglacial?
16	Hillsborough	peat	> 43 000	GSC-1680	45°54.6'	64°39.8'	30	Schroeder and Arsenault, 1978	WMI	boreal; interglacial?
17	Hillsborough	coprolite	37 200 ± 1310	GSC-2469	45°54.6'	64°39.8'	30	Blake, 1983	DRG	mastodon in karst; interglacial?
Magdalen Islands (Îles-de-la-Madeleine)										
18	Portage du Cap	peat	> 35 000	GSC-2313	47°14.75'	61°54.60'	18	Prest et al., 1977	VKP	subtill; littoral; temperate; "interglacial"
19	Bassin	wood	106 400 ± 8400	UQT-183	47°14.30'	61°53.97'	1	Mott and Grant, 1985	RJM	subtill; littoral; temperate; "interglacial"
20	Havre Aubert	wood	> 47 000	GSC-3633	47°13.14'	61°51.86'	4	McNeely, 1988	DRG	subtill; littoral; temperate; "interglacial"
21	Millerand	wood	> 46 000	GSC-3631	47°13.14'	61°59.74'	15	McNeely, 1988	DRG	subtill; littoral
22	South Cape	wood	> 38 000	GSC-3413	47°12.46'	61°58.58'	12	McNeely, 1988	DRG	subtill; littoral; cold, wet "interglacial"
Anticosti Island (Île d'Anticosti)										
23	Baie Bonsecours	shell	36 000 ± 3500	UQ-553	49°28.75'	63°36.83'	1-2	Gratton et al., 1984		glaciomarine drift; Mid-Wisconsinan
24	Rivière à la Chute	shell	29 060 ± 1050	UQ-510	49°24.13'	63°30.83'	116	Gratton et al., 1984		till
25	Rivière à la Chute	shell	30 000 ± 1200	UQ-514	49°23.78'	63°31.28'	84	Gratton et al., 1984		till
			*Dates pertaining to Late Wisconsinan and Holocene events*							
Newfoundland										
26	Flat Pond	shell	12 600 ± 160	GSC-1600	50°24.03'	57°16.15'	115	Grant, 1986b	DRG	littoral; near marine limit; inside end moraine
27	Blue Mountain	shell	12 000 ± 170	GSC-1601	50°27.83'	57°10.80'	82	Grant, 1986b	DRG	littoral; early deglacial, northwest coast
28	Ten Mile Moraine	shell	11 000 ± 160	GSC-1324	51°04.87'	58°42.63'	60	Grant, in press	DRG	in till of end moraine; dates readvance, highlands ice cap
29	St. Anthony	shell	10 700 ± 170	GSC-1334	51°22.28'	55°37.35'	60	Grant, in press	DRG	deepwater; early deglacial, northeast coast
30	St. Pauls	shell	9 230 ± 140	GSC-1630	49°48.75'	57°52.15'	3	Grant, 1986b	IAB	glaciomarine drift, late retreat piedmont glaciers
31	Trout River	shell	12 500 ± 120	GSC-2936	49°28.5'	58°06.5'	70	Brookes, 1974	IAB	moraine-delta foresets; ice marginal
32	Lomond	shell	10 500 ± 300	GSC-1575	49°27.25'	57°45.30'	4	Brookes, 1974	IAB	sublittoral silt; min. deglaciation, Bonne Bay fiord
33	Lark Harbour	shell	12 600 ± 320	GSC-1462	49°06.7'	58°24.8'	37	Brookes, 1974	IAB	littoral, early deglaciation outer Bay of Islands
34	Cox's Cove	shell	12 600 ± 170	GSC-868	49°07'	58°05'	35	Brookes, 1974	IAB	deepwater clay, early deglacial
35	Rope Cove	shell	13 700 ± 340	GSC-2942	48°55'	58°29'	40	Brookes, 1974	IAB	bottomsets of proglacial moraine delta
36	Robinsons Head	shell	13 500 ± 210	GSC-1200	48°15'	58°47'	32	Brookes, 1974	IAB	Late Wisconsinan deglaciation; early emergence
37	Kippens	shell	12 600 ± 140	GSC-2295	48°32.6'	58°36.7'	6	Brookes, 1977b	IAB	glaciomarine drift, late retreat piedmont glaciers
38	Abrahams Cove	shell	13 700 ± 230	GSC-1074	48°33.7'	58°42.9'	40	Brookes, 1974	IAB	climax of Robinsons Head readvance
39	Wreckhouse	shell	13 800 ± 260	GSC-2113	47°42.5'	59°18.6'	6	Brookes, 1977a	IAB	Late Wisconsinan deglaciation; early emergence predates readvance of Codroy lowland glacier
40	Baie Verte	shell	11 520 ± 180	GSC-55	49°54.5'	56°17.0'	48	Dyck and Fyles, 1963	EPH	sublittoral sediment; near marine limit; retreat of local ice cap
41	South Brook	shell	12 000 ± 220	GSC-1733	49°25.50'	56°06.50'	20	Tucker, 1974	DRG	prodelta mud; minimum age of coastal end moraines
42	Pilley's Island	shell	11 900 ± 200	GSC-1505	49°30.9'	55°43.0'	105	Blake, 1983	NOD	glaciomarine drift; early deglacial submergence
43	Leading Tickles	gyttja	13 200 ± 300	GSC-3608	49°28.28'	55°28.39'	105	Blake, 1983	JBM	maximum age of deglaciation of outer coast
44	Bishops Falls	shell	11 600 ± 210	GSC-2134	49°03.08'	55°22.35'	12	Blake, 1983	DRG	early deglacial submergence
45	Conne River	gyttja	11 100 ± 100	GSC-3634	48°14.8'	55°29.6'	60	Blake, 1983	JBM	deglaciation of east-central ice divide area
46	Lawn	gyttja	11 300 ± 120	GSC-3649	46°55.16'	55°36.75'	114	Anderson, 1983	TWA	beginning of late glacial climatic reversal

Table 5.1. Isotopic ages cited in text

Site No.[1]	Locality	Material[2]	Age* (years BP)	Laboratory[3] Dating No.	Location latitude	Location longitude	Elev[4] (m)	Reference	Collector[5]	Stratigraphic Context; Comments
Nova Scotia										
47	Benacadie Point	peat	11 300 ± 90	GSC-2146	45°54.16'	60°43.68'	1	MacNeill, 1969	DRG	subtill; predates Late Wisconsinan readvance
48	East Bay	peat	10 300 ± 150	GSC-1578	46°02.66'	60°22.50'	50	Grant, 1972b	DRG	subtill; predates early Holocene readvance local ice cap
49	Port Hood	wood	11 300 ± 160	GSC-541	46°01.18'	61°34.35'	18	Terasmae, 1974	JT	subcolluvial; suggests climatic deterioration
50	Big Brook	mud	11 000 ± 90	GSC-3378	45°48.4'	61°12.9'	30	Mott et al., 1986	RJM	sublacustrine; predates proglacial lake dammed by readvance
51	Lantz	peat	10 900 ± 90	GSC-3771	44°58.7'	63°29.0'	15	Mott et al., 1986	RRS	subtill; beginning of early Holocene readvance
52	Brookside	peat	11 100 ± 100	GSC-2930	45°24.09'	63°14.35'	53	Mott et al., 1986	GJB	subcolluvium; indicates late-glacial climatic deterioration
53	Joggins	peat	11 400 ± 100	GSC-3550	45°42.85'	64°26.10'	6	Mott et al., 1986	RRS	subsand peat; onset possible climatic deterioration
54	Debert	charcoal	10 585 ± 47	P-(var.)	45°25.5'	63°27'	45	Borns, 1966	HWB	Paleo-indian site; minimum age deglaciation Minas Basin
55	Canoran Lake	gyttja	11 700 ± 160	GSC-1486	44°35.8'	64°34.9'	107	Lowdon and Blake, 1973	JBR	postglacial; minimum age disappearance South Mountain ice cap
56	Gilbert's Cove	seaweed	14 100 ± 200	GSC-1259	44°29.15'	65°57.00'	0.3	Grant, 1971b	DRG	early postglacial; min. age deglaciation outer Bay of Fundy
57	Spencers Island	peri-ostracum	14 300 ± 320*	BETA-12858	45°21.51'	64°42.53'	2	Stea and Wightman, 1987	DMW	proglacial delta; min. age deglaciation
58	Leak Lake	gyttja	12 900 ± 160	GSC-2728	45°26.2'	64°21'	42	Wightman, 1980	DMW	tundra; age kettle formation
New Brunswick										
59	Shippegan Island	shell	12 600 ± 400	GSC-1383	47°44'	64°46.50'	0	Thomas et al., 1973	MLHT	deglaciation of eastern New Brunswick coast
60	Jacquet River	shell	12 500 ± 500	GSC-1557	47°55.3'	66°01.7'	13	Blake, 1983	DH	minimum age, retreat of Chaleur Bay glacier
61	Mispec	shell	14 400 ± 530	GSC-2573	45°13.5'	65°57.2'	37	Bélanger, 1978	JRB	minimum age, deglaciation of Bay of Fundy
62	Saint John	shell	13 100 ± 160	GSC-3557	45°13.5'	66°06.6'	45	Rampton et al., 1984	VNR	approximate age of Saint John end moraine
Prince Edward Island										
63	Miminegash	shell	12 410 ± 170	GSC-101	46°52'	64°14'	10	Prest, 1973	VKP	glacial isostatic regression in progress
64	Tignish Shore	shell	12 670 ± 340	GSC-160	46°58.5'	63°59.7'	8	Prest, 1973	VKP	glacial isostatic regression in progress
65	Iris Station	gyttja	6 600 ± 270	I-GSC-12	45°56.25'	62°40'	-2	Frankel and Crowl, 1961	VKP	temperate; late Holocene subsidence
Gulf of St. Lawrence										
66	Magdalen Shelf	shell	10 200 ± 440	GSC-1528	48°33'	63°10'	-198	Loring and Nota, 1973	DHL	sandy glaciomarine sediments; late ice remnant
Magdalen Islands (Îles-de-la-Madeleine)										
67	Entry Island	peat	11 300 ± 110	GSC-3696	47°16.09'	61°42.74'	3.4	Mott et al., 1986	RJM	probably colluvial; climatic cooling
68	Entry Island	peat	10 600 ± 100	GSC-3699	47°16.09'	61°42.74'	3.4	Mott et al., 1986	RJM	probably colluvial; climatic cooling
Anticosti Island										
69	Rivière du Brick	shell	13 100 ± 150	UQ-512	49°21.47'	63°23.37'	14	Gratton et al., 1984	DG	deep sea; southern extent of Laurentide ice
70	Pointe du Sud-Ouest	shell	13 570 ± 200	UQ-502	49°24.25'	63°34.8'	9	Gratton et al., 1984	DG	glaciomarine; southern extent of Laurentide ice
71	Pointe du Sud-Ouest	shell	14 500 ± 800	UQ-551	49°24.00'	63°34.63'	4	Gratton et al., 1984	DG	glaciomarine; southern extent of Laurentide ice
72	Pointe du Sud-Ouest	shell	11 950 ± 150	UQ-515	49°24.27'	63°34.63'	3	Gratton et al., 1984	DG	glaciomarine; ice proximal
73	Airport	shell	11 600 ± 200	UQ-1063	49°50.08'	64°15.83'	61	Dubois et al., in press	LSP	proglacial delta; correlates with Sainte-Marine readvance
St. Lawrence Estuary										
74	St-Donat	shell	13 360 ± 320	QU-264	48°30.18'	68°15.91'	98-126	Locat, 1977	JL	glaciomarine
75	Bic	shell	9 830 ± 150	QU-270	48°21.23'	68°45.35'	19-45	Locat, 1977	JL	marine; emergence oscillation
76	Bic	shell	9 540 ± 150	GSC-1216	48°22.58'	68°42.41'	15-30	Locat, 1977	JL	marine; emergence oscillation
77	Rivière St-Charles	wood	5 500 ± 170	QU-5	46°51'	71°13.45'	10	Dionne, 1985b	JCD	alluvial; emergence oscillation
78	Rivière St-Charles	wood	5 460 ± 100	BGS-174	46°51'	71°13.45'	10	Dionne, 1985b	JCD	alluvial; emergence oscillation

* Ages are by the conventional ^{14}C method, except 57 which is by AMS (accelerator-mass spectrometer), and 3, 6, 8, 11, 19 which are by Thorium/Uranium disequilibrium.
1 located in Figure 5.7b.
2 wood is usually contained in peat beds; peat is freshwater bog and fen sediment; gyttja is organic lake and pond sediment; shells are marine species.
3 BETA = Beta Analytic, Inc.; BGS = Brock University, Department of Geological Sciences; GSC = Geological Survey of Canada; GX = Geochron, Cambridge, MA.; I = Isotopes Ltd.; P = University of Pennsylvania; QU = Laboratoire de géochronologie, Ministère de l'Énergie et des Ressources du Québec; UQ = Université du Québec à Montréal (Th/U lab).
4 all elevations are in metres above or below ordinary spring tides, not mean sea level.
5 ADV = Anne de Vernal; DG = Denis Gratton; DH = David Honeyman; DHL = D.H. Loring; DMW = Daryl M. Wightman; DRG = Douglas R. Grant; EPH = Eric P. Henderson; GJB = Gerry J. Beke; HWB = Harold W. Borns Jr.; IAB = Ian A. Brookes; JBM = Joyce Brown Macpherson; JBR = John B. Railton; JCD = Jean-Claude Dionne; JL = Jacques Locat; JRB = J. Robert Bélanger; JT = Jaan Terasmae; LSP = Luc St-Pierre; MLHT = Martin L.H. Thomas; NJZ = New Jersey Zinc; NOD = Neil O'Donnell; RJM = Robert J. Mott; RRS = Ralph R. Stea; TWA = Thane W. Anderson; VKP = Victor K. Prest; VNR = Vern N. Rampton; WMI = William Macintosh.
6 "interglacial" means having vegetation or inferred temperature equivalent to any period within the present interglacial, i.e. the last 10 ka; "interstadial" refers to any conditions cooler than that of any interglacial.
7 Date, though finite, is considered minimal because fauna indicates conditions more temperate than at present.

emerged a few metres above its modern counterpart (Fig. 5.9). The platform and littoral sand rise only slightly higher than the shelly muds and in a few places underlie the interglacial peat beds. They represent a paleoshore that is recognized in various parts of the region and commonly underlies the multiple till sequence. In view of its parallelism to modern sea level, and similarity in elevation to dated oxygen isotope stage 5 shorelines in other areas of the world, the shore level is referred to the sea level maximum of the last interglaciation (stage 5e; Grant, 1980a). Its geodynamic and geoidal significance is treated in a later section. Stratigraphically, it and the organic beds serve as markers for the base of the late Quaternary sequence. The near absence of glacial or other deposits beneath the organic beds, or cut by the platform, is puzzling because it requires that the sediments of all previous glacial and nonglacial periods either do not exist or have been misidentified.

Sequence of till sheets

Glacial drift is the dominant surface material in the area. It is thin and discontinuous on rocky upland areas and generally thickens downslope to locally reach depths of 100 m on lowlands. The thickest accumulations occur along the Yarmouth coast, the Mahone Bay area, in Hants and Colchester counties, the Cumberland Lowland, and in Bras D'Or Lake area. In each of these areas, the drift mantle comprises several superposed till sheets of contrasting colour, lithology, texture, and provenance (Fig. 5.8). From the earliest days of study, when Honeyman (1882) began to recognize long-distance transport of distinctive till masses and northern erratics, much effort has been expended to map and analyze the various tills and to link the succession of drifts to the sequence of ice flow phases represented by crossing striations. A complicating factor is the varied bedrock geology which has led a given ice movement to produce several textural and lithological facies along its path.

More or less comparable till sequences are recognized in central Nova Scotia (Stea, 1982b, c; Stea and Finck, 1984), Atlantic coast area (Grant, 1963), Yarmouth area (Grant 1980b; Stea and Grant, 1982), and Cape Breton Island (D.R. Grant, unpublished). From these sequences it is possible to make a generalized description of the distribution and character of major till units.

Lower grey drift

Correlation of the lowest till unit is difficult because exposures are small and few, texture and composition varies, and stratigraphic setting is not always clear. In exposures near Halifax, the lowest till (Hartlen Till of Stea and Fowler, 1979) is grey, silty, and extremely compact. It is predominantly derived from local quartzite and shows manganese staining as evidence of subaerial weathering. Fabric data suggest deposition by onshore ice flow. In the Yarmouth area the lowest till is locally derived, by ice flowing westward off the Southern Uplands, and locally shows subaerial oxidation. These tills may relate to initial ice buildup during the Wisconsinan. Alternatively, if the oxidation is due to subaerial exposure in the last interglaciation, these units must be pre-Sangamonian. Even older till might occur at Milford and Miller Creek where Stea (1982b) found tills beneath the interglacial peats. The karst setting and evidence of glacial thrusting, however, make age assignment based on superposition tenuous.

Early red till

The main drift sheet of the province is a thick, clay-rich red till found along the entire Atlantic coast and extending northward in the Fundy Lowlands (Fig. 5.9). Typically it forms several large fields of drumlins (Grant in Scott, 1976) which may mark ice streams in the glacier (Grant, 1963). The fabric and underlying striations indicate eastward and southeastward transport. Lithologically, it is enriched in Ca and Mg and contains distinctive erratics from the northern part of the province and from New Brunswick. This till has been studied in most detail in the Halifax area where it first attracted Honeyman's attention, and was later termed the Lawrencetown Till (Grant, 1975b). Studies by Nielsen (1976), Podolak and Shilts (1978), Stea and Fowler (1979) and Stea and O'Reilly (1982) document the heavy mineral, lithological, and chemical attributes of the till. Because of its distinctive fine red matrix which was derived from sources in the Permo-Carboniferous redbed terrane, most compositional aspects show strong contrasts with tills derived locally from quartzite, shale, and granite terranes. Grant (in Scott, 1976) argued that the till lithology evolved by a gradual admixture of local components as the bedload travelled as much as 100 km away from its source. In northern Nova Scotia it is the lower till at Joggins (Wickenden, 1941) which Stea et al. (1986) referred to as the McCarron Brook Till. In southern Nova Scotia the red drift contains shell fragments transported from Bay of Fundy and was named the Red Head till by Grant (1980b). On Cape Breton Island, the supposed correlative is the Richmond Till which is similarly fine grained and red regardless of underlying bedrock lithology. The ice sheet that produced it appears to have moved uniformly southeastward across the whole province, including the Cape Breton Highlands except possibly for a small area above 300 m on the northern tip.

In sum, this red clayey till constitutes a prime marker bed in the glacial sequence, and it may record the first main advance (Early Wisconsinan) of the last glaciation. Stea (1982b) and Stea et al. (1986) noted alteration that suggested subaerial exposure of this till; if so, an ice-free period intervened before deposition of overlying tills. Because there is no absolute chronological data, that interval might have been the Sangamon Interglaciation, in which case the till would be Illinoian.

Intermediate age tills

Younger tills cover the red till sheet and occupy the areas between its drumlins. The lower of these is the middle unit of the three-part till sequence present in most areas. The middle unit has been emplaced by a variety of ice movements, most of them stemming from local sources, and has been given several local names.

The common characteristic is that the lithology (dominantly local), the colour (grey), and texture (sandy to silty) contrast with these characteristics of the underlying red clayey till. In central Nova Scotia, Stea and Finck (1984) attributed the middle or Hants Till of the Atlantic Uplands (Stea, 1982b) and the Joggins and Eatonville tills on the Cumberland Lowlands (Stea et al., 1986) to a southward ice movement from the direction of Prince Edward Island or the Magdalen Shelf. It may therefore possibly represent the Acadian Bay Lobe of Goldthwait (1924, p. 79). On Cape Breton Island, this glacial advance may correlate with the south-southeastward ice flood (Grant, 1971c), whereas in Yarmouth area, the middle or Saulnierville Till was emplaced by southwestward flow.

Roughly corresponding to this intermediate interval Grant (1971c) described a strong northward ice flow over Cape Breton Island, presumably from a centre on the shelf, that brought marine shell material onshore. This movement terminated at the flanks of the northern highland plateau. Similarly, Gravenor (1974) postulated an ice dome offshore from Yarmouth to explain northward fabrics in earlier formed drumlins. Stea and Finck (1984) depicted a similar northward movement that crossed Cobequid Highlands from sources to the south. All three movements may have emanated from a single ice divide that became localized off the Atlantic margin of Nova Scotia. Grant and King (1984) suggested that this divide may have formed as a response to calving in Gulf of St. Lawrence. The northward event may therefore be either a reversal of flow within the same ice mass that earlier had moved southward, or a separate glaciation by a new ice sheet.

Middle Wisconsinan interstadial? deposits

There is some evidence for an ice-free interval in the middle part of the last glaciation. In the Yarmouth area, there is subaerial oxidation of the middle till, which locally is overlain by proglacial fluvial beds (Cape Cove Gravel, Grant, 1980b, p. 56). Stea (1982b, 1983) reported weathering of the lower till and lenses of intertill sediments in central Nova Scotia and at Joggins (Stea et al., 1986). MacEachern et al. (1984) described a paleosol that occurs between lower and upper tills in the eastern area. In addition, at several widely separated sites early southeastward trending striations were weathered and iron stained before onshore readvance of ice partly freshened the outcrop. Finally, remains of Pleistocene elephants have been found near Elmsdale (Livingstone, 1951), Antigonish (W. Shaw, St. Francis Xavier University, personal communication, 1984), and on Cape Breton Island near Baddeck and Middle River (Piers, 1915). At the latter site, a femur from the modern floodplain, was dated at 31.9 ka (13 of Table 5.1; Grant and King, 1984). At East Bay, Castle Bay, and Bay St. Lawrence on Cape Breton Island, nonglacial deposits evidently span the interval 80-22 ka based on Th/U ages (e.g. 8 of Table 5.1). Therefore, the hypothesis of Middle Wisconsinan glacial retreat, at least from some extremities of the Nova Scotian peninsula, seems valid.

Young tills

Surface tills vary in texture and provenance. As Stea and Fowler (1978) and Stea and O'Reilly (1982) have shown, most are typically thin, patchy, and lithologically reflect the underlying bedrock. Evidently they were deposited by ice sources located in Nova Scotia, primarily on the Atlantic Uplands and Cape Breton Lowlands. One consequence of this development of local ice caps was reversal of ice flow directions in some areas. For the Yarmouth area Grant (1980b) described these young tills as immature, loose, sandy, and rubbly (Beaver River Till). It is inferred to have been emplaced by weak movement from a South Mountain ice cap. In central Nova Scotia a correlative till, called the Rawdon Till (Stea and Finck, 1984), is attributed to ice draining into Minas Basin from a centre near Antigonish Highlands. For northern Nova Scotia, Stea et al. (1986) recognized the Shulie Lake Till as having been deposited by a late ice stream in Chignecto Bay coming from a source in southeastern New Brunswick. For Cape Breton Island, D.R. Grant (unpublished) postulated late radial movement of a lowlands ice cap that deposited similarly thin, discontinuous tills and built minor morainal features as it shrank.

Little information relative to timing of advance and retreat of Late Wisconsinan ice is available. Radiocarbon dates indicate that ice had retreated from the Minas Basin by 14.3 ka (57 of Table 5.1), from South Mountain by 11.7 ka (55 of Table 5.1) and from the outer Bay of Fundy by 14.1 ka (56 of Table 5.1).

Glaciofluvial and fluvial deposits

Features and deposits produced by meltwater during the last deglaciation are a minor component of surficial deposits that provide an important corroboration of the retreat pattern of the ice masses that produced the surface till. These features indicate that Nova Scotia was characterized by essentially radial flow and by concentric retreat inland from the Atlantic, Bay of Fundy, and Gulf of St. Lawrence, except for western Cobequid Highlands and adjacent lowlands where the last ice advanced into the province from Northumberland Strait area. There was thus free downslope proglacial drainage. The small eskers and minor outwash terraces in some major valleys indicate three main dispersal centres: the South Mountain ice cap (Railton, 1972); the so-called Antigonish ice cap over the eastern counties; and the Bras D'Or ice mass, plus a small carapace on Cape Breton Highlands. These deposits are important sources of granular aggregate, and consequently are the subject of detailed mapping and testing (e.g., Wright, 1985).

Postglacial fluvial deposits are usually associated with the major outwash corridors. The largest areas occur along the Annapolis, Cornwallis, La Have, Shubenacadie, Stewiacke, and Margaree rivers. In their lower courses these streams are aggrading their floodplains in response to modern rise of sea level.

Ice frontal features

Moraines which could serve to mark ice marginal positions during retreat are uncommon. Excluding those built into the sea which are discussed under marine deposits, none are recognized along the mainland Atlantic coast, except for a few scattered small till ridges perpendicular to ice flow. Along the St. Marys Bay coast Grant (1980b, p. 35) described a massive ridge sculpted from older tills by the last ice during a seemingly important stillstand just before 14 ka. A large ice marginal complex extends eastward from Apple River along north side of Cobequid Highlands which Stea et al. (1986) argued is a 12 ka recessional margin, not the Late Wisconsinan glacial limit as Grant (1977b) postulated. On Cape Breton Island, a wide belt of ice contact deposits in the Sydney-Mira Bay area marks a stand of the Bras D'Or-centred ice mass, while small morainal arcs at Lake O'Laws and Whycocomagh may be correlative positions of outlet glaciers. As well, small ice contact ridges at Margaree Harbour and Inverness seem to have been formed where Cape Breton ice abutted a glacier impinging from Prince Edward Island; these may predate the final ice retreat as might extensive kame fields at Cheticamp. End moraines of the Highlands ice cap occur where outlet valley glaciers debouched at Pleasant Bay, Red River, and Ingonish. Younger ridges are seen on the plateau; these and sets of sidehill meltwater channels track the retreat to several small ice centres. None of the moraines has been dated to substantiate various suppositions about the history of deglaciation. The only hard information related to the timing of deglaciation is a number of scattered basal dates from lake and bog sediments which range 10-12 ka and indicate that most areas were ice-free by that time.

A late glacial climatic reversal

At twelve sites relatively young organic sediments are buried by mineral sediment (sites showing Younger Dryas cooling event on Fig. 5.7b; see also Mott et al., 1986). Dates for seven of these (47-53 of Table 5.1) range from 11.4-10.3 ka. The number of sites, their distribution, and narrow age range suggest a sudden cessation of organic production and deposition of rock debris over a large area. The nature of the overlying sediment is critical to the understanding of this event. At several sites the cover is a nondescript diamicton which may be colluvial in origin. At a few places the mantle might be till and hence would suggest a minor readvance of a remnant glacier, or renewed glacierization as Borns (1965, 1966) postulated. Whatever the glacial implications, Mott et al. (1986) attribute the event to an abrupt climatic deterioration caused by southward migration of the Polar Front in the North Atlantic, as was postulated by Ruddiman and McIntyre (1973) to account for the coeval Younger Dryas readvance/cooling in Europe.

Marine deposits

Postglacial marine sediments comprise two sequences: one, a series of deltas and a gravelly regressive mantle produced during late glacial uplift of the Fundy coast; the other, a muddy transgressive wedge deposited during a more recent rise of sea level throughout the region. The broad implications of the age and elevation of these deposits are treated in the section on paleosea levels in the regional interpretation, but some local stratigraphic details are given here.

Raised gravels along Bay of Fundy coasts take the form of wave-washed veneers on till slopes, beach ridges, and massive ice marginal deltas. Of the latter, the more significant examples include those at Sandy Cove (Goldthwait, 1924, p. 100), Margaretsville (Hickox, 1962, p. 32), and a series adjacent to the Minas Basin at Parrsboro termed the Five Islands Formation by Swift and Borns (1967). Deltas at Sandy Cove and Margaretsville were built at the margin of the South Mountain ice cap while ice stood along the coast — a phase estimated to slightly predate 14 ka (Grant, 1980b, p. 452). The Minas Basin deposits are the product of an ice sheet standing along the north flank of Cobequid Highlands which debouched outwash through passes to build an extensive marine terrace and ice contact deltas (Stea et al., 1986). Wightman (1980, p. 371), assigned a tentative age of 14 ka to deglaciation based on a date (total carbon) of 14.2 ka (57 of Table 5.1) for bay-bottom sediment (Amos, 1978) and basal lake sediment dates (e.g., 12.9 ka, 58 of Table 5.1) from kettles on a delta and this is confirmed by Stea and Wightman (1987) (57 of Table 5.1). The gravelly marine beds as a group are thus directly related to the withdrawal of ice masses from Nova Scotia. Broadly speaking the raised marine beds were deposited during the Late Wisconsinan period of marine inundation, termed the De Geer Sea by Lougee (1953), when the Fundy region was depressed by adjacent ice sheets. Postglacial rebound tilted the crust so that the upper limit of the marine sequence rises northward, in the direction of greatest ice load, from zero near Yarmouth to a maximum in Nova Scotia of 37 m on Cape Chignecto (Stea, 1983). The regional significance of the warping pattern is treated in a later section, but it may be noted that as De Geer (1892) initially observed, the absence of marine overlap over most of Nova Scotia points to relatively insignificant ice loads, that is to say, thin and/or restricted glaciers.

The main type of marine deposits is modern estuarine mud (Amherst Silt; Grant, 1980b) which is accumulating in the intertidal zone in estuaries and other quiescent areas as sea level is presently rising. These flat aggradational areas mainly support salt marsh meadows which in many areas have been isolated from the sea by dykes for use as pasture. The most extensive and thickest areas of muds are in upper Bay of Fundy where they attain thicknesses exceeding 50 m. Consisting of finely laminated silt and clay with rare intercalated organic horizons, the beds in many places overlie a terrestrial humus layer or forest bed. Dates on the basal layer, and on levels within the sequence, document the generally steady rise of tide level during the past 5 ka at rates ranging from 1 to 3 m/ka (Grant, 1970). Regional patterns of relative sea level change deduced from these sequences, and interpretations of the geodynamic and tidal factors are found in the section on regional analysis of paleo sea-levels.

Organic deposits

On land, postglacial sedimentation is largely restricted to peat accumulation in bog areas and the infilling of lake basins. Peatlands cover about 3% of the province. The bogs are numerous but small in the hummocky hardrock Atlantic Uplands; larger areas occur on shallow till plains in the southwestern uplands, and on the Cape Breton plateau. Peat sequences have served as gross indicators of major climatic events (Auer, 1930) and have provided a detailed record of climatic variation (Ogden, 1960). The most complete postglacial records are however contained in lake sediments and these have been studied by several detailed coring programs. For Cape Breton Island, Livingstone and Livingstone (1958), Livingstone and Estes (1967), and Livingstone (1968) conducted early work. In southern Nova Scotia, Railton (1972) studied cores in relationship to deglaciation. Their record of postglacial climatic evolution is treated elsewhere in a separate chapter (Anderson et al., 1989). Apart from their stratigraphic value as a record of environmental changes, both lakes and bogs are important ecosystems and resources.

New Brunswick

Systematic study of the Quaternary geology of New Brunswick began over a century ago when Matthew and Chalmers completed a series of surficial geology maps covering most of the area. Their basic observations remain valid today, and Chalmers' concept of local ice caps as the main process that created most of the deposits and features is the essence of current interpretations about glacial history and sedimentary succession. Since his time more than 300 reports on a variety of aspects have appeared. The latest, by Rampton et al. (1984) is the first comprehensive synthesis of the Quaternary of the entire province. It provides a key to the studies cited above and traces the historical development of Quaternary concepts. Based on recent mapping of the entire province and reassessment of previous data, it includes a complete description of surface deposits and geomorphic features, and presents for the first time a reconstruction of the sequence of events. For an in-depth appreciation of New Brunswick Quaternary geology, the reader is referred to Rampton et al. (1984); it forms the basis for this summary outline.

The arrangement and sequence of deposits in New Brunswick (Fig. 5.7) are similar to those found in Nova Scotia owing to the similarity of bedrock and physiography and to an apparently parallel geological evolution.

Preglacial and/or early Quaternary weathered bedrock

Over large areas, primarily in Northern Miramichi Highlands and Caledonian Highlands, weathered bedrock is present in various forms and to varying depths. The alteration ranges from weak mechanical disintegration to pervasive chemical diagenetic transformation. It is significant that these areas lack a strong glacial imprint although locally the regolith is overlain by till (Gauthier, 1980b). Instead, the terrain displays a remarkable fluvial maturity; a deeply entrenched system of dendritic valleys is incised into the pre-Quaternary peneplains. In some parts of the glaciated lowlands, notably along the Saint John Valley, the regolith commonly underlies till.

It could be argued that the mechanically disrupted bedrock in the highlands might be quite young, even postglacial, owing to the vigour of modern frost action at those elevations. However, the saprolite described by Wang et al. (1981) and the large areas of corestones and deep grus (Gauthier, 1980b; Veillette and Nixon, 1982) suggest considerably longer exposure, and a climate more temperate than that of Quaternary interglaciations. The presence of woolsack tors in the grus/corestone terrain of Big Bald Mountain (Fig. 5.10) may give the impression that the rock alteration is equally modern, but it may be argued that the tors are being set in relief today simply because repeated forest fires have permitted accelerated postglacial erosion of the grus leaving the corestones. In forested areas the corestones remain buried and are seen in roadcuts.

The wide occurrence of weathered rock has profound implications for concepts of style of glaciation and thus for practical applications of glacial geology. Even if much of the alteration is assumed to have occurred during nonglacial intervals within the Quaternary, it means that glacial action failed to strip away the debris and was relatively weak and/or infrequent, particularly over the highland areas. Gauthier (1980b) considered the disposition of the northern highland weathering features in relation to the presence of summit erratics and till patches; he postulated that the latest glaciers had covered the highest areas, but that they were cold based, effectively frozen to the substrate, and hence not sliding and eroding. In contrast, Rampton et al. (1984) found this hypothesis impossible to support paleogeographically and, because clear ice marginal features locally bound the weathered tracts, particularly around Caledonian Highlands, they argued that the weathered highland areas lay beyond the limit of several glacial advances — at least the Late Wisconsinan glacial limit (Fig. 5.11a). Although

Figure 5.10. Woolsack tors, Big Bald Mountain, northern New Brunswick. Present in the subsurface over a large area, those forms originate as corestones which are unweathered remnants of deep disintegration of granite. The spheroidal masses have been exposed by recent erosion of the surrounding grus. They testify to minimal Quaternary erosion. 203673-Z

Seaman (1985) presented evidence to refute that interpretation, the cold-based ice hypothesis remains valid (Fig. 5.11b), because it seems to explain better the same relationships in adjacent areas (Gaspésie, Magdalen Islands, and Cape Breton Island).

The presence of this mantle of weathered rock affects the practice of drift prospecting. It means that the geochemical attributes of given source rock areas are not necessarily uniform and that tills can be enriched or depleted in various constituents depending on whether they were derived from weathered or unweathered bedrock. As well, the possibility of cold-based glacier conditions greatly complicates assumptions about the regularity and predictability of dispersal over the province.

Middle Wisconsinan and older nonglacial deposits

Virtually all Quaternary deposits and features are glacial in origin and most of these are considered to date from the Late Wisconsinan substage. Recognition of pre-Late Wisconsinan beds would require that they either underlie Late Wisconsinan deposits, or predate them (yield ^{14}C dates greater than 25 ka). In this category three occurrences are known. Near Sussex, an organic bed, dated >35 ka (14 of Table 5.1) underlies till. It contains a temperate macrofossil assemblage and wood which has amino acid racemization ratios within the range of those from Nova Scotia interglacial organic beds (N.W. Rutter, University of Alberta, Edmonton, personal communication, 1986). Accordingly, it is referred to the Sangamon Interglaciation, as is the subtill marine abrasion platform near Rockport. Secondly, near Hillsborough, remains of a mastodon were discovered in peat and pond mud occupying a sinkhole (Squires, 1966).

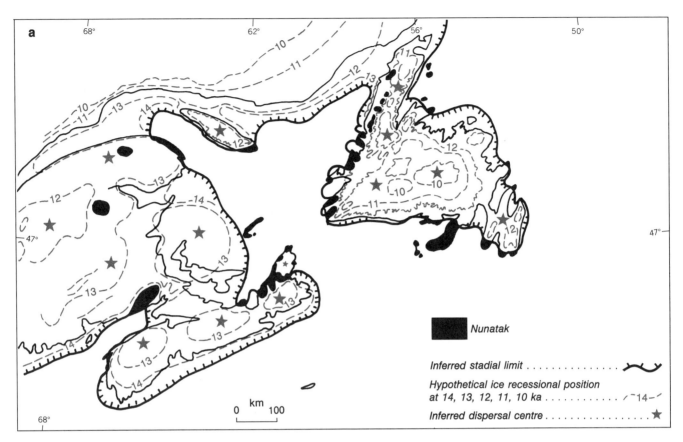

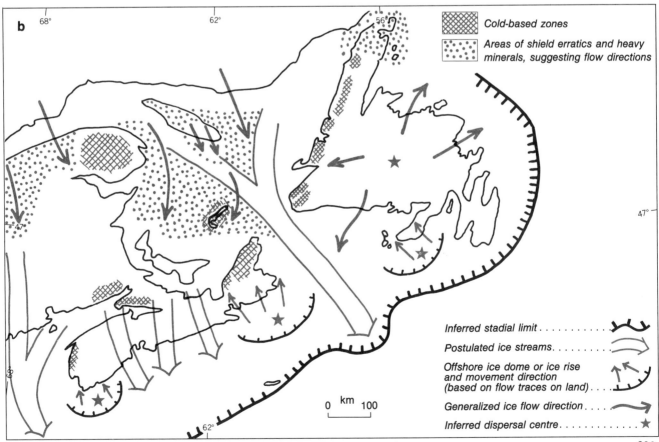

The bone dated 13 600 ± 220 BP (15 of Table 5.1) and the enclosing peat was >43 ka (16 of Table 5.1). These two dates led Schroeder and Arsenault (1978) to conclude that a postglacial elephant fell into a much older bog. Wood from the associated coprolites, however, is 37.2 ka (17 of Table 5.1) so Grant (in Blake, 1983, p. 9) assigned the occurrence to a Middle Wisconsinan interstade. However, pollen in the organic sediments points to an interglacial age (R.J. Mott, Geological Survey of Canada, personal communication, 1987). The only other materials predating glacial units described below are the mantles of colluvial and residual debris that cover surfaces inferred to lie beyond the reach of the last glaciers.

Late Wisconsinan deposits

Layered sequences of superposed till sheets which might substantiate a succession of glacial episodes are virtually unknown. One exception is an area of two till exposures in upper Saint John River valley (Rampton et al., 1984, Fig. 10) where brown till overlies grey till which locally is subaerially oxidized. In general, the surface till constitutes the only till exposed throughout the province and the succession of glacial phases described below is based entirely on a sequence of crosscutting ice flow features, mainly striations.

Texturally the surface till is commonly a loam or sandy loam, owing to the great variety of parent rock types that have been comminuted and mingled. Clast content is greatest over the harder rock terranes. Otherwise no rigorous textural classification and differentiation is yet possible from existing mapping. Soils maps, where available, however, are a good source of textural information because the parent tills are specified. Colour varies with texture and reflects the major source terranes. Lodgment and ablation facies are present; the latter averages more than 10 m in northern areas and is more common in valley bottoms and on divides where ice flow was impeded or stagnant, and a hummocky or ridged deposit resulted.

Geomorphologically the till surface presents the usual variety of morainal landforms. An area of strong fluting and streamlining occurs in the southwest corner of the province where a part of the Maine drumlin field appears; a broad belt sweeps around the south and east flank of Northern Miramichi Highlands, and there is a small area on Chignecto Isthmus. Generally the till surface is broadly rolling to undulating, but hummocky and ribbed tracts occur in some valleys and on some divides as mentioned above. The latter are generally transverse to ice flow as determined from striations. De Geer moraines that are somewhat less regular in form and spacing than is normally understood by the term in Canada occur along Nepisiguit Valley where an ice front is inferred to have retreated in a high-standing sea. Frontal moraines, major or minor, are rare; the largest occur in a belt along the Fundy coast and on the north side of Caledonian Highlands. Others occur along Chaleur Bay, near Plaster Rock, and Lorne. Taken together these features lend substance to inferences of ice flow and recessional positions based on striations and meltwater discharge.

Sequence of glacial events, and lacustrine and marine phases

Ice flows of considerable antiquity are locally recorded in the Bathurst area where onshore and offshore movements were followed by a weathering interval before reglaciation with an eastward sense. The onshore movement is attributed to a dispersal centre on Magdalen Shallows, termed the Escuminac Ice Centre (Fig. 5.11a). The earliest ice flow of wide extent was an eastward movement over the entire area that arrived from sources in the Quebec Appalachians or Maine. Incursion of Laurentide ice was apparently restricted to upper Saint John River valley — the only area where Shield erratics have been introduced (Fig. 5.11b). Its age may be pre-Wisconsinan in that striations from this event are locally oxidized (iron-stained) then crossed by fresh striations. On the other hand, till associated with this movement overlies a possible interglacial peat near Sussex, and striations with this flow sense are inscribed on the supposed interglacial marine platform near Rockport. As well, this movement has the same trajectory as the first main event in Nova Scotia which is probably Early Wisconsinan if not older, as noted above.

After the initial main flood of external ice, local sources became more important. A series of six episodes, of generally decreasing vigour and extent, are distinguished; all are assigned to the Late Wisconsinan because evidence of deglaciation between episodes is lacking. During the first or Chignecto Phase, three ice sources coalesced. To the north, ice over northern Maine and Gaspésie produced a strong Chaleur Bay lobe and a general southward flow over the western part of the province. This flow, however, is considered to have failed to reach above 500 m in the northern highlands which stood as a broad nunatak. A central lowland dome, the Gaspereau Ice Centre, buttressed on the west by northern ice, spread northward to Chaleur Bay and southward to Bay of Fundy. It was abutted on the east by a glacier centred over western Prince Edward Island, which moved over Tracadie Peninsula and Chignecto Isthmus. The Caledonian Highlands were not covered by ice and lay at the margin of a coastal reentrant where relative sea levels reached 48 m, much higher than on adjacent coastal segments covered by later ice. The Chignecto event is estimated by Rampton et al. (1984, p. 44) at 15-18 ka. They link its margin to the Late Wisconsinan maximum limit in northern Nova Scotia postulated by Grant (1977b).

During the following Bantalor Phase, northern sources continued to operate and proglacial Lake Nepisiguit formed in the intermontane valleys of the northern highlands. Gaspereau ice however could not surmount the Tracadie Peninsula, reached low elevations around the flanks of Caledonian Highlands, terminated in the sea at Saint John, and built large moraine segments farther east. The Saint John end moraine is at least 13.1 ka (62 of Table 5.1) and the Bay of Fundy was ice free by 14.4 ka (61 of Table 5.1). Escuminac ice had shifted farther east over Prince Edward Island from whence it continued to push ice into Chignecto Bay.

Figure 5.11. Extent of Late Wisconsinan glaciers: a) "Minimum concept" based on terrestrial morphology and stratigraphy; b) "Maximum concept" based mainly on offshore stratigraphy and theoretical considerations.

The glacial phases that followed are largely recessional configurations of the northern ice mass which dominated the deglacial picture after the Gaspereau and other southern lowland centres had ceased activity. The Millville-Dungarvon phase is an important stillstand based on inferred connections of a series of ice marginal features extending from Chaleur Bay, around the west side of the highlands where small proglacial lakes were imponded at times in several valleys, to lobes in the Tuadook-Renous areas, and thence southwestward across Saint John River to Chiputneticook Lakes. Peripheral crustal depression allowed De Geer Sea to flood far up the valley. The age is estimated at 12.7 ka based on dates on shells from the associated marine submergence. This corresponds in time with the Robinson's Head readvance in Newfoundland (Brookes, 1977b; 37 of Table 5.1) and is equivalent in age to the Piedmont Moraines of western Newfoundland (Grant, 1972a). The ice had retreated from the eastern coast of New Brunswick by 12.6 ka (59 of Table 5.1).

The penultimate event was the Plaster Rock/Chaleur phase when ice cover was shrinking to a centre in the northwestern corner of the province, which was probably a part of a dome of ice extending eastward from northern Maine (Lowell, 1985). Ultimately it functioned independently of Gaspésie ice which continued to feed into Chaleur Bay where marine kame moraines were constructed 12.5-12.2 ka (60 of Table 5.1). The ice in northwestern New Brunswick flowed both northward up Saint John River valley and southward (Gauthier, 1980a). An extension of De Geer Sea, Inland Sea Acadia, flooded northern Saint John Valley as the ice retreated northward. The last remnants of local ice caps may have disintegrated in the St-Quentin area ca. 12.1 ka but a subtill peat dating 11 ka points to a late readvance (R.J. Mott, Geological Survey of Canada, personal communication, 1987). Final changes involved crustal adjustments and regression during the Madawaska Phase when marine waters shallowed, freshened, and ultimately disappeared as the present gradient of the Saint John River became established.

Prince Edward Island

Prest (1973) depicted the distribution of deposits and detailed the sequence of Quaternary events. The Quaternary cover is generally thin, and bedrock is at the surface in several areas. The surface till covers no older deposit and, with the exception of a single occurrence of two superposed drifts in the southeast corner of Prince Edward Island, it seems to be the only drift sheet. It is generally a mantle no more than 3-5 m thick showing textural facies ranging from clayey to sandy depending partly on the local bedrock and whether it was deposited by lodgment or ablation processes. The provenance is local but granite erratics from New Brunswick are abundant in the western end of the island and decrease markedly eastwards. A large network of ice contact stratified drift covers the eastern half of the island. This is thought to have been deposited under dead ice conditions as meltwater winnowed the till surfaces in lower areas. Only a few scattered kames and eskers occur in the western half of the island. During deglacial submergence, marine waters washed the area west of Malpeque Bay leaving a mainly sand veneer with a few gravel beaches. Peat has accumulated in bedrock and morainal depressions and along parts of the coast where rising sea level is causing paludification. Sandy salt-marsh deposits are aggrading in estuaries for the same reason.

The record of Quaternary events on the island is entirely glacial and postglacial — no evidence of preglacial or interglacial intervals is recognized. A sequence of glacier movements has been reconstructed entirely from the cross-cutting relationships of striations, supplemented by dispersal patterns of erratics. The first movement was a west to east invasion by an external ice sheet — evidently of Appalachian origin judging by the eastward-decreasing spread of New Brunswick granite erratics across the province. This event also is recorded in southern New Brunswick and northern Nova Scotia. During a second phase ice moved southward from Gulf of St. Lawrence and may have emplaced the shield erratics which occur sporadically in the northern and eastern parts of the island (Prest in Wanless et al., 1978). Goldthwait (1924) attributed this event to the Acadian Bay Lobe; Rampton et al. (1984) invoked an offshore source, the Escuminac Ice Centre. Later the island was the site of an ice cap that flowed radially but mainly southward to Nova Scotia. During the final phases, the south flank of the ice mass became progressively more directed laterally into the eastern and western ends of Northumberland Strait, as if being drawn down by calving bays encroaching up the deep submarine leads of Shediac Valley and Cape Breton Trough.

Ultimately, marine waters of the deglacial Goldthwait Sea in Gulf of St. Lawrence advanced over the area. Because the western end of the island was depressed more than the eastern end, owing to the relatively greater mass of New Brunswick glaciers, the deglacial shores rebounded more in that area. Consequently marine limit, after modification by subsidence and resubmergence, is now tilted westward from present sea level near the middle of the island to 25 m at the western end. Shell dates show that the glacial isostatic regression was in progress more than 12 ka ago (63, 64 of Table 5.1), and dates on salt marsh and on the underlying submerged forests trace the effects of the subsequent Late Holocene subsidence (Frankel and Crowl, 1961; 65 of Table 5.1; Kranck, 1972). The postglacial climatic evolution has been traced by means of palynological studies of cores through bog and pond sediment (Anderson, 1980).

Newfoundland

Distribution of deposits

The main glacial drift deposits of Newfoundland are associated with three main foci of glacial dispersal: Avalon Peninsula, Northern Peninsula, and the central part of the Newfoundland Uplands (Fig. 5.11a). The pattern of erosion and deposition is evidently the product of radial glacial movement. Thus the central or axial areas of ice dispersal are characterized either by shallow immature drift or regolith on largely unmodified preglacial landforms or by heavy drift accumulations, which locally obscure deep bedrock valleys, and which are generally fluted or ribbed in clear expression of the radial dispersal. Areas of thinner till occur peripheral to the central area; a belt of scoured bedrock is present adjacent to the coast and till is present mainly as scattered crag-and-tail hills.

On high and outlying parts of the island, notably Burin, Hermitage, Bonavista, and Avalon peninsulas, and the summit tablelands of the Long Range Mountains, are glacial terrains that are characterized by distinctly greater maturity of surface form compared to the essentially fresh glacial landscapes which prevail throughout the island and at lower elevations (Fig. 5.12a, b). Erratic blocks from the interior attest to former ice cover of the weathered areas, but morainic topography has been degraded by solifluction, and bedrock outcrops are covered by disintegrated rock. Two different degrees of alteration can be distinguished — the higher and presumably older, termed Weathered Zone C, is at a mature stage of alteration whereby till slopes are uniformly graded, all hollows and hummocks in the drift have been removed, bedrock tracts are buried by in situ formed rubble, and all rock basins and knobs have been obliterated (Fig. 5.12a). At slightly lower elevations, and presumably much younger, is a glacial tract of intermediate character, termed Weathered Zone B; in this zone till surfaces are modified but still carry morainic features, and bedrock shows incipient felsenmeer and grus accumulation. These zones contrast sharply with the lower freshly glaciated areas (Fig. 5.12b). Detailed delineation of the limits of the fresh, mature, and old-age glacial terrains in the western fiordlands (Grant, 1977b; Brookes, 1977a) reveals that they are topographically controlled and that their elevations rise regularly inland.

In two small areas, namely a few small summits of the southern Long Range Mountains above Codroy Valley and Blow Me Down Highlands around Bay of Islands, glacial erratics and features are entirely lacking (Fig. 5.12c; Coleman, 1926; Brookes, 1977a). These lie above, and are sharply truncated by, the most mature glacial terrain. These bedrock areas are in an advanced stage of disintegration and rare tors project from the gentle, featureless slopes. As already noted, these areas are believed to be remnants of pre-Quaternary peneplains.

Glaciofluvial deposits usually occur as esker-kame complexes inland, where they radiate seaward from the main ice-cap centres. Nearer the coast, and in major valleys where meltwater collected during deglaciation, glacial alluvium was deposited as outwash plains, usually transitional to marine beds. Together with the distribution of sidehill meltwater channels, these fluvial features represent the glacial hydraulic system and hence supply data pertinent to the location of the main deglacial ice remnants (Grant, 1974b).

Emerged marine deposits are restricted to a narrow belt along the present shore because the land generally rises steeply away from the coasts. Exceptions are along the western coastal lowlands and around parts of Notre Dame Bay.

Figure 5.12b. Late Wisconsinan glacial terrain, Gros Morne National Park: Typical landscape of areal scouring with rock basins, polished outcrops, and perched erratics. 203673-U

Figure 5.12c. Unglaciated summit, southern Long Range Mountains: Pre-Quaternary peneplain is unmodified by glacial action and large tors occur (foreground). 203673-V

Figure 5.12a. Oldest glacial terrain ("Zone C"), Big Level, Gros Morne National Park: Colluvium and solifluctuated till underlie maturely graded slopes. Rock basins are filled and small tors emerge. Age is estimated at stage 12 (ca. 430 ka). 203673-O

Distal to major meltwater systems, the deposits generally comprise a gravelly littoral facies overlying a silty deepwater facies. Fossils, which occur mainly in areas of carbonate bedrock, have been dated to infer the time of deglaciation and the course of sea level change over the last 14 ka. The bulkiest deposits and those with greatest deglacial history significance are massive kame moraines or ice contact deltas. These were constructed at marine limit at the snouts of ice lobes occupying deep valleys. Their construction evidently represents important interruptions in the general retreat, and some may mark the ice position of the last stadial maximum. These features, together with the upper limit of beaches, wave modified till, and bedrock terrain stripped of loose ablation debris, define the upper limit of postglacial marine submergence which rises from sea level on Avalon Isthmus to 150 m along Strait of Belle Isle. Whether some are related to deglaciations earlier than the last stade, as inferred by Grant (1980a), remains conjectural. Although the details of the pattern of isobases based on elevation of this highest emerged shore have varied in detail from De Geer (1892) to Rogerson (1982), all tend to support the notion of an isostatic up-doming of Newfoundland as if in response to removal of an island-centred glacial complex.

Organic deposits occur widely under sedge and sphagnum bogs in a variety of settings at all elevations, as detailed by Wells and Pollett (1983). Besides constituting an important potential energy resource, bog accumulations have been used to supplement lake sediment cores to reveal postglacial climatic evolution (Macpherson, 1982; Anderson et al., 1989).

Till stratigraphy and geochemistry

Till compositional attributes have not been mapped regionally but have been studied locally in areas of good exposure and in connection with tracing boulder trains or explaining geochemical anomalies. O'Donnell (1973) compared drift dispersal with till geochemistry in an area of unexplained ore-grade erratics around Gullbridge. Alley and Slatt (1975) used geochemistry and distribution of erratics to determine that localized metalliferous erratics came from a window of older till surrounded by barren surface till. In a test of geochemical methods in different settings, Hornbrook et al. (1975) detailed the glacial geology and chemical composition of surface materials in relation to two subcropping orebodies. They found that basal till in drumlinized areas was the best prospecting medium; otherwise stream sediment could be used to narrow the target area. In the central ice divide area, Grant and Tucker (1976) delineated a number of Ni-Cu-Pb-Zn drift anomalies in a region of several divergent ice flow patterns. From these examples, it can be seen that drift prospecting and basic research into mineral dispersal is just beginning in Newfoundland, and in such a heavily drift covered area much more work is necessary if dispersal of indicator materials in drift is to be used as a means of locating ore bodies.

Quaternary stratigraphic studies in Newfoundland are still at the reconnaissance stage. Initially Coleman (1926) recognized unglaciated tracts in the western highlands and divided the Pleistocene sequence into a fresh or 'Wisconsin' age drift and an older weathered 'Jerseyan' drift. Although some of his observations cannot be reproduced, his general twofold scheme remains valid. MacClintock and Twenhofel (1940) studied the deposits around St. Georges Bay and divided the Wisconsinan sequence into two drift sheets and an intervening marine bed. Brookes (1969, 1974) substantiated and elaborated on their findings. In the western highlands Grant (1977b), on the basis of the truncation of older mature glacial terrains by younger ones, recognized three different ages of glacial expansion. The sequence was found to apply also in the Codroy area (Brookes, 1977a) which, in addition, contained summits never glaciated. In Burin Peninsula area, Tucker (1979) and Tucker and McCann (1980), building on the work of Grant (1975c), proposed four glacial phases and a marine interval. Prompted by Widmer's (1950) findings in Hermitage Peninsula area, Leckie (1979) outlined the Quaternary history as including three elements: the activity of a local ice cap; invasion by inland ice; and a sequence of marine submergences. Finally, in the central part of the island around Red Indian Lake and in Deer Lake lowland, Vanderveer and Sparkes (1982) have unravelled the flow sequence and correlated till sheets with sets of striations. All such studies have tended to support the long-held hypothesis that Newfoundland was glaciated largely by its own ice cap complex.

Sequence of events

Most Quaternary deposits and features in Newfoundland pertain to glaciation and these clearly implicate an island-centred glacier complex which expanded several times, but failed to overtop certain high coastal summits. External or Laurentide ice evidently reached only the northern extremity where it attained elevations of 300 m and was confluent with Newfoundland ice perhaps as far south as Port au Choix. Elsewhere on the western coast there are no shield erratics to record Laurentide invasion, and three glacial trimlines slope seaward from the interior. Figure 5.11a shows the disposition of the various subcentres of Newfoundland ice, some of which functioned only during early and late stages of growth and decay. The only other exception to the generally land-based glacier complex flowing seaward is a centre which is inferred to have lain south of Burin Peninsula (Tucker and McCann, 1980), but it may have been an early manifestation of Avalon ice or of a larger ice cap on emergent Grand Banks.

Evidence of interglacial time is fragmentary and largely indirect. Henderson (1972) recognized an emerged intertidal platform that has been related to a stand of sea level 3 to 10 m above present (Grant, 1980a). No associated littoral deposits have been found in Newfoundland but the relationship of a correlative platform elsewhere in Atlantic Canada to tills and other deposits suggests a Sangamonian age. Only one organic record of nonglacial conditions with a temperate climate like the present interglaciation is known; it consists of brackish/marine sediments in a karst depression at Woody Cove on Cabot Strait (Brookes et al., 1982). This proved to be beyond the limit of radiocarbon dating (1 of Table 5.1) and the age is presumed to be Sangamonian; but because the implied sea level is much higher than that recorded regionally by the supposed Sangamonian shore platform, it might date from an earlier interglaciation. The only other evidence of ice-free conditions comes from iron staining on glacially facetted outcrops (Grant, 1975c; Fig. 5.13), and from pedological studies of older glacial tracts (I.A. Brookes, York University, Toronto, personal communication 1983) which have revealed much more strongly rubefied surface soils and paleosols than the 50 cm brown podsol on the youngest drift.

The oldest glaciation, inferred to predate the last interglaciation (oxygen isotope stage 5) on geomorphological grounds (Grant, 1977b; Brookes, 1977a) and on stratigraphic relations (Tucker and McCann, 1980), extended the farthest seaward and reached highest in the western fiordlands. Its elevational limit ranges from 600 m in the northern Long Range Mountains to only 400 m near Cabot Strait. It might date from oxygen isotope stage 12 which is the earliest episode of maximum glaciation recorded offshore (Alam and Piper, 1977).

The next youngest advance, which may also be pre-Wisconsinan, reached within 100 m of the all-time maximum and evidently came from similar sources. A raised shoreline, presumed to be related to the glacial isostatic submergence of this second phase has been recognized by Tucker et al. (1982) in Hermitage and Burin areas. During this, or some other glaciation of intermediate age, ice moved northwards onshore over Burin Peninsula, seemingly from a centre located on the Grand Banks which are thought to have been partly emergent during glaciations. The intermediate glaciation is tentatively assigned to isotope stage 6 (Illinoian) because that was the second most important period of ice activity recorded offshore (Alam and Piper, 1977).

The final glacial phase is of course the best known and is reasonably well dated. Its culmination, believed to represent the Late Wisconsinan stadial maximum, is chronometrically bracketed to the interval 14-13 ka by shell dates from marine sediment above and below till (35, 36, 38, 39 of Table 5.1). Ice moved seaward from interior sources and, on the western lowlands, constructed large piedmont moraines which loop a few kilometres offshore. A conceptualization of its approximate outer boundary appeared in Grant (1977b). In the Burin Peninsula area Tucker (1979) proposed a limit slightly farther inland, and in Strait of Belle Isle Rogerson (1982) proposed an even more restricted extent. Prest (1984) used these limits in portraying his minimum extent of Late Wisconsinan ice. Whatever its precise outer limit, the last ice shrank to a number of discrete centres (Fig. 5.11a) which are mapped from the trend of ice flow indicators, ice marginal meltwater channels, and eskers (Grant, 1974b). These are disposed at a variety of elevations and thus seem to suggest that the major control was the location of main snow-accumulation areas in this low relief maritime setting. The deglacial sea stood high against the retreating glacier and consequently ice contact marine deltas occur either at the mouths or the heads of most major outlet valleys (Jenness, 1960; Tucker, 1974; see 31, 35, 41 of Table 5.1). These probably mark positions of stability reached after a period of calving, although given the paucity of dates, a climatic control cannot be ruled out. An example of such a position of stability is marked by the Ten Mile Lake Moraine (Grant, 1969b) which was built by a readvance culminating in the sea at 11 ka (28, 46 of Table 5.1). As the readvance followed a lengthy retreat by calving, it could represent a surge in response to a steepened ice profile. However, as the age of this readvance correlates well with the Younger Dryas climatic deterioration that affected much of the north Atlantic region and is recorded throughout the region as an interruption of deglacial warming (Anderson, 1983; Macpherson and Anderson, 1985; Mott et al., 1986), the moraine is regarded as a localized glacier response to a cool climate interval. On the Northern Peninsula one morainal belt is thought by Waitt (1981) to date from the period 8-9 ka. Although the pattern of final

Figure 5.13. Intersecting glacial facets, Burin Peninsula, Newfoundland. The outcrop records an ancient glaciation moving southward from the interior, interrupted by a weathering interval when the till and rock surface were oxidized, before the last glaciation which moved westward from offshore and cut a new clean stoss surface. 203673-Q

deglaciation can be mapped from glacial features, and its timing at the outer coast determined by marine shell dates (26, 27, 29, 30, 32-34, 40, 42-45 of Table 5.1), its progress and timing remain obscure, even problematical in light of the surprisingly old lakebottom sediment ages being obtained in the interior (e.g., 45 in Table 5.1). Small moraines are nested in cirques on the western summits but none are dated, and nivation continues to etch the escarpments under present conditions. In addition, fossil rock glaciers occur where ground ice on slopes has mobilized scree and lateral moraines.

Sea level changes, though for the most part related to glacial phases, have attracted special attention and are currently the subject of separate studies. The overall situation is summarized by Grant (1980a). The oldest paleoshore is an emerged littoral wave-cut platform 3-10 m above present tide level that is preserved locally beneath younger tills in southern Newfoundland. Henderson (1968) recognized it on Avalon Peninsula, Brookes (1977a) noted it on Cabot Strait, and Tucker et al. (1982) mapped it on Hermitage and Burin peninsulas. Intra-Wisconsinan glacial age shorelines are inferred only for Hermitage area (Leckie, 1979) and are unknown elsewhere. Most paleoshores are Late Wisconsinan and Holocene in age and relate to the last deglaciation and attendant crustal recovery. The pattern of marine limit, as variously depicted by Grant (1980a) and Rogerson (1982), clearly shows up-doming due to unloading of central parts of the island. Isobases exist for only one younger deglacial level — that of the "Bay of Islands Surface" (Flint, 1940) — and it is concordant with the general trend shown by isobases on the highest level. Grant (in press) correlates the Bay of Islands Surface with the Ten Mile Lake readvance, which is to suggest that the expanding ice cap stabilized sea level gravitationally. Dates on intermediate sea levels are few and scattered and only two sea level curves are available. One for the northern coastal lowland shows a regular emergence continuing essentially to present (Grant, 1972a); one for the central west coast (Brookes et al., 1985) shows that emergence ended about 6 ka ago and that submergence has followed. Other scattered evidence suggests that the southern half of the island is subsiding, presumably due to migration of a collapsing ice marginal crustal bulge.

Deglaciation, crustal unloading, and sea level change have been combined in numerical models (Quinlan and Beaumont, 1982). These show that the observed changes in relative sea level are those that would be expected to follow ice retreat from the limit of Late Wisconsinan glacier cover postulated by Grant (1977b). Thus the glacial limits drawn on the basis of geological evidence are corroborated by theoretical models which link observed sea level change to isostatic crustal rebound.

Gulf of St. Lawrence area

Our knowledge of the Quaternary geology of this part of the region comes mainly from reconnaissance mapping of the bottom sediments and the bottom morphology of this vast inland sea. Information from the Magdalen Islands provides a crucial link between what has been inferred for the southern gulf and the more fully developed history for Nova Scotia and Prince Edward Island. Similarly, data from Anticosti Island extend our knowledge of Laurentide events on the Canadian Shield into the northern part of the Gulf of St. Lawrence. In sum, the general outline of the history of the Gulf of St. Lawrence can now be meshed with the more detailed interpretation of adjacent land areas, such that a coherent picture is emerging.

Submarine areas

Loring and Nota (1973) provided an interpretation of bottom samples collected from a large part of the gulf. Important stratigraphic information was presented by Conolly et al. (1967), while Monahan (1971) and Shearer (1973) presented glacial geological reconstructions based on geomorphological evidence. Bedrock outcrops are rare except near the coast and on part of Magdalen Shallows. Till occurs in the subsurface as a thin mantle over much of the area, underlying postglacial marine mud. Where till is exposed along the south edge of Laurentian Channel, its lithology matches that of the underlying Permo-Carboniferous redbeds. Elsewhere, the seafloor is composed of gravelly sand over the banks and shelves to water depths of 100 m where it merges with the cover of mud that occupies deeper channel areas (see surface materials map in Keen and Williams, 1989). Because coring and seismo-stratigraphic profiling have not been conducted, and because the surface sediment is largely the result of postglacial marine sedimentation and winnowing, little is known about the Quaternary stratigraphy of the region. In fact in most areas it is not even known whether the surface material is a postglacial sediment or whether it is relict or derived from underlying sediments.

The provenance of heavy mineral assemblages has been studied in order to identify transport pathways and to infer agents of deposition of seafloor sand and mud (Loring and Nota, 1969). Three main dispersal fans were noted. The largest, consisting of an amphibole-pyroxene assemblage of Canadian Shield origin, covers the northern gulf, with tongues projecting towards Prince Edward Island and reaching the northern tip of Cape Breton Island (Fig. 5.11b). Assuming that the surficial sands containing this mineral suite were derived from glacially transported debris, a major flood of Laurentide ice is necessary to explain this occurrence as well as the scattered igneous erratics around Magdalen Islands. This movement of Laurentide ice is corroborated by a trail of carbonate debris, which extends from Anticosti Island southeastward over clastic rock terrane. In the southern Gulf of St. Lawrence two locally derived assemblages occur: one around Magdalen Islands is characterized by zircon and the other, north of Prince Edward Island, contains anatase and tourmaline which supports the notion of a late, local ice cap over the island. In addition to the three main fans, there is a suggestion of a metasedimentary heavy mineral suite being introduced from the Chaleur Bay area. This would correspond with terrestrial evidence for a major ice stream flowing eastward from the bay. Finally, Cape Breton Island mineral assemblages are not found offshore. This appears to support the interpretation that the island was reached by northern glaciers, and that its local ice cap did not shed debris into the Gulf of St. Lawrence. Thus, with the exception of the area north of Prince Edward Island, glaciation seems to have involved only a major southward incursion of Laurentide ice; local ice caps from areas adjacent to the gulf apparently had little offshore extent. Information from Anticosti and Magdalen islands and Newfoundland are in harmony with these findings.

Specific information is also available on the positions of glacier margins from the eastern gulf that is consistent with interpretations for the adjacent land areas. Monahan (1971, Fig. 1) depicted several ice frontal ridges. Two are arcuate moraines off the mouths of Bonne Bay and Bay of Islands which functioned as major glacial escapeways for Newfoundland ice. An interlobate moraine trails southward from Banc Beaugé where two trunk glaciers in Esquiman Channel and Anticosti Channel converged. A recessional moraine is inferred for Laurentian Channel, which is similar to that in Cabot Strait, and which Fader et al. (1982) termed the Laurentian Moraine. Shearer (1973) inferred two major ice marginal positions: one attributed to Canadian Shield ice standing on the northside of Esquiman Channel, the other on its southeast margin off Port au Port Peninsula. All these observations support the notion that Laurentide ice did not override Newfoundland but that Laurentide ice and Newfoundland ice were convergent in Strait of Belle Isle and flowed into the Gulf of St. Lawrence.

The timing of ice sheet coverage of the gulf, of ice retreat, and of moraine building cannot be deduced from existing submarine data. Conolly et al. (1967), however, presented stratigraphic data from cores of Laurentian Channel sediment that bear on the question. Their 12 m sequences show a surface grey unit which they say indicates glacial retreat inland of the present coast before 11 ka. A few metres deeper are two very thin, brick-red diamictons with unabraded clasts in a thick glaciomarine drift. The "tills" are considered to be ice-rafted debris dropped in the calving zone of a floating ice sheet that came mainly from the Carboniferous redbed terrane. No till was penetrated. This suggests that the glaciers that sculpted Laurentian Channel probably carved it out to its present size much earlier. The only dated deposit associated with glaciation is a proglacial marine drift 10.2 ka (66 of Table 5.1), which covers glaciomarine sediments, and which is thought to have been derived from Magdalen Shallows bedrock by a glacier advancing northeastwards (Loring and Nota, 1973, p. 126). Perhaps this event, if correctly dated, signals continued activity of a late remnant of the ice centred on or north of Prince Edward.

Magdalen Islands (Îles-de-la-Madeleine)

This small archipelago offers crucial evidence bearing on the style, sequence, and age of glacial and nonglacial events in central Gulf of St. Lawrence. Because of the deep dissection and maturely graded slopes, the lack of clear glacial features, and a peculiar assemblage of surficial materials, the questions of whether and when the islands were ice covered, and what agents produced the deposits have been debated for more than a century. There is no doubt that the islands have an anomalous aspect and ambiguous deposits. Recent work by Prest et al. (1977) and Grant et al. (1985) has brought to light new facts and cleared up several questions, but contradictions and disagreements remain.

As presently perceived, the surficial sequence comprises seven main units: deeply weathered bedrock, an interglacial complex of littoral gravels and terrestrial peats, a till, a problematic substratified diamicton (Demoiselle Drift), local ice contact gravels and an end moraine, colluvium, and late Holocene relict barrier beaches. Maturely weathered bedrock, seen locally on volcanic rocks but not yet analyzed, suggests that the islands have suffered relatively little glacial erosion.

The most important unit is the buried littoral gravels and included peat beds. This deposit, which seems to be a remnant of a beach system similar to the modern tombolos, has been found up to 20 m elevation. The unit comprises two gravel members: the lower is oxidized and locally iron or calcite cemented; the upper is loose, clean, and contains clasts of the lower. These are interpreted as signifying two periods of elevated sea level, separated by a period of regression and subaerial weathering. The lower unit is composed mainly of crystalline rocks from the Canadian Shield. This might indicate a previous Laurentide glaciation (for which a till sheet is lacking) or could merely be due to nonglacial seasonal ice rafting similar to that of today which brings many erratics from the north shore of the Gulf of St. Lawrence to the beaches. These elevated interglacial littoral deposits are tentatively correlated with the raised platform found elsewhere in the region. This in turn leads to the suggestion that the two members correspond with the two sea level maxima of the last interglaciation — stages 5a and 5e.

At six sites, the littoral deposits have organic interbeds which have supplied indeterminant ^{14}C dates (18-22 of Table 5.1). Paleoecology of one at Portage du Cap shows thermophilous forest comparable to the present (Prest et al., 1977), and another at Le Bassin shows a long record of warming from shrub tundra to a forest of oak and pine suggesting a climate warmer than that of today. On the latter bed, Th/U ages range from 89 to 106 ka (Vernal et al., 1986; 19 of Table 5.1). Thus pollen and dating support the assignment of this complex to the Sangamonian Stage (Mott and Grant, 1985).

At all sites the interglacial beds are overlain by diamicton which at three sites is a typical tough compact silty clay lodgment till containing striated stones. Glacial tectonic deformation is further evidence of glacial action. At Portage du Cap the gravel beds have been isoclinally folded, and at Boisville and Des Buttes the peat/gravel assemblage is seen as an allochthon that has been shoved more than 1 km inland to an elevation of 40 m. In addition, bedrock on several islands has been locally plastically crumpled into chevron folds which wrap over the younger marine gravel (Dredge and Grant, 1987). The direction of thrusting and shear is east and southeast regardless of local slope so deformation by nonglacial mass movement appears to be ruled out. Glacial invasion by Laurentide ice is shown by the large crystalline erratics on southern islands (Prest et al., 1977), the crystalline boulder mantle on a central island, and the ice contact deposit that covers much of Île de la Grande Entrée (the Coffin Island end moraine of Alcock, 1941). There is thus abundant evidence of Laurentide ice, like that represented by the suite of Laurentide heavy minerals spread southward over the entire Magdalen Shelf (Loring and Nota, 1973). Perhaps Alcock's moraine marks a recessional position, if not the limit of a younger advance. Based on the glacial tectonic evidence, the Laurentide invasion is post-Sangamonian and possibly correlative with the Early Wisconsinan(?) ice flood that swept southward over Gaspésie, New Brunswick, and Nova Scotia.

Evidence of later events is ambiguous and contradictory. Gravels, peats, and tills are overlain by a reddish brown, sandy, substratified blanket, the Demoiselle Drift, which also mantles much of the rock surface, and which contains subangular clasts mainly of the local volcanic rocks. It has been considered by some to be a till because it contains striated clasts, and by others to be a glaciomarine drift because of the stratification. At Le Bassin it appears to grade into the underlying till and so must be a closely related facies. Perhaps it is a basal melt-out till or possibly it is a submarine ice shelf deposit. Proposal of a marine origin faces the problem that clear evidence for marine submergence is lacking. Unlike other parts of the region where marine inundation is registered by trimlines, beaches, terraces, and gravel lags, no such features occur on Magdalen Islands. A nick point or slope declivity at 30-40 m, often cited as the limit of wave action, varies in elevation and appears to correspond to the contact of soft sandstone with resistant volcanic plugs (Sanschagrin, 1964). Probably it represents an ancient pre-Quaternary(?) degradation level. The Demoiselle Drift covers mature slopes which are graded as if modified by prolonged solifluction and are cut by flat-floored "vallons". These aspects might point to a lengthy periglacial period which possibly corresponds to the time of a period of glaciation elsewhere in the region.

The youngest deposit of Quaternary stratigraphic significance is a thin peat bed dating between 11.3 and 10.6 ka (67, 68 of Table 5.1), which is overlain by sand and a stony mud that superficially resembles Demoiselle Drift but is more likely a younger colluvial deposit. The sequence signifies a late glacial climatic cooling, which correlates with the oscillation ascribed to a Younger Dryas southward incursion of the Polar Front (Mott et al., 1986).

During the late Holocene, sea level has been rising. The transgression constructed barrier beaches which have now linked the islands by tombolos enclosing narrow lagoons. Certain features suggest that a recent small sea level fluctuation is recorded. The tombolos extend from what is now a fossil cliff, and raised or fossil berms are now being truncated, as though sea level was once about a metre higher than present, then fell slightly, and is now rising again.

In sum, the Quaternary record of events includes a warm interglacial period with a higher sea level, a major ice sheet glaciation, a period of colluviation and/or glaciomarine (ice shelf?) sedimentation, a possible moraine-building Laurentide readvance, a Wisconsinan/Holocene climate reversal, and a rising Holocene sea level with one possible recent fluctuation.

Anticosti Island (Île d'Anticosti)

Initial reconnaissance studies led early workers to conclude that the island was completely glaciated during Late Wisconsinan time by the Laurentide Ice Sheet, and that there was a single phase of glacial isostatic submergence to about 76 m after 13 ka (Bolton and Lee, 1960; Prest, 1970). Earlier, however, Fernald (1925) had inferred from botanical associations that the glaciation was more ancient and, using this interpretation, Grant (1977b) speculated that the Late Wisconsinan limit may have lain north of the island. Recent work by Dubois et al. (in press), however, has demonstrated that Anticosti Island was almost completely ice covered during the Late Wisconsinan and has a record of three glacial advances of decreasing extent with time, with two associated glaciomarine or ice contact marine units which apparently span much of the Wisconsinan (Painchaud et al., 1984; St-Pierre et al., 1985; Bigras et al., 1985; St-Pierre et al., 1987). The following summary is based on their work, particularly the latter report.

Three tills containing materials derived from the Canadian Shield mark the recurrent expansion of Laurentide ice. Lithology and carbonate content reveal the influx of crystalline (shield) debris relative to calcareous material of local derivation. The earliest event recorded is a nonglacial marine transgression exceeding 100 m. The oldest, the Rivière à la Patate Till, is grey, compact, silty ($\approx 50\%$ silt), and stony, with a southeastward fabric. Its age is estimated at more than 85 ka based on amino acid ratios in overlying marine sediments. The second, Rivière Jupiter Till, is coarser (34% sand), has a higher carbonate content (72%), and was emplaced by southwestward ice flow. It is laterally transitional to fossiliferous glaciomarine drift which contains shells dated at 36.0 ka (23 of Table 5.1). This marine sediment is interpreted to represent glacial isostatic submergence of at least 84 m that was caused by an advance of Laurentide ice during Middle Wisconsinan time. At that time the glacier limit was located on the south coast and was not farther south at any later time. Shells in the upper part of the till are said to indicate that the advance culminated 29-30 ka (24, 25 of Table 5.1), at which time relative sea level was +84 to +122 m.

Subsequent retreat is recorded by glacial marine and littoral sediments that overlie the till. They signify regression at least to present sea level on the south shore, presumably due to differential rebound. In the interior of the island a significant period of glaciofluvial and fluvial erosion and sedimentation followed deposition of Rivière Jupiter Till. The extent of the erosion and deposition suggests that a residual ice cap may have persisted on the central plateau of the island (Bigras et al., 1985) while most of the island was ice free.

The overlying third Laurentide drift, called Île d'Anticosti Till, has a similar aspect and trajectory to older tills. It extends across the island but terminates at the south coast where it is interbedded with glaciomarine drift that was deposited in a sea that reached +55 m about 15-13 ka (69, 70, 71 of Table 5.1). These dates are taken to signify the age of the Late Wisconsinan stadial maximum in this area and it thus represents the southern extent of Laurentide ice in the northern Gulf of St. Lawrence. That the ice went farther as a grounded glacier in the nearby deep arms of Laurentian Channel (as Gratton et al., 1986 depicted) is considered extremely unlikely. However, a floating ice shelf may have extended farther south, at least over Laurentian Channel. This could explain the Demoiselle Drift on Magdalen Islands and the red stony diamictons in Laurentian Channel (Conolly et al., 1967).

As Laurentide ice retreated off the island between 12 and 11 ka (72, 73 of Table 5.1), relative sea level transgressed to 60 m on the south coast and to 70 m on the north coast (Bigras et al., 1985). As it withdrew, a satellitic ice cap became separated from the ice sheet. It remained active to produce the Sainte-Marie Till, a highly calcareous drift because of its local derivation. A late readvance of the remnant ice cap culminated in the sea and built the Sainte-Marie Moraine. Final retreat brought relative sea level rapidly down to 30 m at 9 ka (at a rate of 0.2 m/ka) then more gradually at 0.025 m/ka to its present level.

In sum, these results provide the first chronostratigraphic evidence of Laurentide glacial activity in northern Gulf of St. Lawrence. The major advances are documented and their ages are closely estimated, and Late Wisconsinan stadial limits have been established and dated. The Anticosti interpretations place constraints on limits of grounded glaciers throughout the gulf and have implication for inferences of glacier extent and age farther south on the continental shelf.

Gaspésie

The Quaternary geology of Gaspésie is crucial to an understanding of the rest of the region because it helps clarify the relationship between the activity of local Appalachian ice caps and the invasion of Laurentide ice. Early studies were topical and local (Bell, 1863; Chalmers, 1887b, 1906; Goldthwait, 1911; Alcock, 1935, 1944; Flint et al., 1942). Interpretations of the whole peninsula by Coleman (1922) and McGerrigle (1952) were turning points because these initiated the ongoing debate about whether the entire area was ice covered; whether the ice was only of Laurentide origin or also of local origin; and over the relative rank, age and duration of each glacial phase. The debate is being resolved by systematic mapping and by detailed stratigraphy which are largely summarized in the stratigraphic framework published by Lebuis and David (1977) and David and Lebuis (1985). The latter interpretation forms the basis of this account, and although it deals mainly with the western half of the peninsula, it is believed to be largely applicable also to the eastern part where mapping is in progress (LaSalle et al., 1985b).

Surficial deposits

Systematic mapping is in progress at this time so it is possible to give only a general overview of the distribution of deposits and features in relation to landscape. According to Gray et al. (1981), the northern coastal fringe consists of rocky coastlands, marine rock platforms, some littoral sediment, and has rare bodies of ice marginal sediment in valley mouths. The broad tablelands forming the northern half of the peninsula are generally rocky, mantled by a thin colluvial cover, and summits above 1100 m have tors, are mantled by felsenmeer blockfields, and have patches of till smoothed by solifluction and cryoturbation. The deep valleys, which dissect these surfaces in a rectangular network, have steep colluvial slopes and narrow floors underlain by glacial and postglacial deposits. Till is most widespread and thickest in two belts crossing the peninsula along the valleys of the Matapédia and Cascapédia rivers. Interfluves in the vicinity of these belts are ice abraded. The eastern half of the peninsula has southward sloping peneplains veneered with regolith, indistinct patches of till, and scattered erratics. A third belt of till and pockets of marine sediment borders Chaleur Bay.

Quaternary sedimentary succession and sequence of events

Lebuis and David (1977) recognized four superposed tills. The stratigraphic sequence begins with gravel overlain by varved sediments which could signal the approach of glaciers. Over this lies the Tamagodi Till which was dispersed initially northeastward and then southeastward, by a local Gaspésie ice cap. Locally it is thoroughly oxidized, so perhaps this till predates the last interglaciation; if not, it has been weathered during a long Wisconsinan interstade. During the early ice cap phase, granite indicator erratics from McGerrigle batholith were dispersed radially.

The main invasion by Laurentide ice is registered by the Langis Till which was spread southward across the waist of the peninsula by ice streaming towards Chaleur Bay. Numerous distinctive indicator lithologies, both Laurentide and Appalachian, permit the ice flow trend to be defined. From numerous lithological and heavy mineral analyses, together with striation measurements, David and Lebuis (1985) concluded that Laurentide ice crossed the western part of the peninsula. Towards the east the till sheets and erratics trains merge with colluvium and weathered rock with scattered erratics.

David and Lebuis (1985) believed that the entire peninsula was ice covered during the Late Wisconsinan. Variations in concentration of erratics, sharp limits of till and glacially abraded terrain against regolith- and felsenmeer-covered surfaces, and the preservation of pre-Quaternary surfaces and soils are attributed to variable temperature conditions at the base of the ice. They postulated that the ice was warm-based over areas that have typical glacial deposits and features (namely the valleys and lowlands) and cold-based over the areas with none. Thus with the ice frozen to its bed, earlier formed surfaces, features, and deposits would be preserved. The frozen-bed state is said to have prevailed in the high areas where ice was thin and on the southern lowlands where ice flow was divergent. Proof that ice did in fact cover these "nonglacial" terrains is in the form of scattered erratics, patches of ablation till, and glacial tectonically deformed regolith (LaSalle et al., 1985b). It is perhaps significant that a cave older than 230 ka has survived glacial unroofing (Roberge et Gascoyne, 1978).

The cold based ice hypothesis thus resolves many of the contradictions about the glacial geology of Gaspésie just as it has explained preservation of old deposits and juxtaposition of freshly glaciated and older surfaces in northern New Brunswick, Cape Breton Island, Newfoundland, and the Magdalen Islands. The difference between this and other areas in the region is the age that is assigned to the extensive, regional ice sheet phase. David and Lebuis (1985) argued for a Late Wisconsinan age, whereas elsewhere it is thought to be Early Wisconsinan. In all likelihood the regional ice sheet phases that are recorded as a major early event in the widely separated areas probably all represent the same event. There are no dates to support age assignment in any areas but there is stratigraphic evidence in areas other than Gaspésie to support an Early Wisconsinan age for this ice. Putting this age controversy to one side, the possibility still has not been ruled out that the ancient weathered summits with no erratics in Gaspésie were not glaciated during the Wisconsinan.

The regional southeastward flow indicators are crossed by northward pointing indicators as far south as the median line of the peninsula. David and Lebuis (1985) referred this movement to a reversal of flow caused by calving in the St. Lawrence estuary. Once the Laurentide ice source was cut off by a calving bay moving up the St. Lawrence estuary (see also Occhietti, 1989), the Gaspésie ice would reorganize into local caps flowing radially to the coast. The Petit Matane Till is the drift unit that records this interval of northward movement to the St. Lawrence shore. It intertongues with Goldthwait Sea marine sediments which date about 13.4 ka (74 of Table 5.1). This part of the local ice cap phase is therefore definitely Late Wisconsinan.

The final phase involved activity of remnant ice caps, one in the McGerrigle Mountains and the other on Murdockville highlands (Chauvin, 1984), and of several small valley glaciers fed by cirques cut into the plateau margins. These produced the thin, bouldery, locally derived Grand-Volume Till.

The pattern of postglacial marine overlap and regression suggests certain possibilities for Late Wisconsinan ice disposition. Cooke (1930, Fig. 3) showed that marine overlap along the St. Lawrence coast decreases in elevation at a rate of 0.35 m/km eastward. This suggests that the ice load increased towards the west. Two higher terraces occur with a similar slope near Cape Gaspé at the eastern tip of the peninsula. Unless the two higher levels date from an earlier glaciation, the three paleoshores indicate a three-step retreat of an ice margin up the St. Lawrence estuary. While scattered dates show that marine withdrawal was under way by 14 ka, the only regression curve comes from just west of the area (Locat, 1977). Initially, emergence was rapid until construction of the Bic Terrace 10-9 ka (75, 76 of Table 5.1). This may represent a sea level stillstand caused by enhanced gravitational attraction due to possible expansion of local and/or Laurentide ice at a time corresponding to formation of the Ten Mile Lake Moraine in Newfoundland. A second oscillation in the generally steady emergence possibly occurred after 5.4 ka (77, 78 of Table 5.1) when relative sea level transgressed from a postglacial low at or below present sea level and reached +10 m to deposit clay on top of a peat bed (Dionne, 1985a). A final small transgression is inferred to have cut a low cliff and terrace (Mitis Terrace) into the clay plain at 4-6 m. Either the 5.4 ka transgression or the event responsible for the Mitis Terrace possibly correlates with the low cliff noted in Magdalen Islands. One of these events might mark the passage of the collapsing glacier-marginal forebulge.

Rock glaciers, a few of which may be still active, are present in the area today (Gray et al., 1981, p. 111). In addition,

Table 5.2. Regional lithostratigraphic correlation chart (modified after Grant and King, 1984) showing some climato-stratigraphic terms that have been proposed for the region, and the tentative correlation with deep sea oxygen isotope chronology and with conventional mid-continent chronostratigraphy.

CHRONOLOGY			LITHOSTRATIGRAPHY								Chrono-stratigraphy
Oxygen isotope stage	ka	Continental slope, seamounts Unit	Scotian Shelf	Nova Scotia Southwest	Nova Scotia Centre	Cape Breton Island	New Brunswick	Magdalen Islands Gulf of St. Lawrence	Newfoundland West Coast	Newfoundland Burin Peninsula	Labrador Torngat Mtns / Shelf	
1	11	A (arctic)	Lahave Clay / Sable Island sand and gravel; Sambro Sand; Scotian Shelf End Moraine		Amherst silt		Saint John end moraine	Entry Id organic bed	Ten Mile Moraine; Robinson's Head, Piedmont and Bradore moraines		Shepard Moraines	HOLOCENE
2	25	B1 (arctic) B2 (subarctic)	(Ice shelf) Emerald Silt	Port Maitland Gravel	diamicton organic beds 10-11 ka Five Islands Fm Rawdon Till		BANTALOR PHASE			Till	Tasiuyak Moraine Saglek Moraine	Late WISCONSINAN
3	50	B3 (arctic) B4 (subarctic)	Scotian Shelf End Moraines Scotian Shelf Drift (Ice sheet)	Beaver River Till Soil Salmon River Sand 38.6? Saulnierville Till	Bennett Bay Till Hants Till	Big Brook Organic 36.2? mastodon 31.9 ka? Shelly Till	Hillsborough mastodon 37.3 ka?		St. Georges Bay Drift Weathered Zone A	Langlade silt ?	Cartwright Moraine	Middle WISCONSINAN
4	62 75	B5 (arctic)		Soil Cape Cove Gravel Red Head Till	Eagles Nest Clay Lawrencetown Till East Milford Till	Soil Richmond Till	CHIGNECTO PHASE	Coffin Island Moraine Demoiselle Drift Till			Saglek Weathered Zone	
5a b c d e	100 125 128	B6 (subarctic, Holocene) B7 (<arctic) ?		ORGANIC BEDS Barton Silt, Salmon River Sand, Sandford Gravel Little Brook Till	ORGANIC BEDS Noel B, Miller Creek, East Milford Hartlen Till? Miller Creek Till ORGANIC BEDS Milford, Addington Forks, Noel A	ORGANIC BEDS Dingwall, Bay St. Lawrence, Leitches Creek B, Mabou, River Inhabitants, Benacadie, Hillsboro Colluvium Till ORGANIC BEDS Green Point, Mabou, East Bay, Leitches Creek A, Big Brook A, raised beach	CALEDONIA PHASE Soil	ORGANIC BEDS Bassin, Havre Aubert, Milleraud, Des Buttes, Boisville raised beach Soil ORGANIC BEDS Portage du Cap, Wolf Island Conglomerate, raised beach	Soil Codroy beds raised beach	Soil Langlade Silt raised beach	Mid-Bank moraines	Early WISCONSINAN SANGAMONIAN
6	150	RED MUD (coldest water)		Bridgewater Conglomerate	Conglomerate	Colluvium Mabou Conglomerate		erratics in lag gravels	Weathered Zone B	Main Brook Till	Shelf Edge Moraine Koroksoak Weathered Zone	Late ILLINOIAN
12 13		RED MUD *P. lacunosa* zone							Weathered Zone C		Torngat Weathered Zone	

References:

1 Ruddiman and McIntyre (1981)
2 Alam and Piper (1977)
3 King (1980)
4 General: Grant (1963), Prest (1977)
  West: Grant (1980b)
  Centre: Stea (1982b)
  CBI: Grant (unpublished)
5 after Rampton and Paradis (1981a,b), Rampton et al. (1984)
6 D.R. Grant, V.K. Prest, R.J. Mott, L.A. Dredge (unpublished data)
7 after Grant (1977), Brookes (1977a), Tucker and McCann (1980)
8 Grant (1980b)
9 Dreimanis and Karrow (1972)

GSC

bodies of permafrost 45-60 m thick have been proven to exist by drilling by Gray and Brown (1979) who place the lower edge of the permafrost at 1100 m. This level approximately corresponds with the lower edge of blockfield occurrence. This raises the possibility that the mature blockfield landscape of the higher areas has been produced by postglacial periglacial activity rather than being part of an old surface that escaped destruction during the last glaciation.

Gaspésie thus provides evidence to support findings elsewhere in the region, namely that there were early and late local ice cap phases and a major Laurentide ice advance. However, two interpretations of the glacial succession persist. One attributes all glacial phases to reorientations of flow regime during a single glacial stage (the Late Wisconsinan) and attributes weathering and terrain maturity to preservation of preexisting features beneath a cold-based ice sheet, while the other holds that weathering and terrain maturity differences delimit areas covered during separate glacial stages. The demonstration that basal ice regimes can explain such fundamental differences in Gaspésie as elsewhere in the region, adds an important new dimension to glaciological interpretation, and to models of ice distribution based on geomorphology.

REGIONAL CORRELATIONS

As described above, composite sequences have been used to develop local Quaternary histories for each area, which in turn were combined to produce the stratigraphic correlation scheme of Grant and King (1984, Table 1). Their table is reproduced here with some modifications as Table 5.2 and provides the framework for the following discussion of Quaternary events.

Pre-Quaternary weathered relicts

In this category may be placed the small patches of deeply rotted rock (saprolite) found on highlands (Wang et al., 1982) and associated with peneplains of Tertiary age (see *Denudation Surfaces*). The maturity of chemical weathering is like that found south of the glacial limit and in present-day warm temperate and subtropical areas. It is therefore reasonable to suppose that these remnants are relicts of pre-Quaternary conditions. The surficial materials on high summits thought to lie above or beyond the reach of Quaternary glaciers are also included as pre-Quaternary weathered relicts. These consist of areas in western Newfoundland and small enclaves on northern Cape Breton Island and the McGerrigle Mountains. The interpretation of these older features is controversial and the question remains whether they exist because they were protected by glaciers frozen to their surfaces or because they were never overridden. In either case, these relicts demonstrate the limited and/or weak action of glaciers in some parts of the region throughout the Quaternary.

Tills and glacial terrains of uncertain pre-Wisconsinan age

Throughout the region there is scattered geomorphic and stratigraphic evidence of ancient glaciations, but much of it is circumstantial and indirect. The levels of glaciation, lying above (or beyond) the limit of fresh Wisconsinan drift, in western Newfoundland, northern New Brunswick, and Gaspésie are examples of terrains that might be pre-Wisconsinan in age. The limits of the Newfoundland "weathered zones" are sharp and inclined towards the sea. They seem to delimit margins of former ice caps moving seaward from Newfoundland. Their presence seems to rule out invasion by Laurentide ice. Of the two "older terrains" on Newfoundland, the highest is estimated on geomorphic grounds to be several hundred thousand years old. Perhaps it correlates with the major glaciation recorded offshore and ascribed to stage 12 (Alam and Piper, 1977). The second most extensive ice sheet coverage is referred to oxygen isotope stage 6 because the terrain is considerably less mature, and because this too was a time of deep glacial erosion of the Gulf of St. Lawrence by grounded glaciers (Alam et al., 1983). This second event may correlate with the glaciation responsible for the eastward flow indicators which cross the entire Maritime Provinces area.

The stratigraphic significance and age of "old" tills found in other areas are unknown. Throughout the region surface drifts locally overlie dissimilar tills which are judged to be much older than the last glaciation. These "old tills" include those that are overlain by a temperate climate organic bed of probable Sangamonian age (Miller Creek, Milford, Addington Forks, and Leitches Creek occurrences) and deeply and thoroughly oxidized tills (Bridgewater Conglomerate, Main Brook Till). The tills underlying organic beds might be Illinoian in age but the others could be older because they appear to have been weathered during several warm periods. Additional candidates for this pre-Wisconsinan class are unoxidized tills which underlie the oldest drift reliably assigned to Wisconsinan time. These may be simply local or advance facies of the overlying till sheet, or may relate to different glaciations. Alam et al. (1983) suggested that at least four major pre-Wisconsinan glaciations are recorded in the deep offshore sediment sequence. The terrestrial Quaternary succession cannot be correlated with this sequence nor can it be used to confirm or deny the validity of these offshore interpretations.

Sangamonian (stage 5) organic beds and raised marine features and deposits

The earliest period for which there is consistent and datable evidence is the last interglaciation. It is represented by a suite of 32 buried organic beds mainly in Nova Scotia and the Magdalen Islands, and one each in New Brunswick and Newfoundland, which formed under climatic conditions that span the range of conditions analogous to those of the Holocene interglaciation. Because glacial deposits separate these organics from Holocene deposits, they are assigned to the last interglaciation. Th/U age estimates support assignment of these units to Sangamonian time, and distinctive pollen assemblages, together with broad groupings of dates, suggest that there were three separate warm intervals during stage 5 (Mott and Grant, 1985). It is noteworthy that plant successions during the last interglaciation at times followed distinctly different lines than during the Holocene. This points to the difficulty of using fossil plant successions to infer certain climatic analogues. Nonetheless the Quaternary sedimentary succession in this area has a well represented interglacial complex which serves as a starting point to define the later glacial sequence.

A second independent reference horizon for separating the late Quaternary from older sequences is the intertidal rock bench and shoreline deposits left by the Sangamonian sea. Because these have regional extent and are often found beneath till along the coast, they provide a distinctive marker bed for correlation. Moreover, the elevation with respect to present tide level (or more precisely the modern intertidal platform) is a means of assessing the amount of crustal movement since it was formed. Its general northward tilt probably reflects the amount of postglacial subsidence yet to occur. Local and regional tectonic tilting, as well as geoidal drift, cannot be ruled out as possible causes of tilting over this long period. For the Atlantic area, at the rate sea level is presently rising, the sea will reoccupy its previous interglacial position within 1500 years. Hence, in terms of sea level, the glacial/crustal cycle will then be complete. The platform has also been used as an indication of postglacial faulting. It is upthrown 14 m by Aspy Fault, Cape Breton Island and is disrupted where it crosses the Cape Ray Fault, Newfoundland (Grant, 1987b, p. 44).

Wisconsinan events

Early Wisconsinan

In Wisconsinan time, as recorded by tills overlying the interglacial beds, there were evidently two major pulses of glaciation. The first was the most extensive. For the Maritimes the major push of ice came from the north with Laurentide ice overtopping Gaspésie and penetrating New Brunswick. Controlled by a regional southeastward gradient, the ice moved uniformly seaward across Nova Scotia and onto the Scotian Shelf, without deflection down the Bay of Fundy. At the same time southward flowing Laurentide ice filled Gulf of St. Lawrence so that Cape Breton Island was virtually covered. This ice was joined by trunk outlet glaciers from the Newfoundland ice cap complex which moved onto the shelf via Cabot Strait and Laurentian Channel. The general extent of grounded glaciers in inland submarine areas (the areas underlain by Carboniferous redbeds) may have been comparable to that in previous maximal glaciations judging by the amount of red mud in offshore sequences that record this phase.

Middle Wisconsinan

At an intermediate stage in the Wisconsinan, important reorganizations of ice dispersal occurred. For a time, glacial domes were localized offshore from Nova Scotia and Newfoundland so that ice flow reversed its former trend and moved inland over Yarmouth area, Cape Breton Island, eastern New Brunswick, and Burin Peninsula. The glaciology of this condition is unclear. Possibly deglaciation of Gulf of St. Lawrence and some adjacent areas, for which there is some terrestrial evidence, caused drawdown and flow reversal of peripheral glaciers. Conditions necessary for offshore ice domes are low glaciation levels, a sea level low enough to have emergent banks on which to construct ice caps, and a retreat of inland glaciers so that ice flow could be reversed. Local evidence that low glaciation levels and low sea level did co-exist is seen in the submerged cirques of Grey Islands, northeastern Newfoundland (Grant, 1986b). One might also speculate that a local ice cap grew on Grand Banks at this time. Such a local cap would better account for the till in that area than hypothesizing that Newfoundland ice pushed far to the east, because there is no evidence that Avalon Peninsula was ever overrun by ice from the west or that it supported other than small ice caps.

The behaviour and retreat chronology of inland ice during the Middle Wisconsinan are recorded offshore on Scotian Shelf where the ice sheet began retreating from outer shelf basins by 46 ka and was replaced by an ice shelf which covered the basins until 32 ka when it dissipated (King and Fader, 1986). This liftoff or retreat across the shelf reached Gulf of Maine and central Laurentian Channel by 40 ka. The demise of buttressing ice caps probably signalled the end of the regional ice sheet phase and marked the inception of the local ice cap complex that is postulated for most areas during the Late Wisconsinan.

Late Wisconsinan

As described for all parts of the region, the surface tills are assigned to a phase of radially spreading local ice caps situated on uplands, lowlands, and even on emergent shelf areas. This distinctive regime is referred to the Late Wisconsinan because in most areas dates between 14 and 10 ka bracket the early recessional stages and the final ice disappearance. On land, there are no dates on the beginning of the local ice cap phase, but Vernal et al. (1986) believed that the East Bay organic sequence extends to as late as 62 ka, and Piper et al. (1989) date the beginning of open water on Scotian Shelf at 32 ka. It is not known when this glacial phase attained its maximum development.

The movements associated with the last glacial maximum, and the resulting tills, have been detailed for a number of areas, but the limit reached remains largely speculative. This is because the limit lies offshore in most places and ice marginal deposits have been erased by subsequent marine transgression. In a few areas, a land-based limit for the last glacial expansion has been recognized on the basis of weathering contrasts, discordant flow indicators, and disjunct marine levels. The Late Wisconsinan glacial limit at the stadial maximum shown in Figure 5.11a is based on that in Grant (1977b) with some important revisions required by subsequent work: Anticosti Island was overridden; Cape Breton Highlands and Burin Peninsula had small ice caps; there were no small nunataks in the Newfoundland Uplands; and more of North Mountain and Cobequid Highlands were overridden. Also shown are inferred ice divides and postulated dispersal centres which have been ascribed to the last or Late Wisconsinan glacial maximum. The portrayal constitutes the so-called "minimum model" of glacial extent (Prest, 1984). It adequately accounts for most observed geomorphic, stratigraphic, and ice flow features. However, its main weaknesses are that there is no direct chronometric control on the proposed limit or in the supposed extraglacial areas, and that it conflicts with most offshore reconstructions which require grounded glaciers in Gulf of Maine and in Gulf of St. Lawrence, including even the deepest parts of Laurentian Channel.

Because there are no radiocarbon constraints, an alternative or "maximum model" can be entertained to accommodate the offshore scenario as well as disagreements in northern New Brunswick and Gaspésie (Fig. 5.11b). This concept, which resembles that depicted by Mayewski et al. (1981), is favoured by those working offshore (King and Fader, 1986) and in adjacent New England (Hughes et al., 1985). It envisages a coherent regional ice mass, contiguous with, if not an integral part of, the Laurentide Ice Sheet, flowing more or less uniformly to the continental margin. This model includes many of the same elements as the minimum model in addition to several distinctive elements and flow regimes which resolve apparent contradictions with the minimum model. Where water depths were too great for grounded glaciers, such as in offshore basins and channels, ice shelves floated across and deposited stratified glaciomarine drift. Where terrestrial ice lifted off to become ice shelves, and where the shelves contacted offshore prominences, grounding line moraines composed of basal and melt-out till were deposited which intertongue with the basinal deposits (King and Fader, 1986). Both sediment types apparently demonstrate the continuity of the ice sheet from land, across basins, to outer banks. On outer banks ice rises may have mushroomed and, during retreat, could have had independent onshore flow trends such as is observed in southern Cape Breton Island. The ice shelves buttressed ice streams which were localized where ice marginal calving bays, such as in Laurentian Channel (Thomas, 1977) and in shelf basins, induced drawdown and accelerated flow. Such "ice currents", as postulated across Nova Scotia (Grant, 1963), would explain the swaths of far travelled drumlin till and the trains of erratics (Scott, 1976). Within the ice sheet cold-based patches are postulated for high areas where ice would have been thin and/or divergent, like Gaspésie (David and Lebuis, 1985), northern New Brunswick (Gauthier, 1980b), and Cape Breton Island (Grant, 1987a, p. 38). Cold-based ice has also been inferred over porous substrates like Cobequid Mountain (Stea et al., 1986). Another potential candidate area for cold-based ice might be the western Newfoundland highlands if the sharp morainal boundaries to the weathered zones and their downslope parallelism are discounted. The main weaknesses of the maximum model are that it cannot accommodate the dated ice margin on Anticosti Island which requires an ice-free Gulf of St. Lawrence during the Late Wisconsinan and it would make all flow events and till sheets, despite the weathering breaks, one age. It thus overlooks much of the onshore stratigraphy and geomorphology. Moreover, if modelling of former ice loads from the resulting crustal deformation has any value, the maximum model is a poorer approximation than the minimum model of the ice necessary to produce the observed changes in postglacial relative sea level (see below).

CRUSTAL MOVEMENTS AND RELATIVE SEA LEVEL CHANGES

Changes of level include local and regional independent movements of both crust and ocean level, on a range of time spans, rates, and amplitudes (Grant, 1980a). These changes give evidence of major earth processes such as epeirogeny, neotectonics, gravitational and oceanographic change, and the extent and behaviour of the last ice sheets. It is now clear that glaciers have been responsible in one way or another for most such changes of level in the Atlantic Appalachian region throughout the Quaternary. It is rarely possible, however, to differentiate between absolute crustal movement and sea level change since the two are closely intertwined.

The account begins with broad scale movements beginning in the early Cenozoic since those may be reasonably assumed to continue as a background to other more recent and transient events. In a similar way, the state of stress in the Earth's crust, as a possible cause of modern faulting and earthquakes, is as much a product of plate tectonics and continental drift as it is due to glacial isostatic crustal deflections and the rising load of seawater on the continental shelf. Therefore, any modern appraisal of delevelling must take the widest view of all factors in order to understand the root cause of present events and to appreciate the continuity of processes.

Long-term Cenozoic movements

Throughout the Cenozoic, continental drift and thermal subsidence have determined regional crustal behaviour. This segment of the Atlantic margin has been sinking at an average rate of 5-10 cm/ka since the Cretaceous in order to accumulate the 3000-8000 m-thick wedge of littoral-neritic sediments that overlie the foundered basement (Fig. 5.3). The subsidence has been interrupted by wide swings of sea level which periodically caused the Cenozoic marine succession to be dissected, even into the Quaternary (King and MacLean, 1976). The fact that no Cenozoic marine beds are emergent, with exception of the last interglacial and the last deglacial isostatic sediments, is evidence of the long-term net subsidence. The present coastal configuration therefore represents the latest regional transgression, such that all Cenozoic peneplains descend below sea level.

Displacements of the last interglacial shore

The most recent clearly documented relative sea level change and possible associated crustal movement is shown by variations in the height of the shoreline ascribed to the maximum of the last interglaciation (Grant, 1980a). Clearly, if there have been no net changes in gravity, crustal position, or oceanographic condition in the last 100 ka, the last interglacial shore should be coincident or parallel to the present shore. Conversely, departures will reflect disequilibrium of some sort. The fossil shore platform thus serves not only as a stratigraphic marker, but also as a geodynamic and geoidal datum plane. Along the outer Atlantic coast, it is found fairly consistently at about 4-6 m above its modern counterpart in four areas, as shown in Figure 5.14: southwestern Nova Scotia (Grant, 1980b); Cape Breton Island (Grant, 1987b, p. 29); Burin Peninsula (Tucker and McCann, 1980); on Avalon Peninsula (Henderson, 1972). The Mic Mac Terrace in St. Lawrence Estuary (Goldthwait, 1911; Dionne, 1963) at 7-9 m is likely correlative. In Bay of Fundy it is not obvious except perhaps at Advocate (Grant, 1986). Along Northumberland Strait and in eastern New Brunswick, it is probably represented by the variably glacially degraded surface up to 10 m elevation that underlies the tills and is seen with beach gravel overlying it in George Bay. The platform is conspicuously absent above tide level along the Nova Scotia coast between Cape Sable and Chedabucto Bay, which perhaps suggests some local negative crustal movement south of Chedabucto Fault. Faulting of the platform is seen in the vicinity of Cape Ray Fault southwestern Newfoundland (Codroy Valley) where steps of several decimetres break the surface. A dramatic example of offset by faulting occurs in northern Cape Breton Island where a 14 m up to south displacement occurs along the Aspy Fault (Neale, 1964; Grant, 1975a). On Magdalen Islands, the interglacial beaches attain at least 20 m elevation, possibly because of salt diapirism or because of an uncompensated ice marginal crustal forebulge. Clearly, if this unique horizon was mapped throughout the region, it could reveal four

CHAPTER 5

major crustal parameters over the last 100 ka: local faulting, uncompensated isostasy, geoidal displacements due to gravitational change, and ocean level displacement due to steric and tidal change.

Late Wisconsinan crustal dislocations and glacial isostatic deflections

Postglacial faults

The third major group of crustal/sea level movement indicators pertains to faulting and warping caused by glaciation. It is not uncommon to find glaciated pavements disrupted by small faults (Fig. 5.15). These have been noted at Halifax and Caledonia Corner by Goldthwait (1924, p. 154), at Saint John by Matthew (1894a, b) and near Yarmouth (Grant, 1980a, p. 31). All are in vertically cleaved slate and may indicate either elastic recovery or localized isostatic rebound during deglaciation. Alternatively it is possibly linked to the moderate northeast-southwest compressional stress field that is revealed by borehole deformation (Plumb and Cox, 1987). In Strait of Belle Isle area similar but larger amplitude displacements in Canadian Shield terrane are noted (Grant, in press).

Large-scale slumps

A possibly related kind of massive bedrock displacement takes the form of large-scale coherent slumps of mountain sides ("sackung") which occur in many places in western Newfoundland. One on the flank of Bonne Bay fiord (Grant, 1974a) has a volume of 10^9 m^3 and may be the largest such failure in Canada; nearly 50 others occur in the surrounding area. Although a few occur on steep granite faces, most are on gentler slopes of high-standing ultrabasic ophiolite thrust sheets. Perhaps the cause is not so much glacial oversteepening as some residual tectonic or isostatic stress in these klippen, or instability because they rest on less competent sedimentary rocks. The leading edge of the blocks eventually becomes so unstable and weakened by tension fissures, that large masses fall as landslides. Where they fall into water, as for example in Western Brook Pond, a tourist attraction, they pose a serious hazard to boat traffic.

Postglacial shore level change

By far the largest and most extensive changes of Quaternary relative sea level position were those that occurred over the last 14 ka as a result of the combined effects of crustal recovery and general eustatic ocean rise. Across the region there are large local differences in the amount and trend of relative sea level (Fig. 5.16). In general, the northern half shows net emergence, and the south net submergence. Thus the north had a higher postglacial sea level phase and shows a suite of raised shorelines (recording the Goldthwait Sea in Gulf of St. Lawrence and De Geer Sea in Bay of Fundy) whereas in the south, all the postglacial paleoshores are now

Figure 5.14. Crustal movements and delevelled paleoshores. Inset A, slope of the geoid above geodetic datum (m); Inset B, smoothed free-air gravity (red = positive; grey = negative); Inset C, profile of differential subsidence; Inset D, 50 year tide-gauge record of rising mean sea level; Inset E, Late Holocene rise of high tide level in Bay of Fundy (after Grant, 1975b).

Figure 5.15. Postglacial faulting, Cape Cove, Nova Scotia. A glaciated outcrop of slate is offset along cleavage planes; striations can be traced across the steps. Movement, which is up to the south and exceeds 1 m, occurred during Wisconsinan time. 203673-W

submerged and the sea is still rising. Shorelines of a given age therefore slope southward from levels as high as +150 m on the North Shore to −115 m on the outer shelf. The general emergence phase, now virtually complete except perhaps for the north shore of Gulf of St. Lawrence, is treated first.

Deglacial emergence phase

An approximation of the general pattern of postglacial crustal rebound is given by the elevation of the highest deglacial paleoshore or "marine limit" above present sea level. In Figure 5.14, marine limit ranges from a high of 150 m at Strait of Belle Isle to 0 m through Nova Scotia, Prince Edward Island, and southern Newfoundland. Between the two, the emergence pattern shows two broad domes over Newfoundland and the Maritimes, with a pronounced reentrant in Gulf of St. Lawrence. In this respect, all such maps from the first (De Geer, 1892) to the most recent (Rogerson, 1982), with the exception of Flint (1940), have shown the same pattern. All authors concur that the two domes reflect the main masses of the Appalachian glacier complex, that the northward gradient is the edge effect of the massive Laurentide Ice Sheet and that the saddle over the Gulf of St. Lawrence indicates minimal ice load there. Indeed, inverse modelling of the glacier distribution from the sea level response by Quinlan and Beaumont (1981, 1982) confirms a two centre Appalachian domain only slightly more extensive than the minimum ice concept (Fig. 5.11a). They do stress, however, that the same sea level pattern would result if the maximum configuration prevailed several thousand years earlier than the 18.0 ka age assumed for the minimum model. In other words, sea level data support the minimum ice concept and exclude the maximum from the last 18.0 ka.

Further numerical modelling with more and better sea level data will likely give more reliable glaciological reconstructions for the regions than will additional fieldwork in the few areas where ice extent can be defined stratigraphically or geomorphologically.

From the zero isobase, the position of sea level at the time of deglaciation descends below present sea level. However, whether the submerged terraces found at depths of −40 m at the mouth of Bay of Fundy (Fader et al., 1977), −110 to −120 m on outer Scotian Shelf (King and MacLean, 1976), −55 m on inner Scotian Shelf, −90 m on western Grand Banks (Fader et al., 1982), −62 to −72 m around Magdalen Islands (Loring and Nota, 1973), and −79 m in eastern Northumberland Strait (Kranck, 1972) are its direct correlatives has not been demonstrated either by tracing or dating. Part of the reason is that marine limit onshore is the last postglacial transgression maximum, whereas the submerged terraces are undated regression minima. This may be the case for the Fundy and Magdalen terraces which lie far below the projected marine limit. Indeed, Quinlan and Beaumont (1981) placed sea level at the time of deglaciation much higher in both areas.

As marine limit is not a synchronous shore, but ranges in age from more than 14 ka in Bay of Fundy to less than 10 ka on the North Shore of Gulf of St. Lawrence, the position of marine limit cannot be used as an indication of the amount or rate of emergence and there is insufficient dating and levelling information to know the shape of any given paleoshore. A possible exception is the "Bay of Islands Surface" (Flint, 1940, p. 1772) which is a prominent rock platform that rises from 0 m in St. Georges Bay to 80 m at Strait of Belle Isle. There it has been correlated with the Ten Mile Lake readvance ca. 10.9 ka and attributed to a sea level stillstand induced gravitationally by the mass increase of the expanding ice cap (Grant, in press).

The pattern of emergence is controlled partly in some areas by variations in ice retreat. Thus delayed deglaciation is believed responsible for the local decrease of marine limit along Minas Basin, Chaleur Bay, some Newfoundland fiords, and north of Anticosti. In the latter case, marine limit is about 80 m lower than elsewhere along the North Shore, and Gratton et al. (1986) postulated that an ice lobe persisted there as late as 9 ka; Quinlan and Beaumont (1982) also inferred a late ice load in the same area.

While it is generally assumed that all submergence features are of Late Wisconsinan age, Rampton et al. (1984) and Tucker et al. (1982) showed that in two areas — Bay of Fundy coast and the Burin-Hermitage shore, respectively — problems in drawing a single coherent trend surface result from the overlap of two different ages of marine transgressions. As the tilt of southwestern Nova Scotia marine limit is not coplanar with that of known age in southern New Brunswick, it too might be older.

Given then that postglacial emergence began at times ranging from 14 ka in De Geer Sea and 9 ka on the North Shore, and reflects gross differences in former ice thickness, a series of local sea level histories (Fig. 5.16) expresses the regional variation. A zonation, like that predicted theoretically by Clarke et al. (1978) is revealed. A northern zone of continuous emergence is represented in Newfoundland, North Shore, and Gaspésie. Emergence, although beginning late, completely overshadowed any other movements and evidently continues slowly. The rest of the region has two opposed trends: early emergence and later resubmergence, with a regression minimum at different times and depths. The central zone, where early emergent movement was greater than later submergence, has both emerged and submerged sequences, and is well recorded in southwestern Newfoundland, Prince Edward Island, and New Brunswick. Possibly the upper St. Lawrence estuary is also in this zone. In the outer zone, through Newfoundland, Nova Scotia, and the continental shelf near or beyond the glacial limit, the initial emergence has been exceeded by the recent resubmergence which is treated separately in the following section.

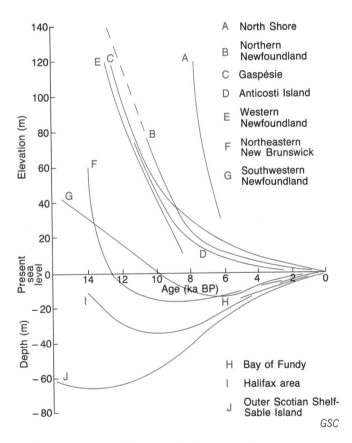

Figure 5.16. Variations in relative sea level change across the region from north to south based mainly on observed shore level displacement curves (A-H), plus two theoretical curves (I, J). Curve A: North Shore, Gulf of St. Lawrence, near Sept-Îles (from data in Dubois, 1980; Dredge, 1983); B: northern Newfoundland (after Grant, 1972a); C: Gaspésie (Locat, 1977); D: western Anticosti Island (Painchaud et al., 1984); E: western Newfoundland (derived from Grant, 1972a; Brookes and Stevens, 1985); F: northeastern New Brunswick (Thomas et al., 1973); G: southwestern Newfoundland (Brookes et al., 1985); H: Bay of Fundy (after Grant, 1975b; Scott and Greenberg, 1983); I, J: Halifax and Sable Island (Quinlan and Beaumont, 1982).

Holocene submergence phase

The modern rise of sea level was among the first Quaternary phenomena to be noted in the region (Dawson, 1856) and it continues to be the subject of intensive study because of its economic impact, and because it is the most sensitive indicator of earth rheology and crustal behaviour. The submergence is recorded by the progressive burial of terrestrial organic deposits by marine organic deposits. Earlier studies (Frankel and Crowl, 1961; Harrison and Lyon, 1963; Grant, 1970) dated mainly humus and forest beds beneath salt marsh in order to determine a maximum age for the first arrival of high tide conditions. This approach showed an average rate of relative rise of 15 cm per century for Atlantic and Gulf coasts, whereas Bay of Fundy submerged twice as fast because of increasing tidal range (see below). At present, the optimal method of determining paleotidal levels to within 15 cm is to core salt marshes and recognize indicative foraminiferal zones (Scott and Medioli, 1978). By this approach, the earlier estimates have been substantiated in some areas and modified in others (Scott et al., 1981); the regression minimum has been detected at -30 m at about 7 ka for the Mahone Bay area (Scott and Medioli, 1982); and the outer shelf has been shown to be submerging much faster than the mainland (Scott et al., 1984). Fluctuations during the general submergence postulated by Harrison and Lyon (1963) are not yet proven, but at least four extensive supratidal peat layers occur in the 4.0 ka tidal sequence at head of Bay of Fundy (Lammers and De Haan, 1980; Noordijk and Pronk, 1981; Grant, 1986). It is unlikely they represent stillstands when crustal lowering ceased; they seem to require regressions of several decimetres which may be tidal or steric, rather than crustal or geoidal in origin.

Referring to the five southern curves in Figure 5.16, the amount and time of inception of the resubmergence is seen to decrease towards the north from a maximum of about -65 m at 14 ka at Sable Island to perhaps -1 m at 5 ka in St. Lawrence estuary. As first postulated by Daly (1920), this progression is the result of a northward migrating, collapsing, ice marginal crustal forebulge. Using similar data for eastern United States, Pardi and Newman (1987) tracked a northward shift of the forebulge of 1500 km in 12 ka. The Maritimes data suggest that the crest of the bulge is now in the northern Gulf of St. Lawrence region. It would explain the fossil cliff seen a few metres above tide level throughout the Gulf area.

Given their longevity, emergence and submergence trends will continue for several centuries, presumably until the regional gravity anomalies are removed (see Inset B, Fig. 5.14). A distinct deficiency of mass persists north of Anticosti Island as a legacy of a Labrador Sector remnant of the Laurentide Ice Sheet, while excess mass characterizes the subsiding areas along the continental margin. At present rates, the subsidence will lower the last interglacial shore back into the intertidal zone within 1.5-2.0 ka, providing there has not been permanent change in crustal or geoidal level in the last 100 ka. In geodynamic terms, it could be another 2.0 ka before the long-term effects of the last glaciation will have ended.

Paleotidal evolution

Bay of Fundy with its anomalous tides presents an additional factor in the recent relative sea level picture. Grant (1970) postulated that as depth/area relations in the bay changed due to emergence at the head and flooding of emergent banks at the mouth, conditions for resonance improved and thereby amplified the tidal range beginning about 6.0 ka. From numerical tidal models, Scott and Greenberg (1983) derived a 7.0 ka date for the start of this phenomenon. The converse might also have occurred, in that microtidal areas such as western Northumberland Strait may have experienced reductions in tidal range as amphidromic points moved to their present position during submergence. The minor oscillation in St. Lawrence estuary (Dionne, 1985a) might also be a tidal feature, if not the result of the migrating bulge.

Recent crustal movement and seismicity

The long term deformational behaviour over the last several millennia is partly reflected in modern lithospheric activity as expressed by earthquakes and stress deformation of boreholes, and as measured by tide gauge movements and geodetic relevelling. The first indication of contemporary crustal movement was tide gauges (Inset D, Fig. 5.14) that showed relative subsidence of 2.5-4.5 m/ka (Grant, 1970). That the change is due to real crustal lowering is well recorded by differential movement along the primary geodetic level line from Halifax to the North American Datum at Pointe-au-Père on the Gaspésie coast. Subsidence since 1909 amounts to more than 30 cm along the Fundy axis (Inset C, Fig. 5.14). Using all the relevelled segments in the Maritimes, Lambert and Vaníček (1979) showed a regional pattern of strong warping, ranging from 4.0 m/ka subsidence in upper Bay of Fundy to 10.0 m/ka uplift in northeastern New Brunswick and Gaspésie (Fig. 5.14).

The strong local uplift is difficult to explain in view of the abundant geomorphic evidence for submergence and subsidence (aggrading salt marshes, erosion of coastal peat bogs, transgressing barriers), but the Fundy subsidence is consistent with that inferred from long-term submergence data. That the belt of subsidence crosscuts the earlier emergence trend may only mean that the migrating forebulge which it possibly represents has a different path than the wave of uplift. Whether this is physically possible is unknown. Vaníček and Nagy (1981) did point out that the trend of the subsidence is only partly coincident with slope of the geoid (Inset A, Fig. 5.14), and hence cannot be entirely attributed to postglacial mass readjustment. In adjacent southeastern Maine, modern localized subsidence and associated high seismicity is attributed to a northwest-southeast direction of principal crustal stress (Anderson et al., 1984).

Earthquakes and rock deformations suggest that the region is under a variety of stresses. According to Burke (1984) major and minor earthquakes in four areas have several origins. Those off Laurentian Channel are attributed to ocean floor faulting, if not to excessive sediment loading. Those in the New Brunswick highlands, especially the 1982 shocks of magnitudes 5.7, 5.4, and 5.1, caused surface thrust faulting and buckling that indicate high horizontal east-west compression (Basham and Adams, 1984). In this case, the surface stress is thought to be localized around a rigid pluton that lies in the regional northeast-southwest compression field which has been revealed by borehole deformation (Plumb and Cox, 1987). Those in the Saint John area may be due to stress concentration along major lines of weakness like the Fundy-Cobequid Fault perhaps because of rapid modern subsidence or regional compression. The postglacial faults in Saint John (Matthew, 1894b) may be a further expression of these same stresses. Lastly, the Moncton group of quakes may be due to subsurface flow or solution of salt.

In sum, it is evident that the northern Appalachian region is in horizontal compression, perhaps because of westward plate drift. Simultaneously the region is being warped by differential postglacial uplift and subsidence, which produce local extension and compression, respectively. These superimposed distortions are nonetheless part of the total stress field and may be sufficient to trigger failure (Quinlan, 1984), especially along lines of weakness such as major faults, fracture zones, and pluton margins. More refined recent sea level histories and fault inventories are needed to quantify the stress-producing mechanisms in the region.

POSTGLACIAL CLIMATE CHANGE

The gradual amelioration of climate which, after deglaciation, produced a succession of vegetation suites as temperature and precipitation varied, has been discussed by Anderson (1985), Mott (1985) and in Chapter 7 of this volume (Anderson et al., 1989) and is not discussed further here. Perhaps the most interesting example of process/response relationships between climate and sedimentation is the sudden cooling about 11-10 ka which is thought to have been caused by a southward shift of the North Atlantic Polar Front (Mott et al., 1986). The cooling correlates with a resurgence of the ice cap in Newfoundland which built the Ten Mile Lake Moraine, and with a devegetation of slopes and consequent colluvial action throughout the region, which caused mineral sediment to bury organic sediments. This impact of a given ocean circulation event on the region's climate further supports the postulated link between oceanic temperature cycles and glacial phases in Atlantic Canada.

PRESENT DAY PROCESSES

Modern conditions are in the continuum of the trends outlined above. Knowledge of their evolution and rates is useful for prognoses of future conditions — discounting anthropogenic interventions. Climate continues in its slightly cooler mode which developed after the optimum about 6 ka (the warming in recent decades may be attributable to the postindustrial era "greenhouse effect"). Landscapes have stabilized and there is minimum slope degradation and erosion, except for incision of older floodplains. Periodic flooding may be increasing in frequency and intensity as forest clear-cutting reduces precipitation-holding capacity. Cryoturbation and solifluction continue at higher elevations, with nivation, mudboils, stone circles, and solifluction terraces, and a few rock glaciers remaining active, and sporadic permafrost existing above a level which descends from 300 m on Cape Breton Island to sea level at Strait of Belle Isle.

Along the coast, rising sea level is causing rapid erosion at rates predicted by the Bruun Rule which holds that recession is a function of sea level rise and land gradient. For example on the non resistant shores around the southern Gulf of St. Lawrence where the coastal gradient is 1:500, the measured rate of long-term shoreline retreat of 150 m/100 a is consistent with the average submergence rate of 3 m/ka.

ECONOMIC CONSIDERATIONS

Knowledge and understanding of Quaternary deposits and processes have numerous practical applications ranging from resource exploitation, land use development, and natural hazards. In essence, any natural condition either creates wealth or consumes it. Optimum land-use policy must therefore take account of certain physical realities. In economic terms, Quaternary geological aspects constitute assets and opportunities which are only partly utilized, as well as liabilities which could be further reduced. In the absence of special studies from which to cite illustrative examples, only a general review of the more important positive and negative impacts can be outlined under three broad headings.

Quaternary deposits as substrates

As bedrock is largely buried in this region, the composition, texture, and thickness of the surficial cover have a bearing on construction, excavation, trafficability, agriculture, forestry, and mining. Most urban centres are situated on till or gravel which, although it presents no stability problems for small structures, normally must be removed from under multi-storey buildings. It may be noted that gravel areas could be further preserved for their resource value; many prime deposits are being forfeited by construction. Many fine textured tills are abnormally weak and are subject to flowage and slumping so care must be taken to determine the physical characteristics of such foundation materials. On the other hand, bouldery tills are difficult to handle, thereby raising excavation costs. On the plus side, areas of deep medium textured drift have been chosen for massive industrial developments such as along Strait of Canso and on Avalon Isthmus because they provide excellent foundation conditions.

Trafficability has become an important consideration for efficient agricultural and forestry operations. Ratings are now given routinely in soils reports for all soils types and the reader is referred to those publications for detailed assessments. In general, the clay tills of Nova Scotia for example, although the best agricultural land, have undesirable characteristics wherever the naturally stabilized slopes are altered.

Surface sediments participate in acid rain ecology to the extent of their natural buffering capacity. Most, however, are already acid and therefore have no reserve capability. An exception are the large areas of calcareous till dispersed widely over Nova Scotia in what otherwise are sensitive bedrock terranes. Unfortunately their distribution is not yet defined for the purpose of impact studies.

As the major surface material, Quaternary sediments are obviously the prime plant medium for forestry and agricultural purposes. Soils mapping provides detailed information in accessible areas, but Quaternary geological mapping, particularly by remote sensing, in Nova Scotia highlands and interior Newfoundland, is a major method of obtaining pertinent data. Given a suitable climate, geological substrate type is the prime determinant for reforestation programs.

Quaternary deposits are a help as well as a hindrance to both mineral exploration and development. As bedrock is largely concealed by overburden, its mineral potential cannot be evaluated by direct observation. Fortunately the same deposits which cover an orebody will usually contain particles eroded from it (although in some circumstances bedrock will be passively till-covered without contributing tell-tale debris). The dispersed mineralized debris thus effectively broadens the surface expression of orebodies in subcrop, and these are then amenable to discovery by interpretation of boulder trains and geochemical anomalies. For this reason, Quaternary mapping emphasizes determination of ice flow sequences, even in areas of abundant outcrop because a large orebody may still be concealed by a small pond or bog, but its glacially transported phase will be available for tracing. Still, many float occurrences remain unexplained. Cost-effective discovery of their sources will depend on knowing which till sheet an occurrence belongs to and its direction of emplacement. Much exploration effort has been ineffective because of unfounded assumptions about flow direction, and because of imperfect stratigraphic schemes.

Areas of thick drift, particularly in Nova Scotia, have been a hindrance to the discovery and exploitation of valuable evaporite minerals (gypsum, potash, celestite) which are major components in the mineral economy. Large economic deposits are deeply mantled by heavy clay till necessitating extensive test drilling and expensive stripping operations.

Quaternary deposits as resources

Sorted aggregates of gravel and sand are essential ingredients for construction. Effective utilization and preservation depend on detailed mapping and resolution of conflicting land use practices. In order of relative quality and quantity, deposits of glaciofluvial outwash, ice contact eskers and kames, and marine deltas and beaches have been intensively mined in urbanized areas, although large reserves at tidewater exist in more remote areas for direct shipping to the eastern megalopolis. Examples of other opportunities for obtaining suitable aggregate are beneficiation of sandy till, especially in Newfoundland, and mining of subtill deposits on Magdalen Islands.

Peat is an example of a Quaternary resource that is abundant in many parts of the area. It could be more extensively utilized as an energy source, an agricultural medium, and a waste-disposal substrate. More information on types of peatlands and the distribution of peat in this region is given in the Chapter 11 of this volume (Tarnocai, 1989).

The groundwater content of surficial materials is becoming increasingly important as acid rain, atmospheric pollution and soil degradation reduce the quality of surface water supplies. At present groundwater is tapped mainly in rural areas by shallow wells, but large food processing operations are entirely dependent on its superior and stable quality. This suggests that urban supplies could be upgraded, augmented, and secured by additions from potential aquifers in buried valleys — yet to be defined and tested.

Quaternary conditions as hazards

The Atlantic region is not a high risk physical environment, but distinct hazards exist, failures occur, and damage is sustained — all of which unnecessarily drain financial resources, reduce output, and interrupt services. Most are avoidable, all can be reduced.

Along transportation corridors, failure of till and rock slopes is a continuing nuisance that is not yet designed into route selection and gradients. Earth flows in certain till types, and rock falls in fractured rock are common and remedial measures are expensive. Coastal clay bodies in western Newfoundland and Chaleur Bay locally fail by slumping of unsupported eroding faces, and by settlement because of overloading. Hitherto unrecognized risk of death and damage from earth flows of waterlogged till on steep slopes points to the need for special attention to this condition. The massive slow-moving valley-wall rock sags (see section on large-scale slumps) seem passive, but those poised on fiord walls could conceivably fail suddenly and create dangerous waves. At the very least, no structures should be built on these deforming terrains.

Rock bursts and postglacial faulting point to ongoing stress release. Certain pyritiferous slates are subject to bacterial expansion. Considerable care should be exercised in siting massive, sensitive installations like power plants, dams, and bridges. Neotectonic mapping is clearly called for.

Seismic risk, based on past events, is considered low for the region but the 1970 and 1982 earthquakes in New Brunswick (Basham and Adams, 1984; Burke, 1984) surprised theoreticians. No doubt there will be future surprises, so attention to alluvial areas where ground accelerations are larger (Burke, 1984, p. 17), and to areas subject to secondary ground failure and to seismic sea waves is warranted.

Shoreline changes continually impact on property values, investment, and production, and locally necessitate expensive remedial measures. This process has yet to be acknowledged as a permanent and predictable component of coastal environmental change, and one that impacts on shoreline engineering.

ACKNOWLEDGMENTS

This chapter has greatly benefited from critical reviews at several stages by various noted experts. Throughout the lengthy process, colleagues L.A. Dredge, V.K. Prest, A.S. Dyke, and D.A. Hodgson (Geological Survey of Canada), were always available for advice and consultation. H.B.S. Cooke (Dalhousie University), R.R. Stea (Nova Scotia Department of Mines and Energy), N.R. Gadd (Geological Survey of Canada), and J.B. Bird (McGill University) reviewed the entire chapter. Several local experts are thanked for their detailed critical comments on separate parts. For Newfoundland, I.A. Brookes (York University), R.J. Rogerson (Memorial University), and D.G. Vanderveer (Newfoundland Department of Mines and Energy) provided incisive comments. The Nova Scotia section was reviewed by R.R. Stea and H.B.S. Cooke, and New Brunswick by V.N. Rampton (Terrain Analysis and Mapping Services Ltd., Carp, Ontario). Sections on Gaspésie and Anticosti Island were brought up to date in the light of recent work by P.P. David (Université de Montréal) and P. LaSalle (Ministère de l'Énergie et des Ressources), and by Q.H.J. Gwyn and J-M. Dubois (Université de Sherbrooke), respectively. B.V. Sanford (Geological Survey of Canada) amended the treatment of bedrock geology and physiography, while D.J.W. Piper (Atlantic Geoscience Centre) reviewed the offshore part.

REFERENCES

Alam, M. and Piper, D.J.W.
1977: Pre-Wisconsin stratigraphy and paleoclimates off Atlantic Canada and its bearing on glaciation in Québec; Géographie physique et Quaternaire, v. 31, p. 15-22.

Alam, M., Piper, D.J.W., and Cooke, H.B.S.
1983: Late Quaternary stratigraphy and paleo-oceanography of the Grand Banks, continental margin, eastern Canada; Boreas, v. 12, p. 253-261.

Alcock, F.J.
1926: Mount Albert map area, Quebec; Geological Survey of Canada, Memoir 144, 75 p.
1935: Geology of Chaleur Bay region; Geological Survey of Canada, Memoir 183, 146 p.
1941: The Magdalen Islands, their geology and mineral deposits; Canadian Institute of Mining and Metallurgy, Transactions, v. 44, p. 623-649.
1944: Further information on glaciation in Gaspé; Royal Society of Canada, Transactions, v. 38, p. 15-21.

Alley, D.W. and Slatt, R.M.
1975: Drift prospecting and glacial geology in the Sheffield Lake-Indian Pond area, north central Newfoundland; Newfoundland Department of Mines and Energy, Mineral Development Division, Report 75-3.

Ambrose, J.W.
1964: Exhumed paleoplains of the Precambrian Shield of North America; American Journal of Science, v. 262, p. 817-857.

Amos, C.L.
1978: The post glacial evolution of the Minas Basin, N.S. A sedimentological interpretation; Journal of Sedimentary Petrology, v. 48, p. 965-982.

Anderson, T.W.
1980: Holocene vegetation and climatic history of Prince Edward Island, Canada; Canadian Journal of Earth Sciences, v. 17, p. 1152-1165.
1983: Preliminary evidence for Late Wisconsinan climatic fluctuations from pollen stratigraphy in Burin Peninsula, Newfoundland; in Current Research, Part B, Geological Survey of Canada, Paper 83-1B, p. 185-188.
1985: Late-Quaternary pollen records from eastern Ontario, Québec and Atlantic Canada; in Pollen Records of Late-Quaternary North American Sediments, V.M. Bryant and R.G. Holloway (ed.); American Association of Stratigraphic Palynologists Foundation, Tulsa, p. 281-326.

Anderson, T.W., Mathewes, R.W., and Schweger, C.E.
1989: Holocene climatic trends in Canada with special reference to the hypsithermal interval; in Chapter 7 of Quaternary Geology of Canada and Greenland, R.J. Fulton (ed.); Geological Survey of Canada, Geology of Canada, no. 1 (also Geological Society of America, The Geology of North America, v. K-1).

Anderson, W.A., Kelley, J.T., Thompson W.B., Borns, H.W., Jr., Sanger, D., Smith, D.C. Tyler, D.A., Anderson, R.S., Bridges, A.E., Crossen, K.J., and Ladd, J.W.
1984: Crustal warping in coastal Maine; Geology, v. 12, p. 677-680.

Antevs, E.
1929: Maps of the Pleistocene glaciations; Geological Society of America, Bulletin, v. 40, p. 631-720.

Auer, V.
1930: Peat bogs in southeastern Canada; Geological Survey of Canada, Memoir 162, 32 p.

Basham, P.W. and Adams, J.
1984: The Miramichi, New Brunswick earthquakes: near-surface thrust faulting in the northern Appalachians; Geoscience Canada, v. 11, p. 115-121.

Bélanger, J.R.
1978: Surficial geology of Saint John, New Brunswick; Geological Survey of Canada, Open File 575, scale 1:50 000.

Bell, R.
1863: On the surficial geology of Gaspé Peninsula; Canadian Naturalist and Geologist, v. 8, p. 175-183.

Bigras, P.J.C., Dubois, J-M.M., and Gwyn, Q.H.J.
1985: Relative sea-level fluctuations during the last 35 000 years, northern Gulf of St. Lawrence (abstract); in Program with Abstracts, Geological Association of Canada, v. 10, p. A5.

Bird, J.B.
1972: The denudational evolution of the Maritime Provinces, Canada; Revue de géographie de Montréal, v. 26, p. 421-432.

Blake, W., Jr.
1983: Geological Survey of Canada, radiocarbon dates XXIII; Geological Survey of Canada, Paper 83-7, 33 p.
1984: Geological Survey of Canada radiocarbon dates XXIV; Geological Survey of Canada, Paper 84-7, 35 p.
1988: Geological Survey of Canada radiocarbon dates XXVII; Geological Survey of Canada, Paper 87-7.

Bolton, T.E. and Lee, L.K.
1960: Post-glacial marine overlap of Anticosti Island, Quebec; Geological Association of Canada, Proceedings, v. 12, p. 67-78.

Borns, H.W., Jr.
1965: Late-glacial ice wedge casts in northern Nova Scotia, Canada; Science, v. 148, p. 1223-1226.
1966: The geography of Paleo-Indian occupation in Nova Scotia; Quaternaria, v. 8, p. 49-57.

Bostock, H.H.
1964: A provisional physiographic map of Canada; Geological Survey of Canada, Paper 64-35, 24 p., Map 13-1964.

Brinsmead, R.A.
1974: Surficial geology and granular resources of Petitcodiac East (21 H/14E); New Brunswick Department of Natural Resources, Mineral Resources Division, Map Plate 74-121.
1979: Granular aggregate resources of Charlo map area (21 O/16); New Brunswick Department of Natural Resources, Mineral Resources Division, Open File 79-10, 79 p.

Brookes, I.A.
1964: The upland surfaces of western Newfoundland; unpublished MSc thesis, McGill University, Montréal, 127 p.
1969: Late-glacial marine overlap in western Newfoundland; Canadian Journal of Earth Sciences, v. 6, p. 1397-1404.
1970: New evidence for an independent Wisconsin-age ice cap over Newfoundland; Canadian Journal of Earth Sciences, v. 7, p. 1374-1382.
1972: The physical geography of the Atlantic Provinces; in Studies in Canadian Geography, A. Macpherson (ed.); University of Toronto Press, 182 p.
1974: Late Wisconsin glaciation of southwestern Newfoundland with special reference to the Stephenville map area; Geological Survey of Canada, Paper 73-40, 31 p.
1977a: Geomorphology and Quaternary geology of Codroy Lowland and adjacent plateaus, southwest Newfoundland; Canadian Journal of Earth Sciences, v. 14, p. 2101-2120.
1977b: Radiocarbon age of Robinson's Head moraine, west Newfoundland, and its significance for postglacial sea level changes; Canadian Journal of Earth Sciences, v. 14, p. 2121-2126.
1982: Ice marks in Newfoundland: A history of ideas; Géographie physique et Quaternaire, v. 36, p. 139-163.

Brookes, I.A. and Stevens, R.K.
1985: Radiocarbon age of rock-boring *Hiatella-arctica* (Linné) and postglacial sea-level change at Cow Head, Newfoundland; Canadian Journal of Earth Sciences, v. 22, p. 136-140.

Brookes, I.A., McAndrews, J.H. and von Bitter, p.H.
1982: Quaternary interglacial and associated deposits in southwest Newfoundland; Canadian Journal of Earth Sciences, v. 19, p. 410-423.

Brookes, I.A., Scott, D.B. and McAndrews, J.H.
1985: Postglacial relative sea-level change, Port au Port area, west Newfoundland; Canadian Journal of Earth Sciences, v. 22, p. 1039-1047.

Burke, K.B.S.
1984: Earthquake activity in the Maritime Provinces; Geoscience Canada, v. 11, p. 16-22.

Chalmers, R.
1881: On the glacial phenomena of the Bay of Chaleur region; Canadian Naturalist, New Series, v. 10, p. 37-54.
1882: On the surface geology of the Baie de Chaleur region; Canadian Naturalist, New Series, v. 10, p. 193-212.
1885: Report on the surface geology of western New Brunswick with special reference to the area included in York and Carleton counties; Geological Survey of Canada, Report of Progress 1882-84, Part XI, 47 p.
1886: Preliminary report on the surface geology of New Brunswick; in Annual Report 1885; Geological Survey of Canada, v. 1, Part GG, 58 p.
1887a: Report to accompany quarter-sheet maps 3 S.E. and 3 S.W. Surface geology. Northern New Brunswick and southeastern Quebec; Geological Survey of Canada, Annual Report for 1886, v. 2, Part M, 39 p.
1887b: On the glaciation and Pleistocene subsidence of northern New Brunswick and southeastern Quebec; Royal Society of Canada Transactions, Section 4, v. 4, p. 139-145.
1888: Report on the surface geology of northeastern New Brunswick; Geological Survey of Canada, Summary Report 1887 and 1888, v. III, Part N, 33 p.
1890: Report on the surface geology of southern New Brunswick; Geological Survey of Canada, Summary Report 1889, v. IV, Part N, 92 p.
1893: Height of the Bay of Fundy coast in the glacial period relative to sea level, as evidenced by marine fossils in the boulder clay at Saint John, New Brunswick; Geological Society of America, Bulletin, v. 4, p. 361-370.
1895: Report on the surface geology of eastern New Brunswick, northwestern Nova Scotia, and a portion of Prince Edward Island; Geological Survey of Canada, Annual Report 1894, v. VII, Part M, 135 p.

1901: Work on the surface geology of northwestern New Brunswick, chiefly in the area of the Grand Falls sheet; Geological Survey of Canada, Summary Report 1900, v. 13, Part A, p. 151-161.

1902: Report on the surface geology shown on the Fredericton and Andover quarter-sheet maps, New Brunswick; Geological Survey of Canada, Annual Report 1899, v. 12, Part M, 41 p.

1906: Surface geology of eastern Québec; Geological Survey of Canada, Annual Report 1904, v. 16, Part A, p. 250-263.

Chauvin, L.
1984: Géologie du Quaternaire et dispersion glaciaire en Gaspésie, région de Mont-Louis — Rivière-Madeleine; Québec Ministère de l'energie et des ressources, ET 83-19, 33 p.

Clark, J.A., Farrell, W.E., and Peltier, W.R.
1978: Global changes in postglacial sea level: A numerical calculation; Quaternary Research, v. 9, p. 265-287.

Clarke, A.H., Jr., Grant, D.R., and Macpherson, E.
1972: The relationship of *Atractodon stonei* (Pilsbry) (Mollusca, Buccinidae) to the Pleistocene stratigraphy and paleoecology of southwestern Nova Scotia; Canadian Journal of Earth Sciences, v. 9, p. 1030-1038.

Coleman, A.P.
1922: Physiography and glacial geology of Gaspé Peninsula, Québec; Geological Survey of Canada, Museum Bulletin 34, 52 p.

1926: The Pleistocene of Newfoundland; Journal of Geology, v. 34, p. 193-223.

Conolly, J.R., Needham, H.D., and Heezen, B.C.
1967: Late Pleistocene and Holocene sedimentation in the Laurentian Channel; Journal of Geology, v. 75, p. 131-147.

Cooke, H.C.
1930: Glacial depression and post-glacial uplift; Part II of Studies of Physiography of the Canadian Shield; Royal Society of Canada, Transactions, v. 24, section IV, p. 51-87.

Daly, R.A.
1902: Physiography of Nova Scotia; Harvard College, Museum of Comparative Zoology, Bulletin 38, p. 73-103.

1920: Oscillation of level in the belts peripheral to the Pleistocene ice-caps; Geological Society of America, Bulletin, v. 31, p. 303-318.

1921: Postglacial warping of Newfoundland and Nova Scotia; American Journal of Science, 5th Series, v. 201, p. 381-391.

David, P.P. and Lebuis, J.
1985: Glacial maximum and deglaciation of western Gaspé, Québec, Canada; Geological Society of America, Special Paper 197, p. 85-109.

Davies, E.H., Akande, S.O., and Zentilli, M.
1984: Early Cretaceous deposits in the Gays River lead-zinc mine, Nova Scotia; in Current Research, Part A, Geological Survey of Canada, Paper 84-1A, p. 353-358.

Dawson, J.W.
1855: Acadian Geology (first edition); Oliver and Boyd, Edinburgh, 388 p.

1856: On a modern submerged forest at Fort Lawrence, N.S.; American Journal of Science, Series 2, v. 21, p. 440-442.

De Geer, G.
1892: On Pleistocene changes of level in eastern North America; Proceedings of the Boston Society Natural History, Bulletin 25, p. 454-477.

Dionne, J-C.
1963: Le problème de la terrasse et de la falaise Mic Mac; Revue canadienne de géographie, v. 17, p. 723-735.

1985a: Evidence of a low sea level in the St. Lawrence estuary during the Holocene (abstract); in Program with Abstracts, Geological Association of Canada, v. 10, p. A14.

1985b: Observations sur le Quaternaire de la rivière Bayer, côte sud de l'estuaire du Saint-Laurent, Québec; Géographie physique et Quaternaire, vol. 39, p. 35-46.

Dredge, L.A.
1983: Surficial geology of the Sept-Iles area, Quebec North Shore; Geological Survey of Canada, Memoir 408, 40 p.

Dredge, L.A. and Grant, D.R.
1987: Glacial deformation of bedrock and sediment, Magdalen Islands and Nova Scotia, Canada: evidence for a regional grounded ice sheet; in Tills and Glaciotectonics, J.J.M. van der Meer (ed.); A.A. Balkema, Rotterdam, p. 183-195.

Dreimanis, A., and Karrow, P.F.
1972: Glacial history of the Great Lakes-St. Lawrence region, the classification of the Wisconsin(an) stage, and its correlation; 24th International Geological Congress, (Montreal), Section 12, Quaternary Geology, p. 5-15.

Dubois, J-M.M.
1980: Environnements quaternaires et évolution postglaciaire d'une zone côtière en émersion en bordure sud du Bouclier canadien: la Moyenne Côte Nord du Saint-Laurent, Québec; thèse de PhD non publiée, Université d'Ottawa, Ottawa, 754 p.

Dubois, J-M.M. and Dionne, J-C.
1985: The Québec North Shore moraine system: A major feature of Late Wisconsin deglaciation; Geological Society of America, Special Paper 197, p. 125-133.

Dubois, J-M.M., Gwyn, Q.H.J., Bigras, P., Gratton, D., Perras, S., et St-Pierre, L.
sous Géologie des formations en surface, l'Île d'Anticosti; Commission
presse: géologique du Canada, Carte 1660A, échelle 1/250 000.

Dyck, W., and Fyles, J.G.
1963: Geological Survey of Canada radiocarbon dates II; Radiocarbon, v. 5, p. 39-55.

Fader, G.B., King, L.H., and Josenhans, H.W.
1982: Surficial geology of the Laurentian Channel and the western Grand Banks of Newfoundland; Geological Survey of Canada, Paper 81-22, 37 p.

Fader, G.B., King, L.H., and MacLean, B.
1977: Surficial geology of the eastern Gulf of Maine and Bay of Fundy; Geological Survey of Canada, Paper 76-17, 23 p.

Fairchild, H.L.
1918: Postglacial uplift in northeastern North America; Geological Society of America, Bulletin, v. 29, p. 187-238.

Fernald, M.L.
1925: The persistence of plants in unglaciated areas of boreal America; American Academy of Arts and Sciences, Memoirs, v. 15, p. 239-311.

Flint, R.F.
1940: Late Quaternary changes of level in western and southern Newfoundland; Geological Society of America, Bulletin, v. 51, p. 1757-1780.

1963: Altitude, lithology, and the Fall Zone in Connecticut; The Journal of Geology, v. 71, p. 683-697.

Flint, R.F., Demorest, M., and Washburn, A.L.
1942: Glaciation of Shickshock Mountains, Gaspé Peninsula; Geological Society of America, Bulletin, v. 53, p. 1211-1230.

Frankel, L. and Crowl, G.H.
1961: Drowned forests along the eastern coast of Prince Edward Island, Canada; Journal of Geology, v. 69, p. 352-357.

Gadd, N.R.
1973: Quaternary geology of southwest New Brunswick, with particular reference to Fredericton area; Geological Survey of Canada, Paper 71-34, 31 p.

Ganong, W.F.
1896: The outlet delta of Lake Utopia, New Brunswick, Natural History Society of New Brunswick, Bulletin, v. 14, p. 410-412.

1913: Notes on the natural history and physiography of New Brunswick; Natural History Society of New Brunswick, Bulletin, v. 30, p. 419-451.

Gauthier, R.C.
1980a: Existence of a central New Brunswick ice cap based on evidence of northwestward moving ice in the Edmunston area, New Brunswick; in Current Research, Part A, Geological Survey of Canada, Paper 80-1A, p. 377-378.

1980b: Decomposed granite, Big Bald Mountain area, New Brunswick; in Current Research, Part B, Geological Survey of Canada, Paper 80-1B, p. 277-282.

1982: Surficial deposits, Northern New Brunswick; Geological Survey of Canada, Open File 856.

Goldthwait, J.W.
1911: The twenty-foot terrace and sea-cliff of the Lower Saint Lawrence; American Journal of Science, v. 32, p. 291-317.

1913: Physiography of Eastern Quebec and the Maritime Provinces; in Geological Survey of Canada, Twelfth International Geological Congress, Excursion Guidebook, no. 1, Part 1, p. 16-24.

1924: Physiography of Nova Scotia; Geological Survey of Canada, Memoir 140, 179 p.

Grant, D.R.
1963: Pebble lithology of the tills of southeast Nova Scotia; unpublished MSc thesis, Dalhousie University, Halifax, Nova Scotia, 235 p.

1969a: Surficial deposits, geomorphic features and late Quaternary history of the terminus of the northern peninsula of Newfoundland and adjacent Quebec-Labrador; Maritime Sediments, v. 5, no. 3, p. 123-125.

1969b: Late Pleistocene re-advance of piedmont glaciers in western Newfoundland: Maritime Sediments, v. 5, no. 3, p. 126-128.

1970: Recent coastal submergence of the Maritime Provinces, Canada; Canadian Journal of Earth Sciences, v. 7, p. 676-689.

1971a: Surficial geology, southwest Cape Breton Island, Nova Scotia; in Report of Activities, Part A, Geological Survey of Canada, Paper 71-1A, p. 161-164.

1971b: Glacial deposits, sea level changes and Pre-Wisconsin deposits in southwest Nova Scotia; in Report of Activities, Part B, Geological Survey of Canada, Paper 71-1B, p. 110-113.

1971c: Glaciation of Cape Breton Island; in Report of Activities, Part B, Geological Survey of Canada, Paper 71-1B, p. 118-120.
1972a: Surficial geology, western Newfoundland; in Report of Activities, Part A, Geological Survey of Canada, Paper 72-1A, p. 157-160.
1972b: Surficial geology of southeast Cape Breton Island; in Report of Activities, Part A, Geological Survey of Canada, Paper 72-1A, p. 160- 163.
1974a: Terrain studies of Cape Breton Island, Nova Scotia and of the Northern Peninsula, Newfoundland; in Report of Activities, Part A, Geological Survey of Canada, Paper 74-1A, p. 241-248.
1974b: Prospecting in Newfoundland and the theory of multiple shrinking ice caps; in Report of Activities, Part B, Geological Survey of Canada, Paper 74-1B, p. 215-216.
1975a: Surficial geology of northern Cape Breton Island; in Report of Activities, Part A, Geological Survey of Canada, Paper 75-1A, p. 407- 408.
1975b: Recent coastal submergence of the Maritime Provinces; Nova Scotian Institute of Science, Proceedings, v. 27, Supplement 3, p. 83-102.
1975c: Glacial features of the Hermitage-Burin Peninsula area, Newfoundland; in Report of Activities, Part C, Geological Survey of Canada, Paper 75- 1C, p. 333-334.
1976: Glacial style and the Quaternary stratigraphic record in the Atlantic Provinces, Canada: in Quaternary Stratigraphy of North America, W.C. Mahaney (ed.), Dowden, Hutchinson and Ross, Inc. Stroudsburg, U.S.A., p. 33-35.
1977a: Altitudinal weathering zones and glacial limits in western Newfoundland, with particular reference to Gros Morne National Park; in Report of Activities, Part A, Geological Survey of Canada, Paper 77-1A, p. 455-463.
1977b: Glacial style and ice limits, the Quaternary stratigraphic record, and changes of land and ocean level in the Atlantic Provinces, Canada; Géographie physique et Quaternaire, v. 31, p. 247-260.
1980a: Quaternary sea-level change in Atlantic Canada as an indication of crustal delevelling; in Earth Rheology, Isostasy and Eustasy, Nils-Axel Mörner (ed.); John Wiley and Sons Ltd., New York, p. 201-214.
1980b: Quaternary stratigraphy of southwestern Nova Scotia; glacial events and sea level changes; Geological Association of Canada, Annual Meeting (Halifax), Field Trip Guidebook, Trip 9, 63 p.
1986a: Glaciers, sediment and sea level — Northern Bay of Fundy, Nova Scotia; 14th Arctic Workshop, Field Trip Guidebook B, 43 p; Geological Survey of Canada, Open File 1323.
1986b: Surficial geology, Port Saunders, Newfoundland; Geological Survey of Canada, Map 1622A, scale 1:250 000.
1987a: Quaternary geology of Nova Scotia and Newfoundland: XII INQUA Congress, Ottawa 1987; Excursion Guide Book A-3/C-3, National Research Council of Canada, Ottawa, NRC 27525, 62 p.
1987b: Glacial advances and sea level changes, southwestern Nova Scotia, Canada; in Centennial Field Guide, Northeastern Section, D.C. Roy (ed.); Geological Society of America, CFG-5, p. 427-432.
in press: Surficial geology, St. Anthony — Blanc-Sablon map areas, Newfoundland and Quebec; Geological Survey of Canada, Memoir.

Grant, D.R. and King, L.H.
1984: A Quaternary stratigraphic framework for the Atlantic Provinces region; in Quaternary Stratigraphy of Canada — A Canadian Contribution to IGCP Project 24, R.J. Fulton (ed.); Geological Survey of Canada, Paper 84-10, p. 173-191.

Grant, D.R. and Tucker, C.M.
1976: Preliminary results of terrain mapping and base-metal analysis of till in the Red Indian Lake and Gander Lake map-areas of central Newfoundland; in Report of Activities, Part A, Geological Survey of Canada, Paper 76-1A, p. 283-285.

Grant, D.R., Prest, V.K., Dredge, L.A., and Mott, R.J.
1985: Lithostratigraphy and Quaternary history, Magdalen Islands, Québec (abstract); in Geological Association of Canada, Program with Abstracts, v. 10, p. A22.

Gratton, D., Gwyn, Q.H.J., et Dubois, J-M.M.
1984: Les paléoenvironnements sédimentaires au Wisconsinien moyen et supérieur, Île d'Anticosti, Golfe du Saint-Laurent, Québec; Géographie physique et Quaternaire, vol. 38, p. 229-242.

Gratton, D., Dubois, J-M.M., Painchaud, A., et Gwyn, Q.H.J.
1986: L'Île d'Anticosti, a-t-elle été récemment englacée?; Geos, v. 15, p. 21-23.

Gravenor, C.P.
1974: The Yarmouth drumlin field, Nova Scotia, Canada; Journal of Glaciology, v. 13, p. 45-54.

Gray, J.T. and Brown, R.J.E.
1979: Permafrost existence and distribution in the Chic-Chocs Mountains, Gaspésie, Québec; Géographie physique et Quaternaire, v. 33, p. 299-316.

Gray, J.T., Boudreau, F., Hétu, B., Labelle, C., Lafrenière, L.B., Payette, S., and Richard, P.
1981: Weathering zones and the problem of glacial limits; Excursion and conference in Gaspésie, Quebec; Canadian Quaternary Association, Papers and Guidebook for Conference and Excursion, 167 p.

Greenwood, B. and Davidson-Arnott, G.D.D.
1972: Quaternary history and sedimentation: A summary and select bibliography; Maritime Sediments, v. 8, p. 88-100.

Gustavson, T.C.
1976: Paleotemperature analysis of the marine Pleistocene of Long Island, New York, and Nantucket Island, Massachusetts; Geological Society of America, Bulletin, v. 87, p. 1-8.

Harrison, W. and Lyon, C.J.
1963: Sea-level and crustal movements along the New England-Acadian shore, 4,500-3,000 B.P.; Journal of Geology, v. 71, p. 96-108.

Henderson, E.P.
1968: Patterned ground in southeastern Newfoundland; Canadian Journal of Earth Sciences, v. 5, p. 1443-1453.
1972: Surficial geology of Avalon Peninsula, Newfoundland; Geological Survey of Canada, Memoir 368, 121 p.

Hickox, C.J., Jr.
1962: Pleistocene geology of the central Annapolis Valley, Nova Scotia; Nova Scotia Department of Mines, Memoir 5, 36 p.

Honeyman, D.C.L.
1882: Nova Scotian geology, superficial; Nova Scotian Institute of Science, Transactions, v. 5, p. 319-331.

Hornbrook, E.H.W., Davenport, P.H., and Grant, D.R.
1975: Regional and detailed geochemical exploration studies in glaciated terrain in Newfoundland; Newfoundland Department of Mines and Energy, Report 75-2, 116 p.

Hughes, T., Borns, H.W., Jr., Fastook, J.L., Kite, J.S., Hyland, M.R., and Lowell, T.V.
1985: Models of glacial reconstruction and deglaciation applied to Maritime Canada and New England; in Late Pleistocene History of Northeastern New England and Adjacent Quebec, H.J.W. Borns, P. LaSalle and W.B. Thompson (ed.); Geological Society of America, Special Paper 197, p. 139-150.

Jansa, L.F. and Wade, J.A.
1975: Paleogeography and sedimentation in the Mesozoic and Cenozoic, southeastern Canada; in Canada's Continental Margins and Offshore Petroleum Exploration, C.J. Yorath, E.R. Parker, and D.F. Glass (ed.); Canadian Society of Petroleum Geologists, Memoir 4, p. 79-102.

Jenness, S.E.
1960: Late Pleistocene glaciation of eastern Newfoundland; Geological Society of America, Bulletin, v. 71, p. 161-180.

Keen, M.J. and Williams G.L. (editors)
1989: Geology of the Continental Margin of Eastern Canada; Geological Survey of Canada, Geology of Canada, no. 2 (also Geological Society of America, The Geology of North America, v. I-1).

Kerr, J.H.
1870: Observations on ice-marks in Newfoundland; Quarterly Journal, Geological Society of London, v. 26, p. 704-705.

King, L.H.
1980: Aspects of regional surficial geology related to site investigation requirements — eastern Canadian Shelf; in Offshore Site Investigation, D.A. Ardus (ed.); Graham and Trotman, p. 37-59.

King, L.H. and Fader, G.B.J.
1986: Wisconsinan glaciation of the Atlantic continental shelf of southeast Canada; Geological Survey of Canada, Bulletin 363, 72 p.

King, L.H. and MacLean, B.
1970: Origin of the outer part of the Laurentian Channel; Canadian Journal of Earth Sciences, v. 7, p. 1470-1484.
1976: Geology of the Scotian Shelf; Geological Survey of Canada, Paper 74-31, 31 p, Map 812 H, scale 1:1 000 000.

King, L.H., MacLean, B., and Gordon, G.F.
1974: Unconformities on the Scotian Shelf; Canadian Journal of Earth Sciences, v. 11, p. 89-100.

Kranck, K.
1972: Geomorphological development and post-Pleistocene sea level changes, Northumberland Strait, Maritime Provinces; Canadian Journal of Earth Sciences, v. 9, p. 835-844.

Lambert, A. and Vaniček, P.
1979: Contemporary crustal movements in Canada; Canadian Journal of Earth Sciences, v. 16, p. 647-668.

Lammers, W. en De Haan, F.A.
1980: De Holocene afzettingen in het gebied van de Tantramar en Aulac, Bay of Fundy, Canada; unpublished PhD thesis, Vrije Universiteit, Amsterdam, 74 p.

LaSalle, p., De Kimpe, C.R., and Laverdière, M.R.
1985a: Sub-till saprolites in southeastern Quebec and adjacent New England: Erosional, stratigraphic, and climatic significance; Geological Society of America, Special Paper 197, p. 13-20.
1985b: Soil development in southern Gaspésie Peninsula in relation to last glaciation events (abstract); in Program with Abstracts, Geological Association of Canada, v. 10, p. A33.

Lawson, A.C.
1890: Note on the pre-Paleozoic surface of the Archean terranes of Canada; Geological Society of America, Bulletin, v. 1, p. 163-174.

Lebuis, J. et David, P.P.
1977: La stratigraphie et les évènements du Quaternaire de la partie occidentale de la Gaspésie, Québec; Géographie physique et Quaternaire, v. 31, p. 275-296.

Leckie, D.A.
1979: Late Quaternary history of the Hermitage area Newfoundland; unpublished MSc thesis, McMaster University, Hamilton, 188 p.

Lee, H.A.
1953: Two phases of till and other glacial problems in the Edmundston-Grand Falls region (New Brunswick, Quebec, and Maine); unpublished PhD thesis, Chicago University, Chicago, Illinois.
1957: Surficial geology of Fredericton, York and Sudbury counties, New Brunswick; Geological Survey of Canada, Paper 56-2, 11 p.

Leggett, R.F.
1980: Glacial geology of Grand Manan Island, New Brunswick; Canadian Journal of Earth Sciences, v. 17, p. 440-452.

Lindroth, C.
1963: The fauna history of Newfoundland illustrated by Carabid beetles; Opuscula Entomologica, Supplement 23, 112 p.

Livingstone, D.A.
1951: A new record for the *Mastodon americanus* Kerr, from Nova Scotia; Nova Scotian Institute of Science, Proceedings and Transactions, v. 22, p. 15-16.
1968: Some interstadial and postglacial pollen diagrams from eastern Canada; Ecological Monographs, v. 38, p. 89-125.

Livingstone, D.A. and Estes, A.H.
1967: A carbon-dated pollen diagram from the Cape Breton plateau, Nova Scotia; Canadian Journal of Botany, v. 45, p. 339-359.

Livingstone, D.A. and Livingstone, B.G.R.
1958: Late-glacial and postglacial vegetation from Gillis Lake in Richmond County, Cape Breton Island, Nova Scotia; American Journal of Science, v. 256, p. 341-359.

Locat, J.
1977: L'émersion des terres dans la région de Baie-des-Sables/Trois-Pistoles, Québec; Géographie physique et Quaternaire, vol. 31, p. 297-306.

Logan, W.E.
1863: Geology of Canada; Geological Survey of Canada, Report of Progress from Commencement to 1863.

Loring, D.H. and Nota, D.J.G.
1969: Mineral dispersal patterns in the Gulf of St. Lawrence; Revue de géographie de Montréal, v. 23, p. 289-305.
1973: Morphology and sediments of Gulf of St. Lawrence; Fisheries Research Board of Canada, Bulletin 182, 147 p.

Lougee, R.J.
1953: A chronology of post-glacial time in eastern North America: Scientific Monthly, v. 76, p. 259-276.

Lowdon, J.A. and Blake, W., Jr.
1973: Geological Survey of Canada radiocarbon dates XIII; Geological Survey of Canada, Paper 73-7, 61 p.

Lowell, T.B.
1985: Late Wisconsin ice-flow reversal and deglaciation, northwestern Maine; in Late Pleistocene History of Northeastern New England and Adjacent Quebec, H.J.W. Borns, P. LaSalle and W.B. Thompson (ed.); Geological Society of America, Special Paper 197, p. 71-83.

Lundqvist, J.
1965: Glacial geology in northeastern Newfoundland; Geologiska Foreningens i Stockholm Forhandlingar, v. 87, p. 285-306.

MacClintock, P. and Twenhofel, W.H.
1940: Wisconsin glaciation of Newfoundland; Geological Society of America, Bulletin, v. 51, p. 1729-1756.

MacEachern, I.J., Stea, R.R., and Rogers, P.J.
1984: Till stratigraphy and gold distribution, Forest Hill Gold District, Nova Scotia; in Current Research, Part A, Geological Survey of Canada, Paper 84-1A, p. 651-654.

MacNeill, R.H.
1969: Some dates relating to the dating of the last major ice sheet in Nova Scotia; Maritime Sediments, v. 5, p. 3.

Macpherson, J.
1982: Postglacial vegetational history of the eastern Avalon Peninsula, Newfoundland, and Holocene climatic change along the eastern Canadian seaboard; Géographie physique et Quaternaire, v. 36, p. 175-196.

Macpherson, J. and Anderson, T.W.
1985: Further evidence of late glacial climatic fluctuations from Newfoundland: pollen stratigraphy from a north coast site; in Current Research, Part B, Geological Survey of Canada, Paper 85-1B, p. 383-390.

Mathews, W.H.
1975: Cenozoic erosion and erosion surfaces in eastern North America; American Journal of Science, v. 275, p. 818-824.

Matthew, G.F.
1872: On the surface geology of New Brunswick; Canadian Naturalist, v. 6, p. 326-328.
1879: Report on the superficial geology of southern New Brunswick; Geological Survey of Canada, Report of Progress 1877-78, Part X, 36 p.
1894a: Movements of the earth's crust at St. John, New Brunswick, in postglacial times; Natural History Society of New Brunswick, Bulletin, no. 12, p. 34-42.
1894b: Post-glacial faults at St. John, N.B.; American Journal of Science, v. 48, p. 501-503.

Mayewski, P.A., Denton, G.H., and Hughes, T.J.
1981: Late Wisconsinan ice sheets of North America; in The Last Great Ice Sheet, G.H. Denton and T.J. Hughes (ed.); John Wiley and Sons, New York, p. 67-178.

McGerrigle, H.W.
1952: Pleistocene glaciation of Gaspé Peninsula (Quebec); Royal Society of Canada, Transactions, v. 46, sec. 4, p. 37-51.
1959: Région de la rivière Madeleine; Ministère des Mines, Québec, Rapport Géologique 77, 52 p.

McIver, N.L.
1972: Cenozoic and Mesozoic stratigraphy of the Nova Scotia Shelf; Canadian Journal of Earth Sciences, v. 9, p. 54-70.

McKeague, J.A., Grant, D.R., Kodama, H., Beke, G.J., and Wang, C.
1983: Properties and genesis of a soil and the underlying gibbsite-bearing saprolite, Cape Breton Island, Canada; Canadian Journal of Earth Sciences, v. 20, p. 37-48.

McNeely, R.N.
1988: Geological Survey of Canada radiocarbon dates XXVIII; Geological Survey of Canada, Paper 88-7.

Miller, P.E.
1973: The depositional environment of the Salmon River section, southwestern Nova Scotia; unpublished BSc thesis, Dalhousie University, Halifax, Nova Scotia, 24 p.

Milne, J.
1874: Notes on the physical features and mineralogy of Newfoundland; Quarterly Journal of the Geological Society of London, v. 30, p. 722-745.
1876: Ice and ice-work in Newfoundland; Geological Magazine, v. 3, p. 303-308.

Monahan, D.
1971: On the morphology of the northeast Gulf of St. Lawrence; Maritime Sediments, v. 7, p. 73-75.

Mott, R.J.
1971: Palynology of a buried organic deposit, River Inhabitants, Cape Breton Island, Nova Scotia; in Report of Activities, Part B, Geological Survey of Canada, Paper 71-1B, p. 123-125.
1985: Late-glacial climatic change in the Maritime Provinces; in Climatic Change in Canada 5: Critical periods in the Quaternary Climatic History of Northern North America, C.R. Harington (ed.); Syllogeus 55, p. 281-300.
1989: Late Pleistocene paleoenvironments in Atlantic Canada; in Chapter 7 of Quaternary Geology of Canada and Greenland, R.J. Fulton (ed.); Geological Survey of Canada, Geology of Canada, no. 1 (also Geological Society of America, The Geology of North America, v. K-1).

Mott, R.J. and Grant, D.R.
1985: Pre-Late Wisconsinan paleoenvironments in Atlantic Canada; Géographie physique et Quaternaire, v. 39, p. 239-254.

Mott, R.J. and Prest, V.K.
1967: Stratigraphy and palynology of buried organic deposits from Cape Breton Island; Canadian Journal of Earth Sciences, v. 4, p. 709-724.

Mott, R.J., Anderson, T.W. and Matthews, J.V., Jr.
1982: Pollen and macrofossil study of an interglacial deposit in Nova Scotia; Géographie physique et Quaternaire, v. 36, p. 197-208.

Mott, R.J., Grant, D.R., Stea, R.R., and Occhietti, S.
1986: Lateglacial climatic oscillation in Atlantic Canada equivalent to the Allerød/Younger Dryas event; Nature, v. 323, no. 6085, p. 247-250.

Murray, A.
1883: Glaciation of Newfoundland; Royal Society of Canada, Proceedings, Transactions, v. 1, p. 55-76.

Murray, R.C.
1955: Directions of glacier motion in south-central Newfoundland; Journal of Geology, v. 63, p. 268-274.

Neale, E.R.W.
1964: Geology, Cape North, Nova Scotia; Geological Survey of Canada, Map 1150A, scale 1:63,360.

Nielsen, E.
1974: A mid-Wisconsinan glacio-marine deposit from Nova Scotia (abstract); in International Symposium on Quaternary Environments, Abstracts Volume, York University, Toronto.
1976: The composition and origin of Wisconsinan till in mainland Nova Scotia; unpublished PhD dissertation, Dalhousie University, Halifax, Nova Scotia, 256 p.

Noordijk, A. en Pronk, T.
1981: De Holocene afzettingen in de dalen van de Missiguash, La Planche, en Nappan, Bay of Fundy, Canada; unpublished PhD thesis, Vrije Universiteit, Amsterdam, 65 p.

Occhietti, S.
1989: Quaternary geology of the St. Lawrence Valley and adjacent Appalachian subregion; in Chapter 4 of Quaternary Geology of Canada and Greenland, R.J. Fulton (ed.); Geological Survey of Canada, Geology of Canada, no. 1 (also Geological Society of America, The Geology of North America v. $\overline{K-1}$).

O'Donnell, N.D.
1973: Glacial indicator trains near Gullbridge, Newfoundland; unpublished MSc thesis, University of Western Ontario, London, 259 p.

Ogden, J.G., III
1960: Recurrence surfaces and pollen stratigraphy of a postglacial raised bog, Kings County, Nova Scotia; American Journal of Science, v. 258, p. 341-353.

Painchaud, A., Dubois, J-M.M., et Gwyn, Q.H.J.
1984: Déglaciation et émersion des terres de l'ouest de l'Île d'Anticosti, golfe du Saint-Laurent, Québec; Géographie physique et Quaternaire, v. 38, p. 93-111.

Pardi, R.B. and Newman, W.S.
1987: Late Quaternary sea levels along the Atlantic coast of North America; Journal of Coastal Research, v. 3, p. 325-330.

Piers, H.
1915: Mastodon remains in Nova Scotia; Nova Scotian Institute of Science, Proceedings, v. 13, p. 163-164.

Piper, D.J.W., Fader, G.B., Josenhans, H., MacLean,B., Mudie, P.J., and Vilks, G.
1989: Quaternary geology; Chapter 10 in Geology of the Continental Margin of Eastern Canada, M.J. Keen and G.L. Williams (ed.);Geological Survey of Canada, Geology of Canada, no. 2 (also Geological Society of America, The Geology of North America, v. $\overline{I-1}$).

Plumb, R.A. and Cox, J.W.
1987: Stress directions in eastern North America determined to 4.5 km from borehole elongation measurements; Journal of Geophysical Research, v. 92, p. 4805-4816.

Podolak, W.E. and Shilts, W.W.
1978: Some physical and chemical properties of till derived from the Meguma Group, southeast Nova Scotia; in Current Research, Part A, Geological Survey of Canada, Paper 78-1A, p. 459-464.

Poole, W.H.
1976: Plate tectonic evolution of the Canadian Appalachian region; in Report of Activities, Part B; Geological Survey of Canada, Paper $\overline{76}$-1B, p. 113- 126.

Prest, V.K.
1970: Quaternary geology of Canada; in Geology and Economic Minerals of Canada, R.J.W. Douglas (ed.); Geological Survey of Canada, Economic Geology Report 1, Fifth Edition, p. 675-764.
1973: Surficial deposits, Prince Edward Island; Geological Survey of Canada, Map 1366A, scale 1:126,720.
1977: General stratigraphic framework of the Quaternary in eastern Canada; Géographie physique et Quaternaire, v. 31, p. 7-14.
1984: The Late Wisconsinan glacier complex; in Quaternary Stratigraphy of Canada — A Canadian Contribution to IGCP Project 24, R.J. Fulton (ed.); Geological Survey of Canada, Paper 84-10, p. 21-36.

Prest, V.K. and Grant, D.R.
1969: Retreat of the last ice sheet from the Maritime Provinces — Gulf of St. Lawrence region; Geological Survey of Canada, Paper 69-33, 15 p.

Prest, V.K., Grant, D.R., Borns, H.W. Jr., Brookes, I.A., MacNeill, R.H., and Ogden, J.G. III
1972: Quaternary geology, geomorphology and hydrogeology of the Atlantic Provinces; 24th International Geological Congress, Guidebook A61-C61, 79 p.

Prest, V.K., Terasmae, J., Matthews, J.V., Jr., and Lichti-Federovich, S.
1977: Late-Quaternary history of Magdalen Islands, Quebec; Maritime Sediments, v. 12, p. 39-59.

Prest, W.H.
1898: Glacial succession in central Lunenburg; Nova Scotian Institute of Science, Proceedings and Transactions, v. 9, p. 158-170.

Quinlan, G.
1984: Postglacial rebound and the focal mechanisms of eastern Canadian earthquakes; Canadian Journal of Earth Sciences, v. 21, p. 1018-1023.

Quinlan, G. and Beaumont, C.
1981: A comparison of observed and theoretical postglacial relative sea level in Atlantic Canada; Canadian Journal of Earth Sciences, v. 18, p. 1146- 1163.
1982: The deglaciation of Atlantic Canada as reconstructed from the postglacial relative sea-level record; Canadian Journal of Earth Sciences, v. 19, p. 2232-2246.

Railton, J.B.
1972: Vegetational and climatic history of southwestern Nova Scotia in relation to a South Mountain ice cap; unpublished PhD thesis, Dalhousie University, 146 p.

Rampton, V.N. and Paradis, S.
1981a: Quaternary geology of Woodstock map area (21J), New Brunswick; New Brunswick Department of Natural Resources, Map Report 81-1, 37 p.
1981b: Quaternary geology of Moncton map area (21I), New Brunswick; New Brunswick Department of Natural Resources, Map Report 81-2, 31 p.
1981c: Quaternary geology of Amherst map area (21H), New Brunswick; New Brunswick Department of Natural Resources, Map Report 81-3, 36 p.

Rampton, V.N., Gauthier, R.C., Thibault, J., and Seaman, A.A.
1984: Quaternary geology of New Brunswick; Geological Survey of Canada, Memoir 416, 77 p.

Roberge, J. et Gascoyne, M.
1978: Premiers résultats de datations dans la grotte de Saint-Elzéar, Gaspésie, Québec; Géographie physique et Quaternaire, vol. 32, p. 281- 287.

Rogerson, R.J.
1981: The tectonic evolution and surface morphology of Newfoundland; in The Natural Environment of Newfoundland, Past and Present, A.G. Macpherson and J.B. Macpherson (ed.); Department of Geology, Memorial University of Newfoundland, p. 24-55.
1982: The glaciation of Newfoundland and Labrador; in Prospecting in Areas of Glaciated Terrain — 1982, P.H. Davenport (ed.); Canadian Institute of Mining and Metallurgy, Geology Division Publication, p. 37-56.

Roland, A.E.
1982: Geological background and physiography of Nova Scotia; Nova Scotian Institute of Science, Halifax, Canada, 311 p.

Ruddiman, W. and McIntyre, A.
1973: Time-transgressive deglacial retreat of polar waters from the North Atlantic; Quaternary Research, v. 3, p. 117-130.
1981: Oceanic mechanisms for amplification of the 23,000-year ice-volume cycle; Science, v. 212, p. 617-627.

St-Pierre, L., Gwyn, Q.H.J., et Dubois, J-M.M.
1985: Dynamique des écoulements glaciaires Wisconsiniens, Île d'Anticosti, golfe du Saint-Laurent (résumé); in Programme et Résumés, Association Géologique du Canada, vol. 10, p. A60.
1987: Lithostratigraphie et dynamique glaciaires au Wisconsinien, Île d'Anticosti, golfe du Saint-Laurent; Journal canadien des sciences de la terre, vol. 24, p. 1847-1858.

Sanford, B.V.
in press:St. Lawrence Platform; Part III in Sedimentary Cover of the Craton: Canada, D.F. Stott and J.D. Aitken (ed.); Geological Survey of Canada, Geology of Canada, no. 6 (also Geological Society of America, The Geology of North America, v. $\overline{D-1}$).

Sanford, B.V. and Grant, G.M.
1976: Physiography, Eastern Canada and adjacent areas; Geological Survey of Canada, Map 1399A (4 sheets), scale 1:2 000 000.

Sanford, B.V., Grant, A.C., Wade, J.A., and Barss, M.S.
1979: Geology of Eastern Canada and adjacent areas; Geological Survey of Canada, Map 1401A, (4 sheets), scale 1:2000 000.

Sanschagrin, R.
1964: Magdalen Islands; Québec Department of Natural Resources, Geological Report 106, 58 p.

Schroeder, J. et Arsenault, S.
1978: Discussion d'un karst dans le gypse d'Hillsborough, Nouveau-Brunswick; Géographie physique et Quaternaire, vol. 32, p. 249-261.

Scott, D.B. and Greenberg, D.A.
1983: Relative sea-level rise and tidal development in the Fundy tidal system; Canadian Journal of Earth Sciences, v. 20, p. 1554-1564.

Scott, D.B. and Medioli, F.S.
1978: Vertical zonations of marsh foraminifera as accurate indicators of former sea levels; Nature, v. 272, p. 528-531.
1982: Micropaleontological documentation for early Holocene fall of relative sea level on the Atlantic coast of Nova Scotia; Geology, v. 10, p. 278- 281.

Scott, D.B., Medioli, F.S., and Duffett, T.E.
1984: Holocene rise of relative sea level at Sable Island, Nova Scotia, Canada; Geology, v. 12, p. 173-176.

Scott, D.B., Williamson, M.A., and Duffett, T.E.
1981: Marsh foraminifera of Prince Edward Island: Their recent distribution and application for former sea level studies; Maritime Sediments and Atlantic Geology, v. 17, p. 98-129.

Scott, J.S.
1976: Geology of Canadian tills; in Glacial Till, R.F. Leggett (ed.); Royal Society of Canada, Special Publication 12, p. 50-66.

Seaman, A.A.
1982a: Granular aggregate resources, McAdam (NTS 21 G/11) and Forest City (NTS 21 G/12); New Brunswick Department of Natural Resources, Mineral Resources Division, Open File 82-14, 101 p.
1982b: The late Quaternary history of the Burtts Corner area, New Brunswick; New Brunswick Department of Natural Resources, Mineral Resources Branch, Open File 83-2.
1985: Glaciation of the northern Miramichi Highlands, New Brunswick (abstract); in Program with Abstracts, Geological Association of Canada, v. 10, p. A55.

Shearer, J.M.
1973: Bedrock and surficial geology of the northern Gulf of St. Lawrence as interpreted from continuous seismic reflection profiles; in Earth Science Symposium on Offshore Eastern Canada, Geological Survey of Canada, Paper 71-23, p. 285-303.

Sparkes, B.G.
1981: Star Lake-Victoria Lake surficial and glacial mapping, Newfoundland; in Current Research, Newfoundland Department of Mines and Energy, Report 81-1, p. 188-191.
1983: Surficial and glacial mapping of the Buchans map-area (12 A/14), Newfoundland; in Current Research, Newfoundland Department of Mines and Energy, Report 83-1, p. 189-191.

Squires, W.A.
1966: The Hillsborough mastodon; The Atlantic Advocate, v. 56, p. 29-32.

Stea, R.R.
1978: Notes on the "Bridgewater Conglomerate" and related deposits along the eastern shore of Nova Scotia; in Report of Activities 1977, Nova Scotia Department of Mines, Report 78-1, p. 15-17.
1980: A study of a succession of tills exposed in a wave-cut drumlin, Meisners Reef, Lunenburg County; in Report of Activities 1979, Nova Scotia Department of Mines, Report 80-1, p. 9-20.
1982a: Pleistocene geology and till geochemistry of south central Nova Scotia (Sheet 6); Nova Scotia Department of Mines and Energy, Map 82-1, scale 1:100 000.
1982b: The properties, correlation and interpretation of Pleistocene sediments in central Nova Scotia; unpublished MSc thesis, Dalhousie University, Halifax, Nova Scotia, 215 p.
1982c: Pleistocene stratigraphy of central Nova Scotia; in Report of Activities 1981, Nova Scotia Department of Mines and Energy, Report 82-1, p. 47-54.
1983: Surficial geology of the western part of Cumberland County, Nova Scotia; in Current Research, Part A, Geological Survey of Canada, Paper 83-1A, p. 197-202.

Stea, R.R. and Brisco, V.J.
1984: Bibliography of the surficial geology of Nova Scotia and the offshore (with radiocarbon date list); Nova Scotia Department of Mines and Energy, Report 84-6, 117 p.

Stea, R.R. and Finck, P.W.
1984: Patterns of glacier movement in Cumberland, Colchester, Hants, and Pictou counties, northern Nova Scotia; in Current Research, Part A, Geological Survey of Canada, Paper 84-1A, p. 477-484.

Stea, R.R. and Fowler, J.
1978: Regional mapping and geochemical reconnaissance of Pleistocene till, Eastern Shore, Nova Scotia; in Report of Activities 1977, Nova Scotia Department of Mines, Report 78-1, p. 5-14.
1979: Minor and trace-element variations in Wisconsinan tills, Eastern Shore region, Nova Scotia; Nova Scotia Department of Mines and Energy, Paper 79-4, 30 p.
1981: Petrology of Lower Cretaceous silica sands at Brazil Lake, Hants County, Nova Scotia; in Report of Activities 1980, Nova Scotia Department of Mines and Energy, Report 81-1, p. 47-64.

Stea, R.R. and Grant, D.R.
1982: Pleistocene geology and till geochemistry of southwestern Nova Scotia (Sheets 7 and 8); Nova Scotia Department of Mines and Energy, Map 82-10 (with explanatory notes).

Stea, R.R. and Hemsworth, D.
1979: Pleistocene stratigraphy of the Miller Creek section, Hants County, Nova Scotia; Nova Scotia Department of Mines and Energy, Paper 79-5, 16 p.

Stea, R.R. and O'Reilly, G.A.
1982: Till geochemistry of the Meguma Terrane in Nova Scotia and its metallogenic implications; in Prospecting in Areas of Glaciated Terrain — 1982, P.H. Davenport (ed.); Canadian Institute of Mining and Metallurgy, Geology Division Publication, p. 82-104.

Stea, R.R. and Wightman, D.M.
1987: Age of the Five Islands Formation, Nova Scotia, and deglaciation of the Bay of Fundy; Quaternary Research, v. 27, p. 211-219.

Stea, R.R., Finck, P.W., and Wightman, D.M.
1986: Quaternary geology and till geochemistry of the western part of Cumberland County, Nova Scotia (Sheet 9); Geological Survey of Canada, Paper 85-17, 58 p.

Stevenson, I.M.
1959: Shubenacadie and Kennetcook Map-areas, Colchester, Hants and Halifax counties, Nova Scotia; Geological Survey of Canada, Memoir 302, 88 p.

Sugden, D.E.
1976: A case against deep erosion of shields by ice sheets; Geology, v. 4, p. 580-582.

Summers, W.F.
1949: Physical geography of the Avalon Peninsula of Newfoundland; unpublished MSc thesis, McGill University, Montréal, Quebec, 204 p.

Swett, K.
1981: A probable sub-Cambrian glacial erosional surface in southern Labrador and eastern Quebec; in Earth's Pre-Pleistocene Glacial Record, M.J. Hambrey and W.B. Harland (ed.); Cambridge University Press, Cambridge, U.K., p. 772-773.

Swift, D.J.P. and Borns, H.W., Jr.
1967: A raised fluviomarine outwash terrace, north shore of the Minas Basin, Nova Scotia; Journal of Geology, v. 75, p. 693-710.

Tarnocai, C.
1989: Peat resources of Canada; in Chapter 11 of Quaternary Geology of Canada and Greenland, R.J. Fulton (ed.); Geological Survey of Canada, Geology of Canada, no. 1 (also Geological Society of America, The Geology of North America, v. K-1).

Terasmae, J.
1974: Deglaciation of Port Hood Island, Nova Scotia; Canadian Journal of Earth Sciences, v. 11, p. 1357-1365.

Thibault, J.
1978: Granular aggregate resources of the Nepisiguit Falls map area (21 P/5); New Brunswick Department of Natural Resources, Mineral Resources Branch, Open File 78-4.
1982: Granular aggregate resources of Rollingdam (NTS 21 G/6) and St. Stephen (NTS 31 G/3); New Brunswick Department of Natural Resources, Mineral Resources Branch, Open File 82-15.

Thomas, M.L.H., Grant, D.R., and Degrace, M.
1973: A late Pleistocene marine shell deposit at Shippegan, New Brunswick; Canadian Journal of Earth Sciences, v. 10, p. 1329-1332.

Thomas, R.H.
1977: Calving bay dynamics and ice sheet retreat up the St. Lawrence valley system; Géographie physique et Quaternaire, v. 31, p. 347-356.

Tucker, C.M.
1974: A series of raised Pleistocene deltas, Halls Bay, Newfoundland; Maritime Sediments, v. 10, p. 1-6.
1976: Quaternary studies in Newfoundland: a short review; Maritime Sediments, v. 12, p. 61-73.
1979: Late Quaternary events on the Burin Peninsula, Newfoundland with reference to the islands of St. Pierre et Miquelon (France); unpublished PhD thesis, McMaster University, Hamilton, Ontario, 282 p.

Tucker, C.M. and McCann, S.B.
1980: Quaternary events on the Burin Peninsula, Newfoundland, and the islands of St. Pierre and Miquelon, France; Canadian Journal of Earth Sciences, v. 17, p. 1462-1479.

Tucker, C.M., Leckie, D.A., and McCann, S.B.
1982: Raised shoreline phenomena and postglacial emergence in south-central Newfoundland; Géographie physique et Quaternaire, v. 36, p. 165-174.

Twenhofel, W.H. and MacClintock, P.
1940: Surface of Newfoundland; Geological Society of America, Bulletin, v. 51, p. 1665-1728.

Van Alstine, R.E.
1948: Geology and mineral deposits of the St. Lawrence area, Burin Peninsula, Newfoundland; Newfoundland Geological Survey, Bulletin 23, 51 p.

Vanderveer, D.G. and Sparkes, B.G.
1982: Regional Quaternary mapping — an aid to mineral exploration; in Prospecting in Areas of Glaciated Terrain 1982, P.H. Davenport (ed.); Canadian Institute of Mining and Metallurgy, Geology Division Publication, p. 284-299.

Vaníček, p. and Nagy, D.
1981: On the compilation of the map of contemporary vertical crustal movements in Canada; Tectonophysics, v. 71, p. 75-86.

Veillette, J.J. and Nixon, F.M.
1982: Saprolite in the Big Bald Mountain area, New Brunswick; in Current Research, Part B, Geological Survey of Canada, Paper 82-1B, p. 63-70.

Vernal de, A. et Mott, R.J.
1986: Palynostratigraphie et paléoenvironnements du Pléistocène supérieur dans la région du lac Bras d'Or, île du Cap-Breton, Nouvelle-Écosse; Journal canadien des sciences de la terre, vol. 23, p. 491-503.

Vernal de, A., Causse, C., Hillaire-Marcel, C., Mott, R.J., and Occhietti, S.
1986: Palynostratigraphy and Th/U ages of upper Pleistocene interglacial and interstadial deposits on Cape Breton Island, eastern Canada; Geology, v. 14, p. 554-557.

Vernal de, A., Richard, P.J.H., et Occhietti, S.
1983: Palynologie et paléoenvironnements du Wisconsinien de la région de la baie Saint-Laurent, île du Cap-Breton; Géographie physique et Quaternaire, vol. 37, p. 307-322.

Vincent, J-S.
1989: Quaternary geology of the southeastern Canadian Shield; in Quaternary Geology of Canada and Greenland, R.J. Fulton (ed.); Geological Survey of Canada, Geology of Canada, no. 1 (also Geological Society of America, The Geology of North America, v. K-1).

Waitt, R.B. Jr.
1981: Radial outflow and unsteady retreat of late Wisconsin to early Holocene icecap in the northern Long Range Upland, Newfoundland; Geological Society of America, Bulletin, v. 92, p. 834-838.

Wang, C., Ross, G.J., Gray, J.T., and Lafrenière, L.B.
1982: Mineralogy and genesis of saprolite and strongly weathered soils in Appalachian region of Canada; Maritime Sediments and Atlantic Geology, v. 18, p. 130-138.

Wang, C., Ross, G.J., and Rees, H.W.
1981: Characteristics of residual soils developed on granite and of the associated pre-Wisconsin landforms in north-central New Brunswick; Canadian Journal of Earth Sciences, v. 18, p. 487-494.

Wanless, R.K., Stevens, R.D., Lachance, G.R., and Delabio, R.N.
1978: Age determinations and geological studies, K-Ar isotopic studies, Report 13; Geological Survey of Canada, Paper 77-2, 60 p.

Wells, E.D. and Pollett, F.C.
1983: Peatlands; in Biogeography and Ecology of the Island of Newfoundland, G.R. South (ed.); Dr. W. Junk Publishers, The Hague, p. 207-265.

Welsted, J.
1976: Post-glacial emergence of the Fundy Coast: An analysis of the evidence; Canadian Geographer, v. 20, no. 4, p. 367-383.

White, W.A.
1972: Deep erosion by continental ice sheets; Geological Society of America, Bulletin, v. 83, p. 1037-1056.

Wickenden, R.T.D.
1941: Glacial deposits of part of northern Nova Scotia; Royal Society of Canada, Transactions, v. 35, p. 143-149.

Widmer, K.
1950: Geology of the Hermitage Bay area, Newfoundland; unpublished PhD thesis, Princeton University, 459 p.

Wightman, D.M.
1980: Late Pleistocene glaciofluvial and glaciomarine sediments on the north side of Minas Basin, Nova Scotia; unpublished PhD thesis, Dalhousie University, Halifax, Nova Scotia, 426 p.

Wightman, D. and Cooke, H.B.S.
1978: Postglacial emergence in Atlantic Canada; Geoscience Canada, v. 5, no. 2, p. 61-65.

Williams, H.
1979: Appalachian Orogen in Canada; Canadian Journal of Earth Sciences, v. 16, p. 792-807.
in press: The Appalachian/Caledonian Region: Canada and Greenland; Geological Survey of Canada, Geology of Canada, no. 5 (also Geological Society of America, The Geology of North America, no. 5, v. F-1).

Wright, W.J.
1985: Aggregate resources, Cape Breton Island; Nova Scotia Department of Mines and Energy, Maps 85-3, 85-4, 85-5, 85-6.

Author's Address
D.R. Grant
Geological Survey of Canada
601 Booth Street
Ottawa, Ontario K1A 0E8

Chapter 6

QUATERNARY GEOLOGY OF THE QUEEN ELIZABETH ISLANDS

Summary

Introduction — *D.A. Hodgson*

Surficial materials — *D.A. Hodgson*

Quaternary stratigraphy and chronology — *D.A. Hodgson*

Northeast Ellesmere Island — *J. England and J. Bednarski*

Queen Elizabeth Islands glaciers — *R.M. Koerner*

Acknowledgments

References

An outlet glacier on the west side of an unnamed ice cap lying between Agassiz and Prince of Wales ice caps in central Ellesmere Island, July 1978. The ice is unusually free of debris, even for a glacier at this latitude where limited basal melting occurs. Debris probably is present at the glacier sole which is obscured by a snowbank. Courtesy of D.A. Hodgson. 204621-I.

Chapter 6

QUATERNARY GEOLOGY OF THE QUEEN ELIZABETH ISLANDS

Chapter Co-ordinator
D.A. Hodgson

SUMMARY

The physiography of the Queen Elizabeth Islands was developed during episodes of subaerial erosion and planation, and relatively recent rifting may have formed interisland channels. Marine erosion has been a relatively minor facet in physiographic development and glacial processes have had only local effect.

The dominant surficial material is weathered and colluviated bedrock, which varies in character depending on the lithology and hardness of the source material. This thin mantle includes the weathered residue of early or middle Quaternary tills. Identifiable tills and other glacial deposits occur only locally and are generally located near margins of existing ice caps, although they are present to some degree on most islands. Offshore and littoral offlap marine deposits mantle low-lying shores but rarely extend up to marine limit. Glacier ice covers one quarter of the land area.

The processes responsible for erosion of plateau surfaces, deposition of locally preserved pre-Quaternary unconsolidated deposits, and division by interisland channels are not understood. Also unknown is whether tectonism postdated the onset of glaciations in the late Tertiary.

Distribution of erratics suggests that continental ice at one time extended onto Prince Patrick Island and at least as far north as Ellef Ringnes Island, and that Greenland ice overran the northeast margin of Ellesmere Island. During these and other times, ice was possibly generated within the northern archipelago either from an ice complex over the eastern islands, or as ice caps on individual islands. The demonstrated extent of the last glaciation is restricted to an extension of up to 60 km of existing ice caps (referred to as the Franklin Ice Complex). A greater, hypothesized expansion which provided contiguous ice caps and shelves in eastern and southern islands has been named the Queen Elizabeth Islands Glacier Complex. An even more extensive pan-archipelago Innuitian Ice Sheet has been hypothesized, partly on the basis of the Holocene emergence record; however, this emergence has also been explained as due to isostatic depression resulting from the composite load of a small local ice complex, and adjacent Laurentide and Greenland ice. Late Wisconsinan Laurentide ice touched southern shores of southwest islands at least once. Nonsynchronous retreat of the ice margins of the last glaciation occurred between ca. 11 and 7 ka.

The most clearly recorded Quaternary event is the Late Pleistocene/early Holocene marine overlap, which occurred up to a few metres above sea level in the extreme west to more than 150 m a.s.l. in the central and eastern islands. Shells related to this event yielded radiocarbon dates of ca. 11.5 ka in the west to ca. 7 ka in the east. There is no certain record of sea level fluctuations between the high late Quaternary stand and the preceding interstade or interglaciation, when levels were as high or higher. Numerous shell radiocarbon dates of 20-45 ka, or older, record this pre-last glaciation event. Low arctic flora and fauna with ages beyond the range of radiocarbon dating indicate warmer conditions than those at present.

Climate was relatively warm during the early Holocene, and temperatures reached a maximum in the middle Holocene. Cooler climate for much of the period since that time is indicated by growth of glaciers, as well as by physical properties of glacier cores.

INTRODUCTION
D.A. Hodgson

The Queen Elizabeth Islands presently are best known among Quaternary geologists for controversy over the extent of the last glaciation (Paterson, 1977a), although the essence of this debate is not widely understood. Glacial studies have dominated Quaternary geology in the far north, as an inevitable extension of their understandable dominance in continental Canada. Despite the current 25% cover of glaciers and the near perennial snowcover elsewhere, however, indicators of past glaciation only occur sparsely on many islands. One explanation to be considered is that low annual temperatures combined with remoteness from major precipitation sources produced glaciers that had little impact on the landscape. Alternatively, sufficient time may have passed since the last extensive glaciations occurred for much of the drift to have been eroded.

Hodgson, D.A. (co-ordinator)
1989: Quaternary geology of the Queen Elizabeth Islands; Chapter 6 in Quaternary Geology of Canada and Greenland, R.J. Fulton (ed.); Geological Survey of Canada, Geology of Canada, no. 1 (also Geological Society of America, The Geology of North America, v. K-1).

Hodgson, D.A.
1989: Introduction (Quaternary geology of the Queen Elizabeth Islands); in Chapter 6 of Quaternary Geology of Canada and Greenland, R.J. Fulton (ed.); Geological Survey of Canada, Geology of Canada, no. 1 (also Geological Society of America, The Geology of North America, v. K-1).

Correlation of the scattered glacial indicators is hindered by fragmentation of the region into numerous islands, which total 420 000 km^2, separated by deep channels, covering about 250 000 km^2 (Fig. 6.1). Fortunately, this physiography combined with glacial isostatic sea level changes has resulted in extensive raised marine deposits, and it is from these that much of the known Quaternary record is drawn. Reconstruction of Quaternary events from other nonglacial materials and landforms, including weathered deposits (chiefly derived from bedrock), fluvial deposits, and ground ice is still at an elementary level. Cores from the relatively thick and static, long-lived ice caps reveal proxy data on last glacial and interglacial climates.

Observations on Quaternary deposits of the Queen Elizabeth Islands, which have been published since the early nineteenth century (Washburn, 1947), anticipate the modern controversy. Whereas estimates of the extent of past glaciation have varied from complete inundation to a cover no greater than at present, sea level change, and in particular Holocene regression, has been readily recognized.

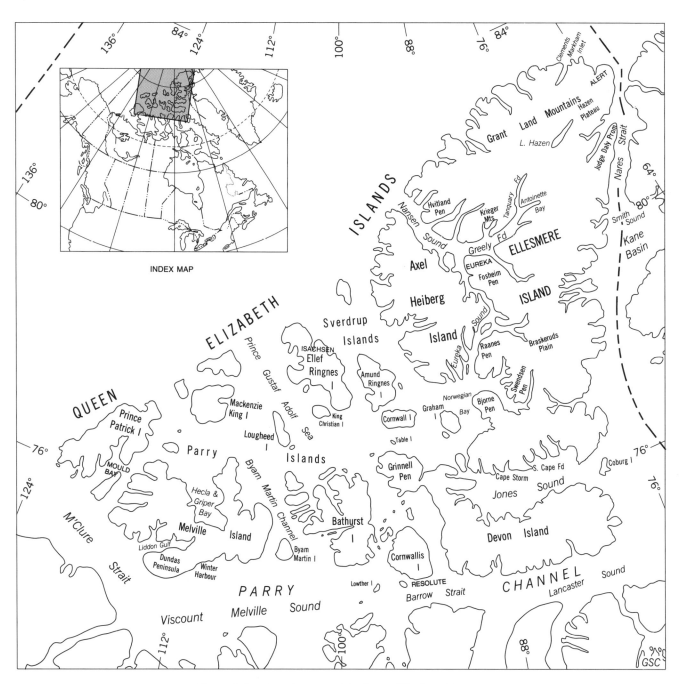

Figure 6.1. Location map of the Queen Elizabeth Islands.

The earlier concept of glaciation of the eastern islands with only local ice caps in central and western islands resulted from the paucity of well preserved glacial depositional and erosional landforms, especially in the central and western islands (Washburn, 1947; Jenness, 1952; Taylor, 1956; Wilson et al., 1958; Bird, 1959; Craig and Fyles, 1960, 1965; Fyles in Jenness, 1962; Boesch, 1963; Roots, 1963; Blake, 1964; Tozer and Thorsteinsson, 1964; Christie, 1967; Hattersley-Smith, 1969a). Subsequently, from the increased body of data on elevation and radiocarbon-dated age of marine limit, the assumption was made that the entire region of isostatic emergence had been subjected to an earlier ice load (Craig and Fyles, 1960; Farrand and Gajda, 1962; Blake, 1964; Andrews, 1970; Prest, 1970; Walcott, 1972). This, together with other factors described later, led Blake (1970, 1972, 1975) to propose that during the last glaciation much of the Queen Elizabeth Islands were covered by the Innuitian Ice Sheet. Blake did not portray ice sheet limits; however, a speculative margin of the last ice complex encompassing all the Queen Elizabeth Islands was drawn by Prest (1969) and Mayewski et al. (1981) and this is commonly equated with Blake's Innuitian Ice Sheet. A plausible model of the topography of such an ice sheet was described by Reeh (1984).

Intensive work by England and others in northern Ellesmere Island (England, 1976a, b, 1978, 1983; Bednarski, 1986; Retelle, 1986a) countered the "maximum" model. England believed that at the last glacial maximum separate ice caps existed on the eastern islands while a higher sea level caused by glacial isostatic depression overlapped adjacent land in the archipelago, with much of the depression due to adjacent Laurentide and Greenland ice sheets. Paterson (1977a) compared the full ice cover and partial ice cover models and drew on ice core studies from Greenland for additional data. Notwithstanding local studies, for the northern archipelago there is still insufficient evidence (Hodgson, 1985) for models of late Quaternary ice cover to be anything more than speculative.

Pre-Quaternary geology

The Queen Elizabeth Islands archipelago is a subplate, bounded by major shear zones, that was broken by rifting in the Late Cretaceous and Tertiary. Some interisland channels occupy grabens (Fig. 6.2; Kerr, 1980, 1981). The islands are largely underlain by the deformed sediments of the Innuitian Province, lying between sediments related to the modern Arctic Ocean basin in the northwest and the stable Arctic Platform flanked by the northernmost Canadian Shield in the southeast; the Arctic Coastal Plain and Arctic Platform are relatively undeformed compared to sediments of the Franklinian Geosyncline, Hazen Trough, and eastern Sverdrup Basin (Fig. 6.2; Trettin, in press).

Common lithologies include shale, anhydrite, siltstone, sandstone, conglomerate, limestone, and dolomite, as well as shield metamorphic rocks and mafic dykes and sills. Clastic sediments vary locally and regionally from uncemented to fully lithified, and on this basis Sverdrup Basin is divided here into an eastern half of generally lithified sediments and a western half of dominantly weakly cemented sediments.

Lithologies that produce useful indicator erratics include the crystalline and metamorphic rocks that only occur on and adjacent to the northernmost Canadian Shield rocks of southeastern Ellesmere and Devon islands, apart from some plutons on northernmost Ellesmere and Axel Heiberg islands, which outcrop sporadically over 250 km^2 (Trettin et al., 1972). Ice could, however, carry shield lithologies from south of Parry Channel, where they outcrop east of 96°W, and from one locality on Victoria Island. Carbonate rocks outcrop widely in the south and east, declining in age, frequency of occurrence and degree of cementation towards the northwest.

Physiography

Taylor (1956), Dunbar and Greenaway (1956), Roots (1963), and Tozer and Thorsteinsson (1964) divided much of the region into physiographic units, which are generalized in Bostock (1970). Figure 6.3, which shows the general physiographic subdivisions, indicates that both elevation and relief decline towards the northwest.

Present topography was shaped by folding and thrusting of basin sediments, by rifting, by subaerial erosion, and to a lesser extent by glacial and marine erosion. The only large body of relatively undisturbed sediments underlies the Arctic Coastal Plain.

To the west of the northernmost shield, successive basins filled with sediments were deformed parallel to the present northeast-southwest topographic grain. While deformational events continued late into the Cenozoic, initial fragmentation by rifting of the northernmost North American Plate into an archipelago possibly had started by the late Paleozoic (Kerr, 1981).

Peneplains and possibly etchplains or pediplains, in many cases exhumed, occur widely but are particularly evident on Melville, central Devon, eastern Axel Heiberg, and western Ellesmere islands (Tozer and Thorsteinsson, 1964; Prest, 1970; Trettin et al. 1972). The oldest and highest erosion surface is recorded by summits on the northernmost shield, which accord with the 1500-1800 m Penny Surface of eastern Baffin Island. This surface is suggested by Bird (1967) to be pre-Proterozoic in age. Bird (1959, 1967) tentatively correlated several lower Phanerozoic denudation surfaces of the southern archipelago and continental mainland with plateaus and accordant summit elevations north of Parry Channel. The youngest peneplain which bridges present interisland channels underlies the Miocene(?) Beaufort Formation in the northwest, and equivalent unnamed beds in eastern Axel Heiberg and western Ellesmere islands (Balkwill and Bustin, 1975; Bustin, 1982). This surface now lies 150-700 m a.s.l.; adjacent channels are 200-500 m deep. Similar flat-lying unconsolidated sediments truncate older strata on eastern Bathurst and southern Ellef Ringnes islands and occur on higher parts of Byam Martin, Cornwall, Graham, and King Christian islands. Deposition clearly predated development of interisland channels (Thorsteinsson and Tozer, 1970).

The relative roles of rifting, fluvial erosion and glacial scouring in formation of interisland channels is not clear. Widespread rifting caused at least local downdrop of present channels (Daae and Rutgers, 1975; Kerr, 1980, 1981) and possibly explains the segments of remarkably straight cliffed coastline locally intersected by hanging valleys (Somerset, southern Devon, southwest Melville islands). Fortier and Morley (1956) suggested that the channels were eroded by rivers flowing from an Arctic/Atlantic watershed, and detailed bathymetry from the divide at the middle of Parry Channel enabled Bornhold et al. (1976) to reconstruct a finely developed drainage system with a pattern characteristic of present day continental basins, undisturbed by glaciation. This, together with the preservation of unconsolidated Tertiary sediments on summits, plateaus, and valley bottoms, indicates that widespread glacial scouring has not occurred in the channels. However studies of the northwest and eastern island margins summarized by Pelletier (1966) and of Nansen Sound by Hattersley-Smith (1969a) suggest deepening and widening of trunk fluvial valleys by glaciers. The role of glaciers in shaping these is indicated by rock sills or morainal barriers which occur at the ocean margins of these troughs. Shoreline processes associated with late Cenozoic and Quaternary sea level fluctuations are likely responsible for cutting gently inclined aprons which extend from 50-100 m to below present sea level around many islands.

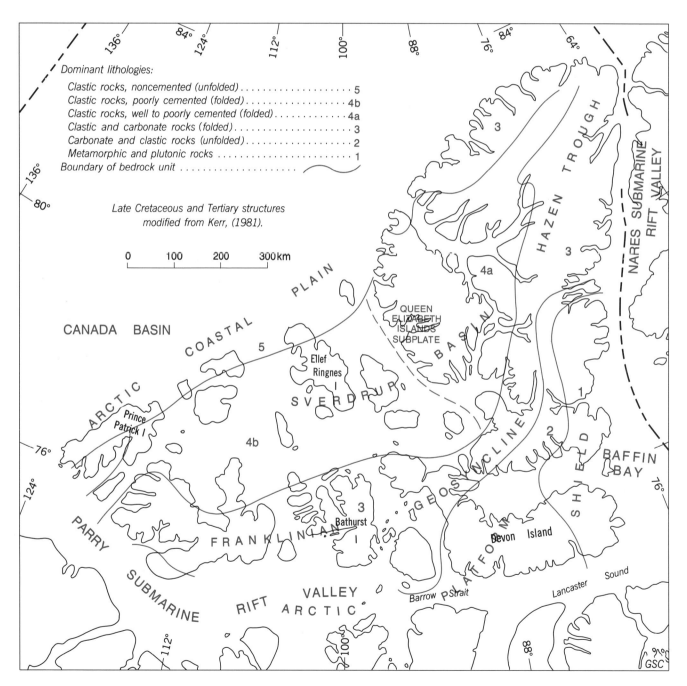

Figure 6.2. Pre-Quaternary geology, structural provinces of Queen Elizabeth Islands, and late Cretaceous and Tertiary structures.

Climate and vegetation

Much of the region lies under the influence of an anticyclonic air mass centred in the central Arctic Ocean, off the northwest islands, whereas the southeast is greatly influenced by cyclonic activity associated with Baffin Bay. The mountains of the eastern islands lie in a transitional zone where in summer the cool northwest air flow is blocked, and lee-side katabatic winds moderate temperatures (Maxwell, 1980; Edlund and Alt, 1989).

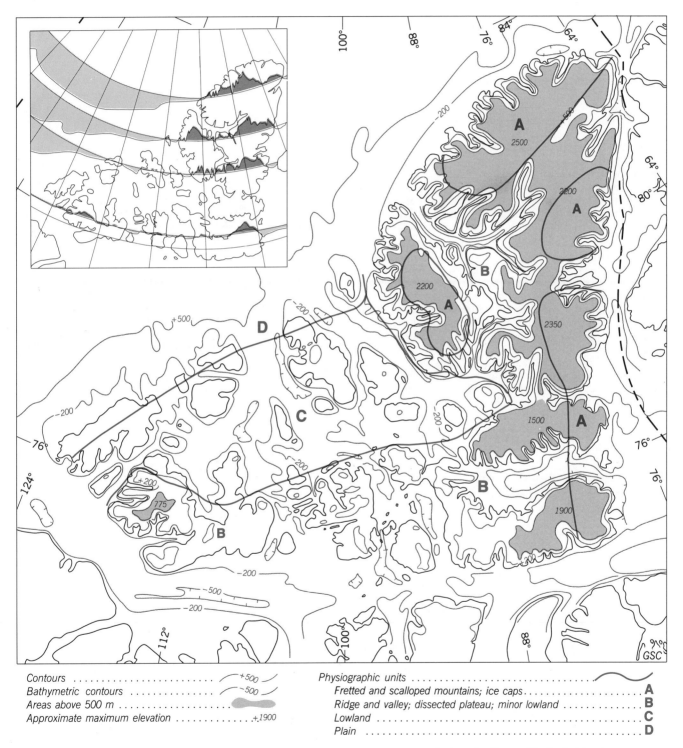

Contours	+500
Bathymetric contours	-500
Areas above 500 m	
Approximate maximum elevation	+1900

Physiographic units	
Fretted and scalloped mountains; ice caps	A
Ridge and valley; dissected plateau; minor lowland	B
Lowland	C
Plain	D

Figure 6.3. Physiographic units and topography (terrestrial and submarine) of the Queen Elizabeth Islands (after GEBCO 5-17, Canadian Hydrographic Service, 1979); inset map shows four profiles across the region (from National Atlas of Canada, 1973).

Annual mean daily temperatures range from -16° to -19°C; the mean daily maximum for the warmest month (July) is 7° to 8.5°C and the extreme maximum 15° to 22°C. All of the northern archipelago, except the southeast, is characterized by extreme aridity with an annual net water balance close to zero (Bovis and Barry, 1974). Total precipitation is 60-160 mm of which about 40% is rain at most stations; however, the greatest 24 hour rains of up to 50 mm can catastrophically unbalance normal geomorphic processes (Cogley and McCann 1976; Hodgson, 1982). It must be noted, however, that weather stations are only in coastal areas and that no climatic data are available for the southeast part of the area.

Permafrost is universally present under land, and has been recorded more than 250 m thick even at the present shoreline (Taylor et al., 1983). The short thaw season and arid climate restrict surface water flow and ponding. Nevertheless, segregated ice is commonly present in the upper few metres of unconsolidated (and, in places, consolidated) deposits, and massive ground ice is present in frost fissures, in rare pingos (Pissart, 1967; Balkwill et al., 1974; Hodgson, 1982), and in palsas in the scattered thin organic cover (Washburn, 1983).

The northernmost limits of a number of major vascular plant families and species (dwarf woody plants, particularly willow and heath, and sedges; Edlund, 1983) presently cross the Queen Elizabeth Islands. They are sensitive indicators of climatic change. Zones of vegetation, impoverished to the northwest, follow a "reversed L" pattern: zonal across western and southern islands and meridional around the mountainous eastern islands which block the northwesterly airflow (Edlund and Alt, 1989). Though subordinate to climatic effects, surficial materials influence species distribution, especially strong alkalinity associated with weathered carbonate rock, strong acidity of some shales, and salinity of fine grained emerged marine deposits (Edlund, 1983). Phytogeography is complicated by the possibility that refugia existed within the eastern islands during the last glaciation (Brassard, 1971).

Geotechnical considerations

Permafrost and the seasonal active layer (Heginbottom, 1989) control land use activities, though effects of thaw are mitigated by low annual precipitation. Regional factors which must be taken into account when assessing geotechnical applications include a general fining of deposits towards the west, and the presence of ice caps in the east. Most reports describing geotechnical conditions are consequences of energy development and transportation feasibility studies (Stangl et al., 1982) and of surficial geology and biophysical mapping projects (Barnett et al., 1977; Hodgson, 1982).

A major concern in assessing the effects of land use is the disturbance of permafrost. The most common sequence of man-initiated disturbance of permafrost is disruption of surface drainage, initiation of thermal erosion and slope failures. Sensitivity of different materials to these processes has been assessed by Babb and Bliss (1974), Barnett et al. (1975), Hodgson (1982), Hodgson and Edlund (1975, 1978), and Kurfurst and Veillette (1977).

A variety of other geomorphological processes not related to man's activities may place the integrity of a structure at risk. Unpredictable fluvial processes that may be damaging can be triggered by heavy rain (Cogley and McCann, 1976) and collapse of ice or snow dams (Ballantyne and McCann, 1980; Heginbottom, 1984). Natural slope failures within the active layer are common (Hodgson, 1982; Heginbottom, 1984; Mathewson and Mayer-Cole, 1984). A large rock failure has been reported only from Cape Hotham, Cornwallis Island (Thorsteinsson, 1958). Other potentially destructive processes include eolian erosion, abrasion or deposition, corrosion related to drainage from acidic rock units (Hodgson, 1982), and sea ice pushing onshore (Taylor and McCann, 1983).

Earthquake epicentres are clustered under Byam Martin Channel and northwest Baffin Bay (Basham et al., 1977). Although no displacement due to seismicity or to differential Holocene uplift has been reported, the possibility of recent tectonic uplift of Lougheed Island has been discussed (Hodgson, 1981).

Resources

Quartzitic sand which could be used as aggregate is widely distributed in fluvial, deltaic, and littoral deposits, and as unconsolidated beds within some Mesozoic and Cenozoic rock formations. Gravel is common in fluvial, glaciofluvial, and high-energy beach deposits of the eastern and southern islands. Rock suitable for crushing has a similar distribution. Aggregate sources are rarer in the central and western islands, where in some locations a lag cover on fine grained sediments has been scraped from large areas to provide for local needs. Few natural reservoirs of unfrozen water exist year round. For many rivers, especially in the west, snowmelt is the most significant water source and this runoff occurs as a brief turbid early summer flow.

SURFICIAL MATERIALS
D.A. Hodgson

Surficial materials of the Queen Elizabeth Islands consist predominantly of weathered or poorly lithified rock but in some areas variable thicknesses of glacial and nonglacial Quaternary sediments are present.

Weathered rock

Much of the surficial material of the Queen Elizabeth Islands is weathered rock derived by physical disaggregation of bedrock. Mechanisms of weathering and transport in this cold arid climate are not well understood; a number of contributing processes are reviewed by French (1976) and Washburn (1980). Although physical disaggregation by hydration shattering is believed by Hudec (1973) to be the dominant weathering process, disaggregation also is promoted by the growth of segregated ice widely present below the frost table.

Hodgson, D.A.
1989: Surficial materials (Queen Elizabeth Islands); in Chapter 6 of Quaternary Geology of Canada and Greenland, R.J. Fulton (ed.); Geological Survey of Canada, Geology of Canada, no. 1 (also Geological Society of America, The Geology of North America, v. K-1).

Thickness of weathered rock, recorded in boreholes, is commonly several metres, which greatly exceeds the present thickness of the active layer (10-80 cm). This raises the question of whether this physical disintegration occurred below the active layer or if the thicker deposits of weathered rock formed when the active layer was thicker. Chemical weathering causes surface etching of carbonate and (especially) evaporite rocks, and Watts (1983) suggested that microfracturing by salt crystallization occurs in crystalline rocks of the southeast, which are influenced by weather systems carrying moisture from Baffin Bay. Aridity elsewhere causes salt efflorescence on the surface of some Cretaceous shales (Christie, 1967). Pedogenic processes rarely develop sufficiently for soil classes to be recognized, though Tarnocai (1976) described frost-churned (Cryosolic) soils from imperfectly or poorly drained areas of Bathurst Island, and Tedrow (1977) identified polar soil groups at sites not mixed by cryoturbation. At drier sites on Ellef Ringnes Island no clear evidence of soil development was found by Foscolos and Kodama (1981) and Hodgson (1982).

Size of clasts in the weathered mantle is largely dependent on the strength of rock cementation. Weakly cemented rock disaggregates to sand and silt whereas strongly cemented rock breaks into pebble to boulder size fragments. Where surface processes are intense, fine particles may accumulate separately from blocks, for example in the centre of frost-sorted cells, and at the base of mass wasting slopes. Fines, largely composed of silt and fine sand-clay size particles, are normally a minor component. Other characteristics of underlying rock which can be clearly recognized in the weathered mantle include alkalinity, shown by the nature of vegetation over the Paleozoic carbonates (Edlund, 1983), and corrosive pH levels found in surficial materials derived from some Sverdrup Basin sediments (Hodgson, 1982). Little data exist on rates of weathering or the absolute age of weathered mantle, and as in other areas peripheral to Late Wisconsinan Laurentide ice (Vincent, 1989), the relationship between extent of ice and degree of bedrock weathering is problematic.

Of particular significance is whether different weathering zones coincide with different ages of former ice coverage, or whether these zones mark boundaries between parts of a single ice mass with differing erosional capacity. Weathering zones have been described from a number of areas on Ellesmere Island (Troelsen, 1952; Fyles in Jenness, 1962; Hattersley-Smith, 1969a) and Axel Heiberg Island (Boesch, 1963). England and Bradley (1978) described at least four weathering zones in northeastern Ellesmere Island which range in age from pre-Quaternary to the last glaciation and do correlate with the age of glaciations. A problem with interpreting weathering zones as marking the limits of previous glaciations was raised by Blake (1978), Watts (1983), and Hodgson (1985) who showed that the last ice sheet did not disturb large areas of weathered mantle while scouring deeply elsewhere.

Established rock stratigraphic formations commonly provide the most convenient and detailed framework for describing weathered rock (Hodgson, 1982). Nevertheless, many mapped formations are composed of different lithologies which cannot be separated on a surficial material map unless they outcrop over a wide area, as for example do the flat-lying sediments of the Arctic Platform (see Hodgson and Vincent, 1984b). Despite redistribution of weathered rock by mass movement and fluvial processes, bedrock unit contacts commonly stand out on airphotos and satellite images, owing to induced variations in vegetation, moisture and microrelief, as well as the albedo of the weathered rock.

Glacial deposits

Much, if not all, of the Queen Elizabeth Islands were glaciated during the Pleistocene, yet glacial deposits are sparse on most of the landmass (Fig. 6.4). This is in marked contrast to islands south of Parry Channel where glacial deposits are relatively abundant (Vincent, 1983; Dyke, 1983). Part of the explanation of this difference may be that Late Pleistocene glacier systems north of the channel differed fundamentally from those to the south, generally leaving either no glacial deposits, or a thin discontinuous till almost indistinguishable from weathered rock. It may also be that thin drift deposits of the Queen Elizabeth Islands are relict, left after subaerial erosion of an archipelago-wide till sheet predating the last glaciation. The ages and the styles of these possible earlier glaciations remain unknown (England, 1985). It is, however, known that the bulk of glacial deposits lie in those parts of the eastern and southern islands proven to have been deglaciated during the early Holocene. Small but prominent glacial deposits flank modern ice sheet margins, which have fluctuated through the Holocene (Blake, 1981b).

Till

Clearly identifiable till generally occurs as a veneer (<2 m), devoid of constructional landforms. Discontinuous lateral moraines of till were left by valley glaciers on Ellesmere and Axel Heiberg islands, whereas associated terminal moraines are largely constructed of glaciofluvial and glaciomarine sediments. End moraines are rare elsewhere.

Till texture is clearly related to upstream rock units where ice flowed overland. Where till was deposited by ice confined within a channel, it commonly incorporates fine grained marine deposits (southwest shore of Greely Fiord, northeast Amund Ringnes Island, south shores of Melville and Byam Martin islands). In some areas adjacent to channels, glacially transported Quaternary shell fragments have been reported to at least 600 m elevation (Sim, 1961).

Erratics in the tills may indicate the general flow direction of the depositing ice. Crystalline erratics, deposited by Greenland ice, are located on northern Ellesmere Island east of Lake Hazen (Christie, 1967; England and Bradley, 1978). On a more local scale, Sverdrup Basin Permo-Carboniferous sandstones and conglomerates, which outcrop as nunataks in the Grant Land Mountains of northern Ellesmere Island, have been deposited by past glacial action over the Hazen Plateau to the southeast. On all the larger central and western islands, erratics from south of Parry Channel are recorded from above the highest known marine limit (Fortier et al., 1963; Tozer and Thorsteinsson, 1964; St-Onge, 1965; J.G. Fyles, Geological Survey of Canada, personal communication, 1986). Furthermore, tills of southern provenance are recognized on southern Melville Island by their markedly calcareous nature (similar to tills on Victoria Island), whereas local rock lithologies are noncalcareous. Oddly enough, however, shield rock erratics have not been reported immediately north of the northernmost shield on the eastern islands, though gneiss and granite erratics do occur up to at least 300 m a.s.l. between the Canadian Shield and southern Eureka Sound (Tozer, 1963), and at higher elevations immediately east of the shield (Christie, 1967).

CHAPTER 6

Glaciofluvial deposits

Esker-like landforms occur on central and western islands whereas eskers are extremely rare in the eastern islands. Linear to slightly sinuous ridges of silty gravel 5 to 20 m or more thick, 50 to 500 m wide, and up to 15 km long occur on Amund Ringnes, Bathurst, Ellef Ringnes, King Christian, Lougheed, and northwest Melville islands (Roots, 1963; Blake, 1964; St-Onge, 1965; Prest et al., 1968; Stott 1969; Hodgson, 1982). No internal structure has been observed, or sense of direction of sediment transport determined.

Sandurs and kame deltas were deposited during still-stands and retreats of major ice lobes or valley glaciers in the eastern islands (Hattersley-Smith and Long, 1967; Hodgson, 1973, 1985; England, 1974). A belt of extensive degraded sandurs in northeast Axel Heiberg Island (Fig. 6.4) is undated, as are massive kames at the distal margin of Dundas Till on Melville Island (Hodgson et al., 1984).

Erosional evidence of meltwater in the form of marginal drainage channels is much more widespread than deposits. Flights of channels are not only well preserved adjacent to

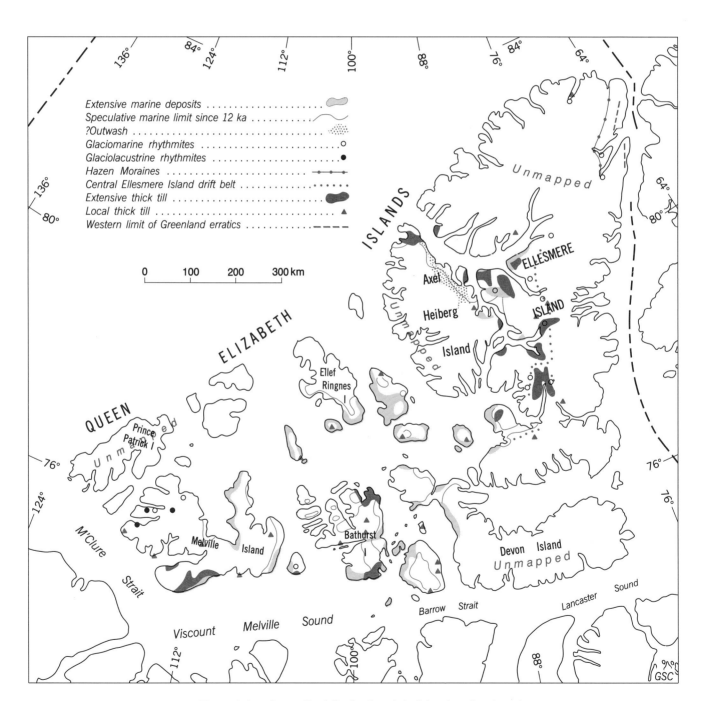

Figure 6.4. Generalized distribution of glacial and marine deposits.

modern ice sheets of the eastern islands, but undated ones are also found to elevations of 600 m adjacent to Eureka Sound, as well as on Devon Island, Cornwallis Island (Edlund, in press), Bathurst Island (Blake, 1964), and Melville Island (Hodgson et al., 1984). In all cases these channels indicate ice retreat inland. Unequivocal meltwater channels have not been reported from the Ringnes and adjacent islands.

Glacial marine deposits

Marine deposits that can be associated with glaciation are few and widely scattered (Fig. 6.4) but where they do occur they are of great significance because radiocarbon dates on any contained molluscs could provide ages for glacial events. Silt or fine sand deposits, locally 50 m thick, consisting of 1-50 cm-thick beds are commonly found abutting or a short distance distal to prominent end moraines or kame deltas in central and northeast Ellesmere Island (Figure 6.4; England, 1974, 1978; Hodgson, 1973, 1985). Banded sediments on western Byam Martin Island were deposited on the distal margin of Laurentide ice whereas isolated rhythmites on Cornwall and northwest Melville islands are likely glacial lacustrine in origin (Hodgson et al., 1984). Marlowe (1968) suggested that laminated sediments below a thin structureless bed in cores from Prince Gustav Adolf Sea and the environs of Lougheed Island resulted from glacial sedimentation into a sea of reduced volume and restricted circulation.

Nonglacial deposits

Marine deposits are the only nonglacial surficial materials of sufficient areal distribution to show on Figure 6.4. Fluvial deposits are relatively abundant but occur as thin separated deposits along river channels. Organic deposits and wind formed features are minor nonglacial elements.

Marine deposits

Marine sediments occur from maximum elevations of only a few metres in the extreme west to over 100 m above sea level on many of the central and eastern islands (Fig. 6.4). Large parts of the islands were submerged during the Late Pleistocene and early Holocene, leaving a thin cover of sediment which was partly removed by subsequent erosion. In general, maximum ages of molluscs from these sediments decrease from ca. 11 ka in the west to ca. 9 to 7 ka in some inner fiords of Ellesmere Island.

Over much of this region marine limits do not correspond with the highest elevation of inorganic littoral sediments. Marine shells commonly are found higher than other identifiable marine deposits from the same event. In other cases, halophytic vegetation may be the only remaining evidence that the sea once stood higher than the level of remaining marine sediments or marine shells (Barnett, 1972). Onset of beach development may not have occurred when the sea stood at its maximum level but may have coincided with a slowing of initially rapid emergence, or with seasonal removal of the land-fast sea ice (Bird, 1967, p. 118). Under present conditions in the northwest islands, year round ice cover, combined with a paucity of coarse material, restricts beach development (McLaren, 1982; Taylor and McCann, 1983). The size of raised beach berms generally increases in a southeastward direction, following the increase in wave energy which is related to the increased length of the summer ice-free period.

The normal sequence of offlap sediments resulting from emergence of this region is silt or clay-silt locally 30 m thick, overlain by sand or gravel littoral deposits, locally more than 10 m thick near deltas. This sequence occurs where the sediment source is of mixed grain size. Where the underlying and inland rock lithology is all coarse or all fine grained, however, all offlap sediments from deep water to shallow water facies may have similar characteristics. Transgressive marine deposits, which might be expected to have been deposited during the isostatic downwarping associated with the last glaciation, and at least locally preserved in unglaciated areas, have not been reported.

Fluvial deposits

Fluvial deposits are most extensive on the coarse grained bedrock of the low-lying northwest islands. In this area networks of wide, flat-floored channels have been developed by the early summer snowmelt. In the higher relief areas of the southern and eastern islands, streams are more confined, emergence has been greater, and terraces and incised channels are common. In general, the relatively short stream lengths combine with high snowmelt to flush finer sediment out of the systems, leaving only coarser material on floodplains and terraces. Many glacial regime rivers are currently constructing sandurs. High Arctic hydrological regimes are discussed by McCann et al. (1972), French (1976), and Washburn (1980).

Prograding (river-dominated) deltas are readily constructed in the low energy shore environment. This process, combined with Holocene emergence, commonly leaves a ribbon of deltaic sediment from marine limit to present shore, which becomes incised by the modern river. The raised delta surface has an unbroken rise inland in areas where shorelines are characterized by low energy. On the other hand, a step-like progression of raised deltas is common where waves and currents influence channel mouths in areas of greater shoreline energy towards the southeast. Composition of these deltas is varied and dependent on sediment supplied by drainage basin, local stream flow characteristics and shoreline energy levels. Drainage is impeded on terraces, which thus become particularly subject to growth and thaw of ice-filled frost fissures. Extreme development of thermokarst terrain on some deltas on Ellef Ringnes and King Christian islands suggests that these deposits are older than Holocene.

Eolian deposits

Wind action is significant, particularly on and adjacent to sandurs and emerged coastal plains (St-Onge, 1965; French, 1976; Pissart et al., 1977; Hodgson, 1982). Evidence of wind erosion is widespread but deposits, such as dunes, occur only adjacent to major sources of sand. Erosional eolian features include desert lag gravel and ventifacts.

Organic deposits

Extensive (square kilometres) though thin (<10 cm) layers of accumulating vegetation are restricted to terrain that is low-lying, has impeded drainage, and contains suitable nutrients for relatively prolific vegetation growth. This includes central Bathurst Island (Polar Bear Pass), the eastern shores of Norwegian Bay, and central Fosheim Peninsula. Smaller areas commonly occur at heads of fiords and bays, even where surrounding terrain is barren of vegetation. Scattered peat deposits several metres thick commonly date less than 9 ka. Peat predating the last glacial stade has also been exposed at a number of sites in the Queen Elizabeth Islands (Blake, 1972, 1974, 1982).

QUATERNARY STRATIGRAPHY AND CHRONOLOGY

D.A. Hodgson

Pleistocene and early Holocene chronostratigraphic studies have been most successful in the northeast and southwest, where extra-archipelago ice left identifiable deposits which interfinger with marine and local glaciogenic deposits. These two areas are described separately from the remainder of the Queen Elizabeth Islands.

Pre-last glaciation

Tundra forest grew in coastal regions of the Arctic Ocean in the Pliocene (Funder et al., 1985; Matthews, 1987). Ice-rafted debris in Pliocene sediments of the Arctic Ocean (Herman and Hopkins, 1980), in raised marine beds in North Greenland (Funder et al., 1985), and in the Atlantic Ocean (Berggren, 1972) indicates inception of glaciers prior to the Quaternary, probably in Alaska, Baffin Island, and Greenland, as well as in Ellesmere Island.

Geological evidence suggests that the glaciers that overran the Queen Elizabeth Islands an unknown number of times during the Quaternary and late Tertiary were thickest and most extensive prior to the last glaciation. Craig and Fyles (1960) suggested that the shield erratics, which are found on most islands and which are the principal evidence for extensive ice covers in the past, were distributed by an Ellesmere-Baffin glacier complex. It is my opinion, however, that the uniform sparsity of these erratics over the entire area points to north flowing continental ice from the distant mainland shield as the source. Vincent et al. (1984) suggested that there was only one advance of continental ice strong enough to cover significant parts of the western Queen Elizabeth Islands and that it was coeval with the Banks Glaciation and probably occurred during the Matuyama Reversed Epoch (early Quaternary). The eastern islands on the other hand may not have been covered by continental ice even during this period of extensive glaciation and instead may have maintained independent though at times coalescent ice caps (Craig and Fyles, 1960, Fig. 1). Northeast Ellesmere Island was invaded more than once by Greenland ice (Christie, 1967; Prest et al., 1968). England and Bednarski (this chapter, 1989) suggest a maximum overlap of 100 km from Nares Strait.

Hodgson, D.A.
1989: Quaternary stratigraphy and chronology (Queen Elizabeth Islands); *in* Chapter 6 of Quaternary Geology of Canada and Greenland, R.J. Fulton (ed.); Geological Survey of Canada, Geology of Canada, no. 1 (*also* Geological Society of America, The Geology of North America, v. K-1).

Last interglaciation or interstade

Numerous collections of emerged marine shells and fewer samples of terrestrial plant material have yielded ^{14}C dates either between 20 and 45 ka or beyond the range of the ^{14}C dating. It is prudent to treat the finite dates as minimums (Blake, 1974). Published amino acid ratios (free fraction) for these samples cluster between 0.15 and 0.3 (Blake, 1980; England et al., 1981; Washburn and Stuiver, 1985). None of these samples were collected between deposits representing sequential glacial advances, or from deposits associated with buried weathering horizons or soils, although some likely indicate a warmer climate than at present.

In several places in situ shell samples were collected from beneath diamicton containing striated clasts. Shells and marine algae from this position on southern Ellesmere Island dated 35-40 ka (Blake, 1980). Thick peat lying in an ice marginal channel and covered by till in south-central Ellesmere Island dated >52 ka (Blake and Matthews, 1979; Blake, 1982).

Shell samples dated 20-45 ka or beyond the range of ^{14}C dating have been collected from the surface or within till or diamicton. Some of these are found well above any known marine limit and may predate glacial transport, or they may be the sole relict deposits of an old submergence. This group includes shells from Cornwallis and adjacent islands from 124 to 205 m a.s.l. (Washburn and Stuiver, 1985, Table 1) and southern Ellesmere Island (Blake, 1975, Table 1). Shells >600 m a.s.l. occur on the northeast shore of Eureka Sound (Sim, 1961; Dyck and Fyles, 1964). Secondly, shells >20 ka occur at or below the Holocene marine limit on northern Ellesmere Island (England et al., 1981), northern Axel Heiberg Island (Dyck and Fyles, 1964, p. 179), south-central Ellesmere Island (Hodgson, 1985), southern Ellesmere and Graham islands (Blake, 1975, Table 1), Devon Island (Barr, 1971; Lowdon and Blake, 1973, p. 39), Cornwallis and adjacent islands (Washburn and Stuiver, 1985, Table 1), Bathurst Island (Blake, 1974, Table 2), and Melville Island (Lowdon and Blake, 1968, p. 241). In most cases the mollusc species are representative of present arctic conditions, though *Mytilus edulis* dated >38 ka from Coburg Island has not been found in the Queen Elizabeth Islands at present (Blake, 1973).

Terrestrial organic deposits, chiefly tundra vegetation preserved in peat deposits, have yielded finite and infinite ^{14}C dates >30 ka (Blake, 1974, Tables 1 and 3; Blake, 1982). Some of these deposits include insect fragments and seeds typical of a middle or low arctic environment (Blake, 1982; Hodgson, 1985, p. 351). Sea level during deposition ranged from as low as present to higher than the Holocene maximum.

In summary, insufficient stratigraphic and chronological data are available to tell if ^{14}C dated organic samples >20 ka were deposited during the last interglaciation, in an ice-free embayment during an interstade of the last glaciation, or in a nonglacial environment which existed throughout the last glaciation.

Last glaciation[1]

Opinion on the nature and extent of the last glaciation has shifted through three phases. Firstly, reconnaissance surveys showed classical glacial indicators largely restricted to

---

[1] Events on southern Melville Island and northern Ellesmere Island are discussed separately, later in this chapter.

the eastern islands (Ellesmere, Axel Heiberg, and Devon islands). It was supposed that these islands were linked by an ice complex originating in and covering the eastern highland rim, while local ice caps possibly existed elsewhere (Wilson et al., 1958; Craig and Fyles, 1960, Fig. 1). Subsequently, the Innuitian Ice Sheet, presumably covering much of the Queen Elizabeth Islands, was proposed by Blake (1964, 1970, 1972, 1975) and portrayed by Prest (1969) chiefly on the basis of the Holocene emergence pattern. The absence of relatively fresh (or for some areas, even any) glacial landforms in the central and western islands was explained by Mayewski et al. (1981) as due to a cold thermal regime at the base of the ice sheet. England (1974, 1976b, 1978, 1983), on the other hand, argued for a spatially restricted last glaciation in northeast Ellesmere Island and northwest Greenland. He proposed that existing upland ice caps in the Queen Elizabeth Islands advanced 10-40 km while ice caps grew on presently unglaciated uplands. Collectively, these

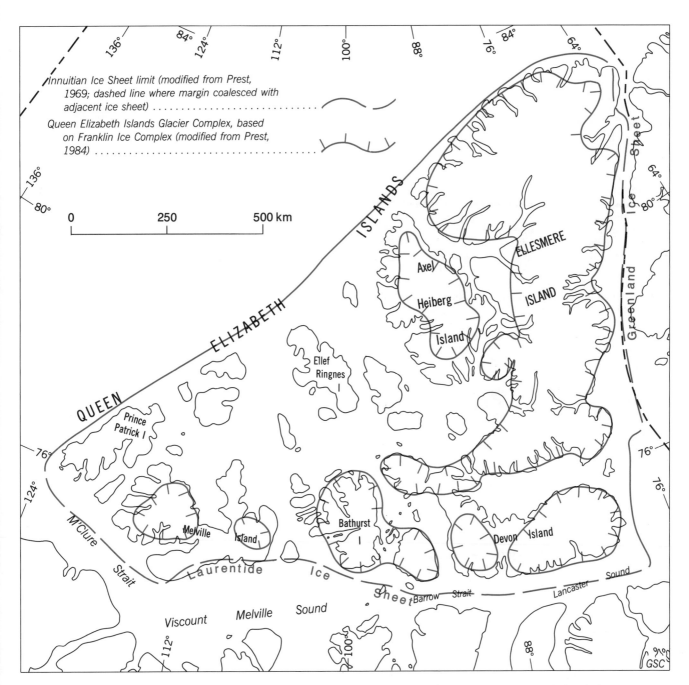

Figure 6.5. Maximum (Innuitian Ice Sheet?) and minimum (Franklin Ice Complex?) extent of glaciation during the last glacial stade in the Queen Elizabeth Islands.

glaciers formed the largely noncontiguous Franklin Ice Complex which left much of the northern archipelago unglaciated. Commentaries on these hypotheses are provided by Andrews and Miller (1976), Paterson (1977a), Boulton (1979), England et al., (1981), and Hodgson (1985). Even though neither of the major protagonists in this controversy has illustrated his proposed models, Figure 6.5 shows the approximate limits of the hypothesized Innuitian Ice Sheet and the Franklin Ice Complex.

Extent of ice cover

The area that has positively been shown to have been ice covered during the last glaciation is restricted to areas adjacent to existing ice caps on northern and central Ellesmere Island, and to an area on the southern coast of Melville Island (Fig. 6.4) overlapped by an ice shelf of Laurentide Ice Sheet origin. Cores through the present Agassiz and Devon ice caps indicate that the areas overlain by these features have been covered continuously by ice for the last 100 ka (Fisher and Koerner, 1983).

A 500 km-long belt of drift (Fig. 6.4) was deposited on west-central Ellesmere Island, 10 to 60 km west of margins of present ice caps, by a fluctuating ice margin between 9 and 7 ka (Hodgson, 1985). This morainal belt is coeval with the Hazen Moraines on northeast Ellesmere Island (England, 1978; this chapter, England and Bednarski, 1989). Shells from glacial marine rhythmites at two locations on the distal side of the drift belt date 8.8 ka and 8.7 ka (Hodgson, 1985). Fiord heads on the proximal side of the drift belt were deglaciated by 7.5 to 6.7 ka. However, unlike the Hazen Moraines, which England believed mark the last glacial maximum, Hodgson inferred that the last glacial limit in central Ellesmere Island lay an unknown distance beyond the drift belt. He also inferred, from evidence of fresh glaciation of the topographically protected Braskeruds Plain (which lies 500 m a.s.l.), that much of central and southern Ellesmere Island and adjacent areas, including Grinnell Peninsula, Devon Island, were glaciated at this time.

For the Krieger Mountains, north of the drift belt, Völk (1980), King (1981) and Barsch et al. (1981) suggested that the last ice limit lies only a few kilometres beyond present ice margins. This is based on morphostratigraphy and on allochthonous plant debris dated ca. 35 ka which occurs in proglacial deposits. Ice marginal landforms at the head of Tanquary Fiord and Antoinette Bay, similar to those of the central Ellesmere drift belt, predate marine sediments dated 6.8 ka (Hattersley-Smith and Long, 1967).

A limited last glaciation ice advance was also proposed for west-central Axel Heiberg Island, where Boesch (1963) found an upper zone of weathered glacial erosional landforms and a lower zone of well preserved landforms. Within the latter zone, Müller (1963) collected molluscs dated 9 ka overlying ice marginal deposits 5 m from the present terminus of Thompson Glacier.

Speculative last ice cover shown in Figure 6.6 is composed of ice caps more coalescent than those outlined by England (1976b). This is similar to the Queen Elizabeth Islands Glacier Complex outlined by Prest (1984). An ice cover prior to ca. 9 ka on Bathurst Island is suggested by the clear evidence of radial ice flow (Blake, 1964) and the difference in age of 1 ka between shells believed to lie at the marine limit in different parts of the island. In addition, the absence of Laurentide ice shelf deposits, which occur on Byam Martin and Melville islands immediately to the west, suggest that a local ice cap covered Bathurst Island. A similar local ice cover is shown on western Melville Island; the occurrence of this is indicated by yet undated outwash and glacial lake and glacial marine deposits.

In contrast to the picture of local ice cover presented above, Blake (1986) suggested that a massive ice stream fed from Ellesmere Island and probably Greenland flowed south through Smith Sound during the last glaciation. The evidence for this is freshness of glacially sculptured terrain supported indirectly by chronological data from lake and raised marine deposits.

Marine submergence

Whatever the extent of ice during the last glaciation, it is clear that towards the end of this event, sea level throughout the northern archipelago was raised in places at least 150 m above present level (Fig. 6.7). Sea and land relationships over the last 15 ka are reviewed by Blake (1976). In the first panarchipelago models of sea level change (Blake, 1970; Walcott, 1970), it was assumed that the oldest marine deposits of finite age immediately postdated deglaciation (Blake, 1970, Fig. 12). Hence, according to this hypothesis, the general pattern of declining age of the highest dated molluscs from ca. 11.5 ka to 9 ka from western to east-central islands records an ice front retreating in this same direction. It was also assumed that greatest emergence (area between Bathurst Island and northern Eureka Sound) occurred where the former ice load was greatest. At present however, last glacial and last high sea level deposits have been found in contact only on Ellesmere Island, on Melville Island (Laurentide ice), and Axel Heiberg Island. This model has been criticized by England (1976b) and Andrews

Figure 6.6. Minimum extent of last ice cover as proposed in this report.

and Miller (1976, p. 13); the latter stated that marine submergence is not necessarily an indicator of an ice cover at every marine limit locality, and England pointed out that the relationship between glacial and marine deposits in the central Queen Elizabeth Islands "has yet to be stratigraphically demonstrated". An additional problem with using marine limit to determine glacial history is that marine sedimentation may have commenced earlier than dated molluscs indicate, in an environment unsuitable for marine organisms (England, 1983).

England (1976b, 1983) developed an alternative model in which the pattern of marine submergence outlined by Blake is explained by the combined effect of the Franklin Ice Complex and Greenland ice to the east. Beyond the ice complex England predicted a raised "full glacial" sea. England (1983) reported evidence of such a stable sea level related to glacial maximum conditions in northern Ellesmere Island but evidence of a correlative sea has not been found elsewhere (Hodgson, 1985).

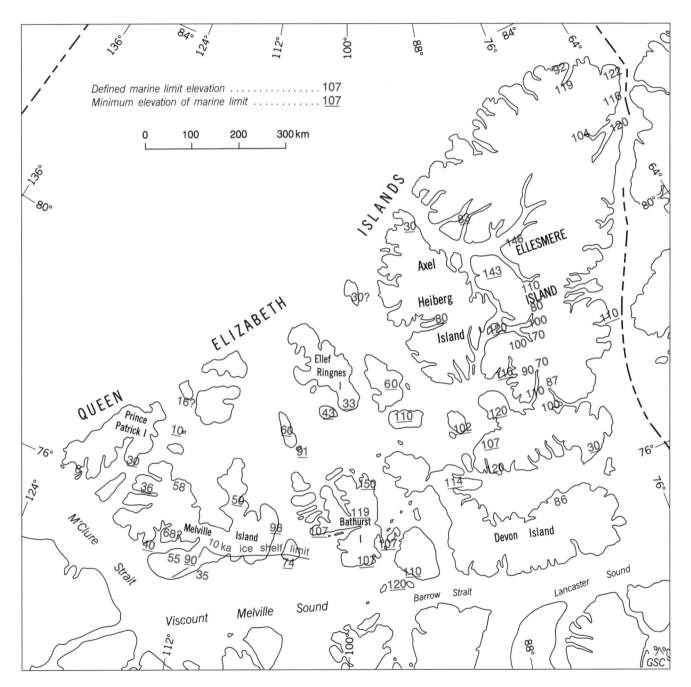

Figure 6.7. Maximum sea level in the last 12 ka, based on ^{14}C dated shells. Possibly ice cover in the northwest throughout the last glacial stage was no greater than that at present, whereas in inner fiords of Ellesmere Island deglaciation and marine overlap occurred as late as 7 ka.

CHAPTER 6

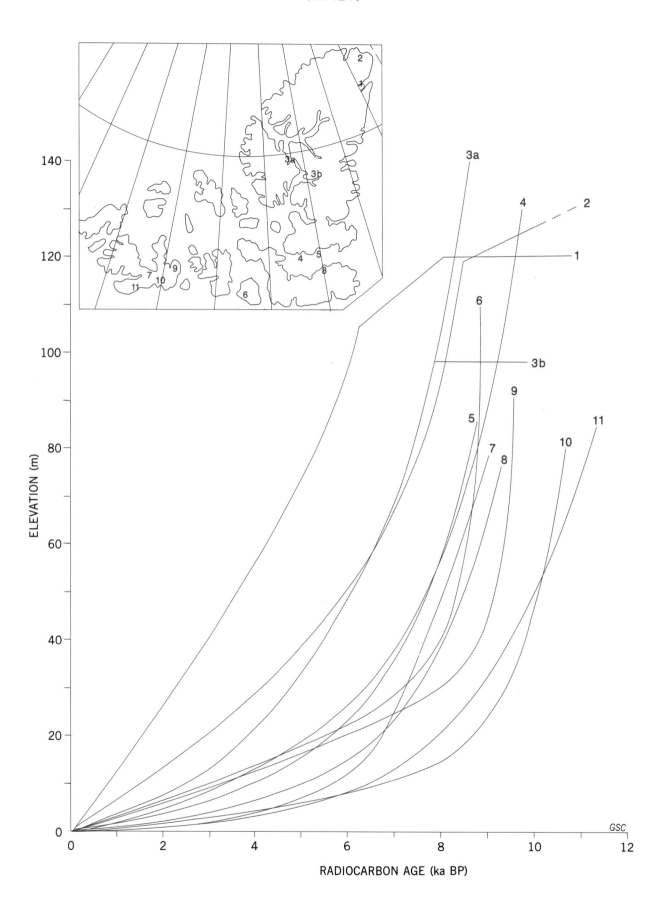

Deglaciation and initial emergence

If the Innuitian Ice Sheet hypothesis is accepted, then the western Queen Elizabeth Islands must have been deglaciated between 11.5 and 10 ka with the eastward receding, partly marine ice front reaching fiord heads of central Ellesmere Island by 7 ka. If ice coverage was restricted to those areas containing tangible evidence of the last glaciation, however, then it would appear that ice retreated from the periphery towards the centre of the main islands. Retreat was completed by ca. 10 ka on western Melville Island, by 8.5 ka on Bathurst Island, and between 9.5 and 6 ka in fiords of Ellesmere Island. In northernmost Ellesmere Island, glacier retreat southward towards the northern Grant Land Mountains commenced ca. 2 ka before the rapid deglaciation of the plateau east of Lake Hazen (Bednarski, 1984). The Viscount Melville Sound Ice Shelf of the Laurentide Ice Sheet overlapped and then retreated from southern Melville Island ca. 10 ka.

Isobases drawn on postglacial shorelines are not portrayed because the distribution of dated shorelines is inadequate to show the influence of the several deglacial ice centres within the Queen Elizabeth Islands, as well as Laurentide and Greenland ice sheets (England, 1976a,b; Dyke, 1984). The known history of rifting and the presence of seismically active areas suggest that tectonic forces may contribute to uplift (Hodgson, 1981).

Holocene

The Holocene of the Queen Elizabeth Islands is characterized by continuing retreat of the ice masses that developed during the last glaciation, shoreline emergence, and several swings in climatic trends. In the following discussion these developments are discussed separately but all are closely tied to the fluctuating change from environmental conditions at the maximum of the last glaciation to those of present.

Marine events

Most sea level data available for the Queen Elizabeth Islands (mainly from the eastern and southern islands) show an initial rapid emergence for ca. 2 ka followed by a steadily declining rate of sea level fall (Fig. 6.8). Exceptions to this are the data from northeast Ellesmere Island where, apparently, sea levels were fairly stable for 2-4 ka and then fell sharply with only a moderate decline in emergence rate as present levels were approached (this chapter, England and Bednarski, 1989).

Figure 6.8. Holocene and Late Pleistocene generalized shoreline emergence curves: (1) northeast and (2) northernmost Ellesmere Island (modified from England, 1983, Fig. 7); (3a) western (Eureka Sound) and (3b) central Ellesmere Island (adapted from Hodgson, 1985, Fig. 8); (4) Cape Storm and (5) South Cape Fiord (after Blake, 1975, Fig. 22 and 25); (6) Cornwallis Island area (modified from Washburn and Stuiver, 1985, Fig. 2); (7) central Melville Island (after Henoch, 1964, Fig. 6); (8) northeast Devon Island (Truelove Inlet, after Barr, 1971, Fig. 5); (9) eastern and (10) southern Melville Island (after McLaren and Barnett, 1978, Fig. 3 and 4); (11) southern Melville Island (adapted from Hodgson et al., 1984, Fig. 14).

The best controlled emergence data are from Cape Storm on southern Ellesmere Island (4 of Fig. 6.8) where more than 50 radiocarbon dates on driftwood, whale bone, and molluscs have been used to trace the 130 m fall of sea level from marine limit to present (Blake, 1975). At this site initial emergence was at a rate of 70 m/ka with over one half of the total emergence occurring in the first 1 ka and then declining to less than 3 m/ka over the last 2.4 ka. The data are sufficient to show that there have been no fluctuations of sea level with amplitude greater than 2 m or with periods of greater than 500 years during the past 9 ka (Walcott, 1972). Concentration of pumice on a 5 ka shoreline is related by Blake (1975) to a slight eustatic sea level rise, to a period of more open water, or a combination of the two.

The presence or absence of driftwood on shorelines has been variously interpreted. Blake (1972) interpreted the arrival of driftwood in the Queen Elizabeth Islands at 8.5-8 ka to indicate that the Innuitian Ice Sheet had broken up, permitting driftwood to be carried into the interisland channels. England (1976b) argued that the sudden arrival of driftwood would not necessarily indicate breakup of any Innuitian Ice Sheet but rather that it could have been due to the breakup of landfast sea ice that excluded driftwood from the area until 8 to 8.5 ka. He went on to relate periods of abundant driftwood deposition to periods of reduced summer sea ice cover; Stewart and England (1983) noted two such periods in northern Ellesmere Island — 6 to 4.2 ka and since 500 BP.

Indicators of recent sea level rise occur at the extremities of the archipelago. Offshore bars have been built on the Arctic Ocean coast of the westernmost islands and off extreme southeast Devon Island. England and Bednarski (this chapter, 1989) suggest that recent submergence has also occurred on the northernmost coast of Ellesmere Island.

Paleoenvironment

Studies of Holocene paleoenvironment have been confined largely to the eastern margin of the archipelago. Data from lake sediment cores (Blake, 1981a; Retelle, 1986b; Smol, 1983; Hyvärinen, 1985), cores in surface peat accumulations (Blake, 1964, 1974; Brassard and Blake, 1978), and ice cap cores (this chapter, Koerner, 1989) indicate a relatively warmer period from at least 9 ka up to ca. 5-4 ka. Blake (1972) noted that driftwood dated 6.5-4.5 ka is particularly abundant, suggesting that this might have been the time of maximum driftwood penetration of interisland channels and hence possibly of maximum warmth. Stewart and England (1983) corroborated this in noting that driftwood penetration on the north coast of Ellesmere Island, along with increased accumulations of terrestrial organic material, occurred between 6 and 4 ka. Regrowth of glaciers and ice shelves occurred in the following cool period (Hattersley-Smith et al., 1955; Hattersley-Smith, 1960; Crary, 1960; Lyons and Mielke, 1973). During the past 4.5 ka there have been intermittent periods of climatic amelioration as indicated by the presence of paleo-Eskimo cultures, particularly Independence I (ca. 4 ka; Knuth, 1967), Independence II (ca. 2.5-2 ka; Knuth, 1966), and Thule cultures (ca. 1-0.7 ka; Hattersley-Smith, 1973; Schledermann, 1980). Driftwood also entered the fiords intermittently during the past 4.5 ka, indicating at least brief intervals of reduced summer sea ice (Blake, 1981b). The lichen kill on high plateaus likely dates from the "Little Ice Age" of the 16th to 19th centuries

CHAPTER 6

(Koerner, 1980; Edlund, 1985). Many large outlet glaciers in the area are encroaching upon or overriding raised marine features which were deposited during deglaciation, indicating that these glaciers are at their maximum positions (England, 1978; Stewart and England, 1983). On the other hand, many glaciers show evidence of recent retreat, probably from the warming of this century (Hattersley-Smith, 1963a).

The present day climate supports flora and microfauna little different to those of the early Holocene. Furthermore, the elevation of the present glacier equilibrium line rarely exceeds 1000 m a.s.l. (Miller et al., 1975). Hence climatic warming would have been insufficient to be a dominant mechanism in the breakup of any hypothesized archipelago-wide ice cover, such as the Innuitian Ice Sheet. This leads to

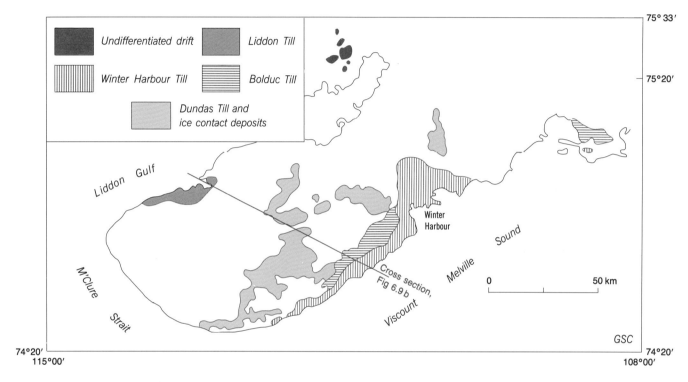

Figure 6.9a. Quaternary tills, southern Melville Island (after Hodgson et al., 1984, Fig. 6).

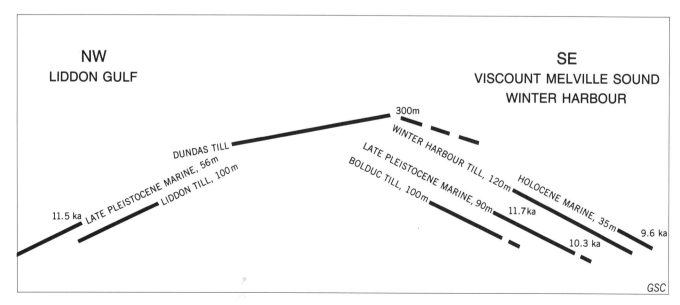

Figure 6.9b. Schematic relationships of till and marine units across Dundas Peninsula, southern Melville Island.

the conclusion that if the Innuitian Ice Sheet did exist, it disintegrated by means of a mechanism other than rise in temperature.

Southern Melville Island

Pre-last glaciation

The oldest recognized evidence of continental ice from the south is Dundas Till and associated ice contact deposits of central Dundas Peninsula which contain shield erratics (Fig. 6.9; Hodgson et al., 1984). The absolute age of this event is unknown, and no former relative sea level has been determined. On the basis of geographical extent, this till is tentatively correlated with till of either the Banks or Thomsen glaciations of Banks Island, lying to the southwest (Vincent, 1989).

Last glaciation or older

Bolduc Till, which includes numerous erratics of southern provenance, was deposited by an advance subsequent to that which deposited Dundas Till and which overlapped southeast Dundas Peninsula to 100 m a.s.l. (Fig. 6.9; Hodgson et al., 1984). Shell fragments on the till surface date >33 ka and may be glacially transported or may date an otherwise unidentified overlapping marine event younger than the till. The oldest marine sediments which definitely overlap the till are dated 11.7 ka.

Liddon Till, on the coast of outer Liddon Gulf, was deposited by continental ice, entering the gulf from M'Clure Strait (Fig. 6.9; Hodgson et al., 1984). The oldest overlapping sediments dated 11.5 ka. Direct correlation between this area and southern Dundas Peninsula is not possible, but on the basis of morphology and elevation this till is more likely coeval with Bolduc Till than with younger Winter Harbour Till.

A marine transgression with a minimum age of 11.7 ka overlapped all shores. Marine deposits related to this event stratigraphically overlie Bolduc and Liddon tills (Fig. 6.9). At ca. 11 ka, sea level in central and northern areas lay at 55 m; a former waterplane recorded by numerous perched deltas.

Last glaciation and postglacial marine events

Winter Harbour Till is the least extensive till on the south shore of Dundas Peninsula (Fig. 6.9). This till is believed to have been deposited by an ice shelf representing the maximum northern extent of Late Wisconsinan Keewatin ice (Hodgson and Vincent, 1984a). The evidence for deposition by an ice shelf includes the presence of a topographically featureless till of similar composition up to an elevation of 120 m a.s.l. over hundreds of kilometres on both northern and southern coasts of Viscount Melville Sound. Winter Harbour Till overlies marine sediments dated 11.3 to 11.5 ka. Hodgson and Vincent (1984a) believed that marine sediments dated 10.3 ka also predate the grounding of the ice shelf. If this is the case, the ice shelf disintegrated a few hundred years later, prior to incursion of the sea by 9.6 ka.

Isostatic recovery appears to have continued while the ice shelf was present because pre-Winter Harbour marine sediments, dated 11.4 ka, lie at 82 m but the postglacial marine limit developed on Winter Harbour, dated slightly older than 9.6 ka, lies at only 35 m a.s.l. The elevation of the postglacial marine limit rises eastwards to >75 m on Byam Martin Island and >120 m on Lowther Island. Holocene emergence in southern Melville Island is smaller though similar in pattern to eastern Melville Island (Henoch, 1964; McLaren and Barnett 1978; Fig. 6.8).

Northeast Ellesmere Island

J. England and J. Bednarski

Northern Ellesmere Island and Greenland support the largest concentration of permanent snow and ice in the Northern Hemisphere. At its closest point, northeast Ellesmere Island is separated by only 40 km from Greenland across Nares Strait, the overall name for the dividing water body (Fig. 6.10); an ice-free plateau of low to moderate relief on northeast Ellesmere Island separates the present day ice sheet from the strait and Greenland.

Northeast Ellesmere Island is composed of two main physiographic regions: the ice-covered Grant Land Mountains rising to 2500 m a.s.l., and the largely ice-free Hazen Plateau to the south (Fig. 6.10). Lake Hazen occupies an elongate trough at the base of the Lake Hazen Fault Zone which abruptly terminates the southern flank of the Grant Land Mountains. The Hazen Plateau (26 000 km^2) is gently rolling and rises from the Lake Hazen basin (150 m a.s.l.) to 1100 m along its southern rim where it is incised by spectacular fiords. The large valleys that cut into the southern rim of the plateau are occupied by prominent, southward-flowing rivers whose drainage predates the tilting of the plateau during the late Tertiary (Christie, 1967; Kerr, 1967).

Older glaciations

Glaciers from both Ellesmere Island and Greenland reached their maximum extents on northeastern Ellesmere Island prior to the last glaciation (Fig. 6.11). The relative ages of advances are indicated by weathering zones. Three weathering zones are recognized along eastern Judge Daly Promontory: (1) extensively weathered, unglaciated summits extending from 1000 to as low as 470 m a.s.l.; (2) a zone of sparse Greenland erratics amongst deeply weathered bedrock whose upper limit descends from 840 m in the north to 470 m in the south and whose lower limit is crosscut by; (3) till and ice-shelf moraines deposited by the maximum advance of the Ellesmere Island ice extending from 370 m to the Holocene marine limit (120 m a.s.l.). Although the maximum Ellesmere Island ice advance crosscuts the zone of Greenland erratics on Judge Daly Promontory, the Greenland moraines 40 km to the north, along western Robeson Channel, have not been overridden. Either these Greenland moraines represent a younger advance in this area or the Ellesmere Island ice did not move into this area. Because the ice-shelf moraines on eastern Judge Daly Promontory have not been overridden by later Greenland ice, it seems unlikely that a younger Greenland advance deposited the moraines along western Robeson Channel.

England, J. and Bednarski, J.
1989: Northeast Ellesmere Island (Quaternary stratigraphy and chronology); in Chapter 6 of Quaternary Geology of Canada and Greenland, R.J. Fulton (ed.); Geological Survey of Canada, Geology of Canada, no. 1 (also Geological Society of America, The Geology of North America, v. K-1).

CHAPTER 6

Therefore Ellesmere Island ice from the Grant Land Mountains probably has not reached Robeson Channel since the maximum advance of the Greenland ice.

Greenland ice

The oldest erratics deposited on the south-central rim of the Hazen Plateau are sparse quartzite cobbles which may be fluvial erratics deposited during the Tertiary (England, 1978). The oldest erratics of definite glacial origin occur on the southwest rim of the Hazen Plateau and eastern Judge Daly Promontory (Fig. 6.11). These erratics are granite and gneissic boulders derived from the Precambrian Shield beneath the Greenland Ice Sheet (Christie, 1967; England and Bradley, 1978). The distribution of these crystalline erratics indicates that Greenland ice penetrated up to 25 km onto the southeast Hazen Plateau during its maximum advance. This extension is only 100 km beyond its present margin in Petermann Fiord and Newman Bay.

Along eastern Judge Daly Promontory, 20 km south of Cape Baird, crystalline erratics of Greenland provenance are incorporated into deeply weathered bedrock on summits up to approximately 800 m a.s.l. bordering Kennedy Channel. The upper limit of crystalline erratics descends to 600 m a.s.l. 20 km inland (to the west) where they terminate below apparently unglaciated summits (England and Bradley, 1978). Some 50 km farther to the southwest, along western Kennedy Channel, the uppermost crystalline erratics descend to 470 m a.s.l. (England et al., 1981). This uppermost profile of erratics deposited by Greenland ice marks the extension of an outlet glacier from Petermann Fiord and it precludes an all pervasive ice ridge over Nares Strait as previously proposed (Dansgaard et al., 1973; Hughes et al., 1977; Mayewski et al., 1981).

The age of the maximum advance of Greenland ice onto northeast Ellesmere Island is unknown; however, the crystalline erratics exhibit moderate weathering (frost shattering and exfoliating desert varnish) and they are commonly incorporated into deeply weathered bedrock, suggesting considerable antiquity. On northeast Judge Daly Promontory Greenland till contains fragmented shells whose amino acid ratios suggest an age of >80 ka (England and Bradley, 1978). Observations at Wrangel Bay, along the northeast rim of the Hazen Plateau, show that at one stage during retreat of Greenland ice a contemporaneous beach, >2 km in length, developed, which now lies 285 m a.s.l. Nearby, a silty diamicton, interpreted as contemporaneous deep water, glaciomarine sediment, contains in situ molluscs dated >32 ka (Retelle, 1986a). This date indicates that the Greenland ice was retreating from its maximum position on northeast Ellesmere Island earlier than 32 ka. Glacial lakes and recessional moraines formed at the margin of Greenland ice in the same area (Retelle, 1986a).

The distribution of the uppermost crystalline erratics on northeast Ellesmere Island is noteworthy because they overtop coastal summits up to 800 m a.s.l. (>1000 above the adjacent seafloor) but descend rapidly southward down Kennedy Channel and terminate only 5 to 25 km inland

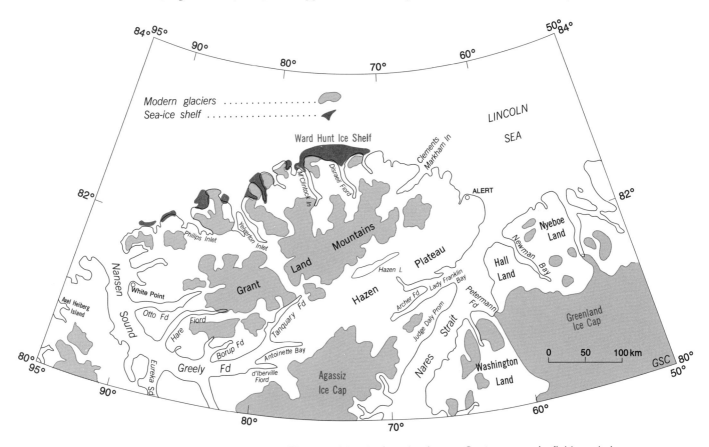

Figure 6.10. Place names on northern Ellesmere Island referred to in text. Contemporary icefields and glaciers are shown; note also sea ice and ice shelves on the northernmost coast.

along the west side of northern Nares Strait (Fig. 6.11). The distribution of erratics indicates a substantial filling of Hall Basin yet considerably less ice in Kennedy Channel, which would have provided the least resistance to ice flow. This suggests that Greenland ice might have deposited the erratics before late Tertiary faulting completely separated Greenland from Ellesmere Island (forming Kennedy Channel). Weidick (1975) suggested that a Greenland Ice Sheet was established in the Late Pliocene (ca. 3.5 Ma) based on the earliest ice-rafting recorded in cores from the North Atlantic (Berggren, 1972), and Herman and Hopkins (1980) reported that the earliest ice rafting occurred before 4.5 million years ago in the Arctic Ocean.

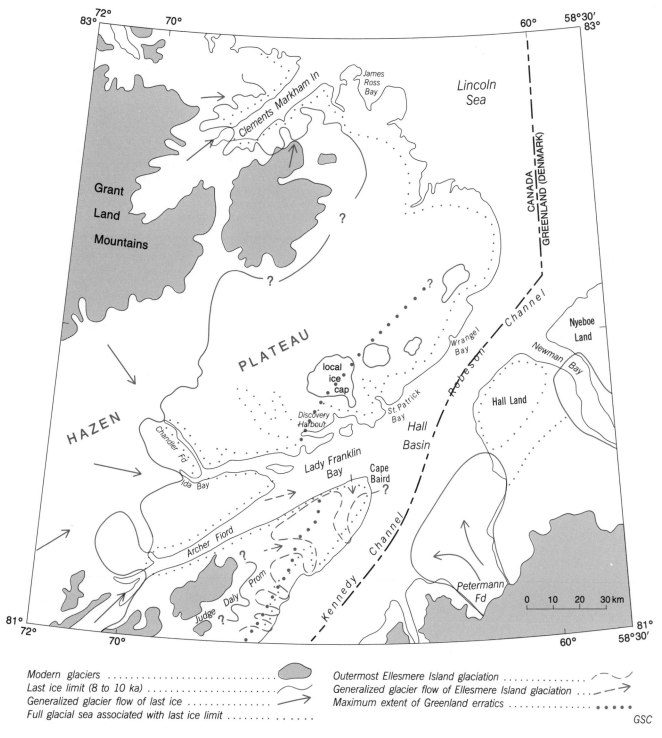

Figure 6.11. Extent of past glaciations on northern Ellesmere Island and adjacent coast of Greenland.

Ellesmere Island ice

The maximum advance of Ellesmere Island ice was characterized by the radial expansion of glaciers from the Grant Land Mountains and the Agassiz Ice Cap (Fig. 6.10). The advance dispersed erratics southward across the Hazen Plateau and into adjacent fiords. Along the south-central flank of the Grant Land Mountains, Hattersley-Smith (1969a) reported erratics up to 640 m a.s.l. at the head of Tanquary Fiord (Fig. 6.11). England (1978) mapped erratics up to 670 to 830 m a.s.l. at the heads of Archer Fiord, Ida Bay, and Discovery Harbour and noted their decline to the east, reflecting the profile of a former trunk glacier reaching outer Lady Franklin Bay. Here, part of the main trunk glacier flowed southward across northern Judge Daly Promontory and deposited till and moraines that crosscut the zone of Greenland erratics at 200-300 m a.s.l. bordering Kennedy Channel (England and Bradley, 1978). During this same interval, glaciers also advanced from a local ice cap on Judge Daly Promontory and reached western Kennedy Channel via Daly River valley and other smaller valleys to the south (Fig. 6.11). These topographically controlled glaciers entered isostatically depressed valleys along western Kennedy Channel where they formed small ice shelves. Relative sea level at the time of ice shelf formation was ca. 162-175 m a.s.l. which is consistent with the estimated water depth required to float these glaciers (England et al., 1978).

The minimum age of the maximum Ellesmere Island ice advance along eastern Judge Daly Promontory is likely beyond the range of ^{14}C dating because shells from subtill and supratill marine deposits have provided dates of 40.3 ka and >39 ka (England, et al., 1981). Amino acid ratios (both D-alloisoleucine to L-isoleucine and aspartic acid) for the same samples provide age estimates of >35 ka. Several other radiocarbon dates from the ice-shelf moraines and ice contact marine deposits along 70 km of eastern Judge Daly Promontory are of similar ages (England et al., 1978, 1981).

Last glaciation

During the last glaciation, ice from the Grant Land Mountains advanced 5 to 40 km to the southeast and formed the Hazen Moraines on the central and southern parts of the Hazen Plateau (England, 1974, 1978). This ice filled the Lake Hazen basin and reached sea level at the head of Chandler Fiord and likely the head of Tanquary Fiord. Glaciers from the northeast end of the Agassiz Ice Cap also reached sea level at the head of Ida bay, and inner Archer Fiord (England, 1978). On central Judge Daly Promontory a prominent, probably correlative moraine system was formed 8 km beyond the present southeast margin of a small ice cap. This moraine system is interpreted as the last ice limit because it abuts older, weathered terrain that extends downvalley (eastward) from this point to the undisturbed Ellesmere Island moraines and sea levels dated >35 ka along western Kennedy Channel (England et al., 1981).

England (1976a, b, 1978) provided evidence that throughout the last glaciation an ice-free corridor existed between northeast Ellesmere Island and northwest Greenland ice (Fig. 6.11). This interpretation differs from that of Greenland workers who envisage the channels and basins between Greenland and Ellesmere Island as being filled with ice at this time (England, 1985, 1987; Bennike et al., 1987; Funder, 1989). The ice-free corridor was part of a peripheral isostatic depression in which the marine limit marks the uppermost extent of a "full glacial sea" (England, 1983). The marine limit in the "full glacial sea" trims either weathered till or higher shorelines formed during previous glaciations. The "full glacial sea" is indicated by: (1) ^{14}C dates on marine fauna that predate any glacial unloading indicating that the sea, rather than glaciers, occupied these areas during maximum glacial time and (2) initial emergence that occurred simultaneously throughout the ice-free corridor, indicating the synchronous decay of a peripheral depression. Relative sea level curves (Fig. 6.12) from the "full glacial sea" are of paleoclimatic interest because they record the complete history of glacial unloading. They are also of geophysical interest because they show the nature of postglacial emergence that one would theoretically expect to find in a peripheral depression.

The limit of the "full glacial sea" is isostatically tilted from 110 m a.s.l. inland of Discovery Harbour, to 120 m a.s.l. at Cape Baird, northernmost Judge Daly Promontory (Fig. 6.13). This upward tilt to the southeast reflects the glacial isostatic dominance of the northwest Greenland ice load during the last glaciation where the apex of the full glacial sea reaches 150 m a.s.l. (see next section). On northeast Ellesmere Island, initial ice retreat and postglacial emergence began slowly between 8.2 and 6.2 ka. After 6.2 ka ice retreat was rapid as indicated by the onset of rapid postglacial emergence. An amelioration of the climate at this time (ca. 6.2 ka) is also suggested by the accumulation of organic debris at Tanquary Fiord and Clements Markham Inlet and by the increased penetration of driftwood into local fiords suggesting reduced summer sea ice (Stewart and England, 1983).

On northern Ellesmere Island Bednarski (1986) located the limit of the last glaciation in Clements Markham Inlet using morphostratigraphy and sea level history (Fig. 6.11). A major glacier advanced some 45 km from the Grant Land Mountains to occupy the head of the inlet until at least 10 ka. Most of the outer inlet remained ice free and was inundated by a high, "full glacial sea" that reached at least 124 m a.s.l. at the glacial limit.

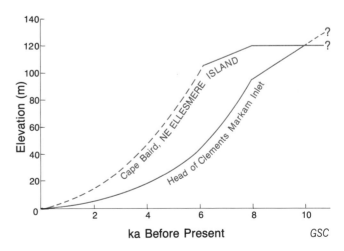

Figure 6.12. Emergence curves showing different histories of glacial isostatic unloading between Clements Markham Inlet and northeast Ellesmere Island / northwest Greenland (recorded here at Cape Baird).

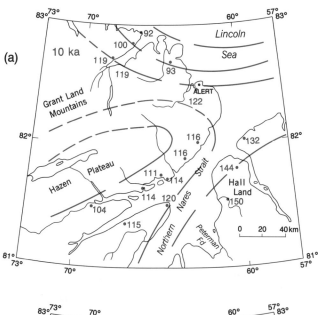
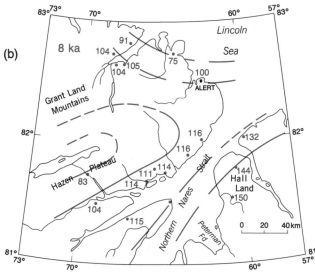
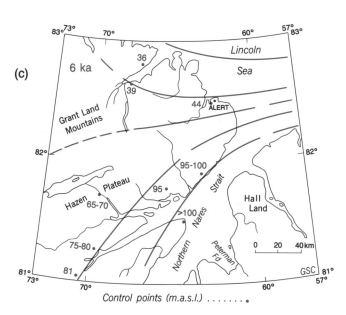

More than 40 radiocarbon dates indicate that small, side glaciers slowly retreated by calving into the "full glacial sea" between 10 and 9 ka. This retreat, however, was minor because the ice was not extensive during the last glaciation, and some outlet glaciers from the Grant Land Mountains were within 6 km of their present margins by 9.7 ka. The main trunk glacier also retreated slowly and it was only 5 km behind its 10 ka limit at 8 ka.

While slow deglaciation progressed in Clements Markham Inlet, the coastline began to slowly emerge from the full glacial sea. Radiocarbon dates on marine shorelines indicate that, distal to the last ice limit, initial emergence from the marine limit occurred simultaneously (10 ka). This limit is now isostatically tilted from 92 m at the mouth of the inlet, to 124 m a.s.l. along the last ice limit near the head of the inlet. Slow deglaciation, coupled with slow emergence, continued until 8 ka. After this, the rate of retreat was rapid so that the entire lowland at the head of Clements Markham Inlet became ice free within 400 years. A sudden termination of the sediment supply to the outer coastline of Clements Markham Inlet between 7 and 5 ka suggests that the adjacent uplands became ice free at this time.

Postglacial emergence and isobases

The magnitude and timing of postglacial emergence on northeast Ellesmere Island are controlled by: (1) the glacial isostatic dominance of the northwest Greenland ice, which influenced most areas southeast of the Grant Land Mountains (England, 1976a, 1982, 1985) and (2) the isostatically independent and earlier unloading of local Ellesmere Island ice on the northern and western sides of the Grant Land Mountains (England, 1983; Bednarski, 1984). These two controls are well illustrated by the regional isobases on the 10, 8, and 6 ka shorelines (Fig. 6.13) and the tilt of shorelines developed at different times (Fig. 6.14).

The isobases on the 10 ka shoreline have several prominent features. First, there is a marked, upward tilt of the shoreline towards northwest Greenland. This indicates that the glacial isostatic loading by the Greenland Ice Sheet extended across Nares Strait dominating most of the Hazen Plateau during the last glaciation (England, 1976a, 1982, 1983, 1985). Second, there is another centre of maximum emergence (former ice loading) over the Grant Land Mountains that is independent of the Greenland ice load. As a result of these two emergence centres on northern Ellesmere Island and northwest Greenland, there is an intervening cell of lower emergence centred over the Lake Hazen basin (Fig. 6.13a, b).

The main features seen in the 10 ka isobases persisted at 8 ka. Because rapid emergence began by 8 ka on the north coast of Ellesmere Island, the 6 ka isobases are considerably lower there than along Nares Strait where little emergence

Figure 6.13. Postglacial isobases drawn on shorelines dated 10 ka, 8 ka, and 6 ka. Note emergence that occurred between 10 and 8 ka in the area of Clements Markham Inlet while the area of northern Nares Strait remained stable. This reflects different histories of glacial unloading also shown in Figure 6.12. Profiles of shorelines along the transect are portrayed in Figure 6.14.

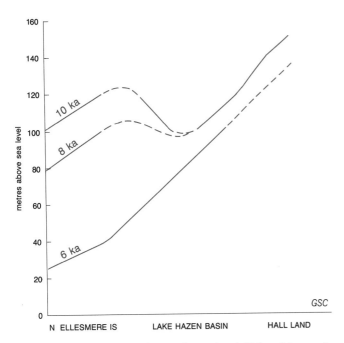

Figure 6.14. Profiles of shorelines dated 10 ka, 8 ka, and 6 ka, drawn along a 300 km transect (cf. Fig. 6.13) from Petermann Fiord to Cape Columbia (just west of mouth of Clements Markham Inlet). Note rapid emergence between 10 and 8 ka on northern Ellesmere Island and slow emergence for the same interval between the Lake Hazen Basin and Hall Land.

occurred between 8 and 6 ka. Rapid glacial unloading along northern Nares Strait did not occur until after 6.2 ka. However, because considerable emergence had occurred by 6 ka farther to the north, the cell of lower emergence around Lake Hazen disappears on the 6 ka isobases (Fig. 6.13c).

The sequence of isobases indicates that initial emergence north of the Grant Land Mountains preceded that on the southeast side by 2 ka. There are two principal reasons why glacial unloading began earlier on northern Ellesmere Island. First, there was probably a dissimilar climatic regime on the north side of the Grant Land Mountains, facing the Arctic Ocean, from that on the southeast side, bordering the Hazen Plateau and Nares Strait. Second, it is also likely that the difference in initial emergence between the north and south sides of the Grant Land Mountains was influenced by the delayed retreat of the Greenland ice which strongly influenced the loading of northeast Ellesmere Island. Such a delay would slow the initial emergence on the south side of the Grant Land Mountains and, at a later date, accentuate rapid emergence along Nares Strait compared to the slower emergence on northern Ellesmere Island. Nonetheless, independent evidence shows that the glacier occupying the head of Chandler Fiord, central Hazen Plateau, was still within 5 km of its last ice limit at 6.2 ka (England, 1983), suggesting that deglaciation was likely delayed both on northeast Ellesmere Island and northwest Greenland until this time.

QUEEN ELIZABETH ISLANDS GLACIERS
R.M. Koerner

Glacier characteristics

Ice covers 108 600 km^2 of the Queen Elizabeth Islands. This amount constitutes 5% of the northern hemisphere's ice cover. Most of the glaciers are located on the higher, mountainous areas of Devon, Ellesmere, and Axel Heiberg islands (Fig. 6.15). The western part of the Arctic Archipelago is almost entirely ice-free with the notable exception of Melville Island, which bears four small, stagnant ice caps on its higher ground.

It is apparent from ice core studies (Paterson et al., 1977) that the large ice caps (e.g., A of Fig. 6.16) are about 100 ka old. They contain ice from the last glaciation in their lower layers and interglacial ice under that. Thus these ice caps are neither products of the Holocene nor solely relicts of the last ice age.

The larger stagnant ice caps like Meighen Ice Cap (Fig. 6.15), where the maximum thickness is of the order of 120 m, began to grow during the climatic deterioration following the Hypsithermal some 4 ka (Koerner 1968). Thinner stagnant ice caps of about 30-40 m maximum thickness (e.g., A of Fig. 6.17) probably began their growth about 1.5 ka ago following a period of warmth that formed an ablation surface on Meighen Ice Cap (Koerner, 1968) and the Ward Hunt Ice Shelf (Lyons and Mielke, 1973). The smallest ice caps, which measure only about 1-2 km in diameter (e.g., B of Fig. 6.17), and which are probably less than 20 m thick, were initiated during the most recent cold period which started about 0.4 ka.

Ice cap surfaces are strongly controlled by the bedrock topography (Fig. 6.16). Ice from large areas of the ice caps flows through outlet glaciers either of the valley or lobate type (B and C, Fig. 6.16, respectively). Elsewhere the ice cap margins lie on rock plateaus at approximately 800 to 1000 m above sea level. Ice thickness varies between about 300 and 800 m on the larger ice caps and between 200 and 500 m on the outlet glaciers (Hattersley-Smith et al., 1969; Paterson and Koerner, 1974; Oswald, 1975; Koerner, 1977a; Narod and Clarke, 1983). Thickness is usually greater on the sides of the ice caps that face water bodies. This effect is most pronounced around Baffin Bay and has been attributed to higher rates of snow accumulation (Koerner, 1977a).

In general, ice velocities in ice caps are less than 20 m/a whereas outlet glaciers move at 20-40 m/a. Summer velocities of some valley glaciers are about twice the mean annual velocity (Iken, 1974) due to penetration of meltwater to the bed. The velocities are low when compared to those in Greenland and Antarctica where the accumulation zones are much larger. Consequently, most of the Queen Elizabeth Islands glaciers are relatively crevasse-free (Sverdrup Glacier, Fig. 6.15, 6.17), and calving at the terminus rarely forms an important term in the mass balance equation. Most of the icebergs endangering the shipping lanes off Newfoundland are from the extremely dynamic Greenland outlet glaciers (Reeh, 1989).

Koerner, R.M.
1989: Queen Elizabeth Islands glaciers; in Chapter 6 of Quaternary Geology of Canada and Greenland, R.J. Fulton (ed.); Geological Survey of Canada, Geology of Canada, no. 1 (also Geological Society of America, The Geology of North America, v. K-1).

Englacial temperatures are dependent on air temperature, geothermal heat flow at the glacier bed, ice thickness, ice velocity, and the amount of accumulation of snow or ablation of ice at the surface. Most of the ice in the Arctic Islands is below freezing (Fig. 6.18) as glacier thickness and air temperatures are low. High ablation rates at the surface of some outlet glaciers, however, leads to melting at the bed (6 of Fig. 6.18). This has important implications as it leads to basal erosion. Evidence of this can be seen in the presence of englacial debris in ice exposed at the snouts of some glaciers.

Ice temperatures do not necessarily decrease with increasing elevation up the glacier. This is partly a consequence of mean annual temperature inversion conditions in the High Arctic (Wilson, 1969). It is also due to melting conditions at the surface in summer. In the firn zone, high up on the ice caps, the latent heat released by refreezing of meltwater is largely retained as the meltwater refreezes in the firn. At lower elevations the meltwater runs off so that latent heat is lost to the glacier. Thus it is not uncommon for the upper regions (accumulation zones) of ice caps to have higher englacial temperatures than the ablation zones lower down (White Glacier, Table 6.1). On the ice caps generally, the change in temperature with elevation at an ice depth of 10 m is much less than the normal adiabatic lapse rate (Devon Ice Cap, Table 6.1).

Snow accumulation varies between high rates of over 44 $g/(cm^2 \cdot a)$ on the slopes facing Baffin Bay to less than 15 $g/(cm^2 \cdot a)$ in the interior parts of northern Ellesmere Island (Koerner, 1979). Of the snowfall in the high accumulation area around Baffin Bay, 17-26% most probably originates directly from the bay (Schriber et al., 1977; Koerner and Russell, 1979). The low snowfall rates in the northern parts of Ellesmere Island are partly due to increased distance from moisture sources but more importantly to the precipitation-shadow effect of the surrounding high land. The elevation of the firn line, which forms the lower limit of a zone permanently covered by new or old snow (firn), is largely

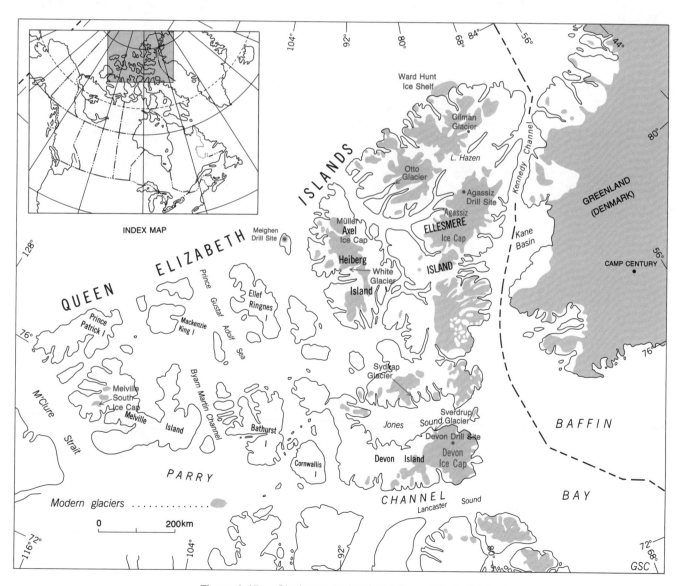

Figure 6.15. Glacier study sites in the Queen Elizabeth Islands.

determined by the snow accumulation rate (Table 6.2). Thus the height is lowest on the slopes facing Baffin Bay. The equilibrium line (positive balance above, negative balance below) lies about 100-300 m below the firn line except in exceptionally warm summers when it may rise slightly above the mean firn line.

Ablation rates at sea level vary from about 25 g/(cm²·a) on Ward Hunt Ice Shelf (Hattersley-Smith and Serson,

Figure 6.16. Landsat image of central Axel Heiberg Island showing the major ice caps. Note the unchannelled outflow of ice to the eastern margins contrasting with flow through valley glaciers on the west side of the main ice cap. This contrast in the modes of flow between one side of an ice cap and the other is common in the Queen Elizabeth Islands. Image 20950-18542, August 29, 1977, Canada Centre for Remote Sensing.

QUATERNARY GEOLOGY — QUEEN ELIZABETH ISLANDS

1970), through 135 g/(cm^2·a) on Sverdrup Glacier (Koerner, 1970), to 290 g/(cm^2·a) on White Glacier (Blatter, 1985). While there is a good inverse correlation between elevation and ablation, there is a poor relationship between latitude and ablation.

Present glacier balance

The mass balance of a glacier is generally measured by setting poles in holes drilled into the ice or firn and measuring how much ice and snow melts or accumulates around them each year. Each pole represents an area of the glacier and

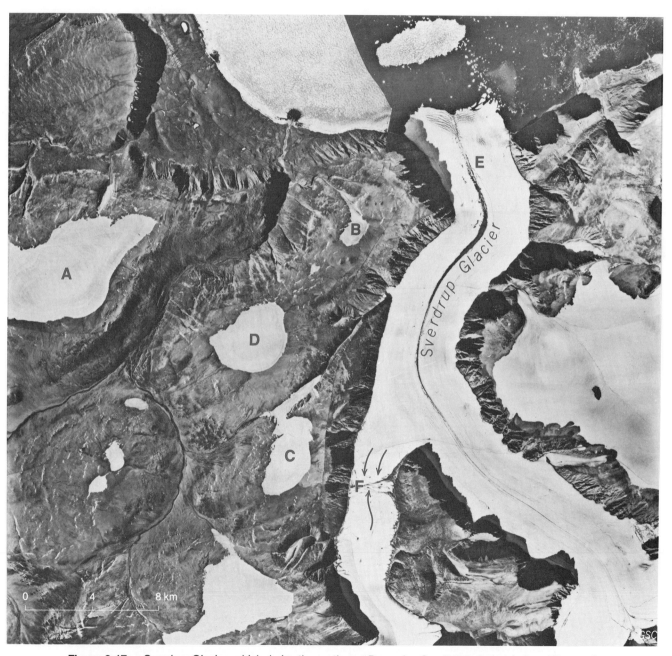

Figure 6.17. Sverdrup Glacier, which drains the northwest Devon Ice Cap (1959 airphoto). A and B are referred to in the text. Note the generally uncrevassed nature of the glacier, the central melt stream disappearing into a moulin at E and the junction of two tributaries flowing towards each other at F. Ice cap C separated into a gully section and an ice cap section in the early 1960s. Ice cap D, now has four small nunataks showing through the ice. NAPL mosaic MG 1888.

CHAPTER 6

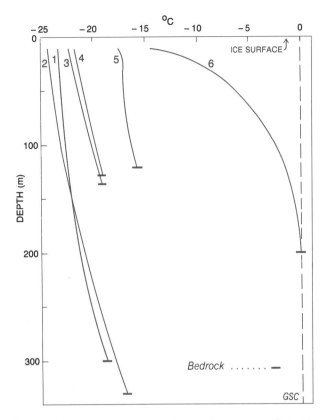

Figure 6.18. Temperature profiles: (1) Devon Ice Cap, elevation 1800 m (Paterson and Clarke, 1978); (2), (3), and (4) Agassiz Ice Cap, elevation 1700 m, 1710 m, and 1715 m, respectively (D.A. Fisher, Geological Survey of Canada, personal communication, 1986); (5) Meighen Ice Cap, elevation 268 m (Paterson, 1968); (6) White Glacier, elevation 300 m (Blatter, 1985).

the glacier balance is the sum of the volumes for each area where accumulation is positive and ablation negative. Calving at the snout of the glaciers generally plays only a minor part in their ablation (Reeh, 1989).

The results of the longest sets of records are shown in Figure 6.19. In each case the 20-25 year balance is slightly negative. The negative balances at the beginning of the records are representative of the warmer conditions of the 1950s. There is no evidence of a warming or cooling trend over the 25 year period of record.

Area changes

A preliminary study of aerial photography taken in the 1940s, 1950s, and 1960s and of Landsat satellite imagery obtained in the 1970s has shown a slight reduction in area of some of the glaciers and small ice caps in the Arctic Islands since the 1940s. The trend is a continuing one.

There are, however, three known cases of substantial marginal change. Otto Glacier in Northern Ellesmere (Fig. 6.15) advanced 2-3 km between 1950 and 1959 and a further 2-3 km between 1959 and 1964 (Hattersley-Smith, 1969b). Good Friday Glacier on Axel Heiberg Island (Fig. 6.16) advanced about 2 km between 1952 and 1959 (Müller, 1969). Both glaciers may be of the surging type. In contrast to these two glaciers, the Sydkap Glacier on southern Ellesmere Island (Fig. 6.15) retreated 6.5 km between 1957 and July, 1974. The retreat is associated with increased crevassing of the surface towards the terminus but the surface of the glacier has not changed in elevation. None of the nearby glaciers have advanced or retreated over the same period and it is not known whether the glacier has retreated from or to a more common position.

Areal changes, however, are not always directly related to volume changes. For example, measurements of ice level change have shown that while the edges of Meighen Ice Cap

Table 6.1. Ice temperatures at different elevations on the Devon Ice Cap and White Glacier[1]

Location	Elevation m a.s.l.	Temperature (°C) at 10 m depth
Devon Ice Cap (SE)	1879	-23.3
Devon Ice Cap (SE)	1858	-21.2
Devon Ice Cap (NW)	1787	-23.5
Devon Ice Cap (SE)	1729	-22.5
Devon Ice Cap (SE)	1609	-20.7
Devon Ice Cap (NW)	1595	-21.0
Devon Ice Cap (SE)	1522	-20.3
Devon Ice Cap (NW)	1317	-19.0
Devon Ice Cap (NW)	500	-14.8
Devon Ice Cap (NW)	305	-14.3
White Glacier	1424	-10.3
White Glacier	867	-14.7
White Glacier	844	-15.7
White Glacier	385	-12.0
White Glacier	206	-12.7
White Glacier	164	-11.4

[1]Blatter, 1985

Table 6.2. Firn line elevations (June, 1974)

Location (cf. Fig. 6.15)	Approximate Latitude °N	Elevation m a.s.l.
Devon Ice Cap (NW)		1280
Devon Ice Cap (SE)	75.5	800
Sydkap Glacier	75.0	920
Manson Icefield	76.5	470-630
Prince of Wales Icefield	77.7 78.7	760-910
Ice cap south of Agassiz Ice Cap	80.0	850-910
Agassiz Ice Cap	80.8	850-940
Muller Ice Cap	79.6 79.8	990-1350

The lower firn line elevations are on slopes facing water bodies (mostly Baffin Bay and its northern extensions). The equilibrium line is generally found 100-300 m below the firn line.

have retreated, the central parts have been thickening. For the 1960-1982 period the surface has been lowered at a mean rate of 100 mm/a near the edge while the centre has thickened at a rate of 90-100 mm/a.

The ice shelves, which lie along the north coast of Ellesmere Island, bordering the Arctic Ocean, are substantially smaller now than they were thirty years ago. The major calving occurred in the early 1960s when 596 km² broke away from the Ward Hunt Ice Shelf (Fig. 6.15), which reduced the ice shelf to about half its former size (Hattersley-Smith, 1963b). Further calving has occurred since then but has been of much smaller extent (Jeffries and Serson, 1983).

Thickness changes

Repeated measurements of depth in surface-to-bedrock boreholes, at the top of the Devon and Agassiz ice caps by the Polar Continental Shelf Project, have not shown any significant changes of thickness in recent years. Repeated gravity measurements on Devon Ice Cap confirm one set of these results. Levelling of two profiles in the ablation zone on Devon Ice Cap in 1965 and 1975 also showed no significant changes in glacier elevation. Arnold (1968), using traditional survey methods, found no changes in elevation in the accumulation zone above Gilman Glacier, Ellesmere Island (Fig. 6.15) for the 1957-1967 period. However, Arnold measured an average surface lowering of 0.17 m/a between 1957-1967 in the ablation zone of Gilman Glacier and of 0.83 m/a between 1960 and 1970 near the terminus of White Glacier (Arnold, 1981).

The unchanging thickness of ice caps in their accumulation zones suggests that accumulation rates have not changed significantly for several decades. This is supported by recent work on ice cores by Paterson and Waddington (1984) who extend this unchanging period to a thousand years. Consequently, the thinning of Gilman and White glaciers and the slight marginal retreat of many glaciers must be due to a recent slight increase in summer melting.

Past glacier balance (ice core record)

Using a thermal drill designed and made by the Cold Regions Research Laboratory of the U.S. Corps of Engineers, seven cores have been extracted from three Queen Elizabeth Islands ice caps (Table 6.3). The information gathered from these cores has provided insight into climatic change and distribution of glaciers over the past 100 ka.

The time scale

The top of Meighen Ice Cap, where the first Polar Continental Shelf Project (PCSP) core was drilled in 1965, suffers ablation in some years. The record is, therefore, discontinuous (Table 6.4). The tops of Devon and Agassiz ice caps, where all subsequent cores have been drilled, may be regarded as zones of continuous accumulation, i.e. annual layers are never lost. The climatic record from these cores is, therefore, a continuous one.

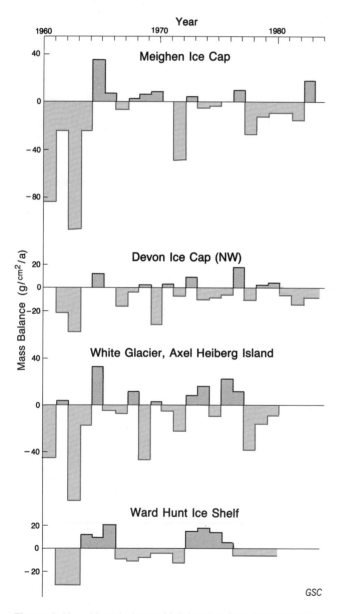

Figure 6.19. Mass balance, Meighen Ice Cap, Devon Ice Cap (northwest side), White Glacier on Axel Heiberg Island (Blatter, 1985), Ward Hunt Ice Shelf (Hattersley-Smith and Serson, 1970; H. Serson, Defence Research Establishment Pacific, personal communication, 1980). Black shading denotes a positive balance and red shading a negative balance.

Table 6.3. Surface-to-bedrock ice cores

Site (cf. Fig. 6.15)	Surface elevation m	Core length m	Year drilled
Meighen Ice Cap	268	121	1965
Devon Ice Cap	1800	300	1972
Devon Ice Cap	1800	300	1973
Agassiz Ice Cap	1700	340	1977
Agassiz Ice Cap	1710	137	1979
Agassiz Ice Cap	1715	127	1984

Note: One core on Devon Ice Cap, drilled in 1971, did not reach bedrock and is not included here.

Table 6.4. Climatic record from the Meighen Ice Cap core

Depth interval (m)	Estimated time interval (years)	Climate and balance at the core site
0	present-80	ablation surface formed during period of negative balance which removed a maximum of 13 m of ice at the core site
0-24	80-390	positive balance of superimposed ice
24-44	390-560	relatively little summer melt; coldest period in ice cap's history; positive balance of superimposed ice and firn
44-54	560-660	core site on a well drained slope; positive balance
54	660-2500/2000	long period of negative balance which reduced the thickness and area of the ice cap
54-116	2500/2000-?	overall positive balance with several short periods of negative balance
116-121	?- ca. 3500	positive balance, with the ice cap covering most of Meighen Island

There are various methods of dating ice cores from such areas. Apart from a theoretical approach to calculate a preliminary time scale, which was used for core cutting and field purposes, four main approaches have been used. The first method is to arrive at a time/depth relationship using the vertical strain rate, which may be measured at various depths in the borehole (Paterson, 1976); however, for this to be accurate it is necessary for the vertical velocities to have remained constant in the past and this may not be the case. Secondly, the decay rate of radioactive isotopes present in the ice, such as ^{32}Si and ^{14}C, may be used to date various levels in the ice as was done in the 1973 borehole on Devon Ice Cap (Paterson et al., 1977). Thirdly, suites of about 12 annual layers are identified at several intervals along the cores by measuring seasonal variations in the concentration of insoluble microparticles or various cations and anions. On the Queen Elizabeth Islands ice caps low accumulation rates and rapid thinning of the annual layers with depth, limit the application of this method to those parts of the core that accumulated over the last 7 ka. Finally, volcanic layers and major climatic events of known age (Fig. 6.20, 6.21) are identified within the cores. The major Icelandic volcanic eruptions, first detected in the Greenland Ice Sheet by Hammer et al. (1978), have also been found in the Queen Elizabeth Islands ice cores (Fisher et al., 1983) and can be used in this context.

The Queen Elizabeth Islands ice cores, with the exception of the core from Meighen Ice Cap, have been dated using a combination of the four methods. The time scales are accurate to about 5% for the last 5 ka of core and to about 10% for ice deposited 5 and 10 ka. For ice older than 10 ka, the time scales have been derived by comparisons between the oxygen isotope and microparticle profiles and those from Greenland cores; however, that time scale is only an approximation (Paterson et al., 1977).

The last 100 ka

The oxygen isotope results from ice older than 10 ka in the core drilled on Agassiz Ice Cap (Fig. 6.15) in 1979 are shown in Figure 6.21. This core is chosen for detailed discussion because it probably contains the most continuous and least disturbed record compiled so far. This is because the drill site is within 200 m of the top of a flow line. The major datum levels are the warming step at about 10 ka (A, Fig. 6.20), the very cold episodes centred at about 20 ka (B), 40 ka (C), and 60 ka (D), and the cooling step at about 70 ka (just below D, Fig. 6.20). All these ages may be substantially in error as will be discussed later. Those parts of the Camp Century (northwest Greenland; Fig. 6.15) and Devon ice cores deposited between 70 and 100 ka have oxygen isotope values higher than those in the Holocene. In the Agassiz Ice

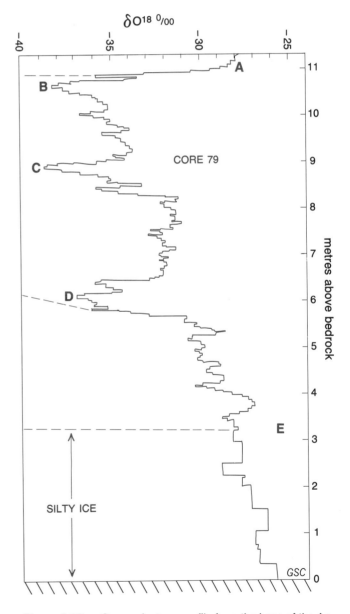

Figure 6.20. Oxygen isotope profile from the base of the Agassiz Ice Cap (1979). The letters are referred to in the text.

Cap cores the values for this interval (D to E, Fig. 6.20) are slightly lower than for the Holocene except for a small section 3.8 m above the bed. Isotope values such as these indicate warmer conditions than can be expected for a glacial period and yet ocean records suggest that sea level was lower than in the Holocene for the 70-100 ka period; this implies early glacial conditions. Paterson et al. (1977) explained this anomaly by considering that this ice is derived from moisture with a different source than the present-day ice. From a study of various components of the ocean record, Ruddiman et al. (1980) concluded that during this period, while the North American ice sheet was growing mainly in its eastern parts, the ocean was still warm. If true, this could mean that the ice caps of the Queen Elizabeth Islands were getting moisture from the same source area as today but in addition were receiving much more "warm" summer snow.

The oxygen isotope curves for the Agassiz ice cap cores were similar to those for cores from Greenland ice cores. Dansgaard et al. (1982), however, changed the original time scale for Camp Century record and adjusted it to fit the deep sea records. Thus what was once the 60-70 ka step (corresponding to the area just below D, Fig. 6.20) is now given an age of 115 ka. The revised time scale simplifies the oxygen isotope profile interpretation in terms of climatic change. Following this interpretation, the Camp Century record, and by comparison the Devon and Agassiz ice cap records as well, have continuously "cold" oxygen isotope values throughout what is identified as the last ice age. The ice with "warmer" isotope values in lower parts of the core (below D to the bed, Fig. 6.20) is, on this time scale, of interglacial origin.

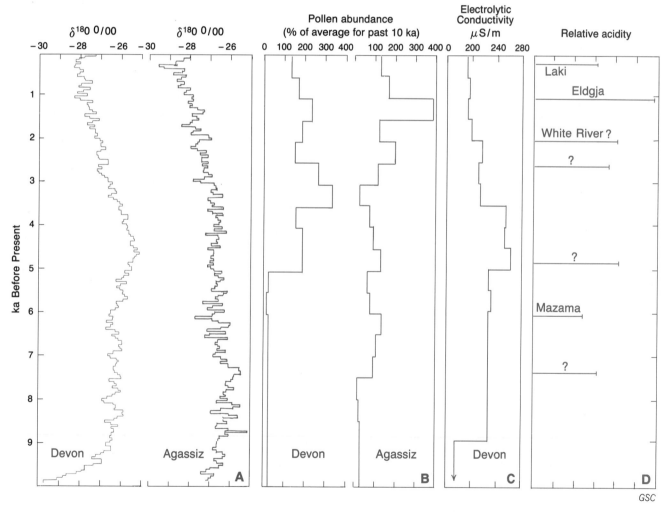

Figure 6.21. A. Oxygen isotope profiles from Devon Ice Cap (1973) and Agassiz Ice Cap (1979). B. Pollen profiles from Agassiz Ice Cap (1977) and Devon Ice Cap (1973), each expressed as percent of the average for the past 10 ka. C. Electrolytic conductivity profile from Devon Ice Cap (1973); the measurements are made on melted ice samples. D. Volcanic fallout levels in the ice: Laki and Eldgja were Icelandic eruptions, White River in the Yukon, and Mazama in the western United States. The volcanic eruptions were determined by acidity measurements made directly on the ice itself using the dry technique of Hammer (1980) and the arbitrary scale is relative hydrogen ion concentration.

The ice with very negative oxygen isotope values (glacial period ice, B to C, Fig. 6.20) has relatively high microparticle and calcium concentrations, indicative of high atmospheric turbidity. The presence of calcium suggests substantial influx to the ice caps of materials from continental shelves exposed during periods of low sea levels when there were extensive ice sheets over North America and Fennoscandia. The same ice also has different rheological properties from the ice above and below (Paterson, 1977b, 1981). In the Devon and Agassiz ice cap cores this ice is relatively "soft" so that borehole closure rates and tilt are high. This has important implications for ice dynamics as it means that Pleistocene ice sheets may have had different dynamic properties from the present ones on which glaciological models are based.

In general, the cores from Devon Ice Cap and the 1977 core from Agassiz Ice Cap have very low dirt concentrations (less than 1 ppm by volume). The lowermost 3 m of all three cores contain a few visible dirt pockets raising the average dirt concentration above 1 ppm (by volume). However, the lowermost 3 m of the core shown in Figure 6.20 (E to the bed) is quite different as it contains substantial amounts of silt-laden ice. Radio-echo sounding indicates that this silty layer is characteristic of the basal ice over much of the bedrock hill underlying the core site and extends for more than 1 km around that hill (Walford and Harper, 1981). The morphology of the grains in this dirty ice is consistent with a history of basal melt followed by freeze-on at the bed of the ice cap. The dirty ice may be older than the last interglacial so that the debris was picked up by freeze-on following pressure melting upstream along a paleo-flowline when the ice was about 200 m thicker. The evidence is not compelling, however, and present work suggests that the ice may be interglacial in origin and the dirt may have been blown onto the ice cap when it was small. In this case, it is unlikely that very much, if any, ice survived the early parts of the last interglacial period in the High Arctic.

The last 10 ka

Oxygen isotope profiles for the last 10 ka for the Devon (1973) and Agassiz (1979) ice cap cores are shown in Figure 6.21a. The Agassiz record shows a continuously warm period from 10 ka to about 4 ka and since then a gradual cooling. The warmest period was from 9 to 7 ka. The Devon record, on the other hand, shows a warm period reaching a peak 5 ka followed by a gradual cooling. Both records show a cold period beginning about 0.3 ka followed by a pronounced warming this century.

Although both the Agassiz and Devon ice cap records suggest a cooling over the last 4-5 ka, the Meighen Ice Cap record (Table 6.4) suggests that there was a lengthy warm period ending 0.6-0.7 ka (Koerner and Paterson, 1974). There is similar evidence for this warm period in the Ward Hunt Ice Shelf (Lyons and Mielke, 1973). Despite these discrepancies, it is probably safe to say that the isotope records show that the past 3 ka have been generally colder than the previous 6 ka and that the period often referred to as the "Little Ice Age" was among the coldest of the past 10 ka.

Pollen grains and spores are found in the snows of the Queen Elizabeth Islands, although only in concentrations of a few grains per litre of melted snow. At present no trends in their areal distribution have been found, although there is a steep gradient between the southern limits of the islands and treeline (Bourgeois et al., 1985). However, there are significant temporal variations in the pollen concentrations in the cores (Fig. 6.21b). There is a clear contrast between pollen drought conditions in the lower parts of the cores representing the Late Pleistocene and the early Holocene and relatively high concentrations in the second half of the Holocene. During the last glaciation, most of the present-day pollen source areas were covered by the ice sheets. Pollen influx to the High Arctic could not restart until deglaciation and recolonization occurred. On Devon Ice Cap there is a 5 ka delay between the oxygen isotope event defining the beginning of the Holocene and a sudden increase in pollen concentations. On Agassiz Ice Cap there is only a 2-3 ka delay, although the initial increase is only to levels already existing in early Holocene ice of the Devon Ice Cap. In the last 5 ka maximum pollen concentrations occurred about 3.0 and 1.0 ka on the Devon Ice Cap and 2.0 and 1.0 ka on the Agassiz Ice Cap. It is interesting that the pollen levels have remained high during the period that the oxygen isotopes suggest is one of cooling climate. Pollen concentrations did peak, however, during a generally recognized period of warmth centred around 1.0 ka.

The geochemical record for the Devon Ice Cap cores (Fig. 6.21c) shows evidence for more open sea-ice conditions during the period of warmth centred around 5 ka. This is deduced from the higher electrolytic conductivity of ice accumulated during this warm period. The high conductivities are attributed to increased concentrations of marine salts deposited on the ice cap during long periods of open water each year in Baffin Bay (Fisher and Koerner, 1981). This is in agreement with Blake (1972) and Stewart and England (1983) who, from studies of ^{14}C ages of driftwood, concluded that there were more open water conditions than usual in the arctic island channels during the Hypsithermal period ending about 4.5 ka.

While there is a dramatic difference between the concentration of microparticles in the Pleistocene ice compared to that in Holocene ice, there are no major trends apparent within the Holocene itself. This suggests that there have been no significant changes of atmospheric turbidity between the various microparticle sources and the drill sites during the Holocene.

Volcanic fallout onto the ice caps can be seen in the cores in the form of highly acidic layers (Fig. 6.21d). Whereas acid fallout following a volcanic eruption lasts for about two years, the seasonal cycles of metals such as calcium, sodium and potassium continue undisturbed, although occasionally at higher levels. Microparticle concentrations, however, rarely show an increase and, even using an electron microscope, ash layers have not been detected. This is true even for highly acidic layers of known volcanic origin (the Laki eruption in Iceland, Fig. 6.21d). Most of the large acid peaks in the cores originate from high northern latitude eruptions, mainly from Iceland and to a lesser extent Alaska and the Yukon.

The last 1 ka

For the period covering the last 1 ka an additional proxy temperature indicator has been used — the percentage of each annual layer of accumulation that has undergone melt to form an ice layer. It is related to summer temperatures when the snow forming that core was at the surface (Koerner, 1977b). The melt records extend to 150 m depth, where microfractures begin to obscure the stratigraphy.

The melt record, together with the curve for oxygen isotopes, shows the cold period of the "Little Ice Age" lasting from about 1570 to 1850 AD with the coldest part beginning in 1650 AD (Fig. 6.22). The records suggest that this period, one of the coldest in the last 10 ka, was 1.5-2.5°C colder than the warm period we have just emerged from. Some five-year periods in the Agassiz Ice Cap record show no summer melt — a phenomena which occurs with the coldest summer in ten today. Alt et al. (1985) have shown how the low temperatures and resultant severe sea-ice conditions hampered the exploration of the Northwest Passage and contributed to the failure of the Franklin expedition. In contrast, parts of this century stand out as unique in terms of the last thousand years, as they contain some of the warmest summers and years of the period.

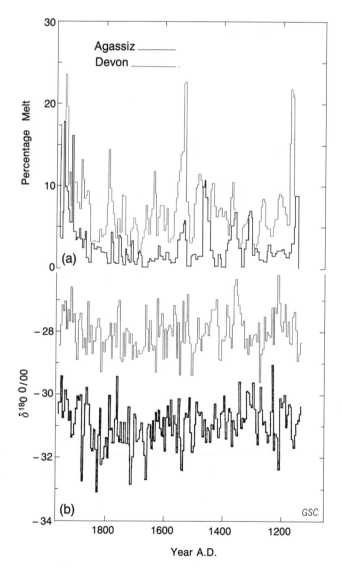

Figure 6.22. Oxygen isotope and percentage melt profiles for Agassiz (black) and Devon (red) ice caps. Percentage melt is the percentage of each annual layer of accumulation that has undergone melt to form an ice layer.

ACKNOWLEDGMENTS

Manuscript maps and field notebooks of J.G. Fyles (Geological Survey of Canada) have provided a wealth of information on surface materials and their origin. This chapter has been reviewed by W. Blake, Jr., N.R. Gadd (Geological Survey of Canada), J.B. Bird (McGill University), J. England (University of Alberta), and W.H. Mathews (University of British Columbia).

REFERENCES

Alt, B.T., Koerner, R.M., Fisher, D.A., and Bourgeois, J.C.
1985: Arctic climate during the Franklin era as deduced from ice cores; in Proceedings of "The Franklin Era in Canadian Arctic History", P. Sutherland (ed.); Archaeological Survey of Canada, Mercury Series, Paper 131, p. 69-92.

Andrews, J.T.
1970: A geomorphic study of postglacial uplift with particular reference to Arctic Canada; Institute of British Geographers, London, Special Publication 2, 156 p.

Andrews, J.T. and Miller, G.H.
1976: Quaternary glacial chronology of the eastern Canadian Arctic: a review and a contribution on amino acid dating of Quaternary molluscs from the Clyde cliffs; in Quaternary Stratigraphy of North America, W.C. Mahaney (ed.); Dowden, Hutchinson and Ross, Inc., Stroudsburg, Pennsylvania, p. 1-32.

Arnold, K.C.
1968: Determination of changes of surface height 1957-1967, of the Gilman Glacier, northern Ellesmere Island, Canada; unpublished MA thesis, McGill University, Montréal, Quebec, 74 p.
1981: Ice ablation measured by stakes and terrestrial photogrammetry — a comparison on the lower part of the White Glacier; Axel Heiberg Island Research Reports, Glaciology, No. 2, McGill University, Montréal, Quebec, 98 p.

Babb, T.A. and Bliss, L.C.
1974: Susceptibility to environmental impact in the Queen Elizabeth Islands; Arctic, v. 27, p. 234-237.

Balkwill, H.R. and Bustin, R.M.
1975: Stratigraphic and structural studies, central Ellesmere Island and eastern Axel Heiberg Island, District of Franklin; in Report of Activities, Part A, Geological Survey of Canada, Paper 75-1A, p. 513-517.

Balkwill, H.R., Roy, K.J., and Hopkins, W.S., Jr.
1974: Glacial features and pingos, Amund Ringnes Island, Arctic Archipelago; Canadian Journal of Earth Sciences, v. 11, p. 1319-1325.

Ballantyne, C.K. and McCann, S.B.
1980: Short lived damming of a high-arctic ice-marginal stream, Ellesmere Island, N.W.T., Canada; Journal of Glaciology, v. 25, p. 487-491.

Barnett, D.M.
1972: Surficial geology and geomorphology of Melville Island, District of Franklin; in Report of Activities, Part A, Geological Survey of Canada, Paper 72-1A, p. 152-153.

Barnett, D.M., Edlund, S.A., and Dredge, L.A.
1977: Terrain characterization and evaluation: An example from eastern Melville Island; Geological Survey of Canada, Paper 76-23, 18 p.

Barnett, D.M., Edlund, S.A., Dredge, L.A., Thomas, D.C., and Prevett, L.S.
1975: Terrain classification and evaluation, eastern Melville Island, N.W.T.; Geological Survey of Canada, Open File 252, 1318 p., 3 maps.

Barr, W.
1971: Postglacial isostatic movement in northeastern Devon Island: a reappraisal; Arctic, v. 24, p. 249-268.

Barsch, D., King, L., and Mausbacher, R.
1981: Glaziologische Beobachtungen an der Stirn des Webber-Gletschers, Borup-Fjord-Gebiet, N-Ellesmere Island, N.W.T., Kanada; Heidelberger Geographische Arbeiten, v. 69, p. 269-284.

Basham, P.W., Forsyth, D.A., and Wetmiller, R.J.
1977: The seismicity of northern Canada; Canadian Journal of Earth Sciences, v. 14, p. 1646-1667.

Bednarski, J.
1984: Glacier fluctuations and sea level history of Clements Markham Inlet, northern Ellesmere Island; unpublished PhD thesis, The University of Alberta, Edmonton, Alberta, 232 p.
1986: Late Quaternary glacial and sea-level events, Clements Markham Inlet, northern Ellesmere Island; Canadian Journal of Earth Sciences, v. 23, p. 1343-1355.

Bennike, O., Dawes, P.R., Funder, S., Kelly, M., and Weidick, A.
1987: The late Quaternary history of Hall Island, northwest Greenland: Discussion; Canadian Journal of Earth Sciences, v. 24, p. 370-374.

Berggren, W.A.
1972: Late Pliocene-Pleistocene glaciation; in Initial Reports of the Deep Sea Drilling Project, v. 12, A.S. Laughton et al. (ed.); U.S. Government Printing Office, p. 953-963.

Bird, J.B.
1959: Recent contributions to the physiography of Northern Canada; Zeitschrift für geomorphologie, Band 3, Heft 2, p. 151-174.
1967: The Physiography of Arctic Canada (with special reference to the area south of Parry Channel); The John Hopkins Press, Baltimore, 336 p.

Blake, W., Jr.
1964: Preliminary account of the glacial history of Bathurst Island, Arctic Archipelago; Geological Survey of Canada, Paper 64-30, 8 p.
1970: Studies of glacial history in Arctic Canada. I. Pumice, radiocarbon dates, and differential postglacial uplift in the eastern Queen Elizabeth Islands; Canadian Journal of Earth Sciences, v. 7, p. 634-664.
1972: Climatic implications of radiocarbon-dated driftwood in the Queen Elizabeth Islands, Arctic Canada; in Climatic Changes in Arctic Areas During the Last Ten-Thousand Years, Y. Vasari, H. Hyvärinen, and S. Hicks (ed.); Proceedings of a symposium held at Oulanka and Kevo, Finland, October 1971, Acta Universitatis Ouluensis, Ser. A, Scientiae Rerum Naturalium, No. 3, Geologica No. 1, p. 77-104.
1973: Former occurrence of *Mytilus edulis* L. on Coburg Island, Arctic Archipelago; Naturaliste canadien, v. 100, p. 51-58.
1974: Studies of glacial history in Arctic Canada. II. Interglacial peat deposits in Bathurst Island; Canadian Journal of Earth Sciences, v. 11, p. 1025-1042.
1975: Radiocarbon age determinations and postglacial emergence at Cape Storm, southern Ellesmere Island, Arctic Canada; Geografiska Annaler, v. 57A, p. 1-71.
1976: Sea and land relations during the last 15 000 years in the Queen Elizabeth Islands, Arctic Archipelago; in Report of Activities, Part B, Geological Survey of Canada, Paper 76-1B, p. 201-207.
1978: Rock weathering forms above Cory Glacier, Ellesmere Island, District of Franklin; in Current Research, Part B, Geological Survey of Canada, Paper 78-1B, p. 207-211.
1980: Mid-Wisconsinan interstadial deposits beneath Holocene beaches, Cape Storm, Ellesmere Island, Arctic Canada (abstract); in Abstracts, AMQUA 6th Biennial Meeting, Orono, Maine, p. 26-27.
1981a: Lake sediment coring along Smith Sound, Ellesmere Island and Greenland; in Current Research, Part A, Geological Survey of Canada, Paper 81-A, p. 191-200.
1981b: Neoglacial fluctuations of glaciers, southeastern Ellesmere Island, Canadian Arctic Archipelago; Geografiska Annaler, v. 63A, p. 201-218.
1982: Terrestrial interstadial deposits, Ellesmere Island, N.W.T., Canada (abstract); in Abstracts, AMQUA 7th Biennial Meeting, Seattle, Washington, p. 73.
1986: Glacial history along Smith Sound, Arctic Canada and Greenland (abstract); in Abstracts, Nordiska Geologmotet 17th Meeting, Helsingfors Universitet, Finland, p. 16.

Blake, W., Jr. and Matthews, J.V., Jr.
1979: New data on an interglacial peat deposit near Makinson Inlet, Ellesmere Island, District of Franklin; in Current Research, Part A, Geological Survey of Canada, Paper 79-1A, p. 157-164.

Blatter, H.
1985: On the thermal regime of Arctic glaciers; Zurcher Geographische Schriften, v. 22, Geographisches Institut, eidgenossische Technische Hochschule Zurich, 107 p.

Boesch, H.
1963: Notes on the geomorphological history; in Axel Heiberg Research Reports, Preliminary Report 1961-62, F. Müller et al. (ed.); McGill University, Montréal, p. 163-167.

Bornhold, B.D., Finlayson, N.M., and Monahan, D.
1976: Submerged drainage patterns in Barrow Strait, Canadian Arctic; Canadian Journal of Earth Sciences, v. 13, p. 305-311.

Bostock, H.H.
1970: Physiographic regions of Canada; Geological Survey of Canada, Map 1245A, scale 1:5 000 000.

Boulton, G.S.
1979: A model of Weischselian glacier variation in the North Atlantic regions; Boreas, v. 8, p. 373-395.

Bourgeois, J.C., Koerner, R.M., and Alt, B.T.
1985: Airborne pollen: a unique air mass tracer, its influx to the Canadian High Arctic; Annals of Glaciology, v. 7, p. 109-116.

Bovis, M.J. and Barry, R.G.
1974: A climatological analysis of north polar desert areas; in Polar Deserts and Modern Man, T.L. Smiley and J.H. Zumberge (ed.); University of Arizona Press, Tucson, p. 23-31.

Brassard, G.R.
1971: The mosses of northern Ellesmere Island, Arctic Canada. I. Ecology and photogeography, with an analysis for the Queen Elizabeth Islands; The Bryologist, v. 74, p. 233-281.

Brassard, G.R. and Blake, W., Jr.
1978: An extensive subfossil deposit of the arctic moss *Aplodon wormskioldii*; Canadian Journal of Botany, v. 56, p. 1852-1859.

Bustin, R.M.
1982: Beaufort Formation, eastern Axel Heiberg Island, Canadian Arctic Archipelago; Canadian Petroleum Geology, Bulletin, v. 30, p. 140-149.

Canadian Hydrographic Service
1979: General bathymetric chart of the oceans; Canadian Hydrographic Service, Sheet 5-17, 5th edition, scale 1:6 000 000.

Christie, R.L.
1967: Reconnaissance of the surficial geology of northeastern Ellesmere Island, Arctic Archipelago; Geological Survey of Canada, Bulletin 138, 50 p.

Cogley, J.G. and McCann, S.B.
1976: An exceptional storm and its effects in the Canadian high Arctic; Arctic and Alpine Research, v. 8, p. 105-110.

Craig, B.G. and Fyles, J.G.
1960: Pleistocene geology of Arctic Canada; Geological Survey of Canada, Paper 60-10, 21 p.
1965: Quaternary of Arctic Canada; in Anthropogene Period in Arctic and Subarctic (Translation); Scientific Research Institute of the Geology of the Arctic, State Geological Committee, U.S.S.R., Moscow, 143, p. 5-33. (Russian with English summary).

Crary, A.P.
1960: Arctic ice island and ice shelf studies, Part II; Arctic, v. 13, p. 32-50.

Daae, H.B. and Rutgers, A.T.C.
1975: Geological history of the Northwest Passage; in Canada's Continental Margins and Offshore Petroleum Exploration, C.J. Yorath, E.R. Parker, and D.J. Glass (ed.); Canadian Society of Petroleum Geologists, Memoir 4, p. 477-500.

Dansgaard, W., Clausen, H.B., Gundestrup, N., Hammer, C.U., Johnsen, S.F., Kristindottir P.M., and Reeh, N.
1982: A new Greenland ice core; Science, v. 218, p. 1273-1277.

Dansgaard, W., Johnsen, S.J., Clausen, H.B., and Gundestrup, N.
1973: Stable isotope glaciology; Meddelelser om Grønland, v. 197, p. 1-53.

Dunbar, M. and Greenaway, K.R.
1956: Arctic Canada from the air; Defence Research Board, Ottawa, Queen's Printer, 541 p.

Dyck, W. and Fyles, J.G.
1964: Geological Survey of Canada radiocarbon dates III; Radiocarbon, v. 6, p. 167-181.

Dyke, A.S.
1983: Quaternary geology of Somerset Island, District of Franklin; Geological Survey of Canada, Memoir 404, 32 p.
1984: Quaternary geology of Boothia Peninsula and northern District of Keewatin, central Canadian Arctic; Geological Survey of Canada, Memoir 407, 26 p.

Edlund, S.A.
1983: Bioclimatic zonation in a High Arctic region: central Queen Elizabeth Islands; in Current Research, Part A, Geological Survey of Canada, Paper 83-1A, p. 381-390.
1985: Lichen-free zones as neoglacial indicators on western Melville Island, District of Franklin; in Current Research, Part A, Geological Survey of Canada, Paper 85-1A, p. 709-712.
in press: Surficial geology of Cornwallis and adjacent islands, District of Franklin, Northwest Territories; Geological Survey of Canada, Paper.

Edlund, S.A. and Alt, B.T.
1989: Regional congruence of vegetation and summer climate patterns in the Queen Elizabeth Islands, Canada; Arctic, v. 42, p. 3-22.

England, J.
1974: The glacial geology of the Archer Fiord/Lady Franklin Bay area, northeastern Ellesmere Island, N.W.T., Canada; unpublished PhD thesis, Department of Geography, University of Colorado, Boulder, Colorado, 234 p.
1976a: Postglacial isobases and uplift curves from the Canadian and Greenland High Arctic; Arctic and Alpine Research, v. 8, p. 61-78.
1976b: Late Quaternary glaciation of the eastern Queen Elizabeth Islands, Northwest Territories, Canada: alternative models; Quaternary Research, v. 6, p. 185-202.
1978: The glacial geology of northeastern Ellesmere Island, Northwest Territories, Canada; Canadian Journal of Earth Sciences, v. 15, p. 603-617.

1982: Postglacial emergence along northern Nares Strait; in Nares Strait and the Drift of Greenland; A Conflict in Plate Tectonics; P.R. Dawes and J.W. Kerr (ed.); Meddelelser om Gronland, Geoscience, v. 8, p. 65-75.
1983: Isostatic adjustments in a full glacial sea; Canadian Journal of Earth Sciences, v. 20, p. 895-917.
1985: The late Quaternary history of Hall Land, northwest Greenland; Canadian Journal of Earth Sciences, v. 22, p. 1394-1408.
1987: The late Quaternary history of Hall Island, northwest Greenland: Reply; Canadian Journal of Earth Sciences, v. 24, p. 374-380.

England, J. and Bradley, R.S.
1978: Past glacial activity in the Canadian High Arctic; Science, v. 200, p. 265-270.

England, J., Bradley, R.S., and Miller, G.H.
1978: Former ice shelves in the Canadian High Arctic; Journal of Glaciology, v. 20, p. 393-404.

England, J., Bradley, R.S., and Stuckernrath, R.
1981: Multiple glaciations and marine transgressions, western Kennedy Channel, Northwest Territories, Canada; Boreas, v. 10, p. 71-89.

Farrand, W.R. and Gajda, R.T.
1962: Isobases on the Wisconsin marine limit in Canada; Geographical Bulletin, v. 17, p. 5-22.

Fisher, D.A. and Koerner, R.M.
1981: Some aspects of climatic change in the High Arctic during the Holocene as deduced from ice cores; in Quaternary Paleoclimate, W.C. Mahaney (ed.); Geo Books, Norwich, p. 249-271.
1983: Ice-core study: a climatic link between the past, present and future; in Climatic Change in Canada 3, C.R. Harington (ed.); National Museum of Natural Sciences, Syllogeus, v. 49, p. 50-69.

Fisher, D.A., Koerner, R.M., Paterson, W.S.B., Dansgaard, W., Gundestrup, N., and Reeh, N.
1983: Effect of wind scouring on climatic records from ice core oxygen-isotope profiles; Nature, v. 301, p. 205-209.

Fortier, Y.O. and Morley, L.W.
1956: Geological unity of the Arctic Islands; Royal Society of Canada, Transactions, v. 50, Ser. 3, p. 3-12.

Fortier, Y.O., Blackadar, R.G., Glenister, B.F., Grenier, H.R., McLaren, D.J., McMillan, N.J., Norris, A.W., Roots, E.F., Souther, J.G., Thorsteinsson, R., and Tozer, E.T.
1963: Geology of the north central part of the Arctic Archipelago, Northwest Territories (Operation Franklin); Geological Survey of Canada, Memoir 320, 671 p.

Foscolos, A.E. and Kodama, H.
1981: Mineralogy and chemistry of arctic desert soils on Ellef Ringnes Island, Arctic Canada; Soil Science Society of America Journal, v. 45, p. 987-993.

French, H.M.
1976: The Periglacial Environment; Longmans, London, 309 p.

Funder, S.
1989: Quaternary geology of North Greenland; in Chapter 13 of Quaternary Geology of Canada and Greenland, R.J. Fulton (ed.); Geological Survey of Canada, Geology of Canada, no. 1 (also Geological Society of America, The Geology of North America, v. K-1).

Funder, S., Abrahamsen, H., Bennike, O., and Feyling-Hanssen, R.W.
1985: Forested arctic: evidence from North Greenland; Geology, v. 13, p. 542-546.

Hammer, C.U.
1980: Acidity of polar ice cores in relation to absolute dating, past volcanism and radio-echoes; Journal of Glaciology, v. 25, p. 359-372.

Hammer, C.U., Clausen, H.B., Dansgaard, W., Gundestrup, N., Johnsen, S.J., and Reeh, N.
1978: Dating of Greenland ice cores by flow models, isotopes, volcanic debris, and continental dust; Journal of Glaciology, v. 20, p. 3-26.

Hattersley-Smith, G.
1960: Some remarks on glaciers and climate in northern Ellesmere Island; Geografiska Annaler, v. 42, p. 45-48.
1963a: Climatic inferences from firn studies in northern Ellesmere Island; Geografiska Annaler, v. 45, p. 139-151.
1963b: The Ward-Hunt Ice Shelf: recent changes in the ice front; Journal of Glaciology, v. 4, p. 415-424.
1969a: Glacial features of Tanquary Fiord and adjoining areas of northern Ellesmere Island, Northwest Territories; Journal of Glaciology, v. 8, p. 23-50.
1969b: Recent observations on the surging Otto Glacier, Ellesmere Island; Canadian Journal of Earth Sciences, v. 6, p. 883-889.
1973: An archaeological site on the north coast of Ellesmere Island; Arctic, v. 26, p. 255-256.

Hattersley-Smith, G. and Long, A.
1967: Postglacial uplift at Tanquary Fiord, northern Ellesmere Island, N.W.T., Arctic, v. 20, p. 255-260.

Hattersley-Smith, G. and Serson, H.
1970: Mass balance of the Ward-Hunt Ice Rise and Ice Shelf: a 10 year record; Journal of Glaciology, v. 9, p. 247-252.

Hattersley-Smith, G., Crary, A.P., and Christie, R.L.
1955: Northern Ellesmere Island, 1953 and 1954; Arctic, v. 8, p. 3-36.

Hattersley-Smith, G., Fuzesy, A., and Evans, S.
1969: Glacier depths in northern Ellesmere Island: airborne radio-echo sounding in 1966; Defence Research Board, Ottawa, DREO Technical Note 69-6, 23 p.

Heginbottom, J.A.
1984: The bursting of a snow dam, Tingmisut Lake, Melville Island, Northwest Territories; in Current Research, Part B, Geological Survey of Canada, Paper 84-1B, p. 187-192.

Heginbottom, J.A. (co-ordinator)
1989: A survey of geomorphic processes in Canada; Chapter 9 in Quaternary Geology of Canada and Greenland, R.J. Fulton (ed.); Geological Survey of Canada, Geology of Canada, no. 1 (also Geological Society of America, The Geology of North America, v. K-1)

Henoch, W.E.S.
1964: Postglacial marine submergence and emergence of Melville Island, N.W.T.; Geographical Bulletin, no. 22, p. 105-126.

Herman, Y. and Hopkins, D.M.
1980: Arctic Ocean climate in Late Cenozoic time; Science, v. 209, p. 557-562.

Hodgson, D.A.
1973: Landscape, and late-glacial history, head of Vendom Fiord, Ellesmere Island; in Report of Activities, Part B, Geological Survey of Canada, Paper 73-1B, p. 129-136.
1981: Surficial geology, Lougheed Island, northwest Arctic Archipelago; in Current Research, Part C, Geological Survey of Canada, Paper 81-1C, p. 27-34.
1982: Surficial materials and geomorphological processes, western Sverdrup and adjacent islands, District of Franklin; Geological Survey of Canada, Paper 81-9, 37 p.
1985: The last glaciation of west-central Ellesmere Island, Arctic Archipelago, Canada; Canadian Journal of Earth Sciences, v. 22, p. 347-368.

Hodgson, D.A. and Edlund, S.A.
1975: Surficial materials and biophysical regions, eastern Queen Elizabeth Islands: Part I, Baumann Fiord (including Bjorne Peninsula, NTS 49 C) and Graham Island (NTS 59 D); Geological Survey of Canada, Open File 265.
1978: Surficial materials and vegetation, Amund Ringnes and Cornwall islands, District of Franklin; Geological Survey of Canada, Open File 541.

Hodgson, D.A. and Vincent, J-S.
1984a: A 10 000 yr. BP extensive ice shelf over Viscount Melville Sound, Arctic Canada; Quaternary Research, v. 22, p. 18-30.
1984b: Surficial geology, central Melville Island, Northwest Territories; Geological Survey of Canada, Map 1583A, scale 1:250 000.

Hodgson, D.A., Vincent, J-S., and Fyles, J.G.
1984: Quaternary geology of central Melville Island, Northwest Territories; Geological Survey of Canda, Paper 83-16, 25 p.

Hudec, P.P.
1973: Weathering of rocks in arctic and subarctic environments; in Proceedings: Symposium on the Geology of the Canadian Arctic, J.D. Aitken and D.J. Glass (ed.); Geological Association of Canada, p. 313-335.

Hughes, T., Denton, G.H., and Grosswald, M.G.
1977: Was there a late-Wurm Arctic Ice Sheet?; Nature, v. 266, p. 596-602.

Hyvärinen, H.
1985: Holocene pollen stratigraphy of Baird Inlet, east-central Ellesmere Island, Arctic Canada; Boreas, v. 14, p. 19-32.

Iken, A.
1974: Velocity fluctuations of an Arctic valley glacier — a study of the White Glacier, Axel Heiberg Island; Axel Heiberg Island Research Reports, Glaciology, No. 5, McGill University, Montreal, 116 p.

Jeffries, M.O. and Serson, H.
1983: Recent changes at the front of Ward Hunt Ice Shelf, Ellesmere Island, N.W.T.; Arctic, v. 36, p. 289-290.

Jenness, J.L.
1952: Problem of glaciation in the western islands of arctic Canada; Geological Society of America, Bulletin, v. 63, p. 939-952.

Jenness, S.E.
1962: Fieldwork, 1961; Geological Survey of Canada, Information Circular No. 5, 82 p.

Kerr, J.W.
1967: Nares submarine rift valley and the relative rotation of north Greenland; Canadian Petroleum Geology, Bulletin, v. 15, p. 483-520.
1980: Structural framework of Lancaster Aulacogen, Arctic Canada; Geological Survey of Canada, Bulletin 319, 24 p.
1981: Evolution of the Canadian Arctic Islands: a transition between the Atlantic and Arctic Oceans; in The Ocean Basin Margin, Volume 5, The Arctic Ocean, A.E.M. Nairn, M. Churkin, Jr., and F.G. Stehli (ed.); Plenum Press, New York, p. 115-199.

King, L.
1981: Studies in glacial history of the area between Oobloyah Bay and Esayoo Bay, northern Ellesmere Island, N.W.T. Canada; in Results of the Heidelberg Ellesmere Island Expedition; Heidelberger Geographische Arbeiten, v. 69, p. 233-267 (English abstract and captions).

Knuth, E.
1966: The ruins of the Musk-ox way; Folk, v. 8, p. 191-219.
1967: Archeology of the Muskox Way; École Pratique des Hautes Études, Contributions du Centre d'Études arctiques et fenno-scandinaves, n°5, 70 p.

Koerner, R.M.
1968: Fabric analysis of a core from the Meighen Ice Cap, Northwest Territories, Canada; Journal of Glaciology, v. 10, p. 421-430.
1970: The mass balance of the Devon Island Ice Cap; Journal of Glaciology, v. 9, p. 325-336.
1977a: Ice thickness measurements and their implications with respect to past and present ice volumes in the Canadian High Arctic ice caps; Canadian Journal of Earth Sciences, v. 14, p. 2697-2705.
1977b: Devon Island Ice Cap: core stratigraphy and paleoclimate; Science, v. 196, p. 15-18.
1979: Accumulation, ablation and oxygen isotope variations on the Queen Elizabeth Island ice caps, Canada; Journal of Glaciology, v. 22, p. 25-41.
1980: The problem of lichen-free zones in Arctic Canada; Arctic and Alpine Research, v. 12, p. 87-94.

Koerner, R.M. and Paterson, W.S.B.
1974: Analysis of a core through the Meighen Ice Cap, Arctic Canada, and its paleoclimatic implications; Quaternary Research, v. 4, p. 253-263.

Koerner, R.M. and Russell, R.D.
1979: $\delta^{18}O$ variations in snow on the Devon Island Ice Cap, Northwest Territories, Canada; Canadian Journal of Earth Sciences, v. 16, p. 1419-1427.

Kurfurst, P.J. and Veillette, J.J.
1977: Geotechnical characterization of terrain units, Bathurst, Cornwallis, Somerset, Prince of Wales and adjacent islands; Geological Survey of Canada, Open File 471.

Lowdon, J.A. and Blake, W., Jr.
1968: Geological Survey of Canada radiocarbon dates VII; Radiocarbon, v. 10, p. 207-245.
1973: Geological Survey of Canada radiocarbon dates XIII; Geological Survey of Canada, Paper 73-7, 61 p.

Lyons, J.B. and Mielke, J.E.
1973: Holocene history of a portion of northernmost Ellesmere Island; Arctic, v. 26, p. 314-323.

Marlowe, J.I.
1968: Sedimentology of the Prince Gustaf Adolf Sea area, District of Franklin; Geological Survey of Canada, Paper 66-29, 83 p.

Mathewson, C.C. and Mayer-Cole, T.A.
1984: Development and runout of a detachment slide, Bracebridge Inlet, Bathurst Island, Northwest Territories, Canada; Association of Engineering Geologists, Bulletin, v. 21, p. 407-424.

Matthews, J.V., Jr.
1987: Plant macrofossils from the Neogene Beaufort Formation on Banks and Meighen islands, District of Franklin; in Current Research, Part A, Geological Survey of Canada, Paper 87-1A, p. 73-87.

Maxwell, J.B.
1980: The Climate of the Canadian Arctic Islands and Adjacent Waters; Canada, Department of Environment, Atmospheric Environment Service, Climatological Studies, No. 30, v. 1, 532 p.

Mayewski, P.A., Denton, G.H., and Hughes, T.J.
1981: Late Wisconsinan ice sheets of North America; in The Last Great Ice Sheets, G.H. Denton and T.J. Hughes (ed.); John Wiley, New York, p. 67-178.

McCann, S.B., Howarth, P.J., and Cogley, J.G.
1972: Fluvial processes in a periglacial environment, Queen Elizabeth Islands, N.W.T., Canada; Institute of British Geographers, Transactions, v. 55, p. 69-82.

McLaren, P.
1982: The coastal geomorphology, sedimentology and processes of eastern Melville and western Byam Martin islands; Geological Survey of Canada, Bulletin 333, 39 p.

McLaren, P. and Barnett, D.M.
1978: Holocene emergence of the south and east coasts of Melville Island, Queen Elizabeth Islands, Northwest Territories, Canada; Arctic, v. 31, p. 415-427.

Miller, G.H., Bradley, R.S., and Andrews, J.T.
1975: The glaciation level and lowest equilibrium line altitude in the high Canadian Arctic: maps and climatic interpretation; Arctic and Alpine Research, v. 7, p. 155-168.

Müller, F.
1963: Radiocarbon dates and notes on the climatic and morphological history; in Axel Heiberg Island Research Reports, Preliminary Report 1961-1962, F. Müller et al. (ed.); McGill University, Montréal, Quebec, p. 169-172.
1969: Was the Good Friday Glacier on Axel Heiberg Island surging?; Canadian Journal of Earth Sciences, v. 6, p. 891-894.

Narod, B.B. and Clarke, G.K.C.
1983: UHF radar system for airborne ice thickness surveys; Canadian Journal of Earth Sciences, v. 20, p. 1073-1086.

National Atlas of Canada
1973: Relief profiles; National Atlas of Canada, Physical Geography Section, Map 3-4, 4th edition; Surveys and Mapping Branch, Canada, Department of Energy, Mines and Resources, scale 1:20 000 000.

Oswald, G.K.A.
1975: Investigation of sub-ice bedrock characteristics by radio-echo sounding; Journal of Glaciology, v. 15, p. 75-87.

Paterson, W.S.B.
1968: A temperature profile through the Meighen Ice Cap, Arctic Canada; International Association of Scientific Hydrology Publication, no. 89, p. 440-449.
1976: Vertical strain-rate measurements in an Arctic ice cap and deductions from them; Journal of Glaciology, v. 17, p. 3-12.
1977a: Extent of the late-Wisconsin glaciation in northwest Greenland and northern Ellesmere Island: a review of the glaciological and geological evidence; Quaternary Research, v. 8, p. 180-190.
1977b: Secondary and tertiary creep of glacier ice as measured by borehole closure rates; Reviews of Geophysics and Space Physics, v. 15, p. 47-55.
1981: The Physics of Glaciers; Pergamon Press, Oxford, 380 p.

Paterson, W.S.B. and Clarke, G.K.
1978: Comparison of theoretical and observed temperature profiles in Devon Island Ice Cap, Canada; Geophysical Journal of the Royal Astronomical Society, v. 55, p. 615-632.

Paterson, W.S.B. and Koerner, R.M.
1974: Radio echo sounding on four ice caps in Arctic Canada; Arctic, v. 27, p. 225-233.

Paterson, W.S.B. and Waddington, E.D.
1984: Past accumulation rates at Camp Century and Devon Island deduced from ice core measurements; Annals of Glaciology, v. 5, p. 222-223.

Paterson, W.S.B., Koerner, R.M., Fisher, D.A., Johnsen, S.J., Clausen, H.R., Dansgaard, W., Bucher, P., and Oschger, H.
1977: An oxygen isotope climatic record from the Devon Island Ice Cap, Arctic Canada; Nature, v. 266, p. 508-511.

Pelletier, B.R.
1966: Development of submarine physiography in the Canadian Arctic and its relation to crustal movements; in Continental Drift, G.D. Garland (ed.); Royal Society of Canada, Special Publication, no. 9, p. 77-101.

Pissart, A.
1967: Les pingos de l'île Prince Patrick (76°N, 120°W); Geographical Bulletin, v. 9, p. 189-217.

Pissart, A., Vincent, J-S., et Edlund, S.A.
1977: Dépôts et phénomènes éoliens sur l'île de Banks, Territoires du Nord-Ouest, Canada; Journal canadien des sciences de la terre, vol. 14, p. 2462-2480.

Prest, V.K.
1969: Retreat of Wisconsin and recent ice in North America; Geological Survey of Canada, Map 1257A, scale 1:5 000 000.
1970: Quaternary geology of Canada; in Geology and Economic Minerals of Canada, R.J.W. Douglas (ed.); Geological Survey of Canada, Economic Geology Report 1, 5th ed., p. 676-764.
1984: Late Wisconsinan glacier complex; Geological Survey of Canada, Map 1584A, scale 1:7 500 000.

Prest, V.K., Grant, D.R., and Rampton, V.N.
1968: Glacial map of Canada; Geological Survey of Canada, Map 1253A, scale 1:5 000 000.

Reeh, N.
1984: Reconstruction of the glacial ice cover of Greenland and the Canadian Arctic islands by three-dimensional perfectly plastic ice sheet modelling; Annals of Glaciology, v. 5, p. 115-122.
1989: Dynamic and climatic history of the Greenland Ice Sheet; Chapter 14 in Quaternary Geology of Canada and Greenland, R.J. Fulton (ed.); Geological Survey of Canada, Geology of Canada, no. 1 (also Geological Society of America, The Geology of North America, v. K-1).

Retelle, M.J.
1986a Glacial geology and Quaternary marine stratigraphy of the Robeson Channel area, northeastern Ellesmere Island, Northwest Territories; Canadian Journal of Earth Sciences, v. 23, p. 1001-1012.
1986b Stratigraphy and sedimentology of coastal lacustrine basins, northeastern Ellesmere Island, N.W.T.; Géographie physique et Quaternaire, v. 40, p. 117-128.

Roots, E.F.
1963: Physiography; in Geology of the north-central part of the Arctic Archipelago, Northwest Territories (Operation Franklin), by Y.O. Fortier et al.; Geological Survey of Canada, Memoir 320, p. 164-179.

Ruddiman, W.F., Mcintyre, A., Niebler-Hunt, V., and Durazzi, J.T.
1980: Oceanic evidence for the mechanism of rapid northern hemisphere glaciation; Quaternary Research, v. 13, p. 13-64.

St-Onge, D.A.
1965: La géomorphologie de l'île Ellef Ringnes, Territoires du Nord-Ouest, Canada; Geographical Branch, Paper 38, 58 p.

Schledermann, P.
1980: Notes on Norse finds from the east coast of Ellesmere Island, N.W.T.; Arctic, v. 33, p. 454-463.

Schriber, G., Stauffer, B., and Müller, F.
1977: $^{18}O/^{16}O$, 2H/1H and 3H measurements on precipitation and air moisture samples from the North Water area; International Association of Scientific Hydrology Publication, No. 118, p. 182-197.

Sim, V.W.
1961: A note on high-level marine shells on Fosheim Peninsula, Ellesmere Island, N.W.T.; Geographical Bulletin, v. 16, p. 120-123.

Smol, J.P.
1983: Paleophycology of a high arctic lake near Cape Herschel, Ellesmere Island; Canadian Journal of Botany, v. 61, p. 2195-2204.

Stangl, K.O., Roggensack, W.D., and Hayley, D.W.
1982: Engineering geology of surficial soils, eastern Melville Island; in Proceedings of the Fourth Canadian Permafrost Conference (The Roger J.E. Brown Memorial Volume), H.M. French (ed.); National Research Council of Canada, Ottawa, p. 136-147.

Stewart, T.G. and England, J.
1983: Holocene sea-ice variations and paleoenvironmental change, northernmost Ellesmere Island, N.W.T., Canada; Arctic and Alpine Research, v. 15, p. 1-17.

Stott, D.F.
1969: Ellef Ringnes Island, Canadian Arctic Archipelago; Geological Survey of Canada, Paper 68-16, 44 p.

Tarnocai, C.
1976: Soils of Bathurst, Cornwallis, and adjacent islands, District of Franklin; in Report of Activities, Part B, Geological Survey of Canada, Paper 76-1B, p. 137-141.

Taylor, A.
1956: Physical geography of the Queen Elizabeth Islands, Canada; 12 volumes, American Geographical Society, New York, N.Y.

Taylor, A., Judge, A., and Desrochers, D.
1983: Shoreline regression: its effect on permafrost and the geothermal regime, Canadian Arctic Archipelago; in Permafrost: Fourth International Conference, Proceedings; National Academy Press, Washington, D.C., p. 1239-1244.

Taylor, R.B. and McCann, S.B.
1983: Coastal depositional landforms in northern Canada; in Shorelines and Isostasy, D.E. Smith and A.G. Dawson (ed.); Institute of British Geographers, Special Publication 16, Academic Press, p. 53-75.

Tedrow, J.C.F.
1977: Soils of the Polar Landscape; Rutgers University Press, New Jersey, 638 p.

Thorsteinsson, R.
1958: Cornwallis and Little Cornwallis islands, District of Franklin, Northwest Territories; Geological Survey of Canada, Memoir 294, 134 p.

Thorsteinsson, R. and Tozer, E.T.
1970: Geology of the Arctic Archipelago; in Geology and Economic Minerals of Canada, R.J.W. Douglas (ed.); Geological Survey of Canada, Economic Geology Report 1, 5th edition, p. 549-590.

Tozer, E.T.
1963: Trold Fiord; in Geology of the north-central part of the Arctic Archipelago, Northwest Territories (Operation Franklin) by Y.O. Fortier et al.; Geological Survey of Canada, Memoir 320, p. 370-380.

Tozer, E.T. and Thorsteinsson, R.
1964: Western Queen Elizabeth Islands, Arctic Archipelago; Geological Survey of Canada, Memoir 332, 242 p.

Trettin, H.P. (editor)
in press: Innuitian Orogen and Arctic Platform: Canada and Greeland; Geological Survey of Canada, Geology of Canada, no. 3 (also Geological Society of America, The Geology of North America, v. E).

Trettin, H.P., Frisch, T.O., Sobczak, L.W., Weber, J.R., Niblett, E.R., Law, L.K., de Laurier, I., and Whitham, K.
1972: The Innuitian Province; in Variations in tectonic styles in Canada, R.A. Price and R.J.W. Douglas (ed.); Geological Association of Canada, Special Paper 11, p. 83-180.

Troelsen, J.C.
1952: Geological investigations in Ellesmere Island; Arctic, v. 5, p. 199-210.

Vincent, J-S.
1983: La géologie quaternaire et la géomorphologie de l'île Banks, Arctique Canadien; Commission géologique du Canada, Mémoire 405, 118 p.
1989: Quaternary geology of the northern Canadian Plains; in Chapter 2 of Quaternary Geology of Canada and Greenland, R.J. Fulton (ed.); Geological Survey of Canada, Geology of Canada, no. 1 (also Geological Society of America, The Geology of North America, v. K-1).

Vincent, J-S., Morris, W.A., and Occhietti, S.
1984: Glacial and nonglacial sediments of Matuyama paleomagnetic age on Banks Island, Canadian Arctic Archipelago; Geology, v. 12, p. 139-142.

Volk, H.R.
1980: Records of emergence around Oobloyah Bay and Neil Peninsula in connection with the Wisconsin deglaciation pattern, Ellesmere Island, Northwest Territories, Canada: a preliminary report; Polarforschung, v. 50, p. 29-44.

Walcott, R.I.
1970: Isostatic response to loading of the crust in Canada; Canadian Journal of Earth Sciences, v. 7, p. 716-726.
1972: Late Quaternary vertical movements in eastern North America: quantitative evidence of glacio-isostatic rebound; Reviews of Geophysics and Space Physics, v. 10, p. 849-884.

Walford, M.E.R. and Harper, M.F.L.
1981: The detailed study of glacier beds using radio-echo techniques; Geophysical Journal of the Astronomical Society, v. 67, p. 487-514.

Washburn, A.L.
1947: Reconnaissance geology of portions of Victoria Island and adjacent regions, Arctic Canada; Geological Society of America, Memoir 22, 142 p.
1980: Geocryology, A Survey of Periglacial Processes and Environments; John Wiley and Sons, New York, 406 p.
1983: Palsas and continuous permafrost; in Permafrost: Fourth International Conference, Proceedings; National Academy Press, Washington, D.C., p. 1372-1377.

Washburn, A.L. and Stuiver, M.
1985: Radiocarbon dates from Cornwallis Island area, Arctic Canada — an interim report; Canadian Journal of Earth Sciences, v. 22, p. 630-636.

Watts, S.H.
1983: Weathering processes and products under arid arctic conditions; Geografiska Annaler, v. 65A, p. 85-98.

Weidick, A.
1975: A review of Quaternary investigations in Greenland; Institute of Polar Studies, Ohio State University, Columbus, Ohio, Report No. 55, 161 p.

Wilson, C.
1969: Climatology of the cold regions, northern hemisphere II; Cold Regions Science and Engineering Monograph 1-A3b, Cold Regions Research and Engineering Laboratory, Hanover, New Hampshire, 158 p.

Wilson, J.T., Falconer, G., Mathews, W.H., and Prest, V.K.
1958: Glacial map of Canada; Geological Association of Canada, scale 1:3 801 600.

Authors' addresses

J. Bednarski
Boreal Institute for Northern Studies
University of Alberta
Edmonton, Alberta
T6G 2E9

J. England
Department of Geography
University of Edmonton
Edmonton, Alberta
T6G 2H4

D.A. Hodgson
Geological Survey of Canada
601 Booth Street
Ottawa, Ontario
K1A 0E8

R.M. Koerner
Geological Survey of Canada
601 Booth Street
Ottawa, Ontario
K1A 0E8

Part 2
APPLIED QUATERNARY GEOLOGY IN CANADA

In this part we return to a point made in the Foreword to this volume: the Quaternary is the geological time period in which we live. In particular, we look at some aspects of the development of the present environment of Canada, noting that this environment is unique in several ways. Man's interactions with this environment, the resources available to him, the geological hazards he is exposed to, and the use he makes of the environment are reviewed.

The discussions are presented in six chapters, each covering one aspect of the subject. The first chapter, on Quaternary paleoenvironments, looks at how the present environment developed in postglacial time and also considers earlier Quaternary environments resulting from cooling and warming of climate. The information is presented in the form of seven paleobotanical case histories, and the emphasis of the chapter is on the development of the present vegetation of Canada and, by inference, the present climate of Canada. One case history, by McAndrews and Boyko-Diakonow, shows the influence of man on the vegetation of a small area of southern Ontario over the last thousand years.

The next chapter, Quaternary geodynamics, is concerned with the pattern and mechanisms of movements of the earth's crust in Canada during the Quaternary. These movements, mainly vertical in nature, are largely the result of differential loading and unloading of the landmass of Canada by the vast ice sheets of the glacial episodes. Estimates of the maximum thickness of the Laurentide Ice Sheet, the main ice sheet to affect Canada during the Wisconsinan, vary from more than 5 km to less than 3 km (N. Reeh, Geological Survey of Greenland, personal communication, 1984). Much of this information on vertical movements of the crust during the Quaternary comes from changes in relative sea level as the crust was depressed by the ice load and then recovered when the load was removed. The picture is complicated by variations in sea level, due largely to the vast amounts of water locked up in the great ice sheets. Worldwide sea level changes of as much as 150 m have occurred during the Quaternary and if the ice now remaining in the Greenland and Antarctic ice sheets were to melt, sea level would rise approximtely 45 m.

The chapter on geomorphic processes examines the processes that sculpt the surface of the Earth and so produce the details of the landscape of Canada as it is today. The controlling effects of physiography, vegetation, and climate are considered. The main suites of geomorphic processes are described, and lesser processes briefly summarized. Several sections of this chapter show that the landscape of Canada is still undergoing adjustment from conditions during the Wisconsinan glacial episode — landsliding is a major process in the development of slopes, and rivers are still adjusting to the drainage disruptions imposed by glaciation. The chapter concludes with a lengthy section on the effects of man; in terms of the quantity of earth materials moved per year, man is more significant than any other geomorphic process.

The chapter on the geochemistry of Quaternary materials reviews the particular legacy of Quaternary glaciation — most of the surficial sediments of Canada are till or are derived almost directly from till; most such sediments are therefore geologically very young. These conditions have had a strong influence on the chemical nature of all the surficial sediments of Canada, and so affect geochemical prospecting procedures and environmental geological conditions, such as sensitivity to acid rain.

The resources which are available to man because of the particular nature of the Quaternary environment are the subject of the next chapter. The specific resources considered are: soil, peat, aggregates, nonmetallic minerals and metallic minerals (primarily gold in placer deposits).

The final chapter looks at ways in which relief, present climate, Quaternary deposits and the legacy of Quaternary processes affect the economic development activities of Canadians. Two sections address problems of geological hazards in the Cordillera and in Eastern Canada: hazards such as rock avalanches, landslides, debris flows, floods, and erosion. For the Prairie region, a broader range of engineering geology topics are described, including problems of construction materials, groundwater supplies, and the stability of open pit mines. The final section reviews the implications of Quaternary geology in the development of urban areas of Canada.

The chapters in Part 2 were written 1984-1986 with some updating in 1987.

J.A. Heginbottom

Chapter 7

QUATERNARY ENVIRONMENTS IN CANADA AS DOCUMENTED BY PALEOBOTANICAL CASE HISTORIES

Summary

Introduction — *J.V. Matthews, Jr. and T.W. Anderson*

Introductory comments on glacial refugia — *J.V. Matthews, Jr. and T.W. Anderson*

The Queen Charlotte Islands refugium: a paleoecological perspective — *R.W. Mathewes*

Paleoecology of the western Canadian ice-free corridor — *C.E. Schweger*

Introductory comments on interglacial environments — *J.V. Matthews, Jr. and T.W. Anderson*

Late Pleistocene paleoenvironments in Atlantic Canada — *R.J. Mott*

Introductory comments on the development of the Canadian biota following ice retreat *J.V. Matthews, Jr. and T.W. Anderson*

History of the boreal forest in Canada — *J.C. Ritchie*

Patterns of post-Wisconsinan plant colonization in Quebec-Labrador — *P.J.H. Richard*

Introductory comments on the paleoenvironmental record of the Holocene *J.V. Matthews, Jr. and T.W. Anderson*

Holocene climatic trends in Canada with special reference to the Hypsithermal interval — *T.W. Anderson, R.W. Mathewes, and C.E. Schweger*

Pollen analysis of varved sediment at Crawford Lake, Ontario: evidence of Indian and European farming — *J.H. McAndrews and M. Boyko-Diakonow*

Acknowledgments

References

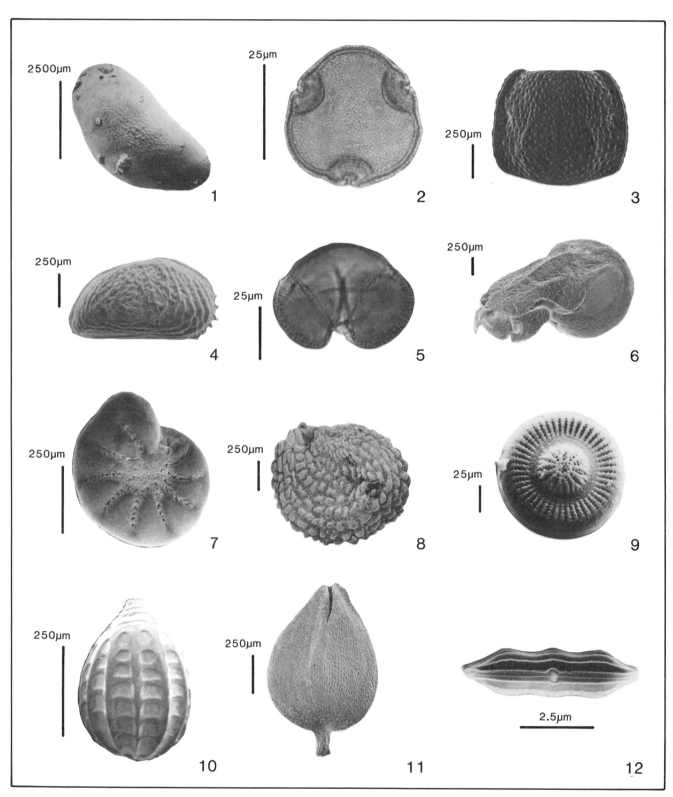

fig. 1. *Candona subtriangulata* (Costracode)
2. *Tilia americana* L. (tree pollen)
3. *Acidota subcarinata* Er.; Staphylinidae (insect)
4. *Heterocyprideis sorbyana* (Jones) right valve
5. *Picea mariana* (Mill.) BSP. (trapollen)
6. *Cleonus* sp.; Curculionidae (insect)
7. *Elphidium* sp. (foraminifera)
8. *Sibbaldia procumbens*
9. *Cyclotella stelligera* Cleve & Grun. (diatom)
10. *Oolina melo* d'Orbifny (foraminifera)
11. *Carex flava* (tree pollen)
12. *Neidium hitchcockii* (Ehr.) Cleve (diatom)

Chapter 7

QUATERNARY ENVIRONMENTS IN CANADA AS DOCUMENTED BY PALEOBOTANICAL CASE HISTORIES

J.V. Matthews, Jr., T.W. Anderson, M. Boyko-Diakonow, R.W. Mathewes,
J.H. McAndrews, R.J. Mott, P.J.H. Richard, J.C. Ritchie, and C.E. Schweger

SUMMARY

This chapter is concerned with the scope and significance of paleoecology-related studies, particularly their application in reconstructing Quaternary environments in Canada, with special reference to seven paleobotanical case histories. Though pollen stratigraphy is the principal evidence that is common to each case history, other important paleoecological data (plant macrofossils, insects and other arthropods, ostracodes, molluscs, diatoms, tree rings, vertebrate and invertebrate fossils) and nonpaleoecological indicators (soil profiles, fire history, historical records) and their relevance to paleoenvironmental research are not overlooked.

The first two case histories present evidence for glacial refugia. Fossil occurrences supported by geological mapping on Queen Charlotte Islands, British Columbia indicate the presence of ice-free areas which possibly served as refugia for certain elements of the biota during the Late Wisconsinan maximum. Similar studies focus on the unglaciated parts of Yukon Territory and the ice-free corridor which is presumed to have existed along parts of the Rocky Mountain Foothills during the Late Wisconsinan maximum. Even though the portal areas to the north and south of the ice-free corridor have certain pre-Wisconsinan fossils in common, the corridor itself lacks critical evidence to indicate that it may have served as a north-south dispersal route.

Records from about twenty buried organic deposits in Atlantic Canada delineate the Late Pleistocene sequence of events and environments there. The Sangamon Interglaciation was characterized by at least three climatic episodes which can be correlated provisionally with the stratigraphic framework for the region and with the deep sea oxygen isotope record.

The Late Wisconsinan and Holocene history of the Canadian flora is the theme of the final four case histories. The first of these deals with the long-term history of the Boreal Forest in western and northwestern Canada and the factors that have contributed to its variability from one region to another. The second of these shows how the post-Wisconsinan vegetation development in Quebec has varied both spatially and temporally. This variability is discussed in terms of the nature of the catchment basins, geomorphological and geographical settings, and climatic conditions.

The final two case histories deal with specific aspects of the Holocene, namely, the Hypsithermal or altithermal interval and man-induced vegetation changes in the late Holocene. The intensity and timing of the Hypsithermal interval, the period of maximum warmth and dryness in early-middle Holocene time, is discussed on a Canada-wide basis. Pollen stratigraphy combined with varve chronology, archeological surveys, and land patenting records in and around Crawford Lake, southern Ontario clearly document two periods of forest disturbance and agriculture, one associated with Indian farming and another with European settlement. Both farming periods left major impacts on the forest succession of the area.

INTRODUCTION

J.V. Matthews, Jr. and T.W. Anderson

Paleoecology: practice and relevance

Our knowledge of Quaternary paleoenvironments in Canada results from research that can be classified broadly as paleoecological. Over the past two decades "paleoecology" has evolved from being no more than an optional consideration in paleontological investigations to a discipline in its own right which considers biological and geological evidence of various types as the means for reconstructing past environments and of documenting environmental and climatic change (Birks and Birks, 1980; Birks, 1981). Few paleoecologists actually employ ecological methods. Instead they function more as paleobiogeographers, documenting past distributional changes of organisms that are preserved as fossils.

Matthews, J.V., Jr., Anderson, T.W., Boyko-Diakonow, M., Mathewes, R.W., McAndrews, J.H., Mott, R.J., Richard, P.J.H., Ritchie, J.C., and Schweger, C.E.
1989: Quaternary environments in Canada as documented by paleobotanical case histories; Chapter 7 in Quaternary Geology of Canada and Greenland, R.J. Fulton (ed.); Geological Survey of Canada, Geology of Canada, no. 1 (also Geological Society of America, The Geology of North America, v. K-1).

Matthews, J.V., Jr. and Anderson, T.W.
1989: Introduction (Quaternary environments in Canada); in Chapter 7 of Quaternary Geology of Canada and Greenland, R.J. Fulton (ed.); Geological Survey of Canada, Geology of Canada, no. 1 (also Geological Society of America, The Geology of North America, v. K-1).

This is the first step to defining the composition of past floras and faunas, an end in itself for some workers, but for most only a part of the process of reconstructing past climatic conditions.

Paleoecology is not without practical value as illustrated by the following examples. First there is the issue of future climate change, especially the potential effects of rising levels of atmospheric CO_2. "General Circulation Models" (better known by the acronym GCMs) predict significant warming, especially in the Canadian Arctic, if current trends in fossil-fuel usage and forest clearance continue into the next century (Menabe and Wetherald, 1980; Schneider, 1984). The warmer climate called for by such models under a doubled concentration of atmospheric CO_2 has no analogue in recorded climatic history. Indeed one must look back to the early Holocene or even the last interglaciation for appropriate analogues of what has been predicted for the next 100 years. Our knowledge of the climate during these earlier periods of warmer climate comes exclusively from interpretation of proxy climate data, much of which originates from various types of paleoecological research.

GCMs yield ambiguous and contradictory projections of precipitation and soil moisture, yet it is future changes of these variables that are likely to have the greatest impact on the Canadian economy and life style. GCMs also deal poorly with the transient responses of CO_2-induced warming. In other words they may predict the equilibrium climatic state under a doubled or quadrupled atmospheric CO_2 level, but not the changes that occur in transit to the equilibrium state (Schneider and Thompson, 1981; Waggoner, 1983; Parry, 1985). For example, at what point in the move to warmer climate would temperature and soil moisture constraints begin to affect distribution of crop and forest plants? Would there occur associated thresholds in incidence of pest insect species? In fact, assuming that higher levels of CO_2 would cause climate warming, can we even assume that such warming would be gradual rather than step-like? Paleoecological research has the potential for providing some answers to these questions and others related to climate futures, and unlike the models, the data generated by paleoecological research relate directly to the variables of vegetation and faunal change that impinge directly on our lives.

Of more immediate concern than the prospect of CO_2-driven climatic change in the next century is short-term (decadal) or background climatic variability. Various forcing functions are involved (Gilliland, 1982; Delcourt et al., 1983), and even though such changes are well documented in the historical record, an evaluation of what causes them depends on reference to longer time series. In order to determine how solar variability affects climate, for example, one must look back in time beyond the start of measured data and into the realm of proxy climate data — the type of data generated by paleoecological research.

Perhaps the greatest value of paleoecology is that it provides a historical perspective of our existing environment. Paleoecological studies often show that former environments lack modern counterparts; in other words, the present environment is unique. Such findings cannot but heighten our appreciation of the complexities, balances, and particularly the vulnerabilities of the existing environment.

Types of paleoecological data

Various types of evidence are used in the investigation of Quaternary environmental history (see Birks and Birks, 1980 for a comprehensive review). The premier approach in most Canadian studies is analysis of fossil pollen because pollen is abundant and well preserved in many types of Quaternary sediments and may be readily isolated and identified. In recent years other types of plant fossils, such as diatoms, silica phytoliths, mosses, and macrofossils such as seeds and fruits, have also been used to reconstruct past Canadian environments or to evaluate the effects of contemporary pollutants such as acid rain (Duthie and Sreenivasa, 1971; Janssens, 1981; Hickman et al., 1984; Delorme et al., 1984; Bombin, 1984). Tree ring analysis or dendrochronology represents another type of paleobotanical data. Tree rings bridge the gap between the first instrumental and historic records and those of prehistoric time. They are useful for dating and documenting climatic change (Fritts, 1982); however, recent Canadian studies (e.g., Parker et al., 1981) show that reading a climate signal directly from variations of tree ring width or wood density is a more complex procedure than has been previously assumed.

Fossil arthropods such as Coleoptera (beetles), chironomid fly larvae, caddisflies (Trichoptera), ostracodes, and Cladocera are proving to be a valuable source of paleoecological data, especially for determining the nature of former local environments, which then can be extrapolated to infer regional climate (Delorme et al., 1977; Williams and Morgan, 1977; Morgan and Morgan, 1980; Williams et al., 1981; Morgan et al., 1983; Delorme and Zoltai, 1984). Chironomid flies have also been used to document anthropogenic changes of water quality in Canadian lakes (Warwick, 1980). Studies such as these are dependent on sound taxonomic knowledge of the present fauna. For this reason, research using fossils of beetles was not feasible until Lindroth (1961-1969) published his monograph of the ground beetles of Canada and Alaska, and Delorme (1968) correctly prefaced his paleoenvironmental research using fossils of freshwater ostracodes with systematic studies of the modern ostracode fauna. Similarly, the emergence of fossils of caddisfly larvae as powerful indicators of former aquatic environments has been realized only after publication of such well illustrated monographs on the contemporary fauna as that of Wiggins (1977).

Shells of freshwater molluscs are also abundant in some types of terrestrial sediments, but in Canada malacological research is still in a nascent phase compared to Europe. Of the few published papers on Canadian fossils some have revealed the same type of Pleistocene distributional changes seen among the Coleoptera (Clarke and Harington, 1978). Bobrowsky (1982) has studied quantitatively the snails associated with archeological sites, pointing the way for use of similar techniques for other Pleistocene molluscan assemblages.

Canadian sites have yielded an abundance of vertebrate fossils (Harington, 1978), and many of these fossils provide insight to past environmental conditions. The Yukon Territory, especially the Dawson area and the Old Crow Basin, is literally a treasure trove of large mammal fossils (Harington, 1977; Morison and Smith, 1987). Recently, such fossils and those in assemblages of comparable age from

Alaska and Siberia have become the centrepiece of a debate on the nature of Late Pleistocene environments in east Beringia (contrast Guthrie, 1982 and Matthews, 1982 with Ritchie, 1984a). The question debated is whether assemblages of fossils composed of large grazing species actually represent former mammalian communities (Matthews, 1982) and, if so, do such communities imply productive steppe-like environments (Guthrie, 1982) or only patchy herb tundra (Ritchie, 1984a).

Although their study has not progressed as far as for the large mammals, small mammal fossils are abundant in some Yukon deposits. Newly discovered fossils of Microtine rodents associated with an Early Pleistocene tephra at one Old Crow (Yukon) locality (Morison and Smith, 1987) promise to provide the long sought-for link between faunas from the southern Canadian Prairies (Klassen, 1989) and those in eastern Siberia (Sher et al., 1979; Repenning, 1980). Small mammal fossils from the Bluefish Cave site in the Northern Yukon (Cinq-Mars, 1979), at the opposite end of the Pleistocene time scale, have yielded new information on the nature of the Late Pleistocene/Holocene transition in east Beringia (Morlan, 1983; Morlan and Cinq-Mars, 1986).

Of course paleoecological research is not the only source of proxy climate and paleoenvironmental data. Prominent in the reconstruction of Canadian environments are soil studies (Sorenson and Knox, 1974; Foscolos et al., 1977) and ice-core research (Koerner, 1989). Historical records of climate and related phenomena (Ingram et al., 1981) are another important data reservoir. The archives of the Hudsons Bay Company contain an astounding quantity of information on the climate of northern Canada during the 19th Century. Wilson (1983) and Ball (1983) have used such records to reconstruct synoptic patterns during particular events, such as the 1816 "year without a summer" that followed the Tambora volcanic eruption (Stothers, 1984).

Canadian paleobotanical case histories

In the last revision of Geology and Economic Minerals of Canada (Douglas, 1970), the entire Quaternary was dealt with in a single chapter (Prest, 1970), a summary that barely touched on paleoenvironmental problems. Much has happened in this realm during the intervening years and this is most clearly indicated by the fact that in the latest bibliographic compilation on Canadian paleoclimatic studies (Hills and Sangster, 1980), 55% of the 376 citations were published after 1968. Obviously, then, a single chapter like this one cannot pretend to be a comprehensive review of Canadian paleoenvironmental studies. Instead, what follows are seven paleobotanical "case histories", each written by one or more specialists, that are intended to illustrate the type and scope of research being conducted in Canada. Admittedly, these case histories are hardly representative of the actual scope of ongoing Canadian paleoenvironmental studies. Only certain regions are discussed (Fig. 7.1), and all of them are heavily biased towards palynology (pollen analysis). A few of the presentations refer in passing to other sources of data, such as plant macrofossils and insect fossils, but because of space constraints, most other useful sources of paleoecological and paleoclimatic data are ignored completely.

The first two case histories (Mathewes, 1989; Schweger, 1989) deal with aspects of glacial refugia (Fig. 7.1). As well,

Schweger's discussion of the ice-free corridor alludes to the immediate preglacial environments of northwestern Canada.

The next case history, by Mott (1989), concerns vegetation and climatic changes in Atlantic Canada (Fig. 7.1) during the Sangamon Interglaciation. It is representative of the type of research being carried out on organic sediments of interglacial and Early Pleistocene age in other areas of Canada. The two case histories by Ritchie (1989) and Richard (1989) are discussions of the sequence of vegetation development in two areas of Canada near the start of the present interglaciation — the Holocene. The case history by Anderson et al. (1989) deals with the peak warming of the present interglacial and how it was expressed differently in different parts of Canada. Such detail is not easily expressed or handled by General Circulation Models, and the Holocene case history shows clearly why any assessment of future climate must be based in part on study of paleoclimates (Schneider, 1986).

The last case history (McAndrews and Boyko-Diakonow, 1989) deals with the latest Holocene, including the historic era. This is possible because of the unique pollen record preserved at Crawford Lake in southern Ontario (Fig. 7.1). In some ways the Crawford Lake series is analogous to a tree ring series since it concerns change on a yearly basis, a rare luxury in Quaternary pollen studies. And like tree ring studies, it blends data derived from normal paleoecological procedures with historical information.

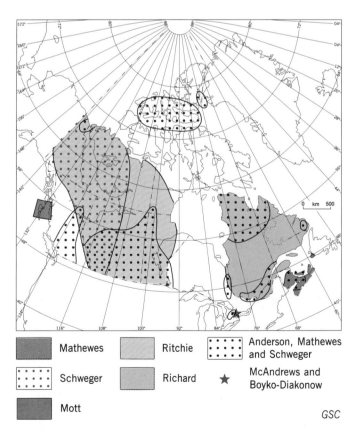

Figure 7.1. Map of Canada showing regions covered by the case histories discussed in this chapter.

INTRODUCTORY COMMENTS ON GLACIAL REFUGIA

J.V. Matthews, Jr. and T.W. Anderson

Accompanying chapters in this volume show that much of Canada has been repeatedly glaciated. As recently as 15 ka, glaciers emanating from various sources and coalescing over broad fronts covered more than 90% of Canada. It is not surprising, therefore, that Canadian paleoecologists and biologists have a special interest in glacial refugia. The one most often discussed is the Beringian refugium, which included parts of Yukon Territory and much of Alaska (Hopkins et al., 1982). Details of the nature of the Beringian refugium, especially its state at the time when *Homo sapiens* first moved into North America, are still debated (Schweger et al., 1982; Colinvaux and West, 1984). But there is no question that a Beringian refugium did exist, probably during each glacial event, and there is ample evidence on the composition of its Late Pleistocene flora and fauna (see various papers in Hopkins et al., 1982).

Such is not the case for many of the smaller refugia that undoubtedly existed at other places on the periphery of the ice sheets (e.g., Newfoundland — Grant, 1977). Many of them have been postulated on geological grounds alone. In other cases, the existing flora and fauna contain a subtle overprint of a former refugial biota, but the actual locus of the refugium and the time of its existence are in doubt. Unambiguous proof of the existence of a glacial refugium are fossils that prove organisms existed at the site when it was surrounded by glaciers or otherwise cut off from other ice-free regions.

The case history by R.W. Mathewes on the Queen Charlotte refugium provides that type of proof. The methodology used for that analysis — detailed geological mapping and reliable isotopic dating combined with pollen and plant macrofossil analyses — illustrates a holistic approach that should be a model for study of other proposed refugia. Macrofossils are of particular importance in the study of refugia, because unlike pollen, they provide the clearest evidence of the species composition of the refugial flora. Although not discussed below, some of the exposures within the Queen Charlotte refugium also contain well preserved beetle fossils which are currently under study (e.g., unpublished Geological Survey of Canada Fossil Arthropod Reports 85-12 and 86-5 by J.V. Matthews, Jr.). Obviously, the Queen Charlotte Islands have considerable potential for future paleoecological research on the refugium question.

The case history described by C.E. Schweger is also concerned with refugia, but only indirectly. It deals primarily with faunal and floral exchange between two large refugia: Beringia in the northwest part of the continent and the unglaciated region of the mid-continent. During the height of glaciation and glacial eustatic lowering of sea level, Alaska and parts of the Yukon (east Beringia) formed the eastern part of the Siberian biogeographic realm (Hopkins, 1973, 1982). In reality, east Beringia was a cul-de-sac, since access to the rest of the North American continent was either blocked by coalesced Cordilleran and Laurentide ice sheets or by less tangible filters such as forest zones (Repenning, 1980). When these ice sheets began to retreat or at times when they were advancing but not yet merged, a narrow neck of unglaciated terrain — the "ice-free corridor" — connected the Beringian and mid-continental unglaciated regions. Although a transitory feature, the ice-free corridor was apparently a route for north-south exchange of organisms that would not be able to cross the boreal barrier that presently extends across Canada. Just as important, the Wisconsinan version of the ice-free corridor was undoubtedly the main route via which late Paleolithic people entered central North America.

Schweger's case history is concerned mostly with the last manifestation of the ice-free corridor. It would be wrong to conclude that earlier iterations of the corridor during pre-Wisconsinan glaciations were necessarily identical to the last one, but evidence of the biota of such earlier corridors is at best indirect and elusive and will remain so until Early Pleistocene faunas from both openings to the corridor are better studied and compared. As noted earlier, we are on the verge of being capable of making such comparisons of Early Pleistocene vertebrate faunas of the northern Yukon and southern Canadian prairies.

THE QUEEN CHARLOTTE ISLANDS REFUGIUM: A PALEOECOLOGICAL PERSPECTIVE

R.W. Mathewes

Introduction

Within the glaciated regions of the world, biologists and biogeographers often theorize about the possible existence of areas that escaped burial by ice and thus served as refugia where plants and animals could have survived the rigours of the Ice Age. Such areas could serve as dispersal centres following deglaciation, and long periods of isolation in small populations could result in the evolution of unique "endemic" forms of plants and animals. Ice-free refugia have been proposed for various parts of Canada and adjacent United States (Scudder, 1979). Coastal refugia have attracted particular attention, both in northern Europe (Dahl, 1955; Funder, 1979) and along the North Pacific Coast of North America, where they have been postulated for Vancouver Island, the Queen Charlotte Islands, and coastal Alaska as far north as Kodiak Island (Heusser, 1960; Karlstrom and Ball, 1969; Fladmark, 1975, 1979). General agreement prevails over the refugial status of some areas such as Kodiak Island, but others, notably the Queen Charlotte Islands, are at the centre of a long-standing controversy between biologists and geologists.

The Queen Charlotte Islands form an archipelago of about 150 large and small islands situated 80 km off the coast of central British Columbia (Fig. 7.2). The biological

---

Matthews, J.V., Jr. and Anderson, T.W.
1989: Introductory comments on glacial refugia; in Chapter 7 of Quaternary Geology of Canada and Greenland, R.J. Fulton (ed.); Geological Survey of Canada, Geology of Canada, no. 1 (also Geological Society of America, The Geology of North America, v. K-1).

Mathewes, R.W.
1989: The Queen Charlotte Islands refugium: a paleoecological perspective; in Chapter 7 of Quaternary Geology of Canada and Greenland, R.J. Fulton (ed.); Geological Survey of Canada, Geology of Canada, no. 1 (also Geological Society of America, The Geology of North America, v. K-1).

evidence suggesting a long period of biotic survival here is varied and intriguing. Among the vascular plants, 13 endemic species or subspecies are recognized (Calder and Taylor, 1968), although most of these were recently found also on northwestern Vancouver Island in another presumed refugium (Pojar, 1980). The bryophyte flora is very unusual, containing at least 14 species that appear to be confined to the Queen Charlotte Islands (Schofield, 1976), including *Wijkia carlottae*, *Seligeria careyana* (Vitt and Schofield, 1976), as well as some undescribed new species.

The liverwort, *Dendrobazzania griffithiana*, is a relict now found only in the Bengal-Bhutan region of Asia and in the Queen Charlottes (Schuster and Schofield, 1982).

The presence of diploid populations of the flowering plant *Saxifraga ferruginea* in the Queen Charlottes, Kodiak Island, and south of the Wisconsin ice margin has also been used to support the refugium hypothesis, for only polyploids occur in glaciated areas (Randhawa and Beamish, 1972).

The fauna is also distinctive, with its own subspecies of mammals and birds (Foster, 1965), including the enigmatic

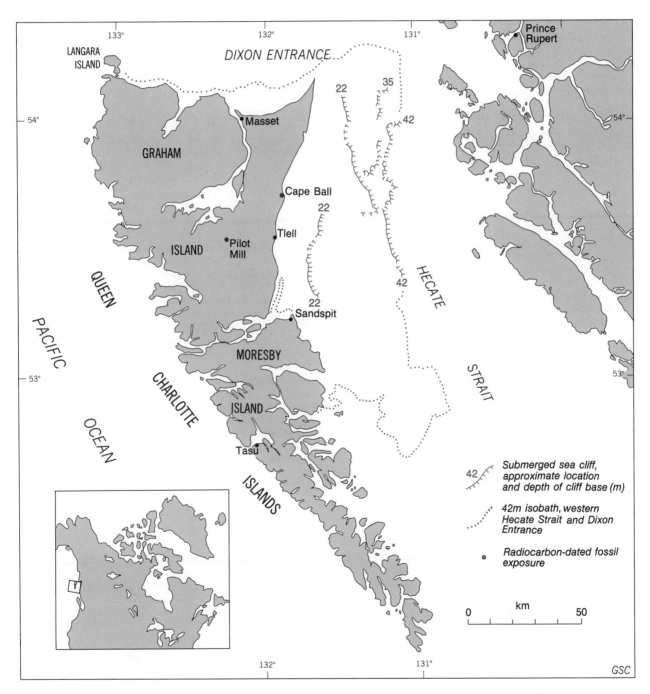

Figure 7.2. Location map of Queen Charlotte Islands showing plant fossil localities and part of Hecate Strait within the 42 m isobath that may have been exposed during the last glacial maximum.

Dawson's caribou (*Rangifer tarandus dawsonii*), whose presumed extinction around 1910 represents an unresolved mystery (Banfield, 1962). Invertebrates of note include a spray-pool amphipod (Bousfield, 1958) and ground beetles of the genus *Nebria*. Originally described from a single known specimen, the endemic species *Nebria charlottae* (Kavanaugh, 1980) was rediscovered in 1981 by Kavanaugh, along with two new presumed endemics. To many biologists, this weight of biological evidence was so compelling that the existence of refugia was taken for granted, so that the major remaining problem was to identify the location(s), size, and number of refugia. Some biologists, however, do not require a refugial hypothesis to account for their data. Moodie and Reimchen (1976) attributed the striking variation in populations of the three-spine stickleback (*Gasterosteus aculeatus*) in lakes of Graham Island to postglacial divergence. Furthermore, Foster (1965) did not require long isolation to account for all the observed vertebrate variations.

Adding to the controversy is the argument proposed by Heusser (1960) and Fladmark (1975, 1979) that a chain of sea level refugia around the north Pacific Coast could have provided a migration route for early humans entering southern North America during the Late Pleistocene. This theory conflicts with the widely held notion for a so-called "ice-free corridor" between the Cordilleran and Laurentide ice masses (see this chapter, Schweger, 1989). If Fladmark is correct, and coastal cultures adapted earlier or as early as continental cultures, then current archeological theories will need revision.

Geologists have been skeptical about these refugia. Since the earliest observations by Dawson (1880), geologists have argued that the islands were totally glaciated by a local ice cap (Sutherland-Brown and Nasmith, 1962; Sutherland-Brown, 1968; Prest, 1969). This ice mass was thought to have met the main Cordilleran ice somewhere in what is now Hecate Strait (Fig. 7.2). Except for acknowledging that a few inhospitable mountain peaks (nunataks) projected above the ice (Sutherland-Brown, 1968), the general consensus has not favoured the existence of a biologically significant refugium, and deglaciation was considered to have occurred at about the same time as the rest of the British Columbia coastline.

A major advance in our knowledge of the glacial and postglacial history of the Queen Charlotte Islands comes from the investigations by Clague (1981) of the seacliff exposures along the northeastern coast of Graham Island. A series of radiocarbon dates from these exposures indicate that the postglacial history of the Cape Ball area (Fig. 7.2) extends back to at least 13 700 ± 100 BP (GSC-3222; Mathewes and Clague, 1982). More recent dates suggest an even greater age. These dates and their refugium implications are the central theme of the following discussion.

Late Pleistocene paleoecology of eastern Graham Island

Prior to Clague's findings, the oldest known terrestrial record of postglacial vegetation history was a peat core from Langara Island (Heusser, 1955), with a basal date of 10 850 ± 800 BP (L-297C; Broecker and Kulp, 1957). This relatively young date did not support the refugium theory, although Heusser (1955) suggested that the diverse plant assemblage of 27 types (indicating presence of forest during deposition of the base of the core) probably did not represent initial postglacial dispersal but instead must have been preceded by an earlier flora.

Pollen and macrofossil analyses of a late glacial peat bed at Cape Ball (Mathewes and Clague, 1982) extend the known history of the terrestrial flora back to 12 400 ± 100 BP (GSC-3112). Samples from the base of the peat contain pollen that represents an open herb-rich tundra assemblage that was invaded first by *Pinus contorta* and then at 11.2 ka by *Picea*. Included in this flora were plants that are now rare elements of the alpine flora (i.e. *Artemisia, Bistorta, Thalictrum*), as well as two species now absent from the islands: *Armeria maritima* and *Polemonium caeruleum*. The only species of Jacob's ladder, *Polemonium*, presently on the islands is *P. pulcherrimum*, a rare plant now growing near sea level on Limestone Island (Mathewes, 1980) and recently also discovered in an alpine area near Tasu (Roemer and Ogilvie, 1983).

Detailed examination of the silty sediments below the peat bed at Cape Ball produced the next major advance in our knowledge of the paleoenvironment. Plant remains screened from a laminated sand and silt unit near the base of the seacliffs produced two radiocarbon dates of 15 400 ± 190 BP (GSC-3319) and 16 000 ± 570 BP (GSC-3370) (Fig. 7.3).

These dates are highly significant because they establish that at least part of Graham Island was ice free during and shortly after the Late Wisconsinan glacial maximum, as defined along the mainland coast (Warner et al., 1982). It is clear, however, that an ice advance did reach Cape Ball prior to deposition of the basal silts, because these fossiliferous sediments are underlain by outwash and till. The age of this glacial event is not known, although radiocarbon dates suggest that the unit immediately below the till is of Middle Wisconsinan age. At the Cinola "Pilot Mill" site, southwest of Cape Ball (Fig. 7.2) a buried peat under a till is dated at between 45 700 ± 970 BP (GSC-3534-2) and 27 500 ± 400 BP (GSC-3530). It is not known if the Pilot Mill till is contemporaneous with the upper till at Cape Ball, but the dates suggest a glacial advance on eastern Graham Island sometime between 27.5 and 16.0 ka. Biological effects of this event include the local extinction of true fir trees (*Abies*) on the Queen Charlotte Islands. Fir was present in the Middle Wisconsinan nonglacial interval (Warner et al., 1984a) but had apparently disappeared by 16.0 ka.

A detailed study of the pollen (Fig. 7.3) and plant macrofossils (Fig. 7.4) covering the period since 16.0 ka is presented by Warner (1984). Together with previous data, his results provide a rare glimpse of the "green patch" that existed while most of British Columbia was locked in the grip of a massive ice mass.

In the oldest sediments, abundant seeds of a "pioneer" flora of *Juncus, Sagina, Rumex*, and *Potamogeton filiformis*, the last named now absent from the Queen Charlottes, are present, along with fragments of *Equisetum*. The high seed concentrations (Fig. 7.4) are typical of weedy colonizing plants, and show that climates were not severe enough to hamper reproduction.

The summary pollen diagram (Fig. 7.3) indicates an early predominance of grass and sedge pollen, along with *Sagina*, Compositae, and a variety of other herbs. These assemblages represent a treeless tundra-like terrain with large areas of exposed mineral soil and weedy herbs along margins of creeks and ponds. Presence of small amounts of tree pollen is likely due to long-distance transport by wind, or perhaps to reworking from older deposits, although the possibility that a few stunted, wind-swept spruce trees were present cannot be discounted. Ponds must have been ice-free for part of the year to allow the continual presence of aquatic plants. Soon after 13 ka, a variety of different aquatic plants appear (Fig. 7.4), including the algae *Chara* and *Nitella*, *Ranunculus aquatilis*, and *Callitriche*. The latter two plants are also well represented by pollen (Fig. 7.3) at this time, as is *Polemonium*. In fact, on the basis of certain pollen characteristics described by Mathewes (1979, 1980), two *Polemonium* pollen types are present.

A shrub phase appears to follow the early herb phase, with the first macrofossil evidence of woody plants appearing by about 12.5 ka. Leaves, bud scales, and seeds of *Salix* were recovered from the silty sediments, accompanied by a striking peak of *Salix* pollen (Fig. 7.3). The leaves are assignable to *Salix reticulata*, now a rare alpine plant in the Queen Charlottes, where it is represented by the subspecies *glabellicarpa*, known outside the islands only on a mountain near Juneau, Alaska. New herbs also appear around this time, including *Ranunculus pygmaeus*, *Selaginella densa*, only recently discovered in the alpine flora (Roemer and Ogilvie, 1983), *Hypericum scouleri* which is not known to occur on Graham Island at present, and *Potentilla pacifica*. Particularly intriguing is a single needle of Sitka spruce (*Picea sitchensis*) at the 250 cm depth. Dating to about 12.4 ka, it predates by at least 1.1 ka the spruce rise higher in the section. The oldest direct evidence of spruce, other than this needle, is wood dated at 11 300 ± 110 BP (GSC-2879) and the spruce pollen content which becomes significant only at about 11.2 ka. The fossil needle tentatively supports the

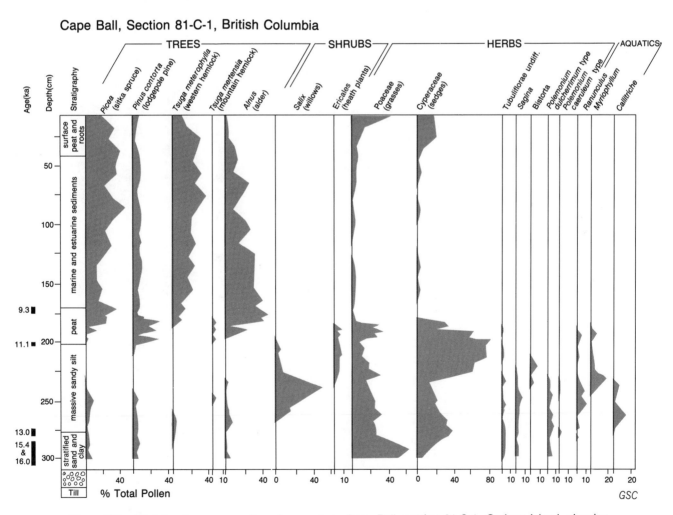

Figure 7.3. Relative frequency pollen diagram from Cape Ball, section 81-C-1, Graham Island, showing changes with time of selected pollen types. Only values of 1% or greater are shown. Radiocarbon dates on diagram are from both the 81-C-1 site and the adjacent 81-C-2 site (after Warner, 1984). The pollen results of this study and others in the text are shown as abbreviated pollen diagrams. The ages or dates (ka) on the left sides of the diagrams are based on ^{14}C dates which are shown in full in the text or in the respective references.

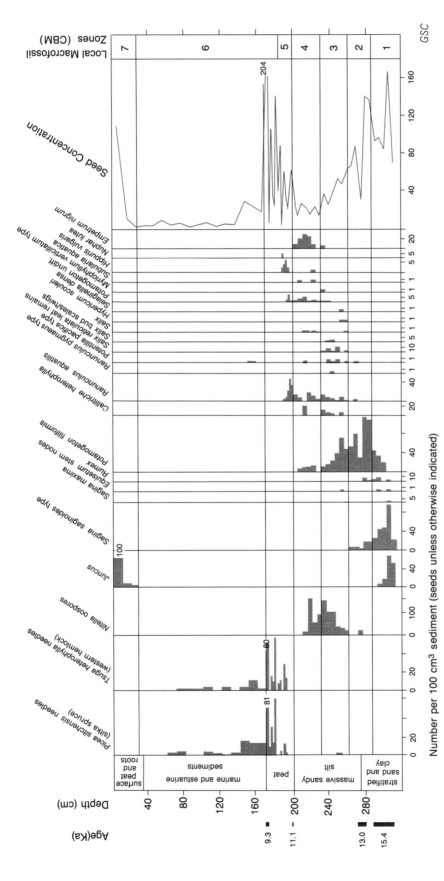

Figure 7.4. Plant macrofossil diagram of Cape Ball, section 81-C-1, Graham Island, showing stratigraphy and radiocarbon dates from this and the adjoining 81-C-2 locality. The total seed concentration curve includes seeds shown as well as those not listed on the diagram (after Warner, 1984).

notion that Sitka spruce survived on the Queen Charlottes during late glacial time. Pollen and abundant macrofossils of *Empetrum* show that prior to the spruce rise, heath plants increased (Fig. 7.3, 7.4), indicating poorer (more acidic) soil conditions. The peat bed at site 81-C-1 was deposited initially in shallow water beginning at 11 100 ± 90 BP (GSC-3337). A forest bed characterized by abundant wood, cones, seeds, and needles (Fig. 7.4) of coniferous trees developed near the end of deposition of the sequence. Between 10.0 ka and 9300 ± 80 BP (GSC-3477), the site was inundated by a marine transgression that deposited marine and estuarine sediments on top of the peat.

Implications for the refugium controversy

Rather than relying on biological inference alone, these new data provide a more solid foundation for hypotheses regarding the nature of glacial refugia on the Queen Charlotte Islands. Significantly, it was the original geological data on sea level changes and new radiocarbon dates that led Clague (1981, p. 22) to suggest "... the possibility that ice cover on the Queen Charlotte Islands during the Fraser Glaciation was not extensive." The 15-16 ka dates and the pollen and macrofossils discussed here show that ice-free areas such as Cape Ball were occupied by a floristically diverse vegetation at the time when the mainland ice mass was at or near its peak extent.

It is important to emphasize that no matter how suggestive this evidence is, it does not prove the existence of a refugium in the strict sense. What would be required for that is evidence of continuous ice-free conditions throughout the last glacial maximum — evidence that we do not yet have. The only conclusion suggested by presently available data is that glacier ice was present in early Late Wisconsinan time (i.e., some time between 27.5 and 16 ka) and that the Cape Ball site was ice free at the time of the Late Wisconsinan maximum on the mainland, that is, after 16 ka (Ryder and Clague, 1989).

Most of Graham Island has yet to be investigated for evidence of refugia. Theoretical considerations for coastal regions (Heusser, 1960; Calder and Taylor, 1968; Fladmark, 1975, 1979) suggest that a variety of sites, ranging from low elevation headlands, interfiord ridges, and offshore islets to high mountain peaks (Dahl, 1955), might have escaped glaciation. The well documented eustatic drop in sea level of around 100 m or more during the last glacial maxima would have exposed much of what is now Hecate Strait (Fladmark, 1975; Warner et al., 1982). Under such conditions and providing it was not scoured by glaciers, the shelf area east of Cape Ball (Fig. 7.2) might have served for a time as a refugium before rising sea level submerged it once again. The Cape Ball peat bed underlies and in part forms an intertidal platform, and one can only wonder if submerged organics occur farther offshore. Perhaps it is in those areas that evidence of plants and animals predating 16 ka will be found. Such offshore sites may also conceal Paleoindian evidence predating 9 ka, the oldest age of primary cultural deposits so far found on the outer coast (Fladmark, 1979). In any case, it is evident from the studies already carried out on Graham Island that the Queen Charlotte Islands will continue to be a fruitful region for study of glacial refugia and the paleoecological evidence of such refugia.

PALEOECOLOGY OF THE WESTERN CANADIAN ICE-FREE CORRIDOR

C.E. Schweger

Introduction

The earliest glacial maps of North America display either a full glacial ice cover from the Atlantic to the Pacific (Wright, 1893) or the existence of a narrow strip of unglaciated terrain separating the Cordilleran and Laurentide ice sheets (Dawson, 1893; Chamberlin, 1894). This proposed narrow unglaciated strip was first described as a "corridor", guiding man's entry into the new world (Antevs, 1945, 1948). It has since come to be known as the "ice-free corridor" (Rutter and Schweger, 1980). Because the corridor would have channelled north-south biotic exchange in North America, its Quaternary history is an issue of priority for a number of disciplines, including vertebrate paleontology (Sher, 1976; Harington, 1980; Repenning, 1980; Guthrie, 1980), ichthyology (Lindsey and Franzin, 1972), invertebrate zoology (Clifford and Bergstrom, 1976), and phytogeography (Packer and Vitt, 1974; Wheeler and Guries, 1982), and it plays a central role in discussions concerning the human colonization of the New World (Haynes, 1971; Reeves, 1983; Martin, 1974).

When the Cordilleran and Laurentide ice masses were joined, biotic exchange ceased and the corridor problem was nonexistent. A "corridor", in the strictest sense, also ceased to exist during interglaciations, such as the present one. Thus the corridor question only applies to narrow windows of Quaternary time, namely those during which Cordilleran and Laurentide ice masses stood near to one another either prior to or after their merger into a single continent-wide ice mass. By definition, then, the ice-free corridor is a variable entity, both in space and time. Its spatial configuration and environments at any one time modulated north-south biotic exchange, but probably in different ways during each major glacial cycle.

Field data from sites ranging from northern Montana (United States) to the mouth of Mackenzie River (Northwest Territories) indicate that Cordilleran and Laurentide ice masses coalesced during the Quaternary. Interbedded tills, trains of erratics, and geomorphology mark a zone of ice fusion along the Rocky Mountain Foothills, then north across Liard River valley (Rutter, 1984). In the north, the western limit of Laurentide ice lay along the eastern edge of the Mackenzie and Richardson mountains, and Cordilleran ice reached its limit in the Yukon along the western edge of the Mackenzie Mountains (Hughes, 1972). In the south, Cordilleran ice extended to or into the Rocky Mountain Foothills and Laurentide ice to the edge of the plains or the Rocky Mountain front (Prest,

---

Schweger, C.E.
1989: Paleoecology of the western Canadian ice-free corridor; in Chapter 7 of Quaternary Geology of Canada and Greenland, R.J. Fulton (ed.); Geological Survey of Canada, Geology of Canada, no. 1 (also Geological Society of America, The Geology of North America, v. K-1).

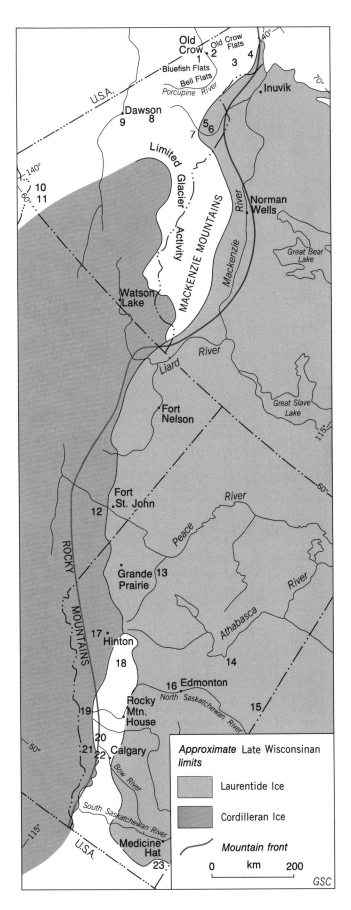

1969). The ice-free corridor opened in the north into an unglaciated Yukon refugium (eastern part of Beringia), while at the southern end it terminated in the Great Plains of southern Alberta and Montana.

There is no question that an ice-free corridor of some form and dimension existed near the beginning and end of each major glaciation, and the evidence of fusion of the Cordilleran-Laurentide glaciers is also clear. However, the chronology of the last such event, and hence of corridor development, is still poorly known. The paleoenvironmental record of earlier corridors is even more cloudy. Mammalian exchange between Siberia and the North American midcontinent via Beringia suggests that earlier corridors existed (Repenning, 1967, 1980), but the mammals say little about the environmental character of these early corridors (Rutter, 1984). For the time being our only recourse is examination of the environmental record at sites near the northern and southern portals of the corridor region.

Yukon: northern portal to the ice-free corridor

The unglaciated region of Yukon Territory is the most eastern part of the Beringian refugium. Much of the Quaternary paleoecology and biogeographic significance of Beringia and northwestern Canada is reviewed elsewhere (Hopkins et al., 1982; Ritchie, 1984a). Controversy and interest surrounding the entrance of early man to North America (Morlan, 1978; Jopling et al., 1981) and resolution of the large-mammal "production paradox" (Schweger et al., 1982) has stimulated paleoecological research in the Yukon. Most attention has focused on the northern Yukon, particularly the Old Crow, Bluefish, and Bell flats (Fig. 7.5), where multidisciplinary studies have dealt with stratigraphy, microfossils, and macrofossils (vertebrate, invertebrate, and plant) at a number of the excellent exposures (Hughes et al., 1989). Paleomagnetic data from some of the exposures and the presence of identified tephras have complemented taphonomic studies (Morlan, 1980) that are in many respects as comprehensive in scope as similar early-man programs in the East Africa (Coppens et al., 1976).

Figure 7.5. Northern and southern sectors of the ice-free corridor (modified from Rutter, 1984). Sites mentioned in text are listed: (1) Twelvemile Bluff (HH-228) on the Porcupine River in Bluefish Flats; (2) CRH-11a on Old Crow River in Old Crow Flats; (3) Polybog, Old Crow Flats; (4) Hanging Lake; (5) Tyrrell Lake, Doll Creek Valley; (6) Lateral Pond, Doll Creek Valley; (7) Hungry Creek Section (HH-72-54) in Bonnet Plume Flats; (8) Chapman Lake; (9) Dawson area; (10) Antifreeze Pond; (11) Silver Creek sections; (12) Charlie Lake archeological site; (13) Watino Sections on Smokey River, Alberta; (14) Lofty Lake, Alberta; (15) Moore Lake; (16) Fort Saskatchewan; (17) Gregg Lake; (18) Fairfax Lake; (19) Goldeye Lake; (20) Vermillion Lakes archeological site; (21) Chalmers Bog; (22) Sibbald Creek archeological site; (23) Robinson paleosol.

Terrestrial fossiliferous sediments have accumulated in the structural basins of northern Yukon since at least the Late Tertiary (Morison and Smith, 1987). The upper part of this sedimentary sequence is exposed in the higher bluffs on Old Crow, Porcupine, and Bell rivers. Although many sections have been studied, the Twelvemile Bluff exposure (site 1, Fig. 7.5) has become the informal type section for the regional stratigraphy and paleoecology. Five fossil pollen assemblage types have been recognized (Lichti-Federovich, 1974) for the six stratigraphic units identified in the section (Hughes, 1969; Stop 30 in Morison and Smith, 1987). Recent paleoecological and stratigraphic work has augmented and clarified the stratigraphy and environmental implications of this important exposure (Fig. 7.6) (see also Fig. 10 in Ritchie 1984a for a slightly different recast of Lichti-Federovich's original data).

Paleomagnetic data, pollen, and macrofossils as well as dates on tephras at Twelvemile Bluff and nearby sections show that Units 1-3 are probably Early Pleistocene in age or older. An in situ forest bed, in Unit 1 contains remains of extinct species of *Picea* and *Larix* as well as *Pinus contorta* and probably *Pinus monticola*, the latter two now limited to central Yukon and southern British Columbia, respectively. This plus presence of significant amounts of *Corylus* (pollen), another taxon not occurring in the Yukon today, points to a warmer and moister climate for the northern Yukon.

Conifer fossils occurring in Unit 2 represent black and white spruce, both of which grow near the site today. Permafrost related structures indicate climate as cold as at present. Pollen of *Picea* and *Pinus* probably represents a large lake that filled the basin when drainage was disrupted by tectonic activity.

Unit 4 is the most complex unit in terms of sediment type and variety of environments represented. The earliest pollen record of tundra vegetation in the Yukon comes from this unit (Lichti-Federovich, 1974). Alluvial sands and gravels of the lower half of the unit contain many cryoturbated silty paleosols and yield shrub-tundra pollen spectra dominated by *Betula*. The upper few metres are silty, with organic zones and several autochthonous peats. At least two horizons of ice-wedge pseudomorphs are evident, attesting to alternating permafrost conditions. Old Crow tephra, an important marker bed in east Beringia which probably fell during stage 5b of the marine oxygen isotope record (Westgate et al., 1983; Schweger and Matthews, 1985), occurs within Unit 4. At the time of the tephra fall, birch-dominated shrub tundra existed over much of northern and central Beringia (Schweger and Matthews, 1985). Evidence of a warmer climate than at present is the pronounced *Picea* pollen peak in Unit 4 (Fig. 7.6) above the tephra and associated fossils of insects and plants that presently do not occur as

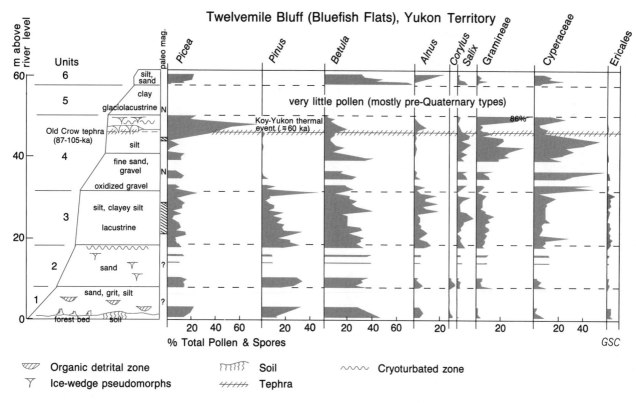

Figure 7.6. Stratigraphy and palynology of the Twelvemile Bluff section, Porcupine River (Bluefish Flats), Yukon Territory based on unpublished data from O.L. Hughes, C.E. Schweger, J.V. Matthews, Jr., S. Lichti-Federovich, and Lichti-Federovich (1974). (See also text and illustrations for Stop 30 in Morison and Smith, 1987). Date on Old Crow tephra from Schweger and Matthews (1985).

far north. This warming, informally designated the Koy-Yukon thermal event (Schweger and Matthews, 1985) and thought to date to the early part of oxygen isotope stage 3 (i.e., about 60 ka), is registered at other sites in east Beringia. A return to tundra conditions is evidenced at the top of Unit 4 (Lichti-Federovich, 1974).

Dated Middle Wisconsinan evidence of tundra and boreal forest is known from exposures on Silver Creek, southern Yukon (Schweger and Janssens, 1980) and Hungry Creek, north-central Yukon (Hughes et al., 1981), respectively. The best evidence of paleoecological change in the Yukon is preserved in lake sediments deposited during the Late Wisconsinan cold cycle. For example, the pollen records from Hanging Lake, northern Yukon (Cwynar, 1982), and Antifreeze Pond, south-western Yukon (Rampton, 1971) (sites 4, 10, Fig. 7.5) extend back to 30 ka. Both studies infer existence of tundra vegetation, akin to modern arctic tundra between 30 and 14.6 ka (Hanging Lake) and lasting until 10 ka (Antifreeze Pond). Following this phase is one dominated by shrub *Betula* (at least in terms of pollen abundance), which undoubtedly reflects late glacial warming. Pollen records from Lateral Pond (Ritchie, 1982) and Chapman Lake (Terasmae and Hughes, 1966) (sites 6, 8, Fig. 7.5) provide further evidence of widespread Late Wisconsinan tundra vegetation in the Yukon.

Macrofossil and palynological evidence points to an absence of boreal forest in eastern Beringia at this time (Hopkins et al., 1981, 1982; Ager, 1983). Spruce, pine, alder, tree birch and aspen, balsam poplar, and many other taxa now typical of the boreal forest or forest-tundra environment in eastern Beringia must have survived in full glacial refugia south of the continental ice sheets (Hopkins et al., 1981). With late glacial warming and the opening of the ice-free corridor they were able to migrate northward, reclaiming the territory occupied during the previous interglacial or interstadial.

This late glacial northward migration scenario explains much of the fossil pollen evidence for Yukon and Northwest Territories. In addition to the pollen sites mentioned above, the records from Polybog (Ovenden, 1982) and Tyrrell Lake (Ritchie, 1982) (sites 3, 5, Fig. 7.5) all show a sharp increase in *Betula* pollen dated as early as 14.6 ka at Hanging Lake. The *Betula*-dominated shrub tundra phase was followed in the early Holocene (e.g., 8.7 ka at Antifreeze Pond and 7.6 ka at Polybog), by a phase during which *Picea* becomes important. Because alder had yet to invade the area, these early Holocene "spruce forests" have no modern counterpart. Contemporary spruce forest first came into existence when alder appeared (6.7 ka at Polybog; 5.7 ka at Antifreeze Pond). Several detailed pollen studies (Ritchie, 1977, 1982, 1984a; Cwynar, 1982) suggest that other taxa such as *Populus, Larix, Myrica,* and *Shepherdia canadensis* may also have been Holocene migrants from the south.

Evidence of migration is not always clear. Setting speculation aside, Ritchie (1984a) has concluded that the radiocarbon chronology of the late glacial Holocene vegetation changes is equivocal when it comes to demonstrating a simple south to north migration for spruce. The earliest spruce pollen records for northwest Canada come from northern Yukon Territory and the Tuktoyaktuk Peninsula, 10.5 ka and 10 ka, respectively. Relying upon the dated pollen records from central Alberta where spruce appeared 10.8 ka (Lichti-Federovich, 1970, 1972), Ritchie (1984a) calculated a south to north migration rate of 1 km per year, which he felt was too fast, even though seed dispersal by river transport could have speeded the process. Seed dispersal by migratory waterfowl may have also been important; however, nothing is known of the extent to which modern fly-ways through western Canada may have been affected by glaciation and ecological changes.

The lack of detailed, well dated pollen studies over such a vast region precludes construction of migrational sequences for other plant taxa at this time (Hopkins et al., 1981; Ritchie, 1984a). Nevertheless, as indicated above, it is clear that migration of plants through the ice-free corridor did occur at the end of the Wisconsinan. The same evidence used to infer plant migration also indicates that a northward flood of species was also characteristic of insects, vertebrates, and other boreal animals. The movement of species through this narrow zone was probably repeated several times during each major glacial cycle, and since the corridor is likely to have served as a filter for some taxa each time it was in existence, the biotic restructuring that occurred after each episode probably formed unique communities having no modern analogues.

Evidence from the northern Yukon points to plant extinctions and range restrictions after the late Tertiary. Extinctions could have conceivably resulted from failure of plant taxa to successfully move southward through the ice-free corridor during onset of glaciations, while more aggressive recolonization of the north by some taxa following deglaciation might have restricted former northern taxa to more southerly ranges. The ice-free corridor probably also served as a temporary biotic filter, in the way that the east-west trending Alps influenced the floral history of Europe during the Tertiary and Pleistocene (van der Hammen et al., 1971).

The unglaciated part of Yukon Territory has long been an important source for Pleistocene mammal remains (Harington, 1970, 1977). Important vertebrate localities are located in the Dawson area (site 9, Fig. 7.5; Harington, 1978), where gold-mining operations involving removal of frozen overburden have yielded an abundance of bones. River erosion, transportation, and hydrological concentration of vertebrate remains have resulted in large collections from modern river bars and Holocene terraces along the Old Crow and Porcupine rivers, northern Yukon. While collections from organic "muck" or alluvium are rich, they lack stratigraphic provenance, thus reducing their paleoecological usefulness and leading to questionable dating attempts, such as the one based on degree of mineral staining (Irving et al., 1977). Detailed analysis of single faunas from controlled stratigraphic contexts greatly increases the evolutionary, biogeographic, and paleoecological significance of vertebrate remains. This requires controlled excavations and screening of sediments from single stratigraphic levels (Morlan, 1980, 1986).

At Old Crow River locality CRH-44 Harington (1978) recovered an "interglacial" fauna, which includes among boreal and arctic taxa fossils of the extinct short-faced skunk (*Brachyprotoma*; Youngman, 1986). Although initially thought to be of interglacial age, perhaps Sangamonian, the 1.2 Ma date on Little Timber tephra (J.A. Westgate, University of Toronto, personal communication, 1984) from approximately the same stratigraphic level warns that this and perhaps some other vertebrate local faunas described by Harington may include taxa of Middle to Early Pleistocene age. The vertebrate fauna from Old Crow locality CRH-11a (site 2, Fig. 7.5) is undoubtedly a mixture of Middle to Late Wisconsinan taxa redeposited in Holocene alluvium. It is an especially noteworthy local fauna because it includes man-modified Wisconsinan bones radiocarbon-dated between 22 and 33 ka (Irving and Harington, 1973; cf. Nelson et al., 1986; Morison and Smith, 1987) as well as a fragment (as yet undated) of a human skull (Irving et al., 1977). Such finds are potentially of great significance to the functioning of the "ice-free corridor" as a portal for entrance of early man into the central part of North America. In this case, however, the significance of the man-related finds is compromised by questionable dating methods (Nelson et al., 1986) and the fact that the fossils, like the majority of vertebrate remains reported in northern Yukon local faunas, are in reworked sediments (Morlan, 1980).

In total, 64 species of Pleistocene vertebrates have been recognized for the unglaciated northern Yukon. The high frequencies of horse, mammoth, and *Bison* remains led Harington (1978) to suggest extensive late Quaternary grasslands. These three taxa constitute the nub of the "productivity paradox" discussed by Hopkins et al. (1982), and, while all Yukon faunas suffer from many of the same problems as those from Old Crow, radiocarbon dating shows that mammoth, horse, *Bison*, and caribou were coeval during the Late Wisconsinan (Matthews, 1982; Morlan and Cinq-Mars, 1982).

Alberta: southern portal to the ice-free corridor

The southern end of the ice-free corridor opens onto the Great Plains of southern Alberta. The Quaternary stratigraphic sequence is best seen at exposures along South Saskatchewan River, near Medicine Hat (Fig. 7.5), where the river bluffs, containing multiple tills as well as alluvium and lacustrine deposits, mark a long record of Quaternary events in western Canada (Stalker and Churcher, 1982; Klassen, 1989). Palynological studies on these deposits have been hindered by poor preservation (R.J. Mott, unpublished Geological Survey of Canada palynological reports). Stalker and Mott (1972) described from the Cypress Hills a Kansan flora from which they reconstructed upland prairies, riparian forests of *Salix*, *Acer negundo*, and *Populus*, and *Picea-Pinus* forests. They suggested that the climate was "probably similar to present or slightly cooler and more moist." Little else is available in the way of pollen evidence with which to characterize the Pleistocene environments of southern Alberta. Even data for the Late Wisconsinan from still farther south into the United States is limited. Clearly, much more work remains to be done before a detailed fossil pollen record is available.

Seventy-two vertebrate taxa have been identified from these sections; the oldest fauna is ascribed to the Nebraskan glacial stage. Large mammals dominate, with the greatest diversity of 31 taxa coming from deposits thought to be of Sangamonian age. Because these mammals were highly mobile and could tolerate a wide range of environmental conditions, their fossils offer only general paleoecological inferences. The variety of species within the equine and bovine groups at southern Alberta sites suggests a persistence of grasslands throughout the Pleistocene. However, Sangamonian faunas contain such presently disjunct taxa as *Rangifer tarandus* (caribou), *Procyon lotor* (raccoon), *Odocoileus virginianus* (white tail deer), and *Hemianchenia* sp. (plains llama). Caribou are known historically from the southern Alberta Rocky Mountains (Soper, 1964); white-tail deer and rarely raccoon are now found along wooded river valley bottoms (Soper, 1964; Banfield, 1974), and the plains llama likely could have found suitable habitat on the cool grasslands of Alberta. Therefore, despite the fact that the assemblage has no modern analogue, it probably represents an environment similar to that of the present.

If the ice-free corridor served to channel northern faunas southward during glaciations, a high degree of faunal similarity between the two regions would be expected. Even though the northern and southern portals of the ice-free corridor are over 2000 km and 15° latitude apart, each has 20 Late Pleistocene mammalian taxa in common (Table 7.1).

It is tempting to suggest that this degree of similarity results from exchange through the ice-free corridor. The majority of these species, however, have great ecological amplitude and are presently found far beyond the portal areas. Some are extinct, allowing us only to speculate on their ecological requirements. Consequently, it is still not clear to what extent the ice-free corridor controlled faunal migrations during glaciations.

Table 7.1. Late Pleistocene mammal fauna common to Medicine Hat, Alberta[1] sections and the Yukon interior[2]

Extant in ice-free corridor today	
Castor canadensis	Canadian beaver
Ondatra zibethicus	muskrat
Erethizon dorsatum	porcupine
Vulpes vulpes	red fox
Canis latrans	coyote
Canis lupus	gray wolf
Lynx canadensis[3]	Canada lynx
Ovis canadensis[3]	mountain sheep
Extant in south today	
Mustela nigripes	black-footed ferret
Cervus canadensis	wapiti
Extant in north today	
Rangifer tarandus[3]	caribou
Extinct forms	
Megalonyx sp.	ground sloth
Brachyprotoma obtusata?	short-faced skunk
Canis dirus	dire wolf
Smilodon fatalis	sabre-toothed cat
Felis leo atrox	Pleistocene lion
Mammuthus primigenius	northern mammoth
Equus scotti	Scott's horse
Camelops hesternus	western camel
Symbos cavifrons	woodland muskoxen

1 adapted from Stalker and Churcher (1982)
2 adapted from Harington (1978), Youngman (1986)
3 known from Rocky Mountains, southern Alberta

Little is known for certain about the southern Alberta portal of the ice-free corridor during the Late Wisconsinan. Ice margins are in dispute (Clayton and Moran, 1982; Moran and Clayton, 1983; Rutter, 1984; Klassen, 1989) and the nearest fossil pollen sites are located over 1000 km farther south. Spruce forests existed to the south during the Late Wisconsinan, and treeless prairie vegetation did not become established there until shortly before the beginning of the Holocene, that is, about 11 ka (Wright, 1984). However, by then spruce forest, complete with its complement of forest insects, had moved northward to colonize areas newly exposed by retreat of glacial ice. It also advanced onto extensive areas of drift underlain by stagnant ice (Ritchie, 1976; Ashworth and Cvancara, 1983). Spruce first appeared at Moore Lake, Alberta (site 15, Fig. 7.5; this chapter, Anderson et al., 1989) at 11.3 ka (Schweger et al., 1981), a time when a pioneer forest dominated by *Populus* existed farther west at Lofty Lake (site 14, Fig. 7.5; Lichti-Federovich, 1970). The earliest evidence for grassland vegetation in southern Alberta comes from the high nonarboreal pollen record and infrared spectra of humic acids from a 10.2 ka paleosol near Robinson (site 23, Fig. 7.5; Westgate, 1972). A southern boreal forest/grassland ecotone may have been located east and presumably north of this site as indicated by pollen analysis of the late glacial Hafichuk site near Moose Jaw, Saskatchewan (Ritchie and DeVries, 1964).

The ice-free corridor

The earliest dated paleoecological record from the ice-free corridor comes from the Watino exposure and similar sites along the Smokey River in north-central Alberta (site 13, Fig. 7.5). At Watino finite dates ranging from 43.5 to 27.4 ka span a Middle Wisconsinan interval characterized by open vegetation with scattered conifers and an invertebrate fauna similar to that of present-day central Alberta (Westgate et al., 1972). Younger events are not recorded at this site.

Lake coring in western Alberta and northeastern British Columbia has resulted in recovery of several cores that date to the beginning of Late Wisconsinan time (White et al., 1979; Holloway et al., 1981), but the dates in some of these are spuriously old due to contamination by coal or other "dead" organics (White, 1983; Hickman et al., 1984). This is a common problem with lakes in the corridor region.

Pollen records whose dates have apparently not been influenced by "dead" organic contamination demonstrate the existence of tundra vegetation in the ice-free corridor during Late Wisconsinan time. For example, the basal pollen zone at Chalmers Bog, southwest of Calgary (site 21, Fig. 7.5) is dated at around 18.3 ka and is dominated by Gramineae, *Artemisia*, and Cyperaceae with a variety of herbs (Mott and Jackson, 1982). Pollen influx values average 350 grains/(cm^2·a), lending support for tundra vegetation at this time. By 8.2 ka pine forests were established at the site. At Yamnuska Bog, another site near Calgary (MacDonald, 1982), a basal zone dominated by *Artemisia*, *Salix*, and *Juniperus* is overlain by a zone (dated at 10 ka) dominated by spruce and pine pollen.

The most complete pollen record from the ice-free corridor is at Goldeye Lake, in the foothills near Nordegg (site 19, Fig. 7.5). Seven radiocarbon dates from the 5.5 m core reveal a pollen record extending back beyond 24 ka. It is divided into four local pollen zones (Fig. 7.7).

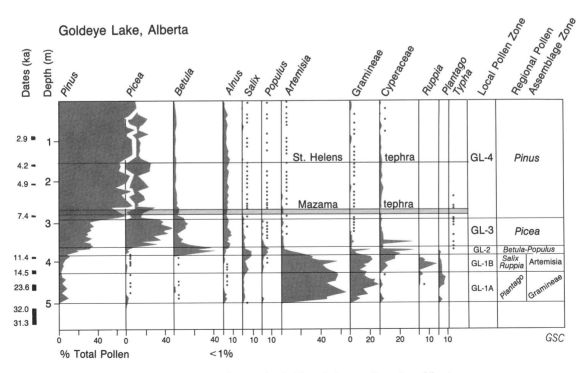

Figure 7.7. Pollen diagram for Goldeye Lake, southwestern Alberta.

The basal zone, GL-1, is dominated by *Artemisia* and Gramineae along with pollen of a large variety of herbs. Dated from >24 to 11.4 ka and displaying pollen influx values that average 37 to 251 grains/(cm$^2 \cdot$a), zone GL-1 represents sparse tundra vegetation. An increase in *Salix* pollen at 15.5 ka, within GL-1, marks the onset of late glacial warming and deglaciation. Pollen zone GL-2 is dominated by *Betula* and *Populus* and is dated at 11.4 to 10.8 ka. *Picea* and *Alnus* dominate between 10.8 and 7.8 ka in zone GL-3, after which a rise of *Pinus* marks the first appearance of the present boreal forest.

The pollen record from Fairfax Lake (site 18, Fig. 7.5), from slightly farther north, reproduces the Goldeye Lake sequence as does the pollen record at Gregg Lake (site 17), which is situated in a glacial meltwater channel that received sediments shortly after ice began retreating from Athabasca River valley.

Thus it is clear that sparse herbaceous tundra existed during the Late Wisconsinan in the ice-free corridor from Chalmers Bog in the south (and presumably farther to the south) to Gregg Lake in the north. Surrounded by glacial ice, which directed strong bora winds through the corridor, and cut off from important moisture sources, the corridor must have possessed a cold, windy, and arid regional climate. Climate began to warm at 15.5 ka. Between 11.5 and 10.5 ka, pioneering forests composed of *Populus* and birch were first established, and these were followed shortly after by boreal forest of the modern type.

By 11.3 ka sediments were accumulating in Moore Lake basin (550 m a.s.l.), eastern Alberta (site 15, Fig. 7.5) (this chapter, Anderson et al., 1989 for pollen and diatom profiles). Spruce-birch forest was growing at Moore Lake at 10.5 ka when Goldeye Lake, 800 m higher in elevation, supported only *Populus*-birch communities. Spruce forests advancing upslope from the east did not appear at Goldeye Lake until nearly a thousand years later.

During the height of the Late Wisconsinan, the ice-free corridor in the Goldeye Lake region was a narrow zone bounded by the Rocky Mountains on the west and the Laurentide ice mass on the east (Fig. 7.8). Deglaciation, starting 15.5 ka, was marked by retreat of the Laurentide ice front, causing the corridor to widen to the east. Near its eastern border, however, this ice-free zone was characterized by large areas of hummocky disintegration moraine and stagnant ice buried beneath a mantle of drift. Biologically productive ponds in the Edmonton area (Fig. 7.5) existed in the stagnant ice region until between 11 and 9 ka (Emerson, 1983). Spruce forest moved onto and across the drift-mantled stagnant ice, spreading rapidly to the northwest, eventually replacing the pioneer vegetation at Goldeye Lake and spreading southward into the Calgary area by 10 ka (MacDonald, 1982). As Laurentide ice continued to thin and stagnate, causing the actual glacial ice barrier to move out of western Canada, the ice-free corridor, as a definable entity, ceased to exist.

Several localities in the corridor region have yielded Rancholabrean (Late Pleistocene) mammal remains. Terrace gravels from Bow River near the Rocky Mountain Front contain a fauna comprised of *Camelops* cf. *hesternus* (camel), *Bison bison antiquus*, *Equus conversidens*, *Mammuthus* sp., *Cervus elaphus* (wapiti), *Rangifer tarandus* (caribou), and *Ovis* sp. (sheep), and dated at 11-10 ka (Churcher, 1968, 1975; Wilson, 1974; Wilson and Churcher, 1978). It is thought to be the type of fauna that would have lived in a region of parkland vegetation (Wilson and Churcher, 1978). Vertebrate remains have also been found in the eastern Peace River region (Churcher and Wilson, 1979). Early postglacial gravels near the Rocky Mountain Front contain *Mammuthus primigenius*, *Equus*, *Bison bison bison*, *Bison bison occidentalis*, *Bison bison athabascae*, a camelid, *Cervus canadensis*, and *Ovis*. These also suggest a parkland-like environment, perhaps similar to the present one in southern and central Alberta.

Only a few of the fossil vertebrate sites date to the Late Wisconsinan maximum. A 22 ka cave fauna from southern Alberta includes *Microtus richardsoni* (southern water vole) and *Lemmus sibiricus* (northern lemming) (Burns, 1980, 1982), presumably reflecting a periglacial environment. Harington (1975) has cited several records of Pleistocene musk oxen (*Symbos*) from Alberta. One specimen from near Fort Saskatchewan (site 16, Fig. 7.5), closes the gap of over 1600 km between *Symbos* remains in eastern Beringia and those south of the Laurentide Ice sheet and "suggests possible links or former routes of movement" through an ice-free corridor. But few vertebrate remains have been dated, making it difficult to say how many mammalian taxa actually occupied and moved through the corridor. And thus we cannot yet say, on the basis of vertebrate fossils alone, that the ice-free corridor was ever suitable for large populations of ungulates. The sparse tundra that existed there under a cold, arid climate probably excluded all but the hardiest species.

The paleoecology of the ice-free corridor, the productivity of its vegetation, and the nature of the mammalian

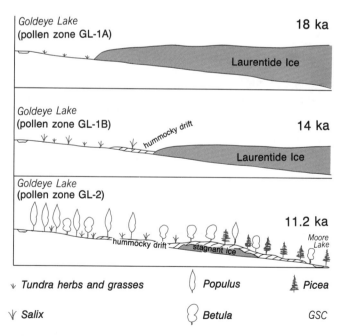

Figure 7.8. Cross-section of the ice-free corridor between Goldeye Lake and Moore Lake (sites 19 and 15, Fig. 7.5), showing presumed landscape characteristics at 18 ka, 14 ka and 11.2 ka (see also Anderson et al., 1989, for a pollen diagram of Moore Lake).

community are of special interest to archeologists. Many consider the corridor as a path used by early human populations, which would have entered the corridor after having passed across the Bering land bridge from Asia and moved through unglaciated east Beringia. Some archeologists have argued that man was in northwestern North America prior to the last development of the corridor. Worked bone, especially mammoth bone flakes and cores, along with antler wedges, a tibia flesher, and evidence of butchery, suggest human activity in the Old Crow region of the northern Yukon (Fig. 7.5) during the Middle Wisconsinan (but see Nelson et al., 1986 for new dates on some of these artifacts). Jopling et al. (1981) inferred a human presence during pre-Sangamonian times, but Morlan (1986) has re-evaluated the evidence and now admits only to the probability of 40 ka man-modified bones (see Stop 28 in Morison and Smith, 1987). In any case, man-modified bones from the Bluefish Cave site near Old Crow (Cinq-Mars, 1979) are conclusive evidence of human hunting activity during the Late Wisconsinan. Significantly the Bluefish Cave evidence predates by several thousand years the oft-cited entrance date of people into North America (Haynes, 1980), which is based on the age of Clovis sites located far south of the corridor.

The Clovis culture, defined on the basis of its distinctive basally fluted projectile points, makes its sudden appearance south of the Laurentide ice just before 11 ka. Haynes (1980) believed that the Clovis people probably migrated to the south through the ice-free corridor. In contrast, a number of archeologists (Bryan, 1969; Bonnichsen and Young, 1980) argued that Clovis originated from pre-existing human populations south of the ice sheets; if so, the Clovis culture must have moved into the area of the ice-free corridor from the south. Two dated sites from the southern part of the corridor appear to support this contention. A rock shelter near Charlie Lake in the vicinity of Fort St. John (Fig. 7.5) has produced bison bone and "Clovis-like" fluted projectile points (Fladmark and Gilbert, 1984) dated as 10.5 ka. The site is somewhat younger than Clovis sites south of the ice sheets. The Vermillion Lakes site (site 20, Fig. 7.5) near Banff, is dated at 11.5-11 ka and contains abundant lithic artifacts and a variety of faunal remains including *Ovis canadensis catclawensis*, a late glacial precursor of Rocky Mountain Big Horn sheep (*Ovis canadensis*) (Fedje, 1984). The Sibbald Creek site in the southern Alberta Foothills (site 22, Fig. 7.5) has also produced fluted points, but unfortunately from an unstratified and undated context (Gryba, 1983). Thus, it is clear that one must remain cautious in attempting to assign a Clovis or fluted point migration vector, whether directed north or south, on the basis of the limited available data.

It has been suggested that people lived in southern Alberta as early as 120 to 80 ka and again at about 40 ka (Stalker, 1977); however, key evidence that appeared to support the "early occupation" are bones of the "Taber Child", which have recently been dated as late Holocene (Brown et al., 1983). Putative pre-Late Wisconsinan flaked chert pebbles that have been cited in support of the 40 ka occupation are not universally accepted as being man-made.

Currently there is still too much uncertainty about the history of the corridor and its environment to warrant further archeological speculation on its probable significance for the history of human beings in North America. Future research efforts designed to improve this situation must focus on the geological problem of documenting in detail Late Wisconsinan changes in the development of the corridor and then seek to demonstrate by appropriate paleoecological means that the ice-free areas were suitable for occupation.

Summary

The ice-free corridor has long been recognized as a feature of the glacial history of western Canada. Its role in paleontological, biological, and archeological dispersal has often been hypothesized but rarely proven. Detailed stratigraphic and paleontological records spanning most of the Pleistocene are known from the contrasting north and south ends of the corridor, but little is known of the early history of the corridor region itself. A better understanding of the Late Wisconsinan history of the corridor is an essential requirement for evaluating its role in the peopling of North America. From studies performed to date, all that is known is that between 24 and 11.4 ka cold, arid tundra vegetation existed in the ice-free corridor from the International Boundary to Athabasca River valley. Woolly mammoth may have been present, lemmings certainly were, perhaps along with horse, caribou, bison, camel, and sheep. At 15.5 ka the botanical record marks the start of deglaciation and by 11.4 ka, coincidental with the earliest archeological record, pioneering forests of birch and poplar or aspen were established in the Foothills region of the corridor.

As Laurentide ice stagnated along the eastern border of the corridor, ice-cored drift was colonized by spruce and poplar trees. Spruce continued to spread, eventually reaching the high Foothills by 11 to 10 ka. As the area of newly deglaciated territory continued to grow, the significance of the ice-free corridor as a spatial and biological entity decreased to zero, thus ending the most recent act of the history of the corridor.

INTRODUCTORY COMMENTS ON INTERGLACIAL ENVIRONMENTS

J.V. Matthews, Jr. and T.W. Anderson

Most of the case histories in this chapter rely on or refer to study of pollen from cores of lake sediments. The following case history represents a different approach in that it deals with fossil pollen from organic sediments exposed in river bluffs, sea coasts, quarries, and gravel pits at various sites in Atlantic Canada.

Pollen from lake cores usually offers the most credible evidence of past regional vegetation and climate. In suitable lake cores it is possible to sample sediments at close intervals and to estimate the rate of pollen deposition per unit area per year, a technique that alleviates some of the statistical problems when pollen fluctuations are expressed in terms of relative abundance (percentages). But sediment focusing and other types of disturbance may make influx data unreliable (Davis and Ford, 1982). In some cases the

---

Matthews, J.V., Jr. and Anderson, T.W.
1989: Introductory comments on interglacial environments; in Chapter 7 of Quaternary Geology of Canada and Greenland, R.J. Fulton (ed.); Geological Survey of Canada, Geology of Canada, no. 1 (also Geological Society of America, The Geology of North America, v. K-1).

core sediments contain too little organic material for radiocarbon dating, a prerequisite for computing pollen influx per annum.

A more serious problem with lake-based studies, especially for Canadian workers, is that because so much of the country was ice-covered during the last glaciation, lacustrine pollen sequences that extend back beyond 14 ka are rare. One might expect that lake basins with appreciably longer records would exist in unglaciated areas such as in east Beringia, and a few are known (Cwynar, 1982); however, most lake basins in northern unglaciated regions have a thermokarst history, leading to a disturbed record. Somewhat similar problems exist with lakes in other unglaciated parts of Canada or even those regions that have been ice-free only since the Middle Wisconsinan (Anderson, 1983). Therefore, to date, Canadian workers do not have a counterpart of the long Grand Pile or Les Echets records from France (Woillard, 1978; DeBeaulieu and Reille, 1984).

To gain information on the environment of the last interglaciation one is forced to resort to sections containing buried organics of that age or buried deposits of lacustrine sediments. However, there are disadvantages to dealing with such evidence. Buried organic deposits usually supply a discontinuous and in places a truncated or severely compressed record of vegetation and faunal change. Also, such sediments are often contaminated by rebedded pollen or contain so much local pollen (usually of aquatic plants) as to mask regional vegetation. On the other hand buried organic deposits also offer advantages: They usually yield an abundance of macrofossils, such as insects and various types of seeds or other plant remains; many represent time periods that are not broached by lake cores; they provide sufficient organic material for various methods of dating; and they can usually be traced and correlated from one section to another. Because of these special characteristics, buried organics can commonly be dated and correlated independently of the fossil evidence, thus avoiding circular arguments. When it is possible to correlate different facies of organics, the fossils from those deposits provide valuable information of several different types of contemporaneous biotopes. In northern areas, organic deposits often help in the identification of structures that signify former development or degradation of permafrost.

Buried organic deposits of presumed interglacial age occur at a number of sites in Atlantic Canada. At many localities they are associated with evidence of former sea level, allowing assessment of the interplay and linkage between marine and terrestrial environments during the last interglacial cycle (Ruddiman and McIntyre, 1979, 1981). Although fossil pollen is the primary type of evidence used

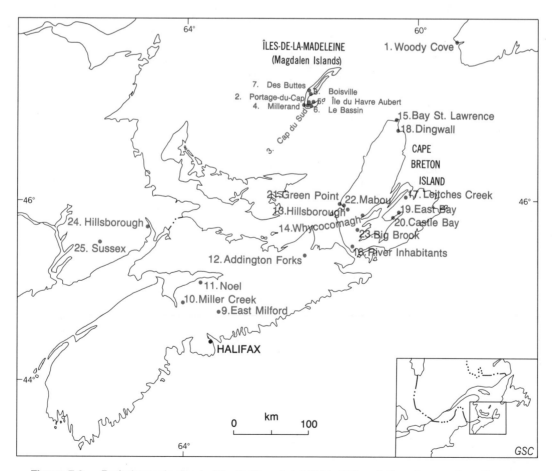

Figure 7.9. Buried organic sites in Atlantic Canada (cf. Table 7.2) and other sites mentioned in text.

in Mott's case history, many of the sections he discusses also contain plant macrofossils and fossil insects (Prest et al., 1976; J.V. Matthews, Jr. and A.V. Morgan, unpublished reports, 1984). Thus, like many of the other case histories in the chapter, the one by Mott is merely a status report on a research programme that is continuing to broaden in its scope.

LATE PLEISTOCENE PALEOENVIRONMENTS IN ATLANTIC CANADA
R.J. Mott

Introduction
Buried organic deposits have been known from Atlantic Canada since Dawson (1855) first noted such a deposit near River Inhabitants on Cape Breton Island. Numerous pre-Late Wisconsinan deposits are now known to occur in the region, the greatest concentration being in Nova Scotia, particularly Cape Breton Island (Fig. 7.9). Several sites have recently been discovered on the Magdalen Islands (Îles-de-la-Madeleine), Quebec; two sites are known from New Brunswick; and one site has been documented from Newfoundland, but as of 1984, buried organic deposits had not been found on Prince Edward Island.

Palynological studies of buried organic deposits in Atlantic Canada began some 30 years ago when L.R. Wilson (in Flint and Rubin, 1955) reported a preliminary palynological analysis of a site near Hillsborough, Nova Scotia.

This was followed by a report on the palynology of four sites on Cape Breton Island (Mott and Prest, 1967). Livingstone (1968) also studied one of these sites, Hillsborough, at about that time. Since these two reports, the tempo of research has increased (Mott, 1971; Prest et al., 1976; Mott et al., 1982; Vernal, 1983; Vernal et al., 1983, 1984; Mott and Grant, 1984, 1985; Vernal and Mott, 1986), and a considerable volume of unpublished data is now also available. This report discusses the results of paleoecological studies conducted to date; attempts to relate the sites to one another and fit them to the regional stratigraphic framework; and characterizes the environmental sequence during the last (Sangamon) interglaciation in Atlantic Canada.

Regional stratigraphic framework
The complex Quaternary stratigraphic framework for the Atlantic Provinces has been outlined by Grant (1989) and Grant and King (1984). It is based on both onshore and offshore sedimentary sequences and has been compared to the deep sea oxygen isotope record, a world standard for correlation of Late Pleistocene climatic events. Quaternary deposits in Atlantic Canada appear to represent three major time intervals: one predating the last interglaciation (stage 6 and earlier); one representing the last interglaciation (stage 5); and one encompassing all of the Wisconsin Glaciation (stages 4, 3, and 2).

Workers in Atlantic Canada consider the last interglaciation, or Sangamonian Stage, to encompass all of oxygen isotope stage 5, a complex period of about 50 ka starting at approximately 128 ka and ending at 75 ka (Grant and King, 1984). Some phases of this period were as warm or warmer than the present; other intervals of time were colder and in some cases even glacial in character.

Collectively the sites in the region have a number of features in common. Nonglacial deposits (organic in many cases) invariably underlie till(s) of the last (Wisconsinan) glaciation, and overlie till or bedrock. In some coastal areas the bedrock forms an elevated platform that presumably was cut during the high sea level stand of isotope stage 5e. A second, lower bench cut into the marine platform but above present sea level is recorded at some sites and may correlate with substage 5c or 5a.

Modern vegetation
Forests of the Atlantic Region vary considerably. Classifications by Rowe (1972) and by Loucks (1962) provide a general picture of the regional forest types (Fig. 7.10).

Both black and white spruce (*Picea mariana* and *P. glauca*) and balsam fir (*Abies balsamea*) characterize the forests of southwestern Newfoundland, the highlands of New Brunswick and Cape Breton Island, the Fundy and Atlantic coasts, and the Magdalen Islands. Ericaceous plants are associated with the conifers in Newfoundland and on the Cape Breton Plateau; in highland areas and on the Magdalen Islands the forests contain abundant white birch (*Betula papyrifera*), and along the Fundy coast red spruce (*Picea rubens*) is a common associate.

Coniferous forests in which red spruce and hemlock (*Tsuga canadensis*) attain their greatest prominence occur throughout the Maritime Lowlands of eastern New Brunswick and Nova Scotia, much of Prince Edward Island, and the lower elevations of peninsular Nova Scotia. Balsam fir, white pine (*Pinus strobus*), and red maple (*Acer rubrum*) are common associates, and beech (*Fagus grandifolia*), sugar maple (*Acer saccharum*), and yellow birch (*Betula alleghaniensis*) are prominent on suitable sites.

Predominantly hardwood forests of sugar maple, yellow birch, and beech with admixtures of balsam fir, white spruce, red spruce, and red maple are common on the uplands of Nova Scotia and New Brunswick and on part of Prince Edward Island. At lower elevations the hardwoods are associated with white pine and hemlock.

Almost pure stands of hardwoods are found only in the central Saint John Valley region, where stands dominated by sugar maple and beech also contain small numbers of white ash (*Fraxinus americana*), ironwood (*Ostrya virginiana*), basswood (*Tilia americana*), and butternut (*Juglans cinerea*).

Chronology
Radiocarbon dates are listed in Table 7.2. Most materials dated are beyond the limit of radiocarbon dating, and the validity of the few finite ones is suspect because (1) other

---

Mott, R.J.
1989: Late Pleistocene paleoenvironments in Atlantic Canada; in Chapter 7 of Quaternary Geology of Canada and Greenland, R.J. Fulton (ed.); Geological Survey of Canada, Geology of Canada, no. 1 (also Geological Society of America, The Geology of North America, v. K-1).

dates from the same deposit are infinite; (2) the suitability of some of the dated material is questionable; or (3) the paleoecological implications of the associated palynological data are inconsistent with the age.

Amino acid analyses were obtained for wood samples from some localities (N.W. Rutter, University of Alberta, Edmonton, personal communication, 1985). In general the aspartic acid ratios fail to provide definitive results. In some cases, the ratios suggest a wide range of ages in contrast to pollen evidence which suggests a short time interval. Wood from one site, interpreted from pollen analysis to represent an interglacial interval, yields ratios similar to those obtained for postglacial wood. Perhaps the only meaningful conclusion that can be drawn is that most of the buried organic deposits represent parts of a lengthy nonglacial interval (Grant and King, 1984).

Thorium/uranium analyses on wood appear to show more promise for clarifying the age of the deposits (Vernal et al., 1984, Causse et al., 1984). Unfortunately, samples from relatively few sites have been analyzed to date (Table 7.2), and the technique must be assessed to ensure the validity of the dates obtained. One serious source of error is caused by leaching of uranium and secondary influx of uranium and thorium. One indication of such secondary contamination is the presence of ^{232}Th. Therefore, those samples containing low amounts of ^{232}Th and high ^{230}Th/^{232}Th ratios (>3.0) are considered the most reliable. Even so, all of the dates should probably be considered minimum ages. The most reliable of them seem to indicate three distinct periods of organic accumulation: one at about 126-100 ka, another around 85 ka, and a third between 60 and 47 ka.

Palynological results

All known sites for which palynological data are available are listed in Table 7.2, but because there are so many sites, only abstracted, pertinent data are given there. Several sites that are typical of various geographical areas or that have spectra representative of distinct intervals are dealt with in more detail below.

Newfoundland. Woody Cove section (site 1, Fig. 7.9) is the only known interglacial site from Newfoundland. A gypsum karst depression exposed in a coastal cliff section near Codroy contains lacustrine and marine sand and clay-silt laminae overlying colluvium and underlying glaciofluvial deposits (Brookes et al.,1982).

Wood of balsam fir (*Abies balsamea*) from an organic horizon in the marine sand has a radiocarbon date of >40 000 BP (I-10203). Palynological evidence indicating a progressive vegetation shift from tundra to boreal forest and back to tundra represents a single climatic cycle. The characteristic boreal forest spectra of *Picea, Abies, Pinus,* and

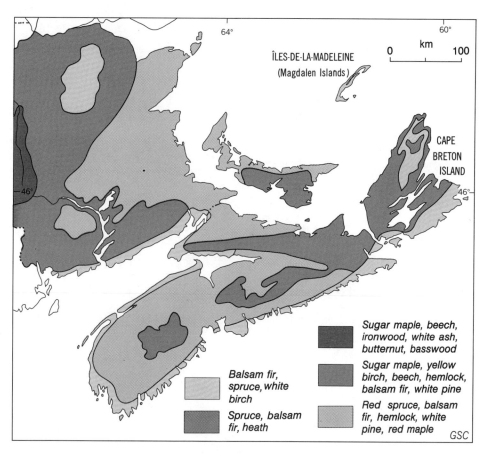

Figure 7.10. Present vegetation zones in Atlantic Canada (after Rowe, 1972 and Loucks, 1962).

Table 7.2. Sites, stratigraphy, dates, and paleoecology in Atlantic Canada

Site Name (cf. Fig. 7.9)	Type of deposit	Stratigraphy	Dates (material dated) ^{14}C and Th/U^*	Paleoecology	Isotopic stage	References
NEWFOUNDLAND						
Woody Cove (site 1)	Coastal cliff (karst)	Lacustrine and marine sand, silt, clay beneath glaciofluvial sediments (details in text)	>40 000 (I-10203; *Abies balsamea*)	Pollen: tundra/boreal/tundra cycle (Fig. 7.11a) Climate: as warm or warmer than present	5e	Brookes et al., 1982
QUEBEC						
Portage-du-Cap (site 2)	Gravel pit	Organic lens interbedded with sand and silt beneath diamicton (details in text)	>35 000 (BGS-259; organic detritus) >38 000 (GSC-2313; *Abies balsamea*)	Pollen: *Picea, Pinus, Abies, Betula, Fagus* (Fig. 7.11d). Macrofossils: plants extant in region Coleoptera and Diatoms: some taxa with more southern modern affinities Climate: as warm or warmer than present	5c	Prest et al., 1976; Lowdon and Blake, 1978
Cap du Sud (site 3)	Coastal cliff	Sand with thin beds of peat over bedrock	>38 000 (GSC-3413; peat)	Pollen: Gramineae, *Betula*, with *Alnus, Salix, Picea, Pinus* Climate: cooler than Portage-du-Cap	5a?	
Millerand (site 4)	Coastal cliff	Organic sediments in sand over bedrock	>46 000 (GSC-3631; wood)	Not studied as of 1984	5a?	
Île du Havre Aubert (site 5)	Coastal cliff	Woody peat over gravel and bedrock beneath till?	>47 000 (GSC-3633; wood)	Not studied as of 1984	5a?	
Le Bassin (site 6)	Lagoon coast	Peat beneath clay and till over laminated silt and sand (details in text)	>46 000 (GSC-3623; *Pinus strobus*) *101 700 + 17 000/-14 200 (UQT-182; *Pinus strobus*) *106 400 + 8400/- 8000 (UQT-183; *Pinus strobus*) *89 400 + 8000/- 7100 (UQT-184; *Pinus strobus*)	Pollen: shrub tundra/boreal coniferous/white pine-hardwood forest (Fig. 7.11f) Macrofossils: wood: *Pinus strobus* Diatoms: freshwater to marine Climate: warmer than at present	5e	
Des Buttes (site 7)	Gravel pit	Organic seams in sand/silt beneath diamicton		Not studied as of 1984	?	
Boisville (site 8)	Gravel pit	Organic seams in sand/silt beneath diamicton		Not studied as of 1984	?	
NOVA SCOTIA						
East Milford (site 9)	Gypsum quarry Karst	Peat, organic clay and clay over diamicton and bedrock beneath tills (details in text)	>33 800 (GSC-33; Wood) >50 000 (GSC-1642; *Larix laricina*) *84 900 + 6500/-6100 (UQT-185; *Abies balsamea*) *84 200 + 11 300/-10 100 (UQT-186; *Abies balsamea*)	Pollen: thermophilous hardwoods, *Fagus/Betula/Abies/Picea, Alnus* (Fig. 7.11e) Macrofossils: seeds, beetles Climate: warm interglacial followed by cooling trend	5c	Prest, 1970; Mott et al., 1982
Miller Creek (site 10)	Gypsum quarry Karst	Peat between clay units over till and beneath till	33 200 ± 2000 (I-3237) >52 000 (GSC-2694; *Pinus banksiana*)	Pollen: *Pinus banks./res..minor Picea*, hardwood taxa	5a?	MacNeill, 1969; Stea and Hemsworth, 1979
Noel Borehole (site 11)	Borehole	20 m redeposited organic silt over till and beneath till		Pollen: uniform boreal spectra of *Picea* and *Pinus* Climate: cool boreal	5a?	R. Stea, personal communication, 1983
Addington Forks (site 12)	Roadcut	Peat band in silt and sand unit between tills (details in text)	33 700 + 2300/-1800 (I-3236) >42 000 (GSC-1598; *Juniperus*)	Pollen: uniform boreal hardwood spectra/*Abies, Picea* (Fig. 7.11g) Climate: warm interglacial declining to cool boreal	5e	MacNeill, 1969; Lowdon and Blake, 1973
Hillsborough (site 13)	Roadcut Karst	Compact peat and organic silty clay over gravels and beneath till	>40 000 (W-157; *Picea*) >51 000 (GSC-570; *Abies balsamea*)	Pollen: *Abies, Alnus/Picea* Climate: Cool boreal	5a?	Flint and Rubin, 1955 Mott and Prest, 1967; Livingstone, 1968
Whycocomagh (site 14)	Roadcut	Organic silt and clay between tills	>44 000 (GSC-290; *Larix laricina*)	Pollen: *Abies/Picea/Alnus* succession Climate: cool boreal	5a	Mott and Prest, 1967
Bay St. Lawrence (site 15)	Coastal cliff	Colluvial gravel sequence with lower peat bed and upper organic silty clay unit on wave-cut bedrock bench (details in text)	>38 800 (GSC-283; *Larix laricina*), 44 200 ± 820 (GSC-3636; *Larix laricina*) >46 000 (GSC-3864; *Picea*) *47 000 + 4700/-4300 (UQT-178; *Larix laricina*)	Pollen: upper silt-tundra and forest tundra taxa with admixed reworked thermophilous hardwoods; Lower peat-herb tundra/shrub-forest tundra/herb tundra cycle (Fig. 7.11h) Microfossils: upper silt-deep water marine diatoms and dinoflagellates Climate: peat-cool boreal; upper silt-tundra	4/3	Mott and Prest, 1967 Guilbault, 1982; Vernal, 1983; Vernal et al., 1983

Table 7.2 (cont'd.)

Site Name (cf. Fig. 7.9)	Type of deposit	Stratigraphy	Dates (material dated) ^{14}C and Th/U*	Paleoecology	Isotopic stage	References
R. Inhabitants (site 16)	River section	Intercalated organic silts and colluvial gravels beneath gravel and till	>49 000 (GSC-1406-2; *Picea*)	Pollen: *Picea, Alnus, Pinus, Betula, Abies* Macrofossils: wood: *Alnus, Picea, Abies* Climate: cool boreal	5a/4?	Dawson, 1855; Grant, 1971; Mott, 1971
Leitches Cr. (site 17)	Borehole	Two organic units separated by till; tills above and below	>52 000 (GSC-2678; peat)	Pollen: *Picea, Abies/Alnus, Betula,* Cyperaceae; both units similar Climate: cool boreal	5a?	R. Stea, personal communication, 1983
Dingwall (site 18)	Coastal cliff	Organic, pebbly silt over till in karst depression; possibly capped by till	32 700 ± 560 (GSC-3381; *Picea*) >39 000 (GSC-3417; *Picea/Larix*) >48 000 (GSC-3541; *Picea*)	Pollen: *Picea,* minor *Alnus* Macrofossils: wood: *Salix* Climate: cool boreal	5a?	
East Bay (site 19)	Coastal cliff	Three organic units interbedded with gravels and beneath till; karst subsidence (details in text)	>50 000 (GSC 3861; *Juniperus*) >49 000 (GSC-3871; *Tsuga canadensis*) >50 000 (GSC-3878; *Picea*) *106 600 + 9600/-8600 (UQT-108; *Pinus strobus*) *86 900 + 6000/-5700 (UQT-109; *Tsuga canadensis*) *60 800 + 5100/-5000 (UQT-179; *Juniperus*) *50 200 ± 5000 (UQT-188; *Picea*) *126 400 + 15 000/-12 800 (UQT-175; wood) *123 400 + 30 000/-23 400 (UQT-176, wood) *62 100 + 5000/-4600 (UQT-177; *Picea*) *98 700 ± 10 500 (UQT-227; wood)	Pollen: (Fig. 7.11b) Unit I-*Pinus strobus/Quercus.* Unit II-*Fagus, Tsuga/Abies/Picea.* Unit III-*Picea, Pinus, Abies,* nonarboreal pollen Macrofossils: wood: Unit I-*Juniperus, Pinus strobus* Unit II-*Tsuga canadensis* Unit III-*Picea* sp. Climate: Unit I-warmer than at present; Unit II-similar to present followed by cooling; Unit III-cool boreal to cold tundra	5e 5c/b 5a/4/3	Occhietti et al., 1984 Vernal et al., 1984
Castle Bay (site 20)	Coastal cliff	Freshwater organic silt over indurated gravel and overlain by laminated, marine, and freshwater silts and deltaic sands, capped by till (details in text)	>42 000 (GSC-1577; organic silt) >52 000 (GSC-1619; *Picea-Larix?*)	Pollen: alternating boreal forest and forest/tundra, spectra progressively more tundra-like upward (Fig. 7.11c) Climate: cool boreal to cold tundra	5a/4 3	Grant, 1972; Lortie et al., 1984; Occhietti et al., 1984; Vernal et al. 1984
Green Point (site 21)	Coastal cliff	Woody organic silt over boulder gravels and beneath till(s)	>53 000 (GSC-3220; *Juniperus*) *117 400 + 10 000/-8800 (UQT-181; *Juniperus*)	Pollen: *Pinus strobus, Quercus,* mixed hardwood spectra Macrofossils: Wood: *Pinus strobus, Juniperus, Fraxinus, Carya* Climate: warm interglacial	5e	Grant and King, 1984
Mabou (site 22)	Roadcut and gravel pit	Peat lenses in silty clay glaciofluvial gravels and till	>53 000 (GSC-3317; *Abies balsamea*)	Pollen: *Alnus, Betula, Picea, Abies,* some non-arboreal pollen Climate: cool boreal	5a?	
Big Brook (site 23)	Gypsum quarry	Complex of two tills with glaciotectonically disturbed organic sediments between and below tills	>49 000 (GSC-3289; *Picea*) 36 200 ± 1280 (GSC-3206; organic silt) >52 000 (GSC-3880; *Picea*)	Pollen: *Picea* or *Alnus-Picea* dominated spectra Climate: cool boreal	5a?	
NEW BRUNSWICK Hillsborough (site 24)	Excavation Karst	Mastodon bones, coprolites, and peat in and beneath clay; details not known	13 600 ± 200 (GSC-1222; bone) >43 000 (GSC-1680; peat) 37 300 ± 1310 (GSC-2469; coprolite) 51 500 ± 1270 (GSC-2469; coprolite matrix)	Pollen: *Picea, Pinus* unlike late glacial and Holocene spectra; similar to spectra from older deposits Climate: cool boreal	?	Grant and King, 1984; Blake, 1983
Sussex (site 25)	Gravel pit	Organic silt seams in base of till overlying gravel; glaciotectonically disturbed	>35 000 (BGS-806; organic silt)	Not studied as of 1984	?	Rampton et al., 1984

Betula (Fig. 7.11) show that climate was as warm or warmer than at present. Brookes et al. (1982) tentatively assigned the interval to the Sangamon Interglaciation and the overlying glaciogenic sediments to the Early Wisconsinan.

Quebec-Magdalen Islands (Îles-de-la-Madeleine). Two important sites are located near one another on Île du Havre Aubert of the Magdalen Islands. Le Bassin section (site 6, Fig. 7.9) at the western end of Le Bassin lagoon shows laminated sand and silt at and below present sea level, overlain by a 20 cm-thick compact layer of peat beneath 20 cm of grey clay with organic seams and about 65 cm of grey silty clay. Yellowish brown sand and purplish red till cap the sequence. Pollen spectra from the basal peat are dominated by *Betula*, nonarboreal pollen, and *Sphagnum* (Fig. 7.11f). These are replaced higher in the unit by *Picea/Betula* and then by *Pinus strobus*/mixed hardwood, the latter extending into the overlying clay unit. Diatoms from the clay (S. Lichti-Federovich, Geological Survey of Canada, personal communication, 1983) imply a marine littoral zone, i.e., higher sea level than at present. Wood from the peat produced a radiocarbon age of >46 000 BP (GSC-3623) and Th/U ages of 101.7 ka (UQT-182), 106.4 ka (UQT-183), and 89.4 ka (UQT-184; Causse and Hillaire-Marcel, 1986). The dates, pollen, and stratigraphic evidence suggest that the site represents the early warming phase of the Sangamon Interglaciation (Mott and Grant, 1985).

Portage-du-Cap section (site 2, Fig. 7.9) is located in a gravel pit at an elevation of 13 m. Organic sediments, dated by radiocarbon analyses of organic detritus at >35 000 BP (BGS-259) and of balsam fir wood at >38 000 BP (GSC-2313), are interbedded with sand, silt, and clay that underlies a red, sandy diamicton (Prest et al., 1976). Macrofossils and diatoms show the organic sediments to be intertidal deposits. Pollen spectra from the organic bed are dominated by *Picea*, *Pinus*, and *Betula*, with lesser amounts of *Quercus*, *Abies*, *Fagus*, and *Alnus*, and low percentages of such thermophilous hardwood genera as *Ulmus*, *Carya*, *Tilia*, and *Acer* (Fig. 7.11d). The pollen data imply a warmer climate than at present, a conclusion supported by diatoms and several types of macrofossils (Prest et al., 1976). Thus the organic horizon undoubtedly formed under interglacial conditions.

Nova Scotia-Mainland. Removal of overburden at the East Milford Gypsum Quarry of the National Gypsum Company has provided several exposures of buried organics (East Milford, site 9, Fig. 7.9). W. Take, in an early study of macrofossils from one karst depression (Prest, 1970), concluded that glacial and interglacial sediments as old as the Kansan Glaciation were present. Wood from the site was dated at >33 800 BP (GSC-33). The original site studied by Take is no longer available, but study and dating of some of the organic material still stored at the Nova Scotia Museum show that the pit contained Holocene materials, thus casting doubt on Take's original interpretation.

Later excavations near the original site exposed a compressed peat overlying organic clay and silt and overlain by till. Tamarack (*Larix* sp.) wood was dated at >50 000 BP (GSC-1642), and balsam fir wood has yielded a Th/U age of 84.9 ka (UQT-185) and 84.2 ka (UQT-186; Causse and Hillaire-Marcel, 1986). A pollen profile (Fig. 7.11e) from the sediment sequence reveals spectra containing abundant thermophilous hardwood taxa in the basal organic sediments, giving way upwards to *Abies*, *Abies/Picea*, and *Alnus*. The pollen stratigraphy and associated macrofossil evidence portrays a deteriorating climate probably associated with the later part of an interglacial interval (Mott et al., 1982).

Continuing quarry operations have exposed additional sections with organic sediments; paleoenvironmental data from these sections replicate the findings of Mott et al. (1982). The palynological and stratigraphic evidence supports the conclusion that only a single episode of organic deposition is represented despite the occurrence of several organic layers at the site (Mott and Grant, 1985).

The Addington Forks site (site 12, Fig. 7.9) contains organic sediments associated with silt and sand between two tills. Wood from the organic unit was originally radiocarbon dated at 33 700 + 2300/-1800 BP (I-3236; MacNeill, 1969), but a more recent date on juniper (*Juniperus* sp.) wood is >42 000 BP (GSC-1598; Lowdon and Blake, 1973). *Quercus*-dominated thermophilous hardwood spectra near the base of the organic unit are succeeded by spectra dominated consecutively by *Abies*, *Abies/Picea/Alnus*, *Picea/Abies/Alnus*, and *Picea* (Fig. 7.11g). Here again, as at the East Milford site, the later part of an interglacial interval appears to be represented (Mott and Grant, 1985).

Nova Scotia-Cape Breton Island. The Bay St. Lawrence site (site 15, Fig. 7.9), a sea cliff on the northern tip of Cape Breton Island near Bay St. Lawrence, shows Quaternary sediments resting on a wave-cut bedrock bench. The sediments comprise a thick sequence of gravel, sand, and silt between a basal sandy and gravelly till-like sediment and an overlying bouldery gravel diamicton, both originally thought to be tills (Mott and Prest, 1967; Prest, 1970). Contained within the sequence are a compact pebbly and sandy peat unit and a lenticular organic silty clay unit. Abundant *Alnus* pollen with minor coniferous taxa and herbs characterize the lower peat unit. The same unit contains tamarack wood that dated >38 000 BP (GSC-283). The upper lenticular clay unit contains marine fossils and pollen assemblages high in *Pinus*, *Picea*, *Betula*, and *Alnus*, with minor amounts of thermophilous hardwood genera (Mott and Prest, 1967).

In a recent detailed study of the site (Vernal, 1983; Vernal et al., 1983), the gravelly diamictons have been interpreted as colluvium rather than till, meaning that there is no definitive evidence of glaciation. A complete vegetation cycle is apparent in pollen assemblages from the lower peat unit (Fig. 7.11h). Herbaceous tundra spectra with abundant Cyperaceae, Gramineae, *Sphagnum*, and *Lycopodium* give way to spectra similar to those reported originally by Mott and Prest (1967), in that they have high percentages of *Alnus crispa* and *Betula* with only small amounts of *Picea*

Figure 7.11. Abbreviated pollen diagrams from selected sites: (a) Woody Cove (after Brookes et al., 1982); (b) East Bay Section (after Vernal et al., 1984); (c) Castle Bay Section (after Vernal et al., 1984); (d) Portage-du-Cap (after Prest et al., 1976); (e) East Milford (after Mott et al., 1982); (f) Le Bassin (after Mott and Grant, 1985); (g) Addington Forks (after Mott and Grant, 1985); (h) Bay St. Lawrence (after Vernal et al., 1983).

QUATERNARY ENVIRONMENTS — PALEOBOTANICAL CASE HISTORIES

a Woody Cove, Newfoundland

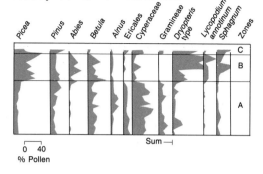

b East Bay Section, Nova Scotia

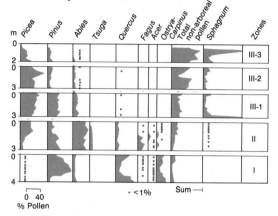

c Castle Bay Section, Unit III, Nova Scotia

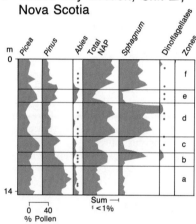

d Portage-du-Cap, Quebec

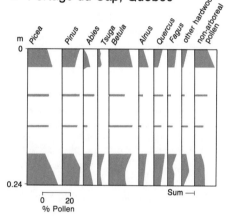

e East Milford, Nova Scotia

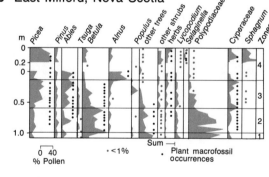

f Le Bassin, Quebec

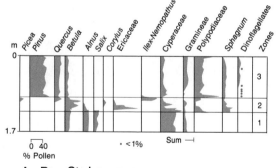

g Addington Forks, Nova Scotia

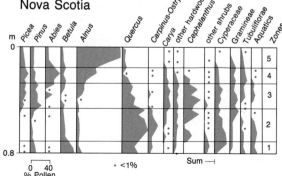

h Bay St. Lawrence, Nova Scotia

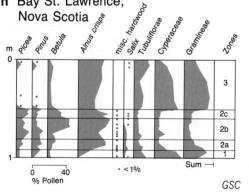

GSC

505

and *Pinus* and even less herbaceous taxa. Herbaceous spectra resembling the lower herbaceous zone complete the sequence (Fig. 7.11h). Successive tundra, boreal forest and tundra environments are indicated. More recent radiocarbon dates on wood from the same organic unit are 44 200 ± 820 BP (GSC-3636) and >46 000 BP (GSC-3864). A Th/U analysis on wood yielded an age of 47.0 ka (UQT-178; Causse and Hillaire-Marcel, 1986).

The upper silty unit contains marine organisms characteristic of a deep water marine environment (Guilbault, 1982), as well as pollen spectra resembling those reported by Mott and Prest (1967). Vernal (1983), however, interpreted the spectra as tundra indicators with an admixture of reworked pollen from an offshore marine deposit representing an older, warmer interval.

At East Bay site (site 19, Fig. 7.9), Quaternary sediments rest on a karstic gypsum bedrock surface. Coarse to medium unstratified basal gravels are overlain by stratified gravels containing lenses of silt, which are in turn capped by till. Included within, underlying the unstratified gravels and usually associated with karst depressions are organic sediments. Compact, steeply dipping peat extending below sea level is exposed at one section. Another section reveals thinly stratified silty clay and interbedded sand and gravel overlying the gypsum bedrock surface. Higher in the section, silty clay containing large logs is overlain by stratified organic silt containing organic lenses. The whole sequence has been deformed, presumably by slumping. Both sections are overlain by gravel and till. Other exposures show beds of more or less deformed organic silt within the unstratified gravel unit.

Detailed study of the East Bay exposures revealed three intervals of organic accumulation, each with a distinctive palynological signature (Occhietti et al., 1984; Vernal et al., 1984; Vernal and Mott, 1986). Palynostratigraphic Unit I, represented by the compact peat, is distinguished by spectra in which *Quercus*, *Carpinus/Ostrya* type, *Tsuga canadensis*, and *Pinus strobus* pollen dominate, along with lesser amounts of a variety of thermophilous hardwood types (Fig. 7.11b). Juniper (*Juniperus*) wood from this unit is dated at >50 000 BP (GSC-3861). Th/U dates on wood from the peat are 126.4 ka (UQT-175), 123.4 ka (UQT-176), 106.6 ka (UQT-108), and 60.8 ka (UQT-179; Causse and Hillaire-Marcel, 1986). The younger dates are considered to be spurious due to discernible contamination. If the older date is accepted, Palynostratigraphic Unit 1 represents isotope substage 5e (Vernal et al., 1984; Vernal and Mott, 1986), with a climate warmer than at present.

Palynostratigraphic Unit II occurs in the woody silty clay and organic silt overlying gypsum bedrock. The basal sediment is dominated by *Abies balsamea* and *Pinus*, with *Fagus grandifolia*, *Tsuga canadensis*, and small amounts of hardwood pollen (Fig. 7.11b). Higher up, *Picea* and *Pinus* dominate. Hemlock (*Tsuga canadensis*) wood produced a radiocarbon age of >49 000 BP (GSC-3871) and a Th/U age of 86.9 ka (UQT-109; Causse and Hillaire-Marcel, 1986). Climatic conditions may have been similar to those at present.

Palynostratigraphic Unit III, based on pollen spectra from the organic silt beds contained in the nonstratified gravels, is characterized by dominance of *Picea*, *Pinus*, and *Abies* pollen and in many cases abundant nonarboreal pollen (Fig. 7.11b). Despite similarities in taxonomic content, the spectra of individual beds are too diverse to discern a trend in the pollen stratigraphy. Spruce wood yielded a radiocarbon date of >50 000 BP (GSC-3878) and Th/U dates of 98.7 ka (UQT-177), 50.2 ka (UQT-188), and 98.7 ka (UQT-227; Causse and Hillaire-Marcel, 1986). Pollen spectra represent conditions colder than at present.

The Castle Bay site (site 20, Fig. 7.9) is a long section on East Bay, Bras D'Or Lake west of the East Bay exposure. Coarse cemented gravel at the base of the section is overlain by 1-2 m of organic silt of freshwater origin. Deltaic sands up to 12 m thick overlie and interfinger with the organic silt in part of the section and are in turn overlain by a coarse gravel unit. Laminated silt and clay, slightly organic and in part strongly contorted, overlie the sand and gravel, and in much of the section directly overlie (with thickness up to 17 m) the basal organic silt. Red till or diamicton caps the laminated silt and clay (Occhietti et al., 1984; Vernal et al., 1984; Vernal and Mott, 1986). Wood from the basal organic unit is dated at >42 000 BP (GSC-1577). Spruce or tamarack wood from higher in the laminated silt and clay is dated at >52 000 BP (GSC-1619), but it may be reworked from an older unit.

Boreal forest-type spectra with high percentages of *Pinus*, *Picea*, and *Abies* occur in the basal organic silt (Vernal et al., 1984). These spectra are succeeded in the overlying laminated sediments by three intervals with forest-tundra spectra, separated by two intervals with boreal-type spectra (Fig. 7.11e). High pollen concentrations throughout the thick laminated silt sequence suggest that it encompasses a considerable length of time. Each of the three vegetation cycles appears to represent colder climate than the preceding one, suggesting onset of glacial conditions. Alternating freshwater and marine diatom assemblages in the laminated silts imply isostatic readjustments of the landscape possibly in response to glacial fluctuations (Lortie et al., 1984).

Paleoenvironmental reconstruction and correlation

The palynological results presented above and in Table 7.2 suggest that the Sangamon Interglaciation in Atlantic Canada comprises three main climate intervals: one in which the vegetation was dominated by thermophilous hardwood genera, a second with thermophilous and coniferous genera, and a third characterized mostly by coniferous genera. The sequence of occurrence and age of these climatic episodes are provisionally discerned by reference to Th/U dates on wood.

Sites such as East Bay, Le Bassin, and Addington Forks indicate that for part of the Sangamon Interglaciation the climate ameliorated to such an extent that thermophilous hardwood forests existed in parts of Atlantic Canada. Oak and white pine were abundant; hardwood taxa such as basswood, hickory, and blue beech/ironwood were plentiful; and maple, beech, and birch were present. Climate must have been distinctly warmer than at present, and the optimum was attained quickly judging by the record at Le Bassin. Stratigraphic relationships and Th/U chronology show that this warm episode occurred early in the Sangamonian between approximately 126 ka and 100 ka.

A second warm interval, dated by Th/U at between 87 to 84 ka (but probably encompassing a much longer time), is characterized by distinctly different palynological spectra.

Its early phases do not appear to be recorded at any of the sites. However, the climate eventually ameliorated to the extent that forests were dominated by beech, birch, and lesser amounts of a variety of other hardwoods (cf., East Milford, Portage-du-Cap, and East Bay). Hickory, basswood, and blue beech/ironwood apparently did not attain their former prominence. Hemlock was an important coniferous taxon, but white pine was of only minor importance. Climatic deterioration at the end of this phase favoured development of boreal forests in which balsam fir, spruce, and alder were the successive dominants.

Amelioration of the climate for a third time, between 62 and 47 ka according to the Th/U dates, is indicated by sections at East Bay, Castle Bay, Bay St. Lawrence, and other locations where conditions favoured proliferation of coniferous forests of spruce and pine and where hardwoods had only a minor role. In fact, a more complex sequence of climatic fluctuations is recorded at some sites, such as Castle Bay, where up to three cycles reflect successively cooler climates, possibly heralding the onset of continental glaciation.

The nature of Sangamonian paleoenvironments in Newfoundland, based on one known deposit with little chronological control, is less well known. Woody Cove section may relate to the warmest climate interval, as the climate has been interpreted to have been warmer than at present even though only boreal forests of balsam fir and spruce predominated during the climate optimum. As can be seen from sites in Nova Scotia, however, the second warm interval was also probably as warm as that at present, and the Woody Cove section may relate to this time.

The record of environmental change as revealed by study of buried organic deposits can be provisionally fitted to the stratigraphic framework outlined by Grant (1989) and Grant and King (1984). Recession of Illinoian glaciers (stage 6) was followed by climatic warming, substage 5e, that culminated in conditions warmer than those at present. Sections that portray this time period are Le Bassin, East Bay I, Addington Forks, and Green Point. The climatic deterioration seen in the Addington Forks sequence may represent the latter part of substage 5e and possibly part or all of 5d as well. This period may include substages 5e to 5c if the Th/U ages are valid (Vernal et al., 1984; Mott and Grant, 1985; Vernal and Mott, 1986). The early part of the next warm period seems to be missing from the record because evidence of a second warming event at East Milford, Portage-du-Cap, and East Bay II appears abruptly. The Mabou site and possibly some others having conifer-dominated assemblages may correlate with the cooling at the end of substage 5c or, judging by the Th/U dates, with substages 5b or 5a (Vernal et al., 1984; Mott and Grant, 1985; Vernal and Mott, 1986).

The third climatic warming event, portrayed at East Bay, Castle Bay, and Bay St. Lawrence (and possibly at Dingwall, Big Brook, River Inhabitants, Miller Creek, some Magdalen Islands sections, and the Noel and Leitches Creek boreholes) apparently failed to attain the thermal level of the others. In fact, it may represent complex fluctuations leading in stepped fashion to ever colder climate. Th/U ages suggest correlation with stages 4 and/or 3 (Vernal et al., 1984; Vernal and Mott, 1986). It is possible, however, that the early part of the third period may relate to substage 5a with progressively cooler cycles correlating with stage 4 and possibly stage 3 (Mott and Grant, 1985).

Despite the numerous sites discovered thus far, the Atlantic Canada paleoenvironmental record of the last interglaciation remains incomplete. Major gaps probably coincide with the periods of colder climate within stage 5, for example, substages 5d and 5b. The beginning of the second warm period also appears to be missing. Is this gap real or have the Th/U dates, which form groups centred around 126 to 100 ka and 87 to 84 ka, led to an incorrect conclusion that two separate warm intervals are represented? An alternative hypothesis could be that East Milford, Portage-du-Cap, and East Bay II sites actually represent the waning phase of the early warm interval represented at East Bay I, Addington Forks, Le Bassin, and Green Point sites, which were assigned to substage 5e.

Obviously, more work is required before the paleoenvironmental history of the Sangamon Interglaciation in Atlantic Canada can be completely unravelled. Additional studies of plant and animal fossils, stratigraphy, and palynology should help to elucidate and embellish the record of this complex and lengthy interval.

INTRODUCTORY COMMENTS ON THE DEVELOPMENT OF THE CANADIAN BIOTA FOLLOWING ICE RETREAT

J.V. Matthews, Jr. and T.W. Anderson

Although the following two case histories (Ritchie, 1989; Richard, 1989), appear by their titles to be quite different, they actually strike at the same problem: the development of Canadian vegetation during and following ice retreat. Many of the issues that they raise, such as differential rates of migration, random associations of species, and the lack of modern analogues for some past communities, have also been discussed in recent reviews (e.g., Wright, 1976, 1984; Davis, 1983). Even though both case histories, like most in this chapter, deal exclusively with vegetational and floristic changes, a similar litany of postglacial events could be cited for arthropods (Morgan and Morgan, 1980) and mammals (Lundelius et al., 1983).

Postglacial revegetation is a theme of much Canadian paleoecological research, which is understandable considering the amount and variety of territory involved (Fig. 7.1). Of course, paleoecologists realize that glacial maps such as that of Prest (1969) are deceptive, for while they show which regions were ice covered, they usually do not portray other glacial effects that tend to retard or enhance biotic recolonization when glaciers retreat. For example, soil is stripped from large areas during a glacial advance, creating, when the ice melts, edaphic conditions that favour unique combinations of pioneer plants. Glaciers also alter drainage systems and are responsible for large water bodies that continue to direct and redirect colonization and migration of plants and animals long after the ice has retreated.

Matthews, J.V., Jr. and Anderson, T.W.
1989: Introductory comments on the development of the Canadian biota following ice retreat; in Chapter 7 of Quaternary Geology of Canada and Greenland, R.J. Fulton (ed.); Geological Survey of Canada, Geology of Canada, no. 1 (also Geological Society of America, The Geology of North America, v. K-1).

Ritchie's case history concerns the development of the boreal forest, one of the major biomes of Canada. But it deals only with the development of the western part of the boreal forest during the present interglaciation — the Holocene. Should we expect the boreal forest of earlier interglaciations to be the same? Is the present boreal forest an exact analogue of past iterations of the biome?

Data from the late Tertiary Beaufort Formation hint at what the precursor of Quaternary boreal forest might have been like. For example, we know that during the Miocene boreal-tundra vegetation occurred near the present latitude of Meighen Island, 2100 km north of its present position. The biota of this early version of the contemporary boreal ecosystem was richer in both tree species and insect species than existing boreal forest (Hills and Matthews, 1974; Hills, 1975; Matthews, 1977). During the Quaternary, the boreal fauna and flora probably suffered taxonomic depletion with each glaciation, much as happened in Europe (van der Hammen et al., 1971).

The location of the boreal forest changed greatly during the Quaternary. As indicated above, it was located as far north as 80°N during the Miocene. In the late Pliocene or early Quaternary it barely reached 70°N (Matthews et al., 1986). During glacial periods, the boreal forest was forced south as far as 40°N (Delcourt et al., 1983; Davis, 1983). Since migration barriers and the migration potential of various boreal species were probably slightly different during each deglacial period, it is reasonable to conclude that no interglacial reconstruction of the boreal forest in Canada was exactly like any other one. Indeed we can easily see in Ritchie's account of the last reconstitution of the boreal forest where subtle differences in previous deglacial events would have led to slightly different types of boreal environment than the present one.

The development of the boreal forest in western Canada has not been so strongly influenced by geographical barriers as was the postglacial vegetation sequence in Quebec and Labrador. There, the presence of large postglacial lakes and seas caused the initial stages of revegetation to vary both spatially and temporally. Also, in Quebec and Labrador glacier ice persisted until much later in the Holocene than in western Canada. In his case history on the Quebec/Labrador revegetation sequence, Richard uses a novel approach to illustrate the importance of these geographical influences. He compares the actual vegetation sequence as represented by the pollen data from numerous lake basins with a hypothetical sequence based on two main variables: latitude and succession. Although Richard concerns himself primarily with plants, there is considerable biogeographical "food-for-thought" in his discussion which should be of interest to a wide spectrum of nonbotanists. For example, entomologists have long been puzzled by the differences in richness of the Arctic Coleoptera fauna west of Hudson Bay compared to that of Arctic Quebec (Campbell, 1980). In recent years paleoecological studies on late glacial sites in Quebec and the eastern United States have shown that northern insects and plants once thought to have survived the last glaciation in unglaciated Beringia (Yukon and Alaska) lived in ice marginal areas of the east during the Late Wisconsinan (Miller and Thompson, 1979; Morgan and Morgan, 1980; Mott et al., 1981). Certain insects were on the very doorstep to Arctic Quebec at 13 to 10 ka yet failed to occupy that region as the ice melted.

Richard's case history suggests a solution to this enigma. He shows that tundra environments, the favoured habitat of the insects in question, were discontinuous in early Holocene Quebec and in addition that these environments differed radically from present tundra, a phenomenon duplicated at some other ice marginal areas in North America (Birks, 1976).

HISTORY OF THE BOREAL FOREST IN CANADA
J.C. Ritchie

Introduction

The boreal forest ecosystem has sufficient cohesiveness in flora, vegetation, and fauna that a comparative analysis of its several patterns of long-term postglacial change might provide a useful case study in paleoecology. What is attempted here is an examination of the nature and timing of long-term Holocene changes in species abundance that have culminated in the modern version of the boreal forests of the western interior of Canada. This vast plant formation is far from uniform, and its only common vegetational element is that it is dominated by white and black spruce (*Picea glauca* and *P. mariana*), aspen (*Populus tremuloides*), balsam poplar (*Populus balsamifera*), and white birch (*Betula papyrifera*), variably associated with jack pine (*Pinus banksiana*), lodgepole pine (*Pinus contorta*), balsam fir (*Abies balsamea*), and larch (*Larix laricina*) (Hare and Ritchie, 1972; Larsen, 1980). The most widely accepted subdivision of the western boreal forest recognized twelve units mappable at a scale of 1:6 000 000; these are distinguished by two primary criteria — prevailing landforms and the particular combinations of tree species (Rowe, 1972).

The data for this comparison are published pollen sequences, with interpretations, from sites in the boreal forest between the Manitoba-Ontario boundary and the western limits of the region — the Cordilleran Foothills zone in the southwest and the arctic and montane forest-tundra and tundra zones in the northwest and north (Fig. 7.12).

Pollen data record with reasonable accuracy long-term changes in vegetation — that is, changes in species composition registered over time intervals of hundreds or thousands of years. On the other hand, few sites in the boreal forest

---

Ritchie, J.C.
1989: History of the boreal forest in Canada; *in* Chapter 7 of Quaternary Geology of Canada and Greenland, R.J. Fulton (ed.); Geological Survey of Canada, Geology of Canada, no. 1 (<u>also</u> Geological Society of America, The Geology of North America, v. K-1).

Figure 7.12. A sketch map of Western Canada showing existing vegetation zones (modified from Rowe, 1972), Laurentide ice limits at 10 ka and 9 ka (after Prest, 1969) and summary pollen diagrams for selected sites: Chalmers Bog (Mott and Jackson, 1982); Twin Tamarack Lake (Ritchie, 1985); Porter Lake, Flin Flon, and Riding Mountain (Ritchie, 1980). The solid line traversing each pollen diagram marks the level when modern vegetation was established.

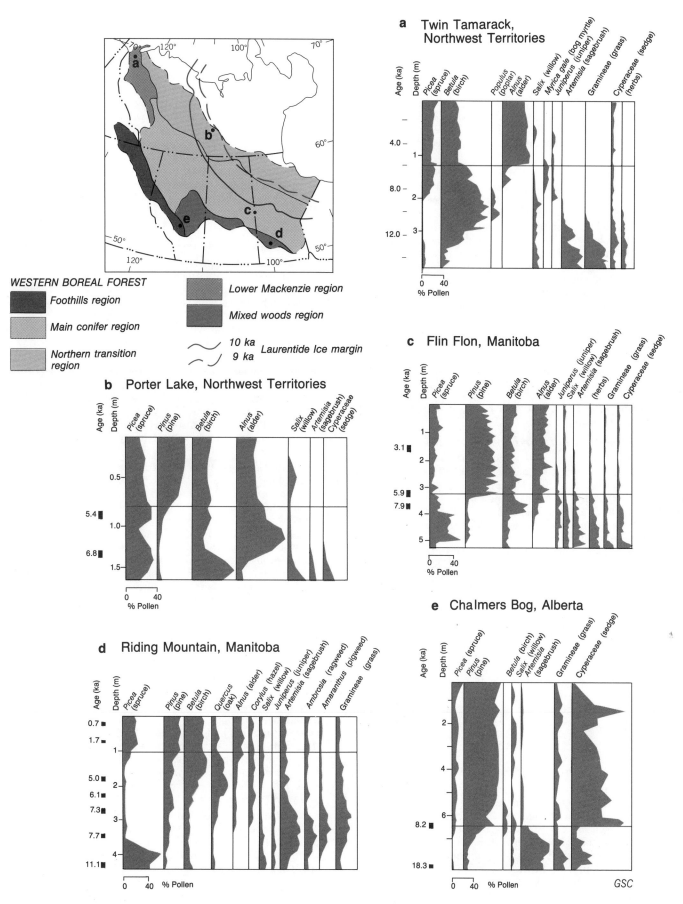

have produced pollen data that record such common and important short-term changes in species abundance and composition as response to fires, widespread insect depredations, or other natural disturbances that have a periodicity of the order of one hundred years. The familiar but partly intractable questions of resolving temporal and spatial scales in pollen data preclude the application of pollen analysis to such problems of vegetation dynamics as gap-phase succession, although progress is being made in this domain using the small-basin approach introduced by Andersen (1978).

A recent review of published records of the late Pleistocene and Holocene pollen stratigraphy of the western interior of Canada shows that about 60 sites have been investigated within the boreal forest region delimited here (Ritchie, 1984b). A significant number of these sites, particularly peat bog sections, are truncated and lack much of the early Holocene and late Pleistocene record, so they are of little use in our present analysis. A partly subjective examination of the roughly 20 remaining sites reveals 5 distinctive patterns of long-term change (Table 7.3). They are described below, followed by a discussion of the factors that might have been responsible for the differences among them. The types are differentiated spatially, so they will be designated accordingly, although the terms chosen have only temporary, practical significance.

Patterns of long-term change in the western boreal forest

Southwest margin or Foothills region

Long reliable records are scarce in this region, but three are similar enough that they might represent a general pattern (see also discussion of Goldeye Lake in this chapter, Schweger, 1989). The Chalmers Bog site from southwest Alberta shows a simple major change at the beginning of the Holocene, from a treeless assemblage that prevailed from 18 ka to roughly 10 ka, to an assemblage dominated by pine and spruce that has remained unaltered to the present day (Fig. 7.12e, based on Mott and Jackson, 1982). The modern boreal forest of the region (the Foothills Forest Section of Rowe, 1972) is dominated by lodgepole pine, associated with white spruce, poplar, and white birch. It was established early in the Holocene and has remained constant in regional composition since that time. A pollen sequence from the nearby Kananaskis Valley (MacDonald, 1982) shows the same sequence, as do the Peace River region sites investigated by White (1983). This is the simplest of the five patterns described here (Table 7.3) in that the pollen record reveals no significant changes in species abundance or composition of the boreal forest since its establishment directly from a tundra precursor at the beginning of the Holocene.

Interior or main conifer region

The vast central tract of the boreal forest is characterized by closed stands of conifers and deciduous angiosperms. The exact composition of the stand mosaics of any locality depends largely on the prevailing landscape-soil regime and the local history of recent fire, insect depredation, or other disturbance factors. The latest significant event in the postglacial development of these forests was the immigration of jack pine, roughly at 6 ka. The small number of sites with complete pollen records is adequately represented by the Flin Flon locality which extends back to before 7-9 ka (Fig. 7.12c). A short-lived treeless episode, when herb and sedge tundra communities prevailed on upland and lowland sites respectively, was followed by an early version of the spruce boreal forest, dominated by spruce with significant open areas occupied by juniper-willow (*Juniperus-Salix*) scrub and patches of herb tundra. White birch immigrated by about 8 ka and the landscape became more or less entirely mantled with closed forests. Finally, at 6 ka, jack pine arrived, and alder (*Alnus*) expanded to its modern proportions, establishing the modern version of the central or main boreal forest (the Northern Coniferous Forest of Rowe, 1972).

Lower Mackenzie region

The uplands adjacent to the Mackenzie River Delta have been investigated relatively intensively, and a consistent pattern of long-term change has emerged from several sites. The site chosen as representative (Fig. 7.12a) is referred to informally as Twin Tamarack Lake, a small shallow pond near Inuvik, Northwest Territories. A full account of the site and its pollen record is available (Ritchie, 1985).

A basal pollen zone, from 14.5 ka to 11.8 ka, is dominated by nonarboreal pollen types, of which sagebrush (*Artemisia*), grass, and willow predominate, associated with a diverse assemblage of herbs. The pollen accumulation rates of all taxa are low. The vegetation reconstructed for this Late Pleistocene episode is a herb tundra on uplands with sedge-grass-willow marsh communities in the lowlands.

Rapid changes in pollen stratigraphy between 11.8 ka and 8.4 ka indicate the replacement of the dwarf birch (*Betula glandulosa*), willow, and herb tundra by a woodland dominated initially by poplar, then by a short-term (500 year) phase with extensive juniper scrub in addition to the poplar groves on uplands, and finally a transition to coniferous woodlands as white and black spruce arrived in the region. The modern boreal woodlands were established two millennia later at about 6 ka, when alder increased abruptly to its modern abundance. The current explanation for some of these changes is that an increase of summer warmth to maximum values at 10 ka initiated the replacement of tundra by woodland. Balsam poplar and aspen, both with highly efficient dispersal mechanisms, moved in rapidly and established dominance in the early Holocene, the former on poorly drained mineral substrata, the latter on uplands. The spruce species, with slower migration rates, arrived about 9 ka and replaced poplar on most sites. Black spruce expanded more slowly than white spruce, reaching its current abundance by about 7 ka when paludification and permafrost aggradation had increased to form extensive mire habitats. White birch, larch and later alder arrived and established the pattern found today in this region. Short-term changes in these northern woodlands, caused mainly by fire, have not been registered in the pollen sequences so far analyzed.

Identical changes have been recorded in the lower Mackenzie region at five other sites with detailed records — M Lake (Ritchie, 1977), SW Lake (Ritchie, 1984b), Kate's Pond (J.C. Ritchie, unpublished), Eildun Lake (Slater, 1978), and Lac Meleze (MacDonald, 1984).

Table 7.3. Patterns of long-term change in the western boreal forest

Foothills Region	Interior/Main Conifer Region	Northwest Transition Region	Lower Mackenzie Region	Mixedwoods Region	ka
					0
				White spruce, white birch, poplars; jack pine on sands; poplar stands common	1
Lodgepole pine dominance, with poplars, white spruce and black spruce; white birch and tamarack of local occurrence	Black spruce and jack pine on uplands; white spruce and poplars on favourable sites; tamarack in mires	Open woodlands, black spruce dominant, with white birch and larch; jack pine rare; white spruce on favourable sites	Black spruce and white birch on uplands; white spruce on alluvium; black spruce and larch on mires; alder common		2
					3
				Oak, white birch, and poplar woodlands; prairie on drier sites	4
					6
	Spruce and white birch dominance with alder and willow scrub	Open spruce, white birch woodlands with alder	White and black spruce white birch and poplar closed woodlands		7
		Spruce - birch woodlands		Prairie on all upland sites; willow and poplar on alluvial sites	8
		Deglaciation			
	Open spruce woodland with juniper, willow, and herbs juniper-birch scrub		Poplar woodlands,		9
Treeless vegetation with abundant grass, sagebush, willows, and herbs			Poplar-willow woodland and scrub	Spruce, juniper, herb willow woodlands	10
	Treeless vegetation,		Poplar-dwarf birch woodland scrub Dwarf birch-herb tundra	**Deglaciation**	
	?herb tundra **Deglaciation**				11
			Herb tundra		12
			Deglaciation		18
Deglaciation					

Northern transition region

A small number of sites along the northern boundary of the boreal forest, in the Northern Transition zone (Rowe, 1972), provide a short, simple record of change in pollen frequencies. A 3-zone pollen diagram (Fig. 7.12b), illustrated here by the Porter Lake site near the eastern extension of Great Slave Lake, Northwest Territories, shows an early spruce-birch phase, a zone delimited by the arrival of alder at 6.7 ka, and the establishment of the modern spectra typical of the contemporary vegetation, which is characterized by the co-dominance of pollen of spruce, birch, pine, and alder. The final event in the development of these subarctic woodlands was the arrival of jack pine at times ranging between 6 and 5 ka. Eastern sites (Thompson, Manitoba and Reindeer Lake, Saskatchewan) show earlier dates for the pine rise than the western localities, suggesting that jack pine might have migrated westwards from an eastern source area, contrary to the opposite notion earlier proposed (Ritchie, 1976). The modern open woodlands, dominated by black spruce on upland till soils, jack pine on sands and bedrock ridges, white spruce on alluvium and eskers, and black spruce and larch on mires, were in place by roughly 5.5 ka. No evidence has been found of a treeless precursor to the woodland phase, probably because the migration rate of spruce and birch was as great or greater than the rate of disintegration of the final stages of the Laurentide Ice Sheet.

Mixedwoods region (southern margin)

Widespread paleosol and pollen evidence indicates that the ecotone between the grassland zone and the southern boreal forest shifted southwards in the late Holocene. Thus modern stands of the boreal forest near this ecotone have short histories of development. This part of the boreal forest is referred to as the Mixedwoods region (Rowe, 1972), which consists of variable mosaics of stands of conifers and deciduous angiosperms. White spruce, white birch, and both poplar species dominate upland sites except those with sandy soils, where jack pine prevails. Black spruce and larch dominate poorly drained, peaty soils.

Well documented examples are found in the Riding Mountain uplands of Manitoba (Fig. 7.12d) where an early Holocene spruce-dominated phase was replaced abruptly by a grassland episode that persisted from 10 to 6 ka. A phase with similar grasslands on well drained sites of southern aspect and with woodlands of aspen, oak (*Quercus*), and white birch on less xeric sites, lasted from 6 to 2.5 ka when, possibly in response to a change to cooler and/or moister summers, black and white spruce, larch, and alder migrated southward to establish the contemporary mixedwoods. A similar pattern was recorded at A Lake in Saskatchewan by Mott (1973), but the sequence has only a single, basal radiocarbon date. The Lofty Lake pollen sequence from a mixedwoods region site in central Alberta (Lichti-Federovich, 1970) resembles the pattern closely, although the prairie period is less fully developed at that site.

Discussion and conclusions

Several interesting points emerge from the above analysis, although none is particularly original in the broader context of ecological succession theory. It is clear that even if one assumes arbitrarily that the western boreal forest is an integrated, definable vegetation formation or biome, its long-term history has varied widely from one region to another (Table 7.3). The factors that have caused this variability in long-term climate change are as follows:

Length of time since deglaciation. This factor has operated differentially, and its expression depends on other influences discussed below. However, it is obvious that the length of time available for vegetation change since deglaciation, roughly proportional to the distance of a site from the centre of glaciation, will have determined some elements in the sequence of long-term changes. For example, in the lower Mackenzie River valley deglaciation occurred early (approximately 14 ka) and several millennia elapsed before the most shade-tolerant, long-lived tree species of upland habitats (white spruce) arrived. As a result, a period of instability in the vegetation prevailed with scrub and poplar woodland phases succeeding each other in response to the maximum early Holocene summer warming. By contrast, the lingering Laurentide Ice Sheet of the central parts of the continent, for example, towards Hudson Bay, persisted into the early Holocene when a climate favorable to tree growth was already established. The result of this circumstance is that by the time land emerged from the ice sheets, spruce had migrated from the south and was available to spread immediately onto the landscape without a preliminary treeless or other stage.

Regional landscape variability. Species abundances in the modern boreal forest vary with landscape pattern. For example, several of the mapped sections of the boreal forest are characterized by the prevalence of a particular landform — such as the existence of sandy outwash of the Athabasca South Section, supporting large jack pine stands to the exclusion of spruce; or the poorly drained terrain of the Manitoba Lowlands supporting extensive black spruce and larch (Rowe, 1972). It is probable, though unconfirmed at present because of lack of sites, that this factor has operated in the past and produced a diversity of long-term successional patterns.

The availability and dispersal of tree populations. The differences between the southwest margin sites and all others is striking and is due primarily to the distributional behaviours of jack pine and lodgepole pine. It appears that all tree taxa, including lodgepole pine, were available to occupy, roughly simultaneously, the Foothills landscapes following an early Holocene climatic warming, with the result that the modern boreal forest composition and general species abundances were established by about 8.5 ka. On the other hand, sites in the central part of the boreal forest and towards the northern boundaries did not achieve regional equilibrium until later in the Holocene, between 6 and 5 ka, due to the relatively slow spread of jack pine northwards. Another striking example of pattern differences that can be ascribed to availability of populations for dispersal is the prominent role of dwarf birch in the Late Pleistocene of the lower Mackenzie Valley sites, while localities in the Foothills region, with comparably long ice-free periods, lacked this shrub in any significant quantity. It has been suggested that dwarf birch spread into the western boreal forest from a source area in Beringia (Ritchie, 1984a).

Site position in relation to ecotones. The short continuous history of the boreal forest at southern sites, lasting only two or three millennia, is a function of their proximity to the forest-grassland ecotone, which migrated southward in response to late Holocene climate change. The corollary, demonstrated by Spear (1983) for a site near Tuktoyaktuk, Northwest Territories, is that the late Holocene movement southwards of the tundra-woodland ecotone produces landscapes dominated by shrub tundra with a relatively short, or at least punctuated, history as a treeless biome.

It is likely, though at present incompletely demonstrated, that in addition to the above factors, different ecotypes have occurred throughout the ranges of some of the important tree species, so that "niche specialization through competitive divergence" (White, 1979) has operated to diversify further the patterns of long-term change.

What emerges is the conclusion that several quite different pathways of vegetation change have occurred leading to the modern boreal forest. These conclusions reinforce the view that the older idea, still propagated in modern text books of biogeography (Pielou, 1979, p. 217), that entire biomes or plant formations migrated en masse in response to climate change, is not applicable to the boreal forest, if to any formation. Indeed the concept of a cohesive boreal forest type is probably invalid, and the most useful concept to depict the boreal forest, past and present, is that of a coenocline (Whittaker, 1975), which implies that the composition and abundance of plant populations vary continuously along environmental gradients.

PATTERNS OF POST-WISCONSINAN PLANT COLONIZATION IN QUEBEC-LABRADOR[1]

P.J.H. Richard

Introduction

Quebec-Labrador is an enormous area (1 825 780 km[2]), with a varied physiography, spreading over 17° of latitude between 45° and 62°N. The number of degree-days over 42°F (5.6°C) ranges from less than 500 in the north to more than 3400 in the south. Present vegetation (Fig. 7.13) varies from tundra in the north and on the tops of some mountains in the Gaspésie area to deciduous forest near Montréal in the southwestern part of Quebec (Grandtner, 1966).

Deglaciation of the region started at about 13 ka or earlier on the coasts of Gaspésie and ended at around 5 ka in northern Quebec (Richard et al., 1982; Occhietti, 1989; Vincent, 1989). Not all land was immediately available for colonization by plants and animals following ice retreat because of temporary inundation of the lowlands by proglacial seas (e.g., Goldthwait, Champlain, Tyrrell, and d'Iberville seas) or vast proglacial lakes (Barlow, Ojibway, and others). Consequently, postglacial revegetation of Quebec-Labrador was widely diachronous and started under a variety of initial conditions.

At any site the time immediately following deglaciation is one of drastic changes of the physical and biotic environments. Under what conditions did late glacial and postglacial plant colonization take place in the different areas? What was the sequence of colonization and replacement of

---

1 Translation of French text.

Richard, P.J.H.
1989: Patterns of post-Wisconsinan plant colonization in Quebec-Labrador; in Chapter 7 of Quaternary Geology of Canada and Greenland, R.J. Fulton (ed.); Geological Survey of Canada, Geology of Canada, no. 1 (also Geological Society of America, The Geology of North America, v. K-1).

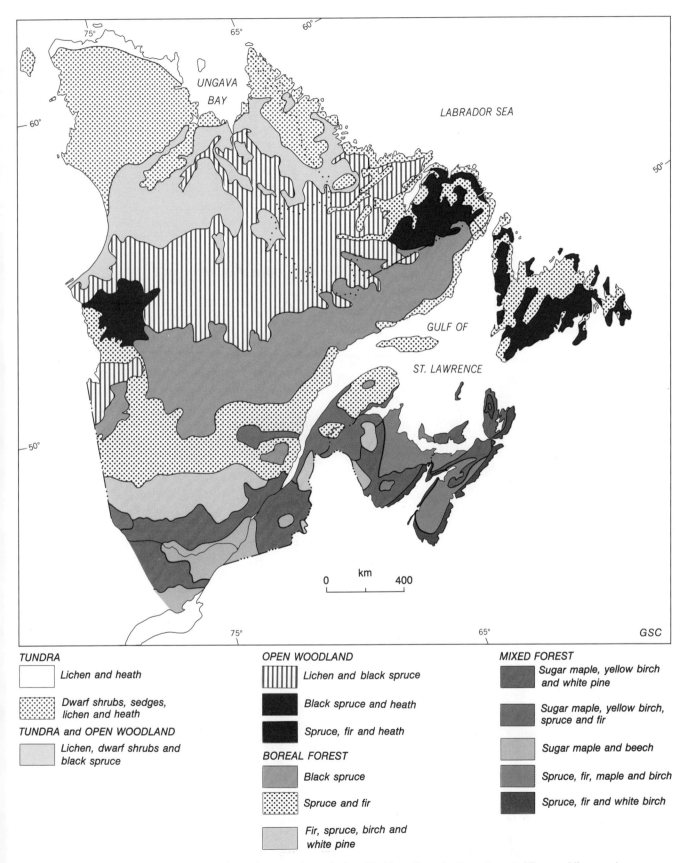

Figure 7.13. Vegetation regions of eastern Canada (modified from Canada, Department of Energy, Mines and Resources, 1973). Forest-tundra boundary changes in New Quebec are based on recent studies by Payette (1983).

CHAPTER 7

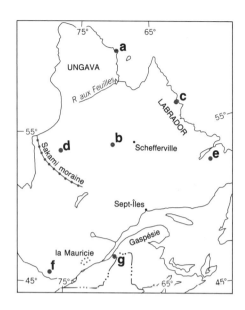

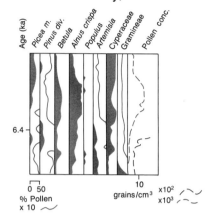

a Heel Cove Valley, Quebec

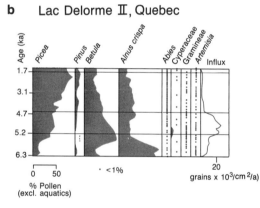

b Lac Delorme II, Quebec

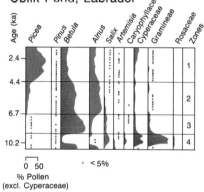

c Ublik Pond, Labrador

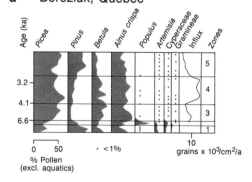

d Bereziuk, Quebec

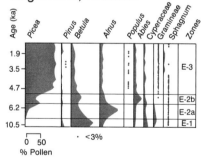

e Eagle Lake, Labrador

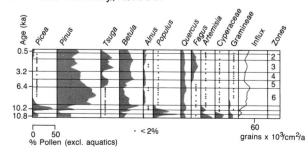

f Lac Ramsay, Quebec

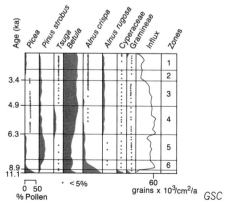

g Lac Colin, Quebec

Table 7.4. Elements of an interpretive model for revegetation of Quebec-Labrador

Standard sequence*	Current zonal equivalent	Pollen assemblage-types
Nonarboreal phase: Absence of trees		
Desert stage	Rocky tundra	Very low pollen concentration
Herbaceous stage	Herb tundra	Herb (grasses, sedges, and others) pollen predominates
Shrubby stage	Shrub tundra	Shrub pollen predominates
Afforestation phase: Presence of trees		
Shrubby stage	Forest-tundra	Shrub and tree (spruce) pollen
Arboreal stage	----	Intolerant tree pollen
Forest phase: Trees dominate, forming more or less closed cover		
Open stage	Taiga	Tree (spruce) and shrub pollen
Closed stage	Closed forest	Tree pollen
Coniferous stage	Spruce forests	Spruce pollen
Mixed stage	Fir forests	Birch with balsam fir pollen
Deciduous stage	Maple forests	Birch or pine with sugar maple pollen

*The elements of the sequence are listed in their chronological order from top to bottom.

plant communities? How do these plant communities compare to the present plant cover? What factors were responsible for the differences? These are the questions asked by the paleophytogeographer interested in the late glacial period in Quebec-Labrador. The answers will not only allow reconstruction of late glacial vegetation landscapes, they will also throw new light on the current distribution of plants or plant communities (disjuncts, relicts). In addition, they will lead to a better understanding of proglacial and periglacial environments, adding to both the geomorphological and climatic points of view for this period.

A sufficient number of Quebec-Labrador sites have now been studied to permit reconstructions for that region on the basis of their pollen and macrofossil content. Abbreviated representative pollen diagrams for seven localities are shown in Figure 7.14. In the following discussions, the palynostratigraphic variability of initial sediments in Quebec-Labrador lakes and bogs will be discussed and interpreted in terms of past vegetation and climate by comparison with a standard sequence (a model) of postglacial vegetation change.

Analogue model of postglacial vegetation change

One approach to the study of Quebec-Labrador revegetation is to compare the reconstructed sequences of vegetation types at a large and widely distributed array of sites with a standard sequence derived from the observation of two phenomena operating presently on the vegetation: latitudinal zonation and succession. The latitudinal zonation of the vegetation reflects a global equilibrium of the floristic stock with the general climate. The units of this latitudinal zonation (the vegetation zones or types) serve as the modern analogues for the reconstructed stages of the vegetation sequence after ice retreat at a given site. This procedure allows the identification of the nonanalogue situations, both in the vegetational composition of the stages and in their relative position in the sequence. There is no *a priori* implication of a corresponding climate involved in such a comparison. Succession is a smaller scale phenomenon that reflects the short-term dynamics of plant communities. The successional events that occur today at the margin of alpine glaciers are dynamic analogues of the revegetation in front of the receding ice. The time involved and the environmental context, both abiotic and biotic, are very different. At such sites, the initial short-term seral stages are dominated by light-demanding, tolerant plants adapted to harsh conditions.

Table 7.4 illustrates the main elements of the model, showing the correspondence between the units of the standard sequence (phases and stages), current plant-zone equivalents, and the basic characteristics of the pollen assemblages that led to the reconstructed stages.

Late glacial palynostratigraphy

Early works in pollen stratigraphy resulted in oversimplified zoning of pollen diagrams, which often masked the diversity of the pollen assemblages. For example, only two zones were attributed to the late glacial period in southern Quebec — an initial herb zone, followed by a *Picea* (spruce) zone (Terasmae, 1969, 1973). The undertaking of many studies, resulting in detailed pollen diagrams, and the advent of modern techniques, like pollen influx measurements, has revealed a greater amount of interregional diversity of late glacial and postglacial pollen assemblages than was evident

Figure 7.14. Abbreviated pollen diagrams for seven sites in Quebec-Labrador and locations of features discussed in text. The Lac Ramsay profile is modified from Mott and Farley-Gill (1981), Lac Colin from Mott (1977), Eagle Lake from Lamb (1980), Ublik Pond from Short and Nichols (1977), Heel Cove Valley from Richard (1977a), Lac Delorme II from Richard et al. (1982), and Bereziuk site from Richard (1979).

before. Pollen studies from Quebec-Labrador have shown that definite series of late glacial pollen assemblages are repeatedly encountered in cores from any one region, a fact which shows that these assemblages are not merely random collections of pollen types (Terasmae, 1976).

Table 7.5 illustrates the variety of pollen assemblages found in the oldest postglacial sediments at various Quebec-Labrador sites. They are named after their dominant or diagnostic taxa. Some of the assemblages shown in the table are missing for particular palynostratigraphic series. Also, the relative abundance of a taxon in an assemblage may be expected to vary from one region to another. Despite such variations, the table clearly shows that recognizable regional assemblages do exist. Furthermore, it is highly probable that the different pollen assemblages within a series reflect a real sequence of plant colonization at a given site.

Except for Labrador, only the earliest assemblages are given in the table. Many of these lack modern analogues. For example, assemblages dominated by *Alnus crispa* or those in which *Populus* and *Abies balsamea* are the main elements have no modern surface pollen counterparts.

Patterns of initial plant colonization

Three major patterns of plant colonization can be distinguished, depending on whether the initial phase is represented by nonarboreal vegetation, by afforestation (a transitional phase between nonarboreal vegetation and forest), or by a forest (Fig. 7.15).

Each pattern includes various modes as far as the age, duration, and composition of the initial plant cover are concerned. Moreover, each presents various types of subsequent evolution in the plant cover. The spatial distribution of these patterns and modes can in general be explained by paleogeographic data.

Sites displaying an initial nonarboreal phase

In its fullest manifestation, this phase comprises an initial desert stage, followed by distinct herbaceous and then shrubby stages (Savoie and Richard, 1979; Labelle and Richard, 1981). The desert stage is characterized by very low pollen concentration and a predominance of allochthonous pollen from distant forested sites. The assemblages may also include retransported preglacial pollen derived from local unconsolidated deposits (J.H. McAndrews, Royal Ontario Museum, personal communication, 1982). The herbaceous stage is generally signalled by a strong (tenfold) increase in pollen concentration (and influx), reflecting the advent of local and regional plant colonization. Cyperaceae, Gramineae and, in some places, *Artemisia* pollen are the

Table 7.5. Palynostratigraphic sequences in Quebec-Labrador

Laurentians, western Appalachians	
Alnus crispa - arboreal *Betula*	youngest
Populus - *Picea* - *Juniperus*	
Shrub *Betula*	
[*Salix* - Cyperaceae - arctic plants]	
Cyperaceae - Gramineae - Cyperaceae	
[*Salix* - Gramineae-Cyperaceae-arctic plants]	
Trees - min PC [arctic plants]	oldest
Eastern James Bay (e.g., Bereziuk)	
Picea - *Alnus crispa*	youngest
Alnus crispa - shrub *Betula*	
Alnus crispa - *Populus* - *Juniperus*	
Pinus divaricata - min PC	oldest
Lac Delorme region (e.g., Delorme II)	
Alnus crispa - *Picea* - *Betula*	youngest
Alnus crispa - *Betula* - *Picea*	
[*Alnus crispa* - min PC]	oldest
Rivière-aux-Feuilles	
Alnus crispa - *Picea*	youngest
Alnus crispa - shrub *Betula*	
Shrub *Betula* - *Alnus crispa*	
trees - min PC	oldest
Arctic Ungava (e.g., Heel Cove)	
Alnus crispa - Cyperaceae - shrub *Betula*	youngest
Salix - *Alnus crispa* - Gramineae	
Gramineae - Cyperaceae - arctic plants	
Trees or *Alnus crispa* - min PC	oldest
Southeast Labrador (e.g., Eagle)	
Picea mariana	youngest
Alnus crispa - *Abies balsamea*	
Alnus crispa - *Picea glauca*	
Betula - Cyperaceae	
Salix - *Betula* - Cyperaceae	
[*Alnus rugosa* - *Betula* - Gramineae]	oldest
Northern Labrador (e.g., Ublik)	
Picea	youngest
Alnus crispa - *Betula* - *Picea*	
Shrub *Betula* - *Alnus crispa*	
[Cyperaceae - Gramineae]	oldest

Min PC = minimal pollen concentration
[ ] = assemblages that are less often encountered

Table 7.6. Arctic-alpine taxa represented in initial postglacial sediments by pollen or macroremains from Quebec-Labrador

Armeria maritima var. *labradorica*	p	m
Arctostaphylos sp.		m
Betula glandulosa	p	m
Carex cf. *C. bigelowii*		m
Carex cf. *C. nardina*		m
Cerastium cf. *C. aplinum*		m
Dryas sp.	p	
Dryas drummondii		m
Dryas integrifolia		m
Draba cf. *D. incana*		m
Koenigia islandica	p	
Lychnis apetala		m
Lychnis alpina/affine		m
Oxyria digyna	p	
Oxytropis maydelliana	p	
Polygonum viviparum	p	
Potentilla cf. *P. nivalis*		m
Ranunculus pedatifidus/sulphureus		m
Ranunculus trichophyllus var. *eradicatus*		m
Salix herbacea	p	
Salix vestita	p	
Saxifraga cf. *S. cernua*	p	
Saxifraga cf. *S. oppositifolia*	p	
Senecio congestus		m
Sibbaldia procumbens		m
Silene acaulis var. *exscapa*		m
Taraxacum cf. *T. pumilum*		m
Tofieldia pusilla	p	

Source: Mott et al., 1981; Larouche, 1979; and unpublished sources.

p = pollen; m = macroremains

Note: Those taxa represented by pollen usually occur in trace amounts, but in some cases, e.g., southern Quebec, they represent 5-25% of the pollen spectra.

dominant components (see Lac Ramsay, Lac Colin, Ublik Pond, and Heel Cove Valley, Fig. 7.14), although most of the diagrams also reveal a diverse assemblage of arctic-alpine pollen types (Richard, 1977b; Mott et al., 1981; Table 7.6). Certain taxa of Caryophyllaceae, Ranunculaceae, Saxifragaceae, Cruciferae, and Rosaceae are especially well represented.

Occasionally the herbaceous stage is preceded by a peak of willow pollen, but in the typical case, the shrub stage follows the herbs and is dominated by dwarf birch (*Betula glandulosa*). Pollen of other shrubs such as Ericaceae, *Shepherdia canadensis*, and green alder (*Alnus crispa*) is consistently present. With the inception of the shrub stage, pollen concentration values show another tenfold rise (see Lac Colin, Fig. 7.14).

These initial pollen assemblages usually come from lake sediments that are low in organic matter; consequently, dating is generally impossible, and precise measurement of the duration of the stages, particularly the desert stage, is not feasible. At several presently forested sites (Richard, 1971; Labelle and Richard, 1981), the duration of the entire nonarboreal stage ranges from a few hundred to a few thousand years.

Inasmuch as the landscape represented by the nonarboreal stages lacked trees and contained taxa currently found in arctic or alpine regions, the term "tundra" is applicable. But tundra also has climatic implications (e.g., cold mean annual temperatures, permafrost) that are not implicit in the pollen data alone. There is also the possibility that the absence of trees may be due to nonclimatic factors, such as a delay in the migration of trees to a site (Richard, 1977a; this chapter, Ritchie, 1989). Further caution against assuming a tundra like that of the present is the fact that few of the taxa represented by pollen or macrofossils are restricted to arctic alpine sites. Instead, many can be found growing today in open habitats (arctic-alpine type) within forest zones.

Except for a site in Laurentide Park, north of Québec City, where the elevation of the terrain caused tundra to persist until 7.5 ka (Richard, 1971), the early tundra phase lasted from about 11.5 to 10 ka (Richard, 1977b). Present data show that the sites with such a diversified nonarboreal initial phase occur in southern Quebec (see Lac Ramsay, Fig. 7.14), in the highlands of the Appalachians, and at the southern edge of the Canadian Shield — all areas outside the region known to have been occupied by postglacial seas. During the 11.5 to 10 ka interval, ice masses in the Appalachians and on the southern edge of the Canadian Shield, near the Champlain and Goldthwait seas, created paleogeographic and paleoclimatic conditions conducive to the spread of this rich tundra within the region bounded by the ice or by proglacial seas and lakes.

Elsewhere in Quebec-Labrador few sites show such a clear early tundra phase. The earliest pollen assemblages from the arctic lakes of Quebec and Labrador are generally more diverse than later assemblages from the same lakes, but most display a lower floristic diversity than the early nonarboreal pollen assemblages at southern Quebec sites (Richard, 1981b). Thus it appears that the late glacial tundra in southern Quebec, the Appalachians, and the edge of the Canadian Shield was floristically richer than today's tundra or the paleotundra of arctic Quebec. This paradox may have been caused by the great abundance of unleached and unconsolidated deposits at the southern sites, a different light regime, and possibly a milder and moister climate.

Earliest pollen assemblages from the coniferous forest zone of Quebec-Labrador are less diverse than those of the south. They are dominated by Cyperaceae and Gramineae as well as shrubs like *Salix*, shrub birch (*Betula glandulosa*), and green alder (*Alnus crispa*) (Wenner, 1947; Morrison, 1970; Jordan, 1975; Short, 1978). In eastern Labrador at sites such as Eagle Lake and Ublik Pond (Fig. 7.14), shrub values reach 80% and persist for long periods, i.e., about 4.5 ka. In southern Labrador, several localities experienced more than 2 ka of such shrubby tundra, before the first arrival of trees at 8 ka (Lamb, 1978, 1980; Engstrom and Hansen, 1985). Farther north, on the Labrador coast, herb tundra preceded shrub tundra, with trees appearing no earlier than 5 ka (Jordan, 1975; Short and Nichols, 1977). Although the earliest pollen assemblages have a lower concentration of arctic plants and are less differentiated into distinct stages than those in southern Quebec, their long duration probably reflects conditions similar to those of the present Arctic.

Elsewhere in Quebec, on the Laurentian Plateau, and in the mountains of the Eastern Townships, initial pollen assemblages, dominated by sedges or shrubs, display more uniformity and lasted for shorter periods (Richard, 1973a, b, 1975a, b; Mott, 1977). They are more similar to the early seral stages of a successional sequence. Some of the assemblages contain pollen of trees that is typical of the next, afforestation phase of the model (see below).

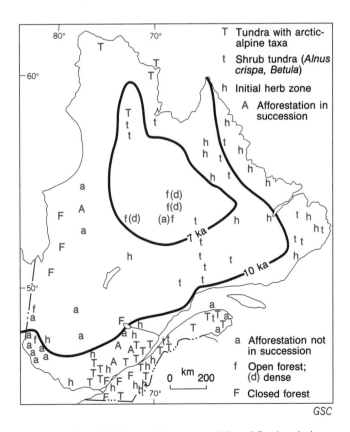

Figure 7.15. Initial vegetation composition of Quebec-Labrador and approximate position of ice margin at 10 and 7 ka.

Subsequent Holocene history of initially nonarboreal sites

The subsequent vegetation history of sites showing an initial nonarboreal stage varies from region to region. In the Ungava region only those sites very close to the current treeline record any northward movement of treeline between 4 and 3.5 ka. Other minor fluctuations, especially in the density of arboreal vegetation, are indicated by macroremains such as stumps found outside the current spruce or larch groves (Payette and Gagnon, 1979). Elsewhere, between 6.2 and 4 ka dense shrub tundra, characterized initially by shrub birch (*Betula glandulosa*) then later by green alder (*Alnus crispa*), replaced the initial herb tundra in diachronous fashion from the Ungava coast towards the interior plateaus.

In general, the Appalachians were characterized by rapid afforestation, with black spruce (*Picea mariana*) dominating, especially in the southwest (Richard, 1975a, 1978; Mott, 1977). In the northeast, balsam fir (*Abies balsamea*) or trembling aspen (*Populus tremuloides*) were the dominants of the earliest arboreal phase. Everywhere, however, common boreal species, especially tamarack (*Larix laricina*) and white birch (*Betula papyrifera*), were parts of the original forests (Labelle and Richard, 1984).

The southern edge of the Canadian Shield underwent slow afforestation. North of the Champlain Sea, tundra was invaded by aspen parkland or riverine aspen groves, which lasted for 500 to 1000 years (Richard, 1977a; Mott and Farley-Gill, 1981). Taiga did not give way to closed forest until about 3 ka after deglaciation in La Mauricie Park and 5 ka after ice retreat in the centre of the Laurentian Highlands (Richard, 1981a). Initially this lag in development of closed forest must have been caused by the biogeographical barrier imposed by the Champlain Sea, but the persistence of a cold climate due to the high elevation of these mountain ranges is undoubtedly the major reason for the lengthy taiga or afforestation period.

On the North Shore (Gulf of St. Lawrence) and in southern Labrador, afforestation extended over several centuries, involving first white spruce, then balsam fir and finally black spruce-white birch communities (Mott, 1976; Lamb, 1980, 1984; Engstrom and Hansen, 1985). Farther north, fir fails to appear and spruce (black?) and larch are the main trees involved in afforestation, forming the open spruce forest that today extends over the inland portions of the area under study, as well as in protected parts of the coast (Jordan, 1975; Short and Nichols, 1977; Lamb, 1982; King, 1984). In southern Labrador trees arrived late (e.g., 8 ka) and in the north even later (4.5 ka); this is probably due to the remoteness of the area from forest refugia and postglacial migration fronts.

A remarkable afforestation phase is evidenced by pollen spectra from sites in Gaspésie, lower St. Lawrence, and the Laurentians between Sept-Îles and Rivière Saint-Maurice. Green alder shows a pronounced pollen maximum, in many places exceeding 50% of the pollen sum, commonly coming at the same time or shortly after pollen maxima for trembling aspen and black spruce. Green alder, therefore, was important for some centuries during colonization of these regions (Richard, 1981a).

Sites with an initial afforestation phase

Large areas of western Quebec (e.g., Abitibi, Témiscamingue, and James Bay areas) had no initial nonarboreal phase, (Richard, 1979, 1980) as shown in the pollen diagram for Lac Louis (Vincent, 1973). Deglaciation in the Témiscamingue area was immediately followed by the Lake Barlow glacial lacustrine episode. Trees were present at least in the form of scattered groves as early as about 10 ka. In fact, the earliest pollen assemblages at some thirty small basins above the glacial lacustrine shoreline limit indicate that the initial vegetation was a mosaic of herbaceous, shrubby and arboreal communities, the latter containing trembling aspen, black spruce, and tamarack.

In the James Bay area, the first sites available for plants were located in a corridor between the Tyrrell Sea and an ice front retreating from the present position of the Sakami Moraine (Vincent, 1989). Trees took more time to spread northward, especially since the entire region to the south had been earlier occupied by proglacial Lake Ojibway. Afforestation is marked at about 6.5 ka by the establishment of aspen parkland, succeeded by black spruce, green alder, and shrub birch to form a taiga with denser tree cover than is typical of the region today. There are still not enough data to indicate whether a similar pattern of initial plant colonization existed for other areas of western Quebec, but it is reasonable to assume that a similar type of postglacial progressive afforestation occurred at other sites on the Canadian Shield.

The most luxuriant "periglacial" plant cover occurred in the heart of Quebec, in the Lac Delorme region, about 100 km west of Schefferville (Fig. 7.14; Richard et al., 1982). Unlike the James Bay area, the earliest pollen assemblages at Lac Delorme are completely dominated by green alder (*Alnus crispa*). The oldest organic sediments contain macroremains as well as high percentages and influx values of green alder pollen. Macroremains of white birch, tamarack, and black spruce are also present. These immigrants were probably also accompanied by aspen and balsam poplar. Thus between 6.2 and 5.6 ka dense luxuriant vegetation, bearing no resemblance to the impoverished taiga of today, covered the landscape in the region of the last outliers of the New Quebec glacier. White birch and balsam fir were most abundant during this period, whereas they are rare there today. Clearly, by 6 ka local ice no longer had a negative effect on the vegetation.

Sites with an initial forest phase

The record of sites at which the oldest sediments portray forested conditions varies according to the nature of the sediment catchment basins, their geomorphological and paleogeographical context, and their geographical location. Compared to the palynostratigraphic model (Table 7.4), the pollen sequences commonly seem to be basally truncated.

The earliest pollen assemblages in the St. Lawrence Lowlands within the boundaries of the Champlain and Goldthwait seas indicate that forests first appeared there between 10 and 7.5 ka. Dominated by boreal species, these forests were quickly invaded by more thermophilous trees, including sugar maple (*Acer saccharum*). By the time of the Lampsilis Lake recession (Brown-Macpherson, 1967; Elson and Elson, 1969), the vegetation of the region was dominated on well drained sites by sugar maple (Richard, 1973c, 1975c; Larouche, 1979; Comtois, 1982). Only the margins of the postglacial seas escaped this pattern of plant colonization. Similar paleogeographical control over the mode of initial plant colonization is evident in the lowlands of Abitibi and Témiscamingue, two areas flooded by proglacial lakes Barlow and Ojibway, and the zone invaded by the Tyrrell Sea (at least in the James Bay area).

Other areas fail to display the typical early postglacial pollen assemblages because the sediment sources are ponds that formed on ice-cored moraine and outwash. In such cases, several thousand years may have elapsed between deglaciation and formation of ponds or lakes. The presence of forest-type pollen assemblages and macroremains at the base of organic sections in such basins indicates that buried ice remnants survived for a long time (2 ka?) beneath forest vegetation at the southern edge of the Laurentians.

Discussion and conclusions

It is apparent from the foregoing that the pattern of post-Wisconsinan plant colonization in Quebec-Labrador was spatially and temporally varied. Initial plant colonization started during the Late Wisconsinan and continued until the middle Holocene. In some areas colonization was retarded by presence of proglacial lakes or seas, but in general, once terrestrial habitats became available, colonization took place immediately but at different rates. For the most part it was unidirectional, that is, without a temporary reversal that might represent an episode of climatic deterioration. Such a late glacial cooling is evident in parts of the Maritime Provinces (Mott et al., 1986), and it may well have occurred in Gaspésie, along the Baie des Chaleurs. There is at present, however, no evidence for similar cooling in the interior of southern Quebec.

South of the current arctic zone, well developed tundra vegetation only occurred in southern Quebec-Labrador. Its character varied greatly from one region to another. In other areas the initial vegetation phase consisted of waves of forest plants. This means that arctic plants now found in northernmost Quebec probably migrated to their present venue via the Labrador coastal area and possibly the now inundated continental shelf. The possibility that such northern areas were colonized from nunatak refugia in the Torngat Mountains has not been convincingly demonstrated (Lamb, 1982; Gangloff, 1983), and the nunatak hypothesis has also been refuted for the alpine flora of the mountains of the Gaspésie (David and Lebuis, 1985). The suggestion of a possible early deglaciation (ca. 20 ka) in Labrador (Vilks and Mudie, 1978) suffers from dating problems, since "old" carbon has undoubtedly contaminated the sediments that have a low total organic content. Available data, therefore, do not support the hypothesis of full glacial plant refugia in Quebec-Labrador. The existence of disjunct arctic or alpine taxa around the Gulf of St. Lawrence is adequately explained by the former existence of a more or less continuous "arctic corridor" at the southern edge of the Wisconsinan ice cap during its maximum extension (Morisset, 1971). For a short time during the Holocene and late glacial such "periglacial" tundra followed the retreating glacial front, leaving arctic or alpine relicts at certain localities.

The case of *Shepherdia canadensis* is a good example of how paleoecological studies may help explain modern plant distribution (Richard, 1974). This plant is a small heliophilous and calcicolous or basiphilous shrub, extending across Canada in essentially boreal habitats. Although it is common in the west, localities in Quebec-Labrador are widely separated, especially in the Canadian Shield; there it grows only in association with calcareous rocks, along marine shorelines and along the Labrador trough (Rousseau, 1974). Pollen data, however, show that this plant was ubiquitous during the afforestation phase. Later it was progressively eliminated by competition from the forest flora taxa and by soil leaching, leaving it isolated at its current growth sites in Quebec-Labrador.

Afforestation was distinctly different in the eastern and western parts of Quebec-Labrador. In the west it was the first stage of colonization; in the southeast, that is, around the Champlain Sea, it followed a protracted shrub-tundra phase. In the west, trees often arrived together, with trembling aspen and jack pine playing an important role. In the east, white spruce, balsam fir, and black spruce arrived in sequence as a result of migration lags, with fir dominating in the first continuous forests. In addition, during afforestation in the east, green alder exhibited a remarkable dominance, a phenomenon that did not occur in the west. These distinctions are probably caused by the maritime effects and the greater rainfall in the east, favouring white spruce and fir. Also, the more northern location of this part of the area precluded occupation by most thermophilous species. In the west, a north-south climatic gradient was established, also eliminating thermophilous trees, although in a more continental context. In the southwest, afforestation took place at a time when a rich forest flora existed close to proglacial lakes and the ice sheet, whereas the east and northeast parts of the region were far removed from the sources of arboreal plants. As a result, trees reached such areas at a later date, and because of the northern location, the flora was more depauperate. For species adapted to the upper latitudes, the climate must also have been a contributing factor in this slow invasion of the land, especially around 9 ka, when the ice sheet was still able to modify the pattern of air mass circulation in the North Shore region. The cooler climate of higher elevations, however, was undoubtedly responsible for the slow afforestation in the Laurentians north of Québec City between 9 and 5 ka. At about 5.5 ka in the central part of Quebec-Labrador, open forests consisting mostly of tamarack and green alder, but denser than present forests in the region, began to invade land abandoned by the last remnants of the New Quebec glacier.

The post-Wisconsinan plant colonization across Quebec-Labrador, as reconstructed from pollen and macroremain analysis of lake and peat bog sediments, is a picture of movement and variety. When compared with the standard sequence of vegetation units (Table 7.4), the data show how expected phases and stages varied from one region to another. Some of these recolonization stages represent vegetation and a flora that have no modern counterparts. Thus variety and uniqueness are prominent characteristics of the late glacial vegetation of Quebec-Labrador. Further pollen work and analyses of plant macroremains should help to clarify the nature and duration of the various postglacial phases and plant migration routes in the area. Geochemical and isotopic studies, as well as other data sources (e.g., fossil insects, diatoms), should provide independent validation for assessment of the climatic conditions under which the postglacial vegetation of Quebec-Labrador developed.

INTRODUCTORY COMMENTS ON THE PALEOENVIRONMENTAL RECORD OF THE HOLOCENE

J.V. Matthews, Jr. and T.W. Anderson

The next two case histories are specifically concerned with the present interglaciation — the Holocene — which, by general agreement, encompasses the last 10 ka. Anderson and his co-authors focus on the nature and timing of the period of peak Holocene warming, a time when climate in most regions of Canada was warmer than at present. What they show is that the early thermal history of the Holocene is not one of synchronous country-wide warming followed by cooling. Rather, peak warmth occurred at different times and was expressed in different ways in different parts of the country — a finding that agrees with a similar conclusion reached in a review of the Hypsithermal interval in the United States (Wright, 1976).

The case history on Holocene climate is also instructive because it is based mainly on the approach of reconstructing vegetation from pollen and only then making the jump to paleoclimatic inference. The other more quantitative and direct approach to climate reconstruction, involving pollen data and the use of transfer functions (Webb and Clark, 1977; Overpeck et al., 1985), is also mentioned below, but as stated there, in the discussion of British Columbia climates, the conclusions reached using transfer functions do not always match those garnered from the more traditional methods of analysis.

The case history by McAndrews and Boyko-Diakonow concerns the last few millenia of the Holocene, including the historical period, at Crawford Lake in southern Ontario. Crawford Lake is unusual because it is one of only a few lakes in northeastern North America that are presently depositing annually laminated (varved) sediments (Boyko, 1973). Rhythmically bedded sediments are found in other lakes, but it is seldom that they can be demonstrated to represent annual deposition so unequivocally as at Crawford Lake. Other lakes in Ontario have been shown to have had phases during which annually bedded sediments accumulated (Cwynar, 1978), but few of them fulfil the special limnic conditions required for such deposition at present.

The advantage of annually layered sediments is that they allow precise calculation of time of deposition by counting rather than by extrapolation between isotopically dated horizons. In the case of Crawford Lake, the annually layered sediments provide a year-by-year chronology for a time period that is too recent to be adequately dated by most isotopic methods; however, with such advantages come problems. Special techniques are required to sample such soft, near-surface sediments without disturbing their yearly signal. McAndrews has been one of the pioneers in the development of a freezing-tube method for sampling such sediments.

The Crawford Lake research discussed below, and especially in Boyko (1973), represents a unique blend of pollen evidence with historical records on plant communities and land clearance. Taken together, these two lines of evidence emphasize the sharp contrast between European agricultural practices and those of an earlier phase of Iroquois agriculture. The authors of the case history allude to the problem of determining whether fluctuations registered in the pollen diagram are due to climate or to succession following abandonment of the Iroquois agricultural plots.

This "succession versus climate" question is a universal one for paleobotanists. The problem is determining the degree to which pollen fluctuations in lake cores reflect climate change, succession and migrational lags, or a complicated interplay of both (Davis, 1981b, 1983; Wright, 1984; this chapter, Richard, 1989). The exquisite time resolution of the Crawford Lake core makes it possible to assess the importance of such subtle variables. In older lake cores, the time resolution is seldom sufficient to reveal a potential succession variable. But pollen studies of a series of lakes from a large region can reveal Holocene migration trends (Davis, 1981b, 1983).

HOLOCENE CLIMATIC TRENDS IN CANADA WITH SPECIAL REFERENCE TO THE HYPSITHERMAL INTERVAL

T.W. Anderson, R.W. Mathewes, and C.E. Schweger

Introduction

The climate of the Holocene has been discussed and evaluated in several recent studies (Wright, 1976; Davis et al., 1980; Ritchie et al., 1983; Webb et al., 1983; Payette, 1984; Dean et al., 1984). The climatic model that has been developed is one of gradual increases in warmth and dryness from early to middle Holocene time, maximum warm and dry conditions in middle Holocene time (Hypsithermal interval, climatic optimum, or thermal maximum), followed by gradual decreases to more or less present conditions in late Holocene time. Superimposed on the general trend, particularly in the Midwest of North America, are two dry pulses separated by a moister interval in middle Holocene time (Dean et al., 1984). The Hypsithermal interval is considered by Wright (1976) to be time-transgressive from region to region.

This conclusion is examined below in the Canadian context. The examples discussed are based on studies of sites in widely spaced areas of Canada (Fig. 7.16), using mostly fossil pollen, but supported in some cases by plant macrofossil, insect, diatom, and stratigraphic data. Collectively they portray regional trends in the Holocene climatic history of Canada, including the differential expression of the Hypsithermal interval.

Matthews, J.V., Jr. and Anderson, T.W.
1989: Introductory comments on the peleoenvironmental record of the Holocene; in Chapter 7 of Quaternary Geology of Canada and Greenland, R.J. Fulton (ed.); Geological Survey of Canada, Geology of Canada, no. 1 (also Geological Society of America, The Geology of North America, v. K-1).

Anderson, T.W., Mathewes, R.W., and Schweger, C.E.
1989: Holocene climatic trends in Canada with special reference to the Hypsithermal interval; in Chapter 7 of Quaternary Geology of Canada and Greenland, R.J. Fulton (ed.); Geological Survey of Canada, Geology of Canada, no. 1 (also Geological Society of America, The Geology of North America, v. K-1).

British Columbia

Early to mid-Holocene

Geological and botanical evidence for early to middle Holocene climate in the Cordillera is complex and in many cases contradictory (Mathewes, 1985). Radiocarbon dates on wood above present timberline in southern British Columbia attest to a period of warmer climate prior to 5.2 ka (Clague, 1981). Some geological evidence, however, points to early Holocene glacial advances during this time, especially in the southern Coast Mountains of southern British Columbia (Clague, 1981). Pollen evidence in southwestern British Columbia and Washington State generally supports the existence of a warmer and/or drier period at some time during the Holocene, although dating varies considerably from one locality to another (Mathewes, 1973; Mathewes and Heusser, 1981; Hebda, 1982).

Coast and coast/interior transition. A dramatic change in pollen stratigraphy occurs between the 10.5 ka and 10 ka levels in many pollen diagrams from south coastal British Columbia and Washington (Heusser, 1960). At Marion Lake (site 1, Fig. 7.16), lodgepole pine, spruce, and both balsam and mountain hemlock decline during this interval (Fig. 7.17a). Douglas fir (*Pseudotsuga menziesii*) appears abruptly, reaching its maximum Holocene values shortly thereafter. Western hemlock appears in low but significant quantities for the first time, and alder, bracken (*Pteridium aquilinum*) and other fern spores begin to increase. Climatic warming may have favoured the expansion of Douglas fir during this interval (Mathewes, 1973; Heusser, 1977), but clear evidence of a subsequent period warmer and drier than the present is not evident at Marion Lake.

A quantitative analysis of climatic trends at Marion Lake (Mathewes and Heusser, 1981) suggests a rapid rise in summer temperatures coupled with declining precipitation between 10.5 and 10 ka, followed by a period of maximum July temperature and minimum precipitation from 10 ka to 7.5 ka (Fig. 7.17a). This informally named "early Holocene xerothermic interval" (Mathewes and Heusser, 1981) supports the qualitative identification of warmer and drier early Holocene conditions at Pinecrest and Squeah lakes in the Yale area (sites 2, 3, Fig. 7.16) near the coast-interior transition (Mathewes and Rouse, 1975), as well as in other coastal sites (Hansen and Easterbrook, 1974; Heusser et al., 1980; Hebda, 1983). Unpublished pollen data on two lake sites recently studied by R.W. Mathewes within the coast-interior ecotone (Blue and Horseshoe Lakes, sites 4, 5, Fig. 7.16) exhibit a pattern similar to that of the Yale area, but in the former xerothermic conditions persist until about 6 ka. At Bear Cove Bog, (site 6, Fig. 7.16) near the present northern limit of Douglas fir on Vancouver Island, the sudden appearance of significant quantities of Douglas fir pollen, accompanied by bracken fern, indicates maximum dry and warm conditions between 8.8 ka and about 7 ka.

A simplified composite pollen diagram for the Yale area in the lower Fraser Canyon reveals qualitative evidence for an early xerothermic interval. The steep, rugged topography of the lower Fraser River Canyon and the "sub-continental" climatic conditions here contrast with the wetter "maritime" conditions of the Marion Lake area to the west (Courtin et al., 1981). Similar distinctions probably existed throughout the Holocene. The presence of high percentages of bracken spores, alder, Douglas fir, and grass pollen signify a period of open forest conditions (Heusser, 1973, 1974, Hebda, 1983). These same characteristics imply higher frequency of fires (McMinn, 1951), which were likely an important reason for forest instability over all of southern British Columbia during early to middle Holocene time.

Reversals in the relative pollen abundance of certain taxa (i.e., western hemlock and possibly balsam fir) assist in defining periods of climatic change in southern British Columbia, especially when the autecological requirements of these taxa are considered. They have low drought tolerances, high "moisture optima", and possess low fire resistance (Minore, 1979). The autecology of Douglas fir is just the opposite; when its pollen is abundant, a warmer, drier climate with the possibility of increased fire frequency is suggested. The clearest example of early Holocene hemlock reversal in British Columbia is the dramatic decline of arboreal pollen from 40% to 11% at Bear Cove Bog on Vancouver Island (Hebda, 1983). Since Douglas fir and bracken peak during this same interval, a warm/dry climate is likely. Similar climatically significant reversals occur in the balsam fir curves at Marion Lake (Fig. 7.17a) and Surprise Lake (Mathewes, 1973).

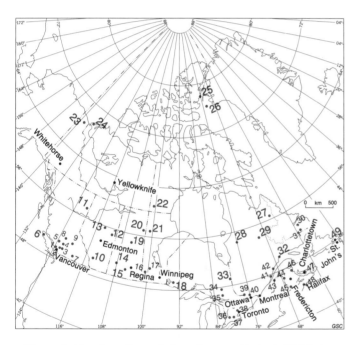

Figure 7.16. Locations of sites discussed in text. (1) Marion and Surprise lakes; (2) Pinecrest Lake; (3) Squeah Lake; (4) Blue Lake; (5) Horseshoe Lake; (6) Bear Cove Bog; (7) Kelowna Bog; (8) Chilhil Lake; (9) Phair Lake; (10) Chalmers Bog; (11) Peace River sites; (12) Moore Lake; (13) Lofty Lake; (14) Herbert site; (15) Hafichuk site; (16) Russell site; (17) Riding Mountain site; (18) Hayes Lake; (19) Cycloid Lake; (20) Reindeer Lake; (21) Lynn Lake; (22) Ennadai Lake; (23) Hanging Lake; (24) Sleet Lake; (25) Baird Inlet Lake; (26) Carey Islands site; (27) Lac des Roches Mountonnées; (28) Bereziuk site; (29) Lac Delorme; (30) Eagle Lake; (31) Moraine Lake; (32) Sept Îles site; (33) Lac Yelle; (34) Jack Lake; (35) Nina Lake; (36) Ellice Bog; (37) Gage Street Bog; (38) Ballycroy Bog; (39) Perch Lake; (40) Lac Ramsay; (41) Lac Marcotte; (42) Lac à l'Ange; (43) Lac Colin; (44) Roulston Lake; (45) Basswood Road Lake; (46) Teagues Lake; (47) MacLaughlin Pond; (48) Silver Lake; (49) Sugarloaf Lake.

CHAPTER 7

a Marion Lake, British Columbia
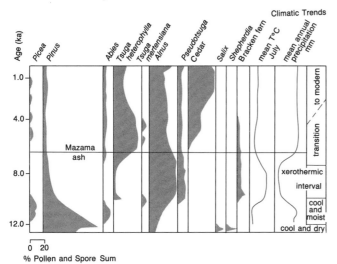

b Moore Lake, Alberta
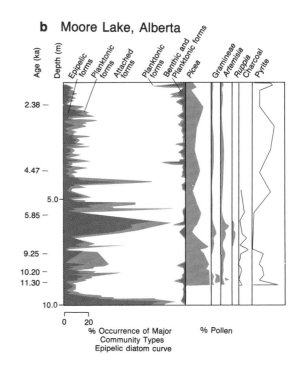

c Sleet Lake, Yukon Territory
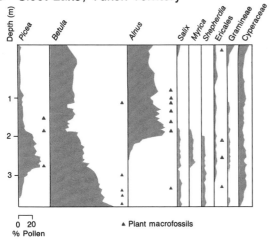

d Lac des Roches Moutonnées, Quebec
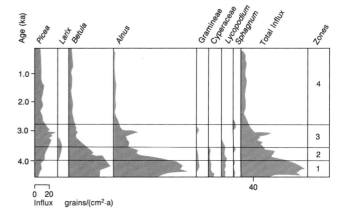

e Baird Inlet Lake, Ellesmere Island
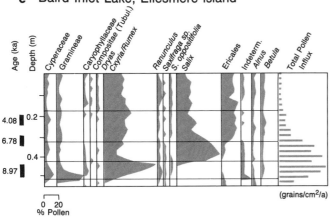

f Ellice Bog, Ontario
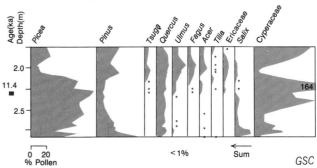

Southern Interior. Hansen (1955) identified a warm, dry period in south-central British Columbia between 7.5 ka and 3.5 ka with a "thermal maximum" about 6.6 ka. Alley (1976) also found evidence of a warm, dry interval in Kelowna Bog (site 7, Fig. 7.16), but occurring there between 8.4 and 6.6 ka. Alley's geomorphic studies corroborate the palynological data, that is, evidence of eolian activity and dune formation in the Okanagan Valley during the early Holocene, followed by dune stabilization. In this region, as on the coast, Mazama volcanic tephra approximately dates the return to moister and cooler conditions.

King (1980) analyzed pollen, plant macrofossils, and aquatic mollusc remains from two lakes, Chilhil and Phair lakes, near Lillooet, British Columbia. Some of the best evidence for early Holocene xerothermic conditions is present at Chilhil Lake (site 8, Fig. 7.16) where pollen and seeds of "mudflat" plants such as Chenopodiaceae and *Rumex*, pulmonate snails, and seeds of shallow water aquatic plants indicate low lake levels between 8.0 and 6.6 ka. This phase is followed by one characterized by a significant reduction of mudflat indicators and hence of increasing lake levels. Phair Lake (site 9, Fig. 7.16) in the same general area, began accumulating sediments just prior to the Mazama eruption. The pollen sequence starts with fibrous shallow water peat with abundant Cyperaceae achenes and cattail (*Typha*) pollen, then gives way at about 5.7 ka to deeper water marls representing a permanent pond with higher water levels.

Hebda (1982) has documented evidence for climatic change on the basis of past changes in grassland distribution in the southern interior. Using the indicator value of non arboreal pollen, especially pollen frequency curves of *Artemisia* and Gramineae, he interpreted the period between 10 ka and 8 ka as a grassland maximum, with climate warmer and drier than present. This was followed, between 8 ka and 4.5 ka, by a "mesic grasslands period" reflecting a general trend to forest expansion and moister climate.

Late Holocene

Coastal region. Although the nature of early to middle Holocene climate is controversial, there is widespread agreement among geologists and paleobotanists that the late Holocene climate in coastal British Columbia was colder and wetter than during the early Holocene (Clague, 1981). The beginning of this cooling trend, however, appears to be time transgressive in the Cordillera. Geological evidence of neoglacial alpine glacier readvances in the Cordillera is fragmentary for the last 5 ka, but the period between 3.1 ka and 2.3 ka is well documented (Clague, 1981).

Paleobotanical indicators suggest that increasingly wet conditions commenced prior to 5 ka at some sites on the coast. For example, the western hemlock and cedar pollen frequencies rise in southwestern British Columbia at about the time of the Mazama ashfall (6.6 ka; Mathewes, 1973; Mathewes and Rouse, 1975; Heusser, 1983). Western hemlock and Sitka spruce frequencies begin their rise shortly after 7 ka at Bear Cove, northern Vancouver Island (Hebda, 1983). Both trends point to cooling and increasing moisture. At Bear Cove this cool, moist interval is followed, starting at about 4-3 ka, by the advent of modern hemlock-red cedar forest. In a similar vein, Hebda and Mathewes (1984) interpreted maximum values for red cedar at many coastal sites during the last 3-2 ka as marking a period of cooling and increased wetness — a period that in an adjacent area may have seen a fall of mean July temperature by 1-2°C (Heusser, 1977).

Paleoclimatic reconstructions relying on pollen-climate transfer functions, however, fail to show either late Holocene cooling or increased wetness. Mathewes and Heusser's (1981) data for Marion Lake (Fig. 7.17) suggest more or less constant conditions since 6 ka. Similarly perplexing is that reconstructed precipitation and temperature curves from a marine core in Saanich Inlet (British Columbia) also show little evidence of change during the Holocene (Heusser, 1983).

Interior. In the Okanagan region (site 7, Fig. 7.16) cooler and moister conditions began about 6 ka (Alley, 1976). Subsequent fluctuations appear to coincide with neoglacial stades. Hebda (1982), on the other hand, reviewed data that indicate moister conditions (although perhaps still warmer than present) occurred between 8 and 6 ka, with a further increase in moisture at approximately 4.5 ka. In southeastern British Columbia, Hazell (1979) showed a change from shrubby vegetation in the early Holocene to closed-canopy forest beginning 7 ka. Although he did not ascribe climatic significance to the appearance of western hemlock pollen at 4 ka, this increase of a moisture-demanding species probably signifies a climatic shift, in keeping with other data. Evidence from northern Washington shows that forests expanded between 5 ka and 4 ka, also probably a response to cooler and moister conditions (Hebda, 1982). This trend stands in sharp contrast to the record at Chalmers bog (site 10, Fig. 7.16; Fig. 7.12a) on the eastern edge of the Cordillera, which shows virtually no vegetation change over the last 6 ka (Mott and Jackson, 1982).

Studies on lake cores from the Peace River area (site 11, Fig. 7.16) in the northern interior region of British Columbia and Alberta (White and Mathewes, 1982; White, 1983) indicate increasingly more moist climate starting about 7 ka. At the height of this phase (5.5-3 ka), permanent ponds and lakes developed on the Alberta Plateau.

Detailed stratigraphic, palynological, and macrofossil studies in Chilhil and Phair lakes, in the Lillooet region (King, 1980) portray a complex post-Mazama ash fall climatic history. A trend to moister climate is indicated by marsh formation at Phair Lake and decreases of mudflat or prairie species at about 7 ka. By 5.7 ka, a permanent waterbody had formed and marl was being deposited. A subsequent change to gyttja at 2 ka possibly signals another rise in water level. Pollen assemblages like those of the present, with reduced grass and sage, appear around 4 ka coinciding with an increase in numbers of pelecypods and decrease of aquatic gastropods, changes perhaps also related to increasing water levels. At Chilhil Lake, the reductions in grass and

Figure 7.17. Abbreviated pollen diagrams for (a) Marion Lake, British Columbia (from Mathewes, 1973); (b) Moore Lake, Alberta; (c) Sleet Lake, Yukon (from Spear, 1983); (d) Lac des Roches Moutonnées, Quebec (from McAndrews and Samson, 1977); (e) Baird Inlet Lake, Ellesmere Island, Northwest Territories (from Hyvarinen, 1985); and (f) Ellice Bog, Ontario (Anderson, 1971).

sage pollen are dated at 4.4 ka but water level changes based on aquatic plant macrofossils and molluscs are not synchronous with those at Phair Lake.

White and Mathewes (1982) and Vance et al. (1983) reviewed other evidence from Alberta and northeastern British Columbia that indicates a period of maximum warmth and aridity prior to about 6 ka followed by different episodes of cooling. Vance et al. (1983) place the end of the Hypsithermal at 4 ka, and the advent of modern forest conditions at 3 ka.

Alberta

Hansen (1949, 1952) was perhaps the first to recognize a warm arid Hypsithermal period in Alberta, while Moss (1955) invoked it as an explanation for the origin of the Peace River grasslands. Recent work, discussed below, shows that the Holocene paleoecological history of Alberta features a 3-4 ka period of warm, dry climate, during which the surface hydrology and vegetation of much of the province underwent significant changes.

Even without pollen evidence, radiocarbon dates on basal sediments from lakes and bogs in central Alberta indicate a complex history of lake formation and water level fluctuations. Lakes with early basal dates are either very deep or are located in the western boreal forest. Those with younger basal ages, e.g., 3-5 ka, are mostly from the climatically controlled aspen parkland ecotone in central Alberta. These findings suggest that a prolonged, severe drought must have characterized the earlier history of some areas of the province.

The pollen records from two basins, one now occupied by Moore Lake and the other by Lofty Lake provide some information on the timing and magnitude of this Holocene arid interval. Moore Lake (site 12, Fig. 7.16; Fig. 7.17b) is situated along the southern boreal forest ecotone in east-central Alberta. It is meromictic, with a maximum water depth of 30 m. A 9 m core of finely laminated sediment has been raised from the deepest part of the lake. A basal date of 11 300 ± 170 BP (GSC-2856) is associated with fossil pollen assemblages dominated by *Populus, Betula*, and *Picea* with trace amounts of *Pinus* (Fig. 7.17b). Apparently a late glacial boreal forest had already colonized this region prior to the development of the Moore Lake basin. Spruce declines upwards in the profile reaching minimum values at 6.8 m. On the other hand, Gramineae and *Artemisia* pollen gradually increase in abundance. *Ruppia* pollen is limited to the interval 6.2 to 8.1 ka, while charcoal fragments reach their greatest frequencies in the dated interval 9250 ± 80 BP (GSC-2858) to 5850 ± 80 BP (GSC-3870) after which they all but disappear in rusty coloured sediments. Spruce becomes more abundant about 5.8 ka, whereas grass and *Artemisia* decline and *Ruppia* disappears from the pollen record. After about 5.8 ka, charcoal fragments decline to low levels while pyrite increases.

The changes in frequency of pollen, charcoal, and pyrite between 9.2 and 5.8 ka can best be explained as a result of regional drought. Grassland vegetation increased at the expense of boreal forest, and, judging from the abundance of charcoal, the spread of grasslands and their maintenance was aided by increased frequency of fire in the drought-stressed bordering forest. The higher evaporation stress on Moore Lake undoubtedly caused an increase in dissolved solids, eventually making the water favourable for *Ruppia occidentalis* (widgeon grass or ditch grass), an aquatic plant presently limited to saline lakes in Alberta and elsewhere. A peak of epipelic diatoms in the 9.2-5.6 ka interval (Fig. 7.17) provides some measure of the extent of water level depression. These diatoms are attached to submerged aquatic vegetation; hence, their dominance indicates that the bottom of Moore Lake was within the phototrophic zone, that is, the water depth had dropped to about 12 m or less, or 18 m below its present level.

Some time before 5.9 ka there was a reversal of the above listed trends. Moore Lake waters freshened, fires were less frequent, grassland gave way to spruce forest, submerged aquatic vegetation died off, planktonic diatoms dominated, and anaerobic conditions returned to the stratified bottom waters (once again resulting in conditions favourable for precipitation of pyrite by anaerobic bacteria). Only an increase in precipitation and/or decrease in summer temperatures could account for these changes.

Lofty Lake, located 100 km west of Moore Lake (site 13, Fig. 7.16), displays a pollen record of increasing prairie influence between 9.2 and about 6 ka (Lichti-Federovich, 1970). Furthermore, rising frequencies of Cyperaceae, *Typha, Myriophyllum*, and *Ruppia* indicate shallow, saline, water conditions. A core from what is presently the deepest part of Lofty Lake shows that the lake largely dried up between about 8.7 and 6.3 ka. At this time, it was reduced to a shallow pond surrounded by a wetland soil, with prairie or open forest on the uplands. Water and core depth calculations indicate at least an 8 m drop in water level. By 6.3 ka lake levels had risen, resulting in algal gyttja being deposited over the paleosol.

Detailed pollen studies of other lakes in central Alberta confirm the pattern shown from Moore and Lofty lakes — the existence in central Alberta of a period of regional aridity, beginning as early as 9.2 ka and terminating by 5.8 ka or slightly later (Lichti-Federovich, 1970; Vance et al., 1983). At the peak of this Holocene arid period deep lakes in central Alberta became shallow saline ponds, some supporting ditch grass (*Ruppia*), while shallow ponds dried to salty pans covered with glasswort (*Salicornia*).

River systems were also affected by these climatic changes. Floodplain occurrences of Mazama ash (dated 6.6 ka) in southern Alberta indicate a period of stability, nondeposition, and soil formation prior to deposition of the ash (Dormaar, 1983; Waters and Rutter, 1983). With a return to a wetter climate, some time after the Mazama ash fall, floodplain aggradation resumed.

Paleohydrological changes during the early to middle Holocene strongly affected the flora and fauna of Alberta. Prehistoric human cultures were probably also influenced. For example, it can be assumed that dwindling surface waters would result in reduction of fish resources, and expansion of grasslands must have affected the supply of woody fuel. Opposing these negative effects would have been the more favourable conditions for production and hunting of large grazing ungulates. A more tangible impact of early Holocene aridity is its effect on the archeological record with respect to site preservation and visibility. The archeological record of Alberta may prove to have a hiatus between 8.5 and 5.5 ka. But if so, it may be due to a bias in the record. Shoreline erosion due to a subsequent rise in lake levels would have destroyed or inundated many archeological sites, while renewal of floodplain development that followed the arid phase probably buried others.

The effects of Holocene aridity are not well demonstrated within the boreal forest, perhaps for ecological reasons elucidated by Ritchie (1981). Evidence has been demonstrated for the existence of arid conditions 8 ka in the Crowsnest Pass region, Alberta (Driver, 1978). Farther north in the Tonquin Pass of Jasper National Park alpine treeline advanced beyond its present position between 8.7 and 4.3 ka (Kearney and Luckman, 1983). This clearly indicates that an increase of temperature was a component of the climatic change that brought on aridity in central and southern Alberta.

Central Canada

For purposes of this discussion central Canada comprises Saskatchewan, Manitoba, and Ontario west of Lake Superior. Southern Manitoba, in particular, was an area of considerable early pollen work by Ritchie (1964, 1967, 1969), Ritchie and Hadden (1975), and Ritchie and Lichti-Federovich (1968). Studies in Saskatchewan were carried out mainly by Ritchie and de Vries (1964), Mott (1973), and Wilson (1984). Ritchie (1976) has summarized the Late Wisconsinan and Holocene vegetation from about 20 pollen profiles from the region. Additional discussion on pollen stratigraphy in this region is also presented in Ritchie (this chapter, 1989).

Early spruce forests gave way to prairie grassland by 10 ka at Herbert (site 14, Fig. 7.16) and by 10.6 ka at Hafichuk (site 15) in southern Saskatchewan, by 10.2 ka at Russell (site 16) in southern Manitoba, and slightly later (ca 9.5 ka) to the north. Grassland is differentiated on the basis of high percentages of herbs, principally *Artemisia*, Chenopodiineae, Gramineae, and Ambrosieae as illustrated in the profile from Riding Mountain (site 17, Fig. 7.16; Fig. 7.12d), Manitoba. North of roughly 52°N, grassland was preceded by a narrow belt of woodland dominated by birch, possibly with aspen. At Hayes Lake (site 18, Fig. 7.16), western Ontario, open, mixed woodland dominated by jack pine and poplar moved into and replaced spruce forest at 9.2 ka; the mixed woodland vegetation persisted until about 3.6 ka and was succeeded by spruce and fir forests (McAndrews, 1982).

The replacement of spruce by grassland in the southern parts of Manitoba and Saskatchewan and by jack pine-dominated woodland in western Ontario is attributed to a warmer and drier climate supported by an increase in frequency of fires. Warm, dry conditions apparently peaked between about 8 ka and 6 ka, resulting in expansion of prairie vegetation beyond its present range and reduction of water levels in some of the more southerly lakes. This interval of maximum warmth correlates with prairie expansion in the Midwest United States and the beginning of the Prairie Peninsula as a discrete floristic area in Illinois (King, 1981).

The incursion of pine (predominantly *Pinus banksiana*), dated at 6 ka at Cycloid Lake (Mott, 1973) and at Reindeer Lake (Ritchie, 1976) (sites 19, 20, Fig. 7.16), is thought to have taken place in response to increased frequency of fires at this time. A subsequent southward shift of the boreal forest and reinvasion of the grasslands by spruce, pine, birch, oak, and alder between about 5.5 and 2 ka is probably a response to a cooler and (or) moister climate.

Northern Canada

Holocene pollen records, vegetation changes, and climatic trends in northwest Canada have been discussed in recent papers by Ritchie (1977, 1982), Ritchie and Hare (1977), Matthews (1980), Cwynar (1982), Ritchie and Cwynar (1982), Ritchie et al. (1983), Spear (1983), and MacDonald (1983). A comprehensive review of data up to approximately 1983 is presented in Ritchie (1984a). The Holocene vegetational history of neighbouring Alaska is reviewed in Ager (1983).

At Sleet Lake (site 24, Fig. 7.16; Fig. 7.17c), on the Tuktoyaktuk Peninsula, Northwest Territories, influx and percentage pollen profiles supplemented by macrofossils provide perhaps the best evidence of an early Holocene warming event in northwest Canada (Spear, 1983, Ritchie, 1984a) Spruce influx increases sharply starting before 10 ka to maximum values at 9 ka, after which influx declines gradually to the present. The prime climatic signal, based mainly on the behaviour of spruce, is one of rapid warming, starting at about 12 ka, followed by a period of warmer-than-present climate, which lasted from approximately 10 to 4.5 ka. Although cooling started as early as 8 ka, it was not until 4.5 ka that the influx of spruce pollen falls to existing levels.

Farther west at Hanging Lake (site 23, Fig. 7.16), northern Yukon Territory (Cwynar, 1982), the early Holocene thermal maximum dates from 11.1 to 8.9 ka. Total pollen influx at this site and at other lakes in the Mackenzie Delta area (Ritchie et al., 1983) rises sharply at 10 ka. This feature, plus occurrence of certain plant macrofossils of indicator plant species, appears to coincide with the early Holocene Milankovitch thermal insolation maximum at 10 ka (Ritchie et al., 1983).

The sequence of Holocene climatic changes recorded in northwest Canada differs from that documented in north-central Canada and the eastern Arctic. Ice persisted in the Keewatin and New Quebec-Labrador regions until the middle Holocene, thereby retarding warming in those regions. Deglaciation was followed immediately at many sites by organic deposition and immigration of trees and shrubs implying that the early Holocene temperature gradient was especially steep in this region. For example, high *Picea* pollen percentages at the base of Lynn and Ennadai lakes (sites 21, 22, Fig. 7.16) in northern Manitoba-Keewatin region date 6.5 ka and 5.8 ka, respectively (Nichols, 1967). *Larix* and *Alnus crispa* dominate the earliest assemblages at Lac Delorme (site 29, Fig. 7.16; Fig. 7.14) in the newly uncovered parts of central New Quebec (Richard et al., 1982). An *Abies* zone precedes abrupt increases in birch and spruce near Sept-Îles (site 32, Fig. 7.16), Quebec (Mott, 1976) and in spruce at Eagle Lake (site 30, Fig. 7.16; Fig. 7.14), southeast Labrador (Lamb, 1980), whereas *Populus* precedes spruce at the Bereziuk site (site 28, Fig. 7.16; Fig. 7.14), west-central Quebec (Richard, 1979). The tree and shrub-dominated assemblages of the early Holocene are especially apparent in the abbreviated pollen diagrams for Lac Delorme, Bereziuk, and Eagle Lake (Fig. 7.14). Spruce percentages generally increase earlier to the southeast and southwest of New Quebec than in New Quebec and Labrador. In contrast with regions farther west, total pollen influx in the New Quebec-Labrador region, represented by a profile from Lac des Roches Moutonnées,

Quebec (McAndrews and Samson, 1977; site 27, Fig. 7.16; Fig. 7.17d) did not peak until about 4 ka. This was a reflection of the northward migration of spruce-dominated forest in response to regional climatic warming.

Other paleoenvironmental data from northern Canada provide additional evidence in support of a middle Holocene Hypsithermal event. Treeline expanded between 5.5 ka and 3.5 ka in Keewatin (Nichols, 1967) and as early as 4.5 ka in New Quebec-Ungava region (Gagnon and Payette, 1981). The presence of podzolic paleosols in and near the forest/tundra ecotone attests to northward migration of forest and hence climatic warming during the middle Holocene. The record of peat development is in some cases also an index of climatic amelioration (Moore and Bellamy, 1974). Dates for maximal peatland development in New Quebec fall mainly in the interval 5.5 to 3.8 ka (Payette, 1984).

Several types of evidence mark the subsequent downturn of climate. Spruce charcoal dating 3.5 ka in a podzol from southwest Keewatin (Sorenson, 1977) is thought to denote a climatic deterioration resulting in destruction by fire of forests which had been rendered incapable of reproducing when cold, dry Arctic air moved south.

Evidence of increased frequencies of fire and eolian activity start at about 5 ka in the forest-tundra region, northwest Quebec (Filion, 1984). Increases in minerogenic sediment represented by profiles of K, Mg, and Ca in Moraine Lake (site 31, Fig. 7.16), Labrador (Engstrom and Wright, 1984), after 4 ka, possibly mark increased erosion under a more open lichen woodland environment.

In the Canadian High Arctic, abundant driftwood on raised beaches, dating to between 6.5 and 4.5 ka, is attributed to the presence of more open water than at any other time in the Holocene (Blake, 1972). These two millennia of reduced ice conditions coincide with a middle Holocene warm interval elsewhere in northern Canada (Blake, 1972). Andrews et al. (1981) estimated that July mean temperatures rose to well above present values for at least the interval 5.5 to 2.5 ka. Accelerated growth of moss peat started at approximately 6.5 ka on Carey Islands (site 26, Fig. 7.16), northwest Greenland somewhat earlier than in adjacent areas (Brassard and Blake, 1978). The end of this phase occurred between 4.5 ka and 4 ka.

Pollen stratigraphy from Baird Inlet Lake (site 25, Fig. 7.16; Fig. 7.17e), east-central Ellesmere Island (Hyvarinen, 1985) shows an early Holocene succession starting with *Oxyria*-dominated assemblages, then a sequence dominated by *Salix* and Ericales, and finally to a late Holocene phase characterized by an increase in bare ground and fell-field indicators. This last phase originates from about 4 ka and is interpreted by Hyvarinen to signify deteriorating climate in response to post-Hypsithermal cooling.

Southeast Canada

This region comprises northeastern and southern Ontario, southern Quebec, and Atlantic Canada and possesses the highest concentration of palynologically studied sites in Canada (Mott, 1975; Mott and Farley-Gill, 1978, 1981; Savoie and Richard, 1979; Anderson, 1980; Richard, 1980, 1981a; Terasmae, 1980; Labelle and Richard, 1981; McAndrews, 1981; Webb et al., 1983; Warner et al., 1984b). The majority of sites are in southern Quebec; fewer sites occur in southern Ontario and in the Maritime Provinces. Areas clearly deficient in pollen stratigraphic sites are central Quebec between the Gulf of St. Lawrence and James Bay, and mainland Newfoundland. Anderson (1985) presented a compilation of all known late Quaternary pollen records and a review of Holocene vegetation and climate history.

Pollen influx variations are an important criterion for deciphering the Holocene climatic history of the region. Figure 7.18 summarizes influx data for fourteen sites extending from northern Ontario to Newfoundland. Total pollen influx rates are low (less than 5000 grains/($cm^2 \cdot a$) prior to 10 ka, but they increase several fold at each site between 10 and 9 ka. The more obvious increases take place at Lac Colin, Lac Ramsay, Basswood Road and Sugarloaf lakes (sites 43, 40, 45, 49, Fig. 7.16; Fig. 7.18a, b). At this time, spruce or spruce and poplar were being replaced by pine-dominated pollen assemblages at Silver Lake (site 48, Fig. 7.16), Basswood Road Lake, and Lac Ramsay, and other lake sites in eastern Ontario, and by pine and birch at Lac Colin and other lakes in Quebec south of St. Lawrence River. During this interval north of St. Lawrence River, shrub-dominated assemblages gave way to spruce and poplar at Marcotte Lake and Lac à l'Ange (sites 41, 42, Fig. 7.16; Fig. 7.18b). These upward shifts in total pollen influx imply an increase in overall vegetation productivity, which is almost certainly correlated with climatic warming. Hence the influx variations mark the beginning of a progression towards Hypsithermal conditions. Maximum influx rates were achieved at most sites especially Lac Colin, Perch Lake (site 39, Fig. 7.16; Fig. 7.18c) and MacLaughlin Pond (site 47, Fig. 7.16; Fig. 7.18a), between about 8 and 4 ka, corresponding with the period of peak warmth.

Pollen records from eastern Ontario and south and south-central Quebec indicate that the vegetation had changed dramatically over this period. Birch populations expanded northward into southern Quebec (Webb et al., 1983), and white pine and birch spread into south-central Quebec. By about 7 ka hemlock and northern hardwoods were widespread. Hemlock declined at approximately 4.8 ka throughout its range (Davis, 1981a) and was replaced mainly by maple, beech, and elm. Hemlock reappeared shortly after 4 ka, but the lower hemlock percentages in most late Holocene pollen diagrams from southeastern Canada indicate that hemlock did not regain its previous level in the mixed forest.

Hypsithermal temperatures in the range 1° to 2°C above modern temperatures in the northern Great Lakes region (McAndrews, 1981) resulted in the expansion of the

Figure 7.18. Total pollen influx (a, b) and spruce (*Picea*) pollen influx (c) profiles for fourteen lakes in eastern Canada. Basswood Road Lake data is from Mott (1975), MacLaughlin Pond from Anderson (1985), Sugarloaf Lake from Macpherson (1982), Silver Lake from Livingstone (1968), Teagues and Roulston lakes (R.J. Mott, unpublished), Lac Marcotte and Lac à l'Ange from Labelle and Richard (1981), Lac Colin from Mott (1977), Lac Ramsay from Mott and Farley-Gill (1981), Perch Lake from Terasmae (1980) and Terasmae and McAtee (1979), Nina and Jack lakes from Liu (1982), and Lac Yelle from Richard (1980).

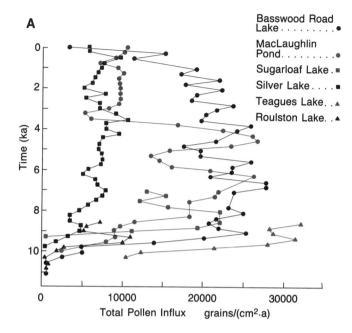

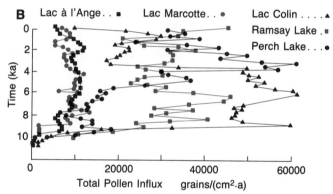

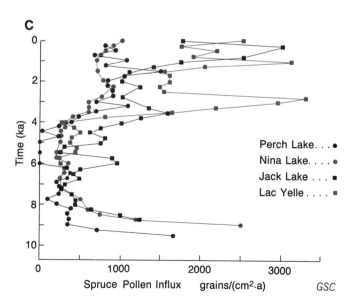

ranges of certain plant taxa and movements of ecotones across forest boundaries. For example, *Pinus strobus* grew about 100 to 120 km north of its present limit (Terasmae and Anderson, 1970; Liu, 1982) and *Thuja occidentalis* expanded its range in northern Ontario between 7 and 5 ka (Liu, 1982). *Nyssa* (most likely *N. sylvatica*) and *Decodon verticillatus* also had northerly extended ranges at this time, coinciding with maximum percentages of white pine (Warner et al., 1984b).

Stratigraphic changes or hiatuses in sediment and pollen records from southwestern Ontario relate to middle Holocene hydrological fluctuations in depositional basins. Transformation from lacustrine marl to peat between about 8 and 7 ka at Gage Street Bog (site 37, Fig. 7.16) at Kitchener, Ontario (Anderson, 1982; Schwert et al., 1985) and at 4.2 ka at Ballycroy Bog (site 38, Fig. 7.16) near Bolton, Ontario (Anderson, 1971) is attributed to lowering groundwater levels at these sites under the influence of a warmer and drier climate during the Hypsithermal interval. In fact, preliminary evidence from Ellice Bog (site 36, Fig. 7.16; Fig. 7.17f), near Stratford, Ontario indicates that peat growth may have slowed down or possibly ceased during the Hypsithermal. The *Pinus* maximum and first peak in *Tsuga* that characterize most lake and bog sites in southwestern Ontario from about 10 to 4 ka (Mott and Farley-Gill, 1978; McAndrews, 1981; Turner et al., 1983) are missing in the pollen stratigraphy at Ellice Bog (Fig 7.17f). *Picea* is replaced immediately by increases in *Tsuga, Acer, Fagus,* and *Ulmus; Pinus*, on the other hand, is insignificant (less than 30% maximum). Cyperaceae and Ericaceae show contrasting trends across this interval (Cyperaceae declines to a minimum and Ericaceae increases to peak values). It is tempting to relate the decline in Cyperaceae to a real decrease or cessation in peat growth due to a reduction of the bog water table at this time. A drier bog surface would have favoured the encroachment of less mesic-loving plants such as heath plants (Ericaceae) and *Salix*. A subsequent rise in water table after about 4 ka may have altered the bog habitat, resulting in the commencement of sedge growth again.

Insect faunal changes documented in nearby Gage Street Bog (Schwert et al., 1985) provide additional evidence in support of a period of maximum warmth in southwestern Ontario. The appearance at approximately 8.4 ka of insect taxa that are presently characteristic of the lowlands of east-central United States marks the onset of warmer-than-present climatic conditions. Inferred mean July temperatures for the area are similar to other estimates postulated for the same period in northern Ontario (McAndrews, 1981; Liu, 1982), New England (Davis et al., 1980), and across the Midwest of North America (Bartlein et al., 1984).

Influx estimates for spruce pollen are given for four sites in the northern Great Lakes-James Bay region (Fig. 17.18c). Spruce influx is generally low between 8 ka and 4 ka and increases at least two-fold at or shortly after 4 ka. These increases are especially pronounced at Nina Lake and Lac Yelle (sites 35, 33, Fig. 7.16). A date of 3450 ± 60 BP (WAT-809) occurs just above the rise in spruce influx at Jack Lake (site 34, Fig. 7.16), whereas at Nina Lake (Fig. 7.18c), the spruce rise is dated at 3960 ± 90 BP (WAT-811; Liu, 1982). The sudden increases in spruce influx may be explained in terms of an increase in the frequency of spruce trees in the vegetation at these times, and presumably reflect a southward shift of the spruce ecotone in response to post-Hypsithermal cooling.

Summary

An overall early Holocene trend to warmer-than-present climatic conditions began prior to 10 ka in northwest Canada and southern coastal British Columbia, from about 10 ka in interior British Columbia and throughout the southern Great Plains, by about 9.2 ka in the northern Great Plains, and between 10 and 9 ka at many sites in southeastern Canada. Peak xerothermic conditions appear to have been achieved earliest in the northwest (ca. 10 ka), between 9 or 8.5 and 7 ka in coastal and interior British Columbia, from about 9 to 6 ka in the Great Plains, and between approximately 8 and 4 ka in southeastern Canada. An abundance of driftwood dating between 6.5 and 4.5 ka in the Arctic Archipelago coincides with a middle Holocene period of maximum warmth throughout northeastern Canada. Climatic deterioration to conditions like those of today was in evidence by about 4.5 ka in British Columbia, the northwest, and central Canada, and at or shortly after 4 ka in southeast Canada, New Quebec-Labrador region, and the eastern Arctic, although gradual cooling and increased precipitation are inferred for some sites prior to 5 ka.

The time transgressive trend of the Hypsithermal interval is apparent especially for the lower boundary, whereas a certain degree of synchroneity prevails for the upper boundary. In view of the often contradictory evidence, the boundaries should perhaps be viewed as transitional in nature on a vegetative, faunal, and climatic basis.

POLLEN ANALYSIS OF VARVED SEDIMENT AT CRAWFORD LAKE, ONTARIO: EVIDENCE OF INDIAN AND EUROPEAN FARMING

J.H. McAndrews and M. Boyko-Diakonow

Introduction

Crawford Lake is part of a conservation area in Ontario administered by the Halton Region Conservation Authority. It is the site of a former Indian village which has been partly reconstructed. A visitor centre is maintained for public education.

Unlike most Canadian lakes, Crawford Lake is depositing annual layers of sediment called varves. These varves provide a precise chronology for fossil pollen analysis and the vegetation history of the area around the lake. Since 1000 AD two intervals of agriculture have been detected: prehistoric Indian farming 1360-1660 (Byrne and McAndrews, 1975; Finlayson and Byrne, 1975) and European farming beginning in 1820 (Boyko, 1973; McAndrews, 1976).

McAndrews, J.H. and Bokyo-Dikonow, M.
1989: Pollen analysis of varved sediment at Crawford Lake, Ontario: evidence of Indian and European farming; in Chapter 7 of Quaternary Geology of Canada and Greenland, R.J. Fulton (ed.); Geological Survey of Canada, Geology of Canada, no. 1 (also Geological Society of America, The Geology of North America, v. K-1).

Geological setting

Crawford Lake is located 65 km southwest of Toronto (Fig. 7.19) at an elevation of 279 m; it has a surface area of 2.4 ka. At its deepest place 24 m of water overlies 4.5 m of Holocene organic sediment that in turn overlies an unknown thickness of glacial sand of Late Pleistocene age. Cliffs of Silurian Amabel dolostone up to 6 m high nearly surround the lake and extend 2 m below the surface. Beneath this massive, flat-lying, erosion-resistant dolostone are the relatively thin bedded limestone, dolostone, sandstone, and shales of the Silurian Clinton and Cataract groups that together are 26 m thick; these overlie the Ordovician Queenston shale. Thus, Crawford Lake lies in a small, deep basin largely eroded from rocks of the Clinton and Cataract groups.

The origin of the Crawford Lake basin is obscure. The oldest sediment, the basal sand, is at least 10.5 ka old, as dated by abundant spruce pollen, but no older than the Ontario lobe glacial advance of the Port Bruce stadial of about 15 ka (Karrow, 1983). During the Port Huron stadial between 14 and 13 ka, the Ontario lobe readvanced westward to the Waterdown Moraine located on top of the Niagara Escarpment 1 km east of Crawford Lake. Meltwater was channelled along the ice front, over the location of Crawford Lake, over a now-abandoned waterfall and through a gorge before terminating in proglacial Lake Warren in the Erie Basin (Chapman and Putman, 1984). A date of 13 660 ± 370 BP (WAT-951) on moss in pond sediment in the channel provides a minimum age for cessation of meltwater flow. One possibility for the formation of Crawford Lake is Late Pleistocene solution of limestone, but this is unlikely because most of the rock that would have to have been removed to form the basin is nearly insoluble dolostone, shale, and sandstone. A second possibility is that a catastrophic rush of water along the spillway hydraulically mined the rock and carried it away downstream. In any event, the lake basin was only partly filled with outwash sediment and must therefore date from the end of meltwater flow or slightly later.

Because the lake is relatively small and deep, it is meromictic — the warm, oxygenated surface water does not

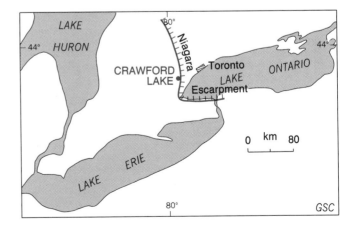

Figure 7.19. Map showing location of Crawford Lake (after Boyko-Diakonow, 1979).

mix seasonally with the relatively dense, cold, and oxygen-poor bottom water. This excludes benthic animals that ordinarily disturb the sediment surface, and thus seasonal pulses of sediment are preserved as distinct layers (Boyko-Diakonow, 1979). In June, when the surface water becomes supersaturated with $CaCO_3$, calcite crystals form, fall to the bottom and collect as a thin, white sediment layer (Ludlam, 1969). In October, deep circulation of oxygen-rich water causes a die off of anaerobic bacteria and the precipitation of pyrite to form a black layer of sediment (Dickman, 1979) on top of the white spring layer. Each couplet of contrasting white and black layers represents a varve that ranges from 0.5 to 2.0 mm thick.

Varves and pollen analysis

In 1971 the upper 83 cm of varved sediment was collected with a freezing tube sampler (Swain, 1973) consisting of a weighted tube, plugged at the base and filled with dry ice. The tube was lowered on a rope into the sediment, allowed to remain in the sediment for 20 minutes and then pulled up. The frozen rind of varved sediment, 2 cm thick, was taken to the laboratory, where the varves were counted and sampled for fossil pollen analysis.

The varves were sampled in units of 5 or 10 years from 1970 back to 1000 AD. Fossil pollen was concentrated by digesting the mud samples with acids; 400 to 1000 pollen grains were identified in each sample. A percentage pollen diagram was constructed from 91 intervals; only the more abundant and significant tree and herb pollen are shown in Figure 7.20.

Indian farming

Two periods of agriculture are marked by relatively abundant herb pollen, particularly those of weedy plants and corn. The period of Indian agriculture begins in 1360 AD with the appearance of corn (*Zea Mays*), weedy grasses and purslane (*Portulaca oleracea*); it ends at 1660.

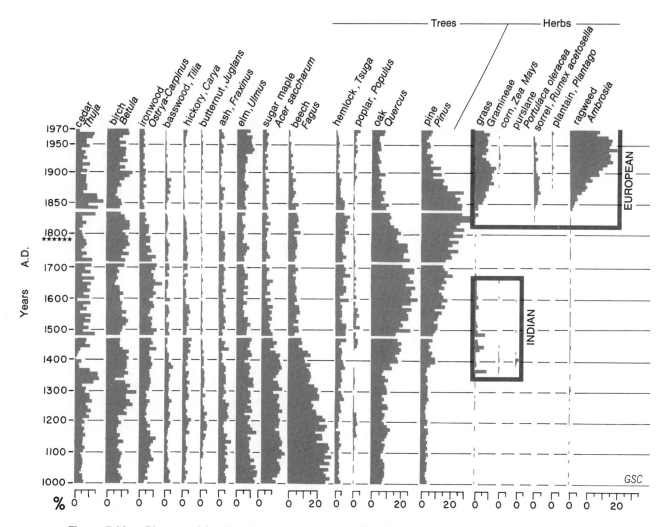

Figure 7.20. Diagram of fossil pollen percentages from Crawford Lake showing Indian farming from 1360-1660 and European farming beginning in 1820. The stars indicate a change in the time scale: five-year intervals above and 10 year intervals below. The intervals 1470-1480, 1710-1720, and 1825-1830 were not analyzed.

Archeological excavations in southern Ontario have shown that agriculture was introduced about 500 AD (Trigger, 1976). The Crawford Lake pollen record suggests that Indians farmed within a few kilometres of the lake for 300 years beginning in 1360. An archeological survey located several village sites, the nearest lying 200 m northwest of the lake. Excavation of this Crawford Lake village placed it in the Middleport substage of the late Ontario Iroquois tradition, about the time of the first appearance of corn and weed pollen. Flotation of village midden soil yielded charred seeds of corn, bean, and purslane, which established that it was a village of farmers. There was evidence of six longhouses, and their dimensions suggest a population of 450 people. Accounts of French explorers and missionaries, beginning in 1615, indicate that villages were occupied for 10 to 20 years before the inhabitants moved, built new villages, and cleared new forest land for corn fields. Further survey and excavation in the vicinity of Crawford Lake has revealed earlier, contemporaneous, and later farming villages.

Indian farming in southern Ontario ceased in 1660 because of the dispersal of the Ontario Iroquois through intertribal warfare. There was virtually no further farming until European farmers arrived in the early 1800s. This agrees with the pollen stratigraphy.

Indian agriculture proceeded by the clearing of the forest through cutting and burning. The pollen diagram shows a decline in climax forest trees, particularly sugar maple and beech. These newly cleared fields were initially fertile, but the fertility declined with continuous cropping, and the fields were abandoned after several years. New forest was cleared for new fields and the cycle repeated. The pollen diagram indicates that these abandoned, weedy fields were not reoccupied by climax maple and beech, but were invaded by poplar, oak (probably red oak), and white pine; because of its greater longevity, pine succeeded the poplar and oak.

European farming

The release of crown land for European settlement (patenting) and subsequent farming followed land surveying. Surveying began in 1790 in townships bordering Lake Ontario and spread northward. Land around Crawford Lake was surveyed in 1813 and 1819, but patenting within 2 km of Crawford Lake ranged from 1822-1864; the lot containing Crawford Lake was patented in 1841 (Boyko, 1973).

The pollen record reflects this gradual forest clearance and farming, beginning in 1820, with the appearance of sorrel pollen. Sorrel (*Rumex acetosella*) pollen peaks in the late 1800s, perhaps reflecting the abundance of this perennial plant in pioneer pastures. The next indicator of farming, grass pollen, begins to proliferate in 1830; some of these pollen grains are relatively large, suggesting cultivated grains such as wheat. Ragweed, a native annual plant, begins to expand in 1840 and peaks in the mid-1900s, probably in response to intense mechanized cultivation associated with corn farming; corn pollen reappears in the record in 1870. The last interesting weed pollen, plantain (mostly *Plantago lanceolata*, an introduced species), also appears in 1870.

With the expansion of weed and cultivated plant pollen, most forest tree pollen declined because of clearing and selective logging. Pine stumps are abundant in the modern forest around the lake, and pine-stump fences border nearby cultivated fields. On the other hand, the pollen record indicates that some trees such as cedar, birch, ironwood (*Ostrya* and *Carpinus*), elm, and poplar held their own or expanded with European farming.

The pollen record indicates that Indian farming was less intense than European farming because of the lack of mechanization and domestic grazing animals. Indian farming did, however, leave a long-term impact upon forest succession, namely the development of a poplar/oak/pine sere followed by pine dominance.

ACKNOWLEDGMENTS

R.W. Mathewes is indebted to B.G. Warner (University of Waterloo) for discussions and use of data from his PhD thesis on the Queen Charlotte Islands. Funding for Mathewes' work was provided by an NSERC operating grant and radiocarbon dates were provided by the Radiocarbon Dating Laboratory, Geological Survey of Canada.

R.J. Mott is grateful to A. de Vernal (Université du Québec à Montréal) for supplying the East Bay and Castle Bay diagrams and for allowing use of her Bay St. Lawrence diagram. J.H. McAndrews (Royal Ontario Museum) permitted reproduction of his Woody Cove diagram. Th/U dates on wood were supplied by C. Causse and C. Hillaire-Marcel (Université du Québec à Montréal). N.W. Rutter (University of Alberta) arranged for the amino acid racemization determinations on wood. H. Jetté (Geological Survey of Canada) ably assisted with pollen analysis and preparation of the manuscript. Comments by J.V. Matthews, Jr. and T.W. Anderson on the original manuscript are greatly appreciated by R.J. Mott.

The original research reported in the case history on the boreal forest was carried out in J.C. Ritchie's laboratory and supported by a NSERC grant A-6320. Ritchie is indebted to Les Cwynar (Memorial University, Cornerbrook) and Glen MacDonald (McMaster University) for useful discussions of boreal forest history and to Kate Hadden (Scarborough College) for technical assistance.

P.J.H. Richard acknowledges the students and staff of Laboratoire de paléobiogéographie et de palynologie, département de Géographie, Université de Montréal who contributed to large parts of the data reported in the case history on the patterns of post-Wisconsinan plant colonization in Quebec-Labrador. The laboratory is supported by NSERC (Canada) and FCAR (Quebec) grants to Richard. Richard is also grateful to J.V. Matthews, Jr., T.W. Anderson, and L.A. Barton (Geological Survey of Canada) for their rewriting of the translation of the original french manuscript.

Anderson and Matthews acknowledge R.J. Mott for providing unpublished pollen data for Roulston Lake and Teagues Lake, New Brunswick. They also thank all of the contributors for their patience in the preparation of this chapter. Finally they are pleased to acknowledge the help of L. A. Barton (Geological Survey of Canada) who was in charge of preliminary technical editing during preparation of the manuscript. They also thank Anne Morgan (University of Waterloo) and H.E. Wright, Jr. (University of Minnesota) who provided critical reviews of the chapter.

REFERENCES

Ager, T.A.
1983: Holocene vegetational history of western United States; in Late Quaternary Environments of the United States, The Holocene, H.E. Wright, Jr. (ed.); University of Minnesota Press, Minneapolis, p. 128-141.

Alley, N.
1976: The palynology and palaeoclimatic significance of a dated core of Holocene peat, Okanagan Valley, southern British Columbia; Canadian Journal of Earth Sciences, v. 13, p. 1131-1144.

Andersen, S.T.
1978: Local and regional vegetational development in eastern Denmark in the Holocene; Geological Survey of Denmark, Yearbook 1976, p. 5-27.

Anderson, T.W.
1971: Postglacial vegetative changes in the Lake Huron-Lake Simcoe District, Ontario, with special reference to Glacial Lake Algonquin; unpublished PhD thesis, Uviversity of Waterloo, Waterloo, Ontario, 246 p.

1980: Holocene vegetation and climatic history of Prince Edward Island, Canada; Canadian Journal of Earth Sciences, v. 17, p. 1152-1165.

1982: Pollen and plant macrofossil analyses on late Quaternary sediments at Kitchener, Ontario; in Current Research, Part A, Geological Survey of Canada, Paper 82-1A, p. 131-136.

1983: Preliminary evidence for late Wisconsinan climatic fluctuations from pollen stratigraphy in Burin Peninsula, Newfoundland; in Current Research, Part B, Geological Survey of Canada, Paper 83-1B, p. 185-188.

1985: Late-Quaternary pollen records from eastern Ontario, Quebec and Atlantic Canada; in Pollen Records of Late Quaternary North American Sediments, V.M. Bryant, Jr. and R.G. Holloway (ed.); American Association of Stratigraphic Palynologists Foundation, p. 281-326.

Anderson, T.W., Mathewes, R.W., and Schweger, C.E.
1989: Holocene climatic trends in Canada with special reference to the Hypsithermal interval; in Chapter 7 of Quaternary Geology of Canada and Greenland, R.J. Fulton (ed.); Geological Survey of Canada, Geology of Canada, no. 1 (also Geological Society of America, The Geology of North America, v. K-1).

Andrews, J.T., Davis, P.T., Mode, W.N., Nichols, H., and Short, S.K.
1981: Relative departures in July temperatures in northern Canada for the past 6000 yr; Nature, v. 289, no. 5794, p. 164-167.

Antevs, E.
1945: Correlation of Wisconsin glacial maxima; American Journal of Science, v. 243A, p. 1-39.

1948: The Great Basin with emphasis on glacial and postglacial times. III. Climatic changes and pre-white man; University of Utah Bulletin, v. 38, p. 168-191.

Ashworth, A.C. and Cvancara, A.M.
1983: Paleoecology of the southern part of the Lake Agassiz Basin; in Glacial Lake Agassiz, J.T. Teller and L. Clayton (ed.); Geological Association of Canada, Special Paper 26, p. 133-156.

Ball,T.F.
1983: Preliminary analysis of early instrumental temperature records from York Factory and Churchill Factory; in Climatic Change in Canada 3, Syllogeus 49, p. 203-219.

Banfield, A.W.F.
1962: The disappearance of the Queen Charlotte Islands' caribou; National Museum of Canada, Contribution to Zoology, Bulletin, v. 185, p. 40-49.

1974: The Mammals of Canada; University of Toronto Press, 438 p.

Bartlein, P.J., Webb, T., III, and Fleri, E.
1984: Holocene climatic change in the northern Midwest: pollen-derived estimates; Quaternary Research, v. 22, p. 361-374.

Birks, H.J.B.
1976: Late Wisconsinan vegetational history at Wolf Creek, central Minnesota; Ecological Monographs, v. 46, p. 395-429.

1981: The use of pollen analysis in the reconstruction of past climates: a review; in Climate and History, T.M.L. Wigley, M.J. Ingram, and G. Farmer (ed.); Cambridge University Press, Cambridge, p. 111-138.

Birks, H.J.B. and Birks, H.H.
1980: Quaternary Paleoecology; Arnold, London, 289 p.

Blake, W., Jr.
1972: Climatic implications of radiocarbon-dated driftwood in the Queen Elizabeth Islands, Arctic Canada; in Climatic Changes in Arctic Areas during the last Ten Thousand Years, Y. Vasari, H. Hyvarinen, and S. Hicks (ed.); Acta Universitatis Ouluensis, Series A, Sciential Rerum Naturalium, no. 3, Geologica no. 1., p. 77-104.

1983: Geological Survey of Canada radiocarbon dates XXIII; Geological Survey of Canada, Paper 83-7, 34 p.

Bobrowsky, P.T.
1982: The quantitative and qualitative significance of fossil and subfossil gastropod remains in archaeology; unpublished MA thesis, Simon Fraser University, Burnaby, B.C., 294 p.

Bombin, M.
1984: On information evolutionary theory, phytoliths, and Late Quaternary ecology of Beringia; unpublished PhD thesis, University of Alberta, Edmonton, Alberta, 152 p.

Bonnichsen, R. and Young, D.
1980: Early technological repertoires: bone to stone; Canadian Journal of Anthropology, v. 1, p. 123-128.

Bousfield, E.L.
1958: Freshwater amphipod crustaceans of glaciated North America; Canadian Field-Naturalist v. 72, p. 55-113.

Boyko, M.
1973: European impact on the vegetation around Crawford Lake in southern Ontario; unpublished MSc thesis, Department of Botany, University of Toronto, 115 p.

Boyko-Diakonow, M.
1979: The laminated sediments of Crawford Lake, southern Ontario, Canada; in Moraines and Varves, Ch. Schluchter (ed.); Balkema, Rotterdam, p. 303-307.

Brassard, G.R. and Blake, W., Jr.
1978: An extensive subfossil deposit of the arctic moss *Aplodon wormskioldii*; Canadian Journal of Botany, v. 56, p. 1852-1859.

Broecker, W.S. and Kulp, J.L.
1957: Lamont natural radiocarbon measurements IV; Science, v. 126, p. 1324-1334.

Brookes, I.A., McAndrews, J.H., and Von Bitter, P.H.
1982: Quaternary interglacial and associated deposits in southwest Newfoundland; Canadian Journal of Earth Sciences, v. 19, p. 410-423.

Brown, R.M., Andrews, H.R., Ball, G.C., Burn, N., Imahori, Y., and Milton, J.C.D.
1983: Accelerator ^{14}C dating of the Taber Child; Canadian Journal of Archaeology, v. 7, p. 233-237

Brown-Macpherson, J.
1967: Raised shorelines and drainage evolution in the Montreal Lowland; Cahiers de géographie du Québec, v. 23, p. 343-360.

Bryan, A.L.
1969: Early man in America and the late Pleistocene chronology of western Canada and Alaska; Current Anthropology, v. 10, p. 339-367.

Burns, J.A.
1980: The brown lemming, *Lemmus sibiricus* (Rodentia, Arvicolidae), in the late Pleistocene of Alberta and its postglacial dispersal; Canadian Journal of Zoology, v. 58, p. 1507-1511.

1982: Water vole *Microtus richardsoni* (Mammalia, Rodentia) from the Late Pleistocene of Alberta; Canadian Journal of Earth Sciences, v. 19, p. 628-631.

Byrne, R. and McAndrews, J.H.
1975: Pre-Columbian purslane (*Portulaca oleracea* L.) in the New World; Nature, v. 253, no. 5494, p. 726-727.

Campbell, M.
1980: Distribution patterns of Coleoptera in eastern Canada; Canadian Entomology, v. 112, p. 1161-1175.

Calder, J.A. and Taylor, R.L.
1968: Flora of the Queen Charlotte Islands, Part 1; Canada Department of Agriculture, Monograph, v. 4, p. 1-659.

Canada Department of Energy, Mines and Resources
1973: The National Atlas of Canada. Vegetation Regions; Map 45-46. Ottawa.

Causse, C. et Hillaire-Marcel, C.
1986: Géochimie des familles U et Th dans la matière organique fossile des dépôts interglaciaires et interstadiaires de l'est et du nord du Canada: potentiel radiochronologique; dans Recherches en cours, Partie B, Commission géologique du Canada, Étude 86-1B, p. 11-18.

Causse, C., Carro, O., et Hillaire-Marcel, C.
1984: Le déséquilibre U/Th dans la matière organique fossile, potentiel radiochronologique (résumé); Association Québécoise pour l'étude du Quaternaire, Cinquième Congrès, Sherbrooke, Québec, 4-7 octobre, 1984, p. 22-23.

Chamberlin, T.C.
1894: Glacial phenomena of North America, p. 724-775; in The Great Ice Age and its Relation to the Antiquity of Man (3rd edition), J. Geike (ed.); Stanford, London, 850 p.

Chapman, L.J. and Putman, D.F.
1984: The physiography of southern Ontario (3rd edition); Ontario Geological Survey, Special Volume 2, 270 p.

Churcher, C.S.
1968: Pleistocene ungulates from the Bow River gravel at Cochrane, Alberta; Canadian Journal of Earth Sciences, v. 5, p. 1467-1488.

1975: Additional evidence of Pleistocene ungulates from the Bow River gravels at Cochrane, Alberta; Canadian Journal of Earth Sciences, v. 12, p. 68-76.

Churcher, C.S. and Wilson, M.
1979: Quaternary mammals from the eastern Peace River District, Alberta; Journal of Paleontology, v. 53, p. 71-76.

Cinq-Mars, J.
1979: Bluefish Cave I: A late Pleistocene eastern Beringian cave deposit in the northern Yukon; Canadian Journal of Archaeology, v. 3, p. 1-32.

Clague, J.J.
1981: Late Quaternary geology and geochronology of British Columbia, Part 2. Summary and discussion of radiocarbon-dated Quaternary history: Geological Survey of Canada, Paper 80-35, p. 1-41.

Clarke, A.H. and Harington, C.R.
1978: Asian freshwater molluscs from Pleistocene deposits in the Old Crow Basin, Yukon Territory; Canadian Journal of Earth Sciences, v. 15, p. 45-51.

Clayton, L. and Moran, S.R.
1982: Chronology of the late Wisconsinan glaciation in middle North America; Quaternary Science Reviews, v. 1, p. 55-82.

Clifford, H.F. and Bergstrom, G.
1976: The blind aquatic isopod *Salmasellus* from a cave spring of the Rocky Mountains' eastern slopes, with comments on a Wisconsin refugium; Canadian Journal of Zoology, v. 54, p. 2028-2032.

Colinvaux, P.A. and West, F.H.
1984: The Beringian ecosystem; The Quarterly Review of Archaeology, p. 10-15.

Comtois, P.
1982: Histoire holocène du climat et de la végétation à Lanoraie, Québec; Journal canadien des sciences de la terre, vol. 19, p. 1938-1952.

Coppens, Y., Howell, F.C., Isaac, G.L., and Leakey, R.E.F. (editors)
1976: Earliest Man and Environments in the Lake Rudolf Basin; University of Chicago Press, Chicago, 656 p.

Courtin, P.J., Klinka, K., Heineman, J., Mitchell, W., and Green, R.N.
1981: Biogeoclimatic Units, Victoria-Vancouver; Sheet NTS 92 SE, British Columbia Ministry of Forests, Vancouver, scale 1:500 000.

Cwynar, L.C.
1978: Recent history of fire and vegetation from laminated sediment of Greenleaf Lake, Algonquin Park, Ontario; Canadian Journal of Botany, v. 56, p. 10-21.
1982: A late-Quaternary vegetation history from Hanging Lake, northern Yukon; Ecological Monographs, v. 52, p. 1-24.

Dahl, E.
1955: Biogeographic and geologic indication of unglaciated areas in Scandinavia during the ice age; Geological Society of America, Bulletin, v. 66, p. 1499-1520.

David, P.P. and Lebuis, J.
1985: Glacial maximum and deglaciation of Western Gaspé, Quebec, Canada; in Late Pleistocene History of Northeastern New England and Adjacent Quebec, H.W. Borns, Jr., P. LaSalle, and W.B Thompson (ed.); Geological Society of America, Special Paper 197, p. 85-109.

Davis, M.B.
1981a: Outbreaks of forest pathogens in Quaternary history: Proceedings, IV International Palynological Conference, v. 3, p. 216-227.
1981b: Quaternary history and the stability of forest communities; in Forest Succession: Concepts and Application, D.C. West, H.H. Shugart, and D.B. Botkin (ed.); Springer-Verlag, New York, p. 132-153.
1983: Quaternary history of deciduous forests of eastern North America and Europe; Annals of the Missouri Botanical Garden, v. 70, p. 550-563.

Davis, M.B. and Ford, M.S.
1982: Sediment focusing in Mirror Lake, New Hampshire; Limnology and Oceanography, v. 27, p. 137-150.

Davis, M.B., Spear, R.W., and Shane, L.C.K.
1980: Holocene climate of New England; Quaternary Research, v. 14, p. 240-250.

Dawson, G.M.
1880: Report on the Queen Charlotte Islands, 1878: Geological Survey of Canada, Report of Progress for 1878-1879, Part B, p. 1-239.

Dawson, J.W.
1855: Acadian Geology (first edition); Oliver and Boyd, Edinburgh, 388 p.
1893: The Canadian ice age: being notes on the Pleistocene geology of Canada, with special reference to the life of the period and its climatic conditions; W.V. Dawson, Montreal.

Dean, W.E., Bradbury, J.P., Anderson, R.Y., and Barnosky, C.W.
1984: The variability of Holocene climate change: evidence from varved lake sediments; Science, v. 226, no. 4679, p. 1191-1194.

DeBeaulieu, J.L. and Reille, M.
1984: A long upper Pleistocene pollen record from Les Echets, near Lyon, France; Boreas, v. 13, p. 111-132.

Delcourt, H.R., Delcourt, D.A., and Webb, T., III
1983: Dynamic plant ecology: the spectrum of vegetational change in space and time; Quaternary Science Reviews, v. 1, p. 153-175.

Delorme, L.D.
1968: Pleistocene freshwater Ostracoda from Yukon, Canada; Canadian Journal of Zoology, v. 46, p. 859-876.

Delorme, L.D. and Zoltai, S.C.
1984: Distribution of an arctic ostracod fauna in space and time; Quaternary Research, v. 21, p. 65-73.

Delorme, L.D., Easterby, S.R., and Duthie, H.C.
1984: Prehistoric pH trends in Kejimkujik Lake, Nova Scotia; Internationale Revue Gesamten Hydrobiologie, v. 69, p. 41-55.

Delorme, L.D., Zoltai, S.C., and Kalas, L.L.
1977: Freshwater shelled invertebrate indicators of paleoclimate in northwestern Canada during Late glacial times; Canadian Journal of Earth Sciences, v. 14, p. 2029-2046.

Dickman, M.D.
1979: A possible varving mechanism for meromictic lakes; Quaternary Research, v. 11, p. 113-124.

Dormaar, J.
1983: Aliphatic carboxylic acids in burned Ah horizons in Alberta, Canada as paleoenvironmental indicators; Canadian Journal of Earth Sciences, v. 20, p. 859-866.

Douglas, R.J.W. (editor)
1970: Geology and Economic Minerals of Canada (5th edition); Geological Survey of Canada, Economic Geology Report no. 1, Part A, 838 p.

Driver, J.C.
1978: Holocene man and environments in the Crowsnest Pass, Alberta; unnpublished PhD thesis, University of Calgary, Calgary, Alberta, 230 p.

Duthie, H.C. and Sreenivasa, M.R.
1971: Evidence for eutrophication of Lake Ontario from the sedimentary diatom succession; Proceedings, 14th Conference of Great Lakes Research, International Association for Great Lakes Research, p. 1-13.

Elson, J.A. and Elson, J.B.
1969: Phases of the Champlain Sea indicated by littoral mollusks; Geological Society of America, Bulletin, v. 70, p. 1596.

Emerson, D.
1983: Late glacial molluscs from Cooking Lake moraine, Alberta, Canada; Canadian Journal of Earth Sciences, v. 20, p. 160-162.

Engstrom, D.R. and Hansen, B.C.S.
1985: Postglacial vegetational change and soil development in southeastern Labrador as inferred from pollen and chemical stratigraphy; Canadian Journal of Botany, v. 63, p. 543-561.

Engstrom, D.R. and Wright, H.E., Jr.
1984: Chemical stratigraphy of lake sediments as a record of environmental change; in Lake Sediments and Environmental History, E.Y. Haworth and J.E.G. Lund (ed.); University of Minnesota Press, Minneapolis, Minnesota, p. 11-67.

Fedje, D.
1984: Archaeological investigations in Banff National Park — 1983; Archaeological Survey of Alberta, Occasional Paper no. 23, p. 77-95.

Filion, L.
1984: A relationship between dunes, fire and climate recorded in the Holocene deposits of Quebec; Nature, v. 309, no. 5968, p. 543-546.

Finlayson, W.D. and Byrne, R.
1975: Investigations of Iroquoian settlement at Crawford Lake, Ontario — a preliminary report; Ontario Archaeology, v. 25, p. 31-36.

Fladmark, K.R.
1975: A paleoecological model for Northwest coast prehistory; National Museum of Man, Mercury Series, Archaeological Survey of Canada, Paper 43, p. 1-328.
1979: Routes: Alternate migration corridors for early man in North America; American Antiquity v. 44, p. 55-69.

Fladmark, J.R. and Gilbert, R.
1984: The paleoindian component at Charlie Lake Cave (abstract); Canadian Archaeological Association, 17th annual meeting, Victoria, p. 5.

Flint, R.F. and Rubin, M.
1955: Radiocarbon dates of pre-Mankato events in eastern and central North America; Science, v. 121, p. 649-658.

Foscolos, A.E., Rutter, N.W., and Hughes, O.L.
1977: The use of pedological studies in interpreting the Quaternary history of central Yukon Territory; Geological Survey of Canada, Bulletin 271, 48 p.

Foster, J.B.
1965: The evolution of the mammals of the Queen Charlotte Islands, British Columbia; British Columbia Provincial Museum, Occasional Paper 14, p. 1-130.

Fritts, H.C.
1982: The climate-growth response; in Climate from Tree Rings, M.K. Hughes, P.M. Kelly, J.R. Pilcher, and V.C. LaMarche, Jr. (ed.); Cambridge University Press, London, p. 33-38.

Funder, S.
1979: Ice-age plant refugia in east Greenland; Palaeogeography, Palaeoclimatology, Palaeoecology, v. 28, p. 279-295.

Gagnon, R. et Payette, S.
1981: Fluctuations Holocène de la limite des forêts de mélèzes, Rivière aux Feuilles, Nouveau-Québec: une analyse macrofossile en milieu tourbeux; Géographie physique et Quaternaire, vol. 35, p. 57-72.

Gangloff, P.
1983: Les fondements géomorphologiques de la théorie des paléonunataks: le cas des monts Torngats; Zeitschrift für Geomorphologie, N.F., v. 47, p. 109-136.

Gilliland, R.L.
1982: Solar, volcanic, and CO_2 forcing of recent climatic changes; Climatic Change, v. 4, p. 111-131.

Grandtner, M.M.
1966: La végétation forestière du Québec méridional; Presses de l'Université Laval, Québec, 216 p.

Grant, D.R.
1971: Surficial geology, southwest Cape Breton Island, Nova Scotia; in Report of Activities, Part A, Geological Survey of Canada, Paper 71-1A, Part A, p. 161-164.
1972: Surficial geology of southeast Cape Breton Island, Nova Scotia; in Report of Activities, Part A, Geological Survey of Canada, Paper 72-1A, p. 160-163.
1977: Glacial style of ice limit, the Quaternary stratigraphic record and changes of land and ocean level in the Atlantic Provinces, Canada; Géographie physique et Quaternaire, v. 31, p. 347-360.
1989: Quaternary geology of the Atlantic Appalachian region of Canada; Chapter 5 in Quaternary Geology of Canada and Greenland, R.J. Fulton (ed.); Geological Survey of Canada, Geology of Canada, no. 1 (also Geological Society of America, The Geology of North America, v. K-1).

Grant, D.R. and King, L.H.
1984: A stratigraphic framework for the Quaternary history of the Atlantic Provinces, Canada; in Quaternary Stratigraphy of Canada — A Canadian Contribution to IGCP Project 24, R.J. Fulton (ed.); Geological Survey of Canada, Paper 84-10, p. 174-191.

Gryba, E.M.
1983: Sibbald Creek: 11,000 years of human use of the Alberta Foothills; Archaeological Survey of Alberta, Occasional Paper 22.

Guilbault, V-P.
1982: The pre-Late Wisconsinan foraminiferal assemblage of Bay St. Lawrence, Cape Breton Island, Nova Scotia; in Current Research, Part C, Geological Survey of Canada, Paper 82-1C, p. 39-43.

Guthrie, R.D.
1980: Bison and man in North America; Canadian Journal of Anthropology, v. 1, p. 55-73.
1982: Mammals of the mammoth Steppe as paleoenvironmental indicators; in Paleoecology of Beringia, D.M. Hopkins, J.V. Matthews, Jr., C.E. Schweger, and S.B. Young (ed.); Academic Press, New York, p. 307-326.

Hansen, H.P.
1949: Postglacial forests in south central Alberta, Canada; American Journal of Botany, v. 36, p. 54-65.
1952: Postglacial forests in the Grande Prairie-Lesser Slave Lake region of Alberta, Canada; Ecology, v. 33, p. 31-40.
1955: Postglacial forests in south central and central British Columbia; American Journal of Science, v. 253, p. 640-658.

Hansen, B.S. and Easterbrook, D.J.
1974: Stratigraphy and palynology of late Quaternary sediments in the Puget Lowland, Washington; Geological Society of America, Bulletin 85, p. 587-602.

Hare, F.K. and Ritchie, J.C.
1972: The boreal bioclimates; Geographical Review, v. 62, p. 333-365.

Harington, C.R.
1970: Ice age mammal research in the Yukon Territory and Alaska; in Early Man and Environments in Northwest North America, R.A. Smith and J.W. Smith (ed.); University of Calgary, Archaeological Association, p. 35-51.
1975: Pleistocene Muskoxen (*Symbos*) from Alberta and British Columbia; Canadian Journal of Earth Sciences, v. 12, p. 903-919.
1977: Pleistocene mammals of the Yukon Territory; unpublished PhD dissertation, University of Alberta, Edmonton, 1059 p.
1978: Quaternary vertebrate faunas of Canada and Alaska and their suggested chronological sequence; National Museums of Canada, Syllogeus Series no. 15, 105 p.
1980: Faunal exchanges between Siberia and North America — evidence from Quaternary land mammal remains in Siberia, Alaska and the Yukon Territory; in The Ice-Free Corridor and Peopling the New World, N.W. Rutter and C.E. Schweger (ed.); Canadian Journal of Anthropology, v. 1, p. 45-50.

Haynes, C.V.
1971: Time, environment, and early man; Arctic Anthropology, v. 8, p. 3-14.
1980: The Clovis Culture; Canadian Journal of Anthropology, v. 1, p. 115-121.

Hazell, S.
1979: Late Quaternary vegetation and climate of Dunbar Valley, British Columbia; unpublished MSc thesis, University of Toronto, Toronto, Ontario.

Hebda, R.J.
1982: Postglacial history of grasslands of southern British Columbia and adjacent regions; in Grassland Ecology and Classification; Symposium Proceedings, A.C. Nicholson, A. McLean, and T.E. Baker (ed.); British Columbia Ministry of Forests, p. 157-191.
1983: Late-glacial and postglacial vegetation history at Bear Cove Bog, northwest Vancouver Island, British Columbia; Canadian Journal of Botany, v. 61, p. 3172-3192.

Hebda, R.J. and Mathewes, R.W.
1984: Holocene history of cedar and native Indian cultures of the North American Pacific Coast; Science, v. 225, no. 4663, p. 711-713.

Heusser, C.J.
1955: Pollen profiles from the Queen Charlotte Islands, British Columbia; Canadian Journal of Botany, v. 33, p. 429-449.
1960: Late-Pleistocene environments of North Pacific North America; American Geographical Society, Special Publication, v. 35, p. 1-308.
1973: Environmental sequence following the Fraser advance of the Juan de Fuca Lake, Washington; Quaternary Research, v. 3, p. 284-306.
1974: Quaternary vegetation, climate, and glaciation of the Hoh River Valley, Washington; Geological Society of America, Bulletin, v. 85, p. 1547-1560.
1977: Quaternary palynology of the Pacific slope of Washington; Quaternary Research, v. 8, p. 282-306.

Heusser, C.J., Heusser, L.E., and Streeter, S.S.
1980: Quaternary temperatures and precipitation for the north-west coast of North America; Nature, v. 286, no. 5774, p. 702-704.

Heusser, L.E.
1983: Palynology and paleoecology of postglacial sediments in an anoxic basin, Saanich Inlet, British Columbia; Canadian Journal of Earth Sciences, v. 20, p. 873-885.

Hickman, M., Schweger, C.E., and Habgood, T.
1984: Lake Wabamun, Alta.: a paleoenvironmental study; Canadian Journal of Botany, v. 62, p. 1438-1465.

Hills, L.V.
1975: Late Tertiary floras of Arctic Canada: An interpretation; Proceedings, Circumpolar Conference on Northern Ecology, I-63-71.

Hills, L.V. and Matthews, J.V., Jr.
1974: A preliminary list of fossils of plants from the Beaufort Formation, Meighen Island, District of Franklin; in Report of Activities, Part B, Geological Survey of Canada, Paper 74-1B, p. 224-226.

Hills, L.V. and Sangster, E.V.
1980: A review of paleobotanical studies dealing with the last 20 000 years, Alaska, Canada and Greenland; in Climatic Change in Canada, C.R. Harington (ed.); National Museums of Canada, Syllogeus Series, v. 26, p. 73-246.

Holloway, R.G., Bryant, V.M., Jr., and Valastro, S.
1981: A 16,000 year pollen record from Lake Wabamun, Alberta, Canada; Palynology, v. 5, p. 195-208.

Hopkins, D.M.
1973: Sea level history in Beringia during the past 250 000 years; Quaternary Research, v. 3, p. 520-540.
1982: Aspects of the paleogeography of Beringia during the late Pleistocene; in Paleoecology of Beringia, D.M. Hopkins, J.V. Matthews, Jr., C.E. Schweger, and S.B. Young (ed.); Academic Press, New York, p. 3-28.

Hopkins, D.M., Matthews, J.V., Jr., Schweger, C.E., and Young, S.B. (editors)
1982: Paleoecology of Beringia; Academic Press, New York.

Hopkins, D.M., Smith, P.A., and Matthews, J.V., Jr.
1981: Dated wood from Alaska and the Yukon: implications for forest refugia in Beringia; Quaternary Research, v. 15, p. 217-249.

Hughes, O.L.
1969: Pleistocene stratigraphy, Porcupine and Old Crow Rivers, Yukon Territory; Geological Survey of Canada, Paper 69-1, p. 209-212.
1972: Surficial geology of northern Yukon Territory and northwestern District of Mackenzie, Northwest Territories; Geological Survey of Canada, Paper 69-36, 11 p.

Hughes, O.L., Harington, C.R., Janssens, J.A., Matthews, J.V., Jr., Morlan, R.E., Rutter, N.W., and Schweger, C.E.
1981: Upper Pleistocene stratigraphy, paleoecology, and archaeology of the northern Yukon interior, eastern Beringia; Arctic, v. 34, p. 329-365.

Hughes, O.L., Rutter, N.W., Matthews, J.V., Jr., and Clague, J.J.
1989: Regional Quaternary stratigraphy and history of unglaciated areas of the Canadian Cordillera; in Chapter 1 of Quaternary Geology of Canada and Greenland, R.J. Fulton (ed.); Geological Survey of Canada, Geology of Canada, no. 1 (also Geological Society of America, The Geology of North America, v. K-1).

Hyvarinen, H.
1985: Holocene pollen stratigraphy of Baird Inlet, east-central Ellesmere Island, arctic Canada; Boreas, v. 14, p. 19-32.

Ingram, M.J., Farmer, G., and Wigley, T.M.L.
1981: Past climates and their impact on man: a review; in Climate and History, T.M.L. Wigley, M.J. Ingram, and G. Farmer (ed.); Cambridge University Press, London, p. 3-49.

Irving, W.N. and Harington, C.R.
1973: Upper Pleistocene radiocarbon-dated artefacts from the northern Yukon; Science, v. 179, no. 4071, p. 335-340.

Irving, W.W., Mayhall, J.T., Melkye, F.J., and Beebe, B.F.
1977: A human mandible in probable association with Pleistocene faunal assemblage in eastern Beringia: a preliminary report; Canadian Journal of Archaeology, v. 1, p. 81-93.

Janssens, J.A.
1981: Subfossil bryophytes in eastern Beringia: their paleoenvironmental and phytogeographical significance; unpublished PhD thesis, University of Alberta, Edmonton, Alberta, 163 p.

Jopling, A.V., Irving, W.N., and Beebe, B.F.
1981: Stratigraphic, sedimentological and faunal evidence for the occurrence of pre-Sangamonian artefacts in northern Yukon; Arctic, v. 34, p. 3-33.

Jordan, R.
1975: Pollen diagrams from Hamilton Inlet, central Labrador, and their environmental implications for the northern Maritime Archaic; Arctic Anthropology, v. 12, p. 92-116.

Karlstrom, T.N.V. and Ball, G.E.
1969: The Kodiak Island Refugium; University of Toronto Press, Toronto, 262 p.

Karrow, P.F.
1983: Quaternary geology of the Hamilton-Cambridge area; Ontario Geological Survey, Open File Report 5429, p. 160.

Kavanaugh, D.H.
1980: Insects of western Canada, with special reference to certain Carabidae (Coleoptera): Present distribution patterns and their origins; Canadian Entomologist, v. 112, p. 1129-1144.

Kearney, M.S. and Luckman, B.H.
1983: Postglacial vegetational history of Tonquin Pass, British Columbia; Canadian Journal of Earth Sciences, v. 20, p. 776-786.

King, G.
1984: Deglaciation and revegetation of Western Labrador and adjacent Quebec; in Program and Abstracts; AMQUA, 8th Biennial Meeting, 13-15 August 1984, Boulder, Colorado, 148 p.

King, J.E.
1981: Late Quaternary vegetational history of Illinois; Ecological Monographs, v. 51, p. 43-62.

King, M.
1980: Palynological and macrofossil analyses of lake sediment from the Lillooet area, British Columbia; unpublished MSc thesis, Department of Biological Sciences, Simon Fraser University, Burnaby, British Columbia.

Klassen, R.W.
1989: Quaternary geology of the southern Interior Plains of Canada; in Chapter 2 of Quaternary Geology of Canada and Greenland, R.J. Fulton (ed.); Geological Survey of Canada, Geology of Canada, no. 1 (also Geological Society of America, The Geology of North America, v. K-1).

Koerner, R.M.
1989: Queen Elizabeth Islands glaciers; in Chapter 6 of Quaternary Geology of Canada and Greenland, R.J. Fulton (ed.); Geological Survey of Canada, Geology of Canada, no. 1 (also Geological Society of America, The Geology of North America, v. K-1).

Labelle, C. et Richard, P.
1981: Végétation tardiglaciaire et postglaciaire au sud-est du parc des Laurentides, Québec; Géographie physique et Quaternaire, vol. 35, p. 345-359.

Labelle, C. et Richard, P.J.H.
1984: Histoire postglaciaire de la végétation dans la région de Mont-Saint-Pierre, Gaspésie; Géographie physique et Quaternaire, vol. 38, p. 257-274.

Lamb, H.F.
1978: Postglacial vegetation change in southeastern Labrador; unpublished MSc thesis, University of Minnesota, Minneapolis, Minnesota.
1980: Late-Quaternary vegetation history of southeastern Labrador; Arctic and Alpine Research, v. 12, p. 117-135.
1982: Late Quaternary vegetational history of the forest-tundra ecotone in north-central Labrador; unpublished PhD thesis, University of Cambridge, Cambridge, England, 195 p.
1984: Modern pollen spectra from Labrador and their use in reconstructing Holocene vegetational history; Journal of Ecology, v. 72, p. 37-59.

Larouche, A.
1979: Étude comparée de l'histoire de la végétation postglaciaire par l'analyse pollinique et macrofossile des sédiments d'un lac et d'une tourbière du Québec; Mémoire de maîtrise (non publié), Université Laval, Québec, 117 p.

Larsen, J.A.
1980: The Boreal Ecosystem; Academic Press, New York, 500 p.

Lichti-Federovich, S.
1970: The pollen stratigraphy of a dated section of Late Pleistocene lake sediment from central Alberta; Canadian Journal of Earth Sciences, v. 7, p. 938-945.
1972: Pollen stratigraphy of a sediment core from Alpen Siding Lake, Alberta; in Report of Activities, Part B, Geological Survey of Canada, Paper 72-1B, p. 113-115.
1974: Palynology of Two Sections of Late Quaternary Sediments from the Porcupine River, Yukon Territory; Geological Survey of Canada, Paper 74-23, 6 p.

Lindroth, C.H.
1961- The ground-beetles of Canada and Alaska, 6 parts; Opuscula
1969: Entomologica, Supplements XX, XXIV, XXIX, XXXIII, XXXIV, 1192 p.

Lindsey, C.C. and Franzin, W.G.
1972: New complexities in zoogeography and taxonomy of the Pygmy Whitefish (*Prosopium coulteri*); Fisheries Research Board of Canada, Journal, v. 29, p. 1772-1775.

Liu, K-B.
1982: Postglacial vegetational history of northern Ontario; unpublished PhD thesis, University of Toronto, Toronto, Ontario, 337 p.

Livingstone, D.A.
1968: Some interstadial and postglacial pollen diagrams from Eastern Canada; Ecological Monographs, v. 38, p. 89-125.

Lortie, G., Vernal, A. de, Mott, R.J., et Occhietti, S.
1984: Sur des alternances de flores diatomifères marines et dulcicoles dans une séquence sédimentaire wisconsinienne au Lac Bras D'Or, Nouvelle-Écosse (résumé); Association Québécoise pour l'Étude du Quaternaire, Cinquième Congrès, Sherbrooke, Québec, 4-7 octobre, 1984, p. 35-37.

Loucks, D.L.
1962: A forest classification for the Maritime Provinces; Proceedings of the Nova Scotia Institute of Science, 25, Part 2, 1959-60, 167 p.

Lowdon, J.A. and Blake, W., Jr.
1973: Geological Survey of Canada radiocarbon dates XIII; Geological Survey of Canada, Paper 73-7, 61 p.
1978: Geological Survey of Canada radiocarbon dates XVIII; Geological Survey of Canada, Paper 78-7, 20 p.

Ludlam, S.D.
1969: Fayetteville Green Lake, New York III. The laminated sediments; Limnology and Oceanography, v. 14, p. 848-857.

Lundelius, E.L., Graham, R.J., Jr., Anderson, El., Guilday, J., Holman, J.A., Steadman, D.W., and Webb, S.D.
1983: Terrestrial vertebrate faunas; in Late Quaternary Environments of the United States: Vol. 1, The Late Pleistocene, S.C. Porter (ed.); University of Minnesota Press, p. 311-353.

MacDonald, G.M.
1982: Late Quaternary paleoenvironments of the Morley Flats and Kananaskis Valley of southwestern Alberta; Canadian Journal of Earth Sciences, v. 19, p. 23-35.
1983: Holocene vegetation history of the upper Natla River area, Northwest Territories, Canada; Arctic and Alpine Research, v. 15, p. 169-180.
1984: Postglacial plant migration and vegetation development in the western Canadian boreal forest; unpublished PhD thesis, University of Toronto, Toronto, Ontario, 261 p.

MacNeill, R.H.
1969: Some dates relating to the dating of the last major ice sheet in Nova Scotia; Maritime Sediments, v. 5, no. 1, p. 3.

Macpherson, J.B.
1982: Postglacial vegetational history of the eastern Avalon Peninsula, Newfoundland, and Holocene climatic change along the eastern Canadian seaboard; Géographie physique et Quaternaire, v. 36, p. 175-196.

Martin, P.S.
1974: The discovery of America; Science, v. 179, no. 4077, p. 969-974.

Mathewes, R.W.
1973: A palynological study of postglacial vegetation changes in the University Research Forest, southwestern British Columbia; Canadian Journal of Botany, v. 51, p. 2085-2103.
1979: Pollen morphology of Pacific Northwestern *Polemonium* species in relation to paleoecology and taxonomy; Canadian Journal of Botany, v. 57, p. 2428-2442.
1980: Pollen evidence for the presence of Tall Jacob's-ladder (*Polemonium caeruleum* L.) on the Queen Charlotte Islands during late-glacial time; Syesis v. 13, p. 105-108.
1985: Paleobotanical evidence for climatic change in southern British Columbia during late-glacial and Holocene time; in Climatic Change in Canada 5, C.R. Harington (ed.); National Museums of Canada, Syllogeus Series, v. 55, p. 397-422.
1989: The Queen Charlotte Islands refugium: a paleoecological perspective; in Chapter 7 of Quaternary Geology of Canada and Greenland, R.J. Fulton (ed.); Geological Survey of Canada, Geology of Canada, no. 1 (also Geological Society of America, The Geology of North America, v. K-1).

Mathewes, R.W. and Clague, J.J.
1982: Stratigraphic relationships and paleoecology of a late-glacial peat bed from the Queen Charlotte Islands, British Columbia; Canadian Journal of Earth Sciences, v. 19, p. 1185-1195.

Mathewes, R.W. and Heusser, L.E.
1981: A 12 000 year palynological record of temperature and precipitation trends in southwestern British Columbia; Canadian Journal of Botany, v. 59, p. 707-710.

Mathewes, R.W. and Rouse, G.E.
1975: Palynology and paleoecology of postglacial sediments from the Lower Fraser River Canyon of British Columbia; Canadian Journal of Earth Sciences, v. 12, p. 745-756.

Matthews, J.V., Jr.
1977: Tertiary Coleoptera fossils from the North American Arctic; Coleopterists Bulletin, v. 31, p. 297-308.
1980: Paleoecology of John Klondike Bog, Fisherman Lake Region, southwest District of Mackenzie; Geological Survey of Canada, Paper 80-22, 12 p.
1982: East Beringia during late Wisconsinan time: a review of the biotic evidence; in Paleoecology of Beringia, D.M. Hopkins, J.V. Matthews, Jr., C.E. Schweger, and S.B. Young (ed.); Academic Press, New York, p. 127-150.

Matthews, J.V., Jr., Mott, R.J., and Vincent, J.-S.
1986: Preglacial and interglacial environments of Banks Island: pollen and macrofossils from Duck Hawk Bluffs and related sites; Géographie physique et Quaternaire, v. 40, p. 279-298.

McAndrews, J.H.
1976: Fossil history of man's impact on the Canadian flora: an example from southern Ontario; Canadian Botanical Association, Bulletin, v. 9, p. 1-5.
1981: Late Quaternary climate of Ontario: temperature trends from the fossil pollen record; in Quaternary Paleoclimate, W.C. Mahaney (ed.); Geo Abstracts, p. 319-333.
1982: Holocene Environment of a Fossil *Bison* from Kenora, Ontario; Ontario Archaeology, v. 37, p. 41-51.

McAndrews, J.H. and Boyko-Daikonow, M.
1989: Pollen analysis of varved sediment at Crawford Lake, Ontario: evidence of Indian and European farming; in Chapter 7 of Quaternary Geology of Canada and Greenland, R.J. Fulton (ed.); Geological Survey of Canada, Geology of Canada, no. 1 (also Geological Society of America, The Geology of North America, v. K-1).

McAndrews, J.H. et Samson, G.
1977: Analyse pollinique et implications archéologiques et géomorphologiques, Lac de la Hutte Sauvage (Mushuauniipi), Nouveau-Québec; Géographie physique et Quaternaire, vol. 31, p. 177-183.

McMinn, R.G.
1951: The vegetation of a burn near Blaney Lake, British Columbia; Ecology, v. 32, p. 135-140.

Menabe, S. and Wetherald, R.T.
1980: On the distribution of climatic change resulting from increase in carbon dioxide content of the atmosphere; Journal of Atmospheric Science, v. 37, p. 99-118.

Miller, N.G. and Thompson, G.G.
1979: Boreal and western North American plants in the Late Pleistocene of Vermont; Journal of the Arnold Arboretum, v. 60, p. 167-218.

Minore, D.
1979: Comparative autecological characteristics of northwestern tree species: a literature review; Pacific NW Forest and Range Experiment Station, USDA Forest Service, General Technical Report PNW-87.

Moodie, G.E.E. and Reimchen, T.E.
1976: Glacial refugia, endemism, and stickleback populations of the Queen Charlotte Islands, British Columbia; Canadian Field-Naturalist, v. 90, p. 471-474.

Moore, P.D. and Bellamy, D.J.
1974: Peatlands; Springer-Verlag, New York, Inc., p. 221.

Moran, S.R. and Clayton, L.
1983: Reply — Chronology of the Late Wisconsinan glaciation in middle North America; Quaternary Science Reviews, v. 1, no. 4, p. 15-22.

Morgan, A.V. and Morgan, A.
1980: Faunal assemblages and distributional shifts of Coleoptera during the late Pleistocene in Canada and the Northern United States; Canadian Entomologist, v. 112, p. 1105-1128.

Morgan, A.V., Morgan, A., Ashworth, A.C., and Matthews, J.V., Jr.
1983: Late Wisconsin fossil beetles in North America; in Late Quaternary Environments of the United States, S.C. Porter (ed.); University of Minnesota Press, Minneapolis, v. 1, p. 354-362.

Morison, S.R. and Smith, C.A.S. (editors)
1987: Quaternary research in Yukon; INQUA 12th International Congress, Ottawa, Canada, Excursion Guide Book A20a, A20b, National Research Council of Canada.

Morisset, P.
1971: Endemism in the vascular plants of the gulf of St. Lawrence region; Naturaliste canadien, v. 18, p. 167-177.

Morlan, R.E.
1978: Early man in northern Yukon Territory: Perspectives as of 1977; in Early Man in America — From a Circum-Pacific Perspective, A.L. Bryan (ed.); Archaeological Researches International, Edmonton, p. 78-95.
1980: Taphonomy and archaeology in the Upper Pleistocene of the northern Yukon Territory: a glimpse of the peopling of the New World; National Museum of Man, Mercury Series, Archaeological Survey of Canada, Paper 94, 398 p.
1983: Counts and estimates of taxonomic abundance in faunal remains: microtine rodents from Bluefish Cave I; Canadian Journal of Archaeology, v. 7, no. 1, p. 61-76.
1986: Pleistocene archaeology in Old Crow Basin: a critical reappraisal; in New Evidence for the Pleistocene Peopling of the Americas, A.L. Bryan (ed.); Center for Study of Early Man, University of Maine, Orono, p. 27-48.

Morlan, R.E. and Cinq-Mars, J.
1982: Ancient Beringians: Human occupation in the Late Pleistocene of Alaska and the Yukon Territory; in Paleoecology of Beringia, D.M. Hopkins, J.V. Matthews, Jr., C.E. Schweger, and S.B. Young (ed.); Academic Press, New York, p. 353-381.
1986: Bluefish Caves and the habitability of the northwestern margin of the Laurentide ice sheet (abstract); Program and Abstracts, 9th AMQUA meeting, Champaign-Urbana, Illinois.

Morrison, A.
1970: Pollen diagrams from interior Labrador; Canadian Journal of Botany, v. 48, p. 1957-1975.

Moss, E.H.
1955: The vegetation of Alberta; The Botanical Review, v. 21, p. 493-567.

Mott, R.J.
1971: Palynology of a buried organic deposit, River Inhabitants, Cape Breton Island, Nova Scotia; in Report of Activities, Part B, Geological Survey of Canada, Paper 71-1B, p. 123-125.
1973: Palynological studies in central Saskatchewan. Pollen stratigraphy from lake sediment sequences; Geological Survey of Canada, Paper 72-49, 18 p.
1975: Palynological studies of lake sediment profiles from southwestern New Brunswick; Canadian Journal of Earth Sciences, v. 12, p. 273-288.
1976: A Holocene pollen profile from the Sept-Iles area, Quebec: Naturaliste canadien, v. 103, p. 457-567.
1977: Late-Pleistocene and Holocene palynology in southeastern Quebec; Géographie physique et Quaternaire, v. 31, p. 139-149.
1989: Late Pleistocene paleoenvironments in Atlantic Canada; in Chapter 7 of Quaternary Geology of Canada and Greenland, R.J. Fulton (ed.); Geological Survey of Canada, Geology of Canada, no. 1 (also Geological Society of America, The Geology of North America, v. K-1).

Mott, R.J. and Farley-Gill, L.D.
1978: A Late-Quaternary pollen profile from Woodstock, Ontario; Canadian Journal of Earth Sciences, v. 15, p. 1101-1111.
1981: Two late Quaternary pollen profiles from Gatineau Park, Quebec; Geological Survey of Canada; Paper 80-31, 10 p.

Mott, R.J. and Grant, D.R.
1984: Pre-late Wisconsinan paleoenvironments in Atlantic Canada (abstract); VIth International Palynological Conference, Calgary, 26 August to 1 September, 1984, p. 11.

1985: Pre-late Wisconsinan paleoenvironments in Atlantic Canada; Géographie physique et Quaternaire, v. 39, p. 239-254.

Mott, R.J. and Jackson, L.E., Jr.
1982: An 18 000 year palynological record from the southern Alberta segment of the classical Wisconsinan "Ice-free Corridor"; Canadian Journal of Earth Sciences, v. 19, p. 504-513.

Mott, R.J. and Prest, V.K.
1967: Stratigraphy and palynology of buried organic deposits from Cape Breton Island, Nova Scotia; Canadian Journal of Earth Sciences, v. 4, p. 709-723.

Mott, R.J., Anderson, T., and Matthews, J.V., Jr.
1981: Late-glacial paleoenvironments of sites bordering the Champlain Sea, based on pollen and macrofossil evidence; in Quaternary Paleoclimate, W.C. Mahaney (ed.); GeoBooks, University East Anglia, Norwich, p. 129-171.

Mott, R.J., Anderson, T.W., and Matthews, J.V., Jr.
1982: Pollen and macrofossil study of an interglacial deposit in Nova Scotia; Géographie physique et Quaternaire, v. 36, p. 197-208.

Mott, R.J., Grant, D.R., Stea, R., and Occhietti, S.
1986: Late-glacial climatic oscillation in Atlantic Canada equivalent to the Allerød/younger Dryas event; Nature, v. 323, no. 6085, p. 247-250.

Nelson, D.E., Morlan, R.E., Vogel, J.S., Southon, J.R., and Harington, C.R.
1986: New dates on northern Yukon artifacts: Holocene not Upper Pleistocene; Science, v. 232, p. 749-751.

Nichols, H.
1967: Central Canadian palynology and its relevance to northwestern Europe in the Late Quaternary Period; Review of Palaeobotany and Palynology, v. 2, p. 231-243.

Occhietti, S.
1989: Quaternary geology of St. Lawrence Valley and adjacent Appalachian subregion; in Chapter 4 of Quaternary Geology of Canada and Greenland, R.J. Fulton (ed.); Geological Survey of Canada, Geology of Canada, no. 1 (also Geological Society of America, The Geology of North America, v. K-1).

Occhietti, S., Mott, R.J., Rutter, R., Vernal, A. de, Lortie, G., DeBoutray, B., Causse, C., et St-Jean, G.
1984: Transition Sangamonien-Wisconsinien et lithostratigraphie au centre de l'Île du Cap Breton, Nouvelle Écosse (résumé); Association Québécoise pour l'Étude du Quaternaire, Cinquième Congrès, Sherbrooke, Québec, 4-7 octobre, 1984, p. 40-41.

Ovenden, L.E.
1982: Vegetation history of a polygonal peatland, northern Yukon; Boreas, v. 11, p. 209-224.

Overpeck, J.T., Webb, T. III, and Prentice, I.C.
1985: Quantitative interpretation of fossil pollen spectra: dissimilarity coefficients and the method of modern analogs; Quaternary Research, v. 23, p. 87-108.

Packer, J.G. and Vitt, D.H.
1974: Mountain Park: a plant refugium in the Canadian Rocky Mountains; Canadian Journal of Botany, v. 52, p. 1393-1409.

Parker, M.L., Jozsa, L.A., Johnson, S.G., and Bramhall, P.A.
1981: Dendrochronological studies on the coasts of James Bay and Hudson Bay; in Climatic Change in Canada 2, C.R. Harington (ed.); National Museums of Canada, Syllogeus, no. 33, p. 129-188.

Parry, M.L.
1985: Estimating the sensitivity of natural ecosystems and agriculture to climatic change — guest editorial; Climatic Change, v. 7, no. 1, p. 1-5.

Payette, S.
1983: The forest tundra and present tree-lines of the northern Quebec-Labrador Peninsula; in Tree-Line Ecology, Proceedings of the Northern Quebec Tree-Line Conference, P. Morisset and S. Payette (ed.); Nordicana, v. 47, p. 3-23.
1984: Peat inception and climatic change in northern Quebec; in Climatic Changes on a Yearly to Millennial Basis, N.-A. Morner and W. Karlin (ed.); D. Reidel Publishing Company, Boston, p. 173-179.

Payette, S. and Gagnon, R.
1979: Tree-line dynamics in Ungava Peninsula, northern Quebec; Holarctic Ecology, v. 2, p. 239-248.

Pielou, E.C.
1979: Biogeography; John Wiley, New York, 351 p.

Pojar, J.
1980: Brooks Peninsula: Possible Pleistocene glacial refugium on northwestern Vancouver Island; Program and Abstracts, Botanical Society of America, Miscellaneous Publication 158, p. 89.

Prest, V.K.
1969: Retreat of Wisconsin and Recent Ice in North America; Geological Survey of Canada, Map 1257A.
1970: Quaternary Geology of Canada; in Geology and Economic Minerals of Canada, R.J.W. Douglas (ed.); Geological Survey of Canada, Economic Geology Report no. 1 (5th edition), p. 676-764.

Prest, V.K., Terasmae, J., Matthews, J.V., Jr., and Lichti-Federovich, S.
1976: Late-Quaternary history of Magdalen Islands, Quebec; Maritime Sediments, v. 12, no. 2, p. 39-59.

Rampton, V.N.
1971: Late Quaternary vegetation and climatic history of the Snag-Klutlan area, southwestern Yukon Territory, Canada; Geological Society of America, Bulletin, v. 82, p. 959-978.

Rampton, V.N., Gauthier, R.C., Thibault, J., and Seaman, A.A.
1984: Quaternary geology of New Brunswick; Geological Survey of Canada, Memoir 416, 77 p.

Randhawa, A.S. and Beamish, K.I.
1972: The distribution of *Saxifraga ferruginea* and the problem of refugia in northwestern North America; Canadian Journal of Botany, v. 50, p. 79-87.

Reeves, B.O.K.
1983: Bergs, barriers and Beringia: reflections on the peopling of the New World; in Quaternary Coastlines and Marine Archaeology, P.M. Masters and N.C. Fleming (ed.); Academic Press, New York, p. 390-411.

Repenning, C.A.
1967: Palearctic-Nearctic mammalian dispersal in the late Cenozoic; in The Bering Land Bridge, D.M. Hopkins (ed.); Standford University Press, p. 288-311.
1980: Faunal exchanges between Siberia and North America; Canadian Journal of Anthropology, v. 1, p. 37-44.

Richard, P.J.H.
1971: Two pollen diagrams from the Québec city area, Canada; Pollen et Spores, v. 13, p. 523-559.
1973a: Histoire postglaciaire comparée de la végétation dans deux localités au nord du parc des Laurentides, Québec; Naturaliste canadien, vol. 100, p. 577-590.
1973b: Histoire postglaciaire comparée de la végétation dans deux localités au sud de la ville de Québec; Naturaliste canadien, vol. 100, p. 591-603.
1973c: Histoire postglaciaire de la végétation dans la région de Saint-Raymond de Portneuf, telle que révélée par l'analyse pollinique d'une tourbière; Naturaliste canadien, vol. 100, p. 561-575.
1974: Présence de *Shepherdia canadensis* (L.) Nutt. dans la région du parc des Laurentides, Québec, au tardiglaciaire; Naturaliste canadien, vol. 101, p. 763-768.
1975a: Contribution à l'histoire postglaciaire de la végétation dans les Cantons-de-l'Est: étude des sites de Weedon et Albion; Cahiers de géographie du Québec, vol. 19, p. 267-284.
1975b: Histoire postglaciaire de la végétation dans la partie centrale du Parc des Laurentides, Québec; Naturaliste canadien, vol. 102, p. 669-681.
1975c: Contribution à l'histoire postglaciaire de la végétation dans la plaine du Saint-Laurent: Lotbinière et Princeville; Revue de géographie de Montréal, vol. 29, p. 95-107.
1977a: Histoire post-wisconsinienne de la végétation du Québec méridional par l'analyse pollinique; Service de la recherche, Direction générale des forêts, Ministère des Terres et Forêts du Québec, Tome 1, 312 p., Tome 2, 142 p.
1977b: Végétation tardiglaciaire au Québec méridional et implications paléoclimatiques; Géographie physique et Quaternaire, vol. 31, p. 161-176.
1978: Histoire tardiglaciaire et postglaciaire de la végétation au Mont Shefford, Québec; Géographie physique et Quaternaire, vol. 32, p. 81-93.
1979: Contribution à l'histoire postglaciaire de la végétation au nord-est de la Jamésie, Nouveau-Québec; Géographie physique et Quaternaire, vol. 33, p. 93-112.
1980: Histoire postglaciaire de la végétation au sud du lac Abitibi, Ontario et Québec; Géographie physique et Quaternaire, vol. 34, p. 77-94.
1981a: Palaeoclimatic significance of the Late-Pleistocene and Holocene pollen record in south-central Quebec, p. 335-360, in Quaternary Paleoclimate, W.C. Mahaney (ed.); Geo Abstracts, University East Anglia, Norwich, 464 p.
1981b: Paléophytogéographie postglaciaire en Ungava, par l'analyse pollinique; Collection Paléo-Québec, no 13, Université du Québec à Montréal, 153 p.
1989: Patterns of post-Wisconsinan plant colonization in Quebec-Labrador; in Chapter 7 of Quaternary Geology of Canada and Greenland, R.J. Fulton (ed.); Geological Survey of Canada, Geology of Canada no. 1 (also Geological Society of America, The Geology of North America, v. K-1).

Richard, P.J.H., Larouche, A., et Bouchard, M.
1982: Âge de la déglaciation finale et histoire postglaciaire de la végétation dans la partie centrale du Nouveau-Québec; Géographie physique et Quaternaire, vol. 36, p. 63-90.

Ritchie, J.C.
1964: Contributions to the Holocene paleoecology of west central Canada 1. The Riding Mountain area; Canadian Journal of Botany, v. 42, p. 181-197.
1967: Holocene vegetation of the northwestern precincts of the Glacial Lake Agassiz Basin; in Life, Land and Water, W.J. Mayer-Oakes (ed.); University of Manitoba Press, Winnipeg, Manitoba, p. 217-229.
1969: Absolute pollen frequencies and carbon-14 age of a section of Holocene lake sediment from the Riding Mountain area of Manitoba; Canadian Journal of Botany, v. 47, p. 1345-1349.
1976: The late-Quaternary vegetational history of the western interior of Canada; Canadian Journal of Botany, v. 54, p. 1793-1818.
1977: The modern and late Quaternary vegetation of the Campbell Dolomite Uplands, near Inuvik, N.W.T.; Ecological Monographs v. 47, p. 401-423.
1980: Towards a Late-Quaternary paleoecology of the ice-free corridor; Canadian Journal of Anthropology, v. 1, p. 15-28.
1981: Problems of interpretation of the pollen stratigraphy of northwest North America, p. 344-391; in Quaternary Paleoclimates, W.C. Mahaney (ed.); Geoabstracts, Norwich, 464 p.
1982: The modern and late-Quaternary vegetation of the Doll Creek area, North Yukon, Canada; New Phytologist, v. 90, p. 563-603.
1984a: Past and Present Vegetation of the Far Northwest of Canada; University of Toronto Press, Toronto, 251 p.
1984b: A Holocene pollen record of boreal forest history from the Travaillant Lake area, Lower Mackenzie River Basin; Canadian Journal of Botany, v. 62, p. 1385-1392.
1985: Late-Quaternary climatic and vegetational change in the Lower Mackenzie Basin, Northwest Canada; Ecology, v. 66, p. 612-621.
1989: History of the boreal forest in Canada; in Chapter 7 of Quaternary Geology of Canada and Greenland, R.J. Fulton (ed.); Geological Survey of Canada, Geology of Canada, no. 1 (also Geological Society of America, The Geology of North America, v. K-1).

Ritchie, J.C. and Cwynar, L.C.
1982: The Late Quaternary vegetation of the north Yukon; in Paleoecology of Beringia, D.M. Hopkins, J.V. Matthews, Jr., C.E. Schweger, and S.B. Young (ed.); Academic Press, New York, p. 113-127.

Ritchie, J.C. and De Vries, B.
1964: Contributions to the Holocene paleoecology of west central Canada — a late glacial deposit from the Missouri Coteau; Canadian Journal of Botany, v. 42, p. 677-692.

Ritchie, J.C. and Hadden, K.A.
1975: Pollen stratigraphy of Holocene sediments from the Grand Rapids area, Manitoba, Canada; Revue paléobotanie et palynologie, v. 19, p. 193-202.

Ritchie, J.C. and Hare, F.K.
1977: Late-Quaternary vegetation and climate near the arctic tree line of northwestern North America; Quaternary Research, v. 1, p. 331-341.

Ritchie, J.C. and Lichti-Federovich, S.
1968: Holocene pollen assemblages from the Tiger Hills, Manitoba; Canadian Journal of Earth Sciences, v. 5, p. 873-880.

Ritchie, J.C., Cwynar, L.C., and Spear, R.W.
1983: Evidence from north-west Canada for an Early Holocene Milankovitch Thermal Maximum; Nature, v. 305, no. 5930, p. 126-128.

Roemer, H.L. and Ogilvie, R.T.
1983: Additions to the flora of the Queen Charlotte Islands on limestone; Canadian Journal of Botany, v. 61, p. 2577-2580.

Rousseau, C.
1974: Géographie floristique du Québec-Labrador; distribution des principales espèces vasculaires; Travaux et documents, Centre d'Études Nordiques, no. 7, Presses Université Laval, Québec, 798 p.

Rowe, J.S.
1972: Forest Regions of Canada; Department of Environment, Canada Forestry Service, Publication 1300, 172 p.

Ruddiman, W.F. and McIntyre, A.
1979: Warmth of the subpolar North Atlantic Ocean during northern hemispheric ice-sheet growth; Science, v. 204, no. 4389, p. 173-175.
1981: Oceanic mechanisms for amplification of the 23 000 year ice-volume cycle; Science, v. 212, no. 4495, p. 617-627.

Rutter, N.W.
1984: Pleistocene history of the western Canadian ice-free corridor; in Quaternary Stratigraphy in Canada — A Canadian Contribution to IGCP Project 24, R.J. Fulton (ed.); Geological Survey of Canada, Paper 84-10, p. 49-56.

Rutter, N.W. and C.E. Schweger (editors)
1980: The ice-free corridor and peopling of the New World; Canadian Journal of Anthropology, v. 1., 139 p.

Ryder, J.M. and Clague, J.J.
1989: Regional Quaternary stratigraphy and history of British Columbia, in Chapter 1 of Quaternary Geology of Canada and Greenland, R.J. Fulton (ed.); Geological Survey of Canada, Geology of Canada, no. 1 (also Geological Society of America, The Geology of North America, v. K-1).

Savoie, L. et Richard, P.J.H.
1979: Paléophytogéographie de l'épisode de Saint-Narcisse dans la région de Sainte-Agathe, Québec; Géographie physique et Quaternaire, vol. 33, p. 175-188.

Schneider, S.H.
1984: On the empirical verification of model-predicted CO_2-induced climatic effects; in Climate Processes and Climate Sensitivity, J.E. Hansen and T. Takahashi (ed.); American Geophysical Union, Geophysical Monographs, v. 29, p. 187-201.
1986: Can modeling of the ancient past verify prediction of future climates — an editorial; Climatic Change, v. 8., p. 117-119.

Schneider, S.J. and Thompson, S.L.
1981: Atmospheric CO_2 and climate: importance of the transient response; Journal of Geophysical Research, v. 86, p. 3135-3147.

Schofield, W.B.
1976: Bryophytes of British Columbia. III. Habitat and distributional information for selected mosses; Syesis, v. 9, p. 317-354.

Schuster, R.M. and Schofield, W.B.
1982: On *Dendrobazzania*, a new genus of Lepidoziaceae (Jungermanniales); The Bryologist, v. 85, p. 231-238.

Schweger, C.E.
1989: Paleoecology of the western Canadian ice-free corridor; in Chapter 7 of Quaternary Geology of Canada and Greenland, R.J. Fulton (ed.); Geological Survey of Canada, Geology of Canada, no. 1 (also Geological Society of America, The Geology of North America, v. K-1).

Schweger, C.E. and Janssens, J.A.
1980: Paleoecology of the Boutellier nonglacial interval, St. Elias Mountains, Yukon Territory, Canada; Arctic and Alpine Research, v. 12, p. 309-317.

Schweger, C.E. and Matthews, J.V., Jr.
1985: Old Crow Tephra and its implications for the synoptic paleoecology/paleoclimatology of Alaska and the Yukon during the mid-Wisconsinan time; Géographie physique et Quaternaire, v. 39, p. 275-290.

Schweger C.E., Habgood, T., and Hickman, M.
1981: Late glacial-Holocene climatic changes of Alberta — the record from lake sediment studies; in The Impacts of Climatic Fluctuations on Alberta's Resources and Environment; Atmospheric Environment Services, Western Region, Environment Canada, Edmonton, Report WAES-1-81, p. 47-60.

Schweger, C.E., Matthews, J.V., Jr., Hopkins, D.M., and Young, S.B.
1982: Paleoecology of Beringia — A Synthesis; in Paleoecology of Beringia, D.M. Hopkins, J.V. Matthews, Jr., C.E. Schweger, and S.B. Young (ed.); Academic Press, New York, p. 425-444.

Schwert, D.P., Anderson, T.W., Morgan, A., Morgan, A.V., and Karrow, P.F.
1985: Changes in Late Quaternary vegetation and insect communities in southwestern Ontario; Quaternary Research, v. 23, p. 205-226.

Scudder, G.E.E.
1979: Present patterns in the fauna and flora of Canada and its insect fauna; in Canada and its Insect Fauna, H.V. Danks (ed.); Memoirs of the Entomological Society of Canada, v. 108, p. 87-179.

Sher, A.V.
1976: The role of Beringian land in the development of Holarctic mammalian fauna in the Late Cenozoic; in Beringia in the Cenozoic Era, V.I. Kontrimavichus (ed.); Academy of Sciences of the USSR, Far Eastern Scientific Center, p. 296-316.

Sher, A.V., Kaplina, T.N., Giterman, R.E., Lozhkin, A.V., Arkhangelov, A.A., Kiselyov, S.V., Kovzentsov, Yu. V., Virina, E.I., and Zazhigin, V.S.
1979: Scientific excursion on problems "late Cenozoic of the Kolyma Lowland"; XIV Pacific Science Congress, Moscow, NAUKA: 116

Short, S.K.
1978: Palynology: a Holocene environmental perspective for archaeology in Labrador-Ungava; Arctic Anthropology, v. 15, p. 9-35.

Short, S.K. and Nichols, H.
1977: Holocene pollen diagrams from subarctic Labrador-Ungava: vegetational history and climatic change; Arctic and Alpine Research, v. 9, p. 265-290.

Slater, D.S.
1978: Late Quaternary pollen diagram from the central Mackenzie corridor area (abstract); Abstracts, American Quaternary Association, 5th Biennial Meeting, Edmonton, p. 176.

Soper, J.D.
1964: The Mammals of Alberta; Hamly Press, Edmonton.

Sorenson, C.J.
1977: Reconstructed Holocene bioclimates; Annals of the Association of American Geographers, v. 67, p. 214-222.

Sorenson, C.J. and Knox, J.C.
1974: Paleosols and paleoclimates related to late Holocene forest/tundra border migrations: Mackenzie and Keewatin, N.W.T., Canada; in International Conference in the Prehistory and Paleoecology of western North American Arctic and Sub-Arctic, S. Raymond and P. Schlederman (ed.); University of Calgary, Archeological Association, Calgary, p. 187-204.

Spear, R.W.
1983: Paleoecological approaches to a study of treeline fluctuation in the Mackenzie Delta region, Northwest Territories, preliminary results; in Tree-line Ecology, Proceedings of the Northern Quebec Tree-Line Conference, P. Morisset and S. Payette (ed.); Nordicana, v. 47, p. 61-72.

Stalker, A.M.
1977: Indications of Wisconsin and earlier man from the southwest Canadian prairies: Annals of New York Academy of Sciences, v. 19, p. 119-136.

Stalker, A.M. and Mott, R.J.
1972: Palynology of the "Kansan" carbonaceous clay unit near Medicine Hat, Alberta; in Report of Activities, Part B, Geological Survey of Canada, Paper 72-1B, p. 117-119.

Stalker, A.M. and Churcher, C.S.
1982: Ice age deposits and animals from the southwestern part of the Great Plains of Canada; Geological Survey of Canada, Miscellaneous Report 31 (wall chart).

Stea, R. and Hemsworth, D.
1979: Pleistocene stratigraphy of the Miller Creek Section, Hants County, Nova Scotia; Nova Scotia Department of Mines and Energy, Paper 79-5, 16 p.

Stothers, R.B.
1984: The great Tambora eruption in 1815 and its aftermath; Science, v. 224, no. 4654, p. 1191-1198.

Sutherland-Brown, A.
1968: Geology of the Queen Charlotte Islands, British Columbia; British Columbia Department of Mines and Petroleum Resources, Bulletin, v. 54, p. 1-226.

Sutherland-Brown, A. and Nasmith, H.
1962: The glaciation of the Queen Charlotte Islands; Canadian Field-Naturalist, v. 76, p. 209-219.

Swain, A.M.
1973: A history of fire and vegetation in northeastern Minnesota as recorded in lake sediments; Quaternary Research, v. 3, p. 383-396.

Terasmae, J.
1969: Quaternary palynology in Quebec: a review and future prospects; Revue de géographie de Montréal, v. 23, p. 281-288.
1973: Notes on late Wisconsin and early Holocene history of vegetation in Canada; Arctic and Alpine Research, v. 5, p. 201-222.
1976: In search of a palynological tundra; Geoscience and Man, v. 15, p. 77-82.
1980: Some problems of Late Wisconsin history and geochronology in southeastern Ontario; Canadian Journal of Earth Sciences, v. 17, p. 361-381.

Terasmae, J. and Anderson, T.W.
1970: Hypsithermal range extension of white pine (Pinus strobus L.) in Quebec, Canada; Canadian Journal of Earth Sciences, v. 7, p. 406-413.

Terasmae, J. and Hughes, O.L.
1966: Late-Wisconsinan chronology and history of vegetation in the Ogilvie Mountains, Yukon Territory, Canada; Palaeobotanist, v. 15, p. 235-242.

Terasmae, J. and McAtee, C.L.
1979: Palynology of Perch Lake sediments: Progress and some potential future developments; in Hydrological and geochemical studies in the Perch Lake Basin: A second report of progress, P.J. Barry (ed.); Atomic Energy of Canada, Chalk River Nuclear Laboratories, Chalk River, p. 279-282.

Trigger, B.G.
1976: The Children of Aataentsic I: A History of the Huron People to 1660; McGill-Queen's University Press, Montreal, 455 p.

Turner, J.V., Fritz, P., Karrow, P.F., and Warner, B.G.
1983: Isotopic and geochemical composition of marl lake waters and implications for radiocarbon dating of marl lake sediments; Canadian Journal of Earth Sciences, v. 20, p. 599-615.

Vance, R.E., Emerson, D., and Habgood, T.
1983: A mid-Holocene record of vegetation change in central Alberta; Canadian Journal of Earth Sciences, v. 20, p. 364-376.

van der Hammen, T., Wijmstra, T.A., and Zagwijn, W.H.
1971: The floral record of the late Cenozoic of Europe; in The Late Cenozoic Glacial Ages, K.K. Turekian (ed.); Yale University Press, New Haven, p. 391-424.

Vernal, A. de
1983: Paléoenvironnements du Wisconsinien par la palynologie dans la région de Baie Saint-Laurent, Île du Cap Breton: Département de Géographie, Université de Montréal, Montréal, Québec, 97 p.

Vernal, A. de et Mott, R.J.
1986: Palynostratigraphie et paléoenvironments du Pléistocène supérieur dans la région du Lac Bras d'Or, Île du Cap-Breton, Nouvelle-Écosse; Journal canadien des sciences de la terre, vol. 23, p. 491-503.

Vernal, A. de, Mott, R.J., Occhietti, S., and Causse, C.
1984: Palynostratigraphie du Pléistocène supérieur dans la région du Lac Bras d'Or, Île du Cap-Breton (Nouvelle Écosse) (résumé); VIth International Palynological Conference, Calgary, 26 August to 1 September, 1984, p. 34.

Vernal, A. de, Richard, P.J.H., et Occhietti, S.
1983: Palynologie et paléoenvironnements du Wisconsinien de la région de la Baie Saint-Laurent, Île du Cap-Breton; Géographie physique et Quaternaire, vol. 37, p. 307-322.

Vilks, G. and Mudie, P.
1978: Early deglaciation of Labrador Shelf; Science, v. 202, no. 4373, p. 1181-1183.

Vincent, J-S.
1973: A palynological study for the Little Clay Belt, northwestern Quebec; Naturaliste Canadien, v. 100, p. 59-70.
1989: Quaternary geology of the southeastern Canadian Shield; in Chapter 3 of Quaternary Geology of Canada and Greenland, R.J. Fulton (ed.); Geological Survey of Canada, Geology of Canada, no. 1 (also Geological Society of America, The Geology of North America, v. K-1).

Vitt, D.H. and Schofield, W.B.
1976: Seligeria careyana, a new species from the Queen Charlotte Islands, Western Canada; The Bryologist, v. 79, p. 231-234.

Waggoner, P.E.
1983: Agriculture and a climate changed by more carbon dioxide; in Board on Atmospheric Sciences and Climate; Commission on Physical Sciences, Mathematics and Resources, National Research Council, National Academy Press, p. 383-418.

Warner, B.G.
1984: Late Quaternary paleoecology of eastern Graham Island, Queen Charlotte Islands, British Columbia, Canada; unpublished PhD thesis, Department of Biological Sciences, Simon Fraser University, Burnaby, British Columbia, 190 p.

Warner, B.G., Clague, J.J., and Mathewes, R.W.
1984a: Geology and paleoecology of a Mid-Wisconsin peat from the Queen Charlotte Islands, British Columbia, Canada; Quaternary Research, v. 21, p. 337-350.

Warner, B.G., Hebda, R.J., and Hann, B.J.
1984b: Postglacial paleoecological history of a cedar swamp, Manitoulin Island, Ontario, Canada; Palaeogeography, Palaeoclimatology, Palaeoecology, v. 45, p. 301-345.

Warner, B.G., Mathewes, R.W., and Clague, J.J.
1982: Ice-free conditions on the Queen Charlotte Islands, British Columbia, at the height of Late Wisconsin glaciation; Science, v. 218, no. 4573, p. 675-677.

Warwick, W.F.
1980: Chironomidae (Diptera) responses to 2800 years of cultural influence: A Palaeolimnological study with special reference to sedimentation, eutrophication, and contamination processes; Canadian Entomologist, v. 112, p. 1193-1238.

Waters, P.L. and Rutter, N.W.
1983: Utilizing paleosols and volcanic ash in correlating Holocene deposits in southern Alberta; in Quaternary Chronologies, W.C. Mahaney (ed.); Geoabstracts Ltd., Norwich, U.K.

Webb, T., III and Clark, D.R.
1977: Calibrating micropaleontological data in climatic terms: a critical review; Annals of the New York Academy of Science, v. 5, no. 288, p. 93-118.

Webb, T., III, Richard, P.J.H., and Mott, R.J.
1983: A mapped history of Holocene vegetation in southern Quebec; National Museums of Canada, Syllogeus Series 49, p. 273-336.

Wenner, C.-G.
1947: Pollen diagrams from Labrador; Geografiska Annaler, v. 29, p. 137-374.

Westgate, J.A.
1972: Cypress Hills; in Quaternary Geology and Geomorphology Between Winnipeg and the Rocky Mountains, N.W. Rutter and E.A. Christiansen (ed.); 24th International Geological Congress, Montréal, Guidebook, Excursion C-22, p. 50-62.

Westgate, J.A., Fritz, P., Matthews, J.V., Jr., Kalas, L., Delorme, L.D., Green, R., and Aario, R.
1972: Geochronology and paleoecology of mid-Wisconsin sediments in west central Alberta, Canada (abstract); 24th International Geological Congress, Montreal, p. 380.

Westgate, J.A., Hamilton, T.D., and Gorton, M.P.
1983: Old Crow Tephra: a new late Pleistocene stratigraphic marker across north-central Alaska and western Yukon Territory; Quaternary Research, v. 19, p. 38-54.

Wheeler, N.C. and Guries, R.P.
1982: Biogeography of Lodgepole pine; Canadian Journal of Botany, v. 60, p. 1805-1814.

White, J.M.
1983: Late Quaternary geochronology and paleoecology of the upper Peace River District, Canada; unpublished PhD thesis, Simon Fraser University, Burnaby, British Columbia, 146 p.

White, J.M. and Mathewes, R.W.
1982: Holocene vegetation and climatic change in the Peace River District, Canada; Canadian Journal of Earth Sciences, v. 19, p. 555-570.

White, J.M., Mathewes, R.W., and Mathews, W.H.
1979: Radiocarbon dates from Boone Lake and their relation to the 'Ice-free Corridor' in the Peace River District of Alberta, Canada; Canadian Journal of Earth Sciences, v. 16, p. 1870-1874.

White, P.S.
1979: Pattern, process and natural disturbance in vegetation; The Botanical Review, v. 45, p. 229-299.

Whittaker, R.M.
1975: Communities and Ecosystems (2nd edition); Macmillan, New York, 385 p.

Wiggins, G.B.
1977: Larvae of the North American Caddisfly Genera (Trichoptera); University of Toronto Press, Toronto, 401 p.

Williams, N.E. and Morgan, A.V.
1977: Fossil caddisflies (Insecta: Trichoptera) from the Don Formation, Toronto, Ontario, and their use in paleoecology; Canadian Journal of Zoology, v. 55, p. 519-527.

Williams, N.E., Westgate, J.A., Williams, D.D., Morgan, A., and Morgan, A.V.
1981: Invertebrate fossils (Insecta: Trichoptera, Diptera, Coleoptera) from the Pleistocene Scarborough Formation at Toronto, Ontario, and their paleoenvironmental significance; Quaternary Research, v. 16, p. 146-166.

Wilson, C.
1983: Some aspects of the calibration of early Canadian temperature records in the Hudson's Bay Company archives: A case study for the summer season, eastern Hudson/James Bay, 1814 to 1821; Climatic Change in Canada 3, Syllogeus 49, p. 144-202.

Wilson, M.A.
1984: Postglacial climatic and vegetation history of the La Ronge area, northern Saskatchewan; Saskatchewan Research Council, no. R-743-2-E-84, 90 p.

Wilson, M.
1974: Fossil *Bison* and artifacts from the Mona Lisa Site, Calgary, Alberta. Part 1: stratigraphy and artifacts; Plains Anthropologist, v. 19, p. 34-45.

Wilson, M. and Churcher, C.S.
1978: Late Pleistocene *Camelops* from the Gallelli Pit, Calgary, Alberta: morphology and geologic setting; Canadian Journal of Earth Sciences, v. 15, p. 729-740.

Woillard, G.
1978: Grand Pile peat bog: a continuous pollen record for the last 140 000 years; Quaternary Research, v. 9, p. 1-21.

Wright, G.F.
1893: Man and the Glacial Period (2nd edition); Paul, London.

Wright, H.E., Jr.
1976: The dynamic nature of Holocene vegetation: a problem in paleoclimatology, biogeography, and stratigraphic nomenclature; Quaternary Research, v. 6, p. 581-596.
1984: Sensitivity and response time of natural systems to climatic change in the late Quaternary; Quaternary Science Review, v. 3, p. 91-131.

Youngman, P.M.
1986: The extinct short-faced skunk *Brachyprotoma obtusata* (Mammalia, Carnivora): first records for Canada and Beringia; Canadian Journal of Earth Sciences, v. 23, p. 419-424.

Authors' addresses

T.W. Anderson
Geological Survey of Canada
601 Booth Street, Ottawa, Ontario
K1A 0E8

M. Boyko-Diakonow
3688 Parkview Street
Penticton, British Columbia
V2A 0H1

R.W. Mathewes
Department of Biological Sciences
Simon Fraser University
Burnaby, British Columbia
V5A 1S6

J.V. Matthews, Jr.
Geological Survey of Canada
601 Booth Street, Ottawa, Ontario
K1A 0E8

J.H. McAndrews
Department of Botany
Royal Ontario Museum
Toronto, Ontario
M5S 2C6

R.J. Mott
Geological Survey of Canada
601 Booth Street, Ottawa, Ontario
K1A 0E8

P.J.H. Richard
Département de géographie
Université de Montréal
Montréal, Québec
H3C 3J7

J.C. Ritchie
Life Sciences, Scarborough College
West Hill, Ontario
M1C 1A4

C.E. Schweger
Department of Anthropology
University of Alberta
Edmonton, Alberta
T6G 2H4

Printed in Canada

Chapter 8

QUATERNARY GEODYNAMICS IN CANADA

Summary

Introduction — *J.T. Andrews*

Nature of the last glaciation in Canada — *J.T. Andrews*

Postglacial emergence and submergence — *J.T. Andrews*

Models of glacial isostasy — *W.R. Peltier*

Acknowledgments

References

Vertebrae and ribs of a bowhead whale at an elevation of 57.5 m on Prescott Island, central Canadian Arctic. A radiocarbon date on these bones gave an age of 9333 ± 145 BP (GSC-2913). Marine limit in this area is at about 107 m and the date on these bones approximates the time the sea reached this island. Courtesy of A.S. Dyke. 204511-V.

Chapter 8

QUATERNARY GEODYNAMICS IN CANADA

John T. Andrews and W.R. Peltier

SUMMARY

This chapter provides a rather modest introduction to the extensive Quaternary data base which currently exists and to the geophysical models which have been developed to interpret these data; it is clear that an active interplay between theory and observation is desirable. The Quaternary geological data can be brought to bear upon several vital geodynamic issues, including the question of the variation of mantle viscosity with depth, the question of the thickness of the continental lithosphere, and the question of the nature of the discontinuities in elastic parameters which are observed seismically to occur at depths of 420 and 670 km in the earth. All of these questions are of importance in attempting to understand the nature of the convective circulation in the mantle which is responsible for continental drift and seafloor spreading; the Quaternary geological record has begun to provide crucial information with which it will be possible in future to further refine our understanding.

This limited space does not allow all areas of Quaternary geodynamic research to be dealt with adequately. Some aspects not dealt with are: (1) the manner in which the isostatic adjustment process may serve to cause earthquakes by the reactivation of slip on old faults located near the ice sheet margin; (2) the manner in which the gravitationally self-consistent model of relative sea level variation may be employed to filter the modern tide gauge data to reveal more clearly the presence of any currently ongoing "eustatic" component of sea level rise due to, for example, the melting of presently existing continental ice sheets; (3) the manner in which certain earth rotation data may also be brought to bear on the question of the viscoelastic stratification of the deep interior, rotation data which also constitute memories of the planet of the last deglaciation event of the current ice age.

Although the data base of Quaternary geology has played an important role in geodynamics since the early work of Haskell and Vening-Meinesz on the isostatic recovery of Fennoscandia, only in the past decade has its unique value been fully realized. These data have begun to play a leading role in helping to decipher the nature of the rheology of the mantle and also to aid in the interpretation of major features in the planetary elastic structure which have been revealed through the analysis of seismic waves.

Much of the data set is now amenable to direct quantitative interpretation through application of the gravitationally self-consistent model for deglaciation induced relative sea level change.

INTRODUCTION
J.T. Andrews

Throughout the Quaternary, Canada has been the site of a large scale natural experiment in geodynamics: the loading and unloading of the crust caused by glaciation and deglaciation. The glaciation of the Canadian northlands may have commenced as early as the Pliocene some 3.4 million years ago. Early glaciations were probably widely spaced but it appears that the periodicity of glaciation shortened during the Quaternary so that during the last approximately 600 ka Canada has been subjected to large scale continental loading and unloading of the crust on time scales of 90 ka or less. The major theme of this chapter is the glacial isostatic response of the Canada landmass to the growth and decay of large ice sheets, notably the Laurentide and Cordilleran, and the Franklin ice complex. Figure 8.1 illustrates the approximate boundaries of these three ice masses during the last glacial advance. These ice sheets were largely coeval although they did not necessarily respond exactly in phase — at least on time scales of 10^3 years. Glacial isostasy has also been linked with faulting and seismicity (Quinlan, 1984). It is feasible that the stresses induced by loading and unloading of large ice sheets could contribute to the reactivation of faults and a consequent increase in seismicity (Clark, 1982a; Quinlan, 1984). Another major aspect of Canadian Quaternary geodynamics which is probably linked with glacial loading is large-scale, negative, free-air gravity anomalies which are apparently spatially associated with the former ice sheets.

Before proceeding further with this chapter it is appropriate to define what we are not discussing and what is being treated elsewhere in the Quaternary geology of Canada and Greenland volume. The specific evidence for Quaternary glacial chronology is presented in Part 1 of this volume in the form of a series of regional chapters. In most cases, these syntheses were not available to us at the time this chapter was written. These include information on marine limit, the elevation and age of shoreline features, and the relationship

---

Andrews, J.T. and Peltier, W.R.
1989: Quaternary geodynamics in Canada; Chapter 8 in Quaternary Geology of Canada and Greenland, R.J. Fulton (ed.); Geological Survey of Canada, Geology of Canada, no. 1 (also Geological Society of America, The Geology of North America, v. K-1).

Andrews, J.T.
1989: Introduction (Quaternary geodynamics in Canada); in Chapter 8 of Quaternary Geology of Canada and Greenland, R.J. Fulton (ed.); Geological Survey of Canada, Geology of Canada, no. 1 (also Geological Society of America, The Geology of North America, v. K-1).

of these to deglaciation. For regions removed from the sea, these regional chapters contain some evidence pertaining to the pattern of shoreline warping around former proglacial lakes which is useful in determining the isostatic response of the crust.

The results of the isostatic response of the crust to glacial unloading has been used for nearly a century to assist in locating the former centre(s) of Canadian ice sheets (cf. Shilts, 1980, 1982; Denton and Hughes, 1981; Dyke et al., 1982b; Peltier and Andrews, 1983). In later sections we evaluate some of the interactions between ice sheets, earth rheology, and relative sea level variations.

During the Quaternary (the last 1.65 Ma) Canada has witnessed other geodynamical processes but these are not discussed here. The Neotectonics volume of the Decade of North American Geology series describes these, including evidence for Holocene faulting and current seismicity in Canada. Hence no attempt is made here to examine the relationship between seismicity and glacial isostasy.

This chapter is organized in three parts. The first part provides a review of our knowledge of the extent, nature, and timing of the last glaciation. The second part outlines the field evidence for postglacial emergence and submergence throughout coastal Canada, and deals with the evidence from the deformation of glacial lake shorelines. The third part presents the results of various modelling exercises that associate the glacial/deglacial history and rheological models of the earth with predicted changes of relative sea level and the extent of the free-air gravity anomaly.

NATURE OF THE LAST GLACIATION IN CANADA

J.T. Andrews

The extent of ice, the variation in ice thickness, and the timing of glaciation and deglaciation across Canada is currently a subject of intense debate. Thus the interpretation of glacial isostatic data does not start with a complete understanding of ice sheet dynamics. Indeed one of the attractions of studying glacial isostasy is that we may be able to deduce glacial history from the pattern of glacial isostatic rebound (Peltier, 1976; Clark, 1980; Quinlan, 1981, 1985; Quinlan and Beaumont, 1982; Wu and Peltier, 1983). The questions that are critical to an interpretation of the glacial isostatic data — ignoring the need to resolve a suitable earth rheology (Peltier, 1981) — are: 1) when did the ice sheets develop; how did they change in extent and thickness throughout the glaciation; and in what manner did they disappear and 2) what was the global history of glaciation and deglaciation over this same time interval. (This also affects the history of sea level movement and deformation over the Canadian landmass (Clark et al., 1978).)

In terms of the interaction between glacial geology and isostasy (both forward and reverse problems (Peltier, 1976; Peltier and Andrews, 1976)) one very important point is the difference between an ice divide and an area of maximum ice thickness (cf. Andrews, 1982). On current ice sheets, such as Greenland (Radok et al., 1982), some ice divides (which control the direction of basal flow) are located over high subglacial topography whereas the thicker ice overlies interior basins. These observations may partly explain spatial disagreements between areas of maximum glacial unloading and ice flow divides (Andrews and Peltier, 1983) and might reconcile some of the conflict on the geometry of the Laurentide Ice Sheet (e.g., Shilts, 1980; Denton and Hughes, 1981; Dyke et al., 1982b).

Glacial isostatic evidence from Canada is nearly entirely restricted to beaches and shoreline features. These data largely document the last deglacial episode. We thus have a restricted view of the isostatic processes associated with the last glacial unloading of the crust and the associated adjustments of the earth's surface. Evidence for the nature of isostatic response during periods of ice sheet growth comes from the eastern coast of outer Baffin Island where an extensive series of wave-cut cliffs face Baffin Bay and Davis Strait (Miller et al., 1977; Nelson, 1981; Brigham, 1983; Mode et al., 1983). In this area the sequence of Quaternary sediments consists of repetitive packages of sediments. Each package is interpreted as a sedimentary response to a cycle of glacial loading and unloading (Fig. 8.2). The last major glacial cycle to directly affect the outer coast of Baffin Island occurred beyond the limits of ^{14}C dating (Løken, 1966; Miller et al., 1977; Dyke et al., 1982a; Andrews and Miller, 1984) and approximate ages for these "old" glacial and glacial marine sediments can be obtained only by U-series dating (Szabo et al., 1981) and amino acid racemization methods (Miller et al., 1977; Miller, 1985). Thus we

Andrews, J.T.
1989: Nature of the last glaciation in Canada; in Chapter 8 of Quaternary Geology of Canada and Greenland, R.J. Fulton (ed.); Geological Survey of Canada, Geology of Canada, no. 1 (also Geological Society of America, The Geology of North America, v. K-1).

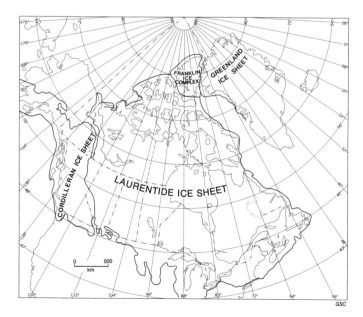

Figure 8.1. The Late Wisconsinan ice sheet complexes of Canada and North America.

have no way of gauging how rapidly the crust reacts to the buildup of an ice load. We do know from the lithofacies on the outer coast of Baffin Island that sea level started to rise some time after the end of an interglaciation and that sea level was rising prior to glacial ice arriving at a site. This has several implications. First, it strongly suggests that the earth's surface was being depressed beyond the ice margin. This provides limits for an appropriate earth model (cf. Brotchie and Sylvester, 1969; Walcott, 1970; Peltier, 1981, 1982, 1984; England, 1983) as it indicates that the crust was flexing in front of the advancing ice margin. Second, the absence of any difference in amino acid racemization ratios between the transgressive and regressive facies (Fig. 8.2) indicates that the complete glacial loading and unloading cycle and the associated sea level rise and fall were complete within a span of approximately 10 ka. This is of similar duration to the cycle of glacial unloading that has been observed throughout Canada (e.g., Walcott, 1972; Blake, 1975; England, 1983).

The question of the extent and timing of glaciation over Canada during the last interglacial/glacial cycle is not well resolved and is discussed at length in the various regional chapters in Part 1 of this volume. It is an extremely important topic for Quaternary geodynamics because without a well constrained history of ice advance and retreat, attempts to reconstruct and/or interpret the ice sheet/earth rheology/relative sea level interactions have been forced to commence with full isostatic compensation at 18 ka (e.g., Peltier, 1974, 1976; Peltier and Andrews, 1976, 1983; Clark et al., 1978). It is clear, however, that in certain parts of Canada (e.g., southern Cordillera and southern Canadian Plains) little or no ice was present during the Middle Wisconsinan and that the extent to which isostatic compensation was achieved during subsequent ice growth can only be inferred.

In the Baffin Sector of the Laurentide Ice Sheet the transgressive deposits associated with the early part of the last glaciation date from marine oxygen isotope stage 5, possibly commencing about 115 ka, whereas in southern Canada the onset of this glaciation was presumably later (70 ka?). Evidence from interior British Columbia (Fulton and Westgate, 1975) indicates that the post-last interglaciation ice advance began prior to about 55 ka. Then there was a major interstade which lasted from >55 ka to about 25 ka, followed by a rapid expansion of ice during the Fraser Glaciation. Evidence for a long interstade during Middle Wisconsinan time comes from many areas within the region covered by the Laurentide Ice Sheet, particularly from the Prairie Provinces and the Great Lakes region. No evidence, however, for a major Middle Wisconsinan interstade of several tens of thousands of years duration has been reported from southern Quebec along the axis of St. Lawrence Valley. In the High Arctic the situation is far from clear. There is evidence for an early glaciation, followed by a Middle Wisconsinan interstade (Blake, 1980), which in turn was succeeded by ice expansion during late glacial time (Blake, 1970, 1975, 1976; England, 1976, 1983; England et al., 1981). There is no agreement, however, on the extent of ice cover during the past 100 ka (see Hodgson, 1989).

An area crucial to understanding the geodynamic response to the last glaciation is the region near the geographic centre of the Laurentide Ice Sheet. A knowledge of the glacial history of this area is necessary before accurate models of ice loading and unloading can be prepared. Hudson Bay and James Bay Lowlands are characterized by

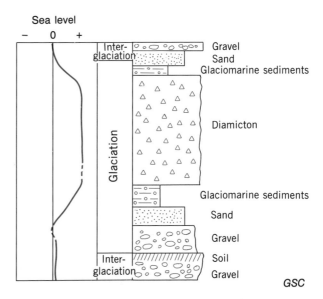

Figure 8.2. Sedimentary sequence (idealized) from the outer east coast of Baffin Island which is associated with the loading and unloading of the coast by the growth and retreat of the Laurentide Ice Sheet (from Mode et al., 1983).

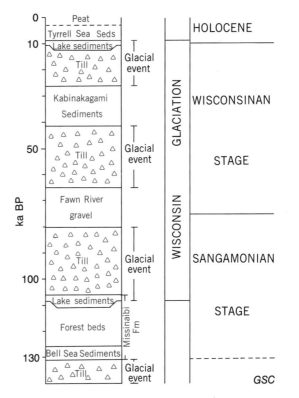

Figure 8.3. Schematic stratigraphy of the Hudson Bay Lowlands and the inferred history of growth and retreat of the Laurentide Ice Sheet over the last 125 ka (from Andrews et al., 1983).

relatively thick sections of Quaternary sediments (e.g., McDonald, 1969, 1971; Skinner, 1973; Shilts, 1982; Andrews et al., 1983). McDonald (1971) originally proposed that the Missinaibi Formation was the correlative of the St. Pierre interstadial beds of southern Quebec and thus envisioned extensive deglaciation of Hudson Bay during the Wisconsin Glaciation. Most workers, however, have assigned the Missinaibi Formation to the "last interglacial" with sediments of the Bell Sea considered to be early last interglaciation (Fig. 8.3). Amino acid ratios of shells from the tills and nonglacial sediments indicated to Andrews et al. (1983; see also comments and discussions by Dyke, 1984 and Andrews et al., 1984) that several populations of shells of different ages were present in the Hudson Bay Lowlands. They calculated ages of the various amino acid groups on the assumption that the sediments of the Bell Sea recorded an interval of deglaciation dated at 130 ka. These age estimates suggest that deglacial events occurred within Hudson Bay about 118, 76, and 35 ka (Fig. 8.3). Near the entrance to Hudson Strait, there is evidence for a Middle Wisconsinan deglacial event (Clark, 1982b, 1984; Miller, 1985). This is in the form of in situ marine shells which have ^{14}C dates between 28 and 42 ka and amino acid ratios only slightly higher than those in early Holocene shells.

These data imply that glaciation of Hudson Bay and the western and southwestern sectors of the Laurentide Ice Sheet occurred at roughly a 40 ka periodicity. Time-dependent models of the glaciation of Canada have been presented by Budd and Smith (1981) and Smith (1984) who used Milankovitch radiation cycles as an input into their glaciological model. Although much of the data input was derived from theoretical considerations it is nevertheless intriguing that they also showed the North American ice sheets waxing and waning on a 40 ka periodicity. Figure 8.3 thus presents a possible framework for the glacial events within Canada over the last 130 ka although it must be stressed that the dating control is minimal.

Peltier (1982) suggested that the deep sea $\delta^{18}O$ record might serve as a surrogate for the history of Wisconsin Glaciation across North America. Such records vary but show a series of oxygen isotope events dominated by a 90 ka cycle, but also with significant periodicities at about 20 and 40 ka. Another possible surrogate for the North American ice sheets might be sea level fluctuations from regions well removed from the ice sheets such as Barbados and New Guinea (Matthews, 1973; Bloom et al., 1974). Although many workers have suggested that the $\delta^{18}O$ record and sea level fluctuations are comparable, and have directly inferred sea level from ^{18}O changes (Shackleton and Opdyke, 1973), Andrews (1982) pointed out that there were large differences between the two. Aharon (1983) reached a similar conclusion with estimated sea levels differing by as much as 70 m between the two methods. Table 8.1 illustrates one aspect of the level of uncertainty in estimates of ice sheet volume on a global scale over the last 133 ka.

There is no agreement on the extent of sea level lowering (i.e., ice sheet volume) during the 18 ka glacial maximum. Peltier and Andrews (1976), for example, suggested that water contained in ice sheets of the Northern Hemisphere was equivalent to about 70-80 m of sea level, whereas in the maximum ice sheet reconstruction of Denton and Hughes (1981), the amount of sea level lowering is almost double that value.

Table 8.1. Estimations of paleosea levels based on (1) elevation of reef crests and uplift calculations in unglaciated areas and (2) $\delta^{18}O$ content of coral from reef crests (from Aharon, 1983)

Age ka	Sea Level (1)	Sea Level (2)
0	0	–
7	4	–65
28.5	–42	–108
40	–39	–84
45	–30	–87
60	–28	–97
85	–13	–32
107	–15	–22
120	+8	–62
133	+5	+4

This section illustrates the extent of our lack of knowledge about the last glaciation in Canada. If we had a complete understanding of the configuration of the ice sheets, and their chronology, then we would be able to rigorously constrain a major element in our modelling of the process of glacial isostasy. However, the ice sheet portion of the modelling (e.g., Peltier, 1974, 1976, 1982; Farrell and Clark, 1976; Peltier and Andrews, 1976, 1983; Quinlan, 1981; Quinlan and Beaumont, 1982) is not well constrained and this imposes limits on what can be said about the geophysics of glacial rebound. These points should be remembered when evaluating the final part of this chapter.

POSTGLACIAL EMERGENCE AND SUBMERGENCE

J.T. Andrews

Observations on the elevation of raised marine and lacustrine deposits and landforms have been features of research in Canada for over a century. This work has included determining marine (or lacustrine) limits, and sea (or lake) levels at specific times, and delimiting synchronous shorelines. Much of the data related to marine shorelines have been compiled in a series of annotated bibliographies (Richards and Fairbridge, 1965; Richards, 1970, 1974; Richards and Shapiro, 1979); in the Richards and Shapiro volume the annotated bibliographies on Canadian shorelines run to 20 pages out of a total global coverage of 230 pages. Similar compilations are not available for lake shoreline data. Paleoshoreline data play an important role in determining the Quaternary geodynamics of Canada but marine limit is probably the single most useful piece of information and it is used as the basic datum from which to discuss glacial isostatic rebound and late glacial history.

Andrews, J.T.
1989: Postglacial emergence and submergence; in Chapter 8 of Quaternary Geology of Canada and Greenland, R.J. Fulton (ed.); Geological Survey of Canada, Geology of Canada, no. 1 (also Geological Society of America, The Geology of North America, v. K-1).

Prior to the advent of radiocarbon dating, little information existed on the timing and rate of glacial isostatic recovery; however, once ^{14}C dating had become accepted, dating the raised beach sequences throughout Canada became a prime research goal. Particularly in the 1960s, but continuing to the present, a focus of research has been the presentation of dated sea level curves from around the coasts of Canada (e.g., Andrews, 1970; Walcott, 1972; Clague, 1975, 1983; Blake, 1976; Dyke, 1979; Hillaire-Marcel and Vincent, 1980). These data have been used by geophysicists to understand the long term relaxation processes of the crust and mantle on time scales of 10^3 to 10^4 years (Walcott, 1970; Peltier, 1974, 1976, 1981, 1982, 1984; Clark, 1980).

Data and errors

Obtaining shoreline related isostatic data involves a number of procedures and interpretations and includes several steps where error and uncertainty may enter the results. The procedure consists of locating the position of a former water plane, measuring the elevation of this paleolevel, and determining the time at which water stood at that specific level. The discussion below relates specifically to marine shoreline features but many of the points apply equally to lacustrine shorelines.

The accuracy and precision of elevation measurements have been discussed by several workers (Bird, 1954; Sim, 1960; Løken, 1962; Andrews, 1970; Blake, 1975; Hillaire-Marcel, 1979; Andrews, 1986). In remote areas there are problems of defining the present reference "sea level". Then there is the question of whether it is low tide, high tide, high high tide, or mean tide that a particular feature relates to. There is a need to make explicit whether an elevation refers to a deposit or landform that was constructed during storm conditions — such as shingle ridges — or represents some other tide level — such as the lip of a delta. These assessments are made more difficult by the probability that tidal ranges have varied as the geometry of a water body changed in response to oscillations in sea level (Grant, 1970; Jardine, 1975; Scott and Greenberg, 1983). This can be critical if excessively high tidal ranges exist, such as in the Bay of Fundy and Ungava Bay.

The determination of former sea levels requires the determination of elevations (above or below present sea level); a level and stadia rod should be used for detailed elevational control (Blake, 1975). Accurate surveying of a former water plane does not, however, get around the difficulty of determining how closely datable organic materials are associated with a specific paleolevel. Shells lying within beach gravels can clearly be reworked from both older deposits offshore or material can be carried to the sea from erosion of pre-existing fossiliferous strata. Wood incorporated into beach gravels can be reworked from older units; articulated marine bivalves represent an in situ horizon within the sediment but can they be unambiguously associated with a specific paleosea-level? If the shells occur in well defined deltaic foresets then the answer is yes, within limits, but if they occur at the base of the section in undifferentiated marine muds then the association with a former sea level is largely a guess. Thus elevational errors consist of an error associated with the surveying technique plus a more poorly defined uncertainty linked to the clarity of the association between the materials being dated and the paleosea-level.

The above discussion has concentrated on the raised marine record; however, in parts of Canada relative sea level changes during the Quaternary have also involved marine transgressions so that paleoshorelines are now submerged. Areas where such evidence is important include much of the Maritimes, much of the coastal area of British Columbia, and the outer margin of the Laurentide Ice Sheet in Arctic Canada, namely the coast of the Beaufort Sea and outer east coast Baffin Island. Evidence for marine transgressions during the Holocene includes submerged forests, saltmarsh, and freshwater peats buried beneath marine muds, and submerged littoral facies and landforms (e.g., Grant, 1970; Clague, 1983; Hill et al., 1985). The study of these deposits and the dating of their association with paleosea-levels is a difficult task because often the information can be obtained only through an elaborate process of marine geophysical surveys and coring.

After an organic sample has been collected and its elevation and stratigraphic context determined, the problem of dating must be faced. Here, relative change of elevation associated with the last glaciation is the main concern so radiocarbon dating is the main technique available. The bulk of ^{14}C dating of sea level related materials in Canada has been carried out in the Geological Survey of Canada (GSC) radiocarbon dating laboratory (Lowdon, 1985). There are several problems associated with ^{14}C dating that must be noted: (1) There may be systematic differences in procedures and in calculations between laboratories. (2) There are differences in the manner in which the standard error about the mean is calculated and reported (as either a one or two standard errors). (3) There are specific differences in the way in which laboratories calculate the ages of marine shells; as an example, the same suite of shells dated by the GSC and the Geochron Laboratory would be reported with a consistent difference of about 410 years — the GSC date being younger (see discussion in Washburn and Stuiver, 1985; Andrews and Miller, 1985). (4) There is not a 1:1 relationship between calendar years and ^{14}C years due to the variation in production of ^{14}C in the atmosphere through time. Over the last 8 ka this correction can amount to nearly 1 ka (Klein et al., 1982). Although it is legitimate to apply such corrections to terrestrial materials, it is not known how the corrections can be applied to marine organisms. Where there are series of comparable dated sets, however (e.g., Blake, 1975 at Cape Storm, Ellesmere Island on whalebone, shell, and driftwood), no consistent differences are found. In contrast, however, there are considerable differences in radiocarbon ages reported on marine shells and the acid-insoluble organic fraction (Fillon et al., 1981; Andrews et al., 1985) of marine muds. This age discrepancy can amount to several thousand years and is attributable to the transportation of "old" carbon (Mudie and Guilbault, 1982).

Thus the dating and reporting of the ages of samples associated with raised or submerged marine/littoral units is not as simple as it might appear. The error quoted on a ^{14}C dated sample does not take into account the geological uncertainty. Despite the litany of problems noted above, smooth, relatively well fitting curves can generally be drawn through sea level dating data (Fig. 8.4).

The advent of mass accelerators for ^{14}C dating purposes may revolutionize paleosea-level studies. This technique enables dates to be obtained on 20-50 mg of material (Litherland, 1980; Andrews et al., 1985; Blake, 1985); thus small sample size may no longer be a limiting factor in dating both raised and submerged deposits.

Glacial lake shorelines

Extensive glacial lakes existed over large areas of North America during the last deglaciation (cf. Prest et al., 1968; Andrews, 1973a; Teller and Clayton, 1983; Karrow and Calkin, 1985). They occurred wherever the ice retreated down the regional slope or where the ice margins blocked natural drainage routes (Fulton and Walcott, 1975). The level of glacial lake shorelines cannot be tied to a worldwide datum in the same manner as paleosea-levels. This means that glacial lake shoreline data cannot be used in exactly the same manner as marine shoreline data, but they provide critical measures of surface deformation in areas far removed from the sea.

By definition, the strike of a deformed water plane parallels regional isobases. Elevational observations along a mappable shoreline that is an isochronous surface, provide a measure of the dip of the deformed shoreline and the rate of change of the deformation along the length of the shoreline. If successively younger shorelines exist then changes through time of these parameters can be looked for. Such techniques have been most often applied in studies of glacial lake shorelines (see Andrews and Barnett, 1972, for references) but a similar approach has been attempted on synchronous raised marine shorelines (Løken, 1962; Andrews, 1970; Dyke, 1983, 1984) with limited success.

Glacial lake shorelines can be dated through radiocarbon assays on associated wood, peat, or shells. The ages of shorelines can also be determined by radiocarbon dating of sediment close to the transition from glacial to lacustrine sedimentation in small lake basins above and below a shoreline. This approach, however, can result in significant errors (e.g., Short, 1981) because of the influx of "old" carbon into sedimentary basins during deglaciation.

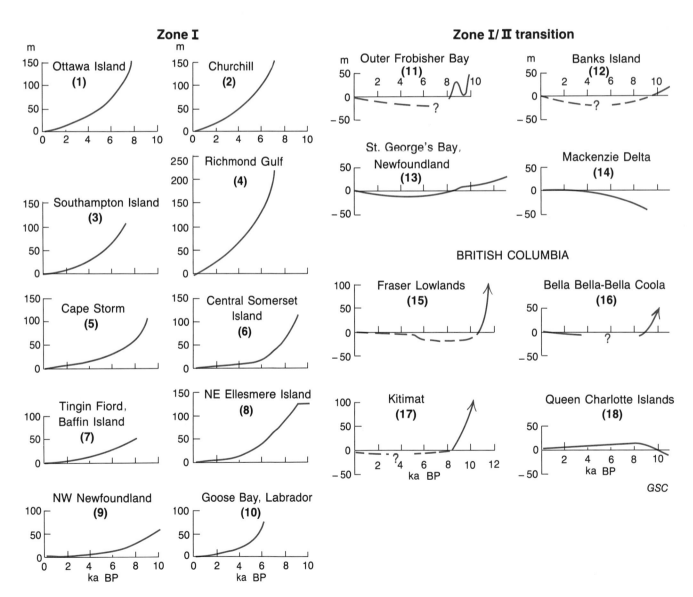

Figure 8.4. Selected relative sea level curves from Canadian sites; the sites are numbered and their locations are shown in Figure 8.5.

QUATERNARY GEODYNAMICS IN CANADA

In areas where glacial lakes existed over several thousands of years (Prest, 1970; Vincent and Hardy, 1979; Teller and Clayton, 1983; Karrow and Calkin, 1985) it is possible to use the shoreline deformation data to discover whether there is evidence for shifts in the direction of glacial isostatic recovery during deglaciation. This is an important question to both the Quaternary geologist interested in the possible changing location of ice divides and the geophysicist concerned with the process of isostatic compensation. At present, however, there is little concrete evidence to document progressive shifts in isobase directions (lake shoreline strikes) during the last 12 ka in Canada.

Figure 8.5 illustrates the direction of dip of glacial lake and marine shorelines andTable 8.2 provides other pertinent data. Although there is reasonable coverage in the southeast, south, and east sectors of the Laurentide Ice Sheet, there is a lack of data from west of Hudson Bay and more measurements would be useful in the southwest and northwest sectors of the former Laurentide Ice Sheet. The intersection of several shoreline dips should indicate either the presence of an area of uplift or at least the former location of a ridge in the ice sheet (Andrews and Barnett, 1972). Evidence from the Great Lakes and glacial Lake Barlow-Ojibway (sites 18, 20, 39, Fig. 8.5) for example indicates that

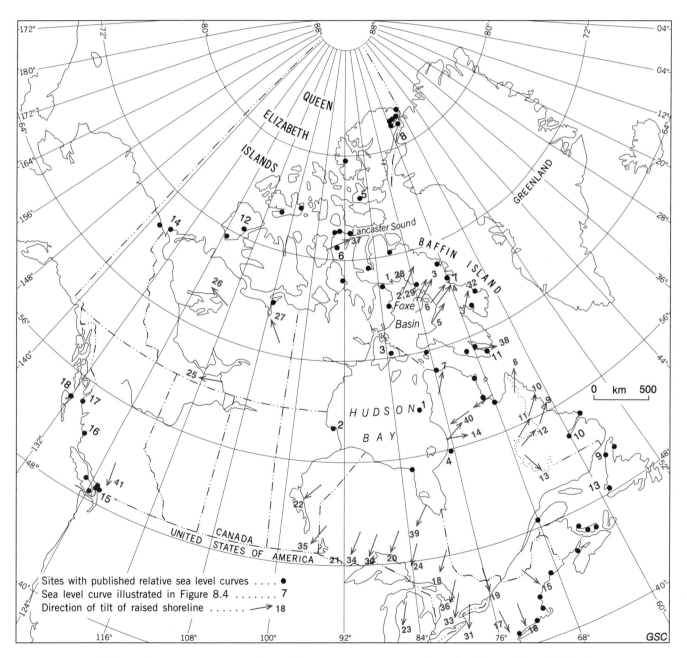

Figure 8.5. Location map of published relative sea level curves — those for sites numbered in black are shown in Figure 8.4. The arrows show direction of tilting of glacial lake and raised marine shorelines. Sites are described in Table 8.2.

the isostatic recovery of the region south and south-southwest of Hudson Bay has been dominated by uplift to the northeast towards the plateaus of central Labrador. The work of Lauriol and Gray (1983) on glacial lakes between Hudson Bay and Ungava Bay (site 40, Fig. 8.5) suggests that the former lake surface is tilted down towards the west, but site 14 shows a tilt in the opposite sense. This latter site is, however, based on marine limit elevations and may not measure a synchronous shoreline deformation. The shorelines from eastern Labrador-Ungava (sites 11, 12) suggest a centre of isostatic recovery towards the centre of the peninsula. Thus there is little evidence for a major centre of recovery located over Hudson Bay, and the evidence from glacial Lake Barlow-Ojibway (site 39) would appear to deny this possibility. An ice dome, however, might have existed over Hudson Bay for a short period of time before continued flow of ice through Hudson Strait led to its destruction and ice flow centres were restricted to ice divides over Labrador and Keewatin. Hillaire-Marcel (1980) has proposed that the postglacial rebound data indicate an early interval of unloading located over Hudson Bay followed by a shift in the locus of uplift to extend across the Labrador plateaus. In my view the strikes of lake shoreline deformations currently available do not support evidence for the migration of a zone of postglacial recovery.

Marine limit

Maps of the highest level reached by the Late Pleistocene/early Holocene seas have been published over many decades. This upper limit of marine processes serves as a fundamental stratigraphic marker in studies of deglaciation. In areas within the maximum limit of the former ice sheets, the marine limit was formed successively as the ice retreated;

Table 8.2. Information on sites where direction of shoreline tilt has been determined. Numbers refer to site locations in Figure 8.5; numbering begins in the northeast and runs clockwise (from Andrews and Barnett, 1972)

Site (cf. Fig. 8.5)	Bearing of tilt ± degrees	Gradient m/km	Approximate age ka	Marine (M) or glacial Lake (GL)
1	215	1.0	7	M
2	229	0.6	7	M
3	230	0.6	7	M
4	230	1.0	8	M
5	225	0.14	7	M
6	232	0.54	8	M
7	205	1.06	7.8	M
8	205	1.0	10 (?)	M
9	235	1.2	10 (?)	M
10	215	0.73	10 (?)	M
11	225	0.44	8.8	GL Naskaupi
12	238	0.44	8.8	GL Naskaupi
13	330	0.35	7.5	GL Wapussakatoo
14	230	0.8	7.7	M
15	315	0.72	12.6	GL Memphremagog
16	350	~1.0	13–12 (?)	GL Sudbury-Concord
17	340	~1.0	13.6 (?)	GL Vermont
18	020	~1.0	9 (?)	GL Barlow-Ojibway
19	345	0.57	11.7	M Champlain
20	045			GL
21	030	0.58	~9.5	GL Agassiz
22	045	0.57	12	GL Agassiz
23	015	~0.76	10.5	GL Algonquin
24	025	~0.08	5	GL Nipissing
25	090		9 (?)	GL McConnell
26	135	0.8	9.5 (?)	GL Great Bear
27	155	<0.3	8.5	M
28	None	No tilt	9.5	GL
29	None	No tilt	3	GL Generator
30	030	0.19	~5	GL Nipissing
31	024	0.93	12.7	GL Whittlesey
32	230	0.50	5	M
33	029			GL Warren
34	027	0.28	12.5	GL Agassiz
35	036	0.25	12.5	GL Agassiz
36	023	0.33	10.5	GL Algonquin
37	060	1.4-.35	9.3	M
38	010	-	10–7	M
39	030		10–9	GL
40	270	-	8–7	M
41	340-350	2.5-1.8	10.5–9	GL British Columbia

thus it is not synchronous and elevations on it do not portray the time-dependent form of glacial rebound. For example, the age of the marine limit associated with the last deglaciation varies across Canada from ca. 14 ka to 5 ka, and the range in age of the highest raised marine deposits, regardless of association with the last deglaciation, ranges between >70 ka to the present (the latter equals no observable marine limit). Nevertheless, the elevation of the marine limit provides an important reference for glacial isostatic recovery, and it points to several problems connected with our understanding of the last glaciation of Canada. Figure 8.6 shows a selection of marine limit elevations; it can be compared with the historical sequence of marine limit isoline maps shown in Figure 8.7. Figure 8.6 also shows the relative sea level zones of Clark et al. (1978). These zones delimit areas of fundamentally different responses to the process of glacial unloading and the addition of meltwater to the global ocean.

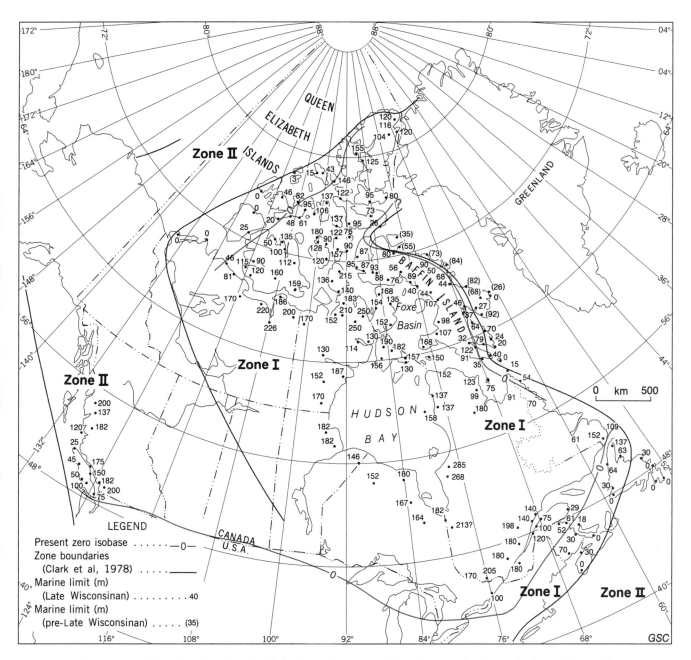

Figure 8.6. Elevation (m) of marine limits from sites around Canada. Map also shows the boundaries of the various sea level response zones of Clark et al. (1978) (see Fig. 8.8) and the location of the present zero (0 m) isobase. Marine limit elevations from Prest et al. (1968), Andrews (1970, 1980), Clague (1981), Dyke (1984), Dredge and Cowan (1989), and Dyke and Dredge (1989).

Marine limits in Canada were once associated exclusively with Late Wisconsinan deglaciation; however, in several areas of eastern and northern Canada the highest marine beaches or sediments predate the Late Pleistocene deglaciation (e.g., Løken, 1966; Miller and Dyke, 1974; Dyke, 1977; England et al., 1978; Andrews, 1980; Vincent, 1982) and final deglaciation is marked by a prominent low-level shoreline that is eroded into the older marine sediments (Andrews, 1989).

The information in Figure 8.6 has not changed dramatically from earlier marine limit maps (see Fig. 8.7). The area of highest observed Holocene rebound is still eastern

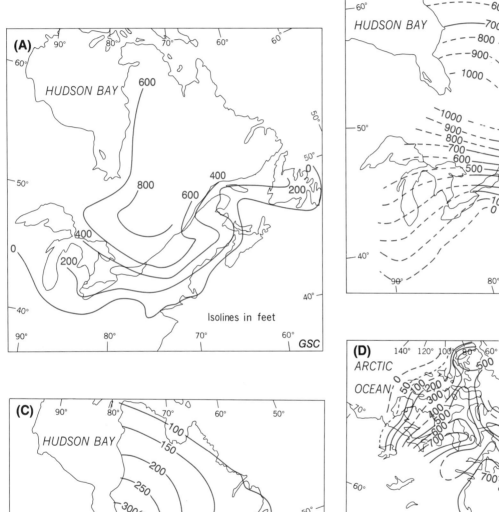

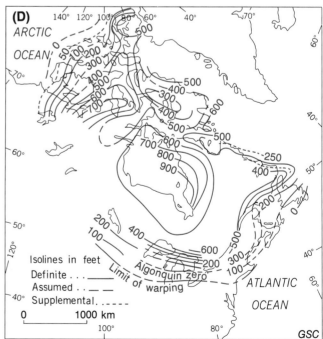

Figure 8.7. Historical sequence of maps showing marine limit isolines for Canada and adjacent parts of the United States (Andrews, 1973b) modified from a) De Geer, 1892; b) Fairchild, 1918; c) Daly, 1934; and d) Farrand and Gajda, 1962.

Hudson Bay where the marine limit has been measured at close to 300 m a.s.l. To the south there is still an area of high marine limit in the interior of St. Lawrence Valley. A major difference between the present record and the map of Farrand and Gajda (1962) is the lower levels of Holocene age along the east coast of Baffin Island and northern Ellesmere Island. Other notable features include the zone of high marine limits along the Arctic coast from Bathurst Inlet to northeast Keewatin, and extending as a ridge northward, and finally the area of high marine limit (>100 m) which extends from Cornwallis Island northeastward to Ellesmere Island (England, 1976, 1983; Washburn and Stuiver, 1985).

An important geodynamical question is how far do these "highs" and "lows" in marine limit elevation reflect differences in ice thickness and how far do they represent local deglacial effects? The simple association of the marine limit with Late Wisconsinan deglaciation is appropriate only within Zones I and possibly parts of the Zone I/II transition (Fig. 8.6). In Zone II and other parts of Zone I/II transition (see next section for discussion of these zones), the "marine limit" represents a Holocene sea level transgression associated with global sea level rise. Two examples of such regions are the extreme outer coast of Cumberland Peninsula, Baffin Island (Dyke, 1977; Andrews, 1980; Quinlan, 1985) and the Queen Charlotte Islands (Clague et al., 1982).

Figure 8.6 shows the zero isobase or the boundary separating coasts that are still undergoing isostatic recovery from those that are submerging (essentially Zone I/Zone I/II transition). This boundary cannot be accurately mapped in many areas because of the lack of data. It is known that much of the eastern coastline of Canada is being submerged. The extent of submergence along the Labrador coast is not well reported but I suggest that the outer coast and islands may be effected by the current rise in global sea level (Gornitz et al., 1982; Barnett, 1983). Much of the outer east coast of Baffin Island is currently being submerged and the boundary runs along the middle sections of the fiords. Submergence of coastal areas has been reported from the Beaufort Sea coast but the position of the boundary along the outer Queen Elizabeth Islands is not well known.

Relative sea level curves

Relative sea level curves are of fundamental importance to both Quaternary studies and to geophysics. Such curves describe the relaxation process related to deglaciation (e.g., Walcott, 1972; Peltier and Andrews, 1976; Clark, 1980) and they can be used to draw isobases on postglacial recovery. About 80 curves for Canada and the adjacent United States (e.g., Thorson, 1981; Bloom, 1977) have been used. Figure 8.4 illustrates some selected curves and Figure 8.5 shows the location of these. They have been drawn by interpolating between data points positioned at 1 ka intervals. Additional curves will be available on a Quaternary paleosea-level map which is currently being prepared by B.R. Pelletier of the Geological Survey of Canada.

Relative sea level curves reflect the combined interaction of glacial rebound and the global rise of sea level associated with the melting of ice sheets. Initially it was postulated that a global eustatic sea level curve existed which could be used to convert relative sea level curves (Fig. 8.4) into actual uplift curves (e.g., Andrews, 1970). However, the works of Walcott, Peltier, and Clark have demonstrated that each point on the earth's surface has a unique response to the combined processes of glacial unloading and sea level rise from ice sheet melting. For this reason we restrict our discussion to relative sea level oscillations.

Figure 8.8 illustrates three simplified relative sea level curves, each typical of curves found in one of the sea level zones of Clark et al. (1978). These show the general movement and timing of sea level change during the past 10 ka. Near the geographic centre of the Laurentide Ice Sheet (Zone I) approximately 220 m of emergence has occurred in about 8 ka (4, Fig. 8.4). On the Ottawa Islands, a short distance farther north (1, Fig. 8.4) about 150 m has occurred in 7.5 ka. Farther north on Southampton Island (3, Fig. 8.4) the work of Bird (1970) suggests that 110 m of emergence has taken place in 7 ka. This figure is similar to that for site 1 and somewhat faster than on western Baffin Island where emergence of 105 m occurred over 6.7 ka (Andrews, 1966). In an intermediate zone around the geographic centre of the Laurentide Ice Sheet deglaciation occurred 12 to 8 ka. In St. Lawrence Valley, between Ottawa

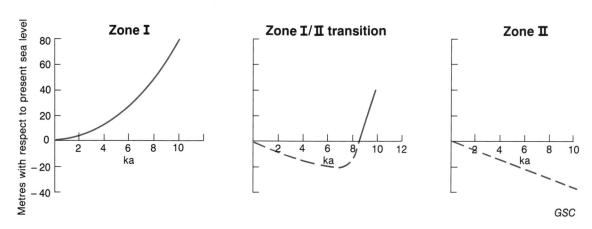

Figure 8.8. Simplified relative sea level curves from the various zones described in Clark et al. (1978) (cf. Fig. 8.5).

200 m of emergence has taken place in the last 12 ka, whereas in the vicinity of Bathurst Inlet, along the Arctic coast, 210 m of emergence is recorded in 8.5 to 9 ka.

Relative sea level curves from the ice marginal zone (i.e. Zone I/II transition, Clark et al., 1978) show differences in the response depending on their location (11-13, Fig. 8.4). In the extreme northwest sea level along the Beaufort coast (site 12) is dominated by submergence (Forbes, 1980; Vincent, 1982; Hill et al., 1985) and represents a similar pattern to that noted for part of the Canadian Maritimes (Grant, 1970). Along the outer coast of eastern Baffin Island the Holocene transgressive limit lies between 0 and 30 m with associated ^{14}C dates of between 9.5 and 10.5 ka (Andrews, 1980; Andrews and Miller, 1984). The last few thousand years, however, have been dominated by submergence (site 11, outer Frobisher Bay). In parts of Newfoundland emergence commenced between 13 and 12 ka (Grant, 1972, 1975, 1980; Brookes, 1969) and amounted to between 30 and 100 m. In many areas the Holocene has been marked by continued submergence (site 13).

Along the coast of British Columbia relative sea level curves (Mathews et al., 1970; Fladmark, 1974; Clague, 1975; Andrews and Retherford, 1978) show a period of initial emergence, followed by submergence/emergence. Present sea level was attained about 9 ka. Since then there has been submergence of a few metres. Even at the fiord heads emergence was replaced by coastal submergence over the last 5 ka. In British Columbia the relatively high marine limits of 200 m occur significantly closer to the outer coast than they do along the eastern and northeastern coasts of Canada (Fig. 8.6). Fulton and Walcott (1975) have suggested that this may be associated with a relatively thin lithosphere in this part of western Canada.

What is probably the most closely dated relative sea level curve in the world comes from Cape Storm, Ellesmere Island in the High Arctic (Blake, 1975). The curve shows the "normal" exponential decrease in sea level as a function of time (5, Fig. 8.4). There is no evidence of reversals in the emergence and Blake (1975) suggested that his data preclude eustatic sea level oscillations of >2 m over the last 8 or so ka. The data suggest that glacial isostatic rebound is still occurring.

Nearly all relative sea level curves from Canada show relatively simple patterns of sea level change dominated by emergence (Zone I), emergence/submergence (Zone I/II), or submergence (Zone II). Examples of these are curves 4, 13, and 14 of Figure 8.4. In the last few years, however, there are increasing hints that sea level variations, at some sites, did not follow these straightforward patterns. For example, Andrews (1980), England (1983, 1985) and Clague (1985) noted evidence for relative sea level stability near the onset of deglaciation (8, Fig. 8.4). England (1983) dealt with the northeast part of Ellesmere Island, only 40-100 km from the coast of northwest Greenland. The sea level curve in Figure 8.4 shows a unique feature in the early part of the record when sea level changed slowly. England (1983) explained this by invoking an ice stillstand between at least 11 and 8 ka which was followed by deglaciation and unloading. In England's interpretation (1976, 1983) the 120 m or so of Holocene rebound reflects unloading of a peripheral depression which was developed by the flexing of the lithosphere away from the margin of the ice sheet. Clague (1985) suggested that sea level remained stable for at least several hundred years around 10 ka near Kitimat, British Columbia.

Other areas where sea level curves do not fit the apparent standard pattern are southern British Columbia, eastern Baffin Island, and Prince of Wales Island. Rapid emergence/submergence/emergence was proposed by Mathews et al. (1970) for southwest British Columbia. The explanation of this sequence of oscillations remains in doubt. On eastern Baffin Island sea level curves from near the fiord heads (e.g., 7 of Fig. 8.4, 8.5) are dominated by emergence (Løken, 1966; Andrews et al., 1970; Dyke, 1979) but on the outer east coast a more complex sea level history is suggested (e.g., 11, Fig. 8.4, 8.5). Andrews (1980) noted that the shoreline diagram from around Clyde Fiord, eastern Baffin Island, suggested a complex sea level history between 10 and 8 ka. Evidence for rapid regressions and transgressions of 10-15 m between 10 and 8 ka have been postulated for outer eastern Baffin Island and northernmost Labrador (Løken, 1962; Nelson, 1978; Andrews and Miller, 1985). The causes of these sudden and relatively large-scale changes in sea level are not well researched. They might be attributed to glacial isostatic adjustments associated with Late Wisconsinan glacial advances, or they might be tectonic in nature. Dyke and Dredge (1989) presented evidence from the northern coast of Prince of Wales Island for a steep and unpredicted fall in the marine limit which they suggest might signify faulting. In Fennoscandia researchers on late Quaternary sea levels have advanced both glacial isostasy (Anundsen and Fjeldskaar, 1983) and tectonics (Mörner, 1980) as mechanisms to explain sharp changes in sea levels. These apparently anomalous changes might also be due to erroneous reconstructions of the history of deglaciation. In the case of the Fennoscandia work, however, these relatively large changes appear to tie in closely with the Younger Dryas glacial event.

This survey of the Late Pleistocene and Holocene emergence and submergence history of the coasts of Canada indicates that there are a variety of responses to the interacting processes of glacial unloading and global sea level rise. Over most of Canada sea level has fallen throughout this period because of the overriding importance of glacial isostatic recovery (i.e., areas in Zone I as shown in Figure 8.6). In areas peripheral to the various ice sheets, however, there have been a more complex set of responses with many areas showing unequivocal evidence of submergence.

Isobase maps

Isobase maps on relative changes of sea level have been presented by several authors over the course of the last two decades. The majority of these have been based on radiocarbon dated sea levels although one has also been based on the occurrence of pumice on beaches in the High Arctic (Blake, 1970). The isobase maps presented by Andrews (1970), Walcott (1972), and Dyke (1974) are largely based on field observations. The contouring (i.e. the interpretation) of the field data has not been identical. Andrews (1970) showed two separate areas of maximum uplift within the borders of the Laurentide Ice Sheet whereas Walcott (1972) showed only one. The latter map led Paterson (1972) to propose a single-ridge model for the Late Wisconsinan ice sheet.

Isobase maps have been complemented in recent years by maps of changes in relative sea level based on glacial isostatic modelling (e.g., Quinlan and Beaumont, 1981; Quinlan, 1985).

The information in Figures 8.4 and 8.5 has been used to produce two isobase maps for North America showing the position of sea level since 7 ka and 2 ka (Fig. 8.9, 8.10). These maps outline the central area of emergence and the peripheral areas of submergence. These two time periods were chosen because sea level data for the 7 ka period has the most extensive spatial coverage of deglaciated Canada, whereas the picture at 2 ka has information pertinent to present rates of land emergence or submergence. The maps are based solely on published relative sea level curves and no attempt has been made to expand this coverage through plotting of individual radiocarbon dated sea level determinations. Contouring the resulting elevation data is not straightforward and different authors will produce different constructs depending on their perceptions of glacial and glacial isostatic history (e.g., Andrews, 1978, who illustrated three different published patterns of isobases for the Queen Elizabeth Islands).

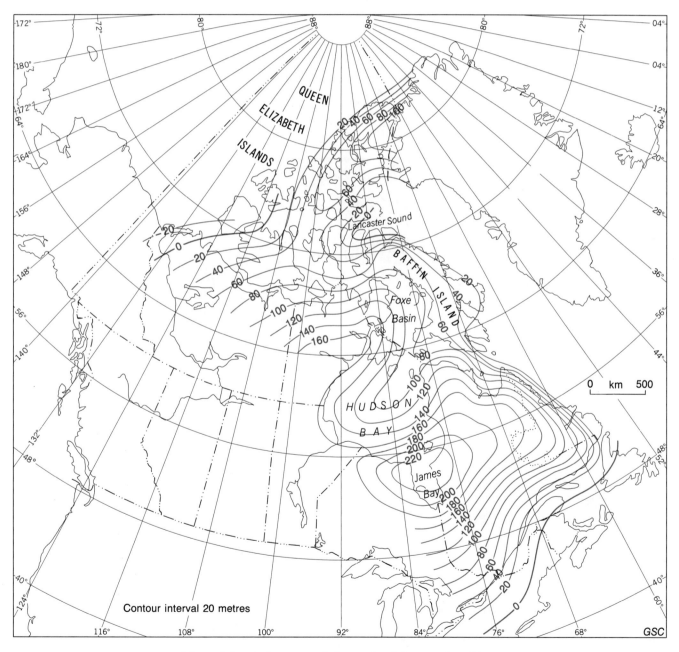

Figure 8.9. Isobases on relative sea level changes since 7 ka.

CHAPTER 8

Relative changes of sea level since 7 ka

Figure 8.9 has been drawn largely on the basis of relative sea level curves and directions of shoreline tilting. The map will differ in detail from an isobase map based on individual radiocarbon dated shoreline sites. Differences between the predicted isobase pattern (Fig. 8.9) and dated individual sites reflect how far this "model" differs from the actual pattern of isostatic unloading. The major features of this reconstruction are: (1) a broad ridge extending from northwest Greenland across into the Queen Elizabeth Islands; (2) a broad ridge with maximum values of 160 m a.s.l. situated inshore and northwest of Hudson Bay; and (3) a broad region of high postglacial emergence located across James Bay. The zero (0 m) isobase cuts northeast-southwest across the Maritime Provinces and down into New England along the southeastern periphery of the former ice sheet. At the opposite side of the ice sheet, the zero isobase is located near the Mackenzie Delta and extends northeast-southwest along the outer Queen Elizabeth Islands. In British Columbia most if not all the coastline was undergoing submergence by 7 ka but data are sparse and no attempt is made to contour the 7 ka sea level in Figure 8.9 in that area.

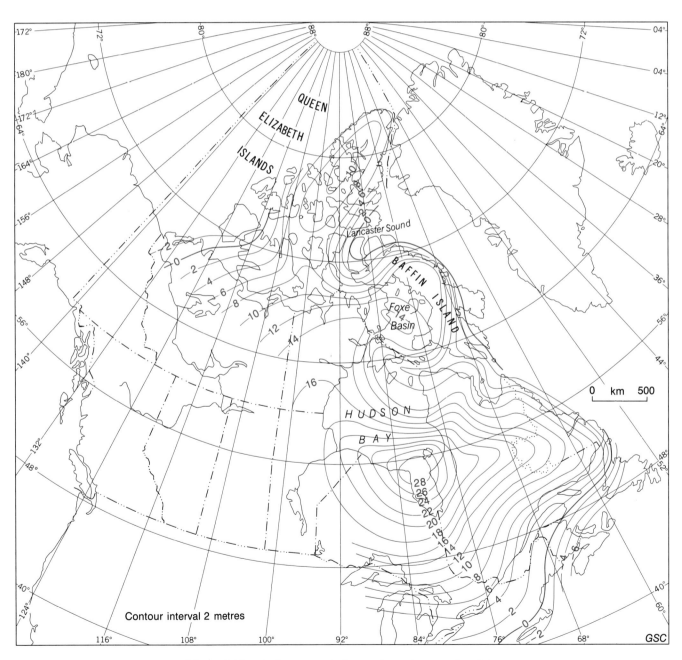

Figure 8.10. Isobases on relative sea level changes since 2 ka.

On the west coast uplift was largely completed by 8 ka and subsequent submergence is generally ascribed to eustatic sea level rise. Other features of note are the inflections of the isobases towards the ice sheet around Lancaster Sound in the north and in toward Hudson Bay. The zero isobase is shown penetrating along Lancaster Sound by 7 ka and it runs southward along the outermost coast of Baffin Island.

Relative changes of sea level since 2 ka

The maximum amount of emergence in the last 2 ka in Canada is ca. 30 m (Fig. 8.10). In gross form Figures 8.9 and 8.10 are similar, although in detail some changes in the pattern of emergence are apparent. Zones of maximum emergence are located over the central Queen Elizabeth Islands, Foxe Basin, and James Bay, whereas significant "lows" occur along Lancaster Sound and Hudson Strait. The zero isobase has not noticeably changed its position. Even though the zero isobase is shown offshore along most of the Labrador coast, wave destruction of paleo-Eskimo sites in northern Labrador suggests that it should be drawn along the outer coast of Labrador, but unfortunately data are not adequate to permit precise location of this isobase.

Neither Figure 8.9 nor Figure 8.10 adequately accounts for the observations of Lauriol and Gray (1983) who stated that lake shorelines between Ungava Bay and Hudson Bay are tilted down towards Hudson Bay. Figure 8.10, however, partly accommodates these data by showing a ridge extending northeast towards Ungava Bay. A local ice centre or late lingering ice might explain the anomaly. Prest (1984, Fig. 1) has suggested that there was an "Ungava Bay" ice centre at some stage in the Late Wisconsinan in order to explain the impounding of glacial Lake Naskaupi between the ice margin and the height of land near the east coast of Labrador (Ives, 1960).

The extensive ridge that is shown extending across Baffin Island from Foxe Basin (Fig. 8.10) might reflect the late deglaciation of Foxe Basin (by 7.7 ka) and the effect of residual ice over Baffin Island. Dyke (1974) inferred a similar pattern in his reconstruction of isobase patterns for the northeastern sector of the ice sheet.

Estimated rates of vertical movement during the Holocene

Relative sea level curves (Fig. 8.4) allow an objective calculation of the rates of sea level rise or fall for the Holocene and latest Pleistocene that give emergence and submergence rates. Prior to about 7 ka there was still extensive ice coverage of parts of northern Canada and thus sea level data earlier than 7 ka are areally restricted. It is for this reason that emergence and submergence based on the rate of sea level change for middle and late Holocene intervals will be discussed but not for the early Holocene when rebound in many areas was still restrained by the presence of ice. Data were taken from published curves (Fig. 8.4) and the change in sea level during a 1 ka interval was expressed as the rate of emergence ($+$) or submergence ($-$) (Fig. 8.11).

The rate of emergence between 8 and 7 ka (Fig. 8.11a) cannot be determined for areas covered by ice at that time. Outside the 8 ka ice margin, where sea level curves on the Ungava coast (Matthews, 1967) and on the Baffin Island side (C.A. Laymon, University of Colorado, unpublished) show extremely rapid relaxations, rates of up to 80 m/ka are estimated. Southeast of the ice sheet margin rates varied between 10 and -10 m/ka and were thus much lower than the estimated values of between 55 and -5 m/ka on the northwestern margin of the ice sheet. Over the Queen Elizabeth Islands a rate of 25 m/ka is estimated for the channels and fiords between Axel Heiberg and Ellesmere islands. It must be pointed out, however, that various reconstructions of emergence in this area either predict a ridge extending westward from Greenland or a discrete centre of uplift (as in Fig. 8.11a).

The rate of change in sea level between 4 ka and 3 ka indicates that over the Queen Elizabeth Islands the maximum emergence was now only 8 m/ka compared to an estimated value of ca. 27 m/ka over James Bay (Fig. 8.11b). A ridge extends northeast from the approximate position of the Keewatin Ice Divide towards Foxe Basin and Baffin Island.

The final map illustrates the calculated rates of emergence over the last thousand years (Fig. 8.11c). This map provides a link with the next section where we discuss the

Table 8.3. Direction and causes of relative sea level changes in British Columbia (Clague et al., 1982)

Age	Queen Charlotte Islands (thin ice)		Inner and middle coasts (thick ice)	
	Relative sea level change	Controlling factors	Relative sea level change	Controlling factors
Late Pleistocene-early Holocene	Sea rising	Eustatic rise[1] Forebulge collapse[1] Tectonic uplift[2]	Sea falling	Isostatic rebound[1] Eustatic rise[2]
middle Holocene	Sea stable to slightly falling	Tectonic uplift[1] Eustatic rise[1] Forebulge collapse?[2]	Sea rising	Eustatic rise[1] Forebulge collapse?[2]
late Holocene	Sea falling	Tectonic uplift[1] Eustatic rise[2]	Sea stable to slightly rising	Tectonic subsidence?[2] Eustatic rise?[2] Forebulge collapse?[2]

[1] Dominant controlling factor
[2] Subordinate factor

CHAPTER 8

evidence for changes in the present elevation of lake and marine sites as forecast from an analysis of tidegauge records. Figure 8.11c provides an estimate for the maximum rate of emergence of around 12 m/ka for the region surrounding James Bay. Within Foxe Basin rates are lower and are estimated to be about 5 m/ka. Continuing submergence at a rate of −1 m/ka is shown for the Maritimes, the eastern coast of Baffin Island, and the northwestern margin of the former Laurentide Ice Sheet.

No attempt was made to show the rates of change for British Columbia. The history of sea level has been discussed at length by Mathews et al. (1970), Clague (1975, 1983), and Clague et al. (1982). Within the time-frame covered by Figure 8.11 most of coastal British Columbia was undergoing a period of slow submergence that followed a period of rapid emergence (see Fig. 8.4, curves from British Columbia). A marked exception occurs in the Queen Charlotte Islands (Fladmark, 1975; Clague et al., 1982;

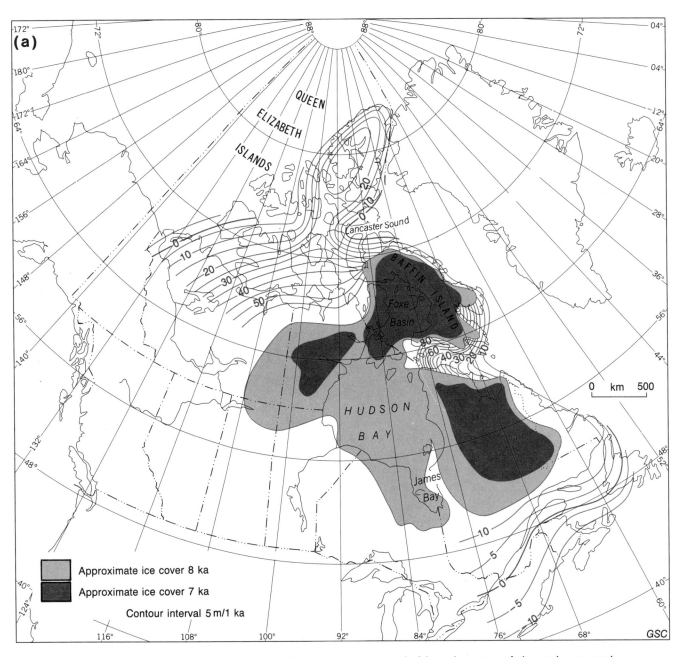

Figure 8.11. Rates (m/ka) of emergence (+) and submergence (−) based on rates of change in sea level which have been determined from relative sea level curves (Fig. 8.4) for (a) 8-7 ka (ice limit sources, Prest, 1969; Bryson, et al., 1969; Dyke, 1974); (b) 4-3 ka; and (c) 1 ka to present.

QUATERNARY GEODYNAMICS IN CANADA

Clague, 1983) where curves (Fig. 8.4) suggest a rise in sea level between 11 and 8.5 ka followed by a slow but steady emergence up until present. Table 8.3 (from Clague et al., 1982) illustrates the suggested controls on sea level history for various regions of British Columbia and suggests that present rates of sea level change are influenced by tectonic uplift or subsidence, forebulge collapse, and eustatic changes.

Present changes in lake and sea levels

The present vertical movements within Canada can be estimated from relevelling of transects, or from an analysis of tide gauge data. Analysis of gauging stations from both maritime and lake sites have been discussed for several decades (e.g., Gutenberg, 1941) and have been used to evaluate the present effects of glacial isostatic unloading. There is currently great interest in estimating the worldwide rise of sea

Figure 8.11. Continued.

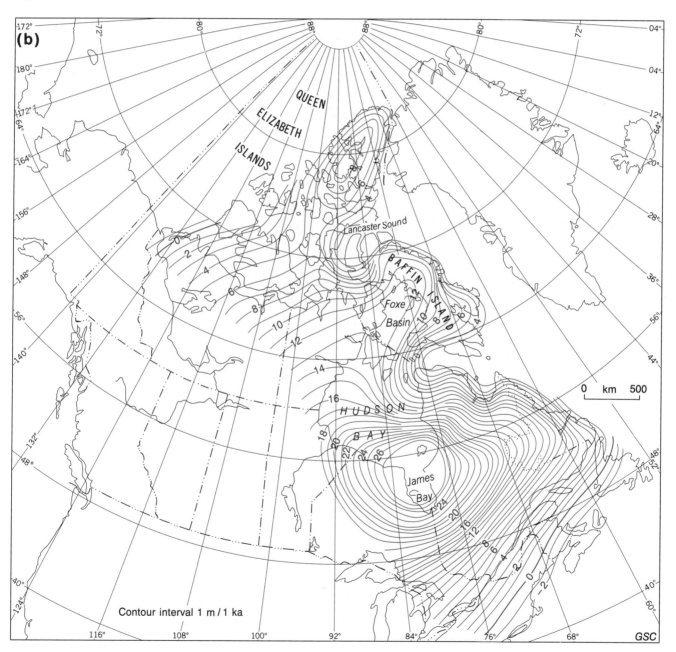

level (Gornitz et al., 1982; Barnett, 1983) and its possible association with climatic warming associated with the atmospheric increase in CO_2 (e.g., Hoffman et al., 1983; National Research Council, 1984) and melting of glaciers and ice shelves. However, the task of distinguishing between eustatic, isostatic, climatic, oceanographic, and tectonic causes of changes in water level at a gauging site (apart from local disturbance) is a major problem. Gutenberg (1941, p. 722-733) presented a thorough account of the difficulties and the methodology of analysis. More recent studies dealing with changes of water level at Canadian gauging stations include Barnett (1968, 1970), Vaníček (1976), Vaníček and Nagy (1980), Quinlan (1981), and Riddihough (1982).

Analysis of tide gauging data is not straightforward, and different methods will lead to different estimates of current vertical motions (e.g., Table 1 in Clague et al., 1982;Table 1 in Riddihough, 1982). These differences are related to: (1) the statistical measure used to compute the trend; (2) whether annual or monthly changes of level are employed; and (3) whether individual sites or site conglomerations are used (Wigen and Stephenson, 1980). In comparing calculated rates, based on the relative sea level curves (e.g., Fig. 8.4), with Maritime Province tide gauge stations Quinlan (1981, p. 382) noted a consistent drift between them which he attributed to modern rise of sea level equivalent to between 1.4 and 2.4 m/ka. It should be noted that

Figure 8.11. Continued.

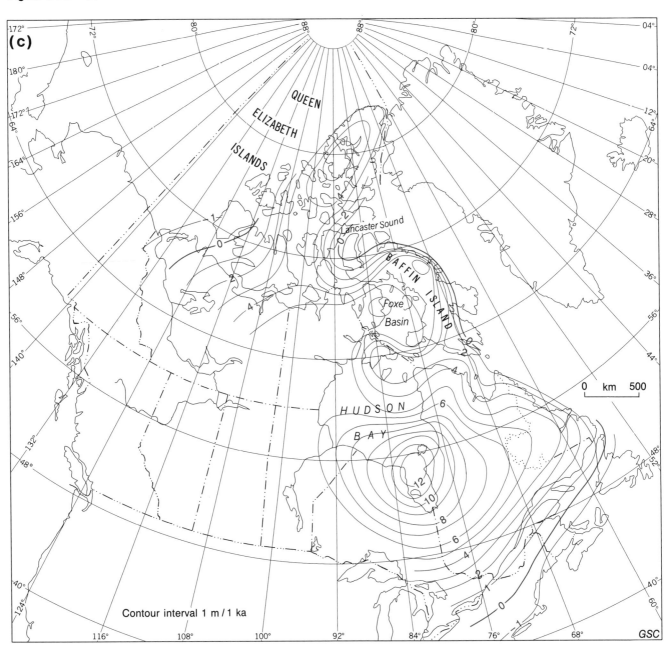

this recent rise of sea level will result in a *relative* decrease in the rate of sea level fall in Zone I areas and an increase in the rate of rise of sea level in Zone II areas.

Figure 8.12 shows a synthesis of the current relative sea level changes. The map indicates much of the outer coast of British Columbia is emerging whereas east of a line running between Victoria and the Strait of Georgia the area is submerging (Riddihough, 1982). Farther east, the broad zone of uplift around the northern Great Lakes is based on lake level data and reflects a strong glacial isostatic component.

Unfortunately there are only a small number of tide gauge stations in the Canadian Arctic (Fig. 8.12). Barnett (1968) examined the tide gauge records from Churchill and determined that sea level was falling at 6.1 m/ka, although this was subsequently corrected to 3.9 m/ka (Barnett, 1970). Vaníček and Nagy (1980) computed the rate of relative sea level change from these tidegauge stations using classical regression analysis. We have examined these records for the period up until 1982. Figure 8.13 plots data from some of the sites as an illustration of the degree of scatter and the trend

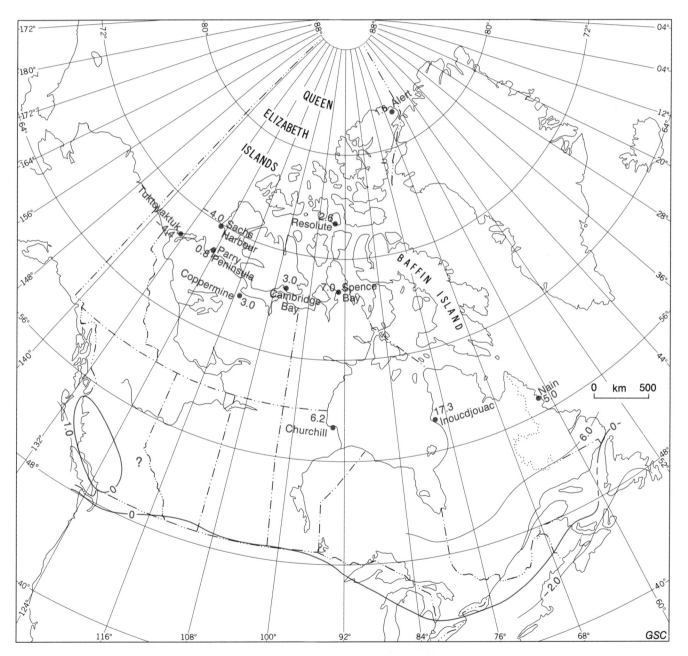

Figure 8.12. Current rates (m/ka) of emergence (+) and submergence (−) as indicated by changes in sea level, uncorrected for any "global" sea level rise, for tide gauge stations in northern Canada (figures obtained from tide gauge data provided by Canadian Hydrographic Service; isolines from Vaníček and Nagy, 1980).

of the resistant line. The resistant line is not influenced by outliers (Velleman and Hoaglin, 1981) and is a more conservative measure of the slope of a relationship.

The suggested trends for the stations agree reasonably well with the estimates of Vaníček and Nagy (1980) although their estimate of 3.5 m/ka for Churchill appears to be incorrect. Our calculation indicates a relative rise of the land (based on annual tide gauge elevations) of 6.2 m/ka (Fig. 8.12, 8.13) which is close to that estimated by Barnett (1968; but see Barnett, 1970). Conventional regression analysis resulted in a rate of 7.6 m/ka. Figure 8.13 shows the resistant line slopes for tide gauge stations in northern Canada that have longer than 10 years of record. These stations provide a potentially important source of information on Canadian neotectonics, although the sites need to be maintained for several more decades in order to obtain more consistent trends.

An intriguing result occurs at Inoucdjouac, on the east side of Hudson Bay (Fig. 8.12). Our results are the average values for the months of October and November, two months which had consistent records with a minimum number of missing values. The trend is for a rise of the station by 17.3 m/ka. Although this is lower than that estimated by Vaníček and Nagy (1980), it is still a high value and is not in keeping with estimates based on relative sea level curves (Fig. 8.11c). This record needs to be confirmed by a continuing analysis of the tide gauge records on a year-to-year basis as well as an enquiry into site conditions. Figure 8.14 plots the association between the tide gauge trends against estimates of emergence/submergence derived from the relative sea level curves (Figs. 8.4, 8.11). In general there is good agreement and this further emphasizes the Inoucdjouac anomaly.

MODELS OF GLACIAL ISOSTASY
W.R. Peltier

The figures presented with this section represent a base of information against which glacial isostatic modelling of the earth's rheology can proceed. However, the links between the field sciences and geophysics are not always clear nor on appropriate scales. In the following section we examine models of glacial isostasy and their implications for the free air gravity anomaly and relative sea levels. The time/thickness model for the Canadian ice sheets which underpins these discussions is based on modifications of the ICE-1 ice sheets model used by Peltier and Andrews (1976). These ice sheets consisted of a single domed Laurentide Ice Sheet centred over Hudson Bay, a Cordilleran Ice Sheet, and limited ice in the High Arctic. The profile of ICE-1 does not conform to some current concepts of the Laurentide Ice Sheet (see regional chapters in Part 1 of this volume and Prest, (1984), but in one examination of this problem Peltier and Andrews (1983) were not able to resolve the problem).

Over the past decade, considerable progress has been made in the development of mathematical models of the physical process of isostatic adjustment. This process influences a variety of observations besides those of relative sea level (as represented by either radiocarbon or tide gauge data) and of glacial lake shoreline tilts, both of which are mentioned in the preceding sections of this chapter. Among the other observations associated with this physical phenomenon are the negative "free air" anomalies in the local vertical component of the gravitational acceleration which are observed over regions that were once ice covered, and certain astronomical observations relating to the earth's rotation. The latter consist of two types of data concerning both the earth's rate of rotation (the so-called *length of day*) and time variations of the position of its rotation pole with respect to the surface geography (the process called *true polar wander*). A recent review of the development of modern theoretical models of glacial isostasy, which began with Peltier (1974), has been presented in Peltier (1982).

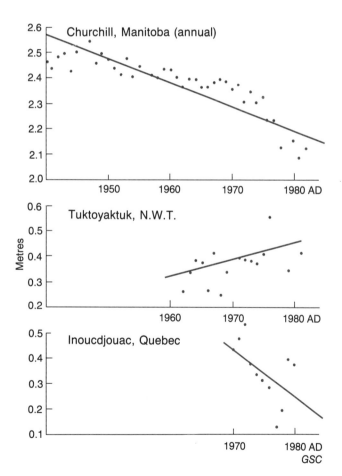

Figure 8.13. Examples of changes in sea level as determined from an examination of tide gauge data from selected stations in northern Canada (the sites are located in Figure 8.12). The trend in the change of sea level has been determined by use of the "resistant line" — a method that reduces the effect of outliers in regression analysis.

Peltier, W.R.
1989: Models of glacial isostasy; in Chapter 8 of Quaternary Geology of Canada and Greenland, R.J. Fulton (ed.); Geological Survey of Canada, Geology of Canada, no. 1 (also Geological Society of America, The Geology of North America, v. K-1).

Relative sea level change and earth models

The main theoretical problem which must be resolved in attempting to predict relative sea level is connected with the complexity of the interaction between the ice sheets, oceans, and solid earth which these variations involve. If the earth were rigid then relative sea level variations forced by an ice sheet disintegration event would be relatively simple. Water produced by ice sheet melting would be added to the ocean basins and sea levels would rise. Even in this simple case, however, the rise of sea level would not be the same everywhere. This is because the surface of the sea must remain a gravitational equipotential surface and because the original equipotential is perturbed nonuniformly by the melting of ice sheets concentrated at specific geographic locations. Meltwater would be distributed over the ocean basins in such a way as to ensure that the sea surface (the geoid) was everywhere on the same equipotential surface. In the rigid earth case the variations in potential are caused entirely by the ice and water loads themselves. This is not the case when the finite elasticity of the planet is accounted for. Since the earth actually does deform under the changing ice and water loads, these deformations, since they redistribute matter in the interior, also contribute to the variation of the gravitational potential and thus require further redistributions of water among the ocean basins in order to ensure that the surface of the sea remains equipotential. If the deformations produced by surface loading were entirely elastic, however, then once the ice sheet disintegration event was complete there would be no further variations of sea level.

The fact that relative sea level (at locations which were once under the Laurentide Ice Sheet) has continued to change since deglaciation was complete between 6 and 7 ka, is direct evidence that the earth is not a simple Hookean elastic solid. Rather, because the time scale over which the surface loads were applied was sufficiently long, the dominant contribution to the load induced deformation is in fact due to an essentially viscous "flow" of earth material under the gravitationally induced stress field. The rate at which earth material flows is governed by its effective viscosity and this parameter can be estimated by fitting an appropriate theory to relative sea level data such as those shown in Figure 8.4. The higher the viscosity of the earth's mantle, the longer will be the memory of the unloading event and the slower will be the relaxation of the viscous component of the deformation following glacial unloading. Clearly this delayed viscous rebound response to ice sheet disintegration further complicates the issue of relative sea level variations forced by ice melting. Because of this time dependent response of the earth, there continues, long after ice sheet melting, a continuous redistribution of meltwater over the ocean basins in such a way as to ensure at all times an ocean surface which is equipotential. At sites which were once covered by ice, however, the dominant contributor to relative sea level variations is the local increase of earth radius which is produced by the viscous inflow of mantle material which was earlier squeezed out from under the ice sheet in the loading phase of the glacial cycle. This inflow of material is responsible for the elevation of the land at sites which were once ice covered and the submergence of the land at sites which were immediately peripheral to the major ice concentrations. These regions were referred to as Zone I and Zone II in the context of the discussion of the data shown in Figure 8.4 (location of zones is shown in Fig. 8.6).

Detailed discussions of the visco-elastic model for the computation of gravitationally self-consistent sea level histories will be found in Peltier (1974, 1976, 1982), Farrell and Clark (1976), Peltier et al. (1978), Clark et al. (1978), and Wu and Peltier (1983). The ability of the model to fit data such as those shown in Figure 8.4 for a number of sites in both Zones I and II of the North American continent will be illustrated. Model fits to the observed relative sea level histories at six sites in Canada are shown in Figure 8.15; at each of these locations the predictions of the relative sea level model are shown for three different visco-elastic structures which differ only with respect to mantle viscosity. The elastic structure is fixed in each case to that of a conventional seismic model, specifically model 1066B of Gilbert and Dziewonski (1975). The lines show fits to the relative sea level data of models which have lower mantle viscosities of 10^{21} Pa·s, 10^{22} Pa·s, and 5×10^{22} Pa·s. In each case the models have upper mantle viscosities (above the 670 km seismic discontinuity) of 10^{21} Pa·s and lithospheric thicknesses of 120.7 km. Inspection of the model fits to the data at the six sites demonstrates that, of the three simple models, the one which best reconciles the observations is that for the model with a uniform viscosity of 10^{21} Pa·s. The model with lower mantle viscosity of 10^{22} Pa·s fits the amplitude of emergence at 6 ka reasonably well but predicts a present day emergence rate which is far too rapid. The model with a further increase of lower mantle viscosity to 5×10^{22} Pa·s predicts the present day rate of uplift correctly but predicts far too little net emergence over the time since 6 ka.

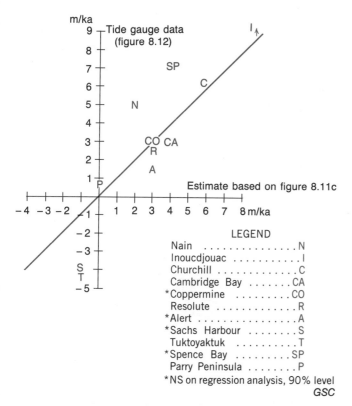

Figure 8.14. Scattergram of the association between rates of sea level change derived from relative sea level curves (Fig. 8.11c) versus determinations from recent tide gauge observations (Fig. 8.12) from northern Canada.

Fits of the same sequence of models to the data for six sites from Zone II are shown in Figure 8.16. Here the fits of the uniform viscosity models to the observations compiled by Bloom (1967) for these sites are rather poor. As recently demonstrated in Wu and Peltier (1983), however, there does exist a systematic misfit of the predictions of the model to the data as a function of position along the coast (e.g., Fig. 8.17). This suggests that some property, to which the Zone I data are not sensitive, has been misspecified. Peltier (1984) has recently shown that an increase of lithospheric thickness from 120.7 km to about 240 km will rectify the observed misfits completely. If this result is correct, it has important geodynamic implications, one of which is that the chemical differences between continents and the material in which they are embedded could persist to much greater

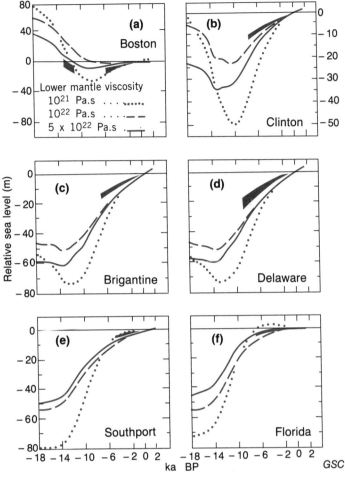

Figure 8.16. Same as Figure 8.15 but for six sites in the peripheral bulge region along the east coast of the United States.

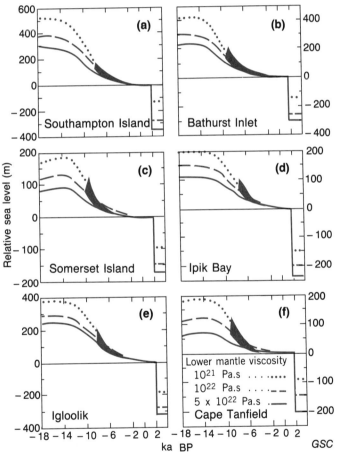

Figure 8.15. Relative sea level data observed and theoretically predicted for six sites which were once covered by the Laurentide Ice Sheet. The ICE-2 deglaciation history is that tabulated in Wu and Peltier (1983). The observations are shown as the shaded red regions while the theoretical predictions are for models that differ from one another only by their lower mantle viscosities. The upper mantle viscosity is held fixed at 1021 Pa·s and the lithospheric thickness at 120.7 km. Uplift is projected through to 2 ka in the future and the remaining uplift for each model is indicated by the horizontal line adjacent to the right margin.

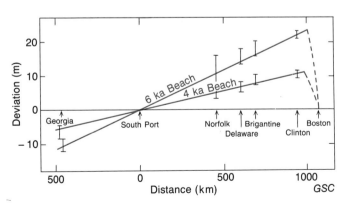

Figure 8.17. Misfit of the theoretical relative sea level predictions for the uniform viscosity model to the observations as a function of distance along the eastern seaboard of the continental United States. Distance is measured positive north from Southport, North Carolina.

depth than is conventionally accepted. The goodness of fit of the thick lithosphere model to the observations is demonstrated explicitly in Table 8.4. This simple example should serve to illustrate the potential of Quaternary geological data in providing constraints upon important geodynamic properties of the earth. The importance of relative sea level in providing direct constraints on the viscosity of the mantle should be equally clear. The magnitude of the viscosity of the mantle is, of course, a crucial ingredient in demonstrating the validity of the thermal convection hypothesis of the

Table 8.4. Observed and predicted relative sea level at selected american eastern seaboard sites. Calculations are based upon the gravitationally self-consistent model with realistic ICE-2 melting chronology and an earth model with a lithospheric thickness of 245 km and 1066B elastic structure

Time (ka)	Southport Observed (m)	Southport Predicted (m)	Virginia Observed (m)	Virginia Predicted (m)	Brigantine Observed (m)	Brigantine Predicted (m)
-8	–	–	-22.0 ± 4.0	-29.52	–	–
-7	–	–	-17.0 ± 3.0	-21.70	–	–
-6	–	–	-13.0 ± 3.0	-14.90	–	–
-5	–	–	-10.0 ± 3.0	-9.63	-10.5 ± 0.6	-11.32
-4	-3.3 ± 0.5	-2.26	-7.0 ± 1.5	-5.95	-7.5 ± 0.5	-7.79
-3	-2.2 ± 0.4	-0.96	-3.0 ± 1.0	-1.45	-5.2 ± 0.5	-4.40
-2	-1.2 ± 0.2	-0.30	-0.5 ± 0.5	-1.78	-3.0 ± 0.4	-2.36
-1	-0.5 ± 0.1	-0.03	-0.4 ± 0.5	-0.68	-1.4 ± 0.4	-0.95
0	0	0	0	0	0	0

Time (ka)	New York Observed (m)	New York Predicted (m)	Clinton Observed (m)	Clinton Predicted (m)	Boston Observed (m)	Boston Predicted (m)
-8	-17.8 ± 3.0	-22.48	–	–	–	–
-7	-16.0 ± 2.4	-19.06	-10.2 ± 0.7	-20.39	–	–
-6	-13.0 ± 2.2	-14.45	-8.4 ± 0.5	-15.5	–	–
-5	-9.4 ± 1.5	-10.02	-6.7 ± 0.5	-10.84	-9.0 ± 3.0	-9.02
-4	-6.4 ± 0.6	-6.56	-4.7 ± 0.5	-7.16	-6.0 ± 3.0	-6.03
-3	-4.4 ± 0.5	-4.00	-3.4 ± 0.5	-4.41	-2.5 ± 2.0	-3.74
-2	-1.2 ± 0.3	-0.88	-1.0 ± 0.2	-0.94	-0.5 ± 0.5	-0.84
0	0	0	0	0	0	0

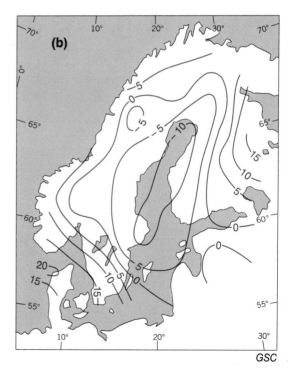

Figure 8.18. Observed free air gravity anomalies (mGal) for (a) northeastern North America and (b) Fennoscandia.

origin of continental drift and seafloor spreading. As demonstrated in Peltier (1980, 1985a), if the viscosity of the mantle were significantly different from the value of 10^{21} Pa·s, required by Quaternary relative sea level data, then the convection hypothesis would not be viable. Since these data provide the only means of directly estimating this parameter, their importance as a cornerstone of modern geodynamics is indisputable.

Free air gravity anomaly

One of the most serious objections which has been raised to the interpretation of the relative sea level record presented in the last section is due to Sir Harold Jeffreys. He has long maintained (e.g., 1972 for a recent review) that the viscous flow interpretation of observations of postglacial relative sea level change was fundamentally incorrect. The basis of his objection has usually been stated in terms of a claimed inability of the viscous flow model (which requires mantle viscosities near 10^{21} Pa·s to fit relative sea level data) to simultaneously reconcile both relative sea level histories and the free air gravity anomalies associated with the main centres of glacial rebound. Figure 8.18 shows the observed free air anomalies over both Canada and Fennoscandia. On the basis of the Fennoscandia data, Jeffreys correctly argued that the free air gravity anomaly predicted by the uniform density and viscosity half space model of Haskell (1936, 1937), with a mantle viscosity of 10^{21} Pa·s, was only 2-3 mGal. He argued that since this is far too small to explain the observed value of about -15 mGal, the viscous flow model must be completely rejected. Jeffreys' position has been that the actual rheology of the earth is such that it is incapable of deforming as a viscous fluid no matter what the time scale of the stress field to which it is subjected. Inspection of Figure 8.18 shows that the same difficulty may well be a problem with attempts to explain the free air anomaly over Canada. This anomaly is much deeper than that over Fennoscandia, with a basin nearing -40 mGal. The geometry of this anomaly is such that it has the form of an ellipse with long axis trending northwest-southeast centred on Hudson Bay. The zero isoline of the free air anomaly passes through the proglacial lakes that ring the Canadian Shield. There can therefore be no doubt that the cause of this anomaly (e.g., Innes et al., 1968; Walcott, 1970) lies in the last deglaciation event.

Recently a theoretical explanation has been advanced as to how an essentially Newtonian viscous model with a mantle viscosity near 10^{21} Pa·s may be able to simultaneously reconcile both relative sea level and free air gravity observations (Peltier, 1982; Peltier and Wu, 1982; Wu and Peltier, 1983). The idea has to do with the elastic component of earth structure and specifically with the variation of density which occurs through the mantle of the earth between 420 and 670 km depth, that is through the so-called "transition region". These two depths are both the locations of rather sharp discontinuities in density and elastic moduli which are conventionally attributed to the occurrence of pressure induced phase transitions at these depths. The boundary at 420 km depth is acknowledged to mark the transition from olivine to spinel, whereas that at 670 km depth is usually attributed to the transition from spinel to a mixture of perovskite and magnesiowustite (e.g., Jeanloz and Thompson, 1983). This interpretation of the 670 km discontinuity, however, is somewhat more controversial.

The physical essence of the new argument which has been advanced to explain both the free air gravity data and the relative sea level data involves the notion that these phase boundaries may be able to induce a buoyant restoring force when they are deflected from their equilibrium levels by surface ice sheet loads. If this occurs then this extra source of internal buoyancy in the model leads to a marked increase of the predicted free air gravity anomaly even when the mantle is taken to have a fairly uniform viscosity of 10^{21} Pa·s as is required by the relative sea level data. In order for

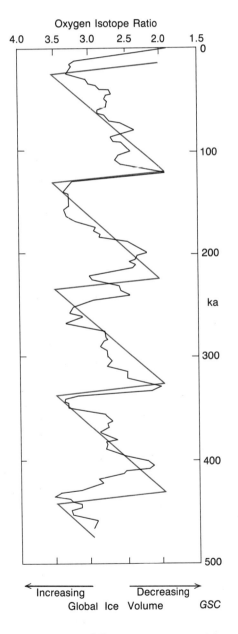

Figure 8.19. A typical $\delta^{18}O$ versus time record from a deep sea core (black curve) and the 10^5 year periodic sawtooth function (red curve) that is employed to approximate the ice volume record for the purpose of computing gravity anomalies.

the 420 and 670 km discontinuities to induce a buoyant restoring force when they are deflected, however, they are required to behave as non-adiabatic discontinuities on the time scale of glacial isostatic adjustment. This is not immediately obvious if these discontinuities are phase boundaries since one might expect that any process which caused such a boundary to be deflected to a different pressure level would simply cause enough material to change phase that the boundary would be kept flat and no buoyancy could be induced. If the boundaries were chemical boundaries, however, then it is perfectly clear that any process which displaced them would be resisted by a buoyant restoring force. On sufficiently short time scales it turns out that even a phase boundary may well act as though it were a chemical discontinuity. The reason for this is that, at least in the approximation that the phase boundary may be treated as univariant, the rate at which the phase boundary can move is determined by the thermal conductivity and this is very low. Analyses in O'Connell (1976) and Mareschal and Gangi (1977) show that if this univariant approximation is correct, then a reconciliation of the relative sea level and free air gravity data is possible to effect.

In order to make accurate predictions of the expected free air gravity anomalies, however, one is obliged to introduce one further complexity and this concerns the detailed history of glaciation and deglaciation. Although this history is not well known for any of the individual locations at which large concentrations of Pleistocene ice were found, there does exist an important data set which provides a strong constraint upon the variation with time of the total volume of the continental ice sheets. This consists of $\delta^{18}O$ versus depth data from sedimentary cores from deep ocean basins. Because such cores preserve stratigraphies of both planktonic and benthic foraminifera, the $\delta^{18}O$ versus depth information contained in the sediment can be employed to unravel approximately the separate effects of temperature and ice volume (Shackleton, 1967). The $\delta^{18}O$ versus depth data are transformed to $\delta^{18}O$ versus time data by locating the depth in the core to a known time horizon, normally the Brunhes-Matuyama magnetic reversal which has an age of approximately 730 ka. An example of these data is shown in Figure 8.19 along with a simple periodic sawtooth function of period 10^5 years. Spectral analysis of such data by Hays et al. (1976) has demonstrated that the apparently dominant 10^5 year cycle in fact contains about 60% of the variability in the $\delta^{18}O$ record.

In computing the free air gravity signal expected at present over a given site of continental deglaciation, it is assumed that the variability of global ice volume represented by the $\delta^{18}O$ record is equally representative of the local ice sheet histories which have contributed to it, at least insofar as the 10^5 year cycle is concerned. Subject to this assumption concerning the glaciation history, explicit predictions may be made of the magnitude of the expected free air gravity anomaly as a function of the remaining parameters of the model and these predictions may be compared to the observed anomalies shown in Figure 8.18. Figure 8.20 illustrates a sequence of model predictions of the peak free air

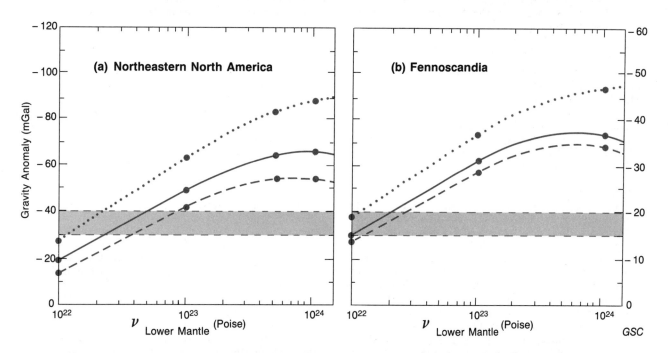

Figure 8.20. Comparison of observed and model predicted gravity anomalies for different lower mantle viscosities: a) northeastern North America and b) Fennoscandia. The shaded areas show the observed peak free air gravity anomalies. The dotted lines are predictions based upon the assumption of initial isostatic equilibrium, the dashed curves include the effect of the load cycle illustrated in Figure 8.19, but the ice sheet radius is held fixed, while the solid curves also include the variation of ice sheet radius as the volume fluctuates. The latter predictions are therefore the most accurate.

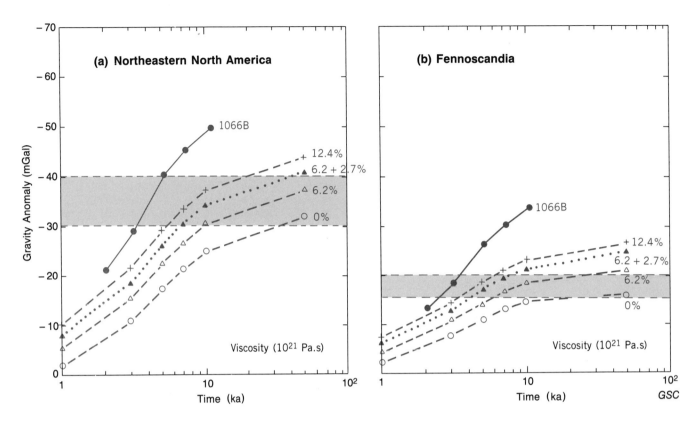

Figure 8.21. Comparison of observed (shaded areas) and model predicted gravity anomalies for a 105 year loading and unloading cycle: a) northeastern North America and b) Fennoscandia. The predicted anomalies are for various models that differ from one another in the amount of internal buoyancy in the mantle. The models are fully described in the text.

anomaly for both northeastern North America and Fennoscandia which explicitly include (solid curves) the full effects of the loading history as this is constrained by the $\delta^{18}O$ data. The computations were performed using spherical cap approximations of the two ice sheets and the density and Lamé constants were taken equal to those in the seismic model 1066B (Gilbert and Dziewonski, 1975). The lithospheric thickness in the earth model was fixed to 120.7 km and the upper mantle viscosity to 10^{21} Pa·s. Only the lower mantle viscosity was varied in the calculation and the predictions of the peak free air anomalies shown in Figure 8.18 have been made as a function of this parameter. It is important to realize that all of the density variation in model 1066B has been treated as nonadiabatic for the purpose of these calculations. When this assumption is made, the analyses then imply a lower mantle viscosity which is close to the upper mantle value of 10^{21} Pa·s. Since such models are also able to satisfy relative sea level data from sites within the centre of uplift, they appear to be able to simultaneously satisfy both the sea level and free air gravity observations.

The crucial physical ingredient of the models which enables them to effect the reconciliation of this long-standing anomaly is the presence of internal buoyancy in the mantle associated with the strong variation of density through the transition region. This is demonstrated explicitly in Figure 8.21 (from Peltier, 1985b); this figure shows model predictions of the peak free air anomaly for both northeastern North America and Fennoscandia which include the full influence of the history of 10^5 year cycles of loading and unloading which have dominated the last 1000 ka of earth history. The elastic earth structures employed for these calculations differ considerably from that of model 1066B (Gilbert and Dziewonski, 1975), however, in that this component of the model has been assumed to consist of a small number of homogeneous spherical shells. Each of the models represented by the dashed curves has a mantle which contains only a single discontinuity (that at 670 km depth), whereas the model whose predictions are shown as the dotted curves contain both a 670 km and a 420 km discontinuity. The density contrasts across the boundaries are shown in per cent adjacent to each curve. For the case of comparison the prediction for the model with a fully nonadiabatic 1066B structure is shown as the solid curve. Inspection of the results of these calculations shows that for both northeastern North America and Fennoscandia the observed free air signals appear to require the presence of some internal buoyancy in the model in order that the free air anomaly be correctly predicted by the same model required by the relative sea level data, that is, one with modest viscosity contrast between the upper and lower mantles.

ACKNOWLEDGMENTS

John Andrews wishes to acknowledge the help provided by Jay Stravers and "Chip" Laymon (University of Colorado), who provided considerable assistance in the gathering of reprints and other necessary data. Harvey Thorleifson (Geological Survey of Canada), was responsible for obtaining the tide-gauge records and developing that source of information. The graduate assistants were provided partly by funds from a contract from the Geological Survey of Canada, and partly from ongoing grants from the National Science Foundation. This chapter has benefited greatly from critical comments of J. Adams, A.S. Dyke, and B.R. Pelletier (Geological Survey of Canada), G. Quinlan (Memorial University), W.H. Mathews (University of British Columbia), and J.B. Bird (McGill University).

REFERENCES

Aharon, P.
1983: 140,000-yr isotope climatic record from raised coral reefs in New Guinea; Nature, v. 304, p. 720-723.

Andrews, J.T.
1966: Pattern of coastal uplift and deglacierization, west Baffin Island, N.W.T.; Geographical Bulletin, v. 8, p. 174-193.
1970: A geomorphological study of post-glacial uplift with particular reference to Arctic Canada; Institute of British Geographers, Special Publication 2, 156 p.
1973a: The Wisconsin Laurentide Ice Sheet: dispersal centers, problems of rates of retreat, and climatic implications; Arctic and Alpine Research, v. 5, p. 185-199.
1973b: Maps of the maximum postglacial marine limit and rebound for the former Laurentide Ice Sheet (*The National Atlas of Canada*); Arctic and Alpine Research, v. 5, p. 41-48.
1978: Sea level history of arctic coasts during the Upper Quaternary: dating, sedimentary sequences, and history; Progress in Physical Geography, v. 2, p. 375-407.
1980: Progress in relative sea level and ice sheet reconstructions, Baffin Island, N.W.T. for the last 125,000 years; in Earth Rheology, Isostasy and Eustasy, N-A. Mörner (ed.); John Wiley and Sons, London, p. 175-200.
1982: On the reconstruction of Pleistocene ice sheets: A review; Quaternary Science Reviews, v. 1, p. 1-30.
1986: Methods in the reconstruction of sea levels in glaciated regions; in Manual of Sea Level Reconstructions, IGCP-61, Geo Abstracts, Norwich, United Kingdom.
1989: Quaternary geology of the northeastern Canadian Shield; in Quaternary Geology of Canada and Greenland, R.J. Fulton (ed.); Geological Survey of Canada, Geology of Canada, no. 1 (also Geological Society of America, The Geology of North America, v. K-1).

Andrews, J.T. and Barnett, D.M.
1972: Analysis of strandline tilt direction in relation to ice centres and postglacial crustal deformation, Laurentide Ice Sheet; Geografiska Annaler, v. 54A, p. 1-11.

Andrews, J.T. and Miller, G.H.
1984: Quaternary glacial and nonglacial correlations in the eastern Canadian Arctic; in Quaternary Stratigraphy of Canada — A Canadian Contribution to IGCP Project 24, R.J. Fulton (ed.); Geological Survey of Canada, Paper 84-10, p. 101-116.
1985: Holocene sea level variations within Frobisher Bay; in Quaternary Environments: Eastern Canadian Arctic, Baffin Bay, and West Greenland, J.T. Andrews (ed.); Allen and Unwin, London, p. 585-607.

Andrews, J.T. and Peltier, W.R.
1983: Glacial geology and glacial isostasy of the Hudson Bay region; in Shorelines and Isostasy, D.E. Smith and A.G. Dawson (ed.); Academic Press, London, p. 286-319.

Andrews, J.T. and Retherford, R.M.
1978: A reconnaissance survey of late Quaternary sea levels, Bella Bella/Bella Coola region, central British Columbia coast; Canadian Journal of Earth Sciences, v. 15, p. 341-350.

Andrews, J.T., Buckley, J.T., and England, J.H.
1970: Late-glacial chronology and glacio-isostatic recovery, Home Bay, east Baffin Island; Geological Society of America, Bulletin, v. 81, p. 1123-1148.

Andrews, J.T., Jull, A.J.T., Donahue, D.J., Short, S.K., and Osterman, L.E.
1985: Sedimentation rates in Baffin Island fiord cores from comparative radiocarbon dates; Canadian Journal of Earth Sciences, v. 22, p. 1827-1834.

Andrews, J.T., Miller, G.H., Vincent, J-S., and Shilts, W.W.
1984: Quaternary correlations in Arctic Canada; in Quaternary Stratigraphy of Canada — A Canadian Contribution to IGCP Project 24, R.J. Fulton, (ed.); Geological Survey of Canada, Paper 84-10, p. 127-136.

Andrews, J.T., Shilts, W.W., and Miller, G.H.
1983: Multiple deglaciations of the Hudson Bay Lowlands, Canada, since deposition of the Missinaibi (last-interglacial?) Formation; Quaternary Research, v. 19, p. 18-37.

Anundsen, K. and Fjeldskaar, W.
1983: Observed and theoretical Late Weichselian shore-level changes related to glacier oscillations at Yrkje, south-west Norway; in Late and Postglacial Oscillation of Glaciers: Glacial and Periglacial Forms, H. Schroeder-Lanz (ed.); A.A. Balkema, Rotterdam, p. 133-170.

Barnett, D.M.
1968: A re-examination and re-interpretation of tide-gauge data for Churchill, Manitoba; Canadian Journal of Earth Sciences, v. 3, p. 77-88.
1970: An amendment and extension of tide gauge data analysis for Churchill, Manitoba; Canadian Journal of Earth Sciences, v. 7, p. 626-627.

Barnett, T.P.
1983: Recent changes in sea level and their possible causes; Climatic Change, v. 5, p. 15-38.

Bird, J.B.
1954: Postglacial marine submergence in central Arctic Canada; Geological Society of America, Bulletin, v. 65, p. 457-464.
1970: The final phase of the Pleistocene ice sheet north of Hudson Bay; Geographica Lodziensia, v. 24, p. 75-89.

Blake, W., Jr.
1970: Studies of glacial history in Arctic Canada. I. Pumice, radiocarbon dates, and differential postglacial uplift in the eastern Queen Elizabeth Islands; Canadian Journal of Earth Sciences, v. 7, p. 634-664.
1975: Radiocarbon age determinations and postglacial emergence at Cape Storm, southern Ellesmere Island; Geografiska Annaler, v. 62A, p. 1-71.
1976: Sea and land relations during the last 15 000 years in the Queen Elizabeth Islands, Arctic Archipelago; in Report of Activities, Part B, Geological Survey of Canada, Paper 76-1B, p. 201-207.
1980: Mid-Wisconsinan interstadial deposits beneath Holocene beaches, Cape Storm, Ellesmere Island, Arctic Canada (abstract); AMQUA 6th Biennial Meeting, Orono, Maine, Abstracts, p. 26-27.
1985: Radiocarbon dating with accelerator mass spectrometry, results from Ellesmere Island, District of Franklin; in Current Research, Part B, Geological Survey of Canada, Paper 85-1B, p. 423-429.

Bloom, A.L.
1967: Pleistocene shorelines: a new test of isostasy; Geological Society of America, Bulletin, v. 78, p. 1477-1493.

Bloom, A.L. (compiler)
1977: Atlas of sea-level curves; International Geological Correlation Program Project 61, Sea level Project, IGCP Secretariat, Division of Earth Sciences, UNESCO, Paris, France.

Bloom, A.L., Broecker, W.S., Chappell, J.M.A., Matthews, R.K., and Mesolella, K-J.
1974: Quaternary sea level fluctuations on a tectonic coast: new ^{230}Th/^{234}U-dates from Huon Peninsula, New Guinea; Quaternary Research, v. 4, p. 185-205.

Brigham, J.K.
1983: Stratigraphy, amino acid geochronology, and correlation of Quaternary sea-level and glacial events, Broughton Island, Arctic Canada; Canadian Journal of Earth Sciences, v. 20, p. 577-598.

Brookes, I.A.
1969: Late-glacial marine overlap in western Newfoundland; Canadian Journal of Earth Sciences, v. 6, p. 1397-1404.

Brotchie, J.F. and Sylvester, R.
1969: On crustal flexure; Journal of Geophysical Research, v. 74, p. 5240-5252.

Bryson, R.A., Wendland, W.M., Ives, J.D., and Andrews, J.T.
1969: Radiocarbon isochrons on the disintegration of the Laurentide Ice Sheet; Arctic and Alpine Research, v. 1, p. 1-14.

Budd, W.F. and Smith, I.N.
1981: The growth and retreat of ice sheets in response to orbital radiation changes; in Sea Level, Ice and Climatic Change, I. Allison (ed.); International Association of Hydrological Sciences, Publication no. 131, p. 369-409.

Clague, J.J.
1975: Late Quaternary sea-level fluctuations, Pacific coast of Canada and adjacent areas; in Report of Activities, Part C, Geological Survey of Canada, Paper 75-1C, p. 17-21.
1981: Late Quaternary geology and geochronology of British Columbia. Part 2: Summary and discussion of radiocarbon-dated Quaternary history; Geological Survey of Canada, Paper 80-35, 41 p.
1983: Glacio-isostatic effects of the Cordilleran ice sheet, British Columbia; in Shorelines and Isostasy, D.E. Smith and A.G. Dawson (ed.); Academic Press, London, p. 285-319.
1985: Deglaciation of the Prince Rupert-Kitimat area, British Columbia; Canadian Journal of Earth Sciences, v. 22, p. 256-265.

Clague, J.J., Harper, J., Jr., Hebda, R.J., and Howes, D.E.
1982: Late Quaternary sea levels and crustal movements, coastal British Columbia; Canadian Journal of Earth Sciences, v. 19, p. 597-618.

Clark, J.A.
1980: The reconstruction of the Laurentide Ice Sheet of North America from sea-level data: Method and preliminary results; Journal of Geophysical Research, v. 85, p. 4307-4323.
1982a: Glacial loading: a cause of natural fracturing and a control of the present stress in regions of high Devonian-shale gas production; Sandia National Laboratories, Albuquerque, New Mexico, SAND-81-2474C, 18 p.

Clark, J.A., Farrell, W.E., and Peltier, W.R.
1978: Global changes in postglacial sea level: a numerical calculation; Quaternary Research, v. 9, p. 265-287.

Clark, P.U.
1982b: Late Quaternary history of the Iron Strand region, Torngat Mountains, northernmost Labrador, Canada (abstract); in Program and Abstracts; American Quaternary Association, Seventh Biennial Conference, Seattle, p. 82.
1984: Glacial geology of the Kangalaksiorvik-Abloviak region, northern Labrador, Canada; unpublished PhD thesis, University of Colorado, Boulder, Colorado, 240 p.

Colvill, A.J.
1982: Glacial landforms at the head of Frobisher Bay, Baffin Island, Canada; unpublished MA thesis, University of Colorado, Boulder, Colorado, 202 p.

Daly, R.A.
1934: The Changing World of the Ice Age; Yale University Press, New Haven, 271 p.

De Geer, G.
1892: On Pleistocene changes of level in eastern North America; Boston Society of Natural History Proceedings, v. 25, p. 454-477.

Denton, G.H. and Hughes, T.J. (ed.)
1981: The Last Great Ice Sheets; John Wiley, New York, 484 p.

Dredge, L.A. and Cowan, W.R.
1989: Quaternary geology of the southwestern Canadian Shield; in Quaternary Geology of Canada and Greenland, R.J. Fulton (ed.); Geological Survey of Canada, Geology of Canada, no. 1 (also Geological Society of America, The Geology of North America, v. K-1).

Dyke, A.S.
1974: Deglacial chronology and uplift history: northeastern sector, Laurentide Ice Sheet; Institute of Arctic and Alpine Research, Occasional Paper 12, 73 p.
1977: Quaternary geomorphology, glacial chronology, and climatic and sea-level history of southwestern Cumberland Peninsula, Baffin Island, Northwest Territories, Canada; unpublished PhD thesis, University of Colorado, Boulder, Colorado, 185 p.
1979: Glacial and sea level history of the southwestern Cumberland Peninsula, Baffin Island, N.W.T., Canada; Arctic and Alpine Research, v. 11, p. 179-202.
1983: Quaternary geology of Somerset Island, District of Franklin; Geological Survey of Canada, Memoir 404, 32 p.
1984: Quaternary geology of Boothia Peninsula and northern District of Keewatin, central Arctic Canada; Geological Survey of Canada, Memoir 407, 26 p.

Dyke, A.S. and Dredge, L.A.
1989: Quaternary geology of the northwestern Canadian Shield; in Quaternary Geology of Canada and Greenland, R.J. Fulton (ed.); Geological Survey of Canada, Geology of Canada, no. 1 (also Geological Society of America, The Geology of North America, v. K-1).

Dyke, A.S., Andrews, J.T., and Miller, G.H.
1982a: Quaternary geology of Cumberland Peninsula, Baffin Island, District of Franklin; Geological Survey of Canada, Memoir 403, 32 p.

Dyke, A.S., Dredge, L.A., and Vincent, J-S.
1982b: Configuration of the Laurentide ice sheet during the Late Wisconsin maximum; Géographie physique et Quaternaire, v. 36, p. 5-14.

England, J.H.
1976: Late Quaternary glaciation of the eastern Queen Elizabeth Islands, N.W.T., Canada: Alternative models; Quaternary Research, v. 6, p. 185-203.
1983: Isostatic adjustments in a full glacial sea; Canadian Journal of Earth Sciences, v. 20, p. 895-917.
1985: The late Quaternary history of Hall Land, northwest Greenland; Canadian Journal of Earth Sciences, v. 22, p. 1394-1408.

England, J.H., Bradley, R.S., and Miller, G.H.
1978: Former ice shelves in the Canadian High Arctic; Journal of Glaciology, v. 20, p. 393-404.

England, J.H., Bradley, R.S., and Stuckenrath, R.
1981: Multiple glaciations and marine transgression, western Kennedy Channel, Northwest Territories, Canada; Boreas, v. 10, p. 71-90.

Fairchild, H.L.
1918: Postglacial uplift of northeastern America; Geological Society of America, Bulletin, v. 29, p. 187-234.

Farrand, W.R. and Gajda, R.T.
1962: Isobases on the Wisconsin marine limit; Geographical Bulletin, v. 17, p. 5-22.

Farrell, W.E. and Clark, J.A.
1976: On postglacial sea level; Geophysical Journal of the Royal Astronomical Society, v. 46, p. 647-667.

Fillon, R.H., Hardy, I.A., Wagner, F.J.E., Andrews, J.T., and Josenhans, H.W.
1981: Labrador shelf: shell and total organic matter — ^{14}C date discrepancies; in Current Research, Part B, Geological Survey of Canada, Paper 81-1B, p. 105-111.

Fladmark, K.R.
1974: A paleoecological model for Northwest prehistory; unpublished PhD dissertation, University of Calgary, Calgary, Alberta.
1975: A paleoecological model for Northwest coast prehistory; National Museum of Man, Mercury series, Archeological Survey of Canada, Paper 43, 328 p.

Forbes, D.L.
1980: Late Quaternary sea level in the southern Beaufort Sea; in Current Research, Part B, Geological Survey of Canada, Paper 80-1B, p. 75-87.

Fulton, R.J. and Walcott, R.I.
1975: Lithospheric flexure as shown by deformation of glacial lake shorelines in southern British Columbia; Geological Society of America, Memoir 142, p. 163-173.

Fulton, R.J. and Westgate, J.A.
1975: Tephrostratigraphy of Olympia Interglacial sediments in south-central British Columbia, Canada; Canadian Journal of Earth Sciences, v. 12, p. 489-502.

Gilbert, F., and Dziewonski, A.M.
1975: An application of normal mode theory to the retrieval of structural parameters and source mechanisms from seismic spectra; Philosophical Transactions of the Royal Society of London, Series A. Mathematical and Physical Sciences, v. 278, no. 1280, p. 187-269.

Gornitz, V., Lebedeff, S., and Hansen, J.
1982: Global sea level trend in the past century; Science, v. 215, no. 4540, p. 1611-1614.

Grant, D.R.
1970: Recent coastal submergence of the Maritime Provinces, Canada; Canadian Journal of Earth Sciences, v. 7, p. 676-689.
1972: Postglacial emergence of northern Newfoundland; in Report of Activities, Part B, Geological Survey of Canada, Paper 72-1B, p. 100-107.
1975: Glacial style and the Quaternary stratigraphic record in the Atlantic Provinces, Canada; in Report of Activities, Part B, Geological Survey of Canada, Paper 75-1B, p. 109.
1980: Quaternary sea-level change in Atlantic Canada as an indication of crustal delevelling; in Earth Rheology, Isostasy, and Eustasy, N-A. Mörner (ed.); John Wiley, New York, p. 201-214.

Gutenberg, B.
1941: Changes in sea level, postglacial uplift, and mobility of the earth's interior; Geological Society of America, Bulletin, v. 52, p. 721-772.

Haskell, N.A.
1936: The motion of a viscous fluid under a surface load; Physics (N.Y.), v. 7, p. 56-61.
1937: The viscosity of the asthenosphere; American Journal of Science, v. 33, p. 22-28.

Hays, J.D., Imbrie, J., and Shackleton, N.J.
1976: Variations in the earth's orbit: Pacemaker of the ice ages; Science, v. 194, p. 1121-1132.

Hill, P.R., Mudie, P.J., Moran, K., and Blasco, S.M.
1985: A sea level curve for the Canadian Beaufort shelf; Canadian Journal of Earth Sciences, v. 22, p. 1383-1393.

Hillaire-Marcel, C.
1979: Les mers post-glaciaires du Québec: quelques aspects; thèse de doctorat, Université Pierre et Marie Curie, Paris VI, vol. 1, 293 p., vol. 2, 249 p.
1980: Les faunes des mers post-glaciaires du Québec: quelques considérations paléoécologiques; Géographie physique et Quaternaire, vol. 34, p. 3-60.

Hillaire-Marcel, C. and Vincent, J-S.
1980: Holocene stratigraphy and sea level changes in southeastern Hudson Bay, Canada; Paleo-Quebec 11, Musée d'Archéologie, Université du Québec à Trois-Rivières, Trois-Rivières, 165 p.

Hodgson, D.A. (co-ordinator)
1989: Quaternary geology of the Queen Elizabeth Islands; in Quaternary Geology of Canada and Greenland, R.J. Fulton (ed.); Geological Survey of Canada, Geology of Canada, no. 1 (also Geological Society of America, The Geology of North America, v. K-1).

Hoffman, J.S., Keyes, D., and Titus, J.G.
1983: Projecting future sea level rise; U.S. Environmental Protection Agency, EPA-230-09-007, 121 p.

Innes, M.J.S., Goodacre, A.K., Weston, A., and Weber, J.R.
1968: Gravity and isostasy in the Hudson Bay region; Publication of the Dominion Observatory, Ottawa, Part 5.

Ives, J.D.
1960: Former ice-dammed lakes and the deglaciation of the middle reaches of the George River, Labrador-Ungava; Geographical Bulletin, v. 14, p. 44-70.

Jardine, W.G.
1975: The determination of former sea levels in areas of large tidal range; in Quaternary Studies, Royal Society of New Zealand, Wellington, p. 163-168.

Jeanloz, R. and Thompson, A.B.
1983: Phase transitions and mantle discontinuities; Reviews of Geophysics and Space Physics, v. 21, p. 51-74.

Jeffreys, H.
1972: Creep in the earth and planets; Tectonophysics, v. 13, p. 569-581.

Karrow, P.F. and Calkin, P.E. (editors)
1985: Quaternary evolution of the Great Lakes; Geological Association of Canada, Special Paper 30, 258 p.

Klein, J., Lerman, J.C., Damon, P.E., and Ralph, E.K.
1982: Calibration of radiocarbon dates; Radiocarbon, v. 24, p. 103-150.

Lauriol, B. and Gray, J.T.
1983: Un lac glaciaire dans la région du lac Minto - Nouveau-Québec; Journal canadien des sciences de la terre, vol. 20, p. 1488-1492.

Litherland, A.E.
1980: Ultrasensitive mass spectrometry with accelerators; Annual Review of Nuclear and Particle Science, v. 30, p. 437-473.

Løken, O.H.
1962: The late glacial and postglacial emergence and deglaciation of northernmost Labrador; Geographical Bulletin, v. 17, p. 23-56.
1966: Baffin Island refugia older than 54 000 years; Science, v. 153, p. 1378-1380.

Lowdon, J.A.
1985: The Geological Survey of Canada radiocarbon dating laboratory; Geological Survey of Canada, Paper 84-24, 19 p.

Mareschal, J-C. and Gangi, A.F.
1977: Equilibrium position of phase boundary under horizontally varying surface loads; Geophysical Journal of the Royal Astronomical Society, v. 49, p. 757-772.

Mathews, W.H., Fyles, J.G., and Nasmith, H.W.
1970: Postglacial crustal movements in southwestern British Columbia and adjacent Washington State; Canadian Journal of Earth Sciences, v. 7, p. 690-702.

Matthews, B.
1967: Late Quaternary land emergence in northern Ungava, Quebec; Arctic, v. 20, p. 176-201.

Matthews, R.K.
1973: Relative elevation of Late Pleistocene high sea level stands: Barbados uplift rates and their implications; Quaternary Research, v. 3, p. 147-153.

McDonald, B.C.
1969: Glacial and interglacial stratigraphy, Hudson Bay Lowland; in Earth Science Symposium on Hudson Bay, P.J. Hood (ed.); Geological Survey of Canada, Paper 68-53, p. 78-99.
1971: Late Quaternary stratigraphy and deglaciation in eastern Canada; in The Late Cenozoic Glacial Ages, K.K. Turekian (ed.); New Haven, Yale University Press, p. 331-354.

Miller, G.H.
1985: Aminostratigraphy of Baffin Island shell-bearing deposits; in Quaternary Environments: Eastern Canadian Arctic, Baffin Bay, and West Greenland, J.T. Andrews (ed.); Allen and Unwin, London.

Miller, G.H. and Dyke, A.S.
1974: Proposed extent of late Wisconsin Laurentide ice on eastern Baffin Island; Geology, v. 2, p. 125-130.

Miller, G.H., Andrews, J.T., and Short, S.K.
1977: The last interglacial-glacial cycle, Clyde Foreland, Baffin Island, N.W.T.: stratigraphy, biostratigraphy and chronology; Canadian Journal of Earth Sciences, v. 14, p. 2824-2857.

Mode, W.N., Nelson, A.R., and Brigham, J.K.
1983: Sedimentologic evidence for Quaternary glaciomarine cyclic sedimentation along eastern Baffin Island, Canada; in Glacial-marine Sedimentation, B.F. Molnia (ed.); Plenum Press, New York, p. 495-534.

Mörner, N-A.
1980: The Fennoscandian Uplift: Geological data and their geodynamical implication; in Earth Rheology, Isostasy and Eustasy, N-A. Mörner (ed.); Wiley and Sons, London, p. 251-284.

Mudie, P.A. and Guilbault, J-P.
1982: Ecostratigraphic and paleomagnetic studies of Late Quaternary sediments on the northeast Newfoundland shelf; in Current Research, Part B, Geological Survey of Canada, Paper 82-1B, p. 107-116.

National Research Council
1984: Environment of West Antarctica: potential CO_2-induced changes; Report of a Workshop, National Academy Press, Washington, D.C., 236 p.

Nelson, A.R.
1978: Quaternary glacial and marine stratigraphy of the Qivitu Peninsula, northern Cumberland Peninsula, Baffin Island, Canada; unpublished PhD dissertation, University of Colorado, Boulder, Colorado, 215 p.
1981: Quaternary glacial and marine stratigraphy of the Qivitu Peninsula, northern Cumberland Peninsula, Baffin Island; Geological Society of America, Bulletin, v. 92, Part I, p. 512-518, Part II, p. 1143-1281.

O'Connell, R.J.
1976: The effects of mantle phase changes on postglacial rebound; Journal of Geophysical Research, v. 81, p. 971-974.

Paterson, W.S.B.
1972: Laurentide Ice Sheet: estimated volumes during the late Wisconsin; Reviews of Geophysics and Space Physics, v. 10, p. 885-917.

Peltier, W.R.
1974: The impulse response of a Maxwell Earth; Reviews of Geophysics and Space Physics, v. 12, p. 649-669.
1976: Glacial isostatic adjustments-II: The inverse problem; Geophysical Journal of the Royal Astronomical Society, v. 46, p. 669-706.
1980: Mantle convection and viscosity; in Physics of the Earth's Interior, A.M. Dziewonski and E. Boschi (ed.); North Holland, Amsterdam, p. 362-431.
1981: Ice age geodynamics; Annual Review of Earth and Planetary Science, v. 9, p. 199-225.
1982: Dynamics of the ice age earth; Advances in Geophysics, v. 24, p. 1-146.
1984: The thickness of the continental lithosphere; Journal of Geophysical Research, v. 89, p. 11303-11306.
1985a: Mantle convection and viscoelasticity; Annual Review of Fluid Mechanics, v. 17, p. 561-608.
1985b: The LAGEOS constraint on deep mantle viscosity: results from a new normal mode method for the inversion of viscoelastic relaxation data; Journal of Geophysical Research, v. 90.

Peltier, W.R. and Andrews, J.T.
1976: Glacial-isostatic adjustment-I. The forward problem; Geophysical Journal of the Royal Astronomical Society, v. 46, p. 605-646.
1983: Glacial geology and glacial isostasy of the Hudson Bay region; in Shorelines and Isostasy, D.I. Smith (ed.); Institute of British Geographers, London, p. 285-319.

Peltier, W.R. and Wu, P.
1982: Mantle phase transitions and the free air gravity anomalies over Fennoscandia and Laurentia; Geophysical Research Letters, v. 9, p. 731-734.

Peltier, W.R., Farrell, W.E., and Clark, J.A.
1978: Glacial isostasy and relative sea level: a global finite element model; Tectonophysics, v. 50, p. 81-110.

Prest, V.K.
1969: Retreat of Wisconsin and Recent ice in North America; Geological Survey of Canada, Map 1257A, scale 1:5 000 000.
1970: Quaternary geology; in Geology and Economic Minerals in Canada, R.J. Douglas (ed.); 5th Edition, Department of Energy, Mines and Resources, Ottawa, Canada, p. 676-764.
1984: The Late Wisconsinan glacier complex; in Quaternary Stratigraphy of Canada — A Canadian Contribution to IGCP Project 24, R.J. Fulton (ed.); Geological Survey of Canada, Paper 84-10, p. 21-36, Map 1584A.

Prest, V.K., Grant, D.R., and Rampton, V.N.
1968: Glacial Map of Canada; Geological Survey of Canada, Map 1253A, scale 1:5 000 000.

Quinlan, G.
1981: Numerical models of postglacial relative sea level change in Atlantic Canada and the eastern Canadian Arctic; unpublished PhD thesis, Department of Oceanography, Dalhousie University, Halifax, Nova Scotia, 499 p.

1984: Postglacial rebound and the focal mechanisms of eastern Canadian earthquakes; Canadian Journal of Earth Sciences, v. 21, p. 1018-1023.

1985: A numerical model of postglacial relative sea level change near Baffin Island; in Quaternary Environments: Eastern Canadian Arctic, Baffin Bay, and West Greenland, J.T. Andrews (ed.); Allen and Unwin, London.

Quinlan, G. and Beaumont, C.
1981: A comparison of observed and theoretical postglacial relative sea level in Atlantic Canada; Canadian Journal of Earth Sciences, v. 18, p. 1146-1163.

1982: The deglaciation of Atlantic Canada as reconstructed from the postglacial relative sea-level record; Canadian Journal of Earth Sciences, v. 19, p. 2232-2248.

Radok, U., Barry, R.G., Jenssen, D., Keen, R.A., Kiladis, G.N., and McInnes, B.
1982: Climatic and physical characteristics of the Greenland Ice Sheet; Cooperative Institute for Research in Environmental Sciences, University of Colorado, Boulder, Colorado, 193 p.

Richards, H.G.
1970: Annotated bibliography of Quaternary shorelines. Supplement 1965-1969; Academy of Natural Sciences, Special Publication 10, Philadelphia, Pennsylvania, 240 p.

1974: Annotated bibliography of Quaternary shorelines. Second supplement 1970-1973; Academy of Natural Sciences, Special Publication 11, Philadelphia, Pennsylvania, 214 p.

Richards, H.G. and Fairbridge, R.W.
1965: Annotated bibliography of Quaternary shorelines (1945-1964); Academy of Natural Sciences, Special Publication 6, Philadelphia, Pennsylvania, 280 p.

Richards, H.G. and Shapiro, E.A.
1979: Annotated bibliography of Quaternary shorelines. Third supplement 1974-1977; GeoAbstracts, Bibliography No. 5, GeoAbstracts Ltd., Norwich, UK, 245 p.

Riddihough, R.P.
1982: Contemporary movements and tectonics on Canada's west coast: a discussion; Tectonophysics, v. 86, p. 319-341.

Scott, D.B. and Greenberg, D.A.
1983: Relative sea-level rise and tidal development in the Fundy tidal system; Canadian Journal of Earth Sciences, v. 20, p. 1554-1564.

Shackleton, N.J.
1967: Oxygen isotope analyses and Pleistocene temperatures re-addressed; Nature, v. 215, p. 15-17.

Shackleton, N.J. and Opdyke, N.D.
1973: Oxygen isotope and paleomagnetic stratigraphy of equatorial Pacific core V28-238: oxygen isotope temperatures and ice volumes on a 105 year and 106 year scale; Quaternary Research, v. 3, p. 35-55.

Shilts, W.W.
1980: Flow pattern in the central North American ice sheet; Nature, v. 286, no. 5770, p. 213-218.

1982: Quaternary evolution of the Hudson/James Bay Region; Naturaliste Canadien, v. 109, p. 309-332.

Short, S.K.
1981: Radiocarbon date list 1, Labrador and northern Quebec, Canada; Institute of Arctic and Alpine Research, University of Colorado, Occasional Paper 36, 33 p.

Sim, V.W.
1960: Maximum postglacial marine submergence in northern Melville Peninsula; Arctic, v. 13, p. 178.

Skinner, R.G.
1973: Quaternary stratigraphy of the Moose River Basin, Ontario; Geological Survey of Canada, Bulletin 225, 77 p.

Smith, I.N.
1984: Numerical modelling of ice masses; unpublished PhD thesis, Meteorology Department, University of Melbourne, Melbourne, Australia.

Szabo, B.J., Miller, G.H., Andrews, J.T., and Stuiver, M.
1981: Comparison of uranium-series, radiocarbon and amino acid data from marine molluscs, Baffin Island, Arctic Canada; Geology, v. 9, p. 451-457.

Teller, J.T. and Clayton L. (ed.)
1983: Glacial Lake Agassiz; Geological Association of Canada, Special Paper 26, 451 p.

Thorson, R.M.
1981: Isostatic effects of the last glaciation in the Puget Lowland, Washington; United States Geological Survey, Open File Report 81-370, 100 p.

Vaníček, P.
1976: Pattern of recent vertical crustal movements in Maritime Canada; Canadian Journal of Earth Sciences, v. 13, p. 661-667.

Vaníček, p. and Nagy, D.
1980: The map of contemporary vertical crustal movement in Canada; EOS, v. 61, p. 145-147.

Velleman, P.F. and Hoaglin, D.C.
1981: Applications, basis, and computing of exploratory data analysis; Duxbury Press, Boston, Massachussets, 354 p.

Vincent, J-S.
1982: The Quaternary history of Banks Island, N.W.T., Canada; Géographie physique et Quaternaire, v. 36, p. 109-232.

Vincent, J-S. and Hardy, L.
1979: The evolution of glacial lakes Barlow and Ojibway, Quebec and Ontario; Geological Survey of Canada, Bulletin 316, 18 p.

Walcott, R.I.
1970: Isostatic response to loading of the crust in Canada; Canadian Journal of Earth Sciences, v. 7, p. 716-726.

1972: Late Quaternary vertical movements in eastern North America; Reviews of Geophysical and Space Physics, v. 10, p. 849-884.

Washburn, A.L. and Stuiver, M.
1985: Radiocarbon dates from Cornwallis Island area, Arctic Canada — an interim report; Canadian Journal of Earth Sciences, v. 22, p. 630-636.

Wigen, S.O. and Stephenson, F.E.
1980: Mean sea level on the Canadian West Coast; in Proceedings 2nd International Symposium on problems related to the definition of North American vertical geodetic networks; Canadian Institute of Surveying, Ottawa, Ontario, p. 105-124.

Wu, p. and Peltier, W.R.
1983: Glacial isostatic adjustment and the free air gravity anomaly as a constraint on deep mantle viscosity; Geophysical Journal of the Royal Astronomical Society, v. 74, p. 377-450.

Authors' addresses

J.T. Andrews
Department of Geological Sciences
and INSTAAR
University of Colorado
Boulder, Colorado 80309

W.R. Peltier
Department of Physics
University of Toronto
Toronto, Ontario
M5S 1A1

Chapter 9

A SURVEY OF GEOMORPHIC PROCESSES IN CANADA

Summary

Introduction — *J.A. Heginbottom*

Climate of Canada — *M.W. Smith*

Physiography of Canada and its effects on geomorphic processes — *H.O. Slaymaker*

Slope processes — *M.A. Carson and M.J. Bovis*

River processes — *T.J. Day*

Cold climate processes — *H.M. French*

Other geomorphic processes

    Geomorphic response to endogenic processes — *S.G. Evans*

    Geomorphology of lakes — *R.A. Klassen*

    Solution processes — *D.C. Ford*

    Eolian processes — *P.P. David*

    Animals as agents of erosion — *P.A. Egginton*

Human society as a geological agent — *K. Hewitt*

Acknowledgments

References

Retrogressive thaw slumps in massive ice-cored terrain near Peninsula Point, southwest of Tuktoyaktuk, Northwest Territories. Scars of earlier stabilized slumps can be seen behind the active slump. Courtesy of J.A. Heginbottom. 204375-S.

Chapter 9

A SURVEY OF GEOMORPHIC PROCESSES IN CANADA

Co-ordinator
J.A. Heginbottom

SUMMARY

Canada is a large northern country with a wide variety of landscapes. Most geomorphic processes are dominated by the special quality of northern climates, with great contrasts in temperature between seasons; large amounts of snow deposited annually and lying undisturbed for months; intensive frost action and widespread permafrost; and peculiar river regimes with a large spring freshet, extensive flooding, and considerable surplus energy for erosion and transportation of debris.

Regionally, the climate varies with latitude and distance from the oceans; the high, north-south alignment of the Cordillera controls air mass movements across the country and so influences the climate on a regional scale. The climate is markedly seasonal, and has undergone considerable change throughout geological time. Recent anthropogenic influences on climate may be affecting geomorphic processes.

Geological history and geomorphic processes control the landscape at local and regional scales. On a continental scale, geological structure and lithology are more significant. Topographic relief, the presence of permafrost, and the nature of soil and vegetation cover combine with geology to provide the details of the physiography of Canada, and the framework of the landscape of Canada.

Major geomorphic processes active in Canada today comprise slope processes, river processes, and cold climate processes.

The land surface of Canada is generally young and so has yet to attain permanent stability with respect to mass failures. Thus slope processes at all scales are active and important agents of landscape modification. Many different types of landslides occur in Canada, due to the range in topographic, geological, climatic, and soil conditions across the country. Even within a region, a variety of landslides occur. In crystalline rock and hard sedimentary rock, slope failures include rockfalls, detachment slides, and rock avalanches. In clay and poorly lithified shale, typical failures include deep-seated rotational landslides, lateral spreading, and earthflows. In unconsolidated regolith, failures comprise shallow translational slides and mobile debris flows. Thawing permafrost slopes are characterized by specific forms of slope failure. Of secondary importance are the processes of soil creep and soil erosion and, locally, wind erosion and subsurface erosion.

The range of climatic and physiographic variation in the landscape of Canada results in different hydrological regimes across the country. These, in turn, produce variations in fluvial transfer and storage of sediment, the intensity of fluvial erosion, and river morphology and processes. There are also long term variations in fluvial sediment transfer arising from postglacial landscape development and, more recently, climatic change and anthropogenic effects.

As the present impact of glaciation is extremely localized, nonglacial or periglacial conditions dominate cold climate processes in Canada. These processes are associated with intense frost action, commonly combined with the presence of permafrost. More than 50% of Canada currently experiences periglacial conditions. The main processes comprise freezing and thawing of the ground, generally coupled with frost heave and ice segregation, and, particularly in permafrost regions, thaw settlement. Ground ice is an important component of the ground in permafrost regions and contributes significantly to landforms and local relief elements. Frost action results in various forms of patterned ground and controls cryogenic weathering processes.

Other geomorphic processes active today are less significant in the magnitude, rate, or extent of their effects. These include the effects of endogenic processes such as vertical crustal movement, the effects of earthquakes, and the effects of volcanic processes. Lakes are numerous and extensive in many parts of Canada; their geomorphology is affected by geological and glacial history, climate, and physiography. Solution processes occur widely but locally, affecting mainly limestone terrains, but also gypsum and salt deposits. Similarly, the effects of eolian processes are widespread but only of local importance; wind was more significant in early postglacial time. Animals act as agents of erosion, directly or indirectly, and, less often but more visibly, as agents of deposition; their significance is commonly underestimated.

Also underestimated are the effects of human society in modifying both the landscape of Canada and natural geomorphic effects within the landscape. Societies create their own, managed landscapes — urban, agricultural, forest land, and mining areas. They affect, control, or promote natural erosion processes and initiate major new, artificial geological cycles and, now, anthropogenic geological regimes. Forest utilization affects the largest areas followed

---

Heginbottom, J.A. (co-ordinator)
1989: A survey of geomorphic processes in Canada; Chapter 9 in Quaternary Geology of Canada and Greenland, R.J. Fulton (ed.); Geological Survey of Canada, Geology of Canada, no. 1 (also Geological Society of America, The Geology of North America, v. K-1).

by croplands and the effects of various agricultural practices. Mining, particularly for construction materials, and the built environment, while less extensive, are more obvious in their effects. In many areas, these modified and artificial processes combine, resulting in even more intensive and accelerating changes. Overall, the continuing effects of human impacts are now a significant component of the earth's geological environment.

INTRODUCTION
J.A. Heginbottom

Canada is one of the largest countries in the world, second only to the USSR in area. A consequence of this great size is that Canada encompasses a wide variety of landscapes, ranging from prairies to polar deserts and from coastal mudflats to high mountains. Landscapes of Canada are the result of the interaction of internal or endogenic geological processes, such as tectonic and volcanic activity, and external, exogenic or geomorphic processes. The former processes contribute to the building of mountains and general increases in the elevation of the land surface. The latter processes are those that wear down, erode, and sculpt the earth's surface and produce the diverse landscapes seen today. These processes, as they occur in Canada today, are the subject of this chapter.

None of the various geomorphic processes to be described are unique to Canada; other northern countries exhibit processes in similar combinations. Scandinavia, Alaska, and the USSR are the major regions in which the geomorphology has affinities with that of Canada. The range and combination of geomorphic processes at work in Canada today are largely a function of the northerness of the country, due to the special qualities of northern climates (Hare, 1980). These special qualities, described in more detail below, include great contrasts in temperatures between the seasons, and between day and night that result in intensive frost action and, in the northern part of the country, widespread permafrost. Furthermore, the generally humid climate, with cold snowy winters and summer storms, produces special river regimes with considerable energy for erosion and transportation of debris.

Geomorphic processes

The processes acting on the surface of the earth are controlled by a number of factors, as noted above. The results of this are that processes vary in both the temporal and spatial frequency of their activity, the efficiency with which they erode or modify the face of the earth, and their significance as perceived by mankind. Man's views of geomorphology are also strongly influenced by his level of social and technological development.

A few geomorphic processes act continuously or quasi-continuously in both time and, within limited regions, in space. Examples include fluvial erosion, sediment transport, and solifluction. More commonly, processes are more or less intermittent in both time and space. Thus, frost action is a diurnally to seasonally active process; some fluvial erosion is flood dependent; and slope wash may occur only during a particular rainstorm event. These are examples of processes that are discontinuous in time. Spatially discontinuous processes include burrowing by animals, wind deflation, or the growth of ice wedges.

For many processes there is a direct relationship between the frequency of action and the magnitude of the events. This relationship is best known for river floods, where the recurrence interval or return period for a flood of a given size can be predicted with a fair degree of reliability, given a reasonably long record of flood levels. A similar relationship exists for other intermittent processes, such as landslides, rockfalls, and earthquakes. From a human perspective, the larger and less frequent events are the ones more likely to pose a direct threat to life and cause extensive property damage. Geological hazards, such as large rockfalls and failure of natural dams, are discussed in Chapter 12 (Jackson, 1989).

As is apparent throughout this volume, the repeated glaciation and deglaciation of Canada has had a profound and overriding effect on the landscape — even in unglaciated areas. The most recent deglaciation occurred between 18 and 6 ka ago (Prest, 1969). During and immediately following deglaciation, in each part of the country, there was a period of intense geomorphological activity, as the newly exposed, unvegetated surface was affected by the full range of geomorphic processes. This "paraglacial" concept (Church and Ryder, 1972) has been discussed by Ryder (1971a, b) with special reference to the development of alluvial fans in south-central British Columbia, and by Jackson et al. (1982) for terrace sediments of Bow River, Alberta. The particular conditions existing during this period of transition from predominantly glacial to dominantly fluvial geomorphic controls may be regarded as effectively unique and, as such, represent the other end of the range of frequencies of activity for geomorphic processes.

Descriptions of the climate and physiography of Canada, as they pertain to geomorphic processes, follow this introduction. After these are three sections discussing the main suites of geomorphic processes presently at work in Canada: slope processes, river processes, and cold climate processes. In turn, these are followed by a section comprising brief accounts of five groups of processes of lesser significance — lesser in the sense of the magnitude or rate of their effects, the proportion of Canada affected by them, or the extent of our knowledge of these processes. The effects of man on geomorphic processes forms the final section of this chapter.

The physiography of Canada, as described here, is used in a broad, classical sense (Bowman, 1911) to include consideration of the vegetation and soil cover. Elsewhere in the volume, the term is used in a narrower sense to refer only to the form of the ground surface.

In several sections of this chapter, many of the examples presented and references cited are drawn from western Canada and from the Cordilleran region in particular. This is partly because the Cordillera constitutes a geologically young and active environment, with volcanic and tectonic mountain building processes active today. This is also an area of economic significance to Canada, with important

---

Heginbottom, J.A.
1989: Introduction (A survey of geomorphic processes in Canada); in Chapter 9 of Quaternary Geology of Canada and Greenland, R.J. Fulton (ed.); Geological Survey of Canada, Geology of Canada, no. 1 (also Geological Society of America, The Geology of North America, v. K-1).

industries locally and major transportation connections between the rest of Canada and the Pacific Coast. In consequence, efforts have been made, and are continuing, to understand geological and geomorphic processes in this region and to assess their significance for man. This topic is discussed in more detail in Chapter 12 (Jackson, 1989).

One major suit of geomorphic processes has been deliberately omitted here. Discussion of all aspects of coastal and shoreline processes is not included here; these topics are presented in the volume on *Geology of the Continental Margin of Eastern Canada* (Keen and Williams, 1989). Present day glacial processes are also excluded, along with any discussion of all the processes involved in the growth and decay of ice sheets and of erosion and sedimentation associated therewith. Much of this material is covered in the first six chapters of this volume.

CLIMATE OF CANADA
M.W. Smith

Much of Canada lies in the sweep of the circumpolar westerly winds, which dominate the atmospheric circulation of the middle latitudes. The climates of these latitudes are characterized by changeability and by the occurrence of a truly cold season, which generally increases in severity as one goes inland, farther away from any maritime influence. For their latitude, however, the climates of Canada are unusually cold, for although the summers bring normal warmth, the winters are long and intensely cold. For example, Ottawa (45°N) is at about the same latitude as the Bordeaux region of France. In July and August, the mean temperature is almost the same (20.0° versus 19.6°C); in January, however, the mean temperature in Ottawa ($-10.9°C$) is 16.1°C colder than in Bordeaux (5.2°C). In fact, the Ottawa winter is colder than that in Leningrad, which is at 60°N. Winnipeg is close to the latitude of Paris, but has a mean frost-free season of only 118 days compared to more than 200 days.

This section presents a brief account of the major factors which produce this unique climatic situation. For a fuller description of climatic conditions across Canada, the reader is referred to Bryson and Hare (1974) and Hare and Thomas (1979).

Effects of physiography

Canada is unusually cold because of the influence of its physiography on air mass movements, in particular the high, north-south alignment of the Cordillera in the west combined with the vast plains of subdued relief to the east (see Bryson and Hare, 1974; this chapter, Slaymaker, 1989). The western mountains present a significant obstacle to the flow of the westerlies and confine maritime influences to a narrow coastal fringe. East of the Cordillera, the annual temperature cycle is not tempered by air mass exchanges with the ocean, and as a consequence the region is subject to extreme continental conditions. This physiographic arrangement is important for yet another reason, in that precipitation, which might have spread far across the open plains to the east, instead falls in prodigious amounts on the west coast mountains leaving the Prairie Provinces, and even some of the interior basins of the Cordillera, dry. Thus, the world's greatest ocean has little direct effect on the climate of most of Canada.

In contrast, the open plains to the east of the Cordillera provide an unobstructed path for great meridional sweeps of arctic and tropical air streams. The climatic regime here is dominated by frequent penetrations of intensely cold arctic air in winter, and warm, often humid, tropical air in summer. These incursions contribute to the marked seasonality, as well as the storminess and changeability that characterize the area east of the Rocky Mountains.

There is another influence of the Cordillera which serves to emphasize the pattern of harsh winter cold in the central and eastern parts of the country. The mountains create a permanent wave in the upper air westerlies, which are diverted northwards as they approach the Pacific coast, and then sweep to the south over the Great Lakes and eastern Canada. This pattern is particularly common in winter, resulting in a high frequency of arctic air flows over central and eastern Canada. Thus, whilst summer east of the Rockies has temperatures that are typical of the latitude, the prolonged and intense winter is unusual, and gives the climate its very marked continental character.

The east coast has a modified continental climate, with frequent incursions of mild Atlantic maritime air associated with mid-latitude cyclones. This reduces the annual range of temperature and spreads precipitation more evenly throughout the year. In the east, also, the Great Lakes markedly influence the climate of their region. When the air is cold but the lakes are unfrozen, the warm open water heats and moistens the cold air flowing across the lakes, producing local precipitation belts along the leeward shores. In summer, the lakes remain relatively cool and serve to moderate somewhat the heat of the continental interior.

Temperature

As one moves northward in Canada, moderating influences are left behind and the climate is never really warm. This is the land of the boreal forest, treeless tundra, and permafrost, and in these environments physical processes proceed at a slow pace. Persistent net outward radiation during the prolonged arctic winter night creates bitterly cold air masses. Mean daily temperature is below $-30°C$ for three to five months in winter throughout the ice-bound Arctic Islands. Outbreaks of frigid arctic air also tend to dominate the winter climate of central and eastern Canada and occasionally reach as far south as Florida. In the arctic summer, heat energy is used to melt the snow and ice, and air temperatures remain relatively low. In the far north the temperature of the warmest month (July) is only around 5°C; farther south, on the mainland, mean July temperatures do rise a little above 10°C.

In the Arctic proper, the cold air circulation pattern keeps out warmer air masses that might otherwise import heat. Farther south, in the boreal zone, the climate is dominated by arctic air streams only in winter; in summer these are replaced by mild westerlies. Because of the southerly

---

Smith, M.W.
1989: Climate of Canada; *in* Chapter 9 of Quaternary Geology of Canada and Greenland, R.J. Fulton (ed.); Geological Survey of Canada, Geology of Canada, no. 1 (*also* Geological Society of America, The Geology of North America, v. K-1).

Table 9.1. Climatic statistics for selected stations

Station	Latitude (N)	Longitude (W)	Mean Annual Temp. (°C)	GDD[1]	Frost-Free Period (days)	FDD[2]	Snow-cover (days)	Annual Total Precip. (mm)
Vancouver	49°11'	123°10'	9.8	2019	212	45	7	1068
Edmonton	53°34'	113°31'	2.8	1516	127	1501	117	447
Saskatoon	52°10'	106°41'	1.6	1618	110	1977	128	353
Winnipeg	49°54'	97°14'	2.3	1791	118	1903	122	535
Toronto	43°40'	79°24'	8.9	2434	192	438	59	790
Toronto/Malton	43°41'	79°38'	7.5	2185	148	616	70	752
Ottawa	45°19'	75°40'	5.8	2069	137	1040	117	851
Montréal	45°30'	73°35'	7.2	2301	183	814	117	999
Montréal/Dorval	45°20'	73°45'	6.5	2165	154	933	114	941
Quebec City	46°48'	71°23'	4.4	1729	132	1153	140	1089
Halifax	44°38'	63°30'	6.8	1708	183	422	63	1381
Churchill	58°45'	94°04'	-7.3	688	81	3791	213	397
Baker Lake	64°18'	96°00'	-12.3	326	61	5206		213
Resolute	74°43'	94°59'	-16.4	36	9	6238	283	136

[1]GDD: Growing degree-days (base temperature +6°C)
[2]FDD: Freezing degree-days (base temperature 0°C)

Source: Hare and Thomas (1979)

sweep of the westerlies as one goes east, the boreal zone extends much farther north in the western part of the country, so that Inuvik (at 68°N) has a mean July temperature close to 14°C, whereas Iqaluit (formerly Frobisher Bay, at about 63°N) reaches only 8°C and Churchill (about 58°N) reaches only 12°C.

In summary, moving eastwards from the coast of British Columbia, with its strikingly mild winters, the severity of winter cold and seasonality increase towards central Canada (Table 9.1). On the west coast, the frost-free season spans well over 200 days, whereas in the southern Prairie Provinces it lasts only 120 days or so. The average period of winter snow cover is less than 10 days on the west coast, but more than 120 days in the Prairies. The severity of winter cold at various locations is indicated by the values for freezing degree-days in Table 9.1 (Fig. 9.1); the value of almost 2000 for Saskatoon, for example, is almost 50 times that for Vancouver.

To the east of the Great Lakes moderating influences appear once again, with winter cold and seasonal temperature contrasts reduced accordingly. Halifax has a frost-free period of 183 days, whilst the city of Toronto has an average value of 192 days; however, nowhere does it reach a duration typical of Pacific Canada. Along with the moderating trend, the period of winter snow cover decreases to only 63 days at Halifax, and freezing degree-day values fall below 500.

In contrast to winter, the variation in summer temperatures across the southern part of the country is small. The growing degree-day values in Table 9.1 are an indication of summer warmth; the range in values (46% of the mean) is far less than for the values of degree-days below freezing (200% of the mean). The highest values occur in southern Ontario, whilst maritime influences on the west coast keep values a little lower there. Values are somewhat lower in the Prairies than in southern Ontario, because of latitude.

The statistics for Churchill, Baker Lake, and Resolute in Table 9.1 reveal the progression towards frigid winters and distinctly cool summers as one moves northward. The frost-free period becomes extremely short (almost nonexistent in fact), the snow cover period correspondingly longer, and the freezing index reaches prodigious values. In the north, the climate clamps an icy padlock on the land for a large part of the year, and the meagre summer warmth brings only a brief respite.

Moisture

The other significant factor of climate is, of course, moisture. In simple terms, Canada can be characterized as being wet on the east and west margins and dry in the Prairie Provinces and the north. Annual precipitation at west coast stations can reach well over 3000 mm, but in eastern Canada the wettest areas receive only 1250 to 1500 mm. Throughout the Prairie Provinces, annual totals are generally below 500 mm. Precipitation decreases as one goes north, and in the Arctic Islands annual totals are generally below 150 mm.

The wet regions are favoured by strong cyclonic disturbances causing uplift of moisture-laden maritime air masses. Such conditions are found both on the west coast and in eastern Canada. In British Columbia, precipitation from moist Pacific air masses is enhanced by uplift over the Coast Mountains. Maximum amounts occur at the highest, most windward locations. On the west coast of Vancouver Island and in the Coast Mountains, mean annual precipitation can approach 5000 mm, and very heavy snows characterize the higher ground everywhere (Hare and Thomas, 1979). The eastern region of Ontario, Quebec, and the Maritimes derives its precipitation from mid-latitude cyclones, which tend to converge on this region, bringing in moist air from the Atlantic Ocean or Gulf of Mexico. Annual amounts generally increase from west to east, and the whole region gets heavy snowfalls (Table 9.1).

In the valleys of southern British Columbia and in the Prairies the climate is dry. This is largely due to the rain-shadow effect of the Cordillera. The Arctic is also dry, because little atmospheric moisture can be held in the cold air, and there are few atmospheric disturbances to provide uplift of air. The surface, however, can be very wet in summer because drainage is impeded by the presence of permafrost.

Seasonality of climate

Seasonal variation is a hallmark of Canadian climates. One key element is the annual freeze-thaw cycle, which is responsible for a special kind of river regime. Each spring there is a spectacular release of water as snow and ice melt and rivers open. During their spring flood, the rivers of Canada have enormous erosion and transporting power. Depending on the region, between 40 and 70% of the annual runoff, and 60 to 90% of the total yearly sediment transport occurs at this time. In summer, too, rivers can occasionally reach spectacular levels following very heavy rains. On the Prairies, for example, the heaviest, but rare rainfalls, only hours in duration, are capable of delivering an amount of water which exceeds one-third the normal total summer precipitation. Falls of this intensity lead to more erosion and transportation of sediment than countless minor storms, although the moderate storms have the greatest effect overall because of their frequency. In the east, the Maritime Provinces, and occasionally Quebec and Ontario, are at times crossed by waning hurricanes, which can also deposit huge rainfall amounts — the record for Halifax being 240 mm in one day.

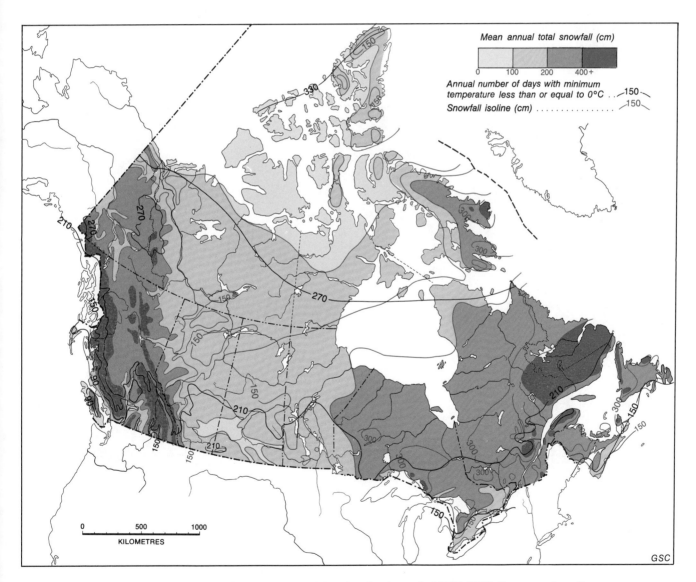

Figure 9.1. The winter climate of Canada. Using data for the period 1951-1980, the maps show the mean annual number of days with minimum temperature of 0°C or below and the mean annual total snowfall (modified from Atmospheric Environment Service, 1987).

Climatic change

Such is a brief view of the average climatic picture of Canada, and for most practical purposes, climate can be considered as essentially stable, allowing for a certain degree of year-to-year variation. On the geological time scale, however, Canada has experienced a wide range of climatic conditions. This will continue to be so, for climatic variation is certain, due to the wide range of natural factors that can influence it. The climate of our lifetimes should not be thought of particularly as "normal" or unchanging, as it is certainly not typical of the last one thousand years, and even less so of the last one million years, which have seen several major ice ages. The last dramatic change in climate initiated the destructive retreat of the vast ice sheet that covered nearly all of Canada at about 18 ka. Since the disappearance of this continental ice there have been several alternating periods of warm and cool climates with the present climate being about 8° to 10°C warmer than at the glacial maximum.

Could Canada, once again, be invaded by glacial ice? Climate history would seem to suggest that this is eventually inevitable. However, whilst the cold winter climate of the present is a fairly close relative of a truly glacial climate, there is no imminent danger of a new continental ice sheet, since this would take thousands of years to develop. In any event, we cannot predict with certainty the course of future climate. This is now, even more the case with the arrival of a new factor on the scene — man.

Since the middle of last century, human activities have substantially increased the amount of carbon dioxide in the

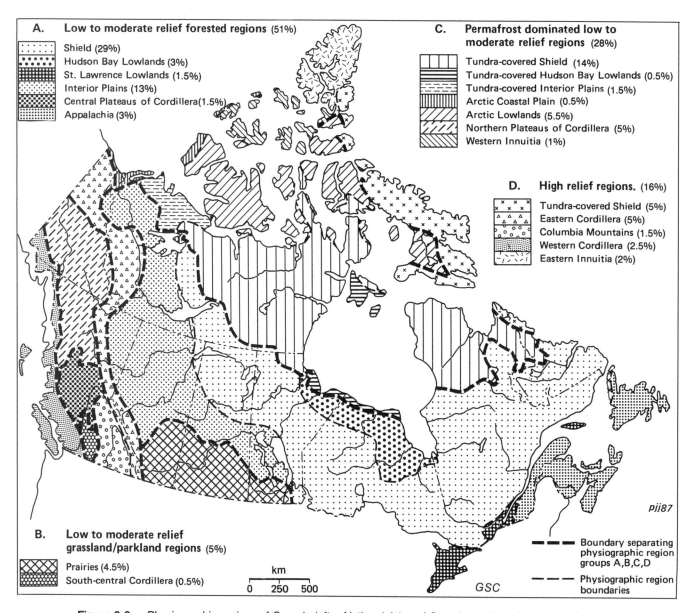

Figure 9.2. Physiographic regions of Canada (after National Atlas of Canada, 1974, using maps of physiographic regions, permafrost, soils, and vegetation).

atmosphere. This has resulted from the combustion of fossil fuels and the world-wide destruction of forests. Although carbon dioxide is only a trace gas in the atmosphere, it plays a critical role in controlling the earth's climate through the greenhouse effect. Theoretical evidence is accumulating that the carbon dioxide increase might well cause major climate change on a global scale, and that increases in global temperature by this means could outweigh natural variations. This possibility is particularly significant to polar latitude countries such as Canada, where we know that temperature changes can be several times greater than the global average. The basic effect of a global warming, concentrated in the polar latitudes, is now generally accepted as the likely consequence of continuing increases in atmospheric carbon dioxide.

There are, indeed, some indications of a beginning trend towards warmer temperatures throughout the Northern Hemisphere since about 1970. After reaching a maximum around 1940, the hemispheric temperature declined to a minimum around 1970, despite any effects of increasing atmospheric carbon dioxide. Since then, it has warmed about half-way to the high value of the 1940s. These changes may be cyclical and the causes have not been firmly established; the increase in carbon dioxide and other trace gases will not be the only major factor affecting our climatic future.

PHYSIOGRAPHY OF CANADA AND ITS EFFECTS ON GEOMORPHIC PROCESSES

H.O. Slaymaker

Physiography concerns the landforms, soils, vegetation, and climate of a region and the interaction between those elements (Bowman, 1911; Holland, 1964; Bostock, 1967; Bird, 1980). When delimiting physiographic regions, earth scientists have given variable weights to geology, landforms, soils, vegetation, and climate; they have used a variety of aggregate measures of these factors; and, when assessing the effects of physiography on geomorphic processes, they have tended to confine attention to local or site scale effects.

At the site scale of analysis the following kinds of geomorphic process are investigated: rockfalls, slides, slumps, earthflows, debris flows, creep, gully development, fluvial erosion, wind erosion, piping, karst development, weathering, and frost action. Two sets of physiographic factors can be analyzed in relation to each of these processes: factors that produce conditional instability or susceptibility to erosion and factors that trigger slope failure or sediment removal at a given time. At the regional scale, it is normally possible to identify only the factors that promote susceptibility to erosion, and consequently terrain mapping has become of increasing importance in assessing the biophysical resources of Canada. A recent example of the value of such an analysis is provided by Evans (1982); he reviewed the distribution of different glacial deposits on mountain slopes in the Cordillera and their association with historical slope failure events. The essential ingredient of local scale analyses is the determination of the precise nature of surficial materials and their geotechnical properties and hence their susceptibility to erosion. Triggering factors, whether meteorological (such as severe rain storms) or geological (such as earthquakes) can only be mapped on a statistical basis. Thus maps can be made showing the probable return frequency for rainstorm events of a given magnitude or the seismic risk for any area.

At a Canada-wide scale, it is not possible to reproduce the resolution of physiographic effects on geomorphic processes that are available at site and regional scales. A physiographic regionalization that is sensitive to geomorphic process must recognize, at least, different lithological groupings, terrain, presence or absence of permafrost, and presence or absence of tree cover. No direct relationship between physiographic and geomorphic process is thereby established, not least because of the enormously variable influences of continental glaciation over Canada's landmass. Twenty physiographic regions defined by geological structure, relief, permafrost, soil, and tree cover boundaries are identified (Fig. 9.2). The individual maps used to define these physiographic regions are derivative from Rowe (1959), Bostock (1967), Brown (1967), and Clayton et al. (1977). Each region is described briefly in the following sections.

Low to moderate relief forested regions

More than 50% of Canada is in these regions (Fig. 9.2); the average rate of geomorphic change is low, a fact that is reflected in the low sediment concentrations and low sediment mass flux in such rivers as the St. Lawrence, Churchill, Nelson, and Saskatchewan. Where sensitive soils are widely distributed and intensive land use changes have been introduced, as in the St. Lawrence Lowlands, considerable local acceleration of geomorphic change has occurred.

Shield. The Canadian Shield is by far the largest of the 20 physiographic regions recognized. It is underlain by predominantly Precambrian gneissic and granitic rocks, with relatively homogeneous chemical composition and rather similar response to geomorphic processes. Linear joint and fault patterns provide the most distinctive bedrock landforms. Variation in surface deposits, consisting mainly of till, glaciofluvial deposits, and glacial lake clay covers, provides the most significant controls on local geomorphic process rates. The soils map (Clayton et al., 1977) indicates large areas of organic soils and gleysols. The region includes most of Quebec, Ontario, much of Manitoba, Saskatchewan, Keewatin, eastern Mackenzie, and southern Labrador.

Hudson Bay Lowland. The Hudson Bay Lowland occupies a structural depression in the Canadian Shield underlain by Paleozoic limestones and dolomites and, locally, Mesozoic clastic rocks. The topography is subdued and thick glacial sediments occur at the surface. In places, Precambrian inliers provide moderate local relief. Much of the region is covered with peatlands, palsas, and peat mounds.

St. Lawrence Lowlands. The St. Lawrence Lowlands are divided into three subregions — western, central, and eastern. The western lowland is surrounded on three sides by Lakes Huron, Erie, and Ontario and is underlain by flat to

---

Slaymaker, H.O.
1989: Physiography of Canada and its effects on geomorphic processes; in Chapter 9 of Quaternary Geology of Canada and Greenland, R.J. Fulton (ed.); Geological Survey of Canada, Geology of Canada, no. 1 (also Geological Society of America, The Geology of North America, v. K-1).

gently dipping Paleozoic strata; the central lowland straddles St. Lawrence River between Ottawa and Québec City and is also underlain by flat-lying Paleozoic strata. The eastern lowland, which includes Anticosti Island, is underlain by southward dipping Paleozoic rocks. The most distinctive controls on contemporary geomorphic activity are the Pleistocene sediments, especially those associated with the late glacial Champlain Sea.

Interior Plains. Northern Alberta and central Saskatchewan and Manitoba plains consist of a series of plateaus underlain by Paleozoic and Mesozoic sediments. Gently undulating to flat surfaces are locally crossed by deeply entrenched and commonly misfit rivers. Extensive areas of peatland and gleysols occur in northern Alberta and Saskatchewan.

Central Plateaus of the Cordillera. The interior plateaus of the Cordillera are underlain in large part by metamorphosed Mesozoic sedimentary and volcanic rocks and by Tertiary volcanic rocks. Rivers deeply dissect the plateaus and steep valley sides are being actively eroded.

Appalachia. The physiography is dominated by a well developed peneplain which slopes to the southeast. Differential erosion has generated a series of subparallel lowlands, highlands, and uplands, which reflect the strike of the highly folded and faulted Paleozoic and Precambrian rocks. Pleistocene glaciation of the whole region differentiates the detailed landforms of this region from the classical Appalachian landscapes of the United States.

Low to moderate relief grassland/parkland regions

According to the Langbein and Schumm (1958) model relating effective precipitation to sediment yield, these regions should have the maximum regional rates of geomorphic change. With the exception of local extreme rates of geomorphic process as in the Red Deer badlands region (Campbell, 1975) where gullying is highly active, available quantitative data do not confirm this prediction.

Prairies. The southern plains of Alberta, Saskatchewan, and Manitoba are underlain by poorly consolidated Mesozoic and Tertiary sediments. Broad relief features are controlled by bedrock relief but on a local scale the geomorphology is developed almost exclusively in glacial drift which is up to 300 m thick. Rivers are, in many places, deeply entrenched in glacial deposits and bedrock. Locally, entrenchment in Cretaceous shales and sands has led to the development of badlands topography. Coulee formation in the valley sides is distinctive (Beaty, 1975).

South-central Cordillera. The lower parts of Fraser, Thompson, and Okanagan valleys enjoy a semi-arid climate which has led to grassland development. The valley systems, entrenched in plateaus, are underlain by extensive glacial lake sediments in which piping processes are well developed (Slaymaker, 1982).

Permafrost-dominated low to moderate relief regions

Some 50% of Canada is underlain by permafrost and much of these regions display the distinctive geomorphic processes resulting in patterned ground, gelifluction lobes, and thermokarst (see this chapter, French, 1989). The processes themselves are most effective as sorting agents in a vertical direction, but the mass export of sediment via regional sediment transfer systems is not great except where local disturbances by man have accelerated erosion.

Tundra-covered shield. Much of central and northern District of Keewatin and the whole of Ungava are included in this physiographic region. Bedrock is similar to the forested shield region, but continuous permafrost underlies the surface and the discontinuous vegetation cover allows strong attack by geomorphic processes. Felsenmeer-covered rock plateaus and patterned ground in unconsolidated materials are ubiquitous.

Tundra-covered Hudson Bay Lowland. This is a discontinuous region of rock deserts, peat-covered tundra and patterned ground and is underlain by Paleozoic limestones and dolomites and thick Quaternary sediments.

Tundra-covered Interior Plains. The Anderson and Horton plains (Bostock, 1967) are underlain by flat-lying Proterozoic, Paleozoic, and Mesozoic strata. In some areas, such as immediately north of Great Bear Lake, glacial drift is thick and the surface is hummocky. In many other areas, the surface is devoid of Quaternary sediments.

Arctic Coastal Plain. The Mackenzie Delta and adjacent Tuktoyaktuk Peninsula is the most studied part of this region (Mackay, 1979a). Its distinctiveness derives in part from the thick accumulation of unconsolidated sediments in which permafrost has developed to a depth of 500 m or more. This region is underlain by semi-consolidated Mesozoic and Tertiary clastic sediments with a Quaternary cover consisting largely of sand, gravel, and silt, with only minor amounts of till. Ice wedges, tundra polygons, and pingos are characteristic landforms. The Yukon coast and the northwestern coasts of Banks, Prince Patrick, and Meighen islands are also included and show massive ground ice.

Arctic Lowlands. The Arctic Lowlands are underlain largely by flat-lying Paleozoic sedimentary rocks, which form somewhat featureless plateaus. Drumlinoid ridges and moraines form important topographic variety. Permafrost is ubiquitous.

Northern Plateaus of the Cordillera. This region consists of structurally complex Paleozoic and Mesozoic sedimentary, igneous, and metamorphic rocks. The presence of permafrost is almost universal, though discontinuous in the south. Even though the area is here classified as a low to moderate relief region, numerous mountain ranges separate the major plateaus.

Western Innuitia. This is a region of low to moderate relief in Mesozoic and Tertiary sandstones and shales. Bedrock directly underlies the surface throughout much of the area. Dissected domes developed on gypsum anhydrite diapirs and ring structures in igneous intrusives are distinctive. The extent of ground ice in the area is not well known.

High relief regions

These regions display the highest rates of regional sediment removal. Because climate, vegetation, and sediment availability vary in a vertical direction, a model of preferred sediment transfer mechanisms based on elevation zones is most appropriate (Fig. 9.3). The elevations of the different sediment transfer zones vary with latitude. The process of redistribution of clastic sediment from high to low elevation zones overshadows the contribution of contemporary weathering of bedrock; with respect to dissolved load export, present breakdown of bedrock seems to be more important than solution of the unconsolidated sediments, because of the short time available for chemical reactions between sediments and soil water.

Tundra-covered shield. Southeastern Ellesmere and eastern Devon islands, much of Baffin Island and the Torngat Mountains fall in this region. Some of the finest alpine scenery in Canada occurs here, and ice caps and glaciers are common, particularly in the northern part of the region. Processes of mass movement and fluvial redistribution of sediments are active, and several large, active sandurs are present (Church, 1972). Permafrost is present throughout the region.

Eastern Cordillera. The eastern Cordillera is composed almost entirely of folded Paleozoic and Mesozoic sedimentary strata and includes glacierized, glaciated, and, in the Yukon, a few unglaciated ranges. The region comprises mainly the Rocky Mountains, Mackenzie Mountains, and the Richardson, British, and Brooks ranges and includes several low relief basins filled with unconsolidated to semi-consolidated Mesozoic, Tertiary, and Quaternary clastic sediments in the Yukon. Permafrost is common everywhere, except the southern Rocky Mountains part of the region. Modern glaciers are small and widely scattered.

Columbia Mountains. Columbia Mountains are underlain by Mesozoic, Paleozoic, and Precambrian metamorphic rocks and granitic intrusions. Glaciers are locally present and snow related processes of nivation and avalanching are characteristic.

Western Cordillera. St. Elias Mountains, at the northern end of the region, include the highest mountain in Canada (Mount Logan, 6050 m). Particularly in the north, this region is heavily glacierized. It consists primarily of crystalline gneisses and granitic rocks. Glacial erosion and snow related processes are active. The Pacific Ocean side of the region has the highest precipitation of Canada (Table 9.1).

Eastern Innuitia. Axel Heiberg Island and most of Ellesmere Island are high relief glacierized areas developed on folded Mesozoic and Paleozoic sedimentary rocks. The presence of polar glaciers is the distinctive feature of this region. Permafrost is ubiquitous

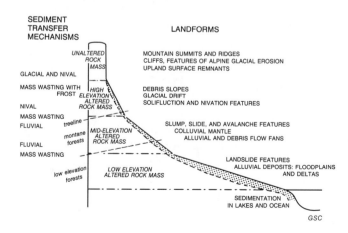

Figure 9.3. Preferred sediment transfer mechanisms as a function of elevation zone (elaborated by Ryder, 1981, after Gilbert, 1973).

SLOPE PROCESSES
M.A. Carson and M.J. Bovis

Sloping land surfaces in Canada, as in other parts of the world, are acted upon by a wide range of processes which cause downslope movement of rock and soil materials. These include various types of landslides, both deep-seated and shallow, both slow moving and rapid; heaving and consolidation of the near surface regolith, producing intermittent creep displacements; splash of surface particles by raindrops; and the entrainment of soil particles by overland flow, sometimes producing "sheet" erosion, more frequently localized in small rills, and occasionally concentrated in gullies.

The relative importance of each process in a given area depends to a large extent on the geomorphic age of the landscape. Slopes that are still intermittently undercut and steepened by streams, or by wave action, are usually dominated by landslide processes. Even after undercutting has ceased, slopes underlain by resistant rocks may take thousands, or even millions of years to flatten to a gradient sufficiently gentle that landslides can no longer occur. Long after undercutting has stopped, rock weathers to produce a regolith of sufficient thickness that the regolith will slide, exposing the underlying rock to further weathering. As slopes flatten towards their ultimate "threshold" angle, below which mass instability will not occur, the time interval between successive slides increases, allowing other, slower acting, processes such as creep, rainsplash, and soil wash to assume greater relative importance in moulding the form of the landscape.

The land surface of Canada is, in general, so young that many of its valley sides, hillslopes, and shoreline bluffs have yet to attain permanent stability with respect to mass failure. Rivers in the lowland areas of the Prairies, Ontario, and Quebec are dissecting land surfaces that, less than 12 ka ago, were lake floors or seabed. Newly emerged or submerged land along the northern coasts and on the shores of the Great Lakes is still adjusting to undercutting by wave action, in some places retreating at rates of more than 1 m/a. In the Cordillera, the transformation of kilometre-high mountain sides into stable slopes is still a long way from completion. Glacially oversteepened bedrock slopes, potentially unstable glacial drift deposits, locally intense precipitation, and a regionally high earthquake frequency conspire to make the Cordilleran region the most geomorphically active area in Canada.

In many parts of Canada, then, landslide processes are more significant than creep and soil erosion effects, at least from the standpoints of rapid and major changes in hillslope form, in addition to the attendant hazards. Accordingly, they are emphasized here. In the concluding part of this section, an attempt is made to assess the temporal and spatial variability of processes, following an overview of landslide and nonlandslide processes in specific geographic regions.

Carson, M.A. and Bovis, M.J.
1989: Slope processes; in Chapter 9 of Quaternary Geology of Canada and Greenland, R.J. Fulton (ed.); Geological Survey of Canada, Geology of Canada, no. 1 (also Geological Society of America, The Geology of North America, v. K-1).

Landslides

The wide range of topographic, geological, climatic, and soil conditions across Canada provides a potential for many different types of landslides. To some extent one can speak of "regional landslide types", along the lines suggested by Mollard (1977). It should be realized, however, that within each region, such as the Prairies or the Cordillera, a wide range of landslide types usually is found. This is well demonstrated for the Cordillera by the recent work of Eisbacher (1979a) and Cruden (1985) (see also Clague, 1989; Evans and Gardner, 1989). Landslides are grouped here primarily according to material type and secondarily according to type of movement: failures in crystalline or hard sedimentary rock, including rockfalls, detachment slides, rock avalanches, and rock topples; failures in clay or poorly lithified shale, including deep-seated rotational failures, lateral spreads, and earthflows; failures in generally unconsolidated regolith, including shallow translational slides, and the much more mobile debris flows; and failures in seasonally thawed ground, principally in areas underlain by permafrost. Some of these types, such as rock avalanches, are localized because of particular topographic and geological controls. Others, such as deep-seated failures in cohesive soils, are more widespread principally because of the common occurrence of steep river or lakeshore bluffs developed in clay or silty clay materials.

Rockslopes

Slopes developed in hard, intact rock theoretically can stand as vertical cliffs thousands of metres high without failing. A simple relationship, stated earlier by Terzaghi (1962) is:

$$H_c = q_u/\gamma \qquad (9\text{-}1)$$

where H_c is the critical height of the slope, in metres, q_u is the unconfined compressive strength of the rock, in N/m² (Pascals), measured by loading a laterally unsupported cylindrical rock specimen to the point of failure, and γ is the unit weight of the rock, in N/m³. For crystalline rock such as granite, typical values are $\gamma = 2.6 \times 10^4$ N/m³, $q_u = 1.0 \times 10^8$ N/m², yielding $H_c = 3850$ m. In nature such spectacular vertical cliffs are rarely seen, since invariably significant defects exist in the rock mass, such as joints, bedding planes, and foliation. These provide planes of weakness along which small to catastrophically large volumes of rock may become detached. Weathering also contributes to the weakening of rock slopes by producing chemical changes along planes of weakness, as well as developing additional fractures, by a variety of chemical and physical processes. In general, the strength of a rockmass depends on its size since the chance of encountering a through-going discontinuity, oriented favourably for slope movement, increases with the volume of the rockmass.

Talus and rockfalls. Fractures and other discontinuities are clearly visible in most natural rock slopes. Under the actions of hydration shattering and repeated freeze-thaw cycles, relatively small volumes of rock are pried loose each year and accumulate as talus. The surface slope of this sheet of angular rock fragments is commonly observed at between 34° and 36°, close to the angle of repose of the material (Carson, 1977a). Innumerable examples of talus slopes still being enlarged by rockfall exist in Canada's mountain areas (Fig. 9.4), especially in the western Cordillera, the Labrador-Ungava region, and the Arctic Archipelago. It would be misleading to suggest that discrete-particle rockfall is the only process contributing to talus development. Snow avalanches are known to be significant contributors of debris, as well as agents of debris redistribution (Luckman, 1978). Avalanches probably contribute also to the concave, basal slope profile observed on many talus slopes (Luckman, 1971). Gardner (1979) cited evidence of debris flow components in talus slopes. These deviations from simple rockfall behaviour tend to occur where distinct "chutes" have developed in the bedrock cliff, along zones of preferential weathering. Once formed, chutes serve as conduits and temporary reservoirs for falling rock fragments, as well as favoured sites for snow accumulation, and hence debris and snow avalanches. These produce a distinct cone-shaped accumulation of talus and other lithic debris, which is more properly termed a debris cone rather than a talus cone (Church et al., 1979). Debris cones are notably flatter (25°-35°) than talus cones or talus sheets produced by rockfall alone.

While rockfall and talus production are not confined to cold regions, they are particularly effective in areas subject to regular freezing and thawing. In arctic, subarctic, and alpine areas, percolation of water from snowmelt or rainfall into the part of the talus below frost table leads to ice accumulation in the voids. Ice may also develop from burial of winter snow by rockfall, rockslide, and debris flow. Interstitial ice can lead to a slow, downslope "flow" of talus as a rock glacier; spectacular examples are reported from the Yukon by Johnson (1978). Some of these contain a "core" of glacier ice, buried by thick accumulations of talus and ablation moraine. Many Canadian rock glaciers show signs of recent movement, judging from the presence of overturned, lichen-covered boulders.

Figure 9.4. Talus slopes developed below Mount Arethusa, near Highwood Pass, Canadian Rockies. Courtesy of J.S. Gardner. 204377C

Detachment slides. Wherever rock slope joints, bedding, or foliation dip downslope, the potential exists for the detachment of entire rock layers as rockslides. The actual volume produced by a single event may vary from a few cubic metres to tens of millions of cubic metres, depending on the slope angle, the dip angle of the rock layering, and the shear strength along such planes of weakness. Detachment of a rock layer by sliding (shear failure) will be possible if the shear stress, produced by the weight of the overlying rockmass, exceeds the shear strength along the plane (Fig. 9.5):

Shear stress at failure = (cohesive strength) + (frictional strength)

$$\tau = c' + (\sigma - u) \tan \phi' \quad (9\text{-}2a)$$

or for the special case where planes of weakness exist parallel to the slope surface (Fig. 9.5), and groundwater flow is also parallel to these planes,

$$\gamma \cdot z \cdot \sin b \cdot \cos b = c' + (\gamma \cdot z \cdot \cos^2 b - \gamma_w \cdot h \cdot \cos^2 b) \tan \phi' \quad (9\text{-}2b)$$

where τ is shear stress, c' is cohesion, σ is normal stress, u is fluid pressure, ϕ' is the angle of shearing resistance on the failure plane, z is the vertical thickness of the rock layer, h is the water depth above the failure plane, b is the dip angle of the plane, assumed roughly parallel to the slope, and γ, γ_w are, respectively, the unit weights of rock and water. The important role of water pressure is to reduce the effective normal pressure along the plane, thus reducing frictional strength; the expression for u in equation 9-2b assumes groundwater flow is parallel to the slope. In many well jointed rock slopes, fluid pressure (u) is probably small, because of efficient drainage of the rockmass. In addition, cohesion may approach zero because of the cumulative effects of weathering, seismic shaking, and very small, previous shear displacements. This leads to the reduced equation:

$$\tan b = \tan \phi' \quad (9\text{-}3)$$

since $\tau/\sigma = \tan \phi'$, giving a limiting slope angle equal to the friction angle. Where it exists, cohesion allows the slope to stand at an angle steeper than the angle of shearing resistance (ϕ') but this creates a potentially unstable slope since, as cohesion (c') tends to zero over time, frictional resistance alone will not support the slope.

Many catastrophic rock slope failures have occurred where the dip angle of the failure plane is approximately equal to or slightly less than the angle of shearing resistance, notably on sedimentary dip slopes in the Rocky Mountains (Cruden, 1976) and in the Mackenzie Mountains of northwest Canada (Eisbacher, 1979b), as well as in failures of metamorphic rock, such as the Hope Slide of 1965 (Fig. 9.6; Mathews and McTaggart, 1978). This points to the negligible cohesion component in many rock slopes, as well as the lack of significant water pressure. Cruden (1982), however, suggested that pore pressure may have contributed to the recent Brazeau rockslide. It has been speculated seismic shaking triggered the Hope Slide, and probably accounts for many of Eisbacher's (1979b) examples, particularly those released from bedding planes inclined 10-15° below the estimated friction angle.

Rock Avalanches. Failure and disintegration of a large rock slab may produce a high velocity rock avalanche, or "sturzstrom" (Hsu, 1975) in which tens of millions of cubic metres of material discharge into a valley in less than a minute. Rock avalanches may run downvalley for several kilometres beyond the originating slope or ride a considerable distance up the opposing valley side, which points to a tremendous store of kinetic energy. Rock avalanches result in major modifications to the landscape, such as huge slide scars and thick deposits, which may impound large lakes. Eisbacher (1979b) mapped almost 50 post-Pleistocene examples in a 75 000 km² tract of northwestern Canada. Joint sets, normal to bedding planes, seem to be a prerequisite for rapid disintegration of the mass into a stream of angular, blocky debris. Various hypotheses have been put forward to account for the long runout distances of sturzstroms on low gradient valley floors. One possibility is momentum transfer from the rear to the front of the debris (Hsu, 1975). In addition, the high rates of shear strain imposed on the basal layers probably force marked dilation of the "flowing" mass, reducing the internal friction angle of the material (Davies, 1982).

At least six major rock avalanches have occurred in the Cordillera in the past 130 years (Table 9.2), an average of about 5 events per century. The obstacles to predicting rock avalanches are clear from equation 9-2a: the material parameters of cohesion (c') and angle of shearing resistance (ϕ') must be known for slopes which have not yet failed, and the likely upper limit of the fluid pressure (u) must be specified, not to mention the effects of future seismic shaking on all the parameters in that equation.

Rock creep. In complete contrast to the preceding catastrophic type of slope behaviour is the slow downslope bending or cambering of rock strata under their own weight on steep mountain slopes. Relatively few examples have been noted to date in the Cordillera (Mollard, 1977; Bovis, 1982), but they may be more widespread than originally thought. Bending movements, which may lead to toppling of near surface rock layers, seem to be favoured by steeply inclined joints or bedding dipping into the slope. Anomalous uphill-facing scarps may be produced by such displacements.

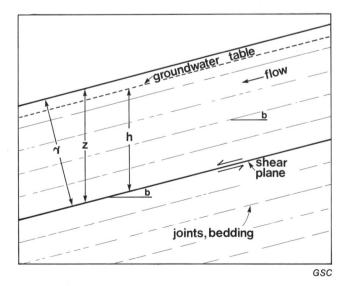

Figure 9.5. Components of a plane translational slide in rock. Refer to equation 9-2 for explanation of terms.

Figure 9.6. Oblique aerial view of the Hope slide of 1965. Province of British Columbia 0445.

Table 9.2. Catastrophic rockslope failures in the Canadian Cordillera during historical times

Year	Event	Geographic Zone	Volume (10^6 m^3)
1855/56	Rubble Creek	Coast Mountains	25
1903	Frank	Rocky Mountains	33
1933	Brazeau Lake	Rocky Mountains	4
1963	Mt. Cayley	Coast Mountains	5
1965	Hope-Princeton	Cascade Mountains	47
1975	Devastation Glacier[1]	Coast Mountains	26

[1] Three failures are known from the Devastation Glacier site during historical times.

Clay and shale slopes

Large areas of Canada are covered by thick deposits of clay or clay-shale of variable strength and age. These include the Cretaceous argillaceous bedrock of the Interior Plains, stiff deposits of Pleistocene till and lacustrine sediments, overconsolidated by the weight of glacial ice, and relatively soft deposits laid down in postglacial lake, estuarine, and marine environments. In all cases, deep-seated landslides may occur. Though instability may take various forms, commonly it is of the rotational type (Fig. 9.7).

Whether or not instability occurs depends partly on the height and steepness of the slope, which affect the "driving torque" of material above the potential failure arc, and partly on the shear strength mobilized along the incipient slip surface. Particularly important in this regard is the groundwater regime in the vicinity of the slope, as this affects the hydraulic head (h) along the failure plane. Equation 9-2a demonstrates how water pressure (u) reduces frictional strength. The only difference here is that fluid pressure is developed in innumerable pore spaces which exist between the clay particles, rather than along discrete joint or bedding planes, as in rock slopes. An increase in water table elevation is the major reason why landslides (of all kinds) occur in periods of prolonged wetness either due to rainfall or, especially in Canada, from spring snowmelt.

Spatial differences in groundwater regime are often significant in promoting local patterns of instability. Pore pressure magnitude is affected not only by water table height, but is related to the direction of seepage: descending flow is associated with lower pressures, enhancing stability; upward seepage involves higher pressures. Upward groundwater discharge is not uncommon at the base of clay slopes and has been shown by LaRochelle et al. (1970) to be a prime cause of deep seated landslides in parts of the St. Lawrence Lowlands. The detailed effects of stratigraphy on the groundwater regime and stability of clay slopes have been outlined by Lafleur and Lefebvre (1980).

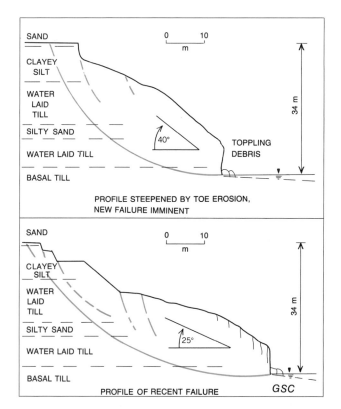

Figure 9.7. Failures of shoreline bluffs, Lake Erie, Ontario; companion profiles showing cyclic pattern of rotational failure, steepening by toe erosion, and new failure (after Quigley et al., 1977).

Even along a single line of clay bluffs, seepage conditions may vary spatially, leading to preferential instability at certain sites and to the production of "scallops" in the cliff line. Spectacular scallops occur on the Scarborough Bluffs of Lake Ontario (Bryan and Price, 1980; Eyles et al., 1985), their initiation apparently associated with favoured sites of seepage and piping. Once formed, scallops are self-reinforcing features, since groundwater seepage converges in the scallops at the expense of the intervening bluff spurs. Because of the decreased water table in the protruding spurs, their stability is increased, and the cliffs may become near-vertical, only succumbing to collapse if actually undermined by wave erosion. Failure of these spurs thus takes the form of collapse of cliff slabs, bounded by joint planes parallel to the cliff face, rather than by deep-seated failure. Once formed, surface drainage is also funnelled into the scallops so that rill wash and gullying often supplement mass movement.

Varying degrees of deep-seated instability along bluffs cut in glacial drift along the north shore of Lake Erie (Quigley et al., 1977) highlight the importance of both cliff height and slope steepness, the latter controlled by the rate of wave erosion. On the lower cliffs (about 27 m high) to the west, deep-seated instability is absent, though in many places the bluffs appear to be on the verge of failure and would collapse if undercut rapidly. Rates of cliff retreat by wave erosion are relatively slow (about 0.6 m/a), which allows ample time for shallow slides and mudflows to regrade slopes. Interestingly, the shallow instability also takes the form of scallops, with gradients of about 37°, separated by much steeper bluff spurs, though in this case there is no evidence that the pattern was caused by prior variability in seepage along the bluffs. Farther east, where bluffs are higher (greater than 40 m), and subject to more rapid retreat (about 2 m/a), periods of bluff oversteepening by wave action do lead to rotational failure. As a consequence, slope steepness is reduced from about 40° to 25° and temporary stability ensues. Subsequent removal of the sediment at the slope toe and further undercutting by waves ultimately lead to another period of oversteepening and failure, the whole cycle taking 10 to 25 years depending on lake levels.

Though the frequency of deep-seated landslides decreases as shallow slides and other processes reduce slope steepness, clay slopes may succumb to deep failure many decades or centuries after the erosion has stopped. Springtime groundwater pressure, for example, varies from year to year; in stable areas where normal water table levels are well below the ground surface, deep-seated failure may occur in an abnormally wet spring if slope height and steepness are great enough. Apart from pore pressure induced changes in shear strength, some clay sediments experience long term strength loss from changes in cohesion and angle of shearing resistance. Two spectacular regional examples of this are discussed below: the postglacial marine clays of eastern Canada, where long term loss of salts from the pore fluid decreases cohesive strength; and the bentonitic clays of the Prairies, where slow creep can lead to a loss of cohesion as well as to a reduction of frictional strength.

Flowslides in marine clay. Though deep-seated landslides in the postglacial marine clay deposits of parts of St. Lawrence and the Ottawa valleys can also take the form of simple rotational failures, frequently an initial deep-seated slope failure spawns a new one immediately behind, and this is repeated, producing a retrogressive "flowslide" that may devastate large areas of flat land behind a valley slope (Crawford, 1968). Recent examples have shown that retrogression is rapid, taking less than a minute to obliterate a few hectares, and only a few hours to produce large scars. Similar flowslides occur in postglacial marine sediments in other recently glaciated parts of the world, such as Scandinavia.

Because of the rapidity of retrogression, and because eye-witness accounts are both few and unreliable, the nature of the processes involved has been a matter of debate. Traditionally, enlargement of the slide area was assumed to take place by repeated rotational failures of the kind just described. But studies of the pattern of disturbance found in several flowslides (Carson, 1977b) show that many involve lateral spreading, with repeated subsidence, producing a multiple horst-graben structure (Fig. 9.8). Subsidence of clay wedges to form grabens requires remoulding of their basal parts under the load of the wedge itself, transforming plastic clay to liquid mud, which is then intruded along the fault zones. Where the clay is weak, widespread remoulding can occur, frequently masking all evidence of the horst-graben pattern, and driving viscous debris far out of the slide area along the valley.

Slide areas in which only small amounts of remoulding take place, preserving the horst-graben structure, are sometimes referred to as "ribbed flowslides". Those where almost complete conversion to fluid mud occurs, often leaving "clean" flow bowls, have been termed *coulées d'argile*. The type examples of these are the South Nation River rib slide

587

Figure 9.8. The 1978 flowslide at Sainte-Madeleine-de-Rigaud, Quebec (from Carson, 1979).

Table 9.3. Flowslides in sensitive sediments of the Ottawa-St.Lawrence valleys during historical times

Year	Location	Area (ha)	Volume (10^6 m^3)
1840	Maskinongé	34	
1894	Saint-Alban	650	460
1898	Saint-Thuribe	35	3
1908	Notre-Dame-de-la-Salette	2	
1910	South Nation River	14	
1924	Rivière St-Maurice	4	
1924	Kenogami	25	2
1945	Yamaska	3	0.1
1951	Rimouski	3	0.2
1951	Rimouski	10	0.8
1955	Hawkesbury	5	0.4
1955	Nicolet	2	0.2
1962	Tounustouc	9	4
1963	Breckenridge	1	0.02
1971	South Nation River	28	6
1971	Saint-Jean-Vianney	28	7

Source: after Mitchell and Markell, 1974

(Ontario) and the Saint-Jean-Vianney earthflow (Quebec), both of which occurred in 1971. Most historic flowslides appear to contain elements of these two basic forms. At Saint-Jean-Vianney, the 28 ha earthflow of 1971 took place within a much larger crater (2000 ha) dated at 400-600 years old.

The prerequisite for deep-seated landslides to degenerate into such retrogressive flowslides is a silty clay sediment which, on disturbance, transforms to a viscous mud. Such sediments are termed "sensitive", with sensitivity defined as the ratio of undrained shear strength in the undisturbed state to that in the remoulded state. The high sensitivity of sediments in some lowland areas of eastern Canada results, to some extent, from high water content. This in turn reflects the flocculated fabric that developed during sedimentation in brackish or saline water, and the general absence of subsequent heavy loads that might have consolidated the sediment more. Where high overburden pressures do develop, cementation appears to have assisted in maintaining the open fabric. The low strength of these sediments, when disturbed, also stems from the weakness of attractive forces among particles. Most of the sediment is rock flour (particles of quartz, feldspars, and micas ground by glacial abrasion to clay size) and shows little of the electrical attraction among particles characteristic of true clay minerals. On the other hand, where salt concentration in the pore fluid remains high (because of incomplete leaching and diffusion of salts since exposure), reflocculation of particles can occur during disturbance, permitting attractive forces to develop and maintain sufficient strength in the sediment to prevent flowage. Significantly, large retrogressive flowslides seem to be confined to areas of the lowlands where salt concentrations have decreased to low levels, less than 2 g/L (Carson, 1981).

Table 9.3 lists some of the largest flowslides to have occurred in historic time. An inventory of flowslide scars in the Lac Saint-Jean area and St. Lawrence Lowlands of Quebec by Lebuis and Rismann (1979), using air photograph interpretation, revealed more than one thousand. Based on those occurring in historic time, the mean recurrence interval of slides is about 2.5 years for those larger than 1 ha, and about 25 years for those larger than 30 ha. Scars of similar slides are abundant along valleys in the lowlands around James Bay, Hudson Bay, Labrador coast, and the North Shore of the Gulf of St. Lawrence. Some have also been noted in postglacial lacustrine sediments of Nouveau-Quebec but have received little attention.

Slides in clay-shale. Landslide terrain similar to the ribbed flow bowls of eastern Canada is found also in the Interior Plains, flanking the South Saskatchewan, Peace, Battle, and other rivers, in valleys which are roughly 100 m in depth. Mollard (1977, p. 38-56) described many of these, noting that some of them affect more than 3 km of valley slope and extend more than 1.5 km from the river edge. Often the valley top material is Quaternary drift, in some places more than 50 m thick. Some of the failures appear to have occurred in this drift throughout their full depth, particularly where it consists of postglacial lacustrine clay (Haug et al., 1977), but most have basal failure surfaces in the underlying Cretaceous bentonitic clay shales.

These prairie landslide movements are slow and ongoing: displacement rates increase "downslope", with a forward translation of 4 m in six years reported at the toe of Little Smoky slide (Fig. 9.9), central Alberta (Thomson and Hayley, 1975). Concurrent with advance of the slope toe, there is subsidence of individual blocks of material behind. At intervals of the order of decades, sufficient subsidence and translation occur in front of the backscarp to permit a new block to fail by slow settlement and the scar to retrogress.

Two important questions about these slides have not been satisfactorily answered. The first is how failure can occur on basal slide planes that are almost horizontal. The second concerns the mechanisms of retrogression.

In connection with the first point, slippage along such gentle gradients requires very small shear strength at depth. Some of the clay shale seems to have been subjected to prolonged, deep weathering during Tertiary times. More important, perhaps, because of the large amounts of smectite clay minerals, the bentonite-rich sediment (drift and bedrock) is prone to softening if sheared, resulting in a cohesion (c') of approximately zero and angle of shearing resistance (ϕ') of less than 8°. Such shearing certainly occurred during glacial periods, with thrusting and brecciation of the uppermost shale, and subsequently in response to release of earth pressure by postglacial cutting of deep valleys (Matheson and Thomson, 1973). Present-day instability contributes further to pressure release along the backscarps of slide areas, with slow deep-seated creep of the next block, leading to a progressive decrease in strength.

So far no satisfactory quantitative analysis of the intermittent retrogression (and ongoing translation) has emerged. The uncertainty as to the precise mode of failure is, in many ways, similar to the confusion that existed regarding rapid retrogression in sensitive sediments just described. Frequently, failure is analyzed in terms of successive slips along curved surfaces (Fig. 9.9), yet field observations repeatedly emphasize horst and graben structures.

Regolith slides and debris flows

Although regolith mantles derived from bedrock weathering are important in unglaciated areas, over most of Canada glaciation has resulted in deep burial or erosion of such mantles. Consequently, the debris mantle which overlies the rock is usually the result of either the accumulation of glacial deposits or postglacial weathering and colluvial action.

Regolith mantled slopes are subject to intermittent slippage of the debris cover, provided that the slope gradient exceeds some threshold value (b_t) which depends on the strength and thickness of the material. Assuming a sliding surface parallel to the slope, equation 9-2 can be rearranged to read:

$$\tan b_t = c'/k + (1 - u/k) \tan \phi' \qquad (9\text{-}4)$$

where c', u, and ϕ' are as previously defined, b_t is the threshold slope angle, and $k = \gamma \cdot z \cdot \cos^2 b$ (or $\gamma \cdot y \cdot \cos b$) is the overburden pressure normal to the potential sliding surface. Regoliths in Canada are susceptible to the development of perched water tables, well above the level of the regional bedrock aquifer, particularly where till of low permeability directly underlies the regolith layer. The cohesion (c') may stem from clay minerals in the regolith or from plant roots that similarly bind the debris together and lock it to the underlying bedrock or overconsolidated till. The angle of shearing resistance (ϕ') is greatly influenced by the angularity, degree of sorting, and packing geometry of regolith particles.

Regoliths lacking both fine grained sediment and deeply penetrating roots have cohesion values roughly equal to zero, in which case the threshold slope gradient is controlled by ϕ', k, and u, all of which can change over time. Weathering of the regolith, for example, leading to particle breakdown, may reduce ϕ', thus contributing to a decrease of b_t, and continuing slippage, over geological time. Whether thickening of the regolith (increase in k) would have the same effect depends on concurrent changes in the pore fluid pressure. For example, if the regolith never develops a saturated layer at its base but does dry out to give pore pressures that are atmospheric ($u = 0$), the value of b_t is identical to ϕ' and is independent of regolith thickness (eq 9-3). The tendency of many talus slopes to stand at the angle of repose, in effect a ϕ' value, illustrates this condition.

In contrast, full saturation of the regolith to the surface by a perched aquifer, combined with $c' = 0$, is approximated by $\tan b_t = 1/2 \tan \phi'$, since the unit weight of water is about half that of saturated soil. In the short-term, buildup of pore pressure during rainstorms or snowmelt, leading to perched aquifers in the regolith, is certainly the main trigger for regolith slides. Whether perched storm water tables become sufficiently high to produce slippage depends on many factors besides rainfall amount. One is the thickness of the regolith: as this increases, the thickness of the perched saturated layer must also increase to maintain a

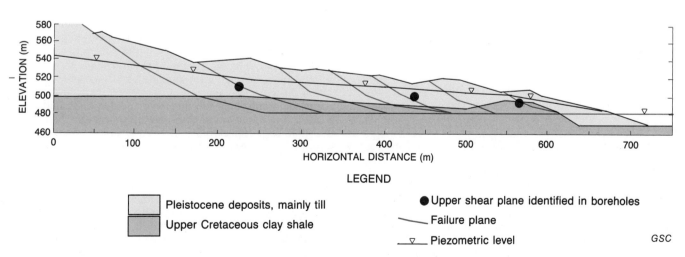

Figure 9.9. Stratigraphic cross-section of the Little Smokey landslide, Alberta (after Thomson and Hayley, 1975).

Figure 9.10. Debris torrents, Coast Mountains, British Columbia. Courtesy of M.J. Bovis. a) Typical debris torrent channel and deposits on a tributary of Lynn Creek, North Vancouver. Note the ravelling of stream banks, which supplies large organic and coarse clastic debris to the channel. b) Recently deposited debris torrent material on a tributary of Ryan River, near Pemberton. Coarse debris is typical of torrents derived from quartz bedrock.

given pore pressure ratio (u/k). Shallow regoliths are therefore more easily destabilized than thick ones by infiltration of water. Soil infiltration capacity, which varies considerably with soil texture, ice content, and water content, also controls the rapidity of pore pressure buildup. Where antecedent moisture content is high, less infiltration is needed to produce a perched, saturated zone. Because of flow divergence on spurs and convergence in linear depressions, shallow landslides commonly occur in topographic depressions, as already noted in the discussion of deep-seated landslides in clay.

Where weathering rates are high, relative to undercutting rates, slopes may quickly flatten to the threshold gradient; as undercutting continues, further slides simply maintain a steady state in which slopes are cut back at the threshold angle. Valley side slopes in the Queenston Shales of southern Ontario provide one example of this. The slopes are still undercut and are essentially straight, mantled with a thin regolith of either mixed shale-clay rubble ($35° < \phi' < 45°$) or more fully weathered clay ($25° < \phi' < 35°$) and stand at angles close to the ϕ' values (Carson, 1975). The true groundwater table in the area is well below the surface and the relatively low permeability of the regolith does not appear to be conducive to development of perched groundwater.

Where the developing regolith does not lose cohesive strength rapidly, slopes can stand well above the ultimate threshold gradient, with slides occurring only infrequently at intervals of hundreds or thousands of years. This time interval reflects the rate of decrease of c'/k as weathering not only diminishes cohesive strength but also increases regolith thickness. A stiff shale, for example, can have as much as 200 kPa of cohesion (c'), decreasing, as it weathers into a clay regolith, to about 10 kPa or less. Simple application of equation 9-4 shows that for a 1 m thick (y) regolith on a 55° slope (with $\phi' = 35°$ and $u/k = \gamma_w/\gamma = 0.5$) the strength c' must drop to 12 kPa for a slide to occur; and for failure on a 35° slope, c' must fall below 6 kPa. Thus as slopes flatten, the time interval between slides increases, and it may take a long time for regolith slopes derived from strong rock to attain permanent stability.

An interesting feature of the quantitative treatment of the stability of cohesive slopes is the role of vegetation. Studies of tree root strength (for example, Wu et al., 1979) have indicated that on forest-covered slopes where roots penetrate to the underlying rock, the cohesion imparted to the regolith can be significantly higher than the natural cohesion of the unforested regolith. Present-day evidence supports the view that clear-cutting of forests can lead to increased instability of regoliths, primarily due to the loss of this cohesive strength as roots decay. On a geological time scale, sequences of deforestation and regrowth, whether due to forest fire or climatic change, would have had comparable effects on the stability of regolith mantled slopes.

Debris flows. The association of regolith slides with periods of rapid snowmelt or intense rainfall often results in the sliding of saturated debris into steep creeks, whose discharge is notably augmented by overland flow and rapid interflow. If the gradient of the creek bed is greater than the threshold slope for stability of fully saturated debris, the saturated mass may accelerate downslope as a channelled, high velocity slurry known as a debris flow. Shear dilation of the debris causes a volume increase and further entrainment of water from the swollen creek. In addition, rocky debris, logs, and wood mulch may be scoured from the creek bed and sides, causing a notable growth in the depth of flow, and therefore velocity and momentum, in a downstream direction. Such rapidly moving, highly mobile masses are termed debris torrents in the western Cordillera (Fig. 9.10). They may be initiated in five ways: by delivery of regolith material from sideslopes as described above; by high storm water discharge in a steep creek causing failure of the bedload by fluid shear; by shock loading of debris from snow avalanche or rockfall impact; by temporary damming of a creek following a debris slide; or by freak events such as "jökulhlaups", which involve the catastrophic drainage of ice-impounded or supraglacial lakes (Jackson, 1979). Recent work has shown that the quantity and stability of accumulated material in the creek bed has to be specified, in addition to the flood discharge (VanDine, 1985). This complex interaction of water and debris makes it difficult to predict the timing and magnitude of events, even when climatic inputs are accurately known (Eisbacher and Clague, 1981).

Considerable attention has been focused on these events in recent years, particularly in view of the inability of culverts, or even bridges, to convey debris flows (Evans, 1982; Hungr et al., 1984). Debris flow activity over postglacial time has led to the accumulation of large debris cones or fans at the mouths of steep mountain creeks, particularly in the Cordillera and the more mountainous parts of the Arctic Archipelago. Impact scars on mature conifers and the presence of deciduous "regrowth" areas on fan surfaces, may provide an indication of the frequency and magnitude of debris flows (Desloges and Gardner, 1981).

Instability of thawing slopes

Although deep-seated slips involving frozen ground have been reported from Mackenzie Valley (McRoberts and Morgenstern, 1974b), instability on most frozen slopes involves only thawed surficial material. Slopes underlain by permafrost and those subject to seasonal freezing are dealt with here.

In ice-poor rock permafrost, thaw on cliffs leads to rockfall and the development of talus, as already noted. The thaw of ice-rich permafrost cliffs in fissile or unconsolidated materials is quite different: instead of talus, soupy debris collects at the cliff base and, if the water content is high enough, flows away from the cliff as a colluvial lobe of viscous or viscoplastic mud. Continued evacuation of debris leads to extensive retreat of permafrost cliffs, in marked contrast to talus development, which acts to mask the parent cliff. Cliffs in ice-rich materials, formed either by wave action along coasts or by stream undercutting, are widespread in the western Arctic and the Arctic Archipelago (Fig. 9.11) and are subject to retreat by ablation each summer. Annual retreat rates of 3 m are not uncommon and, continued over decades, lead to large, amphitheatre-like bowls. Progressive stabilization of the slope occurs when the backscarp becomes so small that overhanging mats of vegetation effectively insulate it from further ablation. There is a superficial resemblance between these bowls and some of the scars produced by quick-clay failures in eastern Canada; both are apparently triggered by an initial failure, which leads to retrogression in the weaker material so exposed. The mechanisms and rates of retrogression are, however, quite different.

On less steep tundra slopes, ice-rich permafrost is commonly insulated from thaw by stable, vegetated regolith. If sudden slippage of regolith occurs, exposure of the permafrost may lead to seasonal retrogression or failure. Alternatively, deeper penetration of thaw into the permafrost itself may induce rapid slippage in which the surface regolith is passively rafted away by failure of the thawed, ice-rich material below. Any disturbance of the insulating vegetation cover may lead to instability as described, but slopes are particularly prone to this type of failure following forest and tundra fires (this chapter, French, 1989). Such rapid failures of the active layer are commonly referred to as skin flows (Fig. 9.12). The mechanisms involved in these cold-climate failures are still not agreed upon, though the regolith involved usually has a water content exceeding its liquid limit. The most intriguing feature of skin flows, however, is not so much the long runout of the wet colluvium as

Figure 9.11. Icy cliff in the headwall of a retrogressive thaw slump near Eureka, Ellesmere Island, Northwest Territories. Massive ground ice extends to within 1 m of the surface. Courtesy of D.G. Harry. 204770.

Figure 9.12. Skin flow slope failure, Caribou Hills, Northwest Territories. Courtesy of J.A. Heginbottom. 185167

the fact that the initial failure can occur on slopes as gentle as 5°. Mass movement on low gradient slopes in cold areas has been known for a long time, usually being attributed to gelifluction — a genetic term denoting slow flowage of regolith. Only recently have the possible mechanisms for such movements been considered in the context of soil mechanics.

The problem of regolith instability on low-gradient slopes may be illustrated by reference to equation 9-4: even with no cohesion and extreme groundwater conditions (that is, water table at the surface with u/k = 0.5), failures should not be able to occur on slopes with gradients less than 1/2 tan ϕ'. Even with clayey regoliths, with ϕ' perhaps as low as 12°, slopes flatter than 6° should be stable. That they frequently are not, in permafrost areas, implies that excess pore pressures, greater than $u = \gamma_w \cdot z \cdot \cos^2 b$, and hence u/k ratios greater than 0.5, must be capable of developing in the saturated, thawing debris. One mechanism that could produce excess pressures in areas of seasonally thawed ground is ice-blocked drainage. If the water table rises (because of recharge from thawing areas behind a slope) while the slope is still frozen, outflow seepage will be partially blocked and artesian pressures may develop. Few field studies have been made of this mechanism.

Thaw-consolidation is another process that could create artesian pressures. In areas where the frozen regolith contains large quantities of segregated ice, the inability of the bulked debris, once thawed, to withstand the force of its own weight means that positive pressures will be set up within the interstitial water as it supports some of the load. The magnitude of the pressure depends on the rate of loading above the thaw plane (and hence on the rate of advance of the plane) in relation to the inherent compressibility of the thawed material and the ease of expulsion of water from the thawed zone. Though thaw-consolidation occurs in non-permafrost areas, it should be particularly effective above a permafrost table, which impedes downward drainage of pore water. In these cases, apart from limited expulsion of water by lateral drainage, self-loading by thaw requires that water pressures in the thaw zone increase until sufficient to expel water upwards to the ground surface. The theory of thaw-consolidation is presented quantitatively by McRoberts and Morgenstern (1974a) who showed that, for compressible sediment of low permeability, failure could be expected on slopes as gentle as 2° to 3°, corresponding to gradients as low as 1/4 tan ϕ' in some cases.

Though thaw-consolidation is probably significant in many landslides in cold areas, there are several aspects of the theory that limit its general validity (Algus, 1986). One is the assumption that the active layer remains fully saturated during thaw. Though this may be true for ice-rich permafrost, sediment that has undergone a volume increase solely due to the phase change of in situ water will not become saturated on thaw until reconsolidation to the prefrozen state and will not necessarily develop excess pore pressures. Secondly, many fine grained permafrost sediments contain a reticulate ice-vein network (Mackay, 1974a), which on thaw may enhance expulsion of porewater from the thaw zone and lead to gross underestimates of permeability based on laboratory tests.

Studies of mudflows in temperate areas have offered an additional mechanism for excess pore fluid pressures that may also be applicable to thawing slopes in periglacial regions: this is the recognition that the pore fluid of remoulded debris is rarely water, but more a dilute slurry formed by admixture of fine particles with water. Because pore pressures are a function of the unit weight of the fluid, conversion of pore water to slurry increases pore pressures. Vallejo (1980) recognized this in silty clay sediments with a reticulate ice-vein pattern: during failure, attrition of the clay blocks between thawed ice veins converted the thaw water to a muddy suspension, hence increasing pore pressure and allowing movement across gentle slopes. The critical question, however, is whether slurry pressures can develop prior to failure, thus initiating movement on low gradient slopes. Evidence is still lacking on this point.

Soil creep and soil erosion

Soil creep, rainsplash, rill and gully erosion, localized subsurface erosion (piping), and wind erosion have so far been ignored in this discussion. The argument has been made that on the actively evolving slopes of the Canadian landscape, they are of secondary importance in comparison to landslides, although in absolute terms all of these processes, as well as mass movement, operate more effectively on steep slopes.

Soil creep

Soil creep is probably the major process affecting vegetated slopes which are not prone to either mass failure or surface erosion. Though many mechanisms can produce creep, the most common is periodic heave of the regolith, normal to the slope, followed by settlement under gravity. Heave-settlement cycles are usually produced by the growth and wastage of ice lenses or needle ice, a process referred to as frost heave (this chapter, French, 1989). If settlement is vertical during the thaw cycle, downslope movement, S, for any point in the regolith is given by:

$$S = H \cdot \tan b \qquad (9\text{-}5)$$

where H is heave and b is the slope angle. Because soil in the upper part of the regolith is affected by volume increase in the material beneath it, creep displacements are greatest near the ground surface; in most cases this produces a velocity profile which is concave downslope. This is usually seen where freezing occurs from the surface downwards (one-sided freezing). Seasonal soil creep due to heave and settlement has been studied in many parts of the world, with surface movements typically of the order 5 mm/a or less, and volumetric rates of 10 cm^3/a per centimetre width (along the contour of the slope). In northern Canada, in the zone of continuous permafrost, creep rates are commonly much higher. Mackay (1981), for example, measured rates of 5 to 10 mm/a on slopes flatter than 7° on Garry Island, and volumetric rates of 7 to 52 cm^3/(cm·a). Interestingly, most of his sites showed greatest movement in the late summer, associated with thaw of an ice-rich base of the active layer that had been formed by upfreezing from the permafrost table. Since autumn freezing also occurs from the surface downwards, this environment experiences two-sided freezing. Soil movement at Mackay's sites tended to be plug-like, rather than decreasing with depth as occurs under one-sided freezing. Some of Mackay's observations were made in terrain classified as "hummocky gelifluction lobes". His findings, together with radiocarbon dates on peat overridden by the moving regolith, indicate that such lobes, sometimes interpreted as the product of relatively recent flowage from skin failures, may be formed by slow frost creep acting over several thousands of years.

Surface soil erosion

Erosion by rainsplash and overland flow is usually only significant on steep colluvial slopes on which arboreal canopy or ground cover is sparse. Local absence of ground cover may be due to fire, logging, construction, or recent mass movement. On a regional scale, sparse vegetation is obviously climatically conditioned, notably in the polar desert areas of the Arctic Archipelago, in the drier parts of the Prairies, and in the semi-arid interior valleys of British Columbia. In the latter case, slope aspect exerts an important influence on local climate and, therefore, on plant cover and soil stability. South-facing slopes typically have an open forest with scattered shrubs and bunchgrass and are more prone to soil erosion (as evidenced by exposed tree roots) than the moister, closed-canopy slopes which face north. This distinction is lost as one moves into the moister Cordilleran valleys where precipitation is greater than about 500 mm/a. Climate also influences soil movement through the character and seasonal distribution of precipitation. Most of the Cordilleran and Prairie regions have a distinct summer maximum of precipitation, delivered usually as short-lived, high intensity rainstorms. This regime, combined with sparse vegetation on slopes of low infiltration rate (clay and shale) may lead to the development of spectacular badlands topography (Fig. 9.13) notably along Red Deer River valley of southeast Alberta (Campbell, 1982). The highly expansive clay soils of this region tend to develop a polygonal network of desiccation cracks, especially on interfluves. The cracks serve as effective "sinks" for overland flow in that, except under heavy rainstorm conditions, a large proportion of incoming precipitation is absorbed by the expanding montmorillonite clays. In existing rills and gullies, however, reworking of shale debris produces a more compact surface with narrower, shallower desiccation cracks than are found on the badlands interfluves. This is then conducive to higher runoff and gully deepening. Periodic widening of gullies occurs by small scale mass failures in the polygonal-fractured clay crust. Subsurface erosion by piping also occurs in this area (Bryan et al., 1978).

Discussion of soil erosion would be incomplete without reference to the Prairies region. Appreciable wind transport of fine soil particles occurred during the 1920s and 1930s as a result of drought conditions. Climatic amelioration, combined with improved farming methods, has allowed substantial restabilization of formerly eroded areas to take place. Contemporary erosion occurs by wind transport as well as by overland flow from snowmelt and summer rainstorms. The spatial variability of erosion rates means that relatively few data exist on primary hillslope erosion, as opposed to suspended load in streams. Recently, isotopic methods have been used to estimate erosion rates over the past 25 years. Fallout of radioactive cesium (^{137}Cs) from atmospheric tests of nuclear devices in the late 1950s and early 1960s was absorbed, presumably uniformly, by surface soil and shows little spatial variation in concentration in soils retaining their native vegetation cover (de Jong et al., 1983). Cultivated areas, however, show a pronounced decrease in ^{137}Cs concentration on slope crests, coupled with an increase in slope-foot areas, indicating a substantial, local reworking of mineral soil by surface runoff and rainsplash processes. Preliminary soil loss figures calculated by this relatively new method are of the order 20 to 60 kg/m^2

Figure 9.13. Oblique aerial view of badlands topography, Red Deer River valley, Alberta. Courtesy of I.A. Campbell.

over the past 25 years, which translates to an average surface lowering rate of 0.5 to 1.5 mm/a. This is substantially lower than the 1 to 8 mm/a of surface lowering reported by Campbell (1982) from a ten year record of erosion on badlands in Alberta.

Though the High Arctic is usually considered to be dominated by cold-climate processes such as gelifluction, there is evidence that normal soil erosion processes may also be important, at least episodically. Heavy rainfall is known to have caused significant erosion in the Queen Elizabeth Islands during the summer of 1973 (Cogley and McCann, 1976). An important finding of this study was that soil erosion was instrumental in exposing ground ice, the subsequent melting of which led to mass movements of the type described above. The geomorphic significance of the storm event may be judged by the fact that suspended sediment concentrations in trunk streams, over the three-day storm period, were nearly three times higher than the levels attained during a major jökulhlaup event which occurred eleven days after the storm runoff peak and which was possibly triggered by the storm runoff itself.

In the coastal Cordillera of British Columbia, the erosional potential of intense, orographically enhanced precipitation falling on steep slopes is largely offset by the interception of rainfall by coniferous forests and by the inhibition of overland flow by a thick layer of litter and decaying logs. Subsurface runoff, particularly through "pipes" created by the decaying of roots, possibly accounts for a large proportion of "quick flow" of water from forested watersheds during rainstorms (deVries and Chow, 1978), though this is still a matter of debate (Sklash et al., 1986). This concentration of subsurface flow undoubtedly leads to erosion of finer material, evidence of which can be seen where logging roads traverse steep colluvial slopes. Although the effects of logging activities on surface soil erosion may be locally severe, particularly along logging roads, most studies have emphasized the importance of shallow regolith slides and debris torrents in the accelerated delivery of sediment to stream channels which normally accompanies timber harvesting (O'Loughlin, 1972; Wilford and Schwab, 1982; this chapter, Hewitt, 1989). Subsurface concentration of water in pipes may in fact serve to accelerate the rise of the perched water table and may augment pore pressures by flow convergence hence increasing the likelihood of debris slides and debris flows.

Variation in rates of hillslope activity

The foregoing brief overview of hillslopes in Canada has dealt with processes which, at the present time, seem to account for most of the geomorphic work. Important temporal changes in the rates of operation of hillslope processes, however, take place, the consequences of which are discussed in this section. Geomorphic activity on hillslopes usually occurs as a series of discrete events, rather than as a continuous downslope movement of material. A seasonal variation in work, driven primarily by climate, is now known to exist over a wide range of processes, from rockfall to soil erosion. The cumulative effect of such action is to leave a distinct morphological imprint on the landscape, which allows inferences about process to be made from hillslope form. On a longer time span, of the order of decades, hillslope events may occur of a magnitude well above the assumed long-term average rate for a given area. Examples of such accelerated activity are the rapid retreat of shoreline bluffs in the Great Lakes region during periods of abnormally high lake levels; very large debris flows triggered by freak rainstorms in the Cordillera; and cycles of thaw-related instability on permafrost cliffs in the Arctic. This behaviour is analogous to the "dynamic metastable equilibrium" proposed by Schumm (1977) for river channel behaviour. For example, the pattern of erosion and deposition of soil by surficial processes such as overland flow may be temporarily disrupted by major erosion events which effect substantial local changes in slope morphology. The significant complication with hillslopes, however, is that a major geomorphic event may change the local rank-ordering of geomorphic processes for a substantial time period. For example, a slope characterized by minor colluvial activity may be transformed by sudden mass failure (caused perhaps by slope undercutting) to such an extent that retrogressive slumping or slope ravelling becomes the dominant agent of denudation as the slope slowly adjusts to a more stable profile.

On a time scale of hundreds to thousands of years, slopes may undergo form and process changes in response to several forcing variables. First, downcutting and in particular lateral migration of river channels both lead to slope steepening and are major causes of slope instability. Climatic changes may, through the intervening variable of plant cover, bring about significant changes in denudation rates. This is amply shown by the rapid rate of fan building in the Cordillera during the early Holocene "paraglacial" period (Ryder, 1971a,b). The importance of late glacial and Neoglacial conditions in the reworking of material on Cordilleran slopes was also emphasized by Slaymaker and McPherson (1977). Preliminary data on the movement of slow earthflows in the Cordillera (Bovis, 1985) suggest that Holocene climatic fluctuations may have been important.

Weathering, leading to progressive fatigue of slope materials, is also important on this longer time scale. Unfortunately, evidence of such progressive strength decrease comes to light only after a detailed, project-related investigation of a hillslope. Finally, seismic shaking must take its toll of hillslopes in the long term. The effects of the 1946 Vancouver Island earthquake are well documented (Mathews, 1979). Large rock avalanches, such as the Hope slide (Fig. 9.6) are suspected of having been seismically triggered (this chapter, Evans, 1989). Whilst the Cordillera has the highest potential for seismically triggered releases, it should be realized that St. Lawrence Valley also has experienced powerful earthquakes within the historical period, some of which have triggered failures of quick-clay material (Desjardins, 1980).

In addition to variation of hillslope activity with time, slope denudation is rarely uniform spatially. For various reasons, preferentially localized denudation leads to the formation of hollows, scallops, bowls, and chutes in otherwise uniform surfaces. Indeed, in some cases, this "dissection" of slopes is sufficiently self-reinforcing that slope processes should be considered not only as sculptors of valley sides, but also of the drainage network itself. Thus while it is correct to think in terms of slope processes acting on the sides of valleys cut by rivers, it is also true that some streams flow in valleys produced, in large part, by slope processes.

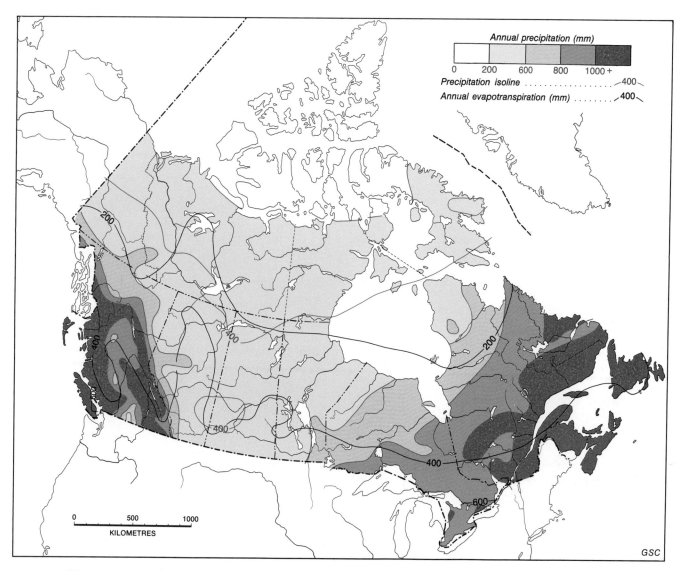

Figure 9.14. Water balance in terms of derived precipitation and evapotranspiration (after Plate 25, Hydrological Atlas of Canada, 1978).

RIVER PROCESSES

T.J. Day

Fluvial sediment transfer and storage are dramatic elements of Canada's landscape. This section presents an overview of the characteristics and tempo of fluvial sediment transfer, its morphological expression, and governing hydrological and geomorphic factors.

Day, T.J.
1989: River processes; in Chapter 9 of Quaternary Geology of Canada and Greenland, R.J. Fulton (ed.); Geological Survey of Canada, Geology of Canada, no. 1 (also Geological Society of America, The Geology of North America, v. K-1).

Hydrological regimes

As presented by Smith (this chapter, 1989) the average climatic picture of Canada shows the manner in which physiographic features, particularly the Cordillera in the west and the vast plains of subdued relief in the east, influence precipitation and the timing and duration of the frost-free season. These features, coupled with evapotranspiration, infiltration, and storage parameters such as interception, soil moisture, and groundwater, determine hydrological regimes. In simple terms, a water balance equation shows that runoff is equal to precipitation minus evapotranspiration plus or minus any change in water storage. The distributions of precipitation and evapotranspiration are shown in Figure 9.14. Major runoff volumes are expected along each coast with lowest volumes expected in the Prairies and arctic coastal plains and lowlands.

Annual runoff, which is the portion of precipitation that reaches stream channels and integrated over a year, is shown in Figure 9.15. Mean annual runoff varies from over 3000 mm on the west coast of Vancouver Island to less than 25 mm in some areas of the Prairies. In terms of percentage coverage of the land surface, about 50% lies between the 100 and 200 mm isolines, with only 0.4% having greater than 3000 mm and 14% having less than 100 mm.

The direction of major drainage and the areas of the contributing basins are shown in Figure 9.16. Within the western Cordillera, drainage is westward either into the Pacific Ocean or, in the case of the Yukon River basin, towards the Bering Strait through Alaska. The Mackenzie River system drains about one fifth of Canada's landmass northward into the Arctic Ocean, including the northeastern portion of the Cordillera, the northern two thirds of the Interior Plains and western portions of the Canadian Shield. The northern portion of the Canadian Shield and the arctic lowlands and coastal plains also drain into the Arctic Ocean. Drainage from the southern Interior Plains and from most of the Canadian Shield and all of Hudson Bay Lowlands is towards Hudson Bay. A good portion of the annual flow is from the Churchill and Saskatchewan river system which drains eastward from the Interior Plains. Systems draining into James Bay are further sites of major annual flow. The other major drainage direction is eastward, along the southeastern portion of the Canadian Shield, through the Great Lakes-St. Lawrence Lowlands towards the Atlantic Ocean. In terms of annual flows by river system, the St. Lawrence carries the largest discharge, followed closely by the Mackenzie. The eastern part of the Canadian Shield in Labrador and the Atlantic Provinces drains directly to the Atlantic Ocean.

Regimes by physiographic region

Cordillera. In this region of mountains, plateaus, and valleys, elevation generally exerts more control on climate than does latitude. In general, precipitation decreases eastward from the Pacific coast, especially in the lee of successive

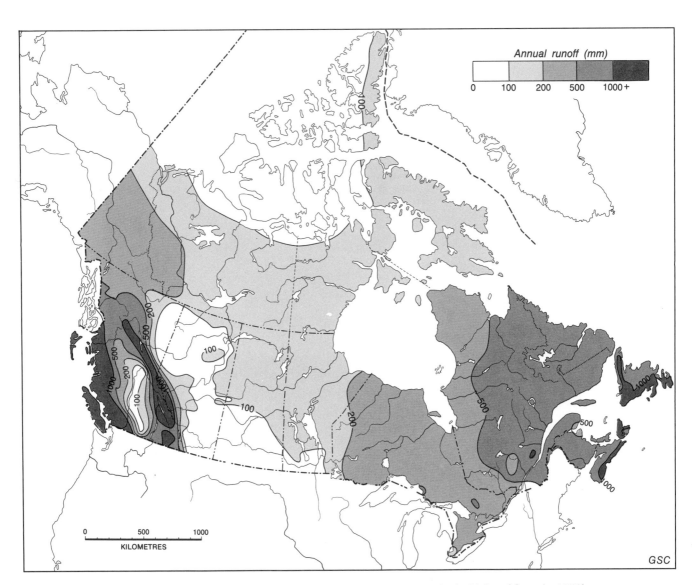

Figure 9.15. Long term average annual runoff (after Plate 24, Hydrological Atlas of Canada, 1978).

mountain ranges. In the western Cordillera, steeply sloping terrain, relatively low evapotranspiration, and rapid snowmelt result in runoff of most precipitation. As shown in Figure 9.17, the seasonal high flow for the major westerly draining river systems occurs in spring or early summer as a result of snowmelt accompanied by rain with some contribution from glacial meltwaters. In coastal areas, however, where winters are mild, the highest runoff occurs in late autumn, the season of heaviest rainfall. Regimes of easterly draining rivers are similarly dominated by a spring snowmelt flood but, because of lower precipitation, runoff can be significantly less. The mean annual discharge from Cordilleran drainage is 24 000 m³/s, which is 23% of the Canadian total.

Interior Plains. The Interior Plains are characterized by low relief and high evaporation rates; discharge accounts for only about one third of the total amount of precipitation — the lowest proportion in Canada. The southern part of the plains is dominated by grasslands (the Prairies) and the north is forest covered. For truly prairie rivers, such as the Assiniboine (Fig. 9.17), annual flow regimes are dominated by snowmelt runoff with peak flow occurring during the

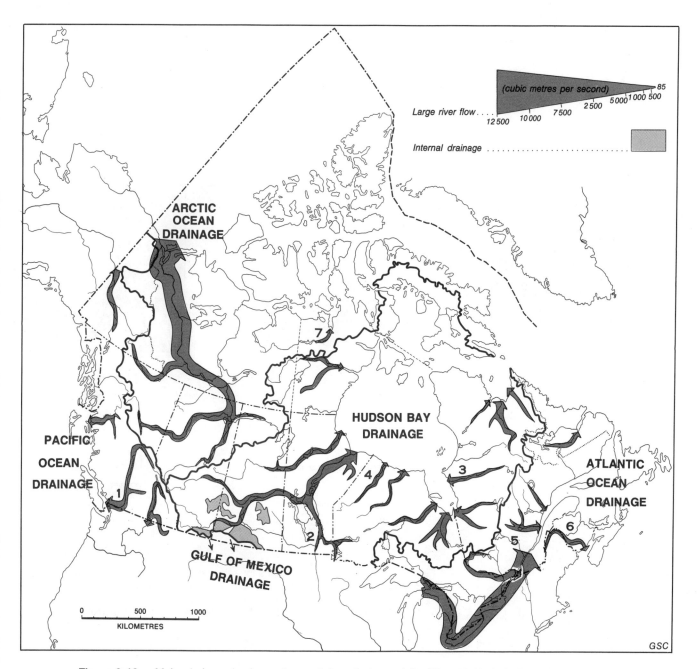

Figure 9.16. Major drainage basins and annual river discharge (after Plate 22, Hydrological Atlas of Canada, 1978). Numbers 1-7 refer to rivers used in Figure 9.17.

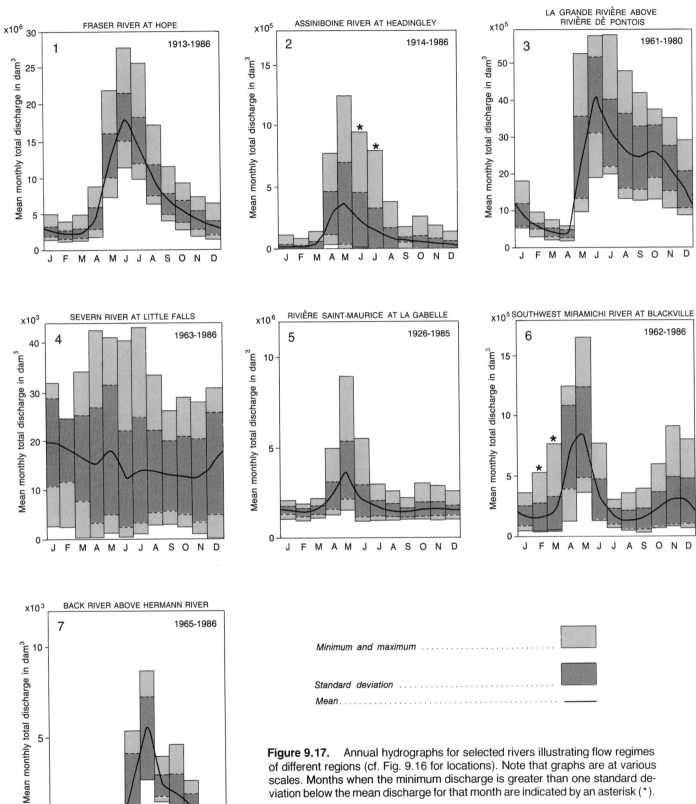

Figure 9.17. Annual hydrographs for selected rivers illustrating flow regimes of different regions (cf. Fig. 9.16 for locations). Note that graphs are at various scales. Months when the minimum discharge is greater than one standard deviation below the mean discharge for that month are indicated by an asterisk (*).

April to July period. Annual variation in flows can be very large. For Saskatchewan River and some of its tributaries, which have their headwaters on the eastern flanks of the Cordillera, prairie and mountain snowmelt peak flows are distinct, with maximum flows generally occurring in June in response to the mountain snowmelt. Again, considerable annual variability occurs particularly for the prairie snowmelt period. Although summer convectional storms may be significant in small eastern slope basins, the peak flows in these streams are still dominated by the snowmelt flood. The major tributaries of Mackenzie River, such as Liard and Peace rivers, rise in the eastern Cordillera and their annual hydrographs are dominated by the mountain snowmelt floods. Extensive peatlands in the central part of the boreal forest zone have high water retention capacities. The mean annual discharge of Mackenzie River is about 9900 m³/s, the second largest in Canada.

Canadian Shield. Annual runoff from the Canadian Shield decreases from 680 mm in the southeast to 160 mm in the northwest (Fig. 9.15). These rates are influenced by the numerous lakes, impervious ground, moderate gradients, and low evapotranspiration (but lake evaporation can be significant during summer months). Depending upon location (Fig. 9.17) maximum runoff occurs between May and August. Annual variability in flows (Fig. 9.17) is generally less than in the Interior Plains. The larger rivers draining into western Hudson Bay have broad flood peaks and less annual variability than those entering from the east.

Hudson Bay Lowlands. Rivers flowing through and those rising in the lowland have flow peaks either in the April-June period when snowmelt dominates or in late summer and early autumn caused by rainfall (Fig. 9.17). Almost 80% of the lowlands is a vast peatland and small lake plain with a high water retention capacity.

Great Lakes-St. Lawrence Lowlands. St. Lawrence River drains this region eastward to the Atlantic Ocean. Most of the main tributaries, such as Ottawa and Saguenay rivers, carry runoff from the Canadian Shield. On the southern banks of St. Lawrence River, tributaries flow from the Appalachian region. Tributary regimes are dominated by the March-May snowmelt with minor peaks occurring in the September-December period in response to local rainfall (Fig. 9.17). Although the St. Lawrence carries the largest annual discharge, less than a third of the region's annual precipitation actually reaches the ocean; the remaining two thirds is lost to evaporation.

Appalachia. The relatively steep and impervious terrain of this region, coupled with moderate river gradients and high precipitation, results in high runoff volumes. Stream discharge accounts for almost 83% of the annual precipitation. This runoff ratio may be inflated because of a negative bias resulting from the siting of most observing stations in valleys and lowland areas. Annual runoff ranges from 860 to 1270 mm in the east and from 510 to 1200 mm in the slightly drier west. Annual runoff maximums occur in the spring (Fig. 9.17) due to snowmelt, while rainfall produces secondary peaks in the fall and winter months.

Arctic Lowlands and Innuitia. Arctic and subarctic nival regimes are characterized by a snowmelt flood in spring and generally low levels of flow throughout the rest of the year, punctuated by periodic rainstorm floods (Fig. 9.17). Such floods may be more severe than those in spring, especially for rivers with headwaters in mountainous regions where heavy summer rainfalls may be orographically enhanced and snowmelt may still continue. On Ellesmere and Baffin islands, along the northeastern boundary of the Innuitian region, proglacial regimes are found. Similar to the proglacial streams in the Cordillera, these rivers do not experience an abrupt spring flood. Their regimes are strongly affected by the nature of the melt season. Discharges continue to rise until late summer as progressively higher zones on the glaciers melt. In the arctic lowlands, peatland regimes associated with poor drainage are common. The large water-retaining capacity of peat and peatland vegetation, as well as its high resistance to runoff, greatly attenuates flood flows.

Fluvial erosion intensity

Denudation on the regional scale is greater in the mountains than in the lowlands; is greater in wet, snowy climates than in dry climates; and is greater where ground ice exists in quantity. Bird (1980, his Table 6-2) presented estimates of regional denudation rates based upon sediment loads in large river basins. Surface lowering in the western Cordillera ranges from 50 to 100 mm/ka. For the arctic lowlands and highlands rates are estimated to be between 10 and 35 mm/ka and 35 and 65 mm/ka, respectively. In the Prairies estimates are 10 to 30 mm/ka. Similar rates of 15 to 35 mm/ka were found for southeastern Canada and 5 to 15 mm/ka for the Canadian Shield. These rates are gross estimates based upon sparse data with short record lengths. Problems with regional denudation rates are that present day sediment transport data can reflect the remobilization of older deposits and the interception of sediment by lakes and floodplains, as well as present denudation rates.

Sediment transport processes and characteristics

Rivers transport the products of erosion and weathering to lower elevations by three primary and very different mechanisms: by suspension, by solution, and by traction as bedload. The characteristics of these mechanisms are different, and the proportion of the load carried by each process varies from region to region in response to variation in topography, geology, and the hydrological regime. In simple terms, fluvial processes are middle and low elevation (Fig. 9.3) processes which transfer glacial, colluvial, and alluvial debris to depositional sites at low elevations. These transfers and deposition (or storage) points are shown schematically in Figure 9.18.

Suspended sediment

With short records and sparse data, attempts to generalize suspended sediment concentrations (Stichling, 1973; Slaymaker and McPherson, 1973) are of limited value. The inherent variability of transport, due to spatial and temporal characteristics of erosion and sediment delivery processes, complicates valid generalizations. Some association, however, can be shown between annual concentrations and yield of suspended matter and physiographic regions. Although indicative of gross spatial characteristics and their variability, mean annual parameters mask important temporal features of suspended sediment transport regimes.

In the Cordilleran region most suspended sediment transport is associated with the spring snowmelt runoff. For example, in lower Fraser River more than 80% of the annual load is carried during the freshet period in late spring and

Table 9.4. Annual loadings of suspended sediment for Fraser River at Mission, B.C.

Year	Suspended Sediment Load (10^3 t)
1966	19 300
1967	26 100
1968	20 900
1969	13 900
1970	11 500
1971	17 500
1972	30 900
1973	12 200
1974	24 900
1975	12 000
1976	24 900
1977	14 500
1978	12 300
1979	15 000
1980	10 900
1981	12 400
1982	25 600
1983	8 090
1984	12 300

Source: Milliman (1980)

early summer. There is a very marked hysteresis in the relationship between suspended sediment and water discharge, with the bulk of the sediment being transported as flows increase and concentrations falling rapidly after peak flow has passed (Milliman, 1980). Annual loadings for Fraser River near the coast are given in Table 9.4. The dominance of the spring-summer snowmelt runoff as the main suspended sediment transport period is seen throughout the Cordillera except along the west coast mountains where runoff from winter rain may carry significant suspended sediment loads.

On the eastern slopes of the Cordillera similar suspended sediment regimes are found. For a very small basin, Nanson (1974) documented three relationships between suspended sediment and discharge. Two were related to the passage of the spring snowmelt and rainfall flood events, and the third to a later, purely snowmelt runoff. Instantaneous concentrations of over 5000 mg/L were measured. Hudson (1983) has shown that almost 100% of the annual suspended sediment load for Elbow River occurs in the May-June snowmelt runoff period. Hudson identified river bank erosion as the source of 25% of the suspended load. Significant hysteresis exists for the transport characteristics of this river as well.

River systems draining the eastern Cordillera are tributary to either the northerly flowing Mackenzie River system or the easterly flowing Saskatchewan system.

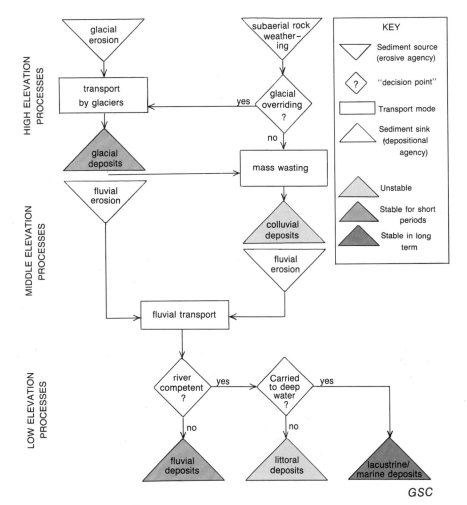

Figure 9.18. Fluvial sediment sources, transfer patterns, and storage sites in the landscape (after Church, 1980).

Consequently, suspended sediment regimes are transferred from the Cordillera into the Interior Plains. For South Saskatchewan River, however, the influence of local prairie snowmelt becomes noticeable near the Alberta-Saskatchewan boundary, where an increase in flow occurs during the March-May period before the arrival of the generally dominant mountain runoff. The suspended sediment regime follows the annual hydrograph with an initial transport period in March-April and then the major event later during May-July. These two separate events produce hysteresis in transport-discharge relationships. Daily sediment concentrations may on occasion be highest during the Interior Plains runoff period, but most of the load passes during the second period and most of this during the rise to peak flow. Although sand can constitute up to 50% of the suspended load, the silt and clay fraction normally constitutes greater than 70%. Saskatchewan River maintains these regime characteristics until it reaches Lake Winnipeg. In the eastern Interior Plains, the silt and clay load increases and, although the highest peak loads are now carried during the Plains runoff period, most of the total load is still transported during dominance of flow by mountain snow melt.

Suspended sediment regimes for truly Interior Plains rivers have only one main event, which for Assiniboine River is during April-May. Again almost 70% of the load is carried during this period. During most of the year, silt and clays constitute over 85% of the suspended material except during high flows when sand can amount to as much as 40%.

The amount of fluvial sediment reaching the Beaufort Sea from Mackenzie River has recently been estimated at about 150×10^6 t/a (Neill and Mollard, 1980; Harper and Penland, Woodward Clyde Consultants, Victoria, B.C., report for Atlantic Geoscience Centre, 1982). Most of this load is transported during the June to October period, and almost all is suspended load. Within Mackenzie River above the delta, there is a sediment discharge peak during August, and a similar but weaker pattern has been found for channels within the delta (Davies, 1975). The suspended materials being transported through the delta are predominantly clay (35-55%) and silt (40-60%) with only minor amounts of sand (less than 5%; Davies, 1975). Most of the fluvial sediment reaching the Mackenzie Delta and Beaufort Sea is derived from the mountains of the eastern Cordillera north of 57°N. Sediment originating from more southern basins in the Cordillera and the Interior Plains is deposited in either major lakes or hydro-electric reservoirs.

On the Canadian Shield sediment concentrations and yields are, as expected, lower than those of the Cordillera and Interior Plains. Krank and Ruffman (1982) estimated that an average of 14×10^6 t of suspended sediment each year is deposited in James Bay by the eastern shore shield rivers. For the western shore of James Bay, they estimated about 27×10^6 t/a is transported by rivers either originating in or passing through the Hudson Bay Lowlands.

In the St. Lawrence Lowlands suspended sediment concentrations and loads are generally higher than in the Canadian Shield. For example, Ongley (1973) determined mean annual sediment loads of 6.3 to 66 t for 52 basins tributary to Lake Ontario. For one basin, Ongley et al. (1981) showed that suspended sediment concentrations displayed an annual pattern of spring highs in response to snowmelt, summer lows, and again higher concentrations in the fall in response to rainfall events. Spring maximum concentrations were measured at 300 mg/L while the summer concentrations were 5 mg/L and less; fall concentrations were similar to those of the spring period.

Julien and Frenette's (1984) investigations of agricultural watersheds on the south shore of St. Lawrence River, within the lowlands, indicate a dominant single transport period during the February to March period with peak transport occurring in April. This single event is in response to snowmelt. Annual loadings of suspended sediment ranged from 40 to 380 t. Frenette and Larinier (1973) have illustrated that the suspended sediment load of St. Lawrence River is more related to the load of its major tributaries than to its own load largely as a consequence of sediment from the upper part of its drainage basin being trapped in the Great Lakes. These authors documented that concentrations downstream of the Ottawa-St. Lawrence confluence range only from 0 to 30 mg/L, with an average annual load of 2.5×10^6 t/a. (This is about one-twentieth the estimated load of the Mackenzie River.) Daily loads showed a maximum of 25 000 t.

Streams and rivers within the Appalachian region generally show a peak transport period in spring in response to snowmelt, low summer transport, then increased transport over the winter months in response to rainfall. For Kennebecasis River in New Brunswick, instantaneous concentrations can reach to over 600 mg/L and daily loads can be in excess of 11 000 t; mean annual transport is about 35×10^3 t.

A common feature of suspended sediment transport is the effect of short duration, high magnitude events. For example, in Highwood River in the eastern Cordillera, 68% of the total annual suspended load for 1975 was transported in one event which lasted only 1% of time the total year. Also loads transported during such events can amount to more sediment movement than the total loads of several other years combined.

Bed load

Reliable data on bed load are sparse; few good field studies have been undertaken, field measurements are difficult, and only limited good quality observational records are available. This is due, in part, to the complex range of factors governing the process. Sediment transported as bed load is most significant in rivers with gravel beds. Gravel bed rivers differ from sand bed rivers in the lack of small scale bed features, sporadic and partial movement of bed material, beds which are inactive most of the time, a wide range of particle sizes, and differences in surface (pavement and armour) and subsurface material sizes. Some results are available, however, to indicate rates and processes.

For example, in a small mountain stream Nanson (1974) indicated that sediment supply to the stream, and hence the amount of bed load being transported, is controlled by the timing and frequency of mass movement events on valley sides. Nanson also identified channel armouring as a further control on rates. Rates ranging from 0.2 to 107 kg/min were measured, with distinct relationships between bed load discharge and water discharge before and after peak discharge, due to supply and bed armouring processes.

Hudson (1983), for Elbow River in the eastern Cordillera, has shown that significant temporal and spatial variations exist in measured bed load rates. He estimated that the measured load amounts to only one quarter of the possible load (based upon hydraulic considerations) due to supply constraints. He also demonstrated a downstream discontinuity in bed load transport as Elbow River passes from the mountains through the foothills (transport capability increases through these reaches) and then through the Interior Plains where transport capacity is significantly less. To accommodate this discontinuity, sediment is stored in waves along the channel. He presented rates of transport, based upon Hollingshead (1968), ranging from 24 to 91 kg/h or 0.6 to 2190 t/day for a measurement site in the lower foothills.

For larger proglacial gravel bed rivers on Baffin Island, Church and Gilbert (1975) determined that bed load amounted to 77 to 90% of the annual load. For short periods of record (one season) Church (1972) computed bed load yields ranging from 29 to 1114 t/km^2 of watershed area for several Baffin Island gravel bed rivers. For Vedder River in British Columbia, Tywoniuk (1977) measured bed load rates ranging from 2.7 to 260 t/day. Jonys' (1976) data for the Vedder are instructive in relation to the effective width of bed transport and the variations of transport rates within this effective width. For still larger gravel rivers in Alberta, Galay (1970) calculated gravel bed load rates for the North Saskatchewan to be 0.46 kg/s per metre width of channel. This rate was for only a short period and was determined from the displacement of large gravel bars.

More recently, on the gravel reach of lower Fraser River, D.G. McLean and M.C. Mannerström (unpublished) have indicated that channel changes are not generally associated with periods of major floods but rather these changes usually develop in response to relatively localized changes in flow alignment. The role of sediment transport in bar growth and main channel deposition is critical to the understanding of the origin of these local morphological changes.

Remarkably little data are available on sand bed load. For the sand bed reach of Fraser River, daily rates of about 450 to 5440 t/day for a discharge range of 2830 to 14 200 m^3/s have been measured (Western Canada Hydraulic Laboratories Ltd., 1978).

Dissolved solids

Canada's surface waters show widely different mineral characteristics from region to region. Coastal rivers generally carry low concentrations of dissolved solids; similar concentrations are recorded for Canadian Shield surface waters, while highly mineralized waters predominate in the Interior Plains region. Most surface waters in Canada fall into the alkaline-earth bicarbonate category (Inland Waters Directorate, 1977). Such waters carry significant calcium and magnesium bicarbonates which originate mainly from limestone, dolomite, and clay. In terms of dissolved load, alkaline-earth bicarbonates range from 11 to 91% of the total, with the content of most major rivers ranging from 50 to 90%. Although the solute load of rivers does not exclusively represent primary denudation of drainage basins, but also includes contributions from atmospheric and other sources (e.g., Singh, 1970; Meybeck, 1983), the effects of these other contributions are not well researched and, in some cases, these effects are estimated to be small (McPherson, 1975; Zerman and Slaymaker, 1978).

In coastal drainage basins of the western Cordillera, median concentrations of dissolved solids are of the order of 50 to 160 mg/L. For the eastern side of the Rocky Mountains, Hudson (1983) showed concentrations ranging from about 120 to 300 mg/L for several small basins. The northwardly flowing Yukon River has a median concentration of about 120 mg/L and is diluted as it flows northward by tributary inflows. In an opposite way, the total dissolved solids carried by Mackenzie River increase northward, primarily as a result of drainage from the limestone and dolomite formations of the Mackenzie and Franklin mountains. The easterly flowing North and South Saskatchewan river systems carry median concentrations of about 200 to 240 mg/L. Concentrations in the North Saskatchewan are usually lower than in those in South Saskatchewan River, due to the inflow of water from the Canadian Shield, but as tributaries from the Interior Plains region enter, concentrations increase. After the North and South Saskatchewan join to become Saskatchewan River, total dissolved solids decrease as more waters from the Shield serve to dilute concentrations. Within the Interior Plains region waters are highly mineralized with median concentrations ranging from 250 to over 1000 mg/L.

Rivers rising in the Canadian Shield have total concentrations of dissolved solids of about 150 mg/L. Rivers flowing across the Canadian Shield show a dilution of solutes, whereas rivers rising on the shield which pass through the Hudson Bay Lowlands show an increase in solute concentrations. In the St. Lawrence Lowlands major rivers rising on the Canadian Shield show increased solute concentrations as waters draining limestone and dolomite terrain enter; within the lowlands solute concentrations are high (median values of about 300 mg/L). Along St. Lawrence River dissolved solids slightly decrease eastward as waters from the Canadian Shield and Appalachians enter. Total concentrations of dissolved solids in Appalachian rivers are low, with median values ranging from about 15 to 50 mg/L.

Although estimates of continental dissolved solid yields are available (e.g., Livingstone, 1963, gives a continental average of 33 t/km^2), little synthesis of available data sources has been undertaken in Canada. Several small and regional scale studies have, however, provided figures on yields of individual basins (e.g., Slaymaker and McPherson, 1973; Ongley, 1973; McPherson, 1975); McPherson (1975) computed annual yields of dissolved solids ranging from 0.31 to 100 t/km^2 for 21 intermediate sized basins (133 to 1427 km^2) draining the eastern slopes of the Cordillera. In 50% of these basins annual yields were less than 17.5 t/km^2. Hudson's (1983) more intensive study of Elbow River in the eastern Cordillera indicated annual yields of 98 t/km^2 for the mountains, 65 t/km^2 for the upper foothills, 54 t/km^2 for the lower foothills, and 25 t/km^2 for the plains. Hudson attributed these low yields to the effects of the low soil temperature of the mountains and foothills. Ongley (1973) determined considerably higher annual yields for 52 basins in the St. Lawrence Lowlands which drain into Lake Ontario; yields ranged from 48 to 267 t/km^2 for basins ranging from 2.6 to 12 941 km^2.

River morphology and processes

Canada's landmass exhibits a diversity of river morphologies ranging from the beaded channels formed in permafrost terrain to the large gravel bed channels of the Canadian Cordillera. A number of factors govern this diversity, including upstream supply of water and sediment; boundary materials; geomorphic setting; geological history and, to a lesser extent, ice regime.

The upstream supply of water and sediment determines the dimensions of a river channel and many of its morphological features. Besides the simple volume of water discharge (to which the main dimensions are scaled), the temporal variability is important. For example, Kellerhals et al. (1972) pointed out the dissimilarity of form (within the same geomorphic setting) between rivers with and without the moderating effect of lake storage. Although sediment load is strongly associated with discharge, in the longer term it reflects the effects of geology, climate, and land use, and changes in these. The complex, and still little understood, relationships among the governing factors of river morphology prevent a simple association of load type and river morphology. The schematic simplification of Mollard (1973), however, is useful in indicating relationships such as the general associations of low sediment supply and predominance of suspended load over bed load, which are characteristics of stable, low gradient rivers exhibiting various patterns of meanders. Conversely, where bed load dominates transport, with large amounts of sediment available to be moved, river channels are unstable and have relatively steep gradients; characteristic morphologies are wandering to braiding. Beyond such general simplification, identification of the morphological effect of sediment supply and mode of transport requires detailed study.

Boundary materials (bed and bank sediments, as well as materials in the vicinity of the channel) are significant in their influence on local channel stability by determining bank strength and erosion thresholds. Of particular significance here is the effect of the composition and stratigraphy of surficial deposits of glacial origin. Thus in many locations in Canada, rivers flowing in coarse glacial materials have heavily armoured or paved beds, which limit bed erosion and "normal" channel development. Also, Church (1983b) has documented the importance of alluvial history and variations in bed and bank materials in understanding the problem of bridge scour along Thompson River in British Columbia. The presence of coarse alluvial gravels over glacial lake silts and clays, as well as the sporadic presence of consolidated tills and glacial lake sediments in this area, contributes to variable scour along the river.

The morphology of a river directly reflects its geomorphic setting: for example, where they cross a delta, channels are generally deep and relatively straight; upstream of a fan, channels commonly meander and flood basins are commonly present, whereas where they cross a fan, channels are generally straight but may be shallow or deeply entrenched. Alluvial plain segments of valleys are generally graded to handle normal discharge and sediment loads and consequently, at times of flood, flow spreads out to reoccupy abandoned channels, fill flood basins and deposit sediment across the floodplain. The relation of a river to its valley and the degree to which the channel is entrenched or confined are important indicators of longer term channel processes. A common feature of many Canadian rivers is their youthfulness, as the landmass has been ice free for only about 12 ka. During this ice-free period, rapid changes in hydrological regimes have occurred resulting in many underfit rivers (Kellerhals et al., 1972; Galay et al., 1973) exhibiting entrenched and confined channels. Geological control of river morphology can be illustrated by examples of bedrock controlled rivers, such as the many rivers on the Canadian Shield that follow the bedrock fracture pattern. On a larger time scale, beyond the effects of glacial and postglacial environments, geological control can be seen in the adjustment of major rivers to conform with major tectonic elements; an example is the eastward flow of Interior Plains rivers away from the Rocky Mountains.

Regional characteristics

The significant diversity of morphological factors and the complex relationships amongst the governing conditions permit only limited regional generalizations. Such generalizations are made more difficult by the fact that few studies have attempted to consider the various possible contributing factors, both historical and present day.

The study by Shaw and Kellerhals (1982) on the major plains rivers in Alberta is a significant exception. These rivers can be divided into three reaches: an upper mountain reach, a central gravel reach, and a lower sand reach. Upon leaving the mountains these rivers follow either preglacial or postglacial valleys, and sometimes the same river (for example the North Saskatchewan) will pass from one type to another. Vertical erosion of river valleys has been the predominant postglacial process, but one that has been complicated by the effects of differential isostatic rebound resulting in decreased gradients to the northeast. Following rebound, the downstream reaches may have entered into an aggrading phase and Shaw and Kellerhals (1982) postulated that the abrupt transition from gravel to sand beds has resulted from these factors.

Within the Cordillera, river morphologies cannot easily be generalized because drainage generally is east-west, whereas the general fabric of the various physiographic units is oriented north-south. Obviously in headwater reaches in the mountains, steep, coarse bed channels are found. Where rivers, such as the Fraser, cross mountain chains, bedrock control dominates reaches and valleys generally follow faults or other tectonic discontinuities. Within the plateau regions of the Cordillera, valleys occupy trenches which in some places may contain large, fiord-like lakes and elsewhere contain thick fills of unconsolidated sediments. Hence two morphological types of reach dominate these rivers — deep valleys excavated in soft sediments and alluvial plains constructed over filled lake basins.

Appalachian rivers are, in a regional sense, similar to those of the Cordillera, but they lack the major mountain source areas and, in general, the tectonic elements controlling the physiography are smaller. On the Canadian Shield, a typical river reach occupies a bedrock controlled valley joining two lakes. Channel gradients are irregular and falls and rapids are common. Relief is generally small, so little sediment is produced today. What sediment can be eroded from the glaciated surface is trapped in the next lake basin, which is never far downstream.

Across Canada and within almost any region a full range of morphologies can be found as rivers adjust to regional or locally dominant governing factors. This diversity is too broad and the reasons too complicated to be presented here. There are, however, several references of note: Kellerhals et al. (1972), Galay (1970), Lewis and McDonald (1973), and Mollard (1973); these reports contain photographs and summary tables of data and information. Also Neill and Mollard (1980) provide information on erosion and sedimentation processes along some northern rivers, while Carter et al. (1987) describe fluvial processes in the Arctic Lowlands.

Long term variations in fluvial sediment transfer

Seasonal and annual characteristics of fluvial sediment transfer have been briefly described in previous sections. Although these demonstrate significant variation, longer term variations — since the last glaciation — can be even more significant. Concepts of periglacial processes, where processes are influenced by the cold temperature associated with the presence of ice, are well accepted (this chapter, French, 1989). In a similar way, proglacial environments significantly affect the pattern and intensity of fluvial processes. The studies of sandurs on Baffin Island by Church (1972) and more recently at Sunwapta Lake, Alberta by Gilbert and Shaw (1981) show the relationships between proglacial environments and sediment movement. However, the key concept relevant to the understanding of many depositional environments and morphologies of many modern rivers is that of paraglacial processes. First introduced by Ryder (1971a), and expanded by Church and Ryder (1972), this concept is based upon the fact that following deglaciation plentiful glacial drift remains in unstable settings. This material is susceptible to movement by mud and debris flows and is moved rapidly downslope where it may be stored temporarily in valleys or become available for fluvial transport.

Within the Interior Plateau of British Columbia, Thompson Valley contains paraglacial fluvial deposits, which may be up to 175 m thick and which consist of early postglacial gravels and sands deposited by large braided streams. There are at least two valley fills of distinct age in Thompson valley, indicating two separate phases of deposition. Anderton (1970) presented evidence that this valley fill possibly accumulated within 1000 years. Similar deposits are found in the Fraser, Similkameen, and Bonaparte valleys of central British Columbia (Church and Ryder, 1972). Further analyzing Baffin Island data, Church and Ryder (1972) demonstrated that outwash plain development between the commencement of ice retreat from valley-head moraines (5.7 ka) until completion of the retreat some 1.4 ka later shows deposition of an order of magnitude larger than fjord sedimentation which began at 4 ka.

More recently, Jackson et al. (1982) presented evidence that paraglacial processes are responsible for the early postglacial (11.5 to 10 ka) aggradation of Bow River on the eastern margin of the Rocky Mountains. During this rapid depositional period Bow River was a braided, gravel bed river, which contrasts significantly with its present quasi-stable, sinuous, single-channel form. Similarly, Smith (1975) demonstrated from delta stratigraphy at Upper Waterfowl Lake, Alberta, that sedimentation rates prior to 6.6 ka were more than twice those recorded since that time.

Besides playing a significant role in providing large volumes of sediment to fluvial systems, the results of this abundant supply can be seen today in the presence of terraces along some of Canada's major rivers (for example, Thompson and Fraser rivers, British Columbia; Bow River, Alberta; and Ottawa River, Ontario) and in the presence of large fans blocking valleys (for example, on Mistaya River, Smith 1975; Smith and Smith 1980; and Thompson River, Ryder, 1971a,b). These major deposits of paraglacial origin have partially filled valleys, have contributed to the formation of lakes, and influence present day river morphologies.

On a much shorter time scale sediment transfer rates also show significant variability. For example, Church (1983a) suggested that the pattern of channel instability of the modern Bella Coola River in the Coast Mountains of the Cordillera most likely results from erosion of Neoglacial moraines (1840 to 1900 AD). Also, Gilbert (1972) demonstrated that sediment yield to Lillooet Lake in British Columbia had almost tripled since 1948. He suggested that this increase resulted from glacial retreat, increased logging, agriculture on the floodplain, and river training, with river training being almost certainly the overwhelmingly dominant factor. Similarly, for portions of lakes Ontario and Erie, Kemp et al. (1974) identified a three fold increase in sedimentation rates since 1930 and 1935, respectively.

COLD CLIMATE PROCESSES

H.M. French

Cold climate processes may be subdivided into those associated with either glacial or nonglacial conditions. The present impact of glacierization is localized, however, and for the most part it is cold nonglacial conditions that dominate the greater part of Canada's landmass. These conditions are commonly referred to as "periglacial".

Although the term periglacial was first applied to areas peripheral to Pleistocene ice sheets and glaciers, modern usage of the term refers to a wide range of cold nonglacial conditions, irrespective of their proximity to glaciers, either in time or space. Conditions referred to as periglacial exist not only in high latitude and tundra regions of northern Canada but also in areas south of treeline such as Mackenzie Valley and central Yukon, and in alpine regions of southern Canada.

Periglacial processes are intimately associated with intense frost action, commonly combined with the presence of permafrost; on this basis, 50% of the land surface of Canada currently experiences periglacial conditions. There are all gradations between environments in which frost-related processes dominate, and where the whole or a major part of the landscape is the result of such processes, and environments in which frost-related processes are subservient to others. Complicating factors are that different lithologies vary in their susceptibility to frost action, and that there is no perfect correlation between areas of intense frost action and areas underlain by permafrost. Moreover, since extensive areas of northern Canada have only recently emerged from beneath Late Wisconsinan ice, the morphology is dominantly glacial, and periglacial processes currently serve only to modify these glaciated landscapes. The "paraglacial" concept, referring to conditions of the glacial-nonglacial transition (Church and Ryder, 1972; this chapter, Day, 1989) may therefore be a more appropriate designation for these environments. However, in those areas of Canada that were either outside the various Pleistocene glacial limits or have experienced protracted nonglacial histories, such as the interior Yukon and northwestern Banks Island, landscapes are probably in equilibrium with current cold climate processes.

French, H.M.
1989: Cold climate processes; in Chapter 9 of Quaternary geology of Canada and Greenland, R.J. Fulton (ed.); Geological Survey of Canada, Geology of Canada, no. 1 (also Geological Society of America, The Geology of North America, v. K-1).

Freezing and thawing of the ground

Central to the operation of cold climate processes is the freezing of the ground surface. This may occur either on a diurnal basis, as in many temperate and subtropical regions, or on a seasonal basis, as in much of Canada. The depth of frost penetration depends mainly upon the intensity of cold, its duration, the thermal and physical properties of the soil and rock, the overlying vegetation, and snow depth. Where the depth of seasonal frost exceeds the depth of thaw during the following summer, a zone of frozen (that is, temperature below 0°C) ground persists throughout the year. This is commonly referred to as permafrost. All three conditions — diurnal frost, seasonal frost, and permafrost — influence the nature, extent, and significance of cold climate processes.

The annual rhythm of freezing and thawing of ground dominates much of northern and central Canada where long cold winters are typical. From a geomorphic viewpoint, the nature and rate of both spring thaw and autumn freeze-back are of interest. The former influences the nature of spring runoff while the latter controls the nature of frost heaving and ice segregation in the soil. Usually, spring thaw occurs quickly and, in areas with only seasonal frost (that is, no permafrost), over three-quarters of the soil thaws during the first four to five weeks in which air temperatures are above 0°C. At Thompson, Manitoba, for example, ground thermal regimes are closely related to snow thickness and density (Brown, 1973; Goodrich, 1982); years of heavy snowfall retard thaw yet prevent deep frost penetration. In areas with permafrost, thaw progression is primarily a function of time, and the depth of thaw depends on the length of the thaw season. Autumn freeze-back is an equally complex process, especially in those regions underlain by permafrost where freezing occurs both downwards from the surface and upwards from the perennially frozen ground beneath. The freeze-back period is much longer and may persist for six to eight weeks. At Inuvik, Northwest Territories, for example, freezing begins in late September and finishes in mid-December (Heginbottom, 1973; Mackay and MacKay, 1976). During the majority of this time the soil remains in a near-isothermal condition, sometimes referred to as the "zero curtain". This phenomenon results from water releasing latent heat as it freezes, thereby retarding the drop in temperatures. Initially, freezing progresses at a slow rate from the surface downwards but then dramatically speeds up at depth. This is because moisture decreases with depth since soil water is initially drawn upwards to the freezing plane, thereby preferentially increasing the latent heat effects in the near surface layers.

The frequency of freeze-thaw cycles in the soil is surprisingly low. Numerous field studies from a variety of localities in northern Canada indicate that at depths of more than about 10 cm only the annual cycle occurs. It is only at the ground surface that freeze-thaw cycles occur with any frequency, and even these fluctuations may be stopped by a significant snowfall.

Frost heave and ice segregation

Intimately associated with ground freezing are the phenomena of frost heaving and ice segregation. These processes occur wherever moisture is present within the soil (Smith, 1986). Frost heaving caused by ice segregation occurs throughout much of Canada. Annual ground displacements of several centimetres with cyclic differential ground pressures of many kilopascals per square centimetre are common. Engineering problems due to these displacements and pressures, together with the adverse effects of accumulations of segregated ice in freezing soil, are widespread and costly.

The process of ice segregation involves complex interrelationships between the ice, an unfrozen liquid phase, and the bulk pore water (Polar Research Board, 1984; Gold, 1985). Complex latent heat of phase change as well as variable interfacial energy between phases are involved. Freezing air temperatures and radiant heat loss from the ground surface create a thermal gradient that induces upward heat flow. Initially, soil water freezes near the ground surface, and then segregated ice crystals form, grow, and ultimately coalesce into lamellar lenses (ice segregation). Ice lens growth is then a consequence of a continuous upward flow of water from below. During this process a balance is achieved between the dissipation of the latent heat of freezing and the upward flow of soil water. The upward displacement of the soil surface (frost heaving) is a consequence of the growth of a series of ice lenses. When displacement is restricted, significant "heaving pressures" can develop.

Both primary (that is, capillary) and secondary heave can be distinguished (Miller, 1972; Williams, 1976; Smith, 1985b). In primary heave, the critical conditions for the growth of segregated ice are:

$$P_i - P_w = (2\sigma/r_{iw}) \cdot (2\sigma/r) \qquad (9\text{-}6)$$

where P_i is pressure of ice, P_w is pressure of water, σ is the surface tension between ice and water, r_{iw} is the radius of the ice-water interface, and r is the radius of the largest continuous pore openings. Secondary heave is not clearly understood but may occur at temperatures below 0°C and at some distance behind the freezing front. Pore water expulsion from an advancing freezing front is another mechanism of ice segregation, especially for massive ice bodies, provided that pore water pressures are adequate to replenish the groundwater that is transformed to ice.

Repeated frost heaving is usually associated with the seasonally frozen layer. Field studies in the Mackenzie Delta region indicate that heave occurs not only during the autumn freeze-back but also during winter when ground temperature is below 0°C (Mackay et al., 1979; Smith, 1985a). Geomorphic indicators of frost heaving include the upthrust of bedrock blocks (Dionne, 1983; Dyke, 1984, 1986), the raising of objects and tilting of stones (Mackay and Burrows, 1979; Mackay 1984b), and the sorting and migration of soil particles (Corte, 1966). Frost heaving presents numerous geotechnical problems in the construction of roads, buildings, and pipelines in cold environments (Carlson, 1982; Hayley, 1982) and in locally induced artificial freezing, such as beneath ice arenas (Leonoff and Lo, 1982).

An interesting but small scale heave phenomenon produced by diurnal freezing at or just beneath the ground surface is the formation of needle ice. The ice crystals grow upwards in the direction of heat loss and can range in length from a few millimetres to several centimetres. Occasionally, they may lift small pebbles or, more commonly, soil particles. Needle ice formation is particularly common in the Cordillera wherever wet, silty soils are present. The thawing and collapse of needle ice is thought significant in frost sorting, frost creep, the differential downslope movement of fine and coarse material, and the origin of certain micropatterned ground forms. The importance of needle ice as a disruptive agent in the soil has probably been underestimated, especially in exposing soils to wind action, deflation, and cryoturbation activity. In some areas it may be responsible for damage to plants when freezing causes vertical mechanical stresses within the root zone (Brink, 1967). In the coastal area of British Columbia, needle ice occurs in oriented stripes (Mackay and Mathews, 1974) and both the presence of shadows and solar radiation have been suggested as explanations; it is not clear whether oriented needle ice is primarily a secondary effect developed by thawing or a freezing effect.

Cryogenic weathering

The weathering of bedrock in cold climates is not fully understood. It is currently believed that a complex of weathering processes, both physical and chemical, operate either independently or in combination.

The most important bedrock weathering process has generally been assumed to be frost wedging, a physical weathering process relying upon repeated freeze-thaw activity. Features attributed to frost wedging include extensive areas of angular bedrock fragments (blockfields), scree and talus deposits, and irregular bedrock outcrops termed tors (Fig. 9.19). All occur widely in Arctic Canada (Bird, 1967; Dyke, 1976; French, 1976). Porous and well bedded sedimentary rocks, such as shales, sandstones, and limestones, are regarded as being especially susceptible to frost wedging. Numerous field studies from both northern Canada and elsewhere (Cooke and Raiche, 1962; Chambers, 1966; Washburn, 1967; Fahey, 1973), however, now demonstrate that not only is the frequency of freeze-thaw cycles limited but also that the required combination of freezing intensity and moisture availability is rarely encountered under natural field conditions (Thorn, 1980). Serious doubts must be cast, therefore, upon the effectiveness of freeze-thaw rock shattering for much of northern Canada.

Several studies have estimated the rate of rockwall recession apparently resulting from frost wedging. For example, in Longyeardalen, Spitzbergen over the last fifty years frost weathering has caused steep rock faces to retreat at a rate of 0.3 mm/a (Jahn, 1976). In northern Canada, rates of cliff recession vary with lithology, but rates as high as 0.6 to 1.0 mm/a have been suggested (French, 1976, his Table 7.4, p. 148).

A recent trend is to regard both hydration weathering, in which the pressure of absorbed water generates forces sufficient to free grains and disintegrate rock, and salt weathering as important processes in cold regions. Their overall significance to cryogenic weathering, however, is still to be agreed upon (Washburn, 1979, p. 73-74; French, 1981). These processes focus attention upon the moisture rather than the temperature requirements for weathering and may be effective without the freezing process being involved. In these processes, physical and chemical effects may combine; for example, field studies undertaken on coarse grained granitic rocks on Ellesmere Island indicate that repeated salt hydration and crystallization following summer precipitation events favour microfracturing and the formation of weathering pits and associated grus (Watts, 1983). Other studies suggest that solutional effects are significant in the High Arctic (Cogley, 1972; Smith, 1972), and permafrost does not necessarily inhibit the development of extensive karst topography (van Everdingen, 1981). Finally, experimental studies in the USSR indicate that many of the widely held assumptions concerning cold climate weathering may need to be re-evaluated. For example, Konishchev (1978) concluded that the resistance of quartz is lower than that of feldspar when subject to freeze and thaw. According to these studies, the ultimate grain size distribution resulting from cryogenic weathering is the reverse to that produced under "normal" (that is, noncryogenic) conditions.

It is not surprising, therefore, that cold climate weathering processes and their associated landforms are currently subject to renewed scrutiny. For example, snow has long been recognized as a potent geomorphic agent. However, the nivation concept itself (St-Onge, 1969) relies upon the assumed effectiveness of freeze-thaw processes in promoting the physical disintegration of bedrock beneath the snowbank. A shift to the hydration hypothesis places greater emphasis upon rock porosity. Many nivation features and tors may be structurally controlled rather than erosional in origin (Dyke, 1983). In the case of the cryoplanation terraces of the Yukon Plateau (Hughes et al., 1972; French et al., 1983), where no obvious structural control exists, their distribution may reflect subtle interactions between rock porosity, available moisture, and prevailing winds.

Permafrost and ground ice

A consequence of the long period of winter cold and the relatively short period of summer thaw in much of northern Canada is the formation of a layer of ground that remains frozen (below 0°C) throughout the year. This perennially frozen ground is termed permafrost. Although essentially a thermal condition, the presence of moisture within permafrost, either in a frozen or unfrozen state, presents problems, as evidenced by frost heave, described above. Soil and rock do not automatically freeze at 0°C, especially if percolating

Figure 9.19. Limestone tors and felsenmeer, Cornwallis Island, Northwest Territories. Courtesy of L.A. Dredge. 204375-Z.

groundwater is highly mineralized or under pressure. As a result, significant quantities of unfrozen pore water may continue to exist at temperatures below 0°C (Williams, 1976; Andersland and Anderson, 1978).

There is a broad zonation of permafrost conditions in Canada according to climate (Fig. 9.20). Zones of continuous and discontinuous permafrost are recognized, in addition to alpine permafrost and subsea permafrost. In total, approximately 50% of Canada's land surface is underlain by permafrost of one sort or another. The southern limit of the zone of continuous permafrost correlates well with the approximate position of the -6 to -8°C mean annual air temperature isotherm, and this relates to the -5°C isotherm of mean annual ground temperatures. The zone of discontinuous permafrost is further subdivided into areas of widespread and sporadic permafrost. At the extreme southern fringes of the zone of discontinuous permafrost, permafrost exists as isolated "islands" beneath peat and other organic sediments.

Ground ice is an important component of permafrost and in areas underlain by unconsolidated sediments, such as the Mackenzie Delta, it may comprise as much as 60% by volume of the upper 1-5 m of permafrost (Pollard and French, 1980). It may also occur in bedrock (French et al., 1986). Although many types of ground ice are recognized (Mackay, 1972; French and Pollard, 1986), pore ice, segregated ice, and wedge ice are the most important, both in terms of volume and widespread occurrence (Fig. 9.21). In general, the volume of ground ice is greatest near to the surface of permafrost where ice wedges, reticulate ice veins, and aggradational ice are all present (Williams, 1968; Mackay, 1974a; Pollard and French, 1980) and the ground is subject to repeated annual thermal changes (Cheng, 1983; Mackay, 1983). The latter induce moisture migration downwards from the active layer above and upwards from the permafrost below. In certain areas of the western Arctic icy sediments and massive beds of segregated ice several tens of

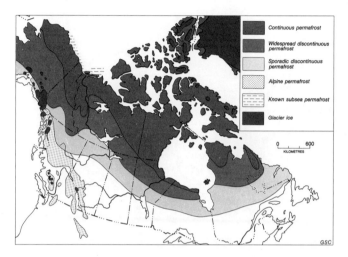

Figure 9.20. Distribution of permafrost in Canada.

Figure 9.21. Deformed massive ground ice of uncertain origin (segregation ice or buried glacier ice), Nicholson Point, Northwest Territories. Deformation of ice and enclosing sediments is presumed to have occurred as a result of overriding by Late Wisconsinan ice. A large, recent ice wedge is visible in the upper part of the section.

Figure 9.22. Large, low centre ice-wedge polygons, Banks Island. The polygons are approximately 20 m across. Courtesy of J-S. Vincent. 204063-B

Figure 9.23. Small peat plateau, northern Manitoba. Courtesy of L.A. Dredge. 203121-N

metres thick are known to exist (Mackay, 1971; Rampton and Mackay, 1971; Mackay et al., 1972). The source of water for such ground ice is of a subterranean rather than subaerial origin. Hence, the growth of many massive ground ice bodies and much segregated ice is associated with the aggradation of permafrost and the Quaternary history of the area (Rampton, 1973; French et al., 1982; Mackay and Matthews, 1983; Burn et al., 1986; Heginbottom and Vincent, 1986).

Permafrost-related processes and landforms

A number of geomorphic processes are directly related to the presence of permafrost and give rise to unique landforms. One of the most widespread of these processes is large-scale thermal contraction cracking of the ground (Mackay, 1986a). This leads to the formation of a polygonal pattern of fissures (Fig. 9.22). The fissures may be as much as 4 to 5 m deep and the polygons may be 15 to 30 m across. If water from melting snow at the ground surface trickles down the fissure and if cracking is repeated at the same locality, a wedge-shaped ice body (ice wedge) forms. In the Mackenzie Delta, detailed field observations indicate that the period of active ice-wedge cracking is concentrated between mid January and mid March (Mackay, 1974b). However, wedges do not crack each year; instead, probably less than 50% of the wedges in any given area crack annually (Mackay, 1975; Harry et al., 1985). As the ice wedges grow wider, peaty ridges usually develop parallel to the fissures and a polygonal, saucer-shaped depression is formed (low-centred polygons). The ice-wedges in such polygons are usually active (that is, the wedges still crack). Many polygons, however, are bun-shaped (high centred) and in these features cracking of the wedge is infrequent. Cracking is not instantaneous but propagates laterally from discrete points, and both upward and downward crack propagation apparently occur (Mackay, 1984a).

Ice wedges may accelerate other geomorphic processes by promoting differential thaw and erosion along the wedge. For example, the catastrophic drainage of a lake may occur via the erosion of ice wedges at its outlet (Harry and French, 1983) while the rapid retreat of coastal cliffs by block slumping in areas with numerous ice-wedge polygons is well known. In the western Canadian Arctic adjacent to the Beaufort Sea, retreat rates of between 1 and 5 m per year have been recorded (Mackay, 1963a; Harry et al., 1983).

Several distinctive landforms result from the aggradation of permafrost and the growth of ground ice. The most well known of these are perennial ice-cored hills termed pingos and perennially frozen peaty mounds termed palsas or peat plateaus (Fig. 9.23). Both are types of frost mounds (Mackay, 1986b). Palsas form in wetlands primarily by ice segregation beneath a peaty organic layer (Zoltai and Tarnocai, 1971, 1975; Kershaw and Gill, 1979). Typically they occur in the discontinuous permafrost zone and are rarely more than 5 m high. Seasonal frost mounds (van Everdingen, 1978, 1982) are sometimes confused with palsas since the two can be morphologically similar. However, these are not the result of ice segregation and instead, possess a core of injection ice (Pollard and French, 1984, 1985). Finally, the various small scale frost mounds commonly observed in tundra regions (French, 1971; Washburn, 1983) are generally regarded as being different from palsas, pingos, and seasonal frost mounds.

Pingos can grow to sizable dimensions, a few exceeding 50 m in height and 300 m in diameter (Fig. 9.24). The ice core of a pingo may be produced by injection of groundwater under pressure from below, by segregation, or by surface water freezing in dilation cracks (Mackay, 1973, 1979a, 1985). Hydrostatic (closed system) type pingos derive their water pressure from pore water expelled by downward permafrost growth. Hydraulic (open system) type pingos derive their water pressure by gravity flow from an upslope source. One of the largest concentrations of closed system pingos, approximately 1350, occurs along the coastal plain of the Mackenzie Delta (Mackay, 1963b); smaller groups occur on Banks Island (French and Dutkiewicz, 1976; Pissart and French, 1976) and western Victoria Island (Fyles, 1963). In most cases, the site of pingo growth is a shallow residual pond left by rapid lake drainage or an abandoned river channel or terrace (Fig. 9.25). In all cases permafrost aggrades into saturated, unfrozen sediments (taliks), causing pore water expulsion. The largest concentration of open system pingos occurs in the unglaciated interior Yukon where more than 700 have been identified (Hughes, 1969). Here, the major requirement is the presence of thin or discontinuous permafrost and the confining of subpermafrost waters. As a consequence, most open system pingos develop in distinct topographic situations, such as valley bottoms or lower valley side slopes.

The degradation of permafrost often involves the melting of ground ice accompanied by local collapse and subsidence of the ground. These processes are termed thermokarst, which is a physical (that is, thermal) process peculiar to permafrost regions (French, 1976, p. 104-133). Since thermokarst merely reflects a disruption in the thermal equilibrium of the permafrost, a range of conditions can initiate it, including changes in regional climate, localized slope instability and erosion, drainage alteration, and either

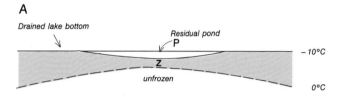

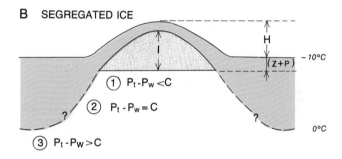

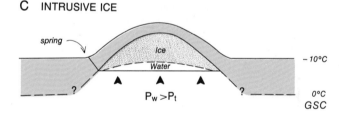

Figure 9.25. Diagrammatic representation of the growth of a pingo in the residual pond of a drained lake (from Mackay, 1979b, 1985).

A. The initial condition: The residual pond has a depth P in the centre, the overburden thickness of the permafrost is Z; permafrost is thinnest beneath the centre of the residual pond.

B. The growth of segregated ice: P_t is the total resistance to heaving and includes lithostatic pressure and resistance to bending of the overburden; P_w is the pore water pressure; and C is the soil constant. Ice lensing is favoured at location 1, beneath the growing pingo; ice lenses and pore ice develop at location 2, beneath the margin of the pingo; and pore water expulsion occurs at locality 3. The values of P_t and P_w change from location 1 to location 3. H and I denote the height of the pingo and the thickness of the ice body, respectively.

C. The formation of a subpingo water lens and the accumulation of intrusive ice: As the pore water pressure exceeds P_t (the total resistance to uplift), a subpingo water lens accumulates and intrusive ice forms by freezing of bulk water. Peripheral failure may result in spring flow.

Figure 9.24. Ibyuk pingo, near Tuktoyaktuk, Northwest Territories. The pingo, one of the largest known, is about 47 m high and 250 m across the base. Courtesy of J-S. Vincent. 204063-A

natural (such as, fire) or man-induced disruptions to the surface vegetation cover (Heginbottom, 1973). Along the western Arctic coastal plain, where alluvial sediments with high ice contents are widespread, thermokarst is believed to be one of the most important processes fashioning the landscape (Rampton, 1973, 1982). Elsewhere, large scale thermokarst phenomena include ground ice slumps (Lamothe and St-Onge, 1961; Mackay, 1966) and thaw lakes (Harry and French, 1983; French and Harry, 1983; this chapter, Klassen, 1989). Some types of thermokarst, such as ground ice slumps, are probably the most rapid erosional agents currently operating in tundra and arctic regions (this chapter, Carson and Bovis, 1989). Where large ice wedges are present within ice-rich silts, striking badlands thermokarst topography can result (French, 1974a). Typical man-induced thermokarst terrain consists of an irregular hummocky and unstable topography of enclosed depressions and standing water bodies (Kerfoot, 1973; French, 1975).

Active layer processes

Between the upper surface of permafrost and the ground surface is the active layer, a zone which thaws each summer and refreezes each autumn. In thermal terms, it is the layer that fluctuates above and below 0°C during the year. In permafrost areas, seasonally frozen and thawed ground can be equated with the active layer. The thickness of this layer varies from as little as 15 to 30 cm in high latitudes to over 1.5 m in subarctic and continental regions. Thickness depends on many factors including the ambient air temperatures; the angle of slope and its orientation; the vegetation cover; the depth, density, and duration of snow cover; the soil and rock type; and ground moisture conditions.

Mass wasting processes that operate in the active layer include frost creep, gelifluction (solifluction), cryoturbation, and certain forms of rapid mass movement (Egginton, 1986). Most patterned ground phenomena occur in the active layer or, where permafrost is absent, in the zone of seasonal frost.

Frost creep is the ratchet-like downslope movement of particles as the result of the frost heaving of the ground and the subsequent settling upon thawing — the heaving being predominately normal to the slope and the settling more nearly vertical. Movement associated with frost creep decreases from the surface downwards and depends upon the frequency of freeze-thaw cycles, angle of slope, moisture available for heave, and frost susceptibility of soil (this chapter, Carson and Bovis, 1989). Gelifluction is a form of solifluction occurring in areas underlain by permafrost. In contrast to gelifluction, solifluction is a more general term for the mass wasting typical of cold subarctic regions not underlain by permafrost. Both processes are faster than soil creep, often of the order of 0.5 to 10 cm/a (French, 1974b; Rampton and Dugal, 1974; Dyke, 1981; Egginton and French, 1985). In Canada, conditions suitable for gelifluction occur in areas where downward percolation of water through the soil is limited by the permafrost table and where the melt of segregated ice lenses provides excess water which reduces internal friction and cohesion in the soil. Particularly favoured sites include areas beneath or downslope of late-lying snowbanks. Features produced by both solifluction and gelifluction include sheets of locally derived surficial materials, tongue-shaped lobes (Fig. 9.26), and alternating stripes of coarse and fine sediment.

Cryoturbation is the term used to refer to the lateral and vertical displacement of soil which accompanies seasonal and/or diurnal freezing and thawing. This process is associated with the formation of patterned ground phenomena, the most common of which is the nonsorted circle or hummock. These are ubiquitous wherever fine grained sediments are present, and are particularly widespread in Mackenzie Valley and Keewatin. At Inuvik, Northwest Territories, for example, hummocks are composed of fine grained, frost-sensitive soils and are typically 1 to 2 m in diameter and 30 to 50 cm high. Beneath the hummock, the late summer frost table is bowl-shaped, and the hummocks grade from those which are completely vegetated (earth hummocks) to those with bare centres (mud hummocks). The mound form has traditionally been attributed to an upward displacement of material resulting from cryostatic (that is, freeze-back) pressures generated in a confined, wet unfrozen pocket in the active layer. The existence of substantial cryostatic pressures in the field, however, has yet to be convincingly demonstrated (Mackay and MacKay, 1976). On more general grounds, it can be argued that the presence of voids in the soil, the occurrence of frost cracks in the winter, and the weakness of the confining soil layers lying above prevent pressures of any magnitude from forming. Moreover, on theoretical grounds, cryostatic pressures should not develop in a frost sensitive soil hummock because ice lensing at the top and/or bottom of the active layer will desiccate the last unfrozen pocket so that the pore water is under tension, not under pressure. Mackay (1979b, 1980) concluded that the upward displacement of material is caused by the freeze and thaw of ice lenses at the top and bottom of the active layer, with a gravity induced, cell-like movement. The latter occurs because the top and bottom of the freeze-thaw zones have opposite curvatures (Fig. 9.27). Evidence of cell-like circulation is deduced from the grain size distribution of the hummocky soils, radiocarbon dating of organic materials intruded into the hummock centres from the sides, and from upward moving tongues of saturated soil observable in late summer (Zoltai and Tarnocai, 1974; Zoltai et al., 1978).

In the Mackenzie Delta area, average rates of movement on hillslopes possessing numerous vegetation-covered hummocks are approximately 1 cm/a (Mackay,

Figure 9.26. The toe of a gelifluction lobe, near Holman, Victoria Island, Northwest Territories. The front of the lobe is about 1 m in height.

1981). Contrary to most other studies of mass wasting, which indicate that movement decreases with depth (Washburn, 1979, p. 200-213), the movement of hummocks is convex downslope or plug-like. Movement progressively buries inter-hummock peat to form a buried organic layer. Most of this movement is attributed to creep which is promoted by the thaw of an ice-rich layer at the base of the active layer. This ice-rich layer is thought to form by upfreezing in winter, and the ice content may be augmented by ice lensing in the summer thaw period.

A somewhat different mechanism has been suggested (Shilts, 1978; Egginton and Shilts, 1978; Dyke and Zoltai, 1980; Egginton and Dyke, 1982) for similar features which occur widely in District of Keewatin but upon poorly sorted sediments (muds) with a significant silt or clay content. These are termed mudboils and consist of round to elongate patches of bare soil 1 to 3 m in diameter. Natural moisture contents are near the liquid limits, so that the muds liquify and flow readily in response to slight changes in moisture content or slight internal or external stresses. When the stresses cannot be relieved by downslope movement, mud may burst through a semi-rigid surface layer to create a mudboil. Driving pressures for such diapirism commonly result from hydrostatic or artesian pressures on slopes and from excess pore water pressures created by rain and/or thawing ice lenses. Deformed structures resulting from liquefaction and density re-adjustments in the active layer have also been described from Banks Island and have been termed periglacial involutions (French, 1986).

Rapid mass movement may also occur within the active layer with the permafrost table acting as a lubricated slip plane and controlling the depth of the failure plane (Fig. 9.28). These skin flows or detachment failures are usually attributed to local conditions of soil moisture saturation and high pore water pressure, commonly the result of thaw consolidation (McRoberts and Morgenstern, 1974a; this chapter, Carson and Bovis, 1989). In some cases, failure takes the form of a mudflow, in others, a distinct slump scar or hollow is formed. In several of the arctic islands, where the active layer is thin and fine grained and ice-rich sediments are widespread, active layer failures are both common and difficult to predict (Barnett et al., 1977; Hodgson, 1982; Stangl et al., 1982). They also occur in the boreal forest zone of the Mackenzie Valley lowlands, especially along river banks and on ice-rich colluvial slopes. It is not uncommon for failures to develop following the destruction of vegetation by forest fire (Zoltai and Pettapiece, 1973).

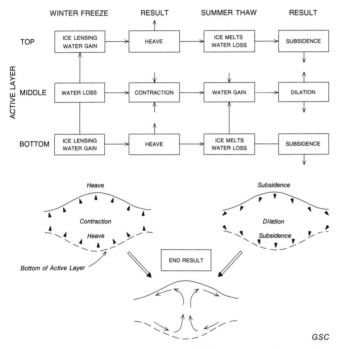

Figure 9.27. The equilibrium model for hummock growth (from Mackay, 1980).

Figure 9.28. Skin flows and multiple mudflows, Masik Valley, southern Banks Island. These rapid mass movements occur in summer and are confined to the active layer.

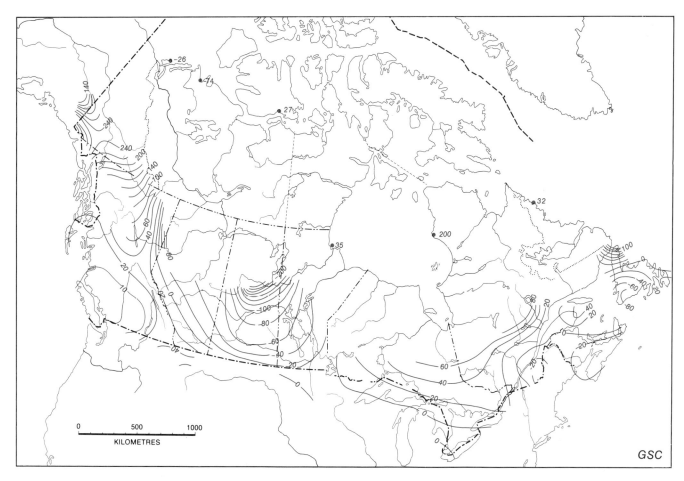

Figure 9.29. General areas of uplift (positive values) and subsidence (negative values) in Canada (after Vaníček and Nagy, 1981).

OTHER GEOMORPHIC PROCESSES
Geomorphic response to endogenic processes
S.G. Evans

Endogenic processes are important controls on the type and rate of exogenic geomorphic processes active in the development of the Canadian landscape. As discussed here these processes include vertical tectonic crustal strain (uplift, tilting, and subsidence), the effects of earthquakes and the modification of existing relief by volcanic activity.

These phenomena may affect the operation of geomorphic processes in several ways. Uplift, subsidence, and tilting result in changes in topography and in regional and local material transport gradients, and contribute to the morphological instability of river systems. Earthquakes result in instantaneous increases in local denudation by mass movement and the subsequent entrainment of large volumes of material into the sediment transfer system. They may also locally increase transport gradients through fault scarp generation. Volcanic activity also produces increases in local transport gradients by the geologically relatively rapid creation of volcanic landforms, which are unstable positive relief features. These become catastrophically unstable during both eruptive and dormant phases. Volcanic activity may also substantially modify energy gradients within a drainage system by the diversion or damming of river channels.

Existing data on endogenic processes in Canada are limited due to the short length of the observation record and the lack of comprehensive studies in remote areas. The effects of endogenic processes are spatially variable in Canada and within the various geological regions. They are most important in the Cordilleran denudation system which, because of its tectonic setting, is subject to rapid uplift, recent volcanism, and frequent earthquakes. A disparity between both the patterns and rates of uplift and denudation, however, is noted. The response of geomorphic systems to the endogenic energy input is a complex function of antecedent conditions, climate, geology, and vegetation. Because of this, comparable endogenic energy inputs may not have comparable geomorphic responses.

Evans, S.G.
1989: Geomorphic response to endogenic processes; *in* Chapter 9 of Quaternary Geology of Canada and Greenland, R.J. Fulton (ed.); Geological Survey of Canada, Geology of Canada, no. 1 (*also* Geological Society of America, The Geology of North America, v. K-1).

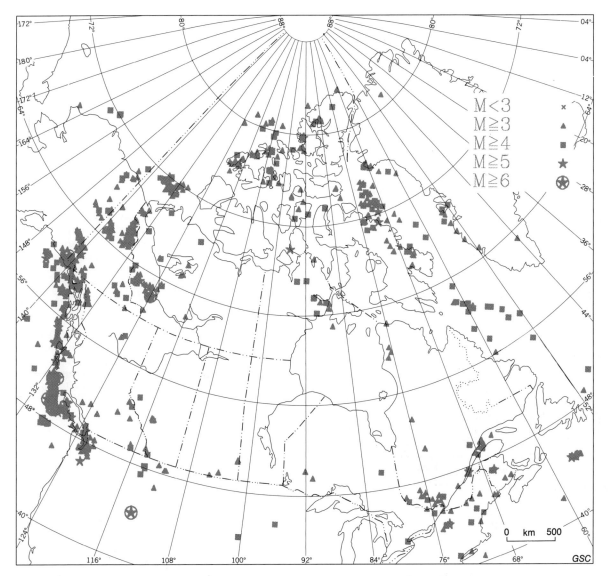

Figure 9.30. Distribution of historic earthquake epicentres in Canada, of magnitude 3 or greater. Courtesy of Geophysics Division, Geological Survey of Canada.

Vertical tectonic crustal movements

Data on present vertical tectonic crustal movements in Canada are limited in space and time. Vaníček and Nagy (1981) presented a partial map of contemporary vertical crustal movements which was constructed on the basis of relevelling surveys. General areas of uplift and subsidence reproduced from this map are given in Fig. 9.29 Note that it is not possible to distinguish vertical crustal movements due to tectonic processes and those due to glacial unloading. The pattern of movement is spatially variable within Canada and within geological regions. Areas of uplift include the Cordillera, the Canadian Shield and parts of the Atlantic Provinces whilst subsidence is seen in the Interior Plains, the southwest rim of the Canadian Shield, and small areas of Atlantic Provinces.

The relationship between uplift and denudation may be examined with reference to the Cordillera. A maximum rate of uplift of 24 mm/a is estimated for the southwest part of the Yukon Territory decreasing to 1 mm/a in southwest British Columbia (Vaníček and Nagy, 1981). These rates may be compared to regional estimates of mean annual denudation made by Slaymaker and McPherson (1977) which were based on total sediment loads of rivers with basin areas in excess of 25 000 km^2. Areas of maximum denudation do not correspond to areas of maximum uplift within the Cordillera and there is an order of magnitude difference between the rate of maximum denudation (less than 1 mm/a) and the rate of maximum uplift (24 mm/a). The existing data, therefore, suggest a considerable disparity between present rates of regional denudation and uplift in the Cordillera, although an unknown amount of this uplift may be due to glacial unloading effects.

With reference to local effects, areas of subsiding coasts may be expected to experience an increase in coastal landslide activity whilst uplift-generated downcutting may be expected to result in an increase in the incidence of landsliding due to slope steepening. Data are not available, however, to confirm these effects.

Earthquakes

The distribution of historic earthquake epicentres in Canada (Fig. 9.30) shows several distinctive clusters, including: the Pacific offshore, southwest British Columbia, the Mackenzie Mountains, the north coast of Baffin Island, and the St. Lawrence Lowlands. Since many areas of high seismic activity are remote from human habitation, the response of exogenic processes to seismic inputs has not been well documented with the exception of the St. Lawrence Lowlands, southwestern British Columbia, and the Nahanni region of the Northwest Territories.

In the St. Lawrence Lowlands, landslides produced by earthquakes have been discussed by Hodgson (1927, 1950), Smith (1962), and Desjardins (1980). By far the most damaging historic event was the 1663 earthquake (MMI = X)[1]. Historical records kept by the Jesuits, and quoted by Hodgson (1950) and Smith (1962), indicate that the earthquake was accompanied by large and extensive landslides in Champlain Sea deposits which devastated areas along Saint-Maurice, Batiscan, and St. Lawrence rivers. Desjardins (1980) obtained radiocarbon dates from large landslide sites on the Saint-Maurice River, near Shawinigan, which bracketed the 1663 event. The village of Les Éboulements, Quebec apparently received its name from landslides associated with the earthquake (Smith, 1962) and a contemporary account reported that the rivers of the region "ran white with mud the following summer" (Hodgson, 1950). The exogenic response to the 1663 seismic event was particularly marked due to the existence of adverse antecedent soil moisture conditions. According to Smith (1962) several comparable earthquakes have occurred in the St. Lawrence Lowlands since 1663 (for example, 1732, 1791, 1860, 1870, 1925, 1935) but significant landsliding has only been reported associated with the 1870 event. This earthquake (MMI = IX) caused a large rockfall at Cape Trinity on Rivière Saguenay (Hodgson, 1950).

In western Canada high magnitude earthquakes have occurred frequently in historic times (Milne, 1956; Milne et al., 1978; Wetmiller et al., 1987). Since they have generally affected remote, uninhabited areas, it has not been possible to document their effect on exogenic processes, apart from isolated landslides (Milne, 1956). The best known is the 1946 Vancouver Island earthquake (M = 7.2)[2]. Landslides (including submarine landslides) were noted at many locations and have been described by Hodgson (1946), Mathews (1979), and Rogers (1980). The landslide "print" of this earthquake is seen in Figure 9.31 and constitutes the most complete record to date of the exogenic response to a Canadian earthquake.

In the northern Cordillera, Eisbacher (1979a,b) and Clague (1981) have noted the association between the location of large rockslides of unknown age and historic earthquake epicentres in the Mackenzie Mountains and St. Elias Mountains, respectively. Although this association would suggest a direct relationship between seismic activity and landslide events, many large historic earthquakes in the vicinity of these mountain ranges are not known to have generated significant large rockslides. Along the Denali Fault, Clague (1979) noted mountain side scarps and "sackung" features which he attributed directly to earthquake shaking.

The October 1985 Nahanni earthquake (M = 6.6) (Wetmiller et al., 1987) triggered a large rock slide (Fig. 9.32) which involved $5-7 \times 10^6$ m³ of Paleozoic limestone (Evans et al., 1987). This is one of few landslides in the Cordillera known to have been triggered by a seismic event.

The 1965 Hope slide (Mathews and McTaggart, 1978) was apparently coincident with a seismic event. The landslide, which had a volume of 47.3×10^6 m³, occurred approximately three hours after a small earthquake (M = 3.2) and coincided within minutes with a second earthquake of similar magnitude. The epicentre of the earthquake was estimated to be directly beneath the landslide site (Mathews and McTaggart, 1978). The fact that the slope had withstood the greater seismic force of several previous historic earthquakes indicates that antecedent conditions within the slope controlled its response to seismic energy inputs or that the characteristics of the earthquake shaking (for example orientation, duration, and frequency) were important.

Volcanic processes

Volcanic processes increase the potential energy within a denudation system through the construction of positive relief features. They also result in the damming or diversion of rivers, as described in the Cordillera by Sutherland-Brown (1969) and Souther (1981). Quaternary volcanic activity in

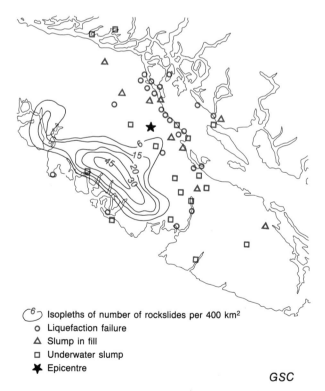

Figure 9.31. Distribution of landslides triggered by the 1946 Vancouver Island earthquake, based on maps in Mathews (1979) and Rogers (1980). Isopleths of rockslide frequency were constructed from the numbers of fresh scars per 400 km² grid square as reported by Mathews (1979).

---

[1] MMI = Maximum Mercalli Intensity (values quoted are for the intensity near the epicentre).

[2] M = Richter Magnitude

Figure 9.32. Aerial photograph of rock slide triggered by the October 1985 Nahanni earthquake. NAPL A27018-266.

Canada is confined to the western and central Cordillera. Over 150 eruptive centres are localized along four main linear belts (Fig. 9.33) — the Wrangell, Stikine, Anahim, and Garibaldi volcanic belts (Souther, 1975). Although no recent eruptions have been documented in the Canadian Cordillera, volcanism was active in the Stikine Volcanic Belt at Aiyansh as recently as about 1760 (Sutherland-Brown, 1969) and at Mt. Edziza, which has erupted several times within the last 2 ka (Souther, 1981).

Volcanic piles are characterized by unstable denudation surfaces that are susceptible to rapid erosion by flowing water and catastrophic mass movement processes. These surfaces are steep slopes which consist of low strength volcanic debris. The slopes are subject to local seismic tremors associated with volcanism and water levels within the slopes may be affected by the thawing of snow or ice caps due to geothermal or climatic processes, as well as by deuteric fluids.

These characteristics give rise to massive slope movements. Examples of rock avalanches or debris flows that have occurred since 1855 in the vicinity of eruptive centres within the Garibaldi volcanic belt have been described by Patton (1976), Mokievsky-Zubok (1977), Moore and Mathews (1978), Clague and Souther (1982), and Evans (1987). The volume involved in these movements varies between 0.5 and 12×10^6 m^3 and as such they represent massive instantaneous mass transfer events during which large volumes of debris are supplied on a recurrent basis to fluvial systems surrounding the volcanic centres. These mass movement events may be expected to change drastically the sediment yield characteristics of the fluvial systems which they affect (e.g., Lehre et al., 1983).

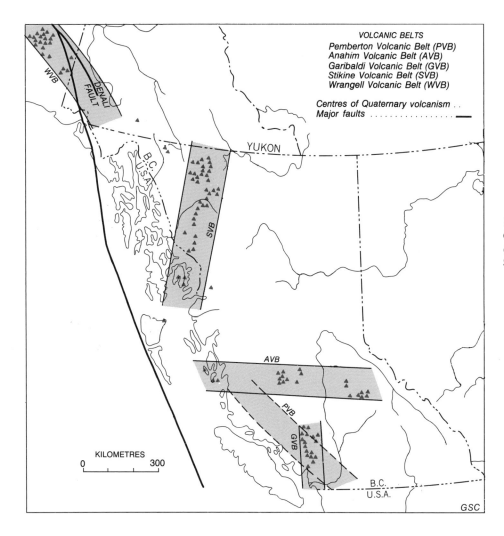

Figure 9.33. Centres of Quaternary volcanism in western Canada (modified after Souther, 1975).

Geomorphology of lakes

R.A. Klassen

The principal physical characteristics of lakes, such as form and bathymetry, shoreline features, composition and availability of sediments, and lake dynamics are strongly influenced by the geological origins and histories of their basins. Most Canadian lakes occupy basins of glacial, thaw, and fluvial origins (Zumberge, 1952; Frey, 1963; Reeves, 1968; Environment Canada, 1973). Other categories include those within structural basins, formed by movements of the earth's crust; mass movement basins created by landslide deposition; volcanic basins; and shoreline basins created in back-beach environments by shoreline deposition.

Bedrock basins of nonglacial origin may have formed originally as a result of ancient fluvial erosion (Ambrose, 1964), solution collapse (Christiansen, 1971), or meteorite impact (Dence et al., 1968). They can retain major aspects of their original morphology, despite glaciation. Other basins are associated with particular rock types or structural features, such as fault zones, that have been preferentially eroded or modified by wind, water or ice; the shape and distribution of lakes that occupy these erosional basins can follow the surface expression of the geological unit or structure.

Lakes that originated by glaciation occupy basins that have been formed or modified by glacier ice through either erosion of bedrock or drift, or through deposition of glacial sediments; in Canada they are pre-eminent in distribution and abundance. The characteristics of lakes in glaciated terrain are varied and depend largely on underlying bedrock types and the kinds of glacial sediments and landforms associated with the lake basins.

Lakes that formed by glacial deposition include those dammed by drift within bedrock basins of whatever origin and those ponded among glacial landforms such as drumlins, ribbed moraine, and hummocky moraine. The orientation and shapes of lakes in glaciated areas are related strongly to the surrounding glacial landforms. Those in deposits formed by actively flowing ice (e.g., flutes, drumlins) are generally elongate parallel to the direction of ice flow and have smooth, straight shorelines. Lakes related to ice

---

Klassen, R.A.
1989: Geomorphology of lakes; in Chapter 9 of Quaternary Geology of Canada and Greenland, R.J. Fulton (ed.); Geological Survey of Canada, Geology of Canada, no. 1 (also Geological Society of America, The Geology of North America, v. K-1).

marginal deposits and features formed transverse to the direction of ice flow (for example ribbed moraine) tend to be elongate perpendicular to the direction of ice flow, irregular in shape, and include numerous elongate islands and shoals. Lakes in areas of disintegration deposits (e.g., hummocky moraine, crevasse fillings) are irregular to complex in orientation and shape, and can include many islands.

Thawing of ground ice in permafrost regions and melting of buried glacier ice, with the collapse of overlying sediments, are analogous processes. The meltout of buried glacier ice results in kettle lakes that are generally small, rounded in shape, and lack preferred orientation. They are commonly developed in areas of ablation drift and within esker systems and outwash plains where they can occur singly or in groups. Kettle lakes characterize large areas of the prairie region. In regions of permafrost, thaw lakes or thermokarst lakes, formed by meltout of ground ice, are common and characterize the Mackenzie Delta (Mackay, 1956, 1963b), parts of southern Baffin Island (Bird, 1967), Banks Island (French and Harry, 1983; Harry and French, 1983), and other areas of extensive, fine grained deposits. Thermokarst lakes are typically round and shallow (less than 3 m deep) and can change configuration and migrate laterally through thermal degradation of their shores. Some have a preferred long axis orientation due to the influence of wind-generated circulation on shoreline erosion (Washburn, 1979, p. 271-273; Carter et al., 1987).

Lake basins originating by fluvial action include those formed in abandoned channels (e.g., oxbow lakes, Mackay, 1963b), by deposition on floodplains (e.g., flood basins), and by construction of fans or other alluvial deposits obstructing a valley (Ashley, 1979). Some lakes occupy abandoned meltwater channels (Last, 1984) and plunge pools (Mackay and Mathews, 1973) that were eroded by glaciofluvial action during deglaciation.

Lakes change with time as a result of (1) accumulation of clastic sediments and organic materials within the basins, (2) shoreline modification by mass movement or by the action of waves or ice, (3) erosion or deposition at outlet sites, (4) change in water inflow/outflow or in the position of the groundwater table, and in areas of permafrost, (5) thermal degradation of ground ice and (6) periglacial processes. Changes in lake levels can result from erosion of outlets, from deposition by landslides, by dam construction, by glacial advance (Clague and Rampton, 1982), and by isostatic rebound. Such changes occur at rates that vary from gradual to rapid and can be periodic. Isostatic rebound has caused the basins of the Great Lakes to tilt resulting in new outlets and higher lake levels during postglacial time (Sly and Thomas, 1974).

Floods, sometimes catastrophic, can result from the breaching of natural dams and the sudden drainage of a lake. Recent examples in Canada have included the failure of a moraine dam at Klattasine Creek, British Columbia, in ca. 1972 (Clague et al., 1985), a snow dam on Melville Island, Northwest Territories in July 1979 (Heginbottom, 1984), a landslide dam in the Squamish River valley in June 1984 (Evans, 1986), and a glacier dam on Noeick River, British Columbia, in October 1984 (Jones et al., 1985). Lake drainage can also result from coastal retreat, changes in erosional base level, the migration of river channels across a floodplain, and in permafrost areas, thermokarst activity and tapping by headward erosion along ice wedges (Harry and French, 1983).

Within lakes, erosion and transport of coarse debris are confined principally to nearshore and shallow water areas, and fine materials are transported towards deeper water in suspension, by turbidity currents and by ice rafting. The surficial geology of the lake basin can influence the rates of shore erosion and clastic sedimentation (Gelinas and Quigley, 1973), as well as lake water chemistry (Ryder, 1964; Coker and Shilts, 1979). Consequently glacial history is an important control of physical and chemical characteristics of lakes and the regional occurrence of some lake water properties.

In addition to wave-cut landforms, distinctive shoreline structures such as ice push ridges, cuspate mounds, and pot holes can be produced by the onshore movement of lake ice driven by wind or by thermal expansion forces (Dionne and Laverdière, 1972; Washburn, 1979, p. 254-257). In arctic regions lake ice can also limit wave action and rates of shore erosion during summer periods (Coakley and Rust, 1968). In regions characterized by permafrost, periglacial features such as rib-and-trough structures and polygonal ground patterns can characterize shallow (<2.5 m) water areas (Dionne, 1974, 1978a; Shilts and Dean, 1975).

In Canada little stratigraphic study has been made of the Quaternary sediments contained within existing lake basins. In general it is assumed sediments predating the last glaciation have been removed or destroyed. Sly and Prior (1984), however, reported the occurrence of glacially overridden sediments in the basin of Lake Ontario. In addition, some large valleys in British Columbia (for example, the Okanagan) contain deposits of several lacustrine fill cycles predating the last glaciation, apparently preserved because ice was not able to sweep the valleys clean to bedrock (Fulton, 1972). The vast majority of Canadian lakes were present at the time of deglaciation and consequently their basins contain glacial, glaciofluvial, and glacial lake sediments related to the last deglacial cycle. In southern parts of the Canadian Shield, where these deposits have been studied, Quaternary stratigraphic successions beneath lakes can be thick (from 10 m to as much as 30 m or more) and complex, and can include deposits of unstratified drift that are overlain by massive to laminated sedimentary sequences deposited in either glacial lakes or a marine environment (Shilts et al., 1976; Klassen and Shilts, 1982; Shilts, 1984). Such thick accumulations of sediment are not necessarily indicated by surficial deposits within the drainage basin. An organic-rich blanket of modern lake sediment over the older sediments represents sedimentary accumulation since deglaciation.

Solution processes

D.C. Ford

In descending order of solubility, the significant karst rocks are salt, gypsum (including anhydrite), limestone, and dolomite. The three latter outcrop over approximately 1.25×10^6 km^2 of Canada, occurring in every geological and

---

Ford, D.C.
1989: Solution processes; in Chapter 9 of Quaternary Geology of Canada and Greenland, R.J. Fulton (ed.); Geological Survey of Canada, Geology of Canada, no. 1 (also Geological Society of America, The Geology of North America, v. K-1).

topographic province. Thick salt is part of the stratigraphic succession beneath 500×10^3 km² of the prairies and significant areas also occur in southern Ontario and the Atlantic Provinces (Fig. 9.34).

Salt and gypsum dissolve by molecular dissociation in water. There is no dynamic threshold for the process and only minor temperature dependence. Solution rates are a function of available runoff or groundwater supply. Salt is so soluble that it does not survive at the surface in Canada. Stoping from dissolution at depth has produced deep breccia pipes in Saskatchewan (Christiansen, 1967); some of which have created shallow surface depressions. It appears that there was a peak of solution activity during Late Wisconsinan deglaciation.

Gypsum karst is best represented by large collapse sinkholes propagated through overlying carbonate rocks. Several thousand occur discontinuously in a belt extending from Wood Buffalo Park and Pine Point through the Franklin Mountains and Colville Hills to the Horton Plain in the Mackenzie Valley (van Everdingen, 1981). The most northerly examples of these are developing in the continuous permafrost zone. A majority are probably postglacial, but reactivated interglacial examples exist and possibly some preglacial ones. Smaller sinkhole karsts with interstadial and interglacial ones occur in Newfoundland and Nova Scotia. Very rugged pinnacle karsts of apparent Holocene age also occur in these provinces.

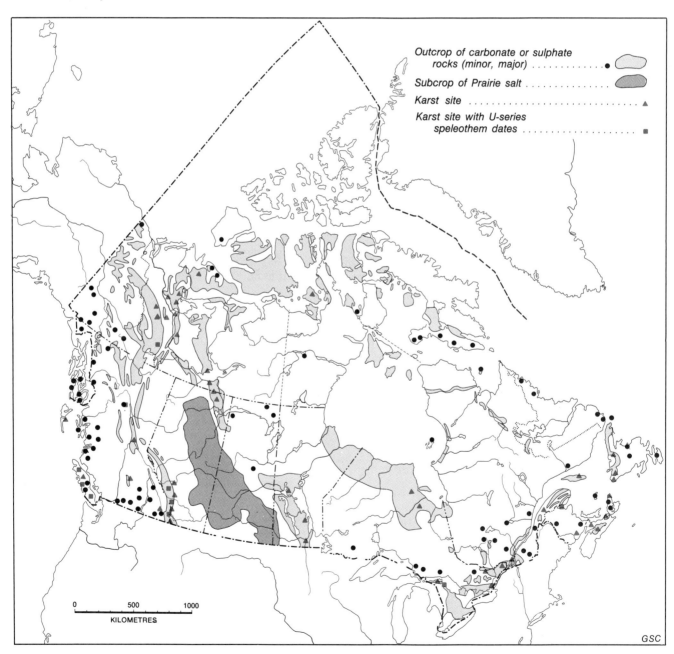

Figure 9.34. Distribution of soluble rocks and karst sites in Canada.

Limestone and dolomite are less soluble. Rates of solution are regulated primarily by supply of carbon dioxide to form carbonic acid. Drake and Ford (1981) have shown that concentrations of dissolved solids in most of Canada are similar to those of warmer humid regions; solution rates are then decided by mean annual runoff. Rates are often lower above the alpine treeline and are notably depleted in some subglacial waters (Ford, 1971). Limestone karst landforms and caves are widespread in mainland Canada, including some of the world's finest examples of alpine and subarctic karst. Development in the northern islands is restricted by permafrost and frost shattering (Bird, 1967). Tracts of dolomite karst are possibly more extensive than in any other nation.

Regional karst development

Rugged, ice-modified sinkhole terrains occur in western Newfoundland. Carbonate platforms dominate the St. Lawrence Lowlands and the Interlake region of Manitoba. They display extensive solution pavements with minor sinkholes and small caves (Ford, 1983a). A majority of features are Holocene but some are older. Speleothems in Saint-Elzéar Cave, Gaspésie are dated to 280 ka (Roberge and Gascoyne, 1978).

There is a great variety of karst in the southern Rocky Mountains: Crowsnest Pass displays preglacial cave systems that are dissected by repeated cirque cutting. Castleguard karst (Banff Park) is a benchland with sinkholes and solution pavement that extends beneath the Columbia Icefield, central portions of which are drained via deep caves. A shallower cave, Castleguard Cave, is Canada's longest explored system. It is more than 1 million years old and postdates the first glaciation of the region (Ford, 1983b).

The greatest density of caves is found in the mountains of northern Vancouver Island. Study is only just beginning; some stalagmites are dated to the last interglaciation (oxygen isotope stage 5e) but no older ones are yet known.

The most remarkable karstlands occur in limestones and dolomites of the eastern Mackenzie Mountains between 61° and 65°N. The only studies are of the Nahanni karst (Brook and Ford, 1978; Schroeder, 1979). This area has been glaciated, but not for more than 300 ka. It displays a labyrinth of solution corridors, large sinkholes, poljes, natural bridges, and towers. There are many affinities to tropical karsts. It demonstrates that rock solution may proceed rapidly in northern interior regions, even where there is discontinuous permafrost today (Fig. 9.35).

Figure 9.35. The Col Karst, Nahanni North Karst, Mackenzie Mountains, Northwest Territories. The Col displays a high density of sinkholes; individuals are up to 40 m deep. Larger karst depressions, 1000 m or more in length, head on either side of the Col. In the background is a cover of shale. 204377-G

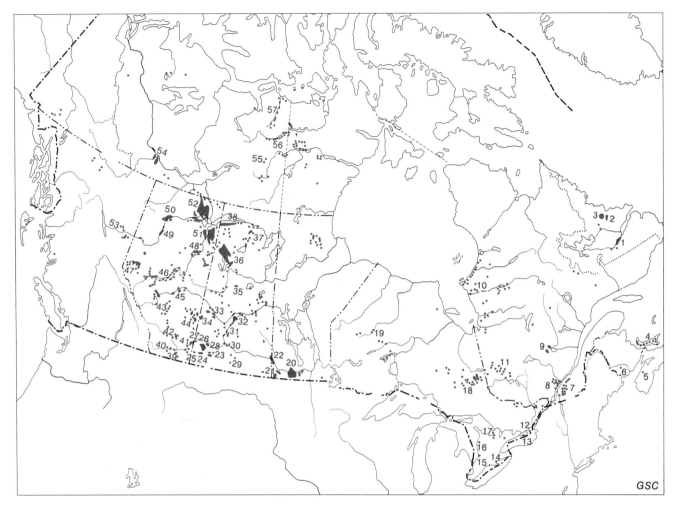

Figure 9.36. Sand dune occurrences in Canada. The number in parentheses following the province name indicates total area of dune occurrences greater than 25 km². The numbered areas are listed by name; the dots generally indicate areas smaller than 25 km². SH signifies recognized "Sand Hills" areas.

Eolian processes

Peter P. David

Eolian processes, varying in nature and intensity through time and in space, have been active in Canada since the retreat of the last ice sheet. Just as the latter was a diachronous process, so was the onset of the early phase of eolian activity in front of the retreating ice. Whereas the early phase was everywhere intense and not dependent so much on climate as on the availability of newly exposed, non-cohesive glacial sediments unprotected by vegetation, in those areas which were deglaciated prior to 10 ka, it did not generally last long because of the prevailing humid climate (David, 1982). Between about 10 ka and 5 ka, the dryer part of the Holocene, eolian activity continued with vigour (David, 1972) until there occurred a progressive stabilization of the eolian features by vegetation cover, caused by a gradual oscillating shift towards the more humid present-day climate. In the presently forested central regions, activity was generally limited to the early phase following deglaciation, whereas in the southern and northern marginal areas it continued into the second, more humid half of the Holocene. In fact, during the last 4 to 5 ka fluctuations in precipitation controlled the recurrence of eolian activities in the southern parkland areas (David, 1971, 1982), while variations in temperature appear to have controlled eolian processes near the northern forest limit (Filion, 1984; Filion and Morrisset, 1983).

The manifestations of former eolian activities in Canada are the wind-abraded bedrock surfaces and faceted pebbles (ventifacts) common in some parts of northern Canada (Tremblay, 1961; David, 1981b; Schreiner et al., 1981; Raupp and Argus, 1982) but less common in the south (Clark and Elson, 1961; David, 1977); the wind-aligned coulees of southern Alberta (Beaty, 1975) and the widespread occurrences of wind-eroded cliffs, both with their associated

David, P.P.
1989: Eolian processes; in Chapter 9 of Quaternary Geology of Canada and Greenland, R.J. Fulton (ed.); Geological Survey of Canada, Geology of Canada, no. 1 (also Geological Society of America, The Geology of North America, v. K-1).

Figure 9.36.

NEWFOUNDLAND (365 km²)
1. Churchill River SH
2. Kanairiktok River SH
3. Harp Lake SH

4. **PRINCE EDWARD ISLAND** (small, sporadic)
   North shore region

5. **NOVA SCOTIA** (small, sporadic)

6. **NEW BRUNSWICK** (small, sporadic)

QUEBEC (264 km²)
7. Drummondville region
8. Trois-Rivières region
9. Lac Saint-Jean region
10. Baie-de-James - Great Whale River regions
11. Val d'Or region

ONTRAIO (329 km²)
   Southern Ontario (small, sporadic)
12. Wolf Island SH
13. Sandbanks SH
14. Simcoe SH
15. Pinery SH
16. Sauble Beach SH
17. Wasaga SH

18. Kirkland Lake - Timmins regions
    McCool Creek SH
    Watabeg Lake SH
    Kennedy Lake SH

19. Western Ontario (small, sporadic)
    Whiteclay Lake SH
    Whitewater Lake SH
    Lonebreast Bay SH

MANITOBA (1693 km²)
20. Brandon SH
21. Oak Lake SH
22. St. Lazare SH

SASKATCHEWAN (6307 km²)
23. Seward and Antelope Lake SH
24. Bigstick Lake and Crane Lake SH
25. Tunstall SH
26. Great SH
27. Burstall and Westerham SH
28. Cramersburg SH
29. South-central Saskatchewan (small, sporadic)
30. Elbow and Birsay SH
31. Dundurn and Pike Lake SH
32. Prince Albert region
    Duck Lake SH
    Holbein SH
    Nisbet Forest SH
33. North Battleford SH
34. Manito Lake SH
35. Emmeline Lake SH
36. Agar Lake and Fortin River SH (sporadic dunes)
37. Cree River region
    Kearns Lake SH
    Lake Athabasca region
38. Athabasca SH

ALBERTA (8007 km²)
39. Pakowki Lake SH
40. Lethbridge region
    Grassy Lake SH
41. Middle and Hilda SH
42. Southwestern Alberta (small, sporadic)
43. Red Deer region (small, sporadic)
44. Wainwright region
    Sounding Lake SH
    Edgerton SH
    Buffalo Park SH
45. Fort Saskatchewan region
    Eastgate SH
    Redwater River SH
    Beaverhill Creek SH
    Ukalta SH
46. Edson - Athabasca regions
    Edson SH
    Windfall region SH
    Whitecourt SH
    Fish Lake SH
    Holmes Crossing SH
    Fort Assiniboine SH
    Nelson Lake SH
    Kilsyth Lake SH
    Chisholm SH
    Bruce Lake SH
    Hondon SH
    Decrene SH
47. Grande Prairie region
    Pinto Creek SH
    Wapiti and Pipestone Creek SH
    Bear River SH
    Grovedale SH
    Ellenwood Lake SH
    Economy Creek SH
    Watino SH
48. Fort McMurry region
    Algar River SH
    Fort McMurry SH
49. Carcajou region
    Cache Creek SH
    Wolverine River SH
50. Fort Vermilion and La Crête SH
51. Lake Athabasca region
    Ronald Lake SH
    Richardson River SH
    Old Fort Bay SH
52. Wood Buffalo Park SH (sporadic dunes)

BRITISH COLUMBIA (96 km²) (small, sporadic)
53. Fort St. John region
    Windy Creek SH
    Halfway River SH

NORTHERN CANADA (665 km²) (small, sporadic)
54. Fort Simpson region
    Fort Simpson SH
    Manners Creek SH
    Trail River SH
55. Thelon River region
56. Back River region
57. Ellice river region

eolian cliff-top deposits (David, 1970, 1972; Bouchard, 1974; Pissart et al., 1977); the common occurrences of sand dune areas throughout Canada (Fig. 9.36; David, 1977; Dionne, 1978b) and the associated loess deposits (Christiansen, 1959; David, 1977, 1981b; Dumanski et al., 1980; Catto, 1983); and, finally, the rare occurrences of volcanic tephra layers associated with other eolian sediments (David, 1970; Westgate, 1970; Dumanski et al., 1980).

At the present time, eolian processes are generally limited to the sporadic mobilization of a few sand dunes (David, 1977, 1979); the continued erosion of exposed cliffs everywhere and of exposed barren land surfaces in the arctic regions (Pissart et al., 1977); the deflation of newly exposed alluvial sediments (Kindle, 1952; Nickling, 1978), the periodic deflation of ploughed fields in the prairies (this chapter, Hewitt, 1989); and the transportation of tephra derived from recent volcanic eruptions (tephra from the recent eruption of Mount St. Helens, May 1981, was carried as far as Montréal, Quebec). An exceptional and highly enigmatic area in regard to its history of eolian activity is the

Athabasca dune area located south of Lake Athabasca in Saskatchewan and Alberta (David, 1977; Abouguendia, 1981; Raupp and Argus, 1982; Carson and MacLean, 1986), where eolian activity, which may have been occurring on a large scale since deglaciation, is not controlled by climate but by local groundwater conditions affecting large barren sand surfaces which show a variety of eolian landforms (Hermesh, 1972; M. Landals, internal report to Alberta Parks, Edmonton, 1978; David, 1981a; Fig. 9.37).

In addition to all the above, important eolian activity occurs during the long winter months when large quantities of snow are remobilized by the wind, directly affecting man's

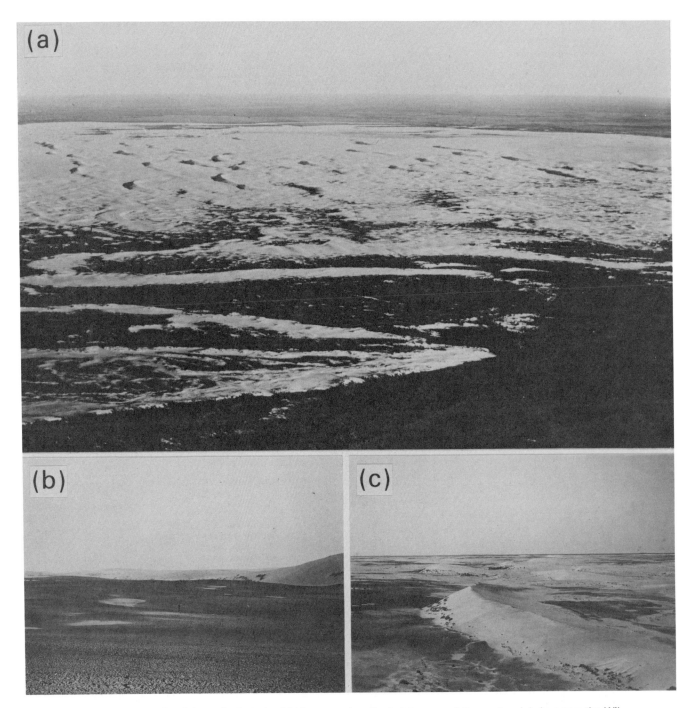

Figure 9.37. Sand dunes in the Lake Athabasca region, Saskatchewan. a) General aerial view over the Williams River sector, looking south. Dune migration is from right to left. b) Deflated segment in front of one of the giant dunes shown in (a), looking south. c) Close up aerial view of the giant dunes shown in (a), looking north. 204377-E, D, F

activity everywhere and indirectly affecting the local spring recharge of groundwater and early spring dune activities in certain areas (David, 1972). Eolian activity during the winter also leads to local accumulations of niveo-eolian deposits (Cailleux, 1972).

Animals as agents of erosion
P.A. Egginton

Almost all geomorphic processes are influenced by biological factors of one kind or another. In this section attention is drawn to a number of animals, other than man, which by various activities promote erosion. Animals may contribute directly to erosion by displacing material downslope, or they may contribute indirectly by altering the surface vegetation cover, exposing surficial materials to different or intensified process activities.

Indirect influence

Regions that are dominated by a single species or level of vegetation may be particularly vulnerable to the influence of animals. Various insects periodically threaten Canadian forests (Hiratsuka et al., 1981; Safranyik et al., 1981) and grasslands. For example, the mountain pine beetle over a few years may kill 90% of the trees in stands that it invades. The rapid loss of a forest cover may cause significant changes in evapotranspiration, snow accumulation, interception, surface erosion, and runoff (e.g., Bethlamy, 1974). This leads to increased sediment yields and problems of slope stability similar to those experienced in areas subject to clearcut logging practices (this chapter, Hewitt, 1989).

A variety of mammals such as bison, deer, geese, hare, and moose, through overgrazing, may destroy the vegetation cover or influence vegetation succession (e.g., Holsworth, 1980; Dionne, 1985). Similarly, various seed eaters, such as the deer mouse, may delay or prohibit "natural" regeneration of a forest (e.g., Williams, 1959; Pank, 1974). In these cases very high animal densities are required for a significant impact; this may occur naturally or result from an imposed reduction in range. Of course, the highest incidence of overgrazing and the attendant problems of increased soil erosion are due to the pasturing of domestic cattle and sheep. Trampling by large animals (for example, caribou, muskoxen, and reindeer) can substantially destroy the surface vegetation cover (Bos, 1967; Pegau, 1970; Bliss and Wein, 1971) exposing previously protected materials. In permafrost regions with high ice contents, this may promote thermokarst development (Mackay, 1970).

Direct influence

The impact may be more direct. Extensive trail systems, such as those developed by caribou, may channel surface runoff, further promoting erosion (Bee and Hall, 1956;

Wiggins and Thomas, 1962). If the surficial materials are "quick", loading and trampling by large animals may induce liquefaction and cause a downslope displacement of the surface muds (Egginton, 1979). On some slopes, terracettes and soil slippage result from repeated traverses by elk or sheep (Stelfax, 1971). The best examples of these occur, however, on slopes extensively grazed by domestic cattle. Terracettes can be produced in a variety of ways (Fairbridge, 1968), but where the terracettes end at a fence it can be stated unequivocally that they are related to the passage of animals. These features provide bare soil for erosion and induce erosion by channelling overland runoff.

A number of animals tunnel or burrow — they move material from depth to the surface, alter soil permeability expose what may be readily eroded materials, and displace material downslope. Grizzly and polar bears, at least in stature, are the giants amongst the earth movers. Anyone walking in grizzly country will have seen the craters excavated by bears digging for ground squirrels; earth is scattered in all directions. Polar and grizzly bear denning areas are equally impressive monuments to bears' earth moving capabilities. On the other hand, the quantity of earth moved by much smaller animals can be prodigious. Darwin (1881), in a classic work, found that earthworms may move more than 18 to 44 t/(ha·a). On Yukon slopes, ground squirrels may displace 18 t/(ha·a) (Price, 1971). Pocket gophers in alpine tundra zones also displace substantial quantities of earth (Stoeker, 1976; Thorn, 1980). Thorn (1978, 1982) suggested that because of the large quantities involved (up to 27.7 t/(ha·a), pocket gophers may be the dominant agent of erosion on vegetated alpine slopes which have a winter snow cover. Voles and moles also may be of significance; in the Ottawa area they raise between 14 and 28 t/(ha·a) (Egginton, 1985). The overturning of surficial materials and the construction of interconnected "macropores" may significantly alter infiltration rates (Beven and Germann, 1982). Green and Askew (1965) reported clay soils which, because they are colonized by ants, have hydraulic conductivities more typical of gravels.

Beaver, because of their distribution and habits, produce one of the most visible impacts on the Canadian land surface. Dams, built of mud and wood, impound several hectares of water for each beaver pond, which ultimately kills the surface vegetation. Where beaver populations are high and multiple dams are built, streams develop a stepped profile (Heede, 1975), water surface slope and velocity are reduced, and the discharge is spread over a longer time. Swamps are formed as the ponds silt up, producing beaver meadows and/or delta-like forms. The failure of beaver dams, which in deep channels may be up to 5 m high, produces localized floods and can cause extensive bed erosion (Retzer et al., 1956).

Generally the role of animals is overlooked by geomorphologists. In fact animals often dislodge or chew on markers and stakes and so interfere with the measurement of geomorphic process and, as van Zon (1981) suggested, they are viewed as nuisances rather than as geomorphological influences. Relatively high densities are required for animals to have a significant influence on rates of erosion. Detailed quantitative studies are generally lacking; however, it is clear that in areas where conditions are favourable, animals are not only significant but can be the dominant agents of erosion.

Egginton, P.A.
1989: Animals as agents of erosion; in Chapter 9 of Quaternary Geology of Canada and Greenland, R.J. Fulton (ed.); Geological Survey of Canada, Geology of Canada no. 1 (also Geological Society of America, The Geology of North America, v. K-1).

HUMAN SOCIETY AS A GEOLOGICAL AGENT

Kenneth Hewitt

Human societies act as geological agents in two substantial ways: they modify the action of natural processes and they deliberately redistribute and artificially transform earth materials. Less obvious, but analogous to the more profound changes in earth history, are chemical and biological changes to the environment made by man.

This section considers the role of human activities in the Quaternary, especially changes induced in the Canadian landscape and habitats. It is intended to compliment and round out the geoscience perspectives of preceding sections; however, a certain shift of emphasis and even style, is difficult to avoid. Why is this?

At the heart of earth science are the description and explanation of natural processes. These planetary conditions are assumed to occur and to obey physical laws with or without human presence. Their analysis is the main way earth science contributes to our understanding of the Canadian environment, albeit with a strong interest in how they enter national social and economic life. Moreover, the revolutionary role of geological ideas for modern thought was to conceive of the earth in terms of processes and especially time spans that dwarf human activities to insignificance. In the Quaternary, the last tiny slice of geological time, the signs of man are sparse and minor until the very end. The phenomena we shall consider are hardly discernable anywhere before about 10 ka. In Canada, most were scarcely present before the late nineteenth century. In addition, the very size of Canada, with extensive areas of the oldest rocks as well as the enormous legacy of Pleistocene glaciations, seems to require the majestic sweep of geological theory. Compelling as that may be, however, it is less useful or even misleading for the present purpose.

If one begins in the places where most Canadians live and economic activity is concentrated, substantial "manmade geology" is found. This is conspicuous in built environments and areas of intensive cultivation, forestry, and mining. Here, the landscape, the materials that surround us and the course of their development are largely shaped by human agency. In fact, our geological role is concentrated in and most directly affects the *habitable* earth: its fertile land surfaces and adjacent layers of atmosphere and hydrosphere. That brings a significance to our subject out of proportion to the overall geological scope. However, we will also discover that, recent as they are, when projected onto geological time scales, the impacts of today's industrialized processes upon landscape and earth materials do assume the magnitude of major "natural" processes.

The evidence from other industrial nations shows that total human removals, transport, and deposition of earth materials approach and even surpass natural processes. In Japan, one of the more geologically active natural environments, Kadomura (1980) found humanly induced erosion has exceeded that of all other surface processes in recent decades. A similar claim has been made for Great Britain since the 1920s (Vita-Finzi, 1977). In many regions, modification of erosion subcycles increased their rates (Table 9.5) by an order of magnitude or more. Data for a full appraisal of the Canadian scene are lacking. Nevertheless, the scale and scope of Canada's urban-industrial activities are not very different from, and in certain resource extractive industries are far larger than Japan's or Britain's. There is also widespread modification of the chemical and physical properties of land surfaces, surface waters, and the atmosphere. Industrial processes now reproduce rock-forming processes of all kinds. The resulting environments built of concrete, brick, plastics, metals, and glass surround an ever larger fraction of the human population and its activities.

Finally, some consequences of human activity might produce rapid and decisive changes in the evolution of the earth. It is believed that the buildup of carbon dioxide in the atmosphere, due to fossil fuel and vegetation burning and oxidation of newly exposed soils, may lead to climatic changes such as those identified in new geological periods and systems of rocks (Cohen, 1986). Acid precipitation caused by human activities threatens us with uncertain, but possibly disastrous, levels of pollution through the release of toxic elements presently bound up in otherwise fairly stable compounds and minerals. It has been argued that nuclear war, with only a modest fraction of the available weapons, could replicate the kind of geological and evolutionary catastrophe popularly identified with the extinction of the dinosaurs (Turco et al., 1983; Royal Society of Canada, 1985). It is thought that risks of comparable changes exist in environmental, chemical, and biological warfare agents (Westing, 1984).

Not surprisingly, therefore, society's role in geological processes often coincides with concern for the whole range of humanly induced changes in the environment. Unlike the objective view that accords with natural processes, here it is difficult to ignore the fact that human impacts ultimately derive from human decisions. In many ways the literature most directly bearing on our subject is in such fields as mining geology, civil engineering, or materials science: work that describes and prescribes how the geosphere is to be modified for human use. The bulk of recent efforts to actually assess the consequences of that use, however, emerge from the wave of environmental concern. It addresses such problem areas as soil erosion or acid rain, damage to "fragile" ecosystems or hazards research (Leggett, 1973; Dasmann, 1976; Environment Canada, 1983). Many relevant studies give as much weight to managerial issues, public awareness, and political responsibility, as to physical processes (Detwyler, 1971; Mitchell and Sewell, 1981). That also creates a different sense of purpose from the mainstream of earth science studies.

Having recognized these issues, it remains important to ask how, where, and on what terms human activities comprise objective geological agents. Thus we may attempt to define them in relation to the kinds of processes and time spans earth science has developed for "natural" conditions. Such would seem to be the appropriate concern of what has been called "anthropo-geomorphology" (Fels, 1965; Nir, 1983). Actually, what we must consider has, rather, the scope of an "anthropogeology", since it involves much more

---

Hewitt, K.
1989: Human society as a geological agent; in Chapter 9 of Quaternary Geology of Canada and Greenland, R.J. Fulton (ed.); Geological Survey of Canada, Geology of Canada, no. 1 (also Geological Society of America, The Geology of North America, v. K-1).

than geomorphic processes. Before turning to Canadian conditions themselves, we may review the scope of this field.

Impacts on natural processes

Human modifications of natural processes are the most extensive and involve the largest net change in the rates of geological processes. The greatest of these effects appears in interference with erosion cycles. Human land uses greatly affect the severity and distribution of erosion by wind, rain, rivers, waves, frost, or mass movement. In some cases, such as damming rivers and the building of groynes along beaches, terraced agriculture, or afforestation, society may impede and slow natural rates of erosion. For the most part, however, human activity has led to much increased rates (Table 9.5). The most widespread consequences have derived from changes in vegetation cover, notably the cutting of forests and burning of grasslands (Thomas, 1956).

In general, a reduced or less continuous vegetation cover increases the erosive potential of rain beat and sheet erosion, gullying and flood waters, wind, groundwater sapping, frost growth, and thaw. Each has relevance for much of southern Canada (Cailleux and Hamelin, 1969). On steeper slopes, especially in mountain lands or on coastal cliffs, in certain susceptible alluvial deposits and over permafrost, the role of rapid or large mass movements is commonly magnified by reduced vegetation cover.

Increases in the rate or severity of erosion are seen especially in increased transport of sediment and also of dissolved solids in rivers, the atmosphere, and seas. Accelerated erosion of farm, forest, or urbanized areas leads to more turbid and chemically enriched surface waters. The use of air, water, and land to dump wastes, or of natural cycles to bear them away, also appears as massive interferences in erosional cycles. It is most popularly identified as "pollution". Finally, all such materials introduced into or forced through natural pathways will be deposited or precipitated in a way similar to natural processes; this is seen in the silting up of waterways, reservoirs, and coastal areas, but also appears as the fall-out of acid rain, buildup of salts in irrigated lands, and of toxic wastes in soil, groundwater, or lakes (Fyfe, 1982).

Artificial geological cycles

The second concern derives from the development by human societies of controlled versions of every component of the geological cycle. There is a "man-made" erosion cycle. It comprises all those "entrainments" of earth materials involved in quarrying, mining, dredging, and construction sites. The "transport" component occurs as earth moving or the shipment of what is extracted from mines. The "deposition" component is seen in built structures, earthworks, landfill, and so forth. Meanwhile, technologies of great diversity simulate other parts of the rock cycle. These artificially produce rock-like substances or transform earth materials to new chemical and physical forms. There are many industrial processes that fuse or gasify rock solids. Others have the same basic characteristics as lithification and igneous and metamorphic processes in nature.

Concrete, cement, and mud brick structures, stucco or grouted foundations are examples of artificial lithification. In a sense, the builtup, paved areas of today constitute "sedimentary facies" of distinctive anthropogenic character and great size. The melting of ores and solidifying of smelted products are artificial "igneous cycles", as is the production of glass or abrasives such as silicon carbide and aluminum oxide. Fired ceramics, work-hardened metals, artificial diamonds, or the aureoles of transformed rock around underground nuclear explosions are artificial "metamorphics".

These could seem to be far-fetched analogies were not the levels of their industrial production and dissemination, and the extent to which most human populations are now surrounded by man-made "rocks", so evident. When the larger — sometimes orders of magnitude larger — wastes and by-products of this artificial rock-making are added, they approach or exceed the scope of natural biogeochemical cycles (Brown, 1970b; Coates, 1973). We must also recognize the chemical and physical changes brought about in agricultural and other lands by artificial fertilizers, biocides, vehicle compaction, warfare, and so forth; and the effects of chemical and radioactive fall-out, oil spills, and dumping on the chemistry of lakes and oceans. By such means, human activity has turned much of the surface of the planet into a more or less artificial one.

Anthropogenic regimes

Most studies relevant to our concerns deal either with the impact and problems of particular activities or impacts upon particular natural processes. As the range and intensity of human action grow, however, it becomes pervasive. There is a shift from environments and landscape dominated by natural conditions to ones that increasingly reflect human agency. The more intensively used and continuously settled areas will come under the sway of human or *anthropogenic regimes*. How do these compare, contrast, and articulate with natural regimes?

When a forest is cleared or when land is irrigated, drained or terraced, the net effect on earth surface processes resembles a more or less drastic climatic change. If the impacts are enduring ones, they assume a similar role to *morphogenetic* controls in nature: the tendency of prevailing climates and vegetation covers to produce characteristic sets of landforms and rates of change (Büdel, 1969).

When slopes are excavated or earthworks and spoil heaps produced or when land is paved over, streams channelled, or shorelines walled, human influence resembles that of distinctive rock types, tectonic events, or volcanism. Here we refer to *structural* controls, the endogenic and constructional forces affecting the earth's surface. Limestones, volcanic cones, and faulting help produce distinctive landforms or sequences of forms, soils, or stream chemistry — and so do humanly reconstructed surface forms and their materials.

Human activities also change the geography and pace of geological development. In many ways these are the more radical departures from nature. The forms of human land use, its settlements and communications, interact with and respond to natural conditions. The result is a very different spatial order, however, from that of either regional geology or climate and natural vegetation. The location of mines, forest industries, or power plants may reflect the natural resource base. But the timing and history of their operations, the scale and type of their solid, liquid, and gaseous products reflect the economic "climates". They respond to technological innovations and other essentially social phenomena. The organized systems for transporting materials

Table 9.5. Observed and estimated quantities of earth materials moved or eroded as a consequence of human activity

WORLD

Ranked estimates of all major land uses (Nir, 1983)
- Agriculture — $156\,800 \times 10^6$ t/a
- Grazing — $50\,000 \times 10^6$ t/a
- Building materials — $10\,000 \times 10^6$ t/a
- Other mining — $5\,000 \times 10^6$ t/a
- Forest clearance — 800×10^6 t/a
- Highways and railroads — 400×10^6 t/a
- Urbanization — 17×10^6 t/a

Total "anthropogeomorphological activities" (Nir, 1983) — $172\,000 \times 10^6$ t/a

Deliveries by rivers to the oceans (Judson, 1968; Nir, 1983)
- Total before human intervention — $9\,300 \times 10^6$ t/a
- Total after human intervention — $24\,000 \times 10^6$ t/a
- Mississippi River, USA — 450×10^6 t/a
- Hwang Ho, China (Yellow River) — $1\,500 \times 10^6$ t/a

Global emissions of airborne particulates (Butler, 1979)
- Natural — $20\,960 \times 10^6$ t/a
- "Man-made" — 269×10^6 t/a

Production of leading minerals (1977) (Nir, 1983)

	Production	Wastes
Coal	$2\,500 \times 10^6$	$2\,500 \times 10^6$ t/a
Brown coal	900×10^6	900×10^6 t/a
Iron	480×10^6	480×10^6 t/a
Copper	2×10^6	400×10^6 t/a
Gold	1	110×10^6 t/a
Bauxite	80×10^6	63×10^6 t/a
Totals	$3\,960 \times 10^6$	$4\,440 \times 10^6$ t/a

CANADA

Agriculture: soil erosion losses by crop types

Crop type	Mean soil erosion loss t/(ha·a)	Range of average soil loss for selected watersheds t/(ha·a)
Southern Ontario: potential sheet erosion (Dickinson and Wall, 1978)		
Horticultural crops (potatoes, tomatoes, etc.)	9.1	6.6 - 12.1
Beans (soy, white)	7.6	5.5 - 9.8
Continuous corn	6.7	2.9 - 11.7
Corn in rotation	3.7	0.9 - 6.9
Tobacco	3.5	2.1 - 4.9
Small grains	3.4	1.5 - 6.9
Meadow in rotation	2.6	0.9 - 5.0
Permanent pasture	0.4	0.1 - 0.8
Woodlands	0.2	0.05 - 0.4
Prince Edward Island (Coote, 1983)		
Potato lands		
-fallow, 7% slope	18.7	
-fallow, 12% slope	19.6	
Potatoes		
-up and down	41.1	
-12% slope	10.0	
Gullied area	65.0	

Soil replacement rate = 3.0 t/(ha·a).

New Brunswick (Coote, 1983)
- Potatoes — 42.0

Table 9.5. (cont.)

Nova Scotia (Coote, 1983)
    Cornfield, 9% slope   26.0

Quebec (near Lac Saint-Jean) (Coote, 1983)
    Hay   < 1.0
    Fallow (loam) 10% slope   > 56.6
    Bare fallow (gravelly loam) 15% slope   28.0
    Pasture (gravelly loam) 15% slope   60

Prairie Provinces (Coote, 1983). Areas of agricultural land subject to severe erosion (mainly wind)
    Moderate to severe 2.5×10^6 ha
    Less severe 3.7×10^6 ha

<u>Mining</u>: estimates of area disturbed by pits and quarries (Marshall, 1982)

All Canada (excluding Territories, 1982)   > 120 000 ha
Eastern Canada (active sites, 1977)   33 072 ha
    -Ontario and Quebec   30 832 ha
    -Ontario   20 916 ha
    -within 8 km of population centres   23 664 ha
Western Canada
    -Saskatchewan   10 926 ha
    -Alberta   9 090 ha
    -Manitoba (within 48 km of Winnipeg)   2 446 ha

<u>Nuclear waste production</u> (Redmond, 1983)
    -low level radioactive wastes (total to year 2000) 1.3×10^6 m³
    -tailings from uranium mills
        -total to 1977 907×10^6 t
        -production in 1978   16 145 t/day

<u>Sanitary Landfill</u> (Levings, 1982; Environment Canada, 1983)
    Canada (1970, estimated) 15.0×10^6 t/a
    Toronto (1981) 1.7×10^6 t

<u>Dredging spoils</u> (Levings, 1982)
    All Canada, marine coastal waters, 1979 19.6×10^6 m³
    Canadian Great Lakes, 1975-79
    - placed in confinement 1.4×10^6 m³/a
    - dumped in open water 0.9×10^6 m³/a
    Toronto Harbour, Leslie Street Spit 16.0×10^6 m³

<u>Airport construction</u> (Nir, 1983; Environment Canada, 1983)
    All Canada, total land used   98 000 ha
    Mirabel
    - total land expropriated   350 ha
    - operational land area   6 900 ha
    - total earth moved   500 000 m³

MISCELLANEOUS NON-CANADIAN (Nir, 1983)

Mining - USA
    - area disturbed by strip and surface mining, 1965 12.5×10^6 ha
    - new openings, per year   65 000 ha
Canals - materials initially moved
    - Panama 675×10^6 m³
    - Suez 110×10^6 m³
Highways - USA
    - all highways - materials excavated $70\,000 \times 10^6$ m³
    - new roads in 1972 - erosional yield 12.3×10^6 t
    - divided highways - sediment yield   1200 t/km²
Urbanization - USA
    - materials eroded from construction sites in Maryland   1 127 t/ha
    - erosion rates, Washington, D.C.
        - rural   400 t/km²
        - urbanized   55 000 t/km²

cut across natural pathways and units. Highways, pipelines, or shipping lanes create networks peculiar to human settlement and commerce. Neither can we ignore large scale rerouting of natural flows, of which interbasin river diversions are a notable development in Canada (Kellerhals et al., 1979).

Anthropogenic regimes also affect the pace of geological development, that is, they modify the types, magnitudes, and rates of given processes in given areas. They make the surface more erodible in one place, less so in another. Activities ranging from cultivation to mining increase the availability of sediment and soluble minerals to natural transport media. Effects, such as those upon slope stability, change the thresholds at which particular processes operate, for example, the quantities of rainfall or wind speeds required to initiate landslide or dune movement.

In such terms, anthropogenic regimes reshape the geological environments and direction of development over much of the earth's surface. In turn, they affect the broader biogeochemical cycles that interact with these surfaces. This applies especially to the places where the majority of mankind lives and works.

The Canadian scene

In keeping with the above discussion, we will examine the Canadian situation in terms of certain dominant types of land use. These approximate anthropogenic regimes as defined above and help suggest a broad, national perspective on the subject. In order of their areal extent, they are forest utilization, croplands, mines and artificial transforming of rock materials, and built environments. The actual intensity and completeness of human control upon earth surface conditions in these "regimes" are inversely related to areal extent. A complicating factor of note involves the reworked or artificial materials introduced into natural transport media, deliberately or inadvertently; these spread the impact of human activities far beyond what is identified in land use patterns.

Forest utilization

Forest clearance has been, perhaps, the single most extensive modification of the world's fertile lands and the precursor to most other human impacts (Darby, 1956; Nir, 1983). The more densely settled and much of the agricultural land of Canada was forested before European settlement and there is still a nation-wide pattern of forest, woodland, and wood-lot clearance to make way for agriculture, highways, power corridors, urbanization, and recreation. Locally it can be a powerful stimulus to erosion; in recent decades that has been counteracted by the areas of marginal farm and grazing land abandoned to forest encroachment. The more substantial and geographically extensive impacts, however, stem from harvesting in forest lands themselves. This is a key component of the entire Canadian economy; it now involves an annual cut of about 800 000 ha (Weetman, 1983, p. 284).

Almost invariably, timber harvesting produces a phase of sharply increased sediment and dissolved solids runoff (Bailey, 1971; Weetman and Webber, 1972; Swanston and Swanston, 1976; Martin and Pierce, 1980). This relates directly to decreased protection of the surface against rainbeat, wind, freeze-thaw and desiccation cycles (Honbeck et al., 1970; Goodell, 1972; Harr et al., 1975). It is associated with increased amounts and rates of runoff and, by far the most important factor, with concentrated erosion where the surface has been scarred, compacted, or cut up for roads and log extraction (Steinbrenner and Gessel, 1955).

The incidence or increase of particular geomorphic processes after logging, notably mass movements, depends upon the habitat as well as harvesting techniques involved. Dramatic increases in rapid mass movements have been reported from mountainous areas (Schwab and Watt, 1981; Evans, 1982). In a study of coastal mountain catchments of southern British Columbia, O'Loughlin (1972) reported a more than four-fold increase in average numbers of landslides and a more than two-fold increase of landslide debris on logged areas as against undisturbed areas (this chapter, Carson and Bovis, 1989). The most extreme case, however, involved 30 times more slides than adjacent unlogged slopes and about 13 500 t/km^2 of debris from clear cut areas, compared to some 280 t/km^2 for uncut areas. Flash floods and outburst floods after logging debris has dammed streams also increase erosion (Toews and Moore, 1982). Along with mass movements they constitute a serious hazard to human populations (Waltham, 1978, p. 65; Eisbacher and Clague, 1981).

Timber extracting practices and silviculture can make a profound difference to the geomorphic impact of logging (Maini and Carlisle, 1974). A high density of logging roads, especially in mountainous areas, tends to have a greater impact on erosion than the logged over area itself (Reid et al., 1981; Rice et al., 1979). It has been shown that selective cutting, the leaving of uncut strips of timber between clear-cut sections, especially along streams and roads, will greatly reduce sediment and dissolved load in streams (Cromack, et al., 1978; Environmental Associates (P.E.I.) Ltd., A report to Executive Council, Charlottetown, P.E.I., 1979; Martin and Pierce, 1980; Weetman, 1983). The industry, however, continues to cite economic arguments for extensive clear cutting; in 1970, clear cutting was involved in 83% of Canadian logging (Maini and Carlisle, 1974).

Erosional consequences of logging are not confined to the year of extraction. They may continue for years and perhaps decades afterwards. The rates of regeneration are critical here and include not only tree-cover itself, but soil and leaf litter, infiltration and moisture retention capacity, and repair of bare spots, deep scars, and access roads. We are only just beginning to understand the role of forest composition and ecology; these may be more decisive factors in the long term (Gordon, 1985). Until recovery is well advanced, runoff and erosion rates are likely to continue well above pre-cut levels. It is a problem further magnified by the limited amount of artificial restocking and rehabilitation of cut-over areas (Weetman, 1981). About 85% of logged areas in Canada are left to regenerate themselves (Maini and Carlisle, 1974) including 17% acknowledged to need silvicultural treatment. Here the most severe of impacts are anticipated.

Forest fires have the same dramatic impacts as clear cutting and in some cases have been linked with forest exploitation activities (Ashworth, 1986, p. 69). It is, of course, difficult to estimate the relative role of forest fires in nature, under pre-European settlement conditions when some deliberate burning occurred (White, 1972) and at present. Certainly, forest industries provide the economic incentive for spending millions of dollars each year to prevent or extinguish forest fires. The access made possible by forest exploitation has aided these endeavours. On the other hand, it can be argued that fire prevention may have ecologically detrimental results.

Other aspects, whose long term effects are hard to define, include the consequences of widespread single-age, single-species stands. There is also the net effect of harvesting upon forest land fertility, upon disease or pest infestations, and of extensive use of biocides to control the latter (Norris, 1981). It is easy to believe that the continuance of present practices will lead to irreversible landscape changes comparable to a marked shift of climate in nature.

In terms of sheer extent, forest impacts are the most substantial in Canada. Of a total area of 9.9×10^6 km^2, some 4.4×10^6 km^2 of the country is defined as forested, of which rather more than half is classed as potentially productive. The amount that is cut by major forest industries in a given year stands at about 800 000 ha. Forest fires consume about the same area in an average year and a comparable area is damaged by diseases and infestations. Losses to small scale cutting for firewood, clearance of woodland for agriculture, urban spread, highways and the like are an unknown quantity, but are probably in the tens of thousands of hectares. In sum, there are drastic changes to the surface environment of roughly 3 or 4% of Canada's mature forest land, and about 1% of the total forested area each year. This means that, at present rates, most of the Canada's forest land and some 20% of the nation's landmass, will undergo these modifications in about one century — a major impact by any standards of environmental change.

Croplands and agricultural practices

About 7.2% of Canada's land is classed as suitable for agriculture and some 5% is in use at present (Coote, 1983). If the amount seems small, its human significance is proportionally the greater. The areas of cultivation have been converted from pre-existing forest and grassland cover. Cultivation practices often simulate the surface conditions of semi-arid or seasonally dry climates, which in nature have the highest climatic rates of erosion (Langbein and Schumm, 1958). Cereal, legume, and vegetable crops are commonly produced in an annual cycle which leaves the ground bare for some weeks or months. This increases the erosive potential of frost heave, rain beat, sheetwash, rill and gully action, and wind. Fall ploughing, which means that ridged and frost heaved or desiccated soil is exposed to the full impact of the spring thaw and rains, greatly enhances soil losses (Anderson and Wenhardt, 1966).

In one of many such studies, it was found that on gently sloping corn plots near Ottawa, Ontario the loss of soil was greatly increased by cultivation practices (Ripley et al., 1961). Moreover, 75% of the soil loss occurred in only 3 out of the 13 years during which the experiment was conducted. In one year alone, 46% of the total loss occurred and of that most soil went out during a single, 60 minute rainstorm in spring when the plot surface was bare and the subsoil was frozen. This is an example of how human impacts can change the scale and significance of natural events and the pace of geological development. Similar observations have been made elsewhere (e.g., Webber, 1964), and indicate how cultivation, like climatic change, will alter the magnitude and frequency relations of erosional processes.

The main agents of erosion on crop lands are, as in nature, water and wind. In general, cultivation increases the amount and intensity of runoff, and hence the erosional power inherent in a given precipitation or snowmelt event. Equal or more significant are the effects of tillage, especially the type of ploughing; where furrows run up and down slopes they provide channels for runoff. Where soil compaction or depletion of fertility occurs, it decreases moisture infiltration and lowers erosional resistance. These matters are the subject of an enormous literature (Nir, 1983, Chapter 3; Coote, 1983).

Bare plots are exposed not only to rain beat but to higher rates of evaporation or sublimation, and hence to desiccation. Bare, dry surfaces are most susceptible to wind erosion. It is a very severe problem where farming has moved into the subhumid to semi-arid lands of the southwestern prairies and the dry interior basins of the western mountains (Anderson, 1975). The problem also occurs, however, in more humid areas not usually associated with wind erosion, including some locally severe examples in eastern Canada (Fitzsimmons and Nickling, 1982). The time of year when soil is left bare takes on special significance; the period between snowmelt and crop emergence in spring is one of exceptional vulnerability if the common windiness of spring is combined with dry weather. In the Prairie Provinces, when soils are freeze-dried — that is, exposed to desiccation in winter — they become much more susceptible to wind erosion than when wet-dried (Anderson and Wenhardt, 1966). If fields are left bare in winter without a stubble or trash cover to help trap snow, there is much more serious risk of wind erosion.

In general, although climate is the broad determinant, cultivation practices can play the main role in releasing or suppressing the erosive effects of precipitation and winds. Sometimes, a seemingly minor and intuitively sensible practice will make a large difference; for example, soil loss from potato fields in New Brunswick was shown to increase sharply with removal of stones (Saini and Grant, 1980), yet no such effect was seen if the stones were crushed and returned to the fields, evidently because the smaller particles still acted as a stone mulch helping resist erosion.

High value and therefore economically attractive crops such as corn and soybeans are especially hard on soil (Dickinson and Wall, 1978). The area under corn in Ontario rose more than three-fold between 1961 and 1976, and soybeans almost as much. At the same time such crops encourage other erosion-inducing trends. Larger fields, fewer hedges or fence rows, clearing of wood lots, heavy multirow equipment, even the speed with which motorized tillage occurs, all contribute to increasing erosion risks. Studies of the most productive farm areas across Canada, including the potato lands of Prince Edward Island, the truck farming of Ontario's Holland Marsh, and Prairie grain fields report high rates of erosion (Mirza and Irwin, 1964; Alberta Environmental Conservation Authority, 1976; Saini and Grant, 1980). The studies commonly quote 15-30 year periods within which productive soil could be wholly lost. According to Coote (1983, p. 231), "...The annual rate at which soil forms in Canada is estimated to be between about 0.25 t/ha (tons per hectare) and just over 1 t/ha. Yet annual erosion rates of 20-25 t/ha are quite common in agricultural land that is suffering rill and gully erosion."

From the volume of literature, soil erosion under agriculture appears to be a widely and continuously researched

CHAPTER 9

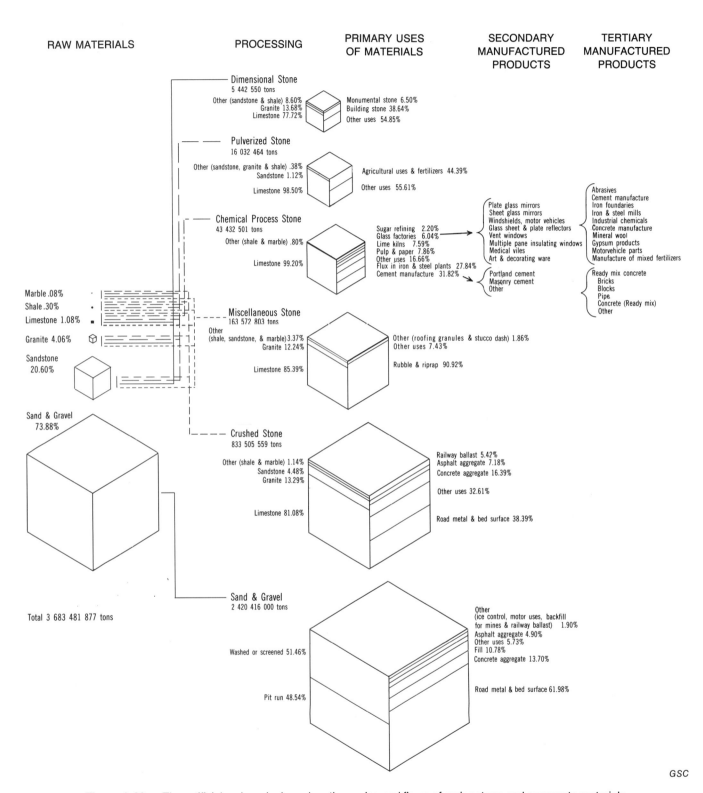

Figure 9.38. The artificial rock cycle, based on the scales and flows of major stone and aggregate materials used commercially in Canada for the eleven years, 1968 through 1979. (Source: Statistics Canada).

concern. Following the disastrous erosional as well as economic events of the 1930s, soil conservation became a high profile, politically favoured issue (Turner, 1955; Anderson, 1975; Warrick and Bowden, 1981). However, two broad observations must be made here. Firstly, no general and nationwide assessment of soil erosion associated with agriculture seems available for Canada. But secondly, half a century after the "dust bowl", a new wave of accelerated erosion on agricultural land became widely in evidence and continues (van Vuurren, 1978; Vontobel, 1980). The main explanation appears to be the highly mechanized, high yielding, artificially sustained monocultures that have become the heart of the farm economy. Unfortunately, industrialized farming tends to hide soil deterioration and loss until they reach an advanced stage. Most soil conservationists appear convinced that present types of crops and levels of productivity could be maintained without serious soil loss if different practices were adopted. Many also recognize, however, that the conditions under which farmers presently work conspire against the investment of the care and resources this requires. Socioeconomic pressures upon farmers in contemporary agribusiness, notably via interest rates, market prices, and government policies, divert concern away from soil conservation, if they do not actually force farmers to trade off long term soil destruction against short-term economic survival (N.E. Stewart and D. Himelman, unpublished report, Prince Edward Island Department of Agriculture, 1975; Ketcheson, 1980).

Mining and the artificial geological cycle

Mining is a leading sector of the Canadian economy and the largest source of exports. In recent decades, Canadians have also become among the highest per capita consumers of extracted materials.

A look at official figures for the thirty leading items shipped from mines and quarries in 1978 indicates about 573×10^6 t (Energy, Mines and Resources, 1979). Of these, the eight largest items accounted for more than 500×10^6 t. Such figures are comparable to the sediment delivery of the world's largest rivers at their mouths. Canadian sand and gravel shipments are nearly double the sediment yield for the whole Mississippi Basin (Milliman and Meade, 1983, p. 2). The "catchment areas" are very different, including the extreme localization of entrainment in pits and quarries; so too are the patterns of movement, transformation, and "deposition". Since stone and aggregates are the most voluminous of the extracted materials, an attempt is made here to define the actual pathways of this artificial geological cycle over a recent decade (Fig. 9.38).

If construction is the main consumer of earth materials, it is worth noting that other mineral cycles commonly involve a larger *hidden* mobilization. It appears in waste piles and removed overburden, tailings and tailings ponds, slag heaps, and gangue materials. In the production of coal, gold, or many other refined materials these "wastes" often amount to vastly greater impacts. In the vicinity of Elliot Lake, Ontario for example, mines generally yield about a tonne of waste rock for each kilogram of uranium produced (Redmond, 1983). By 1981, it was estimated that some 100×10^6 t of tailings had been produced involving disposal areas of some 610 ha. These quantities are expected to increase to 10^9 t in the next three or four decades (Redmond, 1983).

An important feature of industrialized systems, with their huge markets and economies of scale, is not just higher amounts of materials moved, but the use of progressively lower grades of ore and of more remote, large scale sources. Both are much in evidence across Canada.

One consequence of lower grades of ore is that greater volumes will be mined per unit of production. In a study of copper ore, for example, Lasky (1950) showed that as ore grade decreases arithmetically, tonnages required increase geometrically. The relation may be more favourable in other ores, but often is less so. Of even greater significance is the move from subsurface to surface mining that has accompanied this trend. Since individual mines are now much larger so material redistribution is that much greater.

For bulk materials, there is a great incentive to minimize transportation distance. Mine and processing wastes tend to be dumped in the immediate vicinity, creating extreme and dramatic man-made landforms such as in the Thetford Mines area of southern Quebec. Sand and gravel, of relatively high bulk compared to value but used in huge quantities, are sought as close to the point of use as possible. Hence pits are usually within a few hours at most of the major areas of construction. In the early 1960s, Blair (1965) showed that nearly all Toronto's supplies came from within 125 km of the city, most from within 80 km; Hamilton's supplies came from within 45 km. The amount of land for quarries and its concentration around urban centres have major impacts upon landscape. In the absence of considerable rehabilitation efforts, such man-made slopes and depressions have their own distinctive erosional development. McClelland (1983) gives a thorough review of these matters and their management implications.

Canadian shipments of sand and gravel doubled between 1965 and 1979. Not surprisingly, Ontario and Quebec accounted for almost half of all shipments, their roads alone absorbing nearly one third of Canadian consumption (Energy, Mines and Resources, 1979, p. 387). This involves a huge redistributing and resorting of mainly Pleistocene outwash from the Laurentide Ice Sheet.

At the other extreme, we have the growing search for and extraction of high value minerals in remote areas, especially the Canadian North and offshore zones. In 1977 a lead-zinc mine opened on Strathcona Sound, Baffin Island, to produce about 150 000 t/a, and another larger one is in operation at Arvik on Little Cornwallis Island, District of Franklin (Bliss, 1983). The possible and actual environmental effects of oil and gas development in the Arctic have been front page news since the mid 1970s. Generally, most of the North appears a "sensitive" environment in which the disturbance of slopes, vegetation, surface waters, and especially permafrost, lead to increases in erosion and long term surface damage. It is virtually impossible to carry out mining itself and to build its supporting systems of pipelines, roads, or ports without substantial impacts of this kind. Along the coast of the Beaufort Sea, in 1979 and 1980, one of the largest dredging operations in recent years moved 8.8×10^6 m^3 of material to make a harbour in which drillships could overwinter safely (Albery, Pullerits, Dickinson and Associates, 1981).

Such impacts of mining are not, however, new to the Canadian scene. The consequences of vast numbers of placer mines during the Yukon gold rush are still visible on the ground. Gardner (1981, 1986) described the impact of silver mining on erosional processes in the mountainous

Slocan District of British Columbia from 1890 to 1930. Removal of timber for construction, pit props and fuel, forest fires along with roads, flumes and use of packed-down snow for winter transport served to destabilize slopes and create an episode of recurrent large avalanches, floods, and landslides.

The built environment

Construction projects involving the large scale redistribution of earth and rock are also not new to Canada. The eastern section of the Trans-continental (C.N.) Railroad was completed in 1910. It had required the excavation of 44.6×10^6 m^3 of materials, almost two-thirds of it rock (Stevens, 1973, p. 261). The more localized Welland Canal By-Pass Project involved 49.7×10^6 m^3 of excavation (Jackson, 1975, p. 26). Such figures are, of course, dwarfed by the James Bay Project; earth and rock fill for La Grande No. 2 dam alone is about 48.2×10^6 m^3. This is more than moved in either the Hope slide or Frank slide rock avalanches (Table 9.2; Voight, 1978). These projects begin to compare with geological upheavals and are hardly achieved without great environmental change.

If less spectacular individually, the general consequences of urban construction and the communications that link settlements are more widespread and larger in total. Canada's nearly one million kilometres of roads have absorbed the greatest amount of construction materials and involve enormous patterns of earth moving and modification of surface vegetation, drainage, and slopes (Parizek, 1971). Meanwhile, the pace of urbanization, which in southern Ontario raised per capita consumption of construction materials to 17 t/a in the mid 1970s, creates an immense demand for earth materials. It spreads paved, roofed-over, and otherwise drastically modified surfaces at a tremendous rate. If this barely occupies more than 1% of Canada's land area, it is that fraction where some 80% of our population resides.

By means of artificial lithologies, urban areas acquire distinctive surface forms. Their response to sunshine, wind, and rain is distinctive. Much of the paved and roofed-over surface — usually 25-50% in Canadian cities — is impermeable to moisture and designed for rapid runoff. Drainage is channelled and piped, with the channel lengths usually much shorter than natural channels.

The more severe results of urbanization are often seen at and beyond the edge of the built-up area. An interface between paved and open land tends to concentrate erosion on the latter. The impacts of vehicles, whose erosive potential is offset on paved or otherwise protected artificial surfaces, commonly spread beyond the existing built area (Robinson, 1971, Chapter 12). Paving and storm drains sharply increase the volume and peak flows of runoff in surrounding conduits or rivers. Air pollution beyond the city margins may damage or destroy vegetation cover; the case of Sudbury being the extreme but best known for Canada (Pears, 1976).

Interacting and compounding effects

An outstanding feature of recent decades is the concentration of a range of powerful technologies or conflicting land uses and their interaction within a given region, often the same drainage basin or shoreline, sometimes across a single slope. This has received some attention where conflicts of interests arise, such as among logging, mining, recreation, hunting, and highly valued wilderness landscapes. In Canada, however, still more significant is the coincidence of much of the best farmland with the main urban-industrial zones (Gertler and Crowley, 1977). As a consequence, expanding cities and their multiplying "fringe" effects are over-running foodlands. This also compounds the impacts of the two powerful modifiers of erosional conditions — urbanization and agriculture.

These combined impacts are most obvious and extensive in southern Ontario. Here, urbanization, industrial concentration, high production agriculture, and pressures of ex-urban living and recreation are seen in land use change everywhere (Maini and Carlisle, 1974; Yeates, 1975). Erosional consequences can be seen from the shorelines of the Great Lakes to the cottage-crowded shores of small lakes on the Canadian Shield. Along drainage lines there are thousands of kilometres of piped, armoured, or locally modified stream channel; hundreds of small dams and weirs, and several dozen more substantial flood control and flow augmentation dams. Innumerable gullies, washouts, and temporary new deposits can be seen on fields and roadsides every spring. Sand and gravel pits of large size and great number lie just off the major axes of growth and transportation. One result, according to Dickinson and Wall (1976), is that rivers draining to the Great Lakes transport ten times more sediment than would be expected under pre-European settlement conditions. Similar observations can be made of all areas of urbanization from one end of the country to the other — from Victoria to St. John's or Winnipeg to Tuktoyaktuk.

Most of the actual landscapes of Canada include the interaction of human, climatic, and geological conditions. Anthropogenic effects enhance, reduce, or quite transform the role of the latter two. Man-encouraged wind erosion in the dry lands of southern Alberta and Saskatchewan (Anderson, 1975) or artificially increased avalanche activity around mountain passes or winter resorts are processes reflecting natural conditions. The introduction of avalanche defenses or drought-resistant plants are keyed to the natural region involved. The more drastic human modifications, however, serve to expand the geographical range of particular natural processes and stimulate "azonal" processes. Azonal impacts include the patches of severe wind erosion among the humid crop lands of eastern Canada (Fitzsimons and Nickling, 1982) or the badlands that have formed on outcrops of (Ordovician) Queenston Shale after deforestation and attempts at cultivation in southern Ontario.

Standardized forms of human technology may produce erosional changes distinctive to particular settings or common problems across a wide variety of habitats. Highway construction offers examples of both effects (Parizek, 1971). Impacts due to disturbed permafrost and peatland are a distinctive feature in the sub-Arctic and Arctic (Haugen and Brown, 1971; Crawford and Johnston, 1971; Heginbottom 1973; Radforth and Burwash, 1977). So are landslide and avalanche problems in the western Cordillera (Stetham and Shearer, 1980) or dust-plume sources along highways in western dry lands. On the other hand, similar geomorphic results recur at sites from Vancouver Island to Newfoundland; these include gully, creep, and slump damage on cutbanks. Impeded drainage shows up in patches of drowned, dead trees and periodic washouts. Generally enhanced wind action is seen in deadfall swaths where roads cut through forest, and on bald road banks (e.g., de Belle, 1971).

In places where lithology and relief are strongly articulated it is easy to miss or underestimate human influences. The bare or thinly wooded areas of the southern Canadian Shield may be seen as a wilderness shaped wholly by nature and the legacy of the Pleistocene ice sheets. Yet, before it was represented in the classic nature paintings of the Group of Seven, much of it had suffered human impacts. These included logging, deliberate burning and artificially increased forest fire incidence, farming, mining fall-out, and a by no means undisturbed development of secondary forest. Decimation of the beaver population has also had major effects on drainage. Obviously, the vegetation is not pristine. The lack of soil and amount of bare rock outcrop, however, also depend in considerable measure upon artificially accelerated erosion.

Broad landscape influences embrace the main regions of human occupancy. Yet, many of the more severe and distinctive erosional impacts are seen at scattered critical sites and along sensitive boundary or ecotonal areas. Shorelines, mountain foot, and timberline zones are among the natural ecotones which, in Canada, have been identified with marked landscape change. The shorelines of the Great Lakes and coastal zones of the Atlantic, Pacific, and Arctic oceans have long been recognized as suffering environmental damages, among which erosion-related effects of dredging, dumping, and chemical enrichment are common (Environment Canada, 1976; Thomson, 1981; Levings, 1982). The Rocky Mountain Foothills area is a zone of stress, with the growing threat of accelerated erosion and slope instability being central problems (Omara-Ojunga, 1980; Eastern Slopes Publications, 1977). It is these ecotones, along with such ecosystems as wetlands, islands, arctic and alpine tundra, that recent research identifies as "fragile" — that means unusual vulnerability to the impacts of human land uses and technologies. The fragile responses include, if they are not dominated by, marked erosional and related geological change. Some of the fragile zones are in the remote hinterlands, the North, and offshore zones. But others, such as lake and marine shorelines, wetland, and island areas are close to centres of population and sites of highly valued human uses.

Concluding remarks

The work has yet to be done that would provide a reasonably full assessment of these human impacts on the land of Canada. Such evidence as we have been able to draw upon, suggests that these assessments could be an important part of the agenda of earth sciences in coming decades. There seems little doubt that the intensity and diversity of these impacts will continue to grow. It will be useful to conclude by examining some problems that such work encounters.

Firstly, there is the difficulty of deciding where the more critical impacts are occurring, both in a geological and social sense. Our discussions above suggest that landscape change and erosional processes show the more visible and dramatic impacts. Yet these modifications, though locally severe, hardly depart from or exceed the kinds of changes in which the Quaternary abounds. They are also potentially the easiest to survey and assess. However, impacts upon chemical and biotic conditions, especially due to chemical concentration and applications of rarer or novel substances, raise serious problems of study as well as possible consequences. From the evolutionary perspective of historical geology, they resemble the sources of the more profound changes for life and habitats. Change in the chemistry of atmosphere, oceans, or earth's regolith, and in the radiational environment; large and global modification of habitats; the conditions of survival of organisms and the organisms themselves suggest change in the most basic features of the biosphere. Despite several decades of environmental concern, these are not matters readily dealt with by the uniformitarian, long-range ideas that form the mainstream of earth science.

Secondly, it is not easy to feel deep concern in these matters with respect to Canada. The nation is widely perceived to be a young country of relatively small population and vast wilderness areas. The land is seen to abound in as yet untapped natural resources and to be dominated by natural conditions. Despite evidence to the contrary, it is difficult to accept that in the places where Canadians live, work, and relax, and far beyond, large and growing impacts reflect human even more than natural conditions. Yet low population and large area are offset by the pressures of an affluent, mobile society. Its demands for materials and energy are met by advanced technologies and an economy grounded in massive extraction of natural resources.

Here we broach the still more difficult problem, that anthropogenic effects upon geology are inseparable from society's values, expectations, and interests. We anticipate that in nature, the erodible bluffs along lakes Erie or Ontario would recede due to storms, highwater or groundwater sapping, and ice effects in spring. The bluffs along Bow River at Calgary or the North Saskatchewan at Edmonton should be active geomorphic areas without human interference. But, when housing, valuable farm, or recreational land pushes to the edge of such features, erosion assumes a social and economic meaning. Our assessment of both natural and artificial erosion processes carries with it questions of benefit and risk — and beyond that loom questions of blame and responsible management. Indeed, various scholars tend to divide our whole subject into aspects involving either "deliberate" or "inadvertent" human impacts (Brown, 1970a; Vita-Finzi, 1977). All that one need emphasize here is that whatever the context of society's interaction with earth surface processes, the pressure of human interests and perceptions can hardly be avoided.

Finally, we should reflect upon those areas where there has already been substantial study, often with improved practice and managerial action in mind; they include some excellent work. There have been conspicuous local successes in reversing impacts thought to be adverse, and yet the general picture is one of increasing and more novel problems. That is obvious in the case of agriculture. Sixty years of investment in soil erosion research and soil conservation practices have failed to check the overall loss of topsoil. Much the same may be said of forest operations, despite erosion reducing silvicultural methods being known. In the case of more extreme events such as floods, we can also point to billions of dollars of investment in research, forecasting, and control works; yet the toll of damages in North America and world wide, continues to expand. The same may be said of landslides, coastal erosion, and a range of waterborne and airborne pollutants. Thus, we may indeed make the case for more study and improved concepts in the field of anthropogeology. But, to the extent that this will be expected to have practical applications, we can hardly ignore how the continuing thrust of human impacts on the planetary environment seems to overwhelm our knowledge and the improvements we can recommend.

ACKNOWLEDGMENTS

M.J. Church (University of British Columbia), J-C. Dionne (Laval University), and B.R. Pelletier (Geological Survey of Canada) are thanked for reading an earlier draft of the chapter and for providing useful comments. Valuable comments were also received from J.B. Bird (McGill University) and W.H. Mathews (University of British Columbia), each of whom read a draft of the entire volume.

The contributions of Jean Andrey for initial literature research for the last section of the chapter, D. Shrubsole for compiling data for Table 9.5 and Figure 9.38, and P. Schaus for designing Figure 9.38 are acknowledged. J. Gardner and A.G. McLellan (University of Waterloo) read early drafts of the last section in this chapter and provided useful comments.

REFERENCES

Abouguendia, Z.M.
1981: Athabasca sand dunes in Saskatchewan; Mackenzie River Basin Committee, Mackenzie River Basin Study Report, Supplement 7, 335 p., Appendices A-G.

Alberta Environmental Conservation Authority
1976: Proceedings of the public hearings on erosion of land in northwestern Alberta; Alberta Environmental Conservation Authority, Edmonton, Alberta, 8 volumes.

Albery, Pullerits, Dickson and Associates
1981: Report on dredging 1979-80, McKinley Bay and Tuktoyaktuk Harbour, Beaufort Sea; Report, Canadian Marine Drilling Ltd., Calgary, 25 p.

Algus, M.
1986: The development of coastal bluffs in a permafrost environment, Kivitoo Peninsula, Baffin Island, Canada; unpublished PhD thesis, McGill University, Montreal, 300 p.

Ambrose, J.W.
1964: Exhumed paleoplains of the Precambrian Shield of North America; American Journal of Science, v. 262, p. 817-857.

Andersland, O.B. and Anderson, D.M.
1978: Geotechnical Engineering for Cold Regions; McGraw Hill, New York, 566 p.

Anderson, C.H.
1975: A history of soil erosion by wind in the Palliser Triangle of western Canada; Canada Department of Agriculture, Historical Series 8, Ottawa.

Anderson, C.H. and Wenhardt, A.
1966: Soil erodibility, fall and spring; Canadian Journal of Soil Science, v. 46, p. 255-259.

Anderton, L.J.
1970: Quaternary stratigraphy and geomorphology of the lower Thompson River, British Columbia; unpublished MA thesis, University of British Colulmbia, Vancouver.

Ashley, G.M.
1979: Sedimentology of a tidal lake, Pitt Lake, British Columbia, Canada; in Moraines and Varves, Ch. Schluchter (ed.); A.S. Balkima, Rotterdam, p. 327- 345.

Ashworth, W.
1986: The Late Great Lakes: An Environment History; Collins, Toronto, 274 p.

Atmospheric Environment Service
1987: Climatic Atlas Climatique — Canada; Canada Department of the Environment, Atmospheric Environment Service, Canadian Climate Program, Map Series 1 — Temperatures and Degree Days, and Map Series 2 — Precipitation.

Bailey, R.G.
1973: Forest land use implications; in Environmental Geomorphology and Landscape Conservation, Volume III: Non-urban regions, D.R. Coates (ed.), p. 388-413; Dowden, Hutchinson and Ross Inc., Stroudsburg, Pa., Benchmark Papers in Geology, v. 8, 483 p.

Barnett, D.M., Edlund, S.A., and Dredge, L.A.
1977: Terrain characterization and evaluation: an example from eastern Melville Island; Geological Survey of Canada, Paper 76-23, 18 p.

Beaty, C.B.
1975: Coulee alignment and the wind in southern Alberta, Canada; Geological Society of America, Bulletin, v. 86, p. 119-128.

Bee, J.W. and Hall, E.R.
1956: Mammals of northern Alaska; University of Kanasa, Museum of Natural History, Miscellaneous Publication, v. 8, p. 1-309.

Beven, K. and Germann, P.
1982: Macropores and water flow in soils; Water Resources Research, v. 18, p. 1311-1325.

Bethlamy, N.
1974: More streamflow after a bark beetle epidemic; Journal of Hydrology, v. 23, p. 185-189.

Bird, J.B.
1967: The Physiography of Arctic Canada (With Special Reference to the Area South of Parry Channel); Johns Hopkins Press, Baltimore, 336 p.
1980: The Natural Landscapes of Canada: A Study in Regional Earth Science; John Wiley and Sons Canada Ltd., Toronto, 2nd edition, 260 p.

Blair, A.M.
1965: Surface extraction of non-metallic minerals in Ontario southwest of the Frontenac Axis; unpublished PhD thesis, University of Illinois, Urbana, 270 p.

Bliss, L.C.
1983: Modern human impact in the Arctic; in Man's Impact on Vegetation, M.J.A. Werger and I. Ikushima (ed.); W. Junk, The Hague, p. 213-225.

Bliss, L.C. and Wein, R.W.
1971: Changes to the active layer caused by surface disturbance; in Proceedings of a Seminar on the Permafrost Active Layer, 4 and 5 May 1971; National Research Council of Canada, Associate Committee on Geotechnical Research, Technical Memorandum 103, p. 37-47.

Bos, G.
1967: Range types and their utilization by muskox on Nunivak Island, Alaska, reconnaissance study; unpublished MSc thesis, University of Alberta, Edmonton, 124 p.

Bostock, H.S.
1967: Physiographic Regions of Canada; Geological Survey of Canada, Map 1245A.

Bouchard, M.A.
1974: Géologie des dépôts meubles de l'île Herschel, Territoire du Yukon; thèse de Maîtrise non-publiée, Université de Montréal, Montréal, 125 p.

Bovis, M.J.
1982: Uphill-facing (antislope) scarps in the Coast Mountains, southwest British Columbia; Geological Society of America, Bulletin, v. 93, p. 804-812.
1985: Earthflows in the Interior Plateau, southwest British Columbia; Canadian Geotechnical Journal, v. 22, p. 313-334.

Bowman, I.
1911: Forest Physiography; Wiley, New York, 759 p.

Brink, V.C.
1967: Needle ice and seedling establishment in southwestern British Columbia; Canadian Journal of Earth Sciences, v. 47, p. 135-139.

Brook, G.A. and Ford, D.C.
1978: The origin of labyrinth and tower karst and the climatic conditions necessary for their development; Nature, v. 275, no. 5680, p. 493-496.

Brown, E.H.
1970a: Man shapes the earth; Geographical Journal, v. 136, p. 74-84.

Brown, H.
1970b: Human materials production as a process in the biosphere; Scientific American, v. 223, p. 195-208.

Brown, R.J.E.
1967: Permafrost in Canada; Geological Survey of Canada, Map 1246A; National Research Council of Canada, Publication 9769.
1973: Permafrost distribution and relation to environmental factors in the Hudson Bay Lowland; in Proceedings: Symposium on the physical environment of the Hudson Bay Lowlands; University of Guelph, March 30-31, 1973, University of Guelph, Guelph, Ontario, p. 35-68.

Bryan, R.B. and Price, A.G.
1980: Recession of the Scarborough Bluffs, Ontario, Canada; in Coasts Under Stress, A.R. Orme, D.B. Prior, N.P. Psuty, and H.J. Walker (ed.); Zeitschrift für Geomorphologie, Supplementband, v. 34, p. 48-62.

Bryan, R.B., Yair, A., and Hodges, W.K.
1978: Factors affecting the initiation of runoff and piping in Dinosaur Provincial Park badlands, Alberta, Canada; in Field Instrumentation and Geomorphological Problems, O. Slaymaker, A. Rapp, and T. Dunne (ed.); Zeitschrift für Geomorphologie, Supplementband, v. 29, p. 151-168.

Bryson, R.A. and Hare, F.K. (editors)
1974: Climates of North America, World Survey of Climatology, Volume II; Elsevier Scientific Publishing Co., Amsterdam, 420 p.

Büdel, J.
1969: Das system der klima-genetischen geomorphologie; Erdkunde, v. 23, p. 165-182.

Burn, C.R., Michel, F., and Smith, M.W.
1986: Stratigraphic, isotopic and mineralogical evidence for an early Holocene thaw unconformity at Mayo, Yukon Territory; Canadian Journal of Earth Sciences, v. 23, p. 794-803.

Butler, J.D.
1979: Air Pollution Chemistry; Academic Press, New York, 408 p.

Cailleux, A.
1972: Les formes et les dépôts nivéo-éoliens actuels en Antarctique et au Nouveau-Québec; Cahiers de Géographie du Québec, v. 16, p. 377-409.

Cailleux, A. et Hamelin, L.E.
1969: Poste de Baleine (Nouveau-Québec): example de géomorphologie complexe; Revue de Géomorphologie dynamique, v. 3, p. 129-150.

Campbell, I.A.
1975: Stream discharge, suspended sediment and erosion rates in the Red Deer River basin, Alberta; International Association of Scientific Hydrology, Special Publication 122, p. 244-259.
1982: Surface morphology and rates of change during a ten-year period in the Alberta badlands; in Badland Geomorphology and Piping, R.B. Bryan and A. Yair (ed.); Geo Books (Geo Abstracts Ltd.), Norwich, England, p. 221-237.

Carlson, L.E., Ellwood, J.R., Nixon, J.F., and Slusarchuk, W.A.
1982: Field test results of operating a chilled buried pipeline in unfrozen ground; in Proceedings, Fourth Canadian Permafrost Conference, H.M. French (ed.); National Research Council of Canada, Ottawa, NRCC 20124, p. 475-480.

Carson, M.A.
1975: Threshold and characteristic angles of straight slopes; 4th Guelph Symposium on Geomorphology, University of Guelph, Guelph, Ontario, p. 19-33.
1977a: Angles of response, angles of shearing resistance and angles of straight talus slopes; Earth Surface Process, v. 2, p. 363-380.
1977b: On the retrogression of landslides in sensitive muddy sediments; Canadian Geotechnical Journal, v. 14, p. 582-602.
1979: Le glissement de Rigaud de 3 mai 1978: une interprétation du mode de rupture d'après la morphologie de la cicatrice; Géographie physique et Quaternaire, vol. 33, p. 63-92.
1981: Influence of porefluid salinity on instability of sensitive marine clays: a new approach to an old problem; Earth Surface Processes and Landforms, v. 6, p. 499-516.

Carson, M.A. and Bovis, M.J.
1989: Slope processes; in Chapter 9 of Quaternary Geology of Canada and Greenland, R.J. Fulton (ed.); Geological Survey of Canada, Geology of Canada, no. 1 (also Geological Society of America, The Geology of North America, v. K-1).

Carson, M.A. and MacLean, P.A.
1986: Development of hybrid aeolian dunes: the William River dunefield, northwest Saskatchewan, Canada; Canadian Journal of Earth Sciences, v. 23, p. 1974-1990.

Carter, L.D., Heginbottom, J.A., and Woo, M-K.
1987: Arctic Lowlands; Chapter 14 in Geomorphic Systems of North America, W.L. Graf (ed.); Geological Society of America, Centennial Special Volume 2, p. 583-628.

Catto, N.R.
1983: Loess in the Cypress Hills, Alberta; Canadian Journal of Earth Sciences, v. 20, p. 1159-1167.

Chambers, M.J.C.
1966: Investigations of patterned ground at Signy Island, South Orkney Islands: II. Temperature regimes in the active layer; British Antarctic Survey, Bulletin, v. 10, p. 7-83.

Cheng Guodong
1983: The mechanism of repeated segregation for the formation of thick layered ground ice; Cold Regions, Science and Technology, v. 8, p. 57-66.

Christiansen, E.A.
1959: Glacial geology of the Swift Current area, Saskatchewan; Saskatchewan Department of Mineral Resources, Report 32, 62 p.
1967: Collapse structures near Saskatoon, Saskatchewan, Canada; Canadian Journal of Earth Sciences, v. 4, p. 757-767.
1971: Geology of the Crater Lake collapse structure in southeastern Saskatchewan; Canadian Journal of Earth Sciences, v. 8, p. 1505-1513.

Church, M.
1972: Baffin Island sandurs: a study of arctic fluvial processes; Geological Survey of Canada, Bulletin 216, 208 p.
1980: Records of recent geomorphological events; in Timescales in Geomorphology, R.A. Cullingford, D.A. Davidson, and J. Lewis (ed.); Wiley and Sons, Chichester, p. 13-29.
1983a: Pattern of instability in a wandering gravel bed channel; International Association of Sedimentologists, Special Publication, v. 6, p. 169-180.
1983b: The importance of geomorphology in hydraulic engineering; Paper presented to the 6th Canadian Hydrotechnical Conference, June 2 and 3, Ottawa, Canadian Society for Civil Engineering, Hydrotechnical Division, p. 1039-1058.

Church, M. and Gilbert, R.
1975: Proglacial fluvial and lacustrine environments; in Glaciofluvial and Glaciolacustrine Sedimentation, A.V. Jopling and B.C. McDonald, (ed.); Society of Economic Paleontologists and Mineralogists, Special Publication 23, p. 22- 100.

Church, M. and Ryder, J.M.
1972: Paraglacial sedimentation: a consideration of fluvial processes conditioned by glaciation; Geological Society of America, Bulletin, v. 83, p. 3059-3072.

Church, M., Stock, R.F., and Ryder, J.M.
1979: Contemporary sedimentary environments on Baffin Island, N.W.T., Canada; Debris slope accumulations; Arctic and Alpine Research, v. 11, p. 371-402.

Clague, J.J.
1979: The Denali Fault System in southwest Yukon Territory — a geologic hazard?; in Current Research, Part A, Geological Survey of Canada, Paper 79-1A, p. 169- 178.
1981: Landslides in the south end of Kluane Lake, Yukon Territory; Canadian Journal of Earth Sciences, v. 18, p. 959-971.
1989: Economic implications of Quaternary geology in the Cordillera; in Chapter 1 of Quaternary Geology of Canada and Greenland, R.J. Fulton (ed.); Geological Survey of Canada, Geology of Canada, no. 1 (also Geological Society of America, The Geology of North America, v. K-1).

Clague, J.J. and Rampton, V.N.
1982: Neoglacial Lake Alsek; Canadian Journal of Earth Sciences, v. 19, p. 94-117.

Clague, J.J. and Souther, J.G.
1982: The Dusty Creek landslide on Mount Cayley, British Columbia; Canadian Journal of Earth Sciences, v. 19, p. 524-539.

Clague, J.J., Evans, S.G., and Blown, J.G.
1985: A debris flow triggered by the breaching of a moraine-dammed lake, Klattasine Creek, British Columbia; Canadian Journal of Earth Sciences, v. 22, p. 1492-1502.

Clark, T.H. and Elson, J.A.
1961: Ventifacts and eolian sand at Charette, P.Q.; Royal Society of Canada, Transactions, v. 55, Series 3, p. 1-11.

Clayton, J.S., Erlich, W.A., Cann, D.B., Day, J.H., and Marshall, I.B.
1977: Soils of Canada; Canada Department of Agriculture, Ottawa, Publication 1544, 2 volumes and map.

Coakley, J.P. and Rust, B.R.
1968: Sedimentation in an arctic lake; Journal of Sedimentary Petrology, v. 38, p. 1290-1300.

Coates, D.R. (editor)
1973: Environmental Geomorphology and Landscape Conservation, Volume III, Non-urban regions; Dowden, Hutchinson and Ross Inc., Stroudsburg, Pa., Benchmark Papers in Geology, v. 8, 483 p.

Cogley, J.G.
1972: Processes of solution in an arctic limestone terrain; in Polar Geomorphology, R.J. Price and D.E. Sugden (ed.); Institute of British Geographers, Special Publication 4, p. 201-211.

Cogley, J.G. and McCann, S.B.
1976: An exceptional storm and its effects in the Canadian High Arctic; Arctic and Alpine Research, v. 8, p. 105-110.

Cohen, S.J.
1986: Climatic change, population growth and their effects on Great Lakes water supplies; Professional Geographer, v. 38, p. 317-323.

Coker, W.B. and Shilts, W.W.
1979: Lacustrine geochemistry around the north shore of Lake Superior: implications for evaluation of the effects of acid prepiciptation; in Current Research, Part C, Geological Survey of Canada, Paper 79-1C, p. 1-15.

Cooke, F.A. and Raiche, V.G.
1962: Freeze-thaw cycles at Resolute, N.W.T.; Geographical Bulletin, v. 18, p. 64- 78.

Coote, D.R.
1983: Stresses on land under intensive agricultural use; in Stress on Land in Canada; Canada Department of the Environment, Policy Research and Development Branch, Ottawa, p. 228-257.

Cormack, K., Jr., Swanson, F.R., and Grier, C.C.
1978: A comparison of harvesting methods and their impact on soils and environment in the Pacific Northwest; in Forest Soils and Land Use, Proceedings, 5th North American Forest Soils Conference, C.T. Young (ed.); U.S. Department of Agriculture, Washington, D.C., p. 449-476.

Corté, A.E.
1966: Particle sorting by repeated freezing and thawing; Biuletyn Peryglacijalny, v. 15, p. 175-240.

Crawford, C.B.
1968: Quick clays of eastern Canada; Engineering Geology, v. 2, p. 239-265.

Crawford, C.B. and Johnston, G.H.
1971: Construction on permafrost; Canadian Geotechnical Journal, v. 8, p. 236-251.

Cruden, D.M.
1976: Major rock slides in the Rockies; Canadian Geotechnical Journal, v. 13, p. 8-20.
1982: The Brazeau Lake slide, Jasper National Park, Alberta; Canadian Journal of Earth Sciences, v. 19, p. 975-981.
1985: Rock slope movements in the Canadian Cordillera; Canadian Geotechnical Journal, v. 22, p. 528-540.

Darby, H.C.
1956: The clearing of the woodland in Europe; in Man's Role in Changing the Face of the Earth, W.H. Thomas (ed.); University of Chicago Press, Chicago, p. 183-216.

Dasmann, R.F.
1976: Environmental Conservation; Wiley, New York, 4th edition, 427 p.

Darwin, C.R.
1881: The formation of vegetable mould through the action of worms with observations on their habits; published in 1945 as: Darwin and the Earthworms; Faber, London.

David, P.P.
1970: Discovery of Mazama ash in Saskatchewan, Canada; Canadian Journal of Earth Sciences, v. 7, p. 1579-1583.
1971: The Brookdale road section and its significance in the chronological studies of dune activities in the Brandon Sand Hills of Manitoba; Geological Association of Canada, Special Paper No. 9, p. 293-299.
1972: Great Sand Hills, Saskatchewan; in Quaternary geology and geomorphology between Winnipeg and the Rocky Mountains, N.W. Rutter and E.A. Christiansen (ed.); 24th International Geological Congress, Montréal, Guidebook, Excursion C-22 p. 36-50.
1977: Sand dune occurrences in Canada; Canada Department of Indian and Northern Affairs, National Parks Branch, Contract 74-230, Report, 183 p.
1979: Sand dunes in Canada; Geos, Ottawa, Spring issue, p. 12-14.
1981a: The aeolian environment; in Athabasca Sand Dunes in Saskatchewan, Z.M. Abouguendia (ed.), p. B19-B56; Mackenzie River Basin Study Report, Supplement 7, 335 p., Appendices A-G.
1981b: Stabilized dune ridges in northern Saskatchewan; Canadian Journal of Earth Sciences, v. 18, p. 286-310.
1982: Late Pleistocene and Holocene climatic changes based on the eolian stratigraphic record of the Canadian Prairies: an update (abstract); in Program with Abstracts; Geological Association of Canada, Annual Meeting, Winnipeg, Manitoba, v. 7, p. 44.

Davies, K.F.
1975: Mackenzie River input to the Beaufort Sea; Victoria, B.C.: Canada Department of the Environment, Beaufort Sea Technical Report No. 15, 72 p.

Davies, T.R.H.
1982: Spreading of rock avalanche debris by mechanical fluidization; Rock Mechanics, v. 15, p. 9-24.

Day, T.J.
1989: River processes; in Chapter 9 of Quaternary Geology of Canada and Greenland, R.J. Fulton (ed.); Geological Survey of Canada, Geology of Canada, no. 1 (also Geological Society of America, The Geology of North America, v. K-1).

deBelle, G.
1971: Roadside erosion and resource implications in Prince Edward Island; Canada Department of Energy, Mines and Resources, Policy Research and Coordination Branch, Ottawa, 25 p.

de Jong, E., Begg, C.B.M., and Kachanoski, R.G.
1983: Estimates of soil erosion and deposition for some Saskatchewan soils; Canadian Journal of Soil Science, v. 63, p. 607-617.

Dence, M.R., Innes, M.J.S., and Robertson, P.B.
1968: Recent geological and geophysical studies of Canadian craters; in Shock Metamorphism of Natural Materials, B.M. French and N.M. Short (ed.); Mono Book Corporation, Baltimore, Maryland, p. 339-362.

Desjardins, R.
1980: Tremblements de terre et glissements de terrain: corrélation entre des datations au ^{14}C et des données historiques à Shawinigan, Québec; Géographie physique et Quaternaire, vol. 34, p. 359-362.

Desloges, J. and Gardner, J.
1981: Recent chronology of an alpine alluvial fan in southwestern Alberta; Albertan Geographer, no. 17, p. 1-18.

Detwyler, T.R.
1971: Man's Impact on Environment; McGraw-Hill, New York, 414 p.

deVries, J. and Chow, T.L.
1978: Hydrologic behaviour of a forested mountain soil in coastal British Columbia; Water Resources Research, v. 14, p. 935-942.

Dickinson, W.T. and Wall, G.J.
1976: Temporal patterns of erosion and fluvial sedimentation in the Great Lakes Basin; Geoscience Canada, v. 3, p. 158-163.
1978: Soil erosion by water: nature and extent; Notes on Agriculture, v. 14, p. 4-7.

Dionne, J-C.
1974: Cryosols avec triage sur rivage et fond de lacs, Québec central subarctique; La Revue de Géographie de Montréal, vol. 28, p. 323-342.
1978a: Formes et phénomènes périglaciaires en Jamésie, Québec subarctique; Géographie physique et Quaternaire, vol. 32, p. 187-247.
1978b: Dunes et dépôts éoliens en Jamésie et Hudsonie, Québec subarctique; Environnement Canada, Rapport d'information, 36 p.
1983: Frost-heaved bedrock features: a valuable permafrost indicator; Géographie physique et Quaternaire, v. 37, p. 241-251.
1985: Tidal marsh erosion by geese, St. Lawrence Estuary, Québec; Géographie physique et Quaternaire, v. 39, p. 99-105.

Dionne, J-C. and Laverdière, C.
1972: Ice-formed beach features from Lake St. Jean, Quebec; Canadian Journal of Earth Sciences, v. 9, p. 979-990.

Dumanski, J., Pawluk, S., Vucetich, C.G., and Lindsay, J.D.
1980: Pedogenesis and tephrochronolgy of loess derived soils, Hinton, Alberta; Canadian Journal of Earth Sciences, v. 17, p. 52-59.

Drake, J.J. and Ford, D.C.
1981: Karst solution: a global model for groundwater solute concentrations; Japanese Geomorphological Union, Transactions, v. 2, p. 223-230.

Dyke, A.S.
1976: Tors and associated weathering phenomena, Somerset Island, District of Franklin; in Report of Activities, Part B, Geological Survey of Canada, Paper 76-1B, p. 209-216.
1981: Late Holocene solifluction rates and radiocarbon soil ages, central Canadian Arctic; in Current Research, Part C, Geological Survey of Canada, Paper 81-C, p. 17-22.
1983: Quaternary geology of Somerset Island, District of Franklin; Geological Survey of Canada, Memoir 403, 32 p.

Dyke, A.S. and Zoltai, S.C.
1980: Radiocarbon-dated mudboils, central Canadian Arctic; in Current Research, Part B, Geological Survey of Canada, Paper 80-1B, p. 271-275.

Dyke, L.D.
1984: Frost heave of bedrock in permafrost regions; Association of Engineering Geologists, Bulletin, v. 21, p. 389-405.
1986: Frost heaving of bedrock; in Focus: Permafrost Geomorphology, H.M. French (ed.); The Canadian Geographer, v. 30, p. 360-362.

Eastern Slopes Publications
1977: A Policy for Resource Management of the Eastern Slopes; Eastern Slopes Publications, Edmonton.

Egginton, P.A.
1979: Mudboil activity, central District of Keewatin; in Current Research, Part B, Geological Survey of Canada, Paper 79-1B, p. 349-356.
1985: Moles as agents of erosion in the Ottawa area, Ontario; in Current Research, Part A, Geological Survey of Canada, Paper 85-1A, p. 731-733.
1986: Active layer processes; in Focus: Permafrost Geomorphology, H.M. French (ed.); The Canadian Geographer, v. 30, p. 364-365.

Egginton, P.A. and Dyke, L.D.
1982: Density gradients and injection structures in mudboils in central District of Keewatin; in Current Research, Part B, Geological Survey of Canada, Paper 82-1B, p. 173-176.

Egginton, P.A. and French, H.M.
1985: Solifluction and related processes, eastern Banks Island, N.W.T.; Canadian Journal of Earth Sciences, v. 22, p. 1671-1678.

Egginton, P.A. and Shilts, W.W.
1978: Rates of movement associated with mudboils, central District of Keewatin; in Current Research, Part B, Geological Survey of Canada, Paper 78-1B, p. 203-206.

Eisbacher, G.H.
1979a: First-order regionalization of landslide types in the Canadian Cordillera; Geoscience Canada, v. 6, p. 69-79.
1979b: Cliff collapse and rock avalanches (sturzstroms) in the Mackenzie Mountains, northwest Canada; Canadian Geotechnical Journal, v. 16, p. 309-334.

Eisbacher, G.H. and Clague, J.J.
1981: Urban landslides in the vicinity of Vancouver, British Columbia, with special reference to the December 1979 rainstorm; Canadian Geotechnical Journal, v. 18, p. 205-216.

Energy, Mines and Resources, Department of
1979: Canadian Minerals Yearbook; Canada Department of Energy, Mines and Resources, Ottawa, 625 p.

Environment, Department of the
1973: Inventory of Canadian freshwater lakes; Inland Waters Directorate, Water Resources Branch, Ottawa, Publication no. 02KX-KL-327-3-8061, 34 p.
1976: Coastal zone atlas: Ontario-Canada, Great Lakes Shore Damage Survey; Ontario Ministry of Natural Resources, Catalogue no. FS 99-10/1975.
1983: Stress on Land in Canada; Policy Research and Development Branch, Ottawa, Publication no. EN-73-2/6E, 323 p.

Evans, S.G.
1982: Landslides and surficial deposits in urban areas of British Columbia: a review; Canadian Geotechnical Journal, v. 19, p. 269-288.
1986: Landslide damming in the Cordillera of western Canada; in Landslide dams: Processes, Risk and Mitigation: Session Proceedings, Geotechnical Division, American Society of Civil Engineers, Meeting, Seattle, Washington, p. 111-130.
1987: A rock avalanche from the peak of Mount Meager, British Columbia; in Current Research, Part A, Geological Survey of Canada, Paper 87-1A, p. 929-934.
1989: Geomorphic response to endogenic processes; in Chapter 9 of Quaternary Geology of Canada and Greenland, R.J. Fulton (ed.); Geological Survey of Canada, Geology of Canada, no. 1 (also Geological Society of America, The Geology of North America, v. K-1).

Evans, S.G. and Gardner, J.S.
1989: Geological hazards in the Canadian Cordillera; in Chapter 12 of Quaternary Geology of Canada and Greenland, R.J. Fulton (ed.); Geological Survey of Canada, Geology of Canada, no. 1 (also Geological Society of America, The Geology of North America, v. K-1).

Evans, S.G., Aitken, J.D., Wetmiller, R.J., and Morner, R.B.
1987: A rock avalanche triggered by the October 1985 North Nahanni earthquake, District of Mackenzie, N.W.T.; Canadian Journal of Earth Sciences, v. 24, p. 176-184.

Eyles, N., Eyles, E.H., Lau, K., and Clark, B.
1985: Applied sedimentology in an urban environment — the case of Scarborough Bluffs, Ontario; Canada's most intractable erosion problem; Geoscience Canada, v. 12, p. 91-104.

Fairbridge, R.W.
1968: Terracettes, lynchets and 'cattle tracks'; in The Encyclopedia of Geomorphology, R.W. Fairbridge (ed.); Reinhold Book Corporation, New York, p. 13-29.

Fahey, B.D.
1973: An analysis of diurnal freeze thaw and frost heave cycles in the Indian Peaks region of the Colorado Front Range; Arctic and Alpine Research, v. 5, p. 269-281.

Fels, E.
1965: Nochmals: Anthropogene geomorphologie; Petermanns Geographische Mitteiluagen, v. 109, p. 9-15.

Filion, L.
1984: A relationship between dunes, fire and climate recorded in the Holocene deposits of Quebec, Canada; Nature, v. 309, p. 543-546.

Filion, L. and Morrisset, P.
1983: Eolian landforms along the eastern coast of Hudson Bay, northern Québec; in Tree-line Ecology, P. Morrisset and S. Payette (ed.); Proceedings of the Northern Quebec Tree-line Conference, Centre d'Études Nordiques, Université Laval, Québec.

Fitzsimmons, J.G. and Nickling, W.G.
1982: Wind erosion of agricultural soils in southwestern Ontario; University of Guelph, School of Rural Planning and Development, Publication 108, 40 p.

Ford, D.C.
1971: Characteristics of limestone solution in the Southern Rocky Mountains and Selkirk Mountains, Alberta and British Columbia; Canadian Journal of Earth Sciences, v. 8, p. 585-608.
1983a: Effects of glaciations upon karst aquifers in Canada; Journal of Hydrology, v. 61, p. 149-158.
1983b: Castleguard Cave and karst, Columbia Icefields Area, Rocky Mountains of Canada: a symposium; Arctic and Alpine Research, v. 15, p. 425-554.

French, H.M.
1971: Ice-cored mounds and patterned ground, southern Banks Island, western Canadian Arctic; Geografiska Annaler, v. 53A, p. 32-38.
1974a: Active thermokarst processes, eastern Banks Island, western Canadian Arctic; Canadian Journal of Earth Sciences, v. 11, p. 785-794.
1974b: Mass-wasting at Sachs Harbour, Banks Island, NWT, Canada; Arctic and Alpine Research, v. 6, p. 71-80.
1975: Man-induced thermokarst, Sachs Harbour airstrip, Banks Island, Northwest Territories; Canadian Journal of Earth Sciences, v. 12, p. 132-144.
1976: The periglacial environment; London and New York:Longmans Group Limited, 308 p.
1981: Periglacial geomorphology and permafrost; Progress in Physical Geography, v. 5, p. 267-273.
1986: Periglacial involutions and mass displacement structures, Banks Island, Canada; Geografiska Annaler, v. 68A, p. 167-174.
1989: Cold climate processes; in Chapter 9 of Quaternary Geology of Canada and Greenland, R.J. Fulton (ed.); Geological Survey of Canada, Geology of Canada, no. 1 (also Geological Society of America, The Geology of North America, v. K-1).

French, H.M. and Dutkiewicz, L.
1976: Pingos and pingo-like forms, Banks Island, western Canadian Arctic; Biuletyn Peryglacijalny, v. 26, p. 211-222.

French, H.M. and Harry, D.G.
1983: Ground ice conditions and thaw lake evolution, Sachs River lowlands, Banks Island, Canada; in Mesoformen des reliefs in heirtigen Periglacial raumes, H. Poser, and E. Schunke (ed.); Abhandlunger des Akadamie der Wissenschaften in Gottingen, v. 35, p. 70-81.

French, H.M. and Pollard, W.H.
1986: Ground ice investigations, Klondike District, Yukon Territory; Canadian Journal of Earth Sciences, v. 23, p. 550-560.

French, H.M., Bennett, L., and Hayley, D.W.
1986: Ground ice conditions near Rea Point and on Sabine Peninsula, eastern Melville Island; Canadian Journal of Earth Sciences, v. 23, p. 1389-1400.

French, H.M., Harris, S.A., and van Everdingen, R.O.
1983: The Klondike and Dawson City; in Guidebook 3: Permafrost and related features of the Northern Yukon Territory and Mackenzie Delta, Canada; H.M. French and J.A. Heginbottom (ed.), Fourth International Conference on Permafrost and IGU Commission on the Significance of Periglacial Phenomena, (1983), Alaska Division of Geological and Geophysical Surveys, Fairbanks, Alaska.

French, H.M., Harry, D.G., and Clark, M.J.
1982: Ground ice stratigraphy and Late Quaternary events, south-west Banks Island, Canadian Arctic; in Proceedings, Fourth Canadian Permafrost Conference, H.M. French (ed.); National Research Council of Canada, Ottawa, NRCC 20124 p. 81-90.

Frenette, M. and Larinier, M.
1973: Some results of the sediment regime of the St. Lawrence River; in Fluvial Processes and Sedimentation, Hydrology Symposium No. 9, Department of the Environment, Inland Waters Directorate, Ottawa, p.138-157.

Frey, D.G. (editor)
1963: Limnology in North America; Madison, Wisconsin, University of Wisconsin Press, 734 p.

Fulton, R.J.
1972: Stratigraphy of unconsolidated fill and Quaternary development of North Okanagan Valley; in Bedrock topography of the North Okanagan Valley and stratigraphy of the unconsolidated valley fill, British Columbia; Geological Survey of Canada, Paper 72-8, Part B, p. 9-17.

Fyfe, W.S.
1982: Recent sediments: the front line of environmental protection; Geoscience Canada, v. 9, p. 71-73.

Fyles, J.G.
1963: Surficial geology of Victoria and Stefansson Islands, District of Franklin; Geological Survey of Canada, Bulletin 101, 38 p.

Galay, V.J.
1970: Some hydraulic characteristics of coarse-bed rivers; unpublished PhD thesis, University of Alberta, Edmonton.

Galay, V.J., Kellerhals, R., and Bray, D.I.
1973: Diversity of river types in Canada; in Fluvial Processes and Sedimentation, Hydrology Symposium No. 9, Inland Waters Directorate, Canada Department of the Environment, Ottawa, p. 217-250.

Gardner, J.S.
1979: The movement of material on debris slopes in the Canadian Rocky Mountains; Zeitschrift für Geomorphologie, v. 23, p. 45-57.
1981: Snow and silver in Slocan, B.C.: resource-hazard thresholds (abstract); American Association of Geographers, Annual Meeting, Los Angeles, Abstracts, p. 82.
1986: Snow as a resource and hazard in early twentieth century mining, Selkirk Mountains, British Columbia; Canadian Geographer, v. 30, p. 217-228.

Gelinas, P.J. and Quigley, R.M.
1973: The influence of geology on erosion rates along the north shore of Lake Erie; Proceedings of the Sixteenth Conference on Great Lakes Research, International Association of Great Lakes Research, p. 421-430.

Gertler, L.O. and Crowley, R.W.
1977: Changing Canadian Cities: The Next Twenty-five Years; McClelland and Stewart, Toronto, 474 p.

Gilbert, R.E.
1972: Observations on sedimentation at Lillooet Delta, British Columbia; in Mountain Geomorphology: Geomorphological Processes in the Canadian Cordillera, B.C., H.O. Slaymaker and H.J. McPherson (ed.); Tantalus Research Ltd., Vancouver, B.C. Geographical Series No. 14.
1973: Observations of lacustrine sedimentation at Lillooet Lake, British Columbia; unpublished PhD thesis, University of British Columbia, Vancouver, 193 p.

Gilbert, R. and Shaw, J.
1981: Sedimentation in proglacial Sunwapta Lake, Alberta; Canadian Journal of Earth Sciences, v. 18, p. 81-93.

Gold, L.W.
1985: The ice factor in frozen ground; in Field and Theory: Lectures in Geocryology, M. Church and O. Slaymaker (ed.); University of British Columbia Press, Vancouver, p. 74-95.

Goodell, B.C.
1972: Water quantity and flow regime — influences of land use, especially forestry; in Mountain Geomorphology: Geomorphological Processes in the Canadian Cordillera, H.O. Slaymaker and H.J. McPherson (ed.); Tantalus Research Ltd., Vancouver, B.C. Geographical Series No. 14, p. 197-206.

Goodrich, L.
1982: The influence of snow cover on the ground thermal regime; Canadian Geotechnical Journal, v. 19, p. 421-432.

Gordon, A.G.
1985: Budworm! What about the Forest?; in Society of American Foresters, Spruce-Fir Management and Spruce Budworm Region IV; Technical Conference, Ontario Ministry of Natural Resources, Toronto, p. 3-29.

Green, R.D. and Askew, G.P.
1965: Observations on the biological development of macropores in soils of Romney Marsh; Journal of Soil Science, v. 16, p. 342.

Hare, F.K.
1980: Introduction: Canada at large; in The Natural Landscapes of Canada: A Study in Regional Earth Science, by J.B. Bird; 2nd edition, John Wiley and Sons Canada Ltd., Toronto, 260 p.

Hare, F.K. and Thomas, M.K.
1979: Climate Canada; 2nd edition, John Wiley and Sons, Toronto, 230 p.

Harr, R.D., Harper, W.C., Krygier, J.T., and Hseil, F.S.
1975: Changes in storm hydrographs after roadbuilding and clear-cutting in the Oregon Coast Range; Water Resources Research, v. 11, p. 436-444.

Harry, D.G. and French, H.M.
1983: The orientation of thaw lakes, southwest Banks Island, Arctic Canada; Proceedings, Fourth International Conference on Permafrost; National Academy Press, Washington, D.C., p. 456-461.

Harry, D.G., French, H.M., and Clark, M.J.
1983: Coastal conditions and processes, Sachs Harbour, southwest Banks Island, western Canadian Arctic; Zeitschrift für Geomorphologie, Supplement 47, p. 1-26.

Harry, D.G., French, H.M., and Pollard, W.H.
1985: Ice wedges and permafrost conditions near King Point, Beaufort Sea coast, Yukon Territory; in Current Research, Part A, Geological Survey of Canada, Paper 85-1A, p. 111-116.

Haug, M.D., Sauer, E.K., and Fredlund, D.G.
1977: Retrogressive slope failures at Beaver Creek, south of Saskatchewan, Canada; Canadian Geotechnical Journal, v. 14, p. 288-301.

Haugen, R.K. and Brown, J.
1971: Natural and man-induced disturbances of permafrost terrain; in Environmental Geomorphology, D.R. Coates (ed.); State University of New York, Binghampton, N.Y., p. 139-149.

Hayley, D.W.
1982: Application of heat pipes to design of shallow foundations on permafrost; Proceedings, Fourth Canadian Permafrost Conference, H.M. French (ed.); National Research Council of Canada, Ottawa, NRCC 20124 p. 535-544.

Heede, B.H.
1975: Mountain watersheds and dynamic equilibrium; Watershed Management Symposium, ASCE Irrigation and Drainage Division, Logan, Utah, p. 407-420.

Heginbottom, J.A.
1973: Effects of surface disturbance upon permafrost; Environmental-Social Program, Northern Pipelines, Task Force on Northern Oil Development, Report 73-16, 29 p.
1984: The bursting of a snow dam, Tingmisut Lake, Northwest Territories; in Current Research, Part B, Geological Survey of Canada, Paper 84-1B, p. 187-192.

Heginbottom, J.A. and Vincent, J-S. (editors)
1986: Correlation of Quaternary deposits and events around the margin of the Beaufort Sea: contributions from a joint Canadian-American workshop, April 1984; Geological Survey of Canada, Open File 1237, 125 p.

Hermesh, R.
1972: A study of the ecology of the Athabasca sand dunes with emphasis on the phytogenic aspects of dune formation; unpublished MA thesis, University of Saskatchewan, Saskatoon, 158 p.

Hewitt, K.
1989: Human society as a geological agent; in Chapter 9 of Quaternary Geology of Canada and Greenland; R.J. Fulton (ed.); Geological Survey of Canada, Geology of Canada, no. 1 (also Geological Society of America, The Geology of North America, v. K-1).

Hiratsuka, H.F., Cerezke, J., Petty, J., and Still, J.N.
1981: Forest insect and disease conditions in Alberta, Saskatchewan, Manitoba, and the Northwest Territories in 1980 and predictions for 1981; Northern Forest Research Centre, Canadian Forestry Service, Information Report NOR-X-231, 13 p.

Hodgson, D.A.
1982: Surficial materials and geomorphological processes, western Sverdrup and adjacent islands, District of Franklin; Geological Survey of Canada, Paper 81-9, 43 p.

Hodgson, E.A.
1927: The marine clays of Eastern Canada and their relationship to earthquake hazards; Royal Astronomical Society of Canada, Journal, v. 21, p. 257-264.
1946: The British Columbia earthquake, June 23, 1946; Royal Astronomical Society of Canada, Journal, v. 40, p. 285-319.
1950: The Saint Lawrence Earthquake March 1, 1925; Dominion Observatory, Publications, v. 7, no. 10.

Holland, S.S.
1964: Landforms of British Columbia, a physiographic outline; B.C. Department of Mines and Petroleum Resources, Bulletin 48, 138 p., map.

Hollingshead, A.B.
1968: Measurements of the bedload discharge of the Elbow River; unpublished MSc thesis, University of Alberta, Edmonton.

Holsworth, W.N.
1980: Interaction between moose, elk and buffalo in Elk Island National Park, Alberta; unpublished MSc thesis, University of British Colulmbia, Vancouver, 92 p.

Hornbeck, J.W., Pierce, R.S., and Federer, C.A.
1970: Streamflow changes after forest clearance in New England; Water Resources Research, v. 6, p. 1124-1132.

Hsu, K.J.
1975: Catastrophic debris streams (sturzstroms) generated by rockfalls; Geological Society of America, Bulletin, v. 867, p. 129-140.

Hudson, H.R.
1983: Hydrology and sediment transport in the Elbow River Basin, S.W. Alberta; unpublished PhD thesis, University of Alberta, Edmonton.

Hughes, O.L.
1969: Distribution of open system pingos in central Yukon Territory with respect to glacial limits; Geological Survey of Canada, Paper 69-34, 8 p.

Hughes, O.L., Rampton, V.N., and Rutter, N.W.
1972: Quaternary geology and geomorphology, southern and central Yukon; 24th International Geological Congress, Montréal, Guidebook A-11, 59 p.

Hungr, O., Morgan, G.C., and Kellerhals, R.
1984: Quantitative analysis of debris torrent hazards for design of remedial measures; Canadian Geotechnical Journal, v. 21, p. 663-677.

Hydrological Atlas of Canada
1978: Inland Waters Directorate, Canada. Department of Fisheries and Environment, Ottawa.

Inland Waters Directorate
1977: Surface water quality in Canada — an overview; Canada Department of Fisheries and Department of the Environment, Inland Waters Directorate, Water Quality Branch, v. 1, 45 p.

Jackson, J.N.
1975: Welland and the Welland Canal: The Canal By-pass Project; Mika Publishing, Belleville, Ontario, 214 p.

Jackson, L.E., Jr.
1979: A catastrophic glacial outburst flood (jökulhlaup) mechanisms for debris flow generation at the Spiral Tunnels, Kicking Horse River basin, British Columbia; Canadian Geotechnical Journal, v. 16, p. 806-813.

Jackson, L.E., Jr. (co-ordinator)
1989: The influence of the Quaternary geology of Canada on man's environment; Chapter 12 in Quaternary Geology of Canada and Greenland, R.J. Fulton (ed.); Geological Survey of Canada, Geology of Canada no. 1 (also Geological Society of America, The Geology of North America, v. K-1).

Jackson, L.E., Jr., MacDonald, G.M., and Wilson, M.C.
1982: Paraglacial origin for terraced river sediments in Bow Valley, Alberta; Canadian Journal of Earth Sciences, v. 19, p. 2219-2231.

Jahn, A.
1976: Contemporaneous geomorphological processes in Longyeardalen, Vestspitsburgen (Svalbard); Biuletyn Peryglacijalny, v. 26, p. 253-268.

Jones, D.P., Ricker, K.E., Desloges, J.R., and Maxwell, M.
1985: Glacier outburst flood on the Noeick River: the draining of Ape Lake, British Columbia, October 20, 1984; Geological Survey of Canada, Open File 1139, 92 p.

Johnson, P.G.
1978: Rock glacier types and their drainage systems, Grizzly Creek, Yukon Territory; Canadian Journal of Earth Sciences, v. 15, p. 1496-1507.

Jonys, C.K.
1976: Acoustic measurement of sediment transport; Canada Department of the Environment, Inland Waters Directorate, Scientific Series No. 66.

Judson, S.
1968: Erosion of the land or what's happening to our continents; American Scientist, v. 56, no. 4, p. 356-374.

Julien, P. and Frenette, M.
1984: A model for predicting suspended load in northern streams; in Proceedings of Fourth Annual American Geophysical Union; Front Range Branch, Fort Collins, Colorado, p. 119-139.

Kadomura, H.
1980: Erosion by human activities in Japan; Geological Journal, v. 4.2, p. 133-144.

Keen, M.J. and Williams, G.L. (editors)
1989: The Geology of the Continental Margin of Eastern Canada; Geological Survey of Canada, Geology of Canada no. 2 (also Geological Society of America, The Geology of North America, v. I-1).

Kellerhals, R., Church, M., and Davies, L.B.
1979: Morphological effects of interbasin river diversions; Canadian Journal of Civil Engineering, v. 6, p. 18-31.

Kellerhals, R., Neill, C.R., and Bray, D.I.
1972: Hydraulic and geomorphic characteristics of rivers in Alberta; Research Council of Alberta, River Engineering and Surface Hydrology, Report 72-1.

Kemp, A.L.W., Anderson, T.W., Thomas, R.L., and Mudrochova, A.
1974: Sedimentation rates and recent sediment history of lakes Ontario, Erie and Huron; Journal of Sedimentary Petrology, v. 44, p. 207-218.

Kerfoot, D.E.
1973: Thermokarst features produced by man-made disturbances in the tundra terrain; in Research in Polar and Alpine Geomorphology, B.D. Fahey and R.D. Thompson (ed.); Proceedings, 3rd Guelph Symposium on Geomorphology, p. 60-72.

Kershaw, G.P. and Gill, D.
1979: Growth and decay of palsas and peat plateaus in the MacMillan Pass-Tsichu River area, Northwest Territories, Canada; Canadian Journal of Earth Sciences, v. 16, p. 1362-1374.

Ketcheson, J.W.
1980: Long-range effects of intensive cultivation and monoculture on the quality of southern Ontario soils; Canadian Journal of Soil Science, v. 60, p. 403-410.

Kindle, E.D.
1952: Dezadeash map-area, Yukon; Geological Survey of Canada, Memoir 268, 68 p.

Klassen, R.A.
1989: Geomorphology of lakes; in Chapter 9 of Quaternary Geology of Canada and Greenland, R.J. Fulton (ed.); Geological Survey of Canada, Geology of Canada, no. 1 (also Geological Society of America, The Geology of North America, v. K-1).

Klassen, R.A. and Shilts, W.W.
1982: Subbottom profiling of lakes of the Canadian Shield; in Current Research, Part A, Geological Survey of Canada, Paper 82-1A, p. 375-384.

Konishchev, V.N.
1978: Mineral stability in the zone of cryolithogenesis; Proceedings, Third International Conference on Permafrost, Volume 1; National Research Council of Canada, p. 305-311.

Krank, K. and Ruffman, A.
1982: Sedimentation in James Bay; Le Naturaliste Canadien, v. 109, p. 353-361.

La Rochelle, P., Chagnon, J-Y. and Lefebvre, G.
1970: Regional geology and landslides in the marine clay deposits of eastern Canada; Canadian Geotechnical Journal, v. 7, p. 145-156.

Lafleur, J. and Lefevbre, G.
1980: Groundwater regime associated with slope stability in Champlain Clay deposits; Canadian Geotechnical Journal, v. 14, p. 44-54.

Lamothe, C. and St-Onge, D.A.
1961: A note on a periglacial erosional process in the Isachsen area, N.W.T.; Geographical Bulletin, no. 16, p. 104-113.

Langbein, W.B. and Schumm, S.A.
1958: Yield of sediment in relation to mean annual precipitation; American Geophysical Union, Transactions, v. 39, p. 1076-1084.

Lasky, S.G.
1950: Mineral resource appraisal by the U.S. Geological Survey; Colorado School of Mines Quarterly, v. 45, p. 1-27.

Last, W.M.
1984: Sedimentology of playa lakes of the northern Great Plains; Canadian Journal of Earth Sciences, v. 21, p. 107-125.

Lebuis, J. and Rissmann, P.
1979: Earthflows in the Quebec and Shawinigan areas; in Sensitive clays, unstable slopes, corrective works, and slides in the Quebec and Shawinigan area; in Guidebook to Field Trip B-11, J-Y. Chagnon, J. Lebuis, J.D. Allard, and J.M. Robert (ed.); Geological Association of Canada, p. 18-38.

Legget, R.F.
1973: Cities and Geology; McGraw-Hill, New York, 448 p.

Lehre, A.K., Collins, B.D., and Dunne, T.
1983: Post-eruption sediment budget for the North Fork Toutle River Drainage, June 1980-June 1981; Zeitschrift für Geomorphologie, Supplementband, v. 46, p. 143- 163.

Leonoff, C.E. and Lo, R.C.
1982: Solution to frost heave of ice arenas; Proceedings, Fourth Canadian Permafrost Conference H.M. French (ed.); National Research Council of Canada, Ottawa, NRCC 20124 p. 481-486.

Levings, C.D.
1982: The ecological consequences of dredging and dredge spoil dispersal in Canadian waters; Association Committee on Scientific Criteria for Environmental Quality, NRCC 18130, 142 p.

Lewis, C.P. and McDonald, B.C.
1973: Rivers of the Yukon North Slope; in Fluvial Processes and Sedimentation, Hydrology Symposium No. 9; Inland Waters Directorate, Canada Department of the Environment, p. 251-271.

Livingstone, D.A.
1963: Chemical composition of rivers and lakes; U.S. Geological Survey, Professional Paper 708.

Luckman, B.H.
1971: The role of snow avalanches in the evolution of alpine talus slopes; Institute of British Geographers, Special Publication, v. 3, p. 93-110.
1978: Geomorphic work of snow avalanches in the Canadian Rocky Mountains; Arctic and Alpine Research, v. 10, p. 261-276.

Mackay, J.R.
1956: Notes on oriented lakes of the Liverpool Bay area, Northwest Territories; Revue Canadienne de Géographie, v. 10, p. 169-173.
1963a: Notes on the shoreline recession along the coast of the Yukon Territory; Arctic, v. 16, p. 195-197.
1963b: The Mackenzie Delta area, N.W.T.; Canada, Department of Mines and Technical Surveys, Geographical Branch, Memoir 8, 202 p.
1966: Segregated epigenetic ice and slumps in permafrost, Mackenzie Delta area, N.W.T.; Geographical Bulletin, v. 8, p. 59-80.
1970: Disturbances to the tundra and forest tundra environment of the western Arctic; Canadian Geotechnical Journal, v. 7, p. 420-432.

1971: The origin of massive icy beds in permafrost, western Arctic coast; Canadian Journal of Earth Sciences, v. 8, p. 397-422.
1972: The world of underground ice; Association of American Geographers, Annals, v. 62, p. 1-22.
1973: The growth of pingos, western Arctic Coast; Canadian Journal of Earth Sciences, v. 10, p. 979-1004.
1974a: Reticulate ice veins in permafrost, northern Canada; Canadian Geotechnical Journal, v. 11, p. 230-237.
1974b: Ice-wedge cracks, Garry Island, Northwest Territories; Canadian Journal of Earth Sciences, v. 11, p. 1366-1383.
1975: The closing of ice-wedge cracks in permafrost, Garry Island, Northwest Territories; Canadian Journal of Earth Sciences, v. 12, p. 1668-1674.
1979a: An equilibrium model for hummocks (non-sorted circles), Garry Island, Northwest Territories; in Current Research, Part A, Geological Survey of Canada, Paper 79-1A, p. 165-167.
1979b: Pingos of the Tuktoyaktuk Peninsula area, Northwest Territories; Géographie physique et Quaternaire, v. 33, p. 3-61.
1980: The origin of hummocks, western Arctic coast, Canada; Canadian Journal of Earth Sciences, v. 17, p. 996-1006.
1981: Active layer slope movement in a continuous permafrost environment, Garry Island, Northwest Territories, Canada; Canadian Journal of Earth Sciences, v. 18, p. 1666-1680.
1983: Downward water movement into frozen ground, western Arctic coast; Canadian Journal of Earth Sciences, v. 20, p. 120-134.
1984a: The direction of ice-wedge cracking in permafrost: downward or upward; Canadian Journal of Earth Sciences, v. 21, p. 516-524.
1984b: The frost heave of stones in the active layer above permafrost with downward and upward freezing; Arctic and Alpine Research, v. 16, p. 439-446.
1985: Pingo ice of the western Arctic coast, Canada; Canadian Journal of Earth Sciences, v. 22, p. 1452-1464.
1986a: The first 7 years (1978-1985) of ice wedge growth, Illisarvik experimental drained lake site, western Arctic coast; Canadian Journal of Earth Sciences, v. 23, p. 1782-1795.*
1986b: Frost mounds; in Focus: Permafrost Geomorphology, H.M. French (ed.); The Canadian Geographer, v. 30, p. 363-364.

Mackay, J.R. and Burrows, C.
1979: Uplift of objects by an upfreezing ice surface; Canadian Geotechnical Journal, v. 17, p. 609-613.

Mackay, J.R. and MacKay, D.K.
1976: Cryostatic pressures in nonsorted circules (mud hummocks), Inuvik, Northwest Territories; Canadian Journal of Earth Sciences, v. 13, p. 889-897.

Mackay, J.R. and Mathews, W.H.
1973: Geomorphology and Quaternary history of the Mackenzie River Valley near Fort Good Hope, N.W.T., Canada; Canadian Journal of Earth Sciences, v. 10, p. 26-41.
1974: Needle ice striped ground; Arctic and Alpine Research, v. 6, p. 79-84.

Mackay, J.R. and Matthews, J.V., Jr.
1983: Pleistocene ice and sand wedges, Hooper Island, Northwest Territories; Canadian Journal of Earth Sciences, v. 20, p. 1087-1097.

Mackay, J.R., Ostrick, J. and Lewis, C.P., and MacKay, D.K.
1979: Frost heave at ground temperatures below 0°C, Inuvik, Northwest Territories; in Current Research, Part A, Geological Survey of Canada, Paper 79-1A, p. 403-405.

Mackay, J.R., Rampton, V.N., and Fyles, J.G.
1972: Relic Pleistocene permafrost, western Arctic, Canada; Science, v. 176, p. 1321-1323.

Maini, J.S. and Carlisle, A.
1974: Conservation in Canada: A Conspectus; Canada Department of the Environment, Canadian Forestry Service, Publication No. 1340, Ottawa, 441 p.

Marshall, I.B.
1982: Mining, land use and the environment. I. A Canadian overview; Land Use Canada Series No. 22, Lands Directorate, Canada Department of the Environment, Ottawa.

Martin, C.W. and Pierce, R.S.
1980: Clearcutting patterns affect nitrate and calcium in streams of New Hampshire; Journal of Forestry, v. 78, p. 268-272.

Matheson, D.S. and Thomson, S.
1973: Geological implications of valley rebound; Canadian Journal of Earth Sciences, v. 10, p. 961-978.

Mathews, W.H.
1979: Landslides of Central Vancouver Island and the 1946 Earthquake; Seismological Society of America, Bulletin, v. 69, p. 445-450.

Mathews, W.H. and McTaggart, K.C.
1978: Hope rockslides, British Columbia; in Rockslides and Avalanches, v. 1, B. Voight (ed.); Elsevier Scientific Publishing Co., Amsterdam, p. 259-275.

McLelland, A.G.
1983: Pits and quarries — their land impacts and rehabilitation; in Stress on Land in Canada, Canada Department of the Environment, Policy Research and Development Branch, p. 183-226.

McPherson, H.J.
1975: Sediment yield from intermediate-sized stream basins in southern Alberta; Journal of Hydrology, v. 25, p. 243-257.

McRoberts, E.C. and Morgenstern, N.
1974a: The stability of thawing slopes; Canadian Geotechnical Journal, v. 11, p. 447-469.
1974b: Stability of slopes in frozen soil, Mackenzie Valley, N.W.T.; Canadian Geotechnical Journal, v. 11, p. 554-573.

Meybeck, M.
1983: Atmospheric inputs and river transport of dissolved substances (review paper); in Dissolved Loads of Rivers and Surface Water Quantity/Quality Relationships, B.W. Webb (ed.); International Association for Scientific Hydrology, Publication No. 141, p. 173-192.

Miller, R.D.
1972: Freezing and heaving of saturated and unsaturated soils; Highway Research Record, v. 393, p. 1-11.

Milliman, J.D.
1980: Sedimentation in the Fraser River and its estuary, southwestern British Columbia (Canada); Estuarine and Coastal Marine Science, v. 10, p. 609-633.

Milliman, J.D. and Meade, R.H.
1983: World-wide delivery of river sediment to the oceans; Journal of Geology, v. 91, p. 1-21.

Milne, W.G.
1956: Seismic activity in Canada west of the 113° meridian, 1841-1951; Dominion Observatory, Publication v. 18, p. 119-145.

Milne, W.G., Rogers, G.C., Riddihough, R.P., McMechan, G.A., and Hyndman, R.D.
1978: Seismicity of Western Canada; Canadian Journal of Earth Sciences, v. 15, p. 1170-1193.

Mirza, C. and Irwin, R.W.
1964: Determination of subsidence of an organic soil in southern Ontario; Canadian Journal of Soil Science, v. 44, p. 248-253.

Mitchell, B. and Sewell, W.R.D.
1981: Canadian Resource Policies: Problems and Prospects, Methuen, London, 294 p.

Mitchell, R.J. and Markell, A.R.
1974: Flowsliding in sensitive soils; Canadian Geotechnical Journal, v. 11, p. 11-31.

Mokievsky-Zubok, O.
1977: Glacier-caused slide near Pylon Peak, British Columbia; Canadian Journal of Earth Sciences, v. 14, p. 2657-2662.

Mollard, J.D.
1973: Airphoto interpretation of fluvial features; in Fluvial Processes and Sedimentation, Hydrology Symposium No. 9; Department of the Environment, Inland Waters Directorate, Ottawa, p. 341-380.
1977: Regional landslide types in Canada; Geological Society of America, Reviews in Engineering Geology, v. 3, p. 29-56,

Moore, D.P. and Mathews, W.H.
1978: The Rubble Creek landslide, southwestern British Columbia; Canadian Journal of Earth Sciences, v. 15, p. 1039-1052.

Nanson, G.C.
1974: Bedload and suspended-load transport in a small steep, mountain stream; American Journal of Science, v. 274, p. 471-486.

National Atlas of Canada
1974: Surveys and Mapping Branch, Canada Department of Energy, Mines and Resources, Ottawa.

Neill, C.R. and Mollard, J.D.
1980: Examples of erosion and sedimentation processes along some northern Canadian Rivers; International Symposium on River Sedimentation, Beijing, China, Guanghua Press, p. 565-597.

Nickling, W.G.
1978: Eolian sediment transport during dust storms: Slims River Valley, Yukon Territory; Canadian Journal of Earth Sciences, v. 15, p. 1069-1084.

Nir, D.
1983: Man, A Geomorphological Agent: An Introduction to Anthropic Geomorphology; D. Reidel, Dordrecht, 157 p.

Norris, L.A.
1981: The behaviour of herbicides in the forest environment; in Proceedings of the Conference on Weed Control in Forest Management, H.A. Holt and D.C. Fischer (ed.); Purdue University, Lafayette, p. 192-215.

O'Loughlin, C.L.
1972: A preliminary study of landslides in the Coast Mountains of southwestern British Columbia; in Mountain Geomorphology: Geomorphological Processes in the Canadian Cordillera, H.O. Slaymaker and H.J. McPherson (ed.); Tantalus Research Ltd., Vancouver, B.C. Geographical Series No. 14, p. 101-112.

Omara-Ojunga, P.
1980: Resource management in mountainous environments: the case of the East Slopes Region, Bow River Basin, Alberta; unpublished PhD thesis, Department of Geography, University of Waterloo, Waterloo.

Ongley, E.D.
1973: Sediment discharge from Canadian basins into Lake Ontario; Canadian Journal of Earth Sciences, v. 10, p. 146-156.

Ongley, E.D., Bynoe, M.C., and Percival, J.B.
1981: Physical and geochemical characteristics of suspended solids, Wilson Creek, Ontario; Canadian Journal of Earth Sciences, v. 18, p. 1365-1379.

Pank, L.F.
1974: A bibliography on seed-eating mammals and birds that affect forest regeneration; United States Department of the Interior, Fish and Wildlife Service, Special Scientific Report, Wildlife, Number 174, 28 p.

Parizeck, R.R.
1971: Impact of highways on the hydrogeologic environment; in Environmental Geomorphology, D.R. Coates (ed.); State University of New York, Binhampton, N.Y., p. 151-199.

Patton, F.D.
1976: The Devastation Glacier slide, Pemberton, B.C. (abstract); in Program and Abstracts, Geological Association of Canada, Cordillera Section, p. 26-27.

Pears, A.J.
1976: Geomorphic and hydrologic consequences of vegetation destruction, Sudbury, Ontario; Canadian Journal of Earth Sciences, v. 13, p. 1358-1373.

Pegau, R.E.
1970: Effect of reindeer trampling and grazing on lichens; Journal of Range Management, v. 23, p. 95-97.

Pissart, A. and French. H.M.
1976: Pingo investigations, north-central Banks Island, Canadian Arctic; Canadian Journal of Earth Sciences, v. 13, p. 937-946.

Pissart, A., Vincent, J-S., et Edlund, S.A.
1977: Dépôts et phénoménes éoliens sur l'Ile de Banks, Territoires du Nord-Ouest, Canada; Journal canadien des sciences de la terre, vol. 14, p. 2462-2840.

Polar Research Board
1984: Ice segregation and frost heaving; Ad Hoc Study Group on Ice Segregation and Frost Heaving, Polar Research Board, National Research Council, National Academy Press, Washington, D.C., 72 p.

Pollard, W.H. and French, H.M.
1980: A first approximation of the volume of ground ice, Richards Island, Pleistocene Mackenzie Delta, Canada; Canadian Geotechnical Journal, v. 17, p. 509-516.
1984: The groundwater hydraulics of seasonal frost mounds, North Fork Pass, Yukon Territory; Canadian Journal of Earth Sciences, v. 21, p. 1073-1081.
1985: The internal structure and ice crystallography of seasonal frost mounds; Journal of Glaciology, v. 31, p. 157-162.

Prest, V.K.
1969: Retreat of Wisconsin and Recent ice in North America; Geological Survey of Canada, Map l257A, scale 1:5 M.

Price, L.W.
1971: Geomorphic effect of the arctic ground squirrel in an alpine environment; Geografiska Annaler, v. 534, no. 2, p. 100-105.

Quigley, R.M., Gelinas, P.J., Bou, W.T., and Packer, R.W.
1977: Cyclic erosion-instability relationships: Lake Erie north shore bluffs; Canadian Geotechnical Journal, v. 14, p. 310-323.

Radforth, L.M. and Burwash, A.L.
1977: Transportation; in Muskeg and the Northern Environment in Canada, N.W. Radforth and C.O. Brawner (ed.); University of Toronto Press, Toronto, p. 249-263.

Rampton, V.N.
1973: The influence of ground ice and thermokarst upon the geomorphology of the Mackenzie-Beaufort region; in Resarch in Polar and Alpine Geomorphology, B.D. Fahey and R.D. Thompson (ed.); Proceedings, 3rd Guelph Symposium on Geomorphology, p. 43-59.
1982: Quaternary geology of the Yukon Coastal Plain; Geological Survey of Canada, Bulletin 317, 49 p.

Rampton, V.N. and Dugal, J.B.
1974: Quaternary stratigraphy and geomorphic processes on the Arctic coastal plain and adjacent areas, Demarcation Point, Yukon Territory, to Malloch Hill, District of Mackenzie; in Report of Activities, Part A, Geological Survey of Canada, Paper 74-1A, p. 283.

Rampton, V.N. and Mackay, J.R.
1971: Massive ice and icy sediments throughout the Tuktoyaktuk Peninsula, Richards Island, and nearby areas, District of Mackenzie; Geological Survey of Canada, Paper 71-21, 16 p.

Raup, H.M. and Argus, G.W.
1982: The Lake Athabasca sand dunes of northern Saskatchewan and Alberta, Canada. 1. The land and vegetation; National Museums of Canada, Publications in Botany, 12, 96 p.

Redmond, I.G.
1983: Land impact associated with the disposal of radioactive wastes; in Stress on Land in Canada, Canada Department of the Environment, Policy Research and Development Branch, Ottawa, p. 11-32.

Reeves, C.C., Jr.
1968: Introduction to paleolimnology; in Developments in Sedimentology 11; Elsevier Publishing Co., Amsterdam, 228 p.

Reid, L.M., Dunne, T., and Cederholm, C.J.
1981: Application of sediment budget studies to the evaluation of logging road impacts; New Zealand Journal of Hydrology, v. 20, p. 49-62.

Retzer, J.L., Swope, H.M., and Rutherford, W.H.
1956: Suitability of physical factors for beaver management in the Rocky Mountains of Colorado; State of Colorado, Department of Game and Fish, Technical Bulletin, 2, 33 p.

Rice, R.M., Tiley, F.B., and Datzman, P.A.
1979: A watershed's response to logging and roads: South Fork of Caspar Creek, California; U.S. Forest Service Research Paper, SW-146, 20 p.

Ripley, P.O., Kalbfleisch, W., Bourget, S.J., and Cooper, D.S.
1961: Soil erosion by water; Canada Department of Agriculture, Publication 1083.

Roberge, J. et Gascoyne, M.
1978: Premiers résultats de datations dans la Grotte de St.-Elzéar; Géographie physique et Quaternaire, vol. 32, p. 281-287.

Robinson, J.
1971: Highways and our environment; New York, McGraw-Hill, 340 p.

Rogers, G.C.
1980: A documentation of soil failure during the British Columbia earthquake of 23 June 1946; Canadian Geotechnical Journal, v. 17, p. 122-127.

Rowe, J.S.
1959: Forest regions of Canada: Department of Northern Affairs and Natural Resources, Forestry Branch, Bulletin 123.

Royal Society of Canada
1985: Nuclear winter and associated effects: a Canadian appraisal of the environmental impact of nuclear war; Royal Society of Canada, Ottawa, 382 p.

Ryder, J.M.
1971a: The stratigraphy and morphology of para-glacial alluvial fans in south-central British Columbia; Canadian Journal of Earth Sciences, v. 8, p. 279-298.
1971b: Some aspects of the morphometry of paraglacial alluvial fans in south-central British Columbia; Canadian Journal of Earth Sciences, v. 8, p. 1252-1264.
1981: Geomorphology of the southern part of the Coast Mountains of British Columbia; Zeitschrift für Geomorphologie, Supplementband 37, p. 120-147.

Ryder, R.A.
1964: Chemical characteristics of Ontario lakes as related to glacial history; American Fisheries Society, Transactions, v. 93, p. 260-268.

Safranyik, L., Van Sickle, G.A., and Manning, G.H.
1981: Position paper on mountain pine beetle problems with special reference to the Rocky Mountain parks region; Canada Department of the Environment, Canadian Forestry Service Publication, 27 p.

Saini, G.R. and Grant, W.J.
1980: Long-term effects of intensive cultivation on soil quality in the potato-growing areas of New Brunswick (Canada) and Maine (U.S.A.); Canadian Journal of Earth Sciences, v. 60, p. 421-428.

St-Onge, D.A.
1969: Nivation landforms; Geological Survey of Canada, Paper 69-30, 12 p.

Schreiner, B.T., Acton, D.F., and David, P.P.
1981: Geology; in Athabasca sand dunes in Saskatchewan, Z.M. Abouguendia (ed.); p. 8-52, Mackenzie River Basin Committee, Mackenzie River Basin Study Report, Supplement 7, 335 p., Appendices A-G.

Schroeder, J.
1979: Le développement des grottes dans la région du Premier Canyon de la Rivière Nahanni Sud, T.N.O. thèse de PhD non publiée, Université d'Ottawa, Ottawa, 265 p.

Schumm, S.A.
1977: The Fluvial System; Wiley Interscience, New York.
Schwab, J.W. and Watt, W.J.
1981: Logging and soil disturbance on steep slopes in the Quesnel Highlands, Cariboo Forest Region; British Columbia, Ministry of Forests, Research Note 88, 15 p.
Shaw, J. and Kellerhals, R.
1982: The composition of recent alluvial gravel in Alberta river beds; Alberta Geological Survey, Alberta Resarch Council, Bulletin 4.
Shilts, W.W.
1978: Nature and genesis of mudboils, central Keewatin, Canada; Canadian Journal of Earth Sciences, v. 15, p. 1053-1068.
1984: Sonar evidence for postglacial tectonic instability of the Canadian Shield and Appalachians; in Current Research, Part A, Geological Survey of Canada, Paper 84-1A, p. 567-579.
Shilts, W.W. and Dean, W.E.
1975: Permafrost features under arctic lakes, District of Keewatin, Northwest Territories; Canadian Journal of Earth Sciences, v. 12, p. 649-662.
Shilts, W.W., Dean, W.E., and Klassen, R.A.
1976: Physical, chemical and stratigraphic aspects of sedimentation in lake basins of the eastern Arctic Shield; in Report of Activities, Part A, Geological Survey of Canada, Paper 76-1A, p. 245-254.
Singh, T.
1970: Land management practises that affect physical and chemical water quality; in Managing Lands for Water, D.L. Golding (ed.), Canada Department of the Environment, Northern Forest Resarch Centre, Report Nor-X-13, p. 49-72.
Sklash, M.G., Stewart, M.K., and Pearce, A.J.
1986: Storm runoff generation in humid headwater catchments: 2: A case study of hillslope and low-order stream response; Water Resources Research, v. 22, p. 1273-1282.
Slaymaker, H.O.
1982: The occurrence of piping and gullying in the Penticton glaciolacustrine silts, Okanagan Valley, British Columbia; in Badland Geomorphology and Piping, R.B. Bryan and A. Yair (ed.); Geo Abstracts, Norwich, United Kingdom, p. 305-316.
1989: Physiography of Canada and its effects on geomorphic processes; in Chapter 9 of Quaternary Geology of Canada and Greenland, R.J. Fulton (ed.); Geological Survey of Canada, Geology of Canada, no. 1 (also Geological Society of America, The Geology of North America, v. K-1).
Slaymaker, H.O. and McPherson, H.J.
1973: Effects of land use on sediment production; in Fluvial Processes and Sedimentation; Hydrology Syposium No. 9, Ottawa, Canada Department of the Environment, Inland Waters Directorate, p. 158-183.
1977: An overview of geomorphic processes in the Canadian Cordillera; Zeitschrift für Geomorphologie, v. 21, p. 169-186.
Sly, P.G. and Prior, J.W.
1984: Late glacial and postglacial geology in the Lake Ontario Basin: Canadian Journal of Earth Sciences, v. 21, p. 802-821.
Sly, P.G. and Thomas, R.L.
1974: Review of geological research as it relates to an understanding of Great Lakes terminology; Journal of the Fisheries Research Board of Canada, v. 31, p. 795-825.
Smith, D.G. and Smith N.D.
1980: Sedimentation in anastomosed river systems: examples from alluvial valleys near Banff, Alberta; Journal of Sedimentary Petrology, v. 50, p. 1057-1064.
Smith, D.J.
1972: Solution of limestone in an arctic environment; in Polar Geomorphology, R.J. Price and D.E. Sugden (ed.); Institute of British Geographers, Special Publication No. 4, p. 187-200.
Smith, D.M.
1975: Sedimentary environments and late Quaternary history of a 'low-energy' mountain delta; Canadian Journal of Earth Sciences, v. 12, p. 2004-2013.
Smith, M.W.
1985a: Observations of soil freezing and frost heave at Inuvik, Northwest Territories, Canada; Canadian Journal of Earth Sciences, v. 22, p. 283-290.
1985b: Models of soil freezing; in Field and Theory: Lectures in Geocryology, M. Church and O. Slaymaker (ed.); University of British Columbia Press, Vancouver, p. 96-120.
1986: Frost action and soil freezing; in Focus: Permafrost Geomorphology, H.M. French (ed.); The Canadian Geographer, v. 30, p. 353-360.
1989: Climate of Canada; in Chapter 9 of Quaternary Geology of Canada and Greenland, R.J. Fulton (ed.); Geological Survey of Canada, Geology of Canada, no. 1 (also Geological Society of America, The Geology of North America, v. K-1).

Smith, W.E.T.
1962: Earthquakes of eastern Canada and adjacent area (1534-1927); Dominion Observatory, Publications, v. 26, p. 271-301.
Souther, J.G.
1975: Geothermal potential of western Canada; Proceedings, 2nd United Nations Symposium on the Development and Use of Geothermal Resources, v. 1, p. 259-267.
1981: Volcanic hazards in the Stikine region of northwestern British Columbia; Geological Survey of Canada, Open File 770.
Stangl, K., Roggensack, W.D., and Hayley, D.W.
1982: Engineering geology of surficial soils, eastern Melville Island; Proceedings, Fourth Canadian Permafrost Conference, H.M. French (ed.); National Research Council of Canada, Ottawa, NRCC 20124, p. 136-147.
Steinbrenner, E.C. and Gessel, S.P.
1955: The effect of tractor logging on the physical properties of some forest soils in S.W. Washington; Soil Science Society of America, Proceedings, v. 19, p. 372-376.
Stelfax, T.G.
1971: Bighorn sheep in the Canadian Rockies: a history 1800-1970; Canadian Field Naturalist, v. 85, p. 101-122.
Stethem, C.J. and Schaerer, P.A.
1980: Avalanche accidents in Canada. II. A selection of case histories of accidents 1943 to 1978; National Research Council of Canada, Division of Building Research, Paper 926.
Stevens, G.R.
1973: History of the Canadian National Railways; Macmillan, New York, 538 p.
Stichling, W.
1973: Sediment loads in Canadian rivers; in Fluvial Processes and Sedimentation; Hydrology Symposium No. 9, Inland Waters Directorate, Canada Department of the Environment, p. 38-72.
Stoecker, R.
1976: Pocket gopher distribution in relation to snow in the alpine tundra; in Ecological impacts of snowpack augmentation in the San Juan Mountains of Colorado; Colorado State University, Fort Collins, p. 281-287.
Sutherland-Brown, A.
1969: Aiyansh lava flow, British Columbia; Canadian Journal of Earth Sciences, v. 6, p. 1460-1468.
Swanston, O.N. and Swanston, F.J.
1976: Timber harvesting, mass erosion, and steepland forest geomorphology in the Pacific Northwest; in Geomorphology and Engineering, D.R. Coates (ed.); Stroudsbury, Penn, Dowden, Hutchinson and Ross, Inc., Stroudsbury, Pa., p. 199-221.
Terzaghi, K.
1962: Stability of steep slopes on hard unweathered rock; Geotechnique, v. 12, p. 251-270.
Thomas, W.L. (editor)
1956: Man's Role in Changing the Face of the Earth; University of Chicago, Chicago, Illinois, 1193 p.
Thomson, R.E.
1981: Oceanography of the British Columbia coast; Canada Department of Fisheries and Oceans, 291 p.
Thomson, S. and Hayley, D.W.
1975: The Little Smokey Landslide; Canadian Geotechnical Journal, v. 12, p. 379-392.
Thorn, C.E.
1978: A preliminary assessment of the geomorphic role of pocket gophers in the alpine tundra zone of the Colorado Front Range; Geografiska Annaler, v. 60A, p. 181-187.
1980: Alpine bedrock temperatures: an empirical study; Arctic and Alpine Research, v. 12, p. 73-86.
1982: Gopher disturbance: its variability by Braun-Blanquet vegetation units in the Niwot Ridge alpine tundra zone, Colorado Front Range, U.S.A.; Arctic and Alpine Research, v. 14, no. 1, p. 45-51.
Toews, D.A.A. and Moore, M.K.
1982: The effects of streamside logging on large debris on Carnation Creek; British Columbia Ministry of Forests, Vancouver, Land Management Report no. 11.
Tremblay, L.P.
1961: Wind striations in northern Alberta and Saskatchewan, Canada; Geological Survey of America, Bulletin, v. 72, p. 1561-1564.
Turco, R.P., Toon, O.B., Ackerman, T.P., Pollack, J.B., and Sagan, C.
1983: Nuclear winter: global consequences of multiple nuclear explosions; Science, v. 222, p. 1283-1300.
Turner, A.R.
1955: How Saskatchewan dealt with her 'dustbowl'; Geographical Magazine, v. 28, p. 182-192.

Tywoniuk, N.
1977: A study of the bedload transport of a gravel-bed river; unpublished PhD thesis, University of Ottawa, Ottawa.

Vallejo, L.E.
1980: A new approach to the stability analysis of thawing slopes; Canadian Geotechnical Journal, v. 17, p. 607-612.

VanDine, D.F.
1985: Debris flows and debris torrents in the Southern Canadian Cordillera; Canadian Geotechnical Journal, v. 22, p. 44-68.

van Everdingen, R.O.
1978: Frost mounds at Bear Rock, near Fort Norman, Northwest Territories, 1975-1976; Canadian Journal of Earth Sciences, v. 15, p. 263-276.
1981: Morphology, hydrology and hydrochemistry of karst in permafrost terrain near Great Bear Lake, Northwest Territories; National Hydrology Research Institute, Paper 11 (IWD Scientific Series No. 114), 53 p.
1982: Frost blisters of the Bear Rock spring area near Fort Norman, N.W.T.; Arctic, v. 35, p. 243-265.

van Vuuren, W.
1978: Erosion, a growing concern for the future; Notes on Agriculture, v. 4, p. 17-20.

van Zon, H.J.M.
1981: Biological aspects of process measurements; in Geomorphological Techniques, A. Goudie, M. Anderson, T. Burt, J. Lewin, K. Richards, B. Whalley, and P. Worsley (ed.); George Allen and Unwin, London, p. 266-273.

Vaníček, P. and Nagy, D.
1981: On the compilation of the map of contemporary vertical crustal movements in Canada; Tectonophysics, v. 71, p. 75-86.

Vita-Finzi, C.
1977: Physiographic effects of man; Encyclopedia Britannica, Macropaedia, v. 14, p. 429-433.

Voight, B. (editor)
1978: Rockslides and Avalanches. I. Natural Phenomena; Elsevier, New York, 833 p.

Vontobel, R.
1980: Our precious topsoil is wasting away; Canadian Geographic, v. 100, p. 50-59.

Waltham, T.
1978: Catastrophe, the Violent Earth; Crown Publishers, Inc., New York, 170 p.

Warrick, R.A. and Bowden, M.
1981: The changing impacts of drought in the Great Plain; in Great Plains, Perspectives and Prospectus, M. Lawson, and M. Baker (ed.); Center for Great Plains Studies, Lincoln, Nebraska.

Washburn, A.L.
1967: Instrumental observations of mass-wasting in the Mesters Vig district, Northeast Greenland; Meddelelser om Gronland, v. 166, 318 p.
1979: Geocryology: A Survey of Periglacial Processes and Environments; Edward Arnold, London, 406 p.
1983: Palsa and continuous permafrost; Procedings, Fourth International Conference on Permafrost; National Academy Press, Washington, D.C., p. 1372-1377.

Watts, S.H.
1983: Weathering pit formation in bedrock near Cory Glacier, southeastern Ellesmere Island, Northwest Territories; in Current Research, Part A, Geological Survey of Canada, Paper 83-1A, p. 487-491.

Webber, L.R.
1964: Soil physical properties and erosion control; Journal of Soil and Water Conservation, p. 28-30.

Weetman, G.F.
1983: Forestry practices and stress on Canadian forest land; in Stress on Land in Canada, Canada Department of the Environment, Policy Research and Development Branch, Ottawa, p. 259-302.

Weetman, G.F. and Webber, L.R.
1972: The influence of wood harvesting on the nutrient status of two spruce stands; Canadian Journal of Forest Research, v. 2, p. 351-369.

Western Canada Hydraulics Laboratories Ltd.
1978: Analysis of federal sediment survey data taken on the Lower Fraser River; report submitted to Water Survey of Canada, Vancouver, October, 120 p.

Westgate, J.A.
1970: The Quaternary geology of the Edmonton area, Alberta; in Pedology and Quaternray Research, Symposium Proceedings, S. Pawluk (ed.); University of Alberta, Edmonton, p. 129-151.

Westing, A.H. (editor)
1984: Environmental Warfare: A Technical, Legal and Policy Appraisal; Stockholm Peace Research Institute, Taylor and Francis, London, 107 p.

Wetmiller, R.J., Basham, P.W., Weichert, D.H., and Evans, S.G.
1987: The 1985 Nahanni earthquakes: problems for seismic hazard estimates in the northeast Canadian Cordillera; in Earthquake Engineering/Génie Séismique; Proceedings, Fifth Canadian Conference on Earthquake Engineering, p. 695-703.

White, G.H.
1972: A history of fire in North America; in Fire in the Environment, Symposium Proceedings, U.S. Forest Service, Denver, p. 3-11.

Wiggins, I.L. and Thomas, J.H.
1962: A flora of the Alaskan Arctic Slope; Arctic Institute of North America, Special Publication 4, 425 p.

Wilford, D.J. and Schwab, J.W.
1982: Soil mass movements in the Rennell Sound area, Queen Charlotte Islands, British Columbia; Canadian Hydrology Symposium, Associate Committee on Hydrology, National Research Council of Canada, v. 82, p. 521-541.

Williams, O.
1959: Food habits of the deer mouse; Journal of Mammology, v. 40, p. 415-419.

Williams, P.J.
1968: Ice distribution in permafrost profiles; Canadian Journal of Earth Sciences, v. 5, p. 1381-1386.
1976: Volume change in frozen soils; Laurits Bjerrum Memorial Volume, Norwegian Geotechnical Institute, Oslo, Norway, p. 233-246.

Wu, T.M., McKinnell, W.P., and Swanston, D.N.
1979: Strength of tree roots and landslides on Prince of Wales Island, Alaska; Canadian Geotechnical Journal, v. 16, p. 19-33.

Yeates, M.
1975: Main Street: Windsor to Quebec City; MacMillan, Toronto.

Zerman, L.J. and Slaymaker, H.O.
1978: Mass balance model for calculation of ionic input loads in atmospheric fallout and discharge from a mountain basin; Hydrological Sciences Bulletin, v. 23, p. 103-117.

Zoltai, S.C. and Pettapiece, W.W.
1973: Studies of vegetation, landform and permafrost in the Mackenzie Valley: terrain, vegetation and permafrost relationships in the northern part of the Mackenzie Valley and northern Yukon; Environmental-Social Program, Northern Pipelines, Task Force on Northern Oil Development, Report 73-4, p. 105.

Zoltai, S.C. and Tarnocai, C.
1971: Properties of a wooded palsa in northern Manitoba; Arctic and Alpine Research, v. 3, p. 115-129.
1974: Soils and vegetation of hummocky terrain; Environmental-Social Program, Northern Pipelines, Task Force on Northern Oil Development, Report 74-5, 86 p.
1975: Perennially frozen peatlands in the western Arctic and subarctic of Canada; Canadian Journal of Earth Sciences, v. 12, p. 28-43.

Zoltai, S.C., Tarnocai, C., and Pettapiece, W.W.
1978: Age of cryoturbated organic materials in earth hummocks from the Canadian Arctic; in Proceedings of the Third International Conference on Permafrost, Edmonton, Volume 1; National Research Council of Canada, p. 325-331.

Zumberge, J.H.
1952: The lakes of Minnesota, their origin and classification; Minnesota Geological Survey, Bulletin 35, 99 p.

Authors' addresses

M.J. Bovis
Department of Geography
University of British Columbia
Vancouver, B.C.
V6T 1W5

M.A. Carson and Associates
4533 Rithetwood Drive
Victoria, B.C.
V8X 4J5

P.P. David
Department of Geology
University of Montréal
Montréal, Quebec
H3C 3J7

T.J. Day
Water Survey of Canada
Water Resources Branch
Environment Canada
Ottawa, Ontario
K1A 0E7

P.A. Egginton
Geological Survey of Canada
601 Booth Street
Ottawa, Ontario
K1A 0E8

S.G. Evans
Geological Survey of Canada
601 Booth Street
Ottawa, Ontario
K1A 0E8

D.C. Ford
Department of Geography
McMaster University
Hamilton, Ontario
L8S 4K1

H.M. French
Department of Geography and Geology
University of Ottawa
Ottawa, Ontario
K1N 6N5

J.A. Heginbottom
Geological Survey of Canada
601 Booth Street
Ottawa, Ontario
K1A 0E8

K. Hewitt
Department of Geography
Wilfred Laurier University
Waterloo, Ontario
N2L 3C5

R.A. Klassen
Geological Survey of Canada
601 Booth Street
Ottawa, Ontario
K1A 0E8

H.O. Slaymaker
Department of Geography
University of British Columbia
Vancouver, B.C.
V6T 1W5

M.W. Smith
Department of Geography
Carleton University
Ottawa, Ontario
K1S 5B6

Printed in Canada

Chapter 10

TERRAIN GEOCHEMISTRY IN CANADA

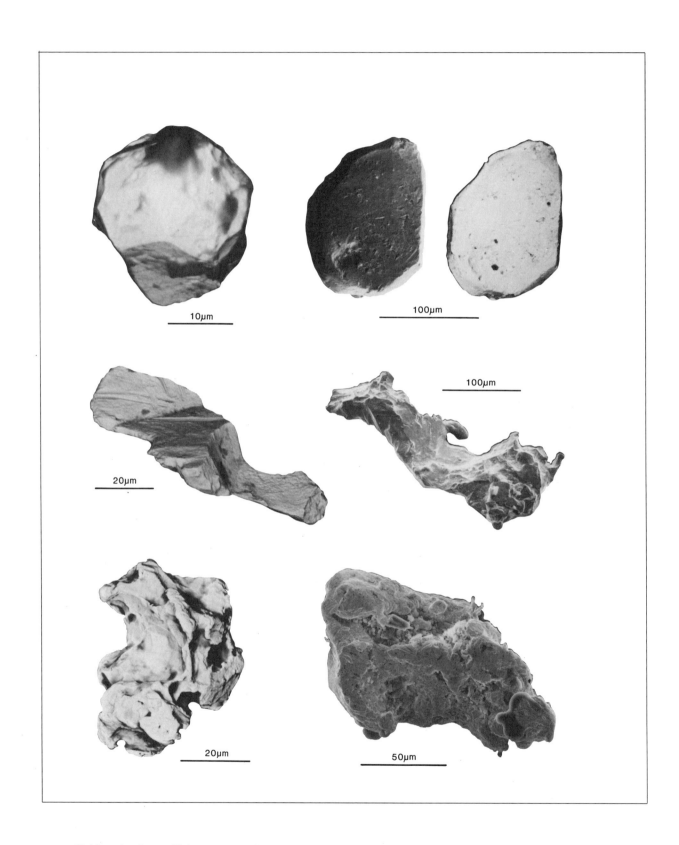

Gold grains from till from several Canadian localities. The upper four retain primary crystal, fracture-filling, and wiry shapes. The lower two show the more common curled and abraded (striated) shapes. Most are silt or fine sand sized as indicated by scale bars. Courtesy of R.N.W. DiLabio.

Chapter 10

TERRAIN GEOCHEMISTRY IN CANADA

R.N.W. DiLabio

SUMMARY

Terrain geochemistry is the study of the geochemistry of surficial materials to solve geological and environmental problems. The areal and stratigraphic collection of geochemical data shows the natural geochemical variation between extremes of impoverishment and enrichment of components in surficial materials. In Canada's glaciated landscape, till is the prime sample type because it is widespread and because it is first-cycle sediment that has a relatively simple history of erosion, transport, and deposition. Trace and minor element data sets exist for till and other media in many areas of the country at reconnaissance and detailed scales of sampling. These data are usable for mineral exploration, for environmental studies, and for studies of Quaternary stratigraphy, mapping, and history.

Drift prospecting is the determination of the provenance of glacial sediments (mainly till) in the search for mineral deposits. This exploration method is based on the premise that debris is usually eroded glacially from a subcropping ore deposit and the debris is deposited in a coherent dispersal train down-ice from the source. Boulder tracing and geochemical analysis of till are the main techniques for mapping dispersal trains.

Recycled till and other glacial sediments, organic detritus, and debris eroded directly from bedrock all contribute to postglacial stream and lake sediments. These sampling media have been used over large regions of the country to characterize the surficial geochemistry of areas where till is unavailable or not economical to collect.

In environmental geology and geomedicine, geochemical data on surficial sediments form the basis for estimating the sensitivity of terrain to acid precipitation and industrial fallout and for identification of areas that are naturally enriched in potentially noxious trace elements or naturally depleted in essential trace nutrients.

Geochemical and lithological information on glacial sediments, mainly till, has been used in the field of classical Quaternary geology to determine ice flow directions, provenance, and stratigraphy.

INTRODUCTION

Terrain geochemistry is defined here as the study of the geochemistry of surficial sediments in order to solve geological and environmental problems. In regions that were glaciated during the Quaternary, geochemical research is hampered by the complexity of the surficial sediments, which are largely allochthonous in relation to the bedrock they overlie. In the context of Canada's almost totally glaciated landscape, the sediments have particular characteristics that influence the selection of sample media, sampling design, and interpretation of data. Foremost among these characteristics is that most of the surficial sediments are till or recycled derivatives of it. Till has distinctive provenance features, is widespread, and is the parent material for most of the other surficial sediments. Data on the geochemistry of till and its derivatives are applicable in mineral exploration, environmental assessments, and Quaternary geology.

Glacial dispersal

Till was produced during the Quaternary by the glacial erosion, transport, and deposition of fresh and weathered unconsolidated sediments and bedrock. It is therefore a geologically young sediment which, at any given site, is not an in

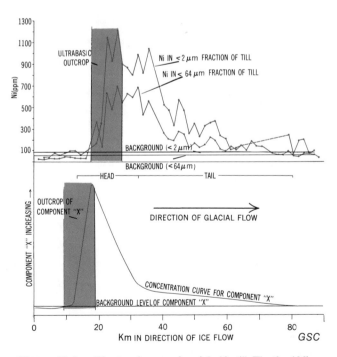

Figure 10.1. Dispersal curves for nickel in till, Thetford Mines area, Quebec. Actual (top) and idealized (bottom) curves show the relationship of the head and tail of a negative exponential curve (after Shilts, 1976).

DiLabio, R.N.W.
1989: Terrain geochemistry in Canada; Chapter 10 in Quaternary Geology of Canada and Greenland, R.J. Fulton (ed.); Geological Survey of Canada, Geology of Canada, no. 1 (also Geological Society of America, The Geology of North America, v. K-1).

CHAPTER 10

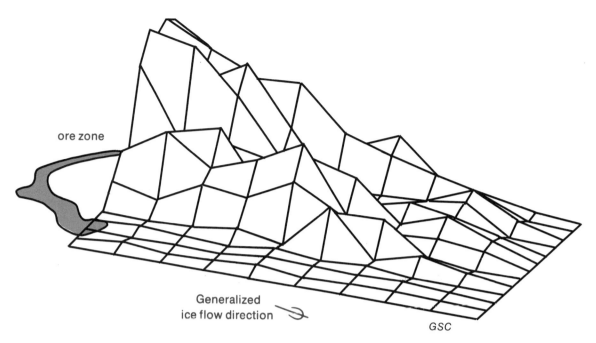

Figure 10.2. Perspective plot of the abundance of ore pebbles in till samples collected down-ice from an ore zone. Grid rectangles represent 30 by 60 m. The highest peak represents 99 per cent (after DiLabio, 1981).

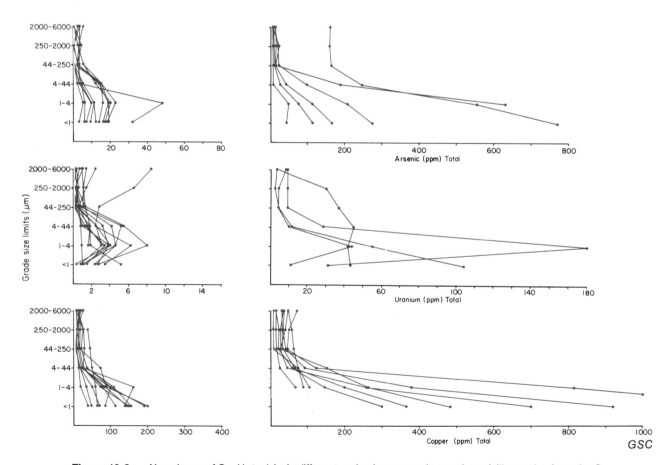

Figure 10.3. Abundance of Cu, U, and As in different grain size ranges in a variety of till samples from the Canadian Shield (after Shilts, 1984). Samples in right-hand figures considered to be anomalous; those on left represent background. Each line represents one sample. Concentrations are from perchloric acid leach and represent "total" metal.

situ weathering product but a lithological summation of source units up-ice from the site. Debris from any size of source unit is dispersed down-ice to produce a ribbon-shaped or flame-shaped dispersal train — a body of till that is enriched in debris from the source relative to the till surrounding the train. Shilts (1976) has shown that a plot of the abundance of glacially dispersed debris vs. distance down-ice approximates a negative exponential curve (Fig. 10.1) in which the concentration of a component reaches a peak near its source and then declines exponentially to background levels down-ice. He called the area of the dispersal train containing the peak the "head", and the area of falling concentrations the "tail". Many dispersal trains also have abrupt lateral edges, with a sharp contrast over a short distance between low concentrations outside the train and high concentrations inside it (Fig. 10.2). The blending of trains derived from different up-ice sources produces the mixed lithology that is a normal feature of till. Most of the individual dispersal trains are not identifiable, however, because they are too small or are composed of rocks and minerals that are not distinctive. The size and shape of a dispersal train are controlled by the orientation of the source relative to ice flow, by the size and erodibility of the source, and by the topography of the source and dispersal areas, which can trap trains in valleys or break them into disjointed segments in rough terrain.

Distinctive components (e.g., gold, tungsten, tin, chromium, niobium, platinum) allow the identification of trains where such components are rare in the rocks of the dispersal area. Shilts (1976) has shown, for example, that chromium-rich minerals derived from ultramafic rocks form distinct dispersal trains because bedrock in most of the dispersal areas is chromium-poor. Similarly, trains of radioactive components can be mapped by airborne radiometric methods. Analytical sensitivity also enhances the distinctiveness of a component; for example, dispersal trains of gold have been identified in which the gold content of the heavy minerals in the till is only about 3 parts per million in contrast to much lower levels (as low as 50 parts per billion) outside the trains, a distinction that is possible only with highly sensitive analytical techniques.

Recycling of sediment

Till that is derived directly by erosion of bedrock is a first-cycle sediment (first derivative of bedrock, Shilts, 1976). Sediments resulting from reworking of till or other unconsolidated sediments are second-cycle sediments; they have

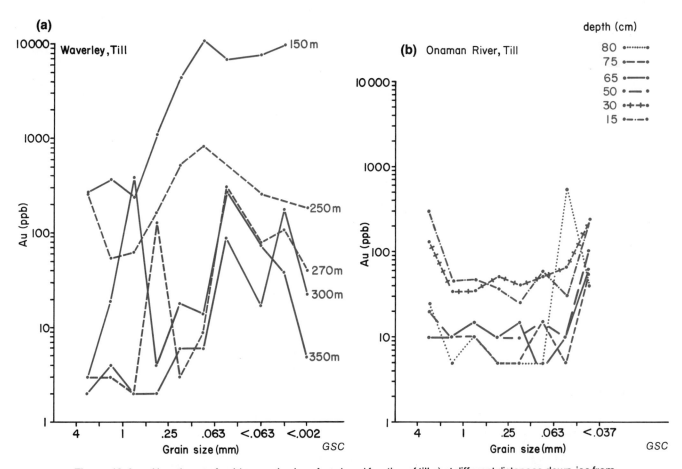

Figure 10.4. Abundance of gold vs. grain size of analyzed fraction of till a) at different distances down-ice from a gold deposit at Waverley, Nova Scotia (after DiLabio, 1982a), b) at varying depths below surface in a dispersal train at Onaman River, Ontario (after DiLabio, 1985).

been subjected to sorting and have undergone an episode of transport in water along a different path from the original. In this way, glaciofluvial gravel and sand represent the coarse fractions and glaciolacustrine silt and clay represent the fine fractions of the till from which they were derived. Because these sediments have travelled along transport paths consisting of two vectors, and have been transported first by ice then by water, it is more difficult to interpret their provenance than it is to trace till to its bedrock source. Postglacial stream and lake sediments are classed as third-cycle sediments because most of them have been recycled from all types of weathered and unweathered glacial and nonglacial sediments and postglacial organic debris and have been transported along a complex path, that has usually consisted of at least three cycles of erosion and deposition between the bedrock source and final site of deposition.

Sampling designs and types of media

Samples used in terrain geochemical surveys are collected usually at or near the surface to cover a specific area, for example, a map area, favourable geological structure, or drainage basin. Samples are collected on a grid pattern or a linear pattern. Recently, increasing use has been made of samples collected stratigraphically using various drilling techniques to produce a three-dimensional data set (Geddes and Kristjansson, 1986). The data derived from a set of samples usually represent the natural or "baseline" geochemical or lithological levels in the sampled surficial material; rarely, the data represent the effects of pollution (artificial contamination).

The types of surficial sediments most commonly used as sampling media are lake sediments, till, and stream sediments. Till is a prime sampling medium because it is first-

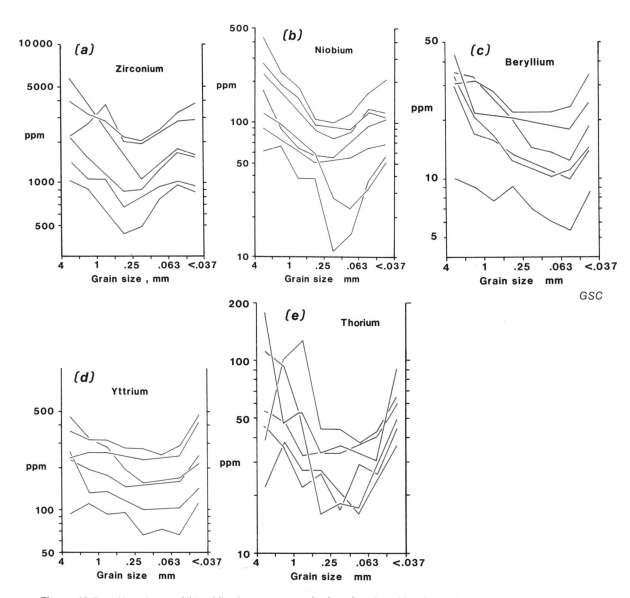

Figure 10.5. Abundance of lithophile elements vs. grain size of analyzed fraction of till from a dispersal train at Lac Brisson, Quebec and Newfoundland. Samples represent anomalous to background sites.

cycle sediment that is widespread. Where till is difficult to collect, for example in the boreal forest, lake sediments have become the prime reconnaissance sampling medium for detecting hydromorphically dispersed trace elements representing the average geochemistry of drainage basins. Stream sediments are most commonly collected in mountainous terrain where they have the highest likelihood of reflecting bedrock lithology because of the active downcutting of high-gradient streams (i.e., they are first-cycle sediments in such an environment).

Several different grain size fractions have been used routinely in the analyses of these media. The <80 mesh (<180 μm) fraction is the traditional analyzed fraction of stream and lake sediments in surveys for base and precious metals and uranium. Surveys using stream sediments and till in the search for deposits containing resistate minerals (e.g., gold, scheelite, cassiterite, platinum) have used sand-sized heavy minerals for analysis. Unstable minerals such as sulphides are seldom recoverable in sand-sized heavy mineral concentrates collected at the surface, but they are routinely recovered where unweathered till samples have been collected from below the surface by drilling.

Experimental work has shown that the geochemically most active fraction of weathered till lies in its finer grain sizes because of the tendency for phyllosilicate minerals and secondary minerals to be enriched in the fine sizes. These phases have a high total surface area and exchange capacity, so they act as scavengers, adsorbing representative portions of the trace metals that are released during weathering of primary, particularly sulphide, minerals. This pattern seems to hold for the base metals and uranium, for which the largest body of data is available (Fig. 10.3). The <2 μm fraction is the best fine grained fraction of weathered till to use when analyzing for many elements because it is more metal-rich than any other fraction. Often, however, the <63 μm (<250 mesh) fraction is analyzed because it is easier to recover.

Similar fractionation experiments on gold-bearing till (DiLabio, 1982a, 1985; Nichol, 1986; Shelp and Nichol, in press) indicate that gold distributions in weathered till are rather complex, being the result of the combined effects of the grain size of glacially comminuted detrital particulate gold, the grain size of gold released from weathering sulphides, the grain size of reprecipitated or adsorbed gold, and the original grain size of the gold at its source. In general, weathered till is richest in gold in its finer size ranges, although coarse sizes (>2 mm, rock fragments) are often auriferous (Fig. 10.4).

In a set of samples from one locality, the lithophile trace elements show a tendency to be enriched in the coarser grain

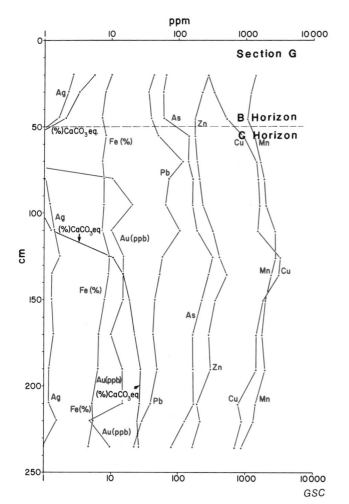

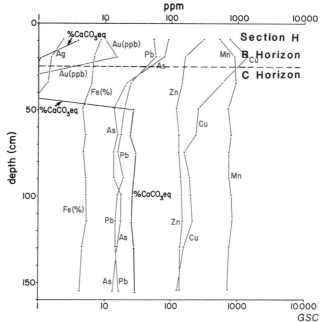

Figure 10.6. Variations in trace element abundances in the <2 μm fraction of the B and C horizons of soil developed on till at Onaman River, Ontario (after DiLabio, 1982b). Gold abundances and per cent CaCO₃ equivalent were determined on the <63 μm fraction. Vertical scale is depth below surface.

size ranges (the size ranges of rock fragments) and to a lesser degree in fine sizes (Fig. 10.5). Because most of these elements are contained in resistate and silicate minerals (pyrochlore, zircon, gittinsite) at this site, the peak of abundance in the finer sizes likely reflects the grain size of the host minerals in bedrock.

Although fractionation results have been collected to guide sampling and analytical programs using till, they have the additional value of showing the original distributions of trace elements in first-cycle sediment, before the sediment itself or the trace elements it contains are recycled into postglacial stream and lake sediments, sample media also commonly used in Canada.

Another important aspect of the weathering of till that influences the geochemistry of till and its derivatives is its behaviour within the solum during postglacial time. Unlike in situ soils in unglaciated terrain, those in glaciated terrain are geologically young, usually <10 ka, and their parent materials are immature allochthonous sediments derived from a variety of sources. Soils developed on till should therefore show a chemistry indicative of their original detrital lithology modified by the effects of trace element redistribution attributable to soil-forming processes. This geochemical modification is illustrated by examining in detail the geochemistry of the B and C horizons of soil profiles developed on till within a dispersal train (Fig. 10.6). In this example, glacial erosion of a mineral occurrence has produced a dispersal train enriched in copper, zinc, silver, and gold (DiLabio, 1982b). The till also contains abundant allochthonous carbonates derived from Paleozoic bedrock 130 km up-ice. The profiles were sampled on the thin edge (section H) and at a thicker part (section G) of the elongated lens of metal-rich till that constitutes the train. This lens of till is part of an apparently homogenous till sheet, most of which is metal-poor. In section H, the dispersal train is restricted to the upper 40 cm of the profile, shown by the restriction of the higher values of the geochemically less mobile lead, arsenic, silver, and gold to that interval. Copper enrichment above the local background value (~100 ppm) deeper in the profile results from pedogenic transport of

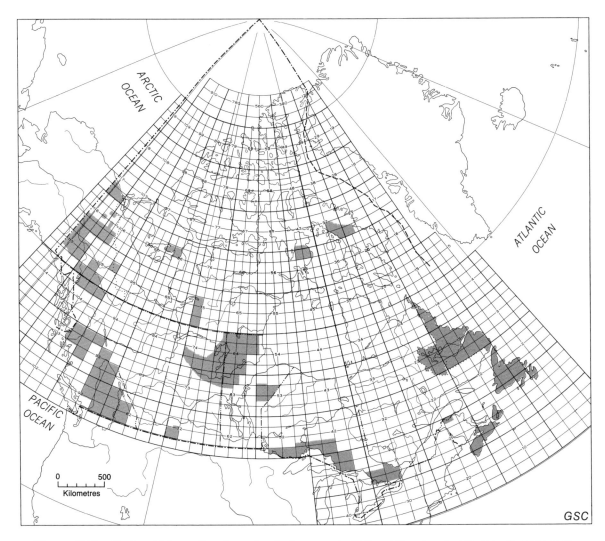

Figure 10.7. Regional lake sediment geochemical surveys to National Geochemical Reconnaissance standards (after Lund, 1987).

copper downward during the postglacial. In section G, the dispersal train makes up at least the full thickness of the exposure (lead, arsenic, silver, and gold data) and pedogenic effects cannot be isolated, even though they must be present.

Types of data and databases

The types of data that exist for significant areas of Canada are mainly trace element and minor element levels in lake sediments, stream sediments, till, and lake waters. Specialized data sets of carbonate levels in till and pebble lithology of till are available for a few regions. The bulk of the data that are available for lake and stream sediments have been collected under the National Geochemical Reconnaissance (NGR) and Uranium Reconnaissance Program (URP) of the Geological Survey of Canada. Figure 10.7 shows the extent of the coverage for these types of programs, including coverage by provincial (e.g., Davenport and Butler, 1983; McConnell, 1984) as well as federal agencies. Figure 10.8 shows areas covered by till geochemical surveys performed by the Geological Survey of Canada and provincial agencies (e.g., Pawluk and Bayrock, 1969; Stea and Fowler, 1979; Stea, 1981; Stea and Grant, 1982). A large number of detailed studies of till have also been made, so several scales of sampling are available.

APPLICATIONS OF TERRAIN GEOCHEMISTRY

Applications to mineral exploration

Drift prospecting

Drift prospecting is defined here as the use of data on the geochemistry and lithology of glacial sediments (mainly till) to identify economically significant components in the sediments and to trace them up-ice to their bedrock source. Drift prospecting research in Canada started with the work of Dreimanis (1956) and has been expanding since the early 1970s (Shilts, 1973b, 1976, 1984). The concept of predictable patterns within dispersal trains (Fig. 10.1, 10.2), when considered during the design of a geochemical exploration program, will influence the choice of sample types, the sampling

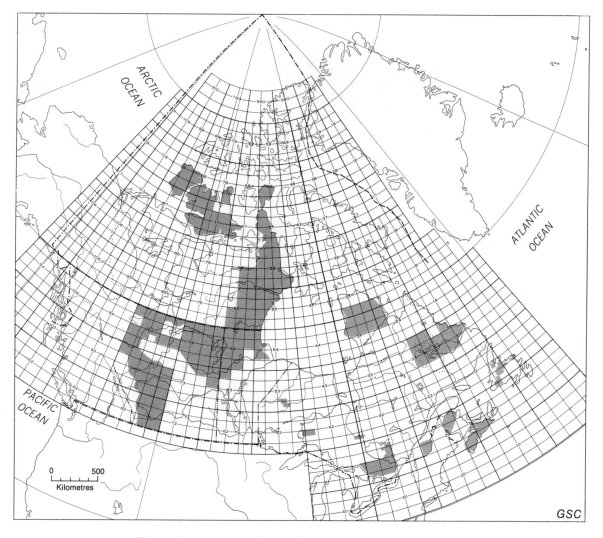

Figure 10.8. Reconnaissance till geochemical surveys in Canada.

plan, the analytical scheme, and the interpretation of the data. Once a dispersal train has been detected, it can be traced relatively easily up-ice to its source because simple clastic dispersal, not more complex processes, is the main process involved in the formation of a train.

Glacial dispersal trains exist at every scale up to hundreds of kilometres in length, such as the train of carbonate-rich till that extends southwards from Hudson Bay and the train of debris derived from Dubawnt Group rocks that extends eastwards into and across northern Hudson Bay (Fig. 10.9, Shilts et al., 1979). Trains of this size are detected only when the area sampled is very large and when the characteristic lithological component of the train is present in adequate amounts and is distinctive against the background rock types in the dispersal area. For drift prospecting purposes, these large trains are significant in that the exotic lithology of the till can mask the lithology and geochemistry of mineralized debris eroded from local sources. Large trains such as these can be detected by "reconnaissance" scale till sampling, in which samples are collected at a density of the order of one sample per 100 km^2.

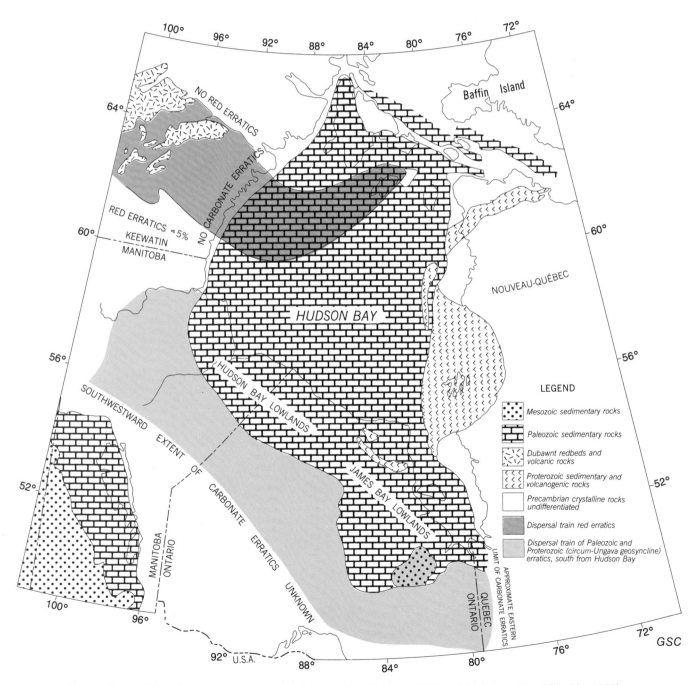

Figure 10.9. Major dispersal trains around Hudson Bay (modified from Shilts, 1982; Kaszycki and DiLabio, 1986).

Smaller dispersal trains derived from individual rock units and distinctive belts of rock are more likely to be detected in the preliminary stages of mineral exploration programs. At this stage of exploration, "local" scale sampling, in which sample density is of the order of one per square kilometre, will suffice to define which parts of a favourable bedrock unit are most metalliferous and may even detect the tails of dispersal trains derived from small mineralized sources. This intermediate-sized dispersal train is illustrated by those derived from the Rankin Inlet-Ennadai belt in the District of Keewatin (Ridler and Shilts, 1974), where local scale sampling (Fig. 10.10) was used to find and map dispersal from the more metalliferous parts of the belt. At the sampling density used there, the large southeastward-trending trains identify areas that should be sampled at a detailed scale to clarify whether the large trains simply represent areas of high background metal levels or are composed of several overlapping small trains derived from areas of mineralized bedrock.

"Detailed" sampling, in which sample density is of the order of one per hectare, is designed to locate the heads of dispersal trains. This is the density of sampling that would normally be carried out in drift prospecting programs tracing trains up-ice or testing geophysical anomalies or favourable geological contacts. Several examples of dispersal trains that have been mapped at a detailed scale (Fig 10.11) illustrate some of the characteristic features of dispersal trains: they are ribbon- or flame-shaped in outline; they have abrupt lateral edges with the surrounding barren till; and the values of the distinctive component within a train decay rapidly down-ice. At this scale of sampling, postglacial mobilization of trace elements in groundwater and soil water may spread the dispersal train downslope, partially obscuring its original shape, which is the result of clastic dispersal.

Most studies of dispersal trains for prospecting purposes have taken place in Fennoscandia and Canada, and the results of several studies are found in Kvalheim (1967), in the proceedings of conferences on prospecting in areas of glaciated terrain (Jones, 1973, 1975; IMM, 1977, 1979, 1984, 1986; Davenport, 1982), in Nichol and Bjorklund (1973), Bradshaw (1975), Shilts (1976, 1984), Minell (1978), Bolviken and Gleeson (1979), and Kauranne (1976). Dispersal trains that have been mapped in Canada include the Steep Rock iron ore train (Dreimanis, 1956); the Kirkland Lake gold train (Lee, 1963); the train of sphalerite-bearing boulders at George Lake, Saskatchewan (Karup-Moller and Brummer, 1970); the train derived from ultramafic rocks at Thetford Mines, Quebec (Shilts, 1973a); the Kidd Creek, Ontario train (Skinner, 1972); the Gullbridge, Newfoundland train (O'Donnell, 1973); the Buchans, Newfoundland train (James and Perkins, 1981); the Icon trains, near Chibougamau, Quebec (DiLabio, 1981); the Currie-Bowman townships train, Ontario (Thompson, 1979); the Mount Pleasant train in New Brunswick (Szabo et al., 1975); the Bathurst Norsemines train (Miller, 1979); the Hopetown, Ontario train (Sinclair, 1979; DiLabio et al., 1982); the Vixen Lake, Saskatchewan train (Geddes, 1982); the Forest Hill train in Nova Scotia (MacEachern and Stea, 1985); the Matachewan barite train in Ontario (Stewart, 1986); the Waddy Lake trains in Saskatchewan (Sopuck et al., 1986); the Golden Pond train at Casa-Berard, Quebec (Sauerbrei et al., in press); the Owl Creek train, near Timmins, Ontario (Bird and Coker, in press); and the Sisson Brook train, New Brunswick (Snow and Coker, in press).

A type of dispersal train that would complicate interpretation of drift prospecting data is the metal-poor or "negative" dispersal train such as the one defined by Klassen and Shilts (1977) in the Baker Lake area of the District of Keewatin (Fig. 10.12). Ice eroded a large volume of metal-poor kaolin-cemented quartz sandstone from the Pitz Lake basin and dispersed it over a wide area towards the southeast. Dilution of the till by this metal-poor debris is clear in the depression of trace element levels in the dispersal area. Similar negative dispersal trains exist down-ice from areas underlain by metal-poor carbonate bedrock and quartz sandstone such as the Hudson Bay Lowland (Fig. 10.9) and the Athabasca Basin. Trace element values that should be considered "high" within a negative train will be significantly lower than those considered "high" values outside the train.

The gravel and sand found in eskers can be considered as the sorted coarse sediment that remains when the mud is washed out of till by subglacial and englacial streams. Esker sediments are thus second-cycle sediments (second derivatives of bedrock of Shilts, 1976) and have been shown to reflect the geochemistry and lithology of the adjacent till where the indicator rocks and minerals survive transport in the esker system (Shilts, 1973b). Because eskers are widely spaced and samples from them represent small areas along their length, they are best used in reconnaissance sampling or in detailed sampling where they cross favourable geological structures. Lee (1965, 1968) demonstrated the effectiveness of sampling eskers in tracing several indicator minerals and rocks to their respective bedrock sources in the Kirkland Lake area, Ontario.

Stream sediment and lake sediment geochemistry

Stream sediment geochemistry, after having been developed in Africa and other unglaciated regions (Levinson, 1980), was introduced to the Canadian scene in the 1950s. The fact that most stream sediments are recycled from a variety of glaciogenic sediments and surficial organic matter and that

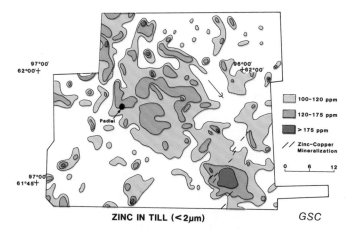

Figure 10.10. Local Zn anomaly from Cu-Zn mineralization superimposed on a regional anomaly in the District of Keewatin; map based on over 1000 till samples collected on a 1.6 km x 1.6 km grid (after Shilts, 1984).

they have been weathered during at least one episode of transport makes them third-cycle sediments (third derivatives of bedrock of Shilts, 1976), and therefore much more difficult than till to trace to bedrock sources. Stream sediments have been used successfully in Canada, however, where drift is thin and where stream gradients are high enough to cause erosion of till and bedrock, such as in mountainous terrain, where the method has been used to explore for deposits of both resistate and unstable minerals (Bradshaw, 1975).

Lake sediment geochemistry has been used extensively to evaluate large areas of terrain at a reconnaissance scale (Fig. 10.7), and to follow up anomalies in individual lake basins. Coker et al. (1979) have reviewed the subject and the following summary is based on their paper.

The application of lake sediment geochemistry to mineral exploration began in the late 1960s. Since that time there has been steady use of the method, particularly in the Canadian Shield, but also in the Cordilleran and Appalachian regions. The success of this method is attributed to the demonstrated ability of lake sediments to reflect the presence of nearby mineralization. In addition, fine grained centre-lake sediments are a homogenous medium that may be sampled relatively economically with a gravity corer or grab sampler lowered from a boat or helicopter.

In the flat-lying, boreal forest terrain characteristic of the southern Canadian Shield, organic matter is common in the lake waters and sediments, and metal-organic interactions control trace metal levels in the sediments. The metals themselves reach the lakes in surface and groundwater and to a lesser extent in detrital mineral grains. The presence of dissolved organic matter can variously enhance trace element mobility, by forming mobile complexes, or retard mobility by direct precipitation of insoluble organic complexes or sulphides, depending on geochemical conditions in the water. Swamps or marshes within the drainage basin may adsorb trace elements, restricting their movement into the lake itself. In contrast, tundra lakes and alpine Cordilleran lakes are fed by snowmelt waters containing very little organic matter. Here, adsorption of metals directly onto clays, rock flour, and hydrous metal oxides and dissolution of detrital mineral grains are the predominant water-sediment interactions.

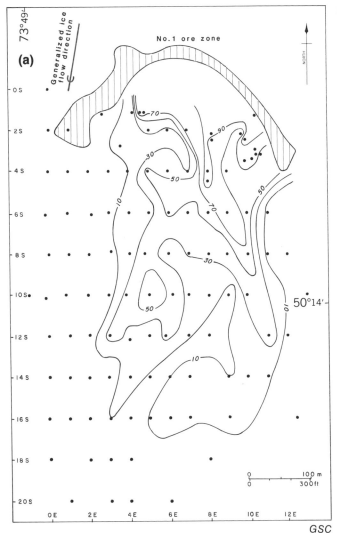

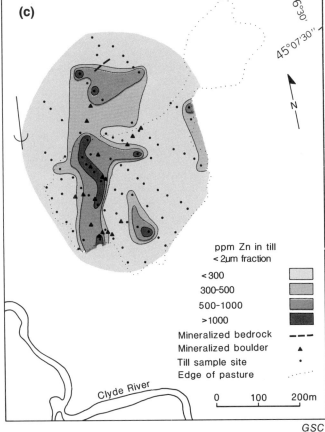

Figure 10.11. Dispersal trains mapped through detailed sampling of till. a) percentage of ore pebbles in till in the Icon train, near Lac Mistassini, Quebec (after DiLabio, 1981); b) trace element levels (ppm) in the <63µm fraction of till in the Lac Brisson train, Quebec and Newfoundland; c) the Hopetown train, near Lanark, Ontario (after DiLabio et al., 1982).

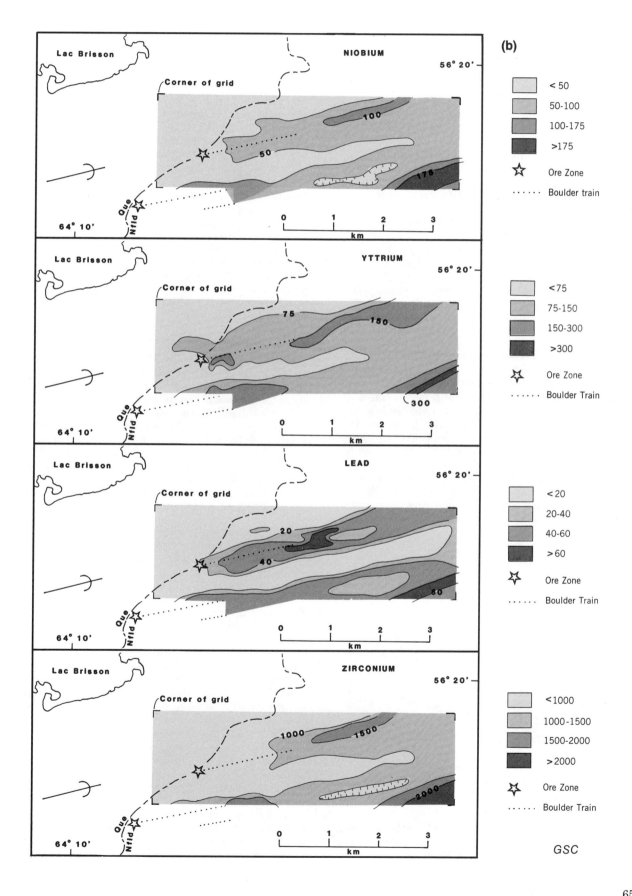

The stratigraphy of lake sediments is similar in most types of lakes in Canada. There is an upper stratum of gelatinous sediment (gyttja), containing a variable quantity of organic material, that has accumulated since deglaciation. The gyttja normally overlies organic-poor late glacial clastic lacustrine sediment. It is invariably present in centres of lakes in the boreal forest and Appalachians and occurs in the centres of most tundra and alpine lakes. The gyttja is thickest and has the highest organic content in boreal forest lakes and is relatively thin, areally restricted, and discontinuous in tundra lakes.

Most trace metals are enriched in the modern organic sediments. This is most probably due to the strength of metal-organic binding and the higher ion-exchange capacity of organic sediments relative to clastic inorganic ones. As a result, the highest and most uniform concentrations of trace metals are found in the modern organic sediment in the deepest central basin of most lakes.

The sampling density of a survey depends on the type and objectives of the survey and the mobility of the trace elements in the surficial environment and their distribution in the rocks in the study area. At the reconnaissance level of sampling (1 sample per 10-20 km²), gross bedrock differences can be discerned and regional trends are clearly outlined, commonly depicting the element associations. As a follow-up to reconnaissance sampling, detailed sampling at one sample per square kilometre or several per lake, at inflows and around the margins, have proved effective in outlining potential economic mineralization.

The interpretation of trace metal levels in lake sediments should normally include assignment of different values for background and anomalous metal levels for large areas of grossly different bedrock and drift lithology. Coker and Dunn (1983) assigned two different sets of values for background and anomalous metal levels in lake sediments in a study in the Athabasca basin, where Athabasca Group rocks and drift derived from them are depleted in some elements (e.g., U) and relatively enriched in others (e.g., As) in comparison to the basement rocks and drift that surround the basin.

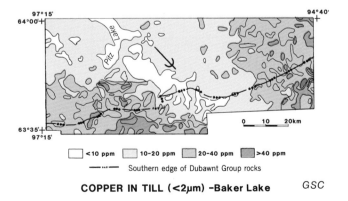

Figure 10.12. Negative Cu anomaly near Baker Lake, District of Keewatin, based on over 2500 till samples collected on a 1.6 km × 1.6 km grid (after Klassen and Shilts, 1977).

Peat geochemistry

A significant number of exploration programs have involved peat geochemistry, usually at a detailed scale on individual small properties where peat bogs cover terrain that is favourable for mineralization (Gunton and Nichol, 1974; Boyle, 1977; DiLabio and Coker, 1982). Reconnaissance peat sampling is usually not done because of the high cost. Bog-covered terrain is most abundant in the boreal forest and covers areas of recessive rock units, swales between hills of glacial sediments, and broad poorly drained lowlands floored with sorted, fine grained sediments. The last type of bog is normally not sampled because of the limited contact between the peat and till or bedrock.

The use of peat as a geochemical sampling medium is based on its known ability to absorb large quantities of metals from surface water and groundwater. Research in Europe and Canada (Szalay, 1964; Usik, 1969; Eriksson, 1973, 1976; Tanskanen, 1976; Larsson, 1976) has indicated that the high cation exchange capacity of peat enables it to absorb metals up to per cent levels (Boyle, 1977; DiLabio and Coker, 1982). Usually, samples from the base of the peat profile and from the margins of bogs will have the highest metal levels and will detect metalliferous till and bedrock beneath and upslope from the bog.

Application to environmental geology

Geochemical and lithological data on glacial and postglacial sediments have been used recently to estimate the sensitivity of terrain to acid precipitation and to identify areas that are naturally enriched in potentially noxious trace elements. For example, Coker and Shilts (1979) have evaluated geochemical data on lake sediments and waters from a large area bordering the north shore of Lake Superior. This study illustrated how data collected for mineral resource evaluation, to aid bedrock mapping, and to delineate natural regional geochemical trends also could be used to detect existing or potential environmental disturbances. They stated that "knowing the areas north of Lake Superior where F, Hg, and As are naturally enriched in lakes (and most probably in associated soils and overburden) allows some predictions to be made about areas where these toxic substances may become dangerous in an "acidified" landscape. Arsenic and mercury in lake sediments and adjacent soils are likely ... in relative equilibrium with their natural environment. Significant decreases in soil and/or water pH may upset this natural equilibrium and could possibly release these and other metals into streams, rivers, lakes, groundwater, etc. in potentially noxious forms, for eventual uptake by vegetation and terrestrial or aquatic organisms" (Coker and Shilts, 1979, p. 12). Other potential problem areas of naturally elevated levels of toxic metals can be outlined by examining the larger existing data bases, such as the National Geochemical Reconnaissance (Geological Survey of Canada, 1983).

In a study designed primarily to identify areas that might be sensitive to acid precipitation, Kettles and Shilts (1987) sampled till and other glacial sediments over an area of 15 000 km² covering the Frontenac Arch of southeastern Ontario. This area lies within the plume of acid precipitation in eastern North America. There, deleterious effects on the landscape would have economic impact on forestry, sport fishery, and tourism. Data on the carbonate content of till (Fig. 10.13a) show dispersal of carbonate debris, derived from Paleozoic limestone bedrock and Proterozoic marble,

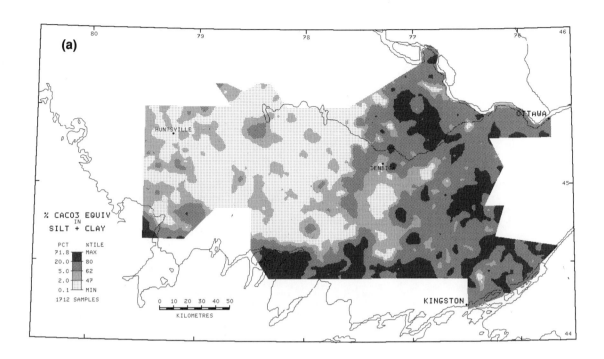
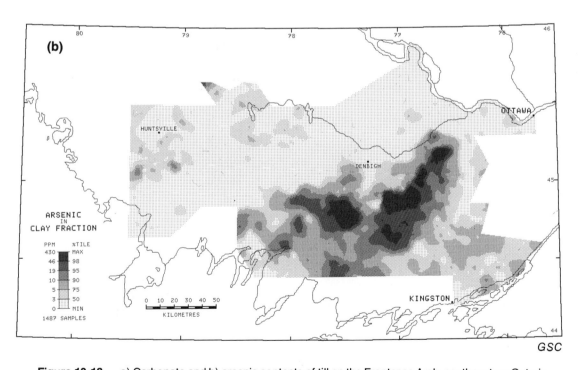

Figure 10.13. a) Carbonate and b) arsenic contents of till on the Frontenac Arch, southeastern Ontario (after Kettles and Shilts, 1987).

onto carbonate-poor crystalline rocks of the Frontenac Arch. The carbonates in the till represent the main acid neutralizing component in the region. Calcareous till covers a much larger area than simply the extent of carbonate bedrock, showing the limitation of estimating the areal extent of land having buffering capacity without considering the extent of glacially transported surficial materials. The arsenic content of the samples (Fig. 10.13b) is one example of data on a potentially noxious element that were mapped in this survey (data on lead, chromium, cadmium, mercury, etc. are given in Kettles and Shilts, 1983). Anomalous arsenic levels follow the Grenville metasedimentary belt in the eastern part of the Frontenac Arch and are also found along the southern edge of the exposed Precambrian bedrock. These anomalous levels highlight areas in which arsenic might be remobilized into surface water and groundwater if soil acidity was increased by acid precipitation. The anomalous levels are also significant in terms of mineral exploration because arsenic is considered as a pathfinder for gold, and consequently, the elevated levels outline areas of known gold occurrences and other land with potential for gold mineralization.

Geochemical data are also used in geomedicine and agriculture to outline areas of natural enrichment or depletion of noxious trace elements or essential micronutrients (Bolviken et al., 1980). These examples show the multiple uses for geochemical data on surficial sediments, uses which might not have been considered when the surveys were originally designed.

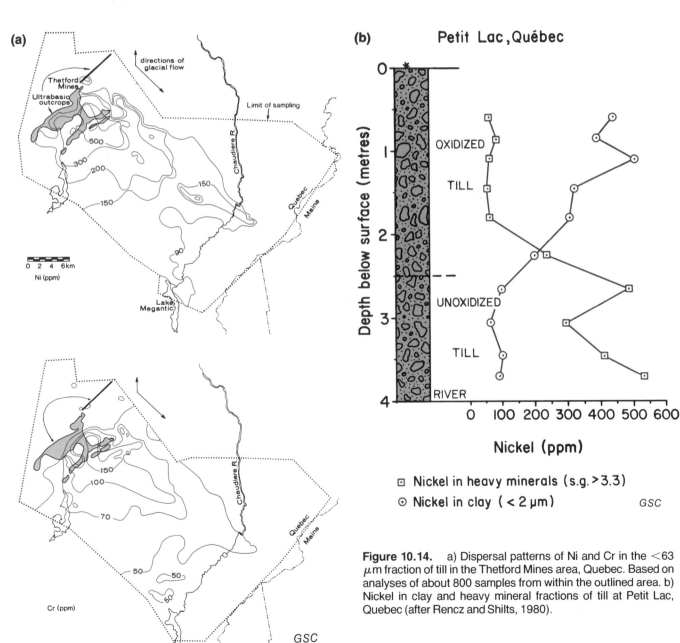

Figure 10.14. a) Dispersal patterns of Ni and Cr in the <63 μm fraction of till in the Thetford Mines area, Quebec. Based on analyses of about 800 samples from within the outlined area. b) Nickel in clay and heavy mineral fractions of till at Petit Lac, Quebec (after Rencz and Shilts, 1980).

Application to Quaternary geology

Data on the geochemistry of glacial sediments have been used to solve problems in Quaternary geology, mainly to determine ice flow directions and provenance, and to differentiate stratigraphic units. The data are useful because their distribution patterns depend on the principle of glacial dispersal and on the predictable shapes of dispersal trains.

Ice-flow directions estimated by measurement of striae are not always the most significant flow directions in terms of drift transport. It has been noted at several sites that till was deposited by movement of ice in a direction different from the ice flow direction indicated by the youngest set of striae. Rencz and Shilts (1980) have shown that a late glacial shift in ice flow direction in St. Lawrence Valley, represented by a widespread set of northward-trending striae, was responsible for minimal glacial transport to the north (Fig. 10.14a). The dispersal train of chromium- and nickel-rich till, extending southeastward from ultrabasic outcrops, indicates that the till was deposited mainly by southeastward-moving ice. Minor northward redirection of debris already in transport southward is visible in the northward distortion of the trace element contours defining the northeastern flank of the train. The redirection is also shown by vertical shifts in chromium and nickel contents of the upper part of the till in the small area of northward distortion of the dispersal train (Fig. 10.14a). A layer of ultrabasic-rich till atop a sequence of regionally "normal" till might be interpreted as a two-till sequence, but knowledge of the redirected ultrabasic-rich debris allows the correct interpretation of the section to be made (Fig. 10.14b). Veillette (1983) has also noted short distances of transport associated with a late glacial redirection of ice flow in the Lac Témiscamingue area of Ontario and Quebec. From lithological data, he showed that the important transport direction for the till matches an earlier set of striae, not the youngest set seen in the region. Not only is a ranking of the importance of different transport directions significant in establishing the glacial history of a region, it also feeds back into mineral exploration, facilitating the tracing of anomalous till geochemistry data up-ice in the proper direction.

Provenance information derived from till geochemistry data has been used in the interpretation of the glacial history of large areas of the central Northwest Territories that have been mapped and sampled at a reconnaissance scale since 1970. Shilts and Cunningham (1977) used maps of till geochemistry data to place constraints on the ice flow directions that may have been possible during the last glaciation in the southern District of Keewatin. Dyke (1984) has interpreted ice flow directions across Boothia Peninsula with the aid of similar data (see also Dyke and Dredge, 1989).

In the field of glacial sedimentology, detailed geochemical data on till has shown subtle vertical and lateral variations in an apparently homogenous till section (Shilts, 1978). This allows the postulation that in some situations a till sheet consists of a set of compositionally distinct layers, each of which was deposited without much mixing with the layers beneath. Strong compositional layering in the debris load of existing glaciers has been noted (DiLabio and Shilts, 1979), suggesting that if it survived postdepositional homogenization, it would produce a till having thin, geochemically distinct compositional layering.

Geochemical data has also been used to characterize individual tills in areas of complex stratigraphy. Trace element contents of fine fractions of till have been shown to reflect slight shifts in provenance (May and Dreimanis, 1973; Shilts, 1976) in sequences of similar tills in the Great Lakes and St. Lawrence basins. Such data are another set of lithological information that can be added to the traditional data sets on pebble lithology, texture, carbonate content, and sand and clay mineralogy (Shilts, 1980, 1981; Vincent, 1983; Dyke, 1984).

ACKNOWLEDGMENTS

This report was improved greatly by the comments provided by W.B. Coker, E.M. Cameron, R.J. Fulton (Geological Survey of Canada), J.B. Bird (McGill University), and W.H. Mathews (University of British Columbia).

REFERENCES

Bird, D.J. and Coker, W.B.
in press: Quaternary stratigraphy and geochemistry at the Owl Creek gold mine, Timmins, Ontario, Canada; Journal of Geochemical Exploration, v. 28, p. 267-284.

Bolviken, B. and Gleeson, C.F.
1979: Focus on the use of soils for geochemical exploration in glaciated terrane; in Geophysics and Geochemistry in the Search for Metallic Ores; P.J. Hood (ed.); Geological Survey of Canada, Economic Geology Report 31, p. 295-326.

Bolviken, B., Ek, J., and Kuusisto, E.
1980: Geochemical data, a basis for geomedical studies; in Geomedical Aspects in Present and Future Research; J. Lag (ed.); Norwegian Academy of Science and Letters, Universitetsforlaget, Oslo, p. 21-33.

Boyle, R.W.
1977: Cupriferous bogs in the Sackville area, New Brunswick, Canada; Journal of Geochemical Exploration, v. 8, p. 495-527.

Bradshaw, P.M.D. (compiler and editor)
1975: Conceptual models in exploration geochemistry — The Canadian Cordillera and Canadian Shield; Journal of Geochemical Exploration, v. 4, p. 1-213.

Coker, W.B., Hornbrook, E.H.W., and Cameron, E.M.
1979: Lake sediment geochemistry applied to mineral exploration; in Geophysics and Geochemistry in the Search for Metallic Ores; P.J. Hood (ed.); Geological Survey of Canada, Economic Geology Report 31, p. 435-478.

Coker, W.B. and Shilts, W.W.
1979: Lacustrine geochemistry around the north shore of Lake Superior: implications for evaluation of the effects of acid precipitation; in Current Research, Part C, Geological Survey of Canada, Paper 79-1C, p. 1-15.

Coker, W.B. and Dunn, C.E.
1983: Lake water and lake sediment geochemistry, NEA/IAEA Athabasca Test Area; in Uranium Exploration in Athabasca Basin, Saskatchewan, Canada, E.M. Cameron (ed.); Geological Survey of Canada, Paper 82-11, p. 117-125.

Davenport, P.H. (editor)
1982: Prospecting in Areas of Glaciated Terrain — 1982; Canadian Institute of Mining and Metallurgy, 339 p.

Davenport, P.H. and Butler, A.J.
1983: Regional Geochemical Surveys; in Current Research, Newfoundland and Labrador Department of Mines and Energy, Paper 83-1, p. 121-125.

DiLabio, R.N.W.
1981: Glacial dispersal of rocks and minerals in the Lac Mistassini — Lac Waconichi area, Quebec, with special reference to the Icon dispersal train; Geological Survey of Canada, Bulletin 323, 46 p.

1982a: Gold and tungsten abundance vs. grain size in till at Waverley, Nova Scotia; in Current Research, Part B, Geological Survey of Canada, Paper 82-1B, p. 57-62.

1982b: Drift prospecting near gold occurrences at Onaman River, Ontario and Oldham, Nova Scotia; in Geology of Canadian Gold Deposits; Canadian Institute of Mining and Metallurgy, Special Volume 24, p. 261-266.

1985: Grain size distribution of gold in weathered and unweathered till; in Current Research, Part A, Geological Survey of Canada, Paper 85-1A, p. 117-122.

DiLabio, R.N.W. and Coker, W.B.
1982: Geochemistry of peat associated with uraniferous bedrock in the Kasmere Lake area, Manitoba; in Prospecting in Areas of Glaciated Terrain — 1982, P.H. Davenport (ed.); Canadian Institute of Mining and Metallurgy, p. 179-194.

DiLabio, R.N.W. and Shilts, W.W.
1979: Composition and dispersal of debris by modern glaciers, Bylot Island, Canada; in Moraines and Varves, C. Schluchter (ed.); A.A. Balkema, Rotterdam, p. 145-155.

DiLabio, R.N.W., Rencz, A.N., and Egginton, P.A.
1982: Biogeochemical expression of a classic dispersal train of metalliferous till near Hopetown, Ontario; Canadian Journal of Earth Sciences, v. 19, p. 2297-2305.

Dreimanis, A.
1956: Steep Rock iron ore boulder train; Geological Association of Canada, Proceedings, Part 1, p. 27-70.

Dyke, A.S.
1984: Quaternary geology of Boothia Peninsula and northern District of Keewatin, central Canadian Arctic; Geological Survey of Canada, Memoir 407, 26 p.

Dyke, A.S. and Dredge, L.A.
1989: Quaternary geology of the northwestern Canadian Shield; in Quaternary Geology of Canada and Greenland, R.J. Fulton (ed.); Geological Survey of Canada, Geology of Canada, no. 1 (also Geological Society of America, The Geology of North America, v. K-1).

Eriksson, K.
1973: Prospecting in an area of central Sweden; in Prospecting in Areas of Glacial Terrain, M.J. Jones (ed.); Institution of Mining and Metallurgy, London, p. 83-86.
1976: Regional prospecting by the use of peat sampling; Journal of Geochemical Exploration, v. 5, p. 387-388.

Geddes, R.S.
1982: The Vixen Lake indicator train, northern Saskatchewan; in Prospecting in Areas of Glaciated Terrain — 1982, P.H. Davenport (ed.); Canadian Institute of Mining and Metallurgy, p. 264-283.

Geddes, R.S. and Kristjansson, F.J.
1986: Quaternary geology of the Hemlo area: constraints on mineral exploration; Canadian Geology Journal of the Canadian Institute of Mining and Metallurgy, v. 1, no. 1, p. 5-8.

Gunton, J.E. and Nichol, I.
1974: Delineation and interpretation of metal dispersion patterns related to mineralization in the Whipsaw Creek area; Canadian Institute of Mining and Metallurgy, v. 67, no. 741, p. 66-75.

IMM (Institution of Mining and Metallurgy)
1977: Prospecting in Areas of Glaciated Terrain 1977; Institution of Mining and Metallurgy, London, 140 p.
1979: Prospecting in Areas of Glaciated Terrain 1979; Institution of Mining and Metallurgy, London, 110 p.
1984: Prospecting in Areas of Glaciated Terrain 1984; Institution of Mining and Metallurgy, London, 232 p.
1986: Prospecting in Areas of Glaciated Terrain 1986; Institution of Mining and Metallurgy, London, 269 p.

James, L.D. and Perkins, E.W.
1981: Glacial dispersion from sulphide mineralization, Buchans area, Newfoundland; in The Buchans Orebodies: Fifty Years of Geology and Mining, E.A. Swanson, D.F. Strong, and J.G. Thurlow (ed.); Geological Association of Canada, Special Paper 22, p. 269-283.

Jones, M.J. (editor)
1973: Prospecting in Areas of Glacial Terrain; Institution of Mining and Metallurgy, London, 138 p.
1975: Prospecting in Areas of Glacial Terrain 1975; Institution of Mining and Metallurgy, London, 154 p.

Karup-Moller, S. and Brummer, J.J.
1970: The George Lake zinc deposit, Wollaston Lake area, Saskatchewan; Economic Geology, v. 65, p. 862-874.

Kaszycki, C.A. and DiLabio, R.N.W.
1986: Surficial geology and till geochemistry, Lynn Lake-Leaf Rapids region, Manitoba; in Current Research, Part B, Geological Survey of Canada, Paper 86-1B, p. 245-256.

Kauranne, L.K.
1976: Conceptual models in exploration geochemistry, Norden, 1975; Journal of Geochemical Exploration, v. 5, p. 173-420.

Kettles, I. and Shilts, W.W.
1983: Reconnaissance geochemical data for till and other surficial sediments, Frontenac Arch and surrounding areas, Ontario; Geological Survey of Canada, Open File 947.
1987: Tills of the Ottawa region; in Quaternary geology of the Ottawa region, Ontario and Quebec, R.J. Fulton (ed.); Geological Survey of Canada, Paper 86-23.

Klassen, R.A. and Shilts, W.W.
1977: Glacial dispersal of uranium in the District of Keewatin, Canada; in Prospecting in Areas of Glaciated Terrain 1977; Institution of Mining and Metallurgy, London, p. 80-88.

Kvalheim, A. (editor)
1967: Geochemical Prospecting in Fennoscandia; Interscience, New York, 350 p.

Larsson, J.O.
1976: Organic stream sediments in regional geochemical prospecting, Precambrian Pajala district, Sweden; Journal of Geochemical Exploration, v. 6, p. 233-249.

Lee, H.A.
1963: Glacial fans in till from the Kirkland Lake fault: a method of gold exploration; Geological Survey of Canada, Paper 63-45, 36 p.
1965: 1. Investigations of eskers for mineral exploration; 2. Buried valleys near Kirkland Lake, Ontario; Geological Survey of Canada, Paper 65-14, 20 p.
1968: An Ontario kimberlite occurrence discovered by application of the glaciofocus method to a study of the Munro esker; Geological Survey of Canada, Paper 68-7, 3 p.

Levinson, A.A.
1980: Introduction to Exploration Geochemistry (second edition); Applied Publishing, Wilmette, Illinois, 924 p.

Lund, N.G.
1987: Index to National Geochemical Reconnaissance Surveys, 1973-1986; Geological Survey of Canada, Map 1661A.

MacEachern, I.J. and Stea, R.R.
1985: The dispersal of gold and related elements in tills and soils at the Forest Hill gold district, Guysborough County, Nova Scotia; Geological Survey of Canada, Paper 85-18, 31 p.

May, R.W. and Dreimanis, A.
1973: Differentiation of glacial tills in southern Ontario, Canada, based on their Cu, Zn, Cr, and Ni geochemistry; in The Wisconsinan Stage, R.F. Black, R.P. Goldthwait, and H.B. Willman (ed.); Geological Society of America, Memoir 136, p. 221-228.

McConnell, J.
1984: Reconnaissance and detailed geochemical surveys for base metals in Labrador; Newfoundland and Labrador Department of Mines and Energy, Report 84-2, 114 p.

Miller, J.K.
1979: Geochemical dispersion over massive sulphides within the continuous permafrost zone, Bathurst Norsemines, Canada; in Prospecting in Areas of Glaciated Terrain 1979; Institution of Mining and Metallurgy, London, p. 101-109.

Minell, H.
1978: Glaciological interpretations of boulder trains for the purpose of prospecting in till; Sveriges Geologiska Undersokning, Serie C, no. 743, 51 p.

Nichol, I.
1986: Geochemical exploration for gold deposits in areas of glaciated overburden: problems and new developments; in Prospecting in Areas of Glaciated Terrain 1986; Institution of Mining and Metallurgy, London, p. 7-16.

Nichol, I. and Bjorklund, A.
1973: Glacial geology as a key to geochemical exploration in areas of glacial overburden with particular reference to Canada; Journal of Geochemical Exploration, v. 2, 133-170.

O'Donnell, N.D.
1973: Glacial indicator trains near Gullbridge, Newfoundland; unpublished MSc thesis, University of Western Ontario, London, Ontario, 259 p.

Pawluk, S. and Bayrock, L.A.
1969: Some characteristics and physical properties of Alberta tills; Research Council of Alberta, Bulletin 26, 72 p.

Rencz, A.N. and Shilts, W.W.
1980: Nickel in soils and vegetation of glaciated terrains; in Nickel in the Environment, J.O. Nriagu (ed.); John Wiley and Sons, p. 151-188.

Ridler, R.H. and Shilts, W.W.
1974: Exploration for Archean polymetallic sulphide deposits in permafrost terrains: an integrated geological/geochemical technique, Kaminak Lake area, District of Keewatin; Geological Survey of Canada, Paper 73-34, 33 p.

Sauerbrei, J.A., Pattison, E.F., and Averill, S.A.
in press: Till sampling in the Casa-Berardi area, Quebec, a case history in orientation and discovery; Journal of Geochemical Exploration, v. 28, p. 297-314.

Shelp, G.S. and Nichol, I.
in press: Dispersion of gold in glacial till associated with gold mineralization in the Canadian Shield; Journal of Geochemical Exploration, v. 28, p. 315-336.

Shilts, W.W.
1973a: Glacial dispersal of rocks, minerals, and trace elements in Wisconsinan till, southeastern Quebec, Canada; in The Wisconsinan Stage, R.F. Black, R.P. Goldthwait, and H.B. Willman (ed.); Geological Society of America, Memoir 136, p. 189-219.
1973b: Drift prospecting; geochemistry of eskers and till in permanently frozen terrain: District of Keewatin; Northwest Territories; Geological Survey of Canada, Paper 72-45, 34 p.
1976: Glacial till and mineral exploration; in Glacial Till, R.F. Legget (ed.); Royal Society of Canada, Special Publication No. 12, p. 205-224.
1978: Detailed sedimentological study of till sheets in a stratigraphic section, Samson River, Quebec; Geological Survey of Canada, Bulletin 285, 30 p.
1980: Geochemical profile of till from Longlac, Ontario to Somerset Island; Canadian Institute of Mining and Metallurgy, Bulletin, v. 73, no. 822, p. 85-94.
1981: Surficial geology of the Lac-Mégantic area, Quebec; Geological Survey of Canada, Memoir 397, 102 p.
1982: Quaternary evolution of the Hudson/James Bay Region; Le Naturaliste Canadien, v. 109, p. 309-332.
1984: Till geochemistry in Finland and Canada; Journal of Geochemical Exploration, v. 21, p. 95-117.

Shilts, W.W. and Cunningham, C.M.
1977: Anomalous uranium concentrations in till north of Baker Lake, District of Keewatin; in Report of Activities, Part B, Geological Survey of Canada, Paper 77-1B, p. 291-292.

Shilts, W.W., Cunningham, C.M., and Kaszycki, C.A.
1979: Keewatin Ice Sheet — re-evaluation of the traditional concept of the Laurentide Ice Sheet; Geology, v. 7, p. 537-541.

Sinclair, I.G.L.
1979: Geochemical investigation of the Clyde River zinc prospect, Lanark County, Ontario; in Geochemical Exploration, 1978, J.R. Watterson and P.K. Theobald (ed.); Association of Exploration Geochemists, p. 487-495.

Skinner, R.G.
1972: Overburden study aids search for ore in Abitibi clay belt; Northern Miner, v. 58, no. 37, p. 62.

Snow, R.J. and Coker, W.B.
in press: Overburden geochemistry related to W-Cu-Mo mineralization at Sisson Brook, New Brunswick, Canada: an example of short and long distance glacial dispersion; Journal of Geochemical Exploration, v. 28, p. 353-368.

Sopuck, V., Schreiner, B.T., and Averill, S.
1986: Drift prospecting for gold in the southeastern Shield of Saskatchewan, Canada; in Prospecting in Areas of Glaciated Terrain 1986; Institution of Mining and Metallurgy, London, p. 217-240.

Stea, R.R.
1981: Pleistocene geology and till geochemistry of south central Nova Scotia; Nova Scotia Department of Mines and Energy, Map 82-1.

Stea, R.R. and Fowler, J.H.
1979: Minor and trace element variations in Wisconsinan tills, eastern shore region, Nova Scotia; Nova Scotia Department of Mines and Energy, Paper 79-4, 30 p.

Stea, R.R. and Grant, D.R.
1982: Pleistocene geology and till geochemistry of southwestern Nova Scotia; Nova Scotia Department of Mines and Energy, Map 82-10.

Stewart, R.A.
1986: Glacial dispersion of barite in till near Matachewan, Ontario; in Prospecting in Areas of Glaciated Terrain 1986; Institution of Mining and Metallurgy, London, p. 261-269.

Szabo, N.L., Govett, G.J.S., and Lajtai, E.Z.
1975: Dispersion trends of elements and indicator pebbles in glacial till around Mt. Pleasant, New Brunswick, Canada; Canadian Journal of Earth Sciences, v. 12, p. 1534-1556.

Szalay, A.
1964: The cation-exchange properties of humic acids and their importance in the geochemical enrichment of $UO_2 2 \pm$ and other cations; Geochimica et Cosmochimica Acta, v. 28, p. 1605-1614.

Tanskanen, H.
1976: Factors affecting the metal contents in peat profiles; Journal of Geochemical Exploration, v. 5, p. 412-414.

Thompson, I.S.
1979: Till prospecting for sulphide ores in the Abitibi clay belt of Ontario; Canadian Institute of Mining and Metallurgy, v. 72, no. 807, p. 65-72.

Usik, L.
1969: Review of geochemical and geobotanical prospecting methods in peatland; Geological Survey of Canada, Paper 68-66, 43 p.

Veillette, J.J.
1983: Les polis glaciaires au Témiscamingue: une chronologie relative; dans Recherches en cours, Partie A, Commission géologique du Canada, Étude 83-1A, p. 187-196.

Vincent, J-S.
1983: La géologie du quaternaire et la géomorphologie de l'Île Banks, Arctique canadien; Commission géologique du Canada, Mémoire 405, 118 p.

Author's address

R.N.W. DiLabio
Geological Survey of Canada
601 Booth Street
Ottawa, Ontario
K1A 0E8

Printed in Canada

Chapter 11

QUATERNARY RESOURCES IN CANADA

Summary

Introduction — *L.E. Jackson, Jr.*

Soils as a resource — *K.W.G. Valentine*

   Soil formation — *D.F. Acton*

   Soils of Canada

     Cordilleran region — *K.W.G. Valentine*

     Interior Plains — *D.F. Acton*

     St. Lawrence Lowlands — *E.W. Presant*

     Atlantic region — *K. Webb*

     Shield region — *D.F. Acton*

     Arctic Canada — *C. Tarnocai*

Peat resources in Canada — *C. Tarnocai*

Aggregate and nonmetallic Quaternary mineral resources — *W.A.D. Edwards*

Placer deposits in Canada — *S.R. Morison*

Acknowledgments

References

a) Wheat grown on a chernozemic soil developed on till (ground moraine) in central Saskatchewan. If it was not for the Quaternary sediment cover, the soil would be a relatively unfertile and difficult to manage solonetzic soil developed on Cretaceous shale. Courtesy of J. Shields. b) Peatland in the Bulmer Lake area, southwestern District of Mackenzie. Extensive peatlands that occur primarily within the Boreal Forest Zone are a vast potential energy resource which has developed in Canada because of the particular Holocene climate. Courtesy of C. Tarnocai. c) Placer gold was the magnet that lead to the settlement of Yukon Territory. The most extensive deposits were found beyond the limit of glaciation. Similar rich deposits were probably present in several other parts of Canada but were destroyed by overriding Quaternary glaciers. Courtesy of D. Norris. d) Gravel pit in a raised delta on the west side of Howe Sound near Fort Mellon, British Columbia. Sand and gravel deposits, related to higher sea and lakes levels and as part of glacial deposits, are an invaluable resoruce in many parts of Canada. Courtesy of R.J. Fulton.

Chapter 11

QUATERNARY RESOURCES IN CANADA

Co-ordinator
L.E. Jackson, Jr.

SUMMARY

Soils are one of Canada's premier Quaternary resources. They sustain agriculture, forestry, and wildlife resources. Soil is the product of the interaction of passive parent material (bedrock or Quaternary sediments) and physiography, and the active elements of climate, flora, and fauna. Most soils in Canada have formed since the end of the last glaciation (18 to 8 ka).

Many of the physical, mineralogical, and chemical properties of soils in Canada are inherited from the parent material. These, in turn, are the results of Quaternary geological events. For example, soils of the Interior Plains are almost stoneless where they have developed on glacial lake sediments, but are stony where they have developed on hummocky moraine or glaciofluvial gravels. Where the covering of glacial sediments has been removed by Holocene erosion, soils are commonly sodic because of the exposure of Cretaceous shales. Former glacial flow patterns may be reflected in the regional distribution of minerals, within the soil — carbonate minerals are an important example. Similar influences of parent material on the physical, mineralogical, and chemical properties of soils can be cited from all regions of Canada.

There is estimated to be more than 100×10^6 ha of peatlands in Canada. Because of great variations in climate and physiographic situations, peatlands occur in a wide range of types. The physical and chemical characteristics of peat materials associated with peatlands depend on their botanical composition and the region in which they were deposited. There are approximately 3×10^{12} m^3 or 335×10^9 t of dry peat in Canada. The energy value of this peat resource is approximately 6.7×10^{21} J. Existing mining techniques are able to handle only the unfrozen, southern peatlands, whose energy equivalent is approximately 2.7×10^{21} J, similar to that of the coal resources of Canada.

Most of Canada's aggregate comes from Quaternary sand and gravel deposits. Crushed stone is locally a significant source of aggregate, however, where sand and gravel are in short supply or long haul distances for sand and gravel offset the higher production costs of crushed stone. Artificial aggregate is used primarily for lightweight concrete. Construction of transportation facilities and urban centres consumes most of Canada's aggregate production. Energy development facilities such as drilling islands, oil sands, plants, and dams are other important consumers. Aggregate depletion has been rapid in urban regions, and shortages of readily accessible sand and gravel deposits have occurred or are predicted for many urban areas. These shortages result in rising aggregate prices, due to longer haul distances, and in substitution by crushed stone, lower grade sources such as till or manufactured aggregate, all of which have higher production costs.

In addition to aggregate, other Quaternary nonmetallic mineral resources are exploited in Canada. For example, silica sand is used by the glass industry and glacial lake sediments are used for the manufacture of bricks and ceramic products. Sodium sulphate, an essential ingredient in the manufacture of Kraft paper, is mined from evaporitic lakes on the prairies. Marl deposits have been exploited by the cement industry and may be used to combat soil acidification.

Placer deposits are accumulations of heavy minerals which have been concentrated through sedimentary processes. A variety of sedimentary processes has been involved in the development of Canadian placer deposits. Placer gold is the most significant mineral in Canada and the only exploitable Quaternary metallic resource of significance. Placer gold deposits of Yukon Territory and British Columbia are the most productive in Canada. Here, placers were deposited in gulches and by braided and meandering streams. In some cases gold concentrations are the result of the fluvial reworking of pre-existing placer deposits, and chemical transport and precipitation. The few small placers in eastern Canada are largely of fluvial and glaciofluvial origin. Low grade deposits of auriferous beach sands locally exist on both east and west coasts.

INTRODUCTION
L.E. Jackson, Jr.

Geological events during and immediately prior to the Quaternary shaped the Canadian landscape and determined the disposition of many of the resources that support Canada's economy and have played significant roles in shaping the history of the country.

Jackson, L.E., Jr. (co-ordinator)
1989: Quaternary resources in Canada; Chapter 11 in Quaternary Geology of Canada and Greenland, R.J. Fulton (ed.); Geological Survey of Canada, Geology of Canada, no. 1 (also Geological Society of America, The Geology of North America, v. K-1).

Jackson, L.E.
1989: Introduction (Quaternary resources in Canada); in Chapter 11 of Quaternary Geology of Canada and Greenland, R.J. Fulton (ed.); Geological Survey of Canada, Geology of Canada, no. 1 (also Geological Society of America, The Geology of North America, v. K-1).

The soils of Canada — one of our ultimate resources — are very much a product of the Quaternary. Although the soil is not usually mined, except for horticultural or landscaping purposes, it is exploited none the less. It is the basis of Canadian agriculture and forestry; its properties dictate in part the hydrology and chemistry of Canada's rivers and lakes. Like mineral aggregate or placer gold, soil is essentially a nonrenewable resource subject to depletion; its losses are due to erosion, salinization, oxidation of organic matter, and burial beneath urban sprawl. Past civilizations have declined and disappeared due to their lack of proper management of this vital resource.

Aggregate resources — sand and gravel deposits — are largely the legacy of past rivers flowing from, beside, and beneath glacial ice. Canada's total aggregate reserves are immense. Around urban areas, however, close and readily available deposits are being rapidly depleted. Urban sprawl has made some aggregate deposits unavailable; local shortages have developed and will develop in some areas. Adequate aggregate surveys and protection of deposits from development will be required to optimize continued use of this resource. Substitution with crushed rock or other manufactured aggregates and hauling of natural aggregates from more distant sources will be the trend in many urban areas. In addition to their exploitation for aggregate, sand and gravel deposits form important reservoirs for groundwater in many parts of Canada.

Other nonmetallic Quaternary mineral resources range from abundant silica sand in sand dunes to sodium sulphate in evaporitic lake basin in the southern Interior Plains.

Peat deposits, which have accumulated since the end of the last glaciation, are found in most regions of Canada. This plentiful and virtually untapped resource could contribute to Canada's energy needs in future years as well as provide feedstock for chemical industries.

The Klondike discovery in Yukon Territory in 1896 triggered one of the most dramatic waves of northern migration and settlement in Canada's history. Auriferous placers in Yukon and elsewhere in Canada continue to be significant sources of this precious metal.

SOILS AS A RESOURCE
K.W.G. Valentine

The soils of Canada have developed from glacial and postglacial sediments under a climatic and biological environment that is largely tundra and boreal forest. Exceptions are the temperate grasslands of the southern Interior Plains and the mixed forests of the southeast.

This interaction of climate, earth materials, and geological events has given Canada soils that are, compared to other parts of the world, relatively young, quite cold and well supplied with nutrients. Apart from parts of Yukon Territory, which were not covered by ice during the Wisconsin Glaciation, no soils are older than about 18 000 years (when ice first retreated). Some, on floodplains, or in the Arctic, are much younger. Soil temperatures are depressed by arctic and continental climates. In fact 40% of Canada's soils contain perennially frozen layers within 2 m of the surface. Chemical and biological weathering is therefore slow. Considerable organic matter accumulates in most topsoils, and mineral nutrients (abundant in such young sediments) are not rapidly lost. Physical weathering, on the other hand, is active, especially where numerous freeze-thaw cycles occur each year. Fire also has a significant, thoroughly underestimated, effect. It has been estimated that 90% of the boreal forest burns regularly, releasing some nutrients from the surface litter, but volatilizing others, and continuously modifying the vegetation.

Covering about 10×10^6 km^2, Canada is the second largest country in the world. Much of our land, however, is too cold, too wet, too dry, too steep, or too rocky for intensive commercial use. Only about one fifth, 2×10^6 km^2, is productive forest land. Even less, about 0.75×10^6 km^2, is good to moderate farmland (with a little more used for grazing). Add the small amount occupied by our economically crucial cities, mines, and transport corridors, and little more than 30% of our land is used intensively. Moreover, climate, economics, politics and social history have combined to concentrate that 30% along our southern border. Nevertheless, it contributes significantly to our gross national product (agriculture, forestry, and trapping combined constitute about 10%), and even more significantly to our exports (forest products provide 18% of the total value of exports, and agriculture about 10%). Neither is the other 70% unused. It sustains many of our native peoples in their traditional lifestyle, it provides habitat for our wildlife, space for our recreation, and a backdrop for our tourist industry.

But the picture is not static. Ever since the first European immigrations, man has changed the landscape and continues to do so. Land was cleared for farming. Trees were cut for ship timber, construction lumber, and later for wood pulp. As a nation we are now becoming more urban, with residential subdivisions and industrial estates spreading onto adjoining farmland. In response to social, economic, and political pressures our land uses have become more intensive. Agricultural crops are fertilized, sprayed, and harvested with specialized machinery. More of our forests (though not an adequate proportion) are replanted instead of being left to reseed naturally. Such activity has increased yields but caused problems. Some soils in the east are becoming more acid, others in the Great Plains are becoming more saline. Wind and water erosion is prevalent in many areas. Continuous cropping is lowering the store of organic matter in topsoils, and heavy machinery is compacting fine-textured subsoils. Climatic change is a further possibility, though one difficult to predict. Higher temperatures due to increasing levels of carbon dioxide in the lower atmosphere could considerably extend our northern limits of agricultural and silvicultural production; however, they could also tax our present methods of dryland cultivation.

Our soils, therefore, form an important part of the natural wealth of this country. But they must be conserved and husbanded carefully if they are to continue to fulfill that role. The following sections describe, in more detail, how they have been formed, what are their major characteristics, and how they vary from one part of the country to another.

Valentine, K.W.G.
1989: Soils as a resource; in Chapter 11 of Quaternary Geology of Canada and Greenland, R.J. Fulton (ed.); Geological Survey of Canada, Geology of Canada, no. 1 (also Geological Society of America, The Geology of North America, v. K-1).

Soil formation

D.F. Acton

Soil formation is the result of the combined effects of the natural factors of parent material, topography and drainage, climate, and vegetation, and the length of time over which these factors have interacted. It also often includes the activities of man. Mineral soils form in material derived from rocks and sediments whereas organic soils form in material derived from plants. Many soil properties are inherited from these materials, and hence, rocks and plants are looked upon as the parent material of soil. The parent materials of most mineral soils in Canada have been moved, sorted, and redeposited by ice, water, or wind, which have provided additional properties for the soil to inherit. The interplay of parent materials, topography and drainage, and climate and vegetation over time are all responsible for changing the "raw" parent material into soil. The intensity of these changes depends upon the type of climate, native vegetation, kind of parent material, and the factor of time. These changes eventually produce the layers or horizons which make up the soil profile. It is not possible, however, to state in absolute time how long it takes to produce a particular kind of soil. Thus, the time factor in soil formation is a combination of the actual length of time and the intensity or speed of the physical, chemical, and biological activities responsible for changing "raw" parent material into a well defined soil profile. In this respect the time factor in soil formation, where a "juvenile" soil may not be young in absolute time, is not identical with geological time, which is concerned primarily with time in a chronological sense.

Dynamics of soil formation

Soils are dynamic bodies. Their genesis is commonly complex and usually never fully understood.

Soil genesis can be viewed as two steps which often merge and overlap: (1) the accumulation or development of the surface sediment to form the parent materials of soils and (2) the differentiation of horizons in the profile. With respect to step 1, the nature of the regolith (rock or sediment) in which horizon differentiation proceeds profoundly affects the rate and direction of changes.

Horizon differentiation (step 2) results from four basic kinds of changes: (1) addition of material, such as the addition of fresh organic residues, to the soil surface; (2) removal of materials, such as soluble salts and decomposed organic constituents, from the soil by leaching; (3) transfer of constituents, such as clay, organic matter, or salts, from one part of the soil profile to another; and (4) transformation of organic matter, soluble salts, carbonates, or sesquioxides from one form to another (Simonson, 1959).

Addition, removal, transfer, and transformation in soils may promote or retard horizon differentiation. These processes are operative in all soils but the relative importance of each is not uniform for all soils. For example, removal of sesquioxides and addition of organic matter occur in all soils, but they are much more important in some soils than in others. It is the balance among individual processes in a given combination that becomes the key to the nature of a soil.

The variety of changes that occurs during the differentiation of horizons in a profile depends upon a host of simpler chemical and physical processes such as hydration, oxidation, solution, leaching, precipitation, and mixing. These simpler and more basic reactions proceed in all soils. They, too, are controlled by factors such as climate, living organisms, parent materials, and topography. The Canadian classification scheme for the soil types resulting from the interplay of these factors may be found in "The Canadian System of Soil Classification" which is summarized in Table 11.1 (Canada Soil Survey Committee, 1978).

Relation of soils to geology

Many soil properties are inherited from the soil parent material. Particle size distribution, mineralogical composition, and age of the material are important considerations in determining the kind of soil that is formed and man's use of the soil. The significance of hydrological and topographical factors will also be discussed from the same perspective.

The amount of clay, or lack of it, in the parent material of the soil has a pronounced influence on the soil's water retention capability thus influencing the type of native vegetative cover and agricultural capability. For example, in semi-arid regions, such as southwestern Saskatchewan and southeastern Alberta, where moisture is often the limiting factor in crop growth, soils formed from parent materials with a high clay content, for example, glacial lake sediments, are more desirable for crop production, in that they have large water-holding capacities. Conversely, parent materials associated with coarse textured, gravelly and sandy soils, for example, glaciofluvial gravels, are of lower value for agricultural purposes, in most cases, because of their smaller water-holding capacity. In humid regions, such as the St. Lawrence Lowlands and the Atlantic provinces, it is the reverse. Clayey soils may be less desirable than sandy soils for agricultural purposes because of slow permeability, poor aeration, or even flooding. These generalizations do not necessarily hold for native vegetation, however. For example, a merchantable tree species may grow more rapidly on a moist, clay-rich, bottomland site than sandy midslope site in a given region.

Particle size distribution also influences the thermal properties of soils. Soil summer temperatures are lower in clay soils due to their high porosity and hence low thermal conductivity. This influence is accentuated by the higher moisture contents usually associated with clay soils.

Mineralogy varies with the particle size distribution of the parent material. Sandy soils tend to be high in quartz, with lesser amounts of feldspars and micas. Surface area of such sediments is small, reaction rates are slow and hence the capacity to exchange ions is small. In contrast, fine grained soils usually have a high content of montmorillonite, kaolinite, or mica and a large surface area that facilitates more rapid weathering and greater exchange of cations between the clay surface and the soil solution. As a consequence, these soils have a far greater capacity to supply plant nutrients.

The nature of the bedrock also affects soil composition as the source of the soil's parent material. In the Interlake

---

Acton, D.F.
1989: Soil formation; in Chapter 11 of Quaternary Geology of Canada and Greenland, R.J. Fulton (ed.); Geological Survey of Canada, Geology of Canada, no. 1 (also Geological Society of America, The Geology of North America, v. K-1).

Table 11.1. Principal characteristics of the nine soil orders in Canada

Order	Dominant features	Natural vegetation	Soil climate	Texture	Parent material Reaction	Drainage
Brunisolic	Soils with brownish subsurface horizons containing weak accumulations of humified organic matter combined with Al and Fe. May have light- or dark-coloured surface horizons	Boreal forest; mixed forest, shrubs, and grass; heath and tundra	Varied	Varied; commonly sandy	Varied	Well to imperfect
Chernozemic	Soils with surface horizons darkened by the accumulation of organic matter from the decomposition of grasses, forbs, and shrubs. May have brownish subsurface horizons.	Grassland or grassland – forest	Cool; subarid to subhumid	Varied; usually not sandy	Base saturated; high Ca + Mg	Well to imperfect
Cryosolic	Mineral or organic soils with permafrost within 2 m of the surface	Tundra subarctic and boreal forest and alpine	Very cold to extremely cold; humid	Varied	Varied	Well to very poor
Gleysolic	Mineral soils that are saturated with water and under reducing conditions for extensive periods of the year	Varied	Mild to moderately cold; aquic	Varied	Varied	Poor to very poor
Luvisolic	Soils with light-coloured horizons near the surface underlain by subsurface horizons in which silicate clay has accumulated	Forest	Mild to very cold; subhumid	Varied; not sandy	Base saturated	Well to imperfect
Organic	Organic (peat, muck, bog) layer usually greater than 40 cm thick that is saturated with water for prolonged periods. Also includes leaf litter over rock or fragmental material	Bog, fen, and forest	Mild to cold; subhumid to perhumid	—	—	Usually poor to very poor
Podzolic	Soils with brown or reddish subsurface horizons containing strong accumulations of humified organic matter combined with Al and Fe. May have a light-coloured surface horizon	Forest or heath	Cool to very cold; humid to perhumid	Varied; usually not clayey	Acid	Well to imperfect
Regosolic	Weakly developed soils lacking distinct soil horizons	Varied	Varied	Varied	Varied	Well to imperfect
Solonetzic	Soils with hard, clay enriched subsurface horizons resulting from deflocculation and translocation of sodium-rich clays. May have a light-coloured surface horizon	Grasses and forbs, in places tree-covered	Cool; subarid to subhumid	Varied; usually not sandy	Base saturated; high Na	Well to imperfect

region of Manitoba, for example, glacial erosion of Paleozoic carbonate bedrock enriched the tills and other surficial deposits with carbonate minerals. This high carbonate content retarded soil development. This has resulted in very thin Chernozemic and Luvisolic soils (Table 11.1). Carbonate mineral content generally decreases as the distance of former glacier travel increases from the carbonate rock source. Consequently, soils become thicker as one moves south from this region down the paleo ice-flow direction. Another example of soil development apparently related to the nature of the parent material is the occurrence of Solonetzic soils in parts of Saskatchewan and Alberta. These soils are found in areas of thin glacial drift and close proximity to Cretaceous shales which are a source of sodium. Sodium ions act as a dispersing agent to the clay. As a result, clay moves readily from the surface A horizon to the subsurface B horizon, leaving grey, leached surfaces and compact, clay-enriched subsoils.

The age of the soil parent material has a marked effect on the kind of soil that has formed. Nearly all of the surficial deposits and associated landscapes in which soils have formed date from the end of the last glaciation, generally 15-8 ka. This relatively short period of soil formation contrasts markedly to the ages of many soils in unglaciated parts of the world. Consequently, many soils in Canada still contain large quantities of soluble salts and carbonates. Even those that have undergone leaching show relatively little alteration of silicates and other minerals.

There are other soils whose origins are more recent. Soils in sand dunes exemplify the effect of climatically induced hydrological changes on soil and sediment stability and susceptibility to wind erosion. In the grassland region of the Prairie provinces, Regosols prevail throughout dune areas, suggesting recent development of these soils. In the forest region, Regosols are succeeded by Brunisols as the eolian sands are stabilized. Even these soils are characterized by youthful features.

The unstable slopes of valley sides and escarpments are commonly characterized by weakly developed Regosols. Here, fluvial erosion and mass movement are more rapid than soil formation.

The influence of near-surface drainage on soil formation has been mentioned previously. Soil development is also affected by groundwater conditions. Poorly drained or Gleysolic soils form in direct response to a high water table. Saline (Solonetzic) soils occur where mineralized groundwater is sufficiently close to the surface to enable salts to rise to the surface by capillary flow. Quaternary stratigraphy can act as a major control in groundwater distribution and hence another aspect of Quaternary geology can play a major role in soil formation.

Soils of Canada

The soils of Canada have been described by Clayton et al. (1977). Table 11.1 summarizes the principal characteristics of the nine soil orders found in Canada. The generalized distribution of these orders across Canada is portrayed in Figure 11.1. The following discussion describes these soils and their parent materials, the major economic enterprises they support, and limitations they pose to land uses.

Cordilleran region
K.W.G. Valentine

The Cordilleran region (Fig. 1, Foreword) forms the western margin of Canada. Viewed simply, it comprises a chain of plateaus or low mountains sandwiched between two much higher mountain complexes — the Rocky, Mackenzie, and Arctic mountains to the east and the Coast, St. Elias, and Insular mountains to the west. It stretches from the 49th Parallel to the Arctic Ocean, and covers most of British Columbia and Yukon Territory, and smaller parts of Northwest Territories and Alberta. All, except portions of the Yukon Plateau in the far north and some summits and headlands, were covered by ice during the Wisconsin Glaciation. Most soils have, therefore, developed in geologically young surficial deposits. In mountains these deposits are till and rubbly colluvium, whose mineralogy and texture depend on the underlying bedrock. In valleys, and on some low plateaus, there are postglacial lacustrine, fluvial, and eolian deposits, whose mineralogy and texture depend on their mode of deposition as well as their source. The climate of the Coast and Insular mountains is wet and relatively mild. The interior plateaus are much drier with a continental range of temperature — warm in summer, cold in winter (extremely so in the north). The Rocky and Mackenzie mountains retain a continental temperature regime but receive more precipitation than the interior plateaus. A boreal forest of spruce and fir covers most of the region. High elevations and latitudes have alpine or arctic tundra of grass, herbs, and dwarf shrubs.

Heavy rain, acidic bedrock, and a dense coniferous forest combine to make Podzolic soils the most common type in the Coast Mountains and islands. These soils are acid, coarse-textured with numerous rock fragments, and generally thin. Their clay minerals are predominantly chlorite, smectite, and mica. Forestry is the most significant commercial landuse. Forest growth in the southern Cordillera is the most rapid in Canada as a result of generous moisture, a relatively long growing season, and nutrient cycling, rather than inherent soil fertility. Steep slopes, abundant rainfall, and thin soils mean tree harvesting and replanting face the added problem of soil erosion and massive landslides.

At the other extreme of productivity are thin and rubbly Brunisolic and Regosolic soils which are present on mountain tops. Short growing seasons and high energy erosive processes retard soil development or remove or bury soil almost as fast as it forms.

Valleys and lowlands of the coastal area, such as those of the Fraser River and Vancouver Island are characterized by more fertile soils. The postglacial sediments, which form the parent materials of these Brunisolic and Gleysolic soils contain abundant mineral nutrients, although their root zones suffer summer moisture deficits in the extreme southwest. These areas are generally used for agriculture, urban sites, and transportation corridors, rather than trees. Small

---

Valentine, K.W.G.
1989: Cordilleran region (Soils of Canada); in Chapter 11 of Quaternary Geology of Canada and Greenland, R.J. Fulton (ed.); Geological Survey of Canada, Geology of Canada, no. 1 (also Geological Society of America, The Geology of North America, v. K-1).

pockets of lowland organic soils support intensive vegetable, fruit, and flower production for neighbouring Vancouver and Victoria, although the organic layers are becoming thinner with continual use.

The Interior has subdued relief of plateaus and low mountains scored by deep river valleys. It is underlain mainly by volcanic and sedimentary rocks including considerable expanses of flat-lying lavas. The predominant surficial material is glacial till, but many basins contain postglacial lacustrine, fluvial, and eolian deposits, including Holocene volcanic ash (notably White River Ash in the southern Yukon, Lerbekmo and Campbell, 1969). The principal clay minerals in subsoils are mica, smectite, kaolinite, and vermiculite. Soil types are Chernozems, under the grasslands of the warmest southern valleys. There, irrigated agriculture produces orchard fruits and vegetables, as well as forage for beef cattle that graze the adjacent forested uplands in summer. The cooler central interior has Luvisolic and Podzolic soils under a forest that is harvested for timber and pulp. The northern interior is dry and very cold and contains little merchantable timber. The soil types are Brunisols, Organic, and perennially frozen Cryosols. Parts of this northern section were not covered by ice in the last glaciation and have deeply weathered soils. Lowland Organic soils regulate water runoff and stream flow through the year. Cryosolic soils, susceptible to erosion when thawed, require special treatment along roads, pipelines, or at drill sites.

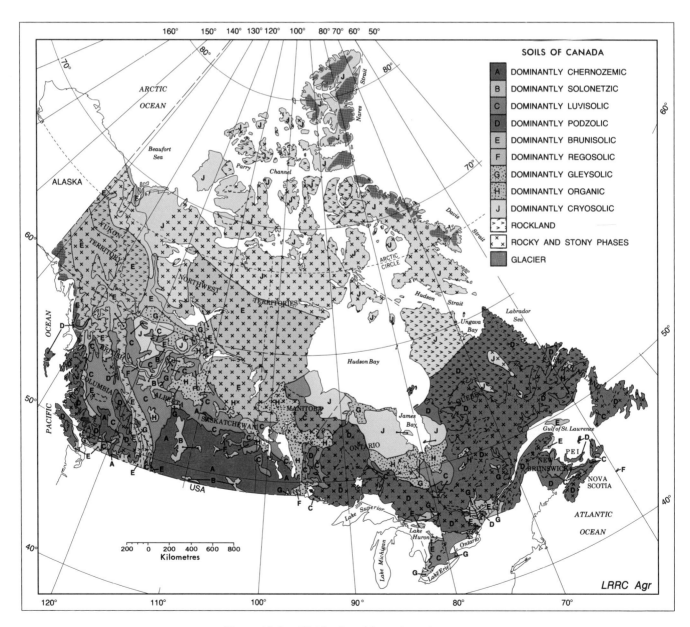

Figure 11.1. Distribution of Canadian soil orders.

The chain of the Rocky, Mackenzie, Ogilvie, and, in the far north, Arctic mountains is formed of strongly folded sedimentary rocks, many of which are calcareous (e.g. limestone in the southern Rockies). Glacial till and, on steeper slopes, rubbly colluvium are the main surficial materials except in the northern mountains which were essentially unglaciated. Soil types are Brunisolic and Podzolic under boreal forest, and Brunisolic, Regosolic, and Cryosolic in alpine and arctic tundra lands.

Interior plains
D.F. Acton

The Interior Plains region of Canada is bound by the Rocky Mountains on the west, the Canadian Shield on the north and east, and the 49th Parallel on the south (Fig. 1, Foreword). It is characterized by soils occurring in climatically induced concentric bands, shown in Figure 11.1, that emanate outward towards cooler and more moist regions from a "warm and dry" core, centred in southwestern Saskatchewan and southeastern Alberta.

Glacial lake silts, clays, and sands are the most widespread parent materials for soils on the Interior Plains (Prest et al., 1968). This is because Laurentide Ice Sheet retreat was generally down the regional slope. Consequently, the ice sheet ponded extensive lakes along its margins; these lakes migrated northeastwards with ice margin retreat. Thus much of the plains are underlain by glacial lake sediments even though glacial lakes probably covered only a small percentage of their area at any one time during deglaciation. Ground and hummocky moraine escaped inundations beneath glacial lakes over large areas. These are the next most common soil parent material following glacial lake sediments. Unlike the lacustrine sediments, these parent materials are predominantly till and contain an appreciable stone content (>2 mm). The finer than 2 mm fraction of these tills is dominantly derived from local bedrock; however, the stones are derived from both local bedrock and gravels and distant Paleozoic sedimentary rocks flanking the Canadian Shield and shield rocks.

Outwash gravels may be locally important as a parent material. They commonly occur within hummocky moraine or as deposits in former proglacial lake spillways. In places glacial deposits of all types are masked by blankets or patchy veneers of loess or dune sands and these are the parent materials of present soils. Complexes of soils are commonly buried within eolian sediments.

Lastly, the Late Mesozoic and early Tertiary bedrock locally may form soil parent material. Soils developed on the latter rocks are commonly Solonetzic.

Brown and Dark Brown Chernozemic soils, formed under mixed prairie grassland, dominate the relatively warm and dry climates in the southern core. Progressing to the north and east, Black and Dark Gray Chernozemic soils associated with moderately cold, subhumid climates and a dominantly fescue prairie-aspen grove vegetation reflect the greater biological productivity of these climatic regions. Farther north and east Gray Luvisolic, Brunisolic, peaty Gleysolic, and Organic soils dominate the cold, subhumid to humid region, with a coniferous forest-dominated vegetation. Such soils reflect much slower decomposition rates of organic matter and much higher weathering rates of mineral fractions as compared to the Chernozemic soils to the south. Finally, through the northern extension of the Interior Plains, Gray Luvisolic, Brunisolic, and Organic Cryosols in the subarctic zone give way to a predominance of Cryosols in the treeless arctic tundra. This represents a marked decrease in biological activity in the subarctic, coinciding with increased occurrences of permafrost.

The Chernozemic soil region described above represents the "heart" of western Canadian agriculture. These soils are richly endowed in a wide range of mineral constituents inherited from the varied assemblage of metamorphic, igneous, and volcanic rocks from the Precambrian Shield, limestone and dolomite of Paleozoic origin, and sedimentary rocks of Cretaceous and Tertiary age — generally well blended by the glacial events of the Pleistocene. As a result of this endowment, there is an abundant supply of minerals essential for plant nutrition. Only rarely do these sediments contain excessive concentrations of minerals that are detrimental to crop production. Physical properties, including clay content and structure, are also usually favourable for crop production.

In addition to this rich mineral endowment contained in the soil's parent material, the geological youthfulness of the surficial deposits and the semi-arid to subhumid Holocene environments have resulted in relatively young soils throughout the region. Most silicate minerals have undergone little alteration and occlusion with amorphous constituents is minimal. Soluble salts and lime are abundant in most soils, they may be excessive in some, however. Topography is also immature, typified by the chaotic character of the hummocky moraine where surface runoff collects locally. While erosion may be severe due to the relative instability of these youthful landscapes, drainage usually has not developed to the point where sediment is removed on a regional basis.

St. Lawrence Lowlands
E.W. Presant

The St. Lawrence Lowlands region covers only 1.5% of the total surface area of Canada (Fig. 1, Foreword); however, its fertile soils and relatively benign climate make it one of the most important agricultural regions in the country. It extends from the southwestern tip of Lake Erie to the Strait of Belle Isle, off Newfoundland's northern tip, but over 75% of the area occurs in southwestern Ontario and in the lowlands of the Ottawa-Montreal area. Flat-lying, gently warped

---

Acton, D.F.
1989: Interior Plains (Soils of Canada); *in* Chapter 11 of Quaternary Geology of Canada and Greenland, R.J. Fulton (ed.); Geological Survey of Canada, Geology of Canada, no. 1 (*also* Geological Society of America, The Geology of North America, v. K-1).

Presant, E.W.
1989: St. Lawrence Lowlands (Soils of Canada); *in* Chapter 11 of Quaternary Geology of Canada and Greenland, R.J. Fulton (ed.); Geological Survey of Canada, Geology of Canada, no. 1 (*also* Geological Society of America, The Geology of North America, v. K-1).

Paleozoic rocks that underlie the region are bounded on the north by the Canadian Shield and on the southeast by the Appalachian Mountains.

The unconsolidated surficial deposits that blanket the Paleozoic strata of the region were deposited during the last glaciation by a complex series of glacial erosional and depositional events culminating in the incursion of the Champlain Sea into Quebec and eastern Ontario. The clay mineralogy of the soils that developed on these relatively youthful deposits reflects this immaturity. Also, the dominant Canadian Shield origin of the glacial materials is indicated by the presence of essentially unweathered mica, chlorite, and vermiculite. The climate is strongly influenced by the close proximity of the Great Lakes and St. Lawrence River, causing milder and wetter conditions than elsewhere in the continental interior.

For the purposes of this discussion the St. Lawrence Lowlands region can be subdivided into the Western, Central, and Eastern lowlands on the basis of soil materials and utilization. The Western Lowland, which comprises most of southwestern Ontario, has fertile soils and favourable climatic conditions for a wide range of field, vegetable, and fruit crops. The soils are developed on glacial till and lacustrine deposits whose topography ranges from steeply sloping moraines and drumlins to flat lacustrine and till plains. Luvisolic soils are predominant. Brunisolic soils are most common in the stony or shallow tills near the Canadian Shield, whereas Gleysolic soils occupy most of the poorly drained areas of lacustrine deposits. The area includes scattered pockets of Organic soils which are used for intensive vegetable production. Many of the soils that are highly suited for crops such as tender fruits are also in great demand for various urban, aggregate, and transportation uses. Soil degradation, in the form of soil erosion on sloping, medium textured soils, or as soil compaction on the heavier silts and clays is also a serious problem.

The Central Lowland, which includes the flat plains bordering Ottawa and St. Lawrence rivers in eastern Ontario and western Quebec, extends into the upper part of the Gulf of St. Lawrence, east of Quebec City. Crop diversity is not as great as in the Western Lowland because of cooler temperatures, even though the soils are just as fertile. There are also many competing nonagricultural demands for land, especially in the vicinities of Montreal and Ottawa. The soils have developed in a variety of glacial deposits, in marine deposits left by the Champlain Sea and in fluvial sands and silts deposited as Ottawa and St. Lawrence rivers reworked older sediments. There are structural and drainage problems associated with the marine clays, and erosion problems on intensively farmed slopes of the glacial deposits. Brunisolic and Gleysolic soils predominate in the Central Lowland; some areas of Podzolic and Organic soils also occur.

The Eastern Lowland extends from the mouth of the St. Lawrence River to the Strait of Belle Isle and includes Anticosti Island and a narrow strip of land along the northwestern coast of Newfoundland. The surface material in this area consists of marine clay, silt and sand, fluvial and glaciofluvial sand and gravel, and till. The Podzolic and Brunisolic soils that developed on these materials are not used extensively for agriculture because of the short growing season and long, harsh winters, but they do support a thriving forestry industry.

Atlantic region

K. Webb

The Atlantic region, on Canada's east coast, is centred around the Gulf of St. Lawrence. The region includes Gaspésie peninsula of southeastern Quebec, the provinces of New Brunswick, Nova Scotia, and Prince Edward Island, and most of the island of Newfoundland (Fig. 1, Foreword). The climate of the region is generally cool and humid but varies from a somewhat mild and wet maritime climate in the southern coastal areas to a more continental climate in the rest of the region. A mixed hardwood-softwood forest covers most of the mainland whereas a boreal coniferous forest predominates at higher elevations in the Gaspésie and highlands of Cape Breton Island. The vegetation of Newfoundland is a mosaic of productive Boreal forest, scrub forest, heath-moss barrens, and extensive peatlands.

Physiographically, the Atlantic region represents the northerly extension of the Appalachian Mountains and is composed of an extensive complex belt of fold mountains consisting of flat-topped rolling highlands and uplands interspersed with deeply entrenched valleys and broad undulating to rolling lowland plains. The highlands and uplands consist of a variety of metasediments, crystalline gneisses, and granitic intrusives; lowlands are underlain mainly by Carboniferous redbeds. All of the region (with the possible exception of one or two small nunataks) was covered by ice during the Pleistocene period and most of the land surface has been scoured and subsequently covered with glacial deposits of varying thickness and composition. It is on these geologically young surficial materials that the region's soils have developed.

Podzolic soils characterize the entire region and typically occur in coarse to medium textured, acid surface materials. Clay minerals are predominantly illite with significant amounts of chlorite and smaller amounts of vermiculite, kaolinite, and montmorillonite.

The Podzolic soils of the highlands and uplands have developed predominantly in shallow, coarse textured, stony, glacial till derived from the underlying igneous and metamorphic bedrock. On gentle slopes, where drainage is impeded, Gleysolic soils occur in association with Podzols. On similar slopes in Newfoundland, peaty Gleysols and Organic soils are developed on extensive shallow peatlands. At high elevations, Regosolic soils are found on frost-shattered bedrock. Cemented and indurated layers commonly occur in the Podzolic soils of the highlands and uplands of Newfoundland and parts of Nova Scotia.

The major land use in the highlands is forestry, recreation, and wildlife habitat. As well, the highlands of the mainland produce maple syrup products and abandoned farmland is used for commercially grown blueberries. The upland soils are similar to the soils of the highlands, and are stony, shallow, and infertile. An exception is along Saint John River valley of western New Brunswick where Podzolic

---

Webb, K.
1989: Atlantic region (Soils of Canada); *in* Chapter 11 of Quaternary Geology of Canada and Greenland, R.J. Fulton (ed.); Geological Survey of Canada, Geology of Canada, no. 1 (also Geological Society of America, The Geology of North America, v. K-1).

soils, developed on rolling terrain in relatively thick, medium textured tills, are used extensively for agriculture. The widespread cropping of potatoes has caused serious erosion of these soils.

On the lowlands of the mainland, Podzolic, Gleysolic, and Luvisolic soils have developed in thick, medium to coarse textured red tills derived from the underlying sedimentary bedrock. Podzolic soils predominate on the sandy glaciofluvial deposits and organic soils occur in the numerous poorly drained areas. Regosolic soils characterize the dyked marshlands surrounding the Bay of Fundy and the alluvial sediments adjacent to rivers and streams.

Forestry and agriculture are the major land uses on the lowlands. The finer textured Luvisols are used for pasture, dairying, and forage production. In Annapolis Valley, the deep, coarse textured Podzols are cultivated for fruits and vegetables, while in other lowland areas these soils are used for cash crops and mixed farming. Small areas of organic soils are used for agriculture in New Brunswick and Newfoundland, and extensive areas of these soils are being considered as potential sources of fuel peat. The major limitations of lowland soils are acidity, excess soil moisture, and shallow, compact subsoils which reduce soil permeability and root growth. Where soils are left bare over winter, as in the areas of potato production in Prince Edward Island, erosion is a major problem.

Shield region
D.F. Acton

The Canadian Shield — a vast region underlain primarily by Precambrian crystalline rock — covers almost half of Canada (Fig. 3.3). The soil descriptions in this section apply to the area south of the arctic treeline. These differ markedly from shield soils north of treeline which are described in the following section on Arctic soils.

Many glacial landform features are evident throughout the Canadian Shield. Hummocky topography, usually consisting of a thin till veneer over crystalline rock, is common but long, sinuous eskers, belts of drumlins and glacial flutings, and areas of ribbed moraine are also prominent features of the shield landscape. Glacial lake, marine, outwash, and organic deposits are widespread in depressions. Raised beaches are common features of the washed landscape around large lakes and areas formerly submerged by the sea. Some of these have been subjected to eolian activity.

The various glacial and organic deposits referred to above form the parent materials of the soils. The glacial deposits are usually sandy or gravelly, with the glacial tills having a high boulder content. Only the glacial lake deposits have an appreciable amount of silt and clay. The mineralogy is dominated by quartz, feldspar, and mica, except in glacial lake basins where montmorillonite and vermiculite are also present in appreciable quantities. Carbonate minerals are notable by their absence from most mineral assemblages.

Soils of the Canadian Shield are predominantly Luvisols, Brunisols, and Podzols. Luvisols are soils in which translocation of clay is the major process. In Brunisols and Podzols, in addition to translocation of clays, alteration of primary minerals occurs and the alteration products can be complexed with organic matter. Brunisolic soils occur where there has been a relatively small amount of alteration and especially where there has been little complexing with organic matter, usually associated with more arid sites. Podzolic soils are those formed under more humid conditions where there is considerable amount of organic-bound amorphous material.

Organic soils occupy the poorly drained low areas in extreme southern portions of the Canadian Shield. Farther north, where the climate is colder, and discontinuous and continuous permafrost is present, organic Cryosols dominate the peatlands.

The vast majority of land in the southern Canadian Shield is forest or peatland so it is natural that the main land use is forestry, hunting, trapping, and recreation. Agricultural development of soils on the Canadian Shield is virtually restricted to the glacial lake clays in northern Ontario, areas of marine clays in the Lac Saint-Jean area and small areas of fine grained outwash, lacustrine sediments, and marine deposits around the southern margin of the shield.

Arctic Canada
C. Tarnocai

Arctic Canada, the area north of the arctic treeline, has mainly Cryosolic soils (Canada Soil Survey Committee, 1978). These soils have developed in a very cold, arctic climate and are associated with near-surface permafrost (Tarnocai, 1978). They are strongly affected by cryogenic processes and are commonly associated with patterned ground such as circles, nets, stripes, steps, and polygons. Cryosolic soils occupy approximately 40% of the area of Canada and are by far the most widely occurring soils in northern Canada.

The Arctic is dominated by soils with marked cryoturbation or mixing — the Turbic Cryosols. In the southern Arctic these soils have a brown B horizon (Brunisolic Turbic Cryosols) much like the Brunisolic soils that occur on coarse textured deposits in this region. As the climate becomes colder, northward, the brown B horizon is less distinct and becomes mixed by cryoturbation, eventually disappearing in the High Arctic as Regosolic Turbic Cryosols become dominant. Gleysolic Turbic Cryosols are found on poorly drained areas. Various types of patterned ground are associated with these Turbic Cryosols.

Static Cryosols, commonly associated with coarse textured material, are the second most commonly occurring group of Cryosols in the Arctic. Little or no cryoturbation

---

Acton, D.F.
1989: Shield region (Soils of Canada); in Chapter 11 of Quaternary Geology of Canada and Greenland, R.J. Fulton (ed.); Geological Survey of Canada, Geology of Canada, no. 1 (also Geological Society of America, The Geology of North America, v. K-1).

Tarnocai, C.
1989: Arctic Canada (Soils of Canada); in Chapter 11 of Quaternary Geology of Canada and Greenland, R.J. Fulton (ed.); Geological Survey of Canada, Geology of Canada, no. 1 (also Geological Society of America, The Geology of North America, v. K-1).

occurs in these soils and patterned ground (except for polygons) is generally lacking. Brunisolic Static Cryosols are common in the southern part of the Arctic, but are less common in the northern part. Unlike the Turbic Cryosols, however, they still occur in the Arctic Islands. Regosolic Static Cryosols are restricted to recent deposits and areas undergoing erosion; Gleysolic Static Cryosols are found on poorly drained areas, mainly on medium and coarse textured materials.

Organic Cryosols associated with peatlands occur mainly in the southern part of the Arctic. The peatland types most commonly encountered are lowland polygons and peat mounds. Organic Cryosols are also found in the Arctic Islands but cover only the small areas associated with strongly eroded, high-centre lowland polygons.

The organic matter content of Turbic and Static Cryosols is high, even in soils developed on the Arctic Islands. A large portion of the organic matter is cryoturbated from the surface and is found in the form of organic layers, intrusions, smears and very fine organic materials mixed with mineral materials.

Soils in the Arctic generally have a high moisture content, especially near the permafrost table. This results in reduced colours and mottles, even in better-drained situations. In the western Arctic, soils associated with earth hummocks show very characteristic mottles and grey colours (Tarnocai and Zoltai, 1978). In the High Arctic, because of the shallowness of the active layer, mottling is uncommon, except in fine textured materials.

Arctic soils have characteristic structures. Vesicular structures (Bunting and Fedoroff, 1973; Zoltai and Woo, 1976; Tarnocai, 1976) are closely related to drainage and texture. Well developed granular structure occurs near the surface of earth hummocks in fine-textured Cryosols, but their lower horizons are massive and structureless, having a high bulk density (Drew and Tedrow, 1962; Zoltai and Tarnocai, 1974). In addition, vertical channels, frost cracks, and banded, granic, orbiculic, and conglomeric fabrics are also found in these soils (Fox and Protz, 1981; Fox, 1983).

Eluvial, or leached, horizons are common in the southern part of the Arctic and have also been observed in the Arctic Islands (McMillan, 1960; Tarnocai, 1976). A surface salt crust is common in the High Arctic (Tedrow et al., 1968; Tarnocai, 1976). A calcareous crust may also be formed on the underside of rocks, both at the surface and within the soil mass.

One of the most striking features of arctic soils is the great amount of ice, in the form of segregated ice crystals, vein ice, ice lenses, ice wedges, and thick, massive ground ice, occurring in the near-surface permafrost (Tarnocai, 1978). Ice content in soil varies and generally increases as soil texture becomes finer. Soils with much ice are sensitive to disturbance and the ice content is one of the major factors controlling their utilization.

The thermal characteristics of arctic soils are radically different from those of soils in temperate climates (Tarnocai, 1984a). Cryosols, because of the shallowness of the active layer and the small amount of stored energy, have very little buffering capacity against a change in atmospheric temperature. In addition, the underlying permafrost acts as a heat sink in summer and fall and continually removes energy from the thawed layer of the soil. Hence, the slightest decrease in air temperature quickly lowers the soil temperature at all depths. This cooling causes the frost table to rise long before the surface freezes (Tarnocai, 1977).

Most arctic soils are under tundra vegetation. Plant density in the tundra community decreases with increasing latitude.

During the last few decades, particularly the 1960s, oil and gas exploration and mining have left scars on the arctic landscape because of improper land use. This was mainly due to a lack of knowledge of the properties of the arctic terrain and how it would react to various land use practices. New land use methods and technology have reduced the damage to the surface soil and thus environmental damage has now been minimized.

PEAT RESOURCES IN CANADA

C. Tarnocai

Peat is an accumulation of variably decomposed and compacted plant debris. Remains of peat-forming mosses and sedges usually dominate most peats although other plant remains and variable quantities of inorganic material may be present. According to the Canadian System of Soil Classification (Canada Soil Survey Committee, 1978), peat has a carbon content greater than 17% by weight. According to the definition used in the fuel peat industry, peat is considered to be an organic material with a maximum ash content of 40% (Monenco Ontario Limited, 1981).

Areas of peat accumulation include a range of biogeomorphic features; they are collectively referred to as peatlands. The main physical requirement for the establishment of peatlands under Canadian climatic regimes is abundant moisture. This can be satisfied by a rainy climate such as that of the west and east coasts of Canada or by the poor drainage conditions and high water tables which occur in permafrost regions or in areas of deranged drainage such as hummocky moraine. Consequently peatlands and deposits can occur and accumulate in most parts of Canada. The most important exceptions are the Interior Plains where aridity has all but prevented the development of peatlands and, to a lesser extent, the High Arctic where aridity and climatic severity limit plant growth in general.

A wide range of values is found in the literature for the total area of peatlands in Canada with the largest being 170×10^6 ha (Kivinen and Pakarinen, 1980), followed by 153×10^6 ha (Zoltai, 1980) and 129×10^6 ha (Muskeg Subcommittee of the NRC, 1977). This range is, in part, the result of varying definitions as to what deposits may be included under the heading of peat. The data concerning peat resources presented in this report are based on a study that evaluated Canadian peatland inventories and established the distribution of peatlands based on these inventories (Tarnocai, 1984b). It should be noted, however, that the figures presented here are still estimates even though they are based on inventories. Large areas of Canada, especially in the North, have only been covered by reconnaissance studies or not at all.

Tarnocai, C.
1989: Peat resources in Canada; in Chapter 11 of Quaternary Geology of Canada and Greenland, R.J. Fulton (ed.); Geological Survey of Canada, Geology of Canada, no. 1 (also Geological Society of America, The Geology of North America, v. K-1).

Table 11.2. Peatland classification according to the wetland classifications by Zoltai et al. (1973) and Tarnocai (1980)

PEATLAND CLASSES			
Bog	Fen	Swamp	Marsh
PEAT LANDFORMS			
Palsa	Northern ribbed	Stream	Estuarine high
Peat mound	Atlantic ribbed	Shore	Estuarine low
Mound	Ladder	Peat margin	Coastal high
Domed	Net	Basin	Coastal low
Polygonal peat plateau	Floating	Flat	Floodplain
Lowland polygon	Stream	Floodplain	Channel
Northern plateau	Collapse		Inactive delta
Atlantic plateau	Palsa		
Collapse	Spring		
Floating	Slope		
Shore	Lowland polygon		
Basin	Horizontal		
Flat	Channel		
String			
Blanket			
Bowl			
Slope			
Veneer			

Table 11.3. Classification of peat materials

Material	Dominant organic constituent(s)	State of decomposition	pH	Van Post[1] value	Rubbed fibre content[2] (%)	Mean bulk density (g/cm^3)	Comments
Sphagnum peat	*Sphagnum* sp.	undecomposed	2.5-4.5	1-3	~60	0.07	entire sphagnum plant readily identifiable
Sedge peat	sedge (*Carex* sp.)	moderately decomposed	4.5-7.0	up to 5-7	8-30	0.11	sedge leaves identifiable with the unaided eye
Brown moss sedge peat	sedge (*Carex* sp.) and brown mosses of the genera *Drepanocladus, Calliergon,* and *Aulacomnium*	undecomposed to moderately decomposed	5.0-7.0	up to 5-7	8-30	0.11	sedge and moss plants identifiable with the unaided eye
Woody-sedge peat	sedge (*Carex* sp.) woody material	moderately decomposed	4.5-7.0	4-6	10-40	0.18	sedge and wood fragments identifiable with unaided eye
Feather moss peat	feather mosses of the genera *Hypnum, Hylocomium,* and *Pleurozium*	moderately decomposed	4.5-6.5	4-7	10-60	0.12	minor amounts of woody material derived from coniferous trees
Sedimentary peat	aquatic plant debris including algae, diatoms, aquatic mosses	highly decomposed and comminuted	4.5-6.5	—	—	0.13 but may be as high as 0.17	few plant fragments visible to unaided eye; plastic and slightly sticky; shrinks to hard clods upon drying; dry clods difficult to rewet
Amorphous peat	variable	highly decomposed	4.5-6.9	5-10	2-8	0.15	plant fragments unidentifiable to the unaided eye

1 The Van Post value rates the relative state of decomposition of vegetative material with 1 being unaltered and 10 being an amorphous residue.
2 Rubbed fibre content is a measure of the resiliency of the fibrous component of the peat.

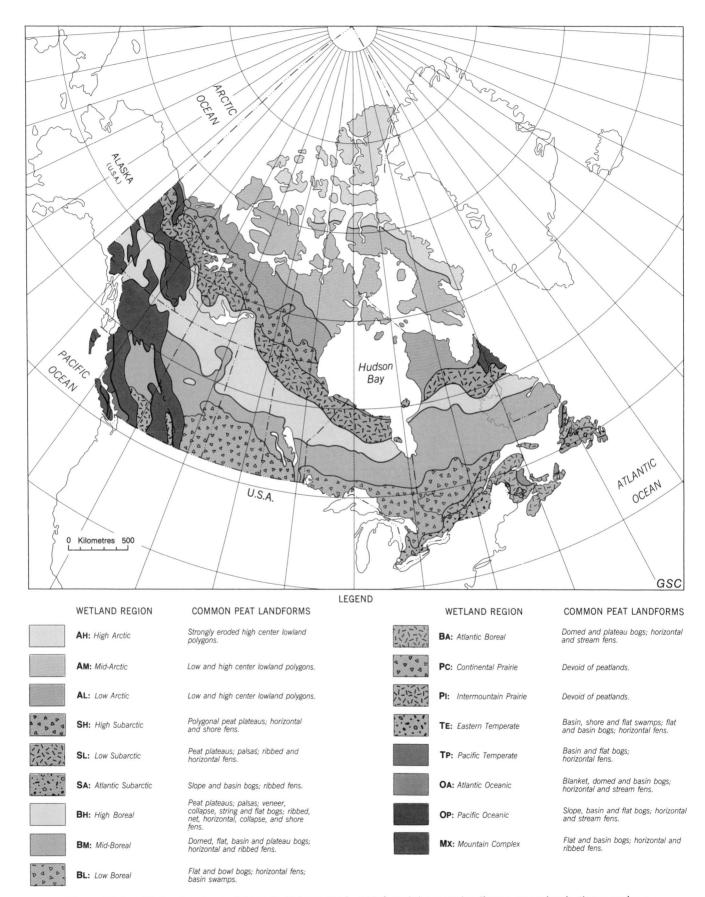

Figure 11.2. Wetland regions of Canada (Adams et al., 1981) and the peat landforms occurring in these regions.

Peatland types

Peatland types in Canada vary with climatic conditions and the physiographic setting. A number of classifications have been proposed for Canadian peatlands. Radforth (1955, 1958) stressed the appearance of muskeg patterns, as seen from the air, with ground information being supplied concerning the vegetation and topography. Others emphasized the use of the peatland vegetation (Jeglum et al., 1974) or the peat landform (Zoltai et al., 1973; Tarnocai, 1980) for classification.

The approach which is most widely accepted and most commonly used in Canada today is a genetically based, wetland classification (Zoltai et al., 1973; Tarnocai, 1980). It includes both peatlands (peat depths greater than 40 cm) and mineral wetlands (peat depths 40 cm or less). There are four general peatland classes: bogs, fens, swamps, and marshes. A description of each class follows:

Bogs. A bog is a peatland which generally has a water table at or near the surface. Its surface is either raised above or level with the surrounding wetlands, and it is virtually unaffected by the nutrient-rich groundwaters from any adjacent mineral soils (ombrotrophic). Hence, the groundwater of the bog is generally acidic and low in nutrients. The dominant peat materials are undecomposed sphagnum and moderately decomposed woody-moss peat underlain, in places, by moderately to well decomposed sedge peat. Bogs may be treed with black spruce or treeless and they are usually covered with sphagnum, feather mosses, and ericaceous shrubs.

Fens. A fen is a peatland with a water table that is commonly at or above its surface. The water is usually nutrient-rich and minerotrophic because of the adjacent mineral soils. The dominant peat materials are shallow to deep, well to moderately decomposed sedge or woody-sedge peat. The vegetation consists primarily of sedges, grasses, reeds, and brown mosses with some shrub cover and, in places, a scanty tree layer.

Swamps. A swamp is a peatland or a mineral wetland with standing or gently flowing water in the form of pools and channels. The water table is commonly at or near the surface. Pronounced water movement occurs from the margins or other mineral sources and hence the waters are nutrient-rich. If peat is present, it is mainly well decomposed woody or amorphous peat underlain, in places, by sedge peat. The vegetation is characterized by a dense tree cover of coniferous or deciduous species and by tall shrubs, herbs, and mosses.

Marshes. A marsh is a mineral wetland or a peatland that is periodically inundated by standing or slowly moving waters. Surface water levels may fluctuate seasonally, with declining levels exposing drawdown zones of matted vegetation or mud flats. The waters are nutrient-rich. The substratum usually consists of mineral material or moderately to well decomposed peat deposits. Marshes characteristically show a zonal or mosaic surface pattern of vegetation, composed of unconsolidated grass and sedge sods, commonly interspersed with channels or pools of open water. Marshes may be bordered by peripheral bands of trees and shrubs but the predominant vegetation consists of a variety of emergent nonwoody plants such as rushes, reeds, reed-grasses, and sedges. Where open water areas occur, a variety of submerged and floating aquatic plants flourish.

These wetland classes are further subdivided according to their landforms. The landform name is based on either

Table 11.4. Peat resources of Canada (from Tarnocai, 1984b)

Provinces and Territories	ha × 10³	Peatland area % of land area within designated areas	% of total Canadian peatlands	Indicated peat volumes m³ × 10³	%	Indicated oven dry weight of peat tonnes × 10⁶	Indicated weight of peat with 50% water content⁺ tonnes × 10⁶	%	Measured peat (Tibbetts and Ismail, 1980) tonnes × 10⁶
Alberta	12 673	20	11	316 822	11	36 118	54 177	11	3
British Columbia	1 289	1	1	38 685	1	4 410	6 615	1	20
Manitoba	20 664	38	19	516 605	17	58 893	88 339	17	103
New Brunswick	120	2	*	4 800	*	466	698	*	103
Newfoundland–Labrador	6 429	17	6	257 160	8	24 945	37 417	8	612
Northwest Territories	25 111	8	23	577 553	19	65 841	98 762	19	-
Nova Scotia	158	3	*	6 320	*	613	920	*	23
Ontario	22 555	25	20	676 653	22	77 138	115 708	23	135
Prince Edward Island	8	1	*	312	*	30	45	*	2
Quebec	11 713	9	11	351 381	12	40 057	60 086	12	64
Saskatchewan	9 309	16	8	232 737	8	26 532	39 798	8	27
Yukon Territory	1 298	3	1	25 968	1	2 960	4 441	1	-
Canada	111 327	12		3 004 996		338 003	507 006		1 092

* less than 1%
⁺ oven dry weight basis

the surface morphology, the surface pattern, the morphology of the basin in which the wetland developed or the associated water bodies. The various peat landforms found in Canada are presented, according to the peatland classes, in Table 11.2 and their geographic distribution is shown in Figure 11.2. They have been described by Tarnocai (1970, 1980) and Zoltai et al. (1973).

Peat materials

Peat materials associated with peatlands are separated according to botanical composition. Thus, the name of the peat material indicates the most common plant material(s) associated with that specific peat. For example, woody-sedge peat is composed dominantly of sedge peat with a subdominant amount of woody peat. A brief description of some common peat materials associated with Canadian peatlands is given in Table 11.3.

Distribution of peat resources

Figures 11.3 and 11.4 show the estimated total areal coverage of peatlands, and the thickness of deposits, and Table 11.4 gives the estimated volume of peat in Canada.

Tibbetts (1983) suggested various categories for estimating the volume of peat, according to the level of confidence in the evaluation. A similar procedure is used in mineral resource evaluation. These categories are:

1. Inferred resources — estimates based solely on satellite data, aerial photographs, and topographical maps and assumed thickness of not more than 1 m.
2. Indicated resources — volume estimated by use of satellite data, aerial reconnaissance, radar, aerial photographs, topographical maps, and at least one measured thickness.

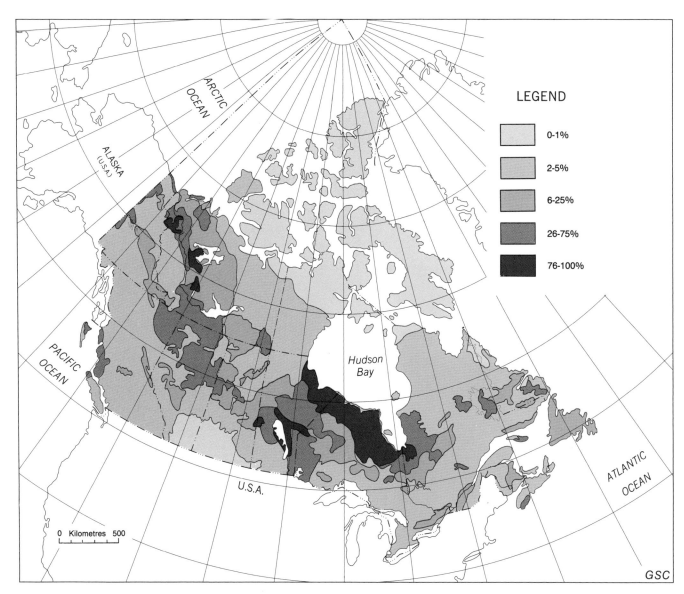

Figure 11.3. Distribution of peatlands in Canada (from Tarnocai, 1984b).

3. Measured resources — volume and quality measured and evaluated over the entire deposit; an example is the New Brunswick inventory (Keys et al., 1982).
4. Reserves — measured resources that can be economically exploited.

Based on these categories the volume estimates given in Table 11.4 would be classified under "indicated resources". In reality, however, these volumes are based mainly on reconnaissance and exploratory surveys and thus many actual measurements were made. Insufficient measurements were made, however, to permit them to be classified as "measured resources".

The weight of peat in the various regions of Canada was determined using the calculated volumes (Table 11.4) and average oven dry bulk density values for sphagnum peat (0.075 g/cm^3), sedge peat (0.119 g/cm^3), and woody peat (0.149 g/cm^3) (Mills, 1974). Using these values, an average bulk density of 0.097 g/cm^3 was derived for the Atlantic Provinces, where the peat deposits are composed mainly of sphagnum and sedge peats. For the rest of the country, where peat deposits are composed chiefly of sphagnum, sedge, and woody peat materials, an average bulk density of 0.114 g/cm^3 was used for calculations. The weights of peat in metric tonnes, based on oven dry weight and 50% water content (on a weight basis) are presented in Table 11.4; also shown are the tonnages of the "measured" peat resources at 50% water content (category 3) from Tibbetts and Ismail (1980), indicating that a small fraction of Canada's peat resource has been measured in detail.

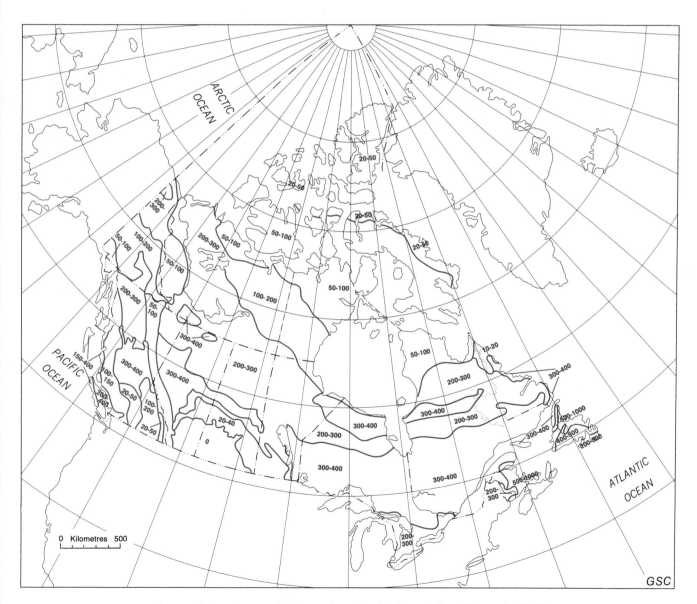

Figure 11.4. Average thickness of peat (cm) in Canada (from Tarnocai, 1984b).

Uses of peat

The following discussion outlines present and potential direct or indirect utilization of peat resources in agriculture and forestry, and as a fuel or a raw material for the chemical industry. The chief value of most of Canada's peatlands, however, is, and probably will remain, as a wildlife habitat and watershed regulator.

The use of peat depends mainly on its properties and quality, the climatic conditions of the area where the peatland is situated, and economic factors.

Peatlands are used in agriculture for growing a variety of crops. Unpublished inventory data from Environment Canada indicate that approximately 280×10^3 ha of peatland are now used for agriculture. These areas are situated mainly in the Atlantic Provinces, Quebec and Ontario, and in the Fraser River delta of British Columbia.

Most peatlands in Canada support nonmerchantable forests. Trials carried out in Newfoundland indicate that fens on slopes are the most suitable peatlands on which to grow merchantable forests. These peatlands are usually the most sheltered, nutrient-rich, and easy to drain (Pollett, 1972). The potential of peatlands for supporting merchantable forests in Canada cannot easily be determined from the limited amount of data presently available. The optimum water table level, soil moisture, and nutrient levels depend on the tree species. Regional variations in climate, topography, and geology are also major considerations in site selection and establishment of future plantations.

Peat is also widely used in Canada for horticultural purposes, mainly as a soil conditioner. The addition of horticultural peat to a soil increases the organic matter content and the ability to hold water and nutrients, and improves aeration. Horticultural peat is used mainly in greenhouses, gardens, and forest nurseries. It is also used in the production of peat pots and as a seed carrier. In addition, Canada is the world's largest exporter of horticultural peat.

Because of its large surface area, porosity, and exchange capacity, peat is often used as a filtration and absorption agent. As a physical filter, it is effective in removing suspended solids from effluent and in reducing the sludge clogging of traditional filters. Peat is also effective in the control of oil spills because it has a high absorbency due to its porous nature. Peat can also be used as a chemical filter because it is effective in removing heavy metals, colour, and toxic materials from industrial effluent.

The use of peat as a fuel dates far into the past (Table 11.5). It can be used for fuel as mined or after it has been upgraded. Peat can be burned in furnaces for heating or in boilers to produce steam in order to generate electricity. It can also be processed into a variety of fuels including coke, synthetic natural gas, and methanol. Methanol so produced can also be used as a feedstock in the chemical industry. Because peat contains about 60% more volatile matter than lignite (Punwani, 1980), it is much easier to convert peat to natural gases and liquids than it is to convert coal. The coke produced from peat is chemically of a higher quality than coal-based coke and can be used as a reducing agent by the chemical and steel industries and as a binder in the production of the iron pellets used by steel mills.

The main characteristics looked for in fuel or energy peats are high degree of humification, high bulk density, relatively low ash content, low content of potential pollutants such as sulphate and mercury, and high calorific value.

Table 11.5, which is based on European data, gives the basic energy values of fuel peat and provides a comparison with various other fuels.

Since a tonne of dry peat has an energy value of approximately 20×10^9 J (Overend, 1981), the total peat resource of Canada has an energy value of 6.7×10^{21} J. For comparison, the energy value of the Canadian oil resources is 80×10^{18} J, of the natural gas resources 145×10^{18} J, of the coal resources 2.2×10^{21} J, and of the oil sands resources 5.8×10^{21} J (Overend, 1983). Unfortunately, with present mining methods it is technically feasible to use only 40% of Canada's peat resources.

Peat mining

Climate is one of the main physical factors determining the feasibility of peat mining operations. Variables such as temperature, precipitation, the amount of sunshine, and the length of the frost-free period are of prime importance, but

Table 11.5. Comparison of the properties of various fuels

Properties of fuels	Heavy fuel oil	Coal (bituminous)	Coal (lignite)	Peat	Wood
C %	83-86	76-87	65-75	50-60	48-50
H %	11.5-12.5	3.5-5.0	4.5-5.5	5.0-6.5	6.0-6.5
O %	1.5-2.5	2.8-11.3	20-30	30-40	38-42
N %	0.2-0.3	0.8-1.2	1-2	1.0-2.5	0.5-2.3
S %	2.0-2.8	1-3	1-3	0.1-0.2	-
Ash content %	0.3	4-10	6-10	2-10	0.4-0.6
Melting point of ash C°	-	1100-1300	1100-1300	1100-1200	1350-1450
Volatiles %	-	10-50	50-60	60-70	75-85
Bulk density kg/m³	920-970	720-880	650-780	300-400	320-640
Effective heat value of dry matter MJ/kg	41.4-41.7	28.5-33.0	20.1-24.3	19.7-21.4	18.4-19.3
Operational moisture %	0.1	3-8	40-60	40-60	30-55
Effective heat value at the lowest operational moisture content MJ/kg	-	27.5-32.0	11.1-13.6	10.5-12.6	12.1-12.7
Effective heat value at the highest operational moisture content MJ/kg	-	26.0-30.2	6.5-8.2	7.5-8.2	6.9-7.3

Source: Monenco Ontario Ltd., 1981

Table 11.6. Sources of aggregate in Canada

Age	British Columbia	Alberta	Saskatchewan	Manitoba	Ontario	Quebec	New Brunswick	Prince Edward Island	Nova Scotia	Newfoundland
Holocene	ALLUVIAL Pumice DELTAIC - fan*	ALLUVIAL* EOLIAN	ALLUVIAL - fan - terrace - plain - bar	EOLIAN			ALLUVIAL - bar - terrace EOLIAN	MARINE - beach		MARINE - beach
Pleistocene (Late Wisconsinan)	GLACIOMARINE - beach GLACIOLACUSTRINE - beach GLACIOFLUVIAL* - spillway - outwash - esker - terrace - delta	GLACIOLACUSTRINE - clay GLACIOFLUVIAL* - outwash	GLACIOLACUSTRINE - clay GLACIOFLUVIAL - outwash - spillway - esker - kame	GLACIOLACUSTRINE - outwash - beach* GLACIOFLUVIAL - esker* - kame* - outwash GLACIAL* - moraine	GLACIOFLUVIAL* - outwash - delta - esker - kame GLACIAL - moraine	GLACIOFLUVIAL* - outwash - delta - esker - kame GLACIAL - moraine	GLACIOMARINE* - apron - delta GLACIOFLUVIAL* - outwash - ice contact		GLACIOFLUVIAL* GLACIAL - till	GLACIOMARINE - terrace - ridges - beach GLACIOFLUVIAL - esker - outwash GLACIAL - till*
(Early Pleistocene)		PREGLACIAL FLUVIAL*								
Tertiary	Basalt	GRAVEL and Conglomerate	GRAVEL and Conglomerate							
Cretaceous	Sandstone									
Jurassic	Diorite									
Triassic										
Permian										
Pennsylvanian										
Mississippian	Amphibolite									
Devonian				Limestone	Limestone	Sandstone				Dolomite
Silurian					Dolomite	Limestone				Volcanics
Ordovician					Limestone	Limestone				Siltstone
Cambrian	Limestone					Dolomite				Shale
	Gneiss			Granite		Granite			Quartzite, Granite	Sandstone
						Andesite				Granite
						Gabbro				

* primary aggregate source
Source: unpublished information supplied by respective provincial mines branches

evapotranspiration and average annual moisture deficiency are also useful indicators of regional suitability for peat mining.

Peat is commercially mined, using either dry or wet methods (Aspinall and Hudak, 1980; Carncross, 1980; and Tomiczek et al., 1982). The dry methods (milled and sod mining techniques) produce peat containing 35 to 50% moisture content. In the dry mining methods, extensive site preparation in the form of draining and grading must be carried out before peat production can begin. Since this method of mining requires field drying, it is greatly affected by meteorological factors. Dry mining methods are most commonly used to mine peat for fuel and horticultural purposes.

The wet mining method produces peat containing 80 to 98% water, depending on the mining equipment used. This high water content peat slurry is then pumped to the plant for dewatering. The wet mining method requires very little site preparation and is used under wet climatic conditions or on peatlands that are difficult to drain.

Present mining techniques are appropriate only for unfrozen peat; unfortunately, approximately 60% of Canadian peatlands are frozen. High ice-content, perennially frozen peatlands cannot be mined with existing techniques; new mining methods will be required to exploit these deposits.

AGGREGATE AND NONMETALLIC QUATERNARY MINERAL RESOURCES

W.A.D. Edwards

Aggregate is hard, inert construction material used with Portland or bituminous cements to form concrete, mortar, and asphalt. It is also used without cement as railroad ballast, in road construction, and in general building construction. Aggregate includes *mineral aggregate* such as sand, gravel, or crushed stone, and *synthetic* or *manufactured aggregate* such as recycled concrete or asphalt, burned shale or clay, and slag. Aggregate is an essential and major commodity in the construction of our transportation systems, urban centres and mega-projects.

In Canada, aggregate is derived predominantly from deposits of sand and gravel although crushed stone may be locally important. Most of the sand and gravel deposits originate as postglacial floodplain or terrace gravels; glaciofluvial outwash sediments including kames, eskers, kame deltas, glacial marine deltas; and preglacial gravels buried in valleys or forming caps on plateaus (Table 11.6). In some cases, material of sand and gravel size is screened from till or derived from beach or eolian deposits. Examples of recent aggregate resource mapping and inventory are given in Table 11.7.

Aggregate from all sources amounts to one of the most valuable nonmetallic mineral materials mined in Canada.

The value of gross production during 1981 and 1982 (Table 11.8) exceeded $500 million annually (Stonehouse, 1984).

Utilization and production of aggregate

Transportation systems (including highway and railroad construction and maintenance) accounted for 70% of the 1980 Canadian consumption of sand and gravel, and 38% of the consumption of crushed stone. In some provinces, such as Prince Edward Island, New Brunswick, and Saskatchewan, transportation projects consume 80% to 90% of total yearly production.

Urban centres provide the major markets for aggregate in many regions. For example, the greater Vancouver/Fraser Lowland area consumes about half of the yearly British Columbia aggregate production; Edmonton and Calgary together use half of Alberta's production; Winnipeg requires up to half of Manitoba's yearly production; and Metropolitan Toronto requires about a quarter of that of Ontario.

According to the Canadian Transport Commission (1978) special or mega-projects may require huge amounts of material. In the Northwest Territories the sand and gravel requirement for oil and gas exploration (camps, wharves, airstrips, roads, and drilling islands) between 1972 and 1976 exceeded 6×10^6 t compared to a total consumption for the Northwest Territories in 1977 of about 2.4×10^6 t. In Alberta, the construction of each new oil sands surface mine is expected to require about 40×10^6 t of aggregate. This is

Table 11.7. Examples of aggregate resource mapping and inventory in Canada

British Columbia	Quebec
Clague and Hicock, 1976	Ministère de l'Énergie et des
Hora and Basham, 1980	Ressources du Québec, 1984
Leaming, 1968	St-Maurice et al., 1984
McCammon, 1977	
Alberta	**New Brunswick**
Edwards, 1979, 1984	Barnett, 1982
Edwards and Hudson, 1984	Barnett et al., 1977
Hudson, 1981	Parise and Barnett, 1981
Shetsen, 1981	
Saskatchewan	**Nova Scotia**
Mollard, 1979	Canadian Transport Commission, 1978
Simpson, 1984	Fowler, 1982
	Fowler and Dickie, 1978
	Wright, 1981
Manitoba	**Newfoundland**
Jones and Bannatyne, 1982	Environmental Geology Section, 1980
Large, 1978	Grant, 1974
Large and Ringrose, 1981	Kirby et al., 1983
Ringrose and Large, 1977	Vanderveer et al., 1981
Ontario	**North West Territories and Yukon**
Cowan, 1977	Mollard, 1982
Peat et al., 1981	Monroe, 1972a, b
Proctor and Redfern Limited, 1974	Thomas and Rampton, 1982
Scott, 1983	Walmsley, 1981

Edwards, W.A.D.
1989: Aggregate and nonmetallic Quaternary mineral resources; *in* Chapter 11 of Quaternary Geology of Canada and Greenland, R.J. Fulton (ed.); Geological Survey of Canada, Geology of Canada, no. 1 (*also* Geological Society of America, The Geology of North America, v. K-1).

Table 11.8. Production and value of Canadian aggregate resources

	Sand and Gravel		Crushed Stone		Lightweight*	
	($t \times 10^3$)	($\$ \times 10^3$)	($t \times 10^3$)	($\$ \times 10^3$)	($t \times 10^3$)	($\$ \times 10^3$)
Total production						
1981	259 661	517 002	81 129	281 895	1 001	20 305
1980	276 452	508 364	97 535	303 262	1 102	25 971
Imports						
1981	1 447	6 084	2 561	16 162		
1980	1 210	4 480	2 457	13 610		
Exports						
1981	319	953	1 758	6 007		
1980	384	924	2 214	6 176		

*Production from domestic raw materials was:
1981 — 520 000t, $7 091 000; 1980 — 624 000t, $10 261 000.

Source: Stonehouse, 1984

Table 11.9. Maximum haul distances for aggregate (1983)

Province	Haul distance (km) and mode
British Columbia	100 (truck) 120 (barge) 200 (rail; railroad ballast, rip-rap)
Alberta	100 (truck)
Saskatchewan	110 (rail)
Manitoba	60 (truck)
Ontario	160 (rail, truck), 150 (ship)
Quebec	100 (truck) 100 (barge, Magdalen Isles)
New Brunswick	100 (truck)
Nova Scotia	110 (truck), 250 (barge)
Prince Edward Island	205 (rail, truck, barge — Nova Scotia) 310 (rail, truck, barge — New Brunswick)
Newfoundland	180 (truck)
Yukon	30 (truck)
Northwest Territories	50 (truck)

Source: unpublished information supplied by respective provincial mines branches, and territorial agencies.

more than the province's total annual consumption. In Quebec the James Bay Project used about 20×10^6 t of aggregate in 1977 and the Montreal Olympics was probably the reason 1975 was a peak year in terms of concrete aggregate consumption.

The predominance of aggregate production from sand and gravel deposits (Table 11.8) is due to the widespread distribution of these unconsolidated, usually glaciofluvial and fluvial, deposits in Canada. Local availability and ease of mining result in lower production costs. The low-unit price and high bulk to weight ratio of aggregate results in transportation commonly being a major cost to the aggregate consumer. Maximum haul distances by province are shown in Table 11.9.

Crushed stone is used as an alternative when its price is competitive, where unconsolidated deposits are unavailable, or where aggregate quality specifications cannot be met from surficial deposits. Lightweight aggregate includes material such as expanded clay, shale, slate, perlite or vermiculite, byproduct slag, and natural pumice. Despite the higher manufacturing costs of lightweight aggregates, they are competitive with sand, gravel, and crushed stone in some applications, particularly for lightweight concrete.

The need for large, locally available supplies of aggregate means that the source, cost, and mode of transportation can differ markedly in different parts of the same province. Significant importation of aggregate from the United States occurs in the greater Vancouver area of British Columbia, and the Windsor-Sarnia area of Ontario; other provinces have also imported some aggregate from the United States. Such import is the result of special local requirements. Aggregate is exported to the United States from all provinces (except Prince Edward Island and Newfoundland). In 1981 over 2×10^6 t of aggregate was exported to the United States and also to Bermuda (Table 11.8).

Outlook for future aggregate supplies

Canada is fortunate in having a large supply of widely distributed unconsolidated granular materials, derived by fluvial and glaciofluvial processes. Even this remarkable inventory of unconsolidated aggregate is diminishing, however, and local shortages are occurring in many areas (Fig. 11.5). Aggregate depletion has resulted in: (1) land use conflicts, which have produced public and special interest group reaction, and government control and legislation; (2) concern about the availability of resource information, resulting in geological and resource evaluation studies; (3) increased use of crushed stone or till as unconsolidated sand and gravel deposits are depleted; (4) increasing costs, due to longer haul distances and the higher costs of processing stone or manufacturing aggregate; and (5) the documentation of resources being depleted through aggregate inventories and forecasts of supply, demand, and use.

Aggregate resources have been and continue to be mapped in all provinces and territories (Fig. 11.5). The normal followup to such studies is an economic assessment including the projected aggregate consumption by the numerous markets in any region and the resultant depletion. Such economic evaluations are available for many of the higher demand areas. A forecast of aggregate depletion to the point of stress on current supply areas in the near future is shown in Figure 11.5.

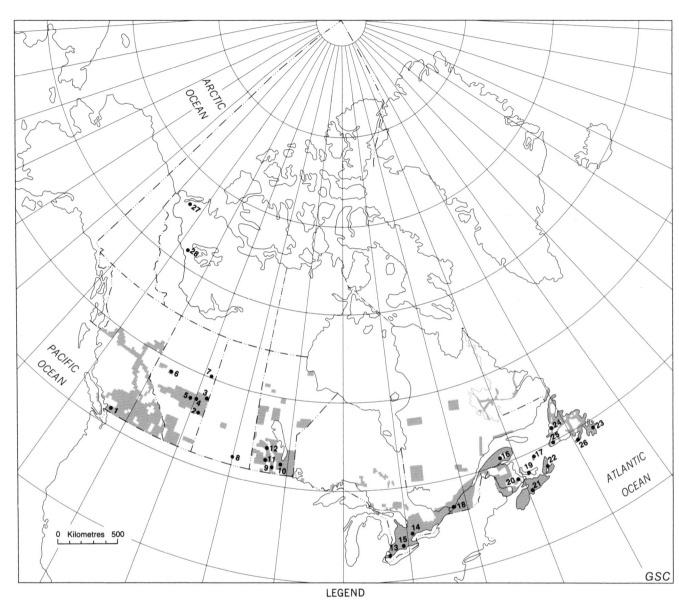

Figure 11.5. Aggregate resource mapping in Canada (light red areas) and areas of stress on aggregate resources.

LEGEND

REGION	EXPECTED OCCURRENCE OF STRESS ON LOCAL RESOURCE
British Columbia	
1 Langley District	current
Maple Ridge District	current
Matsqui District	current
Chilliwack, Abbotsford and Mission Districts	1990
Greater Vancouver	beyond 1990
Alberta	
2 Coronation	current
3 Cold Lake	2025
4 Vegreville	current
5 Edmonton	2020
6 Valleyview	current
7 Fort McMurray	current
Saskatchewan	
8 Weyburn	current
Manitoba	
9 Portage La Prairie	current
10 Winnipeg	2030
11 Brandon	beyond 2030
12 Dauphin	beyond 2000
Ontario	
13 Windsor/Sarnia	current
14 Toronto	current stress: reserve depletion by 2010

REGION	EXPECTED OCCURRENCE OF STRESS ON LOCAL RESOURCE
Ontario	
15 Middlesex, Elgin, Oxford	beyond 2001
Quebec	
16 Gaspe Peninsula	current
17 Magdalen Island	current
18 Montreal	current (sand and gravel)
Prince Edward Island	
19 Prince Edward Island	current (importing)
New Brunswick	
20 Moncton	current
Nova Scotia	
21 Halifax	current (sand)
22 Sydney	current (sand)
Newfoundland	
23 St. John's	current
24 Corner Brook	current
25 Marystown	current
26 Port aux Basques	current
North West Territories	
27 Inuvik	current
28 Normal Wells	current

Other nonmetallic mineral resources

Although aggregate has the largest production value of the nonmetallic Quaternary resources, other important Quaternary mineral resources to be considered include silica sand, clay, peat, marl, and sodium sulphate.

Silica is produced primarily from bedrock sources in Canada. Some Quaternary sources, however, are also being exploited; for example, unconsolidated silica sand is mined in the Bruderheim area, 35 km northeast of Edmonton, for the manufacture of glass fibre insulation. The sand is upgraded from local Holocene dune sand and reserves in the area are adequate for many years.

Clay and shale for the manufacture of brick and other ceramic products also is widely distributed across Canada. Shale and argillaceous sediments of pre-Quaternary age are the major sources of material for brick and pottery; however, some Quaternary and Holocene materials are used. Quaternary deposits of interest to the ceramic industry include stoneless marine and lake sediments, reworked glacial till, interglacial clays and floodplain clays. In Quebec, some brick and tile is made from glacial lake clay and in Ontario and Alberta some clays are mixed with shale or used alone for brick manufacture. Glacial lake clays are used in both Manitoba and Alberta in the production of lightweight aggregate.

Sodium sulphate is used in the manufacture of pulp and paper by the "Kraft" process and in the production of detergents, glass, and other chemicals. Commercial deposits of sodium sulphate are harvested from alkaline lakes and salt flats in southern Saskatchewan and adjacent southeastern Alberta. Concentration of sodium sulphate occurs in undrained basins under semi-arid conditions (Last, 1984). Percolating groundwaters carry dissolved salts into the basins from the surrounding soils, high rates of summer evaporation concentrate the brine to near saturation, and cooler fall temperatures cause crystallization and precipitation of sodium sulphate as the mineral mirabilite, commonly known as Glauber's Salt. Twenty-one deposits in Saskatchewan contain an estimated total of 50×10^6 t of sodium sulphate. Total production in Saskatchewan in 1981 was just over 500 000 t with a value of over $36 million. Alberta's production of sodium sulphate comes from a deposit at Metisko Lake, about 200 km east of Red Deer. Annual capacity is about 75 000 t with production in 1983 valued at $3.9 million. Minor amounts of sodium carbonate, sodium sulphate, and magnesium sulphate have been obtained from several small basins in the semi-arid interior of British Columbia (Barry, 1981).

Marl is a term used to describe a wide range of calcareous deposits from limestones to unconsolidated, highly calcareous clays of marine or fresh water origin. It has been used as a raw material for cement but is most likely to be used in the future as an agricultural liming agent to counteract acidic soils. Nonmarine marl deposits can form when the following conditions are met: presence of a carbonate source in drift or near surface bedrock; leaching of the carbonate by percolating groundwater; transportation of calcium and bicarbonate ions in a permeable aquifer; discharge at the surface as springs, seeps, or in lakes; and precipitation of calcium carbonate. Many small lakes in British Columbia contain deposits of marl which have been used locally for agricultural purposes. In Alberta, marl reserves are estimated at 10×10^6 m^3 with over 4×10^6 m^3 lying in the Half Way Lake deposits north of Edmonton which, until recently were worked for cement production. Numerous marl accumulation sites are reported in Ontario and Quebec. Some of these have been worked commercially. Marine marl deposits include fine grained sediments which are rich in shell fragments, such as those present in coastal British Columbia (Mathews, 1947).

PLACER DEPOSITS IN CANADA
S.R. Morison

Placer deposits are classically defined as accumulations of heavy minerals which have been eroded from lode sources and concentrated through sedimentation processes involving gravity, water, wind, or ice. As is discussed below, however, some placers may form through in place growth of minerals. Boyle (1979) gave an excellent discussion of gold deposits, and this section draws heavily on his descriptions. Placer deposits in Canada range in age from the paleoplacer-type Precambrian uranium deposits of the Blind River-Elliott Lake area to modern gold deposits on Fraser River bars. Although this discussion is concerned primarily with gold and with placer deposits of Quaternary age, some consideration is given to Tertiary placers which in many cases served as sources for Quaternary deposits.

Source and classification of placer gold deposits

The origin of gold concentrated in placers is of economic interest because of the assumption that the source could supply a large and more certain supply of gold than the placer deposit itself. This search for the "mother lode" is based on the premise that placer gold is detrital and the tracing of source areas is dependent on the dynamics and physical characteristics of the concentrating environment. Boyle (1979) summarized possible sources of gold in placers: (1) major lode deposits, (2) scattered, slightly auriferous quartz bodies, (3) trace amounts of gold in various rock types, and (4) pre-existing placers. The first and second sources figure in interpretations of placer gold being strictly of detrital origin; the placer develops when breakdown of the host rock and lodes leaves the stable, heavy gold particles to be concentrated by colluvial and alluvial processes. The third source requires several additional steps between release of gold from the host rock and lodes and concentration of gold particles in the placer: oxidation and destruction of the host minerals (e.g., pyrite) releases trace gold to solutions; the dissolved gold may then be precipitated wherever the solutions are subject to reducing conditions — a favourable site where precipitation might preferentially occur is at an aquitard such as a clay layer or bedrock at the base of coarse sediments.

The relative importance of these various gold sources for placer deposits is not well understood. In many cases there appears to be no relationship between adequate bedrock sources of gold and the amount of placer gold extracted (Boyle, 1979).

Morison, S.R.
1989: Placer deposits in Canada; in Chapter 11 of Quaternary Geology of Canada and Greenland, R.J. Fulton (ed.); Geological Survey of Canada, Geology of Canada, no. 1 (also Geological Society of America, The Geology of North America, v. K-1).

Residual enrichment

Residual enrichment involves the concentration of gold through chemical transport and precipitation. Residual enrichment and supergene migration of gold has been investigated by workers such as: Stokes (1906), Lenher (1912, 1918), Krauskopf (1951), Cloke and Kelly (1964), Kinkel and Lesure (1968), Chernyayev et al. (1969), Goleva et al. (1970), Lesure (1971), Lakin et al. (1971), Boyle et al. (1975), and Boyle (1979). In deposits formed in this manner there is no discernible mechanical sorting, no associated accumulation of heavy minerals, and the gold may occur in sites which it could not reach by fluvial transport (e.g. deep fractures in rock). Examples of morphological types of gold that must have been precipitated as a result of supergene migration are: wire gold, sheet gold (in bedrock fractures), nuggets with concentric structure, and dendritic gold. Precipitation occurs at buried surfaces and in fractures, in bedrock, and in both colluvial and alluvial deposits (which may also contain detrital gold), and commonly appears to be in some way associated with organic materials.

Colluvial placer deposits

Colluvial placers are best developed on terrain where active downslope movement of surface sediments is occurring, generally over a weathered surface, and commonly is associated with primary lode sources.

The concentration processes involve the winnowing and sorting action associated with downward migration of heavy minerals during downslope movement. Factors such as gradient, thickness and type of slope material (i.e., scree, talus, soil creep, solifluction), coefficient of friction, and seasonal frost all affect the grade and type of colluvial placer deposits (Boyle, 1979). Such deposits have been classified according to their position on a slope relative to the primary lode source: eluvial placers are deposits whose outline coincides roughly with the primary ore source; deluvial placers are deposits which extend from primary source to base of slope; proluvial placers are restricted to the base of the slope (Fig. 11.6). Colluvial processes are important during downcutting and provide gold for the development of alluvial placers. It is also important to note that gold associated with colluvial placer deposits is coarser grained proximal to the primary source, and according to Boyle (1979), both the grain size and fineness decrease with distance from the source area.

Alluvial placer deposits

Alluvial placers are those formed in present and past water courses. Their formation is controlled by hydraulic conditions of the stream and the supply of gold and other sediment (Boyle, 1979). Fluvial settings, such as creeks, gulches, meandering and braided streams, and alluvial fans and fan deltas, all have potential for heavy mineral concentration and preservation.

Creek and gulch placer deposits are associated with deeply dissected terrains with highly weathered and colluviated slopes; the valleys are narrow, restricted, and commonly incised. Discharge is seasonal in nature and gradients are moderate to high. The resulting alluvial deposits are generally thin, and the gravelly sediments are poorly sorted, angular, and locally derived. Pay zones are regular, laterally consistent, and on or close to the bedrock surface in the thalweg of the gulch stream bed (Fig. 11.7). The gold is generally coarse grained and the accompanying heavy mineral suite is closely allied to the local bedrock mineralogy (Boyle, 1979).

Stream and river placers are associated with fluvial environments such as meandering and braided streams and alluvial fans; the valleys are commonly wide. In meandering river environments, heavy minerals may accumulate in main channel bedforms such as dunes or as lag deposits. Point

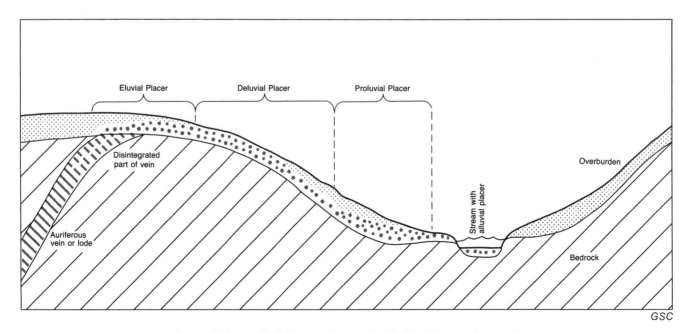

Figure 11.6. Colluvial placer deposit classification (from Boyle, 1979).

bars, on the inside curve of meander scrolls, are also favourable sites for heavy mineral deposition during flood conditions. The distribution of detrital gold deposited in a braided river environment is sporadic and discontinuous (Smith and Minter, 1980). In this environment, favourable areas for heavy mineral concentrations include transverse bars, channel bends, channel junctions, and segments where stream flow is convergent. In addition, entrapment of heavy minerals during aggradation of proximal channel gravels, as either diffuse gravel sheets (Hein and Walker, 1977) or as unit bar forms, is possible. In summary, the pay zone in stream and river placers ranges from linear paystreaks which are parallel to the channel system, to discontinuous concentrations which are not necessarily limited to the base of the alluvial fill.

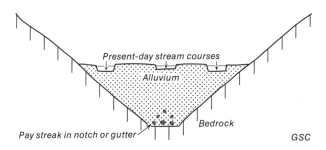

Figure 11.7. Pay streak in a gulch placer deposit (from Boyle, 1979).

Glacial placer deposits

The subglacial erosional characteristics of an advancing ice sheet are such that preglacial gravel deposits in valleys parallel to the direction of ice advance are generally destroyed and incorporated into glacial debris. Ice contact deposits such as kames and eskers, morainal deposits, and proglacial sediments may locally contain low quantities of gold (Boyle, 1979; Morison, 1985). Detrital gold is, however, widely dispersed in glacial deposits so that reconcentration by postglacial streams is generally required before an economic placer deposit is produced (Boyle, 1979).

Beach and nearshore placer deposits

Beach placer deposits are formed on wave-dominated shorelines through the winnowing action associated with longshore currents and wave action. Paystreaks follow strandlines and the gold is generally both fine grained and of high fineness (Boyle, 1979). Nelson and Hopkins (1972) have described detrital gold accumulations in the sediments of the Bering Sea off Nome, Alaska. The concentrations are related to marine regressions and transgressions with associated reworking of auriferous debris deposited on the shelf during glacial intervals.

Stratigraphic and regional distribution of placer gold deposits

The placer deposit types described above occur in five generalized stratigraphic settings in Canada (Table 11.10). It is generally assumed that the gold was released from bedrock

Table 11.10. Stratigraphy and general characteristics of placer gold deposits in Canada

Time Period	Tertiary	Quaternary Pleistocene Preglacial or Nonglacial	Interglacial	Glacial	Holocene
Placer environment and geomorphic location	Buried alluvial sediments in benches above valley floors	Buried alluvial sediments in benches above valley floors; valley fill alluvial sediments; alluvial terraces	Valley fill alluvial sediments; alluvial terraces	Benches of proglacial and ice contact deposits; moraines and drift	Valley bottom alluvial plains and terraces; colluvium and slope deposits; beach and nearshore marine deposits
General sediment characteristics	Mature sediments; well sorted alluvium with a diverse assemblage of sediment types	Locally derived gravel lithology; moderately to well sorted alluvium which is crudely to distinctly stratified	Mixed gravel lithology; moderately to well sorted alluvium which is crudely to distinctly stratified	Regionally derived gravel lithology; variable sorting and stratification depending upon type of glacial drift	Mixed gravel lithology; moderately to well sorted alluvium which is crudely to distinctly stratified; poorly sorted, massive slope deposits; well sorted beach sand
Gold distribution	Greater concentration with depth	Discrete concentrations throughout to pay streaks at base of alluvium	Discrete concentrations throughout to pay streaks at base of alluvium	Dispersed throughout	Discrete concentrations throughout to pay streaks at base of alluvium; pay streaks follow slope morphology and strandline trend
Mining problems	Thick overburden	Thick overburden; variable grade	Variable grade	Low grade	Variable grade and low volume of auriferous sediment
Examples	"White Channel Gravel" of the Klondike area, Yukon Territory	Preglacial fluvial gravels, Clear Creek drainage basin and unglaciated terrain in Yukon Territory	Interglacial stream gravels in Atlin and Cariboo mining districts, British Columbia	Glaciofluvial gravel, Clear Creek drainage basin, Yukon Territory	Valley bottom creek and gulch placers, Clear Creek drainage basin, Colluvial placers in Dublin Gulch area, Yukon Territory; beach placers, Queen Charlotte Islands, British Columbia

by Tertiary weathering or during weathering periods in nonglacial and/or interglacial intervals of the Quaternary. Pliocene and Pleistocene preglacial and/or nonglacial placer alluvium is generally found in benches or terraces buried below overburden, such as glacial drift and nonauriferous alluvium. Interglacial placer alluvium is found as valley bottom fill and terraces in drainage basins that have escaped the effects of glacial scour. During glacial intervals, processes associated with subglacial, ice marginal, and proglacial processes can incorporate gold from bedrock or paleoplacer sources. Holocene placer deposits are found as alluvial plains and terraces, colluvium and slope deposits, and beach and nearshore marine deposits.

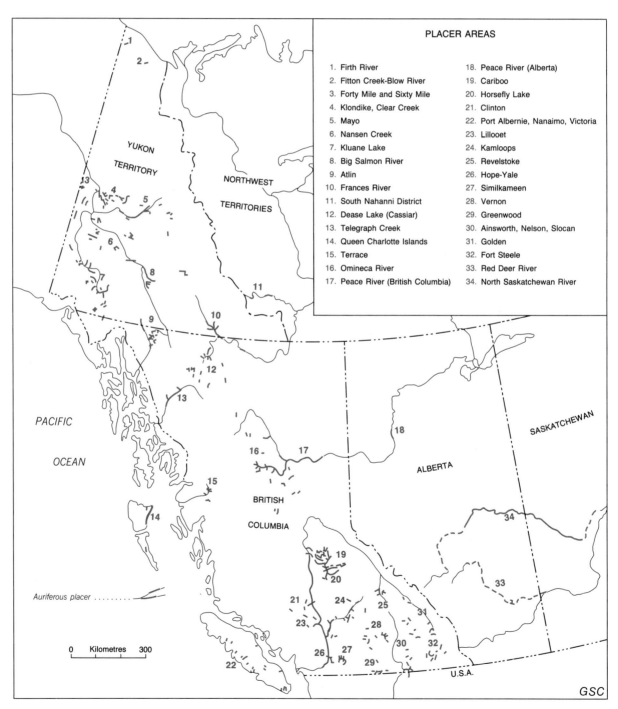

Figure 11.8. Placer gold districts in western Canada (from Boyle, 1979).

Cordillera

Yukon placers have contributed up to 5% of total Canadian gold production, and are the most significant placer deposits in Canada. According to royalty records, Yukon placer deposits produced 11 187 525 fine ounces (347 965 585 g) of gold between 1885 and 1982 (Debicki, 1983). In 1983 and 1984, production from Yukon placer gold deposits was 75 415 and 75 765 fine ounces, respectively (Grapes and Morin, 1983; G. Gilbert, Mineral Resources Directorate, Yukon Territory, personal communication, 1984).

Yukon Territory is unique in Canada in that the west-central and northern regions escaped glaciation (Hughes et al., 1969, 1972) and as a result there is a high degree of placer deposit preservation in these regions. In central and southern regions, however, Quaternary glaciation has eroded and incorporated preglacial placers in valleys oriented parallel to ice flow and buried those in valleys transverse or oblique to ice direction.

Placer mining districts are found throughout central and southern Yukon (Fig. 11.8) and include a variety of placer deposit settings, each with a unique depositional history. For example, in Clear Creek drainage basin (4 of Fig. 11.8; Morison, 1983a, b) placer gold may be found in valley bottom creek and gulch gravel, glacial gravel, and buried preglacial fluvial gravel (Fig. 11.9). In the Dublin Gulch area (5 of Fig. 11.8) small colluvial placers have developed from local bedrock source (Boyle, 1979). Auriferous gravel in the Klondike area (4 of Fig. 11.8) was first described by McConnell (1905, 1907). He divided these deposits into stream gravel, terrace gravel, "White Channel Gravel", and high level river (glacial) gravel (Fig. 11.10). The White Channel deposit (Pliocene?) forms benches 50 to 100 m above present day stream levels and is divided into white and yellow gravel units (McConnell, 1907). White Channel clastic sediment is characterized by a complex assemblage of lithofacies types ranging from laminated silt and clay to massive and disorganized boulder gravel. In a general sense, gold concentration increases with depth in White Channel Gravel (Gleeson, 1970). White Channel sediment and the underlying bedrock may also be argillically

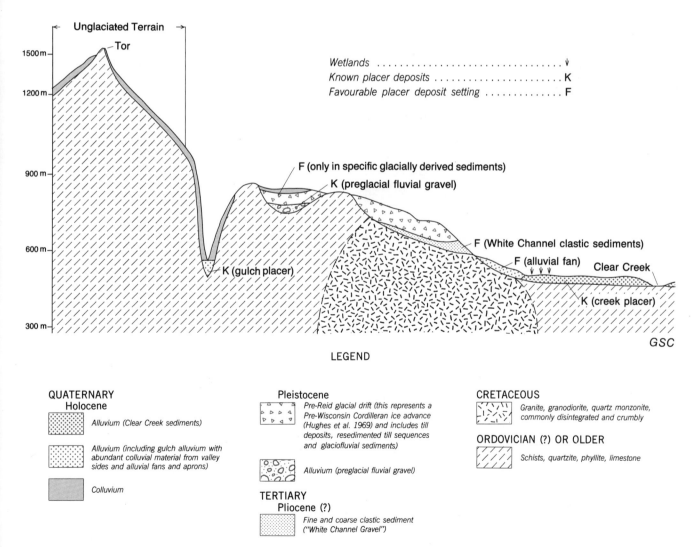

Figure 11.9. Schematic profile of Clear Creek drainage basin showing the settings of different types of placer deposits (from Morison, 1985).

altered. Tempelman-Kluit (1982) was the first to suggest that the alteration processes involved may relate to the apparent gold enrichment in the lower portion of the gravel sequence. Acknowledgment of the relationship between the distribution of gold and both White Channel lithofacies and altered equivalents is necessary to understand the importance of White Channel Gravel as an economic mineral deposit.

Lower Cretaceous (Albian) conglomeratic rocks at McKinnon Creek in the Indian River area of the Yukon may contain up to 3.4 g/t of gold (Lowey, 1984). Traditionally, these rocks have been considered to be paleoplacers (McConnell, 1905). Lowey (1984), however, has introduced evidence which suggests that the gold mineralization may be due to postdepositional hydrothermal activity.

Placer mining districts in British Columbia from 1856 to 1981 produced 4 349 177 fine ounces (163 966 620 g) of gold (Debicki, 1984). Placers in British Columbia include Tertiary gravel and Quaternary interglacial gravel, glacial gravel, and postglacial Holocene gravel (Table 11.10). In the Princeton district (27 of Fig. 11.8), the Similkameen and Tulameen drainage basins are unique in that they have produced commercial quantities of platinum in addition to gold (Raicevic and Cabri, 1976).

Beach placer deposits occur on Graham Island in the Queen Charlotte Islands (Mandy, 1934; 14 of Fig. 11.8). These deposits, which consist mainly of magnetite, are typically low grade and were formed through wave and wind erosion of auriferous glacial sediments (Boyle, 1979).

Plains

The most significant alluvial placer drainage basin on the western plains is North Saskatchewan River (34 of Fig. 11.8; Tyrrell, 1915). In Alberta, additional drainage basins that contain occurrences of placer gold include the McLeod, Athabasca, Peace, Red Deer, Milk, and Redwater rivers (Halferdahl, 1965; Giusti, 1983). Recorded gold production from Alberta placers from 1887 to 1981 was 31 788 fine ounces (Giusti, 1983). In Saskatchewan alluvial placer occurrences are found on North Saskatchewan, Waterhen, and Poplar rivers (Coombe, 1984). The Leaf Lake area in east-central Saskatchewan has yielded limited quantities of gold

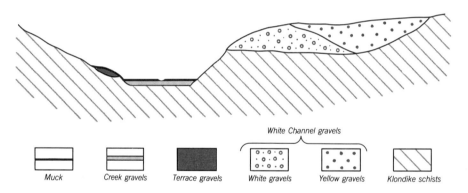

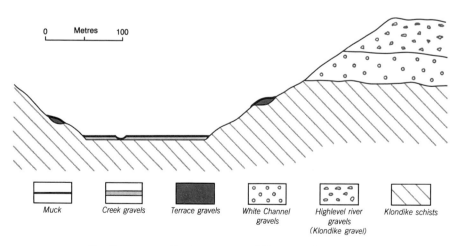

Figure 11.10. Gravel types in the Klondike area (from McConnell, 1905, 1907).

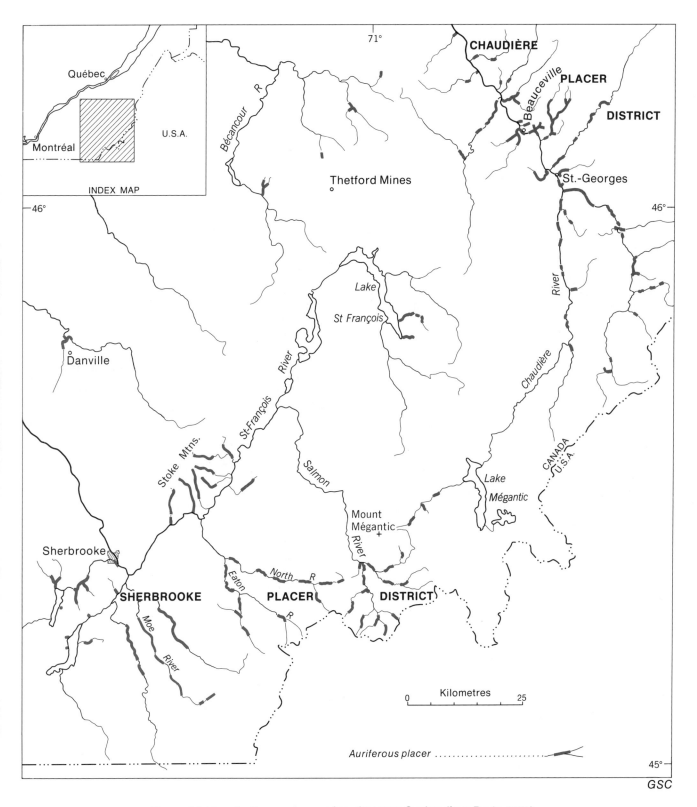

Figure 11.11. Auriferous placers of southeastern Quebec (from Boyle, 1979).

from surrounding gravel deposits (Coombe, 1984). In a general sense, placer gold on the western plains is found in Holocene and modern stream gravel deposits, with the specific source areas not understood.

Central Canada

In Ontario some gold has been obtained from river gravels and glacial sediments in drainage basins such as Vermilion River, Wanapietei River, Meteor Lake, Lake Manitou, Savante Lake, and Grassy River (Boyle, 1979). Glaciofluvial sediments in the Timmins-Kirkland Lake mining area also contain placer gold (Ferguson and Freeman, 1978). The source of the gold in these various occurrences is unknown, but because it is fine grained, it was possibly derived through glacial erosion of gold-bearing lodes. The reader is referred to Ferguson and Freeman (1978) for specific examples of Quaternary float and placer mineral occurrences in Ontario.

Chaudière and Sherbrooke districts in the Eastern Townships are the two main areas where placer gold mining activity has occurred in Quebec (Fig. 11.11; LaSalle, 1980). The Beauceville placer area in the Rivière Chaudière drainage basin was discovered in 1830, and has produced 100 000 ounces of gold (Boyle, 1979). Drainage systems such as Gilbert, Moulin, and des Plantes rivers have produced placer gold in the Chaudière district (Boyle, 1979). The Beauce Placer company dredged Rivière Gilbert from 1959 to 1964, and recovered 4378 ounces of gold and 64 ounces of silver (LaSalle, 1980). The gold of placers of the Appalachian region of Quebec was probably released from host rocks and initially concentrated during Tertiary weathering and dissection. Many of the deposits were destroyed during glaciation, but the varied terrain permitted preservation of placers in valleys oriented transverse to glacier flow. Most of the exploited placers appear to be interglacial and postglacial alluvial sediments which were enriched in gold where the fluvial systems intersected older placers or by concentration of scattered gold dispersed throughout glacial deposits.

Maritimes

In Nova Scotia, beach placer deposits occur where wave action is eroding source rocks of the Meguma Group (Boyle, 1979). An example of this is found at Ovens, Nova Scotia, where approximately 2000 ounces of gold was extracted from auriferous beach sand (Malcolm and Faribault, 1929; Boyle, 1979). Also, offshore from Ovens, gold occurs on the continental shelf in deposits interpreted to be either abandoned stream channels or relict strandlines (Libby, 1969; Boyle, 1979).

ACKNOWLEDGMENTS

This chapter has benefited from critical review by P.B. Hale (Canada Oil and Gas Lands, Department of Energy, Mines and Resources), P.J. Rennie (Canadian Forestry Service), D.W. Scott (Ontario Geological Survey), W.H. Mathews (University of British Columbia), J.B. Bird (McGill University), K. Valentine (Land Resource Research Centre, Agriculture Canada), W.A. Dixon Edwards (Alberta Geological Survey), and R.W. Boyle and B.R. Pelletier (Geological Survey of Canada).

REFERENCES

Adams, G.D., Boissonneau, A.N., Hirvonen, H.E., Mills, G.F., Oswald, E.T., Pettapiece, W.W., Tarnocai, C., Wells, E.D., and Zoltai, S.C.
1981: Wetlands of Canada maps (wetland regions and distribution of wetlands); Ecological Land Classification Series, No.14, Canada, Department of Environment, scale 1:7 500 000.

Aspinall, F. and Hudak, W.
1980: Peat harvesting — state of the art; Symposium, Peat as an energy alternative, Institute of Gas Technology, Arlington, U.S.A., p. 159-173.

Barnett, D.E.
1982: Aggregate resource management in New Brunswick; Canadian Institute of Mining and Metallurgy, Bulletin, v. 75, no. 844, p. 103-109.

Barnett, D.E., Brinsmead, R.A., and Finamore, P.F.
1977: Granular aggregate resources of the Belledune planning district, Gloucester and Restigouche counties; New Brunswick Department of Natural Resources, Mineral Development Branch, Topical Report 77-5.

Barry, G.S.
1981: Sodium sulphate; in Canadian Minerals Yearbook; Canada, Department of Energy, Mines and Resources, Mineral Report 29, p. 443-448.

Boyle, R.W.
1979: The geochemistry of gold and its deposits; Geological Survey of Canada, Bulletin 280, 584 p.

Boyle, R.W., Alexander, W.M., and Aslin, G.E.M.
1975: Some observations on the solubility of gold; Geological Survey of Canada, Paper 75-24, 6 p.

Bunting, B.T. and Fedoroff, N.
1973: Mircromorphological aspects of soils developed in the Canadian high arctic; Soil Microscopy, Proceedings, Fourth International Working Meeting on Soil Micromorphology, G.K. Rutherford (ed.); The Limestone Press, Kingston, Ontario, p. 350-365.

Canada Soil Survey Committee, Subcommittee on Soil Classification
1978: The Canadian system of soil classification; Research Branch, Canada, Department of Agriculture, Publication 1646, Supply and Services Canada, Ottawa, 164 p.

Canadian Transport Commission
1978: The Canada Mineral Aggregate Industry (with emphases on transportation); Traffic and Tariff Studies Division, Ottawa-Hull.

Carncross, C.A.
1980: Wet harvesting of peat; Symposium, Peat as an energy alternative, Institute of Gas Technology, Arlington, U.S.A., p. 175-180.

Chernyayev, A.M., Chernayeva, L.E., Yermeyeva, M.N., and Andreyev, M.I.
1969: Hydrogeochemistry of gold; Geochemistry International 1969, v. 6, p. 348-358.

Clague, J.J. and Hicock, S.R.
1976: Sand and gravel resources of Kitimat, Terrace, and Prince Rupert, British Columbia; in Report of Activities, Part A, Geological Survey of Canada, Paper 76-1A, p. 273-276.

Clayton, J.S., Ehrlich, W.A., Cann, D.B., Day, J.H., and Marshall, I.B.
1977: Soils of Canada; Canada, Canada, Department of Agriculture, Ottawa (2 volumes).

Cloke, P.L. and Kelly, W.C.
1964: Solubility of gold under inorganic supergene conditions; Economic Geology, v. 59, p. 259-270.

Coombe, W.
1984: Gold in Saskatchewan; Saskatchewan Geological Survey, Open File Report 84-1, 134 p.

Cowan, W.R.
1977: Toward the inventory of Ontario's mineral aggregates; Ontario Geological Survey, Miscellaneous Paper 73, 19 p.

Debicki, R.L.
1983: An overview of the placer mining industry in Yukon, 1978 to 1982; in Yukon Placer Mining Industry 1978 to 1982; Canada, Department of Indian and Northern Affairs, Exploration and Geological Services Division, Whitehorse, p. 7-14.

1984: An overview of the placer mining industry in Atlin mining division, 1978 to 1982; British Columbia Ministry of Energy, Mines and Petroleum Resources, Geological Branch, Paper 1984-2, 19 p.

Drew, J.V. and Tedrow, J.C.F.
1962: Arctic soil classification and patterned ground; Arctic, v. 15, p. 109-116.

Edwards, W.A.D.
1979: Sand and gravel deposits in the Canmore Corridor area, Alberta; Alberta Research Council, Earth Sciences Report 79-2, 30 p.
1984: Geology of some gravel deposits in the Edmonton region, Alberta; Canadian Institute of Mining and Metallurgy, Special Volume 29, p. 219-222.

Environmental Geology Section
1980: Summary map of aggregate resource potential, Insular Newfoundland; Inventory of Aggregate Resources Project, Department of Mines and Energy, Government of Newfoundland and Labrador, scale 1:1 000 000.

Ferguson, S.A. and Freeman, E.B.
1978: Ontario occurrences of float, placer gold and other heavy minerals; Ontario Geological Survey, Mineral Deposits Circular 17, 214 p.

Fowler, J.H.
1982: Aggregate resources in Nova Scotia; Canadian Institute of Mining and Metallurgy, Bulletin, v. 75, no. 844, p. 100-102.

Fowler, J.H. and Dickie, G.B.
1978: Sand and gravel occurrence maps; Nova Scotia Department of Mines and Energy, Open File Maps, scale 1:50 000.

Fox, C.A.
1983: Micromorphology of an Orthic Turbic Cryosol — a permafrost soil; in Soil Micromorphology, Volume 2: Soil Genesis, p. Bullock and G.P. Murphy (ed.); A.B. Academic Publishers, Oxford, p. 699-705.

Fox, C.A. and Protz, R.
1981: Definition of fabric distributions to characterize the rearrangement of soil particles in the Turbic Cryosols; Canadian Journal of Soil Science, v. 61, p. 29-34.

Giusti, L.
1983: The distribution, grades and mineralogical composition of gold-bearing placers in Alberta; unpublished MA thesis, University of Alberta, Department of Geology, Edmonton, Alberta, 397 p.

Gleeson, C.F.
1970: Heavy mineral studies in the Klondike area, Yukon Territory; Geological Survey of Canada, Bulletin 173, 63 p.

Goleva, G.A., Krivenkov, V.A., and Gutz, Z.G.
1970: Geochemical trends in the occurrence and migration forms of gold in natural water; Geochemistry International 1970, v. 7, p. 518-529.

Grant, D.R.
1974: Newfoundland, granular resources inventory; Geological Survey of Canada, Open File 194, scale 1:500 000.

Grapes, K.J. and Morin, J.A.
1984: 1983 Yukon mining and exploration overview; in Yukon Exploration and Geology 1983; Canada, Department of Indian and Northern Affairs, Exploration and Geological Services Division, Whitehorse, p. 5-14.

Halferdahl, L.B.
1965: The occurrence of gold in Alberta rivers; Alberta Research Council, Economic Mineral Files, Open File Report 65-11, p.

Hein, F.J. and Walker, R.G.
1977: Bar evolution and development of stratification in the gravelly, braided, Kicking Horse River, British Columbia; Canadian Journal of Earth Sciences, v. 14, p. 562-570.

Hora, Z.D. and Basham, F.C.
1980: Sand and gravel study, 1980, British Columbia Lower Mainland; British Columbia Ministry of Energy, Mines and Petroleum Resources, Paper 1980-10.

Hudson, R.B.
1981: The critical relationship between aggregate and oil sands development (abstract); Canadian Institute of Mining and Metallurgy, Bulletin, v. 74, no. 827, p. 130.

Hughes, O.L., Campbell, R.B., Muller, J.E., and Wheeler, J.O.
1969: Glacial limits and flow patterns, Yukon Territory, south of 65°N latitude; Geological Survey of Canada, Paper 68-34, 9 p.

Hughes, O.L., Rampton, V.N., and Rutter, N.W.
1972: Quaternary geology and geomorphology, southern and central Yukon (Northern Canada); 24th International Geological Congress, Montréal, Excursion A-11, p. 30-34.

Jeglum, J.K., Boissonneau, A.N., and Haavisto, V.F.
1974: Toward a wetland classification for Ontario; Canada, Department of Environment, Canadian Forestry Service, Information Report O-X-215, 54 p.

Jones, C.W. and Bannatyne, B.B.
1982: Evaluation of bedrock for aggregate; Manitoba Mineral Resources Division, Report of Field Activities 1982, p. 100-101.

Keys, D., Gemmell, D.E., and Henderson, R.E.
1982: New Brunswick peat — resources, management, and development potential; Proceedings of a symposium on peat and peatlands; National Research Council of Canada, Atlantic Research Laboratory, Halifax, p. 222-236.

Kinkel, A.R. and Lesure, F.G.
1968: Residual enrichment and supergene migration of gold, southeastern United States; United States Geological Survey, Professional Paper 600-D, p. 174-178.

Kirby, F.T., Ricketts, R.J., and Vanderveer, D.G.
1983: Inventory of aggregate resources in Newfoundland and Labrador-Information report and index maps, to accompany aggregate resource map series 1:250 000 (Open Files Nfld. 1287 and Lab. 602); Government of Newfoundland and Labrador, Department of Mines and Energy, Government of Newfoundland and Labrador, Report 83-2.

Kivinen, E. and Pakarinen, P.
1980: Peatland areas and the proportion of virgin peatland in different countries; Proceedings of the 6th International Peat Congress, Duluth, U.S.A., p. 52-54.

Krauskopf, K.B.
1951: The solubility of gold; Economic Geology, v. 46, p. 858-870.

Lakin et al.
1971: Geochemistry of gold in the weathering cycle; Geochemical exploration proceedings, 3rd International Geochemical Symposium, Canadian Institute of Mining and Metallurgy, Special Volume 11, p. 196.

LaSalle, P.
1980: L'or dans les sédiments meubles: Formation des placers, extraction et occurrences dans le sud-est du Québec; Ministère de l'Énergie et des Ressources du Québec, DPV-745, 26 p.

Large, P.
1978: Sand and gravel in Manitoba; Manitoba Mineral Resources Division, Education Series 78-1, 12 p.

Large, P. and Ringrose, S.
1981: Pitfalls and purpose: The aggregate challenge (abstract); Canadian Institute of Mining and Metallurgy, Bulletin, v. 74, no. 827, p. 110.

Last, W.M.
1984: Sedimentology of playa lakes of the northern Great Plains; Canadian Journal of Earth Sciences, v. 21, p. 107-125.

Leaming, S.F.
1968: Sand and gravel in the Strait of Georgia area; Geological Survey of Canada, Paper 66-60, 149 p.

Lenher, V.
1912: The transportation and depositon of gold in nature; Economic Geology, v. 7, p. 744-750.
1918: Further studies on the deposition of gold in nature; Economic Geology, v. 13, p. 161-184.

Lerbekmo, J.F. and Campbell, F.A.
1969: Distribution, composition and source of the White River Ash, Yukon Territory; Canadian Journal of Earth Sciences, v. 6, p. 109-116.

Lesure, F.G.
1971: Residual enrichment and supergene transport of gold, Calhoun Mine, Lumpkin County, Georgia; Economic Geology, v. 66, p. 178-186.

Libby, F.
1969: Gold in the sea; Sea Frontiers, v. 15, p. 232-241.

Lowey, G.W.
1984: Auriferous conglomerates at McKinnon Creek, west-central Yukon (115 0/11): Palaeoplacer or epithermal mineralization?; in Yukon Exploration and Geology 1983; Canada, Department of Indian and Northern Affairs, Exploration and Geological Services Division, Whitehorse, p. 69-77.

Malcolm, W. and Faribault, E.R.
1929: Gold fields of Nova Scotia; Geological Survey of Canada, Memoir 156, 253 p.

Mandy, J.T.
1934: Gold bearing black-sand deposits of Graham Island, Queen Charlotte Islands; Canadian Institute of Mining and Metallurgy, Transactions, v. 37, p. 563-572.

Mathews, W.H.
1947: Calcareous deposits of the Georgia Strait area; British Columbia Department of Mines, Bulletin 23.

McCammon, J.W.
1977: Surficial geology and sand and gravel deposits of Sunshine Coast, Powell River and Campbell River areas; British Columbia Ministry of Energy, Mines and Petroleum Resources, Bulletin 65.

McConnell, R.G.
1905: Report on the Klondike gold fields; in Annual Report, 1901, v. 14, Part B, Geological Survey of Canada, Publication no. 884, p. 1-71.
1907: Report on gold values in the Klondike high level gravels; Geological Survey of Canada, Publication No.979, 34 p.

McMillan, N.J.
1960: Soils of the Queen Elizabeth Islands (Canadian Arctic); Canadian Journal of Soil Science, v. 11, p. 131-139.

Mills, G.F.
1974: Organic soil parent materials; Proceedings of the Canada Soil Survey Committee Organic Soil Mapping Workshop, Winnipeg, Manitoba, p. 21-43.

Ministère de l'Énergie et des Ressources du Québec
1984: Compilation de la géologie du Quaternaire, région des Appalaches; Ministère de l'Énergie et des Ressources du Québec, DV 84-10, 89 cartes, échelle 1:50 000.

Mollard, J.D.
1979: Granular construction potential Regina — Moose Jaw Region/Saskatoon Region/Prince Albert Region; Saskatchewan Urban Affairs, 3 maps, scale 1:250 000.
1982: Terrain analysis study, North Canol Road, Yukon Territory; J.D. Mollard and Associates Limited, scale 1:20 000.

Monenco Ontario Limited
1981: Evaluation of the potential of peat in Ontario; Ontario Ministry of Natural Resources, Occasional Paper No.7, 193 p.

Monroe, R.L.
1972a: Terrain classification and sensitivity maps, N.W.T. and Yukon (107B, 117A, C, D); Geological Survey of Canada, Open File 120, scale 1:250 000.
1972b: Terrain classification and sensitivity maps, N.W.T. (106 I, J, K, M, N, and O); Geological Survey of Canada, Open File 121, scale 1:250 000.

Morison, S.R.
1983a: Surficial geology of Clear Creek drainage basin, Yukon Territory (NTS sheets 115P, 11, 12, 13, 14); Canada, Department of Indian and Northern Affairs, Open File Release, scale 1:50,000.
1983b: A sedimentologic description of Clear Creek fluviatile sediments (115 P) central Yukon; in Yukon Exploration and Geology 1982; Canada, Department of Indian and Northern Affairs, Exploration and Geological Services Division, Whitehorse, p. 50-54.
1985: Placer deposits of Clear Creek drainage basin (115 P), central Yukon; in Yukon Exploration and Geology 1983; Canada, Department of Indian and Northern Affairs, Exploration and Geological Services Division, Whitehorse, p. 88-93.

Muskeg Subcommittee of the National Research Council of Canada
1977: Muskeg and the Northern Environment in Canada, N.W. Radforth and C.O. Brawner (ed.); University of Toronto Press, Toronto, 399 p.

Nelson, C.H. and Hopkins, D.M.
1972: Sedimentary processes and distribution of particulate gold in the north Bering Sea; United States Geological Survey, Professional Paper 689, p. 1-27.

Overend, R.P.
1981: The peat for energy and chemical R&D program of NRC; Proceedings of the Symposium on Peat: An awakening natural resource, Thunder Bay, Ontario, October 26-28, p. 251-257.
1983: Energy production in the Boreal Forest zone; in Resource and Dynamics of the Boreal Zone, R. Wein, R. Riewe, and I.R. Methuen (ed.); Association of Canadian Universities for Northern Studies, Ottawa, p. 378-396.

Parise, J.C. and Barnett, D.E.
1981: Evaluation of potential crushed stone quarries in New Brunswick; Mineral Development Branch, New Brunswick Department of Natural Resources.

Peat, Marwick and Partners, and M.M. Dillon Ltd.
1981: Mineral aggregate transportation study; Ontario Ministry of Natural Resources, Mineral Resources Branch, Industrial Minerals Background Paper 1, 122 p.

Pollett, F.C.
1972: Nutrient content of peat soils in Newfoundland; Proceedings of the Fourth International Peat Congress, Finland, v. 3, p. 461-468.

Prest, V.K., Grant, D.R., and Rampton, V.N.
1968: Glacial Map of Canada; Geological Survey of Canada, Map 1253A, scale 1:5 000 000.

Proctor and Redfern Limited
1974: Mineral aggregate study, Central Ontario Planning Region; Ontario Ministry of Natural Resources, Mineral Resources Branch, Toronto, 100 p.

Punwani, D.V.
1980: Peat as an energy alternative: An overview; Symposium, Peat as an energy alternative, Institute of Gas Technology, Arlington, U.S.A., p. 1-28.

Radforth, N.W.
1955: Organic terrain organization from the air (altitudes less than 1000 feet); Handbook No.1, Canada, Department of National Defence, Report DR 95, 55 p.
1958: Organic terrain organization from the air (altitudes 1000-5000 feet); Handbook No.2, Canada, Department of National Defence, Report DR 124, 23 p.

Raicevic, D. and Cabri, L.J.
1976: Mineralogy and concentration of Au- and Pt-bearing placers from the Tulameen River area in British Columbia; Canadian Institute of Mining and Metallurgy, Bulletin, v. 69, p. 111-119.

Ringrose, S. and Large, P.
1977: Quaternary geology and gravel resources of the Leaf Rapids Local Government District; Manitoba Mineral Resources Division, Geological Report 77-1, 93 p.

St-Maurice, Y., Paré, C., et Jacob, H.-L.
1984: L'industrie des matériaux de construction au Québec en 1982: État de la situation; Service de l'Économie minérale du Québec, 58 p.

Scott, D.
1983: Structural industrial minerals in Ontario; Ontario Geological Survey, Miscellaneous Paper 14, p. 20-32.

Shetsen, I.
1981: Sand and gravel resources of the Lethbridge Area; Alberta Research Council, Earth Sciences Report 81-4, 41 p.

Simonson, R.W.
1959: Outline of a generalized theory of soil genesis; Soil Science Society of America, Proceedings, v. 23, p. 152-156.

Simpson, M.A.
1984: Aggregate resource maps; Saskatchewan Research Council, Open File, scale 1:50 000.

Stokes, H.N.
1906: Experiments on the solution, transportation, and deposition of copper, silver, and gold; Economic Geology, v. 1, p. 644-650.

Stonehouse, D.H.
1984: Mineral aggregates; in Canadian Minerals Yearbook; Canada, Department of Energy, Mines and Resources, Mineral Report 32, p. 28.1-28.12.

Tarnocai, C.
1970: Classification of peat landforms in Manitoba; Canada, Department of Agriculture, Research Station, Pedology Unit, Winnipeg, 45 p.
1976: Soils of Bathurst, Cornwallis, and adjacent islands, District of Franklin; in Report of Activities, Part B, Geological Survey of Canada, Paper 76-1B, p. 137-141.
1977: Soils of north-central Keewatin; in Report of Activities, Part A, Geological Survey of Canada, Paper 77-1A, p. 61-64.
1978: Distribution of soils in northern Canada and parameters affecting their utilization; 11th International Congress of Soil Science Transactions, Volume 3, Symposia Papers, p. 281-304.
1980: Canadian wetland registry; Proceedings of a Workshop on Canadian Wetlands, C.D.A. Rubec and F.C. Pollett (ed.); Canada, Department of Environment, Ecological Land Classification Series, No.12, p. 9-38.
1984a: Characteristics of soil temperature regimes in the Inuvik area; in Northern Ecology and Resource Management, R. Olson et al. (ed.); The University of Alberta Press, Edmonton, p. 19-37.
1984b: Peat resources of Canada; National Research Council of Canada, Peat Energy Program, NRCC No.24140, 17 p.

Tarnocai, C. and Zoltai, S.C.
1978: Earth hummocks of the Canadian arctic and subarctic; Arctic and Alpine Research, v. 10, p. 581-594.

Tedrow, J.C.F., Bruggemann, P.F., and Walton, G.F.
1968: Soils of Prince Patrick Island; The Arctic Institute of North America, Washington, D.C., Research Paper 44, 82 p.

Tempelman-Kluit, D.J.
1982: White Channel gravel of the Klondike; in Yukon Exploration and Geology 1981; Canada, Department of Indian and Northern Affairs, Exploration and Geological Services Division, Whitehorse, p. 74-79.

Thomas, R.D. and Rampton, V.N.
1982: Surficial geology and geomorphology; North Klondike River (Map 6-1982), Upper Blackstone River (Map 7-1982), Engineer Creek (Map 8-1982), Lower Ogilvie River (Map 9-1982), Moose Lake (Map 10-1982) Rock River (Map 11-1982), Yukon Territory; Geological Survey of Canada, Maps 6-1982 to 10-1982, scale 1:100 000.

Tibbetts, T.E.
1983: The peat forum and the peat for energy and chemicals R&D program; Proceedings of a Peatland Inventory Methodology Workshop, Canada, Department of Agriculture, Land Resource Research Institute; and Canada, Department of Environment, Newfoundland Forest Research Center, Ottawa, p. i-iv.

Tibbetts, T.E. and Ismail, A.
1980: A Canadian approach to peat energy; Symposium, Peat as an energy alternative, Institute of Gas Technology, Arlington, U.S.A., p. 663-677.

Tomiczek, P.W., Phillips, J.D., and Hudak, W.
1982: Potential peat harvesting alternatives; Symposium, Peat as an energy alternative II, Institute of Gas Technology, Arlington, U.S.A., p. 119-133.

Tyrrell, J.B.
1915: Gold on the North Saskatchewan River; Canadian Mining Institute, Transactions, v. 18, p. 160-173.

Vanderveer, D.G., Ricketts, R.J., and Kirby, F.T.
1981: Toward an inventory of aggregate resources in Newfoundland and Labrador (abstract); Canadian Institute of Mining and Metallurgy, Bulletin, v. 74, p. 110.

Walmsley, M.E.
1981: Terrain classification and interpretive maps, Cowley Lake area, Yukon Territory; Pedology Consultants, Vancouver, scale 1:5000.

Wright, W.J.
1981: Aggregate resources of Cape Breton Island; in Mineral Resources Division, Report of Activities, 1981; Nova Scotia Department of Mines and Energy, Report 81-1, p. 23-24.

Zoltai, S.C.
1980: An outline of the wetland regions of Canada; Proceedings of a Workshop on Canadian Wetlands, C.D.A. Rubec and F.C. Pollett (ed.); Canada, Department of Environment, Ecological Land Classification Series, No.12, p. 1-8.

Zoltai, S.C. and Tarnocai, C.
1974: Soils and vegetation of hummocky terrain; Environmental-Social Committee, Northern Pipelines, Task Force on Northern Oil Development, Report 74-5, 86 p.

Zoltai, S.C. and Woo, V.
1976: Soils and vegetation of Somerset and Prince of Wales islands, District of Franklin; in Report of Activities, Part B, Geological Survey of Canada, Paper 76-1B, p. 143-145.

Zoltai, S.C., Pollett, F.C., Jeglum, J.K., and Adams, G.D.
1973: Developing a wetland classification for Canada; Proceedings of the Fourth North American Forest Soils Conference, Québec City, p. 497-511.

Authors' addresses

D.F. Acton
Saskatchewan Soil Survey Unit
Land Resource Research Centre
210 John Mitchell Building
University of Saskatchewan
Saskatoon, Saskatchewan
S7N 0W0

W.A.D. Edwards
Alberta Geological Survey
4445 Calgary Trail South
Edmonton, Alberta
T6H 5R7

L.E. Jackson, Jr.
Geological Survey of Canada
100 West Pender Street
Vancouver, British Columbia
V6B 1R8

S.R. Morison
Department of Indian Affairs and
 Northern Development
200 Range Road
Whitehorse, Yukon
Y1A 3H1

E.W. Presant
Ontario Institute of Pedology
Guelph Agriculture Centre
Box 1030, Guelph, Ontario
N1H 6N1

C. Tarnocai
Land Resource Research Centre
Agriculture Canada
K.W. Neatby Building
Ottawa, Ontario
K1A 0C6

K.W.G. Valentine
Land Resource Research Institute
Agriculture Canada
K.W. Neatby Building
Ottawa, Ontario
K1A 0C6

K. Webb
Canada Soil Survey Unit
Research Branch, Agriculture Canada
Nova Scotia Agriculture College
Truro, Nova Scotia
B2N 5E3

Printed in Canada

Chapter 12

THE INFLUENCE OF THE QUATERNARY GEOLOGY OF CANADA ON MAN'S ENVIRONMENT

Summary

Introduction — *L.E. Jackson, Jr.*

Geological hazards in the Canadian Cordillera — *S.G. Evans and J.S. Gardner*

Engineering geology and land use planning in the Prairie region of Canada — *J.S. Scott*

Geological hazards in Eastern Canada — *J. Locat and J-Y. Chagnon*

Quaternary geology and urban planning in Canada — *O.L. White*

Acknowledgments

References

Landslide in late glacial marine clay at Saint-Jean-Vianney. The landslide, which occurred May 4, 1971, enveloped 40 houses, killing 30 people. Similar slides occur in thick, fine grained Quaternary marine and lacustrine sediments in several parts of Canada. Courtesy of N.R. Gadd.

Chapter 12

THE INFLUENCE OF THE QUATERNARY GEOLOGY OF CANADA ON MAN'S ENVIRONMENT

co-ordinator
L.E. Jackson, Jr.

SUMMARY

Quaternary geology and geomorphology play a major role in controlling the nature and distribution of surface materials and in controlling the processes active at the surface. Consequently Quaternary geology and geomorphology influence the pattern of man's land use activities and are valuable in determining where geological hazards might occur.

High relief, steep slopes, past and present glaciation, seismic activity, and a montane climatic regime make mass wasting processes ubiquitous geological hazards in the Cordillera. These range in increasing magnitude and decreasing frequency from low magnitude rockfall, which may be a daily occurrence during the spring and summer, through debris flow, which may have a recurrence frequency of 20 years or less, to debris avalanches in excess of 10^6 m^3. Only eleven of these latter events have been recorded in the southern Canadian Cordillera during the historic period. Alpine areas have the additional hazard of snow avalanches which have high recurrence frequencies, large volumes, and move at high velocities. All of these hazards can be most economically managed through identification of hazardous areas and avoidance of them. Except for large debris avalanches, protective structures are generally an effective protection if it is necessary to make use of a hazardous area. In the case of avalanches, forecasting and induced release are additional effective management techniques.

Planning and engineering in the Prairie region (southern Interior Plains of Canada) requires an understanding of the Quaternary geology framework and history and of the bedrock formations from which Quaternary sediments were derived. For example, an understanding of the Quaternary stress history, particularly glaciotectonism has proved essential in evaluating the stability of highwalls in Prairie surface mines. Management and conservation of aggregate and groundwater resources require detailed knowledge of Quaternary stratigraphy. Urban sprawl may bury aggregate deposits, and waste storage or disposal sites may pollute aquifers unless accurate three dimensional knowledge of Quaternary stratigraphy is available. There are few geological hazards on the Prairies because of the low relief and aseismicity. Landsliding is a significant problem along some river valleys and flooding is a widespread problem on the Lake Agassiz plain. Expansive clays, derived from Cretaceous shales that underlie the glacial overburden, locally may cause foundation problems.

Geological hazards in eastern Canada result from Quaternary and pre-Quaternary events as well as contemporary tectonism. Depression of the crust by ice sheet loading resulted in the creation of a temporary sea and many temporary or enlarged lakes. These were loci for the deposition of fine sediments and subsequent slope failures in these sediments. Large earthquakes have occurred in the region. Seismic activity is thought to be related to plate interactions and continuing stress release related to glacial isostatic rebound. Flooding and erosion are widespread problems along rivers and lake and sea shores. Flooding results predominantly from the coincidence of heavy rainfall, snowmelt, and the breakup of river ice during the spring. Passive measures including the identification of hazard areas and their avoidance through sound planning are the most effective ways of managing these hazards.

There is an awareness among federal, provincial, and municipal authorities of the utility of employing geological information in urban planning. However, its application and legislation requiring geological investigations and data collection are uneven and of variable quality across Canada. The cost of nonuse of geological information, particularly as it relates to geological hazards, is greatest in urban areas. The high density of construction and high land values in urban areas assure that damage from problems such as differential compaction in foundation substrates will be extremely costly. Where other resources are involved, such as groundwater, alternatives to the application of geological information may not exist.

INTRODUCTION

L.E. Jackson, Jr.

Geology, geomorphology, and engineering geology are important to man's use of the land, whether in the urban setting or in the sparsely populated hinterland of Canada. Study and mapping of the surficial deposits and rock that form the surface and immediate subsurface of the landmass provide information on the earth materials on or within

---

Jackson, L.E., Jr. (co-ordinator)
1989: The influence of the Quaternary geology of Canada on man's environment; Chapter 12 in Quaternary Geology of Canada and Greenland, R.J. Fulton (ed.); Geological Survey of Canada, Geology of Canada, no. 1 (also Geological Society of America, The Geology of North America, v. K-1).

which foundations, roads, and pipelines may be constructed, towns built, or other human activities carried out. Such studies also identify past or presently active geomorphological processes, such as landsliding, fault-related surface displacement, or contamination of aquifers, that may pose hazards or limitations to man's activities. Physical, mineralogical, and engineering testing of surficial deposits and rock assist in the determination of the suitability and probable performance of these materials for the various activities and uses of man.

Establishment of the sequence or stratigraphy of sediments and rocks immediately underlying the surface is essential for age dating and determination of recurrence frequencies of geomorphic processes that may pose hazards to various land uses. Examples include the establishment of past recurrence frequencies of debris flows by counting and dating buried soil horizons on an alluvial fan or the dating of overbank deposits on a low river terrace in order to estimate the past frequencies of large magnitude floods. Stratigraphic investigations also allow further refinement of land use capability by identifying resources such as sand and gravel deposits and aquifers. Studies of observable geomorphic phenomena, such as slope failures or the processes of shoreline erosion, are fundamental to the identification of geological hazards and aid in the design of engineering structures which permit the coexistence of man's works or man himself with potentially hazardous geomorphic settings.

This chapter examines some aspects of the application of geology, geomorphology, and engineering geology to land use planning and man's activities in southern Canada, that is, those parts of the country not extensively affected by permafrost; the problems of permafrost are treated in a separate chapter. Throughout Canada, the level of detail and extent of Quaternary and engineering geological information varies from region to region, partly in response to the specific geological problems characteristic of the various regions. This review of selected issues which affect man's use of the land in Canada today has been prepared from the point of view that the modern geological environment is largely controlled by Quaternary events and deposits.

The chapter is not an exhaustive review of all problems and issues in all areas of Canada, but rather is a series of essays prepared by experts from different regions and fields of specialization which brings out what they conceive to be the main environmental issues related to Quaternary geology and processes. The topics covered are generally representative of the significant concerns of the main regions of southern Canada. Thus, in the Cordillera, an area of relatively recent and rapid uplift, the geological hazards of rock avalanches, rockfalls, debris flows, slumps and slides, and snow avalanches are major concerns. The full extent of the problems and how society should react to them are only now beginning to be assessed. Similarly, the emphasis of the section on eastern Canada is on geological hazards, including earthquakes, landslides, and flood hazards. The section on the Prairie region includes discussion of a broad range of engineering geology issues, including questions of foundations, aggregates, water supply, flooding, and open pit mining. The final section deals with urban Canada and discusses the general nature and availability of the type of Quaternary geology information necessary for the efficient and orderly construction of this environment which is where most Canadians live.

Additional pertinent material is included in many other chapters of this volume. Many of the regional chapters in Part 1 contain sections which bring out the prime land use and environmental problems related to the Quaternary geology of the region. The geomorphic processes associated with the various hazards are described in Chapter 9. Man's use of Quaternary resources is described in Chapter 11 and aspects of Quaternary geology and the chemistry of surficial materials are covered in Chapter 10. Questions of shoreline and coastal erosion and man's use of coastal lands are dealt with in Volume no. 2 of the Geology of Canada series — *Geology of the Continental Margin of Eastern Canada*.

GEOLOGICAL HAZARDS IN THE CANADIAN CORDILLERA

S.G. Evans and J.S. Gardner

Land use planning in the Cordillera is of particular importance owing to the abundance of geological hazards associated with mountainous terrain. A combination of high relief, past and present glaciation, and montane climatic regimes make geological hazards and the planning response to them unique in Canada. Amongst the geological hazards which the planner must take into consideration are snow avalanches, floods and debris flows, low magnitude rockfalls, and high magnitude rock avalanches. These are described here together with a discussion on risk assessment which determines the planning response to these hazards.

Snow avalanches

Snow avalanches and, to a lesser degree, ice avalanches have been persistent hazards to man and his activities in the Cordillera. Although snow avalanches and ice avalanches occur most frequently in the alpine zone, the most damaging events are often those which begin in snow accumulation or source areas above the treeline, traverse lower forested terrain, and run out onto valley floors where most engineered facilities related to human activities are located (Fig. 12.1).

The snow and ice avalanche hazard has been well documented for the Alps of Europe where it has been persistent and damaging over many centuries (Fraser, 1978). Experience with the avalanche hazard in the Canadian Cordillera is relatively recent, beginning with the construction of the Canadian Pacific railway in the 1880s. Snow and ice avalanches have posed a problem for mining (Atwater,

---

Jackson, L.E., Jr.
1989: Introduction (Influence of the Quaternary geology of Canada on man's environment); in Chapter 12 of Quaternary Geology of Canada and Greenland, R.J. Fulton (ed.); Geological Survey of Canada, Geology of Canada, no. 1 (also Geological Society of America, The Geology of North America, v. K-1).

Evans, S.G. and Gardner, J.S.
1989: Geological hazards in the Canadian Cordillera; in Chapter 12 of Quaternary Geology of Canada and Greenland, R.J. Fulton (ed.); Geological Survey of Canada, Geology of Canada, no. 1 (also Geological Society of America, The Geology of North America, v. K-1).

1968; Peck, 1970), transportation (Schaerer, 1962; Freer and Schaerer, 1980), and more recently for urban development and recreation activities (Williamson, 1982).

Avalanches commonly recur in the same locations year after year and in places several times each year. Thus avalanche hazard areas can be identified by terrain and vegetation conditions and on the basis of prior experience or historical record. Avalanches also will occur where they have not occurred before. The early mining literature in British Columbia provides evidence of "green slides" or avalanches occurring in new locations or with greater magnitude than usual in old locations, resulting in the destruction of forest cover.

In terms of loss of life, property damage, and the disruption of human activities in the Cordillera, snow avalanches have been a most significant environmental hazard (Freer and Schaerer, 1980). A few examples will serve to illustrate the dimensions of the hazard.

The Rogers Pass area in the central Selkirk Mountains, is the site of the most serious avalanche hazard to a transportation route in the Cordillera. The hazard to both the Canadian Pacific railway and Trans-Canada Highway, which traverse Rogers Pass, results from a combination of high snowfall amounts and rates, and steep slopes extending directly to the valley floor. The railway, completed through the pass in 1885, was closed by avalanches during the winter

Figure 12.1. Typical avalanche tracks in Kananaskis country of the Rocky Mountains, Alberta. These extend through an elevational range of as much as 1000 m from high in the alpine zone to lower slopes and valley floor. Courtesy of J.S. Gardner.

of 1885-86. In March 1910 a single avalanche killed 62 men. This disaster, plus the long periods of closure, led the Canadian Pacific Railway to construct the Connaught Tunnel under the most dangerous avalanche area in Rogers Pass. Between 1885 and 1916, when operation of the tunnel commenced, over 200 men perished in snow avalanches (Woods and Marsh, 1983). Avalanches continued to plague railway operations despite the construction of snowsheds, the employment of large numbers of men for track clearing, and the eventual use of the rotary snowplow.

In 1956 construction began on the Trans-Canada Highway through Rogers Pass. It was completed in 1962 with careful consideration and planning for avalanche hazard (Schaerer, 1962; McKenzie, 1960). Avalanche hazard, although still a threat to traffic and highway workers, has been reduced considerably through careful location of the highway, the construction of snow sheds, sophisticated weather monitoring and avalanche forecasting, road closures, earth dams, dikes, mounds, a mobile artillery defense, and the use of heavy snow clearance machinery.

Other important avalanche hazard areas are found in northwestern British Columbia along Highway 16 between Terrace and Prince Rupert (Stethem and Schaerer, 1979; Clague, 1978) and on Highway 3 (Salmo-Creston) over Kootenay Pass in south-central British Columbia (Nelson, 1970). In addition snow avalanches are a persistent winter hazard along parts of virtually all transportation routes in the Cordillera and are a factor to be considered in the planning, construction, and maintenance of future routes.

Avalanches have consistently affected mining activities in the Cordillera (Peck, 1970). Many mines have been located in remote, high mountain valleys and at high elevations well above treeline in positions vulnerable to snow avalanches. For example, in February 1965, a severe avalanche occurred at the Granduc Mine near Stewart, British Columbia, which killed 26 men and injured 23 others. In addition, much of the 50 km-long road between the mine and Stewart traverses some of the worst avalanche hazard terrain in the Cordillera. This led to the development of a sophisticated avalanche hazard management program during the operation of the mine.

The other major socio-economic sector affected by snow avalanche hazard is the recreation industry, particularly skiing and mountaineering (Williamson, 1982). A catalogue of recreational avalanche accidents over four decades in Canada is given in Stethem and Schaerer (1979, 1980). Most fatalities and injuries occur outside the controlled downhill ski areas involving backcountry (cross-country) skiers and mountaineers. The advent of helicopter skiing in remote areas of central British Columbia (Selkirk, Monashee, and Caribou mountains) has further escalated the avalanche hazard in the ski industry.

A wide variety of adjustments to the avalanche hazard has been employed in the Cordillera. These include land use management, direct control, and public education. Avalanche forecasting is used in direct control, and relies on understanding the process and its relationship to meteorological, terrain, and vegetation conditions (Perla, 1978). Forecasts are used for public information and decision-making with regard to closures and avalanche control measures. Most direct control measures rely on induced release of snow masses through the use of explosives. Avalanche protection devices such as snowsheds, earth mounds, earth dykes, and diversions are also used as passive measures in avalanche defence (Perla and Martinelli, 1975).

Floods and debris flows

Floods and debris flows take a variety of forms in the mountain environment and it is not unusual for floods to be transformed into debris flows or vice versa (Miles and Kellerhals, 1981; VanDine, 1985).

Flash floods and debris flows have been especially troublesome to road and rail traffic in the Coast and Columbia mountains of British Columbia. Traffic disruptions on the Squamish Highway north of Vancouver, the Trans-Canada Highway, and rail lines in the Fraser Canyon and in the Selkirk and Rocky mountains (Fig. 12.2), and the railway and highway east of Prince Rupert are notable examples. Eisbacher (1979) has discussed the geography of debris flow in the Cordillera, Clague (1978) has published an inventory of potential flood and debris flow areas in Skeena Valley between Prince Rupert and Terrace, and Eisbacher and Clague (1984) have provided an excellent summary describing debris flows in the mountain environment and methods of mitigation. VanDine (1985) has detailed the occurrence and mechanisms of debris flows in the southern Cordillera.

Large quantities of water, the primary ingredient for flash flood and debris flow events, are released in the mountain environment by several mechanisms. Prolonged and heavy rainstorms, both with and without melting snow, are a stimulus throughout the Cordillera (e.g., Evans and Lister, 1984). Intense convective storms are a stimulus in selected locations such as the Front and Western ranges of the Rocky Mountains. Sudden and copious snowmelt in late spring, especially on south and southwest exposures, provides another stimulus. Finally, in glacierized areas, jökulhlaups or catastrophic glacier meltwater outbursts and rupture of moraine-dammed lakes occasionally generate waterfloods and debris flows (Fig. 12.3a; Mathews, 1965; Jackson, 1979; Clague et al., 1985; Blown and Church, 1985).

Large quantities of overland and channel flow frequently mobilize debris. In addition to the volume of discharge, high velocities due to steep gradients and the availability of glacial and colluvial debris in mountain watersheds encourage debris mobilization. The mechanical differences between fluvial transport, as in floods, and grain dispersion transport, as in debris flows, result in different types of deposits and landform. Deposits and landforms not only provide some of the best means for identifying potentially hazardous flood and debris flow locations, but also assist in estimating the magnitude of past events.

One of the best documented debris flow sites in the Cordillera is located in Yoho National Park, British Columbia in the vicinity of Cathedral Mountain and the Spiral Tunnels on the Canadian Pacific Railway. This site and the characteristics of a major debris flow event in September 1978 have been described by Jackson (1979, 1980). The buildup of meltwater in a small glacier on Cathedral Mountain resulted in sudden release of water from beneath the glacier. The water mobilized colluvial and glacial sediments to produce a debris flow which descended some 800 m to the main valley below. This event produced sharp-crested debris levees and debris lobes with reverse sorting characterized by the largest clasts being carried to the outer margins of the deposits. Canadian Pacific Railway records and old photographs indicate that events of similar magnitude occurred in 1962, 1946, and 1925.

J.S. Gardner (unpublished) and Sauchyn et al. (1983) have identified numerous drainage basins which produce debris flows in the southern Rocky Mountains. These basins, usually less than 1 km² in area, characteristically begin in the alpine zone where the necessary water is made available through snowmelt and rainfall. The debris flows then propagate to lower elevations, commonly reaching main valleys, intersecting roads, trails, railways, and campgrounds. Event frequency has been estimated on the basis of dendrochronological, historical, botanical, and other evidence (Sauchyn et al., 1983; Deslodges and Gardner, 1984).

Frequency of events in the Main Ranges is estimated at 1 in 20 years (5% chance per year) whereas small basins in the Front Ranges appear to produce debris flows somewhat less frequently.

Copious rainfall in the Coast Mountains results in debris flows, flash floods, and debris avalanches which are initiated throughout the elevational range from the alpine zone to sea level. Particular problems have been encountered along the eastern shore of Howe Sound, north of Vancouver,

Figure 12.2. Canadian Pacific Railway and Trans-Canada Highway pass beneath the lower slopes of Mount Stephen in Yoho National Park, British Columbia. These routes are subject to low magnitude rockfalls, debris flows, and snow avalanches. Note the presence of debris flow deposits above the railway track. Courtesy of J.S. Gardner.

Figure 12.3. (a) The breached terminal moraine at Nostetuko Lakes, British Columbia. In 1983, the sudden release of about 7×10^6 m³ of water devastated the upper part of Nostetuko River valley. The breaching process was probably initiated when an ice avalanche detached from the toe of the Cumberland Glacier above the lake and generated an overtopping wave. 204047-F (b) Debris flows initiated in steep mountain watersheds are a major geological hazard in the Cordillera. Protective works, such as this check dam on Harvey Creek, have been built at a cost of millions of dollars in the vicinity of some settlements in southwest British Columbia to protect homes from debris flow impact. 204246-P

British Columbia where debris flows caused 12 deaths between 1958 and 1983 (Lister et al., 1984; VanDine, 1985).

Flood and debris flow hazard mitigation can be based primarily on land use control and identification of conditions leading to the events. The morphology and sedimentology of flood and debris flow deposits are well known. Thus identification of potentially hazardous sites is a matter of systematic inventory and mapping. Where hazardous sites cannot be avoided or facilities are already in place, active remedial methods, such as construction of stream channel control works and various types of debris dams and deflectors, can be used to minimize damage (Eisbacher and Clague, 1984). Debris flow control structures have been constructed at considerable cost (Fig. 12.3b) at several locations in southwestern British Columbia. These include deflection dykes at Port Alice (Nasmith and Mercer, 1979) and at Agassiz (Martin et al., 1984), and check dams at several locations along Howe Sound (Hungr et al., 1984; VanDine, 1985).

Low magnitude rockfalls

Low magnitude rockfall involves the free fall and downslope bounding of individual or small groups of rocks. In the mountain environment, low magnitude rockfalls occur on cliffs, scarps, and steep dip slopes and produce ubiquitous talus or scree accumulations characterized by fall-sorting (Gardner, 1971a). In the Cordillera, rockfalls have been studied in a geomorphic context by Gardner (1968, 1970, 1980, 1983) and Luckman (1976) in the Rocky Mountains, and by Gray (1972) in the Ogilvie Mountains, Yukon Territory. Low magnitude rockfall poses a major hazard along highway and railway rock cuts throughout the Cordillera, and substantial reviews have been published on the subject (Peckover and Kerr, 1977; Piteau and Peckover, 1978). Hazardous aspects of rockfall in recreational mountaineering have been discussed by several workers (Gardner, 1971b, 1981; Marsh, 1981).

Detailed studies of alpine rockfall have been conducted by Gardner in the Main Ranges and Front Ranges of the Rocky Mountains. Based on several thousand hours of direct observation of rockfalls in the field, the following generalizations may be made: Low magnitude rockfalls are concentrated during spring thaw and summer. On a diurnal basis, they are most frequent during warm and sunny periods, peaking in the early afternoon hours. They tend to be rare on cool, cloudy days and, on average, are rare between the hours of 2200 and 0700. Intense rainstorms create an exception to these generalizations, resulting in bursts of rockfall activity. Low magnitude rockfalls are concentrated in a spatial or geographic sense. Steep rock slopes with northerly, northeasterly, and easterly exposures, are especially prone to low magnitude rockfall. The spatial and temporal distributions of rockfall occurrences indicate that freeze and thaw processes on susceptible bedrock surfaces are the driving mechanisms.

At lower elevations similar results have been reported by Peckover and Kerr (1977) and Piteau and Peckover (1978). Critical times for low magnitude rockfall hazard to roads, railways, and structures are during thaw periods, particularly thaws of ice and deep-seated frost at the end of the winter period, and during periods of intense rain. Artificially steepened and fractured rock cuts are the most susceptible.

Engineering methods to deal with low magnitude rockfall hazard are reviewed by Peckover and Kerr (1977) and Piteau and Peckover (1978) and include warning signs on roads; stabilization of rock surfaces with bolts anchors, concrete structures, and netting; protective barriers; periodic clearing of loose rock from steep surfaces and of fallen debris from roads and tracks; and location of facilities away from susceptible areas.

Rock avalanches

Major reviews have been recently completed on landslides in the Cordillera (Evans, 1982, 1984; Eisbacher and Clague, 1984; Cruden, 1985). Consequently, this section will use only one landslide type, catastrophic rock avalanches, as an example of factors affecting the planning response to landslide occurrence.

Rock avalanches occur when a detached mass of rock leaves a shear surface, disintegrates, and flows down a valley side slope. Velocities up to 300 km/h may be reached. Before coming to a halt, the debris may move many kilometres down a valley side slope or down the valley itself. Excellent descriptions of typical rock avalanches are found in Eisbacher and Clague (1984). The run-out distance of such landslides appears to be related to the volume of the rock mass involved in the initial detachment (e.g., Scheidegger, 1973). The catastrophic nature of the event, the high velocity of the mass whilst in motion, and the large area covered by the debris make rock avalanches the most potentially damaging of all landslide types.

The distribution of major rock avalanche events in postglacial time in the Cordillera is unknown at both regional and local scales. Workers such as Nasmith (1980), Gardner (1980), and Luckman (1981) have suggested that most rock avalanches in the Cordillera occurred in the immediate postglacial period as a result of glacial oversteepening, the removal of glacial support, and stress-relief processes related to deglaciation. This suggestion, however, is made in the absence of absolute dates. Abele (1974) has stated a similar view with reference to rock avalanches in the European Alps.

Figure 12.4. (a) The northern margin of the Frank Slide (estimated volume 31×10^6 m^3) and the town of Frank, Alberta from Turtle Mountain shortly after the rock avalanche occurred in 1903. 17008

The conclusion that the frequency of landslide was greatest immediately following the last glaciation is contradicted in part by the work of Porter and Orombelli (1981) in the Mount Blanc area of the European Alps, and Whitehouse and Griffiths (1983) in a 10 000 km² area of the Southern Alps of New Zealand. Both these studies utilized absolute dating techniques. Porter and Orombelli (1981) found that most of the rock avalanches for which they had evidence date from the Little Ice Age, the period of climatic deterioration from the 17th to 19th centuries, and that the hazard "is much greater than appreciated". The increased incidence of landslides of all types during the Little Ice Age was also noted by Grove (1972) in western Norway. It appears, therefore, that climatic changes in the Neoglacial period may be a significant factor in the initiation of rock avalanches. In New Zealand, Whitehouse and Griffiths (1983) showed that greatest frequency of rock avalanche was during the last 3 ka but attributed this to 'erosion censoring' and difficulty in identifying older rock avalanche deposits rather than a substantive increase in frequency.

Many rock avalanches are triggered by seismic loading during earthquakes (Keefer, 1984; Evans et al., in press) and their occurrence would therefore partly reflect the magnitude and frequency of these loadings in postglacial time,

Figure 12.4.(b) Oblique aerial photograph of the Hope Slide (estimated volume 47×10^6 m³) shortly after it occurred in 1965. British Columbia Government Airphoto BC (o)445.

unrelated to the timing of Wisconsinan deglaciation or climatic changes.

The time series represented by the distribution of events in postglacial time is, therefore, a result of the superimposition of three effects: (1) the effects of Wisconsinan deglaciation, (2) climatic forcing in the Holocene Neoglacial periods, and (3) seismic forcing. Because of the lack of absolute dating of prehistoric rock avalanche events in any region of the Cordillera, the nature of the time series remains unknown both in general terms and with respect to climatic and seismic forcing. The time series, however, is likely to be nonstationary.

Since 1856, 140 deaths have been caused by four rock avalanche disasters (S.G. Evans and J.J. Clague, unpublished). These deaths amount to 40% of the total deaths due to all types of landslides in the period 1856 to 1983. Four major rock avalanche disasters have occurred since 1900.

Frank, Alberta (April 29, 1903). Approximately 76 people perished when 31×10^6 m³ of fragmented rock descended the eastern flank of Turtle Mountain and struck the southern edge of the town of Frank (Fig. 12.4a). Initial failure occurred on this eastern limb of the Turtle Mountain anticline and involved Paleozoic carbonate rocks. No evidence of a previous occurrence exists at Frank and the effect of coal mining within Turtle Mountain on the stability of the slope has long been the subject of debate (Cruden and Krahn, 1973).

Jane Camp, British Columbia (March 22, 1915). Jane Camp, part of the Britannia Mine complex, located 37 km north of Vancouver, was inundated by a small rock avalanche (estimated volume 1.0×10^5 m³) with the resultant death of approximately 56 people. Tunnelling in fractured Mesozoic volcaniclastic rocks above Jane Camp may have contributed to the landslide. It is unlikely that a previous rock avalanche had occurred at the site (Ramsey, 1967).

Hope, British Columbia (January 9, 1965). This rock avalanche, 17 km east of Hope (Fig. 12.4b), buried four people who were travelling the Hope-Princeton Highway. The landslide involved 47×10^6 m³ of Paleozoic metavolcanics and occurred at the same location as a prehistoric rock avalanche of comparable size. The landslide was initiated by a small earthquake, the epicentre of which coincided with the landslide site (Mathews and McTaggart, 1978).

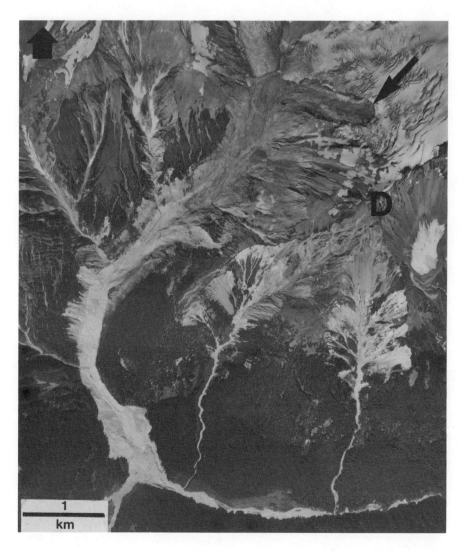

Figure 12.4.(c) Aerial photograph of the 1975 Devastation Glacier rock avalanche - rock debris flow (estimated volume 12×10^6 m³), Mount Meager volcanic complex, Coast Mountains, British Columbia. Source area is indicated by arrow. D = The Devastator. British Columbia Government BC 77117-119.

Devastation Glacier, British Columbia (July 22, 1975). An estimated 12×10^6 m³ of ice and hydrothermally altered volcanic rock broke away from the slopes of The Devastator within the Meager Creek volcanic complex, a part of the Garibaldi Volcanic Belt (Fig. 12.4c). The slide buried four people. Other large landslides of a similar character occurred in the same valley in 1931 and 1947 (Mokievsky-Zubok, 1977).

Table 12.1 lists major rock avalanches known to have occurred in the Canadian Cordillera since 1855. Recent examples include the North Nahanni rock avalanche (Fig. 12.4d) described by Evans et al. (1987) triggered by the October 1985 Nahanni earthquake (estimated volume $5\text{-}7 \times 10^6$ m³) and the 1986 Mount Meager rock avalanche (estimated volume $0.5\text{-}1.0 \times 10^6$) described by Evans (1987). The listing is certain to be incomplete but represents the only basis on which an assessment of the hazard of catastrophic rock avalanches can be made at this time (January 1987).

Damage beyond the limits of landslide debris

In addition to direct damage within the boundaries of the debris, extended and delayed damage may occur as a result of the formation of a lake dammed by rock avalanche debris (Evans, 1986a). Following the damming event, damage may be sustained at upstream sites as a result of flooding during reservoir filling and downstream as a result of the breaching of the landslide dam. The breaching of such dams may be

Figure 12.4.(d) Aerial photograph of North Nahanni rock avalanche (estimated volume $5\text{-}7 \times 10^6$ m³) triggered by the October 1985 Nahanni earthquake, Northwest Territories. NAPL A27018-266.

catastrophic (Evans, 1986b) and may occur days, months, or years after the rock avalanche debris is deposited. Large areas upstream and downstream of the rock avalanche site may thus sustain indirect damage as a result of the rock avalanche. Lakes dammed by rock avalanche debris are common in the Cordillera (Fig. 12.5) as are sites of former lakes that have subsequently drained (Evans, 1986a). There is no documentation, however, of a catastrophic breach similar to that which occurred on Indus River, at Nanga Parbat described by Mason (1929).

Damage beyond the limits of landslide debris may also be caused by destructive landslide-generated waves which occur when landslide debris enters natural or man-made water bodies. Following the Vajont Disaster in the Italian Alps the possibility of such events has been considered in the design of hydroelectric projects in the Cordillera (e.g., Chaudhry et al., 1983).

Figure 12.5. Landslide dam formed of rock avalanche debris in the vicinity of Mount Mason, Coast Mountains, British Columbia. 204047-C

Table 12.1. Major known rock avalanches of the southern Canadian Cordillera since 1855

Rock avalanches	Date	Volume ($\times 10^6$ m^3)	Rock type	Previous occurrences	Source
Rubble Creek, Garibaldi, B.C.	1855 or 1856	25	Quaternary volcanics	yes	Hardy et al., 1978
Frank, Alberta	1903	31	Paleozoic limestone	no	Cruden and Krahn, 1973
Hoodoo Mountain, B.C.	1919	?	Quaternary volcanics	?	Kerr, 1948
Kaouk River, Vancouver Island, B.C.	1928	8	Granite	no	Evans, 1984
Devastation Glacier I, B.C.	1931	~20	Pliocene volcanics	yes	
Brazeau Lake, Alberta	1933	4.5	Paleozoic limestone	no	Cruden, 1982
Mount Colonel Foster, B.C.	1946	0.5	Mesozoic volcanics	no	
Devastation Glacier II, B.C.	1947	~4	Quaternary volcanics	yes	
Pandemonium Creek, B.C.	1959 or 1960	~5	Coast plutonic complex	no	
Dusty Creek, B.C.	1963	5	Quaternary volcanics	yes	Clague and Souther, 1982
Hope, B.C.	1965	47	Paleozoic metavolcanics	yes	Mathews and McTaggart, 1978
Devastation Glacier III, B.C.	1975	12	Quaternary volcanics	yes	
Turbid Creek, B.C.	1984	~5	Quaternary volcanics	yes	
North Nahanni, N.W.T.	1985	~5	Paleozoic limestone	yes	Evans et al., 1987
Mount Meager, B.C.	1986	~1	Quaternary volcanics	no	Evans, 1987
North Creek, B.C.	1986	~0.5	Coast plutonic complex	yes	

Table 12.2. Risk assessment for selected rock avalanche sites in the Cordillera

Site	n[1]	Length of record (years)	Return period (years)	Design life of facility		
				1	100	500
FRANK	1	10 000	10 000	$10^{-4.0}$	$10^{-2.0}$	$10^{-1.3}$
HOPE A	2	10 000	5000	$10^{-3.7}$	$10^{-1.7}$	$10^{-1.0}$
HOPE B	3	10 000	3333	$10^{-3.5}$	$10^{-1.5}$	$10^{-0.8}$
RUBBLE CREEK FAN A	9	9000	1000	$10^{-3.0}$	$10^{-1.0}$	$10^{-0.4}$
RUBBLE CREEK FAN B	3	600	200	$10^{-2.3}$	$10^{-0.4}$	$10^{-0.3}$
DEVASTATION GLACIER/ MEAGER CREEK	4	53	17.66	$10^{-1.2}$	1	1

[1] n = number of encounters for which probability is calculated

Risk assessment for geological hazards

Using the methods of Borgman (1963), an estimate of risk for sites subject to the types of geological hazards discussed in this section is possible. According to Borgman (1963), the encounter probability, E (defined as the probability that an event will occur during the life of a facility), for a site is given by:

$$E = 1 - (1 - 1/T)^L \quad (12\text{--}1)$$

where T = return period and L = design life of the facility. This method has been used for snow avalanche encounter probability by LaChapelle (1966) and is also suggested for rock avalanches by Whitehouse and Griffiths (1983). It may also be used for assessment of rockfall and debris flow sites.

The method is illustrated here with respect to rock avalanches and equation (1) may be applied to selected rock avalanche sites listed in Table 12.1 assuming that the incidence of rock avalanches at a site is statistically independent. This assumption may not be valid if human interference was the landslide trigger or if the occurrence of one rock avalanche produces conditions which increase the probability of a second event, for example, development of an unstable scarp which fails some time later. In Table 12.2 the annual encounter probabilities (L = 1) are given in addition to encounter probabilities for design lives of 100 and 500 years for the following sites; Frank (1st encounter), Hope (2nd and 3rd encounter), Rubble Creek (3rd and 9th encounter), and Devastation Glacier/Meager Creek (4th encounter). Encounter probabilities are expressed for fixed facilities within reach of the debris of previous events.

With respect to Table 12.2 the following points are noted:

1. All annual encounter probabilities are equal to or in excess of 10^{-4}. Starr (1969) proposed an international standard of acceptable annual risk level of about 10^{-6} with respect to "planned activities". Hestnes and Lied (1980) referred to a proposed highest tolerable risk level for dwelling houses in Norway (3×10^{-3} per year) and also to the fact that Norwegian snow avalanche statistics indicate a personal death risk of 6.3×10^{-4} per year.

2. The encounter probability at Frank in 1903 (assuming that mining did not contribute to failure) represents the risk of any site that could be affected by at least one rock avalanche in postglacial time (taken to be 10 ka).

3. The risk of a second encounter at a site where a rock avalanche has previously occurred (e.g., at Hope) reduces the return period to 5 ka but does not dramatically increase the encounter probability in the absence of a clustering effect. Nasmith (1980) has stated that the risk at a site where a catastrophic slide has occurred in the past is of a different order of magnitude (possibly a thousand times greater) when compared to the risk at a location where no catastrophic slide has occurred during the past 10 ka. Data presented in Table 12.2 do not support this view for the risk of a first, second, and third encounter.

4. Similar probabilities calculated for first and second encounters at a site indicate the importance of identifying potential rock avalanche sites in addition to sites where a rock avalanche has already occurred.

5. The high encounter probabilities calculated for the Rubble Creek fan and the Meager Creek site reflect the recurrent nature of the rock avalanches which is associated with Quaternary composite volcanoes (Evans, 1984). The stratigraphy of the rock avalanche deposits in the Rubble Creek fan has been described by Moore and Mathews (1978) and Hardy et al. (1978) and has formed the basis of the two scenarios examined in Table 12.2. In 1973 the British Columbia Supreme Court upheld a decision by the Government of British Columbia to halt development of a mountain village site on the fan on the basis "that there is a sufficient possibility of a catastrophic slide during the life of the community" (Berger, 1973). The encounter probabilities for L = 100 and L = 500 (Table 12.2) appear to support this view.

6. Encounter probabilities are seen to increase substantially with increase in exposure time (L in equation 1). This fact should be considered when planning permanent facilities such as settlements in the Canadian Cordillera (cf. Berger, 1973).

Conclusions

The mountain environment of the Canadian Cordillera has an array of active geomorphic and hydrologic processes that act as natural hazards to human activities. To date, human activity and development in the Cordillera have been relatively limited in relation to development in other mountainous countries. Thus, the destructive impacts of these processes have been relatively minor. Snow avalanches have placed significant constraints on the location and operation of transportation routes, mines, and recreational developments. Debris flows and flash floods have caused damage to roads and buildings in localized parts of the Cordillera particularly in the Coast Mountains. Rock slope instability is widespread throughout the Cordillera. Quaternary composite volcanoes in the southern Coast Mountains have been particularly prone to rock avalanches whereas small scale rockfall is a ubiquitous problem in the vicinity of cliffs and cut slopes along transportation routes. When planning facilities in the Cordillera, consideration should be given to the potential damage that may result from the impoundment of rivers by rock avalanche debris, both during reservoir filling and the possible catastrophic breaching of the dam, and from landslide-generated waves in natural or man-made reservoirs. In dealing with these potentially hazardous processes, earth science information can aid in the planning process by the identification of hazardous sites, the estimation of probable frequencies and magnitudes, and the assessment of risk.

ENGINEERING GEOLOGY AND LAND USE PLANNING IN THE PRAIRIE REGION OF CANADA

J.S. Scott

The Prairie region of Canada, comprising parts of the provinces of Manitoba, Saskatchewan, Alberta, and northeastern British Columbia, is situated between the Precambrian Shield to the northeast and the foothills of the Rocky Mountains to the southwest (Fig. 12.6). Elevations range from above 900 m a.s.l. in the Foothills of Alberta and Cypress Hills of southwest Saskatchewan to less than 100 m in the lowland region of Manitoba. The area has a northeastward regional slope with gradual decreases in elevation except for relatively abrupt declines at the Missouri Coteau and at the Manitoba Escarpment. Drainage is generally northward to the Mackenzie River system and northeastward to Hudson Bay.

The northeastward regional slope has had a profound influence upon the deposits and geological events in the Prairie region. Gravel and sand, derived from the Rocky Mountains, is transported across the Prairies, down the regional slope. Upland remnants of Tertiary gravels, which were carried onto the Prairies from the west, occur in the Cypress Hills area of the south-central Prairies and in several areas of central and western Alberta. In addition, extensive fills of Rocky Mountain source sand and gravel occur in buried valleys throughout the Prairies. During the advance phases of continental glaciation glaciers originating on the Canadian Shield to the northeast moved up the regional slope, diverting drainage to the north and south. During retreat phases, large glacial lakes formed as the ice receded downslope. This interplay between direction of ice advance and retreat, and regional slope resulted in a distinctive complex of spillway valleys, disintegration deposits, and glacial lake sediments. It is the nature and distribution of these deposits and landforms that are of particular importance in planning land use activities.

The Prairie region is underlain by clay shale, siltstone, and sandstone from which are derived more than 80% of the materials making up the clay and silt-rich tills of the region (Scott, 1976). Montmorillonite within the clay fraction of tills imparts low permeability and high plasticity which causes these tills to become sticky when wet. The sand, gravel, and coarser material contained within the Prairie tills is largely derived from Precambrian Shield lithologies, from Paleozoic carbonates that flank the shield, and from carbonates and sandstones derived from the Rocky Mountains.

Quaternary deposits range from a few metres to well over 300 m in thickness (Christiansen, 1970a). The Quaternary succession generally consists of multiple layers of till separated in some localities by stratified deposits. Deposit distribution and stratigraphy are not uniform throughout the region but bear a complex relationship to topography, availability of materials, and glacier dynamics. In parts of Saskatchewan and Alberta the stratigraphic succession of Quaternary deposits is further complicated by ice-thrusting which has resulted in removal and/or repetition of beds including both drift and bedrock (Byers, 1959;

Scott, J.S.
1989: Engineering geology and land use planning in the Prairie region of Canada; in Chapter 12 of Quaternary Geology of Canada and Greenland, R.J. Fulton (ed.); Geological Survey of Canada, Geology of Canada, no. 1 (also Geological Society of America, The Geology of North America, v. K-1).

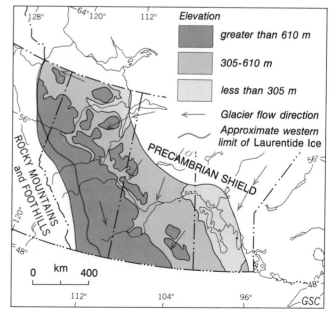

Figure 12.6. Location and general elevation of the Prairie region.

Table 12.3. Quaternary stratigraphic units used in Saskatchewan

Group	Formation	Unit
		Surficial stratified drift
SASKATOON GROUP	BATTLEFORD FORMATION	Intertill deposits Upper till Intertill deposits
	FLORAL FORMATION	Middle till Intertill deposits Lower till Intertill deposits
SUTHERLAND GROUP		Upper till Middle till Lower till
EMPRESS GROUP		Stratified glacial deposits Preglacial deposits in preglacial valleys
	BEDROCK	

Kupsch, 1962; Christiansen and Whitaker, 1976). The relationship among Quaternary stratigraphic units that have been identified in Saskatchewan provides an example of the sequence of deposits that may be encountered in the subsurface (Table 12.3). Further description of the Quaternary geology of the area may be found in Chapter 2 of the regional section (Part 1) of this volume.

The region's geological legacy from the Quaternary Period pervades man's land use and planning activities. The following discussion considers planning and development from geotechnical, groundwater, mining engineering, and waste disposal points of view and illustrates how these considerations have been influenced by Quaternary deposits and processes. The examples used are drawn from the collective experience of the contributors to this section and serve to focus upon the significance of Quaternary geology to the diverse settings and activities that occur throughout the Prairie region of Alberta, Saskatchewan, and Manitoba.

Engineering construction and materials

Planning and construction of engineering works and locating engineering materials involves the collection of a variety of disparate isolated data and using these to develop three-dimensional models from which the nature and reaction of foundation materials can be predicted and which will aid in locating construction materials with specified properties. Without an understanding of the Quaternary geology framework it would not be possible to develop accurate three-dimensional models. Without these, the geotechnical engineer could only react to problems instead of being able to anticipate them. In addition, without this geological input, in some cases, it is possible that faulty construction decisions could be made.

Construction and foundation considerations

From the point of view of construction and foundation considerations three mechanical properties of Quaternary deposits are important: shear strength, compressibility, and hydraulic conductivity. The range of values for these properties is highly variable because of the processes involved in deposition and the complexity of stratigraphy and structure. In terms of these properties, Quaternary deposits can be grouped into four general categories: till, sand and gravel, clay, and alluvium.

Prairie tills are generally considered to be stable foundation materials exhibiting high shearing resistance and low compressibility. Not all tills, however, have the same properties (MacDonald and Sauer, 1970). For example, the youngest till in Saskatchewan is more compressible and less stable than the heavily overconsolidated tills of the Floral Formation (Table 12.3), even though the texture and mineralogy are similar (Sauer, 1974). Excavation in the surface till is generally difficult under poorly drained conditions, whereas road builders have found the tills of the Floral Formation may be excavated without difficulty even below the water table. One till of the Sutherland Group has a high clay content and tends to be unstable on exposed slopes. In addition, the older tills may contain highly fractured layers resulting in values of hydraulic conductivity similar to a uniform sand or poorly sorted gravel (Christiansen, 1968).

Sand and gravel deposits are generally easily excavated and provide stable foundation conditions where they are well drained. Slope stability can, however, be a problem in excavations below water table and high hydraulic heads can be encountered where these materials are capped by till or clay. The presence or absence of these materials can make a significant difference in the feasibility for tunnels or deep open excavations. In a report on the suitability of Quaternary and older materials of the Edmonton area for different land uses, Kathol and McPherson (1975) indicated that the occurrence of sand, gravel and clay is the prime element contributing to unsuitability for deep sewer construction (Fig. 12.7).

Expansive clays, derived from Cretaceous shales by glacial processes, occur extensively throughout the region as glacial lake deposits. These clays shrink and swell with decreasing or increasing moisture content, respectively, thereby imposing stresses upon building foundations and other near surface structures such as sewers and water mains. In addition, surface clays can be highly compressive and, where saturated, create slope stability problems. These problems have been severe in cities such as Edmonton, Regina, and Winnipeg, all of which are located in glacial lake basins. In Winnipeg, pile foundations are commonly used to support medium to heavy structures — a practice that has been recently extended to light structures and residences which previously were commonly supported on reinforced concrete walls resting upon wide footings (Kjartanson, 1983).

Alluvium can exceed 30 m in thickness and range from sand and gravel to organic silts and clays. It occurs in most proglacial (Sauer, 1977) and extraglacial (Christiansen, 1983; Sauer, 1983) meltwater channels and spillways in Saskatchewan. Thick deposits of highly compressible alluvium, which has a very low shear strength, are a major concern for foundation design for bridges, earthfill dams, and roadway embankments. In many sites, artesian groundwater in the alluvium creates severe instability in open excavations or under heavy embankments.

Land use conflicts and protection of aggregate resources

One of the major challenges facing nearly all major cities is the rise in construction costs as haul distances for aggregate resources increase (Edwards, 1989). Commonly, ill-planned development results in a far greater depletion of the gravel resource by pre-emptive land uses than by exhaustion of the resource. Adequate data on the distribution of gravel, both at the surface and within or beneath the Quaternary succession, combined with enlightened planning permit systematic development of the gravel resources in concert with urban expansion. The relationship between granular material resources and development of the city of Calgary provides a useful example.

Gravel resources in Calgary are abundant and conspicuous. Because of the abundance, little heed has been paid to the need to integrate the utilization of the resource with city growth. Many of the older parts of the city have been built over gravel (Fig. 12.8), including the city centre and adjoining areas along Bow and Elbow rivers, as well as other communities such as Montgomery and Bowness. In most of this area, no attempt was made to remove the gravel prior to construction. Extensive reserves of gravel occur as an upland cap, underlying till, on Big Hill and Broadcast Hill. These areas contain estimated reserves of over 200×10^6 m³ but much of the land has already been pre-empted by low density housing and parkland zoning; as a result, huge reserves of gravel have been lost. Now major gravel operations have moved as far as pits east of De Winton, 27 km away, with a substantial increase in haulage costs.

Winnipeg lies in the middle of a vast plain of fine grained glacial lake sediment and might be expected to have aggregate resource problems. Fortunately, a large esker-delta complex lies 16 km northeast of the city. The quality of the aggregate is high because the materials were derived from Paleozoic carbonates and Precambrian crystalline lithologies. The deposit lies far enough from the urban centre

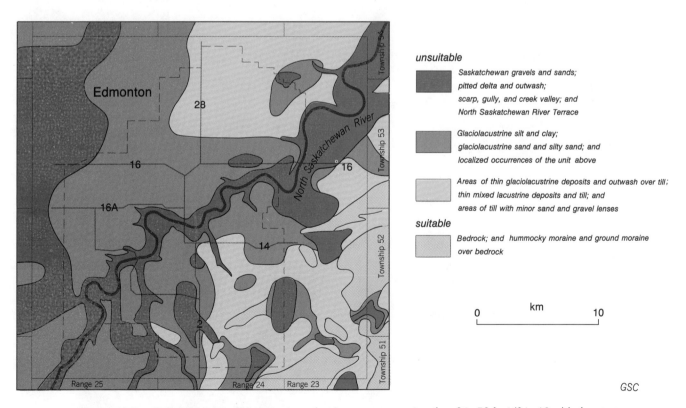

Figure 12.7. Suitability of the Edmonton area for deep sewer construction, 0 to 50 feet (0 to 16 m) below surface (from Kathol and McPherson, 1975).

so that urban sprawl did not pre-empt aggregate extraction before the value of this resource was realized; timely action by various levels of government has conserved the resource and maximum exploitation is proceeding.

The solution to this resource availability and land use conflict has three aspects. First, a community must estimate its aggregate resource needs for the foreseeable future. Next, it must conduct a thorough inventory (using a knowledge of Quaternary geology and process as an aid in predicting location and extent of supplies) to locate the necessary resources. Finally, it must assure that the resource areas are protected in development plans. Special resource management strategies, developed in the provincial planning process, have been adopted to protect the resource from urban sterilization. Manitoba Provincial Land Use Policy No. 13, a regulation under the Planning Act, promulgated in 1981, is an example of legislation designed to ensure conservation of aggregate resources.

Groundwater supply and exploitation

Groundwater is a major source of water supply in rural areas and many smaller urban communities of the Prairie region. Well yields of a few tens of litres per minute may be satisfactory for domestic use and such quantities can be obtained from localized aquifers within till (Meyboom, 1967). The much larger volumes of groundwater required for municipal and industrial use, however, can be obtained only from aquifers having significantly greater areal extent and thickness.

The following description of Quaternary aquifers in Saskatchewan exemplifies the diverse Quaternary geological settings that are of importance as sources of groundwater.

The discussion of exploration, exploitation, and protection of these aquifers illustrates the planning necessary if these resources are to be effectively used and protected. The examples cited in this section are all drawn from Saskatchewan but groundwater conditions and aquifers are similar throughout the Prairie region.

Aquifers

In Saskatchewan potable groundwater supplies are found in aquifers which occur throughout drift deposits and at or near the bedrock surface. They can be grouped into several types: bedrock aquifers, buried valley aquifers, blanket aquifers, intertill aquifers, and surficial aquifers (as defined by Meneley, 1972). Aquifers in Saskatchewan of major areal extent are shown in Figure 12.9; the generalized stratigraphic relationship of the aquifers is shown in Figure 12.10.

Bedrock units which comprise aquifers include carbonate rocks in east-central Saskatchewan but are found primarily within Cretaceous sediments throughout southern Saskatchewan. Some of these aquifers, such as the Swan River aquifer and the Judith River aquifer (where cut by Tyner Valley), are exposed at the bedrock surface. The hydrological evaluation of the groundwater resources of the

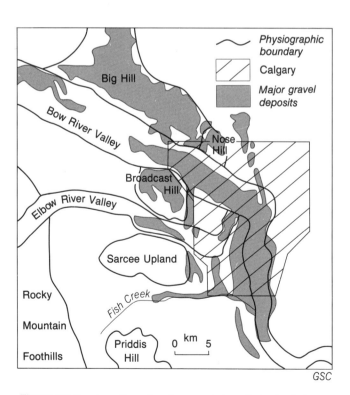

Figure 12.8. Location of major gravel deposits in the Calgary area.

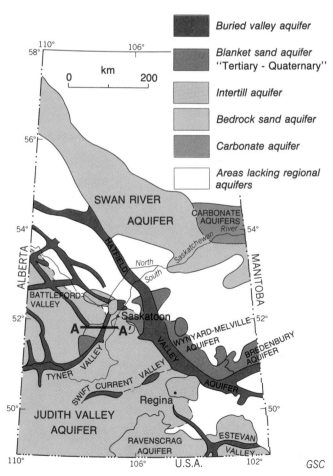

Figure 12.9. Major aquifers in southern Saskatchewan (after Meneley, 1972). Cross-section A-A' is shown in Figure 12.11.

Judith River Formation has been dealt with by Kewen and Schneider (1979) and Whitaker (1982). In southeastern Saskatchewan the Tertiary sand, silt, and clay with coal seams of the Eastend and Ravenscrag Formation are significant aquifers (Meneley, 1983; Fig. 12.9). Other deposits such as the Bredenbury Formation — a sand which grades from fine at the top to gravel at the base — are likely of Tertiary-Quaternary age.

Major aquifers are found within the large buried valleys which occur in southern Saskatchewan (Fig. 12.9). These valleys were formed prior to or during the first glacial advance. The valleys are commonly filled with interstratified gravel, sand, silt, and clay, which may be up to 50 m thick, and form productive aquifers. The largest of these occurs in Hatfield Valley which traverses the entire province. Part of this major valley aquifer system was studied by Maathuis and Schreiner (1982a,b) and Schreiner and Maathuis (1982). Other important aquifers occur in the Battleford, Tyner, Swift Current, Estevan, and smaller buried valleys.

Blanket aquifers consist of large areas of silt, sand, and gravel which lie on the bedrock surface. These aquifers are located along the flanks of major buried valleys such as Hatfield Valley which is bordered by the Wynyard-Melville and Meacham aquifers (Fig. 12.9).

Intertill aquifers are composed of stratified materials consisting primarily of sands but including silts and gravels. Some of these are of interglacial or interstadial age, such as the Tessier aquifer which forms a significant aquifer in the Saskatoon area (Meneley, 1970; SkwaraWoolf, 1980; Fig. 12.11). Other aquifers are formed of layers or lenses of stratified materials within till deposits but these commonly are discontinuous, less extensive, and of lesser importance in terms of water supplies than the intertill aquifers.

In addition to the buried aquifers, surficial aquifers (i.e., aquifers that directly underlie the surface) comprising glaciofluvial, fluvial, alluvial, and lacustrine sediments provide useful water supplies. The main advantages of these are that they are easily accessible and provide good quality water. Disadvantages are that they are easily contaminated, commonly do not produce large quantities of water, and are susceptible to drought.

Delineation of aquifer stratigraphy

Location and evaluation of Quaternary aquifers require delineating the Quaternary stratigraphic framework and understanding of the processes responsible for formation of the aquifer. In Saskatchewan, the stratigraphic framework

Figure 12.10. Schematic illustration of the stratigraphic relationship of the main aquifers and aquitards in southern Saskatchewan.

Lithology	Unit
Sand, silt and clay	SURFICIAL AQUIFER
till	AQUITARD
sand and gravel	INTERTILL AQUIFER
till	AQUITARD
sand and gravel	BURIED VALLEY / BLANKET AQUIFER (BEDROCK SURFACE)
sand and silt	BEDROCK AQUIFER
silt and clay	AQUITARD
sand and silt	BEDROCK AQUIFER
silt and clay	AQUITARD

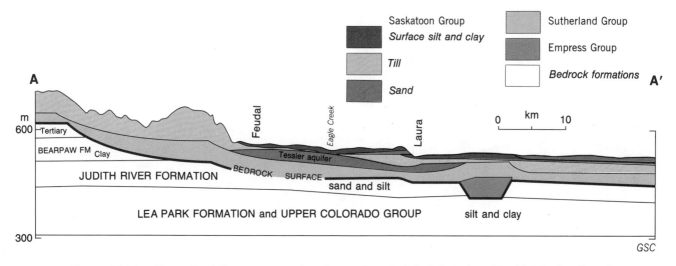

Figure 12.11. Generalized Quaternary stratigraphy southwest of Saskatoon (see Fig. 12.9 for location of cross-section A-A′).

of the Quaternary sediments has been established and sets of criteria have been developed to identify individual stratigraphic units, primarily tills (Christiansen, 1968; Fig. 12.11). This framework, as well as models and processes, have been applied to the delineation of numerous aquifers. Aquifers bear similar relationships to Quaternary geology and bedrock topography in the other Prairie Provinces.

Buried valley aquifers are set into bedrock and generally underlie the Quaternary glacial sequence. The key to defining these is determining the locations and dimensions of the buried valleys and the extent, thickness, and nature of valley fill. It is important to know the stratigraphic succession and continuity of sediments, particularly where a buried valley intersects a bedrock aquifer or where it is overlain by another aquifer, because hydraulic continuity may be developed between aquifers.

Table 12.4. Quaternary stratigraphy of the Sand River area, Alberta

QUATERNARY	
Holocene	
POSTGLACIAL STRATIFIED DEPOSITS	Clay, silt, sand, gravel; undifferentiated eolian, fluvial, and lacustrine deposits
Pleistocene	
GRAND CENTRE FORMATION	
Vilna Member	Clayey diamicton; contains abundant blocks of glacially transported older sediment; very coarse sand fraction is rich in igneous and metamorphic rock fragments; glacial sediment (till)
Kehiwin Lake Member	Sandy diamicton; very coarse sand fraction is rich in igneous, metamorphic, and quartz rock fragments; glacial sediment (till)
Reita Lake Member	Clayey sand diamicton; very coarse sand fraction is rich in igneous and metamorphic rock fragments; glacial sediment (till)
Hilda Lake Member	Clayey diamicton; contains abundant blocks of glacially transported older sediment; very coarse sand fraction is rich in igneous and metamorphic rock fragments; glacial sediment (till)
SAND RIVER FORMATION	Sand and gravelly sand; minor silt and clay; glaciofluvial sediment
MARIE CREEK FORMATION	
Unit 2	Sandy diamicton; very coarse sand fraction rich in carbonate rock fragments; glacial sediment (till)
Unit 1	Clayey diamicton; contains discrete lenses of bedded silt and clay; very coarse sand fraction is rich in carbonate rock fragments; glacial sediment (till)
ETHEL LAKE FORMATION	Silt and clay; minor sand and gravel; predominantly glaciolacustrine sediment
BONNYVILLE FORMATION	
Unit 2	Diamicton; sandy in east two thirds of map area, clayey in west; very coarse sand fraction is rich in quartz fragments; glacial sediment (till)
Unit 1	Clayey diamicton; recognized by very low resistivity response; glacial sediment (till) is overlain by sand and gravel in some places
MURIEL LAKE FORMATION	Sand and gravel; minor silt and clay; glaciofluvial sediment
BRONSON LAKE FORMATION	Clayey diamicton and clay undivided; recognized primarily by very low resistivity response; very coarse sand fraction is rich in quartz and shale bedrock fragments; mixed glacial sediment (till) and glaciolacustrine sediment
EMPRESS FORMATION	
Unit 3	Sand and gravel; contains igneous and metamorphic clasts derived from the Canadian Shield; glaciofluvial sediment
Unit 2	Silt and clay; undivided fluvial and glaciolacustrine sediment
Unit 1	Sand and gravel; contains quartzite and chert clasts derived from the Cordillera; commonly referred to as preglacial Saskatchewan sand and gravel
CRETACEOUS	
BELLY RIVER FORMATION	Grey to greenish grey, thick bedded, feldspath sandstone; grey clayey siltstone, grey and green mudstone; concretionary ironstone beds; nonmarine
LEA PARK FORMATION	Dark grey shale; pale grey glauconitic, silty shale with ironstone concretions; marine

The location and extent of blanket aquifers must be defined, as well as stratigraphy in terms of thickness and relationship to other aquifers, particularly the buried valleys and bedrock aquifers outcropping at the bedrock surface. In many areas the materials of the blanket aquifer may be similar to those of bedrock aquifers, therefore differentiating and defining the two types is commonly based on stratigraphic position.

Quaternary stratigraphy is most important in locating, tracing, and defining intertill aquifers. Correlating till units based on stratigraphic parameters establishes the framework for delineating individual intertill aquifers and provides the basis for determining the interrelationship between various aquifers.

An example where Quaternary stratigraphic data were gathered to solve a major groundwater problem is the Sand River area of east-central Alberta (Andriashek, 1985a, b). This area includes the Cold Lake oil sands deposit. It has been proposed that this heavy oil be exploited by steam injection, a process that requires the availability of large volumes of water and a safe place to dispose of used water. The concerns in this area were that insufficient groundwater was available for both oil sand development and local use and that subsurface disposal of waste water would contaminate community water supplies. Consequently a three-dimensional study of the geometry, thickness, distribution, and lithological properties of Quaternary deposits was undertaken. The results of the study indicate that eight Quaternary formations lie above the heavy-oil deposits (Andriashek, 1985a, b): four till formations (diamictons), likely deposited during four major glaciations; and four formations of stratified sediment, composed of glacial and interglacial deposits (Table 12.4). Two formations, the Empress Group (Whitaker and Christiansen, 1972) and the Muriel Lake Formation, consist of thick, extensive deposits of sand and gravel and are considered to have a high potential for producing significant amounts of groundwater. The Empress Group, at the base of the Quaternary succession, consists of stratified sediment and occupies preglacial valleys (Andriashek and Fenton, 1983; Gold et al., 1983; Table 12.4). The Bronson Lake Formation (till) overlies the Empress Group in many places within the buried valleys. It is very clayey and hence is considered to be an aquitard. The Bronson Lake Formation is overlain by the Muriel Lake Formation, an extensive, generally thick unit of glaciofluvial sand and gravel. The Muriel Lake Formation is in many places coincident with the major buried valleys and also extends out of the valleys. The coarse, granular texture of the Muriel Lake Formation, combined with its extensive distribution makes it a favourable target as a major source of groundwater in the Cold Lake area. The prime value of this study is that careful delineation of Quaternary stratigraphic units has made it possible to define areas where aquitards, which normally separate the major aquifers, have been eroded and thereby allowing hydraulic communication between the aquifers. This information is essential to evaluate the aquifers' potential and possibility of contamination of water within these units. Similar studies should be undertaken in other areas where there is likely to be a major utilization of the aquifers either as a water source or for waste disposal.

Quaternary geology and groundwater quality

Geology influences groundwater quality through the mineralogical composition of geological units and through the geological setting which determines the geological control on groundwater flow. The combination of hydrogeological setting and topography determines the groundwater flow systems, while groundwater chemistry is dependent upon the flow path through the different geological units.

Water in surficial aquifers shows a gradual increase in total dissolved solids with length of travel. Due to the low temperatures, however, and presence mainly of relatively insoluble minerals, such aquifers yield relatively fresh water. Water from these is generally of the calcium bicarbonate type and commonly has a total concentration less than 1000 mg/L. Tills contain relatively higher concentrations of finer grained and more reactive minerals. Consequently a variety of complex, interrelated, geochemical processes involving dissolution of calcite, dolomite, gypsum, and anhydrite, and cation exchange with clays occurs (Freeze and Cherry, 1979). In till waters, sodium, magnesium, calcium, bicarbonate, and sulphate ions commonly occur in major concentrations, and total dissolved solids range from 1000 to 10 000 mg/L. The framework of intertill aquifers consists largely of relatively nonreactive minerals but, because intertill aquifers are generally recharged by downward groundwater flow through tills, the water in intertill aquifers is commonly similar in composition to that in the tills. Where Quaternary aquifers are in hydraulic continuity with older aquifers, groundwater highly charged in dissolved solids may mix with the relatively fresh water of the younger aquifer. Consequently, an understanding of the Quaternary stratigraphic framework and extent of various aquifers is required if variations in water quality are to be understood.

The influence of the geological setting on groundwater quality has been illustrated by Meneley (1970) for the Dalmeny aquifer near Saskatoon. The Dalmeny aquifer is a shallow intertill type occurring at depths of 25 to 45 m below ground surface. It is overlain in part by till and surficial silt and clay, and in part by a relatively thin till layer and a significant surficial sand deposit. The aquifer is recharged by downward movement of groundwater. A distinctive contrast in water chemistry, which can be related to position relative to the overlying materials, can be observed within the aquifer.

Protection of groundwater quality

The handling of waste products from the Cominco Ltd. potash mine, approximately 30 km south of Saskatoon, offers an example of the role of Quaternary geology in containment and disposal of waste material in the Prairie region and the protection of groundwater quality. Wastes from the potash mine are stored at the surface in a waste disposal basin which consists of a brine pond, a decanting pond, and a salt tailings area. Prior to construction of containment facilities, the physical environment of the waste disposal basin and its vicinity was determined (Meneley and Whitaker, 1969). The determination of the physical environment involved the establishment of the geological framework and the determination of hydraulic parameters of relevant geological units. Thus it was possible to identify potential pathways for brine migration and to develop an effective monitoring system for detecting brine movement and evaluating the performance of the waste disposal basin.

The geological setting at the Cominco Ltd. potash waste disposal site is shown schematically in Figure 12.12. Based on this setting the surficial stratified drift and Floral Formation were identified as potential zones for brine migration. There is a potential for seepage through and under the waste disposal basin dykes and for lateral migration of brine within the oxidized zone of the surficial stratified drift towards Rice Lake; however, a clay cut off core at the north edge of the basin, extending through the oxidized surficial stratified drift, and ditches along parts of the basin perimeter make it possible to keep this shallow lateral migration within acceptable limits.

The Floral Formation, which consists of a complex succession of till, silt, clay, sand, and gravel, is considered an aquifer. It was assumed that waste brine would slowly migrate downward through the surficial stratified drift into the Floral Formation aquifer. Due to the low bulk vertical hydraulic conductivity of the surficial stratified drift, it is estimated that downward migration of brine is of the order of several centimetres a year. At this rate it would take several hundreds of years for contamination to reach the Floral Formation aquifer. It is possible, however, that due to dispersion, and possible fracturing, the migration time may be much shorter. Once the brine has reached the Floral Formation aquifer, the principle of hydrodynamic containment can be applied to keep the contamination of this aquifer within acceptable limits.

If the brine reached the Floral Formation aquifer it would possibly continue to migrate downward from this aquifer, through the Sutherland Group, towards the Tyner Valley aquifer. The Sutherland Group consists of a hard, dense, till and presumably has a very low bulk vertical hydraulic conductivity, so that flow through this unit would be slow. In addition, an upward hydraulic gradient exists between the Tyner Valley aquifer and Floral Formation aquifer. The combination of the existing upward hydraulic gradient and low hydraulic conductivity of the Sutherland Group hence would inhibit downward migration of brine from the Floral Formation aquifer.

Similar analyses of Quaternary deposits and of groundwater flow systems should be conducted in the vicinity of all waste disposal sites and sites where hazardous materials that could potentially contaminate groundwater are stored.

Stability of open pit mines

The geology of Quaternary deposits has an important impact on the extraction of resources from older rocks. An example of this impact is the stability of pit walls. Glacial thrusting and deformation of overridden materials has occurred in many parts of the Prairies underlain by Mesozoic bedrock (Slater, 1927; Byers, 1959; Kupsch, 1962; Christiansen and Whitaker, 1976; Sauer, 1978; Fenton and Andriashek, 1978; Babcock et al., 1978; Moran et al., 1980; Fenton, 1983) and exert a strong control on the stability of cut slopes.

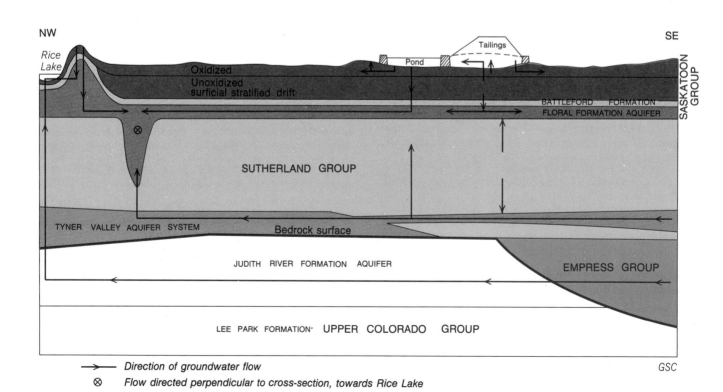

Figure 12.12. Schematic illustration of groundwater flow patterns at Cominco Ltd. mine site, 30 km south of Saskatoon.

Maintaining stability of highwalls is one of the important challenges facing the designer of surface coal mines in the plains of Alberta. In addition to the risks posed to the safety of men and equipment by highwall failures, costs may be incurred in the rehandling of overburden materials, disruption of mining schedules, and the outright loss of potentially recoverable coal. In at least three coal mines currently operating in Alberta, significant highwall failure problems are related to disturbance of the bedrock overburden caused by glaciers (Fenton et al., 1984).

Where glaciotectonic disturbances can be recognized in the planning stages, pit orientation and local changes in steepness of highwalls can be utilized to minimize costs from potential failures. Many, although by no means all, of these features can be identified by their characteristic landform. Others can be detected only by coring or geophysical methods.

An open pit coal mine in the Wabamun Lake area west of Edmonton provides an excellent example of the influence of glacial thrusting of bedrock upon pit wall stability (M.M. Fenton, Alberta Research Council, personal communication, 1983). Bedrock at the mine site includes a six seam coal unit, overlain by sandstone, siltstone, mudstone, and shale of the Paskapoo Formation of Cretaceous age (Green, 1972). Bedrock is overlain by till with local pockets of glacially thrust bedrock. During the last glaciation, ice in this area moved generally southward. The major results of glacial thrusting at the mine site are: (1) local removal of the overburden and, in a few places, the coal seams; (2) deformation of the coal and overburden with little subsequent downglacier transport of the thrust debris; and (3) importation of masses of thrust sediment from an unknown distance upglacier.

A detailed study (M.M. Fenton, personal communication, 1983) was undertaken of a pit wall failure that occurred in Pit 03 of the Wabamun Lake mine. This pit is oriented roughly east-west with the highwall on the south side, the downglacier side, of the pit. The failure began in the eastern part of this highwall in spring and by August had extended westward, almost continuously, over a distance of about 900 m.

Five geological units are recognized in a cross-section south of the pit (Fig. 12.13). Folding predominates in fine grained sediments within the deformed units (1 and 3), and shear planes or faults are present in all sediment types but are particularly extensive in sandstone. Axial planes of folds and surfaces of faults and shears strike east-west and dip northward into the pit thereby contributing to instability of the headwall. Measurement of groundwater levels and piezometric pressure revealed that the formations overlying the relatively impermeable coal seams were essentially saturated and that high pore water pressures existed at the toe of the pit slope. Therefore, the geological data, together with the groundwater data, suggest that the highwall failed because of a combination of high pore water pressure and the presence of a system of inherent planes of weakness related to glacial thrusting. If the presence of the northward dipping planes of weakness had been known before work began, the highwall might have been oriented so that the shear planes did not dip towards the bottom of the cut.

Geological hazards

In comparison with other regions of Canada, the risk of hazards of a geological nature is relatively low in the Prairies. For example, the region lies within zone 0 of the 1970 seismic zoning map of Canada (Heidebrecht et al., 1983). This zonation indicates that the seismic risk to buildings and other engineered structures in contact with the ground is negligible. The topography and geology of the region are such, however, that in some localities both landslides and floods constitute significant hazards; in other localities collapse structures associated with soluble rock in the subsurface are known to occur.

Landslides

In terms of the variety of materials involved and disturbing agents, a multitude of combinations exist that can cause landslides. The causes can be grouped into two main categories: external and internal (Terzaghi, 1950). External causes, such as erosion at the toe of a slope or steepening of the slope by natural or man-made forces, lead to an increase in shear stresses without change in shear resistance of the slope material. Internal causes are those that lead to a decrease in shear resistance of the slope material. Such a decrease is most commonly brought about by an increase in pore water pressure or by a progressive decrease in cohesion of slope material.

The most common locus for slope failures in the Prairie region is along the valleys of large present-day rivers, such as the Peace, the North and South Saskatchewan, and the Red, and along the slopes of proglacial meltwater channels. The geological settings for slope failures can be grouped into three broad categories: (1) clay shale of Cretaceous age, which may be locally disturbed by glacial tectonics (Scott and Brooker, 1968; Thomson and Morgenstern, 1977; Sauer, 1978; Krahn et al., 1979); (2) discontinuities at stratigraphic contacts within the drift (Sauer, 1974, 1979); and (3) surficial stratified drift overlying till deposits (Haug et al., 1977; Lidgren and Sauer, 1982). Each category can be subdivided for analytical purposes on the basis of stratigraphy, structure, geological history, and historical relationships of river dynamics and groundwater regimes. The dynamic factors

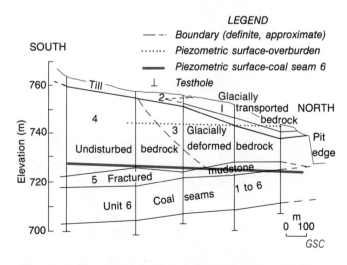

Figure 12.13. Wabamun Lake mine pit: north-south cross-section showing geology and piezometric surfaces in glacially deformed overburden and coal seam 6.

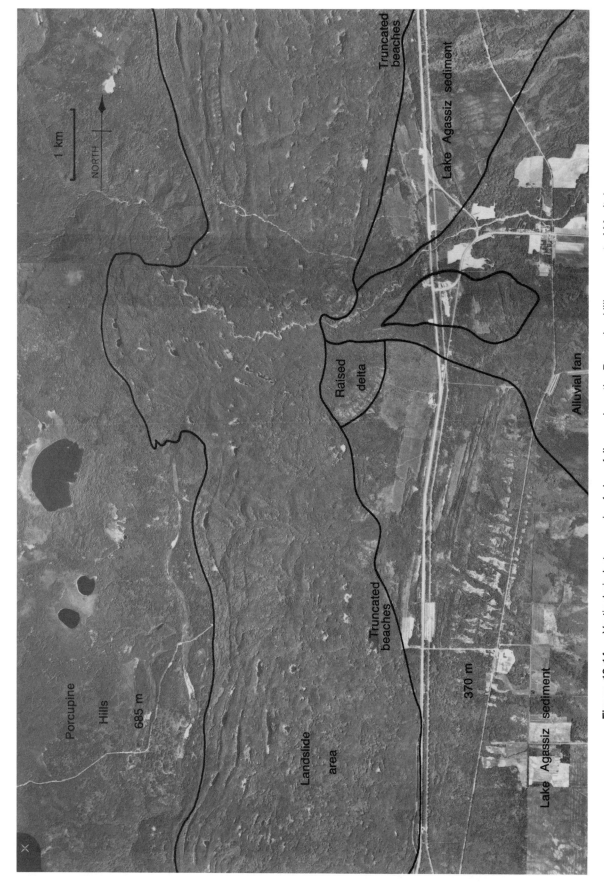

Figure 12.14. Vertical airphoto mosaic of slope failures along the Porcupine Hills segment of Manitoba escarpment. NAPL A24683-189, 191

responsible for many of these landslides are still active today and consequently most landslides areas are presently active (Christiansen, 1983; Sauer, 1983).

One of the most extensive areas of landslides within the region, and perhaps anywhere in Canada, occurs along the Manitoba Escarpment which rises in places 300 m in 10 km. Evidence of slumping is present along much of the escarpment but it is best displayed along the east and northeast side of Porcupine Hills (Wickenden, 1945; Scott and Brooker, 1968; Fig. 12.14). The disturbed zone is estimated to be 200 km^2 and appears to consist of four separate landslides. Individual landslides are typically 5 km wide across the slip planes and 10 km parallel to them. The upper one-third to one-half of each landslide consists of a series of parallel grabens, commonly containing long narrow lakes. The lower parts of the slides are hummocky and comprise mixed black Cretaceous shale and Pleistocene sediments. Cretaceous shales of the Favel and Vermilion River Formations, both containing layers of bentonite, occur along the base of the escarpment. Springs and seepage zones were observed in the toe area of minor slope failures that had occurred along valley walls of creeks that drain the northern part of the slide area (Scott and Brooker, 1968). The age and specific cause of the landslides are not known. The toe of the failure truncates the lower Campbell Beach of glacial Lake Agassiz, however, and thus a maximum age of 9.3 ka can be assigned to the failure. It is possible that the presence of bentonite seams and groundwater discharge may have contributed to instability. In addition, drawdown of water levels associated with the early Holocene draining of Lake Agassiz, may have been a significant triggering factor.

Flood hazards

Spring snowmelt is the main source of runoff and erosion in the Prairie region except for local, intense thunderstorms during summer months. Flooding is not a general problem because of the topographic configuration largely developed in response to Quaternary processes and events. Most large rivers are deeply incised and many areas are dotted with undrained sloughs which maximize groundwater recharge and minimize overland runoff. Risk of flooding, however, occurs in valley bottoms and in the lowlands of Manitoba where Red and Assiniboine rivers drain across the flat-floored Lake Agassiz Basin. Past floods in this basin have caused millions of dollars damage and loss of life. Flood risk potential is greatest during the spring and depends upon winter snow accumulation, rate of snowmelt, and precipitation during the snowmelt period. Governments have spent millions of dollars on the construction and maintenance of dykes, dams, and flood control programs.

The most impressive flood diversion project to date has been the Red River Floodway. It was excavated to carry floodwater around the city of Winnipeg and back into Red River near Lockport. The channel is 56.7 km long and construction involved moving 764×10^6 m^3 of material. The floodway is designed to carry a flow of 850 m^3/s which should protect Winnipeg from floods with a recurrence frequency of 1 in 160 years (Table 12.5). Since the beginning of operational use of the floodway in 1968 flood damage to the City of Winnipeg has been eliminated.

The Lake Agassiz plain has been further protected by a floodway between Assiniboine River and Lake Manitoba. This protects the city of Portage La Prairie, several villages, and a 90 km strip of flat farmland between Portage La Prairie and Winnipeg.

The Federal and Provincial Governments, in 1975, initiated a flood damage reduction program aimed at identifying flood risk areas and discouraging further development in areas subjected to periodic flooding. The program involves mapping of flood risk areas, flood forecasting systems, land use planning, works to control flows, and acquisition of property or easements to control further development in flood risk areas.

Solution collapse

Solution collapse is a potential hazard over wide areas of the Prairies that are underlain by evaporites. For example, southern Saskatchewan is underlain by about 250 000 km^2 of Devonian salt which is up to 200 m thick (Pearson, 1963). Dissolution of these salt beds has been a continuous process since their deposition, resulting in the formation of collapsed structures in overlying beds (Gendzwill and Hajnal, 1971). Crater Lake in southeastern Saskatchewan, the youngest known collapse structure, dates possibly at 13.6 ka (Christiansen, 1971). The lack of abundant collapse structures in the present surface suggests that this process does not pose a significant hazard.

GEOLOGICAL HAZARDS IN CENTRAL AND EASTERN CANADA

Jacques Locat and Jean-Yves Chagnon

Quaternary geology finds similar applications to civil engineering, mining, and hydrology in Ontario, Quebec, and the Atlantic Provinces as it does in the Prairies. Differing physiography and Quaternary histories, especially during the Late Wisconsinan glaciation, however, make for divergent geological hazards between these two regions. It is the management of these hazards that is emphasized in the following discussion of Quaternary geology and planning in eastern Canada.

A geological hazard is defined as a possible event or unpredicted geological condition that is unwanted or undesired. It excludes such things as foundation settlement problems, disposal of toxic wastes, or groundwater supply problems. Natural hazards encountered in Ontario, Quebec, and the Atlantic Provinces are predominantly related to crustal deformation, mass movement, flooding, and erosion. Submarine processes, such as iceberg scour and coastal erosion, are considered in Volume 2 of the Geology of Canada (Keen and Williams, 1989).

Pertinent Quaternary events

In eastern Canada, many geological hazards are directly related to the Quaternary geological history of the area. The preglacial physiography and the processes of glaciation have

---

Locat, J. and Chagnon, J-Y.
1989: Geological hazards in central and eastern Canada; in Chapter 12 of Quaternary Geology of Canada and Greenland, R.J. Fulton (ed.); Geological Survey of Canada, Geology of Canada, no. 1 (also Geological Society of America, The Geology of North America, v. K-1).

controlled the emplacement of the Quaternary deposits and these play a large role in determining the location of many hazards.

The downwarping of the crust, caused by the weight of the Laurentide Ice Sheet, depressed areas such as St. Lawrence and lower Ottawa valleys below sea level so that upon retreat of the ice they were submerged by the sea. Ice damming and differential uplift during glacier retreat resulted in development of many large glacial lakes (e.g., lakes Iroquois, Algonquin, Ojibway). These basins were the loci of fine grained sedimentation. Crustal unloading initiated isostatic uplift (Quinlan, 1984) as evidenced by the emergence above sea level of most coastal areas of eastern Canada, and St. Lawrence and lower Ottawa valleys. Isostatic uplift of lake and marine basins led to stream incision of fine grained marine and lacustrine sediments, which in turn has triggered landsliding (Locat, 1977). Even where these deposits have not been incised, they remain metastable and create problems for the engineering of buildings and road foundations (Smalley, 1979; Quigley, 1980; Torrance, 1983; Locat et al., 1984).

Vertical movement has also caused faulting of Quaternary sediments and a reactivation of ancient vertical faults (Adams, 1981). In some areas, such as the Hudson Bay Lowland (Hardy, 1976; Andrews and Peltier, 1989) crustal uplift continues today.

Tectonic activity and seismicity

Earthquakes are frequent in eastern Canada (Leblanc, 1981; Basham and Adams, 1984). Major ones were noted as early as 1534-1535 (Smith, 1962) and tremors of greater than magnitude 7 on the Richter scale have occurred periodically during the last 300 years. A well known example is the Grand Banks Earthquake of 1929 (Basham and Adams, 1982; Keen et al., 1989) which triggered an enormous landslide on the continental slope off Newfoundland. Table 12.6 lists recent major earthquakes in eastern Canada and Figure 12.15 shows locations of seismic activity. Additional information on zones of intense earthquake activity such as

Table 12.5. Maximum discharge and return period for the fifteen greatest floods on record on Red River at Winnipeg

Date of maximum discharge	Estimated maximum discharge at Redwood Bridge (1000 m^3/s)	Probable return period (years)
1826 May 21	6.37 E	667
1852 May 21	4.67 E	147
1861 May 8	3.54 E	45
1979 May 10	3.00 *	26
1950 May 19	2.93	23
1974 April 24	2.93 *	23
1966 April 14	2.50	14
1970 May 3	2.26 *	10
1882 May 3	2.26	10
1969 May 1	2.17 *	9
1916 April 22	2.02	8
1948 April 30	1.95	7
1956 April 27	1.95	7
1904 April 24	1.87	6
1897 April 27	1.83	

E - Estimated
* computed natural flow without existing control works

Table 12.6. Major earthquakes of eastern Canada

Date	Epicentre	Intensity	Magnitude	Remarks
1534-1535	Les Éboulements	IX-X		
June 1638	St. Lawrence Valley	IX		Near mouth of Saguenay River
Feb. 1663	La Malbaie	X		Greatest ground shaking episode recorded in Quebec
Sept. 1732	Montréal	VIII	5.6-6.0	Resulted in one death and damage or destruction of 300 homes
Dec. 1791	Baie-St-Paul	VIII		
Oct. 1860	Rivière-Ouelle	VIII-IX		Vibrations felt over an area of 1.8 x 10^6 km^2
Oct. 1870	Baie-St-Paul	IX		
28 Feb. 1925	La Malbaie	IX		Caused considerable damage in Québec, Trois-Rivières, and Shawinigan
18 Nov. 1929	Grand Banks		7	
9 Jan. 1982	Miramichi		5.7	

the Mount Tremblant and Miramichi areas are available in Horner et al. (1979) and Basham and Adams (1984), respectively. These earthquakes are expressions of tectonic activity believed to be partly related to plate tectonic movements and possibly also to vertical uplift resulting from the stress relief caused by the melting of the last ice sheet. This strain energy appears to be dissipating in zones of crustal weakness. For example, the Charlevoix area of Quebec was the locus of a meteorite impact 350 Ma ago (Rondot, 1968; Roy and Duberger, 1983). The strain energy imparted by this impact is coupled with the regional tectonic stress to induce in situ principal stress ratios greater than 1 ($\sigma h/\sigma v$, $v<1$) in many areas. These deviatoric stresses can exceed the shear strength of the rock, resulting in popouts (White et al., 1974; Durand and Ballivy, 1974). Compilations of seismicity and neotectonism in eastern Canada have been made by Adams (1981) and Chagnon and Locat (1984), and a new compilation is being prepared by Adams for the neotectonics volume of the Geological Society of America's Geology of North America series.

Mass movement

Many landslides have occurred and will continue to occur in the fine grained sediments of eastern Canada. The most recent major ones are listed in Table 12.7. Loss of life has resulted in many cases from these mass movements. In the last two decades major studies were undertaken to shed more light on the causes of landslides and to attempt some measure of prediction (Chagnon, 1968; Tavenas et al., 1971; Mitchell and Markell, 1974; Quigley, 1980; Lebuis et al., 1983; Locat et al., 1984; Lefebvre, 1984). Natural mass movements are not entirely restricted to Quaternary sediments but also occur locally in rock materials (Dionne, 1969a).

Landslides in soft clay

Landslides of more than one type occur in soft sediments (Fig. 12.16-12.18). Depending on soil properties, deposit stratigraphy, and underlying topography, failures can be a simple block rotation or a large earthflow. Stability analysis of the first rupture can be carried out easily as a routine measure (Lefebvre, 1981), but much remains to be done in the prediction of landslides in time and space (Locat et al., 1984). Mapping techniques for areas prone to landslide failure have been developed for Ontario (Klugman and Chung, 1976) and Quebec (Lebuis et al., 1983). Carson and Bovis (1989) also discuss flowslides in marine clay in another part of this volume.

The major landslides of eastern Canada are found in the so-called sensitive clays located primarily in St. Lawrence, lower Ottawa, and adjacent tributary valleys.

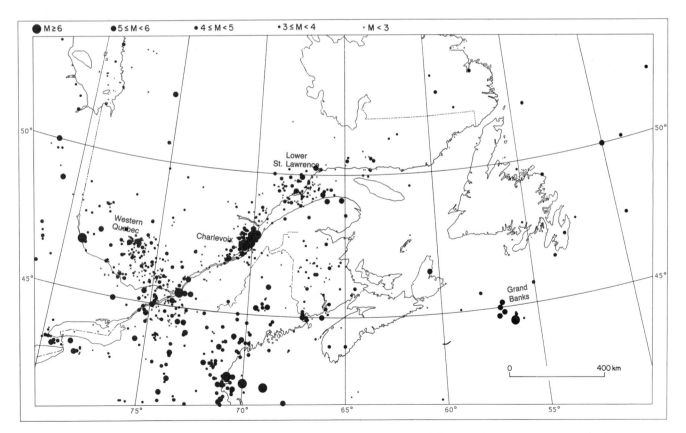

Figure 12.15. Locations of earthquake epicentres and magnitude of earthquakes in eastern Canada and adjacent United States (from Geophysics Division, Geological Survey of Canada).

Quigley (1980) and Torrance (1983) gave a full account of the behaviour of these clays. Torrance (1974), Locat (1982), and Locat and Lefebvre (1986) discussed the role that salts and their leaching play in promoting sensitivity in marine clays. The most sensitive soils are the marine clayey silts (muds) flanking the Canadian Shield. These were deposited from glacial meltwater in brackish waters at a high sedimentation rate. Consequently, these soils are commonly stratified with alternate layers of sandy silts and clayey silts. Stratification aids the leaching process (Donovan and Lajoie, 1979; Locat, 1982) in addition to providing potential zones of high pore pressures. Lowering of the water table by incision related to the glacial isostatic uplift of these areas enhanced leaching of the salts and hence a significant reduction in the liquidity index (or remoulded strength). Mitchell and Klugman (1979) have recently reviewed many aspects of mass instabilities in sensitive clays.

Rock slides and rock avalanches

Rock slides or avalanches are much less frequent than other types of landslides in eastern Canada and are restricted to areas of high relief. An example of a failure of this type occurred near Saint-Fabien-sur-Mer, about 200 km east of Québec, in 1967 (Dionne, 1969a). Many small areas of rock

Table 12.7. Major landslides of eastern Canada

Date	Place	Area or volume	Victims
April 1840	Maskinongé (Co. Maskinongé, Que.)	35 ha	
27 April 1894	St-Alban (Co. Portneuf, Que.)	650 ha	4
Sept. 1895	St-Luc de Vincennes (Co. Champlain, Que.)	2 ha	5
May 1898	St-Thuribe, Riv. Blanche (Co. Portneuf, Que.)	35 ha	1
6 April 1908	N.D.-de-la-Salette (Co. Papineau, Que.)	40 ha	33
April 1925	Portneuf (Co. Portneuf, Que.)	1 ha	
24 July 1935	St-Vallier (Co. Bellechasse, Que.)	6 ha	
Sept. 1938	Ste-Geneviève de Batiscan (Co. Champlain, Que.)	6 ha	
18 May 1945	St-Louis (Co. Richelieu, Que.)	4.5 ha	
12 Nov. 1955	Nicolet (Co. Nicolet, Que.)	190×10^3 m^3	3
23 May 1962	Riv. Toulnustouc (Co. Saguenay, Que.)	3.8×10^6 m^3	8
10 May 1963 / 11 Dec. 1963	St-Joachim-de-Tourelle (Co. Rimouski, Que.)	6.9×10^3 m^3	
13 June 1964	Desbiens (Co. Lac St-Jean, Que.)	24.5×10^3 m^3	4
15 April 1969	Louiseville (Co. Maskinongé, Que.)	?	
4 May 1971	St-Jean-Vianney (Co. Chicoutimi, Que.)	32 ha / 76×10^9 m^3	31
12 May 1972	South Nation River (Ontario)	?	

instability are found along highways of the Gaspésie (A. Drolet, Ministère des Transports du Québec, personal communication, 1979) and parts of the Maritime Provinces. In addition, local rock avalanches have been a persistent and serious problem. For example, between 1836 and 1889 below the scarp of Cap Diamant, in Québec, more than 100 persons were killed in rock avalanches (Chagnon et al., 1979). No general survey, however, has been made of rock slides and avalanches in eastern Canada.

Erosion and flood hazards

Hazards created by the action of both wind and water are included here. Stream and river flooding occurs particularly during spring snowmelt; wind erosion is a problem in some areas (on Iles-de-la-Madeleine, winds of more than 100 km/h are common); wind can also cause flooding along lake or sea shores where the fetch is long enough to permit buildup of storm tides; and, of course, wind and water combine to cause wave erosion in coastal areas.

Erosion and sedimentation

Shoreline erosion is a common problem in areas of Quaternary sediments throughout the Great Lakes-St. Lawrence system. Land loss due to this process is a particularly severe problem along the shores of Lake Erie (Gelinas, 1974), but it is also a persistant problem throughout the Great Lakes basin (Rukavina, 1976, 1978, 1982). Rivers and streams downcutting in fine grained sediments, and waves and floating ice action erode shore areas, causing bank instabilities (e.g., at the delta of the St. Lawrence near Sorel) in wider reaches of St. Lawrence River (Dionne, 1969b, 1970; Troude et al., 1983). Marine coastal areas exposed to the

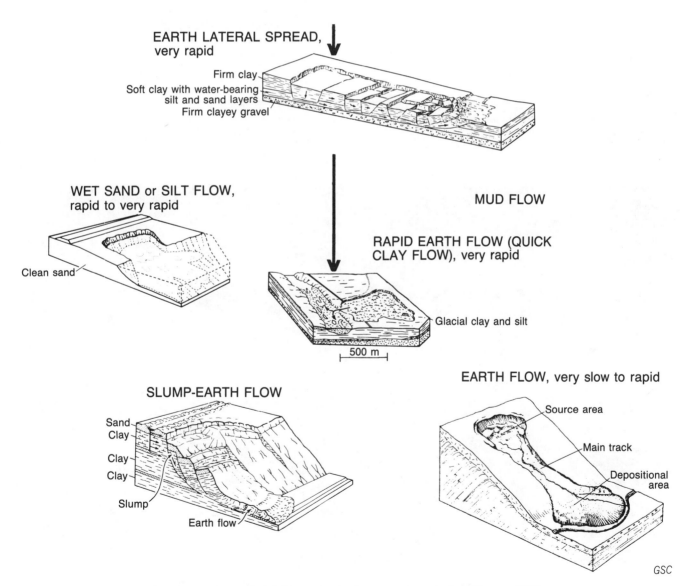

Figure 12.16. Types of landslides occurring in eastern Canada (from Varnes, 1978).

combination of high winds and longshore currents, such as Îles-de-la-Madeleine, Prince Edward Island, and parts of Nova Scotia, suffer extensive erosion and coastal retreat (Owens and Drapeau, 1973). Additional information on coastal erosion, in both marine and fresh water environments, is available in three conference reports (National Research Council, 1979, 1982; McCann, 1980) and is discussed further in a separate volume (no. 2) of this series (Keen and Williams, 1989). In a few areas intense erosion has caused rapid sedimentation in adjacent areas (Greenwood and Davidson-Arnott, 1979; Greenwood and Keay, 1979); for example, at Rivière Pentecôte near Baie-Comeau the intervention of man along the river course has resulted in the siltation of the entire bay which was formerly a deep water port. The material involved is a loose deltaic sand which is common along the north shore of the St. Lawrence estuary and gulf (Dredge and Thom, 1976).

Floods

In local areas of eastern Canada the risk of flood hazard is high. Such areas are subject to periodic torrential rains (occasionally associated with dying hurricanes) and heavy spring rains and snowmelt which commonly occur before breakup of river ice. Floods caused by ice jams on St. Lawrence River during snowmelt are of common occurrence and similar problems plague low-lying areas adjacent to many other rivers and streams. Shipping on the St. Lawrence Seaway helps maintain an open channel and has in part alleviated ice jam problems on St. Lawrence River. Little can be done in other areas prone to this type of flooding, however, other than to control development on floodplains and to attempt to blast ice jams loose when they occur. Small conservation reservoirs built on many streams throughout southern Ontario generally control "normal" flooding but are of little effect during a catastrophic storm, such as that which resulted in 78 deaths through flooding of Holland Marsh north of Toronto in October 1954.

Karst

Many areas of eastern Canada are underlain by limestone and several areas are underlain by evaporites in which solution features can be or have been developed (Roberge, 1977). Karst development is well known on Îles-de-la-Madeleine and in some parts of Nova Scotia and Newfoundland. These underground openings, which are difficult to locate, pose a problem to urban and road development. In the Québec city area alone many kilometres of underground passages can be traced, especially in the Boischatel area (Société québécoise de spéléologie, 1980). At Îles-de-la-Madeleine the dissolution of gypsum is observed at the surface by the presence of depressions or dolines. Caverns are also found in parts of Ontario (Cowell and Ford, 1980). Karst features locally pose engineering problems but as active collapse is not occurring today, karst processes do not appear to constitute a significant hazard.

Figure 12.17. The South Nation River Landslide, May 12, 1972. Courtesy of C. Malcolm Photography, Cornwall.

Figure 12.18. The Moisie River gully flow (Dredge and Thom, 1976). Courtesy of the Iron Ore Company of Canada.

QUATERNARY GEOLOGY AND URBAN PLANNING IN CANADA

Owen L. White

In most Canadian cities the nature and distribution of Quaternary deposits, the topographic grain imposed by Quaternary glaciation and fluvial erosion, and the nature and pattern of ongoing processes control the location of transportation routes, play a role in zoning decisions, determine engineering design and play a major role in determining construction costs. Consequently, for effective planning and efficient and successful development of Canada's cities, Quaternary geological information is essential. Few Canadian urban areas completely lack geological information, although for some areas little information is available. The geological reports and maps that are available for most cities do, however, provide a basic framework from which regional planning and development can be carried out. In many situations the data are not available at a level of detail nor in a format or language that encourages its use by planning authorities.

The value of Quaternary geology information in the planning process is best illustrated by the consequences of its nonuse. For example, an area of hummocky terrain near

---

White, O.L.
1989: Quaternary geology and urban planning in Canada; in Chapter 12 of Quaternary Geology of Canada and Greenland, R.J. Fulton (ed.); Geological Survey of Canada, Geology of Canada, no. 1 (also Geological Society of America, The Geology of North America, v. K-1).

Kitchener, Ontario, was "prepared" for development by modifying the hummocks and filling in the depressions. A residential subdivision was constructed on the new, gently rolling surface with no apparent attention to the variability of the underlying materials, despite the availability of geological maps and aerial photographs. Some of the single and multiple residential units were located on lots underlain by thick organic deposits which had formed in the bottoms of the kettle holes on the original hummocky terrain but which had been covered by a metre or two of fill. Severe differential settlement problems developed on at least five sites within the subdivision; a small apartment building and a single family residence were restored by underpinning; two single family residences required major reconstruction; and one house was eventually demolished. Builder, engineering consultant, mortgagee, and the city all denied responsibility so repair costs and other losses were all borne by the purchasers (Jankowski and McCarthy, 1974a, b).

The following discussion examines Quaternary geological information available in Canadian urban areas and records examples of the manner in which geological information is used in the urban planning process in Canada.

Quaternary geological information

The prime source of Quaternary information is the surficial materials map which shows the distribution of and describes Quaternary geological units. A knowledgeable person can use these and accompanying reports to determine the distribution of different types of surface materials and to predict depth to bedrock and the succession of Quaternary units in the subsurface. One problem, however, is that in few Canadian urban areas have these maps and reports been specifically tailored to the urban developer or planner. Consequently, the planner and developer require a good working knowledge of Quaternary geology before they can obtain maximum benefit from these documents.

Quaternary geological maps in Canadian urban areas

The Vancouver region is one urban area where basic Quaternary geology maps (scale 1:50 000) and a report describing the regional Quaternary geology and discussing its application in construction and with respect to natural hazards are available (Armstrong, 1980a, b, 1984; Armstrong and Hicock, 1980a, b). One specific objective of Armstrong (1984) is to encourage more and better use of geological information by engineers, architects, planners, and contractors as well as the public. In an attempt to further this end, minimum use is made of geological jargon and units are described in terms of engineering properties and the nature of subsurface conditions is described.

In the Toronto area, a number of maps and reports are available. One of the earliest regional Quaternary maps and reports was that of Coleman (1932). In 1980, the Ontario Geological Survey published a 1:100 000 scale compilation of existing map information of the Toronto area supplemented by stratigraphic data obtained from cross-sections along the north-south and east-west Toronto subway routes (Sharpe, 1980). The notes with the map provide a brief description of the Quaternary geology of the area and an extensive bibliography on the local geology. Although this map was prepared with the nongeological user in mind and White (1982) briefly discussed the effect of Toronto geology on underground construction in the city, a description and discussion of the Toronto area geology written for the non-geologist engineer or planner is not available.

A map of the surficial deposits (scale 1:50 000), together with charts and maps showing the engineering characteristics of the geological units and the thickness of the drift deposits, was published for Montréal Island by the Geological Survey of Canada (Prest and Hode-Keyser, 1977). This followed an extensive period of co-operative field and laboratory studies between the City of Montréal and the Federal Government and provides a sound geological framework for urban planning and development.

Even where 1:50 000 scale surficial geology maps are not available, many urban areas are covered by regional maps (i.e., Chapman and Putnam, 1966, scale 1:250 000), which provide a useful guide to Quaternary deposits, or by terrain evaluation maps and reports, prepared with use by planners and other nongeologists in mind, and accompanied by manuals explaining the use of the maps (e.g., Gartner et al., 1981 for Ontario; Ryder and Howes, 1984, for British Columbia).

Agriculture soil survey maps

Soil maps provide data which are of significance in the urban planning and development process. In many areas of Canada, soil survey maps and reports were available many years before a similar coverage of Quaternary geology maps was available. As a result, soil survey maps have been used by planners to provide data on surficial materials.

The City of Edmonton planning staff used soil survey information in 1967 when preparing a review of an area adjoining the city which was subsequently annexed. This report (Bowser et al., 1962) was used even though a surficial geology map published in the same year was available (Bayrock and Hughes, 1962). An interesting question that should be answered if Quaternary geologists want planners to make greater use of their maps is why the planners chose to use the 1:126 720 scale soil map rather than the 1:63 360 scale surface geology map. Perhaps the planners found it easier to translate soil units to surface materials units that they understood than to make the transformation from geological units, or possibly the city chose to use the soil map because it was in colour whereas the geological map was not.

Folios

In some urban areas, folios or collections of maps have been prepared specifically for the use of planners. The information provided in these reports is not meant to replace the need for detailed site investigations but is intended to provide a basis for regional development and to assist in planning more detailed site investigations. Saskatoon, Edmonton, Winnipeg, and Ottawa have such folios.

The Saskatoon folio includes not only maps, cross-sections and discussions of the Quaternary geology and soils, but also includes information on bedrock and climate (Christiansen, 1970b). The folio was prepared "to provide a basic physical framework within which city planning and engineering design can be done more effectively". The surficial geology is presented both at a regional scale (1:250 000) and in more detail (1:63 360) in the central core of the urban area. The physical properties of the subsurface materials are described and related to slope stability problems, the occurrence of groundwater and buried aggregate resources. The excellent Saskatoon example has not been repeated in the same format for any other Canadian urban area.

The Edmonton folio consists of a report with 22 accompanying maps and charts (Kathol and McPherson, 1975). The report describes both the Quaternary and bedrock geology as well as climate, groundwater, and sand and gravel deposits, and shows the application of geology to land use planning by rating the Edmonton terrain in terms of general construction conditions, solid waste disposal sites, slope stability, and deep sewer construction.

The Winnipeg folio (Kjartanson, 1983) is similar in format to the Edmonton report and makes extensive use of geotechnical and water well drill logs to provide the subsurface data. It is somewhat more geotechnical than geological, compared with those for Edmonton and Saskatoon, and is directed more towards engineers rather than planners. Nevertheless it remains, as do the others, a presentation of the surficial materials of the area in terms that should be readily handled by planners and other nongeologists in planning both for regional and city development.

The geoscience information package for the Ottawa region was prepared as an example of the type of information that should be available for urban areas (Bélanger and Harrison, 1980). This publication of the Geological Survey of Canada was an outgrowth of a project to set up automated geological data banks in several centres across Canada (Scott, 1973). The report includes maps showing the distribution of surface materials, depth to bedrock, and drift thickness, and includes general descriptions of important geological features and materials and a discussion of methods of presenting data. Comments are made on specific aspects of materials that are significant from a land use and planning point of view, but the report does not provide rating maps which might be used by planners nor does it supply specific geotechnical data.

Quaternary geological data applied to urban planning

With the increased attention paid to the planning process in urban areas over the past two or three decades and a greater sophistication of applied techniques, there has been an increased use of geological data. Much of the credit for this trend must be given to Dr. R.F. Legget of Ottawa who has consistently championed the cause and value of geology in planning as well as in engineering design and construction. There are many examples of municipalities and major land developers using available geological data or acquiring such data for their own specific requirements. In recent years, however, the biggest changes have been seen in the use of geological data in planning the protection of nonrenewable mineral resources such as sand and gravel, in outlining probable groundwater supplies, in identifying potentially hazardous areas, and in understanding the origins and characteristics of the geological environment.

Geological input
An excellent example of the assembly of geological data for use in the preparation of official plans is seen in the Regional Municipality of Peel, immediately west of Metropolitan Toronto. In 1975 the Regional Government published a physical survey of the region (Regional Municipality of Peel, 1975), partly to assess the effect of the rapid urbanization of the region and partly to determine the priorities for the use of natural resources. The report, prepared by the regional staff with input from consultants, paid major attention to surficial and bedrock geology, soils, and groundwater; identified areas underlain by sand and gravel or potentially useful bedrock resources; and delineated floodplains, areas of poor drainage, and areas susceptible to slope stability problems and erosion, which require special protective attention. Although the regional official plan has not yet been published, the physical survey has, over the years, provided background information for (1) the preparation of the official plans for the three constituent municipalities; (2) the review of specific development sites and of special study areas; and (3) identification of areas where insufficient background data were available and where extra studies were needed to upgrade the data base.

Geological input into the planning process is, of course, essential if the use of sand and gravel deposits is to be planned and managed. Examples of inventories are: in Alberta, Edwards and Hudson (1983), in Newfoundland, Vanderveer (1983), and in Ontario, Scott (1982). The objectives of these inventories are similar: to record the location, quality, and quantity of the sand and gravel deposits and to encourage the wise management of the resource. The inventories are useful in areas where there is a shortage of good quality material, where a shortage could occur in the foreseeable future, or where urban development threatens to pre-empt access to an otherwise exploitable aggregate deposit. In Ontario, where provincial agencies undertake to provide the necessary geological information, the official plan of the municipality should provide for the protection of adequate aggregate resources within the municipality and should "provide a clear and reasonable mechanism to permit the establishment or expansion of pits and quarries...." (Ontario Government, 1986).

Hydrogeological input
Hydrogeological input is significant where urban areas are dependent upon groundwater. These areas require a thorough understanding of the local flow system and all activities in the recharge areas must be monitored. Two examples where this has occurred are at Guelph and the Regional Municipality of Waterloo in Ontario.

At Guelph, the biggest producing wells for the water supply of the city are located at Arkell in a glacial outwash plain in front of the Paris Moraine (Hore, 1975). The city owns the 405 ha of land in the vicinity of the wells and the recharge area, but much of the recharge area lies in another municipality — rural Puslinch Township. On one occasion, when part of the recharge area was offered for sale as a possible sand and gravel extraction site, the city intervened and purchased it.

The Regional Municipality of Waterloo supplies groundwater for the urban communities of Kitchener, Waterloo, and Cambridge. No specific measures are taken to protect the recharge areas as these are all generally located in rural parts of the Municipality and are not under developmental pressure. A depleted gravel pit which was recently acquired for groundwater recharge, however, lies near other gravel extraction properties. These will have to be carefully monitored to ensure there is no detrimental effect on the recharge pit.

The protection of groundwater resources from industrial and waste pollution also calls for major geological input into the planning process and regulatory control. Gartner (1984) reported on the role of geological and hydrogeological information in the planning and selecting of the site for a hazardous waste disposal facility to be operated by the Ontario Waste Management Corporation. Other factors such as transportation, agriculture, and atmospheric and socio-economic conditions are involved, but geology and hydrogeology have been major factors through all phases of the selection process from the initial review of 130 000 km^2, to the selection of the eight candidate sites, and to the detailed investigation of the preferred sites (Ontario Waste Management Corporation, 1985).

Hydrology input

Floodplains present a special problem for planners and developers. Not only are such areas prone to periodic inundation, but channel migration and bank erosion require special protective and training measures and thorough understanding of riverine processes. Examples of ill-advised construction and development in river floodplains are common worldwide, and go back many years before geological and hydrological processes were recognized and understood and before the planning process became accepted in urban areas. At present, in many jurisdictions detailed records are available for flood frequencies, discharges, and attendant sediment loads. Local stream processes have been studied and modelled, and floodplain development is well planned, controlled, and locally protected by engineering works (see for example the discussion of the Red River Floodway by Scott in this chapter).

Geological hazard legislation and urban planning

The direct use of geological data in the planning and regulatory process is commonly seen where unstable slopes are involved (see for example the discussion of geological hazards in the Canadian Cordillera by Gardner and Evans in this chapter). Although the failure of a slope may be related to natural causes (earthquakes or river erosion) or man-made causes (excavation at the toe of a slope or the dumping of fill on the face or brink of a slope), regulation of developmental activities in hazardous areas is usually established in order to minimize injury, loss of life, or physical damage.

Legislation concerning slope stability problems in urban areas varies from province to province. Current legislation in Quebec requires regional municipal councils to map areas of unstable slope conditions, to incorporate consideration for unstable slopes in their official plans, and requires a geotechnical engineering report before a building permit can be issued for construction in an unstable area.

Ontario has not yet adopted legislation but discussions are proceeding on regulations to cover all nonflood hazard lands in the planning process (Ontario, Interministerial Working Group, 1983). An example classification of valley slopes which might be incorporated into future legislation is presented in Klugman and Chung (1976) and Poschmann et al. (1983); this six-fold classification relates specifically to sensitive clay areas.

British Columbia has no overall legislation governing use of hazardous land. The provincial Ministry of Transportation and Highways, however, is responsible for the approval of subdivision development outside of organized territory. If they foresee problems in an area under strong development pressure, the Ministry will initiate studies and withhold permission for development in areas deemed hazardous. This procedure was followed near Penticton in an area underlain by thick silt subject to piping (Miller and Nyland, 1981).

Conclusions

This discussion has pointed out that Quaternary geology is important and has a role to play in urban planning. The unanswered question is why do all urban communities not have active urban geology programs. This question has three probable answers:

1. Methodology: Urban geology methodology has been developed through several successful pilot projects. Specific procedures for recording and presenting urban geology data, however, have not been endorsed by a responsible organization. Consequently, a community starting an urban geology program can not join a functioning, well supported urban geology data network.

2. Jurisdiction and resources: Responsibility for urban planning clearly lies with the local level of government. The local level, however, generally lacks the expertise and money to initiate and conduct an urban geology program.

3. Will to initiate: The final answer is a lack of desire, on the part of the local politicians, to become involved in urban geology programs. Studies that do not solve immediate problems or lead directly to generation of revenue are difficult to sell at a political level. Consequently, there is no push from the local responsible level of government to initiate urban geology programs.

Because well designed urban geology data bases generally are not available, geology is often incorporated into urban planning in a haphazard manner and costly errors are often made. If greater use is to be made of geology in urban planning, geological reports must be written and maps prepared in terms and formats which can be appreciated easily by engineers and planners, every opportunity must be taken to use the mass media to impress on the public the value of geology in solving urban planning problems, and courses in geology must be made mandatory for student engineers and planners.

ACKNOWLEDGMENTS

Contributions to the section on *Engineering geology and land use planning in the Prairie region of Canada* were solicited from individuals, as listed below, in Alberta, Saskatchewan, and Manitoba, who responded generously with their knowledge of the importance and significance of Quaternary geology to land use and planning in their respective provinces. Their contributions, by both text and illustrations, are gratefully acknowledged by the compiler who alone bears the responsibility for such omissions and misrepresentations as may have occurred.

Alberta: M.M. Fenton, L.D. Andriashek, and S.R. Moran (Alberta Research Council). Saskatchewan: E.A.

Christiansen (E.A. Christiansen Consulting Ltd., Saskatoon), J.L. Henry (University of Saskatchewan), H. Maathuis, B.T. Schreiner (Saskatchewan Research Council), and E.K. Sauer (University of Saskatchewan). Manitoba: C. Jones, E. Nielsen, and B. Bailey (Manitoba Department of Energy and Mines).

Although the section on *Quaternary geology and urban planning in Canada* bears only one name, the text includes information obtained through many conversations and discussions with colleagues across the country in all sectors of the community. To each and everyone concerned are extend thanks and appreciation. Nevertheless, the compiler takes the responsibility for all statements made. This contribution is published with the approval of the Director, Ontario Geological Survey, Ministry of Northern Development and Mines.

This overall chapter has benefited from the critical review of W.H. Nasmith (Victoria), J.S. Scott, B.R. Pelletier (Geological Survey of Canada), W.H. Mathews (University of British Columbia), and J.B. Bird (McGill University).

REFERENCES

Abele, G.
1974: Bergsturze in den Alpen; Wissenschaftliche Alpenvereinshefte, no. 25, 230 p.

Adams, J.
1981: Postglacial faulting: a literature survey of occurrences in eastern Canada and comparable glaciated areas; Atomic Energy of Canada Limited, Ontario, Technical Report 142, 63 p.

Andrews, J.T. and Peltier, W.R.
1989: Quaternary geodynamics in Canada; Chapter 8 in Quaternary Geology of Canada and Greenland, R.J. Fulton (ed.); Geological Survey of Canada, Geology of Canada, no. 1 (also Geological Society of America, The Geology of North America, v. K-1).

Andriashek, L.D.
1985a: Quaternary stratigraphy Sand River area; unpublished MSc thesis, Department of Geology, University of Alberta, Edmonton, 368 p.
1985b: Quaternary stratigraphy and surficial geology, Sand River area (73 L); Alberta Research Council, OF 1979-6, Fig. 9-10, EM-181, EM-183, scale 1:200 000.

Andriashek, L.D. and Fenton, M.M.
1983: Drift thickness map and buried valley cross-sections; Alberta Geological Survey, Alberta Research Council, Map.

Armstrong, J.E.
1980a: Surficial geology, Mission, British Columbia; Geological Survey of Canada, Map 1485A, scale 1:50 000.
1980b: Surficial geology, Chilliwack, British Columbia; Geological Survey of Canada Map 1487A, scale 1:50 000.
1984: Environmental and engineering applications of the surficial geology of the Fraser Lowland, British Columbia; Geological Survey of Canada, Paper 83-23, 54 p.

Armstrong, J.E. and Hicock, S.R.
1980a: Surficial geology, New Westminister, British Columbia; Geological Survey of Canada, Map 1484A, scale 1:50 000.
1980b: Surficial geology, Vancouver, British Columbia; Geological Survey of Canada, Map 1486A, scale 1:50 000.

Atwater, M.
1968: The Avalanche Hunters; Macrae Smith, Philadelphia, 235 p.

Babcock, E.A., Fenton, M.M., and Andriashek, L.D.
1978: Shear phenomena in ice-thrust gravels, central Alberta; Canadian Journal of Earth Sciences, v. 15, p. 277-283.

Basham, P.W. and Adams, J.
1982: Earthquake hazards to offshore development on the Eastern Canadian Continental Shelves; in Proceedings of the 2nd Canadian Conference on Marine Geotechnical Engineering, Halifax, Nova Scotia, p. 1-5.
1984: The Miramichi New Brunswick earthquakes: near surface thrust faulting in the Northern Appalachians; Geoscience Canada, v. 11, p. 115-121.

Basham, P.W., Weichart, D.H., and Berry, M.J.
1979: Regional assessment of seismic risk in Eastern Canada; Seismological Society of America, Bulletin, v. 69, p. 1567-1602.

Bayrock, L.A. and Hughes, G.M.
1962: Surficial geology of the Edmonton District, Alberta; Research Council of Alberta, Preliminary Report 62-6, 40 p.

Bélanger, J.R. and Harrison, J.E.
1980: Regional geoscience information: Ottawa-Hull; Geological Survey of Canada, Paper 77-11, 18 p.

Berger, T.B.
1973: Reasons for the judgement of the Honorable Mr. Justice Berger; Supreme Court of British Columbia, report.

Blown, I. and Church, M.
1985: Catastrophic lake drainage within the Homathko River basin, British Columbia; Canadian Geotechnical Journal, v. 22, p. 551-563.

Borgman, L.E.
1963: Risk criteria; Journal of the Waterways and Harbors Division, American Society of Civil Engineers, v. 89, p. 1-35.

Bowser, W.E., Kjearsgaard, A.A., Peters, T.W., and Wells, R.E.
1962: Soil survey of Edmonton sheet (83 H); Alberta Soil Survey, Report 21, 66 p.

Byers, A.R.
1959: Deformation of the Whitemud and Eastend Formations near Claybank, Saskatchewan; Royal Society of Canada, Transactions, v. 53, p. 1-11.

Carson, M.A. and Bovis, M.J.
1989: Slope processes; in Chapter 9 of Quaternary Geology of Canada and Greenland, R.J. Fulton (ed.); Geological Survey of Canada, Geology of Canada, no. 1 (also Geological Society of America, The Geology of North America, v. K-1).

Chagnon, J-Y.
1968: Les coulées d'argile dans la province de Québec; Naturaliste Canadien, vol. 95, p. 1327-1343.

Chagnon, J-Y. et Locat, J.
1984: Étude sur les déformations intraformationnelles dans les sédiments récents (Quaternaire); critères de définition des structures produites par la séismicité; Rapport soumis à la Commission de Contrôle de l'Énergie Atomique du Canada, Ottawa, Rapport GLG-84-40, 168 p.

Chagnon, J-Y., Lebuis, J., Allard, D., and Robert, J.-M.
1979: Sensitive clays, unstable slopes, corrective works and slide in the Québec and Shawinigan area; Geological Association of Canada, Field Trip Guides, B-11, Laval University, Québec, 38 p.

Chapman, L.J. and Putnam, D.F.
1966: The Physiography of Southern Ontario; University of Toronto Press, Toronto, 2nd edition, 386 p.; includes 4 maps, scale 1:250 000.

Chaudhry, M.H., Mercer, A.G., and Cass, D.
1983: Modeling of slide-generated waves in a reservoir; Journal of Hydraulic Engineering, v. 109, p. 1505-1520.

Christiansen, E.A.
1968: Pleistocene stratigraphy of the Saskatoon area, Saskatchewan, Canada; Canadian Journal of Earth Sciences, v. 7, p. 1167-1173.
1970a: Geology and groundwater resources of the Wynyard area (72 P), Saskatchewan; Saskatchewan Research Council, Map No. 10.
1970b: Physical environment of Saskatoon, Canada; Saskatchewan Research Council and National Research Council of Canada, Ottawa, 68 p.
1971: Geology of the Crater Lake collapse structure in southeastern Saskatchewan; Canadian Journal of Earth Sciences, v. 8, p. 1505-1513.
1983: The Denholm Landslide, Saskatchewan, Part I: Geology; Canadian Geotechnical Journal, v. 20, p. 197-207.

Christiansen, E.A. and Whitaker, S.H.
1976: Glacial thrusting of drift and bedrock; in Glacial Till; An Interdisciplinary Study, R.F. Legget (ed.); Royal Society of Canada, Special Publication, no. 12, p. 121-130.

Clague, J.J.
1978: Terrain hazards in the Skeena and Kitimat River basins, British Columbia; in Current Research, Part A, Geological Survey of Canada, Paper 78-1A, p. 183-188.

Clague, J.J. and Souther, J.G.
1982: The Dusty Creek landslide on Mount Cayley, British Columbia; Canadian Journal of Earth Sciences, v. 19, p. 524-539.

Clague, J.J., Evans, S.G., and Blown, I.G.
1985: A debris flow triggered by the breaching of a moraine-dammed lake, Klattasine Creek, British Columbia; Canadian Journal of Earth Sciences, v. 22, p. 1492-1502.

Coleman, A.P.
1932: The Pleistocene of the Toronto Region; Ontario Department of Mines, Annual Report 41, Part 7, 69 p.

Cowell, D.W. and Ford, D.C.
1980: Hydrochemistry of a dolomite karst: the Bruce Peninsula of Ontario; Canadian Journal of Earth Sciences, v. 17, p. 520-526.

Cruden, D.M.
1982: The Brazeau Lake Slide, Jasper National Park, Alberta; Canadian Journal of Earth Sciences, v. 19, p. 975-981.
1985: Rock slope movements in the Canadian Cordillera; Canadian Geotechnical Journal, v. 22, p. 528-540.

Cruden, D.M. and Krahn, J.
1973: A re-examination of the geology of the Frank Slide; Canadian Geotechnical Journal, v. 10, p. 581-591.

Deslodges, J.R. and Gardner, J.S.
1984: Process discharge estimation in ephemeral channels, Canadian Rocky Mountains; Canadian Journal of Earth Sciences, v. 21, p. 1050-1060.

Dionne, J-C.
1969a: Note sur un éboulement à St-Fabien-sur-Mer, Côte sud du St-Laurent; Revue de Géographie de Montréal, vol. 23, p. 55-64.
1969b: Érosion glacielle littorale, estuaire du Saint-Laurent; Revue de Géographie de Montréal, vol. 23, p. 5-20.
1970: Aspects morpho-sédimentologiques du glaciel, en particulier des côtes du St-Laurent; Québec, Laboratoire de Recherche Forestière, Rapport d'Information Q-F-X-9, 324 p.

Donovan, J.J. and Lajoie, G.
1979: Geochemical implications of diagenetic iron sulfide formation in Champlain Sea sediments; Canadian Journal of Earth Sciences, v. 16, p. 575-584.

Dredge, L. and Thom, B.G.
1976: Development of gully-flow near Sept-Îles, Québec; Canadian Journal of Earth Sciences, v. 13, p. 1145-1151.

Durand, J. et Ballivy, G.
1974: Particularités rencontrées dans la région de Montréal résultant de l'arrachement d'écailles de roc par la glaciation; Canadian Geotechnical Journal, vol. 11, p. 302-306.

Edwards, W.A.D.
1989: Aggregate and nonmetallic Quaternary mineral resources; in Chapter 11 of Quaternary Geology of Canada and Greenland, R.J. Fulton (ed.); Geological Survey of Canada, Geology of Canada, no. 1 (also Geological Society of America, The Geology of North America, v. K-1).

Edwards, W.A.D. and Hudson, R.B.
1983: Setting the scene for aggregate resources management in Alberta; in Proceedings, 19th Forum on the Geology of Industrial Minerals, Ontario Geological Survey, Miscellaneous Paper 114, p. 136-143.

Eisbacher, G.H.
1979: First-order regionalization of landslide characteristics in the Canadian Cordillera; Geoscience Canada, v. 6, p. 69-79.

Eisbacher, G.H. and Clague, J.J.
1984: Destructive mass movements in high mountains: hazard and management; Geological Survey of Canada, Paper 84-16, 230 p.

Evans, S.G.
1982: Landslides and surficial deposits in urban areas of British Columbia: A review; Canadian Geotechnical Journal, v. 19, p. 269-288.
1984: The landslide response of tectonic assemblages in the southern Canadian Cordillera; Proceedings, IV International Symposium on Landslides, Toronto, v. 1, p. 495-502.
1986a: Landslide damming in the Cordillera of western Canada; in Proceedings, Symposium on Landslide Dams: Processes, Risks and Mitigation, R.L. Schuster (ed.); American Society of Civil Engineers, Geotechnical Special Publication No. 3, p. 111-130.
1986b: The maximum discharge of outburst floods caused by the breaching of man-made and natural dams; Canadian Geotechnical Journal, v. 23, p. 385-387.
1987: A rock avalanche from the peak of Mount Meager, British Columbia; in Current Research, Part A, Geological Survey of Canada, Paper 87-1A, p. 925-934.

Evans, S.G. and Lister, D.R.
1984: The geomorphic effect of the July 1983 rainstorms in the southern Cordillera and their impact on transportation facilities; in Current Research, Part B, Geological Survey of Canada, Paper 84-1B, p. 223-235.

Evans, S.G., Aitken, J.D., Wetmiller, R.J., and Horner, R.B.
1987: A rock avalanche triggered by the October 1985 North Nahanni Earthquake, District of Mackenzie, N.W.T.; Canadian Journal of Earth Sciences, v. 24, p. 176-184.

Fenton, M.M.
1983: Deformation terrain mid-continent region: Properties, subdivision, recognition (abstract); Geological Society of America, North-Central Section, Program with Abstracts, v. 15, p. 250.

Fenton, M.M. and Andriashek, L.D.
1978: Glaciotectonic features in the Sand River area northeastern Alberta, Canada (abstract); 5th Biennial meeting American Quaternary Association, Edmonton, Alberta, Abstracts, p. 199.

Fenton, M.M., Moell, C.E., Pawlowicz, J.G., Trudell, M.R., and Moran, S.R.
1984: Glaciotectonic deformation and open pit coal mine highwall stability (abstract); in Program with Abstracts, Geological Association of Canada, v. 9, p. 61.

Fraser, C.
1978: Avalanches and Snow Safety; Murray, London, 269 p.

Freer, G.L. and Schaerer, P.A.
1980: Snow-avalanche hazard zoning in British Columbia, Canada; Journal of Glaciology, v. 26, p. 345-354.

Freeze, R.A. and Cherry, J.A.
1979: Groundwater; Prentice-Hall Inc., Englewood Cliffs, New Jersey.

Gardner, J.
1968: Debris slope form and process in the Lake Louise district; unpublished PhD thesis, McGill University, Montréal, 263 p.
1970: Rockfall: A geomorphic process in high mountain terrain; Albertan Geographer, v. 16, p. 16-20.
1971a: Morphology and sediment characteristics of mountain debris slopes in the Lake Louise district; Zeitschrift fur Geomorphologie, v. 15, p. 390-403.
1971b: A note on rockfalls and north faces in the Lake Louise area; American Alpine Journal, v. 7, p. 317-318.
1980: Frequency, magnitude and spatial distribution of mountain rockfalls and rockslides in the Highwood Pass area, Alberta, Canada; in Thresholds in Geomorphology, D.R. Coates and J.D. Vitek (ed.); Allen and Unwin, New York, p. 267-295.
1981: Observations on objective hazards in the mountains; Canadian Alpine Journal, v. 64, p. 52-54.
1983: Rockfall frequency and distribution in the Highwood Pass area, Canadian Rocky Mountains; Zeitschrift fur Geomorphologie, v. 27, p. 311-324.

Gartner, J.F.
1984: Some aspects of engineering geology mapping in Canada; Association of Engineering Geologists, Bulletin, v. 21, p. 269-293.

Gartner, J.F., Mollard, J.D., and Roed, M.A.
1981: Ontario Engineering Geology Terrain Study Users' Manual; Ontario Geological Survey, Northern Ontario Engineering Geology Terrain Study 1, 51 p.

Gélinas, P.J.
1974: Contribution to the study of erosion along the North Shore of Lake Erie; unpublished PhD thesis, Faculty of Graduate Studies, University of Western Ontario, London, 266 p.

Gendzwill, D.J. and Hajnal, Z.
1971: Seismic investigation of the Crater Lake structure in southeastern Saskatchewan; Canadian Journal of Earth Sciences, v. 8, p. 1514-1524.

Gold, C.M., Andriashek, L.D., and Fenton, M.M.
1983: Bedrock topography, Sand River map sheet 73 L; Alberta Geological Survey, Alberta Research Council, Map EM 153, scale 1:250 000.

Gray, J.T
1972: Debris accretion on talus slopes in the Central Yukon Territory; in Mountain Geomorphology, O. Slaymaker and H.J. McPherson (ed.); Tantalus Press, Vancouver, p. 75-84.

Green, R.
1972: Geological map of Alberta; Alberta Research Council, Map 35.

Greenwood, B. and Davidson-Arnott, G.D.
1979: Sedimentation and equilibrium in wave-formed bars: A review and case study; Canadian Journal of Earth Sciences, v. 16, p. 312-332.

Greenwood, B. and Keay, P.A.
1979: Morphology and dynamics of a barrier beach: as study of stability; Canadian Journal of Earth Sciences, v. 16, p. 1533-1546.

Grove, J.M.
1972: The incidence of landslides, avalanches and floods in western Norway during the Little Ice Age; Arctic and Alpine Research, v. 4, p. 131-138.

Hardy, L.
1976: Contribution à l'étude géomorphologique de la portion québécoise des basses-terres de la baie de James; thèse de doctorat non publiée, Département de géographie, Université McGill, Montréal, 264 p.

Hardy, R.M., Morgenstern, N.R., and Patton, F.D.
1978: Report of the Garibaldi Advisory Panel - Part 1; British Columbia Department of Highways, Victoria, British Columbia, 77 p.

Haug, M.D., Sauer, E.K., and Fredlund, D.G.
1977: Retrogressive slope failures at Beaver Creek, south of Saskatoon, Saskatchewan, Canada; Canadian Geotechnical Journal, v. 14, p. 288-301.

Heidebrecht, A.C., Basham, P.W., Rainer, J.H., and Berry, M.J.
1983: Engineering applications of new probabilistic seismic ground-motion maps of Canada; Canadian Journal of Civil Engineering, v. 10, p. 670-680 (NRC/DBR Paper No. 1135).

Hestnes, E. and Lied, K.
1980: Natural-hazard maps for land-use planning in Norway; Journal of Glaciology, v. 26, p. 331-343.

Hore, R.C.
1975: The City of Guelph water supply from the Arkell springs area; in Waterloo '75 Field Trips Guidebook, P.G. Telford (ed.); Geological Association of Canada, p. 315-320.

Horner, R.B., Wetmiller, R.J., and Hasegawa, H.S.
1979: The St. Donat, Québec, earthquake sequence of February 18-23, 1978; Canadian Journal of Earth Sciences, v. 16, p. 1892-1898.

Hungr, O., Morgan, G.C., and Kellerhals, R.
1984: Quantitative analysis of debris torrent hazards for design of remedial measures; Canadian Geotechnical Journal, v. 21, p. 663-677.

Jackson, L.E., Jr.
1979: A catastrophic glacial outburst flood (jökulhlaup) mechanism for debris flow generation at the Spiral Tunnels, Kicking Horse River basin, B.C.; Canadian Geotechnical Journal, v. 16, p. 806-813.
1980: New evidence on the origin of the September 6, 1978 jökulhlaup from Cathedral Glacier, British Columbia; in Current Research, Part A, Geological Survey of Canada, Paper 80-1B, p. 292-294.

Jankowski, C. and McCarthy, E.
1974a: A record of sinking homes, heartaches, frustrations; Kitchener-Waterloo Record, April 17, 1974, p. 17.
1974b: House sinking, owners sue city, finance company; Kitchener-Waterloo Record, December 30, 1974, p. 3.

Kathol, C.P. and McPherson, R.A.
1975: Urban geology of Edmonton; Alberta Research Council, Bulletin 32, 61 p.

Keefer, D.K.
1984: Rock avalanches caused by earthquakes: Source characteristics; Science, v. 223, p. 1288-1290.

Keen, M.J. and Williams, G.L. (editors)
1989: Geology of the Continental Margin of Eastern Canada; Geological Survey of Canada, Geology of Canada, no. 2 (also Geological Society of America, The Geology of North America, v. I-1).

Keen, M.J., Adams, J.E., Moran, K.M., Piper, D.J.W., and Reid, I.
1989: Constraints due to the effects of earthquakes; in Chapter 14 of Geology of the Continental Margin of Eastern Canada, M.J. Keen and G.L. Williams (ed.); Geological Survey of Canada, Geology of Canada, no. 2 (also Geological Society of America, The Geology of North America, v. I-1).

Kerr, F.A.
1948: Lower Atikine and Western Iskut River areas, British Columbia; Geological Survey of Canada, Memoir 246, 94 p.

Kewen, T.J. and Schneider, A.T.
1979: Hydrogeologic evaluation of the Judith River Formation Aquifer in west central Saskatchewan; Saskatchewan Research Council, Geology Division, Report and maps prepared for Saskatchewan Department of Environment, Report G-72-2, 78 p.

Kjartanson, B.
1983: Geological engineering report for urban development of Winnipeg, A. Baracos, D.H. Shields, and B. Kjartanson (ed.); Department of Geological Engineering, University of Manitoba, Winnipeg, 78 p.

Klugman, M.A. and Chung, P.
1976: Slope stability study of the Regional Municipality of Ottawa-Carleton, Ontario, Canada; Ontario Geological Survey, Miscellaneous Paper 68, 13 p., 5 maps.

Krahn, J., Johnson, R.F., Fredlund, D.G., and Clifton, A.W.
1979: A highway cut failure in Cretaceous sediments at Maymont, Saskatchewan; Canadian Geotechnical Journal, v. 16, p. 703-715.

Kupsch, W.O.
1962: Ice-thrust ridges in western Canada; Journal of Geology, v. 70, p. 582-594.

LaChapelle, E.R.
1966: Encounter probabilities for avalanche damage; Alta Avalanche Study Center, Utah, Miscellaneous Report No. 10, 10 p.

Leblanc, G.
1981: A closer look at the September 16, 1732, Montreal Earthquake; Canadian Journal of Earth Sciences, v. 18, p. 539-550.

Lebuis, J., Robert, J.M., and Rissmann, P.
1983: Regional mapping of landslide hazard in Quebec; Symposium on slopes on soft clays, Report no. 17, Swedish Geotechnical Institute, p. 205-262.

Lefebvre, G.
1981: Fourth Canadian Geotechnical Colloquium: strength and slope stability in Canadian soft clay deposits; Canadian Geotechnical Journal, v. 18, p. 420-442.
1984: Geology and slope instability in Canadian sensitive clays; in Proceedings of the 37th Canadian Geotechnical Conference, Toronto, p. 23-24.

Lidgren, R.A. and Sauer, E.
1982: The Gronlid-Crossing, Saskatchewan Provincial Highway No. 6 at the Saskatchewan River; Canadian Geotechnical Journal, v. 19, p. 360-380.

Lister, D.R., Morgan, G.C., VanDine, D.F., and Kerr, J.W.G.
1984: Debris torrents in Howe Sound, British Columbia; in Proceedings, 4th International Symposium on Landslides, Toronto, v. 1, p. 649-654.

Locat, J.
1977: L'émerson des terres dans la région de Baie-des-Sables/Trois-Pistoles; Géographie physique et Quaternaire, vol. 31, p. 297-306.
1982: Contribution à l'étude de la structuration des argiles sensibles de l'est du Canada; thèse de doctorat non publiée, Faculté des sciences appliquées, Département de Génie Civil, Université de Sherbrooke, 512 p.

Locat, J. and Lefebvre, G.
1986: The origin of structuration of the Grande-Baleine marine sediments, Quebec, Canada; Quarterly Journal of Engineering Geology, London, v. 19, p. 365-374.

Locat, J., Demers, B., Lebuis, J., et Rissmann, P.
1984: Prédiction des glissements de terrain: application aux argiles sensibles, rivière Chacoura, Québec, Canada; in Proceedings, 4th International Symposium on Landslides, Toronto, v. 2, p. 549-555.

Luckman, B.H.
1976: Rockfalls and rockfall inventory data: Some observations from Surprise Valley, Jasper National Park, Canada; Earth Surface Processes, v. 1, p. 287-298.
1981: The geomorphology of the Alberta Rocky Mountains: A review and commentary; Zeitschrift für Geomorphologie, Supplement-Band 37, p. 91-119.

Maathuis, H. and Schreiner, B.T.
1982a: Hatfield Valley aquifer system in the Wynyard region, Saskatchewan; prepared for Saskatchewan Department of Environment v. 1, text, appendixes A-E, and maps, Report G-744-7-C-82a, 61 p.
1982b: Hatfield Valley aquifer system in the Waterhen River area (73 K), Saskatchewan; prepared for Saskatchewan Department of Environment, v. 1, text, appendixes A-E, Report G-744-7-C-82b, 33 p.

MacDonald, A.B. and Sauer, E.K.
1970: The engineering significance of Pleistocene stratigraphy in the Saskatoon area, Saskatchewan, Canada; Canadian Geotechnical Journal, v. 7, p. 116-126.

Marsh, D.
1981: Rockfall; Explore Alberta, v. 3, p. 24-30. Martin, D.C., Piteau, D.R., Pearce, R.A., and Hawley, P.M.
1984: Remedial measures for debris flows at the Agassiz Mountain Institution, British Columbia; Canadian Geotechnical Journal, v. 21, p. 505-517.

Mason, K.
1929: Indus flood and Shyok glaciers; Himalayan Journal, v. 1, p. 11-29.

Mathews, W.H.
1965: Two self-dumping ice-dammed lakes in British Columbia; The Geographical Review, v. 55, p. 46-52.

Mathews, W.H. and McTaggart, K.C.
1978: Hope rockslides, British Columbia; in Rockslides and Avalanches, v. 1, B.Voight (ed.); Elsevier Scientific Publishing Co., Amsterdam, p. 259-275.

McCann, S.B. (ed.)
1980: The coastlines of Canada; Geological Survey of Canada, Paper 80-10, 439 p.

McKenzie, W.C.
1960: Avalanche sheds in Roger's Pass, B.C.; B.C. Professional Engineer, v. 11, p. 15-20.

Meneley, W.A.
1970: Groundwater Resources, in Physical environment of Saskatoon, Canada, E.A. Christiansen (ed.); Saskatchewan Research Council and National Research Council of Canada, NRC Publication No. 11378.
1972: Groundwater resources in Saskatchewan; in Water supply for the Saskatchewan Nelson Basin, Appendix 7, Section F, p. 673-723; Saskatchewan-Nelson Basin Board, Ottawa.
1983: Hydrogeology of the Eastend to Ravenscrag Formations in southern Saskatchewan; W.A. Meneley Consultants Ltd., Saskatoon, report prepared for Saskatchewan Department of Environment, 21 p.; available from Saskatchewan Water Corporation Library, Moose Jaw, Saskatchewan.

Meneley, W.A. and Whitaker, S.H.
1969: A monitoring program for the Cominco Limited potash waste disposal basin near Vanscoy, Saskatchewan. Phase II: Environment evaluation, design, and construction of the monitoring system; Saskatchewan Research Council, unpublished report prepared for Cominco Ltd.; available from Saskatchewan Water Corporation Library, Moose Jaw, Saskatchewan.

Meyboom, P.
1967: Interior plains hydrogeological region; in Groundwater in Canada, I.C. Brown (ed.); Geological Survey of Canada, Economic Geology Report No. 24.

Miles, M. and Kellerhals, R.
1981: Some engineering aspects of debris torrents; Proceedings of the 5th Canadian Hydrotechnical Conference, Canadian Society of Civil Engineers, p. 395-420.

Miller, G.E. and Nyland, D.
1981: Geological hazards and urban development of glacio-lacustrine silt deposits in the Penticton area, British Columbia; in Preprints; Canadian Geotechnical Society 34th Annual Meeting, Fredericton, N.B., 20 p.

Mitchell, R.J. and Klugman, M.A.
1979: Mass instabilities in sensitive Canadian soils; Engineering Geology, v. 14, p. 109-134.

Mitchell, R.J. and Markell, A.R.
1974: Flowsliding in sensitive soils; Canadian Geotechnical Journal, v. 11, p. 11-31.

Mokievsky-Zubok, O.
1977: Glacier-caused slide near Pylon Peak, British Columbia; Canadian Journal of Earth Sciences, v. 14, p. 2657-2662.

Moore, D.P. and Mathews, W.H.
1978: The Rubble Creek landslide, southwestern British Columbia; Canadian Journal of Earth Sciences, v. 15, p. 1039-1052.

Moran, S.R., Clayton, L., Hooke, R.LeB., Fenton, M.M., and Andriashek, L.D.
1980: Glacier-bed landforms of the prairie region of North America; Journal of Glaciology, v. 25, p. 457-476.

Nasmith, H.W.
1980: Catastrophic rockslides and urban development; Proceedings of Specialty Conference on Slope Stability Problems in Urban Areas, Canadian Geotechnical Society, Toronto, 8 p.

Nasmith, H.W., and Mercer, A.G.
1979: Design of dykes to protect against debris flows at Port Alice, British Columbia; Canadian Geotechnical Journal, v. 16, p. 748-757.

National Research Council
1979: First Canadian Conference on Marine Geotechnical Engineering, Calgary, Alberta; W.J. Eden (ed.); Canadian Geotechnical Society, 467 p. (available from Bitech Publishers, Vancouver).
1982: Second Canadian Conference on Marine Geotechnical Engineering, Halifax, Nova Scotia; Canadian Geotechnical Society, 225 p. (available from Bitech Publishers, Vancouver).

Nelson, J.W.
1970: Problems caused by avalanches on highways in British Columbia; Ice Engineering and Avalanche Forecasting and Control, National Research Council of Canada, Technical Memorandum No. 98, p. 84-90.

Ontario Government
1986: Policy statement, Mineral Aggregate Resources; Order in Council No. 1249-86, May 9th, 1986.

Ontario, Interministerial Working Group
1983: Non-flood hazard lands-sensitive clay slopes; Report, Interministerial Working Group, Ministries of Municipal Affairs and Housing and Natural Resources, Toronto, 45 p.

Ontario Waste Management Corporation (Toronto)
1985: OWMC Exchange, v. 2, 4 p.

Owens, E.H. and Drapeau, G.
1973: Changes in beach profiles at Chedabucti Bay, Nova Scotia, following large-scale removal of sediments; Canadian Journal of Earth Sciences, v. 10, p. 1226-1232.

Pearson, W.J.
1963: Salt deposits of Canada; Symposium on Salt, Northern Ohio Geological Society, p. 196-239.

Peck, J.W.
1970: Mining vs. avalanches-British Columbia; Ice Engineering and Avalanche Forecasting and Control, National Research Council of Canada, Technical Memorandum No. 98, p. 79-83.

Peckover, F.L. and Kerr, J.W.G.
1977: Treatment and maintenance of rock slopes on transportation routes; Canadian Geotechnical Journal, v. 14, p. 487-507.

Perla, R. (editor)
1978: Avalanche Control, Forecasting and Safety; National Research Council of Canada, Associate Committee on Geotechnical Research, Technical Memorandum No. 120, 301 p.

Perla, R.I. and Martinelli, M.
1975: Avalanche Handbook; United States Department of Agriculture; Agriculture Handbook 489, 238 p.

Piteau, D.R. and Peckover, F.L.
1978: Engineering of rock slopes; in Landslides: Analysis and Control, R.L. Schuster and R.J. Krizek (ed.); National Research Council, Transportation Research Board, Special Report 176, p. 192-228.

Porter, S.C. and Orombelli, G.
1981: Alpine rockfall hazards; American Scientist, v. 69, p. 67-75.

Poschmann, A.S., Klassen, K.E., Klugman, M.A., and Goodings, D.
1983: Slope stability study of the South Nation River and portions of the Ottawa River; Ontario Geological Survey, Miscellaneous Paper 114, 20 p., 2 maps.

Prest, V.K. and Hode-Keyser, J.
1977: Geology and engineering characteristics of surficial deposits, Montreal Island and vicinity, Quebec; Geological Survey of Canada, Paper 75-27, 29 p.

Quigley, R.M.
1980: Geology, mineralogy and geochemistry of Canadian soft soils: a geotechnical perspective; Canadian Geotechnical Journal, v. 17, p. 261-285.

Quinlan, G.
1984: Postglacial rebound and the focal mechanism of eastern Canadian earthquakes; Canadian Journal of Earth Sciences, v. 21, p. 1018-1023.

Ramsey, B.
1967: Britannia — The Story of a Mine; Agency Press Ltd., Vancouver, 177 p.

Regional Municipality of Peel
1975: Physical survey - towards a regional official plan; Planning Department, Regional Municipality of Peel, 116 p., maps.

Roberge, J.
1977: Karste de la Haute-Saumon, Ile D'Anticosti, Québec, modèle de développement d'un karste jeune; Spéléo-Québec, vol. 3-4, p. 19-36.

Rondot, J.
1968: Nouvel impact météoritique fossile? La structure semi-circulaire de Charlevoix; Journal canadien des sciences de la terre, vol. 7, p. 1305-1317.

Roy, D.W. and Duberger, R.
1983: Relations possibles entre la microséismicité récente et l'astroblème de Charlevoix; Journal canadien des sciences de la terre, vol. 20, p. 1613-1618.

Rukavina, N.A.
1976: Workshop on Great Lakes coastal erosion and sedimentation, N.A. Rukavina (ed.); Canada Centre for Inland Waters, Burlington, Ontario, 137 p.
1978: Second Workshop on Great Lakes coastal erosion and sedimentation, N.A. Rukavina (ed.); Canada Centre for Inland Waters, Burlington, Ontario, 118 p.
1982: Third Workshop on Great Lakes coastal erosion and sedimentation, N.A. Rukavina (ed.); Canada Centre for Inland Waters, Burlington, Ontario, 177 p.

Ryder, J.M. and Howes, D.E.
1984: Terrain information, a user's guide to terrain maps in British Columbia; Surveys and Resource Mapping Branch, British Columbia, 16 p.

Sauchyn, J., Gardner, J.S., and Suffling, R.
1983: Evaluation of botanical methods of dating debris flows and debris flow hazard in the Canadian Rocky Mountains; Physical Geography, v. 4, p. 182-201.

Sauer, E.K.
1974: Geotechnical implications of Pleistocene deposits in southern Saskatchewan; Canadian Geotechnical Journal, v. 11, p. 359-373.
1977: A valley crossing in Pleistocene deposits; Engineering Geology, v. 11, p. 1-21.
1978: The engineering significance of glacier ice-thrusting; Canadian Geotechnical Journal, v. 15, p. 457-472.
1979: A slope failure in till at Lebret, Saskatchewan; Canadian Geotechnical Journal, v. 16, p. 242-250.
1983: The Denholm landslide, Saskatchewan. Part II: Analysis; Canadian Geotechnical Journal, v. 20, p. 208-220.

Schaerer, P.A.
1962: The avalanche hazard evaluation and prediction at Roger's Pass; National Research Council of Canada, DBR Paper No. 142, 32 p.

Scheidegger, A.E.
1973: On the prediction of the reach and velocity of catastrophic landslides; Rock Mechanics, v. 5, p. 231-236.

Schreiner, B.T. and Maathuis, H.
1982: Hatfield Valley aquifer system in the Melville region, Saskatchewan; Prepared for Saskatchewan Department of Environment, v. 1, text, appendixes A-E, and maps, Report G-743-3-B-82, 90 p.

Scott, D.W.
1982: Aggregate resources inventory of Ontario; Canadian Mining and Metallurgical Bulletin, v. 75, p. 84-93.

Scott, J.S.
1973: Urban geology in Canada; in Proceedings, National Conference on Urban Engineering Terrain Problems; Associate Committee for Geotechnical Research, Technical Memorandum No. 109, National Research Council, Ottawa, p. 7-19.
1976: Geology of Canadian tills; in Glacial Till, R.F. Legget (ed.); Royal Society of Canada, Special Publication, p. 50-66.

Scott, J.S. and Brooker, E.W.
1968: Geological and engineering aspects of upper Cretaceous shales in western Canada; Geological Survey of Canada, Paper 66-37, 75 p.

Sharpe, D.R.
1980: Quaternary geology of Toronto and surrounding area; Ontario Geological Survey, Preliminary Map P2204, Geological Series, Scale 1:100 000.

SkwaraWoolf, T.
1980: Mammals of the Riddell local fauna (Floral Formation, Pleistocene, Late Rancholabrean), Saskatoon, Canada; Saskatchewan Culture and Youth and Museum of Natural History, Natural History Contributions, No. 2, p. 129.

Slater, G.
1927: Structure of the Mud Buttes and Tit Hills in Alberta; Geological Society of America, Bulletin, v. 38, p. 721-730.

Smalley, I.J.
1979: Sensitive soils and quick clays; Engineering Geology, v. 14, p. 81-217.

Smith, W.E.T
1962: Earthquakes of eastern Canada and adjacent areas, 1534-1927; Publication of the Dominion Observatory, Ottawa, v. 26, p. 271-301.

Société québécoise de spéléologie
1980: Le karste de la plate-forme de Boischatel et le karste barré de La Rédemption, état des connaissances; Livret-guide, ACQUA Congress, Québec, 110 p.

Starr, C.
1969: Social benefit versus technological risk; Science, v. 165, p. 1232-1238.

Stethem, C.J. and Schaerer, P.A.
1979: Avalanche accidents in Canada I: A selection of case histories of accidents, 1955 to 1976; National Research Council of Canada, DBR Paper No. 834, 114 p.
1980: Avalanche accidents in Canada II: A selection of case histories of accidents, 1943-1978; National Research Council of Canada; DBR Paper No. 926, 75 p.

Tavenas, F., Chagnon, J-Y., and Larochelle, P.
1971: The St-Jean-Vianney landslide: Observations and eye-witness accounts; Canadian Geotechnical Journal, v. 7, p. 463-478.

Terzaghi, K.
1950: Mechanics of landslides; in Application of Geology to Engineering Practice, S. Paige (chairman) (Berkey Volume); Geological Society of America, p. 83-123.

Thomson, S. and Morgenstern, N.R.
1977: Factors affecting distribution of landslides along rivers in southern Alberta; Canadian Geotechnical Journal, v. 14, p. 508-523.

Torrance, J.K.
1974: A laboratory investigation of the effect of leaching on the compressibility and shear strength of Norwegian marine clays; Geotechnique, v. 24, p. 155-173.
1983: Towards a general model of quick clay development; Sedimentology, v. 30, p. 547-555.

Troude, J.P., Serodes, J-B., et Elouard, B.
1983: Étude des mécanismes sédimentologiques des zones intertidales de l'estuaire moyen du Saint-Laurent, cas de la batture de Cap-Tourmente; Université Laval, Départment de Génie Civil et Environnement Canada, Direction Générale des Eaux Intérieures, Québec, 117 p.

Vanderveer, D.G.
1983: Aggregates — The often maligned and often forgotten industrial mineral; in Proceedings, l9th Forum on the Geology of Industrial Minerals; Ontario Geological Survey, Miscellaneous Paper 114, p. 65-78.

VanDine, D.F.
1985: Debris flows and debris torrents in the southern Canadian Cordillera; Canadian Geotechnical Journal, v. 22, p. 44-68.

Varnes, D.J.
1978: Slope movement and types and processes; in Landslides: Analysis and Control; Transportation Research Board, National Academy of Sciences, Washington, D.C., Special Report 176, Chapter 2.

Whitaker, S.H.
1982: Groundwater resources at the Judith River Formation in southwestern Saskatchewan; prepared for Saskatchewan Department of Environment; available from Saskatchewan Water Corporation Library, Moose Jaw, Saskatchewan.

Whitaker, S.H. and Christiansen, E.A.
1972: The Empress Group in southern Saskatchewan; Canadian Journal of Earth Sciences, v. 9, p. 353-360.

White, O.L.
1982: Toronto's subsurface geology; in Geology under Cities, Reviews in Engineering Geology, R.F. Legget (ed.); Geological Society of America, v. 5, p. 119-124.

White, O.L., Karrow, P.F., and MacDonald, J.R.
1974: Residual stress relief phenomena in southern Ontario; Proceedings of the Ninth Canadian Rock Mechanics Symposium, Ottawa, p. 323-348.

Whitehouse, I.E. and Griffiths, G.A.
1983: Frequency and hazard of large rock avalanches in the central Southern Alps, New Zealand; Geology, v. 11, p. 331-334.

Wickenden, R.T.D.
1945: Mesozoic stratigraphy of the eastern plains, Manitoba and Saskatchewan; Geological Survey of Canada, Memoir 239, 87 p.

Williamson, D.
1982: A study of avalanche hazard management in the Canadian West; unpublished BSc thesis, Department of Geography, University of Waterloo, Waterloo, 65 p.

Woods, J.G. and Marsh, J.S.
1983: Snow war: An illustrated history of Roger's Pass, Glacier National Park, British Columbia; National and Provincial Parks Association of Canada, Toronto, 52 p.

Authors' addresses

J-Y. Chagnon
Groupe de recherche en géologie de l'ingénieur
Département de géologie
Université Laval
Ste. Foy, Québec
G1K 7P4

S.G. Evans
Geological Survey of Canada
601 Booth Street
Ottawa, Ontario
K1A 0E8

J.S. Gardner
Department of Geography
Faculty of Environmental Sciences
University of Waterloo
Waterloo, Ontario
N2L 3G1

L.E. Jackson, Jr.
Geological Survey of Canada
100 West Pender Street
Vancouver, British Columbia
V6B 1R8

J. Locat
Groupe de recherche en géologie de l'ingénieur
Département de géologie
Université Laval
Ste. Foy, Québec
G1K 7P4

J.S. Scott
Geological Survey of Canada
601 Booth Street
Ottawa, Ontario
K1A 0E8

O.L. White
Ontario Geological Survey
77 Grenville Street
Toronto, Ontario
M7A 1W4

Printed in Canada

Part 3
QUATERNARY GEOLOGY OF GREENLAND

With an area of 2 200 000 km² Greenland is the largest island in the world. The northernmost point, Kap Morris Jesup, reaches 83°40′N, while the southern tip, Kap Farvel, lies 2700 km to the south at 59°45′N. At present 80% of the land area is covered by the Inland Ice filling up a depression extending to approximately 250 m below sea level. The ice-free land makes up a mountainous rim, up to 300 km wide and 3700 m high, bordering on a coastal shelf which attains its maximum width of 300 km off the coast of northeast Greenland.

Earlier reviews of the Quaternary geology of Greenland include works by Weidick (1975a, 1976a). Since their appearance, much new knowledge has been gained, not only by "traditional" studies in the ice-free land, but also by geophysical work on the coastal shelves and on the Inland Ice. These chapters combine, and as far as presently possible, synthesize knowledge gained from these widely different fields of study.

The Quaternary geology of Greenland is presented in two chapters, the first concerns the Quaternary of the ice-free land and the adjacent coastal shelves and the second describes the Greenland Ice Sheet and results of glaciological investigations. The first three sections of Chapter 13 describe the regional Quaternary geology of West Greenland, East Greenland, and North Greenland. These are followed by a summary of recently collected data from the shelf areas adjacent to Greenland, by a section on sea level history, and one on Quaternary marine and terrestrial ecosystems. The chapter closes with a summary of development of climatic, glacial, and oceanographic factors. The Greenland Ice Sheet chapter (14) emphasises the wealth of glaciological and climatological data that have been gathered over the past decade.

Development of Quaternary concepts

As was the case in other Arctic areas, the accretion of scientific knowledge on Greenland has been from its start closely connected with exploration history. The first purely scientific expeditions were in the early 19th Century. For Quaternary geology, a significant event was the publication of works by Rink (1852, 1853), in which for the first time a clear distinction was made between *highland ice,* that is, local ice caps and mountain glaciers in coastal areas, and the so-called Inland Ice, and it was shown that these two types of glaciation were governed by distinctly different dynamics.

These observations immediately caught the attention of "glacialists" in Europe who saw what Rink had not seen, that the Inland Ice served as a long sought model for the ice age glaciation in Europe. As a consequence, a number of scientists visited Greenland in the following years to study the Inland Ice, and were also able to demonstrate that the ice-free land had once been inundated by the ice sheet (e.g., Brown, 1871; Nordenskiöld, 1871; Steenstrup, 1881; Helland, 1876).

Thus by 1875 it was generally accepted that Greenland had also had its ice age, and that the retreat of the ice had been accompanied by an invasion of the sea which, when it regressed, left marine sediments on dry land.

In 1876 "Kommissionen for Videnskabelige Undersøgelser i Grønland" (The Commission for Scientific Research in Greenland) was established, and this scientifically directed government institution at once initiated systematic geological surveying. In the following years such Quaternary "standard information" as distribution of erratics, moraines, and glacial scouring, as well as elevation determinations and collections of fossils from raised marine sediments began to appear.

Initially this work was concentrated in "Danish West Greeenland", that is, West Greenland south of 73°N, but by the end of the century scientific expeditions began to penetrate the remote and inaccessible areas of East and North Greenland, and onto the Inland Ice. The Inland Ice was first traversed by Nansen in 1888, and the last remaining unknown part of Greenland's perimeter in northeast Greenland was surveyed during the Denmark Expedition, 1906-1908.

For the next 50 years information on the Quaternary geology was acquired largely through routine geological mapping which was organized on a large scale. In East Greenland "Dr. Lauge Koch's East Greenland Expeditions" (1926-1958) substituted motor vessels and airplanes for the formerly used umiak and dog sled. In West Greenland, following 1946, mapping was carried out by the newly founded Geological Survey of Greenland, which for the first time employed specialists to deal with the Quaternary deposits. In North Greenland the "Danish Peary Land Expeditions" have been doing similar work intermittently since 1947.

The first attempts to establish a chronlogy for the Quaternary was based on correlation of subfossil marine faunas, and it was more or less implicit that the record covered only the period after the last ice age, since older deposits had been removed by glaciers. This view began to disintegrate with the advent of ^{14}C dating in the 1950s. For the first time a tool was available for accurate dating and correlation of ice margin deposits in all parts of Greenland — and it became clear that the Quaternary record was longer in East and North Greenland than in "classical" West Greenland. The trend towards a longer and more complex Quaternary record was further accentuated in the 1970s with the introduction of new methods for dating and correlation: amino acid analyses of marine bivalves, Th/U dating, micropaleontological correlation, paleomagnetic investigations, and correlation with the continuous time-series developed from deep sea cores in adjacent oceans. Application of these various techniques has made it possible to extend the Quaternary record back to the Pliocene-Pleistocene boundary, although hiata still dominate.

Equally important for the concept of the Quaternary was the glaciological and paleclimatological work carried out on the Inland Ice mainly under the auspices of the "Greenland Ice Sheet Programme" (GISP), an international research project begun after the first successful penetration of the Inland Ice in a deep boring in 1966. Together with evidence emerging from oil exploration on the shelf areas, begun in 1968, this information will eventually allow a detailed reconstruction of the Inland Ice fluctuations at least during the last ice age.

Chapters 13 and 14 were written during 1984 and 1985; final revisions were made in 1987.

S. Funder

Chapter 13

QUATERNARY GEOLOGY OF THE ICE-FREE AREAS AND ADJACENT SHELVES OF GREENLAND

Summary

Introduction — *S. Funder*

Quaternary geology of West Greenland — *S. Funder*

Quaternary geology of East Greenland — *S. Funder*

Quaternary geology of North Greenland — *S. Funder*

Quaternary geology of the shelves adjacent to Greenland — *S. Funder and H.C. Larsen*

Sea level history — *S. Funder*

Paleofaunas and floras — *S. Funder and B. Fredskild*

Development of climate, glaciation, and oceanographic circulation — *S. Funder*

Acknowledgments

References

Eastward view along Øfjord, Scoresby Sund area, East Greenland. Glacial erosion formed the trough in a gently undulating Tertiary surface that was elevated by epeirogenic uplift in the late Tertiary. Courtesy of Geodetic Institue, Copenhagen.

Chapter 13

QUATERNARY GEOLOGY OF THE ICE-FREE AREAS AND ADJACENT SHELVES OF GREENLAND

Co-ordinator
S. Funder

SUMMARY

Greenland, the largest island in the world, has a general bowl shape with peripheral mountainous areas surrounding a central basin that extends below sea level. The Greenland Ice Sheet occupies the central bowl, covers much of the fringing mountains, and in places pushes to the coast where it calves into the sea. Ice-free regions at the fringes of the ice sheet are in most areas mountainous, cut by fiords and contain scattered thin deposits of till and local thick deposits of Quaternary nonglacial sediments of a variety of different ages.

Evidence from shelf areas indicates that an early glaciation of Greenland, which was more extensive than any succeeding one, occurred near the end of the Pliocene (about 2.4 Ma). A younger phase of glaciation (about 1.8 Ma) is recorded near the base of the Kap København Formation and in the Lodin Elv Formation. Sediments above deposits related to this glaciation were deposited under cool temperate conditions. Tree remnants included in these sediments suggest a climate incompatible with existence of an inland ice sheet.

Several areas contain a record of a glaciation that occurred prior to the last interglaciation. This has been referred to as Fiskebanke, Scoresby Sund, and Bliss Bugt glaciations in West, East, and North Greenland, respectively. On the basis of intensity of weathering it is suggested that these three are correlative. The glaciation is tentatively referred to Illinoian. This glaciation was more extensive than subsequent glaciations, and distribution of erratics indicates that ice in coastal areas was thick enough to move independent of the underlying topography.

The last interglaciation is recorded in the Kaffehavn and Langelandselv deposits that contain subarctic marine fossils which are near to or farther north than similar Holocene faunas. No terrestrial interglacial deposits have been found but data from cores of the Inland Ice have been used to suggest that the Greenland Ice Sheet did not exist at the peak of this interglaciation. The marine sediments that overlie the Langelandselv deposits contain an Arctic fauna and are found as much as 100 m above present sea level. These, the Jameson Land marine beds do not contain direct evidence of glaciation but the Arctic fauna indicates cooler conditions than during the preceding interglaciation and isostatic depression caused by a buildup of ice is required to explain their position well above sea level.

The ice-free areas of Greenland record only one main ice advance of Wisconsinan age. This ice advance occurred after 40 ka and is referred to as the Sisimiut, Flakkerhuk, and Independence Fjord glaciations in West, East, and North Greenland, respectively. Off West Greenland, this ice was grounded as far as 30-50 km offshore and locally off East Greenland it extended as far as 200 km. The culmination of this glaciation apparently occurred about 14 ka and large scale oscillations apparently were in phase in all parts of the ice sheet.

Ice retreat possibly began soon after 14 ka and by 10-11 ka the ice margin was located near the present coast. A readvance, possibly caused by increased precipitation, occurred between 10.3 and 9.5 ka in East Greenland. This oscillation, referred to as the Milne Land stade deposited moraines and outwash plains in many fiords. The Taserqat stade of West Greenland is of similar age. By about 7 ka the ice sheet had retreated to its present limits, and between then and 3 ka it lay as much as 10 km inside this position. Readvance after 3 ka brought the ice sheet to a maximum position in the late 19th Century and this was followed by minor retreat to the present limits.

Relative sea level has fallen in all areas since Wisconsinan glaciation largely due to emergence of the land. Maximum uplift (generally up to 100 m but locally as much as 140 m) is recorded near the outer coast with the level decreasing towards the present ice margin. Submergence, possibly caused by the late Holocene expansion of the Greenland Ice Sheet, has been occurring over the past few centuries.

During the early Holocene fell field vegetation was replaced by dwarf shrub heath and subarctic-boreal bivalves began to appear in coastal waters. These suggest that by 7-8 ka temperatures were slightly warmer than at present. The comparatively warmer water fauna and temperate vegetation began disappearing at different times in different areas. Late Holocene cooling apparently commenced in coastal East Greenland about 5 ka, in West Greenland about 4 ka, and in North Greenland about 3.5 ka.

Funder, S. (co-ordinator)
1989: Quaternary geology of the ice-free areas and adjacent shelves of Greenland; Chapter 13 in Quaternary Geology of Canada and Greenland, R.J. Fulton (ed); Geological Survey of Canada, Geology of Canada, no. 1 (also Geological Society of America, The Geology of North America, v. K-1).

INTRODUCTION
S. Funder

The Quaternary geology of the ice-free land is closely associated with the former glacial regime of the Inland Ice, which in turn is a result of the interaction between bedrock geology, physiography, and climate. Consequently this chapter begins with a brief description of these environmental features.

At present the Inland Ice regimes range from cold-polar and dry in the north to subpolar and humid in the south. This pattern has existed for a long time and has played a major role in the creation of different types of landscapes and glacial deposits. These differences make it natural to discuss the Quaternary geology of Greenland in terms of three regions: West Greenland with its characteristic landscapes of heavily abraded bedrock and sparse Quaternary cover; East Greenland with its spectacular fiord landscapes, created by selective glacial erosion, and at least locally thick Quaternary deposits; and North Greenland exhibiting a variety of landscape types, including the thickest accumulations of Quaternary sediments found in Greenland.

The Quaternary has in general terms been an era of erosion for Greenland, and much of the eroded material can now be found on the coastal shelves. The record of the shelves hence forms an important adjunct to the terrestrial Quaternary record. Geological information from these areas is presently emerging as a by-product to oil exploration. A detailed account of the geology of the East Greenland shelf is given elsewhere in this series (Larsen, in press). A summary of the currently emerging Quaternary history of the shelves adjacent to Greenland is also presented in this chapter.

Other sections of this chapter discuss sea level history and the Quaternary development of marine and terrestrial ecosystems, with main emphasis on the Holocene. The final segment is a summary history that considers the development of climate, glaciers, and oceanographic circulation during the Quaternary.

Bedrock geology and physiography
Bedrock geology

The largest part of the country, probably including areas under the Inland Ice, is made up of Precambrian crystalline rocks — gneisses and migmatites with interbedded supracrustal and various plutonic rock types (e.g., Kalsbeek, 1982). Several mobile belts are recognized in this Precambrian Shield which was established as a craton by middle Proterozoic time (Fig. 13.1).

Along the northern and eastern margins of this stable block, sedimentary basins began to develop in mid-late Proterozoic time, and thick piles of sediment accumulated on the subsiding continental crust. In East Greenland, sandstone, pelites, and carbonates accumulated in the basin in Late Proterozoic and Early Paleozoic times and during the Caledonian Orogeny (Early Silurian); these sediments were folded, thrusted, metamorphosed, migmatized, and locally intruded by granite (Henriksen, 1985; Hurst et al., 1985). In North Greenland the sedimentary basin was occupied by a shallow shelf sea to the south, with mainly carbonate sedimentation (Peel, 1985), and a deep water trough to the north, with a sequence of mainly turbiditic sandstone and shale (e.g., Dawes and Peel, 1981; Surlyk and Hurst, 1984; Higgins et al., 1985). During the Innuitian Orogeny (Late Paleozoic) the deep water basin was deformed and the sediments folded and progressively metamorphosed up to amphibolite facies in the north, while the carbonate platform to the south escaped major deformation and metamorphism.

Following the Caledonian Orogeny, Greenland was part of a single North Atlantic landmass, comprising the North American and North Eurasian plates. In East Greenland thick deposits of Devonian and Carboniferous terrestrial sandstones were deposited in intramontane basins and testify to crustal instability and rapid downwasting of the newly formed mountain range (e.g., Birkelund and Perch-Nielsen, 1976). Later, especially in Mesozoic time, a new complex sedimentary basin arose by block faulting along lines parallel to the present coast (Surlyk et al., 1981). This basin was filled with continental and marine-epicontinental sandstone and shale. At right angles to the main direction of faulting, crossfaults with a northwest-southeast strike formed. Much later these determined the location of the spectacular fiords of East Greenland.

The Precambrian Shield in West Greenland also began to fragment in Mesozoic time, and a new sedimentary basin — the West Greenland Basin — was formed in the Late Cretaceous by rifting and faulting parallel to the present coastline. The sediments are exposed on land in the Nugssuaq Embayment as a thick pile of clastic marine and fluvial sandstone and shale but are mainly confined to the shelf (Henderson et al., 1981).

Thus, by Late Mesozoic time the stage was set for a decisive chapter in the history of Greenland — the history of how the country attained its present shape and physiography. This process was closely related to the formation, by active ocean floor spreading, of the Baffin Bay-Labrador Sea in the west and of the Greenland and Norwegian seas in the east. In the early Tertiary, ocean floor spreading was preceded by a period of effusive volcanism, which produced sheets of plateau basalts in the Nugssuaq Embayment of West Greenland (Clarke and Pedersen, 1976) and locally in East Greenland (Larsen, 1980; Brooks and Nielsen, 1982; Larsen and Watt, 1985).

During the ensuing period of ocean floor spreading vast amounts of terrigenous detritus were deposited in the West Greenland shelf area and in a newly formed sedimentary basin extending along the entire East Greenland coast.

Funder, S.
1989: Introduction (Quaternary geology of the ice-free areas and adjacent shelves of Greenland); in Chapter 13 of Quaternary Geology of Canada and Greenland, R.J. Fulton (ed); Geological Survey of Canada, Geology of Canada, no. 1 (also Geological Society of America, The Geology of North America, v. K-1).

Figure 13.1. Geology of Greenland's ice-free land and shelf in relation to neighbouring land areas (modified from Escher and Watt, 1976; shelf geology from Henderson et al., 1981; Larsen, 1983; and H.C. Larsen, unpublished).

QUATERNARY GEOLOGY — ICE-FREE AREAS AND ADJACENT SHELVES OF GREENLAND

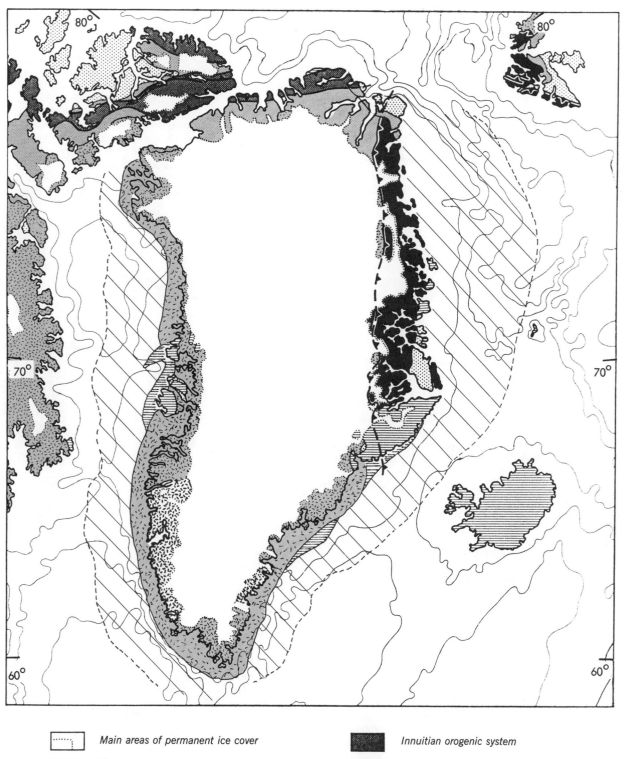

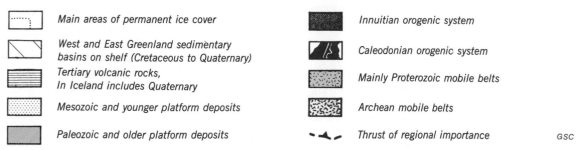

- Main areas of permanent ice cover
- West and East Greenland sedimentary basins on shelf (Cretaceous to Quaternary)
- Tertiary volcanic rocks, In Iceland includes Quaternary
- Mesozoic and younger platform deposits
- Paleozoic and older platform deposits
- Innuitian orogenic system
- Caleodonian orogenic system
- Mainly Proterozoic mobile belts
- Archean mobile belts
- Thrust of regional importance

GSC

Seismic and aeromagnetic investigations have shown that these sediments, which were deposited mainly in the Tertiary and Quaternary, attain thicknesses of 7 km (Larsen, 1980; Henderson et al., 1981), and the record preserved in them provides an indirect record of the development of the physiography of the adjacent land.

Development of preglacial physiography

As noted above, the shaping of Greenland and its preglacial topography is a consequence of ocean floor spreading in adjacent oceans during the Tertiary. While the oceans formed, tectonic adjustments took place along the margins of the continental block, resulting in subsidence of the seaward border and uplift of the land.

Using geomorphological criteria, remnants of ancient uplifted terrain surfaces have been recognized in many parts of Greenland (see summary by Weidick, 1975a). The most extensive of these features is the "initial topography" described from East Greenland by Ahlmann (1941). The "topography" lies between 72° and 76°N, and extends from under the Inland Ice to the coast. It formerly was an undulating plain, but now is preserved as deeply dissected mountain plateaus and summit areas at approximately 2000 m elevation. The plain is thought to have developed during a prolonged period of denudation and fluvial erosion and to have been uplifted in the late Tertiary.

From the structures of early Tertiary plateau basalt and fission track dating of later intrusions in southeast Greenland, Brooks (1979) and Gleadow and Brooks (1979) were able to show that large-scale fluvial erosion was followed by epeirogenic uplift of up to 2.5 km, in Oligocene and Miocene times. Great thicknesses of Upper Oligocene to Upper Miocene terrigenous sediments on the adjacent shelf may be a consequence of this uplift (Larsen, 1984).

West Greenland also has experienced considerable uplift in coastal areas so that deep-crustal rock types on land, border on a thick sequence of mainly Cenozoic fluvial and marine sediments on the adjacent shelf. Precise timing of this uplift is still lacking; however, analyses from boreholes on the shelf indicate initial Paleocene-early Eocene sediment was derived from an extensive drainage system in a deeply weathered land area with mature relief. Late Eocene and Oligocene clastic sediments reflect rejuvenation of the relief by uplift of the land (Henderson et al., 1981).

Hence it appears that the mountains of Greenland — like those elsewhere around the North Atlantic — initially were formed by late Tertiary epeirogenic uplift. The uplift apparently was most intense in the east, causing the asymmetry of drainage. Presently the main drainage divide runs north-south near the eastern ice margin, leaving large parts of the Inland Ice to flow towards the west, while only smaller sectors drain eastwards (see Chapter 14 on the Inland Ice), the development of the eastern drainage pattern in Paleocene to Pliocene times has been described by Larsen (in press).

A major drainage outlet from the Inland Ice is located on the west coast in the Disko Bugt area (Fig. 13.2). In this area early seismic investigations showed the presence of deep subglacial channels extending more than 200 km inland from the ice margin (Holtzscherer and Bauer, 1954). It was suggested by Weidick (1975a, Fig. 1) that these channels were part of an extensive preglacial drainage system. Their presence was later confirmed by airborne radio echo sounding (Gudmandsen, 1978) and by analysis of Landsat satellite imagery of the Inland Ice surface (Thomsen, 1983).

Although the channels in their present form may well be glacially eroded, bathymetric investigations on the shelf give some support to the idea that the largest preglacial drainage outlet of Greenland was located outside the present Disko Bugt. On the shelf off the mouth of this bay a large half circular structure with a diameter of more than 200 km protrudes from the outer shelf margin, and was interpreted as a preglacial fluvial delta cone by Henderson (1975, Fig. 25). Similar, but smaller structures also occur outside other major fiord systems and developed as prograding deltas in the late Tertiary possibly in the Pliocene — at a time when base level was ca. 300 m lower than at present (Sommerhoff, 1979).

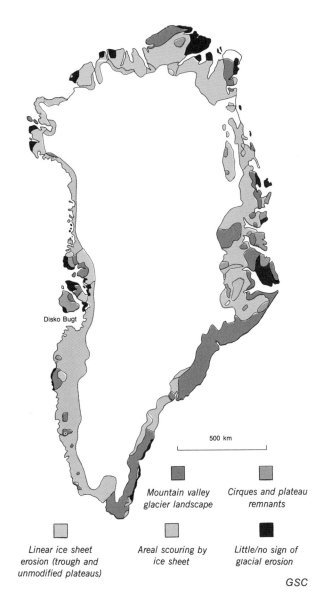

Figure 13.2. Landscapes of glacial erosion in Greenland (from Sugden, 1974).

It therefore seems probable that the main features of the present Inland Ice drainage were inherited from a pre-existing fluvial drainage system with a main outlet in the Disko Bugt area.

Present physiography — an overview

The main agent responsible for forming the Greenland landscape is ice — in the form of ice sheets and glaciers. An excellent survey of landscape types has been provided by

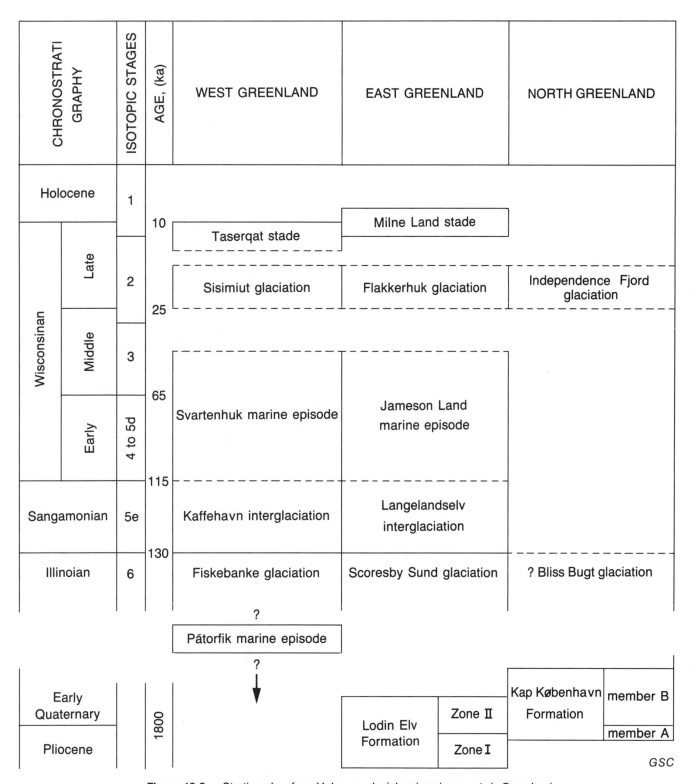

Figure 13.3. Stratigraphy of pre-Holocene glacial and marine events in Greenland.

Sugden (1974; Fig. 13.2). The landscape types are classified from type and intensity of glacial erosion, ranging from areal scouring by an ice sheet to landscapes of selective linear erosion occurring in areas where movement is concentrated in ice streams. Figure 13.2 shows a general trend of decreasing erosion intensity from the south and west towards the north and east. This can be related to changes in bedrock elevation and lithology, and to climate.

Areas with little or no sign of glacial erosion occur especially in East and North Greenland. This type of landscape was thought to reflect polar type glaciation with no basal slip at the ice-bedrock contact (Sugden, 1974). As discussed below, however, stratigraphic evidence seems to show that some of these areas have escaped glaciation for a long time so the "little or no sign of glaciation" may be a factor of time since glaciation rather than style of glaciation. A possible implication from this interpretation is that the general distribution of landscape types is caused not just by conditions during the last glaciation, but by a repetition of this pattern during several glaciations — possibly throughout the Quaternary.

Quaternary stratigraphy and terminology

Modern stratigraphic work in Greenland has aimed at establishing local successions of marine and glacial events. The main units with their predominantly informal names are shown in Figure 13.3, and discussed in the regional sections below. Most of these units have been established within the last decade, as a result of the application of modern dating techniques, and some — the Independence Fjord and Bliss Bugt glaciations — are introduced here for the first time.

For chronostratigraphic correlations the North American, the northwest European, and even the central European terminologies have been used by field workers in Greenland. In the present report correlation is made with the North American chronostratigraphy which is considered to match that of northwest Europe, thus Illinoian is considered equivalent to Saalian, Sangamonian to Eemian, and Wisconsinan to Weichselian. This implies that the Sangamonian in this report is considered equivalent to substage 5e in the isotopic record, and hence has a short duration from ca. 130 to 115 ka (e.g., Shackleton and Opdyke, 1973; Mangerud et al., 1979). Consequently the Wisconsinan has a long duration from ca. 115 to 10 ka.

Table 13.1. Mean temperatures for warmest (w) and coldest (c) months and annual precipitation at five stations in the period 1969-1979 (from Publikationer fra det Meteorologiske Institut).

	Latitude	°C w	°C c	Precip. mm
Station Nord*	81°36'N	3	-34	256
Scoresbysund	70°25'N	3	-20	549
Godhavn*	69°14'N	7	-16	479
Narssarssuaq	61°11'N	10	-11	649
Prins Christian Sund	60°02'N	7	-6	2505

*Observations incomplete

Climate, permafrost, oceanography, and biogeography

This section presents a brief survey of the physical environment in the ice-free land, to serve as a background to the treatment of Quaternary ecosystems and paleoclimates. Details of the Inland Ice climate are given in Chapter 14 dealing with the Inland Ice. A comprehensive treatment of the climate of Greenland has been supplied by Putnins (1970).

The distribution of temperatures and precipitation is here illustrated in a general way by observations from five stations showing climatic gradients from north to south (Station Nord - Prins Christian Sund), from west to east (Godhavn - Scoresbysund), and from the outer coast to an interior fiord area at almost the same latitude (Prins Christian Sund — Narssarssuaq) (Table 13.1 and Fig. 13.4).

Owing to the stabilizing influence of the Inland Ice and the meridional pattern of air circulation, summer temperatures are remarkably uniform from north to south. However, the number of months with mean temperatures above 0°C, a measure of the length of growing season, increases from one in the north to seven in the south. The latitudinal effect is

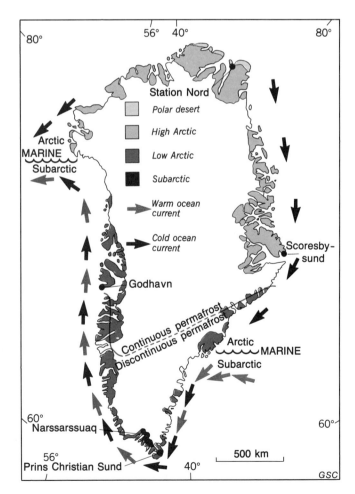

Figure 13.4. Terrestrial and marine biogeographic divisions and permafrost distribution in Greenland (modified from Funder, 1978c; permafrost boundary from Weidick, 1975a).

clearly visible in the biologically less important winter temperatures. Besides the gradient determined by latitude, all areas show a gradient from oceanic climate at the outer coasts to continental climate in the regions bordering on the Inland Ice margin. Precipitation is mainly supplied by maritime air following cyclone tracks along the southeast and west coasts, and is most abundant in the south.

As pointed out by Weidick (1975a), the southern limit for continuous permafrost seems to follow the mean annual temperature isotherm of −5°C and extends farthest south in continental areas close to the Inland Ice margin (Fig. 13.4). To the south of this limit, discontinuous and sporadic permafrost are encountered.

Especially in coastal regions, the climate is strongly influenced by the oceanographic circulation pattern. The main feature is the East Greenland Polar Current carrying cold polar water from the Arctic Ocean south along the east coast. Off southeast Greenland the polar water mixes with warm Atlantic water from the Irminger Current, and the mixture, forming the West Greenland Current, flows around the southern tip and northwards along the West Greenland coast, as far north as 78°N (Fig. 13.4).

Owing to the influence of the East Greenland Polar Current, East Greenland is colder than West Greenland at the same latitude. This asymmetry is apparent in the distribution of biogeographical zones (Fig. 13.4). The warmest summers occur in the interior south, where subarctic vegetation with birch and rowan woods occur in sheltered valleys. The major part of West Greenland is characterized by Low Arctic vegetation with willow and alder copses, while northeast and North Greenland falls within the High Arctic, with vegetation dominated by dwarf shrub heaths composed of dwarf-birch and a number of ericaceous dwarf-bushes. To this scheme may be added a zone of polar desert in coastal regions north of 80°N, where woody plants are absent or extremely rare.

In the shallow marine environment, the boundary between subarctic and arctic water masses is defined by the northernmost occurrence of such subarctic molluscs as *Mytilus edulis*, *Chlamys islandica*, and *Littorina saxatilis*. In northwest Greenland this boundary has fluctuated over several degrees of latitude in this century (see section on *Paleofaunas and floras*).

QUATERNARY GEOLOGY OF WEST GREENLAND

S. Funder

Topography, drainage, and glaciation

West Greenland comprises the rim of ice-free land from Kap Alexander (78°N) to Kap Farvel (60°N) (Fig. 13.5). The rim attains maximum width, approximately 200 km, in the Søndre Strømfjord region, whereas at Melville Bugt to the north it is absent except for a few nunataks.

The dominant landscape type is hilly upland composed of rounded knolls of crystalline bedrock at elevations between 300 and 1500 m. In some areas the hilly uplands abut on areas of distinctly alpine topography with elevations of 2000 m or more. Such areas occur in the south where the mountains extend below the ice cover, and locally between 65° and 67°N. In the latter area, the high mountains near the coast border on a 10 to 30 km-wide coastal strandflat, composed of heavily abraded gneiss at elevations below 300 m.

To the north of Disko Bugt the thick sequence of plateau basalt gives rise to a different type of topography characterized by gently undulating high mountain plateaus dissected by cirques and steep sided valleys which form a complex drainage pattern.

South of 65°N the land areas are dissected by numerous fiord troughs with depths down to 600 m below sea level, the fiords are usually headed by calving glaciers. To the north, up to 68°N, there are fewer fiords and they are headed not by glaciers but by broad valleys carrying meltwater from the Inland Ice margin.

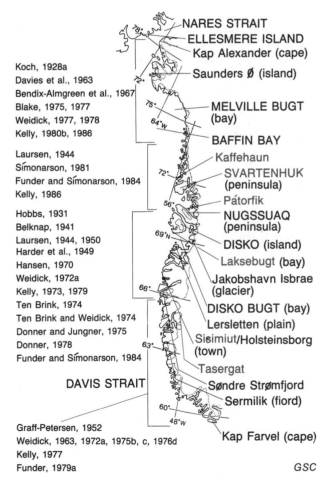

Figure 13.5. Regional Quaternary studies in West Greenland and locality names mentioned in text. Names shown in red have been utilized in stratigraphic terminology.

Funder, S.
1989: Quaternary geology of West Greenland; in Chapter 13 of Quaternary Geology of Canada and Greenland, R.J. Fulton (ed.); Geological Survey of Canada, Geology of Canada, no. 1 (also Geological Society of America, The Geology of North America, v. K-1).

Disko Bugt is the largest inlet in the West Greenland coastline, and separates Disko Island from the "mainland". This inlet provides easy access into Baffin Bay for icebergs from several productive calving glaciers. The largest of these is Jakobshavn Isbrae which has an annual calving production of 25 km^3 water equivalent of ice, and is the single largest outlet from the Inland Ice (Reeh, 1989).

For a distance of some 300 km in Melville Bugt the Inland Ice margin is located close to the present shoreline. In this sector, the products of ablation by calving and melting are discharged directly into Baffin Bay.

As noted above, each general type of topography in this region appears to foster different mechanisms for ablation from the Inland Ice margin. Hence each sector may respond differently to climatic change and these differences are reflected in the history of Holocene deglaciation.

Quaternary studies

West Greenland south of 73°N supports the majority of Greenland's population and is the location of the centres of administration; as a consequence the area has a richer and longer record of scientific research than other parts of Greenland. Systematical geological studies began after 1850 and included work on the Quaternary; this work has mainly been associated with the mapping projects of the Geological Survey of Greenland.

Earlier reviews of the Quaternary geology in this region comprise works by Weidick (1968) and Kelly (1985), emphasizing the Holocene and Late Quaternary stratigraphy. A number of regional studies are listed in Figure 13.5, the results have been compiled by A. Weidick in two Quaternary maps, covering the area from 63° to 71°N (Geological Survey of Greenland, 1974, 1978).

Nature and distribution of Quaternary sediments and landforms

All of West Greenland — excluding only some high mountains near the coast — was covered by the Inland Ice during the Sisimiut glaciation of Late Wisconsinan age. This recent glaciation is clearly recorded by erosion rather than sediment accumulation in most areas. Thick Quaternary deposits are of restricted occurrence and are generally confined to major valleys and lowlands along the coasts.

Till. The most widespread glacial deposits are patches of loose gravelly and sandy diamicton and scattered erratic boulders considered to be melt-out till. These materials form a continuous cover in the interior near the present Inland Ice margin, whereas at the outer coasts glacial deposits are reduced to scattered erratics lying on the abraded bedrock surface. This thin sediment, modified by later slope wash, originated as sparse debris in the last ice body which covered the area, and was lowered down onto the terrain surface during a brief stage of stagnation and melting (e.g., Funder, 1979a). Thicker deposits of melt-out till occur in lateral and terminal moraines and are especially common between 64° and 67°N, where they form north-south trending zones reflecting periods of stillstand or even readvance of the Inland Ice margin in middle Holocene time (Weidick, 1972a; Ten Brink and Weidick, 1974; Ten Brink, 1975).

Whereas melt-out till is widespread, typical lodgment till is rare, occurring as patches on the stoss side of glacially abraded bedrock knolls (Sugden, 1972). The apparent rarity of this sediment may possibly be explained by the absence of fine grained source material for glacial erosion.

Since till deposits in all parts of the region were laid down during the deglacial stages following the Sisimiut glaciation, there is generally no discernible difference in surface freshness. An exception to this occurs in coastal mountains above 1000 m elevation where weathered erratics are associated with autochthonous felsenmeer and tors, showing that these areas were nunataks during the Sisimiut glaciation (Kelly, 1985).

Till clasts and transportation routes. Assessing source areas and transportation routes for the till components is made difficult by the monotony of the bedrock geology, which provides few indicator boulders. In an area of varied bedrock lithology in South Greenland, however, it has been shown that more than 50% of the stones, pebbles, and boulders could be accounted for by known bedrock exposures within 10 km of the site (Funder, 1979a). The tills probably also contain a long distance transport component, although this is rarely recognizable. This has been documented in the area north of Disko Bugt where Steenstrup (1883) found abundant gneiss erratics at the outer coast and on high mountain plateaus, 150 km from their nearest source areas to the east.

Considering that ice movement, possibly throughout the Quaternary, has been from the presently ice covered interior towards the coast, it is noteworthy that only in one area, to the south of Melville Bugt, have exotic erratics been found that cannot be referred to rock types exposed in the narrow rim of land which is now ice free (M. Kelly, University of Lancaster, Lancaster, England, personal communication, 1985).

A peculiar type of till clast, which has been observed especially in the most recent till deposits close to the ice margins, is a calcareous concretion which commonly contains a nucleus of organic matter. An interglacial age has been suggested for some concretions (Bryan, 1954), but it is more likely that the concretions were derived from middle Holocene marine deposits which, in late Holocene time have been overridden and eroded by glaciers (Kelly, 1975, 1980a).

Moraines. Moraines dating from Holocene deglaciation stages occur in all parts of the area, and in their general distribution follow that outlined for till deposits. They are especially abundant in the inland parts of the region which is the classic area for the study of the deglaciation history in Greenland (Weidick, 1968, 1972a; Sugden, 1972; Ten Brink and Weidick, 1974; Ten Brink, 1975).

Moraines generally have developed only along active sectors of the ice margin — lobes and outlet glaciers — while the regionally more extensive passive sectors have created few moraines. This is especially clear in fiord regions where moraines along the fiords testify to the former existence of fiord glaciers, while there are few or no traces of the corresponding ice margins on the interfluves between fiords. The location of moraines along the fiords — at fiord junctions and bends, and at places where the sides change from steep to gentle slopes — indicates that the moraines commonly were formed as an interaction between the glacier and the topography of its bed, rather than in response to climatic change.

From the location and dating of the moraines it appears that the sensitivity of the Inland Ice margin to climatic change in the Holocene was strongly controlled by the ablation mechanism. Consequently, in the Disko Bugt area and the fiords to the south, where ablation took place by calving, the marginal recession was faster by several millennia than it was in the intervening inland areas where melting was the main mechanism of ablation (Weidick, 1984).

Glaciofluvial and fluvial deposits. These sediments cover the floors of all major valleys, occurring as outwash plains and fluvial terraces deposited from braided rivers. The most extensive plains occur in the valleys which now connect the Inland Ice margin with the heads of fiords. Usually the valleys contain several generations of river terraces formed largely by changes in base level caused by glacial isostatic movements. In our century some of these plains have been utilized as air fields, and they are the main transportation routes in the country.

Glaciofluvial sand and gravel also occur as kame terraces along valley sides and — though of more restricted occurrence — as kame and kettle topography on some valley floors. These features, as well as the abundant meltwater channels which may be incised into bedrock, were formed mainly in a proglacial environment. There is sparse evidence for subglacial meltwater: eskers have been observed only in the interior continental areas, and potholes in bedrock are not common.

Marine deposits. Marine sediments ranging from coarse littoral gravel to massive or laminated silt are widespread in the coastal areas, occurring up to 140 m above present sea level — the maximum elevation of Holocene marine limit. Marine sediments commonly occur in a prodeltaic facies adjacent to major raised deltas at the mouths of large valleys. The most extensive marine deposits occur at Lersletten to the south of Disko Bugt. In this area marine silt and sand form a plain between knolls of gneiss and were deposited in early and middle Holocene time in a skaergaard (skerryguard) which was later isostatically raised. These sediments have yielded rich mollusc faunas which were the basis for the first attempts to establish a Quaternary chronology in Greenland (Jensen and Harder, 1910).

Marine sand and silt older than the Sisimiut glaciation occur as scattered erosion remnants, especially in areas north of Disko Bugt. These deposits are especially important for the understanding of the Quaternary chronology, and their ages have been subject to much discussion, as mentioned below.

Periglacial and eolian features. The distribution of frozen ground features closely follows that of continuous permafrost (Fig. 13.4). Thus pingos, ice wedges, and rock glaciers are rare south of 67°N.

The large majority of pingos have been observed in valleys in the area of Mesozoic sediments north of Disko Bugt (Weidick, 1975a). Analyses of gas and water from the pingos have shown a very high content of methane which probably aided the formation of these pingos which are believed to be of the open system type (Henderson, 1969). In addition, rock glaciers seem to be especially common on Disko and Nugssuaq, on steep valley sides (Humlum, 1981).

Palsas and other types of frost mounds are most common in the border area between continuous and discontinuous permafrost. To the south of this area they may have formed in response to neoglacial permafrost expansion

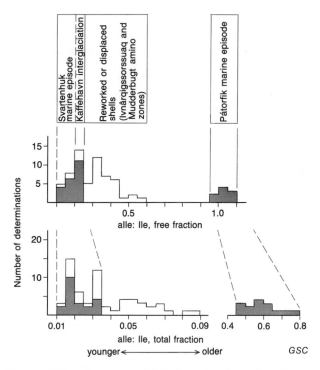

Figure 13.7. Frequency distribution of amino acid ratios on mollusc shells (*Mya truncata* and *Hiatella arctica*) and event stratigraphy in West Greenland. Shaded columns, shells found in situ; open columns, reworked shells or shells from reworked deposits. The ratios belong to 25 samples with 3-6 determinations in each. Total ratios obtained before 1982 have been corrected for laboratory fractionation (Kelly, 1986). Correlation is here based on ratios in the free fraction because ratios in the total fraction show more spread and less consistency, possibly as a result of leaching. Sources: Funder and Simonarson, 1984; Kelly, 1986; stratigraphy adapted from Kelly, 1985.

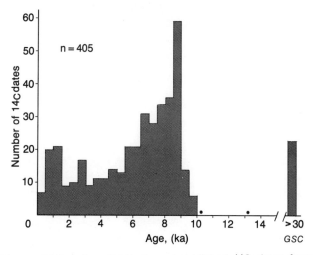

Figure 13.6. Age distribution of published ^{14}C dates from West Greenland, including dates obtained for archeological purposes. Dots denote single dates. Compiled from dating lists in reports of the Geological Survey of Greenland, Radiocarbon, and elsewhere.

(Kelly, 1981), while in the north palsa-like peat mounds seem to date from the Holocene warm period (Brassard and Blake, 1978).

In the interior parts strong winds from the Inland Ice may raise dust storms over the large fluvial plains. Coarser sediments are deposited as small sand dunes in the valleys, while dust settles in vegetated areas nearby to form a thin layer of loess (Böcher, 1949; Hansen, 1970).

Succession of events

The Quaternary stratigraphy of the West Greenland region has been treated, with emphasis on the Holocene, by Weidick (1968). An excellent review of this and more recent work has been prepared by Kelly (1985), who carefully discussed the validity of the data and offered alternative interpretations. The presentation given here relies heavily on Kelly's work — both for concepts and stratigraphic terminology. However, I have chosen simple explanations where there is no compelling reasons for complex ones, and the stratigraphy presented here (Fig. 13.3) is a slightly modified version of Kelly's.

Approximately 400 ^{14}C dates, made mainly on bivalve shells, peat, and gyttja, have been obtained from the region in order to date Holocene and Wisconsinan events (Fig. 13.6). As elsewhere in Greenland, ages in the interval from ca. 10 to 20 ka are rare and may be due to contamination of the dated material. There seems to be no good reason, however, to suspect the age of 13 380 ± 175 BP obtained on marine shells in southern West Greenland (I-7624, Weidick, 1975c).

In recent years amino acid analysis of mollusc shells has become an important tool in the differentiation of older deposits (Fig. 13.7). Shells from 25 samples have been analyzed, and the results have been discussed by Funder and Símonarson (1984) and Kelly (1985, 1986).

As in other historical records the Quaternary stratigraphy in this area shows increasing detail with decreasing age. Owing to the ravages of the ice sheet during the Sisimiut glaciation, however, the record has an abrupt break; there is little evidence for the record before this event, while the time after is known in reasonable detail.

Sea level history, subfossil marine faunas, and vegetation development are treated in a later section which covers all Greenland.

Pátorfik marine beds. The Pátorfik marine beds at the coast of Nugssuaq peninsula (71°N, Fig. 13.8) comprise a 40 m-thick sequence of deltaic and prodeltaic sand, silt, and mud, exposed along 2.5 km of the coast. These sediments contain a rich mollusc fauna which attracted the attention of geologists as early as 1848. The deposits, as well as their long history of widely different dating estimates, have recently been described in detail by Símonarson (1981). The rather peculiar mollusc fauna combined with high amino acid ratios has resulted in an early Quaternary age being assigned to these deposits (Fig. 13.7 and see *Paleofaunas and floras*; Funder and Símonarson, 1984). The sediments probably owe their preservation to a covering bed of lithified talus breccia.

Fiskebanke glaciation and Kaffehavn interglaciation. Subarctic marine molluscs with infinite ^{14}C ages have been

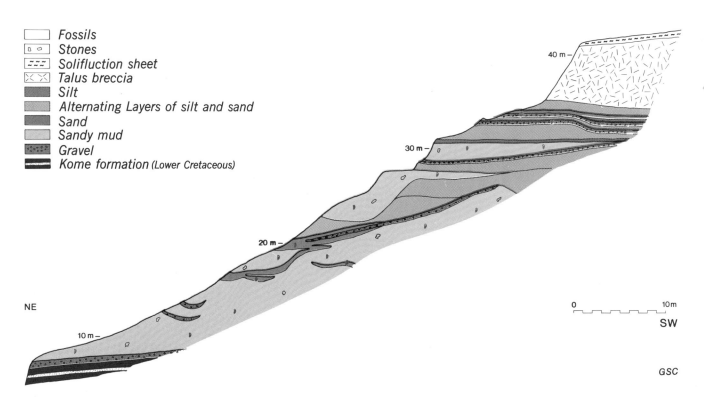

Figure 13.8. Section of the Pátorfik marine beds (from Símonarson, 1981) showing transition from prodeltaic (5-20 m a.s.l.) to deltaic (20-35 m a.s.l.) facies. The marine beds are covered by a younger talus breccia. Mollusc fauna is of boreal-subarctic type throughout the sequence.

found in marine deposits at sites in northern West Greenland and reflect a hydrographic pattern similar to the present (Funder and Símonarson, 1984; Kelly, 1985, 1986). Amino acid analyses and faunal composition support correlation of the deposits with subarctic molluscs in littoral gravel below till in a 7 m section at Kaffehavn. Known from four localities in northern West Greenland, these deposits are referred to the Kaffehavn interglaciation, provisionally correlated with the Sangamonian (Fig. 13.3).

The southernmost of these localities, Laksebugt, shows marine conditions colder than the present, and these deposits have tentatively been correlated with an early deglacial stage of the Kaffehavn interglaciation (Funder and Símonarson, 1984).

On Saunders Ø the subarctic marine fauna is underlain by till (Davies et al., 1963; Blake, 1975), which consequently should predate the Sangamonian and tentatively is referred to the Illinoian.

It is noteworthy that although the deposits at all these sites may well contain significant hiata, none contains positive evidence for more than two periods of extensive glaciation, and only one in the Wisconsinan. In more southerly parts of West Greenland the oldest glaciation phase is denoted by deeply weathered till on high coastal mountains representing the Fiskebanke glaciation, which is also — tentatively — correlated with the Illinoian. It should be pointed out, however, that age control is poor and is based mainly on the Saunders Ø section[1] (Kelly, 1985).

Svartenhuk marine episode. The Early and Middle Wisconsinan record is known from scattered localities in the northern part of the region where ice sheet erosion has been least severe (Fig. 13.9). These occurrences comprise marine fossiliferous sand and silt with sparse arctic molluscs, overlain by till. Some deposits are in situ while others appear to be megaclasts in younger till.

The marine deposits, here referred to as the Svartenhuk marine beds, give ages beyond ^{14}C dating limit or old-finite ^{14}C ages (more than 20 ka), and amino acid data indicate that they belong to the same depositional episode, following the Kaffehavn interglaciation. Sedimentation lasted well into the Wisconsinan (Fig. 13.7), and the episode possibly terminated about 40 ka (Kelly, 1986).

Shells from a number of "reworked" deposits have yielded amino acid ratios that would imply a somewhat older depositional period (Fig. 13.7). Since these deposits are glacially reworked or transported as megaclasts, however, the shells do not necessarily satisfy one of the basic requirements for providing correlatable amino acid ratios — they have not necessarily shared the same thermal history. Therefore, until the ratios can be duplicated from in situ deposits, they should not be included in correlation charts for this area.

Sisimiut glaciation. During the Late Wisconsinan Sisimiut glaciation the Inland Ice expanded and transgressed the present coastline in all parts of the region. The ice masses moved onto the present shelf where their margin may have been located ca. 50 km offshore on the inner shelf (Kelly, 1985). The glacial geomorphology of the shelf shows, as discussed in a later section, that the ice sheet at this time was mainly land based in a period of eustatically lowered sea level (see *Quaternary geology of the shelves adjacent to Greenland*).

A model of the ice flow, developed by Reeh (1984), shows a pattern similar to the present with westward directed flow from a central ice divide located near the present one. Only in the Nares Strait region of northwest Greenland did a more complex flow pattern develop as a result of the

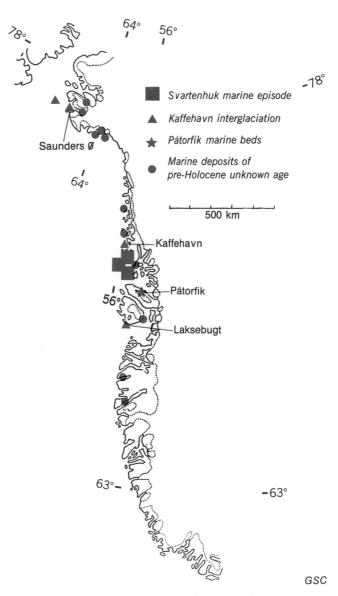

Figure 13.9. Occurrence of pre-Holocene marine deposits in West Greenland (modified from Kelly, 1985).

---

[1] This important locality, and others in the area, was investigated in detail by members of the NORDQUA expedition to the area in 1986, and extensive thermoluminescent dating, amino acid analyses, as well as faunal and lithological work have been carried out. Although the time frame remains, much new information has been gained, some of which alters concepts reported here. Preliminary results from this collective work have been reported by Feyling-Hanssen, Funder, Houmark-Nielsen, Kronborg, Mörner, Reeh and Thomsen, Sejrup and Sorby in Danmarks geologiske Undersøgelse(1988).

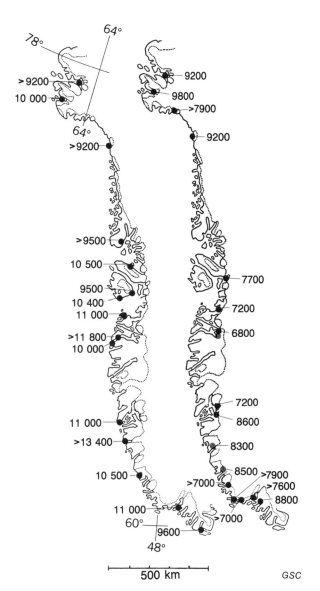

Figure 13.10. Ages for deglaciation at the outer coasts (left), and dates for the attainment of the present state of glaciation (right) (from Kelly, 1985). Deglaciation ages are based on ^{14}C dates of marine molluscs and uncertainties are ±500 years. Ages for the attainment of present glaciation are from ^{14}C dates of marine shells in front of ice margins, and on organic material reworked in younger moraines (red).

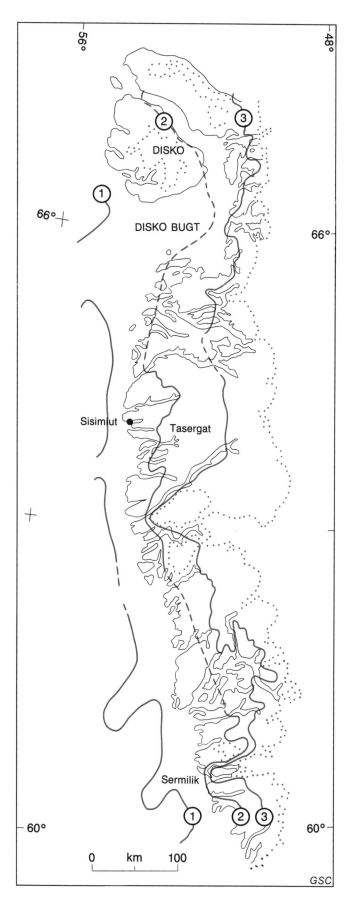

Figure 13.11. West Greenland Inland Ice margins during (1) Sisimiut glaciation (ca. 14 ka), (2) Taserqat stade (ca. 10 ka), (3) and Fjord stade (8 ka). Dotted line shows present ice margin. Sources: Weidick, 1975c, 1984; S. Funder, unpublished; shelf moraines from Figure 13.26.

Table 13.2. Ages and correlation of Holocene moraine systems in West Greenland (adapted from Kelly, 1985).

Age ka BP	Disko Bugt (Weidick, 1968; S. Funder, unpub.)	N. Isortoq N. Stromfjord (M. Kelly, unpub.)	Holsteinsborg S. Stromfjord (Ten Brink and Weidick, 1974)	Godthaabsjford (Weidick, 1975c)	Julianehaab Frederikshaab (M. Kelly unpub.; Weidick, 1963)	Narssaurssuaq
					Sarfa	
-		Isortoq				
10					Eqaluit	Niaqornakasik
		Y Isortoq	Taserqat			
-	Disko stade	Itivdlerssuaq	Avatdleq			
	"Marrait"	Maligia	Old Fjord	Kapisigdlit	Aussivik	Tunugdliarfik
8	(Fjord 2)		Aukua 1 & 2	Young Fjord		
			Umivit			
-	"Tasiussaq" (Fjord 1)	Ugssuit	Mt. Keglen			
6						Narssarssuaq
-						
2						
-			Orkendalen			
0	Historic	Historic	Historic	Historic	Historic	Historic

coalescence of the Inland Ice with an ice sheet over Ellesmere Island. This region is discussed in the section on North Greenland.

There is no dating control for the growth phase of the ice sheet. It is only known that maximum advance occurred after the Svartenhuk marine episode mentioned above. Retreat from the Wisconsinan limit began at or before 13.5 ka, as deduced from ^{14}C dates in the coastal areas (Kelly, 1985). The disintegration occurred in two distinct phases: disintegration of marine parts and melting of land-based parts.

The early phases of deglaciation are poorly known because they took place in areas that are now covered by the sea. It is noteworthy, however, that the radiocarbon dates indicate that the ice margin in all parts of the region was located near the present coastline at 10-11 ka (Fig. 13.10, 13.11). This synchroneity can best be explained by the assumption that the breakup of the ice sheet was triggered by a eustatic sea level rise rendering the shelf-based portions unstable along the entire margin. The ice sheet adjusted to the change in sea level by retreating until the margin was again land based, near the present coastline.

The shelf areas were not the only areas cleared of ice during this phase. Major inlets such as the fiords of South Greenland, Disko Bugt, and the fiords in the northwest, were also deglaciated as shown by ^{14}C dates. Thus by 10-11 ka the ice margin roughly coincided with the present coastline, and it was after this time that the present land was uncovered (Fig. 13.10, 13.11). At Sermilik Fjord, 63°N, a ^{14}C age of 13 380 ± 175 BP has been obtained for marine shells, indicating that in this particular area the ice margin had retreated from the shelf before this time, and remained stationary at the present coast line for the following five millennia (I-7624, Weidick, 1975c).

Holocene. The second stage in the retreat of the Sisimiut glaciation ice sheet involved melting of land-based ice. From the dating of moraines it can be seen that the melting proceeded at different rates in different areas with local climate and topography the main controlling factors.

The fastest retreat took place in the warm and dry areas of West Greenland between 65° and 68°N where a 200 km-wide rim of land was exposed within the next 3-4 ka, even though some regional moraines were built during this time (Table 13.2 and Fig. 13.11). The slowest retreat was in the cool moist climate of Melville Bugt in the north, where the ice margin never melted back from the sea-fronting position it has occupied since 8-9 ka, as implied by dates obtained by Kelly (1980b) and Fredskild (1985b).

In each area the course of deglaciation has been studied from its moraines and other ice marginal features, and moraine stratigraphy has been developed especially in central West Greenland, where individual segments of moraines can commonly be followed for tens of kilometres (Weidick, 1968, 1972a; Sugden, 1972; Ten Brink and Weidick, 1974; Ten Brink, 1975). The results from these areas do not correspond well with those to the north and south (Kelly, 1985), consequently it appears that local topography may have played an important part in moraine formation. In general terms however, two periods of moraine formation can be recognized in most areas, at 8.5-8 ka and around 7.5 ka (Table 13.2; Fig. 13.11).

During deglaciation, local ice caps were isolated on some highland areas. In Disko Bugt the disappearance of the marine Inland Ice tongue was followed, at 9 ka, by a major readvance from a local ice cap on the highlands of Disko Island into the waters of the newly ice-free embayment. It has not been possible to demonstrate any contemporaneous change in the behaviour of the adjacent Inland Ice margin, and this readvance, the Disko stade, probably was a response to increased precipitation in coastal areas (S. Funder, unpublished).

Deglaciation eventually brought the ice margins behind their present locations, as documented by the occurrence of reworked early Holocene marine sediments and shells in young moraines at many sites (Fig. 13.10). This phase, when the Inland Ice sheet margin may have been more than 10 km behind the present one, is dated to the period from ca. 6 to 3 ka (Kelly, 1980a; Weidick, 1985).

Thus the disintegration of the Sisimiut glaciation ice sheet, which lasted for ca. 10 ka, may be envisaged as a transition from a stable full glacial ice sheet in a cold, dry climate to a stable postglacial ice sheet in a mild, humid climate (Weidick, 1975b). Sea level during this same time rose from a low glacial to a high postglacial position.

At ca. 3 ka glaciers began again to advance, culminating with glacier maxima in the 18th, 19th, and early 20th centuries. Moraines from these advances have been dated by comparison with old written and photographic records by Weidick (1959, 1984, 1985) and by lichenometry (Beschel, 1961; Gordon, 1981). The moraines seem to group in three age categories: mid 18th Century, 1850-1890, and ca. 1920. In many areas the ice reached its late Holocene maximum in the 1880-1890 period, partly as a response to very cold conditions 100-200 years earlier (Weidick, 1984).

In the period 1920-1960, probably as a result of high summer temperatures, the ice margin and its outlet glaciers have experienced thinning and retreat, exposing a zone of fresh unvegetated ground moraine; however, ice movements are not synchronous, and while some sectors are retreating, others may be advancing (Fig. 13.12).

QUATERNARY GEOLOGY OF EAST GREENLAND

S. Funder

Topography, drainage, and glaciation

This region comprises the coastal areas between Kap Farvel (59°N) and Ingolf Fjord (80°N; Fig. 13.13) — an area characterized by its multitude of glacier-headed fiords. Based on physiography, the area falls into two subregions: a relatively homogeneous area of high mountain plateaus developed on the Caledonides north of Scoresby Sund and a variable area of alpine topography developed on Tertiary volcanics and Precambrian gneiss to the south (Fig. 13.1).

The difference between these two subregions is well shown by the nature and dimensions of fiords. South of Scoresby Sund, fiords — with some exceptions — are narrow, unbranched, closely spaced and less than 50 km long. North of Scoresby Sund in the "East Greenland fiord zone",

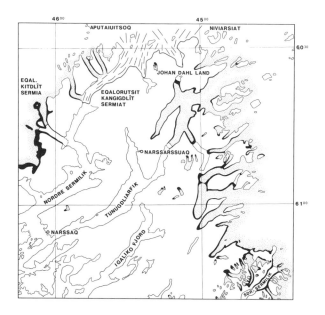

Figure 13.12. Advancing and retreating sectors of the Inland Ice in southwest Greenland (from Weidick, 1984).

Funder, S.
1989: Quaternary geology of East Greenland; in Chapter 13 of Quaternary Geology of Canada and Greenland, R.J. Fulton (ed); Geological Survey of Canada, Geology of Canada, no. 1 (also Geological Society of America, The Geology of North America, v. K-1).

fiords are organized in branching systems, penetrate the land up to a distance of 300 km from the coast, and narrow steep walled interior fiords contrast with the wide inlets at the coast. The major inlets, Scoresby Sund, Kong Oscars Fjord, Foster Bugt, and Gael Hamkes Bugt, have depths that rarely exceed 600 m, whereas the steep sided fiords of the interior are 800-1200 m deep, with a northern hemisphere record of 1500 m attained in Nordvestfjord in the Scoresby Sund fiord system.

In the Caledonides the topography is dominated by gently undulated high mountain plateaus at 1500-2500 m elevation which are commonly covered by small ice caps. In the areas of Mesozoic sedimentary rocks, on the other hand, the terrain is gently sloping from inland elevations of 1200 m to extensive areas of lowland near the outer coasts. The largest of these are Jameson Land peninsula in the southern part of the fiord zone, and Hochstetter Forland in its northern part.

The topographical contrast between the interior and the coastal parts of this region has played a major control on the impact of glaciation; in the interior glacial erosion resulted in the overdeepened fiord troughs, while the coastal areas to the east have suffered little erosion. In these latter areas, thick sequences of sediments have been preserved along the coasts, containing the most complete upper Quaternary record known from Greenland.

To the south of Scoresby Sund bedrock lithology and topography change abruptly. The thick Tertiary plateau basalts between 70° and 68°N form a landscape of arêtes and pinnacles rising over glacier-filled valleys and smaller ice caps which in the west are fused with the Inland Ice. To the south of 68°N, the bedrock is dominated by gneiss, forming alpine landscapes with decreasing elevation towards the south. The highest mountains in Greenland occur in this region, where Gunnbjørn Fjeld reaches 3700 m elevation.

The Inland Ice margin in this region is, in contrast to the west Greenland margin, composed mainly of high elevation passive sectors with large fiord glaciers accounting for most of the ablation. The only ice shelf in Greenland fills parts of the inlet Jøkelbugt at 79°N. It extends for 100 km in a north-south direction, and towards the open ocean it is buttressed against a chain of islands (Koch and Wegener, 1911).

In general terms, the landscapes of East Greenland suggest a different, more passive, and less erosive regime for the Inland Ice than those of West Greenland (Fig. 13.2).

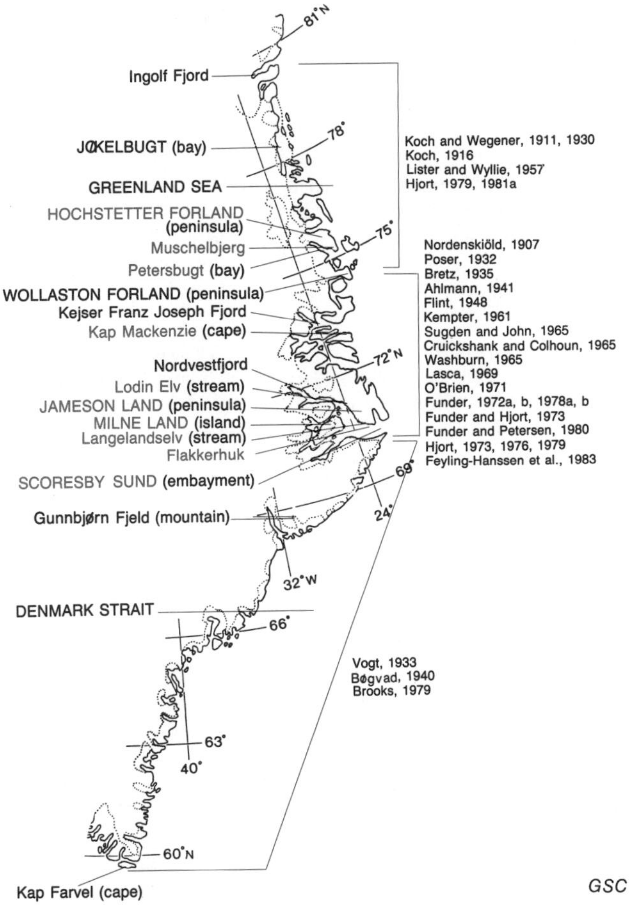

Quaternary studies

The East Greenland region is sparsely populated. Approach by ship has in some periods been difficult, owing to the belt of pack ice drifting along the coast on the East Greenland Polar Current. Although scientific exploration began in the early 19th Century, the first 100 years of research was carried out by expeditions working for a short time in restricted areas, which often were not of their own choosing, but determined by the pack ice situation.

Systematic geological work was initiated in 1926 by Dr. Lauge Koch's "East Greenland Expeditions" which, in the period up to 1958, mapped the whole northern part of the region south to 72°N. Unfortunately, Quaternary geology received only slight attention in this work. In the period 1968-1973 the bedrock of the Scoresby Sund area was mapped by the Geological Survey of Greenland. Some information on the Quaternary of this area is included in the resulting 1:100 000 geological maps. Other major activities include the Mestersvig geomorphic research program directed by A.L. Washburn in the period 1955-1969.

A number of regional Quaternary studies are cited in Figure 13.13. Little information is available for areas south of Scoresby Sund, and the treatment below concentrates on the East Greenland fiord zone.

Nature and distribution of Quaternary sediments and landforms

As noted above, the effects of the last major glaciation phase are not as prominent in this region as in West Greenland. Especially in coastal areas, glacial erosion was slight or absent, and here thick and continuous covers of pre-last glaciation sediments, the Jameson Land marine beds, have locally been preserved.

In the interior parts, on the other hand, the steep topography limits the distribution of loose sediments which are generally found only on valley floors and low lying areas of

Figure 13.13. Regional Quaternary studies in East Greenland and place names mentioned in text. Names shown in red have been utilized in stratigraphic terminology.

abraded bedrock along fiords. The high mountain plateaus, where not ice covered, are characterized by either till or autochtonous felsenmeer with scattered and weathered erratics.

Till. Coarse sandy till is widespread in lowlands along fiords as well as on some high mountain plateaus; usually, however, the occurrence is patchy and discontinuous, broken by bedrock knobs, and the thickness rarely exceeds a few metres. Indeed, the thickest and densest till cover is found on high mountain plateaus.

On the sandstone plateaus of Jameson Land, till and till-like deposits form the terrain surface over extensive areas between 500 and 1100 m elevation. These sediments range from sorted gravel and boulders to matrix-supported sandy and clayey till with a thickness of more than 10 m (Funder, 1972a). The clast fraction has a more than 50% representation of far-travelled crystalline rock types and includes the enigmatic Scolithos quartzite (see below). These deposits are highly weathered (fines removed and surface of clasts disintegrating) and are here referred to the oldest recognizable glaciation phase in the region, the Scoresby Sund glaciation, which in most other areas is recorded only by scattered exotic erratics in autochthonous felsenmeer or regolith.

Weathering differences in tills and in their associated landforms have been observed in all parts of the area and show that the region has been subject to a succession of glaciations, each less extensive than the preceding. Two or three distinct phases of glaciation were recognized by the early field workers (Koch and Wegener, 1930; Bretz, 1935; Flint, 1948; Lister and Wyllie, 1957; Washburn, 1965). In the chronology section below these observations are discussed in the light of recent absolute dating methods.

Till clasts and their transportation routes. The varied bedrock lithology, and the general north-south trend of contacts, allow the assessment of routes of glacial transportation from indicator boulders. Crystalline rock types, and boulders of upper Proterozoic and Paleozoic sedimentary rocks from the interior regions occur as glacial erratics in all parts, and their distribution is compatible with a simple model for ice movement along west-east trajectories.

A distinctive and characteristic erratic which occurs in all parts of the fiord zone from 70° to 78°N is Scolithos quartzite. In spite of its wide distribution as an erratic, however, this rock type has nowhere been observed in situ. Consequently it has been suggested that the boulders come from a bed of upper Proterozoic quartzite now hidden under the Inland Ice (e.g., Haller, 1971). Although this remains the most plausible explanation, it is surprising that a single bed of uniform lithology should extend over a distance of more than 800 km.

Both Scolithos quartzite and other far-travelled clasts have their highest frequency in coastal areas, outside the limits of the Late Wisconsinan Flakkerhuk glaciation, indicating that the dispersal of the boulders took place mainly during an earlier glaciation when an ice sheet expanded over the entire land area, crossing such topographical barriers as the Scoresby Sund basin and the mountains of Liverpool Land (Funder, 1972a).

Moraines. Moraines occur along the fiords and in valleys and are commonly located at topographical breaks, at fiord junctions and bends, and at valley mouths, indicating topographical rather than climatic control in their location.

The most conspicuous moraines are related to the Late Wisconsinan-early Holocene Milne Land stade. In general these are located at the junction between the narrow fiords of the interior and the wide coastal inlets between 70° and 76°N. In the Scoresby Sund area these moraines occur in swarms, and individual ridges may attain a height of 50 m and lengths of 5-10 km. They are composed of coarse sandy till with boulders of more than 10 m diameter on their surface (Sugden and John, 1965; Funder, 1978a; Hjort, 1979, 1981a).

Glaciofluvial and fluvial deposits. Glaciofluvial deposits composed of poorly sorted gravel and sand form outwash plains on valley floors and kame terraces along fiords and valleys, and were deposited mainly during the Holocene phase of glacier retreat. On coastal Jameson Land, sediments and landforms associated with dead ice disintegration — kames, kame and kettle topography, hummocky moraine, and kame deltas — occupy a zone up to 10 km wide along the entire western coast, and reflect the decay of a thin glacier in the adjacent Scoresby Sund basin, after the Flakkerhuk glaciation (S. Funder, 1984, unpublished).

Fluvial sand and gravel form extensive plains deposited from braided rivers in all major valleys. The plains are graded to sea levels higher than the present, and were deposited during the Holocene period of isostatic uplift. In coastal areas the poorly cemented Mesozoic sandstone provided the source for extensive fluvial erosion and large Gilbert type deltas formed. On western Jameson Land these sediments are incised into older marine deposits and form desert-like plains, covering up to 100 km^2.

Marine deposits. Marine deposits comprise prodeltaic, massive or laminated silt, and littoral and deltaic sand and gravel, occurring up to 130 m above sea level (Holocene marine limit in the region). Matrix-supported silty diamicton containing large boulders and resembling a lodgment till occurs in some areas. The setting of these deposits is, however, used to interpret the sediment as marine ice-drop sediment (Nordenskiöld, 1907; Washburn and Stuiver, 1962; Lasca, 1969; Feyling-Hanssen et al., 1983).

Marine sediments generally are restricted to valley mouths and coastal lowlands; however, some low lying valleys and valley systems contain marine sediments as much as 60 km up valley. These valleys were shallow fiords for some millennia after their deglaciation (e.g., Koch, 1916; Bretz, 1935; Cruickshank and Colhoun, 1965; Street, 1977).

Unique to this region are the thick Quaternary sediments which cover coastal lowlands at the mouths of major fiord systems and which have been named Jameson Land marine beds (Funder, 1984). The sediments locally attain a thickness of 100 m. In general they consist of a basal laminated and homogeneous silt with large boulders which grades upwards into stratified sand and silt with small and large scale crossbedding and infilled channels. In situ mollusc shells occur locally throughout the sequence and show, together with the lithology and structures, that the sediments were deposited in shallow water marine and deltaic environments. In most areas the top of the sequence is eroded and disturbed by later glacier overriding. The depositional environment and the events preceding and following deposition of these early Wisconsinan sediments are discussed further in the section *Succession of events.*

Periglacial and eolian features. The whole of the East Greenland fiord zone lies within the zone of continuous permafrost (Fig. 13.4), and cryoturbation structures and frost wedge polygons occur in Quaternary deposits throughout the region (e.g., Poser, 1932; Sørensen, 1935; Washburn, 1965, 1967, 1969). Open system pingos have developed in the fluvial deposits on the floors of major valleys as far south as 71°N on Jameson Land (Müller, 1959; Cruickshank and Colhoun, 1965; O'Brien, 1971). Small turf hummocks are a common feature in mossy meadows, and may be related to palsas (Raup, 1965). Rock glaciers occur on the steep fiord sides of the interior (Funder and Petersen, 1980).

Eolian features occur in the major valleys which, as elsewhere in Greenland, serve as channels for strong winds. The effects of deflation can be seen especially on old fluvial terraces which often have developed a deflation armour consisting of a pebble layer at the surface. The windblown sediments are deposited as a mantle of fine sand and silt in adjacent vegetated areas, and locally small sand dunes have developed on valley floors (e.g., Poser, 1932; Bretz, 1935).

Succession of events

Early relative chronologies for the glacial succession in the East Greenland fiord zone were based on weathering and geomorphological evidence. Most of the early workers agreed that ice coverage during the last ice age was not complete (e.g., Nordenskiöld, 1907; Koch, 1916; Bretz, 1935; Washburn, 1965). More recent stratigraphic work and the application of modern dating methods have confirmed this contention, showing that while complete ice coverage was attained during an earlier stage, Late Wisconsinan glaciers were thin or failed to reach the coastal areas. This has allowed preservation of a more detailed upper Quaternary record in this region than elsewhere in Greenland.

More than 200 ^{14}C dates have been obtained in the region, mainly on marine mollusc shells and lake sediments. The distribution of ages is shown in Figure 13.14, and is similar to that shown by West Greenland samples (Fig. 13.6); however, there is a considerable number of "old-finite" and infinite ages, which come mainly from the Jameson Land marine beds. A few samples have given ages in the interval 11-20 ka. Some of these are total carbon determinations on lake clay and probably should not be considered reliable; others, which are shell dates, may represent mixtures of Holocene and much older shells.

Thirty collections of shells have been analyzed for amino acids ratios and some results have been published by Hjort (1981a) and Funder (1984). All available measurements are plotted in Figure 13.15, which provides a basis for the correlation of pre-Holocene events. Besides the methods outlined above, Th/U dating has been performed on some samples (Hjort, 1981a).

The succession of glacial and nonglacial events is outlined in Figure 13.16, and follows ideas outlined previously (Funder, 1982b, 1984). A slightly different stratigraphy has been presented for Hochstetter Forland in the northern part of the fiord zone (Hjort, 1981b; Hjort and Björck, 1984), and an attempt to correlate this with that of the southern areas is shown in Figure 13.16. This is based primarily on a re-evaluation of the significance of amino acid ratios for absolute dating, as discussed by Funder (1984).

Sea level change, mollusc faunas, and vegetation are treated in later sections.

The Lodin Elv Formation. The Lodin Elv Formation is known only from an isolated erosional remnant on the west coast of Jameson Land. It comprises a 40 m-thick sequence of prodeltaic sand and silt, and silty diamicton (Fig. 13.17). Foraminifer faunas, supported by amino acid ratios for mollusc shells, imply an age at the Pliocene-Pleistocene transition (Feyling-Hanssen et al., 1983). The diamicton is interpreted as an ice-drop sediment and, together with similar evidence from North Greenland, it provides the earliest,

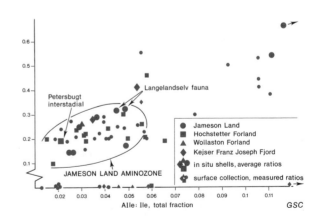

Figure 13.15. Plot of amino acid ratios on East Greenland shells of *Mya truncata* and *Hiatella arctica* (from Funder, 1984). Ratios determined before 1982 are corrected for laboratory fractionation by a factor 0.66; this includes all samples from Hochstetter Forland. Large symbols denote average values from samples collected in situ, and small symbols are single measurements from shells collected from the surface. Samples plotted on abscissa were too small to yield free ratios.

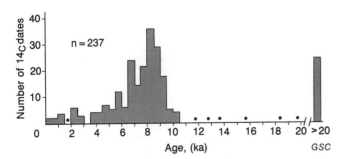

Figure 13.14. Age distribution of ^{14}C dates from East Greenland; dots denote a single date. Compiled from dating lists published in the reports of the Geological Survey of Greenland and Radiocarbon. Ages on marine material are corrected for reservoir effect (Funder, 1978a).

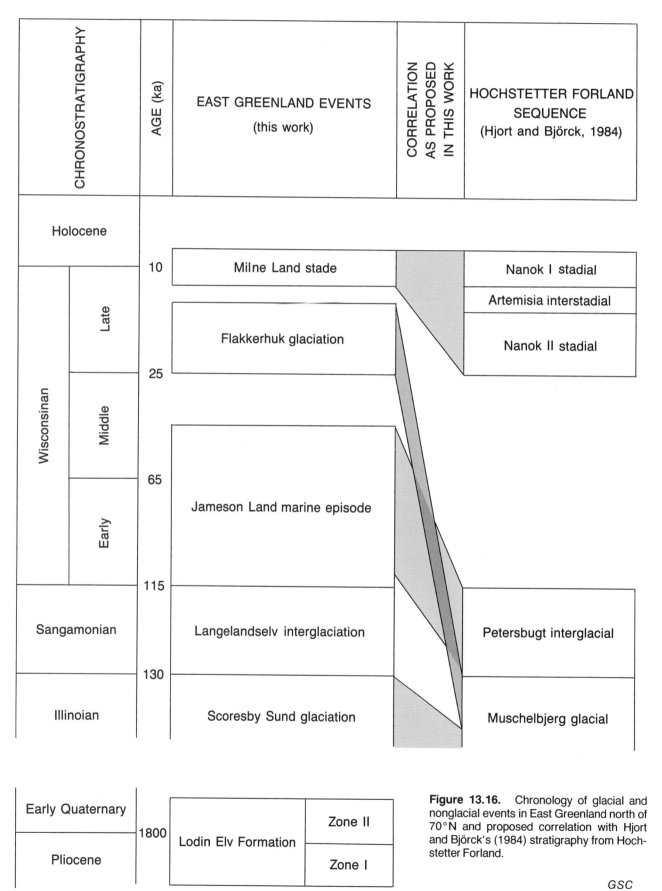

Figure 13.16. Chronology of glacial and nonglacial events in East Greenland north of 70°N and proposed correlation with Hjort and Björck's (1984) stratigraphy from Hochstetter Forland.

albeit circumstantial, evidence for glaciation on land in Greenland.

Scoresby Sund Glaciation. High elevation till beds on Jameson Land, and scattered exotic erratics, notably of Scolithos quartzite, which occur in deeply weathered autochthonous felsenmeer on some mountain plateaus and in coastal lowlands, testify to an old phase of extensive glaciation — the Scoresby Sund Glaciation. The erratics are found in areas characterized by deeply incised V-shaped valleys with gently sloping sides and other features of subaerial rather than glacial erosion (Bretz, 1935). Locally these areas are mantled by a thick regolith and may have sandstone tors (Funder, 1982a). These features indicate a great age for the glaciation that preceded the landscape formation.

Most of these "old" glacial deposits and features were earlier referred to the Kap Mackenzie stade (Funder and Hjort, 1973; Hjort, 1979, 1981a). Recent work by Hjort and Björck (1984), however, has shown that this term is ambiguous, and it is therefore abandoned.

As noted above, the trajectories which can be derived from erratics imply ice sheet movement from west to east over the region, independent of the present topography (e.g., Funder, 1972a). In the coastal areas the ice thickness exceeded 500-1000 m, and the ice margin must have been located on the shelf at a considerable distance from the present coast.

The deposits from the Scoresby Sund glaciation are older than the Langelandselv interglaciation and are provisionally referred to the Illinoian.

Langelandselv interglaciation. Deposits that have been referred to this period are known from only one site on the southwest coast of Jameson Land where they form the base of the Jameson Land marine beds. They consist of sand and gravel exposed 0.5-3 m above sea level in a coastal cliff. The sediments contain a rich and diverse mollusc fauna (see *Paleofaunas and floras*), including some subarctic species which are now absent from the area (Petersen, 1982). The fauna implies greater than Holocene influx of subarctic water to the region and is correlated with isotopic substage 5e in the adjacent Greenland Sea (Funder, 1984). This interglaciation is referred to as the Langelandselv in East Greenland (Fig. 13.16).

Jameson Land marine episode. The thick piles of marine and deltaic sediments constituting the Jameson Land marine beds have been described above. The sediments occur at the mouths of major fiord systems (Fig. 13.18), and amino acid ratios imply that all these subtill marine sediments were deposited during the same nonglacial period (Fig. 13.15; Funder, 1984). These sediments were deposited in a coastal, unglaciated environment following the last interglaciation and contain a sparse mollusc fauna (see *Paleofaunas and floras*). However, although the wide outer fiords remained free of glaciers, the fact that these deposits extend up to 100 m above present sea level indicates that the area was isostatically depressed, and that ice had or was accumulating on adjacent land. The uppermost sediments are generally eroded and the timing for final deposition is not known; ^{14}C dates and amino acid ratios suggest that it was later than 40 ka.

Flakkerhuk glaciation. Evidence for the Late Wisconsinan Flakkerhuk glaciation occurs in the coastal parts of the wide outer fiords, where "medium fresh", and in places continuous till, covers mountain sides. In addition, dead ice disintegration features, such as kames, kame and kettle topography, and meltwater channels, testify to a phase of extensive fiord glaciation, and deposits of the preceding Jameson Land marine episode were overridden and in some areas folded and thrusted (Funder and Petersen, 1980; Funder, 1982a).

An attempt to reconstruct the Inland Ice margin during this stage for the area from Scoresby Sund to the north is shown in Figure 13.18. However, owing to sparsity of ice marginal features (moraines and outwash), it is based on a variety of less reliable evidence, such as the upper limit of till cover, glacial striations, and weathering differences as well as a certain amount of imagination. Although the exact location of the ice margin is in part based on conjecture, it is obvious that the outlet glaciers which filled the outer fiord basins were thin and had low surface gradients. In Scoresby

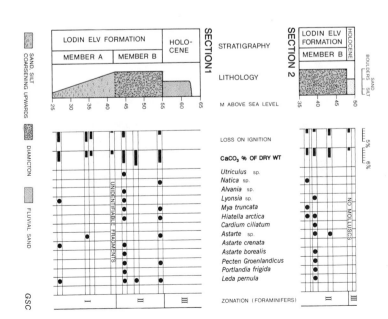

Figure 13.17. Lithology and mollusc fauna of the Lodin Elv Formation, Jameson Land (from Feyling-Hanssen et al., 1983).

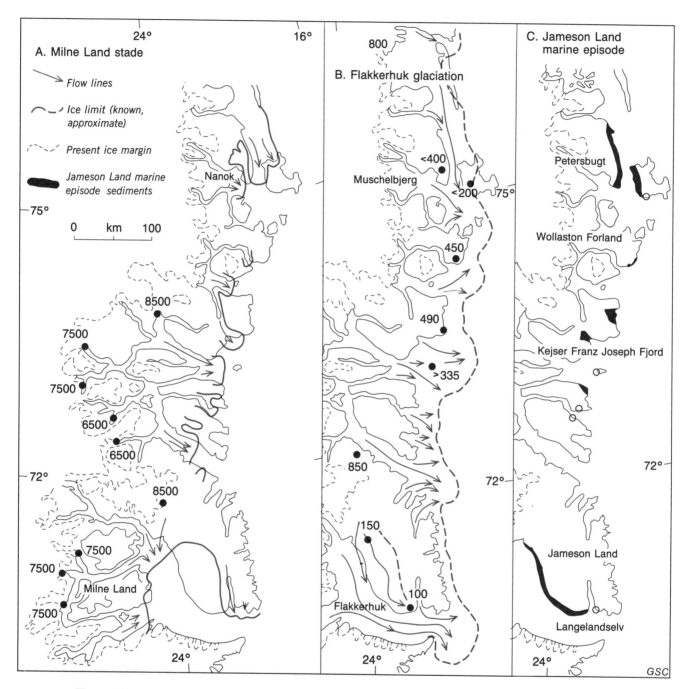

Figure 13.18. Past and present glaciation in the East Greenland fiord zone. A. Known ice limit during Milne Land stade (10 ka), flow lines, and minimum dates for deglaciation of sites at the present ice margin (^{14}C years BP). Sources: Funder and Hjort, 1978; Hjort, 1979; S. Funder, unpublished. B. Conceptual model of Flakkerhuk glaciation (ca. 14 ka), ice distribution, major flow lines, and ice thickness (metres above present sea level), based entirely on land evidence. Sources: Hjort, 1981a; Funder, 1984. Surface elevations based on weathering limits quoted from the regional studies. C. Distribution of sediments from the Jameson Land marine episode (Early and Middle Wisconsinan). Circles show location of shells from same episode, reworked in younger till. Source: Funder, 1984.

Sund the glacier surface declined less than 100 m over a distance of 125 km, indicating that the glacier was floating although it may well have been grounded at its terminus on the shelf. In the fiords to the north it was somewhat thicker. The limit proposed here differs slightly from that given by Hjort and Björck (1984) in that it implies that an ice shelf developed and was buttressed towards the ocean behind a chain of islands at Hochstetter Forland in the north. The lack of features or deposits in the interior regions which can be attributed to this phase implies that the thickness of ice here was not significantly greater than during the later Milne Land stade.

The exact age of the Flakkerhuk glaciation remains uncertain. All that is known with certainty is that it is younger than the Jameson Land marine episode but older than the Milne Land stade. Here it is provisionally assigned a Late Wisconsinan age.

Milne Land stade. Some time after the glacier recession which followed the Flakkerhuk glaciation, the fiord and valley glaciers advanced again. During this phase — the Milne Land stade — the outlet glaciers did not fill the wide outer fiords, but ended at calving fronts located where the steep fiords of the interior joined the wider coastal fiords (Fig. 13.18). The evidence for this advance comprises well preserved lateral moraines and outwash plains. In the Scoresby Sund area in the south and on Hochstetter Forland in the north the moraines are conspicuous and occur in swarms, indicating small glacier fluctuations during the period (Sugden and John, 1965; Funder, 1978a; Hjort, 1979, 1981a). Throughout the region the moraines are correlated by ^{14}C dating of marine molluscs in deposits that are contemporaneous with or postdate the moraines. These results have been discussed by Hjort (1979, 1981a), who showed that the oldest moraines in all parts were formed at ca. 10.3 ka while the youngest moraines were abandoned at ca. 9.5 ka. Funder and Hjort (1973) suggested that this advance was caused by increased precipitation.

Holocene. Following the Milne Land stade the outlet glaciers melted rapidly back through fiords and valleys. The retreat was interrupted by short periods of stillstand and possibly by minor readvance. This is reflected by kame terraces and lateral moraines along the fiords and terminal moraines in some valleys. Both the location of these features, where fiords bend or branch or at breaks in slope, and the lack of synchroneity in neighbouring fiords indicate that the interruptions in retreat were caused by interaction between the glaciers and the topography of their beds, rather than climatic changes (e.g., Funder, 1978a). Topographical differences probably also explain differences in retreat rates in the fiords with average rates varying from 20 to 40 m/a in the Scoresby Sund area for the period between 9.5 and 7.0 ka (Funder, 1972b). After ca. 7 ka the glaciers retreated behind their present margins. This can be seen at some glacier snouts where middle Holocene marine sediment presently is being overridden by glaciers (Hjort and Funder, 1974).

Fresh unvegetated moraines occur a short distance in front of most valley and fiord glaciers in the area and mark the late Holocene maximum of glaciation in the region. Early observations on these moraines have been reviewed by Ahlmann (1941) who, from analogy with other North Atlantic areas, suggested that the moraines were formed at some time in the interval 1750-1850, and a similar conclusion has been reached by Pert (1971) for some glaciers in Stauning Alper (72°N).

QUATERNARY GEOLOGY OF NORTH GREENLAND

S. Funder

Topography, drainage, and glaciation

This large region constitutes the northern cap of Greenland, from Ingolf Fjord (80°N) in the east to Kap Alexander (78°N) in the west (Fig. 13.19). It comprises some of the most extensive stretches of ice-free land in Greenland, and measures from east to west 1200 km, from south to north 500 km, including the world's northernmost land at 83°40′N.

General surveys of the physiography in this remote region have been undertaken by Koch (1928a) and Davies (1972). The topography varies greatly, but in general terms two main types of landscape can be distinguished: dissected mountain plateaus and areas of alpine mountains. The distribution of these two landscape types is conditioned by bedrock geology. Thus plateaus are formed over undeformed Proterozoic and Paleozoic sedimentary rocks (Fig. 13.1) and occur at the southern edge of the area. They are lower than 1000 m elevation and are, especially in Peary Land, dissected by numerous U-shaped valleys. Alpine mountain morphology is developed in the low grade metamorphic turbidites of the North Greenland fold belt which extends along the Arctic Ocean coast of Peary Land. Summits in these mountains rarely exceed 2000 m elevation, and they support numerous small glaciers and ice caps.

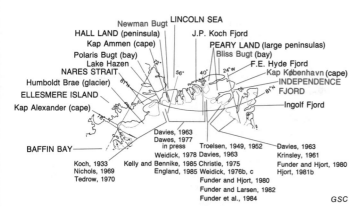

Figure 13.19. Regional Quaternary studies in North Greenland and place names mentioned in text. Names shown in red have been utilized in stratigraphic terminology.

Funder, S.
1989: Quaternary geology of North Greenland; in Chapter 13 of Quaternary Geology of Canada and Greenland, R.J. Fulton (ed.); Geological Survey of Canada, Geology of Canada, no. 1 (also Geological Society of America, The Geology of North America, v. K-1).

To these two landscape types may be added a third which is peculiar to North Greenland — a coastal plain. The 2 to 10 km-wide plain extends for a distance of 250 km along the north and east coasts of Peary Land and is blanketed by Quaternary glacigenic and marine sediments.

The region is divided into segments by some of the longest and widest fiords in Greenland. The fiords drain into the Arctic Ocean and Nares Strait and generally trend in directions perpendicular to the present coastline. The fiords are all headed by glaciers which drain a significant portion of the Inland Ice (Reeh, 1989). The largest single outlet is the Humboldt Gletscher with a width of nearly 100 km. The glaciers are not very active, however, and rates of both accumulation and ablation over this part of the Inland Ice are very low (Reeh, 1989).

When compared to West and East Greenland, it is noteworthy that in North Greenland traces of glacial erosion are confined to fiords and valleys, and traces of "areal scouring" are scarce (Fig. 13.2). This suggests that glaciation over this region has been characterized by cold inert ice, as it is at present.

Quaternary studies

Owing to its inaccessibility, the region has been little visited, and Quaternary observations are still lacking from some areas. Scientific and exploration expeditions began in the western areas in mid-19th Century and the eastern and northern parts became known through the travels of Robert Peary and the Denmark Expedition early in this century. The first regional treatment of the geology of the region was by Koch (1920, 1928a,b) who traversed it from west to east by dog sled in 1921-1922.

Since 1947 "The Danish Peary Land Expeditions" have been working intermittently in the region, and especially in interior areas adjacent to Independence Fjord. The expeditions have covered a wide range of natural sciences, concentrating especially on archeology.

Up until the 1950s travelling was by dog sled during the snow season, and the possibility for making Quaternary observations was limited. Modern helicopter-supported surveys, however, were begun by the United States military in the late 1950s (e.g., Davies, 1972), and in 1978 Geological Survey of Greenland initiated a major geological project, including mapping of the Quaternary geology. Much information from these recent activities is, at the time of writing, available only in the report series of the Geological Survey of Greenland. A number of regional Quaternary studies are listed in Figure 13.19.

Nature and distribution of Quaternary sediments and landforms

In general terms North Greenland has been less affected by glacial erosion than other parts of Greenland and consequently bears some of the thickest and most extensive accumulations of Quaternary sediments. The sediments occur along fiords and coasts, and in the major valleys, but they are especially extensive on the coastal plain of the northeast, and at the mouths of the large fiords of Peary Land.

Till. As elsewhere in Greenland patches of coarse sandy till and scattered glacial erratics are ubiquitous. Till also forms a continuous cover in areas to the south of Independence Fjord and along the north and east coasts as mapped by Davies (1972) and Dawes (1987). The till consists of clayey sandy matrix-supported diamicton with a thickness of up to 2 m. With a smooth surface, draping the underlying bedrock surface, it may cover areas of several hundred square kilometres up to elevations of 400 m above sea level. Appearance and composition suggest that these deposits may be lodgment till, a depositional type which is rare in other parts of Greenland.

Till clasts and their transportation routes. The distribution of erratic boulders shows that the flow of ice during the ice ages has followed routes which are more complex than those outlined for other parts of Greenland.

As recorded by Koch (1928b), the maximum extension of the Inland Ice is reflected by several types of erratics, notably red granite, which is now known to occur at or below the present Inland Ice margin in the south. The boulders occur scattered over the mountain plateaus and at the outer coasts in most parts of the region, as well as on adjacent parts of Ellesmere Island in the Canadian Arctic (Christie, 1967). It was also noted by Koch, however, that these erratics are absent from Peary Land where a local ice cap blocked the Inland Ice. This observation was substantiated by Troelsen (1952), Davies (1963), and Christie (1975), who recorded a limit for Inland Ice erratics running along Independence Fjord some 30 km to the north of its north coast in Peary Land.

Although the erratics encountered in northern and eastern Peary Land generally can be accounted for by short distance transport from local sources, there are a few erratics that require other explanations. Boulders of granite and gneiss occur sparsely on the coastal plain of northeast Peary Land (Koch, 1928b; Dawes, 1970); they have recently been described in detail and discussed by Dawes (1986) who suggested source areas either in southern North Greenland, 200-300 km to the southeast, or in northern Ellesmere Island, twice that distance to the southwest. Both possibilities draw heavy implications for the former glacial regime at the shores of the Arctic Ocean, and further investigations in this remote area are badly needed. Awaiting these investigations, I choose the more conservative of two possibilities — transport from the south — and suggest that the crystalline erratics were distributed over the area at a time when the Inland Ice overrode the entire region, but that later local glaciation has removed or buried them in most areas. In addition, on some small islands off the north coast of Peary Land, acidic volcanics dominate the boulder communities, whereas this rock type is extremely rare in the adjacent coastal areas. The most likely source of these erratics lies on the coast 100 km to the west, implying substantial flow along the north coast of Peary Land (Funder and Larsen, 1982).

Moraines. Lateral and terminal moraines occur abundantly along fiords and in major valleys, especially in interior parts, and are thought to reflect short stillstands or readvances during the Holocene.

In the lowland areas of central Hall Land large lateral moraines have been deposited both below and above former sea level; these moraines, which may be up to 1 km in cross-section, show marked differences in freshness, and their stratigraphical significance has been discussed vividly (e.g., Koch, 1928a; Davies, 1972; England, 1985; Kelly and Bennike, 1985; Dawes, 1987; Bennike et al., 1987).

Particular features of this region are moraine fragments and meltwater channels which are perched on and incised into mountain sides along the outer coast. These features are especially common on the coastal plain of northern Peary Land (Davies, 1963; Funder and Larsen, 1982). The moraines and meltwater channels show that ice once flowed eastward along the outer coast and record the presence of ice shelves on the shores of the Arctic Ocean.

Glacier thrust sediments. On the north side of the mouth of Independence Fjord the coastal lowland is composed of ridges up to 15 km long and 150 m high trending in broad arcs along the fiord. The ridges were interpreted as terminal moraines by Koch (1928a, b), Troelsen (1952), and Davies (1963). However, the ridges are composed not of till, but of marine sediments of the early Quaternary Kap København Formation, and structures in the sediments show that the sediments were dislocated and overridden by a glacier originating in Independence Fjord. Thus the ridges are not moraines, but crests of 100 m-thick rafts of sediments (Funder et al., 1984). This landscape, unique in Greenland, occupies an area of approximately 1000 km² and probably owes its existence to the particular lithology of the sediments of the Kap København Formation.

Glaciolacustrine, glaciofluvial, and fluvial deposits. Glaciolacustrine deposits seem to be more common in this region than elsewhere in Greenland. They occur at valley mouths and in lowlands between fiords, where fiord glaciers have dammed large lakes. The deposits occur also on the coastal plain of northern Peary Land, where the lakes were dammed against ice moving along the coast (Koch, 1928a; Davies, 1963; Kelly and Bennike, 1985). The largest glacial lake basin occupied central parts of Hall Land (Koch, 1928a; Kelly and Bennike, 1985; Dawes, 1987), while other large lakes occurred in the southeastern parts (Davies, 1963; Krinsley, 1961). The ice dammed lakes probably existed during the last glaciation and its waning phases. The sediments are up to 50 m thick and comprise laminated sand and silt, which may include graded bedding, climbing ripples, and thick beds of diamicton. At low elevations the sediments may be conformably overlain by marine sediment, whereas higher up they are commonly covered by glaciofluvial deposits or till.

Glaciofluvial and fluvial deposits form outwash plains and valley trains with several generations of dissected terraces present in all major valleys. Kame terraces in association with lateral moraines flank the valleys, and kames are common on valley floors. The largest kames occur on the till plains described above and may form narrow plateaus, 10×1 km, rising more than 100 m above the surroundings — as in the area at the mouth of F.E. Hyde Fjord. Large kames also occur on the coastal plain of northern Peary Land and form islands off the coast (Funder and Larsen, 1982). These large kames probably reflect the location of crevasse systems at the confluence of major outlet glaciers and smaller tributaries.

Marine deposits. Marine deposits here, as elsewhere in Greenland, comprise massive laminated silt and sand and gravel associated with raised deltas or beaches. Presently, at least in the northern parts, the coasts are permanently ice bound, and no coarse grained beach sediments occur along the shores. In these areas beach ridges composed of gravel and shingle do occur at higher elevations, showing that sea ice conditions in the Arctic Ocean have changed over the Holocene, a contention which is supported by the occurrence of driftwood logs of non-Greenland origin on the raised beach ridges (e.g., Fredskild, 1969; Stewart and England, 1983; England and Bednarski, 1989). The most extensive beach ridge landscape occurs on the coastal plain of eastern Peary Land up to 60 m above sea level.

Radiocarbon dating of mollusc shells in the marine sediments has shown that although most are of Holocene age, older assemblages occur especially in areas near the outer coasts. Of particular note are older marine sediments of the Late Pliocene-early Quaternary Kap København Formation which are exposed in the hilly landscape at the mouth of Independence Fjord (see below).

Periglacial and eolian features. The region lies within the zone of continuous permafrost, and cryoturbation structures and frost wedge polygons occur throughout. In the humid coastal regions solifluction sheets and lobes may occupy extensive areas, especially where the surface sediment consists of silty marine sediments and till (e.g., Nichols, 1969). In the dry interior solifluction is uncommon and frost wedges are often flanked by ridges up to 1 m high (Nichols, 1969); they occur not only in unconsolidated Quaternary sediments, but also in carbonate bedrock (Davies, 1961). Sand wedge polygons and rock glaciers also occur in these areas (Bennike, 1987). Pingos are generally absent although collapsed pingos have been observed at two localities in Peary Land (Bennike, 1983a). Hummocky landscapes produced by thermokarst processes in fluvial and marine sediments are common in all areas.

Although accumulations of windblown sediments are not common, deflation features with ventifacts are common features on fluvial terraces in the dry interior (e.g., Troelsen, 1949; Fristrup, 1952; Nichols, 1969; Bennike, 1987).

Succession of events

As elsewhere in Greenland, thick stratigraphic sections are lacking and the chronology must be pieced together from sections representing short time intervals. The sections are dated and correlated by ^{14}C dating, amino acid parameters, or biostratigraphic correlation (Fig. 13.20).

The stratigraphy falls into two well defined parts: the Pliocene-Pleistocene transition period, and the late Quaternary. Preliminary amino acid results have indicated that marine deposits on the coastal plains of northern and eastern Peary Land may date from the long time interval between these two extremes, however field observations from these remote areas are still too sparse to allow further stratigraphic speculation.

The age distribution of 197 radiocarbon dates obtained in the region is similar to that from other parts of Greenland (Fig. 13.21). The ages have been obtained mainly on mollusc shells, but in this area driftwood has also been an important dating medium, especially in conjunction with archeological studies. The paleo-oceanographic significance of these dates is dealt with below in the section on Quaternary paleoclimates.

The results of amino acid analyses of 25 samples of mollusc shells are shown in Figure 13.22. The method has been especially important for the correlation of outcrops of the Kap København Formation. The results show that for these old samples the total amino acid ratios have a much poorer time resolution than ratios from the free fraction.

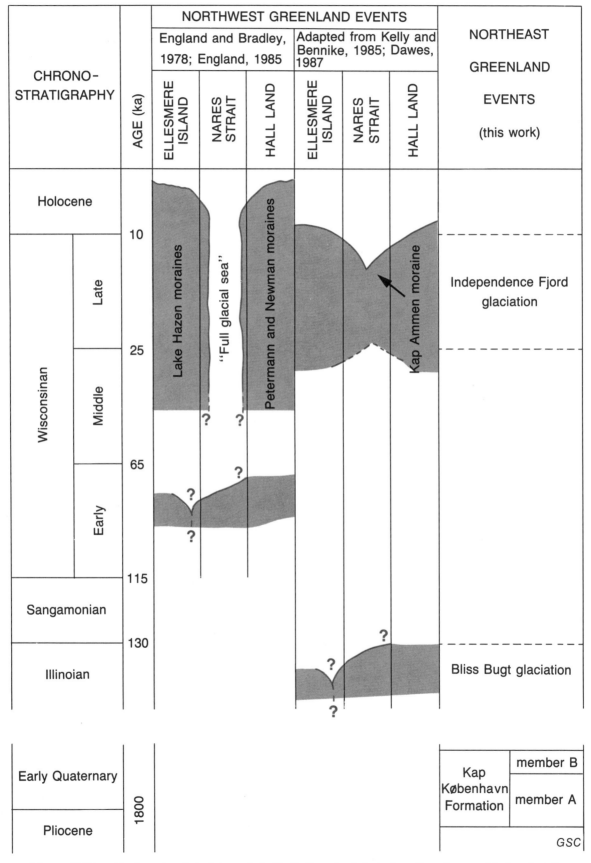

Figure 13.20. Stratigraphy of glacial events in North Greenland, including alternative models for glaciation along Nares Strait. Red shading indicates ice cover.

Kap København Formation

The Kap København Formation, a more than 100 m-thick sequence of unconsolidated marine sand and silt in eastern Peary Land, allows insight into the physical environments of Greenland before the onset of the ice age (Fig. 13.23). The peculiar glacier-tectonized landscape in which the sediments occur has been described above. The sediments have been dated by various methods, the most reliable being analysis of its foraminiferal biostratigraphy which indicates a Late Pliocene and Early Pleistocene age (Funder et al., 1985). In addition, magnetostratigraphical studies in member B have shown reversed polarity, which is correlated with part of the Matuyama Polarity Chron (Abrahamsen and Marcussen, 1986).

A High Arctic marine fauna and pockets of silty diamicton with large boulders suggest that the lower part of the sequence (Upper Pliocene), was deposited in an arctic environment characterized by drifting debris-loaded icebergs (Funder et al., 1984). This, together with the evidence from the contemporaneous Lodin Elv Formation in East Greenland, is the earliest direct evidence of glaciers on land in Greenland. Subsequent to this period cool temperature conditions prevailed, and forest tundra grew in Peary Land — as seen from the trunks, branches, twigs, needles, and cones in the sediments (see section below on paleoflora). An extrapolation from the climatic data suggests that the Inland Ice did not exist at this time.

The sediments were subsequently overridden by glaciers. Sediments correlatable with the Kap København Formation may occur elsewhere along Independence Fjord (Funder et al., 1984), and there is a possibility that other localities containing wood with nonfinite ^{14}C ages may belong to the same period. Such occurrences have been noted by Weidick (1978) and Jepsen (1982).

Bliss Bugt glaciation. Most field workers in the region have, by means of weathering criteria and the distribution of glacial erratics, distinguished two phases of extensive glaciation. During the oldest and most extensive phase the Inland Ice crossed Nares Strait in the west and inundated coastal areas of Ellesmere Island to an elevation of 600 m (Christie, 1967). Precise dating of this event is still wanting. On Ellesmere Island it predates the maximum Ellesmere Island ice advance which, based on amino acid ratios, has tentatively been said to be older than 80 ka (England and Bradley, 1978; England and Bednarski, 1989). Kelly and Bennike (1985) stated that it probably is older than the last interglaciation.

In the northeast, traces of extensive glaciation are few; however, as discussed above, scattered crystalline erratics on the north coast of Peary Land may indicate that the Inland Ice once overrode the North Greenland mountain range pushing northwards into the Arctic Ocean. This somewhat hypothetical event is here termed the Bliss Bugt glaciation, and is tentatively correlated with the maximum extension of the Inland Ice in the west which occurred just before the last integlaciation.

Independence Fjord glaciation. During the subsequent phase of extensive glaciation the ice was thinner and its movement controlled by topography. However, the extent and timing of this glaciation is, especially in the Nares Strait region, a matter of debate. An excellent review of the opposing theories and their field evidence has been provided by Paterson (1977), and even though a considerable amount of field work has been conducted in this area since then, no agreement has been reached. The disagreement concerns the extent of the Middle and Late Wisconsinan and Holocene Greenland Inland Ice and the extent of ice cover over Arctic Canada. The differences of opinion are shown graphically in Figure 13.20.

One school of thought contends that ice sheet margins which were only slightly more advanced than the present remained stable during the period from before 35 to ca. 8 ka, while the Nares Strait was an ice-free corridor partly filled with a "full glacial sea" (England and Bradley, 1978; England, 1982, 1983, 1985; England and Bednarski, 1989).

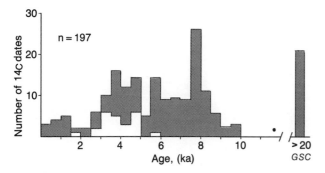

Figure 13.21. Age distribution of ^{14}C dates from North Greenland, including ages obtained for archeological purposes. Grey columns represent ages obtained mainly on bivalve shells, but also on gyttja, peat, wood, and bone; open columns represent ages obtained on driftwood. All ages on marine material are corrected for reservoir effect (Funder, 1982b). Compiled from dating lists in the reports of the Geological Survey of Greenland, Radiocarbon, and elsewhere.

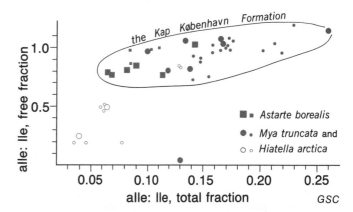

Figure 13.22. Plot of amino acid ratios from North Greenland shells of *Mya truncata* and *Hiatella arctica* (circles) and *Astarte borealis* (squares). Ratios in the total fraction, obtained before 1982 are corrected for laboratory fractionation with a factor 0.66 (open symbols); younger determinations (closed symbols) are uncorrected. Large symbols denote average for determinations on in situ shells from one sample; small symbols represent single determinations on shells collected from the surface. Sample plotted on abscissa too small to yield free fraction ratios. Sources: Funder et al., 1985; S. Funder, unpublished.

Opposed to this view another group of field workers believe that during Late Wisconsinan time Nares Strait was filled by a trunk glacier, more than 500 m thick. This glacier was nourished from both Ellesmere Island and Greenland and flowed northwards into the Lincoln Sea and southwards into Baffin Bay (Fig. 13.24). The disintegration of this ice tongue began before 10 ka. Observations supporting aspects of this theory have recently been published by Blake (1977), Weidick (1978), Christie (1983), Kelly and Bennike (1985), Dawes (1987), and Bennike et al. (1987), and an ice sheet model based on this scenario has been constructed by Reeh (1984). England (1983) could not find evidence for the Nares Strait ice tongue on Ellesmere Island and used apparent stability of relative sea level at the time the ice tongue was reported to have been disintegrating and lack of moraines or till that could be clearly associated with a Late Wisconsinan ice lobe as arguments against its former existence; field data from Hall Land has been used to corroborate the Late Wisconsinan history developed on northern Ellesmere Island (England, 1985). Bennike et al. (1987), however, dispute this interpretation of the Hall Land data.

A discussion of the two theories lies beyond the scope of the present report, however, the nature of glacial landforms along the Greenland shore of Nares Strait, especially in its southern parts, as well as the lack of absolute dates from the putative "full glacial sea" (see Fig. 13.21) would seem to favour the presence of a Late Wisconsinan ice tongue in Nares Strait.

In northeast Greenland a young phase of extensive glaciation is reflected by moraines and glaciofluvial deposits up to elevations of 500 m on mountain sides along the outer coasts. As noted above, these features indicate the existence of ice shelves along the Arctic Ocean coast, nourished from

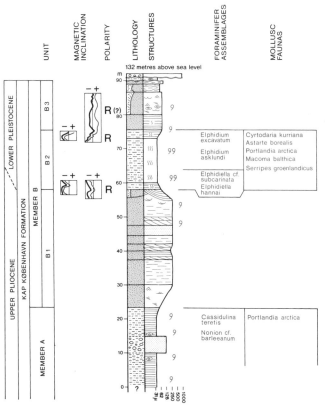

Figure 13.23. Paleomagnetism, stratigraphy, lithology, and marine faunas of the Kap København Formation (from Funder et al., 1985).

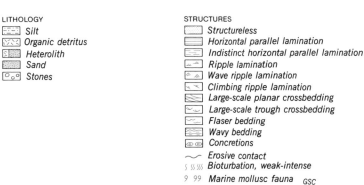

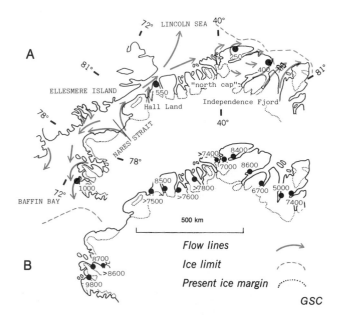

Figure 13.24. Past and present glaciation in North Greenland. A. Conceptual model of Independence Fjord glaciation (Late Wisconsinan), proposed location of ice margins, major flow lines, and ice surface elevation (metres above present sea level). The model is based entirely on land evidence. Sources: Blake, 1977; Funder and Hjort, 1980; Funder and Larsen, 1982; Christie, 1983; Reeh, 1984; Kelly and Bennike, 1985. B. Dates for deglaciation in interior fiords. Sources: ^{14}C dates published by Weidick, 1972a, 1973, 1978; Funder, 1982b; Kelly and Bennike, 1985; Fredskild, 1985b.

outlet glaciers in fiords and valleys. This event is here termed the Independence Fjord glaciation since this large fiord marks the northernmost extension of the Inland Ice. Over the mountains of Peary Land in the north a dynamically independent ice cap, the "north cap", existed at the same time (e.g., Weidick, 1976b, c). The Independence Fjord glaciation, predated the Early Holocene marine transgression into the area and is provisionally referred to the Late Wisconsinan (Fig. 13.20).

Hence the evidence indicates that the Late Wisconsinan ice cover over North Greenland was thin and, especially in the northeast, characterized by extensive ice shelves along the coasts (Fig. 13.24). Erratics on islands off the north coast of Peary Land, indicate that some ice shelves may have been large and may imply that the Lincoln Sea between Canada and Greenland was filled by an ice shelf (Funder and Larsen, 1982).

Holocene. The shrinkage of the Inland Ice and local ice caps is demonstrated by numerous moraines along fiords and valleys and began at ca. 10 ka. At 9 ka the greater part of the large Independence Fjord was free of ice (Bennike, 1987), and at 7 ka most major glaciers seem to have achieved their present extent (Fig. 13.24). By 6 ka at least some glaciers were up to 20 km behind their present front (e.g., Funder and Hjort, 1980; m. Kelly, University of Lancaster, Lancaster, England, and O. Bennike, Geological Museum, University of Copenhagen, personal communication, 1985).

Fresh unvegetated moraines in front of the glaciers testify to later readvance which culminated in historical times.

Radiocarbon dating of reworked organic material in the moraines has given ages between 5 and 0.2 ka (Weidick, 1976b, 1977; Hjort in Hakansson, 1982), and oscillations within the last 50 years have been detailed by Davies and Krinsley (1961) using old maps and especially the detailed descriptions by Koch (1928b) as reference. It was concluded that glaciers generally were retreating in this period, the retreat being most marked in the southwestern areas as a response to a decrease in precipitation.

QUATERNARY GEOLOGY OF THE SHELVES ADJACENT TO GREENLAND

S. Funder and H.C. Larsen

The continental shelf of Greenland attains a maximum width of 300 km in the northeast and narrows to 35 km in the south. In the northwest it is fused with the Canadian shelf (Fig. 13.1). Although glacial erosion has affected some parts and ploughed out deep channels, the area is predominantly one of sediment accumulation, and the thickest and most extensive sequences of Quaternary sediments in Greenland occur here. This has been shown by bathymetric and geophysical surveys, which have been carried out in the past two decades, with a view to the fishing and oil potentials of the shelf. This section reviews the Quaternary information that has emerged as a by-product of this work with relevant publications being listed in Table 13.3. Unfortunately this information is largely limited to areas south of 72°N.

The pre-Quaternary development of the shelves has been treated briefly above (section on *Bedrock geology*).

General morphology of the shelf

Some main features in the physiography of the shelf are shown in Figures 13.25 and 13.26. Traces of ice sheet erosion, that is, glacially abraded bedrock with roches moutonnées-like features on the surface, occur in areas of crystalline bedrock and lavas and are most common off southeast Greenland where they form a zone extending up to 100 km from land (Larsen, 1983; Fig. 13.26). Off West Greenland the zone has a width of 20-30 km, and contains a coast-parallel trough with depths of about 500 m. The trough has developed over the boundary between the Precambrian basement and the younger sedimentary basin to the west and may be fault controlled (e.g., Holtedahl, 1970). Off northern East Greenland, traces of ice sheet erosion seem to be absent (Larsen, 1984) and in general the distribution of glacial erosion features seems to be a continuation of the pattern known from onshore (Fig. 13.2).

On the outer shelf glacial erosion has left its clear mark in deep transverse channels which occur at the mouths of all major drainage outlets (Fig. 13.25). The channels traverse

---

Funder, S. and Larsen, H.C.
1989: Quaternary geology of the shelves adjacent to Greenland; in Chapter 13 of Quaternary Geology of Canada and Greenland, R.J. Fulton (ed.); Geological Survey of Canada, Geology of Canada, no. 1 (also Geological Society of America, The Geology of North America, v. K-1).

Table 13.3. Geophysical and geological investigations on the Greenland shelves with implications for Quaternary geology.

Methods	Reference	Area* (N. lat.)
Bathymetry, bottom sampling	Rvachev, 1963	W 60-68°
Bathymetry (bottom sampling)	Sommerhoff 1973	E 60-64°
Bathymetry	Henderson, 1975	W 59-70°
Bathymetry	Sommerhoff, 1975a,b,1979,1981,1983	E,W 55-68°
Shallow seismic, bathymetry	Johnson et al., 1974	
Shallow seismic, gravimetry, side scan sonar	Brett and Zarudzki, 1979	W 64-70°
Shallow seismic	Roksandić, 1979	W 61-64°
Deep seismic, aeromagnetism	Larsen, 1980,1984	E 61-74°
Shallow seismic, bathymetry side scan sonar	Larsen, 1983	E 65-69°
Drilling	Risum et al., 1980	W 66-68°
Drilling, shallow seismic	Henderson et al., 1981	W 65-68°

* E,W: East and West Greenland shelves

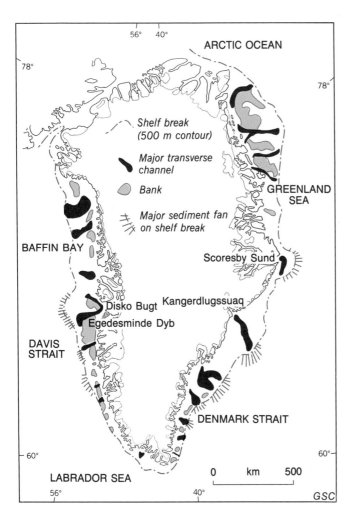

Figure 13.25. Some features in the morphology of the Greenland shelf. Sources: Sommerhoff, 1983; Naval Research Laboratory, Acoustics Division, 1985.

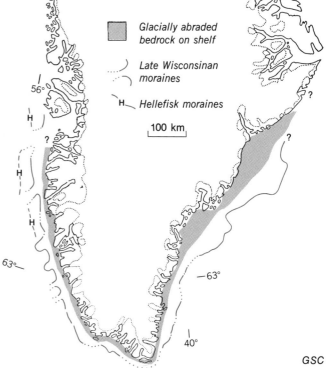

Figure 13.26. Occurrence of glacially sculptured bedrock surfaces and moraines on Greenland's shelf. Sources: Brett and Zarudzki, 1979; Sommerhoff, 1983; Larsen, 1983.

the shelf, usually with a southwards deflection of trend, and they are generally less than 500 m deep although depths exceeding 1000 m have been recorded (Sommerhoff, 1973). Their dimensions may be impressive: thus the channel off Kangerdlugssuaq is 50 km wide and 350 km long (Larsen, 1983), while Egedesminde Dyb at the mouth of Disko Bugt is 250 km long (Zarudzki, 1980). At the mouths of the channels there is usually a marked bulge of the shelf edge, composed of a semicircular sediment fan measuring as much as 200 km in diameter (e.g., at the mouths of Disko Bugt and Scoresbysund). Although the U-shaped channels were eroded by outlet glaciers, most authors speculate that glaciers utilized an older fluvial drainage system over the shelf (Henderson, 1975; Sommerhoff, 1979; Roksandć, 1979).

Seawards of the glacially abraded bedrock are banks which form large areas of relatively level surface at depths of usually 200 m. Off West Greenland between 62° and 69°N, however, the banks are less than 100 m deep and locally may even be less than 10 m. Like their Canadian counterparts, the banks sustain a rich fish life, which has been exploited by man for centuries. Nearly a hundred years ago it was suggested that they were terminal moraines deposited by the Inland Ice during its maximum (Jessen, 1896) and although recent investigations have shown their history to be more complex, their surfaces do seem to be formed of glacial deposits; the outer ice margins probably reached them during one or several glaciations. Seismic investigations show, however, that the banks are composed mainly of Tertiary and Quaternary marine shelf sediments, and they thus must be considered to be areas of sediment accumulation which have been preserved through severe periods of glacial erosion.

Nature and distribution of Quaternary sediments and erosion forms

Although bottom sampling and drilling have given some first hand information on the sediments, most knowledge comes from detailed bathymetry, which provides an accurate picture of the surface morphology, and from acoustic data, which show the internal structures of the sediments. These data indicate that the outer shelves have a continuous cover of Quaternary sediments with thicknesses up to 300 m (as shown in borings, Risum et al., 1980). In general, however, thicknesses are of the order of 100 m, as apparent from seismic profiling (Brett and Zarudzki, 1979; Larsen, 1980, 1984; Larsen, 1983). The base of the Quaternary, as shown in the profiles, is an unconformity believed to have developed by glacial erosion of Tertiary sediments (Larsen, 1983; H.C.Larsen, unpublished).

In spite of the limitations in the geophysical methods, especially in providing lithological information, several types of Quaternary sediments have been recognized.

Till and moraines. Till is identified mainly from its surface characteristics, and in each area only the youngest glacial event is recorded. A till sheet with gently rolling topography covers areas of the East Greenland shelf at 64°N and extends to a distance of 100 km from the coast (Johnson et al., 1975). Hummocky moraine has been recorded by Brett and Zarudzki (1979), occupying 20 km-wide zones behind terminal moraines on the West Greenland shelf. Finally, Roksandć (1979) described 60 m-thick fillings of homogeneous till in transverse channels off southern West Greenland.

Moraines have been observed on the shelves south of 68°N and comprise push moraines on the banks, and calving or grounding line moraines in transverse channels (Sommerhoff, 1973, 1983). The push moraines occur in swarms, up to 4 km wide, and may attain a height of 100 m, although they are generally much lower (Brett and Zarudzki, 1979). On the West Greenland shelf two distinct moraine systems have been recorded (Fig. 13.26). The oldest follows the outer shelf margin between 65° and 66°N (Brett and Zarudzki, 1979) and was termed the Hellefisk moraines by Kelly (1985). The younger moraines follow the inner bank edges some 60 km to the east and flank the transverse channels. These moraines are correlated with moraines on the outer shelf margin in southeast Greenland by Sommerhoff (1981, 1983), who considered them to represent the maximum of Wisconsin Glaciation. This agrees with Kelly (1985) who, from ice thickness considerations, correlated them with the Late Wisconsinan Sisimiut glaciation in West Greenland. The highly lobate nature of the ice margin indicates that the moraines in West Greenland were deposited mainly on land, while off East Greenland the less lobate ice margin was probably grounded in shallow water (Sommerhoff, 1973, 1983).

Stratified deposits and traces of marine erosion. Sediments judged on the basis of surface morphology to be glaciofluvial and glaciomarine in origin appear to form outwash plains sloping gently towards the outer shelf margin on the distal side of moraines (Johnson et al., 1975). Holocene marine sediments apparently occur only sporadically over the shelf, probably owing to strong current activity (Sommerhoff, 1981, 1983). This is in agreement with Rvachev's (1964) observation that the surface sediments usually are inhomogeneous, coarse grained, and contain a component of ice rafted debris. Seismic profiles, however, indicate that some transverse channels may have up to 5 m thick stratified marine sediments on their bottom (Sommerhoff, 1979).

Traces of Holocene marine erosion have been recorded especially on the shallow West Greenland banks and comprise marine abrasion terraces and beach ridge systems eroded into glacigenic deposits. The beach ridges occur down to a depth of 70 m between 62° and 65°N on the West Greenland banks (Sommerhoff, 1975a). They substantiate the theory that the Late Wisconsinan ice margin in this area was essentially land based, and show that eustatic sea level lowering in this area exceeded isostatic subsidence during the last glaciation.

Iceberg scouring. Side scan sonar investigations have shown that iceberg scours, measuring up to 4 km in length and 75 m in width, occur commonly down to depths of 350 m at the mouth of Disko Bugt in West Greenland, and in the Denmark Strait off East Greenland (Brett and Zarudzki, 1979; Larsen, 1983); in the latter area they also occur sporadically down to a depth of 650 m. The depths of scours show that the icebergs were too deep to escape over fiord thresholds, and therefore must have originated in an ice sheet on the shelf. The fresh appearance of the scours is additional evidence of the lack of Holocene marine sedimentation in these areas.

Quaternary history

From its nature, the Quaternary record of the shelf has an even larger element of speculation than that on land, and is based to a large extent on morphological considerations. Hopefully in the future correlation of seismic sections with worldwide sea level history, and with the deep sea record from areas adjacent to the Greenland shelf may provide a firmer age control.

Even though there are great differences in both method and scope between work on land and on the shelf, there seems to be general agreement in the results, and, as on land, the history of the shelf has two chapters — one dealing with the Late Pliocene and early Quaternary, the other with the late Quaternary. This leaves a long time interval in between about which little or nothing is known.

Pliocene-Pleistocene. The large sediment fans which lie at the shelf break are thought by Henderson (1975) and Sommerhoff (1975a, 1979) to have originated at least partly as fluvial cones, at a time before the onset of glaciation. Sea level at that time was 300 m lower than present, relative to the Greenland shelf, and this period of fan formation is tentatively dated to the Late Pliocene and early Quaternary.

Seismic sections from the shelf off northeast Greenland have been used to suggest that this area was entirely covered with land-based ice in the Late Pliocene (H.C. Larsen, unpublished). This is the earliest evidence for an extensive ice sheet over Greenland and corroborates the land record mentioned above that indicates glaciers were present in the Late Pliocene in both East and North Greenland. These observations agree with data from deep sea drilling cores that show substantial amounts of ice rafted detritus and significant oxygen isotope changes began to appear in adjacent parts of the North Atlantic and Baffin Bay in the Late Pliocene (Shackleton et al., 1984; Eldholm, Thiede, Taylor et al., 1987; Srivastava, Arthur, Clement et al., 1987).

Hellefisk moraines. Of the two moraine systems on the West Greenland shelf, the youngest, and innermost, has been correlated with the Late Wisconsinan, but the age of the older Hellefisk moraines remains uncertain. Provisionally these moraines may be correlated with the evidence for old glaciation on land, the Fiskebanke glaciation (see above) and referred to the Illinoian (Kelly, 1985).

During this phase the Inland Ice covered the entire West Greenland shelf, and may even have extended into the oceanic Davis Strait.

Wisconsinan. A much more detailed knowledge has now been obtained on the Late Wisconsinan distribution of ice on the shelves, especially in areas south of 68°N. This is because of the detailed bathymetry carried out by Sommerhoff (1981, 1983). The former ice limits are reflected by terminal moraines on the banks, dated partly by their being the youngest glacial event which affected the shelf, and partly by Kelly's (1985) correlation of the moraines with the Sisimiut glaciation in West Greenland.

The picture which has emerged shows that the extent of glacier coverage of the shelf varied from area to area. Off West Greenland, as noted above, the ice margin was essentially land based on the inner shelf, 30-50 km from the present coast, but large calving outlet glaciers filled the transverse channels, flowing towards the shelf break. On the shelf off southeast Greenland the ice margin was grounded in shallow water; and in the area at 66°N, the ice sheet covered the shelf to a maximum of 200 km from the coast. Farther to the north on the East Greenland shelf, the sparsity of traces of glacial activity agree with the evidence on land indicating that the Flakkerhuk glaciation (see above) was characterized by thin outlet glaciers which had their snouts on the inner shelf. In the extreme north and northeast, however, the land evidence indicates considerable shelf glaciation during the Independence Fjord glaciation (see above). Unfortunately no investigations have been made offshore in these regions.

Holocene. The sporadic occurrence of Holocene sediments has been mentioned above. Apparently marine transgression took place over the shallow West Greenland banks after the last glaciation. The transgression can probably be correlated with the early Holocene transgression in West Greenland (Sommerhoff, 1975a).

SEA LEVEL HISTORY
S. Funder

Raised marine sediments and features of marine erosion are scattered along the coasts in all parts of the country. The large majority of these occurrences are of Holocene age and reflect submergence at the end of the last glaciation and the isostatic rebound which followed retreat of the Inland Ice margin. This section emphasizes the Holocene rebound history and gives a brief survey of the scant information on Sangamonian and Wisconsinan sea levels.

A start was made more than a century ago on collection of elevation determinations of the raised marine features, and these observations were synthesized into a general uplift model by Vogt (1933), who attempted to date the emergence history by comparison with that of Scandinavia. An important step forward came with Washburn and Stuiver's (1962) demonstration of the utility of ^{14}C dated mollusc shells for dating of the uplift and this stimulated uplift studies in all parts of the country. Weidick (1972a) synthesized these results and was able to relate uplift patterns to the deglacial behaviour of the ice margins in different areas. Since the appearance of this important report, much new data have accumulated which has necessitated some revision. Reviews of these data covering West Greenland have been published by Kelly (1979, 1985), and covering East Greenland by Funder (1978a) and Hjort (1979).

Sangamonian sea level

Fossiliferous marine deposits from three areas are thought to be of Sangamonian age. The Langelandselv beds, East Greenland consist of coarse sublittoral sediments, and their lithology and fauna reflect sea level 5-8 m above the present (Funder, 1984). The subarctic fauna on Saunders Ø, provisionally correlated with the Kaffehavn interglaciation of northern West Greenland, reflects sea level slightly higher than 6 m above the present (Blake, 1975). The Laksebugt fauna, West Greenland reflects sea level higher than 55 m above the present and is thought to date from an early stage of the Kaffehavn interglaciation (Funder and Símonarson, 1984).

Funder, S.
1989: Sea level history (Greenland); in Chapter 13 of Quaternary Geology of Canada and Greenland, R.J. Fulton (ed.); Geological Survey of Canada, Geology of Canada, no. 1 (also Geological Society of America, The Geology of North America, v. K-1).

Wisconsinan sea level

The Jameson Land marine beds, East Greenland overlie deposits from the Langelandselv interglaciation and show that this interglaciation ended with submergence of the land areas. The marine beds are up to 100 m thick and are generally shallow water sediments. They therefore reflect a prolonged period of slow subsidence, but the occurrence of cut and fill structures and, locally, erosional unconformities suggests that the submergence was not necessarily continuous. The sediments are glacier eroded and the exact amount of subsidence in this period is unknown. On Jameson Land it probably did not exceed 80 m, which can be compared to a Holocene marine limit in the same area of 65 m. This implies either that Late Wisconsinan ice load was not as great as that which caused the earlier submergence or that deglaciation did not occur until considerable thinning (and uplift) had taken place (S. Funder, unpublished).

As noted above, submarine abrasion terraces and beach ridges at depths down to 70 m imply that parts of the West Greenland shelf was dry land during the Late Wisconsinan. This indicates that in this area at the edge of the shelf, eustatic sea level lowering exceeded isostatic subsidence caused by the ice sheet (Sommerhoff, 1975a).

Marine limits and distribution of uplift

The pattern of emergence following deglaciation is shown in Figure 13.27, based on the elevation of Holocene marine limits. The marine limits are defined as the uppermost occurrence at each site of marine delta terraces, shorelines, or such erosional features as fossil cliffs and washed out boulders, each ideally associated with fossiliferous marine sediments. The last of these demands may be hard to meet, and the decision whether the features are indeed marine or whether they could have been formed in glacier dammed lakes or as kame deltas involves an element of interpretation. Therefore the old, pre-radiocarbon dating literature often has greatly overestimated values for the marine limits. This uncertainty applies also to some of the observation points in Figure 13.27, especially in East Greenland south of 68°N where most areas have not been visited since the 1930s. The data points are not on marine limit at one time but are on marine limit transgressing time. Hence the changes we are seeing are a function of relatively slow deglaciation with the lowest values in each area coming from the first sea level record inscribed after deglaciation. Marine limit elevation shows emergence or submergence since deglaciation and does not represent all movements that occurred during deglaciation. It is controlled largely by the rate of ice retreat and even if the total uplift was large, and if unloading was slow so rebound kept pace with deglaciation, marine limit would be little tilted.

In all regions the maximum visible uplift occurred at or near the outer coasts, and values decrease towards the present ice margin. This may be explained by the later deglaciation of the interior parts which allowed a portion of the rebound to take place while the areas were still ice covered. An additional complicating factor is that submergence, possibly tied to recent expansion of the Greenland Ice Sheet, has been occurring over the last few centuries (see below). Consequently the tilt of the Holocene marine limit near the ice sheet margin is probably being further accented today.

In general the Holocene emergence is less than 100 m relative to present sea level. However, higher maximum values occur locally, reaching approximately 140 m in coastal West Greenland between 66° and 68°N, and about 130 m in East Greenland, in a restricted area in Scoresby Sund, whilst in North Greenland they are variously reported as approximately 120 m or 150 m in Hall Land (Kelly and Bennike, 1985; England, 1985). In the Nares Strait region of western North Greenland the uplift pattern has been interpreted to show the combined effect of both Greenland and Canadian ice (England, 1976, 1982; Weidick, 1978). Very little emergence, 20 m or less, occurred in areas in the northwest, as well as in the south and southeast.

Figure 13.27. Holocene marine limits in Greenland. Lines for 20, 40, 80, and 120 m emergence relative to present sea level. Shaded areas have more than 120 m emergence. Dots mark observation points used. Black dashed line shows proposed maximum extent of Late Wisconsinan glaciation. Sources (clockwise from western North Greenland): Nichols, 1969; Kelly and Bennike, 1985; Funder and Hjort, 1980; Hjort, 1981c; Koch and Wegener, 1911; Hjort, 1979, 1981b; S. Funder, 1978a, unpublished; Brooks, 1979; Bøgvad, 1940; Kelly, 1985.

CHAPTER 13

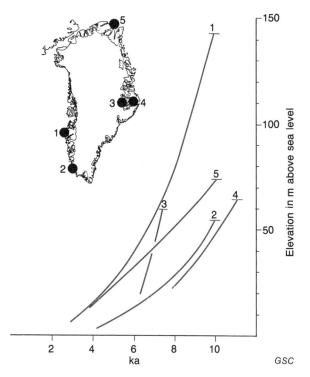

Figure 13.28. Some emergence curves from Greenland; inset map shows locations. Sources: 1 and 2 (Nordre Isotoq and Frederikshaab), Kelly, 1985; 3 and 4 (inner Scoresby Sund and Jameson Land), S. Funder, 1978a, unpublished; 5 (Kap København), Funder and Abrahamsen, 1988.

The pattern of uplift thus shows great variability. In a general way, however, it reflects the amount of land-based ice which was removed in each area in the early Holocene.

Holocene emergence history

Emergence curves show the age and amount of uplift relative to present sea level in restricted land areas. A number of these curves have been constructed from different areas by ^{14}C dating mollusc shells in raised deposits and assessing their depositional water depths. Locally some control has been obtained by dating the onset of organic sedimentation in lakes which became isolated from the sea at various stages during uplift of the land. These results have been summarized by Kelly (1979) and Funder (1978a) from West and East Greenland.

The reliability of the curves has been subject to much discussion, and it is generally agreed that many uncontrollable errors may affect their delineation, especially in the lower parts. Also, errors will generally tend to make the curves too steep. Bearing the uncertainties in mind, however, the curves do give an — albeit smoothed and general — impression of the process of isostatic rebound at the sites.

Figure 13.28 shows some of the more recently constructed emergence curves, which have been selected to show the fluctuations of relative sea level in areas with different amounts of uplift and different deglaciation histories.

In areas near the outer coasts a record of uplift began at the time deglaciation permitted entry of the sea. In most parts of the area, this was more or less simultaneously at 10-11 ka (curves 1, 2, 4 and 5). Initial uplift rates were as high as 50 m/ka (curve 1), but vary proportionally with the total amount of uplift, and were ca. 25 m/ka in areas with moderate uplift (curves 2, 4, and 5). The very high initial uplift rates, about 90 m/ka, postulated by Washburn and Stuiver (1962) must now be considered erroneous, and use of gravitational attraction of the ice sheet as an explanation of high relative sea level (Clark, 1976) seems to be without observational background.

The curves show that the high initial uplift rates gradually decreased. The ensuing period, however, was one of steady emergence, although it may not have been as continuous as shown by the curves. Thus Hjort (1973, 1981c) has interpreted field evidence in East Greenland to show that a eustatically conditioned transgression, which is not reflected in the emergence curves, took place 6 ka ago. Areas adjacent to the present ice margin obviously have a much shorter history of visible emergence, but here also initial uplift rates were high (curve 3).

For an area in western North Greenland, England (1985) has suggested that after a slow start at 8 ka, uplift rates reached a maximum at ca. 6 ka, owing to rapid ice melting at this time; however, there are no Greenland data to support this.

The youngest phase of emergence is least well known. In coastal West Greenland sea level may have been more than 20 m above the present at ca. 4.5 ka (e.g., Kelly, 1979), and in interior northeast Greenland the early inhabitants, 4.0-4.5 ka, lived on beaches which are now 10-20 m above sea level (Knuth, 1966). Obviously emergence in these areas must have continued for some time after this, as indicated also by an observation from interior Disko Bugt, showing that emergence here was operating after 2 ka (S. Funder, unpublished).

The time of the end of emergence is not known, but it probably happened at different times in different areas. At least 1 ka ago emergence had changed to submergence in most areas. This is seen from medieval Norse and eskimo house ruins, which are now below high tide, and from direct measurements of sea level change in the last century; the evidence for subsidence in West Greenland has been summarized by Kelly (1980b). Some areas have subsided 0.5 m in the period 1897-1946, while total submergence in West Greenland may amount to as much as 5-8 m. Similar, but fewer observations have been made in East Greenland, as summarized by Hjort (1981c). Detailed levelling shows subsidence rates of the order of 1 m/ka but submergence changed again to emergence after 1940 (Saxov, 1958, 1961).

Sea level changes within the last century seem to be directly related to the growing and waning of glaciers in the same period (Weidick, 1975b), and it was suggested by Kelly (1980a) that the subsidence may have begun with the onset of glacier readvance ca. 3 ka.

PALEOFAUNAS AND FLORAS

S. Funder and B. Fredskild

For more than a century co-operation with biologists has been an integral part of the Quaternary work in Greenland, and the organic remains found in the sediments were identified and discussed in relation to the present climatic and ecological demands of the species.

This section gives a brief survey of this work dealing with the record of marine invertebrates, the less well known history of vertebrates, and finally the comprehensive results in the field of vegetation history.

Marine invertebrates

The basic concepts for interpreting the subfossil mollusc faunas in terms of past climate and hydrography were established by the marine biologist Ad. S. Jensen (e.g., Jensen, 1917, 1942; Jensen and Harder, 1910; Harder et al., 1949), and refined by Laursen (1944, 1950, 1954). Recently the subfossil molluscs have been interpreted from their association with distinct water masses to show hydrographic rather than climatic changes. This method provides a basis for correlation with paleohydrographic results from deep sea investigations in adjacent oceans (Funder and Símonarson, 1984; Funder, 1984; Kelly, 1986).

While the early work was concerned mainly with Holocene faunas, an accumulating number of ^{14}C dates on shells has gradually substantiated that some faunas were older (e.g., Funder and Hjort, 1973). An important step towards dating the old faunas, especially from the Late Pliocene and early Quaternary, came with Feyling-Hanssen's demonstration of the utility of inner shelf benthic foraminifer assemblages for biostratigraphical correlation (Feyling-Hanssen et al., 1983; Feyling-Hanssen, 1985, 1986).

A review of the subfossil occurrence of a large number of mollusc species has recently been supplied by Símonarson (1981).

Late Pliocene and early Quaternary faunas. Characteristic foraminifers and molluscs from the Lodin Elv and Kap København formations are listed in Figures 13.17 and 13.23.

Foraminifer assemblages dominated by *Cassidulina teretis* and *Elphidiella hannai* in the Kap København Formation correlate with similar assemblages in the Lodin Elv Formation, zone II, and date from the Plio-Pleistocene transition period (Funder et al., 1985; Feyling-Hanssen, 1986).

The mollusc faunas in the two sequences are composed entirely of species which are widespread in Greenland waters at present, although *Portlandia arctica* and *Pecten groenlandicus* are confined to the northern parts, while *Macoma balthica* now is associated with subarctic water, and lives only along the west coast.

The richest subfossil mollusc fauna known from Greenland occurs in the Pátorfik marine beds of West Greenland (Table 13.4 and Fig. 13.8), which have been visited frequently by geologists over the last 100 years (e.g., Símonarson, 1981). Besides one endemic species which is extinct (*Alvania patorfikensis*), the fauna contains some that are now absent, and known only from this locality in Greenland — the gastropod *Oenopota angulosa* and the bivalves *Cyrtodaria siliqua* and *Periploma* sp., which now occur only along the east coast of North America (Símonarson, 1974; Funder and Símonarson, 1984). The bivalve *Panopea norvegica*, previously known only from the Pátorfik beds, has recently also been identified in mid-Holocene marine deposits in West Greenland (Weidick and Funder, unpublished).

Table 13.4. Recent geographical distribution of the molluscs and barnacles at Pátorfik (from Símonarson, 1981)

Species	LUSI-TANIAN	BOREAL			ARCTIC		
		Low	Mid	High	Low	Mid	High
Lepeta (Lepeta) caeca							
Margarites (Margarites) helicinus							
Margarites (Margarites) groenlandicus							
Margarites (Pupillaria) cinereus							
Lacuna (Epheria) vincta							
Lacuna (Epheria) crassior							
Putilla (Parvisetia) globula							
Alvania (Alvania) patorfikensis							
Alvania (Alvania) sp.							
Alvania (Frigidoalvania) janmayeni							
Tachyrhynchus erosus							
Natica (Lunatia) pallida							
Natica (Tectonatica) affinis							
Trophon (Boreotrophon) truncatus							
Trophon (Boreotrophon) clathratus							
Colus sp.							
Neptunea (Neptunea) despecta							
Buccinum undatum							
Buccinum groenlandicum							
Oenopota bicarinata							
Oenopota decussata							
Oenopota trevelliana							
Oenopota angulosa							
Oenopota nobilis							
Amaura candida							
Toledonia limnaeoides							
Diaphana minuta							
Retusa (Retusa) obtusa var. pertenuis							
Spiratella retroversa							
Nucula (Leionucula) tenuis expansa							
Nuculana (Nuculana) pernula							
Portlandia (Yoldiella) lucida							
Portlandia (Yoldiella) lenticula							
Portlandia (Yoldiella) fraterna							
Mytilus (Mytilus) edulis							
Chlamys (Chlamys) islandica							
Palliolum (Delectopecten) groenlandicum							
Tridonta (Tridonta) borealis							
Tridonta (Tridonta) elliptica							
Tridonta (Nicania) montagui							
Axinopsida orbiculata							
Serripes groenlandicus							
Clinocardium ciliatum							
Macoma (Macoma) calcarea							
Hiatella (Hiatella) arctica							
Panopea (Panomya) norvegica							
Cyrtodaria siliqua							
Mya (Mya) truncata							
Mya (Mya) pseudoarenaria							
Lyonsia (Bentholyonsia) arenosa							
Periploma sp.							
Balanus (Balanus) balanus							
Balanus (Balanus) crenatus							
Balanus (Chirona) hameri							

GSC

Funder, S. and Fredskild, B.
1989: Paleofaunas and floras (Greenland); in Chapter 13 of Quaternary Geology of Canada and Greenland, R.J. Fulton (ed.); Geological Survey of Canada, Geology of Canada, no. 1 (also Geological Society of America, The Geology of North America, v. K-1).

Although as noted above, the Late Pliocene and early Quaternary faunas are of modern type, they deviate by their sparse representation of the bivalve species *Mya truncata* and *Hiatella arctica* which dominate almost all younger faunas.

Sangamonian. The fauna from the Langelandselv interglaciation contains upwards of 25 taxa (Table 13.5) and is especially rich in gastropods (Petersen, 1982). It differs from both the present and other known subfossil faunas in East Greenland by containing such subarctic gastropods as *Amaura candida, Lacuna vincta*, and *Trophon clathratus*, as well as the bivalves *Mytilus edulis* and *Chlamys islandica* (Fig. 13.29). The latter two lived in the area for some millennia during the Holocene.

In West Greenland the deposits of the Kaffehavn interglaciation also contain *Mytilus edulis* and *Chlamys islandica* and the subarctic barnacle *Balanus hameri*, showing that conditions were at least as favourable as at present (Funder, 1984; Kelly, 1986).

Wisconsinan. The upper Jameson Land and Svartenhuk marine beds of East and West Greenland contain faunas of low diversity and dominated by species which are now widespread in Arctic waters (Table 13.5), notably *Mya truncata* and *Hiatella arctica*. The faunas also contain, as minor constituents, a few arctic sublittoral species, which occur in shallow waters in the extreme north while in most other areas they occur at depths of 100 m or more. Because of the depths of their habitats, these species are extremely rare in Holocene raised deposits, and their presence in faunas ascribed to the Wisconsinan may indicate that the present layer of warm surface water was thin or absent. This applies to the occurrence of *Astarte crenata* ssp. *crenata* in the

Table 13.5. Molluscs and barnacles in Sangamonian and Wisconsinan faunas in West and East Greenland

	La[1]	Ka[2]	Sv[3]	Jl[4]
BIVALVES				
Mylitus edulis		+	+	
Chlamys islandica	+	+		
Astarte borealis	+		+	+
Astarte montagui	+		+	+
Astarte elliptica	+		+	+
Axinopsis orbiculata	+			
Serripes groenlandicus	+	+	+	+
Clinocardium ciliatum	+	(+)	+	+
Macoma calcarea	+	+	+	
Hiatella arctica	+	+	+	+
Mya truncata	+	+	+	+
Pandora glacialis	+			
Thracia myopsis	+			
Nucula tenuis		(+)	+	
Portlandia arctica		(+)	+	+
Modiolaria nigra			+	
Pecten groenlandicus			+	+
Thyasira flexuosa			+	
Macoma moësta		(+)		
Portlandia intermedia			+	
Astarte crenata ssp. *crenata*				+
Macoma loveni				+
GASTROPODS				
Natica sp.				+
Puncturella noachina	+	+		
Margarita groelandica	+			
Lacuna vincta		+		
Cingula castanea	+			
Turritellopsis acicula	+			
Amauropsis islandica	+			
Lunatia pallida		+		
Lunatia nana	+			
Trophon truncatus	+			
Trophon violacea	+			
Bela simplex		+		
Amaura candida	+			
Amaura alba	+			
Cylichna occulta	+			
Acmaea testudinalis		+		
Cingula arenaria		+		
Lora sp.			+	
BARNACLES				
Balanus balanus	+	+		
Balanus hameri		+		

1 Langelandselv interglaciation, East Greenland
2 Kaffehavn interglaciation, West Greenland (Laksebugt fauna in parentheses)
3 Svartenhuk marine episode, West Greenland
4 Jameson Land marine episode, East Greenland

Sources: Funder and Hjort, 1973; Petersen, 1982; Funder and Símonarson, 1984; Kelly, 1986.

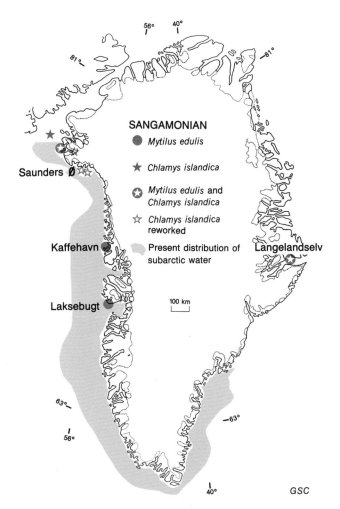

Figure 13.29. Sangamonian occurrences and present distribution of the bivalves *Chlamys islandica* and *Mytilus edulis* in Greenland. Sources: Funder, 1984; Kelly, 1985.

Jameson Land marine beds (Funder, 1984), and possibly also to *Portlandia intermedia* which has been noted by Kelly (1986) in the Svartenhuk marine beds, and *Macoma moesta* in the Laksebugt fauna.

Holocene. Following the deglaciation of the coasts at ca. 10 ka bottom communities similar to those at present invaded the shallow water habitats, as shown by the numerous ^{14}C dated mollusc shells.

The sequence of postglacial faunal change was first worked out by Jensen and Harder (1910), and was based especially on collections from the shell-rich sediments at Orpigssoq, West Greenland. Later, the stratigraphy was further refined (Laursen, 1944, 1950; Harder et al., 1949). However, ^{14}C dates from these sites have induced some modifications (Donner and Jungner, 1975; Kelly, 1979), and the Holocene record now seems to comprise a development from cool to warmer to cool.

Although this succession is found in both West and East Greenland, there are some differences in timing and intensity. The first penetration of warm subarctic water was along the west coast and is reflected by the occurrence of one or both of the species *Mytilus edulis* and *Chlamys islandica* as early as 9 ka, possibly throughout their entire present range (Fig. 13.30). Later, even higher temperatures of the surface water is denoted by the occurrence in central West Greenland of such subarctic-boreal bivalves as *Zirphaea crispata, Anomia squamula, Arctica islandica*, as well as the recently identified *Panopea norvegica* and the gastropods *Emarginula fissura* and *Acmaea virginea* (Fig. 13.30). These species, which are now absent from Greenland

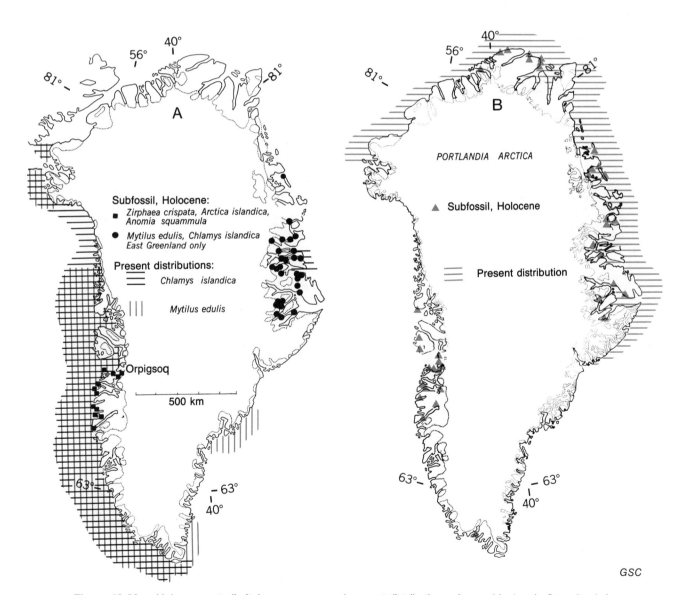

Figure 13.30. Holocene extralimital occurrences and present distributions of some bivalves in Greenland. A, subarctic bivalves in West and East Greenland and B, an arctic species (*Portlandia arctica*). Sources: Jensen, 1942; Hjort and Funder, 1974; Donner and Jungner, 1975; S. Funder, 1978a, unpublished; Kelly and Bennike, 1985; C. Hjort and S. Funder, unpublished.

waters, were locally abundant in the period from ca. 7.5 to shortly after 5 ka, but apparently only in those areas which today have the longest open water season. During this period of "marine optimum", the arctic *Portlandia arctica* disappeared from the West Greenland waters (Fig. 13.31), and it has not later been able to reimmigrate (Jensen, 1942).

East Greenland saw a period of warm water from 8.5 to 5.5 ka, when *Mytilus edulis* and *Chlamys islandica* were abundant in the fiords where they now are extinct or survive only as small relict populations (Fig. 13.30, 13.31; Hjort and Funder, 1974).

In earlier literature the occurrence of the small bivalve *Limatula subauriculata* in raised deposits at the coast of the Arctic Ocean in Greenland and Canada was considered as an indicator for warmer conditions in the extreme north (Laursen, 1954; Stewart and England, 1983). Recently, however, an examination of the shells has shown that they all belong to the species *Limatula hyperborea*, an arctic bathyal species that is extant in the area (S. Funder, unpublished).

Although the marine faunas are known mainly from mollusc shells, more delicate organisms have also been preserved especially in calcareous concretions which occur both in situ in marine sediments and as clasts in young till deposits. The faunas contained in the concretions comprise imprints and skeletal remnants of rich faunas including polychaetes and echinoderms (Gripp, 1932; Weidick, 1972b).

Vertebrate faunas

The sparse finds of Quaternary vertebrate remains in Greenland have been summarized by Bendix-Almgreen (1976). Out of a total of nine taxa only one, reindeer, is terrestrial, while the rest are marine fish and mammals. The reindeer is considered to have immigrated from Ellesmere Island and appeared in North Greenland as early as 8 ka. Since then the populations have fluctuated, both in numbers and distribution, with climatic change as the main factor (Meldgaard, 1986). The best known vertebrate fossils are skeletons of the fish *Mallotus villosus* which occur frequently in concretions, and indeed were the first Greenland fossils to become known to science in the late 18th Century.

Other vertebrate remains have been found mainly at the surface and few are dated, but most are believed to be of Holocene age. An exception to this are the finds of bones from the fish *Gasterosteus aculeatus* (Fredskild and Røen, 1982), and a tooth from the extinct hare genus *Hypolagus* (Funder et al., 1985), both found in the Late Pliocene and early Quaternary Kap København Formation (Fig. 13.23).

This brief survey does not include vertebrate remains found at human occupation sites dating back as far as 4.5 ka. For information on these faunas as well as on human history the reader is referred to the archeological literature.

Paleofloras

Speculations on the age and development of Greenland's vegetation go back more than a century and began with a

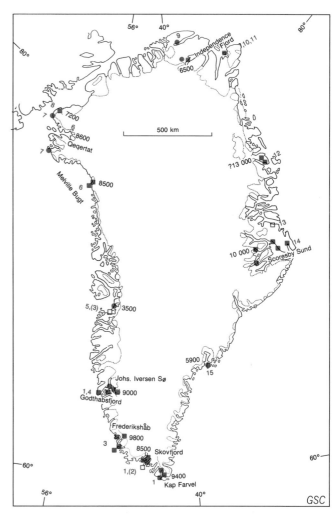

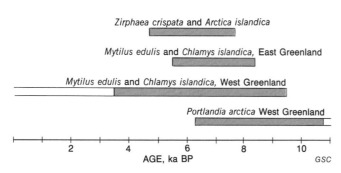

Figure 13.31. Time span of some arctic and subarctic bivalves in Greenland waters. Red shaded areas, presence demonstrated by ^{14}C dates; open areas, presence surmised (adapted from Kelly, 1985).

Figure 13.32. Location of pollen data from Greenland, with oldest ^{14}C age (years BP) in each area. Squares represent lake sediments; circles represent peat. Solid symbols represent pollen diagrams; open squares represent other data. Sources (referenced by number; numbers in parentheses refer to open symbols): 1. Fredskild, 1973; 2. Funder, 1979a; 3. Kelly and Funder, 1974; 4. Iversen, 1934, 1952; Fredskild, 1983a; 5. Fredskild, 1967; 6. Fredskild, 1985b; 7. Malaurie et al., 1972; 8. Blake et al., 1985; 9. Bennike, 1983b; 10. Fredskild, 1969; 11. Funder and Abrahamsen, 1988; 12. Björck and Persson, 1981; 13. Funder, 1979b; 14. Funder, 1978a; 15. Bick, 1978.

sometimes heated discussion about the age of the flora and the possibility for survival of plants in refugia during the ice ages (summarized by Funder, 1979b). The issue was removed from the sphere of speculation with Iversen's (1934, 1952) publication of pollen diagrams from West Greenland, which, although representing only the Holocene, showed that plants that are now dominating components of the vegetation had immigrated, apparently from abroad, in fairly recent times. These results initiated palynological and limnological studies in all parts of the country, which have provided a wealth of detailed information on the Holocene history of terrestrial ecosystems. This history has recently been reviewed, with main emphasis on West Greenland, by Fredskild (1985a).

Pre-Holocene vegetation is known from only one unit, the Kap København Formation in North Greenland, which recently has supplied a detailed picture of the Late Pliocene and early Quaternary vegetation in the extreme Arctic.

This section provides a brief survey of the flora development during this early stage and the Holocene.

Late Pliocene and early Quaternary flora. The shallow water marine sand of the Kap København Formation (Fig. 13.23) contains abundant transported remains of terrestrial plants, including 10-20 cm-thick beds of moss fragments, and trunks, branches, twigs, needles, leaves, cones, fruits, and seeds from trees, bushes, and herbs (Fredskild and Røen, 1982; Mogensen, 1984; Funder et al., 1984, 1985). The plant remains occur throughout member B of the formation (Fig. 13.23) and reflect an environment at the forest-tundra ecotone, consisting of scattered trees growing in heath. The identified tree remains are dominated by *Larix* sp. and *Picea mariana*, with *Thuja occidentalis*, *Taxus* sp., and tree birch as minor constituents. The most common remains among the higher plants, however, come from arctic plants, notably *Dryas* cf. *octopetala* and *Betula nana*.

Table 13.6 shows that the identified taxa from terrestrial flora and fauna are a mixture of boreal and low arctic organisms, unlike any present day environment. The majority of species, however, now live near the forest-tundra transition in Labrador 2500 km to the south, and the composition of the flora reflects cool, moist, boreal, Low Arctic climate, quite different from the present High Arctic continental climate in the area (Funder et al., 1985).

Holocene vegetation history. Pollen diagrams from all parts of the country reflect the development of vegetation since the sites became ice free or emerged from the sea during isostatic rebound. The majority of diagrams have been constructed from analyses of cores of lake sediments, and show regional vegetation changes, while site-related changes appear in diagrams from peat sections (Fig. 13.32).

In most regions the record goes back to ca. 10 ka, only in the far north is it shorter. In East Greenland three lakes have given ^{14}C ages in the interval 12-16 ka (Funder, 1979b; Björck and Persson, 1981). The ages were determined on slightly humic clay and silt, and contamination by inactive carbon from nearby coal deposits cannot be definitively ruled out. Therefore these ages should probably be treated with caution until corroborated by additional evidence.

The general pattern of vegetation development in different parts of Greenland is shown in Figure 13.33, based on analyses of sediments from 25 lake basins.

In its basic outline the development in all areas has three stages: an early stage of "fell field vegetation," that is, herbs and grasses growing scattered on raw minerogenic soil. After a transition period the areas were in the second stage covered by dwarf shrub heaths of varying composition and density. In the south, copses and even woods appeared. Finally, a stage of deterioration occurs in all areas with recurrence of less dense and less diverse plant communities.

The vegetation development at sites in West and East Greenland is shown by the pollen diagrams in Figures 13.34

Table 13.6. Terrestrial flora and fauna from the Kap København Formation and the present climatic distribution of these species

	Boreal	Arctic Low	Arctic High
Betula alba s.l. (tree)	+		
Cirsium sp.[1] (herb)	+		
Cornus stolonifera (bush)	+		
Larix cf. *occidentalis* (tree)	+		
Picea mariana (tree)	+		
Potamogeton natans (herb)	+		
Cristatella mucedo[2] (bryozoan)	+		
Taxus sp. (tree or bush)	+		
Thuja occidentalis (tree)	+		
Bembidion grapi[2] (beetle)	+	(+)	
Dicranum undulatum[3] (moss)	+	(+)	
Schistidium maritimum[3] (moss)	+	(+)	
Botrychium sp.[1] (fern)	+	+	
Gyrinus opacus[2] (beetle)	+	+	
Menyanthes trifoliata (herb)	+	+	
Potamogeton gramineus (herb)	+	+	
Potentilla cf. *palustris* (herb)	+	+	
Alona affinis[2] (daphnia)	+	+	(+)
Alonella excisa[2] (daphnia)	+	+	(+)
Cinclidium stygium[3] (moss)	+	+	(+)
Diphasium sp. (club moss)	+	+	(+)
Empetrum nigrum s.l. (dwarf bush)	+	+	(+)
Gasterosteus aculeatus[2] (fish)	+	+	(+)
Hydroporus sp.[2] (beetle)	+	+	(+)
Lycopodium annotinum[1] (club moss)	+	+	(+)
Ranunculus reptans[2] (herb)	+	+	(+)
Sphagnum sp.[3] (moss)	+	+	(+)
Brassicaceae[1] (herbs)	+	+	+
Calliergon sarmentosum (moss)	+	+	+
Carex spp (sedges)	+	+	+
Chamaenerion sp.[1] (herb)	+	+	+
Caryophyllaceae (herbs)	+	+	+
Cenococcum geophilum[1] (fungus)	+	+	+
Cyrtomnium hymenophylloides[3] (moss)	+	+	+
Distichum sp.[3] (moss)	+	+	+
Distichum flexicaule[3] (moss)	+	+	+
Erigeron tp.[1] (herbs)	+	+	+
Equisetum sp. (horsetail)	+	+	+
Hippuris vulgaris (herb)	+	+	+
Huperzia selago[1] (club moss)	+	+	+
Mnium thomsonii[3] (moss)	+	+	+
Orthothecium sp.[3] (moss)	+	+	+
Pedicularis sp.[1] (herb)	+	+	+
Poaceae (grasses)	+	+	+
Polytrichum sp.[3] (moss)	+	+	+
Salix spp (trees, bushes, dwarf bushes)	+	+	+
Saxifraga spp[1] (herbs)	+	+	+
Scorpidium scorpioides[3] (moss)	+	+	+
Zygnema tp.[1] (alga)	+	+	+
Betula nana (bush)	(+)	+	(+)
Potamogeton filiformis (herb)	(+)	+	(+)
Cystopteris fragilis ssp. *dickieana*[1] (fern)	(+)	+	+
Dryas octopetala s.l. (dwarf bush)	(+)	+	+
Oxyria digyna (herb)	(+)	+	+
Plagiomnium medium ssp. *curvatulum*[3] (moss)	(+)	+	+
Polygonum viviparum tp.[1] (herb)	(+)	+	+
Ledum palustre ssp. *decumbens* (dwarf bush)		+	
Drepanocladus exannulatus[3] (moss)		+	(+)
Cyrtomnium hymenophyllum[3] (moss)		+	+
Papaver sect. *scapiflora* (herbs)		+	+
Vaccinium uliginosum ssp. *microphyllum* (bush)		+	+
Cassiope tetragona (dwarf bush)		(+)	+
Cinclidium arcticum[3] (moss)		(+)	+
Lepidurus arcticus[2] (crustacean)		(+)	+
Ranunculus nivalis/pedatifidus[2] (herbs)		(+)	+
Hypolagus sp.[4] (mammal)		extinct	

(+) restricted occurrence
[1] identified from pollen or spores
[2] from Fredskild and Røen, 1982
[3] identified by G.S. Mogensen (Funder et al., 1984; Mogensen, 1984)
[4] identified by C.A. Repenning (United States Geological Survey, personal communication, 1983)
All others identified by O. Bennike

CHAPTER 13

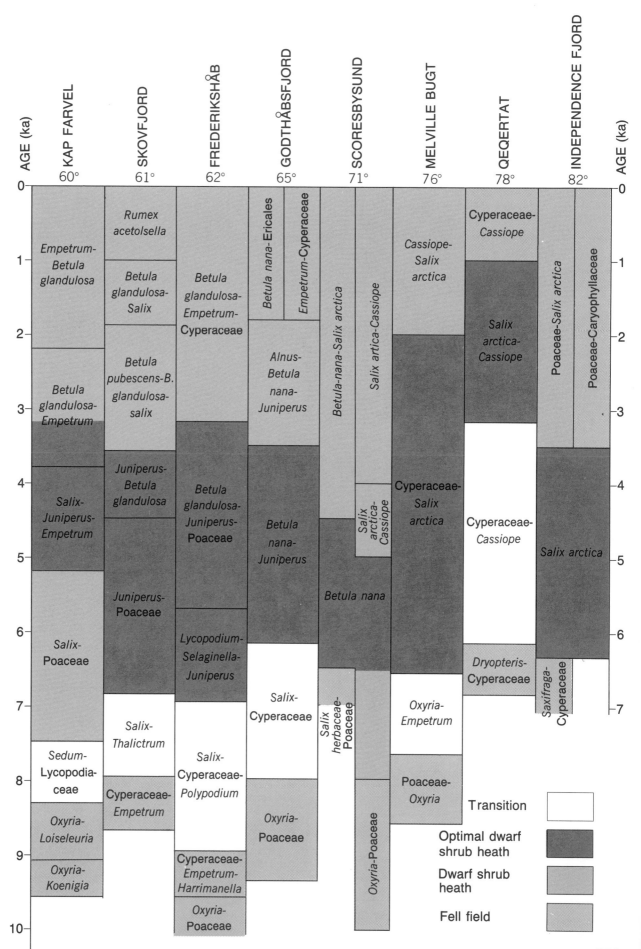

and 13.35. The early fell field stage is characterized by abundant pollen from grass and such herbs as *Oxyria digyna*, *Saxifraga oppositifolia*, and Caryophyllaceae, plants which today occur in all parts of the country and are among the first to invade newly exposed ground. The sparse fell field vegetation is adapted to cope with such problems as cryoturbation, soil erosion during snow melting, and the difficult soil water balance problems caused by a shallow permafrost table — problems which are serious enough but only indirectly related to climate. Certain of the low arctic species that were present, such as *Angelica archangelica*, *Thalictrum alpinum*, and others, show that summer temperatures were similar to or only slightly cooler than present.

This early stage ended at or before 8 ka, in regions south of 75°N, in the north it persisted one or two millennia longer (Fig. 13.33).

At the end of the fell field stage, dwarf shrubs began to dominate mesic sites in the lowlands. Some species had been sparsely present earlier, notably *Empetrum nigrum* and *Vaccinium uliginosum*, whereas many had to immigrate from North America or northern Europe (Fig. 13.36). Thus *Betula nana*, an early European immigrant, first appeared in East Greenland at 8 ka and expanded almost explosively. However, it took well over 1 ka before it became established in West Greenland (Funder, 1979b; Fredskild, 1985a).

The North American immigrants comprise high arctic species which apparently crossed Nares Strait, and low arctic species which seem to have crossed Davis Strait from Labrador in the south. The first group includes *Salix arctica* and *Cassiope tetragona*, two species which now dominate the vegetation in many parts of North Greenland. The second group includes *Betula glandulosa* and *Alnus crispa*. *Alnus* pollen occur frequently in the lake sediments of coastal West Greenland back to 8 ka, this early pollen was thought by Iversen (1952) and Fredskild (1973) to have been transported by winds over Davis Strait from North America, but Kelly and Funder (1974) suggested that it came from native plants. Recently Fredskild (1983a) presented evidence in favour of the first theory and indicated that alder did not immigrate until ca. 4 ka.

In the south the heaths were composed of a great number of shrubs, including the thermophilous *Juniperus communis*, dwarf birches, and several species of ericales. Moist ground was invaded by copses of *Salix glauca* and later *Alnus crispa*, and in sheltered valleys in the extreme south

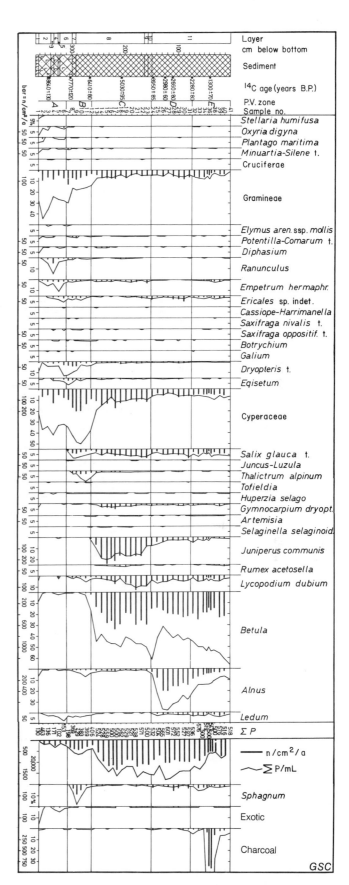

Figure 13.33. Holocene pollen assemblages and vegetation development in Greenland; Godthabsfjord, Scoresby Sund and Independence Fjord in the interior, on left, and coastal area to right (for sources and location see Figure 13.32).

Figure 13.34. Pollen diagram showing development of terrestrial vegetation in a lake in West Greenland (Johannes Iversen SØ, see Fig. 13.32, from Fredskild, 1985a).

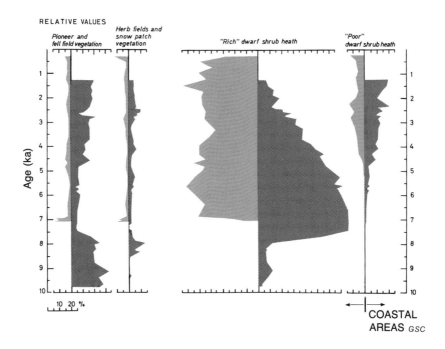

Figure 13.35. Composite pollen diagram showing development of major plant communities in the Scoresby Sund area, East Greenland (from Funder, 1978a). Curves facing left (red) show development in the interior while those facing right (black) show the development in coastal areas.

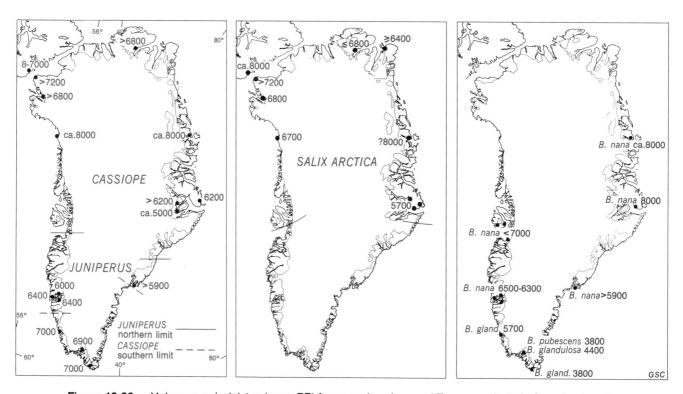

Figure 13.36. Holocene arrival dates (years BP) for some American and European plants in Greenland, and their present distributions (from Fredskild, 1985a,b).

open woods composed of *Betula pubescens* and *Sorbus americana* were formed 2-3 ka ago. In the north the heaths were open and composed of few species, notably *Salix arctica*, *Cassiope tetragona*, and *Vaccinium uliginosum*.

In the southern areas the heath development was associated with the development of stable humic soil. This is reflected in the composition of lake sediments and their content of planktonic algae, showing a gradual shift from eutrophy during the early stage when washed out mineral nutrients were available, to oligotrophy and acidification when the nutrients in the lakes were replaced by acid decomposition products from humic soils (Fredskild, 1983a, b).

By 5-6 ka the heaths in most areas reached their maximum extension and diversity. After this time decline set in, but began at different times in different areas (Fig. 13.33). Deterioration apparently first hit coastal East Greenland, and some millennia later affected West Greenland. In coastal East Greenland and in North Greenland the heaths were reduced to local areas and fell field vegetation again expanded, while in West Greenland the heaths in the interior became dominated by dwarf birches and at the coasts by *Empetrum*, as they are at present (Fig. 13.34, 13.35).

A special facet of the vegetation history deals with the Norse settlers who colonized West Greenland between 60° and 65°N at ca. AD 1000, and for 500 years subsisted in the area on a pasture and hunting economy. Pollen diagrams from peat and lake sediments near the farmsteads have shown that the colonization began with clearing and burning of the willow copse to develop areas for hay production (Iversen, 1934, 1952; Fredskild, 1981). A number of weeds were introduced, and in pollen diagrams from South Greenland a sudden expansion of *Rumex acetosella* indicates the arrival of the colonists. Although peat cutting and overgrazing led to soil erosion in areas close to the farmsteads, the pollen diagrams give no clear explanation to why the colonists finally disappeared.

DEVELOPMENT OF CLIMATE, GLACIATION, AND OCEANOGRAPHIC CIRCULATION

S. Funder

This section gives an outline of late Tertiary and Quaternary paleoenvironments based on past ice distribution, pollen and mollusc records, and sea level change which has been described in the preceding sections. This record is compared with data from two other sources which in the past decades have provided a wealth of information on glaciation, paleohydrography, and paleoclimate: the deep sea and the Inland Ice records.

Funder, S.
1989: Development of climate, glaciation, and oceanographic circulation (Greenland); in Chapter 13 of Quaternary Geology of Canada and Greenland, R.J. Fulton (ed.); Geological Survey of Canada, Geology of Canada, no. 1 (also Geological Society of America, The Geology of North America, v. K-1).

The ice sheet and deep sea records have the advantages over those from ice-free land because they contain continuous, or nearly continuous, depositional records and they contain information which is pertinent for large geographical regions. On the other hand they have the disadvantage that their signals, such as the isotopic composition of deep sea benthic foraminifers, need a sometimes rather opaque element of interpretation before they can be translated into meaningful knowledge.

In recent years a comprehensive literature dealing with deep sea studies has accumulated. The survey presented below is, by necessity, restricted to data that have been obtained in oceans around Greenland, and to information which is relevant for the Greenland land record.

The results from Inland Ice cores have been treated in detail by Reeh (1989). In this section, the record, which spans the period since the last interglaciation, is compared with that from the ice-free land.

One conclusion from the discussion in the preceding sections is that the Quaternary record of Greenland Quaternary apparently falls into two distinct sections: the late Tertiary to early Quaternary, and the late Quaternary (Illinoian to the Holocene; Fig. 13.3). There is hardly any information on the time interval between these two, as will appear also from the presentation below.

Deep sea data

Under the auspices of The Deep Sea Drilling Project and Ocean Drilling Programme, cores have been obtained in Baffin Bay and Labrador Sea (Davies, 1972; Srivastava, Arthur, Clement et al., 1987), in Denmark Strait (Schambach, 1978), and in Greenland-Norwegian seas (White, 1976; Eldholm, Thiede, Taylor et al., 1987). They generally extend back into the Paleogene, and even though their time resolution is too low for detailed Quaternary work, these cores have provided important information on late Tertiary hydrographic and climatic developments and the onset of glaciation on adjacent land.

A considerably larger data base is available in the more numerous piston and trigger cores which have a finer time resolution, but generally go back only some 200 to 300 ka. Presently detailed information is available from approximately 40 stations, in Baffin Bay (Aksu, 1985; de Vernal et al., in press), the Labrador Sea (Fillon, 1985; Aksu and Mudie, 1985), the Denmark Strait and Greenland-Norwegian Seas (Kellogg, 1980), and the Arctic Ocean (Markussen and Thiede, 1985). The results from these studies include analyses of oxygen isotope and faunal composition of foraminifer faunas as well as the chemistry and lithology of the sediments. Dating is usually made by reference to the global isotopic time scale (Shackleton and Opdyke, 1973, 1977), but important marker horizons are provided by two ash layers, dated to ca. 10 and ca. 65 ka, which seem to occur extensively in the North Atlantic and may have been distributed from Icelandic sources by either drifting ice (Ruddiman and Glover, 1972) or by atmospheric transport (Mangerud et al., 1984).

The deep sea cores, and especially the piston and trigger cores, have provided a wealth of information on past glaciation and distribution of water mass types in the oceans around Greenland. This information must be considered when discussing the record on land even though serious problems of correlation remain.

Onset of glaciation

The earliest evidence of an ice sheet over Greenland is an erosional unconformity and glacial sediments which appear in seismic profiles from the East Greenland shelf (see *Quaternary geology of the shelves adjacent to Greenland, Pliocene-Pleistocene*). This was caused by a more extensive glaciation of the shelf than has occurred since. A preliminary age estimate of 2.4 Ma has been placed on this event (H.C. Larsen, Geological Survey of Greenland, Denmark, personal communication, 1986). This is in general agreement with recent results from deep sea drilling indicating a similar age for a phase of major glaciation on land bordering the North Atlantic, even though the onset of glaciation may have been earlier — especially in regions bordering the Arctic Ocean (Clark, 1982; Eldholm, Thiede, Taylor et al., 1987).

A younger glaciation phase is reflected in diamicton, exposed in the Lodin Elv and Kap København formations and biostratigraphically dated to the time of the Pliocene-Pleistocene transition at ca. 1.8 Ma. These sediments have been interpreted as formed by deposition from icebergs (Fig. 13.17, 13.23). Arctic climate at this time is shown also by the marine mollusc faunas in the sediments (see section on paleofaunas). However, the exposures do not reveal whether deposition was related to local glaciers or to a receding ice sheet margin.

The arctic fauna in the Kap København Formation is overlain by sand with remains of boreal-low arctic fauna and flora. This indicates that temperatures rose again, and climate in the extreme north of Greenland was similar to that which now prevails 2500 km to the south (see section on Late Pliocene and early Quaternary flora). Although mountain glaciers may well have existed during this interval, it is inconceivable that the ice sheet could have survived these temperatures. Thus although there is evidence for the existence of an ice sheet and arctic climate as far back as the Late Pliocene, the ice sheet at that time appears to have disappeared during interglaciations.

(?) Illinoian

As noted in the regional chapters, weathered till and scattered erratics on high mountains and at the outer coasts show that all parts of the country saw a period of extensive glaciation before 30-40 ka (based on radiocarbon dates, from younger marine sediment). In West, East, and North Greenland this glaciation phase has been termed the Fiskebanke, Scoresby Sund, Bliss Bugt glaciations, respectively. From an assessment of weathering intensity it is here postulated, as a working hypothesis, that the three are correlative. Based on exposures on Saunders Ø and Jameson Land in northwest and East Greenland this glaciation is tentatively referred to the Illinoian.

During this glaciation the shelf areas were inundated by the ice sheet which in West Greenland extended to the outer shelf break, 100-150 km from the present coast, and formed the Hellefisk moraines. In the northwest the ice sheet crossed Nares Strait and overrode coastal parts of Ellesmere Island in the Canadian Arctic, while in East Greenland it moved independently of the present fiord topography and extended for an unknown distance onto the shelf.

The Laksebugt marine fauna of West Greenland is believed to reflect initial influx of subarctic water after the disappearance of this ice sheet (Table 13.5).

Although attempts at correlation of these somewhat hypothetical events with other records obviously is premature, it can be noted that Fillon (1985) in Labrador Sea-Baffin Bay noted very high input of sand, indicating intensive glaciation of the surrounding land areas, in late isotopic stage 6. This is here considered as equivalent to the Illinoian. In the Greenland-Norwegian seas this time interval is not well represented in the deep sea cores; however, it appears that the region was dominated by cold polar water (Kellogg et al., 1978; Streeter et al., 1982).

Sangamonian

The marine Kaffehavn and Langelandselv deposits of northern West and East Greenland are assigned an interglacial status because their mollusc faunas have a component of subarctic species at sites which are close to or farther north than their modern distributional limit (Fig. 13.29).

For the Greenland-Norwegian seas, Kellogg (1980) found that oceanography during isotopic substage 5e resembled the present, although circulation was slightly more vigorous. During the last 400 ka this type of circulation has been attained only twice: during this substage and in the Holocene. On this basis, the Langelandselv marine beds are correlated with substage 5e, with an age of 125-115 ka (Kellogg et al., 1978). In the Labrador Sea-Baffin Bay substage 5e is also marked by a peak in the abundance of subpolar planktonic foraminifera, but unfortunately the record is obscured by differential solution of the shells. However, it was noted also by Fillon (1985) that substage 5e was the last time when oceanographic conditions in this area were more or less identical with those at present. As a consequence, it is likely also that the Kaffehavn deposits correlate with this substage.

A corollary to this is that Sangamonian marine conditions were slightly warmer than those attained during the Holocene.

No deposits of interglacial terrestrial material have been observed, but as noted by Reeh (1989), conditions on land at that time have been debated. While the Camp Century core seems to contain ice sheet glacier ice formed during the Sangamonian, the evidence from the Dye 3 core in the south is more complicated, and the basal ice here could have been formed in a local alpine glacier, which would mean that the southern ice dome did not exist. Previously this possibility was deemed feasible from an ice budget aspect by Weidick (1975b).

Wisconsinan

The transition from the Sangamonian is reflected in the shift from mollusc faunas with subarctic components to faunas with sublittoral arctic components in the Svartenhuk and Jameson Land marine beds in West and East Greenland (Table 13.5). No glacigenic sediments have been found in these sequences, and amino acid results show that at the outer coasts the Sangamonian was followed by a prolonged period without glaciation, but with cold polar conditions. Somewhere on land, however, ice was accumulating. This can be inferred from the Jameson Land beds, which require downwarp of the East Greenland coast in this period. Eventually, later than 40 ka, the marine sediments were overridden by glaciers that moved onto the shelf.

Thus the record, fragmentary as it is, contains no firm evidence for glaciation beyond the present coastline in Early and Middle Wisconsinan. It implies, rather, that the ice sheet was built up gradually during successive stages until a culmination was reached during the Sisimiut, Flakkerhuk, Independence Fjord glaciations in West, East, and North Greenland, at or before 14 ka. The evidence further implies that large scale oscillations of the ice sheet margins in all parts of the country were in phase; however, the intensity and character of glaciation varied between regions.

The largest growth of the ice sheet took place in the southeast and west where the ice was several hundred kilometres beyond its present margin, and thicknesses measured at least 1000 m at the present coastline. The landscapes developed bear witness to a vigorous glacial regime with ice sliding over the bedrock surface (Fig. 13.2). The ice sheet margin either ended on dry shelf, as in parts of West Greenland, or was grounded in shallow water, as in southeast Greenland (Fig. 13.26).

In the north and in East Greenland north of 69°N, on the other hand, the ice sheet did not exceed 400-600 m in thickness, and mountain areas and coastal lowlands remained ice free throughout the Wisconsinan. Ice movement here was constrained by local topography, as shown both by the landscape and by the distribution of erratics. Ice tongues debouching from fiords spread out on the shelf, but probably did not form a continuous ice margin (Fig. 13.18). In the extreme north, however, the ice tongues probably coalesced to form ice shelves along the coast, and it may be postulated that an extensive ice shelf occupied Lincoln Sea between Greenland and Ellesmere Island (Fig. 13.24). The major source for this ice shelf was a trunk glacier, nourished from both Canada and Greenland, that moved northwards through Nares Strait into the Lincoln Sea, and southwards into northern Baffin Bay.

Retreat from the Wisconsinan glacial limit began before 10 ka, and possibly at or before 14 ka, as indicated by ^{14}C dates from West Greenland. By 10-11 ka the ice margin was located along the West Greenland coastline (Fig. 13.10, 13.11), and on a major topographical threshold in East Greenland where the Milne Land advance of fiord glaciers took place at this time (Fig. 13.18). In West Greenland, the apparent synchroneity in ice margin behaviour during this early stage, which seems to contrast with the variable pattern of Holocene retreat, may indicate that the early phase of break up was caused largely by rising sea level which rendered the shelf-bound portions of the ice margin unstable along the entire coastline.

A comparison between the glaciation record, and the data from neighbouring deep seas shows both agreement and disagreement. Although the significance of this, considering the potential errors in both sets of data, should not be overestimated, several points can be mentioned.

The difference in glacial regimes in the southeast and west, as compared to the north, imply steeper east-west temperature and moisture gradients than the present. This is suggested also by data from deep sea cores which show different oceanographic developments in the Greenland-Norwegian seas to the east from those in Baffin Bay-Labrador Sea to the west.

In the Greenland-Norwegian seas the oxygen isotopic substage 5e/d transition, here considered as the transition from interglacial to glacial conditions at about 115 ka, was marked by abrupt changes to heavier oxygen isotopes in the benthic foraminifera, disappearance of planktonic subpolar foraminifera, and a peak of benthic foraminifera indicative of an episode of year round sea ice cover (Kellogg et al., 1978; Kellogg, 1980; Streeter et al., 1982). These changes were more abrupt here than elsewhere in the North Atlantic and reflect the sudden amputation of the warm water supply from the south, a supply which was not re-established until the very end of the Wisconsinan. Thus the Greenland-Norwegian seas remained cold and nearly isolated from oceanic circulation throughout the later part of the Sangamonian and the Wisconsinan, implying that the climate along Greenland's eastern seaboard was frigid and continental — in agreement with the land evidence for thin, slowly accumulating and rather passive glaciation in East Greenland north of 69°N.

In Baffin Bay-Labrador Sea, on the other hand, the substage 5e/d transition was less dramatic, and subarctic water from the Atlantic appeared intermittently along the west coast of Greenland throughout the Wisconsinan. These warmer water periods alternated with shorter periods of impeded current activity and high arctic dry climate over the region (Fillon, 1985; Aksu, 1985). In general terms the warmer and more humid climate over this side of Greenland agrees with the thicker and more vigorous character of glaciation observed on land. However, the Sisimiut glaciation of West Greenland could here fit into isotope stage 2 when the West Greenland Current had ceased (de Vernal and Hillaire-Marcel, 1987).

There are thus some points of apparent agreement between the land record and deep sea data; although the glaciation record on land in general deviates from the record in the Baffin Bay - Labrador Sea region — as well as from the histories of neighbouring ice sheets in the same period — by its apparent simplicity. Incompleteness of the geological record on land may account for some of the disagreement; however some is undoubtedly real. Thus the amounts of ice involved both in the growth and decay phases of the Greenland Ice Sheet were undoubtedly smaller than those of neighbouring ice sheets, and the most remarkable facet of the Wisconsinan-Holocene history of the Greenland Ice Sheet is probably its stability, situated as it is in the middle of a region of turbulent climate and large scale meridional heat exchange.

Holocene

Wisconsinan-Holocene transition. The Wisconsinan-Holocene transition is a clear signal in records from the ice-free land, the Greenland Ice Sheet, and the deep seas. The signals can generally be explained as the direct or indirect response to rapid climatic warming.

From the ice-free land there is an abrupt increase in the number of ^{14}C dates, determined on marine shells and organic lake sediments (Fig. 13.6, 13.14, 13.21). This implies that isostatic uplift caused by rapid melting of the land-based ice margin had begun. In the deglaciated land areas vegetation was sufficiently dense to allow organic sedimentation in some lake basins.

In East Greenland north of 69°N a major readvance of fiord glaciers — the Milne Land stade — took place at this time, possibly as the result of an increase in precipitation, caused by the disappearance or reduction of summer pack ice along the coasts. In West Greenland the contemporaneous Taserqat stade left moraines along a considerable part of the coastline (Fig. 13.11); however whether they mark a climatically conditioned readvance, or a glaciodynamically conditioned stillstand of the ice sheet margin is uncertain (e.g., Kelly, 1985). The Disko stade, dated to ca. 9 ka, was an expansion of a local ice cap in a coastal area, probably caused by increased precipitation, thus both in the east and the west, ice margins reacted spasmodically to the climatic changes.

The vegetation that spread over the newly ice-free land was one of "fell field"; however, among the scattered herbs were some thermophilous low arctic species reflecting summer temperatures only slightly lower than those at present. Similarly, subarctic molluscs occurred along the coast of West Greenland as early as 9 ka, showing that the warm West Greenland Current was operating at this time, and probably penetrated almost as far into Baffin Bay as it does now (Fig. 13.29).

The deep Inland Ice cores also contain marked changes at the Wisconsinan-Holocene boundary in oxygen isotope composition, in dust content, acidity, and concentration of chemical components (Dansgaard et al., 1984; Hammer et al., 1985; Herron and Langway, 1985; see also Reeh, 1989), and it has been inferred that the change from glacial to interglacial mode may have occurred in a few decades in response to a drastic change in the atmospheric circulation over the North Atlantic and increased precipitation on the Greenland Ice Sheet (Hammer et al., 1986).

In the deep North Atlantic Ocean change is reflected in the distribution of ice rafted ash, increased productivity of planktonic foraminifera, and expansion of subpolar species. These all indicate a northwestwards shift of the oceanic polar front and, by ca. 9 ka, establishment of the present circulation pattern in Baffin Bay-Labrador Sea (e.g., Ruddiman and McIntyre, 1981). According to these results, however, ice age conditions still prevailed along the eastern seaboard. The evidence from all sources thus shows that the period from ca. 11 to 9 ka was one of very rapid environmental change, in Greenland, as it was in northwestern Europe.

Early Holocene. As noted in the regional chapters above, early Holocene was a period of rapid ice margin retreat and by ca. 7 ka the present distribution of ice cover had been attained in all sectors. There was some deviation from this, however, as in the cool/moist climate areas of northern and southern West Greenland the recession had stopped one or two millennia earlier, while the ice margin had probably already retreated behind present limits in the warm/dry areas of central West Greenland (Fig. 13.11). The many moraines formed in all parts of the country during this period probably reflect topographically conditioned stillstand or minor readvance of the destabilized ice margins. After 7 ka the ice margins had retreated as much as 10 km behind their present locations.

In addition, by ca. 7 ka the fell field vegetation on all suitable sites had given way to dwarf shrub heath. South of 76°N the heaths were dense and composed of many species, while the heaths in the north were less diverse (Fig. 13.32, 13.33). From the occurrence of exotic North American pollen, Fredskild (1984) concluded that strong southwesterly winds from Labrador reached southwest Greenland during the early Holocene summers, but failed to penetrate into the Baffin Bay region in the north.

Along the coast of East Greenland subarctic bivalves began to appear 8.5 ka, showing greater than present influx of warm subarctic water to the area at this time. At about 7.5 ka boreal-subarctic molluscs, now absent from Greenland waters, lived in some areas along the West Greenland coast (Fig. 13.30).

Thus 7-8 ka temperatures were slightly higher than the present. This is in agreement with the Greenland Ice Sheet record, in which light isotopic composition of the ice begins about 8.5 ka (Hammer et al., 1986).

Late Holocene. The Holocene warm period is seen to end at different times in different areas. The first area to cool was apparently coastal East Greenland where dense dwarf shrub heaths were transformed into open vegetation, with recurrence of cryoturbation in the soil, at ca. 5 ka (Fig. 13.34). At the same time subarctic molluscs became extinct or had their areas severely reduced in adjacent fiords. In West Greenland signs of deterioration appear in the pollen record at ca. 4 ka (Fig. 13.33; Fredskild, 1984), and at nearly the same time the boreal-subarctic molluscs disappeared from Greenland waters (Fig. 13.31). In coastal areas in the extreme north, high arctic dwarf shrub heath gave way to polar desert at ca. 5 ka (Funder, and Abrahamsen, 1988). For at least another millennium, however, driftwood continued to land on shores which are now permanently ice-bound (Fig. 13.21). This indicates that sea ice conditions were less severe than present, allowing wood-carrying ice floes to drift across the Arctic Ocean (e.g., Stewart and England, 1983).

Eventually, probably ca. 3 ka, the ice margins began to resurge, a resurgence which in most areas seems to have reached a maximum in the late 19th Century. The advance to this position, which is marked by fresh moraines, may be a response to very cold conditions 100-200 years earlier (Weidick, 1984), and the readvance of the ice margins over the last millennium may be responsible for the present subsidence of Greenland's coasts.

ACKNOWLEDGMENTS

The writing of this review has taken well over a year and the authors are indebted to their institutions, the Geological and Botanical Museums at the University of Copenhagen and the Geological Survey of Greenland, for providing both time and necessary facilities. The Geological Survey of Greenland has also given access to unpublished results and internal files, and this report is published with permission from the Director of the Survey. The tedious task of reading and discussing early manuscript chapters has been undertaken with acumen by Ole Bennike (Geological Museum, University of Copenhagen), Weston Blake, Jr. (Geological Survey of Canada), Peter Dawes, Anker Weidick, Niels Henriksen (Geological Survey of Greenland), Christian Hjort (University of Lund), and Michael Kelly (University of Lancaster, England). Their advice — even when not taken — has been much appreciated.

REFERENCES

Abrahamsen, N. and Marcussen, C.
1986: Magnetostratigraphy of the Plio-Pleistocene Kap København Formation, eastern North Greenland; Physics of the Earth and Planetary Interiors, v. 44, p. 53-61.

Ahlmann, H.W.
1941: Studies in north-east Greenland 1939-1940; Geografiska annaler, v. 23, p. 145-209.

Aksu, A.E.
1985: Climatic and oceanographic changes over the past 400 000 years: evidence from deep-sea cores on Baffin Bay and Davis Strait; in Quaternary Environments Eastern Canadian Arctic, Baffin Bay and Western Greenland, J.T. Andrews (ed.); Allen and Unwin, Boston, p. 181-210.

Aksu, A.E. and Mudie, P.J.
1985: Late Quaternary stratigraphy and paleontology of Northwest Labrador Sea; Marine Micropaleontology, v. 9, p. 537-557.

Belknap, R.L.
1941: Physiographic studies in the Holsteinsborg district of southern Greenland; University of Michigan Studies, Scientific Series, no. 6 II, p. 199-255.

Bendix-Almgreen, S.E.
1976: Palaeovertebrate faunas of Greenland; in Geology of Greenland, A. Escher and W.S. Watt (ed.); The Geological Survey of Greenland, Copenhagen, p. 534-573.

Bendix-Almgreen, S.E., Fristrup, B., and Nichols, R.L.
1967: Notes on the geology and geomorphology of the Carey Øer, Northwest Greenland; Meddelelser om Grønland, v. 164, no. 8, 19 p.

Bennike, O.
1983a: Pingos in Peary Land, North Greenland; Geological Society of Denmark, Bulletin, v. 32, p. 1-3.
1983b: Palaeoecological investigations of a Holocene peat deposit from Vølvedal, Peary Land, North Greenland; Grønlands geologiske Undersøgelse, report no. 115, p. 15-20.
1987: Quaternary geology and biology of the Jørgen Brønlund Fjord area, North Greenland; Meddelelser om Grønland, Geoscience, no. 18, 23 p.

Bennike, O., Dawes, P.R., Funder, S., Kelly, M., and Weidick, A.
1987: The late Quaternary history of Hall Land, northwest Greenland: Discussion; Canadian Journal of Earth Sciences, v. 24, p. 370-374. Beschel, R.E.
1961: Dating rock surfaces by lichen growth and its application to glaciology and physiography (lichenometry); in Geology of the Arctic, Volume 2, G.O. Raasch (ed.); Toronto University Press, Toronto, p. 1042-1062.

Bick, H.
1978: A postglacial pollen diagram from Angmagssalik, East Greenland; Meddelelser om Grønland, v. 204, no. 1, 22 p.

Birkelund, T. and Perch-Nielsen, K.
1976: Late Palaeozoic-Mesozoic evolution of central East Greenland; in Geology of Greenland, A. Escher and W.S. Watt (ed.); Geological Survey of Greenland, Copenhagen, p. 302-339.

Björck, S. and Persson, T.
1981: Weichselian and Flandrian biostratigraphy and chronology from Hochstetter Forland, Northeast Greenland; Meddelelser om Grønland, Geoscience, no. 5, 19 p.

Blake, W., Jr.
1975: Glacial geological investigations in north-western Greenland; in Report of Activities, Part A, Geological Survey of Canada, Paper 75-1A, p. 435-439.
1977: Radiocarbon age determinations from the Carey Islands, Northwest Greenland; in Report of Activities, Part A, Geological Survey of Canada, Paper 77-1A, p. 445-454.

Blake, W., Jr., Boucherle, M.M., Smol., J.P., Fredskild, B., and Janssens, J.A.
1985: Holocene lake sediments from Inglefield Land, Northwestern Greenland; 4th International Symposium on Paleolimnology, Ossiach, Austria, p. 11.

Böcher, T.W.
1949: Climate, soil and lakes in continental West Greenland in relation to plant life; Meddelelser om Grønland, v. 147, no. 2, 63 p.

Bøgvad, R.
1940: Quaternary geological observations etc. in south-east and south Greenland; Meddelelser om Grønland, v. 107, no. 3, 42 p.

Brassard, G.R. and Blake, W., Jr.
1978: An extensive subfossil deposit of the arctic moss Aplodon wormskioldii; Canadian Journal of Botany, v. 56, p. 1852-1859.

Brett C.P. and Zarudzki, E.F.K.
1979: Project Westmar, a shallow marine geophysical survey on the West Greenland shelf; Grønlands geologiske Undersøgelse, report no. 87, 27 p.

Bretz, J.H.
1935: Physiographic studies in East Greenland; in The Fjord Region of East Greenland, L.A. Boyd (ed.); American Geographical Society, Special Publications, v. 18, p. 159-266.

Brooks C.K.
1979: Geomorphological observations at Kangerdlugssuaq, East Greenland; Meddelelser om Grønland, Geoscience no. 1, 21 p.

Brooks C.K. and Nielsen, T.F.D.
1982: The Phanerozoic development of the Kangerdlugssuaq area, East Greenland; Meddelelser om Grønland, Geoscience no. 9, 30 p.

Brown, R.
1871: On the physics of Arctic Ice, as explanatory of the glacial remains in Scotland; Quaternary Journal from the Geological Society 1871, p. 671-701.

Bryan, M.S.
1954: Interglacial pollen spectra from Greenland; Danmarks geologiske Undersøglse, 3. raekke, v. 80, p. 65-72.

Christie, R.L.
1967: Reconnaissance of the surficial geology of northeastern Ellesmere Island, Arctic Archipelago; Geological Survey of Canada, Bulletin 138, 50 p.
1975: Glacial features of the Børglum Elv region, eastern North Greenland; Grønlands geologiske Undersøgelse, report no. 75, p. 26-27.
1983: Lithological suites as glacial "tracers", eastern Ellesmere Island, Arctic Archipelago; in Current Research, Part A, Geological Survey of Canada, Paper 83-1A, p. 399-402.

Clarke, D.L.
1982: Origin, nature and world climate effect of Arctic Ocean ice-cover; Nature, v. 300, p. 321-325.

Clark, J.A.
1976: Greenland's rapid postglacial emergence: A result of ice-water gravitational attraction; Geology, v. 4, p. 310-312.

Clarke D.B. and Pedersen, A.K.
1976: Tertiary volcanic province of West Greenland; in Geology of Greenland, A. Escher and W.S. Watt (ed.); Geological Survey of Greenland, Copenhagen, p. 364-385.

Cruickshank J.G. and Colhoun, E.A.
1965: Observations on pingos and other landforms in Schuchertdal, northeast Greenland; Geografiska annaler, v. 47A, p. 224-236.

Danmarks geologiske Undersøgelse
1988: "18. Nordiske Geologiske Vintermøde, København 1988, Abstracts; Danmarks geologiske Undersøgelse, Copenhagen.

Dansgaard, W., Johnsen, S.J., Clausen, H.B., Dahl-Jensen, D., Gundestrup, N., Hammer, C.U., and Oeschger, H.
1984: North Atlantic climatic oscillations revealed by deep Greenland ice cores; Geophysical Monograph no. 29, Maurice Ewing series, v. 5, p. 288-298.

Davies, T.A. (editor)
1972: Initial reports of the Deep Sea Drilling Project, Volume 12; U.S. Government Printing Office, Washington, 1243 p.

Davies, W.E.
1961: Surface features of permafrost in arid areas; in Geology of the Arctic, Volume 2, G.O. Raasch (ed.); Toronto University Press, Toronto, p. 981-987.
1963: Glacial geology of Northern Greenland; Polarforschung, v. 4, p. 94-103.
1972: Landscape of northern Greenland; Cold Regions Research Engineering Laboratory, Special Report no. 164, 55 p.

Davies, W.E. and Krinsley, D.B.
1961: The recent regimen of the ice cap margin in North Greenland; in Symposium of Obergurgl, International Association of Scientific Hydrology, Publication no. 58, p. 119-130.

Davies, W.E., Krinsley, D.B., and Nicol, A.H.
1963: Geology of the North Star Bugt area, Northwest Greenland; Meddelelser om Grønland, v. 162, no. 12, 68 p.

Dawes, P.R.
1970: Quaternary studies in northern Peary Land; Grønlands geologiske Undersøgelse, Report no. 28, p. 15-18.
1977: Geological photo-interpretation of Hall Land: part of the regional topographical-geological mapping of northern Greenland; Grønlands geologiske Undersøgelse, Report no. 85, p. 25-29.
1986: Glacial erratics on the Arctic Ocean margin of North Greenland; implications for an extensive ice-shelf; Geological Society of Denmark, Bulletin, v. 35, p. 59-69.
1987: Topographic and geological maps of Hall Land, North Greenland: Description of a computer supported photogrammetrical research programme for production of new maps and the Lower Paleozoic and surficial geology; Grønlands geologiske Undersøgelse, Bulletin no. 155, 88 p. scale 1:66 500.

Dawes, P.R. and Peel, J.S.
1981: The northern margin of Greenland from Baffin Bay to the Greenland Sea; in The Ocean Basins and Margins, Volume 5, A.E.M. Nairn, M. Churkin, and F.G. Stehli (ed.); Plenum Press, New York, p. 201-264.

Donner, J.
1978: Holocene history of the west coast of Disko, central West Greenland; Geografiska annaler, v. 60A, p. 63-72.
Donner, J. and Jungner, H.
1975: Radiocarbon dating of shells from marine Holocene deposits in the Disko Bugt area, West Greenland; Boreas, v. 4, p. 25-45.
Eldholm, O., Thiede, J., Taylor, E. et al.
1987: Summary and preliminary conclusions, ODP leg 104; Proceedings of the Ocean Drilling Program, Initial reports part A, National Science Foundation, Washington, D.C., v. 104, p. 751-771.
England, J.
1976: Postglacial isobases and uplift curves from the Canadian and Greenland high arctic; Arctic and Alpine Research, v. 8, p. 61-78.
1982: Postglacial emergence along northern Nares Strait; in Nares Strait and the Drift of Greenland: A Conflict in Plate Tectonics, P.R. Dawes and J.W. Kerr (ed.); Meddelelser om Grønland, Geoscience, no. 8, p. 65-75.
1983: Isostatic adjustments in a full glacial sea; Canadian Journal of Earth Sciences, v. 20, p. 895-917.
1985: The late Quaternary history of Hall Land, Northwest Greenland; Canadian Journal of Earth Sciences, v. 23, p. 1-17.
England, J. and Bednarski, J.
1989: Northeast Ellesmere Island (Quaternary stratigraphy and chronology); in Chapter 6 of Quaternary Geology of Canada and Greenland, R.J. Fulton (ed.); Geological Survey of Canada, Geology of Canada, no. 1 (also Geological Society of America, The Geology of North America, v. K-1).
England, J. and Bradley, R.S.
1978: Past glacial activity in the high arctic; Science, v. 200, p. 265-270.
Escher, A. and Watt, W.S. (editors)
1976: Geology of Greenland; The Geological Survey of Greenland; Copenhagen, 603 p.
Feyling-Hanssen, R.W.
1985: Late Cenozoic marine deposits of East Baffin Island and East Greenland: microbiostratigraphy, correlation and age; in Quaternary Environments Eastern Canadian Arctic, Baffin Bay and Western Greenland, J.T. Andrews (ed.); Allen and Unwin, Boston, p. 354-393.
1986: Graensen mellem Tertiaer og Kvartaer i Nordspen og i Arktis, fastlagt og korreleret ved hjaelp af bethoniske foraminiferer; Geological Society of Denmark, Årsskrift 1985, p. 19-33.
Feyling-Hanssen, R.W., Funder, S., and Petersen, K.S.
1983: The Lodin Elv Formation; a Plio-Pleistocene occurrence in Greenland; Geological Society of Denmark, Bulletin, v. 31, p. 81-106.
Fillon, R.H.
1985: Northwest Labrador Sea stratigraphy, sand input and paleooceanography during the last 160 000 years; in Quaternary Environments Eastern Canadian Arctic, Baffin Bay and Western Greenland, J.T. Andrews (ed.); Allen and Unwin, Boston, p. 210-248.
Fillon, R.H. and Aksu, A.E.
1985: Evidence for a subpolar influence in the Labrador Sea and Baffin Bay during isotopic stage 2; in Quaternary Environments Eastern Canadian Arctic, Baffin Bay and Western Greenland, J.T. Andrews (ed.); Allen and Unwin, Boston, p. 248-263.
Flint, R.F.
1948: Glacial geology and geomorphology; in The Coast of Northeast Greenland, L.A. Boyd (ed.); American Geographical Society, Special Publications, v. 30, p. 91-210.
Fredskild, B.
1967: Palaeobotanical investigations at Sermermiut, Jakobshavn, West Greenland; Meddelelser om Grønland, v. 178, no. 4, 54 p.
1969: A postglacial standard pollendiagram from Peary Land, North Greenland; Pollen et Spores, v. 11, p. 573-583.
1973: Studies in the vegetational history of Greenland; Meddelelser om Grønland, v. 198, no. 4, 245 p.
1981: The natural environment of the norse settlers in Greenland; in Proceedings of the International Symposium "Early European Exploitation of the Northern Atlantic 800-1700", Arctic Centre, University of Groningen, Netherlands, p. 27-42.
1983a: The Holocene vegetational development of the Godthåbsfjord area, West Greenland; Meddelelser om Grønland, Geoscience no. 10, 28 p.
1983b: The Holocene development of some low and high arctic Greenland lakes; Hydrobiologia, v. 103, p. 217-224.
1984: Holocene palaeo-winds and climatic changes in West Greenland as indicated by long-distance transported and local pollen in lake sediments; in Climatic Changes on a Yearly to Millenial Basis, N.-A. Mörner and W. Karlén (ed.); D. Reidel, Stuttgart, p. 163-171.
1985a: Holocene pollen records from West Greenland; in Quaternary Environments Eastern Canadian Arctic, Baffin Bay and Western Greenland, J.T. Andrews (ed.); Allen and Unwin, Boston, p. 643-681.
1985b: The Holocene vegetational development of Tugtuligssuaq and Qeqertat, Northwest Greenland; Meddelelser om Grønland, Geoscience no. 14, 20 p.

Fredskild, B. and Røen, U.
1982: Macrofossils from an interglacial peat deposit at Kap København, North Greenland; Boreas, v. 11, p. 181-185.
Fristrup, B.
1952: Winderosion within the Arctic deserts; Geografisk Tidsskrift, v. 52, p. 51-65.
Funder, S.
1972a: Remarks on the Quaternary geology of Jameson Land and adjacent areas, Scoresby Sund, East Greenland; Grønlands geologiske Undersøgelse, report no. 48, p. 93-98.
1972b: Deglaciation of the Scoresby Sund fjord region, north-east Greenland; Institute of British Geographers, Special Publications, v. 4, p. 33-42.
1978a: Holocene stratigraphy and vegetation history in the Scoresby Sund area, East Greenland; Grønlands geologiske Undersøgelse, Bulletin, v. 129, 66 p.
1978b: Glacial flutings in bedrock, an observation in East Greenland; Geological Society of Denmark, Bulletin, v. 27, p. 9-13.
1978c: Holocene (10 000-0 years BP) climates in Greenland, and North Atlantic atmospheric circulation; in Proceedings of the Nordic Symposium on Climatic Changes and Related Problems, Copenhagen 24-28 April 1978; K. Frydendahl (ed.); Danish Meteorological Institute, Climatological Papers no. 4, p. 175-181.
1979a: The Quaternary geology of the Narssaq area, South Greenland; Grønlands geologiske Undersøgelse, Report no. 86, 24 p.
1979b: Ice-age plant refugia in East Greenland; Palaeogeography, Palaeoclimatology, Palaeoecology, v. 28, p. 279-295.
1982a: Planterefugierne i Grønland; Naturens Verden 1982, p. 241-255.
1982b: C-14 dating of samples collected during the 1979 expedition to North Greenland; Grønlands geologiske Undersøgelse, report no. 110, p. 9-14.
1984: Chronology of the last interglacial/glacial cycle in Greenland: first approximation; in Correlation of Quaternary Chronologies, W.C. Mahaney (ed.); Geobooks, Norwich, p. 261-279.
Funder, S. and Abrahamsen, N.
1988: Palynology in a Polar desert, eastern Greenland; Boreas, v.17, p.195-207.
Funder, S. and Hjort, C.
1973: Aspects of the Weichselian chronology in central East Greenland; Boreas, v. 2, p. 69-84.
1978: Weichsel interstadial(er) i Østgrønland. Aldersbestemmelse og geologi. And Isens udbredelse i Østgrønland i sen Weichsel og tidlig Flandern — morfologisk afgraensning og absolut datering; 13. Nordiske Vintermøde (abstract); Geologisk Centralinstitut, Copenhagen, p. 85.
1980: A reconnaissance of the Quaternary geology of eastern North Greenland; Grønlands Geologiske Undersøgelse, Report no. 99, p. 99-105.
Funder, S. and Larsen, O.
1982: Implications of volcanic erratics in Quaternary deposits of North Greenland; Geological Society of Denmark, Bulletin, v. 31, p. 57-61.
Funder S. and Petersen, K.S.
1980: Glacitectonic deformations in East Greenland; Geological Society of Denmark, Bulletin, v. 28, p. 115-122.
Funder, S. and Simonarson, L.A.
1984: Bio- and aminostratigraphy of some Quaternary marine deposits in West Greenland; Canadian Journal of Earth Sciences, v. 21, p. 843-852.
Funder, S., Abrahamsen, N., Bennike, O., and Feyling-Hanssen, R.W.
1985: Forested Arctic: Evidence from North Greenland; Geology, v. 13, p. 542-546.
Funder, S., Bennike, O., Mogensen, G.S., Noe-Nygaard, B., Pedersen, S.A.S., and Petersen, K.S.
1984: The Kap København Formation, a late Cainozoic sedimentary sequence in North Greenland; Grønlands geologiske Undersøgelse, Report, v. 120, p. 9-18.
Geological Survey of Greenland
1974: Quaternary map of Greenland Søndre Strømfjord — Nugssuaq; Copenhagen, scale 1:500 000.
1978: Quaternary map of Greenland Sheet 2, Frederikshåbs Isblink Søndre Strømfjord; Copenhagen, scale 1:500 000.
Gleadow, A. and Brooks, C.K.
1979: Fission track dating, thermal histories and tectonics of igneous intrusions in East Greenland; Contributions to mineralogy and petrology, v. 71, p. 45-60.
Gordon, J.E.
1981: Glacier margin fluctuations during the 19th and 20th centuries in Ikamiut kangerdluarssuat area, West Greenland; Arctic and Alpine Research, v. 13, p. 47-62.
Graff-Petersen, P.
1952: Glacial morphology of the Kuvinilik area; Geological Society of Denmark, Bulletin, v. 12, p. 266-274.

Gripp, K.
1932: Einige besondere Fossilien in Geschieben aus dem Indlandeis Gronlands; Meddelelser om Grønland, v. 91, 11 p.

Gudmandsen, P.
1978: Application of space techniques in solid earth physics; in The Conference on Application of Space Techniques in Navigation, Geodesy, Oceanography and Solid Earth Physics, København 27. September 1978; Dansk Nationalkomité for den Internationale Union for Geodaesi og Geofysik, Charlottenlund, p. 84-94.

Håkansson, S.
1982: University of Lund radiocarbon dates XV; Radiocarbon, v. 24, p. 194-213.

Haller, J.
1971: Geology of the East Greenland Caledonides; Interscience, London, 413 p.

Hammer, C.U., Clausen, H.B., Neftel, A., Kristinsdottir, P., and Johnson, E.
1985: Continuous impurity analyses along the DYE 3 deep core; in Greenland Ice Core: Geophysics, Geochemistry, and the Environment, C.C. Langway, H. Oeschger, and W. Dansgaard (ed.); American Geophysical Union, Monograph no. 33, p. 90-94.

Hammer, C.U., Clausen, H.B., and Tauber, H.
1986: Ice core dating of the Pleistocene/Holocene boundary applied to a calibration of the C-14 scale; Radiocarbon, v. 28, p. 284-291.

Hansen, K.
1970: Geological and geographical investigations in Kong Frederik IX's Land; Meddelelser om Grønland, v. 188, no. 4, 77 p.

Harder, P., Jensen, Ad.S., and Laursen, D.
1949: The marine Quaternary sediments in Disko Bugt; Meddelelser om Grønland, v. 149, no. 1, 85 p.

Helland, A.
1876: Om de isfyldte fjorde og de glaciale dannelser i Nordgrønland; Archiv for Matematik og Natur i Christiania, Oslo, 68 p.

Henderson, G.
1969: Oil and gas prospects in the Cretaceaous-Tertiary basin of West Greenland; Grønlands geologiske Undersøgelse, Report no. 22, 63 p.
1975: New bathymetric maps covering offshore West Greenland 59°-69°30'N; in Offshore Technology Conference in Houston, Texas; Paper OTC 2223, Dallas, p. 761-764; available from 6200 North Central Expressway, Dallas, Texas.

Henderson, G., Schiener, E.J., Risum, J.B., Croxton, C.A., and Andersen, B.B.
1981: The West Greenland Basin; Canadian Society of Petroleum Geologists, Memoir 7, p. 399-429.

Henriksen, N.
1985: The Caledonides of East Greenland 70-76°; in The Caledonide Orogen — Scandinavia and related areas, D.G. Gee and B.A. Sturt (ed.); Wiley and Sons, New York, p. 1095-1113.

Herron, M.M. and Langway, C.C.
1985: Chloride, nitrate, and sulfate in the DYE 3 and Camp Century, Greenland ice cores; in Greenland Ice Core: Geophysics, Geochemistry, and the Environment, C.C. Langway, H. Oeschger, and W. Dansgaard (ed.); American Geophysical Union, Monograph no. 33, p. 77-84.

Higgins, A.K., Soper, N.J., and Friedrichsen, J.D.
1985: North Greenland fold belt in eastern North Greenland; in The Caledonide Orogen — Scandinavia and related areas, D.G. Gee and B.A. Sturt (ed.); Wiley and Sons, New York, p. 1017-1029.

Hjort, C.
1973: The Vega Transgression, a hypsithermal event in central East Greenland; Geological Society of Denmark, Bulletin, v. 22, p. 25-38.
1976: Remarks on the glacial chronology of the Kong Oscars Fjord and Vega Sund districts, central East Greenland, as evidenced by glacial sculpture, striae and similar features; University of Lund, Department of Quaternary Geology, Report no. 9, 25 p.
1979: Glaciation in northern East Greenland during the Late Weichselian and Early Flandrian; Boreas, v. 8, p. 281-296.
1981a: A glacial chronology for northern East Greenland; Boreas, v. 10, p. 259-274.
1981b: Quaternary geology in northeasternmost Greenland; in Geoscience During the Ymer-80 Expedition to the Arctic, v. Schytt, K. Boström, and C. Hjort (ed.); Geologiska Föreningens i Stockholm Förhandlingar, v. 103, p. 109-119.
1981c: Present and middle Flandrian coastal morphology in Northeast Greenland; Norsk geografisk Tidsskrift, v. 35, p. 197-207.

Hjort, C. and Björck, S.
1984: A re-evaluated chronology for northern East Greenland; Geologiska Föreningens i Stockholm Förhandlingar, v. 105, p. 235-243.

Hjort, C. and Funder, S.
1974: The subfossil occurrence of **Mytilus edulis** L. in central East Greenland; Boreas, v. 3, p. 23-33.

Hobbs, W.H.
1931: Loess, pebble bands and boulders from glacial outwash of the Greenland continental glacier; Journal of Geology, v. 39, p. 381-385.

Holtedahl, O.
1970: On the morphology of the West Greenland shelf with general remarks on the "marginal channel" problem; Marine Geology, v. 8, p. 155-172.

Holtzscherer, J-J. and Bauer, A.
1954: Contribution à la connaissance de L'Indlandsis du Groenland; International Association of Scientific Hydrology, Publication no. 39, p. 244-296.

Humlum, O.
1981: Rock glacier types on Disko, Central West Greenland; Geografisk Tidsskrift, v. 82, p. 59-66.

Hurst, J.M., Jepsen, H.F., Kalsbeek, F., McKerrow, W.S., and Peel, J.S.
1985: The geology of the northern extremity of the East Greenland Caledonides; in The Caledonide Orogen — Scandinavia and related areas, D.G. Gee and B.A. Sturt (ed.); Wiley and Sons, New York, p. 1047-1063.

Iversen, J.
1934: Moorgeologische Untersuchungen auf Grönland; Geological Society of Denmark, Bulletin, v. 8, p. 341-358.
1952: Origin of the flora of western Greenland in the light of pollen analysis; Oikos, v. 4:II, p. 85-103.

Jensen, Ad. S.
1917: Quaternary fossils collected by the Danmark Expedition; Meddelelser om Grønland, v. 43, p. 621-632.
1942: Two new West Greenland localities for deposits from the ice age and the post-glacial warm period; Kongelige Danske Videnskabernes Selskab, Biologiske Meddelelser, v. 17, no. 4, 35 p.

Jensen, Ad.S. and Harder, P.
1910: Post-glacial changes of climate in Arctic regions as revealed by investigations on marine deposits; in Postglaziale Klimavernänderungen, 11 Internationale geologische Kongress, Stockholm 1910, p. 399-407.

Jepsen, H.F.
1982: The Bjørnehiet Formation: a faulted preglacial conglomerate, Washington Land, North Greenland; in Nares Strait and the Drift of Greenland: A Conflict in Plate Tectonics, P.R. Dawes and J.W. Kerr (ed.); Meddelelser om Grønland, Geoscience no. 8, p. 55-59.

Jessen, A.
1896: Geologiske Iagttagelser; in Opmaalingsexpeditionen til Julianehaabs Distrikt 1894, C. Moltke (ed.); Meddelelser om Grønland, v. 16, p. 123-170.

Johnson, G.L., Sommerhoff, G., and Egloff, J.
1975: Structure and morphology of the West Reykjanes Basin and the southeast Greenland continental margin; Marine Geology, v. 18, p. 175-196.

Kalsbeek, F.
1982: The evolution of the Precambrian shield of Greenland; Geologische Rundschau, v. 71, p. 38-60.

Kellogg, T.B.
1980: Paleoclimatology and paleo-oceanography of the Norwegian and Greenland seas: glacial-interglacial contrasts; Boreas, v. 9, p. 115-137.

Kellogg, T.B., Duplessy, J.C., and Shackleton, N.J.
1978: Planktonic foraminiferal and oxygen isotopic stratigraphy and palaeoclimatology of Norwegian Sea deep-sea cores; Boreas, v. 7, p. 61-73.

Kelly, M.
1973: Radiocarbon dated shell samples from Nordre Stromfjord, West Greenland; Grønlands geologiske Undersøgelse, Report no. 59, 20 p.
1975: A note on the implications of two radiocarbon dated samples from Qaleragdlit ima, South Greenland; Geological Society of Denmark, Bulletin, v. 24, p. 21-26.
1977: Quaternary geology of the Ivigtut-Nunarssuit region, South-West and South Greenland; Grønlands geologiske Undersøgelse, Report no. 85, p. 64-67.
1979: Comments on the implications of new radiocarbon dates from the Holsteinsborg region, central West Greenland; Grønlands geologiske Undersøgelse, Report no. 95, p. 35-42.
1980a: The status of the Neoglacial in western Greenland; Grønlands geologiske Undersøgelse, Report no. 96, 24 p.
1980b: Preliminary investigations of the Quaternary of the Melville Bugt and Dundas, North-West Greenland; Grønlands geologiske Undersøgelse, Report no. 100, p. 33-38.
1981: Permafrost related features in Holsteinsborg District, West Greenland; Geological Society of Denmark, Bulletin, v. 30, p. 51-56.
1985: A review of the Quaternary geology of western Greenland; in Quaternary Environments Eastern Canadian Arctic, Baffin Bay and Western Greenland, J.T. Andrews (ed.); Allen and Unwin, Boston, p. 461-501.
1986: Quaternary, pre-Holocene, marine events of western Greenland; Grønlands geologiske Undersøgelse, Report no. 131, 23 p.

Kelly, M. and Bennike, O.
1985: Quaternary geology of parts of central and western North Greenland; Grønlands geologiske Undersøgelse, Report no. 126, p. 111-116.

Kelly, M. and Funder, S.
1974: The pollen stratigraphy of late Quaternary lake sediments of South-West Greenland; Grønlands geologiske Undersøgelse, Report no. 64, 26 p.

Kempter, E.
1961: Die Jungpaläozoische Sedimente von Süd Jameson Land; Meddelelser om Grønland, v. 144, no. 1, 120 p.

Knuth, E.
1966: The ruins of the Musk-Ox way; Folk, v. 8, p. 191-219.

Koch, J.P.
1916: Survey of northeast Greenland; Meddelelser om Grønland, v. 46, p. 79-468.

Koch, J.P. and Wegener, A.
1911: Die glaziologischen Beobachtungen der Danmark-Expedition; Meddelelser om Grønland, v. 46, p. 1-76.
1930: Wissenschaftliche Ergebnisse der danischen Expedition nacn Dronning Louise Land und quer über das Inlandeis von Nordgrönland 1912-1913; Meddelelser om Grønland, v. 75, 676 p.

Koch, L.
1920: Stratigraphy of northwest Greenland; Geological Society of Denmark, Bulletin, v. 5, p. 1-78.
1928a: The physiography of North Greenland; in Greenland, Volume 1, M. Vahl, G.C. Amdrup, L. Bob, and Ad.S. Jensen (ed.); C.A. Reitzel, Copenhagen, p. 491-518.
1928b: Contributions to the glaciology of North Greenland; Meddelelser om Grønland, v. 65, no. 2, p. 181-464.
1933: The geology of Inglefield Land; Meddelelser om Grønland, v. 73, no. 2, p. 1-38.

Krinsley, D.B.
1961: Late Pleistocene glaciation in northeast Greenland; in Geology of the Arctic, Volume 2, G.O. Raasch (ed.); Toronto University Press, Toronto, p. 747-751.

Larsen, B.
1983: Geology of the Greenland-Iceland ridge in the Denmark Strait; in Structure and Development of the Greenland-Scotland Ridge, M.H.P. Bott, S. Saxov, M. Talwani, and J. Thiede (ed.); Plenum Press, New York, p. 425-444.

Larsen, H.C.
1980: Geological perspectives of the East Greenland continental margin; Geological Society of Denmark, Bulletin, v. 29, p. 77-101.
1984: Geology of the East Greenland shelf; in Petroleum Geology of the North European Margin, A. Graham and B. Trotman (ed.); Norwegian Petroleum Society, Stavanger, p. 329-339.
in press: Geology of the East Greenland shelf; in Geology of the Arctic Ocean Region, A. Grantz, J. Sweeney, and G.L. Johnson (ed.); Geological Society of America, The Geology of North America, v. L.

Larsen, L.M. and Watt, W.S.
1985: Episodic volcanism during break-up of the North Atlantic: evidence from the East Greenland plateau basalts; Earth and Planetary Science Letters, v. 73, p. 105-116.

Lasca, N.P.
1969: The surficial geology of Skeldal, Mesters Vig, Northeast Greenland; Meddelelser om Grønland, v. 176, no. 3, 56 p.

Laursen, D.
1944: Contributions to the Quaternary geology of northern West Greenland, especially the raised marine deposits; Meddelelser om Grønland, v. 135, no. 8, 123 p.
1950: The stratigraphy of the marine Quaternary deposits in West Greenland; Meddelelser om Grønland, v. 151, no. 1, 142 p.
1954: Emerged Pleistocene marine deposits of Peary Land (North Greenland); Meddelelser om Groønland, v. 127, no. 5, 24 p.

Lister, H. and Wyllie, P.J.
1957: The geomorphology of Dronning Louise Land; Meddelelser om Grønland. v. 158, no. 1, 73 p.

Malaurie, J., Vasari, Y., Hyvärinen, H., Delibrias, G., and Labeyrie, J.
1972: Preliminary remarks on Holocene paleoclimates in the regions of Thule and Inglefield Land, above all since the beginning of our own era; in Climatic Changes in Arctic Areas During the Last Ten Thousand Years, Y. Vasari, H. Hyvärinen, and S. Hicks (ed.); Acta Universitatis Ouluenis, Series A Scientia rerum Naturalium, no. 3, p. 105-136.

Mangerud, J., Sønstergård, E., and Sejrup, H-P.
1979: Correlation of the Eemian (interglacial) Stage and the deep-sea oxygen-isotope stratigraphy; Nature, v. 277, p. 189-192.

Mangerud, J., Lie, S.E., Furnes, H., Kristiansen, I.L., and Lømo, L.
1984: A Younger Dryas ash bed in western Norway, and its possible correlations with tephra in cores from the Norwegian Sea and the North Atlantic; Quaternary Research, v. 21, p. 85-104.

Markussen, B. and Thiede, J.
1985: Late Quaternary sedimentation in the eastern Arctic Basin: stratigraphy and depositional environment; Palaeogeography, Palaeoclimatology, Palaeoecology, v. 50, p. 271-284.

Meldgaard, M.
1986: The Greenland caribou — zoogeography, taxonomy, and population dynamics; Meddelelser om Grønland, Bioscience, v. 20, 88 p.

Mogensen, G.S.
1984: Pliocene or early Pleistocene mosses from Kap København, North Greenland; Lindbergia, v. 10, p. 19-26.

Müller, F.
1959: Beobachtungen über Pingos, Detailuntersuchungen in Ostgrönland und in der Kanadischen Arktis; Meddelelser om Grønland, v. 153, no. 3, 127 p.

Naval Research Laboratory, Acoustics Division
1985: Bathymetry of the Arctic Ocean; Washington, D.C. (map sheet).

Nichols, R.L.
1969: Geomorphology of Inglefield Land, North Greenland; Meddelelser om Grønland, v. 188, no. 1, 109 p.

Nordenskiöld, A.E.
1871: Redogörelse för en expedition till Grönland år 1870; Kungliga Vetenskaps-Akademiens Förhandlingar, 1870, no. 10, p. 923-1082.

Nordenskiöld, O.
1907: On the geology and physical geography of East Greenland; Meddelelser om Grønland, v. 28, p. 151-284.

O'Brien, R.
1971: Observations on pingos and permafrost hydrology in Schuchert Dal, N.E. Greenland; Meddelelser om Grønland, v. 195, no. 1, 19 p.

Paterson, W.S.B.
1977: Extent of the Late-Wisconsinan Glaciation in Northwest Greenland and northern Ellesmere Island: a review of the glaciological and geological evidence; Quaternary Research, v. 8, p. 180-190.

Peel, J.S.
1985: Cambrian-Silurian platform stratigraphy of eastern North Greenland; in The Caledonide Orogen — Scandinavia and related areas, D.G. Gee and B.A. Sturt (ed.); Wiley and Sons, New York, p. 1077-1094.

Pert, G.J.
1971: Some glaciers of the Stauning Alper, Northeast Greenland; Meddelelser om Grønland, v. 188, no. 5, 50 p.

Petersen, K.S.
1982: Attack by predatory gastropods recognized in an interglacial marine molluscan fauna from Jameson Land, East Greenland; Malacologia, v. 22, p. 721-726.

Poser, H.
1932: Einige Untersuchungen zur Morphologie Ostgrönlands; Meddelelser om Grønland, v. 94, no. 5, 55 p.

Putnins, P.
1970: The climate of Greenland; in Climates of the Polar regions, S. Orvig (ed.); Elsevier, Amsterdam, p. 3-128.

Raup, H.M.
1965: The structure and development of turf hummocks in the Mesters Vig district, Northeast Greenland; Meddelelser om Grønland, v. 166, no. 3, 112 p.

Reeh, N.
1984: Reconstruction of the glacial ice covers of Greenland and the Canadian arctic islands by three-dimensional, perfectly plastic ice-sheet modelling; Annals of Glaciology, v. 5, p. 115-121.
1989: Dynamic and climatic history of the Greenland Ice Sheet; Chapter 14 in Quaternary Geology of Canada and Greenland, R.J. Fulton (ed.); Geological Survey of Canada, Geology of Canada, no. 1 (also Geological Society of America, The Geology of North America, v. K-1).

Rink, H.
1852: Om den geographiske Beskaffenhed af de danske Handelsdistrikter i Nordgronland. Udsigt over Nordgrönlands geognosi; Det Kongelige danske Videnskabernes Selskab, Skrifter, 5. Rk. Naturvidenskabelig-mathematiske Afdeling, v. 3, p. 37-98.
1853: On the large continental ice of Greenland, and the origin of icebergs in the Arctic seas; Royal Geographical Society, Journal, v. 23, p. 145-154.

Risum, J.B., Croxton, C.A., and Rolle, F.
1980: Developments in petroleum exploration offshore West Greenland; Grønlands geologiske Undersøgelse, Report no. 100, p. 55-61.

Roksandić, M.M.
1979: Geology of the continental shelf off West Greenland between 61°15'N and 64°00'N; Grønlands geologiske Undersøgelse, Report no. 92, 15 p.

Ruddiman, W.F. and Glover, L.K.
1972: Vertical mixing of ice rafted volcanic ash in North Atlantic sediments; Geological Society of America, Bulletin, v. 83, p. 2817-2836.

Ruddiman, W.F. and McIntyre, A.
1981: The North Atlantic Ocean during the last deglaciation; Palaeogeography, Palaeoclimatology, Palaeoecology, v. 35, p. 145-214.

Rvachev, V.D.
1964: Relief and bottom deposits of the shelf of southwestern Greenland; Deep Sea Research, v. 11, p. 646-653.

Saxov, S.
1958: The uplift of western Greenland; Geological Society of Denmark, Bulletin, v. 13, p. 518-523.
1961: The vertical movement of eastern Greenland (Angmagssalik); Geological Society of Denmark, Bulletin, v. 14, p. 413-416.

Shackleton, N.J. and Opdyke, N.D.
1973: Oxygen isotope and palaeomagnetic stratigraphy of Equatorial Pacific core V28-238: Oxygen isotope temperatures and ice volumes on a 10^5 and 10^6 year scale; Quaternary Research, v. 3, p. 39-55.
1977: Oxygen isotope and paleomagnetic stratigraphy of Pacific core V28-239 late Pliocene to latest Pleistocene; in Investigation of Late Quaternary Paleooceanography and Paleoclimatology, R.M. Cline and J.D. Hays (ed.); Geological Society of America, Memoir 145, p. 449-464.

Shackleton, N.J., Backman, J., Zimmerman, H. et a l.
1984: Oxygen isotope calibration of the onset of ice-rafting and history of glaciation in the North Atlantic region; Nature, v. 307, p. 620-623.

Shambach, J.D. (editor)
1978: Initial reports of the Deep Sea Drilling Project, Volume 49; U.S. Government Printing Office, Washington, 1020 p.

Símonarson, L.A.
1974: Recent Cyrtodaria and its fossil occurrence in Greenland; Geological Society of Denmark, Bulletin, v. 23, p. 65-75.
1981: Upper Pleistocene and Holocene marine deposits and faunas on the north coast of Nugssuaq, West Greenland; Grønlands geologiske Undersøgelse, Bulletin, v. 140, 107 p.

Sommerhoff, G.
1973: Formenschattz and Morphologische Gliederung des südostgrönländischen Schelfgebietes und Kontinentalabhanges; "Meteor" Forschungs-Ergebnisse, v. C15, p. 1-54.
1975a: Glaziale Gestaltung und marine uberformung der Schelfbänke vor SW-Grönland; Polarforschung, v. 45, p. 22-31.
1975b: Versuch einer geomorphologischen Gliederung des südwestgrönländischen Kontinentalabhanges; Polarforschung, v. 45, p. 87-101.
1979: Submarine glazial übertiefte Täler von Südgrönland; Eiszeitalter und Gegenwart, v. 29, p. 201-213.
1981: Geomorphologische Prozesse in der Labrador-und Imingersee. Ein Beitrag zur submarinen Geomorphologie einer subpolaren Meeresregion; Polarforschung, v. 51, p. 175-191.
1983: Untersuchungen zur Geomorphologie des Meeresbodens in der Labrador-und Irmingersee; Munchener geographische Abhandlungen, v. 28, 86 p.

Sørensen, T.
1935: Bodenformen und Pflanzendecke in Nordostgrönland; Meddelelser om Grønland, v. 93, no. 4, p. 27-43.

Srivastava, S.P., Arthur, M., Clement, B. et al.
1987: Proceedings of the Ocean Drilling Program, Initial Reports Part A; National Science Foundation, Washington, D.C., v. 105, 650 p.

Steenstrup, K.J.V.
1881: Bemaerkninger til et geognostisk oversigtskort; Meddelelser om Grønland, v. 2, p. 27-43.
1883: Bidrag til Kjendskab til de geognostiske og geographiske Forhold i en Del af Nord-Grønland; Meddelelser om Grønland, v. 4, p. 173-242.

Stewart, T.G. and England, J.
1983: Holocene sea-ice variations and paleoenvironmental change, northernmost Ellesmere Island, N.W.T., Canada; Arctic and Alpine Research, v. 15, p. 1-17.

Street, A.F.
1977: Deglaciation and marine paleoclimates in Schuchert Dal, Scoresby Sund, East Greenland; Arctic and Alpine Research, v. 9, p. 421-426.

Streeter, S.S., Belanger, P.E., Kellogg, T.B., and Duplessy, J.C.
1982: Late Pleistocene paleo-oceanography of the Norwegian-Greenland Sea: benthic foraminiferal evidence; Quaternary Research, v. 18, p. 72-90.

Sugden, D.E.
1972: Deglaciation and isostasy in the Sukkertoppen Ice Cap area, West Greenland; Arctic and Alpine Research, v. 4, p. 97-117.
1974: Landscapes of glacial erosion in Greenland and their relationship to ice, topographic and bedrock conditions; Institute of British Geographers, Special Publication, no. 7, p. 177-195.

Sugden, D.E. and John, B.S.
1965: The raised marine features of Kjove Land, East Greenland; Geographical Journal, v. 131, part 2, p. 235-247.

Surlyk, F. and Hurst, J.M.
1984: The evolution of the early Paleozoic deep-water basin of North Greenland; Geological Society of America, Bulletin, v. 95, p. 131-154.

Surlyk, F., Clemmensen, L.B., and Larsen, H.C.
1981: Post Paleozoic evolution of the East Greenland continental margin; in Geology of the North Atlantic Borderlands, J.W. Kerr and J. Ferguson (ed.); Canadian Society of Petrolem Geologist, Memoirs, v. 7, p. 611-645.

Tedrow, J.C.F.
1970: Soil investigations in Inglefield Land, Greenland; Meddelelser om Grønland, v. 188, no. 3, p. 1-93.

Ten Brink, N.W.
1974: Glacio-isostasy: new data from West-Greenland and geophysical implications; Geological Society of America, Bulletin, v. 85, p. 219-228.
1975: Holocene history of the Greenland ice sheet based on radiocarbon-dated moraines in West Greenland; Meddelelser om Grønland, v. 202, no. 4, 44 p.

Ten Brink, N.W. and Weidick, A.
1974: Greenland ice sheet history since the last glaciation; Quaternary Research, v. 4, p. 429-440.

Thomsen, H.H.
1983: Satellitdata - et redskab til studier af Indlandsisens randzone i forbindelse med vandkraftundersøgelser; Grønlands geologiske Undersøgelse, Gletscher-hydrologiske meddelelser, v. 83/8, 24 p.

Troelsen, JC.
1949: Contributions to the geology of the area round Jørgen Brønlunds Fjord, Peary Land, North Greenland; Meddelelser om Grønland, v. 149, no. 2, 28 p.
1952: Notes on the Pleistocene geology of Peary Land, North Greenland; Geological Society of Denmark, Bulletin, v. 12, p. 211-220.

Vernal, A. de and Hillaire-Marcel, C.
1987: Paleoenvironments along the eastern Laurentide ice sheet margin and timing of the last ice maximum and retreat; Géographie physique et Quaternaire, v. 41, p. 265-279.

Vernal, A. de, Hillaire-Marcel, C., Aksu, A.E., and Mudie, P.
1987: Palynostratigraphy and chronostratigraphy of Baffin Bay deep sea cores: climatostratigraphic implications; Palaeogeography, Palaeoclimatology, Palaeoecology, v. 62, p. 96-106.

Vogt, T.
1933: Late Quaternary oscillations of level in southeast Greenland; Skrifter om Svalbard og Ishavet, v. 60, 44 p.

Washburn, A.L.
1965: Geomorphic and vegetational studies in the Mesters Vig district, northeast Greenland; Meddelelser om Grønland, v. 166, no. 1, 60 p.
1967: Instrumental observations of mass-wasting in the Mesters Vig district, northeast Greenland; Meddelelser om Grønland, v. 166, no. 4, 296 p.
1969: Weathering, frost action, and patterned ground in the Mesters Vig district, northeast Greenland; Meddelelser om Grønland, v. 176, no. 4, 299 p.

Washburn, A.L. and Stuiver, M.
1962: Radiocarbon-dated postglacial delevelling in northeast Greenland and its implications; Arctic, v. 15, p. 66-73.

Weidick, A.
1959: Glaciation variations in West Greenland in historical time; Meddelelser om Grønland, v. 158, no. 4, 196 p.
1963: Ice margin features in the Julianehab District, South Greenland; Meddelelser om Grønland, v. 165, no. 3, 133 p.
1968: Observations on some Holocene glacier fluctuations in West Greenland; Meddelelser om Grønland, v. 165, no. 6, 202 p.
1972a: Holocene shore-lines and glacial stages in Greenland — an attempt at correlation; Grønlands geologiske Undersøgelse, report no. 41, 39 p.
1972b: Notes on Holocene glacial events in Greenland; in Climatic Changes in Arctic Areas During the Last Ten Thousand Years, Y. Vasari, H. Hyvärinen, and S. Hicks (ed.); Acta Universitatis Ouluenis, Series A Scientia rerum Naturalium no. 3, p. 177-204.
1973: C-14 Dating of survey material performed in 1972; Grønlands geologiske Undersøgelse, Report no. 55, p. 66-75.
1975a: A review of Quaternary investigations in Greenland; Institute of Polar Studies, Report no. 55, 161 p.
1975b: Estimates on the mass balance changes of the Inland Ice since Wisconsin-Weichsel; Grønlands geologiske Undersøgelse, Report no. 68, 21 p.
1975c: Quaternary geology of the area between Frederikshåbs Isblink and Ameralik; Grønlands geologiske Undersøgelse, Report no. 70, 22 p.
1976a: Glaciation and the Quaternary; in Geology of Greenland, A. Escher and W.S. Watt (ed.); The Geological Survey of Greenland, Copenhagen, p. 431-458.

1976b: Quaternary observations in southern Peary Land, North Greenland; Grønlands geologiske Undersøgelse, Report no. 80, p. 15-17.
1976c: Glaciations of northern Greenland - new evidence; Polarforschung v. 46, p. 26-33.
1976d: Observations on the Quaternary geology in the Fiskenaesset area during the summer of 1973; Grønlands geologiske Undersøgelse, Report no. 73, p. 96-100.
1977: A reconnaissance of Quaternary deposits in northern Greenland; Grønlands geologiske Undersøgelse, Report no. 85, p. 21-24.
1978: Comments on radiocarbon dates from northern Greenland made during 1977; Grønlands geologiske Undersøgelse, Report no. 90, p. 124-128.
1984: Studies of glacier behaviour and glacier mass balance in Greenland - a review; Geografiska annaler v. 66A, p. 183-195.
1985: Review of glacier changes in West Greenland; Zeitschrift für Gletscherkunde und Glazialgeologie, Band 21, p. 301-309.

White, S.M. (editor)
1976: Initial reports of the Deep Sea Drilling Project, Volume 38; U.S. Government Printing Office; Washington, 1256 p.

Zarudski, E.F.K.
1980: Interpretation of shallow seismic profiles over the continental shelf in West Greenland between latitudes 64° and 69°30′N; Grønlands geologiske Undersøgelse, Report no. 100, p. 58-61.

Authors' addresses

B. Fredskild
Greenland Botanical Survey
Gothersgade 130
DK-1123 Copenhagen
Denmark

S. Funder
Geologisk Museum
Øster Voldgade 5-7
1350 Copenhagen K
Denmark

H.C. Larsen
Geological Survey of Greenland
Øster Voldgade 10
1350 Copenhagen
Denmark

Chapter 14
DYNAMIC AND CLIMATIC HISTORY OF THE GREENLAND ICE SHEET

Terminus of Daugaard-Jensen Gletscher, an outlet glacier of the Greenland Ice Sheet which terminates at the head of Nordvestfjord, Scoresby Sund, East Greenland. Courtesy of N. Reeh.

Chapter 14

DYNAMIC AND CLIMATIC HISTORY OF THE GREENLAND ICE SHEET

Niels Reeh

SUMMARY

The last few decades have witnessed several major programs aimed at investigating the Greenland Ice Sheet. These programs have greatly improved our knowledge of the ice sheet in terms of topography, ice thickness, ice temperature, flow, and mass balance. It is concluded that presently the Greenland Ice Sheet as a whole is close to a balanced state, the slight thinning of some marginal sectors likely being compensated for by a slight thickening in the central area. The information on the dynamic and climatic history of the ice sheet as retrieved from the deep ice cores drilled at Camp Century and Dye 3, is reviewed; the problems involved in interpreting ice core records and the main results of the studies — derived past temperatures, precipitation rates, records of past volcanism — are presented. Whereas major changes in global climate are unambiguously revealed by the ice core records, changes of smaller amplitude and shorter duration seem to be more difficult to extract, probably due to a rather high noise level combined with disturbances induced by the flow of the ice. To correct for the latter effect, modelling of the ice sheet flow is needed. New important information on the internal deformations of the ice sheet has been obtained from measurements in the deep drill holes and from deformation tests on core ice. This information has significantly improved dynamic models for the ice sheet and thus has helped to interpret the environmental records from the ice cores. Recent work aimed at collecting paleoenvironmental information from the marginal sectors of the ice sheet is also reviewed. Even though the records that can be retrieved from the ice margin are not as detailed as ice core records, studies on the ice margin seem to have a great potential for obtaining an overview on a regional basis of the dynamic and climatic history of the ice sheet.

HISTORY OF EXPLORATION OF THE GREENLAND ICE SHEET

The last few decades have displayed a veritable breakthrough as regards our understanding of the dynamic and climatic history of the Greenland Ice Sheet: the EGIG expeditions (Expédition Glaciologique Internationale au Groenland) from 1957-1968 investigated the dynamics of a profile across the central Greenland Ice Sheet. The American deep drilling program at Camp Century, Northwest Greenland from 1963-1966 (Hansen and Langway, 1966) and the American-Danish-Swiss deep drilling at Dye 3, South Greenland, which was the main objective of the Greenland Ice Sheet Program (GISP) from 1971-1981 (Langway et al., 1985) provided climatic and other geophysical records reaching back in time far into the Wisconsinan and probably even into the previous interglacial, and in addition gave invaluable information about the internal deformations of the ice sheet. Also, as part of the GISP program, the ice thickness distribution was determined by airborne radio echo soundings along a network of profile lines (Gudmandsen, 1978). The Greenland Ice Sheet Program was primarily concerned with investigations in the interior regions of the ice sheet. A new program was initiated in 1984 in the marginal sector of the ice sheet, the main objective of which is to study the dynamics of a major outlet glacier — the Jakobshavn Isbrae — in order to understand the interaction of the ice streams with the dynamically more quiet sectors of the Inland Ice (DPP, NSF, 1983).

In recent years, application of satellite methods has documented that this technique also has large potential for ice sheet investigations. Up to now the most fertile application of satellites has been to obtain accurate surface elevations (Zwally et al., 1983) and ice flow velocities (Drew and Whillans, 1984) of the interior regions of the ice sheet.

The accelerating progress in data collection on the Greenland Ice Sheet in the last few decades owes a lot to the likewise accelerating technical development that has taken place in the same period, in particular the introduction of more advanced transportation equipment for personnel as well as scientific instruments: from skis and dog sledges to small snowmobiles, large weasels capable of pulling several heavy sleighs, to aircraft and satellites.

The great conquests of recent years, however, should not let sink into oblivion the scientific results obtained by explorers and scientists in the first half of this and the final half of the last century, often under the most difficult and dangerous conditions. It is not possible within the scope of this work to give a detailed account of the early history of exploration of the Greenland Ice Sheet; reference is made to Fristrup (1966). Some of the more prominent events and results of early as well as more recent investigations are summarized in Table 14.1.

Reeh, N.
1989: Dynamic and climatic history of the Greenland Ice Sheet; Chapter 14 in Quaternary Geology of Canada and Greenland, R.J. Fulton (ed.); Geological Survey of Canada, Geology of Canada, no. 1 (also Geological Society of America, The Geology of North America, v. K-1).

Table 14.1. Exploration of the Inland Ice

Year	Investigator	Event	References
1848-1868	Henrik Rink	Investigation of West Greenland Ice Sheet margin; first to observe calving rates from outlet glaciers; introduced the name "Indlandisen" (the Inland Ice); presented theory of rapid ice stream motion within more quiet sectors of the Inland Ice.	Rink (1853, 1877, 1889)
1874-1880	A. Helland, K.J.V. Steenstrup, R.R.I. Hammer	Observation of calving front velocities of West Greenland outlet glaciers.	Helland (1876), Steenstrup (1883), Hammer (1883)
1888	Fridtjof Nansen	First crossing of the Greenland Ice Sheet; first observations of elevations and temperatures of interior regions.	Nansen (1890)
1892-1893	Erich von Drygalski	Year round observations of ice motion of an outlet glacier; first observations of annual ablation rates.	Drygalski (1897)
1912	Alfred de Quervain	First observations of snow accumulation rates and densities on an ice sheet traverse.	Quervain and Mercanton (1925)
1912-1913	J.P. Koch, A. Wegener	First wintering on the Greenland Ice Sheet (close to the eastern margin); first drilling (to 24 m) and firn temperature observations.	Koch and Wegener (1930)
1929-1931	A. Wegener	First wintering on central Greenland Ice Sheet; first observations of ice thickness in interior regions; meteorological observations; recording of snow stratigraphy; introduction of motorized vehicles in Inland Ice investigations.	Wegener (1933)
1948-1954	Paul-Emile Victor	Expéditions Polaires Françaises. Systematic ice thickness survey south of 74°N; drilling to 151 m in central Greenland.	Holtzscherer and Bauer (1954)
1952-1955	Carl Benson	Comprehensive stratigraphic studies of snow and firn.	Benson (1962)
1957-1968	EGIG	Expédition Glaciologique Internationale au Groenland. Investigation of the dynamics of a profile across the central Greenland Ice Sheet; investigation of the energy balance at the ice sheet surface.	Ambach (1963), Hofmann (1964), Mälzer (1964), Gerke (1969), Seckel (1977a, b), Hofmann (1974)
1959	S. Epstein, R.P. Sharp	First application of stable isotope methods in firn and ice studies.	Epstein and Sharp (1959)
1963-1966	U.S. Army	First drilling to bedrock through the Greenland Ice Sheet at Camp Century, northwest Greenland; climatic and other geophysical records from ice core analyses.	Hansen and Langway (1966), Dansgaard et al. (1969)
1971-1983	GISP	Greenland Ice Sheet Program. Drilling of numerous firn and ice cores from 10-2038 m length at locations distributed over the ice sheet surface; drilling to bedrock at Dye 3, south Greenland; ice thickness distribution determined by radio echo sounding flights; determination of velocities in interior regions by repeated satellite positioning and borehole inclinometry.	Langway et al. (1985), Drew and Whillans (1984), Gundestrup and Hansen (1984)
1978	Brooks	Ice sheet topography determined by satellite altimetry.	Brooks et al. (1978)

PRESENT CONDITIONS OF THE ICE SHEET

Comprehensive reviews of the physical and climatic characteristics of the Greenland Ice Sheet have been given by Weidick (1975, 1976), Budd et al. (1982), and Radok et al. (1982). Considerable ice sheet data (surface elevation, ice thickness, surface temperature, accumulation rate) collected by various expeditions up to 1965 are listed in Appendix A of Mock and Weeks (1965). These data supplemented by more recent ones, are filed in the World Data Centers for Glaciology: University of Colorado, Boulder, CO 80309, USA; Molodezhnaya 3, Moscow 117 296, USSR; and Scott Polar Research Institute, Lensfield Road, Cambridge, England CB2 1ER.

General description

The Greenland Ice Sheet (often referred to as the Inland Ice) is the largest ice mass in the northern hemisphere, and the second largest in the world, surpassed only by the Antarctic Ice Sheet. It contains about 7% of the world's fresh water. Snow, which accumulates in the low temperature and highly elevated interior of Greenland, is transformed into ice as it slowly moves downwards into the ice sheet and outwards towards the warmer marginal regions. There the ice either melts in the ablation zone or is removed by calving of icebergs (Fig. 14.1).

Areal distribution and ice volume

The Greenland Ice Sheet covers 78% (1 701 300 km²), local glaciers cover 3% (65 500 km²), and ice-free areas account for 19% (408 800 km²) of the land surface (total area 2 175 600 km²) (Weidick, 1985b). The accumulation and ablation areas of the ice sheet constitute 84 and 16% of the total ice sheet area, respectively; the volume has been estimated at 2.4×10^6 km³ of water equivalent (Holtzscherer and Bauer, 1954). Slightly different estimates have been obtained in other studies; Radok et al. (1982), for example, applying a different map base and incorporating recent ice thickness observations obtained by radio echo soundings, gave the total area of the ice sheet as 1.67×10^6 km² and the ice volume as 2.74×10^6 km³ of water equivalent.

Surface topography

The surface elevations of the ice sheet are shown in Figure 14.2a. The general features are a southern and a northern dome with maximum elevations of 2830 and 3205 m, respectively, connected by a long almost horizontal saddle with elevations around 2500 m (Mock, 1976). The main north-south ice divide is shifted to the east with respect to the centre line of the ice sheet, as a reflection of higher elevations of the sub-ice topography to the east than to the west. The saddle around 67°N owes its existence to easy ice drainage to both east and west through the outlet glaciers Sermilik and Jakobshavn Isbrae (Reeh, 1984).

The surface of the ice sheet is generally smooth with average surface slopes typically ranging from 0.005 or less in the interior regions to 0.05 in the marginal zones. Detailed surface profiling, however, has revealed that stationary undulations with wave lengths of the order of 10 km and wave heights ranging between a few metres and 50 m are common features (Mälzer, 1964; Brooks et al., 1978; Zwally et al., 1983; Overgaard and Gundestrup, 1985). The undulations reflect the hills and valleys of the sub-ice topography (e.g., Overgaard and Gundestrup, 1985).

Surface elevations on the Greenland Ice Sheet have been determined by pressure altimetry (Quervain and Mercanton, 1925; Koch and Wegener, 1930; Joset and Holtzscherer, 1954; Paterson, 1955; Benson, 1962; Mock and Ragle, 1963; Gerke, 1969; Overgaard and Gundestrup, 1985), by geometrical levelling (Mälzer, 1964; Seckel, 1977a), by combined airborne barometric altimetry and radar distance measurement (Gudmandsen, 1978), and south of 72°N by satellite radar altimetry (Brooks et al., 1978; Zwally et al., 1983).

Pressure altimetry measurements generally are accurate only within 50-100 m to judge from comparing elevations at the points of intersection of profile lines. As regards the satellite altimetry, the accuracy is anticipated to be approximately 2 m over the smooth, near-horizontal interior parts of the ice sheet, and approximately 15 m over the more steeply sloping and undulating marginal regions (Zwally et al., 1983). Precise elevations at a few locations on the ice sheet surface have been obtained by means of geoceiver positioning (Mock, 1976; Drew and Whillans, 1984; Overgaard and Gundestrup, 1985). The precision of these elevations is estimated to be of the order of decimetres.

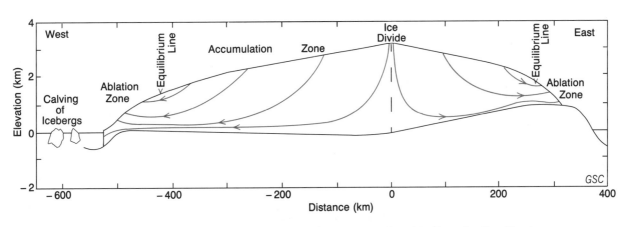

Figure 14.1. Schematic diagram showing flow in a cross-section of the Greenland Ice Sheet.

Ice thickness and sub-ice topography

The ice thickness of the Greenland Ice Sheet is shown in Figure 14.2b. The largest observed ice thickness is 3420 m at 71°42′N, 38°48′W. The average ice thickness has been calculated as 1790 m (Radok et al., 1982).

Moving from the interior regions towards the ice margin, ice thicknesses display a general decrease (Fig. 14.3a). In southern and central Greenland the land-based parts of the ice sheet usually thin out to zero thickness at the margin whereas in northern Greenland the ice sheet commonly terminates in near vertical cliffs up to 30 m high (Koch and Wegener, 1911). Where the ice sheet reaches sea level in deep fiords extending inland from the coast and also along many sectors in Melville Bugt in northwestern Greenland, it terminates in iceberg-producing glacier fronts with thicknesses of typically several hundred metres. The thickest margins of 700-800 m are in the Jakobshavn Isbrae and the Rink Isbrae areas (Carbonnell and Bauer, 1968).

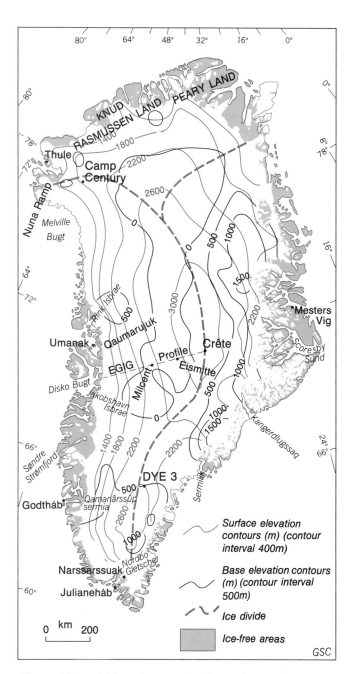
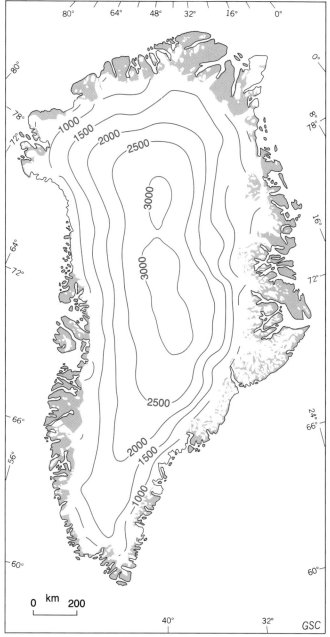

Figure 14.2. (a) Location map showing surface and base elevations (m), and main ice divides of the Greenland Ice Sheet (modified from Radok et al., 1982).

Figure 14.2. (b) Ice thickness (m) on the Greenland Ice Sheet (after Radok et al., 1982).

Sub-ice elevations determined by subtracting ice thicknesses from surface elevations reveal that the sub-stratum south of 66°N is a mountainous region with elevations typically between 500 and 1000 m. Also the eastern marginal zone of the ice sheet rests on a mountainous landscape in places more than 1000 m high. In the central and northwestern regions, on the other hand, the base of the ice sheet is rather level with large areas below present sea level (Fig. 14.2a).

The first measurements of ice thickness in the interior regions of the ice sheet were made by Brockamp et al. (1933) who measured the thickness at Eismitte (71.2°N, 39.9°W) to be 2500-2700 m using seismic methods. The same method was also used by Joset and Holtzscherer who determined ice thicknesses along profiles with a total length of several thousand kilometres south of 74°N (Holtzscherer and Bauer, 1954), and by Bull (1956) on his ice sheet traverse along 77°N. With the introduction of airborne radio-echo soundings during the Greenland Ice Sheet Program (1971-1982), however, ice thicknesses were determined along a network of profile lines covering the entire ice sheet area (Gudmandsen, 1978). In addition, detailed local investigations of ice thickness variations have been performed by means of radio-echo soundings from the surface (Overgaard and Gundestrup, 1985; Jezek et al., 1985).

Temperature

Mean annual ice temperatures of the Greenland Ice Sheet range from 0 to −32°C. Below a depth of about 10 m, the amplitude of the annual temperature cycle is at most a few tenths of a degree. Where melting is negligible the temperature at 10 m depth deviates less than 2°C from the mean annual air temperature (Loewe, 1970). In the accumulation zone the deviation of the temperature at 10 m depth from the mean annual air temperature becomes more and more positive with increasing amount of melt. In the ablation zone, the deviation may decrease again as explained by Koerner (1989).

The distribution of temperatures at 10 m depth over the Greenland Ice Sheet is shown in Figure 14.4a. With increasing depth in the ice sheet, temperatures generally increase due to the geothermal heat flux and to internal heating caused by ice deformation. In some places temperatures reach the pressure melting point at the base of the ice sheet (-2.6°C for 3000 m-thick ice). Two different sets of conditions favour high basal temperatures: 1) large ice thickness combined with a small accumulation rate (central, middle, and northeast Greenland) and 2) rapid motion combined with high surface temperatures (marginal zones in south and west Greenland). This distribution is indicated by basal isotherms which are determined by steady state ice flow modelling (Fig. 14.4b). Even though these temperatures locally may be in error by many degrees due to imperfections of the model applied, the figure illustrates the expected trend of basal ice temperatures from large negative values in the east to higher temperatures and likely melting in large sectors adjacent to the western margin.

Close to the ice margin, local deviations from this pattern are likely to occur: sectors with high, elevated, thin and almost stagnant ice will be cold based, whereas low areas carrying rapidly moving channelled flow will favour basal melting. The thermal regime at the ice sheet base is important in terms of glacial erosion and consequently plays a role in controlling the landforms developed beneath the ice sheet.

Even though temperatures generally increase with depth in the ice sheet, a small negative temperature-depth gradient (decreasing temperature with depth) is commonly observed in the upper layers because of advective transport of cold from higher elevated and consequently generally cooler regions.

Observations of temperatures at 10 m depth in the ice sheet were compiled by Diamond (1960), Bader (1961), Benson (1962), Mock and Weeks (1965), and Radok et al. (1982). Temperatures at greater depths in the ice sheet have been observed at the locations indicated in Figure 14.4a; most of the observed profiles are plotted in Robin (1983, fig. 4.1). Other temperature profiles are presented by Stauffer and Oeschger (1979), Colbeck and Gow (1979), and Gundestrup and Hansen (1984). In a few cases, temperatures were measured from the surface to the base of the ice sheet. At the interior sites where deep drilling was conducted (Camp Century and Dye 3) basal temperatures were about −13°C (Weertman, 1968; Gundestrup and Hansen, 1984). At two locations close to the ice margin in West Greenland, basal temperatures were close to the pressure melting point (Colbeck and Gow, 1979; Stauffer and Oeschger, 1979).

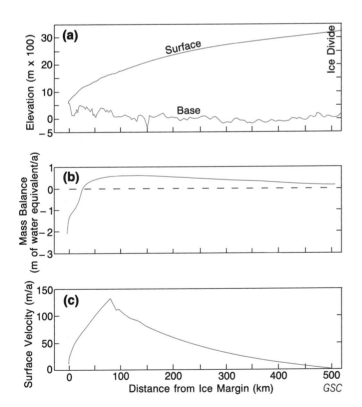

Figure 14.3. EGIG profile (for location see Figure 14.2a) (a) Observed surface and base profiles; (b) observed mass-balance distribution; and (c) observed surface velocity (all modified from Reeh, 1983).

Flow

The general flow pattern of the Inland Ice is away from the northern and southern domes towards the margins. Since flow along the north-south trending main ice divide is very slow, however, the simpler concept of an east-west directed outward flow away from the main divide towards the coastal areas, is not far from reality. Northern Greenland though, is an exception; here the flow has a large northward component.

Ice flow is generally in the direction of the maximum surface slope, that is, normal to the surface elevation contours (Hofmann, 1974; Drew and Whillans, 1984). In the marginal zones, however, where sliding over the base may contribute substantially to the movement (see below), significant deviations are to be expected between the directions of ice flow and maximum surface slope.

Horizontal surface velocities increase from small values in the interior regions to 50-100 m/a near the equilibrium

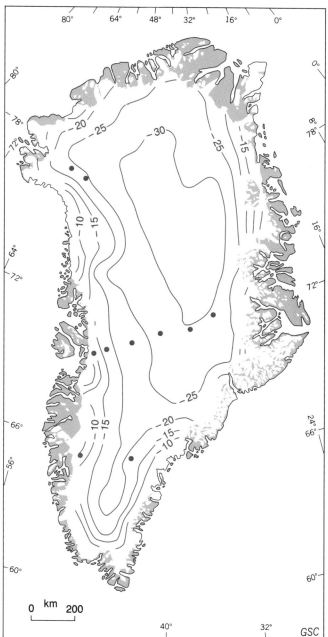

Figure 14.4. (a) Isotherms (°C) at 10 m depth in the Greenland Ice Sheet; locations of temperature profile measurements at >10 m depth are indicated by dots (modified from Radok et al., 1982).

Figure 14.4. (b) Isotherms (°C) at the base of the ice sheet; results obtained by steady state flow-line modelling (after Radok et al., 1982).

line, and from there generally decrease towards the ice margin to values ranging between a few metres per year and about 20 m/a (Fig. 14.3c). This flow pattern applies to the quiet sectors of the ice margin terminating on land where ablation by melting is dominant. As regards the ice margin sectors that feed the large iceberg-producing outlet glaciers terminating in the sea, ice motion accelerates towards the calving fronts to velocities of the order of several metres per day. For example, the calving front region of the Jakobshavn Isbrae exhibits velocities of more than 20 m/day or 7.3 km/a (Carbonnell and Bauer, 1968; Lingle et al., 1981).

Turning to the vertical dimension, velocities generally decrease towards the base of the ice sheet, and approach zero where basal temperatures are below the pressure melting point. Where the basal ice is at the pressure melting point, however, a substantial sliding motion of the ice over the bed may take place. In the floating or almost floating frontal regions of outlet glaciers terminating in deep water, basal friction is absent or greatly reduced. Consequently, the vertical velocity gradient is negligible and basal and surface velocities are essentially equal (block flow).

Horizontal surface velocities have been measured by standard surveying methods at several locations along the ice margin (e.g., Helland, 1876; Hammer, 1883; Ryder, 1889; Drygalski, 1897; Sorge, 1933a; Carlson, 1941; Bauer et al., 1968a; Olesen and Reeh, 1969; Colbeck, 1974; Lingle et al., 1981). Frontal velocities of the outlet glaciers calving into Disko Bugt and Umanaq Bugt were also determined by means of repeated aerial photography (Bauer et al., 1968b; Carbonnell and Bauer, 1968).

Determination of surface velocities in the interior regions of the ice sheet requires special techniques due to the absence of fixed landmarks. The first successful measurements were those of Hofmann (1974), who in 1959 and 1967 repeatedly positioned points in a triangulation network extending right across the ice sheet around 71°N (EGIG profile, fig. 14.2a). In this study the vertical ice motion at the surface was also determined (Seckel, 1977b).

Recently, repeated geoceiver positioning has proved to be a powerful method for measuring velocities in the interior regions of the ice sheet. By this method Drew and Whillans (1984) determined surface velocities in a transect of the Greenland Ice Sheet at 65°N. Velocities increased from a few metres per year at the ice divide to about 50 m/a, 150 km down the western slope of the ice sheet.

Mass balance

The mass balance (budget) is the difference between accumulation (mainly snow in the interior regions) and ablation (melting) and loss by calving of icebergs. The line separating the accumulation area from the ablation area is termed the equilibrium line. A full description of mass balance terms is available in Paterson (1981). Figure 14.3b shows a profile of the mass balance from the central dome of the ice sheet to the western margin (on EGIG profile of fig. 14.2a).

The elevation of the equilibrium line on the western slope of the ice sheet is reported to vary from about 1000 m a.s.l. in the north to 1600-1800 m in the south (Loewe, 1933; Holtzscherer and Bauer, 1954; Benson, 1962; Weidick, 1976, fig. 363). As shown by Braithwaite and Thomsen (1984), however, the year to year variation in equilibrium line elevation is up to 300 m in central West Greenland, corresponding to a shift in horizontal position of more than 30 km. This large year to year variation is one of the main difficulties in assessing the mass balance of the ice sheet; rather than being faced with a well defined line, separating zones of positive and negative mass balance, one finds a wide zone of great ambiguity, where refreezing of meltwater, forming superimposed ice (Schytt, 1955), is an important and complicating factor.

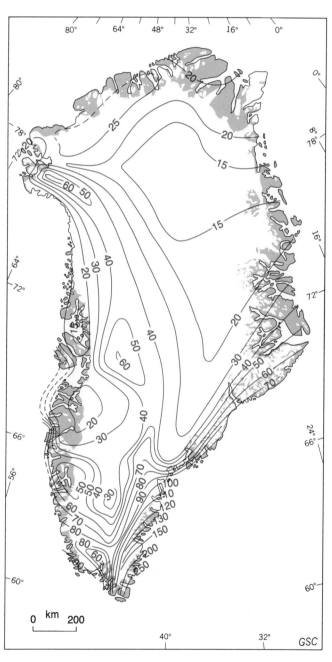

Figure 14.5. Precipitation g/(cm²·a) distribution in Greenland (after Reeh, 1985b).

Precipitation

Figure 14.5 shows distribution of precipitation in Greenland based on accumulation rates in ice-covered areas and recorded precipitation at coastal weather stations. The general trend is a decrease of precipitation from south to north from about 250 g/(cm²·a) in southeastern coastal areas to less than 15 g/(cm²·a) in interior northeast Greenland. Noteworthy are several precipitation minima that occur in West Greenland, in areas lying in the lee of high coastal mountains. Also it is tempting to interpret the precipitation maximum at 70°N, 46°W as being due to relatively easy access of humid air masses passing through the Disko Bugt area to this part of the ice sheet. Similar features are likely to occur also in East Greenland; however, at present this cannot be verified due to insufficient data coverage.

Estimates of the mean accumulation rate over the accumulation area range between 31 and 42.5 g/(cm²·a) (Radok et al., 1982, Table 4.7). Accumulation rates on the ice sheet have been obtained by means of rammsonde profiles (Bull, 1958) and by stratigraphic analyses of pit walls and firn and ice cores. Rotman et al. (1982) derived the distribution of accumulation rate in North Greenland from satellite microwave radiometric data. This method, however, seems to be accurate only in the dry snow zones of the ice sheet.

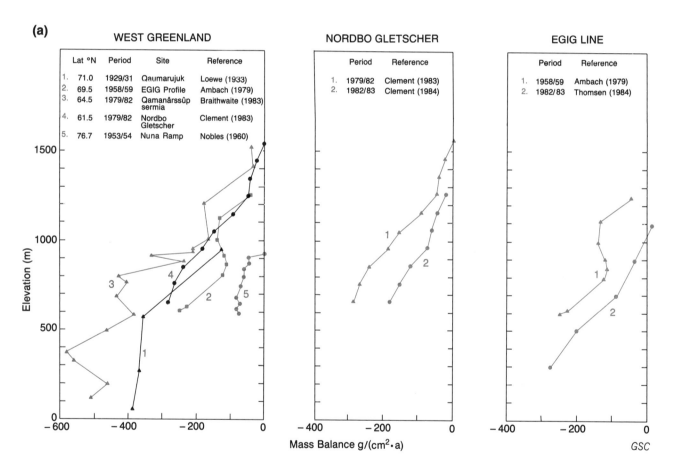

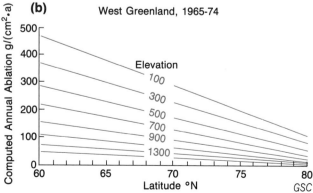

Figure 14.6. (a) Ablation measurements in West Greenland (after Reeh, 1985a) and (b) computed annual ablation for 1965-74 in West Greenland (after Braithwaite, 1980).

The accumulation rate observations have been compiled by Diamond (1960), Benson (1962), Fristrup (1966), Mock (1967), and Radok et al. (1982). Precipitation rates from coastal weather stations have been published by the Danish Meteorological Institute. Summaries of precipitation in coastal areas are given in Blinkenberg (1952), Lysgaard (1969), and Putnins (1970).

According to Benson (1962) the accumulation area of the Greenland Ice Sheet can be subdivided into three facies: the wet snow facies, the percolation facies, and the dry snow facies. Benson also dealt with wind generated features on the snow surface (barchans, ripples, drifts, sastrugi) and the metamorphism of snow to firn and glacier ice (densification theory). Investigations in these fields have also been made by Quervain et al. (1969) and Herron and Langway (1980).

Melt rates

Compared to a fairly complete coverage of accumulation data, observations of ablation rates (melt rates) on the Greenland Ice Sheet are more scanty and are largely restricted to the western marginal zone (Fig. 14.6a).

Estimates of the mean ablation rate throughout the ablation zone ranges between 107 and 130 g/(cm^2·a) (Radok et al., 1982, Table 4.7). However, since the ablation-elevation relationship at a given location can vary greatly from one year to another, these estimates, which are based on only a few years of observations at few locations, should be taken with reservation. A better approach would be to stretch the observations by applying models designed to predict ablation rates on the basis of meteorological records from nearby weather stations (Ambach, 1972; Braithwaite, 1980; fig. 14.6b).

Ablation rates were observed by Drygalski (1897), Quervain and Mercanton (1925), Loewe (1933), Ahlmann (1941), Bauer (in Ambach, 1979), Griffiths (1960), Nobles (1960), and the Geological Survey of Greenland (summarized in Reeh, 1985a).

Closely related to the ablation rate studies, are investigations of the energy balance at the ice sheet surface (Ambach, 1963). Ambach concluded that radiation exchange accounts for about 85% of the energy excess responsible for melting the ice. The study was carried out around 69.5°N in the western marginal zone of the ice sheet.

Calving rates

The most extensive studies of iceberg discharge from the Greenland Ice Sheet are those of Bauer et al. (1968b) and Carbonnell and Bauer (1968), who determined the calving flux from the outflow glaciers calving into Disko Bugt and Umanak Bugt, West Greenland by means of repeated aerial photography. The results of these studies vary by about 10% (82 km^3 of water equivalent per year in 1957 and 93 km^3 of water equivalent per year in 1964).

Weidick and Olesen (1978), based on the investigations mentioned above and scattered information about calving glaciers in southwest Greenland, estimated total iceberg production from West Greenland glaciers south of 71°N to be 97 km^3 of water equivalent per year. Few observations of calving rates are available for East Greenland; for the northern part of Scoresby Sund Olesen and Reeh (1969) reported iceberg production to be 11 km^3 of water equivalent per year

and for the southern part, the calving flux based on observations in 1972 has been estimated at 6.5 km^3 of water equivalent per year (O.B. Olesen and N. Reeh, unpublished). In northwestern and northern Greenland a survey of the major outlet glaciers, which included determination of calving front velocities, has been conducted by Kollmeyer (1980), but the results have not yet been published.

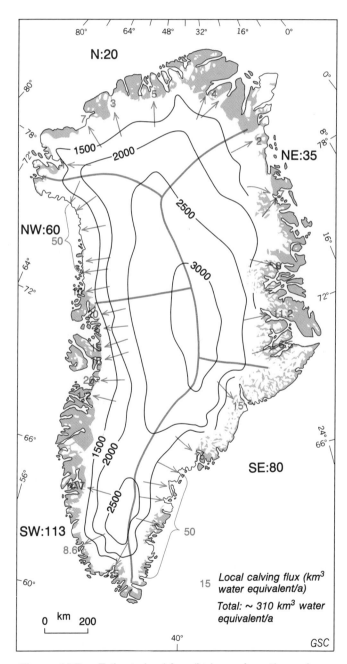

Figure 14.7. Estimated calving discharge from the various sectors of the Greenland Ice Sheet (km^3 of water equivalent per year), based on balanced-state assumption (after Reeh, 1985a). Contours indicate surface elevation (m).

To illustrate the present state of knowledge about calving rates from Greenland glaciers, Reeh (1985a) concluded that out of an estimated total calving flux of about 310 km³ of water equivalent per year, only about 130 km³ of water equivalent per year (about 45%) is based on concrete data such as calving front ice velocities and thicknesses. The distribution of major iceberg producing areas in Greenland, together with observed/estimated calving fluxes, is shown in Figure 14.7.

Total balance of the Greenland Ice Sheet

The discussion of the individual mass balance terms in the previous discussions shows that our present knowledge does not permit accurate assessment of the mass balance of the ice sheet as a whole. Not even the sign of the total balance is presently known. Nevertheless several estimates have been made for the annual gain, loss, and total balance of the Greenland Ice Sheet. They have been summarized by Weidick (1985a) as follows:

Accumulation   500 ± 100 km³ of water equivalent per year
Melting   295 ± 100 km³ of water equivalent per year
Calf ice   205 ± 60 km³ of water equivalent per year

The above estimates suggest that the Greenland Ice Sheet is close to a balanced state under present climatic conditions. This is confirmed by direct observations of the change in surface elevation along the EGIG line in central Greenland (Bauer et al., 1968a; Seckel, 1977a). These investigations show a lowering of the ice sheet surface in the ablation zone of the order of 0.2-0.3 m/a. On the contrary, Seckel's observations in the accumulation zone indicate an increase of surface elevation of about 0.1 m/a for west facing slopes and a decrease of surface elevation of about a few centimetres per year for east facing slopes.

The above mentioned results for the ablation zone agree with an observed average thinning rate of 0.3-0.5 m/a, derived at various sites along the West Greenland ice margin by comparing trimline elevations corresponding to the ice sheet extension around 1880 to present glacier surface elevations (Thomsen, 1983). Also the general retreat in this century of the ice margin in West Greenland (Weidick, 1968; Weidick and Thomsen, 1983) may be taken as evidence for current thinning of this part of the ice sheet.

For the interior regions, the balance has been estimated by flux-divergence considerations based on observed local values of accumulation rate, ice thickness, surface velocities, and surface strain rates (Shumsky, 1965; Mellor, 1968; Mock, 1976); recently Reeh and Gundestrup (1985) applied this method to the ice sheet near the Dye 3 deep drill site in South Greenland. The older estimates indicate large imbalances of the ice sheet (up to more than 1 m/a), in some cases positive in others negative. As discussed by Reeh and Gundestrup (1985), however, the older estimates suffer from lack of precise data and neglect of the difference between surface and depth-averaged values. At Dye 3 precise data are available, and the surface elevation change was calculated at 3 ± 3 cm/a, close to a balanced state. This result agrees with the direct observations farther to the north by Seckel (1977b), referred to above.

ICE CORE AND BOREHOLE ANALYSES

With the drilling of the 1390 m-deep ice core by the U.S. Army, Cold Regions Research and Engineering Laboratory (CRREL) at Camp Century, northwest Greenland in 1964-1966 (Hansen and Langway, 1966; Ueda and Garfield, 1969) and with studies on the ice core from the nearby Site 2 (Langway, 1970), a new era was initiated in the investigation of the Greenland Ice Sheet. The ice core studies demonstrated that the great ice sheets are rich sources of information on past environmental conditions. From 1971-1981 a number of ice cores were drilled from 30 to 400 m depth by the Greenland Ice Sheet Program (GISP) which had a main

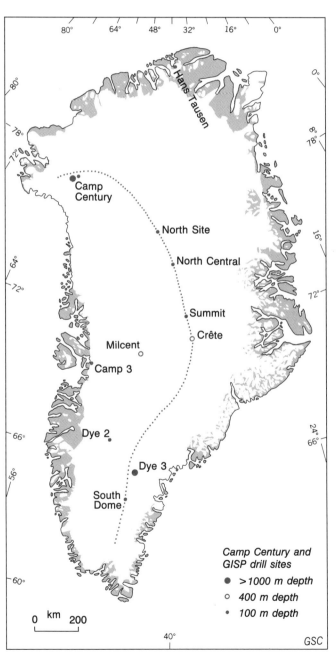

Figure 14.8. Camp Century and Greenland Ice Sheet Program (GISP) drill sites; the dotted curve shows the main ice divide (after Langway et al., 1985).

objective of extracting paleoenvironmental information (Rufli et al., 1976; Johnsen et al., 1980; fig. 14.8). The GISP field activities culminated in 1981, when a 2037 m-deep ice core was retrieved at Dye 3, South Greenland. Extensive stratigraphic, physical, and chemical studies were performed on the ice cores (Langway et al., 1985), providing information in many scientific disciplines, e.g., glaciology, climatology, meteorology, atmospheric chemistry, cosmic physics, geology, and volcanology. Table 14.2 lists the various aspects of ice core analyses and provides references to the studies.

Interpretation of ice core records

Ice cores provide a broad gamut of data useful in deciphering the past environment because all types of fallout, airborne-terrestrial dust and biological material, volcanic debris, sea salts, cosmic particles, isotopes produced by cosmic radiation, and artificially and naturally produced strong acids are incorporated in the snow. Layers of snow containing these contaminants are gradually compressed into solid ice which includes small cavities containing samples of atmospheric air. The layers are buried and sink into the ice sheet under continuous horizontal stretching and vertical thinning, so that they approach zero thickness close to the bottom.

Table 14.2. Information from ice core analyses

Discipline	Topic	Measured quantity	References
Glaciology	Accumulation rates	stable isotopes	Dansgaard et al., 1973
	Dating of internal strata	radioactive isotopes	Dansgaard et al., 1973
		microparticles	Hammer et al., 1978
		conductivity	Hammer et al., 1985
		chemistry	Craigin et al., 1977
	Ice flow and ice flow law	borehole deformations	Gundestrup and Hansen, 1984
		ice crystal size	Herron et al., 1985
		ice crystal fabric	Herron et al., 1985
		ultrasonic velocity	
		deformation of ice samples	Shoji and Langway, 1985
	Past ice sheet elevations	total gas content	Raynaud and Lorius, 1973
	Ice sheet temperatures	temperature profiles	Weertman, 1968
	Transformation of snow into ice	density	Langway, 1970
		crystal growth	Langway, 1970
Climatology	Past and present temperatures	stable isotopes	Dansgaard et al., 1973
		temperature profiles	Robin, 1983
	Past precipitation rates	stable isotopes	Dansgaard et al., 1973
		radioactive isotopes	Dansgaard et al., 1973
		microparticles	Hammer, 1977b
		conductivity	Hammer et al., 1978
		chemistry	Langway et al., 1977
	Past atmospheric circulation	microparticles	Dansgaard et al., 1984
		chemistry	Herron and Langway, 1985
	Glacial climatic cycles	stable isotopes	Dansgaard et al., 1969
Meteorology	Atmospheric circulation pattern	stable isotopes	Fisher and Alt, 1985
	Exchange across the tropopause	radioactive isotopes	Hammer, 1977b
		stratospheric dust	Hammer, 1977b
	Residence time in the atmosphere	radioactive isotopes	Picciotto and Wilgain, 1963
Atmospheric chemistry	Composition changes	composition of entrapped air bubbles	Raynaud and Delmas, 1977
	Change of CO_2 content of the atmosphere	CO_2 content of entrapped air bubbles	Neftel et al., 1982
	Pollution	lead	Murozumi et al., 1969
		fission products	Picciotto and Wilgain, 1963
Cosmic physics	Change in cosmic radiation flux	^{14}C	Beer et al., 1985
		^{32}Si	Clausen, 1973
		^{10}Be	Beer et al., 1985
	Correction of ^{14}C scale	see: Glaciology, dating of internal strata	Hammer et al., 1986
Geology	Sequence of glaciations	stable isotopes	Dansgaard et al., 1971
		impurities	Hammer et al., 1985
	Sub-ice sediments and rocks	debris content	Herron and Langway, 1979
Volcanology	Volcanic activity index	conductivity	Hammer, 1977a
	Dating volcanic eruptions	conductivity	Hammer, 1977a

Table 14.3 Ice core dating methods (after Hammer et al., 1978)

Method	under favourable conditions		Remarks	Typical references
	Time range (years)	Accuracy		
1. **Radioactive decay**				
1.1 ^{3}H and ^{210}Pb	100	10%	amount of ice needed: 1 kg	Theodorsson, 1977
1.2 ^{32}Si and ^{39}Ar	1 000	10%	amount of ice needed: 10^3 kg	Clausen 1973; Oeschger et al., 1976
1.3 ^{14}C	25 000	5-20%	amount of ice needed: $>10^4$ kg	Coachman et al., 1958 Oeschger et al., 1976
1.4 ^{10}Be, ^{26}Al, ^{36}Cl, ^{53}Mn, ^{81}Kr to be developed	>100 000?	?	amount of ice needed: $>10^5$ kg	Oeschger et al., 1976
2. **Glacier dynamics**				
2.1 Theoretical flow models	10 000	3%	steady-state ice sheets	Dansgaard and Johnsen, 1969
2.2 Vertical velocities along boreholes	?	?	steady-state ice sheets	Paterson, 1976
3. **Stratigraphy**				
3.1 Reference horizons				
3.1.1 fallout of fission products	back to 1954	1 year	fallout of bomb-produced debris	Picciotto and Wilgain, 1963
3.1.2 volcanic ash and dust	>200	1 year	dated volcanic eruptions	Gow, 1968
3.1.3 soluble volcanic debris	>200	1 year	dated volcanic eruptions	Hammer, 1977a
3.1.4 radio reflection layers	10 000	?	indirect (imply calibration with dated ice core)	Gudmandsen, 1976
3.1.5 stable isotopes, ^{18}O or ^{2}H	500 000?		drastic δ-shifts (glaciations) indirect	Dansgaard et al., 1973
3.2 Seasonal variations				
3.2.1 classical stratigraphy, density, ice fabrics, deposits of mineral or clay particles	200	10%		Sorge, 1933b
3.2.2 radioactive isotopes	few decades	10%	range and accuracy depend strongly on location for accumulation rates >0.20 m/a ice	Theodorsson, 1977 Hamilton and Langway, 1968; Hammer, 1977b
3.2.3 microparticles	10 000	3-10%		
3.2.4 stable isotopes, ^{18}O or ^{2}H, and trace elements	15 000	1-5%		Dansgaard et al., 1973; Johnsen et al., 1977
3.3 Long-periodic δ^{18}O cycles	100 000	?	implies assumption of periodic δ-cycles due to solar influence	Dansgaard et al., 1971

In the interior regions, where little or no surface melting occurs, the layer sequence is essentially undisturbed (wind erosion may destroy continuity to certain depths). Within a variable distance of the base, shear or even fracture zones due to local irregularities of the basal relief may violate the continuity in the layer sequence. For example, the layer sequence in the 2037 m-long Dye 3 ice core is most likely continuous down to 88 m above the bottom. At this depth a visible silty ice layer containing debris, presumably derived from bedrock, and a step change in all observed ice core constituents have been interpreted as indication of a shear zone (Hammer et al., 1985).

Once ice core information has been extracted by physical or chemical analysis, the next step is to establish a historical record (a time series). This step, which is not at all straightforward involves 1) dating, 2) correction for advective transport due to motion within the ice sheet, 3) correction for possible changes during this motion, 4) sorting out of noise introduced in the deposition phase, and 5) consideration of the difference between ice and air concentrations.

Dating

There are essentially four methods for dating ice core records: 1) stratigraphic dating, 2) dating by radioactive isotopes, 3) theoretical dating by ice flow model calculations, and 4) dating by correlation with other records. Hammer et al. (1978) thoroughly discussed the various dating methods and showed many examples from Greenland (see also Robin, 1983). Table 14.3 (from Hammer et al., 1978) summarizes ice core dating methods. The following describes the methods mentioned above in more detail.

At favourable locations stratigraphic dating can be carried out using seasonally variable parameters such as $\delta^{18}O$, dust, chemistry, or conductivity; this can permit dating of up to 10 000 year-old ice, with an accuracy of about ± 3 a/ka (Hammer et al., 1978, 1986; Dansgaard et al., 1982). In addition, horizons exhibiting distinct physical or chemical characteristics and with a known time of deposition also provide a means for stratigraphically dating ice cores. For example, well dated volcanic events such as the Laki eruption in 1783 and the Eldgja eruption in AD 934-935 can be identified in the ice cores as horizons exhibiting strongly acidic signals (Hammer et al., 1980). Unfortunately in Greenland the detection of volcanic events by acidity measurements is restricted to the Holocene ice because the Wisconsinan ice is alkaline due to the great abundance of calcareous dust deposited in the Late Wisconsinan (Craigin et al., 1977). Another useful marker in Greenland ice cores is the greatly elevated concentration of insoluble microparticles found in Late Wisconsinan ice (Craigin et al., 1977; Thompson, 1977; Hammer et al., 1978, 1985).

The radioactive dating methods are generally less precise, as appears from Table 14.3. In spite of relatively low accuracy, radiocarbon dating in particular becomes useful where stratigraphic methods fail, for example in dating lower parts of an ice core. With recent progress in isotope analysis by accelerator techniques, it is hoped that both range and accuracy of the radiocarbon method can be significantly improved (Bennett et al., 1978). The potential of this method, however, and also of the proposed ^{36}Cl-^{10}Be dating method, discussed by Dansgaard (1981), remains yet to be demonstrated.

Theoretical dating by flow model calculations has in principle no age limit. However, unless supported by information about upstream distributions of accumulation rate and ice thickness, and measurements of velocity and temperature profiles in boreholes, the dating accuracy is in general poor and hard to assess. If sufficient information is available, the precision of theoretical dating is probably better than 10% within a time range of about 10 ka (Reeh et al., 1985). Beyond this age (the end of the last glaciation), dating by ice flow modelling would have to consider the large changes in ice sheet geometry and environmental conditions that occurred at that time. With the present state of knowledge, it is not possible to specify these changes with sufficient accuracy for useful dating. With regard to the pre-Holocene part of the ice core records, therefore, flow model calculations should be considered only as a useful tool, providing references to which the ice core records can be compared.

Dating by correlating characteristic features of the records with similar features of other dated records such as pollen records from peat bogs or deep sea sediment records, are at least as imprecise as dating by using theoretical flow considerations. Several time scales derived by such comparisons have been suggested for one ice core (Dansgaard and Johnsen, 1969; Dansgaard et al., 1982). As discussed by Paterson and Waddington (1984), the same event in the Camp Century $\delta^{18}O$ record has been placed at 70 and 120 ka in different correlation studies.

Correction for advective transport

Several parameters measured on ice cores show large spatial variations, when measured at the surface of the ice sheet. Therefore, the variation in a quantity recorded in an ice core, will be affected by the original spatial variation of that parameter, since the ice within the core did not all originate at the same location. Consequently, in order to extract a series of temporal variations from an ice core record, corrections must be made for the advective transport. This is done by means of ice flow model calculations. The precision of such corrections depends on how well the upstream distributions of the relevant glaciological parameters are known. Also the time span, over which such corrections can be reliably performed, is in most cases limited to the Holocene, since corrections further back in time would involve consideration of the poorly known changes of ice sheet flow pattern and environmental conditions at the end of the last ice age.

On an ice sheet dome or on a horizontal crest there is no horizontal ice motion and consequently no need for corrections for advective transport, at least not for as long as the dome position has remained constant. This is an argument for choosing drill sites on domes or on slightly sloping divides, since at least the more recent parts of the corresponding ice core records should be easier to interpret.

Corrections relating to the snow deposition phase

Generally the concentration of a trace substance in precipitation is different from that in the precipitating air mass (Junge, 1977). Hammer (1985) has discussed the relationship between the concentration of a substance in the air and the concentration in the ice. Hammer (1985) concluded that in areas of high precipitation (more than 10-20 cm of ice equivalent per year) ice cores reflect the local air concentrations of trace substances with only moderate influence from the annual individual precipitation characteristics, and that changes in annual precipitation due to climate changes have little effect on the relationship between local air to local ice concentration.

Since snow precipitation is a discontinuous process, an ice core does not represent a uniform sampling of a time interval (e.g., one year), but rather is made up of subsamples, each representing a precipitation event. This means, for example, that high precipitation seasons are overrepresented at the expense of low precipitation seasons. Therefore, differences between, for example, mean annual $\delta^{18}O$ values measured on an ice core need not reflect differences in the corresponding mean annual 'air' concentration of $\delta^{18}O$. They might as well be due to year-to-year changes in the annual distribution of precipitation, because the $\delta^{18}O$ values exhibit a pronounced annual cycle. In principle, this applies to all ice core constituents showing significant annual cycles.

In addition to the above considerations, local areal inequalities in snow deposition due to drifting on the ice surface will induce a random element (deposition noise) in any ice core record. Fisher et al. (1985), for example, showed that the signal to noise variance ratio of annual layer thickness and $\delta^{18}O$ records from Greenland ice cores ranges between 1 and 4. A similar phenomenon is the systematic removal of snow falling in the winter season, with its generally high storm activity, from the crest regions of an ice sheet and maybe from local crests of surface undulations by wind erosion. As demonstrated by Fisher et al. (1983), this may strongly influence the mean annual values of $\delta^{18}O$ determined from ice cores, particularly in low accumulation areas.

Surface melting is a problem that affects ion concentrations and gas composition. Components like CO_2 and HNO_3 can be highly enriched in ice cores compared to the corresponding mixing ratio in the air, if surface melting takes place (Neftel et al., 1982). Another important matter concerning gas samples affects the 60-70 m-thick permeable firn layer, above the level of pore close off. Air exchange driven by gas diffusion in the pore space, combined with pumping by barometric pressure changes at the surface, may occur within this layer. In Greenland the permeable layer may include up to several hundred years, thus setting a limit to the time resolution one can obtain for gas-composition records derived from ice cores (Stauffer et al., 1985a).

Aging effects

For some parameters changes occur during movement of the ice so that the values of parameters measured from a drill core differ from those of that ice at the point of snow deposition. For example, annual layers of deposited snow become progressively thinner as they are transformed into ice, and later the ice layers change thickness in accordance with the flow induced strain history. Also, diffusion processes, either gas diffusion in the pore space of the upper firn layers or the relatively slower diffusion in the crystal lattice of the solid ice, will smooth gradients and eventually obliterate variations of oxygen isotope ratios (Johnsen, 1977a). The smoothing depends on the accumulation rate. The annual cycles of $\delta^{18}O$, for example, are detectable over a considerable time span if accumulation rates are above 20 cm/a (Johnsen, 1977a).

Changes by chemical reactions during ice transportation are in general not to be expected, since most reactions that might take place have already occurred in the atmosphere. Organic gases (e.g., methane), however, may oxidize during transportation in the ice sheet (Craig and Chou, 1982).

Other places where errors might be introduced are in the actual drilling and sampling procedure. When the ice is brought back into contact with the atmosphere, chemical reactions and contamination are serious dangers for many ice core constituents, and careful storage and handling of the ice cores is needed in order to ensure that the values recorded are true values.

Results

Even though interpretation of ice core records involves certain problems as discussed in the preceding section, analyses of Greenland cores have provided a wealth of new knowledge about the climatic and dynamic history of the Greenland Ice Sheet.

A thorough account of the many aspects of the climatic records extracted from Greenland ice cores, and presentation of data mainly from before 1973, is given by Robin (1983). Some of the more important of these and newer findings, particularly the results from the Dye 3 deep ice core, drilled in 1979-1981 (Langway et al., 1985), are given in the following section.

Temperature and precipitation history

As shown by Dansgaard et al. (1973), a close relationship exists between the ratio of heavy $H_2^{18}O$ to light $H_2^{16}O$ in the surface snow from the Greenland Ice Sheet and the mean annual temperature at the snow deposition site (usually expressed as the relative difference from a standard value in parts per thousand, and designated $\delta^{18}O$; Dansgaard, 1964). This has led to the suggestion that $\delta^{18}O$ provides a qualitative estimate of the atmospheric condensation temperature. As discussed by Dansgaard et al. (1984) and documented by global oxygen isotope models (Joussaume et al., 1984; Fisher and Alt, 1985), however, the $\delta^{18}O$ signal may also reflect changes in moisture source areas controlled for example by atmospheric circulation variations (Lawson et al., 1982) or sea ice extent (Ruddiman and McIntyre, 1981). Furthermore, $\delta^{18}O$ is also influenced by processes such as the ratio of summer to winter precipitation, deposition noise, advective transport, and changes in ice sheet surface elevations.

Major changes in the global climate are unambiguously preserved in the $\delta^{18}O$ records. For example, the transition from glacial to interglacial conditions at 11-10 ka is prominent in Greenland deep ice cores (Fig. 14.9a, b) as well as in records from Arctic Canada (Koerner, 1989) and Antarctica (Dansgaard et al., 1984). In addition, large amplitude oscillations seen in both the Dye 3 and the Camp Century $\delta^{18}O$ records (Fig. 14.9b), referred to Middle and Late Wisconsinan time, most likely reflect abrupt climatic events (Dansgaard et al., 1984). These probably reflect rapid changes in the mean latitude of the polar front, in analogy with the event 13 ka documented in studies of benthic foraminifera in North Atlantic deep sea sediments (Ruddiman and McIntyre, 1981).

In contrast to large amplitude fluctuations, climatic changes of shorter durations seem to be more difficult to extract from the $\delta^{18}O$ records. Robin (1983) compared $\delta^{18}O$ records from different sites on the Greenland Ice Sheet and Arctic Canada, concluding that a warm spell starting around 1930 is almost the only feature common to all records. Robin argued that this may be due to differences in the short term temperature trends at the different sites. An equally likely explanation, however, is that the direct climatic signal in the $\delta^{18}O$ records is buried in a high noise level caused by the variety of processes discussed in the section on interpretation of core records. This is illustrated in Figure 14.9c, which shows the observed $\delta^{18}O$ profile for the upper 650 m of the Dye 3 deep ice core spanning the last 1500 years, and the same profile after correction for upstream effects back to 1150. Clearly, the corrections are of the same order of magnitude as the signal. In spite of this, there may be good correlations between $\delta^{18}O$ records and temperature information. Dansgaard et al. (1975) successfully correlated the Crête $\delta^{18}O$ record with temperature records from England and Iceland over the last millennium. As was pointed out by Robin (1983), however, similar correlations would not be found with the isotopic records from other sites in Greenland. The explanation could be that the records from other Greenland sites are more influenced by upstream effects than the Crête record, which is from the main central ice divide.

In order to reduce the noise and to obtain a geographically more widespread representation, Hammer et al. (1981) combined four Greenland $\delta^{18}O$ profiles into a single isotope index curve which was combined with other records of proxy climatic data to give a Northern Hemisphere temperature index curve (Fig. 14.10). This curve indicates that warm periods occurred in the twentieth, twelfth, and eighth centuries and cold periods in the nineteenth, seventeenth (culmination of the Little Ice Age), and ninth centuries.

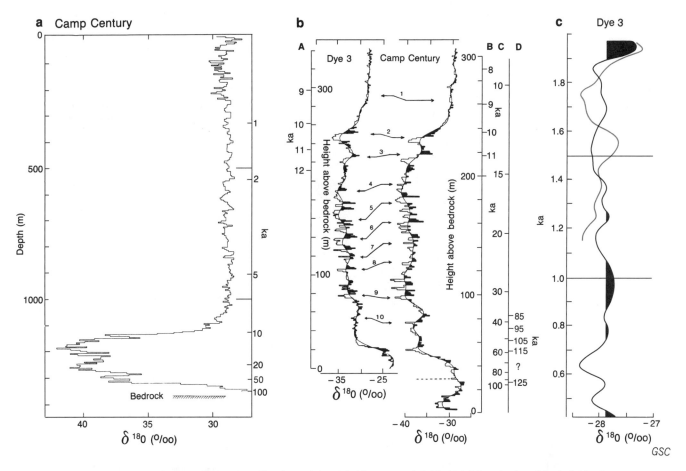

Figure 14.9. Oxygen isotope profiles from Greenland ice cores (cf. Fig. 14.8 for site locations). (a) Isotope profile for Camp Century core (after Robin, 1983). (b) The deepest parts of oxygen isotope profiles for cores from Dye 3 and Camp Century plotted as functions of distance from bedrock (after Dansgaard et al., 1982). The arrows show common features in the two records. Time scales A and B are deduced from simple ice flow modelling (Hammer et al., 1978), time scale C is from Fourrier spectral analysis of the Camp Century record (Dansgaard et al., 1971) and time scale D is from comparison with a deep sea record (Dansgaard et al., 1982). (c) Oxygen isotope profile for the upper 650 m of the Dye 3 ice core, spanning the last 1500 years (black line); same curve after correction for upstream effects back to 1150 AD (red line). The curves are smoothed with a 100 year digital low-pass filter (after Dansgaard et al., 1985a).

Climatic signals in terms of past summer climate are also provided by stratigraphic ice core records of melt features from the Dye 3 area (Hibler and Langway, 1977; Herron et al., 1981). The derived summer temperature index roughly agrees with the Northern Hemisphere temperature index referred to above. In several cases, however, melt feature records obtained from sites less than 1 km apart did not correlate. The reason is probably that the melt feature records contain a substantial noise component of advective character caused by spatial variations of the melt rate over surface undulations.

Information about past temperature conditions in Greenland is also contained in borehole temperature profiles. To extract this information, however, is a complicated procedure involving ice flow modelling and therefore requiring past ice sheet flow patterns and mass balances to be estimated. The farther the temperature profile location is from an ice divide and the further back in time the temperature record is being extended, the greater the complications and the more uncertain the results. A combined study, attempting to extract the climatic temperature signal from measured profiles of $\delta^{18}O$, and total gas content (reflecting the elevation above sea level of the snow deposition site; Raynaud and Lorius, 1973), was performed on the Camp Century deep ice core by Budd and Young (1983). They concluded that the temperature 15 ka was 8°C cooler than at present and that the ice sheet in the Camp Century area by that time was some 600 m thicker than the present. This was contradicted by Reeh (1984), however, who modelled the three-dimensional surface elevation and flow pattern for Late Wisconsinan northwest Greenland ice. This study indicated that ice thickness changed substantially less than that suggested by Budd and Young (1983). As a consequence, the relatively large $\delta^{18}O$ shift at the Wisconsinan/Holocene transition of 11-12‰ compared to an estimated climatic shift of only 6 ± 1‰ (Dansgaard et al., 1985b) could hardly be explained by thickness changes of the ice sheet alone. Dansgaard et al. (1985b) pointed at an anomalously high climatic temperature shift or an anomalously high $\delta^{18}O$ change per degree of warming in northwest Greenland as contributing causes.

Several factors are involved in deriving records of mass balance history from ice core data. Such records can be deduced from the thicknesses of annual layers, identified by means of annual cycles in, for example, $\delta^{18}O$, microparticle content, conductivity, or chemical constituents (Hammer et al., 1978). It is, however, necessary to apply the proper corrections, assessing the postdepositional distortion of the layers (Reeh et al., 1978; Paterson and Waddington, 1984). Modelling of the flow history of the ice sheet, therefore, is necessary to reconstruct the mass balance history. Local variability in accumulation patterns introduces additional noise and also topographically controlled accumulation rate variations must be extracted before any climatic significance can be ascribed to mass balance records derived from ice cores (Reeh et al., 1985).

A study based on observed annual layer thickness profiles along three 400 m-long ice cores from Dye 3, Milcent, and Crête (Reeh et al., 1978) concluded that in central Greenland, accumulation rates have decreased by a few per cent during the last 1500 years. Deviations from this long term trend were generally less than 5% (Fig. 14.11a). The larger deviations of up to 11% found for the Dye 3 area in South Greenland were later shown to be due to topographically controlled upstream variations of the accumulation rate, and therefore not to have climatic significance (Reeh et al., 1985; fig. 14.11b).

Moving back in time, the correction for postdepositional strain history becomes increasingly uncertain. In Greenland two ice core records (Camp Century and Dye 3) reach more than 1500 years back, and at neither of these locations sufficient upstream data have been collected to allow reliable estimates of accumulation rates prior to about 3 ka. As an example of how this uncertainty affects interpretations, the bulge in the measured Dye 3 layer thickness profile in the depth interval 1200-1500 m (Fig. 14.11b), which corresponds to the time interval 3.5-6.0 ka, can be explained in two ways: it may be due to a 20-30% increase in precipitation during the climatic optimum, or equally likely it is caused by straining of the layers due to the ice moving over a subglacier mountain some 30 km upstream from the drill site. Unfortunately, the basal topography is not known in sufficient detail to resolve this important question.

Estimates of precipitation rates in the Wisconsinan have been derived from Greenland deep ice core records.

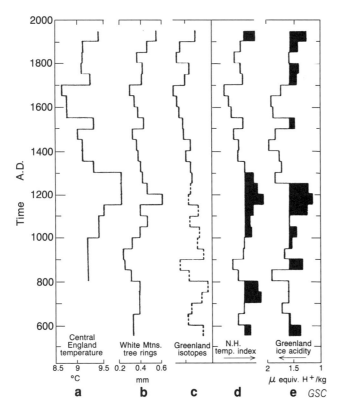

Figure 14.10. Correlation between climatic variations and volcanic activity (after Hammer et al., 1981). Curve d is a composite curve composed of the three indirectly determined records (a-c), with warm periods shown in solid black. Volcanic activity (e) is represented by an acidity profile drawn from data obtained from the Crête core. Lower than normal volcanism corresponds to low acidity and is shown in black.

Hammer et al. (1986) observed an abrupt, more than two-thirds reduction in annual layer thicknesses of the Dye 3 deep ice core at the Holocene/Wisconsinan transition. This is probably due to a similar reduction in accumulation rate in the Late Wisconsinan as compared to early Holocene. Based on comparison of the Camp Century record with lake sediment and deep sea foraminifera records, Dansgaard et al. (1982) suggested up to five times lower than present precipitation rates around 15 ka and relatively high precipitation rates in the periods 125-115, 80-60, and 40-30 ka. These periods correspond to those in which the deep sea records indicate a considerable buildup of continental ice. As discussed by Paterson and Waddington (1984), however, conclusions based on correlation dating of ice cores should be considered tentative.

History of atmospheric gases

As the snow layers deposited at the surface are buried and sink into the ice sheet, they compact to form solid ice which encloses air bubbles. These bubbles contain a sample of atmospheric gases. The past composition of gases in the atmosphere can therefore be measured directly in air bubbles trapped in the ice. The age of the samples, however, is not generally the same as that of the ice in which it is trapped. Because the upper part of the ice is permeable and atmospheric air may move through the upper 60-70 m thick firn layer, the trapped air may be considerably younger than the ice. In Greenland the permeable layer may include up to several hundred years of precipitation layers, thus setting a limit to the time resolution of the gas composition records.

Studies on ice samples from the Camp Century and Dye 3 deep ice cores indicate that the $O_2:N_2$ ratio of the atmosphere has not changed significantly during the last 100 ka (Raynaud and Delmas, 1977; Stauffer et al., 1985a). The CO_2 concentration, however, displays large variations. CO_2 records from both Greenland deep ice cores reveal a large and rapid increase from 180-200 ppm by volume at the end of the Wisconsinan to 260-300 ppm by volume in the Holocene (Neftel et al., 1982; Stauffer et al., 1985b). Also a series of rapid changes of CO_2 concentrations in the

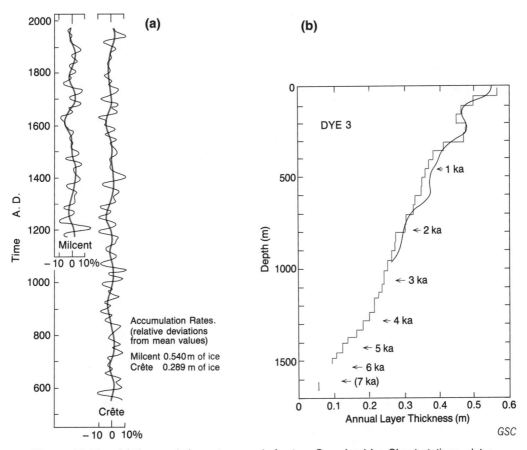

Figure 14.11. (a) Accumulation rate records for two Greenland Ice Sheet stations, determined mainly from seasonal $\delta^{18}O$ cycles (adapted from Reeh et al., 1978). The annual accumulation rate series have been smoothed by digital low-pass filters with cut off periods of 120 years (red curves) and 30 years (black curves). (b) Annual layer thickness profiles along the Dye 3 deep ice core (modified after Dansgaard et al., 1985a). The red curve shows observed thicknesses; the black curve is layer thicknesses as predicted by an ice flow model accounting for the short-distance variations in ice thickness and accumulation rate upstream of the drillsite.

Wisconsinan — 60 ppm by volume in time scales of the order of one century (Stauffer et al., 1984) — are well correlated with climatic changes deduced from stable isotope records (Fig. 14.9b). If these changes reflect changes in the atmospheric concentration, they may be a clue to understanding the changing climate during and at the termination of the last ice age. Artifacts due to CO_2-enriched melt layers and exchange with carbonates in the ice matrix which display a concentration variation corresponding to the variation in CO_2, however, cannot be excluded (Oeschger, 1985).

Also, the alkaline character of the Greenland Wisconsinan ice (Hammer et al., 1985) may influence the CO_2 content of this ice. These points should be clarified before CO_2 concentration variations in the atmosphere, as indicated by ice core records, should be considered a decisive factor in explaining the glacial/interglacial climatic cycles. The few measurements available for ice only a few centuries old suggest that pre-industrial CO_2 concentrations were 260-270 ppm by volume (Wolff and Peel, 1985). Present day air contains about 30% more CO_2.

Until recently, CO_2 was the only gas seriously considered as a mechanism for a possible future global warming, the so-called greenhouse effect. It has now become evident that many other trace gases, for example, methane (CH_4) and chlorofluorocarbons (CCl_3F and CCl_2F_2), currently accumulating in the atmosphere, may contribute significantly to global warming. Studies on Greenland and other polar ice cores suggest a rapid increase of CH_4 beginning about 200 years ago (Craig and Chou, 1982; Rasmussen and Khalil, 1984), and that the chlorofluorocarbons are entirely anthropogenic (Rasmussen and Khalil, 1984).

As mentioned previously, analysis of the total gas content of ice cores has been used to infer past elevations of deposition sites (Raynaud and Lorius, 1973). The method is based on the fact that the barometric pressure and therefore also the gas content of the ice changes with height. There are still unsolved questions, however, regarding the precision of the method (Paterson, 1981, p. 343).

Impurity content

Pre-industrial Greenland Ice Sheet impurities consist of marine, continental, volcanic, stratospheric, and extraterrestrial material (Hammer et al., 1985) and may be classified as follows:

Microparticles, terrestrially derived and mainly clay-like. During the Wisconsinan this material was strongly alkaline; presumably it consisted of Ca-rich dust originating from vast areas of continental shelves, exposed because of low sea level (Craigin et al., 1977; Hammer et al., 1985). The concentration of microparticles in Holocene ice in Greenland is of the order of 0.05 mg/kg of ice, which is about 6 times higher than in Antarctic ice, but is 3 to 70 times lower than in Greenland ice of Wisconsinan age (Fig. 14.12). The extremely high dust content of Greenland ice particularly in the Late Wisconsinan, was probably due to a combination of severe storminess at mid-latitudes and rich sources of aerosols in the areas covered by glacial outwash south of the North American and Eurasian ice sheets and also on the exposed continental shelf areas.

Sea salts, mainly NaCl. The chloride concentration profile along the Dye 3 deep ice core has been interpreted as indicating lower than present precipitation rates over the ice sheet in the Wisconsinan. This, combined with high wind speeds that increased sea salt aerosol concentrations over the oceans resulted in Cl concentrations as much as four to five times higher than in the Holocene (Herron and Langway, 1985).

Volcanically produced strong electrolytes. H_2SO_4, HCl, and HF and salts of these acids. Large volcanic eruptions in the northern hemisphere leave a high acidity signal in the ice due to washout of acids (Fig. 14.13). The volcanic acidity peaks rise over a background level of approximately 1μ equivalent H+ per kg in the Holocene period. This background

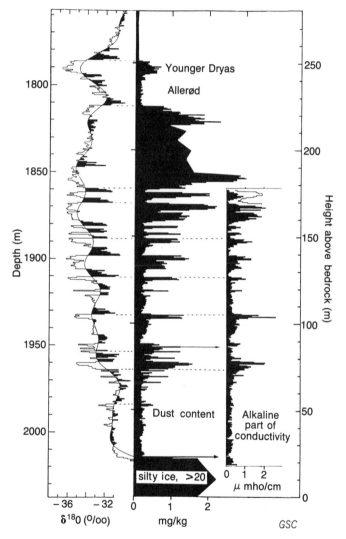

Figure 14.12. Ice age part of the Dye 3 deep ice core showing $\delta^{18}O$, dust concentration, and alkalinity in samples averaging 55 cm in length (after Hammer et al., 1985). The thin curve over the alkalinity is the total electrical conductivity of the melted samples. The difference between alkalinity and electrical conductivity is due to sea salts, nitrates, and volcanic impurities. High dust concentration, high alkalinity, and high sea salt concentrations vary in parallel manner, whereas low $\delta^{18}O$ values correspond to high impurity content. The smooth line through the $\delta^{18}O$ curve represents the isotope values after smoothing with a 24-point digital low-pass filter. The arrows in the dust content curve indicate visible dust layers presumably of bedrock origin.

and even the acid volcanic signals, however, are suppressed in ice deposited in Greenland during most of the Wisconsinan. This ice is alkaline due to the then high alkaline aerosol load that obviously neutralized the atmospheric acids (Hammer et al., 1980, 1985).

Acidity peaks in the ice cores corresponding to historically known volcanic eruptions have been used to check time scales obtained by stratigraphic dating of the ice cores, and on the other hand, have been used to date accurately volcanic events known according to historical tradition (Hammer, 1984) or from tephra chronology (Thorarinsson, 1981; fig. 14.12). Also the impact of volcanism on climate has been studied by means of records derived from ice cores. Hammer et al. (1980) found a significant correlation

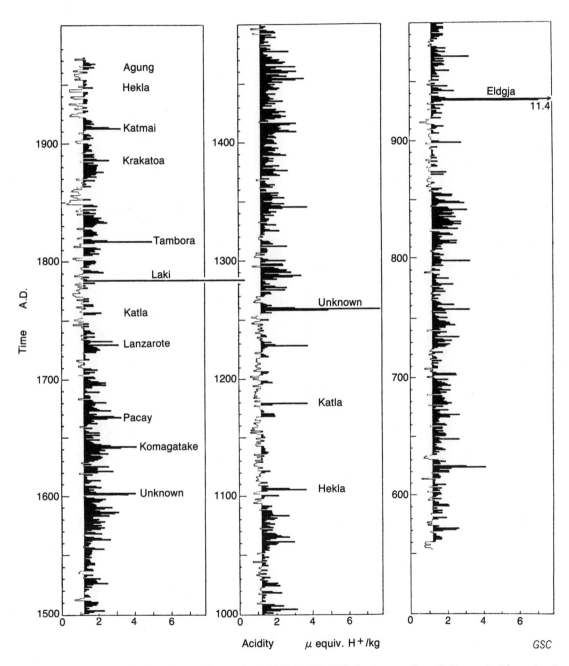

Figure 14.13. Mean acidity of annual layers from 1972 to 553 AD in the ice core from Crête, central Greenland (after Hammer et al., 1980). Acidity above the background (shaded areas), $1.2 \pm 0.1 \mu$ equivalent H^+ / kg ice, is due to fallout of volcanic acids, mainly H_2SO_4, from eruptions north of 20°S. The ice core is dated with an uncertainty of ± 1 year for the past 900 years, increasing to ± 3 years at 553 AD (Hammer et al., 1978), which makes possible the identification of several large eruptions known from historical sources, and the accurate dating of the Icelandic Eldgja eruption.

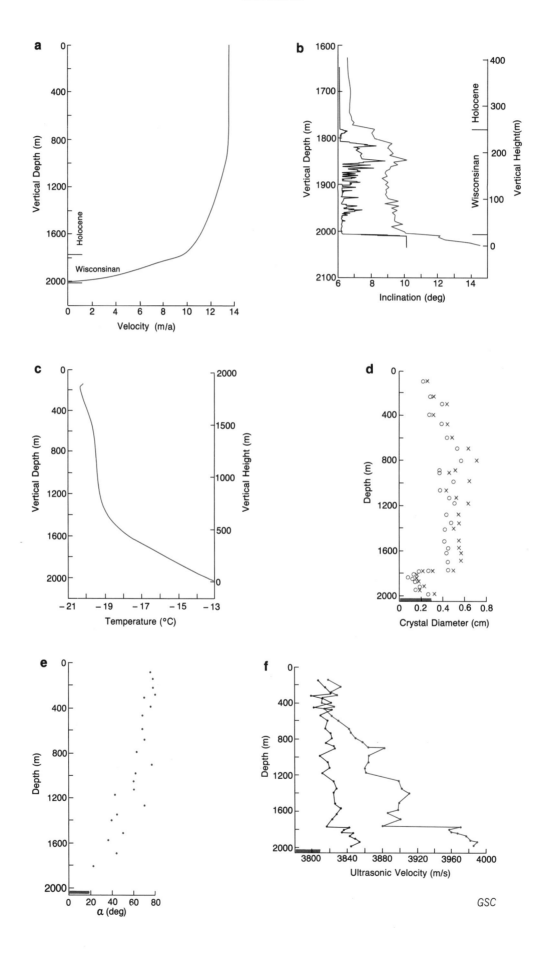

between a northern hemisphere temperature index and Greenland ice acidity in the period AD 550-1950 (Fig. 14.10), indicating that periods of low/high volcanic activity are associated with warm/cold climates.

Radioactive isotopes of different origins. These are in part due to interaction of cosmic radiation with atoms and molecules existing in the upper atmosphere (e.g., ^{3}H, ^{10}Be, ^{14}C, ^{32}Si, and ^{36}Cl). Others are introduced as a result of nuclear bomb tests (e.g., ^{137}Cs, ^{90}Sr), and some are released from the earth's crust (e.g., ^{210}Pb). As mentioned earlier, radioactive isotopes have been used to date ice cores by measuring radioactive decay (e.g., ^{3}H, ^{14}C, ^{32}Si, ^{210}Pb), and by detecting reference horizons of high B-activity (e.g., ^{137}Cs, ^{90}Sr). In addition, variations in ^{10}Be, with a half life of 1.5 Ma, measured on ice cores by means of accelerator based mass-spectrometry technique, have also been used to detect changes in the ^{10}Be production rate and to derive past ice accumulation rates (Beer et al., 1985). It was found that ^{10}Be concentration variations of about 60% in recent ice from the Dye 3 ice core correlated with the 11-year solar cycle and therefore probably reflected variations in production rate. On the other hand, a significant change of the ^{10}Be concentration at the Holocene/Wisconsinan transition by a factor of 2.5-3, most likely reflects a change in ice accumulation rate from relatively low values in the Late Wisconsinan to relatively high values in the Holocene in accordance with the findings of Hammer et al. (1986).

Besides these main classes of impurities, the ice sheet contains various soluble impurities of terrestrial and stratospheric origin (Craigin et al., 1977; Risbo et al., 1981; Herron et al., 1985), and extraterrestrial material (Quervain and Mercanton, 1925; Langway, 1970; Maurette et al., 1986). Further, in the marginal zones and in the near bottom layers of the ice sheet, wind blown particles from the ice free foreland and debris derived from bedrock may occur in high concentrations.

Internal deformations in the ice sheet

Ice core analyses also provide information on the mechanical properties and the structure of the ice which is important for understanding ice deformation. Further, measurements in boreholes enable the internal deformations in the ice sheet to be determined. The horizontal velocity profile (Fig. 14.14a) measured by repeated inclinometer surveys (Fig. 14.14b) of the Dye 3 deep drillhole (Gundestrup and Hansen, 1984) shows a velocity profile with about 75% of the velocity increase taking place in the lower 12% of the ice sheet. This part of the ice sheet is of Wisconsinan age. In the upper half of the ice sheet the velocity is essentially constant. As shown by Dahl-Jensen (1985), the stress and temperature distributions in the ice sheet (Fig. 14.14c) cannot by themselves account for this pronounced nonuniform distribution of the shear rates. The observed velocity profile suggests that ice of Wisconsinan age seems to flow about three times faster than ice of Holocene age, when subjected to the same stress and temperature. The enhanced flow of this ice is believed to be due to high concentrations of dust (Fig. 14.14b) and other impurities, and to small crystal size (Fig. 14.14d; Paterson, 1981, p. 34). The ice fabric does change with depth (Fig. 14.14e), although more gradually than dust content and ice crystal size; Dahl-Jensen (1985) concluded that ice fabric does not seem to affect the deformation rates at Dye 3. Deformation tests on fresh ice core samples from Dye 3 (Shoji and Langway, 1985), on the other hand, indicate an enhancement factor of 8-12 for the Wisconsinan ice at the base of the ice sheet when compared to ice near the upper surface; however, the enhancement factor is only about 4 when compared to ice immediately above the Wisconsinan/Holocene transition. Shoji and Langway (1985) ascribed this difference to the influence of ice fabrics and put the enhancement factor due to fabric at 4, and that due to impurities at 2-3. One of the borehole surveys which Dahl-Jensen used in her calculations, was rather inaccurate (Gundestrup and Hansen, 1984) and a new accurate borehole survey may resolve the apparent inconsistency between the results of borehole and ice sample deformation measurements.

Ultrasonic velocities were also measured on the Dye 3 ice core (Herron et al., 1985; fig. 14.14f). The difference in vertical and horizontal ultrasonic velocities correlate with the ice fabric, and hence can be used as a fast method of measuring this quantity.

ICE MARGIN INVESTIGATIONS

Traditionally, investigations in the marginal zone of the Greenland Ice Sheet have been related to mass balance studies and to studies of ice marginal deposits. More recent ice margin investigations, however, are also concerned with studies of glacier hydrology, glacier dynamics, and paleoclimatology. Dansgaard (1961) measured stable isotope values on icebergs calved from outlet glaciers in West Greenland. By comparing iceberg isotope values to the isotope distribution in the surface layers of the accumulation region feeding the outlet glaciers, the deposition site of the ice in the icebergs could be determined. Moreover, the age of the icebergs was determined by radiocarbon dating of the CO_2 contained in air bubbles entrapped in the ice, thus allowing travel times and average ice flow velocities to be determined. Similarly Robin (1983), used data collected by Raynaud (1976) to link locations with the same observed mean $\delta^{18}O$ values in the accumulation and ablation zones along the EGIG profile in West Greenland. Robin concluded that there is excess ablation in the profile over that for a steady state, an interpretation in agreement with the observed surface lowering in the ablation area (see *Mass balance*). Reeh and Thomsen (1986) and Reeh et al. (1987) also reported on the application of stable isotope methods in the margin of the Greenland Ice Sheet. They used $\delta^{18}O$ as a natural tracer

Figure 14.14. Parameters relating to internal ice sheet deformations obtained by analysis of the Dye 3 deep ice core and bore hole. (a) Horizontal ice velocity versus depth (adapted from Gundestrup and Hansen, 1984). (b) Hole inclination versus depth below 1600 m (red curve); the black curve is a dust index (after Gundestrup and Hansen, 1984). (c) Temperature profile (adapted from Gundestrup and Hansen,1984). (d) Crystal size profile; the circle represents vertical dimension or intercept length and the x represents horizontal dimension (after Herron et al., 1985). (e) Fabric parameter (α), the half-apex angles of the cones containing 90% of the crystal axes (after Herron et al., 1985). (f) Ultrasonic velocity profiles: the black line illustrates the horizontal velocities; the red line represents vertical velocities (after Herron et al., 1985).

to delineate ice sheet meltwater drainage basins. Also they distinguished between different kinds of ice (local ice, superimposed ice, regelation ice, etc.) by the different isotopic ratios and presented $\delta^{18}O$ records dating back into the Wisconsinan obtained from surface sampling of the outermost kilometre of the ice sheet margin.

The background for the ice margin isotope studies follows: As mentioned in an earlier section, $\delta^{18}O$ values from the accumulation zone of the Greenland Ice Sheet show significant temporal and geographic variations. The temporal variations cover time scales from annual to glacial-interglacial cycles, with low/high $\delta^{18}O$ values being related to cool/warm periods of snow deposition. The geographic variations generally reflect the distribution of mean annual temperatures on the ice sheet surface. This causes a positive trend in $\delta^{18}O$ from the high, cold, central regions of the accumulation zone towards the lower, warmer areas near the equilibrium line, and also a positive trend from north to south. In the ablation zone along the ice sheet margin, the ice originally deposited in the inland accumulation area reappears at the surface. The nearer to the ice divide the snow was originally deposited in the inland accumulation area reappears at the surface. The nearer to the ice divide the snow was originally deposited, the closer to the ice margin the corresponding ice will resurface (Fig. 14.1), and the older it will be. Therefore, the isotopic composition of the ice in the ablation zone will display variations, reflecting the temporal and geographic variations in the depositional area. This is illustrated in Figure 14.15, where $\delta^{18}O$ values of samples collected on the West Greenland ice margin are plotted against elevation. The data group around two straight lines, one for the accumulation zone and one for the ablation zone. The linking of points in the accumulation zone with those with similar oxygen isotope ratios in the ablation zones is illustrated by the vertical line connecting the two regression lines.

Oxygen isotope records from the outermost kilometre of the ice sheet margin are displayed in Figure 14.16 along with an interpretation of the main features. The West Greenland record, which is the most detailed, shows that features known from Greenland deep ice core records have apparently been preserved also in the ice margin isotopic records — for example the Allerød/Younger Dryas oscillation at the end of the last ice age. The continuity of the ice margin isotopic records, however, must be questioned due to numerous shear layers and meltwater refreezing occurring in the marginal zone of the ice sheet. The different isotopic levels of the Wisconsinan (pre-Holocene) ice in West Greenland (-40 to $-42‰$, representing the summit region of central Greenland), North Greenland (-46 to $-47‰$) and at Camp Century (-42 to $-43‰$; fig. 14.9), may be taken as evidence that the Wisconsinan ice of the Camp Century and North Greenland records does not originate from a common deposition area in central Greenland, but is of local origin (Reeh et al., 1987).

Recently it has been documented that the melt zone of the Greenland Ice Sheet contains what may be the richest deposits of micrometeorites on the earth surface (Maurette et al., 1986). The micrometeorites falling on the ice sheet surface are carried to the melt zone where meltwater helps to concentrate the particles in supraglacial ponds and slowly flowing streams. The collected extraterrestrial particles are generally identical to tektites found on the ocean floor but a pure glass type has been discovered that has not been seen in deep sea samples. Iron-rich spheres on the other hand, are conspicuously rare in the ice marginal samples.

DYNAMIC MODELS FOR THE GREENLAND ICE SHEET

In the last few decades dynamic modelling of ice sheets has developed from a mainly qualitative description of ice sheet dynamic behaviour into a discipline providing quantitative solutions to specific dynamics problems. An important step in this development is the improved knowledge of the flow properties of glacier ice, manifesting itself in the assessment of upper and lower limits on ice flow law parameters to be used in dynamic modelling of glaciers and ice sheets (Paterson and Budd, 1982). The implementation of numerical models, capable of dealing with the complex conditions of the real world, is also an important basis for this development.

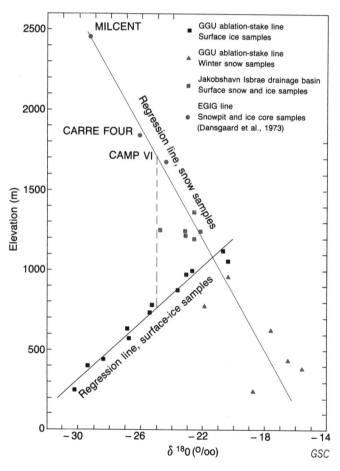

Figure 14.15. Elevation vs. oxygen isotope variation in surface ice and snow samples from the western sector of the Greenland Ice Sheet around 69-70°N (after Reeh and Thomsen, 1986). The vertical line connecting points on the regression lines illustrates the linking of points in the accumulation and ablation zones with same mean surface $\delta^{18}O$ values.

Ice sheet dynamic models have been applied to: 1) explain observed surface topography and motion of ice sheets (e.g., Reeh, 1984; Dahl-Jensen, 1985); 2) obtain an overall picture of the present dynamic and thermal state of the ice sheet (Budd et al., 1982; Radok et al., 1982); 3) provide reference for interpretation of ice core records (e.g., Dansgaard and Johnsen, 1969; Reeh et al., 1985); 4) reconstruct past ice sheet morphology and flow (e.g., Denton and Hughes, 1981); 5) calculate the response of ice sheets to changing environmental conditions (e.g., Budd and Young, 1983; Paterson and Waddington, 1986); and 6) derive climatic history from temperature profiles (e.g., Johnsen, 1977b; Budd and Young, 1983; Dahl-Jensen and Johnsen, 1986).

Here the use of modelling as a means of estimating past ice sheet surface elevations and flow patterns will be mentioned. With inferred older ice marginal positions up to 200 km in advance of the present ice edge in West Greenland during the Late Wisconsinan, Dansgaard et al. (1973) estimated that the elevation of the ice sheet surface in central Greenland must have been about 500 m higher at that time; this estimate was based on two dimensional ice sheet profile theory. Reeh (1984), applying three-dimensional ice sheet modelling, and considering the effects of the irregular basal topography and isostatic depression of the ice sheet base, found the elevation increase to be only 200 m for the same advanced ice marginal positions. This modelling, however, did not consider the effect of the changed climatic conditions in the Late Wisconsinan, that is, decreased accumulation rates and changed temperatures within the ice sheet nor did it account for the fact that the ice sheet in the Late Wisconsinan was probably composed of ice, considerably softer than present-day ice. Reeh (1985c) discussed the influence of these phenomena on ice sheet elevations, concluding that in the Late Wisconsinan, the Greenland Ice Sheet was likely to be just as thin or even thinner than now in spite of its greater geographic extent.

Efforts have also been made to follow in more detail the time response of the ice sheet to changing climate. Jenssen (1977) used a three-dimensional ice sheet model to calculate the future changes in surface elevation and dynamics of the Greenland Ice Sheet. Reeh (1983) used a 2600 year climatic record derived from ice cores to generate an equally long ablation rate record for central West Greenland. This was then used as a forcing function in a model for calculating the response during the last 1000 years of a sector of the West Greenland Ice Sheet margin. The calculated response generally agrees with observed and estimated ice margin fluctuations (Fig. 14.17). The model was also used to predict the ice marginal response for the future decades, and it was concluded that generally the present recession of the ice margin

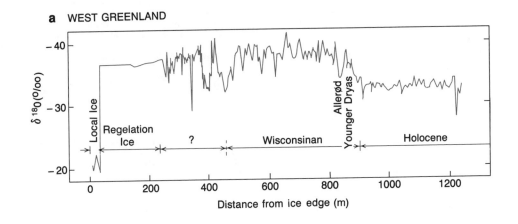

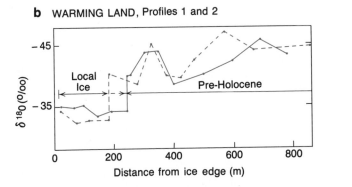

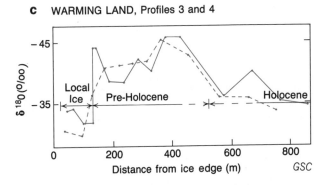

Figure 14.16. Oxygen isotope profiles from the outermost part of the Greenland Ice Sheet margin (after Reeh et al., 1987): (a) Jakobshavn area in West Greenland (69°25'N, 50°15'W); (b) and (c) Warming Land, North Greenland (81°10'N, 52°W).

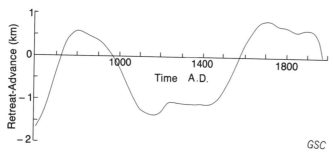

Figure 14.17 Advance and retreat of the front of a glacier lobe from the Greenland Ice Sheet (69°20'N, 50°20'W) in response to ablation variations generated by means of a 2600 year $\delta^{18}O$ record from Dye 3; deviations are from the 1975 ice margin position (modified from Reeh, 1983).

will continue for at least some decades, unless accumulation and/or ablation rates dramatically change in the near future.

THE FUTURE OF THE GREENLAND ICE SHEET

As stated by Weidick (1975), the present situation of glaciation limits compared to the unglaciated (uplifted) relief of Greenland indicates that the ice sheet would not develop under the present climatic conditions. It must be regarded as a glacial relict, owing its present existence to an ice sheet/climate feedback mechanism. A total deglaciation of Greenland in interglacial times has been suggested; however, neither glacial geology nor ice core investigations have proved or disproved this hypothesis. In both of the two Greenland deep ice cores, ice layers close to the bottom which have $\delta^{18}O$ values 3-5‰ higher than present-day surface values (Fig. 14.9b), have been interpreted as ice from the last interglaciation (Sangamon-Eem; Dansgaard et al., 1971, 1985b). In the Camp Century core this isotopically "warm" ice is underlaid by an isotopically "cold" layer which might be interpreted as evidence that the Greenland Ice Sheet survived in northwestern Greenland during the last interglaciation (Weidick, 1975). The Dye 3 core, on the other hand, terminates at the bottom with isotopically warm ice. The $\delta^{18}O$ level of this ice corresponds to the level one would expect if the ice originated from local glaciers that developed at the beginning of the last glaciation in the up to 1500 m-high mountains that would result from uplift of the present subglacier landscape in the Dye 3 area in a deglaciated southern Greenland.

Speculations have also been made on the possible future behaviour of the Greenland Ice Sheet; in particular, the effect of a CO_2 induced climatic change has attracted attention. Bindschadler (1985), for example, estimated that if the present predictions of the magnitude of the CO_2 induced climatic warming hold true, then the mass loss from the Greenland Ice Sheet is likely to increase to between 2 and 4 times the present mass loss. This would make the ice sheet disappear in 2-5 ka. In this estimate, however, a likely effect of climate warming, that is, increased precipitation over the ice sheet, and a likely increase of the fraction of the meltwater that refreezes on the ice sheet, was not considered. Both effects will decrease the rate of disintegration of the ice sheet.

ACKNOWLEDGMENTS

Much of this report was written while I worked at the Geophysical Institute, University of Copenhagen under contract CLI.067 of the European Economic Communities, XII Directorate General. The results on The Geological Survey of Greenland material are published with permission of the Director of the Geological Survey of Greenland. I would like to thank Sven Funder (Geological Museum, University of Copenhagen) for many encouraging discussions during the preparation of the manuscript. I greatly appreciate the critical reviews of a draft of the manuscript by Anker Weidick (Geological Survey of Greenland), H.B. Clausen and C.U. Hammer (Geophysical Institute, Univeristy of Copenhagen), and R.M. Koerner (Geological Survey of Canada).

REFERENCES

Ahlmann, H.W.
1941: Studies in north-east Greenland 1939-1940; Geografiska annaler, v. 23, p. 145-209.

Ambach, W.
1963: Untersuchungen zum Energieumsatz in der Ablations-zone des Grönländischen Inlandeises (Camp IV EGIG, 69°40'05"N, 49°37'58"W); Meddelelser om Grønland, v. 174, no. 4, 311 p.
1972: Zur Schätzung der Eis-ablation im Randgebiet des Grönländischen Inlandeises; Polarforschung, v. 42, no. 1, p. 18-23.
1979: Zur Nettoeisablation in einem Höhenprofil am Grönländischen Inlandsei; Polarforschung, v. 49, no. 11, p. 55-60.

Bader, H.
1961: The Greenland ice sheet; U.S. Army, Corps of Engineers, Cold Regions Research and Engineering Laboratory, Report 1-B2, 18 p.

Bauer, A., Ambach, A., et Schimpp, O.
1968a: Mouvement et variation d'altitude de la zone d'ablation ouest de l'Inlandsis du Groenland entre 1948 et 1959; Meddelelser om Grønland, vol. 174, n° 1, 79 p.

Bauer, A., Baussart, M., Carbonnell, M., Kasser, P., Perroud, P., et Renaud, A.,
1968b: Missions aériennes de reconnaissance au Groenland 1957-1958. Observations aériennes, détermination des vitesses des glaciers vêlant dans Diske Bugt et Umanak Fjord; Meddelelser om Grønland, vol. 173, n° 3, 116 p.

Beer, J., Andree, M., Oeschger, H., Stauffer, B., Balzer, R., Bonani, G., Stoller, C., Suter, M., Wolfli, W., and Finkel, R.C.
1985: ^{10}Be variations in polar ice cores; American Geophysical Union, Geophysical Monograph 33, p. 66-70.

Bennett, C.L., Beukens, R.P., Clover, H.R., Gove, H.E., Liebert, R.B., Litherland, A.E., Purser, K.H., and Sondheim, W.E.
1978: Radiocarbon dating with electrostatic accelerators; dating of milligram samples; Science, v. 201, no. 4343, p. 345-347.

Benson, C.
1962: Stratigraphic studies in the snow and firn of the Greenland ice sheet; U.S. Army, Corps of Engineers, Cold Regions Research and Engineering Laboratory, Research Report 70, 93 p.

Bindschadler, R.A.
1985: Contribution of the Greenland ice cap to changing sea level: present and future; in Glaciers, Ice Sheets, and Sea Level: Effects of a CO_2-Induced Climatic Change; National Academy Press, Washington, D.C. p. 258-266.

Blinkenberg, H.
1952: Vejrforholdene over de grønlandske kystområder; Beretninger vedrørende Grønland, v. 2, 220 p.

Braithwaite, R.J.
1980: Regional modelling of ablation in West Greenland; Grønlands Geologiske Undersøgelse, Report no. 98, 20 p.
1983: Glaciological investigations at Qamanârssûp sermia; Interim report 1982 and Appendix tables; Grønlands Geologiske Undersøgelse, Gletsher-hydrologiske Meddelelser, no. 8314, 34 p.

Braithwaite, R.J. and Thomsen, H.H.
1984: Runoff conditions at Paakitsup Akuliarusersua, Jakobshavn estimated by modelling; Grønlands Geologiske Undersøgelse, Gletscher-hydrologiske Meddelelser, no. 84/3, 22 p.

Brockamp, B., Sorge, E., and Wolcken, K.
1933: Seismik; in Wissenschaftliche Ergebnisse der Deutschen Grönland-Expedition Alfred Wegener 1929 und 1930/31, K. Wegener (ed.); F.A.Brockhaus, Leipzig, 160 p.

Brooks, R.L., Campbell, W.J., Ramseier, R.O., Stanley, H.R., and Zwally, H.J.
1978: Ice sheet topography by satellite altimetry; Nature, v. 274, p. 539-543.

Budd, W.F. and Young, N.W.
1983: Application of modelling techniques to measured profiles of temperatures and isotopes; in The Climatic Record in Polar Ice Sheets, G. de Q. Robin (ed.); Cambridge University Press, p. 150-177.

Budd, W.F., Jacka, T.H., Jenssen, D., Radok, U., and Young, N.W.
1982: Derived physical characteristics of the Greenland ice sheet; The University of Melbourne, Meteorological Department, Australia, Publication no. 23, 103 p.

Bull, C.
1956: Seismic investigations on the northern part of the Greenland ice sheet; in British North Greenland expedition 1952-4: Scientific results; Geographical Journal, v. 122, p. 219-225.
1958: Snow accumulation in North Greenland; Journal of Glaciology, v. 3, no. 24, p. 237-248.

Carbonnell, M. et Bauer, A.
1968: Exploration des couvertures photographiques aériennes répétées du front vêlant dans Disko Bugt et Umanak Fjord Juin-Juillet 1964; Meddelelser om Grønland, v. 173, no. 5, 68 p.

Carlson, W.S.
1941: Report of the northern division of the Fourth University of Michigan Greenland Expedition 1930-31; University of Michigan Studies, Scientific Series, no. 6, p. 61-156.

Clausen, H.B.
1973: Dating of polar ice by ^{32}Si; Journal of Glaciology, v. 12, p. 411-416.

Clement, P.
1983: Glaciologiske undersøgelser i Johan Dahl Land 1982; Grønlands Geologiske Undersøgelse, Gletscher-hydrologiske Meddelelser, no. 83/1, 56 p.
1984: Glaciological activities in Johan Dahl Land area, South Greenland, as a basis for mapping hydropower potential; Grønlands Geologiske Undersøgelse, Report no. 120, p. 113-121.

Coachman, L.K., Hemmingsen, E., Scholander, P.F., Enns, T., and de Vries, H.
1958: Gases in glaciers; Science, v. 127, p. 1288-89.

Colbeck, S.C.
1974: A study of glacier flow for an open-pit mine: An exercise in applied glaciology; Journal of Glaciology, v. 13, no. 64, p. 401-414.

Colbeck, S.C. and Gow, A.J.
1979: The margin of the Greenland ice sheet at Isua; Journal of Glaciology, v. 24, no. 90, p. 155-165.

Craig, H. and Chou, C.C.
1982: Methane: the record in polar ice cores; Geophysical Research Letters, no. 9, p. 1221-1224.

Craigin, J.H., Herron, M.M., Langway, C.C., Jr., and Klouda, G.A.
1977: Interhemispheric comparison of the changes in the composition of atmospheric precipitation during the late Cenozoic era; in Polar Oceans, M.J. Dunbar (ed.); Arctic Institute of North America, Calgary, Alberta, p. 617-631.

Dahl-Jensen, D.
1985: Determination of the flow properties at Dye 3, South Greenland, by bore-hole-tilting measurements and perturbation modelling; Journal of Glaciology, v. 31, p. 92-98.

Dahl-Jensen, D. and Johnsen, S.J.
1986: Palaeotemperatures still exist in the Greenland ice sheet; Nature, v. 320, p. 250-252.

Dansgaard, W.
1961: The isotopic composition of natural waters; Meddelelser om Grønland, v. 165, 120 p.
1964: Stable isotopes in precipitation; Tellus, v. 16, no. 4, p. 436-468.
1981: Ice core studies: dating the past to find the future; Nature, v. 290, p. 360-361.

Dansgaard, W. and Johnsen, S.J.
1969: A flow model and a time scale for the ice core from Camp Century, Greenland; Journal of Glaciology, v. 8, no. 53, p. 215-223.

Dansgaard, W., Clausen, H.B., Dahl-Jensen, D., and Gundestrup, N.
1985a: Climatic history from ice core studies in Greenland. Data correction procedures; Proceedings Symposium on EEC Climatology Programme.

Dansgaard, W., Clausen, H.B., Gundestrup, N., Hammer, C.U., Johnsen, S.J., Kristinsdottir, P.M., and Reeh, N.
1982: A new Greenland deep ice core; Science, v. 218, p. 1273-1277.

Dansgaard, W., Clausen, H.B., Gundestrup, N., Johnsen, S.J., Rygner, C.
1985b: Dating and climatic interpretation of two deep Greenland ice cores; American Geophysical Union, Geophysical Monograph 33, p. 71-76.

Dansgaard, W., Johnsen, S.J., Clausen, H.B., Dahl-Jensen, D., Gundestrup, N., Hammer, C.U., and Oeschger, H.
1984: North Atlantic climatic oscillations revealed by deep Greenland ice cores; in Climate Processes and Climate Sensitivity; American Geophysical Union, Geophysical Monograph 29, p. 288-298.

Dansgaard, W., Johnsen, S.J., Clausen, H.B., and Gundestrup, N.
1973: Stable isotope glaciology; Meddelelser om Grønland, v. 197, no. 2, 53 p.

Dansgaard, W., Johnsen, S.J., Clausen, H.B., and Langway, C.C., Jr.
1971: Climatic record revealed by the Camp Century ice core; in The Late Cenozoic Glacial Ages, K.K. Turekian (ed.); Yale University Press, New Haven, Connecticut, p. 37-56.

Dansgaard, W., Johnsen, S.J., Møller, J., and Langway, C.C.
1969: One thousand centuries of climatic record from Camp Century on the Greenland ice sheet; Science, v. 166, p. 377-381.

Dansgaard, W., Johnsen, S.J., Reeh, N., Gundestup, N., Clausen, H.B., and Hammer, C.U.
1975: Climatic changes, norsemen and modern man; Nature, v. 255, p. 24-28.

Denton, G.H. and Hughes, T.J. (editors)
1981: The Last Great Ice Sheets; John Wiley and Sons, New York, 484 p.

Diamond, M.
1960: Air temperature and precipitation on the Greenland ice sheet; Journal of Glaciology, v. 3, p. 558-567.

DPP, NSF
1983: Workshop on the Jakobshavn Glacier (Greenland), Northwestern University, Evanston, February 8-10, 1983; Division of Polar Programs, National Science Foundation, 92 p.

Drew, A.R. and Whillans, I.M.
1984: Measurement of surface deformation of the Greenland ice sheet by satellite tracking; Annals of Glaciology, v. 5, p. 51-55.

Drygalski, E. von
1897: Grönland Expedition der Gesellschaft für Erdkunde zu Berlin 1891-1893; W.H. Kuhl 1, Berlin, p. 1-556.

Epstein, S. and Sharp, R.P.
1959: Oxygen isotope studies; American Geophysical Union, Transactions, v. 40, p. 81-84.

Fisher, D.A. and Alt, B.T.
1985: A global oxygen isotope model - semi empirical, zonally averaged; Annals of Glaciology, v. 7, p. 117-124.

Fisher, D.A., Koerner, R.M., Paterson, W. S. B., Dansgaard, W., Gundestrup, N., and Reeh, N.
1983: Effect of wind scouring on climatic records from ice-core oxygen-isotope profiles; Nature, v. 301, p. 205-209.

Fisher, D.A., Reeh, N., and Clausen, H.B.
1985: Stratigraphic noise in time series derived from ice cores; Annals of Glaciology, v. 7, p. 76-83.

Fristrup, B.
1966: The Greenland ice cap; Rhodos International Science, Copenhagen, 312 p.

Gerke, K.
1969: Höhenbestimmungen auf dem Grönländischen Inlandeis bei der internationalen glaziologischen Grönland Expedition (EGIG) 1959; Meddelelser om Grønland, v. 173, no. 8, 77 p.

Gow, A.J.
1968: Deep core studies of the accumulation and densification of snow at Byrd Station and Little America V, Antarctica; U.S. Army, Corps of Engineers, Cold Regions Research and Engineering Laboratory, Research Report 197.

Griffiths, T.M.
1960: Glaciological investigations in the TUTO area of Greenland; U.S. Army, Corps of Engineers, Cold Regions Research and Engineering Laboratory, Technical Report 47, 63 p.

Gudmandsen, P.
1976: Studies of ice by means of radio echo soundings; Technical University of Denmark, Laboratory of Electromagnetic Theory, R 162, Lyngby, Denmark.
1978: Application of space techniques in solid earth physics; in The conference on application of space techniques in navigation, geodesy, oceanography and solid earth physics, P.E. Gudmandsen, E. Kejlsø, and C.C. Tscherning (ed.); Den danske Nationalkomite for den Internationale Union for Geodaesi og Geofysik, p. 84-94.

Gundestrup, N.S. and Lyle Hansen, B.
1984: Borehole survey at Dye 3, South Greenland; Journal of Glaciology, v. 30, no. 106, p. 282-288.

Hamilton, W.L. and Langway, C.C., Jr.
1968: A correlation of microparticle concentrations with oxygen isotope ratios in 700-year old Greenland ice; Earth and Planetary Science Letters, v. 3, p. 363-366

Hammer, C.U.
1977a: Past volcanism revealed by Greenland ice sheet impurities; Nature, v. 270, p. 482-486.
1977b: Dating of Greenland ice cores by microparticle concentration analyses; IAHS-AISH, Publication no. 118, p. 297-300.
1984: Traces of Icelandic eruptions in the Greenland ice sheet; Jökull, no. 34, p. 51-65.
1985: The influence on atmospheric composition of volcanic eruptions as derived from ice-core analysis; Annals of Glaciology, v. 7, p. 125-129.

Hammer, C.U., Clausen, H.B., and Dansgaard, W.
1980: Greenland ice sheet evidence of post-glacial volcanism and its climatic impact; Nature, v. 288, p. 230-235.
1981: Past volcanism and climate revealed by Greenland ice cores; Journal of Volcanology and Geothermal Research, v. 11, p. 3-10.

Hammer, C.U., Clausen, H.B., Dansgaard, W., Gundestrup, N., Johnsen, S.J., and Reeh, N.
1978: Dating of Greenland ice cores by flow models, isotopes, volcanic debris and continental dust; Journal of Glaciology, v. 20, no. 82, p. 3-26.

Hammer, C.U., Clausen, H.B., Dansgaard, W., Neftel, A., Kristinsdottir, P., and Johnsen, E.
1985: Continuous impurity analysis along the Dye 3 deep core; American Geophysical Union, Geophysical Monograph 33, p. 90-94.

Hammer, C.U., Clausen, H.B., Tauber, H.
1986: Ice core dating of the Pleistocene/Holocene boundary applied to a calibration of the ^{14}C time scale; Radiocarbon, v. 28, no. 2A, p. 284-291.

Hammer, R.R.I.
1883: Undersøgelser ved Jakobshavn Isfjord og naermeste omegn i Vinteren 1879-80; Meddelelser om Grønland, v. 4, no. 1, p. 1-67.

Hansen, B.L. and Langway, C.C., Jr.
1966: Deep core drilling in ice and ice core analysis at Camp Century, Greenland, 1961-66; Antarctic Journal, v. 1, p. 207-208.

Helland, A.
1876: Om de isfyldte Fjorde og de glaciale Dannelser i Nordgrønland; Archiv for Matematik og Naturvidenskab, v. 1, p. 58-125.

Herron, M. and Langway, C.C., Jr.
1980: Firn densification: an empirical model; Journal of Glaciology, v. 25, no. 93, p. 373-385.
1985: Chloride, nitrate and sulfate in the Dye 3 and Camp Century, Greenland ice cores; American Geophysical Union, Geophysical Monograph 33, p. 77-84.

Herron, M., Herron, S.L., and Langway, C.C.
1981: The climatic signal of ice melt features in southern Greenland; Nature, v. 293, p. 389-391.

Herron, S. and Langway, C.C., Jr.
1979: The debris-laden ice at the bottom of the Greenland ice sheet; Journal of Glaciology, v. 23, no. 89, p. 193-207.

Herron, S.L., Langway, C.C., Jr., and Brugger, K.A.
1985: Ultrasonic velocities and chrystalline anisotropy in the ice from Dye 3, Greenland; American Geophysical Union, Geophysical Monograph 33, p. 23-31.

Hibler, W.D. and Langway, C.C., Jr.
1977: Ice core stratigraphy as a climatic indicator; in Polar Oceans, M.J. Dunbar (ed.); Arctic Institute of North America, Calgary, p. 589-601.

Hofmann, W.
1964: Die Geodätische Lagemessung über das Grönländischen Indlandeis der Internationalen Glaziologischen Grönland-Expedition (EGIG) 1959; Meddelelser om Grønland, v. 173, no. 6, 145 p.
1974: Die internationale glaziologische Grönland-Expedition (EGIG). 2. Die geodätische Lagemessung. Eisbewegung 1959-1967 in den EGIG-profilen; Zeitschrift fur Gletscherkunde und Glazialgeologie, Bd. X, p. 217-224.

Holtzscherer, J.J. and Bauer, A.
1954: Contribution à la connaissance de l'Inlandsis du Groenland; IAHS-AISH Publication no. 39, p. 244-296.

Jenssen, D.
1977: A three-dimensional polar ice-sheet model; Journal of Glaciology, v. 18, no. 80, p. 373-389.

Jezek, K.C., Roeloffs, E.A., and Grieschar, L.
1985: A geophysical survey of subglacial geology around the deep drilling site at Dye 3, Greenland; American Geophysical Union, Geophysical Monograph 33, p. 105-110.

Johnsen, S.J.
1977a: Stable isotope homogenization of polar firn and ice; IAHS-AISH Publication no. 118, p. 210-219.
1977b: Stable isotope profiles compared with temperature profile in firn with historical temperature records; IAHS-AISH Publication no. 118, p. 388-392.

Johnsen, S.J., Dansgaard, W., Clausen, H.B., and Langway, C.C., Jr.
1972: Oxygen isotope profiles through the Antarctic and Greenland ice sheets; Nature, v. 235, p. 429-434.

Johnsen, S.J., Dansgaard, W., Gundestrup, N., Hansen, S.B., Nielsen, J.O., and Reeh, N.
1980: A fast lightweight core drill; Journal of Glaciology, v. 25, p. 169-174.

Joset, M.A. and Holtzscherer, J.J.
1954: Détermination des épaisseurs de l'Inlandsis du Groenland; Rapport Scientifique des Expéditions Polaires Françaises N III 2, Annales de Géophysique, v. 10, no. 4, p. 351-381.

Joussaume, S., Sadourny, R., and Jouzel, J.
1984: A general circulation model of water isotope cycles in the atmosphere; Nature, v. 311, p. 24-29.

Junge, C.E.
1977: Processes responsible for the trace content in precipitation; IAHS-AISH Publication no. 118, p. 63-77.

Koch, J.P. and Wegener, A.
1911: Die glaziologischen Beobachtungen der Danmark-Expedition; Meddelelser om Grønland, v. 46, no. 1, 77 p.
1930: Wissenschaftliche Ergebnisse der Dänischen Expedition nach Dronning Louises-Land und quer über das Inlandeis von Nordgrönland (1912-13 unter Leitung von Hauptmann J.P. Koch); Meddelelser om Grønland, v. 75, 676 p.

Koerner, R. M.
1989: Queen Elizabeth Islands glaciers; in Chapter 6 of Quaternary Geology of Canada and Greenland, R.J. Fulton (ed.); Geological Survey of Canada, Geology of Canada, no. 1 (also Geological Society of America, The Geology of North America, v. K-1).

Kollmeyer, R.C.
1980: West Greenland outlet glaciers: an inventory of the major iceberg producers; Cold Regions Science and Technology, v. 1, p. 175-180.

Langway, C.C., Jr.
1970: Stratigraphical analysis of a deep core from Greenland; Geological Society of America, Special Paper 125, 186 p.

Langway, C.C., Jr., Klouda, G.A., Herron, M.M., and Craigin, J.H.
1977: Seasonal variations of chemical constituents in annual layers of Greenland deep ice deposits; IAHS-AISH Publication no. 118, p. 302-306.

Langway, C.C., Jr., Oeschger, H., Dansgaard, W.
1985: The Greenland Ice Sheet Program in perspective; American Geophysical Union, Geophysical Monograph 33, p. 1-8.

Lawson, M.P., Kuivinen, K.C., and Balling, R.C.
1982: Analysis of the climatic signal in the South Dome, Greenland ice core; Climatic Change, v. 4, p. 375-384.

Lingle, C.S., Hughes, T.J., and Kollmeyer, R.C.
1981: Tidal flexure of Jakobshavn Glacier, West Greenland; Journal of Geophysical Research, v. 86, no. B5, p. 3960-3968.

Loewe, F.
1933: Die Schneepegelbeobachtungen; in Wissenschaftliche Ergebnisse der Deutschen Grönland-Expedition Alfred Wegener 1929 und 1930/31, vol. 1, K. Wegener (ed.); F.A. Brockhaus, Leipzig, p. 153-161.
1970: Screen temperatures and 10 m temperatures; Journal of Glaciology, v. 9, p. 263-628.

Lysgaard, L.
1969: Foreløbig oversigt over Grønlands klima i perioderne 1921-50, 1951-60, og 1961-65; Danske Meteorologiske Institut, Meddelelser no. 21, 35 p.

Mälzer, H.
1964: Das Nivellement über das Grönländische Inlandeis; Meddelelser om Grønland, v. 173, no. 7, 122 p.

Maurette, M., Hammer, C.U., Brownlee, D.E., Reeh, N., and Thomsen, H.H.,
1986: Placers of cosmic dust in the blue lakes of Greenland; Science, v. 233, p. 869-872.

Mellor, M.
1968: The Greenland mass balance. Flux divergence considerations; AIHS-IASH Publication no. 79, p. 275-281.

Mock, S.J.
1967: Calculated patterns of accumulation on the Greenland ice sheet; Journal of Glaciology, v. 6, p. 795-803.
1976: Geodetic positions of borehole sites of the Greenland Ice Sheet Program; U.S. Army, Corps of Engineers, Cold Regions Research and Engineering Laboratory, Report 76-41, 7 p.

Mock, S.J. and Ragle, R.H.
1963: Elevations on the ice sheet of southern Greenland; U.S. Army, Corps of Engineers, Cold Regions Research and Engineering Laboratory, Technical Report 124, 9 p.

Mock, S.J. and Weeks, W.F.
1965: The distribution of ten-meter snow temperatures on the Greenland ice sheet; U.S. Army, Corps of Engineers, Cold Regions Research and Engineering Laboratory, Research Report 170, 44 p.

Murozumi, M., Chow, T.J., and Patterson, C.
1969: Chemical concentrations of pollutant lead aerosols, terrestrial dusts and sea salts in Greenland and Antarctic snow strata; Geochimica et Cosmochimica Acta, v. 33, p. 1247-1294.

Nansen, F.
1890: The First Crossing of Greenland; Longmans, Green, and Co., London.

Neftel, A., Oeschger, H., Schwander, J., Stauffer, B., and Zumbrunn, R.
1982: Ice core sample measurements give atmospheric CO_2 content during the past 40,000 yr; Nature, v. 295, p. 220-223.

Nobles, L.H.
1960: Glaciological investigations, Nunatarssuaq ice ramp, northwestern Greenland; U.S. Army, Corps of Engineers, Cold Regions Research and Engineering Laboratory, Technical Report 66, 57 p.

Oeschger, H.
1985: North Atlantic deep water formation: information from ice cores; NASA Conference Publication 2367, NASA Goddard Space Flight Center, p. 23-27.

Oeschger, H., Stauffer, B., Bucher, P., and Moell, M.
1976: Extraction of trace components from large quantities of ice in bore holes; Journal of Glaciology, v. 17, p. 117-128.

Olesen, O.B. and Reeh, N.
1969: Preliminary report on glacier observations in Nordvestfjord, East Greenland; Grønlands Geologiske Undersøgelse, Rapport no. 21, p. 41-53.

Overgaard, S. and Gundestrup, N.S.
1985: Bedrock topography of the Greenland ice sheet in the Dye 3 area; American Geophysical Union, Geophysical Monograph 33, p. 49-56.

Paterson, W.S.B.
1955: Altitudes on the Inland Ice in North Greenland; Meddelelser om Grønland, v. 137, no. 1, 12 p.
1976: Vertical strain-rate measurements in an arctic ice cap and deductions from them; Journal of Glaciology, v. 17, p. 3-12.
1981: Physics of Glaciers; Pergamon Press, Oxford, New York, Toronto, 380 p.

Paterson, W.S.B. and Budd, W.F.
1982: Flow parameters for ice sheet modelling; Cold Regions Science and Technology, no. 6, p. 175-177.

Paterson. W.S.B. and Waddington, E.D.
1984: Past precipitation rates derived from ice core measurements: methods and data analysis; Reviews of Geophysics and Space Physics, v. 22, no. 2, p. 123-130.
1986: Estimated basal ice temperatures at Crête, Greenland, throughout a glacial cycle; Cold Regions Science and Technology, no. 12, p. 99-102.

Picciotto, E.E. and Wilgain, S.E.
1963: Fission products in Antarctic snow: A reference level for measuring accumulation; Journal of Geophysical Research, v. 68, p. 5965-5972.

Putnins, P.
1970: The Climate of Greenland; in Climates of the Polar Region, S. Orvig (ed.); Elsevier Publishing Company, Amsterdam, p. 3-128.

Quervain, A. de et Mercanton, P.L.
1925: Résultats scientifiques de l'expédition Suisse au Groenland 1912-13; Meddelelser om Grønland, v. 59, no. 5, p. 55-271.

Quervain, M., Brandenberger, F., Renaud, A., Roch, A., and Schneider, R.
1969: Schneekundliche Arbeiten der Internationale Glaziologische Grönland Expedition (Nivologi); Meddelelser om Grønland, v. 177, no. 4, 283 p.

Radok, U., Barry, R.G., Jenssen, D., Keen, R.A., Kiladis, G.N., and McInnes, B.
1982: Climatic and physical characteristics of the Greenland ice sheet; CIRES, University of Colorado, Boulder, 193 p.

Rasmussen, R.A. and Khalil, M.A.K.
1984: Atmospheric methan (CH^4): trends and seasonal cycles; Journal of Geophysical Research, v. 86, p. 9826-9832.

Raynaud, D.
1976: Les inclusions gazeuses dans la glace de glacier; Laboratoire de Glaciologie du Centre National de la Recherche Scientifique, Grenoble, Publication 214.

Raynaud, D. et Delmas, R.
1977: Composition de gaz contenus dans la glace polaire; AIHS-IASH Publication 118, p. 377-381.

Raynaud, D. and Lorius, C.
1973: Climatic implications of total gas content in ice at Camp Century; Nature, v. 243, p. 283-284.

Reeh, N.
1983: Ikke-stationaer beregningsmodel for Indlandsisens randzone; Grønlands Geologiske Undersøgelse, Gletscher-hydrologiske Meddelelser, nr. 83/7, 81 p.
1984: Reconstruction of the glacial ice covers of Greenland and the Canadian Arctic Islands by three-dimensional, perfectly plastic ice-sheet modelling; Annals of Glaciology, v. 5, p. 115-121.
1985a: Greenland ice-sheet mass balance and sea level change; in Glaciers, Ice Sheets and Sea Level: Effects of a CO_2-Induced Climatic Change; National Academy Press, Washington, D.C., p. 155-171.
1985b: Greenland ice sheet mass balance; NASA Conference Publication 2367, NASA Goddard Space Flight Center, p. 45-49.
1985c: Was the Greenland ice sheet thinner in the late Wisconsinan than now?; Nature, v. 317, p. 797-799.

Reeh, N. and Gundestrup, N.S.
1985: Mass balance of the Greenland ice sheet at Dye 3; Journal of Glaciology, v. 31, p. 198-200.

Reeh, N. and Thomsen, H.H.
1986: Stable isotope studies on the Greenland ice-sheet margin; Grønlands Geologiske Undersøgelse, Rapport nr. 130, p. 108-114.

Reeh, N., Clausen, H.B., Dansgaard, W., Gundestrup, N., Hammer, C.U., and Johnsen, S.J.
1978: Secular trends of accumulation rate at three Greenland stations; Journal of Glaciology, v. 20, no. 82, p. 27-30.

Reeh, N., Johnsen, S.J., and Dahl-Jensen, D.
1985: Dating the Dye 3 deep ice core by flow model calculations; American Geophysical Union, Geophysical Monograph 33, p. 57-65.

Reeh, N., Thomsen, H.H., and Clausen, H.B.
1987: The Greenland ice-sheet margin — a mine of ice for paleo-environmental studies; Palaeogeography, Palaeoclimatology, Palaeoecology, no. 58, p. 229-234.

Rink, H.
1853: Om den geografiske Beskaffenhed af de danske Handelsdistrikter i Nordgrønland; Kongelige Danske Videnskabernes Selskab, Skrivelser, 5. Raekke, Naturvidenskabelige-matematiske Afdeling 3, p. 37-98.
1877: Om Indlandsisen og om Frembringelsen af de svømmende Isfjelde; Geografisk Tidsskrift, v. 1, p. 112-119.
1889: Nogle Bemaerkninger om Indlandsisen og Isfjeldenes Oprindelse; Meddelelser om Grønland, v. 8, p. 271-279.

Risbo, T., Clausen, H.B., and Rasmussen, K.L.
1981: Supernovae and nitrate in the Greenland ice sheet; Nature, v. 294, p. 637-639.

Robin, G. de Q.
1983: The climatic record in polar ice sheets; Cambridge University Press, Cambridge, 212 p.

Rotman, S.R., Fisher, A.D., and Staelin, D.H.
1982: Inversion for physical characteristics of snow using passive radiometric observations; Journal of Glaciology, v. 28, p. 179-185.

Ruddiman, W.F. and McIntyre, A.
1981: The North Atlantic Ocean during the last deglaciation; Palaeogeography, Palaeoclimatology, and Palaeoecology, v. 35, p. 145-214.

Rufli, H., Stauffer, B., and Oeschger, H.
1976: Lightweight 50 m core drill for firn and ice; in Proceedings, Symposium on ice-core drilling, J.F. Splettstoesser (ed.); University of Nebraska Press, Lincoln, Nebraska, p. 139-153.

Ryder, C.H.
1889: Undersøgelser af Grønlands Vestkyst fra 72° til 74° N. Br. 1886 og 1887; Meddelelser om Grønland, v. 8, p. 203-270.

Schytt, V.
1955: Glaciological investigations in the Thule Ramp area; U.S. Army, Corps of Engineers, Cold Regions Research and Engineering Laboratory; Technical Report 28, 88 p.

Seckel, H.
1977a: Das geometrische Nivellement über das Grönländische Inlandeis der Gruppe Nivellement A der Internationalen Glaziologischen Grönland Expedition 1967-68. Sommercampagne 1968; Meddelelser om Grønland, v. 187, no. 3, 86 p.
1977b: Höhenänderungen im Grönländischen Inlandeis zwischen 1959 und 1968; Meddelelser om Grønland, v. 187, no. 4, 58 p.

Shoji, H. and Langway, C.C., Jr.
1985: Mechanical properties of fresh ice core from Dye 3, Greenland; American Geophysical Union, Geophysical Monograph 33, p. 39-48.

Shumsky, P.A.
1965: Ob izmenenii massy lednikovogo pokrova v tsentra Grenlandii; Akademiya Nauk SSSR, Doklady, v. 162, no. 2, p. 320-322.

Sorge, E.
1933a: Universal — Dr. Fanck Grönland-Expedition 1932. Umiamako und Rink Gletscher; Deutsche Universal-Film A.G., Berlin, 24 p.
1933b: The scientific results of the Wegener expeditions to Greenland; Geographical Journal, v. 81. p. 333-344.

Stauffer, B. and Oeschger, H.
1979: Temperaturprofile in Bohrlochern am Rande des Grönländischen Inlandeises; Versuchsanstalt fur Wasserbau, Hydrologie und Glaziologie an der EHT Zürich, Mitteilungen Nr. 41, p. 301-313.

Stauffer, B., Hofer, H., Oeschger, H., Schwander, J., and Siegenthaler, U.
1984: Atmospheric CO_2 concentration during the last glaciation; Annals of Glaciology, v. 5, p. 160-164.

Stauffer, B., Neftel, A., Oeschger, H., and Schwander, J.
1985b: CO_2 concentration in air extracted from Greenland ice samples; American Geophysical Union, Geophysical Monograph 33, p. 85-89.

Stauffer, B., Schwander, J., and Oeschger, H.
1985a: Enclosure of air during metamorphosis of dry firn to ice; Annals of Glaciology, v. 7, p. 108-112.

Steenstrup, K.J.V.
1883: Bidrag til Kjendskab til Braererne og Brae-isen i Nord-Grønland; Meddelelser om Grønland, v. 4, 43 p.

Theodorsson, P.
1977: 40-years tritium profiles in a polar and a temperate glacier; IAHS-AISH Publication no. 118, p. 393-398.

Thomsen, H.H.
1983: Glaciologiske undersøgelser ved Pakitsup ilordlia 1982 Ilulissat/Jakobshavn; Grønlands Geologiske Undersøgelse, Gletscher-hydrologiske Meddelelser 83/3, 24 p.
1984: Glaciologiske undersøgelser i Disko Bugt om rådet 1983; Grønlands Geologiske Undersøgelse, Geltscher-hydrologiske Meddelelser, no. 84/1, 31 p.

Thompson, L.G.
1977: Variations in microparticle concentration, size distribution and elemental composition found in Camp Century, Greenland, and Byrd Station, Antarctica, deep ice cores; IAHS-AISH Publication no. 118, p. 351-364.

Thorarinsson, S.
1981: The application of tephrachronology in Iceland; in Proceedings of the NATO Advanced Study Institute "Tephra Studies as a tool in Quaternary Research"; S. Self and R.S.J. Sparks (ed.); Reidel Publishing Company, Dordrecht, p. 109-134.

Ueda, H.T. and Garfield, D.E.
1969: The USA CRREL drill for thermal coring in ice; Journal of Glaciology, v. 8, p. 311-314.

Weertman, J.
1968: Comparison between measured and theoretical temperature profiles of the Camp Century, Greenland, bore hole; Journal of Geophysical Research, v. 73, p. 2691-2700.

Wegener, K. (editor)
1933: Wissenschaftliche Ergebnisse der Deutschen Grönland-Expedition Alfred Wegener 1929 und 1930/31, Band I-III; F.A. Brockhaus, Leipzig.

Weidick, A.
1968: Observations on some Holocene glacier fluctuations in West Greenland; Meddelelser om Grønland, v. 165, no. 6, 202 p.
1975: A review of Quaternary investigations in Greenland; Institute of Polar Studies, The Ohio State University, Report no. 55, 161 p.
1976: Glaciation and the Quaternary of Greenland; in Geology of Greenland, A. Escher and W.S. Watt (ed.); The Geological Survey of Greenland, p. 429-458.
1985a: Review of glacier changes in West Greenland; Zeitschrift fur Gletscherkunde und Glazialgeologie, v. 21, p. 301-309.
1985b: The ice cover of Greenland; Grønlands Geologiske Undersøgelse, Gletscher-hydrologiske Meddelelser nr. 85/4, 8 p.

Weidick, A. and Olesen, O.
1978: Hydrologiske bassiner i Vestgrønland; Grønlands Geologiske Undersøgelse, København, 160 p.

Weidick, A. and Thomsen, H.H.
1983: Lokalgletschere og Indlandsisens rand i forbindelse med udnyttelse af vandkraft i bynaere bassiner; Grønlands Geologiske Undersøgelse, Gletscher-hydrologiske Meddelelser 83/2, 129 p.

Wolff, E.W. and Peel, D.A.
1985: The record of global pollution in polar snow and ice; Nature, v. 313, p. 535-540.

Zwally, H.J., Bindschalder, R.A., Brenner, A.C., Martin, T.V., and Thomas, R.H.
1983: Surface elevation contours of the Greenland and Antarctic ice sheets; Journal of Geophysical Research, v. 88, no. C3, p. 1589-1596.

Author's address

N. Reeh
Alfred Wegener Institute
Bremerhaben, West Germany

Printed in Canada

INDEX

Abernethy Lowland 195
Abitibi lowlands 518
Abitibi Upland 215
Acadian Bay Lobe 410,416
Acton, D.F. 669-671,673,675
Adam Till 233,236
Adelaide Peninsula 194,208
Adirondack Mountains 322,373
Advance
   Battle Mountain advance 74
   Canmore advance 66,68
   Crowfoot advance 74
   Drystone Creek advance 66
   Eisenhower Junction advance 66
   Labuma advance 152,164
   Late Portage Mountain advance 54,66
   Marquette ice advance 161,162,236,242,349
   Obed advance 66
   Two Rivers advance 349
   Waconichi ice advance 272
   Waterton I advance 154
   Waterton II advance 64
   Waterton IV-Hidden Creek advance 65,66
Agassiz Ice Cap 454,462,469,470,471,472
Age
   10 ka to 8.4 ka 208,240,295,556
   11 ka 207,295
   11 ka to 10 ka 208,240,299
   8.4 ka 209,297
   Cape Ball 488,491
   Early Pleistocene 119,152
   Early Wisconsinan 57,61,66,68,125-128,131,155,325,337
      344,345,368,410,415,421,423,426,753
   Holocene 52,54,55,70,72-74,76,108,136,146
      162,236,242,275,291,292,299,421,431,444,452,457
      459,472,508,512,518,550,552,594,750,756
   Illinoian 55,57,61,67,125,129,344,410,419,425
      507,748,753,761,784
   Irvingtonian land mammal age 152
   Kansan Age 152,164,504
   Late Pleistocene 154,550
   Late Wisconsinan 48,50,57,68,72,74,79,108,126,127
      129,131,132,136,157,185,201,204,233,234,275,291
      346,356,402,405,411,412,415,426,449,459,488,491
      494,497,554,750,767
   Little Ice Age 74,179,214,297,457,472,473,708,811
   Matuyama Reversal Epoch 69,109,119452
   Middle Pleistocene 152,154
   Middle Wisconsinan ... 48,70,136,146,155,337,338,344,346,369
      407,411,413,426,488,494,496,545,546,573
   Milankovich thermal maximum 136,525,546,767
   Miocene 119
   Pre-Late Wisconsinan 258
   Pre-Reid 68
   Pre-Sangamonian 49,66,410,498
   Pre-Wisconsinan 119,203,344
   Sangamon Interglaciation ... 121,125,126,152,202,204,226,230
      232,234,236,336,345,356,364,367,412
   Sangamonian 289,338,418,425,495,500,753,784
   Wisconsinan 155,201,289,668,709
   Yarmouthian Age 152,164
Age *see Stade,Interstade,Stage,Interval,Advance,Readvance
Aggregate resources -Quaternary 684-687
Aguanus-Kenamiou Moraine 269
Agutua Moraine 224,243

Aikins Moraine 160
Alaska 24,40,119,120
Alaskan Coastal Plain 129
Albany River 236
Alberta Plateau 140
Albertan Till 43,63,154
Aleutian Trench 26
Alexis River 263
Algoma Great Lakes 249
Algonquin Arch 326
Algonquin Highlands 373
Allan Park kame delta 343
Alle/Ile * see amino acid 235,286
Allegheny Basin 326
Allerod Interval 368
Alliston embayment 344
Amadjuak Lake 184,277
America Plate 25
Amery Till 228,231
Amherst Silt 412
Amino acid racemization 5
Amino acid stratigraphy -definition ... 235,286,544,545
Amund Ringnes Island 449
Amundsen Glaciation 121
Amundsen Gulf 102,127,128,132
Anahim Belt 25,49
Anderson Plain 131
Anderson River 102,128,132
Anderson, T.W. 483-486,498-500,507-508,520-528
Andrews, J.T. 178-189,235,276-301,543-562
Ange-Gardien Till 371
Annapolis River 411
Antarctica Ice Sheet 34,185,218
Antevs, E. 4
Antevs Lake 272
Anticosti Channel 420
Anticosti Island 268,395,403,420,422
Antifreeze Pond 494
Antigonish Highlands 411
Antigonish Ice Cap 411
Appalachian glacier complex 406,582
Appalachian Mountains 321,322
Appalachian Orogen 394,397
Apple River 411
Aquifers 713-723
Arctic Archipelago 119,136-186,445,464,584
Arctic Canada -soils 675-676
Arctic Coastal Plain -geology ... 101,108,120,129,445,582
Arctic Front -bioclimatic boundary 214
Arctic Mountains 21,26
Arctic Ocean 445,452,464
Arctic Platform 445,449
Arctic Red River 130
Argonaut Plain 66
Armstrong, J.E. 4
Arnott Moraine 244
Aspy Fault 426,427
Assiniboine Valley 157,161,162,240
Astrobleme -Charlevoix 351,379
Athabasca Basin 655
Athabasca drumlin field 195
Athabasca Lake 131
Athabasca River 42,43,65,68,692
Athabasca Till 65

823

INDEX

Atlantic Appalachian region
  bedrock geology ... 394,395
  economics ... 432,433
  New Brunswick ... 402
  Newfoundland ... 400
  physiography ... 394-400
  Quaternary history ... 405
  soils ... 674-675
  tectonostratigraphy ... 395
Atlantic region -chronology ... 500
Atlantic uplands ... 403,410,411
Atlin Placers ... 79
Atmospheric carbon dioxide ... 484
Aulitivik Island ... 291
Avalanche disasters ... 709,710
Avalon ice sheet ... 418
Avalon Isthmus ... 418
Avalon Peninsula ... 416,419
Axel Heiberg Islands ... 445,452,453,464,557
Aylen Phase ... 270
Ayr Lake Stade ... 286,288-290,298
Babine Lake ... 52
Back Lowland ... 191,209-210,223
Baffin area ... 178,186,202,210,276,278,279,283,285,298
547,553,557,558,744
Baffin Bay Strait ... 544
Baffin Sector ... 545
Baffinland Stade ... 286,289,290,292
Baie-Trinité Moraine ... 268
Baird Inlet Lake ... 526
Baker Lake ... 202,213
Baker Till ... 121
Ballycroy Bog ... 527
Bank Beauge ... 395
Banks Glaciation ... 119,452,459
Banks Island ... 100-102,104,108,126,127,134,136
Banks Sea ... 119
Barlow-Ojibway Formation ... 233
Barnes Ice Cap ... 184,276,278,279,287,288,297
Barrow Strait ... 191
Barrow Surface ... 191,278
Baseline Till ... 65
Basswood Road Lake ... 526
Bathurst Inlet ... 126,179,189,208,211
Bathurst Island ... 445,457
Bathurst Lowland ... 191,201
Bathurst Norsemines dispersal train ... 655
Bathurst Peninsula ... 108,121,126,128,136
Battle Mountain advance ... 74
Battleford Formation ... 160
Bay d'Espoir ... 400
Bay of Fundy ... 412
Bay of Islands ... 339,400,417,420
Bayfield Valley ... 339
Beams Plateau ... 191
Beauce region ... 326,351
Beauceville ... 694
Beaufort Formation ... 101,445,508
Beaufort Sea area -geomorphic processes ... 21,108,121,128,547
Beaver River Moraine ... 234
Beaver River Till ... 411
Beaverhouse Moraine ... 243
Bécancour Till ... 366,367
Bed load ... 601-602
Bedford Hills-Belair Moraine ... 239,240
Bednarski, J. ... 459-464
Belair Moraine ... 161
Belcher Island ... 185,186,215,237,273
Bell Flat ... 32,68,69,492,493
Bell, R. ... 4
Bell Sea ... 209,235,546
Belles-Amours Moraine ... 268
Belleville phase ... 373
Bengal-Bhutan Asia ... 487
Bering land bridge ... 498
Beringia ... 121,485,486
Bernard River ... 104,121
Bernard Till ... 105,116,119
Bessborough Stage ... 66
Bessette Sediments ... 51,57,70
Big Bald Mountain ... 413
Big Muddy spillway ... 160,161
Big River ... 104,121
Big Sea Sediments ... 121
Big Trout Lake ... 236
Bioclimatic boundary -Arctic Front ... 214
Biostratigraphy -Champlain Sea ... 188,322,349,375
Birch Mountain ... 141
Blanket aquifer ... 717
Bliss Bugt glaciation ... 748,767,784
Bloor Member ... 346
Blow Me Down Highlands ... 417
Blow River ... 125
Bluefish Flat ... 32,68,69,492
Bog -definition ... 679
Bois Blanc Formation ... 343
Bolduc Till ... 127,459
Bonavista Bay ... 399
Bonavista Peninsula ... 416
Bonnet Plume Depression ... 70
Boothia Peninsula ... 189,191,194,198,199,200,202,205,214
245,661
Boreal forest -history in Canada ... 508-512
Bostock, H.S. ... 4
Boulter esker ... 270
Boutellier Nonglacial Interval ... 61,72
Bovis, M.J. ... 583-594
Bow River ... 65,162,576
Bow Valley Till ... 68
Boyko-Diakonow, M. ... 528-530
Braddock Channel ... 160
Bradore Moraine ... 268
Bradtville Drift ... 346
Brampton esker ... 331,343
Bras d'Or ice mass ... 411
Bras d'Or Lakes ... 400,410
Braskeruds Plain ... 454
Breslau Moraine ... 341
Bridgewater Conglomerate ... 406,425
British Columbia
  continental shelf ... 42
  paleoecology ... 70-74
  Quaternary stratigraphy ... 48
Brock Upland ... 102,104,121,126,128,132
Brocket Till ... 154
Bronson Lake Formation ... 719
Brooks Range ... 119
Broughton Island ... 286,288
Bruce Peninsula ... 327
Brunhes-Matuyama Boundary ... 1,120,121,152,567
Brunisolic Turbic Cryosols ... 676
Brunisols ... 671,672
Bruun Rule Definition ... 432
Buchans ... 655
Buckland Glaciation ... 125,129,131,136
Buffalo Lake Till ... 65,68
Bugaboo Glacier ... 74
Bulkley River Valley ... 54
Burin Peninsula ... 398,402,416,418
Burntwood-Knife Interlobate Moraine ... 178,200,204,208,209
214,220,225,238,239,241,243
Byam Martin Island ... 134,445,449,459
Bylot Island ... 276
Cabot Strait ... 419,420,426

INDEX

Caledonian Highlands 413,415
Caledonian Orogeny 744
Cameron Hills 101,141
Cameron Ranch Formation 155
Campbell Phase 131
Canada -natural processes and agents 624-633
Canada -last glaciation 544
Canadian Interior Plains -economic considerations 165-166
Canadian Shield 175-317
Canadian Shield
   * see Northeastern, Northwestern, Southeastern,
   Southwestern Canadian Shield
Canadian Shield -soils 675
Canmore advance 66,68
Canning Till 346
Canyon Ranges 130
Cape Adair transgression 292
Cape Aston 290
Cape Ball 488,491
Cape Breton Highlands 406,410
Cape Breton Island 399,403,410,500,504,506
Cape Breton Island -paleostratigraphy 506
Cape Breton lowlands 411,416
Cape Collinson Interglaciation 119,121
Cape Cove Gravel 406,411
Cape Dalhousie 129
Cape Dyer 298
Cape Hatt 292
Cape Henrietta Maria 290
Cape Henry Kater 247
Cape Ray Fault 426,427
Cape Storm 554
Capilano Sediments 54
Carbon dioxide 484
Carey Islands 526
Caribou Dome 187
Caribou placers 79
Caribou Plateau 49
Caribou River 125,131,141,198
Carson, M.A. 583-594
Cartier-McConnell Moraines 243
Cascade Mountains 26,146
Cascapédia River 423
Case Histories -paleobotany 483-539
Cass Phase 240
Cassiar Mountain 26,40
Catfish Creek Till 324,341,346,347
Caverns 351
Central Arctic Ocean 119
Central Arctic Upland 278
Chagnon, J-V. 723-729
Chain Lake Silt 64
Chaleur Bay 398,403,415
Chalmers Bog 496,497,523
Chalmers, R. 4
Champlain Sea 240,270, 349,363,372,373,374-377,512,518
Chandler Fiord 462
Chantrey Inlet 196,201,209
Chantrey Moraine System 196,208,209,210
Chapleau Moraine 218
Chapman Pond 494
Charlevoix astrobleme 351,379
Chaudière Till 356,369,694
Chedabucto Fault 399,427
Chedabucto-Cobequid Fault 400
Chernozemic 671,672,673
Cherry River Moraine 372
Chesterfield Inlet 211
Chicago outlet 248,350
Chignecto Bay 411
Chignecto Isthmus 415
Chignecto phase 415

Chilcotin Plateau 49
Chippewyn Lake 204
Chiputneticook Lakes 416
Christopher Island Formation 198
Churchill River Till 204,225,237,238,247,256,263
Circum Ungava Geosyncline 185
Clague, J.J. 17-26,34-74,76-83
Clausen Glaciation 121,125,129
Clay and shale slopes -geomorphic processes 586
Clayhurst Stage 66
Clear Hills 141
Clearwater-Mackenzie drainage 161,162
Clements Markham Inlet 462,463
Climate
   Canada 577-581
   moisture 578
   temperature 577-578
Climate change- general circulation models 484,485
Clovis culture 498
Clyde Foreland Formation 262,286,288
Coast Belt 22
Coast Mountains 21,26,32,40,54,58
Coats Island 179,198,239
Cobalt Plain 215
Cobequid Highlands 406,411,412,426,427
Coburg Island 452
Cochrane Formation 208,218,232,234,236,253,265
Cochrane I & II surge 249,272
Cochrane readvance 187,209,210,236,239,244
Cockburn Substage 297
Codroy Valley 417
Coenocline -definition 512
Coffin Island 421
Cold Climate Processes -geomorphic processes 604-611
Cold Lake oil sands 719
Coleman, A.P. 4
Columbia Mountains 26,583
Columbia River 26,34
Colville Moraines 133
Condie Moraine 161
Conestogo Lake 331
Connaught sequence 224,236
Connaught Tunnel 704
Continental Shelf 34
Contwoyto basin 200
Coppermine River 131,133,200,201,208
Coquitlam Drift 54,57
Coquitlam Stade 57
Cordillera
   bedrock geology 22-25
   climate 26-31
   erosion 39
   economic considerations 76-83
   landscape 32-34
   natural hazards 80-83
   paleoecology and paleoclimatology 70-74
   physiography/drainage 26
   Quaternary deposits 34-38
   regional 17-22
   soils 671-672
   tectonic setting 25-26
Cordilleran Ice Sheet 39-42,562
   British Columbia 48-58
   glaciated fringe 63-70
   stratigraphy 48
   Yukon Territory 58-62
Cornwall Island 445
Cornwallis Island 201
Cornwallis River 411
Coronation Gulf 182,191,201,205
Coronation Gulf-Dolphin Strait 133
Corrugated moraine 149

INDEX

Coulées d'argile 587
Coulonge River 269
Cowan, W.R. 178-189,214-249
Cowichan Head Formation 50,52,55,57
Craton 22,394
Crawford Lake 340,485,520,528-530
Cree Lake Moraine 147,161,194,196,242
Crête de coq 379
Crowfoot advance 74
Crowsnest River 63,64
Cryogenic weathering -geomorphic processes 606,610
Cryosols 675,676
Cryoturbation -definition 610
Cumberland Lowlands 410
Cumberland Peninsula 276,283,290,298,301
Currie Township 236,655
Cypress Hills 138,141,144,146,161,495,713
D'Iberville Sea 274,512
Daly River 462
Danish Peary Land Expedition 764
Dark erratics 65,155,185,186,237
Darlingford Moraine 161
David, P.P. 620-623
Davis Strait 544
Dawson Creek 68,484
Dawson G.M. 3,4
Dawson J.W. 3
Day, T.J. 595-604
De Geer moraine 196,201,209,210,220,224,242,254,269,272
                                                     273,279,415
De Geer Sea 412,416,429,430
Debris flows 590
Deep Sea Drilling Project 783
Deer Lake 402,418
Deglaciation 157,239,264,265,497,552,576,618
Demoiselle Drift 421,422
Denali River Fault 25
Dendrochronology 484
Dreimanis, A. 6
Des Moines Lobe 239
Des Plantes River 694
Deserter's Canyon advance 54
Detachment slides -geomorphic processes 585
Devon Ice Cap 454
Devon Islands 445,452,454,464,465,469,471,472
Diamond Jenness Peninsula 105,128,133
DiLabio, R.N.W. 645-663
Dirt Hills Moraine 160
Disko Bugt 746,747,749,750,771,786
Disko Island 750,756
Dispersal trains 655
Dissection Creek 129
Dissolved solids -river processes 602
Ditchfield Moraine 372
Dixon Entrance 34,44
Dog Lake Moraine 236,242
Dolphin Strait 102,132
Don Formation 335,336,345
Donnacona Till 366
Dorset Ridge 395
Double Island 406
Dredge, L.A. 178-249
Drift prospecting 413,418,653-655
Drystone Creek advance 66
Du Lièvre River 269
Dubawnt Group 179,193,196,200,203,214,654
Duck Hawk Bluffs 115,116,121
Duck Mountains 140
Duke River Fault 25
Dummer Moraine 341
Dundas Peninsula 119,127
Dundas Till 450,459

Dundee Formation 329
Dune sands 339
Dunn Peak Moraine 74
Dunwich Drift 346
Duval Moraine 288
Dyer Glaciation 289
Dyke, A.S. 178,214
Dynamic metastable equilibrium -definition 594
Eagle Findlayson Moraine 240
Eagle River 70
Early Holocene 550
Early Pleistocene 1,119,152
Early Portage Mountain advance 54
Early Wisconsinan .. 2,57,61,66,68,125-128,131,155,325,337,344
                                345,368,410,415,421,423,426,753
Earthquakes 614
Earthquakes -Eastern Canada 724
East Coast Sea 127
East Greenland Polar Current 749,757
Eastend Formation 716
Eastern System 26
Eastern Townships 322,694
Eatonville Till 410
Echo Lake Gravels 146,154
Echoing River 231
Eclipse Glaciation 289,290
Eclipse Sound 276
Ecotone 512,521,526,528
Edson Till 65,68
Edwards, W.A.D. 684-687
Eemian 748
Egginton, P.A. 623
Eisenhower Junction advance 66
Ekalugad Moraine 298
Elk River Valley 68
Ellef Ringnes Islands 445,451
Ellesmere Island 445,453,454,457,462,465,553,767
Ellesmere-Baffin glacier complex 452
Ellice Bog 527
Elma Till 341,347
Elmira Moraine 331
Emergence curve 211
Emerson Phase 243
Empress Group 138,719
Endogenic processes -geomorphic processes 612-615
Engineering geology -Prairie region 713-723
England, J. 459-464
Ennadai Lake 214
Eolian landforms 334,379,451,620-622
Erie Interstade 347,371
Erie Lobe 329,340,341,347
Ernst Till 65,68
Erosion -animal agents 623
Erosion -human agents 625,628
Erratics Train Till 65
Escarpment -Ipperwash,Niagara,Ondaga 327,328,329,333
Escuminac Ice Centre 415,416
Esker
    Boulter esker 270
    Brampton esker 331,343
    Guelph esker 343
    Munro esker 220,224
    Norwood esker 331
    Thelon esker 208
Eskimo Lakes 125
Esquiman Channel 420
Eureka Sound 449,450,452
Eurekian Rifting Episode 191
Evans, S.G. 612-615,702-713
Evaporite -solution collapse 723
Evilsmelling Band 146,155
Explorer Plate 25

INDEX

F.E. Hyde Fjord 765
Fairfax Lake 497
Fall Zone peneplain 398
Favel Formation 723
Fawn River 231,233,234,235
Fen 679
Fenelon Falls outlet 349
Fennoscandian Ice Sheet 45
First Canyon Glaciation 63,66,119,125
Fisher River 202
Fiskebanke Glaciation 752,753,784
Five Islands Formation 412
Flakkerhuk Glaciation 758,761,763,772,785
Flat River Till 65
Flat River-Clausen Glaciation 65
Flaxman Member 129
Flint, R.F. 4
Flood hazards 723
Floods and debris flows 704,727
Floral Formation 146,154,714,720
Flowslides -geomorphic processes 587
Fluvial (alluvial) deposits 36,257-258
Fluvial erosion intensity 599
Fluvial sediment transfer 604
Foothills Erratics Train 43,65,68,144,155
Ford, D.C. 617-619
Foreland Belt 22
Forest Hill dispersal train 655
Formation
    Barlow-Ojibway Formation 127,129,233
    Battleford Formation 160
    Beaufort Formation 101,445,508
    Bois Blanc Formation 343
    Bronson Lake Formation 719
    Cameron Ranch Formation 155
    Christopher Island Formation 198
    Clyde Foreland Formation 262,286,288
    Cochrane Formation 208,218,232,234,236,253,265
    Cowichan Head Formation 50,52,55,57
    Don Formation 335,336,345
    Dundee Formation 329
    Favel Formation 723
    Five Islands Formation 412
    Floral Formation 146,154,714,720
    Fort Langley Formation 54
    Gayhurst Formation 363,369,370,371
    Gubik Formation 119,129
    Judith River Formation 716
    Kanguk Formation 115
    Kidluit Formation 121,126,128
    Kipalu Formation 215
    Kittigazuit Formation 121,126,128
    Largs Formation 154
    Lennard Formation 160
    Liard Formation 58
    Lodin Elv Formation 759,767,775,784
    Massawippi Formation 356,368,371
    Missinaibi Formation 185,204,226,228,230-232,258,546
    Mitchell Bluff Formation 146,164
    Muir Point Formation 50,55
    Muriel Lake Formation 719
    Omarralluk Formation 144,215,228,237
    Peel Sound Formation 191,196
    Pierreville Formation 366
    Pitz Formation 198
    Pottery Road Formation 337,345,346
    Rivière aux Vaches Formation 367
    Rosa Formation 154
    Saint-François-du-Lac Formation 367,369
    Scarborough Formation 325,337,343,345,346
    Shell Formation 154
    Souris Formation 138
    Tee Lakes Formation 154

Thelon Formation 196
Thorncliffe Formation 346
Vermilion River Formation 694,723
Wellsch Farm Formation 152
Whidbey Formation 55
Wigwam Creek Formation 232
Woodmore Formation 154
Worth Point Formation 115,116,120,136
Fort Ann phase 373
Fort Langley Formation 54
Fort Nelson Lowland 140,141
Fossil pollen analyses 488-539
Foster Bugt 757
Fox Valley Moraine 160
Foxe Basin 178,182,186,557
Foxe Glaciation 202,205,239,288,290,291,292,294
Foxe Ice 184,208,210,213,238,276,278
Frank Mackie Glacier 74
Franklin Bay Stade 128
Franklin Ice Complex 454,455
Franklin Mountains 26,101,104,125,133,136
Franklinian Geosyncline 445,449
Fraser Glaciation 44,45,48,49,50,51,42,53,70,72,491,545
Fraser Lowland 45,49,50,54
Fraser River 26,32,55,259,265
Frederick House series 224,236
Fredskild, B. 775-783
Free air gravity anomaly 562-569
French, H.M. 604-611
Friday Creek Sediments 233
Frobisher Bay Moraine 277,295,297,298
Frontenac Arch 218,661
Frontenac Axis 215
Frost creep -definition 610
Frost heave 605
Full glacial sea 462
Fulton, R.J. 1-14
Funder, S. 739-786
Fundy Epigeosyncline 394
Fundy Lowlands 410
Fundy-Cobequid Fault 431
Fyles, J.G. 5
Gael Hamkes Bugt 757
Gage Street Bog 527
Galt Moraine 329
Gander Lake 400
Gardner, J.S. 702-713
Garibaldi Belt 25
Garibaldi phase 74
Garry Island Member 128
Gaspereau Ice Centre 415
Gaspésie 403,415,423,512
Gaspésie Ice Cap 423
Gatineau River 269,270
Gayhurst Formation 363,369,370,371
Gelifluction 592,610
General circulation models -climatic change 484,485
Gentilly Till 362,369
Geochemistry
    drift prospecting 413,418
    environmental geology 661
    peat 660
    Quaternary geology 658
    stream and lake sediment 656
    terrain 645-663
Geological hazards
    Flood hazards 723
    Floods and debris flows 704,727
    Risk assessment 712
    Rockfalls 706
    Snow avalanches 702-703
Geological hazards -central and eastern Canada 723-728
Geological hazards -Cordillera 702-713

INDEX

Geomorphic processes 573-644
    Beaufort Sea area 21,108,121,128,547
    Clay and shale slopes 586
    Cold climate processes 604-611
    Cryogenic weathering 606,610
    Debris flows 590
    Detachment slides 585
    Earthquakes 614
    Endogenic processes 612-615
    Flowslides 587
    Frost heave 605
    Gelifluction 592,610
    Hillslope processes 594
    Ice segregation 605
    Karst development 619
    Needle ice formation 605
    Regolith slides 589
    River processes 595-604
    Rock avalanches 585,707,726-727
    Rock creep 585
    Rockslopes 584
    Skin flows 591
    Soil creep 592
    Soil erosion 593
    Solution processes 616-617
    Thaw slopes 591-592
    Vertical tectonic movements 613
    Volcanic processes 615
George Lake -dispersal train 655
Georgian Bay Lobe 327,340,347
Gilbert Glacier 74
Gilbert River 694
Gilbert type deltas 758
Gilman Glacier 469
Gimli Phase 243
Glacial dispersal -terrain geochemistry 649-653
Glacial fluvial deposits 34,145,200,220-222,254-256,331
                                    332,341,342,362,411,450
Glacial isostasy -models 562-568
Glacial lacustrine deposits 35,343,362
Glacial Lake
    Agassiz 131,144,157,161,187,196,200,207,298,210
                          223,232,233,239-243,245,723
    Algonquin 240,325,333,334,339,344,349,724
    Athabasca 144,182,208
    Barlow 189,223,243,249,270,518,519
    Barlow-Ojibway 223,232,233,236,239,243,549,550
    Bigstick 161
    Chambly 373
    Champlain 363
    Chippewa 243
    Cree 196
    Duluth 240,243
    Gayhurst 371
    Grassmere 347
    Hough 240,248,349
    Houghton 243,350
    Hyper Dubawnt 209
    Iosegun 161
    Iroquois 334,343,349,724
    Ivitaruk 127
    Johnston 240
    Kazan 209
    Kincaid 160
    Leduc 161
    Lost Mountain 161
    Lundy 347
    Masik 127
    Mattawaskin 273
    Maumee 333,347
    McConnell 131,162,194,196,200,207,208,243
    McLean 265,267
    Meadow 161,207
    Melfort 161
    Melville 255,263,267,268
    Memphremagog 372
    Minong 240,242,243,248,349
    Minto 273
    Nahanni 65
    Nakina 243
    Nantais 274
    Naskaupi 265,267,557
    O'Laws 411
    Ogilvie 243
    Ojibway 162,187,189,210,223,225,232,236,238,243,249,254
                                255,272,273,275,519,724
    Old Wives 161
    Ostram 243
    Peace 66,68,144,161,207
    Peel 334
    Regina 144,161
    Saltcoats 161
    Schomberg 343,344
    Souris 160,161
    Stanley 243,248,349
    Sultan 243
    Thelon 208
    Unity 161
    Vermont 363
    Warren 325,333,347
    Whittlesey 333,347
    Wollaston 196
Glacial marine deposits 36,201,224-225,257,363,412,451
Glacial maximum model -Queen Elizabeth Islands 445
Glacial refugia 485,486
Glaciation
    Banks Glaciation 119,452,459
    Bliss Bugt Glaciation 748,767,784
    Buckland Glaciation 125,129,131,136
    Clausen Glaciation 121,125,129
    Dyer Glaciation 289
    Eclipse Glaciation 289,290
    First Canyon Glaciation 63,66,119,125
    Fiskebanke Glaciation 752,753,784
    Flakkerhuk Glaciation 758,761,763,772,785
    Flat River-Clausen Glaciation 65
    Foxe Glaciation 202,205,239,288,290,291,292,294
    Fraser Glaciation 44,45,48,49,50,51,42,53,70,72,491,545
    Great Glaciation 43,63,64,154
    Hole-in-the-Wall Glaciation 66
    Hungry Creek Glaciation 69,129,131,136,494
    Jackfish Glaciation 129
    Kansan Glaciation 504
    Klaza Glaciation 58,60
    Kluane Glaciation 48,60
    Komaktorvik Glaciation 259
    Koroksoak Glaciation 259,265
    Langelandselv glaciation 761,776,784
    Macauley Glaciation 48,60,61
    Mason River Glaciation 121,126,128
    McConnell Glaciation 48,58,60,243,270
    Mirror Creek Glaciation 60,61
    Nansen Glaciation 58,60,66,72
    Neoglaciation 74
    Reid Glaciation 58,60,61,65
    Saglek Glaciation 259,263,265
    Scoresby Sund Glaciation 756,758,761,784
    Sisimiut Glaciation 750,751,753,756,771,785
    Thomsen Glaciation 119,121
    Wisconsin Glaciation 126,185,263,500,546,671,771

INDEX

Glacier
    Appalachian glacier complex 406,582
    Frank Mackie Glacier 74
    Gilbert Glacier 74
    Gilman Glacier 469
    Good Friday Glacier 468
    Kaskawulsh Glacier 61
    Otto Glacier 468
    Sydkap Glacier 468
    Thompson Glacier 454,459
    Tiedemann Glacier 74
    White Glacier 465,467,469
Gleysolic Turbic Cryosols 676
Gleysols 671
Gods River 226,230,231,234
Gold Placer deposits
    alluvial 688
    beach and nearshore 689
    classification 687-689
    colluvial 688
    distribution 690
    glacial 689
    residual enrichment 688
Golden Pond -dispersal train 655
Goldeye Lake 496,497
Goldthwait, J.W. 4
Goldthwaite Sea . 188,268-270,363,375,416,423,429,512,518,692
Good Friday Glacier 468
Graham Island 65,66,67,445,488,491
Granby Region 362
Grand Bend beach 339
Grand Lake 400
Grand Marais Moraine 242
Grand River 325,333,339,423
Grand Volume Till 423
Granduc Mine 704
Grandview Hills 131
Grant Land Mountains 449,459,462,463,464
Grant, D.R. 393-440
Grassy River 694
Great Beach 225
Great Bear Lake Ice Lobe 128,133,182,207
Great Bear River 125,130
Great Cordilleran Glacier 154
Great Glaciation 43,63,64,154
Great Lakes 182,236,550
    bedrock geology 326-329
    physiography 329-335
    Quaternary deposits 335-344
    Quaternary history 344-350
Great Slave Lake 144,182,208,511
Great Slave Plain 141
Great Whale River area 245,255
Greely Fiord 449
Green Mountain 351
Greenland 119,191
Greenland
    bedrock geology and physiography 744-748
    biogeography 748,749
    climate 748,749,783
    continental shelves 769
    Holocene emergence history 774
    Quaternary geology 742-792
    sea level history 772
Greenland advance 460
Greenland -East Greenland
    Quaternary deposits 757-763
    Quaternary geology 756-763
Greenland Ice Sheet 40,445,460,463,773,786
Greenland -North Greenland
    Quaternary deposits 764-765
    Quaternary geology 763

Greenland -West Greenland
    Quaternary deposits 750
    Quaternary geology 749
Greenock swamp 334
Gregg Lake 497
Grenville Front 215
Grenville Province 394
Grinnell Ice Cap 297
Grinnell Peninsula 454
Grotte Valerie I & II Interglaciation 121,125
Grus/corestone terrain 413
Gubik Formation 119,129
Guelph esker 343
Gulf of Boothia 182,205,210
Gulf of Maine 395,400
Gulf of St. Lawrence 403,426
Gulf of St. Lawrence -Quaternary History 420
Gullbridge -dispersal train 655
Gunnbjorn Fjeld 757
Gunsight Mountain 119
Hafichuk 496
Halfway River 66
Haliburton Highlands 215
Hall Basin 461
Hall Land 764,768,773
Hall Moraine 295
Hall Peninsula 276,297
Halton Till 341
Hand Hills 141
Hanging Lake 494
Hants Till 410
Hare Indian River 102,125
Hargrave Moraine 242
Harper Creek Moraine 74
Harptree Moraine 160
Harricana Interlobate Moraine 178,214,220,238,240,241
                              243,245,249,255,257,263,264,270,272
Harrowby Sea 121,126
Hartlen Till 410
Hartman Moraine 236,242
Hatfield Valley 716
Hayes Lobe 220,228,234,239,242,243,247
Hazen Moraine 454
Hazen Plateau 449,459,460,462,463,464
Hazen Trough 445
Hecate Lobe 66
Hecate Strait 26,32,44,65,488,491
Heginbottom, J.A. 479-480,576-577
Hellefisk Moraine 771,772,784
Herman Phase 161
Hermitage Peninsula 418
Herschel Island 108,125,129
Hewitt, K. 624-633
Highbury Sediments 50,55
Highlands Ice Cap 411
Highrock Moraine 242
Highwood River 63,64,65
Hillslope processes -geomorphic processes 594
Hochstetter Forland 757,759,763
Hodgson, D.A. 443-459
Hole-in-the-Wall Glaciation 66
Holland marsh 334
Holocene Epoch 1
Holocene 52,54,55,70,72,73,74,76,108,136,146,162
                236,242,275,291,292,299,421,431,444,452,457
                459,472,508,512,518,550,552,594,750,756
Holocene
    paleoenvironment 520
    Greenland Inland Ice 767
    Milne Land stade 758
    xerothermic interval 521
Homathko River 34

INDEX

Hookean elastic solid 563
Hooper Clay 125
Hopetown -dispersal train 655
Horn Plateau 101,104,130,133
Hornaday River 102,132,133
Horton River basin 102,126
Hot spots 25,26
Houghton Phase 248
Hudson Bay Lowland 162,179,182,185,195,203,209,214
218,225,236,238,244,249,545,581,655
Hudson Ice 155,184,218,232,239
Hudson Ice Divide 237,238,270,273
Hudson Strait 178,182,185,187,210,237,245,249,274,295
546,550,557
Hudwin Moraine 242
Hughes, O.L. 58-70
Humber River 339
Humboldt Gletscher 764
Hummingbird Till 65
Hummocky moraine 147
Hungry Creek Glaciation 69,129,131,136,494
Huron Lobe 329,333,340,347,349
Huron Stade 371
Hydrogeology - urban use 729-732
Hydrological regimes -river processes 595-596
Hypsithermal Interval 74,162,520-528
Ibyuk pingo 129
ICE -1 ice sheet model 562
Ice-thrust features 147
Ice cap
    Agassiz ice cap 454,462,469,470,471,472
    Antigonish ice cap 411
    Devon ice cap 454
    Gaspésie ice cap 423
    Grinnell ice cap 297
    Highlands ice cap 411
    Meighen ice cap 464,468,470
    Newfoundland ice cap 402
    South Mountain ice cap 411,412
    Terra Nivea ice cap 297
Ice centre
    Escuminac ice centre 415,416
    Gaspereau ice centre 415
    Patrician ice centre 236
Ice core record 470-472
Ice divide
    definition 184,186,187,544,753
    Hudson Ice Divide 237,238,270,273
    Keewatin Ice Divide 186,195,200,202,203,205,207,208,211
    Kivivic Ice Divide 274
    M'Clintock Ice Divide 133,205,208,209,213
Ice dome -definition 184,411
Ice flow patterns 239,544,658
Ice-free corridor 486,488,491-499
    Alberta 495-496
    Western Canada 491
    Yukon 492-494
Ice frontal features 411
Ice segregation -geomorphic processes 605
Ice sheet
    dynamics 544
    perspective 6-10
Ice shelf
    Ross Ice Shelf 34
    Viscount Melville Sound Ice Shelf 457
    Ward Hunt Ice Shelf 464,466,469,472
Icefield Drift 60
Icon -dispersal train 655
Île de la Grande Entrée 421
Illinoian 1,55,57,61,67,125,129,344,410,419,425,507
748,753,761,784
Independence Fjord 748,767,767,769,772,785

Indian River 692
Industrial mineral resources 687
Ingolf Fjord 756,763
Inland Ice 744,746,755,756,758,764,767
Innerkip Interstadial 337,341
Innuitian Ice Sheet 205,445,453,457,458
Innuitian Orogeny 744
Innuitian Province 445
Insular Belt 22
Insular Mountains 26
Interglaciation -Sangamon 407,410,413,485,500,504,506
Interior Plains 32,63,64,66,99,100,126,137,187,581
Interior Plains -soils 673
Interior Plateau 26,35,40
Interlake region 669
Intermontane Belt 22
Interstade
    Erie Interstade 347,371
    Friday Creek Interstade 233
    Innerkip Interstade 337,341
    Mackinaw Interstade 347,371
    Plum Point Interstade 323,346,371
    Port Talbot I, II Interstade 323,337,338,344,346,371
    Sidney Interstade 338,494
    St. Pierre Interstade 323,337,346,356,359,366,368,369
371,546
    Two Creeks Interstade 240,347,349,368
    Wisconsinan interstade 129,155,776,784
Intertill aquifer 717
Inugsuin Fiord 290
Ipperwash escarpment 328
Irminger Current 749
Iron Strand area 262,265,291
Irvingtonian land mammal age 152
Isobases 213,554
Isortoq Moraine 297
Isostasy
    Free air gravity anomaly 562-569
    Hookean elastic solid 563
    Lithosphere model 565
    Newtonian viscous model 566
    Seismic model 563
    Visco-elastic model 563
    Viscous flow model 566
Isostatic data -definition 547
Ispatinow 196
Ivitak phase 263
Jackfish Creek Till 65,68
Jackfish Creek-Sylvan Lake Episode 65
Jackfish Glaciation 129
Jackson, L.E. Jr. 63-68,667-668,701-702
Jacques Ranges 102
Jakobshavn Isbrae 750
James Bay Lowlands 133,182,215,247,258,545
Jameson Land marine episode 761
Jameson Land peninsula 757,758,759,773
Jerseyan drift 418
Jesse Till 121,127,128
Joggins Till 410
Johnston, W.A. 4
Johnville Till 356,366
Jökulhlaups 82,591,594
Juan de Fuca Plate 25
Juan de Fuca Strait 21,34,54
Judge Daly Promotory 460,462
Judith River Formation 716
Kabinakagami River 234,235
Kaffehavn Interglaciation 752,753,776,784
Kaishak Moraine 127,129,242
Kaminak Lake 198,202,210,211
Kamloops Lake Drift 53
Kanairiktok River 265,268

INDEX

Kangalaksiorvik Fiord 262
Kangalaksiorvik Moraine 265
Kanguk Formation 115
Kansan Age 152,164,504
Kansan Glaciation 504
Kap Alexander 763
Kap Farvel 756
Kap Kobenhavn Formation 119,765,767,775,778,779,784
Kap Mackenzie stade 761
Karrow, P.F. 321-350
Karst development -geomorphic processes 619
Kaskawulsh Glacier 61
Kazan River 202
Kazan Upland 200,215
Keele River 70,130
Keewatin Ice Divide 186,195,200,202,203,205,207,208,211
Keewatin Ice 65,99,184,189,210,459
Keewatin-Mackenzie Plateau 182
Kellett Till 121
Kelly Lake Moraine 133
Kendall Sediments 121,125,128
Kennedy Channel 460
Kidd Creek -dispersal train 655
Kidluit Formation 121,126,128
Killinek Island 265
King Christian Islands 445,451
King William Island 195,201,205,208
Kinojevis Phase 272
Kipalu Formation 215
Kirkfield outlet 240,349
Kirkland Lake -dispersal train 655
Kittigazuit Formation 121,126,128
Kivivic Ice Divide 274
Klassen, R.A. 616-617
Klassen, R.W. 138-166
Klaza Glaciation 58,60
Klondike Plateau 32,64,66,67,68
Kluane Glaciation 48,60
Knife Interlobate Moraine 205
Knife River 233,244
Kodiak Island 486
Koerner, R.M. 464-473
Kogalu aminozone 262,268,289,298
Kogaluk River 265
Komaktorvik Glaciation 259
Komakuk Beach 129
Kong Oscars Fjord 757
Kootenay Pass 704
Kootenay River 34
Korah Phase 243,270
Koroksoak Glaciation 259,265
Koy-Yukon thermal event 494
Kreiger Mountains 454
Kugmallit Bay 128
Kuuijjuarapik 255,272,273
Kwataboahegan River 230,231,234
La Grande region 189,247,253,272,273,275
La Guadeloupe Moraine 372
La Have River 411
Labrador Ice 184,185,187,234,238,245,254,431
Labrador marginal trough 182,253,584
Labrador Plateau 179,265
Labrador Sea 182,744
Labrador Sector 184,186,263
Labuma advance 152,164
Labuma Till 43,63,154
Lac à l'Ange 526
Lac à l'Eau Claire 273
Lac Baskatong 270
Lac Colin 526
Lac Daigle Moraine 269
Lac de la Hutte Sauvage 265

Lac Delorme 274,518
Lac Guillaume-Delisle 273,275
Lac Maskinongé 269
Lac Mistassini 253,255,264,273
Lac Ramsay 526
Lac Saint-Jean 179,269,375
Lac Seul Moraine 242
Lac Stakel 274
Lac Témiscamingue 179,269,270,661
Lac Yelle 527
Laflamme Sea 188,269
Lake Agassiz 131,144,157,161,187,196,200,207,298,210
                        223,232,233,239-243,245,723
Lake Algonquin 240,325,333,334,339,344,349,724
Lake Athabasca 144,182,208
Lake Barlow 189,223,243,249,270,518,519
Lake Barlow-Ojibway 223,232,233,236,239,243,549,550
Lake Bigstick 161
Lake Chambly 373
Lake Champlain 363
Lake Chippewa 243
Lake Cree 196
Lake Duluth 240,243
Lake Erie 339
Lake Gayhurst 371
Lake Grassmere 347
Lake Hazen Fault Zone 459
Lake Hough 240,248,349
Lake Houghton 243,350
Lake Huron 188,215,240,248,327
Lake Huron Basin 240,243,339,347
Lake Hyper Dubawnt 209
Lake Iosegun 161
Lake Iroquois 334,343,349,724
Lake Ivitaruk 127
Lake Johnston 240
Lake Kazan 209
Lake Kincaid 160
Lake Leduc 161
Lake Lost Mountain 161
Lake Lundy 347
Lake Manitou 694
Lake Masik 127
Lake Mattawaskin 273
Lake Maumee 333,347
Lake McConnell 131,162,194,196,200,207,208,243
Lake McLean 265,267
Lake Meadow 161,207
Lake Melfort 161
Lake Melville 255,263,267,268
Lake Memphrémagog 372
Lake Michigan Basin 240,243
Lake Minong 240,242,243,248,349
Lake Minto 273
Lake Nahanni 65
Lake Nakina 243
Lake Nantais 274
Lake Naskaupi 265,267,557
Lake Nipigon 240
Lake O'Laws 411
Lake Ogilvie 243
Lake Ojibway 162,187,189,210,223,225,232,236,238,243
                        249,254,255,272,273,275,519,724
Lake Old Wives 161
Lake Ostram 243
Lake Peace 66,68,144,161,207
Lake Peel 334
Lake Regina 144,161
Lake Saint-François 378
Lake Saint-Louis 378
Lake Saint-Pierre 378
Lake Saltcoats 161

INDEX

Lake Schomberg	343,344
Lake Scugog	334
Lake Simcoe	329,334
Lake Souris	160,161
Lake Stanley	240,248,342,349
Lake Sultan	243
Lake Superior	188,215
Lake Thelon	208
Lake Timiskaming basin	223,236,240
Lake Unity	161
Lake Vermont	363
Lake Warren	325,333,347
Lake Whittlesey	333,347
Lake Winnipeg	162,182,240
Lake Winnipegosis	218
Lake Wollaston	196
Lamoral Till	68
Lampsilis Lake	377,518
Lancaster Sound	202,276,557
Land planning -database	729-732
Land use planning -Prairie region	713-723
Landsdowne House area	236
Landslides	584,707,721,725,726
Langelandselv glaciation	761,776,784
Langis Till	423
Largs Formation	154
Larsen, H.C.	769-772
Late Holocene	523
Late Pleistocene	154,550
Late Portage Mountain advance	54,66
Late Wisconsinan	2,48,50,57,68,72,74,79,108,126,127 129,131,132,136,157,185,201,204,233,234,275,291,346,356 402,405,411,412,415,426,449,459,488,491,497,554,750,767
Late Wisconsinan deglaciation	157,497,552,618
Lateral Pond	494
Laurentian Channel	34,400,420,422,426,427
Laurentian Highlands	188,250,255,351,358,395
Laurentian Moraine	420
Laurentian Region	395,517
Laurentian Valley	327,328,336
Laurentide -nomenclature	184
Laurentide advance	43
Laurentide Ice Sheet	19,35,39,63,99,126,133,178-237,265 356,400,403,418,421,422,423,427,429,431 445,449,457,486,497,511,544,562,673
Laverlochère Moraine	270
Lawrencetown Till	410
Leaf Lake	692
Leda Clay	325
Leinan Moraine	160
Lennard Formation	160
Lennoxville Till	370,371
Lethbridge Moraine	155,157
Liard Formation	58
Liard Lowland	35,58,61
Liard River	42,104,129,491
Liddon Till	459
Limestone Interlobate Moraine	220,243
Limestone Island	488
Lithosphere -Pacific Ocean	22
Lithosphere model -isostasy	565
Lithostratigraphy	
Hudson Bay Lowlands	225-234
Ontario shield	235-239
Little Ice Age	74,179,214,297,457,472,473,708,811
Liverpool Bay	125,126
Lobe	
Acadian Bay Lobe	410,416
Des Moines Lobe	239
Erie Lobe	329,340,341,347
Georgian Bay Lobe	327,340,347
Great Bear Lake Ice Lobe	128,133,182,207
Hayes Lobe	220,228,234,239,242,243,247
Hecate Lobe	66
Huron Lobe	329,333,340,347,349
Ontario Lobe	340,347
Prince Albert Lobe	127,129
Prince of Wales Lobe	127
Puget Lobe	57
Rainy Lobe	239,240,242
Red River Lobe	238,240
Simcoe Lobe	347
Souris Lobe	161
St. Louis Sublobe	218,240
Superior Lobe	236,239
Thesiger Lobe	127
Weyburn Lobe	161
Windigo Lobe	220,224,236
Locat, J.	723-729
Lochart Phase	161,240
Lodin Elv Formation	759,767,775,784
Lofty Lake	496,524
Logan's Line	351
Loks Land Member	262,291
Lonely Creek	58
Long Creek	161
Long Point beach	339
Long Range Mountains	398,416,417,419
Lougheed Island	451
Low, A.P.	4
Lucknow Valley	339
Luvisol	58,66,72,671,672,675
M'Clintock Channel	179,191
M'Clintock Dome	189
M'Clintock Ice Divide	133,205,208,209,213
M'Clure Stade	121,127,128,129,133,136
M'Clure Strait	104,459
MacAlpine Moraine	196,208
Macauley Glaciation	48,60,61
MacDonald, G.M.	70-74
Mackay, J.R.	5
Mackenzie Delta	70-74,101,102,104,125,136,556
Mackenzie Mountains	19,21,40,43,63,65,74,121,125,491 510,585
Mackenzie River	25,34,101,104,125,128,129,130,132,491,713
Mackinaw Interstade	347,371
MacLaughlin Pond	526
Madawaska Phase	416
Magdalen Islands	395,403,415,420,421,423,500,504
Mahone Bay area	410
Main Brook Till	425
Maitland Valley	339
Maktak Fiord	283
Malcolm River	129
Manitoba	146,184
Manitoba Escarpment	140,143,157,240,713,723
Manitoba Plain	140
Manitou-Matamec Moraine	269
Mansel Island	239
Mapleguard Sediments	50
Marcotte Lake	526
Margaree River	411
Marine isotope record	3
Marion Lake	521
Maritime Plain	395
Marks-Mackenzie Moraine	236,242
Marlboro Till	65,68
Marquette ice advance	161,162,236,242,349
Marsh	679
Marsh Creek Till	65
Maryhill Till	341,347
Mason River Glaciation	121,126,128
Massawippi Formation	356,368,371
Matachewan -dispersal train	655

INDEX

Matapédia River 423
Matawin Moraine 359
Matheson Till 236
Mathewes, R.W. 486-491,520-528
Mathews, W.H. 32-34
Mattawa River 240,248
Matthews, J.V. Jr. 68-70,483-486,498-500,507-508,520
Matuyama Reversal Epoch 69,109,119,452
Matuyama Reversed Polarity Chron 119,152,154,767
Maunsell Till 64,154
Maycroft Till 64
Mazama eruption 523
McAndrews, J.H. 528-530
McBeth Fiord 290
McCarron Brook Till 410
McConnell Glaciation 48,58,60,243,270
McDougall Pass 131
McGerrigle Mountains 423,425
McLeod River 692
Meadowcliffe Till 346
Mealey Mountains 250,258,263,267,268
Mécatina Plateau 395
Meek Point Sea 127
Mégantic Moraine 372
Meguma Group 694
Meighen Ice Cap 464,468,470
Melville Bugt 750,756
Melville Hills 102
Melville Island 119,126,127,201,445,449,457,459,464
Melville Peninsula 178,179,276,278
Mercy Till 127
Meta Incognita Peninsula 274,297,299,301
Meteor Lake 694
Metersvig geomorphic research program 757
Mic-Mac terrace 364
Michigan Basin 215,326
Middle Pleistocene 1,152,154
Middle Till Complex 363
Middle Wisconsinan 2,48,70,136,146,155,337,338,344,346
369,407,411,413,426,488,494,496,546,573
Milankovich thermal maximum 136,525,546,767
Milk River 692
Millville-Dungarvon phase 416
Milne Inlet 292
Milne Land stade 763,785
Milverton Moraine 333
Minas Basin 411,412
Mineral exploration applications -drift prospecting 653-655
Mineral exploration examples -dispersal trains 655
Minesing swamp 334
Mingan Ridge 395
Mining activity -Canada natural processes 624-633
Minto Inlet 133
Minto Uplift Province 101
Miocene 119
Miramichi Highlands 415
Mirror Creek Glaciation 60,61
Missinaibi Formation 185,204,226,228,230-232,258,546
Missinaibi Interglacial 194
Missouri Coteau 140,143,147,157,713
Missouri River 138
Mitchell Bluff Formation 146,164
Mokowan Butte 63,66
Montana 32,160,491
Monteregian Hills 364,373
Moore Lake 497,524
Moorhead Phase 240,242
Moose Mountain 141
Moose River 228,230,231,233
Moraine
    Aguanus-Kenamiou Moraine 269
    Aikins Moraine 160
    Arnott Moraine 244
    Baie-Trinité Moraine 268
    Beaver River Moraine 234
    Beaverhouse Moraine 243
    Bedford Hills-Belair Moraine 239,240
    Belair Moraine 161,205
    Belles-Amours Moraine 268
    Bradore Moraine 268
    Breslau Moraine 341
    Burntwood-Knife Interlobate Moraine 178,200,204,208,209
214,220,225,238,239,241,243
    Cartier-McConnell Moraines 243
    Chantrey Moraine System 196,208,209,210
    Cherry River Moraine 372
    Colville Moraines 133
    Condie Moraine 161
    Corrugated moraine 149
    Cree Lake Moraine 147,161,194,196,242
    Darlingford Moraine 161
    De Geer moraine 196,201,209,210,220,224,242,254
269,272,273,279,415
    Dirt Hills Moraine 160
    Ditchfield Moraine 372
    Dog Lake Moraine 236,242
    Dummer Moraine 341
    Dunn Peak Moraine 74
    Duval Moraine 288
    Ekalugad Moraine 298
    Elmira Moraine 331
    Fox Valley Moraine 160
    Frobisher Bay Moraine 277,295,297,298
    Galt Moraine 329
    Grand Marais Moraine 242
    Hall Moraine 295
    Hargrave Moraine 242
    Harper Creek Moraine 74
    Harptree Moraine 160
    Harricana Interlobate Moraine 178,214,220,238,240,241
243,245,249,255,257,263,264,270,272
    Hartman Moraine 236,242
    Hazen Moraine 454
    Hellefisk Moraine 771,772,784
    Highrock Moraine 242
    Hudwin Moraine 242
    Hummocky moraine 147
    Isortoq Moraine 297
    Kaishak Moraine 127,129,242
    Kangalaksiorvik Moraine 265
    Kelly Lake Moraine 133
    Knife Interlobate Moraine 205
    La Guadeloupe Moraine 372
    Lac Daigle Moraine 269
    Lac Seul Moraine 242
    Laurentian Moraine 420
    Laverlochère Moraine 270
    Leinan Moraine 160
    Lethbridge Moraine 155,157
    Limestone Interlobate Moraine 220,243
    MacAlpine Moraine 196,208
    Manitou-Matamec Moraine 269
    Marks-Mackenzie Moraine 236,242
    Matawin Moraine 359
    Mégantic Moraine 372
    Milverton Moraine 333
    Mountain Rapids Moraine 233,237
    Nakina Moraine 243,244
    Nickerson Moraine 239
    Nipigon Moraine 218,242,243
    Noodleook Moraine 265
    Oak Ridges Moraine 327,331,333,334,340,343,346,347,373
    Orangeville Moraine 331
    Paradise Moraine 263,264

INDEX

Paris Moraine	329
Pasquia Moraine	242
Petlura Moraine	161
Piedmont Moraine	416
Pinard Moraine	244
Port Huron Moraine	347
Qu'Appelle Moraine	161
Quebec North Shore Moraine	255,256,268,403
Roulier Moraine	243,270
Sachigo Moraine	220,242
Saint-Antonin Moraine	362,376
St. (Saint-) Narcisse Moraine	249,254,269,270,359,362,363 369,375,379
Sainte-Marie Moraine	422
Sakami Moraine	254,255,272,275,518
Sand Hills Moraines	127
Sandilands Moraine	161
Sebaskachu-Little Drunken Moraine S	268
Sipiwesk-Cantin Moraine	242
Stoke Mountain Moraine	372
Stoughton Moraine	160
Tasiuyak Moraine	265
Ten Mile Lake Moraine	419,423,432
The Pas Moraine	147,157,208,237
Thomson Moraine	160
Tingwick-Ulverton Moraine	372
Tinutyaruik Moraine	263
Tutsieta Lake Moraine	131
Two Loon Moraine	259,262,265
Two Rivers Moraine	240
Waterloo Moraine	331,334
Whitewater Moraine	243
Wyoming Moraine	329,348
Moraine Lake	526
Morgan Bluffs Interglaciation	119
Morison, S.R.	687-694
Mornington Till	341,347
Mott, R.J.	500-507
Moulin River	694
Mount Edziza	49
Mount Logan	26,32
Mount Pleasant -dispersal train	655
Mount Thoresby	265
Mountain Rapids Moraine	233,237
Mountain River	130
Muchalat River Drift	50
Muir Point Formation	50,55
Munro Esker	220,224
Murdockville Highlands	423
Muriel Lake Formation	719
Nachvak Fiord area	262,265
Nachvak phase	263
Nahanni Range	129
Nakina Moraine	243,244
Nansen Glaciation	58,60,66,72
Nansen Sound	446
Nares Strait	452,459,460,461,463,464,753,764,767,768
Naskaupi River basin	268
Natural processes - human agents	624-633
Nebraskan glacial stage	495
Needle ice formation	605
Nelson River	213,218,225,228,234,247
Neoglaciation	74
Nepisiguit Valley	415
Nettiling Lake	277
New Brunswick -Quaternary history	402,412,413
Newfoundland -Quaternary history	400,416
Newfoundland Ice Cap	402
Newfoundland Uplands	395,416
Newtonian viscous model	566
Niagara Escarpment	327,329,333
Nicholson Peninsula	126,128
Nickerson Moraine	239
Nicolet Stade	367
Nina Lake	527
Nipigon Moraine	218,242,243
Nipigon Plain	215
Nipissing Great Lakes II	248,249
Nipissing level	339,340,350
Nissouri Stade	341,346
Nith River	333,343,346
Nonglacial Wisconsinan	155
Noodleook Moraine	265
Nootka transform	25
Nordvestfjord	757
Norman Range	102,133
North American Craton	22
North American Plate	445,744
North Atlantic Polar Front	432
North Bay outlet	240,248,373
North Eurasian Plate	744
North Klondike River	65
North Knife River	193,194,200,204,244,246
North Saskatchewan River	65,68,692
North Shore Gulf of St. Lawrence	398
Northeastern Canadian Shield	
bedrock geology	276-278
climate	278-279
physiography	276-278
Quaternary deposits	279-285
Quaternary history	285-297
sea level	297-301
Northern Canadian Interior Plains	100-137
Arctic Archipelago	119-136
bedrock geology	101
climate	104
economics	137
physiography	101-104
Quaternary deposits	104-119
sea level	136
Northern Miramichi Highlands	413
Northern Peninsula	416,419
Northumberland Strait	395,416,427
Northwestern Canadian Shield	
bedrock geology	189-193
physiography	191-193
Quaternary deposits	193-201
Quaternary history	201-214
Norwood Esker	331
Notre Dame Bay	399,417
Notre Dame Mountains	397
Nova Scotia Upland	395,504
Nugssuaq Embayment	744
Nunatak	488,519
Oak Ridges Moraine	327,331,333,334,340,343,346,347,373
Obed advance	66
Occhietti, S.	321-326,350-379
Ocean Drilling Programme	783
Ochre River	125
Odanak Till	366
Ogilvie Mountains	26,706
Oka Hills	350
Okanagan Centre Drift	50,57
Old Crow Basin	119,484,498
Old Crow Flats	32,61,68,69,70,492,493,494
Oldman River	43,63,64,65
Olduvai Normal-Polarity Subchron	1,152
Olympia Nonglacial Interval	50,52,57,70,72
Omaralluk Formation	144,215,228,237
Omineca Belt	22

INDEX

Omineca Mountains 26,57
Onondaga Escarpment 328
Ontario Island 347
Ontario Lobe 340,347
Opasatika River 230
Opemiska Lake 272
Orangeville Moraine 331
Organic 672,673,674
Organic Cryosols 676
Ottawa Islands 186,238,245,247,553
Ottawa River valley 182,218,240,243,269,349,358,373
Otto Glacier 468
Outlet
    Chicago outlet 248,350
    Fenelon Falls outlet 349
    Kirkfield outlet 240,349
    North Bay outlet 240,248,373
    Port Huron outlet 248,349,350
Owl Creek -dispersal train 655
Pacific Ocean lithosphere 22
Pacific Plate 25
Paleobotany -case histories 483-539
Paleoecology 483-539
    Atlantic Canada 500-507
    British Columbia 70-72
    eastern Cordillera 73-74
    Graham Island 488-491
    Pollen analyses 483-530
    southern Canadian Interior Plains 162-164
    Yukon Territory 72-73
Paleoplain 397,399,417
Paleosols 63,72,157,411,418,496,511
Palsas 750
Paradise Moraine 263,264
Paraglacial concept 576,594
Paris Moraine 329
Parry Channel 445
Parry Peninsula 128
Paskapoo Formation 721
Pasley River 191,194,202
Pasquia Moraine 242
Patorfik marine beds 752,775
Patrician ice centre 236
Payne centre 274
Peace River 42,52,63,66,68,130,140,141,497,692
Peary Land 763,764,765
Peat resources -Canada 676-684
Peatland types 679
Peel Plateau 26,63
Peel River 70,125
Peel Sound Formation 191,196
Pelican Mountains 141
Pelly Mountains 26
Peltier, W.R. 562-568
Pembina spillway 161
Penny Ice Cap 184,279,297
Penny Surface 445
Penokean Hills 215
Perch Lake 526
Permafrost 136,582,607-611
Petit Matane Till 423
Petite Nation River 269
Petlura Moraine 161
Phair Lake 523
Phase
    Aylen Phase 270
    Belleville phase 373
    Campbell Phase 131
    Cass Phase 240
    Chignecto phase 415
    Emerson Phase 243
    Fort Ann phase 373
    Garibaldi phase 74
    Gimli Phase 243
    Glacial Episode II 64
    Glacial Episode III 65
    Herman Phase 161
    Houghton Phase 248
    Ivitak phase 263
    Kinojevis Phase 272
    Korah Phase 243,270
    Lochart Phase 161,240
    Madawaska Phase 416
    Millville-Dungarvon phase 416
    Moorhead Phase 240,242
    Nachvak phase 263
    Nipissing 339,340,350
    Plaster Rock-Chaleur phase 416
    Sheguiandah Phase 270
    Tutsieta Lake Phase 126,128,131,132,136
Physiographic regimes -river processes 596-597
Physiography -Canada 577-583
Piedmont Moraine 416
Pierreville Formation 366
Pilot Mill Till 488
Pinard Moraine 244
Pipun Stage 162
Pitz Formation 198
Placer deposits -Canada 687-694
Placer gold deposits Canada -stratigraphy 689
Placers *see also Gold placers 77
Placers -Beauce region 326
Plains Ice 184,187
Plaster Rock-Chaleur phase 416
Platinum placer deposits 692
Pleistocene Epoch. 1
Plum Point Interstade 323,346,371
Podsols 672,674,675
Pointe Saint-Nicolas Till 366
Polar Continental Shelf Project 469
Polar Front 412,421
Polybog 494
Pop-ups 329
Poplar River 692
Porcupine River 19,68,70,72,131,138,140,493,494
Port Arthur Hills 215
Port au Port Peninsula 420
Port Bruce Stade 341,371
Port Huron Moraine 347
Port Huron outlet 248,349,350
Port Huron Stade 347
Port Stanley Till 341,346,347
Port Talbot I ,II 323,337,338,344,346,371
Portage Mountain 66
Possession Drift 57
Pottery Road Formation 337,345,346
Prairie steps 140
Pre-Late Wisconsinan 258
Pre-Reid 68
Pre-Sangamonian 49,66,410,498
Pre-Wisconsinan 119,203,344
Prelate Ferry Paleosol 157
Presant, E.W. 673-674
Prest, V.K. 5
Prince Albert Lobe 127,129
Prince Albert Peninsula 105,121,126,128,133,134
Prince Edward County Peninsula 329
Prince Edward Island -Quaternary History 395,416
Prince Gustav Adolf Sea 451
Prince of Wales Island 191,195,201,202,205,214,554

INDEX

Prince of Wales Lobe 127
Prince of Wales Lowland 191,194
Prince of Wales Strait 102,128
Prince Rupert 704
Proto-Atlantic Ocean 394
Puget Lobe 57
Puget Lowland 25,40,53
Qivitu Foreland 286,288
Qu'Appelle Moraine 161
Quadra Sand 51,52,57
Quajon Fiord 290
Quaternary 1
Quaternary geodynamics -Canada 543-572
Quaternary resources -Canada 665-697
Quebec North Shore Moraine 255,256,268,403
Queen Charlotte Islands 21,45,65,66,67,486
Queen Charlotte Islands refugium 486-491
Queen Charlotte Ranges 63
Queen Charlotte Sound 44
Queen Charlotte-Fairweather Fault 24,25
Queen Elizabeth Islands
   climate 448
   glaciers 464
   last glaciation 452,453,454,553,556
   last interglaciation 452,553,556
   nonglacial deposits 451
   northeast Ellesmere Island 459-464
   paleoenvironment 457,458
   permafrost 448
   physiography 119,445
   Quaternary history 452
   surficial materials 449
Queen Elizabeth Islands Glacier Complex 454
Queen Maud Gulf 182,191
Queenston Shale 341
Quinn Lake readvance 208
Radiocarbon dating 5
Rainy Lobe 239,240,242
Rankin Inlet-Ennadai belt 655
Rannoch Till 347
Rasmussen Lowland 191,194
Rat River 125,131
Ravenscrag Formation 716
Rawdon Till 411
Red Deer River 692
Red Head Till 410
Red Indian Lake 400,402,418
Red Lake area 236
Red River Lobe 238,240
Redstone River 130
Redwater River 692
Reeh, N. 793-823
Refugium 486-491
Regolith slides -geomorphic processes 589
Regolsolic Turbic Cryosol 676
Regosol 671
Reid Drift 61,72
Reid Glaciation 58,60,61,65
Reid Lakes 60
Remote Peninsula 289
Rennell Sound Fault 25
Richard, P.J.H. 275,512-519
Richards Island 104,126
Richardson Mountains ... 32,35,63,68,73,74,119,125,128,131,491
Richardson River valley 208
Richmond Gulf 185
Richmond Till 410
Riddell Member 146,154
Riding Mountain 140

Rigaud Mountains 350
Risk assessment -geological hazards 712
Ritchie, J.C. 508-512
River morphology 602-603
River processes
   bed load 601-602
   dissolved solids 602
   fluvial erosion intensity 599
   fluvial sediment transfer 604
   geomorphic processes 595-604
   hydrological regimes 595-596
   physiographic regimes 596-597
   river morphology 602-603
   sediment transport 599
   suspended sediment 599-601
Rivière à la Baleine 265
Rivière à la Patate Till 422
Rivière Alluviaq 265
Rivière aux Mélèzes 273,274
Rivière aux Vaches Formation 367
Rivière Broadback 272,273
Rivière Caniapiscau 274,275
Rivière de Rupert 272,273
Rivière Eastmain 273
Rivière George 265
Rivière Jupiter Till 422
Rivière Koroc 265
Rivière Laforge area 273
Rivière Manicouagan 255,269
Rivière Mécatina 269
Rivière Moisie 269
Rivière Nastapoca 273
Rivière Natashquan 269
Rivière Nottaway 259
Rivière Ouareau 270
Rivière Romaine 269
Rivière Saguenay 269
Rivière Saint-Maurice 269
Rivière Sainte-Anne 270
Roaring River Clay 146
Robeson Channel 459,460
Robinson's Head readvance 416
Rock avalanches 585,707,726-727
Rock creep 585
Rockfalls 706
Rockslopes 584
Rocky Mountain Foothills 491,492,497,510
Rocky Mountain Trench 26,50,52,54
Rocky Mountains 21,26,43,63,66,73,150,155,491,706
Rogers Pass 702-703
Rosa Formation 154
Ross Ice Shelf 34
Roulier Moraine 243,270
Running River 125
Rupert Readvance 244
Russell Island 214
Russell Stade 127,131
Rutter, N.W. 58-70
Ryder, J.M. 26-32,48-58,76-76
Saalian 748
Sachigo Moraine 220,242
Sachs Till 127
Sackung 614
Saglek Glaciation 259,263,265
Saint John River 403,413,415,416
Saint-Antonin Moraine 362,376
Saint-François-du-Lac Formation 367,369
Saint-Jean-Vianney 588
Sainte-Marie Moraine 422

836

INDEX

Sakami Moraine 254,255,272,275,518
Salmon River Sand 407
Sampling procedures -terrain geochemistry 649-653
Sand Hills Moraines 127
Sandilands Moraine 161
Sandspit Fault 25
Sangamon Interglaciation 121,125,126,152,202,204,226,230
   232,234,236,336,345,356,364,367,412,407,410,413,485
   500,504,506,507
Sangamon soil 155
Sangamonian 1,289,338,418,425,495,500,753,784
Sangamonian optimum 356,366
Sangamonian sea level 772
Sangamonian Stage 49,50,55,66,154,185,203,225,262
   344,345,421,425,500
Saskatchewan Gravels and Sands 146,152
Saskatchewan Plain 140
Satellite imagery -erosion 184
Sauble Beach 339
Saugeen River 325,335,339
Saulnierville Till 410
Sault Ste. Marie 215,218
Savante Lake 694
Scarborough Formation 325,337,343,345,346
Schefferville area 259,263
Schuyter Point Sea 127,134
Schweger, C.E. 491-499,520-528
Scolithos quartzite 758
Scoresby Sund 756,758,761,784
Scott, J.S. 713-723
Sea
   Banks Sea 119
   Bell Sea 209,235,546
   Champlain Sea 512,518
   D'Iberville Sea 274,512
   De Geer Sea 412,416,429,430
   East Coast Sea 127
   Goldthwaite Sea 188,268-270,363,375,416,423,429
      512,518,692
   Harrowby Sea 121,126
   Labrador Sea 182,744
   Laflamme Sea 188,269
   Meek Point Sea 127
   Prince Gustav Adolf Sea 451
   Schuyter Point Sea 127,134
   Tyrrell Sea 162,188,196,201,209-211,213,225,232,235
      239,244,245,254,272-273
Sea level changes 43-48,136,246,247,298,427,457,546,553
   556-562,772,773
Seal River 191,204,238,245
Sebaskachu-Little Drunken Moraine System 268
Sediment transport 599
Sedimentology of Arctic Fiords Experiment project 279,281
Seismic loading 708
Seismic model 563
Selkirk Mountains 703
Selwyn Mountain 26,40
Semiahmoo Drift 50,52,55
Seminary Till 346
Sermilik Fjord 755
Serpentine Belt 351
Severn Upland 215,231
Shaler Mountains 102,104,121,126,128,133
Shediac Valley 416
Sheguiandah Phase 270
Shell Formation 154
Sherbrooke region 694
Shubenacadie River 411
Shulie Lake Till 411

Shuswap Highlands 74
Sidney Interstade 338,494
Silver Creek 58,61
Silver Lake 526
Simcoe Lobe 347
Simpson Lowland 194,195
Sipiwesk-Cantin Moraine 242
Sisimiut Glaciation 750,751,753,756,771,785
Sisson Brook 655
Sitidgi Lake Stade 131,132,136
Skin flows 591
Slave River 130
Slaymaker, H.O. 581-583
Smith, M.W. 577-581
Snake River Till 68,125
Snow avalanches 702-703
Soil creep 592
Soil erosion 593
Soil formation 669-671
Soil orders in Canada -Table 670
Solonetzic 671,673
Solution processes 616-617
Somerset Island 179,181,191,194,201,205,211,214,445
Souris channel 161
Souris Formation 138
Souris Lobe 161
South Mountain Ice Cap 411,412
South Nahanni River 65,66,119,121,125
South Saskatchewan Gravels 138
South Seal River 200
Southampton Island 178,210,248,278,298,553
Southeastern Canadian Shield
   bedrock geology 249
   physiography 249-253
   Quaternary deposits 253-258
   Quaternary history 258-275
Southern Canadian Interior Plains 138-166
   bedrock geology 138-141
   climate 142-143,164-165
   drainage 141-142
   economics 165-166
   Quaternary deposits 143-162
   Quaternary history 149-162
Southern Indian Lake 237
Southern Uplands 406,410
Southwestern Canadian Shield
   bedrock geology 214-215
   physiography 215-220
   sea level 246-248
Speleothems 32,57,66,67,70,619
Spiral Tunnels 704
St. Elias Mountains 21,26,32,40,58,60,61,73,74
St. Lawrence Lowlands 187,321,322,395,581
   economics 323
   soils 673-674
St. Lawrence Lowlands *see Great Lakes,
St. Lawrence Valley
St. Lawrence Valley 182,242,269,322,351
   bedrock geology 350-351
   physiography 351-355
   Quaternary deposits 355-364
   Quaternary history 364-377
St. Louis Sublobe 218,240
St. Marys River 248
St. (Saint-) Narcisse Moraine 249,254,269,270,359,362,363
   369,375,379
St. Pierre Sediments 356,359,362
St. Pierre interstade .. 323,337,346,356,359,366,368,369,371,546
St. Quentin area 416

837

INDEX

Stade
- Ayr Lake Stade 286,288-290,298
- Baffinland Stade 286,289,290,292
- Coquitlam Stade 57
- Franklin Bay Stade 128
- Huron Stade 371
- Kap Mackenzie Stade 761
- M'Clure Stade 121,127,128,129,133,136
- Milne Land Stade 763,785
- Nicolet Stade 367
- Nissouri Stade 341,346
- Port Bruce Stade 341,371
- Port Huron Stade 347
- Russell Stade 127,131
- Sitidgi Lake Stade 131,132,136
- Sumas Stade 58
- Sunneshine Stade 289
- Taserqat Stade 786
- Toker Point Stade 125,128,129,131,136
- Trois-Rivieres Stade 369
- Vashon Stade 54,57

Stage
- Bessborough Stage 66
- Clayhurst Stage 66
- Cockburn Substage 297
- Eemian Stage 748
- Pipun Stage 162
- Saalian Stage 748
- Sangamonian Stage 49,50,55,66,154,185,203,225, 262,344,345,421,425,500
- Weichselian Stage 748

Stanton Sediments 121,126
Static Cryosols 676
Steep Rock -dispersal train 655
Stewart River 60
Stewiacke River 411
Stikine Belt 25
Stikine Plateau 26,40
Stikine River 49,53
Stirton Till 347
Stoke Mountain Moraine 372
Stoughton Moraine 160
Strandflat 33
Stratford Till 347
Succession 520
Sugarloaf Lake 526
Sumas Drift 54
Sumas Stade 58
Sunneshine Stade 289
Sunnybrook Till 346
Superior Lobe 236,239
Surficial aquifer 717
Suspended sediment 599-601
Sutherland Group 152,154
Sutton Hills 215,225,237,714,720
Sutton Mountains 351
Svartenhuk marine beds 753,777
Sverdrup Basin 445,449,467
Swan Hills 18,141
Swan River 716
Swift Current Channel 160
Sydkap Glacier 468
Sydney-Mira Bay 411
Sylvan Lake Till 65,68
Tadoule Lake 218
Taiga Valley 64
Tamagodi Till 423
Tanquary Fiord 454,462
Tarnocai, C. 675-684
Taserqat Stade 786
Tasiuyak Moraine 265
Tavistock Till 341,347

Tectonics -Eurekian Rifting Episode 191
Tee Lakes Formation 154
Ten Mile Lake Moraine 419,423,432
Ten Mile Lake readavance 419,430
Terra Nivea Ice Cap 297
Terrain geochemistry 645-663
Thames River 325,333,339
Thaw slopes 591-592
The Pas Moraine 147,157,208,237
The Ramparts 34
Thelon basin 200,202
Thelon Esker 208
Thelon Formation 196
Thermokarst 451,499,610
Thesiger Lobe 127
Thetford Mines -dispersal train 655
Thompson Glacier 454,459
Thompson River 52,54
Thomsen Glaciation 119,121
Thomson Moraine 160
Thorncliffe Formation 346
Tiedemann Glacier 74

Till
- Ange-Gardien Till 371
- Beaver River Till 411
- Bécancour Till 366,367
- Canning Till 346
- Catfish Creek Till 324,341,346,347
- Chaudière Till 356,369,694
- Churchill River Till 204,225,237,238,247,256,263
- Definition 143
- Donnacona Till 366
- Dundas Till 450,459
- Eatonville Till 410
- Edson Till 65,68
- Elma Till 341,347
- Ernst Till 65,68
- Erratics Train Till 65
- Flat River Till 65
- Foothills Erratics Train 43,65,68,144,155
- Gentilly Till 362,369
- Grand Volume Till 423
- Halton Till 341
- Hants Till 410
- Hartlen Till 410
- Hummingbird Till 65
- Jackfish Creek Till 65,68
- Jesse Till 121,127,128
- Joggins Till 410
- Johnville Till 356,366
- Kellett Till 121
- Labuma Till 43,63,154
- Lamoral Till 68
- Langis Till 423
- Lawrencetown Till 410
- Lennoxville Till 370,371
- Liddon Till 459
- Main Brook Till 425
- Marlboro Till 65,68
- Marsh Creek Till 65
- Maryhill Till 341,347
- Matheson Till 236
- Maunsell Till 64,154
- Maycroft Till 64
- McCarron Brook Till 410
- Meadowcliffe Till 346
- Mercy Till 127
- Middle Till Complex 363
- Mornington Till 341,347
- Odanak Till 366
- Petit Matane Till 423
- Pilot Mill Till 488

INDEX

Pointe Saint-Nicolas Till 366
Port Stanley Till 341,346,347
Rannoch Till 347
Rawdon Till 411
Red Head Till 410
Richmond Till 410
Rivière à la Patate Till 422
Rivière Jupiter Till 422
Sachs Till 127
Saulnierville Till 410
Seminary Till 346
Shulie Lake Till 411
Stirton Till 347
Stratford Till 347
Sunnybrook Till 346
Sylvan Lake Till 65,68
Tamagodi Till 423
Tavistock Till 341,347
Wartburg Till 341,347
Winter Harbour Till 134,204,459
York Till 345
Till *see also tills identified by name 34,143,340-341,449
Tingwick-Ulverton Moraine 372
Tintina Fault 24
Tintina Trench 26,64,65
Tinutyaruik Moraine 263
Toker Point Stade 125,128,129,131,136
Tom Creek Silt 60,61
Torngat Mountains 250,253,255,258,259,264,265,519
Touchwood Hills 141
Tracadie Peninsula 415
Trent Valley 333,349
Trois-Rivières Stade 369
Trough 182,253,445,584
Tuktoyaktuk Peninsula 104,121,125,126,128,136,494,512,525
Tundra 517,582,583
Turbic Cryosols 675
Turtle Mountain 138
Tutsieta Lake Moraine 131
Tutsieta Lake Phase 126,128,131,132,136
Tuya 33,49
Twelvemile Bluff 493
Two Creeks Interstade 240,347,349,368
Two Loon Moraine 259,262,265
Two Rivers advance 349
Two Rivers Moraine 240
Tyner Valley 720
Tyrell, J.R. 4
Tyrrell Lake 494
Tyrrell Sea 162,188,196,201,209-211,213,225,232,235
                       239,244,245,254,272-273
Ungava Peninsula 179,253,297,512,518
Union Strait 102,132,133
Upham, W. 4
Urban planning -hydrogeology 729-732
Valders readvance 240
Valentine, K.W.G. 668,671-672
Vancouver Island 21,40,52,70
Varve stratification 343,486
Vashon Stade 54,57
Vegetation pattern -Quebec-Labrador 512-519
Vermilion River Formation 694,723
Vertical tectonic movements 613
Victoria Island 100,101,102,104,121,128,129,133,186,196,211
Victoria Lowland 191
Vincent, J-S. 100-137,178-189,249-275,
Visco-elastic model 563
Viscount Melville Sound 104,129,133,134,187,204
Viscount Melville Sound Ice Shelf 457
Viscous flow model 566

Vixen Lake -dispersal train 655
Volcanic processes 615
Wabamun Lake 721
Waconichi ice advance 272
Waddy Lake -dispersal train 655
Wager Bay 213
Wanapietei River 694
Wapawekka Upland 141
Ward Hunt Ice Shelf 464,466,469,472
Wartburg Till 341,347
Wasaga beach 339
Washington 40,42
Waskesiu Upland 141
Waterhen River 692
Waterloo Moraine 331,334
Waterton I advance 154
Waterton II advance 64
Waterton IV-Hidden Creek advance 65,66
Waterton Lakes 63,64,65,68
Watino Interval 157,496
Webb, K. 674-675
Weichselian 748
Wellsch Farm Formation 152
Wernecke Mountains 63
West Greenland Current 749
West River 128
Western Arctic Archipelago 99
Western System 26
Westlynn Drift 49,55
Westwold Sediments 50,57
Weyburn Lobe 161
Whidbey Formation 55
White Channel gravels 32,64,78,691
White Glacier 465,467,469
White, O.L. 729-732
Whitewater Moraine 243
Whycocomagh area 411
Wigwam Creek Formation 232
Wilson, J. Tuzo 5
Windigo Lobe 220,224,236
Wingham-Dundas Valley 328
Winisk River 225,246
Winona Block 25
Winter Harbour Till 134,204,459
Wisconsin Glaciation 126,185,263,500,546,671,771
Wisconsinan 2,155,201,289,668,709
   interstade 129,155,776,784
   sea level 773
Wisconsinan Stade 155
Wisconsinan Stage 50,182,201,233,234
Wisconsinan-Holocene transition 785
Wollaston Peninsula 105,133
Wood Mountain 138,141,150,154
Woodmore Formation 154
Worth Point Formation 115,116,120,136
Wrangell Belt 26
Wrigley Lake 129
Wyoming Moraine 329,347
Yamnuska Bog 496
Yarmouthian Age 152,164
Yeltea Lake 131
York Till 345
Younger Dryas event 412,419,421,554
Yukon Coastal Plain 70,74,105,119,121,125,129
Yukon Plateau 26,40
Yukon refugium 492
Yukon Territory -Chronology 42
Yukon Territory -Quaternary History 19,24,58-60,72
Zero curtain 605